- Main Group metals
- Transition metals, lanthanide series, actinide series
- Metalloids
- Nonmetals, noble gases

	1B	2B	3A	4A	5A	6A	7A	8A	
								2 **He** Helium 4.0026 [G]	1
			5 **B** Boron 10.811 [S]	6 **C** Carbon 12.011 [S]	7 **N** Nitrogen 14.0067 [G]	8 **O** Oxygen 15.9994 [G]	9 **F** Fluorine 18.9984 [G]	10 **Ne** Neon 20.1797 [G]	2
			13 **Al** Aluminum 26.9815 [S]	14 **Si** Silicon 28.0855 [S]	15 **P** Phosphorus 30.9738 [S]	16 **S** Sulfur 32.066 [S]	17 **Cl** Chlorine 35.4527 [G]	18 **Ar** Argon 39.948 [G]	3
28 **Ni** Nickel 58.693 [S]	29 **Cu** Copper 63.546 [S]	30 **Zn** Zinc 65.39 [S]	31 **Ga** Gallium 69.723 [S]	32 **Ge** Germanium 72.61 [S]	33 **As** Arsenic 74.9216 [S]	34 **Se** Selenium 78.96 [S]	35 **Br** Bromine 79.904 [L]	36 **Kr** Krypton 83.80 [G]	4
46 **Pd** Palladium 106.42 [S]	47 **Ag** Silver 107.8682 [S]	48 **Cd** Cadmium 112.411 [S]	49 **In** Indium 114.82 [S]	50 **Sn** Tin 118.710 [S]	51 **Sb** Antimony 121.757 [S]	52 **Te** Tellurium 127.60 [S]	53 **I** Iodine 126.9045 [S]	54 **Xe** Xenon 131.29 [G]	5
78 **Pt** Platinum 195.08 [S]	79 **Au** Gold 196.9665 [S]	80 **Hg** Mercury 200.59 [L]	81 **Tl** Thallium 204.3833 [S]	82 **Pb** Lead 207.2 [S]	83 **Bi** Bismuth 208.9804 [S]	84 **Po** Polonium (209) [S]	85 **At** Astatine (210) [S]	86 **Rn** Radon (222) [G]	6
110* — — (269) [X]	111 — — (272) [X]	112 — — (277) [X]							7

63 **Eu** Europium 151.965 [S]	64 **Gd** Gadolinium 157.25 [S]	65 **Tb** Terbium 158.9253 [S]	66 **Dy** Dysprosium 162.50 [S]	67 **Ho** Holmium 164.9303 [S]	68 **Er** Erbium 167.26 [S]	69 **Tm** Thulium 168.9342 [S]	70 **Yb** Ytterbium 173.04 [S]	71 **Lu** Lutetium 174.967 [S]
95 **Am** Americium (243) [X]	96 **Cm** Curium (247) [X]	97 **Bk** Berkelium (247) [X]	98 **Cf** Californium (251) [X]	99 **Es** Einsteinium (252) [X]	100 **Fm** Fermium (257) [X]	101 **Md** Mendelevium (258) [X]	102 **No** Nobelium (259) [X]	103 **Lr** Lawrencium (260) [X]

CHEMISTRY

& Chemical Reactivity

•

Fourth Edition

JOHN C. KOTZ

SUNY Distinguished Teaching Professor
State University of New York
College at Oneonta

PAUL TREICHEL, JR.

Professor of Chemistry
University of Wisconsin-Madison

SAUNDERS GOLDEN SUNBURST SERIES

Saunders College Publishing

Harcourt Brace College Publishers

Fort Worth Philadelphia San Diego New York Orlando Austin
San Antonio Toronto Montreal London Sydney Tokyo

Publisher: Emily Barrosse
Publisher/Acquisitions Editor: John Vondeling
Product Manager: Pauline Mula
Developmental Editors: Elizabeth C. Rosato and Sarah Fitz-Hugh
Project Editor: Robin C. Bonner
Production Manager: Charlene Squibb
Art Director: Caroline McGowan
Cover Designer: Ruth Hoover

Cover Credit: Ammonium chloride decomposes when heated to gaseous ammonia and hydrogen chloride. These combine, forming the "cloud" seen in the photograph. Photo, C. D. Winters; molecular model, Susan M. Young and Rolin Graphics, Inc.

Printed in the United States of America

CHEMISTRY & CHEMICAL REACTIVITY, fourth edition
0-03-023762-9

Library of Congress Catalog Card Number: 98-85606

8901234567 032 10 9876543

To Katie, David, and Peter

To Isabel, David, and Janet

JOHN C. KOTZ

PAUL TREICHEL, JR.

Contents Overview

Chemical Perspectives and Applications

Historical Perspectives

Current Perspectives

Problem-Solving Tips and Ideas

A Closer Look

Contents

Part 2
The Structure of Atoms and Molecules

Part 3
States of Matter

Part 4
The Control of Chemical Reactions

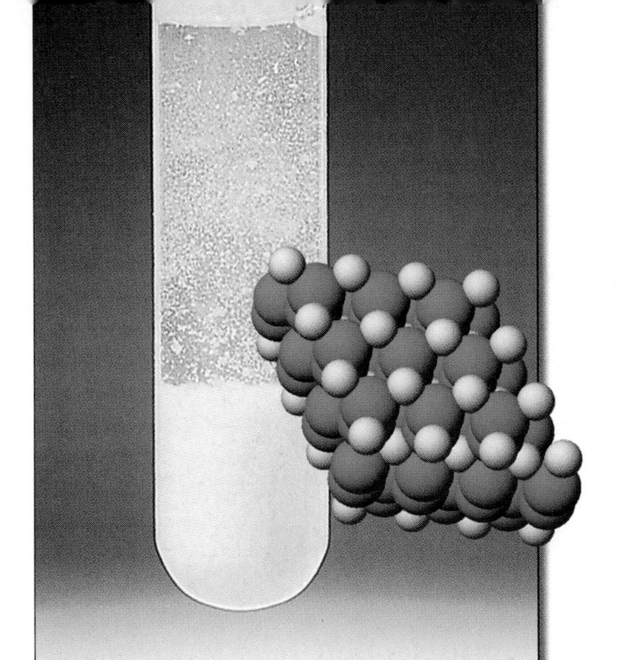

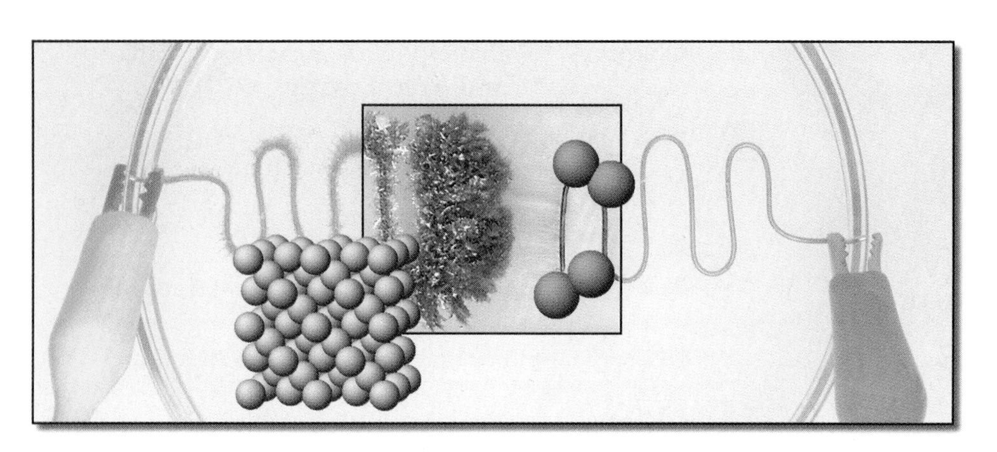

Appendices

Preface

This is the fourth edition of *CHEMISTRY & CHEMICAL REACTIVITY*. The principal theme of the book, beginning with the first edition more than 10 years ago, is to provide a broad overview of the principles of chemistry and the reactivity of the chemical elements and their compounds. This edition, however, brings a new organization to that theme: the close relation between the macroscopic observations we make of chemical and physical changes, the symbols we use to describe those changes, and the way we view those changes at the atomic and molecular levels.

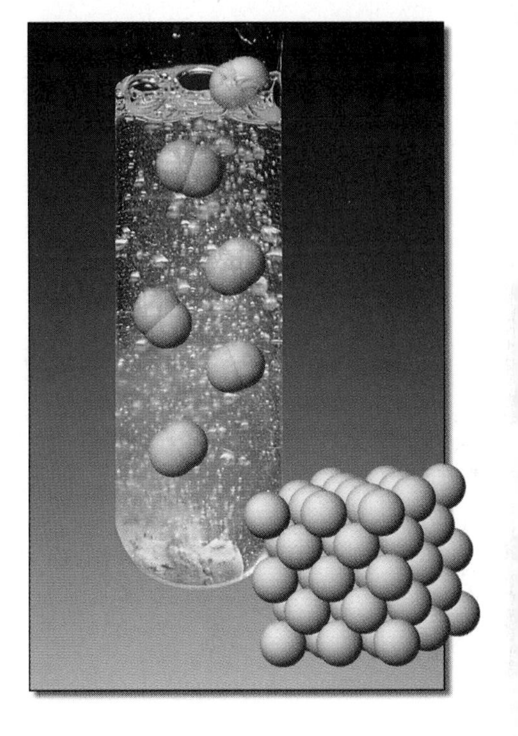

In addition to a fresh viewpoint on chemistry, we want to convey a sense of chemistry as a field that not only has a lively history but also one that is currently dynamic, with important new developments on the horizon. Furthermore, we want to provide some insight into the chemical aspects of the world around us. What materials are important to our economy? What chemical reactions take place in plants and animals and in our environment? What role do chemists play in protecting the environment? By tackling the principles leading to answers to these questions, you can come to a better understanding of nature and to an appreciation for consumer products. Indeed, one of the objectives of this book is to provide the tools needed for you to function as a chemically literate citizen. Learning something of the chemical world is just as important as understanding some basic mathematics and biology—and as important as having an appreciation for fine music and literature.

We are also pleased that this edition will again be offered with the *Saunders Interactive General Chemistry CD-ROM*. Computers have become a more and more powerful way to organize and convey information. The first edition of our CD-ROM has been used by thousands of students worldwide and is the most successful attempt to date to allow students to interact with chemistry. Additional interactivity will be available in an expanded version of the CD-ROM with the incorporation of *ActivChemistry*, software that allows you to design and perform simulated laboratory experiments. Finally, the disks again include portions of the molecular modeling tools from the Oxford Molecular Group. (Models of virtually every chemical compound mentioned in this book are contained in a library on the CD-ROM and can be viewed with the modeling software.) The CD-ROM package—two disks and an accompanying *Workbook*—is available for purchase with the textbook or as a stand-alone product.

The authors of this book became chemists because, simply put, it is exciting to discover new compounds and to find new ways to apply chemical principles. In this book, we hope we have conveyed that sense of enjoyment as well as our awe at what is known about chemistry, and, just as important, what is not known!

AUDIENCE FOR *CHEMISTRY & CHEMICAL REACTIVITY* AND THE *SAUNDERS INTERACTIVE GENERAL CHEMISTRY CD-ROM*

The textbook and CD-ROM are designed for introductory courses in chemistry for students interested in further study in science, whether that science is biology, chemistry, engineering, geology, physics, or related subjects. Our assumption is that students beginning this course have had some preparation in algebra and in general science. Although undeniably helpful, a previous exposure to chemistry is neither assumed nor required.

PHILOSOPHY AND APPROACH OF THE BOOK

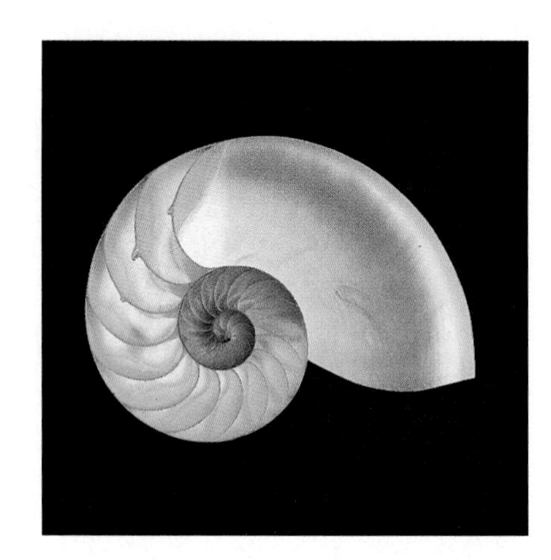

When the first edition of this book was planned, we had two major, but not independent, goals. This edition has these same goals. The first was to construct a book that students would enjoy reading and that would offer, at a reasonable level of rigor, chemistry and chemical principles in a format and organization typical of college and university courses today. Second, we wanted to convey the utility and importance of chemistry by introducing the properties of the elements, their compounds, and their reactions as early as possible and by focusing the discussion as much as possible on these subjects.

A glance at the introductory chemistry texts currently available shows that there is a generally common order of treatment of chemical principles used by educators. With a few minor changes we have followed that order as well. That is not to say that the chapters cannot be used in some other order. For example, although the behavior of gases is often studied early in a chemistry course, the chapter on this topic (Chapter 12) has been placed with chapters on liquids, solids, and solutions because it logically fits with these other topics. It can easily be read and understood, however, after covering only the first four or five chapters of the book.

The discussion of organic chemistry (Chapter 11) is typically left to one of the final chapters in chemistry textbooks. We believe, however, that the importance of organic compounds in biochemistry and in the chemical industry means that we should present that material earlier in the sequence of chapters. Therefore, it follows the chapters on bonding theories because organic chemistry illustrates well the application of models of chemical bonding and molecular structure.

In addition, one of the authors of this text often teaches much of the material on equilibria involving insoluble solids (Chapter 19) before acid-base equilibria (Chapters 17 and 18), and introduces kinetics (Chapter 15) and thermodynamics (Chapter 20) as a unit, after all of the material on equilibria. Although chapters are loosely organized into groups with common themes, we have made every attempt to make individual chapters as independent as possible.

The order of topics in the text was also devised to introduce as early as possible the background required for the laboratory experiments usually done in general chemistry courses. For this reason, chapters on chemical and physical properties, common reaction types, and stoichiometry begin the book. In addition, because an understanding of energy is so important in the study of chemistry, thermochemistry is introduced in Chapter 6.

The American Chemical Society has been urging educators to put "chemistry" back into introductory chemistry courses. As inorganic chemists, we agree wholeheartedly. Therefore, we have tried to describe the elements, their compounds, and their reactions as early and as often as possible in three ways. First, there are numerous color photographs of reactions occurring, of the elements and common compounds, and of common laboratory operations and industrial processes. Furthermore, we have tried to bring material on the properties of elements and compounds into the Exercises and Study Questions as early as possible and to introduce new principles using realistic chemical situations. In general, the descriptive chemistry of the elements has been treated in two ways. First, much has been woven into the book. Second, relevant highlights are given in Chapters 22 and 23 as a capstone to the principles described earlier.

Additionally, special sections called *Current Perspectives* and *A Closer Look* attempt to bring relevance and perspective to a study of chemistry. These include such topics as "Science and Flight 800" (The Nature of Chemistry); "Coral and Broken Bones" (Chapter

1); "Essential Elements" (Chapter 2); "Diet Soda—What's in It?" (Chapter 3); "What to Take for an Upset Stomach" (Chapter 4); "It's in the Bag—Thermodynamics and Consumer Products" (Chapter 6); "Magnetic Resonance Imaging" (Chapter 8); "Why Sweeteners are Sweet" (Chapter 9); "UV Radiation, Skin Damage, and Sunscreens" (Chapter 11); "The Chemistry of Survival" (Chapter 14); "Depletion of Stratospheric Ozone" (Chapter 15); and "Lead Pollution, Ancient and Modern" (Chapter 22).

ORGANIZATION OF THE BOOK

Chemistry & Chemical Reactivity is organized in two ways. First, there are chapters that are especially important in carrying the themes of the book. That is, there are chapters on the *Principles of Reactivity,* and there are others on *Bonding and Molecular Structure.*

The chapters on *Principles of Reactivity* are intended to introduce you to the factors that lead chemical reactions to successfully produce products. Thus, under this topic you will study common types of reactions, the energy involved in reactions, and the factors that affect the speed of a reaction.

The *Principles of Bonding and Molecular Structure* are particularly important. If you page through the book you will notice the abundance of molecular models, most of them drawn by computer. (They were done with programs from Oxford Molecular Group, and we thank them for providing the latest versions for us to use. Portions of this software are included on the CD-ROM and a model of virtually every molecule in the book is included on the CD-ROM.) As described in several places in the book [among them in *An Introduction—The Nature of Chemistry; Computer Molecular Modeling* (Chapter 3); and Chapters 9 through 11], an understanding of molecular structures is one cornerstone of modern chemistry. Using the latest laboratory techniques for uncovering molecular structures, and computer programs that generate revealing portraits of structures, chemists have enormous insight into the ways molecules react.

Second, the book is divided into five parts, each with a grouping of chapters with a common theme.

Part 1: The Basic Tools of Chemistry

Certain basic ideas and methods form the fabric of chemistry, and these are introduced in Part 1. Chapter 1 defines important terms and is a review of units and mathematical methods. Chapters 2 and 3 introduce basic ideas of atoms and molecules, and Chapter 2 introduces one of the most important organizational devices of chemistry, the periodic table. In Chapters 4 and 5 we begin to discuss the principles of chemical reactivity and to introduce the numerical methods used by chemists to extract quantitative information from chemical reactions. Chapter 6 is an introduction to the energy involved in chemical processes.

Part 2: The Structure of Atoms and Molecules

The major goal of this section is to outline the current theories of the arrangement of electrons in atoms and some of the historical developments that led to these ideas (Chapters 7 and 8). With this background, we can understand why atoms and their ions have different chemical and physical properties. This discussion is tied closely to the arrangement of elements in the periodic table so that these properties can be recalled and predictions made. In Chapter 9 we discuss for the first time how the electrons of atoms in a molecule lead to chemical bonding and the properties of these bonds. In addition, we show how to derive the three-dimensional structure of simple molecules. Finally, Chapter 10 considers the major theories of chemical bonding in more detail.

This part of the book is completed with a discussion of organic chemistry (Chapter 11), primarily from a structural point of view. Organic chemistry is such an enormous area of chemistry that we cannot hope to cover it in detail in this book. Therefore, we have focused on compounds of particular importance, including synthetic polymers, and the structures of these materials.

Part 3: States of Matter

The behavior of the three states of matter—gases, liquids, and solids—is described in that order in Chapters 12 and 13. The discussion of liquids and solids is tied to gases through the description of intermolecular forces, with particular attention given to liquid and solid water. Chapter 13 also considers the solid state, an area of chemistry currently undergoing a renaissance. In Chapter 14 we talk about the properties of solutions, intimate mixtures of gases, liquids, and solids.

Part 4: The Control of Chemical Reactions

This section is wholly concerned with the *Principles of Reactivity.* Chapter 15 examines the important question of the rates of chemical processes and the factors controlling these rates. With this in mind, we move to Chapters 16 through 19, a group of chapters that consider chemical reactions at equilibrium. After an introduction to equilibrium in Chapter 16, we highlight the reactions involving acids and bases in water (Chapters 17 and 18) and reactions leading to insoluble salts (Chapter 19). To tie together the discussion of chemical equilibria, we again explore thermodynamics in Chapter 20. As a final topic in this section we describe in Chapter 21 a major class of chemical reactions, those involving the transfer of electrons, and the use of these reactions in cells that produce a voltage.

Part 5: The Chemistry of the Elements and Their Compounds

Although the chemistry of the various elements has been described throughout the book to this point, Part 5 considers this topic in a more systematic way. Chapter 22 is devoted to the chemistry of the representative elements, whereas Chapter 23 is a discussion of the transition elements and their compounds. Finally, Chapter 24 is a brief discussion of nuclear chemistry.

PHILOSOPHY AND APPROACH OF THE *SAUNDERS INTERACTIVE GENERAL CHEMISTRY CD-ROM*

The CD-ROM was designed to take advantage of what computers do best: allow the user to *interact* with information. Therefore, our goal was to produce "an interactive movie about the book." The material in each chapter is presented in a series of "screens," each of which presents an idea or concept and allows the user to interact with the information in some manner—by seeing video of a reaction in progress, by changing a variable in a chemical experiment and watching what happens to the system, or by listening to important tips and ideas about ways to understand a concept or to solve a problem. In addition, you will see practicing chemists describe how the topic of the chapter applies to their work.

Version 2.5 of the CD-ROM will accompany the fourth edition of *CHEMISTRY & CHEMICAL REACTIVITY*. Its core is identical to Versions 2.0 and 2.1. No significant alterations have been made in the material, with the exception of the addition of *ActivChemistry*. This software package, which allows the student to design and perform simulated chemistry laboratory experiments, has been integrated into the screens of the previous version of the CD-ROM.

The CD-ROM is a complete learning environment for chemistry. In addition to the interactive presentation, which covers virtually the entire field of chemistry, it offers programs for plotting information, molecular modeling software from Oxford Molecular Group, and a library of the hundreds of molecular structures mentioned in this book, structures that can be viewed with this software.

NEW TO THIS EDITION

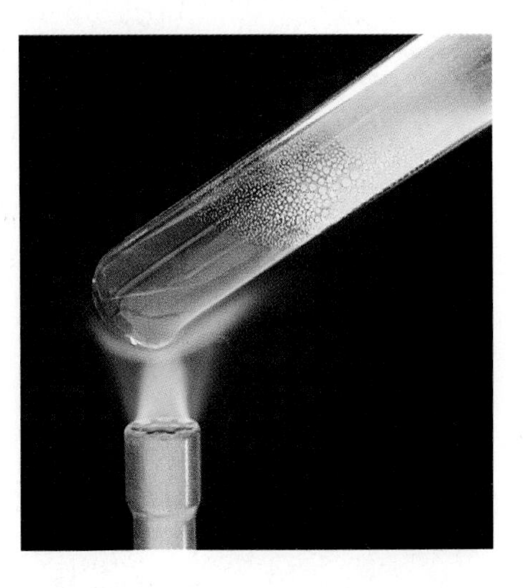

When we finish an edition of this book, we wonder what we can possibly do to change and improve the book. The answer is always that much can be done, and many changes have been made for this edition.

- Chapters 1 through 5 have been reorganized and completely rewritten. The flow of topics in Chapter 1 (*Matter and Measurement*) has been improved.
- Discussion of the concept of the "mole" has been consolidated in Chapter 3 (*Molecules and Compounds*).
- The core concept of chemical stoichiometry has been introduced in Chapter 4 (*Chemical Equations and Stoichiometry*) rather than in Chapter 5, as in the third edition.
- The material on types of chemical reactions has been reorganized in Chapter 5 (*Reactions in Aqueous Solution*). Particular attention has been paid to helping students learn to write net ionic equations.
- The discussion of ionic bonding has been consolidated in Chapter 9.
- The chapter on organic chemistry (Chapter 11) has been rewritten and reorganized to give even greater emphasis to chemical bonding and structure.
- The nature of the solution process (Chapter 14) has been rewritten to give more attention to the role of entropy.
- The important subject of catalysis has been moved forward in Chapter 15 (*Chemical Kinetics*) as an introduction to reaction mechanisms.

FEATURES AND LEARNING AIDS IN THE BOOK

Icons

This icon alerts the student to discussions that are focused on the relationship of macroscopic observations, the symbolic representation of those ideas, and their explanation at the particulate (atomic and molecular) level.

Marginal annotations tie the text to the appropriate section of the *Saunders Interactive General Chemistry CD-ROM.*

Molecular models in the book were created with software from the Oxford Molecular Group. This icon reminds the student that the models in that section are found on the *Saunders Interactive General Chemistry CD-ROM* and/or the *Saunders World Wide Web Site* (http://www.saunderscollege.com). Models of essentially all molecules, as well as ionic, molecular, network, and metallic solids, are found on the CD-ROM or on the Web site.

WWW This icon points out useful sites on the World Wide Web.

This icon points to a place earlier in the book where a relevant principle is described.

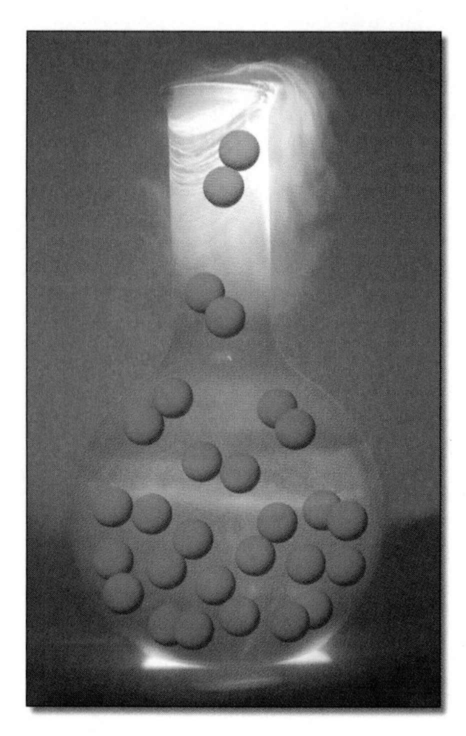

Problem Solving

Worked Examples and Solved Problems

Several hundred worked-out Examples serve as models for solving end-of-chapter problems. The detailed solutions are developed using the technique of dimensional analysis, and answers are highlighted. Examples are followed by similar Exercises, the solutions to which are in Appendix N.

Problem-Solving Flow Diagrams

Scattered throughout the book and in the Examples are problem-solving flow diagrams. These diagrams help organize the information contained in the problems.

End-of-Chapter Study Questions

The end-of-chapter Study Questions, some of which are illustrated with photographs and molecular models, include review questions, questions classified by type, and general questions. The classified problems are in matched pairs. Even-numbered questions have a bold-faced number, and the solution is given in Appendix O and in the *Student Solutions Manual.*

Conceptual Questions

These questions ask the student to think through the solution to a question or problem. Calculations are generally not involved.

Challenging Questions

Students can try their newly acquired skills on questions that involve more detailed calculations or an even deeper understanding.

Summary Questions

Summary questions tie together the concepts from the current chapter with those in previous chapters.

Questions from the Interactive General Chemistry CD-ROM

The workbook that accompanies the *Saunders Interactive General Chemistry CD-ROM* contains hundreds of questions that focus the student's attention on understanding chemistry. Some of these questions have been organized and rewritten to fit the topics in this book. Virtually all require the student to draw conclusions by observing an experiment or chemical process.

Problem-Solving Tips and Ideas

From years of teaching chemistry, we have found that students make certain errors in solving problems and that they have specific difficulties. These "Tips and Ideas" pass on our experience to the student.

Essays on History and Current Issues

Historical Perspective

These essays, about the lives of important scientists, both living and dead, provide some insight into the historical background of chemistry. They are intended to help the student see the human side of chemistry and to learn how ideas and concepts develop.

Current Perspectives

Scattered about the book are a number of essays that describe the applications of chemistry in the world today, from the current controversy over banning chlorine to the uses of buckyballs in medicine.

Illustrations

Full-Color Photography and Molecular Models

Over 600 full-color photos are included, each chosen specifically for this book. These photos illustrate common elements, compounds, and minerals as well as reactions and other processes in progress.

Molecular Models

Wherever appropriate we have included computer-generated models of the molecules involved in chemical reactions or that are being illustrated in a photograph. (See page xxix for the color scheme used in these models.) Most of these models are found on the *Saunders Interactive General Chemistry CD-ROM* and/or on the *Saunders Chemistry Web Site* (at http://www.saunderscollege.com). Software on the CD-ROM and software that can be downloaded from the World Wide Web allow the student to rotate these models, change the type of model (ball and stick or space-filling), and measure bond angles and bond distances. (See page xx for more on the *Saunders Chemistry Web Site.*)

Art

The art for this edition is almost entirely new. Color has been used to make the diagrams as attractive and meaningful as possible. In addition, color is used pedagogically.

- In the periodic tables in the text, main group metals are shown in red, transition metals in yellow, metalloids in gray, and nonmetals in green.
- A common color scheme is used in the illustration of molecular models (see page xxix of this text).

Other Features

- Each chapter ends with *Chapter Highlights*, a summary of the important concepts, equations, and key terms.
- Boxes titled "A Closer Look" delve deeper into topics closely related to the subject being discussed. They are meant to provide a more detailed discussion of a subject or a look at ideas used in chemical research.
- Appendices at the back of the book include a review of mathematical methods, a table of conversion factors, important constants, and a glossary of terms in the combined index/glossary. Inside the back cover are short tables of useful constants, a listing of the more useful data tables in the book, and a list of the "top 25" chemicals.

SUPPORTING MATERIALS FOR STUDENTS

Saunders Interactive General Chemistry CD-ROM, Version 2.5 with ActivChemistry, a multimedia companion to *CHEMISTRY & CHEMICAL REACTIVITY,* was originally designed by John Kotz and William Vining, University of Massachusetts, and produced by Archipelago Productions. Divided into chapters, which closely follow the organization of *CHEMISTRY & CHEMICAL REACTIVITY,* the CD-ROM presents ideas and concepts with which the user can interact. One can watch a reaction in progress, change a variable in an experiment and experience the result, follow stepwise solutions to problems, explore the periodic table, and listen to tips and suggestions on problem solving and understanding concepts. The CD-ROM includes original graphics, over 100 video clips of chemical experiments, which are enhanced by sound and narration, and several hundred molecular models and animations.

The CD-ROM also includes molecular modeling software from Oxford Molecular Group that can be used to view hundreds of models, rotate the models for a fuller understanding of their structures, and measure bond lengths and bond angles.

The CD-ROM has been used by thousands of students worldwide since its introduction in 1996. It can be purchased as a package with the textbook or as a stand-alone product.

The **World Wide Web Site** includes molecular models, on-line activities, problem-solving, quizzing, and information about chemistry, to name just a few of the things that can be found at http://www.saunderscollege.com. Tap into the Math Review section to brush up on math skills.

Pocket Guide by John DeKorte, Glendale Community College, contains useful summaries of each text section, as well as helpful problem-solving reminders and tips.

Student Study Guide by Paul Hunter, Michigan State University, has been designed around the key objectives of the book. It provides section summaries, review questions and answers, study questions and problems with answers, crossword puzzles, and a sample test at the end of each of the five parts of the book.

Student Solutions Manual by Alton Banks, North Carolina State University, contains detailed solutions to the even-numbered, end-of-chapter Study Questions.

Student Lecture Outline by Ronald Ragsdale, University of Utah, is an aid in organizing the material in the text.

SUPPORTING MATERIALS FOR INSTRUCTORS

Instructor's Resource Manual by Susan Young, Hartwick College, suggests alternative organizations of the course, classroom demonstrations, and worked-out solutions to odd-numbered, end-of-chapter Study Questions.

PowerPoint™ Slide Presentation comprises several years of class lectures by author John Kotz. Hundreds of slides that cover the entire year of introductory chemistry have been created for lecture presentations. They use the full power of PowerPoint™ and incorporate videos, animations, and photos from the *Saunders Interactive General Chemistry CD-ROM.* Professors can customize their lecture presentations by adding their own slides or by deleting or changing existing slides. The PowerPoint™ files are available on the Saunders Chemistry World Wide Web Site.

The **World Wide Web Site** also includes an Instructor's Section that gives professors access to lecture notes and chapter-by-chapter outlines, among other things. The PowerPoint™ slides are available in this section.

The **1999 Instructor's Resource CD-ROM** is a dynamic lecture tool containing imagery from Saunders 1999 chemistry titles. It can be used in conjunction with commercial presentation software such as PowerPoint™, Persuasion™, and Podium™. The CD-ROM is for both Macintosh and Windows platforms.

The **Overhead Transparency Set** is a collection of 150 full-color transparencies with sizable labels for viewing in large lecture halls. The illustrations chosen are those most often used in the classroom, and most are marked with an icon in the Instructor's Annotated Edition of the text.

A **Test Bank** by Ronald O. Ragsdale, University of Utah, contains over 1100 new, multiple-choice questions and numerous fill-in questions for each chapter.

ExaMaster+™ Computerized Test Bank is the software version of the printed Test Bank. Instructors can create thousands of questions in a multiple-choice format. A command reformats the multiple-choice questions into short-answer questions. Problems can be added or modified, and graphics can be added. ExaMaster can be used to record and graph student grades. Available in both Macintosh and Windows formats.

Chemical Principles in the Laboratory by Emil Slowinski and Wayne Wolsey of Macalester College, and William Masterton of the University of Connecticut, provides detailed directions and study assignments. The manual contains 42 experiments that have been thoroughly class-tested and selected with regard to safety and cost. An Instructor's Manual provides lists of equipment and chemicals needed for each experiment.

CalTech Chemistry Animation Project (CAP) is a set of six video units that cover the chemical topics of Atomic Orbitals, Valence Shell Electron Pair Repulsion Theory, Crystals and Unit Cells, Molecular Orbitals in Diatomic Molecules, Periodic Trends, and Hybridization and Resonance.

Periodic Table Videodisc: Reactions of the Elements by Alton Banks, North Carolina State University, features still and live footage of the elements, their uses, and their reactions with air, water, acids, and bases. Available to qualified adopters. Also available in CD-ROM format through JCE:Software, Chemistry Department, University of Wisconsin, Madison, WI 53706, (800) 991-5534.

Shakhashiri Chemical Demonstration Videotapes feature Bassam Shakhashiri of the University of Wisconsin-Madison performing 50 three- to five-minute chemical demonstrations. An accompanying Manual describes each demonstration and includes discussion questions.

Other Supporting Materials

The marginal notations in the Instructor's Annotated Edition of the book list a number of resources. Sources of some of these materials are:

Videodiscs

Chem Demos I and II and *The World of Chemistry—Selected Demonstrations and Animations,* Volumes I and II, are available from JCE:Software, Chemistry Department, University of Wisconsin, Madison, WI 53706, (800) 991-5534.

Chemical Demonstrations

B. Z. Shakhashiri, ***Chemical Demonstrations—A Handbook for Teachers of Chemistry,*** University of Wisconsin Press, Madison, WI. Four volumes have been published.

L. R. Summerlin and J. L. Ealy, Jr., ***Chemical Demonstrations—A Sourcebook for Teachers,*** Volume 1, American Chemical Society, Washington, D.C., 1988.

L. R. Summerlin, C. L. Borgsford, and J. B. Ealy, ***Chemical Demonstrations—A Sourcebook for Teachers,*** Volume 2, American Chemical Society, Washington, D. C., 1988.

Saunders College Publishing may provide complimentary instructional aids and supplements or supplement packages to those adopters qualified under our adoption policy. Please contact your sales representative for more information. If as an adopter or potential user you receive supplements you do not need, please return them to your sales representative or send them to:

Attn: Returns Department
Troy Warehouse
465 South Lincoln Drive
Troy, MO 63379

Acknowledgments

Preparing the fourth edition of *Chemistry & Chemical Reactivity* took almost eighteen months of continuous effort. However, as in our work on the first three editions, we have had the support and encouragement of our families and of some wonderful friends, colleagues, and students.

SAUNDERS COLLEGE PUBLISHING

The editorial staff of Saunders College Publishing has once again been extraordinarily helpful. The project has benefitted from their good humor, friendship, and dedication. Much of the credit goes to our Publisher, John Vondeling. We have worked with John for many years and have become fast friends. His support and confidence are greatly appreciated. Not only does he understand publishing, but he knows a thing or two about a good trout stream.

The Developmental Editors for this edition were Beth Rosato and Sarah Fitz-Hugh. In addition to being very pleasant colleagues, they kept the project organized and focused. In addition, they were trusted friends and confidants. We greatly appreciate their efforts to make this a successful book.

Our Project Editor for this edition was Robin Bonner. Her attention to detail, her energy, and her enthusiasm for the project will surely help make this edition as successful as previous editions.

No book can be successful without proper marketing. Pauline Mula was new to this edition, and she was a delight to work with. She was knowledgeable about the market and worked tirelessly to bring the book to everyone's attention.

Caroline McGowan was in charge of the art and design program at Saunders College Publishing. Her assistance was invaluable in helping us learn how to produce art for the book, in helping to ensure that we had chosen appropriate materials, and in designing the final product. All agree it is a beautiful book.

Our team at Saunders College Publishing is completed with Emily Barrosse, Vice President and Publisher, and Charlene Squibb, Production Manager. We appreciate their patience, creativity, and organizational skills.

PHOTOGRAPHY AND ART

Most of the color photographs for this edition were again beautifully done by Charles D. Winters of Oneonta. He produced dozens of new images for this book, often under great deadline pressure. His work gets better and better, and he has added a new dimension by using software tools to combine photographs with molecular models, often with stunningly beautiful results. We have worked with Charlie for some years and have become close friends. We listen to his jokes, both new and old—and always forget them. When we finish the book, we look forward to a kayaking trip.

We are very pleased to have Susan Young back in Oneonta, New York, where she is now on the faculty at Hartwick College. Susan produced the molecular models for the book using software from the Oxford Molecular Group and worked with the authors and Charlie Winters to design the models for the composite illustrations. In addition, she read the manuscript, the galley and page proofs, and generally provided good advice. We simply could not have done this project without Susan's help, creativity, good humor, and energy.

OXFORD MOLECULAR GROUP

Several years ago CAChe Scientific, Inc. made a grant of a Molecular Modeling Worksystem to author John Kotz. This software has been used heavily by students at Oneonta in general and inorganic chemistry courses. More recently, the Oxford Molecular Group made a gift of Personal CAChe systems to John Kotz and Charles Winters. These systems were used to prepare the molecular models for this book and the CD-ROM. The people at CAChe—especially George Purvis and Leo Brown—have been extremely helpful, and we wish to acknowledge their support with gratitude.

OTHERS

Bill Vining, a former student of John Kotz and now a member of the Chemistry Department at the University of Massachusetts, has contributed creatively to this and the previous edition. Bill and John Kotz collaborated on the CD-ROM and wrote the *Workbook* that accompanies the CD-ROM set. Questions on the CD-ROM for this book have been drawn from the *Workbook,* and we wish to acknowledge Bill's contribution.

Publishing a book is a complicated process, and a large team of people is needed to carry out the task. At least one more member of our team deserves special thanks: Katie Kotz kept the work in Oneonta, New York, and Madison, Wisconsin—photography, text preparation, and photo research—organized. Her organizational skills and her expertise in maintaining a large database of information have been invaluable. In addition, she has been the wonderful wife of one of the authors for 37 years.

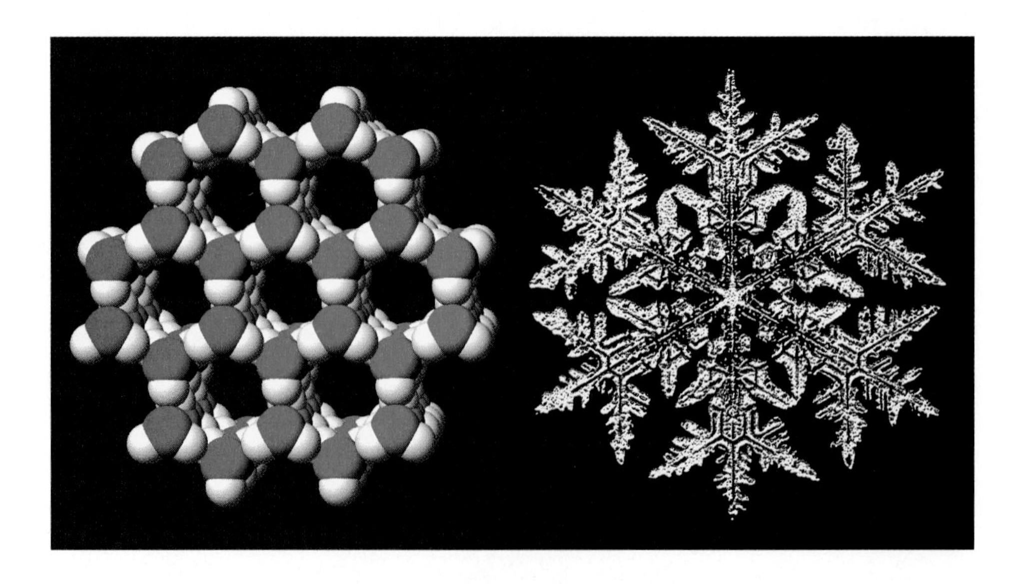

Reviewers

We believe the success of any book is due in no small way to the quality of the reviewers of the manuscript. The reviewers of previous editions made important contributions that are still part of this book. Reviewers of the fourth edition continued that tradition. We wish to acknowledge with gratitude the efforts of all those listed below. Several should be noted in particular, however. Gary Riley and Steve Landers worked all of the problems in the book. We want to add a special thanks to John DeKorte, who played a major role in the final development of the manuscript.

Alton Banks
North Carolina State University

John DeKorte
Glendale Community College

L. Peter Gold
Pennsylvania State University

Paul Hunter
Michigan State University

Donald Kleinfelter
University of Tennessee—Knoxville

Michael E. Lipschutz
Purdue University

Mark E. Noble
University of Louisville

Lee G. Pedersen
University of North Carolina, Chapel Hill

Ronald O. Ragsdale
University of Utah

Wayne Tikkanen
California State University, Los Angeles

John B. Vincent
University of Alabama

Reviewers of the Third Edition

Lester Andrews
University of Virginia

Jeffrey R. Appling
Clemson University

Caroline L. Ayers
East Carolina University

Cindy A. Burkhardt
Radford University

Michael P. Castellani
Marshall University

Geoffrey Davies
Northeastern University

Randall Davy
Liberty College

Dan Decious
California State University, Sacramento

John DeKorte
Glendale Community College

Karen E. Eichstadt
Ohio University

Kevin Grundy
Dalhousie University

Mary Gurnee
Embry-Riddle Aeronautical University

Suzanne Harris
University of Wyoming

Jerry P. Jasinski
Keene State College

Martin Kellerman
California Polytechnical State University, San Luis Obispo

Christine S. Kerr
Montgomery College, Rockville Campus

D. Whitney King
Colby College

Donald Kleinfelter
The University of Tennessee—Knoxville

Robert M. Kren
The University of Michigan—Flint

Joan Lebsack
Fullerton College

David E. Marx
University of Scranton

William H. Myers
University of Richmond

Frank A. Palocsay
James Madison University

Pete Poston
Western Oregon State College

Robert A. Pribush
Butler University

Nancy C. Reitz
American River College

Vic Shanbhag
Mississippi State University

Alka Shukla
Houston Community College—Southeast Branch

Saul I. Shupack
Villanova University

William E. Stanclift
Northern Virginia Community College

Conrad Stanitski
The University of Central Arkansas

Juan F. Villa
Herbert H. Lehman College of The City University of New York

Wayne Wesolowski
Illinois Benedictine College

Christopher J. Willis
The University of Western Ontario

Laura Yeakel
Henry Ford Community College

Reviewers of the Second Edition

Tom Baer
University of North Carolina at Chapel Hill

Muriel Bishop
Clemson University

Edward Booker
Texas Southern University

Donald Clemens
East Carolina University

Michael Davis
University of Texas, El Paso

Carl Ewig
Vanderbilt University

Russell Grimes
University of Virginia

Anthony W. Harmon
University of Tennessee—Martin

Alan S. Heyn
Montgomery College

Lisa Hibbard
Spelman College

Mary Hickey
Henry Ford Community College

William Jensen
South Dakota State University

Ronald Johnson
Emory University

Stanley Johnson
Orange Coast College

Lenore Kelly
Louisiana Tech University

Paul Loeffler
Sam Houston State University

Brian McGuire
Northeast Missouri State University

Jerry Mills
Texas Tech University

Mark Noble
University of Louisville

Lee G. Pedersen
University of North Carolina, Chapel Hill

Chester Pinkham
Tri-State University

Steve Ruis
American River College

Jerry Sarquis
Miami University

Steven Strauss
Colorado State University

Larry C. Thompson
University of Minnesota, Duluth

Milt Wieder
Metropolitan State College

Reviewers of the First Edition

Bruce Ault
University of Cincinnati

Alton Banks
Southwest Texas State University

O.T. Beachley
SUNY—Buffalo

Jon M. Bellama
University of Maryland

James M. Burlitch
Cornell University

Geoffrey Davies
Northeastern University

Glen Dirreen
University of Wisconsin

John M. DeKorte
Northern Arizona University

Darrell Eyman
University of Iowa

Lawrence Hall
Vanderbilt University

James D. Heinrich
Southwestern College

Forrest C. Hentz
North Carolina State

Marc Kasner
Montclair State College

Philip Keller
University of Arizona

Herbert C. Moser
Kansas State University

John Parson
Ohio State University

Lee G. Pedersen
University of North Carolina

Harry E. Pence
SUNY—Oneonta

Charles Perrino
California State University (Hayward)

Elroy Post
University of Wisconsin—Oshkosh

Ronald Ragsdale
University of Utah

Eugene Rochow
Harvard University

Steven Russo
Indiana University

Charles W.J. Scaife
Union College

George H. Schenk
Wayne State University

Peter Sheridan
Colgate University

Kenneth Spitzer
Washington State University

Donald D. Titus
Temple University

Charles A. Trapp
University of Louisville

Trina Valencich
California State University, Los Angeles

A Note to Students

ON THE SYSTEM OF SYMBOLS AND COLORS USED IN THIS BOOK

Modern chemical evaluation of our world and its physical changes involves three levels of observation and representation. This book centers on the close interrelationship of these levels—macroscopic, symbolic, and particulate—to which the book's hundreds of photos paired with molecular models bear witness.

No matter what the level of your awareness of chemistry, you are going to experience chemical phenomena at the **macroscopic level.** That is, you see chemical processes with your eyes. To be able to report those experiences to one another, chemists use symbols; that is, we represent chemical phenomena at the **symbolic level.** Finally, to understand what has happened, we describe observations in terms of particles called atoms and molecules. That is, we represent chemistry at the **particulate level.**

This special icon will denote instances where we wish to draw particular attention to the relation between the macroscopic, symbolic, and particulate views of a chemical process.

These three types of representation can be thought of as the points of a triangular matrix—discussions of chemistry occur using a combination of each of the three forms of representation. *This is an important organizing principle, and we shall use it throughout this book.* A special icon will be placed in the margin to call attention to the different types of representations.

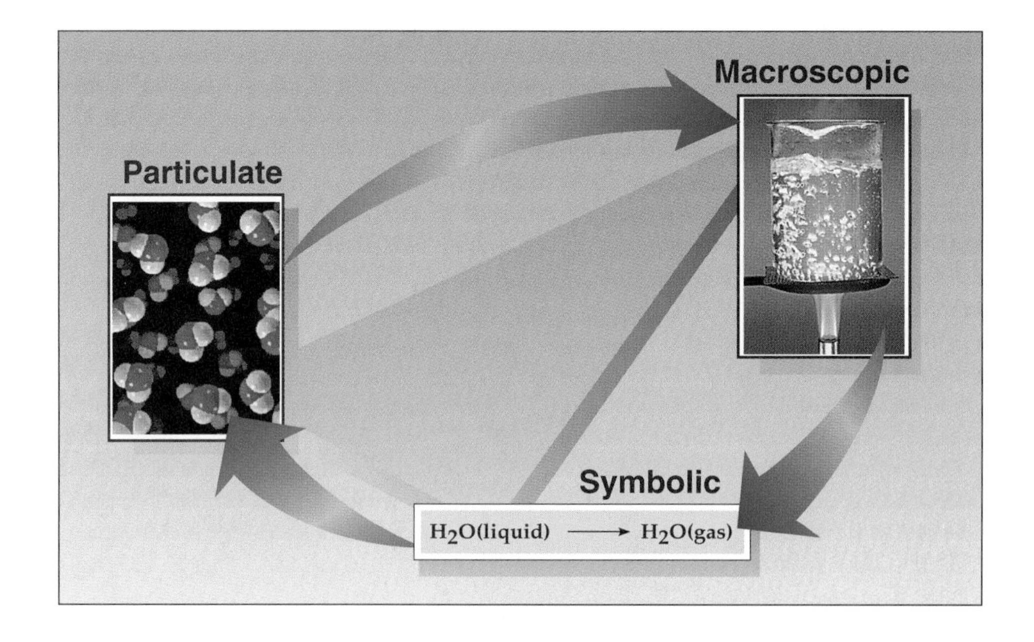

Color

Chemistry makes standard use of color to represent particular elements and groups of elements. In the periodic tables used in this book, the main group metals are shown in red, the transition metals in yellow, the metalloids in gray, and the nonmetals in green (see inside the front cover).

We have also used a common color scheme in computer-generated molecular models, as illustrated here.

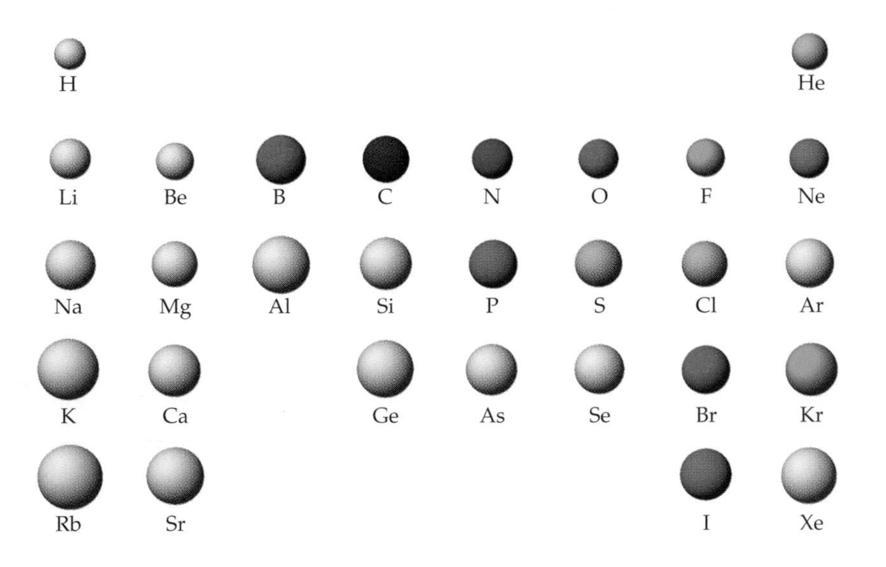

A Note to Faculty and Students Who Will Use This Book

There are almost as many ways to teach chemistry properly as there are faculty members in the chemistry departments across North America. We make no pretense that we have found the best way of organizing chemistry, of explaining a principle, or of solving a problem. Further, even though we have put our best efforts into creating an error-free book, errors will surely be found. Therefore, if you find a way to explain something more clearly, a better way to demonstrate a reaction or a principle, or errors in our discussions, we hope you will feel free to write to us or to someone at Saunders College Publishing; they will see that we get your letter. Many specific things can be corrected in subsequent printings of the book, and more general ideas can be incorporated in any editions of this text that may follow. And don't be surprised if we send you an email to find out more about your ideas.

A final word to students using the book. We believe chemistry is a challenging, exciting, and worthwhile area of study, and we hope you agree. However, like anything worthwhile, it does take some work to understand the subject. As you work through the book, just remember the encouraging words of Dr. Seuss from his book "Oh, the Places You'll Go" (Theodor S. Geisel and Audrey S. Geisel, Random House, New York, 1990):

> "Onward up many
> a frightening creek
> though your arms may get sore
> and your sneakers may leak."

JOHN C. KOTZ
KotzJC@oneonta.edu

PAUL M. TREICHEL
Treichel@chem.wisc.edu

About the Authors

John C. Kotz received his Ph.D. from Cornell University in 1964 and currently teaches chemistry at the SUNY College at Oneonta, where he was promoted to University Distinguished Teaching Professor in 1986. In 1979, he was Fulbright Lecturer and Research Scholar in Lisbon, Portugal. He has received the National Catalyst Award in Chemical Education from the Chemical Manufacturers Association. Dr. Kotz has been an editor of *ChemMatters* magazine and is on the board of editors of the *Journal of Chemical Education: Software.* He is also the co-author of two inorganic chemistry textbooks and another Saunders introductory general chemistry text, *The Chemical World: Concepts and Applications.* Dr. Kotz welcomes any questions or comments you may have and can be reached at his email: KotzJC@oneonta.edu.

Paul Treichel, Jr. received his B.S. from the University of Wisconsin-Madison in 1958 and a Ph.D. from Harvard University in 1962. After a year of postdoctoral study at Queen Mary College in London, he assumed a faculty position at the University of Wisconsin-Madison, where he has taught general chemistry and inorganic chemistry for 32 years and served as Department Chair from 1986 to 1996. Dr. Treichel's research in organometallic chemistry, aided by 75 graduate and undergraduate students, has resulted in the publication of more than 160 articles in scientific journals. Dr. Treichel would like to receive your comments or questions. He may be reached at Treichel@chem.wisc.edu.

Introduction: The Nature of Chemistry

Many plants contain chemicals that have medicinal use. The bark of white willows, for example, contains salicylic acid, a chemical compound related to aspirin, a model of which is shown here. *(Photo, C. D. Winters; model, S. M. Young)*

See Roald Hoffmann and Vivian Torrance: *Chemistry Imagined, Reflections on Science,* Washington, D.C., Smithsonian Institution Press, 1993.

Chemistry is about change. It was once only about changing one substance into another—wood and oil burn, grape juice turns into wine, and cinnabar, a red mineral from the earth, changes into shiny quicksilver. It is still about change, but now chemists focus on the change of one molecule into another (Figure 1).

Although chemistry is endlessly fascinating—at least to chemists—why should you study chemistry? Each person probably has a different answer, but many of you may be taking this chemistry course because someone else has decided it is an important part of preparing for a particular career. Chemistry is especially useful because it is central to our understanding of disciplines as diverse as biology, geology, materials science, medicine, physics, and many branches of engineering. In addition, chemistry plays a major role in our economy; chemistry and chemicals affect our daily lives in a wide variety of ways. Furthermore, a course in chemistry can help you see how a scientist thinks about the world and how to solve problems. The knowledge and skills developed in such a course will benefit you in many career paths and will help you become a better informed citizen in a world that is becoming technologically more complex—and more interesting. Therefore, to begin your study of chemistry, this chapter discusses some fundamental ideas used by chemists. It also introduces you to a problem confronting chemists, and indeed all citizens, almost every day—the weighing of the benefits of a discovery or a practice against its risks to individuals and society.

(a)

(b)

(c)

Figure 1 Chemistry is the study of atoms and molecules, and their transformations. Here metallic aluminum and orange-brown, liquid bromine (a) combine with each other (b), and are transformed into white, solid aluminum bromide (c). (*Photos, C. D. Winters; models, S. M. Young*)

TOO MUCH OF A GOOD THING?

The earth's air temperature has risen 0.3 to 0.6 °C over the past century. Many climate experts believe this is evidence that human beings are perturbing the earth's climate, primarily from the burning of fossil fuels. Combustion of wood, coal, and oil produces carbon dioxide, CO_2, which is called a "greenhouse" gas because it helps trap the earth's heat within the atmosphere. Because carbon dioxide levels have risen 30% since preindustrial times, climate experts say that global warming and carbon dioxide levels are connected.

Experiments have been done in which iron is added to the oceans because iron promotes the growth of phytoplankton, which in turn live on atmospheric carbon dioxide. But, do we really want to tinker with the ecology of the world's oceans? See R. Monastersky: "Iron versus the greenhouse." *Science News*, Sept. 30, 1995, p. 220.

The realization that global warming is likely to bring unwanted ecological changes has led some scientists to consider ways to remove some CO_2 from the atmosphere. We know that carbon dioxide is taken up by green plants and turned into the carbon-based molecules of the plant kingdom. So, one way to decrease atmospheric carbon dioxide might be to promote the growth of green plants. Not surprisingly, this experiment was done recently, but, as we will see shortly, the results were not what was expected.

Farmers have known for centuries that putting bone chips, manure, and wood ashes on crop lands will boost plant yields. Only relatively recently, however, was it known that these provided such important nutrients as phosphorus, potassium, and nitrogen. "Natural fertilizers" were used to provide these nutrients well into the 19th century, but agriculturists realized that the world's supply of "fixed" nitrogen was not keeping up with the demand. There could be a worldwide crisis in agriculture if something was not done.

By 1909 Fritz Haber, a German chemist, had accomplished the incredible feat of using nitrogen in the air to produce ammonia, NH_3. Using the Haber process, approximately 17 million tons of ammonia are now made annually in the United States, slightly less than that produced by China, the global leader. About one-third of this is applied directly to the soil as fertilizer (Figure 2), and about half is turned into other nitrogen-containing fertilizers. Intensive use of manufactured fertilizers has had an enormous impact on the world's economy and is largely responsible for the "green revolution" of the past few decades.

Nitrogen-based fertilizers have fueled the green revolution, but, like all other beneficial advances, they are not without problems. Researchers in Minnesota and Canada recently found that, in the short term, nitrogen-containing fertilizers spur plant growth and cause plants to use more atmospheric carbon dioxide to incorporate, or "fix," more carbon in their tissues. The long-term problem is that added nitrogen encourages fast-growing "invasive species," or weeds, that are less efficient at fixing carbon. Furthermore, the level of nitrate salts in the soil from fertilizers and natural processes increases, and these can end up in rivers, streams, and lakes and contribute to poor water quality (Figure 3). In fact, a panel of experts in biogeochemistry has stated recently that synthetic fertilizers, farm crops, and the burning of coal and oil have doubled the rate at which atmospheric nitrogen is fixed or converted to a form that can be used by living things. The result is that environmental disruption caused by a planetary overload of nitrogen is emerging as a new worldwide concern.

***Figure* 2 Ammonia gas is "drilled" into the soil of a farm field.** Most of the ammonia manufactured in the world is used as fertilizer because ammonia supplies the nitrogen needed by green plants. (*Arthur C. Smith III, from Grant Heilman Photography*)

Although much of the world's ammonia is used directly for fertilizer, a significant amount is used to produce ammonium nitrate, NH_4NO_3. Although most of this is also used as a fertilizer, about one quarter is used in explosives. That this compound could be used as an explosive was illustrated dramatically on April 16, 1947, when a ship being loaded with fertilizer blew up in Texas City, Texas,

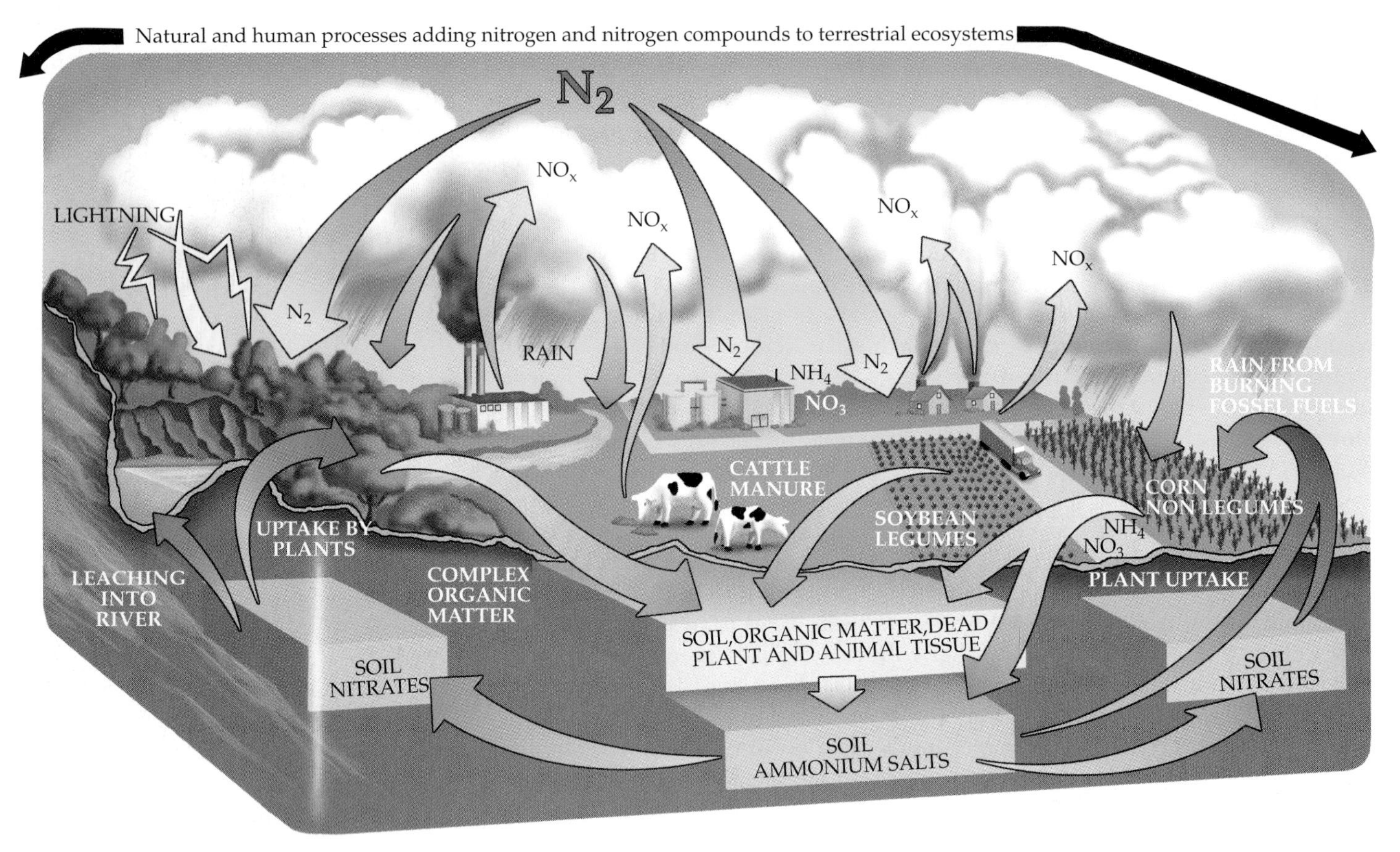

Figure 3 Nitrogen circulation in the atmosphere and soils. The nitrogen in our environment is mostly in the form of gaseous nitrogen, N_2. It is turned into usable forms, such as ammonium and nitrate salts, by natural chemical processes, largely by nitrogen-fixing plants like legumes. Such plants have bacteria in their root nodes that can use atmospheric N_2. Nitrogen taken up by plants is recycled through soils by decay, and some is returned to the atmosphere. Humans have increased the supply of usable nitrogen by raising more and more cattle, planting nitrogen-fixing crops, and using fertilizers. Excess nitrogen-containing compounds leach out of the soil, however, into ground water and can damage water supplies, rivers, and coastal wetlands.

killing 576 people and injuring nearly 5000 more! Ammonium nitrate had long been used as a component of dynamite, according to the original formula developed by Alfred Nobel. The Texas ship disaster, however, led engineers to realize that ammonium nitrate alone could be used as an effective blasting agent.

Today, a mixture of ammonium nitrate and fuel oil is used in 95% of all commercial blasting for mining and road building in the United States. Such explosives can unfortunately be put to less peaceful uses, however. They were used, for example, in the bombing of the World Trade Center in New York in 1994 and the federal building in Oklahoma City in 1995.

Chemistry and Ecology

The ecology of our planet is complex and endlessly fascinating. It is clear that chemistry plays an intimate role in processes that occur within the living cell and on a global level (see Figure 3). To better understand these processes, we all need to understand some of the basic concepts of chemistry. We also need to know enough about chemistry to realize what is not known or understood. For example, it is not completely understood how simple molecules of nitrogen, N_2, the most abundant component of the air we breathe, are converted, or "fixed," by plants to form ammonia.

As we help you study the basic concepts of chemistry, we hope we can also give you an appreciation not only for what is known but also for what is not yet understood. Chemistry is a dynamic subject, extending to the very frontier of knowledge. Chemists know enough about the fundamental chemistry of the elements and important compounds, and about atomic and molecular structure, to be able to address truly important questions in areas such as environmental chemistry, oceanography, geochemistry, materials chemistry, and biochemistry.

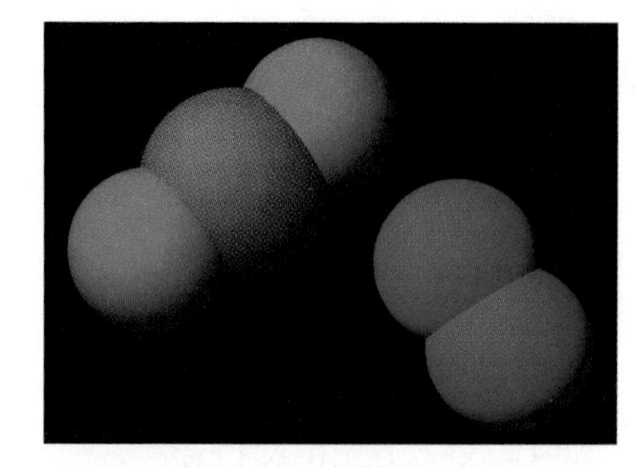

Models of two important molecules of the atmosphere: carbon dioxide and nitrogen. Here the carbon atom is gray, oxygen is red, and nitrogen is blue.

SCIENCE AND ITS METHODS

Among the nitrogen-containing substances in living systems is DNA—deoxyribonucleic acid—a very large molecule that contains nitrogen, carbon, hydrogen, oxygen, and phosphorus and that is found in the nucleus of every cell in your body. The way that the basic structure of DNA was uncovered, and its use in criminal and other legal issues, illustrate the way science works.

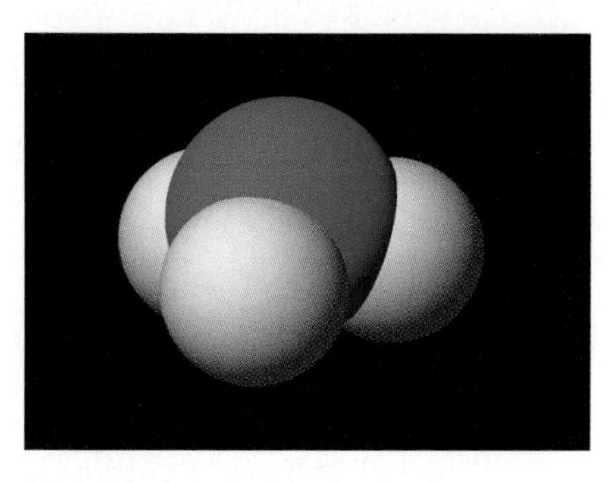

A model of ammonia, a molecule composed of nitrogen (blue) and hydrogen (white) atoms.

Hypotheses, Laws, and Theories

As scientists we study questions of our own choosing or ones that someone else poses in the hope of finding an answer or of discovering some useful information. Although it may not seem so as you begin to study chemistry, it is easy to conceive an idea to study. The goal should be to state clearly a problem that is worth studying and that is also narrow enough in scope that a useful conclusion may be reached. James Watson and Francis Crick did just that in the 1950s when they set out to understand the structure of DNA, a story well told in Watson's book *The Double Helix* and outlined on pages 6 and 7.

Having posed a reasonable question, you should first look at the work that others have already done so that you have some notion of the most productive direction to take. Watson and Crick did this by reading papers published by Linus Pauling, among others, and by talking with Maurice Wilkins and Rosalind Franklin, colleagues who were also studying DNA. They then proceeded to the next step, forming a **hypothesis,** a tentative explanation or prediction of experimental observations.

After formulating one or more hypotheses, scientists perform experiments that are designed to give results that may confirm some hypotheses and invalidate others. In chemistry this usually requires that both quantitative and qualitative information be collected. **Quantitative** information usually means numerical data, such as the temperature at which a chemical substance melts. **Qualitative** information, in contrast, consists of nonnumerical observations, such as the color of a substance or its physical appearance. Watson and Crick built physical models of parts of DNA to see how the atoms would fit together and if the arrangement matched the experimental data they obtained from Wilkins and Franklin.

Richard Feynmann demonstrated the scientific method in his brilliantly simple experiment that showed the likely cause of the explosion of the Space Shuttle *Challenger* in 1986. See Screen 2 of the Introductory Chapter of the *Saunders Interactive General Chemistry CD-ROM.*

Historical Perspective

The Double Helix

The structure of DNA was uncovered by James D. Watson, Francis Crick, and Maurice Wilkins, who shared the 1962 Nobel Prize in medicine and physiology for their work. It was one of the most important scientific discoveries of this century and has opened the way to rapid progress in molecular biology.

The tale of the discovery of DNA's structure has been told by Watson in his book *The Double Helix.* When he was a graduate student at Indiana University, Watson had an interest in the gene and said he hoped that its biological role might be solved "without my learning any chemistry." Later, however, he and Crick found out just how useful chemistry can be when they used a classic text on organic compounds, and their imaginations, to make one of the major discoveries of this century.

Solving important problems requires teamwork among scientists of many kinds, so chemists often work with colleagues around the world. Therefore, Watson went to Cambridge University in England in the fall of 1951. There he first met

James D. Watson (*left*) and Francis Crick (*right*), together with Maurice Wilkins, received the Nobel Prize in medicine and physiology in 1962. (*A. Barrington Brown/Science Source/Photo Researchers, Inc.*)

Francis Crick, who, Watson said, talked louder and faster than anyone else. Crick shared Watson's belief in the fundamental importance of DNA, and they soon learned that Maurice Wilkins and Rosalind Franklin at King's College in London were using the technique of x-ray crystallography in the hope of learning more about DNA's structure. Watson and

Rosalind Franklin of King's College, London. Her work was important to the model of DNA developed by Watson and Crick. She died in 1958 at the age of 37, however, and so did not share in the Nobel Prize because the prizes are never awarded posthumously. Although the relation between Watson, Crick, and Franklin was initially strained, Watson later said that "we later came to appreciate . . . the struggles the intelligent woman faces to be accepted by a scientific world which often regards women as mere diversions from serious thinking." (*Vittorio Luzzati/Centre National de Génétique Molèculaire*)

Crick believed that understanding the structure was crucial to understanding genetics. To solve the structural problem, however, they needed experimental data,

---

Preliminary experiments usually lead to the revision and extension of the original hypothesis. This in turn creates the need for new experiments. After a number of experiments have been done, and the results checked to ensure they are reproducible, a pattern of behavior or results may emerge. At this point it may be possible to summarize the observations in the form of a general rule. Finally, after numerous experiments by many scientists over an extended period of time, the original hypothesis may become a **law**—a concise verbal or mathematical statement of a relation that seems always to be the same under the same conditions.

We base much of what we do in science on laws because they help us predict what may occur under a new set of circumstances. For example, we know from experience that if we allow the chemical element sodium to come in contact with the chemical compound water, a violent reaction occurs and several new chemical compounds are formed (Figure 4). But the result of an experi-

***Figure* 4 The metallic element sodium reacts vigorously with water.** It is a law that the metals lithium, sodium, and potassium, among others, react vigorously with water. To explain this law, a chemist would refer to current theories of the arrangement of electrons in the atom. (*C. D. Winters*)

The threads of a drill or screw twist along the axis of a helix, and some plants also climb by sending out tendrils that twist helically. (*C. D. Winters*)

and the type of information they needed could come from the experiments at King's College. Franklin had experimental data, but, to Watson and Crick, she seemed unwilling to share it. To compound the problem, the experiments were difficult to perform, and Watson and Crick had neither the expertise nor the equipment to do them. And finally, it seemed to Watson and Crick that the King's College scientists did not really appreciate the significance of their work. So Watson and Crick were faced with a dilemma: How could they convince the King's College group to share their data, and, more generally, was it ethical to work on a problem that others had claimed as theirs. "The English sense of fair play would not allow Francis to move in on Maurice's problem," said Watson.

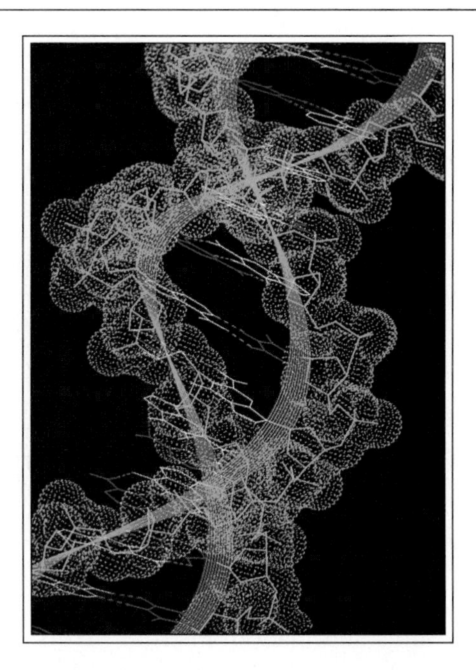

A model of DNA. The chains, which consist of atoms of oxygen and phosphorus, together with sugar molecules, twist in a double helix. The chains are joined by molecules called nucleotide bases. (*Will & Deni McIntyre/Photo Researchers, Inc.*)

Watson and Crick knew from the beginning that the overall structure of DNA was a helix, that is, the molecule twisted in space like the threads of a screw or the strands of a grapevine. They also knew what chemical elements it contained and roughly how they were grouped together. What they did not know was the detailed structure of the helix. Exactly how were the atoms arranged? By the spring of 1953 they had the answer. DNA is a double helix of chains, each chain consisting of oxygen and phosphorus atoms along with sugar molecules. Attached to each chain are a series of molecules called nucleotide bases, which are made up of atoms of carbon, nitrogen, hydrogen, and oxygen. The structure of DNA is a double helix because the nucleotides of one chain line up with the nucleotides of the second chain in a particular way. The two chains are held together by the forces of attraction between matched pairs of bases. As you read through this book, you will see the basic principles on which this structure is based.

ment might be different from what is expected based on a general rule. When that happens, a chemist gets excited because experiments that do not follow the known rules of chemistry are the most interesting. We know that understanding the exceptions almost invariably gives new insights.

Once enough reproducible experiments have been conducted, and experimental results have been generalized as a law, it may be possible to conceive a theory to explain the observation. A **theory is** a unifying principle that explains a body of facts and the laws based on them. It is capable of suggesting new hypotheses. Theories abound, not only in the sciences, but also in other disciplines like economics and sociology. In chemistry, excellent examples of theories are those developed to account for chemical bonding (Chapters 9–11). It is a fact that atoms are held together, or bonded to one another, as in a molecule such as ammonia or water. But how and why does this happen? Several theories are currently used in chemistry to answer these questions, but are they correct? What are their limits? Can the theories be improved, or are completely new theories necessary? Laws summarize the facts of nature and rarely change. Theories are inventions of the human mind. Theories can and do change as new facts are uncovered.

Goals of Science

See G. Gale: *Theory of Science: An Introduction to the History, Logic, and Philosophy of Science,* New York, McGraw-Hill, 1979.

The sciences, including chemistry, have two goals. The first of these is prediction and control. This is the reason we do experiments and seek generalities. We want to be able to predict what may occur under a given set of circumstances. We also want to know how we might control the outcome of a chemical reaction or process.

The second goal is explanation and understanding. We know, for example, that a group of elements that includes lithium, sodium, and potassium (see Figure 4) will react vigorously with water. But why should this be true? And why is this extreme reactivity unique to these elements? To explain and understand this, we turn to theories such as those developed in Chapters 7 and 8.

The Importance of Serendipity

People who work outside of science usually have the idea that science is an intensely logical field. They picture white-coated chemists moving logically from hypothesis to experiment and then to laws and theories without human emotion or foibles. This is a great simplification! Watson and Crick worked many months and made numerous errors before they understood DNA's structure.

For examples of serendipity in chemistry, see the sidebar "The Art of Science" on Screen 2 in the Introductory Chapter of the *Saunders Interactive General Chemistry CD-ROM.*

Often, scientific results and understanding arise quite by accident, otherwise known as **serendipity.** Creativity and insight are needed to transform a fortunate accident into useful and exciting results, wonderful examples of which are the recent discovery of the cancer drug cisplatin by Barnett Rosenberg, the discovery of penicillin by Alexander Fleming (1881–1955) in 1928, or of iodine by Bernard Courtois (1777–1838).

Dilemmas and Integrity in Science

You may think research in science is straightforward: Do an experiment, draw a conclusion. But, research is seldom that easy. Frustrations and disappointments are common enough, and results can be inconclusive. Complicated experiments often contain some level of uncertainty, and spurious or contradictory data can be collected. For example, suppose you do an experiment expecting to find a direct relation between two experimental quantities. You collect six data sets. When plotted, four of the sets lie on a straight line, but two others lie far away from the line. Should you ignore the last two points? Or should you do more experiments when you know the time they take will mean someone else could publish first and thus get the credit for a new scientific principle? Or consider that the two points not on the line might indicate that your original hypothesis is wrong, and that you will have to abandon a favorite idea you have worked on for a year. Scientists have a responsibility to remain objective in the face of these difficulties, but it is sometimes hard to do.

It is important to remember that scientists are human and therefore subject to the same moral pressures and dilemmas as any other person. To help ensure integrity in science, some simple principles have emerged over time that guide scientific practice: Experimental results should be reproducible, conclusions should be reasonable and unbiased, credit should be given where it is due, and results should be reported in sufficient detail that they can be used or reproduced by others.

CURRENT PERSPECTIVES

Science and Flight 800

The unraveling of the mystery of the fate of TWA Flight 800, which crashed on its way from New York to Paris on July 17, 1996, is an excellent illustration of the scientific method and how it can affect all of our lives. Investigators of the crash made observations, formed hypotheses, tested their hypotheses against the information they had collected, and even did experiments before beginning to reach some conclusions. (This theme was expressed in a letter to the editor of the *New York Times* from Gary H. Posner, August 25, 1996.)

The Boeing 747 exploded in flight shortly after taking off. The cockpit voice recorder recorded an abrupt sound a fraction of a second before it stopped working, and the front portion of the plane broke away from the rest. At first it seemed that a bomb or even a missile had brought the plane down. Traces of explosives were indeed found, but it turned out the plane had been used in a test for dogs being trained to find explosives, and traces of explosives had remained on the plane from that test. Gradually, attention centered on the fuel tank under the main cabin in the center of the fuselage. The tank contained only a small amount of fuel, and it was initially thought that a bomb in the cabin overhead had ignited the fuel. But the data did not support that hypothesis either. Then it was suggested that the fuel in the tank was set off by a spark from some source. But what was the temperature of the fuel at the time? Was the mixture of fuel and air sufficient to explode? What was the source of an igniting spark? At the time this book is being written, the answer is not yet known, and there is some concern that it may never be known. (This is not unusual. It is also sometimes true in scientific studies in chemistry or any of the other sciences.) Nonetheless, careful investigation, using the same approaches that any scientist takes in studying nature, has been used by the investigators of the crash of Flight 800.

A Boeing 747 landing in Vancouver, British Columbia. (*David Nunk, Science Photo Library, Photo Researchers, Inc.*)

REPRESENTING CHEMISTRY

You may be studying general chemistry because you want to become a chemist, biochemist, or chemical engineer, and chemistry will be the focus of your profession. Others of you will become physicians or will be involved in some other way in the health professions. Here, too, chemistry is important although the reasons will not always be obvious. Finally, others of you will enter a wide variety of professions in which the immediate applications of chemistry are not obvious but will nonetheless influence your lives. No matter what the level of your awareness of chemistry, you are going to experience chemical phenomena at the **macroscopic level.** That is, you see chemical processes with your eyes. To be able to report those experiences to one another, chemists use symbols; that is, we represent chemical phenomena at the **symbolic level.** Finally, to understand what has happened, we describe observations in terms of particles called atoms and molecules. That is, we represent chemistry at the **particulate level.**

These three types of representation can be thought of as the points of a triangular matrix—discussions of chemistry occur using a combination of each of the three forms of representation (Figure 5). *This is an important organizing principle, and we shall use it throughout this book.* A special icon will be placed in the margin to call attention to the different types of representations.

This special icon will denote instances where we wish to draw particular attention to the relation between the macroscopic, symbolic, and particulate views of a chemical process.

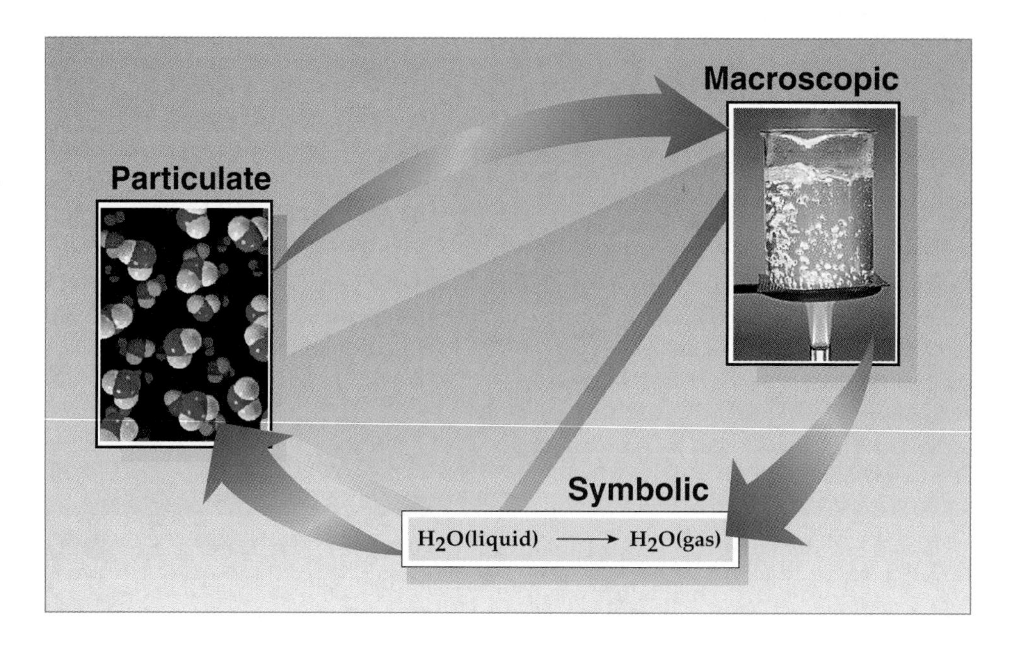

Figure 5 Chemistry can be represented as a triangular matrix of concepts. We observe chemical processes on the macroscopic scale and then write symbols to represent those observations. To understand or illustrate those processes we try to view or imagine what has occurred at the particulate—atomic and molecular—level. (*Photo, C. D. Winters; model, Roy Tasker/University of Western Sydney, Australia*)

Chemistry at the Macroscopic Level

How do you prepare pasta in the kitchen? You first warm the water until it just begins to boil, then drop in a pinch of salt and stir until it dissolves. Next, drop in a few milliliters of oil so the pieces of pasta don't stick to one another. You have just done chemistry at the macroscopic level. You saw water undergo a change in its state from liquid to vapor, and you observed that salt (sodium chloride) can form a solution with water. Finally, you observed that oil does not dissolve in water but floats on the surface of water because oil is less dense than water and does not form a homogeneous mixture with water.

Doing chemistry in the laboratory is much like cooking. You add a solid to water, or mix two solutions and observe something occur, or you observe a change in some substance, such as the boiling of water (see Figure 5).

Symbolism in Chemistry

Musicians can interpret the notes on a page of sheet music differently from what the composer intended. This can happen in chemistry as well, even though an observation is expressed in standard symbols.

See the first screen in the Introductory Chapter of the *Saunders Interactive General Chemistry CD-ROM* for an animation of solids, liquids, and gases as applied to water. Model building is also discussed.

So that a piece of music can be played by others, the composer expresses the music in standard symbols—musical notes—that are understood by all musicians. In the same way, chemists convey observations to one another using standard symbols as a shorthand. For example, we could indicate the evaporation of water by

$$H_2O(\text{liquid}) \longrightarrow H_2O(\text{gas})$$

Here H and O are the symbols for the chemical elements hydrogen and oxygen, respectively, and water molecules are composed of H and O atoms in a 2-to-1 ratio. The arrow connecting H_2O(liquid) and H_2O(gas) means that liquid water changes to water vapor.

Symbols are widely used for convenience. For example, the symbol mph stands for "miles per hour," or a rate of progress. Changing m to k in mph (to obtain kph, or "kilometers per hour") leads to a different value. A speed of 50

Historical Perspective

Serendipity

In the 1960s Barnett Rosenberg set out to study a problem that had interested him for some time—the effect of electric fields on living cells—but the results of the experiment were quite different from his expectations. He and his students had placed an aqueous suspension of live *Escherichia coli* bacteria in an electric field between supposedly inert platinum plates. Much to their surprise they found that cell growth was affected; cell division had stopped! After careful experimentation, the effect on cell division was traced to tiny amounts of a compound now called cisplatin, which was produced by reaction of the platinum with electrically charged chemical species in the water.

Much to the benefit of cancer chemotherapy, Rosenberg recognized that these laboratory results had wider implications, and subsequent experiments led to compounds now used to treat cancer. He recently said that the use of cisplatin has meant that "Testicular cancer went from a disease that normally killed about 80% percent of the patients, to one which is close to 95% curable. This is probably the most exciting development in the treatment of cancers that we have had in the past 20 years. It is now the treatment of first choice in ovarian, bladder, and osteogenic sarcoma [bone] cancers as well."

Dr. Barnett Rosenberg, Head of the Barros Research Institute in East Lansing, Michigan. (*Doug Elbinger*)

mph is about 82 kph. The symbols you use in chemistry also have precise meanings and must be used carefully. Changing the H to K in H_2O (to get K_2O), for example, produces the formula for a very different substance.

Chemistry at the Particulate Level

In one drop of liquid water there are many millions of H_2O molecules in close proximity (Figure 6). All are moving rapidly, but each is strongly influenced by its neighbors. As energy is transferred to the drop in the form of heat, the molecules move even more rapidly and begin to separate from one another. Eventually, after sufficient energy has been added, the molecules are more widely separated and have even less influence on one another. The water is now described as a vapor.

We have just described the process of a change in the physical state of water at the particulate or molecular level. We cannot directly see atoms and molecules with our eyes, or even with an optical microscope. Nonetheless, special instruments and many, many experiments have led to the knowledge of the composition and structure of water and to the nature of water in its solid, liquid, and gaseous forms. To understand chemical phenomena, it is very helpful to imagine what is happening at the atomic and molecular level, that is, at the particulate level. We shall use many such "pictures," and many are found animated on the *Saunders Interactive General Chemistry CD-ROM.*

To ensure that you become familiar with the importance of molecular models, illustrations of molecular models occur throughout this book. In addition, if you have access to a computer you can view and interact with models at the World Wide Web site for the book. Finally, if your computer has a CD-ROM drive, then you can also use the library of molecular structures on the *Saunders Interactive General Chemistry CD-ROM.* See the Preface for more details.

Model Building, Science, and Molecular Structures

One way to solve a problem is to build a model, something that we think mimics reality. This can be a mathematical model for the production and income of

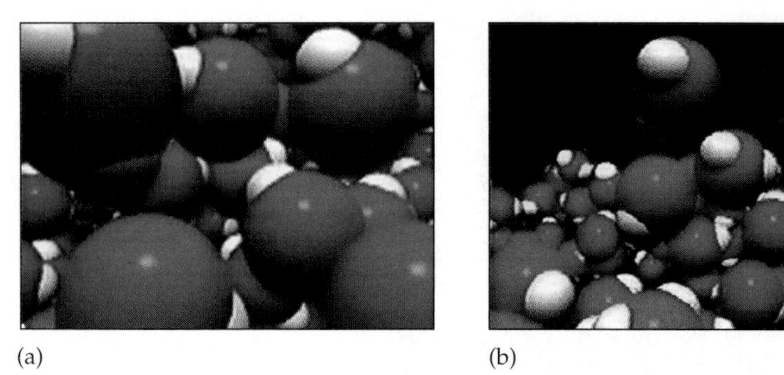

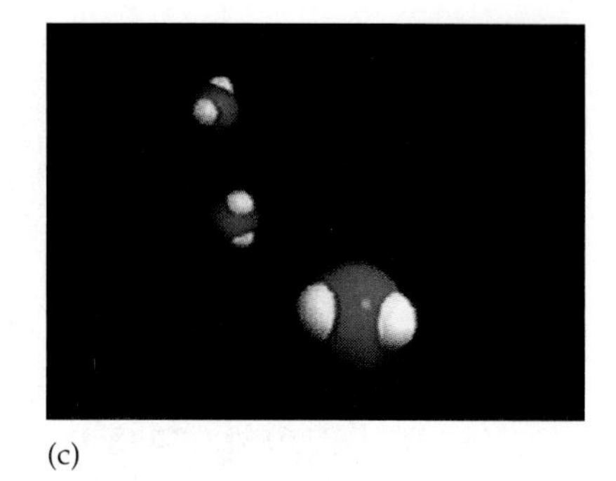

(a) (b) (c)

Figure 6 The change of liquid water to water vapor at the particulate level. (a) Water molecules are relatively close to one another and are in constant motion. (b) As energy is transferred to the water, some molecules begin to break away and become separated more widely. (c) When sufficient energy has been transferred, the molecules are widely separated and still moving rapidly. The substance is now considered a vapor. (*From an animation by Roy Tasker/University of Western Sydney, Australia*)

a small business, the model an architect constructs to see how the rooms in a house fit together, or the model a chemist might make of a molecule using paper or plastic or constructed using a computer (as in Figures 6 and 7).

James Watson and Francis Crick used models when solving the problem of the structure of DNA (pages 6–7 and Figure 7). First, they built physical models of the pieces of DNA. They then used their chemical knowledge and intuition to see how these pieces best fit together to create a molecule with properties that matched experimental results. Watson said that he and Crick thought that "model building represented a serious approach to science, not the easy resort of slackers who wanted to avoid the hard work necessitated by an honest scientific career."

The powerful x-ray techniques that provided the experimental proof of the Watson-Crick DNA model have been used extensively over the past 40 years to explore the shapes, or structures, of thousands of molecules. Scientists now have well-defined ideas of molecular structure, and, just as importantly, our understanding of the relation between molecular structure and chemical behavior is increasing. Indeed, the importance of molecular structure cannot be understated. DNA is a helix because of the geometry of the linkages between the chemical elements in its backbone. It is a double helix because of the way that molecules from one helical chain fit together with those from the other chain (see Figure 7). And it is able to carry genetic information precisely because of the geometries of the molecules involved.

A knowledge of molecular structure is now regarded as fundamental to progress in chemistry. Indeed, it is so important that pharmaceutical and chemical companies, for example, model chemical reactions using computer-generated structures such as those in Figure 7 before doing laboratory work.

RISKS AND BENEFITS

The risks of smoking are clear: the annual death rate is 40 per million people. But did you know that the death rate for high school football players is about 10 per million? In comparison, the death rate caused by asbestos in schools is

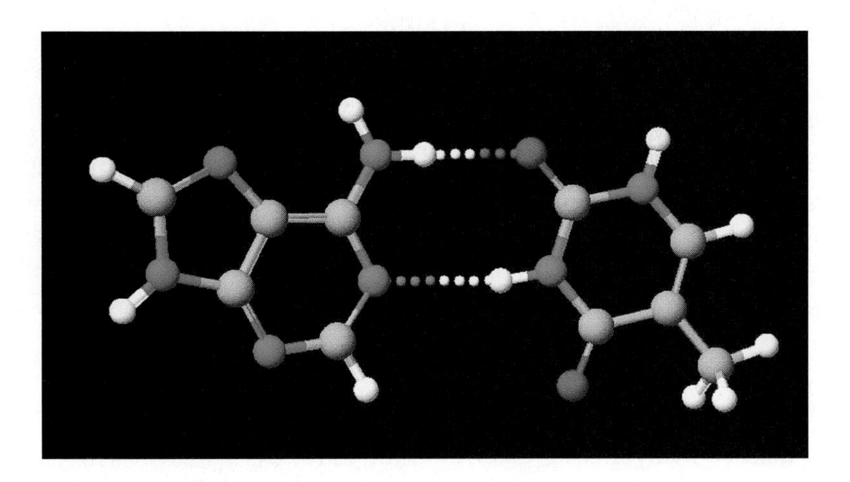

Figure 7 Four different molecules bridge the two strands of the "backbone" of DNA (see page 7). Two of them (adenine, *left*, and thymine, *right*) are shown here. Because of their geometry, these two molecules fit tightly together, providing one of the many linkages that hold together the two strands of the DNA backbone. (The atoms in these computer-generated molecular structures are labeled by color: hydrogen = white; carbon = gray; oxygen = red; and nitrogen = blue.)

estimated to be in the range of 0.005 to 0.093 per million. At such a low rate, is it worth the cost of removing the asbestos from schools? Or is it worth the risk to play football?

Weighing risks and benefits is not only an issue for society at large but is also meaningful to you and your family. How do you make decisions about a perceived risk? Common sense will carry you a long way, but some questions can be answered only if you have a background in the sciences, including chemistry. An example is the Clean Air Act of 1970 and current suggestions for changes to it.

The Clean Air Act

The Clean Air Act, passed in 1970, set standards for air quality. For example the act specified that air is considered safe if it contains no more than 0.12 ppm (part per million) of ozone, O_3. Also, airborne particles should be no larger than 10 μm (1 micrometer = 1×10^{-6} m) in diameter. To achieve these standards, over $450 billion has been spent in the United States since 1970, and enormous success has been achieved. Pollution levels have dropped nearly 30% on average. Even greater gains have been made in reducing the levels of some serious pollutants such as lead and carbon monoxide.

A part per million (ppm) means we can find 1 gram of a substance in 1 million grams of material. That's about a tenth of a drop in a bucket.

In spite of these improvements, the Environmental Protection Agency (EPA) believes that further improvements can still be made (Figure 8) and so has proposed new, much tighter air quality standards for ground-level ozone and particulate matter. The controversy that erupted over the proposed new standards illustrates the ongoing process of balancing risks and benefits in our society.

The current standards of the Clean Air Act were set because they represented the threshold of danger. With recent advances in chemistry and other sciences, however, the head of the EPA decided that new scientific evidence compelled her to tighten standards and that the benefits would outweigh the costs. (The EPA initially estimated the cost of new regulations at close to $8.5 billion, whereas the annual benefit could be as much as $120 billion.) The benefits are largely the money saved because people, especially children, would not become ill and require medical care. The EPA has said, for example, that "Asthma among our children has increased 118% between 1980 and 1993. It is the single greatest cause of childhood hospital admissions and among the leading causes of missed days of school."

For a discussion of epidemiology and the connection between diet, lifestyle, and environmental factors, see G. Taubes: "Epidemiology faces its limits," *Science,* Vol. 269, July 14, 1995, pp. 164–169.

Figure 8 The Clean Air Act of 1970 has led to much cleaner air in the United States. Nonetheless, considerable improvement can still be made, and the EPA has proposed tightening standards. (*Sue McCartney/Photo Researchers, Inc.*)

The EPA actions illustrate important new disciplines in our modern society. First is **risk assessment,** the process of bringing together information from many disciplines—such as chemistry, environmental science, and epidemiology—to understand an issue such as the effect of ozone on the environment. Second is **risk management,** the setting of standards or writing of regulations to minimize risk and maximize benefits. Risk management involves science, ethics, economics, and other matters that are part of governmental, political, and social interactions.

In the case of the Clean Air Act, controversy has arisen over both risk assessment and risk management. Critics claim that health improvements would be slight if ozone standards are tightened, and the benefits claimed for the new standards for particles rest on shaky scientific assumptions. Furthermore, such bodies as the President's Council of Economic Advisors estimate that the costs could be many times greater than the EPA has estimated.

Analysis and Detection

The EPA decided to tighten air quality standards because of advances in the sciences. As you study chemistry you will learn methods of detecting and analyzing substances, and it will become apparent that special methods are needed to detect very low levels of a substance. Chemists, however, are inventing ever more sensitive methods of measurement. For example, in 1960 mercury could be detected at a concentration of 1 ppm, in 1970 the detection limit was 1 ppb (part per billion), and by 1980 the limit had dropped to 1 ppt (part per trillion).

The ability to detect lower and lower levels of environmental pollutants has an important effect. A few decades ago, when food, air, or water was tested, toxic substances were often not found. Today, though, most materials contain detectable amounts of numerous toxic substances. The news media reports these findings, and the public becomes alarmed. Demands are made for regulation or corrective action, which is often interpreted as a demand for reduction of the pollutants below the detection limit. Two problems arise from such demands. First, although we expect that chemistry will push detection limits lower and lower, measuring zero in a chemical analysis will always be impossible. Second, it is not clear that minuscule amounts of toxic substances are a sufficient enough health threat that costly responses are required.

Risks in Your Own Life

When you size up the risks in your own life, such as being diagnosed with cancer, keep in mind that epidemiologists have found only a few environmental agents that are strongly linked to that disease. These are cigarette smoke, alcohol, ionizing radiation, a few drugs, a handful of occupational carcinogens, and perhaps three viruses (hepatitis-B, human T-cell leukemia, and human papillomavirus). Every year, though, dozens of papers are published in the scientific literature that report potential new environmental causes of cancer. Experts say, however, that you should take these new reports seriously only if there is at least a twofold increased risk. For example, using a mouthwash high in alcohol has been reported to increase the risk of mouth cancer 1.5 times. Another study states that if you consume olive oil once a day or less, the risk of having breast cancer increases by only a factor of 1.25. The problem is that this report was contradicted by a study that showed that the breast cancer risk was reduced by 25%! Risk analysis and management is difficult; it is not an exact science.

Risks, Benefits, and Chemistry

Why study chemistry? The reasons are clear. You will be called upon to make many decisions in your life for your own good or for the good of those in your community—whether that be your neighborhood or the world. An understanding of the nature of risk, of science in general, and of chemistry in particular, can only serve to help in these decisions.

READINGS ABOUT SCIENCE

You will find the following list of books about science both interesting and informative:

- Rachel Carson: *Silent Spring,* New York, Houghton Mifflin, 1962.
- Richard Feynman: *What Do You Care What Other People Think?* W.W. Norton and Company, New York, 1988; and *Surely You're Joking, Mr. Feynman,* New York, W.W. Norton and Company, 1985.
- Roald Hoffmann and Vivian Torrance: *Chemistry Imagined, Reflections on Science,* Smithsonian Institution Press, Washington, D.C., 1993.
- Thomas S. Kuhn: *The Structure of Scientific Revolutions,* Chicago, The University of Chicago Press, 1970.
- Primo Levi: *The Periodic Table,* New York, Schocken Books, 1984.
- Sharon D. McGrayne: *Nobel Prize Women in Science,* New York, Birch Lane Press, 1993.
- Lewis Thomas: *The Lives of a Cell,* New York, Penguin Books, 1978.
- J. D. Watson: *The Double Helix, A Personal Account of the Discovery of the Structure of DNA,* New York, Atheneum, 1968.

CHAPTER HIGHLIGHTS

Having studied this chapter you should

- Understand the general methods of science.
- Understand the meaning of the terms **hypothesis, law,** and **theory.**
- Be aware of the difference between **quantitative** and **qualitative** information.
- Appreciate the goals of science and the importance of **serendipity.**
- Be aware of the moral dilemmas that can occur in science.
- Begin to think of chemistry in terms of its main conceptual viewpoints: **macroscopic, symbolic,** and **particulate** representations.
- Appreciate the importance of models and model building in science.
- Understand the ways in which risks can be assessed and managed and how risks and benefits can be weighed.

STUDY QUESTIONS

These questions are meant to help you review the topics covered in this chapter. Answers to the questions with a boldfaced number are given in Appendix N and in the Student Solutions Manual.

GENERAL QUESTIONS

1. What are some of the risks and benefits of nitrogen?
2. The first section of the chapter is titled "Too Much of a Good Thing." What does this mean?
3. When studying some metals in the laboratory, you observe that a piece of lead, which melts at 328 °C, is dull gray in color and bends easily. Which part of this information is qualitative and which part is quantitative?
4. A piece of copper metal, whose mass is 15 grams, is shiny and conducts electricity. Which part of this information is qualitative and which part is quantitative?
5. The air in an automobile tire exerts a pressure because the molecules of air (oxygen and nitrogen, among others) are moving about very rapidly inside the tire and are colliding with the tire walls. What part of this statement expresses a fundamental law of nature and which part is a theory? (See Gases and Their Properties, Chapter 12.)
6. Are supply-side economics and creationism laws or theories?
7. You will study the structures of molecules in detail beginning in Chapter 9. Examine the structures of carbon dioxide and ammonia on page 5. In your own words, describe these structures. What atom is in the "center" of the molecule? Are the molecules straight, bent, flat, or pyramidal?
8. Water is identified by the symbol H_2O. We know that the O atom is located between the two H atoms, and that the molecule is bent. Sketch a picture of the structure of this molecule.
9. Name three risks that you assumed today (such as riding a bicycle, smoking, etc.). In your estimation, which was the most risky?
10. How does the discussion of the use and overuse of nitrogen illustrate goals of science?

INTERACTIVE GENERAL CHEMISTRY CD-ROM

The Saunders Interactive General Chemistry CD-ROM *was designed to accompany* Chemistry & Chemical Reactivity. *The CD-ROM can be used as a complete chemistry course or as a supplement to the book. All of the topics in the book are included on the CD-ROM, but the latter further illustrates and amplifies the book by using videos and animations. All of the chapters in the book include references to the CD-ROM and questions that enable you to better appreciate the topics described in the book.*

Screen 1 The Nature of Chemistry

The first screen of the introductory chapter of the CD-ROM includes various descriptions of chemistry and its role in our world.

(a) Watch the first video on this screen (*left side*). How is chemistry defined?
(b) The second of the five videos describes models and modeling and includes an animation of liquid water. Describe the nature of liquid water.

Screen 2 Science and Its Methods

On January 28, 1986, the Space Shuttle *Challenger* exploded on liftoff. The accident was thoroughly investigated, and, during public hearings, the physicist Richard Feynmann performed a simple experiment that illustrated the probable cause of that accident. Describe that experiment and how it illustrates the scientific method.

Molecular Modeling

The CD-ROM includes molecular-modeling software from CAChe/Oxford Molecular. (See the manual accompanying the CD-ROM for instructions on using this software.)

(a) Find the model of carbon dioxide. (The model is labeled "CO_2," and is in the folder marked "Inorganic.") Examine this model and describe the structure in your own words.
(b) Find the models of ammonia and water (in the folder marked "Inorganic," labeled NH_3 and water, respectively). Examine these models and describe the structure in your own words.
(c) Figure 7 shows the bases adenine and thymine, important components of DNA. Use the CAChe/Oxford molecular-modeling software on the CD-ROM to inspect adenine further. (See the folder marked "Biochem," and then the folder marked "Bases.") Describe the structure in your own words. Is it flat or bent or puckered in some way? Are there rings of atoms? If so, how many atoms are there in a ring? What atoms are included in adenine?

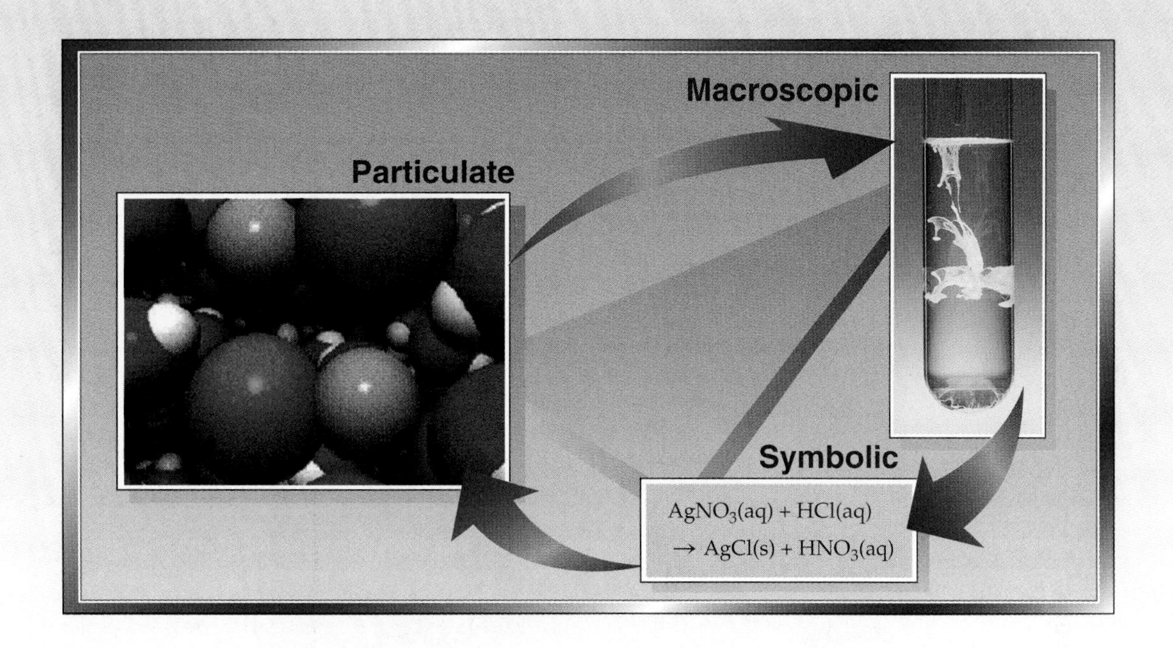

Part 1

The Basic Tools of Chemistry

Chemistry is the study of the chemical elements, the compounds they form, and their transformations. You are just beginning a tour of the properties of elements such as lithium, carbon, nitrogen, iron, zinc, and mercury. Compounds such as water, ammonia, and sweet sucrose are on our tour as well. We are going to ask why the chemical elements have the properties they have and why they can combine to form compounds. We also want to know why molecules have the properties and shapes they do and what holds them together. Why are diamonds hard? Why is water such a unique substance? It is these and hundreds of other questions that make up the fabric of chemistry.

To even begin answering the kinds of questions we have posed, we will have to lay a foundation of principles. Thus, Chapter 1 defines some important terms and is a review of units and numerical methods. Chapter 2 introduces you to atoms and elements, and Chapter 3 is an introduction to molecules and compounds. It is here that we begin one of the major themes of this book—the connection between the structures of molecules and their properties. In Chapter 4 we lay out the principles of the quantitative study of chemical reactions, and with Chapter 5 we outline the fundamental types of chemical reactions. With Chapter 5 we also truly begin to emphasize what you see in the figure on this page, the connection between what we can see, the macroscopic, how we write down our observations as symbols, and how we imagine that these changes occur on an atomic and a molecular scale. Finally, with Chapter 6 we examine the energy changes that occur during a chemical or physical change and what these changes in energy mean. ● *(Left, R. Tasker/University of Western Sydney, Australia; right, C. D. Winters)*

Chapter 1

Matter and Measurement

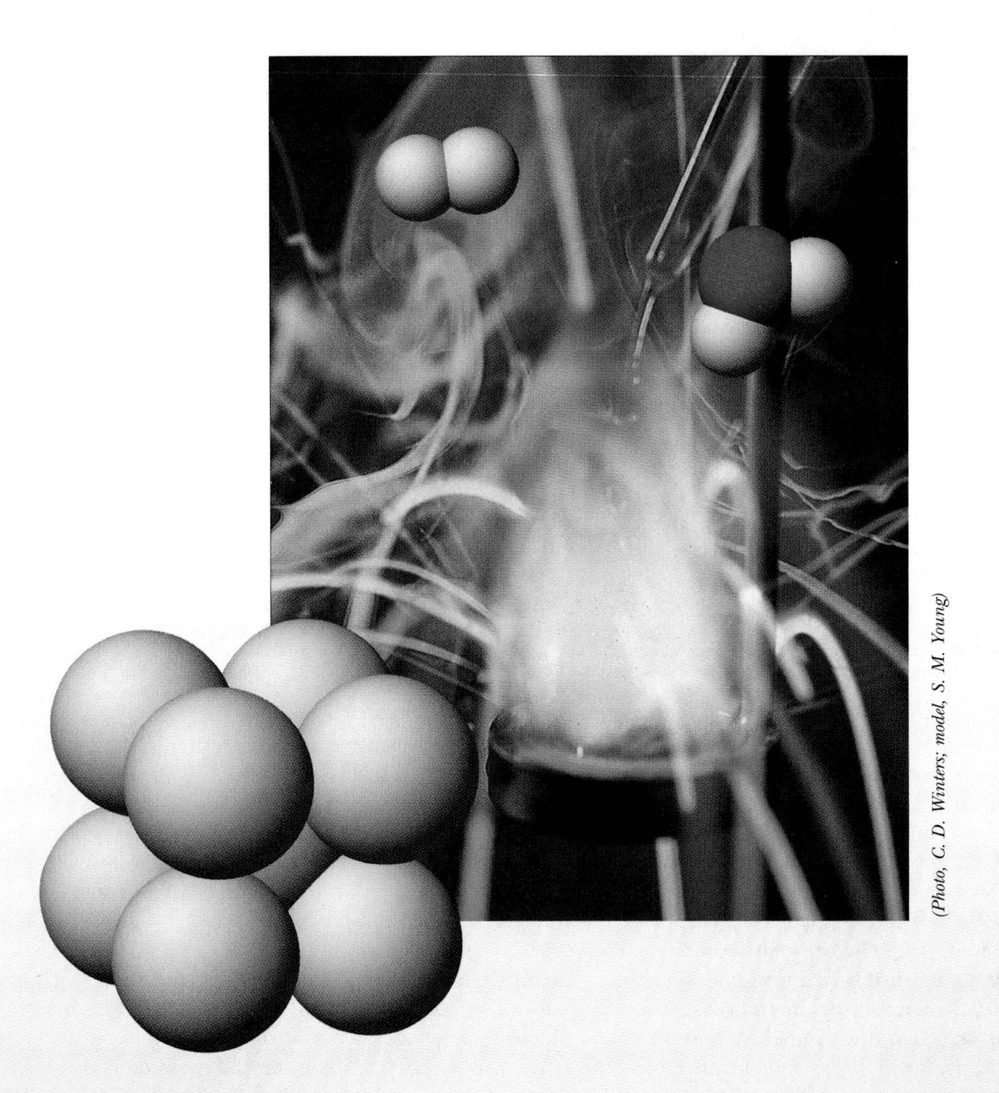

(Photo, C. D. Winters; model, S. M. Young)

A Chemical Puzzler

The chemical element potassium reacts with water. What states of matter are illustrated? What types of changes are observed? What qualitative observations can be made? (For a video of this reaction see the *Saunders Interactive General Chemistry CD-ROM*, Screen 8.15.)

Imagine a tall glass filled with a clear liquid. Sunlight from a nearby window causes the liquid to sparkle, and the glass is cool to the touch. A drink of water would certainly taste good, but should you take a sip? If the glass were sitting in your kitchen you would probably say "yes" without further thought. But what if this scene occurred in a chemical laboratory? How would you know that the glass held water? Or, perhaps we should pose a more chemical question: How would you prove this liquid is water?

How could you prove that the clear, colorless liquid is water? Think about this as you read the chapter, and then see Study Question 66 at the end of the chapter.

We usually think of the water we drink as being pure, but this is not strictly true. In some instances material may be suspended in it or bubbles of gases such as oxygen may be visible to the eye. Some tap water has a slight color from dissolved iron. In fact, drinking water is almost always a mixture of substances, some dissolved and some not. As with any mixture, we could ask many questions. What are the components of the mixture—dust particles, bubbles of oxygen, dissolved sodium, calcium or iron salts—and what are their relative amounts? How can these substances be separated from one another, and how are the properties of one substance changed when it is mixed with another?

These examples represent some of the problems chemists face everyday. Chemists seek answers to questions by making observations and doing experiments. They also draw on known information and create new information with an organized approach. This chapter takes up some basic topics that will help you begin to see how chemists think about matter.

1.1 CLASSIFYING MATTER

A chemist looks at a glass of drinking water and sees a liquid. The liquid could be pure water, a chemical compound. More likely, the liquid really is a homogeneous mixture of water and dissolved salts, that is, a solution. Or, it is possible our water sample is a heterogeneous mixture, with solids suspended in the liquid. These all represent ways to classify matter, and the ways that we think about it are represented in Figure 1.1.

See the *Saunders Interactive General Chemistry CD-ROM,* Screen 1.3, States of Matter.

States of Matter

An easily observed and very useful property of matter is its **state,** or **phase** (Figure 1.2), that is, whether a substance is a solid, liquid, or gas at room temperature or at some other temperature. A **solid** can be recognized because it has a rigid shape and a fixed volume that changes very little as temperature and pressure change. Like solids, **liquids** have a fixed volume, but a liquid is fluid—it takes on the shape of its container and has no definite form of its own. **Gases** are fluid also, but the volume of a gas is not fixed; rather, it is determined by the size of the container. The volume of a gas varies with temperature and pressure.

Figure 1.2 illustrates the connection between the macroscopic and particulate worlds. For example, bromine is an orange-brown liquid or vapor that consists of particles (molecules) in rapid motion.

At low temperatures, virtually all matter is found in the solid state. As the temperature is raised, though, solids generally melt to form liquids. Eventually, if the temperature is raised high enough, liquids evaporate to form gases. Volume changes accompany changes in state. For a given mass of material, there is usually a small increase in volume on melting—water being a very significant exception—and then a large increase in volume upon evaporation.

Water is an exception to the rule that the solid phase has a smaller volume than an equal mass of the liquid. Ice floats on liquid water because the density of ice is less than that of the liquid. See Section 1.4.

Figure 1.1 A scheme for the classification of matter.

Matter (may be solid, liquid, or gas): anything that occupies space and has mass

Heterogeneous matter: nonuniform composition

Physically separable into

Homogeneous matter: uniform composition throughout

Pure substances: fixed composition; cannot be further purified

Physically separable into

Solutions: homogeneous mixtures; uniform compositions that may vary widely

Compounds: elements united in fixed ratios

Chemically separable into

Combine chemically to form

Elements: cannot be subdivided by chemical or physical changes

Kinetic-Molecular Theory

The **kinetic-molecular theory** of matter helps us interpret the properties of solids, liquids, and gases (see Figure 1.2). According to this theory, all matter consists of extremely tiny particles (atoms and molecules), which are in constant

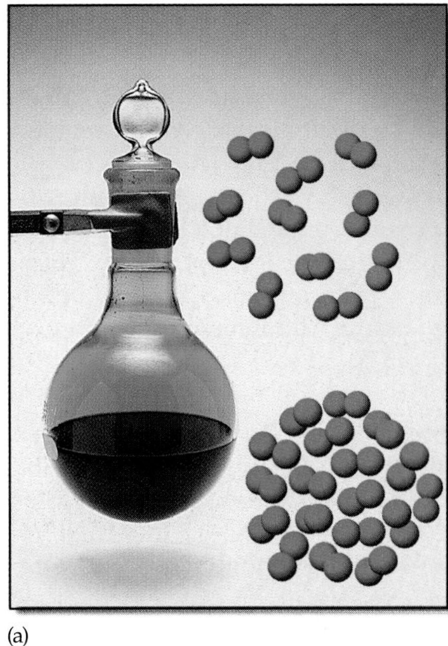

(a)

(b)

Figure 1.2 **The three states of matter.** (a) In the gas phase, atoms or molecules (here, of bromine) move rapidly over distances larger than the sizes of the atoms or molecules themselves. Although the particles collide with one another often, there is little interaction between them. In the liquid state, however, the atoms or molecules are much closer together and they interact with one another. Motion of the particles is still very evident, although the particles move over only very small distances. (b) The particles in a solid (here, iron) are even closer together and almost totally restricted to specific locations. (*Photo, C. D. Winters; model, S. M. Young*)

motion. In solids these particles are packed closely together, usually in a regular array. The particles vibrate back and forth about their average positions, but seldom does a particle in a solid squeeze past its immediate neighbors to come into contact with a new set of particles.

In liquids or gases, atoms or molecules are arranged at random rather than in the regular patterns found in solids. Liquids and gases are fluid because the particles are not confined to specific locations and can move past one another. Under normal conditions, the particles in a gas are far apart.

Gas molecules move extremely rapidly because they are not constrained by their neighbors. In air at room temperature, for example, the average nitrogen molecule is going faster than 450 m/s (meters per second), or over 1000 mph. Molecules of a gas fly about, colliding with one another and with the container walls. This random motion allows gas molecules to fill their container, so the volume of the gas sample is the volume of the container.

One further aspect of the kinetic-molecular theory is that the higher the temperature the faster the particles move. The particles' energy of motion (called the **kinetic energy**) acts to overcome the forces of attraction between the particles. A solid melts to form a liquid when the temperature of the solid is raised to the point at which the particles vibrate fast enough and far enough to push one another out of the way and move out of their regularly spaced positions. As the temperature increases even more, the particles move even faster until finally they can escape the clutches of their comrades and the substance becomes a gas. *Increasing temperature corresponds to faster and faster motions of atoms and molecules,* a general rule you will find useful in many future discussions.

See the *Saunders Interactive General Chemistry CD-ROM,* Screen 1.7, for an animation of solids, liquids, and gases and a discussion of kinetic-molecular theory.

Matter at the Macroscopic Level and Particulate Level

The characteristic properties of gases, liquids, and solids just described are observed by the unaided human senses. They are determined using samples of matter large enough to be seen, measured, and handled. Using samples of finite and measurable size, we can also determine, for example, the color of a substance, whether it dissolves in water, or whether it conducts electricity or reacts with oxygen. Observations and manipulation take place in the macroscopic world of chemistry (Figure 1.3). It is the world of experiments and of observations. It is most likely the world that you think of when you think of chemistry.

See the *Saunders Interactive General Chemistry CD-ROM,* Screen 1.4, for a visualization of the macroscopic and particulate levels of chemistry.

Now let us move to the level of atoms and molecules, a world of chemistry that we cannot see. Take a macroscopic sample of material and divide it, again and again, past the point where the amount of sample can be seen by the naked eye, past the point where it can be seen using an optical microscope. Eventually, in this process of subdividing you will reach that level of individual particles that make up all of matter. You have reached what chemists refer to as the **submicroscopic,** or **particulate,** world of chemistry (see Figure 1.3). This is the world of atoms and molecules.

Figure 1.3 shows the connection between the macroscopic, symbolic, and particulate views of matter.

As described on page 9, chemists are interested in the structure of matter at the particulate level. Atoms and molecules cannot be "seen" in the same way that one views the macroscopic world of chemistry, but they are no less real to chemists. Chemists imagine what atoms must look like and how they might fit together to form molecules. They create models to represent atoms and molecules and use these models to think about chemistry and to explain the observations that they have made about the macroscopic world.

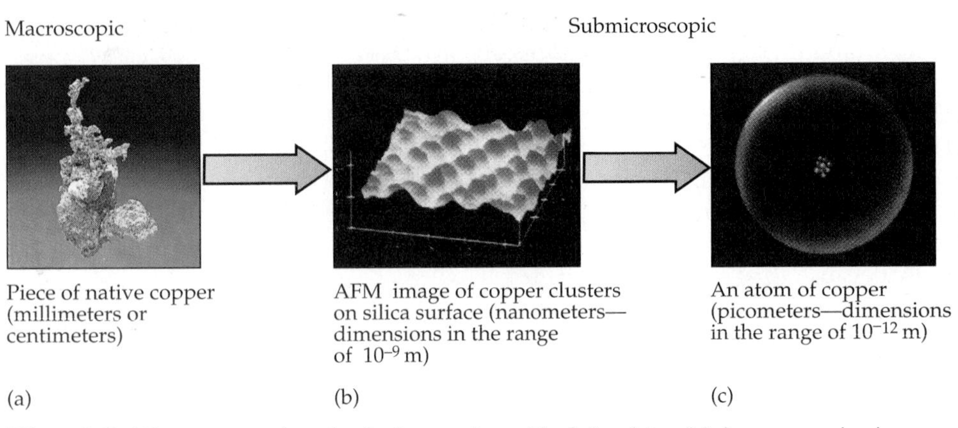

Figure 1.3 The macroscopic and submicroscopic world of chemistry. (a) A macroscopic piece of copper metal. This is native copper, a piece of naturally occurring elemental copper. (b) Scanning tunneling (STM) and atomic force microscopy (AFM) are powerful methods of probing the surface of solids with atomic resolution. They can reveal the submicroscopic world of atoms. Here the figure shows an AFM topographic image of the copper atoms on the surface of silica. The image is 1.70 nm square, and the rows of atoms are separated by about 0.44 nm. (c) Representation of an atom of copper. (*a, C. D. Winters; b, Reproduced, with permission, from X. Xu, S. M. Vesecky, and D. W. Goodman, Science, Vol. 258, p. 788, 1992*)

Figure 1.4 Iron pyrite (right). This compound, composed of atoms of iron and sulfur, can form large cubic crystals that reflect the arrangement of the atoms deep inside the crystal. In the model of the solid at the atomic level, iron atoms are silver and sulfur atoms are yellow. (*Photo, C. D. Winters; model, S. M. Young*)

It has been said that chemists carry out experiments at the macroscopic level, but they think about chemistry at the particulate level (Figure 1.4). This is a useful perspective that will help you as you study chemistry. Indeed, one of our goals is to help you realize this connection.

Pure Substances

See the *Saunders Interactive General Chemistry CD-ROM,* Screen 1.13, Mixtures and Pure Substances.

Let us think again about our glass of drinking water. How would you tell whether this sample is pure water or a mixture of substances? Begin by making a few simple observations. Is solid material floating in the liquid? Does the liquid have an odor or unexpected taste or color?

Every pure substance has two features. First, it has a set of unique properties by which it can be recognized; all samples of the substance have these characteristic properties. Pure water, for example, is colorless, odorless, and certainly does not contain suspended solids. If you wanted to identify the substance conclusively as water, you would have to examine its properties more carefully and compare them against the known properties of pure water. The sample may be a colorless liquid, but this is not conclusive evidence because many clear, colorless liquids are known. To make a definite identification, you need to look at other properties. Melting point and boiling point serve the purpose well here. If you could show that the substance melts at 0 °C and boils at 100 °C at atmospheric pressure, you can be certain it is water. No other substance melts and boils at precisely these temperatures.

Many pure substances occur naturally. Gold, diamonds, sulfur, and many minerals occur naturally in very pure form, but these substances are special cases. We live in a world of mixtures—all living things, the air and food we depend on, and many products of technology are mixtures.

The second feature of a pure substance is that it cannot be separated into two different species by any physical technique. If this were true, our sample would be classified as a mixture.

(a)

(b)

(c)

Figure 1.5 **Mixtures.** (a) A chocolate chip cookie is a heterogeneous mixture. (b) A sample of blood may look homogeneous, but it is in fact a heterogeneous mixture of liquids and suspended particles. (c) A homogeneous solution, here of salt in water. (The model shows that salt consists of separate particles, called ions, in water.) (*a, C. D. Winters; b, Ken Edward/Science Source/Photo Researchers; c, photo, C. D. Winters and model, S. M. Young*)

Mixtures: Homogeneous and Heterogeneous

A chocolate chip cookie is obviously a mixture, because one can see that the chocolate chips are different from the surrounding material (Figure 1.5). A mixture in which the uneven texture of the material can be detected is called a **heterogeneous mixture.** Heterogeneous mixtures may appear completely uniform but on closer examination are not. Blood, for example, may not look heterogeneous until you examine it under a microscope and red and white blood cells are revealed (see Figure 1.5b). Milk appears smooth in texture to the unaided eye, but magnification reveals fat and protein globules within the liquid. In a heterogeneous mixture the properties in one region are different from those in another region, and the properties differ from sample to sample.

Homogeneous matter is completely uniform at the macroscopic level and consists of one or more substances in the same phase (see Figure 1.5c). No amount of optical magnification reveals a homogeneous mixture to have different properties in one region than in another, because heterogeneity exists only at the atomic or molecular level, where the individual particles are too small to be seen with ordinary light.

The properties of a homogeneous mixture are the same everywhere in the sample. Such mixtures are often called **solutions,** and common examples include air (mostly a mixture of nitrogen and oxygen gases), gasoline (a mixture of carbon- and hydrogen-containing compounds called hydrocarbons), urine (a mixture of water, urea, and salts excreted by your body), and soft drinks.

When a mixture is separated into its pure components, the components are said to be **purified** (see Figure 1.1). Most efforts at separation are not complete in a single step, however, and repetition almost always will give an increasingly pure substance. For example, iron can be separated from a heterogeneous mixture of iron and sulfur by repeatedly stirring the mixture with a magnet (Figure 1.6). When the mixture is stirred the first time and the magnet is removed, much of the iron is removed with it, leaving the sulfur in a higher state of purity. After one stirring, however, the sulfur may still have a dirty appearance due to a small amount of iron that remains. Repeated stirrings with the magnet, or perhaps the use of a very strong magnet, finally leave a bright yellow sample of sulfur

Figure 1.6 **Purifying sulfur.** The iron chips in a mixture of two chemical elements, iron and sulfur, may be removed by stirring the heterogeneous mixture with a magnet. (*C. D. Winters*)

that apparently cannot be purified further, at least by this technique. This purification process uses a property of the mixture, its color, to measure the extent of purification. (The color depends on the relative quantities of iron and sulfur in the mixture.) When the bright yellow color is obtained, all the iron is assumed to have been removed and the sulfur is considered purified.

Homogeneous and heterogeneous mixtures (Exercise 1.1). Solutions to this and all other Exercises in the book are found in Appendix N. (*C. D. Winters*)

Exercise 1.1 Mixtures and Pure Substances

The photo in the margin shows two mixtures. Which one is homogeneous and which is heterogeneous? Which one is a solution?

1.2 ELEMENTS AND ATOMS

Pure water can be decomposed to hydrogen and oxygen by passing an electric current through it (Figure 1.7). Substances like hydrogen and oxygen that are composed of only one type of atom are classified as elements (Figure 1.8). Currently 112 elements are known. Of these, only about 90 are found in nature; the remainder have been created by scientists. The name and a symbol for each element is listed in the table at the front of the book and inside the back cover. Carbon (C), sulfur (S), iron (Fe), copper (Cu), silver (Ag), tin (Sn), gold (Au), mercury (Hg), and lead (Pb) were known in relatively pure form to the early Greeks and Romans and to the alchemists of ancient China, the Arab world, and medieval Europe. However, many others—such as aluminum (Al), silicon (Si), iodine (I), and helium (He)—were not discovered until the 18th and 19th centuries. Finally, artificial elements, such as technetium (Tc), plutonium (Pu), and americium (Am), were made in the 20th century using the techniques of modern physics.

Be sure to notice that only the first letter of an element's symbol is capitalized. For example, cobalt is Co and not CO. The notation CO represents the chemical compound carbon monoxide.

Figure 1.7 Electrolyzing water. When an electric current is passed through water, the elements hydrogen and oxygen are formed. The macroscopic observation is that liquid water generates two gases, hydrogen and oxygen. Here the water is represented as a three-particle, or three-atom, molecule consisting of two H atoms and one O atom. Hydrogen gas consists of two atoms of H; oxygen consists of two atoms of O. Thus, we write the symbols H_2O, H_2, and O_2. (*Photo, C. D. Winters; model, S. M. Young*)

Figure 1.8 The chemical elements copper, aluminum, mercury, and sulfur (clockwise from bottom left). All can be distinguished by their color. In addition, three of the elements—copper, aluminum, and sulfur—are solids, whereas mercury is a liquid. (*C. D. Winters*)

Many elements have names and symbols with Latin or Greek origins, but more recently discovered elements have been named for their place of discovery or for a person or place of significance (Table 1.1).

The table inside the front cover of this book, in which the symbol and other information for each element are enclosed in a box, is called the **periodic table.** We will describe this important tool of chemistry in more detail beginning in Chapter 2.

See the *Saunders Interactive General Chemistry CD-ROM,* Screen 1.5, Elements and Atoms.

Table 1.1 • THE NAMES OF SOME CHEMICAL ELEMENTS

Element	Symbol	Date of Discovery	Discoverer	Derivation of Name or Symbol
Berkelium	Bk	1950	G. T. Seaborg S. G. Thompson A. Ghiorso (U.S.)	Berkeley, California was the site of Seaborg's laboratory.
Copper	Cu	Ancient		Latin, *cuprum*, copper, or *cyprium*, from Cyprus.
Lead	Pb	Ancient		Latin, *plumbum*, lead, meaning heavy.
Oxygen	O	1774	J. Priestley (G. B.) K. W. Scheele (Sweden)	French, *oxygene*, generator of acid, derived from the Greek, *oxy* and *genes* meaning acid forming (because oxygen was thought to be part of all acids).

Current Perspectives

Coral Chemistry and Broken Bones

Some years ago one of the authors of this book took his family on a vacation to a tropical island where they sailed, snorkeled among the fish and corals, and walked along beautiful beaches. While surfing one day, however, his son was tossed by a wave and suffered a severely broken neck. Fortunately, he was not paralyzed, so he was flown to a hospital where surgeons repaired a broken vertebra. The repair was made by grafting a piece of his hipbone onto the broken vertebra, a successful but very painful procedure.

The young man with the broken neck was not alone. Every year Americans spend several billion dollars on plates and pins for broken bones and for dental implants and for reconstructive devices such as hip and knee implants. Nearly 500,000 bone grafts and 260,000 hip operations are performed in the U. S. annually. A recent estimate is that hip fractures alone cost the United States $10 billion each year. A new type of bone cement may help to lower some of these costs as well as reduce pain and suffering.

Photo of coral. (*Susan Blanchet/ Dembinsky Photo Associates*)

Over 10 years ago a graduate student at the University of California got an idea for making bone cement. He was studying coral, living substances in tropical waters that form their skeletons from chemical elements found in the seas: calcium, carbon, sodium, phosphorus, and oxygen. Coral skeletons are chemically very similar to human bones. Perhaps, the student thought, the right combination of ingredients, based on the elements found in corals, could lead to bone formation in humans. He was right. A combination of chemicals found in almost any laboratory, and in many foods or other natural substances, cures into a substance with the same structure as bone. So, the student formed a company, and he and others spent several years finding just the right combination of ingredients to create a paste that can be injected into the bony area to be repaired. Upon injection into the body, the paste hardens within about 10 minutes, and within 12 hours it has achieved 85% to 90% of its final strength. (See *Science*, March 24, 1995, Vol. 267, p. 1796; and *Chemical & Engineering News*, August 25, 1997, pp. 27–32.)

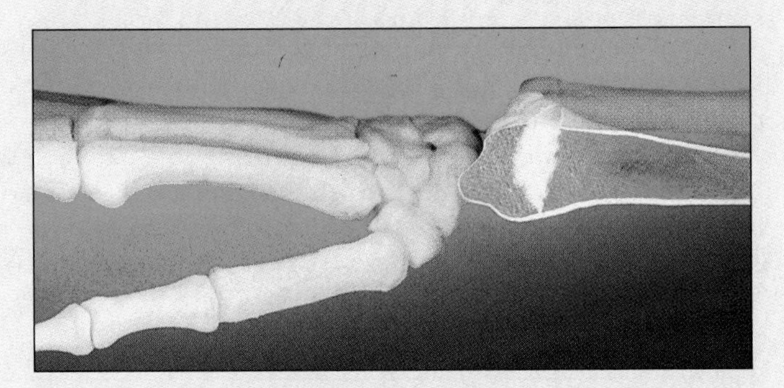

A distal radius (wrist) fracture repaired with bone cement (white area). (*Norian Corp., Cupertino, CA 95014*)

An **atom** is the smallest particle of an element that retains the chemical properties (see Section 1.5) of that element. Modern chemistry is based on an understanding and exploration of nature at the atomic level, and we have much more to say about atoms and atomic properties in Chapters 2, 7, and 8, in particular.

Exercise 1.2 Elements

Using the periodic table inside the front cover of this book,

(a) find the names of the elements with symbols Na, Cl, and Cr.

(b) find the symbols for the elements zinc, nickel, and potassium.

1.3 COMPOUNDS AND MOLECULES

See the *Saunders Interactive General Chemistry CD-ROM,* Screen 1.6, Compounds and Molecules.

Pure substances like sugar, salt, and water, which are composed of two or more different elements, are referred to as chemical compounds. Even though only 112 elements are known, there appears to be no limit to the number of compounds that can be made from those elements. More than 12 million compounds are now known, with about a half million added to the list each year.

When elements become part of a compound, their original properties, such as color, hardness, and melting point, are replaced by the characteristic properties of the compound. Consider ordinary salt (sodium chloride), which is composed of two elements.

- Sodium is a shiny metal that reacts violently with water.
- Chlorine is a light yellow gas that has a distinctive, suffocating odor and is a powerful irritant to lungs and other tissues.

Salt, on the other hand, is a white, crystalline solid with properties completely unlike those of the two elements from which it is made (Figure 1.9).

It is important to make a careful distinction between a mixture of elements and a compound of two or more elements. Pure metallic iron and yellow, powdered sulfur can be mixed in varying proportions. In the chemical compound known as iron pyrite (see Figure 1.4), however, this kind of variation cannot occur. Not only does iron pyrite exhibit properties peculiar to itself and different from those of either iron or sulfur, or a mixture of these elements, but it has a definite percentage composition by weight (46.55% Fe and 53.45% S, or 46.55 g of Fe and 53.45 g of S in 100.00 g of sample). Thus, two major differences ex-

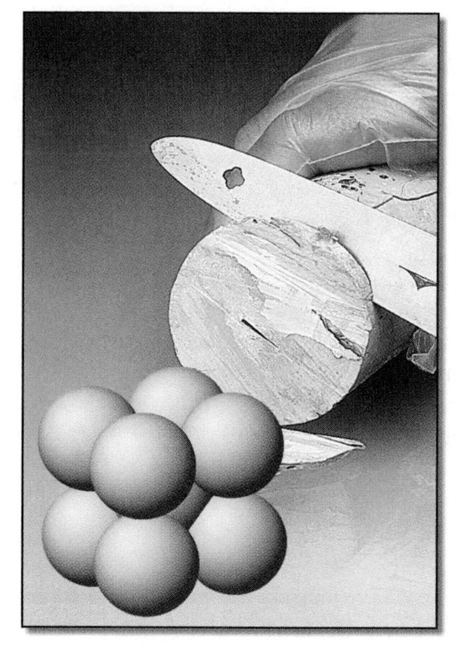

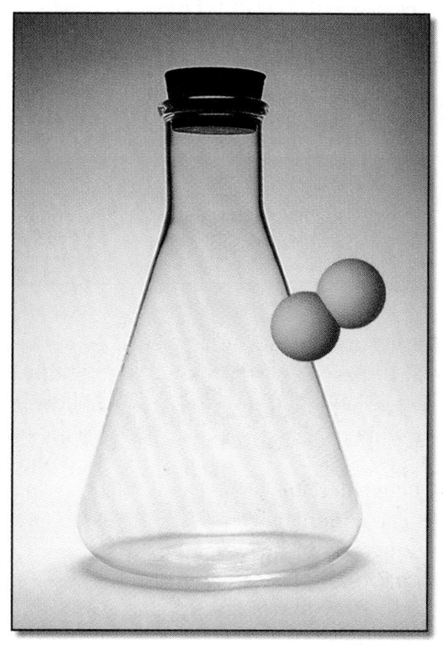

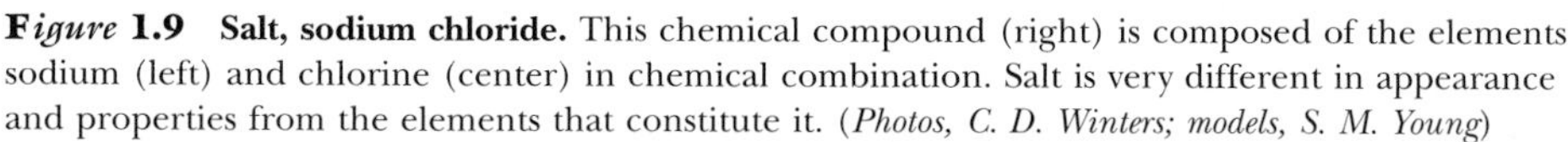

***Figure* 1.9 Salt, sodium chloride.** This chemical compound (right) is composed of the elements sodium (left) and chlorine (center) in chemical combination. Salt is very different in appearance and properties from the elements that constitute it. (*Photos, C. D. Winters; models, S. M. Young*)

ist between mixtures and pure compounds: Compounds have distinctly different properties from their parent elements, and compounds have a definite percentage composition (by mass) of their combining elements.

A reminder that many of the molecular models in the book are also on the *Saunders Interactive General Chemistry CD-ROM* or are found in the World Wide Web site for this book. See http://www.saunderscollege.com.

Some compounds—such as salt, NaCl—are composed of **ions,** which are electrically charged atoms or groups of atoms. (We describe such compounds beginning in Chapter 3.) Other compounds—such as water and sugar—consist of **molecules,** which are the smallest, discrete units that retain the chemical characteristics of the compound. The composition of any compound can be represented by its formula. You already know the formula for water, H_2O. The symbol for hydrogen, H, is followed by a subscript "2" to indicate that two atoms of hydrogen occur in a single water molecule. Because the symbol for oxygen appears in the formula without a subscript, only a single oxygen atom occurs in the molecule.

1.4 PHYSICAL PROPERTIES

See the *Saunders Interactive General Chemistry CD-ROM,* Screen 1.2, Physical Properties of Matter.

See the *Saunders Interactive General Chemistry CD-ROM,* Screen 1.8, Density, and Screen 1.9, Density on the Submicroscopic Scale. In the "toolbox" you will find an extensive database of properties of the elements, one of them the density of the element.

Your friends recognize you by your physical appearance: your height and weight and the color of your eyes and hair. The same is true of chemical substances. When you are cooking in the kitchen, you can distinguish sugar from water because you know that sugar consists of small white, solid particles, whereas water is a colorless liquid. Maple syrup for your pancakes is also a liquid, but it comes in light and dark colors and is much more viscous (pours more slowly) than water. Properties such as these, which can be observed and measured without changing the composition of a substance, are called **physical properties.** The chemical elements copper, aluminum, mercury, and sulfur, for example, clearly differ in color, appearance, and physical state (see Figure 1.8). Physical properties allow us to classify and identify substances of the material world. Table 1.2

Table 1.2 • Some Physical Properties

Property	Comment
Color	
State of matter	Is it a solid, liquid, or gas?
Melting point	The temperature at which a solid melts
Boiling point	The temperature at which a liquid boils
Heat of vaporization	The heat required to change a liquid to a vapor
Heat of fusion	The heat required to change a solid to a liquid
Density	Mass per unit volume
Solubility	Mass of substance that can dissolve in a given volume of water or other solvent
Metallic character	
Electric conductivity	
Conductivity of heat	
Magnetic properties	
Shape of crystals of a solid	
Malleability	The ease with which a solid can be deformed
Ductility	The ease with which a solid can be drawn into a wire
Viscosity	The susceptibility of a liquid to flow

lists some physical properties of matter that chemists commonly use. A few of these are discussed in more detail in this chapter, and others are taken up later in the text.

Exercise 1.3 Physical Properties

Identify as many physical properties in Table 1.2 as you can for the following common substances: (a) iron, (b) water, (c) table salt (whose chemical name is sodium chloride), and (d) oxygen.

Density

Density, the ratio of the mass of an object to its volume, is a physical property that is useful for identifying substances.

Your body mass index (BMI) is related to density (BMI = body mass/body surface area). Because muscle is denser than fat, a fit person will have a higher BMI than an unfit person of the same size. What is your BMI? (A healthy adult should have a BMI of about 20–25.)

$$\text{Density} = \frac{\text{mass}}{\text{volume}}$$

Your brain unconsciously uses the density of an object you want to pick up by estimating volume visually and preparing your muscles to lift the expected mass. For example, the person shown in Figure 1.10 is holding objects of similar size or volume but of widely different masses and therefore of widely differing densities. By looking at the appearance of the solids—their physical properties—can you tell which one has the higher density?

You've heard the expression "Get the lead out!" to mean hurry up. Why does this apply? Because lead has a very high density, 11.35 g/cm^3 (11.35 grams per cubic centimeter). That is, 1.00 cm^3 of lead has a mass of 11.35 g. In contrast, titanium, a metal used for its corrosion resistance, has a much lower density, only 4.5 g/cm^3. Although pieces of these metals could look quite similar, you could tell the difference between them by measuring their densities.

The density of pure water at 25 °C is 0.997 g/cm^3.

***Figure* 1.10** **A student holding two different solids.** From your experience, and the appearance of the solids, can you tell which one is denser? *(C. D. Winters)*

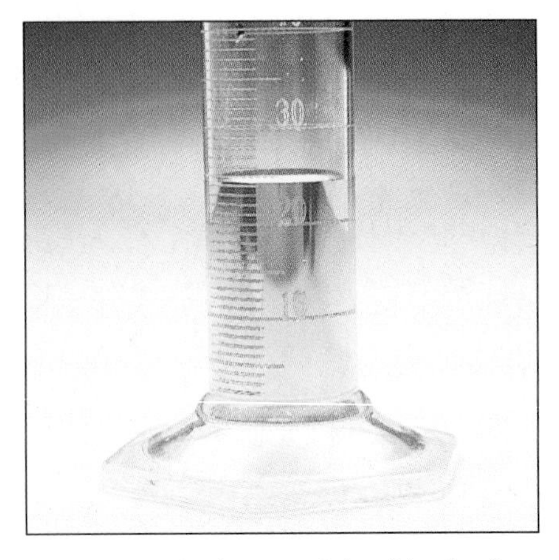

Graduated cylinder containing 24 mL of mercury. See text for a calculation using this information. (*C. D. Winters*)

If any two of three quantities—mass, volume, and density—are known for a sample of matter, the third can be calculated. Density equals mass divided by volume; therefore

$$\text{Volume} \times \text{density} = \text{volume } (\cancel{\text{cm}^3}) \times \frac{\text{mass (g)}}{\text{volume } (\cancel{\text{cm}^3})} = \text{mass (g)}$$

You can use this approach to find the mass of 24 cm^3 [or 24 mL (milliliters)] of mercury in the graduated cylinder in the photo. A handbook of information for chemistry lists the density of mercury as 13.534 g/cm^3 (at 20 °C).

$$24\ \cancel{\text{cm}^3} \cdot \frac{13.534\ \text{g}}{1\ \cancel{\text{cm}^3}} = 320\ \text{g}$$

Therefore, this very small volume of mercury has a large mass (about 3/4 of a pound using the English system of measurements).

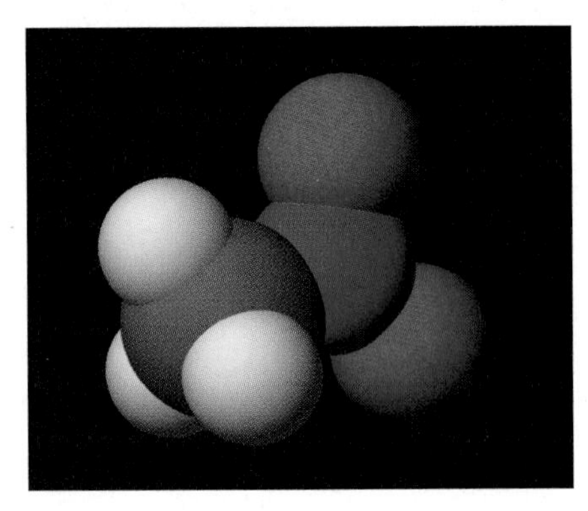

Nitromethane, the chemical compound referred to in Example 1.1, is used as a fuel in race cars.

Example 1.1 Density

Liquid nitromethane is used as a fuel for rockets and race cars. It has a density of 1.13 g/cm^3. What is the mass, in grams, of 56 L (or 56,000 cm^3) of the fuel?

Solution You know the density and volume of the sample. Because density is the ratio of the mass of a sample to its volume (density = mass/volume), then mass = volume × density. Therefore, to find the sample's mass, multiply the volume by the density.

$$56{,}000\ \cancel{\text{cm}^3} \cdot \frac{1.13\ \text{g}}{1\ \cancel{\text{cm}^3}} = 63{,}000\ \text{g}$$

The answer is 63,000 g, or about 140 pounds (to use a measurement common in the United States). Finally, be sure to notice that units of cm^3 cancel in the calculation.

Exercise 1.4 Density

The density of dry air is 1.12×10^{-3} g/cm^3. What volume of air, in cubic centimeters, has a mass of 15.5 g? (Answer = 1.38×10^4 cm^3)

Temperature

See the *Saunders Interactive General Chemistry CD-ROM*, Screen 1.10, Temperature.

Another useful physical property of pure elements and compounds is the temperature at which the solid melts (its melting point) or the liquid boils (its boiling point). **Temperature** is the property of matter that determines whether heat energy can be transferred from one body to another and the direction of that transfer: Heat energy transfers spontaneously only from a hotter object to a cooler one.

The number that represents an object's temperature depends on the unit chosen for the measurement. Three scales for temperature measurement are in common use today: **Fahrenheit, Celsius,** and **Kelvin** units (Figure 1.11). The

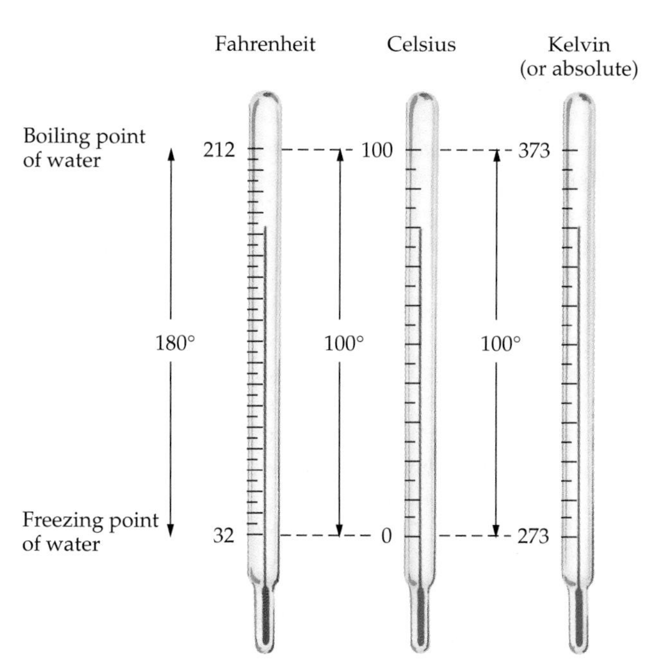

Figure 1.11 A comparison of Fahrenheit, Celsius, and Kelvin temperature scales. The reference, or starting point, for the Kelvin scale is absolute zero (0 K = −273.15 °C), which has been shown theoretically to be the lowest possible temperature. Note that the abbreviation K for the kelvin unit is used *without* the degree sign (°). Also note that 1 °C = 1 K = (9/5) °F.

Celsius scale is generally used for measurements in the laboratory. When calculations incorporate temperature data, however, the data generally must be expressed in kelvins.

The Celsius and Fahrenheit Temperature Scales

In the United States everyday temperatures are reported using the Fahrenheit scale, but the Celsius scale is used in most other countries and in scientific notation. The latter scale was first suggested by Anders Celsius (1701–1744), a Swedish astronomer, and like the Fahrenheit scale it is based on the properties of water. The size of the Celsius degree is defined by assigning zero as the freezing point of pure water (0 °C) and 100 as its boiling point (100 °C).

The size of the Fahrenheit degree is equally arbitrary. Gabriel Fahrenheit (1686–1736), a German physicist, defined 0 °F as the freezing point of a solution in which he had dissolved the maximum amount of salt (because this was the lowest temperature he could reproduce reliably), and he intended 100 °F to be the normal human body temperature (but this turned out to be 98.6 °F). Today, the reference points are set at 32 °F and 212 °F, the freezing and boiling points of pure water, respectively. The number of units between these points is 180 Fahrenheit degrees (see Figure 1.11).

If the Celsius and Fahrenheit units are compared, the Celsius degree is seen to be 1.8 times larger than the Fahrenheit degree. It takes only 5 Celsius degrees to cover the same temperature range as 9 Fahrenheit degrees, and this relationship can be used to convert a temperature on one scale to a temperature on the other.

To help you think in terms of the Celsius scale, it is useful to know that water freezes at 0 °C and boils at 100 °C, a comfortable room temperature is about 22 °C (72 °F), your average body temperature is 37 °C (98.6 °F), and the hottest water you could put your hand into without serious burns is about 60 °C (140 °F).

William Thomson, also known as Lord Kelvin, was the professor of natural philosophy at the University in Glasgow, Scotland, from 1846 to 1899. He was best known then for his work on heat and work, from which came the concept of the absolute temperature scale. He is also remembered, however, for his part in a famous 19th century debate over the age of the earth. (*E. F. Smith Collection/Van Pelt Library/University of Pennsylvania*)

The Kelvin Temperature Scale

Winter temperatures in many places can easily drop below 0 °C, that is, to temperatures given by negative numbers on the Celsius scale. In the laboratory, much colder temperatures can be achieved easily, and the temperatures are given by even larger negative numbers. There is a limit to how low the temperature can go, however, and hundreds of experiments have found that the limiting temperature is −273.15 °C (or −459.67 °F).

William Thomson, known as Lord Kelvin (1824–1907), first suggested a temperature scale that does not use negative numbers. Kelvin's scale, now adopted as the international standard for science, uses the same size degree as the Celsius scale, but it takes the lowest possible temperature as its zero, a point called **absolute zero.** Because *kelvin and Celsius units are the same size,* the freezing point of water is reached 273.15 degrees above the starting point; that is, 0 °C is the same as 273.15 kelvins, or 273.15 K. Temperatures in Celsius units are readily converted to kelvins, and vice versa, using the relation

$$t(\mathrm{K}) = \frac{1\ \mathrm{K}}{1\ {}^{\circ}\mathrm{C}}(t\ {}^{\circ}\mathrm{C} + 273.15\ {}^{\circ}\mathrm{C})$$

Thus, a common room temperature of 23.5 °C is

$$t(\mathrm{K}) = \frac{1\ \mathrm{K}}{1\ {}^{\circ}\mathrm{C}}(23.5\ {}^{\circ}\mathrm{C} + 273.15\ {}^{\circ}\mathrm{C}) = 296.7\ \mathrm{K}$$

Finally, notice that the degree symbol (°) is not used with Kelvin temperatures. The name of the unit on this scale is the kelvin (not capitalized), and such temperatures are designated with a capital K.

Exercise 1.5 Temperature Changes

Liquid nitrogen boils at 77 K. What is this temperature in Celsius degrees? (Answer = −196 °C)

Temperature (°C)	Density of Water (g/cm^3)
0 (ice)	0.917
0 (liq water)	0.99984
2	0.99994
4	0.99997
10	0.99970
20	0.99820
100	0.95836

Temperature Dependence of Physical Properties

The temperature of a sample of matter often affects the numerical values of its properties. Density is a particularly important example. Although the change in density of water relative to temperature seems quite small, it profoundly affects our environment. For example, as the water in a lake cools, the density of the water increases, and the denser water sinks (Figure 1.12). This continues until the lake water reaches 3.98 °C, the temperature at which water has its maximum density (0.999973 g/cm^3). If the temperature drops further, the density decreases slightly, and this colder water floats on top of water at 3.98 °C. This leads to the seasonal "turnover" of water observed in lakes. As water cools and sinks deeper in a lake, it carries with it dissolved oxygen, which replenishes the oxygen lost in the decay of plants and used in the metabolism of fish and aquatic life. Turnover is also essential in recycling nutrients from decayed plants and animals back into the ecosystem.

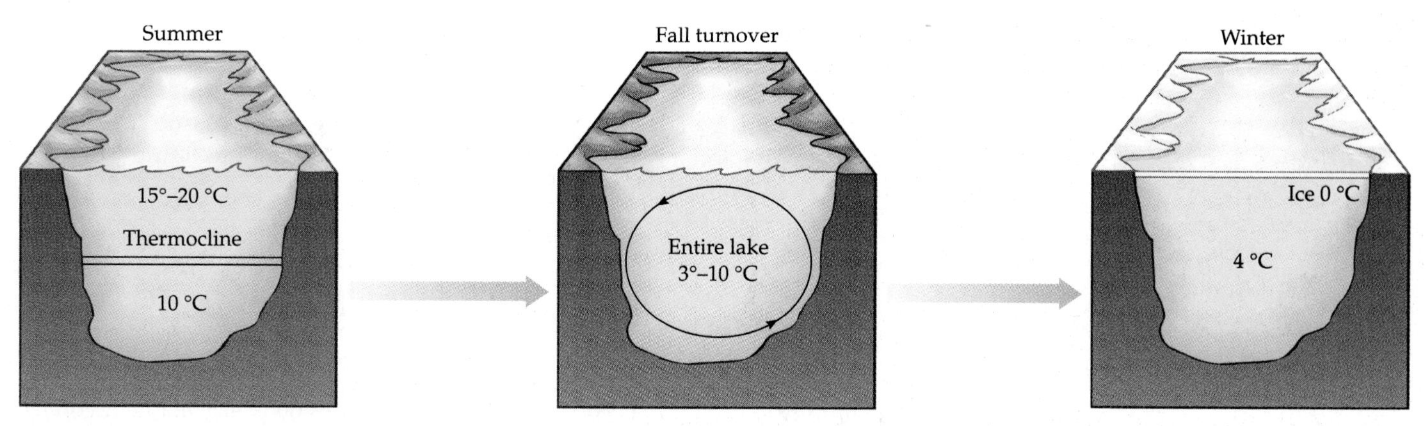

Figure 1.12 The density of water. Although the density of water changes to a very small extent with temperature, the change is sufficient to have profound environmental effects. As water in a lake cools down its density increases, and the denser water sinks. This continues until the lake water reaches 3.98 °C, the temperature at which water has its maximum density. If the temperature drops further the density decreases slightly, and this less dense water rises. The importance of this is that it leads to the seasonal "turnover" of water in lakes. The density of ice is much less than that of water at any temperature, so ice floats on liquid water. This means that lakes and rivers don't freeze solid, no matter how cold the winter gets.

If water cools still further, solid ice forms. The density of ice is much less than of water, so ice floats on water.

Because the density of liquids changes with temperature, you should always report the temperature when you make a volume measurement in the laboratory. [This is also the reason that laboratory glassware that is used to make accurate volume measurements always specifies the temperature (Figure 1.13).]

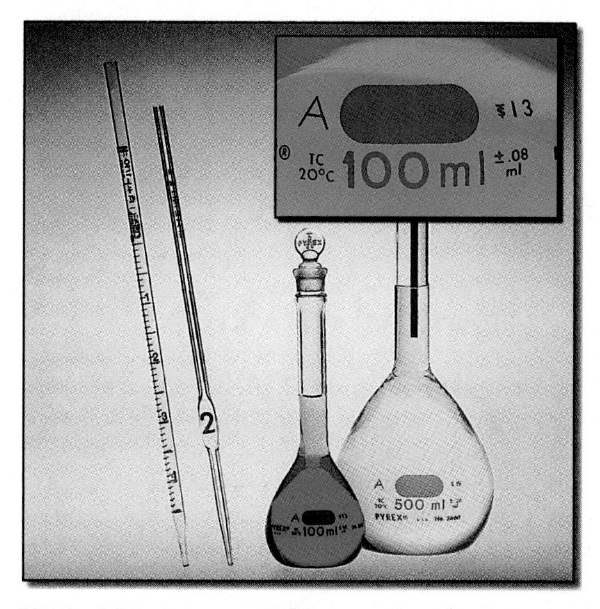

Figure 1.13 Laboratory glassware and temperature. These pieces of laboratory glassware will contain exactly the volume specified only if the temperature is as marked. *(C. D. Winters)*

Exercise 1.6 Density and Temperature

Carbon tetrachloride, CCl_4, a liquid compound commonly found in the laboratory, has a density of 1.58 g/cm^3. If you place a piece of a plastic soda bottle (d = 1.37 g/cm^3) and a piece of aluminum (d = 2.70 g/cm^3) in liquid CCl_4, will the plastic or aluminum float or sink?

Problem-Solving Tips and Ideas

1.1 Finding Data

All the information you may need to solve a problem in this book may not be presented in the problem. For example, although the density of the liquid was given in Example 1.1, we could have left it out and assumed you would (a) recognize that you needed density to convert a volume to a mass and (b) know where to find the information. The Appendices of this book contain a wealth of information, and even more is available on the *Saunders Interactive General Chemistry CD-ROM.* Various handbooks of information are also available in most libraries, and among the best are the *Handbook of Chemistry and Physics* (CRC Press) and *Lange's Handbook of Chemistry* (McGraw-Hill). Another excellent source of data is produced by the National Institutes for Standards and Technology (http://www.nist.org).

Extensive and Intensive Properties

Extensive properties depend on the amount of a substance present. Thus, the mass and volume of each sample of an element in Figure 1.8 are extensive properties. In contrast, **intensive properties** are those that do not depend on the amount of substance. A sample of ice will melt at 0 °C, no matter whether you have an ice cube or an iceberg.

Density is an intensive property, which seems curious at first because it is the quotient of two extensive properties. On reflection, however, you recognize that the density of gold, for example (19.3 g/cm^3), does not depend on the size of the sample. Whether you have a flake of pure gold, or a solid gold ring, both have the same ratio of mass to volume.

Figure **1.14 A physical property used to distinguish compounds.** Aspirin and naphthalene are both white solids. You can tell them apart by, among other things, a difference in physical properties. At the temperature of boiling water, naphthalene (melting point 80.2 °C) is a liquid, whereas aspirin (melting point, 135 °C) is a solid. (*C. D. Winters*)

1.5 PHYSICAL AND CHEMICAL CHANGE

Chemistry is about change. Changes involving physical properties are called **physical changes.** In a physical change the chemical identity of a substance is preserved even though it may have changed its physical state or the gross size and shape of its pieces. An example of a physical change is the melting of a solid, and the temperature at which this occurs is often so characteristic that it can be used to identify the solid. For example, naphthalene (used as a moth repellent) and aspirin resemble each other in outward appearance (both are white, powdery solids), but naphthalene melts at 80.2 °C, whereas aspirin melts at 135 °C (Figure 1.14).

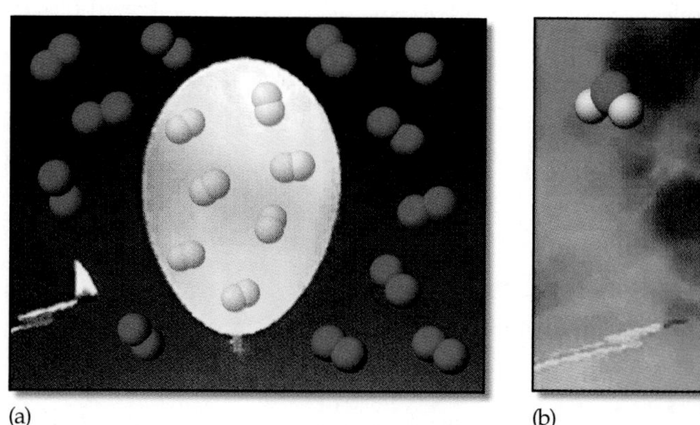
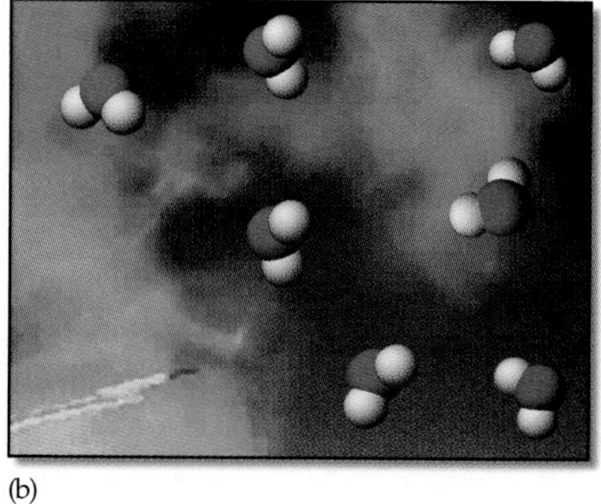

(a) (b)

Figure 1.15 The explosion of a hydrogen/oxygen mixture. (a) A balloon filled with molecules of hydrogen gas, H_2, and surrounded by molecules of oxygen, O_2, in the air. (The balloon floats in air because gaseous H_2 is less dense than air.) (b) When H_2 and O_2 mix, and the mixture is ignited, it explodes, producing water, H_2O. (*Photos are from a video on Screen 1.11 of the Saunders Interactive General Chemistry CD-ROM; models, S. M. Young*)

Now suppose a lighted candle is brought up to a balloon filled with hydrogen gas (Figure 1.15). When the heat causes the skin of the balloon to rupture, the hydrogen mixes with the oxygen in the air, and the heat of the candle sets off an explosion. This is an example of a **chemical change,** or **chemical reaction,** because one or more substances (the **reactants**) are transformed into one or more different substances (the **products**).

See the *Saunders Interactive General Chemistry CD-ROM,* Screen 1.11, Chemical Change, and Screen 1.12, Chemical Change on the Molecular Scale.

At the molecular, or particulate, level a chemical change produces a new arrangement of atoms without a gain or loss in the number of atoms of each kind. The molecules present after the reaction are different from those present before the reaction. Here hydrogen and oxygen molecules react to form water molecules, which can be represented as

Recall that H atoms are white and O atoms are red in molecular models. See the color chart in the Preface.

$$2\ H_2(g) + O_2(g) \longrightarrow 2\ H_2O$$

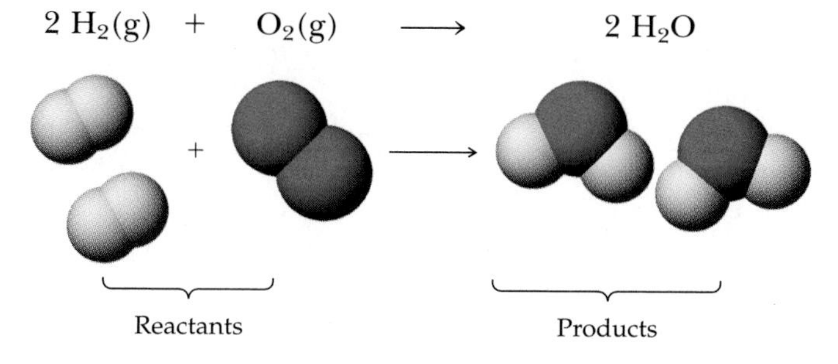

Reaction of hydrogen and oxygen to give water.

This symbolic representation of the change is called a **chemical equation** because it shows that the substances on the left produce, or "give," the substances on the right. The equation suggests that if there are four atoms of H and 2 atoms of O before reaction, the same number must be present after reaction, albeit in a different arrangement.

In contrast with a chemical change, a physical change does not result in a new chemical substance. The molecules present before and after the change are the same, but their arrangement relative to one another (farther apart in a gas, closer together in a solid, for example) is different.

Finally, physical changes and chemical changes are often accompanied by transfer of energy. The reaction of hydrogen and oxygen to give water (see Figure 1.15) transfers a tremendous amount of energy (in the form of heat and light) to its surroundings.

Exercise 1.7 Chemical Reactions and Physical Changes

When camping in the mountains, you boil a pot of water on a campfire. What physical and chemical changes take place in this process?

1.6 UNITS OF MEASUREMENT

See the *Saunders Interactive General Chemistry CD-ROM,* Screen 1.15, Units of Measurement, and Screen 1.16, The Metric System.

Doing chemistry requires observing chemical reactions and physical changes. In the laboratory you mix two solutions and see a beautiful golden yellow solid form, and, because the solid is denser than water, the solid drops to the bottom of the test tube. The color and appearance of the substances involved, if heat was involved, or if the process occurred very quickly or very slowly, are qualitative observations. No measurements and numbers were involved.

To understand a chemical reaction more completely, chemists usually make **quantitative** measurements. These involve numerical information. For example, if two compounds react with each other, how much product forms? Do we have to heat the reactants to make the change occur, and if so, how much heat is required and for how long?

In chemistry, measurements involve time, mass, volume, and distance, among other things. The image of copper atoms on a surface in Figure 1.3 is 1.7 nm (nanometers) square. A nanometer (nm) is equivalent to 10^{-9} m (meter), a very small dimension to engineers, a large one to chemists, but a common one to biologists—a bacterium is about 1000 nm in length, whereas a typical molecule is only about 0.1 nm across.

To understand much of science and engineering, it is essential to be familiar with the dimensions of the objects being studied, the units being used, and the relation between them. The scientific community has chosen a modified version of the **metric system** as the standard system for recording and reporting measurements. This is a decimal system, in which all of the units are expressed as powers of 10 times some basic unit. The resulting system, as applied internationally in science, is called the Système International d'Unités (International System of Units), abbreviated SI.

SI Units

In SI, all units are derived from base units, some of which are listed in Table 1.3. Larger and smaller quantities are expressed by using an appropriate prefix with the base unit (Table 1.4). For instance, highway distances are given in kilometers, in which 1 km (kilometer) is exactly 1000, or 10^3, m (meters). In chemistry, length is most often given in subdivisions of the meter, such as centimeters (cm) or millimeters (mm). The prefix "centi-" means 1/100, so 1 centimeter is 1/100 of a meter ($1 \text{ cm} = 1 \times 10^{-2}$ m); 1 millimeter is 1/1000 of a meter

Table 1.3 • SOME SI BASE UNITS

Measurement	Name of Unit	Abbreviation
Mass	kilogram	kg
Length	meter	m
Time	second	s
Temperature	kelvin	K
Amount of substance	mole	mol
Electric current	ampere	A

(1 mm = 1×10^{-3} m). On the atomic scale, dimensions are often given in nanometers (nm) (1 nm = 1×10^{-9} m) or picometers (pm) (1 pm = 1×10^{-12} m).

Several conversion factors that allow you to convert between SI and non-SI units are given in Appendix C and inside the back cover of the book.

1.7 USING NUMERICAL INFORMATION

See the *Saunders Interactive General Chemistry CD-ROM,* Screen 1.17, Using Numerical Information.

As part of your course in chemistry you will prepare materials in the laboratory that will require you to make some calculations. You will collect numerical data and use those data to calculate a result, or you will look for correlations among the pieces of data. All these activities mean you have to be comfortable handling numerical information. This section describes some common calculations and proper ways to handle quantitative information.

Suppose you are given a rectangular piece of aluminum and are asked to find its density in units of grams per cubic centimeter. Because density is the ratio of mass to volume, you need to measure the mass and determine the volume of the piece. The data in the margin were collected in the laboratory. To find the volume of the aluminum sample in cubic centimeters, you multiply its length by its width and its thickness. First, however, all the measurements must have the same units, meaning that the thickness must be converted to centimeters.

$$3.1 \text{ mm} \cdot \frac{1 \text{ cm}}{10 \text{ mm}} = 0.31 \text{ cm}$$

DETERMINING THE DENSITY OF ALUMINUM

Measurement	Data Collected
Mass of aluminum	13.56 g
Length	6.45 cm
Width	2.50 cm
Thickness	3.1 mm

Table 1.4 • SELECTED PREFIXES USED IN THE METRIC SYSTEM

Prefix	Abbreviation	Meaning	Example
mega-	M	10^6	1 megaton = 1×10^6 tons
kilo-	k	10^3	1 kilogram (kg) = 1×10^3 g
deci-	d	10^{-1}	1 decimeter (dm) = 1×10^{-1} m
centi-	c	10^{-2}	1 centimeter (cm) = 1×10^{-2} m
milli-	m	10^{-3}	1 millimeter (mm) = 1×10^{-3} m
micro-	μ	10^{-6}	1 micrometer (mm) = 1×10^{-6} m
nano-	n	10^{-9}	1 nanometer (nm) = 1×10^{-9} m
pico-	p	10^{-12}	1 picometer (pm) = 1×10^{-12} m

With all the dimensions of the piece of aluminum in the same unit, its volume and density can be calculated:

$$\text{Length} \times \text{width} \times \text{thickness} = \text{volume}$$

$$6.45 \text{ cm} \cdot 2.50 \text{ cm} \cdot 0.31 \text{ cm} = 5.0 \text{ cm}^3$$

$$\text{Density} = \frac{13.56 \text{ g}}{5.0 \text{ cm}^3} = 2.7 \text{ g/cm}^3$$

See Appendix A for more information on using numerical information.

In a conversion factor (such as 1 cm/10 mm), the top and bottom terms are equivalent quantities expressed in different units. Here 1 cm is equivalent to 10 mm.

You will find that *dimensional analysis,* an approach explored further in Appendix A and used throughout the book, is very useful in solving problems such as the density calculation. Using this approach to change 3.1 mm to its equivalent in centimeters, for example, you multiply the number you wish to convert (3.1 mm) by a factor called a **conversion factor** (here 1 cm/10 mm) to produce a result in the desired unit (0.31 cm). Units are handled like numbers, and because the unit "mm" was in both the numerator and the denominator, dividing one by the other leaves a quotient of 1. We often say the units "canceled out." Here this leaves the answer in centimeters, the desired unit.

A conversion factor expresses the equivalence of a measurement in two different units (1 cm ≡ 10 mm; 1 g ≡ 1000 mg; 12 eggs ≡ 1 dozen). Because the numerator and denominator describe the same distance or quantity, the factor is equivalent to the number 1. Therefore, multiplication by this factor does not change the measured distance or quantity, only its units. A conversion factor is always written so that it has the form "new units divided by units to be converted."

$$\text{Number} \cdot (\cancel{\text{unit}}) \times \left[\frac{\text{new unit}}{\cancel{\text{unit to be converted}}}\right] = \text{new number} \cdot \text{new unit}$$

↑ Quantity to express in new units — ↑ Conversion factor — ↑ Quantity now expressed in new units

This way the units in the denominator "cancel" the units of the original data, leaving the desired units. In the previous problem, units of millimeters canceled, leaving the result in centimeters.

Problem-Solving Tips and Ideas

1.2 Using Scientific Notation

The number 0.001 or 1/1000 is written as 1×10^{-3}. This notation—called scientific, or exponential, notation—is used throughout the book and is explained further in Appendix A. Scientific notation makes it easier to handle very large or very small numbers.

Make sure you know how to use your calculator to solve problems with exponential numbers. When entering a number such as 1.23×10^{-4} into your calculator, you first enter 1.23 and then press a key marked EE or EXP (or something very similar). This enters the " × 10" portion of the notation for you. You then complete the entry by keying in the exponent of the number, −4. (To change the exponent from +4 to −4, you need to press the "+/−" key.)

A common error made by students is to enter 1.23, then press the multiply key (×) and then key in 10 before finishing by pressing EE or EXP followed by −4. This gives you an entry that is 10 times too large. Try this! Experiment with your calculator so you are sure you are entering data correctly.

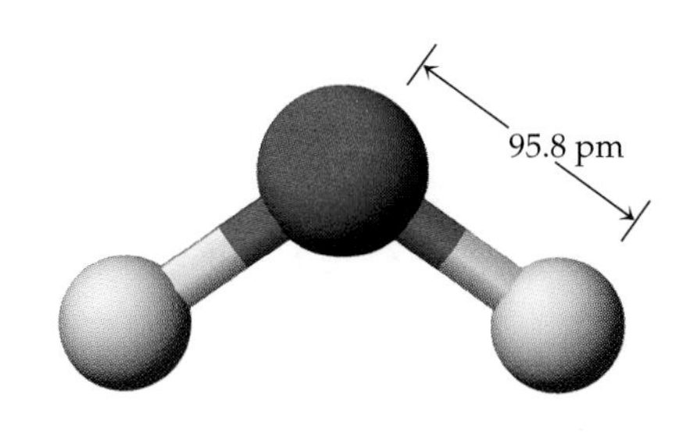

Example 1.2 Distances on the Molecular Scale

An oxygen atom is connected to two H atoms in water. The distance between O and H is 95.8 pm. What is this distance in meters? In nanometers (nm)?

Solution You can solve this problem by knowing the conversion factor between the units in the information you are given (picometers) and the units of the desired outcome (meters or nanometers). There is no direct conversion from nanometers to picometers given in our tables, but relationships are listed between meters and picometers and between meters and nanometers (Table 1.4; 1 pm = 1×10^{-12} m and 1 nm = 1×10^{-9} m). Therefore, we first convert picometers to meters and then convert meters to nanometers.

$$\text{Picometers} \xrightarrow{\times \frac{\text{m}}{\text{pm}}} \text{Meters} \xrightarrow{\times \frac{\text{nm}}{\text{m}}} \text{Nanometers}$$

$$95.8\ \text{pm} \cdot \frac{1 \times 10^{-12}\ \text{m}}{1\ \cancel{\text{pm}}} = 95.8 \times 10^{-12}\ \text{m} = 9.58 \times 10^{-11}\ \text{m}$$

$$9.58 \times 10^{-11}\ \cancel{\text{m}} \cdot \frac{1\ \text{nm}}{1 \times 10^{-9}\ \cancel{\text{m}}} = 9.58 \times 10^{-2}\ \text{nm (or 0.0958 nm)}$$

Notice how units cancel to leave an answer whose unit is that of the numerator of the conversion factor.

Exercise 1.8 Interconverting Units of Length

The pages of a typical textbook are 25.3 cm long and 21.6 cm wide. What is each length in meters? In millimeters? What is the area of a page in square centimeters? In square meters? (Answer = 0.253 m; 253 mm; 546 cm^2; 0.0546 m^2)

Exercise 1.9 Density

A platinum sheet is 2.50 cm square and has a mass of 1.656 g. The density of platinum is 21.45 g/cm^3. What is the thickness of the platinum sheet, in millimeters? (Answer = 0.124 mm)

Chemists often handle chemicals in glassware such as beakers, flasks, pipets, graduated cylinders, and burets, which are marked in volume units (Figure 1.16). The SI unit of volume is the cubic meter (m^3), which is too large for everyday laboratory use. For example, if you used cubic meters, the volume of a common laboratory beaker would be 0.0006 m^3 and you would often work with volumes of chemicals in the range of 0.001 m^3 or less in the laboratory.

A number such as 0.001 m^3 is inconvenient for routine use, so we usually use the unit called the **liter,** symbolized by **L.** A cube with sides equal to 10 cm (0.1 m) has a volume of 10 cm × 10 cm × 10 cm = 1000 cm^3 (or 0.001 m^3). This is defined as 1 liter.

$$1\ \text{liter (L)} = 1000\ \text{cm}^3$$

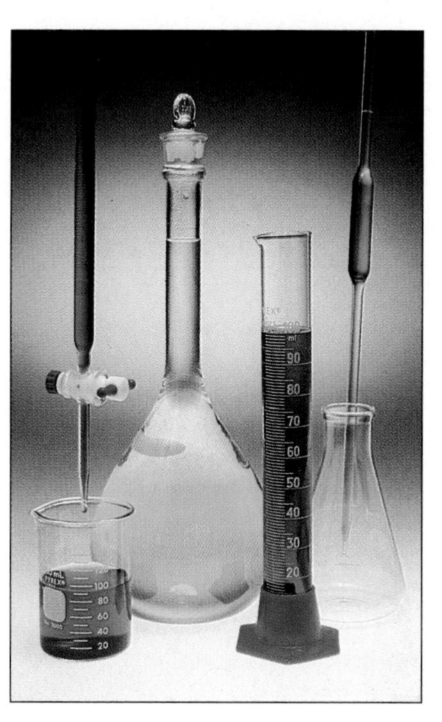

***Figure* 1.16 Some common laboratory glassware**. Volumes are marked in units of milliliters (mL). Remember that 1 mL is equivalent to 1 cm^3. (*C. D. Winters*)

The liter is a convenient unit to use in the laboratory, as is the milliliter (mL). Because there are 1000 mL and 1000 cm^3 in a liter, this means that

$$1\ cm^3 = 0.001\ L = 1\ \text{milliliter}\ (1\ mL)$$

Because the liter is exactly equal to 1000 cm^3, the terms *milliliter and cubic centimeter* (or "cc") *are interchangeable.* Therefore, a flask that contains exactly 125 mL has a volume of 125 cm^3 or one-eighth of a liter.

$$125\ \cancel{cm^3} \cdot \frac{1\ L}{1000\ \cancel{cm^3}} = 0.125\ L$$

The cubic decimeter (dm^3) is not widely used in the United States, but it is common in the rest of the world. A length of 10 cm is called a decimeter (dm) because it is 1/10 of a meter. Because a cube 10 cm on a side defines a volume of one liter, a liter is equivalent to a cubic decimeter: 1 L = 1 dm^3. Products in Europe and other parts of the world are often sold by the cubic decimeter.

Example 1.3 Units of Volume

A laboratory beaker has a volume of 0.6 L. What is its volume in cubic centimeters and milliliters?

Solution The relation between liters and cubic centimeters is 1 L = 1000 cm^3. Therefore, you should multiply 0.6 L by the conversion factor (1000 cm^3/L). The units of L cancel to leave an answer with units of cm^3.

$$0.6\ \cancel{L} \cdot \frac{1000\ cm^3}{1\ \cancel{L}} = 600\ cm^3$$

Because cubic centimeters and milliliters are equivalent, we can also say the volume of the beaker is 600 mL.

Exercise 1.10 Volume

(a) A standard wine bottle has a volume of 750 mL. How many liters does this represent?

(b) One U.S. gallon is equivalent to 3.7865 L. How many liters are in a 2.0-quart carton of milk? (Remember there are 4 quarts in a gallon.) How many cubic decimeters? (Answer = 1.9 L = 1.9 dm^3)

To determine the density of a piece of aluminum, we needed to know its mass. The **mass** of a body is the fundamental measure of the amount of matter in that body, and the SI unit of mass is the kilogram (kg). Smaller masses are expressed in grams (g) or milligrams (mg) (Table 1.4).

$$1\ kg = 1000\ g$$
$$1\ g = 1000\ mg$$

In the United States the English system of mass measurement is most commonly used, so masses are often given in pounds; 1 lb (pound) is equivalent to 453.59237 g. This means that a mass of 1.00 kg is equivalent to 2.20 lb, or about the mass of a quart of milk. But let us say you are living in Portugal and you have bought 750 g of strawberries. How many kilograms does this represent and how many pounds?

$$750 \not{g} \cdot \frac{1 \text{ kg}}{1000 \not{g}} = 0.75 \text{ kg}$$

$$750 \not{g} \cdot \frac{1 \text{ lb}}{454 \not{g}} = 1.7 \text{ lb}$$

Example 1.4 Mass in Kilograms, Grams, and Milligrams

A new U.S. penny has a mass of 2.49 g. Express this mass in kilograms and milligrams.

Solution Here the relation between the unit of the desired answer and the unit of the information given is 1 kg = 1000 g and 1000 mg = 1 g. Therefore, multiply the mass in grams by a conversion factor that has the form "(units for answer/units of information given)."

$$2.49 \not{g} \cdot \frac{1 \text{ kg}}{1000 \not{g}} = 2.49 \times 10^{-3} \text{ kg}$$

$$2.49 \not{g} \cdot \frac{1000 \text{ mg}}{1 \not{g}} = 2.49 \times 10^{3} \text{ mg}$$

Example 1.5 Density in Different Units

Oceanographers often express the density of sea water in units of kilograms per cubic meter. If the density of sea water is 1.025 g/cm^3 at 15 °C, what is its density in kilograms per cubic meter?

Solution To simplify this problem, let us break it into two steps. We shall first change grams to kilograms and then convert cubic centimeters to cubic meters. The density is the ratio of the mass in kilograms to the volume in cubic meters.

$$1.025 \not{g} \cdot \frac{1 \text{ kg}}{1000 \not{g}} = 1.025 \times 10^{-3} \text{ kg}$$

No direct conversion factor is available in one of our tables for changing units of cubic centimeters to cubic meters. We can find one, however, by cubing (raising to the third power) the relation between the meter and the centimeter.

$$1 \text{ cm}^3 \cdot \left(\frac{1 \text{ m}}{100 \text{ cm}}\right)^3 = 1 \not{\text{cm}^3} \cdot \left(\frac{1 \text{ m}^3}{1 \times 10^6 \not{\text{cm}^3}}\right) = 1 \times 10^{-6} \text{ m}^3$$

Therefore, the density of sea water is

$$\text{Density} = \frac{1.025 \times 10^{-3} \text{ kg}}{1 \times 10^{-6} \text{ m}^3} = 1.025 \times 10^{3} \text{ kg/m}^3$$

Current Perspectives

The Gimli Glider and the Importance of Units

On July 23, 1983, a new Boeing 767 was flying smoothly at 26,000 ft from Montreal to Edmonton as Air Canada Flight 143. A warning buzzer sounded in the cockpit. It appeared a fuel pump had failed. And then another warning buzzer! Another fuel pump problem. The pilots immediately changed course for Winnipeg, the closest large airport, but more warning buzzers indicated a fueling problem. Both of the mammoth jet engines of the plane shut down. One of the world's largest planes was now a glider, not because the fuel pumps had failed but because the plane had run out of fuel!

How did this modern jet airplane, equipped with the latest technology, run out of fuel? A simple mistake had been made in calculating the amount of fuel required for the flight.

Like all Boeing 767s, this plane had a sophisticated fuel gauge, but it was not working properly. The plane was still allowed to fly, however, because there is an alternative method of determining the quantity of fuel in the tanks. Mechanics can use a stick, much like the oil dipstick in an automobile engine, to measure the fuel level in each of the three tanks. The mechanics in Montreal read the dipsticks, which were calibrated in centimeters, and translated those readings to a volume in liters. According to this, the plane had a total of 7682 L of fuel.

Pilots always calculate fuel quantities in units of mass because they need to know the total mass of the plane before trying to take off. Air Canada pilots had always calculated the quantity of fuel in their planes in *pounds*. In accord with a plan to convert to metric units nationwide, however, the new 767's fuel consumption was given in *kilograms*. The pilots knew that 22,300 kg of fuel was required for the trip. If 7682 L of fuel remained in the tanks, how much had to be added? All they had to do was convert 7682 L to a mass in kilograms by multiplying by the fuel's density, calculate the mass to be added, and then convert that mass to the volume of fuel to be added, again using the fuel density.

The First Officer of the plane asked one of the mechanics for the factor necessary to do the volume-to-mass conversion, and the mechanic replied "1.77." Using that, the First Officer and the mechanics did the following calculation:

$$\text{Fuel remaining} = 7682\ \text{L} \cdot 1.77 = 13{,}597\ \text{kg}$$

$$\text{Fuel to be added} = 22{,}300\ \text{kg req'd} - 13{,}597\ \text{kg available} = 8703\ \text{kg}$$

$$\text{Volume to be added} = 8703\ \text{kg}/1.77 = 4916\ \text{L}$$

But later calculations show they should have added 20,163 L of fuel. Only about one fourth of the required fuel was added! Why? Because no one thought about the *units* of the number 1.77. They realized later that 1.77 has the unit of *pounds* per liter and not *kilograms* per liter. (See Study Question 75 at the end of this chapter.)

Out of fuel, the plane could not make it to Winnipeg, so controllers directed them to Gimli, to a small airport abandoned by the Royal Canadian Air Force. After gliding for almost 30 minutes, the plane approached the Gimli runway. The runway, however, had been converted to a race course for cars, and a race was being held that day. Furthermore, a steel barrier had been erected across the runway. Nonetheless, the pilot managed to touch down very near the end of the runway and, as the plane sped down the concrete strip, the nose wheel collapsed and several tires blew. The plane skidded to a stop just before the barrier, and the passengers and crew rushed down the emergency chutes. The Gimli glider had made it! And somewhere an aircraft mechanic is paying more attention to units on numbers.

After running out of fuel, Air Canada Flight 143 glided 29 minutes before landing on an abandoned airstrip at Gimli, Manitoba, near Winnipeg. (*AP/Wide World Photos*)

Exercise 1.11 Mass and Density

(a) The amount of vitamin C in a tablet is 5.00×10^2 mg. How many grams is this? How many kilograms?

(b) The density of iron is 7.874 g/cm^3. What is the mass of iron, in kilograms, in a slab that is 4.5 m long, 1.5 m wide, and 25 cm thick? (Answer = 1.3×10^4 kg)

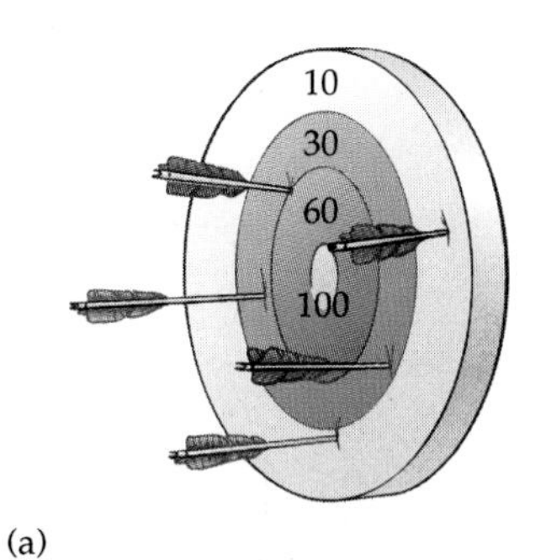

(a)

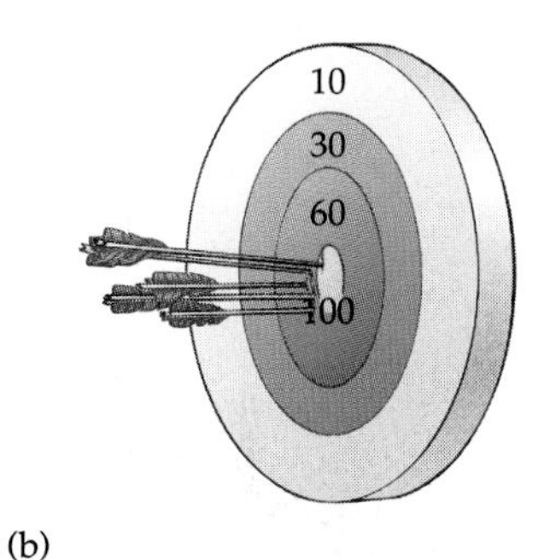

(b)

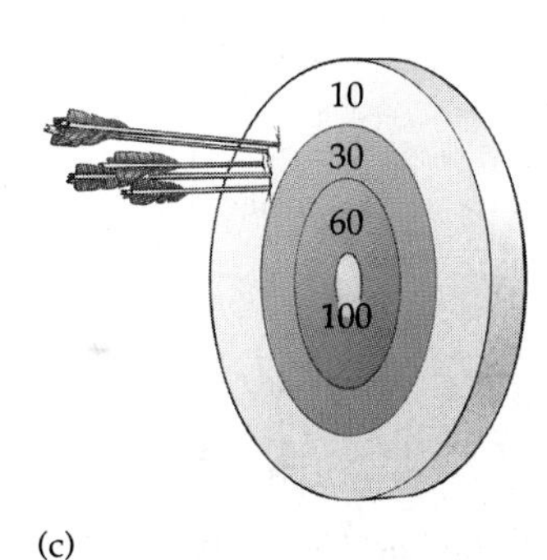

(c)

Figure 1.17 Precision and accuracy. (a) Poor precision and poor accuracy. (b) Good precision and good accuracy. (c) Good precision and poor accuracy.

Precision, Accuracy, and Experimental Error

The **precision** of a measurement indicates how well several determinations of the same quantity agree. This is illustrated by the results of shooting arrows at a bull's eye (Figure 1.17). In part (a) the arrows are scattered all over the target; the archer was apparently not very skillful (or shot the arrows from a long distance away from the target), and the precision of their placement on the target is low. In Figure 1.17b the arrows are all clustered together, indicating much better consistency on the part of the archer, that is, greater precision.

Accuracy is the agreement of a measurement with the accepted value of the quantity. Figure 1.17b also shows that our archer was quite accurate—the average of all shots is very close to the accepted position, namely the bull's eye.

Figure 1.17c illustrates that it is possible to be precise without being accurate—the archer has consistently missed the bull's eye, although all the arrows are clustered precisely around one point on the target. This third case is like an experiment with some flaw (either in its design or in a measuring device) that causes all results to differ from the correct value by the same amount.

The precision of a measurement is often expressed by the **average deviation** (see Example 1.6). That is, we calculate the difference between each experimental result and the average result. These differences, each expressed as a *positive* quantity, are averaged, and the results of the experiment are reported as the average value plus or minus (±) the average deviation.

Example 1.6 Precision and Accuracy

Student A makes four measurements of the diameter of a coin using a precision tool called a micrometer. Student B measures the same coin using a simple plastic ruler. They report the following results:

Student A	Student B
28.246 mm	27.9 mm
28.244	28.0
28.246	27.8
28.248	28.1

Calculate the average value and the average deviation for each set of data.

Solution The average for each set of data is obtained by summing the four values and dividing by 4.

Student A	Deviation from Average	Student B	Deviation from Average
28.246 mm	0.000	27.9 mm	0.1
28.244	0.002	28.0	0.0
28.246	0.000	27.8	0.2
28.248	0.002	28.1	0.1
Av. = 28.246	Av. = 0.001	Av. = 28.0	Av. = 0.1

Student A would report the experimental results as 28.246 ± 0.001, whereas B would report 28.0 ± 0.1. Student A's results are more precise.

The standard kilogram in a laboratory in Australia. (*Courtesy Csiro Div. of Applied Physics, National Measurement Laboratory, Australia*)

Who determines what is "accepted" so we know when a result is accurate? In the case of mass, for example, there is a standard kilogram mass kept by the national laboratories in several countries. In the United States this is the National Institutes for Standards and Technology (NIST). The balance used in your laboratory was certified by its manufacturer against standard masses from NIST, for example. In general, however, scientists establish as the standard of accuracy any result that has been determined reproducibly in many different laboratories. The masses of the chemical elements are one example.

If you are measuring a quantity in the laboratory, you often report the **error** in the result, the difference between your result and the accepted value,

$$\text{Error} = \text{experimentally determined value} - \text{accepted value}$$

or you report the **percent error.**

$$\text{Percent error} = \frac{\text{error}}{\text{accepted value}} \times 100\%$$

Example 1.7 Error

Suppose the coin described in Example 1.6 has an "accepted" diameter of 28.054 mm. What are the experimental errors of students A and B?

Solution

Student A: Error = 28.246 mm − 28.054 mm = +0.192

Student B: Error = 28.0 mm − 28.1 mm = −0.1

Student A's results were more precise, but student B's results were more accurate.

Exercise 1.12 Precision, Accuracy, and Error

Two students measured the freezing point of an unknown liquid, one using an ordinary laboratory thermometer calibrated in 0.1 °C units. The other student used a thermometer certified by NIST and calibrated in 0.01 °C units. Their results were as follows:

Student A: −0.3 °C; 0.2 °C; 0.0 °C; and −0.3 °C

Student B: 273.13 K; 273.17 K; 273.15 K; and 273.19 K

Calculate the average value and average deviation for each student. Knowing that the unknown liquid was water, calculate the error for each student. Which student has the more precise values? Which has the smaller error?

Significant Figures

In most experiments several kinds of measurements must be made, and some can be made more precisely than others. It is common sense that a calculated result can be no more precise than the least precise piece of information that went into the calculation. This is where the rules for "significant figures" come in. As an example let us consider again the calculation of the density of aluminum described on page 37. The mass (13.56 g) and dimensions (6.45 cm long, 2.50 cm wide, and 0.31 cm thick) were determined by standard techniques. All these numbers have two digits to the right of the decimal, but they have different numbers of significant figures.

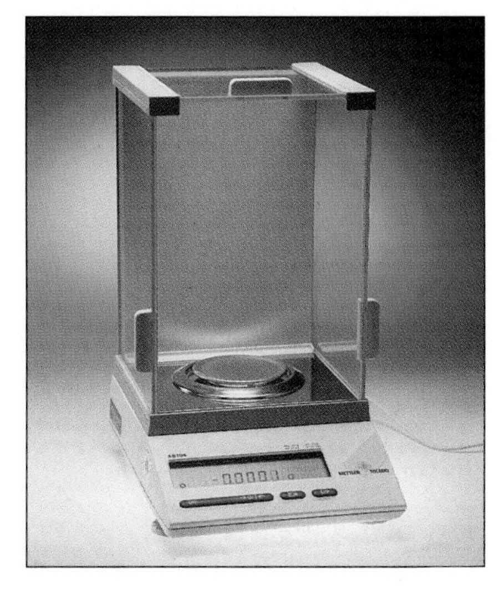

Standard laboratory balances can determine the mass of an object to the nearest milligram. Thus, an object may have a mass of 13.456 g (13456 mg, five significant figures), 0.123 g (123 mg, three significant figures), or 0.072 g (72 mg, two significant figures).

Measurement	Data Collected	Significant Figures
Mass of aluminum	13.56 g	4
Length	6.45 cm	3
Width	2.50 cm	3
Thickness	0.31 cm	2

The quantity 0.31 cm has two significant figures. That is, the 3 in 0.31 is exactly right, but the 1 is not known exactly. The thickness could have been as small as 0.30 or as large as 0.32. Unless indicated otherwise, the standard convention used in science is that the final digit in a number is uncertain to the extent of ± 1. In general, *in a number that represents a scientific measurement, the last digit to the right is taken to be inexact, but all digits farther to the left are assumed to be exact.*

When the data on the piece of aluminum are combined, the calculated density is 2.7 g/cm^3, a number with two significant figures (page 38). This follows from the previous statement that the calculated result can be no more precise than the least precise piece of information.

When doing calculations using measured quantities, you must follow some simple rules so that the results reflect the precision of all the measurements that go into the calculations. These rules are laid out in the box titled ***Guidelines for Determining Significant Figures*** (p. 47).

One last word on significant figures and calculations. In working problems with a hand-held calculator, you should do the calculation with all the digits allowed by the calculator and round off only at the end of the calculation. Rounding off in the middle can introduce errors. *When showing multistep calculations in this book we shall give the answer to each step to the correct number of significant figures. However, more significant figures than are required are carried in the calculator to the end of the calculation.* Example 1.9 shows this procedure.

If your answer to a problem in this book does not quite agree with the answers in the back of the book, this may be the result of rounding errors. *Remember:* Only round off at the end of a calculation.

Example 1.8 Significant Figures

An example of a calculation you will do later in the book (Chapter 12) is

$$\text{Volume of gas (L)} = \frac{(0.120)(0.08206)(273.15 + 23)}{(230/760.0)}$$

Calculate the final answer to the correct number of significant figures.

Solution As a first step, we analyze each number in the equation.

Number	Number of Significant Figures	Comments
0.120	3	The trailing 0 is significant. See Rule 1 (1.20×10^{-1}).
0.08206	4	The 0 to the right of the decimal is not significant. See Rule 1 (8.206×10^{-2}).
273.15 + 23 = 296	3	23 has no decimal places, so the sum can have none. See Rule 2.
230/760.0 = 0.30	2	230 has two significant figures because the last zero is not significant (2.3×10^{2}). In contrast, there is a decimal point in 760.0, so there are four significant digits. The quotient must have two significant digits. See Rules 1 and 3.

The analysis shows that one of the pieces of information is known to only two significant figures. Therefore, the answer must be volume of gas = 9.6 L.

Exercise 1.13 Significant Figures

(a) What are the sum and product of 11.19 and 0.054?
(b) What is the result of the calculation?

$$x = \frac{(110.2 - 57)}{(0.021 + 0.00115)}$$

(Answers: **(a)** = 11.24, 0.60; **(b)** = 2.4×10^{3})

1.8 PROBLEM SOLVING

Chemistry is a quantitative science. We make measurements, collect data, and then search for patterns in those data. We might calculate the quantity of a substance required for a chemical reaction or, knowing the density of a metal and its identity, calculate the size of its atoms. Some of these calculations can be quite complex, so the example and exercise that follow illustrate how to work through a situation involving more than one or two steps. As you do this, think about the problem-solving strategy in the *Problem-Solving Tips and Ideas 1.3* box (p. 49).

Guidelines for Determining Significant Figures

Rule 1. To determine the number of significant figures in a measurement, read the number from left to right and count all digits, starting with the first digit that is *not* zero.

Example	Number of Significant Figures
1.23	3
0.00123 g	3; the zeros to the left of the 1 simply locate the decimal point. To avoid confusion, write numbers of this type in scientific notation; thus, $0.00123 = 1.23 \times 10^{-3}$.
2.0 and 0.020 g	Both have two significant digits. When a number is greater than 1, *all zeros to the right of the decimal point are significant.* For a number less than 1, only zeros to the right of the first significant digit are significant.
100 g	1; in numbers that do not contain a decimal point, "trailing" zeros may or may not be significant. The practice followed in this book is to include a decimal point if the zeros are significant. Thus, 100. is used to represent three significant digits, whereas 100 has only 1. To avoid confusion, an alternative method is to always write numbers in scientific notation because *all digits are significant when written in scientific notation.* Thus, 1.00×10^2 has three significant digits, whereas 1×10^2 has only 1.
100 cm/m	Infinite number of significant digits. This is a defined quantity.
$\pi = 3.1415926$	The value of π is known to a greater number of significant figures than you will ever use in a calculation.

Rule 2. When adding or subtracting numbers, the number of decimal places in the answer should be equal to the number of decimal places in the number with the fewest places.

0.12	2 decimal places	2 significant figures
1.9	1 decimal place	2 significant figures
10.925	3 decimal places	5 significant figures
12.945	3 decimal places	5 significant figures

The sum should be reported as 12.9, a number with one decimal place, because 1.9 has only one decimal place.

Rule 3. In multiplication or division, the number of significant figures in the answer should be the same as that in the quantity with the fewest significant figures.

$$\frac{0.01208}{0.0236} = 0.512 \text{ or, in scientific notation, } 5.12 \times 10^{-1}$$

Because 0.0236 has only three significant digits and 0.01208 has four, the answer should have three significant digits.

Rule 4. When a number is rounded off, the last digit to be retained is increased by one only if the following digit is 5 or greater.

Full Number	Number Rounded to Three Significant Digits
12.696	12.7
16.349	16.3
18.35	18.4
18.351	18.4

Example 1.9 Problem Solving

A mineral oil has a density of 0.875 g/cm^3. Suppose you spread 0.75 g of this oil over the surface of water in large beaker, which has a diameter of 21.6 cm. How thick is the oil layer? Express the thickness in centimeters.

Solution We often begin solving such problems by sketching a picture of the situation. Looking at this sketch, we recognize that the solution to our problem is to find the volume of the oil on the water. If we know the volume, then we can find the thickness because

$$\text{Volume of oil layer} = \text{thickness of layer} \times \text{area of oil layer}$$

So, we need two things: (a) the volume of the oil layer and (b) the area of the layer.

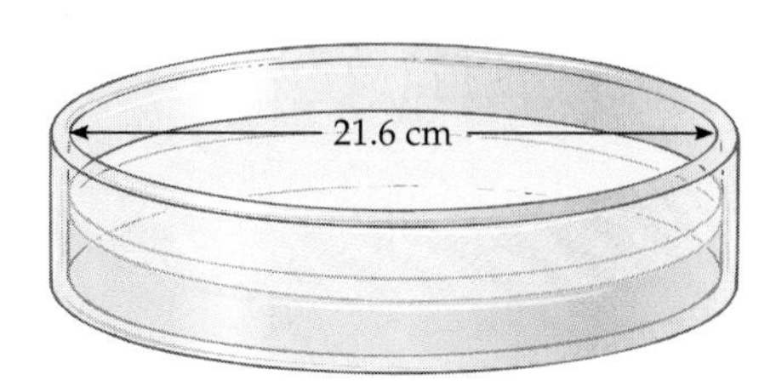

See Example 1.9.

(a) Volume of oil layer

The mass of the oil layer is known, so combining the mass of oil with its density gives us the volume of the oil used.

$$0.75\ \text{g} \cdot \frac{1\ \text{cm}^3}{0.875\ \text{g}} = 0.86\ \text{cm}^3$$

The calculator shows 0.857143. . . . Notice that the quotient should have two significant figures because 0.75 has only two significant figures, so the result of this step is 0.86 cm^3.

(b) Area of oil layer

Scientific calculators have the value of π programmed into the calculator. There should be a key labeled π.

The oil is spread over a circular surface, whose area is given by the

$$\text{Area} = \pi \cdot (\text{radius})^2$$

In this case, the radius is 10.8 cm, so

$$\text{Area of oil layer} = (3.142)(10.8\ \text{cm})^2 = 366\ \text{cm}^2$$

The calculator shows 366.435. . . . The answer to this step, however, should have only three significant figures because 10.8 has three.

(c) Thickness of oil layer

$$\text{Thickness of layer} = \frac{\text{Volume of layer}}{\text{Area of layer}} = \frac{0.86\ \text{cm}^3}{366\ \text{cm}^2} = 0.0023\ \text{cm}$$

In scientific notation, the answer is 2.3×10^{-3} cm. It has two significant figures.

Remember: When doing multistep calculations in this book we shall show the answer to each step to the correct number of significant figures. All digits are carried in the calculator to the end of the calculation, however. The final answer is then given in the correct number of significant figures. A feature of many calculators that solves this problem is that they will display the answer as a preset number of significant figures.

***Exercise* 1.14** Problem Solving

A particular paint has a density of 0.914 g/cm^3. You need to a cover a wall that is 7.6 m long and 2.74 m high with a paint layer 0.13 mm thick. What volume of paint (in liters) is required? What mass (in grams) will the paint layer have? (Answer = 2.7 L; 2.5×10^3 g)

The Concept of "Percent"

Chemists often express the composition of matter in terms of **percent.** For example, we know that 88.81% of a given mass of water is oxygen, that sucrose is 42.11% carbon or that a 5¢ coin, a nickel, is only about 25% nickel (the rest is copper). Because "percent" is a widely used concept in chemistry, it is worth thinking about it while we are discussing units.

You are familiar with "percent" from looking for a bargain at the local mall or from paying sales taxes. For example, paying a sales tax of 6.00% means that

Problem-Solving Tips and Ideas

1.3 Problem-Solving Strategies

Now that you have seen solutions for several numerical problems, it is an appropriate time to comment on the strategy of problem solving.

Step 1. *Define the problem.* Read the question carefully. What key principles are involved? What information is necessary, and what is there just to place the question in the context of chemistry? Organize the information to see what is necessary and see relationships among the data given to you. Try writing the information down in a table form. If it is numerical information, be sure to include units.

One of the greatest difficulties for a student in introductory chemistry is picturing what is being asked for. Try sketching a picture of the situation involved. For example, sketch a picture of the piece of aluminum whose density we wanted to calculate, and put the dimensions on the drawing.

Step 2. *Develop a plan.* Have you done a problem of this type before? If not, perhaps the problem is really just a combination of several easier ones you have seen before. Break it down into those simpler components. Try reasoning backward from the units of the answer. What data do you need to find an answer in those units?

Step 3. *Execute the plan.* Carefully write down each step of the problem, being sure to keep track of the units on numbers. (Do the units cancel to give you the answer in the desired units?) Don't skip steps. Don't do anything but the simplest steps in your head. Students often say they got a problem wrong because they "made a stupid mistake." Your instructor—and book authors—make them, too, and it is usually because they don't take the time to write down the steps of the problem clearly.

Step 4. *Is the answer reasonable?* As a final check when doing any calculation, ask yourself if the answer is reasonable. Let us say you are asked to convert 100. yd to a distance in meters. Using dimensional analysis, and some well-known factors for converting from the English system to the metric system, we have

$$100.\text{ yards} \cdot \frac{3\text{ ft}}{1\text{ yd}} \cdot \frac{12\text{ in.}}{1\text{ ft}} \cdot \frac{2.54\text{ cm}}{1\text{ in.}} \cdot \frac{1\text{ m}}{100\text{ cm}} = 91.4\text{ m}$$

You should recognize that a distance of 91.4 m is about right. Because a meter is a little more than 3 ft, the distance should be a little less than 100 m. In the first step, if you divided instead of multiplied by 3, the final answer would be a little more than 10 m. This would be equivalent to only about 30 ft, and you know a 100-yd football field is longer than that.

you pay 6.00 dollars per 100 dollars of things purchased. Therefore, if you buy a new shirt for $31.50, the tax is $1.89.

$$31.50\ \cancel{\$\text{ spent}} \cdot \frac{6.00\ \$\text{ tax}}{100\ \cancel{\$\text{ spent}}} = 1.89\ \$\text{ tax}$$

Two important points should be made here. First, notice that the problem—one that you have probably done in your head many times—was solved using units.

We used the conversion factor "\$ tax/\$100 spent," and the unit "\$ spent" cancels out and leaves units of "\$ tax." Second, the word "percent" tells us what the conversion factor must be because *per cent* is a shortened version of the Latin phrase *per centum,* meaning *in 100.* The conversion factor in a percent calculation is always the value of the percent divided by 100.

Example 1.10 Using Percent

Battery plates in lead storage batteries (the type used in automobiles) are made from a mixture of two chemical elements: lead (Pb, 94.0%) and antimony (Sb, 6.0%). If a piece of a battery plate has a mass of 25.0 g, what masses of lead and antimony (in grams) are present?

Solution Let us first solve for the mass of lead, using the known percentage of lead. The plate is 94.0% lead, which means that 94.0 g of lead is present in every 100. g of plate.

$$25.0 \text{ g } \cancel{\text{battery plate}} \cdot \frac{94.0 \text{ g lead}}{100. \text{ g } \cancel{\text{battery plate}}} = 23.5 \text{ g lead}$$

We know that the plate contains only lead and antimony, so

$$25.0 \text{ g plate} = 23.5 \text{ g lead} + x \text{ g antimony}$$

Solving for x, we find that the mass of antimony is 25.0 g − 23.5 g = 1.5 g of antimony. Of course, we could obtain the same result from the calculation

$$25.0 \text{ g } \cancel{\text{battery plate}} \cdot \frac{6.0 \text{ g antimony}}{100. \text{ g } \cancel{\text{battery plate}}} = 1.5 \text{ g antimony}$$

Notice that the answer has only two significant figures, because there are only two in 6.0 g of antimony.

Exercise 1.15 Using Percent

Fertilizer for household plants is labeled "15–30–15." This means the fertilizer contains 15% nitrogen measured as N, 30% phosphorus measured as P_2O_5, and 15% potassium measured as K_2O. If you use 15 g of fertilizer, what mass of N and what mass of P_2O_5 are you using?

CHAPTER HIGHLIGHTS

Having studied this chapter, you should be able to

- Recognize the different **states** of matter (**solids, liquids,** and **gases**) and give their characteristics (Section 1.1).
- Understand the basic ideas of the **kinetic-molecular theory** (Section 1.1).
- Understand the difference between matter represented at the **macroscopic** level and at the particulate level.
- Appreciate the difference between pure substances and mixtures and the difference between **homogeneous** and **heterogeneous** mixtures (Section 1.1).
- Identify the name or symbol for an element, given its symbol or name (Section 1.2).

- Use the terms **atom, element, molecule,** and **compound** correctly (Sections 1.2 and 1.3).
- Define **physical properties** of matter and give some examples (Section 1.4).
- Use **density** as a way to connect the volume and mass of a substance (Sections 1.4 and 1.7). The **density** of a substance is defined as the ratio of its mass to its volume.

$$\text{Density} = \frac{\text{mass}}{\text{volume}}$$

In chemistry density is usually in units of grams per cubic centimeter.

- Convert between temperatures on the **Celsius** and **Kelvin** scales (Section 1.4).
- Understand the difference between **extensive** and **intensive properties** and give examples (Section 1.4).
- Explain the difference between **chemical** and **physical change** (Section 1.5).
- Recognize and know how to use the prefixes that modify the sizes of metric units (Section 1.6).
- Begin using dimensional analysis to carry out unit conversions and other calculations (Section 1.7 and Appendix A).
- Know the difference between **precision** and **accuracy** and how to calculate percent error (Section 1.7).
- Understand the use of **significant figures** (Section 1.7).
- Recognize and use percent in calculations (Section 1.7).

STUDY QUESTIONS

Answers to questions with boldface numbers appear in Appendix O.

REVIEW QUESTIONS

1. The mineral fluorite contains the elements calcium and fluorine. What are the symbols of these elements? How would you describe the shape of the fluorite crystals in the photo? What can this tell us about the arrangement of the atoms inside the crystal?

The mineral fluorite. (*C. D. Winters*)

2. What are the states of matter and how do they differ from one another?
3. Small chips of iron are mixed with sand (see photo below). Is this a homogeneous or heterogeneous mixture? Suggest a way to separate the iron and sand from each other.

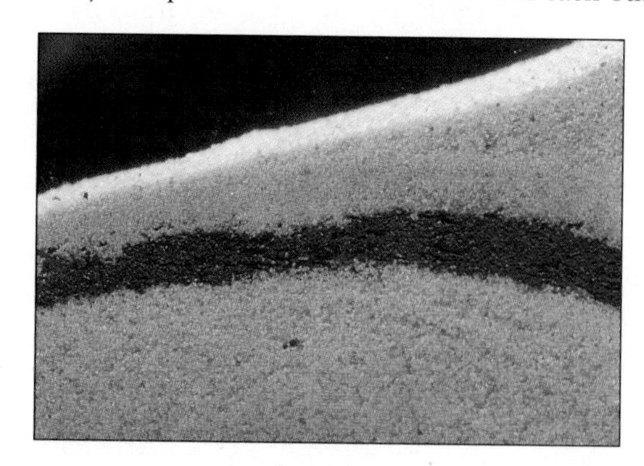

Chips of iron mixed with sand. (*C. D. Winters*)

4. The photo following shows copper balls, immersed in water, floating on top of mercury. What are the liquids and solids in this photo? Which substance is densest? Which is least dense?

Water, copper, and mercury. (*C. D. Winters*)

5. What is the difference between the terms "compound" and "molecule"? Use these words in a sentence.
6. In each case, decide if each underlined property is a physical or chemical property:
   (a) The normal color of sulfur is yellow.
   (b) Iron is transformed into rust in the presence of air and water.
   (c) Dynamite can explode.
   (d) The density of uranium metal is 19.05 g/cm^3.
   (e) Aluminum metal melts at 933 K.
   (f) Chlorophyll, a plant pigment, is green.
7. In each case, decide if the change is a chemical or physical change.
   (a) A cup of household bleach changes the color of your favorite T-shirt from purple to pink.
   (b) Water vapor in your breath condenses in the air on a cold day.
   (c) Plants use carbon dioxide from the air to make sugar.
   (d) Butter melts when placed in the sun.
8. Give the number of significant figures in each of the following numbers:
   (a) 9.87 (c) 0.00823
   (b) 1050 (d) 1.67×10^{-6}
9. In the photo, you see a gemstone called aquamarine. It is surrounded by samples of two of the elements of which it is composed, aluminum and silicon, in the form of foil and powder, respectively. (The other element combined in aquamarine is oxygen.)

Aquamarine is the bluish crystal. It is surrounded by aluminum foil and silicon powder. (*C. D. Winters*)

   (a) What are the symbols of the three elements that combine to make aquamarine?
   (b) Based on the photo, describe some of the physical properties of the elements and the mineral. Are any the same? Are any properties different?
10. In the photo you see a quartz crystal. If you studied the crystal, you might observe that it is colorless and clear, and has a density of 2.65 g/cm^3, a mass of 2.5 g, and a length of 4.6 cm. Which of these observations are qualitative and which are quantitative? Which of the observations is extensive and which is intensive?

A sample of quartz called a Herkimer "diamond." (*C. D. Winters*)

11. In Figure 1.2 you see a piece of iron and in the drawing a representation of the internal structure of iron. Which is the macroscopic view and which is the particulate view? How are the macroscopic and particulate views related?

NUMERICAL AND OTHER QUESTIONS

Density

12. Ethylene glycol, $C_2H_6O_2$, is a liquid that is the base of the antifreeze you use in your car's radiator. It has a density of 1.1135 g/cm^3 at 20 °C. If you need 500. mL of this liquid, what mass of the compound, in grams, is required?
13. A piece of silver metal has a mass of 2.365 g. If the density of silver is 10.5 g/cm^3, what is the volume of the silver?
14. Water has a density at 25 °C of 0.997 g/cm^3. If you have exactly 500 mL of water at 25 °C, what is its mass in grams? In kilograms?
15. A chemist needs 2.00 g of a liquid compound. (a) What volume of the compound is necessary if the density of the liquid is 0.718 g/cm^3? (b) If the compound costs $2.41 per milliliter, what is the cost of 2.00 g?
16. A sample of unknown metal is placed in a graduated cylinder containing water. The mass of the sample is 37.5 g, and the water levels before and after adding the sample to the cylinder are as shown in the figure. Which metal in the following list is most likely the sample (*d* is the density of the metal)?

(a) Mg, $d = 1.74\ g/cm^3$
(b) Fe, $d = 7.87\ g/cm^3$
(c) Ag, $d = 10.5\ g/cm^3$
(d) Al, $d = 2.70\ g/cm^3$
(e) Cu, $d = 8.96\ g/cm^3$
(f) Pb, $d = 11.3\ g/cm^3$

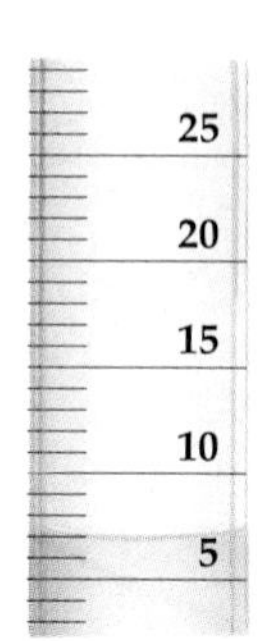

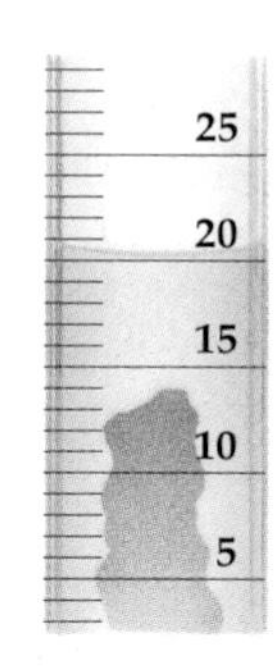

Graduated cylinders with unknown metal (*right*).

17. Iron pyrite is often called "fool's gold" because it looks like gold (see Figure 1.4). Assume you have a piece of a solid that looks like gold, but you believe it to be fool's gold. The sample has a mass of 23.5 g. When the sample is lowered into the water in a graduated cylinder (see Study Question 16), the water level rises from 47.5 mL to 52.2 mL. Is the sample fool's gold ($d = 5.00\ g/cm^3$) or gold ($d = 19.3\ g/cm^3$)?
18. The "cup" is a volume widely used by cooks in the United States. One cup is equivalent to 237 mL. If 1 cup of olive oil has a mass of 205 g, what is the density of the oil (in grams per cubic centimeter)?
19. Peanut oil has a density of $0.92\ g/cm^3$. If a recipe calls for 1 cup of peanut oil (1 cup = 237 mL), what mass of peanut oil (in grams) are you using?

Temperature

20. Many laboratories use 25 °C as a standard temperature. What is this temperature in kelvins?
21. The temperature on the surface of the sun is 5.5×10^3 °C. What is this temperature in kelvins?
22. Make the following temperature conversions:

	°C	K
(a)	16	———
(b)	———	370
(c)	−40	———

23. Make the following temperature conversions:

	°C	K
(a)	———	77
(b)	60.	———
(c)	———	1450

24. The element gallium has a melting point of 29.8 °C. If you held a sample of gallium in your hand, should it melt? Explain briefly.
25. Neon, a gaseous element used in neon signs, has a melting point of −248.6 °C and a boiling point of −246.1 °C. Express these temperatures in kelvins.

Elements and Atoms

26. Give the name of each of the following elements:
(a) C (b) Na (c) Cl (d) P (e) Mg (f) Ca
27. Give the name of each of the following elements:
(a) Mn (b) Cu (c) Xe (d) K (e) F (f) Cr (g) As (h) Fe
28. Give the symbol for each of the following elements:
(a) lithium (b) titanium (c) iron (d) silicon (e) lead (f) zinc
29. Give the symbol for each of the following elements:
(a) silver (b) aluminum (c) plutonium (d) cadmium (e) tin (f) barium (g) krypton (h) palladium

Units and Unit Conversions

30. A race covers a distance of 1500 m. What is this distance in kilometers? In centimeters?
31. The average lead pencil, new and unused, is 19 cm long. What is its length in millimeters? In meters?
32. A standard U. S. postage stamp is 2.5 cm long and 2.1 cm wide. What is the area of the stamp in square centimeters? In square meters?
33. A compact disk has a diameter of 11.8 cm. What is the surface area of the disk in square centimeters? In square meters?
34. A typical laboratory beaker has a volume of 250. mL. What is its volume in cubic centimeters? In liters? In cubic meters?
35. Some soft drinks are sold in bottles with a volume of 1.5 L. What is this volume in milliliters? In cubic centimeters?
36. A book has a mass of 2.52 kg. What is this mass in grams?
37. A new U. S. quarter has a mass of 5.63 g. What is its mass in kilograms? In milligrams?
38. A standard sheet of notebook paper is 8.5 × 11 in. What are these dimensions in centimeters? What is the area of the paper in square centimeters?
39. Suppose your bedroom is 18 ft long, 15 ft wide, and the distance from floor to ceiling is 8 ft, 6 in. You need to know the volume of the room in metric units for some scientific calculations. What is the room's volume in cubic meters? In liters?

Accuracy, Precision, and Error

40. You and your lab partner are asked to determine the density of an aluminum bar. The mass is known accurately (to four significant figures). To measure its volume you use two methods: A and B. You use a simple metric ruler and find the results in A. Your partner uses a precision micrometer, and obtains the results in B.

Method A (g/cm^3)	**Method B** (g/cm^3)
2.2	2.703
2.3	2.701
2.7	2.705
2.4	5.811

The accepted density of aluminum is 2.702 g/cm^3.

(a) Calculate the average density and average deviation of the individual values. Should all the experimental results be included in your calculations? If not, justify any omissions.

(b) Calculate the error for each method's average value.

(c) Which method's average value is more precise? Which method is more accurate?

41. The accepted value of the melting point of pure aspirin is 135 °C. Trying to verify that value, you obtain 134 °C, 136 °C, 133 °C, and 138 °C in four separate trials. Your partner finds 138 °C, 137 °C, 138 °C, and 138 °C.

(a) Calculate the average value and average deviation for you and your partner.

(b) Calculate the error for each set of data.

(c) Which of you is more precise? More accurate?

Significant Figures

42. What is the volume, in cubic centimeters, of a rectangular backpack whose dimensions are 22.86 cm × 38.0 cm × 76 cm?

43. Carry out the following calculation, and report the answer in the correct number of significant figures: 0.000523 × 0.0263 × 263.28.

44. Carry out the following calculation, and report the answer in the correct number of significant figures.

$$(1.68)(7.847)\left(\frac{1.0000}{55.85}\right)$$

45. Carry out the following calculation, and report the answer in the correct number of significant figures.

$$(0.0546)\left[\frac{23.56 - 1.4}{1.345 \times 10^3}\right]$$

46. Solve the equation for n, and report the answer in the correct number of significant figures.

$$\left(\frac{43.7}{760.0}\right)(125) = n(0.082057)(298.2)$$

47. Solve the equation for V, and report the answer in the correct number of significant figures.

$$\left(\frac{456}{760.0}\right)V = (0.000328)(0.082057)(273.15 + 19.6)$$

GENERAL QUESTIONS

48. Diamond has a density of 3.513 g/cm^3. The mass of diamonds is often measured in "carats," 1 carat equaling 0.200 g. What is the volume (in cubic centimeters) of a 1.50-carat diamond?

49. Molecular distances are usually given in nanometers (1 nm = 1×10^{-9} m) or in picometers (1 pm = 1×10^{-12} m). However, a commonly used unit is the angstrom, where 1 Å = 1×10^{-10} m. (The angstrom unit is not an SI unit.) If the distance between the Pt atom and the N atom in the cancer chemotherapy drug cisplatin is 1.97 Å, what is this distance in nanometers? In picometers?

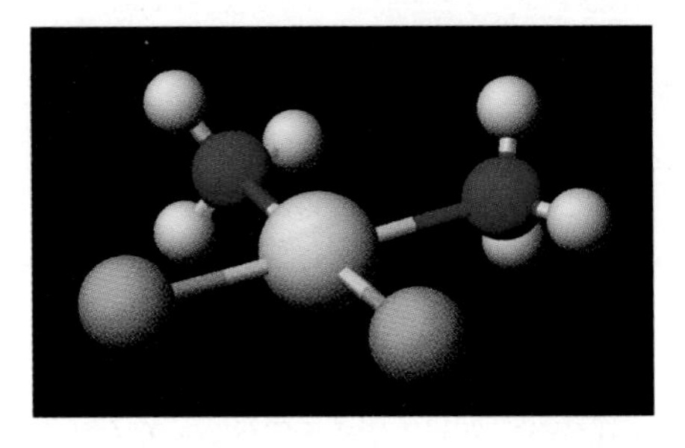

A model of the cancer chemotherapy agent cisplatin. The molecule has a Pt atom in the center. Two Cl atoms and two ammonia (NH_3) molecules are attached.

50. The separation between carbon atoms in diamond is 0.154 nm. (a) What is their separation in meters? (b) What is the carbon atom separation in angstrom units (where 1 Å = 1×10^{-10} m)?

A cut diamond with the internal structure illustrated. The separation between C atoms in the crystal is 0.154 nm.

51. At 25 °C the density of water is 0.997 g/cm^3, whereas the density of ice at −10 °C is 0.917 g/cm^3.

(a) If a soft drink can (volume = 250. mL) is filled completely with pure water and then frozen at −10 °C, what volume does the solid occupy?

(b) Can the ice be contained within the can?

52. An ancient gold coin is 2.2 cm in diameter and 3.0 mm thick. It is a cylinder for which volume = (π) (radius)^2(thickness). If the density of gold is 19.3 g/cm^3, what is the mass of the coin in grams? Assume a price of gold of $410/troy ounce. How much is the gold in the coin worth? (1 troy ounce = 31.10 g; π = 3.14159)

53. The smallest repeating unit of a crystal of common salt is a cube with an edge length of 0.563 nm. What is the volume of this cube in cubic nanometers? In cubic centimeters?

A crystal of salt with a model of the internal structure of salt. The length of an edge of the smallest repeating unit is 0.563 nm. (*Photo, C. D. Winters; model, S. M. Young*)

54. You can identify a metal by carefully determining its density *d*. An unknown piece of metal, with a mass of 29.454 g, is 2.35 cm long, 1.34 cm wide, and 1.05 cm thick. Which of the following is the element?

(a) Nickel, d = 8.91 g/cm^3

(b) Titanium, d = 4.50 g/cm^3

(c) Zinc, d = 7.14 g/cm^3

(d) Tin, d = 7.23 g/cm^3

55. A red blood cell has a diameter of 7.5 μm (micrometer). What is this dimension in (a) meters, (b) nanometers, and (c) angstrom units (where 1 Å = 1 × 10^{-10} m)?

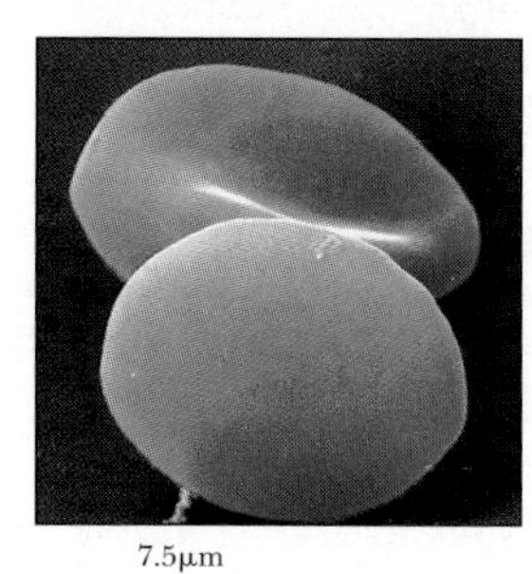

A red blood cell has a diameter of 7.5 μm. (*Meckes/Ottawa/Photo Researchers, Inc.*)

56. Which occupies a larger volume, 600 g of water (with a density of 0.995 g/cm^3) or 600 g of lead (with a density of 11.34 g/cm^3)?

57. You have a 100.0-mL graduated cylinder containing 50.0 mL of water. You drop a 154-g piece of brass (density = 8.56 g/cm^3) into the water. How high does the water rise in the graduated cylinder?

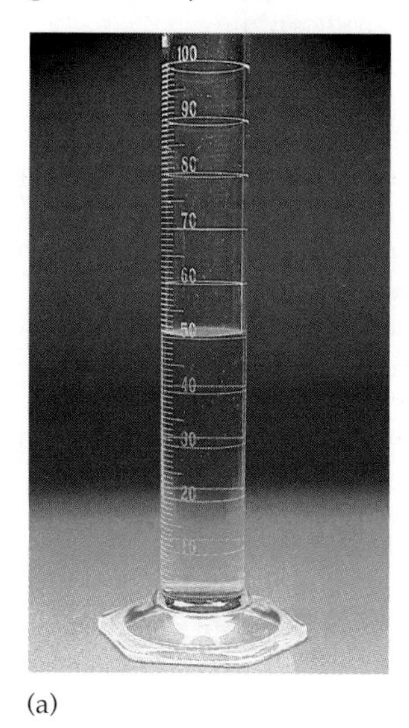

(a)

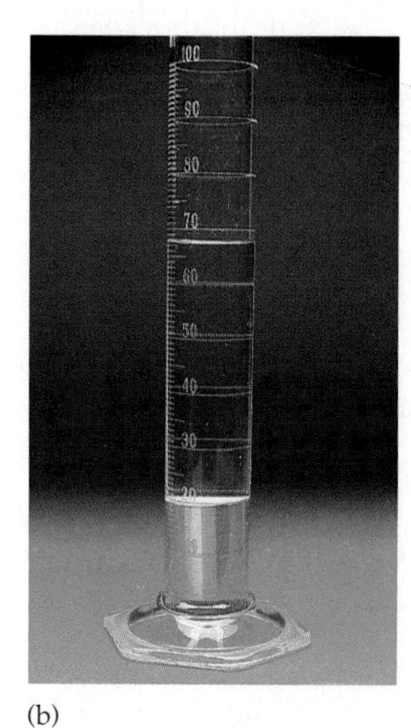

(b)

(a) A graduated cylinder with 50.0 mL of water. (b) A piece of brass is added to the cylinder. (*C. D. Winters*)

58. The density of pure water is given at various temperatures.

t (°C)	d (g/cm^3)
4	0.99997
15	0.99913
25	0.99707
35	0.99406

Suppose your laboratory partner tells you the density of water at 20 °C is 0.99910 g/cm^3. Is this a reasonable number? Why or why not?

59. When you heat popcorn, it pops because it loses water explosively. Assume a kernel of corn, weighing 0.125 g, weighs only 0.106 g after popping. What percent of its mass did the kernel lose on popping? Popcorn is sold by the pound. Using 0.125 g as the average mass of a popcorn kernel, how many kernels are there in a pound of popcorn?

60. The platinum-containing cancer drug cisplatin contains 65.0% platinum. If you have 1.53 g of the compound, how many grams of platinum can be recovered from this sample?

61. The solder once used by plumbers to fasten copper pipes together consists of 67% lead and 33% tin. What is the mass (in grams) of lead in a 1.00-lb block of solder? What is the mass (in grams) of tin (1 lb = 453.59 g)?

62. The density of a solution of sulfuric acid is 1.285 g/cm^3, and it is 38.08% acid by mass. What volume of the acid solution (in milliliters) do you need to supply 125 g of sulfuric acid?

63. The anesthetic procaine hydrochloride is often used to deaden pain during dental surgery. The compound is packaged as a 10.% solution (by mass; $d = 1.0$ g/ml) in water. If your dentist injects 0.50 mL of the solution, how many milligrams of procaine hydrochloride are injected?

CONCEPTUAL QUESTIONS

64. Make a drawing, based on the kinetic-molecular theory and the ideas about atoms and molecules presented in this chapter, of the arrangement of particles in each of the cases listed here. For each case draw ten particles of each substance. It is acceptable for your diagram to be two-dimensional. Represent each atom as a circle and distinguish each different kind of atom by shading its circle.
(a) A sample of solid iron (which consists of iron atoms)
(b) A sample of *liquid* water (which consists of H_2O molecules)
(c) A sample of pure water *vapor*
(d) A homogeneous mixture of water vapor and helium gas (which consists of helium atoms)
(e) A heterogeneous mixture consisting of liquid water and solid aluminum; show a region of the sample that includes both substances.
(f) A sample of brass (which is a homogeneous mixture of copper and zinc)

65. You are given a sample of a silvery metal. What information would you seek to prove that the metal is silver?

66. Suggest a way to determine if the colorless liquid in a beaker is water. If it is water, does it contain dissolved salt?

67. Describe an experimental method that can be used to determine the density of an irregularly shaped piece of metal.

68. Three liquids of different densities are mixed. Because they are not miscible (do not form a homogeneous solution with one another) they form discrete layers, one on top of the other. Sketch the result of mixing carbon tetrachloride (CCl_4, $d = 1.58$ g/cm^3), mercury ($d = 13.546$ g/cm^3), and water ($d = 1.00$ g/cm^3).

69. Diabetes can alter the density of urine, and so urine density can be used as a diagnostic tool. Diabetics can excrete too much sugar or excrete too much water. What do you predict will happen to the density of urine under each of these conditions?

70. A copper-colored metal is found to conduct an electric current. Can you say with certainty that it is copper? Why or why not? Suggest additional information that could provide unequivocal confirmation that the metal is copper.

71. What experiment can you use to
(a) Separate salt from water?
(b) Separate iron filings from small pieces of lead?
(c) Separate elemental sulfur from sugar?

CHALLENGING QUESTIONS

72. The aluminum in a package containing 75 ft^2 of kitchen foil weighs approximately 12 oz. Aluminum has a density of 2.70 g/cm^3. What is the approximate thickness of the aluminum foil in millimeters? (1 oz = 28.4 g.)

73. The fluoridation of city water supplies has been practiced in the United States for several decades because it is believed that fluoride prevents tooth decay, especially in young children. This is done by continuously adding sodium fluoride to water as it comes from a reservoir. Assume you live in a medium-sized city of 150,000 people and that each person uses 660 L (175 gal) of water per day. How many kilograms of sodium fluoride must be added to the water supply each year (365 days) to have the required fluoride concentration of 1 ppm (part per million), that is, 1 kilogram of fluoride per million kilograms of water? (Sodium fluoride is 45.0% fluoride, and water has a density of 1.00 g/cm^3.)

74. Copper has a density of 8.94 g/cm^3. If a factory has an ingot of copper that has a mass of 57 kg (125 lb), and the ingot is drawn into wire with a diameter of 9.50 mm, what length of wire (in meters) can be produced?

75. The Gimli Glider was a Boeing 767 that ran out of fuel. Read the description on page 42, and then verify that the ground crew should have had added 20,163 L of fuel (and not 4916 L). The crucial piece of information is the density of fuel. The crew used 1.77, but they did not recognize that the units were *pounds* per liter. To solve this problem you need to first find the fuel density in units of *kilograms* per liter (1 lb = 453.6 g).

76. About two centuries ago, Benjamin Franklin showed that 1 teaspoon of oil would cover about 0.5 acre of still water. If you know that 1.0×10^4 m^2 = 2.47 acres, and that there are approximately 5 cm^3 in a teaspoon, what is the thickness of the layer of oil? How might this thickness be related to the sizes of molecules?

77. Automobile batteries are filled with sulfuric acid. What is the mass of the acid (in grams) in 500. mL of the battery acid solution if the density of the solution is 1.285 g/cm^3 and if the solution is 38.08% sulfuric acid by mass?

THE CHEMICAL PUZZLER

78. The chapter-opening photo shows the chemical element potassium reacting with water.
(a) What states of matter are illustrated?
(b) What types of changes are observed?
(c) What qualitative observations can be made?
(d) The structure of potassium shown on the photo is an internal view of potassium metal. It shows potassium atoms "stacked together" as closely as possible. Describe the structure in your own words.

INTERACTIVE GENERAL CHEMISTRY CD-ROM

Screen 1.2 Physical Properties of Matter

(a) In the photo using mercury and water as an example of differences in density, how do you know which beaker contains mercury and which contains water?
(b) How could you distinguish gold from silver? A ruby from a diamond?

Screen 1.3 Physical Properties

The photo was taken by freezing bromine in the freezing compartment of an ordinary home refrigerator and then moving the sample to the studio as quickly as possible. What does this tell you about the freezing point of bromine? Within what temperature range might bromine melt?

Screen 1.5 Chemical Elements

Five different elements are pictured on this screen. Under "normal" conditions, which ones are solid? Which ones are liquid? Which ones are gases?

Screen 1.6 Compounds and Molecules

Bring up the structure of the caffeine molecule by clicking on the red arrow next to the photo of solid caffeine. Each "ball" in this "ball-and-stick" model of the molecule represents an atom. (The "sticks" are chemical bonds, the "glue" that holds the molecule together.) The carbon atoms are gray, the hydrogen atoms are white, the oxygen atoms are red, and the nitrogen atoms are blue.
(a) How many total atoms are in the caffeine molecule?
(b) How many carbon atoms? How many hydrogen atoms? How many oxygen atoms? And how many nitrogen atoms?
(c) Describe any features of this molecule that you think are interesting or important.

Screen 1.7 The Kinetic-Molecular Theory

Describe at least two differences among solids, liquids, and gases.

Screen 1.8 Density

Explain why there is a difference in the density of the brick and the Styrofoam block shown on this screen, considering their structure on the molecular scale.

Screen 1.8 Density

The densities of the elements are given in the *Periodic Table* database. (Click on the "toolbox" icon.)
(a) What is the density of aluminum at 25 °C?
(b) What is the density of uranium at 25 °C?
(c) Look up the densities of lead, platinum, and mercury at 25 °C. Which is the densest? Which is the least dense?

Screen 1.11 Chemical Change

The video on this screen shows the reaction of hydrogen gas (in the balloon) with oxygen.
(a) What is the source of the oxygen for this reaction?
(b) Why is the hydrogen-filled balloon floating in air?
(c) The narrator states that energy is released. What evidence is there for this energy release? In what form is this energy?

Screen 1.12 Chemical Changes on the Molecular Scale

Examine the video of the reaction of elemental phosphorus and chlorine and the animation of this reaction. (*Note:* The animation is meant only to illustrate the form in which P and Cl are found before and after reaction.)
(a) How many P_4 molecules are there in the beginning?
(b) How many Cl_2 molecules are there in the beginning?
(c) How many PCl_3 molecules are formed?

Screen 1.13 Mixtures and Pure Substances and Screen 1.14 Separation of Mixtures

Describe how you might separate cobalt(II) hydroxide from the solution in the middle tube and how you could separate dissolved potassium chromate from water in the right-hand test tube.

Screen 1.15 Units of Measurement

The speedometer in your car is probably marked off in two scales (such as the one on this screen). One scale is in units of miles per hour (white numerals on this screen) and the other is kilometers per hour (yellow numerals on this screen).
(a) Does either scale use SI units? How should the speedometer be marked if SI units were used?
(b) Use the speedometer to arrive at the approximate relation between kilometers and miles. That is, how many kilometers are there, approximately, per mile?

Screen 1.18 Return to the Puzzler

How is metallic iron separated from breakfast cereal? Can you think of an application of this technique in the recycling of household waste?

Chapter 2

Atoms and Elements

(C. D. Winters)

A Chemical Puzzler

There are 112 chemical elements, but only a few are important in commerce or biology. Several of these chemical elements are pictured above. How many can you identify by looking at their physical state, their color, or other physical properties?

Primo Levi was an Italian writer who told his life's story in the beautiful book *The Periodic Table.* Each element in the periodic table reminds him of part of his passage through life, but he reserves a special place for carbon.

Levi imagines what may happen to a single atom of carbon. He says that

> Our character lies for hundreds of millions of years, bound to three atoms of oxygen and one of calcium, in the form of limestone. One day, though, the limestone was sent to a kiln where it was roasted until it separated from the calcium. . . . Still firmly clinging to two of its three former oxygen companions, [the carbon atom] issued from the chimney and took the path of the air. It was caught by the wind, flung down on earth, lifted ten kilometers high. It was breathed in by a falcon, descending into its precipitous lungs, but did not penetrate its rich blood and was expelled. . . . It traveled on the wind for eight years . . . on the sea and among the clouds, over forests, deserts, and limitless expanses of ice; then it stumbled into capture and the organic adventure. Carbon is a singular element: it is the only element that can bind itself in long chains without a great expense of energy, and for life on earth . . . precisely long chains are required. Therefore, carbon is the key element of living substances, but its promotion, its entry into the living world, is not easy and must follow an obligatory, intricate path, which has been clarified only in recent years. If the elaboration of carbon were not a common daily occurrence, on the scale of billions of tons a week, whenever the green of a leaf appears, it would by full right deserve to be called a miracle.

When limestone, calcium carbonate, is heated or "roasted" it decomposes to calcium oxide and carbon dioxide.

$$CaCO_3(solid) \longrightarrow CaO(solid) + CO_2(gas)$$

To understand such stories as Levi's eloquent description of the chemistry of carbon, we turn next to a description of the nature of atoms and elements. To appreciate the current view of the atom, however, it is helpful first to learn how these ideas have developed over the centuries of human history.

2.1 ORIGINS OF ATOMIC THEORY

In Chapter 1 atoms were described as tiny spheres in constant motion. This simple model goes back to the Greek philosopher Democritus (460–370 BC), who reasoned that if a piece of matter, such as gold, were divided into smaller and smaller pieces, one would ultimately arrive at a tiny particle of gold that could not be divided further but that still retains the properties of gold. He used the word "atom," which literally means "uncuttable," to describe this undividable, ultimate particle of matter. According to Aristotle (384–322 BC), Democritus taught that atoms are hard, move spontaneously, and link to one another by some "hook-and-eye" connection. Epicurus (341–271 BC) attributed mass to atoms, and about 100 BC Asklepiades introduced the idea of clusters of atoms, corresponding to what we now call molecules.

See the *Saunders Interactive General Chemistry CD-ROM,* Screen 2.3, Origins of Atomic Theory and Screen 2.4, The Discovery of Atomic Structure.

Democritus used his concept of atoms to explain the properties of substances. For example, the high density and softness of lead can be interpreted if lead atoms are packed together like marbles in a box and move easily past one another. Iron, on the other hand, is less dense but harder than lead, and Democritus argued that iron atoms might be shaped like corkscrews so that they would entangle in a rigid but relatively lightweight structure. He also explained in a simple way other common observations, such as drying clothes, the appearance of moisture on the outside of a vessel of cold water, how an odor moves

Historical Perspective

Primo Levi (1919–1987)—Chemist and Writer

Levi was born in Turin, Italy, in 1919. He was trained as a chemist, and worked in the field of paint chemistry for 30 years. His knowledge and love of chemistry made him the writer he was and also saved his life.

During World War II he was a member of the anti-Fascist resistance in Italy. He was arrested, however, and sent to a death camp, Auschwitz, in 1944. He survived the camp because the Nazis found his skills as a chemist useful. The experience, however, left him with an "absolute need to write." His memoirs of life in the death camps "are considered a triumph of lucid intelligence over modern barbarism." Perhaps his best known book is *The Periodic Table* (Schocken Books, New York, 1984). In this elegantly written book Levi uses the properties of various elements to describe his life and work. Levi also wrote *Survival in Auschwitz, The Reawakening, The Drowned and the Saved, The Monkey's Wrench,* and *Other People's Trades.*

Primo Levi. (*Giansanti/Sygma*)

through a room, and how crystals grow from a solution. To explain the macroscopic events he saw, he imagined atoms scattering or collecting together as needed. Atomic theory has been built on the assumptions of Democritus: the properties of matter that we can see are explained by the properties and behavior of atoms that we cannot see.

Plato (428–348 BC) and Aristotle argued against the existence of atoms, and their ideas prevailed for centuries. Most of those in the mainstream of enlightened thought rejected or remained ignorant of the atomic theory proposed by Democritus, though a few well-known scientists did refer to atoms. Galileo Galilei (1564–1642) reasoned that the appearance of a new substance through chemical change involved a rearrangement of parts too small to be seen, and Francis Bacon (1561–1626) speculated that heat might be a form of motion of small particles. Robert Boyle (1627–1691) was guided in his work on gases and other aspects of chemistry by what he called his "corpuscular philosophy." None of these people, however, provided detailed, quantitative explanations of physical and chemical facts in terms of atomic theory.

Exercise 2.1 Atomic Theory

Use the idea that matter consists of atoms or molecules to interpret each of the following observations. Describe what the atoms or molecules are doing and how that explains what happens.

1. Wet clothes hung on a line eventually become dry.
2. Moisture appears on the outside of a glass of ice water.
3. Crystals of solid sugar dissolve in water.
4. Sugar dissolves faster in hot water than in cold water.

John Dalton and His Atomic Theory

In 1803, John Dalton (1766-1844) forcefully revived the idea of atoms. Dalton linked the existence of elements, which cannot be decomposed chemically, to the idea of atoms, which are indivisible. Compounds, which can be broken down into two or more new substances, must contain two or more different kinds of atoms. Dalton went further to say that each kind of atom must have its own properties—especially a characteristic mass. This idea allowed his theory to account quantitatively for the masses of different elements that combine chemically to form compounds. Thus, Dalton's ideas could be used to interpret known chemical facts, and to do so quantitatively. The postulates of his atomic theory are that

See the *Saunders Interactive General Chemistry CD-ROM,* Screen 2.5, The Dalton Atomic Theory.

1. All matter is made of atoms. These indivisible and indestructible objects are the ultimate chemical particles.
2. All atoms of a given element are identical, both in mass and in properties. Atoms of different elements have different masses and different properties.
3. Compounds are formed by combination of two or more different kinds of atoms. Atoms combine in the ratio of small whole numbers, for example, one atom of A with one atom of B, or two atoms of A with one atom of B.
4. Atoms are the units of chemical change. A chemical reaction involves only combination, separation, or rearrangement of atoms. Atoms are not created, destroyed, divided into parts, or converted into other kinds of atoms during a chemical reaction.

John Dalton (*Oesper Collection in the History of Chemistry/University of Cincinnati*)

John Dalton's ideas were accepted by the scientific community because they helped early scientists understand two important scientific laws: the law of the conservation of matter and the law of constant composition. Both grew out of careful experimental studies involving the masses of elements and compounds.

The **law of conservation of matter** had been formulated a few years earlier by Antoine Lavoisier (1743–1794). He carefully determined the masses of reactants and products in a series of chemical reactions and found the sum of the masses of the reactants always equaled the sum of the masses of the products. This result can be understood from Dalton's second and fourth postulates. If each kind of atom has a characteristic mass, and if exactly the same number of each kind of atom exists before and after a reaction, the total masses before and after must also be the same.

See the brief biography of the great chemist Antoine Lavoisier in Chapter 4.

The French chemist Joseph Louis Proust (1754–1826) formulated the **law of constant composition** (also called the **law of definite proportions**) from his analysis of minerals. He determined that the ratio of masses of the elements in pure samples of a given compound did not vary, regardless of the sample's origin. Dalton's second and third postulates rationalize this result, which must arise if the ratio of different atoms in a compound is fixed and if each kind of atom has a characteristic mass. For example, one of the compounds formed from carbon and oxygen always has a one-third greater mass of oxygen than carbon. This would be true if this compound contained one carbon atom for each oxygen atom—giving the formula CO—and if the mass of one oxygen atom was one-third greater than the mass of a carbon atom. Using this kind of reasoning, Dalton was able to determine ultimately which atoms were heavier than others and how much of one element could be expected to combine with another element in a compound.

Dalton's theory was valuable because it explained existing facts, but he went further and proposed a new law, the **law of multiple proportions.** This states that if two elements form two different compounds, the mass ratio of the elements making up one compound is a whole number multiple of the mass ratio of the elements in the second compound. Consider the two oxides of carbon: carbon monoxide (CO) and carbon dioxide (CO_2). By careful laboratory measurements, it is possible to show that the ratio of the mass of oxygen to the mass of carbon in CO is 1.3333 (mass of O/mass of C = 1.3333), and the ratio of the masses of O to C in CO_2 is 2.6666 (mass of O/mass of C = 2.6666). The mass ratio for CO_2 is exactly twice the value found for CO.

Dalton's atomic theory was important because it suggested a new law and stimulated Dalton and his contemporaries to do a great deal more research, thereby contributing to scientific progress. A good theory not only accounts for existing knowledge but also stimulates the search for new knowledge. Although it was not until the 1860s that a consistent set of relative masses of the atoms was agreed on, Dalton's idea that the masses of atoms are crucial to quantitative chemistry was accepted from the early 1800s on.

2.2 PROTONS, NEUTRONS, AND ELECTRONS: A HISTORICAL PERSPECTIVE

See the *Saunders Interactive General Chemistry CD-ROM,* Screens 2.6, Electricity and Electric Charge.

Although Dalton's atomic theory said nothing about Democritus's idea that atoms have structure, we know now they do. This knowledge is important because it gives us insights into how and why atoms stick together to form molecules. Democritus's idea that atoms are held together by "hooks and eyes" is not accepted, but his fundamental notion that we can better understand how elements behave if we know about the structures of atoms turned out to be correct. The next few sections describe how scientists arrived at our current understanding of the structure of atoms.

Electricity

Electricity is involved in many of the experiments from which the theory of atomic structure was derived. The fact that objects can bear an electric charge was first observed by the ancient Egyptians, who noted that amber, when rubbed with wool or silk, attracted small objects. You can observe the same thing when you comb your hair on a dry day—your hair is attracted to the comb. A bolt of lightning or the shock you get when touching a doorknob results when an electric charge moves from one place to another.

Two types of electric charge had been discovered by the time of Benjamin Franklin (1706–1790), the American statesman and inventor. He named them positive (+) and negative (−), because they appear as opposites and can neutralize each other. Experiments show that *like charges repel each other and unlike charges attract each other.* Franklin also concluded that charge is balanced: if a negative charge appears somewhere, a positive charge of the same size must appear somewhere else. The fact that a charge builds up when one substance is rubbed over another implies that the rubbing separates positive and negative charges (Figure 2.1). By the 19th century it was understood that positive and negative charges are somehow associated with matter—perhaps with atoms.

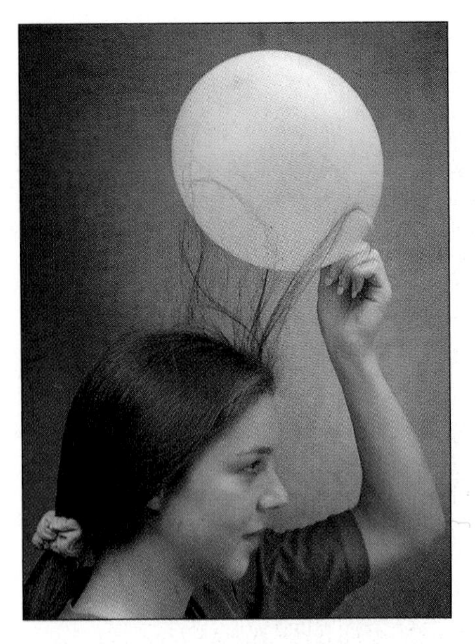

Figure 2.1 Static electricity. If you brush your hair, or comb it, or rub it with a balloon, a static electric charge appears on the surface of the balloon. Unlike electric charges attract each other and like electric charges repel each other. *(C. D. Winters)*

Radioactivity

In 1896, the French physicist, Henri Becquerel (1852–1908), discovered that a uranium ore emitted rays that could darken a photographic plate, even though the plate was covered by black paper to protect it from being exposed by light rays. In 1898, Marie Curie (1867–1934) and coworkers isolated polonium and radium, which also emitted the same kind of rays, and in 1899 she suggested that atoms of certain substances disintegrate when they emit these unusual rays. She named this phenomenon **radioactivity,** and substances that display this property are said to be radioactive.

See the *Saunders Interactive General Chemistry CD-ROM,* Screen 2.7, Evidence for Subatomic Particles, Radioactivity.

Early experiments identified three kinds of radiation: alpha (α), beta (β), and gamma rays (γ). These behave differently when passed between electrically charged plates, as shown in Figure 2.2. Alpha and β rays are deflected, but γ rays pass straight through. This implies that α and β rays are electrically charged particles, because charges are attracted or repelled by the charged plates. Even though an α particle was found to have an electric charge (+2) twice as large as that of a β particle (−1), α particles are deflected less, which implies that α particles must be heavier than β particles. Gamma rays have no detectable charge or mass; they behave like light rays.

Marie Curie's suggestion that atoms disintegrate contradicts Dalton's idea that atoms are indivisible, and requires an extension of Dalton's theory. If atoms can break apart, there must be something smaller than an atom; that is, atomic structure must involve subatomic particles.

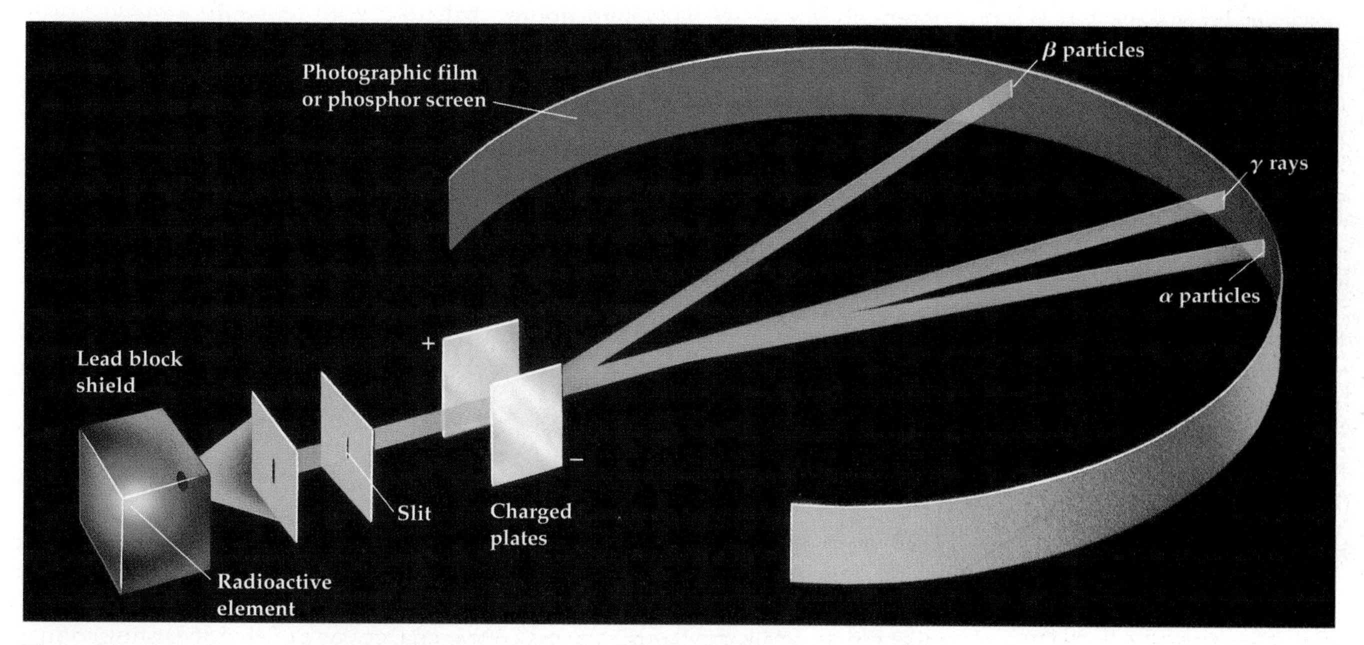

***Figure* 2.2 Separation of subatomic particles.** Alpha (α), beta (β), and gamma (γ) rays from a radioactive element are separated by passing them through electrically charged plates. Positively charged α particles are attracted to the negative plate, and negative β particles are attracted to the positive plate. (Note that the heavier α particles are deflected less than the lighter β particles.) Gamma rays have no electric charge and pass undeflected between the charged plates.

Historical Perspective

Marie Curie (1867–1934)

She was born Marya Sklodovska in Poland, but her family called her Maya. When she later lived in France she was Marie, and today she is often referred to as Madame Curie.

She was the daughter of school teachers and grew up in a household full of books. When she graduated from high school with honors, she wanted to attend the University of Warsaw, but the University did not accept women. She could not go to Paris to study either because neither she nor her family had enough money. She and her sister Bronya, however, worked out a plan to help each other. Marya would work to support Bronya while Bronya studied medicine in Paris, and then Bronya would work to support Marya. Their plan worked, and Marya enrolled at the Sorbonne in Paris in November 1891. She lived extremely frugally, spending almost nothing on luxuries or even on food. On several occasions she fainted because she worked constantly and lived only on a few cherries and radishes. But she successfully completed her degree, finishing first in her class. Almost exactly 15 years later Marie Curie became the first woman ever hired to teach at that university.

By 1896, several physicists, especially Henri Becquerel, had been studying the curious radiation emanating from uranium. Marie Curie was thinking about possible topics for her doctoral thesis and had read about this work. Because her French physicist husband, Pierre (1859–1906), was working at the École Supérieure de Physique et de Chimie de Paris, she was given a laboratory there—but she had to supply her own materials. She was interested in whether other elements would emit such radiation. After Pierre Curie had developed a highly sensitive way of detecting the radiation from a radioactive source, Marie Curie tested every substance she could find for radioactivity. One of her first findings was to confirm Becquerel's observation that uranium metal itself was radioactive and that the degree to which a uranium-bearing sample was radioactive depended on the percentage of uranium present. When she tested pitchblende, a common ore containing uranium and other metals (e.g., lead, bismuth, and copper), she was astonished to find that it was even more radioactive than pure uranium. There was only one explanation: pitchblende contained an element more radioactive than uranium.

Marie Curie with her daughter Irène and husband Pierre. (*Mutter Museum, Philadelphia College of Physicians*)

In 1898, the Curies published their work on the separation of pitchblende into the various metals that it contained. Using techniques of qualitative analysis—techniques that are so effective that they are still in use today in the introductory chemistry laboratory—they separated a new element from the pitchblende. The Curies stated that "We therefore think that the substance that we have extracted from pitchblende contains a metal previously unknown, a neighbor of bismuth in its analytical properties. If the existence of the new metal is confirmed, we propose to call it polonium, after the native land of one of us." They had discovered the next element after bismuth in the periodic table. Further investigation of the pitchblende by the Curies uncovered another new, highly radioactive element, radium.

In 1903, Becquerel and Marie and Pierre Curie shared the Nobel Prize in physics for their discovery of "spontaneous radioactivity." She was awarded a second Nobel Prize in 1911, this one in chemistry, for the discovery of radium and polonium. One of her daughters, Irène, married Frédéric Joliot, and they shared the 1935 Nobel Prize in chemistry for their discovery of artificial radioactivity.

Electrons

See the *Saunders Interactive General Chemistry CD-ROM,* Screen 2.8, Electrons.

Passing an electric current through a solution of a compound—a technique called electrolysis—can cause a chemical reaction (Figure 2.3), such as plating gold or silver onto another metal or the production of chlorine from sodium chloride. In 1833, the British scientist Michael Faraday (1791–1867) showed that the same quantity of electric current caused different quantities of different metals to deposit, and postulated that those quantities were related to the relative masses of the atoms of those elements. Such experiments were interpreted to mean that, just as an atom is the fundamental particle of an element, a funda-

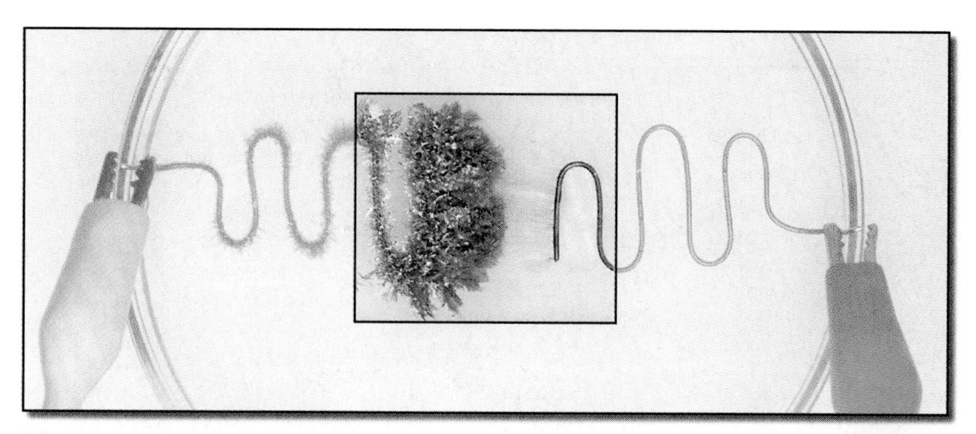

Figure 2.3 Electrolysis. A chemical reaction can be induced when an electric current is passed through a substance (electrolysis). Here a current is passed through a solution of tin chloride in water. (The current enters the solution through the wires or electrodes at the left and right, which are attached to a battery.) At one electrode (*right*) the element chlorine is produced, whereas elemental tin is produced at the other electrode (*left*). Such experiments by Michael Faraday in the 19th century led directly to the discovery of the electron. (*C. D. Winters*)

mental particle of electricity must exist. This "particle" of electricity was given the name **electron.**

Further evidence that atoms are composed of smaller particles came from experiments with cathode ray tubes. These are glass tubes from which most of the air has been removed and that have a piece of metal called an electrode sealed into each end (Figure 2.4). When a sufficiently high voltage is applied to the electrodes, a "cathode ray" flows from the negative electrode (cathode) to the positive electrode (anode). Experiments showed that cathode rays travel in straight lines, cause gases to glow, can heat metal objects red hot, can be deflected by a magnetic field, and are attracted toward positively charged plates. When cathode rays strike a fluorescent screen, light is given off in a series of tiny flashes. We can understand all these observations if a cathode ray is assumed to be a beam of negatively charged particles.

You are already familiar with cathode rays. Television pictures and the images on a computer monitor are formed by the deflection of cathode rays by electrically charged plates inside the tube. This principle was first exploited by Sir Joseph John Thomson (1856–1940) to prove experimentally the existence of the electron. In his experiments he applied electric and magnetic fields simultaneously to a beam of cathode rays (Figure 2.5). By balancing the effect of the electric field against that of the magnetic field and using basic laws of electricity and magnetism, Thomson calculated the ratio of charge to mass for the particles in the beam. He was not able to determine either charge or mass independently, however. But, because cathode rays have a negative charge, Thomson suggested the particles were the same as the electrons associated with Faraday's experiments. Furthermore, he obtained the same charge-to-mass ratio in experiments using 20 different metals as cathodes and several different gases. These results suggested that electrons are present in all kinds of matter and that they presumably exist in atoms of all elements.

It remained for the American physicist Robert Andrews Millikan (1868–1953) to measure the charge on an electron and thereby enable scien-

Discharge tubes and cathode rays are the forerunners of modern television tubes. See "The Body Electric" by S. Devons in *The Sciences,* March/April 1997, pp. 26–31.

Computer monitors are often called CRTs because they are cathode-ray tubes.

Thomson announced the discovery of the electron on April 30, 1897. For a very brief account of the discovery of the electron, proton, and neutron, and the origins of the names, see B. M. Peake: *Journal of Chemical Education,* 1989, Vol. 66, p. 738, 1989.

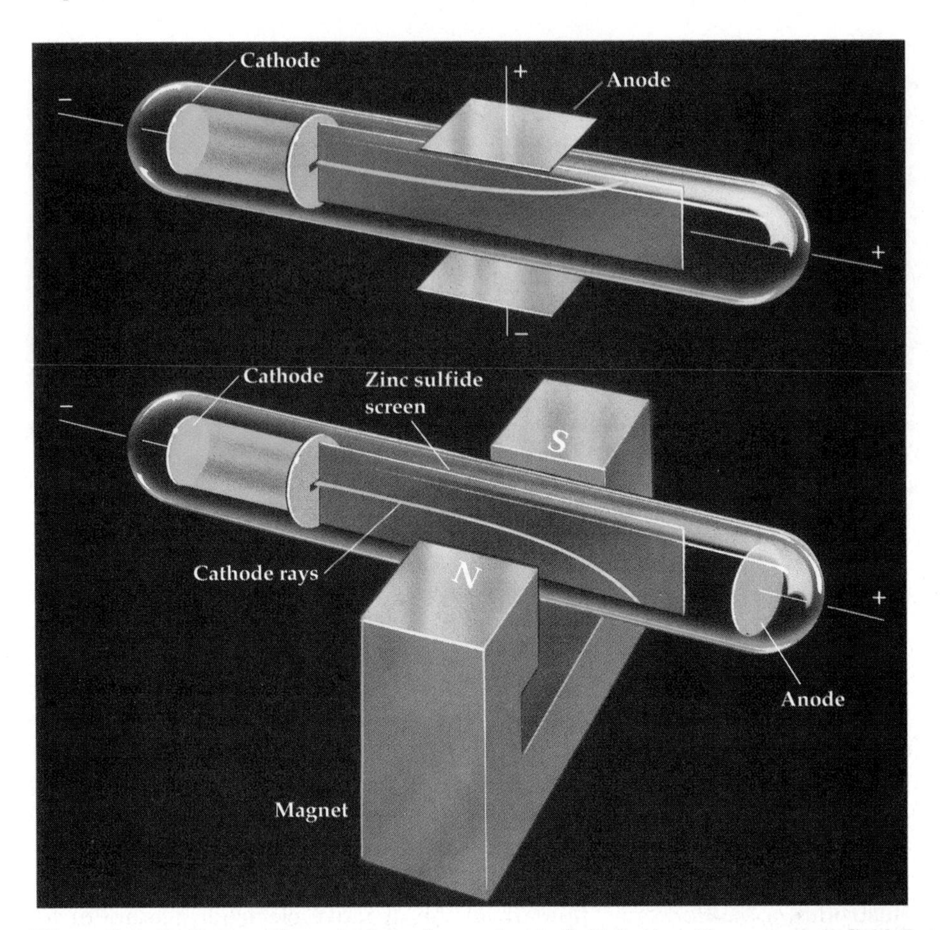

***Figure* 2.4 Deflection of a cathode ray by an electric field (*top*) and by a magnetic field (*bottom*).** When an external electric field is applied, the cathode ray is deflected toward the positive pole. When a magnetic field is applied, the cathode ray is deflected into a curved path. In both cases, the curvature is related to the mass and velocity of the particles of the cathode rays and the magnitude of the field.

tists to calculate its mass. His apparatus is sketched in Figure 2.6. Tiny droplets of oil were sprayed into a chamber. As they settled slowly through the air, the droplets were exposed to x-rays, which caused air molecules to transfer electrons to them. Millikan used a small telescope to observe individual droplets. If the electric charge on the plates above and below the droplets is adjusted, the electrostatic attractive force pulling a droplet upward is just balanced by the force of gravity pulling the droplet downward. From the equations describing these forces, Millikan calculated the charge on the droplet. Different droplets had different charges, but Millikan found that each was a whole-number multiple of the same smaller charge, 1.60×10^{-19} C (where C represents the coulomb, the SI unit of electric charge; Appendix D). Millikan assumed this to be the fundamental unit of charge, the charge on an electron. Because the charge-to-mass ratio of the electron was known, the mass of an electron could be calculated. The currently accepted value for the electron mass is 9.109389×10^{-28} g, and the currently accepted value of the electronic charge is $-1.60217733 \times 10^{-19}$ C. When talking about the properties of fundamental particles, we always express

See the *Saunders Interactive General Chemistry CD-ROM,* Screen 2.9, Mass of the Electron.

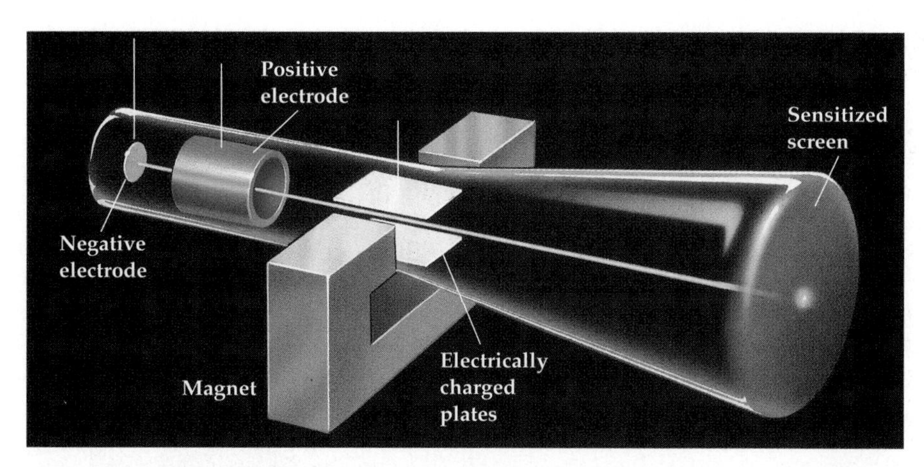

Figure 2.5 Thomson's experiment to measure the charge-to-mass ratio of the electron. A beam of electrons (cathode rays) passes through an electric field and a magnetic field. The experiment is arranged so that the electric field causes the beam of electrons to be deflected in one direction, and the magnetic field deflects the beam in the opposite direction. By balancing the effects of these fields, the charge-to-mass ratio of the electron can be determined.

charge relative to the charge on the electron, which is given the value of -1.

Additional experiments showed that cathode rays had the same properties as the β particles emitted by radioactive elements, providing further evidence that the electron is a fundamental particle of matter.

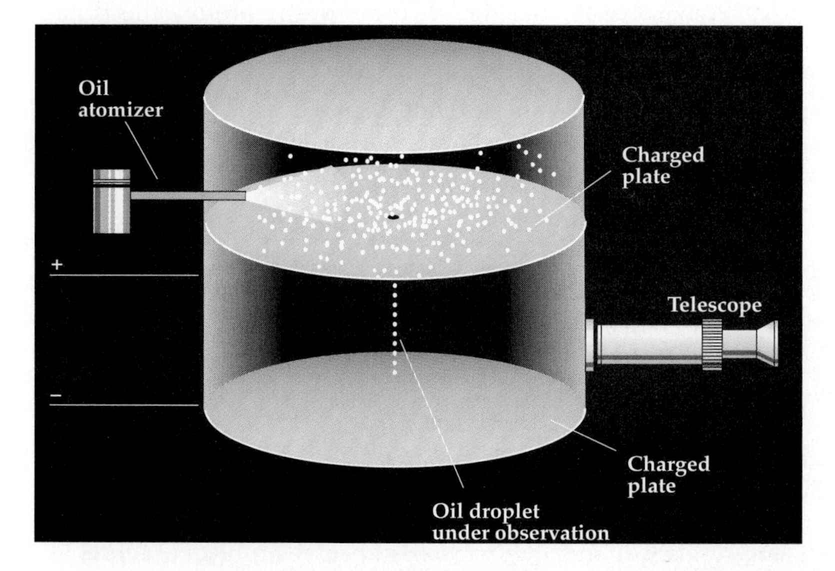

Figure 2.6 Millikan oil drop experiment. A fine mist of oil drops is introduced into one chamber. The gas molecules in the chamber are ionized (split into electrons and a positive fragment) by a beam of x-rays. The electrons adhere to the oil drops, some droplets having one electron, some two, and so on. These negatively charged droplets fall under the force of gravity into the region between the electrically charged plates. By carefully adjusting the voltage on the plates, the force of gravity on the droplet is exactly counterbalanced by the attraction of the negative droplet to the upper, positively charged plate. Analysis of these forces led to a value for the charge on the electron.

Protons

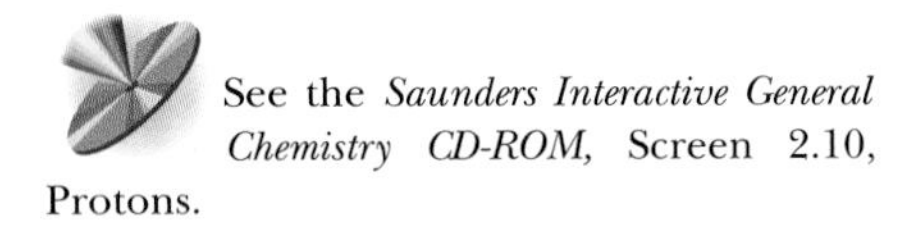

See the *Saunders Interactive General Chemistry CD-ROM,* Screen 2.10, Protons.

The first experimental evidence of a fundamental *positive* particle came from the study of "canal rays" (Figure 2.7), which were observed in a special cathode-ray tube with a perforated cathode. When high voltage is applied to the tube, cathode rays are observed. On the other side of the perforated cathode, however, a different kind of ray is seen. Because these rays are attracted toward a negatively charged plate, they must be composed of positively charged particles.

Each gas used in the tube gives a different charge-to-mass ratio for the positively charged particles (unlike cathode rays, which are the same no matter what gas is used). When hydrogen gas is used, the largest charge-to-mass ratio is obtained, suggesting that hydrogen provides positive particles with the smallest mass. These were considered to be the fundamental, positively charged particles of atomic structure and were later called **protons** (from a Greek word meaning "the primary one") by Ernest Rutherford.

The mass of a proton is known from experiment to be 1.672623×10^{-24} g. The relative charge on the proton, equal in size but opposite in sign to the charge on the electron, is +1.

Neutrons

See the *Saunders Interactive General Chemistry CD-ROM,* Screen 2.12, Neutrons.

Because atoms normally have no charge, the number of protons must equal the number of electrons in an atom. Most atoms, however, have masses greater than would be predicted on the basis of only protons and electrons, indicating that uncharged particles must also be present. Because this third type of particle has no charge, the usual methods of detecting particles could not be used. Nonetheless, in 1932, many years after the discovery of the proton, the British physicist James Chadwick (1891–1974) devised a clever experiment that produced these expected neutral particles and then detected them by having them knock hydrogen ions, a detectable species, out of paraffin. This particle, now known as the **neutron,** has no electric charge and has a mass of $1.6749286 \times 10^{-24}$ g, slightly greater than the mass of a proton.

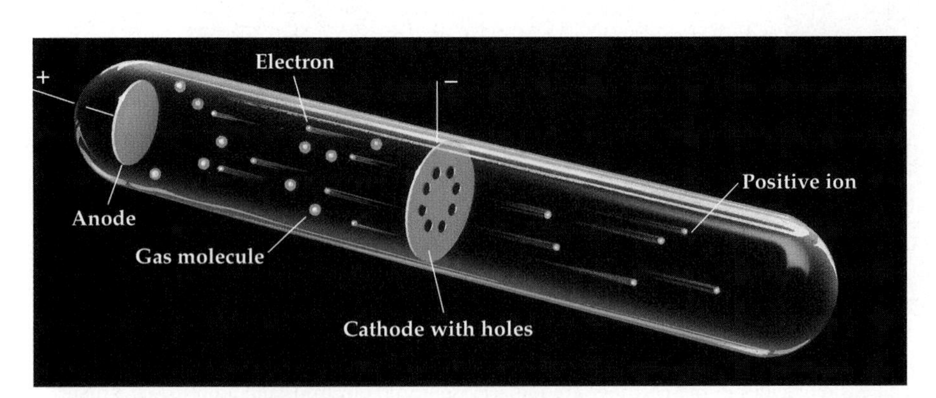

F*igure* 2.7 Electrons collide with gas molecules in this cathode-ray tube with a perforated cathode. Positively charged fragments of these molecules are produced, and these are attracted to the negatively charged, perforated cathode. Some positive fragments pass through the holes and form a beam, or "ray," of positively charged particles. Like cathode rays, positive rays (or "canal rays") are deflected by electric and magnetic fields but much less so than for a given value of the field, because positive particles are much heavier.

The Nucleus of the Atom

J. J. Thomson had supposed that an atom was a uniform sphere of positively charged matter within which thousands of electrons circulated in coplanar rings. Thomson and his students thought the only question was the number of electrons circulating within this sphere. To test this model experimentally, they directed a beam of electrons at a very thin metal foil. They expected that as the beam passed through the foil, the electrons of the beam would encounter the very large number of electrons within the atoms, and the negative charges would repel each other. A tiny deflection of the beam from its straight path should be observed with each encounter, the size of the total deflection related to the number of electrons in the atom. A deflection was in fact observed, but it was much smaller than expected, forcing Thomson and his students to revise their estimate of the number of electrons, but not their model of the atom.

See the *Saunders Interactive General Chemistry CD-ROM,* Screen 2.11, The Nucleus of the Atom.

About 1910, Ernest Rutherford (1871–1937) decided to test Thomson's model further. Rutherford had discovered earlier that α rays (see Figure 2.3) consisted of positively charged particles having the same mass as helium atoms. He reasoned that, if Thomson's atomic model were correct, a beam of such massive particles would be deflected very little as it passed through the atoms in a very thin sheet of gold foil. Rutherford's associate, Hans Geiger (1882–1945), and a student, Ernst Marsden, set up the apparatus diagrammed in Figure 2.8 and observed what happened when α particles hit the foil. Most passed almost straight through, but Geiger and Marsden were amazed to find that a few α particles were deflected at large angles, and some came almost straight back! Rutherford later described this unexpected result by saying, "It was about as credible as if you had fired a 15-inch [artillery] shell at a piece of paper and it came back and hit you."

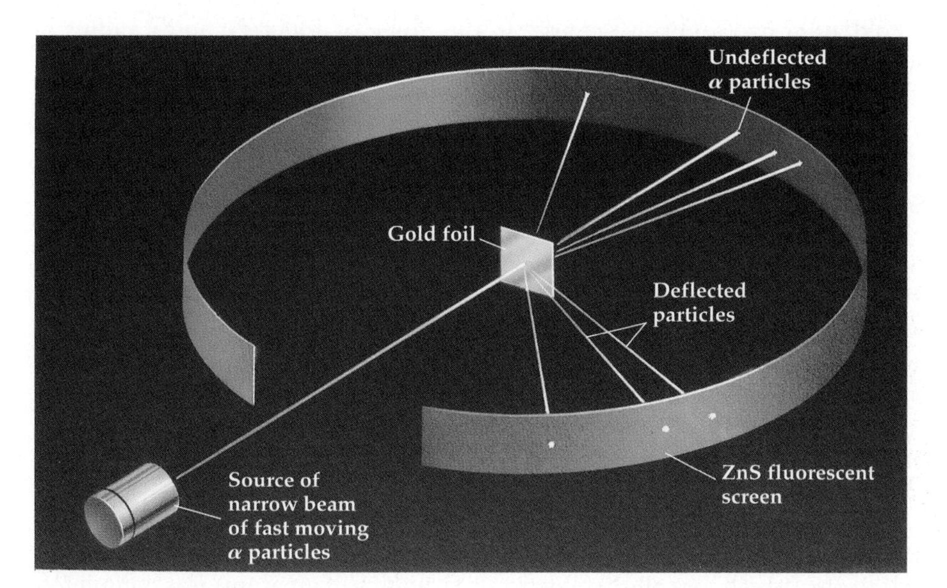

Figure 2.8 The experimental arrangement of the Rutherford experiment. A beam of positively charged α particles was directed at a very thin gold foil. A luminescent screen coated with zinc sulfide (ZnS) was used to detect particles passing through. Most of the particles passed through the foil, but some were deflected from their path. A few were even deflected backward.

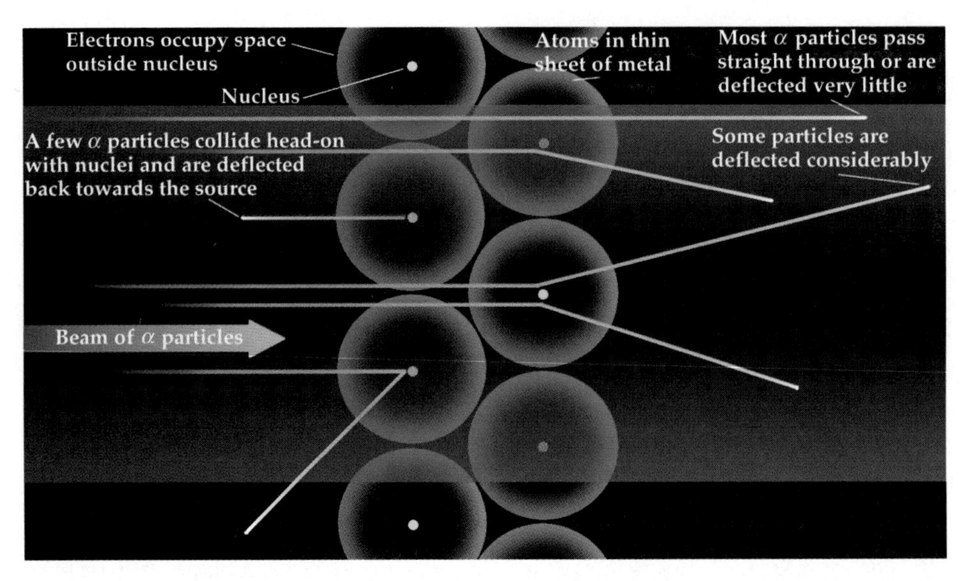

Figure 2.9 Rutherford's interpretation of the results of an experiment done by Geiger and Marsden.

Atoms are extremely small; the radius of the typical atom is between 30 and 300 pm (3×10^{-11} m to 3×10^{-10} m). To get a feeling for the incredible smallness of an atom, consider that one teaspoon of water (about 1 cm^3) contains about three times as many atoms as the Atlantic Ocean contains teaspoons of water.

The only way to account for this was to discard Thomson's model and to conclude that all of the positive charge and most of the mass of the atom is concentrated in a very small volume. Rutherford called this tiny core of the atom the **nucleus** (Figure 2.9). The electrons occupy the rest of the space in the atom. From their results Rutherford, Geiger, and Marsden calculated that the nucleus of a gold atom had a positive charge in the range of 100 ± 20 and a radius of about 10^{-12} cm. The currently accepted values are 79 for the charge and about 10^{-13} cm for the radius.

2.3 ATOMIC STRUCTURE

All atoms have neutrons in the nucleus (usually at least equal to the number of protons), except for the hydrogen atom, which has only a single proton in its nucleus.

The experiments of Thomson, Rutherford, and others led to the now-accepted model of the atom (Figure 2.10). Three primary particles—protons, neutrons, and electrons—make up all atoms. The model places protons and neutrons in a very small nucleus, which means that the nucleus contains all the positive charge and almost all the mass of an atom. Negatively charged electrons surround the nucleus and occupy most of the volume. Atoms have no net charge, so *the number of electrons outside the nucleus equals the number of protons in the nucleus.*

Although experiments in the early part of the 20th century established that electrons occupied the space outside the nucleus, the detailed arrangement of the electrons (which we call the electronic structure of an atom) was completely unknown to Rutherford and other scientists at that time. As we shall see in Chapter 8, the electronic structure of atoms is of particular importance in chemistry because this determines how atoms react.

Exercise 2.2 Describing Atoms

We know now the radius of the nucleus is about 0.001 pm, and the radius of an atom is, say, 100 pm. If an atom were a macroscopic object with a radius of 100 m, it would approximately fill a football stadium. What would be the radius of the nucleus of such an atom? Can you think of an object that is about that size?

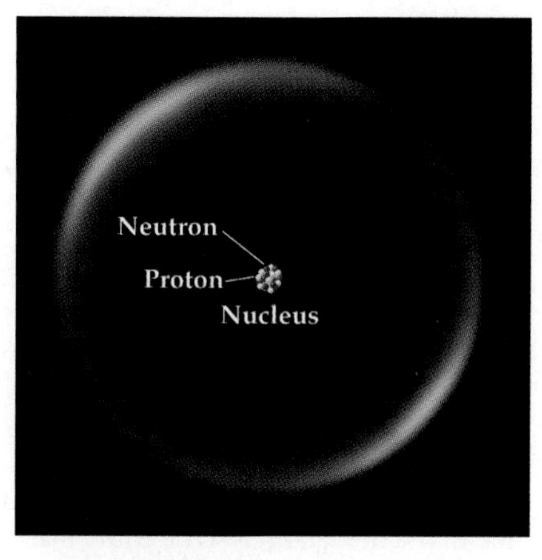

Figure 2.10 The structure of the atom. All atoms consist of one or more protons (positively charged) and usually as many neutrons (no charge) packed into an extremely small nucleus. Electrons (negatively charged) are arranged in space as a "cloud" around the nucleus. In an electrically neutral atom the number of electrons equals the number of protons.

2.4 ATOMIC NUMBER AND THE MASS OF AN ATOM

All atoms of an element have the same number of protons in the nucleus. Hydrogen is the simplest element, with one nuclear proton. A helium atom has two protons, while a lithium atom has three and a beryllium atom has four. The most recently discovered element, not yet named, has 112 protons in its nucleus.

Currently known elements are listed in the periodic table inside the front cover of the book. The integer number at the top of the box for each element is its atomic number. The **atomic number,** generally given the symbol **Z,** is the number of protons in the nucleus of an atom of that element. A sodium atom, for example, has a nucleus containing 11 protons, so its atomic number is 11. A uranium atom has 92 nuclear protons and $Z = 92$.

What is the mass of an individual atom? Chemists in the 18th and 19th centuries recognized that careful experiments could give *relative* **atomic masses.** Such experiments showed, for example, that the mass of a single oxygen atom is 1.33 times the mass of a single carbon atom or that a calcium atom is 2.5 times the mass of an oxygen atom.

Using mass ratios is cumbersome when doing quantitative chemistry, so a number that represents the mass of an atom is preferable. To create such a set of numbers, one could assign a specific mass to a given element and then calculate the masses of other elements from this value using mass ratios. After trying several standards, scientists settled on the current one, carbon. A carbon atom with six protons and six neutrons in the nucleus is assigned a mass value of exactly 12.000. Oxygen, which has 1.3333 times the mass of carbon, then becomes 16.000 on the atomic mass scale. Atomic masses of other elements were assigned from other data in a similar manner.

Because experimentally determined atomic masses are actually ratios, these numbers do not have units. It is useful, however, to have a unit of mass when describing masses of atoms. For this reason the **amu,** which stands for **atomic mass unit,** has been introduced. *One atomic mass unit, amu, is 1/12th of the mass of an atom of carbon with six protons and six neutrons.* Thus, such a carbon atom has a mass of 12.000 amu, whereas an oxygen atom with eight protons and eight neutrons in the nucleus has a mass of 16.000 amu.

Protons and neutrons have masses very close to 1 amu (Table 2.1). The electron, in contrast, has a mass that is only about 1/2000th of this value. Because proton and neutron masses are so close to 1 amu, the mass of an atom can be *estimated* if the number of neutrons and protons is known. The sum of the number of protons and neutrons is called the **mass number** of the particular atom and is given the symbol A. For example, a sodium atom has 11 protons and 12 neutrons in its nucleus, so its mass number is 23. The most common atom of

Historical Perspective

Ernest Rutherford (1871–1937)

Lord Rutherford, one of the most interesting people in the history of science, was born in New Zealand in 1871 but went to Cambridge University in England to pursue his Ph.D. in physics in 1895. His original interest was in a phenomenon that we now call radio waves, and he apparently hoped to make his fortune in the field, largely so he could marry his fiancée back in New Zealand. However, his professor at Cambridge, J. J. Thomson, convinced him to work on the newly discovered phenomenon of radioactivity. It was at Cambridge that he discovered α and β radiation, but he moved to McGill University in Canada in 1899. At McGill he did further experiments to prove that α radiation is composed of helium nuclei and that β radiation consists of electrons. For this work he received the Nobel Prize in chemistry in 1908.

At McGill, Rutherford was fortunate to have a talented student, Frederick Soddy (1877–1956), work with him. The two studied a radioactive gas coming from the radioactive element, thorium. Their experiments showed that the gas was argon, which meant that they had made the first observation of the spontaneous disintegration of a radioactive element, one of the great discoveries of 20th-century physics.

Ernest Rutherford (*Oesper Collection in the History of Chemistry/University of Cincinnati*)

In 1903, Rutherford and his young wife visited with Pierre and Marie Curie in Paris, on the very day that Madame Curie received her doctorate in physics (see page 64). There was of course a celebration, and that evening while the party was in the garden of the Curie's home, Pierre Curie brought out a tube coated with a phosphor and containing a large quantity of radium in solution. The phosphor glowed brilliantly from the radiation given off by the radium. Rutherford later said the light was so bright he could clearly see Pierre Curie's hands were "in a very inflamed and painful state due to exposure to radium rays."

In 1907, Rutherford moved from Canada to Manchester University in England, and it was there that he performed the experiments that gave us the modern view of the atom. In 1919, he moved back to Cambridge and assumed the position formerly held by J. J. Thomson. Not only was Rutherford responsible for very important work in physics and chemistry, but he also guided the work of no fewer than ten future recipients of the Nobel Prize. Element 104, rutherfordium (Rf), was recently named in his honor.

uranium has 92 protons and 146 neutrons, so $A = 238$. With this information, we can symbolize an atom of known composition by the notation

$$\text{mass number} \longrightarrow \quad {}^{A}_{Z}\text{X} \longleftarrow \text{element symbol}$$
$$\text{atomic number} \longrightarrow$$

(where the subscript Z is optional because the element symbol tells us what the atomic number must be). For example, the atoms described previously have the

Table 2.1 • Properties of Subatomic Particles*

	Mass			
Particle	**Grams**	**amu**	**Charge**	**Symbol**
Electron	9.109389×10^{-28}	0.0005485799	−1	${}^{0}_{-1}\text{e}$
Proton	1.672623×10^{-24}	1.007276	+1	${}^{1}_{1}\text{p}$
Neutron	1.674929×10^{-24}	1.008665	0	${}^{1}_{0}\text{n}$

* These constants and others in the book are taken from E. R. Cohen and B. N. Taylor: "Fundamental Physical Constants," *Physics Today*, Vol. 40, pp. BG11-BG15, 1987.

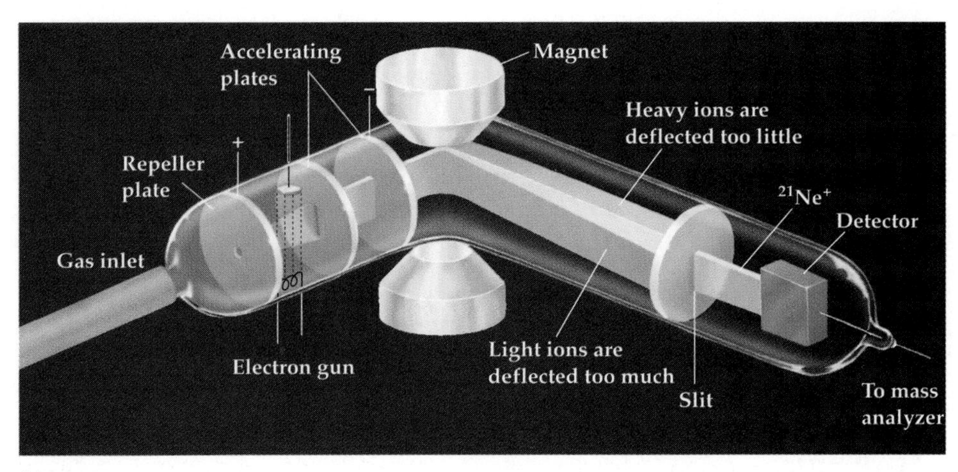

(a)

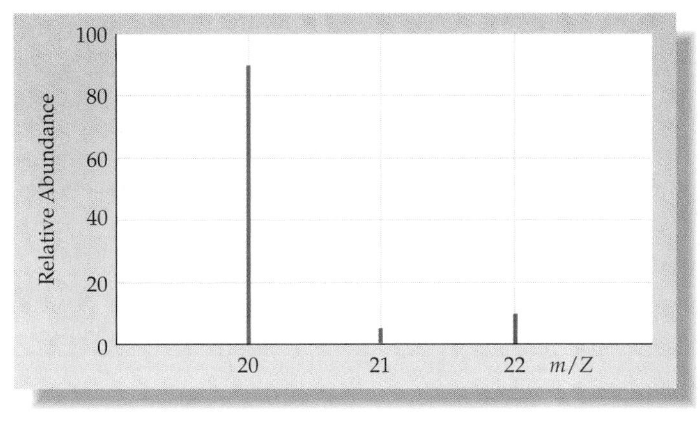

(b)

***Figure* 2.11 Simplified diagram of an electron ionization mass spectrometer.** (a) A sample is introduced as a vapor into the ionization chamber. There it is bombarded with high-energy electrons that strip electrons from the molecules of the sample. The resulting positive ions are accelerated by a series of negatively charged accelerator plates into an analyzing chamber. This chamber is in a magnetic field, which is perpendicular to the direction of the ion beam. The magnetic field causes the ion beam to curve (see Figure 2.4). The radius of curvature depends on the charge of the ion (Z), its mass (m), and on the accelerating voltage and the strength of the magnetic field. (b) A mass spectrum is a plot of relative ion abundance versus the (m/Z) ratio. Here ions of ^{21}Ne are focused on the detector, whereas beams of ions of ^{20}Ne and ^{22}Ne (of lighter or heavier mass) experience greater or lesser curvature, respectively, and so fail to be detected. By scanning the mass range, however, a spectrum of masses is observed.

symbols $^{23}_{11}Na$ or $^{238}_{92}U$, or just ^{23}Na or ^{238}U. In words, we say "sodium-23" or "uranium-238."

Masses of atoms have been determined experimentally using mass spectrometers (Figure 2.11). Although the atomic mass approximately equals the mass number, the actual mass is not an integral number (except for ^{12}C, which is exactly 12 by definition). For example, the mass of an iron atom with 32 neutrons, ^{58}Fe, is 57.9333 amu, slightly less than the mass number.

Because the amu is a unit of mass, it can be related to other units of mass using conversion factors; that is, 1 amu $= 1.661 \times 10^{-24}$ g.

An obvious thing to do is to check the experimental mass of an atom by adding up all the masses of the protons, neutrons, and electrons in the atom to see if the experimental mass is obtained. This sum is always found to be slightly greater than the actual mass, however. This is not a mistake. Rather, the difference (sometimes called the "mass defect") is related to the energy binding the particles of the nucleus together; this is discussed in more detail in Chapter 24.

Example 2.1 Atomic Composition

How many neutrons are in an atom of platinum with a mass number of 195?

Solution Platinum has the symbol Pt, and its atomic number (shown in the periodic table inside the front cover) is 78. An atom of platinum, therefore, has 78 protons in the nucleus (and 78 electrons arranged outside the nucleus). The mass number of the atom is the sum of the number of protons and neutrons in the nucleus. Therefore,

$$\text{Mass number} = 195 = \text{number of protons} + \text{number of neutrons}$$
$$= 78 + \text{number of neutrons}$$
$$\text{Number of neutrons} = 195 - 78 = 117$$

Example 2.2 Atomic Mass

What is the mass number of a fluorine atom with ten neutrons? Such an atom has a mass of 18.9984 amu. What is its mass in grams?

Solution All fluorine atoms have nine protons in the nucleus. For an atom with ten neutrons, the mass number A would be 19. The atom's symbol is $^{19}_{9}F$.

To calculate the mass of one atom, use the conversion factor that relates amu to mass in grams.

$$\text{Mass of } ^{19}F \text{ atom} = 18.9984 \text{ amu} \cdot (1.661 \times 10^{-24} \text{ g/amu})$$
$$\text{Mass of } ^{19}F \text{ atom} = 3.156 \times 10^{-23} \text{ g}$$

Exercise 2.3 Atomic Composition

1. What is the mass number of a copper atom with 34 neutrons?
2. A copper atom with 34 neutrons has a mass of 62.9396 amu. What is its mass in grams?
3. How many protons, neutrons, and electrons are in a $^{59}_{28}Ni$ atom?

2.5 ISOTOPES

See the *Saunders Interactive General Chemistry CD-ROM,* Screen 2.13, Isotopes.

If we examine a natural sample of an element using a mass spectrometer (see Figure 2.11) we find, in most cases, atoms with several masses. Take boron, for example. The mass spectrum of boron shows the presence of two different atoms, one with a mass of about 10 and the second with a mass of about 11. That is, there are two types of boron atoms: ^{10}B and ^{11}B. Atoms with the same atomic number but different mass numbers are called **isotopes.**

Frederick Soddy, Ernest Rutherford's assistant, coined the word "isotope" to describe the different forms of the same element.

All atoms of the same element have the same number of protons, five in the case of boron. To have different masses, isotopes must have *different numbers of neutrons.* The nucleus of a ^{10}B atom contains five protons and five neutrons, whereas the nucleus of a ^{11}B atom contains five protons and six neutrons.

Most elements have at least two stable (nonradioactive) isotopes, but a few have only one isotope (aluminum, fluorine, and phosphorus, for example).

Table 2.2 • MASSES OF ISOTOPES OF SOME ELEMENTS

Element	Symbol	Atomic Mass	Mass Number (amu)	Isotopic Mass (amu)	Natural Abundance (%)
Hydrogen	H	1.00794	1	1.0078	99.985
	D		2	2.0141	0.015
	T*		3	3.0161	≈0
Boron	B	10.811	10	10.0129	19.91
			11	11.0093	80.09
Nitrogen	N	14.00674	14	14.0031	99.63
			15	15.0001	0.37
Oxygen	O	15.9994	16	15.9949	99.759
			17	16.9993	0.037
			18	17.9992	0.204
Magnesium	Mg	24.305	24	23.9850	78.99
			25	24.9858	10.00
			26	25.9826	11.01

*Radioactive

Figure 2.12 Ice made from "heavy water." Water containing ordinary hydrogen (^1_1H, protium) forms a solid that is less dense (d = 0.917 g/cm^3 at 0 °C) than liquid H_2O (d = 0.997 g/cm^3 at 25 °C) and so floats in the liquid. (Water is unique in this regard. The solid phase of virtually all other substances sinks in the liquid phase of that substance.) Similarly, "heavy ice" (D_2O) floats in "heavy water." D_2O-ice is denser than H_2O, however, so cubes made of D_2O sink in the liquid phase of H_2O. (The solid at the bottom of the glass in this photograph is D_2O.) (*C. D. Winters*)

Conversely, some other elements have many isotopes (tin, for example, has ten stable isotopes). Generally, we refer to a particular isotope by giving its mass number (for example, uranium-238, ^{238}U), but the isotopes of hydrogen are so important that they have special names and symbols. Hydrogen atoms all have one proton. When that is the only nuclear particle, the isotope is called *protium,* or more usually "hydrogen." When one neutron is also present, the isotope ^2_1H is called *deuterium,* or "heavy hydrogen" (symbol = D) (Table 2.2). The nucleus of radioactive hydrogen, ^3_1H, or *tritium* (symbol = T), contains one proton and two neutrons.

The substitution of one isotope of an element for another of that element in a compound can have interesting consequences (Figure 2.12). This is especially true when deuterium is substituted for hydrogen because the mass of deuterium is double that of hydrogen.

Isotope Abundance

A mass spectrometer can be used to obtain not only accurate masses of an element's isotopes but also their percent abundance, where **percent abundance** is defined as the percentage of atoms of each kind in a sample.

$$\text{Percent abundance} = \frac{\text{number of atoms of a given isotope}}{\text{total number of atoms of all isotopes of that element}} \times 100\%$$

A mass spectrum of neon isotopes is illustrated in Figure 2.11. The spectrum reflects the fact that the element has three isotopes with abundances ^{20}Ne = 90.48%, ^{21}Ne = 0.27%, and ^{22}Ne = 9.22%.

Consider the two isotopes of boron. A mass spectrometer allows us to measure the masses of the boron-10 and boron-11 isotopes as 10.0129 and 11.0093, respectively. The boron-10 isotope has an abundance of 19.91%; that of boron-11 is 80.09%. This means that if you could count out 10,000 boron atoms from an "average" natural sample, 1991 of them would be boron-10 atoms and 8009 of them would be boron-11 atoms.

Example 2.3 Isotopes

Silver has two isotopes, one with 60 neutrons and the other with 62 neutrons. What are the mass numbers and symbols of these isotopes?

Solution Silver has an atomic number of 47, so it has 47 protons in the nucleus. The two isotopes, therefore, have mass numbers of

$$\text{Isotope 1: } A = 47 \text{ protons} + 60 \text{ neutrons} = 107$$
$$\text{Isotope 2: } A = 47 \text{ protons} + 62 \text{ neutrons} = 109$$

The first isotope has a symbol $^{107}_{47}Ag$ and the second is $^{109}_{47}Ag$.

Exercise 2.4 Isotopes

Silicon has three isotopes with 14, 15, and 16 neutrons, respectively. What are the mass numbers and symbols of these three isotopes?

2.6 ATOMIC WEIGHT

See the *Saunders Interactive General Chemistry CD-ROM,* Screen 2.15, Atomic Mass.

Chemists usually use the term "atomic weight" of an element rather than "atomic mass." Although the quantity is more properly called a "mass" than a "weight," the term "atomic weight" is so commonly used that it has become accepted.

Because a sample of boron has some atoms with a mass of 10.0129 amu and others with a mass of 11.0093 amu, the average atomic mass must be somewhere between these values. This average mass of a representative sample of atoms, expressed in atomic mass units, is called the **atomic mass** or **atomic weight.** For boron, the atomic mass is 10.81 amu, as is shown by the following calculation, in which the mass of each isotope is multiplied by its percent abundance divided by 100:

$$\text{Atomic mass} = \left(\frac{19.91}{100}\right) \cdot 10.0129 \text{ amu} + \left(\frac{80.09}{100}\right) \cdot 11.0093 \text{ amu} = 10.81 \text{ amu}$$

In general, the average atomic mass can be calculated using the equation

$$\text{Average atomic mass} = \left(\frac{\%\text{ abundance of isotope 1}}{100}\right)(\text{mass of isotope 1}) + \left(\frac{\%\text{ abundance of isotope 2}}{100}\right)(\text{mass of isotope 2}) + \cdots$$

The periodic table entry for copper.

29 ⟵ atomic number

Cu ⟵ symbol

63.546 ⟵ atomic weight

The atomic mass of each stable element has been determined by experiment, and these masses appear in the periodic table in the front of the book. In the periodic table, each element's box contains the atomic number, the element symbol, and the atomic mass. For unstable (radioactive), synthetic elements, the mass number of the most stable isotope is given in parentheses.

Example 2.4 Calculating Average Atomic Mass from Isotopic Abundances

Bromine (used to make silver bromide, the important component of photographic film) has two naturally occurring isotopes, one with a mass of 78.918336 amu and a percent abundance of 50.69%. The other isotope, of mass 80.916289, has a percent abundance of 49.31%. Calculate the atomic mass of bromine.

Solution The atomic mass of any element is the average of the masses of all the isotopes in a representative sample. To calculate the atomic mass, multiply the mass of each isotope by its percent abundance divided by 100. (The percent abundance divided by 100 is called the *fractional abundance.*)

For bromine, the calculation is as follows:

Average atomic mass of bromine

$$= \text{atomic mass}$$
$$= (0.5069)(78.918336 \text{ amu}) + (0.4931)(80.916289 \text{ amu})$$
$$= 79.90 \text{ amu}$$

Example 2.5 Calculating Isotopic Abundances

Antimony, Sb, has two stable isotopes with experimentally determined masses of 120.904 amu (^{121}Sb) and 122.904 amu (^{123}Sb). What are the relative abundances of these isotopes?

Solution The atomic mass of antimony is 121.757 amu. Because the atomic mass is closer to 121 amu than to 123 amu, this must mean that the lighter isotope (^{121}Sb) is the more abundant. To calculate the actual abundances, we must contend with the fact that there are two unknown, but related, quantities.

$$\text{Atomic mass} = 121.757 \text{ amu} = (\text{fractional abundance of } ^{121}\text{Sb})(120.904 \text{ amu}) + (\text{fractional abundance of } ^{123}\text{Sb})(122.904 \text{ amu})$$

or

$$121.757 \text{ amu} = x(120.904 \text{ amu}) + y(122.904 \text{ amu})$$

Recall from algebra that if you have two unknowns, you need two equations to be able to solve for the unknowns. We have the first of them. The second comes from the fact that the sum of the fractional abundances must equal 1. That is,

$$\text{Fractional abundance of } ^{121}\text{Sb} + \text{fractional abundance of } ^{123}\text{Sb} = 1$$

or

$$x + y = 1$$

and so

$$y = \text{fractional abundance of } ^{123}\text{Sb} = 1 - x$$

Making the substitution for y, we have

$$121.757 \text{ amu} = x(120.904 \text{ amu}) + (1 - x)(122.904 \text{ amu})$$
$$= (120.904 \text{ amu})x + 122.904 \text{ amu} - (122.904 \text{ amu})x$$
$$121.757 \text{ amu} - 122.904 \text{ amu} = (120.904 \text{ amu} - 122.904 \text{ amu})x$$
$$x = \frac{121.757 \cancel{\text{amu}} - 122.904 \cancel{\text{amu}}}{120.904 \cancel{\text{amu}} - 122.904 \cancel{\text{amu}}} = 0.5735$$

HISTORICAL PERSPECTIVE

Dmitri Ivanovitch Mendeleev (1834–1907)

Mendeleev was born in Tobolsk, Siberia, but was educated in St. Petersburg where he lived virtually all his life. He taught at St. Petersburg University and there wrote books and published his concept of chemical periodicity.

It is interesting that Mendeleev did little else with chemical periodicity after his initial articles. He went on to other interests, among them studying the natural resources of Russia and their commercial applications. In 1876, he visited the United States to study the fledgling oil industry and was much impressed with the industry but not with the country. He found Americans uninterested in science, and he felt the country carried on the worst features of European civilization.

Dmitri Ivanovitch Mendeleev *(Oesper Collection in the History of Chemistry/University of Cincinnati)*

By the end of the 19th century, political unrest was growing in Russia, and Mendeleev lost his position at the university. He was appointed Chief of the Chamber of Weights and Measures for Russia, however, and established an inspection system for guaranteeing the honesty of weights and measures used in Russian commerce.

All pictures of Mendeleev show him with long hair. He made it a rule to cut his hair only once a year, in the spring, whether he had to appear at an important occasion or not.

The fractional abundance of ^{121}Sb is 0.5735. This means the fractional abundance of ^{123}Sb is $1 - x = 0.4265$. Translating these to percent abundances (in four significant figures), we have

$$^{121}Sb = 57.35\% \quad \text{and} \quad ^{123}Sb = 42.65\%$$

This result confirms our earlier estimate that the lighter isotope was the more abundant of the two.

Exercise 2.5 Calculating Atomic Mass

Verify that the atomic mass of chlorine is 35.45 amu, given the following information:

$$^{35}Cl \text{ mass} = 34.96885 \text{ amu; percent abundance} = 75.77\%$$
$$^{37}Cl, \text{ mass} = 36.96590 \text{ amu; percent abundance} = 24.23\%$$

2.7 THE PERIODIC TABLE

See the *Saunders Interactive General Chemistry CD-ROM,* Screen 2.16, The Periodic Table, and Screen 2.17, Chemical Periodicity. There is also an interactive periodic table (with a glossary of terms and a database of element properties). Click on the "toolbox" icon.

The periodic table of elements in the front of the book is one of the most useful tools in chemistry. Not only does it contain a wealth of information, but it can be used to organize many of the ideas of chemistry. It is important therefore that you be familiar with its main features and terminology.

Features of the Periodic Table

The elements are arranged so that those with similar chemical and physical properties lie in vertical columns called **groups.** The table commonly used in the

United States has groups numbered 1 through 8, with each number followed by a letter: A or B. Using this system, chemists often designate the A groups as **main group elements** and B groups as **transition elements.** The horizontal rows of the table are called **periods,** and they are numbered beginning with 1 for the period containing only H and He. For example, sodium, Na, is in Group 1A and is the first element in the third period. Mercury, Hg, is in Group 2B and in the sixth period.

There are two primarily different ways to designate the groups in the periodic table. In one, the groups are numbered 1 through 18 from left to right. In the other, the main group elements are labeled as Groups 1A–8A, whereas the transition elements are labeled as Groups 1B–8B. References to elements in the book shall use the A/B system, the predominant system in the United States.

The periodic table can be divided into several regions according to the properties of the elements. On the table inside the front cover of this book, elements that behave as metals are indicated in orange or yellow, those that are nonmetals are indicated in green, and elements called metalloids are in gray. Elements gradually become less metallic as one moves from left to right across a period, and the metalloids lie along the metal-nonmetal boundary. Some elements are shown in Figure 2.13 on pages 80 and 81.

You are probably familiar with many properties of **metals** from everyday experience. Metals are solids (except for mercury, page 81), conduct electricity, are usually ductile (can be drawn into wires) and malleable (can be rolled into sheets), and can form alloys (solutions of one or more metals in another metal). Iron (Fe) and aluminum (Al) are used in automobile parts because of their ductility, malleability, and low cost relative to other metals. Copper (Cu) is used in electric wiring because it conducts electricity better than most metals. Chromium (Cr) is plated onto automobile parts, not only because its metallic luster makes cars look better but also because chrome-plating protects the underlying metal from reacting with oxygen in the air.

Nonmetals have a wide variety of properties. Some are solids; bromine is a liquid, and a few, like nitrogen and oxygen, are gases at room temperature. With the exception of graphite, nonmetals do not conduct electricity, which is one of the main features that distinguishes them from metals. All nonmetals lie to the right of a diagonal line that stretches from B to Te in the periodic table.

Some of the elements next to the diagonal line have properties that make them difficult to classify as a metal or nonmetal. Because of this, chemists have come to call them metalloids or, sometimes, semimetals. You should know, however, that there is often disagreement among chemists, not only about what a metalloid is but also what elements fit this category. We shall define a **metalloid** as an element that has some of the physical characteristics of a metal but some of the chemical characteristics of a nonmetal, and we shall include only B, Si, Ge, As, Sb, and Te in the category. This may not be a completely satisfying distinction, but it does reflect the ambiguity in the behavior of these elements. Antimony, for example, conducts electricity as well as many elements that are truly metals. Its chemistry, however, resembles that of a nonmetal such as phosphorus.

Historical Development of the Periodic Table

Although the arrangement of the elements in the periodic table can now be understood on the basis of atomic structure, the table was originally developed from many, many experimental observations of the chemical and physical properties of elements and is the result of the ideas of a number of chemists in the 18th and 19th centuries.

The historical development of the periodic table illustrates the way chemistry has developed: experimental observations led to empirical correlations of

Figure 2.13 **Some common elements.** (*C. D. Winters*)

Group 1A: lithium (Li)

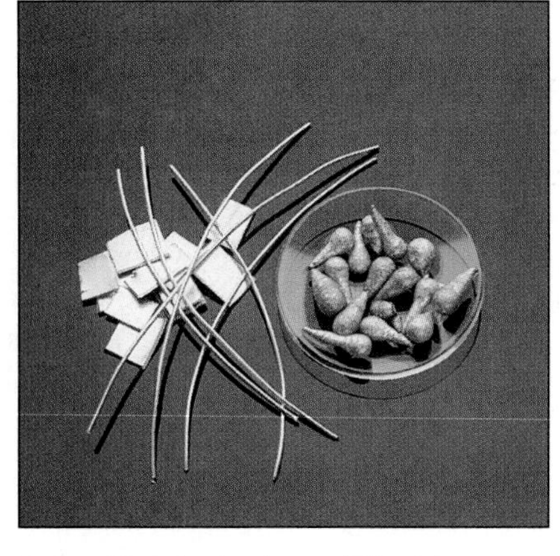

Group 3A: left, aluminum (Al); right, indium (In)

Group 5A: phosphorus (P)

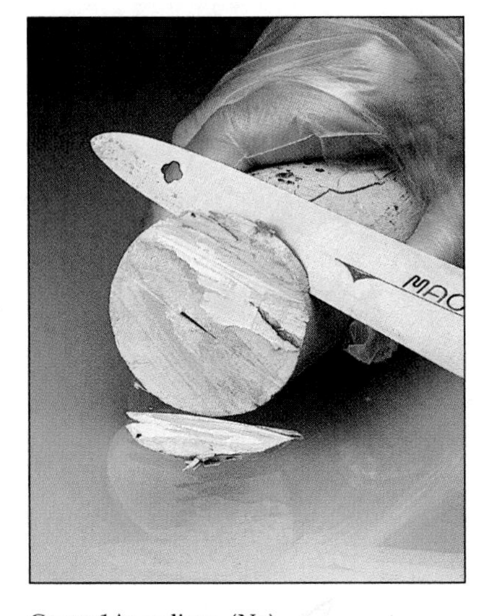

Group1A: sodium (Na)

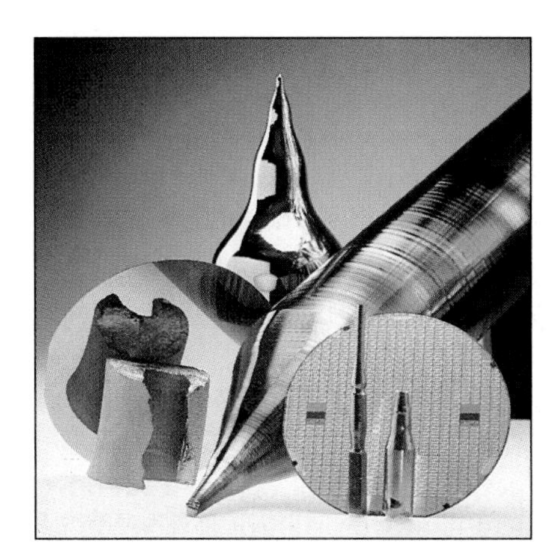

Group 4A: silicon (Si)

Group 4A: left, carbon (C); top, lead (Pb); right, tin (Sn); bottom, silicon (Si)

Group 1A: Potassium (K)

Group 6A: sulfur (S)

Group 6A: selenium (Se)

Group 5A: Nitrogen (N_2)

Group 7A: Bromine (Br_2) (left); iodine (I_2) (right)

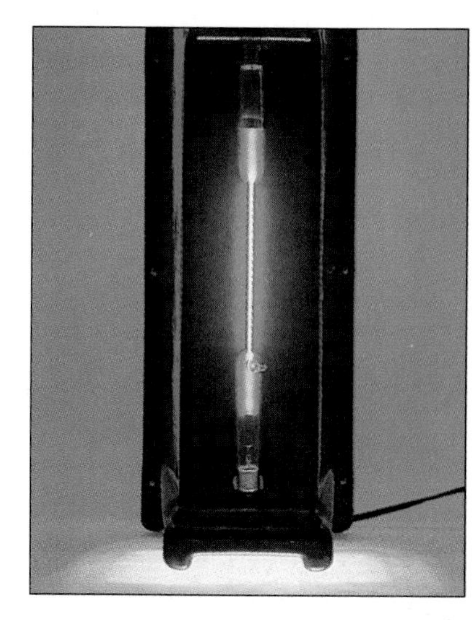

Group 8A: neon (Ne)

Fourth-period transition metals:
left to right, Ti, V, Cr, Mn, Fe, Co, Ni, Cu

Group 1B: copper (Cu)

Group 1B: silver (ag)

Group 1B: gold (Au)

Group 2B: left, zinc (Zn); right, mercury (Hg)

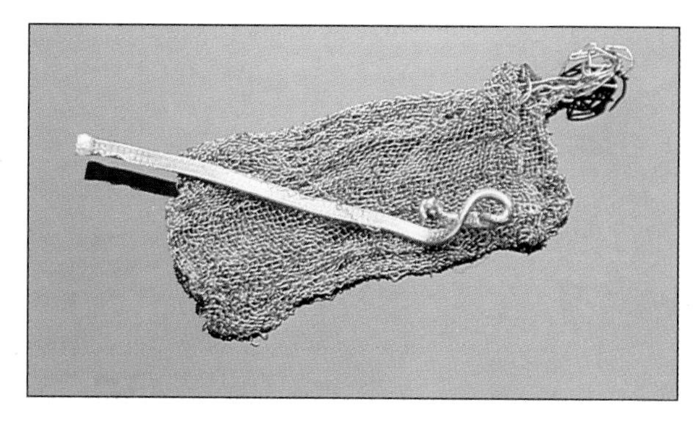

Group 8B: platinum (Pt)

In addition to Mendeleev, many others contributed to the notion of the periodicity of properties and suggested some form of a periodic table: A. E. B. de Chancourtois (1820–1886); John Newlands (1837–1898); and most especially Lothar Meyer (1830–1895).

Tellurium and iodine are not the only elements for which the element with a smaller atomic mass occurs later in the periodic table. Other reversed pairs in the modern periodic table include argon (Ar)–potassium (K) and cobalt (Co)–nickel (Ni). Can you find others?

properties and then to the prediction of results of further experiments. Once those predictions were tested, a theory could then be developed.

On the evening of February 17, 1869, at the University of St. Petersburg in Russia, a 35-year-old professor of general chemistry, Dmitri Ivanovitch Mendeleev (1834–1907) was writing a chapter of his soon-to-be-famous textbook on chemistry. He had the properties of each element written on a separate card. While shuffling the cards trying to gather his thoughts before writing his manuscript, he realized that, if the elements were arranged in order of increasing atomic mass, certain properties were repeated several times! That is, he saw that a "periodicity" occurred in the properties of elements (Figure 2.14), and soon thereafter he summarized this in a table (Figure 2.15), a predecessor to the modern periodic tables we use today. Mendeleev built the table by lining up the elements in a horizontal row in order of increasing atomic mass. Every time he came to an element with properties similar to one already in the row, he started a new row. The columns, then, contained elements with similar properties.

The most important feature of Mendeleev's table—and a mark of his genius and daring—was that he left empty spaces to retain the rationale of an ordered arrangement based on the periodic reoccurrence of similar properties. He deduced that these spaces would be filled by as-yet undiscovered elements. For example, in order of increasing atomic mass were copper (Cu), zinc (Zn), and then arsenic (As). If arsenic had been placed next to zinc, arsenic would have fallen under aluminum (Al). But arsenic forms compounds similar to those formed by phosphorus (P) and antimony (Sb), not aluminum. Arsenic, therefore, belonged in a position under phosphorus, and Mendeleev reasoned there must be two as-yet undiscovered elements whose atomic masses were between zinc and arsenic. The two missing elements were soon discovered: gallium (Ga) in 1875 and germanium (Ge) in 1886. In later years other gaps in Mendeleev's periodic table were filled as other predicted elements were discovered.

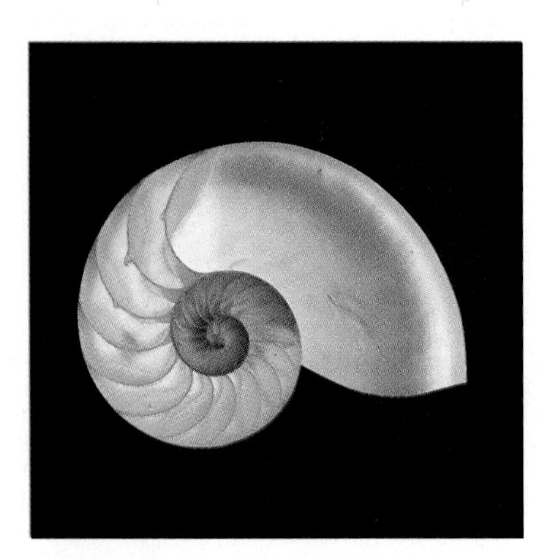

F*igure* 2.14 One half of a shell of a chambered nautilus. As the animal builds its shell, the pattern of the shell repeats itself with each revolution, an illustration of periodicity. (*C. D. Winters*)

TABELLE II

REIHEN	GRUPPE I. — R^2O	GRUPPE II. — RO	GRUPPE III. — R^2O^3	GRUPPE IV. RH^4 RO^2	GRUPPE V. RH^3 R^2O^5	GRUPPE VI. RH^2 RO^3	GRUPPE VII. RH R^2O^7	GRUPPE VIII. — RO^4
1	H = 1							
2	Li = 7	Be = 9,4	B = 11	C = 12	N = 14	O = 16	F = 19	
3	Na = 23	Mg = 24	Al = 27,3	Si = 28	P = 31	S = 32	Cl = 35,5	
4	K = 39	Ca = 40	— = 44	Ti = 48	V = 51	Cr = 52	Mn = 55	Fe = 56, Co = 59, Ni = 59, Cu = 63.
5	(Cu = 63)	Zn = 65	— = 68	— = 72	As = 75	Se = 78	Br = 80	
6	Rb = 85	Sr = 87	?Yt = 88	Zr = 90	Nb = 94	Mo = 96	— = 100	Ru = 104, Rh = 104, Pd = 106, Ag = 108.
7	(Ag = 108)	Cd = 112	In = 113	Sn = 118	Sb = 122	Te = 125	J = 127	
8	Cs = 133	Ba = 137	?Di = 138	?Ce = 140	—	—	—	— — — —
9	(—)	—	—	—	—	—	—	
10	—	—	?Er = 178	?La = 180	Ta = 182	W = 184	—	Os = 195, Ir = 197, Pt = 198, Au = 199.
11	(Au = 199)	Hg = 200	Tl = 204	Pb = 207	Bi = 208	—	—	
12	—	—	—	Th = 231	—	U = 240	—	— — — —

F*igure* 2.15 Dmitri Mendeleev's 1872 periodic table. The spaces marked with blank lines represent elements that Mendeleev deduced existed but that were unknown at the time; he left places for them in the table. The symbols at the top of the columns (e.g., R^2O and RH^4) are molecular formulas written in the style of the 19th century.

Not only did Mendeleev predict the existence of then unknown elements, but he also thought that he could detect inaccurate atomic masses. Given the chemical methods of that day, such inaccuracies were expected. For example, the elements tellurium (Te) and iodine (I) were assigned atomic masses of 128 and 127, respectively, but chemical similarities meant that he had to place Te in the same group with sulfur (S, Group 6A) and I in the same group as chlorine (Cl, Group 7A). While he was correct in his placement of these elements, time has proved him wrong on his mass prediction. We know now that the atomic mass of tellurium is indeed slightly greater than that of iodine.

The ordering of Te and I suggests that atomic mass is not the property that governs periodicity, but that another closely related property does. This was identified in 1913 by H. G. J. Moseley (1887–1915), a young English scientist working with Ernest Rutherford. Moseley bombarded many different metals with electrons in a cathode-ray tube and observed the x-rays emitted by the metals. Most importantly, he found that the wavelengths of x-rays emitted by a particular element are related in a precise way to the *atomic number* of that element.

Based on his experiments, Moseley realized that other atomic properties may be similarly related to atomic number and not, as Mendeleev had believed, to atomic mass. Indeed, if the elements are arranged in order of increasing atomic number, the defects in the Mendeleev table are corrected. That is, the **law of chemical periodicity** should be restated as "the properties of the elements are periodic functions of atomic number."

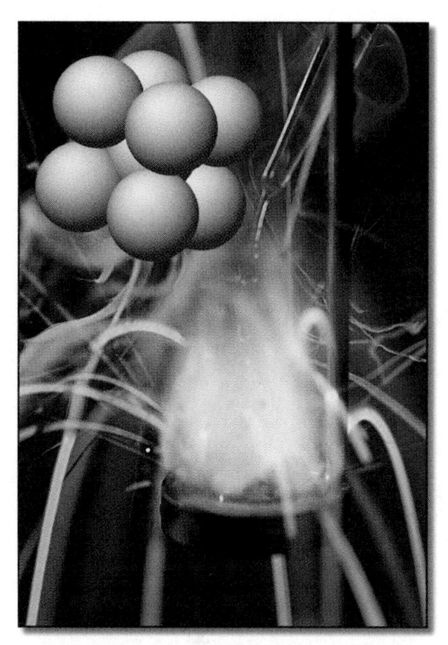

Figure 2.16 The alkali metals. The alkali metals in the solid state consist of metal atoms tightly packed into an array called a "lattice" (see also pp. 6 and 18). When an alkali metal such as potassium is treated with water, a very vigorous reaction occurs, giving an alkaline solution and hydrogen gas, which burns in air. (*Photo, C. D. Winters; model, S. M. Young*)

2.8 THE ELEMENTS, THEIR CHEMISTRY, AND THE PERIODIC TABLE

The vertical columns, or groups, of the periodic table contain elements having similar chemical and physical properties, and several groups of elements have distinctive names that are useful to know.

WWW There are several excellent sources of information on the elements on the World Wide Web. Search for *Web Elements* or begin with the Saunders web site at http://www.saunderscollege.com.

Group 1A: H and the Alkali Metals Li, Na, K, Rb, Cs, Fr

Elements in the leftmost column, **Group 1A,** are known as the **alkali metals.** Except for hydrogen, all are metals and are solids at room temperature (see Figure 2.13). In contrast, hydrogen, a gas under normal conditions, consists of diatomic, or "two-atom," molecules.

The metals of Group 1A are all very reactive. For example, they react with water to produce hydrogen and alkaline solutions (Figure 2.16). Because of their reactivity, these metals are only found in nature combined in compounds, never as the free element. A compound of sodium, sodium chloride (NaCl), has played an important role in history, because it is a fundamental part of the diet of humans and animals. Potassium compounds are important plant nutrients. In addition to their similar reactivities, all these metallic elements form compounds with oxygen that have formulas like A_2O, where A represents the alkali metal: Li_2O, Na_2O, K_2O, Rb_2O, Cs_2O. Hydrogen also forms a compound having the same general formula: water, H_2O. This similarity of formulas was important to Mendeleev when he set up the periodic table (see Figure 2.15).

The word "alkali" comes from the Arabic language; ancient Arabian chemists discovered that ashes of certain plants, which they called *al-qali,* gave water solutions that felt slippery and burned the skin. These ashes contain compounds of Group 1A elements that produce alkaline (basic) solutions.

Table 2.3 • THE 10 MOST ABUNDANT ELEMENTS IN THE EARTH'S CRUST

Rank	Element	Abundance (ppm)*
1	Oxygen	474,000
2	Silicon	277,000
3	Aluminum	82,000
4	Iron	41,000
5	Calcium	41,000
6	Sodium	23,000
7	Magnesium	23,000
8	Potassium	21,000
9	Titanium	5,600
10	Hydrogen	1,520

*ppm = g per 1000 kg

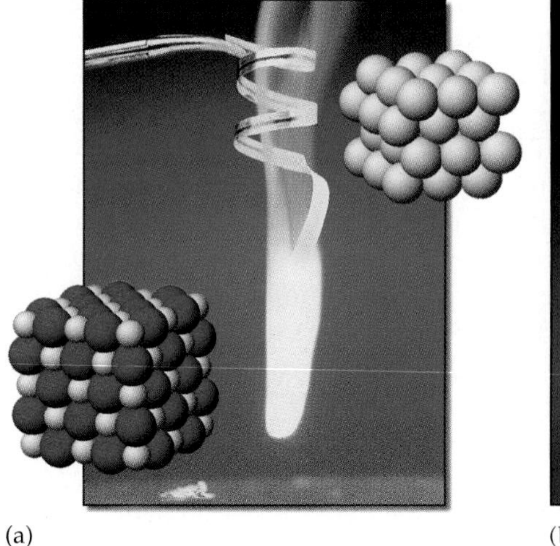
(a)

(b)

Figure 2.17 **The alkaline earth metals.** (a) The structure of magnesium (and other metals) consists of tightly packed atoms. When heated in air, magnesium burns to give magnesium oxide, a white solid in which Mg and O are now packed together in a regular array. (b) Some common calcium-containing substances: calcite (the clear crystal); a seashell; limestone; and an over-the-counter remedy for excess stomach acid. *(Photos, C. D. Winters; models, S. M. Young)*

Group 2A, the Alkaline Earth Metals: Be, Mg, Ca, Sr, Ba, Ra

Na and CaO

The second group in the periodic table, **Group 2A,** is composed entirely of metals that occur naturally only in compounds (Figure 2.17). Except for beryllium (Be), these elements also react with water to produce alkaline solutions, and most of their oxides (such as lime, CaO) form alkaline solutions; hence, they are known as the **alkaline earth elements.** Magnesium (Mg) and calcium (Ca) are the seventh and fifth most abundant elements in the earth's crust, respectively (Table 2.3). Calcium is especially well known. It is one of the important elements in teeth and bones, and it occurs in vast limestone deposits. Calcium carbonate ($CaCO_3$) is the chief constituent of limestone and of corals, sea shells, marble, and chalk (see Figure 2.17b). Radium (Ra), the heaviest alkaline earth element, is radioactive and is used to treat some cancers by radiation. When alkaline earth metals combine with oxygen (see Figure 2.17a), the general formula of the product is EO, where E is one of the alkaline earth elements; for example, magnesium forms MgO and calcium gives CaO.

The Transition Elements

Following Group 2A is a series of so-called **transition elements** that fill the fourth through the seventh periods in the center of the periodic table. All are metals (see Figure 2.13; Figure 2.18), and 13 of them are in the top 30 elements in terms of abundance in the earth's crust. Some, like iron (Fe), are very abundant in nature (Table 2.4). Most occur naturally in combination with other elements, but a few—silver (Ag), gold (Au), and platinum (Pt)—are much less reactive and so can be found in nature as pure elements.

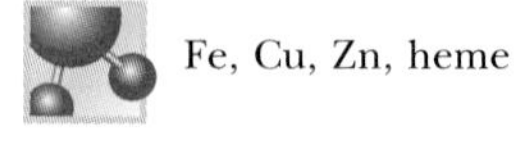
Fe, Cu, Zn, heme

Virtually all of the transition elements have commercial uses (see Figure 2.18). They are used as structural materials (iron, titanium, chromium, copper);

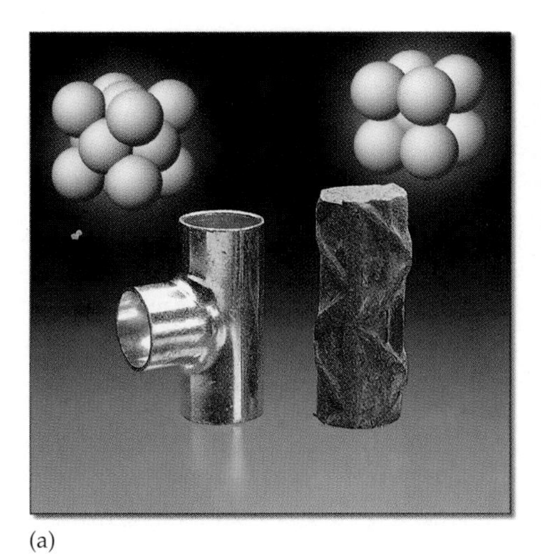

(a)

(b)

***Figure* 2.18** **Transition metals.** (a) All transition metals are solid (except for mercury, Hg, which is a liquid). Several arrangements of metal atoms are typical, as in iron and copper. (b) Transition metals are found in many consumer products, such as paint (titanium and chromium), wiring and plumbing (copper), structural steel (iron), and jewelry (gold, silver, platinum). *(Photos, C. D. Winters; models, S. M. Young)*

***Table* 2.4** • Relative Abundance of the 10 Most Abundant Transition Elements in the Earth's Crust

Rank	Element	Abundance (ppm)*
4	Iron	41,000
9	Titanium	5,600
12	Manganese	950
18	Zirconium	190
19	Vanadium	160
21	Chromium	100
23	Nickel	80
24	Zinc	75
25	Cerium	68
26	Copper	50

*ppm = g per 1000 kg

in paints (titanium, chromium); in the catalytic converters in automobile exhaust systems (platinum and rhodium); in coins (copper, nickel, zinc); and in batteries (manganese, nickel, cadmium, mercury).

A number of the transition elements play important biological roles as well. For example, iron, the fourth most abundant element in the earth's crust (see Table 2.4), is the central element in the chemistry of hemoglobin, the oxygen-carrying component of blood, and in the closely related cytochromes, the electron transfer agents in living systems (Figure 2.19).

Two rows at the very bottom of the table accommodate the **lanthanides** [the series of elements between the elements lanthanum (57) and hafnium (72)]

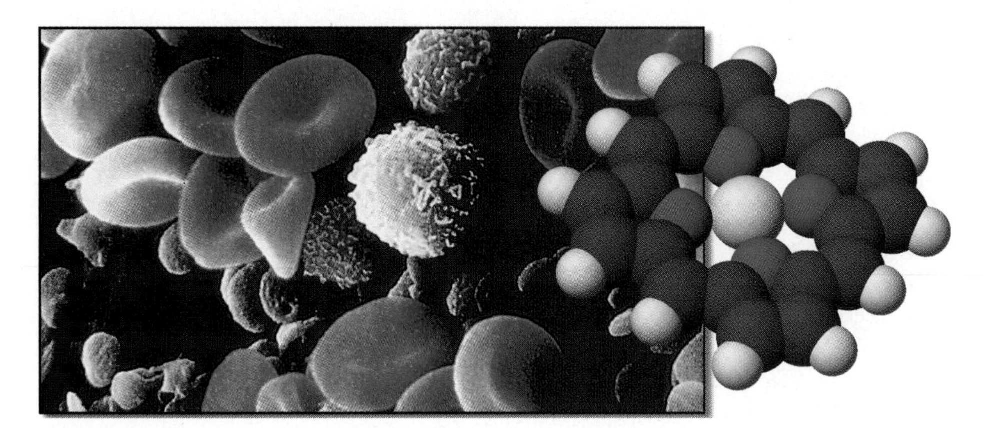

***Figure* 2.19** **The structure of the heme portion of hemoglobin.** Blood cells contain hemoglobin, a protein that contains iron as a central element. The iron is bound up in the heme group, shown here in the form of a molecular model. *(Photo, Ken Edwards/Science Source/Photo Researchers, Inc.; model, S. M. Young; model added by C. D. Winters)*

and the **actinides** [the series of elements between actinium (89) and rutherfordium (104)]. Some lanthanide compounds are used in color television picture tubes, uranium (U) is the fuel for atomic power plants, and americium (Am) is used in smoke detectors.

Group 3A: B, Al, Ga, In, Tl

Al

Group 3A contains one element of great importance, aluminum (Figure 2.20). This element and three others (Ga, In, Tl) are metals, whereas boron (B) is a metalloid. Aluminum (Al) is the most abundant metal in the earth's crust at 8.2% by mass (or 82000 ppm). It is exceeded in abundance only by the nonmetals oxygen and silicon. These three elements are found combined in clays and other common minerals.

Boron occurs in the mineral borax, which is mined in Death Valley, California, for use as a cleaning agent, antiseptic, and flux. In the late 19th century the mineral was hauled out of the valley in wagons drawn by 20 mules, hence the name of a popular washing powder. Gallium, indium, and thallium are much less familiar because they have fewer important uses. All of these elements form oxygen compounds with the general formula X_2O_3, such as Al_2O_3.

Group 4A: C, Si, Ge, Sn, Pb

Thus far all of the elements we have described, except boron, have been metals. Beginning with **Group 4A,** however, the groups contain more and more nonmetals. There is a nonmetal, carbon (C), two metalloids, silicon (Si) and germanium (Ge), and two metals, tin (Sn) and lead (Pb) in Group 4A. Because of

(a)

(b)

***Figure* 2.20 Group 3A elements.** (a) Aluminum metal consists of atoms tightly packed into a solid array. (b) Aluminum is very abundant in the earth's crust and finds many commercial uses as the metal as well as in aluminum sulfate, which is used in water purification. It is found in all clays. Boron is perhaps best known in the form of borax, a natural compound used in soap. It is also a component of borosilicate glass, the glass used for laboratory glassware. *(Photos, C. D. Winters; model, S. M. Young)*

the change from nonmetallic to metallic behavior, more variation occurs in the properties of the elements of this group than in most others. All form oxygen compounds having the general formula XO_2, however. An example is carbon dioxide, CO_2, normally found as a colorless gas, and silicon dioxide, SiO_2, a solid whose best-known form is quartz or common sand.

Carbon is the basis for the great variety of chemical compounds that make up living things. On earth it is found in the atmosphere as CO_2, in carbonates like limestone (see Figure 2.17b), and in coal, petroleum, and natural gas—the fossil fuels.

One of the most interesting aspects of the chemistry of the nonmetals is that a particular element can often exist in several different and very distinct forms, called **allotropes.** Carbon has at least three allotropes, the best known of which are graphite and diamond (Figure 2.21). In graphite the atoms of carbon are arranged in flat sheets of interconnected, hexagonal rings in which each carbon atom is connected to three others in the same layer. These sheets of rings cling only weakly to one another, so one layer can slip easily over another. This explains why graphite is soft, is a good lubricant, and works in your pencil. (Pencil "lead" is not the element lead; it is a composite of clay and graphite that leaves a trail of graphite on the page as you write.) In addition, chemists and engineers have been able to form sheets of graphite into fibers and from them make extraordinarily strong composite materials now used to make tennis rackets, fishing rods, and the masts and booms of sail boats.

Graphite, diamond, CO_2, SiO_2, and C_{60}

The carbon atoms in diamond are also arranged in six-sided rings, but each carbon atom is connected to four others, and these surround the central carbon atom at the corners of a tetrahedron (see Figure 2.21b). The atoms are connected throughout the solid as in a graphite layer, but the carbon atom rings in diamond cannot be flat. This structure causes diamonds to be extremely hard, denser than graphite ($d = 3.51$ g/cm^3 for diamond and $d = 2.22$ g/cm^3 for graphite), and chemically less reactive than graphite. Diamonds are excellent

(a)

(b)

Figure **2.21 Two of the allotropes of carbon.** (a) Graphite consists of layers of carbon atoms, linked in six-member rings. (b) Diamond has an extended array of linked carbon atoms. In this case they are also arranged in six-member rings, but the rings are not flat because each C atom is surrounded tetrahedrally by four other C atoms. *(Photos, C. D. Winters; models, S. M. Young)*

Current Perspectives

Buckyball Chemistry

In the 1970s radio astronomers first detected molecules such as cyanoacetylene, HC_3N, in outer space. Harold Kroto, a chemist at the University of Sussex in England, was using radio-frequency radiation to study the structure of simple molecules, and he realized that he could use his expertise to work with astronomers in searching for other molecules in space. He contacted Takeshi Oka and astronomers at the Canadian National Research Council, and Kroto and that group soon detected expanded versions of HC_3N with even more carbon atoms. At about that time a now-famous red giant star was located in which even more carbon-containing molecules were subsequently discovered.

Early in the 1980s Kroto attended a spectroscopy meeting in Austin, Texas, where he met Robert Curl of Rice University. Kroto was quite interested in talking with Curl because Curl and Richard Smalley, also of Rice University, had built an instrument to look for clusters of atoms under extreme conditions, the kind of conditions that might exist in the vicinity of a star. In 1985, Curl and Smalley invited Kroto to come to Rice to use their instrument to look for new species that might exist in vaporized carbon. Kroto said he was in Houston within three days of being invited. Together with several graduate students, Kroto spent a week or so looking at the products of vaporizing graphite with a laser beam. Much to their amazement, their instrument detected a molecule that contained 60 carbon atoms, C_{60}, as the most prominent product (with C_{70} another prominent species). Being chemists, the first question they asked themselves was what structure this thing might have. Both Smalley and Kroto assumed from the start that these new allotropes of carbon must be pieces of flat, graphite-like sheets. But what was magic about 60 or even 70 carbons? The answer came when they discarded the notion of flat surfaces and realized that they were "closed" molecules based on rings of five and six carbon atoms. Smalley built a structure of rings cut out of paper, and quickly saw that a structure looking just like a soccer ball was the most logical candidate. They gave it the rather cumbersome name of buckminsterfullerene after the American engineer, R. Buckminster Fuller, who had invented the geodesic dome, which their new-found structure resembled. Chemists shortened the name to "buckyball."

The next major problem to solve was to prepare enough C_{60} to be able to prove the correctness of the proposed structure and to see what the chemistry of such molecules might be. Nothing happened until 1990 when teams at the University of Arizona and the Max Planck Institute in Germany discovered a way to make buckyballs by the pound instead of in nanogram quantities.

Richard Smalley has said that "To a chemist [the discovery of buckyballs is] like Christmas." The fullerenes have extraordinary properties, and dozens of uses have been proposed: microscopic ball bearings, lightweight batteries, new lubricants, new plastics, and antitumor therapy for cancer patients (by enclosing a radioactive atom within the cage). Very recently chemists in Italy attached a buckyball to a nicotine molecule and showed that it was active against a wide range of pathogens, including *Mycobacterium avium,* an emerging species of bacteria associated with HIV infections.

Finally, the circle has been closed on Kroto's idea that interesting new molecules can be found in outer space. Jeffrey Bada, a geochemist at the Scripps Institution of Oceanography in La Jolla, California, described buckyballs that may have come from outside our solar system. Buckyballs found in a 1.8 billion-year-old meteorite in Canada contained helium atoms trapped in the cage. The ratio of the ^3He and ^4He isotopes for the trapped atoms was quite different from that in our solar system or on earth. It clearly implied that the He atoms trapped in C_{60} came from far outside our solar system and so, too, did the buckyballs.

Harold Kroto said recently that the fullerene story is "a shining example of how fundamental science can lead to important discoveries." He also said that the "most important discoveries are serendipities, because they are things you don't expect."

Because Curl, Kroto, and Smalley opened up a new area of fundamental chemistry, and an area of considerable promise, they received the Nobel Prize for Chemistry in 1996.

For more information see R. F. Curl and R. E. Smalley, "Fullerenes," *Scientific American,* October, 1991, p. 54; and H. Aldersey-Williams, *The Most Beautiful Molecule.* New York, John Wiley & Sons, 1995.

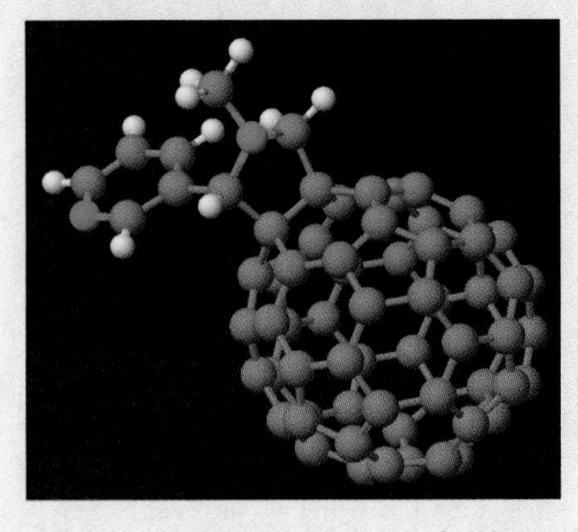

C_{60} has recently been joined to a nicotine molecule to produce a compound that is active against many pathogens.

Figure 2.22 Buckyballs. A member of the family called buckminsterfullerenes, C_{60}, is an allotrope of carbon. Sixty carbon atoms are arranged in a spherical cage that resembles a hollow soccer ball. Notice that each six-member ring is part of three other six-member rings and three five-member rings. Chemists call this compound a "buckyball." *(Photo, C. D. Winters; model, S. M. Young)*

conductors of heat. This property, combined with their hardness, means that they can be used on the tips of drills and other cutting tools. The demand for industrial diamonds is so great that over 60 tons of diamonds are made synthetically each year.

One of the most exciting developments in chemistry in the last decade has been the discovery that it is possible to grow thin films of diamond. Such films are very useful because the surface of another material can be coated with an extraordinarily hard film that conducts heat but not electricity. Some scientists believe that the development of diamond thin films is potentially the greatest advance in materials science since the invention of plastic.

Diamond and graphite have been known for centuries. In the late 1980s, however, another form of carbon was identified as a component of black soot, the stuff that collects when carbon-containing materials are burned in a deficiency of oxygen. This substance is made up of molecules with 60 carbon atoms arranged as a spherical "cage" of carbon atoms (Figure 2.22). If you look carefully, you may recognize that it resembles a hollow soccer ball; the surface is made up of five- and six-member rings. The shape also reminded its discoverers of a structure called a geodesic dome invented several decades ago by the innovative American philosopher and engineer, R. Buckminster Fuller. The official name of the allotrope is therefore buckminsterfullerene, but chemists often call C_{60} molecules "buckyballs."

Si and SiO_2

Silicon is the basis of many minerals such as quartz and beautiful gemstones like amethyst (Figure 2.23). Tin and lead have been known for centuries be-

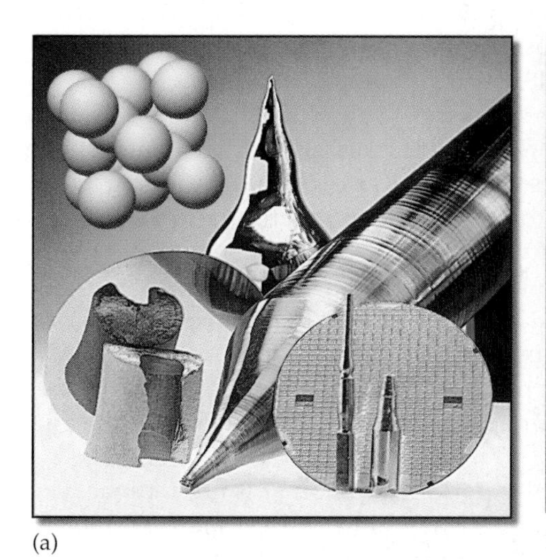

(a)

(b)

Figure 2.23 Silicon. (a) Elemental silicon has the same structure at the molecular level as the diamond allotrope of carbon. Notice that the silicon atoms are tetrahedrally arranged. (b) Silicon exists in nature largely as silicon dioxide, SiO_2, in the form of quartz and common sand. Here the silicon atoms are surrounded tetrahedrally by oxygen atoms. The O atoms link silicon atoms, and the structure consists of rings of six Si atoms linked by O atoms. *(Photos, C. D. Winters; models, S. M. Young)*

H_2														N_2	O_2 O_3	F_2	
3 Li	4 Be											5 B	6 C			Cl_2	
11 Na	12 Mg											13 Al	14 Si	15 P	16 S		
19 K	20 Ca	21 Sc	22 Ti	23 V	24 Cr	25 Mn	26 Fe	27 Co	28 Ni	29 Cu	30 Zn	31 Ga	32 Ge	33 As	34 Se	Br_2	
37 Rb	38 Sr	39 Y	40 Zr	41 Nb	42 Mo	43 Tc	44 Ru	45 Rh	46 Pd	47 Ag	48 Cd	49 In	50 Sn	51 Sb	52 Te	I_2	
55 Cs	56 Ba	57 La	72 Hf	73 Ta	74 W	75 Re	76 Os	77 Ir	78 Pt	79 Au	80 Hg	81 Tl	82 Pb	83 Bi	84 Po	85 At	86 Rn
87 Fr	88 Ra	89 Ac															

***Figure* 2.24 Diatomic molecules.** Seven of the elements in the periodic table exist as diatomic, or two-atom, molecules. Ozone, an allotrope of oxygen, has three O atoms in the molecule.

cause they are easily smelted from their ores. Tin alloyed with copper makes bronze, which was used for centuries in utensils and weapons. Lead has been used in water pipes and paint, even though the element is quite toxic to humans. Indeed, one theory for the fall of the Roman Empire contends that the Romans suffered from lead poisoning from the lead pipes they used to supply their water. In fact, the word "plumbing" comes from the Latin word for lead, *plumbum,* a name that is also the origin of its symbol, Pb.

Group 5A: N, P, As, Sb, Bi

N_2 and P_4

Nitrogen, in the form of N_2 (Figure 2.24), makes up about three-fourths of earth's atmosphere and accounts for nearly all the nitrogen at the earth's surface. As described beginning on page 3, the element is essential to life, so ways of fixing atmospheric nitrogen (forming compounds from the element) have been sought for at least a century. Nature accomplishes this easily in plants, but severe methods (high temperatures, for example) must be used in the laboratory to cause N_2 to react with other elements.

Phosphorus is also essential to life as an important constituent in bones and teeth. The element glows in the dark if it is in the air, and its name, based on Greek words meaning "light-bearing," reflects this. This element also has several allotropes, the most important being white and red phosphorus. White phosphorus ignites spontaneously in air and so is normally stored under water (Figure 2.25). It consists of four phosphorus atoms, P_4, linked together and arrayed at the corners of a tetrahedron. The red form is slightly different in that the tetrahedra of P atoms are linked one to another.

***Figure* 2.25 White and red phosphorus.** A molecular model of white phosphorus shows that P atoms are located at the corners of a tetrahedron. *(Photo, C. D. Winters; model, S. M. Young)*

Both forms of phosphorus are used commercially. White phosphorus is oxidized to form P_4O_{10}, which reacts with water to form pure phosphoric acid, H_3PO_4. This acid is then used in food products such as soft drinks. Red phosphorus also reacts with oxygen in the air and so is used in the striking strips on match books.

Bismuth is the heaviest element in the periodic table that is not radioactive; all elements with atomic numbers greater than bismuth's (83) emit α, β, or γ rays. In contrast to the other elements of **Group 5A,** bismuth is a metal. All the Group 5A elements form oxygen- or sulfur-containing compounds with the general formula E_2O_3 or E_2S_3. Examples include the brilliant yellow mineral orpiment (As_2S_3) and black stibnite (Sb_2S_3); the latter was used as a cosmetic by women in ancient societies.

Group 6A: O, S, Se, Te, Po

Group 6A begins with oxygen, which constitutes about 20% of earth's atmosphere and which combines readily with most other elements. Most of the energy that powers life on earth is derived from reactions in which oxygen combines with other substances.

Sulfur has been known in elemental form since ancient times as brimstone or "burning stone" (Figure 2.26). Sulfur, selenium, and tellurium are referred to collectively as chalcogens (from the Greek word, *khalkos,* for copper) because copper ores contain them. Their compounds are foul-smelling and poisonous; nevertheless sulfur and selenium are essential components of the human diet. By far the most important compound of sulfur is sulfuric acid, the compound the chemical industry manufactures in the largest amount (see the inside back cover of this book).

As in Group 5A, the second- and third-period elements have very different structures. Like nitrogen, oxygen is also a diatomic molecule (see Figure 2.24). Unlike nitrogen, however, oxygen has an allotrope, the well-known compound ozone, O_3. For reasons we shall explore in Chapter 9, the molecule has a bent structure. Sulfur, which can be found in nature as a yellow powder or crystalline substance, has many allotropes. The most common allotrope consists of eight-member, crown-shaped rings of sulfur atoms (see Figure 2.26).

Polonium was isolated in 1898 by Marie and Pierre Curie, who separated it from tons of a uranium-containing ore and named it for Madame Curie's native country, Poland.

Figure 2.26 Sulfur. The most common allotrope of sulfur consists of eight-member, crown-shaped rings. *(Photo, C. D. Winters; model, S. M. Young)*

Group 7A, Halogens: F, Cl, Br, I, At

At the far right of the periodic table are two groups composed entirely of nonmetals. Within each group the elements are quite similar, but they are completely different from elements in the other group. The **Group 7A** elements—fluorine, chlorine, bromine, and iodine—all exist as diatomic molecules (see Figure 2.24). All combine violently with alkali metals to form salts such as table salt, NaCl, for example (Figures 1.9 and 2.27). The name for this group, the

Be sure to notice the spelling of fluorine (u before o). A common misspelling is "flourine," which would be pronounced "flower-ene."

NaCl, S_8

(a)

(b)

Figure 2.27 The reaction of sodium with chlorine. Like all alkali metals, sodium reacts vigorously with halogens. (a) A flask containing yellow chlorine gas (Cl_2). (b) A piece of sodium was placed in the flask, and the sodium reacted with the chlorine to give table salt (sodium chloride, NaCl). The reaction evolves energy in the form of heat and light. (See also Figure 1.9.) *(Photo, C. D. Winters; models, S. M. Young)*

CURRENT PERSPECTIVES

Essential Elements

As our knowledge of biochemistry—the chemistry of living systems—increases, we learn more and more about **essential elements.** These are so important to life that a deficiency in any one will result in either death or severe developmental abnormalities. No other element can take the place of an essential element.

Of the 112 known elements, only about 20 or so are "essential," and these can be divided roughly into two categories: *bulk* elements and *trace* elements. Bulk elements include main group elements (C, H, N, O, P, S) that are used predominantly in the construction of proteins and other organic cell components. Also included are mineral ions such as Na^+, K^+, Mg^{2+}, Ca^{2+}, Cl^-, and HPO_4^{2-} that serve as electrolytes.

Foods rich in essential elements. (*C. D. Winters*)

A number of metal ions are required only in trace amounts. The ions interact with specific biological molecules and activate or regulate their function by changing their structure or by serving as catalytic centers.

Vitamins are essential biological organic nutrients, and the essential elements are their inorganic counterparts. Unlike the vitamins, essential elements cannot be made by living organisms but must be present in the environment. It is therefore not surprising that the relative quantities of essential elements in the human body (Table A) reflect in many ways the relative abundances of the known elements in the earth's oceans and, to a lesser extent, in the crust. Because of this, there has been some speculation that life began close to water or in water.

Four elements (O, C, H, and N) account for most of the mass of the human body. About 99% of the structure of the body is built of elements up to atomic number 21. Exceptions are Li, Be, Al, and the rare gases. It is important to point out, though, that lithium compounds are used to treat manic depression, and aluminum has been found in the brain cells of persons who have died of Alzheimer's disease. The remaining 1% of the essential elements come largely from the first series of transition elements. In fact, their importance is out of all proportion with their abundance in the body, as iron, zinc, and copper are among the most important elements.

halogens, comes from the Greek words *hals,* meaning "salt," and *genes,* for "forming." The halogens react with many other metals to form salts, and they also combine with most nonmetals. They are among the most reactive of elements.

Group 8A, Noble Gases: He, Ne, Ar, Kr, Xe, Rn

The **Group 8A** elements—helium, neon, argon, krypton, xenon, and radon—are the least reactive elements. All are gases, and none is very abundant on earth or in earth's atmosphere. Because of this, they were not discovered until the end of the 19th century. Helium, the second most abundant element in the universe after hydrogen, was detected in the sun in 1868 by analysis of the solar spectrum. (The name of the element comes from the Greek word for the sun, *helios.*) It was not found on earth until 1895, however. Until 1962, when a compound of xenon was first prepared, it was believed that none of these elements would combine chemically with any other element. This led to the name **noble gases** for this group, a term meant to denote their general lack of reactivity. For the same reason they are sometimes called the **inert gases** or, because of their low abundance, the **rare gases.**

Much of the 3 or 4 grams of iron in the body is used in hemoglobin, the substance responsible for carrying oxygen to the cells of the body. Iron deficiency is marked by fatigue, infections, and mouth inflammation. The average person also contains about 2 g of zinc. A deficiency of this element will be noticed by loss of appetite, failure to grow, and skin changes.

Your body also has about 75 mg of copper, about a third of which is found in the muscles and the remainder in other tissues. It is involved in many biological functions, so a deficiency shows up in a variety of ways: anemia, degeneration of the nervous system, impaired immunity, and defects in hair color and structure.

Although many elements are essential to health and can be obtained from many common foods (Table B), an overabundance can be detrimental. The optimum amount of selenium, for example, is in the range of 50 to 200 μg/day, whereas 200 to 1000 μg is toxic. Anything over 1000 μg is lethal.

Table A • RELATIVE AMOUNTS OF ESSENTIAL ELEMENTS IN THE HUMAN BODY

Element	Percent by Mass
Oxygen	65
Carbon	18
Hydrogen	10
Nitrogen	3
Calcium	1.5
Phosphorus	1.2
Potassium, sulfur, chlorine	0.2
Sodium	0.1
Magnesium	0.05
Iron, cobalt, copper, zinc, iodine	< 0.05
Selenium, fluorine	< 0.01

Table B • SOURCES OF SOME ESSENTIAL ELEMENTS

Element	Source	mg/100 g
Iron	Brewer's yeast	17.3
	Eggs	2.3
Zinc	Brazil nuts	4.2
	Chicken	2.6
Copper	Oysters	13.7
	Brazil nuts	2.3
Calcium	Swiss cheese	925
	Whole milk	118
	Broccoli	103
Selenium	Butter	0.15
	Cider vinegar	0.09

Because of its low density, helium gas is used in lighter-than-air craft such as blimps. Neon and argon are used in advertising signs. Radon is radioactive and the cause of indoor air pollution problems when it seeps out of the ground into buildings.

Exercise 2.6 The Periodic Table

How many elements are in the third period of the periodic table? Give the name and symbol of each. Tell whether each element in the period is a metal, metalloid, or nonmetal.

CHAPTER HIGHLIGHTS

Having studied this chapter, you should be able to

- Explain the historical development of the atomic theory and identify some of the scientists who made important contributions (Sections 2.1–2.2).

- Describe **electrons, protons,** and **neutrons,** and the general structure of the atom (Section 2.3).
- Define **isotope** and give the **mass number** and number of neutrons for a specific isotope (Section 2.4).
- Calculate the atomic mass of an element from isotopic abundances and masses (Section 2.4).
- Explain the difference between the **atomic number** and **atomic mass** of an element (Sections 2.3–2.5).
- Identify the periodic table location of **groups, periods, metals, metalloids, nonmetals, alkali metals, alkaline earth metals, halogens, noble gases,** and the **transition elements** (Section 2.6).

STUDY QUESTIONS

REVIEW QUESTIONS

1. Which of John Dalton's ideas, first put forth early in the 19th century, are not true according to our present views of atomic theory?
2. What are the three fundamental particles from which atoms are built? What are their electric charges? Which of these particles constitute the nucleus of an atom? Which is the least massive particle of the three?
3. Define "atomic mass unit (amu)."
4. What is the difference between the mass number and the atomic number of an atom?
5. What did the discovery of radioactivity reveal about the structure of atoms?
6. What is the relationship between the work of J. J. Thomson and that of Robert Millikan? How was Ernest Rutherford's work related to J. J. Thomson's?
7. If the nucleus of an atom were the size of a medium-sized orange (let us say with a diameter of about 6 cm), what would be the diameter of the atom?
8. The volcanic eruption of Mt. St. Helens in the state of Washington in 1980 produced a considerable quantity of a radioactive element in the gaseous state. The element has atomic number 86. What are the symbol and name of this element?
9. Titanium and thallium have symbols that are easily confused with each other. Give the symbol, atomic number, atomic weight, and group and period number of each. Are they metals, metalloids, or nonmetals?
10. Lithium has two stable isotopes: ^{6}Li and ^{7}Li. One of them has an abundance of 92.5%, and the other has an abundance of 7.5%. Knowing that the atomic mass of lithium is 6.941, which is the more abundant isotope?
11. Primo Levi (page 60) said of zinc that "it is not an element which says much to the imagination, it is gray and its salts are colorless, it is not toxic, nor does it produce striking chromatic reactions; in short, it is a boring metal." From this, and from reading this chapter, make a list of the properties of zinc. For example, include in your list the position of the element in the periodic table, and tell how many electrons and protons an atom of zinc has. What is its atomic number and atomic mass? Zinc certainly is important in our economy. Can you think of any uses of the element? Check your dictionary, a reference book such as *The Handbook of Chemistry and Physics,* or the World Wide Web.
12. What was incorrect about Mendeleev's original concept of the periodic table? What is the modern "law of chemical periodicity," and how is this related to Mendeleev's ideas?
13. What is the difference between a group and a period in the periodic table?
14. Name and give symbols for (a) three elements that are metals, (b) four elements that are nonmetals, (c) and two elements that are metalloids. In each case locate the element in the periodic table by giving the group and period in which the element is found.
15. Name and give symbols for three transition metals in the fourth period. Look up each of your choices on the World Wide Web and make a list of the uses that you find.
16. Name three transition elements, a halogen, a noble gas, and an alkali metal.
17. Name two halogens. Look up each of your choices in a dictionary (or the World Wide Web), and make a list of their properties and uses.
18. Name an element that was first discovered by Madame Curie. Give its name, symbol, and atomic number. Use a dictionary to find the origin of the name of this element.
19. Which of the nonmetallic elements have allotropes? Name those elements and describe the allotropes of each.

NUMERICAL AND OTHER QUESTIONS

The Composition of Atoms

20. Give the mass number of each of the following atoms: (a) sodium with 12 neutrons, (b) titanium with 26 neutrons, and (c) germanium with 40 neutrons.

21. Give the mass number of (a) an iron atom with 30 neutrons, (b) an americium atom with 148 neutrons, and (c) a tungsten atom with 110 neutrons.

22. Give the complete symbol ($^{A}_{Z}X$) for each of the following atoms: (a) potassium with 20 neutrons, (b) argon with 21 neutrons, and (c) cobalt with 33 neutrons.

23. Give the complete symbol ($^{A}_{Z}X$) for each of the following atoms: (a) nitrogen with 8 neutrons, (b) zinc with 34 neutrons, and (c) xenon with 75 neutrons.

24. How many electrons, protons, and neutrons are there in an atom of (a) magnesium-24, ^{24}Mg; (b) tin-119, ^{119}Sn; and (c) plutonium-244, ^{244}Pu?

25. How many electrons, protons, and neutrons are there in an atom of (a) carbon-13, ^{13}C; (b) copper-63, ^{63}Cu; and (c) bismuth-205, ^{205}Bi?

26. Fill in the blanks in the table (one column per element).

Symbol	^{58}Ni	^{33}S	______	______
Number of protons	______	______	10	______
Number of neutrons	______	______	10	30
Number of electrons in the neutral atom	______	______	______	25
Name of element	______	______	______	______

27. Fill in the blanks in the table (one column per element).

Symbol	^{65}Cu	^{86}Kr	______	______
Number of protons	______	______	78	______
Number of neutrons	______	______	117	46
Number of electrons in the neutral atom	______	______	______	36
Name of element	______	______	______	______

Isotopes

28. The synthetic radioactive element technetium is used in many medical studies. Give the number of electrons, protons, and neutrons in an atom of technetium-99.

29. Radioactive americium-241 is used in household smoke detectors and in bone mineral analysis. Give the number of electrons, protons, and neutrons in an atom of americium-241.

30. Cobalt has three radioactive isotopes used in medical studies. Atoms of these isotopes have 30, 31, and 33 neutrons, respectively. Give the symbol for each of these isotopes.

31. Which of the following are isotopes of element X, the atomic number for which is 9: $^{19}_{9}X$, $^{20}_{9}X$, $^{9}_{18}X$, and $^{21}_{9}X$?

Atomic Mass

32. Verify that the atomic mass of lithium is 6.94 amu, given the following information:

^{6}Li, mass = 6.015121 amu; percent abundance = 7.50%

^{7}Li, mass = 7.016003 amu; percent abundance = 92.50%

33. Verify that the atomic mass of magnesium is 24.31 amu, given the following information:

^{24}Mg, mass = 23.985042 amu; percent abundance = 78.99%

^{25}Mg, mass = 24.985837 amu; percent abundance = 10.00%

^{26}Mg, mass = 25.982593 amu; percent abundance = 11.01%

34. Gallium has two naturally occurring isotopes, ^{69}Ga and ^{71}Ga, with masses of 68.9257 amu and 70.9249 amu, respectively. Calculate the percent abundances of these isotopes of gallium.

35. Copper has two stable isotopes, ^{63}Cu and ^{65}Cu, with masses of 62.939598 amu and 64.927793 amu, respectively. Calculate the percent abundances of these isotopes of copper.

36. Thallium has two stable isotopes, ^{203}Tl and ^{205}Tl. Which is the more abundant of the two?

37. Strontium has four stable isotopes. Strontium-84 has a very low natural abundance, but ^{86}Sr, ^{87}Sr, and ^{88}Sr are all reasonably abundant. Which of these more abundant isotopes predominates?

The Periodic Table

38. How many elements occur in Group 5A of the periodic table? Give the name and symbol of each of these elements. Tell whether each is a metal, nonmetal, or metalloid.

39. How many elements occur in the fourth period of the periodic table? Give the name and symbol of each. Tell if each is a metal, nonmetal, or metalloid.

40. How many periods of the periodic table have 8 elements, how many have 18 elements, and how many have 32 elements?

41. How many elements occur in the seventh period? What is the name given to the majority of these elements and what well-known property characterizes them?

42. Based on the formulas of other oxides of Group 2A elements, what is the formula for the oxide of barium?

43. Salt, sodium chloride, has the formula NaCl. What is the formula for the fluoride and chloride of potassium?

GENERAL QUESTIONS

44. Potassium has three naturally occurring isotopes (^{39}K, ^{40}K, and ^{41}K), but ^{40}K has a very low natural abundance. Which of the other two is the more abundant? Briefly explain your answer.

45. The elements of Group 4A can combine with halogens to give compounds such as carbon tetrachloride, CCl_4. Give the formulas for the compounds formed between Cl and the other Group 4A elements: silicon and germanium.

46. Crossword Puzzle: In the 2 × 2 crossword shown here, each letter must be correct four ways: horizontally, vertically, diagonally, and by itself. Instead of words, use symbols of elements. When the puzzle is complete, the four spaces will contain the overlapping symbols of 10 elements. There is only one correct solution.

1	2
3	4

Horizontal

1–2: Two-letter symbol for a metal used in ancient times.
3–4: Two-letter symbol for a metal that burns in air and is found in Group 5A.

Vertical

1–3: Two-letter symbol for a metalloid.
2–4: Two-letter symbol for a metal used in U.S. coins.

Single squares: all one-letter symbols

1. A colorful nonmetal.
2. Colorless gaseous nonmetal.
3. An element that makes fireworks green.
4. An element that has medicinal uses.

Diagonal

1–4: Two-letter symbol for an element used in electronics.
2–3: Two-letter symbol for a metal used with Zr to make wires for superconducting magnets.

This puzzle first appeared in *Chemical and Engineering News*, December 14, 1987 (p. 86) (submitted by S. J. Cyvin) and in *Chem Matters*, October, 1988.

47. The chart shown here is a plot of the logarithm of the relative abundance of elements 1 through 36 in the solar system. (The abundances are given on a scale that gives silicon a relative abundance of 1×10^6 [the logarithm of which is 6].)

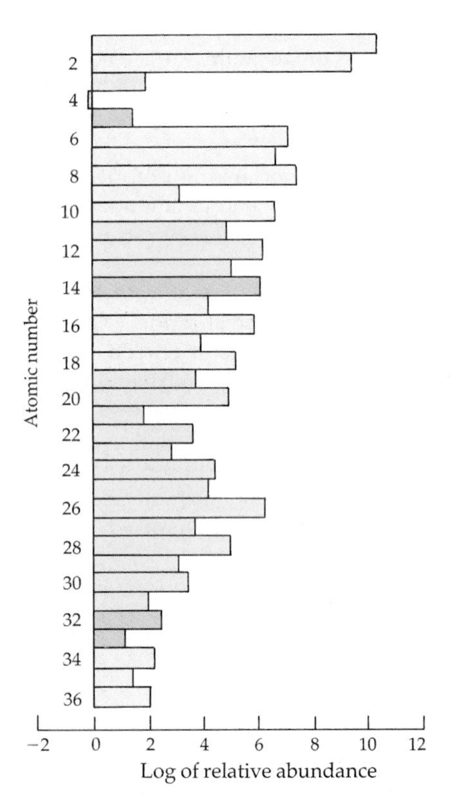

The relative abundances of elements 1–36 in the solar system.

(a) What is the most abundant metal?
(b) What is the most abundant nonmetal?
(c) What is the most abundant metalloid?
(d) Which of the transition elements is most abundant?
(e) What halogens are included on this plot and which is the most abundant?

48. The plot shows the variation in density with atomic number for the first 36 elements. Use this plot to answer the following questions:

(a) What three elements in this series have the highest density. What is their approximate density? Are these elements metals or nonmetals?
(b) Which element in the second period has the largest density? Which element in the third period has the largest density? What do these two elements have in common?
(c) Some elements have densities so low that they do not show up on the plot. What elements are these? What property do they have in common?

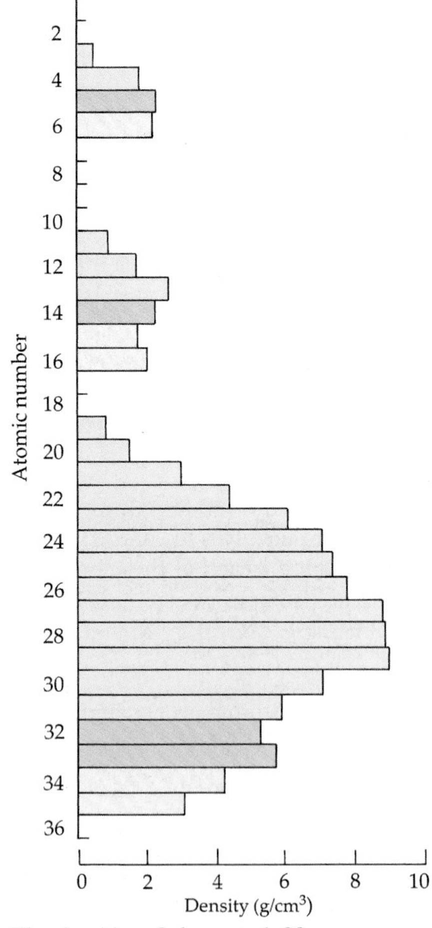

The densities of elements 1–36.

49. Draw a picture showing the approximate positions of all protons, electrons, and neutrons in an atom of helium-4. Make

certain that your diagram indicates both the number and position of each type of particle.

50. Reviewing the periodic table.
   (a) Name an element in Group 2A.
   (b) Name an element in the third period.
   (c) What element is in the second period in Group 4A?
   (d) What element is in the third period in Group 6A?
   (e) What halogen is in the fifth period?
   (f) What alkaline earth element is in the third period?
   (g) What noble gas element is in the fourth period?
   (h) Name the nonmetal in Group 6A and the third period.
   (i) Name a metalloid in the fourth period.

51. Reviewing the periodic table.
   (a) Name an element in Group 2B.
   (b) Name an element in the fifth period.
   (c) What element is in the sixth period in Group 4A?
   (d) What element is in the third period in Group 6A?
   (e) What alkali metal is in the third period?
   (f) What noble gas element is in the fifth period?
   (g) Name the element in Group 6A and the fourth period. Is this a metal, nonmetal, or metalloid?
   (h) Name a metalloid in Group 5A.

CONCEPTUAL QUESTIONS

52. Calcium reacts with air to form a compound in which the ratio of calcium to oxygen atoms is 1 : 1; that is, the formula is CaO. Based on this fact, predict the formulas for compounds formed between oxygen and the other Group 2A elements.

53. Compare the formulas of the compounds formed between O and Li, Be, Al, and C. Can you see a pattern in these formulas? Describe this pattern. Is there any relation of the pattern to the periodic group of the element combining with oxygen?

54. Again consider the plot of log abundance versus atomic number in Study Question 47. Can you uncover any relation between abundance and atomic number? Is there any difference between elements of even atomic number and those of odd atomic number?

55. The photo here depicts what happens when magnesium and calcium are placed in water. Based on their relative reactivities, what might you expect to see when barium, another Group 2A element, is placed in water? Give the period in which each element (Mg, Ca, and Ba) is found. What correlation do you think you might find between the reactivity of these elements and their position in the periodic table?

Magnesium (*left*) and calcium (*right*) in water. (*C. D. Winters*)

56. Answer the following questions using the figures shown here. (The circles represent atoms. Atoms of different elements are depicted by different colors.)

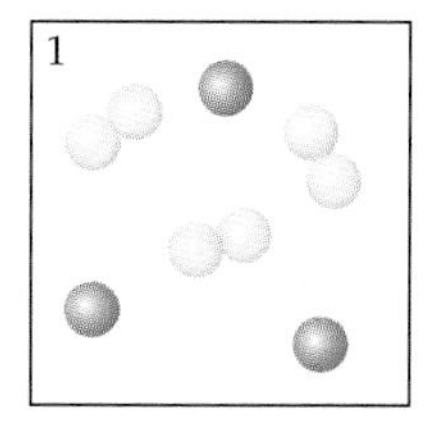

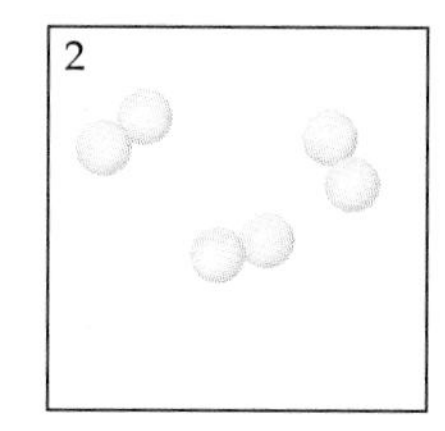

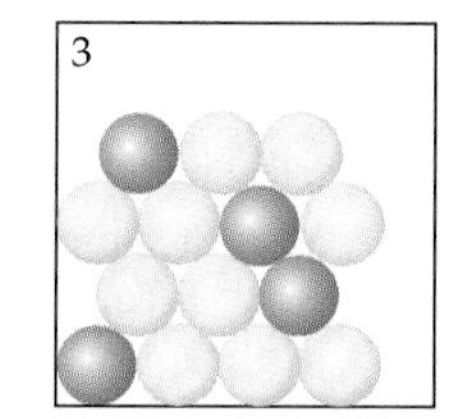

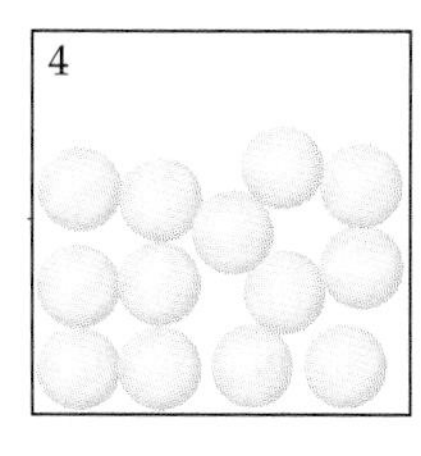

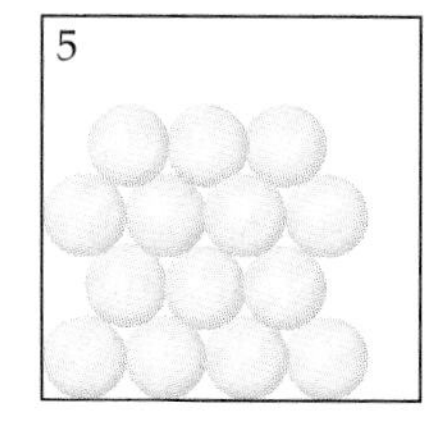

   (a) Which represents the particulate level view of a sample of a solid element such as iron?
   (b) Which represents the particulate level view of a sample of a liquid such as mercury?
   (c) Which represents the particulate level view of a sample of oxygen gas?
   (d) Which represents the particulate level view of solid bronze—a mixture of copper and zinc?
   (e) Which represents the particulate level view of a sample of a mixture of nitrogen gas and neon gas?

57. Sketch a representation of the particulate level view of solid iodine in the manner of the drawings in Study Question 56.

STRUCTURAL CHEMISTRY

Models of many of the elements mentioned in this chapter are available at the Saunders site on the World Wide Web or are on the *Interactive General Chemistry CD-ROM.* The RasMol, Chime, or CAChe model viewers allow you to measure angles in molecules and distances between atoms. Use these tools to answer the questions that follow.

58. Examine the model of ozone, O_3. What is the distance between neighboring O atoms in this molecule? Are the two distances the same or different? What is the O—O—O angle?

59. Examine the models of the three carbon allotropes.
   (a) Measure the C—C—C angles in graphite and diamond. What are these angles? Are they the same or different?
   (b) Are the C—C distances shorter or longer in graphite than in diamond?
   (c) Examine the C_{60} model. Are the C—C distances all the same or do they vary with the type of carbon ring?

60. Compare the structures of diamond and elemental silicon. Describe the similarities and differences.

61. Examine the structures of elemental aluminum and copper.
    (a) Describe the similarities and differences.
    (b) Measure the distance along one edge of the cube in each structure. Which has the greater distance? Is there any correlation with the position of the element in the periodic table?

CHALLENGING QUESTIONS

62. When a sample of phosphorus burns in air, the compound P_4O_{10} forms. Assume one experiment showed that 0.744 g of phosphorus formed 1.704 g of P_4O_{10}. Use this information to determine the ratio of the atomic masses of phosphorus and oxygen (mass P/ mass O). If the atomic mass of oxygen is assumed to be 16.000 amu, calculate the atomic mass of phosphorus.

63. Consider an atom of ^{64}Zn.
    (a) Calculate the density of the nucleus in grams per cubic centimeter, knowing that the nuclear radius is 4.8×10^{-6} nm and the mass of the ^{64}Zn atom is 1.06×10^{-22} g. Recall that the volume of a sphere is $(4/3)\pi r^3$.
    (b) Calculate the density of the space occupied by the electrons in the zinc atom, given that the atomic radius is 0.125 nm and the electron mass is 9.11×10^{-28} g.
    (c) Having calculated the densities above, what statement can you make about the relative densities of the parts of the atom?

64. The data in the table were collected in a Millikan oil drop experiment.

Oil Drop	Measured Charge on Drop (C)
1	1.59×10^{-19}
2	11.1×10^{-19}
3	9.54×10^{-19}
4	15.9×10^{-19}
5	6.36×10^{-19}

    (a) Use these data to calculate the charge on the electron (in coulombs).
    (b) How many electrons have accumulated on each oil drop?
    (c) The accepted value of the electron charge is 1.60×10^{-19}. Calculate the average deviation and error for the data in the table.

65. The mass ratios of oxygen to magnesium, O/Mg, determined for different samples of an oxide of magnesium are 1.60 g/2.43 g, 0.658 g/1.00 g, and 2.29 g/3.48 g. Do these ratios confirm the law of constant composition?

INTERACTIVE GENERAL CHEMISTRY CD-ROM

Screen 2.6 Electricity and Electric Charge

(a) What is an important principle of operation of the electroscope?
(b) How do the attraction and repulsion of electric charges apply to the structure of the atom? Go back to Screen 2.2 and think again about the structure of the atom.

Screen 2.8 Electrons

(a) How could you demonstrate that an electron is a negatively charged particle?
(b) What property of the electron was measured in Thomson's experiment?

Screen 2.10 Protons

(a) How are positively charged particles generated in the "canal-ray" experiment?
(b) Why does hydrogen have the largest charge-to-mass ratio of all the elements studied?

Screen 2.11 The Nucleus of the Atom

(a) Why are so few particles deflected as they pass through the foil?
(b) Why are even fewer particles deflected back in the direction from which they came?

Screen 2.17 Chemical Periodicity

(a) After viewing the videos of the reactions of lithium, sodium, and potassium with water, how would you describe the relative speeds of the reactions?
(b) Give a detailed description of the reaction of sodium with water.
(c) Speculate on the reason that sodium forms a little ball when it reacts with water.
(d) Why do you think the sodium ball scoots around on top of the water?

Chapter 3

Molecules and Compounds

(C. D. Winters)

A Chemical Puzzler

Stir a red compound with the formula $CoCl_2 \cdot 6\ H_2O$ into a beaker of water to make a solution with a faint red color. Take some of this solution and write your message on a piece of paper. When the paper dries, the words are no longer visible. You have written a secret message! But warm the paper above a candle flame, and the words appear again. What is happening? What kind of chemical compounds are involved? Does this process happen with other compounds? Are the ingredients of our magic ink—hydrated cobalt(II) chloride and water—essential to the process or would others work?

Models of the molecules illustrated or described in this chapter are available on the *Saunders Interactive General Chemistry CD-ROM* or at the Saunders College Publishing site on the World Wide Web (http://www.saunderscollege.com). A key to atom colors in the models appears in the Preface.

Nature is full of beautiful things. A field of flowers in the spring time. Roses in a garden. A bowlful of ripe strawberries. Or shiny red apples. Why do we find them appealing? It is often because they have pleasing shapes or delicious flavors. But it also might be because of their beautiful colors. Roses, strawberries, and apples all owe their color to various chemical compounds. The compound pelargonidin, for example, is responsible for the red color of common geraniums and strawberries.

But many compounds, like water, are colorless. Why do some compounds have specific colors while others are colorless? The simple answer relates the color of a compound to the wavelengths of light it absorbs from the visible spectrum of light. (The color you see is the light that is *not* absorbed as light passes through a sample or that is reflected from its surface.) Chemists want to know more than that, however. Why do compounds absorb light at all? And what happens when they do? To answer these questions, we need to know about the structure or shape of molecules and how those shapes are determined, a task we return to often in this book.

This chapter is the beginning of your examination of the major types of chemical compounds and their properties. Chemists have used modern techniques to study hundreds of thousands of compounds, and many can now be made in the laboratory in large quantities and can be used in hundreds of ways. An understanding of their composition and structure is just one part of an understanding of their physical and chemical properties.

3.1 MOLECULES AND COMPOUNDS

A chemical formula is "a collection of element symbols with right subscripts giving the relative numbers of atoms of different kinds of elements in the entity in question." P. Block, W. H. Powell, and W. C. Fernelius, *Inorganic Chemistry Nomenclature*, p. 16, Washington, D. C., American Chemical Society.

Compounds are pure substances that can be decomposed into two or more *different* pure substances. When sucrose, common sugar, is mixed with sulfuric acid,

A model of the molecule pelargonidin, the compound that gives the red color to geraniums. *(Photo, C. D. Winters; model, S. M. Young)*

***Figure* 3.1 The decomposition of a compound into its respective elements.** Sucrose, $C_{12}H_{22}O_{11}$, can be decomposed to elemental carbon (*the black solid*) and water (*seen as steam emerging from the beaker*) by treating sugar with concentrated sulfuric acid. *(C. D. Winters)*

a chemical reaction occurs to form two other substances: black, elemental carbon and colorless water, another compound (Figure 3.1).

$$1 \text{ molecule of } C_{12}H_{22}O_{11} \longrightarrow 12 \text{ atoms of C} + 11 \text{ molecules of } H_2O$$

To describe this chemical change (or chemical reaction) on paper, the composition of each compound has been represented by a symbol or **formula,** a shorthand way of expressing the number of atoms of each type in one molecule, the smallest unit of the compound. Sucrose, for example, is represented by the formula $C_{12}H_{22}O_{11}$, indicating that each molecule contains 12 C atoms, 22 H atoms, and 11 O atoms.

A striking feature of compounds is that the characteristics of the constituent elements are lost. Shiny, metallic aluminum foil reacts violently with the element bromine, a foul-smelling, red-orange liquid, to produce a solid, white compound Al_2Br_6 (Figure 1, page 2). The formula of this compound shows that eight atoms occur per molecule: two atoms of aluminum and six atoms of bromine.

There are often several ways to write the formulas of compounds. For example, the formula of ethanol, common alcohol, can be represented as C_2H_6O. This simple molecular formula provides information on the composition of ethanol molecules—two carbon atoms, six hydrogen atoms, and one oxygen atom occur per molecule—but it gives us no structural information. This structural information is important because another molecule, dimethyl ether, has exactly the same composition but is completely different in its structure and chemical and physical properties. For this reason it is useful to try to write formulas to convey some information about a compound's structure.

In order to provide some of this information, it is useful to write formulas in which certain atoms are grouped together. This tells us more about how atoms

The formula of sucrose can be written as $C_{12}(H_2O)_{11}$, making it seem that the compound is a "hydrate" of carbon. This is the origin of the name *carbohydrate.*

See the *Saunders Interactive General Chemistry CD-ROM,* Screen 4.2 for a video of the aluminum-bromine reaction.

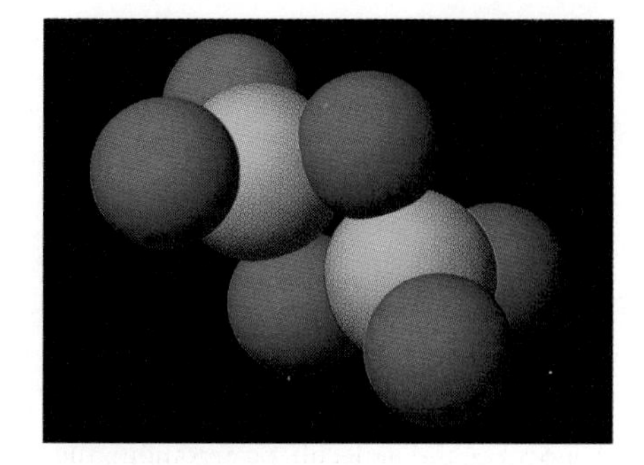

A molecule of aluminum bromide has two Al atoms and six Br atoms for a total of eight atoms.

See the *Saunders Interactive General Chemistry CD-ROM,* Screens 3.3 and 3.4 for a description of molecular formulas.

are connected and identifies important parts of the molecule. Such formulas are called **structural formulas.**

C_2H_6O — molecular formula of ethanol

CH_3CH_2OH — structural formula of ethanol

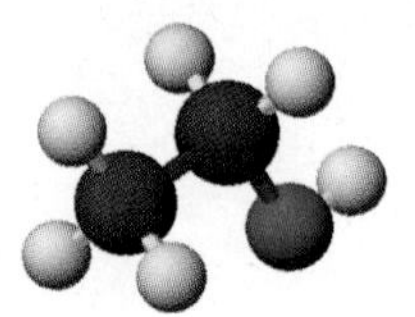

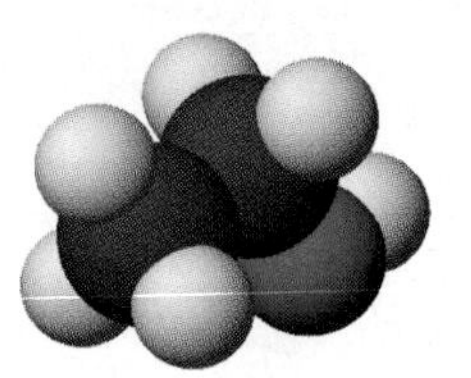

Ball-and-stick and space-filling models of ethanol.

C_2H_6O — molecular formula of dimethyl ether

CH_3OCH_3 — structural formula of dimethyl ether

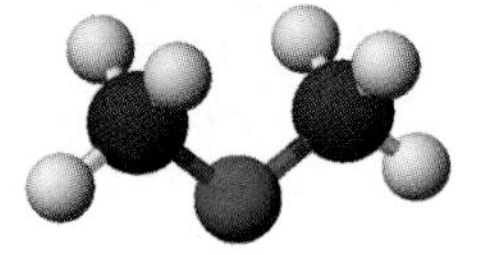

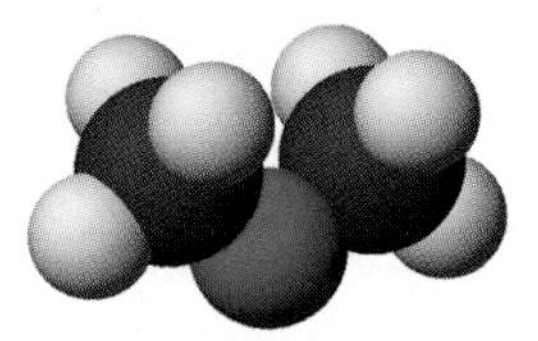

Ball-and-stick and space-filling models of dimethyl ether.

Writing structural formulas is valuable because it lets us categorize compounds. Ethanol is just one of an important group of compounds called *alcohols.* Other common examples are methanol (CH_3OH) and propanol ($CH_3CH_2CH_2OH$). By writing structural formulas it is possible to see a similarity in these compounds: all contain an OH group. The OH group is one of many types of **functional groups** that can occur in organic compounds. Such groups are responsible for the characteristic chemistry of the class of compounds and can be related to some of their physical properties. Thus, writing structural formulas rather than molecular formulas makes sense. You can, however, determine the molecular formula of a compound from its structural formula just by counting up the atoms.

You will often see simple structural formulas that present an even higher level of structural detail, showing how all the atoms are attached within a molecule. Several examples are given here. The lines between atoms represent **chemical bonds,** the forces that hold atoms together in molecules.

STRUCTURAL FORMULAS

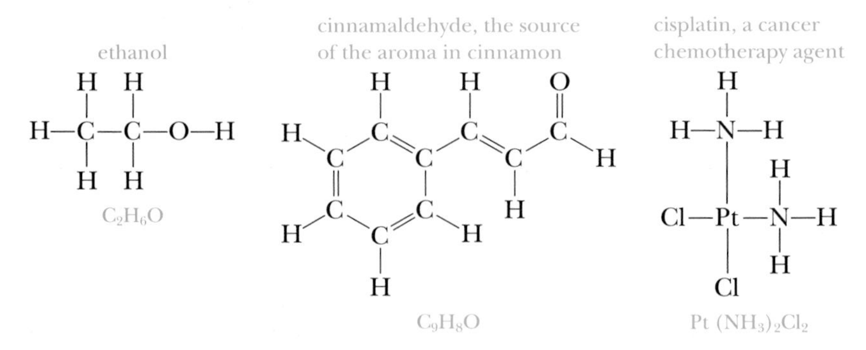

Example 3.1 Molecular Formulas

The acrylonitrile molecule is the building block for acrylic plastics (such as Orlon and Acrilan). Its structural formula is shown here. What is the molecular formula for acrylonitrile?

```
    H   H
    |   |
H—C=C—C≡N
```

Solution Acrylonitrile has three C atoms, three H atoms, and one N atom. Therefore, its molecular formula is C_3H_3N.

Exercise 3.1 Molecular Formulas

The styrene molecule is the building block for polystyrene, the material familiar in drinking cups, building insulation, and packing material. Its structural formula is shown here. What is the molecular formula for styrene?

```
        H       H
        |       |
H     C       C       H
  \ /   \\  /   \\  /
   C      C       C
   ||     |       |
   C      C       H
  / \   //  \
H     C       H
      |
      H
```

styrene

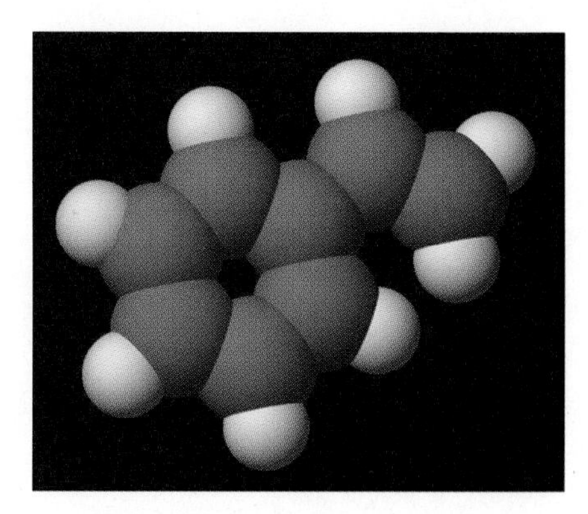

A molecular model of styrene. C atoms are gray and H atoms are white.

3.2 MOLECULAR MODELS

Visualizing the structures of molecules is important to chemists. One reason is that molecular structures are often very beautiful in the same sense that art is beautiful. The intriguing patterns of symmetry in a molecule's structure capture the imagination, much as symmetry is a part of some kinds of art (Figure 3.2). For example, there is something intrinsically attractive about the regular hexagonal shapes in graphite or diamond (see Figure 2.21) or the pattern created as water molecules are assembled in ice (Figure 3.3).

More important, however, is the fact that the molecular structure of a substance is often essential in explaining its physical and chemical properties. Two well-known features of ice are easily related to its molecular structure. The first is the shapes of ice crystals. The sixfold symmetry of the macroscopic crystals also appears at the particulate level, in the form of six-sided rings involving hydrogen and oxygen atoms. The second is water's unique property of being less dense when solid than it is when liquid. This fact, which has enormous consequences for earth's climate, results from the fact that molecules of water are not packed together tightly.

Most molecules are three-dimensional, so it is often difficult to draw their structures on paper. Certain conventions have been developed, however, that help represent three-dimensional structures on two-dimensional surfaces. Simple

Figure 3.2 *(Bruce Beasley, Gnomonos/Bank America Corporation Art Collection, San Francisco)*

Figure 3.3 Ice. Snowflakes are six-sided structures, reflecting the underlying structure of ice. Ice consists of six-sided rings formed by water molecules, in which each side consists of two O atoms and an H atom. *(Mehau Kulyk/Science Photo Library/Photo Researchers, Inc.; model by S. M. Young added by C. D. Winters)*

perspective drawings are often used. Solid lines (—) are used to represent bonds in the plane of the paper, a solid wedge (▼) is used to represent a bond extending out from the plane of the paper, and a dashed line (---) is used for a bond behind the paper. Using this convention, methane would be drawn as illustrated in Figure 3.4(a).

Molecular models are a very useful way of representing molecular structure (see Figure 3.4). Created from wood or plastic, these models give an even bet-

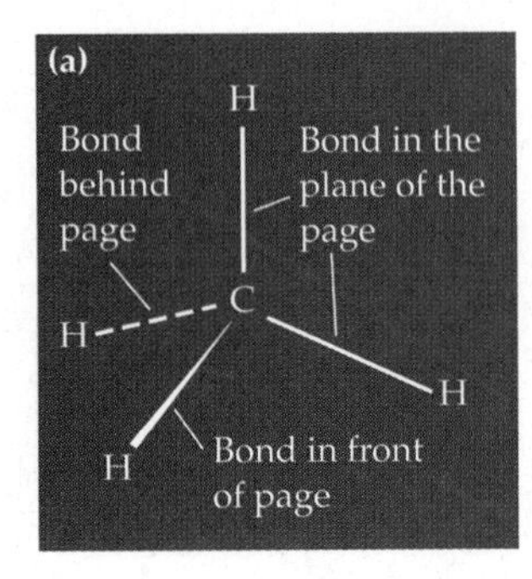

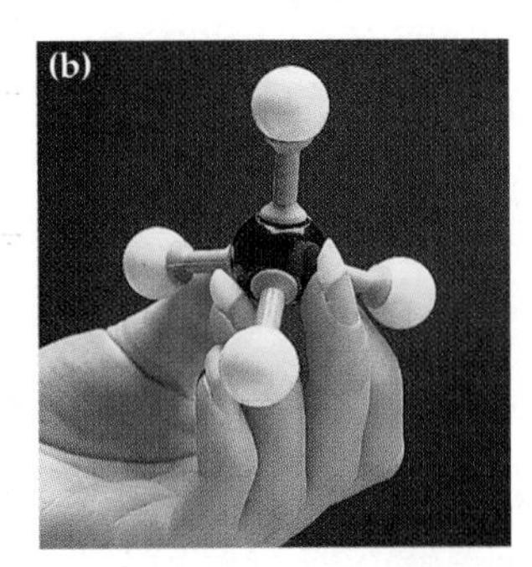

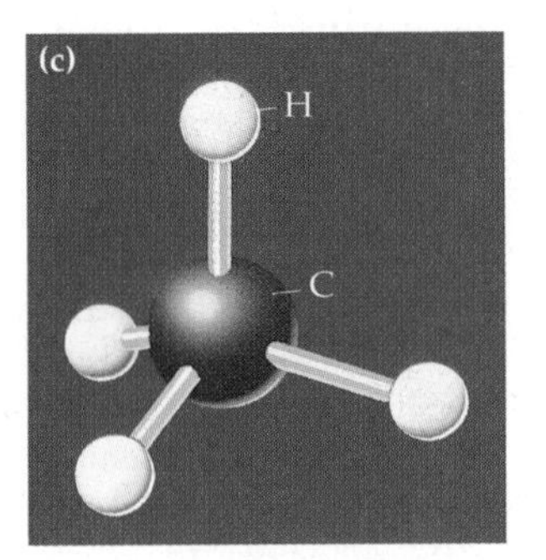

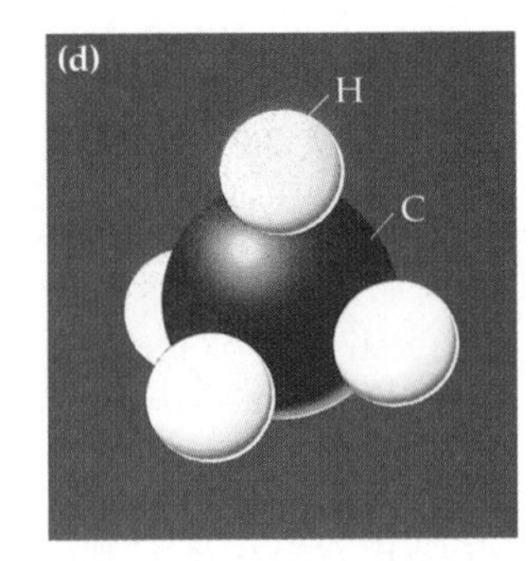

Figure 3.4 Representing molecular models. The simple molecule methane, CH_4, can be represented by (a) lines and symbols on a flat surface or (b) a plastic model can be built. The molecule can also be represented by a model drawn by computer as a (c) ball-and-stick model or a (d) space-filling model. All four types are used in this book. The black ball represents a C atom, and the H atoms are represented by white balls. *(b, C. D. Winters)*

A CLOSER LOOK — *Computer Resources for Molecular Modeling*

Computer programs for molecular modeling are an important recent development. With the availability of relatively low-cost, high-powered computers, the use of molecular-modeling programs has become fairly common. Although the computer screen is two-dimensional, the perspective drawings obtained from molecular-modeling programs are usually quite good. In addition, most programs offer an option to rotate the molecule on the computer screen to allow the viewer to see the structure from any desired angle. Both ball-and-stick and space-filling representations can be portrayed by computer modeling. Most of the drawings in this book were prepared with the CAChe/Oxford Molecular modeling software. The *Saunders Interactive General Chemistry CD-ROM* that accompanies this book contains a portion of the CAChe/Oxford Molecular software that you can use to visualize models of a large number of substances. The Saunders College Publishing Company Homepage on the World Wide Web (http://www.saunderscollege.com) contains a link to RasMol and Chime, molecular visualization software. Models of many of the compounds mentioned in this book are on the CD-ROM or are available at the Saunders site on the World Wide Web. You can visualize these molecules using the software on the CD-ROM, or, if you download RasMol or Chime and configure your browser properly, you can download files from the Internet that allow you to visualize these models on your own computer.

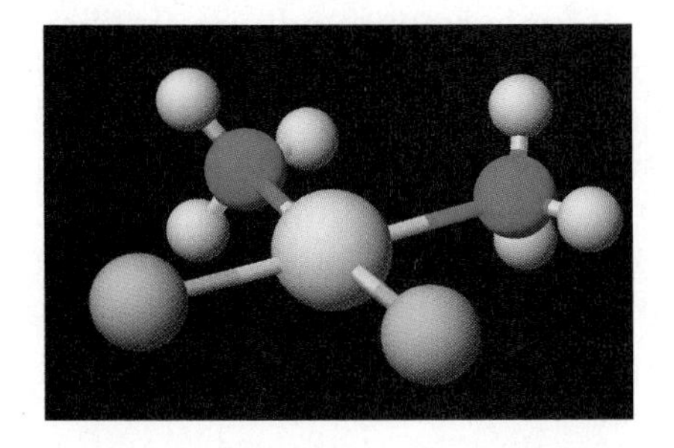

A model of cisplatin, $Pt(NH_3)_2Cl_2$, as viewed using the CAChe/Oxford Molecular software on the *Saunders Interactive General Chemistry CD-ROM.*

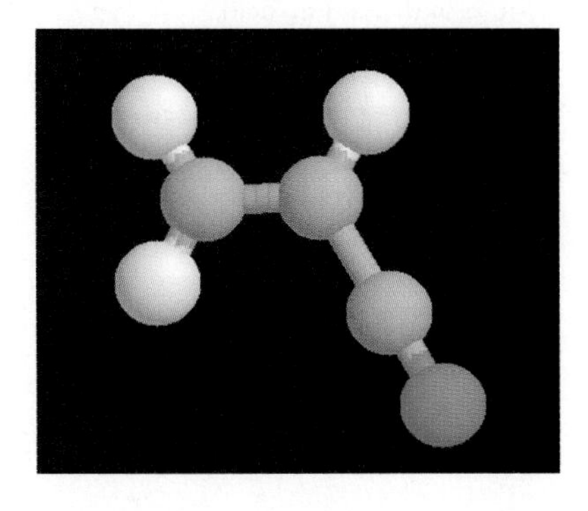

A model of acrylonitrile, C_3H_3N, viewed using RasMol, a program available on the Internet.

ter sense of three-dimensional structure. Molecular models can be held in the hand and rotated to put all parts of the molecule in view. There are several kinds of molecular models. One is called the *ball-and-stick model.* Spheres, usually in different colors, represent the atoms, and the sticks represent the bonds holding them together. These models make it easy to see how atoms are attached to one another.

Molecules can also be represented using *space-filling models.* These models are more realistic because they are a better representation of actual sizes of atoms and molecules. A disadvantage of space-filling models is that atoms can often be hidden from view.

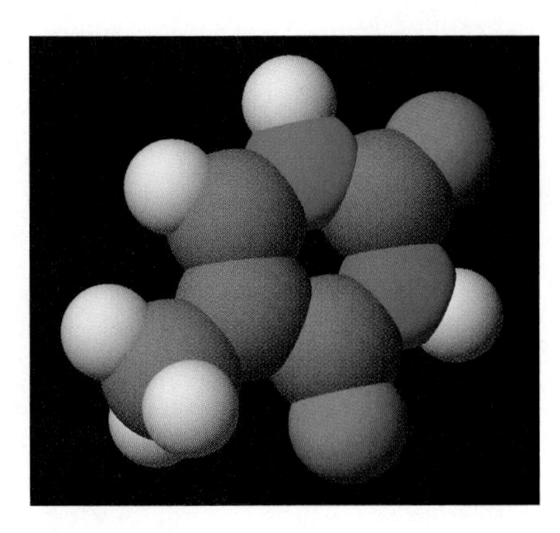

A space-filling model of thymine, one of the bases in DNA.

Example 3.2 Writing Molecular Formulas

Write the molecular formula of thymine, whose structure is given in the margin. The color codes for the models are: carbon atoms = gray; hydrogen atoms = white; nitrogen atoms = blue; and oxygen atoms = red.

Solution Thymine has five C atoms, six H atoms, two N atoms, and two O atoms, giving a formula of $C_5H_6N_2O_2$.

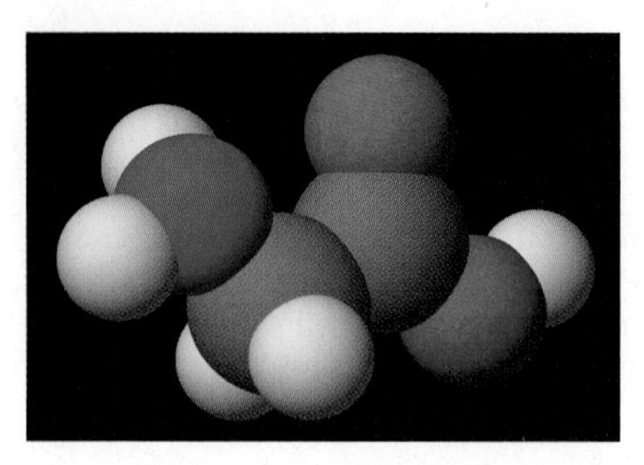

Glycine, an essential amino acid.

Exercise 3.2 Formulas of Molecules

Glycine, whose molecular model is illustrated here, is an important amino acid and a constituent of many living things. What is its molecular formula? See Example 3.2 and the Preface for the color coding of the model.

3.3 IONS

Thus far we have described molecular compounds, that is, compounds that consist of discrete molecules at the particulate level. **Ionic compounds** are another major class of compounds. As their name implies, they consist of **ions,** atoms or groups of atoms that bear a positive or negative electric charge. Many important and familiar compounds are composed of ions. Table salt, or sodium chloride (NaCl), and alum $[KAl(SO_4)_2 \cdot 12\ H_2O]$, which is used in making paper and pickles, are just two. To recognize ionic compounds, and to be able to write formulas for these compounds, it is important to know the formulas and charges of common ions. You also need to know their names and how to name the compounds they form.

Metal atoms generally lose one or more electrons in the course of reaction to form ions with a positive charge, ions commonly called **cations.** (The name is pronounced "cat′-ion.") A lithium atom is electrically neutral because it has three protons and three electrons (Figure 3.5). The atom loses one of its electrons to form a lithium ion. The ion has a net charge of 1+ because it now has one more proton (three) than electrons (two). We symbolize the resulting lithium cation as Li^+. When two or more electrons are lost by an atom, the number is written with the charge sign. For example, the symbol for a calcium ion, an ion with 20 protons and 18 electrons, is Ca^{2+}.

In contrast with metals, nonmetals frequently *gain* one or more electrons to form ions with a negative charge in the course of their reactions. Such ions are called **anions.** (The name is pronounced "ann′-ion.") Figure 3.5 depicts a fluorine atom with nine protons and nine electrons gaining an electron to form a fluoride ion (F^-), which has nine protons and ten electrons.

Charges on Monatomic Ions

Monatomic ions are single atoms that have lost or gained electrons. Typical charges on such ions are indicated in the periodic table in Figure 3.6. Be sure to notice that *metals of Groups 1A–3A form positive ions with a charge equal to the group number of the metal.*

Group	Metal Atom	Electrons Lost	Metal Ion
1A	Na (11 protons, 11 electrons)	1	Na^+ (11 protons, 10 electrons)
2A	Ca (20 protons, 20 electrons)	2	Ca^{2+} (20 protons, 18 electrons)
3A	Al (13 protons, 13 electrons)	3	Al^{3+} (13 protons, 10 electrons)

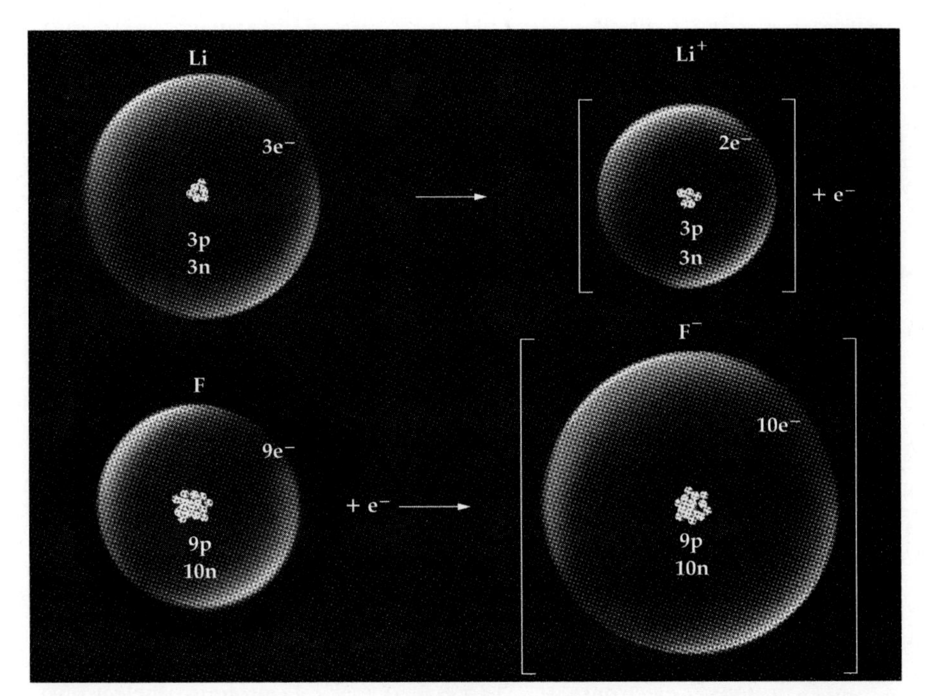

Figure 3.5 Ions. A lithium atom is electrically neutral because the number of positive charges (three protons) and negative charges (three electrons) are the same. When it loses one electron, it has one more positive charge than negative charge, so it has a net charge of 1+. We symbolize the resulting lithium cation as Li^+. A fluorine atom is similarly electrically neutral, having nine protons and nine electrons. Because it is a nonmetal it gains electrons. As an F^- anion, it has one more electron than it has protons, so it has a net charge of 1−.

1A	2A	3B	4B	5B	6B	7B	8B	8B	8B	1B	2B	3A	4A	5A	6A	7A	8
H^+																H^-	
Li^+														N^{3-}	O^{2-}	F^-	
Na^+	Mg^{2+}											Al^{3+}		P^{3-}	S^{2-}	Cl^-	
K^+	Ca^{2+}		Ti^{4+}		Cr^{2+} Cr^{3+}	Mn^{2+}	Fe^{2+} Fe^{3+}	Co^{2+} Co^{3+}	Ni^{2+}	Cu^+ Cu^{2+}	Zn^{2+}				Se^{2-}	Br^-	
Rb^+	Sr^{2+}									Ag^+	Cd^{2+}		Sn^{2+}		Te^{2-}	I^-	
Cs^+	Ba^{2+}										Hg_2^{2+} Hg^{2+}		Pb^{2+}	Bi^{3+}			

Figure 3.6 Charges on some common monatomic cations and anions. Note that metals usually form cations. For main group metals *(light orange)*, the positive charge is given by the group number. Nonmetals *(green)* generally form anions with a negative charge equal to 8 minus the group number. For transition metals *(yellow)*, a simple prediction of the charge is not possible as most form ions of different charges. A good rule of thumb, however, is that first-row transition metals typically form 2+ or 3+ ions.

It is extremely important that you know the ions commonly formed by the elements shown in Figure 3.6 as well as the common polyatomic ions given in Table 3.1 (page 110).

Transition metals also form cations. Unlike main group elements, though, no easily predictable pattern of behavior occurs for transition metal cations. In addition, many of these metals form several different ions. An iron-containing compound, for example, may contain either Fe^{2+} or Fe^{3+} ions. Indeed, 2+ *and 3+ ions are typical of many transition metals* (see Figure 3.6).

Group	Metal Atom	Electrons Lost	Metal Ion
7B	Mn (25 protons, 25 electrons)	2	Mn^{2+} (25 protons, 23 electrons)
8B	Fe (26 protons, 26 electrons)	2	Fe^{2+} (26 protons, 24 electrons)
8B	Fe (26 protons, 26 electrons)	3	Fe^{3+} (26 protons, 23 electrons)

The Hale-Bopp comet that passed through the solar system in the spring of 1997 had a tail composed of sodium ions. The broad tail pointing upward in the photo is the dust trail of the comet. The straighter tail, pointing up to the left in the photo, is a tail of sodium ions, Na^+. *(Pekka Parviainen/Science Photo Library/Photo Researchers, Inc.)*

Nonmetals often form ions with a negative charge equal to 8 minus the group number of the element. For example, N is in Group 5A and so has a charge of 3− because an atom of N can *gain* three electrons.

Group	Nonmetal Atom	Electrons Gained	Nonmetal Ion
5A	N (7 protons, 7 electrons)	3 (= 8 − 5)	N^{3-} (7 protons, 10 electrons)
6A	S (16 protons, 16 electrons)	2 (= 8 − 6)	S^{2-} (16 protons, 18 electrons)
7A	Br (35 protons, 35 electrons)	1 (= 8 − 7)	Br^- (35 protons, 36 electrons)

Notice that hydrogen appears at two locations in Figure 3.6; the H atom can either lose or gain electrons, depending on the other atoms it encounters.

Lose: H (1 proton, 1 electron) $\longrightarrow$ H^+ (1 proton, 0 electrons) + e^-

Gain: H (1 proton, 1 electron) + e^- $\longrightarrow$ H^- (1 proton, 2 electrons)

Finally, the noble gases lose or gain electrons only in special cases because these elements are extremely stable chemically.

Valence Electrons

As illustrated in Figure 3.6, the metals of Groups 1A, 2A, and 3A form ions having 1+, 2+, and 3+ charges; that is, their atoms have lost one, two, or three electrons, respectively. In each case *the number of electrons remaining on the ions is the same as the number of electrons in an atom of a noble gas.* For example, Mg^{2+} has ten electrons, the same number as in an atom of Ne, the noble gas preceding magnesium in the periodic table.

An atom of a nonmetal at the right in the periodic table must lose a great many electrons to achieve the same number as a noble gas atom of lower atomic number. (For instance, Cl would have to lose seven electrons to have the same number of electrons as Ne.) If an atom of a nonmetal were to gain a few electrons, however, it would have the same number as a noble gas atom of higher atomic number. For example, an oxygen atom has eight electrons. By gaining two per atom it forms O^{2-}, which has ten electrons, the same number as Ne. Because the noble gases are in periodic Group 8A, eight minus the group number gives the number of electrons gained and hence the negative charge on the ion. As illustrated in Figure 3.6, *ions having the same number of electrons as a noble gas atom are especially favored in chemical compounds.*

The American chemist G. N. Lewis (1875–1946) devised the idea that electrons in atoms might be arranged in shells around the nucleus. Each shell could hold a characteristic number of electrons, and only those electrons in the outermost shell were involved when one atom combined with another. These outermost electrons came to be known as **valence electrons.** For example, a magnesium atom has two valence electrons (two more electrons than a neon atom), and, when these valence electrons are lost, a Mg^{2+} ion forms. An oxygen atom has six valence electrons (six more than a helium atom) and has room for two more (which would give it the same number as a neon atom); thus, an O^{2-} ion is reasonable.

Example 3.3 Predicting Ion Charges

Predict the charges on ions of aluminum and sulfur.

Solution Aluminum is a metal in Group 3A of the periodic table, so it is predicted to lose three electrons to give the Al^{3+} cation.

$$Al \longrightarrow Al^{3+} + 3e^-$$

Sulfur is a nonmetal in Group 6A, so it is predicted to gain electrons to give an anion. The number of electrons gained is $8 - 6 = 2$. Therefore,

$$S + 2e^- \longrightarrow S^{2-}$$

Exercise 3.3 Predicting Ion Charges

Predict formulas for monatomic ions formed from (1) K, (2) Se, (3) Ba, and (4) Cs.

Polyatomic Ions

A **polyatomic ion** contains two or more atoms, and the aggregate bears an electric charge. For example, carbonate ion, CO_3^{2-}, is a common polyatomic anion. It consists of one C atom and three O atoms with two units of negative charge. Another common polyatomic cation is NH_4^+, the ammonium ion. In this case, four H atoms surround an N atom, and the group bears a 1+ charge. It is extremely important to know the names, formulas, and charges of the common polyatomic ions listed in Table 3.1. Some are also illustrated in Figure 3.7.

See the *Saunders Interactive General Chemistry CD-ROM,* Screen 3.8 for a description of polyatomic ions.

3.4 IONIC COMPOUNDS

Ionic compounds are those in which the component particles are ions. All compounds, whether molecular or ionic, are electrically neutral, that is, there is no net charge. Because ionic compounds are composed of ions, the numbers of

Table 3.1 • NAMES AND COMPOSITION OF SOME COMMON POLYATOMIC IONS

Formula	Name
CATION: Positive Ion	
NH_4^+	ammonium ion
ANIONS: Negative Ions	
Based on a Group 4A element	
CN^-	cyanide ion
$CH_3CO_2^-$	acetate ion
CO_3^{2-}	carbonate ion
HCO_3^-	hydrogen carbonate ion (or bicarbonate ion)
Based on a Group 5A element	
NO_2^-	nitrite ion
NO_3^-	nitrate ion
PO_4^{3-}	phosphate ion
HPO_4^{2-}	hydrogen phosphate ion
$H_2PO_4^-$	dihydrogen phosphate ion
Based on a Group 6A element	
OH^-	hydroxide ion
SO_3^{2-}	sulfite ion
SO_4^{2-}	sulfate ion
HSO_4^-	hydrogen sulfate ion (or bisulfate ion)
Based on a Group 7A element	
ClO^-	hypochlorite ion
ClO_2^-	chlorite ion
ClO_3^-	chlorate ion
ClO_4^-	perchlorate ion
Based on a transition metal	
CrO_4^{2-}	chromate ion
$Cr_2O_7^{2-}$	dichromate ion
MnO_4^-	permanganate ion

positive and negative ions must be set so that the positive and negative charges balance. Sodium chloride is a simple example. Because the sodium ion has a 1+ charge (Na^+) and the chloride ion has a 1− charge (Cl^-) these ions must be present in a 1-to-1 ratio. Thus, the formula for sodium chloride is NaCl (see Figure 2.28).

Let us take the compound formed from aluminum ions (Al^{3+}) and oxide ions (O^{2-}) as a second example (Figure 3.8). Here the ions have positive and negative charges that are of different absolute size. If combined in a 1-to-1 ratio, the charges would not balance. To have a compound with the same number of positive and negative charges, therefore, two Al^{3+} ions [total charge is $6+ = 2 \times (3+)$] must combine with three O^{2-} ions total charge is $6- = 3 \times (2-)$] to give a formula of Al_2O_3.

Aluminum is a metal in Group 3A and so loses three electrons to form the Al^{3+} cation. Oxygen is a nonmetal in Group 6A and so gains two electrons to form an O^{2-} anion. Notice that the charge on the cation is the subscript on the anion (and vice versa).

$$Al^{3+} O^{2-} \longrightarrow Al_2O_3$$

This often works well, but be careful. The subscripts in $Ti^{4+} + O^{2-}$ are reduced to the simplest ratio (1 Ti to 2 O and not 2 Ti to 4 O)

$$Ti^{4+} + O^{2-} \longrightarrow TiO_2$$

(a) (b) (c)

Figure 3.7 Common polyatomic ions. (a) The carbonate ion, CO_3^{2-}, is combined with calcium ion, Ca^{2+}, in calcium carbonate. (b) The ammonium ion, NH_4^+, is combined with the dichromate ion, $Cr_2O_7^{2-}$, in ammonium dichromate. (c) The sulfite ion, SO_3^{2-}, is combined with sodium ion, Na^+, in sodium sulfite, Na_2SO_3. *(Photos, C. D. Winters; models, S. M. Young)*

Calcium ion has a 2+ charge because the metal is a member of Group 2A. It can combine with a variety of anions to form ionic compounds such as those in the table shown here.

Ion Combination	Compound	Overall Charge on Compound
$Ca^{2+} + 2\ Cl^-$	$CaCl_2$	$(2+) + 2(1-) = 0$
$Ca^{2+} + CO_3^{2-}$	$CaCO_3$	$(2+) + (2-) = 0$
$3\ Ca^{2+} + 2\ PO_4^{3-}$	$Ca_3(PO_4)_2$	$3 \times (2+) + 2 \times (3-) = 0$

In writing all these formulas, it is conventional that the *symbol of the cation is always given first, followed by the symbol for the anion.* Also notice the use of parentheses when there is more than one polyatomic ion.

Figure 3.8 Aluminum oxide. Aluminum and oxygen form white aluminum oxide, Al_2O_3. This compound is the basis of gemstones such as rubies. In this case chromium(III) ions, Cr^{3+}, replace some of the Al^{3+} ions, giving the gem its color. Aluminum oxide also forms a microscopic coating on aluminum metal and protects the underlying metal from corrosion. Finally, the compound is very hard and is the material used on "sand" paper. *(C. D. Winters)*

Example 3.4 Ionic Compounds

For each of the following ionic compounds, write the symbols for the ions present and give the number of each: (1) $MgBr_2$, (2) Li_2CO_3, and (3) $Fe_2(SO_4)_3$.

Solution

1. $MgBr_2$ is composed of one Mg^{2+} ion and two Br^- ions. When a halogen such as bromine is combined only with a metal, you can assume the halogen is an anion with a charge of 1−. Magnesium is a metal in Group 2A and always has a charge of 2+ in its compounds.

2. Li_2CO_3 is composed of two lithium ions, Li^+, and one carbonate ion, CO_3^{2-}. Li is a Group 1A element and so always has a 1+ charge in its compounds. Because the two 1+ charges neutralize the negative charge of the carbonate ion, the latter must be 2−.
3. $Fe_2(SO_4)_3$ contains two Fe^{3+} ions and three sulfate ions, SO_4^{2-}. The way to rationalize this is to recall that sulfate is 2−. Because three sulfate ions are present (with a total charge of 6−), the two iron cations must have a total charge of 6+. This is possible only if each iron cation has a charge of 3+.

Example 3.5 Writing Formulas for Compounds Formed from Ions

Write formulas for ionic compounds composed of an aluminum cation and each of the following anions: (1) fluoride ion, (2) sulfide ion, and (3) nitrate ion.

Solution First, the aluminum cation is predicted to have a charge of 3+ because Al is a metal in Group 3A.

1. Fluorine is a Group 7A element and so its charge is predicted to be 1− (from $8 - 7 = 1$). Therefore, we need 3 F^- ions to combine with one Al^{3+}. The formula of the compound is AlF_3.
2. Sulfur is a nonmetal in Group 6A and so forms a 2− anion. Thus, we need to combine two Al^{3+} ions [total charge is $6+ = 2 \times (3+)$] with three S^{2-} ions [total charge is $6- = 3 \times (2-)$]. The compound has the formula Al_2S_3.
3. The nitrate ion has the formula NO_3^- (see Table 3.1). The answer is therefore similar to the AlF_3 case, and the compound has the formula $Al(NO_3)_3$. Here we place parentheses around the NO_3 portion of the formula to show that three polyatomic NO_3^- ions are involved.

Problem-Solving Tips and Ideas

3.1 Formulas for Ions and Ionic Compounds

Writing formulas for ionic compounds takes practice, and it requires that you know the formulas and charges of the most common ions. The charges on monatomic ions are often evident from the position of the element in the periodic table, but you simply have to remember the formulas and charges of polyatomic ions, especially the most common ones such as nitrate, sulfate, carbonate, phosphate, and acetate.

If you cannot remember the formula of a polyatomic ion, or if you encounter an ion you have not seen before, you may be able to figure out its formula from the formula and the name of one of its compounds. For example, suppose you are told that $NaCHO_2$ is sodium formate. You know that the sodium ion is Na^+, so the formate ion must be the remaining portion of the compound and it must have a charge of 1− to balance the 1+ charge on the sodium ion. Thus, the formate ion must be CHO_2^-.

Finally, when writing the formulas of ions, you *must* include the charge on the ion. Writing Na when you mean sodium ion is incorrect. There is a vast difference in the properties of the element sodium (Na) and those of its ions (Na^+).

Exercise 3.4 Ionic Compounds

1. Give the number and identity of the constituent ions in each of the following ionic compounds: (a) NaF, (b) $Cu(NO_3)_2$, and (c) $NaCH_3CO_2$.
2. Iron is a transition metal and so can form ions with at least two different charges. Write the formulas of the compounds formed between two different iron cations and chloride ion.
3. Write the formulas of all of the neutral ionic compounds that can be formed by combining the cations Na^+ and Ba^{2+} with the anions S^{2-} and PO_4^{3-}.

Problem-Solving Tips and Ideas

3.2 Is a Compound Ionic?

Students often ask how one knows a compound is ionic. No method works all of the time, but here are some useful guidelines.

1. Most metal-containing compounds are ionic. So, if a metal atom appears in the formula of a compound, a very good "first guess" is that it is ionic. (There are interesting exceptions, but few come up in introductory chemistry.) It is helpful in this regard to recall trends in metallic behavior (see Figure 3.6): all elements to the left of a diagonal line running from boron to tellurium in the periodic table are metallic.
2. If there is no metal in the formula, it is likely that the compound is not ionic. The exceptions here are compounds composed of polyatomic ions based on nonmetals (e.g., NH_4Cl or NH_4NO_3).
3. Learn to recognize the formulas of polyatomic ions (see Table 3.1) because they combine to form ionic compounds. Chemists write the formula of ammonium nitrate as NH_4NO_3 (and not as $N_2H_4O_3$) to alert others to the fact that it is an ionic compound composed of the common polyatomic ions NH_4^+ and NO_3^-.

As an example of these guidelines, you can be sure that Mg^{2+} with Br^- and K^+ with S^{2-} are combinations that are suitable for ionic compound formation. On the other hand, the compound BCl_3, formed from two nonmetals, B and Cl, is not considered ionic.

Ionic Compounds and Coulomb's law

What is the "glue" that causes ions of opposite electric charge to be held together, to form an orderly arrangement of ions called an ionic compound (Figure 3.9)? As described in Section 2.2, when a substance with a negative electric charge is brought near a substance with a positive electric charge, a force of attraction occurs between them. Similarly, there is a force of *repulsion* when two negatively charged substances, or two positively charged substances, are brought together. These forces are called **electrostatic forces.** The force of attraction between a positive and a negative ion is given by **Coulomb's law,**

$$\text{Force of attraction} = k\frac{(n^+e)(n^-e)}{d^2}$$

where n^+ is the number of charges on the positive ion (e.g, three for Al^{3+}), n^- is the number of charges on the negative ion (e.g, two for O^{2-}), e is the charge

See the *Saunders Interactive General Chemistry CD-ROM,* Screen 3.9 for a discussion of Coulomb's law and an "interactive tool" illustrating the law. Coulomb's law is fundamental to understanding many things in chemistry. We shall refer to it often.

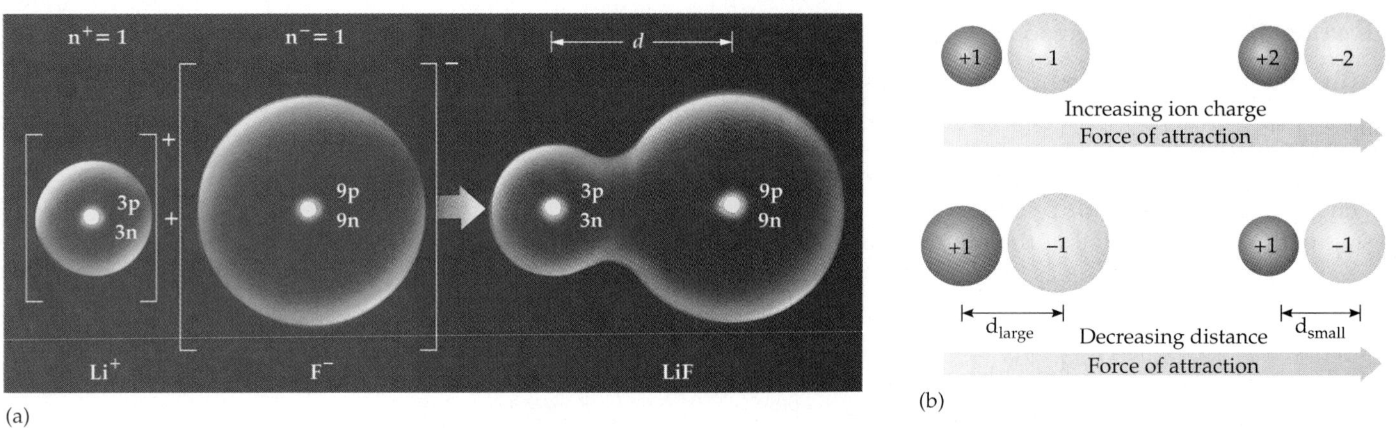

Figure 3.9 Coulomb's law and electrostatic forces. (a) Ions such as Li^+ and F^- are held together by an electrostatic force. Here a lithium ion (Li^+, with three protons and two electrons) is attracted to a fluoride ion (F^-, with nine protons and ten electrons). (b) Forces of attraction between ions of opposite charge increase with increasing ion charge and with decreasing distance.

on the electron (1.602×10^{-19} C), d is the distance between the centers of the ions (see Figure 3.9), and k is a proportionality constant. The equation shows that *the force of attraction between oppositely charged ions increases as their charges increase and as the distance, d, between the ions becomes smaller.*

The strength of electrostatic forces directly influences the properties of compounds formed from ions. Because the force of attraction increases as the charges on the ions increase, the attraction between ions with charges of 2+ and 2− is greater than between ions with 1+ and 1− charges (see Figure 3.9). The closer the centers of the ions can come to one another, the greater the force of attraction. This means that smaller ions of opposite charge will attract one another more strongly than larger ions.

A crystal of sodium chloride consists of an extended lattice of sodium ions and chloride ions. *(Photo, C. D. Winters; model, S. M. Young)*

Crystal Lattices and Properties of Ionic Compounds

Ionic compounds are not molecules in the sense that water consists of molecules of H_2O, nor do they consist of simple pairs or small groupings of positive and negative ions. Rather, the ions are generally arranged in an extended three-dimensional network called a **crystal lattice.** Sodium chloride and pyrite (Figure 1.4) are excellent examples.

Careful examination of the crystal lattice of sodium chloride should reveal that each sodium ion is surrounded by six chloride ions, and each chloride ion is surrounded by six sodium ions. This arrangement extends for many, many ions in every direction. No discrete NaCl molecules exist.

Although they do not consist of molecules, ionic compounds do have well-defined ratios of one kind of ion to another, and these are represented by their formulas. For example, NaCl represents the simplest ratio of sodium ions to chloride ions in the lattice, namely 1:1. The combination of one Na^+ ion and one Cl^- ion is referred to as a **formula unit** of sodium chloride.

Compounds made up of ions have characteristic properties that can be understood in terms of the charges of the ions and their arrangement in the lattice. Because each ion is surrounded by many oppositely charged nearest neighbors, each ion is held fairly tightly in its allotted location. At room temperature each ion can move a bit around its average position, but much more energy must be added before an ion can move fast enough and far enough to escape the attraction of neighboring ions. Thus, if enough energy is added, the lattice structure collapses, and the substance melts. As the coulombic force of attraction increases, ever more energy—and higher and higher temperatures—are required to cause melting. Thus, Al_2O_3, a solid composed of Al^{3+} and O^{2-} ions (Figure 3.8), melts at a much higher temperature (2072 °C) than NaCl (801 °C), a solid composed of Na^+ and Cl^- ions.

The high melting points of ionic compounds are useful. For example, white Al_2O_3 is used in fire bricks, ceramics, and other materials that must withstand high temperatures.

3.5 NAMES OF COMPOUNDS

Chemists try to use precise language, but assigning clear, unambiguous names to compounds has always been a problem. It is a problem that continues today as new, ever more complicated molecules are discovered. Nonetheless, rules do exist for naming the kinds of compounds you will read about in this text, and you should study them thoroughly.

Naming Ionic Compounds

The name of an ionic compound is built from the names of the positive and negative ions in the compound. The name of the positive ion is given first, followed by the name of the negative ion.

See the *Saunders Interactive General Chemistry CD-ROM,* Screen 3.13 for more on naming ionic compounds.

Naming Positive Ions

With a few exceptions (such as NH_4^+), the positive ions described in this text are metal ions. Positive ions are named by the following rules:

1. For a monatomic positive ion, that is, a metal cation, the name is that of the metal plus the word "ion." For example, we have already referred to Al^{3+} as the aluminum ion.
2. Some cases occur, especially in the transition series, in which a metal can form more than one type of positive ion. The most common practice is to indicate the charge of the ion by a Roman numeral in parentheses immediately following the ion's name. For example, Co^{2+} is the cobalt(II) ion, and Co^{3+} is the cobalt(III) ion. [An older naming system for ions uses the ending "-ous" for the ion of lower charge and "-ic" for the ion of higher charge. For example, there are cobaltous (Co^{2+}) and cobaltic (Co^{3+}) ions, and ferrous (Fe^{2+}) and ferric (Fe^{3+}) ions. We do not use this system in this book, but some chemical manufacturers continue to use it.]

Another cation that you will see on occasion is Hg_2^{2+}, the name of which is the mercury(I) ion. The reason for the Roman numeral (I) is that the ion is composed of two Hg^+ ions bonded together.

Finally, you will encounter the ammonium cation, NH_4^+, many times in this book, in the laboratory, and in your environment. Do not confuse the *ammonium ion* with the neutral *ammonia molecule,* NH_3.

Naming Negative Ions

Two types of negative ions must be considered: those having only one atom (*monatomic*) and those having several atoms (*polyatomic*).

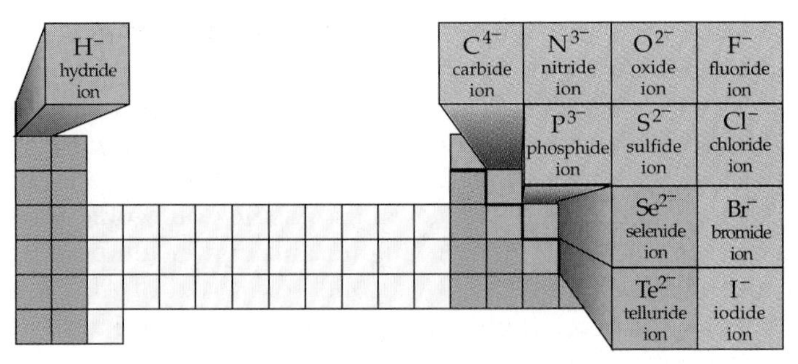

Figure 3.10 Names and charges of some common monatomic ions of the nonmetals.

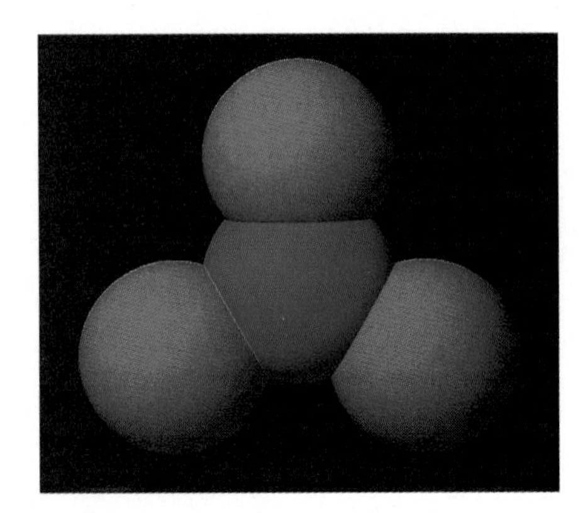

A model of the nitrate ion, NO_3^-.

1. A *monatomic negative ion* is named by adding *-ide* to the stem of the name of the nonmetal element from which the ion is derived (Figure 3.10). As a group, the anions of the Group 7A elements, the halogens, are called **halide ions.**
2. *Polyatomic negative ions* are quite common, especially those containing oxygen (called *oxoanions*). The names of some of the most common oxoanions are given in Table 3.1. Although most of these names must simply be learned, some guidelines can help. For example, consider the following pairs of ions:

   NO_3^- is the nitrate ion, whereas NO_2^- is the nitrite ion

   SO_4^{2-} is the sulfate ion, whereas SO_3^{2-} is the sulfite ion

The oxoanion with the *greater number of oxygen atoms* is given the suffix *-ate*, and the oxoanion with the *smaller number of oxygen atoms* has the suffix *-ite*. For a series of oxoanions with more than two members, the ion with the largest number of oxygen atoms has the prefix *per-* and the suffix *-ate*. The ion with the smallest number of oxygen atoms has the prefix *hypo-* and the suffix *-ite*. The oxoanions containing chlorine are good examples.

ClO_4^-	*per*chlor*ate* ion
ClO_3^-	chlor*ate* ion
ClO_2^-	chlor*ite* ion
ClO^-	*hypo*chlor*ite* ion

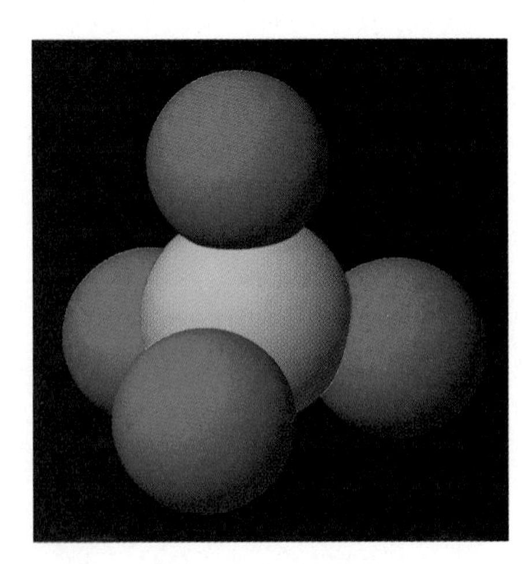

A model of the sulfate ion, SO_4^{2-}.

Oxoanions that contain hydrogen are named by adding the word "hydrogen" before the name of the oxoanion. If two hydrogens are in the compound, we say dihydrogen. Many of these hydrogen-containing oxoanions have common names that are so often used that you should know them, too. For example, the hydrogen carbonate ion, HCO_3^-, is often called the bicarbonate ion.

Ion	Systematic Name	Common Name
HPO_4^{2-}	hydrogen phosphate ion	
$H_2PO_4^-$	dihydrogen phosphate ion	
HCO_3^-	hydrogen carbonate ion	bicarbonate ion
HSO_4^-	hydrogen sulfate ion	bisulfate ion
HSO_3^-	hydrogen sulfite ion	bisulfite ion

Figure 3.11 Some common ionic compounds. Clockwise from the top they are: A box of salt (sodium chloride, NaCl), a clear crystal of calcite (calcium carbonate, $CaCO_3$), a pile of $CoCl_2 \cdot 6H_2O$ [cobalt(II) chloride hexahydrate], and an octahedral crystal of fluorite (calcium fluoride, CaF_2). *(C. D. Winters)*

Naming Ionic Compounds

When naming ionic compounds, the positive ion name is given first, followed by the name of the negative ion. Some examples are given in the table shown here, and others are shown in Figure 3.11.

Ionic Compound	Ions Involved	Name
$CaBr_2$	Ca^{2+} and 2 Br^-	calcium bromide
$NaHSO_4$	Na^+ and HSO_4^-	sodium hydrogen sulfate
$(NH_4)_2CO_3$	2 NH_4^+ and CO_3^{2-}	ammonium carbonate
$Mg(OH)_2$	Mg^{2+} and 2 OH^-	magnesium hydroxide
$TiCl_2$	Ti^{2+} and 2 Cl^-	titanium(II) chloride
Co_2O_3	2 Co^{3+} and 3 O^{2-}	cobalt(III) oxide

Exercise 3.5 Names and Formulas of Ionic Compounds

1. Give the formula for each of the following ionic compounds. Use Table 3.1 and Figure 3.10.
   (a) ammonium nitrate **(d)** vanadium(III) oxide
   (b) cobalt(II) sulfate **(e)** barium acetate
   (c) nickel(II) cyanide **(f)** calcium hypochlorite
2. Name the following ionic compounds:
   (a) $MgBr_2$ **(d)** $KMnO_4$
   (b) Li_2CO_3 **(e)** $(NH_4)_2S$
   (c) $KHSO_3$ **(f)** $CuCl$ and $CuCl_2$

Naming Binary Compounds of the Nonmetals

See the *Saunders Interactive General Chemistry CD-ROM,* Screens 3.5 and 3.6 for a description of binary compounds of the nonmetals.

Thus far we have described naming ions and ionic compounds. Another kind of compound comes from the combination of two nonmetals and is composed of molecules. These "two-element," or **binary,** compounds of nonmetals, can also be named in a systematic way.

Hydrogen forms binary compounds with all the nonmetals (except the noble gases). For compounds of oxygen, sulfur, and the halogens, the H atom is generally written first in the formula and is named first. The other nonmetal is named as if it were a negative ion.

Compound	Name
HF	hydrogen fluoride
HCl	hydrogen chloride
H_2S	hydrogen sulfide

Take note of the way formulas of binary nonmetal compounds are written. Simple hydrocarbons such as methane and ethane have formulas written with H following C, and the formulas of ammonia and hydrazine have H following N. Water and the hydrogen halides, however, have the H atom preceding O or the halogen atom. Tradition is the only explanation for such oddities in chemistry.

Virtually all binary, nonmetal compounds are based on a combination of elements from Groups 4A–7A with one another or with hydrogen. The formula is generally written by putting the elements in order of increasing group number. When naming the compound, the number of atoms of a given type in the compound is designated with a prefix, such as "di-," "tri-," "tetra-," "penta-," and so on.

Compound	Systematic Name
NF_3	nitrogen trifluoride
NO	nitrogen monoxide
NO_2	nitrogen dioxide
N_2O	dinitrogen monoxide
N_2O_4	dinitrogen tetraoxide
PCl_3	phosphorus trichloride
PCl_5	phosphorus pentachloride
SF_6	sulfur hexafluoride
S_2F_{10}	disulfur decafluoride

Finally, many of the binary compounds of nonmetals were discovered years ago and have names so common that they continue to be used. These names must simply be learned.

Compounds such as methane and butane belong to a class of hydrocarbons called alkanes (see Chapter 11).

Compound	Common Name
CH_4	methane
C_2H_6	ethane
C_3H_8	propane
C_4H_{10}	butane
NH_3	ammonia
N_2H_4	hydrazine
PH_3	phosphine
NO	nitric oxide
N_2O	nitrous oxide ("laughing gas")
H_2O	water

Exercise 3.6 Naming Compounds

1. Give the formula for each of the following binary, nonmetal compounds:
   (a) carbon dioxide (d) boron trifluoride
   (b) phosphorus triiodide (e) dioxygen difluoride
   (c) sulfur dichloride (f) xenon trioxide
2. Name the following binary, nonmetal compounds:
   (a) N_2F_4 (c) SF_4 (e) P_4O_{10}
   (b) HBr (d) BCl_3 (f) ClF_3

3.6 ATOMS, MOLECULES, AND THE MOLE

Atoms and the Mole

One of the most exciting parts of chemistry is the discovery of some new substance. Chemistry is also a quantitative science, however. When two chemicals react with each other, you want to know how many atoms or molecules of each are used so that a formula of the product can be established. This means we need some method of counting atoms, no matter how small they are. That is, we must discover a way of connecting the macroscopic world, the world we can see, with the particulate world of atoms and molecules. The solution to this problem is to define a convenient unit of matter that contains a known number of particles. The chemical counting unit that has come into use is the **mole.**

See the *Saunders Interactive General Chemistry CD-ROM,* Screens 2.18 and 2.19 for a discussion of the mole and moles of elements.

The word "mole" was apparently introduced about 1896 by the German chemist Wilhelm Ostwald (1853–1932), who derived the term from the Latin word *moles,* meaning a "heap" or "pile." The mole, whose symbol is mol, is the SI base unit for measuring an *amount of substance* (see Table 1.3) and is defined as follows:

The mole is the chemist's six-pack or dozen; it is a counting unit. Many objects come in similar counting units. Shoes, socks, and gloves are sold by the pair, soft drinks by the six-pack, and eggs and donuts by the dozen. Atoms and molecules are counted by the mole.

> A **mole** is the amount of substance that contains as many elementary entities (atoms, molecules, or other particles) as there are atoms in exactly 12 g of the carbon-12 isotope.

The key to understanding the concept of the mole is that *one mole always contains the same number of particles, no matter what the substance.* One mole of sodium contains the same number of atoms as one mole of iron. One mole of aspirin contains the same number of molecules as one mole of water. But how many particles? Many, many experiments over the years have established that number as

$$1 \text{ mole} = 6.022136736 \times 10^{23} \text{ particles}$$

This value is commonly known as **Avogadro's number** in honor of Amedeo Avogadro, an Italian lawyer and physicist (1776–1856) who conceived the basic idea (but never determined the number). Understand that there is nothing "special" about the value. It is fixed by the definition of the mole as exactly 12 g of carbon-12. If one mole of carbon were defined to have some other mass, then Avogadro's number would have a different value. It is interesting that the num-

Avogadro's number is known to 10 significant figures ($6.022136736 \times 10^{23}$), which is not unusual precision for many physical constants. This constant, and others in this book, are taken from E. R. Cohen and B. N. Taylor, "The Fundamental Physical Constants," *Physics Today,* Vol. 40, pp. BG11–BG15, 1987.

Current Perspectives

A Diet Soda—What's in It?

Have you ever read the label on a food product and wondered what all those chemicals are? Most of the compounds listed are naturally occurring and have been added to enhance flavor or to preserve the taste and quality of the product. The label on one popular diet soda indicates that it contains, in decreasing concentration, "carbonated water, caramel color, aspartame, phosphoric acid, potassium benzoate (to protect taste), natural flavors, citric acid, and caffeine." Elsewhere a label says "Phenylketonurics: Contains phenylalanine." What are some of these compounds and why the warning label?

Aspartame and the warning about phenylalanine are connected. Aspartame—sold under the brand name Nutrasweet—is made by chemically combining two naturally occurring amino acids, aspartic acid and phenylalanine. The product is about 200 times sweeter than sugar. Phenylalanine is an essential amino acid, meaning that we cannot live without it. A small number of people, however, cannot tolerate an excess of phenylalanine because they lack an enzyme that metabolizes the acid. People who suffer from phenylketonuria must avoid foods containing aspartame or Nutrasweet, as well as foods such as steak, eggs, cheese, yogurt, and chocolate.

Potassium benzoate is an ionic compound derived from naturally occurring benzoic acid, $C_6H_5CO_2H$. Most berries contain this acid or the anion that comes from removing an H^+ ion from the $—CO_2H$ group.

Citric acid is a colorless, crystalline solid that dissolves readily in water and has a strongly acidic taste. It is of course present in citrus fruits, most especially lemons, and is made by fermentation of carbohydrates. It is used in foods not only as a flavoring agent but also as an antioxidant.

As a final aside, notice that aspartic acid, phenylalanine, benzoic acid, and citric acid all possess a common structural feature: a $—CO_2H$ group, which is the group that makes them acids.

For more information, see *The Consumer's Good Chemical Guide* by John Emsley, W. H. Freeman, New York, 1994.

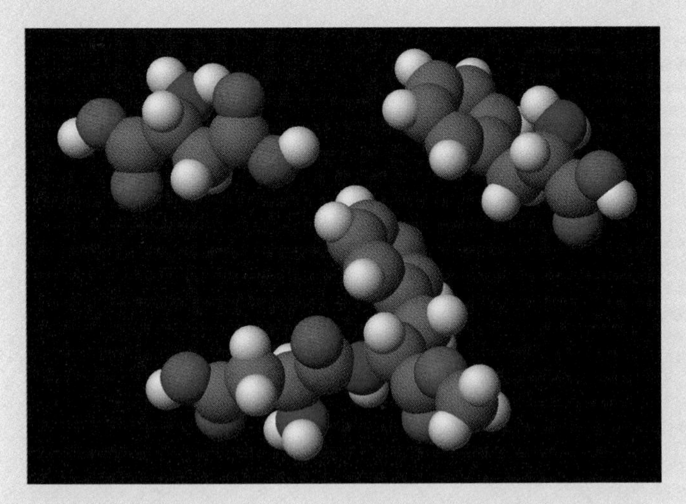

Models of aspartic acid *(above left)*, phenylalanine *(above right)*, and aspartame *(below)*.

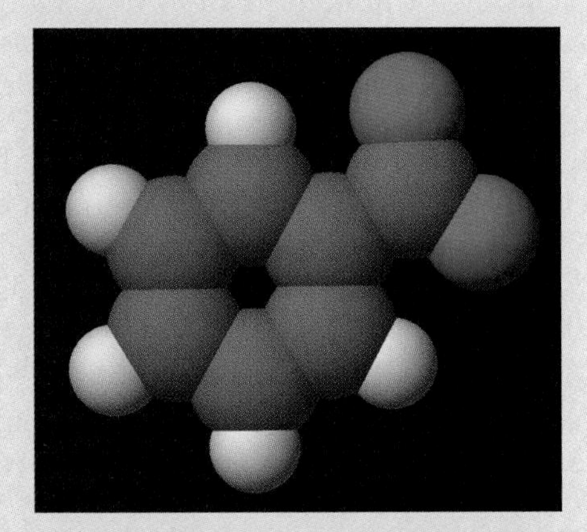

Model of the benzoate ion, $C_6H_5CO_2^-$.

Model of citric acid, $C_6H_8O_7$.

ber was just revised (from 6.022045×10^{23}) in the 1980s to the new value as a result of new and better measurements. Even the most fundamental values and ideas of science are under constant scrutiny.

As you saw in Section 2.4, the atomic mass scale is a relative scale, with the ^{12}C atom chosen as the standard. The masses of all other atoms have been established by experiment and placed on this scale. For example, experiments

Historical Perspective

Amedeo Avogadro (1776–1856) and His Number

Lorenzo Romano Amedeo Carlo Avogadro, an Italian nobleman, was educated as a lawyer and practiced the profession for many years. In about 1800, however, he turned to science and was the first professor in Italy in mathematical physics. In 1811, he first suggested the hypothesis, which we now call a law, that "equal volumes of gases under the same conditions have equal numbers of molecules." From this eventually came the concept of the mole. Avogadro's ideas were not accepted during his lifetime, and scientists were not convinced until the Italian chemist, Stanislao Cannizzaro (1826–1910), described experiments proving them at the great chemical conference in Karlsruhe, Germany, in 1860.

One of the great difficulties presented by Avogadro's number is comprehending its size. It may help to write it out in full as

$$6.022 \times 10^{23} = 602{,}200{,}000{,}000{,}000{,}000{,}000{,}000$$

or as

$$602{,}200 \times 1 \text{ million} \times 1 \text{ million} \times 1 \text{ million}$$

But, think of it this way: If you had Avogadro's number of unpopped popcorn kernels and poured them over the continental United States, the country would be covered in popcorn to a depth of 9 miles! Or, if you divided one mole of pennies equally among every man, woman, and child in the United States, each person could pay off the national debt (currently \$5.5 trillion, or 5.5×10^{12}) and still have \$17 trillion left over for an ice cream cone or two.

How is Avogadro's number determined? It is clearly not possible to count all of the atoms in a mole. (If a computer counted 10 million atoms per second, it would take about 2 billion years to count all of the atoms in a mole.) At least four experimental methods can be used, one of them being the measurement of the dimensions of the smallest repeating unit in a solid element (Chapter 13).

Amedeo Avogadro, conte di Quaregna. *(E. F. Smith Collection/VanPelt Library/University of Pennsylvania)*

show that a ^{16}O atom is 1.33 times heavier than a ^{12}C atom or that an ^{19}F atom is 1.58 times heavier than a ^{12}C atom. Because a mole of carbon-12 has a mass of exactly 12 g and contains 6.0221367×10^{23} atoms, and because a mole of one kind of atom always contains the same number of particles as a mole of another kind of atom, this means a mole of ^{16}O atoms has a mass in grams 1.33 times 12.0 g, or 16.0 g. Similarly, a mole of ^{19}F atoms is 1.58 times greater than 12.0 g of carbon, or 19.0 g.

Moles of Atoms: The Molar Mass

The mass in grams of one mole of atoms of any element (6.0221367×10^{23} atoms of that element) is the **molar mass** of that element. Molar mass is conventionally abbreviated with a capital italicized M and is expressed in units of grams per mole (g/mol). For elements *molar mass is an amount in grams numerically equal to the atomic mass in atomic mass units.* Using sodium and lead as examples,

$$\begin{aligned}\text{Molar mass of sodium (Na)} &= \text{mass of exactly 1 mol of Na atoms} \\ &= 22.99 \text{ g/mol} \\ &= \text{mass of } 6.022 \times 10^{23} \text{ Na atoms}\end{aligned}$$

$$\begin{aligned}\text{Molar mass of lead (Pb)} &= \text{mass of exactly 1 mol of Pb atoms} \\ &= 207.2 \text{ g/mol} \\ &= \text{mass of } 6.022 \times 10^{23} \text{ Pb atoms}\end{aligned}$$

The relative physical sizes of 1-mol quantities of some common elements are shown in Figure 3.12. Although each of these "piles of atoms" has a different volume and different mass, each contains 6.022×10^{23} atoms.

Figure 3.12 One-mole quantities of common elements. Clockwise from the top left: copper beads (63.546 g); aluminum foil (26.982 g); lead shot (207.2 g); magnesium chips (24.305 g); chromium (51.996 g); and sulfur powder (32.066 g). Four of the samples are in 50-mL beakers. (*C. D. Winters*)

The mole concept is the cornerstone of quantitative chemistry. It is essential to be able to convert from moles to mass and from mass to moles. Dimensional analysis, which is described in Section 1.7 and in Appendix A, shows that this can be done in the following way:

MASS ⟷ MOLES CONVERSION

Moles to Mass	**Mass to Moles**
$\text{Moles} \cdot \dfrac{\text{grams}}{1\ \text{mol}} = \text{grams}$	$\text{Grams} \cdot \dfrac{1\ \text{mol}}{\text{grams}} = \text{moles}$
↑ molar mass	↑ 1/molar mass

For example, suppose you wish to use 0.35 mol of aluminum. What mass, in grams, of aluminum is this? Using the molar mass of aluminum (27.0 g/mol), you find that 9.5 g of aluminum is required.

$$0.35\ \cancel{\text{mol Al}} \cdot \frac{27.0\ \text{g Al}}{1\ \cancel{\text{mol Al}}} = 9.5\ \text{g Al}$$

Look at the periodic table in the front of the book and notice that some atomic masses are known to more significant figures and decimal places than others. When using molar masses in a calculation, the convention followed in this book is to use one more significant figure in the molar mass than in any of the other data. For example, if you weigh out 16.5 g of carbon, you use 12.01 g/mol for the molar mass of C to find the moles of carbon present.

$$16.5\ \cancel{\text{g C}} \cdot \frac{1\ \text{mol C}}{12.01\ \cancel{\text{g C}}} = 1.37\ \text{mol C}$$

↑ Note that four significant figures are used in the molar mass, but there are only three in the sample mass.

Using one more significant figure means the accuracy of the molar mass is greater than the other numbers and does not limit the precision of the result.

Example 3.6 Mass to Moles

How many moles are represented by 125 g of silicon, an element used in semiconductors?

Solution Use the periodic table in the front of the book to find the molar mass of silicon (28.09 g/mol). Convert the mass of silicon to its equivalent in moles.

$$125\ \cancel{\text{g Si}} \cdot \frac{1\ \text{mol Si}}{28.09\ \cancel{\text{g Si}}} = \boxed{4.45\ \text{mol Si}}$$

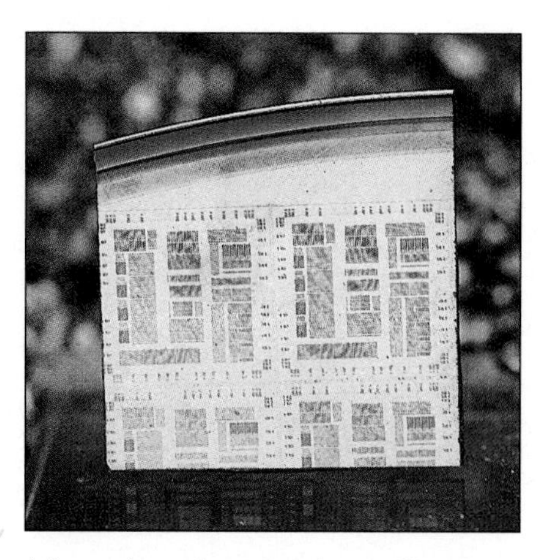

A large piece of a wafer of pure silicon. The tiny rectangles on the wafer are electronic circuits etched into the surface of the wafer. See Example 3.6. *(C. D. Winters)*

Example 3.7 Moles to Mass

What mass, in grams, is equivalent to 2.50 mol of lead (Pb)?

Solution For a conversion between mass and moles, you always need the molar mass, which is 207.2 g/mol in the case of lead. Thus, the number of grams of lead in 2.50 mol is

$$2.50\ \cancel{\text{mol Pb}} \cdot \frac{207.2\ \text{g Pb}}{1\ \cancel{\text{mol Pb}}} = \boxed{518\ \text{g Pb}}$$

The photograph shows 518 g of lead shot in a 150-mL beaker.

This student is holding a 150-mL beaker containing 2.50 mol of lead. See Example 3.7. *(C. D. Winters)*

Example 3.8 Mole Calculation

The graduated cylinder in the photograph contains 32.0 cm^3 of mercury. If the density of mercury at 25 °C is 13.534 g/cm^3, how many moles of mercury are in the cylinder? How many atoms of mercury are there?

Solution Volume and moles are not directly connected (see *Problem-Solving Tips and Ideas: 3.3,* page 124). You must therefore first use the density to convert the volume to a mass, and then derive the quantity of mercury, in moles, from the mass. Finally, the number of atoms is obtained from the number of moles.

$$\boxed{\text{Volume, cm}^3} \xrightarrow[\text{use density}]{\times \frac{\text{g}}{\text{cm}^3}} \boxed{\text{Mass, g}} \xrightarrow[\text{use molar mass}]{\times \frac{\text{mol}}{\text{g}}} \boxed{\text{Moles}} \xrightarrow[\text{use Avogadro's number}]{\times \frac{\text{atoms}}{\text{mol}}} \boxed{\text{Atoms}}$$

The mass of mercury is found to be equivalent to 433 g.

$$32.0\ \cancel{\text{cm}^3} \cdot \frac{13.534\ \text{g}}{1\ \cancel{\text{cm}^3}} = 433\ \text{g}$$

Knowing the mass, you can now find the quantity in moles.

$$433\ \cancel{\text{g Hg}} \cdot \frac{1\ \text{mol Hg}}{200.6\ \cancel{\text{g Hg}}} = \boxed{2.16\ \text{mol Hg}}$$

Finally, because you know the relation between atoms and moles (Avogadro's number), you can find the number of atoms in the sample.

$$2.16\ \cancel{\text{mol Hg}} \cdot \frac{6.022 \times 10^{23}\ \text{atoms}}{1\ \cancel{\text{mol Hg}}} = \boxed{1.30 \times 10^{24}\ \text{atoms Hg}}$$

A graduated cylinder containing 32.0 cm^3 of mercury. See Example 3.8. *(C. D. Winters)*

Problem-Solving Tips and Ideas

3.3 More on Unit Conversions

In Example 3.8 you wanted to find the quantity of mercury, in moles, in a given volume of the element, and then you wanted to know the number of atoms of mercury in that volume. You know that the mass of an element is directly related to the number of moles represented by that mass, and the number of atoms can be calculated once the number of moles is known.

$$\text{Mass, g} \xrightarrow[\text{use molar mass}]{\times \frac{\text{mol}}{\text{g}}} \text{Moles} \xrightarrow[\text{use Avogadro's number}]{\times \frac{\text{atoms}}{\text{mol}}} \text{Atoms}$$

The problem here was that you do not have the mass of mercury, only its volume. This is the reason you need the density of the substance, that is, the relation between volume and mass. Therefore, you first need to convert milliliters of mercury into grams of mercury using density as the conversion factor.

$$\text{Volume, cm}^3 \xrightarrow[\text{use density}]{\times \frac{\text{g}}{\text{cm}^3}} \text{Mass, g}$$

As a final point, note that all three of the conversions in Example 3.8 and the other problems we have done have involved calculations of the form

$$\text{Given data} \cdot \frac{\text{Desired units}}{\text{Units of given data}} = \text{Answer in desired units}$$

conversion factor

Exercise 3.7 Mass/Mole Conversions

1. What is the mass, in grams, of 1.5 mol of silicon?
2. How many moles are represented by 454 g of sulfur? How many atoms?

Exercise 3.8 Atoms

The density of gold, Au, is 19.32 g/cm^3. What is the volume (in cubic centimeters) of a piece of gold that contains 2.6×10^{24} atoms? If the piece of metal is a square with a thickness of 0.10 cm, what is the length, in centimeters, of one side of the piece?

Molecules, Compounds, and the Mole

See the *Saunders Interactive General Chemistry CD-ROM,* Screen 3.15 for a discussion of moles of compounds and Screen 3.16 for "Using Molar Mass."

The formula of a compound tells you the type of atoms or ions in the compound and the relative number of each. For example, in one molecule of methane, CH_4, one atom of C combines with four atoms of H. But suppose you have Avogadro's number of C atoms (6.022×10^{23}) combined with the proper number of H atoms. The compound's formula informs us that four times as many

H atoms are required (24.09×10^{23} H atoms) to give Avogadro's number of CH_4 molecules. What masses of atoms are combined, and what is the mass of this many CH_4 molecules?

C	4 H	CH_4
6.022×10^{23} C atoms	$4 \times 6.022 \times 10^{23}$ H atoms	6.022×10^{23} CH_4 molecules
= 1.000 mol of C	= 4.000 mol of H atoms	= 1.000 mol of CH_4 molecules
= 12.01 g of C atoms	= 4.032 g of H atoms	= 16.04 g of CH_4 molecules

Because we know the number of moles of C and H atoms, we know the masses of carbon and hydrogen that combine to form CH_4. It follows from the law of the conservation of mass that the mass of CH_4 is the sum of these masses (Section 2.1). That is, 1 mol of CH_4 has a mass equivalent to the mass of 1 mol of C atoms (12.01 g) plus 4 mol of H atoms (4.032 g). This is the **molar mass, *M*,** of CH_4. Chemists often use the term "molecular weight" for molar mass.

Just as chemists often use the term "atomic weight" when they mean "atomic mass," they often refer to the molecular weight of a compound. Although this term is very commonly used, we shall generally use molar mass in this book.

MOLAR AND MOLECULAR MASSES

Element or Compound	Molar Mass (g/mol)	Average Mass of One Molecule* (g/molecule)
O_2	32.00	5.314×10^{-23}
P_4	123.9	2.057×10^{-22}
NH_3	17.03	2.828×10^{-23}
H_2O	18.02	2.992×10^{-23}
CH_2Cl_2	84.93	1.410×10^{-22}

*See text, page 126, for the calculation of the mass of one molecule.

Ionic compounds such as NaCl do not exist as individual molecules. Thus, no *molecular* formula can be given; rather, one can only write the simplest formula that shows the relative number of each kind of atom in a sample. Nonetheless, we talk about the molar mass of such compounds and calculate it from the simplest formula. To differentiate substances like NaCl that do not contain molecules, however, chemists sometimes refer to their *formula weight* instead of their molar mass or molecular weight.

Figure 3.13 is a photograph of 1-mol quantities of several common compounds. To find the molar mass of any compound you need only add up the atomic masses for each element in one formula unit. As an example, let us find the molar mass of aspirin, $C_9H_8O_4$. In one formula unit there are nine carbon atoms, eight hydrogen atoms, and four oxygen atoms, which add up to 180.2 g/mol of aspirin.

$$9 \text{ mol of C/mole of } C_9H_8O_4 = 9 \cancel{\text{mol C}} \cdot \frac{12.01 \text{ g C}}{1 \cancel{\text{mol C}}} = 108.1 \text{ g C}$$

$$8 \text{ mol of H/mole of } C_9H_8O_4 = 8 \cancel{\text{mol H}} \cdot \frac{1.008 \text{ g H}}{1 \cancel{\text{mol H}}} = 8.064 \text{ g H}$$

$$4 \text{ mol of O/mol of } C_9H_8O_4 = 4 \cancel{\text{mol O}} \cdot \frac{16.00 \text{ g O}}{1 \cancel{\text{mol O}}} = 64.00 \text{ g O}$$

$$\text{Molar mass of } C_9H_8O_4 = 180.2 \text{ g}$$

Figure 3.13 One-mole quantities of some compounds. Clockwise from front right, they are: NaCl, white (M = 58.44 g/mol); water, H_2O (M = 18.02 g/mol); aspirin (M = 180.2 g/mol); and $NiCl_2 \cdot 6H_2O$, green (M = 237.7 g/mol). *(C. D. Winters)*

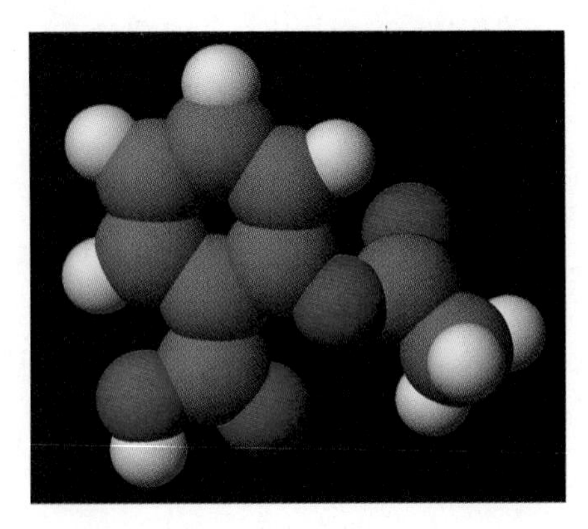

A molecular model of aspirin, $C_9H_8O_4$. As in all other all models carbon atoms are gray, hydrogen atoms are white, and oxygen atoms are red.

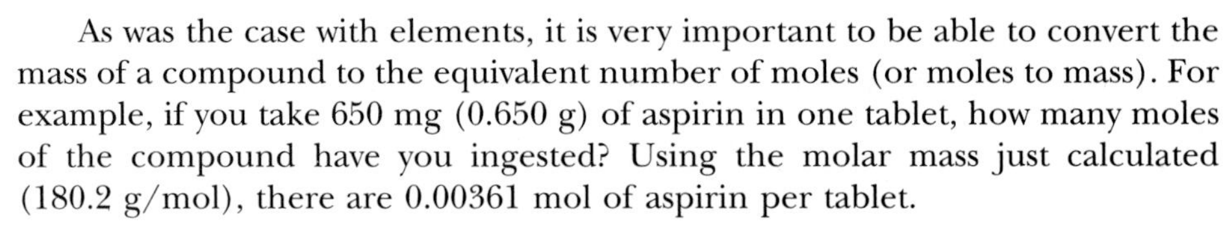

As was the case with elements, it is very important to be able to convert the mass of a compound to the equivalent number of moles (or moles to mass). For example, if you take 650 mg (0.650 g) of aspirin in one tablet, how many moles of the compound have you ingested? Using the molar mass just calculated (180.2 g/mol), there are 0.00361 mol of aspirin per tablet.

$$0.650\ \cancel{\text{g aspirin}} \cdot \frac{1\ \text{mol aspirin}}{180.2\ \cancel{\text{g aspirin}}} = 0.00361\ \text{mol aspirin}$$

The molar mass of a compound is the mass in grams of Avogadro's number of molecules (or of Avogadro's number of formula units of an ionic compound) (see Figure 3.13). With this knowledge, it is possible to determine the number of molecules in any sample from its mass or even to determine the average mass of one molecule. For example, the number of aspirin molecules in one tablet is

$$0.00361\ \cancel{\text{mol aspirin}} \cdot \frac{6.022 \times 10^{23}\ \text{molecules}}{1\ \cancel{\text{mol aspirin}}} = 2.17 \times 10^{21}\ \text{aspirin molecules}$$

and the mass of one molecule is

$$\frac{180.2\ \text{g}\ \cancel{\text{aspirin}}}{1\ \cancel{\text{mol aspirin}}} \cdot \frac{1\ \cancel{\text{mol}}}{6.022 \times 10^{23}\ \text{molecules}} = 2.99 \times 10^{-22}\ \text{g/molecule}$$

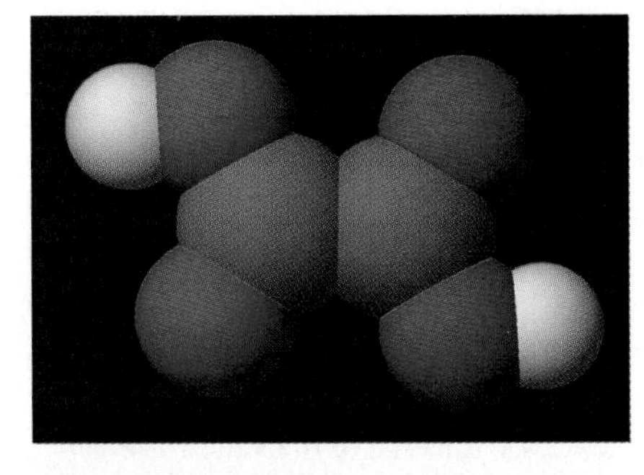

Oxalic acid, $H_2C_2O_4$, is a common, naturally occurring acid. It is found in appreciable concentrations in leafy green plants, such as spinach and rhubarb. It can also combine with calcium ions to form calcium oxalate, CaC_2O_4. This insoluble compound can form in your body and is the source of kidney stones.

Example 3.9 Molar Mass and Moles

You have 16.5 g of the common compound oxalic acid, $H_2C_2O_4$.

1. How many moles are represented by this mass of oxalic acid?
2. How many molecules of oxalic acid are in 16.5 g?
3. How many atoms of carbon are in 16.5 g of oxalic acid?
4. What is the mass of one molecule of oxalic acid?

Solution The first step in any problem involving the conversion of mass and moles is to find the molar mass of the compound in question. Then we can perform the other calculations as outlined by the following scheme:

$$\text{Mass, g} \xrightarrow[\text{use molar mass}]{\times\ \frac{\text{mol}}{\text{g}}} \text{Moles} \xrightarrow[\text{use Avogadro's number}]{\times\ \frac{\text{molecules}}{\text{mol}}} \text{Molecules} \xrightarrow[\text{use chemical formula}]{\times\ \frac{\text{C atoms}}{\text{molecule}}} \text{Number of C atoms}$$

Step 1. *Calculate molar mass:*

$$2\ \text{mol of C/mol of}\ H_2C_2O_4 = 2\ \cancel{\text{mol C}} \cdot \frac{12.01\ \text{g C}}{1\ \cancel{\text{mol C}}} = 24.02\ \text{g C}$$

$$2\ \text{mol of H/mol of}\ H_2C_2O_4 = 2\ \cancel{\text{mol H}} \cdot \frac{1.008\ \text{g H}}{1\ \cancel{\text{mol H}}} = 2.016\ \text{g H}$$

$$4\ \text{mol of O/mol of}\ H_2C_2O_4 = 4\ \cancel{\text{mol O}} \cdot \frac{16.00\ \text{g O}}{1\ \cancel{\text{mol O}}} = 64.00\ \text{g O}$$

$$\text{Molar mass of}\ H_2C_2O_4 = 90.04$$

Step 2. *Calculate number of moles:* The molar mass expressed in units of grams per mole is the conversion factor in all mass-to-mole conversions.

$$16.5 \text{ g} \cdot \frac{1 \text{ mol}}{90.04 \text{ g}} = 0.183 \text{ mol oxalic acid}$$

Step 3. *Calculate number of carbon atoms:* To find the number of C atoms in the acid sample, we first find the number of oxalic acid molecules. Here we convert 0.183 mol of $H_2C_2O_4$ to the number of molecules of oxalic acid in that quantity of compound by using Avogadro's number.

$$0.183 \text{ mol} \cdot \frac{6.022 \times 10^{23} \text{ molecules}}{1 \text{ mol}} = 1.10 \times 10^{23} \text{ molecules}$$

You know that each molecule contains two carbon atoms, so you can find the number of carbon atoms in 16.5 g of the acid.

$$1.10 \times 10^{23} \text{ molecules} \cdot \frac{2 \text{ C atoms}}{1 \text{ molecule}} = 2.20 \times 10^{23} \text{ C atoms}$$

Step 4. *Calculate the mass of one molecule:* The units of the desired answer are grams per molecule, which indicates that you should multiply the starting unit of molar mass (grams per mole) by 1/Avogadro's number (mole/molecule), so that the unit "mol" cancels.

$$\frac{90.04}{1 \text{ mol}} \cdot \frac{1 \text{ mol}}{6.022 \times 10^{23} \text{ molecules}} = 1.495 \times 10^{-22} \text{ g/molecule}$$

Exercise 3.9 Molar Mass and Moles-to-Mass Conversions

1. Calculate the molar mass of (a) limestone, $CaCO_3$, and (b) the amino acid cysteine, $HSCH_2CH(NH_2)CO_2H$.
2. If you have 454 g of $CaCO_3$, how many moles does this represent?
3. In order to have 2.50×10^{-3} mol of cysteine, how many grams must you have?

A molecular model of the amino acid cysteine. The S atom is a yellow sphere. See Exercise 3.9.

3.7 DESCRIBING COMPOUND FORMULAS

Given a sample of an unknown compound, how can its formula be determined? The answer lies in chemical analysis, a major branch of chemistry that deals with the determination of molecular formulas and structures.

Percent Composition

See the *Saunders Interactive General Chemistry CD-ROM,* Screen 3.17 for "Percent Composition."

According to the law of constant composition, any sample of a pure compound always consists of the same elements combined in the same proportion by mass. Thus, it seems that molecular composition can be expressed in at least two ways: (1) in terms of the number of atoms of each type per molecule or (2) in terms of the mass of each element per mole of compound. Actually, there is at least one more way of expressing molecular composition, a method derived from method 2. Composition can be given by the mass of each element in the compound relative to the total mass of the compound—that is, in terms of the mass percent of each element, or **percent composition** by mass. For ammonia, this is

$$\text{Mass percent N in } NH_3 = \frac{\text{mass of N in 1 mol of } NH_3}{\text{mass of 1 mol of } NH_3}$$
$$= \frac{14.01 \text{ g N}}{17.03 \text{ g } NH_3} \times 100\%$$
$$= 82.27\% \text{ (or } 82.27 \text{ g N}/100.0 \text{ g } NH_3)$$

$$\text{Mass percent of H in } NH_3 = \frac{\text{mass of H in 1 mol of } NH_3}{\text{mass of 1 mol of } NH_3}$$
$$= \frac{3(1.008) \text{ g H}}{17.03 \text{ g } NH_3} \times 100\%$$
$$= 17.76\% \text{ H (or } 17.76 \text{ g H}/100.0 \text{ g } NH_3)$$

Exercise 3.10 Percent Composition

Express the composition of each of the following compounds in terms of the mass of each element in 1.00 mol of compound and the mass percent of each element:

1. NaCl, sodium chloride
2. C_6H_{14}, hexane
3. $(NH_4)_2SO_4$, ammonium sulfate

Empirical and Molecular Formulas

See the *Saunders Interactive General Chemistry CD-ROM,* Screens 3.18 and 3.19 for "Determining Empirical Formulas" and "Determining Molecular Formulas."

Ways of expressing molecular composition:

1. A formula giving the number of atoms of each element per molecule
2. Mass of each element per mole of compound
3. Mass of each element per 100 g of compound (percent composition)

Now let us consider the *reverse* of the procedure just described and use relative mass or percent composition data to find a molecular formula. Suppose you know the identity of the elements in a sample and have determined the mass of each element in a given mass of compound (the percent composition) by chemical analysis. You can then calculate the relative number of moles of each element in one mole of compound and from this the relative number of atoms of each element in the compound. For example, for a compound composed of atoms of A and B, the steps from percent composition to a formula are

Convert weight percent to mass and then to moles

$$\left.\begin{array}{l}\% \text{ A} \longrightarrow \text{g A} \longrightarrow x \text{ mol A} \\ \% \text{ B} \longrightarrow \text{g B} \longrightarrow y \text{ mol B}\end{array}\right\} \xrightarrow[\text{mole ratio}]{\text{find}} \frac{x \text{ mol A}}{y \text{ mol B}} \xrightarrow[\text{ratio gives formula}]{} A_xB_y$$

Consider hydrazine, a close relative of ammonia that is used to remove dissolved oxygen in hot water heating systems and to remove metal ions from polluted water. The mass percentages in a sample of hydrazine are 87.42% N and 12.58% H. Taking a 100.00-g sample of hydrazine, the percent composition data tell us that the sample contains 87.42 g of N and 12.58 g of H. The number of moles of each element in the 100.00-g sample is therefore

$$87.42\ \text{g N} \cdot \frac{1\ \text{mol N}}{14.007\ \text{g N}} = 6.241\ \text{mol N}$$

$$12.58\ \text{g H} \cdot \frac{1\ \text{mol H}}{1.0079\ \text{g H}} = 12.48\ \text{mol H}$$

Now we can use the number of moles of each element in 100.00-g of sample to find the number of moles of one element relative to the other. For hydrazine, this ratio is 2 mol of H to 1 mol of N,

When finding the mole ratio of two elements, always divide the larger number by the smaller one. For example, if you have 6 mol of A and 2 mol of B, dividing 6 by 2 will give the ratio of moles of element A to 1 mol of B. That is, the simplest formula would be A_3B.

$$\frac{12.48\ \text{mol H}}{6.241\ \text{mol N}} = \frac{2.000\ \text{mol H}}{1.000\ \text{mol N}} \longrightarrow NH_2$$

showing that there are 2 mol of H atoms for every mole of N atoms in hydrazine. Thus, in one molecule two atoms of H occur for every atom of N, that is, the simplest atom ratio is represented by the formula NH_2.

Percent composition data allow you to calculate the atom ratios in the compound. A *molecular* formula, however, must convey *two* pieces of information: (1) the *relative number* of atoms of each element in a molecule (the atom ratios) and (2) *the total number* of atoms in the molecule. For hydrazine you know now that twice as many H atoms occur as N atoms. This means the molecular formula could be NH_2, but because percent composition data give only the *simplest possible ratio* of atoms in a molecule, the formula could be N_2H_4, N_3H_6, N_4H_8, or any other with a 1-to-2 ratio of N to H. A formula such as NH_2, in which the atom ratio is the simplest possible, is called the **empirical formula.** In contrast, the **molecular formula** shows the *true number* of atoms of each kind in a molecule; it can always be derived by multiplying the empirical formula by a whole number.

To determine the molecular formula, the molar mass must be obtained from experiment. For example, experiments show that the molar mass of hydrazine is 32.0 g/mol, twice the formula mass of NH_2, which is 16.0 g/mol. This must mean that the molecular formula of hydrazine is two times the empirical formula of NH_2, that is, N_2H_4.

As another example of the usefulness of percent composition data, let us say that you collected the following information in the laboratory for the compound isooctane, the compound used as the standard for determining the octane rating of a fuel: % carbon = 84.12; % hydrogen = 15.88; molar mass = 114.2 g/mol. Using these data, you want to calculate the empirical and molecular formulas for the compound. The percent composition data tell you that 84.12 g of C and 15.88 g of H occur in a 100.0-g sample. You can therefore find the number of moles of each element in this sample.

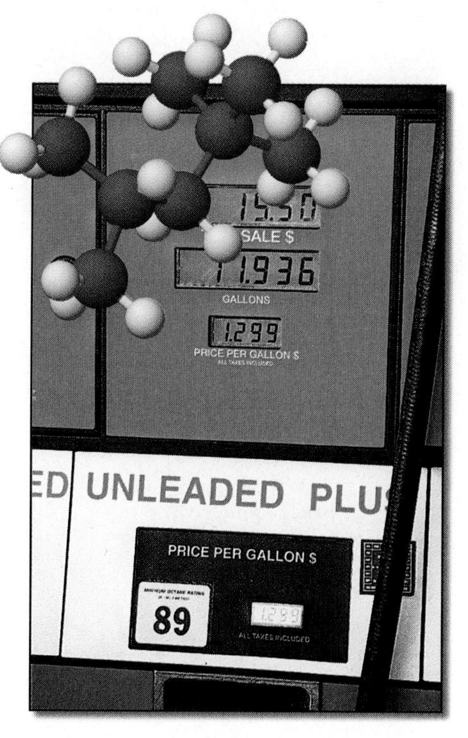

Isooctane is the compound that provides the standard for the octane rating of gasolines. Its molecular formula is C_8H_{18}, whereas its empirical formula is C_4H_9. *(Photo, C. D. Winters; model, S. M. Young)*

$$84.12\ \text{g C} \cdot 1\ \frac{1\ \text{mol C}}{12.011\ \text{g C}} = 7.004\ \text{mol C}$$

$$15.88\ \text{g H} \cdot \frac{1\ \text{mol H}}{1.0079\ \text{g H}} = 15.76\ \text{mol H}$$

This means that, in any sample of isooctane, the ratio of moles of H to C is

$$\text{Mole ratio} = \frac{15.76\ \text{mol H}}{7.004\ \text{mol C}} = \frac{2.250\ \text{mol H}}{1.000\ \text{mol C}}$$

Now your task is to turn this decimal fraction into a whole-number ratio of H to C. To do this, recognize that 2.25 is the same as $2\frac{1}{4}$ or 9/4. Therefore, the ratio of C to H is

$$\text{Mole ratio} = \frac{2.25 \text{ mol H}}{1.00 \text{ mol C}} = \frac{2\frac{1}{4} \text{ mol H}}{1 \text{ mol C}} = \frac{9/4 \text{ mol H}}{1 \text{ mol C}} = \frac{9 \text{ mol H}}{4 \text{ mol C}}$$

and you know now that nine H atoms occur for every four C atoms in isooctane. Thus, the simplest, or *empirical, formula* is C_4H_9. If C_4H_9 were the molecular formula, the molar mass would be 57.12 g/mol. Your experiments gave the actual molar mass as 114.2 g/mol, however, twice the value for the empirical formula.

$$\frac{114.2 \text{ g/mol for isooctane}}{57.12 \text{ g/mol of } C_4H_9} = 2.00 \text{ mol } C_4H_9 \text{ per mol of isooctane}$$

The molecular formula is therefore $(C_4H_9)_2$, or C_8H_{18}.

Eugenol, $C_{10}H_{12}O_2$, is a principal component of oil of cloves. *(Photo, C. D. Winters; model, S. M. Young)*

Example 3.10 Calculating a Formula from Percent Composition

Eugenol is the active component of oil of cloves. It has a molar mass of 164.2 g/mol and is 73.14% C and 7.37% H; the remainder is oxygen. What are the empirical and molecular formulas for eugenol?

Solution This problem can be solved by following the "map" in the preceding text. First, find the number of moles of carbon and hydrogen in a 100.00-g sample of eugenol.

$$73.14 \text{ g C} \cdot \frac{1 \text{ mol C}}{12.011 \text{ g C}} = 6.089 \text{ mol C}$$

$$7.37 \text{ g H} \cdot \frac{1 \text{ mol H}}{1.008 \text{ g H}} = 7.31 \text{ mol H}$$

But what about oxygen? Because the total sample mass is 100.00 g, this means that the mass of oxygen is

$$100.00 \text{ g} = 73.14 \text{ g C} + 7.37 \text{ g H} + \text{mass of O}$$
$$\text{Mass of O} = 100.00 \text{ g} - (73.14 \text{ g C} + 7.37 \text{ g H})$$
$$= 19.49 \text{ g O}$$

Now we can calculate moles of O as well.

$$19.49 \text{ g O} \cdot \frac{1 \text{ mol O}}{15.999 \text{ g O}} = 1.218 \text{ mol O}$$

To find the mole ratio, the best approach is to base the ratios on the element with the smallest number of moles present. In this case it is oxygen. Therefore,

$$\frac{\text{Mol C}}{\text{Mol O}} = \frac{6.089 \text{ mol C}}{1.218 \text{ mol O}} = \frac{4.999 \text{ mol C}}{1.000 \text{ mol O}} = \frac{5 \text{ mol C}}{1 \text{ mol O}}$$

$$\frac{\text{Mol H}}{\text{Mol O}} = \frac{7.31 \text{ mol H}}{1.218 \text{ mol O}} = \frac{6.00 \text{ mol H}}{1.00 \text{ mol O}} = \frac{6 \text{ mol H}}{1 \text{ mol O}}$$

Now we know there are 5 mol of C and 6 mol of H to 1 mol of O, so the empirical formula is C_5H_6O. The experimentally determined molar mass is 164.2 g/mol. This is *twice* as large as the mass for C_5H_6O (82.1 g/unit).

$$\frac{164.2 \text{ g/mol for eugenol}}{82.10 \text{ g/mol of } C_5H_6O} = 2.00 \text{ mol } C_5H_6O/\text{mol of eugenol}$$

Therefore, the molecular formula is $C_{10}H_{12}O_2$.

Exercise 3.11 Empirical and Molecular Formulas

Isoprene is a volatile, fragrant liquid that polymerizes to form natural rubber. It has 88.17% carbon and 11.83% hydrogen. Its molar mass is 68.11 g/mol. What are its empirical and molecular formulas?

Problem-Solving Tips and Ideas

3.4 Finding Empirical and Molecular Formulas

- The experimental data available to find a formula may be in the form of percent composition or the masses of elements combined in some mass of compound. No matter what the starting point, the first step is always to convert masses of elements to moles.
- Be sure to use at least three significant figures when calculating empirical formulas. Fewer significant figures usually gives a misleading result.
- When finding atom ratios, it is easiest to divide the larger number of moles by the smaller one.
- Empirical and molecular formulas often differ for molecular compounds. In contrast, the formula of an ionic compound is generally the same as its empirical formula.
- To determine the molecular formula of a compound after calculating the empirical formula, the molar mass must be obtained by some experimental method.

Exercise 3.12 Empirical and Molecular Formulas

Propylene glycol has been suggested as an alternative to ethylene glycol, which is used as antifreeze. Ethylene glycol is toxic to animals that accidentally drink antifreeze, but propylene glycol is not. Propylene glycol has 47.35% C and 10.60% H. The remainder is oxygen. If the molar mass is 76.10 g/mol, what are the empirical and molecular formulas of propylene glycol?

Determining Formulas

The empirical formula of a compound can be calculated if the percent composition of the compound has been experimentally determined. Then, if the molar mass has also been determined by any of a number of experimental methods, the empirical formula can be converted to a molecular formula.

Figure 3.14 Preparing a compound of copper and sulfur. *(left)* A weighed quantity of copper is combined with an excess of sulfur. *(center)* The mixture is heated strongly in a crucible to cause the reaction between copper and sulfur and to evaporate any unreacted sulfur. *(right)* The product of the reaction is a black copper sulfide. *(C. D. Winters)*

Copper metal and yellow, powdered sulfur react to form a copper sulfide (Figure 3.14)

$$\text{Cu metal} + \text{S(s)} \xrightarrow{\text{will give}} \text{Cu}_x\text{S}_y$$

The Cu-to-S ratio is unknown. To determine this, we can heat a weighed quantity of copper with sulfur in a crucible. The quantity of S used is more than is required to react with all of the copper. After Cu_xS_y has been formed, continued heating will cause the excess sulfur to vaporize, leaving pure Cu_xS_y.

Suppose you collect the following data in your laboratory experiment to determine the formula of Cu_xS_y.

Mass of crucible	19.732 g
Mass of crucible plus copper	27.304 g
Mass of crucible plus Cu_xS_y after heating	29.214 g

The first step is to find what masses of copper and sulfur have combined. Next, convert those masses to moles of the elements, and then find the ratio of Cu to S to give the formula

Mass of copper and Mass of sulfur —use molar mass→ Moles of copper and Moles of sulfur → Mole ratio, Cu/S = empirical formula

1. Mass of copper:

Mass of crucible + Cu	27.304 g
Mass of crucible	− 19.732 g
Mass of Cu	7.572 g

Mass of sulfur:

Mass of crucible + Cu_xS_y	29.214 g
Mass of crucible + Cu	27.304 g
Mass of S	1.910 g

2. Moles of copper and sulfur

$$7.572 \text{ g Cu} \cdot \frac{1 \text{ mol Cu}}{63.546 \text{ g Cu}} = 0.1192 \text{ mol Cu}$$

$$1.910 \text{ g S} \cdot \frac{1 \text{ mol S}}{32.066 \text{ g S}} = 0.05956 \text{ mol S}$$

3. Ratio of moles of copper and sulfur

$$\frac{\text{Mol Cu}}{\text{Mol S}} = \frac{0.1192 \text{ mol Cu}}{0.05956 \text{ mol S}} = \frac{2.001 \text{ mol Cu}}{1.000 \text{ mol S}} = \frac{2 \text{ mol Cu}}{1 \text{ mol S}}$$

This calculation shows there are twice as many moles of copper in the compound as moles of S, so there are twice as many atoms of copper as atoms of sulfur per formula unit. The empirical formula is Cu_2S.

Example 3.11 Determining the Formula of a Metal Halide

Tin metal and purple iodine will react to give an orange tin iodide, Sn_xI_y.

$$\text{Sn metal} + I_2 \longrightarrow Sn_xI_y$$

The experiment is set up so that the quantity of Sn in the original mixture of elements is far in excess of that needed to react with all the I_2 present. After all the iodine has reacted and the tin-iodine compound has been formed, unreacted solid tin metal can be separated from the mixture. The following data were collected in an experiment to find the values of x and y in Sn_xI_y.

Mass of tin (Sn) in the original mixture	1.056 g
Mass of iodine (I_2) in the original mixture	1.947 g
Mass of tin (Sn) recovered after reaction	0.601 g

Solution The first step is to find the mass of Sn that combined with 1.947 g of I_2. These masses are then converted to moles of Sn and I, and finally to the ratio of moles of Sn and I in the empirical formula.

Mass of Sn in the original mixture	1.056 g
Mass of Sn recovered after the reaction	−0.601 g
Mass of Sn consumed in the reaction	0.455 g

You now know that 0.455 g of Sn combined with 1.947 g I_2. Next, calculate the moles of each element that reacted to form Sn_xI_y.

$$0.455 \text{ g Sn} \cdot \frac{1 \text{ mol Sn}}{118.7 \text{ g Sn}} = 3.83 \times 10^{-3} \text{ mol Sn}$$

$$1.947 \text{ g } I_2 \cdot \frac{1 \text{ mol } I_2}{253.81 \text{ g } I_2} = 7.671 \times 10^{-3} \text{ mol } I_2$$

At this point recognize that iodine combines with tin as I atoms. Therefore, the moles of I combined with tin is

$$7.671 \times 10^{-3} \text{ mol } I_2 \cdot \frac{2 \text{ mol I}}{1 \text{ mol } I_2} = 15.34 \times 10^{-3} \text{ mol I}$$

Now we can calculate the ratio of Sn and I that combined.

$$\frac{15.34 \times 10^{-3} \text{ mol 1}}{3.83 \times 10^{-3} \text{ mol Sn}} = \frac{4.00 \text{ mol I}}{1.00 \text{ mol Sn}}$$

The ratio of atoms of I to atoms of Sn is 4:1, which gives the empirical formula SnI_4. More experimental data are needed to find the molecular formula, but such experiments would show the molecular formula is the same as the empirical formula in this case.

Exercise 3.13 Determining the Formula of a Binary Oxide

Analysis shows that 0.586 g of potassium can combine with 0.480 g of O_2 gas to give a white solid with a formula of K_xO_y. What is the formula of the white solid?

Problem-Solving Tips and Ideas

3.5 Finding Empirical and Molecular Formulas

The calculations in the text and in Examples 3.10 and 3.11 illustrate only a few of the ways of using chemical information to determine formulas. You should always focus on using the data to

1. Find the number of moles of atoms of the elements combined in a given mass of compound and then
2. Find the ratio of those moles

The data you are given may be mass percentages, masses of elements in a given amount of compound, or experimental information from a chemical reaction.

Using the Formulas of Compounds

Once the formula of a compound is known, you can use the information in a variety of ways. One of those is to place the atoms in three-dimensional space and define the structure of the compound. This is an important part of chemistry because structural knowledge leads to insight into reactivity, a subject we shall return to often in this text. Another use of molecular formulas is in quantitative chemistry. For example, because we know the formula of the product of the reaction of aluminum and bromine (Figure 1, page 2), we know how much bromine is required to react completely with the aluminum. In other words, knowing the formula of a compound, we know how much product can be derived from it in a chemical reaction. We shall illustrate that idea here but will devote much of the next two chapters to a more complete exploration of the possibilities.

Example 3.12 Using Chemical Formulas

What mass of copper(I) sulfide, Cu_2S, may be obtained from 2.00 kg of copper (see Figure 3.14)?

Solution Once a formula is known, the molar mass of the compound, a vital piece of information, can be calculated. Here it is the link between the mass of copper and the mass of Cu_2S produced. Our plan for solving this problem is first to find the moles of copper in 2.00 kg of compound. Next, the formula Cu_2S tells us that two moles of Cu occur in each mole of Cu_2S. Moles of Cu_2S are therefore linked to the mass of copper.

$$\text{Mass Cu, g} \xrightarrow{\times \frac{\text{mol}}{\text{g Cu}}} \text{Mol Cu} \xrightarrow{\text{Relate moles Cu to moles } Cu_2S \text{ using the formula}} \text{Mol } Cu_2S \xrightarrow{\times \frac{\text{g } Cu_2S}{\text{mol}}} \text{Mass } Cu_2S\text{, g}$$

$$2.00 \text{ kg Cu} \cdot \frac{1000 \text{ g Cu}}{1 \text{ kg Cu}} \cdot \frac{1 \text{ mol Cu}}{63.55 \text{ g Cu}} = 31.5 \text{ mol Cu}$$

$$31.5 \text{ mol Cu} \cdot \frac{1 \text{ mol } Cu_2S}{2 \text{ mol Cu}} \cdot \frac{159.2 \text{ g } Cu_2S}{1 \text{ mol } Cu_2S} = 2510 \text{ g } Cu_2S$$

This is very similar to the approach we shall use to study weight relations in chemical reactions beginning in Chapter 4. There is an alternative in this case, however. Because the formula is known, we know the weight percent of copper in copper(I) sulfide. This tells us directly the number of grams of copper in 100 g of Cu_2S, so we can use this relationship to convert grams of copper to grams of copper(I) sulfide. The weight percent of Cu in Cu_2S is 79.83%. Therefore,

$$2.00 \times 10^3 \text{ g Cu} \cdot \frac{100.0 \text{ g } Cu_2S}{79.83 \text{ g Cu}} = 2510 \text{ g } Cu_2S$$

Either approach is effective and acceptable.

Exercise 3.14 Using a Formula

What mass of iron(III) oxide contains 50.0 g of iron?

3.8 HYDRATED COMPOUNDS

See the *Saunders Interactive General Chemistry CD-ROM,* Screen 3.14 for "Hydrated Compounds."

If ionic compounds are prepared in water solution and then isolated as solids, the crystals often have molecules of water trapped in the lattice. Compounds in which molecules of water are associated with the ions of the compound are called **hydrated compounds.** The beautiful green nickel(II) compound in Figure 3.13, for example, has a formula that is conventionally written as $NiCl_2 \cdot 6\ H_2O$. The dot between $NiCl_2$ and $6\ H_2O$ indicates that 6 mol of water are associated with every mole of $NiCl_2$; it is equivalent to writing the formula as $NiCl_2(H_2O)_6$. The name of the compound, nickel(II) chloride hexahydrate, reflects the presence of 6 mol of water. The molar mass of $NiCl_2 \cdot 6\ H_2O$ is 129.6 g/mol (for $NiCl_2$) plus 108.1 g/mol (for $6\ H_2O$), or 237.7 g/mol.

Hydrated compounds are quite common. The walls of your home may be covered with wallboard, or "plaster board" (Figure 3.15) These sheets contain hydrated calcium sulfate, or gypsum ($CaSO_4 \cdot 2\ H_2O$), as well as unhydrated $CaSO_4$, sandwiched between paper. Gypsum is a mineral that can be mined. Now, however, it is more commonly a byproduct in the manufacture of hydrofluoric

(a)

(b)

***Figure* 3.15 Hydrated compounds.** (a) Wallboard, or "plaster board," consists of gypsum sandwiched between sheets of heavy paper. Gypsum is hydrated calcium sulfate, $CaSO_4 \cdot 2\ H_2O$. (b) When water molecules are associated with cobalt(II) chloride, $CoCl_2$, the hydrated compound, $CoCl_2 \cdot 6\ H_2O$, is red. When it is heated, however, the compound turns deep blue (called "cobalt blue") as water is lost and the structure of the compound changes. *(C. D. Winters)*

acid and phosphoric acid, or it also comes from cleaning sulfur dioxide from the exhaust gases of power plants.

If gypsum is heated between 120 and 180 °C, the water is partly driven off to give $CaSO_4 \cdot \frac{1}{2}\ H_2O$, a compound commonly called "plaster of Paris." If you have ever broken an arm or leg and had to have a cast, the cast may have been made of this compound. It is an effective casting material because, when added to water, it forms a thick slurry that can be poured into a mold or spread out over a part of the body. As it takes on more water, the material increases in volume and forms a hard, inflexible solid. These properties also make plaster of Paris a useful material to artists, because the expanding compound fills a mold completely and makes a high-quality reproduction.

Hydrated cobalt(II) chloride is the deep red solid in Figure 3.15b. When heated it turns purple and then deep blue as it loses water to form anhydrous $CoCl_2$; "anhydrous" means a substance without water. On exposure to moist air, anhydrous $CoCl_2$ takes up water and is converted back into the red hydrated compound. It is this property that allows crystals of the blue compound to be used as a humidity indicator. You may have seen them in a small bag packed with a piece of electronic equipment. The compound also makes a good "invisible ink." A solution of cobalt(II) chloride in water is red, but if you write on paper with the solution it cannot be seen. When the paper is warmed, however, the cobalt compound dehydrates to give the deep blue anhydrous compound, and the writing becomes visible.

There is no simple way to predict how much water will be present in a hydrated compound, so it must be determined experimentally. Such an experiment may involve heating the hydrated material so that all the water is released from the solid and evaporated (Figure 3.16). Only the anhydrous compound is left. The formula of hydrated copper(II) sulfate, commonly known as "blue vitriol," is determined in this manner in Example 3.13.

Example 3.13 Determining the Formula of a Hydrated Compound

Suppose you want to know the value of x in blue, hydrated copper(II) sulfate, $CuSO_4 \cdot x\,H_2O$, that is, the number of water molecules for each unit of $CuSO_4$. In the laboratory you weigh out 1.023 g of the solid. After heating the solid thoroughly in a porcelain crucible (see Figure 3.16), 0.654 g of nearly white, anhydrous copper(II) sulfate, $CuSO_4$, remains.

$$1.023\text{ g } CuSO_4 \cdot x\,H_2O + \text{heat} \longrightarrow 0.654\text{ g } CuSO_4 + ?\text{ g } H_2O$$

Solution The law of the conservation of matter dictates that the mass of the original, hydrated compound must equal the sum of the mass of the water driven away and the mass of the anhydrous compound left behind.

Mass of hydrated compound	1.023 g
−Mass of anhydrous compound, $CuSO_4$	−0.654 g
Mass of water	0.369 g

Because you want to know the number of moles of H_2O that occur for each mole of $CuSO_4$, the next step is to convert the masses of these compounds to moles.

$$0.369\text{ g } H_2O \cdot \frac{1\text{ mol } H_2O}{18.02\text{ g } H_2O} = 0.0205\text{ mol } H_2O$$

$$0.654\text{ g } CuSO_4 \cdot \frac{1\text{ mol } CuSO_4}{159.6\text{ g } CuSO_4} = 0.00410\text{ mol } CuSO_4$$

The value of x is determined from the mole ratio.

$$\frac{0.0205\text{ mol } H_2O}{0.00410\text{ mol } CuSO_4} = \frac{5.00\text{ mol } H_2O}{1.00\text{ mol } CuSO_4}$$

This tells us the water-to-$CuSO_4$ ratio is 5:1, so the formula of the hydrated compound is $CuSO_4 \cdot 5\ H_2O$, and its name is copper(II) sulfate pentahydrate.

Figure **3.16 Determining the formula of a hydrated compound.** The formula of a hydrated compound can be determined by heating a weighed sample enough to cause the compound to release its water of hydration. Knowing the mass of the hydrated compound before heating, and the mass of the anhydrous compound after heating, gives the mass of water in the original sample. *(C. D. Winters)*

Exercise 3.15 Determining the Formula of a Hydrated Compound

Naturally occurring hydrated copper(II) chloride is called eriochalcite. When heated to 100 °C, the compound is dehydrated, that is, the water of hydration is lost. If 0.235 g of $CuCl_2 \cdot x\,H_2O$ gives 0.185 g of $CuCl_2$ on heating, what is the value of x?

CHAPTER HIGHLIGHTS

Having studied this chapter, you should be able to

- Interpret the meaning of **molecular formulas** and **structural formulas** (Section 3.1).
- Recognize that metal atoms commonly lose one or more electrons to form positive ions (cations), and nonmetal atoms often gain electrons to form negative ions (anions) (see Figure 3.5).

- Recognize that the charge on a metal cation (other than the transition metals) is equal to the group number in which the element is found in the periodic table (M^{n+}, n = Group number) (Section 3.3). Transition metal cations are often 2+ or 3+, but other charges are observed.)
- Recognize that the negative charge on a single-atom or monatomic anion, X^{n-}, is given by $n = 8 -$ Group number (Section 3.3).
- Give the names or formulas of **polyatomic ions**, knowing their formulas or names, respectively (Table 3.1 and Section 3.3).
- Write the formulas for **ionic compounds** by combining ions in the proper ratio to give no overall charge (Section 3.4).
- Understand the importance of **Coulomb's law.** Recall from Section 2.2 that **electrostatic forces** are responsible for the attraction or repulsion of charged species (ions). The magnitude of the force is given by Coulomb's law, which states that the force of attraction between oppositely charged species increases with electric charge and with decreasing distance between the species (Section 3.3).
- Name ionic compounds and simple binary compounds of the nonmetals (Section 3.5).
- Understand that the **molar mass** of an element or a compound (often called the atomic or molecular weight, respectively) is the mass in grams of Avogadro's number of atoms of an element or of molecules of a compound (Section 3.6). For ionic compounds, which do not consist of individual molecules, we often refer to the sum of atomic masses as the formula mass (or formula weight).
- Calculate the molar mass of a compound from its formula and a table of atomic weights (Section 3.6).
- Calculate the number of moles of an element or compound that are represented by a given mass, and vice versa (Section 3.6).
- Express molecular composition in terms of **percent composition** (Section 3.7).
- Use percent composition to determine the **empirical formula** of a compound (Section 3.7).
- Use formulas in chemical calculations (Section 3.7).
- Use experimental data to calculate the number of water molecules in a **hydrated compound** (Section 3.8).

STUDY QUESTIONS

REVIEW QUESTIONS

1. A model of the pelargonidin molecule is illustrated on page 100. How many O atoms occur in one molecule? What is the formula of this molecule?
2. A model of cisplatin is given on page 105. How many N atoms are in one molecule? How many H atoms occur in one molecule? How many H atoms are in 1 mol of the compound? What is its molar mass?
3. The molecule illustrated here is methanol. Using Figure 3.4 as a guide, decide which atoms are in the plane of the paper, which lie above the plane, and which lie below. Sketch a ball-and-stick model. If available to you, go to the Saunders site on the World Wide Web (or the *Saunders Interactive General Chemistry CD-ROM*) and find the model of methanol.

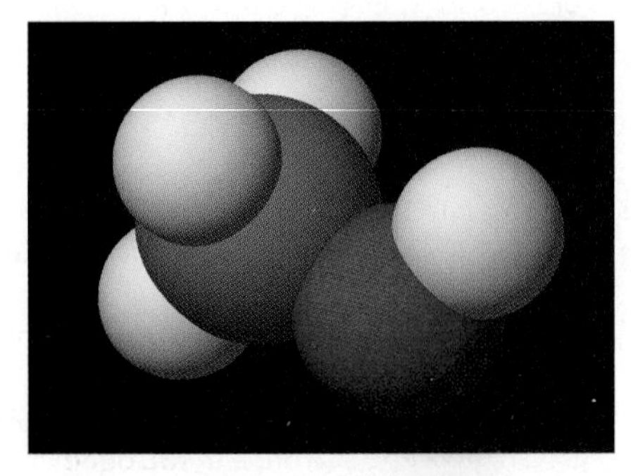

A molecular model of methanol.

4. How many electrons are in a strontium atom (Sr)? Does an atom of Sr gain or lose electrons when forming an ion? How many electrons are gained or lost by the atom? When Sr forms an ion, the ion has the same number of electrons as which one of the noble gases?
5. The compound $(NH_4)_2SO_4$ consists of two polyatomic ions. What are the names and electric charges of these ions? What is the molar mass of this compound?
6. The structure of vanillin is shown here.
   (a) Explain why 40 g of vanillin represents less mass than 3.0×10^{23} molecules of the compound.
   (b) What feature of the molecule is shared with molecules such as styrene (page 103) and graphite (Figure 2.21)?

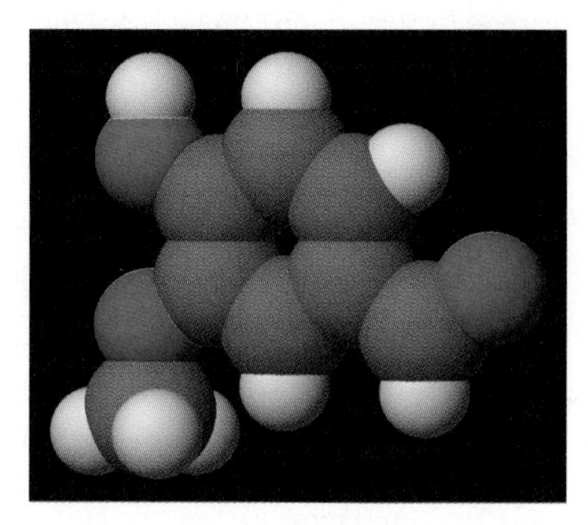

A molecular model of vanillin.

7. Knowing that the formula of sodium borate is Na_3BO_3, give the formula and charge of the borate ion. Is the borate ion a cation or an anion?
8. Which has the larger mass, 0.5 mol of Na or 0.5 mol of Si?
9. Which has the larger mass, 10 atoms of iron or 10 atoms of calcium?
10. What is the difference between an empirical and a molecular formula? Use the compound ethane, C_2H_6, to illustrate your answer.
11. Which compound has the larger weight percentage of oxygen, water or methanol (CH_3OH)?

NUMERICAL AND OTHER QUESTIONS

Molecular Formulas

12. Write the molecular formula of each of the following compounds:
    (a) One of a series of compounds called the boron hydrides has 4 boron atoms and 10 hydrogen atoms per molecule.
    (b) Vitamin C, ascorbic acid, has 6 carbon atoms, 8 hydrogen atoms, and 6 oxygen atoms per molecule.
    (c) A molecule of aspartame, an artificial sweetener, has 14 carbon atoms, 18 hydrogen atoms, 2 nitrogen atoms, and 5 oxygen atoms.
13. Write the molecular formula for each of the following compounds:
    (a) Isoprene, the building block of rubber, has 5 carbon atoms and 8 hydrogen atoms per molecule.
    (b) A molecule of saccharin, an artificial sweetener, has 7 carbon atoms, 5 hydrogen atoms, 3 oxygen atoms, and 1 each of nitrogen and sulfur per molecule.
    (c) A molecule of camphor has 10 carbon atoms, 16 hydrogen atoms, and 1 oxygen atom per molecule.
14. Give the total number of atoms of each element in one formula unit of each of the following compounds:
    (a) CaC_2O_4
    (b) $C_6H_5CHCH_2$
    (c) $Cu_2CO_3(OH)_2$
    (d) $Co(NH_3)_6Cl_3$
    (e) $K_4Fe(CN)_6$
15. Give the total number of atoms of each element in each of the following molecules.
    (a) $Co_2(CO)_8$
    (b) $HO_2CCH_2CH_2CO_2H$, succinic acid
    (c) $CH_3CH(NH_2)CO_2H$, alanine, an amino acid
    (d) $C_6H_2CH_3(NO_2)_3$, TNT, an explosive
16. Write a molecular formula for the following two organic acids:

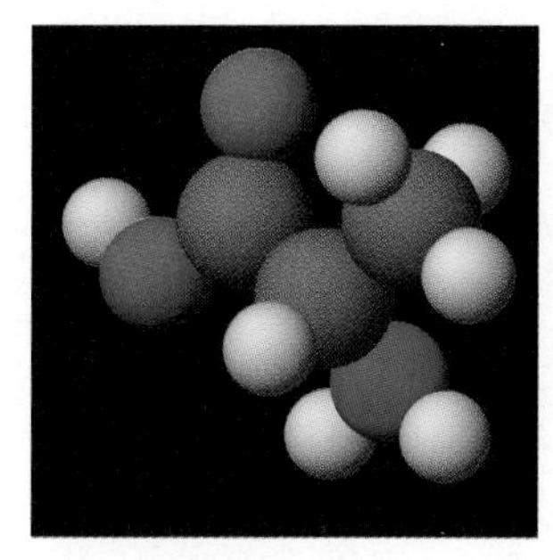

A molecular model of alanine.

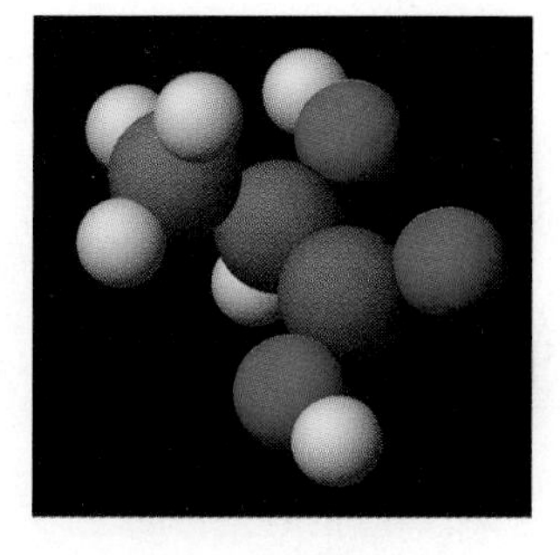

A molecular model of lactic acid.

17. Write a molecular formula for each of the following molecules:
    (a) Acetaminophen is an analgesic (found in such over-the-counter drugs as Tylenol).

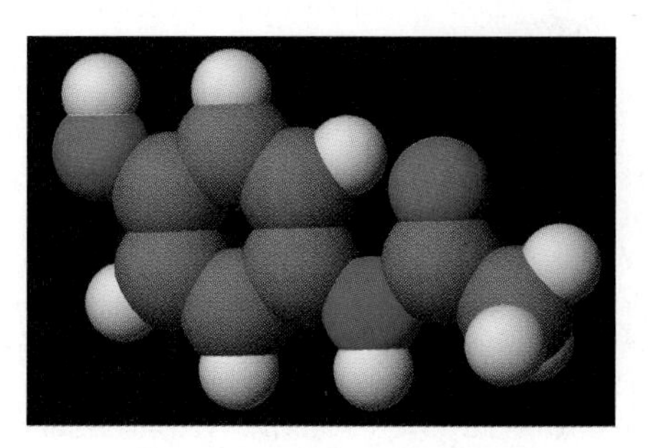

A molecular model of acetaminophen.

(b) Dimethyl terephthalate is one of the two molecules used to make the polymers Dacron and Mylar.

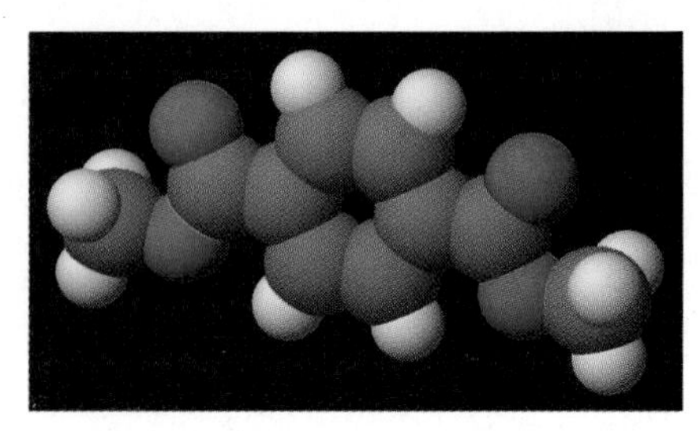

A molecular model of dimethyl terephthalate.

Molecular Models

18. A ball-and-stick model of sulfuric acid is illustrated here. What is the formula for sulfuric acid? Describe the structure of the molecule. Is it flat? That is, are all the atoms in the plane of the paper? (Color code: sulfur atoms are yellow; oxygen atoms are red; and hydrogen atoms are white.)

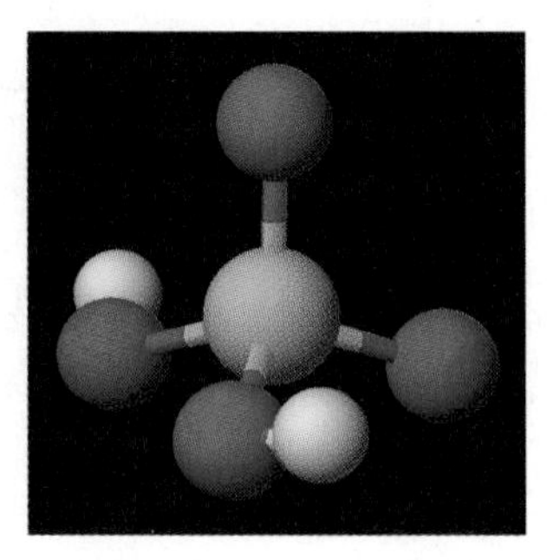

A molecular model of sulfuric acid.

19. A ball-and-stick model of toluene is illustrated here. What is its formula? Describe the structure of the molecule. Is it flat or is only a portion of it flat? (Color code: carbon atoms are gray and hydrogen atoms are white.)

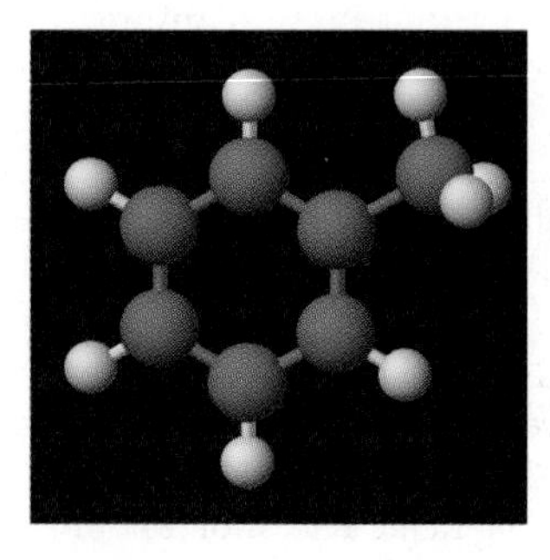

A molecular model of toluene.

Ions and Ion Charges

20. What charges are most commonly observed for monatomic ions of the following elements?
(a) magnesium (c) iron
(b) zinc (d) gallium

21. What charges are most commonly observed for monatomic ions of the following elements?
(a) selenium (c) nickel
(b) fluorine (d) nitrogen

22. Give the symbol, including the correct charge, for each of the following ions:
(a) strontium ion (e) sulfide ion
(b) titanium(IV) ion (f) perchlorate ion
(c) aluminum ion (g) cobalt(II) ion
(d) hydrogen carbonate ion (h) ammonium ion

23. Give the symbol, including the correct charge, for each of the following ions:
(a) permanganate ion (d) sulfate ion
(b) nitrite ion (e) phosphate ion
(c) dihydrogen phosphate ion (f) sulfite ion

24. When potassium becomes a monatomic ion, how many electrons does it lose or gain? What noble gas atom has the same number of electrons as a potassium ion?

25. When oxygen and sulfur become monatomic ions, how many electrons do they lose or gain? What noble gas atom has the same number of electrons as an oxygen ion? What noble gas atom has the same number of electrons as a sulfur ion? How do O and S resemble each other in their behavior?

Ionic Compounds

26. Predict the charges of the ions in an ionic compound containing barium and bromine.

27. What are the charges of the ions in an ionic compound containing nickel(II) and fluoride ions?

28. For each of the following compounds, give the formula, charge, and number of each ion that makes up the compound:
(a) K_2S (c) $(NH_4)_3PO_4$ (e) $KMnO_4$
(b) $NiSO_4$ (d) $Ca(ClO)_2$

29. For each of the following compounds, give the formula, charge, and number of each ion that makes up the compound:
(a) $Ca(CH_3CO_2)_2$ (c) $Al(OH)_3$ (e) $CuCO_3$
(b) $Co_2(SO_4)_3$ (d) KH_2PO_4

30. Cobalt is a transition metal and forms ions with at least two different charges. Write the formulas for the two different cobalt oxides.

31. Platinum is a transition element and forms Pt^{2+} and Pt^{4+} ions. Write the formulas for the compounds of each of these ions with (a) chloride ions and (b) sulfide ions.

32. Which of the following are correct formulas for ionic compounds? For those that are not, give the correct formula.
(a) AlCl (c) Ga_2O_3
(b) NaF_2 (d) MgS

33. Which of the following are correct formulas for ionic compounds? For those that are not, give the correct formula.
(a) Ca_2O (c) Fe_2O_5
(b) $SrCl_2$ (d) Li_2O

Coulomb's Law

34. Sodium ion, Na^+, forms ionic compounds with fluoride, F^-, and iodide, I^-. You know that the radii of these ions are: $Na^+ = 116$ pm; $F^- = 119$ pm; and $I^- = 206$ pm. In which ionic compound, NaF or NaI, are the forces of attraction between cation and anion stronger? Explain your answer.

35. Consider the two ionic compounds NaCl and CaO. In which compound are the cation-anion attractive forces stronger? Explain your answer.

Naming Ionic Compounds

36. Name each of the following ionic compounds:
(a) K_2S (c) $(NH_4)_3PO_4$
(b) $NiSO_4$ (d) $Ca(ClO)_2$

37. Name each of the following ionic compounds:
(a) $Ca(CH_3CO_2)_2$ (c) $Al(OH)_3$
(b) $Co_2(SO_4)_3$ (d) KH_2PO_4

38. Give the formula for each of the following ionic compounds:
(a) ammonium carbonate
(b) calcium iodide
(c) copper(II) bromide
(d) aluminum phosphate
(e) silver(I) acetate

39. Give the formula for each of the following ionic compounds:
(a) calcium hydrogen carbonate
(b) potassium permanganate
(c) magnesium perchlorate
(d) potassium hydrogen phosphate
(e) sodium sulfite

40. Write the formulas for all the ionic compounds that can be made by combining each of the cations with each of the anions listed here. Name each compound formed.

Cations	Anions
K^+	CO_3^{2-}
Ba^{2+}	Br^-
NH_4^+	NO_3^-

41. Write the formulas for all the ionic compounds that can be made by combining each of the cations with each of the anions listed here. Name each compound formed.

Cations	Anions
Co^{2+}	SO_4^{2-}
Mg^{2+}	PO_4^{3-}
Li^+	S^{2-}

Naming Binary, Nonmetal Compounds

42. Give the name for each of the following binary, nonionic compounds:
(a) NF_3 (c) BBr_3
(b) HI (d) PF_5

43. Give the name for each of the following binary, nonionic compounds:
(a) N_2O_3 (c) OF_2
(b) P_4S_3 (d) XeF_4

44. Give the formula for each of the following nonmetal compounds:
(a) sulfur dichloride
(b) dinitrogen pentaoxide
(c) silicon tetrachloride
(d) diboron trioxide (commonly called boric oxide)

45. Give the formula for each of the following nonmetal compounds:
(a) bromine trifluoride
(b) xenon difluoride
(c) hydrazine
(d) diphosphorus tetrafluoride

Atoms, Elements, and the Mole

46. Calculate the mass, in grams, of
(a) 2.5 mol of boron (c) 0.015 mol of oxygen
(b) 1.25×10^{-3} mol of iron (d) 653 mol of helium

47. Calculate the mass, in grams, of
(a) 6.03 mol of gold (c) 0.045 mol of uranium
(b) 15.6 mol of Ne (d) 3.63×10^{-4} mol of Pu

48. Calculate the number of moles represented by each of the following:
(a) 127.08 g of Cu (c) 5.0 mg of americium
(b) 0.012 g of potassium (d) 6.75 g of Al

49. Calculate the number of moles represented by each of the following:
(a) 16.0 g of Na (c) 0.0034 g of platinum
(b) 0.876 g of arsenic (d) 0.983 g of Xe

50. Superman comes from the planet Krypton. If you have 0.00789 g of the gaseous element krypton how many moles does this represent? How many atoms?

51. The recommended daily allowance (RDA) of iron in your diet is 15 mg. How many moles is this? How many atoms?

52. What is the average mass of one copper atom?

53. What is the average mass of one atom of titanium?

54. In an experiment, you need 0.125 mol of sodium metal. Sodium can be cut easily with a knife (Figure 2.13), so if you cut out a block of sodium, what should the volume of the block be in cubic centimeters? If you cut a perfect cube, what is the length of the edge of the cube? (The density of sodium is 0.968 g/cm^3.)

55. Dilithium is the fuel for the Starship *Enterprise.* Because its density is quite low, however, you need a large space to store a large mass. To estimate the volume required, we shall use the element lithium. If you need 256 mol for an interplanetary trip, what must the volume of a piece of lithium

be? If the piece of lithium is a cube, what is the dimension of an edge of the cube? (The density for the element lithium is 0.534 g/cm^3 at 20 °C.)

Molecules, Compounds, and the Mole

56. Calculate the molar mass of each of the following compounds:
(a) Fe_2O_3, iron(III) oxide
(b) BCl_3, boron trichloride
(c) $Ni(NO_3)_2 \cdot 6\ H_2O$, nickel(II) nitrate hexahydrate
(d) $C_6H_8O_6$, ascorbic acid (vitamin C)

57. Calculate the molar mass of each of the following compounds:
(a) $Fe(C_6H_{11}O_7)_2$, iron(II) gluconate, a dietary supplement
(b) $CH_3CH_2CH_2CH_2SH$, butanethiol, has a skunk-like odor
(c) $C_{20}H_{24}N_2O_2$, quinine, used as an antimalarial drug

58. How many moles are represented by 1.00 g of each of the following compounds?
(a) C_3H_7OH, propanol, rubbing alcohol
(b) $C_{11}H_{16}O_2$, an antioxidant in foods, also known as BHA (butylated hydroxyanisole)
(c) $MgSO_4 \cdot 7\ H_2O$, magnesium sulfate heptahydrate (Epsom salt)

59. Assume you have 0.250 g of each of the following compounds. How many moles of each are present?
(a) $C_9H_8O_4$, aspirin
(b) $C_{14}H_{10}O_4$, benzoyl peroxide, used in acne medications
(c) $Pt(NH_3)_2Cl_2$, cisplatin, a cancer chemotherapy agent

60. Acetonitrile, CH_3CN, was found in the coma of Comet Hale-Bopp in 1997 (page 108). If you have 2.50 kg of acetonitrile, how many moles of the compound are present?

61. Acetone, $(CH_3)_2CO$, is an important industrial solvent. In 1996, 1260 million kg of this organic compound was produced. How many moles does this represent?

62. An Alka-Seltzer tablet contains 324 mg of aspirin ($C_9H_8O_4$), 1904 mg of $NaHCO_3$, and 1000. mg of citric acid ($C_6H_8O_7$). (The last two compounds react with each other to provide the "fizz," bubbles of CO_2, when the tablet is put into water.)
(a) Calculate the number of moles of each substance in the tablet.
(b) If you take one tablet, how many molecules of aspirin are you consuming?

63. Sulfur trioxide, SO_3, is made industrially in enormous quantities by combining oxygen and sulfur dioxide, SO_2. The trioxide is not usually isolated but is converted to sulfuric acid. If you have 1.00 kg of sulfur trioxide, how many moles does this represent? How many molecules? How many sulfur atoms? How many oxygen atoms?

Percent Composition

64. Calculate the mass percent of each element in the following compounds.
(a) PbS, lead(II) sulfide, galena
(b) C_3H_8, propane, a hydrocarbon fuel
(c) NH_4NO_3, ammonium nitrate, a fertilizer

65. Calculate the mass percent of each element in the following compounds:
(a) $MgCO_3$, magnesium carbonate
(b) C_6H_5OH, phenol, an organic compound used in some cleaners
(c) $CoCl_2 \cdot 6\ H_2O$, a beautiful red compound (see Figure 3.11).

66. Capsaicin, the compound that gives the hot taste to chili peppers, has the formula $C_{18}H_{27}NO_3$.
(a) Calculate the molar mass.
(b) If you eat 55 mg of capsaicin, how many moles have you consumed?
(c) Calculate the mass percent of each element in the compound.
(d) How many milligrams of carbon are there in 55 mg of capsaicin?

67. The copper-containing compound $Cu(NH_3)_4SO_4 \cdot H_2O$ is a beautiful blue solid. Calculate the molar mass of the compound and the mass percent of each element. What is the mass in grams of copper and of water in 10.5 g of the compound?

Empirical and Molecular Formulas

68. Succinic acid occurs in fungi and lichens. Its empirical formula is $C_2H_3O_2$ and its molar mass is 118.1 g/mol. What is its molecular formula?

69. A compound used in analytical chemistry, dimethylglyoxime, has the empirical formula C_2H_4NO. If its molar mass is 116.1 g/mol, what is the molecular formula of the compound?

70. Acetylene is a colorless gas used as a fuel in welding torches, among other things. It is 92.26% C and 7.74% H. Its molar mass is 26.02 g/mol. Calculate the empirical and molecular formulas of acetylene.

71. A large family of boron-hydrogen compounds, the boron hydrides, have the general formula B_xH_y. (All react with air and burn or explode.) One member of this family contains 88.5% B; the remainder is hydrogen. Which of the following is its empirical formula: BH_2, BH_3, B_2H_5, B_5H_7, or B_5H_{11}?

72. Nitrogen and oxygen form an extensive series of oxides with the general formula N_xO_y. One of them, a blue solid, contains 36.84% N. What is the empirical formula of this oxide?

73. Cumene is a hydrocarbon, a compound composed only of C and H. It is 89.94% carbon, and the molar mass is 120.2 g/mol. What are the empirical and molecular formulas of cumene?

74. Mandelic acid is an organic acid composed of carbon (63.15%), hydrogen (5.30%), and oxygen (31.55%). Its molar mass is 152.14 g/mol. Determine the empirical and molecular formulas of the acid.

75. Nicotine, a poisonous compound found in tobacco leaves,

is 74.0% C, 8.65% H, and 17.35% N. Its molar mass is 162 g/mol. What are the empirical and molecular formulas of nicotine?

76. Cacodyl, a compound containing arsenic, was reported in 1842 by the German chemist Robert Wilhelm Bunsen. It has an almost intolerable garlic-like odor. Its molar mass is 210 g/mol, and it is 22.88% C, 5.76% H, and 71.36% As. Determine its empirical and molecular formulas.

77. The action of bacteria on meat and fish produces a compound called cadaverine. As its name and origin imply, it stinks! (It is also present in bad breath and adds to the odor of urine.) It is 58.77% C, 13.81% H, and 27.40% N. Its molar mass is 102.2 g/mol. Determine the molecular formula of cadaverine.

78. The "alum" used in cooking is potassium aluminum sulfate hydrate, $KAl(SO_4)_2 \cdot x\,H_2O$. To find the value of *x*, you can heat a sample of the compound to drive off all of the water and leave only $KAl(SO_4)_2$. Assume you heat 4.74 g of the hydrated compound and that the sample loses 2.16 g of water. What is the value of *x*?

79. If "Epsom salt," $MgSO_4 \cdot x\,H_2O$, is heated to 250 °C, all the water of hydration is lost. On heating a 1.687-g sample of the hydrate, 0.824 g of $MgSO_4$ remains. How many molecules of water occur per formula unit of $MgSO_4$?

80. A new compound containing xenon and fluorine was isolated by shining sunlight on a mixture of Xe (0.526 g) and F_2 gas. If you isolate 0.678 g of the new compound, what is its empirical formula?

81. Elemental sulfur (1.256 g) is combined with fluorine, F_2, to give a compound with the formula SF_x, a very stable, colorless gas. If you have isolated 5.722 g of SF_x, what is the value of *x*?

Using Formulas

82. What mass of lead(II) sulfide (the mineral galena) contains 2.00 kg of lead?

83. The mineral ilmenite, $FeTiO_3$, is a source of titanium. What quantity of ilmenite, in grams, is required if you wish to obtain 750 g of titanium?

84. Elemental phosphorus is made by heating calcium phosphate with carbon and sand in an electric furnace. What quantity of calcium phosphate, in kilograms, must be used to produce 15.0 kg of phosphorus?

85. Chromium is obtained by heating chromium(III) oxide with carbon. If you want to produce 850 kg of chromium metal, what quantity of Cr_2O_3 (in kilograms) is required?

GENERAL QUESTIONS

86. A drop of water has a volume of about 0.05 mL. How many molecules of water are in a drop of water? (Assume water has a density of 1.00 g/cm^3.)

87. Write the molecular formula and calculate the molar mass for each of the molecules shown here. Which has the larger percentage of nitrogen? Of carbon?

(a) Trinitrotoluene (TNT)

CH_3 | O_2N–C=C(–C–)–C–NO_2 ; H–C=C(–NO_2)–C–H

(b) Serine, an essential amino acid

$$HO{-}\underset{H}{\overset{H}{C}}{-}\underset{NH_2}{\overset{H}{C}}{-}CO_2H$$

88. Your doctor has diagnosed you as being anemic, that is, as having too little iron in your blood. At the drugstore you find two iron-containing dietary supplements, one with iron(II) sulfate, $FeSO_4$, and the other with iron(II) gluconate, $Fe(C_6H_{11}O_7)_2$. If you take 100. mg of each compound, which delivers more iron?

89. Malic acid, an organic acid found in apples, contains C, H, and O in the following ratios: $C_1H_{1.50}O_{1.25}$. What is the empirical formula of malic acid?

90. Spinach is high in iron (2 mg/90-g serving). It is also a source of the oxalate ion, $C_2O_4^{2-}$, however, which combines with iron ions to form iron oxalate, a substance that prevents your body from absorbing the iron. Analysis of a 0.109-g sample of iron oxalate shows that it contains 17.46% iron. What is the empirical formula of the compound?

91. Write the molecular formula and calculate the molar mass for the molecule shown here. What are the weight percentages of carbon, hydrogen, and sulfur in the compound? How many moles does a 10.0-g sample of the compound represent? How many grams of sulfur are in 10.0 g of the compound? (The molecule is called dimethyl sulfoxide, and it is commonly used as a solvent.)

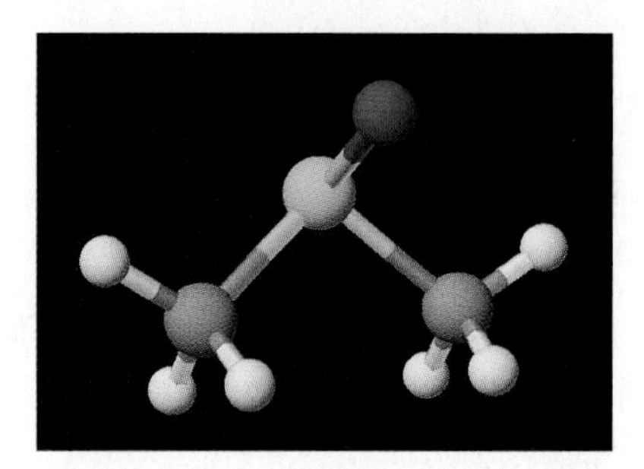

A ball-and-stick model of dimethyl sulfoxide.

92. The molecule shown here, β-D-ribose, is a member of the family of compounds called carbohydrates. If 10.0 g of β-D-ribose is decomposed to carbon and water, how many grams of water is obtained?

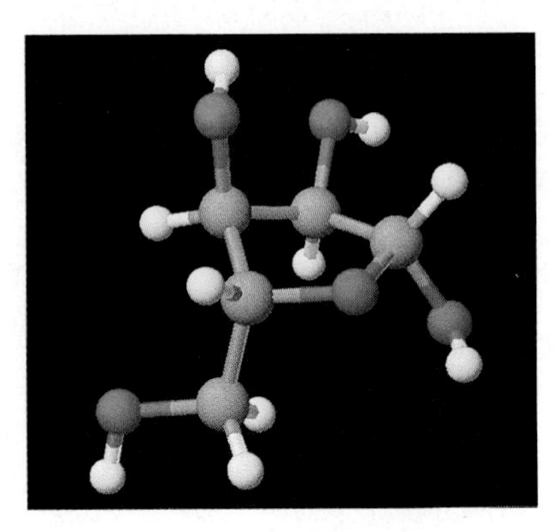

A ball-and-stick model of β-D-ribose.

93. Saccharin is over 300 times sweeter than sugar. It was first made in 1897, a time when it was common practice for chemists to record the taste of any new substances they synthesized.
   (a) Write the molecular formula for the compound.
   (b) If you ingest 125 mg of saccharin, how many moles is this?
   (c) What mass of sulfur is contained in 125 mg of saccharin?

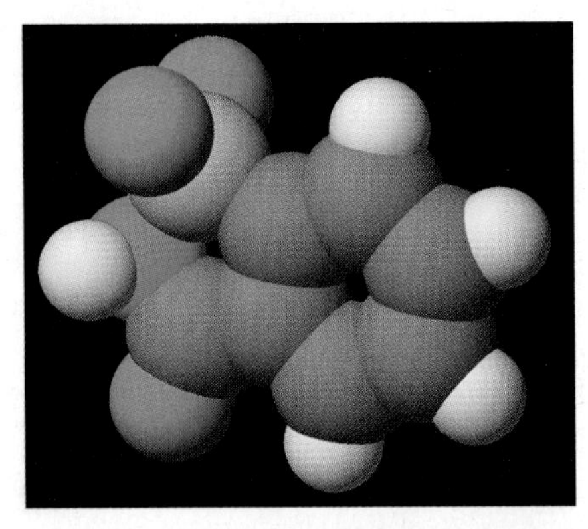

A space-filling model of saccharin.

94. Which of the following pairs of elements are likely to form ionic compounds when they react with each other? Write appropriate formulas for the ionic compounds you expect to form, and give the name of each.
   (a) chlorine and bromine
   (b) nitrogen and bromine
   (c) lithium and sulfur
   (d) indium and oxygen
   (e) sodium and argon
   (f) sulfur and bromine
   (g) magnesium and fluorine

95. Name each of the following compounds, and tell which ones are best described as ionic:
   (a) ClF_3
   (b) NCl_3
   (c) $CaSO_4$
   (d) $Mg(NO_3)_2$
   (e) XeF_4
   (f) OF_2
   (g) NaI
   (h) Al_2S_3
   (i) PCl_5
   (j) K_3PO_4

96. Write the formula for each of the following compounds, and tell which ones are best described as ionic:
   (a) sodium hypochlorite
   (b) boron triiodide
   (c) aluminum perchlorate
   (d) calcium acetate
   (e) potassium permanganate
   (f) ammonium sulfite
   (g) potassium dihydrogen phosphate
   (h) disulfur dichloride
   (i) chlorine trifluoride
   (j) phosphorus trifluoride

97. Fluorocarbonyl hypofluorite is composed of 14.6% C, 39.0% O, and 46.3% F. If the molar mass of the compound is 82 g/mol, determine the empirical and molecular formulas of the compound.

98. Azulene, a beautiful blue hydrocarbon, has 93.71% C and a molar mass of 128.16 g/mol. What are the empirical and molecular formulas of azulene?

99. A major oil company has used a gasoline additive called MMT to boost the octane rating of its gasoline. What is the empirical formula of MMT if it is 49.5% C, 3.2% H, 22.0% O, and 25.2% Mn?

100. Direct reaction of iodine (I_2) and chlorine (Cl_2) produces an iodine chloride, I_xCl_y, a bright yellow solid. If you completely used up 0.678 g of iodine, and produced 1.246 g of I_xCl_y, what is the empirical formula of the compound? A later experiment showed the molar mass of I_xCl_y was 467 g/mol. What is the molecular formula of the compound?

101. Pepto-Bismol, which helps provide soothing relief for an upset stomach, contains 300. mg of bismuth subsalicylate, $C_{21}H_{15}Bi_3O_{12}$, per tablet. If you take two tablets for your stomach distress, how many moles of the "active ingredient" are you taking? How many grams of Bi are you consuming in two tablets?

102. Iron pyrite, often called "fool's gold," has the formula FeS_2 (see photo). If you could convert 15.8 kg of iron pyrite to iron metal, how many kilograms of the metal do you obtain?

A sample of iron pyrite, or "fool's gold." Notice the cubic nature of the crystals. (See also Figure 1.4) (*C. D. Winters*)

103. Stibnite, Sb_2S_3, is a dark gray mineral from which antimony metal is obtained. If you have 1.00 kg of an ore that contains 10.6% antimony, what mass of Sb_2S_3 (in grams) is in the ore?

104. Which of the following is impossible?
(a) Silver foil that is 1.2×10^{-4} m thick.
(b) A sample of potassium that contains 1.784×10^{24} atoms
(c) A gold coin of mass 1.23×10^{-3} kg
(d) 3.43×10^{-27} mol of S_8

105. Transition metals can combine with carbon monoxide (CO) to form compounds such as $Fe(CO)_5$ and $Co_2(CO)_8$. Assume that you combine 0.125 g of nickel with CO and isolate 0.364 g of $Ni(CO)_x$. What is the value of x?

106. A metal M forms a compound with the formula MCl_4. If the compound is 74.75% chlorine, what is the identity of M?

107. The mass of 2.50 mol of a compound with the formula ECl_4, in which E is a nonmetallic element, is 385 g. What is the molar mass of ECl_4? What is the identity of E?

108. In a reaction 2.04 g of vanadium combines with 1.93 g of sulfur to give a pure compound. What is the empirical formula of the product?

109. The weight percent of oxygen in an oxide that has the formula MO_2 is 15.2%. What is the molar mass of this compound? What element or elements are possible for M?

CONCEPTUAL QUESTIONS

110. An ionic compound can dissolve in water because the cations and anions are attracted to water molecules. The drawing here shows how a cation and a water molecule, which has a negatively charged O atom and positively charged H atoms, can interact. Which of the following cations should be most strongly attracted to water: Na^+, Mg^{2+}, or Al^{3+}? Explain briefly.

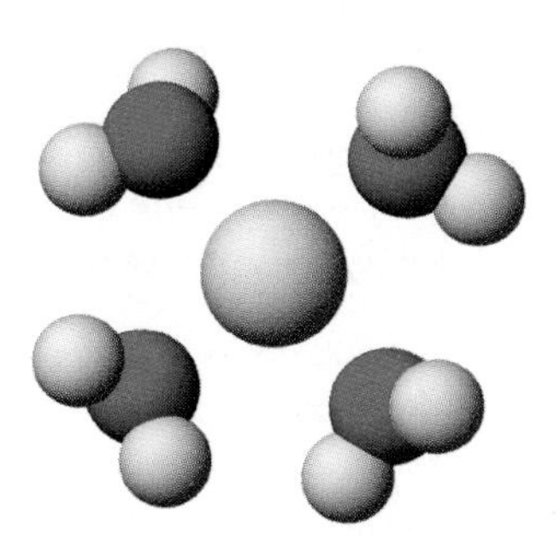

Water molecules interacting with a metal cation, such as Mg^{2+}.

111. When analyzed, an unknown compound gave these experimental results: C, 54.0%; H, 6.00%; and O, 40.0%. Four different students used these values to calculate the empirical formulas shown here. Which answer is correct? Why did some students not get the correct answer?
(a) $C_4H_5O_2$ (c) $C_7H_{10}O_4$
(b) $C_5H_7O_3$ (d) $C_9H_{12}O_5$

112. Two general chemistry students working together in the lab weigh out 0.832 g of $CaCl_2 \cdot 2\ H_2O$ into a crucible. After heating the sample for a short time and allowing the crucible to cool, these students determine that the sample has a mass of 0.739 g. They then do a quick calculation. On the basis of this calculation, what should they do next?
(a) Congratulate themselves on a job well done.
(b) Assume the bottle of $CaCl_2 \cdot 2\ H_2O$ was mislabeled; it actually contained something different.
(c) Heat the crucible again, and then reweigh it.

STRUCTURAL CHEMISTRY

Models of many of the molecules mentioned in this chapter, and in many others, are available at the Saunders site on the World Wide Web or are on the *Saunders Interactive General Chemistry CD-ROM.* Use these to answer the questions that follow:

113. Examine a model of benzene, C_6H_6.
(a) Describe its structure. What are its important features?
(b) Measure the H—C—C and C—C—C angles. Are these the same or different?
(c) Examine the models of graphite and diamond (Figure 2.21). Is the benzene structure related to either of these?

114. Examine the models of the organic acids formic acid, HCO_2H, acetic acid, CH_3CO_2H, and oxalic acid, $H_2C_2O_4$.
(a) What common feature do these molecules share?
(b) Measure the O—C—O angle in each molecule. Are these the same or different?

115. Examine the models of ethane, ethylene, and acetylene.
(a) Give the formula of each.
(b) Measure the H—C—C angle in each. Are they the same or different?
(c) Measure the distance between C atoms. Are they the same or different? Is there any relation to the formula?

CHALLENGING PROBLEMS

116. The elements A and Z combine to produce two different compounds: A_2Z_3 and AZ_2. If 0.15 mole of A_2Z_3 has a mass of 15.9 g and 0.15 mole of AZ_2 has a mass of 9.3 g, what are the atomic masses of A and Z?

117. Although carbon-12 is now used as the standard for atomic masses, this has not always been the case. Early attempts at classification used hydrogen as the standard, with the mass of hydrogen equal to 1.0000 amu. Later attempts defined atomic masses using oxygen (with a mass of 16.0000 amu). In each instance, the atomic masses of the other elements were defined relative to these masses.
(a) If H = 1.0000 amu was used as a standard for atomic masses, what would the atomic mass of oxygen be? What would be the value of Avogadro's number under these circumstances?
(b) Assuming the standard is O = 16.0000 amu, determine the value for the atomic mass of hydrogen and the value of Avogadro's number. (To answer this question, you need more precise data on current atomic masses: H, 1.00797 and O, 15.9994.)

118. Most standard analytical balances can measure accurately to the nearest 0.0001 g. Assume you have weighed out a 2.0000-g sample of carbon. How many atoms are contained in this sample? Assuming the indicated accuracy of the measurement, what is the largest number of atoms that can be present in the sample?

119. An estimation of the radius of a lead atom:
    (a) You are given a cube of lead that is 1.000 cm on each side. The density of lead is 11.35 g/cm^3. How many atoms of lead are contained in the sample?
    (b) Atoms are spherical; therefore, the lead atoms in this sample cannot fill all the available space. As an approximation, assume that 60% of the space of the cube is filled with spherical lead atoms. Calculate the volume of one lead atom from this information. From the calculated volume (V), and the formula $V = 4/3(\pi r^3)$, estimate the radius (r) of a lead atom.

Summary Questions

120. A piece of nickel foil, 0.550 mm thick and 1.25 cm square, is allowed to react with fluorine, F_2, to give a nickel fluoride. (a) How many moles of nickel foil were used? (The density of nickel is 8.908 g/cm^3.) (b) If you isolate 1.261 g of the nickel fluoride, what is its formula? (c) What is its name?

121. Uranium is used as a fuel, primarily in the form of uranium(IV) oxide, in nuclear power plants. This question considers some uranium chemistry.
    (a) A small sample of uranium metal (0.169 g) is heated to between 800 and 900 °C in air to give 0.199 g of a dark green oxide, U_xO_y. How many moles of uranium metal were used? What is the empirical formula of the oxide, U_xO_y? What is the name of the oxide? How many moles of U_xO_y must have been obtained?
    (b) The naturally occurring isotopes of uranium are ^{234}U, ^{235}U, and ^{238}U. Which is the most abundant?
    (c) If the hydrated compound $UO_2(NO_3)_2 \cdot z\ H_2O$ is heated gently, the water of hydration is lost. If you have 0.865 g of the hydrated compound and obtain 0.679 g of $UO_2(NO_3)_2$ on heating, how many molecules of water of hydration are in each formula unit of the original compound? (The oxide U_xO_y is obtained if the hydrate is heated to temperatures over 800 °C in the air.)

INTERACTIVE GENERAL CHEMISTRY CD-ROM

Screen 3.9 Exploring Coulomb's Law

Place the movable sphere (an ion) about 2 cm to the left of the stationary sphere (another ion). Give the left sphere a charge of 1+ and the right sphere a charge of 1−. (Use the red arrows to raise or lower the ion charge as appropriate.) The gray arrows are pointing at one another, indicating that the ions are attracted to one another. The size of the arrow indicates the magnitude of the force of attraction.

(a) Give one of the spheres a charge of 0. What do you observe?
(b) Raise the charge on one ion by 1 (e.g., increase the cation charge from 1+ to 2+). Now what do you observe about the magnitude of the interaction between the ions?
(c) Raise the ion charge by yet another unit. (The cation is now 3+, and the anion is 1−, for example.) What do you observe about the magnitude of the attraction? What is the relation between ion charge and the magnitude of the attraction between ions?
(d) Now give one ion a charge of 2+ and the other ion a charge of 2−. How does the force of attraction compare with the force of attraction when one ion is 1+ and the other is 1− ?
(e) Give one ion a charge of 2+ and the other a charge of 2 −, and place them about 2 cm apart. Move the left ion until the ions are about 4 cm apart (this is the limit of movement of the left ion). Compare the magnitude of the force of attraction at the two positions.
    (i) Move the ions so they are about 1 cm apart. Compare the magnitude of forces with the previous positions of the ion.
    (ii) What can you conclude about the relation between the ion positions and the magnitude of their attraction?
(f) Finally, give each ion the same charge, say 2+. What do you observe?
(g) Summarize the relation between ion charge, the distance between ions, and the forces of attraction and repulsion.

Screen 3.10 Ionic Compounds

Examine the video of the reaction of sodium with chlorine.

(a) The sodium metal for the reaction is removed from a beaker, where chips of the metal are clearly seen under a clear, colorless liquid. Is this liquid water? Explain. (Refer to Screen 2.17.)
(b) What happens to the sodium in the course of the reaction? Does each sodium atom gain or lose one or more electrons? If so, how many electrons? What is the final form of the sodium in sodium chloride?
(c) Is energy involved in this reaction? Is there evidence for the evolution of energy? What is that evidence?

Screens 3.10 and 3.11 Ionic Compounds

Complete the table by placing symbols, formulas, and names in the blanks.

Cation	Anion	Name	Formula
______	______	ammonium bromide	______
Ba^{2+}	______	______	BaS
______	Cl^-	iron(II) chloride	______
______	F^-	______	PbF_2
Al^{3+}	CO_3^{2-}	______	______
______	______	iron(III) oxide	______
______	______	______	$LiClO_4$
______	______	aluminum phosphate	______
______	Br^-	lithium bromide	______
______	______	______	$Ba(NO_3)_2$
Al^{3+}	______	aluminum oxide	______
______	______	iron(III) carbonate	______

Chapter 4

Chemical Equations and Stoichiometry

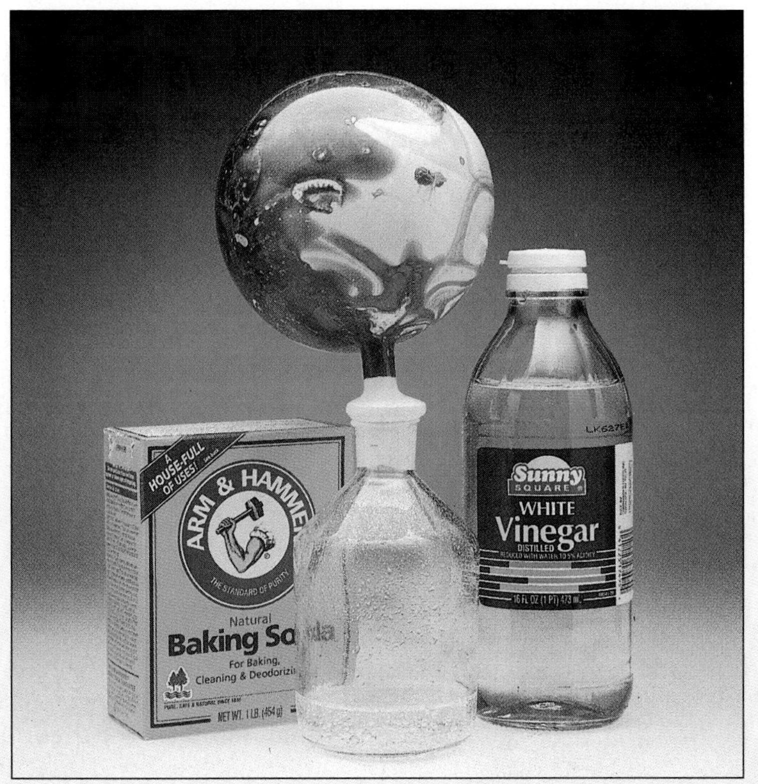

(C. D. Winters)

A Chemical Puzzler

Vinegar (an aqueous solution of acetic acid) and baking soda (sodium hydrogen carbonate) react to produce sodium acetate, water, and carbon dioxide gas.

$$CH_3CO_2H(aq) + NaHCO_3(aq) \longrightarrow NaCH_3CO_2(aq) + H_2O(\ell) + CO_2(g)$$

If you take a handful of baking soda, put it in a glass, and drop in vinegar, carbon dioxide bubbles out of the mixture. As you drop in more and more vinegar, more and more CO_2 gas is evolved. Eventually, though, evolution of CO_2 stops. More vinegar does not lead to more CO_2 gas. Why?

When you think about chemistry you think of chemical reactions. The medieval image of a chemist as an alchemist who mixes chemicals to produce an explosion or fire lingers on. But of course there is much more. The mere act of reading this sentence involves an untold number of chemical reactions in your body. Indeed, every activity of living things depend on carefully regulated chemical reactions. The reaction of glucose with oxygen to produce carbon dioxide and water is the fundamental energy-producing reaction in the human body:

$$C_6H_{12}O_6(s) + 6\ O_2(g) \rightarrow 6\ CO_2(g) + 6\ H_2O(\ell)$$

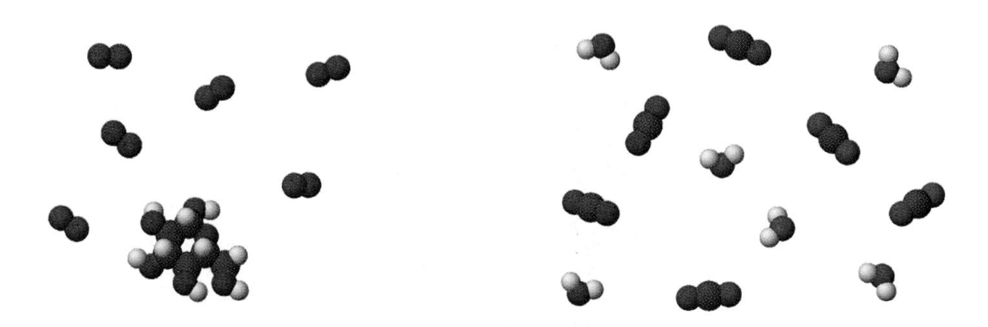

Our objective in this chapter is to begin the subject of **stoichiometry,** the quantitative study of chemical reactions. This will help us determine, for example, what quantity of oxygen is required for complete combustion of a given quantity of glucose or what mass of carbon dioxide can be obtained. Stoichiometry is a part of chemistry that is fundamental to much of what chemists, chemical engineers, biochemists, molecular biologists, geochemists, and many others do.

4.1 CHEMICAL EQUATIONS

You have already seen several examples of chemical reactions: sugar and acid (Figure 3.1), aluminum and bromine (Figure 1, p. 2), and sodium and chlorine (Figure 1.9).

To create the photograph in Figure 4.1, we directed a stream of chlorine gas (Cl_2) onto solid phosphorus (P_4). The mixture burst into flame, and the chemical reaction produced liquid phosphorus trichloride, PCl_3. We can depict this using the balanced chemical equation shown opposite, top, which shows the relative amounts of reactants (the substances combined in the reaction) and products (the substances produced). The physical states of the reactants and products are also often indicated in an equation. The symbol (s) indicates a solid, (g) a gas, and (ℓ) a liquid. What the equation does not show are the conditions of the experiment or if any energy (in the form of heat or light) is involved. Lastly, a chemical equation does not tell you if the reaction happens very quickly or if it takes 100 years.

See the *Saunders Interactive General Chemistry CD-ROM,* Screens 4.1–4.3, for chemical reactions and the principle of matter conservation.

In the 18th century, the great French scientist Antoine Lavoisier introduced the law of conservation of matter, which later became part of Dalton's atomic theory (Section 2.1). Lavoisier showed that matter can neither be created nor destroyed. This means that if you use 10 g of reactants, then, if the reaction is complete, you must end up with 10 g of products. Combined with Dalton's atomic theory, this also means that if 1000 atoms of a particular element react, then those 1000 atoms must appear in the products in some fashion. When applied

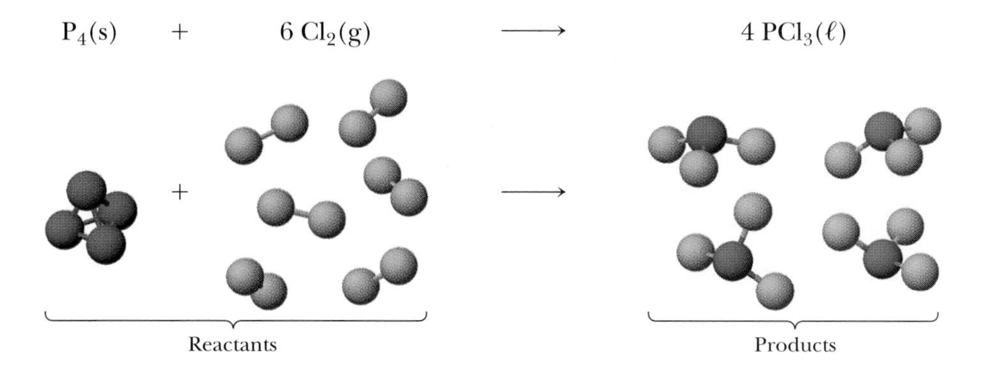

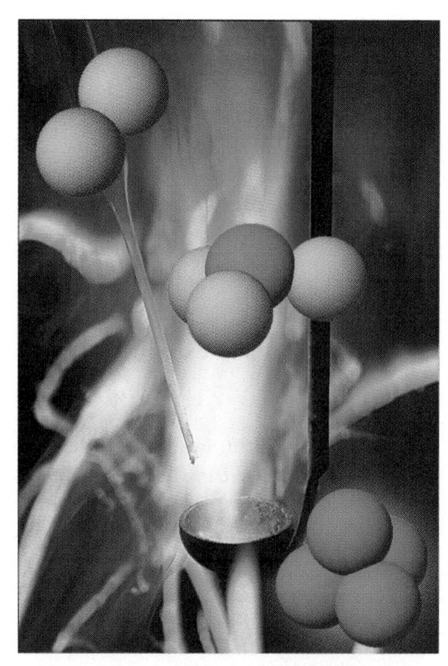
Figure 4.1 A chemical reaction. Phosphorus (P_4) burns in chlorine gas (Cl_2) to produce liquid phosphorus trichloride, PCl_3. (*Photo, C. D. Winters; models, S. M. Young*)

to the reaction of phosphorus and chlorine, the conservation of matter means that 1 molecule of phosphorus (with 4 phosphorus atoms) and 6 diatomic molecules of Cl_2 (with 12 atoms of Cl) are required to produce four molecules of PCl_3. Because each PCl_3 molecule contains 1 P atom and 3 Cl atoms, the four PCl_3 molecules account for 4 P atoms and 12 Cl atoms in the product.

The numbers in front of each formula in a *balanced* chemical equation reflect the principle of the conservation of matter. Review the equation for the reaction of phosphorus and chlorine, and consider the balanced equation for the reaction of aluminum and bromine (Figure 1, page 2).

$$2\ Al(s) + 3\ Br_2(\ell) \longrightarrow Al_2Br_6(s)$$

The number in front of each chemical formula can be read as a number of atoms and molecules (2 atoms of Al and 3 molecules of Br_2 form 1 molecule of Al_2Br_6). It can refer equally as well to number of moles. Thus, the balanced equation for the reaction of aluminum and bromine tells us that 2 mol of solid aluminum reacts with 3 mol of liquid bromine to produce 1 mol of solid Al_2Br_6. The relationship between the quantities of chemical reactants and products is called **stoichiometry** (pronounced "stoy-key-AHM-uh-tree"), and the coefficients in a balanced equation are the **stoichiometric coefficients.**

Balanced chemical equations are fundamentally important in depicting the outcome of chemical reactions and in the quantitative understanding of chemistry. Thus, this chapter has two important goals:

- Learning to balance simple chemical equations (Section 4.2)
- Learning how to apply the conservation of matter to the determination of the quantity of substance produced from given quantities of reactants or the quantity of reactants required to produce a given quantity of product (Sections 4.3 and 4.4)

Exercise 4.1 Chemical Reactions

The reaction of iron with oxygen is shown in Figure 4.2. The equation for the reaction is

$$4\ Fe(s) + 3\ O_2(g) \longrightarrow 2\ Fe_2O_3(s)$$

1. What are the stoichiometric coefficients in this equation?
2. If you were to use 8000 atoms of Fe, how many molecules of O_2 are required to consume the iron completely?

Figure 4.2 Iron reacting with oxygen. A stream of powdered iron is squirted into the flame of a Bunsen burner. The tiny particles of iron react with oxygen to form the iron oxide Fe_2O_3. The flame and the energy released in the reaction heat the particles to incandescence. You have probably seen this reaction before, in Fourth of July fireworks. (*C. D. Winters*)

Historical Perspective

Antoine Laurent Lavoisier (1743–1794)

On Monday, August 7, 1774, the Englishman Joseph Priestley (1733–1804) became the first person to isolate oxygen. He heated solid mercury(II) oxide, HgO, causing the oxide to decompose to mercury and oxygen.

$$2\ HgO(s) \longrightarrow 2\ Hg(\ell) + O_2(g)$$

Priestley did not immediately understand the significance of the discovery, but he mentioned it to the French chemist Antoine Lavoisier in October, 1774. One

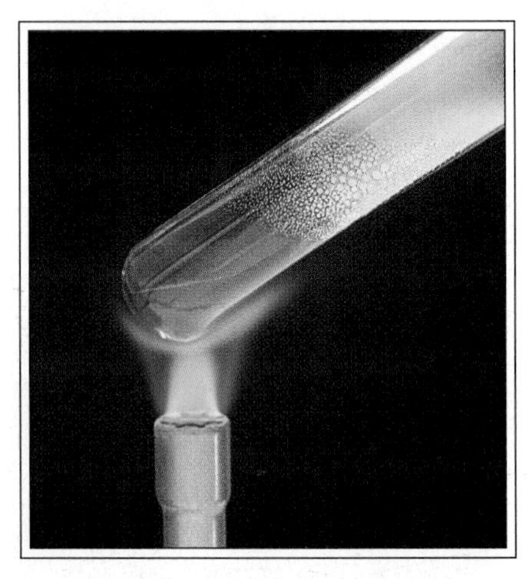

The decomposition of red mercury(II) oxide to give metallic mercury and oxygen. The mercury is seen as a film on the surface of the test tube. (*C. D. Winters*)

Lavoisier with his wife, painted in 1788 by Jacques Louis David. (*© 1986 The Metropolitan Museum of Art, purchase, Charles Wrightsman Gift, honor of Everett Fahy, 1977*)

of Lavoisier's contributions to science was his recognition of the importance of exact scientific measurements and of carefully planned experiments, and he applied these methods to the study of oxygen. From this work he came to believe Priestley's gas was present in all acids and so he named it "oxygen," from the Greek words meaning "to form an acid." In addition, Lavoisier observed that the heat produced by a guinea pig when exhaling a given amount of carbon dioxide is similar to the quantity of heat produced by burning carbon to give the same amount of carbon dioxide. From this and other experiments he concluded that "Respiration is a combustion, slow it is true, but otherwise perfectly similar to that of charcoal." Although he did not understand the details of the process, this was an important step in the development of biochemistry.

Lavoisier was a prodigious scientist and the principles of naming chemical substances that he introduced are still in use today. Furthermore, he wrote a textbook in which he for the first time applied the principles of the conservation of matter to chemistry and used the idea to write early versions of chemical equations.

Because Lavoisier was an aristocrat, he came under suspicion during the Reign of Terror of the French Revolution. He was an investor in the Ferme Générale, the infamous tax-collecting organization in 18th-century France. Tobacco was a monopoly product of the Ferme Générale, and it was common to cheat the purchaser by adding water to the tobacco, a practice that Lavoisier opposed. Nonetheless, because of his involvement with the Ferme, his career was cut short by the guillotine on May 8, 1794, on the charge of "adding water to the people's tobacco." *

*For an account of Lavoisier's life and his friendship with Benjamin Franklin, see Stephen Jay Gould: "The passion of Antoine Lavoisier." *Bully for Brontosaurus,* New York, Norton, 1991.

4.2 BALANCING CHEMICAL EQUATIONS

A chemical equation for a reaction must be balanced before useful quantitative information can be obtained about the reaction. Balancing an equation ensures that the same number of atoms of each element appear on both sides of the equation. Many chemical equations can be balanced by trial and error, although some will involve more trial than others.

***Figure* 4.3 Magnesium burning in air.** A piece of magnesium ribbon burns in air to give the white solid, magnesium oxide, MgO. (*Photo, C. D. Winters; model, S. M. Young*)

One general class of chemical reactions is the reaction of metals or nonmetals with oxygen to give oxides of the general formula M_xO_y. For example, iron can react with oxygen to give iron(III) oxide (Figure 4.2),

$$4\ Fe(s) + 3\ O_2(g) \longrightarrow 2\ Fe_2O_3(s)$$

magnesium gives magnesium oxide (Figure 4.3),

$$2\ Mg(s) + O_2(s) \longrightarrow 2\ MgO(s)$$

and phosphorus, P_4, reacts vigorously with oxygen to give tetraphosphorus decaoxide, P_4O_{10} (Figure 4.4):

$$P_4(s) + 5\ O_2(g) \longrightarrow P_4O_{10}(s)$$

These equations are balanced as we have written them because the same number of metal or phosphorus atoms and oxygen atoms occur on each side of the equation.

The **combustion,** or burning, of a fuel in oxygen is accompanied by the evolution of heat and light. You are familiar with combustion reactions such as the "combustion" of glucose in your body (page 148) or the burning of octane, C_8H_{18}, a component of gasoline, in an automobile engine:

$$2\ C_8H_{18}(g) + 25\ O_2(g) \longrightarrow 16\ CO_2(g) + 18\ H_2O(\ell)$$

In all combustion reactions involving oxygen, some or all the elements in the reactants end up as compounds containing oxygen, that is, as oxides. For hydrocarbons (compounds containing only C and H) the products of complete combustion are always carbon dioxide and water.

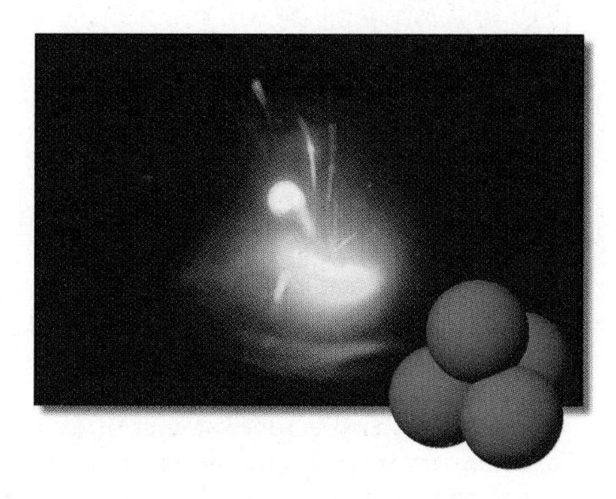

***Figure* 4.4 White phosphorus reacts with oxygen.** In the presence of excess oxygen, the product is tetraphosphorus decaoxide. To prevent oxidation, white phosphorus, P_4, is usually stored underwater (Figure 1.4). (*Photo, C. D. Winters; model, S. M. Young*)

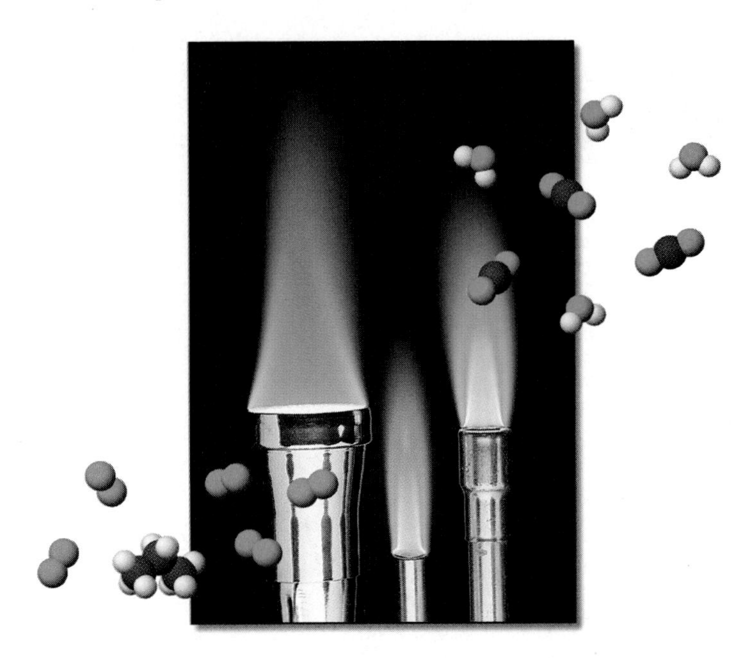

Figure 4.5 A combustion reaction. Here propane burns in air to give carbon dioxide and water. The balanced chemical equation informs us that one molecule of propane will combine with five molecules of oxygen to give three molecules of carbon dioxide and four molecules of water. (*Photo, C. D. Winters; model, S. M. Young*)

As an example of equation balancing, let us write the balanced equation for the complete "combustion" of propane, C_3H_8.

Step 1. *Write the correct formulas for the reactants and products.*

$$C_3H_8(g) + O_2(g) \xrightarrow{\text{unbalanced equation}} CO_2(g) + H_2O(\ell)$$

Here propane and oxygen are the reactants, and carbon dioxide and water are the products (Figure 4.5).

Step 2. *Balance the number of C atoms.* In combustion reactions it is usually best to balance the carbon atoms first and leave the oxygen atoms to the end (because the oxygen atoms are not all found in one compound). In this case three carbon atoms are in the reactants, so three must occur in the products. Three CO_2 molecules are therefore required on the right side:

$$C_3H_8(g) + O_2(g) \longrightarrow 3\ CO_2(g) + H_2O(\ell)$$

Step 3. *Balance the number of H atoms.* Eight H atoms are in the reactants. Each molecule of water has two hydrogen atoms, so four molecules of water account for the required eight hydrogen atoms on the right side:

$$C_3H_8(g) + O_2(g) \longrightarrow 3\ CO_2(g) + 4\ H_2O(\ell)$$

Step 4. *Balance the number of O atoms.* Ten oxygen atoms are on the right side ($3 \times 2 = 6$ in CO_2 plus $4 \times 1 = 4$ in water). Therefore, five O_2 molecules are needed to supply the required ten oxygen atoms:

$$C_3H_8(g) + 5\ O_2(g) \longrightarrow 3\ CO_2(g) + 4\ H_2O(\ell)$$

Step 5. *Verify that the number of atoms of each element is balanced.* The equation shows three carbon atoms, eight hydrogen atoms, and ten oxygen atoms on each side.

The balancing of equations by trial and error is successful if you are organized in your approach. *All equations that have the correct formulas for all reactants and products can be balanced.* Finally, it is important to understand that subscripts in the formulas of reactants and products cannot be changed to balance equations. These subscripts identify these substances, and changing them changes the identity of the substance. For example, you cannot change CO_2 to CO to balance an equation because carbon monoxide, CO, and carbon dioxide, CO_2, are different compounds.

Example 4.1 Balancing the Equation for a Combustion Reaction

See the *Saunders Interactive General Chemistry CD-ROM,* Screen 4.4, Balancing Chemical Equations.

Write the balanced equation for the combustion of butane, C_4H_{10}.

Solution

Step 1. *Write the correct formulas for reactants and products.* As is always the case for compounds containing only carbon and hydrogen, the products of complete combustion are CO_2 and H_2O. Therefore, the unbalanced equation for the combustion is

$$C_4H_{10}(g) + O_2(g) \longrightarrow CO_2(g) + H_2O(\ell)$$

Step 2. *Balance the C atoms.* Four carbon atoms in butane require the production of four CO_2 molecules:

$$C_4H_{10}(g) + O_2(g) \longrightarrow 4\ CO_2(g) + H_2O(\ell)$$

Step 3. *Balance the H atoms.* There are ten hydrogen atoms on the left, so five molecules of H_2O, each having two hydrogen atoms, are required on the right:

$$C_4H_{10}(g) + O_2(g) \longrightarrow 4\ CO_2(g) + 5\ H_2O(\ell)$$

Step 4. *Balance the O atoms.* As the reaction stands after Step 3, there are two oxygen atoms on the left side and 13 on the right ($4 \times 2 = 8$ in CO_2 plus $5 \times 1 = 5$ in H_2O). That is, there is an even number of O atoms on the left and an odd number on the right. Because there cannot be an odd number of O atoms on the left (O atoms are paired in O_2 molecules), multiply each coefficient on both sides of the equation by two so that an even number of oxygen atoms (26) now occurs on the right side:

$$2\ C_4H_{10}(g) + \_\_\ O_2(g) \longrightarrow 8\ CO_2(g) + 10\ H_2O(\ell)$$

Now the oxygen atoms can be balanced by having 13 O_2 molecules on the left side of the equation:

$$2\ C_4H_{10}(g) + 13\ O_2(g) \longrightarrow 8\ CO_2(g) + 10\ H_2O(\ell)$$

Step 5. *Verify the result.* We see that 8 carbon atoms, 20 hydrogen atoms, and 26 oxygen atoms occur on each side of the equation.

Exercise 4.2 Balancing the Equation for a Combustion Reaction

1. Pentane, C_5H_{12}, can burn completely in air to give carbon dioxide and water. Write a balanced equation for this combustion reaction.
2. Write a balanced chemical equation for the complete combustion of tetraethyllead, $Pb(C_2H_5)_4$ (which was used until recently as a gasoline additive). The products of combustion are $PbO(s)$, $H_2O(\ell)$, and $CO_2(g)$.

4.3 MASS RELATIONSHIPS IN CHEMICAL REACTIONS: STOICHIOMETRY

See the *Saunders Interactive General Chemistry CD-ROM,* Screens 5.1–5.3, for a discussion of stoichiometry.

A balanced chemical equation shows the quantitative relationship between reactants and products in a chemical reaction. Let us apply this concept to the reaction of phosphorus and chlorine (see Figure 4.1).

$$P_4(s) + 6\ Cl_2(g) \longrightarrow 4\ PCl_3(\ell)$$

Suppose you use 1.00 mol of phosphorus (P_4, 124 g/mol) in this reaction. The balanced equation shows that 6.00 mol, or 425 g, of Cl_2 must be used and that 4.00 mol, or 549 g, of PCl_3 can be produced.

Reactants			Product
$P_4(s)$	$6\ Cl_2(g)$	$\longrightarrow$	$4\ PCl_3(\ell)$
1 molecule	6 molecules	$\longrightarrow$	4 molecules
1.00 mol	6.00 mol	$\longrightarrow$	4.00 mol
124 g	425 g	$\longrightarrow$	549 g (= 124 g + 425 g)

The balanced equation for the reaction of phosphorus and chlorine applies no matter how much P_4 is used. If 0.0100 mol of P_4 (1.24 g) is used, then 0.0600 mol of Cl_2 (4.25 g) is required, and 0.0400 mol of PCl_3 (5.49 g) can form. You can confirm by experiment that if 1.24 g of P_4 and 4.25 g of Cl_2 are used, then 5.49 g (= 1.24 g + 4.25 g) of PCl_3 can be produced.

Following this line of reasoning further, suppose you have a piece of phosphorus with a mass of 1.45 g. What mass of Cl_2 is required if all the phosphorus is to react? The following procedure leads to the solution.

Step 1. *Write the balanced equation* (using correct formulas for reactants and products). This is always the first step when dealing with chemical reactions.

$$P_4(s) + 6\ Cl_2(g) \longrightarrow 4\ PCl_3(\ell)$$

Step 2. *Calculate moles from masses.* From the mass of P_4, calculate the number of moles of P_4 available. This must be done because the balanced equation shows mole relationships, not mass relationships.

$$1.45\ \text{g}\ P_4 \cdot \frac{1\ \text{mol}\ P_4}{123.9\ \text{g}\ P_4} = 0.0117\ \text{mol}\ P_4$$

Step 3. *Use a stoichiometric factor.* The number of moles available of one reactant (P_4) must be related to the number of moles required of the other reactant (Cl_2).

$$0.0117\ \text{mol}\ P_4 \cdot \frac{6\ \text{mol}\ Cl_2\ \text{required}}{1\ \text{mol}\ P_4\ \text{available}} = 0.0702\ \text{mol}\ Cl_2\ \text{required}$$

↑ **a stoichiometric factor** (from the balanced equation)

To perform this calculation the number of moles of phosphorus available has been multiplied by a **stoichiometric factor,** a *mole ratio factor relating moles of the required reactant to moles of the other reactant.* The stoichiometric factor comes *directly* from the coefficients in the balanced chemical equation. This is the reason you must balance chemical equations before proceeding with calculations. Here the calculation shows that 0.0702 mol of Cl_2 is required to react with all the available phosphorus.

Step 4. *Calculate mass from moles.* From the number of moles of Cl_2 calculated in Step 3, you can now calculate the mass of Cl_2 required.

$$0.0702 \text{ mol Cl}_2 \cdot \frac{70.91 \text{ g Cl}_2}{1 \text{ mol Cl}_2} = 4.98 \text{ g Cl}_2$$

Because the objective of this example was to find the mass of Cl_2 required, the problem is solved.

You may also want to know the mass of PCl_3 that can be produced in the reaction of 1.45 g of phosphorus with the required mass of chlorine (4.98 g) when the reaction goes to completion, that is, when all of at least one of the reactants has been used completely. Because matter is conserved, the answer can be obtained by adding the masses of P_4 and Cl_2 used (giving 1.45 g + 4.98 g = 6.43 g of PCl_3 produced). Alternatively, Steps 3 and 4 can be repeated, but with the appropriate stoichiometric factor and molar mass.

Step 3′. *Use a stoichiometric factor.* Relate the number of moles of available P_4 to the number of moles of PCl_3 that can be produced.

$$0.0117 \text{ mol P}_4 \cdot \frac{4 \text{ mol PCl}_3 \text{ produced}}{1 \text{ mol P}_4 \text{ available}} = 0.0468 \text{ mol PCl}_3 \text{ produced}$$

↑

a stoichiometric factor (from the balanced equation)

Step 4′. *Calculate mass from moles.* Convert moles of PCl_3 produced to a mass in grams.

$$0.0468 \text{ mol PCl}_3 \text{ produced} \cdot \frac{137.3 \text{ g PCl}_3}{1 \text{ mol PCl}_3} = 6.43 \text{ g PCl}_3$$

Example 4.2 Mass Relations in Reactions

Propane, C_3H_8, can be used conveniently as a fuel in your home, car, or barbecue grill because it is easily liquefied and transported. If 454 g (1.00 lb) of propane is burned, what mass of oxygen (in grams) is required for complete combustion, and what masses of carbon dioxide and water (in grams) are formed?

Solution

Step 1. *Remember that the first step must always be to write a balanced equation.*

$$C_3H_8(g) + 5\ O_2(g) \longrightarrow 3\ CO_2(g) + 4\ H_2O(\ell)$$

Having balanced the equation, you can perform the stoichiometric calculations.

Problem-Solving Tips and Ideas 4.1 suggests that you proceed in the following way:

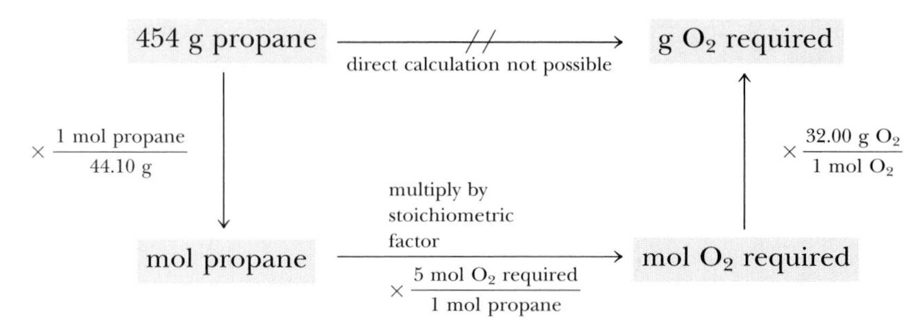

A direct calculation of the mass of O_2 required from the mass of propane is not possible. Instead, first find the moles of propane available, then relate this to moles of O_2 required using the stoichiometric factor. Finally, find the mass of O_2 required from the moles of O_2.

Step 2. *Convert the mass of propane to moles.*

$$454 \text{ g } C_3H_8 \cdot \frac{1 \text{ mol } C_3H_8}{44.10 \text{ g } C_3H_8} = 10.3 \text{ mol } C_3H_8$$

Step 3. *Use the stoichiometric factor to calculate moles of O_2 required.*

$$10.3 \text{ mol } C_3H_8 \cdot \frac{5 \text{ mol } O_2 \text{ required}}{1 \text{ mol } C_3H_8 \text{ available}} = 51.5 \text{ mol } O_2 \text{ required}$$

Step 4. *Convert the number of moles of O_2 required to mass in grams.*

$$51.5 \text{ mol } O_2 \text{ required} \cdot \frac{32.00 \text{ g } O_2}{1 \text{ mol } O_2} = 1650 \text{ g } O_2 \text{ required}$$

Repeat Steps 3 and 4 to find the mass of CO_2 produced in the combustion. First, relate the number of moles of C_3H_8 available to the number of moles of CO_2 produced by using a stoichiometric factor.

$$10.3 \text{ mol } C_3H_8 \text{ available} \cdot \frac{3 \text{ mol } CO_2 \text{ produced}}{1 \text{ mol } C_3H_8 \text{ available}} = 30.9 \text{ mol } CO_2 \text{ produced}$$

Then convert the number of moles of CO_2 produced to the mass in grams.

$$30.9 \text{ mol } CO_2 \cdot \frac{44.00 \text{ g } CO_2}{1 \text{ mol } CO_2} = 1360 \text{ g } CO_2$$

Now, how can you find the mass of H_2O produced? You could go through Steps 3 and 4 again. It is easier, however, to recognize that the total mass of the reactants

$$454 \text{ g } C_3H_8 + 1650 \text{ g } O_2 = 2104 \text{ g of reactants}$$

must be the same as the total mass of products. The mass of water that can be produced is therefore

$$\text{Total mass of products} = 2104 \text{ g} = 1360 \text{ g } CO_2 \text{ produced} + ? \text{ g } H_2O$$

$$\text{Mass of } H_2O \text{ produced} = 744 \text{ g}$$

Exercise 4.3 Weight Relations in Chemical Reactions

What mass of carbon, in grams, can be consumed by 454 g of O_2 in a combustion to give carbon monoxide? What mass of CO is produced?

$$2\ C(s) + O_2(g) \longrightarrow 2\ CO(g)$$

Problem-Solving Tips and Ideas

4.1 Stoichiometry Calculations

You are asked to determine what mass of product can be formed from a given mass of reactant. Keep in mind that it is not possible to calculate the mass of product in a single step. Instead, you must follow a route such as that illustrated here for the reaction of a reactant A to give the product B according to an equation such as $x\,A \longrightarrow y\,B$. Here the mass of reactant A is converted to moles of A. Then, using the stoichiometric factor, you find moles of B. Finally, the mass of B is obtained by multiplying moles of B by its molar mass.

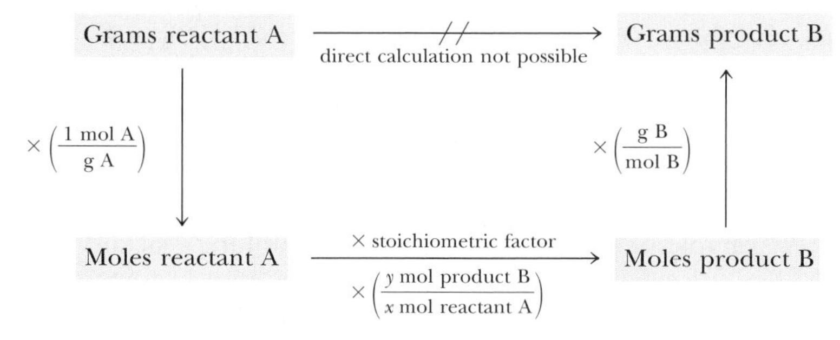

As you practice working with chemical stoichiometry, remember that you will always use a stoichiometric factor at some point.

4.4 REACTIONS IN WHICH ONE REACTANT IS PRESENT IN LIMITED SUPPLY

See the *Saunders Interactive General Chemistry CD-ROM,* Screen 5.4 (Reactions Controlled by the Supply of One Reactant) and Screen 5.5 (Limiting Reactants, The Details).

When carrying out a reaction a chemist, or nature for that matter, rarely supplies the reactants in the exact stoichiometric ratio. Because the goal of a reaction is to produce the largest possible quantity of a useful compound from a given quantity of starting material, it is often the case that a large excess of one reactant is supplied to ensure that the more expensive reactant is completely converted to the desired product. As an example, consider the preparation of cisplatin, $Pt(NH_3)_2Cl_2$, a compound used to treat certain cancers (see "Historical Perspective: Serendipity," page 11):

$$\underset{}{(NH_4)_2PtCl_4(s)} + \underset{\text{ammonia}}{2\ NH_3(aq)} \longrightarrow 2\ NH_4Cl(aq) + \underset{\text{cisplatin}}{Pt(NH_3)_2Cl_2(s)}$$

In this case, it makes sense to combine the more expensive chemical $(NH_4)_2PtCl_4$ (roughly $100 per gram) with a much greater amount of the less expensive chem-

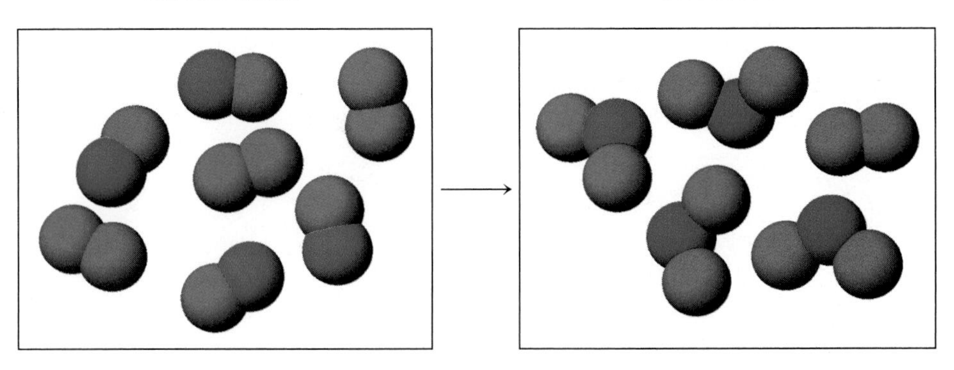

Figure 4.6 A reaction involving a limiting reactant. The reaction occurring is $2\ NO(g) + O_2(g) \longrightarrow 2\ NO_2(g)$. Here 4 NO molecules are present before reaction with 3 O_2 molecules. The 4 NO molecules require only 2 O_2 molecules, however, so 1 O_2 molecule remains after reaction. Which is the limiting reactant?

ical NH_3 (only pennies per gram) than is called for by the balanced equation. Thus, on completion of the reaction, all the $(NH_4)_2PtCl_4$ has been converted to product, and some NH_3 remains. How much $Pt(NH_3)_2Cl_2$ is formed? It depends on the amount of $(NH_4)_2PtCl_4$ present at the start, not on the amount of NH_3, because more NH_3 is present than was required by stoichiometry. A compound such as $(NH_4)_2PtCl_4$ in this example is called the **limiting reactant** because its amount determines, or limits, the amount of product formed (Figure 4.6).

Another example of a reaction that involves a limiting reactant is the manufacture of the pure silicon that is used in computer chips or solar cells (Figure 4.7). The final step in the process is the reduction of purified, liquid silicon tetrachloride with very pure magnesium to give pure elemental silicon.

$$SiCl_4(\ell) + 2\ Mg(s) \longrightarrow Si(s) + 2\ MgCl_2(s)$$

Suppose that 225 g $SiCl_4$ is mixed with 225 g of Mg. Are these reactants mixed in the correct stoichiometric ratio or is one of them in short supply? That is, will one of them limit the quantity of silicon that can be produced? If so, how

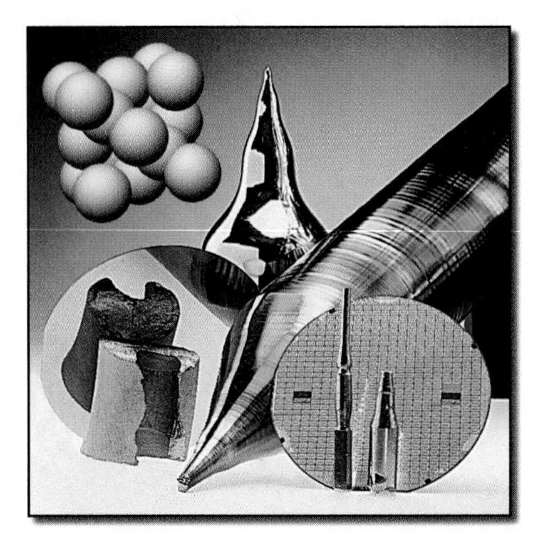

Figure 4.7 Silicon. The manufacture of silicon begins with sand, SiO_2. This is converted to $SiCl_4$ in a chemical reaction, and the $SiCl_4$ is then converted to pure, elemental silicon. (*Photo, C. D. Winters; model, S. M. Young*)

much silicon can be formed if the reaction goes to completion? And how much of the excess reactant is left over when the maximum amount of silicon has been formed?

Because the quantities of both starting materials are given, the first step in answering these questions involves finding the number of moles of each.

$$225 \text{ g } SiCl_4 \cdot \frac{1 \text{ mol } SiCl_4}{169.9 \text{ g } SiCl_4} = 1.32 \text{ mol } SiCl_4 \text{ available}$$

$$225 \text{ g Mg} \cdot \frac{1 \text{ mol Mg}}{24.31 \text{ g Mg}} = 9.26 \text{ mol Mg available}$$

Are these reactants present in the correct stoichiometric ratio as given by the balanced equation?

$$\text{Mole ratio of reactants required by balanced equation} = \frac{2 \text{ mol Mg}}{1 \text{ mol } SiCl_4}$$

$$\text{Mole ratio of reactants actually available} = \frac{9.26 \text{ mol Mg}}{1.32 \text{ mol } SiCl_4} = \frac{7.02 \text{ mol Mg}}{1.00 \text{ mol } SiCl_4}$$

Dividing the moles of Mg available by moles of $SiCl_4$ available shows that the ratio is much larger than the (2 mol Mg/1 mol $SiCl_4$) ratio required by the balanced equation. This means more magnesium is available than is needed to react with all the available $SiCl_4$. Conversely, this means not enough $SiCl_4$ is present to react with all the available Mg. Therefore, *$SiCl_4$ is the limiting reactant.*

Now that $SiCl_4$ has been shown to be the limiting reactant, we can calculate the mass of product expected based on the quantity of $SiCl_4$ available.

$$1.32 \text{ mol } SiCl_4 \cdot \frac{1 \text{ mol Si}}{1 \text{ mol } SiCl_4} \cdot \frac{28.09 \text{ g Si}}{1 \text{ mol Si}} = 37.1 \text{ g Si}$$

Magnesium is the "excess reactant" because more than enough magnesium is available to react with the available silicon tetrachloride, $SiCl_4$. Now we can calculate the quantity of magnesium that remains after all the $SiCl_4$ has been used. To do this, we first need to know the quantity of magnesium that is required to consume all the limiting reactant, $SiCl_4$.

$$1.32 \text{ mol } SiCl_4 \cdot \frac{2 \text{ mol Mg required}}{1 \text{ mol } SiCl_4 \text{ available}} = 2.64 \text{ mol Mg required}$$

Because 9.26 mol of magnesium is available, the number of moles of excess magnesium can be calculated,

$$\begin{aligned}\text{Excess Mg} &= 9.26 \text{ mol Mg available} - 2.64 \text{ mol Mg consumed} \\ &= 6.62 \text{ mol Mg remain}\end{aligned}$$

and then converted to a mass.

$$6.62 \text{ mol Mg} \cdot \frac{24.31 \text{ g Mg}}{1 \text{ mol Mg}} = 161 \text{ g Mg in excess of that required}$$

Finally, because 161 g of magnesium is left over, this means that 225 g − 161 g = 64 g of magnesium has been consumed.

Figure 4.8 A car that uses methanol in its fuel system. Methanol is a very versatile substance. In this van methanol is converted to hydrogen, which is then combined with oxygen in a fuel cell, a type of battery. The fuel cell generates electric energy that runs the car. (*Ballard Power Systems, Inc.*)

Example 4.3 A Reaction Involving a Limiting Reactant

Methanol, CH_3OH, an excellent fuel (Figure 4.8), can be made by the reaction of carbon monoxide and hydrogen.

$$CO(g) + 2\ H_2(g) \longrightarrow \underset{\text{methanol}}{CH_3OH(\ell)}$$

Suppose 356 g of CO is mixed with 65.0 g of H_2. Which is the limiting reactant? What is the maximum mass of methanol that can be formed? What mass of the excess reactant remains after the limiting reactant has been consumed?

Solution As a first step, find the number of moles of each reactant.

$$\text{Moles of CO} = 356\ \text{g CO} \cdot \frac{1\ \text{mol CO}}{28.01\ \text{g CO}} = 12.7\ \text{mol CO}$$

$$\text{Moles of } H_2 = 65.0\ \text{g } H_2 \cdot \frac{1\ \text{mol } H_2}{2.016\ \text{g } H_2} = 32.2\ \text{mol } H_2$$

Are these reactants, in the quantities available, present in a perfect stoichiometric ratio?

$$\frac{\text{Mol } H_2 \text{ available}}{\text{Mol CO available}} = \frac{32.2\ \text{mol } H_2}{12.7\ \text{mol CO}} = \frac{2.54\ \text{mol } H_2}{1.00\ \text{mol CO}}$$

The required mole ratio is 2 mol of H_2 to 1 mole of CO. Clearly more hydrogen is available than is required. It follows then that not enough CO is present to use up all of the hydrogen. *CO is the limiting reactant.*

What is the maximum quantity of CH_3OH that can be formed? This calculation is based on the quantity of limiting reactant available.

$$12.7\ \text{mol CO} \cdot \frac{1\ \text{mol } CH_3OH \text{ formed}}{1\ \text{mol CO available}} \cdot \frac{32.04\ \text{g } CH_3OH}{1\ \text{mol } CH_3OH} = 407\ \text{g } CH_3OH \text{ formed}$$

What quantity of H_2 remains when all the CO has been converted to product? First, we must find the quantity of H_2 required to react with all the CO.

$$12.7\ \text{mol CO} \cdot \frac{2\ \text{mol } H_2}{1\ \text{mol CO}} = 25.4\ \text{mol } H_2 \text{ required}$$

Because 32.2 mol of H_2 is available, but only 25.4 mol is required by the limiting reactant, 6.8 mol of H_2 is in excess. This is equivalent to 14 g of H_2.

$$6.8\ \text{mol } H_2 \cdot \frac{2.02\ \text{g } H_2}{1\ \text{mol } H_2} = 14\ \text{g } H_2 \text{ remain}$$

Example 4.4 Stoichiometry and Limiting Reactants

One way to discover the stoichiometry of a reaction is to add varying amounts of one reactant to a fixed amount of another. Suppose you carry out a series of

Figure 4.9 The zinc-iodine reaction. Metallic zinc, in the form of a powder, and solid, purple iodine (*left*) react vigorously to produce zinc iodide, ZnI_2 (*right*). The purple vapor is excess iodine that is vaporized by the heat of the reaction. Models show the structures of zinc, iodide, and zinc iodide. (*Photos, C. D. Winters; models, S. M. Young*)

experiments in which you add different amounts of zinc to a fixed mass of iodine (Figure 4.9). The results are plotted as mass of product versus mass of zinc.

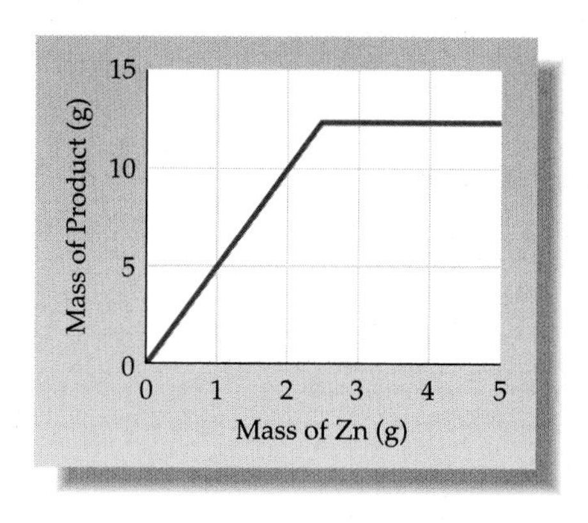

(a) What mass of iodine is used when 2.50 g of Zn is consumed in the reaction?
(b) What is the mole ratio of I_2 to Zn at the point where 2.50 g of Zn is consumed?
(c) What is the empirical formula of the product of the reaction?
(d) What is the balanced equation for the reaction?
(e) When 3 g of Zn is used, which is the limiting reactant, Zn or I_2?

Solution If we work through the first two questions, we can discover the stoichiometry of the reaction.

(a) From the graph you can see that 12.2 g of product is produced when 2.50 g of Zn is consumed. This means that (12.20 g − 2.50 g = 9.70 g) of I_2 must have combined with 2.50 g of Zn.

(b) The quantity of zinc and iodine, in moles, can be calculated.

$$2.50 \text{ g Zn} \cdot \frac{1 \text{ mol Zn}}{65.39 \text{ g Zn}} = 0.0382 \text{ mol Zn}$$

$$9.70 \text{ g I}_2 \cdot \frac{1 \text{ mol I}_2}{253.8 \text{ g I}_2} = 0.0382 \text{ mol I}_2$$

Therefore, the I_2/Zn ratio is 1:1.

(c) Because the number of moles of Zn and I_2 combining is the same, the formula for the product must be ZnI_2. (This formula is expected because zinc is a Group 2B element and typically forms a 2+ ion, whereas iodine is invariably I^- in ionic compounds.)

(d) The balanced equation for the reaction.

$$\text{Zn(s)} + \text{I}_2\text{(s)} \longrightarrow \text{ZnI}_2\text{(s)}$$

(e) Up to 2.50 g of Zn, there is sufficient I_2 present to consume the zinc completely, as indicated by the fact that the quantity of product increases to this point. If more than 2.50 g of zinc is added, the quantity of product is a constant. This must mean that iodine is now the limiting reactant. There is more than enough zinc present, but not enough iodine.

Exercise 4.4 A Reaction Involving a Limiting Reactant

You have 20.0 g of elemental sulfur, S_8, and 160. g of O_2. Which is the limiting reactant in the combustion of S_8 in oxygen to give SO_2 gas? What amount of which reactant (in moles) is left after complete reaction? What mass of SO_2, in grams, is formed in the complete reaction? (*Hint:* Remember first to write the balanced chemical equation for the reaction.)

Exercise 4.5 A Reaction Involving a Limiting Reactant

Titanium tetrachloride, $TiCl_4$, is an important industrial chemical. For example, TiO_2, the material used as a white pigment in paper and paints, can be made from it. Titanium tetrachloride can be made by combining titanium-containing ore (which is often impure TiO_2) with carbon and chlorine.

$$TiO_2(s) + 2\ Cl_2(g) + C(s) \longrightarrow TiCl_4(\ell) + CO_2(g)$$

If one begins with 125 g each of Cl_2 and C, but plenty of TiO_2-containing ore, which is the limiting reactant in this reaction? What quantity of $TiCl_4$, in grams, can be produced?

Problem-Solving Tips and Ideas

4.2 More on Reactions Involving a Limiting Reactant

In the preceding examples involving a limiting reactant, we first calculated moles of each reactant. Next, we compared the ratio of moles available to the stoichiometric factor from the balanced equation. This identified the limiting reactant.

There is another method that some of our students find works well: They calculate the mass of product expected based on each reactant. The limiting reactant is that reactant that gives the smallest quantity of product. For example, refer to the $SiCl_4$ reaction with Mg on page 158. To confirm that $SiCl_4$ is the limiting reactant, calculate the quantity of elemental silicon that can be formed starting with (a) 1.32 mol of $SiCl_4$ and unlimited magnesium or (b) with 9.26 mol of Mg and unlimited $SiCl_4$.

1. Quantity of Si produced from 1.32 mol $SiCl_4$ and unlimited Mg

$$1.32\ \cancel{\text{mol SiCl}_4}\ \text{available} \cdot \frac{1\ \cancel{\text{mol Si}}}{1\ \cancel{\text{mol SiCl}_4}} \cdot \frac{28.09\ \text{g Si}}{1\ \cancel{\text{mol Si}}} = 37.1\ \text{g Si}$$

2. Quantity of Si produced from 9.26 mol Mg and unlimited $SiCl_4$

$$9.26\ \cancel{\text{mol Mg}}\ \text{available} \cdot \frac{1\ \cancel{\text{mol Si}}}{2\ \cancel{\text{mol Mg}}} \cdot \frac{28.09\ \text{g Si}}{1\ \cancel{\text{mol Si}}} = 1.30 \times 10^2\ \text{g Si}$$

Comparing the quantities of silicon produced shows that the available $SiCl_4$ is capable of producing less silicon (37.1 g) than the available Mg (130 g). This confirms the conclusion that silicon tetrachloride, $SiCl_4$, is the limiting reactant.

As a final note, you will find this approach easier to use when there are more than two reactants, each present initially in some designated quantity.

4.5 PERCENT YIELD

The maximum quantity of product that can be obtained from a chemical reaction is the **theoretical yield.** Invariably some waste occurs during the isolation and purification of products, however; no matter how good a chemist you are, you will invariably "lose" small quantities of material along the way. For this reason, the **actual yield** of a compound—the quantity of material you actually obtain in the laboratory or chemical plant—is likely to be less than the theoretical yield. The efficiency of a chemical reaction and the techniques used to obtain the desired compound in pure form can be evaluated by calculating the ratio of the actual yield to the theoretical yield. We call the result the **percent yield** (Figure 4.10).

See the *Saunders Interactive General Chemistry CD-ROM,* Screen 5.6, Percent Yield.

$$\text{Percent yield} = \frac{\text{actual yield}}{\text{theoretical yield}} \times 100\%$$

***Figure* 4.10 Percent yield.** We began with 20 popcorn kernels and found that only 16 of them popped. The percent yield from our "reaction" was (16/20) × 100%, or 80%. (*C. D. Winters*)

Suppose you made aspirin in the laboratory by the following reaction:

$$\underset{\text{salicylic acid}}{C_7H_6O_3(s)} + \underset{\text{acetic anhydride}}{C_4H_6O_3(\ell)} \longrightarrow \underset{\text{aspirin}}{C_9H_8O_4(s)} + \underset{\text{acetic acid}}{CH_3CO_2H(\ell)}$$

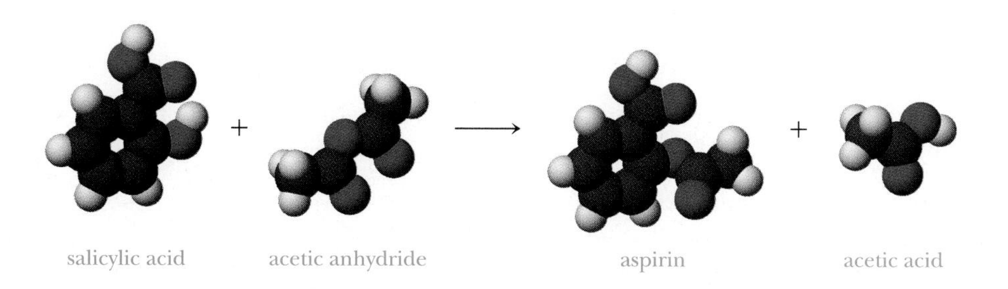

and that you began with 14.4 g of salicylic acid and a stoichiometric excess of acetic anhydride. This means salicylic acid is the limiting reactant. If you obtain 6.26 g of aspirin, what is the percent yield of this product? The first step is to find the number of moles of the limiting reactant, salicylic acid ($C_7H_6O_3$).

$$14.4 \text{ g } \cancel{C_7H_6O_3} \cdot \frac{1 \text{ mol } C_7H_6O_3}{138.1 \text{ g } \cancel{C_7H_6O_3}} = 0.104 \text{ mol } C_7H_6O_3$$

Next, use the stoichiometric factor from the balanced equation to find the number of moles of aspirin expected based on the limiting reactant, $C_7H_6O_3$.

$$0.104 \text{ mol } \cancel{C_7H_6O_3} \cdot \frac{1 \text{ mol aspirin}}{1 \text{ mol } \cancel{C_7H_6O_3}} = 0.104 \text{ mol aspirin}$$

The maximum quantity of aspirin that can be produced—the theoretical yield—is 0.104 mol. Because the quantity you measure in the laboratory is the mass of the product, it is customary to express the theoretical yield as a mass in grams.

$$0.104 \text{ } \cancel{\text{mol aspirin}} \cdot \frac{180.2 \text{ g aspirin}}{1 \text{ } \cancel{\text{mol aspirin}}} = 18.8 \text{ g aspirin}$$

Finally, with the actual yield known to be only 6.26 g, the percent yield of aspirin can be calculated.

$$\text{Percent yield} = \frac{6.26 \text{ g aspirin actually isolated}}{18.8 \text{ g aspirin expected}} \times 100\% = 33.3\% \text{ yield}$$

When a chemist makes a new compound or carries out a reaction, the percent yield is usually reported. This is useful information because it gives other chemists some idea of the quantity of product that can be reasonably expected from the reaction. But be aware that it is often difficult to obtain yields near 100%. Often expensive or time-consuming procedures are required to keep track of every drop of liquid or crumb of solid. In addition, many reactions simply do not go completely to products (as you may know from popping a cup of popcorn kernels). Finally, more complicated reactions can produce more than one set of products from a given set of reactants, which means the yield of the desired product is lower than anticipated. For all these reasons reaction yields are often less than 100% and sometimes *much* lower than that.

Exercise 4.6 Percent Yield

Methanol, CH_3OH, can be burned in oxygen to provide energy, or it can be decomposed to form hydrogen gas, which can then be used as a fuel (see Figure 4.8).

$$CH_3OH(\ell) \longrightarrow 2\ H_2(g) + CO(g)$$

If you decompose 125 g of methanol, what is the theoretical yield of hydrogen? If you obtained only 13.6 g of hydrogen, what is the percent yield of the gas?

See the *Saunders Interactive General Chemistry CD-ROM,* Screen 5.7, Using Stoichiometry: Chemical Analysis of a Mixture.

4.6 CHEMICAL EQUATIONS AND CHEMICAL ANALYSIS

One of the most exciting recent findings in biochemistry and molecular biology is the role of the simple molecule NO in biochemistry. A major challenge in biochemistry has therefore been to find ways to detect and measure tiny amounts of NO. This is the kind of problem addressed by analytical chemists. People in this field use their creativity to identify substances as well as to measure the quantities of components of mixtures. Analytical chemistry is often done now using instrumental methods (Figure 4.11), but classical chemical reactions and stoichiometry still play a central role.

Figure 4.11 A modern analytical instrument: a gas chromatograph mass spectrometer. (*C. D. Winters*)

Analysis of a Mixture

The analysis of mixtures is often challenging. It can take a great deal of imagination to figure out how to use chemistry to determine what, and how much, is in the mixture. All analyses, however, depend on one basic idea: the quantity of a reactant or product is directly related to the quantity of an unknown reactant. Suppose the quantity of the amino acid glycine in a sample is unknown.

Current Perspectives

What to Take for an Upset Stomach

Stoichiometry lies at the heart of chemistry—and of your everyday life. Treating an upset stomach is a good example.

As soon as food reaches your stomach, acidic gastric juices are released by the glands in the mucous lining of the stomach. The high acidity, due to dissolved hydrochloric acid (HCl), is needed for the enzyme pepsin to catalyze the digestion of proteins in food. When you eat too much food, or when your stomach is irritated by very spicy food, your stomach responds by producing even more acid, and you can feel discomfort. Heartburn is a frequent symptom, and it can be relieved with an antacid. The reaction of milk of magnesia is typical of many antacids.

$$Mg(OH)_2(s) + 2\ HCl(aq) \longrightarrow MgCl_2(aq) + 2\ H_2O(\ell)$$

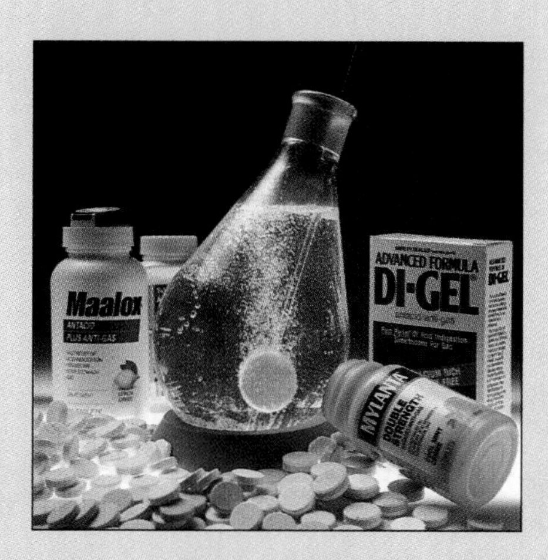

A variety of commercial antacids contain sodium bicarbonate, magnesium hydroxide, aluminum hydroxide, or calcium carbonate. The photo also shows the reaction of an Alka-Seltzer tablet. Notice that one product is CO_2, which comes from the reaction of the sodium bicarbonate and the citric acid in the tablet. (*C. D. Winters*)

One mole of magnesium hydroxide is required to remove 2 mol of hydrochloric acid. Antacids related to this are Amphogel [$Al(OH)_3$] and Maalox [a combination of $Mg(OH)_2$ and $Al(OH)_3$]. In the case of aluminum hydroxide, each mole of $Al(OH)_3$ can remove *three* moles of acid.

$$Al(OH)_3(s) + 3\ HCl(aq) \longrightarrow AlCl_3(aq) + 6\ H_2O(\ell)$$

Which of the hydroxides—$Mg(OH)_2$ or $Al(OH)_3$—do you suppose neutralizes more acid per gram of the hydroxide? (See Study Questions 44 and 45 at the end of this chapter.)

Tums and Di-Gel represent another type of antacid. The "active ingredient" in these is ordinary calcium carbonate, $CaCO_3$, a compound present in enormous quantities on the earth in the form of seashells, corals, and limestone. Calcium carbonate reacts with acids to produce water and carbon dioxide, as well as a calcium salt.

$$CaCO_3(s) + 2\ HCl(aq) \longrightarrow CaCl_2(aq) + CO_2(g) + H_2O(\ell)$$

The reaction of calcium carbonate with HCl is closely related to the reaction you see when an Alka-Seltzer tablet is dropped into water. In this case sodium hydrogen carbonate, commonly called baking soda, reacts with a naturally occurring acid, citric acid ($C_6H_8O_7$, page 120).

$$3\ NaHCO_3(aq) + C_6H_8O_7(aq) \longrightarrow Na_3C_6H_5O_7(aq) + 3\ CO_2(g) + 3\ H_2O(\ell)$$

You know this reaction occurs because you see carbon dioxide bubbling out of the water. The reaction of an Alka-Seltzer tablet is very similar to the one you see in the *Chemical Puzzler* (page 179).

The reactions we have described represent several classes of chemical reactions. Can you see a common theme? Think about this as you read Chapter 5.

It is known, however, that glycine reacts readily and completely with sodium hydroxide.

$$\underset{\text{glycine}}{H_2NCH_2CO_2H(aq)} + NaOH(aq) \longrightarrow H_2NCH_2CO_2Na(aq) + H_2O(\ell)$$

If the exact quantity of sodium hydroxide used in the reaction is measured, then the quantity of glycine present is also known.

Another example of this approach relates the quantity of a known reaction *product* to the quantity of a reactant. Suppose you have a white powder that is a mixture of magnesium oxide (MgO) and magnesium carbonate ($MgCO_3$), and you are asked to find the weight percent of $MgCO_3$ in the mixture. Many metal carbonates decompose on heating to give metal oxides and carbon dioxide. For magnesium carbonate the reaction is

$$MgCO_3(s) \longrightarrow MgO(s) + CO_2(g)$$

Here the quantity of a product, CO_2 gas, is directly related to the quantity of $MgCO_3$ in the sample. The solid left after heating the mixture consists only of MgO, and its mass is the sum of the mass of the MgO that was originally in the mixture, plus the MgO remaining from the $MgCO_3$ decomposition. Thus, the difference in mass of the solid before and after heating gives the mass of CO_2 evolved and, using stoichiometric relationships, the mass of $MgCO_3$ in the original mixture can be determined.

Assume you have 1.599 g of a mixture of MgO and $MgCO_3$ and that heating evolves CO_2 and leaves 1.294 g of MgO. What is the weight percent of $MgCO_3$ in the original mixture? Let us first find the mass of CO_2 evolved.

Mass of mixture ($MgO + MgCO_3$)	=	1.599 g
−Mass after heating (pure MgO)	=	−1.294 g
Mass of CO_2	=	0.305 g

Next, the mass of CO_2 can be converted to moles of CO_2.

$$0.305 \text{ g } CO_2 \cdot \frac{1 \text{ mol } CO_2}{44.01 \text{ g } CO_2} = 0.00693 \text{ mol } CO_2$$

The balanced equation for the decomposition of $MgCO_3$ shows that, for every mole of CO_2 evolved on heating, a mole of $MgCO_3$ existed in the mixture. Therefore, there must have been 0.00693 mol of $MgCO_3$ in the mixture, and the mass of $MgCO_3$ in the mixture was

$$0.00693 \text{ mol } CO_2 \cdot \frac{1 \text{ mol } MgCO_3}{1 \text{ mol } CO_2} \cdot \frac{84.31 \text{ g } MgCO_3}{1 \text{ mol } MgCO_3} = 0.584 \text{ g } MgCO_3$$

This means the weight percent of magnesium carbonate in the mixture was

$$\frac{\text{Mass of } MgCO_3}{\text{Sample mass}} \cdot 100\% = \frac{0.584 \text{ g } MgCO_3}{1.599 \text{ g sample}} \cdot 100\% = 36.5\% \text{ } MgCO_3$$

Example 4.5 Analysis of a Mixture

Titanium(IV) oxide is the predominant white pigment in the world. (It is used in paper, paint, and ink, for example.) Therefore, it is important to have a method of analyzing for the amount of this material in mixtures with other substances. One way to do it is to combine a mixture containing TiO_2 with bromine trifluoride.

$$3 \text{ } TiO_2(s) + 4 \text{ } BrF_3(\ell) \longrightarrow 3 \text{ } TiF_4(s) + 2 \text{ } Br_2(\ell) + 3 \text{ } O_2(g)$$

Oxygen gas is evolved quantitatively, it can be captured readily, and its mass can be determined. Suppose 2.367 g of a TiO_2-containing sample evolves 0.143 g of O_2. What is the weight percent of TiO_2 in the sample?

Solution Let us first calculate the quantity of O_2 evolved, in moles.

$$0.143\ \cancel{g\ O_2} \cdot \frac{1\ mol\ O_2}{32.00\ \cancel{g\ O_2}} = 0.00447\ mol\ O_2$$

Now we can calculate the quantity of TiO_2 present.

$$0.00447\ \cancel{mol\ O_2} \cdot \frac{3\ \cancel{mol\ TiO_2}}{3\ \cancel{mol\ O_2}} \cdot \frac{79.88\ g\ TiO_2}{1\ \cancel{mol\ TiO_2}} = 0.357\ g\ TiO_2$$

Finally, the weight percent of TiO_2 can be calculated.

$$\text{Weight \% of } TiO_2 = \frac{0.357\ g\ TiO_2}{2.367\ g\ sample} \cdot 100\% = 15.1\%\ TiO_2$$

Exercise 4.7 Analysis of a Mixture

You have 2.357 g of a mixture of $BaCl_2$ and $BaCl_2 \cdot 2H_2O$. If experiment shows that the mixture has a mass of only 2.108 g after heating to drive off all the water of hydration in $BaCl_2 \cdot 2H_2O$, what is the weight percent of $BaCl_2 \cdot 2H_2O$ in the original mixture?

Determining the Formula of a Compound

The empirical formula of a compound can be determined if the percent composition of the compound is known (Section 3.7). But where do the percent composition data come from? Various methods are used, and many depend on reactions that decompose the unknown but pure compound into known products. Assuming the reaction products can be isolated in pure form, the masses and the number of moles of each can be determined. Then, the moles of each product can be related to the number of moles of each element in the original compound. One method that works well for compounds that burn in oxygen is *analysis by combustion.* Each element (except oxygen) in the compound combines with oxygen to produce the appropriate oxide.

See the *Saunders Interactive General Chemistry CD-ROM,* Screen 5.8, Using Stoichiometry: Determination of an Empirical Formula.

Consider an analysis of the hydrocarbon, methane, as an example of combustion analysis. A balanced equation for the combustion of methane shows that every mole of carbon in the original compound is converted to a mole of CO_2. Every mole of hydrogen atoms in the original compound gives *half* a mole of H_2O. (Here the 4 moles of H atoms in CH_4 give 2 moles of H_2O.)

$$CH_4(g) + 2\ O_2(g) \longrightarrow CO_2(g) + 2\ H_2O(g)$$

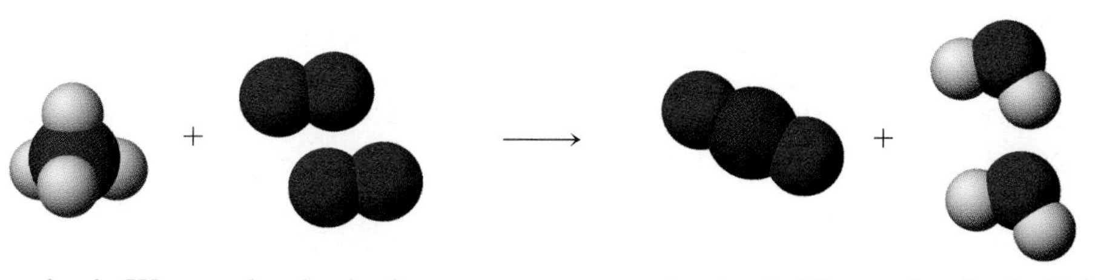

1 molecule CH_4 2 molecules O_2 1 molecule CO_2 2 molecules H_2O

The carbon dioxide and water produced by the combustion are gases that can be separated and their masses determined as illustrated in Figure 4.12. These

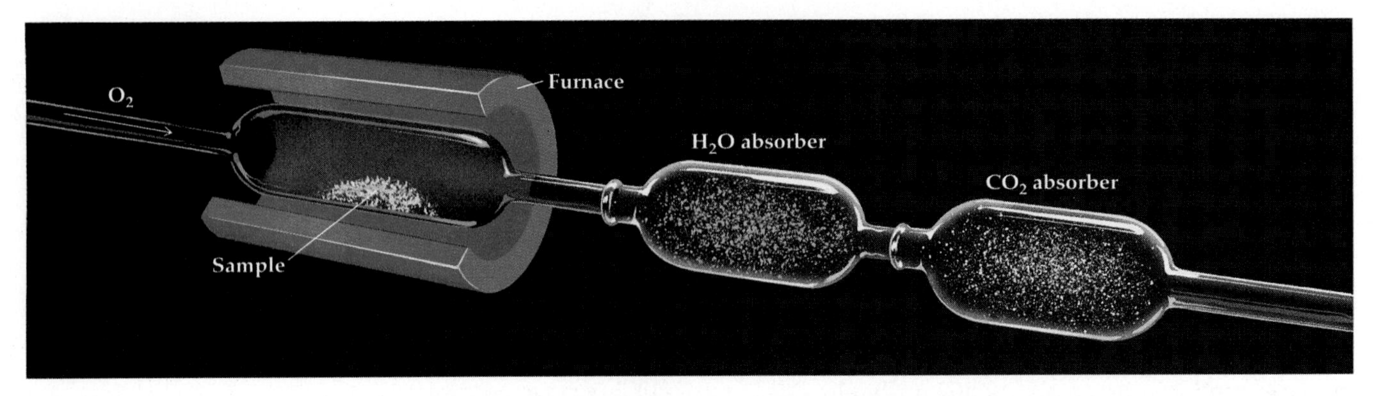

Figure 4.12 Combustion analysis of a hydrocarbon. If a compound containing C and H is burned in oxygen, CO_2 and H_2O are formed, and the mass of each can be determined. The H_2O is absorbed by magnesium perchlorate, and the CO_2 is absorbed by finely divided NaOH supported on asbestos. The mass of each absorbent before and after combustion gives the masses of CO_2 and H_2O. Only a few milligrams of a combustible compound are needed for analysis.

masses can then be converted to the moles of C and H in CO_2 and H_2O, respectively, and the ratio of the moles of C and H in a sample of the original compound can be found. This ratio gives the empirical formula:

$$C_xH_y \xrightarrow{\text{burn in } O_2} \text{g } CO_2 \xrightarrow{\times \frac{1 \text{ mol } CO_2}{44.01 \text{ g}}} \text{mol } CO_2 \xrightarrow{\times \frac{1 \text{ mol C}}{1 \text{ mol } CO_2}} \frac{\text{mol C}}{\text{mol H}} \rightarrow \text{Empirical formula}$$

$$C_xH_y \longrightarrow \text{g } H_2O \xrightarrow{\times \frac{1 \text{ mol } H_2O}{18.02 \text{ g}}} \text{mol } H_2O \xrightarrow{\times \frac{2 \text{ mol H}}{1 \text{ mol } H_2O}} \frac{\text{mol C}}{\text{mol H}}$$

A key concept in using this procedure is mass balance, which goes back to the reason that we always work with balanced chemical equations. That is, for every mole of H_2O observed from combustion, there must have been *two* moles of H atoms in the unknown carbon-hydrogen (or carbon-hydrogen-oxygen) compound. Similarly, for every mole of CO_2 observed, there must have been one mole of carbon in the unknown compound.

Example 4.6 Using Combustion Analysis to Determine the Formula of a Hydrocarbon

When 1.125 g of a liquid hydrocarbon, C_xH_y, was burned in an apparatus like that in Figure 4.12, 3.447 g of CO_2 and 1.647 g of H_2O were produced. The molar mass of the compound was found to be 86.2 g/mol in a separate experiment. Determine the empirical and molecular formulas for the unknown hydrocarbon, C_xH_y.

Solution The first step is to calculate the moles of CO_2 and H_2O isolated from the combustion:

$$3.447 \text{ g } CO_2 \cdot \frac{1 \text{ mol } CO_2}{44.010 \text{ g } CO_2} = 0.07832 \text{ mol } CO_2$$

$$1.647 \text{ g } H_2O \cdot \frac{1 \text{ mol } H_2O}{18.015 \text{ g } H_2O} = 0.09142 \text{ mol } H_2O$$

The previous discussion explains that for every mole of CO_2 isolated, a mole of C must have existed in the compound C_xH_y.

$$0.07832 \cancel{\text{mol } CO_2} \cdot \frac{1 \text{ mol C in } C_xH_y}{1 \cancel{\text{mol } CO_2}} = 0.07832 \text{ mol C in } C_xH_y$$

and that 2 mol of H must have existed in the compound for every mole of H_2O isolated.

$$0.09142 \cancel{\text{mol } H_2O} \cdot \frac{2 \text{ mol H in } C_xH_y}{1 \cancel{\text{mol } H_2O}} = 0.1828 \text{ mol H in } C_xH_y$$

The original 1.125-g sample of compound therefore contained 0.07832 mol of C and 0.1828 mol of H. To determine the empirical formula of C_xH_y we find the ratio of moles of H to moles of C. (Here we use techniques outlined in Section 3.7.)

$$\frac{0.1828 \text{ mol H}}{0.07832 \text{ mol C}} = \frac{2.335 \text{ mol H}}{1.000 \text{ mol C}} = \frac{(2\frac{1}{3} \text{ mol H})}{1 \text{ mol C}} = \frac{7/3 \text{ mol H}}{1 \text{ mol C}} = \frac{7 \text{ mol H}}{3 \text{ mol C}}$$

The empirical formula of the hydrocarbon is therefore C_3H_7.

In a separate experiment, the molar mass was determined to be 86.2 g/mol. Comparing this with the molar mass calculated for the empirical formula,

$$\frac{\text{Molar mass from experiment}}{\text{Molar mass for } C_3H_7} = \frac{86.2 \text{ g/mol}}{43.1 \text{ g/mol}} = \frac{2}{1}$$

we find that the molecular formula is twice the empirical formula. That is, the molecular formula is $(C_3H_7)_2$, or C_6H_{14}.

As an aside, the determination of the molecular formula does not end the problem for a chemist. As is sometimes the case, the formula C_6H_{14} is appropriate for several distinctly different molecules. Two of the five molecules having this formula are shown here:

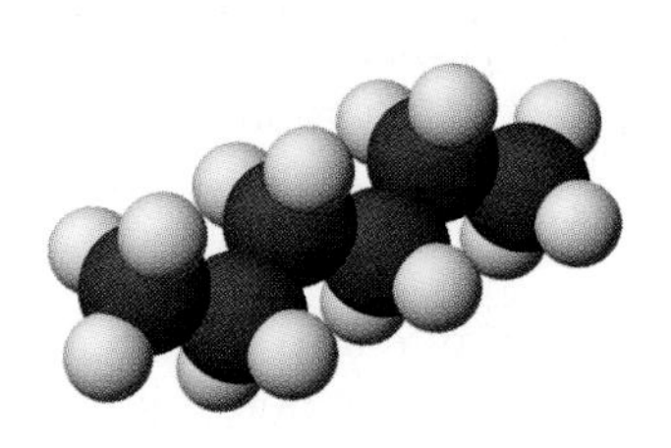

hexane
Boiling point = 68.73 °C

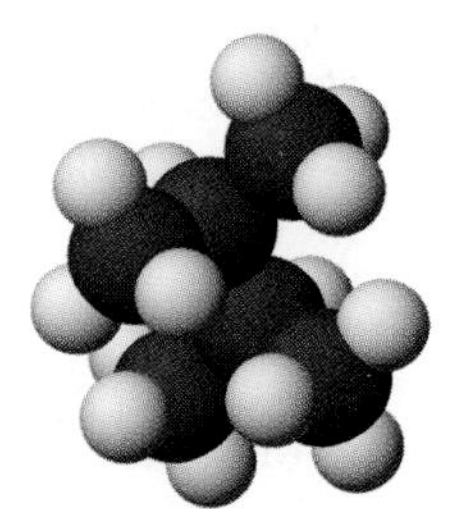

2, 3-dimethylbutane
Boiling point = 57.98 °C

To decide finally the identity of the unknown compound, more laboratory experiments are still necessary.

Exercise 4.8 Determining the Empirical and Molecular Formulas for a Hydrocarbon

A 0.523-g sample of the unknown compound C_xH_y was burned in air to give 1.612 g of CO_2 and 0.7425 g of H_2O. A separate experiment gave a molar mass for C_xH_y of 114 g/mol. Determine the empirical and molecular formulas for the hydrocarbon.

In Example 4.6 the moles of C and H found in the combustion products of the unknown compound were used directly to find the carbon-hydrogen ratio for the unknown compound. Alternatively, the moles of C and H could have been converted to the mass of each element and then to a weight percentage of the original 1.125-g sample. Using the data in Example 4.6, we have

$$0.1828 \cancel{\text{mol H}} \text{ in } C_xH_y \cdot \frac{1.0079 \text{ g H}}{1 \cancel{\text{mol H}}} = 0.1842 \text{ g H}$$

$$0.07832 \cancel{\text{mol C}} \text{ in } C_xH_y \cdot \frac{12.011 \text{ g C}}{1 \cancel{\text{mol C}}} = 0.9407 \text{ g C}$$

This leads to weight percentages of C and H in the 1.125-g sample of C_xH_y of 83.62% and 16.38%, respectively. These weight percentages can then be used to find the empirical formula by the method given in Section 3.7. Be sure to notice, though, that a calculation of the weight percentages is not necessary when information from an experiment is available that directly gives the moles of each element in a given sample mass. *Any method that gives the moles of each element in a given sample will lead to an empirical formula.*

Example 4.7 Determining an Empirical Formula

Suppose you isolate an acid from clover leaves and know that it contains only the elements C, H, and O. Heating 0.513 g of the acid in oxygen produces 0.501 g of CO_2 and 0.103 g of H_2O.

$$\underset{0.513\text{ g}}{C_xH_yO_z(s)} + \text{some } O_2(g) \longrightarrow \underset{0.501\text{ g}}{x\, CO_2(g)} + \underset{0.103\text{ g}}{y/2\; H_2O(g)}$$

What is the empirical formula of the acid $C_xH_yO_z$? Given that another experiment has shown that the molar mass of the acid is 90.04 g/mol, what is its molecular formula?

Solution The difference between this example and the situation in Example 4.6 is that the moles of O in the sample must also be determined. Unfortunately, this cannot be calculated from the masses of CO_2 and H_2O because the oxygen in CO_2 and the H_2O comes not only from $C_xH_yO_z$ but also from the O_2 used in combustion. Therefore, we shall pursue the following strategy:

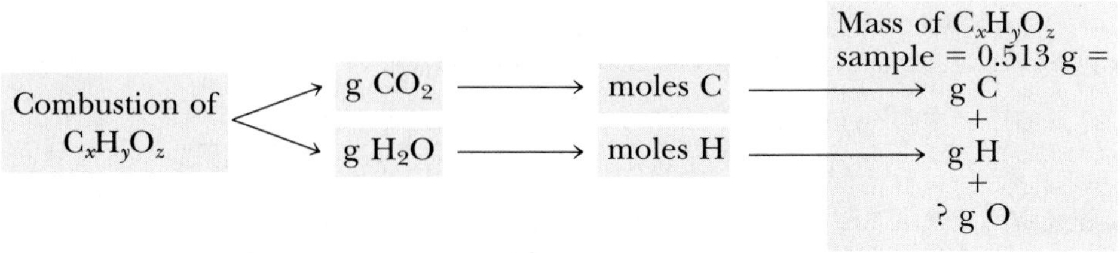

Combustion of $C_xH_yO_z$ → g CO_2 → moles C → Mass of $C_xH_yO_z$ sample = 0.513 g = g C + g H + ? g O

Combustion of $C_xH_yO_z$ → g H_2O → moles H → g H

The masses of CO_2 and H_2O from combustion can be used to find the masses of C and H in the 0.513-g sample of $C_xH_yO_z$. Because the masses of C, H, and

O must add up to 0.513 g, the mass of O can be found by subtraction. With the masses of all elements known, the number of moles of each can be calculated and the ratio of moles of each in $C_xH_yO_z$ determined.

The first step is to convert the masses of CO_2 and H_2O to moles.

$$0.501 \text{ g } CO_2 \cdot \frac{1 \text{ mol } CO_2}{44.010 \text{ g } CO_2} = 0.0114 \text{ mol } CO_2$$

$$0.103 \text{ g } H_2O \cdot \frac{1 \text{ mol } H_2O}{18.015 \text{ g } H_2O} = 0.00572 \text{ mol } H_2O$$

The moles of CO_2 and H_2O can now be converted to the masses of C and H that were in the original compound.

$$0.0114 \text{ mol } CO_2 \cdot \frac{1 \text{ mol C}}{1 \text{ mol } CO_2} \cdot \frac{12.01 \text{ g C}}{1 \text{ mol C}} = 0.137 \text{ g C in } CO_2 \text{ and formerly in the acid sample}$$

$$0.00572 \text{ mol } H_2O \cdot \frac{2 \text{ mol H}}{1 \text{ mol } H_2O} \cdot \frac{1.008 \text{ g H}}{1 \text{ mol H}} = 0.0115 \text{ g H in } H_2O \text{ and formerly in the acid sample}$$

These calculations reveal that the 0.513-g sample contains 0.137 g of C and 0.0115 g of H; the remaining mass, 0.365 g, must be oxygen.

$$0.513\text{-g sample} - 0.137 \text{ g C} - 0.0115 \text{ g H} = 0.365 \text{ g O}$$

To find the empirical formula of the unknown acid, you need only find the number of moles of each element in the sample.

$$0.137 \text{ g C} \cdot \frac{1 \text{ mol C}}{12.01 \text{ g C}} = 0.0114 \text{ mol C}$$

$$0.0115 \text{ g H} \cdot \frac{1 \text{ mol H}}{1.008 \text{ g H}} = 0.0114 \text{ mol H}$$

$$0.365 \text{ g O} \cdot \frac{1 \text{ mol O}}{16.00 \text{ g O}} = 0.0228 \text{ mol O}$$

Then, to find the mole ratio of elements, divide the number of moles of each element by the smallest number of moles.

$$\frac{0.0114 \text{ mol H}}{0.0114 \text{ mol C}} = \frac{1.00 \text{ mol H}}{1.00 \text{ mol C}} \quad \text{and} \quad \frac{0.0228 \text{ mol O}}{0.0114 \text{ mol C}} = \frac{2.00 \text{ mol O}}{1.00 \text{ mol H}}$$

The mole ratios show that, for every C atom in the molecule, one H atom and two O atoms occur. The empirical formula of the acid is therefore CHO_2, and the molar mass of the empirical formula unit is 45.02 g/mol.

Finally, to determine the molecular formula, the experimental molar mass of the compound and the molar mass of the empirical formula are compared.

$$\frac{90.04 \text{ g/mol unknown acid}}{45.02 \text{ g/mol } CHO_2} = \frac{2.000 \text{ mol } CHO_2}{1.000 \text{ mol unknown acid}}$$

Thus, the molecular formula of the acid is twice the empirical formula, that is, $C_2H_2O_4$. (Writing the formula in the manner customary for acids, we have $H_2C_2O_4$).

The compound investigated in Example 4.7 is oxalic acid. This simple acid is widely distributed as the potassium and calcium salts in the leaves, roots, and rhizomes of various plants. It also occurs in human and animal urine, and calcium oxalate is a major constituent of kidney stones. (*Photo, C. D. Winters; model, S. M. Young*)

Exercise 4.9 Formula Determination from Combustion Analysis

Vitamin C is composed of C, H, and O. Determine the empirical formula of vitamin C from the following data: burning 0.400 g of solid vitamin C in pure oxygen gives 0.600 g of CO_2 and 0.163 g of H_2O.

Exercise 4.10 Formula Determination from Combustion Analysis

A compound is composed of only C, H, and Cr. When 0.178 g of the compound is burned in air, the products are CO_2 (0.452 g), H_2O (0.0924 g), and Cr_2O_3.

$$\underset{0.178\text{ g}}{C_xH_yCr_z(s)} + \text{some } O_2(g) \longrightarrow \underset{0.452\text{ g}}{x\, CO_2(g)} + \underset{0.0924\text{ g}}{y/2\, H_2O(g)} + z/2\, Cr_2O_3(s)$$

What is the empirical formula of the compound?

CHAPTER HIGHLIGHTS

Now that you have studied this chapter, you should be able to

- Interpret information conveyed by a **balanced chemical equation** (Section 4.1).
- Balance simple chemical equations (Section 4.2).
- Calculate the mass of one reactant or product from the mass of another reactant or product by using the balanced chemical equation (Section 4.3). The following general scheme is used for a reaction such as $x\text{ A} \longrightarrow y\text{ B}$.

$$\text{Mass A} \xrightarrow[\times\left(\frac{1\text{ mol A}}{\text{g A}}\right)]{} \text{Moles A} \xrightarrow[\times\left(\frac{y\text{ moles B}}{x\text{ moles A}}\right)]{\times\text{ Stoichiometric factor}} \text{Moles B} \xrightarrow[\times\left(\frac{\text{g B}}{1\text{ mol B}}\right)]{} \text{Mass B}$$

- Determine which of two reactants is the **limiting reactant** (Section 4.4).
- Explain the differences among **actual yield, theoretical yield,** and **percent yield,** and calculate percent yield (Section 4.5).

$$\text{Percent yield} = \frac{\text{Actual yield (g)}}{\text{Theoretical yield (g)}} \times 100\%$$

- Use stoichiometry principles to analyze a mixture or to find the empirical formula of an unknown compound (Section 4.6).

STUDY QUESTIONS

REVIEW QUESTIONS

1. What information is provided by a balanced chemical equation?
2. Which of the following represents a balanced chemical equation for the production of ammonia?
   (a) $N_2(g) + H_2(g) \longrightarrow NH_3(g)$
   (b) $N_2(g) + H_2(g) \longrightarrow 2\, NH_3(g)$
   (c) $N_2(g) + 3\, H_2(g) \longrightarrow 2\, NH_3(g)$
3. In the reaction of aluminum and bromine,

$$2\, Al(s) + 3\, Br_2(\ell) \longrightarrow Al_2Br_6(s)$$

(Figure 1, page 2), how many molecules of Br_2 do you need for complete reaction if you have 2000 atoms of Al? How many molecules of Al_2Br_6 do you obtain from the reaction?

4. Given that you have 3 mol of N_2 and wish to calculate the

quantity of NH_3 produced in the reaction of N_2 and H_2 (see Study Question 2), what stoichiometric factor would you use in the calculation?

5. Suppose you combine a weighed quantity of zinc metal with excess iodine to obtain zinc iodide.

$$Zn(s) + I_2(s) \longrightarrow ZnI_2(s)$$

Explain how to calculate the theoretical yield for this reaction. What experimental information do you need to collect in order to calculate the percent yield?

Balancing Equations

6. Balance the following equations:
(a) $Cr(s) + O_2(g) \longrightarrow Cr_2O_3(s)$
(b) $Cu_2S(s) + O_2(g) \longrightarrow Cu(s) + SO_2(g)$
(c) $C_6H_5CH_3(\ell) + O_2(g) \longrightarrow H_2O(\ell) + CO_2(g)$

7. Balance the following equations:
(a) $Cr(s) + Cl_2(g) \longrightarrow CrCl_3(s)$
(b) $SiO_2(s) + C(s) \longrightarrow Si(s) + CO(g)$
(c) $Fe(s) + H_2O(g) \longrightarrow Fe_3O_4(s) + H_2(g)$

8. Balance the following equations and name the reaction products:
(a) $Fe_2O_3(s) + Mg(s) \longrightarrow MgO(s) + Fe(s)$
(b) $AlCl_3(s) + H_2O(\ell) \longrightarrow Al(OH)_3(s) + HCl(aq)$
(c) $NaNO_3(s) + H_2SO_4(\ell) \longrightarrow Na_2SO_4(s) + HNO_3(g)$
(d) $NiCO_3(s) + HNO_3(aq) \longrightarrow Ni(NO_3)_2(aq) + CO_2(g) + H_2O(\ell)$

9. Balance the following equations and name the reaction products:
(a) $SO_2(g) + HF(\ell) \longrightarrow SF_4(g) + H_2O(\ell)$
(b) $NH_3(aq) + O_2(aq) \longrightarrow NO(g) + H_2O(\ell)$
(c) $BF_3(g) + H_2O(\ell) \longrightarrow HF(aq) + H_3BO_3(aq)$

10. Balance the following equations:
(a) The synthesis of urea, a common fertilizer

$$CO_2(g) + NH_3(g) \longrightarrow NH_2CONH_2(s) + H_2O(\ell)$$

(b) Reactions used to make uranium(VI) fluoride for the enrichment of natural uranium

$$UO_2(s) + HF(aq) \longrightarrow UF_4(s) + H_2O(\ell)$$
$$UF_4(s) + F_2(g) \longrightarrow UF_6(s)$$

(c) The reaction to make titanium(IV) chloride, which is then converted to titanium metal

$$TiO_2(s) + Cl_2(g) + C(s) \longrightarrow TiCl_4(\ell) + CO(g)$$
$$TiCl_4(\ell) + Mg(s) \longrightarrow Ti(s) + MgCl_2(s)$$

11. Balance the following equations:
(a) Reaction to produce "superphosphate" fertilizer

$$Ca_3(PO_4)_2(s) + H_2SO_4(aq) \longrightarrow Ca(H_2PO_4)_2(aq) + CaSO_4(s)$$

(b) Reaction to produce diborane, B_2H_6

$$NaBH_4(s) + H_2SO_4(aq) \longrightarrow B_2H_6(g) + H_2(g) + Na_2SO_4(aq)$$

(c) Reaction to produce tungsten metal from tungsten(VI) oxide

$$WO_3(s) + H_2(g) \longrightarrow W(s) + H_2O(\ell)$$

(d) Preparation of platinum

$$(NH_4)_2PtCl_6(s) \longrightarrow NH_4Cl(s) + Pt(s) + Cl_2(g)$$

General Stoichiometry

12. Aluminum reacts with oxygen to give aluminum oxide.

$$4\ Al(s) + 3\ O_2(g) \longrightarrow 2\ Al_2O_3(s)$$

If you have 6.0 mol of Al, how many moles of O_2 are needed for complete reaction? What mass of Al_2O_3, in grams, can be produced?

13. Several brands of antacid tablets use aluminum hydroxide to neutralize excess acid.

$$Al(OH)_3(s) + 3\ HCl(aq) \longrightarrow AlCl_3(aq) + 3\ H_2O(\ell)$$

What quantity of HCl, in grams, can a tablet with 0.750 g of $Al(OH)_3$ consume? What quantity of water is produced?

14. Suppose 16.04 g of methane, CH_4, is burned in oxygen.
(a) What are the products of the reaction?
(b) What is the balanced equation for the reaction?
(c) What mass of O_2, in grams, is required for complete combustion of methane?
(d) What is the total mass of products expected from 16.04 g of methane?

15. If 10.0 g of carbon is combined with an exact, stoichiometric amount of oxygen (26.6 g) to produce carbon dioxide, what mass, in grams, of CO_2 can be obtained? That is, what is the theoretical yield of CO_2?

16. Like many metals, aluminum reacts with a halogen to give a metal halide (see Figure 1, page 2).

$$2\ Al(s) + 3\ Br_2(\ell) \longrightarrow Al_2Br_6(s)$$

What quantity of Br_2, in grams, is required for complete reaction with 2.56 g of Al? What mass of white, solid Al_2Br_6 is expected?

17. The equation for one of the reactions in the process of reducing iron ore to the metal is

$$Fe_2O_3(s) + 3\ CO(g) \longrightarrow 2\ Fe(s) + 3\ CO_2(g)$$

(a) What is the maximum mass of iron, in grams, that can be obtained from 454 g (1.00 lb) of iron(III) oxide?
(b) What mass of CO is required to reduce the iron(III) oxide to iron metal?

18. Iron reacts with oxygen to give iron(III) oxide, Fe_2O_3.
(a) Write a balanced equation for the reaction.
(b) If an ordinary iron nail (assumed to be pure iron) has a mass of 2.68 g, what mass (in grams) of Fe_2O_3 does it produce if the nail is converted completely to this oxide?
(c) What mass of O_2 (in grams) is required for the reaction?

19. Burning coal and oil in a power plant produces pollutants

such as sulfur dioxide, SO_2. The sulfur-containing compound can be removed from other waste gases, however, by the following reaction:

$$2\ SO_2(g) + 2\ CaCO_3(s) + O_2(g) \longrightarrow 2\ CaSO_4(s) + 2\ CO_2(g)$$

(a) Name the compounds involved in the reaction.
(b) What mass of $CaCO_3$ is required to remove 155 g of SO_2?
(c) What mass of $CaSO_4$ is formed when 155 g of SO_2 is consumed completely?

20. The metabolic disorder diabetes causes a buildup of acetone (CH_3COCH_3) in the blood of untreated victims. Acetone, a volatile compound, is exhaled, giving the breath of untreated diabetics a distinctive odor. The acetone is produced by a breakdown of fats in a series of reactions. The equation for the last step is

$$CH_3COCH_2CO_2H \longrightarrow CH_3COCH_3 + CO_2$$

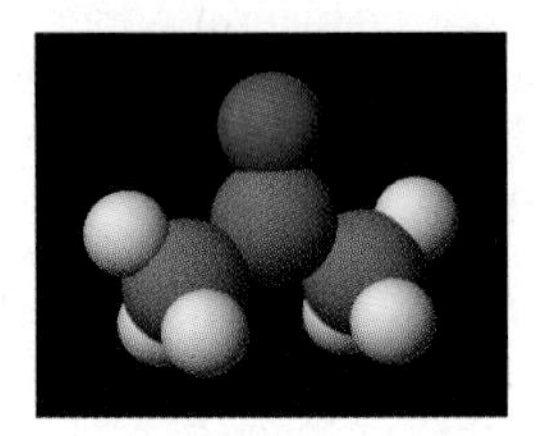

Acetone, CH_3COCH_3, is present in the breath of people suffering from diabetes.

What quantity of acetone can be produced from 125 mg of acetoacetic acid ($CH_3COCH_2CO_2H$)?

21. Your body deals with excess nitrogen by excreting it in the form of urea, NH_2CONH_2. The reaction producing it is the combination of arginine ($C_6H_{14}N_4O_2$) with water to give urea and ornithine ($C_5H_{12}N_2O_2$).

$$C_6H_{14}N_4O_2 + H_2O \longrightarrow NH_2CONH_2 + C_5H_{12}N_2O_2$$

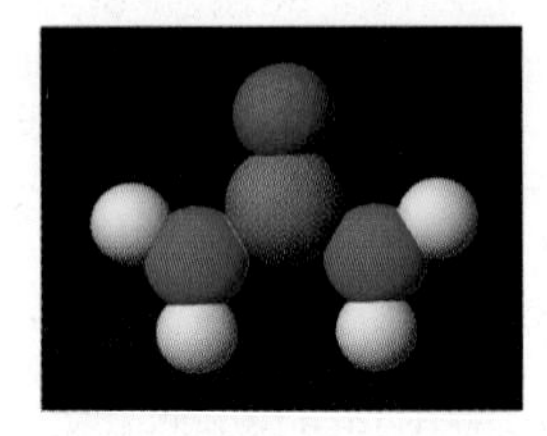

Your body excretes excess nitrogen as urea, NH_2CONH_2.

If you excrete 95 mg of urea, what quantity of arginine must have been used? What quantity of ornithine must have been produced?

Limiting Reactants

22. Aluminum chloride, $AlCl_3$, is an inexpensive reagent used in many industrial processes. It is made by treating scrap aluminum with chlorine according to the following balanced equation:

$$2\ Al(s) + 3\ Cl_2(g) \longrightarrow 2\ AlCl_3(s)$$

(a) Which reactant is limiting if 2.70 g of Al and 4.05 g of Cl_2 are mixed?
(b) What mass of $AlCl_3$ can be produced from 2.70 g of Al and 4.05 g of Cl_2?
(c) What mass of the excess reactant remains when the reaction is completed?

23. The reaction of methane and water is one way to prepare hydrogen:

$$CH_4(g) + H_2O(g) \longrightarrow CO(g) + 3\ H_2(g)$$

If you begin with 995 g of CH_4 and 2510 g of water, what is the maximum possible yield of H_2?

24. Methanol, CH_3OH, is a clean-burning, easily handled fuel (Figure 4.8). It can be made by the direct reaction of CO and H_2 (obtained from heating coal with steam):

$$CO(g) + 2\ H_2(g) \longrightarrow CH_3OH(\ell)$$

Starting with a mixture of 12.0 g of H_2 and 74.5 g of CO, which is the limiting reactant? What mass of the excess reactant (in grams) remains after reaction is complete? What is the theoretical yield of methanol?

25. Disulfur dichloride, S_2Cl_2, is used to vulcanize rubber. It can be made by treating molten sulfur with gaseous chlorine:

$$S_8(\ell) + 4\ Cl_2(g) \longrightarrow 4\ S_2Cl_2(\ell)$$

Starting with a mixture of 32.0 g of sulfur and 71.0 g of Cl_2, which is the limiting reactant? What mass of S_2Cl_2 (in grams) can be produced? What mass of the excess reactant remains when the limiting reactant is consumed?

26. Ammonia gas can be prepared by the reaction of a metal oxide such as calcium oxide with ammonium chloride.

$$CaO(s) + 2\ NH_4Cl(s) \longrightarrow 2\ NH_3(g) + H_2O(g) + CaCl_2(s)$$

If 112 g of CaO and 224 g of NH_4Cl are mixed, what is the maximum possible yield of NH_3? What mass of the excess reactant remains after the maximum amount of ammonia has been formed?

27. Aspirin ($C_9H_8O_4$) is produced by the reaction of salicylic acid ($C_7H_6O_3$) and acetic anhydride ($C_4H_6O_3$) (page 163).

$$C_7H_6O_3(s) + C_4H_6O_3(\ell) \longrightarrow C_9H_8O_4(s) + CH_3CO_2H(aq)$$

If you mix 100. g of each of the reactants, what is the maximum mass of aspirin that can be obtained?

Percent Yield

28. Ammonia gas can be prepared by the following reaction:

$$CaO(s) + 2\ NH_4Cl(s) \longrightarrow 2\ NH_3(g) + H_2O(g) + CaCl_2(s)$$

If 103 g of ammonia is obtained, but the theoretical yield is 136 g, what is the percent yield of this gas?

29. Diborane, B_2H_6, is a valuable compound in the synthesis of new organic compounds. One of several ways this boron compound can be made is by the reaction

$$2\ NaBH_4(s) + I_2(s) \longrightarrow B_2H_6(g) + 2\ NaI(s) + H_2(g)$$

Suppose you use 1.203 g of $NaBH_4$ with an excess of iodine and obtain 0.295 g of B_2H_6. What is the percent yield of B_2H_6?

30. The reaction of zinc and chlorine has been used as the basis of a car battery.

$$Zn(s) + Cl_2(g) \longrightarrow ZnCl_2(s)$$

What is the theoretical yield of $ZnCl_2$ if 35.5 g of zinc is allowed to react with excess chlorine? If only 65.2 g of zinc chloride is obtained, what is the percent yield of the compound?

31. Disulfur dichloride, which has a revolting smell, can be prepared by directly combining S_8 and Cl_2, but it can also be made by the following reaction:

$$3\ SCl_2(\ell) + 4\ NaF(s) \longrightarrow SF_4(g) + S_2Cl_2(\ell) + 4\ NaCl(s)$$

Assume you begin with 5.23 g of SCl_2 and excess NaF. What is the theoretical yield of S_2Cl_2? If only 1.19 g of S_2Cl_2 is obtained, what is the percent yield of the compound?

Chemical Analysis

32. A mixture of $CuSO_4$ and $CuSO_4 \cdot 5\ H_2O$ has a mass of 1.245 g, but, after heating to drive off all the water, the mass is only 0.832 g. What is the weight percent of $CuSO_4 \cdot 5\ H_2O$ in the mixture? (See Figure 3.16.)

33. A sample of limestone and other soil materials is heated, and the limestone decomposes to give calcium oxide and carbon dioxide. A 1.506-g sample of limestone-containing material gives 0.711 g of CaO, in addition to gaseous CO_2, after being heated at a high temperature. What is the weight percent of $CaCO_3$ in the original sample?

34. A 1.25-g sample contains some of the very reactive compound $Al(C_6H_5)_3$. On treating the compound with aqueous HCl, 0.951 g of C_6H_6 is obtained.

$$Al(C_6H_5)_3(s) + 3\ HCl(aq) \longrightarrow AlCl_3(aq) + 3\ C_6H_6(\ell)$$

Assuming that $Al(C_6H_5)_3$ was converted completely to products, what is the weight percent of $Al(C_6H_5)_3$ in the original 1.25-g sample?

35. Bromine trifluoride reacts with metal oxides to evolve oxygen quantitatively. For example,

$$3\ TiO_2(s) + 4\ BrF_3(\ell) \longrightarrow 3\ TiF_4(s) + 2\ Br_2(\ell) + 3\ O_2(g)$$

Suppose you wish to use this reaction to determine the weight percent of TiO_2 in a sample of ore. To do this the O_2 gas from the reaction is collected. If 2.367 g of the TiO_2-containing ore evolves 0.143 g of O_2, what is the weight percent of TiO_2 in the sample?

Determination of Empirical Formulas

36. Styrene, the building block of polystyrene, is a hydrocarbon, a compound consisting only of C and H. If 0.438 g of styrene is burned in oxygen and produces 1.481 g of CO_2 and 0.303 g of H_2O, what is the empirical formula of styrene?

37. Mesitylene is a liquid hydrocarbon. If 0.115 g of the compound is burned in oxygen to give 0.379 g of CO_2 and 0.1035 g of H_2O, what is the empirical formula of mesitylene?

38. Menthol, from *oil of mint,* has a characteristic cool taste. The compound contains only C, H, and O. If 95.6 mg of menthol burns completely in O_2, and gives 269 mg of CO_2 and 110 mg of H_2O, what is the empirical formula of menthol?

39. Quinone, a chemical used in the dye industry and in photography, is an organic compound containing only C, H, and O. What is the empirical formula of the compound if 0.105 g of the compound gives 0.257 g of CO_2 and 0.0350 g of H_2O when burned completely in oxygen?

40. Silicon and hydrogen form a series of compounds with the general formula Si_xH_y. To find the formula of one of them, a 6.22-g sample of the compound is burned in oxygen. On doing so, all of the Si is converted to 11.64 g of SiO_2 and all of the H to 6.980 g of H_2O. What is the empirical formula of the silicon compound?

41. To find the formula of a compound composed of iron and carbon monoxide, $Fe_x(CO)_y$, the compound is burned in pure oxygen, a reaction that proceeds according to the following *unbalanced* equation.

$$Fe_x(CO)_y(s) + O_2(g) \longrightarrow Fe_2O_3(s) + CO_2(g)$$

If you burn 1.959 g of $Fe_x(CO)_y$ and obtain 0.799 g of Fe_2O_3 and 2.200 g of CO_2, what is the empirical formula of $Fe_x(CO)_y$?

GENERAL QUESTIONS

42. Glucose, $C_6H_{12}O_6$, burns in air to give CO_2 and water.

$$C_6H_{12}O_6(s) + 6\ O_2(g) \longrightarrow 6\ CO_2(g) + 6\ H_2O(\ell)$$

If you burn 125 mg of glucose, what quantity of oxygen is required for complete combustion? What quantity of CO_2 is produced? What quantity of water?

43. Nitrogen gas can be prepared in the laboratory by the reaction of ammonia with copper(II) oxide according to the following unbalanced equation.

$$NH_3(g) + CuO(s) \longrightarrow N_2(g) + Cu(s) + H_2O(g)$$

If 26.3 g of gaseous NH_3 is passed over an excess of solid CuO, what mass of N_2, in grams, can be obtained?

44. Several antacids contain aluminum hydroxide, $Al(OH)_3$, which reacts with stomach acid according to the equation

$$Al(OH)_3(s) + 3\ HCl(aq) \longrightarrow AlCl_3(aq) + 3\ H_2O(\ell)$$

Assume that 15.5 g of $Al(OH)_3$ is combined with 20.5 g of HCl.

(a) Which reactant is in excess and which is the limiting reactant?

(b) What mass of water, in grams, can be formed?

45. Some over-the-counter antacids contain aluminum hydroxide (Study Question 44), whereas others contain $CaCO_3$. (See *Current Perspectives*, page 165.)

$$CaCO_3(s) + 2\,HCl(aq) \longrightarrow CaCl_2(aq) + CO_2(g) + H_2O(\ell)$$

If a tablet contains 750 mg of $Al(OH)_3$, and another contains 750 mg of $CaCO_3$, which consumes the greater mass of HCl?

46. Many metals react with halogens to give metal halides. For example, iron gives iron(III) chloride, $FeCl_3$, on reaction with chlorine gas.

The reaction of iron and chlorine. When hot steel wool is plunged into a flask filled with yellow chlorine gas, the steel wool glows brightly, and a cloud of brown iron(III) chloride forms. (*C. D. Winters*)

(a) Write the balanced chemical equation for the reaction.
(b) Beginning with 10.0 g of iron, what mass of Cl_2, in grams, is required for complete reaction? What quantity of $FeCl_3$, in moles and in grams, can be produced?
(c) If only 18.5 g of $FeCl_3$ is obtained, what is the percent yield?

47. Aluminum bromide, a valuable laboratory chemical, is made by the direct reaction of the elements

$$2\,Al(s) + 3\,Br_2(\ell) \longrightarrow Al_2Br_6(s)$$

What is the theoretical yield in grams of Al_2Br_6 if 25.0 mL of liquid bromine (density = 3.10 g/mL) and 12.5 g aluminum metal are used? Is any aluminum or bromine left over when the reaction has gone to completion? If so, what mass of which reactant remains?

48. Some metal halides react with water to produce the metal oxide and the appropriate hydrogen halide (see photo). For example,

$$TiCl_4(\ell) + 2\,H_2O(\ell) \longrightarrow TiO_2(s) + 4\,HCl(g)$$

Titanium tetrachloride, $TiCl_4$, is a liquid at room temperature. When exposed to the air, it forms a dense fog of titanium(IV) oxide. (See Study Question 48.) (*C. D. Winters*)

(a) Name the four compounds involved in this reaction.
(b) If you begin with 14.0 mL of $TiCl_4$ (d = 1.73 g/mL), what mass of water, in grams, is required for complete reaction?
(c) What mass of each product is expected?

49. Boron forms an extensive series of compounds with hydrogen, all with the general formula B_xH_y.

$$B_xH_y(s) + \text{excess } O_2(g) \longrightarrow B_2O_3(s) + H_2O(g)$$

If 0.148 g of B_xH_y gives 0.422 g of B_2O_3 when burned in excess O_2, what is the empirical formula of B_xH_y?

CONCEPTUAL QUESTIONS

50. Chlorine and iodine are mixed in a reaction flask. The mixture is represented here with I atoms as red spheres and Cl atoms as green spheres. What will the contents look like after the reaction has produced ICl_3?

$$I_2(g) + 3\,Cl_2(g) \longrightarrow 2\,ICl_3(g)$$

Original reaction mixture

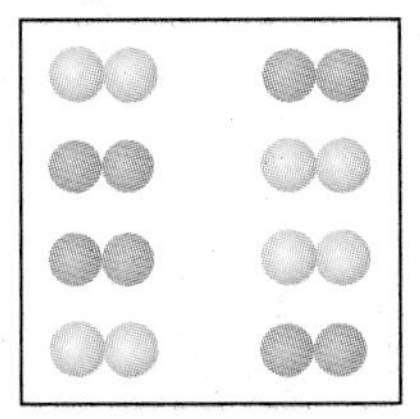

Possible product mixtures

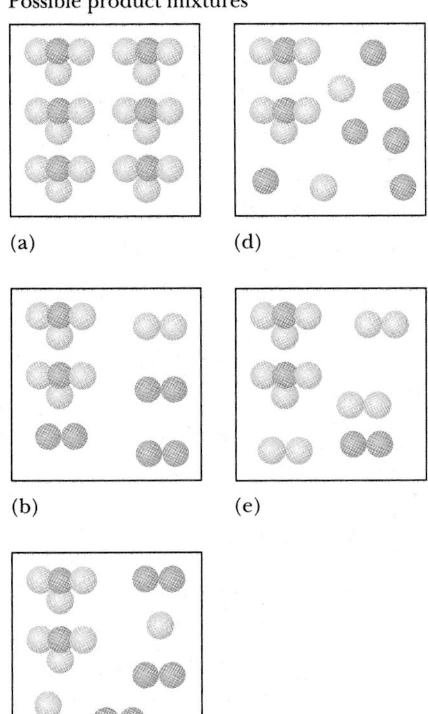

(c)

51. Nitrogen monoxide, NO, and O_2 react to give NO_2. (See Figure 4.6.) The contents of flasks A and B are mixed in a reaction flask. Using open and solid circles for atoms, show the situation in the reaction flask after reaction has occurred. (Represent N atoms by ●, and O atoms by ○.)

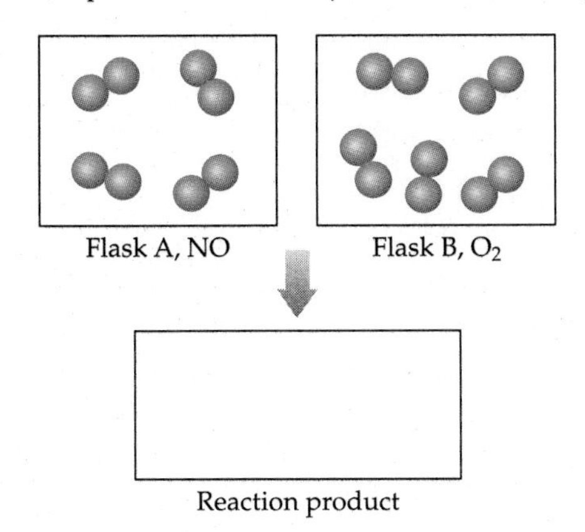

52. The *Chemical Puzzler* involves the reaction with baking soda and acetic acid. Explain why the evolution of CO_2 gas eventually ceases as more and more vinegar is added to a $NaHCO_3$ sample. Suppose you began with 5.0 g of $NaHCO_3$. What quantity of acetic acid would you have to add to just get to the point where CO_2 evolution ceases?

53. Suppose you want to prepare a sample of tin(IV) iodide, SnI_4 (page 133). You weigh out 0.945 g of Sn and 1.834 g of I_2. After mixing in an appropriate solvent, the distinctive color of iodine fades away, signaling the completion of the reaction. The orange solid product, SnI_4, is collected on a filter. The solid has a mass of 1.935 g.
    (a) Which is the limiting reactant, Sn or I_2?
    (b) What is the theoretical yield of SnI_4?
    (c) What is the percent yield of SnI_4?
    (d) Give at least two reasons that the actual yield is not equal to the theoretical yield.

CHALLENGING QUESTIONS

54. In an experiment 1.056 g of a metal carbonate, containing an unknown metal M, is heated to give the metal oxide and 0.376 g CO_2.

$$MCO_3(s) + \text{heat} \longrightarrow MO(s) + CO_2(g)$$

What is the identity of the metal M?
(a) M = Ni (c) M = Zn
(b) M = Cu (d) M = Ba

55. An unknown metal reacts with oxygen to give the metal oxide, MO_2. Identify the metal based on this information:

Mass of metal = 0.356 g

Mass of sample after converting metal completely to oxide = 0.452 g

56. Titanium(IV) oxide, TiO_2, is heated in hydrogen gas to give water and a new titanium oxide, Ti_xO_y. If 1.598 g of TiO_2 produces 1.438 g of Ti_xO_y, what is the formula of the new oxide?

57. The elements silver, molybdenum, and sulfur combine to form Ag_2MoS_4. What is the maximum mass of Ag_2MoS_4 that can be obtained if 8.63 g of silver, 3.36 g of molybdenum, and 4.81 g of sulfur are combined?

58. Balance the following equations.
(a) A method for dissolving silver metal in the presence of metallic gold:

$$Ag(s) + H_2SO_4(aq) \longrightarrow Ag_2SO_4(s) + SO_2(g) + H_2O(\ell)$$

(b) A method for dissolving gold:

$$Au(s) + HNO_3(aq) + HCl(aq) \longrightarrow HAuCl_4(aq) + NO(g) + H_2O(\ell)$$

(c) Dissolving zinc in aqueous sodium hydroxide:

$$Zn(s) + NaOH(aq) + H_2O(\ell) \longrightarrow Na_2Zn(OH)_4(aq) + H_2(g)$$

59. A mixture of sodium carbonate and sodium hydrogen carbonate is treated with aqueous hydrochloric acid. The unbalanced equations for the resulting reactions are

$$Na_2CO_3(s) + HCl(aq) \longrightarrow NaCl(aq) + CO_2(g) + H_2O(\ell)$$
$$NaHCO_3(aq) + HCl(aq) \longrightarrow NaCl(aq) + CO_2(g) + H_2O(\ell)$$

You treat 9.540 g of a $Na_2CO_3/NaHCO_3$ mixture with an excess of aqueous HCl and isolate 9.355 g of NaCl. What is the weight percent of each substance in the mixture?

Summary Question

60. A weighed sample of iron (Fe) is added to liquid bromine (Br_2) and allowed to react completely. The reaction produces a single product, which can be isolated and weighed. The experiment was repeated a number of times with *different masses of iron but with the same mass of bromine.* The results are summarized in the graph.

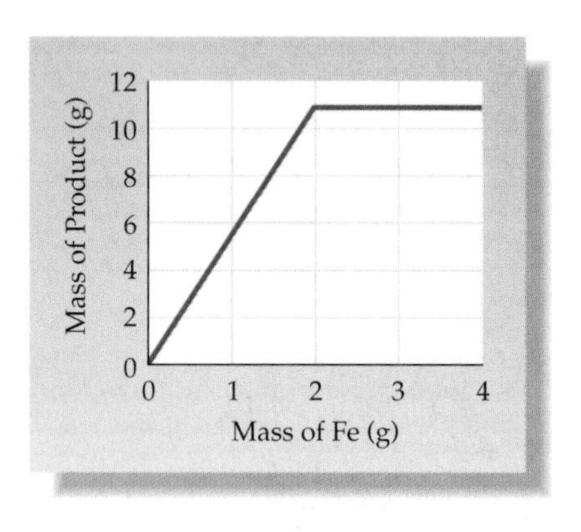

(a) What mass of Br_2 is used when the reaction consumes 2.0 g of Fe?
(b) What is the mole ratio of Br_2 to Fe in the reaction?
(c) What is the empirical formula of the product?
(d) Write the balanced chemical equation for the reaction of iron and bromine.
(e) What is the name of the reaction product?
(f) Which statement or statements best describe the experiments summarized by the graph?
 (i) When 1.00 g of Fe is added to the Br_2, Fe is the limiting reagent.
 (ii) When 3.50 g of Fe is added to the Br_2, there is an excess of Br_2.
 (iii) When 2.50 g of Fe is added to the Br_2, both reactants are used up completely.
 (iv) When 2.00 g of Fe is added to the Br_2, 10.0 g of product is formed. The percent yield must therefore be 20.0%.

INTERACTIVE GENERAL CHEMISTRY CD-ROM

Screen 4.3 The Law of Conservation of Matter

Observe the decomposition of mercury(II) oxide and the accompanying animation.

(a) What visual evidence is there for the decomposition of the oxide?
(b) What happens to the oxygen evolved in the decomposition?
(c) In the animation, how many molecules of oxygen are evolved? How many atoms of mercury? How many "molecules" of HgO must have decomposed? Does this agree with the balanced chemical equation for the decomposition?

Screen 4.4 Balancing Chemical Equations

(a) *Reactions that form oxides:* Observe the video of the reaction of elemental phosphorus. Here the phosphorus is removed from a beaker full of water and placed on a laboratory spoon in the air. (The phosphorus used here is called "white" phosphorus, even though its color is often closer to yellow.)
 (1) Is elemental phosphorus a solid, liquid, or gas?
 (2) What can you conclude about the relative tendency of phosphorus to react with water and air? Why is the element stored under water?
(b) *Combustion reactions:* Observe the animation of the reaction of propane and oxygen. How does this animation illustrate the principle of the conservation of matter?

Screen 5.4 Reactions Controlled by the Supply of One Reactant

Examine the problem associated with this screen (Screen 5.4PR). Which of the diagrams at the right is correct? Why?

Screen 5.5 Limiting Reactants

Examine the series of photos of the reaction between varying amounts of Zn metal and a constant amount of hydrochloric acid.

(a) What gas inflates the balloons in the demonstration?
(b) Why are the volumes of gas in the first two balloons greater than the volume in the balloon on the right?
(c) In which flask is Zn metal the limiting reactant and in which flask is HCl(aq) the limiting reactant? Explain.

Chapter 5

Reactions in Aqueous Solution

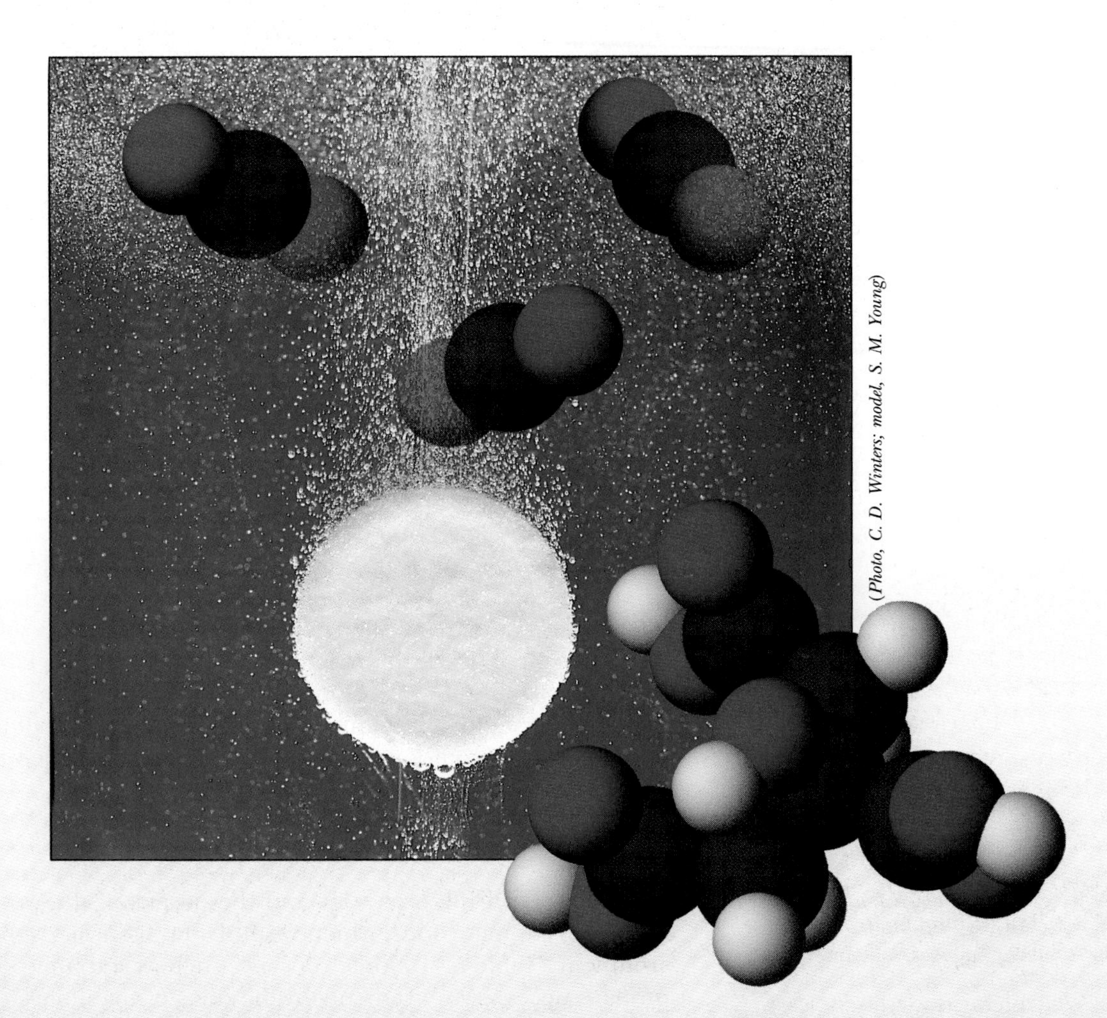
(Photo, C. D. Winters; model, S. M. Young)

A Chemical Puzzler

Alka-Seltzer is a well-known remedy for stomach acidity. A tablet contains citric acid ($C_6H_8O_7$) and sodium hydrogen carbonate ($NaHCO_3$). What is the chemistry of these tablets? What kind of chemical reaction produces the bubbles you see when you drop a tablet into a glass of water?

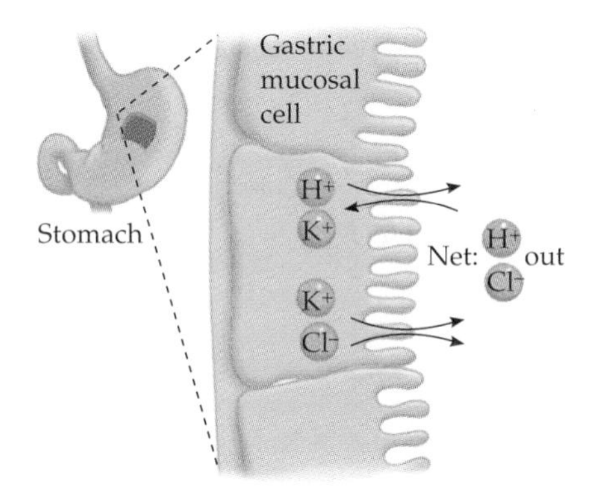

Ions in biochemistry. Ions in aqueous solution are important in many processes in living systems. Large differences in K^+ concentrations inside and outside cells, for example, are used by your body to cause nerve impulses.

Professor Ronald Breslow of Columbia University once said that, "among the sciences, chemistry is unique because it expands what is known." We do this by making new substances using chemical reactions, many of which occur in water, that is, in **aqueous solution.** For example, two of the most important products made by industry are sodium hydroxide (NaOH) and chlorine (Cl_2), and both are made by passing an electric current through an aqueous solution of salt (NaCl). But just as importantly, we turn food into energy, we breathe, we see, and we move, all because of chemical reactions, most of them occurring in aqueous solution.

This chapter, which introduces you to the chemistry of aqueous solutions, has several objectives:

- To understand the nature of ionic substances when dissolved in water
- To be able to predict the water solubility of ionic compounds
- To show that only a few general types of reactions occur in aqueous solution
- To learn how to handle quantitative aspects of reactions in aqueous solution

5.1 PROPERTIES OF COMPOUNDS IN AQUEOUS SOLUTION

A **solution** is a homogeneous mixture of two or more substances, in which one is generally considered the **solvent,** the medium in which another substance—the **solute**—is dissolved (Section 1.1). Chemical reactions in plants and animals occur largely among substances dissolved in water as a solvent, that is, in aqueous solutions. Many of the reactions you will see in the laboratory are also done in aqueous solution. Therefore, to understand such reactions, it is important first to understand something about the behavior of compounds in water. The focus here is on compounds that produce ions in aqueous solution.

Ions in Aqueous Solution: Electrolytes

See the *Saunders Interactive General Chemistry CD-ROM,* Screen 4.5, Compounds in Aqueous Solution.

The water you drink every day and the oceans of the world contain small concentrations of many ions, which result from dissolving solid materials in the environment (Table 5.1). Some of these, such as sodium, potassium, and calcium ions, are necessary ingredients in the diets of mammals.

Dissolving an ionic solid involves separating each ion from the oppositely charged ions that surround it in the solid state (Figure 5.1). Water is especially good at dissolving ionic compounds because each water molecule has a positively charged end and a negatively charged end. Therefore, a water molecule can attract a positive ion to its negative end, or it can attract a negative ion to its positive end. When an ionic compound dissolves in water, each negative ion becomes surrounded by water molecules with their positive ends pointing toward it, and each positive ion becomes surrounded by the negative ends of several water molecules.

The water-encased ions coming from a dissolved ionic compound are free to move about in solution. Under normal conditions, the movement of ions is random, and the cations and anions from a dissolved ionic compound are dis-

Table 5.1 • SOME CATIONS AND ANIONS IN LIVING CELLS AND THEIR ENVIRONMENT*

Element	Dissolved Species	Sea Water	*Valonia*[†]	Red-Blood Cells	Blood Plasma
Chlorine	Cl^-	550	50	50	100
Sodium	Na^+	460	80	11	160
Magnesium	Mg^{2+}	52	50	2.5	2
Calcium	Ca^{2+}	10	1.5	10^{-4}	2
Potassium	K^+	10	400	92	10
Carbon	HCO_3^-, CO_3^{2-}	30	≈10	≈10	30
Phosphorus	HPO_4^{2-}	< 1	5	3	≈3

*Data taken from J. J. R. Fraústo da Silva and R. J. P. Williams: *The Biological Chemistry of the Elements*, Oxford, Clarendon Press, 1991. Concentrations are given in millimoles per liter. (A millimole is 1/1000 of a mole.)

[†]*Valonia* are single-celled algae that live in sea water.

persed uniformly throughout the solution. If two **electrodes** (conductors of electricity such as copper wire) are placed in the solution and connected to a battery, however, cations migrate through the solution to the negative electrode and anions move to the positive electrode. If a light bulb is inserted into the circuit as in Figure 5.2, the bulb lights, showing that the circuit is complete. Compounds whose aqueous solutions conduct electricity are called **electrolytes,** and all ionic compounds that are soluble in water are electrolytes.

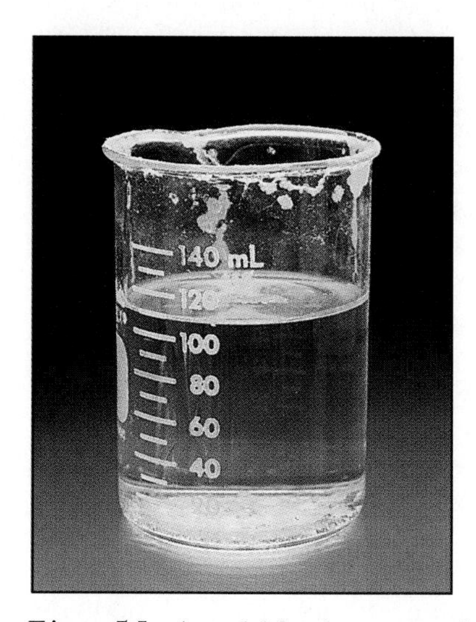

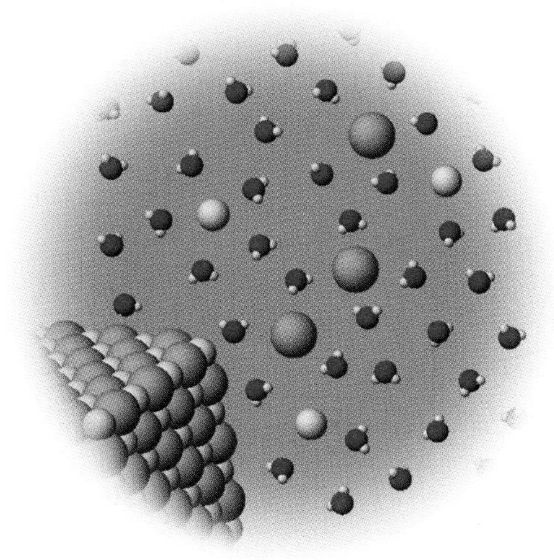

***Figure* 5.1 A model for the process of dissolving NaCl in water.** The ions of the crystal dissociate to give Na^+ (silver) and Cl^- (yellow) ions in aqueous (water) solution. These ions, sheathed in water molecules, are free to move about. Such solutions conduct electricity, so the dissolved substance is called an electrolyte. (*Photo, C. D. Winters; model, S. M. Young*)

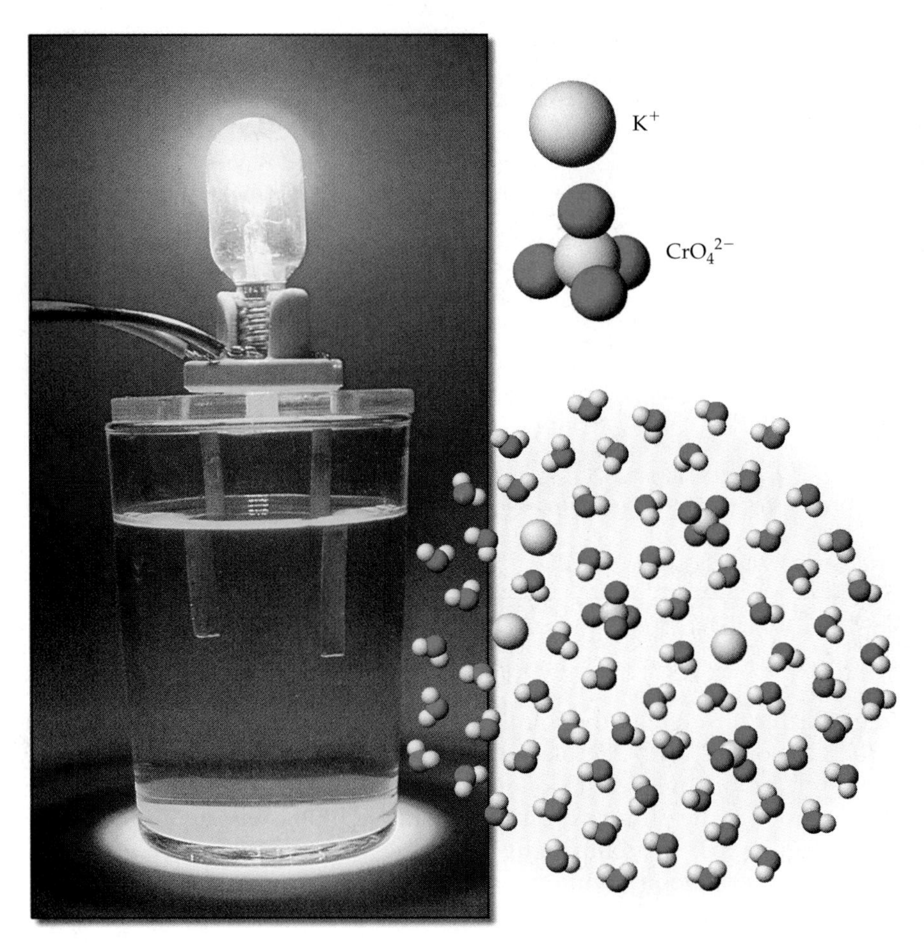

Figure 5.2 Electrolytes. Pure water does not conduct electricity. When an electrolyte is dissolved in water, however, ions exist in solution and are free to move about. Here electrodes attached to a battery have been inserted into a solution of K_2CrO_4, potassium chromate. The light bulb included in the circuit glows brightly, indicating that the electric circuit has been completed. This is experimental evidence that potassium chromate, like all ionic compounds that dissolve in water, is a strong electrolyte. In water, K_2CrO_4 has dissociated into its ions, K^+ and CrO_4^{2-}. Also shown is a model of the complete dissociation of an ionic compound in aqueous solution to form ions. (*Photo, C. D. Winters; model, S. M. Young*)

Types of Electrolytes

Electrolytes can be classified as strong or weak. When sodium chloride and many other ionic compounds dissolve in water, the ions separate or dissociate completely. For every mole of NaCl that dissolves, 1 mol of Na^+ and 1 mol of Cl^- ions enter the solution (Figure 5.1).

The idea that salts such as NaCl dissociate completely to give only ions in solution is a simplification. In fact, there can be a measurable concentration of species such as NaCl(aq), species called "ion pairs."

$$NaCl(aq) \equiv Na^+(aq) + Cl^-(aq)$$

$$100\% \text{ Dissociation} \equiv \text{strong electrolyte}$$

Because ions are in high concentration in solution, the solution is a good conductor of electricity. Substances whose solutions conduct well are **strong electrolytes** (see Figure 5.2).

Other substances produce only a small concentration of ions when they dissolve, and so are poor conductors of electricity; they are known as **weak elec-**

Figure 5.3 Electrolytes and nonelectrolytes. (a) An aqueous solution of a weak electrolyte, acetic acid, conducts electricity poorly, as indicated by the observation that the light bulb in the electric circuit glows only weakly. Also shown is a model of the behavior of a weak electrolyte. The dissolved compound forms relatively few ions in solution. (b) Ethanol dissolved in water does not conduct electricity and so is a nonelectrolyte. The light bulb does not light when attached to a battery. (*Photos, C. D. Winters; models, S. M. Young*)

trolytes (Figure 5.3). For example, when acetic acid—an important ingredient in vinegar—dissolves in water, only a few percent of the molecules are ionized to produce a cation and an anion. Only a few molecules in every 100 molecules of acetic acid are ionized to form acetate and hydrogen ions.

$$CH_3CO_2H(aq) \longrightarrow CH_3CO_2^-(aq) + H^+(aq)$$

acetic acid < 5% ionization = weak electrolyte — acetate ion — hydrogen ion

Many other substances dissolve in water but do not ionize. These are called **nonelectrolytes** because their solutions do not conduct electricity (see Figure

5.3). Nonelectrolytes are generally molecular compounds; examples include sugar ($C_{12}H_{22}O_{11}$), ethanol (CH_3CH_2OH), and antifreeze (ethylene glycol, $HOCH_2CH_2OH$).

Exercise 5.1 Electrolytes

Epsom salt, $MgSO_4 \cdot 7\ H_2O$, is sold in drugstores and used, as a solution in water, for various medical purposes. Methanol, CH_3OH, is dissolved in gasoline in the winter in colder climates to prevent the formation of ice in automobile fuel lines. Which of these compounds is an electrolyte and which is a nonelectrolyte?

Solubility of Ionic Compounds in Water

See the *Saunders Interactive General Chemistry CD-ROM,* Screen 4.6, for a discussion of the solubility of ionic compounds in water.

Not all ionic compounds dissolve in water. Many dissolve only to a small extent, and still others are essentially insoluble. Fortunately, we can make some general statements about which types of ionic compounds are water soluble.

Figure 5.4 lists broad guidelines that help predict whether a particular ionic compound will be soluble in water. For example, sodium nitrate, $NaNO_3$, con-

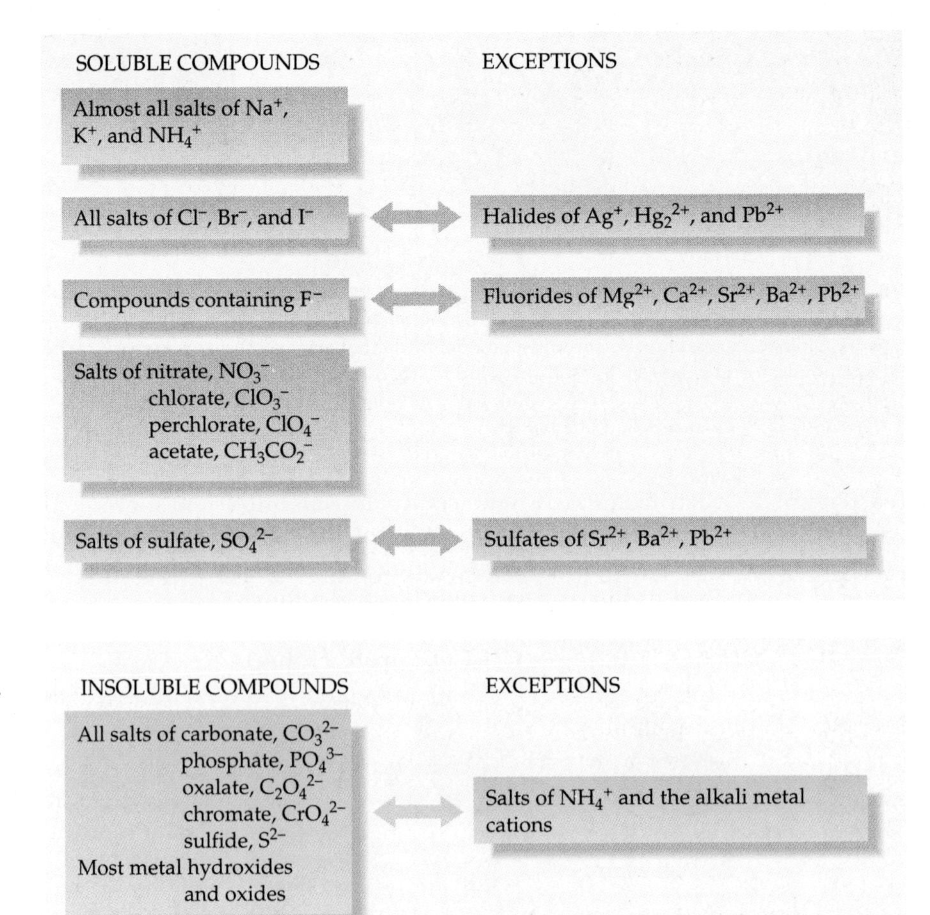

SOLUBLE COMPOUNDS
EXCEPTIONS
Almost all salts of Na^+, K^+, and NH_4^+
All salts of Cl^-, Br^-, and I^-
Halides of Ag^+, Hg_2^{2+}, and Pb^{2+}
Compounds containing F^-
Fluorides of Mg^{2+}, Ca^{2+}, Sr^{2+}, Ba^{2+}, Pb^{2+}
Salts of nitrate, NO_3^-
chlorate, ClO_3^-
perchlorate, ClO_4^-
acetate, $CH_3CO_2^-$
Salts of sulfate, SO_4^{2-}
Sulfates of Sr^{2+}, Ba^{2+}, Pb^{2+}
INSOLUBLE COMPOUNDS
EXCEPTIONS
All salts of carbonate, CO_3^{2-}
phosphate, PO_4^{3-}
oxalate, $C_2O_4^{2-}$
chromate, CrO_4^{2-}
sulfide, S^{2-}
Most metal hydroxides and oxides
Salts of NH_4^+ and the alkali metal cations

Figure 5.4 Guidelines to predict the solubility of ionic compounds. If a compound contains one of the ions in the column to the left in the top chart, the compound is predicted to be at least *moderately* soluble in water. There are a few exceptions, and these are noted at the right. Most ionic compounds formed by the anions listed at the bottom of the chart are *poorly* soluble (with exceptions such as compounds with NH_4^+ and the alkali metal cations).

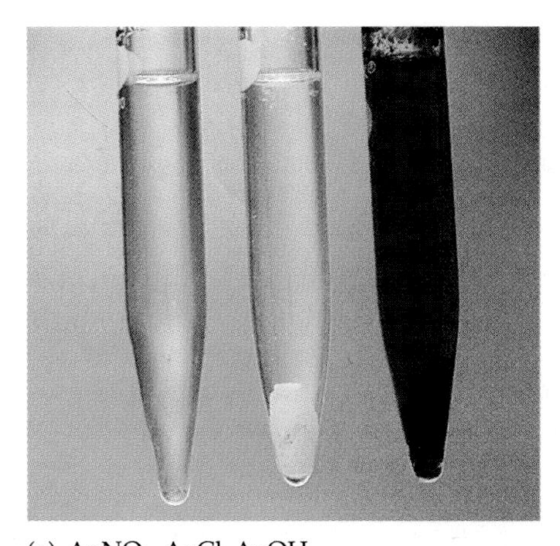

(a) $AgNO_3$, AgCl, AgOH

(b) $(NH_4)_2S$, CdS, Sb_2S_3, PbS

(c) NaOH, $Ca(OH)_2$, $Fe(OH)_3$, $Ni(OH)_2$

***Figure* 5.5 Illustration of the solubility guidelines in Figure 5.4.** (a) Ionic compounds containing Cl^- and NO_3^- are water-soluble. The few exceptions include AgCl. (b) Metal sulfides such as CdS, Sb_2S_3, and PbS, are almost always insoluble. Exceptions are $(NH_4)_2S$ and Na_2S, where the presence of the NH_4^+ and Na^+ cations leads to solubility. (c) Although a few hydroxides such as NaOH and KOH are water-soluble, many are not. (*C. D. Winters*)

tains both an alkali metal cation, Na^+, and the nitrate anion, NO_3^-. According to Figure 5.4, the presence of either of these ions ensures that the compound is soluble in water. Because *ionic compounds that dissolve in water are electrolytes,* an aqueous solution of $AgNO_3$ consists of the separated ions $Ag^+(aq)$ and $NO_3^-(aq)$ (Figure 5.5a). On the other hand, calcium hydroxide is poorly soluble in water. For example, if a spoonful of $Ca(OH)_2$ is placed in 100 mL of water, only 0.17 g, or 0.0023 mol, will dissolve at 10 °C (Figure 5.5c).

Milk of magnesia, $Mg(OH)_2$, a common antacid, is a good example of a metal hydroxide that is nearly insoluble in water.

$$0.0023 \text{ mol } Ca(OH)_2 \text{ dissolves in 100 mL water at 10 °C} \longrightarrow 0.0023 \text{ mol } Ca^{2+}(aq) + 2 \times 0.0023 \text{ mol } OH^-(aq)$$

When solid $Ca(OH)_2$ is placed in water, nearly all of it remains as the solid, and a heterogeneous mixture is the result. This is generally true of all metal hydroxides except those containing an alkali metal cation (see Figure 5.5c).

Observations such as those shown in Figure 5.5 were used to create the solubility guidelines in Figure 5.4. Remember that an ionic compound will be moderately soluble in water if it contains at least one of the ions listed in the *Soluble Compounds* column of Figure 5.4.

***Example* 5.1** Solubility Guidelines

Predict whether each of the following ionic compounds is likely to be water-soluble. If a compound is soluble, list the ions existing in solution.

1. KCl **2.** $MgCO_3$ **3.** MnO_2 **4.** CaI_2

Solution You must first recognize the cation and anion involved and then decide the probable water solubility.

1. KCl is composed of K^+ and Cl^- ions. According to Figure 5.4, the presence of either of these ions means that the compound is likely to be soluble in water.

$$\text{KCl in water} \longrightarrow K^+(aq) + Cl^-(aq)$$

Indeed, its actual solubility is about 35 g in 100 mL of water at 20 °C, and the solution consists of K^+ and Cl^- ions.

2. Magnesium carbonate is composed of Mg^{2+} and CO_3^{2-} ions. Mg^{2+} is in the alkaline earth group, a group that often forms water-insoluble compounds. The carbonate ion usually gives insoluble compounds (see Figure 5.4), unless combined with an ion like Na^+ or NH_4^+. $MgCO_3$ is therefore predicted to be insoluble in water. (The actual solubility of $MgCO_3$ is less than 0.2 g/100 mL of water.)
3. Manganese(IV) oxide is composed of Mn^{4+} and O^{2-} ions. Again, Figure 5.4 suggests that oxides are soluble only when O^{2-} is combined with an alkali metal ion; Mn^{4+} is a transition metal ion, so MnO_2 is insoluble.
4. Calcium iodide is composed of Ca^{2+} and I^- ions. According to Figure 5.4 almost all iodides are soluble in water, so CaI_2 is water-soluble and produces calcium and iodide ions in water.

$$CaI_2 \text{ in water} \longrightarrow Ca^{2+}(aq) + 2\,I^-(aq)$$

Notice that the compound gives two I^- ions on dissolving in water.

When an ionic compound with halide ions dissolves in water, the halide ions are released into aqueous solution. Thus, $BaCl_2$ produces one Ba^{2+} and two Cl^- ions (and *not* Cl_2 or Cl_2^{2-} ions).

Exercise 5.2 Solubility of Ionic Compounds

Predict whether each of the following ionic compounds is likely to be soluble in water. If soluble, write the formulas for the ions present in aqueous solution.

1. KNO_3
2. $CaCl_2$
3. CuO
4. $NaCH_3CO_2$

5.2 PRECIPITATION REACTIONS

See the *Saunders Interactive General Chemistry CD-ROM,* Screen 4.11, for a discussion of the major types of reactions in aqueous solution.

A ***precipitation reaction*** *produces an insoluble product, a* ***precipitate.*** The reactants in such reactions are generally water-soluble ionic compounds. When these are dissolved in water, they dissociate to give the appropriate cations and anions. If one of the cations can form an insoluble compound with one of the anions, a precipitation reaction occurs. For example, silver nitrate and potassium chloride, both water-soluble ionic compounds, form insoluble silver chloride and soluble potassium nitrate (Figure 5.6).

$$AgNO_3(aq) + KCl(aq) \longrightarrow AgCl(s) + KNO_3(aq)$$

Reactants	Products
$Ag^+(aq) + NO_3^-(aq)$	Insoluble AgCl
$K^+(aq) + Cl^-(aq)$	$K^+(aq) + NO_3^-(aq)$

Notice that, when two ionic compounds react to form a solid precipitate, they do so by *exchanging* ions. See Section 5.6. In the lead(II) nitrate + potassium chromate reaction, for example, lead(II) ions exchange nitrate for chromate ions, and potassium ions exchange chromate ions for nitrate ions.

$Pb^{2+} + NO_3^-$

$K^+ + CrO_4^{2-}$

Many precipitation reactions are possible because many combinations of positive and negative ions give insoluble substances (see Figure 5.4). For example, because most chromates are insoluble, lead(II) chromate is easily precipitated by the reaction of a water-soluble lead(II) compound with a water-soluble chromate compound (Figure 5.7a).

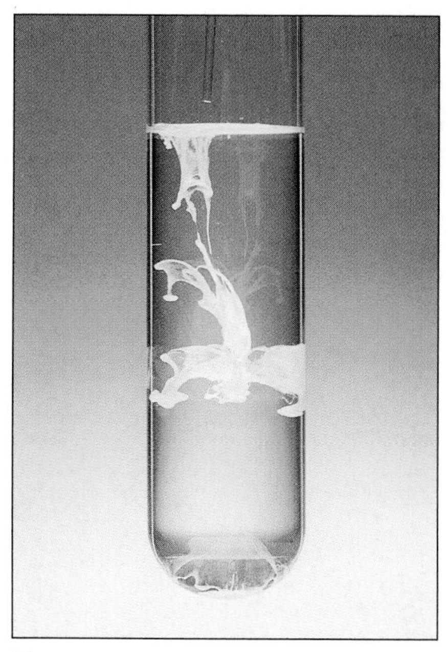

(a)

(b)

(c)

(d)

***Figure* 5.6 Precipitation of silver chloride.** (a) Adding aqueous silver nitrate to potassium chloride produces white, insoluble silver chloride. In (b) through (d) you see a model of the process. In (b) the silver (Ag^+, gray) and chloride (Cl^-, green) ions are widely separated by water molecules. When Ag^+ and Cl^- ions approach, they form ion pairs (c). As more and more silver and chloride ions come together, a precipitate of solid AgCl forms (d). (*Photo, C. D. Winters; models from an animation by Roy Tasker, University of Western Sydney, Australia. See Screen 19.7 of the Saunders Interactive General Chemistry CD-ROM*)

$$Pb(NO_3)_2(aq) + K_2CrO_4(aq) \longrightarrow PbCrO_4(s) + 2\ KNO_3(aq)$$

Reactants	Products
$Pb^{2+}(aq) + 2\ NO_3^-(aq)$	Insoluble $PbCrO_4$
$2\ K^+(aq) + CrO_4^{2-}(aq)$	$2\ K^+(aq) + 2\ NO_3^-(aq)$

Almost all metal sulfides are insoluble in water (Figure 5.7b). If a soluble metal compound in nature comes in contact with a source of sulfide ions (say from a volcano, a natural gas pocket in the earth, or a "black smoker" in the ocean, page 190), the metal sulfide precipitates.

$$Pb(NO_3)_2(aq) + (NH_4)_2S(aq) \longrightarrow PbS(s) + 2\ NH_4NO_3(aq)$$

Reactants	Products
$Pb^{2+}(aq) + 2\ NO_3^-(aq)$	Insoluble PbS
$2\ NH_4^+(aq) + S^{2-}(aq)$	$2\ NH_4^+(aq) + 2\ NO_3^-(aq)$

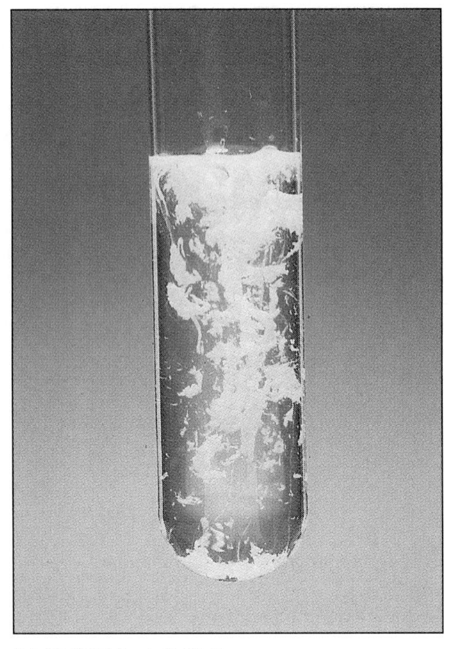

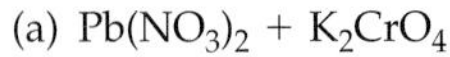

(a) $Pb(NO_3)_2 + K_2CrO_4$

(b) $Pb(NO_3)_2 + (NH_4)_2S$

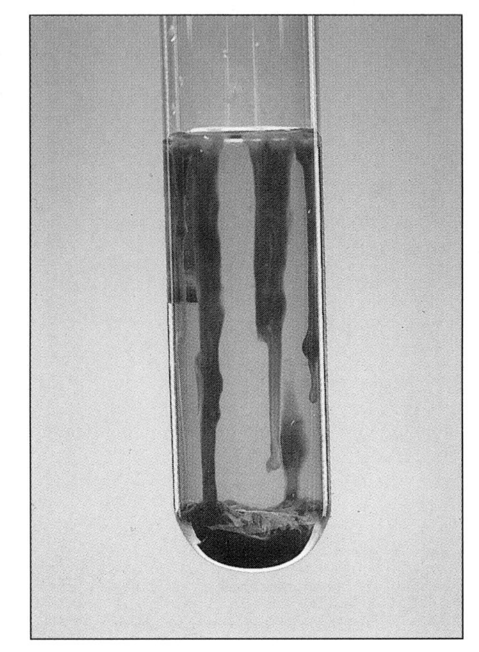

(c) $FeCl_3 + NaOH$

***Figure* 5.7 Precipitation reactions.** (a) Mixing solutions of lead(II) nitrate and potassium chromate gives yellow, insoluble lead(II) chromate and soluble potassium nitrate. Lead(II) chromate is commonly known as chrome yellow and has been used as a pigment in paint. You may know it as "school bus yellow." (b) Mixing solutions of lead(II) nitrate and ammonium sulfide gives black, insoluble lead(II) sulfide and soluble ammonium nitrate. (c) Mixing solutions of iron(III) chloride and sodium hydroxide gives rust-colored, insoluble iron(III) hydroxide and soluble sodium chloride. (*C. D. Winters*)

Sulfide ion, S^{2-}, does not exist in aqueous solution. The actual ion present under most conditions is HS^-.

In fact, this is how many metal sulfur-containing minerals such as iron pyrite (FeS_2, see Figure 1.4) are believed to have been formed in nature.

Finally, with the exception of the alkali metals, all metals form insoluble hydroxides. Thus, water-soluble iron(III) chloride and sodium hydroxide react to give insoluble iron(III) hydroxide (Figure 5.7c).

$$FeCl_3(aq) + 3\,NaOH(aq) \longrightarrow Fe(OH)_3(s) + 3\,NaCl(aq)$$

Reactants	Products
$Fe^{3+}(aq) + 3\,Cl^-(aq)$	Insoluble $Fe(OH)_3$
$3\,Na^+(aq) + 3\,OH^-(aq)$	$3\,Na^+(aq) + 3\,Cl^-(aq)$

Example 5.2 Writing the Equation for a Precipitation Reaction

Is an insoluble product formed when aqueous solutions of potassium chromate and silver nitrate are mixed? If so, write the balanced equation.

Solution First, let us decide what ions are formed when potassium chromate and silver nitrate are placed in water. Both are water-soluble, so the ions Ag^+,

NO_3^-, K^+, and CrO_4^{2-} are released into solution when the compounds are dissolved.

$$AgNO_3(aq) \equiv Ag^+(aq) + NO_3^-(aq)$$
$$K_2CrO_4(aq) \equiv 2\,K^+(aq) + CrO_4^{2-}(aq)$$

Here Ag^+ could combine with CrO_4^{2-}, and K^+ could combine with NO_3^-. The former combination, Ag_2CrO_4, is an insoluble compound, whereas KNO_3 is soluble in water. Thus, the balanced equation for the reaction of silver nitrate and potassium chromate is (Figure 5.8)

$$2\,AgNO_3(aq) + K_2CrO_4(aq) \longrightarrow Ag_2CrO_4(s) + 2\,KNO_3(aq)$$

Figure 5.8 Precipitation reaction. Water-soluble silver nitrate and potassium chromate react to form red, insoluble silver chromate and soluble potassium nitrate. See Example 5.2. (*C. D. Winters*)

***Exercise* 5.3** Precipitation Reactions

Does a precipitation reaction occur in any of the following cases?

1. Sodium carbonate is mixed with nickel(II) chloride.
2. Sodium carbonate is mixed with potassium nitrate.
3. Aluminum chloride is mixed with potassium hydroxide.

Write a balanced chemical equation for the precipitation reactions that occur.

Net Ionic Equations

Let us look at the reaction of silver nitrate and potassium chloride again. Solutions containing the ions Ag^+, NO_3^-, K^+, and Cl^- are mixed (see Figure 5.6). Insoluble AgCl precipitates, and the ions K^+ and NO_3^- remain in solution.

$$\underbrace{Ag^+(aq) + NO_3^-(aq) + K^+(aq) + Cl^-(aq)}_{\text{before reaction}} \longrightarrow \underbrace{AgCl(s) + K^+(aq) + NO_3^-(aq)}_{\text{after reaction}}$$

The K^+ and NO_3^- ions are present in solution before and after reaction and so appear on both the reactant and product sides of the balanced chemical equation. Such ions are often called **spectator ions** because they are not involved in the *net* reaction process; they only "look on" from the sidelines. Little chemical or stoichiometric information is lost if the equation is written without them, and so we can simplify the equation to

$$Ag^+(aq) + Cl^-(aq) \longrightarrow AgCl(s)$$

The balanced equation that results from leaving out the spectator ions is the ***net ionic equation*** *for the reaction.* Only the aqueous ions and nonelectrolytes (which can be insoluble compounds or soluble, molecular compounds such as sugar) that participate in a chemical reaction need to be included in the net ionic equation.

See the *Saunders Interactive General Chemistry CD-ROM,* Screen 4.10, Equations for Reactions in Aqueous Solution: Net Ionic Equations.

Leaving out the spectator ions does not imply that K^+ and NO_3^- ions are totally unimportant in the $AgNO_3$ + KCl reaction. Indeed, Ag^+ and Cl^- ions cannot exist alone in solution; a negative ion of some kind must be present to balance the positive ion charge of Ag^+, for example. Any anion will do, however, as long as it forms water-soluble compounds with Ag^+ and K^+. Thus, we could have used ClO_4^- and combined $AgClO_4$ and KCl. The net ionic equation would have been the same.

Current Perspectives

Black Smokers and the Origin of Life

Francis Crick (page 6) once wrote "The origin of life appears almost a miracle, so many are the conditions which would have had to be satisfied to get it going." That does not mean, though, that chemists and biologists have not tried to find those conditions.

Charles Darwin thought that life might have begun when simple molecules combined, under the right conditions of light, heat, and electrical energy, to produce ones of greater and greater complexity. At the end of the 19th century the Swedish chemist Svante Arrhenius theorized that microbes floating throughout the universe served as the "seeds of life" on earth, the panspermia hypothesis.

Darwin's idea lives on in experiments such as those done by Stanley Miller in 1953. Attempting to recreate what was thought to be the atmosphere of the primeval earth, he filled a flask with the gases methane, ammonia, and hydrogen and added a bit of water. A discharge of electricity acted like lightning in the mixture. The inside of the flask was soon covered with a reddish slime, a mixture that turned out to be rich in amino acids, the building blocks of proteins. Chemists thought they would soon know in more detail how living organisms began their development—but it was not to be. As Miller said recently, "The problem of the origin of life has turned out to be much more difficult than I, and most other people, envisioned."

Many more theories have been advanced to account for the origin of life. (See J. Horgan: "In the beginning . . . ," *Scientific American,* February 1991, pp. 117–125.) The most recent arises from a discovery made in 1977. Scientists were exploring the juncture of two of the tectonic plates that form the earth's surface in the Pacific Ocean. There, on the ocean's floor, they found thermal springs gushing a hot, black soup of minerals. Water seeping into cracks in the thin surface along the ridge is superheated to between 300 and

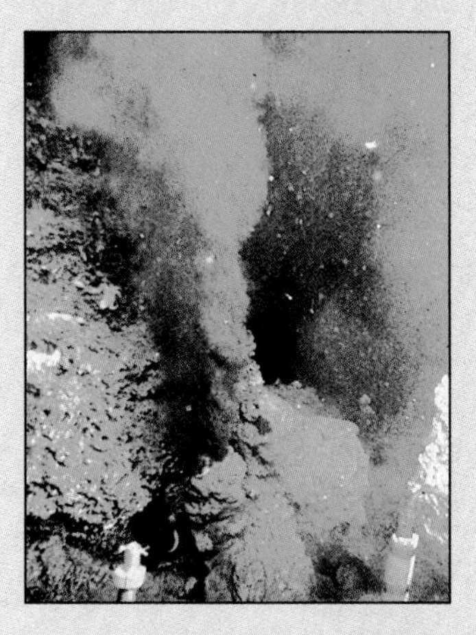

A "black smoker" in the east Pacific rise. *(Dudley Foster/Woods Hole Oceanographic Institution)*

Finally, notice that there will always be a conservation of charge as well as mass in a balanced chemical equation. Thus, in the $Ag^+ + Br^-$ net ionic equation, the cation and anion charges balance to give a net electric charge on the left of 0, the same as the charge on AgBr(s) on the right.

Example 5.3 Writing and Balancing Net Ionic Equations

Write a balanced, net ionic equation for the reaction of $AgNO_3$ and $CaCl_2$ to give AgCl and $Ca(NO_3)_2$.

Solution

Step 1. First write the complete, balanced equation using the correct formulas for reactants and products.

$$2\,AgNO_3 + CaCl_2 \longrightarrow 2\,AgCl + Ca(NO_3)_2$$

Step 2. Decide on the solubility of each compound using Figure 5.4. One general guideline is that nitrates are almost always soluble, so $AgNO_3$ and $Ca(NO_3)_2$

400 °C by the magma of the earth's core that is close to the surface. This superhot water dissolves minerals in the crust and provides conditions for the conversion of sulfates in seawater to hydrogen sulfide, H_2S. When this hot water, now laden with dissolved minerals and rich in sulfides, gushes through the surface, it cools, and metal sulfides, such as those of copper, manganese, iron, zinc, and nickel, precipitate.

$$H_2S(aq) + Cu^{2+}(aq) \longrightarrow CuS(s) + 2\,H^+(aq)$$

Many metal sulfides are black, thus giving the appearance of a black "smoke" coming from the earth, and the vents have been called "black smokers." The smoke settles around the edges of the vent, and eventually forms a "chimney" of precipitated minerals. (See R. A. Lutz and R. M. Haymon: *National Geographic,* Vol. 186, No. 5, 1994, pp. 115–126, for more information.)

Scientists were amazed to discover that the vents were surrounded by peculiar, primitive animals living in the hot, sulfide-rich environment. Because smokers are under hundreds of feet of water, and sunlight does not penetrate to these depths, the animals have developed a way to live without the energy from sunlight. It is currently believed they derive the energy needed to make the organic compounds on which they depend from the reaction of oxygen with hydrogen sulfide or the hydrogen sulfide ion, HS^-.

$$HS^-(aq) + 2\,O_2(g) \longrightarrow HSO_4^-(aq) + \text{energy}$$

Most recently a German lawyer and scientist, G. Wächtershäuser, has suggested that the exotic environment of black smokers is favorable to the construction of complex molecules from simple ones. He and Claudia Huber published experiments showing that metal sulfides such as iron pyrite, which exists in smokers or volcanoes and is ubiquitous on earth, can act as a catalyst to promote the formation of a compound with a C—C bond from molecules present in the gases from smokers. (C. Huber and G. Wächtershäuser: *Science,* Vol. 276, April 11, 1997, p. 245.) This is an important result for several reasons. First, simple molecules with carbon-carbon bonds have been prepared under conditions similar to a black smoker. Second, the product of the reaction reacts with water to form biochemically important acetic acid. Third, the reaction is much like that known to occur in primitive organisms called *Archaebacteria.*

$$2\,CH_3SH + CO \longrightarrow CH_3CO(SCH_3) + H_2S$$

This reaction, done by Huber and Wächtershäuser, shows that organic molecules with C—C bonds can be prepared under conditions resembling those in black smokers deep in the ocean.

are water-soluble. Chloride salts are also usually water-soluble although AgCl is an exception. We can therefore write

$$2\,AgNO_3(aq) + CaCl_2(aq) \longrightarrow 2\,AgCl(s) + Ca(NO_3)_2(aq)$$

Step 3. Recognizing that all soluble ionic compounds dissociate to form ions in aqueous solution, we have

$$AgNO_3(aq) \equiv Ag^+(aq) + NO_3^-(aq)$$
$$CaCl_2(aq) \equiv Ca^{2+}(aq) + 2\,Cl^-(aq)$$
$$Ca(NO_3)_2(aq) \equiv Ca^{2+}(aq) + 2\,NO_3^-(aq)$$

This results in the following ionic equation:

$$2\,Ag^+(aq) + 2\,NO_3^-(aq) + Ca^{2+}(aq) + 2\,Cl^-(aq) \longrightarrow 2\,AgCl(s) + Ca^{2+}(aq) + 2\,NO_3^-(aq)$$

Step 4. Several spectator ions occur in the ionic equation (Ca^{2+} and NO_3^-), so these are eliminated to give the following net ionic equation.

$$2\,Ag^+(aq) + 2\,Cl^-(aq) \longrightarrow 2\,AgCl(s)$$

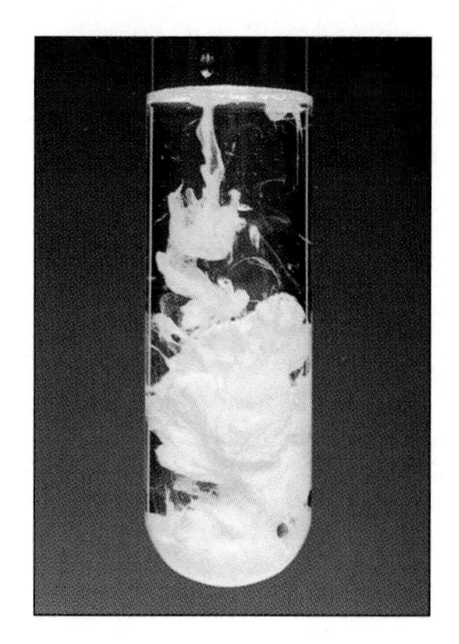

Precipitation reaction. The reaction of silver nitrate and calcium chloride produces insoluble silver chloride and water-soluble calcium nitrate. See Example 5.3. (*C. D. Winters*)

Notice that each species in the net equation is preceded by a coefficient of 2. Therefore, the equation can be simplified by dividing through by 2 to give

$$Ag^+(aq) + Cl^-(aq) \longrightarrow AgCl(s)$$

Step 5. Finally, notice that the sum of ion charges is the same on both sides of the equation. On the left, 1+ and 1− give zero; on the right the electric charge on AgCl is also zero.

Exercise 5.4 Net Ionic Equations

Write balanced net ionic equations for each of the following reactions:

1. $BaCl_2(aq) + Na_2SO_4(aq) \longrightarrow BaSO_4(s) + NaCl(aq)$
2. Iron(III) chloride is mixed with potassium hydroxide to give iron(III) hydroxide and potassium chloride. See Figure 5.7c.
3. Solutions of lead(II) nitrate and potassium chloride are mixed to give lead(II) chloride and potassium nitrate.

5.3 ACIDS AND BASES

See the *Saunders Interactive General Chemistry CD-ROM,* Screens 4.7, 4.8, and 4.11, for a discussion of acids and bases and their reactions.

Acids and bases, two very important classes of compounds, have a number of properties in common. Solutions of acids or bases, for example, can change the colors of vegetable pigments in specific ways. In a reaction you may have seen, acids change the color of litmus, a dye derived from certain lichens, from blue to red. And when you add lemon to tea, the citric acid in the lemon reacts with compounds in the tea and changes their color.

Bases also affect pigments. For example, they can cause phenolphthalein, a colorless compound sometimes used in laxatives, to turn pink (Figure 5.9a). If an acid has made litmus paper turn red, adding a base reverses the effect, making the litmus blue again. Thus, acids and bases seem to be opposites. A base can neutralize the effect of an acid, and an acid can neutralize the effect of a base.

Acids have other characteristic properties. They taste sour, produce bubbles of CO_2 gas when added to limestone, and dissolve many metals, producing hydrogen gas (H_2) at the same time (Figure 5.9c). Although tasting substances is *never* done in a chemistry laboratory, you have probably experienced the sour taste of citric acid, which is commonly found in fruits and is added to candies and soft drinks. Bases, in contrast, have a bitter taste. Also, bases often cause metal ions to form insoluble compounds that precipitate from solution (see Figure 5.7c). Such precipitates can be made to dissolve by adding an acid, another case in which an acid counteracts a property of a base.

Acids

The properties of acids can be interpreted in terms of a common feature of acid molecules, and a different feature can explain the properties of bases. *An* ***acid*** *is any substance that, when dissolved in pure water, increases the concentration of* ***hydrogen ions, H^+(aq),*** *in the water.* One of the most common acids is hydrogen

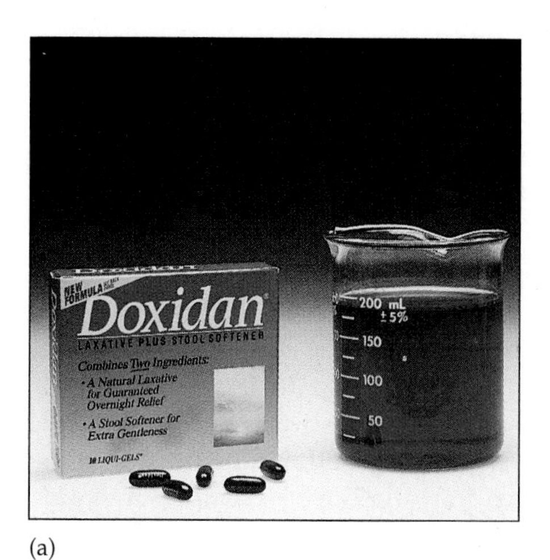

(a)

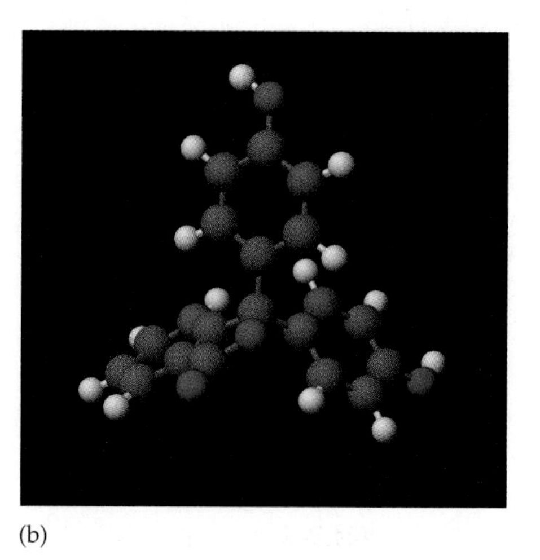

(b)

(c)

***Figure* 5.9 Some properties of acids and bases.** (a) Most metals dissolve in acids such as hydrochloric or sulfuric acid. (b) Some commercially available laxatives contain the dye phenolphthalein. The compound turns pink in basic solution (a) but is colorless in acid. (Try this yourself. Crush a laxative tablet in water and add the solution to household ammonia, a base, and then to vinegar, which contains acetic acid.) (c) A metal reacting with acid. (*Photo, C. D. Winters; model, S. M. Young*)

chloride, which ionizes in water to form a hydrogen ion, $H^+(aq)$, and a chloride ion, $Cl^-(aq)$.

$$HCl(aq) \longrightarrow H^+(aq) + Cl^-(aq)$$

hydrochloric acid
strong electrolyte
= 100% ionized

Because it is completely converted to ions in aqueous solution, HCl is a **strong acid** (and a strong electrolyte). See Table 5.2 for other common acids.

***Table* 5.2 • Common Acids and Bases**

Strong Acids (Strong Electrolytes)		**Strong Bases (Strong Electrolytes)**	
HCl	hydrochloric acid	LiOH	lithium hydroxide
HBr	hydrobromic acid	NaOH	sodium hydroxide
HI	hydroiodic acid	KOH	potassium hydroxide
HNO_3	nitric acid		
$HClO_4$	perchloric acid		
H_2SO_4	sulfuric acid		
Weak Acids (Weak Electrolytes)*		**Weak Base (Weak Electrolyte)**	
H_3PO_4	phosphoric acid	NH_3	ammonia
CH_3CO_2H	acetic acid		
H_2CO_3	carbonic acid		

*These are representative of hundreds of weak acids. See, for example, citric acid.

Models of the acids in Table 5.2 (and NH_3) and others in this chapter are found in the Models folder on the *Saunders Interactive General Chemistry CD-ROM* and the Web site for this book.

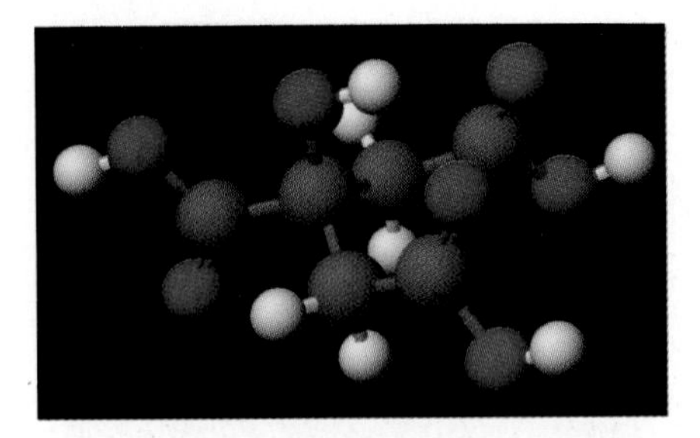

Acids. How do you know when an acid or base is strong or weak? The common strong acids and bases are those listed in Table 5.2. There are many additional weak acids and bases, however, and many are natural substances. Citric acid, $H_3C_6H_5O_7$, is among them. Other naturally occurring weak acids include tartaric acid (found in grapes), aspirin (page 1), and oxalic acid (page 171). Weak bases include nicotine and caffeine.

Many acids, such as sulfuric acid, can provide more than 1 mol of H^+ per mole of acid.

$$\underset{\text{sulfuric acid, 100\% ionized}}{H_2SO_4(aq)} \longrightarrow \underset{\text{hydrogen ion}}{H^+(aq)} + \underset{\text{hydrogen sulfate ion}}{HSO_4^-(aq)}$$

$$\underset{\text{hydrogen sulfate ion, < 100\% ionized}}{HSO_4^-(aq)} \longrightarrow \underset{\text{hydrogen ion}}{H^+(aq)} + \underset{\text{sulfate ion}}{SO_4^{2-}(aq)}$$

The first ionization reaction is essentially complete, so sulfuric acid is considered a strong acid (and, thus, a strong electrolyte). The hydrogen sulfate ion (HSO_4^-), however, like acetic acid, is only partially ionized in aqueous solution. Both the hydrogen sulfate ion and acetic acid are therefore classified as **weak acids.**

Bases

A ***base*** *is a substance that increases the concentration of* ***hydroxide ion, OH^-(aq),*** *when dissolved in pure water.* The property that aqueous bases have in common is the formation of OH^-(aq). Compounds that contain hydroxide ions, such as sodium or potassium hydroxide, are obvious bases. As water-soluble ionic compounds they are strong bases (and strong electrolytes).

$$\underset{\text{sodium hydroxide, soluble base; strong electrolyte = 100\% dissociated}}{NaOH(s)} \longrightarrow Na^+(aq) + \underset{\text{hydroxide ion}}{OH^-(aq)}$$

Ammonia, NH_3, also a very common base, does not have an OH^- ion as part of its formula. Instead, it produces the OH^- ion on reaction with water.

$$\underset{\text{ammonia, base; weak electrolyte < 100\% ionized}}{NH_3(aq)} + H_2O(\ell) \longrightarrow \underset{\text{ammonium ion}}{NH_4^+(aq)} + \underset{\text{hydroxide ion}}{OH^-(aq)}$$

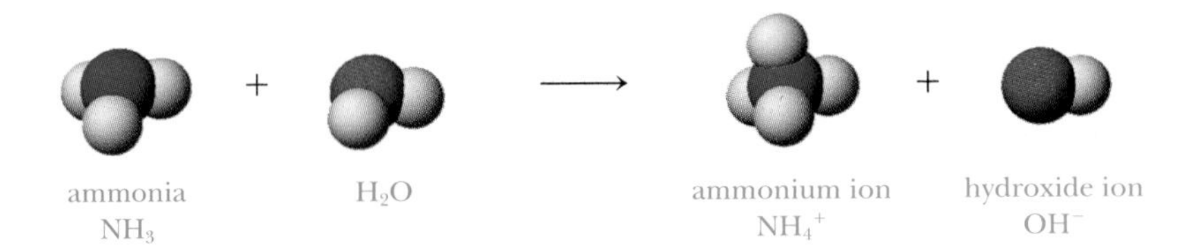

Only a small concentration of ions is present, so ammonia is a **weak base** (and a weak electrolyte).

Exercise 5.5 Acids and Bases

1. What ions are produced when perchloric acid dissolves in water?
2. Barium hydroxide is not very soluble in water. What little does dissolve, however, is dissociated. What ions are produced?

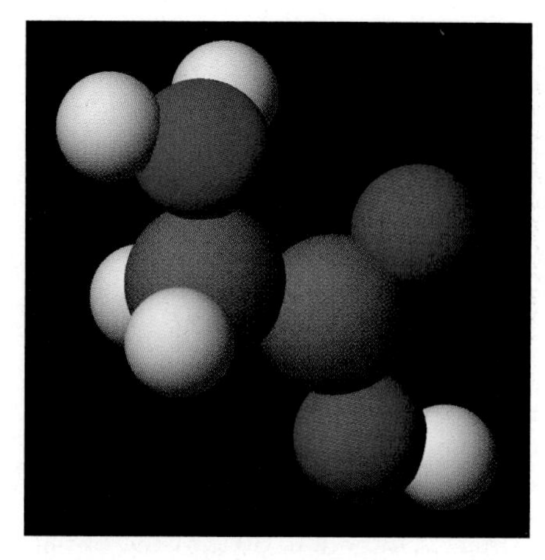

The structure of glycine, $NH_2CH_2CO_2H$. Ammonia is the basis of a vast family of compounds, many of biological importance. For example, the essential amino acid glycine can be considered a derivative of ammonia, where one H atom has been replaced by $—CH_2CO_2H$.

Oxides of Nonmetals and Metals

Each acid shown in Table 5.2 has one or more H atoms in the molecular formula that can be released to form H^+ ions in water. There are, however, less obvious compounds that form acidic solutions. These are oxides of nonmetals, such as carbon dioxide and sulfur trioxide, which have no H atoms but which react with water to produce H^+ ions. Carbon dioxide, for example, dissolves in water to a small extent, and some of the dissolved molecules react with water to form the weak acid, carbonic acid. This acid then ionizes to a small extent to form the hydrogen ion, H^+, and the bicarbonate ion, HCO_3^-,

$$CO_2(g) \xrightarrow{+H_2O} H_2CO_3(aq) \longrightarrow HCO_3^-(aq) + H^+(aq)$$

This reaction is very important in our environment. Carbon dioxide is normally found in small amounts in the atmosphere, so rainwater is always slightly acidic. Oxides like CO_2 that can react with water to produce H^+ ions are known as **acidic oxides.**

See the *Saunders Interactive General Chemistry CD-ROM,* Screen 4.9, for a discussion of acidic and basic oxides.

Oxides of sulfur and nitrogen are present in significant amounts in polluted air and can lead ultimately to acids and other pollutants. For example, sulfur dioxide, SO_2, from human and natural sources can react with oxygen to give sulfur trioxide, SO_3, which then forms sulfuric acid with water. Nitrogen dioxide, NO_2, reacts with water to give nitric and nitrous acids.

$$\underbrace{2\,SO_2(g)}_{\text{from burning fossil fuels and from volcanoes}} + O_2(g) \longrightarrow 2\,SO_3(g)$$

$$SO_3(g) + H_2O(\ell) \longrightarrow \underbrace{H_2SO_4(aq)}_{\text{sulfuric acid}}$$

$$2\,NO_2(g) + H_2O(\ell) \longrightarrow \underbrace{HNO_3(aq)}_{\text{nitric acid}} + \underbrace{HNO_2(aq)}_{\text{nitrous acid}}$$

These reactions are the origin of acid rain from the burning of fossil fuels such as coal and gasoline in the United States, Canada, and other industrialized countries. The gaseous oxides mix with water and other chemicals in the troposphere, and the rain that falls is more acidic than if it contained only dissolved CO_2. When the rain falls on areas that cannot easily tolerate this greater than normal acidity, such as the northeastern parts of the United States and the eastern provinces of Canada, serious environmental problems can occur.

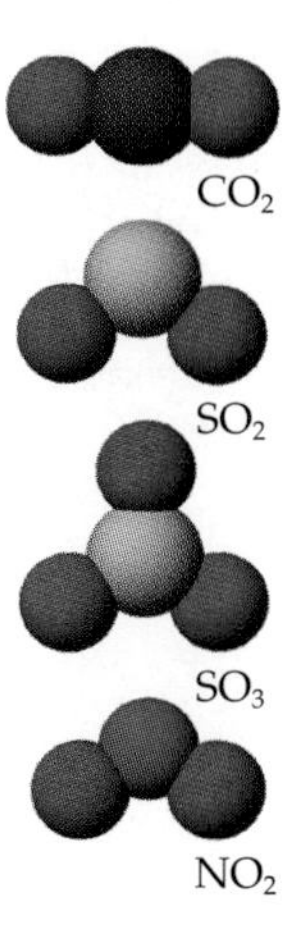

Some common nonmetal oxides that form acids in water.

Oxides of metals can give basic solutions if they dissolve appreciably in water. Perhaps the best known example is calcium oxide, CaO, often called *lime,*

Current Perspectives

Sulfuric Acid—A Gauge of Business Conditions

Sulfuric acid is a colorless, syrupy liquid with a density of 1.84 g/mL and a boiling point of 270 °C. It has several desirable properties that have led to its widespread use: it is generally less expensive to produce than other acids, is a strong acid, can be handled in steel containers, reacts readily with many organic compounds to produce useful products, and reacts readily with lime (CaO), the least expensive and most readily available base.

The acid is made by the contact process. The first step is combustion of sulfur in air to give sulfur dioxide.

$$\frac{1}{8}\,S_8(s) + O_2(g) \longrightarrow SO_2(g)$$

This gas is then combined with more oxygen, in the presence of a catalyst, to give sulfur trioxide,

$$SO_2(g) + \frac{1}{2}\,O_2(g) \longrightarrow SO_3(g)$$

which can give sulfuric acid when absorbed in water.

$$SO_3(g) + H_2O(\ell) \longrightarrow H_2SO_4(aq)$$

For some years sulfuric acid has been the chemical produced in the largest quantity in the United States (and in many other industrialized countries). About 43 billion kilograms (43 million metric tons) was made in the United States in 1995. Some economists have said this production is a measure of a nation's industrial strength or general business conditions.

Currently, over two thirds of the production is used in the phosphate fertilizer industry, which makes "superphosphate" fertilizer by treating phosphate rock with sulfuric acid.

$$Ca_{10}F_2(PO_4)_6(s) + 7\,H_2SO_4(aq) + 3\,H_2O(\ell) \longrightarrow 3\,Ca[H_2PO_4]_2 \cdot H_2O(s) + 7\,CaSO_4(s) + 2\,HF(g)$$

The remainder is used to make pigments, explosives, alcohol, pulp and paper, detergents, and as a component for storage batteries.

Sulfuric acid is not just an industrial product, however. It is also excreted by sea slugs to defend themselves.

Sulfur is found in very pure form in underground deposits along the coast of the United States in the Gulf of Mexico. It is recovered by pumping superheated steam into the sulfur beds to melt the sulfur. The molten sulfur is brought to the surface by means of compressed air. (*Farrel Grehan/Photo Researchers, Inc.*)

A sea slug excretes sulfuric acid in self-defense. (*Sharksong/M. Kazmers/Dembinsky Photo Associates*)

or *quicklime.* This metal oxide reacts with water to give calcium hydroxide, commonly called *slaked lime.* The latter compound, although not very soluble in water (0.17 g/100 g H_2O at 10 °C), is widely used in industry as a base because it is inexpensive.

$$\underset{\text{lime}}{CaO(s)} + H_2O(\ell) \longrightarrow \underset{\text{slaked lime}}{Ca(OH)_2(s)}$$

A CLOSER LOOK — H^+ *Ions in Water*

The H^+ ion is a hydrogen atom that has lost its electron. Only the nucleus, a proton, remains. Because a proton is only about 1/100,000 as large as the average atom or ion, water molecules can approach very closely, and the proton and the water molecules are strongly attracted. The H^+ ion in water is better represented by the combination of H^+ and H_2O, $\mathbf{H_3O^+}$, an ion called the **hydronium ion.** Experiments also show, however, that other forms of the ion exist in water, one example being $[H_3O(H_2O)_3]^+$. Thus, the ionization of HCl actually provides a solution containing the hydrogen ion surrounded by some number of water molecules. We shall often use H^+ (aq) in this text to indicate the presence of hydronium and similar ions, but be aware that the real situation is much more complex.

$$HCl(aq) + H_2O(\ell) \longrightarrow H_3O^+(aq) + Cl^-(aq)$$

hydrochloric acid
strong electrolyte
= 100% ionized

hydronium ion

Oxides like CaO that react with water to produce OH^- ions are known as **basic oxides.** Almost 19 billion kg of lime was produced in the United States in 1995 for use in the metals and construction industries, in sewage and pollution control, in water treatment, and in agriculture.

Exercise 5.6 Acidic and Basic Oxides

For each of the following, indicate whether you expect an acidic or basic solution when the compound dissolves in water:

1. SeO_2 **2.** MgO **3.** P_4O_{10}

5.4 REACTIONS OF ACIDS AND BASES

In general, acids react with strong bases in aqueous solution to produce a salt and water. For example (Figure 5.10),

$$\underset{\text{hydrochloric acid}}{HCl(aq)} + \underset{\text{sodium hydroxide}}{NaOH(aq)} \longrightarrow \underset{\text{sodium chloride}}{NaCl(aq)} + \underset{\text{water}}{H_2O(\ell)}$$

The word "salt" has come into the language of chemistry to describe any ionic compound whose cation comes from a base (here Na^+ from NaOH) and whose anion comes from an acid (here Cl^- from HCl). Reaction of any of the acids listed in Table 5.2 with any of the bases listed there produces a salt and water. Reactions of acids with weak bases like ammonia, however, produce only salts; see Example 5.4.

Hydrochloric acid and sodium hydroxide are strong electrolytes in water (see Table 5.2), so the complete ionic equation for the reaction of HCl(aq) and NaOH(aq) should be written as

$$\underbrace{H^+(aq) + Cl^-(aq)}_{\text{from HCl(aq)}} + \underbrace{Na^+(aq) + OH^-(aq)}_{\text{from NaOH(aq)}} \longrightarrow \underbrace{Na^+(aq) + Cl^-(aq)}_{\text{salt}} + \underset{\text{water}}{H_2O(\ell)}$$

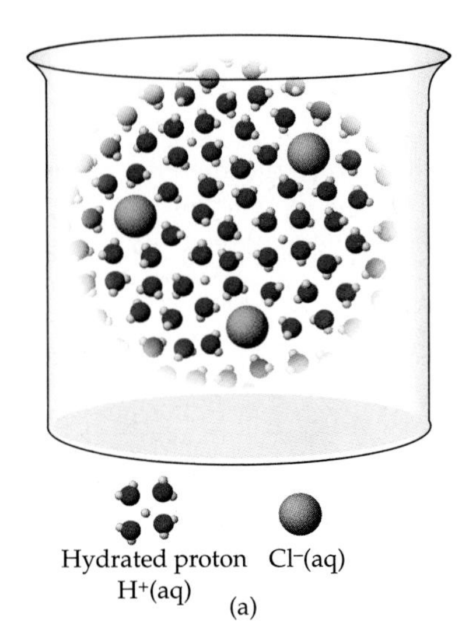

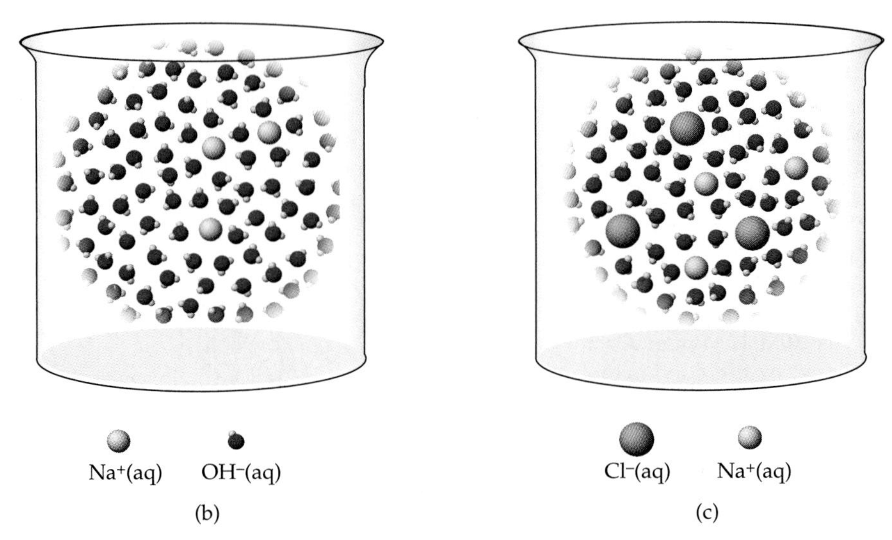

Figure 5.10 An acid-base reaction. (a, b) Beakers of HCl(aq) and NaOH(aq). The acid and base consist of ions in aqueous solution. (c) On mixing, the acid and base react to give a salt and water.

Because Na^+ and Cl^- ions appear on both sides of the equation, the net ionic equation is simply the combination of the ions H^+ and OH^- to give water.

$$H^+(aq) + OH^-(aq) \longrightarrow H_2O(\ell)$$

All acid-base reactions do not lead to a "neutral" solution. (See Chapter 18.) The term "neutralization" is generally used, however, when referring to acid-base reactions.

Indeed, *this is the net ionic equation for the reaction between any* ***strong acid*** *and any* ***strong base.*** Reactions between strong acids and strong bases are called **neutralization reactions** because, on completion of the reaction, the solution is neutral, neither acidic nor basic. The other ions (the cation of the base and the anion of the acid) remain unchanged. If the water is evaporated, however, the cation and anion form a solid salt. In the example above, NaCl can be obtained, whereas nitric acid, HNO_3, and NaOH give the salt sodium nitrate, $NaNO_3$.

$$HNO_3(aq) + NaOH(aq) \longrightarrow NaNO_3(aq) + H_2O(\ell)$$

Calcium oxide, lime, is inexpensive and is used in waste and pollution control. Indeed, one of its major uses is in "scrubbing" sulfur oxides from the exhaust gases of power plants fueled by coal and oil. The oxides of sulfur dissolve in water to produce acids (page 195), and these acids can react with a base. Lime produces the base calcium hydroxide when added to water. This mixture is sprayed into the exhaust stack of the power plant, where it reacts with sulfur-containing acids. One such reaction is

$$Ca(OH)_2(s) + H_2SO_4(aq) \longrightarrow CaSO_4 \cdot 2\,H_2O(s)$$

The compound $CaSO_4 \cdot 2\,H_2O$, hydrated calcium sulfate, is found in the earth as the mineral gypsum. Assuming the gypsum from a coal-burning power plant is not contaminated with compounds based on "heavy" metals (e.g., Hg) or other pollutants, it is environmentally acceptable to put it into the earth.

Example 5.4 Ammonia as a Base

Write the balanced overall equation and the net ionic equation for the reaction of aqueous ammonia and nitric acid to produce the salt ammonium nitrate.

Solution The overall, balanced equation for the reaction is

$$\underset{\text{ammonia}}{NH_3(aq)} + \underset{\text{nitric acid}}{HNO_3(aq)} \longrightarrow \underset{\text{ammonium nitrate}}{NH_4NO_3(aq)}$$

Ammonia is a base because it reacts with water, to a *very* small extent, to produce hydroxide ions (Figure 5.11).

$$NH_3(aq) + H_2O(\ell) \longrightarrow NH_4^+(aq) + OH^-(aq)$$

Nitric acid is a strong acid and produces hydrogen and nitrate ions. It is therefore written as $H^+(aq) + NO_3^-(aq)$. The OH^- ions produced by NH_3 react with the H^+ ions produced by HNO_3 to form water.

$$OH^-(aq) + H^+(aq) + NO_3^-(aq) \longrightarrow H_2O(\ell) + NO_3^-(aq)$$

Adding these two equations gives the balanced, net ionic equation for the reaction.

$$NH_3(aq) + H_2O(\ell) \longrightarrow NH_4^+(aq) + OH^-(aq)$$
$$OH^-(aq) + H^+(aq) + NO_3^-(aq) \longrightarrow H_2O(\ell) + NO_3^-(aq)$$

$$NH_3(aq) + H^+(aq) \longrightarrow NH_4^+(aq)$$

Notice that when the equations were summed, OH^-, NO_3^-, and H_2O are eliminated. The net result is that the net ionic equation for ammonia reacting with a strong acid is the transfer of H^+ ion from the acid to NH_3.

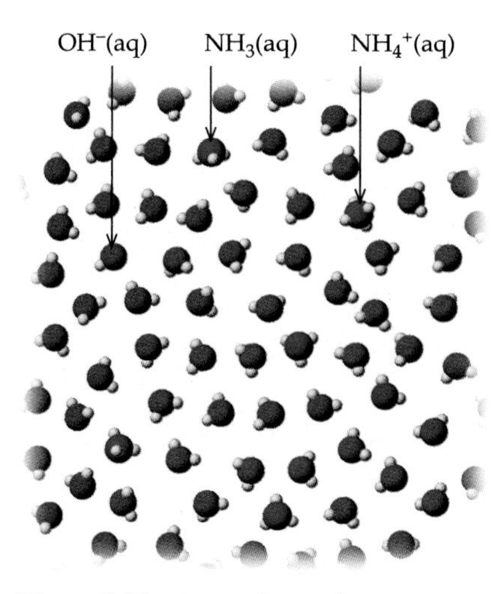

Figure 5.11 Ammonia as a base. Ammonia is a weak base because *very* few NH_3 molecules react with H_2O to produce OH^- and NH_4^+ ions in aqueous solution.

Example 5.5 Acid-Base Reactions

Write the balanced, overall equation and the net ionic equation for the reaction of calcium hydroxide with acetic acid.

Solution Like many acid-base reactions, $Ca(OH)_2$ and CH_3CO_2H produce a salt and water.

$$\underset{\text{calcium hydroxide}}{Ca(OH)_2(s)} + \underset{\text{acetic acid}}{2\ CH_3CO_2H(aq)} \longrightarrow \underset{\text{calcium acetate}}{Ca(CH_3CO_2)_2(aq)} + 2\ H_2O(\ell)$$

Here the salt is calcium acetate, which consists of one calcium ion, Ca^{2+}, and two acetate ions, $CH_3CO_2^-$. Notice that calcium hydroxide supplies 2 mol of OH^- ions per mole of $Ca(OH)_2$ and acetic acid supplies only 1 mol of H^+ ions per mole of the acid.

To write the net ionic equation we first recognize (a) that $Ca(OH)_2$ is not very soluble in water and (b) that acetic acid is a weak electrolyte. Therefore, we write $Ca(OH)_2$ as a solid reactant and acetic acid as a nonionized compound. The product, calcium acetate, is a soluble ionic compound, and so it consists of Ca^{2+} and $CH_3CO_2^-$ ions in water. This means the net ionic equation is not very different from the overall equation.

$$Ca(OH)_2(s) + 2\ CH_3CO_2H(aq) \longrightarrow Ca^{2+}(aq) + 2\ CH_3CO_2^-(aq) + 2\ H_2O(\ell)$$

Exercise 5.7 Acid-Base Reactions

Write the balanced, overall equation and the net ionic equation for the reaction of magnesium hydroxide with hydrochloric acid.

5.5 GAS-FORMING REACTIONS

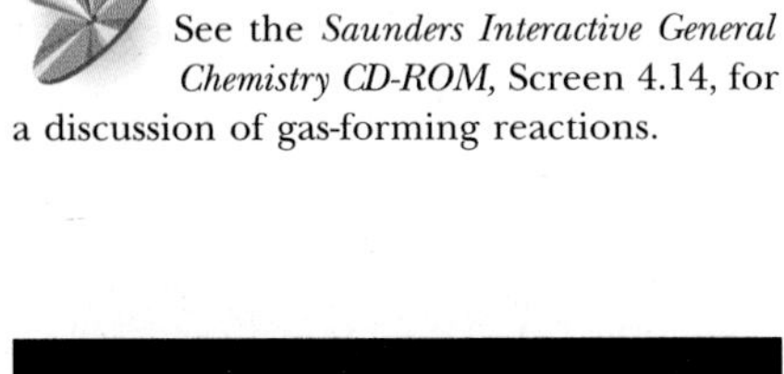

See the *Saunders Interactive General Chemistry CD-ROM,* Screen 4.14, for a discussion of gas-forming reactions.

Have you ever made biscuits or muffins? As you bake the dough, it rises in the oven (Figure 5.12). But what makes it rise? A gas-forming reaction occurs between an acid of some type and baking soda, sodium hydrogen carbonate (bicarbonate of soda, $NaHCO_3$). One possible acid used is tartaric acid, an acid found in many foods, so a typical reaction would be

$$\underset{\text{tartaric acid}}{C_4H_6O_6(aq)} + \underset{\text{hydrogen carbonate ion}}{HCO_3^-(aq)} \longrightarrow \underset{\text{tartrate ion}}{C_4H_5O_6^-(aq)} + H_2O(\ell) + CO_2(g)$$

Figure 5.12 Biscuits rise because of a gas-forming reaction. The acid and sodium bicarbonate in baking powder produce carbon dioxide gas. The acid used in many baking powders is $CaHPO_4$, but $NaAl(SO_4)_2$ is also common. (The aluminum-containing compound forms an acidic solution when placed in water. See Chapter 17.) (*C. D. Winters*)

In dry baking powder the acid and $NaHCO_3$ are kept apart by using starch as a filler. When mixed into the moist batter, however, the acid and the hydrogen carbonate ion react to produce CO_2, and the dough rises.

Several types of chemical reactions lead to gas formation (Table 5.3), but the most common is CO_2 formation. All metal carbonates (and bicarbonates) react with acids to produce carbonic acid, H_2CO_3, which in turn decomposes to carbon dioxide and water (Figure 5.13).

$$CaCO_3(s) + 2\,HCl(aq) \longrightarrow CaCl_2(aq) + H_2CO_3(aq)$$

$$H_2CO_3(aq) \longrightarrow H_2O(\ell) + CO_2(g)$$

A salt and H_2CO_3 are always the products from an acid and a metal carbonate. Carbonic acid is unstable, however, and is rapidly converted to water and CO_2 gas. If the reaction is done in an open beaker, most of the gas bubbles out of the solution.

The *Chemical Puzzler* at the beginning of the chapter described a reaction between Alka-Seltzer and water. The bubbles that arise when the tablet is dropped into water come from the reaction of an acid (citric acid) with the hydrogen

Table 5.3 • Gas-Forming Reactions

Metal carbonate or bicarbonate + acid $\longrightarrow$ metal salt + $CO_2(g)$ + H_2O
$Na_2CO_3(aq) + 2\,HCl(aq) \longrightarrow 2\,NaCl(aq) + CO_2(g) + H_2O(\ell)$
Metal sulfide + acid $\longrightarrow$ metal salt + $H_2S(g)$
$Na_2S(aq) + 2\,HCl(aq) \longrightarrow 2\,NaCl(aq) + H_2S(g)$
Metal sulfite + acid $\longrightarrow$ metal salt + $SO_2(g)$ + $H_2O(\ell)$
$Na_2SO_3(aq) + 2\,HCl(aq) \longrightarrow 2\,NaCl(aq) + SO_2(g) + H_2O(\ell)$
Ammonium salt + strong base $\longrightarrow$ metal salt + $NH_3(g)$ + H_2O
$NH_4Cl(aq) + NaOH(aq) \longrightarrow NaCl(aq) + NH_3(g) + H_2O(\ell)$

carbonate ion (or bicarbonate ion). (See also *Current Perspectives: What to Take for an Upset Stomach,* page 165.)

$$\underset{\text{citric acid}}{H_3C_6H_5O_7(aq)} + \underset{\text{hydrogen carbonate ion}}{HCO_3^-(aq)} \longrightarrow \underset{\text{dihydrogen citrate ion}}{H_2C_6H_5O_7^-(aq)} + H_2O(\ell) + CO_2(g)$$

Figure 5.13 Metal carbonates react with acids to give a salt, carbon dioxide gas, and water. Here a piece of coral, which is mostly $CaCO_3$, is reacting with hydrochloric acid. (*C. D. Winters*)

Example 5.6 Gas-Forming Reactions

Write a balanced equation for the reaction that occurs when nickel(II) carbonate is treated with sulfuric acid.

Solution As described in the text, when a metal carbonate (or metal hydrogen carbonate) is treated with an acid, the products are water and CO_2, as well as a metal salt. The anion of the metal salt is the anion from the acid. Here the metal salt must be nickel(II) sulfate.

$$NiCO_3(s) + H_2SO_4(aq) \longrightarrow NiSO_4(aq) + H_2O(\ell) + CO_2(g)$$

Exercise 5.8 A Gas-Forming Reaction

1. Barium carbonate, $BaCO_3$, is widely used in the brick, ceramic, glass, and chemical manufacturing industries. Write a balanced equation that shows what happens when barium carbonate is treated with nitric acid. Give the name of each of the reaction products.
2. Write a balanced equation for the reaction of ammonium sulfate with sodium hydroxide.

5.6 ORGANIZING REACTIONS IN AQUEOUS SOLUTION

One goal of this chapter is to explore the most common types of reactions that can occur in aqueous solution. This helps you decide, for example, what may occur when an Alka-Seltzer tablet is dropped into water. This section summarizes or organizes our discussions thus far.

Chemists are interested in reactions in aqueous solution not only because they provide one way to make useful products, but also because these kinds of reactions occur on the earth and in plants and animals. Because such reactions are important, it is useful to look for common reaction patterns to see what their "driving forces" might be. How can you know when you mix two chemicals in water that they will combine to produce one or more new compounds?

To organize what you have learned about reactions in aqueous solution, recognize that a common theme ties all the reactions we have described together: in each case *the ions of the reactants changed partners.*

$$A^+B^- + C^+D^- \longrightarrow A^+D^- + C^+B^-$$

Thus, we might refer to them as **exchange reactions,** which gives us a good way of predicting the products of precipitation, acid-base, and gas-forming reactions.

Precipitation Reactions (see Figure 5.7): Ions combine in solution to form an insoluble reaction product.

Overall Equation

$$Pb(NO_3)_2(aq) + 2\,KI(aq) \longrightarrow PbI_2(s) + 2\,KNO_3(aq)$$

Net Ionic Equation

$$Pb^{2+}(aq) + 2\,I^-(aq) \longrightarrow PbI_2(s)$$

Acid-Base Reactions: The cation of the base and the anion of the acid form a salt. For strong acids and bases, water is also a product.

Overall Equation

$$HNO_3(aq) + KOH(aq) \longrightarrow KNO_3(aq) + HOH(\ell)$$

Net Ionic Equation

$$H^+(aq) + OH^-(aq) \longrightarrow H_2O(\ell)$$

Gas-Forming Reactions (see Figure 5.13): The most common examples involve metal carbonates and acids but others exist (see Table 5.3). One product is carbonic acid, H_2CO_3, most of which decomposes to H_2O and CO_2. Carbon dioxide is the gas in the bubbles you see during this reaction.

Overall Equation

$$NiCO_3(s) + 2\,HNO_3(aq) \longrightarrow Ni(NO_3)_2(aq) + H_2CO_3(aq)$$
$$H_2CO_3(aq) \longrightarrow CO_2(g) + H_2O(\ell)$$

Net Ionic Equation

$$NiCO_3(s) + 2\,H^+(aq) \longrightarrow Ni^{2+}(aq) + CO_2(g) + H_2O(\ell)$$

Oxidation-Reduction Reactions: A fourth reaction type, **oxidation-reduction reactions,** also occurs in water. These reactions, which are described in more detail in Section 5.7, involve the transfer of electrons from one substance to another rather than the exchange of partners.

Overall Equation

$$Cu(s) + 2\,AgNO_3(aq) \longrightarrow Cu(NO_3)_2(aq) + 2\,Ag(s)$$

Net Ionic Equation

$$Cu(s) + 2\,Ag^+(aq) \longrightarrow Cu^{2+}(aq) + 2\,Ag(s)$$

In summary, four common "driving forces" are responsible for reactions in aqueous solution.

Reaction Type	Driving Force
Precipitation	Formation of an insoluble compound
Acid-base; neutralization	Formation of a salt and water; proton transfer
Gas-forming	Evolution of a water-insoluble gas such as CO_2
Oxidation-reduction	Electron transfer

Acid-base and oxidation-reduction reactions have a similar feature. Acid-base reactions involve the transfer of a proton (H^+) from acid to base. Oxidation-reduction reactions (Section 5.7) involve the transfer of electrons from reducing agent to oxidizing agent.

These types of reactions are usually easy to recognize, but keep in mind that a reaction may have more than one driving force. For example, barium carbonate reacts readily with sulfuric acid to give barium sulfate, carbon dioxide, and water, a reaction that is both a precipitation and a gas-forming reaction.

$$BaCO_3(s) + H_2SO_4(aq) \longrightarrow BaSO_4(s) + H_2O(\ell) + CO_2(g)$$

Exercise 5.9 Classifying Reactions

Classify each of the following reactions as a precipitation, acid-base reaction, or gas-forming reaction. Predict the products of the reaction, and then balance the completed equation.

1. $CuCO_3(s) + H_2SO_4(aq) \longrightarrow$
2. $Ba(OH)_2(s) + HNO_3(aq) \longrightarrow$
3. $ZnCl_2(aq) + (NH_4)_2S(aq) \longrightarrow$

Problem-Solving Tips and Ideas

5.1 Net Ionic Equations

Net ionic equations are commonly used for chemical reactions in aqueous solution because they describe as accurately as possible the actual chemical species involved in a reaction. Now that you have reviewed the common types of chemical reactions that occur in aqueous solution, we can summarize some features of net ionic equations.

To write net ionic equations we must know what compounds exist as ions in solution.

1. Strong acids, soluble strong bases, and soluble salts exist as ions in solution. Examples include the acids HCl and HNO_3, a base such as NaOH, and salts such as NaCl and $MgCl_2$ (see Figure 5.4 and Table 5.2).
2. All other species should be represented in molecular form. Weak acids exist in solutions primarily as molecules. Even though their structures are made up of ions, insoluble salts such as $CaCO_3(s)$ or insoluble bases such as $Mg(OH)_2(s)$ should not be written in ionic form.

The best way to approach writing net ionic equations is to follow precisely a set of steps.

1. Write a complete, balanced, molecular equation. Indicate the state of each substance (aq, s, ℓ, g).
2. Next rewrite the whole equation, writing all strong acids, strong soluble bases, and soluble salts as ions. (Look carefully at anything labeled with an "(aq)" suffix. Select from those all strong acids, soluble salts, and soluble bases, and rewrite them as ions.)
3. Finally, you will likely see that some of the ions remain unchanged in the reaction (the ions that appear in the equation both as reactants or products). This means that they are not changed in the reaction and that they are not part of the chemistry that is going on; we call them *spectator ions.* Thus, you can cancel them from each side of the equation.

Here are three general net ionic equations it is helpful to remember:

- The reaction between any strong acid and any soluble strong base. The equation for all such reactions is $H^+(aq) + OH^-(aq) \longrightarrow H_2O(\ell)$.

- The reaction of any weak acid (HX, such as HCN, HF, HOCl, CH_3CO_2H) and a soluble strong base; the equation is $HX + OH^-(aq) \longrightarrow H_2O(\ell) + X^-(aq)$.
- The reaction of ammonia with any acid HX: $NH_3(aq) + HX(aq) \longrightarrow NH_4^+(aq) + X^-(aq)$.

Finally, an important hint: Like molecular equations, net ionic equations must be balanced. The same number of atoms appears on each side of the arrow. But there is an additional requirement: The sum of the ion charges on the two sides must also be equal.

5.7 OXIDATION-REDUCTION REACTIONS

See the *Saunders Interactive General Chemistry CD-ROM,* Screens 4.15 and 4.16, for a discussion of oxidation-reduction reactions.

The terms "oxidation" and "reduction" come from reactions that have been known for centuries. Ancient civilizations learned how to change metal oxides and sulfides to the metal, that is, how to reduce ore to the metal. A modern example is the reduction of iron oxide with carbon monoxide to give iron metal (Figure 5.14). Here carbon monoxide removes oxygen from iron(III) oxide, so the iron(III) oxide is said to have been reduced.

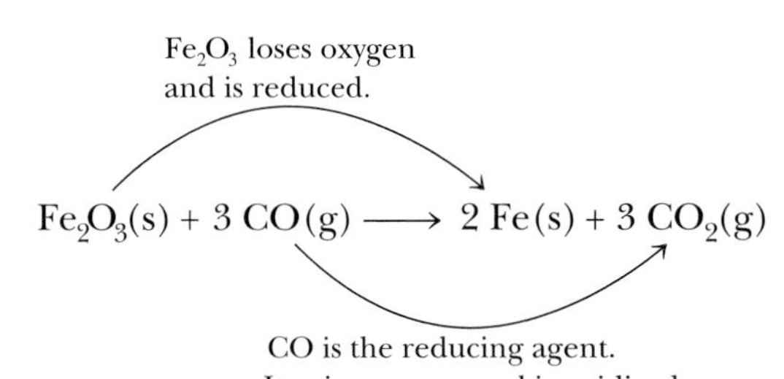
Fe_2O_3 loses oxygen and is reduced.
$Fe_2O_3(s) + 3\ CO(g) \longrightarrow 2\ Fe(s) + 3\ CO_2(g)$
CO is the reducing agent.
It gains oxygen and is oxidized.

In this reaction carbon monoxide is the agent that brings about the reduction of iron ore to iron metal, so carbon monoxide is called the **reducing agent.**

When Fe_2O_3 is reduced by carbon monoxide, oxygen is removed from the iron ore and added to the carbon monoxide, which is "oxidized" by the addition of oxygen to give carbon dioxide. *Any process in which oxygen is added to another substance is an oxidation.* This too is a process known for centuries, and Figures 4.2 through 4.5 show more examples. In the reaction with magnesium, for example (see Figure 4.3), oxygen is the **oxidizing agent** because it is the agent responsible for the oxidation.

Figure 5.14 Reduction of iron oxide. Iron ore is reduced to metallic iron with carbon or carbon monoxide in a blast furnace, a process done on a massive scale worldwide. (*Rosenfeld Images LTD/Science Photo Library/Photo Researchers, Inc.*)

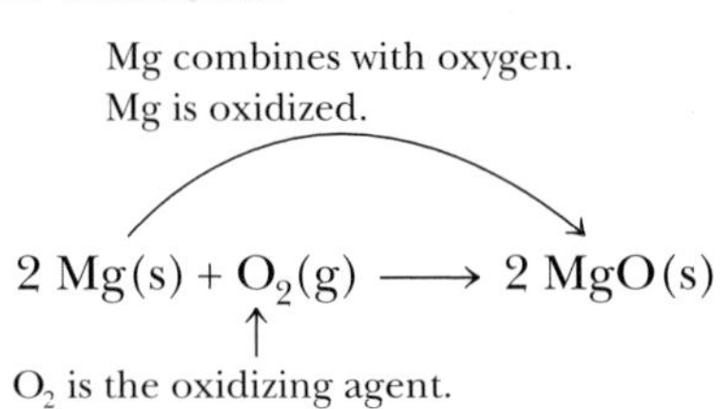
Mg combines with oxygen.
Mg is oxidized.
$2\ Mg(s) + O_2(g) \longrightarrow 2\ MgO(s)$
O_2 is the oxidizing agent.

The experimental observations outlined previously point to several important conclusions:

- If one substance is oxidized, another substance in the same reaction *must* be reduced. For this reason, such reactions are often called oxidation-reduction reactions, or **redox reactions** for short.
- The reducing agent is itself oxidized, and the oxidizing agent is reduced.
- Oxidation is the opposite of reduction. For example, the removal of oxygen is reduction and the addition of oxygen is oxidation.

Redox Reactions and Electron Transfer

Not all redox reactions involve oxygen. *All oxidation and reduction reactions, however, do involve transfer of electrons between substances.* When a substance accepts electrons, it is said to be **reduced** because there is a reduction in the electric charge on an atom of the substance. In the net ionic equation shown here, positively charged Ag^+ is reduced to uncharged Ag(s) on accepting electrons from copper metal.

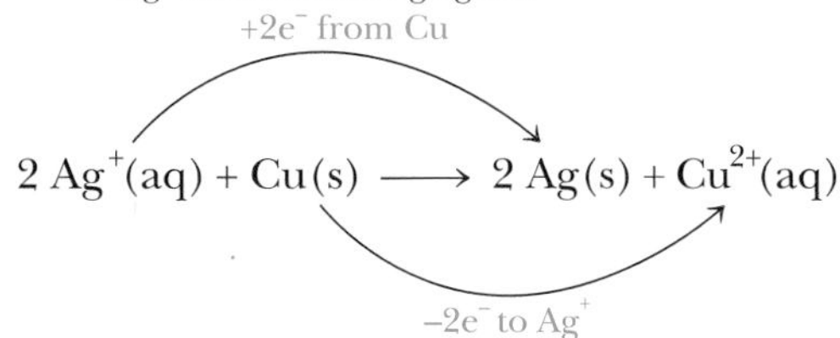
Ag^+ accepts electrons from Cu and is reduced to Ag;
Ag^+ is the oxidizing agent.
$+2e^-$ from Cu
$2\,Ag^+(aq) + Cu(s) \longrightarrow 2\,Ag(s) + Cu^{2+}(aq)$
$-2e^-$ to Ag^+
Cu donates electrons to Ag^+ and is oxidized to Cu^{2+};
Cu is the reducing agent.

Because copper metal supplies the electrons and causes the Ag^+ ion to be reduced, Cu is called the reducing agent (Figure 5.15).

When a substance *loses electrons,* the positive charge on an atom of the substance increases. The substance is said to have been **oxidized.** In our example, copper metal releases electrons on going to Cu^{2+}; because its positive charge has increased, it is said to have been oxidized. For this to happen, something must be available to take the electrons offered by the copper. In this case, Ag^+ is the electron acceptor, and its charge is reduced to zero in silver metal. Therefore, Ag^+ is the "agent" that causes Cu metal to be oxidized, so Ag^+ is the oxidizing agent. In every oxidation-reduction reaction, one reactant is reduced

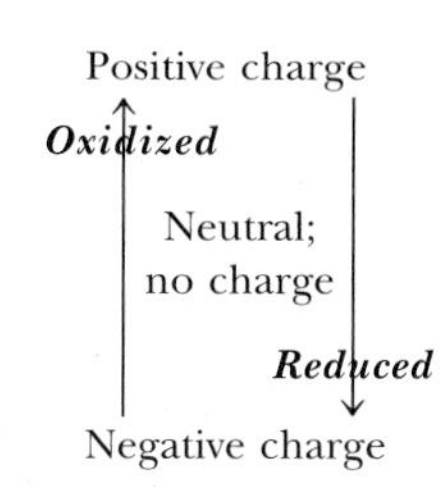
Positive charge
Oxidized
Neutral;
no charge
Reduced
Negative charge

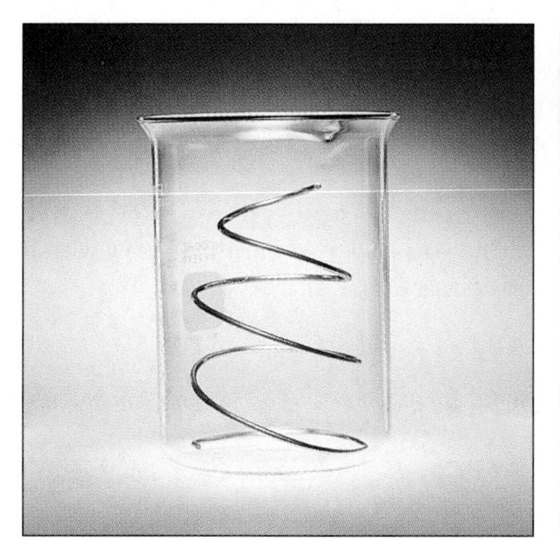
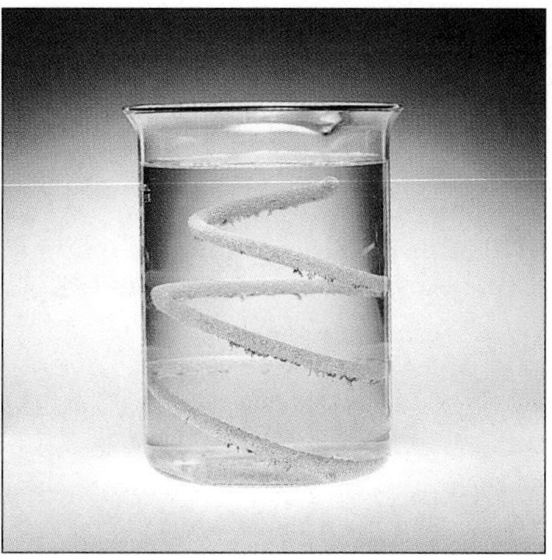
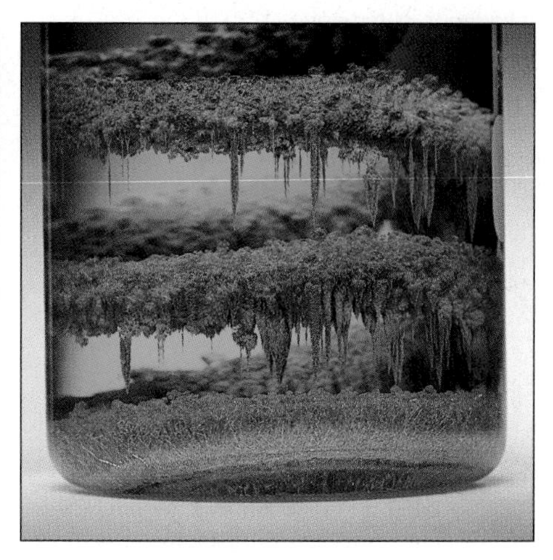

Figure 5.15 The oxidation of copper metal by silver ions. A clean piece of copper wire is placed in a solution of silver nitrate, $AgNO_3$. With time, the copper reduces Ag^+ ions to silver metal crystals, and the copper metal is oxidized to copper ions, Cu^{2+}. The blue color of the solution is due to the presence of aqueous copper(II) ions. (*C. D. Winters*)

(and is therefore the oxidizing agent) and one reactant is oxidized (and is therefore the reducing agent). In summary,

The redox reaction $X + Y \longrightarrow X^{n+} + Y^{n-}$ can be divided into two parts:

$X \longrightarrow X^{n+} + ne^-$: X is oxidized, losing n electrons (to Y) to form X^{n+}. X is the reducing agent in the process.

$Y + ne^- \longrightarrow Y^{n-}$: Y is reduced, gaining n electrons (supplied by X) to form Y^{n-}. Y is the oxidizing agent in the process.

In the reaction of magnesium and oxygen (see Figure 4.3), oxygen is the oxidizing agent because it gains electrons (four electrons per molecule) on going to the oxide ion.

Mg releases $2e^-$ per atom;
Mg is oxidized and is the reducing agent.

$$2\,Mg(s) + O_2(g) \longrightarrow 2\,[Mg^{2+}, O^{2-}]$$

O_2 gains $4e^-$ per molecule;
O_2 is reduced and is the oxidizing agent.

In the same reaction, magnesium is the reducing agent because it releases two electrons per atom on forming the Mg^{2+} ion. All redox reactions can be analyzed in a similar manner.

Oxidation Numbers

See the *Saunders Interactive General Chemistry CD-ROM,* Screen 4.17, Oxidation Numbers.

Except for monatomic ions, the oxidation number for an atom does *not* represent the actual electric charge on that atom. See Chapter 9 for more on atom charges in molecules.

How can you tell an oxidation-reduction reaction when you see one? How can you tell which substance has gained or lost electrons and so decide which one is the oxidizing or reducing agent? The answer is to *look for a change in the oxidation number of an element in the course of the reaction.* The **oxidation number** of an atom in a molecule or ion is defined as the electric charge an atom has, or *appears* to have, as determined by some guidelines for assigning oxidation numbers (see *Guidelines for Determining Oxidation Numbers*).

The reason for learning about oxidation numbers at this point is to be able to identify which reactions are oxidation-reduction processes and to know which is the oxidizing agent and which is the reducing agent in a reaction.

Example 5.7 Determining Oxidation Numbers

Charges on ions are written as (number, sign) (Cu^{2+}), whereas oxidation numbers are written as (sign, number). For example, the oxidation number of the Cu^{2+} ion is +2.

Determine the oxidation number of the indicated element in each of the following compounds or ions:

1. Aluminum in aluminum oxide, Al_2O_3
2. Sulfur in sulfuric acid, H_2SO_4
3. Manganese in the permanganate ion, MnO_4^-
4. Each Cr atom in the dichromate ion, $Cr_2O_7^{2-}$

Guidelines for Determining Oxidation Numbers

1. **Each atom in a pure element has an oxidation number of zero.** The oxidation number of Cu in metallic copper, as well as for each atom in I_2 or S_8, is 0.
2. **For ions consisting of a single atom, the oxidation number is equal to the charge on the ion.** Elements of Periodic Groups 1A–3A form monatomic ions with a positive charge and an oxidation number equal to the group number. Magnesium therefore forms Mg^{2+}, and its oxidation number is +2. (See Section 3.3.)
3. **Fluorine is always −1 in compounds with other elements.**
4. **Cl, Br, and I are always −1 in compounds *except* when combined with oxygen and fluorine.** This means that Cl has an oxidation number of −1 in NaCl (in which Na is +1, as predicted by the fact that it is an element of Group 1A). In the ion ClO^-, however, the Cl atom has an oxidation number of +1 (and O has an oxidation number of −2; see Guideline 5).
5. **The oxidation number of H is +1 and of O is −2 in most compounds.** Although this statement applies to many, many compounds, a few important exceptions occur.
   - When H forms a binary compound with a metal, the metal forms a positive ion and H becomes a hydride ion, H^-. Thus, in CaH_2 the oxidation number of Ca is +2 (equal to the group number) and that of H is −1.
   - Oxygen can have an oxidation number of −1 in a class of compounds called *peroxides,* compounds based on the O_2^{2-} ion. For example, in H_2O_2, hydrogen peroxide, H is assigned its usual oxidation number of +1, and so O is −1.
6. **The algebraic sum of the oxidation numbers in a neutral compound must be zero; in a polyatomic ion, the sum must be equal to the ion charge.** Examples of this rule are the previously mentioned compounds and others found in Example 5.7.

Solution

1. Al_2O_3 is a neutral compound. Assuming that O has its "normal" oxidation number of −2, the oxidation number of Al must be +3, as predicted by its position in the periodic table.

$$Al_2O_3 = (2\,Al^{3+})(3\,O^{2-})$$

$$\begin{aligned}\text{Net charge} &= 0\\ &= \text{oxidation number of Al} + \text{oxidation number of O}\\ &= 2(+3) + 3(-2)\end{aligned}$$

2. H_2SO_4 has an overall charge of 0. If the oxygen atoms each have an oxidation number of −2 and that of the H atoms is +1, then the oxidation number of S must be +6.

$$H_2SO_4 = (2\,H^+)(S^{6+})(4\,O^{2-})$$

$$\begin{aligned}\text{Net charge} &= 0\\ &= \text{sum of oxidation numbers for H atoms} + \text{oxidation number of S} + \text{sum of oxidation numbers for O atoms}\\ &= 2(+1) + (+6) + 4(-2)\end{aligned}$$

3. The permanganate ion, MnO_4^-, has an overall charge of 1−. Because this compound is not a peroxide, O is assigned an oxidation number of −2, which means that Mn has an oxidation number of +7.

$$MnO_4^- = [(Mn^{7+})(4\,O^{2-})]^-$$

$$\begin{aligned}\text{Net charge} &= -1\\ &= \text{oxidation number of Mn} + \text{sum of oxidation numbers for O atoms}\\ &= (+7) + 4(-2)\end{aligned}$$

Dichromate ion, $Cr_2O_7^{2-}$

Be sure to notice in each of the cases in Example 5.7 that the O atom is first assumed to have an oxidation number of −2 and H is +1. Oxidation numbers of the remaining elements can then be found.

4. Compounds containing dichromate ions, $Cr_2O_7^{2-}$, are widely used in the laboratory. Because the net charge on the ion is 2−, and O is assigned an oxidation number of −2, each Cr atom must have an oxidation number of +6.

$$Cr_2O_7^{2-} = [(2\ Cr^{6+})(7\ O^{2-})]^{2-}$$

$$\begin{aligned} \text{Net charge} &= -2 \\ &= \text{sum of oxidation numbers for Cr atoms + sum of oxidation numbers for O atoms} \\ &= 2(+6) + 7(-2) \end{aligned}$$

Exercise 5.10 Oxidation Numbers

Assign an oxidation number to the underlined atom in each of the following molecules or ions.

1. $\underline{Fe}_2O_3$ **2.** $H_3\underline{P}O_4$ **3.** $\underline{C}O_3^{2-}$ **4.** $\underline{N}O_2^{+}$

Recognizing Oxidation-Reduction Reactions

See the *Saunders Interactive General Chemistry CD-ROM,* Screen 4.18, Recognizing Oxidation-Reduction Reactions.

Having learned some guidelines for determining oxidation numbers, you can tell which reactions can be classified as oxidation-reduction and which must be of some other type. In many cases, however, it will be obvious that the reaction is an oxidation-reduction because it involves a well-known oxidizing or reducing agent (Table 5.4).

Like oxygen, O_2, the halogens (F_2, Cl_2, Br_2, and I_2) are always oxidizing agents in their reactions with metals and nonmetals. An example is the reaction of chlorine with sodium metal (see Figure 2.27).

Na releases $1e^-$ per atom.
Oxidation number increases.
Na is oxidized and is the reducing agent.

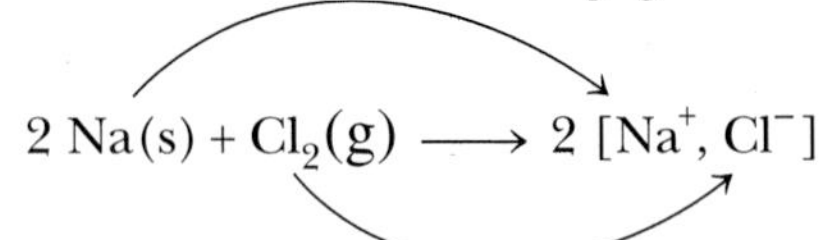

Cl_2 gains $2e^-$ per molecule.
Oxidation number decreases by 1 per Cl atom.
Cl_2 is reduced and is the oxidizing agent.

Chlorine ends up as Cl^-, having acquired two electrons (from two Na atoms) per Cl_2 molecule. Thus, the oxidation number of each Cl atom has decreased from 0 to −1. This means Cl_2 has been reduced and so is the oxidizing agent.

Chlorine gas is widely used as an oxidizing agent to treat water and sewage. For example, it can remove hydrogen sulfide, H_2S, from drinking water by oxidizing the sulfide to insoluble, elemental sulfur. (Hydrogen sulfide has a characteristic "rotten egg" odor; it comes from the decay of organic matter or underground mineral deposits.)

$$8\ Cl_2(g) + 8\ H_2S(aq) \longrightarrow S_8(s) + 16\ HCl(aq)$$

Table 5.4 • Common Oxidizing and Reducing Agents

Oxidizing Agent	Reaction Product	Reducing Agent	Reaction Product
O_2, oxygen	O^{2-}, oxide ion or O combined in H_2O	H_2, hydrogen	$H^+(aq)$, hydrogen ion or H combined in H_2O
Halogen, F_2, Cl_2, Br_2, or I_2	Halide ion, F^-, Cl^-, Br^-, or I^-	M, metals such as Na, K, Fe, and Al	M^{n+}, metal ions such as Na^+, K^+, Fe^{2+} or Fe^{3+}, and Al^{3+}
HNO_3, nitric acid	Nitrogen oxides* such as NO and NO_2	C, carbon (used to reduce metal oxides)	CO and CO_2
$Cr_2O_7^{2-}$, dichromate ion	Cr^{3+}, chromium(III) ion (in acid solution)		
MnO_4^-, permanganate ion	Mn^{2+}, manganese(II) ion (in acid solution)		

*NO is produced with dilute HNO_3, whereas NO_2 is a product of concentrated acid.

Figure 5.16 illustrates the chemistry of another excellent oxidizing agent, nitric acid, HNO_3. Here the acid oxidizes copper metal to give copper(II) nitrate, and the nitrate ion is reduced to the brown gas NO_2. The net ionic equation for the reaction is

$$\underset{\text{reducing agent}}{Cu(s)} + \underset{\text{oxidizing agent}}{2\,NO_3^-(aq)} + 4\,H^+(aq) \longrightarrow Cu^{2+}(aq) + 2\,NO_2(g) + 2\,H_2O(\ell)$$

Nitrogen has been reduced from +5 (in the NO_3^- ion) to +4 (in NO_2); therefore, the nitrate ion in acid solution is an oxidizing agent. Copper metal is clearly the reducing agent in this reaction; each metal atom has given up two electrons to produce the Cu^{2+} ion.

In the reactions of sodium with chlorine and copper with nitric acid, the metals are oxidized. This is typical of many metals, which are generally good reducing agents. Indeed, the alkali and alkaline earth metals are especially good reducing agents. Another example of this is the reaction of potassium with water. Here potassium reduces the hydrogen in water to H_2 gas (page 18).

$$\underset{\text{reducing agent}}{2\,K(s)} + \underset{\text{oxidizing agent}}{2\,H_2O(\ell)} \longrightarrow 2\,KOH(aq) + H_2(g)$$

Aluminum metal is yet another a good reducing agent and is capable of reducing iron(III) oxide to iron metal in a reaction called the *thermite reaction.*

$$\underset{\text{oxidizing agent}}{Fe_2O_3(s)} + \underset{\text{reducing agent}}{2\,Al(s)} \longrightarrow 2\,Fe(s) + Al_2O_3(s)$$

Such a large quantity of heat is evolved in the reaction that the iron is produced in the molten state (Figure 5.17).

***Figure* 5.16 The reaction of copper with nitric acid.** Copper (a reducing agent) reacts vigorously with concentrated nitric acid, an oxidizing agent, to give the brown gas NO_2 and a deep green solution of copper(II) nitrate. Notice that nitric acid, a common acid, can be both an oxidizing agent and an acid. (*C. D. Winters*)

(a)

(b)

(c)

Figure 5.17 The thermite reaction. After the reaction is started with a fuse of burning magnesium wire (a), iron(III) oxide reacts with aluminum powder to give aluminum oxide and iron metal (b). This reaction generates so much heat that iron metal is produced in the molten state (c). (*C. D. Winters*)

Hundreds of compounds are good oxidizing and reducing agents and, when mixed, undergo reaction. For this reason you should be aware that it is not a good idea to mix a strong oxidizing agent with a strong reducing agent; a violent reaction, even an explosion, may take place. This is a reason that chemicals are no longer stored on shelves in alphabetical order. This can be unsafe, because such an ordering may place a strong oxidizing agent next to a strong reducing agent.

Finally, Table 5.5, as well as Table 5.4, may help you organize your thinking as you look for oxidation-reduction reactions and use their terminology.

Table 5.5 • Recognizing Oxidation-Reduction Reactions

	Oxidation	Reduction
In terms of oxidation number	Increase in oxidation number of an atom	Decrease in oxidation number of an atom
In terms of electrons	Loss of electrons by an atom	Gain of electrons by an atom
In terms of oxygen	Gain of one or more atoms	Loss of one or more atoms

Example 5.8 Oxidation-Reduction Reactions

In each reaction, decide which atom is undergoing a change in oxidation number, and then identify the oxidizing agent and the reducing agent. Match the reactions with one of the categories in Table 5.4:

1. $5\,Fe^{2+}(aq) + MnO_4^-(aq) + 8\,H^+(aq) \longrightarrow 5\,Fe^{3+}(aq) + Mn^{2+}(aq) + 4\,H_2O(\ell)$
2. $H_2(g) + CuO(s) \longrightarrow Cu(s) + H_2O(g)$

Solution

1. *The reaction of the iron(II) ion with permanganate ion (Figure 5.18).* The permanganate ion is described as an oxidizing agent in Table 5.4. This is because the Mn oxidation number in MnO_4^- is +7, and it decreases to +2 in the product, the Mn^{2+} ion. Because Mn has been reduced, the MnO_4^- ion is the oxidizing agent.

   The oxidation number of iron has increased from +2 to +3, so the Fe^{2+} ion has lost electrons upon being oxidized to Fe^{3+} (see Table 5.5). This means the Fe^{2+} ion is the reducing agent.
2. *The reaction of hydrogen gas with copper(II) oxide (Figure 5.19).* H_2 gas is a common reducing agent, widely used in the laboratory and in industry. In these reactions of hydrogen, its oxidation number changes from 0 (in H_2) to +1 (in H_2O). Copper(II) oxide is the substance reduced here because the oxidation number of copper changes from +2 in CuO to 0 in copper metal (and CuO has lost an O atom). So, hydrogen is the reducing agent, and copper(II) oxide is the oxidizing agent.

Figure 5.18 The reaction of iron(II) ion and permanganate ion. The reaction of purple permanganate ion (MnO_4^-, the oxidizing agent) with the iron(II) ion (Fe^{2+}, the reducing agent) in acidified aqueous solution gives the nearly colorless manganese(II) ion (Mn^{2+}) and the iron(III) ion (Fe^{3+}). (*C. D. Winters*)

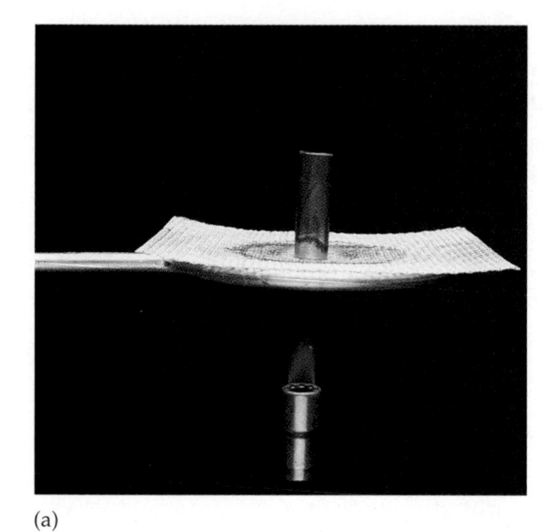

(a)

(b)

Figure 5.19 Reduction of copper oxide with hydrogen. (a) A piece of copper has been heated in air to form a film of black copper(II) oxide on the surface. (b) When the hot copper metal, with its film of CuO, is placed in a stream of hydrogen gas (from the yellow tank at the rear), the oxide is reduced to copper metal, and water forms as the byproduct. (*C. D. Winters*)

Example 5.9 Types of Reactions

Classify each of the following reactions as (a) precipitation, (b) acid-base, (c) gas-forming, or (d) oxidation-reduction.

1. $2\ HNO_3(aq) + Ca(OH)_2(s) \longrightarrow Ca(NO_3)_2(aq) + 2\ H_2O(\ell)$
2. $SO_4^{2-}(aq) + 2\ CH_2O(aq) + 2\ H^+(aq) \longrightarrow H_2S(aq) + 2\ CO_2(g) + 2\ H_2O(\ell)$

Solution One of the best ways to differentiate reactions is to look first for a reaction that could involve oxidation-reduction, that is, a reaction in which the oxidation number of some element changes. This is the case only in reaction 2. Writing this reaction with the oxidation number of each element indicated,

$$\underset{+6,-2}{SO_4^{2-}(aq)} + \underset{0,+1,-2}{2\ CH_2O(aq)} + \underset{+1}{2\ H^+(aq)} \longrightarrow \underset{+1,-2}{H_2S(aq)} + \underset{+4,-2}{2\ CO_2(g)} + \underset{+1,-2}{2\ H_2O(\ell)}$$

you see that the oxidation number of S changes from +6 to −2, and that of C changes from 0 to +4. Therefore, sulfate, SO_4^{2-}, has been reduced (and is the oxidizing agent), and CH_2O has been oxidized (and is the reducing agent).

No changes occur in oxidation number for the elements in reaction 1.

$$\underset{+1,+5,-2}{HNO_3(aq)} + \underset{+2,-2,+1}{Ca(OH)_2(s)} \longrightarrow \underset{+2,\ +5,-2}{Ca(NO_3)_2(aq)} + \underset{+1,-2}{2\ H_2O(\ell)}$$

Instead, reaction 1 is classified as an acid-base reaction. Here an acid (nitric acid, HNO_3) reacts with a base (calcium hydroxide, $Ca(OH)_2$) to give a salt (calcium nitrate, $Ca(NO_3)_2$) and water.

Exercise 5.11 Oxidation-Reduction Reactions

The following reaction occurs in a device for testing the breath for the presence of ethanol. Identify the oxidizing and reducing agents and the substance oxidized and the substance reduced (Figure 5.20).

$$\underset{\text{ethanol}}{3\ C_2H_5OH(aq)} + \underset{\text{dichromate ion; orange-red}}{2\ Cr_2O_7^{2-}(aq)} + 16\ H^+(aq) \longrightarrow \underset{\text{acetic acid}}{3\ CH_3CO_2H(aq)} + \underset{\text{chromium(III) ion; green}}{4\ Cr^{3+}(aq)} + 11\ H_2O(\ell)$$

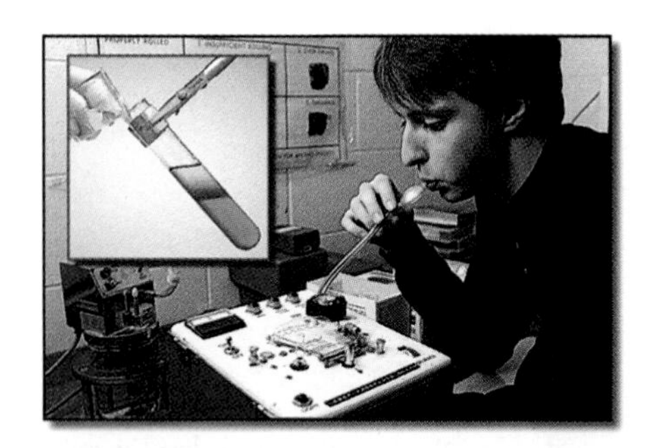

Figure 5.20 The redox reaction of ethanol and dichromate ion is the basis of the test used in a Breathalyzer. In the inset photo ethanol is poured into a solution of orange-red dichromate ion, which is converted to the green chromium(III) ion. (*C. D. Winters*)

Exercise 5.12 Oxidation-Reduction Reactions

Decide which of the following reactions are oxidation-reduction reactions. In each case explain your choice and identify the oxidizing and reducing agents.

1. $NaOH(aq) + HNO_3(aq) \longrightarrow NaNO_3(aq) + H_2O(\ell)$
2. $Cu(s) + Cl_2(g) \longrightarrow CuCl_2(s)$
3. $Na_2CO_3(aq) + 2\ HClO_4(aq) \longrightarrow CO_2(g) + H_2O(\ell) + 2\ NaClO_4(aq)$
4. $2\ S_2O_3^{2-}(aq) + I_2(aq) \longrightarrow S_4O_6^{2-}(aq) + 2\ I^-(aq)$

A CLOSER LOOK *Balancing Net Ionic Equations for Redox Reactions*

Many of the equations for oxidation–reduction reactions presented in this section are net ionic equations and all are balanced. Not only do the same number of atoms of each kind appear on both sides of the equation, but, just as importantly, the net electric charge is the same on both sides. For example, in the reaction of copper metal with silver ions, the total charge on both sides of the equation is +2.

$$Cu(s) + 2\,Ag^{+}(aq) \longrightarrow Cu^{2+}(aq) + 2\,Ag(s)$$

An approach to balancing many redox reactions takes advantage of the fact that such reactions occur by electron transfer. For example, on page 205 you learned that, in the Cu/Ag^{+} reaction, copper supplies two electrons ($2e^{-}$) and a silver ion acquires an electron (e^{-}) to become a silver atom.

Reducing agent; oxidation:

$$Cu(s) \longrightarrow Cu^{2+}(aq) + 2e^{-}$$

Oxidizing agent; reduction:

$$Ag^{+}(aq) + e^{-} \longrightarrow Ag(s)$$

These two equations represent the functioning of the reducing and oxidizing agents, respectively. Each of them is separately balanced for number of atoms and for electric charge. These two equations can be combined to give the balanced equation for the overall reaction by recognizing that, if one copper atom can supply two electrons, then *two* silver ions must be available to accept those electrons. Therefore, we have

$$\begin{array}{l} Cu(s) \longrightarrow Cu^{2+}(aq) + 2e^{-} \\ 2[Ag^{+}(aq) + e^{-} \longrightarrow Ag(s)] \\ \hline Cu(s) + 2\,Ag^{+}(aq) \longrightarrow Cu^{2+}(aq) + 2\,Ag(s) \end{array}$$

This is a very general approach for balancing net ionic equations for redox reactions. You will, however, have to learn some additional techniques for balancing equations for reactions that occur in an acidic or basic solution. This will be considered in Chapter 21 in the context of the discussion of electrochemistry.

5.8 MEASURING CONCENTRATIONS OF COMPOUNDS IN SOLUTION

Most kinds of chemical studies require quantitative measurements, and this includes experiments involving aqueous solutions. To accomplish this, we continue to use balanced equations and moles, but we measure volumes of solution rather than masses of solids, liquids, or gases. Solution concentration expressed as molarity relates the volume of solution in liters or milliliters to the amount of substance in moles.

Solution Concentration: Molarity

See the *Saunders Interactive General Chemistry CD-ROM,* Screen 5.11, Preparing Solutions of Known Concentration.

The concept of concentration is useful in many contexts. For example, about 4,900,000 people live in Wisconsin, and the state has a land area of roughly 56,000 square miles; therefore, the average concentration of people is about 88 per square mile. In chemistry the amount of solute dissolved in a given volume of solution, the **concentration** of the solution, can be found in the same way. Solution concentration, c, is usually reported as *moles of solute per liter of solution;* this is called the **molarity** of the solution.

Be sure to notice that the volume of the solution must be in units of liters when expressing and using concentrations in terms of molarity.

$$\text{Concentration } (c_{\text{molarity}}) = \frac{\text{Quantity of solute (mol)}}{\text{Volume of solution (L)}}$$

For example, if 58.4 g, or 1.00 mol, of NaCl is dissolved in enough water to give a total solution volume of 1.00 L, the concentration, c, is 1.00 mol/L, or

1.00 *molar*. This is often abbreviated as 1.00 M, where the capital M stands for "moles per liter."

$$c_{\text{molarity}} = 1.00 \text{ M} = [\text{NaCl}]$$

Another common notation is to place the formula of the compound in square brackets; this implies that the concentration of the solute in moles of compound per liter of solution is being specified. Finally, note that chemists use the terms "moles per liter" and "molar" interchangeably.

It is important to notice that molarity refers to moles of solute per liter of solution (and not to liters of solvent). If one liter of water is added to one mole of a solid compound, the final volume probably will not be exactly one liter, and the final concentration will not be exactly one molar (Figure 5.21). Therefore, when making solutions of a given molarity, it is almost always the case that we dissolve the solute in a volume of solvent smaller than the desired volume of solution, then make up the final volume with more solvent.

Figure 5.21 Preparing a 0.100 M solution of copper sulfate. To make a 0.100 M solution of $CuSO_4$, 25.0 g or 0.100 mol of $CuSO_4 \cdot 5\,H_2O$ (the blue crystalline solid) was placed in a 1.00-L volumetric flask. In this photo, exactly 1.00 L of water was measured out and slowly added to the volumetric flask. When enough water had been added so that the solution volume was exactly 1.00 L, approximately 8 mL of water (the quantity in the small graduated cylinder) was left over. This emphasizes that molar concentrations are defined as moles of solute per liter of solution and not per liter of water or other solvent. (*C. D. Winters*)

Potassium permanganate, $KMnO_4$, which was used at one time as a germicide in the treatment of burns, is a shiny, purple-black solid that dissolves readily in water to give deep purple solutions. Suppose 0.435 g of $KMnO_4$ has been dissolved in enough water to give 250. mL of solution (Figure 5.22). What is the molar concentration of $KMnO_4$? The first step is to convert the mass of material to moles.

$$0.435 \text{ g } \cancel{\text{KMnO}_4} \cdot \frac{1 \text{ mol KMnO}_4}{158.0 \text{ g } \cancel{\text{KMnO}_4}} = 0.00275 \text{ mol KMnO}_4$$

Now that the number of moles of substance is known, this can be combined with the volume of solution—*which must be in liters*—to give the molarity. Because 250. mL is equivalent to 0.250 L,

$$\text{Molarity of KMnO}_4 = \frac{0.00275 \text{ mol KMnO}_4}{0.250 \text{ L solution}} = 0.0110 \text{ M}$$

The $KMnO_4$ concentration is 0.0110 molar, or 0.0110 M. This is useful information, but it is often equally useful to know the concentration of each type of ion in a solution. Like all soluble ionic compounds, $KMnO_4$ dissociates completely into its ions, K^+ and MnO_4^-, when dissolved in water.

$$\underset{\text{100\% dissociation}}{\text{KMnO}_4(\text{aq})} \longrightarrow \text{K}^+(\text{aq}) + \text{MnO}_4^-(\text{aq})$$

One mole of $KMnO_4$ provides 1 mol of K^+ and 1 mol of MnO_4^-. Accordingly, 0.0110 M $KMnO_4$ gives a concentration of K^+ in the solution of 0.0110 M; similarly, the concentration of MnO_4^- is also 0.0110 M.

Another example of ion concentrations is provided by the dissociation of an ionic compound such as $CoCl_2$ (Figure 5.23).

$$\underset{\text{100\% dissociation}}{\text{CoCl}_2(\text{aq})} \longrightarrow \text{Co}^{2+}(\text{aq}) + 2\,\text{Cl}^-(\text{aq})$$

If 0.1 mol of $CoCl_2$ is dissolved in enough water to make 1 L of solution, the concentration of the cobalt(II) ion is $[Co^{2+}] = 0.1$ M. Because the compound dissociates to provide 2 mol of Cl^- ions for each mole of $CoCl_2$, however, the concentration of chloride ion, $[Cl^-]$, is 0.2 M.

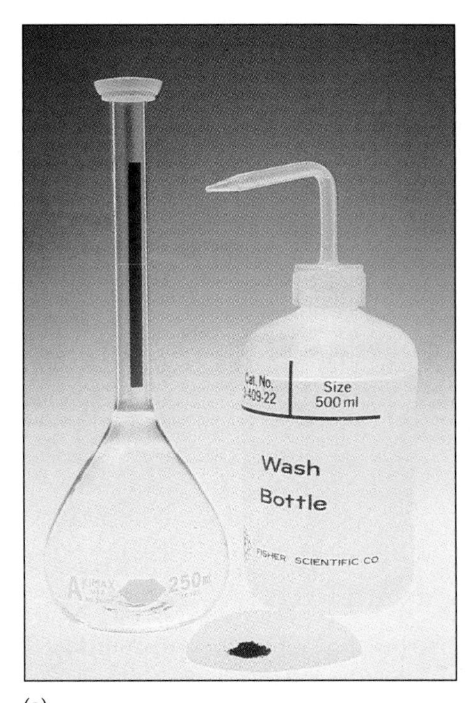

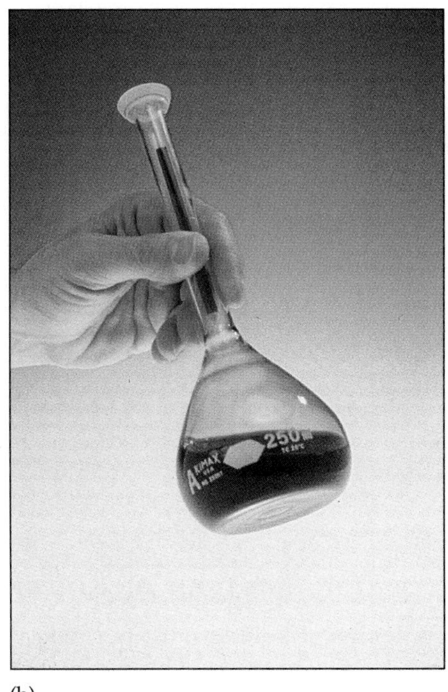

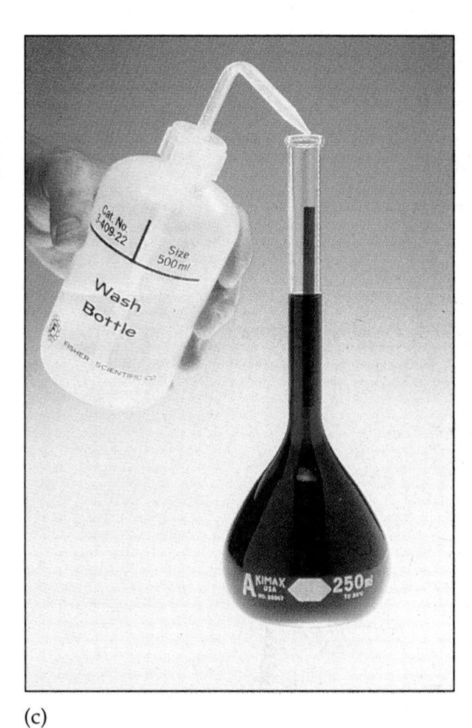

(a) (b) (c)

Figure 5.22 Making a solution. (a) A 0.0110 M solution of $KMnO_4$ is made by adding enough water to 0.435 g of $KMnO_4$ to make 0.250 L of solution. (b) To ensure the correct solution volume, the $KMnO_4$ is placed in a volumetric flask and dissolved in a small amount of water. (c) After dissolving is complete, sufficient water is added to fill the flask to the mark on the neck. The flask now contains 0.250 L of solution. (*C. D. Winters*)

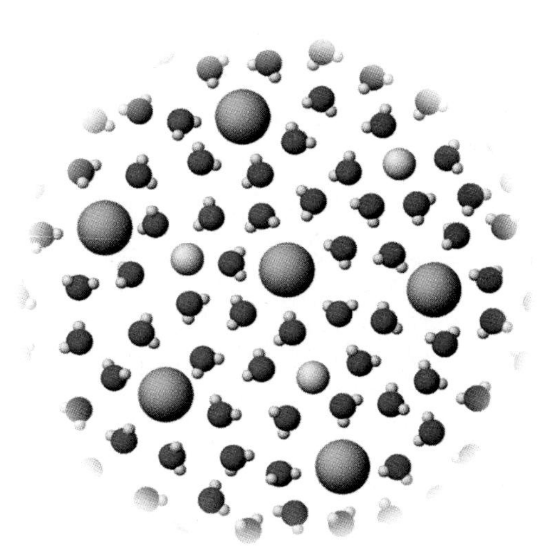

Figure 5.23 Dissolving $CoCl_2 \cdot 6\,H_2O$ in water. When this salt is dissolved, each unit of the compound provides one cobalt(II) ion and two chloride ions. (*Photo, C. D. Winters; model, S. M. Young*)

See the *Saunders Interactive General Chemistry CD-ROM,* Screen 5.10, Solution Concentration: Molarity. Also see on Screen 5.10, A Closer Look—Ion Concentrations.

Exercise 5.13 Solution Molarity

Sodium bicarbonate, $NaHCO_3$, is used in baking powder formulations, in fire extinguishers, and in the manufacture of plastics and ceramics, among other things. If 26.3 g of the compound is dissolved in enough water to make 200. mL of solution, what is the molar concentration of $NaHCO_3$?

Exercise 5.14 Ion Concentrations in Solution

Sodium sulfate, Na_2SO_4, is a strong electrolyte when dissolved in water. Assume enough has been dissolved so that $[Na_2SO_4] = 0.500$ M. State the concentration of each ion.

Preparing Solutions of Known Concentration

A task chemists often must perform is preparing a given volume of solution of known concentration. The problem is to find out what mass of solute to use.

Combining a Weighed Solute with the Solvent

Suppose you wish to prepare 2.00 L of a 1.50 M solution of Na_2CO_3. You have some solid Na_2CO_3 and distilled water. You also have a 2.00-L volumetric flask, a special flask with a line marked on its neck (see Figure 5.22). If the flask is filled with a solution to this line (at 20 °C), it contains precisely the volume of solution specified. To make the solution, you must weigh the necessary quantity of Na_2CO_3 as accurately as possible, carefully place all the solid in the volumetric flask, and then add water to dissolve the solid. After the solid has dissolved completely, more water is added to bring the solution volume to 2.00 L. The solution then has the desired concentration and the volume specified.

But what mass of Na_2CO_3 is required to make 2.00 L of 1.50 M Na_2CO_3? First, calculate the number of moles of substance required,

$$2.00\ \cancel{\text{L}} \cdot \frac{1.50\ \text{mol Na}_2\text{CO}_3}{1.00\ \cancel{\text{L}}\ \text{solution}} = 3.00\ \text{mol Na}_2\text{CO}_3\ \text{required}$$

and then the mass in grams.

$$3.00\ \cancel{\text{mol Na}_2\text{CO}_3} \cdot \frac{106.0\ \text{g Na}_2\text{CO}_3}{1\ \cancel{\text{mol Na}_2\text{CO}_3}} = 318\ \text{g Na}_2\text{CO}_3$$

Thus, to prepare the desired solution, you should dissolve 318 g of Na_2CO_3 in enough water to make 2.00 L of solution.

Exercise 5.15 Preparing Solutions of Known Concentration

An experiment in your laboratory requires 250. mL of a 0.0200 M solution of $AgNO_3$. You are given solid $AgNO_3$, distilled water, and a 250.-mL volumetric flask. Describe how to make up the required solution.

Diluting a More Concentrated Solution

Making a sodium carbonate solution as just described illustrates the most common way to create a solution of known concentration. Another method, however, is to *begin with a concentrated solution and add water to make it more dilute until the desired concentration is reached.* Many of the solutions prepared for your laboratory course are probably made by this dilution method. It is more efficient to store a few liters of a concentrated solution and then add water to make it into many liters of a dilute solution.

See the *Saunders Interactive General Chemistry CD-ROM,* Screen 5.12, Preparing Solutions of Known Concentration—Solution by Dilution.

As an example of the dilution method, suppose you need 1.00 L of 0.0025 M potassium dichromate, $K_2Cr_2O_7$, for use in chemical analysis. You have available a few liters of 0.100 M $K_2Cr_2O_7$ and some distilled water and appropriate glassware. How can you make the required 0.0025 M solution? The approach is to take some of the more concentrated $K_2Cr_2O_7$ solution, put it in a flask, and then add water until the $K_2Cr_2O_7$ is contained in a larger volume of water, that is, until it is less concentrated (or more dilute).

A practical example of making a solution by dilution is mixing frozen concentrated orange juice with water to make orange juice of the right concentration to drink.

The problem is to make a specified quantity of solution of known concentration. If the volume and concentration are known, then the number of moles of solute is also known. The number of moles of $K_2Cr_2O_7$ that must be in the final dilute solution is therefore

$$\text{Amount of } K_2Cr_2O_7 \text{ in final solution} = 1.00 \text{ L} \cdot 0.0025 \text{ mol/L} = 0.0025 \text{ mol } K_2Cr_2O_7$$

A more concentrated solution containing this number of moles of $K_2Cr_2O_7$ must be placed in the flask and then diluted to a total volume of 1.00 L. The volume of 0.100 M $K_2Cr_2O_7$ that contains the required number of moles is 25 mL.

$$0.0025 \cancel{\text{mol } K_2Cr_2O_7} \cdot \frac{1.00 \text{ L}}{0.100 \cancel{\text{mol } K_2Cr_2O_7}} = 0.025 \text{ L, or 25 mL}$$

Thus, to prepare 1.00 L of 0.0025 M $K_2Cr_2O_7$, place 25 mL of 0.100 M $K_2Cr_2O_7$ in a 1.00-L flask and add water until a volume of 1.00 L is reached (Figure 5.24).

The direction that one can prepare a solution by adding water to a more concentrated solution is correct *except* for sulfuric acid solutions. When mixing water and sulfuric acid, the resulting solution becomes quite warm. If water is added to concentrated sulfuric acid, so much heat is evolved that the solution may boil over or splash and burn someone nearby. To avoid this, chemists always add concentrated sulfuric acid *to water* when making a dilute acid solution.

(a)

(b)

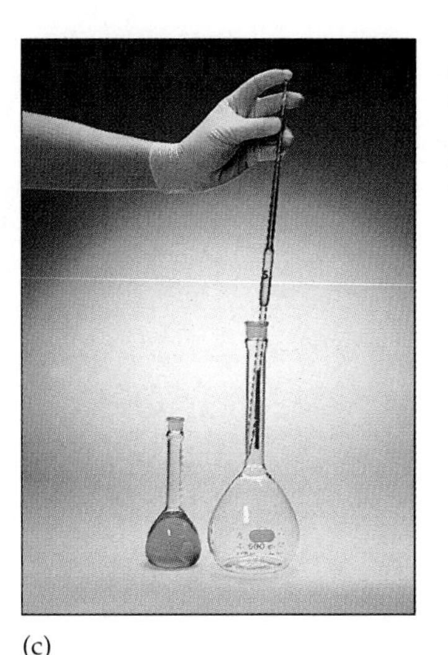

(c)

(d)

Figure 5.24 Making a solution by dilution. (a) Equipment required. (b) A 5.00-mL sample of 0.100 M $K_2Cr_2O_7$ is withdrawn from a flask using a volumetric pipet. (c) The 5.0-mL sample is transferred to a 500.-mL volumetric flask. (d) The 500.-mL flask is filled with distilled water to the mark on the neck, and the concentration of the now diluted solution is 0.00100 M. (*C. D. Winters*)

Problem-Solving Tips and Ideas

5.2 Moles, Volume, and Molarity

Another look at the discussion in this section suggests a useful form of the definition of molarity. That is,

$$\text{Quantity of solute (moles)} = c_{\text{molarity of solute}}\ (\text{mol/L}) \cdot \text{volume of solution (L)}$$
$$= c_{\text{molarity}} \cdot V$$

As you practice working with solutions, you will find this form of the equation to be quite convenient.

Problem-Solving Tips and Ideas

5.3 Preparing a Solution by Dilution

A second look at the preparation of the $K_2Cr_2O_7$ solution suggests a simple way to remember how to do these calculations. The central idea is that the number of moles of $K_2Cr_2O_7$ in the final, dilute solution has to be equal to the number of moles of $K_2Cr_2O_7$ taken from the more concentrated solution. If c is the concentration (molarity) and V is the volume (and the subscripts d and c identify the dilute and concentrated solutions, respectively), the number of moles of solute in either solution can be calculated as follows:

$$\text{Amount of } K_2Cr_2O_7 \text{ in the final dilute solution} = c_dV_d = 0.0025 \text{ mol}$$
$$\text{Amount of } K_2Cr_2O_7 \text{ taken from the more concentrated solution} = c_cV_c = 0.0025 \text{ mol}$$

Because both cV products are equal to the same number of moles, we can use the following equation:

$$c_cV_c = c_dV_d$$
Moles of reagent in concentrated solution = moles of reagent in dilute solution

This equation is valid for all cases in which a more concentrated solution is used to make a more dilute one. It can be used to find, for example, the molarity of the dilute solution, c_d, from values of c_c, V_c, and V_d.

Students are sometimes tempted to use the equation above in stoichiometry problems when relating two solutions (Section 5.9). The equation applies *only* in cases for which the stoichiometric coefficient is 1/1.

Example 5.10 Preparing a Solution by Dilution

You are doing an experiment to find the quantity of iron in a vitamin pill, and you need a standard solution of iron(III) ion. The procedure suggests the standard can be prepared by placing 1.00 mL of 0.236 M iron(III) nitrate in a volumetric flask and diluting to exactly 100.0 mL. What is the concentration of the diluted iron(III) solution?

A CLOSER LOOK *Ions and Life*

When we began the discussion of ions, we noted that ions in aqueous solution are common in our environment and in living systems (see Table 5.1). But it is much more interesting than that. Notice, for example, that seawater contains very high concentrations of sodium and chloride ions, but red blood cells or blood plasma do not (see Table 5.1). On the other hand, K^+ ions are present in a low concentration in sea water, but in much higher concentration in blood and in an especially high concentration in *Valonia* algae in sea water.

One example of the role of common ions in living systems is in respiration. Carbon dioxide from respiration enters red blood cells and is converted to HCO_3^- and H^+. A chloride-bicarbonate ion exchanger protein allows the HCO_3^- ion to move out of the cell into the blood plasma and a Cl^- ion to move in, thus maintaining the charge balance in the cell. The HCO_3^- ion is transported to the lungs in the blood plasma, where the ion is again exchanged for a Cl^- ion in a red blood cell in the lungs. There the HCO_3^- ion is converted back to CO_2, which diffuses out of the cell and is exhaled.

Living systems have various mechanisms to maintain constant ion concentrations. If these are disrupted, disease or a malfunction can result. In fact, a natural bacterial antibiotic, valinomycin, works by changing the K^+ ion concentration in cells. Valinomycin is a relatively large organic molecule that can encapsulate a K^+ ion in a cell, neutralize its charge, and carry it out of the cell. This kills the cell by disrupting the normal cell processes.

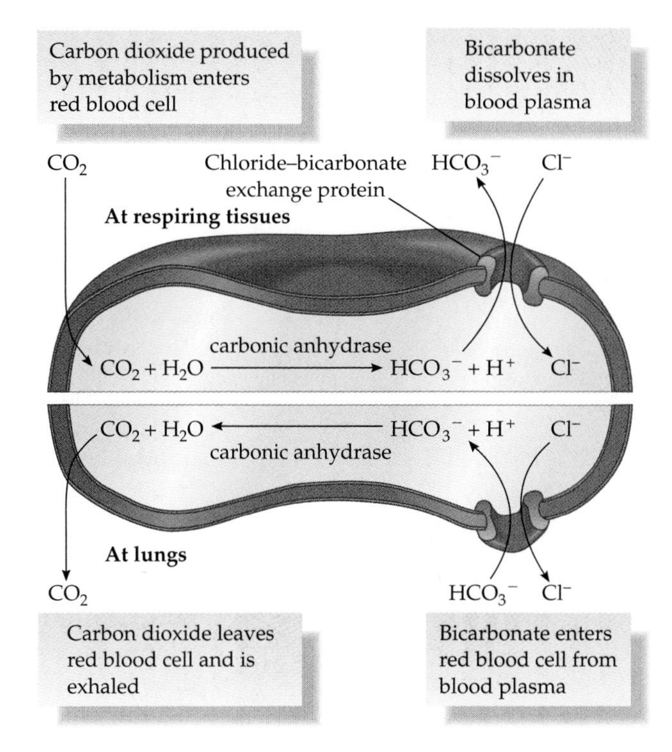

Many other ions are important in biological systems. In addition to those listed in Table 5.1, there are cations of metals such as manganese, iron, cobalt, nickel, copper, and zinc.

For more information see S. J. Lippard and J. M. Berg: *Principles of Bioinorganic Chemistry,* Mill Valley, CA, University Science Books, 1995.

Solution You can find the number of moles of iron(III) ion in the more concentrated iron solution from its volume and molarity.

$$c_M \cdot V = 0.236 \text{ mol/L} \cdot 1.00 \times 10^{-3} \text{ L} = 2.36 \times 10^{-4} \text{ mol of iron(III)}$$

This number of moles is also found in the 100.0 mL of dilute solution. The new concentration is therefore

$$\frac{2.36 \times 10^{-4} \text{ mol iron(III)}}{0.100 \text{ L}} = 2.36 \times 10^{-3} \text{ M}$$

The solution was diluted by a factor of 100, so the new concentration is 100 times smaller than the original concentration.

Exercise 5.16 Preparing a Solution by Dilution

An experiment calls for you to use 250. mL of 1.00 M NaOH, but you are given a large bottle of 2.00 M NaOH. Describe how to make the 1.00 M NaOH in the desired volume.

Exercise 5.17 Preparing a Solution by Dilution

In one of your laboratory experiments you are given a solution of $CuSO_4$ that has a concentration of 0.15 M. If you mix 6.0 mL of this solution with enough water to have a total volume of 10.0 mL, what is the concentration of $CuSO_4$ in this new solution?

5.9 STOICHIOMETRY OF REACTIONS IN AQUEOUS SOLUTION

General Solution Stoichiometry

See the *Saunders Interactive General Chemistry CD-ROM,* Screen 5.13, Stoichiometry of Reactions in Solution.

A common type of reaction is that between a metal carbonate and an aqueous acid to give a salt and CO_2 gas, the type of reaction that occurs when you take a popular remedy for upset stomach (Figure 5.25) (see Section 5.5).

$$\underset{\text{metal carbonate}}{CaCO_3(s)} + \underset{\text{acid}}{2\,HCl(aq)} \longrightarrow \underset{\text{salt}}{CaCl_2(aq)} + \underset{\text{water}}{H_2O(\ell)} + \underset{\text{carbon dioxide}}{CO_2(g)}$$

Figure 5.25 A commercial remedy for excess stomach acid. The tablet contains calcium carbonate, which reacts with hydrochloric acid, the acid present in the digestive system. The most obvious product is CO_2 gas. (*C. D. Winters*)

Suppose we want to know what mass of $CaCO_3$ is required to react completely with 25 mL of 0.750 M HCl. This can be solved in the same way as all the stoichiometry problems you have seen so far, except that the amount of one reactant is given in volume and concentration units instead of as a mass in grams. The first step is to find the quantity of HCl in moles,

$$0.025\,\cancel{\text{L HCl}} \cdot \frac{0.750 \text{ mol HCl}}{1\,\cancel{\text{L HCl}}} = 0.019 \text{ mol HCl}$$

and then to relate this to the quantity of $CaCO_3$ required.

$$0.019\,\cancel{\text{mol HCl}} \cdot \frac{1 \text{ mol } CaCO_3}{2\,\cancel{\text{mol HCl}}} = 0.0094 \text{ mol } CaCO_3$$

Finally, moles of $CaCO_3$ are converted to a mass in grams.

$$0.0094\,\cancel{\text{mol } CaCO_3} \cdot \frac{100.\text{ g } CaCO_3}{1\,\cancel{\text{mol } CaCO_3}} = 0.94 \text{ g } CaCO_3$$

Chemists do such calculations many times in the course of their work in research and product development. If you follow the general scheme outlined in *Problem-Solving Tips and Ideas 5.3,* and pay attention to the units on the numbers, you can successfully carry out any kind of stoichiometry calculations involving concentrations.

Example 5.11 Stoichiometry of a Reaction in Solution

Metallic zinc reacts with aqueous solutions of acids such as HCl, as do many other metals (see Figure 5.9).

$$Zn(s) + 2\,HCl(aq) \longrightarrow ZnCl_2(aq) + H_2(g)$$

Such reactions are often used to produce hydrogen gas in a laboratory. If you have 10.0 g of zinc, what volume of 2.50 M HCl (in milliliters) would be required to convert the zinc completely to zinc chloride?

Solution The balanced equation for the reaction shows that 2 mol of HCl is required for each mole of zinc. Once the quantity of zinc available is known, the quantity of HCl required can be calculated. The volume of solution required is determined from that.

Begin by calculating the quantity of zinc available.

$$10.0 \text{ g Zn} \cdot \frac{1.00 \text{ mol Zn}}{65.39 \text{ g Zn}} = 0.153 \text{ mol Zn}$$

Next, use the stoichiometric factor "2 mol HCl/1 mol Zn" to find the quantity of HCl required.

$$0.153 \text{ mol Zn} \cdot \frac{2 \text{ mol HCl}}{1 \text{ mol Zn}} = 0.306 \text{ mol HCl}$$

Finally, knowing the quantity of HCl required and the concentration of the acid, you can calculate the volume of acid required.

$$0.0306 \text{ mol HCl} \cdot \frac{1.00 \text{ mol solution}}{2.50 \text{ mol HCl}} = 0.122 \text{ L of HCl solution, or } 122 \text{ mL}$$

Because the answer is needed in milliliters, as a final step you convert liters to milliliters.

PROBLEM-SOLVING TIPS AND IDEAS

5.4 Stoichiometry Calculations Involving Solutions

In *Problem-Solving Tips and Ideas, 4.1,* you learned about a general approach to stoichiometry problems. We can now modify that scheme for a reaction such as $x\text{A}+y\text{B} \longrightarrow$ products for reactions occurring in solution.

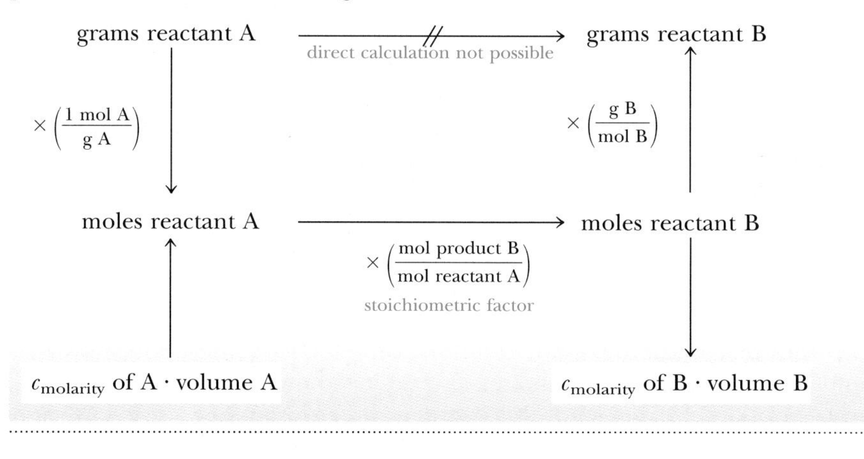

Exercise 5.18 Solution Stoichiometry

Metal carbonates and aqueous acids react to give a salt and CO_2 gas.

$$Na_2CO_3(aq) + 2\ HCl(aq) \longrightarrow 2\ NaCl(aq) + H_2O(\ell) + CO_2(g)$$

If you combine 50.0 mL of 0.450 M HCl and an excess of Na_2CO_3, what mass of NaCl (in grams) is produced?

Titrations

If you know (1) the balanced equation for a reaction and (2) the exact quantity of one of the reactants, you can then calculate the quantity of any other substance consumed or produced in the reaction. This is the essence of any technique of **quantitative chemical analysis,** the determination of the quantity of a given constituent in a mixture.

Qualitative analysis is the determination of the identity of the constituents of a mixture.

An acid isolated from clover leaves was shown in Example 4.7 to have the molecular formula $H_2C_2O_4$. This compound, called oxalic acid, is commercially important in manufacturing paint and textiles, in metal treatment, and in photography. Suppose you are asked to analyze a sample of oxalic acid to ascertain its purity. Because the compound is an acid, it reacts with a base such as sodium hydroxide (see Section 5.4)

$$H_2C_2O_4(aq) + 2\ NaOH(aq) \longrightarrow Na_2C_2O_4(aq) + 2\ H_2O(\ell)$$

The main use of the oxalate ion is to remove calcium ions from aqueous solutions as insoluble calcium oxalate.

$$Ca^{2+}(aq) + H_2C_2O_4(aq) \longrightarrow CaC_2O_4(s) + 2\ H^+(aq)$$

Unfortunately, this reaction can also occur in your body and can lead to kidney stones.

See the *Saunders Interactive General Chemistry CD-ROM,* Screen 5.14, Titrations, and Screen 5.15, Titration Simulation.

You can use this reaction to determine the quantity of oxalic acid present in a given mass of sample if the following conditions are met:

- You can determine when the amount of sodium hydroxide added is just enough to react with all the oxalic acid present in solution.
- You know the volume of the sodium hydroxide solution added at the point of complete reaction.
- You know the concentration of the sodium hydroxide solution.

These conditions are fulfilled in a **titration,** a procedure illustrated in the series of photographs in Figure 5.26. The solution containing oxalic acid is placed in a flask along with an acid-base indicator. An **indicator** is a dye that changes color when the reaction used for analysis is complete. In this case, the dye is colorless in acid solution but pink in basic solution (see Figure 5.9). Aqueous sodium hydroxide of accurately known concentration is placed in a buret. (The buret in Figure 5.26 has a volume of 50.0 mL and is calibrated in 0.1-mL divisions.) As the sodium hydroxide in the buret is added slowly to the acid solution in the flask, the acid reacts with the base according to the net ionic equation

$$H_2C_2O_4(aq) + 2\ OH^-(aq) \longrightarrow C_2O_4{}^{2-}(aq) + 2\ H_2O(\ell)$$

As long as some acid is present in solution, all the base supplied from the buret is consumed, and the indicator remains colorless. At some point, however, the number of moles of OH^- added exactly equals the number of moles of H^+ that can be supplied by the acid. This is called the **equivalence point.** As soon as the slightest excess of base has been added beyond the equivalence point, the solution becomes basic, and the indicator changes color (Figure 5.27).

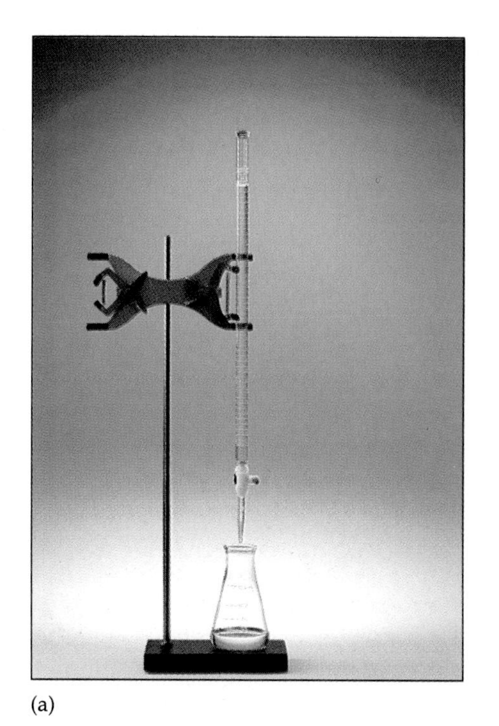

(a)

(b)

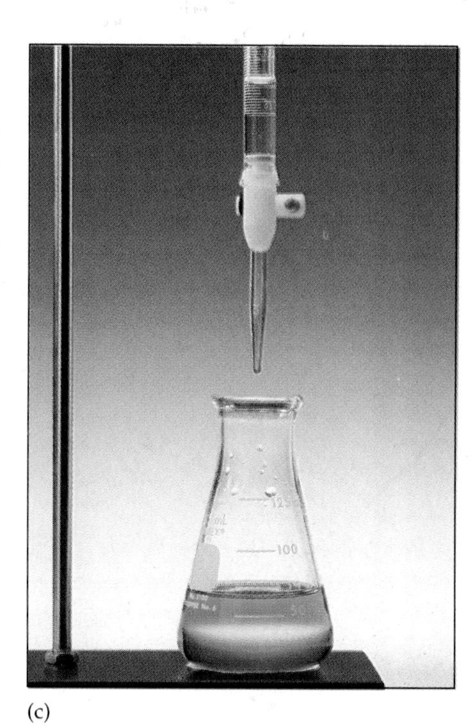

(c)

Figure 5.26 Titration of an acid in aqueous solution with a base. (a) A buret, a volumetric measuring device calibrated in divisions of 0.1 mL, is filled with an aqueous solution of a base of known concentration. (b) Base is added slowly from the buret to the solution. (c) A change in the color of an indicator signals the equivalence point. (The indicator used here is phenolphthalein. See page 193.) (*C. D. Winters*)

Figure 5.27 Acid-base indicator. The juice of a red cabbage could be used as an acid-base indicator because it turns color when the acidity of a solution changes. When the solution is highly acidic, the juice gives the solution a red color. As the solution becomes less acid (more basic), the color changes from red to violet to yellow. (*C. D. Winters*)

When the equivalence point has been reached in a titration, the volume of base added since the beginning of the titration can be determined by reading the calibrated buret. From this volume and the concentration of the base, the number of moles of base used can be found:

Moles of base added = concentration of base (mol/L) · volume of base (L)

Then, using the stoichiometric factor from the balanced equation, the number of moles of base added is related to the number of moles of acid present in the original sample.

Example 5.12 Acid-Base Titration

Suppose you dissolve a 1.034-g sample of impure oxalic acid in some water, add an acid-base indicator, and titrate it with 0.485 M NaOH. The sample requires 34.47 mL of the NaOH solution to reach the equivalence point. What is the mass of oxalic acid and what is its mass percent in the sample?

Solution The objective—to find the mass of oxalic acid from a knowledge of the volume and concentration of the NaOH solution—can be reached using the scheme in *Problem-Solving Tips and Ideas 5.4.* First, calculate the number of moles of NaOH used in the reaction from the volume (in liters) and concentration (in moles per liter) of the NaOH solution.

$$0.03447\,\cancel{\text{L}} \cdot \frac{0.485\text{ mol NaOH}}{1\,\cancel{\text{L}}} = 0.0167\text{ mol NaOH}$$

The balanced, net ionic equation for the reaction shows that 1 mol of oxalic acid requires 2 mol of sodium hydroxide.

$$H_2C_2O_4(aq) + 2\,OH^-(aq) \longrightarrow C_2O_4^{2-}(aq) + 2\,H_2O(\ell)$$

This is the stoichiometric factor required for the calculation of moles of oxalic acid, and it gives the number of moles of oxalic acid present.

$$0.0167\,\cancel{\text{mol NaOH}} \cdot \frac{1\text{ mol }H_2C_2O_4}{2\,\cancel{\text{mol NaOH}}} = 0.00836\text{ mol }H_2C_2O_4$$

The mass of oxalic acid is found from the number of moles of the acid.

$$0.00836\,\cancel{\text{mol }H_2C_2O_4} \cdot \frac{90.04\text{ g }H_2C_2O_4}{1\,\cancel{\text{mol }H_2C_2O_4}} = 0.753\text{ g }H_2C_2O_4$$

This mass of oxalic acid represents 72.8% of the total sample mass.

$$\frac{0.753\text{ g }H_2C_2O_4}{1.034\text{ g sample}} \cdot 100\% = 72.8\%\ H_2C_2O_4$$

In Example 5.12 the concentration of the base used in the titration was given. In real life this usually has to be found by a prior measurement. The procedure by which the concentration of an analytical reagent is determined accurately is called **standardization,** and there are two general approaches.

One approach is to weigh accurately a sample of a pure, solid acid or base (known as a **primary standard**) and then titrate this sample with a solution of the base or acid to be standardized (Example 5.13). Another approach to standardizing a solution is to titrate it with another solution that is already standardized (Exercise 5.20). This is often done with standard solutions purchased from chemical supply companies.

You might think that solid NaOH would make a good primary standard. Solid NaOH is difficult to weigh accurately, however, because it is deliquescent, that is, it absorbs water from humid air. Furthermore, its solutions readily take up CO_2 from the air, thus changing the concentration of OH^-:

$$2\,NaOH(aq) + CO_2(g) \longrightarrow Na_2CO_3(aq) + H_2O(\ell)$$

Example 5.13 Standardizing an Acid by Titration

Solutions of acids such as HCl can be standardized by titrating a base such as Na_2CO_3, a solid that can be obtained in pure form, that can be weighed accurately, and that reacts completely with an acid. Suppose that 0.263 g of Na_2CO_3 requires 28.35 mL of aqueous HCl for titration to the equivalence point. What is the molar concentration of the HCl?

Solution As usual, the balanced equation for the reaction is written first.

$$Na_2CO_3(aq) + 2\,HCl(aq) \longrightarrow 2\,NaCl(aq) + H_2O(\ell) + CO_2(g)$$

Next, the mass of Na_2CO_3 used as the standard is converted to moles.

$$0.263\ \cancel{g\ Na_2CO_3} \cdot \frac{1.00\ mol\ Na_2CO_3}{106.0\ \cancel{g\ Na_2CO_3}} = 0.00248\ mol\ Na_2CO_3$$

Knowing the number of moles of Na_2CO_3, the number of moles of HCl in the solution can be calculated by using the appropriate stoichiometric factor.

$$0.00248\ \cancel{mol\ Na_2CO_3} \cdot \frac{2\ mol\ HCl\ required}{1\ \cancel{mol\ Na_2CO_3}\ available} = 0.00496\ mol\ HCl$$

The 28.35-mL (or 0.02835 L) sample of aqueous HCl therefore contains 0.00496 mol of HCl, and so the concentration of the HCl solution is 0.175 M.

$$HCl\ concentration = \frac{0.00496\ mol\ HCl}{0.02835\ L} = 0.175\ M$$

Exercise 5.19 Acid-Base Titration

A 25.0-mL sample of vinegar requires 28.33 mL of a 0.953 M solution of NaOH for titration to the equivalence point. What mass (in grams) of acetic acid is in the vinegar sample, and what is the concentration of acetic acid in the vinegar?

$$\underset{\text{acetic acid}}{CH_3CO_2H(aq)} + NaOH(aq) \longrightarrow \underset{\text{sodium acetate}}{NaCH_3CO_2(aq)} + H_2O(\ell)$$

Exercise 5.20 Standardization of a Base

Hydrochloric acid, HCl, can be purchased from chemical supply houses in a solution with a concentration of 0.100 M, and such a solution can be used to standardize the solution of a base. If titrating 25.00 mL of a sodium hydroxide solution to the equivalence point requires 29.67 mL of 0.100 M HCl, what is the concentration of the base?

Acid-base titrations are extremely useful for quantitative chemical analysis, but you should not get the impression that a titration can only be done with an acid reacting with a base. Oxidation-reduction reactions (see Section 5.7) lend themselves very well to chemical analysis by titration. Many of these reactions go rapidly to completion in aqueous solution, and methods exist for finding their equivalence points.

Example 5.14 Using an Oxidation-Reduction Reaction in a Titration

Suppose you wish to analyze an iron ore for its iron content. The iron in a sample of solid ore can be converted quantitatively to the iron(II) ion, Fe^{2+}, in aqueous solution, and this solution can then be titrated with aqueous potassium permanganate, $KMnO_4$. The balanced net ionic equation for the reaction occurring in the course of this titration is

$$\underset{\substack{\text{purple}\\ \text{oxidizing agent}}}{MnO_4^-(aq)} + \underset{\substack{\text{colorless}\\ \text{reducing agent}}}{5\,Fe^{2+}(aq)} + \underset{\text{colorless}}{8\,H^+(aq)} \longrightarrow \underset{\text{colorless}}{Mn^{2+}(aq)} + \underset{\text{pale yellow}}{5\,Fe^{3+}(aq)} + 4\,H_2O(\ell)$$

This is a useful analytical reaction, because it is easy to detect when all the iron(II) has reacted (see Figures 5.18 and 5.28). The MnO_4^- ion is deep purple, but when it reacts with Fe^{2+} the color disappears because the reaction product, the Mn^{2+} ion, is colorless. Thus, as $KMnO_4$ is added from a buret, the purple color disappears as the solutions mix. When all the Fe^{2+} has been converted to Fe^{3+}, any additional $KMnO_4$ will give the solution a permanent purple color. Therefore, $KMnO_4$ solution is added from the buret until the initially colorless, Fe^{2+}-containing solution just turns a faint purple color, the signal that the equivalence point has been reached.

(a)

(b)

(c)

Figure **5.28 Titration involving an oxidation-reduction reaction.** (a) Here a solution of Fe^{2+} ion (a reducing agent) is titrated with aqueous $KMnO_4$ (the oxidizing agent; in the buret). (b) As $KMnO_4$ is added to the solution, the iron(II) ion is oxidized and the deep purple $KMnO_4$ solution is reduced. (c) The products (Fe^{2+} and Mn^{2+}) are nearly colorless. Just past the equivalence point, when a slight excess of $KMnO_4$ has been added, the solution takes on a faint purple color. (*C. D. Winters*)

A 1.026-g sample of iron-containing ore requires 24.35 mL of 0.0195 M $KMnO_4$ to reach the equivalence point. What is the mass percent of iron in the ore?

Solution Because the volume and molar concentration of the $KMnO_4$ solution are known, the number of moles of $KMnO_4$ used in the titration can be calculated. Remembering first to change the volume to liters, we have

$$0.02435\ \text{L} \cdot \frac{0.0195\ \text{mol}}{1\ \text{L}} = 0.000475\ \text{mol KMnO}_4$$

Based on the balanced chemical equation for the reaction of $KMnO_4$ with Fe^{2+}, the number of moles of iron(II) that were present in the solution (and therefore the number of moles of iron in the ore sample) can be calculated.

$$0.000475\ \text{mol KMnO}_4 \cdot \frac{5\ \text{mol Fe}^{2+}}{1\ \text{mol KMnO}_4} = 0.00237\ \text{mol Fe}^{2+}$$

The mass of iron is calculated from this,

$$0.00237\ \text{mol Fe} \cdot \frac{55.85\ \text{g Fe}}{1\ \text{mol Fe}} = 0.133\ \text{g iron (Fe)}$$

and, finally, the mass percent of iron in the ore can be calculated.

$$\frac{0.133\ \text{g iron}}{1.026\ \text{g sample}} \cdot 100\% = 12.9\%\ \text{iron}$$

Exercise 5.21 Using an Oxidation-Reduction Reaction in a Titration

Vitamin C, ascorbic acid, has the formula $C_6H_8O_6$. It is a reducing agent in addition to being an acid. One way to determine the ascorbic acid content of a vitamin pill is to react the ascorbic acid with excess iodine,

$$C_6H_8O_6(aq) + I_2(aq) \longrightarrow C_6H_6O_6(aq) + 2\ H^+(aq) + 2\ I^-(aq)$$

and then titrate the iodine that did *not* react with the ascorbic acid with sodium thiosulfate. The balanced, net ionic equation for the reaction occurring in the course of the titration is

$$I_2(aq) + 2\ S_2O_3{}^{2-}(aq) \longrightarrow 2\ I^-(aq) + S_4O_6{}^{2-}(aq)$$

Suppose 50.00 mL of 0.0520 M I_2 was added to the sample containing ascorbic acid. After the reaction was complete, the I_2 not used in this reaction required 20.30 mL of 0.196 M $Na_2S_2O_3$ for titration to the equivalence point. Calculate the mass of ascorbic acid in the unknown sample.

Oxidation-reduction reactions, as well as acid-base reactions, are clearly useful in chemical analysis. No matter what type of reaction is used in analysis, however, you must remember that *before you use any reaction as a quantitative analytical method, you must know the balanced chemical equation* (or at least the stoichiometric relation between the key reactants and products) *to know the necessary stoichiometric factors.*

CHAPTER HIGHLIGHTS

Having studied this chapter, you should be able to

- Explain the differences between **electrolytes** and **nonelectrolytes** (Section 5.1).
- Predict the solubility of ionic compounds in water (Section 5.1).
- Recognize what ions are formed when an ionic compound dissolves in water (Sections 5.1–5.3).
- Predict the products of **precipitation reactions** (Section 5.2), the formation of an insoluble reaction product by the exchange of anions between the cations of the reactants.

$$Pb(NO_3)_2(aq) + 2\,KI(aq) \longrightarrow PbI_2(s) + 2\,KNO_3(aq)$$

- Write **net ionic equations** and show how to arrive at such an equation for a given reaction (Sections 5.2 and 5.6).
- Recognize common **acids** and **bases** and understand their behavior in aqueous solution (Section 5.3 and Table 5.2).
- Predict the products of **acid-base reactions** involving common acids and strong bases (Section 5.4).

$$HNO_3(aq) + KOH(aq) \longrightarrow \underset{\text{potassium nitrate, a salt}}{KNO_3(aq)} + H_2O(\ell)$$

- Understand that the net ionic equation for the reaction of a strong acid with a strong base is $H^+(aq) + OH^-(aq) \longrightarrow H_2O(\ell)$ (Section 5.4).
- Predict the products of **gas-forming reactions** (Section 5.5), the most common of which are those between a metal carbonate and an acid.

$$NiCO_3(s) + 2\,HNO_3(aq) \longrightarrow Ni(NO_3)_2(aq) + CO_2(g) + H_2O(\ell)$$

- Use the ideas developed in Sections 5.2–5.7 as an aid in recognizing four of the common types of reactions that occur in aqueous solution and write balanced equations for such reactions (Section 5.6).

Reaction Type	Driving Force
Precipitation	Formation of an insoluble compound
Acid-strong base; neutralization	Formation of a salt and water
Gas-forming	Evolution of a water-insoluble gas such as CO_2
Oxidation-reduction	Transfer of electrons

- Understand that an **oxidation number** for an element in a compound is the electric charge an atom has, or appears to have, when the electrons of the compound are counted according to a set of "Guidelines" (Section 5.7).
- Calculate oxidation numbers for elements in a compound (Section 5.7).
- Recognize **oxidation-reduction reactions** (often called **redox** reactions) (Section 5.7 and Tables 5.3 and 5.4).
- Explain **molarity,** and calculate molarity values (Section 5.8). The concentration of a solute in a solution in units of molarity is

$$\text{Concentration } (c_{\text{molarity}}) = \frac{\text{Quantity of solute (mol)}}{\text{Volume of solution (L)}}$$

A useful form of this equation is "moles solute = $c_{\text{molarity}} \cdot$ liters of solution."

- Describe how to prepare a solution of a given molarity from the solute and water or by dilution from a more concentrated solution (Section 5.8). When preparing a solution by diluting a more concentrated solution, the number of moles of solute remains the same. The product of the concentration and volume of the more concentrated solution (c) must therefore be the same as that for the diluted solution (d).

$$c_c \cdot V_c = c_d \cdot V_d$$

- Solve stoichiometry problems using solution concentrations (Section 5.9).
- Explain how a **titration** is carried out, explain **standardization,** and calculate concentrations or amounts of reactants from titration data (Section 5.9).

STUDY QUESTIONS

REVIEW QUESTIONS

1. Define the terms solution, solvent, and solute.
2. Find one example in the chapter of each of the following reaction types: acid-base, precipitation, gas-forming, and oxidation-reduction reactions. Describe how each reaction exemplifies its class. Name the reactants and products of each reaction.
3. Find two examples in the chapter of an acid-base reaction. Write the balanced equation for each, and name the reactants and products.
4. What is an electrolyte? How can you differentiate experimentally between a weak and a strong electrolyte? Give an example of each.
5. Name two acids that are strong electrolytes and one that is a weak electrolyte. Name two bases that are strong electrolytes and one that is a weak electrolyte.
6. Which of the following nickel(II) salts are soluble in water and which are insoluble: $Ni(NO_3)_2$, $NiCO_3$, $Ni_3(PO_4)_2$, and $NiCl_2$?
7. Name the spectator ion or ions in the reaction of nitric acid and magnesium hydroxide and write the net ionic equation:

$$2\,H^+(aq) + 2\,NO_3^-(aq) + Mg(OH)_2(s) \longrightarrow 2\,H_2O(\ell) + Mg^{2+}(aq) + 2\,NO_3^-(aq)$$

What type of reaction is this?

8. Name the water-insoluble product in each reaction:
   (a) $CuCl_2(aq) + H_2S(aq) \longrightarrow CuS + 2\,HCl$
   (b) $CaCl_2(aq) + K_2CO_3(aq) \longrightarrow 2\,KCl + CaCO_3$
   (c) $AgNO_3(aq) + NaI(aq) \longrightarrow AgI + NaNO_3$
9. Find two examples of precipitation reactions in the chapter. Write balanced equations for these reactions and name the reactants and products.
10. Find two examples of gas-forming reactions in the chapter. Write balanced equations for these reactions and name the reactants and products.
11. Oxidation-reduction reactions:
    (a) Explain the difference between oxidation and reduction. Give an example of each.
    (b) Explain the difference between an oxidizing agent and a reducing agent. Give an example of each.
12. Bromine is obtained from seawater by the following reaction:

$$Cl_2(g) + 2\,NaBr(aq) \longrightarrow 2\,NaCl(aq) + Br_2(\ell)$$

What has been oxidized? What has been reduced? Name the oxidizing and reducing agents.

13. Identify each of the following substances as an oxidizing or reducing agent: HNO_3, Na, Cl_2, O_2, CO, and $KMnO_4$.
14. You have 0.500 mol of KCl, some distilled water, and a 250.-mL volumetric flask. Describe how you would make a 0.500 M solution of KCl.
15. Which contains the greater mass of solute: 1 L of 0.1 M NaCl or 1 L of 0.06 M Na_2CO_3?
16. Which solution contains the larger concentration of ions: 0.20 M $BaCl_2$ or 0.25 M NaCl?
17. Titrations
    (a) What is the equivalence point in a titration?
    (b) What is the function of an indicator?

PROPERTIES OF AQUEOUS SOLUTIONS

Solubility of Compounds

18. Which compound or compounds in each of the following groups is (are) expected to be soluble in water?
    (a) FeO, $FeCl_2$, and $FeCO_3$
    (b) AgI, Ag_3PO_4, and $AgNO_3$
    (c) Li_2CO_3, NaCl, and $KMnO_4$
19. Which compound or compounds in each of the following groups is (are) expected to be soluble in water?
    (a) $PbSO_4$, $Pb(NO_3)_2$, and $PbCO_3$
    (b) Na_2SO_4, $NaClO_4$, $NaCH_3CO_2$
    (c) AgBr, KBr, Al_2Br_6
20. Give the formula for
    (a) A soluble compound containing the chloride ion
    (b) An insoluble hydroxide
    (c) An insoluble carbonate
    (d) A soluble nitrate-containing compound

21. Give the formula for
   (a) A soluble compound containing the acetate ion
   (b) An insoluble sulfide
   (c) A soluble hydroxide
   (d) An insoluble chloride
22. Each of the following compounds is water-soluble. What ions are produced in water?
   (a) KOH (c) $NaNO_3$
   (b) K_2SO_4 (d) $(NH_4)_2SO_4$
23. Each of the following compounds is water-soluble. What ions are produced in water?
   (a) KI (c) $KHSO_4$
   (b) $Mg(CH_3CO_2)_2$ (d) KCN
24. Decide whether each of the following is water-soluble or not. If soluble, tell what ions are produced.
   (a) Na_2CO_3 (c) NiS
   (b) $CuSO_4$ (d) $CaBr_2$
25. Decide whether each of the following is water-soluble or not. If soluble, tell what ions are produced.
   (a) $BaCl_2$ (c) $Pb(NO_3)_2$
   (b) $Cr(NO_3)_3$ (d) $BaSO_4$
26. Name two water-soluble compounds containing Cu^{2+}. Name two water-insoluble Cu^{2+} compounds.
27. Name two water-soluble compounds containing Ba^{2+}. Name two water-insoluble Ba^{2+} compounds.

TYPES OF CHEMICAL REACTIONS

Precipitation Reactions

28. Balance the equation for the following precipitation reaction, and then write the net ionic equation. Indicate the state of each species (s, ℓ, aq, or g).

$$CdCl_2 + NaOH \longrightarrow Cd(OH)_2 + NaCl$$

29. Balance the equation for the following precipitation reaction, and then write the net ionic equation. Indicate the state of each species (s, ℓ, aq, or g).

$$Ni(NO_3)_2 + Na_2CO_3 \longrightarrow NiCO_3 + NaNO_3$$

30. Predict the products of each precipitation reaction, and then balance the completed equation.
   (a) $NiCl_2(aq) + (NH_4)_2S(aq) \longrightarrow$
   (b) $Mn(NO_3)_2(aq) + Na_3PO_4(aq) \longrightarrow$
31. Predict the products of each precipitation reaction, and then balance the completed equation.
   (a) $Pb(NO_3)_2(aq) + KBr(aq) \longrightarrow$
   (b) $Ca(NO_3)_2(aq) + KF(aq) \longrightarrow$
   (c) $Ca(NO_3)_2(aq) + Na_2C_2O_4(aq) \longrightarrow$
32. Write an overall, balanced equation for the precipitation reaction that occurs when aqueous lead(II) nitrate is mixed with an aqueous solution of potassium hydroxide. Name each reactant and product.
33. Write an overall, balanced equation for the precipitation reaction that occurs when aqueous copper(II) nitrate is mixed with an aqueous solution of sodium carbonate. Name each reactant and product.

Writing Net Ionic Equations

34. Balance each of the following equations, and then write the net ionic equation:
   (a) $(NH_4)_2CO_3(aq) + Cu(NO_3)_2(aq) \longrightarrow CuCO_3(s) + NH_4NO_3(aq)$
   (b) $Pb(OH)_2(s) + HCl(aq) \longrightarrow PbCl_2(s) + H_2O(\ell)$
   (c) $BaCO_3(s) + HCl(aq) \longrightarrow BaCl_2(aq) + H_2O(\ell) + CO_2(g)$
35. Balance each of the following equations, and then write the net ionic equation:
   (a) $Zn(s) + HCl(aq) \longrightarrow H_2(g) + ZnCl_2(aq)$
   (b) $Mg(OH)_2(s) + HCl(aq) \longrightarrow MgCl_2(aq) + H_2O(\ell)$
   (c) $HNO_3(aq) + CaCO_3(s) \longrightarrow Ca(NO_3)_2(aq) + H_2O(\ell) + CO_2(g)$
36. Balance each of the following equations, and then write the net ionic equation. Show states for all reactants and products (s, ℓ, g, aq).
   (a) $ZnCl_2 + KOH \longrightarrow KCl + Zn(OH)_2$
   (b) $AgNO_3 + KI \longrightarrow AgI + KNO_3$
   (c) $NaOH + FeCl_2 \longrightarrow Fe(OH)_2 + NaCl$
37. Balance each of the following equations, and then write the net ionic equation. Show states for all reactants and products (s, ℓ, g, aq).
   (a) $Ba(OH)_2 + HNO_3 \longrightarrow Ba(NO_3)_2 + H_2O$
   (b) $BaCl_2 + Na_2CO_3 \longrightarrow BaCO_3 + NaCl$
   (c) $Na_3PO_4 + Ni(NO_3)_2 \longrightarrow Ni_3(PO_4)_2 + NaNO_3$

Acid-Base Reactions

38. Write a balanced equation for the ionization of nitric acid in water.
39. Write a balanced equation for the ionization of perchloric acid in water.
40. Oxalic acid, which is found in certain plants, can provide two hydrogen ions in water. Write balanced equations (like those for sulfuric acid on page 194) to show how oxalic acid, $H_2C_2O_4$, can supply one and then a second H^+ ion.
41. Phosphoric acid can supply one, two, or three H^+ ions in aqueous solution. Write balanced equations (like those for sulfuric acid on page 194) to show this successive loss of hydrogen ions.
42. Write a balanced equation for reaction of the basic oxide, magnesium oxide, with water.
43. Write a balanced equation for the reaction of sulfur trioxide with water.
44. Complete and balance the following acid-base reactions. Name the reactants and products.
   (a) $CH_3CO_2H(aq) + Mg(OH)_2(s) \longrightarrow$
   (b) $HClO_4(aq) + NH_3(aq) \longrightarrow$
45. Complete and balance the following acid-base reactions. Name the reactants and products.
   (a) $H_3PO_4(aq) + KOH(aq) \longrightarrow$
   (b) $H_2C_2O_4(aq) + Ca(OH)_2(s) \longrightarrow$
   ($H_2C_2O_4$ is oxalic acid, an acid capable of donating two H^+ ions.)
46. Write a balanced equation for the reaction of barium hy-

droxide with nitric acid to give barium nitrate, a compound used in pyrotechnics such as green flares.

47. Aluminum is obtained from bauxite, which is not a specific mineral but a name applied to a mixture of minerals. One of those minerals, which can dissolve in acids, is gibbsite, $Al(OH)_3$. Write a balanced equation for the reaction of gibbsite with sulfuric acid.

Gas-Forming Reactions

48. Many minerals are metal carbonates, and siderite is a mineral that consists largely of iron(II) carbonate. Write an overall, balanced equation for the reaction of the mineral with nitric acid, and name each reactant and product.

49. The beautiful red mineral rhodochrosite is manganese(II) carbonate. Write an overall, balanced equation for the reaction of the mineral with hydrochloric acid. Name each reactant and product.

The red mineral is rhodochrosite, which is nearly pure manganese(II) carbonate. (*C. D. Winters*)

Types of Reactions in Aqueous Solution

50. Balance these reactions and then classify each one as a precipitation, an acid-base reaction, or a gas-forming reaction.
(a) $Ba(OH)_2(s) + HCl(aq) \longrightarrow BaCl_2(aq) + H_2O(\ell)$
(b) $HNO_3(aq) + CoCO_3(s) \longrightarrow Co(NO_3)_2(aq) + H_2O(\ell) + CO_2(g)$
(c) $Na_3PO_4(aq) + Cu(NO_3)_2(aq) \longrightarrow Cu_3(PO_4)_2(s) + NaNO_3(aq)$

51. Balance these reactions and then classify each one as a precipitation, an acid-base reaction, or a gas-forming reaction.
(a) $K_2CO_3(aq) + Cu(NO_3)_2(aq) \longrightarrow CuCO_3(s) + KNO_3(aq)$
(b) $Pb(NO_3)_2(aq) + HCl(aq) \longrightarrow PbCl_2(s) + HNO_3(aq)$
(c) $MgCO_3(s) + HCl(aq) \longrightarrow MgCl_2(aq) + H_2O(\ell) + CO_2(g)$

52. Balance these reactions and then classify each one as a precipitation, an acid-base reaction, or a gas-forming reaction. Show states for the products (s, ℓ, g, aq) and then balance the completed equation. Write the net ionic equation.
(a) $MnCl_2(aq) + Na_2S(aq) \longrightarrow MnS + NaCl$
(b) $K_2CO_3(aq) + ZnCl_2(aq) \longrightarrow ZnCO_3 + KCl$
(c) $K_2CO_3(aq) + HClO_4(aq) \longrightarrow KClO_4 + CO_2 + H_2O$

53. Balance these reactions and then classify each one as a precipitation, an acid-base reaction, or a gas-forming reaction. Write the net ionic equation.
(a) $Fe(OH)_3(s) + HNO_3(aq) \longrightarrow Fe(NO_3)_3 + H_2O$
(b) $FeCO_3(s) + HNO_3(aq) \longrightarrow Fe(NO_3)_2 + CO_2 + H_2O$
(c) $FeCl_2(aq) + (NH_4)_2S(aq) \longrightarrow FeS + NH_4Cl$
(d) $Fe(NO_3)_2(aq) + Na_2CO_3(aq) \longrightarrow FeCO_3 + NaNO_3$

Oxidation-Reduction Reactions

54. Determine the oxidation number of each element in the following ions or compounds:
(a) BrO_3^- (d) CaH_2
(b) $C_2O_4^{2-}$ (e) H_4SiO_4
(c) F_2 (f) SO_4^{2-}

55. Determine the oxidation number of each element in the following ions or compounds:
(a) SF_6 (d) N_2O_4
(b) H_3AsO_4 (e) PCl_4^+
(c) UO_2^+ (f) XeO_4^{2-}

56. Which of the following reactions is (are) oxidation-reduction reactions? Explain your answer in each case. Classify the remaining reactions.
(a) $Zn(s) + 2\,NO_3^-(aq) + 4\,H^+(aq) \longrightarrow Zn^{2+}(aq) + 2\,NO_2(g) + 2\,H_2O(\ell)$
(b) $Zn(OH)_2(s) + H_2SO_4(aq) \longrightarrow ZnSO_4(aq) + 2\,H_2O(\ell)$
(c) $Ca(s) + 2\,H_2O(\ell) \longrightarrow Ca(OH)_2(s) + H_2(g)$

57. Which of the following reactions is (are) oxidation-reduction reactions? Explain your answer briefly. Classify the remaining reactions.
(a) $CdCl_2(aq) + Na_2S(aq) \longrightarrow CdS(s) + 2\,NaCl(aq)$
(b) $2\,Ca(s) + O_2(g) \longrightarrow 2\,CaO(s)$
(c) $Ca(OH)_2(s) + 2\,HCl(aq) \longrightarrow CaCl_2(aq) + 2\,H_2O(\ell)$

58. In each of the following reactions, decide which reactant is oxidized and which is reduced. Designate the oxidizing agent and reducing agent.
(a) $2\,Mg(s) + O_2(g) \longrightarrow 2\,MgO(s)$
(b) $C_2H_4(g) + 3\,O_2(g) \longrightarrow 2\,CO_2(g) + 2\,H_2O(g)$
(c) $Si(s) + 2\,Cl_2(g) \longrightarrow SiCl_4(\ell)$

59. In each of the following reactions, decide which reactant is oxidized and which is reduced. Designate the oxidizing agent and reducing agent.
(a) $Ca(s) + 2\,HCl(aq) \longrightarrow CaCl_2(aq) + H_2(g)$
(b) $Cr_2O_7^{2-}(aq) + 3\,Sn^{2+}(aq) + 14\,H^+(aq) \longrightarrow 2\,Cr^{3+}(aq) + 3\,Sn^{4+}(aq) + 7\,H_2O(\ell)$
(c) $FeS(s) + 3\,NO_3^-(aq) + 4\,H^+(aq) \longrightarrow 3\,NO(g) + SO_4^{2-}(aq) + Fe^{3+}(aq) + 2\,H_2O(\ell)$

SOLUTION CONCENTRATION

60. If 6.73 g of Na_2CO_3 is dissolved in enough water to make 250. mL of solution, what is the molarity of the sodium carbonate? What are the molar concentrations of the Na^+ and CO_3^{2-} ions?

61. Some potassium dichromate ($K_2Cr_2O_7$), 2.335 g, is dissolved in enough water to make exactly 500. mL of solution. What is the molarity of the potassium dichromate? What are the molar concentrations of the K^+ and $Cr_2O_7^{2-}$ ions?

62. What is the mass, in grams, of solute in 250. mL of a 0.0125 M solution of $KMnO_4$?

63. What is the mass, in grams, of solute in 125 mL of a 1.023×10^{-3} M solution of Na_3PO_4? What are the molar concentrations of the Na^+ and PO_4^{3-} ions?

64. What volume of 0.123 M NaOH, in milliliters, contains 25.0 g of NaOH?

65. What volume of 2.06 M $KMnO_4$, in liters, contains 322 g of solute?

66. If 4.00 mL of 0.0250 M $CuSO_4$ is diluted to 10.0 mL with pure water, what is the molarity of copper(II) sulfate in the diluted solution?

67. If you dilute 25.0 mL of 1.50 M hydrochloric acid to 500. mL, what is the molar concentration of the dilute acid?

68. If you need 1.00 L of 0.125 M H_2SO_4, which of the following methods would you use to prepare this solution?
(a) Dilute 20.8 mL of 6.00 M H_2SO_4 to a volume of 1.00 L.
(b) Add 950. mL of water to 50.0 mL of 3.00 M H_2SO_4.

69. If you need 300. mL of 0.500 M $K_2Cr_2O_7$, which of the following methods would you use to prepare this solution?
(a) Add 30.0 mL of 1.50 M $K_2Cr_2O_7$ to 270. mL of water.
(b) Dilute 250. mL of 0.600 M $K_2Cr_2O_7$ to a volume of 300. mL.

70. For each solution, identify the ions that exist in aqueous solution, and specify the concentration of each.
(a) 0.25 M $(NH_4)_2SO_4$
(b) 0.056 M HNO_3
(c) 0.123 M Na_2CO_3
(d) 0.00124 M $KClO_4$

71. For each solution, identify the ions that exist in aqueous solution, and specify the concentration of each.
(a) 0.12 M $BaCl_2$
(b) 0.0125 M $CuSO_4$
(c) 0.146 M $AlCl_3$
(d) 0.500 M $K_2Cr_2O_7$

STOICHIOMETRY OF REACTIONS IN SOLUTION

72. What volume of 0.125 M HNO_3, in milliliters, is required to react completely with 1.30 g of $Ba(OH)_2$?

$$2\ HNO_3(aq) + Ba(OH)_2(s) \longrightarrow Ba(NO_3)_2(aq) + 2\ H_2O(\ell)$$

73. What mass of Na_2CO_3, in grams, is required for complete reaction with 25.0 mL of 0.155 M HNO_3?

$$Na_2CO_3(aq) + 2\ HNO_3(aq) \longrightarrow 2\ NaNO_3(aq) + CO_2(g) + H_2O(\ell)$$

74. One of the most important industrial processes in our economy is the electrolysis of brine solutions (aqueous solutions of NaCl). When an electric current is passed through an aqueous solution of salt, the NaCl and water produce $H_2(g)$, $Cl_2(g)$, and NaOH—all valuable industrial chemicals.

$$2\ NaCl(aq) + 2\ H_2O(\ell) \longrightarrow H_2(g) + Cl_2(g) + 2\ NaOH(aq)$$

What mass of NaOH can be formed from 10.0 L of 0.15 M NaCl? What mass of chlorine can be obtained?

75. Diborane, B_2H_6, is a useful reagent in organic chemistry. One of the several ways it can be prepared is by the following reaction:

$$2\ NaBH_4(aq) + H_2SO_4(aq) \longrightarrow 2\ H_2(g) + Na_2SO_4(aq) + B_2H_6(g)$$

What volume of 0.0875 M H_2SO_4, in milliliters, should be used to consume completely 1.35 g of $NaBH_4$? What mass of B_2H_6 can be obtained?

76. In the photographic developing process, silver bromide is dissolved by adding sodium thiosulfate:

$$AgBr(s) + 2\ Na_2S_2O_3(aq) \longrightarrow Na_3Ag(S_2O_3)_2(aq) + NaBr(aq)$$

If you want to dissolve 0.250 g of AgBr, what volume of 0.0138 M $Na_2S_2O_3$, in milliliters, should be used?

(a) (b)

Silver chemistry. (a) A precipitate of AgBr formed by adding $AgNO_3(aq)$ to KBr(aq). (b) On adding $Na_2S_2O_3(aq)$, sodium thiosulfate, the solid AgBr dissolves. (*C. D. Winters*)

77. Hydrazine, N_2H_4, a base like ammonia, can react with an acid such as sulfuric acid.

$$2\ N_2H_4(aq) + H_2SO_4(aq) \longrightarrow 2\ N_2H_5^+(aq) + SO_4^{2-}(aq)$$

What mass of hydrazine reacts with 250. mL of 0.225 M H_2SO_4?

78. What volume of 0.750 M $Pb(NO_3)_2$, in milliliters, is required to react completely with 1.00 L of 2.25 M NaCl solution? The balanced equation is

$$Pb(NO_3)_2(aq) + 2\,NaCl(aq) \longrightarrow PbCl_2(s) + 2\,NaNO_3(aq)$$

79. What volume, in milliliters, of 0.512 M NaOH is required to react completely with 25.0 mL of 0.234 M H_2SO_4?

80. You place 2.56 g of $CaCO_3$ in a beaker containing 250. mL of 0.125 M HCl (Figure 5.13). When the reaction has ceased,

$$CaCO_3(s) + 2\,HCl(aq) \longrightarrow CaCl_2(aq) + CO_2(g) + H_2O(\ell)$$

does any calcium carbonate remain? Explain your reasoning. What mass of $CaCl_2$ can be produced?

81. You can dissolve an aluminum soft drink can in an aqueous base such as potassium hydroxide.

$$2\,Al(s) + 2\,KOH(aq) + 6\,H_2O(\ell) \longrightarrow 2\,KAl(OH)_4(aq) + 3\,H_2(g)$$

If you place 2.05 g of aluminum in a beaker with 125 mL of 1.25 M KOH, will any aluminum remain? What mass of $KAl(OH)_4$ is produced?

TITRATIONS

82. What volume of 0.812 M HCl, in milliliters, is required to titrate 1.33 g of NaOH to the equivalence point?

$$NaOH(aq) + HCl(aq) \longrightarrow NaCl(aq) + H_2O(\ell)$$

83. If 32.45 mL of HCl is used to titrate 2.050 g of Na_2CO_3 according to the following equation, what is the molarity of the HCl?

$$Na_2CO_3(aq) + 2\,HCl(aq) \longrightarrow 2\,NaCl(aq) + CO_2(g) + H_2O(\ell)$$

84. What volume of 0.955 M HCl, in milliliters, is needed to titrate 2.152 g of Na_2CO_3 to the equivalence point?

$$Na_2CO_3(aq) + 2\,HCl(aq) \longrightarrow 2\,NaCl(aq) + CO_2(g) + H_2O(\ell)$$

85. Potassium hydrogen phthalate, $KHC_8H_4O_4$, is used to standardize solutions of bases. The acidic anion reacts with strong bases (such as NaOH or KOH) according to the following net ionic equation:

$$HC_8H_4O_4^-(aq) + OH^-(aq) \longrightarrow C_8H_4O_4^{2-}(aq) + H_2O(\ell)$$

If a 0.902-g sample of potassium hydrogen phthalate is dissolved in water and titrated to the equivalence point with 26.45 mL of NaOH, what is the molarity of the NaOH?

86. A noncarbonated soft drink contains an unknown amount of citric acid, $H_3C_6H_5O_7$. If 100. mL of the soft drink requires 33.51 mL of 0.0102 M NaOH to neutralize the citric acid completely, how many grams of citric acid does the soft drink contain per 100. mL? The reaction of citric acid and NaOH is

$$H_3C_6H_5O_7(aq) + 3\,NaOH(aq) \longrightarrow Na_3C_6H_5O_7(aq) + 3\,H_2O(\ell)$$

87. You are given a 4.554-g sample that is a mixture of oxalic acid, $H_2C_2O_4$, and another solid that does not react with sodium hydroxide. If 29.58 mL of 0.550 M NaOH is required to titrate the oxalic acid in the 4.554-g sample to the equivalence point, what is the weight percent of oxalic acid in the mixture? Oxalic acid and NaOH react according to the equation

$$H_2C_2O_4(aq) + 2\,NaOH(aq) \longrightarrow Na_2C_2O_4(aq) + 2\,H_2O(\ell)$$

88. An unknown solid acid is either citric acid or tartaric acid. To determine which acid you have, you titrate a sample of the solid with NaOH. The appropriate reactions are:

Citric acid:

$$H_3C_6H_5O_7(aq) + 3\,NaOH(aq) \longrightarrow Na_3C_6H_5O_7(aq) + 3\,H_2O(\ell)$$

Tartaric acid:

$$H_2C_4H_4O_6(aq) + 2\,NaOH(aq) \longrightarrow Na_2C_4H_4O_6(aq) + 2\,H_2O(\ell)$$

You find that a 0.956-g sample requires 29.1 mL of 0.513 M NaOH for titration to the equivalence point. What is the unknown acid?

89. You have 0.954 g of an unknown acid, H_2A, which reacts with NaOH according to the balanced equation

$$H_2A(aq) + 2\,NaOH(aq) \longrightarrow Na_2A(aq) + 2\,H_2O(\ell)$$

If 36.04 mL of 0.509 M NaOH is required to titrate the acid to the equivalence point, what is the molar mass of the acid?

90. Vitamin C is the simple compound $C_6H_8O_6$. Besides being an acid, it is also a reducing agent. One method for determining the amount of vitamin C in a sample is therefore to titrate it with a solution of bromine, Br_2, an oxidizing agent.

$$C_6H_8O_6(aq) + Br_2(aq) \longrightarrow 2\,HBr(aq) + C_6H_6O_6(aq)$$

Suppose a 1.00-g "chewable" vitamin C tablet requires 27.85 mL of 0.102 M Br_2 for titration to the equivalence point. How many grams of vitamin C are in the tablet?

91. To analyze an iron-containing compound, you convert all the iron to Fe^{2+} in aqueous solution and then titrate the solution with aqueous $KMnO_4$ according to the following balanced, net ionic equation:

$$MnO_4^-(aq) + 5\,Fe^{2+}(aq) + 8\,H^+(aq) \longrightarrow Mn^{2+}(aq) + 5\,Fe^{3+}(aq) + 4\,H_2O(\ell)$$

If a 0.598-g sample of the iron-containing compound requires 22.25 mL of 0.0123 M $KMnO_4$ for titration to the equivalence point, what is the weight percent of iron in the compound?

GENERAL QUESTIONS

92. The mineral dolomite contains magnesium carbonate. Write the net ionic equation for the reaction of magnesium carbonate and hydrochloric acid and name the spectator ions. What type of reaction is this?

$$MgCO_3(s) + 2\ HCl(aq) \longrightarrow CO_2(g) + MgCl_2(aq) + H_2O(\ell)$$

93. Mg metal reacts readily with HNO_3, and the following compounds are all involved:

$$Mg(s) + HNO_3(aq) \longrightarrow Mg(NO_3)_2(aq) + NO_2(g) + H_2O(\ell)$$

(a) Balance the equation for the reaction.
(b) Name each compound.
(c) Write the net ionic equation for the reaction.
(d) What are the oxidizing and reducing agents?

94. Ammonium sulfide, $(NH_4)_2S$, reacts with $Hg(NO_3)_2$ to give HgS and NH_4NO_3.
(a) Write the overall balanced equation for the reaction. Indicate the state (s or aq) for each compound.
(b) Name each compound.
(c) What type of reaction is this?

95. What species (atoms, molecules, or ions) are present in an aqueous solution of each of the following compounds?
(a) NH_3 (c) NaOH
(b) CH_3CO_2H (d) HBr

96. The *Chemical Puzzler* on page 179 shows the reaction of citric acid with sodium bicarbonate. The unbalanced equation is

$$H_3C_6H_5O_7(aq) + NaHCO_3(aq) \longrightarrow Na_3C_6H_5O_7(aq) + CO_2(g) + H_2O(\ell)$$

(a) What type of reaction is this?
(b) If an Alka-Seltzer tablet contains exactly 100 mg of citric acid, how many milligrams of sodium hydrogen carbonate must it also contain?

97. In the *Chemical Puzzler* for Chapter 4 sodium hydrogen carbonate reacts with acetic acid.

$$NaHCO_3(aq) + CH_3CO_2H(aq) \longrightarrow NaCH_3CO_2(aq) + CO_2(g) + H_2O(\ell)$$

Suppose you have 125 mL of 0.15 M acetic acid. Is CO_2 gas still being evolved after you have added 15.0 g of $NaHCO_3$? Explain your reasoning.

98. What are the concentrations of ions in a solution made by diluting 10.0 mL of 2.56 M HCl with water to obtain 250. mL of solution?

99. One-half liter (500. mL) of 2.50 M HCl is mixed with 250. mL of 3.75 M HCl. Assuming the total solution volume after mixing is 750. mL, what is the concentration of hydrochloric acid in the resulting solution?

100. Sodium thiosulfate, $Na_2S_2O_3$, is used as a "fixer" in black-and-white photography. Assume you have a bottle of sodium thiosulfate and want to determine its purity. The thiosulfate ion can be oxidized with I_2 according to the equation

$$I_2(aq) + 2\ S_2O_3^{2-}(aq) \longrightarrow 2\ I^-(aq) + S_4O_6^{2-}(aq)$$

This reaction occurs rapidly and quantitatively, and a simple method exists for observing when the reaction has reached the equivalence point. It can be used, therefore, as a method of analysis by titration. If you use 40.21 mL of 0.246 M I_2 in a titration, what is the weight percent of $Na_2S_2O_3$ in a 3.232-g sample of impure material?

101. The following reaction can be used to prepare iodine in the laboratory. (See photos.)

$$2\ NaI(s) + 2\ H_2SO_4(aq) + MnO_2(s) \longrightarrow Na_2SO_4(aq) + MnSO_4(aq) + I_2(g) + 2\ H_2O(\ell)$$

(a) Determine the oxidation number of each atom in the equation.
(b) What is the oxidizing agent and what has been oxidized? What is the reducing agent and what has been reduced?
(c) What quantity of iodine can be obtained if 20.0 g of NaI is mixed with 10.0 g of MnO_2 (and a stoichiometric excess of sulfuric acid)?

Preparation of iodine. A mixture of sodium iodide and manganese(IV) oxide was placed in a flask in a hood (*left*). On adding concentrated sulfuric acid (*right*), brown gaseous I_2 was evolved. (*C. D. Winters*)

CONCEPTUAL QUESTIONS

102. Explain how to prepare zinc chloride by (a) an acid-base reaction, (b) a gas-forming reaction, and (c) an oxidation-reduction reaction. The available starting materials are $ZnCO_3$, HCl, Cl_2, HNO_3, $Zn(OH)_2$, NaCl, $Zn(NO_3)_2$, and Zn. (See Question 105 for a discussion of compound preparation.)

103. Sketch a picture of an aqueous solution of barium chloride. Use circles to show the different ions that exist in aqueous solution. Be sure to show them in the correct stoichiometric ratio.

104. Two students titrate different samples of the same solution of HCl using 0.100 M NaOH solution and phenolphthalein indicator (see Figure 5.26). The first student pipets 20.0 mL of the HCl solution into a flask, adds 20 mL of distilled water and a few drops of phenolphthalein solution, and titrates until a lasting pink color appears. The second student pipets 20.0 mL of the HCl solution into a flask, adds 60 mL of distilled water, and a few drops of phenolphthalein solution, and titrates to the first lasting pink color. Each student correctly calculates the molarity of a HCl solution. The second student's result will be:
(a) Four times less than the first student's
(b) Four times more than the first student's
(c) Two times less than the first student's
(d) Two times more than the first student's
(e) The same as the first student's

105. The types of reactions described in this chapter can be used to prepare compounds. For example, insoluble barium chromate can be made by a precipitation reaction involving the soluble compounds $BaCl_2$ and K_2CrO_4.

$$BaCl_2(aq) + K_2CrO_4(aq) \longrightarrow BaCrO_4(s) + 2\,KCl(aq)$$

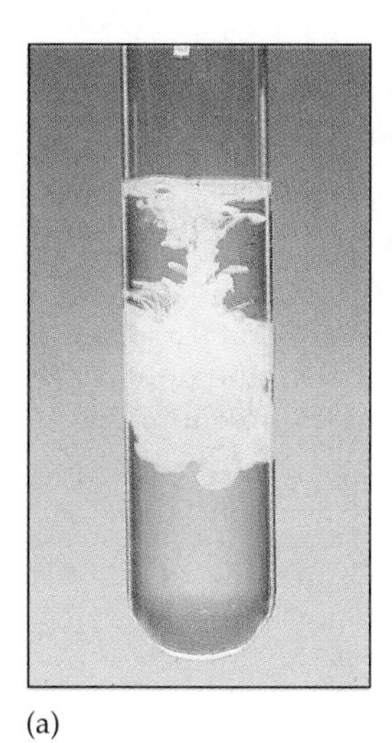

(a) (b)

Preparation of barium chromate. (a) An aqueous solution of barium chloride is added from a dropper to a solution of potassium chromate, K_2CrO_4. (b) Barium chromate, $BaCrO_4$, precipitates and is separated from the solution by collecting it on a filter paper. (*C. D. Winters*)

The product, $BaCrO_4$, can be separated from dissolved products or reactants by filtering the product mixture. The insoluble $BaCrO_4$ is trapped on the filter paper, and aqueous KCl passes through the paper. Suggest a precipitation reaction and a gas-forming reaction by which barium sulfate can be made.

CHALLENGING PROBLEMS

106. You have moved into a new house with a beautiful, heart-shaped hot tub. You need to know the volume of water in the tub, but, owing to the tub's irregular shape, it is not simple to determine its dimensions and calculate the volume. Instead, you solve the problem by stirring in a solution of a dye (1.0 g of methylene blue, $C_{16}H_{18}ClN_3S$, in 50.0 mL of water). After the dye has mixed with the water in the tub, you take a sample of the water. Using an instrument such as a spectrophotometer, you determine that the concentration of the dye in the tub is 4.1×10^{-8} M. What is the volume of water in the tub?

107. Gold can be dissolved from gold-bearing rock by treating the rock with sodium cyanide in the presence of oxygen.

$$4\,Au(s) + 8\,NaCN(aq) + O_2(g) + 2\,H_2O(\ell) \longrightarrow 4\,NaAu(CN)_2(aq) + 4\,NaOH(aq)$$

If you have exactly one metric ton (1 metric ton = 1000 kg) of gold-bearing rock, what volume of 0.075 M NaCN, in liters, do you need to extract the gold if the rock is 0.019% gold?

108. Suppose you mix 25.0 mL of 0.234 M $FeCl_3$ solution with 42.5 mL of 0.453 M NaOH. What is the maximum mass, in grams, of $Fe(OH)_3$ that precipitates? Which reactant is in excess? What is the molar concentration of the excess reactant remaining in solution after the maximum mass of $Fe(OH)_3$ has been precipitated?

109. You wish to determine the weight percent of copper in a copper-containing alloy. After dissolving a sample of an alloy in acid, an excess of KI is added, and the Cu^{2+} and I^- ions undergo the reaction

$$2\,Cu^{2+}(aq) + 5\,I^-(aq) \longrightarrow 2\,CuI(s) + I_3^-(aq)$$

The liberated I_3^- is titrated with sodium thiosulfate according to the equation

$$I_3^-(aq) + 2\,S_2O_3^{2-}(aq) \longrightarrow S_4O_6^{2-}(aq) + 3\,I^-(aq)$$

If 26.32 mL of 0.101 M $Na_2S_2O_3$ is required for titration to the equivalence point, what is the weight percent of Cu in 0.251 g of the alloy?

110. Cobalt(III) ion forms many compounds with ammonia. To find the formula of one of these compounds, you titrate the NH_3 in the compound with standardized acid.

$$Co(NH_3)_xCl_3(aq) + x\,HCl(aq) \longrightarrow x\,NH_4^+(aq) + Co^{3+}(aq) + (x + 3)\,Cl^-(aq)$$

Assume that 23.63 mL of 1.500 M HCl is used to titrate 1.580 g of $Co(NH_3)_xCl_3$. What is the value of x?

111. You have done a laboratory experiment in which you obtained a new iron-containing compound. Its formula could be one of two possibilities: $K[Fe(C_2O_4)_2(H_2O)_2]$ or $K_3[Fe(C_2O_4)_3]$. (In each of these, the $C_2O_4^{2-}$ ion is the oxalate ion.) To find which is correct, you dissolve 1.356 g of the compound in acid, which converts the oxalate ion to

oxalic acid,

$$2\,H^+(aq) + C_2O_4^{2-}(aq) \longrightarrow H_2C_2O_4(aq)$$

and then titrate the oxalic acid with potassium permanganate. The balanced, net ionic equation for the titration is

$$5\,H_2C_2O_4(aq) + 2\,MnO_4^-(aq) + 6\,H^+(aq) \longrightarrow 2\,Mn^{2+}(aq) + 10\,CO_2(g) + 8\,H_2O(\ell)$$

The titration requires 34.50 mL of 0.108 M $KMnO_4$. What is the correct formula of the iron-containing compound?

Summary Question

112. The cancer chemotherapy drug cisplatin, $Pt(NH_3)_2Cl_2$, can be made by reacting $(NH_4)_2PtCl_4$ with ammonia in aqueous solution. Besides cisplatin, the other product is NH_4Cl.
(a) Write a balanced equation for this reaction.
(b) To obtain 12.50 g of cisplatin, what mass of $(NH_4)_2PtCl_4$ is required? What volume of 0.125 M NH_3 is required?
(c) Cisplatin can react with the organic compound pyridine, C_5H_5N, to form a new compound.

$$Pt(NH_3)_2Cl_2(aq) + x\,C_5H_5N(aq) \longrightarrow Pt(NH_3)_2Cl_2(C_5H_5N)_x(s)$$

Suppose you treat 0.150 g of cisplatin with what you believe is an excess of liquid pyridine (1.50 mL; d = 0.979 g/mL). When the reaction is complete, you can find out how much pyridine was not used by titrating the solution with standardized HCl. If 37.0 mL of 0.475 M HCl is required to titrate the excess pyridine,

$$C_5H_5N(aq) + HCl(aq) \longrightarrow C_5H_5NH^+(aq) + Cl^-(aq)$$

what is the formula of the unknown compound $Pt(NH_3)_2Cl_2(C_5H_5N)_x$?

INTERACTIVE GENERAL CHEMISTRY CD-ROM

Screen 4.11 Types of Reaction

After reviewing each reaction on this screen, go back to Screen 4.1SB, the *Chemical Puzzler.* Identify each reaction in the *Puzzler* as one of the four types of reactions.

Screen 5.9 Solutions

(a) Describe what happens as $KMnO_4$ is added to water. Is the mixture homogeneous or heterogeneous?
(b) In the animation, the K^+ ions are shown as yellow spheres, whereas the MnO_4^- ions are a collection of blue and smaller red spheres.
  (1) Describe what you see as the animation proceeds. How does an ionic solid dissolve in water? What forces are at work that allow the solid to dissolve?
  (2) Water molecules are attached to both positive and negative ions. What does this tell you about the nature of water?
  (3) How does the orientation of the water molecules differ when they encounter a K^+ ion as compared with a MnO_4^- ion? What does this tell you about the nature of the water molecule? That is, describe how electric charge is distributed in the water molecule.

Screen 5.10SB A Closer Look—Ion Concentration in Solution

In the animation the green spheres represent negative ions (say Br^-), and the gray spheres represent positive ions (say Mg^{2+}).
(a) What is the ratio of positive to negative ions? What is the formula of the salt?
(b) Refer to Screen 5.9 and the nature of water molecules. If you had not been told the green spheres were negative ions, how could you have figured this out?

Screen 5.11 Preparing Solutions of Known Concentration (1)

(a) Sufficient water was added to hydrated nickel(II) chloride to make exactly 250 mL of a 0.140 M solution. What is the concentration of the Ni^{2+} ion? Of the Cl^- ion?
(b) Why is it important to shake the volumetric flask thoroughly before using the solution you have made?
(c) Examine the glassware on Screen 5.11SB. Which piece of glassware is least accurate?

Screen 5.12 Preparing Solutions of Known Concentration (2)

Why is the color of the diluted solution K_2CrO_4 less intense than the color of the original, stock solution?

Screen 5.14 Titrations

Observe the video of a titration of oxalic acid with aqueous NaOH.
(a) What is the approximate volume of NaOH solution used?
(b) If the NaOH solution had a concentration of 0.10 M, what mass of oxalic acid must have been present in the flask?

Screen 5.15 Titration Simulation

The experiment on this screen simulates the technique of titration. You can add aqueous sodium hydroxide (0.0886 M) from the buret in small increments to 25.0 mL of aqueous HCl in the flask. The indicator, an organic dye, should turn color just at the point at which the acid has been consumed completely by the base. Use the simulation to determine the quantity of 0.0886 M NaOH required for titration of your HCl solution to the equivalence point. (*Note:* You may redo the titration using the same solution as many times as you like. If you leave this screen and then return to it later, however, a different unknown is chosen at random.)

$$HCl(aq) + NaOH(aq) \longrightarrow NaCl(aq) + H_2O(\ell)$$

(a) What quantity of 0.0886 M NaOH is required? How many moles of NaOH were used in the titration?
(b) How many moles of HCl were in the flask?
(c) What was the concentration of the HCl solution you titrated? (When you have finished your calculations, check the follow-up text to see if you are correct.)

Chapter 6

Principles of Reactivity: Energy and Chemical Reactions

(C. D. Winters)

A Chemical Puzzler

Peanuts and peanut oil are organic materials and burn in air. How many burning peanuts does it take to provide the energy to boil a cup of water? How could you figure out how much thermal energy can be derived from a peanut or a handful of peanuts? (For an answer see Screen 6.19 of the *Saunders Interactive General Chemistry CD-ROM.*)

Some chemical reactions begin as soon as the reactants come into contact and continue until at least one reactant (the limiting reactant) is completely consumed. If a candy Gummi Bear, which is mostly sugar, is dropped into hot, molten potassium chlorate, the sugar and $KClO_3$ react vigorously (Figure 6.1).

$$\underset{\text{sugar, reducing agent}}{C_6H_{12}O_6(s)} + \underset{\text{potassium chlorate, oxidizing agent}}{4\ KClO_3(\ell)} \longrightarrow 6\ CO_2(g) + 6\ H_2O(g) + 4\ KCl(s)$$

If aluminum and bromine are combined, they also react rapidly to give aluminum bromide and to evolve considerable energy (see Figure 1, page 2). Many other reactions happen much more slowly, but reactants are still converted almost completely to products. An example is the rusting of iron at room temperature. Given many years' time, a piece of iron exposed to air will be converted completely to iron(III) oxide; that is, it will rust away. Still other reactions are so slow at room temperature that even a lifetime is not enough to observe a measurable change. An example is the oxidation of gasoline, which burns rapidly at high temperatures but can be stored for long periods in direct contact with air—so long as the temperature is not raised by a spark or flame. Nevertheless, chemists are convinced that if one could wait long enough, gasoline and oxygen would be converted to CO_2 and H_2O.

Two questions about chemical reactions are important: Will a reaction occur under a given set of conditions to give predominantly products, and, if it does, how fast will it go? In this chapter we want to examine some aspects of the first question: How can we predict when a reaction will occur? We shall postpone discussing the factors that determine the speed of a reaction until Chapter 15.

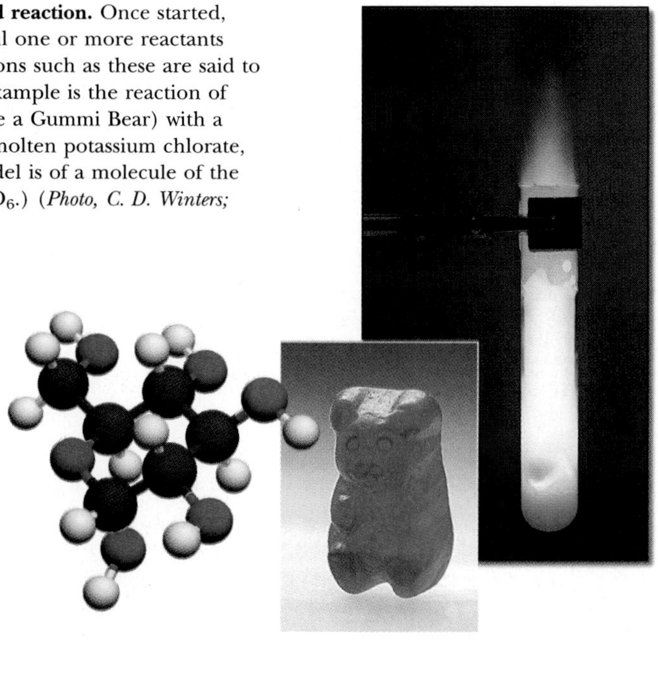

Figure 6.1 Product-favored reaction. Once started, many reactions continue until one or more reactants have been consumed. Reactions such as these are said to be "product-favored." One example is the reaction of sugar (a reducing agent, here a Gummi Bear) with a good oxidizing agent (here molten potassium chlorate, $KClO_3$). (The molecular model is of a molecule of the simple sugar glucose, $C_6H_{12}O_6$.) *(Photo, C. D. Winters; model, S. M. Young)*

6.1 TYPES OF CHEMICAL REACTIONS AND THERMODYNAMICS

Let us begin by looking at two familiar chemical reactions. The first is the reaction of gaseous hydrogen and oxygen to form water (see Figure 1.15), a reaction described by the equation

$$2\ H_2(g) + O_2(g) \longrightarrow 2\ H_2O(\ell)$$

A characteristic of this reaction is that, once initiated by a flame or spark, it proceeds until at least one reactant (H_2 or O_2) has been totally consumed. Conversion of the reactants to the product (H_2O) is also accompanied by the evolution of energy in the form of heat and light. Based on the favorability of this reaction to give products, we categorize it as **product-favored.**

One way to prepare hydrogen gas is to pass an electric current through water, converting it to hydrogen and oxygen. The equation for this reaction is the reverse of the previous equation:

$$2\ H_2O(\ell) \longrightarrow 2\ H_2(g) + O_2(g)$$

In contrast to the combination of hydrogen and oxygen, this reaction proceeds *only* if energy is supplied. This reaction will be sustained only as long as energy is supplied. Because of this, we refer to the reaction as **reactant-favored.**

In the context of product-favored and reactant-favored reactions, we can see that it is possible to write the reverse of any equation by simply changing the direction of the arrow. For example, sodium reacts with chlorine in a product-favored reaction (see Figure 2.27); writing the equation in the opposite direction results in a reactant-favored reaction:

Product-favored reaction: $2\ Na(s) + Cl_2(g) \longrightarrow 2\ NaCl(s)$

Reactant-favored reaction: $2\ NaCl(s) \longrightarrow 2\ Na(s) + Cl_2(g)$

As in the example of the hydrogen-oxygen reaction, the reactant-favored reaction does not occur unless there is outside intervention.

What do we mean by outside intervention? Usually it is some flow of energy. For example, at very high temperatures, small but significant quantities of NO can be formed from air:

$$N_2(g) + O_2(g) \xrightarrow{+\text{ energy}} 2\ NO(g)$$

Such high-temperature conditions are found in power plants and automobile engines, and a large number of such sources can produce enough NO and other nitrogen oxides to cause significant air pollution problems. In the case of the decomposition of NaCl, only by using electric energy can molten salt be converted to sodium and chlorine. In each case energy can cause a reactant-favored system to produce products.

Energy is a central idea when describing a reaction as product- or reactant-favored. For this reason it is useful to know something about energy and its interactions with matter. The most common of these interactions is the transfer of energy as heat or thermal energy when chemical reactions occur. Transfer of heat is a major theme of **thermodynamics,** the science of heat and work—the subject of this chapter.

See the *Saunders Interactive General Chemistry CD-ROM,* Screen 6.2, Product-Favored Systems, and Screen 6.3, Control of Chemical Reactions—Thermodynamics and Kinetics.

In Chapter 5 you learned that product-favored reactions include ones that

- Produce a precipitate
- Produce a salt and water from an acid and base
- Produce a gas such as CO_2

or that

- Are redox reactions such as metals reacting with oxygen or the halogens

Nitrogen oxides, labeled NO_x, are significant components of polluted air.

Energy transfer in chemical reactions is one factor in predicting when reactions are product-favored. This is further described in Chapter 20.

Thermodynamics is also related to the problem of energy use in your home and to ways of conserving energy, to the question of the recycling of materials, and to the problem of current and future energy use in our economy. We shall spend considerable time describing energy transfer and the principles of thermodynamics as applied to simple chemical systems, but keep in mind the practical applications. We shall turn to a few of these at the end of the chapter.

6.2 ENERGY: ITS FORMS AND UNITS

See the *Saunders Interactive General Chemistry CD-ROM,* Screen 6.4, Energy, and Screen 6.5, Forms of Energy.

Energy is defined as the capacity to do work, and work is something you experience all the time. You do some work against the force of gravity if you carry yourself and hiking equipment up a mountain. You can do this work because you have the energy or capacity to do so, the energy having been provided by the food you have eaten. Food energy is chemical energy—energy stored in chemical compounds and released when the compounds undergo the chemical reactions of metabolism.

Energy can be classified as kinetic or potential. An object has **kinetic energy** because it is moving. Examples are:

- *Thermal energy* of atoms, molecules, or ions in motion at the submicroscopic level—All matter has thermal energy because, according to the kinetic-molecular theory, the submicroscopic particles of matter are in constant motion (see Section 1.1).
- *Mechanical energy* of a macroscopic object like a moving baseball or automobile
- *Electric energy* of electrons moving through a conductor
- *Sound,* which corresponds to compression and expansion of the spaces between molecules

Potential energy is energy that results from an object's position. Examples are:

- *Chemical potential energy* resulting from attractions among electrons and atomic nuclei in molecules
- *Gravitational energy,* such as that of a ball held well above the floor or that of water at the top of a waterfall (Figure 6.2)
- *Electrostatic energy* such as that of positive and negative ions a small distance apart (see Figure 6.2)

Potential energy is stored energy—it can be converted into kinetic energy. For example, as water droplets fall over a waterfall, the potential energy of the water is converted into kinetic energy, and the drops move faster and faster. Similarly, kinetic energy can be converted to potential energy: the kinetic energy of falling water can turn a turbine to produce electricity (Figure 6.2).

Temperature, Heat, and the Conservation of Energy

See the *Saunders Interactive General Chemistry CD-ROM,* Screen 6.5, A Closer Look: Heat, Hotness, and Thermometers.

When you stand on a diving board, poised to dive into a swimming pool, you have considerable potential energy because of your position above the water. Once you jump off the board, some of that potential energy is converted pro-

(a) (b)

Figure 6.2 Energy. (a) Water at the top of a water wheel represents stored, or potential, energy. As it flows over the wheel, its energy is converted to mechanical energy. (b) Lightning results from electrostatic energy. The discharge converts this form of potential energy into radiant and thermal energy. (*a, Bruce Roberts/Photo Researchers, Inc.; b, Kent Wood/Photo Researchers, Inc.*)

gressively into kinetic energy (Figure 6.3), which depends on your mass (m) and velocity (v):

$$\text{Kinetic energy of a body in motion} = \frac{1}{2}mv^2 = \frac{1}{2}(\text{mass}) \cdot (\text{velocity})^2$$

During the dive, your mass is constant, whereas the force of gravity accelerates your body to move faster and faster as you fall, so your velocity and kinetic energy increase. This happens at the expense of potential energy. At the moment you hit the water, your velocity is abruptly reduced, and much of your kinetic energy is converted to mechanical energy of the water, which splashes as your body moves it aside by doing work on it. Eventually you float on the surface, and the water becomes still again. If you could see them, however, you would find that the water molecules were moving a little faster in the vicinity of your dive; that is, the temperature of the water would be a little higher.

This series of energy conversions, from potential to kinetic, and from the energy of motion to heat energy, illustrates the **law of energy conservation,** which states that energy can neither be created nor destroyed—*the total energy of the universe is constant.* This law summarizes the results of a great many experiments in which heat, work, and other forms of energy transfer have been measured and the total energy found to be the same before and after an event.

The law of energy conservation is the reason we are careful not to say, for example, that the energy of oil is used up when it is burned. What has been con-

Figure 6.3 The law of energy conservation. During the course of a dive, the diver's potential energy is converted to kinetic energy and finally to thermal energy, illustrating the law of energy conservation. (*J. Williamson/Photo Researchers, Inc.*)

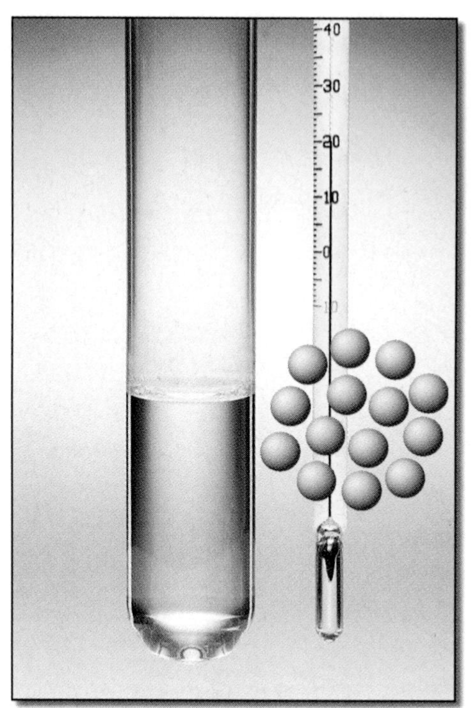

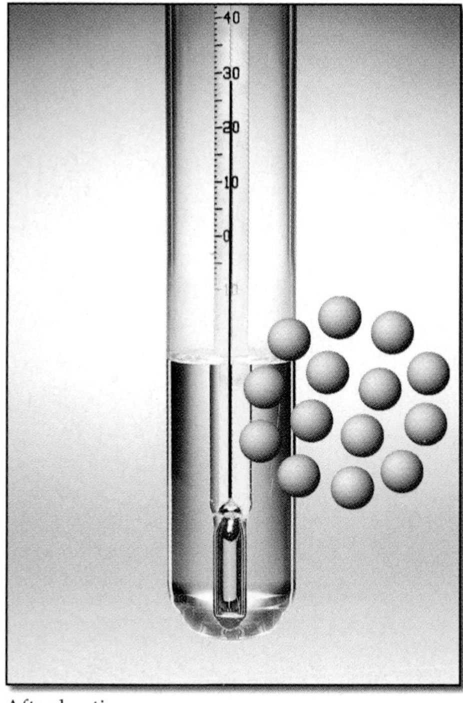

Before heating After heating

***Figure* 6.4 Measuring temperature.** The mercury in a thermometer expands as heat is transferred to the liquid metal from water. As the liquid metal warms, the mercury atoms move more rapidly and are slightly farther apart (as indicated by a decrease in density as the temperature increases). (*Photos, C. D. Winters; models, S. M. Young*)

sumed is an energy resource: the oil's capacity to transfer heat energy to its surroundings when it is burned. If oil is burned in a power plant, its chemical energy can be changed to an equal quantity of energy in other forms. These are mainly electricity, which can be very useful, and thermal energy in the gases going up the smokestack. It is not the oil's energy, but rather its ability to store energy and release it in a form that people can use that has been used up.

Notice also that in our diving example the potential energy the diver originally had on the diving board ended up heating the water. Thermal energy, or heat, is related to temperature, but heat is not the same as temperature. Transferring energy by heating an object increases the object's temperature, and the temperature increase can be measured with a thermometer. For example, Figure 6.4 shows a thermometer containing mercury. When the thermometer is placed in hot water, heat is transferred from the water to the thermometer. The increased energy of the mercury atoms means they are moving about more rapidly, which slightly increases the volume of the spaces between the atoms. Consequently the mercury expands (the density declines, as is the case for most substances when heated), and the upper end of the column of mercury rises higher in the thermometer tube.

Heat is not a substance that is contained in a body. Rather, it is observed as the interaction of a body with its surroundings.

Energy transfer, as heat, also occurs when two objects at different temperatures are brought into contact. *Energy always transfers spontaneously from the hotter to the cooler object.* For example, a piece of metal being heated in a Bunsen burner flame and a beaker containing cold water (Figure 6.5, *left*) are two objects with different temperatures. When the hot metal is plunged into the cold water

Figure 6.5 Energy transfer. Water in a beaker is warmed when a hotter object (a brass bar) is plunged into the water. Heat is transferred from the hotter metal bar to the cooler water (and enough heat is transferred that the water immediately around the hot metal boils). Eventually the bar and the water reach the same temperature (at which point they achieve thermal equilibrium). *(C. D. Winters)*

(Figure 6.5, *right*), heat is transferred from the metal to the water until the two objects reach the same temperature. The quantity of heat may be sufficient to raise the temperature of the water to 100 °C immediately around the metal, shown as boiling water.

Two important aspects of thermal energy should be understood:

- The more energy an atom has, the faster it moves.
- The total thermal energy in an object is the sum of the individual energies of all the atoms or molecules in that object.

The average speed of motion of atoms and molecules is related to the temperature. *The total thermal energy depends on temperature, the type of atoms or molecules, and the number of them in a sample.* For a given substance, thermal energy depends on temperature and the amount of substance. Thus a cup of steaming coffee may contain less thermal energy than a bathtub full of warm water, even though the coffee is at a higher temperature.

Under most circumstances, objects in a given region, such as your room, are at about the same temperature. If an object such as a cup of coffee or tea is much hotter than this, it transfers energy by heating the rest of the objects in your room until the hot coffee cools off (and your room warms up a bit). If an object, say a cold glass of ice water, is much cooler than its surroundings in your room, heat is transferred to the water from everything else until it warms up (and your room cools off a little). Because the total amount of material in your room is much greater than that in a cup of coffee or tea or a glass of ice water, the room temperature changes very little. Heat transfer occurs until everything is at the same temperature.

Exercise 6.1 Energy

A battery stores chemical potential energy. Into what types of kinetic energy can this potential energy be converted?

A CLOSER LOOK

Why Doesn't the Heat in a Room Cause Your Cup of Coffee to Boil?

It is interesting (and useful) to think about *why* the heat in a room doesn't cause a cup of cold coffee to boil. According to the law of energy conservation, it could just as well happen that energy could be transferred from the rest of your room to a hot cup of coffee. The coffee would then get hotter and hotter, eventually boiling. But we know from experience that this never happens. There is a *directionality* in heat transfer: *energy always transfers from a hotter object to a colder one,* never the reverse. This directionality corresponds to a spreading out of energy over the greatest possible number of atoms or molecules. Whether the relatively small number of molecules in a hot cup of coffee cools by transferring energy to a large number of atoms and molecules surrounding the cup, or the large number of particles in the surrounding environment heats a glass of ice water by transferring some of their energy to relatively few molecules in the glass, the end result is that the thermal energy is spread evenly over the maximum number of molecules. Concentrating energy in only a few particles at the expense of many, or even concentrating energy over a large number of particles at the expense of a few, is highly unlikely and is never observed on a macroscopic scale.

The idea that energy is spread over as many atoms and molecules as possible can be used to help us predict directionality in the conversion of chemical reactants to products, that is, to say whether a mixture of substances will be product-favored or reactant-favored. We shall return explicitly to this idea in Chapter 20.

Energy Units

See the *Saunders Interactive General Chemistry CD-ROM,* Screen 6.6, Energy Units and Biography: James P. Joule.

Units for measuring energy were originally designed to measure heat. A **calorie (cal)** was originally defined as the quantity of energy (transferred by heating) that is required to raise the temperature of 1.00 g of pure liquid water by 1.00 degree Celsius, from 14.5 °C to 15.5 °C. This represents a very small quantity of heat, and because we often work with larger quantities of matter, the **kilocalorie (kcal)** is used instead. The kilocalorie is equivalent to 1000 calories.

Most of us tend to think of heat as measured in calories, probably because we hear about dieting or read breakfast cereal boxes. The "calorie" referred to here, though, is the dietary Calorie (Cal), with a capital C, a unit equivalent to the kilocalorie. Thus, a breakfast cereal that gives you 100 Calories of nutritional energy per serving provides 100 kcal or 100×10^3 calories (calories with a small c).

1 kilocalorie = 1 kcal = 1000 calories
1 calorie = 4.184 joules

Currently most chemists use the **joule (J),** the SI unit of energy (Figure 6.6). The joule is preferred as a unit of heat energy because it is related directly to the units used in the calculation of mechanical energy (i.e., potential and kinetic energy). If a 2.0-kg object (about 4 lb) is moving with a velocity of 1.0 m/s (roughly 2 mph), the kinetic energy is

$$\text{Kinetic energy} = \frac{1}{2}mv^2 = \frac{1}{2}(2.0\text{ kg})(1.0\text{ m/s})^2 = 1.0\text{ kg}\cdot\text{m}^2/\text{s}^2 = 1.0\text{ J}$$

To give you some feeling for joules, suppose you are holding a six-pack of soft drink cans, each full of liquid. You drop the six-pack on your foot. Although you probably do not take time to calculate the kinetic energy at the moment of impact, it is about a calorie or two, that is, about 4 to 10 J.

Like the calorie, the joule can be inconveniently small as a unit for many purposes in chemistry and other sciences, and so the **kilojoule (kJ),** which is equivalent to 1000 joules, is often used. For example, burning peanuts may produce 40,000 J or more per gram of peanut oil. Chemists find it more convenient to express large numbers such as these in kilojoules, that is, as 40 kJ/g in this case.

***Figure* 6.6 Energy content of foods.** In many countries that use standardized SI units, food energy is also measured in joules. This can of diet soda from Australia is said to have an energy content of only 1 J! (*C. D. Winters*)

***Example* 6.1** Using Energy Units

The average solar energy received by a horizontal surface in Madison, Wisconsin in the summer is $2.3 \times 10^7 \text{ J/m}^2 \cdot \text{day}$. If a home in Madison has a horizontal roof that measures 10. m by 25 m, what quantity of energy, measured in kilojoules per day, strikes the roof?

Solution Because we know the quantity of energy received per square meter, we need only find the number of square meters of roof.

$$\text{Roof area} = (10.\text{ m})(25\text{ m}) = 250\text{ m}^2$$

$$\text{Energy received} = 250\text{ m}^2 \cdot \frac{2.3 \times 10^7\text{ J}}{\text{m}^2 \cdot \text{day}} \cdot \frac{1\text{ kJ}}{1000\text{ J}} = 5.8 \times 10^6\text{ kJ/day}$$

As an aside, it is also interesting to find the energy received in kilowatt-hours, the energy unit used by your local electric utility. Knowing that 1 kwh = 3.61×10^6 J and

$$5.8 \times 10^9\text{ J} \cdot \frac{1\text{ kwh}}{3.61 \times 10^6\text{ J}} = 1.6 \times 10^3\text{ kwh}$$

we find that the roof receives about 1600 kwh of energy. This is equivalent to the energy from several 100-watt light bulbs burning for 24 hours a day for a year.

***Exercise* 6.2** Energy Units

1. A serving of a breakfast cereal (with skim milk) provides 250 Cal. What is this energy in joules?
2. The energy used by a light bulb of *W* watts for *s* seconds is *Ws* joules. If a 75-W bulb burns for 3.0 h, how many joules of energy have been used?
3. A nonsugar sweetener provides 16 kJ of nutritional energy per serving. What is this energy in kilocalories?

6.3 SPECIFIC HEAT CAPACITY AND THERMAL ENERGY TRANSFER

A cup of water placed on a lighted lab burner takes a few minutes before the water boils. But heating a bathtub full of water with a Bunsen burner would take a long time! The difference is their *heat capacity*. **Heat capacity** is defined as the heat required to produce a given temperature change in some substance. A large object has a larger heat capacity than a small one of the same material.

See the *Saunders Interactive General Chemistry CD-ROM,* Screen 6.7, Heat Capacity, and Screen 6.8, Heat Capacity of Pure Substances.

Historical Perspective

James P. Joule (1818–1889) and Benjamin Thompson (1753–1814)

The joule is named for James Joule, the son of a wealthy brewer in Manchester, England, and a student of John Dalton (Chapter 1). The Joule family wealth, and a workshop in the brewery, gave James Joule the opportunity to pursue scientific studies. One of the "hot" topics of the day was the relation between heat and mechanical energy. Scientists of the time held the view that heat was a massless fluid called caloric. Joule began the process of disproving the concept of the caloric fluid by doing very precise experiments. He showed that work could be converted quantitatively to heat. This was experimental evidence of the concept of the mechanical equivalence of heat, that heat and work can be interconverted and that heat is not a massless fluid.

Joule's work was actually preceded by experiments done by one of the most interesting men to come from the early United States. Benjamin Thompson was born in Woburn, Massachusetts, in 1753, the son of a poor farmer. He was given the title Count Rumford by the King of Bavaria in 1792, and he died a wealthy and honored man in France, having been a spy, a diplomat, a social experimenter, a scientist, and a ladies' man. In Munich, Germany, there is a statue of him and a street named for him. Harvard University still has a Rumford professorship of physics, and a Rumford medal is awarded today by Britain's Royal Society for scientific research on heat and light.

Benjamin Thompson was clearly a very bright young man, and early on he impressed the royal governor of New Hampshire. In 1774, Thompson was made a major in the state's royalist forces even though he had no military training. Because of his royalist sympathies, however, he was soon in trouble, and he escaped to Boston and then to London in 1776. In London he befriended the right people, and was soon named Under Secretary of State for the Colonies. Because his duties involved providing men and supplies for the British forces in America, he grew rich through bribes. But it was also in London that he showed his interests in science by studying the design of cannons and the formulation of gun powder, neither of which were well understood at the time.

James Prescott Joule (1818–1889). *(Oesper Collection in the History of Chemistry/University of Cincinnati)*

After the American Revolution, Thompson moved to the continent, settling in Bavaria. There he instituted a unique system to care for the poor, built the famous English Garden in Munich, and created a candle so consistent in its light level that it became the international standard for measuring "candle power." He became a nutritional expert, stressing the potato, and concocted a soup still known as *Rumfordsuppe.* He also invented a more efficient fireplace, the modern kitchen range and convection oven, and the first double boiler and pressure cooker. What is important here, however, are the experiments he did on heat. When visiting a cannon-boring factory, he noticed that the cannon barrels were hot, and the bore-hole shavings were even hotter. This had been observed for centuries, but Thompson wanted to know more—what caused the heat and how it was passed along. He set up controlled experiments, and—well before Joule's important experiments—convinced himself that heat could not be a substance.

Thompson eventually returned to London where he was involved again with the Royal Institution and recruited the first professors, most notably a young chemist Humphry Davy (1778–1829), who went on to great fame.

On his way back to Bavaria in 1801, Thompson stopped in Paris where he was acclaimed by Napoleon and elected to the French Academy, only the second American to be elected (after Thomas Jefferson). There he also met Madame Lavoisier, the widow of Antoine Lavoisier (page 150). They bought a splendid house with her money and were married in 1805. He described her first as an "incarnation of goodness," but by about 1808 he called her a "female dragon." They divorced in 1809 and Thompson-Rumford moved to Auteuil, France, where he died in August, 1814.

The apparatus used by Thompson to study heat and work. He showed that, by drilling a hole in a metal ball, heat was transferred to water and could warm the water. *(Richard Howard)*

Heat capacity is awkward to use because it depends on the size of an object. Thus, chemists generally use **specific heat capacity, *C*.** This is the heat required to produce a given temperature change *per gram* of material. Mathematically, we express this as

Heat capacities are also often expressed on a *per mole* basis. The *molar heat capacity* is the heat required to raise the temperature of one mole of a substance by one degree Celsius.

$$\text{Specific heat capacity } (C) = \frac{\text{quantity of heat supplied}}{(\text{mass of object}) \cdot (\text{temperature change})} \quad [6.1]$$

$$C(\text{J/g} \cdot \text{K}) = \frac{q(\text{J})}{[m(\text{g})] \cdot [\Delta T(\text{K})]}$$

where the *quantity of heat transferred by heating* is symbolized by q. The symbol Δ (the capital Greek letter delta) means "change in." In chemistry we use heat in units of joules, mass in grams, and temperature in kelvins. The specific heat capacity of a substance, which is often just called its *specific heat*, is therefore expressed in units of joules per gram per kelvin (J/g · K).

Specific heat is determined by experiment. For example, experiments show that 59.8 J is required to change the temperature of 25.0 g of ethylene glycol (a compound used as antifreeze in automobile engines) by 1.00 K. This means the specific heat capacity of the compound is

$$\text{Specific heat capacity of ethylene glycol} = \frac{59.8\ \text{J}}{(25.0\ \text{g})(1.00\ \text{K})} = 2.39\ \text{J/g} \cdot \text{K}$$

The specific heat capacities of many substances have been determined experimentally, and a few values are listed in Table 6.1. Notice that water has one

Table 6.1 • Specific Heat Capacity Values for Some Elements, Compounds, and Common Solids

Substance	Name	Specific Heat Capacity (J/g · K)
Elements		
Al	aluminum	0.902
C	graphite	0.720
Fe	iron	0.451
Cu	copper	0.385
Au	gold	0.128
Compounds		
$NH_3(\ell)$	ammonia	4.70
$H_2O(\ell)$	water (liquid)	4.184
$C_2H_5OH(\ell)$	ethanol	2.46
$HOCH_2CH_2OH(\ell)$	ethylene glycol (antifreeze)	2.39
$H_2O(s)$	water (ice)	2.06
Common Solids		
wood		1.76
cement		0.88
glass		0.84
granite		0.79

of the highest known values of heat capacity. For water, the heat capacity is 4.184 J/g · K, whereas it is only about 0.45 J/g · K for iron or 0.84 J/g · K for common glass. This means it takes about nine times as much heat to raise the temperature of a gram of water 1 K as it does for a gram of iron. Or, about five times as much heat is required to increase the temperature of a gram of water by the same amount as a gram of glass. The high specific heat capacity of water also means that a considerable quantity of thermal energy must be transferred *out* of the substance before it cools down appreciably. For example, 1.00 g of water must give up 4.18 J of thermal energy to cool 1.00 K, whereas 1.0 g of glass gives up only 0.84 J or about one fifth as much thermal energy as water for the same temperature change.

The very high specific heat capacity of water is important to life on earth. A great deal of energy must be absorbed by a large body of water to raise its temperature just a degree or so. Conversely, a great deal of energy must be lost before the temperature of the water drops by more than a degree. Thus, a lake can store an enormous quantity of energy, and bodies of water have a profound influence on our weather.

The rate at which an object cools is determined by its thermal conductivity. For most metals, low heat capacity means a higher thermal conductivity.

The greater the heat capacity of a substance, and the larger its mass, the more thermal energy the substance can store. This is important in many ways. For example, you might wrap some bread in aluminum foil and heat it in an oven. You know you can remove the foil with your fingers after taking the bread from the oven, even though the bread is very hot. The aluminum foil you used has a low mass, and the metal has a low specific heat, so when you touch the hot foil, only a small amount of heat will be transferred to your fingers. This is also the reason a chain of fast-food restaurants warns you that the filling of an apple pie can be much warmer than the paper wrapper or the pie crust (Figure 6.7).

The wide variation in specific heat capacities of substances can be used to distinguish them from one another, just as density can.

If you know the specific heat capacity and mass of a substance, as well as its temperature change, you can calculate the heat transferred to or from that substance. Conversely, you can calculate the temperature change that should occur when a given quantity of heat is transferred to or from a sample of known mass. A convenient equation for such calculations is Equation 6.2, a rearranged version of Equation 6.1:

$$\text{Heat transferred} = q = (\text{specific heat in J/g} \cdot \text{K})(\text{mass in g})(\Delta T \text{ in kelvins}) \quad [6.2]$$

Figure 6.7 A practical example of specific heat. The filling of an apple pie has a higher specific heat than the pie crust or the wrapper. (Notice the warning on the wrapper.) (*C. D. Winters*)

For example, the thermal energy required to warm a 10.0-g piece of copper from 25 °C (298 K) to 325 °C (598 K) is

$$q = \left(\frac{0.385\,\text{J}}{\text{g} \cdot \text{K}}\right)(10.0\,\text{g})(598\,\text{K} - 298\,\text{K}) = 1160\,\text{J}$$

To this point, we have emphasized only the quantity of heat transferred. Equation 6.2, however, allows us to calculate not only the quantity of heat but also know the direction in which it is transferred. When ΔT is determined as in the previous calculation, that is, as

$$\Delta T = \text{final temperature} - \text{initial temperature}$$

it has an algebraic sign: positive (+) for T increase ($T_f > T_i$) and negative (−) for a decrease ($T_f < T_i$). If the temperature of the substance increases, ΔT has a positive (+) sign and so does q. This means heat was transferred *to* the substance (as in the example of heating a piece of copper bar). The opposite case, a decrease in the temperature of the substance, means ΔT has a negative (−) sign and so does q; heat was transferred *from* the substance.

Let us find the quantity of heat energy transferred from a cup of hot tea to your body and the surrounding air if the temperature of the tea held in your hand drops from 60.0 °C (333.2 K) to 37.0 °C (310.2 K) (normal body temperature). Assume the tea has a mass of 250. g and that its specific heat capacity is the same as water. The quantity of heat transferred can be obtained from Equation 6.2.

$$q = \text{heat transferred} = \frac{4.184\ \text{J}}{\text{g} \cdot \text{K}}(250.\ \text{g})(\underset{\text{final temp.}}{310.2\ \text{K}} - \underset{\text{initial temp.}}{333.2\ \text{K}})$$

$$= -24.1 \times 10^3\ \text{J} = -24.1\ \text{kJ}$$

Notice that the final answer has a *negative* value. In this case heat energy is transferred *from* the tea *to* the surroundings, and the temperature of the tea decreases.

See the *Saunders Interactive General Chemistry CD-ROM,* Screen 6.10, Heat Transfer Between Substances.

Example 6.2 Using Specific Heat Capacity

A lake has a surface area of $2.6 \times 10^6\ \text{m}^2$ (about 1 square mile) and an average depth of 10. m. What quantity of heat (in kilojoules) must be transferred to the lake water to raise the temperature by 1.0 °C? Assume the density of the water is $1.0\ \text{g/cm}^3$.

Solution To calculate the heat, we must know the mass of the lake water. Therefore, let us first calculate the volume of water and then find the mass using the density of water. With the mass known, the quantity of heat required can be found.

$$\text{Volume of water, m}^3 \xrightarrow{\times \frac{10^6\ \text{cm}^3}{1\ \text{m}^3}} \text{Volume of water, cm}^3 \xrightarrow{\times \frac{\text{g}}{\text{cm}^3}} \text{Mass of water, g} \xrightarrow{\text{Equation 6.2}} \text{Quantity of heat, kJ}$$

$$\text{Volume of water} = (2.6 \times 10^6\ \text{m}^2)(10.\ \text{m})\left(\frac{100\ \text{cm}}{1\ \text{m}}\right)^3 = 2.6 \times 10^{13}\ \text{cm}^3$$

$$\text{Mass of water} = 2.6 \times 10^{13}\text{cm}^3 \cdot \frac{1.0\ \text{g}}{\text{cm}^3} = 2.6 \times 10^{13}\ \text{g}$$

With the mass known, the quantity of heat required can be found using Equation 6.2, where ΔT is 1.0 °C, or 1.0 K:

$$q = (2.6 \times 10^{13}\ \text{g})(4.184\ \text{J/g} \cdot \text{K})(1.0\ \text{K}) = 1.1 \times 10^{14}\ \text{J, or } 1.1 \times 10^{11}\ \text{kJ}$$

Example 6.3 Determining a Specific Heat Capacity

Suppose a 55.0-g piece of metal was heated in boiling water to 99.8 °C and then dropped into water in an insulated beaker (as in Figure 6.5). There is 225 mL of water (density = 1.00 g/mL) in the beaker, and its temperature before the

A CLOSER LOOK — *Sign Conventions in Energy Calculations*

Whenever you take the difference between two quantities in chemistry, you should *always subtract the initial quantity from the final quantity.* A consequence of this convention is that the algebraic sign of the result indicates an increase (+) or a decrease (−) in the quantity for the substance being studied. This is an important point, as you will see in other chapters of this book.

Thus far, we have described temperature changes and the direction of heat transfer. The table summarizes the conventions used.

Be sure to understand that the sign of q is just a "signal" to tell the direction of heat transfer. Heat itself cannot be negative; it is simply a quantity of energy. As an example, consider your bank account. Assume you have \$260 in your account ($A_{initial}$), and after a withdrawal you have \$200 (A_{final}). The cash flow is thus

$$\text{Cash flow} = A_{final} - A_{initial} = \$200 - \$260 = -\$60$$

The negative sign on the \$60 indicates that a withdrawal has been made; the cash itself is not a negative quantity. Thus, *when we talk about heat we use an unsigned number.* When we want to indicate the direction of transfer in a process, however, we attach a negative sign (heat transferred from the substance) or a positive sign (heat transferred into the substance) to the value of q.

ΔT of Object	Sign of ΔT	Sign of q	Direction of Heat Transfer
Increase	+	+	heat transferred into object
Decrease	−	−	heat transferred out of object

metal was dropped in was 21.0 °C. The final temperature of the metal and water is 23.1 °C. What is the specific heat of the metal? Assume no heat transfers through the walls of the beaker or to the atmosphere.*

Solution Here are the most important aspects of this problem:

- The water and the metal bar end up at the same temperature (T_{final} is the same for both).
- Because of the principle of energy conservation, the thermal energy transferred into the water on warming up and the thermal energy transferred out of the metal in cooling down are numerically equal.
- q_{metal} has a negative value because its temperature dropped as heat was transferred out of the metal.
- Conversely, q_{water} has a positive value because its temperature increased as heat was transferred into the water. Therefore, $q_{metal} = -q_{water}$.

Using the heat capacity of water from Table 6.1 (and converting Celsius temperatures to kelvins), we have

$$q_{metal} = -q_{water}$$
$$(55.0\ \text{g})(C_{metal})(296.3\ \text{K} - 373.0\ \text{K}) = -(225\ \text{g})(4.184\ \text{J/g}\cdot\text{K})(296.3\ \text{K} - 294.2\ \text{K})$$

where $T_{initial}$ for the metal is 99.8 °C (373.0 K), $T_{initial}$ for the water is 21.0 °C (294.2 K), and T_{final} for both metal and water is 23.1 °C (296.3 K). Solving this for C_{metal}, we find

$$(-4220\ \text{g}\cdot\text{K})(C_{metal}) = -1977\ \text{J}$$
$$C_{metal} = 0.469\ \text{J/g}\cdot\text{K}$$

*To simplify the situation, we also assume the heat required to warm the walls of the beaker from 21 °C to 23.1 °C is negligible. This factor is taken into account in Section 6.10, however.

Exercise 6.3 Using Specific Heat Capacity

If 24.1 kJ is used to warm a piece of aluminum with a mass of 250. g, what is the final temperature of the aluminum if its initial temperature is 5.0 °C? The specific heat capacity of aluminum is 0.902 J/g · K.

Problem-Solving Tips and Ideas

6.1 Calculating ΔT

Notice that specific heat values are given in units of joules per gram per kelvin (J/g · K), where the temperature unit is the kelvin. *Virtually all calculations that involve temperature in chemistry are expressed in kelvins.* In our calculations, however, we could have used Celsius temperatures. The reason for this is that the *size* of a kelvin degree and a Celsius degree is the same, and so the *difference* between two temperatures is the same on both scales. For example, the difference between the boiling and freezing points of water is

$$\Delta T_{\text{Celsius}} = 100\ ^\circ\text{C} - 0\ ^\circ\text{C} = 100 \text{ Celsius degrees}$$

$$\Delta T_{\text{Kelvin}} = 373\text{ K} - 273\text{ K} = 100 \text{ kelvins}$$

So, in calculations that ask you to find the *difference* in temperatures, you can use either Celsius or Kelvin temperatures. If, however, a calculation calls for you simply to use a specific temperature, you *must* use the temperature in kelvins.

6.2 Units of Specific Heat Capacity

Specific heat values are given in this book in units of joules per gram per kelvin (J/g · K). Often, however, heat capacity values are given in handbooks (such as the *CRC Handbook of Chemistry and Physics)* as *molar heat capacities.* For example, liquid water has a heat capacity of 4.184 J/g · K or 75.4 J/mol · K. These values are related as follows:

$$4.184 \frac{\text{J}}{\text{g} \cdot \text{K}} \cdot \frac{18.02\text{ g}}{1\text{ mol}} = 75.40 \frac{\text{J}}{\text{mol} \cdot \text{K}}$$

6.3 Using the Concept of Energy Conservation

In Example 6.3 you learned that the thermal energy transferred from a hot piece of metal was equal to the thermal energy transferred to the water. We expressed this as $q_{\text{metal}} = -q_{\text{water}}$, where the negative sign reflects the fact that q_{metal} has a negative value ($q_{\text{metal}} = -9820\text{ J}$). Another way to look at this is to write the equation as $q_{\text{metal}} + q_{\text{water}} = 0$. This tells us the net value of the energy transferred must be zero; energy must be conserved. You may find this a useful point of view when working problems involving thermal energy transfer between objects.

Exercise 6.4 Determining Specific Heat Capacity

A 15.5-g piece of chromium, heated to 100.0 °C, is dropped into 55.5 g of water at 16.5 °C. The final temperature of the metal and the water is 18.9 °C. What is the specific heat capacity of chromium?

Exercise 6.5 Heat Transfer Between Substances

A piece of iron (400. g) is heated in a flame and then dropped into a beaker containing 1000. g of water. The original temperature of the water was 20.0 °C, but it is 32.8 °C after the iron bar is dropped in and both have come to the same temperature. What was the original temperature of the hot iron bar? (Assume no heat is lost to the beaker or surrounding air.)

6.4 ENERGY AND CHANGES OF STATE

See the *Saunders Interactive General Chemistry CD-ROM,* Screen 6.11, Heat Associated with Phase Changes.

So far we have described transfers of energy between objects as a result of temperature differences. But energy is also transferred when matter is transformed from one form to another in the course of a physical or chemical change.

When a solid melts, its atoms, molecules, or ions move about vigorously enough to break free of the constraints imposed by their neighbors in the solid. When a liquid boils, particles move much farther apart from one another (Figure 1.6). In both cases attractive forces among the particles must be overcome, which

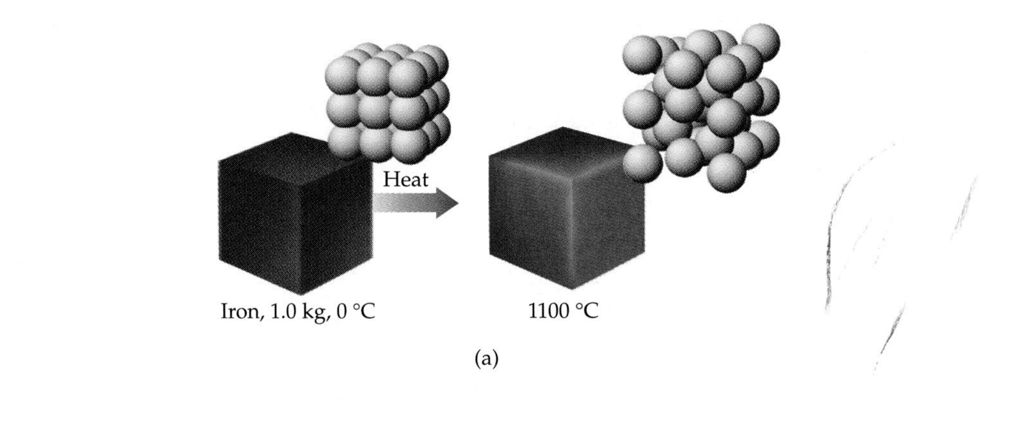

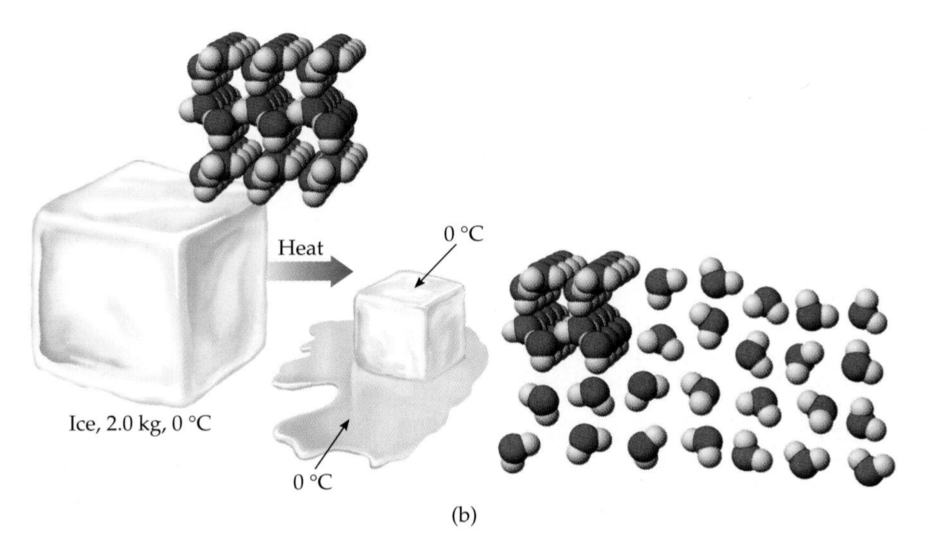

Figure 6.8 Heat transfer and phase change. Heat transfer to an object can lead to a temperature change or a phase change (or both). Here the same quantity of heat (500 kJ) has been transferred to a 1.00-kg block of iron at 0 °C as has been transferred to a 2.00-kg block of ice at 0 °C. The temperature of the iron block has increased by 1100 K and makes the block red hot. On the other hand, this quantity of heat has led only to the melting of 1.5 kg of ice, and the ice and melted water are still at 0 °C.

requires an input of energy. This is the reason heat is associated with changes of state, which *always take place at constant temperature.*

Fusion means melting. When a fuse in an electric circuit melts, it breaks the circuit and prevents current from flowing.

The quantity of heat required to melt ice at 0 °C is 333 J/g and is called the **heat of fusion** of the ice. You see in Figure 6.8 that the same quantity of heat (500. kJ) required to raise the temperature of a 1.00-kg block of iron from 0 °C to 1100 °C, at which it glows red hot, only leads to the melting of 1.50 kg of ice, at which point the ice and the melted water both have a temperature of 0 °C.

$$500.\ \text{kJ} \cdot \frac{1000\ \text{J}}{\text{kJ}} \cdot \frac{1\ \text{g ice melted at 0 °C}}{333.5\ \text{J}} = 1.50 \times 10^3\ \text{g ice melted}$$

Heats of fusion and vaporization are often expressed in units of joules per mole (J/mol). For example, for water
heat of fusion = 333.5 J/g or 6.009 kJ/mol
heat of vaporization = 2256 J/g or 40.66 kJ/mol

The quantity of heat required to convert liquid water to vapor, called the **heat of vaporization,** is 2256 J/g at 100 °C. What quantity of water can be vaporized at 100 °C if 500. kJ of heat is transferred to water at 100 °C? Only 222 g!

$$500.\ \text{kJ} \cdot \frac{1000\ \text{J}}{\text{kJ}} \cdot \frac{1\ \text{g ice vaporized}}{2256\ \text{J}} = 222\ \text{g liquid water vaporized}$$

Figure 6.9 illustrates the quantity of heat absorbed and the consequent temperature change as 500. g of water is warmed from −50 °C to 200 °C. First, the temperature of the ice increases as heat is added. On reaching 0 °C, however, *the temperature remains constant as sufficient heat is absorbed to melt the ice to liquid water.* When all the ice has melted, the liquid absorbs heat and is warmed to 100 °C, the boiling point of water. The temperature is again constant as heat is absorbed to convert the liquid completely to vapor, that is, to steam. Any further heat absorbed heats the steam to a still higher temperature. The heat absorbed at each step is calculated in Example 6.4.

Example 6.4 Energy and Changes of State

Calculate the quantity of heat involved in each step shown in Figure 6.9. That is, calculate the heat required to convert 500. g of ice at −50.0 °C to steam at

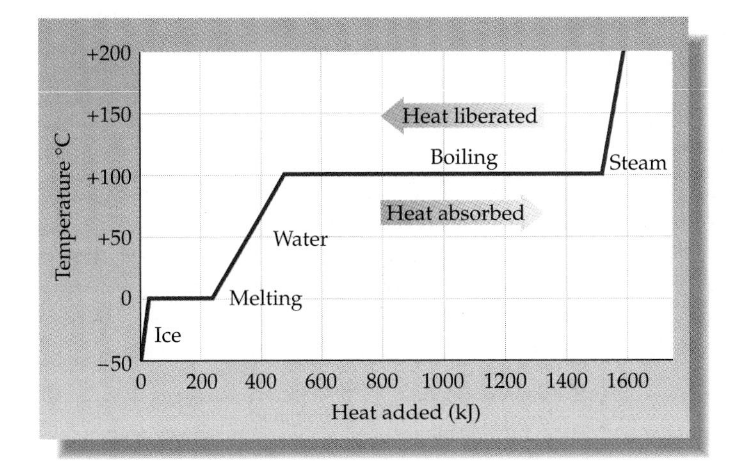

***Figure* 6.9** **Heat transfer and temperature change.** A graph illustrating the quantity of heat absorbed and the consequent temperature change as 500. g of water is warmed from −50 °C to 200 °C. The graph also illustrates the heat evolved when steam at 200 °C is cooled to −50 °C.

200 °C. The required specific heat capacities are

State	Specific Heat Capacity (J/g · K)*
Ice	2.1
Water	4.2
Steam	2.0

Solution The quantity of heat required to warm ice at −50.0 °C to 0.0 °C is calculated by the methods of Section 6.3:

$$q \text{ (to warm ice to 0.0 °C)} = (2.1\ \text{J/g}\cdot\text{K})(500.\ \text{g})(273.2\ \text{K} - 223.2\ \text{K}) = 5.3 \times 10^{4}\ \text{J}$$

Next, the ice is melted at 0.0 °C to water at that temperature:

$$q \text{ (to melt ice)} = 500.\ \text{g} \cdot \frac{333.5\ \text{J}}{\text{g ice}} = 1.67 \times 10^{5}\ \text{J}$$

After melting the ice at 0.0 °C, the water is heated from 0.0 °C to 100.0 °C:

$$q \text{ (to warm water to 100.0 °C)} = (4.2\ \text{J/g}\cdot\text{K})(500.\ \text{g})(373.2\ \text{K} - 273.2\ \text{K}) = 2.1 \times 10^{5}\ \text{J}$$

The water at 100.0 °C is evaporated to steam at 100.0 °C:

$$q \text{ (to evaporate water)} = 500.\ \text{g} \cdot \frac{2256\ \text{J}}{\text{g water}} = 1.13 \times 10^{6}\ \text{J}$$

Steam is heated from 100.0 °C to 200.0 °C:

$$q \text{ (to heat steam to 200.0 °C)} = (2.0\ \text{J/g}\cdot\text{K})(500.\ \text{g})(473.2\ \text{K} - 373.2\ \text{K}) = 1.0 \times 10^{5}\ \text{J}$$

The total thermal energy required is the sum of the thermal energy required in each step.

$$q_{total} = 1.7 \times 10^{6}\ \text{J}, \quad \text{or } 1700\ \text{kJ}$$

Approximately 1700 kJ of thermal energy must be absorbed by the 500-g sample to convert it from ice at −50.0 °C to steam at 200.0 °C. If the steam is to be converted back to ice, 1700 kJ of thermal energy will be liberated. The heat liberated, q, equals −1700 kJ.

Exercise 6.6 Changes of State

What quantity of heat must be absorbed to warm 25.0 g of liquid methanol, CH_3OH, from 25.0 °C to its boiling point (64.6 °C) and then to evaporate the methanol completely at that temperature? The specific heat of liquid methanol is 2.53 J/g · K. The heat of vaporization of methanol is 2.00×10^{3} J/g.

*Specific heat varies with temperature. The values listed are average values for normal temperature ranges.

Exercise 6.7 Using a Change of State to Determine Specific Heat Capacity

An "ice calorimeter" can be used to determine the heat capacity of a metal. A piece of hot metal is dropped into a weighed quantity of ice. The quantity of heat given up by the metal can be determined from the amount of ice melted. Suppose a piece of metal with a mass of 9.85 g is heated to 100.0 °C and then dropped onto ice. When the metal's temperature has dropped to 0.0 °C, it is found that 1.32 g of ice has been melted to water at 0.0 °C. What is the heat capacity of the metal?

6.5 ENTHALPY

Carbon dioxide undergoes a change of state from solid to gas, a process called sublimation, at −78 °C:

$$CO_2(\text{solid}, -78\ °C) \xrightarrow[+\text{ heat}]{} CO_2(\text{gas}, -78\ °C)$$

Figure 6.10 A system absorbing heat at constant volume. When heat is absorbed by solid CO_2 in a closed container, the pressure in the container increases as solid is converted to gas. Because no mechanical connection exists between the system and its surroundings, however, no work is done on the surroundings by the system. Therefore, q_v = heat absorbed at constant volume = ΔE. (*C. D. Winters*)

To describe the change from solid to gas in thermodynamic terms, we first need to extend the earlier discussion of heat as energy transferred between objects. In thermodynamics, one of the "objects" is usually of primary concern. This object, which may be a substance involved in a change of state or substances undergoing a reaction, is called the **system.** The other object is called the **surroundings** and includes everything outside the system that can exchange energy with the system. In Figures 6.10 and 6.11, a CO_2 sample (both solid and vapor) is the *system* and interacts with its *surroundings*—the flask or plastic bag, the table on which they rest, and the air in the room surrounding the sample (and by extension the rest of the universe). A system may be contained within an actual physical boundary, such as a flask or a cell in your body. Alternatively, the boundary may be purely imaginary. For example, you could study the solar system within its surroundings, the rest of the galaxy. In any case, the system can always be defined precisely.

Suppose our system consists of solid and gaseous CO_2 at −78 °C, that the surroundings are also at −78 °C, and that 1 mol CO_2(solid) is transformed into 1 mol CO_2(gas). Just as the potential energy of a ball is greater when it is farther above the earth, which attracts it, the potential energy of a CO_2 molecule is greater when it is farther from another CO_2 molecule, again because an attractive force must be overcome when the two are separated. Thus, the energy of a mole of $CO_2(g)$ is greater than the energy of a mole of $CO_2(s)$. Where does this energy come from? That depends on the circumstances.

See the *Saunders Interactive General Chemistry CD-ROM,* Screen 6.12, Energy Changes in Chemical Processes, and Screen 6.13, The First Law of Thermodynamics.

Let us begin with the case in which the $CO_2(s)$ and $CO_2(g)$ are inside a *rigid container* so that the total volume cannot change (see Figure 6.10). The energy needed for one molecule to escape from solid to gas can be supplied by neighboring molecules, which consequently vibrate a little less about their position in the crystal lattice. As more and more molecules escape, the motion of remaining solid-state molecules diminishes more and more, which would correspond to a lowering of the temperature. Atoms and molecules in the surroundings (the rigid container) are in contact with CO_2 molecules in the solid, however, and energy exchange can occur. Because the temperature of the solid CO_2 is lower

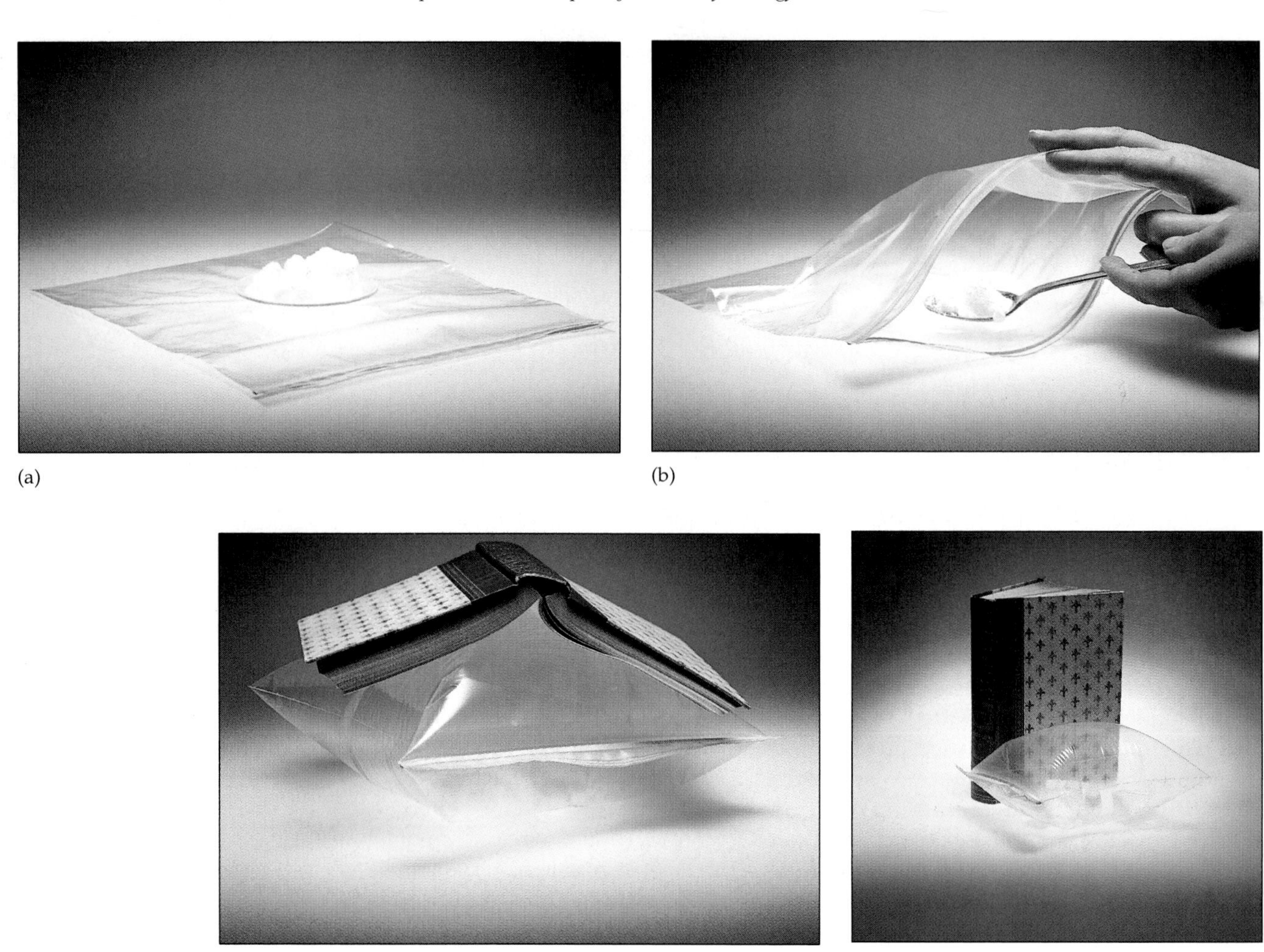

(a) (b) (c) (d)

Figure 6.11 Changes of state for CO_2. (a) Pieces of dry ice and an empty plastic bag. (b) Crushed dry ice has been placed in the plastic bag. (c) The bag has been closed. Some dry ice has sublimed, converting solid CO_2 at −78 °C to gaseous CO_2 at −78 °C. The gas has filled the bag and has done work by raising a book that has been placed on the bag. (d) A similar experiment but without the book. The expanding CO_2 in this case has pushed aside the air that formerly occupied the space taken up by the inflated bag. Pushing aside the atmosphere requires work, just as raising a book does. Work must always be done to push aside the atmosphere by any reaction whose products occupy greater volume than the reactants. The heat absorbed by the system under these constant pressure conditions (q_p) is identified as the enthalpy change, ΔH. *(C. D. Winters)*

than the temperature of the surroundings, energy transfers into the solid CO_2 from the surroundings. This heat transfer from the surroundings must exactly equal the energy required to overcome attractive forces as solid CO_2 sublimes.

What we have just described is transfer of energy as heat from the surroundings to the system while the process $CO_2(s) \longrightarrow CO_2(g)$ takes place. When heat is transferred into a system, the process is said to be **endothermic.** In contrast, the reverse process, converting CO_2 gas to solid CO_2 (dry ice), is

The prefix *exo-* means "out" and *endo-* means "in."

an **exothermic process;** heat is transferred out of the sample of CO_2 to the surroundings when some of the $CO_2(g)$ condenses to a solid. In summary,

Phase Change	Direction of Heat Transfer	Sign of q_{system}	Type of Change
Solid CO_2 $\longrightarrow$ CO_2 Gas	Surroundings $\longrightarrow$ System	Positive	Endothermic
CO_2 Gas $\longrightarrow$ Solid CO_2	System $\longrightarrow$ Surroundings	Negative	Exothermic

When the endothermic process $CO_2(s) \longrightarrow CO_2(g)$ occurs inside a *rigid* container (see Figure 6.10), the only exchange of energy is the heat transferred into the system. Thus, by conservation of energy, *the heat absorbed at constant volume, q_v,* must equal the *change* in the energy of the system, ΔE.

The subscript v on q_v indicates that the quantity of heat has been determined at constant volume.

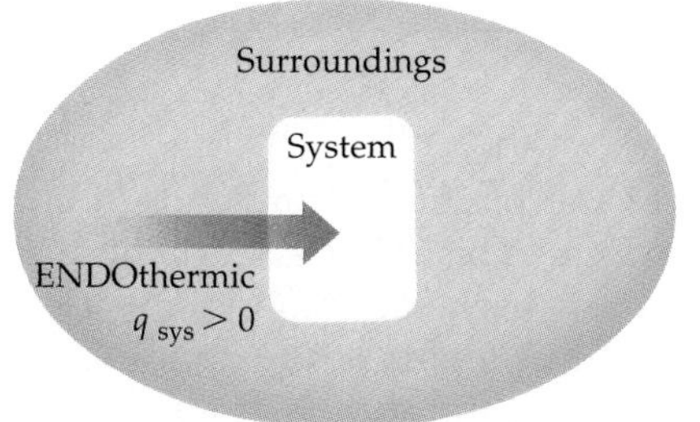

ENDOthermic: energy transferred from surroundings to system

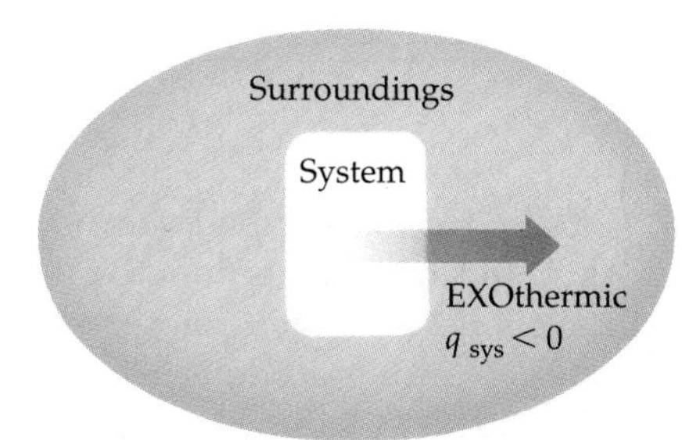

EXOthermic: energy transferred from system to surroundings

If the process does not occur in a rigid container the situation is a bit more complicated, because two energy transfers occur, not just one. As shown in Figure 6.11c, $CO_2(s)$ vaporizing into a flexible container can do some work by raising a book above a table top. Not only is energy transferred into the system as heat, but energy is also transferred out of the system as work. Work is done whenever something is caused to move against an opposing force. In this case the gas has to work against the weight of the book to raise it.

Because energy is now transferred in two ways between system and surroundings, it is no longer true that $\Delta E = q$. Energy transferred in or out of the system by work, symbolized by w, must also be taken into account. The change in energy of the system is now

$$\Delta E = q + w \qquad [6.3]$$

Notice that w represents energy transferred, so it has a direction and a sign, just like q. The work w has a negative value if the system expends energy as work, and w has a positive value if the system receives energy in the form of work. For the process $CO_2(s) \longrightarrow CO_2(g)$, w is negative because work is done on the surroundings when the atmosphere is pushed back.

That is, the energy of the system is changed by the quantity of energy transferred as heat and by the quantity of energy transferred by work. This is a statement of the **first law of thermodynamics,** also called the **law of energy conservation.** Another way of saying the same thing is that the *total amount of energy in the universe is constant.* Energy may be transferred as work or heat, but no energy can be lost, nor can heat or work be obtained from nothing. *All the energy transferred between a system and its surroundings must be accounted for as heat and work.* (This is true as long as no other energy transfer, such as the radiant energy of a glowing light bulb, is involved.) The first law is important in all aspects of our lives; it plays a central role in our models of weather, in designing a power plant, or in understanding why diet and exercise together lead to weight loss.

A CLOSER LOOK

Internal Energy

In chemistry, changes in internal energy, ΔE, that accompany a chemical or physical change are often the focus of attention. The internal energy in any given chemical system can be viewed as the sum of the potential and kinetic energy quantities associated with the system. Potential energy is the energy associated with the attractive and repulsive forces between all the nuclei and electrons in the system. This represents the energy associated with bonds in molecules, forces between ions, and also the forces between molecules in the liquid and solid state. Kinetic energy is the energy of motion of the atoms, ions, and molecules in the system; it is the sum of the energies associated with their translations, vibrations, and rotations.

Consider what happens when a chemical reaction takes place. A reaction rearranges the electrons and nuclei, so the potential energy of the products is different from that of the reactants. Recall, however, that energy, like mass, is conserved in a reaction. If the potential energy associated with the products is lower than that of the reactants, then either energy is transferred out of the system into the surroundings or the energy of the system is retained in the system as kinetic energy. In the latter case, the system becomes hotter. Conversely, if the internal energy of the products is higher than the energy of the reactants, then energy must either be added to the system or the energy needed is gained at the expense of the kinetic energy of the molecules in the system. In this case the temperature of the system decreases.

Even if the book had not been on top of the plastic bag in Figure 6.11c, work would have been done by the expanding gas. This is because whenever a gas expands into the atmosphere it has to push back the atmosphere itself. Instead of raising a book, the expanding gas moves a part of the atmosphere. For the process shown in Figure 6.11d, the energy transferred from the surroundings has two effects: (1) it overcomes the forces holding the molecules together in the solid state at −78 °C, and (2) it does work on the atmosphere as the gas expands. The energy that allowed the system to perform work on its surroundings had to come from somewhere, and that "somewhere" was the energy, q, transferred as heat to the system from its surroundings. The system simply converted part of that heat into work. Therefore, the first law of thermodynamics tells us that the energy change, ΔE, for the CO_2 system must be less than q. The reason for this is that heat is transferred into the system from its surroundings (q has a positive sign) and work is done by the system on the surroundings (w has a negative sign). That is,

Systems that convert heat into work are called heat engines: an example is the engine in an automobile, which converts heat resulting from the combustion of fuel into work to move the car forward.

$$\Delta E = q + w = (q\text{, positive number}) + (w\text{, negative number})$$

and so

$$\Delta E < q$$

In plants and animals, as well as in the laboratory, reactions usually occur at constant pressure. *The heat transferred into (or out of) a system at constant pressure,* q_p, equals a quantity called the **enthalpy change,** symbolized by ΔH ("delta H"). Thus,

See the *Saunders Interactive General Chemistry CD-ROM,* Screen 6.14, Enthalpy Change and ΔH.

$$\Delta H = q_p = \text{heat transferred to or from a system at constant pressure}$$

The subscript p on q_p indicates that heat has been transferred at constant pressure.

Because $\Delta E = q + w$, then at constant pressure $\Delta E = \Delta H + w$, showing that ΔH accounts for all the energy transferred except the amount that does the work of pushing back the atmosphere (which is usually small compared with ΔH). Even if pressure is not constant,

$$\Delta H = (\text{enthalpy of the system at the end of a process}) - (\text{enthalpy of the system at the start of the process}) = H_{final} - H_{initial} \quad [6.4]$$

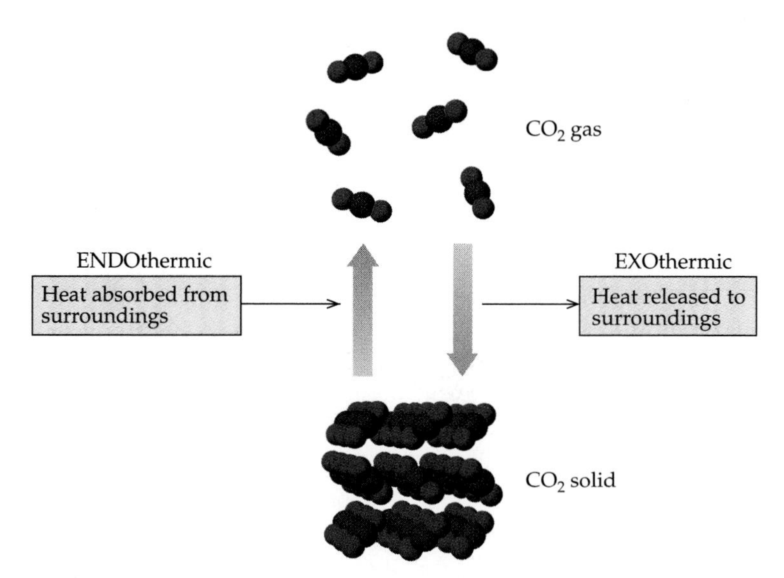

Figure 6.12 Enthalpy change. Enthalpy diagram for interconversion of solid CO_2 and CO_2 gas at constant pressure. Energy in the form of heat must be absorbed when solid CO_2 becomes a gas. The energy breaks the relatively weak forces between CO_2 molecules. If CO_2 gas is to re-form the solid, heat energy is released to the surroundings as the attractive forces between CO_2 molecules come into play.

If the enthalpy of the final system is greater than that of the initial system (as when solid CO_2 changes to CO_2 vapor at −78 °C), the enthalpy has increased (left side of Figure 6.12), and the process has a positive ΔH; q_p must also be positive and the process is endothermic. Conversely, if the enthalpy of the final system is less than that of the initial system, heat has transferred out of the system, and ΔH is negative; the process is exothermic (right side of Figure 6.12).

Another endothermic process is the evaporation of water to water vapor at 25 °C. The evaporation of one mole of water requires 44 kJ:

$$H_2O(\ell) \xrightarrow{+44.0\text{ kJ}} H_2O(g) \qquad \Delta H_{vaporization} = +44.0 \text{ kJ}$$

Change is endothermic.
44.0 kJ of heat energy is transferred from surroundings to the system (liquid H_2O).

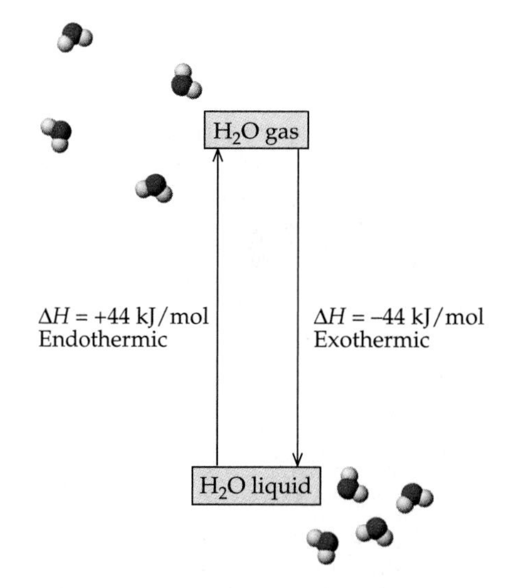

Figure 6.13 Change of state for water. For liquid water to change to water vapor (at 25 °C), the system must absorb 44.0 kJ/mol from the surroundings. When water vapor condenses to liquid water, 44.0 kJ/mol are transferred from the water to the surroundings.

But what about water vapor condensing to form liquid again? If 44.0 kJ of heat energy is required to break the attractions between H_2O molecules in a mole of the liquid so they can move into the gas phase, the same quantity of energy (44.0 kJ/mol) is regained when the molecules in the vapor condense to form the liquid (Figure 6.13). Condensation of water is exothermic; 44.0 kJ/mol is transferred from the water to the surroundings when a mole of H_2O molecules condenses to liquid.

$$H_2O(g) \xrightarrow{-44.0\text{ kJ}} H_2O(\ell) \qquad \Delta H_{condensation} = -44.0 \text{ kJ}$$

Change is exothermic.
44.0 kJ of heat energy is transferred to surroundings from the system (H_2O vapor).

Some key ideas that apply to *all* of thermodynamics may be summarized here:

- When heat transfer occurs (at constant pressure) from a system to its surroundings, as when a vapor condenses to a liquid or solid, the process is exothermic with respect to the system, and ΔH (q_p) has a negative value. Conversely, when heat is absorbed from the surroundings, as when a solid or liquid changes to a vapor, the process is endothermic with respect to the system, and ΔH (q_p) has a positive value.

- For changes that are the reverse of each other, the ΔH values are numerically the same, but their signs are opposite. Thus, for evaporation of water, $\Delta H = +44.0$ kJ/mol, whereas for the condensation of water $\Delta H = -44.0$ kJ/mol.
- The change in energy or enthalpy is directly proportional to the quantity of material undergoing a change. If 2 mol of water is evaporated, twice as much heat energy, or 88.0 kJ, is required.
- The value of ΔH is always associated with a balanced equation for which the coefficients are read as moles, so that the equation shows the macroscopic amount of material to which the value of ΔH applies. Thus, for the evaporation of 2 mol of water

$$2\ H_2O(\ell) \longrightarrow 2\ H_2O(g) \qquad \Delta H = +88.0\ \text{kJ}$$

The enthalpy of vaporization depends on the temperature at which it is measured. For water, $\Delta H_{vaporization}$ is 43.98 kJ/mol at 25 °C but 40.65 kJ/mol at the boiling point.

SUMMARY: ΔH and Phase Changes

Liquid $\longrightarrow$ Solid + heat
Vapor $\longrightarrow$ Liquid + heat
ΔH = negative number; change is **exo**thermic.

Liquid + heat $\longrightarrow$ Vapor
Solid + heat $\longrightarrow$ Liquid
ΔH = positive number; change is **endo**thermic.

As an interesting footnote to this discussion, let us calculate the energy transferred to the surroundings when water vapor in the air condenses to give rain in a thunderstorm. Suppose that an inch of rain falls over a square mile of ground so that 6.6×10^{10} g of water has fallen (a density of 1.0 g/cm^3 has been assumed). The heat of vaporization of water at 25 °C is 44.0 kJ/mol. The quantity of heat transferred to the surroundings from water vapor condensation is therefore

$$6.6 \times 10^{10}\ \text{g water} \cdot \frac{1\ \text{mol}}{18.0\ \text{g}} \cdot \frac{44.0\ \text{kJ}}{12\ \text{mol}} = 1.6 \times 10^{11}\ \text{kJ}$$

This is about the same as the heat released when about 35 million kilograms (about 38,000 tons) of dynamite explodes! (The explosion of 1000 tons of dynamite is equivalent to 4.2×10^9 kJ.) This huge number tells you how much energy is "stored" in water vapor and why we think of storms as such great forces of energy in nature.

One reason water works to put out a fire is that the energy derived from combustion is used to vaporize the water and is therefore not available to cause further combustion. (*Jim Regan/Dembinsky Photo Associates/© 1997*)

Example 6.5 Changes of State and ΔH

The enthalpy change for the vaporization of methanol, CH_3OH, at 25 °C is 37.43 kJ/mol. What quantity of heat energy must be used to vaporize 25.0 g of methanol at 25 °C?

Solution To use the enthalpy of vaporization, we must first convert the mass of the compound to moles of compound:

$$25.0\ \text{g}\ CH_3OH \cdot \frac{1\ \text{mol}}{32.04\ \text{g}\ CH_3OH} = 0.780\ \text{mol}\ CH_3OH$$

Now the enthalpy of vaporization can be used to find the heat required:

$$0.780\ \text{mol}\ CH_3OH \cdot \frac{37.43\ \text{kJ}}{1\ \text{mol}\ CH_3OH} = 29.2\ \text{kJ}$$

Exercise 6.8 Changes of State and ΔH

The enthalpy change for the sublimation of solid iodine is 62.4 kJ/mol.

$$I_2(s) \longrightarrow I_2(g) \qquad \Delta H = 62.4 \text{ kJ}$$

1. What quantity of heat energy must be used to sublime 10.0 g of solid iodine?
2. If 3.45 g of iodine vapor condenses to solid iodine, what quantity of energy is involved? Is the process exothermic or endothermic?

6.6 ENTHALPY CHANGES FOR CHEMICAL REACTIONS

See the *Saunders Interactive General Chemistry CD-ROM,* Screen 6.15, Enthalpy Changes for Chemical Reactions.

The enthalpy change can be determined for any physical or chemical change. For a chemical reaction the products represent the "final system" and the reactants the "initial system." Therefore,

$$\Delta H = H_{\text{products}} - H_{\text{reactants}} \qquad [6.5]$$

Like changes of state, chemical reactions can be exothermic or endothermic. Many of the reactions you have seen so far in photographs in this book [such as the Al + Br_2 reaction (Figure 1, page 2) and the reactions of elements with air (Figures 1.15 and 4.2–4.4)] are exothermic. To learn more about enthalpy changes that accompany chemical reactions, consider the decomposition of a mole of water vapor to its elements.

$$H_2O(g) \xrightarrow{+241.8 \text{ kJ}} H_2(g) + \tfrac{1}{2}\, O_2(g) \qquad \Delta H = +241.8 \text{ kJ}$$

Change is endothermic; ΔH is positive.
Decomposition of 1 mol of water vapor requires 241.8 kJ of energy to be transferred in from the surroundings.

Note that to write an equation for the decomposition of 1 mol of H_2O, it is necessary to use a fractional coefficient for O_2. This is acceptable in thermochemistry because coefficients are always taken to mean moles and not molecules.

The left side of Figure 6.14 shows that the enthalpy of the products of this reaction is greater than that of the reactants. Water vapor would have to absorb 241.8 kJ of energy (at constant pressure) from the surroundings as it decomposes to its elements. That is, the enthalpy change for the endothermic decomposition of water vapor is $\Delta H = +241.8$ kJ/mol.

Now consider the opposite reaction, the combination of hydrogen and oxygen to form water (see Figure 1.15). The quantity of heat energy involved in this oxidation-reduction reaction (see Figure 6.14)

$$H_2(g) + \tfrac{1}{2}\, O_2(g) \xrightarrow{-241.8 \text{ kJ}} H_2O(g) \qquad \Delta H = -241.8 \text{ kJ}$$

Change is exothermic; ΔH is negative.
Formation of 1 mol of water vapor transfers 241.8 kJ of energy to the surroundings.

is the same as for the decomposition reaction except that the reaction of the elements is exothermic. That is, $\Delta H = -241.8$ kJ/mol of water vapor formed.

As in the case of phase changes, some key ideas about enthalpy changes for reactions can be summarized:

- ΔH has a negative value when heat is evolved or transferred (at constant pressure) to the surroundings by an exothermic reaction. ΔH has a posi-

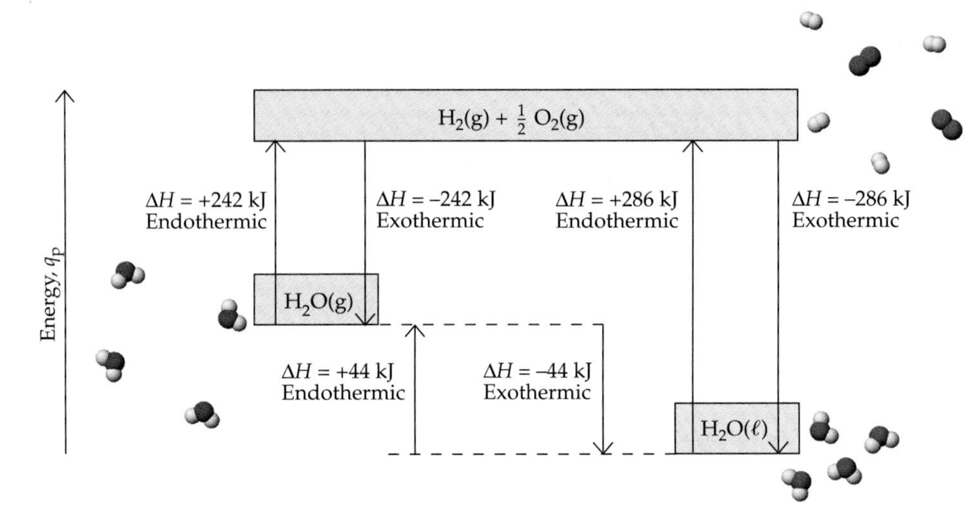

Figure 6.14 Enthalpy change for water. Enthalpy diagram for the interconversion of water vapor, $H_2O(g)$, liquid water, $H_2O(\ell)$, and the elements H_2 and O_2.

SUMMARY: Relationship of Reaction Heat and Enthalpy Change

$$\text{Reactant} \longrightarrow \text{Product + heat}$$

ΔH = negative number; reaction is exothermic.

$$\text{Reactant + heat} \longrightarrow \text{Product}$$

ΔH = positive number; reaction is endothermic.

tive value when heat is absorbed from the surroundings by an endothermic reaction.

- ΔH values are numerically the same, but the signs are opposite, for chemical reactions that are the reverse of each other.
- The change in energy or enthalpy is directly proportional to the quantity of material undergoing a change.
- In thermodynamics, the value of ΔH is always associated with a balanced equation for which the coefficients are read as moles, so that the equation shows the macroscopic amount of material to which the value of ΔH applies. Thus, for the decomposition of 2 mol of water vapor

$$2\ H_2O(g) \longrightarrow 2\ H_2(g) + O_2(g) \qquad \Delta H = +483.6\ \text{kJ}$$

Because energy is transferred when a substance undergoes a change of state, the quantity of energy associated with a chemical reaction must depend on the physical state (solid, liquid, or gas) of the reactants and products. For example, the decomposition of 1 mol of *liquid* water to H_2 and O_2

$$H_2O(\ell) \longrightarrow H_2(g) + \tfrac{1}{2}O_2(g) \qquad \Delta H = +285.8\ \text{kJ}$$

requires more energy than the decomposition of 1 mol of water vapor, as shown in Figure 6.14. The enthalpy of vaporization must be added to the enthalpy change for the decomposition of $H_2O(g)$ to give the value for the decomposition of $H_2O(\ell)$.

Enthalpy changes for reactions have many practical applications. For instance, when enthalpies of combustion are known, the quantity of heat transferred by the combustion of a given mass of a fuel such as propane, C_3H_8, can be calculated. Suppose you want to know how much heat is provided by burning 454 g (1 lb) of propane gas in a furnace. The exothermic reaction that occurs is

$$C_3H_8(g) + 5\ O_2(g) \longrightarrow 3\ CO_2(g) + 4\ H_2O(\ell) \qquad \Delta H = -2220\ \text{kJ}$$

and the enthalpy change is $\Delta H = -2220$ kJ *per mole of propane burned.* How much heat is transferred to the surroundings by burning 454 g of C_3H_8 gas? The first step is to find the number of moles of propane present in the sample:

$$454 \text{ g} \cdot \frac{1 \text{ mol propane}}{44.10 \text{ g}} = 10.3 \text{ mol propane}$$

Then you can multiply by the amount of heat transferred per mole of gas:

$$10.3 \text{ mol propane} \cdot \frac{2220 \text{ kJ evolved}}{1.00 \text{ mol propane}} = 22{,}900 \text{ kJ}$$

$$= \text{total heat evolved by 454 g of propane}$$

This is a substantial quantity of energy when you take into account that when your body completely "burns" 454 g of milk, only about 1400 kJ is evolved, in part because milk is largely water.

Exercise 6.9 Heat Energy Calculation

What quantity of heat energy is required to decompose 12.6 g of liquid water to the elements?

$$H_2O(\ell) \longrightarrow H_2(g) + \tfrac{1}{2}O_2(g) \qquad \Delta H = +285.8 \text{ kJ}$$

6.7 HESS'S LAW

See the *Saunders Interactive General Chemistry CD-ROM,* Screen 6.16, Hess's Law.

It is often important to know how much heat is transferred as a chemical process occurs. First, the direction of heat transfer is a clue that helps predict in which direction a chemical reaction will proceed because, at room temperature, most exothermic reactions are product-favored. Second, ΔH can be used to calculate the heat obtainable when a fuel is burned, for example, as was done in Section 6.6. Third, when reactions are carried out on a larger scale, as in a chemical plant that produces sulfuric acid or ethylene as a raw material for plastics, the surroundings must have enough cooling capacity to prevent an exothermic reaction from overheating and possibly damaging the plant. We therefore want to know ΔH values for as many reactions as possible. For many reactions direct experimental measurements can be made using a device called a *calorimeter* (Section 6.10), but for many other reactions this is not a simple task. Besides, it would be very time-consuming to measure values for every conceivable reaction, and it would take a great deal of space to tabulate these values. Fortunately there is a better way. It is based on the fact that *mass and energy are conserved in chemical reactions.*

Energy conservation is the basis of **Hess's law,** which states that, *if a reaction is the sum of two or more other reactions, then ΔH for the overall process must be the sum of the ΔH values of the constituent reactions.* For example, as illustrated in the preceding section (see Figure 6.14) for the decomposition of liquid water into the elements $H_2(g)$ and $O_2(g)$ (with all substances at 25 °C), the two successive changes are (1) the vaporization of liquid water and (2) the decomposition of water vapor to the elements (with all substances at 25 °C). The equation and

ΔH value for the overall process can be found by adding the equations and ΔH values for the two steps:

(1)	$H_2O(\ell) \longrightarrow H_2O(g)$	$\Delta H_1 = +44.0$ kJ
(2)	$H_2O(g) \longrightarrow H_2(g) + \frac{1}{2} O_2(g)$	$\Delta H_2 = +241.8$ kJ
(1) + (2)	$H_2O(\ell) \longrightarrow H_2(g) + \frac{1}{2} O_2(g)$	$\Delta H_{net} = +285.8$ kJ

Here, $H_2O(g)$ is a product of the first reaction and a reactant in the second. Thus, as in adding two algebraic equations in which the same quantity or term appears on both sides of the equation, $H_2O(g)$ can be canceled out. The net result is an equation for the overall reaction and its associated enthalpy change.

Hess's law is useful because it enables you to find the enthalpy change for a reaction or change in state that cannot be measured conveniently. Suppose you want to know the enthalpy change for the formation of carbon monoxide, CO, from the elements, solid carbon (as graphite) and oxygen gas:

$$C(s) + \frac{1}{2} O_2(g) \longrightarrow CO(g) \qquad \Delta H = ?$$

Experimentally this is not easy to do; it is difficult to keep the carbon from burning completely to carbon dioxide. The way to solve this is to recognize that experiments can be done to measure the enthalpy change for the conversion of carbon to CO_2 and for the conversion of CO to CO_2:

$$C(s) + O_2(g) \longrightarrow CO_2(g) \qquad \Delta H = -393.5 \text{ kJ}$$
$$CO(g) + \frac{1}{2} O_2(g) \longrightarrow CO_2(g) \qquad \Delta H = -283.0 \text{ kJ}$$

The formation of CO is the first of two steps in going from carbon to carbon dioxide (Figure 6.15) ($\Delta H_1 = ?$). The enthalpy changes for the second step (ΔH_2) and for the net reaction (ΔH_{net}) are known. Because (ΔH_{net} is the sum of the ΔH values for the first and second steps, we can solve for the enthalpy change for the first step (ΔH_1):

Step 1	$C(s) + \frac{1}{2} O_2(g) \longrightarrow CO(g)$	$\Delta H_1 = ?$
Step 2	$CO(g) + \frac{1}{2} O_2(g) \longrightarrow CO_2(g)$	$\Delta H_2 = -283.0$ kJ
Net reaction	$C(s) + O_2(g) \longrightarrow CO_2(g)$	$\Delta H_{net} = -393.5$ kJ

$$\Delta H_{net} = \Delta H_1 + \Delta H_2$$
$$-393.5 \text{ kJ} = \Delta H_1 + (-283.0 \text{ kJ})$$
$$\Delta H_1 = -110.5 \text{ kJ}$$

Example 6.6 Using Hess's Law

Suppose you want to know the enthalpy change for the formation of methane, CH_4, from solid carbon (as graphite) and hydrogen gas:

$$C(s) + 2\ H_2(g) \longrightarrow CH_4(g) \qquad \Delta H = ?$$

The enthalpy change for the direct combination of the elements would be extremely difficult to measure in the laboratory. We can, however, measure enthalpy changes for the combustion of the elements and methane in oxygen (see Section 6.10).

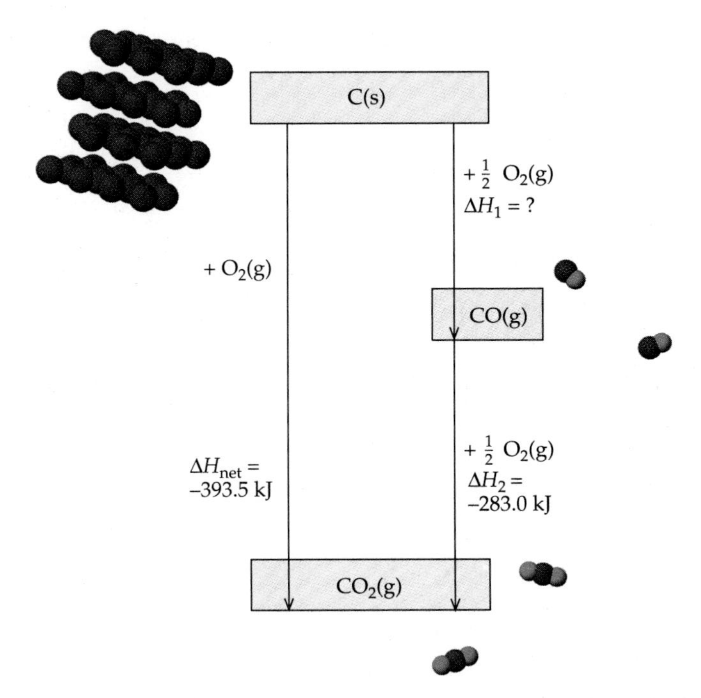

Figure 6.15 Formation of CO_2 from carbon. One mole of carbon dioxide can be formed from carbon and oxygen in a direct reaction ($\Delta H_{net} = -393.5$ kJ) or in two steps. Hess's law states that the sum of the enthalpy changes for the two steps must equal that for the direct reaction. That is, $\Delta H_1 + \Delta H_2 = \Delta H_{net}$. Knowing any two of these values allows us to calculate the third.

	Reaction	ΔH (kJ)
(1)	$C(s) + O_2(g) \longrightarrow CO_2(g)$	−393.5
(2)	$H_2(g) + \frac{1}{2} O_2(g) \longrightarrow H_2O(\ell)$	−285.8
(3)	$CH_4(g) + 2\ O_2(g) \longrightarrow CO_2(g) + 2\ H_2O(\ell)$	−890.3

Use these equations to obtain ΔH for the formation of methane from its elements.

Solution The three reactions (1, 2, and 3), as they are now written, cannot be added together to obtain the equation for the formation of CH_4 from its elements. CH_4 should be a product, but it is a reactant in Equation (3). The solution to this is to reverse Equation (3). At the same time, the sign of ΔH for the reaction is reversed. If a reaction is exothermic in one direction (the combustion of methane evolves energy), its reverse must be endothermic:

$$(3')\quad CO_2(g) + 2\ H_2O(\ell) \longrightarrow CH_4(g) + 2\ O_2(g) \qquad \Delta H = -\Delta H_3 = +890.3 \text{ kJ}$$

Next, we see that 2 mol of $H_2O(\ell)$ is required as a reactant in Equation (3′), whereas Equation (2) is written for only 1 mol of water. We therefore multiply the stoichiometric coefficients in Equation (2) by 2 and multiply the value of ΔH by 2.

$$2 \times (2)\quad 2\ H_2(g) + O_2(g) \longrightarrow 2\ H_2O(\ell) \qquad 2\ \Delta H_2 = 2(-285.8 \text{ kJ}) = -571.6 \text{ kJ}$$

With these modifications, we can add the three equations to give the equation for the formation of methane from its elements.

	Reaction	ΔH (kJ)
(1)	$C(s) + O_2(g) \longrightarrow CO_2(g)$	−393.5
2 × *(2)*	$2\ H_2(g) + O_2(g) \longrightarrow 2\ H_2O(\ell)$	−571.6
(3)′	$CO_2(g) + 2\ H_2O(\ell) \longrightarrow CH_4(g) + 2\ O_2(g)$	+890.3
Net	$C(s) + 2\ H_2(g) \longrightarrow CH_4(g)$	−74.8

The solution to this problem is shown diagrammatically in Figure 6.16.

Exercise 6.10 Using Hess's Law

What is the enthalpy change for the formation of ethane, C_2H_6, from elemental carbon and hydrogen?

$$2\ C(s) + 3\ H_2(g) \longrightarrow C_2H_6(g) \qquad \Delta H = ?$$

Use the reactions representing the enthalpy changes for the formation of $H_2O(\ell)$ and $CO_2(g)$ from Example 6.6 together with the experimentally determined ΔH value for the combustion of ethane.

$$C_2H_6(g) + \tfrac{7}{2}\ O_2(g) \longrightarrow 2\ CO_2(g) + 3\ H_2O(\ell) \qquad \Delta H = -1559.7\ \text{kJ}$$

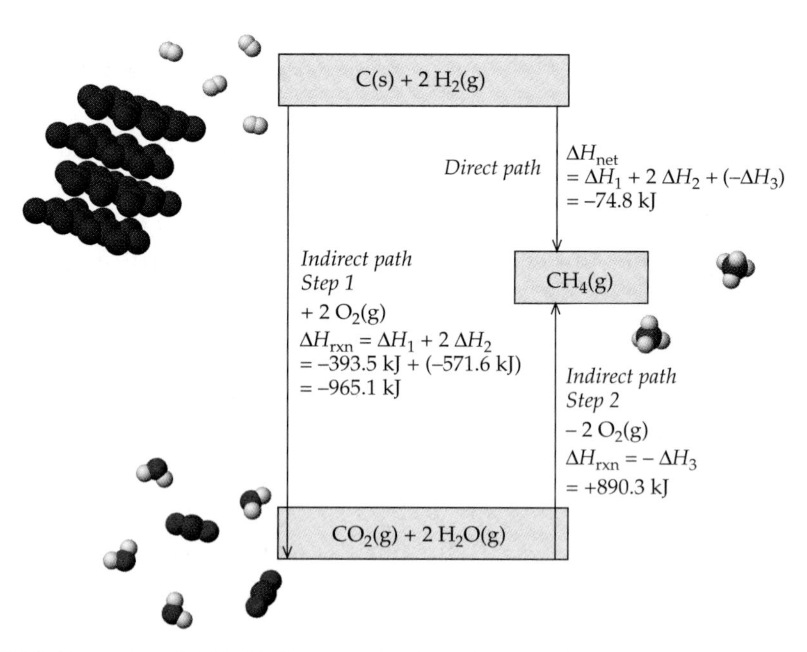

Figure 6.16 Formation of CH_4. Methane may be formed directly from the elements. Alternatively, the carbon may be burned to $CO_2(g)$ and the hydrogen burned to water. Then CO_2 and H_2O are combined to form CH_4 (and the O_2 used in the combustion is returned). Hess's law states that the enthalpy change for the direct reaction (ΔH_{net}) is the sum of the enthalpy changes along the alternative path, $\Delta H_{net} = \Delta H_1 + 2\ \Delta H_2 + (-\Delta H_3)$. See Example 6.6 for a more complete designation of the ΔH values.

Exercise 6.11 Using Hess's Law

Lead has been known and used for centuries. To obtain the metal, lead(II) sulfide (PbS, in the form of a common mineral called galena) is first roasted in air to form lead(II) oxide (PbO).

$$PbS(s) + \frac{3}{2}\,O_2(g) \longrightarrow PbO(s) + SO_2(g) \qquad \Delta H = -413.7\text{ kJ}$$

and then lead(II) oxide is reduced with carbon to the metal.

$$PbO(s) + C(s) \longrightarrow Pb(s) + CO(g) \qquad \Delta H = +106.8\text{ kJ}$$

What is the enthalpy change for the following reaction?

$$PbS(s) + C(s) + \frac{3}{2}\,O_2(g) \longrightarrow Pb(s) + CO(g) + SO_2(g)$$

Is the reaction exothermic or endothermic? How much energy, in kilojoules, is required (or evolved) when 454 g (1.00 lb) of PbS is converted to lead?

6.8 STATE FUNCTIONS

Hess's law works because enthalpy is a **state function,** a quantity whose value is determined only by the state of the system. *The enthalpy change for a chemical or physical change does not depend on the path you choose from the initial conditions to the final conditions.* No matter how you go from reactants to products in a reaction, the net heat evolved or required (at constant pressure) is always the same. This point is illustrated in Figure 6.15 for the formation of CO_2 from its elements and in Figure 6.16 for the formation of methane. For the reaction

$$C(s) + 2\,H_2(g) \longrightarrow CH_4(g) \qquad \Delta H = -74.8\text{ kJ}$$

the enthalpy change for the direct reaction ($\Delta H_{direct} = -74.8$ kJ) is the same as the sum of the enthalpy changes along the indirect pathway that goes through carbon dioxide and water.

Many commonly measured quantities, such as the pressure of a gas, the volume of a gas or liquid, the temperature of a substance, and the size of your bank account, are state functions. You could have arrived at a current bank balance of $25 by having deposited $25, or you could have deposited $100 and then withdrawn $75. The volume of a balloon is also a state function. You can blow up a balloon to a large volume and then let some air out to arrive at the desired volume. Alternatively, you can blow up the balloon in stages, adding tiny amounts of air at each stage. The final volume does not depend on how you got there. For bank balances and balloons, there are an infinite number of ways to arrive at the final state, but the final value depends only on the size of the bank balance or the balloon, and not on the path taken from the initial to the final state.

In principle there is an absolute enthalpy for the reactants ($H_{initial} = H_{reactants}$) and for the products ($H_{final} = H_{products}$) of a reaction. The difference between these enthalpies is the change for the system.

$$\Delta H_{reaction} = H_{final} - H_{initial} = H_{products} - H_{reactants}$$

Because the reaction starts and finishes at the same place no matter what pathway is chosen, $\Delta H_{reaction}$ must always be independent of pathway. Unlike volume, temperature, pressure, energy, or a bank balance, however, the absolute enthalpy of a substance is not usually determined, only its *change* in a chemical or physical process. The thermal energy evolved or required in a chemical process, for example, is a reflection of the *difference* in enthalpy between the reactants and products. We determine only whether the enthalpy of the products is greater than or less than that of the reactants by some amount.

6.9 STANDARD ENTHALPIES OF FORMATION

See the *Saunders Interactive General Chemistry CD-ROM,* Screen 6.17, Standard Enthalpy of Formation.

In 1982, the International Union of Pure and Applied Chemistry chose a pressure of 1 bar for the standard state. This pressure is very close to the average atmospheric pressure observed near sea level (1 bar = 0.98692 standard atmosphere). See Chapter 12.

It is common to use the term *heat of reaction* interchangeably with *enthalpy of reaction.* Understand that it is only the heat of reaction at **constant pressure,** q_p, that is equivalent to the enthalpy change.

Hess's law makes it possible to tabulate ΔH values for a few reactions and to derive a great many other ΔH values by adding together appropriate ones as described in Section 6.8. Because ΔH values depend on temperature and pressure, it is necessary to specify both of these when a table of data is made. Usually a pressure of 1 bar and a temperature of 25 °C are specified, although other conditions of temperature can be used. The **standard state** of an element or a compound is the most stable form of the substance in the physical state in which it exists at 1 bar and the specified temperature. Thus, at 25 °C and a pressure of 1 bar the standard state for hydrogen is the gaseous state, $H_2(g)$, and for sodium chloride it is the solid state, NaCl(s). For carbon, which can exist in at least three solid states at 1 bar and 25 °C, the most stable form, graphite, is selected as the standard.

When a reaction occurs with all the reactants and products in their standard states, the observed or calculated enthalpy change is known as the **standard enthalpy change of reaction, $\Delta H°$,** where the superscript ° indicates standard conditions. All the reactions discussed to this point have followed this convention, so all the ΔH values should have ° attached.

The standard enthalpy change of a reaction for the formation of 1 mol of a compound directly from its elements is called the **standard molar enthalpy of formation, ΔH_f°,** where the subscript *f* indicates that 1 mol of the compound in question has been *formed* in its standard state *from its elements,* also in their standard states. Some of the reactions already discussed define standard molar enthalpies of formation:

$$H_2(g) + \tfrac{1}{2}\,O_2(g) \longrightarrow H_2O(\ell) \qquad \Delta H_f^\circ = -285.8 \text{ kJ/mol}$$
$$C(s) + 2\,H_2(g) \longrightarrow CH_4(g) \qquad \Delta H_f^\circ = -74.8 \text{ kJ/mol}$$

As another example, the following equation shows that 277.7 kJ is evolved if graphite, the standard state form for carbon, is combined with gaseous hydrogen and oxygen to form 1 mol of ethanol at 298 K:

$$2\,C(\text{graphite}) + 3\,H_2(g) + \tfrac{1}{2}\,O_2(g) \longrightarrow C_2H_5OH(\ell) \qquad \Delta H_f^\circ = -277.7 \text{ kJ/mol}$$

Finally, it is important to understand that a standard molar enthalpy of formation (ΔH_f°) is just a special case of an enthalpy change for a reaction. In contrast with the three previous reactions, the enthalpy change for the following reaction is *not* an enthalpy of formation; calcium carbonate has been formed from other compounds, not directly from its elements.

$$CaO(s) + CO_2(g) \longrightarrow CaCO_3(s) \qquad \Delta H^\circ_{rxn} = -178.3 \text{ kJ}$$

Therefore, the enthalpy change is given the more general symbol of ΔH°_{rxn}. The enthalpy change for the following reaction is also *not* a standard enthalpy of formation.

$$P_4(s) + 6\ Cl_2(g) \longrightarrow 4\ PCl_3(\ell) \qquad \Delta H^\circ_{rxn} = -1278.8\ kJ$$

Here a compound has been formed from its elements, but more than 1 mol of the compound has been formed, and so $\Delta H^\circ_{rxn} = 4\ \Delta H^\circ_f$ for $PCl_3(\ell)$.

Table 6.2 and Appendix L list values of ΔH°_f, obtained from the National Institute for Standards and Technology (NIST), for many other compounds. Be sure to notice that *standard enthalpies of formation for the elements in their standard states are zero.* The reason for this is that forming an element in its standard state from the same element in its standard state involves no chemical or physical change.

To distinguish between enthalpies of formation and enthalpy changes for other kinds of reactions, we use the symbols ΔH°_f and ΔH°_{rxn}, respectively. The subscript rxn is short for "reaction."

WWW Thermodynamic data, including enthalpies of formation, can be obtained from the NIST site on the World Wide Web: http://webbook.nist.gov.

Most ΔH°_f values are negative because the process of forming most compounds from their elements is exothermic. Recall that we previously stated that at room temperature most exothermic reactions are product-favored. Thus, forming compounds from their elements (under standard conditions) is generally a product-favored process. Reactions of oxygen with metals are usually exothermic and product-favored:

$$2\ Al(s) + \tfrac{3}{2}\ O_2(g) \longrightarrow Al_2O_3(s) \qquad \Delta H^\circ_f = -1675.7\ kJ$$

and carbon usually forms stable compounds with other elements such as hydrogen and oxygen:

$$C(s) + O_2(g) \longrightarrow CO_2(g) \qquad \Delta H^\circ_f = -393.5\ kJ$$

$$2\ C(s) + 3\ H_2(g) + \tfrac{1}{2}\ O_2(g) \longrightarrow C_2H_5OH(\ell) \qquad \Delta H^\circ_f = -277.7\ kJ$$

One important exception to our general observation is ethyne (commonly called acetylene), a compound with a strongly endothermic standard enthalpy of formation:

$$2\ C(s) + H_2(g) \longrightarrow C_2H_2(g) \qquad \Delta H^\circ_f = +226.7\ kJ$$

Compounds with positive enthalpies of formation are good reservoirs of stored chemical potential energy. Thus, acetylene, for example, is a good fuel and is the starting point for manufacturing many other carbon-containing compounds.

Acetylene, C_2H_2, has a positive standard enthalpy of formation (+226.7 kJ/mol). It is a good reservoir of stored chemical energy. When it burns in pure oxygen, it produces a large quantity of heat and so is used in welding metal. (*C. D. Winters*)

Example 6.7 Writing Equations to Define Enthalpies of Formation

The standard enthalpy of formation of gaseous ammonia is −46.11 kJ/mol. Write the balanced equation for which the enthalpy of reaction is −46.11 kJ.

Solution The equation must show the formation of 1 mol of $NH_3(g)$ from the elements in their standard states; both N_2 and H_2 are gases at 25 °C and 1 bar. The correct equation is therefore

$$\tfrac{1}{2}\ N_2(g) + \tfrac{3}{2}\ H_2(g) \longrightarrow NH_3(g) \qquad \Delta H^\circ_{rxn} = \Delta H^\circ_f = -46.1\ kJ$$

Table 6.2 • Selected Standard Molar Enthalpies of Formation at 298 K

Substance	Name	Standard Molar Enthalpy of Formation (kJ/mol)
$Al_2O_3(s)$	aluminum oxide	−1675.7
$BaCO_3(s)$	barium carbonate	−1216.3
$CaCO_3(s)$	calcium carbonate	−1206.9
$CaO(s)$	calcium oxide	−635.1
$CCl_4(\ell)$	carbon tetrachloride	−135.4
$CH_4(g)$	methane	−74.8
$CH_3OH(\ell)$	methanol	−238.7
$C_2H_5OH(\ell)$	ethanol	−277.7
$CO(g)$	carbon monoxide	−110.5
$CO_2(g)$	carbon dioxide	−393.5
$C_2H_2(g)$	ethyne (acetylene)	+226.7
$C_2H_4(g)$	ethene (ethylene)	+52.3
$C_2H_6(g)$	ethane	−84.7
$C_3H_8(g)$	propane	−103.8
$C_4H_{10}(g)$	butane	−125.6
$CuSO_4(s)$	copper(II) sulfate	−771.4
$H_2O(g)$	water vapor	−241.8
$H_2O(\ell)$	liquid water	−285.8
$HF(g)$	hydrogen fluoride	−271.1
$HCl(g)$	hydrogen chloride	−92.3
$HBr(g)$	hydrogen bromide	−36.4
$HI(g)$	hydrogen iodide	+26.5
$KF(s)$	potassium fluoride	−567.3
$KCl(s)$	potassium chloride	−436.7
$KBr(s)$	potassium bromide	−393.8
$MgO(s)$	magnesium oxide	−601.7
$MgSO_4(s)$	magnesium sulfate	−1284.9
$Mg(OH)_2(s)$	magnesium hydroxide	−924.5
$NaF(s)$	sodium fluoride	−573.6
$NaCl(s)$	sodium chloride	−411.2
$NaBr(s)$	sodium bromide	−361.1
$NaI(s)$	sodium iodide	−287.8
$NH_3(g)$	ammonia	−46.1
$NO(g)$	nitrogen monoxide	+90.3
$NO_2(g)$	nitrogen dioxide	+33.2
$PCl_3(\ell)$	phosphorus trichloride	−319.7
$PCl_5(s)$	phosphorus pentachloride	−443.5
$SiO_2(s)$	silicon dioxide (quartz)	−910.9
$SnCl_2(s)$	tin(II) chloride	−325.1
$SnCl_4(\ell)$	tin(IV) chloride	−511.3
$SO_2(g)$	sulfur dioxide	−296.8
$SO_3(g)$	sulfur trioxide	−395.7

Source: From http://webbook.nist.gov/

Exercise 6.12 Writing Equations to Define Enthalpies of Formation

Write balanced equations to define the formation of methanol and copper sulfate from their respective elements. Give the value of the standard molar enthalpy of formation for each (see Table 6.2). What is the value of ΔH°_{rxn} if 1.5 mol of methanol is formed from its constituent elements?

Standard enthalpies of formation are very useful. For example, you can find the standard enthalpy change for any reaction if the enthalpies of formation for all the reactants and products are known. Suppose you are a chemical engineer and want to know how much heat is required to decompose calcium carbonate (limestone) to calcium oxide (lime) and carbon dioxide, with all substances at standard conditions:

$$CaCO_3(s) \longrightarrow CaO(s) + CO_2(g) \qquad \Delta H^\circ_{rxn} = ?$$

To do this, you find the following enthalpies of formation in a table such as Table 6.2 or Appendix L:

Compound	ΔH°_f (kJ/mol)
$CaCO_3(s)$	−1206.9
$CaO(s)$	−635.1
$CO_2(g)$	−393.5

and then use Equation 6.6 to find the standard enthalpy change for the reaction, ΔH°_{rxn}.

$$\text{Enthalpy change for a reaction} = \Delta H^\circ_{rxn} = \sum[\Delta H^\circ_f(\text{products})] - \sum[\Delta H^\circ_f(\text{reactants})] \qquad [6.6]$$

In Equation 6.6 the symbol Σ (the Greek letter sigma) means to "take the sum." Thus, to find ΔH°_{rxn} you add up the molar enthalpies of formation of the products and subtract from this the sum of the molar enthalpies of formation of the reactants. Applying this to the decomposition of limestone, we have

$$\begin{aligned}\Delta H^\circ_{rxn} &= \Delta H^\circ_f\,[CaO(s)] + \Delta H^\circ_f\,[CO_2(g)] - \Delta H^\circ_f\,[CaCO_3(s)] \\ &= 1\text{ mol }(-635.1\text{ kJ/mol}) + 1\text{ mol }(-393.5\text{ kJ/mol}) \\ &\quad - 1\text{ mol }(-1206.9\text{ kJ/mol}) \\ &= +178.3\text{ kJ}\end{aligned}$$

The decomposition of limestone to lime and CO_2 is endothermic; energy is required to carry out the process.

Example 6.8 Using Enthalpies of Formation

Nitroglycerin is a powerful explosive, giving four different gases when detonated:

$$2\ C_3H_5(NO_3)_3(\ell) \longrightarrow 3\ N_2(g) + \tfrac{1}{2}\ O_2(g) + 6\ CO_2(g) + 5\ H_2O(g)$$

A CLOSER LOOK Hess's Law and Equation 6.6

Equation 6.6 is a convenient way to apply Hess's law when the enthalpies of formation of *all* the reactants and products are known. Let us look again at the decomposition of calcium carbonate,

$$CaCO_3(s) \longrightarrow CaO(s) + CO_2(g) \qquad \Delta H^\circ_{rxn} = ?$$

and think about an alternative route from the reactant to the products. We can imagine the reaction as occurring by first breaking up $CaCO_3$ into its elements (Ca, C, and O_2) and then recombining the elements in a different way to produce CO_2 and CaO.

The enthalpy change for each step in this process is known. Step 1 is the reverse of the equation for the formation of $CaCO_3$ from the elements, so the enthalpy change is the negative of the enthalpy of formation of $CaCO_3$. For the other two steps, ΔH°_{rxn} is the same as the enthalpy of formation. Notice that the sum of reactions 1, 2, and 3 gives the equation for the net reaction. Most importantly, notice that ΔH°_{rxn} for the net reaction is the sum of the enthalpy changes for each step. Here

$$\Delta H^\circ_{rxn} = \Delta H^\circ_f[CaO(s)] + \Delta H^\circ_f[CO_2(g)] - \Delta H^\circ_f[CaCO_3(s)]$$

This is exactly the result given by applying Equation 6.6. The enthalpy change for the reaction is indeed the sum of the enthalpies of formation of the products minus that of the reactant.

	Reaction	ΔH°_{rxn}
Step 1	$CaCO_3(s) \longrightarrow Ca(s) + C(s) + \frac{3}{2} O_2(g)$	$-\Delta H^\circ_f\,[CaCO_3(s)] = -(-1206.9\text{ kJ})$
Step 2	$C(s) + O_2(g) \longrightarrow CO_2(g)$	$\Delta H^\circ_f\,[CO_2(g)] = -393.5\text{ kJ}$
Step 3	$Ca(s) + \frac{1}{2} O_2(g) \longrightarrow CaO(s)$	$\Delta H^\circ_f\,[CaO(s)] = -635.1\text{ kJ}$
Net	$CaCO_3(s) \longrightarrow CaO(s) + CO_2(g)$	$\Delta H^\circ_{rxn} = +178.3\text{ kJ}$

Given that the enthalpy of formation of nitroglycerin, ΔH°_f, is -364 kJ/mol, and consulting Table 6.2 for the enthalpies for the other compounds, calculate the enthalpy change when 10.0 g of nitroglycerin is detonated.

Solution To solve this problem, we need the standard molar enthalpies of formation of the products. Two are elements in their standard states (N_2 and O_2), so their values of ΔH°_f are zero. From Table 6.2, we have

$$\Delta H^\circ_f[CO_2(g)] = -393.5\text{ kJ/mol} \qquad \text{and} \qquad \Delta H^\circ_f\,[H_2O(g)] = -241.8\text{ kJ/mol}$$

The enthalpy change for the reaction can now be found using Equation 6.6:

$$\begin{aligned}\Delta H^\circ_{rxn} &= 6\text{ mol}\cdot\Delta H^\circ_f\,[CO_2(g)] + 5\text{ mol}\cdot\Delta H^\circ_f\,[H_2O(g)] \\ &\quad - 2\text{ mol}\cdot\Delta H^\circ_f\,[C_3H_5(NO_3)_3(\ell)] \\ &= 6\text{ mol }(-393.5\text{ kJ/mol}) + 5\text{ mol }(-241.8\text{ kJ/mol}) \\ &\quad - 2\text{ mol }(-364\text{ kJ/mol}) \\ &= -2842\text{ kJ}\end{aligned}$$

Based on the enthalpy change for the explosion of 2 mol of nitroglycerin, the heat liberated by this exothermic reaction can be calculated when only 10.0 g of nitroglycerin is used:

$$10.0\text{ g nitroglycerin}\cdot\frac{1\text{ mol nitroglycerin}}{227.1\text{ g nitroglycerin}} = 0.0440\text{ mol nitroglycerin}$$

$$q = 0.0440\text{ mol nitroglycerin}\cdot\frac{-2842\text{ kJ}}{2\text{ mol nitroglycerin}} = -62.6\text{ kJ}$$

Exercise 6.13 Using Enthalpies of Formation

Benzene, C_6H_6, is an important hydrocarbon. Calculate its enthalpy of combustion; that is, find the value of ΔH° for the following reaction using Equation 6.6.

$$C_6H_6(\ell) + \tfrac{15}{2}\,O_2(g) \longrightarrow 6\,CO_2(g) + 3\,H_2O(\ell)$$

The enthalpy of formation of benzene, $\Delta H_f^\circ\ [C_6H_6(\ell)]$, is +49.0 kJ/mol. Use Table 6.2 for any other values you may need.

6.10 DETERMINING ENTHALPIES OF REACTION

Calorimetry

See the *Saunders Interactive General Chemistry CD-ROM,* Screen 6.18, Measuring Heats of Reaction—Calorimetry.

The heat evolved by a chemical reaction can be determined by a technique called **calorimetry.** When finding heats of combustion or the caloric value of foods, the measurement is often done in a *combustion calorimeter* (Figure 6.17). A weighed sample of a combustible solid or liquid is placed in a dish that is encased in a "bomb," a cylinder about the size of a large fruit juice can with thick steel walls and ends. The bomb is then placed in a water-filled container with

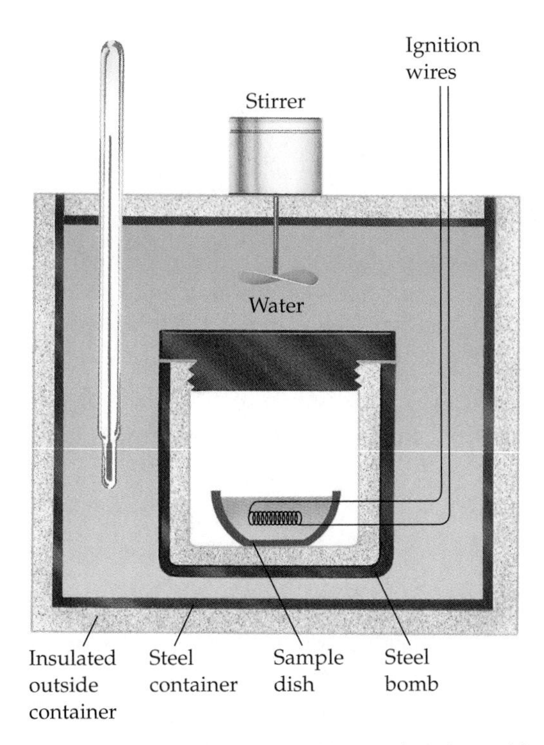

Figure 6.17 A combustion calorimeter. A combustible sample is burned in pure oxygen in a steel "bomb." The heat generated by the reaction warms the bomb and the water surrounding it. By measuring the temperature increase, the heat evolved by the reaction can be determined.

well-insulated walls. After filling the bomb with pure oxygen, the mixture of oxygen and sample is ignited, usually by an electric spark. The heat generated when the sample burns warms the bomb and the water around it, with both coming to the same temperature. In this configuration, the oxygen and the compound represent the *system,* and the bomb and water around it are the *surroundings.* From the law of energy conservation, we can say that

Heat transferred from the system = heat transferred into the surroundings

Heat evolved by the reaction = heat absorbed by water and bomb

$$q_{reaction} = -(q_{water} + q_{bomb})$$

where $q_{reaction}$ has a negative value because the combustion reaction is exothermic. The temperature change of the water, which is also equal to the change for the bomb, is measured. Using these experimental measurements, the total quantity of heat absorbed by the water and the bomb ($q_{water} + q_{bomb}$) can be calculated from the heat capacities of the bomb and the water. According to the previous equation, this total gives the heat evolved by combustion of the compound.

Because the bomb is rigid, the heat transfer is measured at constant volume and is therefore equivalent to ΔE, the change in internal energy. As explained earlier (see Section 6.5), $\Delta E = q_v$, whereas the change in enthalpy is the heat evolved or required at constant pressure, that is, $\Delta H = q_p$. Because ΔE and ΔH are related in a relatively simple way, however, ΔH values can be calculated from ΔE values found in bomb calorimetry experiments.

Example 6.9 Determining the Enthalpy Change for a Reaction by Calorimetry

The heat capacity of a bomb calorimeter, C_{bomb}, can be found in a separate experiment by measuring ΔT produced by burning a measured mass of a compound with a known ΔH of combustion.

Octane, C_8H_{18}, a primary constituent of gasoline, burns in air:

$$C_8H_{18}(\ell) + \tfrac{25}{2}\,O_2(g) \longrightarrow 8\,CO_2(g) + 9\,H_2O(\ell)$$

Suppose that a 1.00-g sample of octane is burned in a calorimeter that contains 1.20 kg of water. The temperature of the water and the bomb rises from 25.00 °C (298.15 K) to 33.20 °C (306.35 K). If the heat capacity of the bomb, C_{bomb}, is known to be 837 J/K, calculate the heat transferred in the combustion of the 1.00-g sample of C_8H_{18}.

Solution The heat that appears as a rise in temperature of the water surrounding the bomb is calculated using Equation 6.2.

$$\begin{aligned} q_{water} &= (\text{specific heat of water})(m_{water})(\Delta T) \\ &= (4.184\ \text{J/g}\cdot\text{K})(1.20 \times 10^3\ \text{g})(306.35\ \text{K} - 298.15\ \text{K}) \\ &= +41.2 \times 10^3\ \text{J} \end{aligned}$$

The heat released by the reaction appears as a rise in temperature of the bomb and is calculated from the heat capacity of the bomb (C_{bomb}, units of joules per kelvin) and the temperature change, ΔT.

$$\begin{aligned} q_{bomb} &= C_{bomb} \cdot \Delta T \\ &= (837\ \text{J/K})(306.35\ \text{K} - 298.15\ \text{K}) \\ &= +6.86 \times 10^3\ \text{J} \end{aligned}$$

The total heat transferred by the reaction to its surroundings is equal to the negative of the sum of q_{water} and q_{bomb}. Thus,

$$\text{Total heat transferred by 1.00 g of octane} = q = -(41.2 \times 10^3\ \text{J} + 6.86 \times 10^3\ \text{J})$$
$$q = -48.1 \times 10^3\ \text{J, or } -48.1\ \text{kJ}$$

The experiment shows that 48.1 kJ of heat is evolved per gram of octane burned. Because the molar mass of octane is 114.2 g/mol, the heat transferred per mole is

$$\text{Heat transferred per mole} = (-48.1\ \text{kJ/g})(114.2\ \text{g/mol}) = -5.49 \times 10^3\ \text{kJ/mol}$$

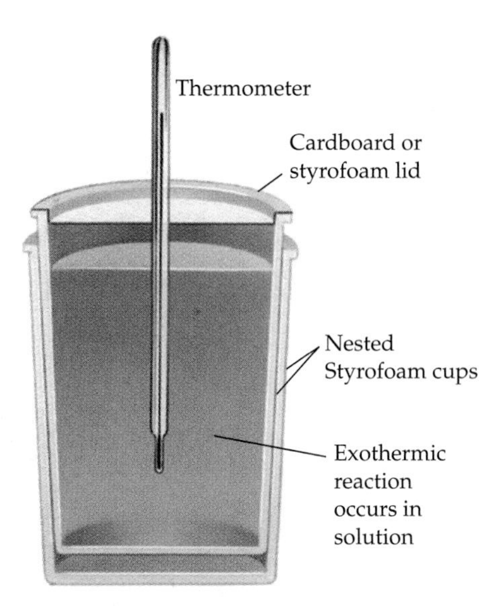

Figure 6.18 A coffee-cup calorimeter. Two Styrofoam coffee cups are placed one inside the other and covered with a lid. The exothermic reaction occurs in the aqueous solution in the innermost container, and the cups provide enough insulation so that the heat is confined to the solution. The change in temperature of the reacting solution can be measured reasonably well with a standard laboratory thermometer.

Exercise 6.14 Determining the Heat Released by a Reaction

A 1.00-g sample of ordinary table sugar (sucrose, $C_{12}H_{22}O_{11}$) is burned in a combustion calorimeter. The temperature of 1.50×10^3 g of water in the calorimeter rises from 25.00 °C to 27.32 °C. If the heat capacity of the bomb is 837 J/K and the heat capacity of the water is 4.184 J/g · K, calculate **(a)** the heat evolved per gram of sucrose and **(b)** the heat evolved per mole of sucrose.

A bomb calorimeter is not always convenient to use, especially in the introductory chemistry laboratory. We can study reactions other than combustions, however, in a "coffee-cup calorimeter" (Figure 6.18). Not only is it a simple, inexpensive device, but it operates at constant pressure. This means the heat evolved in a reaction in a coffee-cup calorimeter (q_p) is a measure of a reaction enthalpy (because $q_p = \Delta H$).

Example 6.10 Using a Coffee-Cup Calorimeter

Suppose you place 0.500 g of magnesium chips in a coffee-cup calorimeter and then add 100.0 mL of 1.00 M HCl. The reaction that occurs is

$$Mg(s) + 2\ HCl(aq) \longrightarrow H_2(g) + MgCl_2(aq)$$

The temperature of the solution increases from 22.2 °C to 44.8 °C. What is the enthalpy change for this reaction per mole of Mg? (Assume the heat capacity of the solution is 4.20 J/g · K and that the density of the HCl solution is 1.00 g/mL.)

The specific heat capacity of a metal can be determined by measuring the temperature change produced when a heated piece of metal is placed in a "coffee-cup calorimeter." See Example 6.3.

Solution Given the mass of solution, its heat capacity, and ΔT, we can find the heat produced in the reaction. (The mass of the solution is approximately the mass of the 100.0 mL of HCl plus the mass of magnesium; the mass is 100.5 g in this case.)

$$q = (100.5\ \text{g})(4.20\ \text{J/g} \cdot \text{K})(318.0\ \text{K} - 295.4\ \text{K}) = 9.54 \times 10^3\ \text{J}$$

This quantity of heat is produced by the reaction of 0.500 g of Mg. The amount produced by the reaction of 1.00 mol of Mg is

$$\text{Heat produced per mole} = \frac{9.54 \times 10^3\ \text{J}}{0.500\ \text{g}} \cdot \frac{24.31\ \text{g}}{1\ \text{mol}}$$
$$= 4.64 \times 10^5\ \text{J/mol, or } 464\ \text{kJ/mol}$$

This means the enthalpy change for the reaction of magnesium with aqueous HCl is $\Delta H = -464$ kJ/mol. Notice that we have included a negative sign because the reaction is clearly exothermic.*

Exercise 6.15 Using a Coffee-Cup Calorimeter

Assume you mix 200. mL of 0.400 M HCl with 200. mL of 0.400 M NaOH in a coffee-cup calorimeter. The temperature of the solutions before mixing was 25.10 °C; after mixing and allowing the reaction to occur, the temperature is 26.60 °C. What is the molar enthalpy of neutralization of the acid? (Assume that the densities of all solutions are 1.00 g/mL and their specific heat capacities are 4.20 J/g · K.)

6.11 APPLICATIONS OF THERMODYNAMICS

Thermodynamics is the science of the transfer of energy as heat and work. As such it is related intimately to problems of energy use in our economy.

Available Energy Resources

What resources are currently available to supply energy for human use? Historically, the first was *biomass,* material produced by living organisms that contains significant quantities of chemical potential energy. Biomass, mostly wood, still provides one-third of the energy resources for some developing countries. *Fossil fuels*—petroleum, coal, and natural gas—made the Industrial Revolution possible and provide most of the energy resources used in the industrialized world today. For 200 years coal was the principal fuel for industrializing nations, only to be replaced by petroleum and natural gas around the middle of this century. Petroleum and natural gas are easier to handle and cleaner to use, but the price of petroleum shifts unpredictably with changing international conditions, and a significant amount of the world's petroleum is produced by nations that lack political stability.

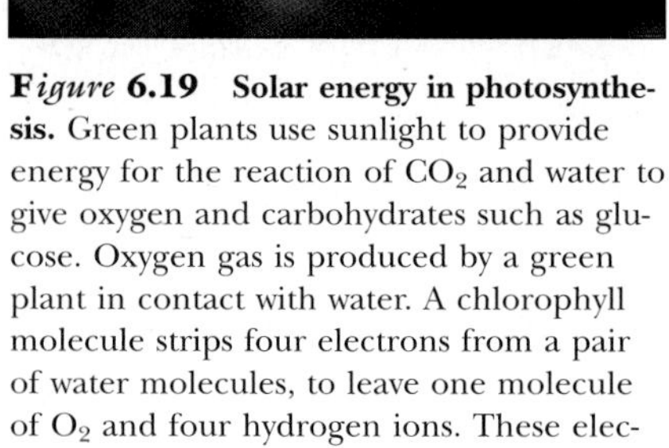

Figure 6.19 Solar energy in photosynthesis. Green plants use sunlight to provide energy for the reaction of CO_2 and water to give oxygen and carbohydrates such as glucose. Oxygen gas is produced by a green plant in contact with water. A chlorophyll molecule strips four electrons from a pair of water molecules, to leave one molecule of O_2 and four hydrogen ions. These electrons are then used in other chemical processes.

$$2\,H_2O(\ell) \longrightarrow O_2(g) + 4\,H^+(aq) + 4\,e^-$$

(*C. D. Winters*)

Great quantities of coal are available, though, and attention is shifting back toward coal for energy and as a new source for many of the chemicals currently obtained from petroleum and natural gas.

Relatively small but still significant quantities of energy are now supplied by *hydroelectric* sources and *nuclear* energy (Chapter 24), and still smaller quantities are provided in the form of direct *solar* energy, *geothermal* energy, and *wind* and ocean currents. These primary energy resources may be converted to electricity, a secondary resource, which is a form of energy we find particularly useful.

*When a ΔH value is stated for a chemical reaction, it is understood that the reactants and products are at the same temperature. In a calorimetry experiment, however, this condition is never met. In a coffee-cup calorimeter (unlike the bomb calorimeter) virtually all the energy produced remains within the system. Thus, the problem is to find how much heat would have to be transferred out of the system to keep its temperature unchanged. That amount of heat is precisely what was obtained when ΔH was calculated in this example. The negative sign is consistent with the transfer of heat out of the system.

(a)

(b)

(c)

Figure 6.20 Solar energy. (a) Some pocket calculators are solar-powered. The small window at the top of the calculator allows light onto the photovoltaic cell. (b) Spacecraft are powered by solar energy collectors. (c) A grid of photovoltaic cells in the Nevada desert. (*a, C. D. Winters; b, © 1997 California Institute of Technology/U. S. Government Sponsorship Ack. NAS7-1260; b, Ales S. MacLean/Landslides*)

Perhaps our most important resource is *solar energy*. Every year the earth's surface receives about ten times as much energy from sunlight as is contained in all the known reserves of coal, oil, natural gas, and uranium combined. This is equivalent to about 15,000 times the world's annual consumption by humans.

See W. Hoagland: "Solar energy," *Scientific American*, p. 170, September 1995; and N. S. Lewis: "Artificial Photosynthesis," *American Scientist*, Vol. 83, p. 534, November-December, 1995.

Solar energy is of course already available to us through biomass as a result of photosynthesis (Figure 6.19); fossil fuels that were formed from biomass as a result of changes that required millions of years; wind and water that store kinetic and potential energy; and direct absorption of solar energy by our bodies and the materials around us. Now, however, there is also considerable interest in using *photovoltaic cells* capable of transforming solar energy directly into electric energy. Such devices are now used in spacecraft and in pocket calculators, and they are also being developed for large-scale commercial use (Figure 6.20).

Photosynthesis in plants uses only about 1% to 3% of the sunlight falling on the plant. Research on photoelectric devices, however, has led to devices that use 20% to 30% of the light. Such high efficiencies are necessary to offset the high cost of making the devices, which are usually based on highly purified silicon.

Using sunlight to produce electricity is feasible in sunny locations such as the southwestern United States. An alternative is to use sunlight (or wind power, for example) when available to produce a storable fuel. One example is artificial photosynthesis, where sunlight is used to drive the reactant-favored decomposition of water into its elements (Figure 6.21):

$$2\ H_2O(\ell) \longrightarrow 2\ H_2(g) + O_2(g)$$

Recent experiments have improved the efficiency and stability of such devices to the point where they may be commercially useful. The hydrogen produced

Figure 6.21 Artificial photosynthesis. A photoelectrochemical cell can generate electricity or produce a chemical fuel. Here a light-sensitive electrode immersed in water and an electrolyte generates hydrogen gas. (*Nathan Lewis/California Institute of Technology*)

Table 6.3 • Energy Released by Combustion of Some Substances

Substance	Energy Released (kJ/g)
hydrogen [to give $H_2O(\ell)$]	142
gasoline	48
crude petroleum	43
typical animal fat	38
coal	29.3
charcoal	29
paper	20
dry biomass	16
air-dried wood or dung	15

Source: Data taken in part from J. Harte: *Consider a Spherical Cow,* University Science Books, Mill Valley, CA, 1988, p. 242.

Figure 6.22 These hydrogen fuel cells are used for transportation purposes. (*Ballard Power Systems*)

could be burned in a power plant or automobile, or the hydrogen and oxygen can be recombined when needed in a fuel cell to generate electricity, as outlined in the following section.

The Hydrogen Economy

Hydrogen gas can be generated in a number of ways (among them by artificial photosynthesis; see Figure 6.21), and the hydrogen can then be used as a fuel. Combustion of hydrogen provides far more energy per gram than any other fuel listed in Table 6.3.

Hydrogen (H_2) is combined with oxygen to provide some of the power to lift the Space Shuttle into orbit. The cigar-shaped tank holds liquid H_2 and liquid O_2. (*NASA*)

Currently the most visible use of hydrogen as a fuel is in the Space Shuttle. The cigar-shaped tank strapped to the shuttle contains 1.46×10^6 L (385,265 gal) of hydrogen and 5.43×10^5 L (143,351 gal) of oxygen. Just a few minutes into the flight, the fuel is exhausted, and the hydrogen and oxygen have been converted completely to nonpolluting water.

The Space Shuttle also uses "fuel cells" to produce electricity from the product-favored reaction of hydrogen and oxygen to give water. Indeed, intense research and development are proceeding around the world on fuel cell technology (Figure 6.22). All of the major automobile manufacturers are involved because hydrogen/oxygen fuel cells would evolve no polluting gases, only water.

Currently, two major barriers exist to the use of hydrogen as an alternative to petroleum. First, an inexpensive way to make H_2 must be found that avoids the use of fossil fuels, which are now the major starting material for the production of hydrogen. Using electricity to break water down to its elements (a process called electrolysis; see Chapter 21) works well, but the electricity needed to do this is currently too expensive to produce hydrogen at a reasonable cost. A current goal of solar energy research is to develop a cell that would provide the electric energy needed for water electrolysis at a low price.

A CLOSER LOOK *It's in the Bag—Thermodynamics and Consumer Products*

Does your elbow hurt from too much tennis? Are your hands or feet cold after a day of skiing? Pull out a cold pack or a hot pack. How do they work?

Cold packs generally contain the salt ammonium nitrate and water in two separate packages. When the inner bag containing water is broken, ammonium nitrate, which is readily soluble in water (see Figure 5.4), dissolves. The dissolving process, however, is strongly endothermic, and heat is transferred from the surroundings—your elbow—to the system—the dissolving salt.

$$NH_4NO_3(s) \longrightarrow NH_4^+(aq) + NO_3^-(aq) \quad \Delta H^\circ_{rxn} = +25.69 \text{ kJ/mol}$$

Hot packs come in several varieties, one of which uses sodium acetate, $NaCH_3CO_2$. When the salt crystallizes from water, heat is evolved and is transferred to the surroundings.

$$Na^+(aq) + CH_3CO_2^-(aq) + 3\,H_2O(\ell) \longrightarrow NaCH_3CO_2 \cdot 3\,H_2O(s)$$
$$\Delta H^\circ_{rxn} = -37.86 \text{ kJ/mol}$$

Another variety of hot pack uses an oxidation-reduction reaction.

$$4\,Fe(s) + 3\,O_2(g) \longrightarrow 2\,Fe_2O_3(s)$$
$$\Delta H^\circ_{rxn} = 2\,\Delta H^\circ_f[Fe_2O_3(s)] = -1648.4 \text{ kJ/mol}$$

This very exothermic reaction allows the hand warmer to maintain a temperature of 57 °C to 69 °C for several hours if oxygen flow is somewhat restricted, as by keeping the warmer in a glove or pocket.

See G. Marsella: "Hot & cold packs," *Chem Matters,* p. 7, February 1987.

Cold packs absorb heat as ammonium nitrate dissolves in water. (*C. D. Winters*)

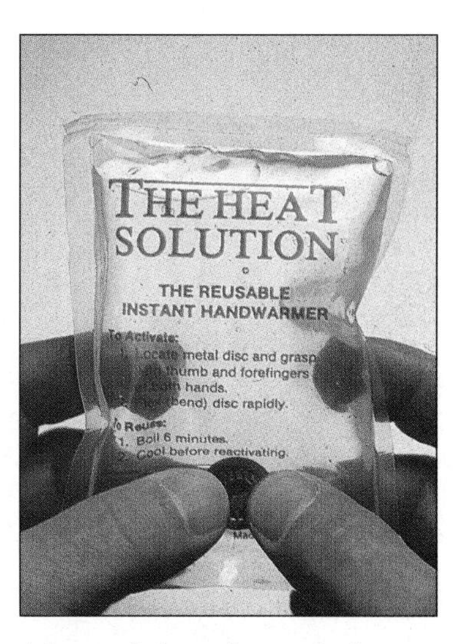

A hot pack that relies on the heat evolved by the crystallization of sodium acetate. (*C. D. Winters*)

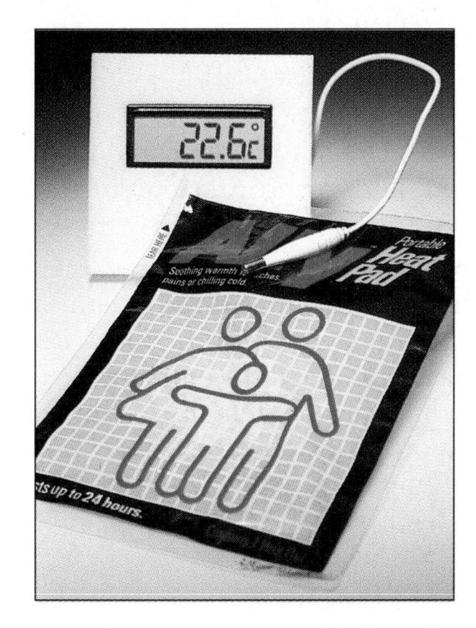

A hand warmer uses the oxidation of iron as the source of heat. (*C. D. Winters*)

The second barrier to using hydrogen as a fuel is the need for a means of convenient storage. The space program has demonstrated that hydrogen can be stored relatively easily and safely as a liquid, although not cheaply because cold temperatures and high pressures are involved. The problem is that this may not be appropriate for a home or car. These uses will require alternative methods of storage, one of which is to store the H_2 as a simple compound with a metal. This compound—a metal hydride—would release the H_2 as a gas when heated.

CHAPTER HIGHLIGHTS

When you have finished studying this chapter, you should be able to

- Understand the terms **reactant-** and **product-favored** and the definition of **thermodynamics** (Section 6.1).
- Describe the various forms of energy and the nature of heat and thermal energy transfer (Section 6.2).
- Know the difference between **kinetic** and **potential energy** (Section 6.2).
- Use the most common energy unit, the **joule** (Section 6.2)
- Use **specific heat capacity** and the sign conventions for heat transfer (Section 6.3).

  The amount of thermal energy transferred into or out of an object can be determined by using the equation

  $$q = (\text{mass of object in g})(\text{heat capacity in J/g} \cdot \text{K})(\Delta T \text{ in °C or K}) \quad [6.2]$$

  When q is negative, thermal energy has been transferred out of the object. When q is positive, thermal energy has been transferred into the object.

- Use **heat of fusion** and **heat of vaporization** to find the quantity of thermal energy involved in changes of state (Section 6.4).
- Recognize and use the language of thermodynamics: the **system** and its **surroundings; exothermic** and **endothermic** reactions; the **first law of thermodynamics** (the law of energy conservation); and **enthalpy** changes (Section 6.5).
- Understand the basis of the **first law of thermodynamics** (Section 6.5).

  When energy is transferred to or from a system, the change in energy for the system is given by $\Delta E = q + w$. Here q is the quantity of thermal energy transferred between the system and its surroundings; similarly, w is the work transferred. Both q and w are positive when heat or work, respectively, are transferred from the surroundings to the system.

- Recognize that when a process is carried out under constant pressure conditions, the heat transferred is equivalent to the enthalpy change, ΔH; that is, $q_p = \Delta H$ (where the subscript p means constant pressure) (Section 6.5).
- Understand that if the enthalpy change for a process is negative, heat has transferred from a system to its surroundings, and the process is exothermic (Section 6.5). Conversely, if ΔH is positive, heat has transferred from the surroundings to a system, and the process is endothermic.
- Use the fact that ΔH for a reaction is proportional to the quantity of material present (Sections 6.5–6.10).
- Apply **Hess's law** to find the enthalpy change for a reaction (Sections 6.7 and 6.10).
- Recognize various state functions, quantities whose value is determined only by the state of the system and not by the pathway by which that state was achieved (Section 6.8).
- Write a balanced chemical equation that defines the **standard molar enthalpy of formation**, ΔH_f°, for a compound (Section 6.9) and use Equation 6.6 to calculate the enthalpy change for a reaction, ΔH_{rxn}°.

$$\text{Enthalpy change for a reaction} = \Delta H_{rxn}^\circ = \sum[\Delta H_f^\circ(\text{products})] - \sum[\Delta H_f^\circ(\text{reactants})] \quad [6.6]$$

where Σ (the Greek letter sigma) means to "take the sum."

- Describe how to measure the quantity of heat energy transferred in a reaction by using **calorimetry** (Section 6.10).
- Discuss uses of energy in our economy and problems and opportunities for developing new energy resources (Section 6.11).

STUDY QUESTIONS

REVIEW QUESTIONS

1. You pick up a six-pack of soft drinks from the floor, but it slips from your hand and smashes into your foot. Comment on the work and energy involved in this sequence. What forms of energy are involved and at what stages of the process?
2. Based on your experience, when ice melts to liquid water is the process exothermic or endothermic? When liquid water freezes to ice at 0 °C is this exothermic or endothermic?
3. What is the first law of thermodynamics? Explain in words and use a mathematical expression.
4. For each of the following, define a system and its surroundings and give the direction of heat transfer:
   (a) Methane is burning in a gas furnace in your home.
   (b) Water drops, sitting on your skin after a dip in a swimming pool, evaporate.
   (c) Water, originally at 25 °C, is placed in the freezing compartment of a refrigerator.
   (d) Two chemicals are mixed in a flask sitting on a laboratory bench. A reaction occurs, and heat is evolved.
5. What is the value of the standard enthalpy of formation for any element under standard conditions?
6. A house is made of wood and glass. Assuming an equal amount of sunshine falls on a wooden wall and a piece of clean glass (of equal mass), which warms more? Explain briefly.
7. Criticize the following statements:
   (a) An enthalpy of formation refers to a reaction in which 1 mol of one or more reactants produces some quantity of product.
   (b) The standard enthalpy of formation of O_2 as a gas at 25 °C and a pressure of 1 bar is 15.0 kJ/mol.
   (c) The thermal energy transferred from 10.0 g of ice as it melts is $q = +6$ kJ.
8. Is the following reaction predicted to favor the reactants or products? Explain your answer briefly.

$$Mg(s) + \tfrac{1}{2}\,O_2(g) \longrightarrow MgO(s) \qquad \Delta H_f^\circ = -601.70 \text{ kJ}$$

9. Which of the following is or is not a state function?
   (a) The volume of a balloon
   (b) The time it takes to drive from your home to your college or university
   (c) The temperature of the water in a coffee cup
   (d) The potential energy of a ball held in your hand

NUMERICAL QUESTIONS

Energy Units

10. If you are on a diet that calls for eating no more than 1200 Cal/day, how many joules would this be?

11. A 2-in. piece of a two-layer chocolate cake with frosting provides 1670 kJ of energy. What is this in Calories?

12. Your home loses heat in the winter through doors and windows and any poorly insulated walls. A sliding glass door (6 ft × $6\frac{1}{2}$ ft with $\frac{1}{2}$ in. of insulating glass) allows 1.0×10^6 J/h to pass through the glass if the inside temperature is 22 °C (72 °F) and the outside temperature is 0 °C (32 °F). What quantity of heat, expressed in kilojoules, is lost per day? If 1 kwh of energy is equal to 3.60×10^6 J, how many kilowatt-hours of energy are lost per day through the door?

13. A parking lot in Los Angeles, California receives an average of 2.6×10^7 J of solar energy per square meter per day in the summer. If the parking lot is 325 m long and 50.0 m wide, what is the total quantity of energy striking the area per day?

Specific Heat

14. The specific heat of nickel is 0.445 J/g · K. How much heat is required to heat a 168-g piece of nickel from −15.2 °C to +23.6 °C?

15. 74.8 J of heat is required to raise the temperature of 18.69 g of silver from 10.0 °C to 27.0 °C. What is the specific heat of silver?

16. Which requires more heat to warm from 22 °C to 85 °C, 50.0 g of water or 200. g of aluminum metal?

17. You hold a gram of copper in one hand and a gram of aluminum in the other. Each metal was originally at 0 °C. (Both metals are in the shape of a little ball that fits into your hand.) If they both take up heat at the same rate, which reaches your body temperature first?

18. How much heat energy (in kilojoules) is required to heat all the aluminum in a roll of aluminum foil (485 g) from room temperature (25 °C) to the temperature of a hot oven (255 °C)?

19. A 20.0-g piece of aluminum at 0.0 °C is dropped into a beaker of water. The temperature of the water drops from 90.0 °C to 75.0 °C. What quantity of heat energy did the piece of aluminum absorb?

20. Ethylene glycol, $HOCH_2CH_2OH$, is often used as an antifreeze in automobile cooling systems. Which requires more heat energy to warm from 25.0 °C to 100. °C, 250 g of water or 250 g of ethylene glycol?

21. Which requires more energy to warm from 0.0 °C to 29.5 °C, a 5.00-kg piece of granite or 5.00 kg of cement?

22. A 192-g piece of copper is heated to 100.0 °C in a boiling water bath and then dropped into a beaker containing 750. mL of water (density = 1.00 g/cm^3) at 4.0 °C. What is the final temperature of the copper and water after they come to thermal equilibrium? (The specific heat of copper is 0.385 J/g · K)

23. A 13.8-g piece of zinc was heated to 98.9 °C in boiling water and then dropped into 15.00 mL of water at 25.0 °C. When the water and metal come to thermal equilibrium, what is the temperature of the system? (The specific heat of zinc is 0.388 J/g · K)

24. When 182 g of gold at some temperature is added to 22.1 g of water at a temperature of 25.0 °C, the final temperature of the resulting mixture is 27.5 °C. If the specific heat of gold is 0.128 J/g · K, what was the initial temperature of the gold sample?

25. When 108 g of water at a temperature of 22.5 °C is mixed with 65.1 g of water at an unknown temperature, the final temperature of the resulting mixture is 47.9 °C. What was the temperature of the second sample of water?

26. A 150.0-g sample of a metal at 80.0 °C is placed in 150.0 g of water at 20.0 °C. The temperature of the final system (metal and water) is 23.3 °C. What is the specific heat of the metal?

27. A 237-g piece of molybdenum, initially at 100.0 °C, is dropped into 244 g of water at 10.0 °C. When the system comes to thermal equilibrium, the temperature is 15.3 °C. What is the specific heat of the metal?

Changes of State

28. The heat energy required to melt 1.00 g of ice at 0 °C is 333 J. If one ice cube has a mass of 62.0 g, and a tray contains 16 ice cubes, what quantity of energy is required to melt a tray of ice cubes at 0 °C?

29. Chloromethane, CH_3Cl, is used as a topical anesthetic. The temperature at which CH_3Cl liquid turns into a vapor (the boiling point) is −24.09 °C. What quantity of heat must be absorbed by the liquid to convert 150. g of liquid to a vapor at −24.09 °C? The heat of vaporization of CH_3Cl is 21.40 kJ/mol.

30. Calculate the quantity of heat required to convert the water in five ice cubes (60.1 g) from $H_2O(s)$ at 0.0 °C to $H_2O(g)$ at 100.0 °C. The heat of fusion of ice at 0 °C is 333 J/g; the heat of vaporization of liquid water at 100 °C is 2260 J/g.

31. Mercury, with a freezing point of −38.8 °C, is the only metal that is liquid at room temperature. How much heat energy (in joules) must be released by mercury if 1.00 mL of the metal is cooled from room temperature (23.0 °C) to −38.8 °C and then frozen to a solid? (The density of mercury is 13.6 g/cm^3. Its specific heat is 0.140 J/g · K and its heat of fusion is 11.4 J/g.)

32. How much heat energy (in joules) is required to raise the temperature of 454 g of tin (1.00 lb) from room temperature (25.0 °C) to its melting point, 231.9 °C, and then melt the tin at that temperature? The specific heat of tin is 0.227 J/g · K, and the metal requires 59.2 J/g to convert the solid to a liquid.

33. Ethanol, C_2H_5OH, boils at 78.29 °C. How much heat energy (in joules) is required to heat 1.00 kg of this liquid from 20.0 °C to the boiling point and then change the liquid completely to a vapor at that temperature? (The specific heat of liquid ethanol is 2.44 J/g · K and its enthalpy of vaporization is 38.56 kJ/mol.)

Enthalpy

34. Nitrogen monoxide has recently been found to be involved in a wide range of biological processes. The gas reacts with oxygen to give brown NO_2 gas.

$$2\,NO(g) + O_2(g) \longrightarrow 2\,NO_2(g) \qquad \Delta H^\circ_{rxn} = -114.1\text{ kJ}$$

Is the reaction endothermic or exothermic? If 1.25 g of NO is converted completely to NO_2, what quantity of heat is absorbed or evolved?

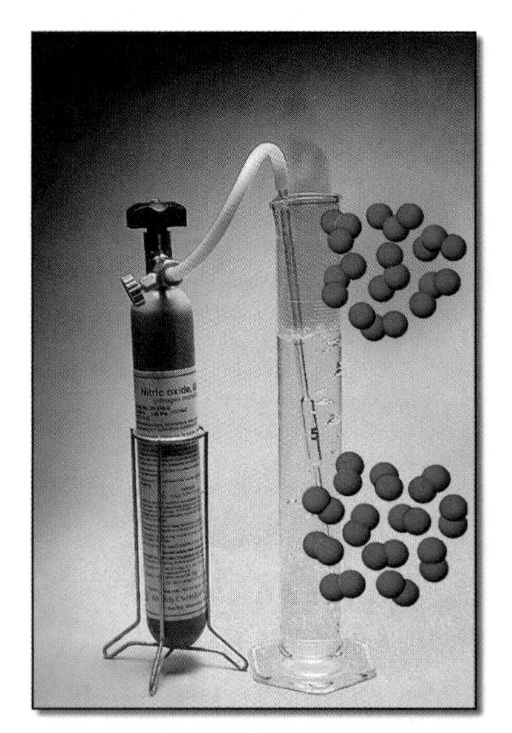

The gas NO is stored in a tank. When it is bubbled through water, the gas in the bubbles is colorless. However, as soon as the NO bubbles enter the atmosphere, colorless NO is oxidized to brown NO_2. *(Photo, C. D. Winters; model, S. M. Young)*

35. Calcium carbide, CaC_2, is manufactured by reducing lime (CaO) with carbon at a high temperature. (The carbide is

used to make acetylene, an industrially important organic chemical. See page 269.)

$$CaO(s) + 3\,C(s) \longrightarrow CaC_2(s) + CO(g) \qquad \Delta H^\circ_{rxn} = +464.8 \text{ kJ}$$

Is the reaction endothermic or exothermic? If 10.0 g of CaO is allowed to react with an excess of carbon, what quantity of heat is absorbed or evolved by the reaction?

36. Isooctane (2,2,4-trimethylpentane, page 129) burns in air to give water and carbon dioxide.

$$2\,C_8H_{18}(\ell) + 25\,O_2(g) \longrightarrow 16\,CO_2(g) + 18\,H_2O(\ell)$$
$$\Delta H^\circ_{rxn} = -10{,}922 \text{ kJ}$$

Is the combustion exothermic or endothermic? If you burn 1.00 L of the hydrocarbon (density = 0.6878 g/mL), what quantity of heat is involved?

37. Acetic acid, CH_3CO_2H, is made by the reaction of methanol and carbon monoxide.

$$CH_3OH(\ell) + CO(g) \longrightarrow CH_3CO_2H(\ell)$$
$$\Delta H^\circ_{rxn} = -355.9 \text{ kJ}$$

Is the reaction exothermic or endothermic? If you produce 1.00 L of the acid (density = 1.044 g/mL), what quantity of heat is involved?

38. Methanol, CH_3OH, is a possible automobile fuel. Energy is evolved in a combustion reaction with O_2.

$$2\,CH_3OH(g) + 3\,O_2(g) \longrightarrow 2\,CO_2(g) + 4\,H_2O(\ell)$$

A 0.0466-g sample of methanol evolves 1110 J when burned at constant pressure. What is the enthalpy change, ΔH°_{rxn}, for the reaction? What is the enthalpy change per mole of methanol (often called the molar heat of combustion)?

39. White phosphorus, P_4, ignites in air to produce heat, light, and P_4O_{10} (see Figure 4.4).

$$P_4(s) + 5\,O_2(g) \longrightarrow P_4O_{10}(s)$$

If 3.56 g of P_4 is burned, 37.4 kJ of heat is evolved at constant pressure. What is the enthalpy change for the combustion of 1.00 mol of P_4?

Hess's Law

40. You need to know the enthalpy change for the formation of benzene, C_6H_6, a value that is not available directly from any data tables.

$$6\,C(\text{graphite}) + 3\,H_2(g) \longrightarrow C_6H_6(\ell) \qquad \Delta H^\circ_f = ?$$

Use the procedure described in Example 6.6 and the enthalpy change for the combustion of benzene, which was determined experimentally, to calculate the standard molar enthalpy of formation of benzene.

$$2\,C_6H_6(\ell) + 15\,O_2(g) \longrightarrow 12\,CO_2(g) + 6\,H_2O(\ell)$$
$$\Delta H^\circ_{rxn} = -6534.8 \text{ kJ}$$

41. Calculate the enthalpy change, ΔH°, for the formation of 1.00 mol of strontium carbonate (the material that gives the red color in fireworks) from its elements.

$$Sr(s) + C(\text{graphite}) + \tfrac{3}{2}\,O_2(g) \longrightarrow SrCO_3(s)$$

The experimental information available is:

$$Sr(s) + \tfrac{1}{2}\,O_2(g) \longrightarrow SrO(s) \qquad \Delta H^\circ_f = -592 \text{ kJ}$$
$$SrO(s) + CO_2(g) \longrightarrow SrCO_3(s) \qquad \Delta H^\circ_{rxn} = -234 \text{ kJ}$$
$$C(\text{graphite}) + O_2(g) \longrightarrow CO_2(g) \qquad \Delta H^\circ_{rxn} = -394 \text{ kJ}$$

42. Using the following reactions, find the enthalpy change for the formation of PbO(s) from lead metal and oxygen gas.

$$Pb(s) + CO(g) \longrightarrow PbO(s) + C(s) \qquad \Delta H^\circ_{rxn} = -106.8 \text{ kJ}$$
$$2\,C(s) + O_2(g) \longrightarrow 2\,CO(g) \qquad \Delta H^\circ_{rxn} = -221.0 \text{ kJ}$$

If 250 g of lead reacts with oxygen to form lead(II) oxide, what quantity of heat (in kilojoules) is absorbed or evolved?

43. You wish to know the enthalpy change for the formation of liquid PCl_3 from the elements.

$$P_4(s) + 6\,Cl_2(g) \longrightarrow 4\,PCl_3(\ell) \qquad \Delta H^\circ_f = ?$$

This reaction cannot be carried out directly. Instead, the enthalpy change for the reaction of phosphorus and chlorine to give phosphorus pentachloride can be determined experimentally

$$P_4(s) + 10\,Cl_2(g) \longrightarrow 4\,PCl_5(s) \qquad \Delta H^\circ_{rxn} = -1774.0 \text{ kJ}$$

The enthalpy change for the reaction of phosphorus trichloride with more chlorine to give phosphorus pentachloride can also be measured.

$$PCl_3(\ell) + Cl_2(g) \longrightarrow PCl_5(s) \qquad \Delta H^\circ_{rxn} = -123.8 \text{ kJ}$$

Use this to calculate the enthalpy change for the formation of 1.00 mol of $PCl_3(\ell)$ from phosphorus and chlorine.

Standard Enthalpies of Formation

44. The standard molar enthalpy of formation of solid chromium(III) oxide is −1139.7 kJ/mol. Write the balanced equation for which the enthalpy of reaction is −1139.7 kJ.

45. The molar enthalpy of formation of methanol, $CH_3OH(\ell)$, is −238.7 kJ/mol. Write the balanced equation for which the enthalpy of reaction is −238.7 kJ.

46. The molar enthalpy of formation of glucose, $C_6H_{12}O_6(s)$, is −1273.3 kJ/mol.
(a) Is the formation of glucose from its elements exothermic or endothermic?
(b) Write a balanced equation depicting the formation of glucose from its elements and for which the enthalpy of reaction is −1273.3 kJ.

47. The molar enthalpy of formation of liquid pyridine, C_5H_5N, is +100.2 kJ/mol.
(a) Write the balanced equation for the formation of pyridine from its elements, that is, for the reaction for which the enthalpy of reaction is +100.2 kJ.
(b) Write the balanced equation for which the enthalpy change is +200.4 kJ.

48. Enthalpy changes have been determined experimentally for the following reactions:

$$Pb(s) + 2\,Cl_2(g) \longrightarrow PbCl_4(\ell) \qquad \Delta H_f^\circ = -329.3 \text{ kJ}$$
$$PbCl_2(s) + Cl_2(g) \longrightarrow PbCl_4(\ell) \qquad \Delta H_{rxn}^\circ = +30.1 \text{ kJ}$$

What is the enthalpy change for the reaction of lead with chlorine to give lead(II) chloride?

49. The enthalpy of formation of solid barium oxide, BaO, is −553.5 kJ/mol, and the enthalpy of formation of barium peroxide, BaO_2, is −634.3 kJ/mol. Calculate the enthalpy change for the reaction

$$BaO(s) + \tfrac{1}{2}\,O_2(g) \longrightarrow BaO_2(s)$$

50. An important step in the production of sulfuric acid is

$$SO_2(g) + \tfrac{1}{2}\,O_2(g) \longrightarrow SO_3(g)$$

It is also a key reaction in the formation of acid rain, beginning with the air pollutant SO_2. Using the data in Table 6.2, calculate the enthalpy change for the reaction. Is the reaction exothermic or endothermic?

51. In photosynthesis, the sun's energy brings about the combination of CO_2 and H_2O to form O_2 and a carbon-containing compound. In its simplest form, the reaction is

$$6\,CO_2(g) + 6\,H_2O(\ell) \longrightarrow C_6H_{12}O_6(s) + 6\,O_2(g)$$

Using the enthalpies of formation in Table 6.2 and $\Delta H_f^\circ[C_6H_{12}O_6(s)] = -1273.3$ kJ/mol, (a) calculate the enthalpy of reaction and (b) decide if the reaction is exothermic or endothermic.

52. The first step in the production of nitric acid from ammonia involves the oxidation of NH_3.

$$4\,NH_3(g) + 5\,O_2(g) \longrightarrow 4\,NO(g) + 6\,H_2O(g)$$

(a) Use the information in Table 6.2 or Appendix L to find the enthalpy change for this reaction. Is the reaction exothermic or endothermic?
(b) What quantity of heat is evolved or absorbed if 10.0 g of NH_3 is oxidized?

53. The Romans used calcium oxide (CaO) as mortar in stone structures. The CaO was mixed with water to give $Ca(OH)_2$, which reacted slowly with CO_2 in the air to give limestone.

$$Ca(OH)_2(s) + CO_2(g) \longrightarrow CaCO_3(s) + H_2O(g)$$

(a) Calculate the enthalpy change for this reaction.
(b) What quantity of heat is evolved or absorbed if 1.00 kg of $Ca(OH)_2$ is allowed to react with a stoichiometric amount of CO_2?

54. Pure metals can often be prepared by reducing the metal oxide with hydrogen gas. For example,

$$WO_3(s) + 3\,H_2(g) \longrightarrow W(s) + 3\,H_2O(\ell)$$

(a) Calculate the enthalpy change for this reaction. (ΔH_f° for $WO_3(s)$ is −842.9 kJ/mol.)
(b) What quantity of heat is evolved or absorbed if 1.00 g of WO_3 is allowed to react with an excess of hydrogen gas?

55. Iron(III) oxide ("rust") can be produced from iron and oxygen in a sequence of reactions that can be written as

$$2\,Fe(s) + 6\,H_2O(\ell) \longrightarrow 2\,Fe(OH)_3(s) + 3\,H_2(g)$$
$$2\,Fe(OH)_3(s) \longrightarrow Fe_2O_3(s) + 3\,H_2O(\ell)$$
$$3\,H_2(g) + \tfrac{3}{2}\,O_2(g) \longrightarrow 3\,H_2O(\ell)$$

(a) What is the enthalpy change for each step?
(b) What is the enthalpy change for the overall process, the reaction of iron with oxygen to give iron(III) oxide?
(c) What quantity of heat is evolved or absorbed if 1.25 g of iron is converted to iron(III) oxide? (Values for ΔH_f° can be found in Table 6.2 and Appendix L. In addition, $\Delta H_f^\circ[Fe(OH)_3(s)] = -696.5$ kJ/mol.)

56. Naphthalene, $C_{10}H_8$, is burned in a calorimeter to find the enthalpy change for the combustion reaction.

$$C_{10}H_8(s) + 12\,O_2(g) \longrightarrow 10\,CO_2(g) + 4\,H_2O(\ell)$$

If $\Delta H_{rxn}^\circ = -5156.1$ kJ, what is the molar enthalpy of formation of naphthalene?

57. Styrene, C_8H_8, is burned in a calorimeter to find the enthalpy change for the combustion reaction.

$$C_8H_8(\ell) + 10\,O_2(g) \longrightarrow 8\,CO_2(g) + 4\,H_2O(\ell)$$

If $\Delta H_{rxn}^\circ = -4395.0$ kJ, what is the molar enthalpy of formation of styrene?

Calorimetry

58. How much heat energy (in kilojoules) is evolved by a reaction in a bomb calorimeter (see Figure 6.17) in which the temperature of the bomb and water increases from 19.50 °C to 22.83 °C? The bomb has a heat capacity of 650. J/K; the calorimeter contains 320. g of water.

59. Sulfur (2.56 g) is burned in a bomb calorimeter with excess $O_2(g)$. The temperature increases from 21.25 °C to 26.72 °C. The bomb has a heat capacity of 923 J/K, and the calorimeter contains 815 g of water. Calculate the heat evolved, per mole of SO_2 formed, in the course of the reaction

$$S_8(s) + 8\,O_2(g) \longrightarrow 8\,SO_2(g)$$

Sulfur burns in oxygen, with a bright blue flame, to give sulfur dioxide gas, SO_2. (*C. D. Winters*)

60. You can find the amount of heat evolved in the combustion of carbon by carrying out the reaction in a combustion calorimeter. Suppose you burn 0.300 g of C(graphite) in an excess of $O_2(g)$ to give $CO_2(g)$.

$$C(graphite) + O_2(g) \longrightarrow CO_2(g)$$

The temperature of the calorimeter, which contains 775 g of water, increases from 25.00 °C to 27.38 °C. The heat capacity of the bomb is 893 J/K. What quantity of heat is evolved per mole of carbon?

61. Suppose you burn 1.500 g of benzoic acid in a combustion calorimeter and find that the temperature of the calorimeter increases from 22.50 °C to 31.69 °C. The calorimeter contains 775 g of water, and the bomb has a heat capacity of 893 J/K. How much heat is evolved per mole of benzoic acid?

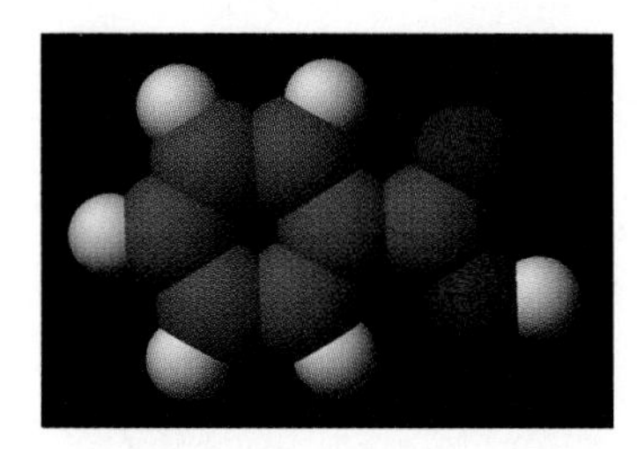

Benzoic acid, $C_7H_6O_2$, occurs naturally in many berries.

62. Assume you mix 100.0 mL of 0.200 M CsOH with 50.0 mL of 0.400 M HCl in a coffee-cup calorimeter. The following reaction occurs.

$$CsOH(aq) + HCl(aq) \longrightarrow CsCl(aq) + H_2O(\ell)$$

The temperature of both solutions before mixing was 22.50 °C, and it rises to 24.28 °C after the acid-base reaction. What is the enthalpy of the reaction per mole of CsOH? Assume the densities of the solutions are all 1.00 g/mL and the heat capacities of the solutions are 4.2 J/g · K.

63. Suppose you mix 125 mL of 0.250 M CsOH with 50.0 mL of 0.625 M HF in a coffee-cup calorimeter. The temperature of both solutions was 21.50 °C, and it rises to 24.40 °C after the reaction.

$$CsOH(aq) + HF(aq) \longrightarrow CsF(aq) + H_2O(\ell)$$

What is the enthalpy of reaction per mole of CsOH? Compare the result of this experiment with the one in Study Question 62. Does the reaction enthalpy depend on the fact that HCl is a strong acid (completely ionized in aqueous solution), whereas HF is a weak acid (incompletely ionized in aqueous solution)? Assume the densities of the solutions are all 1.00 g/mL and the heat capacities of the solutions are 4.2 J/g · K.

64. An "ice calorimeter" can be used to determine the specific heat of a metal. A piece of hot metal is dropped into a weighed quantity of ice. The quantity of heat given up by the metal can be determined from the amount of ice melted. Suppose you heat a 50.0-g piece of metal to 99.8 °C and then drop it onto ice. When the metal's temperature has dropped to 0.0 °C, it is found that 7.33 g of ice has melted. What is the specific heat of the metal?

65. A 9.36-g piece of platinum is heated to 98.6 °C in a boiling water bath and then dropped onto ice. When the metal's temperature has dropped to 0.0 °C, it is found that 0.37 g of ice has melted. What is the specific heat of platinum?

GENERAL QUESTIONS

66. Which gives up more heat on cooling from 50 °C to 10 °C, 50.0 g of water or 100. g of ethanol (specific heat of ethanol = 2.46 J/g · K)?

67. Assume you place 100. g each of copper and gold, each originally at 25 °C, in a boiling water bath at 100 °C. If each metal takes up heat at the same rate (the number of joules of heat absorbed per minute is the same), which piece of metal reaches 100 °C first?

68. The meals-ready-to-eat (MREs) in the military can be heated on a flameless heater. Assume the reaction in the heater is

$$Mg(s) + 2\,H_2O(aq) \longrightarrow Mg(OH)_2(s) + H_2(g)$$

Calculate the enthalpy change under standard conditions (in joules) for this reaction. What quantity of magnesium is needed to supply the heat required to warm 25 mL of water (d = 1.00 g/mL) from 25 °C to 85 °C?

69. The combustion of gaseous diborane, B_2H_6, proceeds according to the equation

$$B_2H_6(g) + 3\,O_2(g) \longrightarrow B_2O_3(s) + 3\,H_2O(g)$$

and 1941 kJ of heat energy is liberated per mole of $B_2H_6(g)$ (at constant pressure). Calculate the molar enthalpy of formation of $B_2H_6(g)$ using this information, the data in Table 6.2, and the fact that ΔH_f° for $B_2O_3(s)$ is −1271.9 kJ/mol.

70. Given the following information,

$$N_2H_4(\ell) + O_2(g) \longrightarrow N_2(g) + 2\,H_2O(g) \qquad \Delta H^\circ_{rxn} = -534\text{ kJ}$$

and the data in Table 6.2, calculate the molar enthalpy of formation for liquid hydrazine, N_2H_4.

71. The following are two of the key reactions in the processing of uranium for use as fuel in nuclear power plants:

$$UO_2(s) + 4\,HF(g) \longrightarrow UF_4(s) + 2\,H_2O(g)$$
$$UF_4(s) + F_2(g) \longrightarrow UF_6(g)$$

(a) Calculate the enthalpy change, ΔH°_{rxn}, for each reaction using the data in Table 6.2, Appendix L, and the following:

Compound	ΔH_f° (kJ/mol)
$UO_2(s)$	−1085
$UF_4(s)$	−1914
$UF_6(g)$	−2147

(b) Calculate the enthalpy change for the overall conversion of UO_2 to UF_6.

$$UO_2(s) + 4\,HF(g) + F_2(g) \longrightarrow UF_6(g) + 2\,H_2O(g)$$

72. The combination of coke and steam produces a mixture called coal gas, which can be used as a fuel or as a starting material for other reactions. The equation for the production of coal gas is

$$C(s) + H_2O(g) \longrightarrow CO(g) + H_2(g)$$

Determine the standard enthalpy change for this reaction. What quantity of heat is involved if 1.0 metric ton (1000.0 kg) of carbon is converted to coal gas?

73. One method of producing H_2 on a large scale is the chemical cycle shown here.

Step 1: $SO_2(g) + 2\,H_2O(g) + Br_2(g) \longrightarrow H_2SO_4(\ell) + 2\,HBr(g)$

Step 2: $H_2SO_4(\ell) \longrightarrow H_2O(g) + SO_2(g) + \frac{1}{2}\,O_2(g)$

Step 3: $2\,HBr(g) \longrightarrow H_2(g) + Br_2(g)$

Using the table of standard enthalpies of formation in Appendix L, calculate ΔH°_{rxn} for each step. What is the equation for the overall process and what is its enthalpy change? Is the overall process exothermic or endothermic?

74. Ammonium nitrate decomposes exothermically to N_2O and water (see photo).

$$NH_4NO_3(s) \longrightarrow N_2O(g) + 2\,H_2O(g)$$

(a) If the enthalpy of formation of $N_2O(g)$ is 82.1 kJ/mol, how much heat is evolved (at constant pressure and under standard conditions)?

(b) If 8.00 kg of ammonium nitrate decomposes, what quantity of heat is evolved (at constant pressure and under standard conditions)?

The decomposition of ammonium nitrate was the explosive charge that damaged New York's World Trade Center in 1993 and destroyed the Federal Building in Oklahoma City in 1995. (*C. D. Winters*)

75. Camping stoves are fueled by propane (C_3H_8), butane [$C_4H_{10}(g)$, $\Delta H^\circ_f = -125.6$ kJ/mol)], gasoline, or ethanol (C_2H_5OH). Calculate the heat of combustion per gram of each of these fuels. (Assume that gasoline is represented by isooctane, $C_8H_{18}(\ell)$, with $\Delta H^\circ_f = -259.2$ kJ/mol.) Do you notice any great differences among these fuels? Are these differences related to their composition? (See T. Smith: "Camping stoves, *Chem Matters,* page 7, April, 1992.)

76. The "combustion" of glucose to CO_2 and H_2O is the fundamental energy-producing reaction in the body. Although combustion actually occurs in a series of steps, the net reaction can be represented as

$$C_6H_{12}O_6(s) + 6\,O_2(g) \longrightarrow 6\,CO_2(g) + 6\,H_2O(\ell)$$

Another important energy-producing carbohydrate is fruit-sugar, or fructose, whose formula is also $C_6H_{12}O_6$. Which produces the great amount of energy per gram, glucose or fructose?

Compound	**ΔH°_f (kJ/mol)**
Glucose	−1273.3
Fructose	−1265.6

77. Hydrazine and 1,1-dimethylhydrazine both react spontaneously with O_2 and can be used as rocket fuels (see photo).

$$\underset{\text{hydrazine}}{N_2H_4(\ell)} + O_2(g) \longrightarrow N_2(g) + 2\,H_2O(g)$$

$$\underset{\text{1,1-dimethylhydrazine}}{N_2H_2(CH_3)_2(\ell)} + 4\,O_2(g) \longrightarrow 2\,CO_2(g) + 4\,H_2O(g) + N_2(g)$$

The molar enthalpy of formation of liquid hydrazine is +50.6 kJ/mol, and that of liquid dimethylhydrazine is +48.9 kJ/mol. By doing appropriate calculations, decide whether the reaction of hydrazine or dimethylhydrazine with oxygen gives more heat per gram (at constant pressure). (Other enthalpy of formation data can be obtained from Table 6.2.)

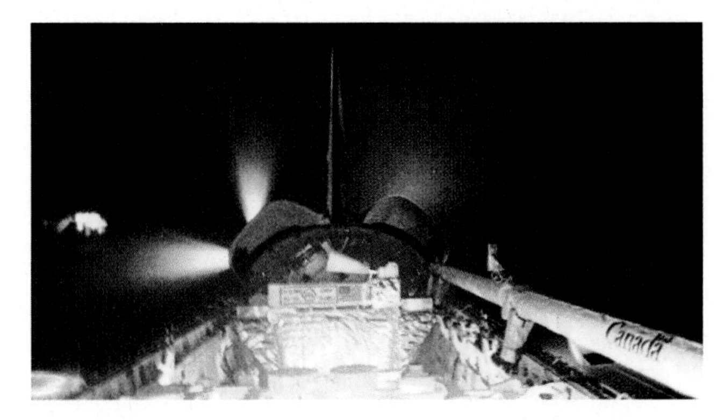

A control rocket on the Space Shuttle uses hydrazine (N_2H_4) as the fuel. (*NASA*)

78. A piece of metal with a mass of 27.3 g was heated to 98.90 °C and then dropped into 15.0 g of water at 25.00 °C. The final temperature of the system is 29.87 °C. What is the specific heat of the metal?

79. In the United States engineers express energy in British thermal units. (1 Btu, which is equivalent to 1055 J, is the amount of heat required to raise the temperature of 1 lb of water from 59.5 °C to 60.5 °C.) If the average solar energy striking a horizontal surface in Washington, D.C., in June is 2080 Btu/ft^2 per day, what is the energy in joules per square meter?

CONCEPTUAL QUESTIONS

80. Calculate the molar heat capacity, J/mol · K, for the four metals in Table 6.1. What observation can you make about these values? Are they widely different or very similar? Using this information can you calculate the specific heat in J/g · K for silver? (The correct value for silver is 0.236 J/g · K.)

81. Look at the enthalpies of formation of metal oxides in Table 6.2 or Appendix L and comment on your observations. Are oxidations of metals generally endothermic or exothermic?

82. How could you determine the enthalpy change for the conversion of graphite to diamond, a reaction for which the enthalpy change cannot be measured easily in the laboratory?

$$\text{C(graphite)} \longrightarrow \text{C(diamond)} \qquad \Delta H^\circ_{rxn} = +1.90 \text{ kJ}$$

Among the experimental information you have is the enthalpy change for the combustion of diamond.

$$\text{C(diamond)} + O_2(g) \longrightarrow CO_2(g) \qquad \Delta H^\circ_{rxn} = -395.4 \text{ kJ}$$

83. Listed in the table are the specific heat capacities of a variety of metals. Suppose you want to work out a method of predicting the specific heat of a metal for which an experimental value is not available. Prepare a plot of the heat capacities in the table versus the atomic weight of the metal. Does any relation exist between specific heat and atomic weight? Use this relation to predict the specific heat of platinum. (The specific heat for platinum is given in the literature as 0.133 J/g · K.) How good is the agreement between the predicted and actual values?

Metal	Specific Heat (J/g · K)
Chromium	0.450
Gold	0.128
Iron	0.451
Lead	0.127
Silver	0.236
Tin	0.227
Titanium	0.522

84. Suppose you are attending summer school and are living in a very old dormitory. The day is oppressively hot. There is no air conditioner, and you can't open the windows of your room because they are stuck shut from layers of paint. There is a refrigerator in the room, however. In a stroke of genius you open the door of the refrigerator, and cool air cascades out. The relief does not last long, though. Soon the refrigerator motor and condenser begin to run, and not long thereafter the room is hotter than it was before. Why did the room warm up?

85. You want to determine the value for the enthalpy of formation of $CaSO_4(s)$.

$$Ca(s) + \tfrac{1}{8}\, S_8(s) + 2\, O_2(g) \longrightarrow CaSO_4(s)$$

The reaction cannot be done directly. You know, however, that both calcium and sulfur react with oxygen to produce oxides in reactions that can be studied calorimetrically. You also know that the basic oxide CaO reacts with the acidic oxide SO_3 to produce $CaSO_4(s)$. Outline a method for determining ΔH°_f for $CaSO_4(s)$ and identify the information that must be collected by experiment. Using information in Table 6.2 confirm that ΔH°_f for $CaSO_4(s) = -1434.5$ kJ/mol.

CHALLENGING QUESTIONS

86. A commercial product called "Instant Car Kooler" contains 10% by weight ethanol, C_2H_5OH, and 90% by weight water. If the interior of your car is overheated, you spray the "Kooler" inside and the interior is cooled. It works because thermal energy must be transferred to the alcohol and water from the warm air. To drop the air temperature from 55 °C to 25 °C requires air to give up 3.6 kJ. How many grams of the ethanol-water mixture must be used to absorb this heat? (The enthalpy of vaporization for ethanol is 850 J/g and for water it is 2360 J/g.) See D. Robson: "Car cooler," *Chem Matters,* p. 11, February, 1993.)

The "Car Kooler" uses the heat of vaporization of ethanol and water to cool the interior of a car on a hot day. See Question 86. (*C. D. Winters*)

87. Suppose you want to heat the air in your house with natural gas (CH_4). Assume your house has 275 m^2 (about 2800 ft^2) of floor area and that the ceilings are 2.50 m from the floors. The air in the house has a molar heat capacity of 29.1 J/mol · K. (The number of moles of air in the house can be found by assuming that the average molar mass of air is 28.9 g/mol and that the density of air at these temperatures is about 1.22 g/L.) How much methane do you have to burn to heat the air from 15.0 °C to 22.0 °C?

88. Methanol, CH_3OH, a compound that can be made relatively inexpensively from coal, is a promising substitute for gasoline. The alcohol has a smaller energy content than gasoline, but, with its higher octane rating, it burns more efficiently than gasoline in combustion engines. (It also has the added advantage of contributing to a lesser degree to some air pollutants.) Compare the heat of combustion per gram of CH_3OH and C_8H_{18} (isooctane), the latter being representative of the compounds in gasoline. ($\Delta H_f^\circ =$ -259.2 kJ/mol for isooctane.)

89. The heat of fusion of ice is 333 J/g. Suppose that three 45-g ice cubes at 0 °C are dropped into 500. mL of tea initially at 20.0 °C. How much ice melts? Assume the tea is weak enough that its specific heat is the same as that of pure water.

90. Isomers are molecules with the same elemental composition but a different atomic arrangement. Three ways of arranging the atoms in a compound with the formula C_4H_8 are shown. The double bond between carbon atoms occurs at a different location, or the arrangement of atoms around the double bond is different. (See structures shown at the bottom of the page.) The enthalpy of combustion of each isomer can be determined using a calorimeter.

(a) Draw an enthalpy diagram such as that in Figure 6.16.

(b) What is the enthalpy change for the conversion of *cis*-2-butene to *trans*-2-butene?

(c) Knowing that the enthalpy of formation of 1-butene is -20.5 kJ/mol, calculate the enthalpy of formation values for both *cis*- and *trans*-2-butene.

91. Suppose you add 100.0 g of water at 60.0 °C to 100.0 g of ice at 0.00 °C. Some of the ice melts and cools the warm water to 0.00 °C. When the ice and water mixture has come to a uniform temperature of 0 °C, how much ice has melted?

92. The standard molar enthalpy of formation, ΔH_f°, of diborane, $B_2H_6(g)$, cannot be determined directly because the compound cannot be prepared by the reaction of boron and hydrogen. It can be calculated, however, using the following reactions:

$$4\,B(s) + 3\,O_2(g) \longrightarrow 2\,B_2O_3(s) \quad \Delta H^\circ_{rxn} = -2543.8 \text{ kJ}$$

$$2\,H_2(g) + O_2(g) \longrightarrow 2\,H_2O(g) \quad \Delta H^\circ_{rxn} = -484 \text{ kJ}$$

$$B_2H_6(g) + 3\,O_2(g) \longrightarrow B_2O_3(s) + 3\,H_2O(g) \quad \Delta H^\circ_{rxn} = -2032.9 \text{ kJ}$$

Calculate ΔH_f° for $B_2H_6(g)$.

93. The standard molar enthalpy of formation, ΔH_f°, of $CS_2(g)$ cannot be determined directly because the compound cannot be prepared by the reaction of carbon and sulfur. It can be calculated, however, using the value for the enthalpy change for the combustion of $CS_2(g)$ to $CO_2(g)$ and $SO_2(g)$ [$\Delta H^\circ_{rxn} = -1104.5$ kJ] and the molar enthalpies of formation of $CO_2(g)$ and $SO_2(g)$. Calculate ΔH_f° for $CS_2(g)$.

94. Predict ΔH° for the production of chloroform, $CHCl_3$, from methane and chlorine.

$$CH_4(g) + 3\,Cl_2(g) \longrightarrow 3\,HCl(g) + CHCl_3(\ell)$$

Use the heat of combustion of methane,

$$CH_4(g) + 2\,O_2(g) \longrightarrow 2\,H_2O(\ell) + CO_2(g) \quad \Delta H^\circ_{rxn} = -890.4 \text{ kJ}$$

the enthalpy of decomposition of HCl(g),

$$2\,HCl(g) \longrightarrow H_2(g) + Cl_2(g) \quad \Delta H^\circ_{rxn} = +184.6 \text{ kJ}$$

the enthalpy of formation of $CO_2(g)$,

$$C(\text{graphite}) + O_2(g) \longrightarrow CO_2(g) \quad \Delta H_f^\circ = -393.5 \text{ kJ}$$

the enthalpy of formation of water,

$$H_2(g) + \tfrac{1}{2}\,O_2(g) \longrightarrow H_2O(\ell) \quad \Delta H_f^\circ = -285.8 \text{ kJ}$$

and the enthalpy of formation of $CHCl_3(g)$ (Appendix L).

$\Delta H_{combustion} = -2687.5$ kJ/mol

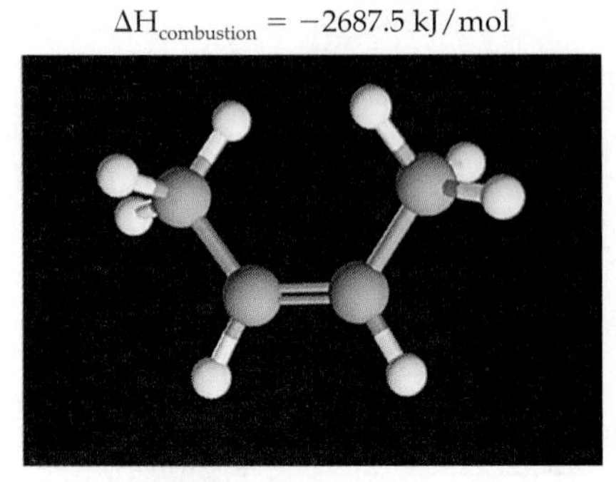

cis-2-butene

$\Delta H_{combustion} = -2684.2$ kJ/mol

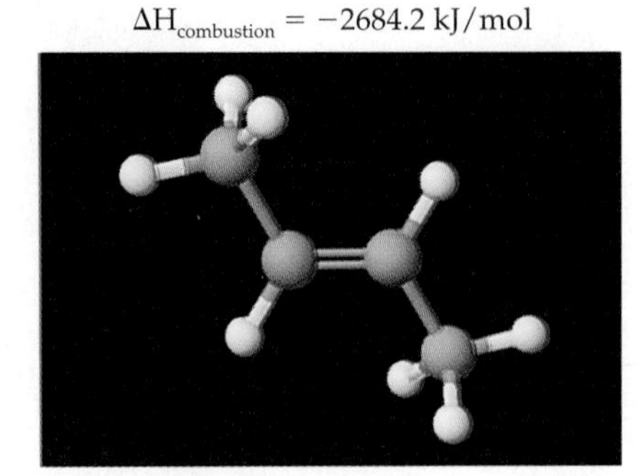

trans-2-butene

$\Delta H_{combustion} = -2696.7$ kJ/mol

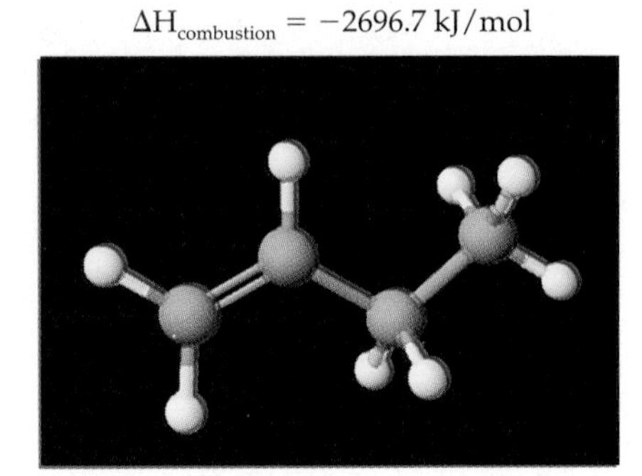

1-butene

Models of three isomers of C_4H_8. See Study Question 90.

Summary Question

95. Sulfur dioxide, SO_2, is a major pollutant in our industrial society, and it is often found in wine.

(a) In wine making, SO_2 is commonly added to kill microorganisms in the grape juice when it is put into vats before fermentation. Furthermore, it is used to neutralize byproducts of the fermentation process, enhance wine flavor, and prevent oxidation. Wine usually contains 80 to 150 parts per million (ppm) SO_2 (1 ppm = 1 g of SO_2/1 × 10^6 g of wine). The United States produced 4.40 × 10^8 gal of wine in 1987. Assuming the density of wine is 1.00 g/cm^3 and that the wine contains 100. ppm of SO_2, how many grams and how many moles of SO_2 were contained in this wine?

(b) When SO_2 is given off by an oil- or coal-burning power plant, it can be trapped by reacting it with MgO in air to form $MgSO_4$.

$$MgO(s) + SO_2(g) + \tfrac{1}{2}O_2(g) \longrightarrow MgSO_4(s)$$

If 21 million tons of SO_2 is given off by coal-burning power plants each year, how much MgO must be supplied to remove all of this SO_2 (1 ton = 9.08 × 10^5 g)? How much $MgSO_4$ is produced?

(c) If ΔH_f° for $MgSO_4$(s) is −2817.5 kJ/mol, how much heat (at constant pressure) is evolved or absorbed per mole of $MgSO_4$ by the reaction in part (b)?

(d) Sulfuric acid comes from the oxidation of sulfur, first to SO_2 and then to SO_3. The SO_3 is then absorbed by water to make H_2SO_4.

$$S(s) + O_2(g) \longrightarrow SO_2(g) \qquad \Delta H_f^\circ = -296.8\ \text{kJ}$$

$$SO_2(g) + \tfrac{1}{2}O_2(g) \longrightarrow SO_3(g) \qquad \Delta H_f^\circ = -98.9\ \text{kJ}$$

$$SO_3(g) + H_2O(\text{in 98\% } H_2SO_4) \longrightarrow H_2SO_4(\ell) \qquad \Delta H_{rxn}^\circ = -130.0\ \text{kJ}$$

The typical plant produces 750 tons of H_2SO_4 per day (1 ton = 9.08 × 10^5 g). Calculate the amount of heat produced by the plant per day.

INTERACTIVE GENERAL CHEMISTRY CD-ROM

Screen 6.2 Product-Favored Systems

(a) Reactions that largely convert reactants to products are considered product-favored systems. Based on your experience, predict if each of the following systems could be designated as product- or reactant-favored. Examine your prediction by going to the next screen.
(1) Rusting of iron
(2) Reaction of a diamond with O_2 at room temperature
(3) Combustion of gasoline
(4) The decomposition of sand, SiO_2, to elemental Si and O_2

(b) What makes us believe the oxidation of the Gummi Bear is product-favored? What do you suppose would happen if you put a cookie in the molten potassium chlorate? What about a piece of celery?

Screen 6.3 Thermodynamics and Kinetics

Chemists say that substances are thermodynamically stable or kinetically stable or both. Describe the difference between the kinetic stability of a substance and its thermodynamic stability. Give examples.

Screen 6.8 Heat Capacity of Pure Substances

This screen and the next one (6.9) take up the concept of specific heat capacity. The "tool" on Screen 6.8 allows you to do an experiment. Note that heat is added to each substance at the same rate, 50 J/s.

(a) What effect does heating the blocks for a longer time have on the final temperature?

(b) What is the effect of the mass of the blocks on the observed temperature change?

(c) Use this experiment to calculate the specific heat capacities of wood, copper, and glass. Do your answers agree with the "official" specific heat capacities for these substances? (For the values, click on the "Summary" button.)

(d) How does this "experiment" verify the relationship between heat transferred and the mass, specific heat, and temperature change of a substance? Explain fully.

(e) If water had been included in the "experiment" you just did, what would its final temperature be if you had heated 5.0 g for 5.0 s? Is this temperature lower or higher than the temperatures you observed for the wood, copper, or glass under these conditions? Explain.

Screen 6.10 Heat Transfer Between Substances

(a) Explain what happens in terms of molecular motions when a hotter object comes into contact with a cooler one.

(b) What does it mean when we say that two objects have come to thermal equilibrium?

Screen 6.11 Heat Associated with Phase Changes

(a) Describe what happens after 1000 J of heat energy is added to 10.0 g of ice at 0 °C and to 10.0 g of iron at 0 °C.

(b) Describe what would have happened to the 10.0 g of ice if 3500 J had been added. What would the temperature of the system be at this point?

Screen 6.14 Enthalpy Change and ΔH

(a) Is the process observed upon adding heat energy to solid CO_2 an endothermic or exothermic process? Why is it endothermic or exothermic?
(b) Is there an increase in enthalpy (positive enthalpy change, $+\Delta H$) or a decrease (negative enthalpy change, $-\Delta H$) when gaseous CO_2 condenses to form a solid?

Screen 6.18 Calorimetry

Use the calorimeter "tool" to determine the temperature change and total heat transferred for three different masses of benzoic acid. What is the relationship between the mass of compound and these two outcomes? Use this to estimate ΔT and total energy if 100 mg of benzoic acid is used in the calorimeter.

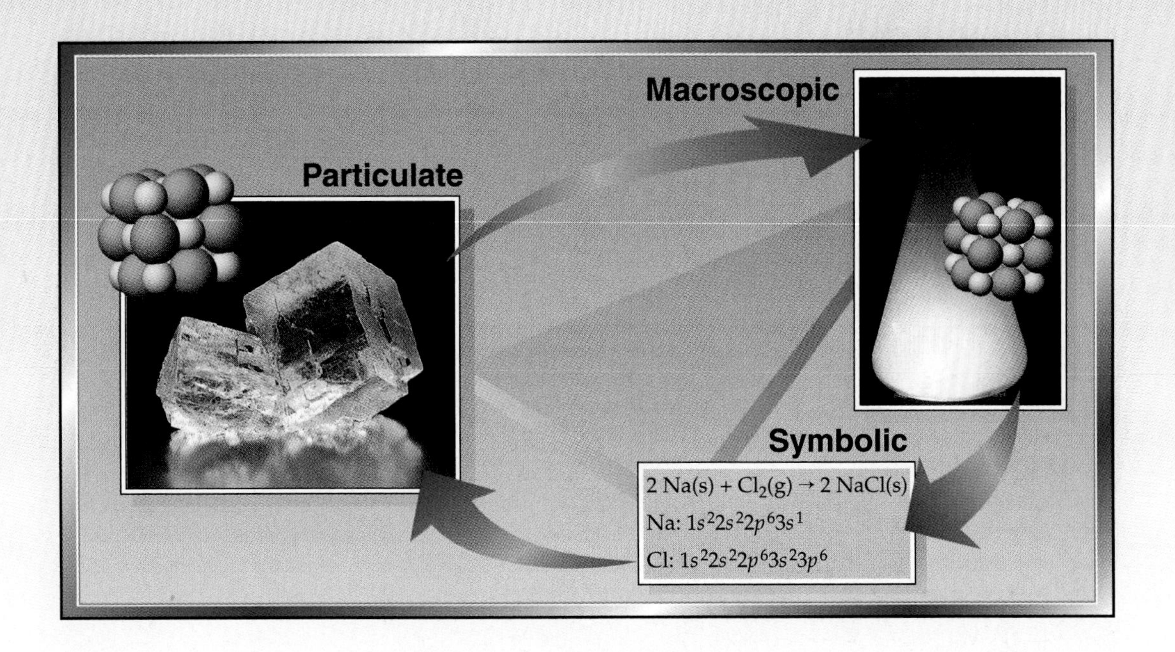

Part 2

The Structure of Atoms and Molecules

To understand the physical and chemical properties of the elements and their compounds, it is important to understand the underlying structure of atoms. Thus, our first goal in this section is to outline the current theories of the arrangement of electrons in atoms and some of the important historical developments that led to these theories (Chapters 7 and 8). With these ideas, you can then understand why atoms and their ions vary in size and in their ability to gain or lose electrons. So that these variations in properties can be remembered, and so that predictions can be made, we shall tie this discussion closely to the arrangement of elements in the periodic table (Chapter 8).

With an understanding of atomic structure, it is possible to see how atoms can join to form molecules. In Chapter 9 we shall first consider how the electrons of the atoms of the molecule are divided into various bonding and nonbonding or lone-pair electrons. We can then show how one can derive the three-dimensional structure of simple molecules. With the structure known, various properties of molecules can then be explained, a topic pursued throughout the book. Chapter 10 considers the major theories of molecular bonding in more detail, and Chapter 11 extends the subject of bonding and molecular structure to the important field of organic chemistry. ● *(Photos, C. D. Winters; models, S. M. Young)*

Chapter 7

Atomic Structure

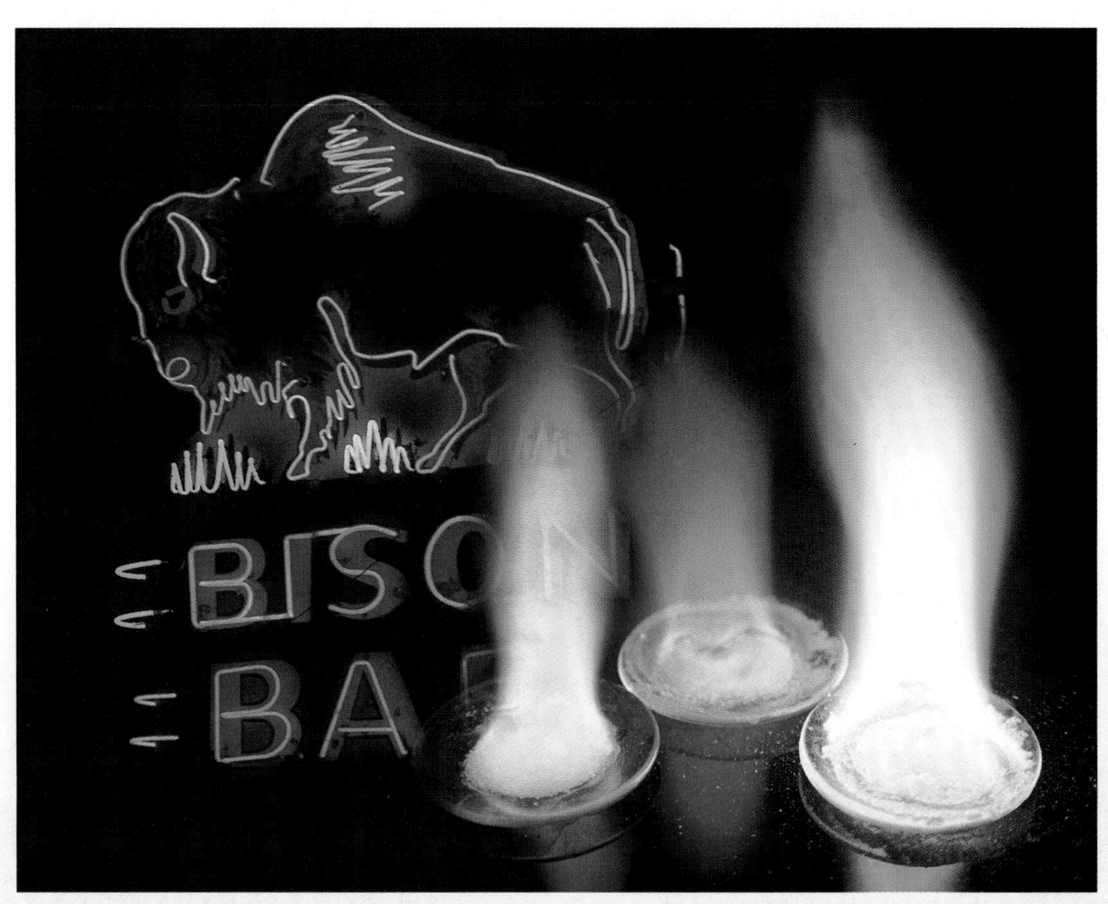

(Burning salts, C. D. Winters; neon sign, Dean Krakel II/Photo Researchers, Inc.)

A Chemical Puzzler

Colorful "neon" signs flash in a rainbow of colors, and similar colors are observed when you put salts such as NaCl, $SrCl_2$, and $CaCl_2$ in a fire. What do these observations have in common? What is the origin of the colors?

The colors of beautiful gemstones or of fireworks bursting in the night sky excite all of us. Have you ever wondered how these colors are produced? Part of the answer can be found in this chapter, and this will lead us to answers to still other questions that have intrigued chemists for much of this century. We shall see that these particular colors arise from ions of various salts used in producing fireworks or from metal ions incorporated naturally into the minerals we consider to be gemstones.

We know that chemical elements that exhibit similar properties are found in the same column of the periodic table. But why should this be so? The discovery of the electron, proton, and neutron (Section 2.2) prompted scientists to look for relationships between atomic structure and chemical behavior. As early as 1902 Gilbert N. Lewis (1875–1946) hit upon the idea that electrons in atoms might be arranged in shells, starting close to the nucleus and building outward. Lewis explained the similarity of chemical properties for elements in a given group by assuming that all the elements of that group have the same number of electrons in the outer shell. These are the **valence electrons,** first introduced in Section 3.3.

Lewis's model of the atom raises a number of questions. Where are the electrons located? Do they have different energies? Does any experimental evidence support this model? These questions were the reason for many of the experimental and theoretical studies that began around 1900 and continue to this day. This chapter and the next one outline the important results so far.

7.1 ELECTROMAGNETIC RADIATION

We are all familiar with water waves, and you may also know that some properties of radiation such as visible light can be described with the ideas of wave motion. These ideas came from the experiments of physicists in the 19th century, among them a Scot, James Clerk Maxwell (1831–1879). In 1864, he developed an elegant mathematical theory to describe all forms of radiation in terms of oscillating, or wave-like, electric and magnetic fields in space (Figure 7.1). Hence, radiation, such as light, microwaves, television and radio signals, and x-rays, is collectively called **electromagnetic radiation.**

See the *Saunders Interactive General Chemistry CD-ROM,* Screen 3.1, Electromagnetic Radiation.

James Clerk Maxwell is regarded by many as the greatest physicist between the times of Newton and Einstein.

Wave forms are illustrated with water waves in Figure 7.2. The **wavelength** of a wave is the distance between successive crests, or high points (or between successive troughs or low points). This distance can be given in meters, nanometers, or whatever unit is convenient. The symbol for wavelength is the Greek letter λ (lambda).

Waves can also be characterized by their **frequency,** symbolized by the Greek letter ν (nu). For a wave passing some point in space, the frequency is equal to the number of complete waves or cycles passing the point in that amount of time. Thus, we usually refer to the frequency as the number of cycles that pass per second (Figure 7.2). The unit for frequency is often written as s^{-1} (standing for 1 per second, 1/s or s^{-1}) and is now called the **hertz.**

If you enjoy water sports, you are familiar with the height of waves. In more scientific terms, the maximum height of a wave, as measured from the axis of propagation of the wave, is called the **amplitude.** In Figure 7.2, notice that the wave has zero amplitude at certain intervals along the wave. Points of zero am-

Maxwell developed the theory of electromagnetic radiation, but its existence was proved by the German experimentalist Heinrich Hertz (1857–1894).

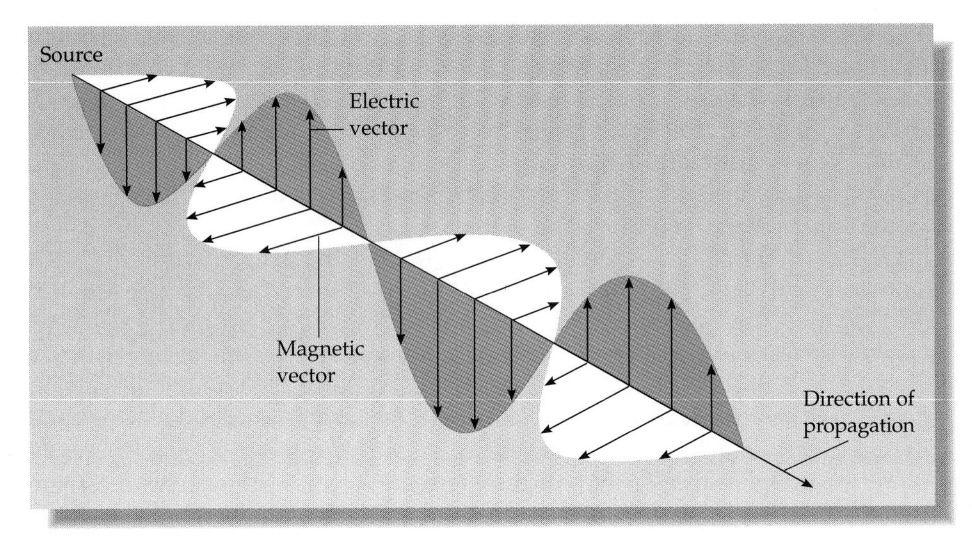

Figure 7.1 Electromagnetic radiation. In the 1860s James Clerk Maxwell developed the currently accepted theory that all forms of radiation are propagated through space as vibrating electric and magnetic fields, the fields being at right angles to one another. Each of the fields is described by a sine wave (because of the mathematical function describing the wave). Such oscillating fields emanate from vibrating charges in a source such as a light bulb or radio antenna.

The concept of nodes is important in understanding atomic orbitals (Section 7.6).

plitude, called **nodes,** always occur at intervals of $\lambda/2$ for waves such as those illustrated in Figure 7.2.

Finally, the speed of a moving wave is an important factor. As an analogy, consider cars in a traffic jam traveling bumper to bumper. If each car is 16 ft long, and if a car passes you every 4 s (that is, the frequency is 1 per 4 seconds, or 1/4 s^{-1}), then the traffic is "moving" at the speed of (16 ft)(1/4 s^{-1}), or

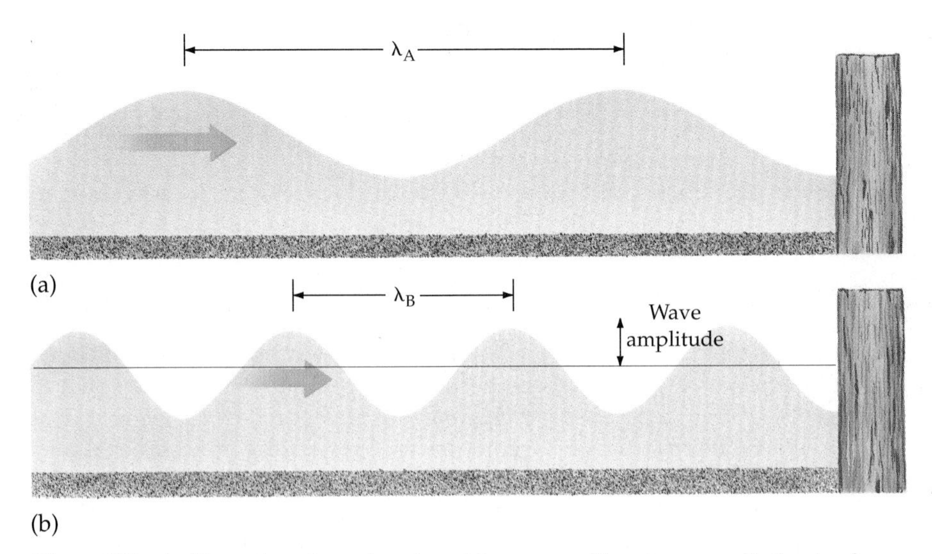

Figure 7.2 An illustration of wavelength and frequency with water waves. Both sets of waves are moving forward toward the post and are traveling with the same velocity. Wave (A) has a longer wavelength than wave (B) ($\lambda_A > \lambda_B$). Wave (A) has a lower frequency than (B) ($\nu_A < \nu_B$), because the number of times per second wave (A) hits the post is less than that of wave (B).

4 ft/s. Multiplying length times frequency gives the speed for any periodic motion, including a wave:

$$\text{Wavelength (m)} \times \text{frequency } (s^{-1}) = \text{velocity } (m \cdot s^{-1})$$

This equation also applies to electromagnetic radiation, where the product of the wavelength and frequency is equal to the speed of light, c:

$$\lambda \cdot \nu = c = 2.99792458 \times 10^8 \text{ m} \cdot \text{s}^{-1} \quad [7.1]$$

The speed of visible light and all other forms of electromagnetic radiation in a vacuum is a constant, c ($= 2.99792458 \times 10^8$ m · s^{-1}; approximately 186,000 miles/s). Given this value, and knowing the wavelength of a light wave, you can readily calculate the frequency, and vice versa. For example, what is the frequency of orange light, which has a wavelength of 625 nm? Because the speed of light is expressed in meters per second, the wavelength in nanometers must be changed to meters before substituting into Equation 7.1 (and use c to four significant figures):

The velocity of light passing through a substance (air, glass, water, etc.) depends on the chemical constitution of the substance and the wavelength of the light. This is the basis for using a glass prism to disperse light and is the explanation for rainbows. Sound velocity is also dependent on the material through which it passes.

$$625 \text{ nm} \cdot \frac{1 \times 10^{-9} \text{ m}}{1 \text{ nm}} = 6.25 \times 10^{-7} \text{ m}$$

$$\nu = \frac{c}{\lambda} = \frac{2.998 \times 10^8 \text{ m} \cdot \text{s}^{-1}}{6.25 \times 10^{-7} \text{ m}} = 4.80 \times 10^{14} \text{ s}^{-1}$$

We are bathed constantly in electromagnetic radiation, including the radiation you can see, visible light. As you know, visible light really consists of a spectrum of colors, ranging from red light at the long-wavelength end of the spectrum to violet light at the short-wavelength end (Figure 7.3). Visible light is,

You can remember the colors of visible light, *in order of decreasing wavelength,* by the famous mnemonic phrase **ROY G BIV,** which stands for red, orange, yellow, green, blue, indigo, and violet.

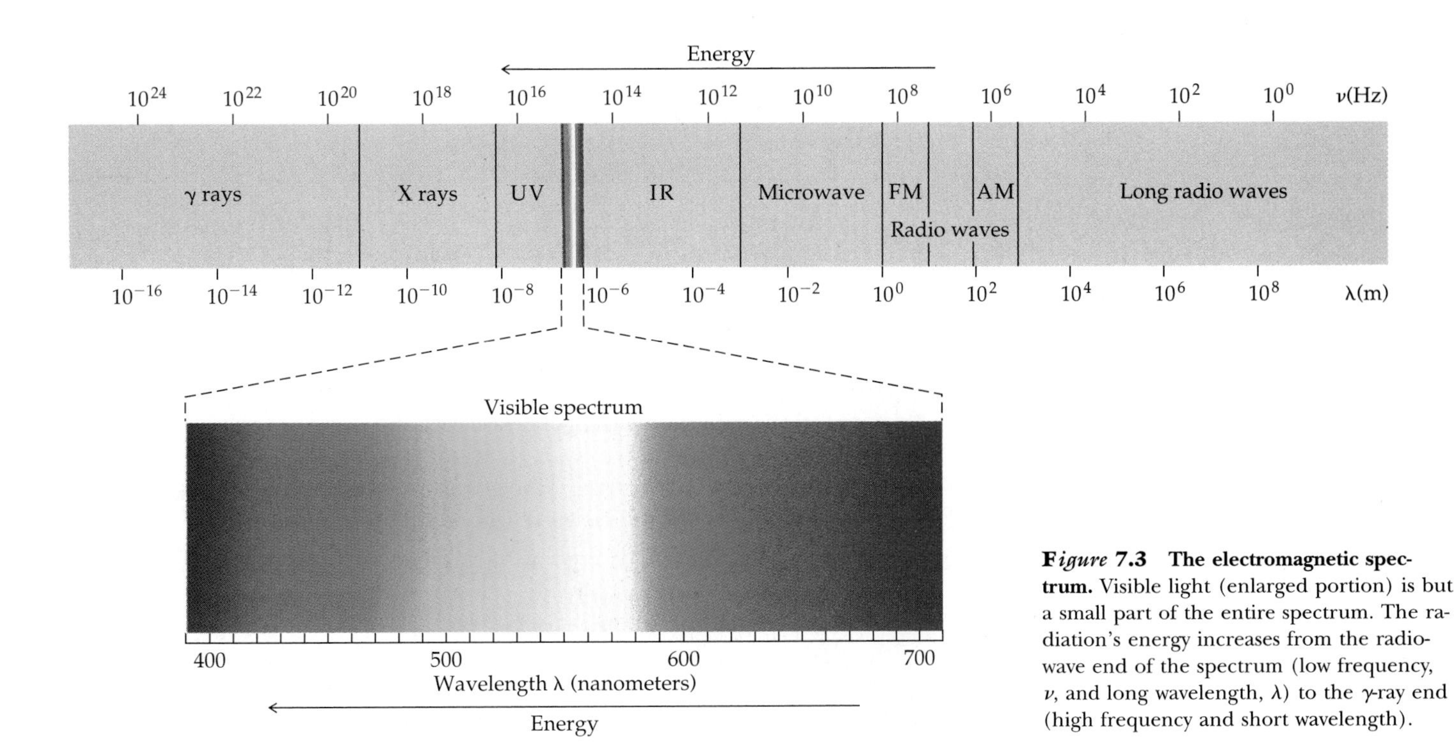

***Figure* 7.3 The electromagnetic spectrum.** Visible light (enlarged portion) is but a small part of the entire spectrum. The radiation's energy increases from the radio-wave end of the spectrum (low frequency, ν, and long wavelength, λ) to the γ-ray end (high frequency and short wavelength).

See the *Saunders Interactive General Chemistry CD-ROM,* Screen 7.4, The Electromagnetic Spectrum.

however, only a small portion of the total electromagnetic spectrum. Ultraviolet (UV) radiation, the radiation that can lead to sunburn, has wavelengths shorter than those of visible light; x-rays and γ rays, the latter emitted in the process of radioactive disintegration of some atoms, have even shorter wavelengths. At longer wavelengths than visible light, we first encounter infrared radiation (IR), the type that is sensed as heat. Longer still is the wavelength of the radiation that is used in a microwave oven and in television and radio transmissions.

Example 7.1 Wavelength-Frequency Conversions

The frequency of the radiation used in all microwave ovens sold in the United States is 2.45 GHz. (The unit GHz stands for "gigahertz;" 1 GHz is a billion cycles per second, or 10^9 s^{-1}.) What is the wavelength (in meters) of this radiation? How much longer or shorter is this than the wavelength of orange light (625 nm)?

Solution The wavelength of microwave radiation in meters can be calculated directly from Equation 7.1:

$$\lambda = \frac{c}{\nu} = \frac{2.998 \times 10^8 \text{ m} \cdot \cancel{\text{s}^{-1}}}{2.45 \times 10^9 \cancel{\text{s}^{-1}}} = 0.122 \text{ m}$$

$$\frac{\lambda \text{ (microwaves)}}{\lambda \text{ (orange light)}} = \frac{0.122 \cancel{\text{m}}}{6.25 \times 10^{-7} \cancel{\text{m}}} = 195{,}000$$

A comparison of the wavelengths of microwave radiation and orange light shows that the microwaves have wavelengths almost 200,000 times longer than the wavelength of orange light.

Exercise 7.1 Radiation, Wavelength, and Frequency

1. Which color in the visible spectrum has the highest frequency? Which has the lowest frequency?
2. Is the frequency of the radiation used in a microwave oven higher or lower than that from your favorite FM radio station (91.7 MHz) (where MHz = megahertz = 10^6 s^{-1})? (see Figure 7.3)
3. Is the wavelength of x-rays longer or shorter than that of ultraviolet light?

The wave motion we have described so far is that of traveling waves. Another type of wave motion, called **standing,** or **stationary, waves,** is relevant to modern atomic theory. If you tie down a string at both ends, as you would the string of a guitar, and pluck it, the string vibrates as a standing wave (Figure 7.4). Several important points should be noted about standing waves:

- A standing wave is characterized by having two or more points of no movement; that is, the wave amplitude is zero at the *nodes.* As with traveling waves, the distance between consecutive nodes is always $\lambda/2$.
- In the first of the vibrations illustrated in Figure 7.4, the distance between the ends of the string, a, is $\lambda/2$. In the second vibration the string length equals one complete wavelength, or $2(\lambda/2)$. In the third, the string length

Figure 7.4 Standing waves. In the first wave, the end-to-end distance is $(1/2)\lambda$, in the second wave it is λ, and in the third wave it is $(3/2)\lambda$.

is $(3/2)\lambda$, or $3(\lambda/2)$. Could the distance between the ends of a standing wave vibration ever be $(3/4)\lambda$, or $3/2(\lambda/2)$? For standing waves, *only certain wavelengths are possible.* Because the ends of a standing wave must be nodes, the only allowed vibrations are those in which $a = n(\lambda/2)$ [where a is the distance from one end, or "boundary," to the other, and n is an integer (1, 2, 3, . . .)]. This is an example of *quantization,* a concept we turn to next.

Exercise 7.2 Standing Waves

The line shown here is 10 cm long.

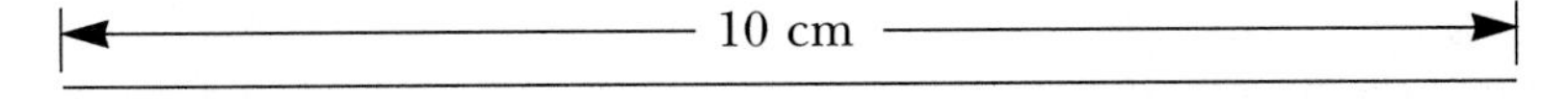

Using this line,

1. Draw a standing wave with one node between the ends. What is the wavelength of this wave?
2. Draw a standing wave with two nodes between the ends. What is its wavelength?
3. If the wavelength of the standing wave is 5 cm, how many waves fit within the boundaries? How many nodes are there?

7.2 PLANCK, EINSTEIN, ENERGY, AND PHOTONS

Planck's Equation

If you heat a piece of metal, it emits electromagnetic radiation, with wavelengths that depend on temperature. At first its color is a dull red. At higher temperatures, the red color brightens, and at still higher temperatures the redness turns to a brilliant white light. For example, the heating element of a toaster becomes "red hot," and the filament of a light bulb glows "white hot."

Your eyes detect the radiation from a piece of heated metal that occurs in the visible region of the electromagnetic spectrum. These are not the only wavelengths of the light emitted by the metal, though. Radiation is also emitted with wavelengths both shorter (in the ultraviolet) and longer (in the infrared) than those of visible light. That is, a spectrum of electromagnetic radiation is emitted (Figure 7.5), with some wavelengths of light more intense than others. As the metal is heated, the maximum in the curve of light intensity versus wavelength is shifted more and more to the ultraviolet region, so the color of the glowing object shifts from red to yellow and finally to white hot.

At the end of the 19th century, scientists were trying to explain the relationship between the intensity and wavelength for radiation given off by heated objects. All attempts were unsuccessful, however. Theories available at the time predicted that the intensity of radiation should increase continuously with decreasing wavelength; that is, there should be no maximum in the intensity versus wavelength graph as shown in Figure 7.5. This perplexing situation became known as the "ultraviolet catastrophe." Clearly, classical physics did not provide a satisfactory explanation, and a new way to look at matter and energy was needed.

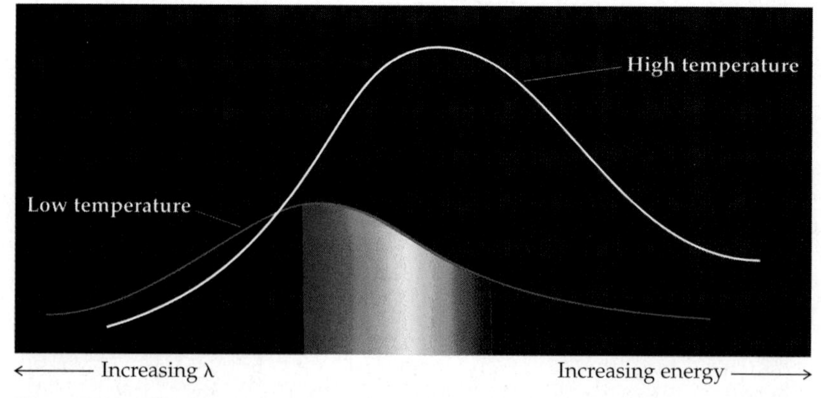

Figure 7.5 The spectrum of the radiation given off by a heated body. The red line is the spectrum for a "red hot" object, since the wavelength of maximum intensity occurs at the wavelength of red light. As the temperature of the object increases, the color of the object becomes more orange and then yellow, as the wavelength of the highest intensity radiation moves toward higher frequencies (toward the ultraviolet). At very high temperatures, the object is "white hot," and there is a comparable intensity of radiation at all wavelengths in the visible spectrum. (Stars are often referred to as "red giants" or "white dwarfs," a reference to their temperatures and relative sizes.)

Historical Perspective

Max Planck (1858–1947)

Max Karl Ernst Ludwig Planck was raised in Munich, Germany, where his father was an important professor at the University. When still in his teens Planck decided to become a physicist in spite of the advice of the head of the physics department at Munich that "The important discoveries [in physics] have been made. It is hardly worth entering physics anymore." Fortunately, Planck did not take this advice, and he worked for a while at the University of Munich before going to Berlin to study thermodynamics.

Max Planck (*E. F. Smith Collection/VanPelt Library/U. of Pennsylvania*)

His interest in thermodynamics led him eventually to consider the ultraviolet catastrophe and to his revolutionary hypothesis. The discovery was announced two weeks before Christmas in 1900, and he was awarded the Nobel Prize in physics in 1918 for this work. Einstein later said it was a longing to find harmony and order in nature, a "hunger in his soul," that spurred Planck on.

In 1900, the German physicist, Max Planck (1858-1947), offered an explanation. Following classical theory, he assumed that vibrating atoms in a heated object give rise to the emitted electromagnetic radiation. He also introduced an important new assumption, however, that vibrations were quantized; that is, only vibrations of specific energies could occur. As you will see as we continue to develop atomic theory, *quantization* is a key concept in modern theory.

To relate the energy and frequency of radiaton, Planck also introduced a very important equation, now called Planck's equation, that relates energy and

(a)

(b)

Infrared radiation has longer wavelengths than light of the visible region of the spectrum. (a) A catalytic converter is used to make wood stoves burn more efficiently and evolve fewer pollutants. It glows red hot when heated, and infrared radiation is experienced as heat. (b) A photo of the San Francisco Bay area taken from a satellite. The film responds to wavelengths in the infrared region. (a: *C. D. Winters;* b: *Earth Satellite Corp./Photo Researchers, Inc.*)

See the *Saunders Interactive General Chemistry CD-ROM,* Screen 7.5, Planck's Equation.

frequency of radiation:

$$E = h \cdot \nu \qquad [7.2]$$

Energy of radiation = (Planck's constant) · (frequency of radiation)

The proportionality constant h is called **Planck's constant,** in his honor. It has the value $6.6260755 \times 10^{-34}$ J · s.

Now, assume as Planck did that there must be a *distribution* of vibrations of atoms in an object—some atoms are vibrating at a high frequency, some at a low frequency, but most with some intermediate frequency. The few atoms with high-frequency vibrations are responsible for some of the light, as are those few with low-frequency vibrations. Most of the light must come, however, from the majority of the atoms that have intermediate vibrational frequencies. That is, a spectrum of light should be emitted with a maximum intensity at some wavelength, in accord with experiment. The ultraviolet catastrophe was solved.

The switch that opens automatic doors is frequently a photoelectric device. *(C. D. Winters)*

Einstein and the Photoelectric Effect

As almost always occurs, the explanation of a fundamental phenomenon—such as the spectrum of light from a hot object—leads to another fundamental discovery. A few years after Planck's work, Albert Einstein (1879-1955) incorporated Planck's ideas into an explanation of the photoelectric effect.

The **photoelectric effect** occurs when light strikes the surface of a metal and electrons are ejected (Figure 7.6). The electrons ejected from the photocathode by the light move to the positively charged anode, and current flows in the cell. Because light causes current to flow, a photoelectric cell therefore acts as a light-activated switch. The automatic door openers in stores and elevators often work this way.

Experiments with photoelectric cells show that electrons are ejected from the surface only if the frequency of the light is high enough. If lower frequency light is used, no effect is observed, regardless of the light's intensity (brightness). If the frequency is above the minimum, however, increasing the light intensity causes a higher current to flow because more and more electrons are ejected.

Einstein decided the experimental observations could be explained by combining Planck's relation between energy and frequency of radiation with a new idea: that light could be described not only as having wave-like properties but also as having quantized, particle-like properties. Einstein assumed these massless "particles," now called **photons,** are packets of energy. The energy of each photon is proportional to the frequency of the radiation, as given by Planck's law (Equation 7.2).

The photoelectric effect can be explained readily using Einstein's proposal. It is easy to imagine that a high-energy particle would have to bump into an atom to cause the atom to lose an electron. It is also reasonable to accept the idea that an electron can be torn away from the atom only if some minimum amount of energy is used. If electromagnetic radiation can also be thought of as a stream of photons, as Einstein said, then the greater the intensity of light, the more photons there are. It then follows that the atoms of a metal surface do not lose electrons when the metal is bombarded by millions of photons if no individual photon has enough energy to remove an electron from an atom. Once the critical minimum energy (i.e., minimum light frequency) is exceeded, the

Einstein received the Nobel Prize in physics in 1921 for "services to theoretical physics, especially for the discovery of the law of the photoelectric effect."

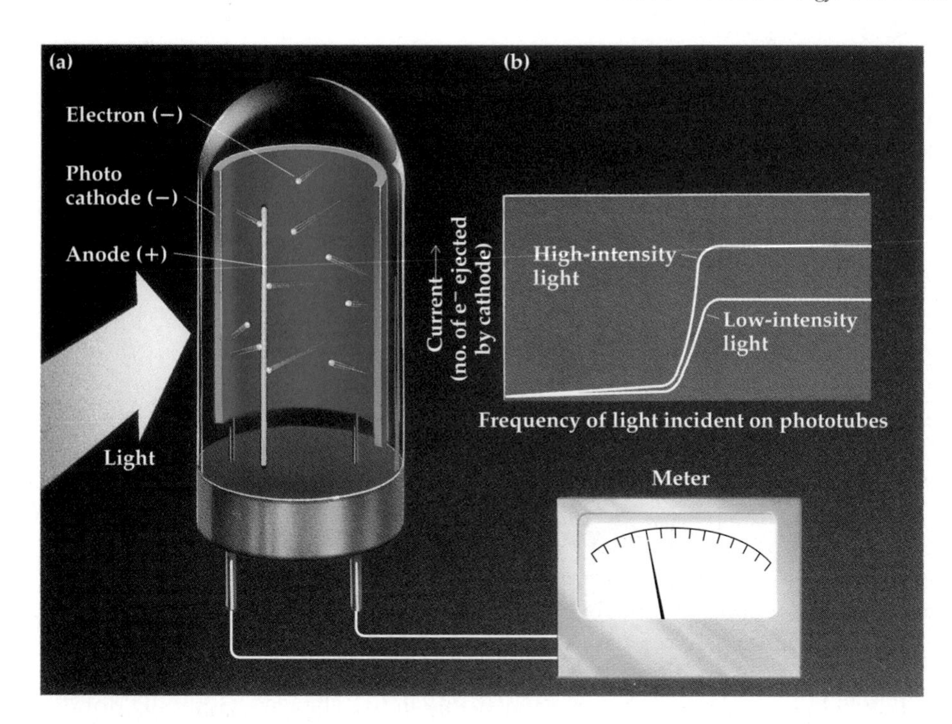

***Figure* 7.6 Photoelectric effect.** (a) A photocell operates by the photoelectric effect. The main part of the cell is a photosensitive cathode. This is a material, usually a metal, that ejects electrons if struck by photons of light of sufficient energy. The ejected electrons move to the anode and a current flows in the cell. Such a device can be used as a switch in electric circuits. (b) No current is observed until the critical frequency is reached. If higher intensity light is used, that is, light with a higher photon density, the only effect is to cause more electrons to be released from the surface; the onset of current is observed at the same frequency as with lower intensity light. When light of higher frequency than the minimum is used, the excess energy of the photon simply makes the electron escape the atom with greater velocity.

energy content of each photon is sufficient to displace an electron from a metal atom. The greater the number of photons with this energy that strike the surface, the greater the number of electrons dislodged. Thus, the connection is made between light intensity and the number of electrons ejected.

Cesium is often used in "electric eye" devices such as automatic door openers because all visible wavelengths of light cause its atoms to emit electrons.

Energy and Chemistry: Using Planck's Equation

Compact disc players use lasers that emit red light with a wavelength of 685 nm. What is the energy of one photon of this light? What is the energy of a mole of photons of the light? To answer this, first convert the wavelength to the frequency of the radiation and then use this to calculate the energy per photon. Finally, the energy of a mole of photons is obtained by multiplying the energy per photon by Avogadro's number:

$$\lambda,\text{ nm} \xrightarrow[\times \frac{10^{-9}\text{ m}}{\text{nm}}]{} \lambda,\text{ m} \xrightarrow[\nu = c/\lambda]{} \nu,\text{ s}^{-1} \xrightarrow[E = h\cdot\nu]{} E,\text{ J/photon} \xrightarrow[\times\text{ Avogadro's number}]{} E,\text{ J/mol}$$

$$685\ \text{nm}\ (10^{-9}\ \text{m/nm}) = 6.85 \times 10^{-7}\ \text{m}$$

$$\nu = \frac{2.998 \times 10^{8}\ \text{m}\cdot\text{s}^{-1}}{6.85 \times 10^{-7}\ \text{m}} = 4.38 \times 10^{14}\ \text{s}^{-1}$$

$$\begin{aligned} E &= h\nu \\ &= (6.626 \times 10^{-34}\ \text{J}\cdot\text{s/photon})(4.38 \times 10^{14}\ \text{s}^{-1}) = 2.90 \times 10^{-19}\ \text{J/photon} \\ &= (2.90 \times 10^{-19}\ \text{J/photon})(6.022 \times 10^{23}\ \text{photons/mol}) = 1.75 \times 10^{5}\ \text{J/mol} \end{aligned}$$

The energy of a mole of photons of red light is equivalent to 175 kJ, and a mole of photons of blue light (λ = 400 nm) has an energy of about 300 kJ. These energies are in a range that can affect the bonds between atoms. It should not be surprising therefore that light can cause chemical processes to occur. For example, you may have seen cases in which sunlight causes paint or dye to fade or cloth to decompose.

The previous calculation shows that, as the frequency of radiation increases, the energy of the radiation also increases (see Figure 7.3). Similarly, the energy increases as the wavelength of radiation decreases:

Energy (E) increases as frequency (ν) increases.

$$E = h \cdot \nu = \frac{h \cdot c}{\lambda}$$

Energy (E) increases as wavelength (λ) decreases.

Therefore, photons of ultraviolet radiation—with wavelengths shorter than those of visible light—have higher energy than visible light. Because visible light has enough energy to affect bonds between atoms, it is obvious that ultraviolet light does as well. That is the reason that ultraviolet radiation can cause a sunburn. In contrast, photons of infrared radiation—with wavelengths longer than those of visible light—have lower energy than visible light. They are generally not energetic enough to cause chemical reactions, but they can affect the vibrations of molecules. We sense infrared radiation as heat, such as the heat given off by a glowing burner on an electric stove.

Example 7.2 Calculating Photon Energies

Compare the energy of a mole of photons of red light from a laser (175 kJ/mol) with the energy of a mole of photons of x-radiation having a wavelength of 2.36 nm. Which has the greater energy? By what factor is one greater than the other?

Solution This calculation follows the scheme outlined in the text. First, we express the wavelength in meters:

$$2.36\ \text{nm} \cdot (1 \times 10^{-9}\ \text{m/nm}) = 2.36 \times 10^{-9}\ \text{m}$$

Next, calculate the frequency of the radiation:

$$\text{Frequency}\ (\nu) = \frac{c}{\lambda} = \frac{2.998 \times 10^{8}\ \text{m} \cdot \text{s}^{-1}}{2.36 \times 10^{-9}\ \text{m}} = 1.27 \times 10^{17}\ \text{s}^{-1}$$

Now the energy can be calculated from Planck's equation,

$$E = h\nu$$
$$= (6.626 \times 10^{-34}\ \text{J} \cdot \text{s/photon})(1.27 \times 10^{17}\ \text{s}^{-1}) = 8.42 \times 10^{-17}\ \text{J/photon}$$

(Alternatively, the wavelength in meters can be converted directly to energy using the equation $E = hc/\lambda$.) Finally, we can calculate the energy per mole of photons.

$$E = (8.42 \times 10^{-17}\ \text{J/photon}) \cdot (6.022 \times 10^{23}\ \text{photons/mol}) = 5.07 \times 10^{7}\ \text{J/mol}$$

The energy of a mole of x-ray photons, equivalent to 5.07×10^4 kJ/mol, is much larger—by a factor of 290—than the energy of a mole of photons of red light.

$$\frac{E \text{ of x-ray photons}}{E \text{ of red light photons}} = \frac{5.07 \times 10^4 \text{ kJ/mol}}{175 \text{ kJ/mol}} = 290.$$

Exercise 7.3 Photon Energies

Compare the energy of a mole of photons of blue light (4.00×10^2 nm) with the energy of a mole of photons of microwave radiation having a frequency of 2.45 GHz (1 GHz = 10^9 s^{-1}). Which has the greater energy? By what factor is one greater than the other?

7.3 ATOMIC LINE SPECTRA AND NIELS BOHR

Atomic Line Spectra

The final piece of experimental information that played a major role in developing the modern view of atomic structure is the observation of the light emitted by atoms after they absorb extra energy. The spectrum of white light, such as that from an incandescent light or from the sun, is the rainbow display of colors shown in Figure 7.7. Such a spectrum, consisting of light of all wavelengths, is called a **continuous spectrum.**

See the *Saunders Interactive General Chemistry CD-ROM,* Screen 7.6, Atomic Line Spectra.

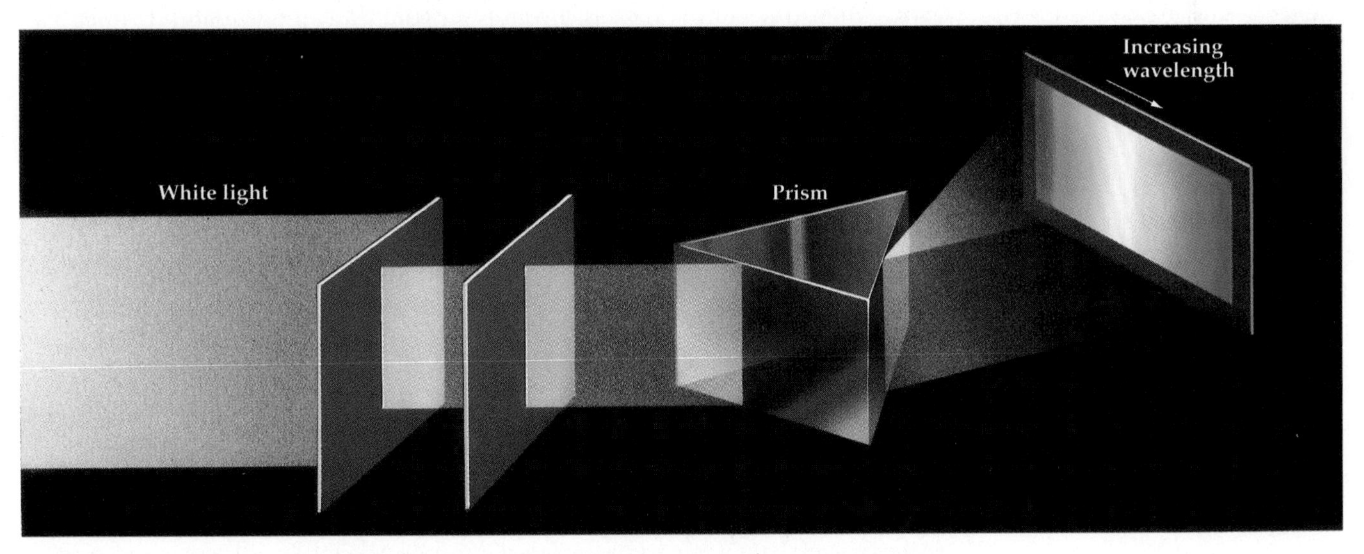

Figure **7.7 A spectrum of white light, produced by refraction in a prism.** The light is observed with a spectroscope or spectrometer by passing the light through a narrow slit to isolate a thin beam, or line, of light. The beam is then passed through a device (a prism or, in modern instruments, a diffraction grating) that separates the light into its component wavelengths. See also Figure 7.3.

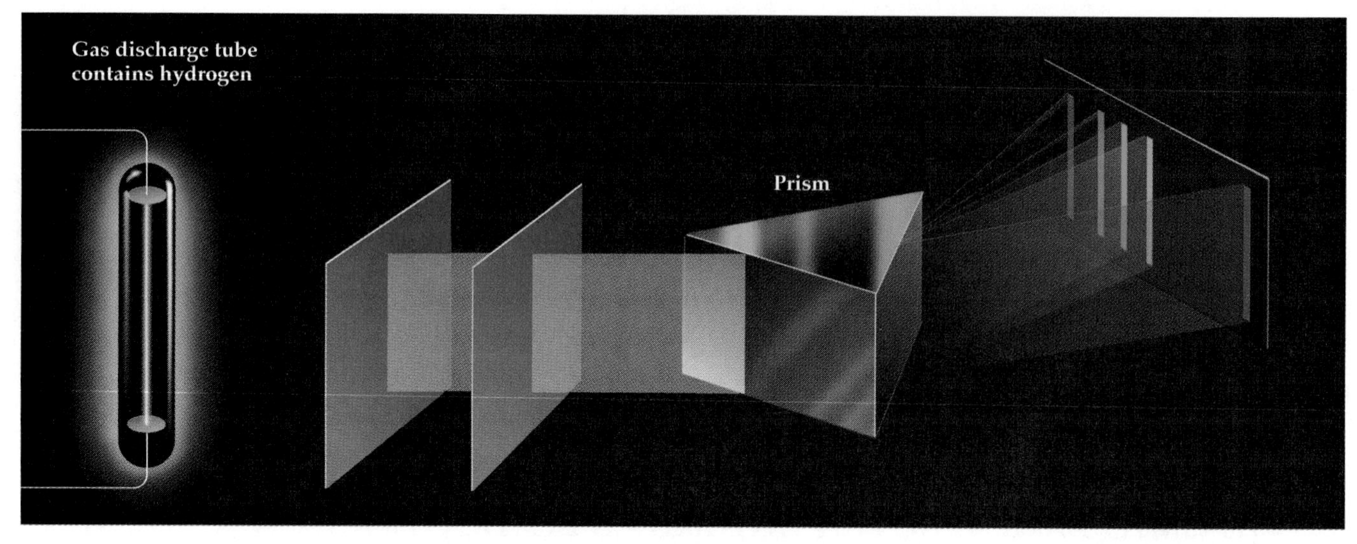

Figure 7.8 The line emission spectrum of hydrogen. The emitted light is passed through a series of slits to create a narrow beam of light, which is then separated into its component wavelengths by a prism. A photographic plate or an instrument detects the separate wavelengths as individual lines. Hence, the name "line spectrum" for the light emitted by a glowing gas.

Gas discharge signs are often referred to as neon signs even though many do not contain that gas. Colors are emitted by excited atoms of noble gases in general: neon, reddish orange; argon, blue; helium, yellowish white. A helium-argon mixture emits an orange light, and a neon-argon mixture gives a dark lavender color. Mercury vapor, used in fluorescent lights, is also added to gas mixtures to obtain a wide range of colors.

If a high voltage is applied to atoms of an element in the gas phase at low pressure, the atoms absorb energy and are said to be "excited." The excited atoms emit light. This light is different, however, from that emitted by a heated object. Instead of producing a continuous spectrum of wavelengths, excited atoms in the gas phase emit only certain wavelengths of light. We know this because when this light is passed through a prism, only a few colored "lines" are seen. This is called a **line emission spectrum** (Figure 7.8). A familiar example of this is the light from a neon advertising sign, in which excited neon atoms emit orange-red light (*Chemical Puzzler*, page 292).

Spectra of visible light emitted by excited atoms of hydrogen, mercury, and neon are shown in Figure 7.9. Every element has a unique line spectrum. Indeed, the characteristic lines in the emission spectrum of an element can be used in chemical analysis, especially in metallurgy, both to identify the element and to determine how much of it is present.

One of the goals of scientists in the late 19th century was to explain why gaseous atoms emitted light of only certain frequencies. Attempts were made to find a mathematical relationship among the observed frequencies. It is always significant if experimental data can be related by a simple equation because a "regular" pattern of information implies a logical explanation for the observations. The first step in this direction came from Johann Balmer (1825–1898) and later Johannes Rydberg (1854-1919). They developed an equation—now called the **Rydberg equation**—from which it was possible to calculate the wavelength of the red, green, and blue lines in the visible emission spectrum of hydrogen atoms (Figure 7.9).

$$\frac{1}{\lambda} = R\left(\frac{1}{2^2} - \frac{1}{n^2}\right) \qquad n > 2 \qquad [7.3]$$

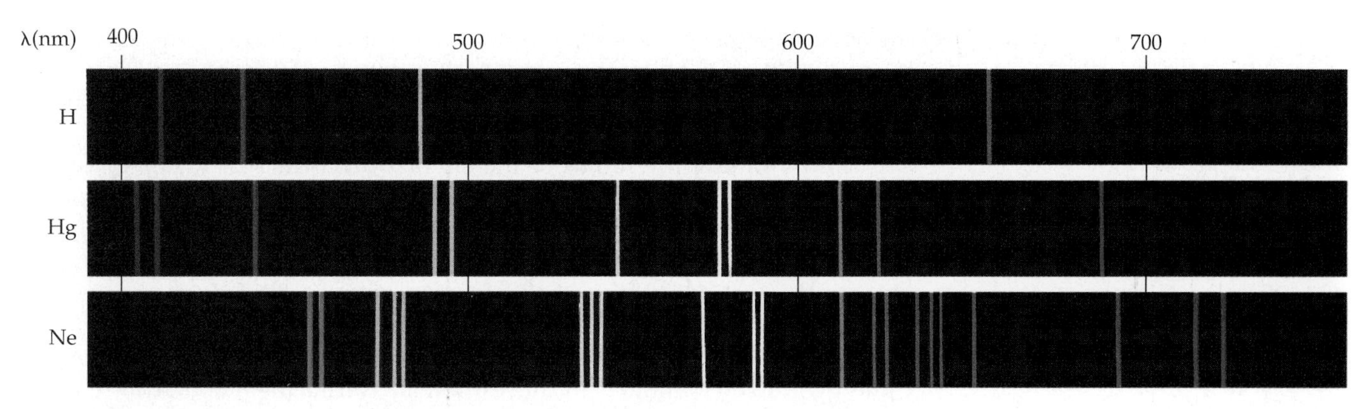

Figure 7.9 Line emission spectra of hydrogen, mercury, and neon. Excited gaseous elements produce characteristic spectra that can be used to identify the element as well as to determine how much element is present in a sample.

In this equation n is an integer, and R, now called the **Rydberg constant,** has the value 1.0974×10^7 m^{-1}. If $n = 3$, the wavelength of the red line in the hydrogen spectrum is obtained (6.563×10^{-7} m, or 656.3 nm). If $n = 4$, the wavelength for the green line is obtained, and $n = 5$ gives the blue line. This group of visible lines in the spectrum of hydrogen atoms (and others for which $n = 6$, 7, 8, and so on) is now called the **Balmer series.**

The symbol m^{-1} means (1/meters), just the same as s^{-1} means 1/seconds.

The Bohr Model of the Hydrogen Atom

Niels Bohr, a Danish physicist, provided the first connection between the spectra of excited atoms and the quantum ideas of Planck and Einstein. From Rutherford's work (Section 2.2), it was known that electrons are arranged in space outside of the atom's nucleus. For Bohr the simplest model of a hydrogen atom was a planetary model, that is, that the electron moved in a circular orbit around the nucleus just as the planets revolve about the sun. In proposing this, however, he had to contradict the laws of classical physics. According to the theories at the time, a charged electron moving in the positive electric field of the nucleus should lose energy. Eventually the electron would have to crash into the nucleus, much in the same way a satellite eventually crashes into the earth as the satellite loses energy by "rubbing up against" the earth's atmosphere. This is clearly not the case; if it were so, matter would eventually be destroyed.

See the *Saunders Interactive General Chemistry CD-ROM,* Screen 7.7, Bohr's Model of the Hydrogen Atom.

To solve the contradiction with the laws of classical physics, Bohr introduced the notion that the electron orbiting the nucleus could occupy only certain orbits or energy levels in which it was stable. That is, the energy of the electron in the atom was "quantized." By combining this quantization postulate with the laws of motion from classical physics, Bohr showed that the potential energy possessed by the single electron in the nth orbit of the H atom is given by the simple equation

$$\text{Energy of the } n\text{th level} = E_n = -\frac{Rhc}{n^2} \qquad [7.4]$$

The *potential energy* of the electron is calculated by Equation 7.4.

R is a proportionality constant, h is Planck's constant, and c is the velocity of light. Each allowed orbit was assigned a value of n, a unitless integer having values of 1, 2, 3, and so on (but not fractional values). This integer is now known as the **principal quantum number** for the electron.

In **Bohr's model** the radius of the circular orbits increases as n increases. As illustrated by Example 7.3, another consequence of the model and Equation 7.4 is that the energy of the electron becomes less negative as the value of n increases. The orbit of lowest or most negative energy, with $n = 1$, is closest to the nucleus, and the electron of the hydrogen atom is normally in this energy level. An atom with its electrons in the lowest possible energy levels is said to be in its **ground state.**

Energy must be supplied to move the electron farther away from the nucleus because the positive nucleus and the negative electron attract each other. When the electron of a hydrogen atom occupies an orbit with n greater than 1, the atom has more energy than in its ground state (the energy is less negative) and is said to be in an **excited state.** The energies of the ground and several excited states are calculated in Example 7.3.

Example 7.3 Energies of the Ground and Excited States of the H Atom

Use Equation 7.4 to calculate the energies of the $n = 1$ and $n = 2$ states of the hydrogen atom in joules per atom and in kilojoules per mole. The values needed to use this equation are $R = 1.097 \times 10^7\ \text{m}^{-1}$, $h = 6.626 \times 10^{-34}\ \text{J}\cdot\text{s}$, and $c = 2.998 \times 10^8\ \text{m/s}$.

Take note of the values of Rhc (2.179×10^{-18} J/atom and 1312 J/mol). We will use these in problems later in the chapter.

Solution When $n = 1$, the energy of an electron in a single H atom is

$$E_1 = -\frac{Rhc}{n^2} = -\frac{Rhc}{1^2} = -Rhc$$

$$= -Rhc = (1.097 \times 10^7\ \text{m}^{-1})(6.626 \times 10^{-34}\ \text{J}\cdot\text{s})(2.998 \times 10^8\ \text{m/s})$$

$$= -2.179 \times 10^{-18}\ \text{J/atom}$$

When $n = 2$, the energy is

$$E_2 = -\frac{Rhc}{2^2} = -\frac{E_1}{4} = -\frac{2.179 \times 10^{-18}\ \text{J/atom}}{4} = -5.448 \times 10^{-19}\ \text{J/atom}$$

The conversion from joules per atom to kilojoules per mole is done using Avogadro's number and the relation 1 kJ = 1000 J.

$$E_1 = \left(\frac{-2.179 \times 10^{-18}\ \text{J}}{\text{atom}}\right)\left(\frac{6.022 \times 10^{23}\ \text{atoms}}{\text{mol}}\right)\left(\frac{1\ \text{kJ}}{1000\ \text{J}}\right) = -1312\ \text{kJ/mol}$$

When chemists say "the energy is lower," they mean the value of E is more negative.

Finally, because $E_2 = E_1/4$, we calculate E_2 to be -328.0 kJ/mol.

Notice that the calculated energies are both negative, with E_1 more negative than E_2. This arises because Bohr's equation reflects the fact that the energy of attraction between oppositely charged bodies (an electron and an atomic nucleus) depends on their charge and the distance between them. From Coulomb's law (Section 3.4), we know that the closer the electron is to the nucleus, the greater the force of attraction. As a result the value of E is more negative as the distance becomes smaller, and chemists or physicists say that the energy is therefore lower (meaning more negative).

Exercise 7.4 Electron Energies

Calculate the energy of the $n = 3$ state of the H atom in (a) joules per atom and (b) kilojoules per mole.

You can think of the energy levels in the Bohr model as the rungs of a ladder climbing out of the basement of an "atomic building" (where the energy of the H atom is -2.18×10^{-18} J/atom; see Example 7.3) to the ground level (where the energy is 0) (Figure 7.10). Each step represents a quantized energy level; as you climb the ladder, you can stop on any rung but not between them. Unlike the rungs of a real ladder, however, Bohr's energy levels get closer and closer together as n increases.

A major assumption of Bohr's theory was that an electron in an atom would remain in its lowest energy level unless disturbed. Energy is absorbed or evolved if the electron changes from one energy level to another, and it is this idea that allowed Bohr to explain the spectra of excited gases.

When the H atom electron has $n = 1$, and so is in its ground state, the energy is a large negative value. As we "climb the ladder" to the $n = 2$ level, the electron is less strongly attracted to the nucleus, and the energy of an $n = 2$ electron is less negative. Therefore, to move an electron in the $n = 1$ state to the $n = 2$ state, the atom must absorb energy, just as energy must be expended in climbing a ladder. Scientists say that the electron must be excited (Figure 7.11).

Using Bohr's equation we can calculate the amount of energy required to carry the H atom from the ground state to its first excited state ($n = 2$). As you learned in Chapter 6, the difference in energy between two states is

$$\Delta E = E_{\text{final state}} - E_{\text{initial state}}$$

Because E_{final} has $n = 2$, and E_{initial} has $n = 1$, ΔE is

$$\Delta E = E_2 - E_1 = \left(-\frac{Rhc}{2^2}\right) - \left(-\frac{Rhc}{1^2}\right)$$

$$= \left(\frac{3}{4}\right)Rhc = 0.75\ Rhc = 0.75\ (1312\ \text{kJ/mol})$$

$$= 984\ \text{kJ/mol of H atoms}$$

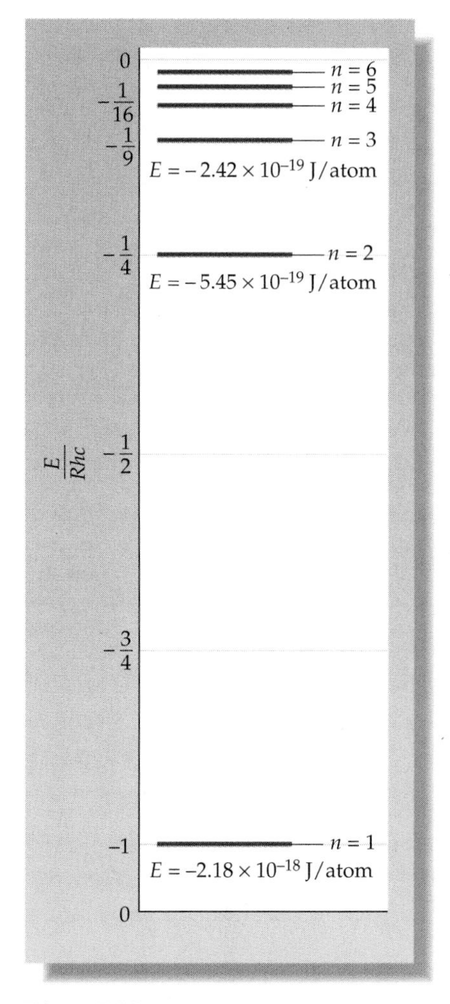

Figure 7.10 Energy levels in the Bohr model. The energies of the electron in a hydrogen atom depend on the value of the principal quantum number n ($E_n = -Rhc/n^2$). Energies are given in joules per atom. Notice that the difference between successive energy states becomes smaller as n becomes larger.

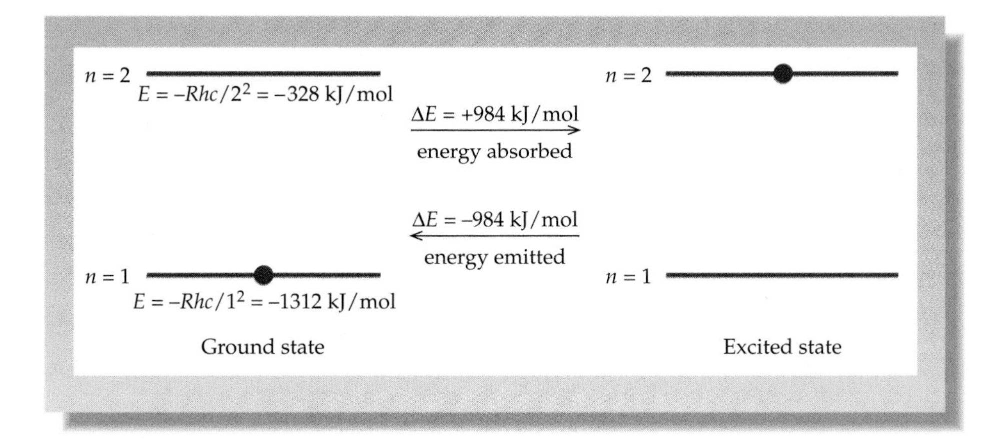

Figure 7.11 Absorption of energy by the atom as the electron moves to an excited state. Absorption ($\Delta E > 0$) and emission ($\Delta E < 0$) of energy by the electron (represented by •) in the H atom moving from the $n = 1$ state (the ground state) to the $n = 2$ state (the excited state) and back again.

Historical Perspective

Niels Bohr (1885–1962)

Niels Bohr was born in Copenhagen, Denmark. He earned a Ph.D. in physics in Copenhagen in 1911 and then went to work first with J. J. Thomson in Cambridge, England, and later with Ernest Rutherford in Manchester, England.* It was there that he began to develop his theory of atomic structure and his explanation of atomic spectra. (For this work he received the Nobel Prize in physics in 1922.) After working with Rutherford for a very short time, Bohr returned to Copenhagen, where he eventually became the director of the Institute for Theoretical Physics. Many young physicists carried on their work in this Institute, and seven of them later received Nobel Prizes for their studies in chemistry and physics. Among them were such well-known scientists as Werner Heisenberg, Wolfgang Pauli, and Linus Pauling. Element 107 was recently named bohrium (Bh) in honor of Bohr and his work.

Niels Bohr *(E. F. Smith Collection/Van Pelt Library/University of Pennsylvania)*

*Rutherford's contributions to chemistry and physics were discussed in Section 2.2. Bohr's life and his work are described by B. L. Haendler: "Presenting the Bohr atom." *Journal of Chemical Education,* Vol. 59, p. 372, 1982; R. Moore: *Niels Bohr,* New York, Alfred Knopf, Inc., 1966; and A. Pais: *Niels Bohr's Times,* New York, Oxford University Press, 1991.

where we used the energy of the $n = 1$ state, as calculated in Example 7.3, for the value of $-Rhc$. The amount of energy that must be absorbed by the atom so that an electron can move from the first to the second energy state is $0.75Rhc$, no more and no less. If $0.7Rhc$ or $0.8Rhc$ is provided, no transition between states is possible. *Energy levels in the H atom are quantized,* with the consequence that only certain amounts of energy may be absorbed or emitted.

Moving an electron from a state of low n to one of higher n is an endothermic process; energy is absorbed, and the sign of the value of ΔE is positive. The opposite process, an electron "falling" from a level of higher n to one of lower n, therefore emits energy. For example, for a transition from $n = 2$ to $n = 1$,

$$\begin{aligned}\Delta E &= E_{\text{final state}} - E_{\text{initial state}} \\ &= E_1 - E_2 = \left(-\frac{Rhc}{1^2}\right) - \left(-\frac{Rhc}{2^2}\right) \\ &= -\left(\frac{3}{4}\right) Rhc\end{aligned}$$

The negative sign indicates energy is evolved; that is, 984 kJ must be *evolved* or *emitted* per mole of H atoms.

Depending on how much energy is added to a collection of H atoms, some atoms have their electrons excited from the $n = 1$ to the $n = 2$ or 3 or higher states. After absorbing energy, these electrons naturally move back down to lower levels (either directly or in a series of steps to $n = 1$) and release the energy the atom originally absorbed. That is, they *emit energy* in the process, and the energy is observed as light. *This is the source of the lines observed in the emission spectrum of H atoms,* and the same basic explanation holds for spectra of atoms of other elements.

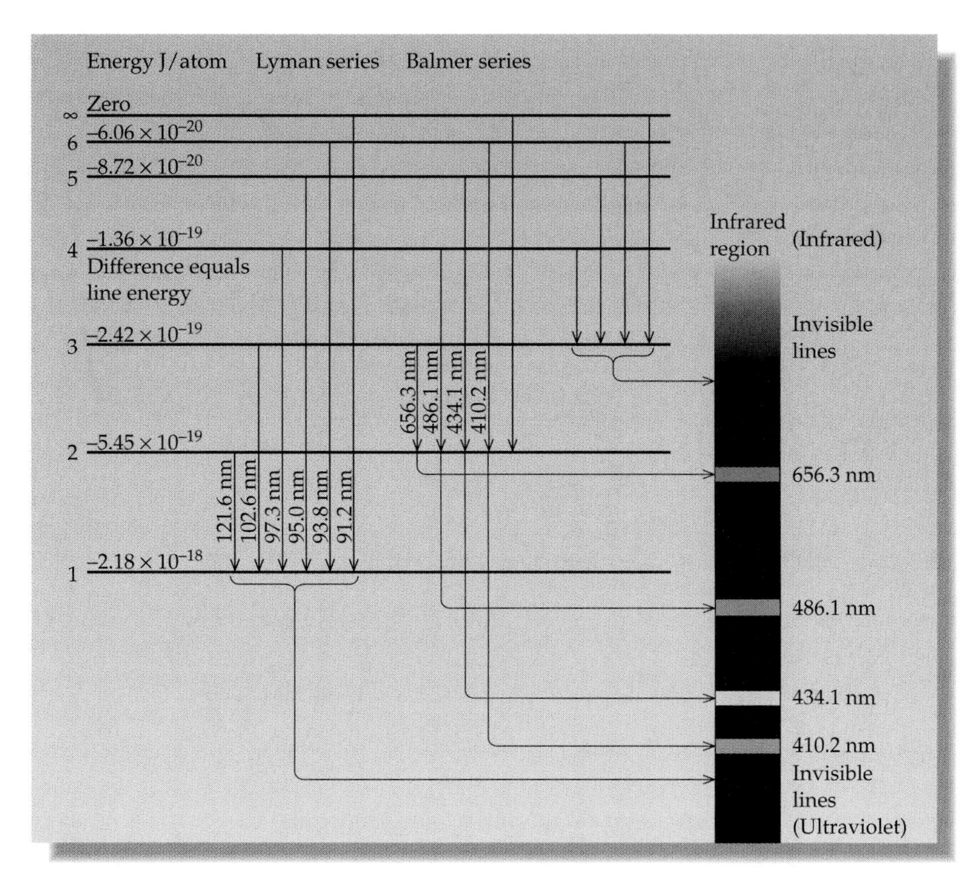

Figure 7.12 Some of the electronic transitions that can occur in an excited H atom. The lines in the ultraviolet region result from transitions to the $n = 1$ level. (This series of lines is called the Lyman series.) Transitions from levels with values of n greater than 2 to $n = 2$ occur in the visible region (Balmer series; see Figure 7.8). Lines in the infrared region result from transitions from levels with n greater than 3 or 4 to the $n = 3$ or 4 levels. (Only the series ending at $n = 3$ is illustrated.)

For hydrogen, the series of emission lines having energies in the ultraviolet region (called the **Lyman series,** Figure 7.12) comes from electrons moving from states with n greater than 1 to that with $n = 1$. The series of lines that have energies in the visible region—the Balmer series—arises from electrons moving from states with $n = 3$ or greater to the lower state with $n = 2$.

In summary, we now recognize that *the origin of atomic spectra is the movement of electrons between quantized energy states.* If an electron is excited from a lower energy state to a higher one, then energy is absorbed. On the other hand, if an electron moves from a higher energy state to a lower one, energy is emitted. If the energy is emitted as electromagnetic radiation, an emission line is observed. The energy of a given emission line in excited hydrogen atoms is

$$\Delta E = E_{\text{final}} - E_{\text{initial}}$$

$$= -Rhc\left(\frac{1}{n^2_{\text{final}}} - \frac{1}{n^2_{\text{initial}}}\right) \quad [7.5]$$

where Rhc is 1312 kJ/mol.

When Bohr's paper describing his ideas was first published in 1913, Einstein declared it to be "one of the great discoveries."

Bohr was able to use his model of the atom to calculate the wavelengths of the lines in the hydrogen spectrum, and there is excellent agreement between the calculated and experimental values. He had tied the unseen (the interior of the atom) to the seen (the observable lines in the hydrogen spectrum)—a fantastic achievement! In addition, he introduced the concept of energy quantization in describing atomic structure, a concept that is still an important part of modern science.

As mentioned previously, agreement between theory and experiment is taken as evidence that the theoretical model is valid. It soon became apparent, however, that a flaw existed in Bohr's theory. It explained only the spectrum of H atoms and of other systems having one electron (such as He^+). Furthermore, *the idea that the electron moves about the nucleus with a path of fixed radius, like that of the planets about the sun, is now thought to be very naive.*

Example 7.4 Energies of Emission Lines for Excited Atoms

Calculate the wavelength of the green line in the visible spectrum of excited H atoms using the Bohr theory.

Solution The green line is the second most energetic line in the visible spectrum of hydrogen (Figure 7.12) and arises from electrons moving from $n = 4$ to $n = 2$. Using Equation 7.5 where $n_{\text{final}} = 2$ and $n_{\text{initial}} = 4$, we have

$$\Delta E = -Rhc\left(\frac{1}{2^2} - \frac{1}{4^2}\right) = -Rhc(0.1875)$$

In the preceding text, we found that Rhc is 1312 kJ/mol, so the $n = 4$ to $n = 2$ transition involves an energy change of

$$\Delta E = -(1312 \text{ kJ/mol})(0.1875) = -246.0 \text{ kJ/mol}$$

The wavelength can now be calculated from the equation $E_{\text{photon}} = h\nu = hc/\lambda$:

$$\lambda = \frac{hc}{E_{\text{photon}}} = \frac{(6.626 \times 10^{-34} \text{ J} \cdot \text{s/photon})(6.022 \times 10^{23} \text{ photons/mol})(2.998 \times 10^{8} \text{ m} \cdot \text{s}^{-1})}{(246.0 \text{ kJ/mol})(10^3 \text{ J/kJ})}$$

$$= 4.863 \times 10^{-7} \text{ m}$$

$$= 4.863 \times 10^{-7} \text{ m } (10^9 \text{ nm/m}) = 486.3 \text{ nm}$$

The experimental value is 486.1 nm (see Figure 7.12). This represents excellent agreement between experiment and theory.

Exercise 7.5 Energy of an Atomic Spectral Line

The Lyman series of spectral lines for the H atom occurs in the ultraviolet region. They arise from transitions from higher levels down to $n = 1$. Calculate the frequency and wavelength of the least energetic line in this series.

A CLOSER LOOK — *Experimental Evidence for Bohr's Theory*

Niels Bohr's model of the hydrogen atom was powerful because it could reproduce experimentally observed line spectra. But there was additional experimental confirmation.

If the electron in the hydrogen atom is moved from the ground state, where $n = 1$, to the energy level, where $n =$ infinity, the electron is considered to have been removed from the atom. That is, the atom has been ionized.

$H^+(g)$	$n = \infty$	$E = 0$ kJ/mol
	$\Delta E = +Rhc = +1312$ kJ/mol	
$H(g)$	$n = 1$	$E = -1312$ kJ/mol

$$H(g) \longrightarrow H^+(g) + e^-$$

We can calculate the energy for this process from Equation 7.5 where $n_{final} = \infty$ and $n_{initial} = 1$.

$$\Delta E = -Rhc\left(\frac{1}{n^2_{final}} - \frac{1}{n^2_{initial}}\right)$$

$$= -Rhc\left(\frac{1}{\infty^2} - \frac{1}{1^2}\right) = +Rhc$$

Because $Rhc = 1312$ kJ/mol, the energy to move an electron from $n = 1$ to $n = \infty$ is 1312 kJ/mol of H atoms. We now call this the **ionization energy** of the atom (Section 8.6) and can measure it in the laboratory. The experimental value is known to be 1312 kJ/mol, in exact agreement with the result calculated from Bohr's theory!

7.4 THE WAVE PROPERTIES OF THE ELECTRON

Einstein used the photoelectric effect to demonstrate that light, usually thought of as having wave properties, can also be thought about in terms of particles, or massless photons. This fact was pondered by Louis Victor de Broglie (1892–1987). If light can be considered as having both wave and particle properties, would matter behave similarly? That is, could a tiny object such as an electron, normally considered a particle, also exhibit wave properties in some circumstances? In 1925, de Broglie proposed that a free electron of mass m moving with a velocity v should have an associated wavelength given by the equation

$$\lambda = \frac{h}{mv} \qquad [7.6]$$

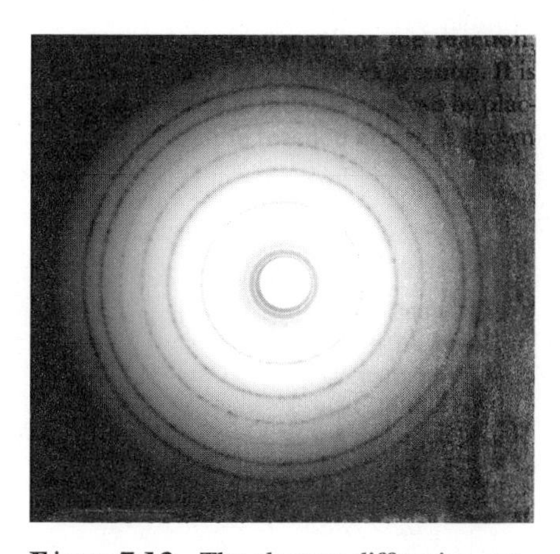

***Figure* 7.13 The electron diffraction pattern obtained for magnesium oxide.** *(R. K. Bohn, Department of Chemistry, University of Connecticut)*

This idea was revolutionary because it linked the particle properties of the electron (m and v) with a wave property (λ). Experimental proof was soon produced. In 1927, C. J. Davisson (1881–1958) and L. H. Germer (1896–1971), working at the Bell Telephone Laboratories in New Jersey, found that a beam of electrons was diffracted like light waves by the atoms of a thin sheet of metal foil and that de Broglie's relation was followed quantitatively (Figure 7.13). Because diffraction is an effect readily explained by the wave properties of light, it followed that electrons also can be described as waves under some circumstances.

Recall that $h = 6.6260755 \times 10^{-34}$ J · s, a very small number.

De Broglie's equation suggests that any moving particle has an associated wavelength. If λ is to be large enough to measure, however, the product of m and v must be very small because h is so small. For example, a 114-g baseball traveling at 110 mph has a large mv product (5.6 kg · m/s) and therefore the incredibly small wavelength of 1.2×10^{-34} m! This tiny value is essentially meaningless because it cannot be measured with any instrument now available. This means we will never assign wave properties to a baseball or any other massive object. It is only possible to observe wave-like properties for particles of extremely small mass, such as protons, electrons, and neutrons.

See the *Saunders Interactive General Chemistry CD-ROM,* Screen 7.8, Wave Properties of the Electron.

Louis Victor de Broglie (1892–1987) (*Oesper Collection in the History of Chemistry/University of Cincinnati*).

Example 7.5 Using De Broglie's Equation

Calculate the wavelength associated with an electron of mass $m = 9.109 \times 10^{-28}$ g traveling at 40.0% of the velocity of light.

Solution First, consider the units involved. Wavelength is calculated from h/mv, where h is Planck's constant expressed as joules times seconds (J · s). As discussed in Chapter 6, $1\ \text{J} = 1\ \text{kg} \cdot \text{m}^2/\text{s}^2$. Therefore, the mass must be used in kilograms and velocity in meters per second:

$$\text{Electron mass} = 9.109 \times 10^{-31}\ \text{kg}$$

$$\begin{aligned}\text{Electron velocity (40.0\% of light velocity)} &= (0.400)(2.998 \times 10^8\ \text{m} \cdot \text{s}^{-1}) \\ &= 1.20 \times 10^8\ \text{m} \cdot \text{s}^{-1}\end{aligned}$$

Substituting these values into de Broglie's equation, we have

$$\lambda = \frac{h}{mv} = \frac{6.626 \times 10^{-34}\ (\cancel{\text{kg}} \cdot \cancel{\text{m}^2}/\cancel{\text{s}^2})(\cancel{\text{s}})}{(9.109 \times 10^{-31}\ \cancel{\text{kg}})(1.20 \times 10^8\ \cancel{\text{m}} \cdot \cancel{\text{s}^{-1}})} = 6.06 \times 10^{-12}\ \text{m}$$

In nanometers, the wavelength is $(6.06 \times 10^{-12}\ \cancel{\text{m}})(1.00 \times 10^9\ \text{nm}/\cancel{\text{m}}) = 6.06 \times 10^{-3}$ nm. This wavelength is only about 1/20 of the diameter of the H atom.

Exercise 7.6 De Broglie's Equation

Calculate the wavelength associated with a neutron having a mass of 1.675×10^{-24} g and a kinetic energy of 6.21×10^{-21} J. (Recall that the kinetic energy of a moving particle is $E = mv^2/2$.)

7.5 THE WAVE MECHANICAL VIEW OF THE ATOM

In Copenhagen, Denmark, after World War I, Niels Bohr assembled a group of physicists who set out to derive a comprehensive theory for the behavior of electrons in atoms from the viewpoint of the electron as a particle. Erwin Schrödinger (1887–1961), an Austrian, independently worked toward the same goal, but he used de Broglie's hypothesis that an electron in an atom could be described by equations for wave motion. Although both Bohr and Schrödinger were successful in predicting some aspects of electron behavior, Schrödinger's approach gave correct results for some properties for which Bohr's failed. For this reason theoreticians today primarily use Schrödinger's concept. In any event, the general theoretical approach to understanding atomic behavior, developed by Bohr, Schrödinger, and their associates, has come to be called **quantum mechanics** or **wave mechanics.**

The Uncertainty Principle

See the *Saunders Interactive General Chemistry CD-ROM,* Screen 7.9, Heisenberg's Uncertainty Principle.

Before you can appreciate Schrödinger's model for the behavior of electrons in atoms, you should know about a great debate that raged in physics early in the 20th century. De Broglie's suggestion that an electron can be described as hav-

A CLOSER LOOK

Seeing Atoms and Molecules

Optical microscopes have been tremendously important in the development of science. Unfortunately, their use is limited by the wavelength of visible light. Using light with a 550-nm wavelength, for example, it is possible to barely resolve (distinguish between) objects that are 0.005 mm apart—good for looking at algae and bacteria, but not for looking at atoms and molecules.

The realization that electrons had wave properties led to the development of the electron microscope and a much higher degree of resolution. Using electrons with a wavelength of 6.06×10^{-3} nm for example (a value much smaller than wavelengths of visible light) gives an improvement of about 10^5 in resolution. With an electron beam instead of a light beam it is possible to distinguish objects separated by only about 50 nm. Although this is in the range of dimensions of some large biomolecules, they can still only be seen as shapeless blobs. The technique is still about an order of magnitude (ten times) too low in resolving power to get useful molecular information.

Because the resolving power of a microscope is related to the wavelength of the radiation used, it might seem logical that simply going to shorter and shorter wavelengths would ultimately make it possible to "see" individual atoms. Unfortunately this is not so, and it is the Heisenberg uncertainty principle that gets in the way. A shorter wavelength means a higher frequency and higher energy, and the energy of the radiation perturbs the particle being studied. A simple picture is that the photon of radiation is so energetic that when it strikes the particle to be imaged it imparts sufficient energy to move it from its position. In effect, with an accurate energy value of the photon, the position becomes imprecise. So, this route to actually seeing atoms is doomed to failure. A totally different approach is needed.

This new approach has been provided by a collection of experiments that go by the name *scanning probe microscopies.* The field developed from research in the early 1980s by Gerd Binnig (1947–) and Heinrich Rohrer (1933–) from the IBM-Zurich Research Laboratory in Ruschlikon, Switzerland. Binnig and Rohrer received the Nobel Prize in physics in 1986 for their invention of the scanning tunneling microscope (STM). In the STM a sharp needle-like probe (the tip having atomic scale dimensions) is positioned within a few atomic diameters from a surface and a voltage applied to cause a current to flow between the probe and the surface. The current is a function of the distance between probe and surface. As the probe moves horizontally along the surface, its height controlled by an intricate feedback circuit. The signal recorded defines the topography of the surface, a series of hills representing atoms, and valleys representing spaces between them. Thus it is possible to "see" individual atoms.

Modifying the technique allows the operator to push atoms around into specific places on the surface. This has led workers in the field to create some interesting pictures such as that in the figure. Workers in the field anticipate that this technique has the potential to create surfaces that are a tailored to specific uses.

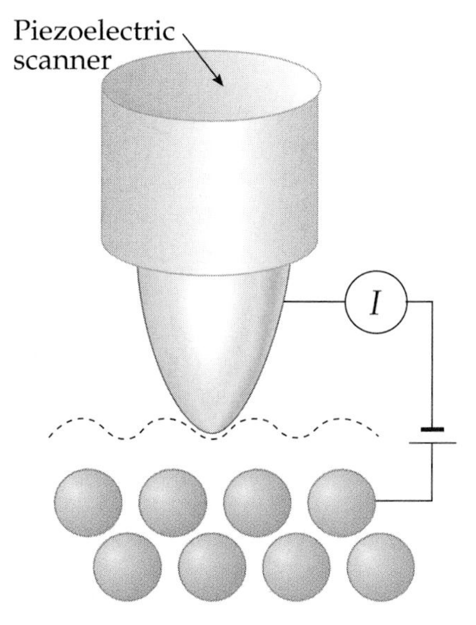

Schematic view of STM probe. The tip is often made of W, Pt, or a Pt-Ir alloy and is prepared so that it ends in a single atom. The spatial arrangement of atoms on a surface is revealed by the variation in electron density as the probe tip moves across the surface. The movement of the tip is controlled with a piezoelectric scanner. Piezoelectric materials expand or contract when a voltage is applied to them, so applying a voltage to the scanner causes it to move in any x, y, or z direction. (Quartz, for example, is a piezoelectric crystal; it is commonly used in watches and the strikers of lighters.)

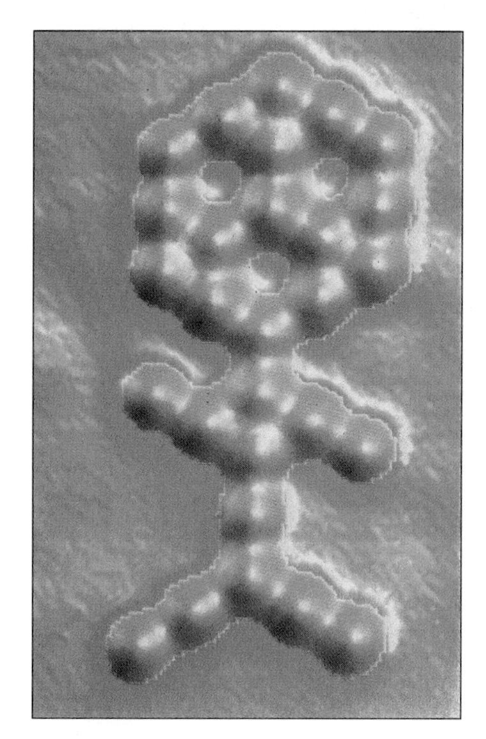

This figure, titled "Molecular Man," was created from 28 individual molecules of carbon monoxide by P. Zeppenfeld, a visiting scientist at IBM. Each of the gold-colored peaks is an image of a CO molecule on a platinum surface. The C atom of each CO is attached to the surface, and the O atom is pointing up toward the viewer. *(IBM Corporation, Research Division, Almaden Research Center)*

Erwin Schrödinger (1887–1961) was born in Vienna, Austria, and studied at the University there. Following service in World War I as an artillery officer, he became a professor of physics at various universities; in 1928 he succeeded Max Planck as professor of theoretical physics at the University of Berlin. He shared the Nobel Prize in physics (with Paul Dirac) in 1933. (*Oesper Collection in the History of Chemistry/University of Cincinnati*)

See the *Saunders Interactive General Chemistry CD-ROM,* Screen 7.10, Schrödinger's Equation and Wave Functions.

ing wave properties was confirmed by experiment (see Section 7.4). J. J. Thomson's experiment to measure the charge-to-mass ratio of an electron showed the particle-like nature of the electron (Figure 2.5). But how can an electron be both a particle and a wave? One can only conclude that the electron has dual properties. The result of a given experiment can be described either by the physics of waves or of particles; no single experiment can be done to show the electron behaves simultaneously as a wave *and* a particle!

What does this wave–particle duality have to do with electrons in atoms? Werner Heisenberg (1901–1976) and Max Born (1882–1970) provided the answer. To explain the behavior of electrons in atoms, it seems most reasonable to assume electrons have wave properties. If this is the case, Heisenberg concluded that it was impossible to fix both the position of an electron in an atom and its energy with any degree of certainty. If we attempt to determine either the location or the energy accurately, then the other is uncertain. In contrast, this is not the case in the world around you. For example, you can determine, with considerable accuracy, the energy of a moving car and its location.

Based on this idea, which we now call **Heisenberg's uncertainty principle,** Max Born proposed that the results of quantum mechanics should be interpreted as follows: if we choose to know the energy of an electron in an atom with only a small uncertainty, then we must accept a correspondingly large uncertainty about its position in the space about the atom's nucleus. In practical terms, this means the only thing we can do is calculate the likelihood, or **probability,** of finding an electron with a given energy within a given space. We shall return to this viewpoint in the next section.

Schrödinger's Model of the Hydrogen Atom and Wave Functions

Schrödinger's model of the hydrogen atom was based on the premise that the electron can be described as a matter wave and not as a tiny particle orbiting the nucleus. Unlike Bohr's model, Schrödinger's approach resulted in an equation that is complex and mathematically difficult to solve except in simple cases. We need not be concerned with the mathematics, but the solutions to the equation—called **wave functions**—are chemically important. If we can understand the implications of these wave functions, then we will understand the modern view of the atom.

Wave functions, which are symbolized by the Greek letter ψ (psi), characterize the electron as a **matter wave.** The following important points can be made:

1. Only certain vibrations, called standing waves (see Figure 7.4), can be observed in a vibrating string. Similarly, the behavior of the electron in the atom is best described as a standing wave. Although electron motion is not as easily visualized as a vibrating string, still *only certain wave functions are allowed.*
2. Each wave function ψ corresponds to an allowed energy value for the electron. This is like Bohr's result for the hydrogen atom ($E_n = -Rhc/n^2$). In the Bohr theory, an energy can be calculated for each value of n. In Schrödinger's model, each wave function is associated with an energy, E_n.
3. Taken together, points 1 and 2 say that *the energy of the electron is quantized.* The concept of quantization enters Schrödinger's theory naturally with the basic

A CLOSER LOOK Heisenberg's Uncertainty Principle

Heisenberg's uncertainty principle is expressed mathematically as

$$\Delta x \cdot \Delta(mv) > h$$

The uncertainty in the position of an electron in an atom (Δx) multiplied by the uncertainty in its momentum (Δmv) (which is related to its energy) must be larger than Planck's constant (h). This tells us that if we wish to know the momentum (or energy) of a very small object like an electron with great certainty ($\Delta mv \rightarrow 0$) then we must accept a great uncertainty in its position ($\Delta x \rightarrow \infty$).

Let us calculate the uncertainty in the position of an electron ($m = 9.11 \times 10^{-28}$ g) moving at 1.20×10^8 m/s (about 40% of the velocity of light). Assume the uncertainty in velocity is 0.100%. From Heisenberg's equation, we find $\Delta x > h/\Delta(mv)$.

$$\Delta x > \frac{(6.626 \times 10^{-34}\ \text{kg} \cdot \text{m}^2/\text{s}^2)(\text{s})}{(9.11 \times 10^{-31}\ \text{kg})(1.20 \times 10^8\ \text{m/s})(0.00100)}$$

$$> 6.06 \times 10^{-9}\ \text{m}$$

This uncertainty is about 6 nm, a very large distance considering that distances on the atomic and molecular scale are measured in nanometers.

Now let us compare the electron with an automobile. What is the position uncertainty for a car ($m = 1.00 \times 10^3$ kg) moving at 26.7 ± 0.0450 m/s (about 60 mph)?

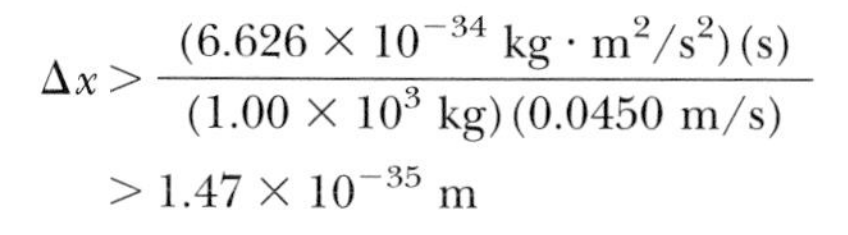

$$\Delta x > \frac{(6.626 \times 10^{-34}\ \text{kg} \cdot \text{m}^2/\text{s}^2)(\text{s})}{(1.00 \times 10^3\ \text{kg})(0.0450\ \text{m/s})}$$

$$> 1.47 \times 10^{-35}\ \text{m}$$

This is such a small uncertainty that it cannot be measured by current instruments. This means we can know the position accurately of a large, moving object such as a car.

Werner Heisenberg (1901–1976) and sons. Heisenberg earned a Ph.D. in theoretical physics at the University of Munich in 1923 and then studied with Max Born and later Niels Bohr. He received the Nobel Prize in physics in 1932. *(Emilio Segré Visual Archives, American Institute of Physics)*

assumption of an electron as a matter wave. This is in contrast with Bohr's theory in which quantization was imposed as a postulate at the start.

4. Each wave function ψ can be visualized in terms of the ideas of probability. Theoreticians have found that the *square* of ψ is related to the probability of finding the electron within a given region of space. Scientists refer to this as the **electron density** in a given region.

By describing the electron as a wave, the idea of a precise location for the electron has no meaning.

Be sure to understand that the theory does not predict the exact position of the electron. Because Schrödinger's theory chooses to define the energy of the electron precisely, Heisenberg's uncertainty principle tells us this must result in a large uncertainty in electron position. This is why we can only describe the *probability* of the electron being at a certain point in space when in a given energy state.

5. The matter waves for the allowed energy states of the electron are called **orbitals.** We shall say more about orbitals in a moment.
6. To solve Schrödinger's equation for an electron in three-dimensional space, three integer numbers—the quantum numbers n, ℓ, and m_ℓ—must be introduced. These quantum numbers may have only certain combinations of values, as outlined below. We shall use these combinations to define the energy states and orbitals available to the electron.

Quantum Numbers

See the *Saunders Interactive General Chemistry CD-ROM,* Screen 7.11, Shells, Subshells, and Orbitals, and Screen 7.12, Quantum Numbers and Orbitals.

In a three-dimensional space, three numbers are required to describe the location of an object. For the wave description of the electron in an atom, this requirement leads to the existence of three **quantum numbers n, ℓ,** and **m_ℓ.** Before looking into the meanings of the three quantum numbers, it is important to say that

- The quantum numbers n, ℓ, and m_ℓ are all integers, but their values cannot be selected randomly.
- The three quantum numbers (and their values) are not parameters that scientists dreamed up. Instead, when the behavior of the electron in the hydrogen atom is described mathematically as a matter wave, the quantum numbers are a natural consequence of that theory.

n, *the Principal Quantum Number = 1, 2, 3, . . .*

The electron energy in an atom with more than one electron depends on n *and* the other quantum numbers ℓ and m_ℓ.

The principal quantum number n can have any integer value from 1 to infinity. As the name implies, it is the most important quantum number because the value of n is the primary factor in determining the energy of an electron. Indeed, *for the hydrogen atom* (with its single electron) *the energy of the electron varies only with the value of* n and is given by the same equation derived by Bohr for the H atom: $E_n = -Rhc/n^2$. The value of n is also a measure of the size of an orbital having that value of n: the greater the value of n, the larger the electron's orbital.

Each electron is labeled according to its value of n. In atoms having more than one electron, two or more electrons may have the same n value. These electrons are then said to be in the same **electron shell.**

ℓ, *the Angular Momentum Quantum Number = 0, 1, 2, 3, . . . ,* n − *1*

The electrons of a given shell can be grouped into **subshells,** each subshell characterized by a different value of the quantum number ℓ and by a characteristic shape. For the nth shell, n different subshells are possible, each subshell corresponding to one of the n different values of ℓ. *Each value of ℓ corresponds to a different orbital shape,* or orbital type.

The value of n limits the number of subshells possible for the nth shell because ℓ can be no larger than $n - 1$. Thus, for $n = 1$, ℓ must equal 0 and only 0. Because ℓ has only one value when $n = 1$, only one subshell is possible for an electron assigned to $n = 1$. When $n = 2$, ℓ can be either 0 or 1. Because two values of ℓ are now possible, there are two subshells in the $n = 2$ electron shell.

The values of ℓ are usually coded by letters according to the following scheme:

Value of ℓ	Corresponding Subshell Label
0	*s*
1	*p*
2	*d*
3	*f*

For example, a subshell with a label of $\ell = 1$ is called a "*p* subshell," and an orbital found in that subshell is called a "*p* orbital." Conversely, an electron assigned to a *p* subshell has an ℓ value of 1.

As an aside, it is interesting to take note of the origin of the letters used to designate subshells. Early studies of the emission spectra of elements other than hydrogen found more lines than could be explained by Bohr's theory. Scientists studying the spectrum of sodium atoms, for example, found four different types of lines, which they labeled *sharp, principal, diffuse,* and *fundamental.* To account for the additional lines, they believed a theory of the atom was needed that included energy subshells for electrons. When the notion of subshells arose from Schrödinger's model, the subshells were labeled as *s, p, d,* and *f.*

No known elements have electrons assigned to orbitals with ℓ greater than 3 in the ground state. However, $\ell = 4$ would correspond to *g* orbitals, $\ell = 5$ to *h* orbitals, and so on. Such orbitals are important to consider for an excited-state atom.

m_ℓ, *the Magnetic Quantum Number* = $0, \pm 1, \pm 2, \pm 3, \ldots, \pm \ell$

The magnetic quantum number, m_ℓ, specifies to which orbital within a subshell the electron is assigned. *Orbitals in a given subshell differ only in their orientation in space, not in their shape.*

The value of ℓ limits the integer values assigned to m_ℓ: m_ℓ can range from $+\ell$ to $-\ell$ with 0 included. For example, when $\ell = 2$, m_ℓ has five values: +2, +1, 0, −1, and −2. The number of values of m_ℓ for a given subshell ($= 2\ell + 1$) specifies the number of orientations that exist for the orbitals of that subshell and thus the number of orbitals in the subshell.

Useful Information from Quantum Numbers

The three quantum numbers introduced thus far are a kind of "electronic zip-code." They tell us to which shell an electron is assigned (n), to which subshell within the shell (ℓ), and that the electron is assigned to an orbital in that subshell (m_ℓ). Their allowed values are summarized in Table 7.1.

Electrons in atoms are assigned to orbitals, which are grouped into subshells. Depending on the value of n, one or more subshells constitute an electron shell.

When $n = 1$ the value of ℓ can only be 0, and so m_ℓ must also have a value of 0. This means that, in the electron shell closest to the nucleus, only one subshell exists, and that subshell consists of only a single orbital. This orbital is labeled "1*s*," the "1" conveying the value of n and "*s*" telling you that $\ell = 0$. *When* $\ell = 0$, *an s orbital is indicated, and only one s orbital can occur in a given electron shell.*

When $n = 2$, ℓ can have two values (0 and 1), so two subshells or two types of orbitals occur in the second shell. One of these is the 2*s* subshell ($n = 2$ and $\ell = 0$), and the other is the 2*p* subshell ($n = 2$ and $\ell = 1$). Because the values of m_ℓ can be +1, 0, and −1 when $\ell = 1$, three *p*-type orbitals exist. Because all three have $\ell = 1$, they all have the same shape, but the different m_ℓ values tell us they differ in their orientation in space. In summary, when $\ell = 1$, *p* orbitals are indicated, and three of them always occur. The converse is true as well: for *p* orbitals in any shell, ℓ is 1.

Electron subshells are labeled by first giving the value of n and then the value of ℓ in the form of its letter code. For $n = 1$ and $\ell = 0$, for example, the label is 1*s*.

When $n = 3$, three subshells, or orbital types, are possible for an electron because ℓ has the values 0, 1, and 2. Because you see ℓ values of 0 and 1 again, you know that two of the subshells within the $n = 3$ shell are 3*s* (one orbital) and 3*p* (three orbitals). The third subshell is *d*, indicated by $\ell = 2$. Because m_ℓ has five values (+2, +1, 0, −1, and −2) when $\ell = 2$, five *d* orbitals (no more and no less) occur in the $\ell = 2$ subshell. Thus, whenever an electron shell has $n = 3$ or greater, one of the ℓ values will be 2, indicating a set of five *nd* orbitals.

Besides *s, p,* and *d* orbitals, we occasionally need to refer to *f* electron orbitals, that is, orbitals for which $\ell = 3$. Seven such orbitals exist because seven values of m_ℓ are possible when $\ell = 3$ (+3, +2, +1, 0, −1, −2, and −3).

Table 7.1 • SUMMARY OF THE QUANTUM NUMBERS, THEIR INTERRELATIONSHIPS, AND THE ORBITAL INFORMATION CONVEYED

Principal Quantum Number **Symbol = n** **Values = 1, 2, 3, . . .** **(Orbital Size, Energy)**	**Angular Momentum Quantum Number** **Symbol = ℓ** **Values = 0 . . . $n-1$** **(Orbital Shape)**	**Magnetic Quantum Number** **Symbol = m_ℓ** **Values = $+\ell$ . . . 0 . . . $-\ell$** **(Orbital Orientations = Number of Orbitals in Subshell)**	**Number and Type of Orbitals in the Subshell** **(Number of Orbitals in Shell = number of values of $m_\ell = 2\ell + 1 = n^2$)**
1	0	0	one 1*s* orbital (one orbital of one type in the $n = 1$ shell)
2	0 1	0 +1,0,−1	one 2*s* orbital three 2*p* orbitals (four orbitals of two types in the $n = 2$ shell)
3	0 1 2	0 +1,0,−1 +2,+1,0,−1,−2	one 3*s* orbital three 3*p* orbitals five 3*d* orbitals (nine orbitals of three types in the $n = 3$ shell)
4	0 1 2 3	0 +1,0, −1 +2,+1,0,−1,−2 +3,+2, +1,0,−1,−2,−3	one 4*s* orbital three 4*p* orbitals five 4*d* orbitals seven 4*f* orbitals (16 orbitals of four types in the $n = 4$ shell)

Exercise 7.7 Using Quantum Numbers

Complete the following statements:

1. When $n = 2$, the values of ℓ can be ______ and ______.
2. When $\ell = 1$, the values of m_ℓ can be ______, ______, and ______, and the subshell has the letter label ______.
3. When $\ell = 2$, the subshell is called a ______ subshell.
4. When a subshell is labeled *s*, the value of ℓ is ______ and m_ℓ has the value ______.
5. When a subshell is labeled *p*, ______ orbitals occur within the subshell.
6. When a subshell is labeled *f*, there are ______ values of m_ℓ, and ______ orbitals occur within the subshell.

7.6 THE SHAPES OF ATOMIC ORBITALS

See the *Saunders Interactive General Chemistry CD-ROM,* Screen 7.13, Shapes of Atomic Orbitals.

The chemistry of an element and of its compounds is determined by the electrons of the element's atoms, particularly the electrons with the highest value of *n*, which are often called valence electrons (Section 3.3). The types of orbitals to which these electrons are assigned is also important, so we turn now to the question of orbital shape and orientation.

s Orbitals

When an electron has $\ell = 0$, we often say the electron is assigned to, or "occupies," an *s* orbital. But what does this mean? What is an *s* orbital? What does it look like? To answer these questions, we begin with the wave function for an electron with $n = 1$ and $\ell = 0$, that is, with a 1*s* orbital. If we assume for the moment that the electron is a tiny particle and not a matter wave, and if we could photograph the 1*s* electron at 1-s intervals for a few thousand seconds, the composite picture would resemble the drawing in Figure 7.14a. This resembles a cloud of dots, so chemists refer to such representations of electron orbitals as **electron cloud pictures.**

The fact that the density of dots is greater close to the nucleus (the electron cloud is denser close to the nucleus) indicates that the electron is most often found near the nucleus (or, conversely, it is less likely to be found farther away). Putting this in the language of quantum mechanics, we say the *electron density* is greater closer to the nucleus or that the *greatest probability* of finding the electron is in a tiny volume of space around the nucleus. Conversely, the electron density falls off on moving away from the nucleus; it is less probable that the electron is farther away. The "thinning" of the electron cloud at increasing distance, shown by the decreasing density of dots in Figure 7.14a, is illustrated in a different way in Figure 7.14b. Here we plotted the *square* of the wave function for the electron in a 1*s* orbital as a function of the distance of the electron from the nucleus. The units of ψ^2 at each point are 1/volume, so the numbers on the vertical axis of this plot represent the probability of finding the electron in each cubic nanometer, for example, at a given distance from the nucleus. For this reason, ψ^2 is called the **probability density.** For the 1*s* orbital, ψ^2 is very

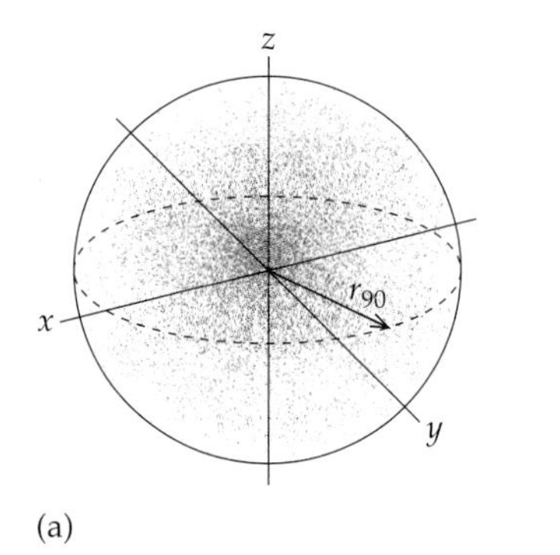

(a)

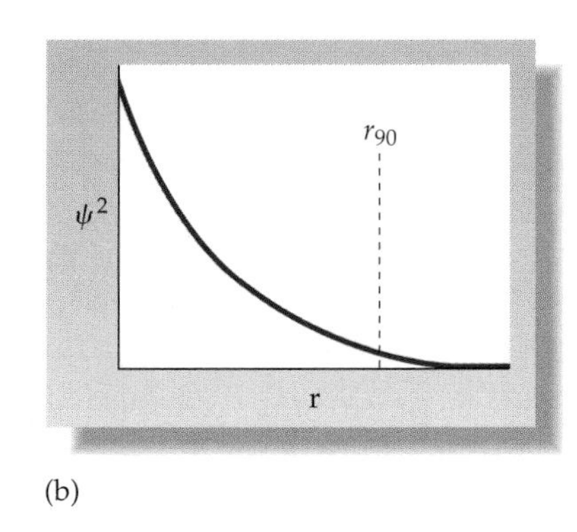

(b)

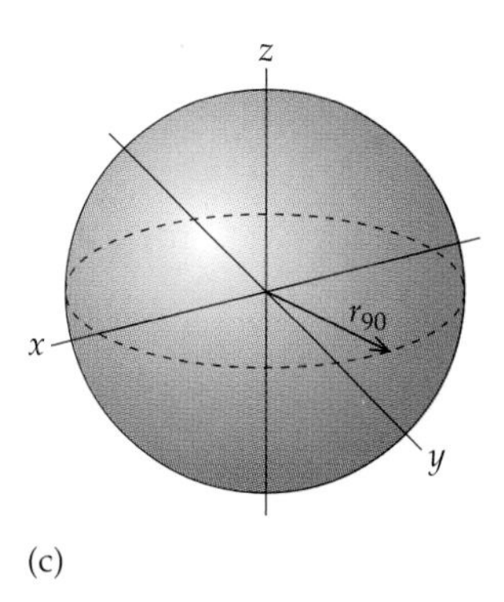

(c)

Figure **7.14 Different views of a 1*s* ($n = 1$ and $\ell = 0$) orbital.** (a) Dot picture of an electron in a 1*s* orbital. Each dot represents the position of the electron at a different instant in time. Note that the dots cluster closest to the nucleus. r_{90} is the radius of a sphere within which the electron is found 90% of the time. (b) A plot of the probability density as a function of distance for a one-electron atom with a 1*s* electron wave. (c) The surface of the sphere within which the electron is found 90% of the time for a 1*s* orbital. This surface is often called a "boundary surface." (A 90% surface was chosen arbitrarily. If the choice was the surface within which the electron is found 50% of the time, the sphere would be considerably smaller.)

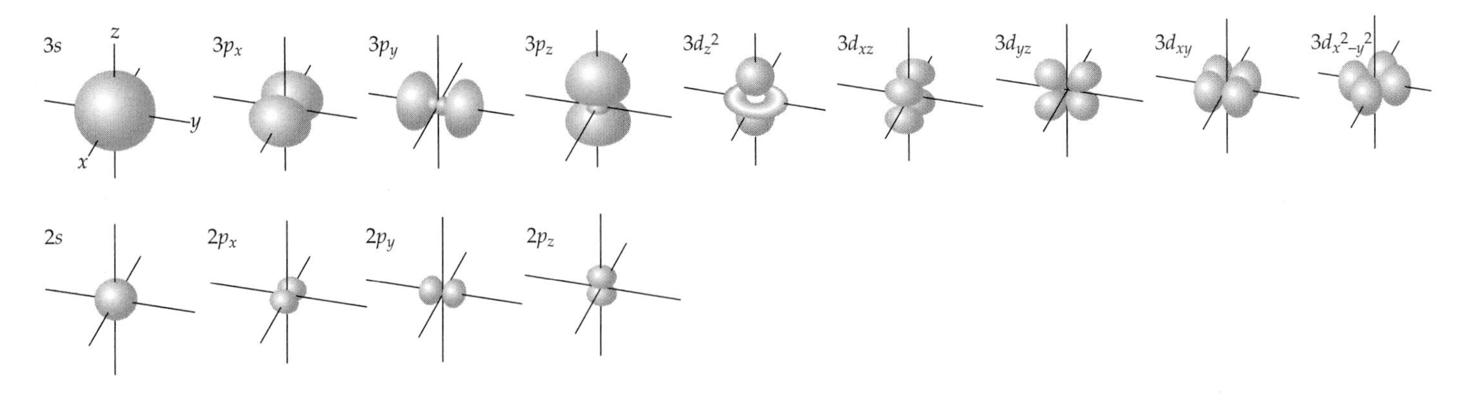

Figure 7.15 Atomic orbitals. Boundary surface diagrams for electron densities of 1*s*, 2*s*, 2*p*, 3*s*, 3*p*, and 3*d* orbitals for a hydrogen atom. For the *p* orbitals the subscript letter on the orbital notation indicates the cartesian axis along which the orbital lies. The plane passing through the nucleus (perpendicular to this axis) is called a planar node ($\ell = 1$). The *d* orbitals all have two planar nodes ($\ell = 2$). See the text for a description of the notation used to differentiate the *d* orbitals.

high for points immediately around the nucleus, but it drops off rapidly as the distance from the nucleus increases. Notice that the probability approaches but never reaches zero, even at very large distances.

For the 1*s* orbital, Figure 7.14a shows the electron is most likely found within a sphere with the nucleus at the center. No matter in which direction you proceed from the nucleus, the probability of finding an electron along a line in that direction drops off (Figure 7.14b). *The 1s orbital is spherical in shape.*

The visual image of Figure 7.14a is that of a cloud whose density is small at large distances from the center; there is no sharp boundary beyond which the electron is never found. The *s* and other orbitals, however, are often depicted as having a sharp boundary surface (Figure 7.14c), largely because it is easier to draw such pictures. To arrive at the diagram in Figure 7.14c we drew a sphere about the nucleus in such a way that the chance of finding the electron somewhere inside is 90%.

Many misconceptions exist about pictures such as Figure 7.14c. Understand that this surface is not real. The nucleus is surrounded by an "electron cloud" and not an impenetrable surface "containing" the electron. The electron is not distributed evenly throughout the volume enclosed by the surface; instead, it is most likely found nearer the nucleus.

From an analysis of ψ for *s* orbitals of different *n* values, we can arrive at the important conclusion that *all* orbitals labeled *s* are spherical in shape. In every case, *s* electrons can be found in a very small volume of space immediately around the nucleus. One important difference between *s* orbitals of different *n*, however, is that *the size of s orbitals increases as n increases* (Figure 7.15). Thus, the 1*s* orbital is more compact than the 2*s* orbital, which is more compact than the 3*s* orbital.

p Orbitals

Atomic orbitals for which $\ell = 1$ are called *p* orbitals and all have the same basic shape. *All p orbitals have one imaginary plane that slices through the nucleus and divides the region of electron density in half* (Figures 7.15 and 7.16).

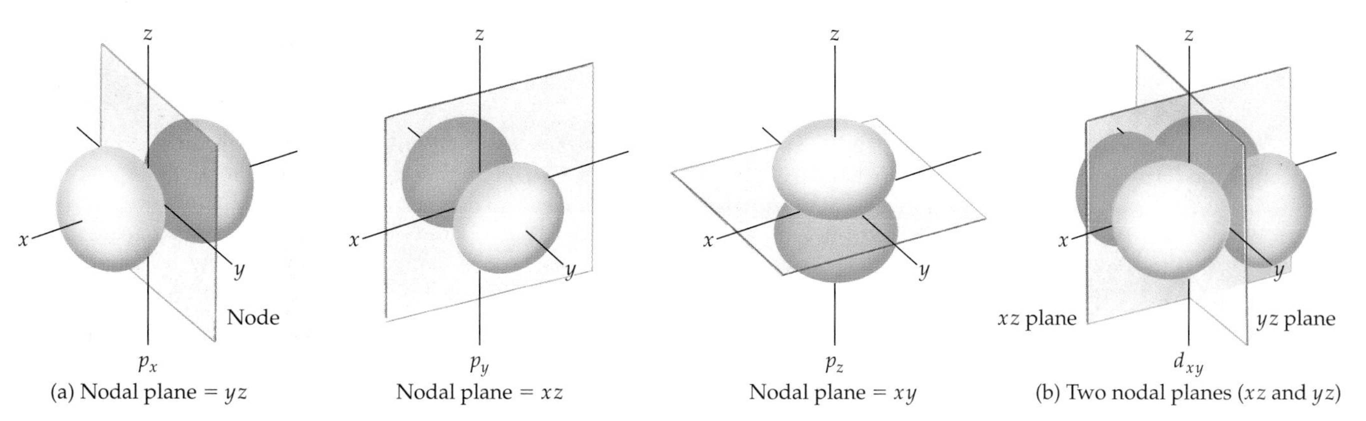

Figure 7.16 Nodal planes in *p* and *d* orbitals. (a) The three *p* orbitals, each having one nodal plane (because $\ell = 1$). (b) The d_{xy} orbital. Like all five *d* orbitals it has two nodal planes (because $\ell = 2$). Here the nodal planes are the *xz*- and *yz*-planes, so the regions of electron density lie in the *xy*-plane and between the *x*- and *y*-axes.

The imaginary plane slicing through the nucleus is called a **nodal plane,** a planar surface on which there is zero probability of finding the electron. The electron can never be found in the nodal plane; the regions of electron density lie on either side of the nucleus. This means that, unlike for *s* orbitals, for *p* orbitals there is no likelihood of finding the electron at the nucleus. A plot of electron probability (ψ^2) versus distance would start at zero at the nucleus, rise to a maximum, and then drop off at still greater distances. If you enclose 90% of the electron density within a surface, the view in Figure 7.16 is appropriate. The electron cloud has a shape that resembles a weight lifter's "dumbbell," and so chemists often describe *p* orbitals as having dumbbell shapes.

According to Table 7.1, when $\ell = 1$, then m_ℓ can only be +1, 0, or −1. That is, three orientations are possible for $\ell = 1$ or *p* orbitals, depending on the location of the nodal plane. There are three mutually perpendicular directions in space (*x*, *y*, and *z*), and the *p* orbitals are commonly visualized as lying along those directions (with the nodal plane perpendicular to the axis). The orbitals are labeled according to the axis along which they lie (p_x, p_y, or p_z) (see Figure 7.16).

The amplitude of the electron wave is zero on a nodal plane.

d Orbitals

The value of ℓ is equal to the number of nodal planes that slice through the nucleus. Thus, *s* orbitals, for which $\ell = 0$, have no nodal planes, and *p* orbitals, for which $\ell = 1$, have one planar nodal surface. It follows that the five *d* orbitals, for which $\ell = 2$, have two nodal surfaces, which results in four regions of electron density. The d_{xy} orbital, for example, lies in the *xy*-plane and the two nodal planes are the *xz*- and *yz*-planes (see Figure 7.16). Two other orbitals, d_{xz} and d_{yz}, lie in planes defined by the *xz*- and *yz*-axes, respectively, and also have two, mutually perpendicular nodal planes.

Of the two remaining *d* orbitals, the $d_{x^2-y^2}$ orbital is easier to visualize. Like the d_{xy} orbital, the $d_{x^2-y^2}$ orbital results from two vertical planes slicing the electron density into quarters. Now, however, the planes bisect the *x*- and *y*-axes, so the regions of electron density lie along the *x*- and *y*-axes.

A CLOSER LOOK Atomic Orbitals

An understanding of atomic orbitals is so crucial to modern chemistry that they are worth a closer look.

We have repeatedly made the point that an electron orbital is best thought of as a matter wave. Like all waves, they have crests, troughs, and nodes (see Figure 7.2). For a 1*s* orbital, the wave crests at the nucleus (Figure A and Figure 7.15), and the wave amplitude declines rapidly on moving away from the nucleus. That is, the algebraic sign of the wave function is everywhere positive. For a 2*s* orbital, we begin to see what you think of as a wave. There is a crest at the nucleus, but, on moving away from the nucleus, the value of ψ^2 decreases rapidly to zero (a node), and rises again before approaching zero at greater distances. The 3*s* orbital adds yet one more crest to the wave. (What do you suppose a 4*s* orbital would look like?) That is, as *n* increases, the wave function becomes zero one or more times on moving away from the nucleus. These points where $\psi = 0$ are **nodes.** Because such a node is always the same distance from the nucleus for any direction we move away (recall that *s* orbitals are all spherical), we call these **spherical nodes** (Figure B). (The number of spherical nodes for an *s*

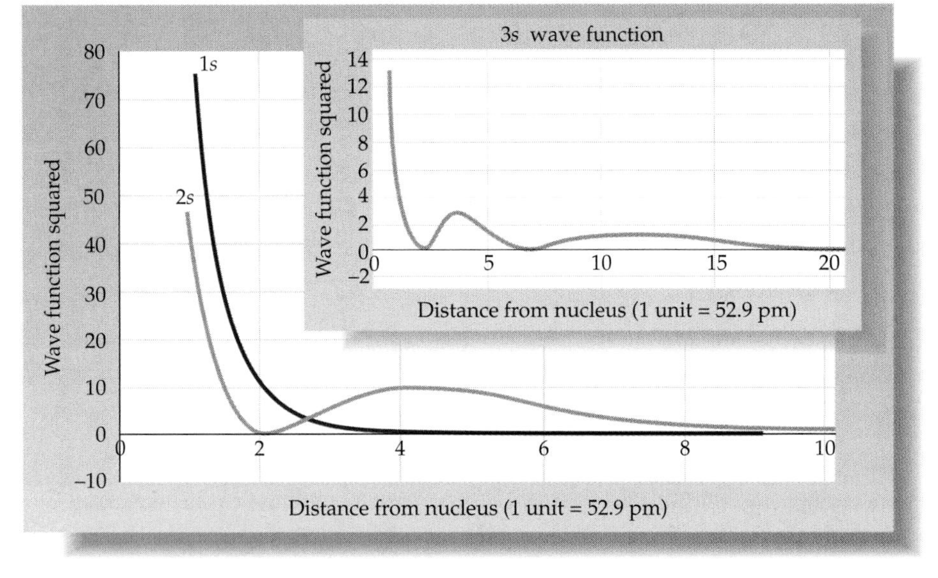

Figure A Plots of the wave function for 1*s*, 2*s*, and 3*s* atomic orbitals (for the H atom). The points in space where the function is zero are called spherical nodes.

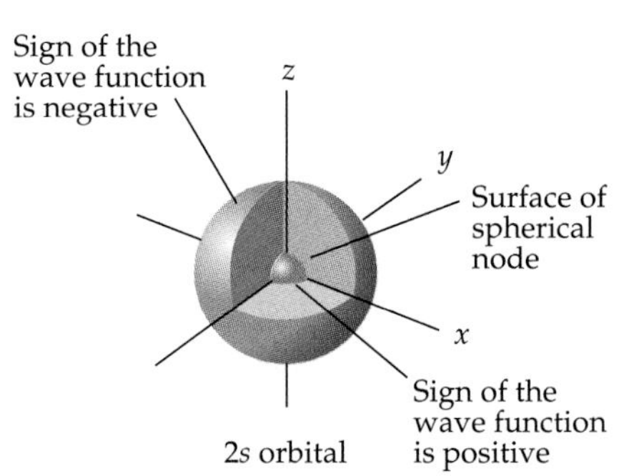

Figure B A 2*s* orbital (for the H atom) showing the spherical nodes.

Nodal planes occur for all *p*, *d*, and *f* orbitals. In some cases, however, the "plane" is not flat and is better called a nodal surface.

The final *d* orbital, d_{z^2}, has two main regions of electron density along the *z*-axis, but a "donut" of electron density also occurs in the *xy*-plane. This orbital has two nodal surfaces, but the surfaces are not flat. Think of an ice cream cone sitting with its tip at the nucleus. One of the electron clouds along the *z*-axis sits inside the cone. If you have another cone pointing in the opposite direction from the first cone, again with its tip at the nucleus, another region of electron density fits inside this second cone. The region outside both cones defines the remaining, donut-shaped region of electron density.

f Orbitals

The seven *f* orbitals all have $\ell = 3$, meaning that three nodal surfaces slice through the nucleus. This makes these orbitals less easily visualized, but one example is illustrated in Figure 7.17.

orbital is $n - 1$.) The best analogy to this picture is that an orbital resembles an onion that has layer on layer of white or pink matter separated by a thin space.

The picture is slightly different for a *p* orbital. Recall that *p* orbitals all have a nodal plane slicing through the nucleus. Thus, the wave crests a slight distance from the nucleus (Figure C). For a *p* orbital, however, the sign of the wave function is positive on one side of the nucleus and negative on the other. Again, we visualize the orbital as a wave with a node at the nucleus (Figure C). (Like *s* orbitals, *p* orbitals of *n* greater than 2 have spherical nodes, their number increasing with *n*.) When drawing three-dimensional pictures of *p* orbitals, they are often represented as in Figure D, where the two different colors represent the different signs of the wave function.

This picture of orbitals is very important in our later discussions of chemical bonding, especially the molecular orbital theory in Chapter 10.

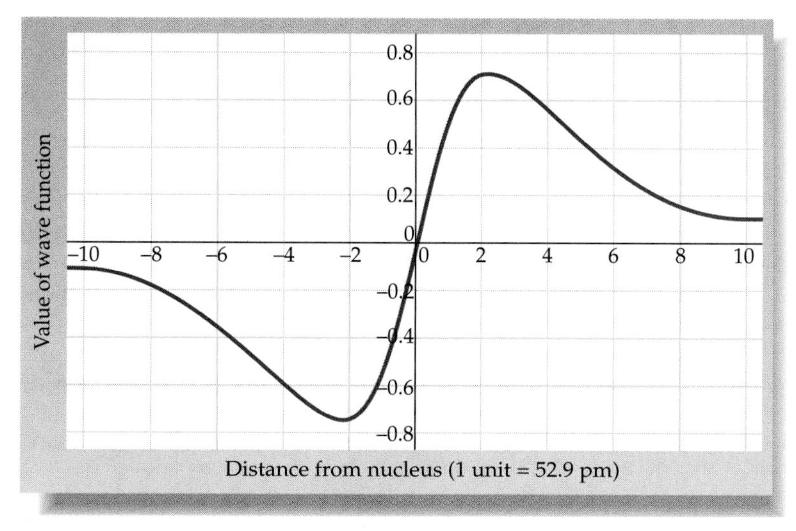

Figure C A *p* orbital has a node at the nucleus. The algebraic sign of the wave functions is positive on one side and negative on the other.

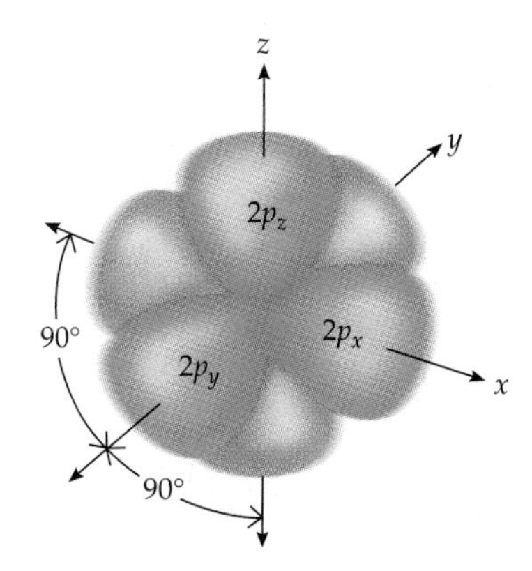

Figure D The 2*p* orbitals, showing that the sign of the wave function is different on opposite sides of the nucleus. Like all *p* orbitals, all of these orbitals have a planar node slicing through the nucleus.

Exercise 7.8 Orbital Shapes

1. What are the *n* and ℓ values for each of the following orbitals: 6*s*, 4*p*, 5*d*, and 4*f*?
2. How many nodal planes exist for a 4*p* orbital? For a 6*d* orbital?

Orbital Shapes and Chemistry

We close with some questions to ponder. When an element is part of a molecule, are the orbitals the same? Do they have the same shapes? What does the shape of orbitals have to do with the chemistry of an element? These are questions we take up in the rest of the book, but a few answers are in order here.

Schrödinger's wave equation can be solved exactly for the hydrogen atom but not for heavier atoms or their ions. Nonetheless, chemists make the as-

Figure 7.17 One of the seven possible *f* orbitals. Notice the presence of three nodal planes as required by an orbital with $\ell = 3$. (*C. D. Winters*)

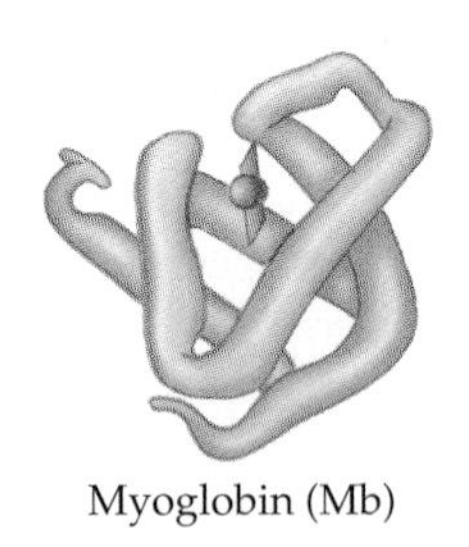
Myoglobin (Mb)

(a)

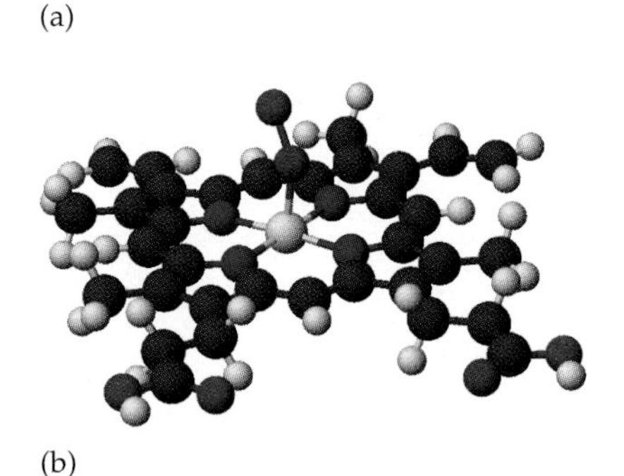

(b)

F*igure* 7.18 Myoglobin, the O_2-binding protein of muscle. (a) The overall structure of myoglobin. The heme group with its iron(II) ion is the square object with the enclosed sphere. The amino acid chain is depicted as a ribbon. (b) The structure of the heme group. It consists of an iron(II) ion bonded to the N atoms of a so-called porphyrin group. When O_2 binds to the iron ion in the heme group it does so by using one of the two O atoms. The other O atom is tipped away from the vertical.

sumption that orbitals in other atoms are hydrogen-like, even when those atoms are part of a molecule. At least this is the best way chemists have found to interpret experimental information on how molecules react.

An example of the importance of thinking in terms of orbitals is the interaction of oxygen molecules with the hemoglobin in your blood or with myoglobin, the oxygen-storage protein in muscle. Myoglobin is the cause of the characteristic red color of muscle, and the muscles of deep-diving mammals such as seals and whales are especially rich in this protein.

Myoglobin, though complex, is the simpler molecule of the two. It consists of an iron-bearing heme group encased in a long chain (called a polypeptide) that consists of 153 amino acid units (Figure 7.18). (Hemoglobin is similar, but it has four heme units, each enclosed in an amino acid chain.) The central iron ion is surrounded by four nitrogen atoms of the heme group, and, in a fifth position, by an N atom of a neighboring molecule, a histidine. The iron is not quite in the plane of the four N atoms, however. It lies 0.055 nm above the plane. As an O_2 molecule approaches, the iron ion is pulled back toward the plane as it binds to the O_2 molecule. Interestingly, the O_2 molecule does not bind to the iron in a "head-on" or in a "sideways" manner. Rather, it is tipped away from the vertical position by about 30°. All of these observations ultimately can be rationalized in terms of the orbitals on the Fe^{2+} ion, which we assume to be like those of an isolated Fe^{2+} ion, and of the O_2 molecule.

Chemistry is the study of molecules and their transformations. By thinking about the orbitals of the atoms in molecules, and by making the simple assumption that they resemble those of the hydrogen atom, we can understand much of the chemistry of even complex systems such as those in plants and animals.

CHAPTER HIGHLIGHTS

When you have finished studying this chapter, you should be able to

- Use the terms **wavelength, frequency, amplitude,** and **node** (Section 7.1).
- Use Equation 7.1 ($\lambda \cdot \nu = c$), which states that the product of the wavelength (λ) and frequency (ν) of electromagnetic radiation is equal to the speed of light (c).
- Recall the relative wavelength (or frequency) of the various regions in the spectrum of electromagnetic radiation (Figure 7.3).
- Use the fact that the energy of a **photon,** a massless particle of radiation, is given by **Planck's equation** (Equation 7.2),

$$E = h\nu \quad [7.2]$$

  where h is Planck's constant (6.626×10^{-34} J · s). This is an extension of Planck's idea that energy at the atomic level is quantized (Section 7.2).
- Describe the **Bohr model** of the atom, how it can account for the emission line spectra of excited atoms, and the limitations of the model (Section 7.3).
- Understand that, in the Bohr model of the H atom, the electron can occupy only certain energy levels, each with an energy given by Equation 7.4

$$E_n = -\frac{Rhc}{n^2} \quad [7.4]$$

where R is the Rydberg constant, h is Planck's constant, c is the velocity of light, and n is an integer equal to or greater than 1 ($Rhc = 2.179 \times 10^{-18}$ J/atom or 1312 kJ/mol) (Section 7.3).

- Recall that Bohr postulated that if an electron moves to another state, the amount of energy absorbed or emitted in the process is equal to the difference in energy between the two states (Section 7.3).
- Understand that in the modern view of the atom electrons are described by the physics of waves (Section 7.4). The wavelength of an electron or any subatomic particle is given by de Broglie's equation (7.6),

$$\lambda = \frac{h}{mv}$$

where m and v are the mass and velocity of the particle, respectively.

- Recognize the significance of **quantum mechanics** in describing the modern view of atomic structure (Section 7.5).
- Understand that an **orbital** for an electron in an atom corresponds to an allowed energy of that electron. [The energy of the electron in the H atom can be calculated by the same equation derived by Bohr (Equation 7.4).]
- Understand that the position of the electron is not known with certainty; only the **probability** of the electron being within a given region of space can be calculated. This is the interpretation of the quantum mechanical model and embodies the postulate called the **Heisenberg uncertainty principle.**
- Describe the allowed energy states of the electron in an atom using three quantum numbers n, ℓ, and m_ℓ (Section 7.5).

Quantum Number	Name	Values	Orbital Property Described
n	principal	1, 2, 3, . . . , ∞	Orbital size, energy, number of subshells, and number of orbital types in the shell
ℓ	angular momentum	0, 1, 2, . . . , $n-1$	Orbital shape, number of orbitals in a subshell
m_ℓ	magnetic	$+\ell$ . . . 0 . . . , $-\ell$	Orbital orientation

- Describe the shapes of the orbitals (Section 7.6).

Value of ℓ	Orbital Label	Nodal Planes	Orbital Shape
0	*s*	0	Spherical
1	*p*	1	"Dumbbell" shape; two regions of electron density
2	*d*	2	Four regions of electron density
3	*f*	3	Eight regions of electron density

STUDY QUESTIONS

REVIEW QUESTIONS

1. Our modern view of atomic structure was developed through many experiments. Name at least three of these experiments, their outcomes, and the person most associated with that experiment. (You may have to review Chapter 2.)
2. Give the equation for each of the following important mathematical relations in this chapter:
   (a) The relationship among wavelength, frequency, and speed of radiation
   (b) The relation between energy and frequency of radiation

(c) The energy of an electron in a given energy state of the H atom

3. State Planck's equation in words and as a mathematical equation.
4. Name the colors of visible light beginning with that of highest energy.
5. Draw a picture of a standing wave, and use this to define the terms "wavelength," "amplitude," and "node."
6. Which of the following best describes the importance of the photoelectric effect as explained by Einstein?
   (a) Light is electromagnetic radiation.
   (b) The intensity of a light beam is related to its frequency.
   (c) Light can be thought of as consisting of massless particles whose energy is given by Planck's equation, $E = h\nu$.
7. What is a photon? Explain how the photoelectric effect implies the existence of photons.
8. What were two major assumptions of Bohr's theory of atomic structure?
9. In what region of the electromagnetic spectrum is the Lyman series of lines found? The Balmer series?
10. Light is given off by a sodium- or mercury-containing streetlight when the atoms are excited in some way. The light you see arises for which of the following reasons:
    (a) Electrons moving from a given energy level to one of higher n
    (b) Electrons being removed from the atom, thereby creating a metal cation
    (c) Electrons moving from a given level to one of lower n
    (d) Electrons whizzing about the nucleus in an absolute frenzy
11. What is incorrect about the Bohr model of the atom?
12. What is Heisenberg's uncertainty principle? Explain how it applies to our modern view of atomic structure.
13. How do we interpret the physical meaning of the square of the wave function? What are the units of ψ^2?
14. What are the three quantum numbers used to describe an orbital? What property of an orbital is described by each quantum number? Specify the rules that govern the values of each quantum number.
15. Give the number of nodal surfaces for each orbital type: *s*, *p*, *d*, and *f*.
16. What is the maximum number of *s* orbitals found in a given electron shell? The maximum number of *p* orbitals? Of *d* orbitals? Of *f* orbitals?
17. Match the values of ℓ shown in the table with orbital type (*s*, *p*, *d*, or *f*).

ℓ Value	Orbital Type
3	______
0	______
1	______
2	______

18. Sketch a picture of the 90% boundary surface of an *s* orbital and the p_x orbital. Be sure the latter drawing shows why the *p* orbital is labeled p_x and not p_y, for example.
19. Complete the following table.

Orbital Type	Number of Orbitals in a Given Subshell	Number of Nodal Planes
s	______	______
p	______	______
d	______	______
f	______	______

NUMERICAL AND OTHER QUESTIONS

Electromagnetic Radiation

20. The colors of the visible spectrum, and the wavelengths corresponding to the colors, are given in Figure 7.3.
   (a) What colors of light involve less energy than green light?
   (b) Which color of light has photons of greater energy, yellow or blue?
   (c) Which color of light has the greater frequency, blue or green?

21. The regions of the electromagnetic spectrum are given in Figure 7.3. Answer the following questions based on this figure:
   (a) Which type of radiation involves less energy, x-rays or microwaves?
   (b) Which radiation has the higher frequency, radar or red light?
   (c) Which radiation has the longer wavelength, ultraviolet or infrared light?

22. An FM radio station has a frequency of 88.9 MHz (1 MHz = 10^6 Hz, or cycles per second). What is the wavelength of this radiation in meters?

23. The U.S. Navy has a system for communicating with submerged submarines. The system uses radio waves with a frequency of 76 s^{-1}. What is the wavelength of this radiation in meters? In miles?

24. Violet light has a wavelength of about 410 nm. What is its frequency? Calculate the energy of one photon of violet light. What is the energy of 1.0 mol of violet photons? Compare the energy of photons of violet light with those of red light. Which is more energetic and by what factor? (See Example 7.2.)

25. Green light has a wavelength of approximately 5.0×10^2 nm. What is the frequency of this light? What is the energy in joules of one photon of green light? What is the energy in joules of 1.0 mol of photons of green light?

26. The most prominent line in the line spectrum of aluminum is found at 396.15 nm. What is the frequency of this line? What is the energy of one photon with this wavelength? Of 1.00 mol of these photons?

27. The most prominent line in the spectrum of magnesium is 285.2 nm. Others are found at 383.8 and 518.4 nm. In what region of the electromagnetic spectrum are these lines found? Which is the most energetic line? What is the energy of 1 mol of photons of the most energetic line? How much more energetic is a photon of this light compared with a photon associated with the least energetic line?

28. The most prominent line in the spectrum of mercury is found at 253.652 nm. Other lines are found at 365.015 nm, 404.656 nm, 435.833 nm, and 1013.975 nm.
   (a) Which of these lines represents the most energetic light?
   (b) What is the frequency of the most prominent line? What is the energy of one photon with this wavelength?
   (c) Are any of these lines mentioned previously found in the spectrum of mercury shown in Figure 7.9? What color or colors are these lines?

29. The most prominent line in the spectrum of neon is found at 865.438 nm. Other lines are found at 837.761 nm, 878.062 nm, 878.375 nm, and 1885.387 nm.
   (a) Which of these lines represents the most energetic light?
   (b) What is the frequency of the most prominent line? What is the energy of one photon with this wavelength?
   (c) Are any of the lines mentioned here found in the spectrum of neon shown in Figure 7.9? What color or colors are these lines?

30. Place the following types of radiation in order of increasing energy per photon:
   (a) yellow light from a sodium lamp
   (b) x-rays from an instrument in a dentist's office
   (c) microwaves in a microwave oven
   (d) your favorite FM music station at 91.7 MHz

31. Place the following types of radiation in order of increasing energy per photon:
   (a) radar signals
   (b) radiation within a microwave oven
   (c) γ rays from a nuclear reaction
   (d) red light from a neon sign
   (e) ultraviolet radiation from a sun lamp

Photoelectric Effect

32. To cause a cesium atom on a metal surface to lose an electron, an energy of 2.0×10^2 kJ/mol is required. Calculate the longest possible wavelength of light that can ionize a cesium atom. What is the region of the electromagnetic spectrum in which this radiation is found?

33. Assume you are an engineer designing a space probe to land on a distant planet. You wish to use a switch that works by the photoelectric effect. The metal you wish to use in your device requires 6.7×10^{-19} J/atom to remove an electron. You know that the atmosphere of the planet on which your device must work filters out all wavelengths of light less than 540 nm. Will your device work on the planet in question? Why or why not?

Atomic Spectra and the Bohr Atom

34. Consider only transitions involving the following energy levels for the H atom.

______ $n = 5$
______ $n = 4$
______ $n = 3$
______ $n = 2$
______ $n = 1$

   (a) How many emission lines are possible, considering only the five quantum levels?
   (b) Photons of the lowest frequency are emitted in a transition from the level with $n =$ __ to a level with $n =$ __.
   (c) The emission line having the shortest wavelength corresponds to a transition from the level with $n =$ __ to the level with $n =$ __.
   (d) The emission line having the highest energy corresponds to a transition from the level with $n =$ __ to the level with $n =$ __.

35. Consider only transitions involving the following energy levels for the hydrogen atom.

______ $n = 4$
______ $n = 3$
______ $n = 2$
______ $n = 1$

   (a) How many emission lines are possible, considering only the four quantum levels?
   (b) Photons of the highest energy are emitted in a transition from the level with $n =$ __ to a level with the $n =$ __.
   (c) The emission line having the longest wavelength corresponds to a transition from the level with $n =$ __ to the level with $n =$ __.

36. If energy is absorbed by a hydrogen atom in its ground state, the atom is excited to a higher energy state. For example, the excitation of an electron from the level with $n = 1$ to the level with $n = 3$ requires radiation with a wavelength of 102.6 nm. Which of the following transitions would require radiation of *longer wavelength* than this?
   (a) $n = 2$ to $n = 4$
   (b) $n = 1$ to $n = 4$
   (c) $n = 1$ to $n = 5$
   (d) $n = 3$ to $n = 5$

37. The energy emitted when an electron moves from a higher energy state to one of lower energy in any atom can be observed as electromagnetic radiation.
   (a) Which involves the emission of less energy in the H atom, an electron moving from $n = 4$ to $n = 2$ or an electron moving from $n = 3$ to $n = 2$?
   (b) Which involves the emission of the greater energy in the H atom, an electron changing from $n = 4$ to $n = 1$ or an electron changing from $n = 5$ to $n = 2$? Explain fully.

38. Calculate the wavelength of light emitted when an electron changes from $n = 3$ to $n = 1$ in the H atom. In what region of the spectrum is this radiation found?

39. A line in the Balmer series of emission lines of excited H

atoms has a wavelength of 410.2 nm. This series originates from electrons changing from high energy levels to the $n = 2$ level. To account for the 410.2 nm line, what is the value of n for the initial level?

40. An electron moves from the $n = 5$ to the $n = 1$ quantum level and emits a photon with an energy of 2.093×10^{-18} J. How much energy must the atom absorb to move an electron from $n = 1$ to $n = 5$?

41. Excited H atoms have many emission lines. One series of lines, called the Pfund series, occurs in the infrared region. It results when an electron changes from higher levels to a level with $n = 5$. Calculate the wavelength and frequency of the lowest energy line of this series.

De Broglie and Matter Waves

42. An electron moves with a velocity of 2.5×10^8 cm/s. What is its wavelength?

43. A beam of electrons ($m = 9.11 \times 10^{-31}$ kg/electron) has an average speed of 1.3×10^8 m/s. What is the wavelength corresponding to the average speed?

44. Calculate the wavelength (in nanometers) associated with a 1.0×10^2-g golf ball moving at 30. m/s (about 67 mph). How fast must the ball travel to have a wavelength of 5.6×10^{-3} nm?

45. A rifle bullet (mass = 1.50 g) is moving with a velocity of 7.00×10^2 mph. What is the wavelength associated with this bullet?

Quantum Mechanics

46. Complete the following table:

Atomic Property	Quantum Number
orbital size	______
relative orbital orientation	______
orbital shape	______

47. An orbital is designated 2*s*, and another orbital is designated 4*p*. Which is the larger orbital? Which has more planar nodes?

48. Answer the following questions:
(a) When $n = 4$, what are the possible values of ℓ?
(b) When ℓ is 2, what are the possible values of m_ℓ?
(c) For a 4*s* orbital, what are the possible values of n, ℓ, and m_ℓ?
(d) For a 4*f* orbital, what are the possible values of n, ℓ, and m_ℓ?

49. Answer the following questions:
(a) When $n = 4$, $\ell = 2$, and $m_\ell = -1$, to what orbital type does this refer? (Give the orbital label, such as 1*s*.)
(b) How many orbitals occur in the $n = 5$ electron shell? How many subshells? What are the letter labels of the subshells?
(c) If a subshell is labeled *f*, how many orbitals occur in the subshell? What are the values of m_ℓ?

50. A possible excited state of the H atom has the electron in a 4*p* orbital. List all possible sets of quantum numbers n, ℓ, and m_ℓ for this electron.

51. A possible excited state for the H atom has an electron in a 5*d* orbital. List all possible sets of quantum numbers n, ℓ, and m_ℓ for this electron.

52. How many subshells occur in the electron shell with the principal quantum number $n = 4$?

53. How many subshells occur in the electron shell with the principal quantum number $n = 5$?

54. Explain briefly why each of the following is not a possible set of quantum numbers for an electron in an atom.
(a) $n = 2$, $\ell = 2$, $m_\ell = 0$
(b) $n = 3$, $\ell = 0$, $m_\ell = -2$
(c) $n = 6$, $\ell = 0$, $m_\ell = 1$

55. Which of the following represent valid sets of quantum numbers? For a set that is invalid, explain briefly why it is not correct.
(a) $n = 3$, $\ell = 3$, $m_\ell = 0$
(b) $n = 2$, $\ell = 1$, $m_\ell = 0$
(c) $n = 6$, $\ell = 5$, $m_\ell = -1$
(d) $n = 4$, $\ell = 3$, $m_\ell = -4$

56. What is the maximum number of orbitals that can be identified by each of the following sets of quantum numbers? When "none" is the correct answer, explain your reasoning.
(a) $n = 3$, $\ell = 0$, $m_\ell = +1$
(b) $n = 5$, $\ell = 1$
(c) $n = 7$, $\ell = 5$
(d) $n = 4$, $\ell = 2$, $m_\ell = -2$

57. What is the maximum number of orbitals that can be identified by each of the following sets of quantum numbers? When "none" is the correct answer, explain your reasoning.
(a) $n = 4$, $\ell = 3$
(b) $n = 5$
(c) $n = 2$, $\ell = 2$
(d) $n = 3$, $\ell = 1$, $m_\ell = -1$

58. How many planar nodes are associated with each of the following orbitals? (a) 2*s*; (b) 5*d*; and (c) 5*f*.

59. How many planar nodes are associated with each of the following atomic orbitals? (a) 4*f*; (b) 2*p*; and (c) 6*s*.

60. State which of the following orbitals cannot exist according to the quantum theory: 2*s*, 2*d*, 3*p*, 3*f*, 4*f*, and 5*s*. Briefly explain your answers.

61. State which of the following orbitals can exist and which cannot according to the quantum theory: 3*p*, 4*s*, 2*f*, and 1*p*. Briefly explain your answers.

62. Write a complete set of quantum numbers (n, ℓ, and m_ℓ) that quantum theory allows for each of the following orbitals: (a) 2*p*, (b) 3*d*, and (c) 4*f*.

63. Write a complete set of quantum numbers (n, ℓ, and m_ℓ) that quantum theory allows for each of the following orbitals: (a) 5*f*, (b) 4*d*, and (c) 2*s*.

64. A given orbital is labeled by the magnetic quantum number $m_\ell = -1$. This could not be a (an)
(a) *f* orbital (c) *p* orbital
(b) *d* orbital (d) *s* orbital

65. A particular orbital has $n = 4$, $\ell = 2$, and $m_\ell = -2$. This orbital must be: (a) $3p$, (b) $4p$, (c) $5d$, or (d) $4d$.

GENERAL QUESTIONS

66. Radiation in the ultraviolet region of the electromagnetic spectrum is quite energetic. It is this radiation that causes dyes to fade and your skin to burn. If you are bombarded with 1.00 mol of photons with a wavelength of 305. nm, what amount of energy (in kilojoules per mole of photons) are you being subjected to?

67. An AM radio station broadcasts at a frequency of 6.00×10^2 KHz (KHz = 1 kilohertz = $1000\ s^{-1}$). What is the wavelength of this signal in meters? What is the energy of one photon of this frequency? Compare with the energy of a photon of red light with $\lambda = 685$ nm (page 301).

68. Exposure to high doses of microwaves can cause damage. Estimate how many photons, with $\lambda = 12$ cm, must be absorbed to raise the temperature of your eye by 3.0 °C. Assume the mass of an eye is 11 g and its heat capacity is 4.0 J/g · K.

69. An advertising sign gives off red light and green light.
   (a) Which light has the higher energy photons?
   (b) One of the colors has a wavelength of 680 nm and the other has a wavelength of 500 nm. Identify which color has which wavelength.
   (c) Which light has the higher frequency?

70. When *Sojouner* landed on Mars in 1997, the planet was approximately 7.8×10^7 km from the earth. How long did it take for the television picture signal to reach earth from Mars?

71. Assume your eyes receive a signal consisting of blue light, $\lambda = 470$ nm. The energy of the signal is 2.50×10^{-14} J. How many photons reach your eyes?

72. If sufficient energy is absorbed by an atom, an electron can be lost by the atom and a positive ion formed. The amount of energy required is called the ionization energy. In the H atom, the ionization energy is that required to change the electron from $n = 1$ to $n =$ infinity. (See *A Closer Look: Experimental Evidence for Bohr's Theory,* page 311.) Calculate the ionization energy for He^+ ion. Is the ionization energy of He^+ more or less than that of H? (Bohr's theory applies to He^+ because it, like the H atom, has a single electron. The electron energy, however is now given by $E = -Z^2Rhc/n^2$, where Z is the atomic number of helium.)

73. What is the shortest wavelength photon an excited H atom can emit? Explain briefly.

74. Hydrogen atoms absorb energy so that the electrons are excited to the $n = 7$ energy level. Electrons then undergo these transitions, among others: (a) $n = 7 \rightarrow n = 1$; (b) $n = 7 \rightarrow n = 6$; and (c) $n = 2 \rightarrow n = 1$. Which transition produces a photon with (i) the smallest energy; (ii) the highest frequency; (iii) the shortest wavelength?

75. The wave function for a $2p$ orbital (in the H atom) is plotted against the distance from the nucleus in *A Closer Look: Atomic Orbitals.* What is the approximate distance from the nucleus at which the maximum probability density is reached?

76. Rank the following orbitals in the H atom in order of increasing energy: $3s$, $2s$, $2p$, $4s$, $3p$, $1s$, and $3d$.

77. How many orbitals in an atom can have the following quantum number or designation?
   (a) $3p$ (d) $6d$ (g) $n = 5$
   (b) $4p$ (e) $5d$ (h) $7s$
   (c) $4p_x$ (f) $5f$

78. Answer the following questions as a summary quiz on the chapter.
   (a) The quantum number n describes the ______ of an atomic orbital.
   (b) The shape of an atomic orbital is given by the quantum number ______.
   (c) A photon of orange light has ______ (less or more) energy than a photon of yellow light.
   (d) The maximum number of orbitals that may be associated with the set of quantum numbers $n = 4$ and $\ell = 3$ is ______.
   (e) The maximum number of orbitals that may be associated with the quantum number set $n = 3$, $\ell = 2$, and $m_\ell = -2$ is ______.
   (f) Label each of the following orbital pictures with the appropriate letter:

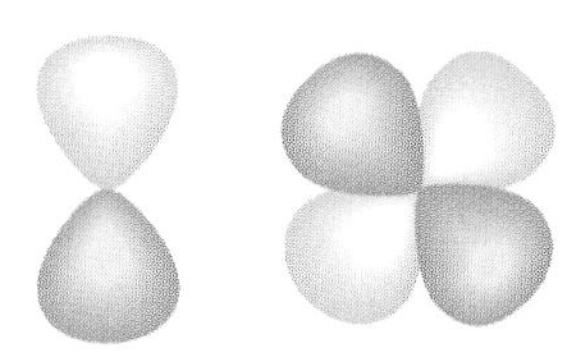

   (g) When $n = 5$, the possible values of ℓ are ______.
   (h) The maximum number of orbitals that can be assigned to the $n = 4$ shell is ______.

79. Answer the following questions as a review of this chapter:
   (a) The quantum number n describes the ______ of an atomic orbital and the quantum number ℓ describes its ______.
   (b) When $n = 3$, the possible values of ℓ are ______.
   (c) What type of orbital corresponds to $\ell = 3$? ______
   (d) For a $4d$ orbital, the value of n is ______, the value of ℓ is ______, and a possible value of m_ℓ is ______.
   (e) Each drawing represents a type of atomic orbital. Give the letter designation for the orbital, its value of ℓ, and specify the number of nodal planes.

letter = ______ ______ ______
ℓ value = ______ ______ ______
nodal planes = ______ ______ ______

(f) An atomic orbital with three nodal planes is ______.
(g) Which of the following orbitals cannot exist according to modern quantum theory: 2*s*, 3*p*, 2*d*, 3*f*, 5*p*, 6*p*?
(h) Which of the following is *not* a valid set of quantum numbers?

n	ℓ	m_ℓ
3	2	1
2	1	2
4	3	0

(i) What is the maximum number of orbitals that can be associated with each of the following sets of quantum numbers? (One possible answer is "none.")
(i) $n = 2$ and $\ell = 1$
(ii) $n = 3$
(iii) $n = 3$ and $\ell = 3$
(iv) $n = 2$, $\ell = 1$, and $m_\ell = 0$

CONCEPTUAL QUESTIONS

80. Suppose you live in a different universe where a different set of quantum numbers is required to describe the atoms of that universe. These quantum numbers have the following rules:

N, principal	1, 2, 3, . . . , ∞
L, orbital	$= N$
M, magnetic	−1, 0, +1

How many orbitals are there altogether in the first three electron shells?

81. Bohr pictured the electrons of the atom as being located in definite orbits about the nucleus, just as the planets orbit the sun. Criticize this model in view of the quantum mechanical model.

82. What does the wave–particle duality mean? What are its implications in our modern view of atomic structure?

83. In what way does Bohr's model of the atom violate the uncertainty principle?

84. Which of these are observable?
(a) position of electron in H atom
(b) frequency of radiation emitted by H atoms
(c) path of electron in H atom
(d) wave motion of electrons
(e) diffraction patterns produced by electrons
(f) diffraction patterns produced by light
(g) energy required to remove electrons from H atoms
(h) an atom
(i) a molecule
(j) a water wave

85. In principle, is it possible to determine
(a) the energy of an electron in the H atom with high precision and accuracy
(b) the position of a high-speed electron with high precision and accuracy
(c) at the same time, both the position and energy of a high-speed electron with high precision and accuracy.

86. The Chemical Puzzler on page 291 shows that light of different colors can come from a "neon" sign or from certain salts when placed in a burning organic liquid. ("Neon" signs are glass tubes filled with neon, argon, and other gases, and the gases are excited by an electric current. They are very similar in this regard to common fluorescent lights, although the color of the latter comes from the phosphor that coats the inside of the tube.) What do these two sources of colored light have in common? How is the light generated in each case?

CHALLENGING QUESTIONS

87. Assume an electron is assigned to the 1*s* orbital in the H atom. Is the electron density zero at a distance of 0.40 nm from the nucleus? (See *A Closer Look: Atomic Orbitals.*)

88. Is the electron density of a $2p_z$ electron at a point on the *x*-axis 0.05 nm from the nucleus higher, lower, or the same when compared with the same distance on the *z*-axis?

89. Cobalt-60 is a radioactive isotope used in medicine for the treatment of certain cancers. It produces β particles and γ rays, the latter having energies of 1.173 and 1.332 MeV. (1 MeV = 1 million electron volts and 1 eV = 9.6485×10^4 J/mol.) What are the wavelength and frequency of a γ-ray photon with an energy of 1.173 MeV?

90. Imagine the nucleus of an H atom is located at the origin (the zero point) of an *x, y, z* graph.
(a) Assume you move along the *x*-axis to a distance *d* away from the nucleus; the probability of finding the 1*s* electron at $x = d$ is, say, 0.01. Is the probability of finding the electron at $y = d$ greater than, less than, or equal to 0.01?
(b) The probability of finding a $2p_x$ electron at $x = d$ is, say, 0.001. Is the probability of finding the electron at $y = d$ greater than, less than, or equal to 0.001?

Summary Question

91. Technetium is not found naturally on earth; it must be synthesized in the laboratory. Nonetheless, because it is radioactive it has valuable medical uses. For example, the element in the form of sodium pertechnetate ($NaTcO_4$) is used in imaging studies of the brain, thyroid, and salivary glands and in renal blood flow studies, among other things.
(a) In what group and period of the periodic table is the element found?
(b) The valence electrons of technetium are found in the 5*s* and 4*d* subshells. What is a set of quantum numbers (n, ℓ, and m_ℓ) for one of the electrons of the 5*s* subshell?
(c) Technetium emits a γ ray with an energy of 0.141 MeV. (1 MeV = 1 million electron volts where 1 eV = 9.6485 $\times 10^4$ J/mol.) What are the wavelength and frequency of a γ-ray photon with an energy of 0.141 MeV?

(d) To make $NaTcO_4$, the metal is dissolved in nitric acid

$$7\ HNO_3(aq) + Tc(s) \longrightarrow HTcO_4(aq) + 7\ NO_2(g) + 3\ H_2O(\ell)$$

and the product, $HTcO_4$, is treated with NaOH to make $NaTcO_4$.

(i) Write a balanced equation for the reaction of $HTcO_4$ with NaOH.

(ii) If you begin with 4.5 mg of Tc metal, how much $NaTcO_4$ can be made? What mass of NaOH (in grams) is required to convert all of the $HTcO_4$ into $NaTcO_4$?

INTERACTIVE GENERAL CHEMISTRY CD-ROM

Screen 7.4 The Electromagnetic Spectrum

(a) Use the spectrum "tool" to answer the following questions:
1. As you move the slider from the blue region to the red region, what happens to the wavelength of the light? To its frequency?
2. Place the slider somewhere in the blue region of the spectrum. What is the approximate wavelength of this radiation? In nanometers? In meters?
3. Place the slider somewhere in the orange region of the spectrum. What is the approximate wavelength of this radiation? In nanometers? In meters?
4. Which color of light in the visible spectrum has the longest wavelength? The shortest wavelength?

(b) Which color of light in the visible spectrum has the highest frequency?

(c) Is the frequency of radiation used in a microwave oven higher or lower than that of your favorite FM radio station? (Assume the station broadcasts at a frequency of 91.7 MHz where MHz = $10^6\ s^{-1}$)

(d) Is the wavelength of x-rays shorter or longer than that of ultraviolet light?

Screen 7.6 Atomic Line Spectra

(a) Why is the spectrum of light in the animation on this screen called a "line" spectrum? Explain fully.

(b) Examine the Balmer equation. As n increases from 3 to higher values, what happens to the wavelength of the radiation emitted by the excited atom? Does the wavelength become longer or shorter? How does this observation relate to the experimental observation in part (a)?

Screen 7.7 Bohr's Model of the Hydrogen Atom

(a) How can we detect the movement of an electron from a higher energy level in an atom to a lower level?

(b) What does it mean when an electron is moved to an energy level with n = infinity?

(c) How could you detect the movement of an electron from an energy level, say $n = 2$, to a higher level, say $n = 4$? What experiment would you do?

(d) Look at the video of the electric pickle on Screen 7.7P. Why does the pickle "glow" when an electric current is applied and why is it thought that the light given off is yellow?

Screen 7.11 Shells, Subshells, and Orbitals

(a) An example of an orbital is $4p_z$. In what shell is the electron located? In what subshell? In what orbital within the subshell?

(b) What is the relation between the value of the quantum number n and the number of subshells in a given shell?

(c) Is there a subshell with four orbitals?

Screen 7.13 Shapes of Atomic Orbitals

(a) Examine the $1s$, $2s$, and $3s$ orbitals. What is their general shape? How do these differ from one another? What are their similarities?

(b) Examine the $2p$ and $3p$ orbitals. What is their general shape? How do these differ from one another? What are their similarities?

(c) Examine the $3d$ orbitals. How would you describe their general shape?

(d) What is the relation between ℓ and the number of planar nodes for an orbital?

Chapter 8

Atomic Electron Configurations and Chemical Periodicity

(Photo, C. D. Winters; model, S. M. Young)

A CHEMICAL PUZZLER

In the photograph hot steel wool is plunged into a flask containing chlorine gas. Very quickly you see a brown cloud of iron(III) chloride in the flask.

$$2\ Fe(s) + 3\ Cl_2(g) \longrightarrow 2\ FeCl_3(s)$$

This compound, a model of which is shown here, is ionic, being composed of iron(III) ions (Fe^{3+}) and chloride ions (Cl^-). The oxidation–reduction reaction that forms the compound requires that electrons be transferred from iron to chlorine. Why is iron such a good reducing agent? Why is chlorine such a good oxidizing agent? Why is solid $FeCl_3$ magnetic, that is, why is it attracted to a magnet? Also, why do the iron(II) and iron(III) cations in the mineral magnetite, Fe_3O_4, make this mineral ferromagnetic?

We developed the modern view of the atom in the previous chapter. That model suggests that electrons can be arranged in shells in the space around the nucleus; each shell is distinguished by the quantum number n and consists of one or more subshells. Each subshell is described by the quantum number ℓ and contains one or more orbitals. The picture that emerges is a satisfying one that reflects the order of the natural world around us.

The model developed in Chapter 7 accurately describes atoms or ions such as H and He^+ that have a single electron. To be useful, though, a model must be applicable to atoms with more than one electron, that is, to all the other known elements. Because the chemical properties of atoms depend on their electronic structure—the number and arrangement of electrons in the atom—one objective of this chapter is to develop a workable picture of the electronic structure of elements other than hydrogen.

Another objective of this chapter is to explore some of the physical properties of atoms, among them the ease with which atoms lose or gain electrons to form ions and the sizes of atoms and ions. These properties are directly related to the arrangement of electrons in atoms and thus to the chemistry of the elements and their compounds.

8.1 ELECTRON SPIN

Three quantum numbers (n, ℓ, and m_ℓ) allow us to define the orbital for an electron. To describe an electron in a multielectron atom completely, however, one more quantum number, the **electron spin magnetic quantum number, m_s,** is required.

See the *Saunders Interactive General Chemistry CD-ROM,* Screen 8.2, Electron Spin, and Screen 8.3, Spinning Electrons and Magnetism.

Around 1920, theoretical chemists realized that, because electrons interact with a magnetic field, there must be one more concept to describe the electronic structure of atoms. It was soon verified experimentally that the electron behaves as though it has a spin, just as the earth has a spin. To understand this property and its relation to atomic structure, we should understand something of the general phenomenon of magnetism.

Magnetism

The needle of a compass at a given location on the earth always points in a given direction, no matter how the compass is moved. The needle is a magnet, such as a piece of iron. In 1600, William Gilbert (1544–1603) concluded that the earth is also a large spherical magnet giving rise to a magnetic field that surrounds the planet (Figure 8.1). The compass needle is "drawn," or "attracted," into this magnetic field, one end of the needle pointing approximately to the earth's geographic North Pole. Thus, we say the end of the compass needle pointing north is the magnet's "magnetic north pole" or simply its "north pole," designated N. The other end of the needle is its "south pole," designated S.

Identical magnetic poles (N–N or S–S) repel each other, and opposite poles (N–S) attract (Figure 8.2). Because the magnetic north pole of the compass needle points to the earth's geographic North Pole, this must mean that this pole is actually the earth's magnetic south pole.

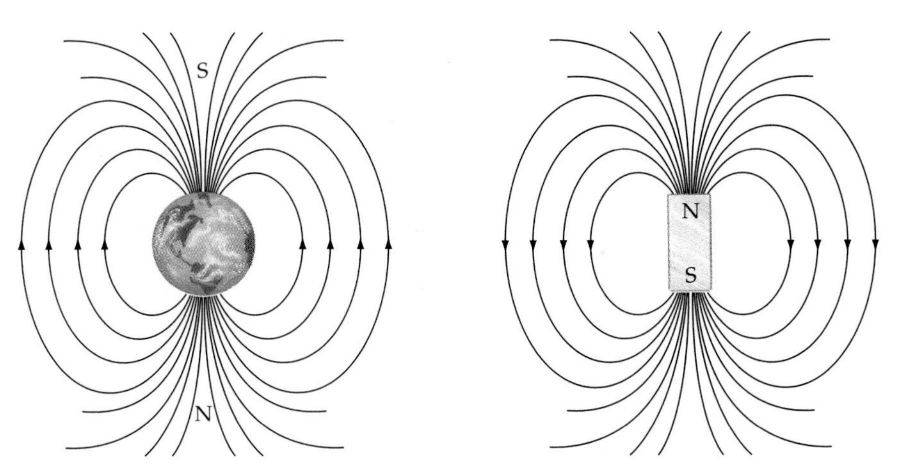

Figure 8.1 The magnetic fields of the earth and of a bar magnet. The magnetic field comes out of one pole, arbitrarily called the "north magnetic pole," N, and loops toward the "south magnetic pole," S. The geographic North Pole of the earth, named before the introduction of the term "magnetic pole," is the magnetic south pole.

Liquid oxygen (boiling point 90.2 K) clings to the poles of a strong magnet. The element is paramagnetic because it has unpaired electrons. For more information, see Chapter 10 and Figure 10.20.

Paramagnetism and Unpaired Electrons

Most substances—chalk, sea salt, cloth—are slightly repelled by a strong magnet. They are said to be **diamagnetic.** In contrast, many metals and other compounds are attracted to a magnetic field. Such substances are generally called **paramagnetic,** and the magnitude of the effect can be determined with an apparatus such as that illustrated in Figure 8.3.

The magnetism of most paramagnetic materials is so weak that you can only observe it in the presence of a strong magnetic field. For example, the oxygen we breathe is paramagnetic; it sticks to the poles of a strong magnet. It does not cling to very weak magnets, however, such as those in the gadgets you stick on

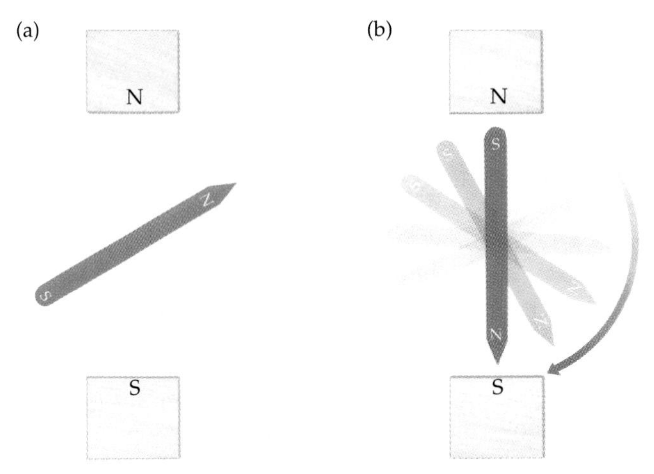

Figure 8.2 A magnetic needle in a strong magnetic field. (a) The repulsion of similar poles and (b) the attraction of opposite poles cause the needle to line up as shown, the N pole of the needle facing the S pole of the magnet.

the door of your refrigerator. Other materials are so strongly magnetic, however, that you readily observe their effect. Examples include the mineral magnetite, Fe_3O_4, and the alloy called "Alnico" (for Al, Ni, and Co). Often referred to as **ferromagnetic** materials, these substances are used to make the little magnetic devices you use in your home.

Paramagnetism and ferromagnetism arise from electron spins. An electron in an atom has the magnetic properties expected for a spinning, charged particle. What is important to us is the relation of that property to the arrangement of electrons in atoms. Experiments have shown that, if you place an atom with a single unpaired electron in a magnetic field, only two orientations are possible for the electron spin. That is, *electron spin is quantized.* One orientation is associated with a spin quantum number value of $m_s = +\frac{1}{2}$ and the other with an m_s value of $-\frac{1}{2}$ (Figure 8.4).

When one electron is assigned to an orbital in an atom, the electron's spin orientation can take either value of m_s. We observe experimentally that hydrogen atoms, each of which has a single electron, are paramagnetic; when an external magnetic field is applied, the electron magnets align with the field—like the needle of a compass—and experience an attractive force. In helium, two electrons are assigned to the same 1*s* orbital, and we can confirm by experiment that *helium is diamagnetic.* To account for this observation, we assume that the two electrons assigned to the same orbital have opposite spin orientations; we say that their spins are **paired.** This means the magnetic field of one electron is "canceled out" by the magnetic field of the second of opposite spin.

In summary, *paramagnetism occurs in substances in which the constituent ions or atoms contain unpaired electrons.* Atoms in which all electrons are paired with partners of opposite spin are diamagnetic. This explanation opens the way to understanding the electron configurations of atoms with more than one electron.

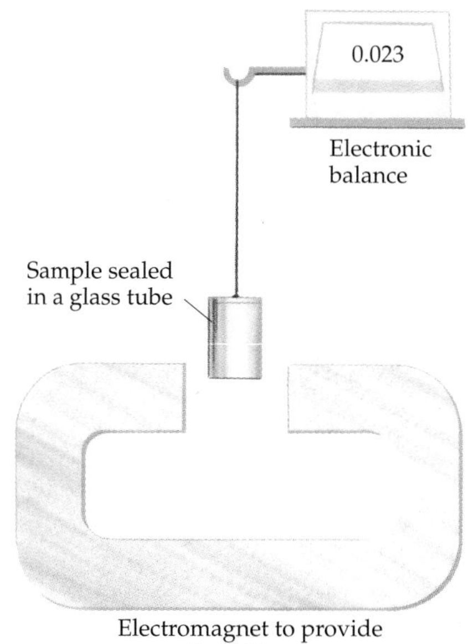

Figure 8.3 A magnetic balance used to measure the magnetic properties of a sample. The sample is first weighed with the electromagnet turned off. The magnet is then turned on and the sample reweighed. If the substance is paramagnetic, the sample is drawn into the magnetic field and the *apparent* weight increases.

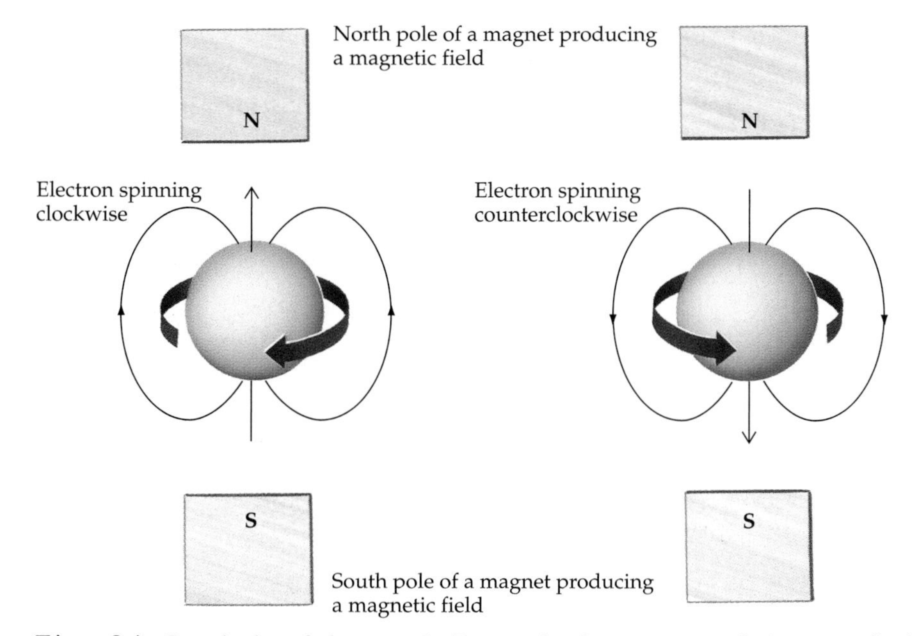

Figure 8.4 Quantization of electron spin. Because the electron acts as a "micromagnet" relative to a magnetic field, only two spins are possible. Any other orientation is forbidden. We therefore can say that the spin of an electron is quantized.

A CLOSER LOOK *Paramagnetism and Ferromagnetism*

Magnetic materials are relatively common, and many are important in our economy. Examples are the magnets in the little objects you stick on refrigerator doors. Magnets are also found in stereo speakers and in telephone handsets, and magnetic oxides are used in recording tapes and computer disks.

The magnetic materials we use are ferromagnetic. The magnetic effect of ferromagnetic materials is much larger than for paramagnetic ones. Ferromagnetism occurs when the spins of unpaired electrons in a cluster of atoms in the solid (called a *domain*) align themselves in the same direction. Only the metals of the iron, cobalt, and nickel subgroups, and a few other metals such as neodymium, exhibit this property. They are also unique in that, once the domains are aligned in a magnetic field, the metal is permanently magnetized.

Many alloys exhibit greater ferromagnetism than do the pure metals themselves. One example is alnico (an alloy of aluminum, nickel and cobalt), and another is an alloy of neodymium, iron, and boron.

Audio and video tapes are plastics coated with crystals of ferromagnetic Fe_2O_3, CrO_2, or other metal oxide. The recording head uses an electromagnetic field to create a varying magnetic field based on signals from a microphone. This magnetizes the tape as it passes through the head, the strength and direction of magnetization varying with the frequency of the sound to be recorded. When the tape is played back, the magnetic field of the moving tape induces a current, which is amplified and sent to the speakers.

Ferromagnetism is temperature-dependent. Substances gain stability by aligning electron spins within micro-

Many common consumer products contain magnets. The round cylinder in the foreground is a cow magnet. It is "fed" to a cow and in the cow's stomach it attracts magnetic junk that the cow may eat and prevents the junk from traveling any farther in the cow's digestive system. (*C. D. Winters*)

8.2 THE PAULI EXCLUSION PRINCIPLE

See the *Saunders Interactive General Chemistry CD-ROM*, Screen 8.4, The Pauli Exclusion Principle.

To make the quantum theory consistent with experiment, the Austrian physicist Wolfgang Pauli (1900–1958) stated in 1925 his **exclusion principle:** *No two electrons in an atom can have the same set of four quantum numbers* (n, ℓ, m_ℓ, and m_s). This principle leads to yet another important conclusion: *No atomic orbital can contain more than two electrons.*

The $1s$ orbital of the H atom has the set of quantum numbers $n = 1$, $\ell = 0$, and $m_\ell = 0$. No other set is possible. If an electron is in this orbital, the electron spin direction must also be specified. Let us represent an orbital by a "box" and the electron by an arrow. A representation of the H atom is then as follows, in which the "electron spin arrow" is shown arbitrarily as pointing upward:

Electron in $1s$ orbital = [↑] ($1s$) Quantum number set $n = 1,\ \ell = 0,\ m_\ell = 0,\ m_s = +\frac{1}{2}$

Orbitals are not literally things or boxes in which electrons are placed. It is not conceptually correct to talk about electrons in orbitals or occupying orbitals, although it is commonly done for the sake of simplicity.

If only one electron is in a given orbital, the electron spin arrow may point either up or down. Thus, an equally valid description would be

Electron in $1s$ orbital = [↓] ($1s$) Quantum number set $n = 1,\ \ell = 0,\ m_\ell = 0,\ m_s = -\frac{1}{2}$

The two preceding "orbital box" diagrams are equally appropriate for the H atom in its ground state (with no magnetic field): one electron is in the $1s$

scopic domains, and domains can be aligned by placing the material in an external magnetic field. At higher temperatures, however, the atoms vibrate more and more within their places in the solid, and the spins can become randomly oriented. The temperature beyond which the energy is sufficient to overcome the aligning forces is called the *Curie temperature,* T_{Curie}. A substance can be ferromagnetic only below this temperature.

Material	T_{Curie} **(K)**
Iron	1043
Cobalt	1388
Nickel	627
Gadolinium	293

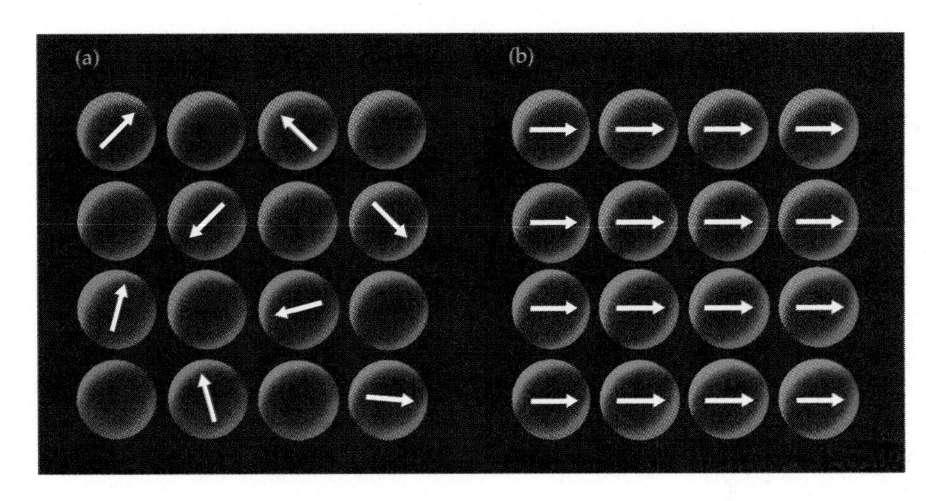

(a) Paramagnetism: the centers (atoms or ions) with magnetic moments are not aligned unless the substance is in a magnetic field. (b) Ferromagnetism: the spins of unpaired electrons in a cluster of atoms or ions align in the same direction.

orbital. For a helium atom, which has two electrons, both electrons are assigned to the $1s$ orbital. From the Pauli principle, you know that each electron must have a different set of quantum numbers, so the orbital box picture now is:

$$\text{Two electrons in } 1s \text{ orbital} = \overset{1s}{\boxed{\uparrow\downarrow}}$$

This electron has $n = 1$, $\ell = 0$, $m_\ell = 0$, $m_s = -\frac{1}{2}$

This electron has $n = 1$, $\ell = 0$, $m_\ell = 0$, $m_s = +\frac{1}{2}$

Each of the two electrons in the $1s$ orbital of an He atom has a *different set of the four quantum numbers.* The first three numbers of a set tell you this is a $1s$ orbital. There are only two choices for the fourth number, $m_s = +\frac{1}{2}$ or $-\frac{1}{2}$. Thus, *the 1s orbital, and any other atomic orbital, can be occupied by no more than two electrons, and these two electrons must have opposite spin directions.* The consequence is that the helium atom is diamagnetic, as experimentally observed.

The $n = 1$ electron shell in any atom can accommodate no more than two electrons. But what about the $n = 2$ shell? There are $n^2 = 4$ orbitals in the $n = 2$ shell: one s orbital and three p orbitals (Table 8.1). Because each orbital can be occupied by two electrons and no more, the $2s$ orbital is assigned to as many as two electrons, and the three $2p$ orbitals accommodate as many as six electrons, for a maximum of eight electrons with $n = 2$. This analysis is carried further for the electron shells normally observed in the known elements in Table 8.1.

The number of orbitals in the nth electron shell is n^2, and the maximum number of electrons in the shell is $2n^2$ because each orbital can be assigned to two electrons.

The results expressed in this table were predicted by the Schrödinger theory and have been confirmed by experiment.

Note that n subshells occur in the nth shell.

Table 8.1 • NUMBER OF ELECTRONS ACCOMMODATED IN ELECTRON SHELLS AND SUBSHELLS WITH $n = 1$ TO 6

Electron Shell (n)	Subshells Available	Orbitals Available ($2\ell + 1$)	Number of Electrons Possible in Subshell [$2(2\ell + 1)$]	Maximum Electrons Possible for nth Shell ($2n^2$)
1	s	1	2	2
2	s	1	2	8
	p	3	6	
3	s	1	2	18
	p	3	6	
	d	5	10	
4	s	1	2	32
	p	3	6	
	d	5	10	
	f	7	14	
5	s	1	2	50
	p	3	6	
	d	5	10	
	f	7	14	
	g*	9	18	
6	s	1	2	72
	p	3	6	
	d	5	10	
	f	7	14	
	g*	9	18	
	h*	11	22	

*These orbitals are not used in the ground state of any known element.

8.3 ATOMIC SUBSHELL ENERGIES AND ELECTRON ASSIGNMENTS

See the *Saunders Interactive General Chemistry CD-ROM,* Screen 8.5, Atomic Subshell Energies.

Our goal is to understand the distribution of electrons in atoms with many electrons. These atoms can be "built" by assigning electrons to shells (defined by the quantum number n) of higher and higher energy. Within a given shell, electrons are assigned to orbitals within subshells (which are defined by the quantum number ℓ) of successively higher energy. Electrons are assigned in such a way that the total energy of the atom is as low as possible.

Experimental Evidence for Electron Configurations

Before learning the details of electron configurations, it is important to understand that our picture of the arrangement of electrons in atoms comes from experiment. The wave model of the atom is a reasonably successful attempt to rationalize these observations.

In Section 3.3 you learned that positive ions are formed by removing one or more electrons from an atom. The energy required in this process is called the **ionization energy** (see Section 8.6 for more details).

The ionization energy for the H atom was calculated from the Bohr theory on page 311.

First ionization energy (899.4 kJ/mol) $$Be(g) \xrightarrow{+\text{ energy}} Be^{+}(g) + e^{-}$$

Second ionization energy (1757.1 kJ/mol) $$Be^{+}(g) \xrightarrow{+\text{ energy}} Be^{2+}(g) + e^{-}$$

The second ionization energy is always larger than the first because the second electron is removed from a positively charged ion, whereas the first electron is removed from a neutral atom.

Ionization energies such as the ones for beryllium can be obtained experimentally. We can plot these data as the ratio of the second to the first ionization energy of an element versus atomic number (Figure 8.5). (Beryllium, for example, has a ratio of 1757.1/899.4 = 1.95.) This plot provides excellent evidence for the arrangement of electrons in shells and subshells of differing energy. It also shows the underlying periodicity of properties that characterizes the chemical elements.

- Lithium, sodium, and potassium each begin a new row of the periodic table, and you see that each has a much larger ratio of ionization energies than the other elements. We can conclude from this experimental observation that one electron (the one removed in the first ionization step) occupies a new electron shell, n, whose energy is much higher than the

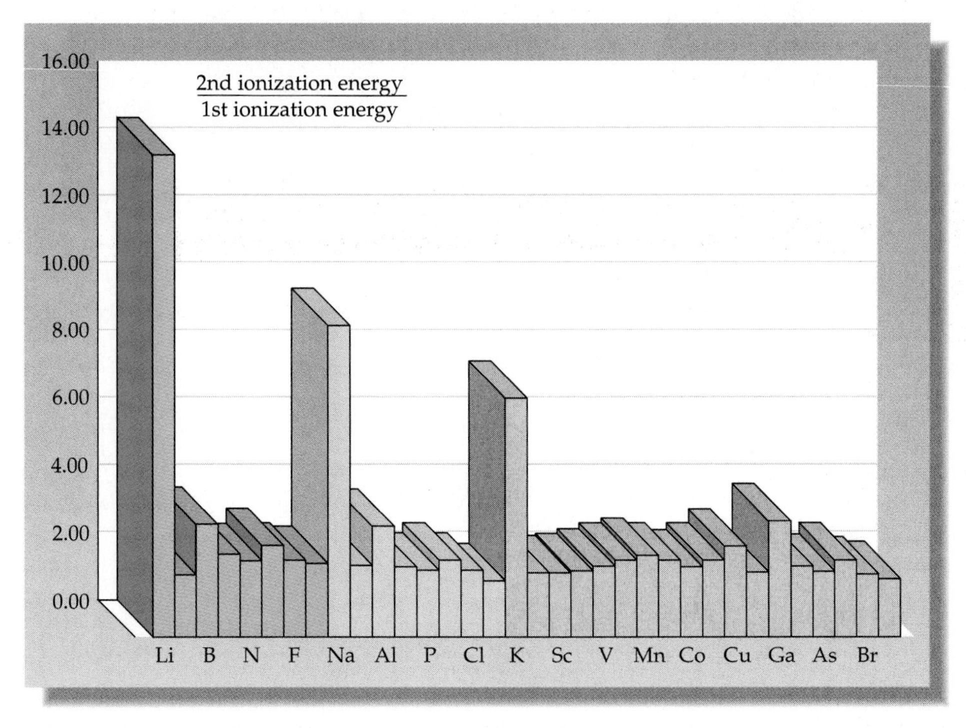

Figure 8.5 Ionization energy ratios versus atomic number. A plot of the ratio of the second and the first ionization energies versus atomic number for the first 36 elements.

CURRENT PERSPECTIVES

Quantized Spins and Magnetic Resonance Imaging

Just as electrons have a spin, so do the nuclei of atoms. In the case of hydrogen atoms, the single proton of the nucleus spins on its axis. For most heavier atoms such as carbon, the atomic nucleus includes both protons and neutrons, but the entire entity has a spin. This is important, because the nuclear spin allows scientists to detect these atoms in molecules and to learn something about their chemical environment.

The technique used to detect the spins of atomic nuclei is known as **nuclear magnetic resonance (NMR).** It is one of the most powerful methods currently available to determine molecular structures. About 20 years ago it was adapted as a diagnostic procedure in medicine where it is called **magnetic resonance imaging, MRI.**

Just as electron spin is quantized, so too is nuclear spin. As with electrons, the H atom nucleus can spin in either of two directions. If the H atom is placed in a strong, external magnetic field, however, the spinning nuclear magnet can align itself with the external field. So, if a sample of ethanol (C_2H_5OH), for example, is placed in a strong magnetic field, a slight excess of the H atom nuclei is aligned with the lines of force of the field.

These nuclei are slightly lower in energy than those with spins opposed to the field. The magnetic resonance technique depends on the fact that energy in the radio-frequency region can be absorbed by the sample and cause the nuclear spins aligned with the field to go out of alignment, that is, to move

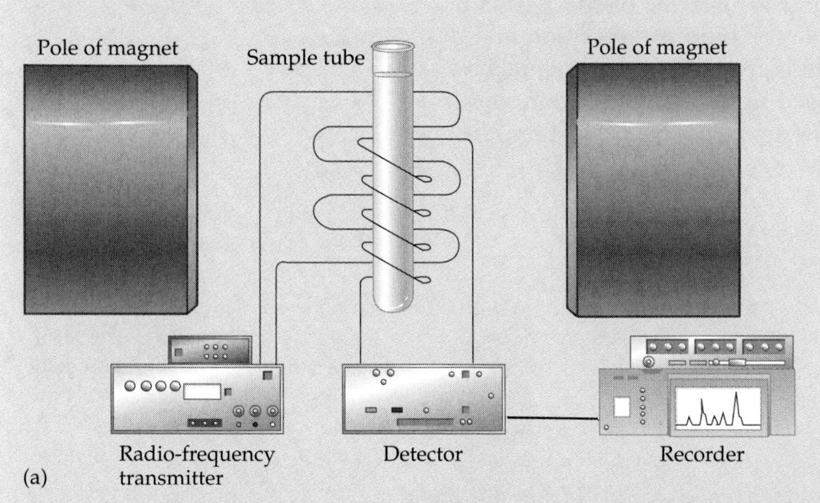

(a)

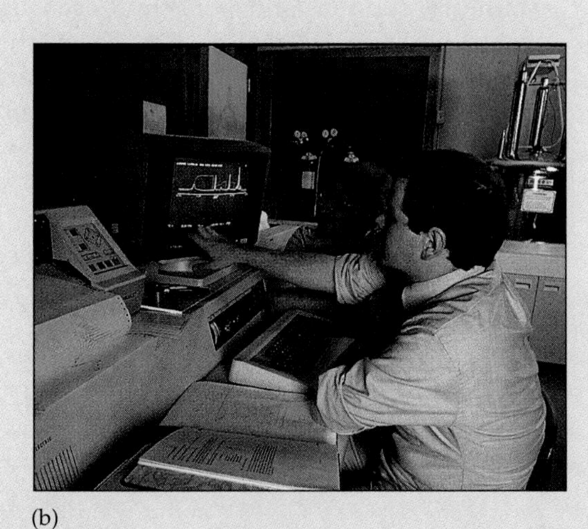
(b)

The nuclear magnetic resonance experiment. (a) A schematic diagram of an NMR spectrometer. (b) A modern NMR. The spectrum seen on the computer screen is observed after placing the sample between the poles of a strong magnet. (*b, James Prince/Photo Researchers, Inc.*)

shell used by the electrons of the elements in the previous row. The second ionization energy is much larger, mainly because the second electron must be removed from the $n - 1$ shell, and this shell lies much lower in energy than the nth shell.

First ionization energy (513.3 kJ/mol) $\quad Li(g) \longrightarrow Li^{+}(g) + e^{-}$ from $n = 2$ shell

Second ionization energy (7298.0 kJ/mol) $\quad Li^{+}(g) \longrightarrow Li^{2+}(g) + e^{-}$ from $n = 1$ shell

- Except for the alkali metals, most elements have about the same ionization energy ratio (values range from about 1.7 to 3). We can rationalize this by assuming that both electrons being removed from the elements

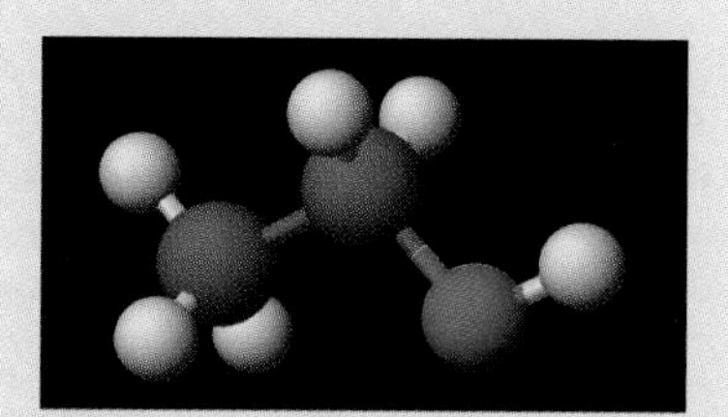

Ethanol, C_2H_5OH.

to a higher energy state. This absorption of energy is detected by the instrument.

The most important aspect of the magnetic resonance experiment is that the difference in energy between the two different spin states depends on the location of H atoms in the molecule. In the case of ethanol, the three CH_3 protons are different from the two CH_2 protons, and both sets are different from the OH proton. These three different sets of H atoms absorb radiation of slightly different energies, and the instrument detects these differences. A student familiar with the technique can quickly distinguish the three different proton environments in the molecule.

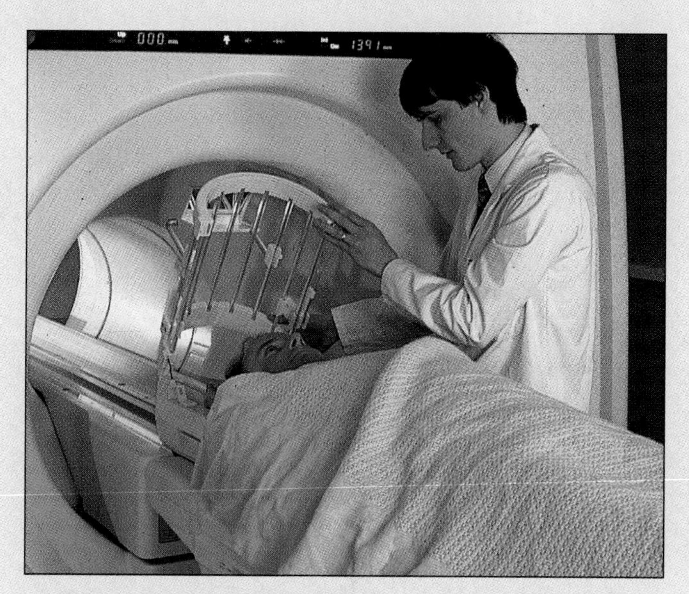

A magnetic resonance imaging (MRI) instrument. The patient is placed inside the large magnet and the tissues to be examined are irradiated with radio-frequency radiation. (*Simon Fraser/Dept. of Neuroradiology, Newcastle General Hospital/Science Photo Library/Photo Researchers, Inc.*)

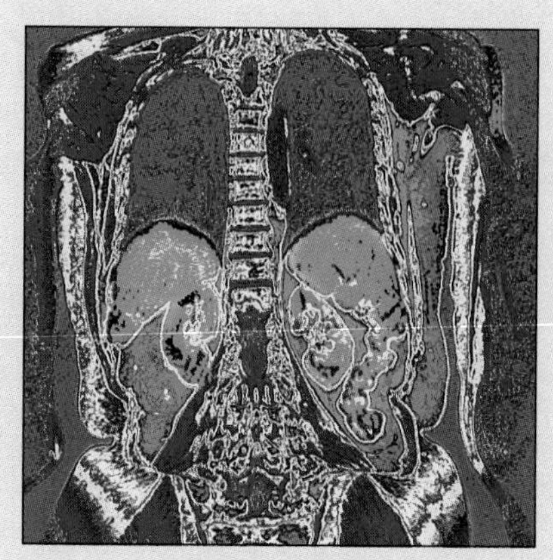

Scan of thorax and abdomen of woman, age 58 years, seen in posterior view. (*Simon Fraser/Royal Victoria Infirmary, Newcastle upon Tyne/Science Photo Library/Photo Researchers, Inc.*)

The MRI experiment is closely related to the experiment described for ethanol. Hydrogen is abundant in the human body in the form of water and in numerous organic molecules. In an MRI experiment, the patient is placed in a very strong magnetic field, and the tissues being examined are irradiated with pulses of radio-frequency radiation. The radio-frequency pulses can be focused at precise locations deep within the tissue and moved in a controlled fashion.

The MRI image is produced by detecting how fast the excited nuclei "relax" from the higher energy state to the lower energy state. The "relaxation time" depends on the type of tissue. When the tissue is scanned, the H atoms in different portions of the body show different relaxation times, and an accurate "image" of the tissue is built up.

The MRI scan gives accurate information on soft tissue—muscle, cartilage, and internal organs—which is unavailable from x-ray scans. MRI is also a noninvasive procedure, and nuclear magnetic resonance fields and radio frequency radiation are not harmful to the body.

beryllium to neon, for example, are in the same electron shell.

- The ratios for boron and aluminum are slightly larger than those of the other elements in their row in the periodic table (except for the relevant alkali metal). This indicates that a subshell has been completed at Be and Mg. Boron and aluminum are elements in which the second electron removed by ionization is in a new subshell that has a slightly higher energy than the subshell used by the electrons of Be and Mg. Therefore, the first electron in boron, for example, is removed from one subshell in the $n = 2$ shell, and the second electron comes from a lower energy subshell within the $n = 2$ shell.

Let us now see how our experimental observations, and our explanations, agree with the currently accepted model of atomic structure.

Order of Subshell Energies and Assignments

Quantum theory predicts that the energy of the H atom, with a single electron, depends only on the value of n ($E = -Rhc/n^2$, Equation 7.4). For heavier atoms, however, the situation is more complex. The ionization energy plot in Figure 8.5 and the experimentally determined order of subshell energies in Figure 8.6 show that the subshell energies of multielectron atoms depend on both n and ℓ. The subshells with $n = 3$, for example, have different energies; for a given atom they are in the order $3s < 3p < 3d$.

The subshell energy order in Figure 8.6 and the actual electron configurations of the elements lead to two general rules that help us predict the electron configurations of elements in the gas phase:

- Electrons are assigned to subshells in order of increasing "$n + \ell$" value.
- For two subshells with the same value of "$n + \ell$," electrons are assigned first to the subshell of lower n.

These rules mean, for example, that electrons are assigned to the $2s$ subshell ($n + \ell = 2 + 0 = 2$) before the $2p$ subshell ($n + \ell = 2 + 1 = 3$), or that they are assigned in the order $3s$ ($n + \ell = 3 + 0 = 3$) before $3p$ ($n + \ell = 3 + 1 = 4$) before $3d$ ($n + \ell = 3 + 2 = 5$). It also means that electrons fill the $4s$ subshell ($n + \ell = 4$) before filling the $3d$ subshell ($n + \ell = 5$). This filling order, which is summarized in Figure 8.7, has been amply verified by experiment.

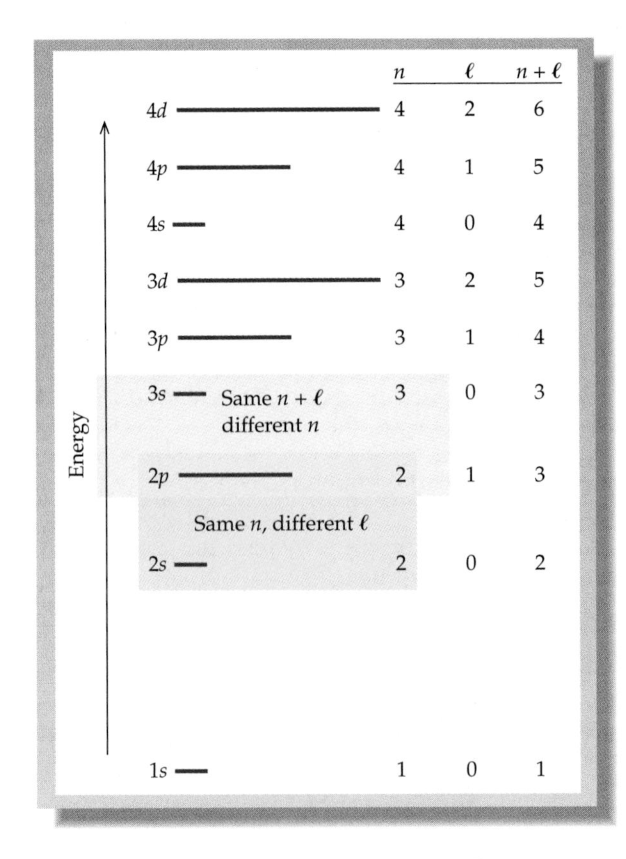

Figure 8.6 Experimentally determined order of subshell energies. In a multielectron atom, energies of electron shells increase with increasing n, and subshell energies increase with increasing ℓ. (*The energy axis is not to scale.*) The energy gaps between subshells for a given shell become smaller as n increases.

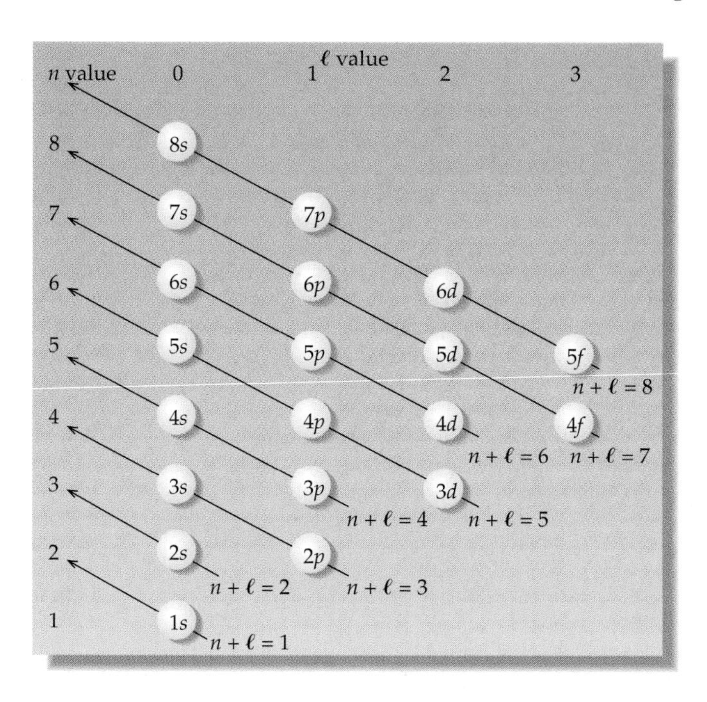

Figure 8.7 Subshell filling order. Subshells in atoms are filled in order of increasing $n + \ell$. When several subshells have the same value of $n + \ell$, the subshells are filled in order of increasing n. To use the diagram begin at 1*s* and follow the arrows of increasing $n + \ell$. (Thus, the order of filling is $1s \rightarrow 2s \rightarrow 2p \rightarrow 3s \rightarrow 3p \rightarrow 4s \rightarrow 3d$, and so on.)

Exercise 8.1 Order of Subshell Assignments

Using the "$n + \ell$" rules discussed previously, you can generally predict the order of subshell assignments (the electron filling order) for a multielectron atom. To which of the following subshells should an electron be assigned first?

1. 4*s* or 4*p*
2. 5*d* or 6*s*
3. 4*f* or 5*s*

Effective Nuclear Charge, Z*

See the *Saunders Interactive General Chemistry CD-ROM,* Screen 8.6, Effective Nuclear Charge, *Z**.

The order in which electrons are assigned to subshells in an atom, and many atomic properties, can be rationalized by the concept of **effective nuclear charge (Z*)**. This is the nuclear charge experienced by a particular electron in a multielectron atom, as modified by the presence of the other electrons.

In the hydrogen atom, with only one electron, the 2*s* and 2*p* subshells have the same energy. However, for lithium, an atom with three electrons, the presence of the 1*s* electrons causes the 2*s* subshell to lie lower in energy than the 2*p* subshell. Why should this be true? This question can be answered in part by referring to Figure 8.8.

Notice in Figure 8.8b that the region in which a 2*s* electron is most likely found lies almost completely *within* the region of probability for a 2*p* electron. Just as importantly, these orbitals *penetrate* the region occupied by the 1*s* electrons, the 2*s* penetration being greater than for 2*p*. That is, an electron in a 2*s* subshell has a much higher probability of being very close to the nucleus than an electron in a 2*p* subshell. Thus, the third electron in lithium occupies the 2*s* subshell, rather than the 2*p* subshell, because the 2*s*–2*p* probability difference near the nucleus leads to a lower energy for a 2*s* than for a 2*p* electron.

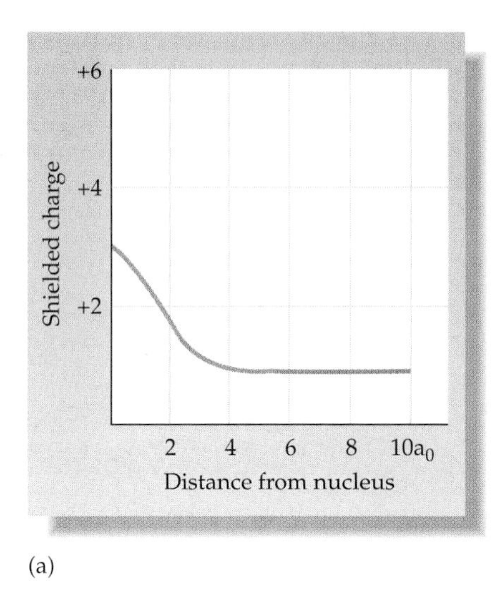

(a)

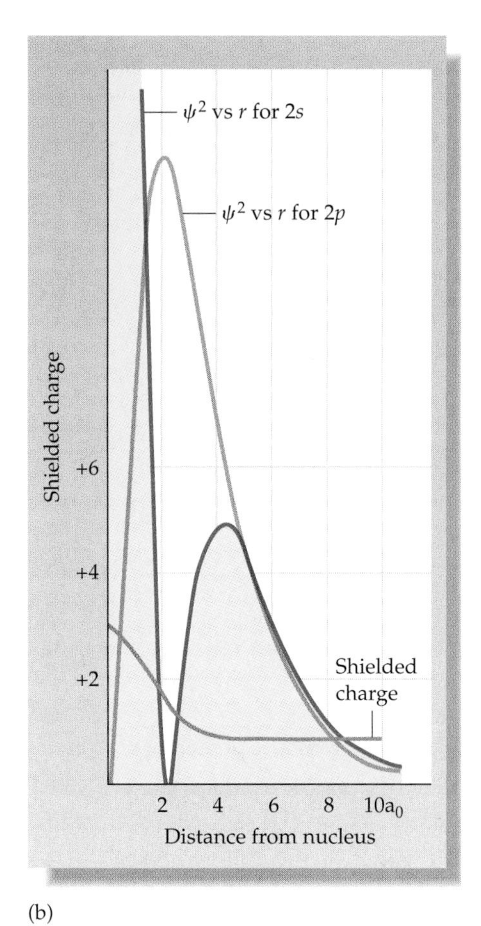

(b)

Figure 8.8 Shielded and effective nuclear charge. (a) The shielded nuclear charge is the positive charge an electron "feels" at a given distance from the nucleus. The shielded charge is shown for the third electron in the Li atom. When the third electron is placed near the nucleus, it feels a charge of 3+. As the electron moves away from the nucleus, however, the positive nuclear charge is screened, or shielded, by the negative 1*s* electrons, and the third electron feels a lower nuclear charge. When the third electron is outside the region occupied by the 1s electrons, the shielded charge felt by the third electron is +1. (b) The red curve is the probability curve (ψ^2) for a 2*s* orbital, whereas the green curve is for the 2*p* orbital. Superimposed on these is the plot for the shielded charge (blue). This shows that the 2*s* orbital allows the third electron to spend more time where the shielded charge is large than a 2*p* orbital does. (Distance from the nucleus is given in units of a_0 [atomic units], where 1 a_0 = 0.0529 nm.) (See page 322.)

Lithium, with a nucleus containing three protons, has two electrons in the 1*s* subshell. When the third electron is almost at the nucleus (regardless of its subshell type), Figure 8.8a shows that this electron experiences a charge of 3+. The positive charge experienced by this third electron drops off rapidly, however, as it moves away from the nucleus; the charge finally reaches 1+. The reason for this drop is that the negative charges of the two electrons in the 1*s* subshell shield, or screen, the effect of the positive nuclear charge, the screening effect becoming greater as the third electron moves outside the region occupied by the 1*s* electrons. The charge felt by any electron at any distance from the nucleus in an atom is called the *shielded nuclear charge.*

Figure 8.8b shows that an electron in a 2*s* subshell has a higher probability than an electron in the 2*p* subshell of being within the region of space where the shielded nuclear charge is larger than about 2+. Thus, *on average,* an electron in a 2*s* subshell experiences a higher shielded nuclear charge than an electron in a 2*p* subshell. The *average shielded charge* felt by the electron is called the **effective nuclear charge, Z^*.** For lithium, the 2*s* electron experiences a larger Z^* than does a 2*p* electron. This means a 2*s* electron is more strongly attracted to the nucleus than a 2*p* electron, and the lithium atom has an electron configuration with a pair of electrons in the 1*s* subshell and a single 2*s* electron. Identical arguments can generally be used to rationalize the electron configurations in Table 8.2.

Effective nuclear charge, Z^*, is important because it can be used to rationalize electron configurations (Section 8.4) and the observed properties of atoms and their ions (Section 8.6).

Table 8.2 • ELECTRON CONFIGURATIONS OF ATOMS IN THE GROUND STATE

Z	Element	Configuration	Z	Element	Configuration	Z	Element	Configuration
1	H	$1s^1$	38	Sr	$[Kr]5s^2$	75	Re	$[Xe]4f^{14}5d^5 6s^2$
2	He	$1s^2$	39	Y	$[Kr]4d^1 5s^2$	76	Os	$[Xe]4f^{14}5d^6 6s^2$
3	Li	$[He]2s^1$	40	Zr	$[Kr]4d^2 5s^2$	77	Ir	$[Xe]4f^{14}5d^7 6s^2$
4	Be	$[He]2s^2$	41	Nb	$[Kr]4d^4 5s^1$	78	Pt	$[Xe]4f^{14}5d^9 6s^1$
5	B	$[He]2s^2 2p^1$	42	Mo	$[Kr]4d^5 5s^1$	79	Au	$[Xe]4f^{14}5d^{10}6s^1$
6	C	$[He]2s^2 2p^2$	43	Tc	$[Kr]4d^5 5s^2$	80	Hg	$[Xe]4f^{14}5d^{10}6s^2$
7	N	$[He]2s^2 2p^3$	44	Ru	$[Kr]4d^7 5s^1$	81	Tl	$[Xe]4f^{14}5d^{10}6s^2 6p^1$
8	O	$[He]2s^2 2p^4$	45	Rh	$[Kr]4d^8 5s^1$	82	Pb	$[Xe]4f^{14}5d^{10}6s^2 6p^2$
9	F	$[He]2s^2 2p^5$	46	Pd	$[Kr]4d^{10}$	83	Bi	$[Xe]4f^{14}5d^{10}6s^2 6p^3$
10	Ne	$[He]2s^2 2p^6$	47	Ag	$[Kr]4d^{10}5s^1$	84	Po	$[Xe]4f^{14}5d^{10}6s^2 6p^4$
11	Na	$[Ne]3s^1$	48	Cd	$[Kr]4d^{10}5s^2$	85	At	$[Xe]4f^{14}5d^{10}6s^2 6p^5$
12	Mg	$[Ne]3s^2$	49	In	$[Kr]4d^{10}5s^2 5p^1$	86	Rn	$[Xe]4f^{14}5d^{10}6s^2 6p^6$
13	Al	$[Ne]3s^2 3p^1$	50	Sn	$[Kr]4d^{10}5s^2 5p^2$	87	Fr	$[Rn]7s^1$
14	Si	$[Ne]3s^2 3p^2$	51	Sb	$[Kr]4d^{10}5s^2 5p^3$	88	Ra	$[Rn]7s^2$
15	P	$[Ne]3s^2 3p^3$	52	Te	$[Kr]4d^{10}5s^2 5p^4$	89	Ac	$[Rn]6d^1 7s^2$
16	S	$[Ne]3s^2 3p^4$	53	I	$[Kr]4d^{10}5s^2 5p^5$	90	Th	$[Rn]6d^2 7s^2$
17	Cl	$[Ne]3s^2 3p^5$	54	Xe	$[Kr]4d^{10}5s^2 5p^6$	91	Pa	$[Rn]5f^2 6d^1 7s^2$
18	Ar	$[Ne]3s^2 3p^6$	55	Cs	$[Xe]6s^1$	92	U	$[Rn]5f^3 6d^1 7s^2$
19	K	$[Ar]4s^1$	56	Ba	$[Xe]6s^2$	93	Np	$[Rn]5f^4 6d^1 7s^2$
20	Ca	$[Ar]4s^2$	57	La	$[Xe]5d^1 6s^2$	94	Pu	$[Rn]5f^6 7s^2$
21	Sc	$[Ar]3d^1 4s^2$	58	Ce	$[Xe]4f^1 5d^1 6s^2$	95	Am	$[Rn]5f^7 7s^2$
22	Ti	$[Ar]3d^2 4s^2$	59	Pr	$[Xe]4f^3 6s^2$	96	Cm	$[Rn]5f^7 6d^1 7s^2$
23	V	$[Ar]3d^3 4s^2$	60	Nd	$[Xe]4f^4 6s^2$	97	Bk	$[Rn]5f^9 7s^2$
24	Cr	$[Ar]3d^5 4s^1$	61	Pm	$[Xe]4f^5 6s^2$	98	Cf	$[Rn]5f^{10}7s^2$
25	Mn	$[Ar]3d^5 4s^2$	62	Sm	$[Xe]4f^6 6s^2$	99	Es	$[Rn]5f^{11}7s^2$
26	Fe	$[Ar]3d^6 4s^2$	63	Eu	$[Xe]4f^7 6s^2$	100	Fm	$[Rn]5f^{12}7s^2$
27	Co	$[Ar]3d^7 4s^2$	64	Gd	$[Xe]4f^7 5d^1 6s^2$	101	Md	$[Rn]5f^{13}7s^2$
28	Ni	$[Ar]3d^8 4s^2$	65	Tb	$[Xe]4f^9 6s^2$	102	No	$[Rn]5f^{14}7s^2$
29	Cu	$[Ar]3d^{10}4s^1$	66	Dy	$[Xe]4f^{10}6s^2$	103	Lr	$[Rn]5f^{14}6d^1 7s^2$
30	Zn	$[Ar]3d^{10}4s^2$	67	Ho	$[Xe]4f^{11}6s^2$	104	Rf	$[Rn]5f^{14}6d^2 7s^2$
31	Ga	$[Ar]3d^{10}4s^2 4p^1$	68	Er	$[Xe]4f^{12}6s^2$	105	Db	$[Rn]5f^{14}6d^3 7s^2$
32	Ge	$[Ar]3d^{10}4s^2 4p^2$	69	Tm	$[Xe]4f^{13}6s^2$	106	Sg	$[Rn]5f^{14}6d^4 7s^2$
33	As	$[Ar]3d^{10}4s^2 4p^3$	70	Yb	$[Xe]4f^{14}6s^2$	107	Bh	$[Rn]5f^{14}6d^5 7s^2$
34	Se	$[Ar]3d^{10}4s^2 4p^4$	71	Lu	$[Xe]4f^{14}5d^1 6s^2$	108	Hs	$[Rn]5f^{14}6d^6 7s^2$
35	Br	$[Ar]3d^{10}4s^2 4p^5$	72	Hf	$[Xe]4f^{14}5d^2 6s^2$	109	Mt	$[Rn]5f^{14}6d^7 7s^2$
36	Kr	$[Ar]3d^{10}4s^2 4p^6$	73	Ta	$[Xe]4f^{14}5d^3 6s^2$			
37	Rb	$[Kr]5s^1$	74	W	$[Xe]4f^{14}5d^4 6s^2$			

8.4 ATOMIC ELECTRON CONFIGURATIONS

The configurations of the elements up to element 109 are given in Table 8.2. These are the ground state electron configurations, where electrons are found in the shells, subshells, and orbitals that result in the lowest energy for the atom. In general, the guiding principle in assigning electrons to available orbitals is to do so in order of increasing "$n + \ell$" (see Figure 8.6). Our emphasis, however, will be to connect the configurations of the elements with their position in the periodic table because this will ultimately allow us to organize a large number of chemical facts.

See the *Saunders Interactive General Chemistry CD-ROM,* Screen 8.7, Atomic Electron Configurations.

Electron Configurations of the Main Group Elements

Hydrogen, the first element in the periodic table, has one electron in a 1*s* orbital. One way to depict its electron configuration is with the **orbital box diagram** used earlier, but an alternative and more frequently used method is the **spectroscopic notation.** Using the latter method, the electron configuration of H is $1s^1$, or "one ess one."

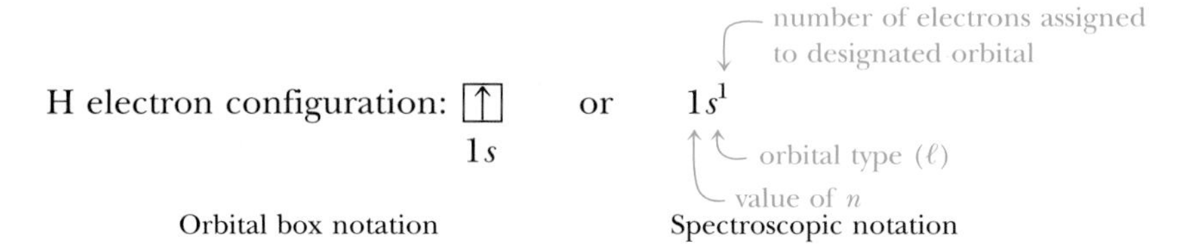

We can use either method to describe the configurations of the other elements. Electron configurations for the first ten elements using both methods are illustrated in Table 8.3.

Following hydrogen and helium, lithium (Group 1A), with three electrons, is the first element in the second period of the periodic table. The first two electrons are in the 1*s* subshell ($n + \ell = 1$), and the third electron must be in the $n = 2$ shell. According to the energy level diagram in Figure 8.6, that electron must be in the 2*s* subshell ($n + \ell = 2$). The spectroscopic notation, $1s^22s^1$, is read "one ess two, two ess one."

The position of lithium in the periodic table tells you its configuration immediately. All the elements of Group 1A (and 1B) have one electron assigned to an *s* orbital of the *n*th shell, for which *n* is the number of the period in which the element is found (Figure 8.9). For example, potassium is the first element in the $n = 4$ row (the fourth period), so potassium has the electron configura-

Table 8.3 • Electron Configurations of Elements with $Z = 1$ to 10

	Electron Configuration	ℓ $n + \ell$	**1s** 0 1	**2s** 0 2	**2p** 1 3
H	$1s^1$		[↑]	[ ]	[ ][ ][ ]
He	$1s^2$		[↑↓]	[ ]	[ ][ ][ ]
Li	$1s^22s^1$		[↑↓]	[↑]	[ ][ ][ ]
Be	$1s^22s^2$		[↑↓]	[↑↓]	[ ][ ][ ]
B	$1s^22s^22p^1$		[↑↓]	[↑↓]	[↑][ ][ ]
C	$1s^22s^22p^2$		[↑↓]	[↑↓]	[↑][↑][ ]
N	$1s^22s^22p^3$		[↑↓]	[↑↓]	[↑][↑][↑]
O	$1s^22s^22p^4$		[↑↓]	[↑↓]	[↑↓][↑][↑]
F	$1s^22s^22p^5$		[↑↓]	[↑↓]	[↑↓][↑↓][↑]
Ne	$1s^22s^22p^6$		[↑↓]	[↑↓]	[↑↓][↑↓][↑↓]

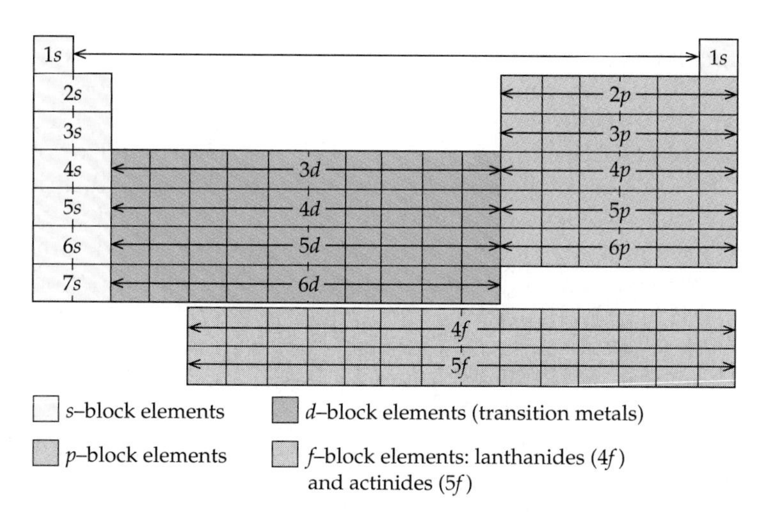

Figure 8.9 Electron configurations and the periodic table. The outermost electrons of an element are assigned to the indicated orbitals. See Table 8.2.

tion of the element preceding it in the table (Ar) *plus* a final electron assigned to the 4*s* orbital.

Beryllium, in Group 2A, has two electrons in the 1*s* orbital plus two additional electrons. Figure 8.6 shows that the 2*s* orbital is appropriate, so the configuration of Be is $1s^2 2s^2$. *All elements of Group 2A have electron configurations of* [electrons of preceding noble gas]ns^2, where *n* is the period in which the element is found in the periodic table. Because all the elements of Group 1A have the configuration ns^1, and those in Group 2A have ns^2, these elements are called ***s*-block elements.**

At boron (Group 3A) you first encounter an element in the block of elements on the right side of the periodic table. Because 1*s* and 2*s* orbitals are filled in a boron atom, the fifth electron must be assigned to a 2*p* orbital. In fact, all the elements from Group 3A through Group 8A have electrons in *p* orbitals, so these elements are sometimes called the ***p*-block elements.** All have the general configuration ns^2np^x, where *x* varies from one to six (and is equal to the group number −2).

Carbon (Group 4A) is the second element in the *p* block, so a second electron is assigned to the 2*p* orbitals. For carbon to be in its lowest energy or *ground state*, this electron *must* be assigned to either of the remaining *p* orbitals, and it must have the same spin direction as the first *p* electron.

In general, when electrons are assigned to *p*, *d*, or *f* orbitals, each successive electron is assigned to a different orbital of the subshell, and each electron has the same spin as the previous one; this pattern proceeds until the subshell is half full. Additional electrons must be assigned to half-filled orbitals. This procedure follows **Hund's rule,** which states that *the most stable arrangement of electrons is that with the maximum number of unpaired electrons,* all with the same spin direction. This arrangement makes the total energy of an atom as low as possible.

Hund's rule has been amply verified by experiments using equipment such as that depicted in Figure 8.3.

Notice that carbon is the second element in the *p* block of elements, so there must be two *p* electrons (besides the two 2*s* electrons already present in the $n = 2$ shell). Because carbon is in the second period of the table, the *p* orbitals involved are 2*p*. Thus, you can immediately write the carbon electron configuration by referring to the periodic table: starting at H and moving from left to right across the successive periods, you write $1s^2$ to reach the end of period 1, and then $2s^2$ and finally $2p^2$ to bring the electron count to six. Carbon is in Group 4A of the periodic table because it has four electrons in the $n = 2$ shell.

Nitrogen (Group 5A) has three electrons, all with the same spin, in three different $2p$ orbitals. Oxygen (Group 6A) has yet another $2p$ electron. Two of the six electrons in oxygen's outer shell are assigned to the $2s$ orbital, and, as it is the fourth element in the p block, the other four electrons are assigned to $2p$ orbitals. This means the fourth $2p$ electron must pair up with one already present. It makes no difference to which orbital this electron is assigned (the $2p$ orbitals all have the same energy), but *it must have a spin opposite to the other electron already assigned to that orbital* (see Table 8.3) so that each electron has a different set of quantum numbers (the Pauli exclusion principle).

Fluorine (Group 7A) has seven electrons in the $n = 2$ shell. Two of these electrons occupy the $2s$ subshell, and the remaining five electrons occupy the $2p$ subshell. All halogen atoms have a similar configuration, ns^2np^5, where n is again the period in which the element is located.

Like all the other elements in Group 8A, neon is a noble gas. All Group 8A elements (except helium) have eight electrons in the shell of highest n value, so all have the configuration ns^2np^6, where n is the period in which the element is found. That is, all the noble gases have filled ns and np subshells. As you will see, the nearly complete chemical inertness of the noble gases correlates with this electron configuration.

The next element after neon is sodium, and with it a new period is begun (recall Figure 8.5). Because sodium is the first element with $n = 3$, the added electron must be assigned to the $3s$ orbital. (Remember that all elements in Group 1A have the ns^1 configuration.) Thus, the complete electron configuration of sodium is that of neon (the preceding electron) plus one $3s$ electron.

Na: $1s^22s^22p^63s^1$ or $[Ne]3s^1$

↑

noble gas notation represents core electrons

We have written the electron configuration for sodium in two ways, one in an abbreviated form that uses the **noble gas notation.** The arrangement preceding the $3s$ electron is that of the noble gas neon, so, instead of writing out "$1s^22s^22p^6$," the completed electron shells are represented by placing the symbol of the corresponding noble gas in brackets.

The electrons included in the noble gas notation are often referred to as the **core electrons** of the atom. Not only is it a time-saving way to write electron configurations, but it also conveys the idea that the core electrons can generally be ignored when considering the chemistry of an element. The electrons beyond the core electrons, the $3s^1$ electron in the case of sodium, are the **valence electrons,** the electrons that determine the chemical properties of an element (Section 3.3).

Example 8.1 Electron Configurations

Give the electron configuration of silicon, using the spectroscopic notation and the noble gas notation.

Solution Silicon, element 14, is the fourth element in the third period ($n = 3$), and it is in the p block. Therefore, the last four electrons in the atom have

1 H																	2 He
3 Li	4 Be											5 B			8 O	9 F	10 Ne
11 Na	12 Mg											13 Al	14 Si		16 S	17 Cl	18 Ar
19 K	20 Ca	21 Sc	22 Ti	23 V	24 Cr	25 Mn	26 Fe	27 Co	28 Ni	29 Cu	30 Zn	31 Ga	32 Ge	33 As	34 Se	35 Br	36 Kr
37 Rb	38 Sr	39 Y	40 Zr	41 Nb	42 Mo	43 Tc	44 Ru	45 Rh	46 Pd	47 Ag	48 Cd	49 In	50 Sn	51 Sb	52 Te	53 I	54 Xe
55 Cs	56 Ba	57 La	72 Hf	73 Ta	74 W	75 Re	76 Os	77 Ir	78 Pt	79 Au	80 Hg	81 Tl	82 Pb	83 Bi	84 Po	85 At	86 Rn
87 Fr	88 Ra	89 Ac	104 Rf	105 Db													

the configuration $3s^23p^2$. These are preceded by the completed shells $n = 1$ and $n = 2$, the electron arrangement for Ne. Therefore, the electron configuration of silicon is

$$1s^22s^22p^63s^23p^2 \quad \text{or} \quad [\text{Ne}]3s^23p^2$$

Example 8.2 Electron Configurations

Give the electron configuration of sulfur, using the spectroscopic, noble gas, and orbital box notations.

Solution Sulfur, element 16, is the sixth element in the third period ($n = 3$), and it is in the p block. The last six electrons assigned to the atom, therefore, have the configuration $3s^23p^4$. These are preceded by the completed shells $n = 1$ and $n = 2$, the electron arrangement for Ne. Therefore, the electron configuration of sulfur is

Spectroscopic notation: $1s^22s^22p^63s^23p^4$

Noble gas notation: $[\text{Ne}]3s^23p^4$

Orbital box notation: [Ne] [↑↓] (3s) [↑↓][↑][↑] (3p)

Example 8.3 Electron Configurations and Quantum Numbers

Write the electron configuration for Al using the noble gas notation, and give a set of quantum numbers for each of the electrons with $n = 3$ (the valence electrons).

Solution Aluminum is the third element in the third period. It therefore has three electrons with $n = 3$, and because Al is in the p block of elements, two of the electrons are assigned to $3s$ and the remaining electron is assigned to $3p$. The element is preceded by the noble gas neon, so the electron configuration is $[\text{Ne}]3s^23p^1$. Using a box notation, the configuration is

Aluminum configuration: [Ne] [↑↓] (3s) [↑][ ][ ] (3p)

The possible sets of quantum numbers for the two $3s$ electrons are

	n	ℓ	m_ℓ	m_s
For ↑	3	0	0	$+\frac{1}{2}$
For ↓	3	0	0	$-\frac{1}{2}$

and for the single $3p$ electron, one of six possible sets would be $n = 3$, $\ell = 1$, $m_\ell = +1$, and $m_s = +\frac{1}{2}$.

Exercise 8.2 Spectroscopic Notation, Orbital Box Diagrams, and Quantum Numbers

1. What element has the configuration $1s^22s^22p^63s^23p^5$?
2. Using the spectroscopic notation and a box diagram, show the electron configuration of chlorine.
3. Write one possible set of quantum numbers for the valence electrons of calcium.

Electron Configurations for the Transition Elements

The names "lanthanides" and "actinides" derive from the name of the first element in these series of elements.

The elements of the fourth through the seventh periods must also use *d* or *f* subshells to accommodate electrons (see Figure 8.9 and Table 8.2). Elements whose atoms are filling *d* subshells are often referred to as the **transition elements.** Those for which *f* subshells are filling are sometimes called the *inner transition elements* or, more usually, the **lanthanides** (filling 4*f* orbitals) and **actinides** (filling 5*f* orbitals).

According to Figure 8.9, which is based on experimentally determined electron configurations, the transition elements are always preceded by two *s*-block elements. Accordingly, scandium, the first transition element, has the configuration $[Ar]3d^14s^2$, and titanium follows with $[Ar]3d^24s^2$ (Table 8.4).

The expected configuration of the chromium atom is $[Ar]3d^44s^2$. The actual configuration, however, has one electron assigned to each of the six available $3d$ and $4s$ orbitals: $[Ar]3d^54s^1$. This is explained by assuming that the $4s$ and the $3d$ orbitals have approximately the same energy at this point, thus giving rise to six orbitals of nearly the same energy. Each of the six valence electrons of chromium is assigned a separate orbital. This occurrence illustrates the fact that there are occasionally minor differences between the predicted and actual configurations. These have little or no effect on the chemistry of the element, however.

Following chromium, atoms of manganese, iron, and nickel have configurations that are expected on the basis of Figure 8.7. Copper, however, is slightly different from expected. This Group 1B element has a single electron in the $4s$ orbital, as implied by its group number, and the remaining ten electrons beyond the argon core are assigned to the $3d$ orbitals. Zinc ends the first transition series. This Group 2B element has two electrons assigned to $4s$, and the $3d$ orbitals are completely filled with ten electrons.

Table 8.4 • ORBITAL BOX DIAGRAMS FOR THE ELEMENTS CA THROUGH ZN

		3*d*	4*s*
Ca	$[Ar]4s^2$	[ ][ ][ ][ ][ ]	[↑↓]
Sc	$[Ar]3d^14s^2$	[↑][ ][ ][ ][ ]	[↑↓]
Ti	$[Ar]3d^24s^2$	[↑][↑][ ][ ][ ]	[↑↓]
V	$[Ar]3d^34s^2$	[↑][↑][↑][ ][ ]	[↑↓]
Cr*	$[Ar]3d^54s^1$	[↑][↑][↑][↑][↑]	[↑]
Mn	$[Ar]3d^54s^2$	[↑][↑][↑][↑][↑]	[↑↓]
Fe	$[Ar]3d^64s^2$	[↑↓][↑][↑][↑][↑]	[↑↓]
Co	$[Ar]3d^74s^2$	[↑↓][↑↓][↑][↑][↑]	[↑↓]
Ni	$[Ar]3d^84s^2$	[↑↓][↑↓][↑↓][↑][↑]	[↑↓]
Cu*	$[Ar]3d^{10}4s^1$	[↑↓][↑↓][↑↓][↑↓][↑↓]	[↑]
Zn	$[Ar]3d^{10}4s^2$	[↑↓][↑↓][↑↓][↑↓][↑↓]	[↑↓]

*These configurations do not follow the "$n + \ell$" rule.

The fifth period ($n = 5$) follows the pattern of the fourth period with minor variations. The sixth period, however, includes the **lanthanide series** beginning with lanthanum, La. As the first element in the *d*-block, lanthanum has the configuration $[Xe]5d^16s^2$. The next element, cerium (Ce), is set out in a separate row at the bottom of the periodic table, and it is with these elements that electrons are first assigned to *f* orbitals. This means the configuration of cerium is $[Xe]4f^15d^16s^2$. Moving across the lanthanide series, the pattern continues with some variation, with 14 electrons being assigned to the seven 5*f* orbitals in lutetium, Lu ($[Xe]4f^{14}5d^16s^2$) (see Table 8.2).

When you have completed this section, you should be able to depict the electron configuration of any element in the *s* and *p* blocks *using the periodic table as a guide.* Regarding the prediction of electron configurations for atoms of elements in the *d* and *f* blocks, you have seen that minor differences can occur between actual (see Table 8.2) and predicted configurations. As you shall see, however, these "anomalies" have little effect on the chemical behavior of the elements.

Example 8.4 Electron Configurations of the Transition Elements

Using the spectroscopic and noble gas notations, give electron configurations for technetium (Tc) and osmium (Os). Base your answer on the positions of the elements in the periodic table. That is, for each element, find the preceding noble gas and then note the number of *s*, *d*, and *f* electrons that lead from the noble gas to the element.

Solution Proceeding along the periodic table, we come to the noble gas krypton, Kr, at the end of the $n = 4$ row before arriving at Tc (element 43) in the fifth period. After the 36 electrons of Kr are assigned, 7 electrons remain. According to the periodic table, two of these seven electrons are in the $5s$ orbital, and the remaining five are in $4d$ orbitals. Therefore, the *technetium* configuration is $[Kr]4d^5 5s^2$.

Osmium is a sixth-period element and the 22nd element following the noble gas xenon. Of the 22 electrons to be added after the xenon core, 2 are of the $6s$ type (Cs and Ba), 14 are of the $4f$ type (Ce through Lu), and the remaining 6 are of the $5d$ type (La through Os). Thus, the *osmium* configuration is $[Xe]4f^{14}5d^6 6s^2$.

In writing electron configurations, we follow the convention of writing the orbitals in order of increasing n. For a given n, the subshells are listed in order of increasing ℓ. Note that this is not always the filling order for the transition and inner transition elements.

Exercise 8.3 Electron Configurations

Using the periodic table and without looking at Table 8.2, write electron configurations for the following elements: (a) P, (b) Zn, (c) Zr, (d) In, (e) Pb, and (f) U. Use the spectroscopic and noble gas notations. When you have finished, check your answers with Table 8.2.

8.5 ELECTRON CONFIGURATIONS OF IONS

See the *Saunders Interactive General Chemistry CD-ROM,* Screen 8.8, Electron Configurations in Ions.

A great deal of the chemistry of the elements is that of their ions. To form a cation from a neutral atom, the general rule is that one or more electrons are removed from the electron shell of highest n. If there is a choice of subshell within the nth shell, the electron or electrons of maximum ℓ are removed. Thus, a sodium ion is formed by removing the $3s^1$ electron from the Na atom,

$$\text{Na: } [1s^2 2s^2 2p^6 3s^1] \longrightarrow \text{Na}^+\text{: } [1s^2 2s^2 2p^6] + \text{e}^-$$

and Al^{3+} is formed by removing two $3s$ electrons and one $3p$ electron from an aluminum atom:

$$\text{Al: } [1s^2 2s^2 2p^6 3s^2 3p^1] \longrightarrow \text{Al}^{3+}\text{: } [1s^2 2s^2 2p^6] + 3\ \text{e}^-$$

The same general rule applies to transition metal atoms. This means the titanium(II) cation has the configuration $[Ar]3d^2$, for example,

$$\text{Ti: } [\text{Ar}]3d^2 4s^2 \longrightarrow \text{Ti}^{2+}\text{: } [\text{Ar}]3d^2 + 2\ \text{e}^-$$

and the iron(II) and iron(III) cations have the configuration $[Ar]3d^6$ and $[Ar]3d^5$, respectively.

All common transition metal cations have electron configurations of the general type [noble gas core]$(n - 1)d^x$. It is very important to remember this because the chemical and physical properties of transition metal cations are determined by the presence of electrons in d orbitals.

Atoms and ions with unpaired electrons are paramagnetic, that is, they are capable of being attracted to a magnetic field (see Section 8.1). Paramagnetism is important here because it provides experimental evidence that transition metal ions with charges of 2+ or greater have no ns electrons. For example, the Fe^{2+} ion is paramagnetic to the extent of four unpaired electrons, and the Fe^{3+} ion has five unpaired electrons. If three $3d$ electrons had been removed instead to form Fe^{3+}, the ion would still be paramagnetic but only to the extent of three unpaired electrons.

Example 8.5 Configurations of Transition Metal Ions

Give the electron configuration for copper, Cu, and for its 1+ and 2+ ions. Are either of these ions paramagnetic? How many unpaired electrons does each have?

Solution As illustrated in Table 8.4, copper has only one electron in the 4*s* orbital and ten electrons in 3*d* orbitals:

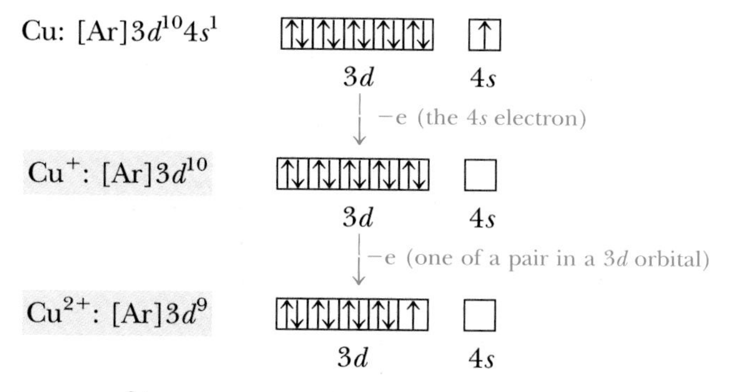

Copper(II) ions (Cu^{2+}) have one unpaired electron and so should be paramagnetic. In contrast, Cu^+ has no unpaired electrons, so the ion and its compounds are diamagnetic.

Exercise 8.4 Metal Ion Configurations

Depict electron configurations for V^{2+}, V^{3+}, and Co^{3+}. Use orbital box diagrams and the noble gas notation. Are any of the ions paramagnetic? If so, give the number of unpaired electrons.

8.6 ATOMIC PROPERTIES AND PERIODIC TRENDS

Reasonably accurate values for the atomic weights of the elements became available in the latter half of the 19th century, and Lothar Meyer (1830–1895) and Dmitri Mendeleev (1834–1907) used them to create the first versions of modern periodic tables (see Section 2.7). Both believed that the chemical and physical properties of the elements were periodic functions of atomic mass.

It is important for us to remember that neither Meyer nor Mendeleev had any knowledge of atomic structure to guide their efforts. Their conclusions were based entirely on experimental observations of chemical and physical properties. Experiments with electrons in cathode-ray tubes were in their infancy in the 19th century; protons and neutrons had not yet been discovered. As a result, neither of these now-famous scientists could explain why groups of elements had similar properties nor why there was a periodic behavior. Now our knowledge of the electronic structure of elements provides this understanding. We realize that *similarities in properties of the elements are the result of similar valence shell electron configurations.*

An objective of this section is to begin to show how atomic electron configurations are related to some of the physical properties of the elements and why those properties change in a reasonably predictable manner when moving down groups and across periods. This background should make the periodic table an even more useful tool in your study of chemistry. With an understanding of electron configurations and their relation to properties, you should be able to organize and predict chemical and physical properties of the elements and their compounds. We will concentrate on physical properties in this section and then look at chemical behavior in Section 8.7.

Atomic Size

See the *Saunders Interactive General Chemistry CD-ROM,* Screen 8.9, Atomic Properties and Periodic Trends, and Screen 8.10, Atomic Size.

Atoms are not like billiard balls. The wave theory of the atom informs us that atoms are not hard shells in which the electrons are contained. An electron orbital has no sharp boundary beyond which the electron never strays (see Figure 7.14). How then can we define the size of an atom? There are actually several ways, and to some extent they give slightly different results.

One of the simplest and most useful ways to define atomic size involves bond lengths. Let us take a simple diatomic molecule such as Cl_2. The radius of a Cl atom can be defined experimentally by dividing the distance between the centers of the two atoms in a Cl_2 molecule by 2. This distance is 200 pm, so the radius of one Cl atom—called the **covalent radius**—is 100 pm. Similarly, the C—C distance in diamond is 154 pm, so a radius of 77 pm can be assigned to carbon. To test these estimates, we can add them together to estimate the distance between Cl and C in CCl_4. The predicted distance of 177 pm agrees well with the experimentally measured C—Cl distance of 176 pm.

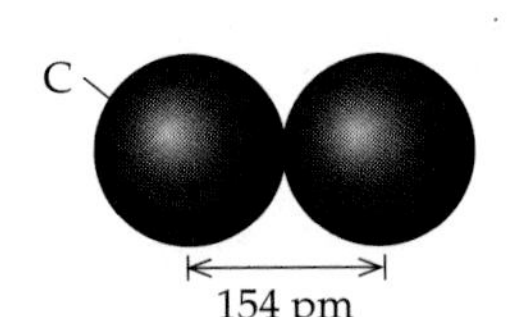

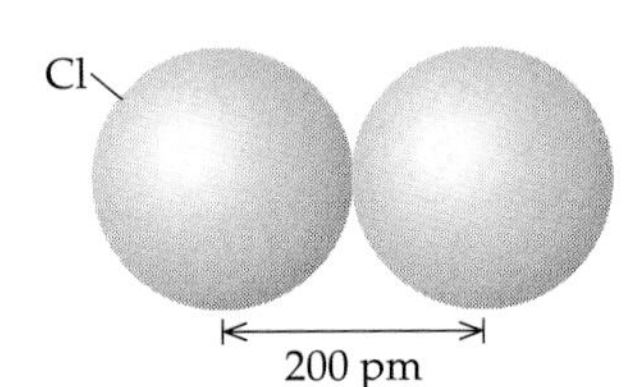

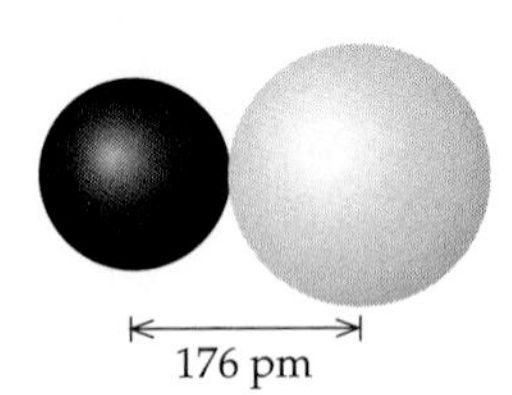

The sum of the atomic radii of C and Cl provides a good estimate of the C—Cl distance in a molecule having such a bond.

This approach to determining radii for atoms will of course only apply if molecular compounds of the element exist. For elements that do not form such compounds, radii are estimated based on calculated electron distributions such as those illustrated on page 322.

A reasonable set of atomic radii has been assembled (Figure 8.10), and some interesting periodic trends are seen immediately. *For the main group elements, atomic radii increase going down a group in the periodic table and decrease going across a period.* These trends reflect two important effects:

- The size of an atom is determined by the outermost electrons. In going from the top to the bottom of a group in the periodic table, the outermost electrons are assigned to orbitals with higher and higher values of the principal quantum number, n. The underlying electrons require some space, so the electrons of the outer shell must be farther from the nucleus.
- For main group elements of a given period, the principal quantum number, n, of the valence orbitals is the same. Going from one element to the next across a period involves adding a proton to the nucleus and an electron to the outer shell. In each step, the effective nuclear charge, Z^*, increases slightly because the effect of each additional proton is more important than the effect of an additional electron. The result is that attraction between the nucleus and electrons increases, and atomic radius decreases.

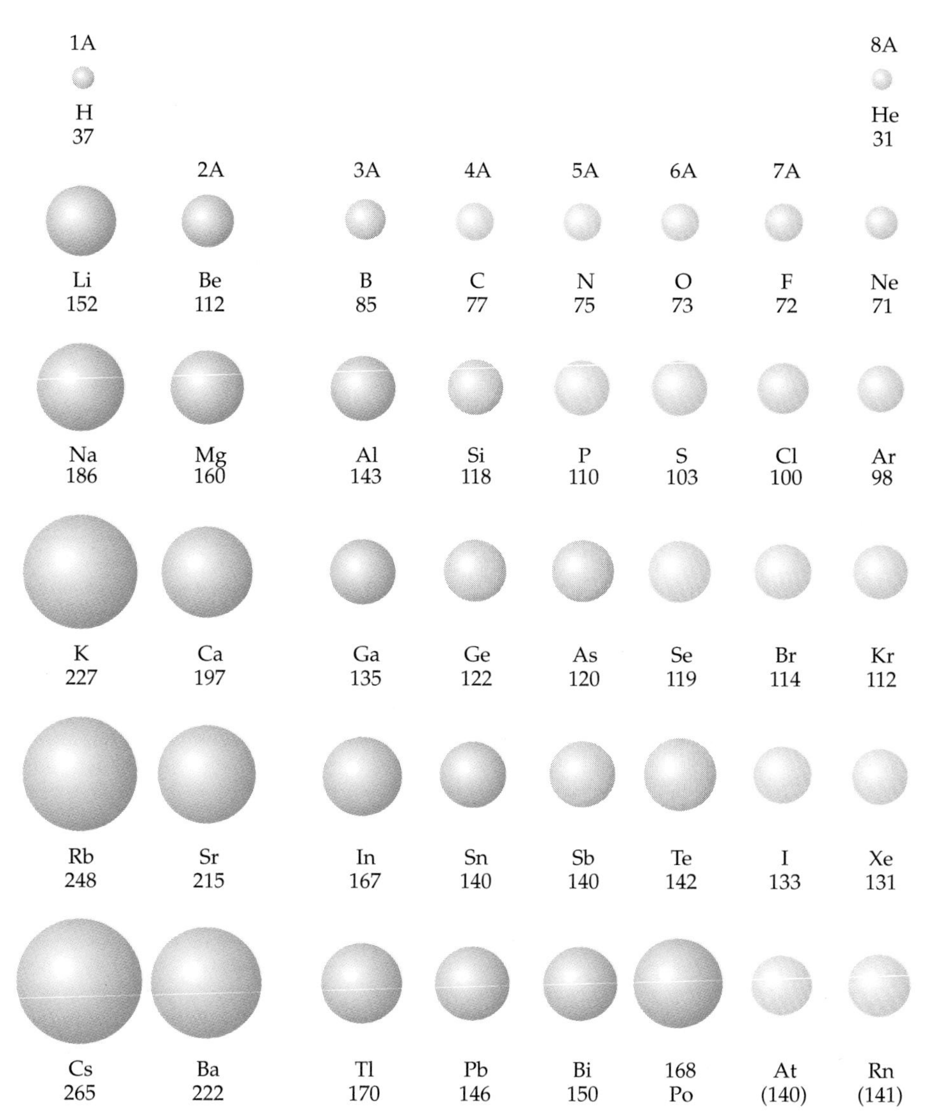

***Figure* 8.10 Atomic radii in picometers for the main group elements.** (You are cautioned that there are numerous tabulations of atomic and covalent radii, and they can vary considerably. This variation comes about because several methods are used to determine the radii of atoms, and the different methods often give different values. For example, tabulations of the atomic radius of N vary from 54.9 pm to 92 pm and O varies from 60.4 to 73 pm.) ($1 \text{ pm} = 1 \times 10^{-12}$ m.)

The periodic trend in the atomic radii of transition metal atoms (Figure 8.11) is somewhat different from that for main group elements. Going from left to right across a given period, the radii initially decrease across the first few elements. The sizes of the elements in the middle of a transition series change very little, whereas a small increase in size occurs at the end of the series. The overall effect can be explained by realizing that the variations in electron configuration across the period are occurring mostly in the $(n-1)d$ subshell. The size of the atom, however, is determined largely by electrons in the outermost shell, that is, by the electrons of the ns subshell. In the first transition series, for example, the outer shell contains the $4s$ electrons, but electrons are being added

When an atom is part of a molecule, the measured radius is the covalent radius, which refers to the role it plays in forming a chemical bond. The radii in Figure 8.10 for some of the nonmetals are covalent radii.

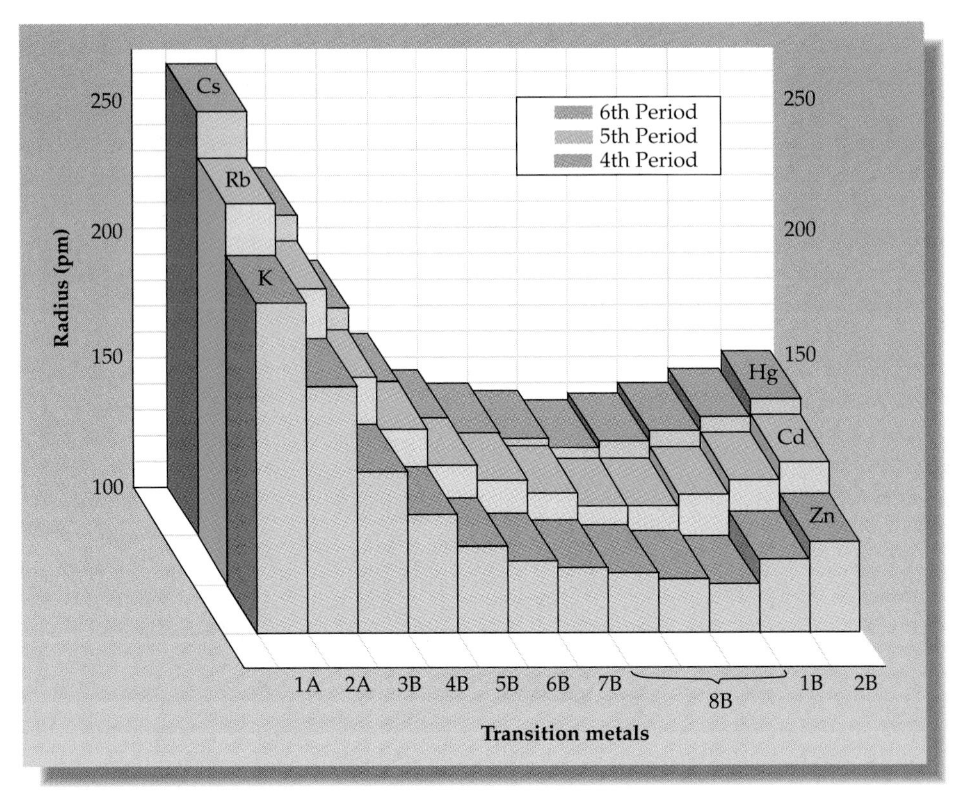

Figure 8.11 Trends in atomic radii. Atomic radii of the Group 1A and 2A metals and the transition metals of the fourth, fifth, and sixth periods.

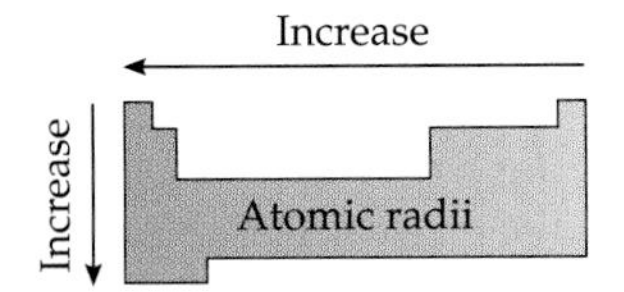

General trends in atomic radii of *s*- and *p*-block elements with position in the periodic table.

to the $3d$ orbitals across the series. The effect of the increased nuclear charge, Z, as one moves from left to right is mostly cancelled out by increased electron–electron repulsion among d electrons. The increase in size for the Group 1B and 2B elements at the end of the series reflects the increased importance of electron–electron repulsion as the d subshell is filled.

Exercise 8.5 Periodic Trends in Atomic Radii

Place the three elements Al, C, and Si in order of increasing atomic radius.

Exercise 8.6 Estimating Atom–Atom Distances

1. Using Figure 8.10, estimate the H—O and H—S distances in H_2O and H_2S, respectively.
2. If the interatomic distance in Br_2 is 228 pm, what is the radius of Br? Using this value, and that for Cl shown previously, estimate the distance between atoms in BrCl.

Ionization Energy

See the *Saunders Interactive General Chemistry CD-ROM,* Screen 8.9, Atomic Properties and Periodic Trends, and Screen 8.12, Ionization Energy.

Ionization energy, the energy required to remove an electron from an atom in the gas phase, was introduced in Section 8.3 as one type of experimental evidence supporting the shell structure of the atom:

$$\text{Atom in ground state(g)} \longrightarrow \text{Atom}^+\text{(g)} + e^-$$
$$\Delta H \equiv \text{ionization energy, } IE$$

Recall from the discussion of the H atom in Section 7.3 that the process of ionization involves moving an electron from a given electron shell to a position a large distance from the atom, that is, to n = infinity. To do this, energy must be supplied to overcome the attraction of the nuclear charge. In accord with thermodynamic convention, the sign of the ionization energy is always positive.

Each atom (except H) can have a series of ionization energies, because more than one electron can always be removed. For example, the first three ionization energies of magnesium are

$$\underset{1s^22s^22p^63s^2}{Mg(g)} \longrightarrow \underset{1s^22s^22p^63s^1}{Mg^+(g)} + e^- \qquad IE(1) = 738 \text{ kJ/mol}$$
$$\underset{1s^22s^22p^63s^1}{Mg^+(g)} \longrightarrow \underset{1s^22s^22p^63s^0}{Mg^{2+}(g)} + e^- \qquad IE(2) = 1451 \text{ kJ/mol}$$
$$\underset{1s^22s^22p^6}{Mg^{2+}(g)} \longrightarrow \underset{1s^22s^22p^5}{Mg^{3+}(g)} + e^- \qquad IE(3) = 7733 \text{ kJ/mol}$$

The very great difference in the second and third ionization energies for Mg is further experimental evidence for the existence of electronic shells in atoms. (See Figure 8.5 and this discussion on pages 339–340.)

Notice that removing each subsequent electron requires more and more energy, and the jump from the second [$IE(2)$] to the third [$IE(3)$] ionization energy is particularly large. Removing an electron from an atom increases the attractive force between the positively charged nucleus and the remaining electrons, and the ionization energy increases. The remaining electrons of the product ion have a lower (or more negative) energy due to the increased attractive forces, and more energy is thus required to remove additional electrons.

The increase in energy required becomes especially great when removing an electron from a shell of lower n. In the preceding magnesium series, the outer electron in Mg^+(g) has the same n and ℓ values as the electron removed in the first step, so the second ionization energy [$IE(2)$] is larger simply because of increased attractive forces. When the third electron is removed, however, the outer electron is a $2p$ electron, that is, it has a lower value of n than the electron removed in the second step; we have dipped into a lower energy electron shell, and the ionization energy [$IE(3)$] increases greatly.

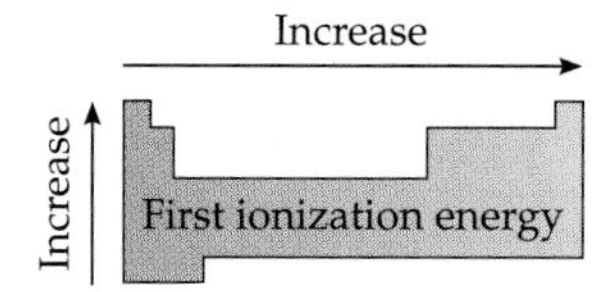

General trends in first ionization energies of A-group elements.

For main group (s- and p-block) elements, *first ionization energies generally increase across a period and decrease down a group* (Figure 8.12 and Appendix F). The trend *across a period* is rationalized by assuming that effective nuclear charge, Z^*, increases with increasing atomic number. Not only does this mean that the atomic radius decreases, but the energy required to remove an electron increases. The general trend *down a group* is rationalized by saying the increase in size makes it easier to remove an electron.

The trend in ionization energies across a given period is not smooth, particularly in the second period. Variations are seen where the value of ℓ changes on going from s-block to p-block elements, from Be to B, for example. The $2p$ electrons are slightly higher in energy than the $2s$ electrons, and so the ionization for B is lower than for Be.

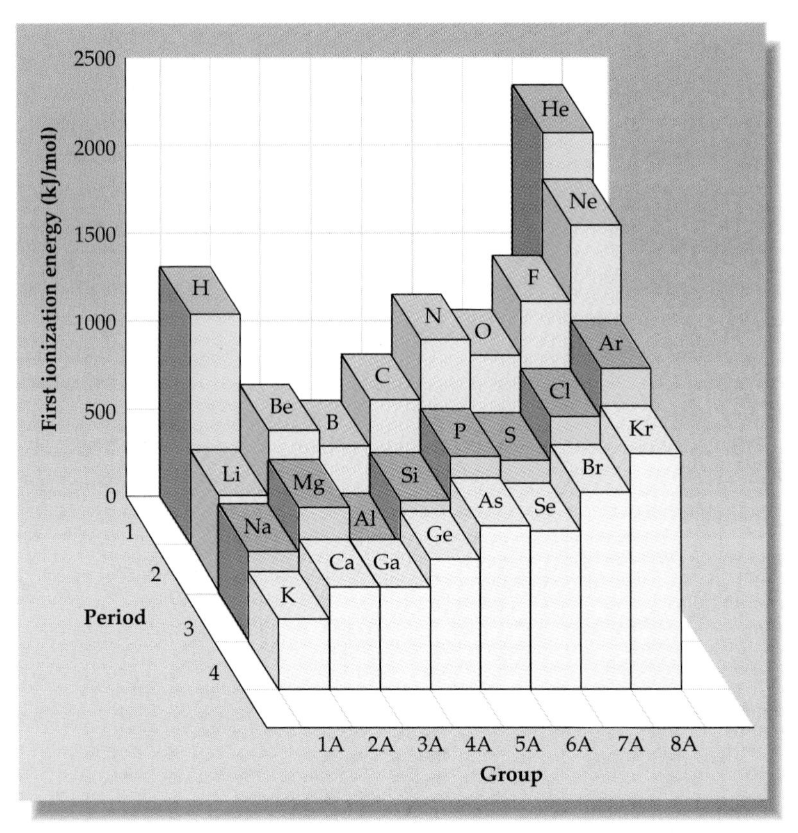

***Figure* 8.12 First ionization energies of the main group elements of the first four periods.** (For data on all the elements see Appendix F.)

Ionization energies represent a balance between electron–nuclear attraction (which depends on Z) and electron–electron repulsion.

On leaving B and moving on to C and then N, the effective nuclear charge increases, which again means an increase in ionization energy. Another dip to lower ionization energy occurs on passing from Group 5A to Group 6A. This is especially noticeable in the second period (N and O). No change occurs in either n or ℓ, but electron–electron repulsions increase for the following reason. In Groups 3A–5A, electrons are assigned to separate p orbitals (p_x, p_y, and p_z). Beginning in Group 6A, however, two electrons are assigned to the same p orbital. The fourth p electron shares an orbital with another electron and thus experiences greater repulsion than it would if it had been assigned to an orbital of its own:

O (oxygen atom)			O^+ (oxygen cation) + e^-	
[Ne][↑↓]	[↑↓][↑][↑]	$\xrightarrow{+1314\ \text{kJ/mol}}$	[Ne][↑↓]	[↑][↑][↑]
$2s$	$2p$		$2s$	$2p$

The greater repulsion experienced by the fourth $2p$ electron makes it easier to remove, and each of the remaining p electrons has an orbital of its own. The normal trend resumes on going from O to F to Ne, however, as the increase in Z^* and decrease in size overcome the effect of pairing electrons in the $2p$ subshell.

Electron Affinity

Some atoms have an affinity, or "liking," for electrons and can acquire one or more electrons to form a negative ion. The **electron affinity,** *EA*, of an atom is defined as the change in energy when the anion of the element loses an electron in the gas phase (Figure 8.13 and Appendix F).

$$A^-(g) \longrightarrow A(g) + e^-(g) \qquad \Delta E \equiv \text{electron affinity, } EA$$

See the *Saunders Interactive General Chemistry CD-ROM,* Screen 8.9, Atomic Properties and Periodic Trends, and Screen 8.13, Electron Affinity.

For example, the electron affinity of fluorine is +328 kJ/mol because the anion, F^-, is so stable. Boron has a much lower electron affinity for an electron, as indicated by its *EA* value of +26.7 kJ/mol. Energy is required to remove the electron from the B^- anion, but the amount of energy is less than 10% of the energy needed to remove an electron from F^-.

Electron affinity and ionization energy are similar in that they represent the energy required to remove an electron. It is therefore not surprising that periodic trends in electron affinity are closely related to those for ionization energy. An element with a high ionization energy generally has a high affinity for an electron. Thus, the values of *EA* become more positive on moving across a pe-

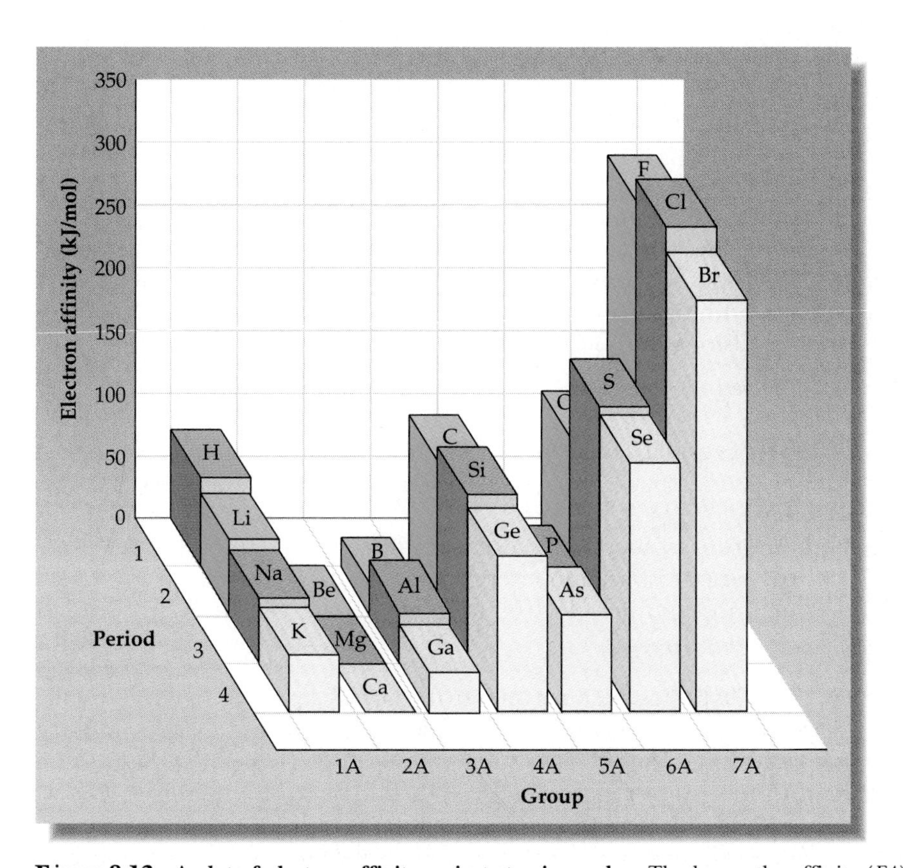

Figure 8.13 A plot of electron affinity against atomic number. The larger the affinity (*EA*) for an electron, the more positive the value. For numerical data, see Appendix F. (Data were taken from H. Hotop and W. C. Lineberger: "Binding energies of atomic negative ions," *Journal of Physical Chemistry*, Reference Data, Vol. 14, p. 731, 1985.)

riod (see Figure 8.13). The effective nuclear charge of the atom is increasing, thus increasing the attraction for an additional electron (and making it more difficult to ionize the atom). One result is that the nonmetals generally have much more positive values of *EA* than the metals. This of course agrees with our chemical experience, which tells us that metals generally do not form negative ions and that elements have an increasing tendency to form anions as we proceed across a period.

The value of *EA* for Be is not measurable because the Be^- anion does not exist. Most tables of *EA* values assign a value of 0.

A severe drop in *EA* is also seen at P and other Group 5A elements on moving across a period. The *EA* of P is much lower than expected.

The trend to higher electron affinity values across a period is not smooth. For example, a beryllium anion, Be^-, is not stable because the added electron must be assigned to a higher energy subshell (2*p*) than the valence electrons (2*s*) (see Figure 8.6). Nitrogen atoms also have no affinity for electrons. Here an electron pair must be formed when an N atom acquires an electron. Significant electron–electron repulsions occur in an N^- ion, making the ion much less stable. The increase in Z^* on going from carbon to nitrogen cannot overcome the effect of these electron–electron repulsions.

The noble gases are not included in our discussion of electron affinity. They have no affinity for electrons because any additional electron must be added to the next higher quantum shell. The higher Z* of the noble gases is not sufficient to overcome this effect.

Electron affinity has been the subject of confusion in introductory chemistry for some years. For a clarifying discussion see J. C. Wheeler: "Electron affinities of the alkaline earth metals and the sign convention for electron affinity." *Journal of Chemical Education,* Vol. 74, pp. 123–127, 1997. Numerical values for *EA* are given in Appendix F.

On descending a group of the periodic table the affinity for an electron generally declines. Electrons are added farther and farther from the nucleus, so the attractive force between the nucleus and electrons decreases. Figure 8.13 shows that this is the case for Cl and Br or P and As, for example. However, the affinity of the F atom for an electron is lower than that of Cl (*EA* for F is less positive than *EA* for Cl); the same phenomenon is observed in Groups 3A through 6A as well. One explanation is that significant electron–electron repulsions occur in the F^- ion, which make the ion less stable. Adding an electron to the seven already present in the $n = 2$ shell of the small F atom leads to considerable repulsion between electrons. Chlorine has a larger atomic volume than fluorine, so adding an electron does not result in such significant electron–electron repulsions in the Cl^- anion.

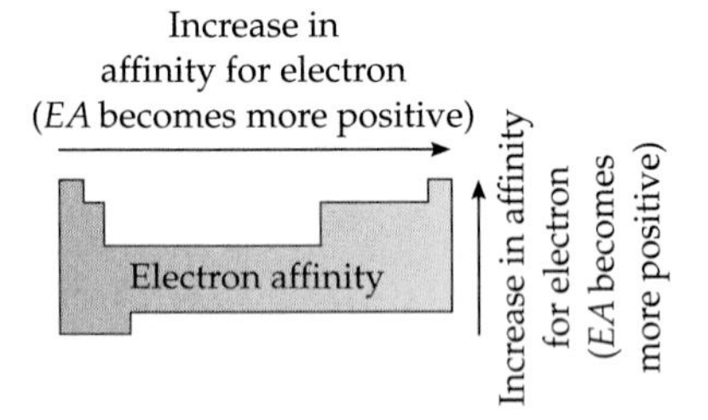

General trends in electron affinities of A-group elements. Exceptions occur at Groups 2A and 5A.

Example 8.6 Periodic Trends

Compare the three elements C, O, and Si.

1. Place them in order of increasing atomic radius.
2. Which has the largest ionization energy?
3. Which has the most positive electron affinity?

Solution

1. *Atomic size.* Atomic radius declines on moving across a period, so oxygen must have a smaller radius than carbon. However, radius increases down a periodic group. Because C and Si are in the same group (Group 4A), Si must be larger than C. In order of increasing size, the elements are therefore O < C < Si.
2. *Ionization energy (IE).* Ionization energy generally increases across a period and decreases down a group; a large decrease in *IE* occurs from the second- to the third-period elements. Thus, the trend in ionization energies should be Si < C < O.

3. *Electron affinity (EA).* Electron affinity values generally become more positive across a period and less positive down a group. Therefore, the *EA* for O should be more positive than the *EA* for C. That is, O (*EA* = +141.0 kJ/mol) has a greater affinity for an electron than does C (*EA* = +121.9 kJ/mol).

It is very interesting to see that the *EA* for Si is more positive (+133.6 kJ/mol) than the *EA* of C (+121.85 kJ/mol). The reason for this is the effect of electron–electron repulsions; such repulsions are larger in the small C^- ion than in the larger Si^- ion.

Exercise 8.7 Periodic Trends

Compare the three elements B, Al, and C.

1. Place the three elements in order of increasing atomic radius.
2. Rank the elements in order of increasing ionization energy. (Try to do this without looking at Figure 8.12; then compare your estimates with the graph.)
3. Which element is expected to have the most positive electron affinity value?

Ion Sizes

Having considered the energies involved in forming positive and negative ions, let us now look at the periodic trends in ionic radii.

Figure 8.14 shows clearly that the periodic trends in the sizes of a few common ions are the same as those for neutral atoms: positive or negative ions of

See the *Saunders Interactive General Chemistry CD-ROM,* Screen 8.9, Atomic Properties and Periodic Trends, and Screen 8.11, Ion Sizes.

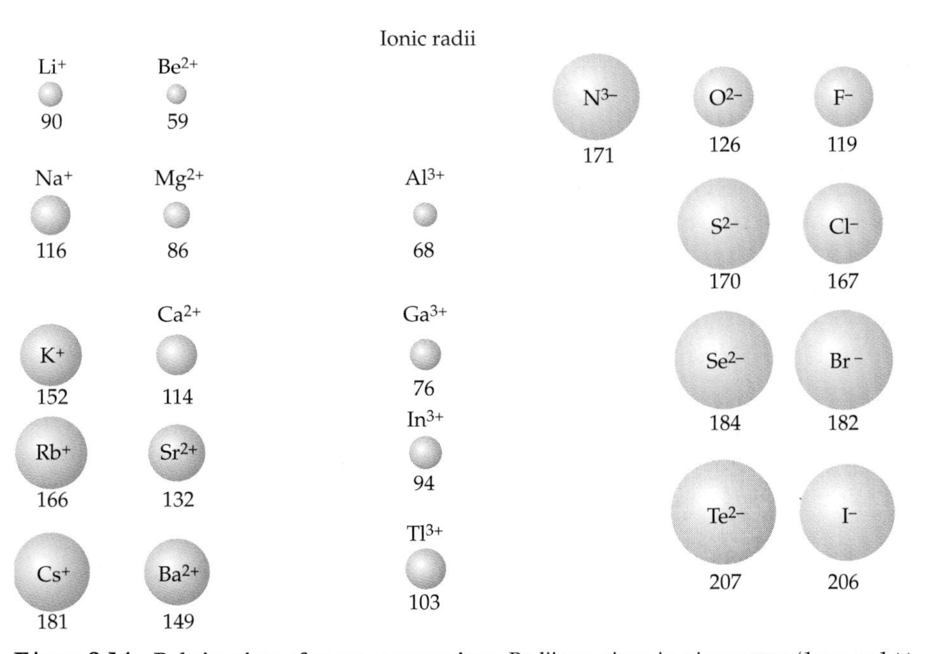

***Figure* 8.14** **Relative sizes of some common ions.** Radii are given in picometers (1 pm = 1×10^{-12} m).

the same group increase in size when descending the group. Pause for a moment, however, and compare Figure 8.14 with Figure 8.10. When an electron is removed from an atom to form a cation, the size shrinks considerably; *the radius of a cation is always smaller than that of the atom from which it is derived.* For example, the radius of Li is 152 pm, whereas that for Li^+ is only 90 pm. This is understandable because when an electron is removed the attractive force of three protons is now exerted on only two electrons, and the remaining electrons contract toward the nucleus. The decrease in ion size is especially great when the last electron of a particular shell is removed, as is the case for Li. The loss of the 2*s* electron from Li leaves Li^+ with no electrons in the $n = 2$ shell.

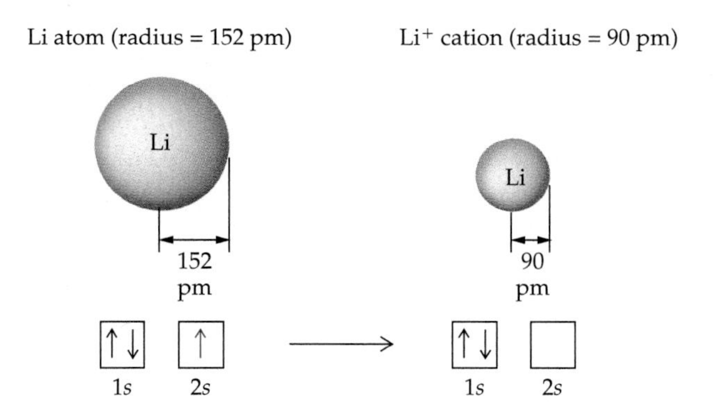

The shrinkage will also be great when two or more electrons are removed, as for Al^{3+} in which it is over 50%:

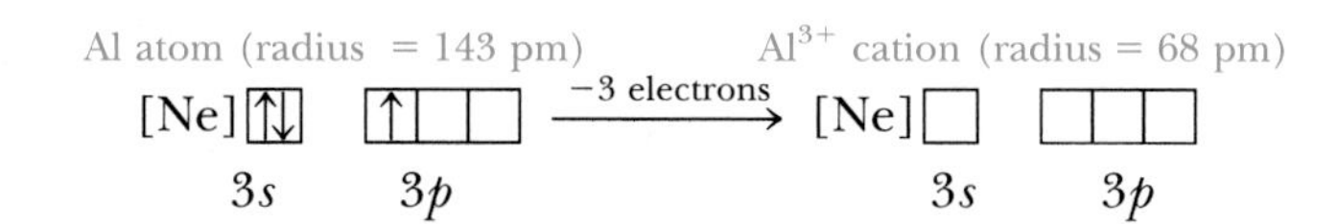

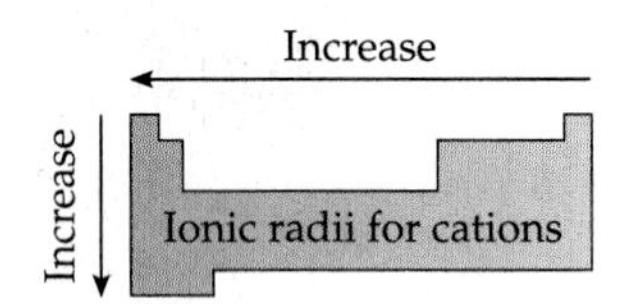

General trends in the positive ion radii of A-group elements.

You can also see by comparing Figures 8.10 and 8.14 that *anions are always larger than the atoms from which they are derived.* Here the argument is the opposite of that used to explain positive ion radii. The F atom, for example, has nine protons and nine electrons. On forming the anion, the nuclear charge is still 9+, but now ten electrons are in the anion. The F^- ion is much larger than the F atom because of increased electron–electron repulsions.

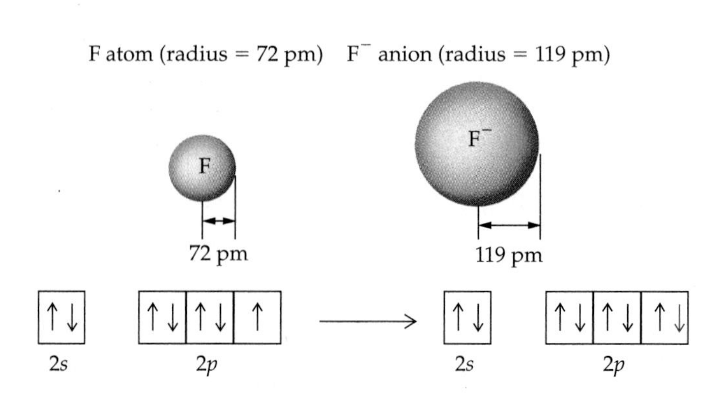

Finally, it is useful to compare the sizes of *isoelectronic ions* across the periodic table. For example, consider O^{2-}, F^-, Na^+, and Mg^{2+}:

Ion	O^{2-}	F^-	Na^+	Mg^{2+}
Ionic radius (pm)	126	119	116	86
Number of nuclear protons	8	9	11	12
Number of electrons	10	10	10	10

All these ions have a total of ten electrons. The O^{2-} ion, however, has only 8 protons in its nucleus to attract these electrons, whereas F^- has 9, Na^+ has 11, and Mg^{2+} has 12. As the number of protons increases in a series of isoelectronic ions, the balance in electron–proton attraction and electron–electron repulsion shifts in favor of attraction, and the radius decreases. As you can see in Figure 8.14, this is generally true for all isoelectronic series of ions.

Exercise 8.8 Ion Sizes

What is the trend in sizes of the ions N^{3-}, O^{2-}, and F^-? Briefly explain why this trend exists.

8.7 PERIODIC TRENDS AND CHEMICAL PROPERTIES

Atomic and ionic radii, ionization energies, and electron affinities are properties associated with atoms, that is, information related to the particulate level of chemistry. We have frequently turned our attention to the particulate level to understand the characteristic properties of elements and compounds. It is reasonable to expect, therefore, that knowledge of these properties will be useful as we continue to explore the chemistry of the elements. Indeed, we use these properties of atoms in many examples as we look more deeply into chemical changes.

Ionization and the Formation of Ionic Compounds

As related earlier, the periodic table was created by grouping together elements that have similar chemical properties. For example, alkali metals characteristically form compounds in which the metal is in the form of a 1+ ion, such as Li^+, Na^+ or K^+. Thus, the reaction between sodium and chlorine gives the ionic compound, NaCl (composed of Na^+ and Cl^- ions) (Figures 1.9 and 8.15), and potassium and water react to form an aqueous solution of KOH (a solution containing the hydrated ions K^+(aq) and OH^-(aq)) (page 18).

Figure 8.15 Sodium metal and chlorine gas react to form NaCl and not compounds with other formulas such as Na_2Cl or $NaCl_2$. The outcome can be explained in part based on ionization energies and electron affinities. See also Figure 1.9 and Chapter 9, Section 9.1. (*C. D. Winters*)

$$2\,\text{Na(s)} + \text{Cl}_2\text{(g)} \longrightarrow 2\,\text{NaCl(s)}$$
$$2\,\text{K(s)} + 2\,\text{H}_2\text{O}(\ell) \longrightarrow 2\,\text{K}^+\text{(aq)} + 2\,\text{OH}^-\text{(aq)} + \text{H}_2\text{(g)}$$

Both of these observations agree with the fact that alkali metals have electron configurations of the type [noble gas core]ns^1 and have low ionization energies.

You might have wondered, however, why sodium doesn't form $NaCl_2$ when it reacts with chlorine, or why potassium gives KOH and not a compound with a different formula such as $K(OH)_2$ when this element reacts with water. Reflecting on the discussion of ionization energies gives us an answer. Reactions generally provide the most stable product, and formation of a Na^{2+} or K^{2+} ion is clearly a very unfavorable process because of the large energy required to remove one of the core electrons. Although the first ionization energy of the Group 1A elements is low, removing a second electron from the metal requires a great deal of energy because this electron must come from the atom's core electrons. In general, formation of a 1+ alkali metal ion is likely, but formation of a 2+ ion of any of the alkali metals in any chemical reaction is energetically unfavorable. For this reason we can make the general statement that *main group metals generally form cations with an electron configuration equivalent to that of the nearest noble gas.*

The addition of more than one electron to an atom is always endothermic, and so much so that the addition of two electrons is endothermic overall. This applies to common anions such as O^{2-} and S^{2-}.

Why isn't Na_2Cl another possible product from the sodium and chlorine reaction? This formula would imply the compound is formed from Na^+ and Cl^{2-} ions. Chlorine atoms have a relatively high electron affinity, but only for the addition of one electron. Adding two electrons per atom means the second must enter the next higher quantum shell, a shell of much higher energy, and an anion such as Cl^{2-} is simply not stable.

The Na_2Cl example leads us to another general statement: *When reacting with a reducing agent, nonmetals generally acquire enough electrons to form an anion with the electron configuration of the next, higher noble gas.* To add additional electrons beyond that produces an unstable ion. This is the reason that chemists often speak of "the tendency of atoms to achieve the noble gas configuration," and this is one reason the noble gases such as He, Ne, and Ar are not reactive. Finally, chemists also speak of "completing an electron octet" because all noble gases have the configuration ns^2np^6 (plus any $(n-1)d^{10}$ and $(n-2)f^{14}$ electrons, as appropriate). The noble gas configuration, with a complete octet of electrons, thus represents a stable configuration for both positive and negative ions in their chemical compounds. This is a feature that we shall discuss further in the next chapter.

We can use similar logic to rationalize results of other reactions. As one goes across the periodic table, from left to right, ionization energies increase. We have seen that Group 1A and 2A elements form ionic compounds, an observation directly related to the low ionization energies for these elements. Ionization energies for elements toward the middle and right side of a period, however, are sufficiently large that cation formation is not favorable. Thus, we do not expect to encounter ionic compounds containing carbon; instead we find carbon *sharing* electrons with other elements in compounds like CO_2 and CCl_4. At the right side of the second period, oxygen and fluorine much prefer taking on electrons than giving them up; these elements have high ionization energies and relatively large electron affinities. Thus, oxygen and fluorine form anions and not cations when they react.

Atomic and Ionic Radii and Similarities Among Transition Metals

Earlier, we called attention to the relatively small variation of atomic radii among the transition elements. This is especially evident for the elements in Group 8B, and it has interesting consequences. Consider cobalt and nickel for instance. These metals have very similar properties such as density (both have densities of 8.9 g/cm^3) and melting points (both metals melt just under 1500 °C). Both occur in meteors, and the elements have important magnetic properties. In the earth's crust, cobalt and nickel usually occur together, a fact necessitating special separation techniques in the metallurgy of these elements. Even the origin of the names for cobalt (German *kobald*, goblin or evil spirit) and nickel (German for satan or "old Nick") suggest a similarity between these elements.

For the main group elements, the size of metal atoms increases in a regular fashion going from the top to the bottom in a periodic group (see Figure 8.11). There is also an increase in atomic radius in going from the first transition series to the second in each group. A similar increase is not seen going from the second to the third transition series of transition metals, however; indeed, there is a very close match in atomic radii for elements in the second and third series and in the same group (such as Ru and Os or Pd and Pt) (see Figure 8.11). The reason for this is that, in building the periodic table, 14 additional elements, the lanthanide elements, are inserted just preceding the third transition series. There is a very gradual decrease in size across the lanthanides (owing to the increase in Z^*). The consequence of having very similar atomic radii is that elements of the same group in the second and third transition series are very similar chemically. They form similar kinds of compounds, and often occur together in ores in the earth's crust. The elements Ru, Os, Rh, Ir, Pd, and Pt represent one particularly striking example of this. These metals are collectively known as the *platinum metals* because they occur together in nature. Other elements that show up together in the earth's crust are silver and gold, molybdenum (Mo) and tungsten (W), and niobium (Nb) and tantalum (Ta).

The term **lanthanide contraction** is commonly used to describe the decrease in size across the series of lanthanide elements.

Exercise 8.9 Energies and Compound Formation

Give a plausible explanation for the observation that magnesium and chlorine give $MgCl_2$ and not $MgCl_3$.

CHAPTER HIGHLIGHTS

When you have finished studying this chapter, you should be able to

- Classify substances as **paramagnetic** (attracted to a magnetic field; characterized by unpaired electron spins) or **diamagnetic** (not magnetic) (Section 8.1).
- Recognize that each electron in an atom has a different set of the four quantum numbers, n, ℓ, m_ℓ, and m_s, where m_s, the spin quantum number, has values of $+\frac{1}{2}$ or $-\frac{1}{2}$ (Section 8.2).
- Understand that the **Pauli exclusion principle** leads to the conclusion that no atomic orbital can be assigned more than two electrons and that the two electrons in an orbital must have opposite spins (different values of m_s) (Section 8.2).
- Using the periodic table as a guide, depict electron configurations of the elements and monatomic ions using an **orbital box notation** or a **spectroscopic notation.** (In both cases, configurations can be abbreviated with the **noble gas notation**) (Sections 8.3 and 8.4).

- Recognize that electrons are assigned to the subshells of an atom in order of increasing subshell energy. In the H atom the subshell energies increase with increasing n, but, in a many-electron atom, the energies depend on both n and ℓ (see Figure 8.6).
- When assigning electrons to atomic orbitals, apply the Pauli exclusion principle and **Hund's rule** (Sections 8.3 and 8.4).
- Predict how properties of atoms—size, **ionization energy** (*IE*), and **electron affinity** (*EA*)—change on moving down a group or across a period of the periodic table (Section 8.7).
  The general periodic trends for these properties are
  (a) Atomic size decreases across a period and increases down a group.
  (b) *IE* increases across a period and decreases down a group.
  (c) *EA* increases generally across a period and decreases down a group.
- Recognize the role that atom and ion size, ionization energy, and electron affinity play in the chemistry of the elements (Section 8.7).

STUDY QUESTIONS

REVIEW QUESTIONS

1. Give the four quantum numbers, specify their allowed values, and tell what property of the electron they describe.
2. What is the Pauli exclusion principle?
3. Using lithium as an example, show the two methods of depicting electron configurations (orbital box diagram and spectroscopic notation).
4. What is Hund's rule? Give an example of its use.
5. What is the noble gas notation? Write an electron configuration using this notation.
6. Name an element of Group 3A. What does the group designation tell you about the electron configuration of the element?
7. Name an element of Group 6B. What does the group designation tell you about the electron configuration of the element?
8. What element is located in the fourth period in Group 4A? What does the element's location tell you about its electron configuration?
9. Describe what happens to atomic size, ionization energy, and electron affinity when proceeding across a period and down a group.

NUMERICAL AND OTHER QUESTIONS

Writing Electron Configurations

10. What are the electron configurations for Al and P? Write these configurations using both the spectroscopic notation and orbital box diagrams. Describe the relation of the atom's electron configuration to its position in the periodic table.

11. What are the electron configurations for Mg and S? Write these configurations using both the spectroscopic notation and orbital box diagrams. Describe the relation of the atom's electron configuration to its position in the periodic table.

12. Using the spectroscopic notation, give the electron configuration of vanadium, V, an element found in some brown and red algae and some toadstools. Compare your answer with Table 8.2.

13. Using the spectroscopic notation, write the electron configurations for atoms of chromium and iron.

14. Depict the electron configuration of an arsenic atom using the spectroscopic and noble gas notations. (A deficiency of arsenic can impair growth in animals even though larger amounts are poisonous.)

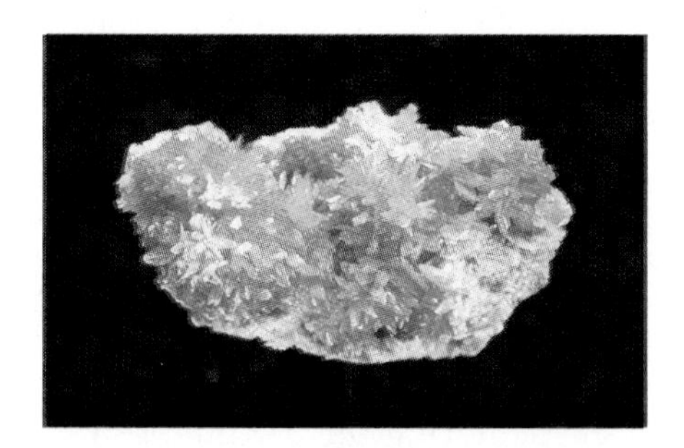

The yellow mineral orpiment contains arsenic. Its formula is As_2S_3. (*C. D. Winters*)

15. Give the electron configuration for the noble gas element krypton using the spectroscopic notation. (The element ranks seventh in abundance of the gases in the earth's atmosphere.)

16. Using the spectroscopic and noble gas notations, write electron configurations for atoms of the following elements and then check your answers with Table 8.2:
(a) Strontium, Sr. This element is named for a town in Scotland.

(b) Zirconium, Zr. The metal is exceptionally resistant to corrosion and so has important industrial applications. Moon rocks show a surprisingly high zirconium content compared with rocks on earth.
(c) Rhodium, Rh. This metal is used in jewelry and in catalysts in industry.
(d) Tin, Sn. The metal was used in the ancient world. Alloys of tin (solder, bronze, and pewter) are important.

17. Use the noble gas and spectroscopic notations to predict electron configurations for the following metals of the third transition series:
(a) Tantalum, Ta. The metal and its alloys resist corrosion and are often used in surgical and dental tools.
(b) Platinum, Pt. This metal was used by pre-Columbian Indians in jewelry. It does not oxidize in air, no matter how high the temperature. It is therefore used by contemporary scientists to coat missile nose cones and in jet engine fuel nozzles.

18. The lanthanides, once called rare earths, are now only "medium rare." All can be purchased for reasonable prices. Using the noble gas and spectroscopic notations, predict reasonable electron configurations for the following elements:
(a) Samarium, Sm. This lanthanide is used in magnetic materials.
(b) Ytterbium, Yb. This element was named for the village of Ytterby in Sweden where a mineral source of the element was found.

19. The actinide americium, Am, is a radioactive element that has found use in home smoke detectors. Depict its electron configuration using the noble gas and spectroscopic notations.

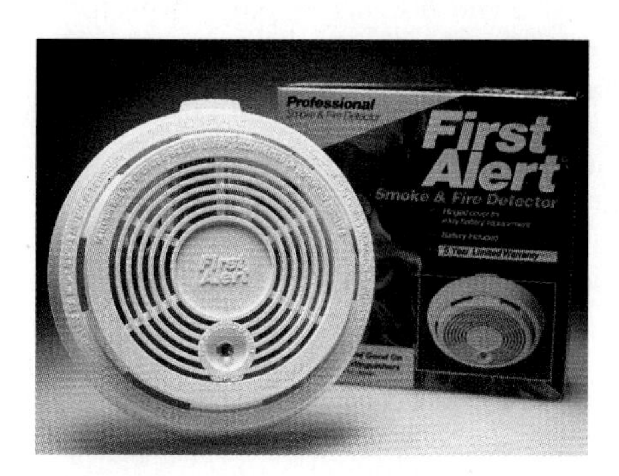

Smoke detectors contain the radioactive element americium, Am. (*C. D. Winters*)

20. Predict reasonable electron configurations for the following elements of the actinide series of elements. Use the noble gas and spectroscopic notations.
(a) Plutonium, Pu. The element is best known as a byproduct of nuclear power plant operation.
(b) Curium, Cm. This actinide was named for Madame Curie (page 64).

21. The name rutherfordium, Rf, has been given to element 104 to honor the physicist Ernest Rutherford (page 72). Depict its electron configuration using the spectroscopic and noble gas notations.

22. Using orbital box diagrams, depict the electron configurations of the following ions: (a) Mg^{2+}, (b) K^+, (c) Cl^-, and (d) O^{2-}.

23. Using orbital box diagrams, depict the electron configurations of the following ions: (a) Na^+, (b) Al^{3+}, (c) Ge^{2+}, and (d) F^-.

24. Using orbital box diagrams and the noble gas notation, depict the electron configurations of (a) V, (b) V^{2+}, and (c) V^{5+}. Are any of the ions paramagnetic?

25. Using orbital box diagrams and the noble gas notation, depict the electron configurations of (a) Ti, (b) Ti^{2+}, and (c) Ti^{4+}. Are any of the ions paramagnetic?

26. Manganese is involved in a number of biological systems. Manganese-depositing bacteria, for example, have been known for many years.
(a) Depict the electron configuration of this element using the noble gas notation and an orbital box diagram.
(b) Using an orbital box diagram, show the electrons beyond those of the preceding noble gas for the 2+ ion.
(c) Is the 2+ ion paramagnetic?
(d) How many unpaired electrons does the Mn^{2+} ion have?

27. Copper plays an important biochemical role, particularly in electron transfer. It exists as both 1+ and 2+ ions. Using orbital box diagrams and the noble gas notation, show electron configurations of these ions. Are either of these ions paramagnetic?

28. Ruthenium, whose compounds are used as catalysts in chemical reactions, has an electron configuration that does not fit the expected pattern (see Table 8.2).
(a) Based on its position in the periodic table, depict the electron configuration of Ru using the noble gas and spectroscopic notations. How does your predicted configuration differ from the actual configuration in Table 8.2?
(b) Using an orbital box notation and the noble gas notation, depict the electron configuration of the ruthenium(III) ion. Can you arrive at the same ion configuration from either the actual or the predicted configuration of the element?

29. Platinum(II) ion is the central ion in cisplatin, $Pt(NH_3)_2Cl_2$, a cancer chemotherapy agent. Platinum has an electron configuration that does not fit the expected pattern (see Table 8.2).
(a) Based on its position in the periodic table, depict the electron configuration of Pt using the noble gas and spectroscopic notations. How does your predicted configuration differ from the actual configuration in Table 8.2?
(b) Using an orbital box notation and the noble gas notation, depict the electron configuration of the plat-

inum(II) ion. Can you arrive at the same ion configuration from either the actual or the predicted configuration of the element?

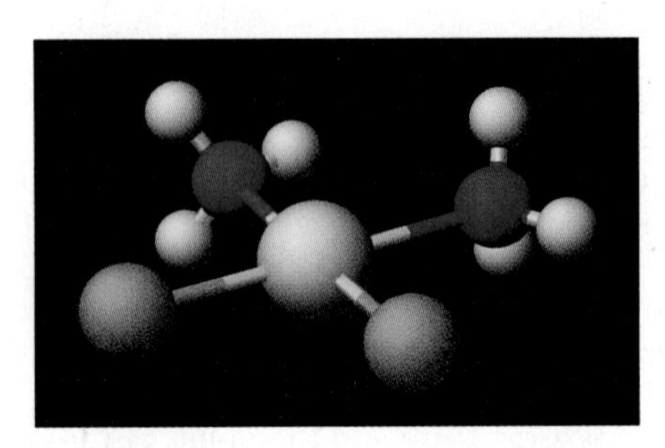

The cancer chemotherapy agent cisplatin has a Pt^{2+} ion in the center of the molecule.

30. Using an orbital box diagram and the noble gas notation show the electron configuration of uranium and of the uranium(IV) ion. Is either of these paramagnetic?

31. The rare earth elements, or lanthanides, commonly exist as 3+ ions. Using an orbital box diagram and the noble gas notation, show the electron configurations of the following:
(a) Ce and Ce^{3+} (cerium)
(b) Ho and Ho^{3+} (holmium)

32. How many unpaired electrons does Ti^{2+} have? The Mn^{3+} ion? Is either of these paramagnetic?

33. Classify the following atoms or ions as paramagnetic or diamagnetic: Al, Al^{3+}, Mg, Co, Co^{3+}.

34. Are any of the 2+ ions of the elements Ti through Zn diamagnetic? Which 2+ ion has the greatest number of unpaired electrons?

35. Two elements in the first transition series (Sc through Zn) have four unpaired electrons in their 2+ ions. What elements fit this description?

Electron Configurations and Quantum Numbers

36. Depict the electron configuration for magnesium using the orbital box and noble gas notations. Give a complete set of four quantum numbers for each of the electrons beyond those of the preceding noble gas.

37. Depict the electron configuration for phosphorus using the orbital box and noble gas notations. Give one possible set of four quantum numbers for each of the electrons beyond those of the preceding noble gas.

38. Using an orbital box diagram and noble gas notation, show the electron configuration of gallium, Ga. Give a set of quantum numbers for the highest energy electron.

39. Using an orbital box diagram and the noble gas notation, show the electron configuration of titanium. Give one possible set of four quantum numbers for each of the electrons beyond those of the preceding noble gas.

40. What is the maximum number of electrons that can be associated with each of the following sets of quantum numbers? In one case, the answer is "none." Explain why this is true.
(a) $n = 4$, $\ell = 3$
(b) $n = 6$, $\ell = 1$, $m_\ell = -1$
(c) $n = 3$, $\ell = 3$, $m_\ell = -3$
(d) $n = 2$, $\ell = 1$, $m_\ell = 1$, $m_s = +\frac{1}{2}$

41. What is the maximum number of electrons that can be identified with each of the following sets of quantum numbers? In some cases, the answer may be "none." In such cases, explain why "none" is the correct answer.
(a) $n = 2$ and $\ell = 1$
(b) $n = 3$
(c) $n = 3$ and $\ell = 2$
(d) $n = 4$, $\ell = 1$, $m_\ell = -1$, and $m_s = -\frac{1}{2}$
(e) $n = 5$, $\ell = 0$, $m_\ell = +1$

42. Explain briefly why each of the following is not a possible set of quantum numbers for an electron in an atom. In each case, change the incorrect value (or values) in some way to make the set valid.
(a) $n = 4$, $\ell = 2$, $m_\ell = 0$, $m_s = 0$
(b) $n = 3$, $\ell = 1$, $m_\ell = -3$, $m_s = -\frac{1}{2}$
(c) $n = 3$, $\ell = 3$, $m_\ell = -1$, $m_s = +\frac{1}{2}$

43. Explain briefly why each of the following is not a possible set of quantum numbers for an electron in an atom. In each case, change the incorrect value (or values) in some way to make the set valid.
(a) $n = 2$, $\ell = 2$, $m_\ell = 0$, $m_s = +\frac{1}{2}$
(b) $n = 2$, $\ell = 1$, $m_\ell = -1$, $m_s = 0$
(c) $n = 3$, $\ell = 1$, $m_\ell = +1$, $m_s = +\frac{1}{2}$

Periodic Properties

44. Estimate the Xe—F bond distance in XeF_2 from the information in Figure 8.10. (Known Xe—F distances are in the range of 190 pm.)

45. Use the data in Figure 8.10 to estimate E—Cl bond distances when E is a Group 5A element.

46. Arrange the following elements in order of increasing size: Al, B, C, K, and Na. (Try doing it without looking at Figure 8.10, then check yourself by looking up the necessary atomic radii.)

47. Arrange the following elements in order of increasing size: Ca, Rb, P, Ge, and Sr. (Try doing it without looking at Figure 8.10, then check yourself by looking up the necessary atomic radii.)

48. Select the atom or ion in each pair that has the larger radius.
(a) Cl or Cl^-
(b) Al or O
(c) In or I

49. Select the atom or ion in each pair that has the larger radius.
(a) Cs or Rb
(b) O^{2-} or O
(c) Br or As

50. Which of the following groups of elements is arranged correctly in order of increasing ionization energy?
(a) C < Si < Li < Ne (c) Li < Si < C < Ne
(b) Ne < Si < C < Li (d) Ne < C < Si < Li

51. Arrange the following atoms in the order of increasing ionization energy: F, Al, P, and Mg.

52. Arrange the following atoms in the order of increasing ionization energy: Li, K, C, and N.

53. Arrange the following atoms in the order of increasing ionization energy: Si, K, As, and Ca.

54. Compare the elements Na, Mg, O, and P.
(a) Which has the largest atomic radius?
(b) Which has the most positive electron affinity?
(c) Place the elements in order of increasing ionization energy.

55. Compare the elements B, Al, C, and Si.
(a) Which has the most metallic character?
(b) Which has the largest atomic radius?
(c) Which has the most positive electron affinity?
(d) Place the three elements B, Al, and C in order of increasing first ionization energy.

56. Periodic trends. Explain each answer briefly.
(a) Place the following elements in order of increasing ionization energy: F, O, and S.
(b) Which has the largest ionization energy: O, S, or Se?
(c) Which has the most positive electron affinity: Se, Cl, or Br?
(d) Which has the largest radius: O^{2-}, F^-, or F?

57. Periodic trends. Explain each answer briefly.
(a) Rank the following in order of increasing atomic radius: O, S, and F.
(b) Which has the largest ionization energy: P, Si, S, or Se?
(c) Place the following in order of increasing radius: Ne, O^{2-}, N^{3-}, or F^-.
(d) Place the following in order of increasing ionization energy: Cs, Sr, Ba.

GENERAL QUESTIONS

58. A neutral atom has two electrons with $n = 1$, eight electrons with $n = 2$, eight electrons with $n = 3$, and two electrons with $n = 4$. Assuming this element is in its ground state, supply the following information:
(a) Atomic number
(b) Total number of *s* electrons
(c) Total number of *p* electrons
(d) Total number of *d* electrons
(e) Whether the element is a metal, metalloid, or nonmetal

59. Element 109, now named meitnerium (for the Austrian–Swedish physicist, Lise Meitner [1878–1968]), was produced in August, 1982, by a team at Germany's Institute for Heavy Ion Research. Depict its electron configuration using the spectroscopic and noble gas notations. Name another element found in the same group as 109.

60. Which of the following is *not* an allowable set of quantum numbers? Explain your answer briefly.

	n	ℓ	m_ℓ	m_s
(a)	2	0	0	$-\frac{1}{2}$
(b)	1	1	0	$+\frac{1}{2}$
(c)	2	1	−1	$-\frac{1}{2}$
(d)	4	3	+2	$-\frac{1}{2}$

61. A possible excited state for the H atom has an electron in a 4*p* orbital. List all possible sets of quantum numbers (n, ℓ, m_ℓ, and m_s) for this electron.

62. How many complete electron shells are in element 71?

63. How many complete electron subshells are there in copper?

64. Name the element corresponding to each characteristic below:
(a) The element with the electron configuration $1s^22s^22p^63s^23p^3$
(b) The element with the smallest atomic radius in the alkaline earth group
(c) The element with the largest ionization energy in Group 5A
(d) The element whose 2+ ion has the configuration $[Kr]4d^5$
(e) The element with the most positive electron affinity in Group 7A
(f) The element whose electron configuration is $[Ar]3d^{10}4s^2$

65. Rank the following in order of increasing ionization energy: Zn, Ca, Ca^{2+}, and Cl^-. Briefly explain your answer.

66. Answer the questions below about the elements A and B, which have the electron configurations shown.

$$A = [Kr]5s^1 \qquad B = [Ar]3d^{10}4s^24p^4$$

(a) Is element A a metal, nonmetal, or metalloid?
(b) Which element has the greater ionization energy?
(c) Which element has the more positive electron affinity?
(d) Which element has the larger atomic radius?

67. Answer the following questions about the elements with the electron configurations shown here:

$$A = [Ar]4s^2 \qquad B = [Ar]3d^{10}4s^24p^5$$

(a) Is element A a metal, metalloid, or nonmetal?
(b) Is element B a metal, metalloid, or nonmetal?
(c) Which element is expected to have the larger ionization energy?
(d) Which element has the smaller atomic radius?

68. Which of the following ions are unlikely to be found in a chemical compound: Cs^+, In^{4+}, Fe^{6+}, Te^{2-}, Sn^{5+}, and I^-? Explain briefly.

69. Place the following elements and ions in order of decreasing size: K^+, Cl^-, S^{2-}, and Ca^{2+}.

70. Answer each of the following questions:
(a) Of the elements S, Se, and Cl, which has the largest atomic radius?
(b) Which is larger, Cl or Cl^-?
(c) Which should have the largest difference between the first and second ionization energy: Si, Na, P, or Mg?
(d) Which has the largest ionization energy: N, P, or As?
(e) Which of the following has the largest radius: Xe, O^{2-}, N^{3-}, or F^-?

71. The following are isoelectronic species: Cl^-, Ar, and K^+. Rank them in order of increasing (a) size, (b) ionization energy, and (c) electron affinity.

72. Compare the elements Na, B, Al, and C with regard to the following properties:
(a) Which has the largest atomic radius?
(b) Which has the largest electron affinity?
(c) Place the elements in order of increasing ionization energy.

73. Two elements in the second transition series (Y through Cd) have four unpaired electrons in their 3+ ions. What elements fit this description?

74. The configuration for an element is given here.

[Ar] [↑↓][↑↓][↑][↑][↑] [↑↓]
3*d* 4*s*

(a) What is the identity of the element with this configuration?
(b) Is a sample of the element paramagnetic or diamagnetic?
(c) How many unpaired electrons does a 3+ ion of this element have?

75. The configuration of an element is given here.

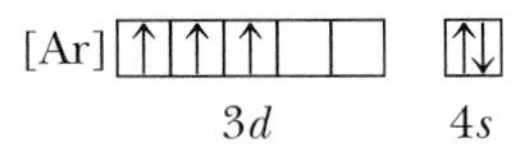

(a) What is the identity of the element?
(b) In what group and period is the element found?
(c) Is the element a nonmetal, a main group element, a transition element, a lanthanide element, or an actinide element?
(d) Is the element diamagnetic or paramagnetic? If paramagnetic, how many unpaired electrons are there?
(e) Write a complete set of quantum numbers (n, ℓ, m_ℓ, m_s) for each of the electrons.
(f) If two electrons are removed to form the 2+ ion, what two electrons are removed? Is the ion diamagnetic or paramagnetic?

CONCEPTUAL QUESTIONS

76. Explain why the sizes of atoms change when proceeding across a period of the periodic table.

77. Explain how the ionization energy of atoms changes and why the change occurs when proceeding down a group of the periodic table.

78. Explain why the sizes of transition metal atoms do not change greatly across a period.

79. Write electron configurations to show the first two ionization processes for potassium. Explain why the second ionization energy is much greater than the first.

80. Predict which of the following elements has the greatest difference between the first and second ionization energy: C, Li, N, and Be. Explain your answer.

81. Why is the radius of Li^+ so much smaller than the radius of Li? Why is the radius of F^- so much larger than the radius of F?

82. Which ions in the following list are likely to be found in chemical compounds: K^{2+}, Cs^+, Al^{4+}, F^{2-}, and Se^{2-}? Do any of these ions have a noble gas configuration?

83. What arguments would you use to convince another student in general chemistry that MgO consists of the ions Mg^{2+} and O^{2-} and not the ions Mg^+ and O^-? What experiments could be done to provide some evidence that the correct formulation of magnesium oxide is $Mg^{2+}O^{2-}$?

84. Explain why the first ionization energy of Ca is greater than that of K, whereas the second ionization energy of Ca is lower than the second ionization energy of K.

85. In general, as you move across a periodic table, the electron affinity of the elements becomes more positive. One exception to this trend, however, is the large decrease in affinity for an electron when going from Group 4A elements to those in Group 5A. Explain why this decrease occurs.

86. Explain why the reaction of calcium and fluorine does *not* have the balanced equation

$$2\ Ca(s) + 3\ F_2(g) \longrightarrow 2\ CaF_3(s)$$

87. Explain why lithium does not reduce argon to form LiAr, a compound having Li^+ and Ar^- ions.

CHALLENGING QUESTIONS

88. Using your knowledge of the trends in element sizes on going across the periodic table, explain briefly why the density of the elements increases from K through V.

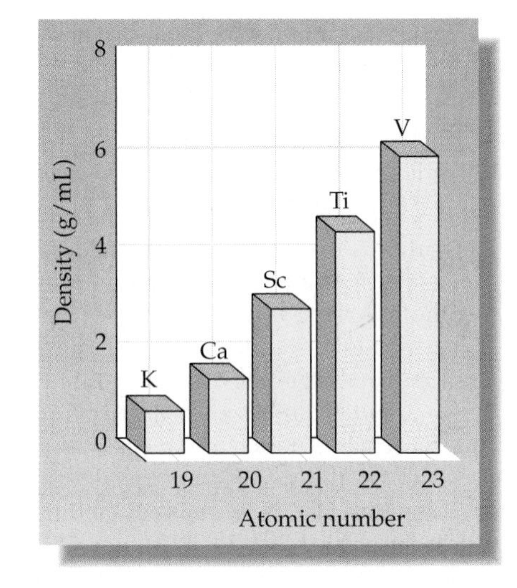

89. The reaction of cobalt metal with HCl gives $CoCl_2$, while the reaction with nitric acid gives $Co(NO_3)_3$. Using the magnetic behavior of these compounds, describe how to tell that neither $CoCl_3$ nor $Co(NO_3)_2$ is a reaction product.

90. The ionization energies for the removal of the first electron in Si, P, S, and Cl are as listed in the table. Briefly rationalize this trend.

Element	First Ionization Energy (kJ/mol)
Si	780
P	1060
S	1005
Cl	1255

91. Suppose that elements on planet Suco have the same periodicity as elements on Earth. Also suppose the planet Suco has a periodic table of elements. Given below is a part of Suco's table, and an incomplete table of atomic radii. Decide on the most likely radius for each element.

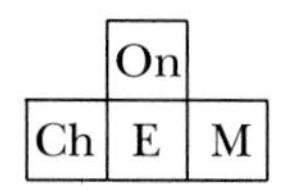

Element	Atomic Radius (pm)
______	90
______	120
______	140
______	180

Summary Question

92. When sulfur dioxide reacts with chlorine, the products are thionyl chloride, $OSCl_2$, and dichlorine oxide, Cl_2O.

$$SO_2(g) + 2\ Cl_2(g) \longrightarrow OSCl_2(g) + Cl_2O(g)$$

(a) Give the electron configuration for an atom of sulfur using the orbital box notation. Do not use the noble gas notation.

(b) Using the configuration given in part (a), write a set of quantum numbers for the highest energy electron in a sulfur atom.

(c) What element involved in this reaction (O, S, Cl) should have the smallest ionization energy? The smallest radius?

(d) Which should be smaller, the sulfide ion, S^{2-}, or a sulfur atom, S?

(e) If you want to make 675 g of $OSCl_2$, how many grams of Cl_2 are required?

(f) If you use 10.0 g of SO_2 and 20.0 g of Cl_2, what is the theoretical yield of $OSCl_2$?

(g) ΔH°_{rxn} for the reaction of SO_2 and Cl_2 is +164.6 kJ/mol of $OSCl_2$ produced. Using this value, and the standard heats of formation of Cl_2O (80.3 kJ/mol) and SO_2 (−296.8 kJ/mol), calculate the standard molar enthalpy of formation of $OSCl_2$.

INTERACTIVE GENERAL CHEMISTRY CD-ROM

Screens 8.2 and 8.3 Spinning Electrons and Magnetism

Describe an experiment that would allow you to detect the presence of unpaired electrons in a molecule.

Screen 8.4 The Pauli Exclusion Principle

(a) What is the difference among a 1*s* orbital, a 2*s* orbital, and a 3*s* orbital?

(b) How do the 2*p* orbitals differ from a 2*s* orbital?

Screen 8.5 Atomic Subshell Energies

How do subshell energies differ on going from hydrogen to elements with more than one electron?

*Screen 8.6 Effective Nuclear Charge, Z**

The animation shows the effective nuclear charge experienced by the 2*s* electron of Li as it penetrates closer to the nucleus. (The distance is measured in units of a_0, Bohr radii, which is equivalent to 0.0529 nm.) Why is the effective nuclear charge felt by this electron only 1+ beyond about $2a_0$, whereas it increases rapidly to 3+ as the electron approaches the nucleus?

Screen 8.10 Atomic Properties and Periodic Trends: Size

If a C atom is attached, or "bonded," to a Cl atom, the calculated distance between the atom is the sum of their radii. Calculate the distance between two C atoms, between C and H, and between C and O. Then use models in the *Molecular Models* folder to examine these distances in ethanol, C_2H_5OH. (See the directions for using these models in the Appendix of this manual. Look particularly for the method of measuring bond distances. Note that the distances given on these models are in Ångstrom units, where 1 Å = 100 pm.) How do the calculated and "observed" distances compare?

Screen 8.12 Atomic Properties and Periodic Trends: Ionization Energy

The difference between the first and second ionization energies of magnesium is about 700 kJ/mol. Why is the difference between the second and third ionization energies over 6000 kJ? What implications does this have for the structure of atoms in general? What implications does this have for the chemistry of magnesium?

Screen 8.14 Properties of Transition Elements

Why are the transition elements of the sixth period so much more dense than the corresponding metals in the fourth and fifth periods? (For a plot of element densities, see the Periodic Table in the "Toolbox.")

Screen 8.15 Chemical Reactions and Periodic Properties (1)

Examine the reactions of lithium, sodium, and potassium with water. Is there a correlation between the relative reactivities of these metals and any of their atomic properties?

Chapter 9

Bonding and Molecular Structure: Fundamental Concepts

(C. D. Winters)

A Chemical Puzzler

Why is sugar sweet? Why is a sugar substitute such as aspartame, a model of which is shown here, 200 times sweeter than sucrose?

Scientists have long known that the key to interpreting the properties of a chemical substance is first to recognize and understand its structure and bonding. *Structure* refers to the way atoms are arranged in space, and *bonding* defines the forces that hold adjacent atoms together. The *Chemical Puzzler* asks why sugar and aspartame are sweet. The answer is that the sweetness is directly related to the molecule's structure. Just how it is related will become more evident as you learn more about the topics of structure and bonding.

Roald Hoffmann (1937–), the Polish-born American chemist, received the 1981 Nobel Prize in chemistry for his work on chemical bonding, the force that binds atoms together to form molecules. Hoffmann has said that "Chemistry is the science of molecules and their transformations. It is the science not so much of the . . . elements, but of the variety of molecules that can be built from them."

The goal of the next three chapters is to figure out how atoms are arranged in chemical compounds and what holds them together. At the same time, we want to begin to show you how to relate the structure and bonding in a molecule to its chemical and physical properties.

Our study of structure and bonding begins with molecules and ions composed of several atoms, and then progresses to larger molecules. Eventually, we want to be able to evaluate substances composed of many atoms. One of the important principles that will come from the initial discussion is that *the bonding and structural characteristics for individual atoms are very similar from molecule to molecule.* This consistency allows us to develop a group of principles that are applicable to all molecules, both small and large.

Knowing a compound's structure and understanding the bonding in that compound are prerequisites to explaining its chemical properties. This applies to simple and complicated molecules alike. For example, we need to know the structure and understand the bonding in water to explain why a collection of such small molecules has an extraordinarily high boiling point, 100 °C. Likewise, knowing about the structure and bonding in common table salt, NaCl, helps explain the high melting point of this ionic compound (801 °C).

You may not think that the boiling point of H_2O, 100 °C, is particularly high. But compare this value to the boiling point of H_2S, a close relative because sulfur is in the same periodic group. Hydrogen sulfide (a gas that smells like rotten eggs) has a boiling point of −60.7 °C, over 160 °C lower than water.

9.1 VALENCE ELECTRONS

The concept of valence electrons was first introduced by the American chemist G. N. Lewis early in this century. The electrons in an atom can be divided into two groups: valence electrons and core electrons (Section 3.3). **Valence electrons** are those in the outermost shell of an atom that determine the chemical properties of the element. The remaining electrons are the **core electrons;** they are not involved in chemical behavior. Chemical reactions involve the loss, gain, or rearrangement of valence electrons.

See the *Saunders Interactive General Chemistry CD-ROM,* Screen 9.2, Valence Electrons.

For main group elements (elements of the A groups in the periodic table), the valence electrons are the *s* and *p* electrons in the outermost shell (Table 9.1). All electrons in inner shells are core electrons. In addition, any electrons in *filled d* subshells are also core electrons. A useful guideline for *main group elements* is that *the number of valence electrons is equal to the group number.* The fact that all elements in a periodic group have the same number of valence electrons accounts for the similarity of chemical properties among members of the group.

Valence electrons for transition elements include the electrons in the *ns* and $(n - 1)d$ orbitals (see Table 9.1). The remaining electrons are core electrons. As was the case with main group elements, the valence electrons for transition metals determine the chemical properties of these elements.

Table 9.1 • CORE AND VALENCE ELECTRONS FOR SEVERAL COMMON ELEMENTS

Element	Periodic Group	Core Electrons	Valence Electrons	Total Configuration
Main Group Elements				
Na	1A	$1s^2 2s^2 2p^6$ = [Ne]	$3s^1$	$[\text{Ne}]3s^1$
Si	4A	$1s^2 2s^2 2p^6$ = [Ne]	$3s^2 3p^2$	$[\text{Ne}]3s^2 3p^2$
As	5A	$1s^2 2s^2 2p^6 3s^2 3p^6 3d^{10}$ = $[\text{Ar}]3d^{10}$	$4s^2 4p^3$	$[\text{Ar}]3d^{10} 4s^2 4p^3$
Transition Elements				
Ti	4B	$1s^2 2s^2 2p^6 3s^2 3p^6$ = [Ar]	$3d^2 4s^2$	$[\text{Ar}]3d^2 4s^2$
Co	8B	[Ar]	$3d^7 4s^2$	$[\text{Ar}]3d^7 4s^2$
Mo	6B	[Kr]	$4d^5 5s^1$	$[\text{Kr}]4d^5 5s^1$

Lewis Symbols for Atoms

Lewis electron dot symbols are variously called Lewis symbols, Lewis dot symbols, or electron dot symbols.

G. N. Lewis also introduced a useful way to represent electrons in the valence shell of an atom. The element's symbol is chosen to represent the atomic nucleus together with the core electrons. Up to four valence electrons, represented by dots, are placed around the symbol one at a time; then, if any electrons remain, they are paired with the ones already there. Chemists now refer to these pictures as **Lewis electron dot symbols.** Lewis symbols for elements of the second and third periods are shown in Table 9.2.

Arranging the electrons around an atom in four groups allows the valence shell to accommodate a maximum of four pairs of electrons. Because this represents eight electrons in all, this is referred to as an **octet** of electrons. As will be evident in later discussions, information about many compounds of the main group elements can be organized based on an octet of electrons around each atom. Hydrogen is an exception, however; *when bonded to another element, hydrogen has two electrons in its valence shell.*

An octet of electrons surrounding an atom is regarded as a stable configuration. This idea derives from the observation that the noble gases (He, Ne, Ar, Kr, Xe, and Rn) demonstrate a notable lack of reactivity. All but helium among these elements have an octet configuration. Helium, neon, and argon do not undergo any chemical reactions, and the other noble gases have very limited chemical reactivity. Because chemical reactions involve changes in the outermost (valence) shell, the limited reactivity of the noble gases is taken as evidence of

Table 9.2 • LEWIS DOT SYMBOLS FOR MAIN GROUP ATOMS

1A ns^1	2A ns^2	3A ns^2np^1	4A ns^2np^2	5A ns^2np^3	6A ns^2np^4	7A ns^2np^5	8A ns^2np^6
Li·	·Be·	·B·	·C·	·N·	:O·	:F·	:Ne:
Na·	·Mg·	·Al·	·Si·	·P·	:S·	:Cl·	:Ar:

Historical Perspective

Gilbert Newton Lewis (1875–1946)

G. N. Lewis introduced the theory of the shared electron-pair chemical bond in a paper published in the *Journal of the American Chemical Society* in 1916. This theory revolutionized chemistry, and it is to honor his contribution that electron dot structures are now known as Lewis structures. Lewis also made major contributions in the field of thermodynamics, and his research included studies on isotopes and on the interaction of light with substances. Later, in Section 17.11, you will encounter Lewis acid–base theory, an extension of his bonding theory applied to acids and bases.

G. N. Lewis *(Oesper Collection in the History of Chemistry/University of Cincinnati)*

G. N. Lewis was born in Massachusetts but raised in Nebraska. He began an academic career after earning his B.A. and Ph.D. at Harvard University. In 1912, he was appointed chairman of the Chemistry Department at the University of California, Berkeley, and remained at Berkeley the rest of his life. Lewis felt that a chemistry department should teach and advance fundamental chemistry, and he was not only a productive researcher but also a teacher who profoundly influenced his students. Among his ideas was the use of problem sets in teaching, an idea still in use today.

the stability of their noble gas (ns^2np^6) electron configuration. Hydrogen, which in its compounds has two electrons in its valence shell, obeys the spirit of this rule by matching the electron configuration of He.

Example 9.1 Valence Electrons

Give the number of valence electrons for Ca and Se. Draw the Lewis electron dot symbol for each element.

Solution For main group elements the number of valence electrons equals the group number. Thus, calcium has two valence electrons, and selenium has six. Dots representing electrons are placed around the element symbol one at a time until there are four electrons. Subsequent electrons are paired with those already present:

$\cdot\text{Ca}\cdot$	$\cdot\ddot{\underset{\cdot\cdot}{\text{Se}}}\cdot$
calcium	selenium

Exercise 9.1 Valence Electrons

Give the number of valence electrons for Ba, As, and Br. Draw the Lewis dot symbol for each of these elements.

9.2 CHEMICAL BOND FORMATION

When a chemical reaction occurs between two atoms, their valence electrons are reorganized so that a net attractive force—a **chemical bond**—occurs between atoms. There are two general types of bonds, ionic and covalent, and their formation can be depicted using Lewis symbols.

See the *Saunders Interactive General Chemistry CD-ROM,* Screen 9.3, Chemical Bond Formation.

(a) (b)

Figure 9.1 Formation of ionic compounds. (a) The reaction of elemental sodium and chlorine to give sodium chloride. (b) The reaction of elemental calcium and oxygen to give calcium oxide. Notice that both reactions are quite exothermic, as reflected by the very negative molar enthalpies of formation of the reaction products. *(C. D. Winters)*

The force of attraction between a positive and negative ion is a coulombic force. See the discussion of Coulomb's law and forces of attraction in Section 3.4.

An **ionic bond** forms when *one or more valence electrons is transferred from one atom to another*, creating positive and negative ions. When sodium and chlorine react (Figure 9.1), an electron is transferred from a sodium atom to a chlorine atom to form Na^+ and Cl^-.

$$\mathrm{Na}\cdot + \cdot\underset{\cdot\cdot}{\ddot{\mathrm{Cl}}}: \longrightarrow \left[\mathrm{Na}\cdot \curvearrowright \cdot\underset{\cdot\cdot}{\ddot{\mathrm{Cl}}}:\right] \longrightarrow \left[\mathrm{Na}^+ \quad :\underset{\cdot\cdot}{\ddot{\mathrm{Cl}}}:^-\right]$$

Metal atom | Nonmetal atom | Electron transfer from reducing agent to oxidizing agent | Ionic compound. Ions have noble gas electron configurations.

The "bond" is the attractive force between the positive and negative ions.

Covalent bonding, in contrast, *involves sharing of valence electrons between atoms.* Two chlorine atoms, for example, share a pair of electrons, one electron from each atom, to form a covalent bond.

$$:\underset{\cdot\cdot}{\ddot{\mathrm{Cl}}}\cdot + \cdot\underset{\cdot\cdot}{\ddot{\mathrm{Cl}}}: \longrightarrow :\underset{\cdot\cdot}{\ddot{\mathrm{Cl}}}:\underset{\cdot\cdot}{\ddot{\mathrm{Cl}}}:$$

It is useful to reflect on the differences in the Lewis electron dot structure representations for ionic and covalent bonding. In both processes, unpaired electrons in the reactants are paired up. Both processes give products in which each atom is surrounded by eight electrons (an octet). The position of the electron pair between the two bonded atoms differs significantly, however. In a chlorine molecule (Cl_2), the electron pair is found *midway* between the two atoms and is shared equally by them. In contrast, the electron pair in sodium chloride has become part of the valence shell of chlorine. The consequences of these differences on the properties of a substance are enormous. The ionic compound, NaCl, is a solid with a high melting point (801 °C), whereas the covalent molecular species, Cl_2, is a gas that liquefies at −34.04 °C and solidifies at −101.5 °C.

As bonding is described in greater detail, you will discover that the two types of bonding—complete electron transfer and the equal sharing of electrons—are extreme cases. Generally, in real compounds electrons are only partially transferred (or, if you like, shared unequally). Not surprisingly, this results in a gradation of properties for the compounds as well.

9.3 BONDING IN IONIC COMPOUNDS

Metallic sodium reacts vigorously with gaseous chlorine to give sodium chloride (see Figure 9.1a), and calcium metal and oxygen react to give calcium oxide (Figure 9.1b). In each case, the product is an ionic compound: NaCl contains Na^+ and Cl^- ions, whereas CaO is composed of Ca^{2+} and O^{2-} ions.

See Section 8.7 for a discussion of the relationship between the position of an element in the periodic table and the types of compounds it can form.

$$Na(s) + \tfrac{1}{2}\,Cl_2(g) \longrightarrow NaCl(s) \qquad \Delta H° = -411.2 \text{ kJ/mol}$$
$$Ca(s) + \tfrac{1}{2}\,O_2(g) \longrightarrow CaO(s) \qquad \Delta H° = -635.09 \text{ kJ/mol}$$

These reactions, which are examples of the general chemical behavior for elements in these periodic groups, can be understood based on the atomic properties described in Chapter 8. The alkali and alkaline earth metals, elements immediately following Group 8A, have low ionization energies. Thus, relatively little energy is required to remove valence electrons from these elements to form cations with a noble gas configuration. In contrast, elements immediately preceding Group 8A (the halogens and the Group 6A elements) have high electron affinities. These elements typically form anions by adding electrons, in most instances giving an ion with the electron configuration equivalent to that of the next noble gas.

The tendency to achieve a noble gas configuration by gain or loss of electrons is an important observation in the chemistry of main group elements. An underlying feature, one that is even more important in guiding the chemistry of the elements and determining the structure of compounds, however, is the favorable energetics of ionic compound formation. In general,

- A chemical reaction occurs when the products of the reaction have a lower potential energy than the reactants. (Such reactions were described as *product-favored reactions* in Chapter 6.)
- The structure of a compound, either ionic or covalent, is the one having the lowest potential energy, that is, the greatest thermodynamic stability.

Energy of Ion Pair Formation

To understand ionic compounds we want to look further at the energetics of their formation. We do this in two steps: first considering the combination of a single cation and anion to form an ion pair, and then looking at the formation of one mole of an ionic compound.

The discussion of the energetics of NaCl is an application of Hess's law (Section 6.7).

The Lewis electron dot representation is a useful starting point to analyze the energy involved in the reaction of a sodium atom with a chlorine atom to form a [Na^+, Cl^-] ion pair. The overall energy of this reaction can be thought of as the sum of three individual steps: (1) the ionization of a sodium atom (the energy of this process is the *ionization energy* of this element), (2) the addition of an electron to a chlorine atom (the negative of the *electron affinity* of the atom), and (3) the pairing of the two ions to form an *ion pair*. The energy of the third process is related directly to the attractive forces between the ions.

$$\begin{array}{ll} Na(g) \longrightarrow Na^+(g) + e^- & IE_{Na} = +502.0 \text{ kJ/mol} \\ Cl(g) + e^- \longrightarrow Cl^-(g) & -EA_{Cl} = -349.0 \text{ kJ/mol} \\ Na^+(g) + Cl^-(g) \longrightarrow NaCl(g) & E_{\text{ion pair}} = -552.0 \text{ kJ/mol} \\ \hline Na(g) + Cl(g) \longrightarrow NaCl(g) & \Delta E = -399.0 \text{ kJ/mol} \end{array}$$

Although it is possible to use the noble gas configuration as a guideline to predict what happens in the reaction of Na and Cl_2, a statement such as "the driving force for formation of Na^+ and Cl^- is to allow these elements to achieve a noble gas configuration" is, at best, misleading.

Examination of the energies for these three steps reveals that the formation of the two ions individually ($IE + EA = +153.0$ kJ) is actually an *endo*thermic process. The formation of Na^+ and Cl^- ions would not be favorable were it not for the fact that ion pair formation is a very exothermic process ($E_{\text{ion pair}}$ has a large negative value). The energy of ion pair formation greatly outweighs the first two steps, so the overall reaction is very favorable on energetic grounds.

The energy associated with the formation of an ion pair, $E_{\text{ion pair}}$, cannot be measured directly, but it can be calculated. This process uses a mathematical equation derived from Coulomb's law, the law describing the force of attraction between ions of opposite charge (page 113):

$$E_{\text{ion pair}} = C\frac{(n^+e)(n^-e)}{d}$$

The energy of the electrostatic attraction follows the sign conventions established in Chapter 6. A negative sign indicates evolution of energy. A lower energy for the system (a more negative value) means the system is more stable.

The symbol C is a constant, d represents the distance between the ion centers, n is the number of positive (n^+) or negative (n^-) charges on an ion, and e is the charge on an electron. Thus, n^+e and n^-e are, respectively, the electronic charges of the cation and anion. Because the two charges have the opposite sign, the energy value is negative. The important feature of this equation is that the energy of attraction between ions of opposite charge depends on two factors.

- *The magnitude of the ion charges.* The higher the ion charges, the greater the attraction, so $E_{\text{ion pair}}$ has a larger negative value. For example, the attraction between Ca^{2+} and O^{2-} ions will be about four times larger [(2+) × (2−)] than the attraction between Na^+ and Cl^- ions, and the energy will be more negative by a factor of about 4.
- *The distance between the ions.* This is an inverse relationship because, as the distance between ions becomes greater (d becomes larger), the attractive force between the ions declines and the energy is less negative. The distance is determined by the sizes of the ions (Figure 8.14).

The two factors affecting the interaction between cations and anions were described qualitatively on page 113.

The effect of ion size is apparent in a plot of $E_{\text{ion pair}}$ for the alkali metal halides (Figure 9.2). The variations are regular and predictable based on periodic trends in the sizes of ions. For example, values of $E_{\text{ion pair}}$ for the chlorides become progressively more negative going from KCl to NaCl to LiCl because the alkali metal ion radii decrease in the order $K^+ > Na^+ > Li^+$ (and so d decreases in this order). The smaller the positive ion, the more negative the value of $E_{\text{ion pair}}$. Similarly, the value of $E_{\text{ion pair}}$ for the halides of a given alkali metal ion becomes more negative with a decrease in size of the halide ion.

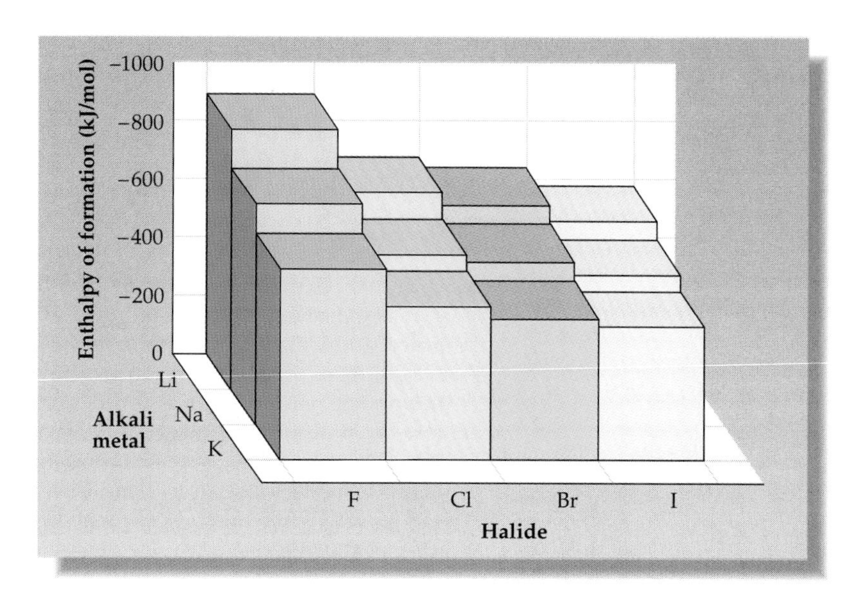

Figure 9.2 Energy of ion pair formation. $E_{\text{ion pair}}$ is illustrated for the formation of the alkali metal halides, MX(g), from the ions $M^+(g) + X^-(g)$. The energy was calculated from the equation $E = (138{,}900\ n^+n^-/d)$ kJ/mol, where n is the ionic charge and d is the interionic distance in picometers (pm).

Lattice Energy

Although ion pair formation is useful for illustrating periodic trends, it is more realistic to think about ionic compounds as they exist under normal conditions. Ionic compounds are solids, and their structures contain positive and negative ions arranged in a three-dimensional lattice. There are no ion pairs in these structures. Ball-and-stick and space-filling models of a small segment of a NaCl lattice are pictured in Figure 9.3. In crystalline NaCl, each Na^+ cation is surrounded by six Cl^- anions, and six Na^+ ions are nearest neighbors to each Cl^-.

Attractive forces between ions in an ionic compound are large, a conclusion anticipated because ionic compounds melt at very high temperatures. A great deal of energy must be supplied to overcome the attractive forces and disrupt the arrangement of ions in the crystal.

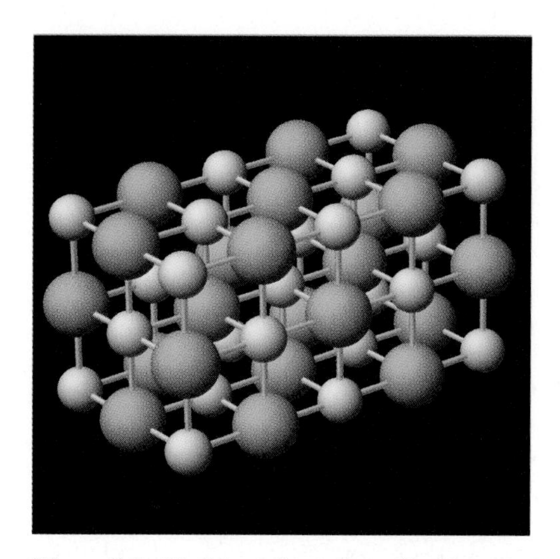

Figure 9.3 Models of the sodium chloride crystal lattice. *(left)* A ball-and-stick model. The "sticks" in the model are there to help identify the locations of the atoms. *(right)* A space-filling model. These models represent only a small portion of the lattice. Ideally, it extends infinitely in all directions. Sodium ions are silver and chloride ions are yellow.

A CLOSER LOOK

Using a Born-Haber Cycle to Calculate Lattice Energies

Lattice energy cannot be measured directly, but it is possible to calculate it from measurable thermochemical quantities. This approach uses a Born-Haber cycle, so named for Max Born (1882–1970) and Fritz Haber (1868–1934), two German scientists prominent earlier in this century. It is an application of Hess's law because the energies involved in one pathway from reactants to products (ΔH_f°) is the sum of the energies involved in another pathway (Steps 1–5). The procedure is illustrated by a calculation of the lattice energy of solid sodium chloride, that is, the energy involved in Step 5. The overall process of lattice formation is first broken into a series of steps:

$Na^+(g)$ + $Cl^-(g)$
↑ *Step 3* ↑ *Step 4*
$Na(g)$ + $Cl(g)$ *Step 5*
↑ *Step 1* ↑ *Step 2*
$Na(s) + \frac{1}{2} Cl_2(g) \xrightarrow{\Delta H_f^\circ} NaCl(s)$

Steps 1 through 4 involve formation of the $Na^+(g)$ and $Cl^-(g)$ ions from the elements. The enthalpy of each step can be measured in an experiment. Step 5 is the formation of the solid ionic lattice, and the energy involved is the lattice energy. ΔH_f° is the standard molar enthalpy of formation of NaCl(s), obtained by calorimetry. The enthalpy values for each step are related by the equation:

$$\Delta H_f^\circ = \Delta H_{Step\ 1} + \Delta H_{Step\ 2} + \Delta H_{Step\ 3} + \Delta H_{Step\ 4} + \Delta H_{Step\ 5}$$

The value for Step 5 can be calculated using this equation because all other values are known.

Step 1. Enthalpy of formation of Na(g) = +107.32 kJ/mol (Appendix L)

Step 2. Enthalpy of formation of Cl(g) = +121.68 kJ/mol (Appendix L)

Step 3. Ionization energy for Na(g) = +496 kJ/mol (Appendix F)

Step 4. −Electron affinity for Cl(g) = −349 kJ/mol (Appendix F)

ΔH_f° Standard heat of formation of NaCl = −411. kJ/mol (Appendix L)

This means the lattice energy is

Step 5. Formation of NaCl(s) from the ions in the gas phase = $E_{lattice}$ = −786 kJ/mol

In this calculation, Steps 1, 2, and 3 are endothermic. Energy is required to vaporize and ionize sodium atoms, and to separate the Cl_2 molecule into two chlorine atoms. The ΔH values for Steps 4 and 5 are both negative; energy is evolved when a chlorine atom acquires an electron and when 1 mol of $Na^+(g)$ and 1 mol of $Cl^-(g)$ ions come together to form the solid crystalline lattice.

See the *Saunders Interactive General Chemistry CD-ROM*, Screens 8.17 and 8.18, Lattice Energy.

Designating the reactants as gaseous ions is meant to indicate that the reactants are so far apart that they experience no attractive forces from other ions.

For ionic compounds, a quantity called the lattice energy is a measure of bonding energy in a crystalline compound. The **lattice energy,** $E_{lattice}$, is defined as the energy of formation of one mole of a solid crystalline ionic compound when ions in the gas phase combine:

$$Na^+(g) + Cl^-(g) \longrightarrow NaCl(s) \qquad E_{lattice} = -786\ kJ/mol$$

The lattice energy for an ionic compound results from the attraction between the cations and anions *in a crystal.* It is not possible to measure lattice energy directly; the reaction that describes this is not one that can be carried out in a laboratory. It is possible to *calculate* lattice energies, however. The math is a bit tedious, though, because it is necessary to include all possible interactions between cations and anions in the crystal. The largest attractive forces are those between neighboring ions of opposite charges. Attractive forces between ions of opposite charge that are farther apart, as well as the forces of repulsion between ions of the same charge in the lattice, must also be included in the calculation.

Fortunately, values of the lattice energy can be determined from measurable thermodynamic parameters. These calculations are usually carried out using enthalpy values for the various steps, yielding an enthalpy value ($\Delta H_{lattice}$).

Internal energy ($E_{lattice}$) and the enthalpy change ($\Delta H_{lattice}$) are related through the first law of thermodynamics (Chapter 6). Determination of the lattice enthalpy of NaCl by this method is illustrated in "A Closer Look: Using a Born-Haber Cycle to Calculate Lattice Energies."

Like values for $E_{ion\ pair}$, the values of $\Delta H_{lattice}$ vary predictably with the charge on the ions. The value of $\Delta H_{lattice}$ for MgO (−4050 kJ/mol), for example, is about four times more negative than the value for NaF (−926 kJ/mol) because the charges on the Mg^{2+} and O^{2-} ions are twice as large as those on Na^+ and F^- ions. The effect of ion size on lattice energy is also predictable: a lattice built from smaller ions generally leads to a more negative value for the lattice energy. Data in Table 9.3 illustrate this point. For the alkali metal halides, for example, the lattice energy for a lithium compound is generally more negative than that for a potassium compound.

Table 9.3 • LATTICE ENERGIES OF SOME IONIC COMPOUNDS*

Compound	$\Delta H_{lattice}$ (kJ/mol)
LiF	−1037
LiCl	−852
LiBr	−815
LiI	−761
NaF	−926
NaCl	−786
NaBr	−752
NaI	−702
KF	−821
KCl	−717
KBr	−689
KI	−649

*D. Cubicciotti: "Lattice energies of the alkali halides and electron affinities of the halogens." *Journal of Chemical Physics*, Vol. 31, p. 1646, 1959.

Exercise 9.2 Using Lattice Energies

Calculate the molar enthalpy of formation, ΔH_f°, of solid sodium iodide using the approach outlined in "A Closer Look: Using a Born-Haber Cycle to Calculate Lattice Energies." The required data can be found in Appendices F and L and in Table 9.3.

Why Compounds Such as $NaCl_2$ and NaNe Don't Exist

Lattice energies are useful in understanding why certain products form—or do not form—in chemical reactions. Let us revisit the question of why a compound such as $NaCl_2$, in which sodium is present as the Na^{2+} ion, is unlikely (Section 8.7). The formation of Na^{2+} (with the configuration $1s^22s^22p^5$) would require the loss of *two electrons* from sodium. Because the second electron must be removed from the $n = 2$ shell, formation of Na^{2+} requires a substantial amount of energy. The total energy for Step 3 in the Born-Haber cycle in *A Closer Look* is the sum of the first and second ionization energies (496 kJ + 4562 kJ). Making the reasonable assumption that the lattice energy for $NaCl_2$ is at least double that of NaCl (the increase coming because the cation charge has doubled, and the size of Na^{2+} is less than that of Na^+), we would estimate a very positive value for ΔH_f of $NaCl_2$ (about + 3000 kJ/mol). A positive value of ΔH_f means that formation of $NaCl_2$ from Na(s) and Cl_2(g) is unfavorable.

Because sodium is a good reducing agent, why doesn't it reduce neon to Ne^- and form NaNe? Again, we can think about this question in terms of the Born-Haber cycle that was given for NaCl (see *A Closer Look,* p. 380). Put Ne in place of Cl_2. Neon exists in the form of atoms, so $\Delta H_{step\ 2}$ is not required. In addition, neon's affinity for an electron should be extremely low. This is because the additional electron has to be placed in the next higher electron shell, and the $n = 3$ shell is much higher in energy than the $n = 2$ shell.

$$\text{Ne}\ (1s^22s^22p^6) + e^- \longrightarrow \text{Ne}^-\ (1s^22s^22p^63s^1) \qquad \Delta H \gg 0$$

This means $\Delta H_{step\ 4}$ should be quite endothermic. The lattice energy of NaNe is not expected to be large enough to overcome this and other endothermic steps, so an overall positive enthalpy change is again expected. The formation of NaNe is energetically unfavorable.

9.4 COVALENT BONDING

See the *Saunders Interactive General Chemistry CD-ROM,* Screen 9.3, Chemical Bond Formation.

The remainder of this chapter is concerned with covalent bonding, in which electron pairs are shared between bonded atoms. Examples of molecules having covalent bonds include gases in our atmosphere (O_2, N_2, H_2O, and CO_2) and common fuels (CH_4). Covalent bonding also serves to hold the atoms together in many common ions such as CO_3^{2-}, CN^-, NH_4^+, NO_3^-, and PO_4^{3-}. We will develop the basic principles of structure and bonding using as examples molecules and ions made up of only a few atoms, but the same principles apply to much larger molecules from aspirin to DNA to zingerone (a compound isolated from ginger root).

The molecules and ions just mentioned are composed entirely of nonmetallic atoms. A point that needs special emphasis is that, in molecules or ions made up *only* of nonmetallic atoms, the atoms will be attached by covalent bonds. Conversely, the presence of a metal in a formula is a signal that the compound is likely to be ionic.

Lewis Electron Dot Structures

See the *Saunders Interactive General Chemistry CD-ROM,* Screens 9.4 and 9.5, Lewis Structures.

In a simple description of covalent bonding, a bond results when one or more electron pairs are shared between two atoms. Thus, the electron pair bond between the two atoms of an H_2 molecule is represented by a pair of dots or, alternatively, a line.

H : H or H—H

Such a representation is called a Lewis electron dot structure or just a **Lewis structure.**

Simple Lewis structures can be drawn starting with Lewis dot symbols for atoms and arranging the valence electrons to form bonds. To create the Lewis structure for F_2, for example, we start with the Lewis dot symbol for a fluorine atom. Fluorine, an element in Group 7A, has seven valence electrons. The Lewis symbol for an F atom has a single unpaired electron along with three electron pairs. In F_2, the single electrons, one on each F atom, pair up in the covalent bond.

:F· + ·F: ⟶ :F:F: or :F—F:

Lone pair of electrons

Shared or bonding electron pair

Lone pairs can be important in a structure. Being in the same electron shell as the bonding electrons, they can influence molecular shape. See Section 9.7.

In the Lewis dot structure for F_2, two types of electrons occur in the valence shell of each atom. The pair of electrons in the F—F bond is the bonding pair, or **bond pair.** The other six pairs in F_2 reside on single atoms and are called **lone pairs.** Because they are not involved in bonding, they are also called **nonbonding electrons.**

Carbon dioxide, CO_2, and dinitrogen, N_2, are important examples of molecules in which two atoms are multiply bonded, that is, they share more than one electron pair.

O=C=O :N≡N:

In carbon dioxide, the carbon atom shares two pairs of electrons with each oxygen and so is linked to each O atom by a **double bond.** The valence shell of each oxygen atom in CO_2 has two bonding pairs and two lone pairs. In dinitrogen, the two nitrogen atoms share three pairs of electrons, so they are linked by a **triple bond.** In addition, each N atom has a single lone pair.

The Octet Rule

An important observation can be made about the molecules you have seen so far: each atom in the molecule (except hydrogen) has a share in four pairs of electrons, so each has achieved a *noble gas configuration. Each atom is surrounded by an octet of electrons.* (Hydrogen typically forms a bond to only one other atom, resulting in two electrons in its valence shell.) If we accept this as a guideline, then a triple bond is necessary in dinitrogen in order to have an octet around each nitrogen atom. The carbon atom and both oxygen atoms in CO_2 achieve the octet configuration by forming double bonds.

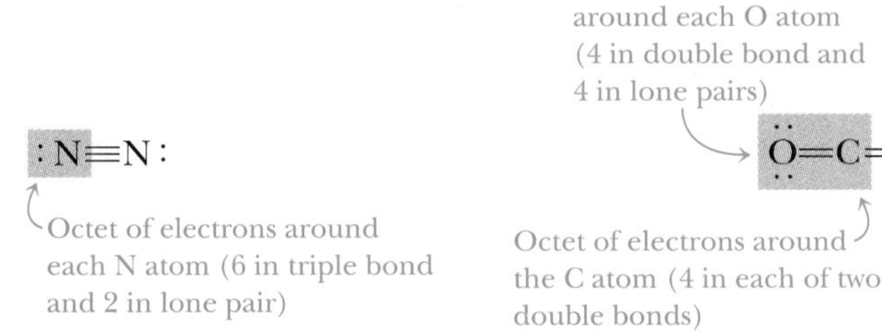

The tendency of a covalent species to have a structure in which eight electrons surround each atom is known as the **octet rule.** Actually, the octet rule is more a useful *guideline* than a rule. It directs you to seek a Lewis structure for covalent compounds in which each atom has eight electrons in its valence shell (or two in the case of hydrogen). A Lewis structure in which each atom achieves an octet is likely to be correct. If an atom in a proposed Lewis structure does not follow the octet rule, you should probably doubt the structure's validity. If a structure obeying the octet rule cannot be written, then it is possible an incorrect formula has been assigned to the compound.

Although the octet rule is widely applicable, there are exceptions. Fortunately, many will be obvious, such as when there are more than four bonds to an element or when there is an odd number of electrons.

There is a systematic approach to constructing Lewis structures of molecules and ions. Let us take formaldehyde, CH_2O, as an example.

1. *Determine the arrangement of atoms within a molecule.* The central atom is *usually* the one with the lowest electron affinity. In CH_2O the central atom is C.

```
H
   C   O
H
```

There is a better criterion than electron affinity for choosing the central atom. That is, it is usually the atom of lowest electronegativity (see Section 9.5.)

You will come to recognize that certain elements often appear as the center atom, among them carbon, nitrogen, phosphorus, and sulfur. Halogens are often terminal atoms forming a single bond to one other atom, but they can be the central atom when combined with O in the oxoacids (such as $HClO_4$). Oxygen is the central atom in water, but in conjunction with carbon, nitrogen, phosphorus, and the halogens it is usually a terminal atom. Hydrogen is always a terminal atom because it typically bonds to only one other atom.

With simple compounds, the first atom in a formula is often the central atom (e.g., SO_2, NH_4^+, NO_3^-). This is not always a reliable predictor, however. Notable exceptions include water (H_2O) and most common acids (HNO_3, H_2SO_4), in which the acidic hydrogen is usually written first.

2. *Determine the total number of valence electrons in the molecule or ion.* In a neutral molecule this number will be the sum of the valence electrons for each atom. For an anion, add a number of electrons equal to the negative charge; for a cation, subtract the number of electrons equal to the positive charge. The number of valence electron pairs will be half the total number of valence electrons. For CH_2O,

$$\text{Valence electrons} = 12 \text{ electrons (6 electron pairs)}$$
$$= 4 \text{ for C} + (2 \times 1 \text{ for two H atoms}) + 6 \text{ for O}$$

H atoms are never central atoms because they can form only one bond.

3. *Place one pair of electrons between each pair of bonded atoms to form a single bond.*

Single bond

H
C—O
H

Here three electron pairs are used to make three single bonds. Three pairs of electrons remain.

4. *Use any remaining pairs as lone pairs around each terminal atom (except H) so that each atom is surrounded by eight electrons.* If there are electrons left over, assign them to the central atom. If the central atom is an element in the third or higher period, it can have more than eight electrons.

Single bond — Lone pair

H
C—Ö:
H

Here all six pairs have been assigned, but the C atom has a share in only three pairs.

5. *If the central atom has fewer than eight electrons at this point, move one or more of the lone pairs on the terminal atoms into a position intermediate between the center and the terminal atom to form multiple bonds.*

H, C—Ö:, H → Move lone pair to create double bond and satisfy octet → H, C=O, H

Single bond — Lone pair — Two shared pairs; double bond

As a general rule double or triple bonds are formed *only* when one or both of the atoms involved is C, N, O, or S. That is, bonds such as C=C, C=N, C=O, and S=O will be encountered frequently.

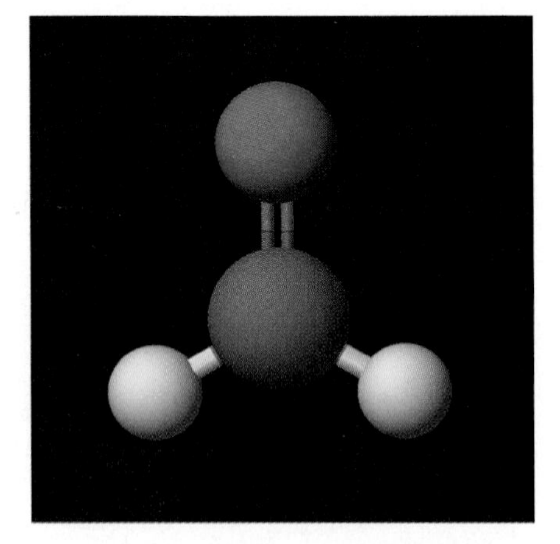

The structure of CH_2O, formaldehyde. The C—H bonds are single bonds, whereas the C=O bond is a double bond.

Example 9.2 Drawing Lewis Structures

Draw Lewis structures for ammonia (NH_3), the hypochlorite ion (ClO^-), the nitronium ion (NO_2^+), and the phosphate ion (PO_4^{3-}).

Solution for NH_3

1. *Central atom.* Hydrogen atoms are always terminal atoms, so nitrogen must be the central atom in the molecule.
2. *Valence electrons.* The total is 8 [4 valence pairs = 5 (for N) + 3 (1 for each H)].

3. *Form single covalent bonds between each atom.* This uses three of the four pairs available.

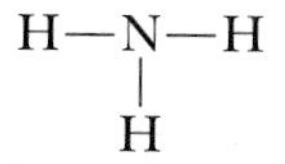

```
H—N—H
  |
  H
```

4. *Place the remaining pair of electrons on the central atom.*

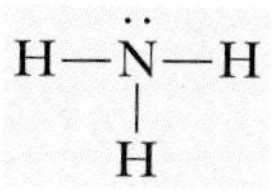

```
  ··
H—N—H
  |
  H
```

Each H atom has a share in one pair of electrons as required, and the central N atom has achieved an octet configuration with four electron pairs. No further steps are required; this is the correct Lewis structure.

Solution for ClO^- ion

1. With two atoms, there is no "central" atom.
2. Valence electrons = 14 (7 valence pairs)
   = 7 (for Cl) + 6 (for O) + 1 (for the negative charge on the ion)
3. One electron pair is used in the Cl—O bond: Cl—O.
4. Distribute the six remaining electron pairs around the "terminal" atoms.

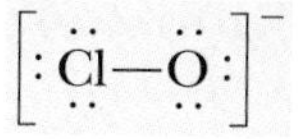

```
[ ··   ·· ]⁻
[:Cl—O:]
[ ··   ·· ]
```

5. As no electrons remain to be assigned and both atoms have an octet, this is the correct Lewis structure.

Solution for the NO_2^+ ion

1. Nitrogen is the center atom, because its electron affinity is lower than that of oxygen.
2. Valence electrons = 16 (8 valence pairs)
   = 5 (for N) + 12 (six for each O) − 1 (for the positive charge)
3. Two electron pairs form the single bonds from the nitrogen to each oxygen:

O—N—O

4. Distribute the remaining six pairs of electrons on the terminal O atoms:

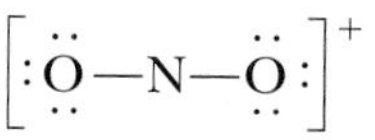

```
[ ··      ·· ]⁺
[:O—N—O:]
[ ··      ·· ]
```

5. The central nitrogen atom is two electron pairs short of an octet. Thus, a lone pair of electrons on each oxygen atom is converted to a bonding electron pair to give two N=O double bonds. Each atom in the ion now has four electron pairs. Nitrogen has four bonding pairs, and each oxygen atom has two lone pairs and shares two bond pairs.

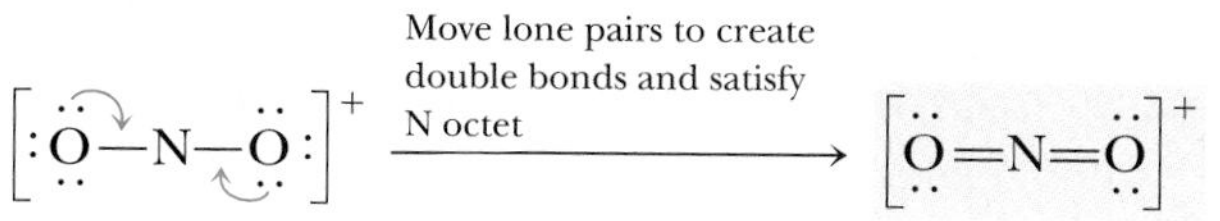

$[:\ddot{O}—N—\ddot{O}:]^+$ —(Move lone pairs to create double bonds and satisfy N octet)→ $[\ddot{O}=N=\ddot{O}]^+$

Solution for the PO_4^{3-} ion

1. Phosphorus, the atom of lower electron affinity, is the center atom.
2. Valence electron pairs = 32 (16 electron pairs)
   = 5 (for P) + 24 (six each for four oxygens) + 3 (for the negative charge)
3. Use four electron pairs for four single bonds linking phosphorus to each oxygen:

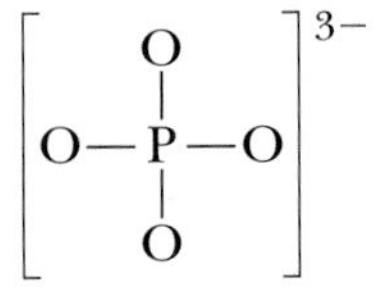

```
[     O     ]3−
      |
[ O — P — O ]
      |
[     O     ]
```

4. Place the remaining 12 electron pairs, 3 pairs at a time, around the four O atoms.

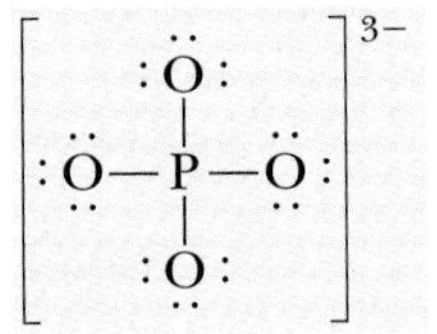

```
[     ..      ]3−
[    :O:      ]
[ ..  |   ..  ]
[:O — P — O:  ]
[ ..  |   ..  ]
[    :O:      ]
[     ..      ]
```

5. No further adjustments are needed as all of the electrons have been accounted four and each atom now has an octet.

Exercise 9.3 Drawing Lewis Structures

Draw Lewis structures for NH_4^+, CO, SCN^-, and SO_4^{2-}.

Predicting Lewis Structures

Lewis structures are useful in gaining a perspective on the structure and chemistry of a molecule or ion. The guidelines for drawing Lewis structures are helpful in this regard, but chemists also rely on patterns of bonding in related molecules. You can quickly draw a Lewis structure if you recognize some of these patterns. Three groups of covalent species are described as examples of this approach.

Hydrogen Compounds

Some common compounds and ions formed from second-period nonmetallic elements and hydrogen are shown in Table 9.4. Their Lewis structures illustrate the fact that the Lewis symbol for an element is a useful guide in determining the number of bonds formed by the element. For example, if there is no charge, nitrogen has five valence electrons. Two electrons occur as a lone pair; the other three occur as unpaired electrons. To reach an octet, it is necessary to pair each of the unpaired electrons with an electron from another atom. Thus, N is pre-

Table 9.4 • Common Hydrogen-Containing Compounds and Ions of the Second-Period Elements

Group 4A	Group 5A	Group 6A	Group 7A
CH_4 methane 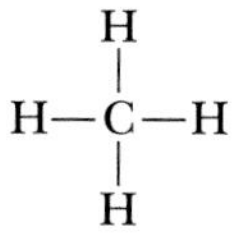	NH_3 ammonia 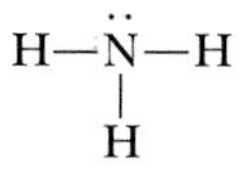	H_2O water 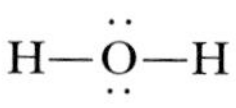	HF hydrogen fluoride 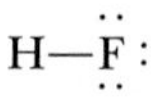
C_2H_6 ethane 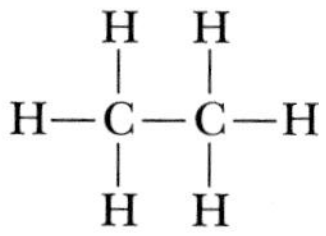	N_2H_4 hydrazine 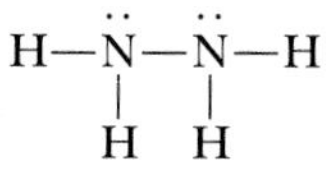	H_2O_2 hydrogen peroxide 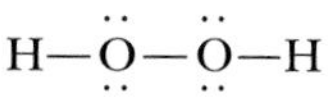	
C_2H_4 ethylene 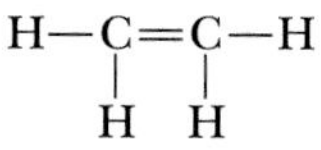	NH_4^+ ammonium ion 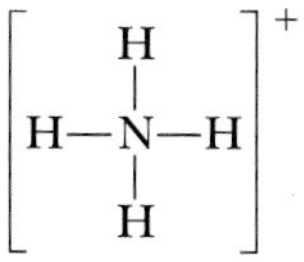	H_3O^+ hydronium ion 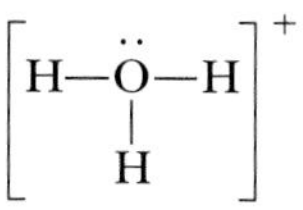	
C_2H_2 acetylene 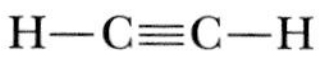	NH_2^- amide ion 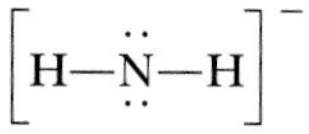	OH^- hydroxide ion 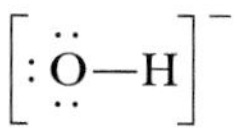	

dicted to form three bonds in uncharged molecules, and this is indeed the case. Similarly, carbon is expected to form four bonds, oxygen two, and fluorine one.

Group 4A Group 5A Group 6A Group 7A

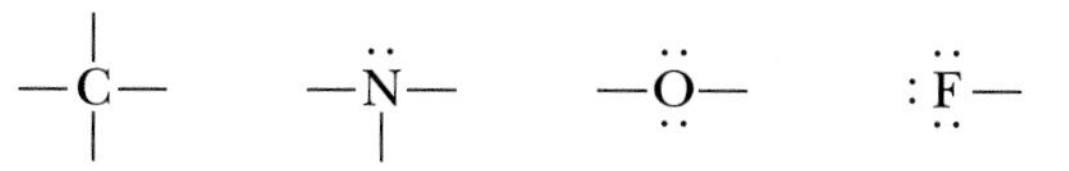

The same ideas apply to Lewis structures for ions. To create the Lewis structure for NH_4^+, for example, we need to bond four H atoms to the central N. One way to build the correct Lewis structure for NH_4^+ is to start with the Lewis symbol for nitrogen (with five electrons), and then subtract one electron, giving N^+ with four unpaired electrons. The N^+ ion can then form four bonds, one to each hydrogen. Similarly, an N^- ion can form only two bonds (and have two lone pairs), an O^+ ion will form three bonds, and an O^- ion will form only one bond.

Group 5A Ions Group 6A Ions

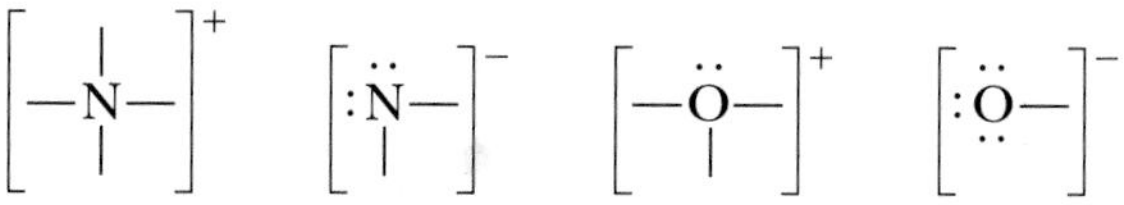

Hydrocarbons are compounds formed from carbon and hydrogen, and the first two members of the series called the *alkanes* are CH_4 and C_2H_6 (see Table 9.4). What is the Lewis structure of the third member of the series, propane, C_3H_8? We can rely on the idea that the atoms in this species each bond in a predictable way. Carbon is expected to form four bonds, and hydrogen can bond to only one other atom. The only arrangement of atoms that meets these criteria has three atoms of carbon linked together by carbon–carbon single bonds.

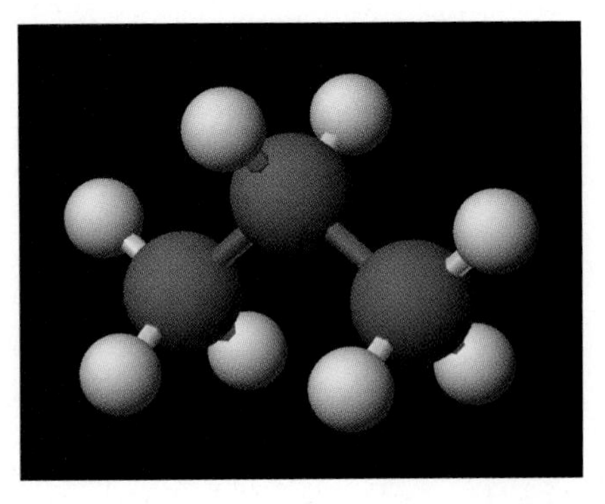

Model of propane, C_3H_8. Notice that the molecule is not linear. The carbon chain is bent. See Section 9.7, Molecular Shapes.

The remaining positions around the carbon atoms are filled in with hydrogens—three hydrogen atoms on the end carbons and two on the middle carbon:

```
    H  H  H
    |  |  |
 H—C—C—C—H
    |  |  |
    H  H  H
```

propane, C_3H_8

Example 9.3 Predicting Lewis Structures

Draw Lewis electron dot structures for CCl_4 and NF_3.

Solution One way to answer this is to recognize that the formulas are similar to CH_4 and NH_3 except that H atoms have been replaced by halogen atoms. Recall that carbon is expected to form four bonds and nitrogen three bonds to give an octet of electrons. In addition, halogen atoms have seven valence electrons, so both Cl and F can attain an octet by forming one covalent bond, just as hydrogen does.

```
       ..
      :Cl:
   ..  |  ..              ..  ..  ..
  :Cl—C—Cl:              :F—N—F:
   ..  |  ..              ..  |   ..
      :Cl:                   :F:
       ..                     ..
```

carbon tetrachloride — nitrogen trifluoride

As a check, count the number of valence electrons for each molecule and verify that all are present.

CCl_4: Valence electrons = 4 for C + 4 × 7 (for Cl) = 32 electrons

NF_3: Valence electrons = 5 for N + 3 × 7 (for F) = 26 electrons

Exercise 9.4 Predicting Lewis Structures

Predict Lewis structures for methanol, H_3COH and hydroxylamine, H_2NOH. (*Hint:* The formulas of these compounds are written to guide you in choosing the correct arrangement of atoms.)

Oxo Acids and Their Anions

Lewis structures of common acids and their anions are illustrated in Table 9.5. In the absence of water these acids are covalently bonded molecular compounds (a conclusion that we should draw because all elements in the formula are nonmetals). Nitric acid, for example, is a colorless liquid with a boiling point of 83 °C. In aqueous solution, however, HNO_3, H_2SO_4, and $HClO_4$ are completely ionized to give a hydrogen ion and the appropriate anion. A Lewis structure for the nitrate ion, for example, can be created using the guidelines on page 383, and the result is a structure with two N—O single bonds and one N=O double

Table 9.5 • Lewis Structures of Common Oxoacids and Their Anions

Acid	Anion
HNO_3 nitric acid	NO_3^- nitrate ion
H_3PO_4 phosphoric acid	PO_4^{3-} phosphate ion
H_2SO_4 sulfuric acid	HSO_4^- hydrogen sulfate ion
	SO_4^{2-} sulfate ion
$HClO_4$ perchloric acid	ClO_4^- perchlorate ion
HOCl hypochlorous acid	OCl^- hypochlorite ion

Anions play a significant biological role. Phosphate ions are important in energy storage and control of cellular function. Sulfate ions maintain charge neutrality in biological systems. Some anions, however, are detrimental to health. For example, another anion based on a Group 5A element, the arsenate ion, AsO_4^{3-}, is very toxic. Arsenate can substitute for the phosphate ion in some reactions and cause those reactions to take a different course. The metal-based chromate ion, CrO_4^{2-}, is very carcinogenic; in the body it is taken up by the sulfate uptake pathway but leads to different reactions taking place.

bond. To form nitric acid, a hydrogen ion is attached to one of the O atoms that has a single bond to the central N (Figure 9.4).

$$[NO_3]^- \underset{-H^+}{\overset{+H^+}{\rightleftharpoons}} HNO_3$$

nitrate ion   nitric acid

A characteristic property of acids in aqueous solution is their ability to donate a hydrogen ion (H^+). The NO_3^- anion is formed when the acid, HNO_3, loses a hydrogen ion. The H^+ ion separates from the acid by breaking the H—O bond, the electrons of the bond staying with the O atom. As a result, HNO_3 and NO_3^- have the same number of electrons, 24, and their structures are closely related.

See Section 5.3 for a discussion of acids in aqueous solution.

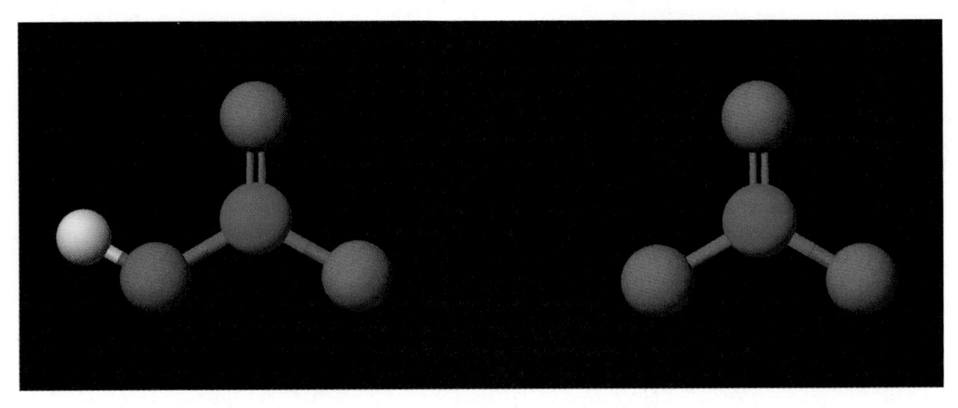

***Figure* 9.4 Models of nitric acid, HNO_3, and nitrate ion, NO_3^-.** The formula for the acid is sometimes written $HONO_2$ in order to give specific information about the arrangement of atoms. The nitrate ion is derived from nitric acid by breaking the H—O bond. The H atom leaves as a hydrogen ion, H^+.

***Exercise* 9.5** Lewis Structures of Acids and their Anions

Draw a Lewis structure for the anion $H_2PO_4^-$, derived from phosphoric acid.

Isoelectronic Species

In what way are NO^+, N_2, CO, and CN^- similar? There are several trivial similarities (they are composed of two atoms, for example), but the most important similarity is that they have the same total number of valence electrons, 10, and thus have the same Lewis structure. The two atoms in each molecule or ion are linked with a triple bond. With three bonding pairs and one lone pair, each atom has an octet configuration.

$$[:N\equiv O:]^+ \quad :N\equiv N: \quad :C\equiv O: \quad [:C\equiv N:]^-$$

Molecules and ions having the *same number of valence electrons and the same Lewis structures* are said to be **isoelectronic** (Table 9.6). You will find it helpful to think in terms of isoelectronic molecules and ions because this is another way to see relationships among common chemical substances.

The term **isostructural** is often used in conjunction with isoelectronic species. Species that are isostructural have the same structure. For example, the PO_4^{3-}, SO_4^{2-}, and ClO_4^- ions in Table 9.6 all have four oxygens bonded to the central atom.

Can you give the formula for another species that is isoelectronic with CN^-? One way to approach this problem is to realize that the isoelectronic molecules and ions illustrated in the previous example are combinations of C^-, N, and O^+, each of which has five electrons in the valence shell (one lone pair and three single electrons, which contribute to forming the triple bond). The only combinations not yet considered are O_2^{2+} and C_2^{2-}. The former is unknown, but the C_2^{2-} is the acetylide ion, a known species.

There are both similarities and important differences in chemical properties of isoelectronic species. For example, both carbon monoxide, CO, and cyanide ion, CN^-, are very toxic, which results from the fact that they can bind to the iron of hemoglobin in blood and block the uptake of oxygen. They are different, though, in their acid–base chemistry. In aqueous solution cyanide ion readily adds H^+ to form hydrogen cyanide, whereas CO does not do so. The isoelectronic species Cl_2 and ClO^- provide a similar example. Attachment of H^+ to OCl^- forms hypochlorous acid, HOCl. In contrast, Cl_2 does not add a proton.

Table 9.6 • Some Common Isoelectronic Molecules and Ions

Formulas	Representative Lewis Structure
BH_4^-, CH_4, NH_4^+	[H | H—B—H | H]$^-$
NH_3, H_3O^+	H—N̈—H | H
CO_2, OCN^-, SCN^-, N_2O, NO_2^+, COS, CS_2	:Ö=C=Ö:
CO_3^{2-}, NO_3^-	[:Ö—N=O: | :Ö:]$^-$
PO_4^{3-}, SO_4^{2-}, ClO_4^-	[:Ö: | :Ö—P—Ö: | :Ö:]$^{3-}$

Problem-Solving Tips and Ideas

9.1 Useful Ideas to Consider When Drawing Lewis Electron Dot Structures

- The octet rule is the most useful guideline when drawing Lewis structures.
- Carbon forms four bonds (four single bonds; two single bonds and one double bond; or one single bond and 1 triple bond). In uncharged species, nitrogen forms three bonds and oxygen two. Hydrogen typically forms only one bond to another atom.
- When multiple bonds are formed, both of the atoms involved are usually among the following: C, N, O, and S. Oxygen has the ability to form multiple bonds with a variety of elements. Carbon forms many compounds having multiple bonds to another carbon or to N or O.
- Nonmetals may form single, double, and triple bonds but never quadruple bonds.
- Always account for single bonds and lone pairs before forming multiple bonds.
- Be alert for the possibility the molecule or ion you are working on is isoelectronic with a species you have seen before.

Exercise 9.6 Identifying Isoelectronic Species

Identify a common molecular (uncharged) species that is isoelectronic with nitrite ion, NO_2^-. Identify a common ion that is isoelectronic with HF.

Resonance

Ozone, O_3, an unstable, blue, diamagnetic gas with a characteristic pungent odor, protects the earth and its inhabitants from intense ultraviolet radiation from the sun. An important feature of its structure is that the two oxygen–oxygen bonds are the same length, suggesting that the two oxygen–oxygen bonds are equivalent. That is, equal O—O bond lengths imply an equal number of bond pairs in each O—O bond. Using the guidelines for drawing Lewis structures, however, you might come to a different conclusion. There are two possible ways of writing the Lewis structure for the molecule:

Alternative ways of creating the Lewis structure of ozone

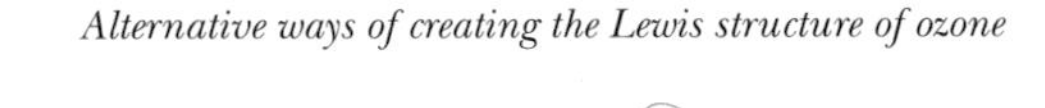

Double bond on the right :Ö—Ö—Ö: ⟶ :Ö—Ö=O:

Double bond on the left :Ö—Ö—Ö: ⟶ :O=Ö—Ö:

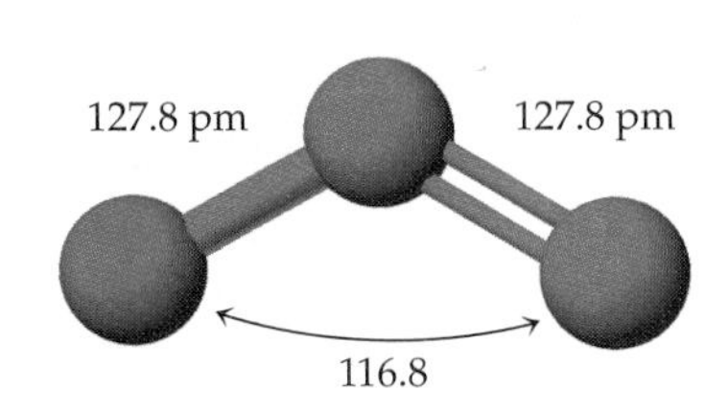

Ozone, O_3, is a bent molecule with oxygen–oxygen bonds of the same length.

These structures are equivalent in that each has a double bond on one side of the central oxygen atom and a single bond on the other side. If either were the actual structure of ozone, one bond should be shorter (O=O) than the other (O—O). The actual structure of ozone shows this is not the case. The inescapable conclusion is that these Lewis structures do not correctly represent the bonding in ozone.

See the *Saunders Interactive General Chemistry CD-ROM,* Screen 9.6, Resonance Structures.

Linus Pauling proposed the **theory of resonance** to reconcile the problem. *Resonance structures are a way to represent bonding in a molecule or ion when a single Lewis structure fails to describe accurately the actual electronic structure.* The alternative structures shown for ozone are called **contributing structures** or **resonance structures.** They have identical patterns of bonding and equal energy. The actual structure of this molecule is a *composite,* or **resonance hybrid,** of the equivalent contributing structures. This is a reasonable conclusion because we see that the O—O bonds both have a length of 127.8 pm, intermediate between the average length of an O=O double bond (121 pm) and an O—O single bond (132 pm).

It is conventional to connect resonance structures with double-headed arrows, ⟷, to indicate that the actual structure is a composite of these structures. One structure can be formed from the other by moving a lone pair of electrons to form a bond, and allowing a bond pair of electrons in a neighboring bond to become a lone pair.

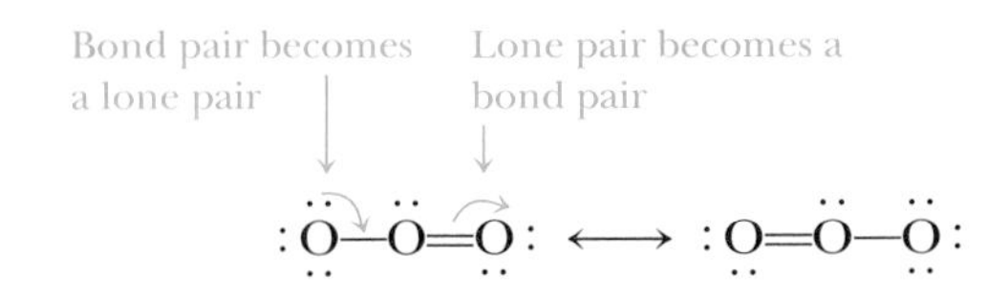

Benzene is a classic example of the use of resonance to represent a structure. The benzene molecule is a six-member ring of carbon atoms with six equivalent carbon–carbon bonds (and a hydrogen atom attached to each carbon atom). The carbon–carbon bonds are 144 pm long, intermediate between the average length of a C=C double bond (134 pm) and a C—C single bond (154 pm).

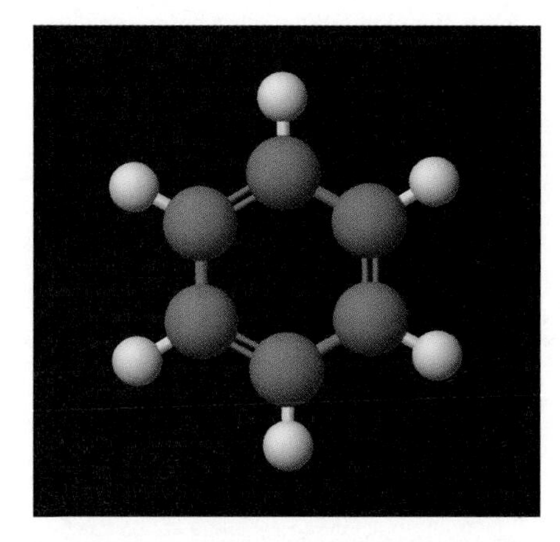

Benzene has a six-member carbon ring with carbon–carbon bonds of equal length.

Two resonance structures can be written for the molecule that differ only in double bond placement. A composite of these two structures, however, will lead to a molecule with six equivalent carbon–carbon bonds.

Let us apply the concepts of resonance to describe bonding in the carbonate ion, CO_3^{2-}, an anion with 24 valence electrons (12 pairs) (and isoelectronic with the nitrate ion).

$$\left[:\ddot{O}{=}C{-}\ddot{O}:\right]^{2-} \longleftrightarrow \left[:\ddot{O}{-}C{-}\ddot{O}:\right]^{2-} \longleftrightarrow \left[:\ddot{O}{-}C{=}\ddot{O}:\right]^{2-}$$

Three equivalent structures can be drawn for this ion, differing only in the location of the C=O double bond. This fits the classical situation for resonance, so it is appropriate to conclude that no single structure correctly describes this ion. Instead, the actual structure is the average of the three structures, in good agreement with experimental results. In the CO_3^{2-} ion, all three carbon–oxygen bond distances are 129 pm, intermediate between C—O single bond (143 pm) and C=O double bond (122 pm) distances.

In aqueous solution, a hydrogen ion can be attached to the carbonate ion to give the hydrogen carbonate, or bicarbonate, ion. This ion can be described as a resonance hybrid of two Lewis structures.

$$\left[:\ddot{O}{=}C{-}\ddot{O}:\ (:\ddot{O}{-}H)\right]^{-} \longleftrightarrow \left[:\ddot{O}{-}C{=}\ddot{O}:\ (:\ddot{O}{-}H)\right]^{-}$$

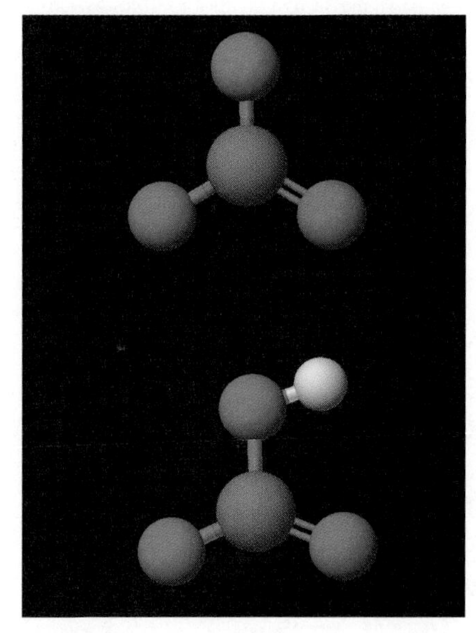

(Top) The carbonate ion, CO_3^{2-}, has three carbon–oxygen bonds of equal length. *(Bottom)* The bicarbonate ion, HCO_3^-, has two resonance structures.

Problem-Solving Tips and Ideas

9.2 Resonance Structures

- Resonance is a means of representing the bonding when a single Lewis structure fails to give an accurate picture.
- The atoms must have the same arrangement in space in all resonance structures. Moving the atoms around or attaching them in a different fashion creates a different compound.
- Resonance structures differ only in the assignment of electron-pair positions, never atom positions.
- Resonance structures differ in the number of bond pairs between a given pair of atoms.
- Even though the formal process of converting one resonance structure to another moves electrons about, resonance is not meant to indicate the motion of electrons.
- The actual structure of a molecule is a composite of the resonance structures.

Example 9.4 Drawing Resonance Structures

Draw resonance structures for the nitrite ion, NO_2^-. Are the N—O bonds single, double, or intermediate in value?

Solution Nitrogen is the center atom in the nitrite ion, which has a total of 18 valence electrons (9 pairs).

Valence electrons = 5 (for the N atom) + 12 (6 for each O atom) + 1 (for negative charge)

After forming N—O single bonds, and distributing lone pairs on the terminal O atoms, a pair remains, which is placed on the central N atom.

$$[:\ddot{\underset{..}{O}}—\ddot{N}—\ddot{\underset{..}{O}}:]^-$$

Both the choice of the name "resonance" and use of an arrow as a symbol to link resonance structures are somewhat unfortunate. An arrow seems to imply that a change is occurring, and the term resonance has the connotation of vibrating or alternating back and forth between different forms. Neither view is correct. Resonance is simply a way of representing a structure.

To complete the octet of electrons about the N atom, form an N=O double bond.

$$[:\underset{..}{O}=\ddot{N}—\ddot{\underset{..}{O}}:]^- \longleftrightarrow [:\ddot{\underset{..}{O}}—\ddot{N}=\underset{..}{O}:]^-$$

Because there are two ways to do this, two equivalent structures can be drawn, and the actual structure must be a resonance hybrid of these two structures. The nitrogen–oxygen bonds are neither single nor double bonds but have an intermediate value.

Exercise 9.7 Drawing Resonance Structures

Draw resonance structures for the nitrate ion, NO_3^-. Sketch a plausible Lewis dot structure for nitric acid, HNO_3.

Exceptions to the Octet Rule

Although a majority of covalent chemical compounds and ions obey the octet rule, there are exceptions. These include molecules and ions that have fewer than four pairs of electrons, those that have more than four pairs, and those that have an odd number of electrons. Only one group of compounds, with boron the central atom, will be mentioned in the first category. In the second category are compounds formed from elements in the third period and below. These elements often are found to exceed an octet in their compounds, especially when bonded to oxygen or one of the halogens. Very few stable molecules or ions with an odd number of electrons are known, and only two examples, both oxides of nitrogen, will be mentioned.

Compounds in Which an Atom Has Fewer Than Eight Valence Electrons

Boron, a nonmetal in Group 3A, has three valence electrons and so is expected to form three covalent bonds with other nonmetallic elements. This results in a valence shell for boron in its compounds with only six electrons, two short of an octet. Many boron compounds of this type are known, including the boron trihalides (BF_3, BCl_3, BBr_3, and BI_3) and boric acid, $B(OH)_3$.

```
    ..     ..              ..        H
                                     |
  : F — B — F :          : O — B — O :
    ..  |  ..              ..  |     ..
                               |
      : F :                H : O — H
        ..                     ..

  boron trifluoride        boric acid
```

Boron chemistry also includes many oxides with more complex structures. For example, the common mineral borax has the formula $Na_2B_4O_5(OH)_4 \cdot 8\ H_2O$ (Figure 9.5). Two of the boron atoms in the $[B_4O_5(OH)_4]^{2-}$ anion have three covalent bonds, and two boron atoms have four bonds:

```
[H                                 ]2−
   \..
    O
    ..\          :O:  ..  H
       B          |  /O
       | \ ..     | / ..
       |   O — B
       |   ..     \
       O — B       :O:
      ..  / \        |
         O   :O:     |
        /..      \   |
       H          B
                   \..
                    O
                    ..\
                       H
```

You have already seen examples of both situations: three bonds to boron (BF_3) or four bonds to boron (BH_4^-). Each oxygen atom in the borax anion is bonded to two other atoms, and the OH groups resemble OH groups in other molecules. This structure illustrates an important observation: *even in complicated molecules, individual atoms have an arrangement of bonds similar to that in simple compounds and to that predicted by basic principles.* This point is crucial to understanding complicated molecular structures.

Boron compounds such as BF_3 that are two electrons short of an octet can be quite reactive. The boron atom can accommodate a fourth electron pair, but only when that pair is provided by another atom. In general, molecules or ions with lone pairs can fulfill this role. Ammonia, for example, reacts with BF_3 to form $H_3N \rightarrow BF_3$. The bond between the B and N atoms in this compound uses an electron pair that originated on the N atom. The reaction of an F^- ion with BF_3 to form BF_4^- is another example.

```
                ..                          ..
      H       : F :               H       : F :
      |         |    ..           |         |    ..
  H — N :  +    B — F :  ---->  H — N  →   B — F :
      |         |    ..           |         |    ..
      H       : F :               H       : F :
                ..                          ..
                             Coordinate covalent bond
```

If the bonding pair of electrons originates on one of the bonded atoms, the bond is called a **coordinate covalent bond.** In Lewis structures, a coordinate covalent bond is often designated by an arrow that points away from the atom donating the electron pair.

Figure 9.5 The mineral borax. The boron is contained in an anion, $[B_4O_5(OH)_4]^{2-}$, which has the structure shown. Notice that two of the boron atoms (silver) have four bonds and thus a completed octet. The other two, however, have only three bonds and so are short a pair of electrons. *(C. D. Winters)*

See the *Saunders Interactive General Chemistry CD-ROM,* Screen 9.7, Electron-Deficient Compounds.

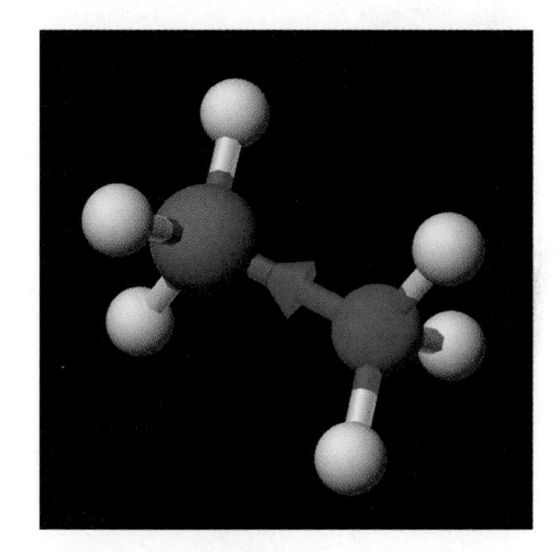

Ammonia reacts with boron trifluoride by forming an N $\rightarrow$ B coordinate covalent bond. The electron pair in this bond originated as the lone pair on the N atom of ammonia (blue).

Table 9.7 • Lewis Structures in Which the Central Atom Exceeds an Octet

Group 4A	Group 5A	Group 6A	Group 7A	Group 8A
SiF_5^-	PF_5	SF_4	ClF_3	XeF_2
SiF_6^{2-}	PF_6^-	SF_6	BrF_5	XeF_4

Compounds in Which an Atom Has More Than Eight Valence Electrons

Elements in the third or higher periods often form compounds and ions in which the central element is surrounded by more than four valence electron pairs (Table 9.7). With most compounds and ions in this category, the central atom is bonded to fluorine, chlorine, or oxygen.

It is often obvious from the formula of a compound that an octet around an atom has been exceeded. As an example, consider sulfur hexafluoride, SF_6, a gas formed by the reaction of elemental sulfur and excess fluorine. Sulfur is the central atom in this compound, and fluorine typically bonds to only one other atom with a single electron pair bond (as in HF and CF_4). Six S—F bonds are required in SF_6, meaning there will be six electron pairs in the valence shell of the sulfur atom.

More than four groups bonded to a central atom is a reliable signal that there are more than eight electrons around a central atom. In compounds with four or fewer bonds an octet may be exceeded, but such a prediction is not possible. Consider two examples from Table 9.7: SF_4 and XeF_2. The central atom in both molecules has five electron pairs in its valence shell.

Compounds of xenon are among the more interesting entries in Table 9.7. When the authors of this book were undergraduates we were taught that the noble gases did not form chemical compounds. Then several noble gas compounds were discovered in the early 1960s. One of the more intriguing compounds is XeF_2, in part because of the simplicity of its synthesis. Xenon difluoride can be made by placing a flask containing xenon gas and fluorine gas in the sunlight. After several weeks, large crystals of colorless XeF_2 fill the flask.

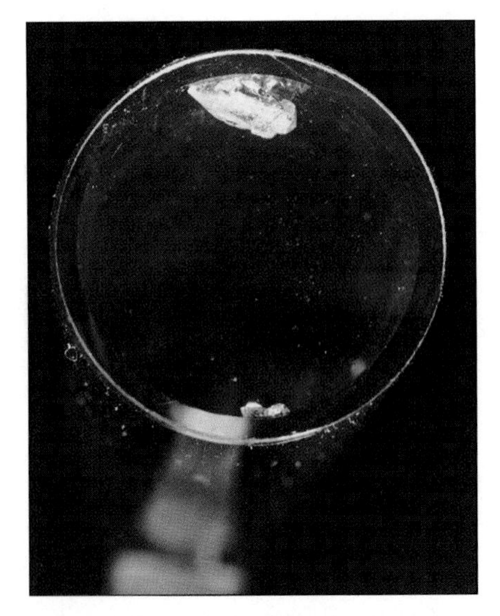

Xenon difluoride. The xenon atom is surrounded by five pairs of electrons, two bond pairs and three lone pairs. See Table 9.7. *(Argonne National Laboratory)*

A useful observation is that *only elements of the third and higher periods in the periodic table form compounds in which an octet is exceeded.* Second-period elements (B, C, N, O, F) are restricted to a maximum of eight electrons in their compounds. For example, nitrogen forms compounds such as NH_3, NH_4^+, and NF_3, but NF_5 is unknown. Phosphorus, the third-period element just below nitrogen in the periodic table, forms many compounds similar to nitrogen (PH_3, PH_4^+,

PF_3), but it also readily accommodates five or six valence electron pairs in compounds such as PF_5 or in ions such as PF_6^-. Arsenic, antimony, and bismuth, the elements below phosphorus in Group 5A, resemble phosphorus in their behavior.

The traditional explanation for the contrasting behavior of second- and third-period elements centers on the number of orbitals in the valence shell of an atom. Second-period elements have four valence orbitals (one 2*s* and three 2*p* orbitals). Two electrons per orbital result in a total of eight electrons being accommodated around an atom. For elements in the third and higher periods, the *d* orbitals in the outer shell are traditionally included among valence orbitals for the elements. Thus, for phosphorus, the 3*d* orbitals are included with the 3*s* and 3*p* orbitals as valence orbitals. The extra orbitals provide the element with an opportunity to accommodate up to 12 electrons.

Example 9.5 Lewis Structures in Which the Central Atom Has More Than Eight Electrons

Sketch the Lewis structure of the $[ClF_4]^-$ ion.

Solution According to the guidelines on page 383, we would proceed as follows:

1. The Cl atom is the central atom.
2. This ion has 36 valence electrons (= 7 for Cl + 4 × (7 for F) + 1 for ion charge) or 18 pairs.
3. Draw the ion with four single covalent Cl—F bonds.

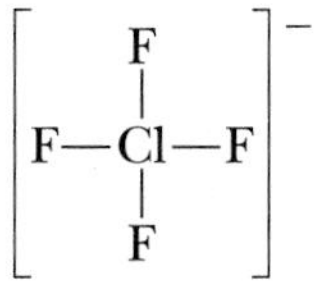

```
[     F     ]−
      |
[ F—Cl—F ]
      |
[     F     ]
```

4. Place lone pairs on the terminal atoms. Because two electron pairs remain after placing lone pairs on the four F atoms, and because we know that Cl can accommodate more than four pairs, these two pairs are placed on the central Cl atom.

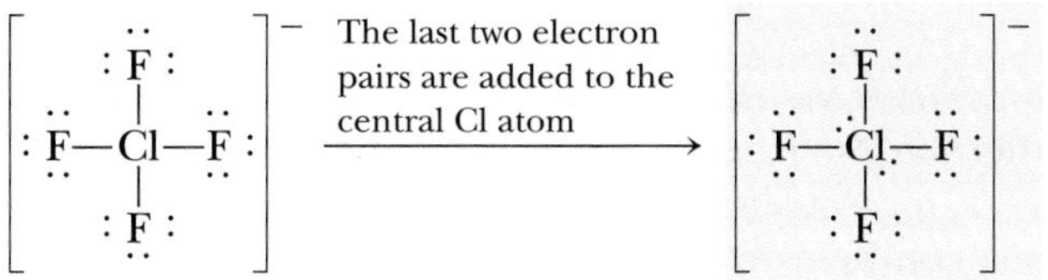

```
[     ..      ]−                              [      ..       ]−
[    :F:      ]                               [     :F:       ]
[ ..  |  ..   ]   The last two electron       [  ..  .|   ..  ]
[:F—Cl—F:]   pairs are added to the      [:F—Cl—F:]
[ ..  |  ..   ]   central Cl atom  ——→      [  ..  |.  ..  ]
[    :F:      ]                               [     :F:       ]
[     ..      ]                               [      ..       ]
```

Exercise 9.8 Lewis Structures in Which the Central Atom Has More Than Eight Electrons

Sketch the Lewis structures for $[ClF_2]^+$ and $[ClF_2]^-$. How many lone pairs and bond pairs surround the Cl atom in each ion?

See the *Saunders Interactive General Chemistry CD-ROM*, Screen 9.8, Free Radicals.

Molecules with an Odd Number of Electrons

Two nitrogen oxides—NO, with 11 valence electrons, and NO_2, with 17 valence electrons—are among a small group of stable molecules with an odd number of electrons. Because they have an odd number of electrons it is impossible to draw a structure obeying the octet rule; at least one electron must be unpaired.

Even though NO_2 does not obey the octet rule, an electron dot structure can be written that approximates the bonding in the molecule. This Lewis structure places the unpaired electron on nitrogen. Two resonance structures imply that the nitrogen–oxygen bonds are equivalent, as observed experimentally.

$$O{=}\dot{N}{-}O \longleftrightarrow O{-}\dot{N}{=}O$$

Experimental evidence for NO indicates that the bonding between N and O is intermediate between a double and a triple bond. It is not possible to write a Lewis structure for NO that is in accord with the properties of this substance, so a different theory is needed to understand bonding in this molecule. We shall return to this compound when molecular orbital theory is introduced in Chapter 10.

The two nitrogen oxides, NO and NO_2, are members of a class of chemical substances called free radicals. **Free radicals** are chemical species—both atoms and molecules—with an unpaired electron. How do these unpaired electrons affect reactivity? Free atoms such as H· and Cl· are very reactive and readily combine with other atoms to give molecules such as H_2, Cl_2, and HCl. We therefore expect free radical molecules to be more reactive than molecules with paired electrons, and most are. A free radical either combines with another free radical to form a molecule in which the electrons are paired, or it reacts with other molecules to produce new free radicals. These kinds of reactions are central to the formation of air pollutants. For example, when small amounts of NO and NO_2 are released from vehicle exhausts, the NO_2 decomposes in the presence of sunlight to give NO and O.

$$NO_2 + \text{sunlight} \longrightarrow \cdot\ddot{N}{=}\ddot{O} + \cdot\ddot{O}\cdot$$

$$\cdot\ddot{O}\cdot + O_2(g) \longrightarrow O_3$$

The free O atom then reacts with O_2 in the air to give ozone, O_3, an air pollutant at ground level that affects the respiratory system.

The two nitrogen oxides, NO and NO_2, are unique in that they can be isolated and neither has the extreme reactivity of most free radicals. When cooled, however, two NO_2 molecules join or "dimerize" to form colorless N_2O_4; the unpaired electrons combine to form an N—N bond in N_2O_4. Even though this bond is weak, the reaction is easily observed in the laboratory (Figure 9.6).

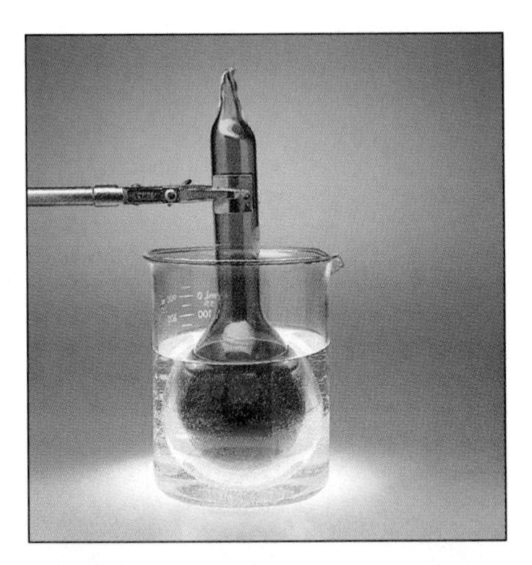

Figure 9.6 Free radical chemistry. When cooled, the brown gas NO_2, a free radical, forms N_2O_4, a molecule with an N—N single bond. The coupling of two free radicals is a common type of chemical reactivity. Because two identical free radicals come together, the product is called a dimer, and the process is called a dimerization. *(C. D. Winters)*

9.5 BOND PROPERTIES

We have described covalent bonds as single, double, or triple bonds and have referred several times to the lengths of bonds. Let us explore their relationship and, in particular, learn what bond lengths can teach us about the electronic structure of a molecule.

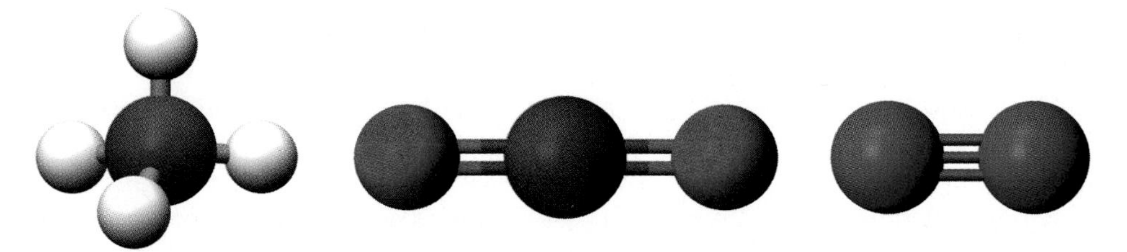

***Figure* 9.7** **Bond order.** The four C—H bonds in methane each have a bond order of 1. The two C═O bonds of CO_2 each have a bond order of 2, whereas the nitrogen–nitrogen bond of N_2 has an order of 3.

Bond Order

See the *Saunders Interactive General Chemistry CD-ROM,* Screen 9.9, Bond Properties.

The **order of a bond** is the number of bonding electron pairs shared by two atoms in a molecule (Figure 9.7). You will encounter bond orders of 1, 2, and 3, as well as fractional bond orders.

In a first-order bond, there is only a single covalent bond between a pair of atoms. Examples are the bonds in molecules such as H_2, F_2, NH_3, and CH_4. The bond order is 2 when two electron pairs are shared between atoms. The C═O bond in CO_2 and the C═C bond in ethylene, C_2H_4, are bonds of order 2. The bond order is 3 when two atoms are connected by three bonds. Examples include the carbon–oxygen bond in carbon monoxide, CO, the nitrogen–nitrogen bond in N_2, and the carbon–nitrogen bond in the cyanide ion, CN^-.

A *fractional bond order* is possible in molecules and ions having resonance structures. For example, what is the bond order for each oxygen–oxygen bond in O_3? Each resonance structure of O_3 has one O—O single bond and one O═O double bond, for a total of three shared bonding pairs accounting for two oxygen–oxygen links. If we define the bond order between any bonded pair of atoms as

$$\text{Bond order} = \frac{\text{number of shared pairs linking X and Y}}{\text{number of X} - \text{Y links in the molecule or ion}}$$

then the order is seen to be 3/2, or 1.5, for ozone.

Bond order = 1
Bond order = 2
One resonance structure
Average bond order for the resonance hybrid = 3/2, or 1.5

Bond Length

Bond length is the distance between the nuclei of two bonded atoms. Bond lengths are largely determined by the sizes of the atoms (Section 8.6). For given elements, the order of the bond determines the final value of the distance.

Table 9.8 lists average bond lengths for a number of common chemical bonds. It is important to recognize that these are average values. Neighboring parts of a molecule can affect the length of a particular bond. For example, Table 9.8 specifies that the average C—H bond has a length of 110 pm. In methane, CH_4, the actual bond length is 109.4 pm, whereas it is only 105.9 pm

Current Perspectives

NO, A Small Molecule with a Big Role

Small molecules such as H_2, O_2, H_2O, CO, and CO_2 are among the most important molecules commercially, environmentally, and biologically. Imagine the surprise of chemists and biologists when it was discovered a few years ago that nitrogen monoxide (nitric oxide, NO), which was widely considered toxic, also has an important biological role.

Nitric oxide is a colorless, paramagnetic gas that is moderately soluble in water. In the laboratory, it can be synthesized by the reduction of nitrite ion with iodide ion:

$$KNO_2(aq) + KI(aq) + H_2SO_4(aq) \longrightarrow NO(g) + K_2SO_4(aq) + H_2O(\ell) + \tfrac{1}{2} I_2(aq)$$

The formation of NO from the elements is unfavorable ($\Delta H_f^\circ = 90.2$ kJ/mol). Nevertheless, small quantities of this compound form from nitrogen and oxygen at high temperatures. Conditions in an internal combustion engine are favorable for this to happen.

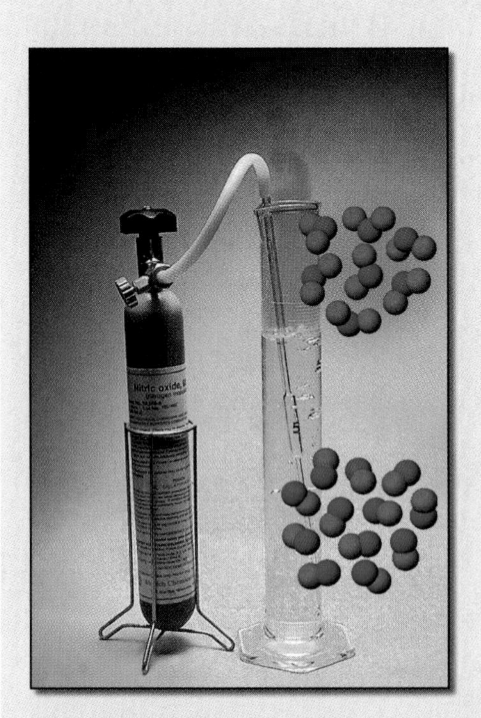

The colorless gas NO is bubbled into water from a high-pressure tank. When the gas emerges into the air, the NO reacts rapidly with O_2 to give brown-orange NO_2 gas. *(C. D. Winters)*

Nitric oxide reacts rapidly with O_2 to form the reddish brown gas NO_2.

$$2\ NO(\text{colorless gas}) + O_2(g) \longrightarrow 2\ NO_2(\text{brown gas})$$

The result is that NO (and compounds such as NO_2 and HNO_3 arising from reactions of NO with O_2 and H_2O) are among pollutants from automobile engines.

Imagine everyone's surprise when it was learned a few years ago that NO is synthesized in a biological process by animals as diverse as barnacles, fruit flies, horseshoe crabs, chickens, trout, and humans. As reported in *Chemical & Engineering News* in late 1993, "[NO] plays a role, often as a biological messenger, in an astonishing range of physiological processes in humans and other animals. Its expanding range of functions already include neurotransmission, blood clotting, blood pressure control, and a role in the immune system's ability to kill tumor cells and intracellular parasites."

Nitroglycerin is an oily, colorless liquid. It is so unstable that, if it is dropped or struck in any way, it explodes, forming CO_2, H_2O, and various gases containing nitrogen and oxygen. (Normally, nitroglycerin is dissolved in some absorbent material such as clay, and the result is dynamite.) In spite of this instability, nitroglycerin has been used for almost 90 years by physicians to regulate blood pressure, particularly in the treatment of chest pain (angina). It works by dilating blood vessels, which decreases the pressure and increases blood flow. The mechanism by which it led to muscle relaxation was only very recently discovered, and chemists have found that it involves NO.

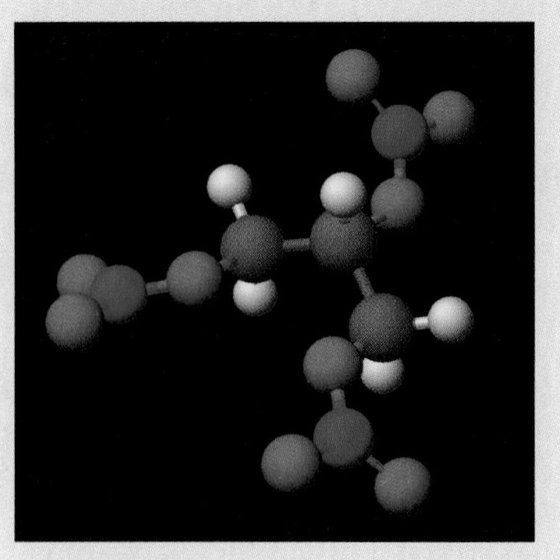

A molecular model of nitroglycerin.

Table 9.8 • Some Approximate Single and Multiple Bond Lengths*

Single Bond Lengths

	Group										
	1A	4A	5A	6A	7A	4A	5A	6A	7A	7A	7A
	H	C	N	O	F	Si	P	S	Cl	Br	I
H	74	110	98	94	92	145	138	132	127	142	161
C		154	147	143	141	194	187	181	176	191	210
N			140	136	134	187	180	174	169	184	203
O				132	130	183	176	170	165	180	199
F					128	181	174	168	163	178	197
Si						234	227	221	216	231	250
P							220	214	209	224	243
S								208	203	218	237
Cl									200	213	232
Br										228	247
I											266

Multiple Bond Lengths

C=C	134	C≡C	121
C=N	127	C≡N	115
C=O	122	C≡O	113
N=O	115	N≡O	108

*In picometers (pm); 1 pm = 10^{-12} m.

long in acetylene, H—C≡C—H. Variations as great as 10% from the average values listed in Table 9.8 are possible.

Because atom sizes vary in a fairly smooth way with the position of the element in the periodic table (Figure 8.10), predictions of trends in bond length can be quickly made. For example, the H—X distance in the hydrogen halides increases in the order predicted by the relative sizes of the halogens: H—F < H—Cl < H—Br < H—I. Likewise, bonds involving carbon and another element in a given period decrease going from left to right, in a predictable fashion; for example, C—C > C—N > C—O > C—F. Trends involving multiple bonds are similar. A C=O bond is shorter than a C=S bond, and a C=N bond is shorter than a C=C bond.

The effect of bond order is evident when bonds between the same two atoms are compared. For example, the bonds become shorter as the bond order increases in the series C—O, C=O, and C≡O:

Bond	C—O	C=O	C≡O
Bond Order	1	2	3
Bond Length (pm)	143	122	113

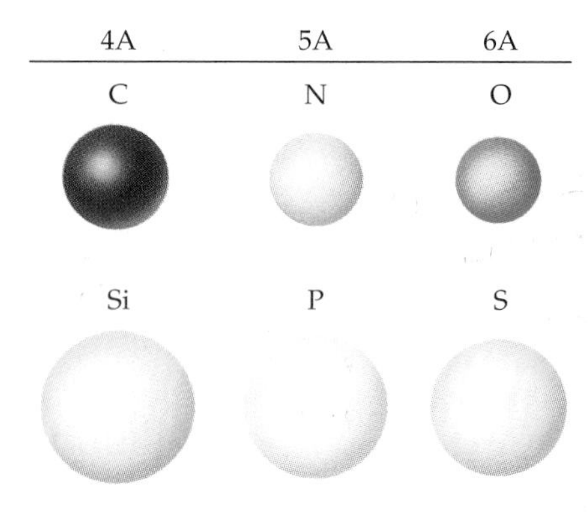

Relative sizes of some atoms of Groups 4A, 5A, and 6A.

Adding a second bond to a C—O single bond to make a C=O shortens the bond by 21 pm. Adding yet another bond results in a further 9-pm reduction in bond length from C=O to C≡O. In general, double bonds are shorter than single bonds between the same set of atoms, and triple bonds between those same atoms are shorter still.

The carbonate ion, CO_3^{2-}, has three equivalent resonance structures. It has a CO bond order of 1.33 (or 4/3) because four electron pairs link the central carbon with the three oxygen atoms. The CO bond distance (129 pm) is intermediate between a C—O single bond (143 pm) and a C=O double bond (122 pm).

Bond order = 2
Bond order = 1
Bond order = 1
Average bond order = 4/3, or 1.33

Exercise 9.9 Bond Order and Bond Length

1. Give the bond order of each of the following bonds and arrange them in order of decreasing bond distance: C=N, C≡N, and C—N.
2. Draw resonance structures for NO_2^-. What is the NO bond order in this ion? Consult Table 9.8 for N—O and N=O bond lengths. Compare these with the NO bond length in NO_2^- (124 pm). Account for any differences you observe.

Bond Energy

See the *Saunders Interactive General Chemistry CD-ROM*, Screen 9.10, Bond Energy.

The **bond dissociation energy,** symbolized by D, is the enthalpy change for breaking a bond in a molecule with the reactants and products in the gas phase under standard conditions.

$$\text{Molecule (g)} \underset{\text{energy released} = -D}{\overset{\text{energy supplied} = D}{\rightleftharpoons}} \text{Molecular fragments (g)}$$

Suppose you wish to break the carbon–carbon bonds in ethane ($H_3C—CH_3$), ethylene ($H_2C=CH_2$), and acetylene (HC≡CH), for which the bond orders are 1, 2, and 3, respectively. For the same reason that the ethane C—C bond is the longest of the series, and the acetylene C≡C bond is the shortest, bond breaking requires the least energy for ethane and the most energy for acetylene.

$$H_3C{-}CH_3(g) \longrightarrow H_3C(g) + CH_3(g) \qquad \Delta H = D = +376 \text{ kJ}$$
$$H_2C{=}CH_2(g) \longrightarrow H_2C(g) + CH_2(g) \qquad \Delta H = D = +720 \text{ kJ}$$
$$HC{\equiv}CH(g) \longrightarrow HC(g) + CH(g) \qquad \Delta H = D = +962 \text{ kJ}$$

Because D represents the energy transferred to the molecule from its surroundings, D has a positive value, and *the process of breaking bonds in a molecule is always endothermic.*

The amount of energy supplied to break carbon–carbon bonds must be the same as the amount of energy released when the same bonds form. *The formation of bonds from atoms or radicals in the gas phase is always exothermic.* This means, for example, that ΔH for the formation of $H_3C—CH_3$ from two $CH_3(g)$ radicals is −376 kJ/mol.

$$H_3C(g) + CH_3(g) \longrightarrow H_3C—CH_3(g) \qquad \Delta H = -D = -376 \text{ kJ}$$

Some experimental bond energies are tabulated in Table 9.9. It is important to understand that these are *average* bond energies. For example, the average bond dissociation energy of a C—H bond is 413 kJ/mol. This value may vary as much as 30 to 40 kJ/mol from molecule to molecule, however, just as bond lengths vary from one molecule to another.

In reactions between molecules, bonds in the reactants are broken and new bonds are formed as the products form. If the total energy released when new bonds are formed exceeds the energy required to break the original bonds, the overall reaction is exothermic. If the opposite is true, then the overall reaction is endothermic. Let us see how this works in practice.

Table 9.9 • SOME AVERAGE SINGLE- AND MULTIPLE-BOND ENERGIES (kJ/mol)*

Single Bonds

	H	C	N	O	F	Si	P	S	Cl	Br	I
H	436	413	391	463	565	318	322	347	432	366	299
C		346	305	358	485			272	339	285	213
N			163	201	283				192		
O				146		452	335		218	201	201
F					155	565	490	284	253	249	278
Si						222		293	381	310	234
P							201		326		184
S								226	255		
Cl									242	216	208
Br										193	175
I											151

Multiple Bonds

N=N	418	C=C	602
N≡N	945	C≡C	835
C=N	615	C=O	732
C≡N	887	C≡O	1072
O=O (in O_2)	498		

*Taken from I. Klotz and R. M. Rosenberg: *Chemical Thermodynamics,* 5th ed., p. 55. New York, John Wiley, 1994, and J. E. Huheey, E. A. Keiter, and R. L. Keiter, *Inorganic Chemistry,* 4th ed., New York, Harper Collins, 1993, Table E.1.

Vegetable oils can be converted to solid fats by a reaction called a *hydrogenation,* the addition of hydrogen. A simple example of this kind of reaction is the conversion of the hydrocarbon propene to propane:

$$\underset{\text{propene}}{\mathrm{H_3C{-}CH{=}CH_2(g)}} + \mathrm{H{-}H(g)} \longrightarrow \underset{\text{propane}}{\mathrm{H_3C{-}CH_2{-}CH_3(g)}}$$

If we knew the standard enthalpies of formation of propene and propane, we could use the procedure described in Chapter 6 to find the enthalpy change for the reaction. Suppose, though, the values were not available. In such a case we can use bond energies to *estimate* the enthalpy change for the reaction. The first step is to examine the reactants and product to see what bonds are broken and what bonds are formed. For the reaction in question one C—C bond and six C—H bonds are *not* changed. Thus we need only to focus on the bonds that have been affected:

Bonds broken: 1 mol of C=C bonds and 1 mol of H—H bonds

$$\mathrm{H_3C{-}CH{=}CH_2(g)} + \mathrm{H{-}H(g)}$$

Energy required = 602 kJ for C=C bonds + 436 kJ for H—H bonds = 1038 kJ

Bonds formed: 1 mol of C—C bonds and 2 mol of C—H bonds

$$\mathrm{H_3C{-}CH_2{-}CH_3(g)}$$

Energy evolved = 346 kJ for C—C bonds + 2 mol (413 kJ/mol for C—H bonds) = 1172 kJ

The enthalpy change for any reaction can be estimated using the equation

$$\Delta H^\circ_{\mathrm{rxn}} = \sum D(\text{bonds broken}) - \sum D(\text{bonds formed})$$

This equation tells you to multiply the bond energy for each bond broken by the number of bonds of that type, and then add all of them. Then, multiply the bond energy for each bond formed by the number of bonds of that type, and add all of these. Because bond formation is exothermic, the quantity "Σ D(bonds formed)" is subtracted from the quantity "Σ D(bonds broken)." Therefore, for the hydrogenation reaction, we have

$$\Delta H^\circ_{\mathrm{rxn}} = 1038\ \mathrm{kJ} - 1172\ \mathrm{kJ} = -134\ \mathrm{kJ}$$

and the overall reaction is exothermic. (Using enthalpies of formation for propene and propane, we calculate $\Delta H^\circ_{\mathrm{rxn}} = -123.8$ kJ, indicating that bond energy calculations can give acceptable results in many cases.)

Example 9.6 Using Bond Energies

Acetone, a common industrial solvent, can be converted to isopropanol, rubbing alcohol, by hydrogenation:

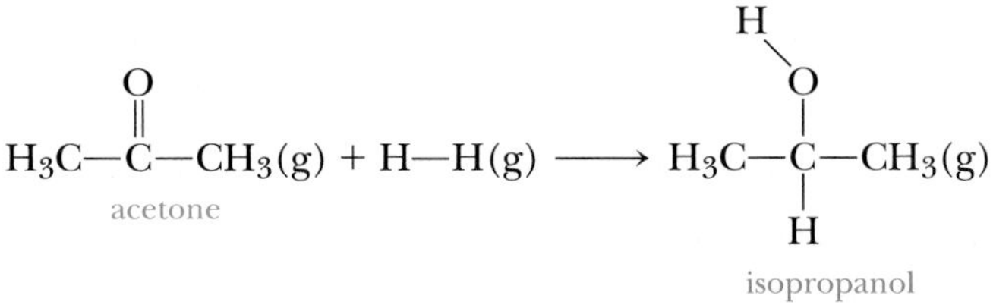

$$H_3C-\overset{\overset{\displaystyle O}{\|}}{C}-CH_3(g) + H-H(g) \longrightarrow H_3C-\overset{\overset{\displaystyle H\diagdown O}{|}}{\underset{\underset{\displaystyle H}{|}}{C}}-CH_3(g)$$

acetone isopropanol

Calculate the enthalpy change for the reaction using bond energies.

Solution The first step is to examine the reactants and product to see what bonds are broken and what bonds are formed. For this reaction the two C—C bonds and six C—H bonds are not changed. We therefore need only focus on the bonds that have been broken in the reactants or formed in the products.

Bonds broken: 1 mol of C=O bonds and 1 mol of H—H bonds

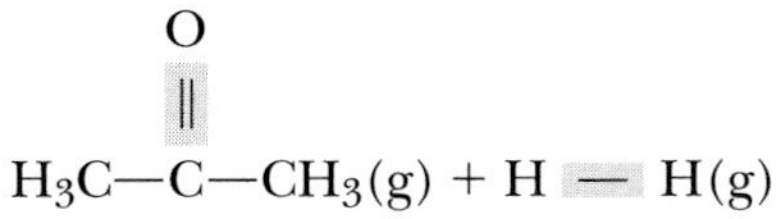

$$H_3C-\overset{\overset{\displaystyle O}{\|}}{C}-CH_3(g) + H-H(g)$$

Energy required = 732 kJ for C=O bonds + 436 kJ for H—H bonds = 1168 kJ

Bonds formed:

1 mol of C—H bonds, 1 mol of C—O bonds, and 1 mol of O—H bonds

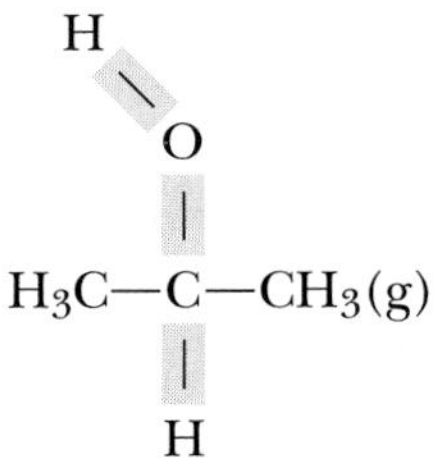

$$H_3C-\overset{\overset{\displaystyle H\diagdown O}{|}}{\underset{\underset{\displaystyle H}{|}}{C}}-CH_3(g)$$

Energy evolved = 413 kJ for C—H + 358 kJ for C—O + 463 kJ for O—H = 1234 kJ

$$\Delta H^\circ_{rxn} = \sum D(\text{bonds broken}) + \sum D(\text{bonds made})$$
$$\Delta H^\circ_{rxn} = 1168\text{ kJ} - 1234\text{ kJ} = -66\text{ kJ}$$

The overall reaction is predicted to be exothermic.

Exercise 9.10 Using Bond Energies

Using the bond energies in Table 9.9, estimate the heat of combustion of gaseous methane, CH_4. That is, estimate ΔH°_{rxn} for the reaction of methane with O_2 to give water vapor and carbon dioxide gas.

9.6 CHARGE DISTRIBUTION IN COVALENT COMPOUNDS

Lewis structures generally provide a fairly good picture of bonding in a covalently bonded molecule or ion. It is possible to "fine tune" this picture, however, to get a more precise description of the distribution of the electrons. In turn, this will provide further insight into the chemical and physical properties of covalent molecules.

Closer analysis of bonding in covalent species reveals that the valence electrons are not distributed among the atoms as evenly as Lewis structures might suggest. Some atoms may have a slight negative charge and others a slight positive charge. There are at least two causes for this. First, when forming bonds, some atoms contribute more electrons than they get back. Second, the electron pair or pairs in a given bond may be drawn more strongly toward one atom than the other. The result is that some atoms have **partial charges,** and the way these are distributed in the molecule is called its **charge distribution.**

The distribution of charges is important in determining the way a molecule reacts. Consider, for example, a diatomic (two-atom) molecule in which one atom is partially positive and the other is partially negative. Such a feature is bound to affect the properties of the molecule. In the solid state, for example, the molecules could be expected to line up with the positive end of one molecule near the negative end of another. The "intermolecular," or "between molecule," force of attraction would be enhanced by the attraction of opposite charges, and properties of the substance that are related to intermolecular forces such as boiling point would be affected.

*Inter*molecular means "between molecules"; *intra*molecular means within a single molecule.

The "location" of the charge in a molecule or ion is also important. The positive or negative charge will influence, among other things, the site at which reaction occurs between the molecule or ion and another chemical species. For example, does a positive H^+ ion attach itself to the O or the Cl of OCl^-? Is the product HOCl or HClO? It is reasonable to expect H^+ to attach to the more negatively charged atom. How can we predict which this will be?

Formal Charges on Atoms

See the *Saunders Interactive General Chemistry CD-ROM,* Screen 9.12, Oxidation Numbers, and Screen 9.13, Formal Charge.

The **formal charge** for an atom in a molecule or ion is the charge calculated for that atom based on the Lewis structure of the molecule or ion using the equation

$$\text{Formal charge of an atom in a molecule or ion} = \text{group number of the atom} - \left[\text{number of lone pair electrons} + \frac{1}{2}\,(\text{number of bonding electrons})\right]$$

The group number represents the number of valence electrons in a free atom. The number of electrons assigned by the Lewis structure to an atom in a molecule or ion is calculated by the term within the square brackets. The difference between the two is the formal charge. An atom will be positive if it "contributes" more electrons to bonding than it "gets back" when in a molecule. The atom will be negative if the opposite is true. There are two important assumptions in this equation. First, lone pairs are assumed to belong to the element on which they reside in the Lewis structure. Second, bond pairs are assumed to be shared equally between the bonded atoms. (The factor of 1/2 divides the bonding electrons equally between the atoms linked by the bond.)

The sum of the formal charges on the atoms in a molecule or ion always equals its net charge. Consider the hydroxide ion. Oxygen is in Group 6A and so has six valence electrons. In the hydroxide ion, however, oxygen can lay claim to seven electrons (six lone pair electrons and one bonding electron), and so the atom has a formal charge of 1−. It has "formally" gained an electron as part of the hydroxide ion.

Formal charge = 1− = 6 − [6 + (1/2)(2)]

$$[:\ddot{\underset{..}{O}}-H]^-$$

Formal charge = 0 = 1 − [0 + (1/2)(2)]

The formal charge on the hydrogen atom in OH^- is zero. (Thus, we have 1− for oxygen and 0 for hydrogen, which equals the net charge of 1− for the ion.) An important conclusion we can draw from this is that, if an H^+ ion approaches an OH^- ion, it should attach itself to the O atom. This of course should lead to water, as is indeed observed.

Formal charges can be calculated for more complicated species such as the nitrate ion. Using one of the resonance structures for the ion, we find that the central N atom has a formal charge of 1+, and the singly bonded O atoms are both 1−. The doubly bonded O atom has no charge. The net charge for the ion is 1−.

Formal charge = 0 = 6 − [4 + (1/2)(4)]

$$\left[\begin{array}{c} :\ddot{O}: \\ \| \\ :\ddot{\underset{..}{O}}-N-\ddot{\underset{..}{O}}: \end{array}\right]^-$$

Formal charge = 1+ = 5 − [0 + (1/2)(8)]

Formal charge = 1− = 6 − [6 + (1/2)(2)]

Is this a reasonable representation of the charge distribution for the nitrate ion? The answer is no, but not because of errors in the formal charge calculation. The problem is that the actual structure of the nitrate ion is not this structure; it is a hybrid of three equivalent resonance structures. Because the three oxygen atoms in NO_3^- are equivalent, we would not expect the charge on one oxygen atom to be different from the other two. This can be resolved, however, if the formal charges on the oxygens are averaged to give an average formal charge of (2/3)−. Summing the charges on the three oxygen atoms and the 1+ charge on the nitrogen atom gives 1−, the charge on the ion.

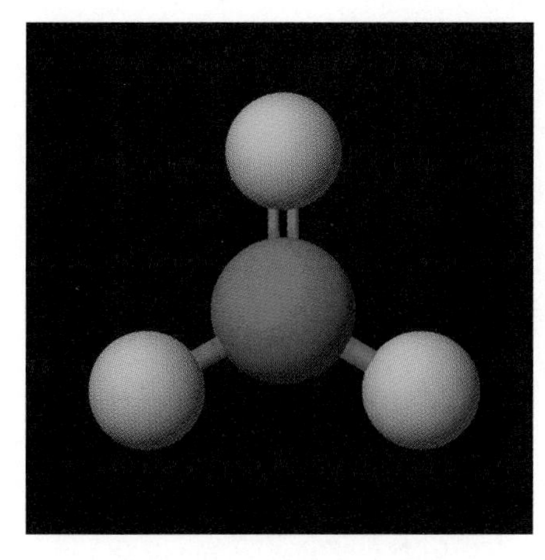

This computer model of the nitrate ion illustrates the formal charges on the N and O atoms of the ion. A red sphere indicates a positive charge and a yellow sphere a negative charge. The relative sizes of spheres reflect the relative charges. Notice especially that the partial charges on the O atoms are the same (the sphere sizes are the same).

In the resonance structures for O_3, CO_3^{2-}, and NO_3^-, all the possible resonance structures are equally likely; they are "equivalent" structures. The molecule or ion therefore has a symmetrical distribution of electrons over all the atoms involved—that is, its electronic structure consists of an equal "mixture," or "hybrid," of the resonance structures.

Example 9.7 Calculating Formal Charges

Calculate formal charges for the atoms in NH_4^+ and in one resonance structure of CO_3^{2-}.

Solution The first step is always to write the Lewis structure for the molecule or ion. Only then can you apply the formula for calculating the formal charges.

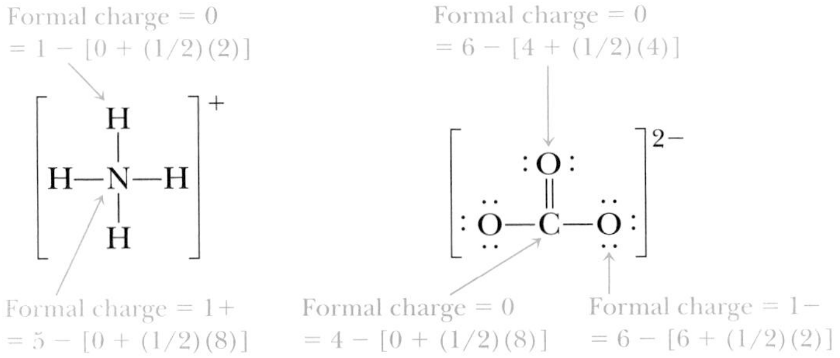

Formal charge = 0
= 1 − [0 + (1/2)(2)]

Formal charge = 0
= 6 − [4 + (1/2)(4)]

Formal charge = 1+
= 5 − [0 + (1/2)(8)]

Formal charge = 0
= 4 − [0 + (1/2)(8)]

Formal charge = 1−
= 6 − [6 + (1/2)(2)]

In each case notice that the sum of the atom formal charges is the charge on the ion. In the carbonate ion, which has three resonance structures, the *average* charge on the O atoms is (2/3)−.

Exercise 9.11 Calculating Formal Charges

Calculate formal charges on each atom in **(1)** CN^- and **(2)** SO_3.

Bond Polarity and Electronegativity

The models used to represent covalent and ionic bonding are the extreme situations in bonding. Pure covalent bonding, in which atoms share an electron pair equally, occurs *only* when two identical atoms are bonded. When two dissimilar atoms form a covalent bond, the electron pair will be unequally shared. The result is a **polar covalent bond,** a bond in which the two atoms have residual or partial charges (Figure 9.8).

Why are bonds polar? Because not all atoms hold onto their valence electrons with the same force, nor do atoms take on additional electrons with equal ease. Recall from the discussion of atom properties that different elements have different values of ionization energy and electron affinity (Section 8.6). These differences in behavior for free atoms carry over to atoms in molecules.

If a bond pair is not equally shared between atoms, the bonding electrons are nearer one of the atoms. The atom toward which the pair is displaced has a larger "share" of the electron pair and thus acquires a partial negative charge. At the same time, the atom at the other end of the bond is depleted in elec-

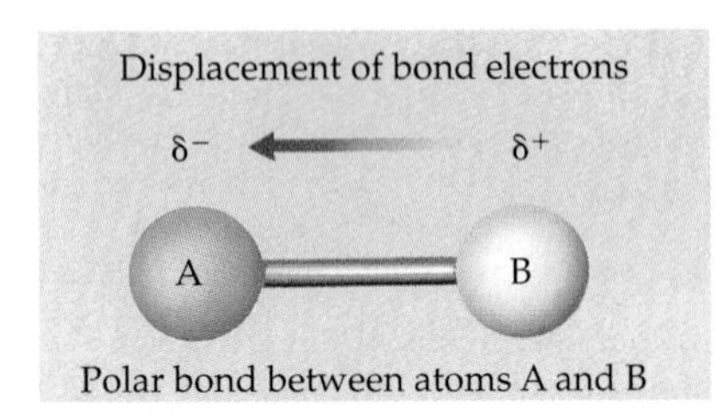

Figure 9.8 A polar covalent bond. Element A has a larger share of the bonding electrons and element B has the smaller share. The result is that A has a partial negative charge (δ−), and B has a partial positive charge (δ+).

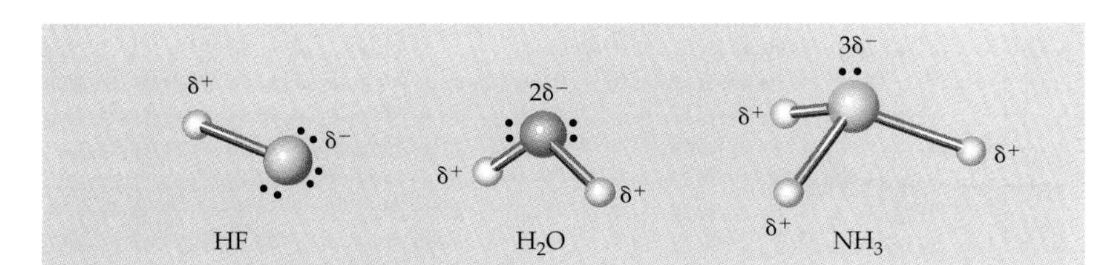

Figure 9.9 Three simple molecules with polar covalent bonds. In each case F, O, and N are more electronegative than H. See Figure 9.10.

trons and acquires a partial positive charge. The bond between the two atoms has a positive end and a negative end; that is, it has negative and positive poles and the bond is called a **polar bond** (having "poles"). The term **dipolar** (having two poles) is also used.

In ionic compounds, displacement of the bonding pair to one of the two atoms is essentially complete, and + and − symbols are written alongside the atom symbols in the Lewis drawings. For a **polar covalent bond** the polarity is indicated by writing the symbols δ+ and δ− alongside the atom symbols, where δ (the Greek letter "delta") stands for a *partial* charge. Hydrogen fluoride, water, and ammonia are three simple molecules having polar, covalent bonds (Figure 9.9). If there is no net displacement of the bond electron pair, the bond is **nonpolar covalent.**

See the *Saunders Interactive General Chemistry CD-ROM,* Screen 9.11, Bond Polarity and Electronegativity.

With so many atoms to use in covalent bond formation, it is not surprising that bonds between atoms can fall anywhere in a continuum from pure ionic to pure covalent. The range of chemical bonding can be represented schematically as

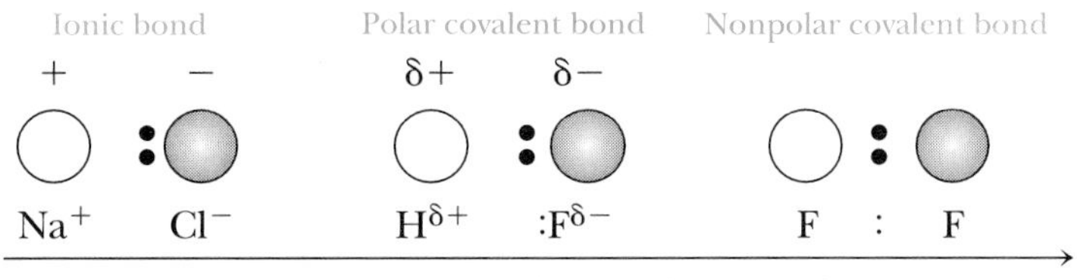

In the 1930s, Linus Pauling proposed a parameter called atom electronegativity that allows us to decide if a bond is polar, which atom of the bond is negative and which is positive, and if one bond is more polar than another. The **electronegativity,** χ, of an atom is defined as a measure of *the ability of an atom in a molecule to attract electrons to itself.*

At first glance, electronegativity and electron affinity may seem very similar, but they are not. *Electronegativity is a parameter that applies only to atoms in molecules.* As described in Section 8.6, *electron affinity is a measurable energy quantity that refers to isolated atoms.*

Values of electronegativity are given in Figure 9.10. Several features and periodic trends are apparent. The element with the largest electronegativity is fluorine; it is assigned a value of $\chi = 4.0$. The element with the smallest value is the alkali metal francium. Electronegativities generally increase from left to right across a period and decrease down a group. This is the opposite of the trend observed for metallic character. Metals typically have low values of electronegativity, ranging from slightly less than 1 to about 2. Electronegativity values for the metalloids are around 2, whereas nonmetals have values greater than 2. No values are given for the noble gases because only xenon and krypton form compounds.

Legend	
<1.0	2.0 – 2.4
1.0 – 1.4	2.5 – 2.9
1.5 – 1.9	3.0 – 4.0

1 **H** 2.1																
3 **Li** 1.0	4 **Be** 1.5											5 **B** 2.0	6 **C** 2.5	7 **N** 3.0	8 **O** 3.5	9 **F** 4.0
11 **Na** 1.0	12 **Mg** 1.2											13 **Al** 1.5	14 **Si** 1.8	15 **P** 2.1	16 **S** 2.5	17 **Cl** 3.0
19 **K** 0.9	20 **Ca** 1.0	21 **Sc** 1.3	22 **Ti** 1.4	23 **V** 1.5	24 **Cr** 1.6	25 **Mn** 1.6	26 **Fe** 1.7	27 **Co** 1.7	28 **Ni** 1.8	29 **Cu** 1.8	30 **Zn** 1.6	31 **Ga** 1.7	32 **Ge** 1.9	33 **As** 2.1	34 **Se** 2.4	35 **Br** 2.8
37 **Rb** 0.9	38 **Sr** 1.0	39 **Y** 1.2	40 **Zr** 1.3	41 **Nb** 1.5	42 **Mo** 1.6	43 **Tc** 1.7	44 **Ru** 1.8	45 **Rh** 1.8	46 **Pd** 1.8	47 **Ag** 1.6	48 **Cd** 1.6	49 **In** 1.6	50 **Sn** 1.8	51 **Sb** 1.9	52 **Te** 2.1	53 **I** 2.5
55 **Cs** 0.8	56 **Ba** 1.0	57 **La** 1.1	72 **Hf** 1.3	73 **Ta** 1.4	74 **W** 1.5	75 **Re** 1.7	76 **Os** 1.9	77 **Ir** 1.9	78 **Pt** 1.8	79 **Au** 1.9	80 **Hg** 1.7	81 **Tl** 1.6	82 **Pb** 1.7	83 **Bi** 1.8	84 **Po** 1.9	85 **At** 2.1
87 **Fr** 0.8	88 **Ra** 1.0	89 **Ac** 1.1														

***Figure* 9.10 Electronegativity values for the elements.** Trends for electronegativities are the opposite of the trends defining metallic character. Nonmetals have high values of electronegativity, the metalloids have intermediate values, and the metals have low values.

There will be a large *difference* in electronegativities if atoms from the left- and right-hand sides of the periodic table form a chemical compound. For cesium fluoride, for example, the difference in electronegativity values, $\Delta\chi$, is 3.2 [= 4.0 (for F) − 0.8 (for Cs)]. The bond is ionic, Cs is positive (Cs^+) and F negative (F^-). In contrast, the electronegativity difference between H and Cl in HCl is only 0.9 [= 3.0 (for Cl) − 2.1 (for H)]. We conclude that HCl must be a covalent species, as expected for a compound formed from two nonmetals. The H—Cl bond is polar, however, with hydrogen being the positive end of the molecule and chlorine the negative end ($H^{\delta+}$—$Cl^{\delta-}$).

Predicting trends in bond polarity in groups of related compounds is possible using values of electronegativity. Among the hydrogen halides, for example, the trend in polarity is HF ($\Delta\chi = 1.9$) > HCl ($\Delta\chi = 0.9$) > HBr ($\Delta\chi = 0.7$) > HI ($\Delta\chi = 0.4$).

Example 9.8 Estimating Bond Polarities

For each of the following bond pairs, decide which is the more polar and indicate the negative and positive poles.

1. Li—F and Li—Cl **2.** Si—O and P—P **3.** C═O and C═S

Solution

1. Li and F are on opposite sides of the periodic table, with F in the extreme upper right corner (χ for Li = 1.0 and χ for F = 4.0). Similarly, Li and Cl are on opposite sides of the table, but Cl is below F in the periodic table (χ for Cl = 3.0) and is therefore less electronegative than F. The difference in electronegativity for LiF is 3.0, for LiCl 2.0. Both bonds are expected to be strongly polar (ionic), with Li positive and the halide atom negative, but the Li—F bond will be more polar than the Li—Cl bond.

Historical Perspective

Linus Pauling (1901–1994)

Linus Pauling was born in Portland, Oregon, in 1901, the son of a druggist. He earned a B.Sc. degree in chemical engineering from Oregon State College in 1922 and completed his Ph.D. in chemistry at the California Institute of Technology in 1925. Before joining Cal Tech as a faculty member, he traveled to Europe where he worked briefly with Erwin Schrödinger and Niels Bohr (Chapter 7). In chemistry he is best known for his work on chemical bonding. Indeed, his book on *The Nature of the Chemical Bond* has influenced several generations of scientists. It was for his pioneering studies of bonding that he was awarded the Nobel Prize in chemistry in 1954. Shortly after World War II, Pauling and his wife began a crusade to limit nuclear weapons, a crusade that came to fruition in the Limited Test Ban Treaty of 1963. For this effort, Pauling was awarded the 1963 Nobel Prize for peace. Never before had any person received two unshared Nobel Prizes.

Linus Pauling. *(Oesper Collection in the History of Chemistry/University of Cincinnati)*

2. Because the bond is between two atoms of the same kind, the P—P bond is nonpolar. Silicon is in Group 4A and the third period, whereas O is in Group 6A and the second period. Consequently, O has a greater electronegativity (3.5) than Si (1.8), so the bond is highly polar ($\Delta\chi = 1.7$), with O the more negative atom.
3. Oxygen lies above sulfur in the periodic table, so oxygen is more electronegative than S. This means the C—O bond is more polar than the C—S bond. For the C—O bond, O is the more negative atom. The value of $\Delta\chi$ (1.0) for CO indicates a moderately polar bond.

Exercise 9.12 Bond Polarity

For each of the following pairs of bonds, decide which is the more polar. For each polar bond, indicate the positive and negative poles. First make your prediction from the relative atom positions in the periodic table; then check your prediction by calculating $\Delta\chi$.

1. H—F and H—I **2.** B—C and B—F **3.** C—Si and C—S

Combining Formal Charge and Bond Polarity

Using formal charge calculations alone to locate the site of a charge in an ion can lead to results that are counterintuitive. The ion BF_4^- illustrates this point. Boron has a formal charge of 1− in this ion, whereas the formal charge calculated for the fluorine atoms is 0. This is not logical: fluorine is the more electronegative atom so the negative charge should reside on F and not on B.

The way to resolve this dilemma is to consider electronegativity in conjunction with formal charge. Based on the electronegativity difference between fluorine and boron ($\Delta\chi = 2.0$) the B—F bonds are expected to be polar, with fluorine being the negative end of the bond, $B^{\delta+}—F^{\delta-}$. So, in this instance, predictions based on electronegativity and formal charge work in opposite directions. The formal charge calculation places the negative charge on boron, but the electronegativity difference says that the charge on boron ends up being distributed onto the fluorine atoms. In effect the charge is "spread out" over the molecule.

Pauling proposed an important idea that applies to the problem with BF_4^- and to all other molecules: the **electroneutrality principle.** This declares that the electrons in a molecule are distributed in such a way that the charges on the atoms are as close to zero as possible. Furthermore, if there is a negative charge, it should be placed on the most electronegative atom. Similarly, a positive charge should be on the least electronegative atom. For BF_4^-, this means the negative charge should not be located on the boron atom alone but spread out over the more electronegative F atoms.

Considering the concepts of electronegativity and formal charge together can also help to decide which of several resonance structures is the more important. For example, when drawing the Lewis structure for CO_2, A is the logical one to draw. But what is wrong with B, in which each atom also has an octet of electrons?

Formal charges 0 0 0 1+ 0 1−

Resonance structures

$$\ddot{\underset{\cdot\cdot}{O}}=C=\ddot{\underset{\cdot\cdot}{O}} \longleftrightarrow :O\equiv C-\ddot{\underset{\cdot\cdot}{O}}:$$

A B

For structure A, each atom has a formal charge of 0, a favorable situation. In B, however, one oxygen atom has a formal charge of 1+ and the other has 1−. This is contrary to the principle of electroneutrality. In addition, B places a positive charge on a very electronegative atom. Thus we can conclude that structure B is of little consequence.

Now use what you have learned with CO_2 to decide which of the three possible resonance structures for the OCN^- ion is the most reasonable. Formal charges for each atom are given on top of the element's symbol.

Example of formal charge calculation: For resonance form *C* for OCN^-, we have

$$O = 6 - [2 + (1/2)(6)] = 1+$$

$$C = 4 - [0 + (1/2)(8)] = 0$$

$$N = 5 - [6 + (1/2)(2)] = 2-$$

Sum of formal charges = 1− = charge on the ion

1− 0 0 0 0 1− 1+ 0 2−

$$\left[:\ddot{\underset{\cdot\cdot}{O}}-C\equiv N:\right]^- \longleftrightarrow \left[\ddot{\underset{\cdot\cdot}{O}}=C=\ddot{\underset{\cdot\cdot}{N}}\right]^- \longleftrightarrow \left[:O\equiv C-\ddot{\underset{\cdot\cdot}{N}}:\right]^-$$

A B C

Structure C will clearly not contribute significantly to the overall electronic structure of the ion. It has a 2− formal charge on the N atom and a 1+ formal charge on the O atom, whereas the formal charges in the other structures are 0, 1+, or 1−. Of structures A and B, A is the more significant because the negative charge is placed on the most electronegative atom. We predict, therefore, that the carbon–nitrogen bond will resemble a triple bond.

The result for OCN^- also allows us to predict that protonation of the ion will lead to HOCN and not HNCO. That is, an H^+ ion will add to the more negative oxygen atom.

1+ 1− 0 0

$$H^+ + :\ddot{\underset{\cdot\cdot}{O}}-C\equiv N: \longrightarrow H-\ddot{\underset{\cdot\cdot}{O}}-C\equiv N:$$

Example 9.9 Calculating Formal Charges

Boron-containing compounds often have a boron atom with only three bonds (and no lone pairs). Why not form a double bond with a terminal atom to complete the boron octet? To answer this, consider the resonance structures of H_2BF and calculate the atoms' formal charges. Are the bonds polar in H_2BF? If so, which is the more negative atom?

Solution The two possible structures for H_2BF are illustrated here with the calculated formal charges on the B and F atoms. (H has a 0 formal charge.)

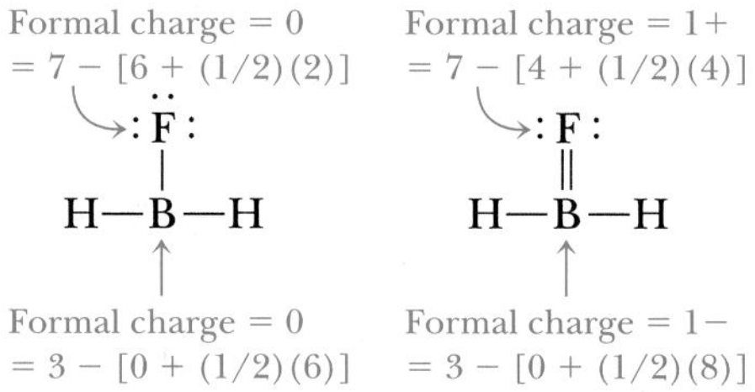

Formal charge = 0
= 7 − [6 + (1/2)(2)]
:F̤: (single bond) H—B—H
Formal charge = 0
= 3 − [0 + (1/2)(6)]

Formal charge = 1+
= 7 − [4 + (1/2)(4)]
:F: (double bond) H—B—H
Formal charge = 1−
= 3 − [0 + (1/2)(8)]

The structure on the left is preferred because all atoms have a zero formal charge, and the very electronegative F atom does not have a charge of 1+.

Both F ($\chi = 4.0$) and H ($\chi = 2.1$) are more electronegative than B ($\chi = 2.0$). All the bonds are polar, the B—F bond being much more polar than the B—H bonds.

Bond polarities in H_2BF

$\overset{\longmapsto}{\text{B—F}}$ ($\delta+$ $\delta-$) $\overset{\longmapsto}{\text{B—H}}$ ($\delta+$ $\delta-$)

Exercise 9.13 Formal Charge, Bond Polarity, and Electronegativity

Consider all possible resonance structures for SO_2. What are the formal charges on each atom in each resonance structure? What are the bond polarities? Do they agree with the formal charges?

9.7 MOLECULAR SHAPES

We can now turn to one of the most important results of drawing Lewis dot structures, the prediction of the three-dimensional geometry of molecules and ions. Because the physical and chemical properties of compounds are tied to their structures, the importance of this subject cannot be overstated. Indeed, this is the reason we have used models of molecular structures throughout the text. Now we can proceed to develop and understand the concepts on which structural models are created.

The **valence shell electron-pair repulsion (VSEPR) model,** devised by Ronald J. Gillespie (1924–) and Ronald S. Nyholm (1917–1971), is a reliable method for predicting the shapes of covalent molecules and polyatomic ions. The model is based on the idea that *bond and lone electron pairs in the valence shell*

See the *Saunders Interactive General Chemistry CD-ROM,* Screens 9.14, 9.15, and 9.16, Molecular Shape and VSEPR.

The structures of the molecules and ions in this chapter are found on the *Saunders Interactive General Chemistry CD-ROM* or at http://www.saunderscollege.com.

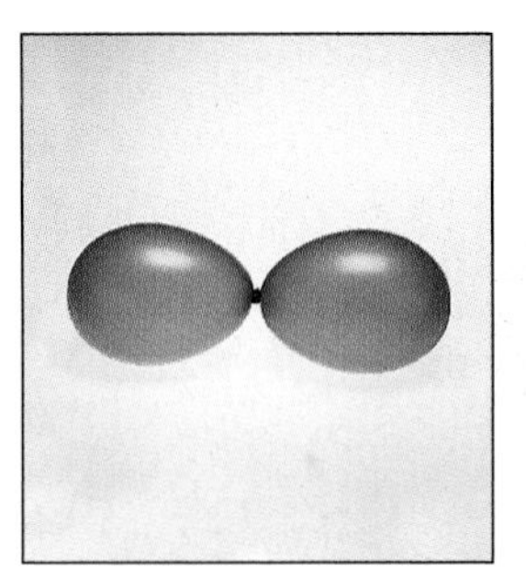

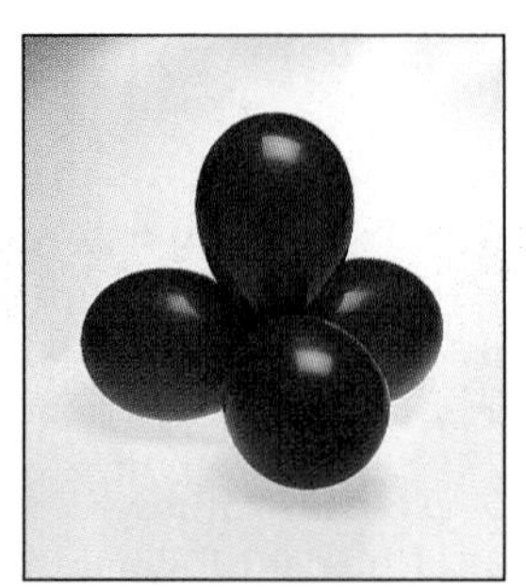

Figure 9.11 Balloon models of electron pair geometries for two to six electron pairs. If two to six balloons of similar size and shape are tied together, they will naturally assume the arrangements shown. These pictures illustrate the predictions of VSEPR. *(C. D. Winters)*

of an element repel each other and seek to be as far apart as possible. The positions assumed by the valence electrons of an atom thus define the angles between bonds to surrounding atoms. VSEPR is remarkably successful in predicting structures of molecules and ions of main group elements; however, it is not generally used to predict structures of molecules based on transition metal atoms.

To get a sense of how valence shell electron pairs repel and determine structure, blow up several balloons to a similar size. Imagine that each balloon's volume represents a repulsive force that prevents other balloons from occupying the same space. When two, three, four, five, or six balloons are tied together at a central point (representing the nucleus and core electrons of a central atom), the balloons naturally form the shapes shown in Figure 9.11. These geometric arrangements minimize interactions between the balloons.

Central Atoms Surrounded Only by Bond Pairs

Let us now apply the VSEPR theory to molecules and ions. The simplest application is to species in which all the electron pairs around the central atom are in single covalent bonds. Figure 9.12 illustrates the geometries predicted by the VSEPR model for molecules that contain only single covalent bonds (and no lone pairs). The general formulas are represented as AX_n, where A is the central atom and n is the number of X groups bonded to it.

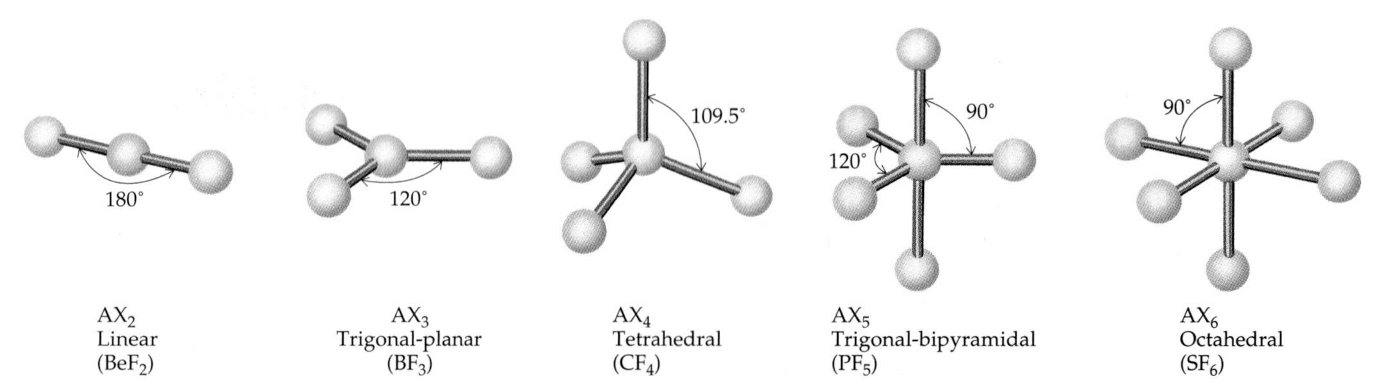

Figure 9.12 Various geometries predicted by VSEPR. Geometries predicted by VSEPR for molecules that contain only single covalent bonds around the central atom.

A CLOSER LOOK Atom Partial Charges

Chemists have long been interested in an assessment of the charges on atoms in molecules, and mathematical methods have been devised that give chemically reasonable values of these charges. Examples of the results of these calculations are included on the *Saunders Interactive General Chemistry CD-ROM* (see the folder marked PARTCHRG in the *Models* folder).

Consider formaldehyde (CH_2O), chloroform ($CHCl_3$), and carbonate ion (CO_3^{-}) as examples of such calculations. In the models shown here the atom sizes are scaled and colored according to their partial charges. Atoms colored yellow have a negative charge and those colored red have a positive charge. The larger the sphere, the larger its partial charge.

In formaldehyde, the oxygen is negative, and the carbon and hydrogens are positive. The calculated values for CH_2O are H = +0.09, C = +0.16, and O = −0.33. These seem reasonable because differences in electronegativity suggest that the O atom in CH_2O should bear a slight negative charge instead of 0, its formal charge.

In chloroform, most of the positive charge resides on the hydrogen atom, the negative charge is on the carbon and chlorine atoms. Calculations for $CHCl_3$ give H = +0.16, C = −0.036, Cl = −0.041. Again electronegativity values suggest the H atom should be positively charged, and not 0, as implied by the formal charge.

In both CH_2O and $CHCl_3$, the sum of the charges is zero, as expected, because there is no net charge on either molecule.

Finally, look at the carbonate ion. The important observation here is that the O atoms all bear the same partial negative charge. The 2− charge for the ion is spread out over the three O atoms, as resonance structures (page 393) are meant to imply.

The partial charges calculated by the CAChe Scientific/Oxford Molecular approach are chemically more reasonable than the formal charges. Such calculations certainly help chemists to make better predictions about the location and size of charges in molecules.

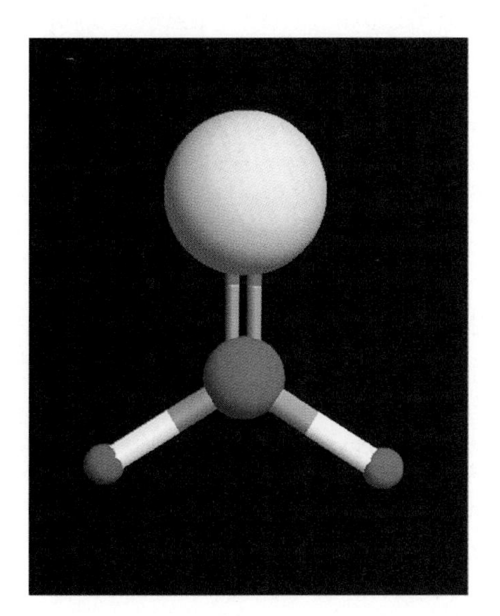

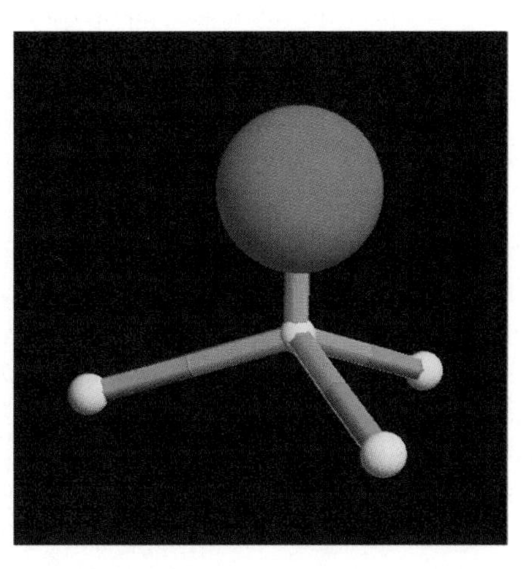

Computer-generated representations of formaldehyde, CH_2O *(left)*, chloroform, $CHCl_3$ *(center)*, and carbonate ion *(right)*. Red designates an atom with a positive charge, and yellow designates one with a negative charge. The relative sizes represent the relative magnitude of the charges.

The **linear** geometry for two bond pairs and the **trigonal-planar** geometry for three bond pairs involve a central atom that does not have an octet of electrons (see Section 9.3). The central atom in a **tetrahedral** molecule obeys the octet rule with four bond pairs. The central atoms in **trigonal-bipyramidal** and **octahedral** molecules have five and six bonding pairs, respectively, and are expected only when the central atom is an element in Period 3 or higher of the periodic table. Because there are many molecules that fit into these categories you should be thoroughly familiar with these geometries.

The word "triangular" is often used in place of "trigonal." Both refer to objects at the corners of a triangle.

Example 9.10 Predicting Molecular Shapes

Predict the shape of silicon tetrachloride, $SiCl_4$.

Solution The first step in predicting the shape of a molecule or ion is to draw its Lewis structure. The Lewis structure does not need to be drawn in any particular way; it only needs to convey the fact that there are four single covalent bonds to silicon. A tetrahedral structure is predicted for the $SiCl_4$ molecule, with Cl—Si—Cl bond angles of 109.5°. This prediction agrees with the actual structure for $SiCl_4$.

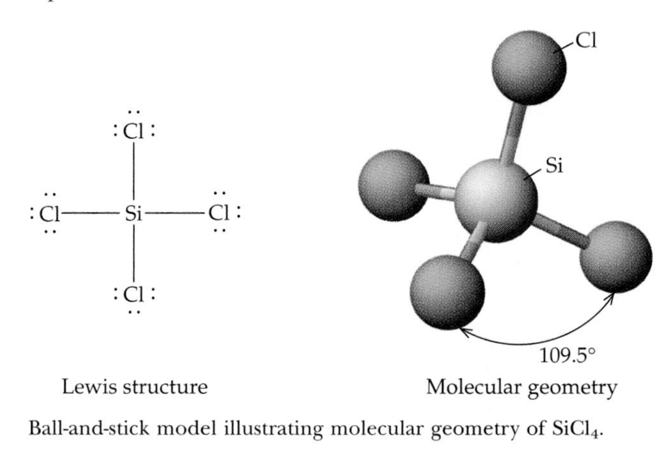

Ball-and-stick model illustrating molecular geometry of $SiCl_4$.

Exercise 9.14 Predicting Molecular Shapes

What is the shape of the dichloromethane molecule? The formula for dichloromethane is CH_2Cl_2. Predict the Cl—C—Cl bond angle.

Central Atoms with Bond Pairs and Lone Pairs

To see how *lone pairs* affect the geometry of the molecule or polyatomic ion, return to the balloon models in Figure 9.11. Recall that the balloons represented *all* the electron pairs in the valence shell. The balloon model therefore predicts the "electron-pair geometry" rather than the "molecular geometry." The **electron-pair geometry** is the geometry taken up by *all* the valence electron pairs around a central atom, whereas the **molecular geometry** involves the arrangement in space of the central atom and the atoms directly attached to it. It is important to recognize that *lone pairs of electrons on the central atom occupy spatial positions even though their location is not included in the verbal description of the shape of the molecule or ion.*

Let us use the VSEPR model to predict the molecular geometry and bond angles in the NH_3 molecule, which has a lone pair on the central atom. First, draw the Lewis structure and count the total number of electron pairs around the central nitrogen atom. There are four pairs of electrons in the nitrogen valence shell, so the *electron-pair geometry* is predicted to be tetrahedral. Draw a tetrahedron with nitrogen as the central atom and the three bond pairs represented by lines. The lone pair is drawn here to indicate its spatial position in the tetrahedron. The *molecular geometry* is described as a *trigonal pyramid.* The nitrogen atom is at the apex of the pyramid, and the three hydrogen atoms form the trigonal base.

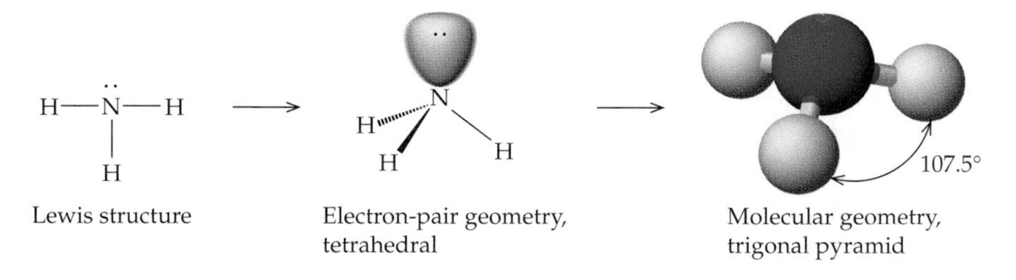

Effect of Lone Pairs on Bond Angles

Because the electron-pair geometry in NH_3 is tetrahedral, we would expect the H—N—H bond angle to be 109.5°. The experimentally determined bond angles in NH_3, however, are 107.5°. This is close to the tetrahedral angle but not exactly that value. This highlights the fact that VSEPR is not a precise theory; it can only predict the approximate geometry. Small variations in geometry (e.g., bond angles a few degrees different from predicted) are actually quite commonplace and often arise because there is a difference between the spatial requirements of lone pairs and bond pairs. Lone pairs of electrons seem to occupy a larger volume than bonding pairs. A rationale for this is that bond pairs are drawn into the bond region between atoms by the strong attractive forces of protons in two nuclei and are therefore relatively compact; we can think of them as "skinny." For a lone pair, however, only one nucleus attracts the electron pair. Because a single nuclear charge is not as effective in attracting the lone pair electrons, lone pairs are considered "fat." The increased volume of lone pairs causes bond pairs to squeeze closer together. Another way to view this is to consider the relative strength of repulsions, which are in the order

Lone pair–lone pair > lone pair–bond pair > bond pair–bond pair

Gillespie and Nyholm recognized the importance of the different spatial requirements of lone pairs and bond pairs and included this as part of their VSEPR model. For example, they used the VSEPR model to predict variations in the bond angles in the series of molecules CH_4, NH_3, and H_2O. Notice in Figure 9.13 that the bond angles decrease in the series CH_4, NH_3, and H_2O as the number of lone pairs on the central atom increases.

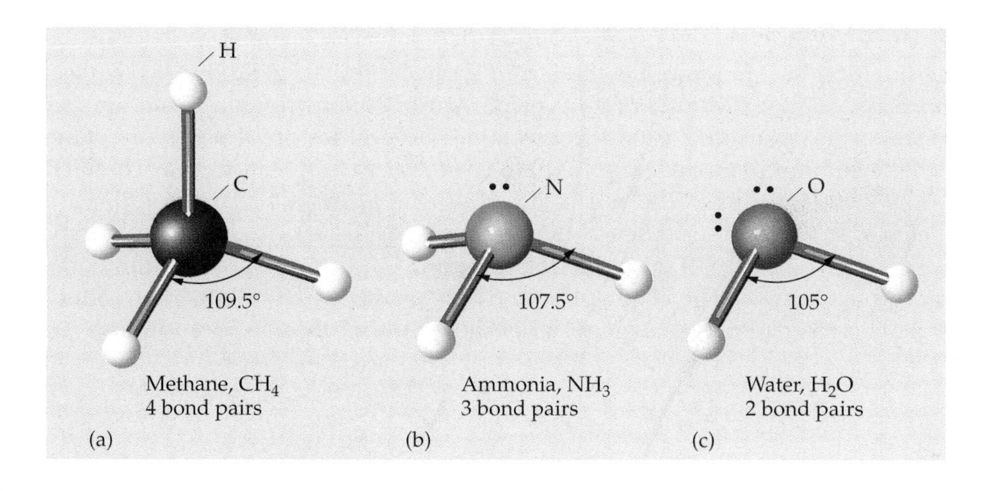

***Figure* 9.13 The geometries of methane, ammonia, and water.** All have four electron pairs around the central atom, so all have a tetrahedral electron-pair geometry. (a) Methane has four bond pairs and so has a tetrahedral molecular shape. (b) Ammonia has three bond pairs and one lone pair, so it has a trigonal-pyramidal molecular shape. (c) Water has two bond pairs and two lone pairs, so it has a bent, or angular, molecular shape. The decrease in bond angles in the series can be explained by the larger spatial requirements of the lone pairs, which squeeze the bond pairs closer together.

Example 9.11 Finding the Shapes of Molecules

What are the molecular shapes of H_3O^+ and ClF_2^+?

Solution

1. The Lewis structure of the hydronium ion, H_3O^+, shows that the oxygen atom is surrounded by four electron pairs, so the electron-pair geometry is tetrahedral. Because three of the four pairs are used to bond terminal atoms, the central O atom and the three H atoms form a trigonal-pyramidal molecular shape like NH_3.

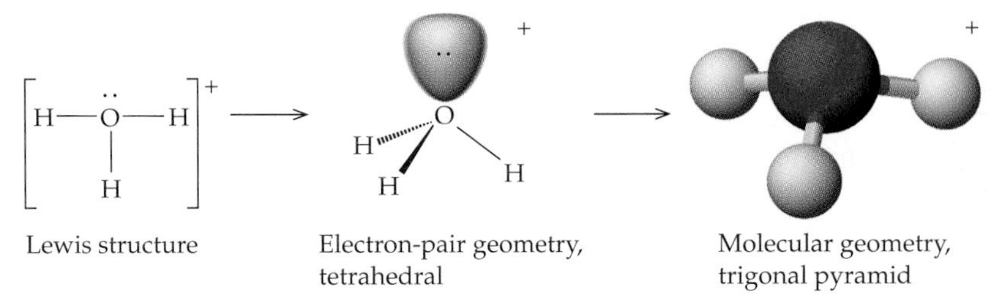

2. Chlorine is the central atom in ClF_2^+. It is surrounded by four electron pairs, so the electron-pair geometry around chlorine is tetrahedral. Because only two of the four pairs are bonding pairs, the ion has a bent geometry.

Lewis structure | Electron-pair geometry, tetrahedral | Molecular geometry, bent or angular

Exercise 9.15 VSEPR and Molecular Shape

Give the electron-pair geometry and molecular shape for BF_3 and BF_4^-. What is the effect on the molecular geometry of adding an F^- ion to BF_3 to give BF_4^-?

Central Atoms with More Than Four Valence Electron Pairs

The situation becomes more complicated if the central atom has five or six electron pairs, some of which are lone pairs. A trigonal-bipyramidal structure has two sets of positions that are not equivalent. The positions in the trigonal plane lie in the equator of an imaginary sphere around the central atom and are called the *equatorial* positions. The north and south poles in this representation are called the *axial* positions. Each equatorial position has two neighboring groups (the axial groups) at 90°, and the axial positions have three groups (the equatorial groups) at 90°. The result is that the lone pairs, which are fatter than bonding pairs, prefer to occupy equatorial positions rather than axial positions.

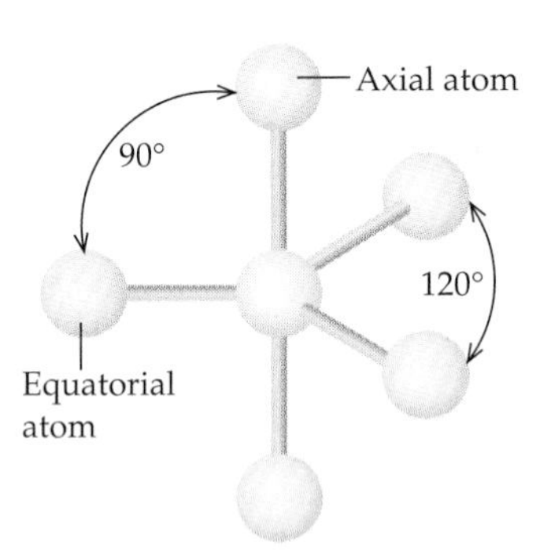

The trigonal bipyramid showing the axial and equatorial atoms. The angles between atoms in the equator are 120°. The angles between equatorial and axial atoms are 90°.

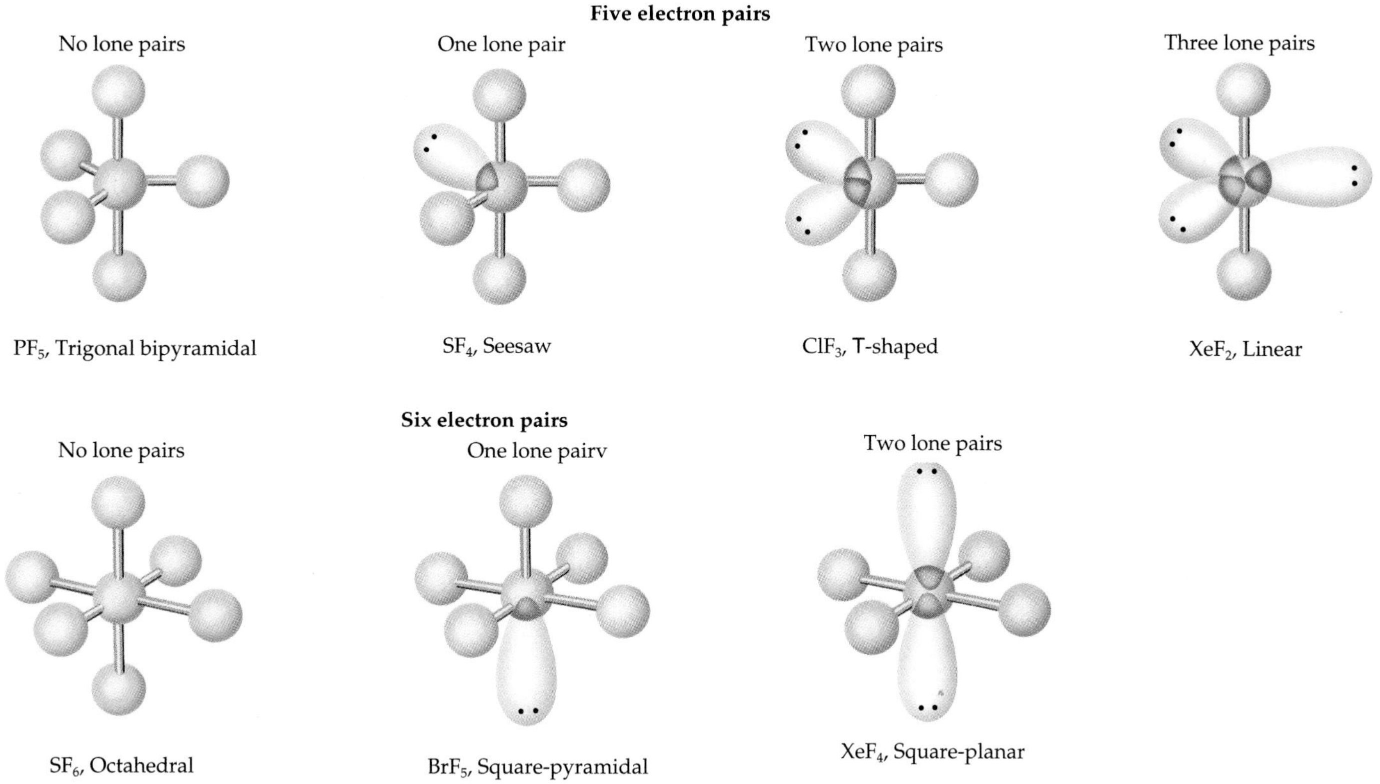

Five electron pairs

No lone pairs — PF_5, Trigonal bipyramidal

One lone pair — SF_4, Seesaw

Two lone pairs — ClF_3, T-shaped

Three lone pairs — XeF_2, Linear

Six electron pairs

No lone pairs — SF_6, Octahedral

One lone pairv — BrF_5, Square-pyramidal

Two lone pairs — XeF_4, Square-planar

***Figure* 9.14** Electron-pair geometries and molecular shapes for molecules and ions with five *(top)* or six *(bottom)* electron pairs around the central atom.

The entries in the top line of Figure 9.14 show species having a total of five valence electron pairs, with zero, one, two, and three lone pairs being present. In SF_4, with one lone pair, the molecule assumes a seesaw shape with the lone pair in one of the equatorial positions. The ClF_3 molecule has three bond pairs and two lone pairs. The two lone pairs in ClF_3 are in equatorial positions; two bond pairs are axial and the third is in the equatorial plane, so the molecular geometry is T-shaped. The third molecule shown is XeF_2. Here, all three equatorial positions are occupied by lone pairs so the molecular geometry is linear.

The geometry assumed by six electron pairs is octahedral (see Figure 9.14), and all the angles at adjacent positions are 90°. Unlike the trigonal-bipyramid, the octahedron has no distinct axial and equatorial positions; all positions are the same. Therefore, if the molecule has one lone pair, as in BrF_5, it makes no difference which position it occupies. The lone pair is often drawn in the top or bottom position to make it easier to visualize the molecular geometry, which in this case is square-pyramidal. If two pairs of the electrons in an octahedral arrangement are lone pairs, they seek to be as far apart as possible. The result is a square-planar molecule, as illustrated by XeF_4.

Example 9.12 Predicting Molecular Shape

What is the shape of the ICl_4^- ion?

Solution A Lewis structure for the ICl_4^- ion shows that the central iodine atom has six electron pairs in its valence shell. Two of these are lone pairs. Placing the lone pairs on opposite sides leaves the four chlorine atoms in a square-planar geometry.

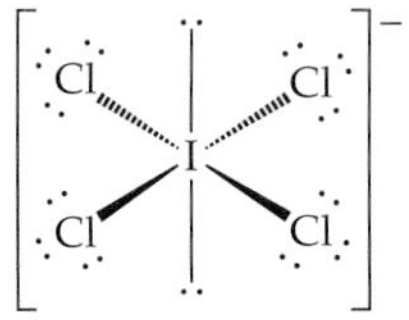

90°

Electron-pair geometry, octahedron

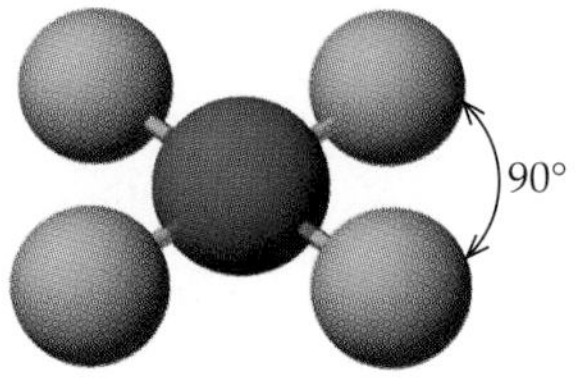

Molecular geometry, square planar

Exercise 9.16 Predicting Molecular Shape

Draw the Lewis structure for ICl_2^- and then decide on the geometry of the ion.

Multiple Bonds and Molecular Geometry

Double and triple bonds involve more electron pairs than single bonds, but this does not affect the overall molecular shape. Electron pairs involved in a multiple bond are all shared between the same two nuclei and therefore occupy the same region of space. Because they must remain in that region, two electron pairs in a double bond (or three in a triple bond) are like a single balloon, rather than two or three balloons. All electron pairs in a multiple bond count as one bond and contribute to molecular geometry the same as a single bond does. For example, the carbon atom in CO_2 has no lone pairs and participates in two double bonds. Each double bond counts as one for the purpose of predicting geometry, so the structure of CO_2 is linear.

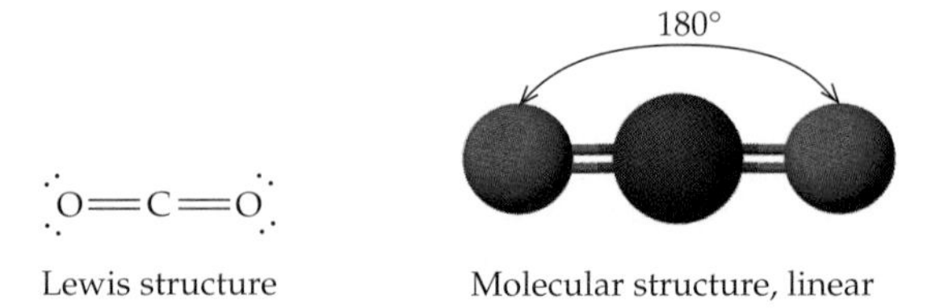

Lewis structure

Molecular structure, linear

When resonance structures are possible, the geometry can be predicted from any of the Lewis resonance structures or from the resonance hybrid structure.

For example, the geometry of the CO_3^{2-} ion is predicted to be trigonal-planar because the carbon atom has three sets of bonds and no lone pairs.

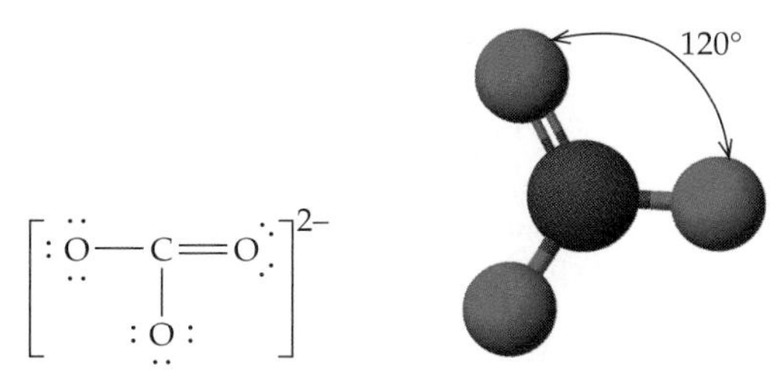

Lewis structure, one resonance structure

Molecular structure, trigonal planar

The NO_2^- ion also has a trigonal-planar electron-pair geometry. Because there is a lone pair on the central nitrogen atom, and two bonds in the other two positions, the geometry of the ion is angular or bent.

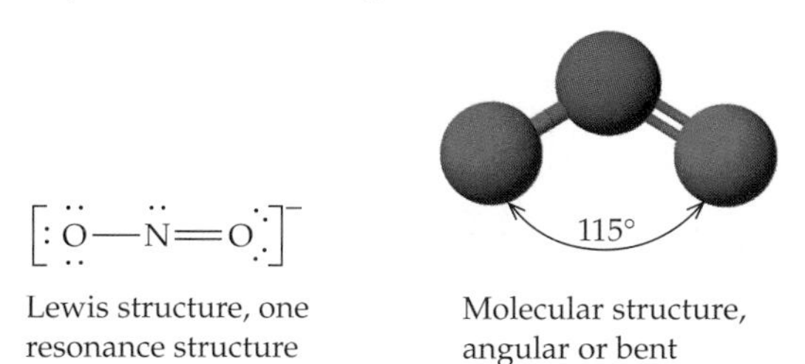

Lewis structure, one resonance structure

Molecular structure, angular or bent

The techniques just outlined can be used to find the geometries of much more complicated molecules. Consider, for example, cysteine, one of the natural amino acids.

cysteine, $HSCH_2CH(NH_2)CO_2H$

Four pairs of electrons occur around the S, N, C_1, and C_2 atoms, so the electron-pair geometry around each is tetrahedral. Thus, the H—S—C and H—N—H angles are predicted to be approximately 109°. The O atom in the grouping C—O—H also is surrounded by four pairs, and so this angle is likewise approximately 109°. Finally, the angle made by O—C_3—O is 120° because the electron-pair geometry around C_3 is planar and trigonal. A computer-generated model is given in Figure 9.15.

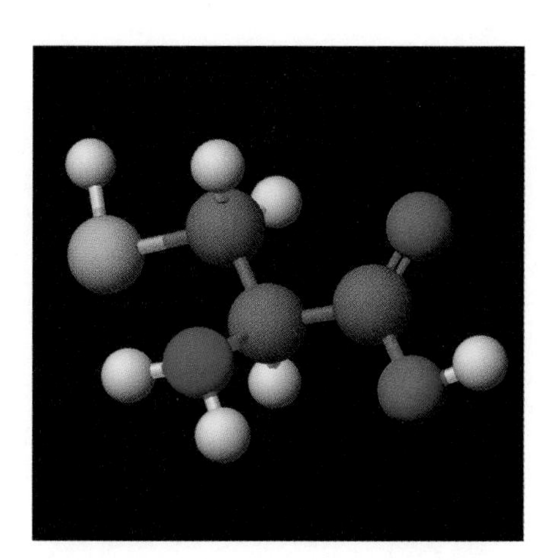

Figure **9.15 Geometry of cysteine.** The structure of cysteine, $HSCH_2CH(NH_2)CO_2H$.

Example 9.13 Finding the Shapes of Molecules and Ions

What are the shapes of the nitrate ion, NO_3^-, and $XeOF_4$?

Solution

1. The NO_3^- ion and CO_3^{2-} ion are isoelectronic. Thus, like the carbonate ion, the electron-pair geometry and molecular shape of NO_3^- are trigonal-planar.

Lewis structure, one resonance structure

Molecular structure, planar trigonal

2. The $XeOF_4$ molecule has a Lewis structure with a total of six electron pairs about the central Xe atom, one of which is a lone pair. It has a square-pyramidal molecular structure. (Two structures are possible based on the position occupied by the oxygen, but there is no way to predict which one is correct. The actual structure is the one shown, with the oxygen in the apex of the square pyramid.)

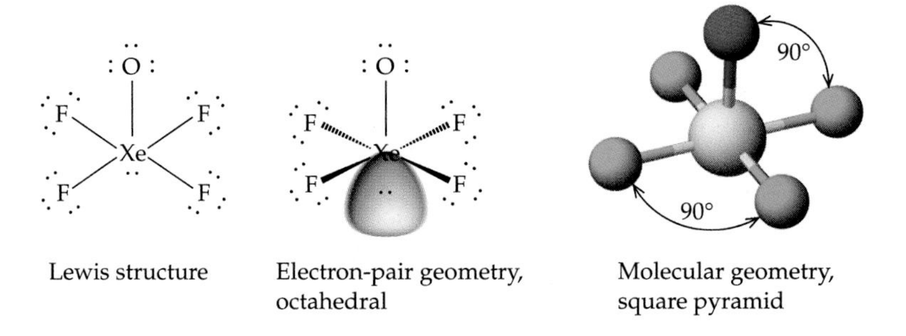

Lewis structure

Electron-pair geometry, octahedral

Molecular geometry, square pyramid

Exercise 9.17 Determining Molecular Shapes

Use Lewis structures and the VSEPR model to determine the electron-pair and molecular geometries for **1.** the phosphate ion, PO_4^{3-}; **2.** the sulfite ion, SO_3^{2-}; and **3.** IF_5.

Problem-Solving Tips and Ideas

9.3 Determining Molecular Structure

The shape of virtually any molecule or ion containing main group elements can be described by thinking through the following steps:

1. Draw the Lewis structure.
2. Determine the total number of single bond pairs (including multiple bonds counted as single pairs).
3. Pick the appropriate electron-pair geometry, and then choose the molecular shape that matches the total number of single bond pairs and lone pairs.
4. Predict the bond angles, remembering that lone pairs occupy more volume than bonding pairs do.

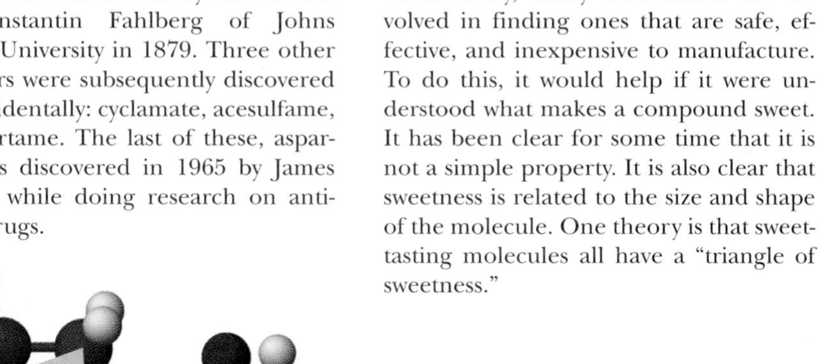

A CLOSER LOOK *Return to the Chemical Puzzler—Why Are Sweeteners Sweet?*

The first artificial sweetener truly safe for human consumption was saccharin, which was discovered by Ira Remsen and Constantin Fahlberg of Johns Hopkins University in 1879. Three other sweeteners were subsequently discovered quite accidentally: cyclamate, acesulfame, and aspartame. The last of these, aspartame, was discovered in 1965 by James Schlatter while doing research on anticancer drugs.

Given the enormous quantities of sugar and artificial sweeteners used in the world today, many laboratories are involved in finding ones that are safe, effective, and inexpensive to manufacture. To do this, it would help if it were understood what makes a compound sweet. It has been clear for some time that it is not a simple property. It is also clear that sweetness is related to the size and shape of the molecule. One theory is that sweet-tasting molecules all have a "triangle of sweetness."

The receptors in our taste buds are composed of proteins, and these have the ability to form chemical bonds with a group such as C=O in another molecule. It also appears that a sweet molecule must have a group such as NH or NH_2, and finally it must have a third site such as a CH_2 or CH_3. All three sites are within a triangle with sides each about 300 to 500 pm long. At least one possible "sweetness triangle" is marked on a model of aspartame. The only way that aspartame can fulfill these criteria is to have the right kinds of atoms the correct distance apart and with the correct geometry around each one. Molecular size and shape are indeed the keys to sweetness!

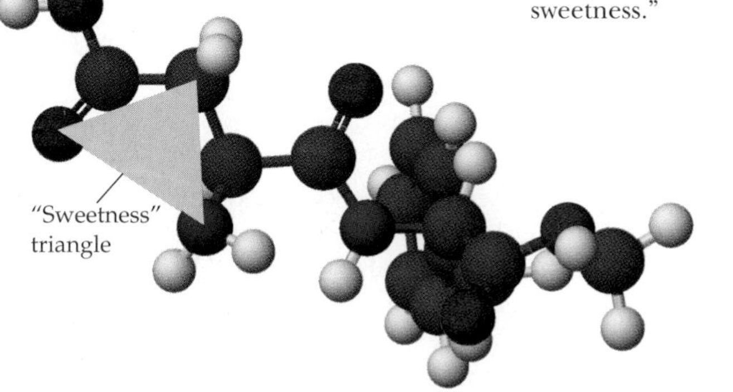

A model of the aspartame molecule showing one possible location for a "triangle of sweetness." See R. H. Mazur, J. M. Schlatter, and A. H. Goldkamp: "Structure-taste relationships of some dipeptides." *Journal of the American Chemical Society*, Vol. 91, p. 2684, 1969.

9.8 MOLECULAR POLARITY

The term "polar" was used in Section 9.6 to describe a bond in which one atom had a partial positive charge and the other a partial negative charge. Because most molecules have at least some polar bonds, molecules can also be polar. In a polar molecule, electron density accumulates toward one side of the molecule, giving that side a negative charge, $\delta-$, and leaving the other side with a positive charge of equal value, $\delta+$.

See the *Saunders Interactive General Chemistry CD-ROM*, Screen 9.17, Molecular Polarity.

Before describing the factors that determine whether a molecule is polar, let us look at the experimental measurement of the polarity of a molecule. Polar molecules experience a force in an electric field that tends to align them with the field (Figure 9.16). When the electric field is created by a pair of oppositely charged plates, the positive end of each molecule is attracted to the negative plate, and the negative end is attracted to the positive plate. The extent to which the molecules line up with the field depends on their **dipole moment,** μ, which is defined as the product of the magnitude of the partial charges ($\delta+$ and $\delta-$) and the distance by which they are separated. The SI unit of the dipole moment is the coulomb-meter, but dipole moments have traditionally been given using a derived unit called the debye (D; 1 D = 3.34×10^{-30} C · m). Dipole moments are determined experimentally, and typical values are listed in Table 9.10.

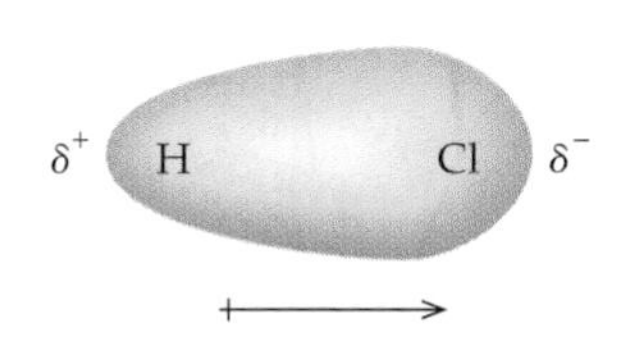

In a polar molecule, the electrons are displaced toward one side. To indicate the direction of molecular polarity an arrow is drawn with the arrowhead pointing toward the negative end, and a plus sign is placed at the positive end of the molecule.

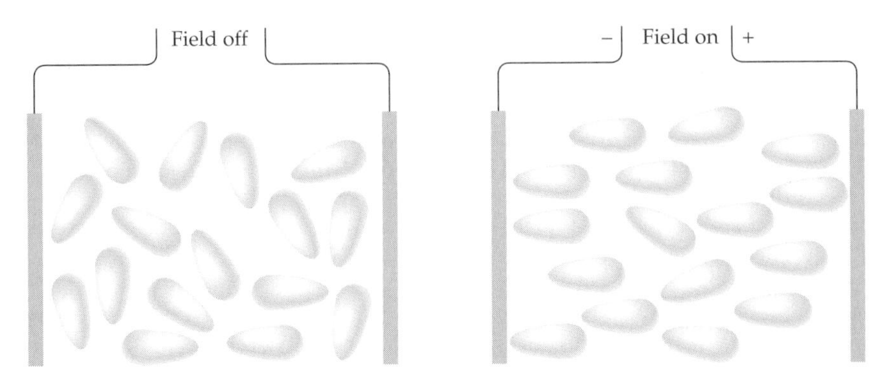

Figure 9.16 Polar molecules in an electric field. When placed between charged plates in an electric field, polar molecules experience a force that tends to align them with the field. The negative end of the molecules is drawn to the positive plate, and vice versa. This property affects the capacitance of the plates (their ability to hold a charge) and provides a way to measure experimentally the magnitude of the dipole.

Peter Debye (1884–1966). The commonly used unit of dipole moments is named in honor of Peter Debye. He was born in The Netherlands but became a professor of chemistry at Cornell University in 1940. He received a Nobel Prize in chemistry in 1936 for his work on polar molecules and x-ray diffraction. He made other contributions, however, particularly to our understanding of electrolyte solutions. *(Rare Book & Manuscript Collections, Carl A. Kroch Library, Cornell University)*

The force of attraction between the negative end of one polar molecule and the positive end of another (called a dipole–dipole force and discussed in Section 13.2) affects the properties of polar compounds. Intermolecular forces (forces between molecules) influence the temperatures at which a liquid freezes or boils, for example. These forces will also help determine whether a liquid dissolves certain gases or solids or whether it mixes with other liquids, and whether it adheres to glass or other solids.

To predict if a molecule is polar, we need to consider if the molecule has polar bonds and how these bonds are positioned relative to one another. Diatomic molecules composed of two atoms with different electronegativities are always po-

Table 9.10 • Dipole Moments of Selected Molecules

Molecule (AB)	Moment (μ, D)	Geometry	Molecule (AB_2)	Moment (μ, D)	Geometry
HF	1.78	linear	H_2O	1.85	bent
HCl	1.07	linear	H_2S	0.95	bent
HBr	0.79	linear	SO_2	1.62	bent
HI	0.38	linear	CO_2	0	linear
H_2	0	linear			
Molecule (AB_3)	**Moment (μ, D)**	**Geometry**	**Molecule (AB_4)**	**Moment (μ, D)**	**Geometry**
NH_3	1.47	trigonal-pyramidal	CH_4	0	tetrahedral
NF_3	0.23	trigonal-pyramidal	CH_3Cl	1.92	tetrahedral
BF_3	0	trigonal-planar	CH_2Cl_2	1.60	tetrahedral
			$CHCl_3$	1.04	tetrahedral
			CCl_4	0	tetrahedral

lar (see Table 9.10); there is one bond and the molecule has a positive and a negative end. But what happens with a molecule with three or more atoms, in which there are two or more polar bonds? Let us look at a series of molecules with stoichiometry AB_2, AB_3, and AB_4, evaluating how the choice of substituent groups (B) and molecular geometry influence the molecular polarity.

Consider first a linear triatomic molecule such as carbon dioxide, CO_2. Here each C—O bond is polar, with the oxygen atom the negative end of the bond dipole. The terminal atoms are at the same distance from the C atom, they both have the same $\delta-$ charge, and they are symmetrically arranged around the central C atom. Therefore, CO_2 has no molecular dipole, even though each bond is polar. This is analogous to a tug-of-war in which the people at opposite ends of the rope are pulling with equal force.

In contrast, water is a bent triatomic molecule. Because O has a larger electronegativity ($\chi = 3.5$) than H ($\chi = 2.1$), each of the O—H bonds is polar, with the H atoms having the same $\delta+$ charge and oxygen having a negative charge ($\delta-$) (Figure 9.17). Electron density accumulates on the O side of the molecule, making the molecule electrically "lopsided" and therefore polar ($\mu = 1.85$ D).

In trigonal-planar BF_3, the B—F bonds are highly polar because F is much more electronegative than B (χ of B = 2.0 and χ of F = 4.0). The molecule is nonpolar, however, because the three terminal F atoms have the same $\delta-$ charge, are the same distance from the boron atom, and are arranged symmetrically around the central boron atom. In contrast, the planar-trigonal molecule phosgene is polar (Cl_2CO, $\mu = 1.17$ D). Here the angles are all about 120°, so the O and Cl atoms are symmetrically arranged around the C atom. The electronegativities of the three atoms in the molecule differ, however: $\chi(O) > \chi(Cl) > \chi(C)$. There is therefore a net displacement of electron density away from the center of the molecule, mostly toward the O atom.

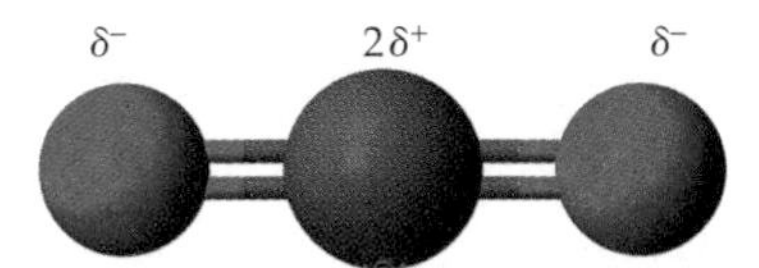

No net dipole moment

CO_2

Carbon dioxide is not a polar molecule. The displacement of electron density on one side of the molecule is equal but opposite to that on the other side.

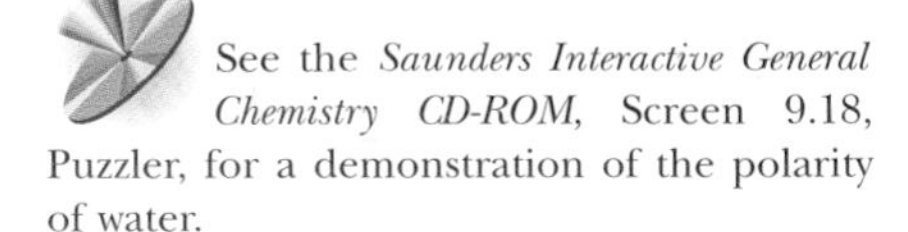

See the *Saunders Interactive General Chemistry CD-ROM*, Screen 9.18, Puzzler, for a demonstration of the polarity of water.

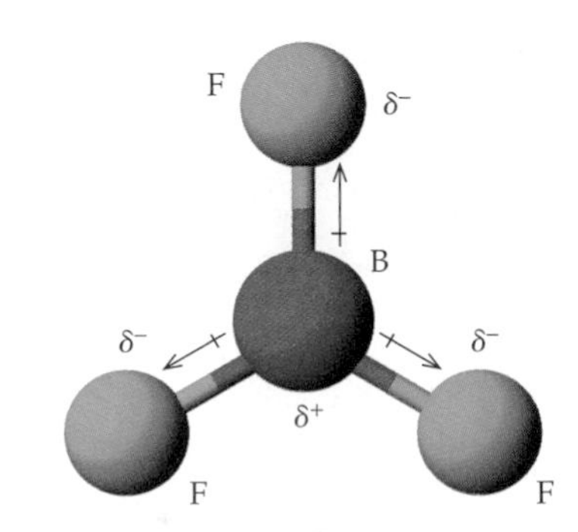

BF_3, No net dipole

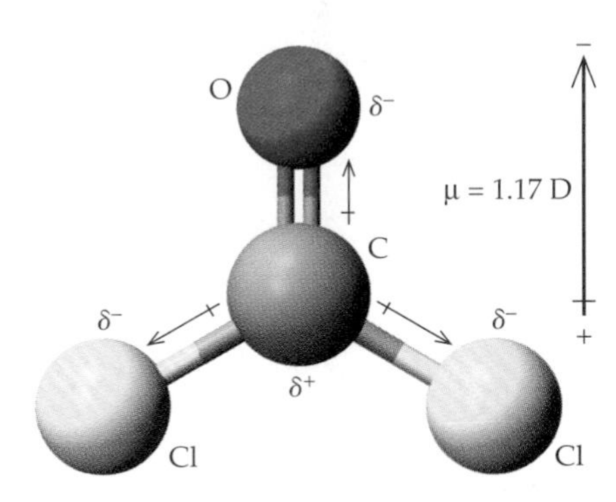

Cl_2CO, Net dipole

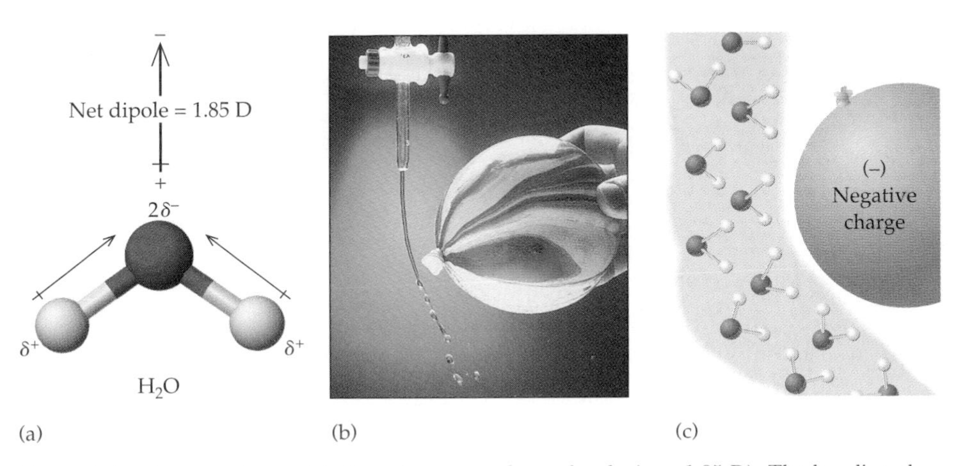

Figure 9.17 Water polarity. (a) Water is a very polar molecule ($\mu = 1.85$ D). The bonding electrons are displaced away from the H atoms and toward the O atom. You may recognize that the molecular dipole of water is the vector sum of the two bond dipoles. Vectors are quantities that have both magnitude and direction. (b) A simple experiment for showing the polarity of water: Rub a balloon on your hair or clothing to induce a static charge on the balloon's surface. Hold the balloon near a thin stream of water and observe the water stream being attracted to the balloon. (c) An illustration of the alignment of polar water molecules with an electrically charged surface. (See the *Saunders Interactive General Chemistry CD-ROM*, Screen 9.18) *(C. D. Winters)*

Ammonia, like BF_3, has AB_3 stoichiometry. In contrast to BF_3, however, NH_3 is polar because the polar N—H bonds are not arranged symmetrically. In general, all trigonal-pyramidal molecules will be polar.

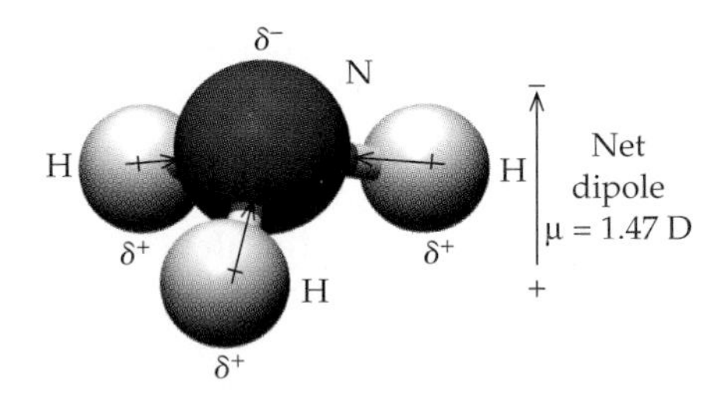

Trigonal-pyramidal ammonia, NH_3, is a polar molecule.

Molecules like carbon tetrachloride, CCl_4, and methane, CH_4, are nonpolar owing to their symmetrical, tetrahedral structures. The four atoms bonded to C have the same partial charge and are the same distance from the C atom. Tetrahedral molecules with both Cl and H atoms ($CHCl_3$, CH_2Cl_2, and CH_3Cl) are polar, however. The electronegativity for H atoms (2.1) is less than that of Cl atoms (3.0), and the carbon–hydrogen distance is different from the carbon–chlorine distances. Because of their electronegativity, the Cl atoms are on the more negative side of the molecule. This means the positive end of the molecular dipole is toward the H atom, and the negative end toward the Cl atoms.

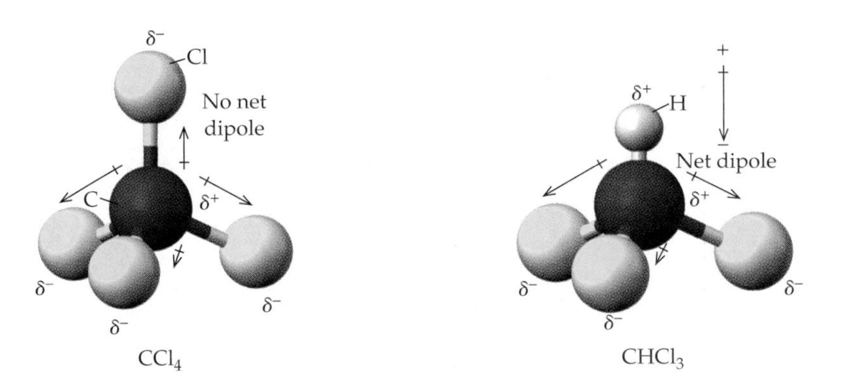

To summarize this discussion of molecular polarity, look again at Figure 9.12. These are sketches of molecules of the type CT_n where C is the central atom and T is a terminal atom. You can predict that a molecule CT_n will *not* be polar, regardless of whether the C—T bonds are polar, if

- All the terminal atoms (or groups), T, are identical, and
- All the T atoms (or groups) are arranged symmetrically around the central atom, C, in the geometries shown.

On the other hand, if one of the T atoms (or groups) is different in the structures in Figure 9.12, or if one of the T positions is occupied by a lone pair, the molecule will be polar.

CURRENT PERSPECTIVES

Cookin' with Polar Molecules

Microwave ovens are common appliances in homes, dorm rooms, and offices. They work because water is polar.

Microwaves are generated in a magnetron, a device invented during World War II for antiaircraft radar. The magnetron is a hollow cylinder with irregular walls, a rod-like cathode in the center, and a strong magnet positioned with north and south poles at opposite ends of the cylinder. An electric current flows from the cathode (which is electrically heated to help free electrons) across the air space to the cylinder wall that serves as the anode. As electrons begin to traverse this passage, the magnetic field forces them to move in circles around the cathode. The circular acceleration of the charged electrons creates electromagnetic waves. The magnetron in ovens is designed to produce microwaves with a frequency of 2.45 gigahertz (GHz). The microwaves flow through a pipe-like guide to the stirrer, which looks like a fan but acts to reflect the microwaves in many directions.

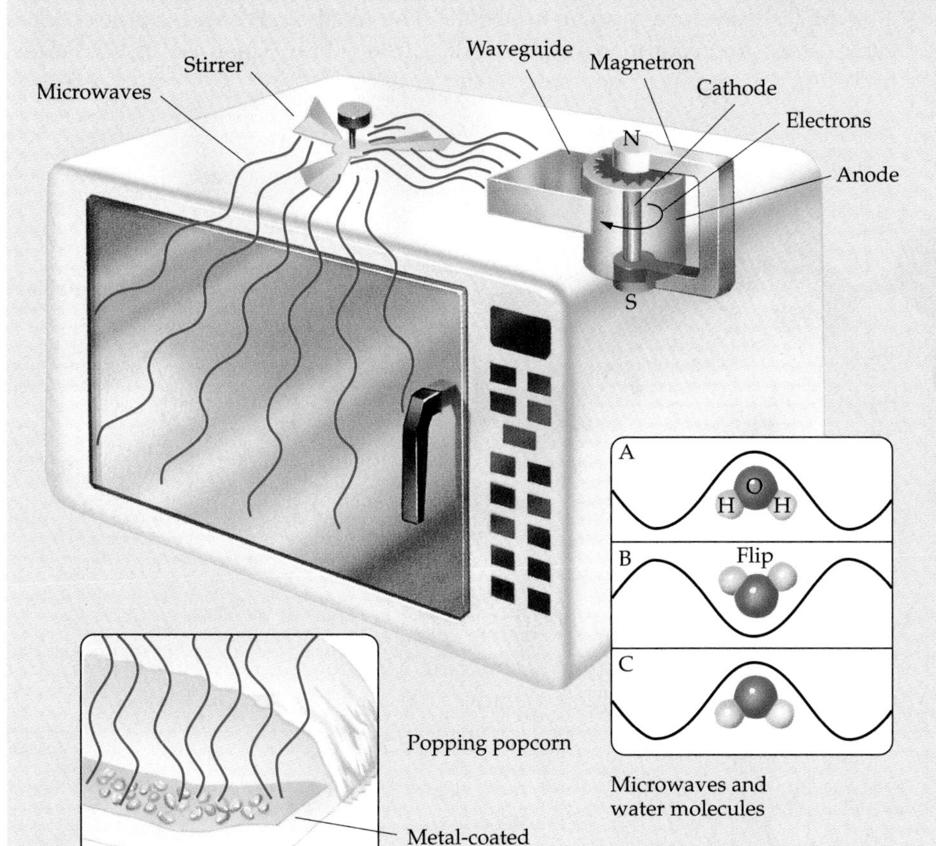

Microwaves and water molecules

Microwaves bounce off the metal walls of the oven and strike the food from many angles. They pass through glass or plastic dishes with no effect. Because electromagnetic radiation consists of oscillating electric (and magnetic) fields (Figure 7.1), however, microwaves can affect mobile, charged particles such as dissolved ions or polar molecules. As each wave crest approaches a polar molecule, the molecule turns to align with the wave and continues turning over or rotating as the trough of the wave passes. Water is the most common polar molecule in food. The microwaves have a frequency of 2.45 GHz because this is close to the optimum rate of rotation of H_2O molecules. The friction from the rotating water molecules heats the surrounding food.

Food is generally not heated above 90 °C in a microwave oven because, as water boils away, the source of heat transfer is lost. Popcorn, however, must be heated in oil above 200 °C, though oil is not heated by microwaves as effectively as water. The food technologists who developed microwave popcorn solved this problem by adding a piece of metal foil (or metal-coated plastic film) to a paper bag. Microwaves are reflected by large metal surfaces but may be absorbed by small metal objects. The microwaves induce electric currents to flow back and forth through the metal, which quickly heats the foil above 200 °C.

Popcorn kernels contain starch, protein, and water sealed with a tight hull. Surrounded by hot oil, the kernel heats up quickly, and the water within rises far above the usual boiling point of water because the sealed hull keeps the contents under pressure. The pressurized steam transforms the starch grains into hot, gelatinized globules. When the hull finally ruptures—at 175 °C and a pressure of about 9 atm—the expanding steam inflates the starch into the fluffy white form we love to eat with salt and butter.

Adapted from J. Emsley: "Microwave chemistry," *Chem Matters,* December 1993. Used with permission.

Example 9.14 Molecular Polarity

Are nitrogen trifluoride (NF_3), dichloromethane (CH_2Cl_2), and sulfur tetrafluoride (SF_4) polar or nonpolar? If polar, indicate the negative and positive sides of the molecule.

Solution

1. NF_3 has the same pyramidal structure as NH_3. Because F is more electronegative than N, each bond is polar, the more negative end being the F atom. This means that the NF_3 molecule as a whole is polar.
2. In CH_2Cl_2 the electronegativities are in the order Cl (3.0) > C(2.5) > H (2.1). This means the bonds are polar, H(δ+)—C(δ−) and C(δ+)—Cl(δ−), with a net displacement of electron density away from the H atoms and toward the Cl atoms. Although the electron-pair geometry around the C atom is tetrahedral, the polar bonds cannot be totally symmetrical in their arrangement. The molecule must be polar, with the negative end toward the two Cl atoms and the positive end toward the two H atoms.
3. Sulfur tetrafluoride, SF_4, has an electron-pair geometry of a trigonal bipyramid. Because the lone pair occupies one of the positions, the S—F bonds are not arranged symmetrically. Furthermore, the S—F bonds are highly polar, the bond dipole having F as the negative end. (χ for S is 2.5 and χ for F is 4.0.) SF_4 is therefore a polar molecule. The axial S—F bond dipoles cancel each other; they point in opposite directions. The equatorial S—F bonds, however, both point to one side of the molecule.

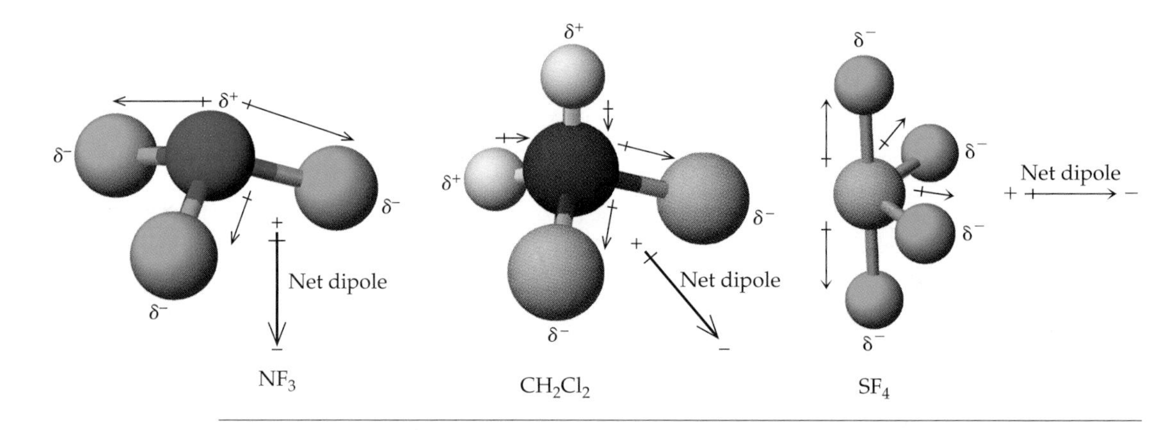

Exercise 9.18 Molecular Polarity

For each of the following molecules, decide whether the molecule is polar and which side is positive and which negative: $BFCl_2$, NH_2Cl, and SCl_2.

CHAPTER HIGHLIGHTS

When you finish this chapter you should be able to:

- Describe the basic forms of chemical bonding—**ionic** and **covalent**—and the differences between them (Section 9.1).

- Predict from the formula whether a compound is ionic or covalent, based on whether a metal is part of the formula (Section 9.1).
- Define the terms **valence electron** and **core electron,** and write **Lewis symbols** for atoms (Section 9.1).
- Describe the basic ideas of **ionic bonding** and how such bonds are affected by the sizes and charges of the ions (Section 9.3).
- Understand **lattice energy** and know how lattice energies are calculated (Born-Haber cycle); recognize trends in lattice energy and how melting points of ionic compounds are correlated with lattice energy (Section 9.3).
- Draw **Lewis structures** for covalent compounds and ions (Section 9.4).
- Understand and apply the **octet rule;** recognize exceptions to the octet rule (Section 9.4).
- Write **resonance structures,** understand what resonance means, and how and when to use this means of representing bonding (Section 9.4).
- Define and predict trends in **bond order, bond length,** and **bond dissociation energy** (Section 9.5).
- Use bond dissociation energies, D, in calculations (Section 9.5).
- Calculate **formal charges** for atoms in a molecule based on the Lewis structure (Section 9.6).
- Define **electronegativity** and understand how it is used to describe the unequal sharing of electrons between atoms in a bond (Section 9.6).
- Combine formal charge and electronegativity to gain a perspective on the **charge distribution** in covalent molecules and ions (Section 9.6).
- Predict the shape or geometry of covalent molecules and ions of main group elements using **VSEPR theory** (Section 9.7). Table 9.11 shows a summary of the relation between valence electron pairs, electron-pair and molecular geometry, and molecular polarity.
- Understand why some molecules are **polar** whereas others are nonpolar (Section 9.8). See Tables 9.10 and 9.11.
- Predict the polarity of a molecule (Section 9.8).

Table 9.11 • Summary of Molecular Shapes and Molecular Polarity

Valence Electron Pairs	Electron-Pair Geometry	Number of Bond Pairs	Number of Lone Pairs	Molecular Geometry	Molecular Dipole?*	Examples
2	linear	2	0	linear	no	$BeCl_2$
3	trigonal planar	3	0	trigonal planar	no	BF_3, BCl_3
		2	1	bent (V-shaped)	yes	$SnCl_2(g)$
4	tetrahedral	4	0	tetrahedral	no	CH_4, BF_4^-
		3	1	trigonal pyramidal	yes	NH_3, PF_3
		2	2	bent (V-shaped)	yes	H_2O, SCl_2
5	trigonal bipyramid	5	0	trigonal bipyramidal	no	PF_5
		4	1	seesaw	yes	SF_4
		3	2	T-shaped	yes	ClF_3
		2	3	linear	no	XeF_2, I_3^-
6	octahedral	6	0	octahedral	no	SF_6, PF_6^-
		5	1	square pyramidal	yes	ClF_5
		4	2	square-planar	no	XeF_4

*For molecules of the CT_n, where the T atoms are identical.

STUDY QUESTIONS

Note: *Models of many of the molecules and ions described in this chapter, and used as examples in Study Questions, are on the* Saunders Interactive General Chemistry CD-ROM *or on the World Wide Web page for this book (http://www.saunderscollege.com).*

REVIEW QUESTIONS

1. Give the number of valence electrons for Li, Sc, Zn, Si, and Cl.
2. Give Lewis symbols for K, Mg, S, and Ar.
3. Describe the formation of KF from K and F atoms using Lewis symbols. Is bonding in KF ionic or covalent?
4. Predict whether the following compounds are ionic or covalent: KI, MgS, CS_2, P_4O_{10}.
5. Define lattice energy. Which should have the more negative lattice energy, LiF or CsF? Explain.
6. Which of the following compounds is not likely to exist: $CaCl_2$ or $CaCl_4$? Explain.
7. Refer to Table 9.4 to answer the following questions:
   (a) How many lone pairs and how many bond pairs are there in the amide ion and in hydrogen fluoride?
   (b) How many single bonds and how many double bonds are in N_2H_4 and in C_2H_4?
8. In boron compounds the B atom often is not surrounded by four valence electron pairs. Illustrate this with BCl_3. Show how the molecule can achieve an octet configuration by forming a coordinate covalent bond with ammonia (NH_3).
9. Which of the following compounds or ions do *not* have an octet of electrons surrounding the central atom: SeF_4, SiF_4, BF_4^-, BrF_4^-, XeF_4?
10. Which of the following are odd-electron molecules or ions: NO_2, SF_4, NH_3, SO_3, O_2^-?
11. Why is a single Lewis structure for benzene not an accurate representation of the structure of benzene?
12. Give the bond order of each bond in acetylene, H—C≡C—H, and phosgene, Cl_2CO.
13. Consider the following resonance structures for the formate ion, HCO_2^-. What is the average C—O bond order in the ion?

$$\left[:\ddot{O}=\overset{H}{\overset{|}{C}}-\ddot{\underset{..}{O}}:\right]^- \longleftrightarrow \left[:\ddot{\underset{..}{O}}-\overset{H}{\overset{|}{C}}=\ddot{O}:\right]^-$$

14. Determine the N—O bond order in the nitrate ion, NO_3^-.
15. Consider a series of molecules in which carbon is bonded by single bonds to atoms of second-period elements: C—O, C—F, C—N, C—C, and C—B. Place these bonds in order of increasing bond length.
16. Define "bond dissociation energy." Does the enthalpy change for a bond-breaking reaction [e.g., C—H(g) ⟶ C(g) + H(g)] always have a positive sign, always have a negative sign, or may the sign vary? Explain briefly.
17. What is the relationship between bond order, bond length, and bond energy for a series of related bonds, say carbon–nitrogen bonds?
18. To estimate the enthalpy change for the reaction

$$O_2(g) + 2\ H_2(g) \longrightarrow 2\ H_2O(g)$$

what bond energies do you need? Outline the calculation, being careful to show correct algebraic signs.
19. Define and give an example of a polar covalent bond. Give an example of a nonpolar bond.
20. How are electronegativity and electron affinity different?
21. Describe the trends in electronegativity in the periodic table.
22. What is the principle of electroneutrality? Use this rule to exclude a possible resonance structure of CO_2.
23. What is the VSEPR theory? What is the physical basis of the theory?
24. What is the difference between the electron-pair geometry and the molecular geometry of a molecule? Use the water molecule as an example in your discussion.
25. What is the molecular geometry for each of the following:

$$H-\ddot{\underset{..}{X}}: \qquad H-\ddot{\underset{..}{X}}-H \qquad H-\overset{H}{\overset{|}{\underset{..}{X}}}-H \qquad H-\overset{H}{\overset{|}{\underset{\underset{H}{|}}{X}}}-H$$

Estimate the H—X—H bond angle for each.
26. If you have four electron pairs around a central atom, how can you have a pyramidal molecule? A bent molecule? What bond angles are predicted in each case?
27. Does SO_2 have a dipole moment? If so, what is the direction of the net dipole in SO_2?

NUMERICAL AND OTHER QUESTIONS

Valence Electrons

28. Give the periodic group number and number of valence electrons for each of the following atoms:
(a) N (d) Mg
(b) B (e) F
(c) Na (f) S
29. Give the periodic group number and number of valence electrons for each of the following atoms:
(a) C (d) Si
(b) Cl (e) Se
(c) Ne (f) Al

The Octet Rule

30. For each of the A groups of the periodic table, give the number of bonds an element is expected to form if it obeys the octet rule.

31. Which of the following elements are capable of forming compounds in which the indicated atom has more than four valence electron pairs? That is, which can form compounds with five or six valence shell electron pairs?
(a) C (d) F (g) Se
(b) P (e) Cl (h) Sn
(c) O (f) B

Ionic Compounds

32. Which compound has the most negative energy of ion pair formation? Which has the least negative value?
(a) NaCl (b) MgS (c) MgF_2

33. Which of the following compounds are not likely to exist: MgCl, $ScCl_3$, BaF_3, CsKr, Na_2O? Explain your choices.

34. Place the following compounds in order of increasing lattice energy (from least negative to most negative): LiI, NaF, CaO, RbI.

35. Calculate the molar enthalpy of formation, ΔH_f°, of solid lithium fluoride using the approach outlined in "A Closer Look: Using a Born-Haber Cycle to Calculate Lattice Energies." The required data can be found in Appendices F and L. In addition, ΔH_f° [Li(g)] = 159.37 kJ/mol.

36. To melt an ionic solid, energy must be supplied to disrupt the forces between ions so the regular array of ions collapses, loses order, and becomes a liquid. If the distance between the anion and cation in a crystalline solid is decreased (but ion charges remain the same), should the melting point decrease or increase? Explain.

37. Which compound in each of the following pairs should have the higher melting point? (See Study Question 36.)
(a) NaCl or RbCl
(b) BaO or MgO
(c) NaCl or MgS

Lewis Electron Dot Structures

38. Draw Lewis structures for the following molecules or ions:
(a) NF_3 (c) HOBr
(b) ClO_3^- (d) SO_3^{2-}

39. Draw Lewis structures for the following molecules or ions:
(a) CS_2 (c) NO_2^-
(b) BF_4^- (d) Cl_2SO

40. Draw Lewis structures for the following molecules:
(a) $CHClF_2$, one of the many chlorofluorocarbons (CFCs) that are no longer being used because of the environmental problems they cause.
(b) Acetic acid, CH_3CO_2H
(c) Acetonitrile, CH_3CN
(d) Methanol, CH_3OH

41. Draw Lewis structures for each of the following molecules:
(a) Tetrafluoroethylene, C_2F_4, the molecule from which Teflon is made
(b) Vinyl chloride, $H_2C{=}CHCl$, the molecule from which PVC plastics are made
(c) Acrylonitrile, $H_2C{=}CHCN$, the molecule from which materials such as Orlon are made
(d) Methyltrichlorosilane, CH_3SiCl_3, a compound used in manufacturing "silicone" polymers

42. Show all possible resonance structures for each of the following molecules or ions:
(a) SO_2
(b) NO_2^-
(c) SCN^-

43. Show all possible resonance structures for each of the following molecules or ions:
(a) Nitrate ion, NO_3^-
(b) Nitric acid, HNO_3
(c) Nitrous oxide (laughing gas), N_2O

44. Draw Lewis structures for each of the following molecules or ions:
(a) BrF_3 (c) XeO_2F_2
(b) I_3^- (d) XeF_3^+

45. Draw Lewis structures for each of the following molecules or ions:
(a) BrF_5 (c) IBr_2^-
(b) IF_3 (d) BrF_2^+

Bond Properties

46. Give the number of bonds for each of the following molecules or ions. Give the bond order for each bond.
(a) H_2CO (c) NO_2^+
(b) SO_3^{2-} (d) NOCl

47. Give the number of bonds for each of the following molecules or ions. Give the bond order for each bond.
(a) CN^- (c) SO_3
(b) CH_3CN (d) $CH_3CH{=}CH_2$

48. In each pair of bonds, predict which is shorter.
(a) B—Cl or Ga—Cl (c) P—S or P—O
(b) Sn—O or C—O (d) C=O or C=N

49. In each pair of bonds, predict which is shorter.
(a) Si—N or Si—O
(b) Si—O or C—O
(c) C—F or C—Br
(d) The C—N bond or the C≡N bond in $H_2NCH_2C{\equiv}N$.

50. Consider the carbon–oxygen bond in formaldehyde (CH_2O) and carbon monoxide (CO). In which molecule is the CO bond shorter? In which molecule is the CO bond stronger?

51. Compare the nitrogen–nitrogen bond in hydrazine, H_2NNH_2, with that in "laughing gas," N_2O. In which molecule is the nitrogen–nitrogen bond shorter? In which is the bond stronger?

52. Consider the nitrogen–oxygen bond lengths in NO_2^+, NO_2^-, and NO_3^-. In which ion is the bond predicted to be longest? Which is predicted to be the shortest? Explain briefly.

53. Compare the carbon–oxygen bond lengths in the formate ion, HCO_2^-, in methanol, CH_3OH, and in the carbonate ion, CO_3^{2-}. In which species is the carbon–oxygen bond predicted to be longest? In which is it predicted to be shortest? Explain briefly.

Bond Energies and Reaction Enthalpies

54. Hydrogenation reactions, the addition of H_2 to a molecule, are widely used in industry to transform one compound into another. For example, the molecule called 1-butene (C_4H_8, a member of a general class of compounds called "alkenes" because of the C=C double bond) is converted to butane (C_4H_{10}, called an "alkane" because there are only C—H bonds and C—C single bonds) by addition of H_2.

```
   H  H  H  H
   |  |  |  |
H—C—C—C=C—H(g) + H2(g) ⟶
   |  |
   H  H

                         H  H  H  H
                         |  |  |  |
                      H—C—C—C—C—H (g)
                         |  |  |  |
                         H  H  H  H
```

Use the bond energies of Table 9.9 to estimate the enthalpy change for this hydrogenation reaction.

55. In principle, dinitrogen monoxide, N_2O, can decompose to nitrogen and oxygen gas:

$$2\ N_2O(g) \longrightarrow 2\ N_2(g) + O_2(g)$$

Use bond energies to estimate the enthalpy change for this reaction.

56. Phosgene, Cl_2CO, is a highly toxic gas that was used as a weapon in World War I. Using the bond energies of Table 9.9, estimate the enthalpy change for the reaction of carbon monoxide and chlorine to produce phosgene.

$$CO(g) + Cl_2(g) \longrightarrow Cl_2CO(g)$$

57. The equation for the combustion of gaseous methanol is:

$$2\ CH_3OH(g) + 3\ O_2(g) \longrightarrow 2\ CO_2(g) + 4\ H_2O(g)$$

Using the bond energies in Table 9.9, estimate the enthalpy change for this reaction. What is the heat of combustion of one mole of gaseous methanol?

58. The compound oxygen difluoride is quite unstable, giving oxygen and HF on reaction with water:

$$OF_2(g) + H_2O(g) \longrightarrow O_2(g) + 2\ HF(g) \qquad \Delta H^\circ_{rxn} = -318\ kJ$$

Using bond energies, calculate the bond dissociation energy of the O—F bond in OF_2.

59. O atoms can combine with ozone to form oxygen:

$$O_3(g) + O(g) \longrightarrow 2\ O_2(g) \qquad \Delta H^\circ_{rxn} = -394\ kJ$$

Using ΔH°_{rxn} and the bond energy data in Table 9.9, estimate the bond energy for the oxygen–oxygen bond in ozone, O_3. How does your estimate compare with the energies of an O—O single bond and an O=O double bond? Does the oxygen–oxygen bond energy in ozone correlate with its bond order?

Formal Charges

60. Determine the formal charge on each atom in each of the following molecules or ions:
(a) N_2H_4 (c) BH_4^-
(b) PO_4^{3-} (d) NH_2OH

61. Determine the formal charge on each atom in each of the following molecules or ions:
(a) SCO (c) O_3
(b) HCO_2^- (formate ion) (d) HCO_2H (formic acid)

62. Determine the formal charge on each atom in the following molecules and ions:
(a) NO_2^+ (c) NF_3
(b) NO_2^- (d) HNO_3

63. Determine the formal charge on each atom in the following molecules and ions:
(a) SO_2 (c) SO_2Cl_2
(b) $SOCl_2$ (d) FSO_3^-

Electronegativity and Bond Polarity

64. In each pair of bonds, indicate the more polar bond and use an arrow to show the direction of polarity in each bond.
(a) C—O and C—N (c) B—O and B—S
(b) P—Br and P—Cl (d) B—F and B—I

65. For each of the bonds listed below, tell which atom is the more negatively charged
(a) C—N (c) C—Br
(b) C—H (d) S—O

66. Acrolein, C_3H_4O, is the starting material for certain plastics.

```
   H  H  H
   |  |  |  ..
H—C=C—C=O
            ..
```

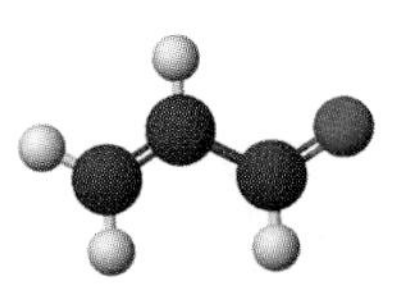

(a) Which bonds in the molecule are polar and which are nonpolar?
(b) Which is the most polar bond in the molecule? Which is the more negative atom of this bond?

67. Urea, $(NH_2)_2CO$, is used in plastics and fertilizers. It is also the primary nitrogen-containing substance excreted by humans.

```
        :O:
  ..    ||    ..
H—N—C—N—H
   |          |
   H          H
```

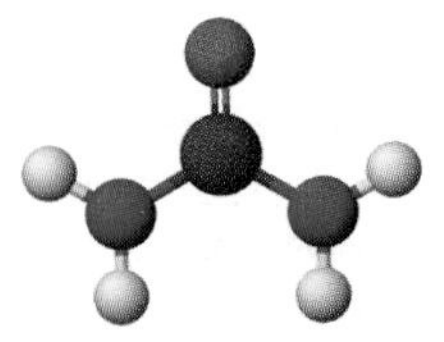

(a) Which bonds in the molecule are polar and which are nonpolar?
(b) Which is the most polar bond in the molecule? Which atom is the negative end of the bond dipole?

Charge Distribution in Molecules

68. Considering both formal charges and bond polarities, predict on which atom or atoms the negative charge resides in the following anions:
(a) BF_4^- (c) OH^-
(b) BH_4^- (d) $CH_3CO_2^-$

69. Considering both formal charge and bond polarities, predict on which atom or atoms the positive charge resides in the following cations:
(a) H_3O^+ (c) NO_2^+
(b) NH_4^+ (d) NF_4^+

70. Three resonance structures are possible for dinitrogen oxide, N_2O.
(a) Draw the three resonance structures.
(b) Calculate the formal charge on each atom in each resonance structure.
(c) Based on formal charges and electronegativity, predict which resonance structure is the most reasonable.

71. Nitric acid, HNO_3, has three resonance structures. One of them, however, contributes much less to the resonance hybrid than the other two. Sketch the three resonance structures and assign a formal charge to each atom. Which one of your structures is the least important?

72. Two resonance structures are possible for NO_2^-. Draw these structures and then find the formal charge on each atom in each resonance structure.

73. Compare the electron dot structures of the carbonate (CO_3^{2-}) and borate (BO_3^{3-}) ions.
(a) Are these ions isoelectronic?
(b) What are the formal atom charges in each ion?
(c) How many resonance structures does each ion have?

74. Draw an electron dot structure for the cyanide ion, CN^-. In aqueous solution this ion interacts with H^+ to form the acid. Should the acid formula be written as HCN or CNH?

75. Draw the electron dot structure for the sulfite ion, SO_3^{2-}. In aqueous solution the ion interacts with H^+. Does H^+ attach itself to the S atom or the O atom of SO_3^{2-}?

Molecular Geometry

76. Draw the Lewis structure for each of the following molecules or ions. Describe the electron-pair geometry and the molecular geometry.
(a) NH_2Cl
(b) Cl_2O (O is the central atom)
(c) SCN^-
(d) HOF

77. Draw the Lewis structure for each of the following molecules or ions. Describe the electron-pair geometry and the molecular geometry.
(a) ClF_2^+ (c) PO_4^{3-}
(b) $SnCl_3^-$ (d) CS_2

78. The following molecules or ions all have two oxygen atoms attached to a central atom. Draw the Lewis structure for each one and then describe the electron-pair geometry and the molecular geometry. Comment on similarities and differences in the series.
(a) CO_2 (b) NO_2^- (c) O_3 (d) ClO_2^-

79. The following molecules or ions all have three oxygen atoms attached to a central atom. Draw the Lewis structure for each one and then describe the electron-pair geometry and the molecular geometry. Comment on similarities and differences in the series.
(a) NO_3^- (b) CO_3^{2-} (c) SO_3^{2-} (d) ClO_3^-

80. Draw the Lewis structure of each of the following molecules or ions. Describe the electron-pair geometry and the molecular geometry.
(a) ClF_2^- (b) ClF_3 (c) ClF_4^- (d) ClF_5

81. Draw the Lewis structure of each of the following molecules or ions. Describe the electron-pair geometry and the molecular geometry.
(a) SiF_6^{2-} (b) PF_5 (c) SF_4 (d) XeF_4

82. Give approximate values for the indicated bond angles.
(a) O—S—O in SO_2
(b) F—B—F angle in BF_3
(c) H—C—H (angle 1) and C—C≡N (angle 2) in acetonitrile

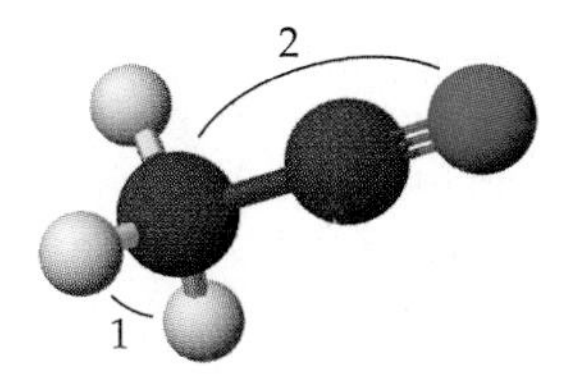

(d) Cl—C—Cl in Cl_2CO

83. Give approximate values for the indicated bond angles.
(a) Cl—S—Cl in SCl_2
(b) N—N—O in N_2O
(c) Bond angles in acetamide

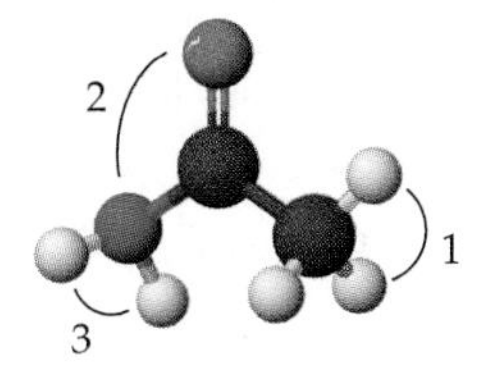

84. Phenylalanine is one of the natural amino acids and is a "breakdown" product of aspartame (p. 372 and 423). Estimate the values of the indicated angles in the amino acid. Explain why the —CH_2—CH(NH_2)—CO_2H chain is not linear.

85. Acetylacetone has the structure shown here. Estimate the values of the indicated angles.

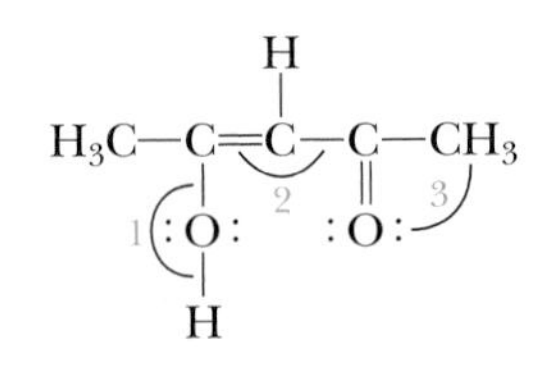

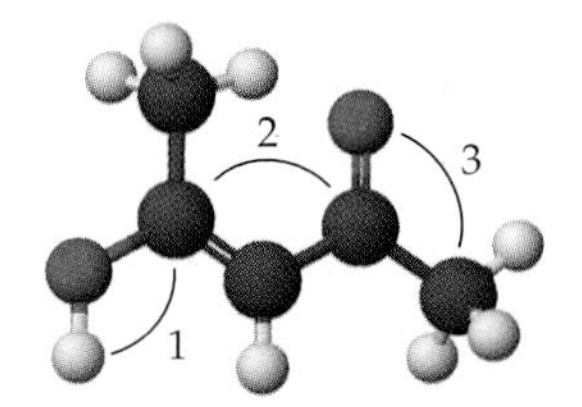

86. Give approximate values for the indicated bond angles.
(a) F—Se—F angles in SeF_4
(b) The O—S—F and the F—S—F bond angles (two are possible for each) in OSF_4 (the O atom is in an equatorial position)
(c) F—Br—F angles in BrF_5
(d) F—P—F angles in PF_6^-

87. Give approximate values for the indicated bond angles.
(a) F—S—F angles in SF_6
(b) F—Xe—F angle in XeF_2
(c) F—Cl—F angle in ClF_2^-
(d) I—I—I angle in I_3^-

88. Which has the greater O—N—O bond angle, NO_2^- or NO_2^+? Explain your answer briefly.

89. Compare the F—Cl—F angles in ClF_2^+ and ClF_2^-. Using Lewis structures determine the approximate bond angle in each ion. Explain which ion has the greater bond angle and why.

Molecular Polarity

90. Consider the following molecules:
(a) H_2O (c) CO_2 (e) CCl_4
(b) NH_3 (d) ClF
(i) In which compound are the bonds most polar?
(ii) Which compounds in the list are not polar?
(iii) Which atom in ClF is more negatively charged?

91. Consider the following molecules:
(a) CH_4 (c) BF_3
(b) NCl_3 (d) CS_2
(i) Which compound has bonds with the greatest degree of polarity?
(ii) Which compounds in the list are not polar?

92. Which of the following molecules is (are) polar? For each polar molecule indicate the direction of polarity, that is, which is the negative and which is the positive end of the molecule.
(a) $BeCl_2$ (c) CH_3Cl
(b) HBF_2 (d) SO_3

93. Which of the following molecules is (are) not polar? In which molecule are there bonds with the greatest degree of polarity?
(a) CO
(b) BCl_3
(c) CF_4
(d) PCl_3
(e) GeH_4

GENERAL QUESTIONS

94. Draw Lewis structures (and resonance structures where appropriate) for the following molecules and ions. What similarities and differences are there in this series?
(a) CO_2
(b) N_3^-
(c) OCN^-

95. What are the orders of the N—O bonds in NO_2^- and NO_2^+? The nitrogen-oxygen bond length in one of these ions is 110 pm and in the other 124 pm. Which bond length corresponds to which ion? Explain briefly.

96. Acrylonitrile, C_3H_3N, is the building block of the synthetic fiber Orlon.

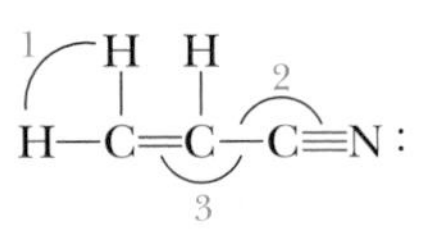

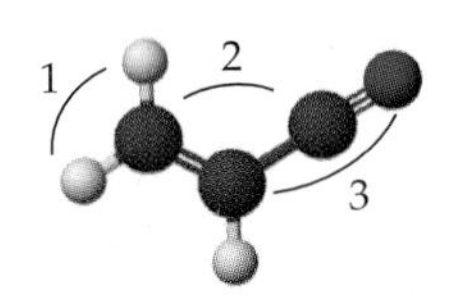

(a) Give the approximate values of angles 1, 2, and 3.
(b) Which is the shorter carbon–carbon bond?
(c) Which is the stronger carbon–carbon bond?
(d) Which is the most polar bond?

97. Vanillin is the flavoring agent in vanilla extract and in vanilla ice cream. Its structure is shown here:

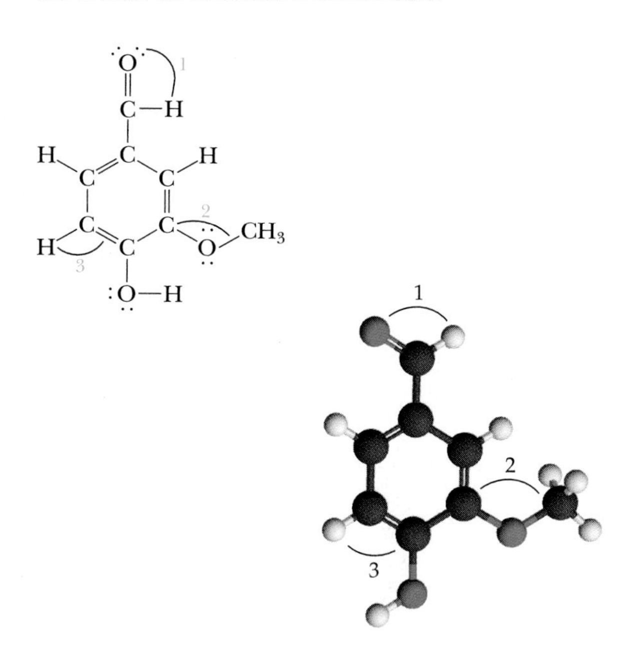

(a) Give values for the three bond angles indicated.
(b) Indicate the shortest carbon–oxygen bond in the molecule.
(c) Indicate the most polar bond in the molecule.

98. The following molecules or ions have fluorine atoms attached to a central atom from Groups 3A through 7A. Draw the Lewis structure for each one and then describe the electron-pair geometry and the molecular geometry. Comment on similarities and differences in the series.
(a) BF_3
(b) CF_4
(c) PF_3
(d) OF_2
(e) HF

99. The formula for nitryl chloride is $ClNO_2$. Draw the Lewis structure for the molecule, including all resonance structures. Describe the electron-pair and molecular geometries, and give values for all bond angles.

100. Given that the spatial requirement of a lone pair is much greater than that of a bond pair, explain why
(a) XeF_2 has a linear molecular structure and not a bent one.
(b) ClF_3 has a T-shaped structure and not a trigonal planar one.

101. In 1962 Watson and Crick received the Nobel Prize for physiology or medicine for their simple but elegant model for the "heredity molecule" DNA (see page 6). The key to their structure (the "double helix") was an understanding of the geometry and bonding capabilities of nitrogen-containing bases such as the guanine molecule shown here.

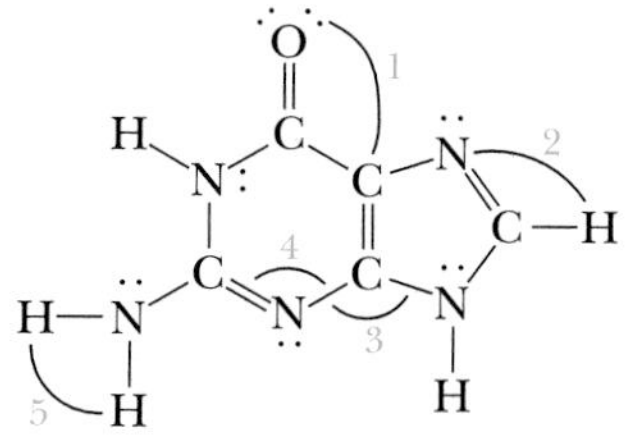

(a) Give approximate values for the indicated bond angles.
(b) Which are the most polar bonds in the molecule?

102. Use the bond energies in Table 9.9 to calculate the enthalpy change for the decomposition of urea (Study Question 67) to hydrazine, $H_2N—NH_2$, and carbon monoxide. (Assume all compounds are in the gas phase.)

103. A model of the formate ion, CHO_2^-, is illustrated here in two ways: a ball-and-stick model and a model in which atoms are colored and sized according to their relative charges. (See A Closer Look: Atom Partial Charges.) Using arguments based on resonance structures, formal charges, and electronegativities, explain why the model showing calculated partial charges is reasonable.

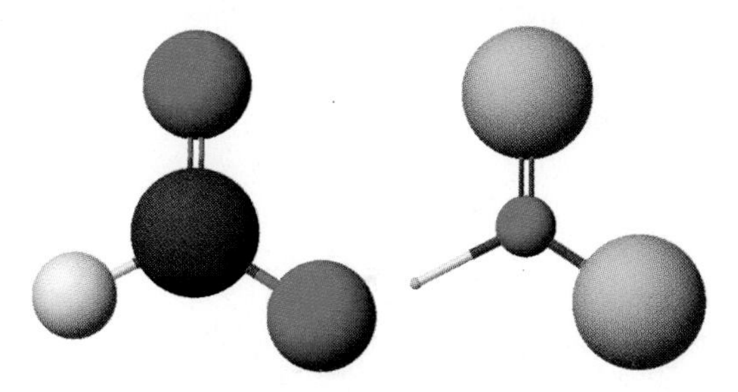

CONCEPTUAL QUESTIONS

104. The cyanate ion, NCO^-, has the least electronegative atom, C, in the center. The very unstable fulminate ion, CNO^-, has the same formula, but the N atom is in the center.
(a) Draw the three possible resonance structures of CNO^-.
(b) On the basis of formal charges, decide on the resonance structure with the most reasonable distribution of charge.
(c) Mercury fulminate is so unstable it is used in blasting caps. Can you offer an explanation for this instability? (*Hint:* Are the formal charges in any resonance structure reasonable in view of the relative electronegativities of the atoms?)

105. The effect of bond order is evident when bonds between the same two atoms are compared. For example, the bonds become shorter as the bond order increases in the series C—O, C=O, and C≡O.

Bond	C—O	C=O	C≡O
Bond order	1	2	3
Bond length (pm)	143	122	113

Adding a second bond to the single bond in C—O shortens the bond by only 21 pm on going to C═O. The third bond results in a further 9-pm reduction in bond length from C═O to C≡O. Suggest why the bond shortening is much less than expected based on the fact that the bond order has doubled and then tripled.

106. Amides are an important class of organic molecules. They are usually drawn as sketched here, but another resonance structure is possible.

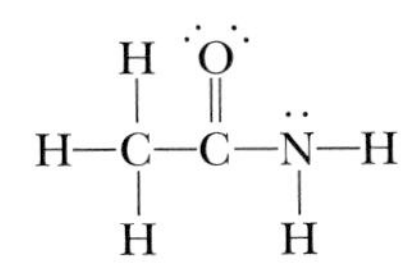

Draw that structure, and then suggest why it is usually not pictured.

107. Examine the trends in lattice energy in Table 9.3. The value of the lattice energy becomes somewhat more negative on going from NaI to NaBr to NaCl. Why is the lattice energy of NaF so much more negative?

CHALLENGING QUESTIONS

108. The molecule shown here, 2-furylmethanethiol, is responsible for the odor of coffee:

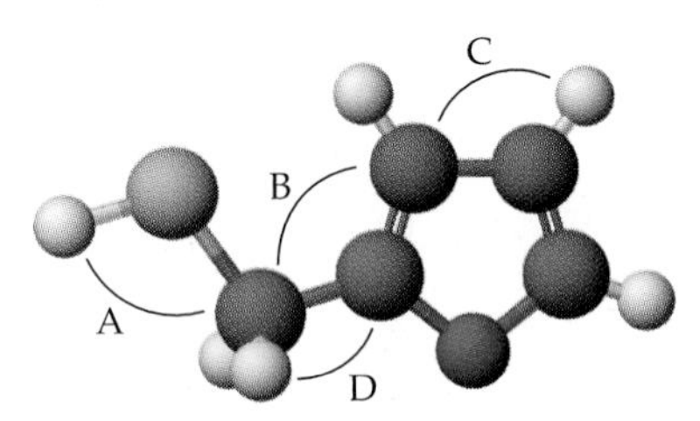

(a) What are the formal charges on the S and O atoms?
(b) Give approximate values of angles A, B, C, and D.
(c) Which are the shorter carbon–carbon bonds in the molecule?
(d) Which bond in this molecule is the most polar?
(e) Which is the most electronegative atom in the molecule?
(f) The molecular model makes it clear that the four C atoms of the ring are all in a plane. Is the O atom in that same plane (making the five-member ring planar), or is the O atom bent above or below the plane?
(g) Is the molecule as a whole polar or nonpolar?

109. Dihydroxyacetone is the basis of quick-tanning lotions. (It reacts with the amino acids in the upper layer of skin and colors them brown in a reaction similar to that occurring when food is browned.)

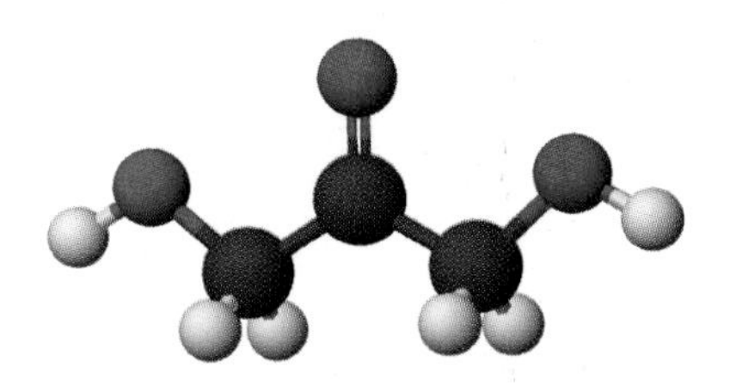

(a) Supposing you can make this compound by treating acetone with oxygen, estimate the enthalpy change for the following reaction (which is assumed to occur in the gas phase) using bond energies. Is the reaction exothermic or endothermic?

```
   H  O  H                         H  O  H
   |  ‖  |                         |  ‖  |
H—C—C—C—H + O2  ⟶  H—O—C—C—C—O—H
   |     |                         |     |
   H     H                         H     H
   acetone                         dihydroxyacetone
```

(b) Is acetone polar?
(c) Positive H atoms can sometimes be removed (as H^+) from molecules with strong bases (which is in part what happens in the tanning reaction). Which H atoms are the most positive in dihydroxyacetone?

Summary Question

110. Chlorine trifluoride, ClF_3, is one of the most reactive compounds known. It reacts violently with many substances generally thought to be inert and was used in incendiary bombs in World War II. It can be made by heating Cl_2 and F_2 in a closed container.
(a) Write a balanced equation to depict the reaction of Cl_2 and F_2 to give ClF_3.
(b) If you mix 0.71 g of Cl_2 with 1.00 g of F_2, what is the theoretical yield of ClF_3?
(c) Draw the electron dot structure of ClF_3.
(d) What is the electron-pair geometry and molecular geometry for ClF_3?

INTERACTIVE GENERAL CHEMISTRY CD-ROM

Screen 9.1 Chemical Puzzler

Why does the balloon cause the stream of water to bend as it passes the balloon?

Screen 9.3 Chemical Bond Formation

(a) What are the major coulombic interactions between two atoms?
(b) What are the three major types of chemical bonds?
(c) What is the difference between an ionic bond and a covalent bond?
(d) What is the difference in carbon–carbon bonding in ethane, ethene (ethylene), and acetylene?

Screen 9.7 Electron-Deficient Compounds

(a) When BF_3 and NH_3 form a coordinate covalent bond, do either of the reactants change their molecular geometry? What is the F—B—F angle in BF_3 and in $F_3B—NH_3$? (See the file "BF3NH3.CSF" in the *Molecular Models* folder.)
(b) Is a reaction between BF_3 and H_2O possible? Explain briefly. (*Hint:* In what way is water similar or dissimilar to ammonia in terms of their Lewis dot structures?)

Screen 9.8 Free Radicals

(a) What experimental observation suggests that NO_2 molecules are combining to form N_2O_4?
(b) Describe one medical use of NO.

Screen 9.9 Bond Properties

(a) Examine the animation of bond breaking on Screen 9.9. As two bonded atoms move apart, and the bond eventually breaks, what happens to the energy of the system?
(b) In the animation of bond energy, the energy increases as the bond distance becomes less than 74 pm. Why does the energy increase?
(c) Locate the molecules in the table shown here in the *Molecular Models* folder and measure the carbon–carbon bond length. Complete the table. (See Appendix A of the *Workbook* that accompanies the CD-ROM to learn how to measure bond distances. Note that they are given in angstrom units, where 1 Å = 100 pm.)

Formula	**Measured Bond Distance (Å)**	**Bond Order**
ethane, C_2H_6		
butane, C_4H_{10}		
ethylene, C_2H_4		
acetylene, C_2H_2		
benzene, C_6H_6		

Screen 9.10 Bond Energy

When H_2 and Cl_2 react to form HCl, all in the gas phase, what bonds are broken and what bonds are made? Is the reaction exothermic or endothermic?

Screen 9.16 Molecular Shape

Locate the following molecules in the *Molecular Models* folder. In each case, measure unique bond angles and bond lengths and use them to label a sketch of the molecule.
(a) Tylenol (*Drugs* folder)
(b) ClF_3 (*Inorganic* folder)
(c) ethylene glycol (*Organic Alcohols* folder)

Screen 9.17 Molecular Polarity

Use the *Molecular Polarity* tool on this screen to explore the polarity of molecules.
(a) Is BF_3 a polar molecule? Describe what happens to the molecular polarity as F is replaced by H on BF_3. Does the polarity change as F is replaced by H? What happens when two F atoms are replaced by H?
(b) Is $BeCl_2$ a polar molecule? Describe what happens to the molecular polarity when Cl is replaced by Br.

Chapter 10

Bonding and Molecular Structure: Orbital Hybridization and Molecular Orbitals

(C. D. Winters)

A Chemical Puzzler

About 20% of earth's atmosphere is made up of dioxygen, O_2. As a gas O_2 is colorless, but if cooled to −183 °C, the gas condenses to a pale blue liquid (at one atmosphere pressure). A blue color is unusual, but what makes O_2 unique is its paramagnetism. That is, when poured between the poles of a magnet liquid oxygen remains suspended there. Paramagnetism indicates the presence of unpaired electrons, so we can conclude that O_2 has one or more unpaired electrons.

Dioxygen is one of a very few molecules for which it is not possible to draw an acceptable Lewis structure. Lewis structures for this molecule show either a double bond between oxygen atoms (as required based on bond lengths) or the presence of two unpaired electrons (as required by the paramagnetism), but not both. The problem is to find a new approach to chemical bonding that more accurately describes the bonding in this molecule but that also applies to many thousands of other molecules as well.

Just how are molecules held together? How can there be two distinctly different molecules with the same formula, such as C_4H_8? How is a change in a molecule's structure connected with your ability to see? Why is oxygen paramagnetic and how is this connected with bonding in the molecule? These are just a few of some fundamental and interesting questions that require us to take a more advanced look at bonding.

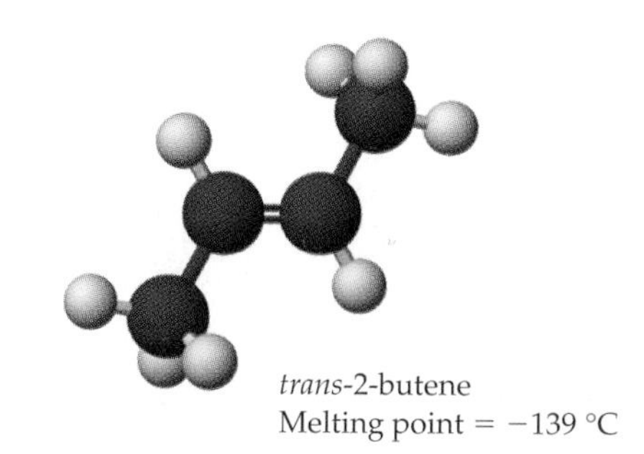

trans-2-butene
Melting point = −139 °C

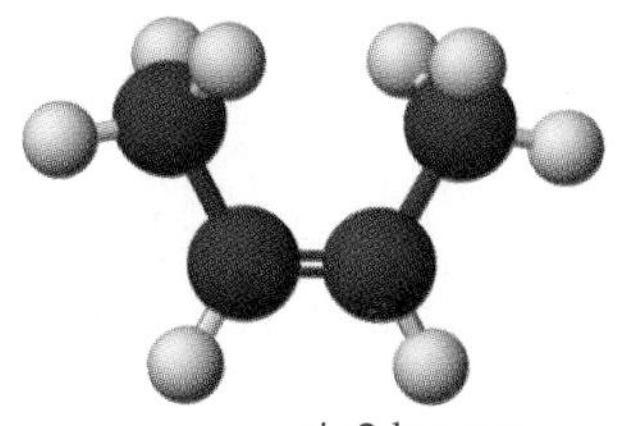

cis-2-butene
Melting point = −106 °C

Two different molecules with the same formula, C_4H_8. How can bonding theory help us understand why such molecules can exist?

10.1 ORBITALS AND BONDING THEORIES

Orbitals, both atomic and molecular, are the focus of this chapter. The quantum mechanical model for the atom, the most successful way to explain the properties of atoms that chemists have yet devised, describes electrons in atoms as waves. An atomic orbital has a specific energy related to electrostatic forces: the attractive force on an electron in that orbital due to the positively charged atomic nucleus, and the repulsive forces on the electron due to the other electrons in the atom. If the energy of the orbital is known accurately, however, its position is known less well (the Heisenberg uncertainty principle), so we think of orbitals as regions in space in which there is a high probability of finding the electron (Figures 7.14 and 7.15).

From Chapter 9 you know that electrons are responsible for the bonds in molecules. It seems reasonable, therefore, that any model describing electrons in molecules would adopt the same premise that is successful for describing electrons in atoms, namely, that electrons in molecules reside in molecular orbitals. Such an approach requires describing electrons in molecules as waves, their energies defined accurately but their location defined less precisely.

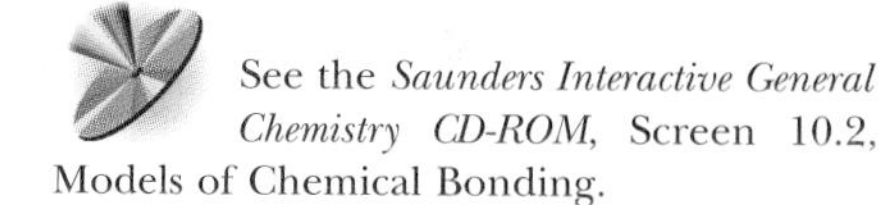

See the *Saunders Interactive General Chemistry CD-ROM,* Screen 10.2, Models of Chemical Bonding.

There are two common approaches to rationalizing chemical bonding based on orbitals: the **valence bond (VB) theory** and the **molecular orbital (MO) theory.** The former was developed largely by Linus Pauling (page 411) and the latter by another American chemist, Robert S. Mulliken (1896–1986). The valence bond approach is closely tied to Lewis's idea of bonding electron pairs between atoms and lone pairs of electrons localized on a particular atom. In contrast, Mulliken's approach was to derive molecular orbitals that are "spread out" or *delocalized,* over the molecule. The atomic orbitals of the atoms in the molecule combine to form a set of orbitals that are the property of the molecule, and the electrons of the molecule are distributed within these orbitals.

Robert Mulliken was awarded the 1966 Nobel Prize in chemistry for the development of molecular orbital theory. Linus Pauling (see page 411) received the 1954 Prize for his work on the nature of the chemical bond.

Why are two theories used? Isn't one more correct than the other? Actually, both give good descriptions of the bonding in molecules and polyatomic ions, but they are used for different purposes. Valence bond theory is generally the method of choice to provide a qualitative, visual picture of molecular structure and bonding. This theory is particularly useful for molecules made up of many atoms. In contrast, molecular orbital theory is used when a more quantitative picture of bonding is needed. Furthermore, valence bond theory provides a good description of the bonding of molecules in their ground, or lowest, energy state. On the other hand, MO theory is essential if we want to describe molecules in higher energy excited states. Among other things, this is important in explaining the colors of compounds. Finally, for a few molecules such as NO and O_2, MO theory is the only way to describe their bonding accurately.

10.2 VALENCE BOND THEORY

The Orbital Overlap Model of Bonding

See the *Saunders Interactive General Chemistry CD-ROM,* Screen 10.3, Valence Bond Theory.

What happens if two atoms at an infinite distance apart are brought together to form a bond? This process is often illustrated with H_2 because, with two electrons and two nuclei, this is the simplest covalent compound (Figure 10.1). Initially, when two hydrogen atoms are widely separated, they do not interact. If the atoms move closer together, however, the electron on one atom begins to experience an attraction to the positive charge of the nucleus of the other atom (Figure 10.2). Because of the attractive forces, the electron clouds on the atoms distort as the electron of one atom is drawn toward the nucleus of the second atom. As a result of these forces of attraction, the potential energy of the system is lowered. Calculations show that when the distance between the H atoms is 74 pm, the potential energy reaches a minimum and the H_2 molecule is most stable. Significantly, 74 pm corresponds to the actual bond distance in the H_2 molecule, a value that can be measured experimentally.

Individual hydrogen atoms each have a single electron. In H_2 the two electrons pair up to form the bond. There is a net stabilization, representing the ex-

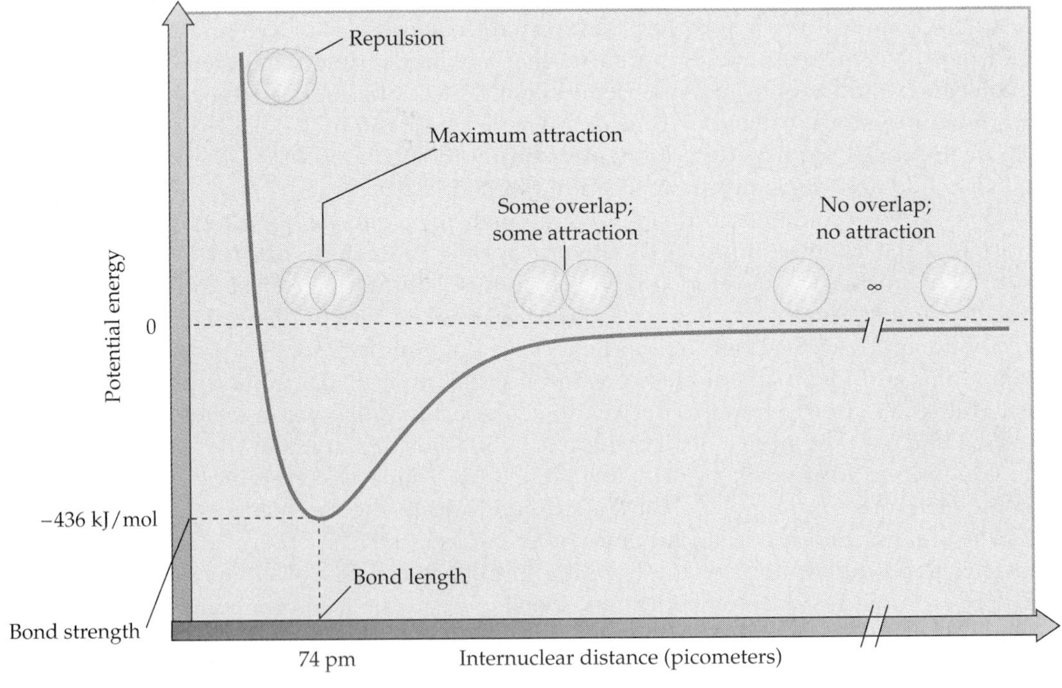

Figure 10.1 Potential energy change during H—H bond formation from isolated hydrogen atoms. The lowest energy is reached at 74 pm, where there is overlap of 1*s* orbitals. At greater distances the overlap is less, and the bond is weaker. At H—H distances less than 74 pm, repulsions between the nuclei and the electrons of the two atoms increase rapidly, and the potential energy curve rises steeply. Thus, an H_2 molecule is expected to be less stable when the distance between the atoms is very small.

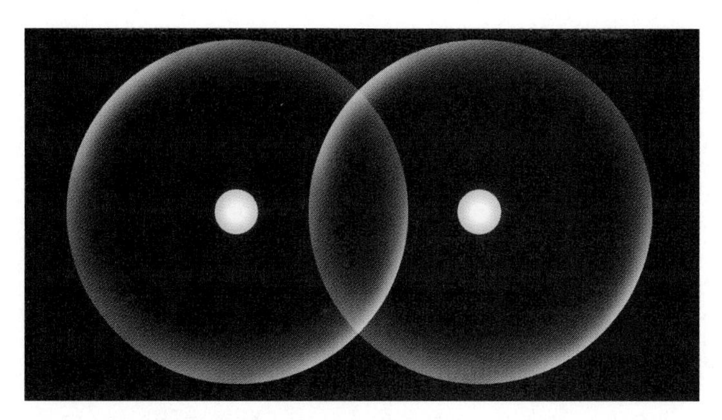

Figure 10.2 The formation of a covalent bond between two H atoms. One pair of electrons (one electron from each atom) moves into the internuclear region and is attracted to both H atom nuclei. It is this mutual attraction for two (or sometimes four or six) electrons by two nuclei that leads to covalent bond formation.

tent to which the energies of the two electrons are lowered from their value in the free atoms. The net stabilization (the extent by which the potential energy was lowered) can be calculated, and the calculated value approximates the experimentally determined value (called the *bond energy*). Agreement between theory and experiment on both bond distance and energy is evidence that this theoretical approach has merit.

Bond formation is depicted in Figures 10.1 and 10.2 as occurring when the electron clouds on the two atoms interpenetrate, or **overlap.** This overlap increases the probability of finding the electrons in the region of space between the two nuclei. *The idea that bonds are formed by overlap of orbitals is the basis for valence bond theory.*

When the single covalent bond is formed in H_2, the electron cloud of each atom is distorted in a way that gives the electrons a higher probability of being found in the region of space between the two hydrogen atoms. This makes sense because this distortion results in the electrons being situated so they can be attracted equally to the two positively charged nuclei. Placing the electrons between the nuclei also matches the Lewis electron dot model.

The covalent bond that arises from the overlap of two *s* orbitals, one from each of two atoms as in H_2, is called a **sigma (σ) bond.** The electron density of a sigma bond is greatest along the axis of the bond.

The main points of the valence bond approach to bonding are

1. Orbitals overlap to form a bond between two atoms (see Figures 10.1 and 10.2).
2. Two electrons, of opposite spin, can be accommodated in the overlapping orbitals. Usually one electron is supplied by each of the two bonded atoms.
3. Because of orbital overlap, the bonding electrons have a higher probability of being found within a region of space influenced by both nuclei. Both electrons are simultaneously attracted to both nuclei (see Figure 10.2). This is the rationalization for Lewis structures in which the electron pair is placed between the two atoms.

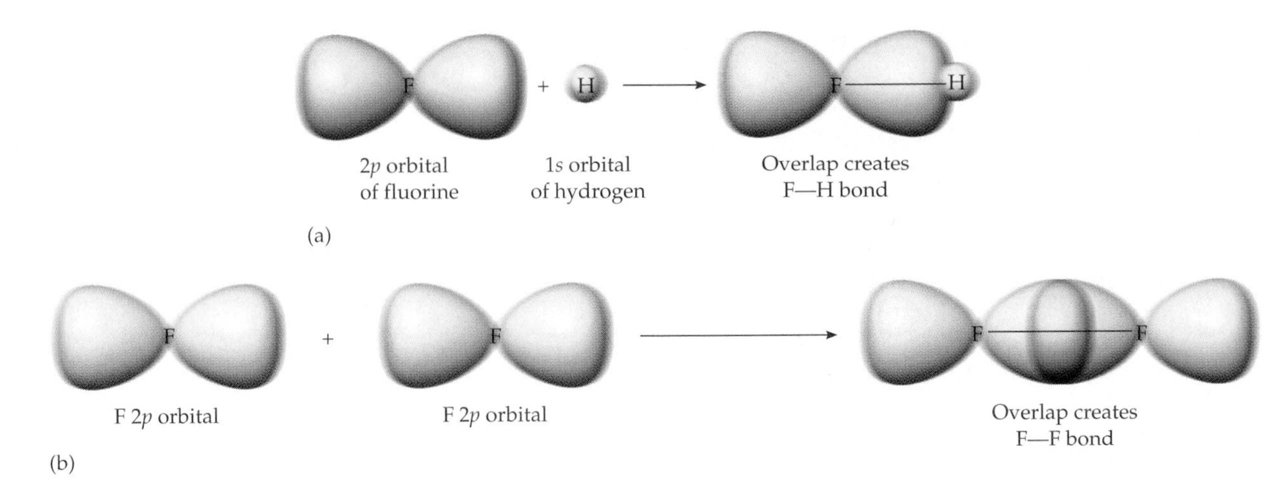

Figure 10.3 Covalent bond formation in HF and F_2. (a) Overlap of hydrogen 1*s* and fluorine 2*p* orbitals to form the sigma (σ) bond in HF. (b) Overlap of 2*p* orbitals on two fluorine atoms forming the sigma (σ) bond in F_2.

What happens for elements below hydrogen? In the Lewis structure of HF, for example, a bonding electron pair is placed between H and F, and three lone pairs of electrons are depicted as localized on the F atom. To use an orbital approach, look at the valence shell electrons and orbitals for each atom that will overlap. The hydrogen atom will use its 1*s* orbital in bond formation. The electron configuration of fluorine is $1s^2 2s^2 2p^5$, and the unpaired electron for this atom is assigned to one of the 2*p* orbitals. A sigma bond results from overlap of the hydrogen 1*s* and the fluorine 2*p* orbital (Figure 10.3).

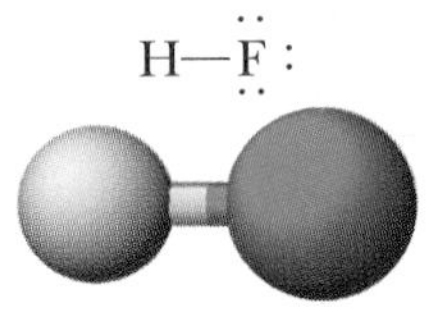

A Lewis structure and a molecular model of hydrogen fluoride, HF.

Formation of the H—F bond is similar to formation of an H—H bond. A hydrogen atom approaches a fluorine atom along the axis containing the 2*p* orbital with a single electron (see Figure 10.3). The orbitals (1*s* on H and 2*p* on F) distort as each atomic nucleus influences the electron and orbital of the other atom. Still closer together, the 1*s* and 2*p* orbitals overlap, and the two electrons pair up to give a σ bond. There is an optimum distance (92 pm) at which the energy is lowest that corresponds to the bond distance in HF. The net stabilization achieved in this process is the energy for the H—F bond.

The remaining electrons on the fluorine atom (two electrons in the 2*s* orbital and four electrons in the other two 2*p* orbitals) are not involved in bonding. They are nonbonding electrons, the lone pairs associated with this element in the Lewis structure.

Extension of this model gives a description of bonding in F_2. The 2*p* orbitals on the two atoms overlap, and the single electron from each atom is paired in the resulting σ bond (Figure 10.3b). The 2*s* and the 2*p* electrons not involved in the bond are the lone pairs on each atom.

Hybridization Involving *s* and *p* Atomic Orbitals

See the *Saunders Interactive General Chemistry CD-ROM*, Screen 10.4, Hybrid Orbitals; Screen 10.5, Sigma Bonding; and Screen 10.6, Determining Hybrid Orbitals.

The simple picture using orbital overlap to describe bonding in H_2, HF, and F_2 works well, but we run into difficulty when molecules with more atoms are considered. For example, a Lewis dot structure of methane, CH_4, shows four C—H covalent bonds. VSEPR theory predicts, and experiments confirm, that the electron-pair geometry of the C atom in CH_4 is tetrahedral, with an angle of 109.5°

between the bond pairs. The hydrogens are identical in this structure. This means that four equivalent bonding electron pairs occur around the C atom. An orbital picture of the bonds should convey both the geometry and the fact that all C—H bonds are the same.

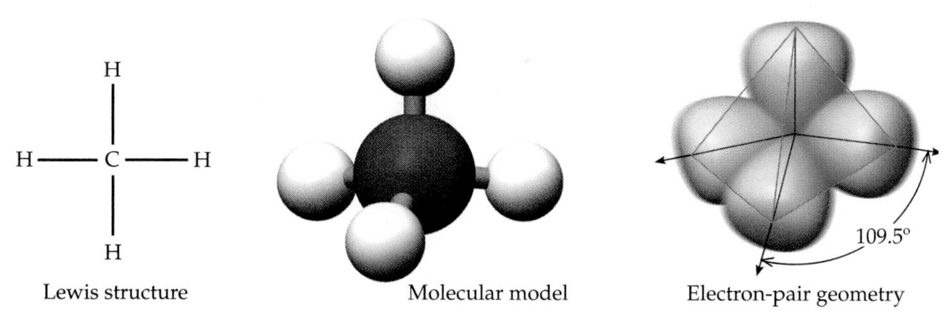

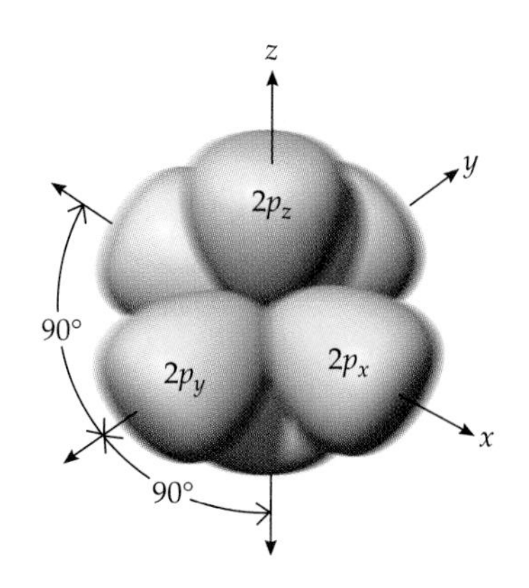

Figure 10.4 The 2p orbitals. The $2p_x$, $2p_y$, and $2p_z$ orbitals are perpendicular to one another. That is, the angle between them is 90°.

If we apply the orbital overlap model used for H_2 and F_2 without modification to describe the bonding in CH_4, a problem arises. The three orbitals for the $2p$ valence electrons of carbon are at right angles, 90° (Figure 10.4), and fail to match the tetrahedral angle of 109.5°. The spherical $2s$ orbital could bond in any direction. Secondarily, in its ground state ($1s^22s^22p^2$), a carbon atom has only two unpaired electrons, not the four that are needed to allow formation of four bonds.

The valence shell electron configuration of carbon.

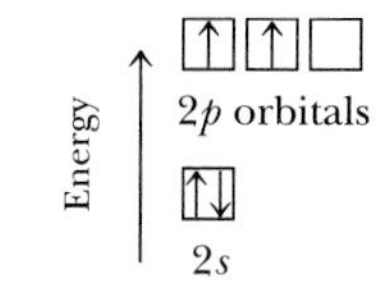

To describe the bonding in methane and other molecules, Linus Pauling proposed the theory of **orbital hybridization** (Figure 10.5). He proposed that a new set of orbitals, called **hybrid orbitals,** could be created by mixing s, p, and/or d atomic orbitals on an atom. Using an appropriate set of atomic orbitals, it is possible to create a new set of orbitals—hybrid orbitals—having an orientation in space that matches the geometry of a compound. The number of hybrid orbitals is the same as the number of atomic orbitals used in their construction.

Be sure to notice that *four* atomic orbitals produce *four* hybrid orbitals. The number of atomic orbitals used is always the same as the number of hybrid orbitals produced.

Figure 10.5 Hybridization. Atomic orbitals can mix, or hybridize, to form hybrid orbitals. An analogy is mixing two different colors *(left)* to produce a third color, which is a "hybrid" of the original colors *(center)*. After mixing there are still two beakers *(right)*, each containing the same volume of solution as before, but the color is a "hybrid" color. *(C. D. Winters)*

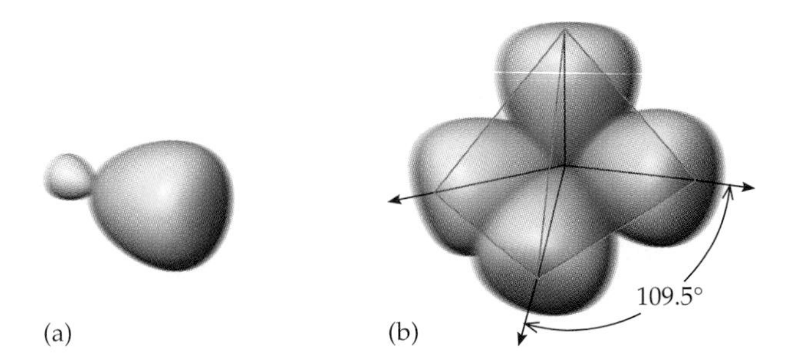

Figure 10.6 **sp^3 hybrid orbitals.** (a) Representation of a single sp^3 hybrid orbital showing the two regions of electron density. (b) The four sp^3 hybrid orbitals are directed at the corners of a tetrahedron, the angle between them being 109.5°.

In the case of methane, four orbitals directed to the corners of a tetrahedron are needed to match the molecular geometry. By mixing the four valence shell orbitals, the 2*s* and all three of the 2*p* orbitals on carbon, a new set of four hybrid orbitals is created that has tetrahedral geometry (Figures 10.6 and 10.7). Each of the four hybrid orbitals is labeled sp^3 to indicate the atomic orbital combination (an *s* orbital and three *p* orbitals) from which they are derived. The four sp^3 orbitals have an identical shape, and the angle between them is 109.5°, the tetrahedral angle.

The energy level diagram in Figure 10.7 illustrates the fact that the four sp^3 hybrid orbitals all have the same energy, which is the weighted average of the parent *s* and *p* orbital energies. Because the orbitals have the same energy, electrons are assigned according to Hund's rule (Section 8.4).

The four valence electrons of the C atom in CH_4 are placed singly in the sp^3 orbitals. Each C—H bond is then formed by overlap of one of the carbon sp^3 hybrid orbitals with the 1*s* orbital of a hydrogen; one electron from the C atom is paired with an electron from an H atom.

Hybridization is an idealized model, which is why it is presented as a qualitative picture. In reality, small deviations in bond angles from the predicted values occur frequently. The value of 107.5° in ammonia is close to 109.5°, and far different from any other choices (such as the unhybridized angle, 90°).

The hybrid orbital model can be applied to many other molecules such as ammonia, NH_3. The Lewis structure for ammonia shows that there are four electron pairs in the valence shell of nitrogen: three bond pairs and a lone pair. VSEPR theory predicts a tetrahedral electron-pair geometry and a trigonal-pyramidal molecular geometry. Structure evidence is a close match to prediction; the H—N—H bond angles are 107.5° in this molecule.

Based on the electron-pair geometry of NH_3, we predict sp^3 hybridization for the N atom. The lone pair is assigned to one of the hybrid orbitals, and each of the other three hybrid orbitals is occupied by a single electron. Overlap of

C atom

↑ ↑ □

2*p* orbitals

Energy

↑↓

2*s*

C atom in CH_4

four sp^3 hybrid orbitals

→ ↑ ↑ ↑ ↑

Electrons available to form σ bonds

Energy level diagram

(a)

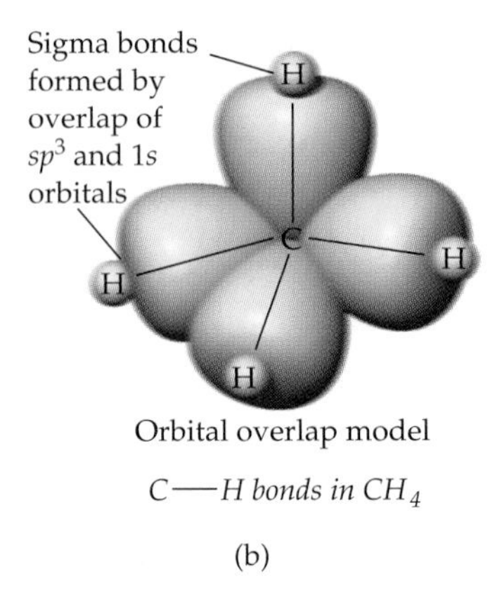

C—H bonds in CH_4

(b)

Figure 10.7 **Bonding in methane.** (a) An energy level diagram showing the result of orbital hybridization. (b) Four sigma bonds are formed by overlap of H atom 1*s* orbitals with C atom sp^3 hybrid orbitals.

H—N̈—H
|
H

Lewis structure

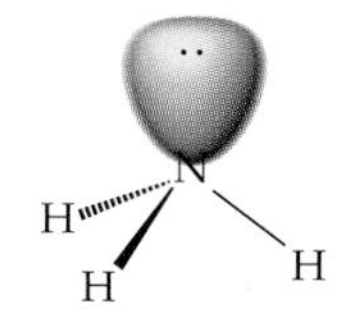

Electron-pair geometry, tetrahedral

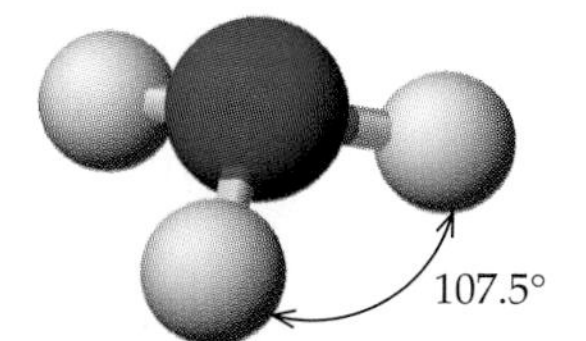

Molecular geometry, trigonal pyramid

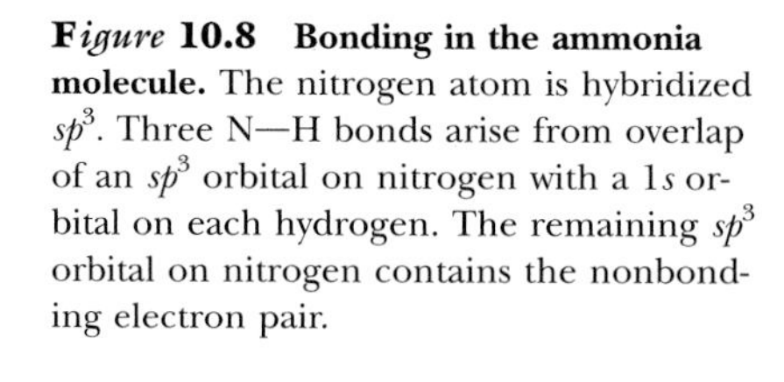

Figure 10.8 Bonding in the ammonia molecule. The nitrogen atom is hybridized sp^3. Three N—H bonds arise from overlap of an sp^3 orbital on nitrogen with a 1*s* orbital on each hydrogen. The remaining sp^3 orbital on nitrogen contains the nonbonding electron pair.

four sp^3 hybrid orbitals

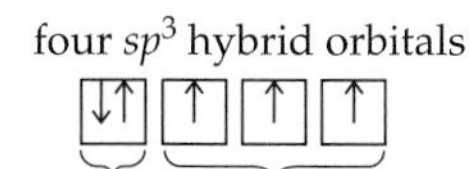

Lone pair electrons — Electrons available to form σ bonds

Electron configuration for N atom in NH_3

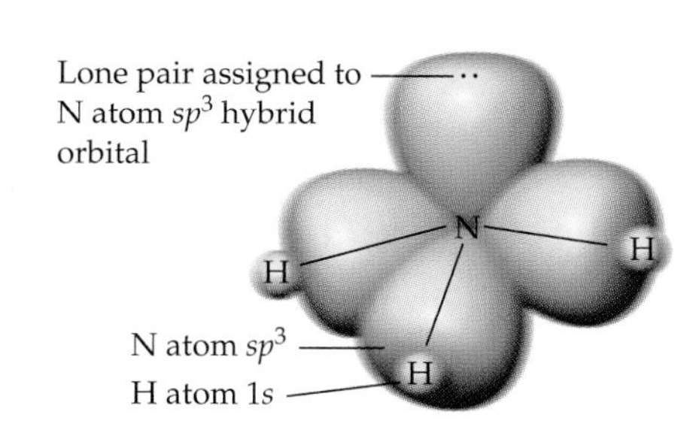

Bonding in NH_3

each of the singly occupied, sp^3 hybrid orbitals with a 1*s* orbital for hydrogen, and pairing of the electrons, creates the N—H bonds (Figure 10.8).

The hybrid orbital model applies to the most important substance on our planet, water. The oxygen atom has two bonding pairs and two lone pairs in its valence shell, and the H—O—H angle is 104.5°. (This value is less than the tetrahedral angle, but close enough that a tetrahedral arrangement for the valence electrons for oxygen can be used in a qualitative picture.) Four sp^3 hybrid orbitals are created from the 2*s* and 2*p* atomic orbitals of oxygen. Two of these sp^3 orbitals are occupied by unpaired electrons and are used to form O—H bonds. Lone pairs occupy the other two orbitals (Figure 10.9).

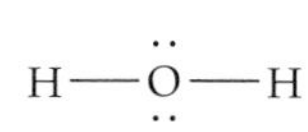

Lewis structure

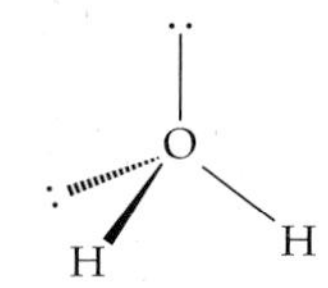

Electron-pair geometry, tetrahedral

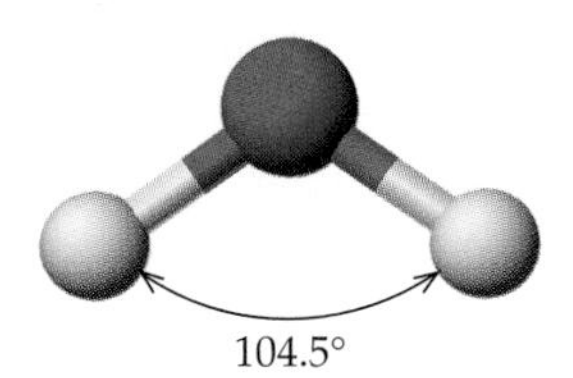

Molecular geometry, bent or angular

four sp^3 hybrid orbitals

Lone pair electrons — Electrons available for σ bonding

Electron configuration for O atom in H_2O

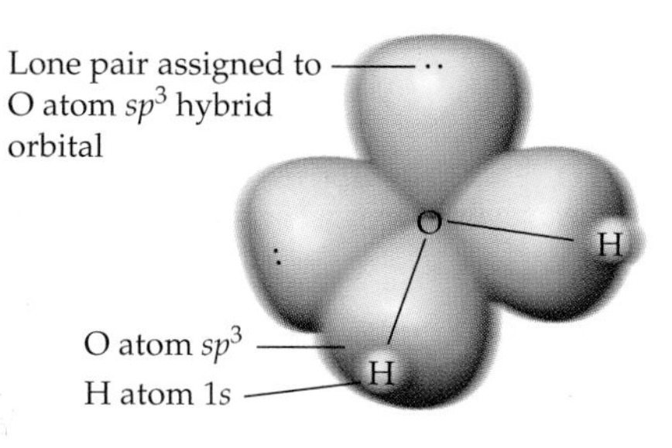

Bonding in H_2O

Figure 10.9 Bonding in H_2O. The oxygen atom is sp^3-hybridized. Two O—H bonds are formed by overlap of sp^3 orbitals on oxygen with hydrogen 1*s* orbitals; the nonbonding electrons on oxygen occupy the other two sp^3 orbitals.

Example 10.1 Valence Bond Description of Bonding

Describe the bonding in ethane, C_2H_6, using valence bond theory.

Solution First, draw the Lewis structure and predict the molecular geometry at both carbon atoms. Next, assign a hybridization to these atoms. Finally, describe covalent bonds that arise based on orbital overlap, and place electron pairs in their proper locations.

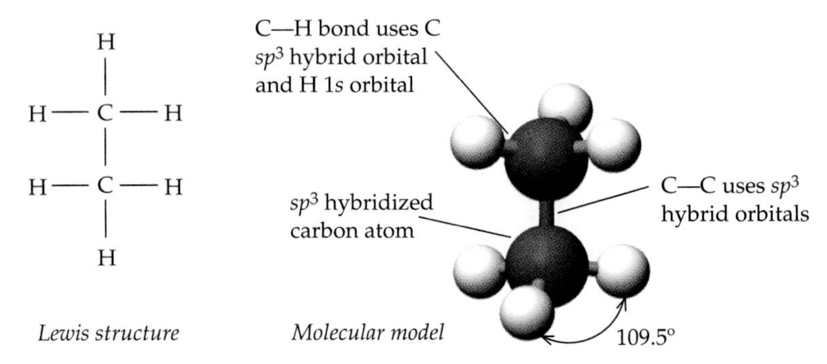

Each carbon atom has an octet configuration, sharing electron pairs with three hydrogen atoms and with the other carbon atom. The electron pairs around carbon have tetrahedral geometry, so carbon is assigned sp^3 hybridization. The C—C bond is formed by overlap of sp^3 orbitals on each C atom, and each of the C—H bonds is formed by overlap of an sp^3 orbital on carbon with a hydrogen 1*s* orbital.

Example 10.2 Valence Bond Description of Bonding

Describe bonding in the molecule CH_3OH using valence bond theory.

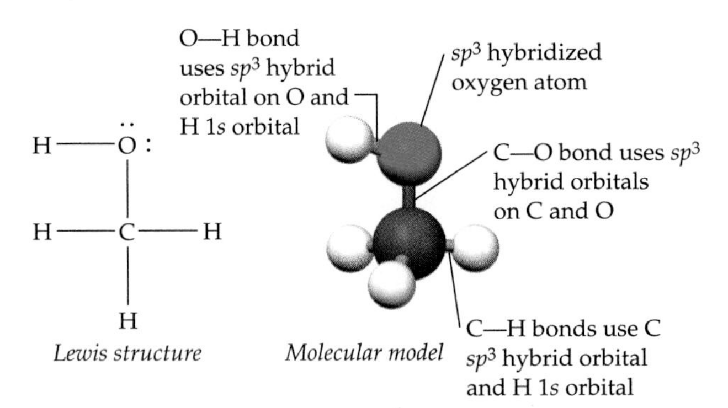

Solution A Lewis structure of CH_3OH implies that the electron-pair geometry around both the C and O atoms is tetrahedral. This geometry is confirmed by experiment. Thus, we may assign sp^3 hybridization to each atom, and the C—O bond is formed by overlap of sp^3 orbitals on these atoms. The C—H bonds are formed by overlap of a carbon sp^3 orbital with a hydrogen 1*s* orbital, and the O—H bond is formed by overlap of an oxygen sp^3 orbital with the hydrogen 1*s* orbital. Two lone pairs on oxygen occupy the remaining sp^3 orbitals.

Exercise 10.1 Valence Bond Description of Bonding

Use valence bond theory to describe the bonding in the hydronium ion, H_3O^+, and methylamine, CH_3NH_2.

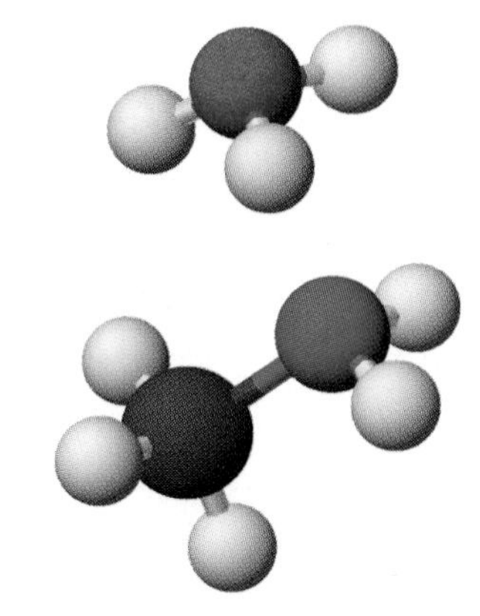

Models of hydronium ion, H_3O^+, and methylamine, CH_3NH_2.

Example 10.2 shows how to predict the structure and bonding in a complicated molecule by looking at each atom separately. This is an important principle that is essential when dealing with molecules made up of many atoms.

Linear and trigonal-planar geometries are also commonly encountered in molecules and ions. For example, BF_3 and other boron halides are trigonal-planar, as are a number of other species, such as NO_3^- and CO_3^{2-}. The carbon atoms in ethylene, $CH_2{=}CH_2$, are also trigonal-planar, and the electron-pair geometry of O_3 and NO_2^- is trigonal-planar. Linear molecules and ions include CO_2, N_2O, OCN^-, and acetylene, H—C≡C—H. Different hybridization schemes are required to describe the bonding in these molecules.

A **trigonal-planar electron-pair geometry** requires a central atom with three hybrid orbitals in a plane, 120° apart. Three hybrid orbitals mean that three atomic orbitals must be combined, and the combination of an *s* orbital with two *p* orbitals is appropriate. If p_x and p_y are the orbitals used in hybrid orbital formation, the **three hybrid sp^2 orbitals** will lie in the *xy*-plane. The p_z orbital not used to form these hybrid orbitals is perpendicular to the plane containing the three sp^2 orbitals (Figure 10.10).

Boron trifluoride has a trigonal-planar electron-pair and molecular geometry. Each boron–fluorine bond in this compound results from overlap of an sp^2 orbital on boron with a *p* orbital on fluorine. Notice that the p_z orbital, which is not used to form the sp^2 hybrid orbitals, is not occupied by electrons. Because the boron atom is not surrounded by an octet of electrons, it seeks a pair of electrons elsewhere. One source is an electron-rich molecule such as ammonia, NH_3, and a coordinate covalent bond can form as described on page 395.

For a molecule in which the central atom has a **linear electron-pair geometry** two hybrid orbitals, 180° apart, are required. One *s* and one *p* orbital can

The hybrid ordital "tool" on Screen 10.6 of the *Saunders Interactive General Chemistry CD-ROM* enables you to "build" hybrid orbitals from various combinations of atomic orbitals. See also the animation of sp^3 hybrid ordital formation on Screen 10.4.

The orientation of the sp^2 hybrid orbitals in space is significant. The fact that the *p* orbital is at right angles to the plane containing the sp^2 orbitals will be important when describing compounds with double bonds.

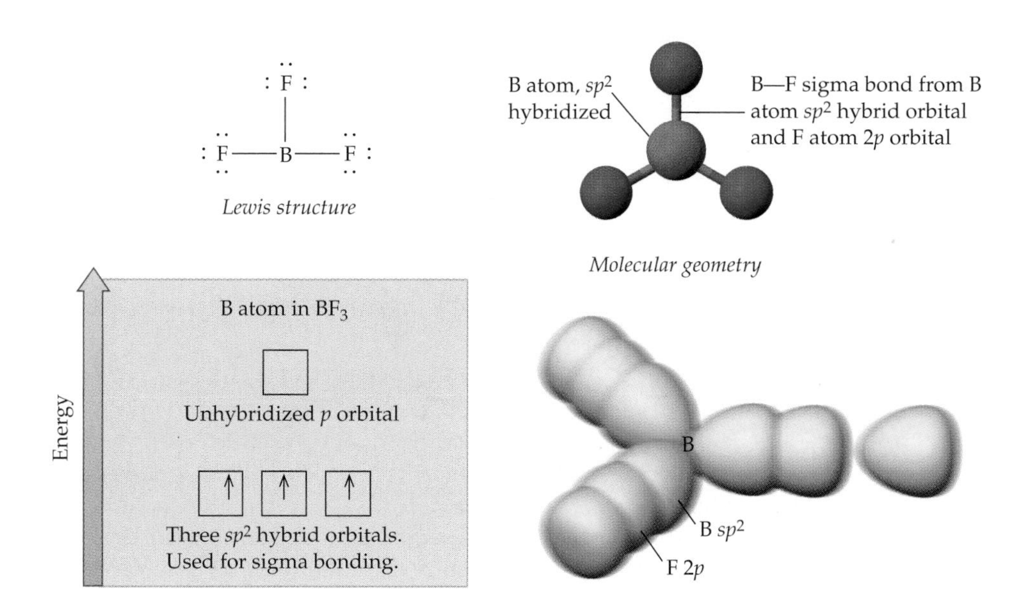

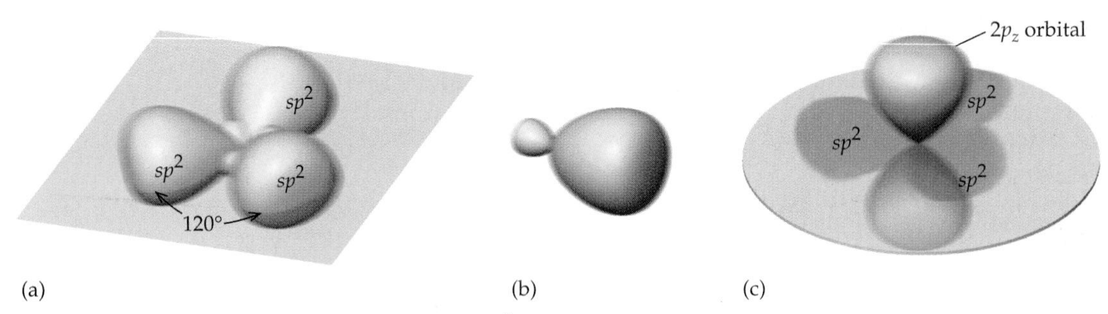

Figure 10.10 Three sp^2 hybrid orbitals. (a) Hybridization of the s, p_x, and p_y atomic orbitals generates three sp^2 orbitals located in the xy-plane. The sp^2 orbitals form angles of 120°. (b) An individual sp^2 hybrid orbital. (c) The unhybridized p_z orbital is perpendicular to the plane containing the sp^2 hybrid orbitals.

be hybridized to form **two *sp* hybrid orbitals** (Figure 10.11). If the p_x orbital is used, then the *sp* orbitals are oriented along the x-axis. The p_y and p_z orbitals are perpendicular to this axis.

Beryllium dichloride, $BeCl_2$, is a solid under ordinary conditions. When it is heated to over 520 °C, however, it vaporizes to give $BeCl_2$ vapor. In the gas phase, $BeCl_2$ is a linear molecule, so *sp* hybridization is appropriate for the beryllium atom. Combining beryllium's $2s$ and $2p_x$ orbitals gives the two *sp* hybrid orbitals that lie along the x-axis. Each Be—Cl bond arises by overlap of an *sp* hybrid orbital on beryllium with a $3p$ orbital on chlorine. In this molecule, there are only two electron pairs around the beryllium atom, so the p_y and p_z orbitals are not occupied.

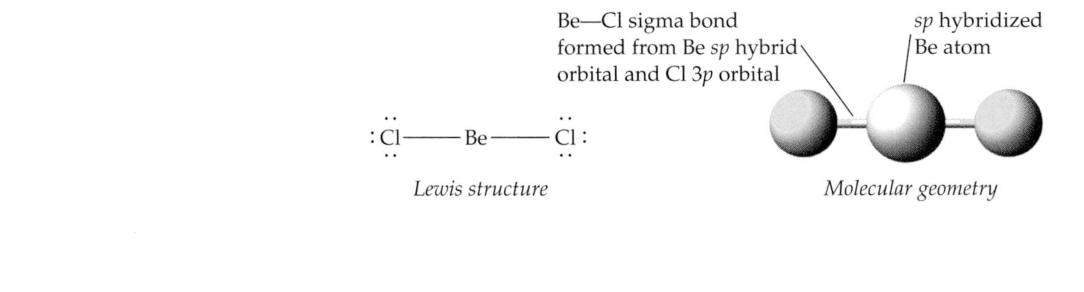

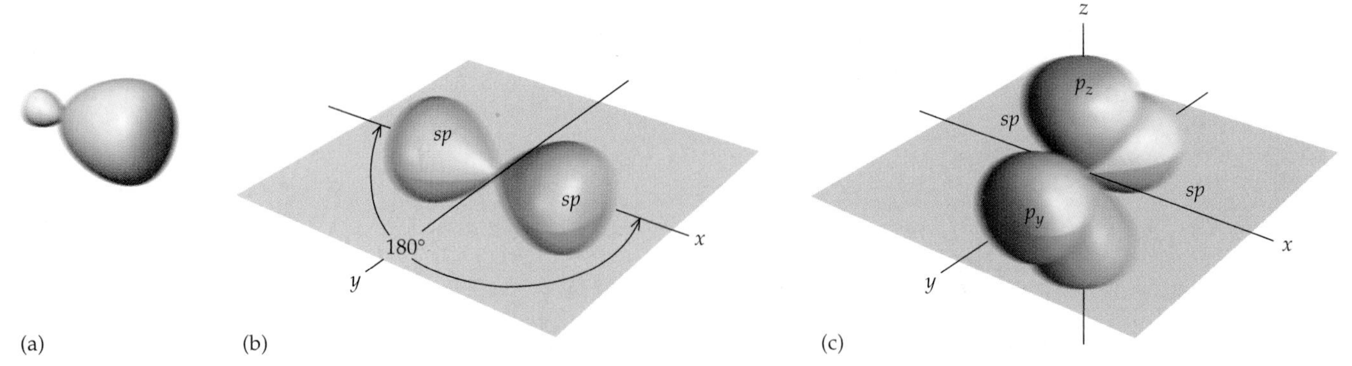

Figure 10.11 *sp* hybrid orbitals. (a) A single *sp* hybrid orbital. (b) Two *sp* hybrid orbitals lie 180° apart. (c) Because only one p orbital is incorporated in the hybrid orbital, two p orbitals remain. These are perpendicular to each other and to the axis along which the two *sp* hybrid orbitals lie.

Hybridization Involving *s*, *p*, and *d* Atomic Orbitals

A basic assumption of Pauling's valence bond theory is that *the number of hybrid orbitals equals the number of valence orbitals used in their creation.* This means the maximum number of hybrid orbitals that can be created from the *s* and *p* orbitals for an atom is four.

How then should we deal with compounds like PF_5 or SF_6, compounds with more than four electron pairs in their valence shell? To describe five or six bonds requires the central atom to have five or six hybrid orbitals, which must be created from five or six atomic orbitals. This is possible if additional atomic orbitals from the *d* subshell are used in hybrid orbital formation. The *d* orbitals are considered to be valence shell orbitals for main group elements of the third and higher periods.

To accommodate six electron pairs in the valence shell of an element, six sp^3d^2 hybrid orbitals can be created from the *s*, three *p*, and two *d* orbitals. The six sp^3d^2 hybrid orbitals are directed in positive and negative directions along the *x*-, *y*- and *z*-axis (Figure 10.12). Thus, they are oriented to accommodate the electron pairs for a compound that has an octahedral electron-pair geometry. Five coordination and trigonal bipyramid geometry are matched to sp^3d hybridization. One *s*, three *p*, and one *d* orbital combine to produce five sp^3d hybrid orbitals.

The hybrid orbital sets for two through six pairs are summarized in Figure 10.12.

Example 10.3 Hybridization Involving *d* Orbitals

Describe the bonding in PF_5 using valence bond theory.

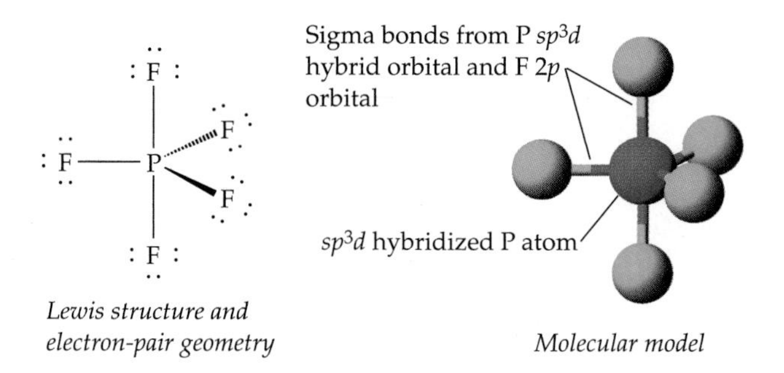

Lewis structure and electron-pair geometry

Molecular model

Solution The first step is to establish the electron-pair and molecular geometry of PF_5. Here the P atom is surrounded by five electron pairs, so PF_5 has a trigonal-bipyramidal electron-pair and molecular geometry. Five covalent bonds must point to the corners of a trigonal bipyramid. The P atom must have a single electron in each of five hybrid orbitals, and the hybrid scheme sp^3d should be used. Each of the five P—F bonds is created by the overlap of one of phosphorus's sp^3d hybrid orbitals with a fluorine *p* orbital.

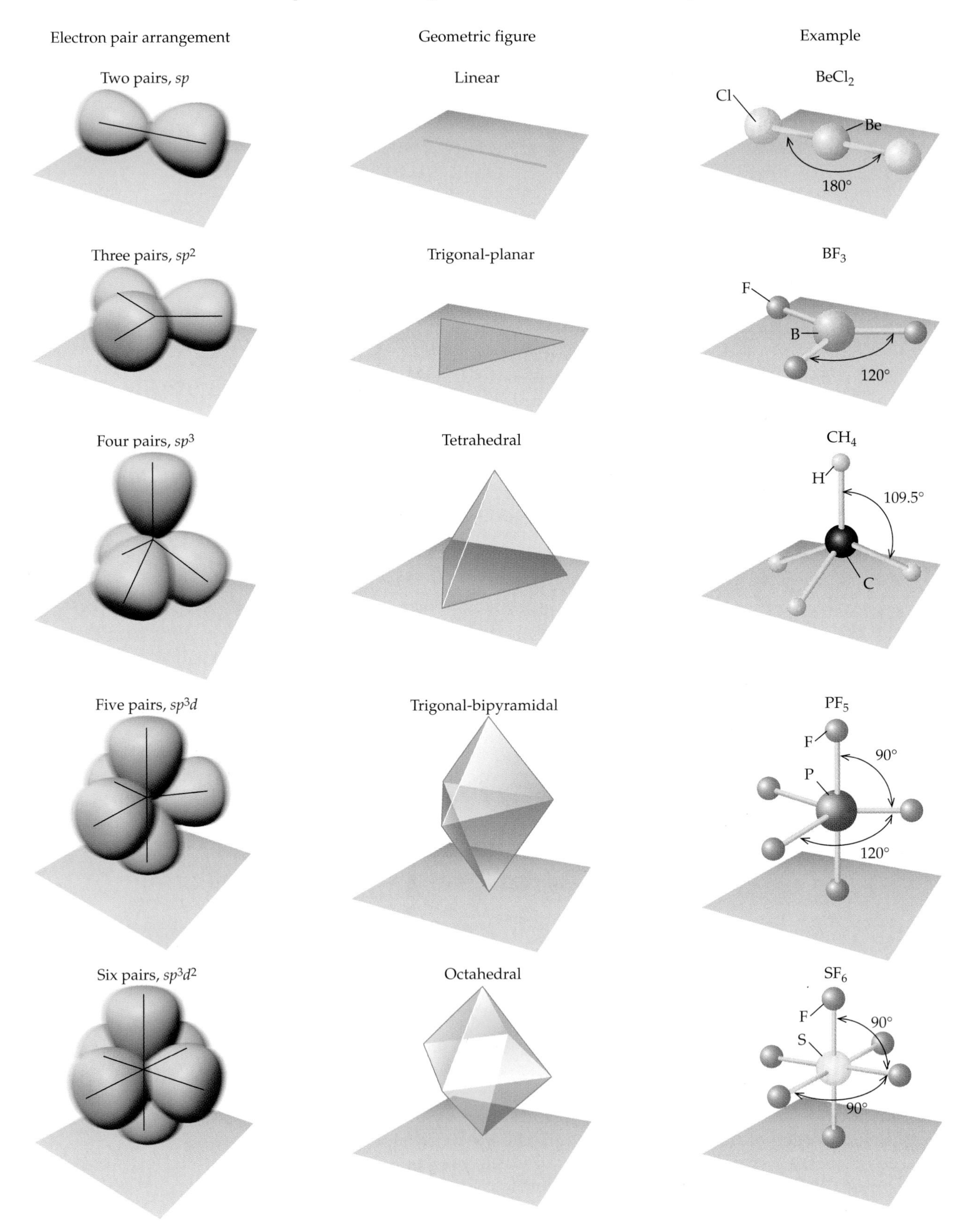

Electron pair arrangement
Geometric figure
Example
Two pairs, sp
Linear
BeCl2
Cl
Be
180°
Three pairs, sp2
Trigonal-planar
BF3
F
B
120°
Four pairs, sp3
Tetrahedral
CH4
H
109.5°
C
Five pairs, sp3d
Trigonal-bipyramidal
PF5
F
P
90°
120°
Six pairs, sp3d2
Octahedral
SF6
F
S
90°
90°

Example 10.4 Recognizing Hybridization

Identify the hybridization of the central atom in the following compounds and ions:

(a) CF_4 (d) SF_3^+
(b) SF_4 (e) I_3^-
(c) $PO_2F_2^-$ (f) SO_4^{2-}

Solution The hybrid orbitals used by a central atom are determined by the electron-pair geometry. Thus, to answer this question we first need to write the Lewis structure and predict the electron-pair geometry.

Following the procedures in Chapter 9 the Lewis structures for four of these molecules or ions can be written as follows:

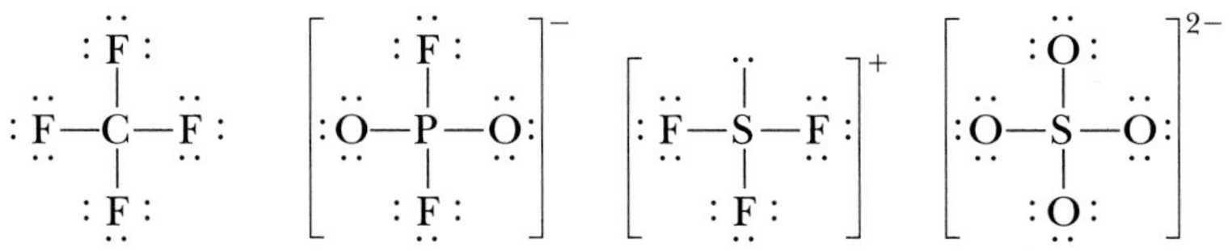

Four electron pairs surround the center atom in CF_4, $PO_2F_2^-$, SF_3^+, and SO_4^{2-}, and the electron-pair geometry for these atoms is tetrahedral. Thus, sp^3 hybridization for the central atom is used to describe the bonding. For I_3^- and SF_4, five pairs of electrons are in the valence shell of the center atom. For these, sp^3d hybridization is appropriate for the central I or S atom.

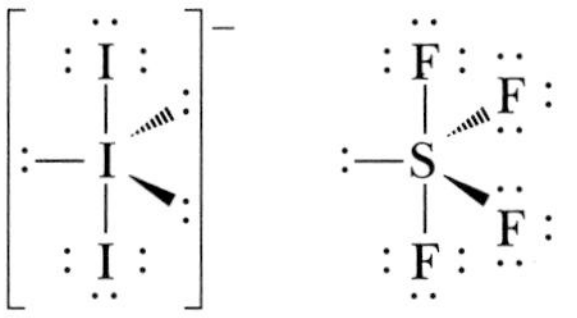

Exercise 10.2 Hybridization Involving *d* Orbitals

Describe the bonding in XeF_4 using hybrid orbitals. Remember to consider first the Lewis structure, then the electron-pair geometry (based on VSEPR theory), and then the molecular shape.

Exercise 10.3 Recognizing Hybridization

Identify the hybridization of the central atom in the following compounds and ions:

(a) BH_4^- (d) ClF_3
(b) SF_5^- (e) BCl_3
(c) OSF_4 (f) XeO_6^{4-}

◀**Figure 10.12** **(at left) The geometry of the hybrid orbital sets for two to six valence shell electron pairs.** In forming a hybrid orbital set, the *s* orbital is always used, plus as many *p* orbitals (and *d* orbitals) as required to give the necessary number of sigma-bonding and lone pair orbitals.

PROBLEM-SOLVING TIPS AND IDEAS

10.1 Hybridization and Hybrid Orbitals

1. Hybridization reconciles the electron-pair geometry with the orbital overlap criterion of bonding. A statement such as "the atom is tetrahedral because it is sp^3-hybridized" is backward. That the electron-pair geometry around the atom is tetrahedral is a fact. Hybridization is one way to rationalize that fact.
2. Hybridization is assigned only after a structure is known. Figure 10.12, which relates structure to hybridization, will be useful.
3. The number of hybrid orbitals always equals the number of atomic orbitals used to create the hybrid orbitals.

Multiple Bonds

See the *Saunders Interactive General Chemistry CD-ROM*, Screen 10.7, Multiple Bonding.

According to valence bond theory, bond formation requires that two orbitals on adjacent atoms overlap. In a single bond, one electron is contributed by each atom; the resulting electron pair forms the bond. Many molecules have two or three bonds between pairs of atoms, however. A double bond requires *two* sets of overlapping orbitals and *two* electron pairs; for a triple bond, *three* sets of atomic orbitals are required, each set accommodating a pair of electrons.

Double Bonds

Consider ethylene, C_2H_4, one of the more common molecules with a double bond. The molecular structure of ethylene places all six atoms in a plane, with H—C—H and H—C—C angles that are approximately 120°. Each carbon atom has trigonal-planar geometry, so sp^2 hybridization is assumed for these atoms. Thus, the model of bonding in ethylene starts with each carbon atom having three sp^2 hybrid orbitals in the molecular plane and an unhybridized p orbital perpendicular to that plane (Figure 10.13). Because each carbon atom is involved in four bonds, a single unpaired electron is placed in each of these orbitals.

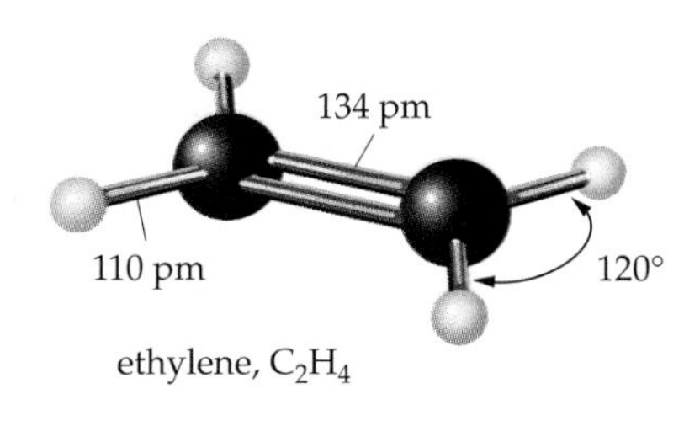

ethylene, C_2H_4

Now we can visualize the C—H bonds, which arise from overlap of sp^2 orbitals on carbon with hydrogen 1s orbitals. After accounting for the C—H bonds, one sp^2 orbital on each carbon atom remains. These orbitals point toward each other and overlap to form one of the bonds linking the carbon atoms (Figure 10.13a). This leaves only one other orbital on each carbon, an unhybridized p orbital, to be used to create the second bond between carbon atoms in C_2H_4.

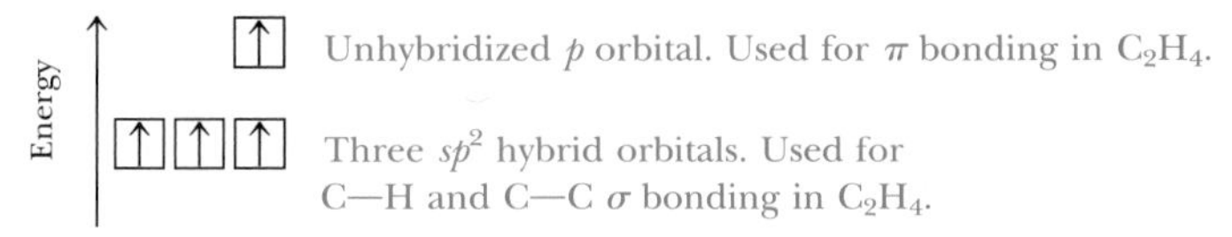

If they are aligned correctly, the p orbitals on the two carbons can overlap, allowing the electrons in these orbitals to be paired. The overlap does not occur directly along the C—C axis, however. Instead, the arrangement compels these orbitals to overlap sideways, and the electron pair occupies an orbital with electron density above and below the plane containing the six atoms (Figure 10.13b).

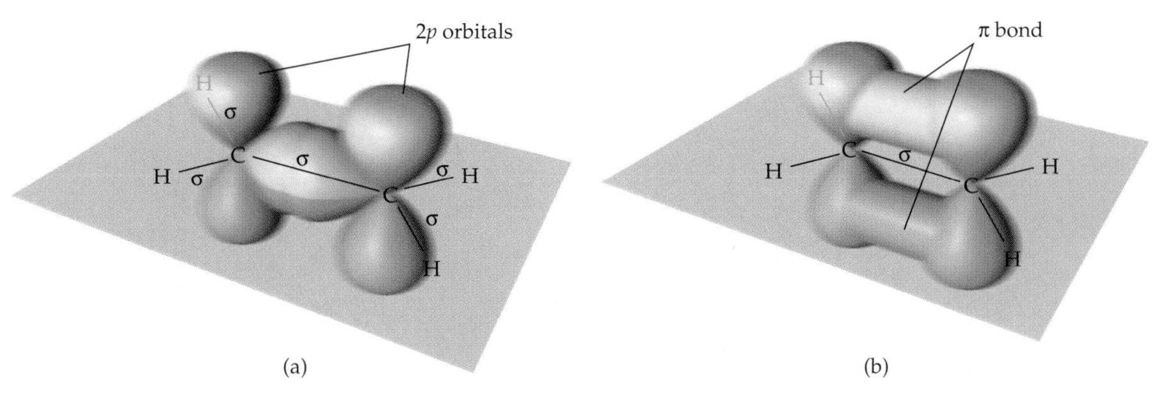

Figure 10.13 The valence bond model of bonding in ethylene, C_2H_4. Each C atom is assumed to be sp^2-hybridized. (a) The C—H bonds are formed by overlap of C atom sp^2 hybrid orbitals with H atom 1*s* orbitals. The σ bond between C atoms arises from overlap of sp^2 orbitals. (b) The carbon–carbon π bond is formed by overlap of an unhybridized 2*p* orbital on each atom.

This description results in two types of bonds in C_2H_4. The C—H and C—C bonds that arise from the overlap of atomic orbitals so that the bonding electrons lie *along* the bond axis are called **sigma (σ) bonds.** A bond formed by sideways overlap of *p* atomic orbitals is called a **pi (π) bond.** In a π bond, the overlap region is above and below the internuclear axis, and the electron density of the π bond is above and below the bond axis (Figure 10.13b and Figure 10.14).

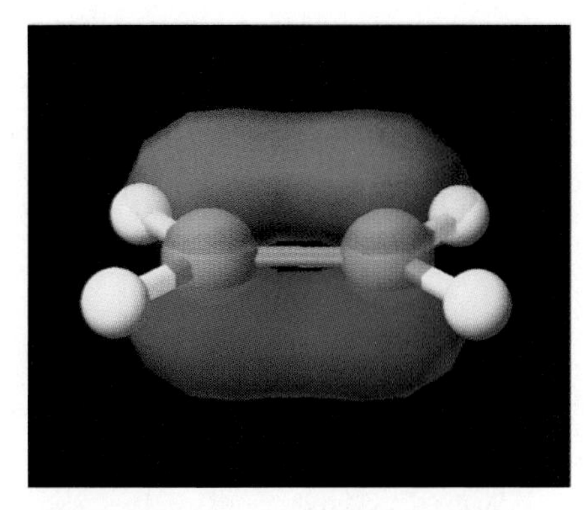

Figure 10.14 A computer-generated model of the π bond in ethylene. Note the lack of electron density along the carbon–carbon axis. The electron density of a π bond lies above and below the σ bond axis. *(Susan Young)*

Be sure to notice that a π bond can form *only* if the *p* orbitals on the adjacent atoms line up. That is, they must be perpendicular to the plane of the molecule and parallel to one another. This happens only if the sp^2 orbitals of both carbon atoms are in the same plane. Thus, *the formation of a π bond requires that this molecule be planar.*

Double bonds between carbon and oxygen, sulfur, or nitrogen are quite common. To illustrate this, consider formaldehyde, CH_2O, in which a carbon–oxygen π bond occurs (Figure 10.15). A trigonal-planar electron-pair geometry indicates sp^2 hybridization for the C atom. The σ bonds from carbon to the O atom and the two H atoms form by overlap of sp^2 hybrid orbitals with half-filled orbitals from the oxygen and two hydrogen atoms. An unhybridized *p* orbital on carbon is oriented perpendicular to the molecular plane (just as for the carbon atoms of C_2H_4). This *p* orbital is available for π bonding, this time with an oxygen orbital.

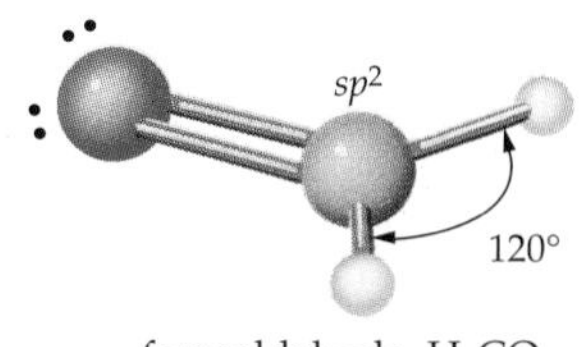

formaldehyde, H_2CO

What orbitals on oxygen are used in this model? The approach in Figure 10.15 assumes sp^2 hybridization for oxygen.* This uses one O atom sp^2 orbital in σ bond formation, leaving two sp^2 orbitals to hold lone pairs. The remaining *p* orbital on the O atom is the participant in the π bond.

*A second equally valid approach is to use unhybridized orbitals on oxygen in bonding. If unhybridized oxygen is assumed, the two *p* orbitals that are oriented at right angles and contain a single electron are used to create the σ and π bonds. The argument favoring hybridization for oxygen is that this would add consistency to the valence bond approach; because hybridization is required for some atoms it makes sense to use it for all. The objection is that hybridization was introduced only to explain the molecular geometry. The oxygen is bonded to only one other atom, so there is no geometry to explain; hybridization does not add anything to the explanation, and it could be regarded as an additional complication.

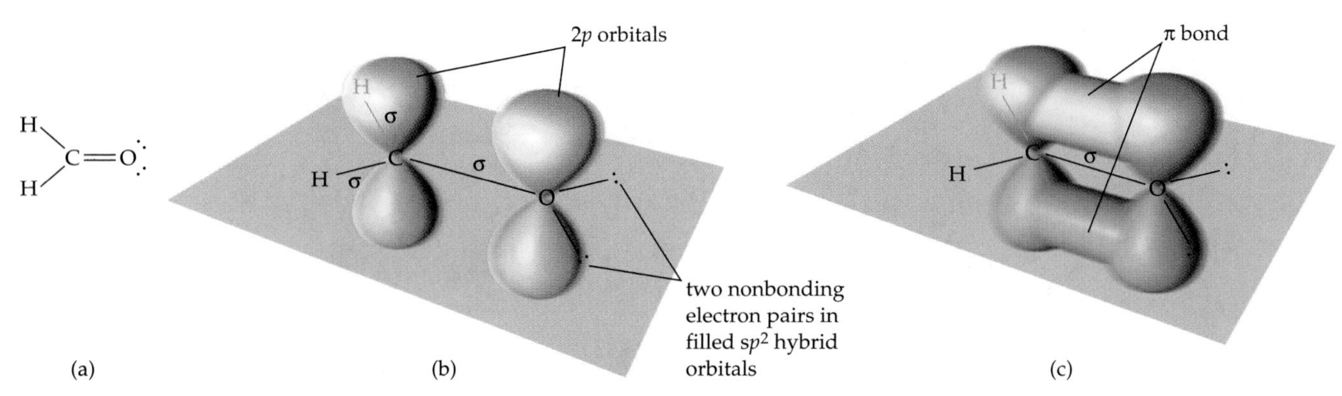

Figure 10.15 Valence bond description of bonding in formaldehyde, CH_2O. (a) Lewis dot structrure of formaldehyde, CH_2O. (b) The carbon atom is assigned sp^2 hybridization. The C—H bonds are formed by overlap of the 1*s* orbital of hydrogen with sp^2 orbitals on carbon. The σ bond between carbon and oxygen arises from the overlap of an sp^2 orbital on carbon and an sp^3 orbital on oxygen. (c) The C—O π bond comes from the side-by-side overlap of *p* orbitals on the two atoms.

Example 10.5 Bonding in Acetic Acid

Using valence bond theory, describe the bonding in acetic acid, CH_3CO_2H, the important ingredient in vinegar.

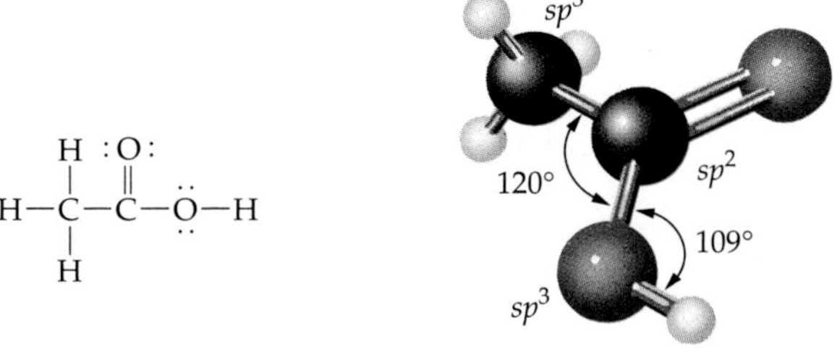

Solution The problem may be approached stepwise, and we first look at the σ bond framework. The carbon atom of the CH_3 group has tetrahedral electron-pair geometry, which means that it is sp^3 hybridized. Three sp^3 orbitals are used to form the C—H bonds. The fourth sp^3 orbital is used to bond to the adjacent carbon atom. This carbon atom (chemists refer to it as the "carbonyl carbon" because it is part of the carbonyl functional group) has a trigonal-planar electron-pair geometry; it must be sp^2-hybridized. The C—C bond is formed using one of these orbitals, and the other two sp^2 orbitals are used to form the σ bonds to the two oxygens. The oxygen of the O—H group has four electron pairs; it must be tetrahedral and sp^3-hybridized. Thus, this O atom uses two sp^3 orbitals to bond to the adjacent carbon and the hydrogen, and two sp^3 orbitals to accommodate the two lone pairs.

Finally, the carbon–oxygen double bond can be described exactly as in the CH_2O molecule. Both the C and O atoms are assumed to be sp^2-hybridized, and the unhybridized *p* orbital remaining on each atom is used to form the carbon–oxygen π bond.

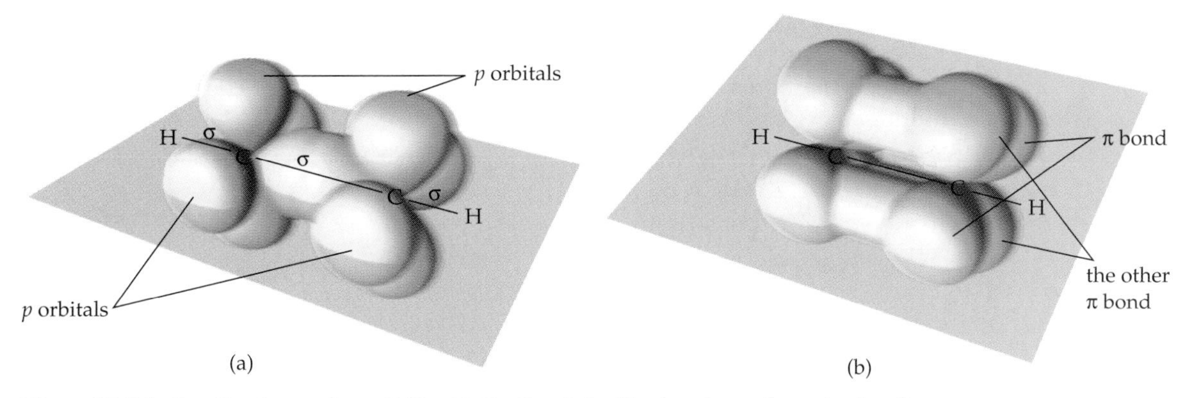

Figure 10.16 Bonding in acetylene, C_2H_2. (a) C—C and C—H σ bonds are formed using C atom *sp* hybrid orbitals. (b) Two π bonds are formed from overlap of two unhybridized *p* orbitals on each C atom.

Exercise 10.4 Bonding in Acetone

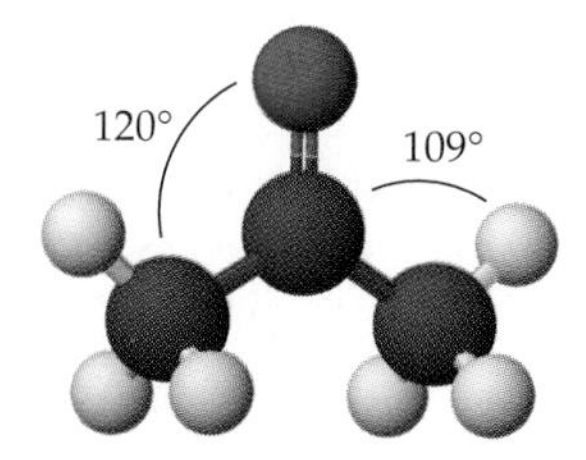

A molecular model of acetone, CH_3COCH_3.

Use valence bond theory to describe the bonding in acetone, CH_3COCH_3.

Triple Bonds

Acetylene, H—C≡C—H, is an example of a molecule with a triple bond. The molecular geometry shows that the four atoms lie in a straight line with the H—C—C angles of 180°, implying the carbon atom is *sp*-hybridized (Figure 10.16). For each carbon atom, there are two *sp* orbitals, one directed toward hydrogen and used to create the C—H σ bond, and the second directed toward the other carbon and used to create a σ bond between the two carbon atoms. Two unhybridized *p* orbitals remain on each carbon, and they are oriented so that it is possible to form *two* π bonds in C_2H_2.

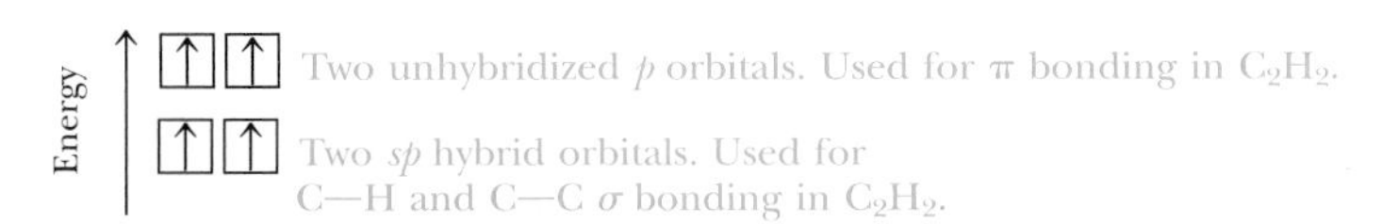

These π bonds are perpendicular to the molecular axis and perpendicular to each other. Three electrons on each carbon atom are paired to form the triple bond consisting of a σ bond and two π bonds (Figure 10.17).

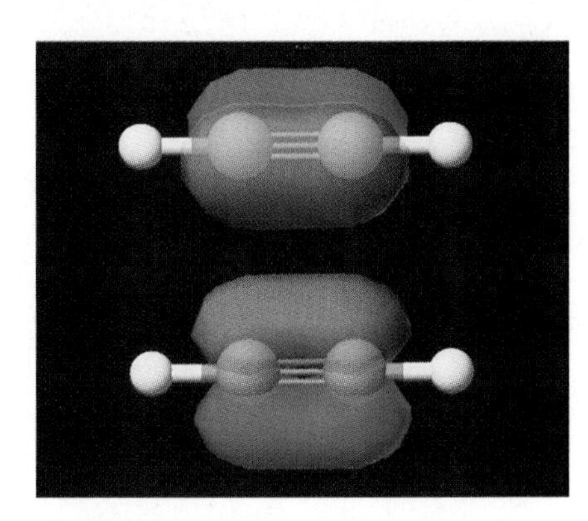

Figure 10.17 Computer-generated views of the two π bonds in acetylene. The π bonds lie perpendicular to each other. The π bond in the bottom structure is in the plane of the paper, whereas the π bond in the top structure is perpendicular to the plane of the paper. *(Susan Young)*

Example 10.6 The C≡O Triple Bond in Carbon Monoxide

Describe the bonding in carbon monoxide, CO, using the orbital hybridization model.

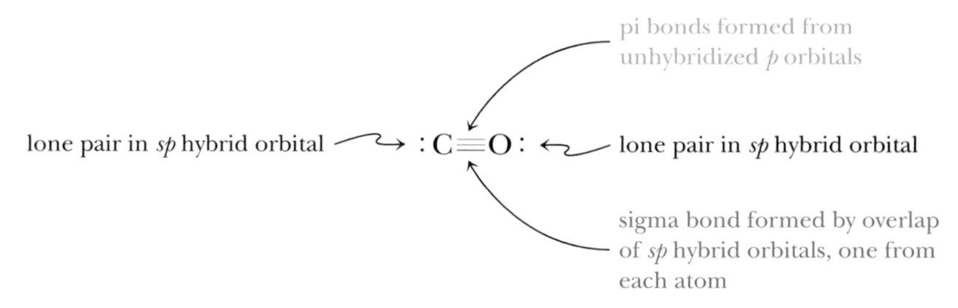

Solution The Lewis structure of carbon monoxide, :C≡O:, informs us that it is useful to consider both the carbon and the oxygen as *sp*-hybridized. For electron "bookkeeping" purposes, one of the hybrid orbitals on each atom is assigned initially one electron, which is then used for σ bond formation. The other *sp* hybrid orbital on each atom is then assigned to a nonbonding electron pair. Two electrons occur in unhybridized *p* orbitals on each atom; they are used to form the two π bonds. The orbital picture describing the triple bond of CO is similar to that used for acetylene, HC≡CH.

Exercise 10.5 Triple Bonds Between Atoms

Describe the bonding in a nitrogen molecule, N_2.

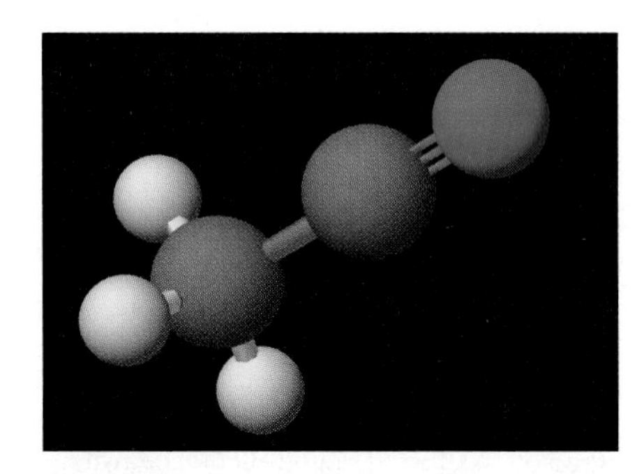

A molecular model of acetonitrile, CH_3CN. See Exercise 10.6.

Exercise 10.6 Bonding and Hybridization

Estimate values for the H—C—H, H—C—C, and C—C—N angles in acetonitrile, $CH_3C{\equiv}N$. Indicate the hybridization of both carbon atoms and the nitrogen atom, and analyze the bonding using valence bond theory.

There are two additional, useful points concerning π bonds. Notice that π bonds do not occur without the bonded atoms also being joined by a σ bond. Thus, a double bond always consists of a σ bond and a π bond. Similarly, a triple bond always consists of a σ bond and *two* π bonds. Because a π bond is formed from *p* atomic orbitals, one on each of two atoms, a π bond may form only if unhybridized *p* orbitals remain on the bonded atoms. If a Lewis structure shows multiple bonds, the atoms involved must therefore be either sp^2- or *sp*-hybridized.

Carbon–Carbon Double Bonds: *Cis* and *Trans* Isomers

In contrast to a double bond, there is free rotation about a single bond. See Screen 10.8 of the *Saunders Interactive General Chemistry CD-ROM* for an animation that contrasts the two types of bonds.

Ethylene, C_2H_4, is a planar molecule. This geometry allows the unhybridized *p* orbitals on the two carbon atoms to line up and form a π bond (see Figure 10.13). Let us speculate on what would happen if one end of the ethylene molecule is twisted relative to the other end (Figure 10.18). This action would distort the molecule away from planarity, and the p orbitals would rotate out of alignment. Rotation would diminish the extent of overlap of these orbitals, and if a twist of 90° is achieved, the two p orbitals would no longer overlap; the π bond would be broken. However, so much energy is required to break this bond (about 260 kJ/mol) that rotation around a C=C bond is not expected to occur at room temperatures.

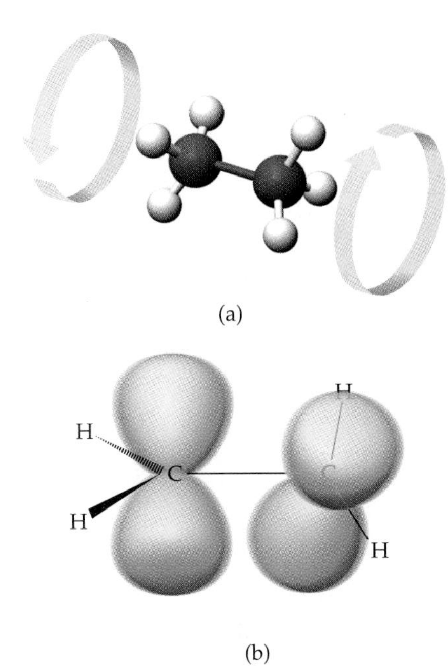

Figure 10.18 Rotation around bonds. (a) Free rotation can occur around the axis of a single (σ) bond. (b) In contrast, rotation is severely restricted around double bonds because doing so would break the π bond, a process generally requiring a great deal of energy.

A consequence of restricted rotation is that isomers occur for many compounds containing a C=C bond. **Isomers** are compounds that have the same formula but different structures. In this case, the two isomeric compounds differ with respect to the orientation of the groups attached to the carbons of the double bond. Two of the compounds with the formula $C_2H_2Cl_2$, *cis-* and *trans-*1,2-dichloroethylene, are isomers. They resemble ethylene, except that two hydrogen atoms have been replaced by chlorine atoms. Because rotation around the C=C double bond does *not* occur, the *cis* compound cannot rearrange to the *trans* compound under ordinary conditions. Each compound can be obtained separately, and each has its own identity. *Cis*-1,2-dichloroethylene boils at 60.3 °C, whereas *trans*-1,2-dichloroethylene boils at 47.5 °C.

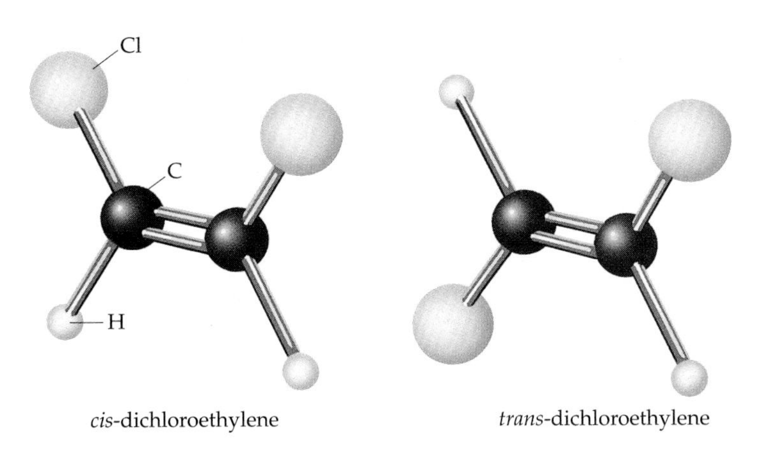

Cis- and *trans*-1,2-dichloroethylene are called *stereoisomers*. In stereoisomers, the atoms are connected in the same order but have different spatial arrangements. Another type of isomer is illustrated by 1,1-dichloroethylene. This compound has the same composition ($C_2H_2Cl_2$), but the atoms are connected in a different way. Such isomers are called *structural isomers*.

Although *cis* and *trans* isomers do not interconvert at ordinary temperatures, they may do so at higher temperatures. According to the kinetic theory of matter (Section 1.1), molecules in the gas and liquid phases in particular move rapidly and often collide with one another. Molecules also constantly flex or vibrate along or around the σ bonds holding them together. If the temperature is sufficiently high, rotation can occur about a carbon–carbon double bond. It may also occur under other special conditions, such as when the molecule absorbs light energy. Indeed, this specific situation is found to occur in the physiological process that allows us to see (see *A Closer Look: Chemical Bonds and Vision*).

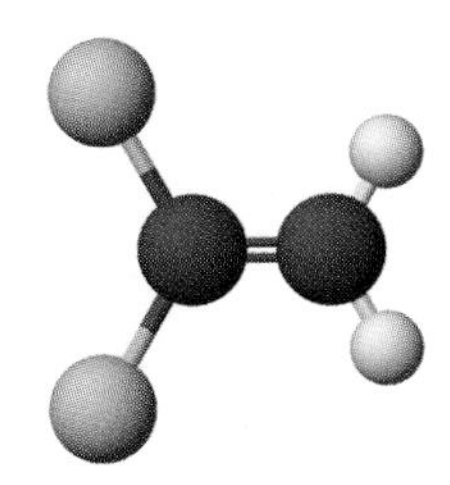

Benzene: Structure and Bonding

Benzene, C_6H_6, is the simplest member of a large group of substances known as *aromatic* compounds, a historical reference to their odor. It occupies a pivotal place in the history and practice of chemistry.

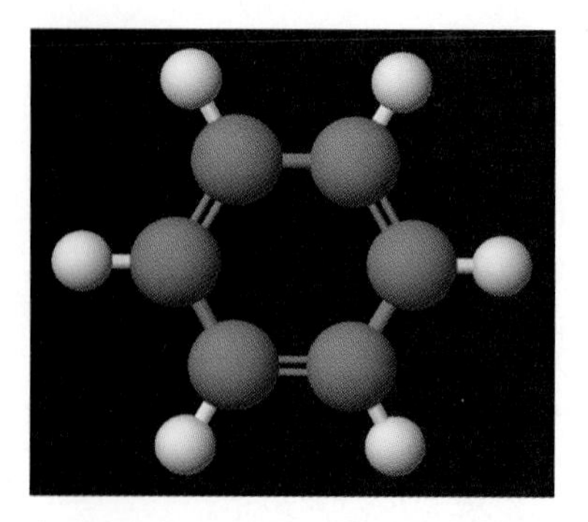

A molecular model of benzene, C_6H_6.

A molecular model of the amino acid tyrosine. Notice that a benzene ring is an important part of its structure, as it is in hundreds of other molecules.

Because the π bond in benzene extends totally around the ring, benzene's structure is commonly written as a six-member ring with a circle in the middle. The circle is meant to convey the idea that the six π electrons are delocalized over the six carbon atoms.

To 19th-century chemists benzene was a perplexing substance with an unknown structure. Based on its chemical reactions, however, August Kekulé (1829–1896) suggested that the molecule has a planar, symmetrical ring structure. We know now he was correct. The ring is flat, and all the carbon–carbon bonds are the same length, 139 pm, a distance intermediate between the average single bond (154 pm) and double bond (134 pm) lengths. If we assume the molecule has two resonance structures with alternating double bonds, this would rationalize the observed structure. The C—C bond order in C_6H_6(1.5) is the average of a single and a double bond.

Resonance structures or *Resonance hybrid*

Understanding the bonding in benzene is important because it is the basis of an enormous number of chemical compounds, the essential amino acid tyrosine being one example. Let us assume the trigonal-planar carbon atoms have sp^2 hybridization. Each C—H bond is formed by overlap of an sp^2 orbital of a carbon atom with a 1*s* orbital of hydrogen, and the C—C σ bonds arise by overlap of sp^2 orbitals on adjacent carbon atoms. After accounting for the σ bonding, an unhybridized *p* orbital remains on each C atom, and each is occupied by a single electron (Figure 10.19). These six orbitals and six electrons form three π bonds. Because all carbon–carbon bond lengths are the same, each *p* orbital overlaps equally well with the *p* orbitals of both adjacent carbons, and the π interaction is unbroken around the six-member ring.

The orbital picture of benzene underscores an important point. The basis of valence bond theory, that a bond is described as a pair of electrons between two atoms, does not work well for the π electrons in benzene—nor does it work whenever resonance is used to describe a structure. However, molecular orbital theory does give us a better view, and that is the subject of the next section.

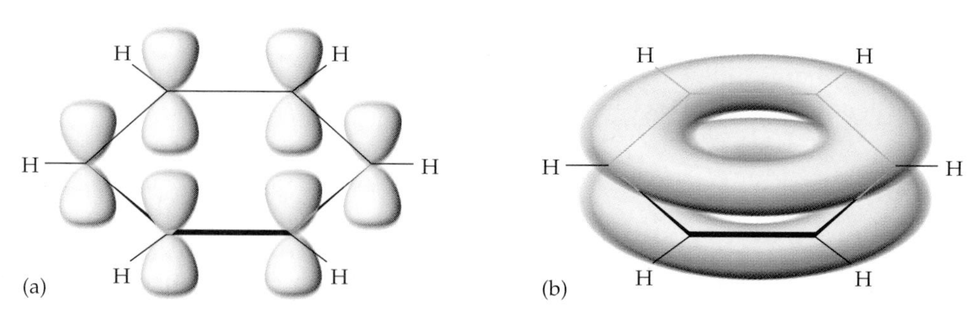

Figure 10.19 Bonding in the benzene molecule. The sigma framework of C—C and C—H bonds is based on sp^2-hybridized C atoms. (a) After accounting for σ bonding, an unhybridized *p* orbital remains on each C atom (b). These overlap to form the π bonds that form a continuous π electron cloud above and below the plane of the ring.

A CLOSER LOOK — *Chemical Bonds and Vision*

Rotation around a double bond occurs in the reactions that allow you to see. A yellow-orange compound, β-carotene, the natural coloring agent in carrots, breaks down in your liver to produce vitamin A, also called retinol. Retinol is oxidized to 11-*trans*-retinal, which isomerizes to 11-*cis*-retinal. The *cis* isomer reacts with the protein opsin in the eye to give the pigment rhodopsin. This light-sensitive combination absorbs intensely in the blue-green region of the visible spectrum. Light striking the pigment triggers rotation around a carbon—carbon double bond, transforming rhodopsin into metarhodopsin II. This change in molecular shape causes a nerve impulse to be sent to your brain, and you perceive a visual image.

Eventually metarhodopsin II reacts chemically to produce 11-*trans*-retinal, and the cycle of chemical changes begins again. Decomposition of metarhodopsin II is not as rapid as its formation, however, and an image formed on the retina persists for a tenth of a second or so. This persistence of vision allows you to perceive movies and videos as continuously moving images, even though they actually consist of separate pictures, each captured on a piece of film or tape for a thirtieth of a second.

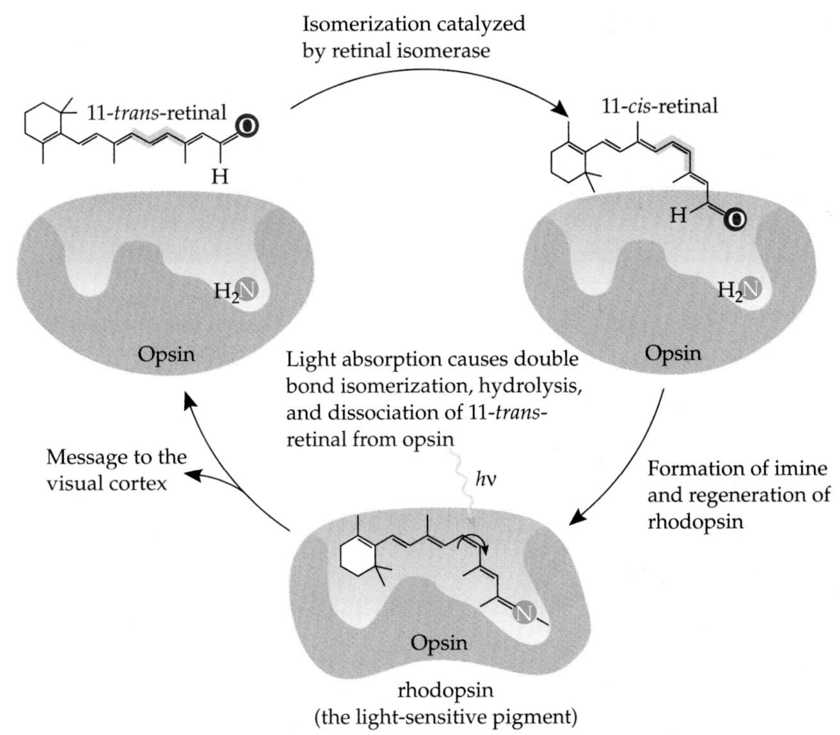

The primary chemical reaction of vision, occurring in the red cells of the eyes, is absorption of light by rhodopsin, followed by isomerization of a carbon–carbon double bond from a *cis* configuration to a *trans* configuration.

10.3 MOLECULAR ORBITAL THEORY

Molecular orbital (MO) theory is an alternative way to view orbitals in molecules. In contrast to the localized bond and lone pair electrons of valence bond theory, MO theory assumes that pure *s* and *p* atomic orbitals of the atoms in the molecule combine to produce orbitals that are spread out, or delocalized, over several atoms or even over an entire molecule. These new orbitals are called **molecular orbitals.**

One reason for learning about the MO concept is that it correctly predicts the electronic structures of certain molecules that do not follow the electron-pairing assumptions of the Lewis approach. The most common example is the O_2 molecule. The rules of Chapter 9 would guide you to draw the electron dot structure of O_2 with all the electrons paired, which fails to explain its paramagnetism (Figure 10.20). The molecular orbital approach can account for this prop-

This dot structure of O_2 was constructed using the guidelines in Chapter 9. It fails, however, to explain the observed paramagnetism of the molecule, which requires the presence of unpaired electrons.

O=O

Figure 10.20 Liquid oxygen. Oxygen in the liquid state is paramagnetic and clings to the poles of a magnet *(right)*. Oxygen gas condenses to a liquid at −183 °C *(left)*. Notice that liquid oxygen is very pale blue *(middle)*. See also the Chemical Puzzler, page 438. *(C. D. Winters)*

erty, but valence bond theory cannot. To see how MO theory can be used to describe the bonding in O_2 and other diatomic molecules, we shall first describe four principles used to develop the theory.

Principles of Molecular Orbital Theory

In MO theory we begin with a given arrangement of atoms in the molecule at the known bond distances. We then determine the *sets* of molecular orbitals that can result from combining all the available orbitals on all the constituent atoms. These molecular orbitals more or less encompass all the atoms of the molecule, and the valence electrons for all the atoms in the molecule are assigned to the molecular orbitals. Just as with orbitals in atoms, electrons are assigned according to the Pauli principle and Hund's rule (Sections 8.2 and 8.4).

The **first principle of molecular orbital theory** is that *the total number of molecular orbitals produced is always equal to the total number of atomic orbitals contributed by the atoms that have combined.* To see the consequence of this orbital conservation principle, let us consider the H_2 molecule.

Bonding and Antibonding Molecular Orbitals in H_2

Orbitals are characterized as waves; therefore, a way to view molecular orbital formation is to assume that two electron waves, one from each atom, interfere with each other. The interference can be constructive, giving a bonding MO, or destructive, giving an antibonding MO.

Molecular orbital theory specifies that when the $1s$ orbitals of two hydrogen atoms overlap, *two* molecular orbitals result from the addition and subtraction of the overlapping orbitals. In the molecular orbital resulting from *addition* of the atomic orbitals, the $1s$ regions of electron density add together, leading to an increased probability that electrons will reside in the bond region between the two nuclei (Figure 10.21). This is called a **bonding molecular orbital** and is the same as that described as a chemical bond by valence bond theory. It is also a σ orbital because the region of electron probability lies directly along the bond axis. This molecular orbital is labeled σ_{1s}, the subscript $1s$ indicating that $1s$ atomic orbitals were used to create the molecular orbital.

The other molecular orbital is constructed by *subtracting* one atomic orbital from the other (see Figure 10.21). When this happens, the probability of finding an electron between the nuclei in the molecular orbital is reduced, and the probability of finding the electron in other regions is higher. Without signifi-

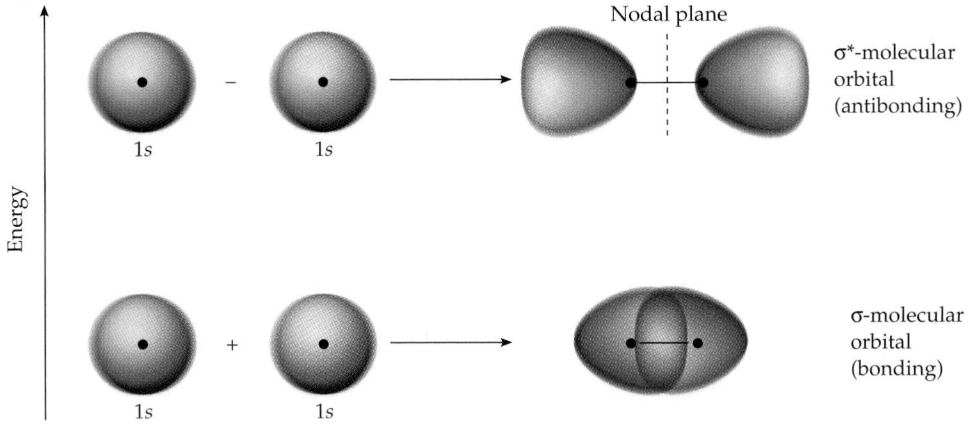

Figure 10.21 Molecular orbitals. Bonding and antibonding σ molecular orbitals are formed from two $1s$ atomic orbitals on adjacent atoms. Notice the presence of a node in the antibonding orbital. (A node is a plane on which there is zero probability of finding an electron.)

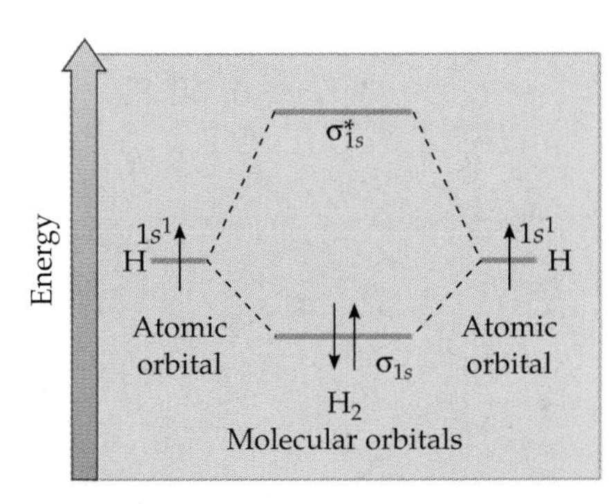

Figure 10.22 A molecular orbital energy level diagram for hydrogen, H_2. Bonding and antibonding molecular orbitals are created from two $1s$ atomic orbitals, one from each H atom. The two electrons of H_2 are placed in the σ_{1s} orbital, the lowest energy molecular orbital.

cant electron density between them, the nuclei repel one another. This type of orbital is called an **antibonding molecular orbital.** Because it is also a σ orbital, it is labeled σ^*_{1s}. The asterisk signifies that it is antibonding. *Antibonding orbitals have no counterpart in valence bond theory.*

A **second principle of molecular orbital theory** is that *the bonding molecular orbital is lower in energy than the parent orbitals, and the antibonding orbital is higher in energy* (Figure 10.22). This means that the energy of a group of atoms is lower than the energy of the separated atoms when electrons are assigned to bonding molecular orbitals. Chemists say the system is "stabilized," and chemical bonds are formed. Conversely, the system is "destabilized" when electrons are assigned to antibonding orbitals because the energy of the system is higher than that of the atoms themselves.

Antibonding orbitals have energies higher than the average energy of the atomic orbitals from which they were created. Their upward energy displacement is slightly greater than the downward displacement of the bonding orbitals. A consequence is that the average energy of molecular orbitals is slightly higher than the average energy of the parent atomic orbitals.

A **third principle of molecular orbital theory** is that the *electrons of the molecule are assigned to orbitals of successively higher energy* according to the Pauli exclusion principle and Hund's rule. This is analogous to the procedure for building up electronic structures of atoms. Thus, electrons occupy the lowest energy orbitals available, and when two electrons are assigned to an orbital, their spins are paired. Because the energy of the electrons in the bonding orbital of H_2 is lower than that of either parent $1s$ electron (see Figure 10.22), the H_2 molecule is stable. We write the electron configuration of H_2 as $(\sigma_{1s})^2$.

What would happen if we try to combine two helium atoms to form dihelium, He_2? Both He atoms have a $1s$ valence orbital that can be added and subtracted to produce the same kind of molecular orbitals as in H_2. Unlike H_2, however, four electrons need to be assigned to these orbitals (Figure 10.23). The pair of electrons in the σ_{1s} orbital stabilizes He_2. The two electrons in σ^*_{1s}, however, destabilize the He_2 molecule. The energy decrease from the electrons in the σ_{1s}-bonding molecular orbital is offset by the energy increase due to the electrons in the σ^*_{1s}-antibonding molecular orbital. Thus, molecular orbital theory predicts that He_2 has no net stability; two He atoms have no tendency to combine. This confirms what we already know, that elemental helium exists in the form of single atoms and not as a diatomic molecule.

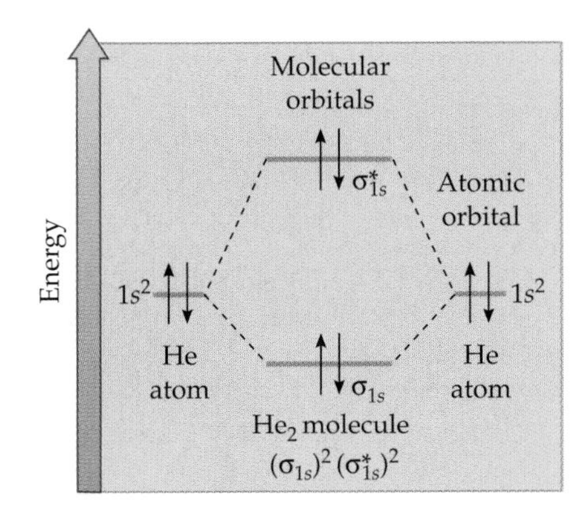

Figure 10.23 A molecular orbital energy level diagram for the dihelium molecule, He_2. This molecule does not exist. In such a molecule, both the bonding (σ_{1s}) and antibonding orbitals (σ^*_{1s}) would be fully occupied.

Bond Order

Bond order was defined in Chapter 9 as the net number of bonding electron pairs linking a pair of atoms. This same concept can be applied directly to molecular orbital theory, but now bond order is defined as

$$\text{Bond order} = \frac{1}{2}(\text{number of electrons in bonding MOs} - \text{number of electrons in antibonding MOs})$$

In the H_2 molecule, there are two electrons in a bonding orbital and none in an antibonding orbital, so H_2 has a bond order of 1. In contrast, in He_2 the stabilizing effect of the σ_{1s} pair is canceled by the destabilizing effect of the σ^*_{1s} pair, so the bond order is 0.

Fractional bond orders are possible. Consider the ion He_2^+. Its molecular orbital electron configuration is $(\sigma_{1s})^2(\sigma^*_{1s})^1$. In this ion, there are two electrons in a bonding molecular orbital, but only one in an antibonding orbital. MO theory predicts that He_2^+ should have a bond order of 0.5, that is, a weak bond should exist between helium atoms in such a species. Interestingly, this ion has been identified in the gas phase, using special experimental techniques.*

Example 10.7 Molecular Orbitals and Bond Order

Write the electron configuration of the H_2^- ion in molecular orbital terms. What is the bond order of the ion?

Solution This ion has three electrons (one each from the H atoms plus 1 for the negative charge). Therefore, its electronic configuration is $(\sigma_{1s})^2(\sigma^*_{1s})^1$, identical with the configuration for He_2^+. This means H_2^- also has a net bond order of 0.5. The ion is predicted to exist under special circumstances.

Exercise 10.7 Molecular Orbitals and Bond Order

What is the electron configuration of the H_2^+ ion? Compare the bond order of this ion with He_2^+ and H_2^-. Do you expect H_2^+ to exist?

Molecular Orbitals of Li_2 and Be_2

A **fourth principle of molecular orbital theory** is that *atomic orbitals combine to form molecular orbitals most effectively when the atomic orbitals are of similar energy.* This principle becomes important when we move past He_2 to Li_2, dilithium, and heavier molecules still.

Is it really true that the Starship *Enterprise* is fueled with dilithium crystals?

*The experiments involve supplying energy sufficient to ionize atoms of gaseous helium and then looking for spectroscopic evidence of transient (short-lived) species that form when He^+ and He combine. This is one of several techniques that are used to observe species that are unstable and cannot be isolated. Enough is known about He_2^+ from theory and experiment that we can safely predict that this ion will not exist in solution or in the solid state in the form of a salt.

A lithium atom has electrons in two orbitals of the s type ($1s$ and $2s$), so a $1s \pm 2s$ combination is theoretically possible. Because the $1s$ and $2s$ orbitals are quite different in energy, however, this interaction cannot make an important contribution. Thus, the molecular orbitals come only from $1s \pm 1s$ and $2s \pm 2s$ interactions (Figure 10.24). This means the molecular orbital electron configuration of dilithium, Li_2, is $(\sigma_{1s})^2(\sigma^*_{1s})^2(\sigma_{2s})^2$. The bonding effect of the σ_{1s} electrons is canceled by the antibonding effect of the σ^*_{1s} electrons, so these pairs make no net contribution to bonding in Li_2. Bonding in Li_2 is due to the electron pair assigned to the σ_{2s} orbital, and the bond order is 1.

The fact that the σ_{1s} and σ^*_{1s} electron pairs of Li_2 make no net contribution to bonding is exactly what you observed in drawing electron dot structures in Chapter 9: core electrons are ignored. In molecular orbital terms, core electrons are assigned to bonding and antibonding molecular orbitals that offset one another.

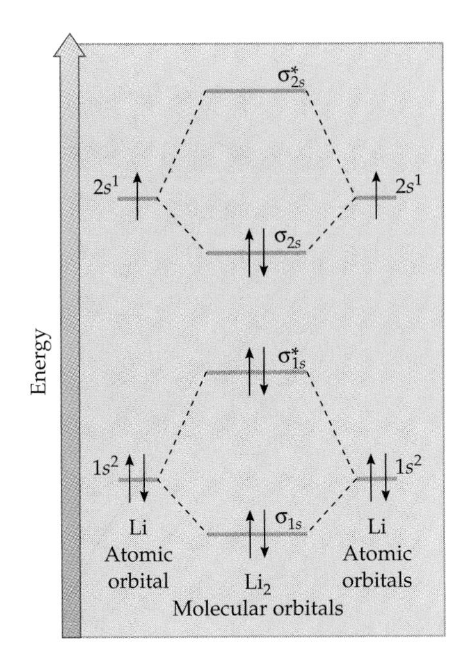

Figure 10.24 Energy level diagram for the combination of two Li atoms with $1s$ and $2s$ atomic orbitals. Notice that the molecular orbitals are created by combining orbitals of similar energies. The electron configuration is shown for Li_2.

A diberyllium molecule, Be_2, is not expected to exist. Its electron configuration is

Be_2 MO Configuration: [core electrons]$(\sigma_{2s})^2(\sigma^*_{2s})^2$

Bond order: $\frac{1}{2}$ (2 bonding electrons − 2 antibonding electrons) = 0

The effects of σ_{2s} and σ^*_{2s} electrons cancel, and there is no net bonding. This species does not exist.

Example 10.8 Molecular Orbitals in Diatomic Molecules

Be_2 does not exist. But what about the Be_2^+ ion? Describe its electron configuration in molecular orbital terms and give the net bond order. Do you expect the ion to exist?

Solution The Be_2^+ ion has only seven electrons (in contrast to eight for Be_2), of which four are core electrons. (The core electrons are assigned to σ_{1s} and σ^*_{1s} molecular orbitals). The remaining three electrons are assigned to the σ_{2s} and σ^*_{2s} molecular orbitals (see Figure 10.24), so the MO electron configuration is [core electrons]$(\sigma_{2s})^2(\sigma^*_{2s})^1$. These means the net bond order is 0.5, and so Be_2^+ is predicted to exist.

Exercise 10.8 Molecular Orbitals in Diatomic Molecules

Could the anion Li_2^- exist? What is the ion's bond order?

Molecular Orbitals from Atomic *p* Orbitals

With the principles of molecular orbital theory in place, we are ready to account for bonding in such important homonuclear diatomic molecules as N_2, O_2, and F_2. First, however, we need to see what types of molecular orbitals form when elements have both s and p valence orbitals. Three types of interactions are possible for two atoms that both have s and p orbitals. Sigma-bonding and anti-

A molecule such as H_2 or Li_2, formed from two identical atoms, is called a **homonuclear diatomic molecule.**

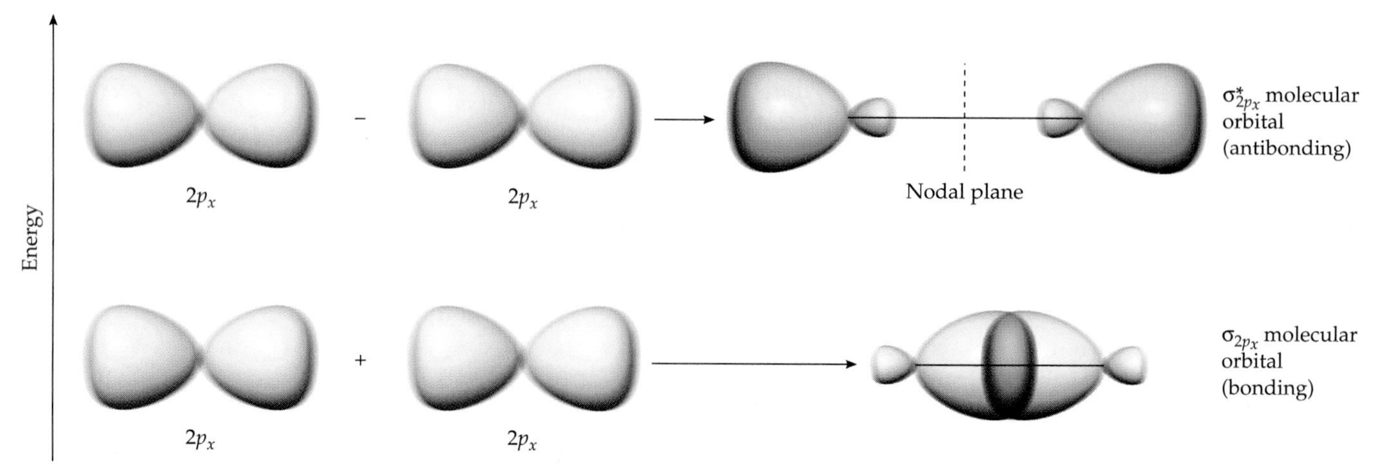

Figure 10.25 Molecular orbitals from p atomic orbitals. Sigma-bonding (σ_{2p}) and antibonding (σ^*_{2p}) molecular orbitals arise from overlap of 2p orbitals. Each orbital can accommodate two electrons. The p orbitals in electron shells of higher n give molecular orbitals of the same basic shape.

bonding molecular orbitals are formed by s orbitals interacting as in Figure 10.21. Similarly, it is possible for a p orbital on one atom to interact with a p orbital on the other atom in a head-to-head fashion to produce a pair of σ-bonding and σ^*-antibonding molecular orbitals (Figure 10.25). And finally, each atom has two p orbitals in planes perpendicular to the σ bond connecting the two atoms. These p orbitals can interact sideways to give pi-bonding and antibonding molecular orbitals (Figure 10.26). The two p orbitals on each atom produce *two* pi-bonding molecular orbitals (π_p) and *two* pi-antibonding molecular orbitals (π^*_p).

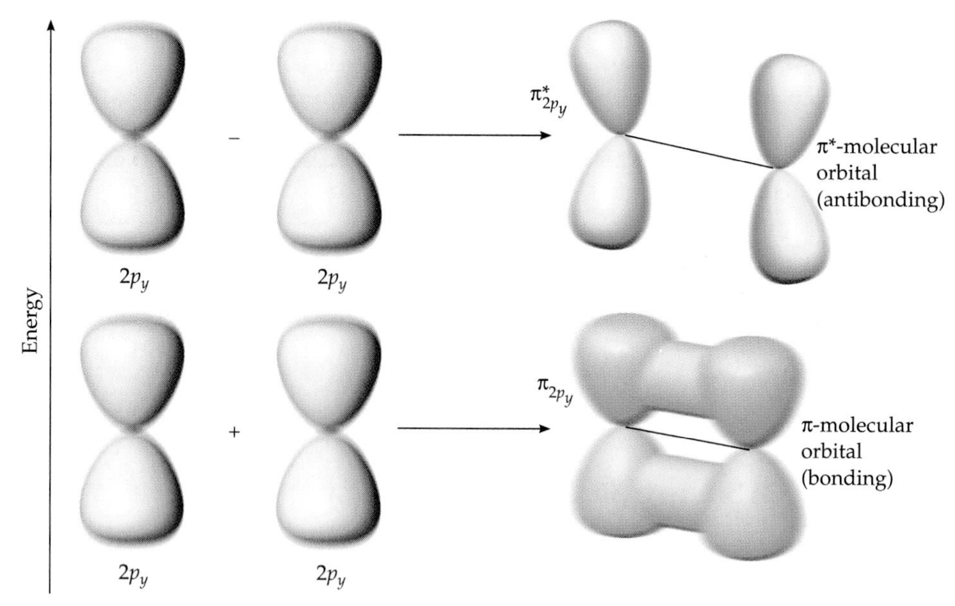

Figure 10.26 Formation of π molecular orbitals. Sideways overlap of atomic 2p orbitals that lie in the same direction in space gives rise to pi-bonding (π_{2p}) and antibonding (π^*_{2p}) molecular orbitals. The p orbitals in shells of higher n give molecular orbitals of the same basic shape.

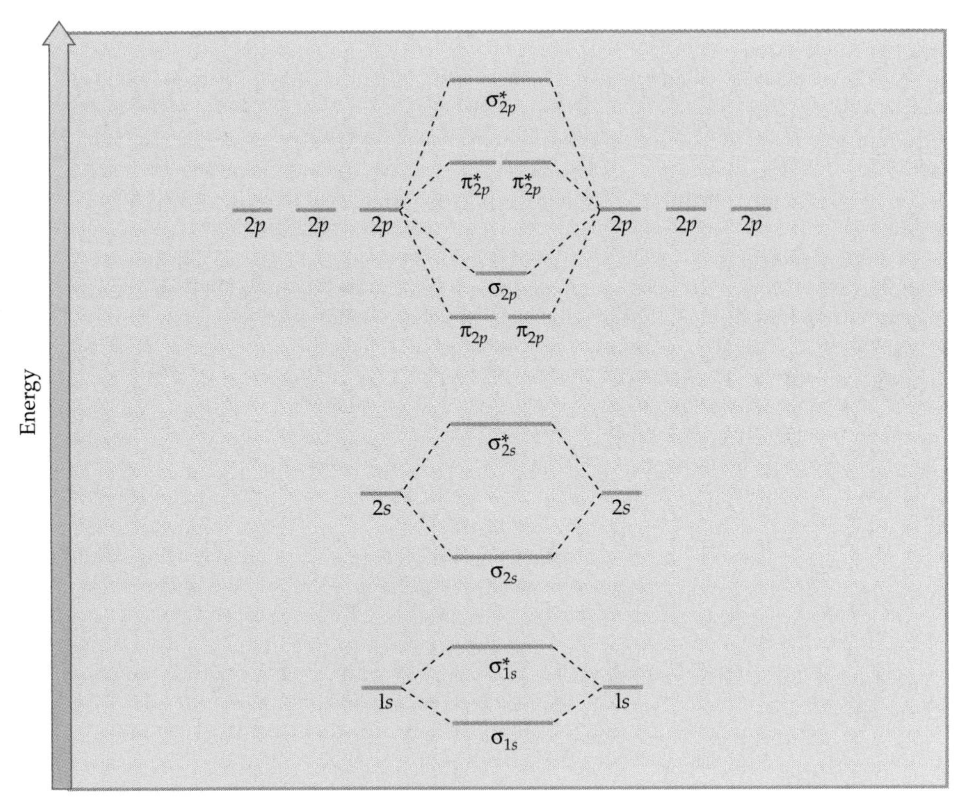

Figure 10.27 Homonuclear diatomic molecules. The molecular orbital energy level diagram for second-period homonuclear diatomic molecules.

Electron Configurations for Homonuclear Molecules for Boron Through Fluorine

Orbital interactions in a second-period, homonuclear, diatomic molecule lead to the energy level diagram in Figure 10.27. Electron assignments can be made using this diagram, and the results for the diatomic molecules B_2 through F_2 are tabulated in Table 10.1, which has two noteworthy features.

First, notice the correlation between the electron configurations and the bond orders, bond lengths, and bond energies at the bottom of Table 10.1. As the bond order between a pair of atoms increases, the energy required to break the bond increases, and the bond distance decreases. Dinitrogen, N_2, with a bond order of 3, has the largest bond energy and shortest bond distance.

Second, notice the configuration for dioxygen, O_2. Dioxygen has 12 valence electrons (6 from each atom), so it has the molecular orbital configuration

O_2 MO Configuration: [core electrons]$(\sigma_{2s})^2(\sigma^*_{2s})^2(\pi_{2p})^4(\sigma_{2p})^2(\pi^*_{2p})^2$

This configuration leads to a bond order of 2 as required by experiment, and it specifies two unpaired electrons (in π^*_{2p} molecular orbitals). Thus, molecular theory succeeds where valence bond theory failed. MO theory explains both the observed bond order and the paramagnetic behavior of O_2.

Table 10.1 • MOLECULAR ORBITAL OCCUPATIONS AND PHYSICAL DATA FOR HOMONUCLEAR DIATOMIC MOLECULES OF SECOND-PERIOD ELEMENTS

	B_2	C_2	N_2	O_2	F_2
σ^*_{2p}	[ ]	[ ]	[ ]	[ ]	[ ]
π^*_{2p}	[ ][ ]	[ ][ ]	[ ][ ]	[↑][↑]	[↑↓][↑↓]
σ_{2p}	[ ]	[ ]	[↑↓]	[↑↓]	[↑↓]
π_{2p}	[↑][↑]	[↑↓][↑↓]	[↑↓][↑↓]	[↑↓][↑↓]	[↑↓][↑↓]
σ^*_{2s}	[↑↓]	[↑↓]	[↑↓]	[↑↓]	[↑↓]
σ_{2s}	[↑↓]	[↑↓]	[↑↓]	[↑↓]	[↑↓]
Bond order	One	Two	Three	Two	One
Bond-dissociation energy (kJ/mol)	290	620	945	498	155
Bond distance (pm)	159	131	110	121	143
Observed magnetic behavior (paramagnetic or diamagnetic)	Para	Dia	Dia	Para	Dia

Example 10.9 Electron Configuration for a Homonuclear Diatomic Ion

When potassium reacts with O_2, potassium superoxide, KO_2, is one of the products. This is an ionic compound, and the anion is the superoxide ion, O_2^-. Write the molecular orbital electron configuration for the ion. Predict its bond order and magnetic behavior.

Solution Use the energy level diagram of Figure 10.26 to generate the configuration of this ion.

O_2^- MO Configuration: [core electrons]$(\sigma_{2s})^2(\sigma^*_{2s})^2(\pi_{2p})^4(\sigma_{2p})^2(\pi^*_{2p})^3$

The ion is predicted to be paramagnetic to the extent of one unpaired electron, a prediction confirmed by experiment. The bond order is 1.5, because there are eight bonding electrons and five antibonding electrons. The bond order for O_2^- is lower than O_2, so we predict the O—O bond length in O_2^- should be longer than the oxygen–oxygen bond length in O_2. The superoxide ion has an O—O bond length of 134 pm, whereas the bond length in O_2 is 121 pm.

You should quickly spot the fact that the superoxide ion (O_2^-), contains an odd number of electrons. This is a third species (in addition to NO and O_2) for which it is not possible to write a Lewis structure that accurately represents the bonding.

A CLOSER LOOK *Molecular Orbitals for Compounds Formed from* p*-Block Elements*

Several features of the molecular orbital energy level diagram in Figure 10.27 might be described in more detail.

- The bonding and antibonding σ orbitals from $2s$ interactions are lower in energy than the σ and π MOs from $2p$ interactions. The reason is that $2s$ orbitals have a lower energy than $2p$ orbitals in the separated atoms.
- The separation of bonding and antibonding orbitals is greater for σ_{2p} than for π_{2p}. This happens because p orbitals overlap to a greater extent when they are oriented head to head (to give σ_{2p} MOs) than when they are side by side (to give π_{2p} MOs). The greater the orbital overlap, the greater the stabilization of the bonding MO and the greater the destabilization of the antibonding MO.

Figure 10.27 shows an energy ordering of molecular orbitals that you might not have expected, but there are reasons for this. A more sophisticated approach takes into account the "mixing" of s and p atomic orbitals, which have similar energies. This causes the σ_{2s} and σ^*_{2s} molecular orbitals to be lower in energy than otherwise expected, and the σ_{2p} and σ^*_{2p} orbitals to be pushed up in energy. This is the reason the energy lowering and raising for the σ_{2s} and σ^*_{2s} orbitals (and for the σ_{2p} and σ^*_{2p} orbitals) in Figure 10.27 is not symmetrical with respect to the $2s$ and $2p$ atomic orbital energies.

Another refinement concerning Figure 10.27 is that, because s and p orbital mixing is important only for B_2, C_2, and N_2, the figure applies strictly only to these molecules. For O_2 and F_2, σ_{2p} is lower in energy than π_{2p}. Nonetheless, Figure 10.27 gives the correct bond order and magnetic behavior for these two molecules.

Electron Configurations for Heteronuclear Diatomic Molecules

The molecules NO, CO, and ClF are examples of simple diatomic molecules formed from two elements of different kinds. These compounds are called heteronuclear diatomic molecules. MO descriptions for heteronuclear diatomic molecules generally resemble those for homonuclear diatomic molecules. As a consequence, an energy level diagram like Figure 10.27 can be used to judge the bond order and magnetic behavior for heteronuclear diatomics.

Let us do this for nitrogen monoxide, NO. Nitrogen monoxide has 11 molecular valence electrons. If these are assigned to the MOs for a homonuclear diatomic molecule, the molecular electron configuration is

NO MO Configuration: [core electrons] $(\sigma_{2s})^2(\sigma^*_{2s})^2(\pi_{2p})^4(\sigma_{2p})^2(\pi^*_{2p})^1$

The net bond order is 2.5, in accord with bond length information. The single, unpaired electron is assigned to the π^*_{2p} molecular orbital. The molecule is paramagnetic, as predicted for a molecule with an odd number of electrons.

Exercise 10.9 Molecular Electron Configurations

The cations O_2^+ and N_2^+ are important components of earth's upper atmosphere. Write the electron configuration of O_2^+. Predict its bond order and magnetic behavior.

Resonance and MO Theory

Ozone, O_3, is a simple triatomic molecule with equal oxygen–oxygen bond lengths. Equal X—O bond lengths are also observed in other molecules and ions, such as SO_2, NO_2^-, and HCO_2^-. Valence bond theory introduced resonance to rationalize the equivalent bonding to the oxygen atoms in these structures. MO theory provides another view of this problem.

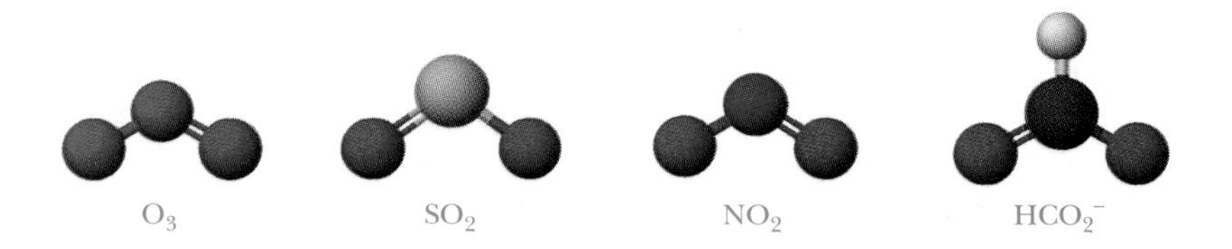

To visualize the bonding in ozone, let us begin by assuming that all three O atoms are sp^2-hybridized. The central atom uses its sp^2 hybrid orbitals to form two σ bonds and to accommodate a lone pair. The terminal atoms use their sp^2 hybrids to form one σ bond and to accommodate two lone pairs. In all, the lone pairs and bonding pairs in the σ framework of O_3 account for seven of the nine valence electron pairs in O_3.

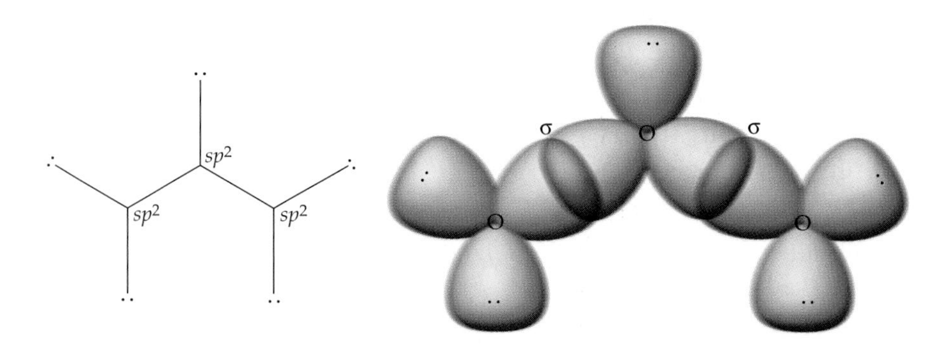

The π bond in ozone arises from the two remaining pairs (Figure 10.28). Because we have assumed each oxygen atom in O_3 is sp^2-hybridized, an unhybridized p orbital perpendicular to the O_3 plane remains on each of the three oxygen atoms. The orbitals are in the correct orientation to form π bonds. A principle of MO theory is that the number of molecular orbitals must equal the number of atomic orbitals. Thus, the three $2p$ atomic orbitals must be combined in a way that forms three molecular orbitals.

One π_p MO for ozone is a bonding orbital because the three p orbitals are "in phase" across the molecule. Another π_p MO is antibonding because the central atomic orbital is "out of phase" with the terminal atom p orbitals. The third π_p MO is nonbonding because the middle p orbital does not participate in the MO. The bonding π_p MO is filled by a pair of electrons, which is delocalized, or "spread over," the molecule, just as the resonance hybrid implies. The nonbonding orbital is also occupied, but the electrons in this orbital are concentrated near the two terminal oxygens. As the name implies, electrons in this molecular orbital neither help nor hinder the bonding in the molecule. The π bond order of O_3 is 0.5, since one bond pair is spread over two O—O linkages. Because the σ bond order is 1.0 and the π bond order is 0.5, the net oxygen–oxygen bond order is 1.5, the same value given by valence bond theory.

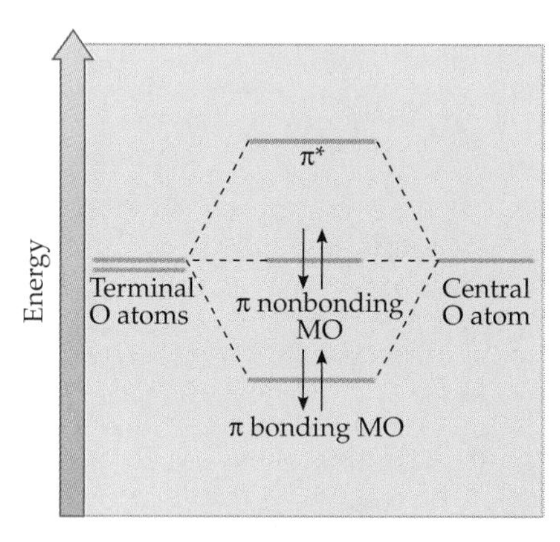

MO energy level diagram for the π bonds in ozone.

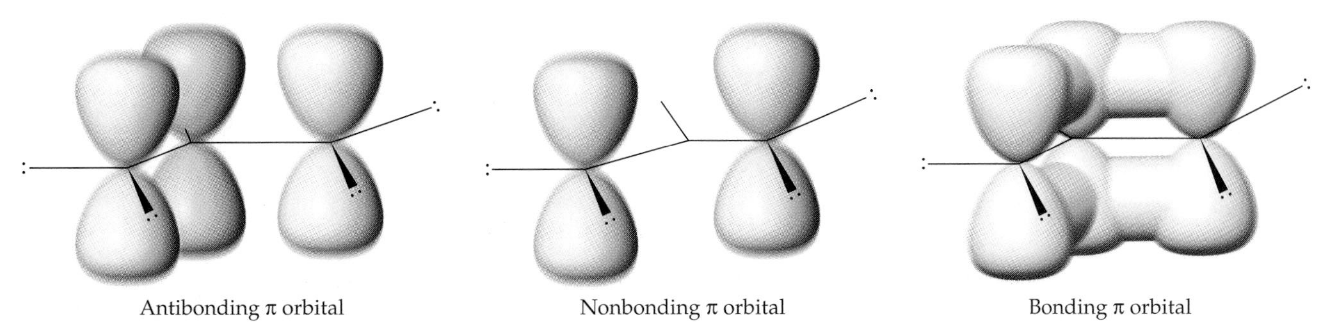

Figure 10.28 Pi-bonding in ozone, O_3. Each O atom in O_3 is sp^2-hybridized. The three $2p$ orbitals, one on each atom, are used to create the three π molecular orbitals. Two pairs of electrons are assigned to the orbitals: one pair in the bonding orbital and one pair in the nonbonding orbital. The π bond order is 0.5, one bonding pair spread across two bonds.

The observation that two of the π molecular orbitals for ozone extend over three atoms illustrates an important point regarding molecular orbital theory: orbitals can extend beyond two atoms. In valence bond theory, all representations for bonding were based on being able to localize pairs of electrons in bonds between two atoms. To further illustrate the MO approach, look again at benzene. On page 458 the point was made that the π electrons in this molecule were spread out over all six carbon atoms. We can now see how the same case can be made with MO theory. Six p orbitals contribute to the π system. Based on the premise that there must be the same number of molecular orbitals as atomic orbitals, there must be six π molecular orbitals in benzene. An energy level diagram for benzene shows that the six p electrons reside in the three lowest energy (bonding) molecular orbitals (Figure 10.29).

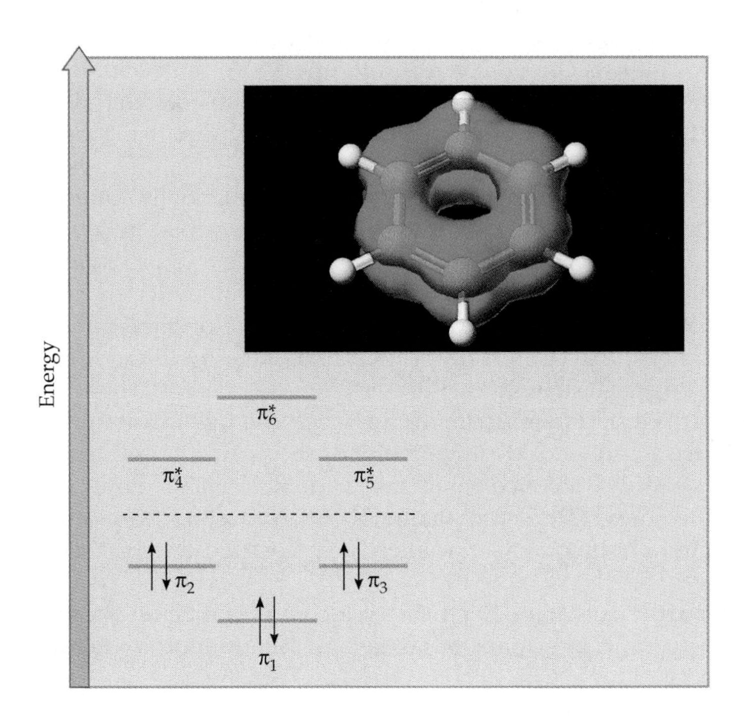

Figure 10.29 Molecular orbital energy level diagram for benzene. Because there are six electrons in the six unhybridized p orbitals, six π molecular orbitals can be formed. Three are bonding and three are antibonding. The three bonding molecular orbitals accommodate the six π electrons. Also illustrated is one of the molecular orbitals as calculated by computer. (All the benzene π molecular orbitals are in the molecular-modeling directory on the *Saunders Interactive General Chemistry CD-ROM.*)

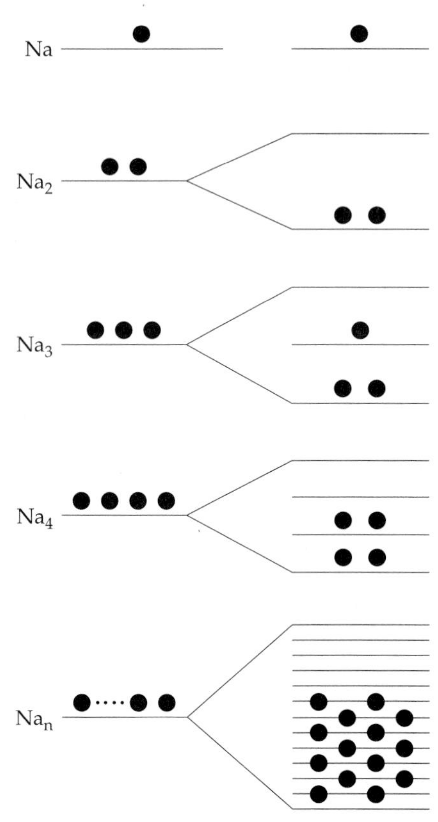

Figure 10.30 Bands of molecular orbitals in a metal crystal. As more and more atoms with the same valence orbitals are added, the number of molecular orbitals grows until the orbitals are so close in energy that they merge into a band.

10.4 METALS AND SEMICONDUCTORS

The simple molecular orbital model used to describe diatomic molecules can be extended to describe the properties of metals and semiconductors.

Conductors, Insulators, and Band Theory

Metal crystals can be viewed as "supermolecules" held together by delocalized bonds formed from the atomic orbitals of all the atoms in a crystal. Even the tiniest piece of metal contains a very large number of atoms, and an even larger number of orbitals is available to form molecular orbitals. If four lithium atoms, for example, are in a group, and each Li atom contributes one $2s$ and three $2p$ orbitals to metallic bonding, 16 molecular orbitals are possible. Four hundred Li atoms taken together would lead to 1600 molecular orbitals, and in a single crystal containing a mole of lithium atoms, there would be $4 \times (6.022 \times 10^{23})$ molecular orbitals.

In a metal, the molecular orbitals spread out over many atoms and blend into a band of molecular orbitals, the energies of which are closely spaced within a range of energies (Figure 10.30). The band is composed of as many orbitals as there are contributing atomic orbitals, and each orbital can accommodate two electrons of opposite spin. The idea that the molecular orbitals of the band of energy levels are spread out, or *delocalized*, over all the atoms in a piece of metal accounts for the bonding in metallic solids. This theory of metallic bonding is called **band theory.**

In a metal, the band of energy levels is only partially filled; there are not enough electrons to fill all of the orbitals (Figure 10.30). Electrons fill the lowest energy molecular orbitals, but the lowest energy for a system (with all electrons in orbitals with the lowest possible energy) is reached only at 0 K. The highest filled level at 0 K is called the **Fermi level** (Figure 10.31). A small input of energy (for example, raising the temperature above 0 K), can cause electrons to move from the filled portion of the band to the unfilled portion. For each electron promoted, two singly occupied levels result, one above the Fermi level and one below. It is the movement of electrons in singly occupied states close to the Fermi level in the presence of an applied electric field that is responsible for the electrical conductivity of metals.

Because the band of unfilled energy levels in a metal is essentially continuous, that is, the energy gaps between orbitals are extremely small, a metal can absorb energy of nearly any wavelength. When light causes an electron in a metal to move to a higher energy state, the now-excited system can immediately emit a photon of the same energy, and the electron returns to the original energy level. It is this rapid and efficient reemission of light that makes polished metal surfaces reflective and appear lustrous.

A metal is characterized by a band structure in which the highest occupied band, called the **valence band,** is only partially filled. In contrast an electrical **insulator** has a completely filled valence band, and the next available empty levels are at much higher energy. The consequence of this is that the promotion of an electron to a higher energy level is not likely, and the solid does not conduct electricity.

Diamonds are electrical insulators. Each carbon atom in a diamond is surrounded by four other carbon atoms at the corners of a tetrahedron (Figure

2.21). In the valence bond model of bonding, each atom is viewed as being sp^3 hybridized, forming four localized carbon—carbon bonds with other carbon atoms. An alternative view, however, is that the orbitals of each carbon atom form molecular orbitals that are delocalized over the solid, representing bonding extending over the whole solid. That is, a band model can also be applied when discussing this solid. In this picture the levels are split into two bands, a *filled* valence band of bonding molecular orbitals and a higher-energy *unfilled* band of antibonding orbitals called the **conduction band** (for a reason that will become apparent in the following discussion). The energy difference separating the two bands is called the **band gap** (Figure 10.32).

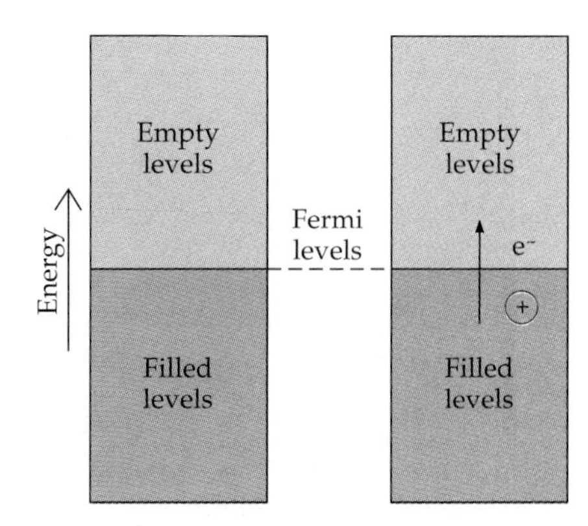

Figure 10.31 The partially filled band of "molecular orbitals" in a metal. (*Left*) The highest filled level is referred to as the Fermi level. The molecular orbitals are delocalized over the metal, a fact accounting for the bonding in metals. Electrons can be promoted from the filled levels to empty levels by the input of a modest amount of energy. (*Right*) The electrons are freer to move in the now partially filled levels; this property accounts for the electrical conductivity of metals.

Semiconductors

The Group 4A elements silicon and germanium have structures similar to diamond, that is, these solids contain tetrahedral silicon or germanium atoms linked by covalent bonds. In contrast to diamond, however, silicon and germanium are semiconductors. **Semiconductors** are materials that are able to conduct small quantities of current. They are much poorer conductors than metals, but they are not insulators. Why are these three structurally similar substances—diamond, silicon, and germanium—different in their conductivity?

The answer lies in the size of the band gap. If an electron is promoted from the valence band to the conduction band, singly occupied states in the valence and conduction bands are formed. As with metals, such a state would allow for electrical conductivity. The band gap in diamond is large, however, on the order of 500 kJ/mol, reflecting the strong bonds between carbon atoms. Semiconductors have band gaps in the range of 50 to 300 kJ/mol. The band gap narrows as one descends Group 4A, and the band gaps of silicon and germanium fall in this range. At least a few electrons can be promoted into the conduction band with the input of only modest amounts of energy, and electrical conduction can occur.

Pure silicon and germanium belong to a class of materials called **intrinsic** semiconductors. The usual view of such materials is that the promotion of an electron from the valence band to the conduction band creates a positive hole in the valence band (Figure 10.33). The semiconductor carries charge because the electrons in the conduction band migrate in one direction and the positive holes in the valence band migrate in the opposite direction. Positive holes "move" because an electron from an adjacent level can move into the hole, thus creating a fresh "hole."

In intrinsic semiconductors, the number of electrons in the conduction band is entirely governed by the temperature and the magnitude of the band gap. The smaller the band gap, the smaller the energy required to promote a significant number of electrons. As the temperature increases, a larger number of electrons can be promoted into the conduction band.

In contrast to intrinsic semiconductors are materials known as **extrinsic** semiconductors. The conductivity of these materials is controlled by adding small numbers of atoms of different kinds of impurities called *dopants*. For example, suppose a few silicon atoms in the silicon lattice are replaced by aluminum atoms (or atoms of some other Group 3A element). Aluminum has only three valence electrons, whereas silicon has four. Four Si—Al bonds are created per aluminum atom in the lattice, but these bonds must be deficient in electrons. According

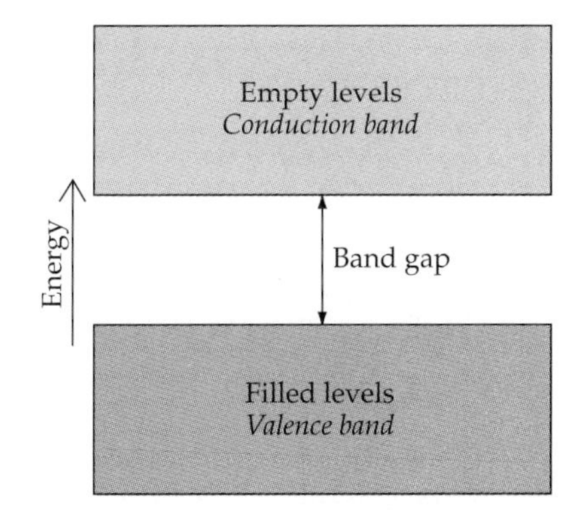

Figure 10.32 Band theory applied to semiconductors and insulators. In contrast to metals, the band of filled levels (the valence band) is separated from the band of empty levels (the conduction band) by a band gap. The band gap can range from just a few kJ/mol to 500 kJ/mol or more.

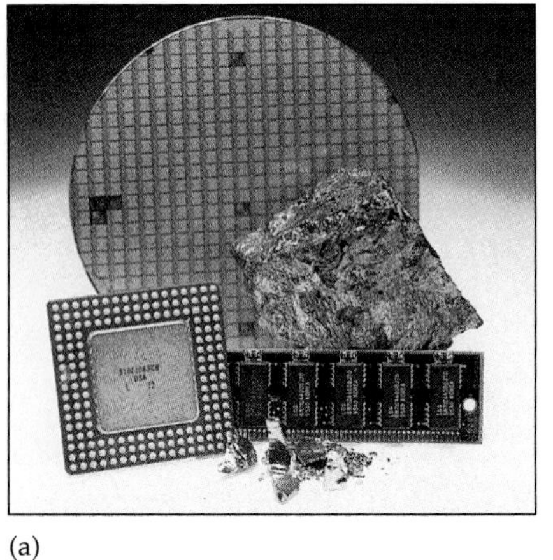

(a)

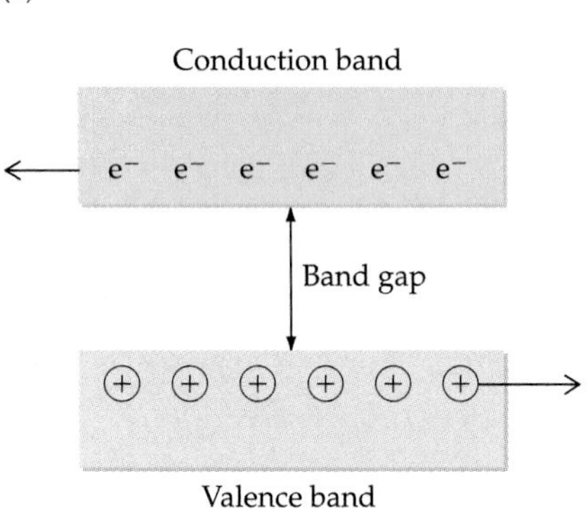

(b)

Figure 10.33 Semiconductors. (a) Various forms of elemental silicon. (b) Positive and negative charge carriers in an intrinsic semiconductor. *(a, C. D. Winters)*

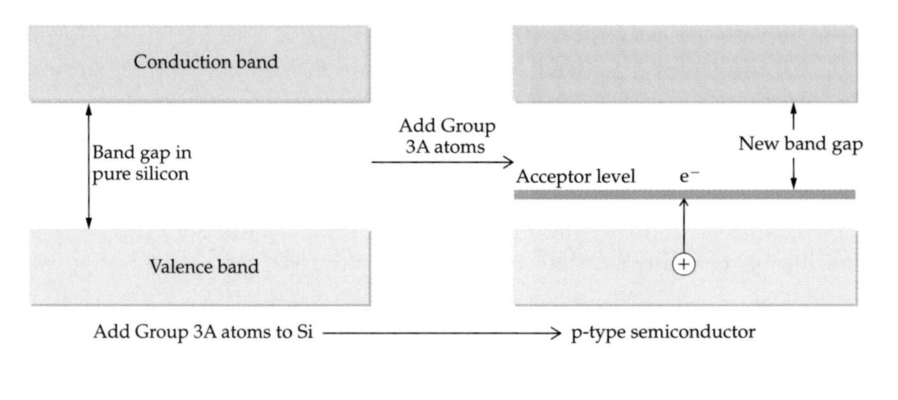

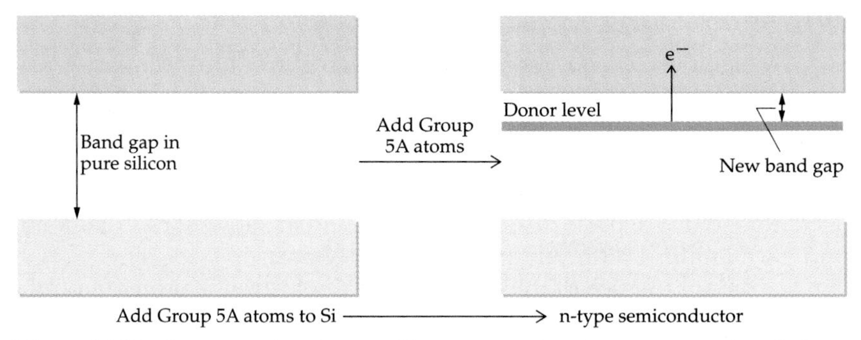

Figure 10.34 Doped semiconductors. *Top:* p-type semiconductors are produced when a 3A element is used as a dopant in a silicon lattice. *Bottom:* n-type semiconductors are produced when a Group 5A element is used as a dopant in a silicon lattice.

to band theory, the Si—Al bonds form a discrete band at an energy level higher than the valence band. This level is referred to as an *acceptor level* because it can accept electrons. The gap between the valence band and the acceptor level is usually quite small, so electrons can be promoted readily to the acceptor level. The positive holes created in the valence band are able to move about under the influence of an electric potential. Because positive holes are created in an aluminum-doped semiconductor, this is called a *p-type semiconductor* (Figure 10.34a).

Suppose phosphorus atoms are incorporated into the silicon lattice instead of aluminum atoms. The material is still a semiconductor, but it now has *extra* electrons because phosphorus has one more valence electron than silicon. Semiconductors doped in this manner have a discrete, *filled* donor level that is just below the conduction band. Electrons are promoted readily to the conduction band, and electrons in the conduction band carry the charge. Such a material is a negative charge carrier and is called an *n-type semiconductor* (Figure 10.34b).

CHAPTER HIGHLIGHTS

Having studied this chapter, you should be able to

- Describe the main features of **valence bond theory** and **molecular orbital theory,** the two commonly used theories for covalent bonding (Section 10.1).

- Recognize that the premise for valence bond theory is that bonding results from the **overlap** of atomic orbitals. By virtue of the overlap of orbitals, electrons are concentrated (or localized) between two atoms (Section 10.2).
- Distinguish how **sigma (σ)** and **pi (π) bonds** arise. For σ bonding, orbitals overlap in a head-to-head fashion, concentrating electrons along the bond axis. Sideways overlap of *p* atomic orbitals results in π bond formation, with electrons above and below the molecular plane (Section 10.2).
- Use the concept of hybridization to rationalize molecular structure (Section 10.2).

Hybrid Orbitals	Atomic Orbitals Used	Number of Hybrid Orbitals	Electron-Pair Geometry
sp	$s + p$	2	Linear
sp^2	$s + p + p$	3	Trigonal-planar
sp^3	$s + p + p + p$	4	Tetrahedral
sp^3d	$s + p + p + p + d$	5	Trigonal bipyramid
sp^3d^2	$s + p + p + p + d + d$	6	Octahedral

- Understand molecular orbital theory (Section 10.3), in which atomic orbitals are combined to form **bonding orbitals, nonbonding orbitals,** or **antibonding orbitals** that are delocalized over several atoms. In this description, the electrons of the molecule or ion are assigned to the orbitals beginning with the one at lowest energy, according to the Pauli exclusion principle and Hund's rule. The bond order, using MO theory, is 1/2(number of bonding electrons − number of antibonding electrons).
- Appreciate **band theory** and how it applies to solids, especially metals (Section 10.4).
- Understand the difference between a conductor of electricity, a **semiconductor,** and an **insulator** (Section 10.4).
- Recognize how dopants such as Group 3A and Group 5A elements affect semiconductor properties (Section 10.4).

STUDY QUESTIONS

Models of many of the molecules and ions in the following Study Questions are available on the Saunders Interactive General Chemistry CD-ROM *or on the Internet at http://www.saunderscollege.com.*

REVIEW QUESTIONS

1. What is the difference between a sigma (σ) and a pi (π) bond?
2. What is the maximum number of hybrid orbitals a carbon atom may form? What is the minimum number? Explain briefly.
3. What are the approximate angles between regions of electron density in sp, sp^2, and sp^3 hybrid orbital sets?
4. For each of the following electron-pair geometries, tell what hybrid orbital set is used: tetrahedral, linear, trigonal-planar, octahedral, and trigonal-bipyramidal.
5. If an atom is *sp*-hybridized, how many pure *p* orbitals remain on the atom? How many pi (π) bonds can the atom form?
6. Consider the three fluorides, BF_4^-, SiF_4, and SF_4.
   (a) Identify a molecule that is isoelectronic with BF_4^-.
   (b) Are SiF_4 and SF_4 isoelectronic?
   (c) Are the central atom hybrid orbitals the same or different in these three species?
7. What is the maximum number of hybrid orbitals a third-period element, say sulfur, can form? Explain briefly.
8. Give an example of a molecule with a central atom that has more than four valence electron pairs. Tell what hybrid orbitals are used by the atom.
9. What is one important difference between molecular orbital theory and valence bond theory?
10. Describe four principles of molecular orbital theory.

11. Sketch a picture of the bonding and antibonding molecular orbitals of H_2 and describe how they differ.
12. What is the connection between bond order, bond length, and bond energy? Use ethane (C_2H_6), ethylene (C_2H_4), and acetylene (C_2H_2) as examples.
13. When is it desirable to use MO theory rather than valence bond theory?
14. What is meant by the terms "localized" and "delocalized" as they pertain to the two bonding theories?
15. How do valence bond theory and molecular orbital theory differ in their explanation of the bond order of 1.5 for ozone?

Valence Bond Theory

16. Draw the Lewis structure for NF_3. What are its electron-pair and molecular geometries? Describe the bonding in the molecule in terms of hybrid orbitals.

17. Draw the Lewis structure for the ClF_2^+ ion. What are its electron-pair and molecular geometries? Describe the bonding in the molecule in terms of hybrid orbitals.

18. Draw the Lewis structure for chloroform, $CHCl_3$. What are its electron-pair and molecular geometries? Describe the bonding in the molecule in terms of hybrid orbitals.

19. Draw the Lewis structure for BeF_4^{2-}. What are its electron-pair and molecular geometries? Describe the bonding in the ion in terms of hybrid orbitals.

20. What hybrid orbital set is used by the underlined atom in each of the following molecules or ions?
(a) $\underline{B}Br_3$ (c) $\underline{C}HCl_3$
(b) $\underline{C}O_2$ (d) $\underline{C}H_2O$

21. What hybrid orbital set is used by the underlined atom in each of the following molecules or ions?
(a) $\underline{C}Se_2$ (c) $\underline{C}O_3^{2-}$
(b) $\underline{S}O_2$ (d) $\underline{N}H_4^+$

22. Describe the hybrid orbital set used by each of the indicated atoms in the molecules below:
(a) The carbon atoms and the oxygen atom in dimethyl ether, CH_3OCH_3
(b) Each carbon atom in propene

```
        H
        |
  H3C—C=CH2
```

(c) The two carbon atoms and the nitrogen atom in the amino acid glycine.

```
     H  H :O:
     |  |  ||   ..
  H—N—C—C—O—H
     ..  |     ..
        H
```

23. Give the hybrid orbital set used by each of the underlined atoms in the following molecules

(a)
```
     H :O: H
     |  ||  |
  H—N—C—N—H
     ..  ‾  ..
```
(underlined: N, C)

(b)
```
        H  H  H
        |  |  |   ..
  H3C—C=C—C=O
                  ..
```
(underlined: $H_3\underline{C}$, first C, last C, O)

(c)
```
      H  H
      |  |
  H—C=C—C≡N:
```
(underlined: first C, last C)

24. Draw the Lewis structure and then specify the electron-pair and molecular geometries for each of the following molecules or ions. Identify the hybridization of the central atom.
(a) SeF_6 (c) ICl_2^-
(b) SeF_4 (d) XeF_4

25. Draw the Lewis structure and then specify the electron-pair and molecular geometries for each of the following molecules or ions. Identify the hybridization of the central atom.
(a) $XeOF_4$ (c) $SeCl_4$
(b) BrF_5 (d) central Br in Br_3^-

26. Draw Lewis structures of HSO_3F and SO_3F^-. What is the number of valence electrons in each species? What is the molecular geometry and hybridization for the sulfur atom in each species?

27. Draw Lewis structures of the acid HPO_2F_2 and its anion $PO_2F_2^-$. What is the number of valence electrons in each species? What is the molecular geometry and hybridization for the phosphorus atom in each species?

28. What is the hybridization of the carbon atom in phosgene, Cl_2CO? Give a complete description of the σ and π bonding in this molecule.

29. What is the hybridization of the sulfur atom in sulfuryl fluoride, SO_2F_2?

Molecular Orbital Theory

30. The hydrogen molecular ion, H_2^+, can be detected spectroscopically. Write the electron configuration of the ion in molecular orbital terms. What is the bond order of the ion? Is the hydrogen–hydrogen bond stronger or weaker in H_2^+ than in H_2?

31. Give the electron configurations for the ions Li_2^+ and Li_2^- in molecular orbital terms. Compare the lithium–lithium bond order in the Li_2 molecule with its ions.

32. Calcium carbide, CaC_2, contains the acetylide ion, C_2^{2-}. Sketch the molecular orbital energy level diagram for the ion. How many net σ and π bonds does the ion have? What is the carbon–carbon bond order? How has the bond order changed on adding electrons to C_2 to obtain C_2^{2-}? Is the C_2^{2-} ion paramagnetic?

33. Oxygen, O_2, can acquire one or two electrons to give O_2^- (superoxide ion) or O_2^{2-} (peroxide ion). Write the electron configuration for the ions in molecular orbital terms, and then compare them with the O_2 molecule on the following basis: (a) magnetic character, (b) net number of σ and π bonds, (c) bond order, and (d) oxygen–oxygen bond length.

34. We can assume for simplicity that the energy level diagram for homonuclear diatomic molecules (Figure 10.27) can be applied to heteronuclear diatomics such as CO.

(a) Write the electron configuration for carbon monoxide, CO.
(b) What is the highest energy, occupied molecular orbital? (Chemists call this the HOMO.)
(c) Is the molecule diamagnetic or paramagnetic?
(d) What is the net number of σ and π bonds? What is its bond order?

35. The nitrosyl ion, NO^+, has an interesting chemistry.
(a) Is NO^+ diamagnetic or paramagnetic? If paramagnetic, how many unpaired electrons does it have?
(b) Assume the molecular orbital diagram for a homonuclear diatomic molecule (Figure 10.27) applies to NO^+. What is the highest energy molecular orbital occupied by electrons?
(c) What is the nitrogen–oxygen bond order?
(d) If NO is ionized to form NO^+, is the nitrogen–oxygen bond stronger or weaker than in NO?

GENERAL QUESTIONS

36. Describe the hybrid orbital set used by sulfur in each of the following molecules or ions:
(a) SO_2 (c) SO_3^{2-}
(b) SO_3 (d) SO_4^{2-}
Does the S atom in all these species use the same hybrid orbitals?

37. Sketch the Lewis structures of ClF_2^+ and ClF_2^-. What are the electron-pair and molecular geometries of each ion? What hybrid orbital set is used by Cl in each ion?

38. What is the hybridization of the nitrogen atom in the nitrate ion, NO_3^-? Describe the orbitals involved in the formation of a nitrogen–oxygen double bond.

39. What is the hybridization of the N atom in NO^+? Describe the orbitals involved in the formation of the multiple bond in this ion.

40. Acrolein, a component of photochemical smog, has a pungent odor and irritates eyes and mucous membranes.

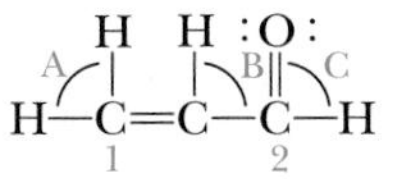

(a) What are the hybridizations of carbon atoms 1 and 2?
(b) What are the approximate values of angles A, B, and C?

41. The organic compound below is a member of a class known as oximes:

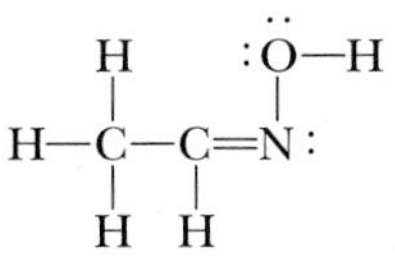

(a) What are the hybridizations of the two C atoms and of the N atom?
(b) What is the approximate C—N—O angle?

42. The compound sketched below is acetylsalicylic acid, better known by its common name, aspirin:

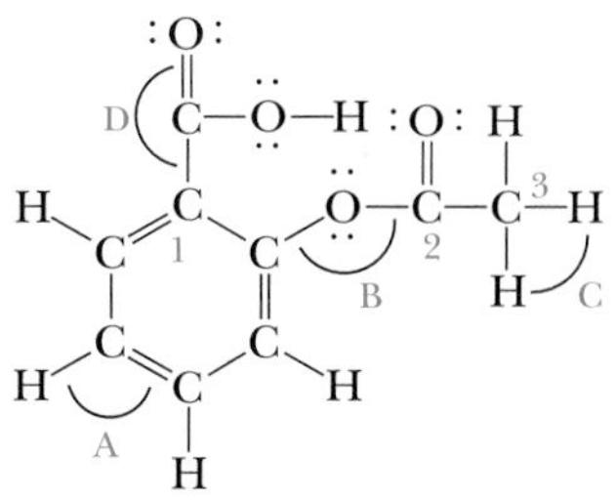

(a) What are the approximate values of the angles marked A, B, C, and D?
(b) What hybrid orbitals are used by carbon atoms 1, 2, and 3?

43. Lactic acid is a natural compound found in sour milk.

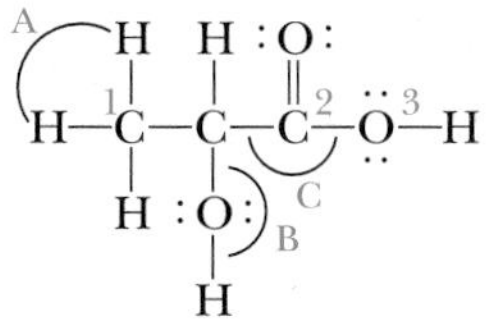

(a) How many π bonds occur in lactic acid? How many σ bonds?
(b) Describe the hybridization of each atom 1 through 3.
(c) Which CO bond is the shortest in the molecule?
(d) What are the approximate values of the bond angles A through C?

44. Histamine is found in normal body tissues and in blood and has the following structure:

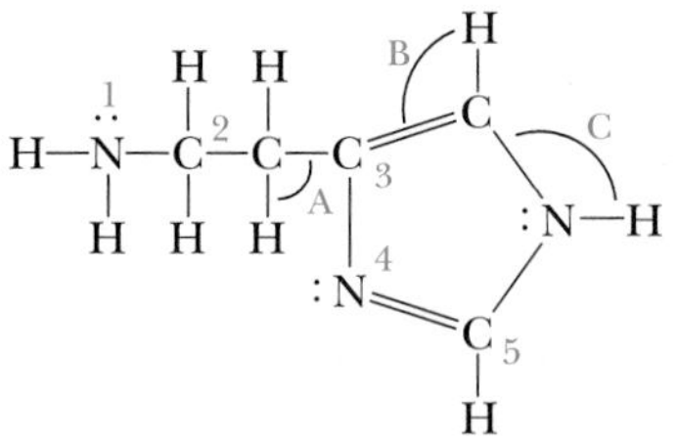

(a) Describe the hybridizations of atoms 1 through 5.
(b) What are the approximate values of the bond angles A through C?

45. Boron trifluoride, BF_3, can accept a pair of electrons from another molecule to form a coordinate covalent bond, as in the following reaction with ammonia:

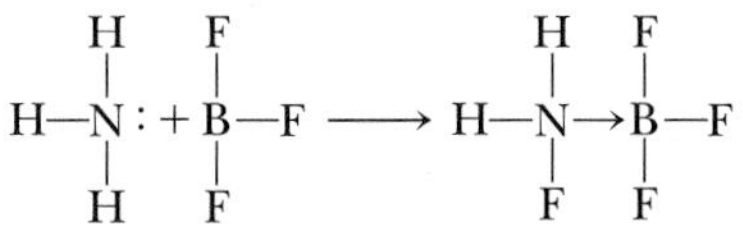

(a) What is the geometry about the boron atom in BF_3? In $H_3N—BF_3$?
(b) What is the hybridization of the boron atom in the two compounds?
(c) Does the boron atom hybridization change on formation of the coordinate covalent bond?

46. The simple valence bond picture of O_2 does not agree with the molecular orbital view. Compare these two theories with regard to the peroxide ion, O_2^{2-}. Do they lead to the same magnetic character and bond order?

47. Nitrogen, N_2, can ionize to form N_2^+ or add an electron to give N_2^-. Using molecular orbital theory, compare these species with regard to (a) their magnetic character, (b) net number of π bonds, (c) bond order, (d) bond length, and (e) bond strength.

48. The ammonium ion is important in chemistry. Discuss any changes in hybridization and bond angles that may occur when ammonia combines with a proton to form the ammonium ion.

$$H^+ + NH_3 \longrightarrow NH_4^+$$

49. Antimony pentafluoride reacts with HF according to the equation

$$2\ HF + SbF_5 \longrightarrow [H_2F]^+[SbF_6]^-$$

(a) What is the hybridization of the Sb atom in the reactant and product?
(b) Draw a Lewis structure for H_2F^+. What is the geometry of H_2F^+? What is the hybridization of F in H_2F^+?

50. Iodine and oxygen form a complex series of ions, among them IO_4^- and IO_5^{3-}. Draw Lewis structures for these ions and specify their electron-pair geometries and the shapes of the ions. What is the hybridization of the I atom in these ions?

51. Xenon forms well-characterized compounds. Two xenon–oxygen compounds are XeO_3 and XeO_4. Draw Lewis structures of each of these compounds and give their electron-pair and molecular geometries. What are the hybrid orbital sets used by xenon in these two oxides?

52. Which of the homonuclear, diatomic molecules of the second-period elements (from Li_2 to Ne_2) are paramagnetic? Which ones have a bond order of 1? Which ones have a bond order of 2? What diatomic molecule has the highest bond order?

53. Which of the following molecules or molecule ions should be paramagnetic? What is the highest occupied molecular orbital (HOMO) in each one? Assume the molecular orbital diagram in Figure 10.27 applies to all of them.
(a) NO (d) Ne_2^+
(b) OF^- (e) CN
(c) O_2^{2-}

54. The sulfamate ion, $H_2N—SO_3^-$, can be thought of as having been formed from the amide ion, NH_2^-, and sulfur trioxide, SO_3.
(a) Sketch a structure for the sulfamate ion. Include estimates of bond angles.
(b) What changes in hybridization do you expect for N and S in the course of the reaction $NH_2^- + SO_3 \longrightarrow H_2N—SO_3^-$?

55. In many chemical reactions, atom hybridization changes. In each of the following reactions, tell what change, if any, occurs to the underlined atom.
(a) $H_2\underline{C}{=}CH_2 + Cl_2 \longrightarrow ClCH_2CH_2Cl$
(b) $\underline{P}(CH_3)_3 + I_2 \longrightarrow P(CH_3)_3I_2$
(c) $\underline{Xe}F_2 + F_2 \longrightarrow XeF_4$
(d) $\underline{Sn}Cl_4 + 2\ Cl^- \longrightarrow SnCl_6^{2-}$

CHALLENGING QUESTIONS

56. The CN molecule has been found in interstellar space. Assuming the electronic structure of the molecule can be described using the molecular orbital energy level diagram in Figure 10.27, answer the following questions.
(a) What is the highest energy molecular orbital to which an electron or electrons is (are) assigned?
(b) What is the bond order of the molecule?
(c) How many net σ bonds are there? How many net π bonds?
(d) Is the molecule paramagnetic or diamagnetic?

57. The molecule CH_3NCS, the structure for which is shown here, is used as a pesticide.

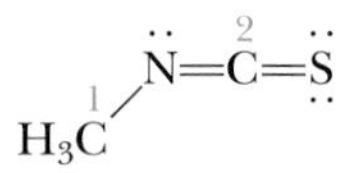

$$H_3\overset{1}{C}—\ddot{N}{=}\overset{2}{C}{=}\ddot{S}$$

(a) Describe the hybrid orbital sets used by carbon 1, carbon 2, and the N atom.
(b) What is the C—N=C bond angle? The N=C=S bond angle?
(c) The C—N=C=S framework is planar. Assume these atoms lie in the plane of the paper and that this is identified as the *xy*-plane. Now consider the orbitals (s, p_x, p_y, and p_z) on carbon atom 2. Which of these orbitals are involved in hybrid orbital formation? Which of these orbitals are used in π bond formation with the neighboring N and S atoms?

58. Amphetamine is a stimulant. Replacing one H atom on the NH_2, or amine, group with CH_3 gives methamphetamine, a particularly dangerous drug commonly known as "speed."

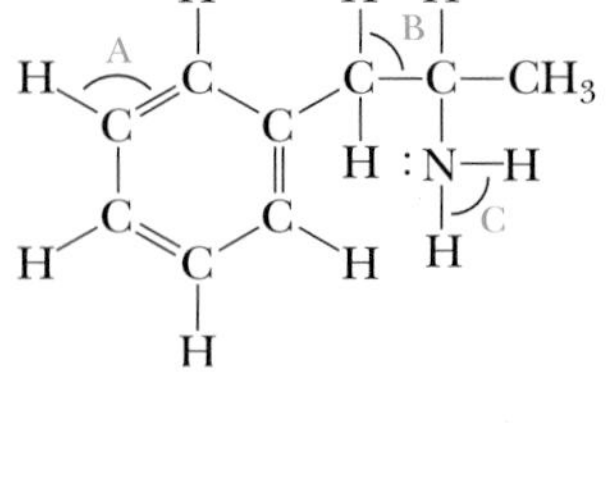

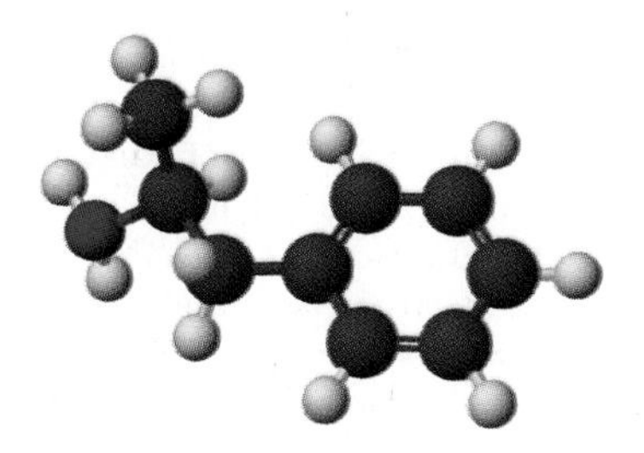

Model of amphetamine.

(a) What are the hybrid orbitals used by the C atoms of the C_6 ring, by the C atoms of the side chain, and by the N atom?
(b) Give approximate values for the bond angles A, B, and C.
(c) How many σ bonds and π bonds are in the molecule?
(d) Is the molecule polar or nonpolar?
(e) Amphetamine reacts readily with a proton (H^+) in aqueous solution. Where does this proton attach to the molecule?

59. Menthol is used in soaps, perfumes, and foods. It is present in the common herb mint, and it can be prepared from turpentine.
(a) What are the hybridizations used by the C atoms in the molecule?
(b) What is the approximate C—O—H bond angle?
(c) Is the molecule polar or nonpolar?
(d) Is the six-member carbon ring planar or nonplanar?

```
         CH3    OH
          |      |
          CH     CH
    H3C /   \CH /   \CH2
            |        |
           H2C     CH
              \C /   \CH3
               H2
```

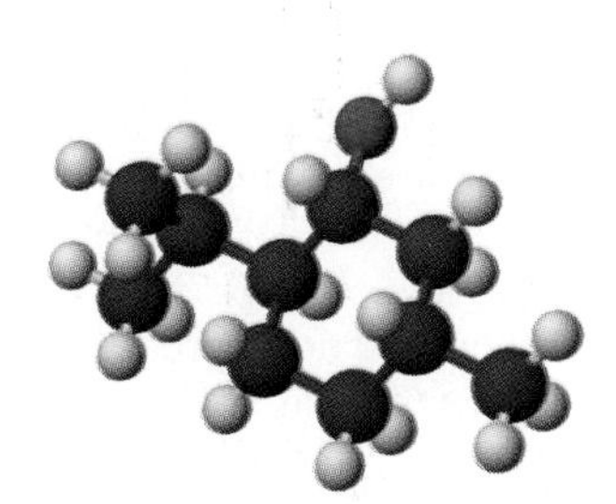

Model of menthol.

60. When two amino acids react with each other, they form a linkage called an amide group, or a peptide link. (If more linkages are added, a protein, or polypeptide is formed.)
(a) What are the hybridizations of the C and N atoms in the peptide linkage?
(b) Is the structure illustrated the only resonance structure possible for the peptide linkage? If another resonance structure is possible, compare it with the one shown. Decide which is the more important structure.
(c) The computer-generated structure shown here, which contains a peptide linkage, shows that the linkage is flat. This is an important feature of proteins. Speculate on reasons that the CO—NH linkage is planar.

```
       H :O:                  H :O:
   ..  |  ||     ..       ..  |  ||     ..
H—N—C—C—O—H  +  H—N—C—C—O—H
   |   |         ..       |   |         ..
   H   H                  H   H

                  | −H2O
                  ↓
          H   :O:         H :O:
    ..    |    ||  ..     |  ||    ..
  H—N—C——C—N——C—C—O—H
    |     |        |      |        ..
    H     H        H      H
                   └ Peptide linkage
```

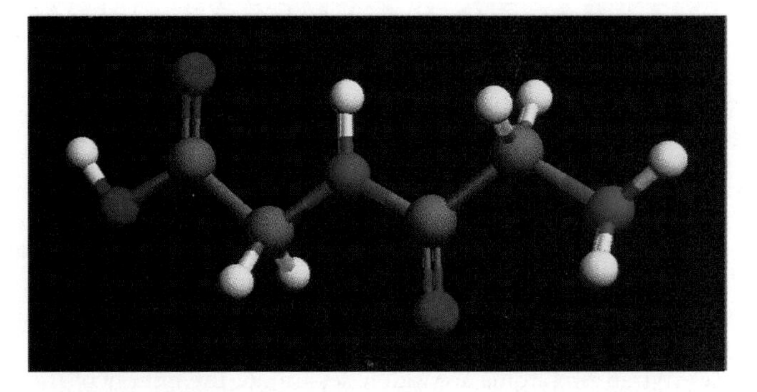

Model of dipeptide.

61. The elements of the second period from boron to oxygen form compounds of the type $X_nE{-}EX_n$, where X can be H or a halogen. Sketch possible molecular structures for B_2F_4, C_2H_4, N_2H_4, and O_2H_2. Give the hybridizations of E in each molecule and specify approximate X—E—E bond angles.

Summary Question

62. The compound whose structure is shown here is acetylacetone. It exists in two forms: one called the enol form and the other called the keto form.

```
           H                                  H
           |                                  |
H3C—C=C—C—CH3  ⇌  H3C—C—C—C—CH3
      |      ||                       ||  |  ||
      O      O                        O   H  O
      |
      H
     enol form                          keta form
```

The molecule reacts with OH^- to form an anion $[CH_3COCHCOCH_3]^-$ (often abbreviated $acac^-$ for acetylacetonate ion). One of the most interesting aspects of this anion is that one or more of them can react with a transition metal cation to give very stable, highly colored compounds (some of which are illustrated in the photo).

Acetylacetone complexes of (*left* to *right*) cobalt, chromium, and iron. *(C. D. Winters)*

(a) Are the keto and enol forms of acetylacetone contributing resonance forms? Explain your answer.
(b) What is the hybridization of each atom (except H) in the enol form? What changes in hybridization occur when this is transformed into the keto form?
(c) What is the electron-pair geometry and molecular geometry around each C atom in the keto and enol forms? What changes in geometry occur when the keto form changes to the enol form?
(d) Draw three possible resonance structures for the acac$^-$ ion.
(e) If you wanted to prepare 15.0 g of deep red $Cr(C_5H_7O_2)_3$ using the following reaction,

$$CrCl_3 + 3\ C_5H_8O_2 + 3\ NaOH \longrightarrow Cr(C_5H_7O_2)_3 + 3\ H_2O + 3\ NaCl$$

what mass of the other reactants is needed?

INTERACTIVE GENERAL CHEMISTRY CD-ROM

Screen 10.3 Valence Bond Theory

(a) This screen describes attractive and repulsive forces that occur when two atoms come near each other. What must be true about the relative strengths of those attractive and repulsive forces if a covalent bond is to form?
(b) It is stated that for a bond to form, orbitals on adjacent atoms must overlap, and each pair of overlapping orbitals will contain two electrons. Explain why neon does not form a diatomic molecule, Ne_2, whereas fluorine forms F_2.
(c) When two atoms are widely separated, the energy of the system is defined as zero. As the atoms approach one another, the energy drops, reaches a minimum, and then increases as they approach still more closely. Explain these observations.

Screen 10.6 Determining Hybrid Orbitals

Examine the *Hybrid Orbitals* tool on this screen. Use this tool to systematically combine atomic orbitals to form hybrid atomic orbitals.

(a) What is the relationship between the number of hybrid orbitals produced and the number of atomic orbitals used to create them?
(b) Do hybrid atomic orbitals form between different p orbitals without involving s orbitals?
(c) What is the relationship between the energy of hybrid atomic orbitals and the atomic orbitals from which they are formed?
(d) Compare the shapes of the hybrid orbitals formed from an s orbital and a p_x orbital with the hybrid atomic orbitals formed from an s orbital and a p_z orbital.
(e) Compare the shape of the hybrid orbitals formed from s, p_x, and p_y orbitals with the hybrid atomic orbitals formed from s, p_x, and p_z orbitals.

Screen 10.7 Multiple Bonding

After examining the bonding in ethylene on the main screen go to the *Closer Look* auxiliary screen.

(a) Explain why the allene molecule is not flat. That is, explain why the CH_2 groups at opposite ends do not lie in the same plane.
(b) Based on the hybrid orbital model, explain why benzene is a planar, symmetrical molecule.
(c) What are the hybrid orbitals used by the three C atoms of allyl alcohol?

$$\mathrm{H_2\overset{3}{C}{=}\overset{2}{C}H{-}\overset{1}{C}H_2{-}\ddot{\underset{\cdot\cdot}{O}}{-}H}$$

Screen 10.8 Molecular Fluxionality

(a) Observe the animations of the rotations of *trans*-2-butene and butane about their carbon–carbon bonds. As one end of *trans*-2-butene rotates relative to the other, the energy increases greatly (from 27 kJ/mol to 233 kJ/mol) and then drops to 30 kJ/mol when the rotation has produced *cis*-2-butene. In contrast, the rotation of the butane molecule requires much less energy for rotation (only about 60 kJ/mol). When butane has reached the halfway point in its rotation, the energy has reached a maximum. Why does *trans*-2-butene require so much more energy to rotate about the central carbon–carbon bond than does butane?
(b) The structure of propene, C_3H_6, is pictured here. Which carbon–hydrogen group (CH_3 or CH_2) can rotate freely with respect to the rest of the molecule?

$$H_3C{-}\overset{\displaystyle H}{\overset{|}{C}}{=}CH_2$$

(c) Can the two CH_2 fragments of allene (see Screen 10.7SB) rotate with respect to each other? Briefly explain why or why not.

Chapter **11**

Bonding and Molecular Structure: Carbon—More Than Just Another Element

(Richard Hutchings/Photo Researchers, Inc.)

A Chemical Puzzler

Do you enjoy chewing bubble gum and blowing the largest possible bubble? Why is it possible to blow large bubbles from bubble gum but not regular gum? What is the chemical compound in bubble gum that allows you to do this? How is it made, and is this compound used in anything else?

This chapter is largely concerned with bonding and structure in organic compounds. Models of many of the molecules mentioned in this chapter are found on the *Saunders Interactive General Chemsitry CD-ROM* or on the World Wide Web site for this book at http://www.saunderscollege.com.

The vast majority of known chemical compounds are built from carbon, and life on this planet is founded on a remarkable collection of substances based on this element. Carbon-based compounds are so well studied that a whole subdiscipline, organic chemistry, has been built around them.

What about carbon makes it unique among the known elements? We already know some things about carbon: it is a nonmetal so it forms covalent bonds to other elements, it has four valence electrons and thus is expected to form four bonds, and it is able to bond to many other elements in a variety of ways. Notably, carbon is able to bond to other carbon atoms to create compounds with many carbon atoms linked together in large multiatom molecules. What better way to continue our discussion of bonding and structure than to look at some of the most important examples of covalent molecular compounds, those based on carbon?

11.1 THE UNIQUENESS OF CARBON

See the *Saunders Interactive General Chemistry CD-ROM,* Screen 11.2, Carbon–Carbon Bonds, and Screen 6.3, Control of Chemical Reactions: Thermodynamics and Kinetics.

That carbon compounds are so numerous, and that their structures have such diversity, is evidence that carbon is unusual among the known elements. Let us examine the properties of this element further to understand why this is so. Carbon has several exceptional features: the number and diversity of ways that it can form bonds to other elements, the strength of bonds between carbon and other elements, and finally, the fact that most reactions of carbon compounds are slow. This last feature, often called **kinetic stability,** refers to the fact that a compound can exist because reactions it might undergo are so slow that, in effect, they do not occur or occur only very slowly.

The Diversity of Bonds Formed by Carbon

With four valence electrons, carbon needs to form four bonds to reach an octet configuration around the atom. In contrast, nitrogen and boron form three bonds, oxygen two, and hydrogen and the halogens one, in nonionic compounds of these elements. Forming four single bonds means that carbon can bond to four different elements, or it can bond to up to four other carbon atoms, each of which can bond to other atoms. The result is an incredible diversity of formulas. Carbon is also unique in the multiplicity of ways it bonds to other elements—with single, double, and triple bonds—and the result is several different geometries around carbon (Table 11.1).

The tetrahedral geometry associated with carbon atoms having four single bonds is especially noteworthy because it leads to the three-dimensionality of its compounds. It is usually possible to join several carbon atoms together in a variety of ways, and when many carbon atoms are linked together the three-dimensionality of the structure can become very complex. Drawing and visualizing three-dimensional structures is very important to understanding the structures of these compounds.

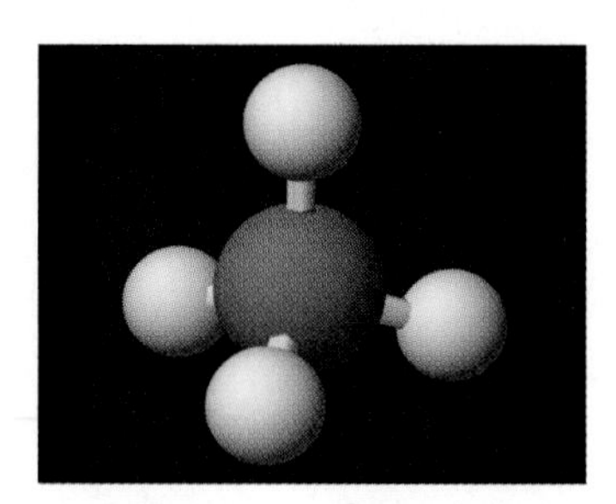

Methane, CH_4, has a tetrahedral molecular geometry.

Joining two carbon atoms together by a double bond creates the possibility of an added structural feature—*cis* and *trans* isomers (page 456). *Cis* and *trans* isomers arise in compounds with carbon–carbon double bonds, such as *cis-* and *trans*-2-butene, because structural rigidity is imposed by the π bond. Rotation of one end of the double bond would require breaking the π bond.

Table 11.1 • BONDING TO CARBON

Description	Lewis Structure	Geometry at Carbon	Hybridization	Examples
four single bonds	(tetrahedral C)	tetrahedral	sp^3	CH_4, CCl_4
two single and one double bond	>C=	trigonal planar	sp^2	C_2H_4, C_6H_6, H_2CO
two double bonds	=C=	linear	sp	CO_2, OCN^-
one triple bond and one single bond	—C≡	linear	sp	C_2H_2, HCN

Another way that carbon enters into bonding is as part of the very stable six-member ring system in benzene. A single Lewis structure does not correctly represent the structure and bonding, so resonance is introduced (page 393).

Resonance structures ⟷ or Resonance hybrid

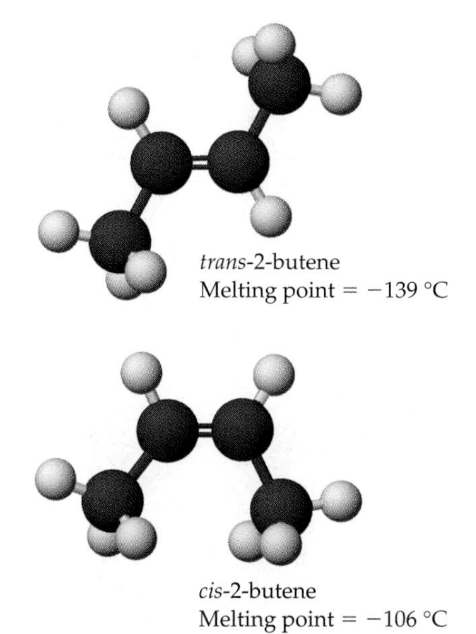

To illustrate the structural diversity of carbon we need look no further than the element itself, in the form of its allotropes: graphite, diamonds, and buckyballs (see Figures 2.21, 2.22, and 11.1). The structure of graphite consists of many layers of carbon atoms. Each carbon atom within a layer is trigonal-planar and bonded to three other carbon atoms, thus forming an extended network of six-member carbon rings. The rings are planar to give each carbon atom the correct bond angles (120°) so the carbon atoms are assumed to be sp^2-hybridized.

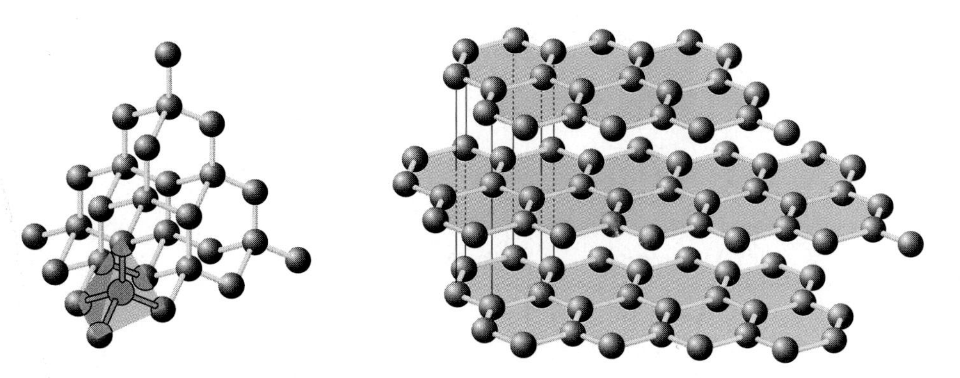

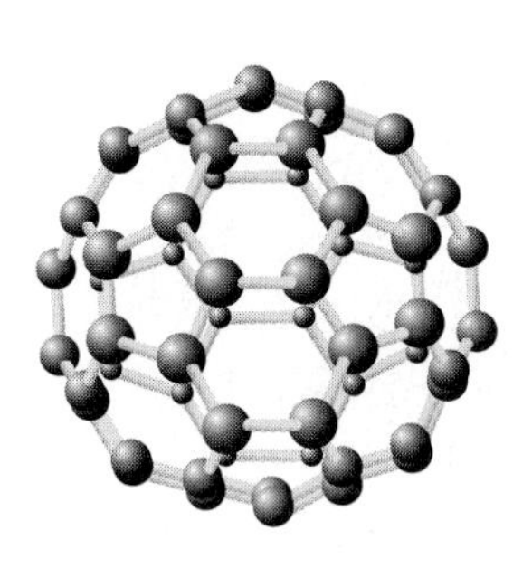

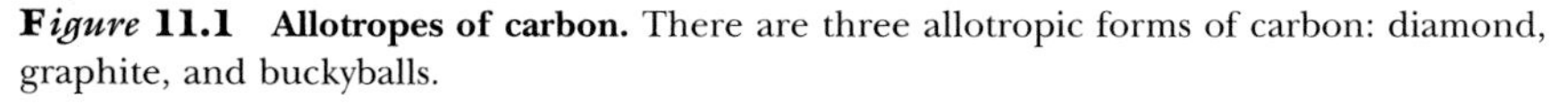
Figure 11.1 **Allotropes of carbon.** There are three allotropic forms of carbon: diamond, graphite, and buckyballs.

CURRENT PERSPECTIVES

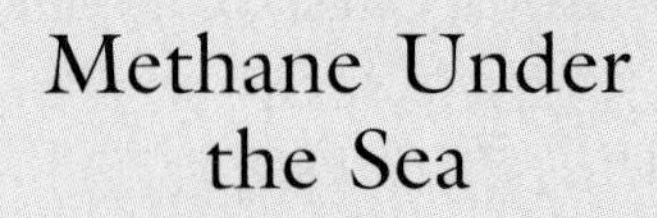

Methane Under the Sea

Methane, CH_4, is the lightest carbon-containing molecule on earth, and also one of the most important molecules. This hydrocarbon is a colorless, odorless, flammable gas under ordinary conditions. Its molecules have the tetrahedral shape that is characteristic of carbon when surrounded by four other atoms.

You probably know it in the form of natural gas, which consists of 70% to 95% methane. It is found deep under the earth's surface in the middle of the United States and in Pennsylvania and Alaska. You might also know methane as *swamp gas* or *marsh gas*. This is formed by bacteria working on organic matter in an environment where the oxygen concentration is low. These conditions occur in sedimentary layers in coastal waters and in marshes. The gas can escape if the sediment layer is thin, and you see it as bubbles in a marsh. It may be trapped under thicker sediment layers, however. Over time it could be squeezed into porous rocks such as sandstone or limestone, and here it waits to be tapped by drilling into the rock layer.

When natural gas pipelines were laid across states and across the continent, pipeline operators soon found that, unless water was carefully kept out of the line, chunks of *methane hydrate* would form and clog the pipes. In methane hydrate the hydrocarbon is trapped in cavities formed by water molecules arranged in an ice-like structure. When methane hydrate melts, the volume of gas released (at normal pressure and temperature) is about 165 times larger than the volume of the hydrate. This would mean that a 3-L sample of methane hydrate would release enough methane to fill a refrigerator.

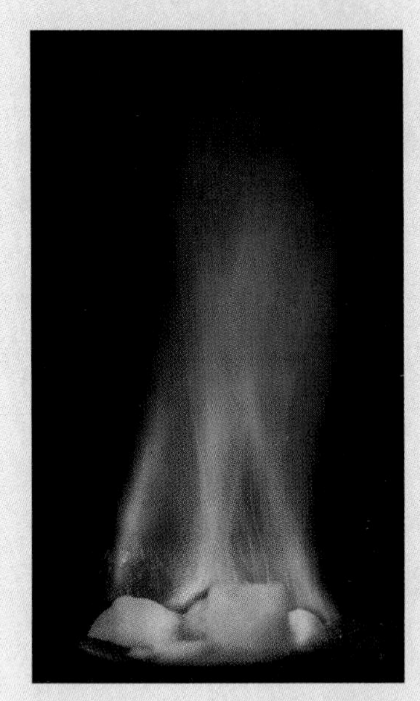

Methane in a sample of methane hydrate burns off. *(John Pinkston and Laura Stern/U.S. Geological Survey/Science News 11/9/96)*

If methane hydrate can form in a pipeline, is it found in nature? In May 1970, oceanographers, drilling into the sea bed off the coast of South Carolina, pulled up samples of the hydrate that fizzed and oozed out of the drill casing. Since then, methane hydrate has been found in many parts of the oceans as well as under permafrost in the Arctic. Indeed, it is estimated that 1.5×10^{13} tons of methane hydrate is buried under the sea floor around the world. In fact, the energy content of this gas may surpass that of the known fossil fuel reserves by as much as a factor of 2!

There is great interest in methane hydrate, not only because of its interesting molecular structure, but also because of the economic potential of the material. In addition, there is speculation that it may be found on the moons in the outer part of our solar system.

More recently, Charles Fisher, a physiological ecologist at Pennsylvania State University, made a surprising find. He usually studies tube worms that live on ocean bottoms. In July 1997, however, when he and his colleagues moved their submersible close to a methane hydrate outcropping in the Gulf of Mexico, they saw the outcropping was crawling with pastel pink worms. The worms turned out to be flat, segmented polychaetes, but of an unknown species. Fisher said the worms must be living off the bacteria that feed on the methane hydrate, but many questions remain.

Finally, there is new evidence that methane hydrate was involved in a burst of evolutionary changes on the earth about 55 million years ago. What if some cataclysmic event triggered a release of methane from the sea bed? Because methane, like carbon dioxide, is a greenhouse gas, an increased atmospheric concentration could lead to an increase in the surface temperature of the earth, an increase that could bring on significant environmental changes.

A colony of worms on an outcropping of methane hydrate in the Gulf of Mexico. *(Charles Fisher, The Pennsylvania State University)*

Carbon–carbon π bonds extend throughout a layer, but no formal carbon–carbon bonds occur between atoms of different layers; thus, the layers are weakly attracted and can slide readily over one another. This is the reason graphite is a lubricant and the most important ingredient in pencil "lead."

In contrast, each carbon atom in diamond is surrounded by four other carbon atoms (see Figures 2.21 and 11.1). The carbon atoms have tetrahedral geometry and are sp^3-hybridized. The structure extends in three dimensions throughout the solid, forming a rigid network of six-member carbon rings. The six atoms in each ring are not in a plane, thus allowing the angles to match the tetrahedral angle, 109.5°. This bonding arrangement makes diamond extremely hard and chemically unreactive. Breaking or distorting a diamond requires breaking carbon–carbon bonds.

Review hybrid orbitals and bonding in Chapter 10, Section 10.1.

One of the most intriguing discoveries of the past decade is a family of compounds that represent a new allotropic form of carbon. The best known of these compounds, often called a buckyball, has the formula C_{60} (see Figures 2.22 and 11.1). Since its discovery, many related compounds of carbon have been discovered, including other cagelike structures with formulas like C_{70} and C_{240}, as well as totally different arrangements in which the carbon atoms form into long tubes (Figure 11.2). In buckyballs and related molecular forms, each carbon

A group of researchers at the Washington University School of Medicine (St. Louis) reported in 1997 that, when injected into mice, a water-soluble form of buckyballs can ameliorate the worst effects of neurodegenerative diseases such as Lou Gehrig's disease.

(a)

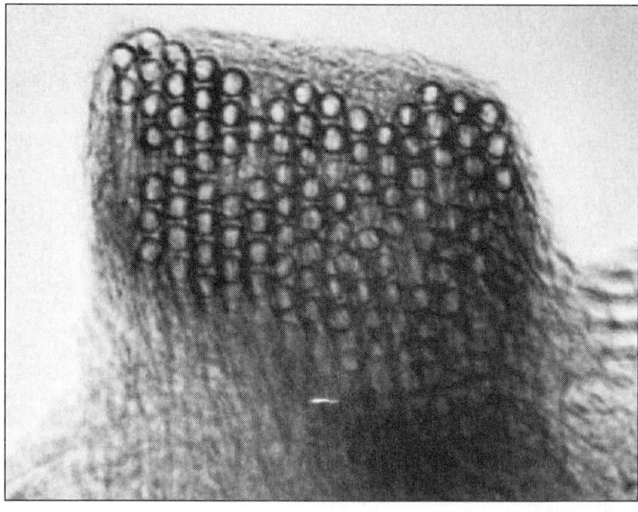

(b)

Figure 11.2 **$C_{1,000,000}$ and beyond.** Not long after the discovery of C_{60} molecules, S. Iijima of NEC Corporation discovered long tubes of carbon atoms in the debris from an arc discharge between carbon electrodes. (a) These tubes can be thought of as buckyballs that have been cut in half and then stitched together with a belt of C_6 rings. Because of their tiny dimensions, they have come to be called "nanotubes." A nanotube long enough to stretch from earth to the moon could be rolled up inside a poppyseed. Here Alex Zettl of Lawrence Berkeley Laboratory holds a model of a carbon nanotube. (b) A bundle of carbon nanotubes, each about 1.4 nm in diameter. The bundle is 10–20 nm thick. Note that the tubes are packed in a triangular arrangement (or, alternatively, six tubes are arranged hexagonally around a central tube). (For other images see http://cnst.rice.edu. For more information see M. W. Browne: "The Next Electronic Breakthrough: Carbon Atoms," *New York Times,* February 17, 1998, page F1; B. I. Yakobson and R. E. Smalley: "Fullerene nanotubes." *American Scientist,* Volume 85, pages 324–337, July–August 1997.) (*a, Lawrence Berkeley Laboratory; b, P. Nikolaev, Rice University, Center for Nanoscale Science and Technology*)

For information on buckyballs and nanotubes, see http://cnst.rice.edu.images

atom has three carbon atom neighbors. It is close to sp^2 hybridization but not quite in a planar arrangement. The carbon atoms are arranged in five- or six-member rings, and the structure closes on itself. In a buckyball, a sphere of 60 carbon atoms results. As with benzene and graphite, a single Lewis structure does not suffice to describe a molecule of C_{60}; several resonance structures can be drawn that place double bonds in various positions to give four bonds to each carbon. This arrangement of atoms in buckyballs and related compounds was so unexpected that chemists are certain the molecule will have interesting applications (see page 88).

Stability of Carbon Compounds

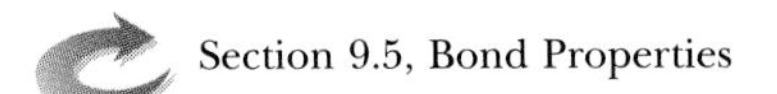
Section 9.5, Bond Properties

The Importance of Bond Dissociation Energy

If a covalent bond is weak, the kinetic energy associated with thermal motion at room temperature can be sufficient to break the bond. Molecular collisions in gases, liquids, and solutions can provide enough energy to cause the molecule to fall apart. Absorption of light energy can also lead to bond breaking. If a molecule is to be stable, strong bonds between atoms are therefore required. Bonds between carbon atoms, and between carbon and most other atoms, are quite stable. The average C—C bond energy is 346 kJ/mol; the C—H bond energy is 413 kJ/mol. Multiple bonds have even higher energies; the energy of the C=C double bond is 602 kJ/mol; the C≡C bond energy is 835 kJ/mol. Contrast the single bond energies in carbon compounds with the bond energies for compounds of silicon, the next element in Group 4A. Bond energies in silicon compounds are at least one-third less strong than those in analogous carbon compounds. Compare, for example, the Si—H bond energy (318 kJ/mol) with that for the C—H bond (413 kJ/mol) and the Si—Si bond energy (222 kJ/mol) with that for the C—C bond (346 kJ/mol).

The consequence of high bond energies is that most carbon compounds possess exceptional thermal stability. For the most part, organic compounds are not degraded under normal conditions; they will fall apart only if heated to a high temperature.

Kinetic Stability of Carbon Compounds

The kinetic stability of its compounds is the third feature that has led to an extensive and sophisticated carbon chemistry. Kinetic stability means a compound can exist because, even though a thermodynamically favored reaction pathway is available to it, reactions leading to its decomposition are slow.

See the *Saunders Interactive General Chemistry CD-ROM,* Screen 6.3. Kinetic stability is not limited to organic compounds. Reactions of most metals with oxygen are highly exothermic, but they also tend to be very slow under ambient conditions.

Kinetic stability is not a rare or unusual feature in chemistry. Perhaps without realizing it, you have encountered this already for organic compounds with respect to their stability toward oxidation by O_2. Because earth's atmosphere is oxidizing (it is about 20% O_2), virtually all organic compounds can be oxidized in air to CO_2 and H_2O. These are favorable reactions that usually evolve a great amount of energy. Nonetheless, compounds like methane, alcohols, and fats and oils coexist with O_2 under normal conditions, and reaction occurs only when initiated by a spark or the addition of heat.

The kinetic stability of organic compounds is an important factor in the rest of organic chemistry as well. One example, the polymerization of ethylene, illustrates this point. Ethylene, C_2H_4, is the simplest example of a hydrocarbon

CURRENT PERSPECTIVES

UV Radiation, Skin Damage, and Sunscreens

Most of us are well aware of the effects of exposure to the sun. A bright red sunburn results, and over a long term, permanent skin damage can occur. Most of this is the result of the damage to organic molecules caused by ultraviolet (UV) radiation from the sun.

UV radiation is broken into three categories: UVA (315–400 nm), UVB (290–315 nm), and UVC (100–290 nm). UVC radiation is dangerous, but most of it is absorbed by earth's ozone layer. UVB light is responsible for your sunburn. Tanning occurs when the light strikes your skin and activates the melanocytes in the skin so that they produce melanin. UVA light also produces damage such as the alteration of connective tissue in the dermis.

Calculation of the energy associated with UV light and a comparison of this value with bond energy values will give you a sense of the problem. As an example, let us calculate the energy (in the units kJ/mol) of UVB light of wave length 300 nm (3.00×10^{-7} m). (See Section 7.1.)

$$E = \frac{(6.63 \times 10^{-34}\ \mathrm{J/s})(3.00 \times 10^{8}\ \mathrm{m/s})}{(3.00 \times 10^{-7}\ \mathrm{m})}$$

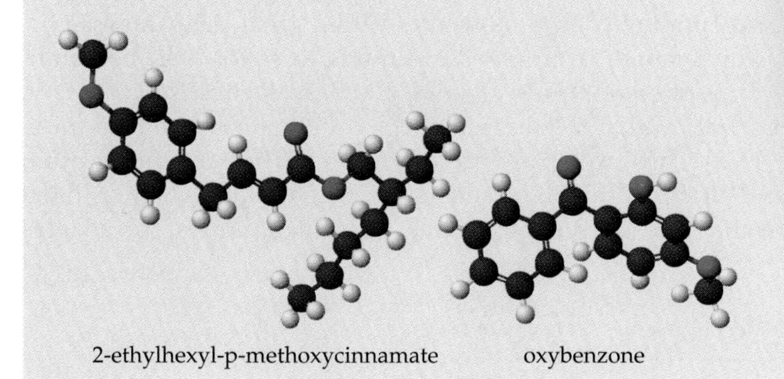

This is in the range of the energy necessary to break C—C or C—H bonds, and there certainly is sufficient energy to break C—S and C—N bonds that are present in proteins.

Various manufacturers have developed mixtures of compounds that protect skin from UVA and UVB radiation. These sunscreens are given "sun protection factor" (SPF) labels that indicate how long the user can stay in the sun without burning. All sunscreens produced by Coppertone, for example, contain 2-ethylhexyl-*p*-methoxycinnamate and oxybenzone. These molecules absorb UV radiation, preventing it from reaching your skin.

See C. Walters, A. Keeney, C. T. Wigal, C. R. Johnston, and R. D. Cornelius: "The spectroscopic analysis and modeling of sunscreens." *Journal of Chemical Education*, Vol. 74, page 99, 1997.

A sunscreen can block some UV radiation from reaching the skin and prevent damage to skin from chemical changes, some of which involve bond-breaking. *(Bachmann/Photo Researchers, Inc.)*

with a C=C double bond. The interesting thing about ethylene is that it is unstable. One reaction it should undergo is polymerization to form polyethylene, the most important plastic manufactured today. That is, π electrons should become σ-bonding electrons in forming bonds from one C_2H_4 unit to another, eventually creating a polymer, a long chain of small molecules linked to form a giant molecule. Using bond energies, we can estimate the enthalpy change in this process. The double bond between carbon atoms in ethylene is broken at the cost of +602 kJ/mol. It is replaced by two single bonds, with a net energy evolution of $2 \times -346 = -692$ kJ/mol. Thus, the net enthalpy change, ΔH_{rxn}, for this reaction is estimated to be −90 kJ/mol, making this reaction quite exothermic. In spite of this, ethylene doesn't polymerize readily. Ethylene can be prepared, stored, and used in various chemical reactions. Once begun, however, the polymerization process is rapid.

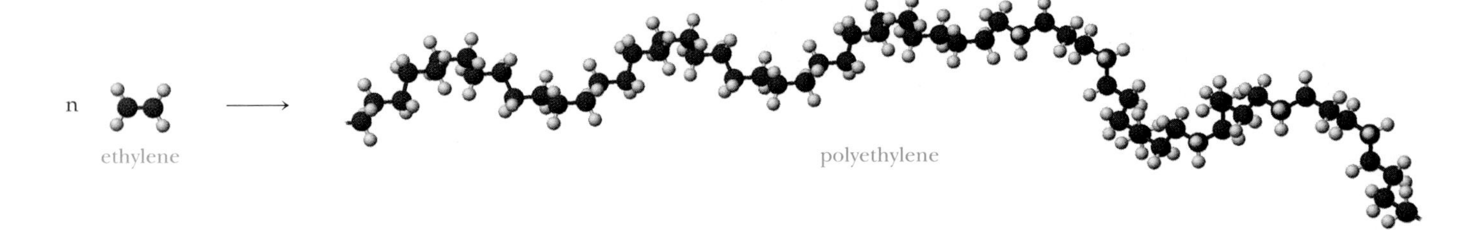

11.2 HYDROCARBONS

Organic chemistry is an important and fascinating subject that covers a vast intellectual territory. This chapter is not an exhaustive study of the field. Many types of compounds are not mentioned, and very few reactions are described. For a more in-depth analysis of this subject, consult books such as W. H. Brown and C. S. Foote: *Organic Chemistry*, 2nd edition. Saunders College Publishing, Philadelphia, 1997.

See the *Saunders Interactive General Chemistry CD-ROM,* Screen 11.3, Hydrocarbons.

Nomenclature for organic compounds is outlined in Appendix E.

Organic chemistry is a huge field. Many students who continue to study chemistry will take one or two semesters of organic chemistry, which most chemists will say is barely an introduction to the subject. Given the breadth of the subject, we can therefore only provide a background in one critical area in this book: the structure of organic compounds. Recognizing types of compounds and being able to visualize their structures are key prerequisites to understanding organic chemistry and its importance in commerce and in living things.

In the course of this chapter we shall consider four general types of compounds. **Hydrocarbons,** compounds constructed of carbon and hydrogen, are the first type. Next, we shall examine compounds with C—O or C—N bonds (alcohols, ethers, and amines). Finally, we shall look at compounds containing a carbonyl (C=O) group (aldehydes, ketones, and acids). Following this, we will consider polymers, exceptionally large molecules with a great many atoms.

Hydrocarbons, which range from gases to liquids to solids, can be subdivided into several groups of compounds: alkanes, alkenes, alkynes, and aromatic compounds, a categorization that relates to the types of chemical bonds in these compounds (Table 11.2). Two properties are shared by all hydrocarbons: none are water-soluble, and all are flammable (and so are potential fuels). We will begin with compounds that have carbon atoms with four single bonds: the alkanes and cycloalkanes.

Table 11.2 • Some Types of Hydrocarbons

Type of Hydrocarbon	Characteristic Features	Example
alkanes	general formula C_nH_{2n+2} C—C single bonds all C atoms surrounded by four single bonds	CH_4, methane C_2H_6, ethane
cyclic alkanes	general formula C_nH_{2n}	C_6H_{12}, cyclohexane
alkenes	general formula C_nH_{2n} C=C double bond	$H_2C{=}CH_2$, ethylene
alkynes	general formula C_nH_{2n-2}	$HC{\equiv}CH$, acetylene
aromatics	rings of C atoms with π bonding extending over all six C atoms	benzene, C_6H_6

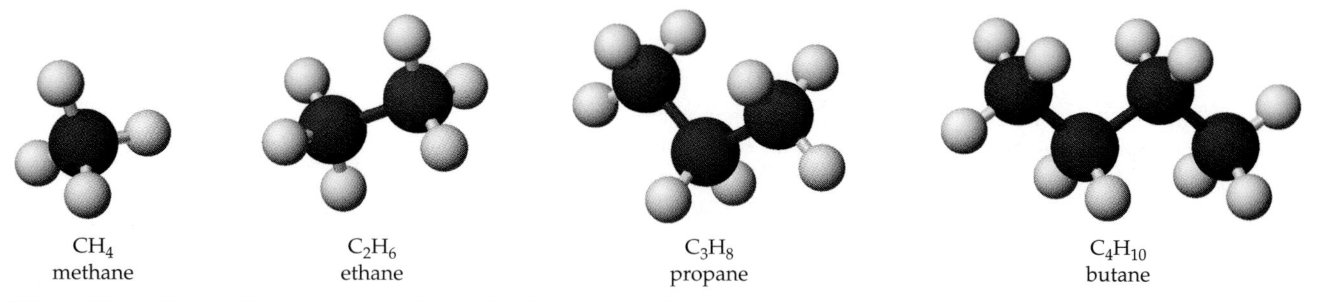

Figure 11.3 **Alkanes.** The lowest molecular weight alkanes are methane, ethane, propane, and butane.

Alkanes

Alkanes have the general formula C_nH_{2n+2}, with n taking integer values. Formulas of specific compounds can be generated from the general formula: CH_4 (methane), C_2H_6 (ethane), C_3H_8 (propane), C_4H_{10} (butane), C_5H_{12} (pentane), and so on (Figure 11.3).

Methane has four hydrogen atoms arranged tetrahedrally around a single carbon. Replacing a hydrogen atom in methane by a $—CH_3$ group gives ethane. If an H atom of ethane is replaced by yet another $—CH_3$ group, propane results. Butane is derived from propane by replacing an H atom of one of the chain-ending carbon atoms with a $—CH_3$ group.

Using the general formula for alkanes, it is possible to create formulas for an infinite number of alkanes. Formulas do not hint at the structural diversity of the alkanes, however. With all the alkanes after propane, it is possible to arrange the carbon atoms in several ways to form **structural isomers.** These compounds have the same formula, but the atoms are connected in a different order.

There are two structural isomers for C_4H_{10},

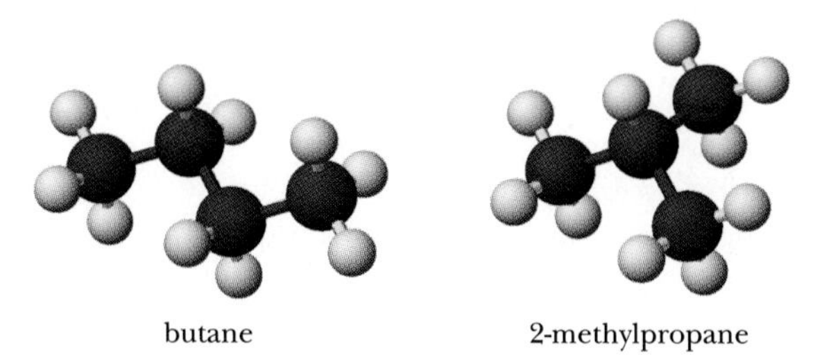

three isomers for C_5H_{12},

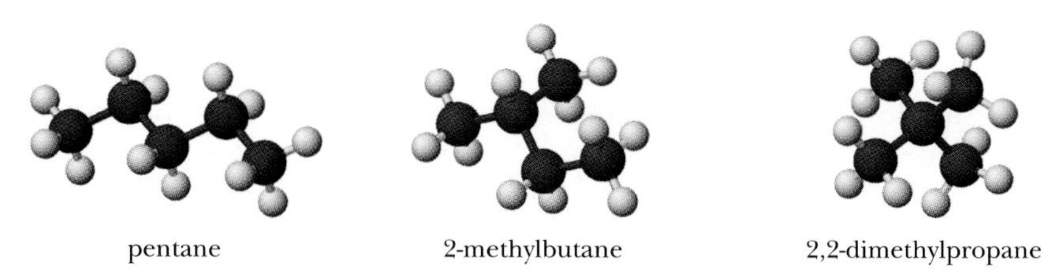

and five for C_6H_{14}. As the number of carbon atoms in an alkane increases, the number of possible structural isomers rapidly increases; there are 9 isomers for C_7H_{16}, 18 for C_8H_{18}, 75 for $C_{10}H_{22}$, and 366,319 structural isomers for $C_{20}H_{42}$.

A CLOSER LOOK Writing Formulas and Drawing Structures

You learned in Chapter 3 that there are various ways of presenting structures (page 102), and it is appropriate to return to that point again as we look at organic compounds. Consider butane, for example. We can represent this molecule in many ways:

1. *Molecular formula:* C_4H_{10}. This type of formula gives information only on molecular composition.
2. *Condensed formula:* $CH_3CH_2CH_2CH_3$. This method of writing the formula gives some information on the way atoms are connected.
3. *Structural formula:* You will recognize this drawing as the Lewis structure. An elaboration on the condensed formula in (2), this representation defines more clearly how each atom is connected, but it fails to describe the three-dimensionality resulting from tetrahedral carbon.

```
    H  H  H  H
    |  |  |  |
H—C—C—C—C—H
    |  |  |  |
    H  H  H  H
```

4. *Structural formula:* This method provides a more accurate representation of (3), conveying the three-dimensional structure of the molecule. To make the drawing appear three-dimensional, wedges and dashed lines are often used for bonds above and below the plane of the paper, respectively. The drawing correctly shows that the carbon atoms are in a zigzag arrangement and not in a straight line.

```
    H         H   H
H   ¦          \ /        H
 \ C           C        /
  /  \        /  \     C—H
 H    C            
     / \             |
    H   H            H
```

Structural formula

5. *Computer-drawn ball-and-stick and space-filling models:* These models afford a better three-dimensional view than the formula in (4).

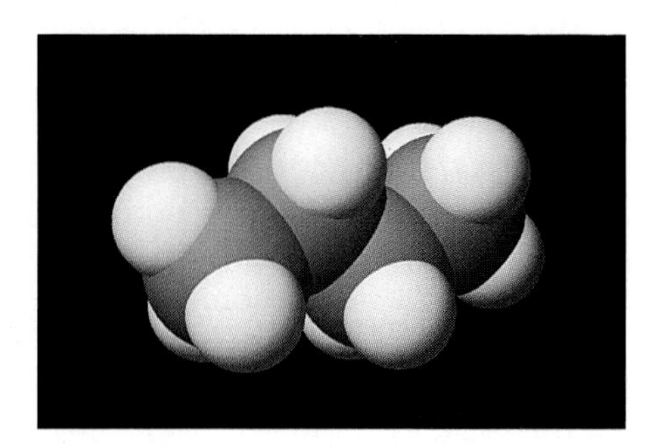

(Not surprisingly, chemists have isolated and characterized only a small fraction of these isomers.)

How does one go about creating structures for all the possible isomers from a given formula? When more than two or three isomers are possible, it is necessary to approach the problem in an organized way. One approach is illustrated in Example 11.1. Before working out this example, however, bear in mind the following points:

- Each alkane is built up with a framework of tetrahedral carbon atoms, and each carbon must have four single bonds. The three-dimensionality conveyed by tetrahedral geometry is a crucial aspect in the structures of these species.
- A hallmark of carbon chemistry is that carbon atoms can be linked together in many different ways. The approach taken here will be to create a framework of carbon atoms and then fill the remaining positions around carbon with H atoms to give each atom four bonds.
- Free rotation occurs around carbon—carbon single bonds. As atoms are assembled to form the skeleton of an alkane, the emphasis is on how carbon atoms are attached to one another.

Example 11.1 Drawing Structural Isomers of Alkanes

Draw structures of the five isomers of C_6H_{14}. Using Appendix E, name each of the structures.

Solution One logical approach is to focus first on the different frameworks that can be built from six carbon atoms. Having created a carbon framework, fill hydrogen atoms into the structure so that each carbon has four bonds.

Step 1. Placing six carbon atoms in a chain gives the framework for the first isomer. Although we have drawn the atoms in a straight line, remember that the bond angles around carbon are 109.5°. Also, remember that free rotation occurs around C—C single bonds. Now fill in hydrogen atoms: three on the carbons on the ends of the chain, two on each of the carbons in the middle. You have created the first isomer, hexane.

```
                              H   H   H   H   H   H
                              |   |   |   |   |   |
C—C—C—C—C—C  ⟶  H—C—C—C—C—C—C—H
                              |   |   |   |   |   |
                              H   H   H   H   H   H
```

carbon framework of hexane — hexane

The names for these alkanes are derived from the systematic nomenclature described in Appendix E.

Step 2. Next draw a chain of five carbon atoms, then add the sixth carbon atom to one of the carbons in the middle of the chain. (Adding it to a carbon at the end of the chain gives a six-carbon chain, the same carbon framework drawn in step 1.) Two different carbon frameworks can be built from the five-carbon chain, depending on whether the sixth carbon is linked to the 2 or 3 position. For each of these frameworks, fill in the hydrogens. The second and third isomers are named 2-methylpentane and 3-methylpentane, respectively.

```
                           H
                           |
                         H—C—H
   C                     H |   H   H   H
   |                     | |   |   |   |
C—C—C—C—C  ⟶  H—C—C—C—C—C—H
                         |     |   |   |
                         H     H   H   H
```

2-methylpentane

carbon framework of methylpentane isomers

```
                               H
                               |
                             H—C—H
      C                  H   H |   H   H
      |                  |   | |   |   |
C—C—C—C—C  ⟶  H—C—C—C—C—C—H
                         |   |   |   |   |
                         H   H   H   H   H
```

3-methylpentane

Step 3. Draw a chain of four carbon atoms. Add in the two remaining carbons, again being careful not to extend the chain length. Two different structures are possible, one with one carbon each in the 2 and 3 positions and another with both extra carbon atoms in the 2 position. Fill in the 14 hydrogens. You have now drawn the fourth and fifth isomers: 2,3-dimethylbutane and 2,2-dimethylbutane.

```
                                        H
                                        |
                                      H—C—H
          C                       H   H   |   H
          |                       |   |   |   |
    C—C—C—C           ⟶     H—C—C—C—C—H
      |                           |   |   |   |
      C                           H   |   H   H
                                    H—C—H
                                      |
                                      H
```

carbon atom frameworks for dimethylbutane isomers

2,3-dimethylbutane

```
                               H
                               |
                             H—C—H
      C                      H   |   H   H
      |                      |   |   |   |
    C—C—C—C    ⟶      H—C—C—C—C—H
      |                      |   |   |   |
      C                      H   |   H   H
                             H—C—H
                               |
                               H
```

2,2-dimethylbutane

Step 4. Should we look for structures in which the longest chain is three carbon atoms? Try it, but you will see that it is not possible to add the three remaining carbons to a three-carbon chain without creating one of the carbon chains already drawn in a previous step. Thus, we have completed the analysis, with five isomers of this compound being identified.

Exercise 11.1 Drawing Structural Isomers of Alkanes

Draw and name the nine isomers having the formula C_7H_{16}. (*Hint:* There is one structure with a seven-carbon chain, two with six-carbon chains, five in which the longest chain has five carbons, and one with a four-carbon chain.)

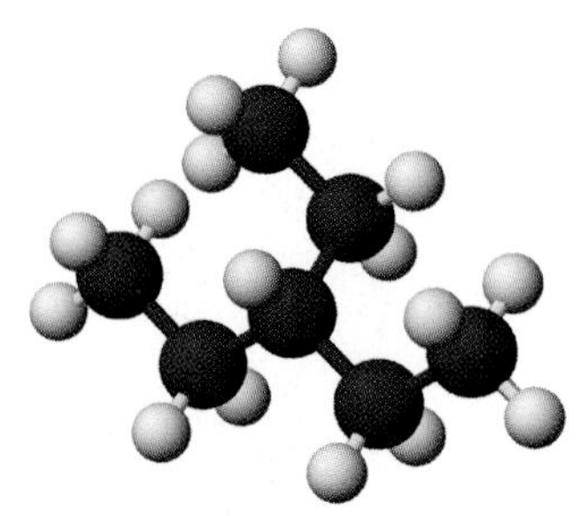

2-ethylpentane, one of the isomers with the formula C_7H_{16}.

Cycloalkanes, C_nH_{2n}

Cycloalkanes are constructed of tetrahedral carbon atoms (each with two H atoms attached) with the carbon framework in the form of a ring. For example, to create the structure of cyclohexane, C_6H_{12}, from hexane, C_6H_{14}, remove one hydrogen atom from each of the C atoms at the ends of the carbon chain in hexane, bend the carbon atoms in a circle, and form a bond between these two carbons. Thus, the formula for a cycloalkane will have two hydrogens less than an alkane containing the same number of carbon atoms.

hexane, C_6H_{14}

Remove 2 H atoms from ends of C_6 chain and form ring

cyclohexane, C_6H_{12}

Cyclohexane, with a ring with six —CH_2 groups, is a well-known cycloalkane. Its six carbon atoms do not lie in a plane. If the carbon atoms were in the form of a regular hexagon with all carbon atoms in one plane, the C—C—C bond angles would be 120°. To have a tetrahedral bond angle of 109.5°, the ring has to pucker. Two structures, the "chair" and "boat" forms, are possible. By partial rotation of several bonds, the two forms can be interconverted. The more stable structure is the chair form because this allows the hydrogen atoms to get as far apart as possible. A side view of cyclohexane reveals that two sets of hydrogen atoms occur in this molecule. Six hydrogen atoms, called the **equatorial** hydrogens, lie in a plane around the carbon ring. The other six hydrogens are positioned above and below the plane and are called **axial** hydrogens. Flexing the ring (a rotation around the C—C single bonds) causes the axial and equatorial hydrogens to exchange positions.

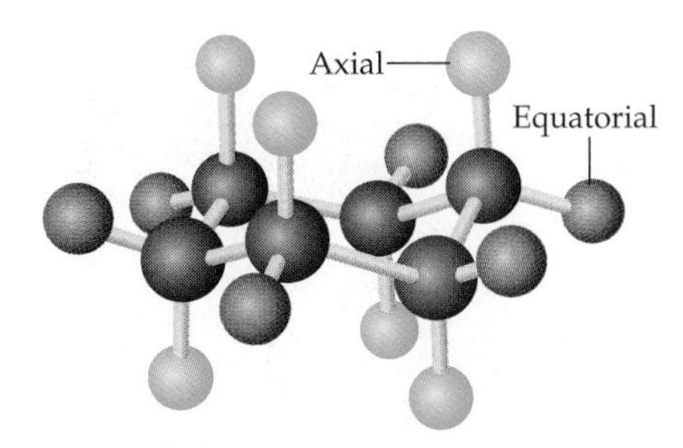

Axial and equatorial positions on cyclohexane

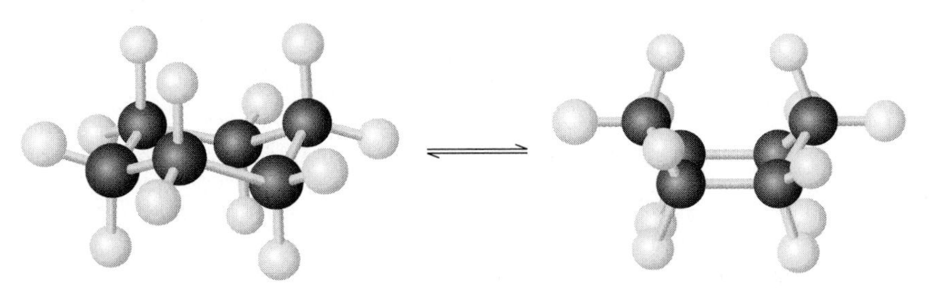

Chair conformation

Boat conformation

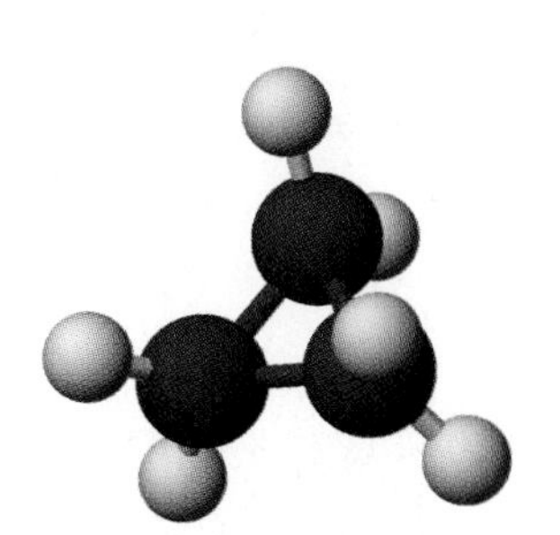

cyclopropane, C_3H_6

Many cycloalkanes exist. Cyclopentane, C_5H_{10}, has a nearly planar five-member ring. The angles of a regular pentagon are 110°, very nearly the tetrahedral angle. A slight distortion from planarity allows hydrogen atoms on adjacent carbon atoms to get a little farther apart. Interestingly, cyclobutane and cyclopropane are also known, although the bond angles in these species are much less than 109.5°. These examples are often called *strained hydrocarbons,* because an unfavorable geometry is imposed around carbon. One of the features of strained hydrocarbons is that the C—C bonds are weaker and the molecules readily undergo ring-opening reactions that relieve the bond angle strain.

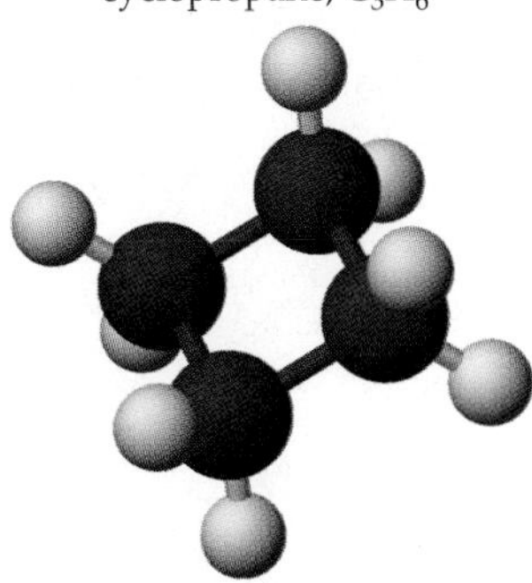

cyclobutane, C_4H_8

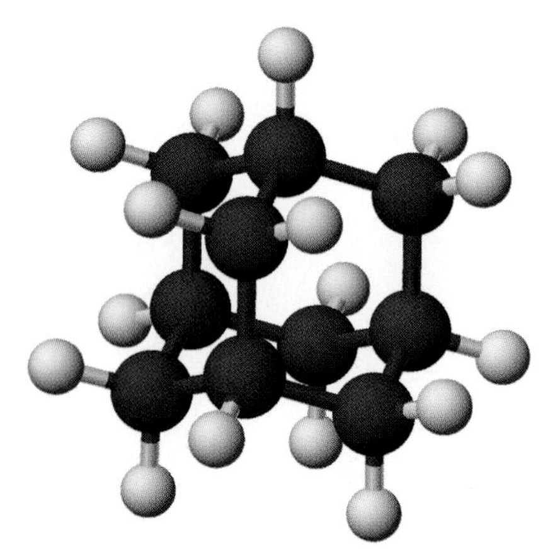

Adamantane, $C_{10}H_{16}$

Finally, to further illustrate the diversity of carbon in forming molecular species, compounds with two, three, or more rings are possible as well. Adamantane, $C_{10}H_{16}$, a tricyclic alkane, is interesting because the carbon skeleton represents a small piece of the diamond structure.

Properties of Alkanes

Table 11.3 lists the lower molecular weight, unbranched alkanes. Methane, ethane, propane, and butane are gases at standard temperature and pressure. Higher molecular weight compounds are liquids or solids. An increase in melting point and boiling point with molecular weight is a general phenomenon that reflects the increased forces of attraction between molecules (Chapter 13).

You already know about alkanes in a nonscientific context because several, such as methane, ethane, propane, and butane, are common fuels. Gasoline and kerosene, and also fuel oils and lubricating oils, are mixtures of various alkanes. White mineral oil is also a mixture of alkanes, as is paraffin wax.

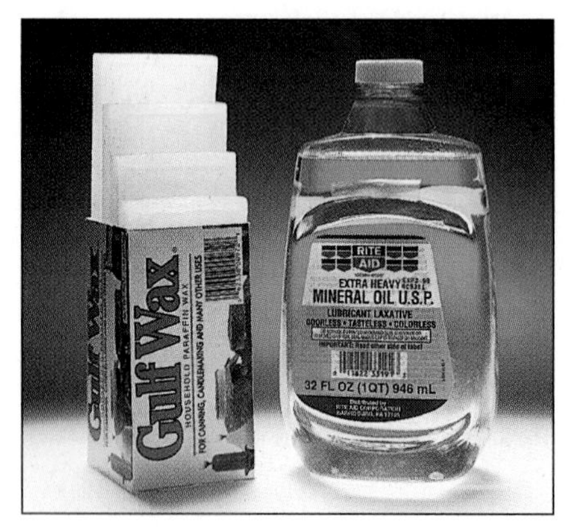

Paraffin wax and mineral oil are mixtures of alkanes. *(C. D. Winters)*

As pure substances, the alkanes are colorless. (The colors seen in petroleum products are due to additives.) The gases and liquids have noticeable but not unpleasant odors, and all of these substances are insoluble in water. (Water-insolubility is typical of compounds that have low polarity. Low polarity is expected because the electronegativity of carbon ($\chi = 2.5$) and hydrogen ($\chi = 2.1$) is not greatly different.)

The chemical reactivity of the alkanes is limited. They all burn readily in air, however, to give CO_2 and H_2O in very exothermic reactions. This is of course the reason they are widely used as fuels.

$$\begin{array}{lcccc} & CH_4(g) + 2\ O_2(g) & \longrightarrow & CO_2(g) + & 2\ H_2O(\ell) \\ \Delta H_f^\circ\ (\text{kJ/mol}): & -74.8 \qquad 0 & & -393.5 & 2(-285.8) \end{array}$$

$$\Delta H^\circ_{rxn} = \sum \Delta H_f^\circ\,(\text{products}) - \sum \Delta H_f^\circ\,(\text{reactants}) = -890.3\ \text{kJ}$$

Table 11.3 • SELECTED HYDROCARBONS OF THE ALKANE FAMILY, C_nH_{2n+2}*

Name	Molecular Formula	State at Room Temperature
methane	CH_4	
ethane	C_2H_6	gas
propane	C_3H_8	
butane	C_4H_{10}	
pentane	C_5H_{12} (pent- = 5)	
hexane	C_6H_{14} (hex- = 6)	
heptane	C_7H_{16} (hept- = 7)	liquid
octane	C_8H_{18} (oct- = 8)	
nonane	C_9H_{20} (non- = 9)	
decane	$C_{10}H_{22}$ (dec- = 10)	
octadecane	$C_{18}H_{38}$ (octadec- = 18)	solid
eicosane	$C_{20}H_{42}$ (eicos- = 20)	

*This table lists only selected alkanes. Liquid compounds with 11 to 16 C atoms are also known, and other solid alkanes include $C_{17}H_{36}$ and $C_{19}H_{40}$.

The hydrogen atoms of an alkane can be replaced by chlorine atoms on reaction with Cl_2. This is formally an oxidation because Cl_2, like O_2, is a strong oxidizing agent. These reactions, which can be initiated by ultraviolet radiation, are free radical reactions; highly reactive Cl atoms are formed from Cl_2 under UV radiation. Reaction of methane with Cl_2 under these conditions proceeds in a series of steps, eventually yielding CCl_4, commonly known as carbon tetrachloride. (HCl is the other product of these reactions.)

$$CH_4 \xrightarrow[UV]{Cl_2} CH_3Cl \xrightarrow[UV]{Cl_2} CH_2Cl_2 \xrightarrow[UV]{Cl_2} CHCl_3 \xrightarrow[UV]{Cl_2} CCl_4$$

	CH_3Cl	CH_2Cl_2	$CHCl_3$	CCl_4
Systematic name	chloromethane	dichloromethane	trichloromethane	tetrachloromethane
Common name	methyl chloride	methylene chloride	chloroform	carbon tetrachloride

The last three compounds have been used as solvents, although less frequently now because of their toxicity. Carbon tetrachloride was also once widely used as a dry cleaning fluid and, because it does not burn, in fire extinguishers.

11.3 ALKENES AND ALKYNES

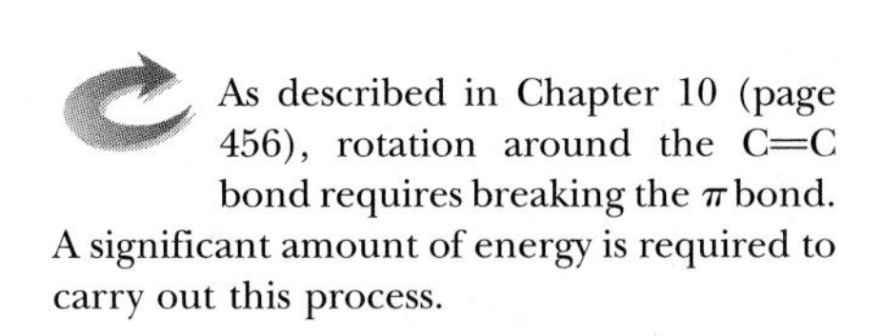

As described in Chapter 10 (page 456), rotation around the C=C bond requires breaking the π bond. A significant amount of energy is required to carry out this process.

The abundance and diversity of compounds seen with the alkanes are repeated with **alkenes.** The presence of a C=C double bond adds a further structural dimension, however, because an additional type of isomerism is now possible.

Isomerism is one of the most interesting aspects of molecular structure. **Structural isomers** are molecules with the same molecular formula but different bonding arrangements of atoms. You have already encountered structural isomers in compounds like butane and 2-methylpropane (page 487). **Stereoisomerism** is a second type of isomerism. Here the atom-to-atom bonding sequence is the same, but the atoms differ in their arrangement in space. One form of stereoisomerism is **geometrical isomerism,** in which the atoms making up a molecule are arranged in different geometrical relationships. *Cis*-2-butene and *trans*-2-butene are geometrical isomers (page 481).

A second type of stereoisomerism is optical isomerism, which arises when a molecule and its mirror image do not superimpose. See Chapter 23.

The general formula for alkenes is C_nH_{2n}. The first two members of the series of alkenes are ethylene, C_2H_4 (also called ethene), and propene, C_3H_6; only a single structure can be drawn for these compounds. As with the alkanes, the occurrence of isomers begins with the four-carbon species. There are four alkene isomers having the formula C_4H_8, and two of them, both named 2-butene, are stereoisomers of each other. This pair are structural isomers of 1-butene and 2-methylpropene.

See the *Saunders Interactive General Chemistry CD-ROM,* Screen 11.3, Hydrocarbons, for a discussion of alkenes and alkynes.

1-butene	2-methylpropene (isobutene)	*cis*-2-butene	*trans*-2-butene
$(H)(H)C{=}C(CH_2CH_3)(H)$	$(H)(H)C{=}C(CH_3)(CH_3)$	$(H_3C)(H)C{=}C(CH_3)(H)$	$(H)(H_3C)C{=}C(CH_3)(H)$

Each of the four butene isomers is a distinct compound that can be isolated (Table 11.4).

Table 11.4 • Properties of Butene Isomers

Name	Boiling Point	Melting Point	Dipole Moment (D)	ΔH_f° (g) (kJ/mol)
1-butene	−6.26 °C	−185.4 °C	———	−20.5
2-methylpropene	−6.95 °C	−140.4 °C	0.503	−37.5
cis-2-butene	3.71 °C	−138.9 °C	0.253	−29.7
trans-2-butene	0.88 °C	−105.5 °C	0	−33.0

Example 11.2 Determining Structural Isomers of Alkenes from a Formula

Draw structures for the six possible alkene isomers with the formula C_5H_{10}.

Solution A procedure that involved drawing the carbon skeleton and then adding hydrogen atoms served well when drawing structures of alkanes, and a similar approach can be used here. It will be necessary to put one double bond into the framework and to be alert for *cis–trans* isomerism.

1. A five-carbon chain with one double bond can be constructed in two ways. One gives rise to *cis–trans* isomers.

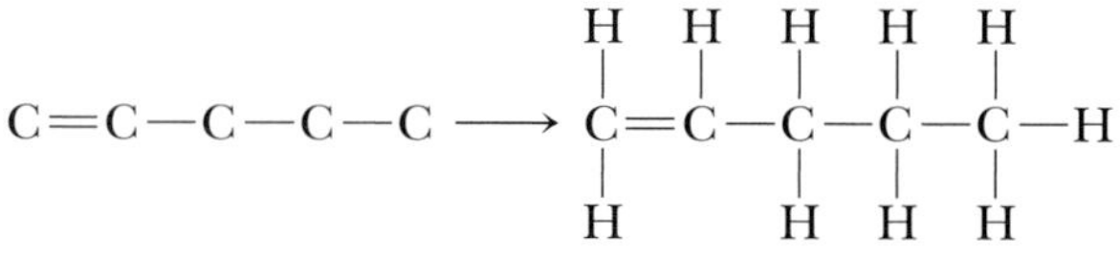

$$C{=}C{-}C{-}C{-}C \longrightarrow H_2C{=}CH{-}CH_2{-}CH_2{-}CH_3$$

1-pentene

The alkene isomers described are shown on the Web site for this book at http://www.saunderscollege.com.

$$C{-}C{=}C{-}C{-}C$$

H H / C=C / H₃C CH₂CH₃ — *cis*-2-pentene: $H(H_3C)C{=}C(H)CH_2CH_3$

cis-2-pentene

H CH₂CH₃ / C=C / H₃C H — $H(H_3C)C{=}C(CH_2CH_3)H$

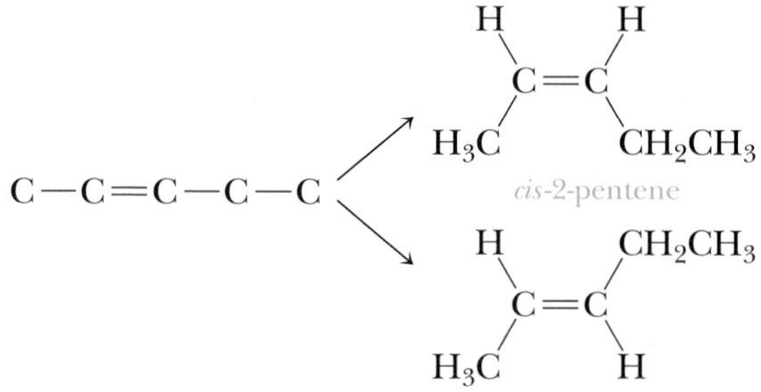

trans-2-pentene

2. Now draw the possible four-carbon chains containing a double bond (analogous to 1-butene and *cis*- and *trans*-2-butene). Add the fifth carbon atom to either the 2 or 3 position. When all possible combinations are found, fill in the hydrogen atoms. This results in three more structures:

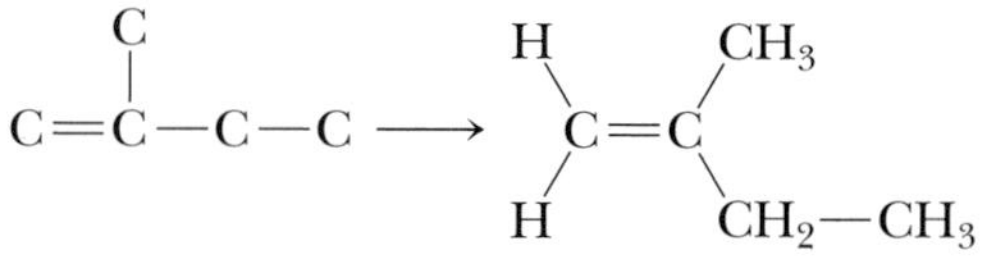

$$C{=}C(C){-}C{-}C \longrightarrow H_2C{=}C(CH_3){-}CH_2{-}CH_3$$

2-methyl-1-butene

$$\mathrm{C{=}C{-}\underset{}{\overset{\displaystyle C}{\overset{|}{C}}}{-}C} \longrightarrow \mathrm{(H)(H)C{=}C(H)(CH(CH_3){-}CH_3)}$$

3-methyl-1-butene

$$\mathrm{C{-}C{=}\overset{\displaystyle C}{\overset{|}{C}}{-}C} \longrightarrow \mathrm{(H)(H_3C)C{=}C(CH_3)(CH_3)}$$

2-methyl-2-butene

Exercise 11.2 Determining Structural Isomers of Alkenes from a Formula

There are 17 possible alkene isomers with the formula C_6H_{12}. Draw structures of the five isomers in which the longest chain has six carbon atoms. (There are also eight isomers in which the longest chain has five carbon atoms, and four isomers in which the longest chain has four carbon atoms. How many can you find?)

At this point it should be apparent that the number of alkenes and the complexity of their structures will rival the alkanes. A further step leads to cycloalkenes, such as cyclohexene. Another direction leads to compounds with two double bonds (*dienes*) such as butadiene, and then to bicyclic compounds, and to *polyenes.* Many alkenes are natural products (Figure 11.4).

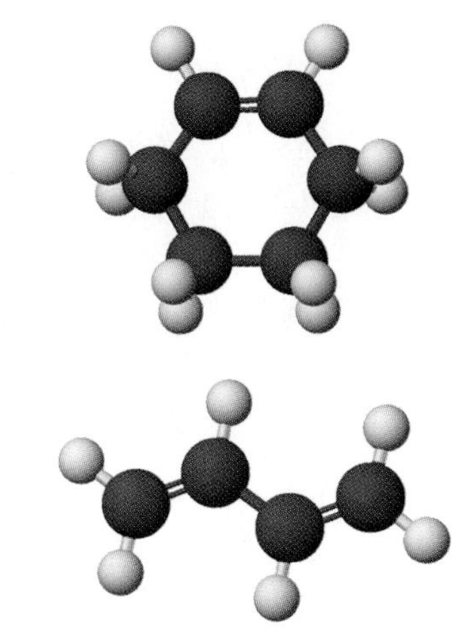

(top) Cyclohexene, C_6H_{10}, and *(bottom)* 1,3-butadiene, C_4H_6.

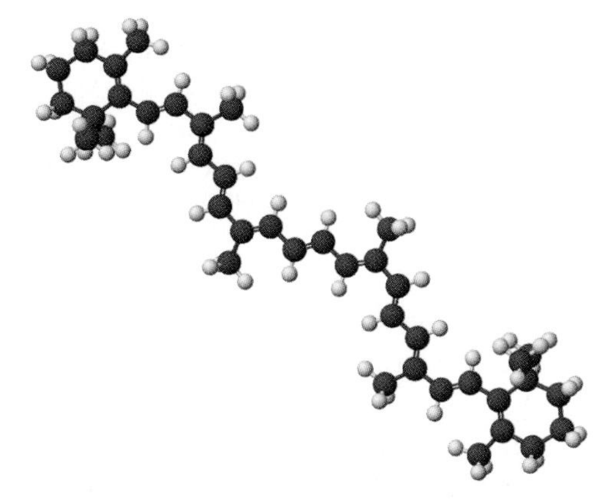

Figure 11.4 Carotene, a naturally occurring compound with eleven C=C bonds. The π electrons can be excited by visible light in the blue-violet region of the spectrum, and so carotene appears orange-yellow to the observer. Carotene or carotene-like molecules are partnered with chlorophyll in nature in the role of assisting in the harvesting of sunlight. Green leaves have a high concentration of carotene. In autumn, green chlorophyll molecules are destroyed and the yellows and reds of carotene and related molecules are seen. The red color of tomatoes comes from a molecule very closely related to carotene. As a tomato ripens, its chlorophyll disintegrates and the green color is replaced by the red of the carotene-like molecule. *(Photo, C. D. Winters; model, S. M. Young)*

Table 11.5 • Some Simple Alkynes

Structure	Systematic Name	Common Name	BP (°C)
H—C≡C—H	ethyne	acetylene	−75
$CH_3C{\equiv}CH$	propyne	methylacetylene	−23
$CH_3CH_2C{\equiv}CH$	1-butyne	ethylacetylene	9
$CH_3C{\equiv}CCH_3$	2-butyne	dimethylacetylene	27

Figure 11.5 **Acetylene.** The reaction of acetylene with oxygen produces a very high temperature. Oxy-acetylene torches, used in welding, take advantage of this fact. *(C. D. Winters)*

Alkynes (C_nH_{2n-2}) will seem fairly simple compared with alkenes. There are fewer isomers and less structural complexity in this group. Table 11.5 lists the four alkynes that have four or fewer carbon atoms.

Properties of Alkenes and Alkynes

Like the alkanes, alkenes and alkynes are generally colorless species. The low-molecular-weight compounds are gases; compounds with higher molecular weights are liquids and solids at STP. They can be oxidized by O_2 to CO_2 and H_2O (Figure 11.5). Because of the C=C bond, however, alkenes have a much more extensive reactivity than alkanes. We have already mentioned one type of reactivity, polymerization, to form molecules with long chains of singly bonded carbons (page 486) and describe it further in Section 11.7.

Addition Reactions of Alkenes and Alkynes

Alkenes are often called olefins. Polymers formed by polymerization of alkenes are referred to as polyolefins.

Compounds with double and triple bonds between the carbon atoms are often referred to as unsaturated, which gives you an insight into their chemical behavior. Carbon is capable of bonding to a maximum of four other atoms (as it does in methane and other alkanes); **unsaturated** refers to the fact that the carbon atoms of the double bond in alkenes are bonded to *fewer* than four other groups. This means that the alkene can use the π electrons of the double bond and *add* other atoms such as Br by σ bond formation:

$$H_2C{=}CH_2 + Br_2 \longrightarrow H{-}CHBr{-}CHBr{-}H$$

1,2-dibromoethane

The addition of different molecules to C=C double bonds to give a **saturated** compound (one with four atoms bonded to each carbon) is illustrated by a general equation:

$$H_2C{=}CH_2 + X{-}Y \longrightarrow H{-}CHX{-}CHY{-}H$$

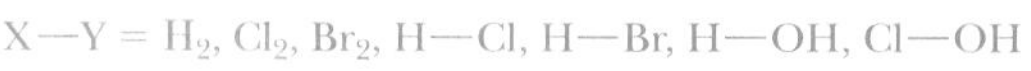

Synthesis of many organic compounds is accomplished by this route. For example, alcohols are formed by adding water to the double bond. Compounds with triple bonds react in a similar fashion:

$$\text{H—C}\equiv\text{C—H} + 2\ \text{Cl}_2 \longrightarrow \text{Cl}_2\text{CHCHCl}_2$$

If the reagent added to a double bond is hydrogen (X—Y = H_2), the reaction is called **hydrogenation,** and the product is an alkane. Hydrogenation is usually a very slow reaction, but it can be speeded up in the presence of a catalyst, often a specially prepared form of a metal, such as platinum, palladium, and rhodium.

A substance that causes a reaction to occur at a faster rate is called a *catalyst.* We will talk more about catalysts when rates of reactions are discussed more fully in Chapter 15.

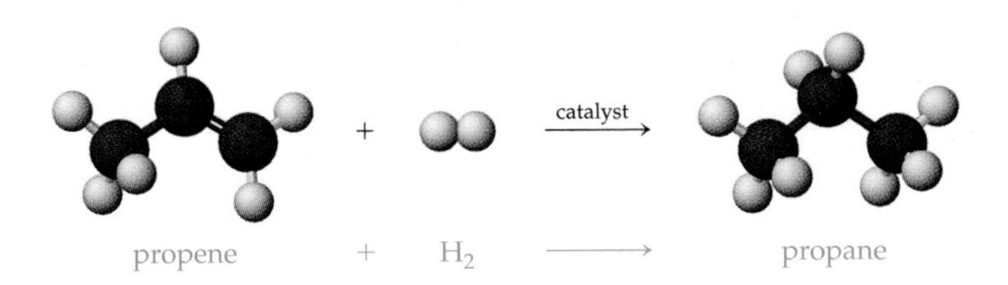

You may have heard the term hydrogenation because certain foods you eat are sometimes "partially hydrogenated." One brand of crackers has a label that says "Made with 100% pure vegetable shortening . . . (partially hydrogenated soybean oil with hydrogenated cottonseed oil)." As described in Section 11.6, hydrogenation makes oils less susceptible to spoilage, among other things.

Exercise 11.3 Reactions of Alkenes

1. Draw the structure of the compound obtained from the reaction of HBr with ethylene.
2. Draw the structure for the compound that comes from the reaction of Br_2 with *cis*-2-butene. Give a reasonable name for this compound.

Aromatic Compounds

Benzene, C_6H_6, is a key molecule in chemistry. It is the simplest **aromatic** compound, a class of compounds so named because they have a significant, and usually not unpleasant, odor. Other members of this class include toluene and naphthalene. A source of many aromatic compounds is coal tar, the tarry material that often accompanies soft coal (Table 11.6).

Guidelines for naming organic compounds are given in Appendix E.

See the *Saunders Interactive General Chemistry CD-ROM,* Screen 11.3, Hydrocarbons, for a discussion of aromatics.

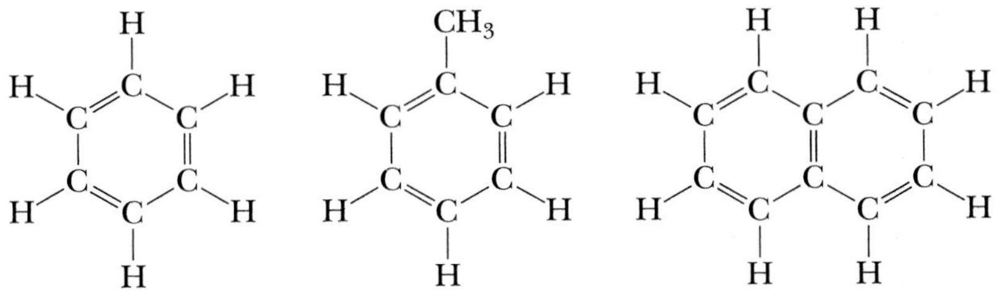

benzene toluene naphthalene

Table 11.6 • Some Aromatic Compounds from Coal Tar

Name	Formula	Boiling Point (°C)	Melting Point (°C)
benzene	C_6H_6	80	+6
toluene	$C_6H_5CH_3$	111	−95
o-xylene	1,2-$C_6H_4(CH_3)_2$	144	−27
m-xylene	1,3-$C_6H_4(CH_3)_2$	139	−54
p-xylene	1,4-$C_6H_4(CH_3)_2$	138	+13
naphthalene	$C_{10}H_8$	218	+80

Review structure and bonding in benzene in Chapter 10 (page 457).

Benzene occupies a pivotal place in the history and practice of chemistry. Michael Faraday discovered this compound in 1825 as a byproduct of illuminating gas, itself produced by heating coal. The structure for benzene was later proposed by August Kekulé (1829–1896). Benzene is an important industrial chemical, usually about 15th on the list of the top 50 chemicals produced annually in the United States. It is used as a solvent and is also the starting point for making thousands of different compounds by replacing the H atoms of the ring.

Toluene was originally obtained from Tolu balsam, which is derived from the pleasant-smelling gum of a South American tree, *Toluifera balsamum.* This balsam has been used in cough syrups and perfumes. Naphthalene is one of the ingredients used in "moth balls," although *p*-dichlorobenzene is now more commonly used. Aspartame (page 423) and another artificial sweetener, saccharin, are also benzene derivatives.

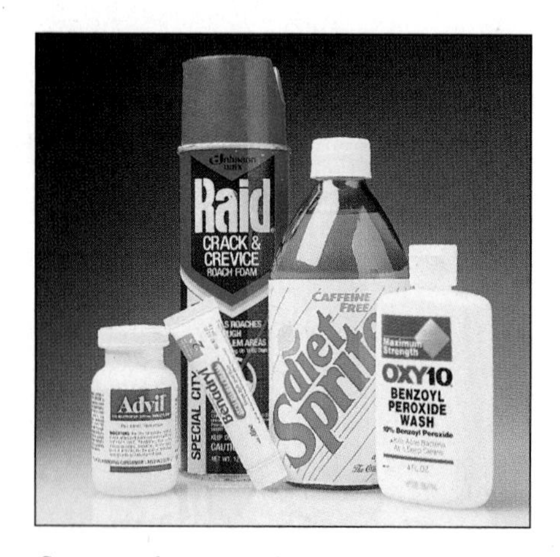

Some products containing compounds based on benzene. Examples include sodium benzoate in soft drinks (page 120), ibuprofen in Advil, and benzoyl peroxide in Oxy-10. *(C. D. Winters)*

The Structure of Benzene

The formula of benzene suggested to 19th-century chemists that this compound should be unsaturated, but, if viewed this way, its chemistry was perplexing. Whereas an alkene readily adds Br_2, for example, benzene does not do so under similar conditions. Instead, benzene reacts with Br_2 at higher temperatures in the presence of other chemicals such as $FeBr_3$ to give a product in which a Br atom has been substituted for an H atom.

Substitution reaction

$$\text{benzene } (C_6H_6) \xrightarrow{Br_2/FeBr_3} \text{bromobenzene } (C_6H_5Br)$$

The story of Kekulé is told in A. J. Ihde; *The Development of Modern Chemistry.* Dover Publications, New York, 1984.

The observation that only one monosubstitution product was obtained for a given reaction was one of the pieces of evidence used by Kekulé to suggest that benzene has a planar and symmetrical six-member ring structure in which all the carbon–carbon and carbon–hydrogen bonds are equivalent (Figure 11.6). Kekulé also suggested (in 1872) that the molecule could be represented by some combination of two structures, which we now call resonance structures.

As mentioned in Chapter 10 (page 458), the structure of benzene has equivalent carbon–carbon bonds, 139 pm long, intermediate between the length of a C—C single bond (154 pm) and a C═C double bond (134 pm). In valence bond terms, its structure is represented as a resonance hybrid (page 393). To save space and time, however, chemists often represent the ring using a shorthand notation.

or simply

Representations of benzene, C_6H_6

Figure 11.6 Benzene structure. An image from a scanning tunneling microscope confirms the six-member ring structure. In this image each ring appears as a raised bump with a slight depression in the middle. The fuzziness of the image is due to the motion of the π-bonding electrons of the C_6 ring. *(Courtesy IBM Corporation Almaden Research Center)*

Substitution Reactions of Aromatic Compounds

Instead of the addition reactions typical of compounds with carbon–carbon double bonds such as alkenes, benzene and other aromatic compounds undergo substitution reactions (Figure 11.7). For example, it is possible to substitute a halogen atom, a nitro group, or an alkyl or other hydrocarbon grouping for one or more of the hydrogen atoms of benzene.

A vast array of compounds that are derivatives of benzene exists in which one or more hydrogens are replaced with other atoms or groups of atoms. Chlorobenzene and nitrobenzene, toluene, *m*-xylene, styrene, and phenol are common examples. Isomers can arise when two or more substituents are attached to a benzene ring.

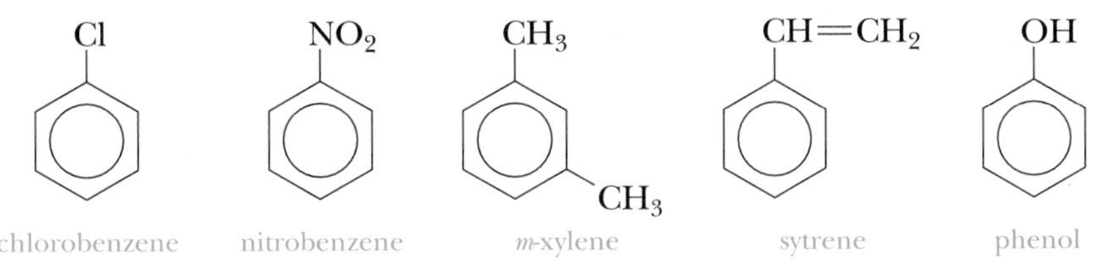

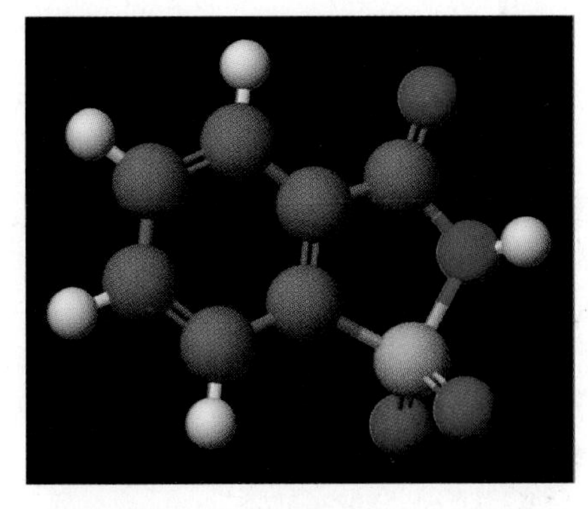

Saccharin, a benzene derivative, is over 300 times sweeter than sugar. At one time it was the most important artificial sweetener in foods, but it has been supplanted by aspartame. Saccharin was discovered by accident in 1879 by Ira Remsen and his student, C. Fahlberg, of Johns Hopkins University.

A systematic nomenclature for derivatives of benzene involves naming substituent groups and identifying their position on the ring by numbering the six carbon atoms (Appendix E). Some common names, which are based on an older

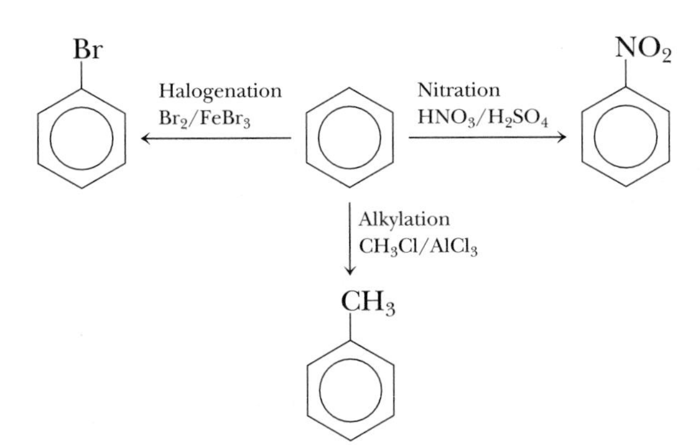

Figure 11.7 Some reactions of benzene. Notice that all are substitution reactions and not addition reactions. Notice also that a combination of reagents is often needed. The substitution of H by $—CH_3$ is promoted by the presence of $AlCl_3$, and $FeBr_3$ assists in the replacement of H by Br.

naming scheme, are also regularly used. This scheme identified isomers of disubstituted benzenes with the prefixes *ortho* (*o*-, carbons adjacent in the benzene ring), *meta* (*m*-,), and *para* (*p*-, carbons on opposite sides of the ring).

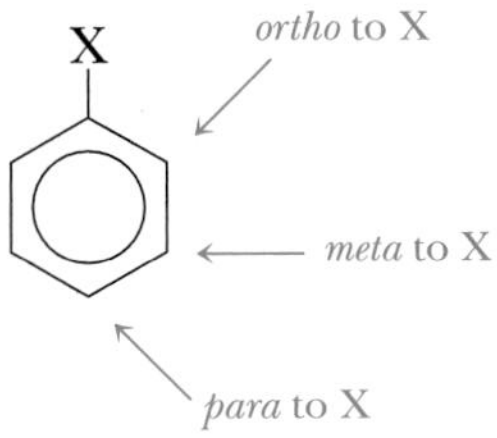

X
ortho to X
meta to X
para to X

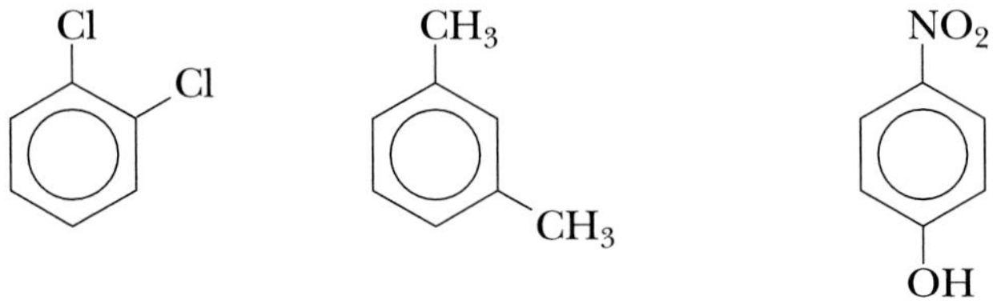

Systematic name	1,2-dichlorobenzene	1,3-dimethylbenzene	1-hydroxy-4-nitrobenzene
Common name	*o*-dichlorobenzene	*m*-xylene	*p*-nitrophenol

Example 11.3 Isomers of Substituted Benzenes

How many isomers are possible for trisubstituted benzenes? Draw and name the isomers of $C_6H_3Cl_3$.

Solution The three isomers of $C_6H_3Cl_3$ are shown here. They are named as derivatives of benzene by specifying the number of substituent groups by the prefix "tri-", the name of the substituent, and the positions of the three groups around the six-member ring.

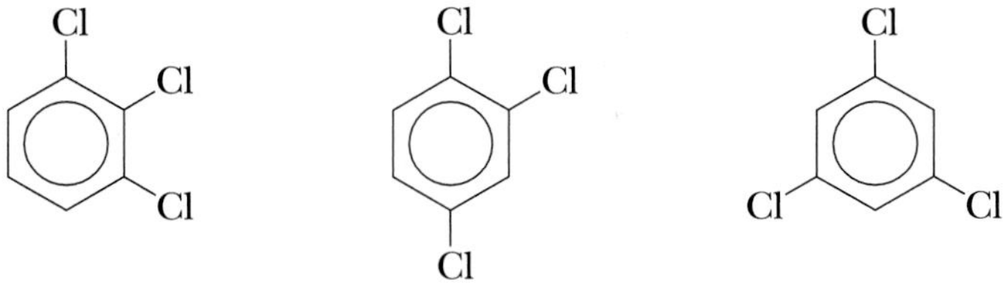

1,2,3-trichlorobenzene 1,2,4-trichlorobenzene 1,3,5-trichlorobenzene

Exercise 11.4 Isomers of Substituted Benzenes

Draw a structure for the compound *p*-dichlorobenzene—one of the two aromatic compounds used in moth balls. What is the systematic name for this compound?

Properties of Aromatic Compounds

Benzene is a colorless liquid, and simple substituted benzenes are liquids or solids at STP. Their properties are typical of hydrocarbons. They are insoluble

A CLOSER LOOK *Petroleum Chemistry*

Much of current chemical technology is based on petroleum. Burning of fuels derived from petroleum provides by far the largest amount of energy needed in the industrial world (Section 6.11). Petroleum and natural gas are also the chemical raw materials for a great many of the things that we take for granted. The petrochemical industry, one of the largest parts of our nation's economy, produces from petroleum the starting materials used in the manufacture of plastics, rubber, and a vast array of other compounds.

Petroleum as it is pumped out of the ground is a complex mixture whose composition varies greatly with source. The primary components of petroleum are alkanes, but, to varying degrees, nitrogen and sulfur-containing compounds are also present. Aromatics may be present, but alkenes and alkynes are *not* components.

An early step in the refining process involves distillation (Chapter 14), a process in which the crude mixture is separated into a series of fractions based on boiling point: a gaseous fraction (mostly alkanes with one to four carbon atoms; this fraction is often burned off), gasoline, kerosene, and fuel oils. After distillation, considerable material, in the form of semisolid, tarlike residues, is likely to remain.

The petrochemical industry seeks to maximize amounts of the higher valued fractions of petroleum and to make specific compounds for which there is a particular need. This means carrying out chemistry on the raw materials on a huge scale. One of the processes to which petroleum is subjected is known as *cracking*. At very high temperatures, bond breaking or "cracking" can occur, and longer chain hydrocarbons fragment to smaller molecular units. These highly complicated reactions are carried out in the presence of a wide array of catalysts, materials that speed up reactions and direct them toward specific products. Among the important products of cracking are ethylene and other alkenes, the important raw materials for the formation of materials such as polyethylene. Ethylene and propylene (propene) are among the top dozen chemicals produced in the United States. Cracking also produces gaseous hydrogen, another widely used raw material in the chemical industry.

In petroleum refining, other kinds of reactions are also important. All are run at elevated temperature and in the presence of specific catalysts. There are isomerization reactions in which the carbon skeleton of an alkane rearranges to a new isomeric species, and reformation processes in which smaller molecules combine to form new molecules. Each process is directed to a specific goal, such as increasing the proportion of branched-chain hydrocarbons in gasoline to obtain higher octane ratings.

A great amount of chemical research has gone into developing and understanding these processes. The petroleum industry is a major employer of chemists and chemical engineers, as research continues to seek ways to improve production of existing products and develop new products.

A modern petroleum refinery. *(Courtesy Marathon Ashland Petroleum)*

in water but soluble in nonpolar solvents. Of course, aromatic hydrocarbons are oxidizable to CO_2 and H_2O.

One of the more important features of benzene is an unusual stability that is associated with the unique π bonding in this molecule. Because resonance is usually used when describing the bonding in benzene, this extra stability is termed **resonance stabilization.** It can be illustrated by comparing the energy evolved in the hydrogenation of benzene with the energy expected for hydrogenation of a compound with three double bonds. The energy evolved in the hydrogenation of three double bonds ($3\ H_2 + 3\ H_2C{=}CH_2 \longrightarrow 3\ C_2H_6$) can be estimated from bond energies (page 402).

$$\begin{aligned}\Delta H_{rxn} &= \sum D(\text{bonds broken}) - \sum D(\text{bonds formed})\\ &= [3\ D(\text{C=C}) + 3\ D(\text{H—H})] - [6\ D(\text{C—H}) + 3\ D(\text{C—C})]\\ &= [3\ \text{mol} \times 602\ \text{kJ/mol} + 3\ \text{mol} \times 436\ \text{kJ/mol}] -\\ &\qquad [6\ \text{mol} \times 413\ \text{kJ/mol} + 3\ \text{mol} \times 346\ \text{kJ/mol}]\\ &= -402\ \text{kJ}\end{aligned}$$

The actual energy of hydrogenation of benzene can be calculated from heats of formation, based on the reaction $C_6H_6(g) + 3\ H_2(g) \longrightarrow C_6H_{12}(g)$. (Using ΔH°_f values for the gaseous reactants and product is done because bond energy calculations are based on gaseous reagents.)

$$\begin{aligned}\Delta H^\circ_{rxn} &= \Delta H^\circ_f\,(\text{products}) - \Delta H^\circ_f\,(\text{reactants})\\ &= \Delta H^\circ_f\,[C_6H_{12}(g)] - \{\Delta H^\circ_f\,[C_6H_6(g)] + 3\ \Delta H^\circ_f\,[H_2(g)]\}\\ &= 1\ \text{mol}\ (-123.4\ \text{kJ/mol}) - [1\ \text{mol}\ (+82.8\ \text{kJ/mol}) + 3\ \text{mol}\ (0)]\\ &= -206.2\ \text{kJ}\end{aligned}$$

Hydrogenation of benzene is less exothermic (by 196 kJ) than adding H_2 to three separate double bonds. This difference is attributed to the added stability given to benzene by resonance.

11.4 ALCOHOLS AND AMINES

Other types of organic compounds arise as elements other than carbon and hydrogen are introduced. Two elements in particular, oxygen and nitrogen, add a rich dimension to carbon chemistry.

See the *Saunders Interactive General Chemistry CD-ROM,* Screen 11.5, Functional Groups, for a discussion of alcohols and amines. See also Screen 11.6, Reactions of Alcohols.

The conventional approach to organizing carbon chemistry considers compounds containing other elements as derivatives of hydrocarbons. Formulas (and structures) are represented by substituting one or more hydrogens in a hydrocarbon molecule by a **functional group,** an atom or group of atoms attached to a carbon atom in the hydrocarbon. Formulas of hydrocarbon derivatives are then written as R—X, in which R is a hydrocarbon lacking a hydrogen atom, and X is the functional group that has replaced the hydrogen in the structures. The chemical and physical properties of the hydrocarbon derivatives are a blend of the properties associated with hydrocarbons and the group that has been substituted for hydrogen.

Table 11.7 identifies a series of substituent groups and the families of organic compounds resulting from their attachment to an alkane.

Alcohols

If one of the hydrogen atoms of an alkane is replaced by a hydroxyl (—OH) group, the result is an **alcohol,** ROH. Methanol, CH_3OH, and ethanol, CH_3CH_2OH, are the most important of the simple alcohols, but other alcohols are also commercially important (Table 11.8). Notice that several have more than one OH functional group.

Table 11.7 • COMMON FUNCTIONAL GROUPS AND DERIVATIVES OF ALKANES

Functional Group*	General Formula	Class of Compound	Examples
F, Cl, Br, I	RF, RCl, RBr, RI	haloalkane	CH_3CH_2Cl, chloroethane
OH	ROH	alcohol	C_2H_5OH, ethanol
OR′	ROR′	ether	$(C_2H_5)_2O$, diethyl ether
NH_2†	RNH_2	(primary) amine	$CH_3CH_2NH_2$, ethyl amine
CHO	RCHO	aldehyde	CH_3CHO methanal (acetaldehyde)
COR′	RCOR′	ketone	CH_3COCH_3, 2-propanone (acetone)
CO_2H	RCO_2H	organic acid	CH_3CO_2H, ethanoic acid (acetic acid)
CO_2R'	RCO_2R'	ester	$CH_3CO_2CH_3$, methyl acetate
$CONH_2$	$RCONH_2$	amide	CH_3CONH_2, acetamide

*R and R′ can be the same or different hydrocarbon groups.
†Secondary amines (R_2NH) and tertiary amines (R_3N) are also possible, see discussion in text.

More than 5.1×10^8 kg of methanol is produced in the United States annually. Most is used to make formaldehyde (H_2CO) and acetic acid (CH_3CO_2H), both of which are components of polymers and are important chemicals in their own right. Methanol is also used as a solvent, as a de-icer in gasoline, and as a fuel in high-powered racing cars. Methanol is often called "wood alcohol" because it was originally produced by heating wood in the absence of air. It is found in low concentration in new wine, where it contributes to the odor, or "bouquet." Like ethanol, methanol causes intoxication, but it is poisonous largely because the human body converts it to formic acid (HCO_2H) and formaldehyde (H_2CO). These compounds attack the cells of the retina in the eye, leading to permanent blindness.

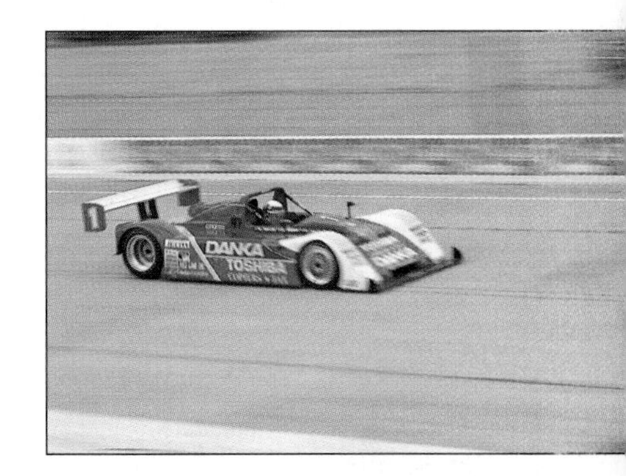

Methanol, CH_3OH, is used as the fuel in cars of the type that race in Indianapolis. *(David Young/Tom Stack & Associates)*

Ethanol is the "alcohol" of alcoholic beverages, in which it is formed by the anaerobic (without air) fermentation of sugar. For many years, industrial alcohol, which is used as a solvent and as a starting material for the synthesis of other

Table 11.8 • SOME IMPORTANT ALCOHOLS

Condensed Formula	BP (°C)	Systematic Name	Common Name	Use
CH_3OH	65.0	methanol	methyl alcohol	fuel, gasoline additive, making formaldehyde
CH_3CH_2OH	78.5	ethanol	ethyl alcohol	beverages, gasoline additive, solvent
$CH_3CH_2CH_2OH$	97.4	1-propanol	propyl alcohol	industrial solvent
$CH_3CH(OH)CH_3$	82.4	2-propanol	isopropyl alcohol	rubbing alcohol
$HOCH_2CH_2OH$	198	1,2-ethanediol	ethylene glycol	antifreeze
$HOCH_2CH(OH)CH_2OH$	290	1,2,3-propanetriol	glycerol (glycerin)	moisturizer in consumer products

compounds, was made by fermentation. In the last several decades, however, it has become cheaper to make ethanol from petroleum byproducts, specifically by the *addition* of water to ethylene.

$$\underset{\text{ethylene}}{H_2C{=}CH_2}(g) + H_2O(g) \xrightarrow{\text{catalyst}} \underset{\text{ethanol}}{H{-}CH_2{-}CH_2{-}OH}(\ell)$$

Ethylene glycol and glycerol are common alcohols having two or more —OH groups. Ethylene glycol is used as antifreeze in automobiles. Glycerol's most common use is as a softener in soaps and lotions. It is also a raw material for the preparation of nitroglycerin (Figure 11.8).

	$H{-}CH(OH){-}CH(OH){-}H$	$H{-}CH(OH){-}CH(OH){-}CH(OH){-}H$
Systematic name	1,2-ethanediol	1,2,3-propanetriol
Common name	ethylene glycol	glycerol or glycerin

It should not be surprising that there are a great many alcohols. Structures for alcohols can be created by replacing a hydrogen atom on the hydrocarbon with an —OH group, but the position of replacement makes a difference. Thus, 1-propanol, $CH_3CH_2CH_2OH$, and 2-propanol, $CH_3CH(OH)CH_3$, are two different alcohols derived from propane.

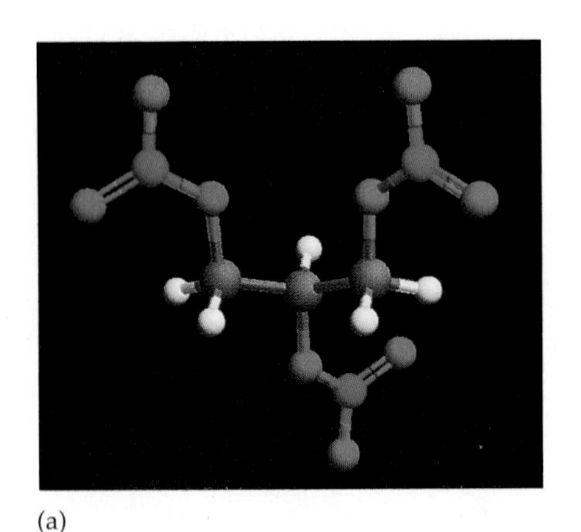

(a)

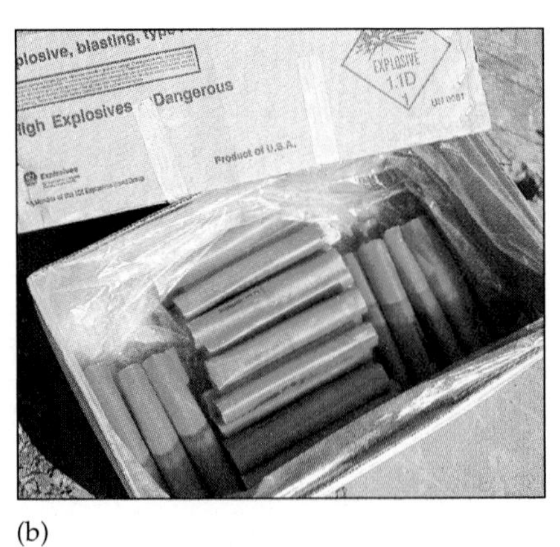

(b)

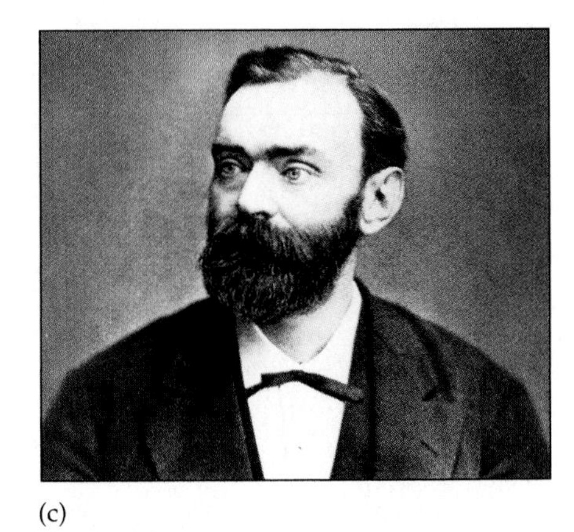

(c)

Figure 11.8 Nitroglycerin. (a) Concentrated nitric acid and glycerin react to form an oily, highly unstable compound, nitroglycerin, $C_3H_5(ONO_2)_3$. (b) Nitroglycerin is more stable if absorbed onto an inert solid, a combination called dynamite. (c) The fortune of Alfred Nobel (1833–1896), built on the manufacture of dynamite, now funds the Nobel Prizes. *(a, model, S. M. Young; b, C. D. Winters; c, The Bettmann Archive)*

Example 11.4 Structural Isomers of Alcohols

How many different alcohols are derivatives of pentane? Draw structures and name each alcohol.

Solution The pentane molecule, C_5H_{12}, has a five-carbon chain. Three different alcohols are possible, depending on whether the —OH group is placed on the first, second, or third carbon atom in the chain. (The fourth and fifth positions are identical to the second and first positions in the chain, respectively.) Alcohols are named as derivatives of the alkane (pentane) by replacing the -e at the end with -ol and indicating the position of the —OH group by a numerical prefix (Appendix E.)

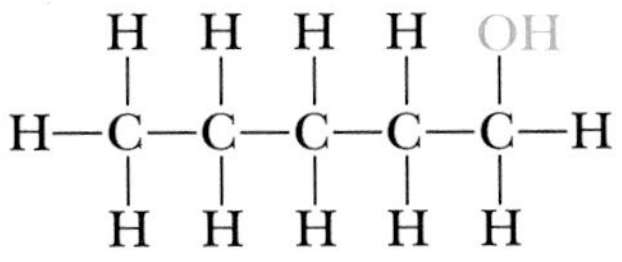

```
     H  H  H  H  OH
     |  |  |  |  |
  H—C—C—C—C—C—H
     |  |  |  |  |
     H  H  H  H  H
```

1-pentanol

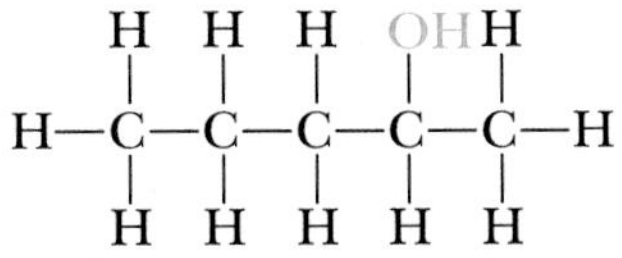

```
     H  H  H  OH H
     |  |  |  |  |
  H—C—C—C—C—C—H
     |  |  |  |  |
     H  H  H  H  H
```

2-pentanol

```
     H  H  OH H  H
     |  |  |  |  |
  H—C—C—C—C—C—H
     |  |  |  |  |
     H  H  H  H  H
```

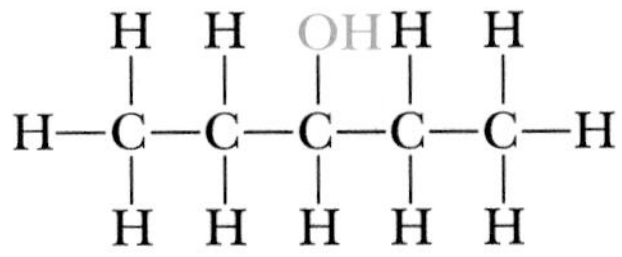

3-pentanol

Exercise 11.5 Structural Isomers of Alcohols

Draw the structures and name the four isomers with the formula C_4H_9OH. (Two have the carbon framework of butane, and two are derived from 2-methylpropane.)

Properties of Alcohols

Methane, CH_4, is a gas (boiling point, −161 °C) with almost no solubility in water. Methanol, on the other hand, is a liquid that is *miscible* with water in all proportions. The boiling point of methanol, 65 °C, is 226 °C higher than the boiling point of methane. What a difference the addition of a single atom into the structure can make in the properties of simple molecules!

Miscible means "soluble in all proportions." Thus, methanol is soluble in all proportions in water.

The formula and structure of alcohols can be described as derived from a hydrocarbon, replacing H with OH. We can approach the structure from another direction, however. Consider alcohols as related to water with one of the H atoms of H_2O replaced by an organic group. If a methyl group is substituted for one of the hydrogens of water, we get the structure of methanol. Ethanol has a —C_2H_5 (ethyl) group, and propanol a —C_3H_7 (propyl) group in place of

CURRENT PERSPECTIVES

New Life for Natural Gas

Natural gas, which is mostly methane, is currently plentiful and cheap. The world's known and projected reserves of natural gas are 2.5×10^{17} L, the energy equivalent of 1.5×10^{12} barrels of oil, although much of it occurs in areas such as Southeast Asia or in the northern reaches of Canada that are far removed from the centers of fuel consumption.

Natural gas is widely used as an energy source; unfortunately, it is also widely wasted. Methane is often simply burned or "flared off" when it comes out of the ground with oil, a substance that is currently more useful. The limitation is transportation. Methane can be carried as a gas in pipelines, but this is expensive if the distances are great. It can also be liquefied and carried by ship, but this procedure risks a fiery catastrophe. Wasting methane could be avoided if it were possible to convert it at low cost, where it is found, to a more readily transportable liquid such as methanol, CH_3OH.

It has been known for some time that methane can be converted to carbon monoxide and hydrogen, and this mixture of gases can readily be turned into methanol in another step.

$$CH_4(g) + \frac{1}{2} O_2(g) \longrightarrow CO(g) + 2\ H_2(g)$$
$$CO(g) + 2\ H_2(g) \longrightarrow CH_3OH(\ell)$$

Unfortunately, the first step is an energy-intensive process that requires a high temperature (>900 °C). In 1998, however, Catalytica, Inc. announced that methane can be converted to methanol with greater than a 70% yield by using a platinum compound as a catalyst in the presence of sulfuric acid.

$$CH_4(g) + \frac{1}{2} O_2(g) \xrightarrow{\text{Pt compound, } H_2SO_4,\ 200\ °C} CH_3OH(\ell)$$

Just as exciting is another discovery regarding methane. Chemical engineers at the University of Minnesota have found that methane can be converted to CO and H_2 under very mild conditions. The photograph shown here illustrates what happens when a mixture of methane and oxygen is passed at room temperature through a heated, sponge-like ceramic disk coated with platinum or rhodium. Rather than oxidizing the methane all the way to water and carbon dioxide, the process produces a hot mixture of CO and H_2, which can be converted in good yield to methanol.

For more information see R. A. Periana, D. J. Taube, S. Gamble, H. Taube, T. Satoh, and H. Fujii: "Platinum catalysts for the high-yield oxidation of methane to a methanol derivative." *Science,* Vol. 280, pp. 560–564, April 24, 1998.

Methane flowing through a catalyst. *(Schmidt/University of Minnesota)*

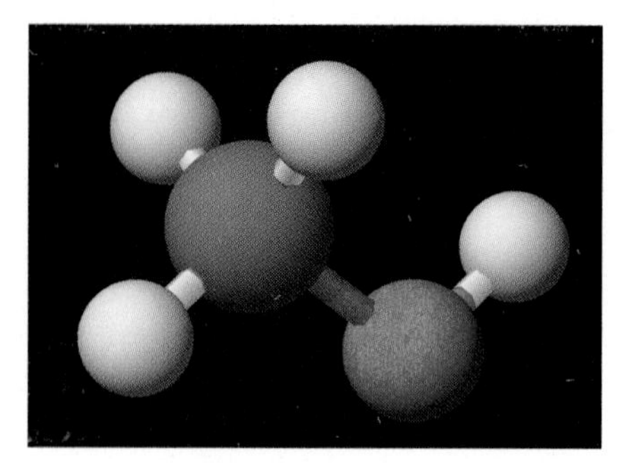

Methanol, CH_3OH, is also known as "wood alcohol."

one of the hydrogens of water. This perspective will also help us understand the properties of alcohols.

The two parts of methanol, the $—CH_3$ group and the —OH group, contribute to its properties. Methanol will burn, a property associated with hydrocarbons. On the other hand, its boiling point is more like that of water. The temperature at which a substance boils is related to the forces of attraction between molecules, the *intermolecular forces:* the stronger the attractive, intermolecular forces in a sample, the higher the boiling point. These forces are particularly strong in water, partly a result of the polarity of the molecule (Section 9.8). Methanol is also a polar molecule, and it is the polar —OH group that leads to methanol's high boiling point. In contrast, methane is nonpolar and its low boiling point reflects low intermolecular forces.

It is also possible to explain the differences in the solubility of methane and methanol in water. The solubility of methanol is conferred by the waterlike character of the polar —OH portion of the molecule. Lacking this polarity, methane has low water-solubility.

Looking at alcohols generally, important trends in properties emerge. As the size of the organic group increases, the boiling point rises, a general trend in families of similar compounds. However, the solubility in water in this series decreases. Methanol and ethanol are completely miscible in water, whereas 1-propanol is only moderately water-soluble and 1-butanol is even less soluble. With an increase in the size of the hydrocarbon group, the organic group (the nonpolar part of the molecule) has become a larger fraction of the molecule, and properties associated with nonpolarity begin to dominate. Space-filling models show that in methanol, the polar and nonpolar parts of the molecule are approximately similar in size, but in 1-butanol the —OH group is less than 20% of the molecule. The molecule is less like water and more "organic."

The intermolecular forces of attraction of compounds with hydrogen attached to a highly electronegative atom, like O, N, or F, are so exceptional that this is accorded a special name: hydrogen bonding. We will discuss hydrogen bonding further in Chapter 13.

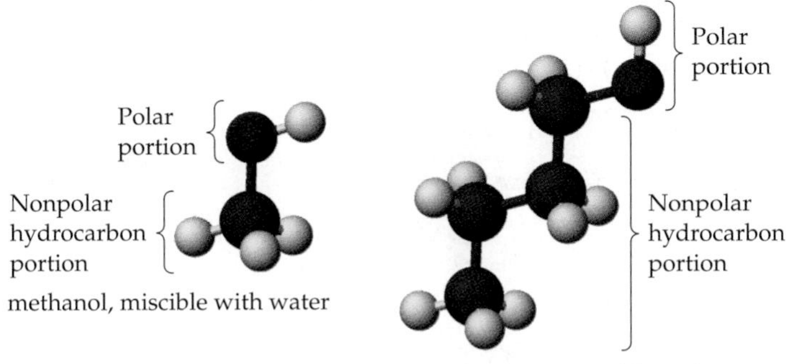

Attaching more than one —OH group to a hydrocarbon framework counteracts this effect. Ethylene glycol, a two-carbon alcohol with two —OH groups, is miscible with water as is glycerol, $HOCH_2CH(OH)CH_2OH$.

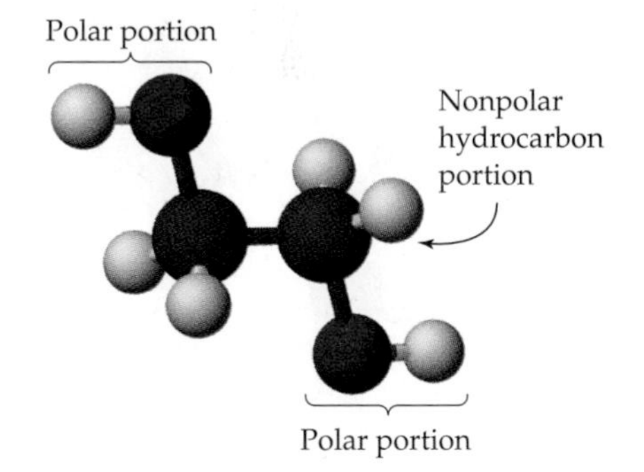

There are similarities in the chemistry of alcohols and water. Like water, alcohols can donate a proton to strong bases to form an anion. If water donates a proton, a hydroxide ion results; an alcohol produces an *alkoxide ion,* RO^-. Alkoxide ions are also formed when alcohols react with alkali metals:

$$C_2H_5OH(\ell) + OH^-(aq) \longrightarrow C_2H_5O^-(aq) + H_2O(\ell)$$

$$ROH(\ell) + Na(s) \longrightarrow NaOR(s) + \tfrac{1}{2}\,H_2(g)$$

Strong acids will protonate alcohols, giving ions of the formula ROH_2^+ (analogous to H_3O^+). Sulfuric acid, however, a compound with a strong affinity for water, converts alcohols to compounds belonging to a different class of organic compound, ethers:

$$\underset{\text{ethanol}}{2\,C_2H_5OH(\ell)} \xrightarrow{H_2SO_4} \underset{\text{diethyl ether}}{C_2H_5OC_2H_5(\ell)} + H_2O(\ell)$$

Ethers have the general formula R—O—R. The best known ether is diethyl ether, $CH_3CH_2OCH_2CH_3$. Lacking an —OH group, the properties of ethers sharply contrast with alcohols. Diethyl ether, for example, has a lower boiling point (34.5 °C) than ethanol (CH_3CH_2OH) (78.3 °C) and is only slightly soluble in water.

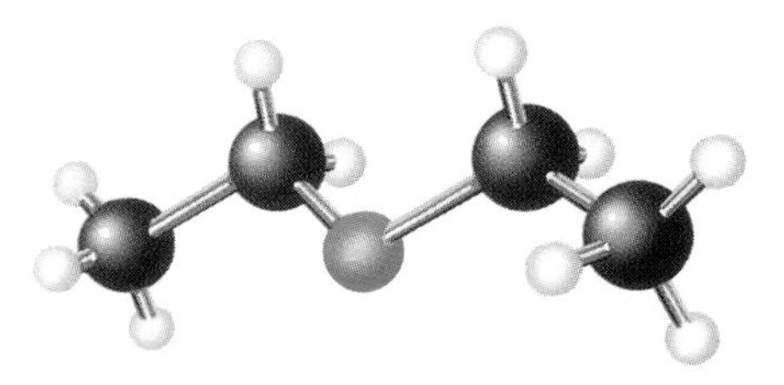

Diethyl ether, $CH_3CH_2OCH_2CH_3$. The compound can be thought of as derived from water by replacing the H atoms of H_2O with ethyl groups, CH_2CH_3.

Amines

It is often convenient to think about water and ammonia as similar molecules: they are both the simplest hydrogen compounds of adjacent second-period elements and are both polar. They also exhibit some similar chemistry, such as protonation (giving H_3O^+ and NH_4^+) and deprotonation (giving OH^- and NH_2^-).

The comparison of water and ammonia can be extended to alcohols and amines. Alcohols have formulas related to water in which one hydrogen is replaced with an organic group (R—OH). In organic **amines,** one (or more) hydrogen atoms of NH_3 is replaced with an organic group. Amine structures are similar to ammonia's structure; that is, the geometry about the N atom is trigonal pyramidal.

Amines are categorized by the number of organic substituents as *primary* (one organic group), *secondary* (two organic groups), or *tertiary* (three organic groups). As examples, consider the three amines with methyl groups: CH_3NH_2, $(CH_3)_2NH$, and $(CH_3)_3N$.

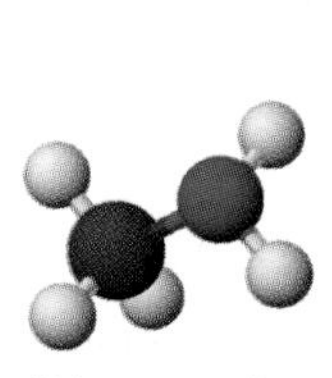

Primary amine
methylamine

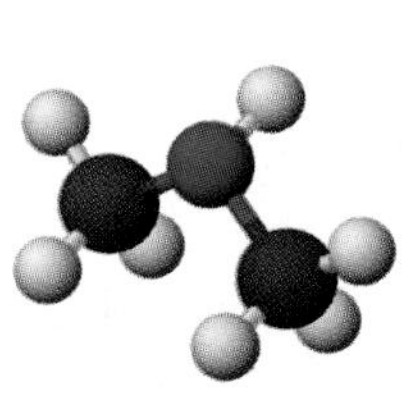

Secondary amine
dimethylamine

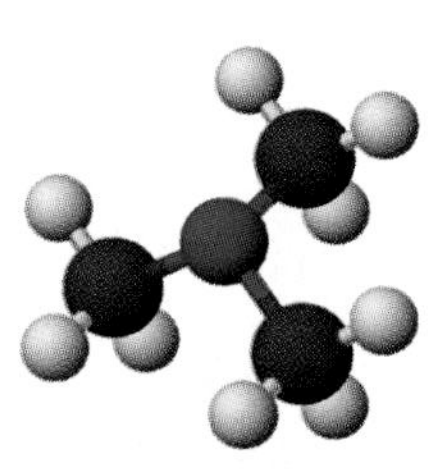

Tertiary amine
trimethylamine

Properties of Amines

Amines usually have offensive odors. You know what the odor is if you have ever smelled decaying fish. Two appropriately named amines, putrescine and cadaverine, add to the odor of urine and are present in bad breath. (Notice that both these compounds have two amino groups.)

$H_2NCH_2CH_2CH_2CH_2NH_2$	$H_2NCH_2CH_2CH_2CH_2CH_2NH_2$
putrescine (1,4-butanediamine)	cadaverine 1,5-pentanediamine

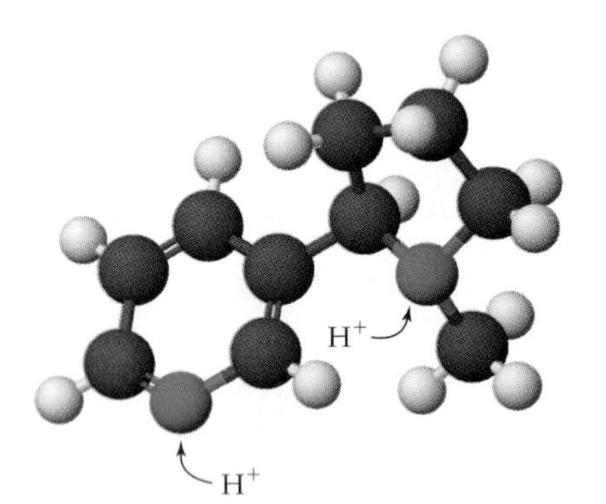

Two nitrogen atoms in the nicotine molecule can be protonated. It is in this form that nicotine is normally found. The protons can be removed, however, by treating it with a base. This "free-base" form is much more poisonous and addictive. See J. F. Pankow: *Environmental Science & Technology,* August 1997.

The smallest amines are water-soluble, but most amines are not. All amines are bases, however, and react with acids to give salts, many of which are water-soluble. As with ammonia, these reactions involve adding H^+ to the lone pair of electrons on the N atom:

$$(CH_3)_3N(aq) + HCl(aq) \longrightarrow [(CH_3)_3NH^+]Cl^-(aq)$$

The fact that an amine can be protonated, and the proton can be removed again by treating with a base, has practical and physiological importance. Nicotine in cigarettes is normally found in the protonated form. (This water-soluble form is often used in insecticides.) Adding a base such as ammonia removes the H^+ ion to leave nicotine in its "free-base" form.

$$NicH_2^{2+}(aq) + 2\ NH_3(aq) \longrightarrow Nic(aq) + 2\ NH_4^+(aq)$$

In this form, nicotine is much more readily absorbed by the skin and mucous membranes so the compound is a much more potent poison.

11.5 COMPOUNDS WITH A CARBONYL GROUP

Formaldehyde (H_2CO), acetic acid (CH_3CO_2H), and acetone (CH_3COCH_3) are among the organic compounds referred to in previous examples in this text. These compounds have a common structural feature: each contains a trigonal-planar carbon atom doubly bonded to an oxygen. The C=O group is called the **carbonyl group,** and all of these compounds are members of a large class of compounds called **carbonyl compounds.**

See the *Saunders Interactive General Chemistry CD-ROM,* Screen 11.5, Functional Groups, for a discussion of aldehydes, ketones, carboxylic acids, and esters.

Carbonyl group formaldehyde acetic acid acetone

Generally, carbonyl compounds are separated into subcategories based on the groups attached to the carbonyl carbon. In this section, we will examine five groups of carbonyl compounds: *aldehydes* (RCHO), *ketones* (RCOR′), organic acids (also called *carboxylic acids*) (RCO_2H), *esters* (RCO_2R'), and *amides* ($RCONR_2'$).

Aldehydes have an organic group (—R) and an H atom attached to a carbonyl group. Ketones have two —R groups attached to the carbonyl carbon; they may be the same groups, as in acetone, or different groups. The carbonyl group in organic acids is attached to an —R group and an —OH group. Esters have —R and —OR′ groups attached to the carbonyl, whereas amides have an —R group and an amino group (—NH_2, —NHR, —NR_2) bonded to the carbonyl carbon. Aldehydes, ketones, and carboxylic acids are oxidation products of alcohols and, indeed, are commonly made by this route. Esters and amides are considered to be derivatives of acids:

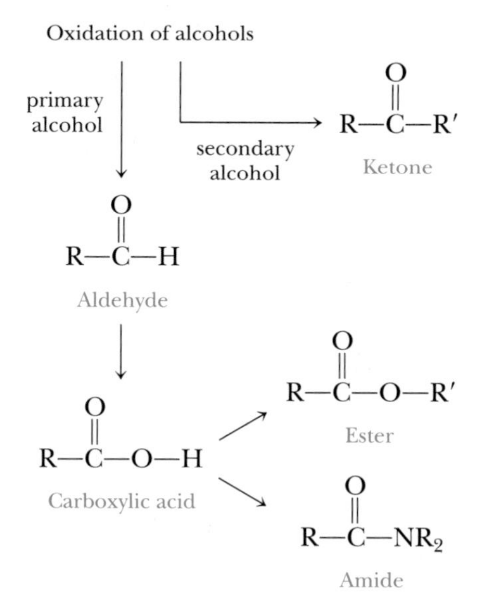

Because the carbonyl group is the center of a trigonal-planar arrangement of atoms, the central carbon atom is assigned sp^2 hybridization. The carbon–oxygen σ and π bonds are described in Section 10.1.

The product obtained on oxidation of an alcohol depends on the alcohol's structure, which is classified according to the number of carbon atoms bonded to the C atom bearing the —OH group. *Primary alcohols* have one carbon and two hydrogen atoms attached, whereas *secondary alcohols* have two carbons and one hydrogen atom attached. *Tertiary alcohols* have three carbon atoms attached to the C atom bearing the —OH group.

The oxidation of a *primary alcohol* occurs in two steps. It is first oxidized to an aldehyde and then in a second step to a carboxylic acid:

$$\underset{\text{Primary alcohol}}{R{-}CH_2{-}OH} \xrightarrow[\text{agent}]{\text{oxidizing}} \underset{\text{Aldehyde}}{R{-}\overset{\overset{O}{\|}}{C}{-}H} \xrightarrow[\text{agent}]{\text{oxidizing}} \underset{\text{Carboxylic acid}}{R{-}\overset{\overset{O}{\|}}{C}{-}OH}$$

The air oxidation of ethanol in wine produces wine vinegar. Because acids have a sour taste, the word "vinegar" means "sour wine" (from the French *vin aigre*)

$$\underset{\text{ethanol}}{H{-}\overset{H}{\underset{H}{C}}{-}\overset{H}{\underset{H}{C}}{-}OH(\ell)} + \underset{\text{Oxidizing agent}}{O_2(g)} \longrightarrow \underset{\text{acetic acid}}{H{-}\overset{H}{\underset{H}{C}}{-}\overset{\overset{O}{\|}}{C}{-}OH(\ell)} + H_2O(\ell)$$

In contrast, the oxidation of a *secondary alcohol* produces a ketone:

$$\underset{\text{Secondary alcohol}}{R{-}\overset{OH}{\underset{H}{C}}{-}R'} \xrightarrow[\text{agent}]{\text{oxidizing}} \underset{\text{Ketone}}{R{-}\overset{\overset{O}{\|}}{C}{-}R'}$$

(—R and —R′ are organic groups. They may be the same or different.)

The oxidizing agents used for these reactions are common reagents such as $KMnO_4$ and $K_2Cr_2O_7$ (Table 5.4).

Alcohol oxidation is used in a device to test one's breath for the presence of ethanol (Figures 5.20 and 11.9). Ethanol is oxidized by yellow-orange potassium dichromate, which in turn is reduced to the green-blue chromium(III) ion. Thus, a color change indicates the presence of the alcohol.

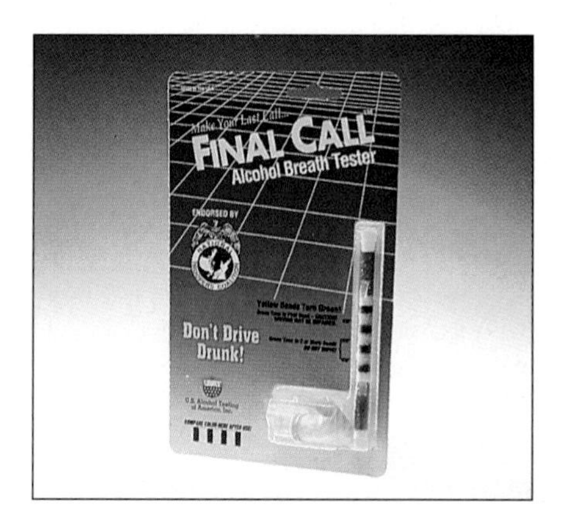

Figure 11.9 Alcohol tester. This device for testing breath for the presence of ethanol relies on the oxidation of the alcohol. If present, ethanol is oxidized by potassium dichromate, $K_2Cr_2O_7$, to acetaldehyde, and then to acetic acid. The yellow-orange dichromate ion is reduced to green Cr^{3+} (aq), the color change indicating that ethanol was present. *(C. D. Winters)*

Aldehydes and Ketones

Aldehydes and ketones have pleasant odors and are often used as the basis of fragrances. Benzaldehyde is responsible for the odor of almonds and cherries, cinnamaldehyde is found in the bark of the cinnamon tree, and the ketone *p*-hydroxyphenyl-2-butanone is responsible for the odor of ripe raspberries (a favorite of the authors of this book). Table 11.9 lists several simple aldehydes and ketones.

The odors of almonds and cinnamon are due to aldehydes, but the odor of fresh raspberries comes from a ketone. *(C. D. Winters)*

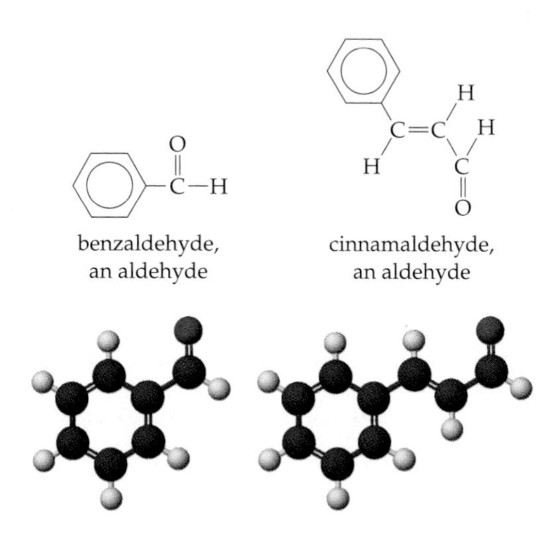

Aldehydes and ketones are the oxidation products of primary and secondary alcohols, respectively. It is therefore reasonable that primary and secondary alcohols can be formed from appropriate carbonyl compounds by the reverse of

Table 11.9 • SIMPLE ALDEHYDES AND KETONES

Structure	Common Name	Systematic Name	BP (°C)
$HC(=O)H$	formaldehyde	methanal	−21
$CH_3C(=O)H$	acetaldehyde	ethanal	21
$CH_3C(=O)CH_3$	acetone	propanone	56
$CH_3C(=O)CH_2CH_3$	methyl ethyl ketone	butanone	80
$CH_3CH_2C(=O)CH_2CH_3$	diethyl ketone	3-pentanone	102

Reagents such as $NaBH_4$, sodium borohydride, are called "hydrides" because the hydrogen is the more electronegative element and thus has a negative partial charge.

an oxidation, that is, by reduction. Commonly used reagents for such reductions are $NaBH_4$ or $LiAlH_4$, although H_2 is used in industrial scale preparations.

$$\underset{\text{Aldehyde}}{R{-}\overset{\overset{\displaystyle O}{\|}}{C}{-}H} \xrightarrow{NaBH_4 \text{ or } LiAlH_4} \underset{\text{Primary alcohol}}{R{-}\overset{\overset{\displaystyle OH}{|}}{\underset{\underset{\displaystyle H}{|}}{C}}{-}H}$$

$$\underset{\text{Ketone}}{R{-}\overset{\overset{\displaystyle O}{\|}}{C}{-}R} \xrightarrow{NaBH_4 \text{ or } LiAlH_4} \underset{\text{Secondary alcohol}}{R{-}\overset{\overset{\displaystyle OH}{|}}{\underset{\underset{\displaystyle H}{|}}{C}}{-}R}$$

***Exercise* 11.6** Aldehydes and Ketones

Draw structural formulas for 2-pentanone. Draw structures for a ketone and two aldehydes that are isomers of 2-pentanone, and name each of these compounds.

***Exercise* 11.7** Aldehydes and Ketones

Draw structures and name the aldehyde or ketone that is formed upon oxidation of the following alcohols:
1. 1-butanol, **2.** 2-butanol, **3.** 2-methyl-1-propanol. Are the three compounds isomers?

Carboxylic Acids

Acetic acid is the most common and most important carboxylic acid. **Carboxylic acids** are the end product of alcohol oxidation, and for years acetic acid was made by oxidizing ethanol produced by fermentation. Now, however, acetic acid is generally made by combining carbon monoxide and methanol in the presence of a catalyst:

$$\underset{\text{methanol}}{CH_3OH(\ell)} + CO(g) \longrightarrow \underset{\text{acetic acid}}{CH_3CO_2H(\ell)}$$

About 1.5×10^9 kg of acetic acid is produced annually in the United States. It is widely used to make plastics and synthetic fibers, as a fungicide, and as the starting material for preparing dietary supplements.

Many organic acids are found naturally (Figure 11.10 and Table 11.10). Acids are recognizable by their sour taste, and you will encounter them in common foods: citric acid in fruits, acetic acid in vinegar, and tartaric acid in grapes are just three examples. Formic acid is a component of the venom of stinging ants, and oxalic acid occurs in significant concentrations in leafy green plants such as rhubarb and spinach (page 171). Lactic acid is particularly widespread in nature, being found in sour milk, pickles, and sauerkraut, among other foods. Benzoic acid is found in most berries.

***Figure* 11.10 Acetic acid in bread.** Acetic acid is produced in bread when leavened with the yeast *Saccharomyces exigus.* Additionally a group of bacteria, *Lactobacillus sanfrancisco,* contribute to the flavor of sourdough bread. These bacteria metabolize the sugar maltose, excreting acetic acid and lactic acid, $CH_3CH(OH)CO_2H$, thereby giving the bread its unique sour taste. *(C. D. Winters)*

Table 11.10 • SOME OTHER NATURALLY OCCURRING CARBOXYLIC ACIDS

Name	Structure	Natural Source
benzoic acid	$C_6H_5{-}CO_2H$	berries
citric acid	$HO_2C{-}CH_2{-}C(OH)(CO_2H){-}CH_2{-}CO_2H$	citrus fruits
lactic acid	$CH_3{-}CH(OH){-}CO_2H$	sour milk
malic acid	$HO_2C{-}CH_2{-}CH(OH){-}CO_2H$	apples
oleic acid	$CH_3(CH_2)_7{-}CH{=}CH{-}(CH_2)_7{-}CO_2H$	vegetable oils
oxalic acid	$HO_2C{-}CO_2H$	rhubarb, spinach, cabbage, tomatoes
stearic acid	$CH_3(CH_2)_{16}{-}CO_2H$	animal fats
tartaric acid	$HO_2C{-}CH(OH){-}CH(OH){-}CO_2H$	grape juice, wine

Carboxylic acids often have common names derived from the source of the acid (Table 11.11). Because formic acid is found in ants, its name comes from the Latin word for ant *(formica)*. Butyric acid gives rancid butter its unpleasant odor, and the name is related to the Latin word for butter *(butyrum)*. The names for caproic (C_6), caprylic (C_8), and capric (C_{10}) acids are derived from the Latin word for goat *(capra)*, because these acids give goats their characteristic, and unpleasant, odor. As seen in Table 11.11, the systematic names of acids are formed

Table 11.11 • SOME SIMPLE CARBOXYLIC ACIDS

Structure	Common Name	Systematic Name	BP (°C)
$HC(=O)OH$	formic acid	methanoic acid	101
$CH_3C(=O)OH$	acetic acid	ethanoic acid	118
$CH_3CH_2C(=O)OH$	propionic acid	propanoic acid	141
$CH_3(CH_2)_2C(=O)OH$	butyric acid	butanoic acid	163
$CH_3(CH_2)_3C(=O)OH$	valeric acid	pentanoic acid	187

A CLOSER LOOK *Muscle Cramps and Hangovers*

Fresh milk can become contaminated easily with bacteria that metabolize the milk sugar lactose and excrete lactic acid, $C_3H_6O_3$. The acid causes the droplets of fat in the milk to coalesce, and the milk curdles. The manufacture of yogurt and cheese is a controlled version of this process.

When you exercise, your body uses glucose as an energy source. Glucose metabolism produces pyruvic acid, CH_3COCO_2H, which is burned aerobically (in the presence of O_2) to give carbon dioxide and water. If you need a sudden burst of energy, as in a sprint race, your muscles may be short of oxygen, so the pyruvic acid is produced anaerobically (without oxygen) to give energy and lactic acid. It is the build-up of this acid in your muscles that makes them hurt after exercise.

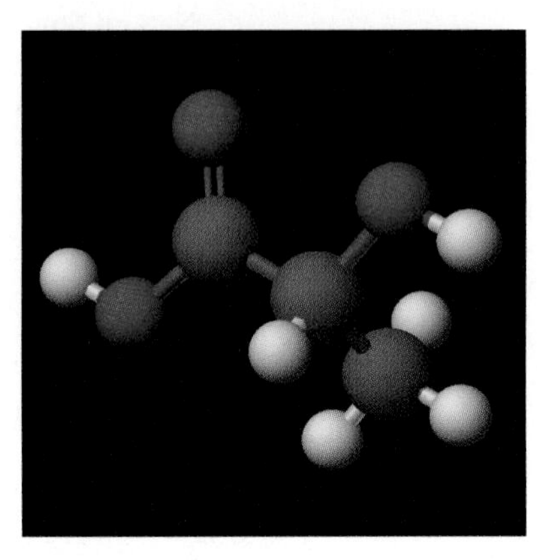

A molecular model of lactic acid, $C_3H_6O_3$.

Lactic acid is produced in your body as part of normal metabolism, and it is removed from your body by your liver. If you drink an alcoholic beverage, your liver also metabolizes the alcohol. If you have had too much to drink, however, lactic acid metabolism may not be efficient, allowing the lactic acid to build up in your body and creating a feeling of fatigue.

See Trevor Smith: "Distance running." *Chem Matters,* Vol. 7, No. 1, page 4, February 1989.

by dropping the "-e" on the name of the corresponding alkane and adding "-oic" (and the word "acid").

Because of electronegativity differences we expect the two O atoms of the carboxylic acid group to be slightly negatively charged, and the H atom of the —OH group to be positively charged (Figure 11.11). This distribution of charges has several important implications:

- The polar acetic acid molecule dissolves readily in water, which you already know because vinegar is an aqueous solution of acetic acid. (Organic acids with larger organic groups are less soluble, however.)
- The hydrogen of the —OH group is the acidic hydrogen (see Figure 11.11). The interaction with water produces the hydronium ion and a carboxylate ion. As noted in Chapter 5, acetic acid is a weak acid in water, as are most organic acids.

Esters

Carboxylic acids (RCO_2H) react with alcohols ($R'OH$) in the presence of a strong acid to form **esters** (RCO_2R'). This reaction is often called **esterification:**

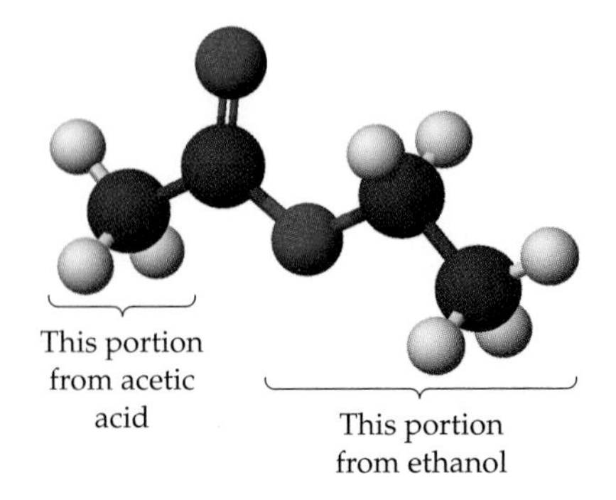

ethylacetate

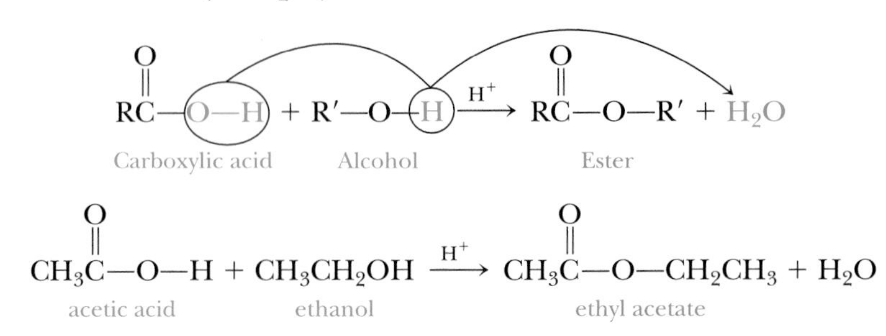

Table 11.12 • Some Acids, Alcohols, and Their Esters

Acid	Alcohol	Ester	Odor of Ester
CH_3CO_2H acetic acid	$CH_3\overset{CH_3}{\overset{\vert}{C}}HCH_2CH_2OH$ 3-methyl-1-butanol	$CH_3\overset{O}{\overset{\Vert}{C}}OCH_2CH_2\overset{CH_3}{\overset{\vert}{C}}HCH_3$ 3-methylbutyl acetate	banana
$CH_3CH_2CH_2CO_2H$ butanoic acid	$CH_3CH_2CH_2CH_2OH$ 1-butanol	$CH_3CH_2CH_2\overset{O}{\overset{\Vert}{C}}OCH_2CH_2CH_2CH_3$ butyl butanoate	pineapple
$CH_3CH_2CH_2CO_2H$ butanoic acid	C_6H_5—CH_2OH benzyl alcohol	$CH_3CH_2CH_2\overset{O}{\overset{\Vert}{C}}OCH_2$—$C_6H_5$ benzyl butanoate	rose

Table 11.12 lists a few common esters and the acid and alcohol from which they are formed. The two-part name of an ester is given by (1) the name of the alkyl group from the alcohol and (2) the name of the carboxylate group derived from the acid. For example, acetic acid and ethanol (commonly called ethyl alcohol) combine to give the ester ethyl acetate.

Perhaps the most important reaction of esters is their **hydrolysis** (literally, a reaction with water), a reaction that is the reverse of the formation of the ester. The reaction, generally done in the presence of a base such as NaOH, produces the alcohol and a salt of the carboxylic acid:

$$\underset{\text{Ester}}{R\overset{O}{\overset{\Vert}{C}}\text{—}O\text{—}R'} + NaOH \xrightarrow[\text{in water}]{\text{heat}} \underset{\text{Carboxylate salt}}{R\overset{O}{\overset{\Vert}{C}}\text{—}O^-Na^+} + \underset{\text{Alcohol}}{R'OH}$$

$$\underset{\text{ethyl acetate}}{CH_3\overset{O}{\overset{\Vert}{C}}\text{—}O\text{—}CH_2CH_3} + NaOH \xrightarrow[\text{in water}]{\text{heat}} \underset{\text{sodium acetate}}{CH_3\overset{O}{\overset{\Vert}{C}}\text{—}O^-Na^+} + \underset{\text{ethanol}}{CH_3CH_2OH}$$

The carboxylic acid can be recovered from this process if the sodium salt is treated with a strong acid such as HCl:

$$\underset{\text{sodium acetate}}{CH_3\overset{O}{\overset{\Vert}{C}}\text{—}O^-Na^+(aq)} + HCl(aq) \longrightarrow \underset{\text{acetic acid}}{CH_3\overset{O}{\overset{\Vert}{C}}\text{—}OH(aq)} + NaCl(aq)$$

Unlike the acids from which they are derived, esters often have pleasant odors (see Table 11.12). Typical examples are methyl salicylate, or "oil of wintergreen," and benzyl acetate. Methyl salicylate is derived from salicylic acid, the parent compound of aspirin.

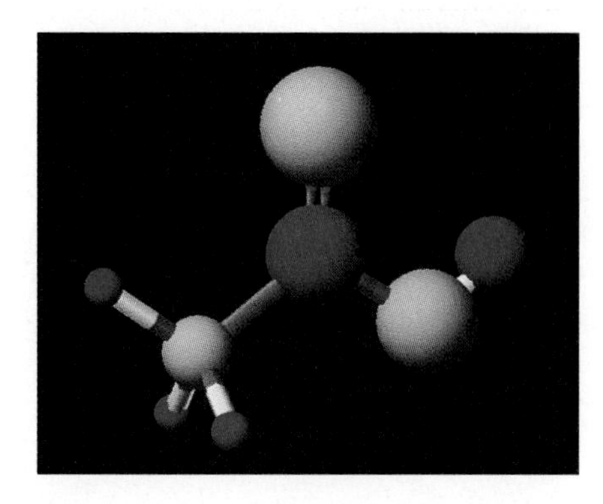

***Figure* 11.11** **Acetic acid, CH_3CO_2H.** This computer-generated model shows the calculated partial charges on the atoms. A yellow sphere indicates a negative charge, and a red sphere indicates a positive charge. The sizes of the spheres reflect the relative magnitude of charge. Notice that the H atom of the —OH group is positive. It is this hydrogen atom that can interact with water to form a hydronium ion.

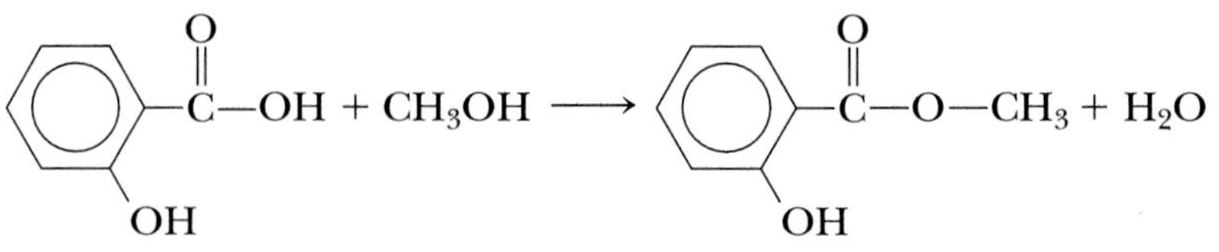

$$\text{salicylic acid } (o\text{-HO}C_6H_4\text{—C(=O)—OH}) + CH_3OH \longrightarrow \text{methyl salicylate } (o\text{-HO}C_6H_4\text{—C(=O)—O—}CH_3) + H_2O$$

salicylic acid   methanol   methyl salicylate (oil of wintergreen)

Benzyl acetate, the active component of "oil of jasmine," is derived from the reaction of benzyl alcohol ($C_6H_5CH_2OH$) and acetic acid. The chemicals are inexpensive, so synthetic jasmine is a common fragrance in less expensive perfumes and toiletries.

Exercise 11.8 Esters

Draw the structure and name the ester formed from each of the following reactions:

1. propanoic acid and methanol
2. butanoic acid and 1-butanol
3. hexanoic acid and ethanol

Exercise 11.9 Esters

Draw the structure and name the acid and alcohol from which the following esters are derived:

1. propyl acetate
2. 3-methylpentyl benzoate
3. ethyl salicylate

Amides

In the same way that alcohols and amines are related, so are esters and **amides** related. An ester has an organic group (—R) and an alkoxide (—OR′) group attached to the carbonyl group (—C═O). Amides also have an organic group attached to the carbonyl group as well as an amino group (—NH_2, —NHR′, or —NR′R″).

The amide grouping is particularly important in some synthetic polymers (Section 11.6) and in many naturally occurring compounds, especially proteins (page 528). You have also seen it in saccharin (page 499). The compound *N*-acetyl-*p*-aminophenol, an analgesic known by the generic name acetaminophen and sold under the names Tylenol, Datril, and Momentum, among others, is an amide. Use of this compound as an analgesic was apparently discovered by accident when a common organic compound called acetanilide (like acetaminophen but without the —OH group) was mistakenly put into a prescription for a patient. Acetanilide acts as an analgesic, but it can be toxic. An —OH group *para* to the amide group makes the compound nontoxic.

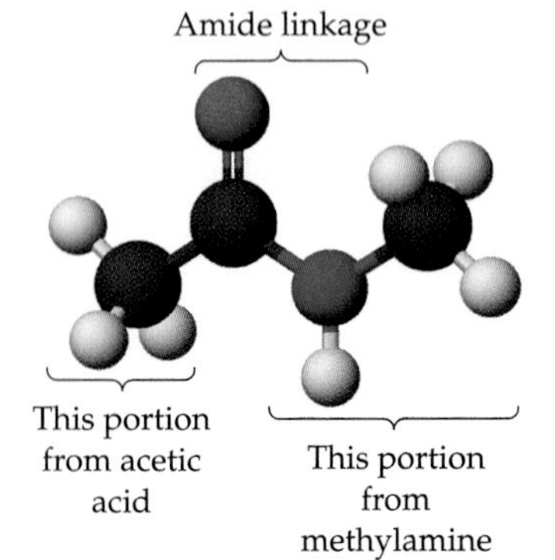

N-Methylacetamide, an example of an amide.

CURRENT PERSPECTIVES

Aspirin is 100 Years Old!

Aspirin is one of the most successful proprietary drugs ever made. Americans swallow more than 50 million aspirin tablets a day, mostly for its pain-relieving (analgesic) effects. Aspirin is also said to ward off heart disease and thrombosis (blood clots), and it has even been suggested as a possible treatment for certain cancers and for senile dementia.

Hippocrates (c. 460–370 BC), the ancient Greek physician, recommended an infusion of willow bark to ease the pain of childbirth. It was not until the 19th century that an Italian chemist, Raffaele Piria, isolated salicylic acid, the active compound, from the bark. Soon thereafter, it was found that the acid could be extracted from a wild flower *Spiraea ulmaria.* It is from the name of this plant that the name "aspirin" (a + spiraea) is derived.

Hippocrates's willow bark extract—salicylic acid—is an analgesic, but it is also very irritating to the stomach lining. It was therefore an important advance when Felix Hoffmann and Henrich Dreser of Bayer Chemicals in Germany found, in 1897, that a derivative of salicylic acid, acetylsalicylic acid, was also a useful drug and had fewer side effects. It is to this compound that we now give the name "aspirin." Acetylsalicylic acid, however, slowly reverts to salicylic acid (and acetic acid) in the presence of moisture; therefore, if you smell the characteristic odor of acetic acid in an old bottle of aspirin tablets, do not use them.

Aspirin is sold under a variety of trade names, such as Anacin, Ecotrin, Excedrin, and Alka-Seltzer. The latter is a combination of aspirin with citric acid and sodium bicarbonate. Sodium bicarbonate is a base, and reacts with the acid to produce a sodium salt, a form of aspirin that is water-soluble and quicker acting.

Besides aspirin, several other popular, over-the-counter painkillers are available. Ibuprofen (Advil) and acetaminophen (Tylenol) are two of the most popular.

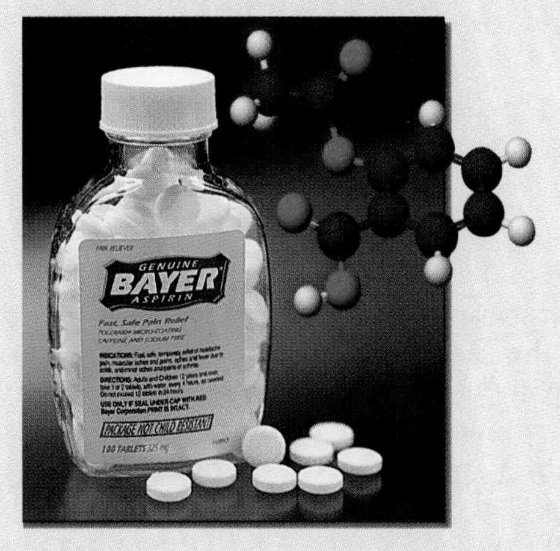

Acetylsalicylic acid is commonly known as aspirin. *(Photo, C. D. Winters; model, S. M. Young)*

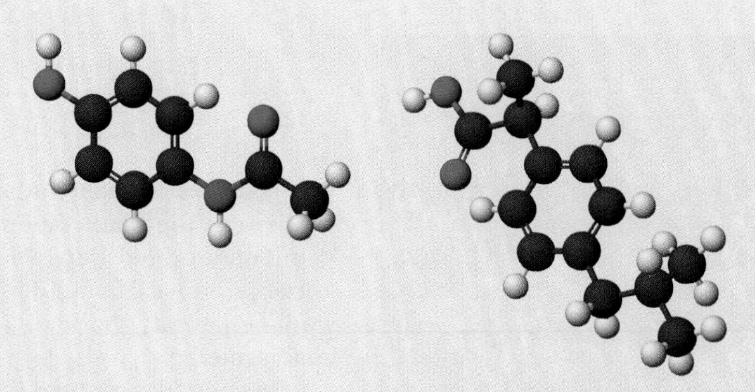

Acetaminophen *(left)* and ibuprofen *(right)*. The latter is often known by one of its trade names, Tylenol.

11.6 FATS AND OILS

Fats and **oils** are triesters made from glycerol (1,2,3-propanetriol) and long-chain carboxylic acids called "fatty acids." The —R groups of the acids, which can be the same or different within the same molecule, can be saturated or unsaturated, that is, they may contain one or more carbon–carbon double bonds.

See the *Saunders Interactive General Chemistry CD-ROM,* Screen 11.7, Functional Groups (2), for a discussion of fats and oils.

Table 11.13 • COMMON FATTY ACIDS

Name	Number of C Atoms	Formula
Saturated Acids		
butanoic	C_4	$CH_3CH_2CH_2CO_2H$
lauric	C_{12}	$CH_3(CH_2)_{10}CO_2H$
myristic	C_{14}	$CH_3(CH_2)_{12}CO_2H$
palmitic	C_{16}	$CH_3(CH_2)_{14}CO_2H$
stearic	C_{18}	$CH_3(CH_2)_{16}CO_2H$
Unsaturated Acids		
oleic	C_{18}	$CH_3(CH_2)_7CH{=}CH(CH_2)_7CO_2H$
linolenic	C_{18}	$CH_3CH_2CH{=}CHCH_2CH{=}CHCH_2CH{=}CH(CH_2)_7CO_2H$

The word "fat" is used generically, like "alcohol" or "acid." All fats are triesters of glycerol, but they are often subdivided into two types depending on whether they are solids (fats) or liquids (oils).

A fatty acid —R group may be unsaturated, monounsaturated, or polyunsaturated, depending on whether one or more double bonds is present. Some common fatty acids are given in Table 11.13:

$$\begin{array}{l} CH_2{-}O{-}H \\ | \\ CH{-}O{-}H \\ | \\ CH_2{-}O{-}H \end{array} + 3\,R\overset{\overset{\large O}{\|}}{C}{-}O{-}H \longrightarrow \begin{array}{l} CH_2{-}O{-}\overset{\overset{\large O}{\|}}{C}R \\ | \\ CH{-}O{-}\overset{\overset{\large O}{\|}}{C}R \\ | \\ CH_2{-}O{-}\overset{\overset{\large O}{\|}}{C}R \end{array} + 3\,H_2O$$

Glycerol   Fatty acid   Fat or oil

Plant seeds often contain oils that serve as a food source for the developing embryo of the plant. About 75% of the carboxylic acid groups in olive oil are monounsaturated fatty acids, principally oleic acid. Linolenic acid, which has three carbon–carbon double bonds, is found in corn, cottonseeds, and soybeans and is especially abundant in rapeseeds. Rapeseed oil is used in cooking oils and margarine.

In general, saturated fatty acids are present in fats (solids), and unsaturated fatty acids are found in oils (liquids). With only single bonds, the hydrocarbon chain in a saturated fatty acid is flexible. The chains can roll up into little balls, making the triester molecules a compact package. These can then pack close to one another, resulting in a solid. In contrast, the presence of a double bond in a fatty acid means the chain is less flexible. It cannot bend around a double bond, so triesters with unsaturated groups are less compact (Figure 11.12). The result is an oil.

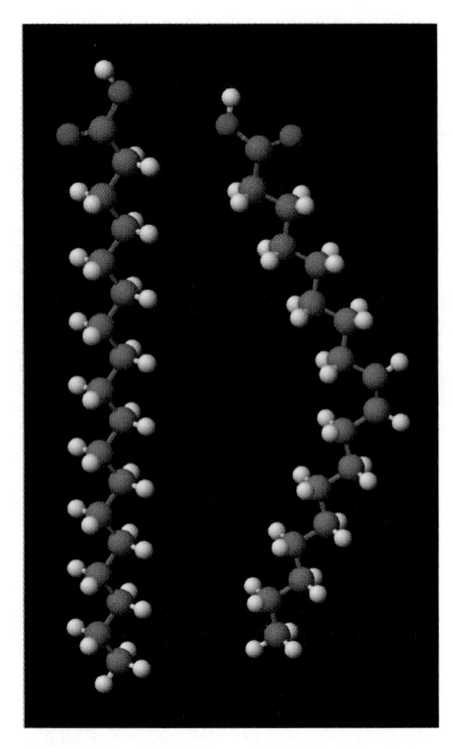

Figure 11.12 **Fatty acids.** Stearic acid *(left)* is saturated, whereas oleic acid *(right)* is unsaturated.

The fatty acids in a glycerol triester have an interesting effect on chocolate. Although most fats contain a complex mixture of fatty acids, cocoa butter, the major component of chocolate, is a relatively homogeneous material. One of the fatty acids in cocoa butter is oleic acid and the other two are either stearic acid or palmitic acid. A consequence of this homogeneity is that the solid has a reasonably sharp melting point at about 34 °C, just below body temperature. When you put a piece of chocolate in your mouth, it melts. Melting requires energy and although it is a very subtle effect, you may notice a cooling effect on your tongue as heat energy is transferred from your body to the chocolate.

Food companies often hydrogenate oils to reduce unsaturation. Although dietary arguments favor unsaturated fats over saturated fats, there are several other reasons saturated fats are used in foods. Double bonds in unsaturated fatty acids are reactive and are attacked by oxygen. Unpleasant odors and flavors develop when an oil is oxidized. Also, hydrogenating an oil makes it less liquid. Food processors often want a solid fat to improve the qualities of the food (e.g., if liquid vegetable oil is used in a cake icing, the icing could slide off the cake). Rather than use animal fat, which also contains cholesterol, manufacturers turn to a hydrogenated or partially hydrogenated oil.

Hydrolysis of esters in the presence of a base breaks the ester down into its component parts, an alcohol and a salt of the acid. This reaction is often called **saponification,** which means "soap making." The treatment of a fat or oil gives glycerol and a salt of a long-chain carboxylic acid such as sodium stearate:

$$
\begin{array}{l}
CH_2{-}O{-}\overset{\overset{\displaystyle O}{\|}}{C}R \\
| \\
CH{-}O{-}\overset{\overset{\displaystyle O}{\|}}{C}R \\
| \\
CH_2{-}O{-}\overset{\overset{\displaystyle O}{\|}}{C}R
\end{array}
+ 3\,NaOH \longrightarrow
\begin{array}{l}
CH_2{-}O{-}H \\
| \\
CH{-}O{-}H \\
| \\
CH_2{-}O{-}H
\end{array}
+ 3\,R\overset{\overset{\displaystyle O}{\|}}{C}{-}O^-Na^+
$$

glyceryl stearate
(a fat)
$R = {-}(CH_2)_{16}CH_3$

glycerol

sodium stearate
(a soap)

Compounds such as sodium stearate that have an ionic end and a long-chain hydrocarbon end are soaps. The ionic end allows them to interact with water, and the hydrocarbon end enables them to mix with oily or greasy substances.

***Exercise* 11.10** Fats and Oils

Draw the structure of glyceryl tripalmitate. When this triester is saponified, what are the products?

11.7 SYNTHETIC ORGANIC POLYMERS

We now turn from molecules built from just a few atoms to large molecules that contain many atoms. These can be either synthetic materials called polymers or naturally occurring molecules such as proteins or nucleic acids. Although these materials have widely varying compositions, their structures and properties are understandable based on the principles we developed for small molecules.

Classifying Polymers

The word "polymer" means "many parts" (from the Greek, *poly* and *meros*). Polymers are giant molecules made by chemically joining many small molecules called **monomers.** Polymer molecular weights range from thousands to millions.

Table 11.14 • PRODUCTION OF SYNTHETIC POLYMERS (IN MILLIONS OF KILOGRAMS)

Plastics

Thermosetting resins, e.g., polyester (3117)

Thermoplastic resins, e.g., high-density polyethylene (23,846)

Synthetic rubber (elastomers)

Styrene butadiene and others (1993)

Synthetic fibers

Nylon and others (4221)

TOTAL (33,177)

Extensive use of synthetic polymers is a fairly recent development. A few synthetic polymers (Bakelite, rayon, and celluloid) were made early in the 20th century, but most of the products you are familiar with originated in the second half of the 20th century. In 1976, synthetic polymers outstripped steel as the most widely used material in the United States. The average production of synthetic polymers in the United States now approaches 150 kg per person annually.

The polymer industry classifies polymers in several different ways. One is their response to heating. **Thermoplastics** (like polyethylene) soften and flow when they are heated and harden when they are cooled. **Thermosetting** plastics (such as Formica) are initially soft but set to a solid when heated and cannot be resoftened.

Another classification depends on the end use of the polymer. The classification as **plastics, fibers, elastomers, coatings,** and **adhesives** is the basis for production data (Table 11.14) published regularly in *Chemical & Engineering News,* an official publication of the American Chemical Society.

A more chemical approach to polymer classification is based on their method of synthesis. **Addition polymers** are made by directly adding monomer units together. **Condensation polymers** are made by combining monomer units and splitting out a small molecule, often water.

Addition Polymers

See the *Saunders Interactive General Chemistry CD-ROM,* Screen 11.9, Synthetic Organic Polymers (1), Addition Polymerization.

Polyethylene, polystyrene, and polyvinylchloride (PVC) are common addition polymers (Figure 11.13). These are built by “adding together” simple alkenes such as ethylene ($CH_2{=}CH_2$), styrene ($C_6H_5CH{=}CH_2$), and vinyl chloride ($CH_2{=}CHCl$). Many other addition polymers are also known (Table 11.15). These polymers, all derived from alkenes, have widely varying properties and uses.

(a)

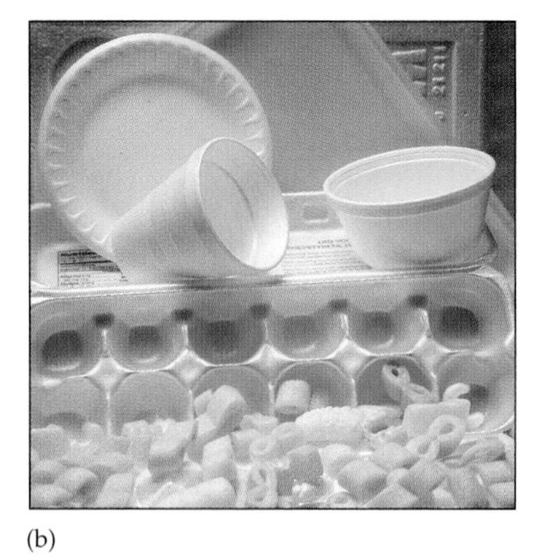

(b)

(c)

Figure 11.13 **Common consumer products.** (a) Packaging materials from high-density polyethylene; (b) from polystyrene; and (c) from polyvinylchloride. Recycling information is provided on most plastics (usually molded into the bottom of the bottle). High-density polyethylene is designated with a “2” inside a triangular symbol and the letters “HDPE.” PVC is designated with a “3” inside a triangular symbol with the letter “V” below. *(C. D. Winters)*

Table 11.15 • Ethylene Derivatives That Undergo Addition Polymerization

Formula	Monomer Common Name	Polymer Name (Trade Names)	Uses	U.S. Polymer Production (Metric tons/Yr)*
$H_2C{=}CH_2$	ethylene	polyethylene (polythene)	squeeze bottles, bags, films, toys and molded objects, electric insulation	7 million
$H_2C{=}CH(CH_3)$	propylene	polypropylene (Vectra, Herculon)	bottles, films, indoor–outdoor carpets	1.2 million
$H_2C{=}CHCl$	vinyl chloride	poly(vinyl chloride) (PVC)	floor tile, raincoats, pipe	1.6 million
$H_2C{=}CH(CN)$	acrylonitrile	polyacrylonitrile (Orlon, Acrilan)	rugs, fabrics	0.5 million
$H_2C{=}CH(C_6H_5)$	styrene	polystyrene (styrene, Styrofoam, Styron)	food and drink coolers, building material insulation	0.9 million
$H_2C{=}CH{-}O{-}C(=O){-}CH_3$	vinyl acetate	poly(vinyl acetate) (PVA)	latex paint, adhesives, textile coatings	200,000
$H_2C{=}C(CH_3){-}C(=O){-}O{-}CH_3$	methyl methacrylate	poly(methyl methacrylate) (Plexiglas, Lucite)	high-quality transparent objects, latex paints, contact lenses	200,000
$F_2C{=}CF_2$	tetrafluoroethylene	polytetrafluoroethylene (Teflon)	gaskets, insulation, bearings, pan coatings	6,000

*One metric ton = 1000 kg

Polyethylene film is produced by extruding the molten plastic through a ringlike gap and inflating the film like a balloon. *(The Stock Market)*

Polyethylene and Other Polyolefins

Polyethylene is by far the production leader among addition polymers. Ethylene (C_2H_4), the monomer from which polyethylene is made, is a product of petroleum refining and one of the top four or five chemicals produced in the United States. When ethylene is heated to between 100 and 250 °C at a pressure of 1000 to 3000 atm in the presence of a catalyst, polymers with molecular weights up to several million are formed. The reaction can be expressed as a balanced chemical equation:

$$n\,H_2C{=}CH_2 \longrightarrow \left(\begin{array}{cc} H & H \\ | & | \\ -C- & C- \\ | & | \\ H & H \end{array}\right)_n$$

ethylene polyethylene

The abbreviated formula of the reaction product, $+CH_2CH_2+_n$, shows that polyethylene is a chain of carbon atoms, each bearing two hydrogens. The chain length for polyethylene can be very long. A polymer with a molecular weight of one million would contain almost 36,000 ethylene molecules linked together.

Polyethylenes formed under various pressures and catalytic conditions have different properties, a result of different molecular structures. For example, when chromium oxide is used as a catalyst, the product is almost exclusively a linear chain (Figure 11.14a). If ethylene is heated to 230 °C at high pressure, however, irregular branching occurs. Still other conditions lead to cross-linked polyethylene, in which different chains are linked together (Figure 11.14b and c).

The high-molecular-weight chains of linear polyethylene pack closely together and result in a material with a density of 0.97 g/mL. This material, referred to as high-density polyethylene (HDPE), is hard and tough, which makes it suitable for things like milk bottles. If the polyethylene chain contains

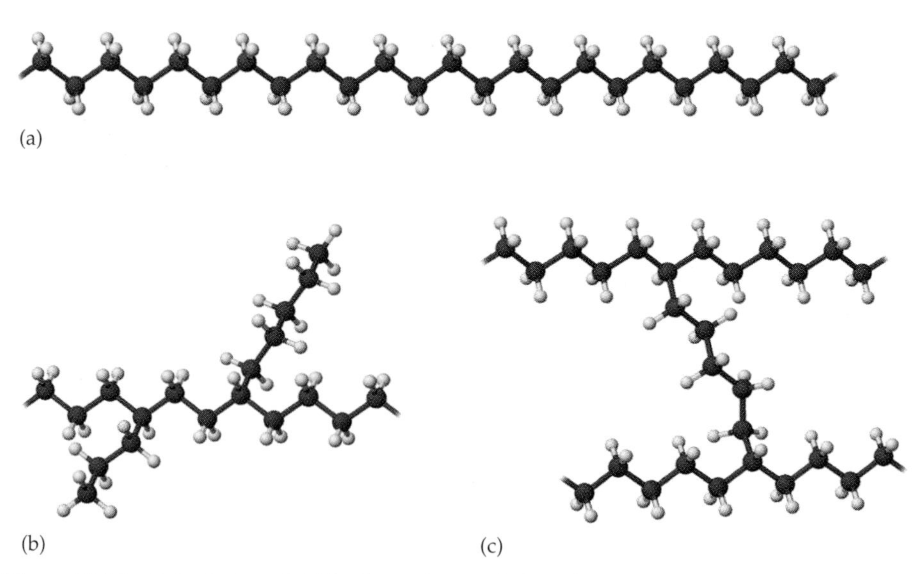

Figure 11.14 Polyethylene. (a) The linear form (HDPE). (b) Branched chains make low-density polyethylene (LDPE). (c) Cross-linked polyethylene (CLPE).

branches, however, the chains cannot pack as closely together, and a lower density material (0.92 g/mL) known as low-density polyethylene (LDPE) results. This material is softer and more flexible than HDPE and is used in such things as sandwich bags. Linking up the polymer chains in cross-linked polyethylene (CLPE) causes the material to be even more rigid and inflexible. Plastic bottle caps are often made of CLPE.

Polymers formed from substituted ethylenes ($CH_2{=}CHX$) have a range of properties and uses (see Table 11.15). Often the properties are predictable based on structure. Polymers without polar substituent groups, like polystyrene, often dissolve in organic solvents, a property useful for some types of fabrication.

Polyvinyl alcohol, $-(CH_2CHOH)-_n$, is a polymer with an affinity for water but little affinity for nonpolar solvents, which is not surprising based on the large number of polar —OH groups that can interact with water (Figure 11.15). Vinyl alcohol itself is not a stable compound (it isomerizes to CH_3CHO), so polyvinyl alcohol cannot be made from this compound. Instead, it is made by hydrolyzing the ester groups in polyvinyl acetate, $-(CH_2CHOCOCH_3)-_n$.

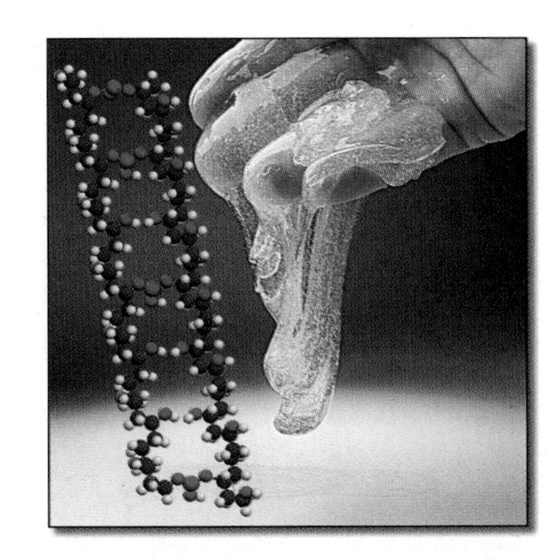

Figure 11.15 Slime. When boric acid, $B(OH)_3$, is added to an aqueous suspension of polyvinyl alcohol, $(CH_2CHOH)_n$, the mixture becomes very viscous. This is because boric acid reacts with the —OH groups on the polymer chain, causing cross-linking to occur. (The model shows an idealized structure of the polymer.) *(Photo, C. D. Winters; model, S. M. Young)*

Solubility in water or organic solvents can be a liability. The many uses of polytetrafluoroethylene [Teflon, $-(CF_2CF_2)-_n$] are a consequence of the fact that it does not interact with water or organic solvents.

Polystyrene $-(CH_2CHC_6H_5)-_n$, with $n = 5700$, is a clear, hard, colorless solid that can be molded easily at 250 °C (Figure 11.16). You are probably more familiar with the very light, foamlike material (Styrofoam) used widely for food and beverage containers and for home insulation. Styrofoam is produced by a process called "expansion molding." Polystyrene beads containing 4% to 7% of a low-boiling liquid like pentane are placed in a mold and heated with steam or hot air. Heat causes the solvent to vaporize and create a foam in the molten polymer that expands to fill the shape of the mold.

Natural and Synthetic Rubber

Natural rubber was first introduced into Europe in 1740, but it remained a curiosity until 1823 when Charles Mackintosh invented a way of using it to water-

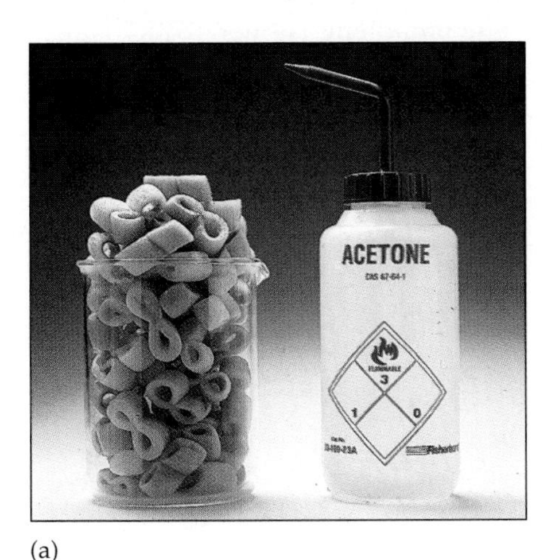

(a)

(b)

Figure 11.16 Polystyrene $-(CH_2CHC_6H_5)-_n$. (a) The polymer is a clear, hard, colorless solid, but it may be more familiar as a light, foamlike material called Styrofoam. (b) Styrofoam has no polar groups and thus dissolves well in organic solvents such as acetone. *(C. D. Winters)*

Figure 11.17 Natural rubber. The sap that comes from the rubber tree is a natural polymer of isoprene. All the linkages in the carbon chain are *cis*. When natural rubber is heated strongly in the absence of air it smells of isoprene, an observation that provided a clue that rubber is composed of this building block. *(Photo, Kevin Schafer/ Tom Stack & Associates; model, S. M. Young)*

Natural rubber has good elasticity but poor grip, so no natural rubber is used in automobile tire treads. Instead, treads are blended of SBR and polybutadiene. Aircraft tires are 100% natural rubber.

proof cotton cloth. The mackintosh, as rain coats are still sometimes called, became extremely popular despite major problems: natural rubber is notably weak and thermoplastic, soft and tacky when warm but brittle at low temperatures. In 1839, after five years of work on natural rubber, the American inventor Charles Goodyear (1800–1860) discovered that heating gum rubber with sulfur produces a material that is elastic, water-repellent, resilient, and no longer sticky.

Rubber is a naturally occurring polymer, the monomers of which are molecules of 2-methyl-1,3-butadiene, commonly called *isoprene.* In natural rubber, isoprene monomers are linked together through carbon atoms 1 and 4, that is, through the end carbon atoms of the C_4 chain (Figure 11.17). This leaves a double bond between carbon atoms 2 and 3. In natural rubber, all these double bonds have a *cis* configuration.

In vulcanized rubber, the material that Goodyear discovered, the polymer chains of natural rubber are cross-linked by short chains of sulfur atoms. Sulfur cross-linking helps to align the polymer chains so the material does not undergo a permanent change when stretched, and it springs back when the stress is removed. Substances that behave this way are called **elastomers.**

With a knowledge of the composition and structure of natural rubber, chemists began searching for ways to make synthetic rubber. When they first tried to make synthetic rubber by linking isoprene monomers together, however, what they made was sticky and useless. The problem was that synthetic procedures gave a mixture of *cis* and *trans* polyisoprene. In 1955, however, chemists at the Goodyear and Firestone companies discovered special catalysts to prepare the all -*cis* polymer (see Figure 11.17). This synthetic material, structurally identical to natural rubber, can now be manufactured cheaply, and over 8.0×10^8 kg of synthetic polyisoprene is produced annually in the United States.

Other kinds of polymers add to the repertoire of elastomeric materials now available. Polybutadiene, for example, is currently used in the production of tires, hoses, and belts. Almost 5.0×10^8 kg of polybutadiene was made in the United States in 1993.

Some elastomers called **copolymers** are formed by polymerization of two (or more) different monomers. A copolymer of styrene and butadiene, made with a 1-to-3 ratio of these raw materials, is the most important synthetic rubber now made; more than about 1.3×10^9 kg of styrene-butadiene rubber (SBR) is produced each year in the United States for making tires.

$$3n\ H_2C{=}CH{-}CH{=}CH_2 + n\ H_2C{=}CH(C_6H_5)$$

1,3-butadiene styrene

$$\downarrow$$

$$\left(\!-CH_2{-}CH{=}CH{-}CH_2{-}CH_2{-}CH{=}CH{-}CH_2{-}CH(C_6H_5){-}CH_2{-}CH{=}CH{-}CH_2-\!\right)_n$$

sytrene-butadiene rubber (SBR)

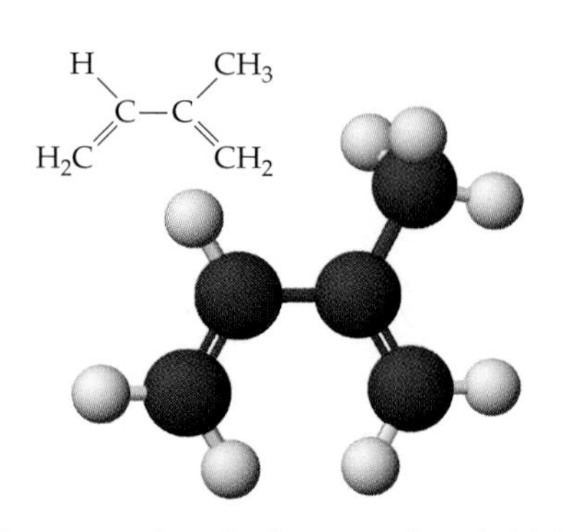

Isoprene, otherwise known as 2-methyl-1,3-butadiene

The covering on a golf ball consists of gutta-percha rubber. In this hard material, produced from the sap of a certain kind of rubber tree, all the double bonds in polyisoprene have *trans* configurations. *(C. D. Winters)*

And a little is left over each year to make bubble gum. The stretchiness of gum once came from natural rubber, but SBR is now used to help you blow bubbles.

The use of SBR in bubble gum was described in G. B. C. Marsella; "Bubble gum." *Chem Matters,* Vol. 12, pp. 10–12, October, 1994.

Condensation Polymers

A chemical reaction in which two molecules react by splitting out, or eliminating, a small molecule is called a **condensation reaction.** A reaction of an alcohol with a carboxylic acid to give an ester is an example. A polymer can be formed in a condensation reaction if two different reactant molecules, each containing *two* functional groups, are used. This is the route used to form polyesters and polyamides, two important types of condensation polymers.

See the *Saunders Interactive General Chemistry CD-ROM,* Screen 11.10, Synthetic Organic Polymers (2), Condensation Polymerization.

Polyesters

Terephthalic acid has two carboxylic acid groups, and ethylene glycol has two alcohol groups. When mixed, the acid and alcohol functional groups at both ends of these molecules can react to form ester linkages, splitting out water. The result is a polymer called polyethylene terephthalate (PET). The multiple ester linkages in the product make this substance a **polyester:**

Polyester "fleece" has recently become very popular for clothing. (See H. Espren, "Fleeced," *New York Times Sunday Magazine,* February 15, 1998, p. 20.)

$$n\,\mathrm{HO\overset{\overset{O}{\|}}{C}\text{—}C_6H_4\text{—}\overset{\overset{O}{\|}}{C}OH} + n\,\mathrm{HOCH_2CH_2OH} \longrightarrow$$

terephthalic acid   ethylene glycol

$$\left(\mathrm{O\overset{\overset{O}{\|}}{C}\text{—}C_6H_4\text{—}\overset{\overset{O}{\|}}{C}OCH_2CH_2O}\right)_n + \mathrm{H_2O}$$

polyethylene terephthalate (PET)

More than about 2×10^9 kg of PET is produced in the United States each year for many uses. For example, bottles made of PET are used for soft drinks. (These bottles have the recycling symbol 1 within a triangle and the letters PETE.)

Figure 11.18 Polyesters. Polyethylene terephthalate is used to make soda bottles and clothing. Film made of Mylar, also a polyester, is used to make recording tape as well as balloons. Because the film has very tiny pores, it can be used for helium-filled balloons; the atoms of gaseous helium move through the pores very slowly. *(C. D. Winters)*

Polyester textile fibers are marketed under such names as Dacron and Terylene. A polyester film, Mylar, has unusual strength and can be rolled into sheets one-thirtieth the thickness of a human hair. Magnetically coated films of this polyester are used to make audio and video tapes (Figure 11.18).

The inert, nontoxic, noninflammatory, and non–blood-clotting properties of Dacron polymers make Dacron tubing an excellent substitute for human blood vessels in heart bypass operations, and Dacron sheets are sometimes used as temporary skin for burn victims.

Polyamides

In 1928, the Du Pont Company embarked on a basic research program headed by Dr. Wallace Carothers (1896–1937). Carothers, who left a faculty position at Harvard to go to Du Pont, was interested in high-molecular-weight compounds, such as rubbers, proteins, and resins. In 1935, his research yielded nylon-6,6 (Figure 11.19), a polyamide prepared from adipoyl chloride, a derivative of adipic acid (a diacid), and hexamethylenediamine, a diamine:

$$n\,\text{Cl}-\overset{\overset{\displaystyle\text{O}}{\|}}{\text{C}}-(\text{CH}_2)_4-\overset{\overset{\displaystyle\text{O}}{\|}}{\text{C}}-\text{Cl} + n\,\text{H}_2\text{N}-(\text{CH}_2)_6-\text{NH}_2 \longrightarrow$$

adipoyl chloride　　　　hexamethylenediamine

$$\left(\overset{\overset{\displaystyle\text{O}}{\|}}{\text{C}}-(\text{CH}_2)_4-\overset{\overset{\displaystyle\text{O}}{\|}}{\text{C}}-\underset{\underset{\displaystyle\text{H}}{|}}{\text{N}}-(\text{CH}_2)_6-\underset{\underset{\displaystyle\text{H}}{|}}{\text{N}}\right)_n + n\,\text{HCl}$$

Amide link

Figure 11.19 Nylon-6,6. Hexamethylenediamine is dissolved in water (bottom layer), and adipoyl chloride (a derivative of adipic acid) is dissolved in hexane (top layer). The two compounds react at the interface between the layers to form nylon, which is being wound onto a stirring rod. *(C. D. Winters)*

Nylon can be extruded easily into fibers that are stronger than natural fibers and chemically more inert. The discovery of nylon jolted the American textile industry at a critical time. Natural fibers were not meeting 20th-century needs. Silk was expensive and not durable, wool was scratchy, linen crushed easily, and cotton did not have a high-fashion image. Perhaps the most identifiable use for the new fiber, however, was in nylon stockings. The first public sale of nylon hosiery took place on October 24, 1939, in Wilmington, Delaware (the site of Du Pont's main office.) The commercial use of nylon soon ended with the start of World War II, however, when all nylon production was diverted to making parachutes and other military gear. It was not until about 1952 that nylon would be able to fulfill the needs of the textile industry.

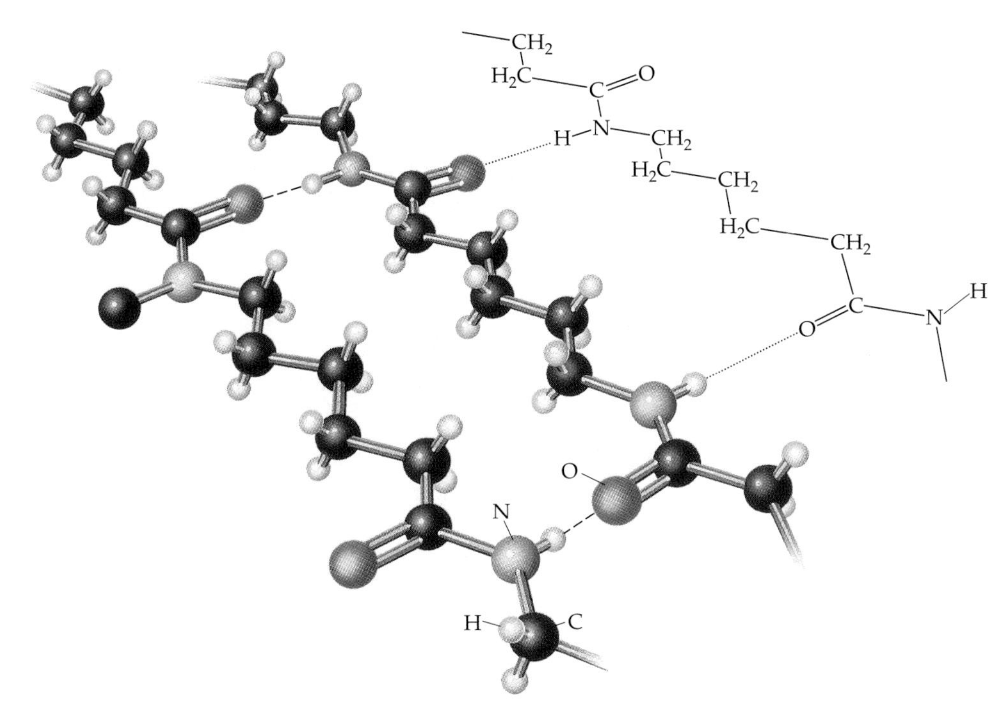

Figure 11.20 Bonding between polyamide chains. Carbonyl oxygen atoms with a partial negative charge on one chain interact with an amine hydrogen with a partial positive charge on a neighboring chain. (This form of bonding, called hydrogen bonding, is described in more detail in Chapter 13.)

Figure 11.20 illustrates why nylon makes such a good fiber. To have good tensile strength (the ability to resist tearing) the polymer chains should be able to attract one another, though not so strongly that the plastic cannot be initially extended to form fibers. Ordinary covalent bonds between the chains (cross-linking) would be too strong. Instead, cross-linking occurs by a somewhat weaker form of bonding (called *hydrogen bonding*) between the hydrogens of N—H groups on one chain and carbonyl oxygens on another chain. The polarities of the N($\delta-$)—H($\delta+$) group and the C($\delta+$)=O($\delta-$) group lead to attractive forces between the polymer chains of the desired magnitude.

Example 11.5 Condensation Polymers

Write the repeating unit of the condensation polymer obtained by combining $HO_2CCH_2CH_2CO_2H$ (succinic acid) and $H_2NCH_2CH_2NH_2$ (1,2-ethylenediamine).

Solution A condensation polymer composed of a diacid and a diamine forms by loss of water between monomers to give an amide bond. The repeating unit is therefore

Amide linkage

$$\left(\overset{\overset{\displaystyle O}{\|}}{C}CH_2CH_2\overset{\overset{\displaystyle O}{\|}}{C}-\underset{\underset{\displaystyle H}{|}}{N}CH_2CH_2\underset{\underset{\displaystyle H}{|}}{N} \right)_n$$

A CLOSER LOOK *Proteins—Natural Polymers*

Nature creates polymers of all types. There are polyamide polymers known as **proteins;** there are **polysaccharides** such as cellulose, a polymer of glucose; and there are **nucleic acids,** polymers consisting of alternating sugar molecules and phosphate ions, with other molecules attached to the polymer chain.

Proteins are important in a variety of ways. As *enzymes* they serve as catalysts in biological synthesis and degradation reactions. *Hormones* serve a regulatory role, and as *antibodies* they protect us from disease. The protein *hemoglobin* transports oxygen in the body. Finally, they are major constituents of cellular and intracellular membranes, skin, hair, and muscle.

Proteins can be divided into two classes: *simple* and *conjugated.* Simple proteins consist only of amino acids, whereas conjugated proteins may contain some other groups in addition to amino acids. Hemoglobin, in which the iron-containing heme group is the site of oxygen-binding (page 85), is an example of the latter type.

Proteins are polymers constructed from α-amino acids, compounds with the general formula $H_2NCHRCO_2H$. (The α in the formula designates the position of the amino group on the carbon adjacent to the carbonyl.) There are 26 naturally occurring amino acids, differing only in the identity of the organic group labeled R in the molecule. Glycine, with R = H, is the simplest amino acid:

$H{-}C(H)(NH_2){-}CO_2H$

glycine

$H_3C{-}C(H)(NH_2){-}CO_2H$

alanine

$HOCH_2{-}C(H)(NH_2){-}CO_2H$

serine

The natural amino acids are subdivided into two groups, depending on whether the R group is nonpolar (as in glycine or alanine) or polar (as in serine).

To form a protein, a water molecule is eliminated between two amino acids in a condensation reaction to form an amide linkage.

$HOCH_2$, H, H_2N, O, H, O

serine

O, H_2N, H, O, H, CH_3

alanine

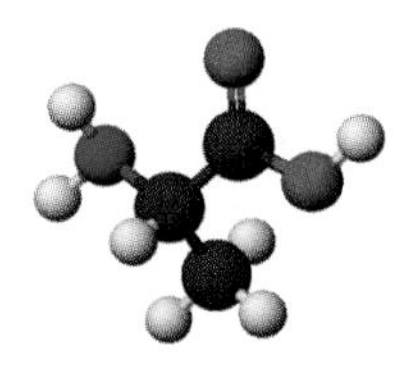

$HOCH_2$, H, H_2N, O, H, N, H, O, O, H, CH_3

Peptide bond

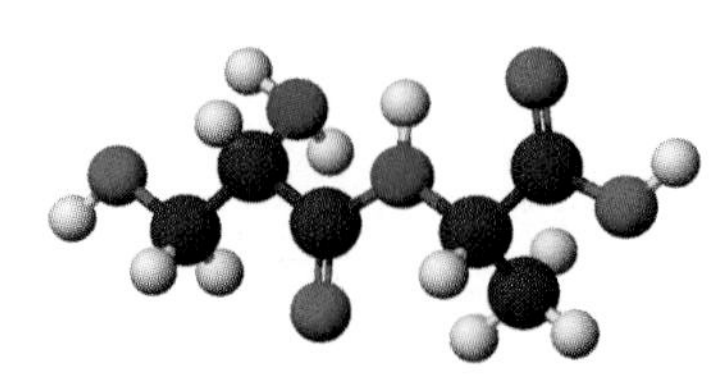

This linkage is also called a **peptide** bond, and linking two amino acids gives a *dipeptide.* Either end of the dipeptide can react with another amino acid to give a tripeptide, which in turn can condense with another amino acid, and so on. When several dozen amino acids have condensed, the polymer is generally called a **polypeptide.** Proteins are biological macromolecules with molar masses of 5000 g/mol or greater, consisting of one or more polypeptide chains. The common protein insulin has 51 amino acid units with two linked chains. In contrast, human hemoglobin contains four protein chains, two identical ones with 141 amino acids and another two, also identical, with 146 amino acids. Considering the ways that 26 amino acids could be put together, it is remarkable that the human body contains only about 100,000 different proteins.

Each protein contains a specific sequence of amino acids. This is called the **primary structure** of the protein. Equally important, though, is the protein's three-dimensional structure (often referred to as its *secondary* and *tertiary structure*). Among the questions one can ask is if the protein is a straight chain or folded in some manner. Let us start by looking at the arrangement of atoms around a peptide bond, and we see some surprising geometry. The bond angles around the N atoms are close to 120°, although we might have predicted angles closer to 109° because the N is surrounded by three bond pairs and one lone pair. It is evident that resonance structure B is an important contributing structure.

A: $:\ddot{O}{=}C{-}\ddot{N}$, with αC, αC, H ⟷ B: $:\ddot{O}^{-\delta}{-}C{=}N^{+\delta}$, with C, C, H; angles 120.0°, 118.5°, 121.5°

A B

The implication of this is that the peptide linkage is nearly planar and that the polypeptide chain can bend significantly only around the α or CHR carbon atoms.

Because of the nature of the peptide bond, Linus Pauling proposed that two stable "folding patterns" are possible for polypeptide chains: the α helix and the β-pleated sheet. In an α helix, the chain is coiled into a spiral. In this manner a positively charged N—H hydrogen on one amino acid is brought into proximity to

$$\overset{-\delta}{\text{O}}\text{—C=N—}\overset{+\delta}{\text{H}}\cdots\overset{-\delta}{\text{O}}\text{—C=N—}\overset{+\delta}{\text{H}}$$

Hydrogen bond

a negatively charged C=O oxygen atom on another amino acid some distance away on the chain. The attraction between the two—called a *hydrogen bond*—stabilizes the structure.

The alternative to the α helix is the β-pleated sheet. This is essentially the structure adopted by nylon and accounts for the elasticity of the polymer (see Figure 11.20).

Protein chemistry is an enormous area of chemistry and biology. For more information see (a) Chapter 13 for a discussion of hydrogen bonding; (b) the *Saunders Interactive General Chemistry CD-ROM,* Screen 8 in the Introductory Chapter for Pauling's description of his model for the α helix and Screen 11.8 for a description of amino acids; and (c) textbooks such as W. H. Brown and C. S. Foote: *Organic Chemistry,* Chapter 27, Saunders College Publishing, 2e, 1997, Philadelphia.

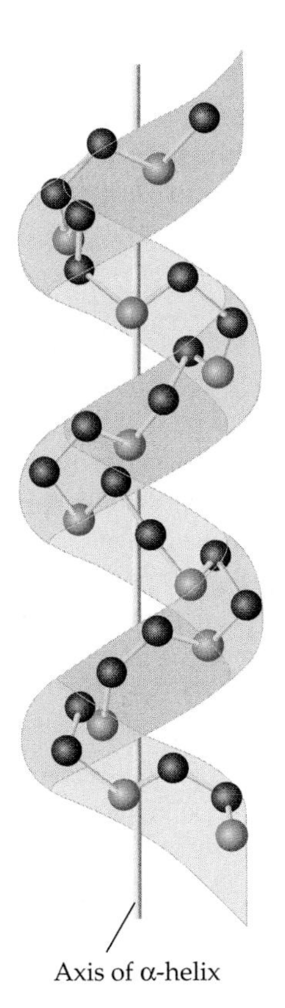

Axis of α-helix

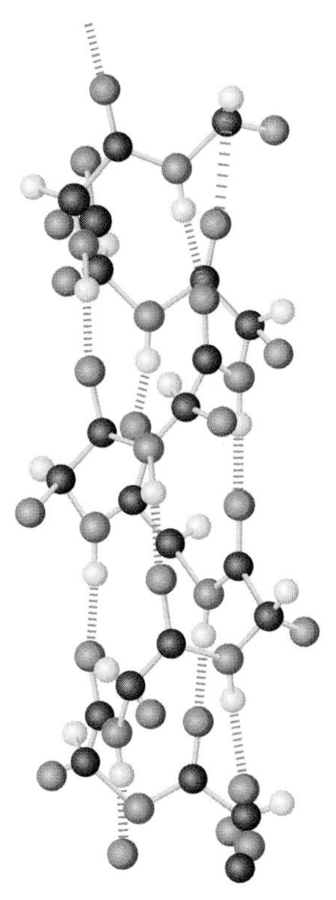

Hydrogen bonds stabilize the helix structure

Composite materials made from graphite fibers and synthetic polymers have found many uses, such as aircraft parts, golf clubs, and fishing poles. Here a rainbow trout is landed with a graphite rod in a river in Australia. *(Bill Bachman/Photo Researchers, Inc.)*

Exercise 11.11 Condensation Polymers

Draw the structure of the repeating unit in the condensation polymer obtained from the reaction of propylene glycol with maleic acid:

```
     H3C  H               O   H   H   O
      |   |               ‖   |   |   ‖
  HO—C—C—OH  +  HO—C—C═C—C—OH
      |   |
      H   H
 propylene glycol          maleic acid
```

(A closely related material is combined with glass fibers [fiberglass] to make hulls for small boats and automobile panels and parts.) Is this a polyamide or a polyester?

New Polymer Materials

Few plastics are used today without some modification. For example, body panels for the Saturn and Corvette automobiles are made of **reinforced plastics,** which contain fibers in a polymer matrix (see Exercise 11.11). These materials are referred to as **composites.** The strongest geometry for a solid is a wire or fiber, and the use of a polymer matrix prevents the fiber from buckling or bending. As a result, reinforced plastics are stronger than steel. In addition, the composites have a low density, from 1.5 to 2.25 g/cm^3, compared with 2.7 g/cm^3 for aluminum, 7.9 g/cm^3 for steel, and 2.5 g/cm^3 for concrete. Wood is the only structural material with a lower density. The low density, high strength, and resistance to corrosion of composites favor use of these materials.

Glass fibers currently account for more than 90% of the fibrous material used to reinforce plastics. Glass is inexpensive and glass fibers possess high strength, low density, good chemical resistance, and good insulating properties. In principle, any polymer can be used as the matrix material. Polyesters are the number one polymer matrix at the present time, and glass-reinforced polymer composites are commonly used for structural applications, such as boat hulls, airplanes, and automobile body panels.

Other fibers and polymers have been used, and the trend is toward increased utilization of composites in automobiles and aircraft. One of the best examples are the **graphite composites** used in the construction of aircraft parts and in a number of sporting goods such as golfclub shafts, tennis racquets, fishing rods, and skis. Carbon or graphite fibers are made from polyacrylonitrile, PAN (see Table 11.15). The PAN fibers are first oxidized around 200 °C and then carbonized at temperatures higher than 1000 °C. This process makes long fibers that are about 90% carbon. Further treatment at temperatures as high as 3000 °C yields fibers that are about 99% carbon. Carbon fiber composites can replace metals because the composites are strong and can be conductive. They also provide strength without the problem of metal fatigue and do not corrode. These composites will find much wider use when the price decreases.

CHAPTER HIGHLIGHTS

When you have finished studying this chapter you should be able to

- Recognize features leading to the diversity of carbon compounds: structure and bonding to carbon, strength of bonds to carbon, and **kinetic stability** of carbon compounds (Section 11.1).
- Use bond energies to predict the energy of a reaction (Section 11.1).
- Draw structural formulas, identify **isomers,** and name branched and unbranched **alkanes** (C_nH_{2n+2}) (Section 11.2).
- Draw structures of **cycloalkanes** (Section 11.2).
- Draw structures, identify isomers including *cis–trans* isomers, and name **alkenes** (Section 11.2).
- Draw structures for benzene and substituted benzenes, and describe the bonding in benzene (Section 11.2).
- Recognize, draw structures, and name **alcohols, amines, aldehydes, ketones, carboxylic acids, esters,** and **amides** (Sections 11.3–11.5).
- Understand the structures of **fats** and **oils** and their properties (saturated and unsaturated fats and oils; **saponification** reactions) (Section 11.6).
- Relate common properties of organic compounds (solubility, boiling point) to structure.
- Recognize chemical properties of the classes of organic compounds (Sections 11.2–11.6).
- Define important terms in polymer chemistry: **thermosetting, thermoplastic, elastomer** (Section 11.7).
- Write equations for the formation of **addition** and **condensation polymers** (Section 11.7).

STUDY QUESTIONS

Naming of organic compounds is described in Appendix E. Structures of many of the compounds used in these questions are found on the Saunders Interactive General Chemistry CD-ROM *or on the Saunders Web site at http://www.saunderscollege.com.*

REVIEW QUESTIONS

Alkanes

1. What is the name of the straight (unbranched) chain alkane with the formula C_7H_{16}?
2. What is the molecular formula for an alkane with 12 carbon atoms?
3. Which of the following is an alkane? Which could be a cycloalkane?
   (a) C_2H_4 (c) $C_{14}H_{30}$
   (b) C_5H_{10} (d) C_7H_8
4. Isooctane, 2,2,4-trimethylpentane is one of the possible structural isomers for the alkane C_8H_{18}. Draw the structure of this isomer, and draw and name structures of two other isomers of C_8H_{18} in which the longest carbon chain is five atoms.
5. Give the systematic name for the following alkane:

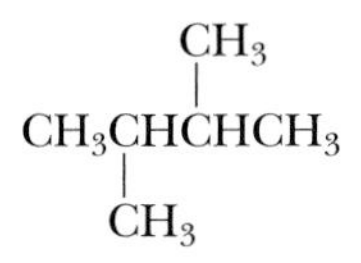

6. Give the systematic name for the following alkane. Draw a structural isomer of the compound and give its name.

CH_3
$CH_3CHCH_2CH_2CHCH_3$
CH_2CH_3

7. Draw the structure of each of the following compounds:
   (a) 2,3-dimethylhexane
   (b) 2,3-dimethyloctane
   (c) 3-ethylheptane
   (d) 2-methyl-3-ethylhexane

8. Draw a structure for cycloheptane. Is the seven-member ring planar? Explain your answer.

9. Draw a structure of cyclohexane in a perspective that allows you to distinguish between axial and equatorial hydrogens.

Alkenes and Alkynes

10. Draw structures for the *cis* and *trans* isomers of 3-methyl-2 hexene.

11. What structural requirement is necessary for an alkene to have *cis* and *trans* isomers? Can *cis* and *trans* isomers exist for an alkyne?

12. A hydrocarbon with the formula C_5H_{10} can be either an alkene or a cycloalkane.
(a) Draw a structure for each of the isomers possible for C_5H_{10}, assuming it is an alkene. Six isomers are possible. Give the systematic name of each isomer you have drawn.
(b) Draw a structure for a cycloalkane having the formula C_5H_{10}.

13. There are five alkenes with the formula C_7H_{14} that have a seven-carbon atom chain. Draw their structures and name them.

14. Draw the structure and give the systematic name for the products of the following reactions:
(a) $CH_3CH{=}CH_2 + Br_2 \longrightarrow$
(b) $CH_3CH_2CH{=}CHCH_3 + H_2 \longrightarrow$

15. Draw the structure and give the systematic name for the products of the following reactions:

(a) 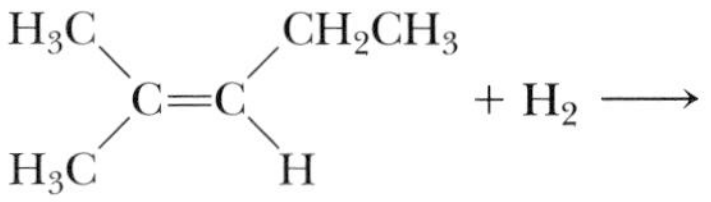 $(H_3C)_2C{=}C(H)(CH_2CH_3) + H_2 \longrightarrow$

H₃C and H₃C on one carbon; CH₂CH₃ and H on the other.

(b) $CH_3C{\equiv}CCH_2CH_3 + 2\ Br_2 \longrightarrow$

16. The compound $CH_3CH_2CH_2CH_2Br$ is a product of addition of HBr to an alkene. Identify the alkene.

17. The compound 2,3-dibromo-2-methylhexane is formed by addition of Br_2 to an alkene. Identify the alkene.

Benzene and Aromatic Compounds

18. Draw structural formulas for the following compounds:
(a) *m*-dichlorobenzene (alternatively called 1,3-dichlorobenzene)
(b) *p*-bromotoluene (alternatively called 1-bromo-4 methylbenzene)

19. Give the systematic name for each of the following compounds:

(a) 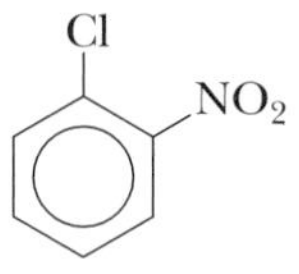 benzene ring with Cl and NO_2 on adjacent carbons (b) 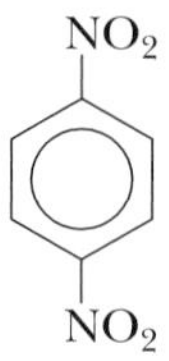 benzene ring with NO_2 at top and NO_2 at bottom (c) 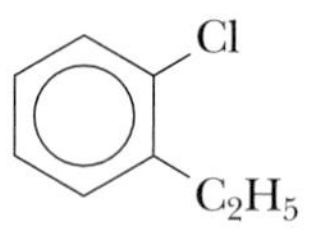 benzene ring with Cl and C_2H_5 on adjacent carbons

20. Show how to prepare ethylbenzene from benzene and an appropriate ethyl derivative.

21. Draw possible resonance structures for naphthalene, $C_{10}H_8$.

Alcohols and Amines

22. Give the systematic name for each of the following alcohols, and tell if each is a primary, secondary, or tertiary alcohol:
(a) $CH_3CH_2CH_2OH$
(b) $CH_3CH_2CH_2CH_2OH$
(c) 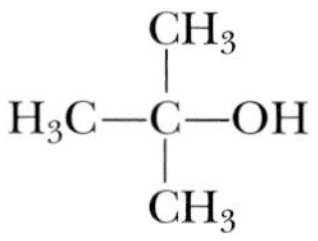 $H_3C{-}C(CH_3)_2{-}OH$ (d) 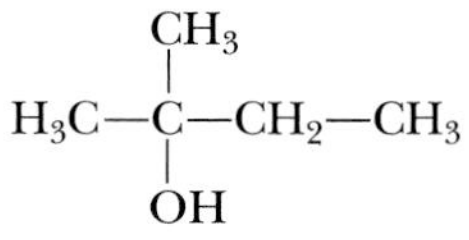 $H_3C{-}C(CH_3)(OH){-}CH_2{-}CH_3$

23. Draw structural formulas for the following alcohols, and tell if each is primary, secondary or tertiary:
(a) 1-butanol
(b) 2-butanol
(c) 3,3-dimethyl-2-butanol
(d) 3,3-dimethyl-1-butanol

24. Give formulas and structures of the following amines:
(a) ethylamine
(b) dipropylamine
(c) butyldimethylamine

25. Name the following amines
(a) $CH_3CH_2CH_2NH_2$
(b) $(CH_3)_3N$
(c) $(CH_3)(C_2H_5)NH$

26. Draw structural formulas for all the alcohols with the formula $C_4H_{10}O$. Give the systematic name of each.

27. Draw structural formulas for all amines with the formula $C_4H_9NH_2$.

28. Complete the following equations:
(a) $CH_3CH_2OH\ (+\ H_2SO_4/180\ °C) \longrightarrow$
(b) $CH_3CH_2CH_2OH + Na \longrightarrow$

29. Describe how to prepare each of the following compounds from an alcohol and other appropriate reagents:
(a) dibutyl ether
(b) CH_3CHO

Compounds with a Carbonyl Group

30. Draw structural formulas for (a) 2-pentanone, (b) hexanal, and (c) pentanoic acid.

31. Give systematic names for each of the following compounds:
(a) 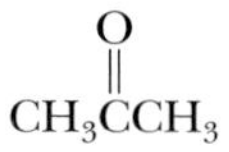 $CH_3C({=}O)CH_3$
(b)  $CH_3CH_2CH_2CH({=}O)$
(c)  $CH_3C({=}O)CH_2CH_2CH_3$

32. Give systematic names for each of the following compounds:
(a) 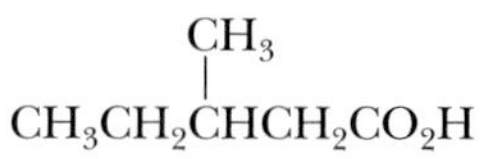 $CH_3CH_2CH(CH_3)CH_2CO_2H$

(b) 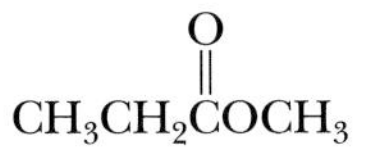 $CH_3CH_2\overset{O}{\overset{\|}{C}}OCH_3$

(c) $CH_3\overset{O}{\overset{\|}{C}}OCH_2CH_2CH_2CH_3$

(d) 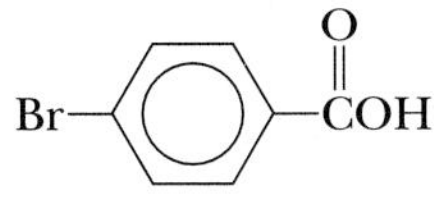 Br—C_6H_4—$\overset{O}{\overset{\|}{C}}OH$

33. Draw structural formulas for the following acids and esters:
   (a) 2-methylhexanoic acid
   (b) pentyl butanoate (which has the odor of apricots)
   (c) octyl acetate (which has the odor of oranges)

34. Give the structural formula and systematic name for the product, if any, from each of the following reactions:
   (a) pentanal and $KMnO_4$
   (b) pentanal and $LiAlH_4$
   (c) 2-octanone and $LiAlH_4$
   (d) 2-octanone and $KMnO_4$

35. Describe how to prepare 2-pentanol beginning with the appropriate ketone.

36. Describe how to prepare propyl propanoate beginning with 1-propanol.

37. Give the name and structure of the product of the reaction of benzoic acid and 2-propanol.

38. Draw structural formulas and give the names for the products of the following reaction:

$$CH_3\overset{O}{\overset{\|}{C}}OCH_2CH_2CH_2CH_3 + NaOH$$

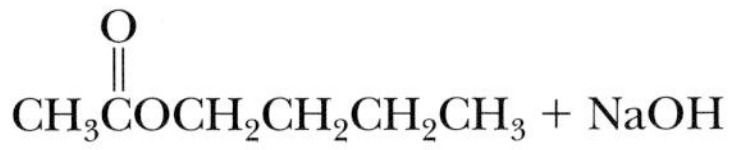

39. Draw structural formulas and give the names for the products of the following reaction:

$$C_6H_5-\overset{O}{\overset{\|}{C}}-O-CH(CH_3)_2 + NaOH$$

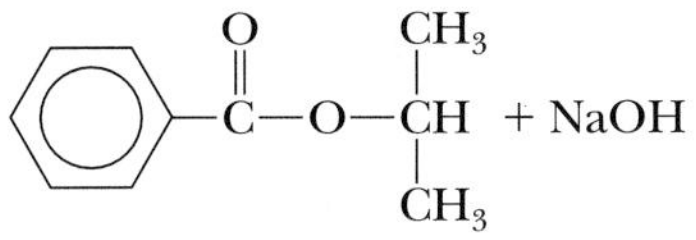

Polymers

40. Polyvinyl acetate is the binder in water-based paints.
   (a) Write an equation for its formation from vinyl acetate.
   (b) Show a portion of this polymer with three monomer units.
   (c) Describe how to make polyvinyl alcohol from polyvinyl acetate.

41. Polychloroprene is made from a chlorinated butadiene pictured here.

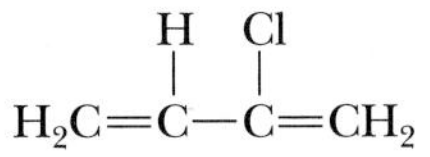

$$H_2C=\overset{H}{\overset{|}{C}}-\overset{Cl}{\overset{|}{C}}=CH_2$$

Monomer for polychloroprene

One of the names under which it is sold is neoprene.

   (a) Write an equation showing the formation of polychloroprene from the monomer.
   (b) Draw the structure of polychloroprene with at least three monomer units.

42. Saran is a copolymer of 1,1-dichloroethene and chloroethene (vinyl chloride). Draw a possible structure for this polymer.

43. The structure of methyl methacrylate is given in Table 11.15. Draw a polymethyl methacrylate (PMMA) polymer that has four monomer units. Is the polymer chain flat or does it twist in some manner? (PMMA has excellent optical properties and is used to make hard contact lenses.)

GENERAL QUESTIONS

44. Draw the structure of each of the following compounds:
   (a) 2,2-dimethylpentane
   (b) 3,3-diethylpentane
   (c) 2-methyl-3-ethylpentane
   (d) 3-propylhexane

45. In addition to the structural isomerism of alkanes, and the *cis–trans* isomerism of some alkenes, other types of isomers are found in organic chemistry. For example, dimethyl ether (CH_3OCH_3) and ethanol (CH_3CH_2OH) are isomers; they have the same molecular formula but their chemical functionality is quite different.
   (a) Draw all of the isomers possible for C_3H_8O. Give the systematic name of each and tell into what class of compound it fits.
   (b) Draw the structural formula for an aldehyde and a ketone with the molecular formula C_4H_8O. Give the systematic name of each.

46. Draw structural formulas for possible isomers of the dichlorinated propane, $C_3H_6Cl_2$. Name each compound.

47. Draw structural formulas for possible isomers of the formula C_3H_6ClBr, and name each isomer.

48. Give structural formulas and systematic names for the three structural isomers of trimethylbenzene, $C_6H_3(CH_3)_3$.

49. Give structural formulas and systematic names for possible isomers of dichlorobenzene, $C_6H_4Cl_2$.

50. Voodoo lilies depend on carrion beetles for pollination. Carrion beetles are attracted to dead animals, and because dead and putrefying animals give off the horrible smelling amine cadaverine, the lily likewise releases cadaverine (and the closely related compound putrescine) (page 508). A decarboxylase enzyme converts the naturally occurring amino acid lysine to cadaverine.

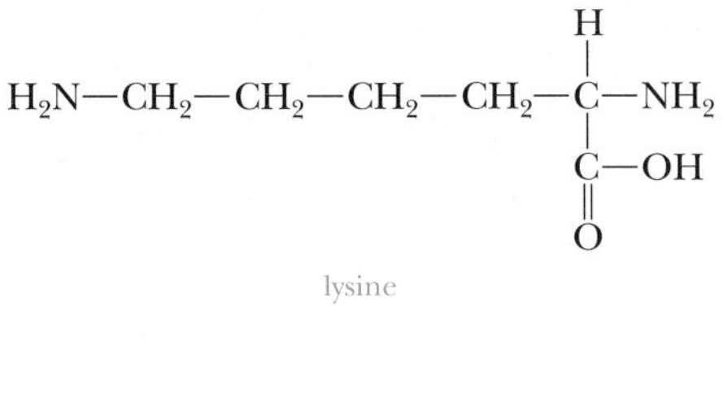

$$H_2N-CH_2-CH_2-CH_2-CH_2-\overset{H}{\overset{|}{C}}(-NH_2)-\overset{O}{\overset{\|}{C}}-OH$$

lysine

What group of atoms must be replaced in lysine to make cadaverine? (Lysine is essential to human nutrition but is not synthesized in the human body.)

51. In the text it is noted that benzoic acid occurs in many berries. When humans eat berries, however, benzoic acid is converted to hippuric acid in the body by reaction with the amino acid glycine.

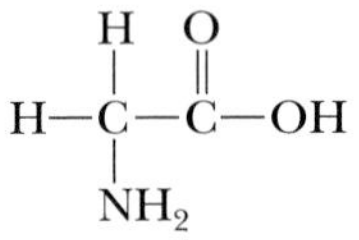

H–C(H)(NH₂)–C(=O)–OH

glycine

Draw the structure of hippuric acid, knowing that it is an amide formed by reaction of the carboxylic acid group of benzoic acid and the amino group of glycine. Why is hippuric acid referred to as an acid?

52. Kevlar is a polyamide made from *p*-phenylenediamine and terephthalic acid. (It is used to make bullet-proof vests, among other things.) Draw the repeating unit of the Kevlar polymer.

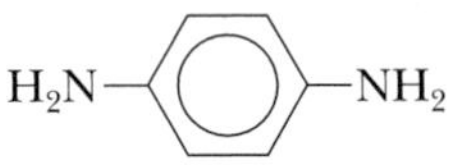

H_2N–C_6H_4–NH_2 HOC(=O)–C_6H_4–C(=O)OH

p-phenylenediamine    terephthalic acid

53. A well-known company selling outdoor clothing has recently introduced jackets made of recycled polyethylene terephthalate (PET), the principal material in many soft drink bottles. Another company makes new PET fibers by treating recycled bottles with methanol to give the diester dimethylterephthalate and ethylene glycol and then repolymerizes these compounds to give new PET. Write a chemical equation to show how the reaction of PET with methanol can give dimethylterephthalate and ethylene glycol.

These students are wearing jackets made of recycled PET soda bottles. *(C. D. Winters)*

54. Draw the structure of glyceryl trilaurate. When this triester is saponified, what are the products?

55. Write a chemical equation describing the reaction between glycerol and stearic acid to give glyceryl tristearate.

CONCEPTUAL QUESTIONS

56. One of the resonance structures for pyridine is illustrated here. Draw another resonance structure for the molecule. Comment on the relationship between this compound and benzene.

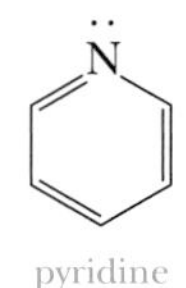

pyridine

57. Write balanced equations for the combustion of ethane and ethanol. Which compound has the more negative enthalpy change for combustion? If ethanol is assumed to be partially oxidized ethane, what effect does this have on the heat of combustion?

58. Describe a simple chemical test to tell the difference between $CH_3CH_2CH_2CH{=}CH_2$ and its isomer cyclopentane.

59. Describe a simple chemical test to tell the difference between 2-propanol and its isomer methyl ethyl ether.

60. Consider *cis*- and *trans*-2-butene and compare these compounds with butane.

(a) *Trans*-2-butene can be converted to *cis*-2-butene by twisting the molecule around the C=C bond. A considerable expenditure of energy is required, however (approximately 264 kJ/mol).

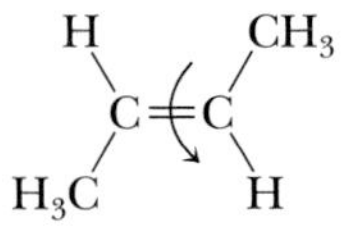

H(H₃C)C=C(CH₃)H

Explain why energy is required for this bond rotation.

(b) It was stated in the text that free rotation occurs around carbon–carbon single bonds. Thus, one end of the butane molecule can rotate relative to the other end around the bond joining the middle two C atoms.

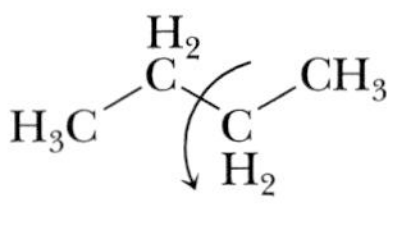

H_3C–CH_2–CH_2–CH_3

An expenditure of energy is in fact required (about 3 kJ/mol), although much smaller than in the case of *trans*-2 butene. Inspect the space-filling model of butane on page 488, and suggest a reason for the fact that rotation around the middle C—C bond requires some energy.

61. Plastics make up about 20% of the volume of landfills. There is, therefore, considerable interest in reusing or recycling these materials. To identify common plastics, a set of universal symbols is now used, five of which are illustrated here.

They symbolize low- and high-density polyethylene, polyvinyl chloride, polypropylene, and polyethylene terephthalate.

PETE

HDPE

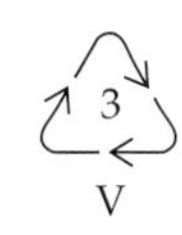

V

LDPE

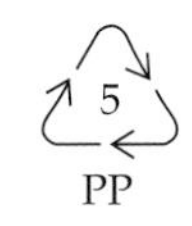

PP

(a) Tell which symbol belongs to which type of plastic.
(b) Find an item in the grocery or drug store made from each of these plastics.
(c) Some properties of several plastics are listed in the table. Based on this information, describe how to separate samples of these plastics from one another.

Plastic	Density (g/cm^3)	Melting Point (°C)
Polypropylene	0.92	170
High-density polyethylene	0.97	135
Poly(ethylene terephthalate)	1.34–1.39	245

Summary Question

62. Maleic acid is prepared by the catalytic oxidation of benzene. It is a dicarboxylic acid, that is, it has two carboxylic acid groups.
(a) Combustion of 0.125 g of the acid gives 0.190 g of CO_2 and 0.0388 g of H_2O. Calculate the empirical formula of the acid.
(b) A 0.261-g sample of the acid requires 34.60 mL of 0.130 M NaOH for complete titration (so that the H^+ ions from both carboxylic acid groups are used). What is the molecular formula of the acid?
(c) Draw a Lewis structure for the acid.
(d) Describe the hybridization used by the C atoms.
(e) What are the bond angles around each C atom?

INTERACTIVE GENERAL CHEMISTRY CD-ROM

Screen 11.2 Carbon–Carbon Bonds

(a) Benzene, C_6H_6, is a planar molecule (Screen 10.7SB). Here we see that another six-carbon cyclic molecule, cyclohexane (C_6H_{12}), is not planar.
1. Contrast the carbon atom hybridization in these two molecules.
2. Why is π electron delocalization possible in benzene?
3. Why is cyclohexane not planar with π electrons delocalized over the ring? Use valence bond theory (with hybrid atomic orbitals) to explain this difference.

(b) What is the main difference between carbon and another Group 4A element, silicon, that explains why carbon can form long chains of atoms, whereas silicon forms only very short chains?

Screen 11.4 Hydrocarbons and Addition Reactions

(a) Why are addition reactions termed "addition?"
(b) Alkenes can be hydrogenated to form alkanes. Alkynes can be hydrogenated as well. What class or classes of hydrocarbons can be formed from the hydrogenation of an alkyne?
(c) Alkanes do not undergo addition reactions. Why not?

Screen 11.5 Functional Groups

Describe the hybrid orbitals used by the indicated atoms:
(a) The O atom in an alcohol
(b) The C═O carbon in an aldehyde
(c) The C═O carbon in a carboxylic acid
(d) The C—O—C oxygen atom in an ester
(e) The N atom in an amine
(f) The N and C═O carbon atoms of an amide

Screen 11.6 Reactions of Alcohols

(a) What is the difference between a substitution reaction and an elimination reaction?
(b) Compare the elimination reaction shown on this screen with the hydrogenation reaction shown on Screen 11.4. In what ways are they similar or dissimilar?

Screen 11.7 Fats and Oils

(a) What type of reaction is used to make a fat or oil from glycerol and a fatty acid: addition, substitution, or elimination?
(b) What is the primary structural difference between fats and oils? What types of functional groups do each contain? What

class of hydrocarbon fragments does each contain?

(c) What structural feature of oil molecules prevents them from coiling up on themselves as fat molecules do?

Screen 11.8 Amino Acids and Proteins

(a) What two functional groups do amino acids contain?

(b) The functional group in a peptide, —NHCO—, is planar. Use hybrid orbitals and the concept of resonance to explain why this can be the case.

(c) What type of reaction is the condensation reaction shown in the animation on this screen: addition, substitution, or elimination?

(d) The peptide shown in the animation is composed of three amino acids. Could this "tripeptide" be further reacted with one or more additional amino acid molecules?

(e) Draw the tripeptide that can be formed by linking together three amino acids called alanine:

```
     H   H  :O:
     |   |   ‖   ..
 H—N—C—C—O—H
   ..    |       ..
        CH3
```

alanine

Screen 11.9 Addition Polymerization

(a) What is the primary structural feature of the molecules used to form addition polymers?

(b) Consider the animation of a polymerization reaction shown on this screen. The polymer made here has a chain of 14 carbon atoms. Could the chain have been shorter than this or longer than this?

(c) What controls the length of the polymer chains formed?

(d) Can the addition polymerization reaction be classified as one of the reaction types we studied earlier: addition, substitution, or elimination?

Screen 11.10 Condensation Polymerization

(a) What is the primary structural feature necessary for a molecule to be useful in a condensation polymerization reaction?

(b) Describe the appearance of the nylon being made in this video.

(c) What does the designation "6,6" mean in nylon-6,6?

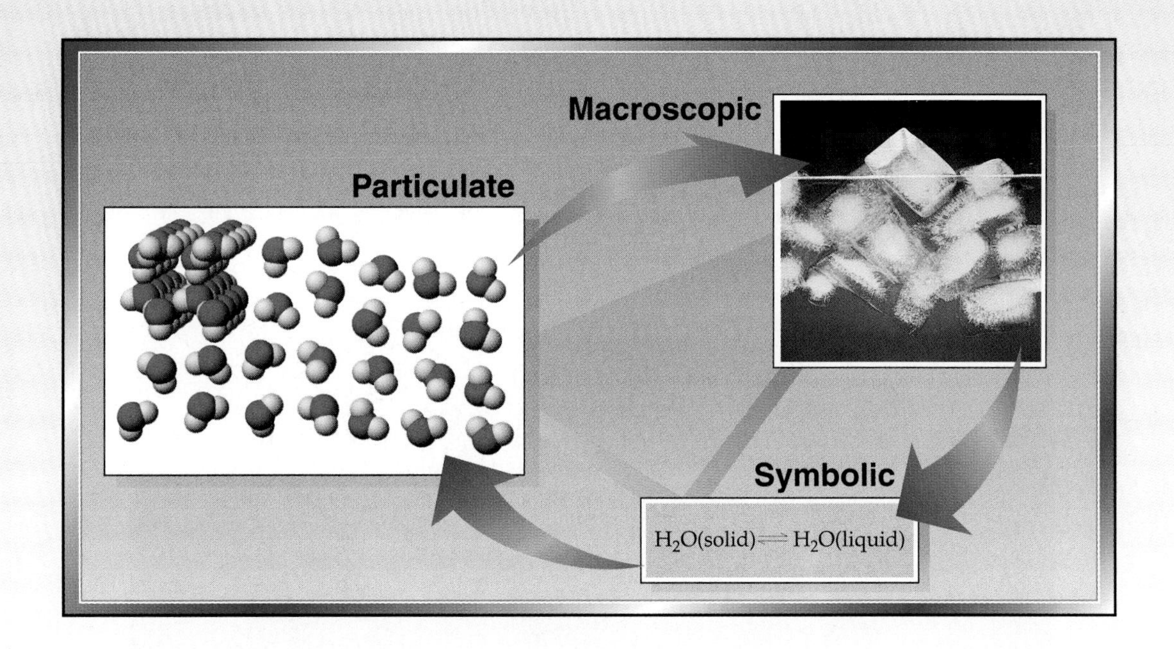

Part 3

States of Matter

The chemical elements and their compounds can exist as gases, liquids, or solutions. The study of gases (Chapter 12) is useful because their behavior can be explained and predicted on the basis of simple mathematical models. When the forces between gaseous atoms or molecules become significant, liquids or solids can form (Chapter 13). Intermolecular forces are examined in Chapter 13, with particular attention given to the forces between water molecules in liquid and solid water. In contrast with liquids, solids often have beautifully regular structures, and we can relate their structures to their molecular formulas. Finally, gases, liquids, and solids can form intimate mixtures or solutions (Chapter 14). Examples are the carbon dioxide that gives the bubbles to champagne, the salt in the sea, or the mixtures of elements that we know as semiconductors in our computers and calculators. • *(Photo, C. D. Winters; model, S. M. Young)*

Chapter 12

Gases and Their Properties

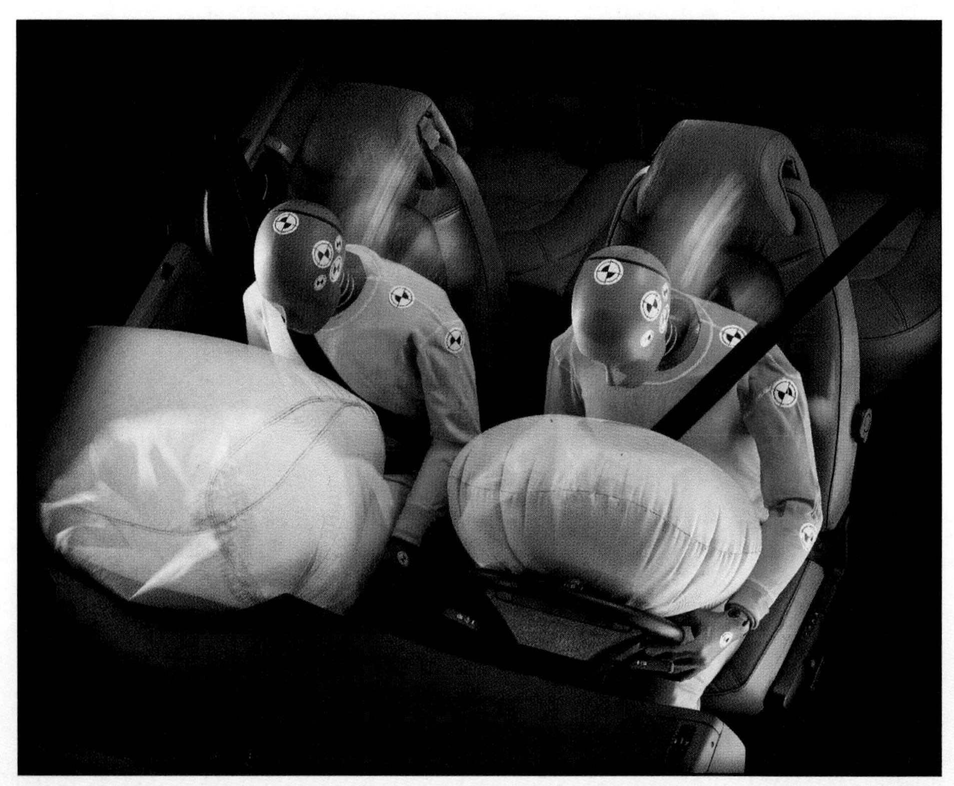

(Saab Cars, USA, Inc.)

A Chemical Puzzler

Automobile air bags, or "supplemental restraint systems (SRS)," depend on some basic chemistry. When a device in your car senses that the car has decelerated very rapidly, your seat belts are tightened, a chemical explodes inside a bag hidden in the steering wheel or dashboard (or both), and the bag is rapidly inflated. Your life is saved by the expansion of a gas. What is the chemistry of air bags? What gas is involved? Does the operation of the air bag depend on temperature? If the air in the car is very cold, as on a winter day in Wisconsin, is the bag effective? What if the accident happens high on a mountain road where the pressure of the atmosphere is much less than at sea level? Is the bag more or less effective?

The Chemical Puzzler poses a number of questions about the operation of the air bag, a safety device now common in most automobiles. In the event of an accident, the air bag is rapidly inflated. Although several chemical reactions occur, the main gas-producing reaction is

$$2\ NaN_3(s) \longrightarrow 2\ Na(s) + 3\ N_2(g)$$

The nitrogen fills the air bag, and your forward momentum is slowed by your seat belt and cushioned by the air bag. If your air bag has a volume of 65 L, how much sodium azide, NaN_3, is required to generate that much gas? Does the amount of sodium azide required to produce 65 L of gas depend on the temperature of the gas? These are two of many questions you may have about how this device works. A number of them can be answered by a study of the properties of gases.

Aside from understanding automobile air bags, there are at least three reasons for studying gases. First, some common elements and compounds exist in the gaseous state under normal conditions of pressure and temperature (Table 12.1). Furthermore, many common liquids can be vaporized, and the properties of these vapors are important. Second, our gaseous atmosphere provides one means of transferring energy and material throughout the globe, and it is the source of life-giving chemicals.

The third reason for studying gases is among the most compelling, however. Of the three states of matter, gases are by far the simplest when viewed at the molecular level and, as a result, gas behavior is well understood. It is possible to describe the properties of gases qualitatively in terms of the behavior of the molecules that make up the gas. Even more impressive, it is possible to describe the properties of gases quantitatively using simple mathematical models. One objective of scientists is to develop precise models of natural phenomena, and a study of gas behavior will introduce you to this approach.

A **gas** is a substance that is normally in the gaseous state at ordinary pressures and temperatures. A **vapor** is the gaseous form of a substance that is normally a liquid or solid at ordinary pressures and temperatures. Thus, we often speak of helium gas and water vapor.

12.1 THE PROPERTIES OF GASES

To describe gases, chemists have learned that only four quantities are needed: pressure, volume, temperature, and the quantity of gas. Let us examine the first of these, the concept of pressure and its units.

Gas Pressure

If you blow up a balloon, the rubber skin stretches and becomes taut. But too much air will cause the rubber skin to break and the air to escape with explosive force. The tightness of the balloon's skin is caused by the force of the gas molecules striking the surface inside the balloon. The force per unit area is the pressure of the gas.

Atmospheric pressure can be measured with a barometer, which can be made by filling a tube with a liquid and inverting it in a dish containing the same liquid. Figure 12.1 shows a mercury-filled barometer. At sea level, the height of the mercury column is about 760 mm above the surface of the mercury in the dish. The pressure exerted by a column of mercury is balanced by the pressure at the bottom of a column of air above the dish—a column of gas that extends to the top of the atmosphere. Pressure is usually reported in units of **millimeters of**

Table 12.1 • Some Common Gaseous Elements and Compounds (at 1 atm pressure and 25 °C)

He	CO
Ne	CO_2
Ar	CH_4
Kr	C_2H_4
Xe	C_3H_8
H_2	HF
O_2	HCl
O_3(ozone)	HBr
N_2	HI
F_2	NO
Cl_2	N_2O
SO_2	H_2S

A simple hand pump can create a vacuum at the top of a well casing. Atmospheric pressure on the water at the bottom of a well causes the water to rise upward. When the water reaches the surface, it flows out. This kind of pump is able to raise water about 10 m, the height of a column of water supported by atmospheric pressure. *(C. D. Winters)*

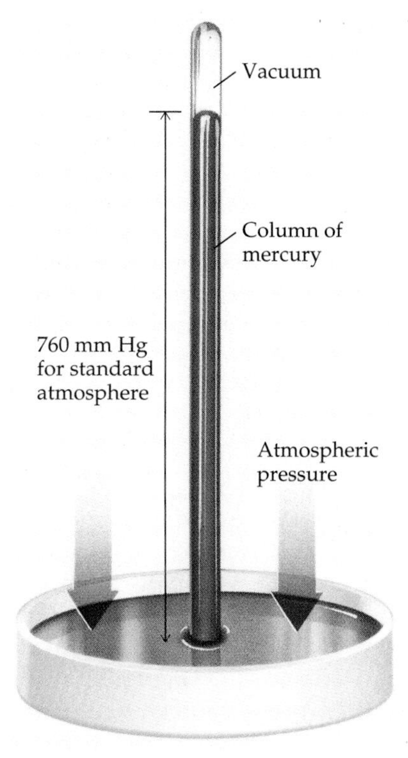

Figure 12.1 A mercury barometer. The pressure of the atmosphere on the surface of the mercury in the dish is balanced by the downward pressure exerted by the column of mercury.

mercury (mm Hg), a unit sometimes called the *torr* in honor of Evangelista Torricelli (1608–1647), who invented the mercury barometer in 1643.

Pressures are also reported as standard atmospheres (atm), a unit defined as follows:

$$1 \text{ standard atmosphere} = 1 \text{ atm} = 760 \text{ mm Hg (exactly)}$$

See the *Saunders Interactive General Chemistry CD-ROM,* Screen 12.2, Gas Pressure.

The SI unit of pressure is the **pascal (Pa),** named for the French mathematician and philosopher Blaise Pascal (1623–1662). It is the only pressure unit that is defined directly in terms of force per unit area:

$$1 \text{ pascal (Pa)} = 1 \text{ newton/meter}^2$$

Because this is a very small unit compared with ordinary pressures, the unit kilopascal (kPa) is more often used. The relationship among these four units is

$$1 \text{ atm} = 760 \text{ mm Hg} = 101.325 \times 10^3 \text{ Pa} = 101.325 \text{ kPa}$$

The *newton* is the SI unit of force.

$$1 \text{ N} = 1 \text{ kg} \cdot \text{m/s}^2$$
$$1 \text{ J} = 1 \text{ N} \cdot \text{m}$$

See page 244.

Finally, atmospheric pressures are sometimes reported in the unit called the **bar,** where 1 bar = 100,000 Pa. Therefore,

$$1 \text{ atm} = 1.013 \text{ bar}$$

The thermodynamic data in Chapter 6 and in Appendix L are given for gas pressures of 1 bar.

A CLOSER LOOK *Measuring Gas Pressure*

Pressure is defined as the force exerted on an object divided by the area over which the force is exerted:

$$\text{Pressure} = \frac{\text{force}}{\text{area}}$$

This book, for example, weighs more than 4 lb and has an area of 82 in.^{2}, so it exerts a pressure of about 0.05 lb/in.^{2} when it lies flat on a surface. (In metric units, the pressure is about 3 g/cm^2.)

Consider the pressure that the column of mercury exerts on the mercury in the dish in a barometer shown in Figure 12.1. This pressure is, by definition, the mass of the mercury column divided by the cross-sectional area of the tube. The mass of mercury is the volume times the density of mercury, 13.53 g/cm^3 (at 25 °C), and the cylinder of mercury has a volume equal to the cross-sectional area of the tube times the height of the column. Combining these relationships in a mathematical expression gives

$$\text{Pressure} = \frac{\text{mass of mercury column}}{\text{cross-sectional area}}$$

$$= \frac{\text{density} \cdot \text{cross-sectional area} \cdot \text{column height}}{\text{cross-sectional area}}$$

$$= \text{density} \cdot \text{column height}$$

The pressure exerted by the column of mercury, which is proportional to the height of the column and the density of mercury, exactly balances the atmospheric pressure outside the beaker. Thus, it is possible to measure atmospheric pressure (or the pressure of any other gas) by measuring the height of the column of mercury it can support.

Mercury is the liquid of choice for barometers because of its high density, but any liquid can be used. The height of the column of liquid in a barometer depends on the density of the liquid. A barometer filled with water would be over 10 m in height. The water column is about 13.6 times as high as a column of mercury because mercury's density (13.53 g/cm^3) is about 13.6 times that of water (density = 0.997 g/cm^3) (at 25 °C).

In the laboratory, gas pressures are often measured with a U-tube manometer. The closed side of a mercury-filled, U-shaped glass tube is evacuated. No gas remains to exert pressure on the mercury surface on that side. The other side is open to the gas whose pressure is to be measured. The gas pressure in mm Hg is read directly as the difference in mercury levels in the closed and open sides.

You may have used a tire gauge to check the pressure in your car or bike tires. This gauge indicates the pressure in pounds per square inch (psi), and some newer ones give the pressure in kilopascals as well. The reading on the scale refers to the pressure *in excess of atmospheric pressure.* (A flat tire is not a vacuum; it contains air at atmospheric pressure.) A tire containing 3.4 atm of air has a gauge pressure of 2.4 atm; the tire gauge would show 35 psi (2.4 atm × 14.7 psi).

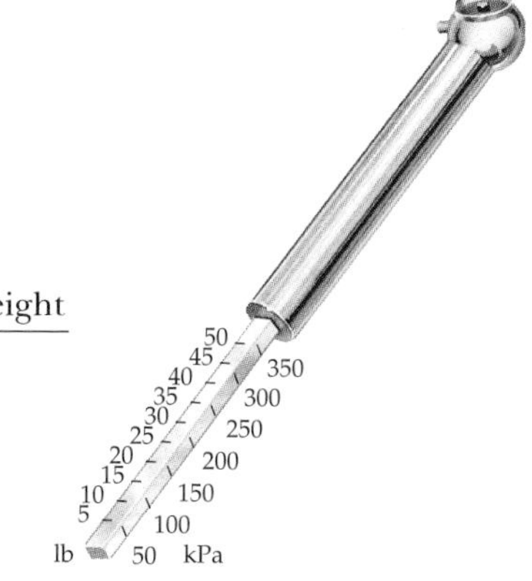

A tire gauge. The "lb" scale measures pressure in pounds per square inch (psi).

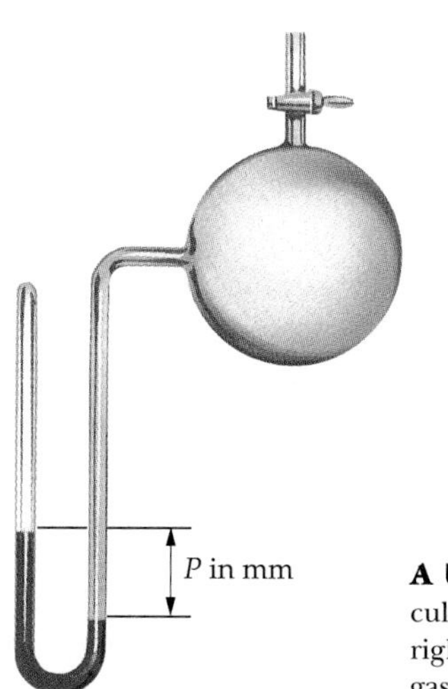

A U-tube manometer. The left side is evacuated; no gas molecules are pressing on the liquid (usually mercury) surface. On the right side, molecules of the sample gas are exerting pressure. The gas pressure is measured by the difference in liquid levels.

Example 12.1 Pressure Unit Conversions

Convert a pressure of 635 mm Hg into its corresponding value in units of atmospheres (atm), bars, and kilopascals (kPa).

Solution The relation between millimeters of mercury and atmospheres is 1 atm = 760 mm Hg. Notice that the given pressure is less than 760 mm Hg, that is, less than 1 atm:

$$635 \text{ mm Hg} \left(\frac{1.00 \text{ atm}}{760. \text{ mm Hg}}\right) = 0.836 \text{ atm}$$

Pressure Conversions:
1 atm = 760 mm Hg (exactly)
= 760 torr (after Torricelli)
= 101.3 kilopascals (kPa)
= 1.013 bar
= 14.7 lb/in.^{2} (psi)
and
1 bar = 1×10^5 Pa (exactly)

The relationship between atmospheres and bars is 1 atm = 1.013 bar. We have

$$0.836 \text{ atm} \left(\frac{1.013 \text{ bar}}{1 \text{ atm}}\right) = 0.847 \text{ bar}$$

The factor relating units of millimeters of mercury and kilopascals is 101.325 kPa = 760 mm Hg. Therefore,

$$635 \text{ mm Hg} \left(\frac{101.325 \text{ kPa}}{760. \text{ mm Hg}}\right) = 84.7 \text{ kPa}$$

Exercise 12.2 Pressure Unit Conversions

Rank the following pressures in decreasing order of magnitude (largest first, smallest last): 75 kPa, 250 mm Hg, 0.83 bar, and 0.63 atm.

12.2 GAS LAWS: THE EXPERIMENTAL BASIS

See the *Saunders Interactive General Chemistry CD-ROM,* Screen 12.3, The Gas Laws.

Experimentation in the 17th and 18th centuries led to three gas laws that provide the basis for understanding gas behavior.

The Compressibility of Gases: Boyle's Law

When you pump up the tires of your bicycle, the pump squeezes the air into a smaller volume (Figure 12.2). This property of a gas is called its **compressibility.** While studying the compressibility of gases, Robert Boyle (1627–1691) observed that the volume of a fixed amount of gas at a given temperature is inversely proportional to the pressure exerted by the gas. All gases behave in this manner, and we now refer to this relationship as **Boyle's law.**

Figure **12.2 A bicycle pump.** This works by compressing air into a smaller volume. You experience Boyle's law because you can feel the increasing pressure of the gas as you press down on the plunger. *(C. D. Winters)*

Boyle's law can be demonstrated in many ways (Figures 12.3 and 12.4). In Figure 12.3 a hypodermic syringe is filled with air and sealed. When pressure is applied to the movable plunger of the syringe, the air inside is compressed. As the pressure increases, the gas volume decreases. When the pressure of the gas in the syringe is plotted as a function of $1/V$, a straight line results. This type of plot demonstrates that the pressure and volume of the gas are inversely proportional; that is, they change in opposite directions. Mathematically, we can write this in two ways (where the symbol $\propto$ means "proportional to"):

$$P \propto \frac{1}{V} \quad \text{or equivalently} \quad V \propto \frac{1}{P}$$

Both show that, for a given quantity of gas at a fixed temperature, the gas volume decreases if the pressure increases. Conversely, if the pressure is lowered, then the gas volume increases.

Boyle's experimentally determined relationship can be put into a useful mathematical form as follows. When two quantities are proportional to each other, they can be equated if a proportionality constant, here called C_B, is introduced. Thus,

$$P = C_B \cdot \frac{1}{V} \quad \text{or} \quad PV = C_B \qquad [12.1]$$

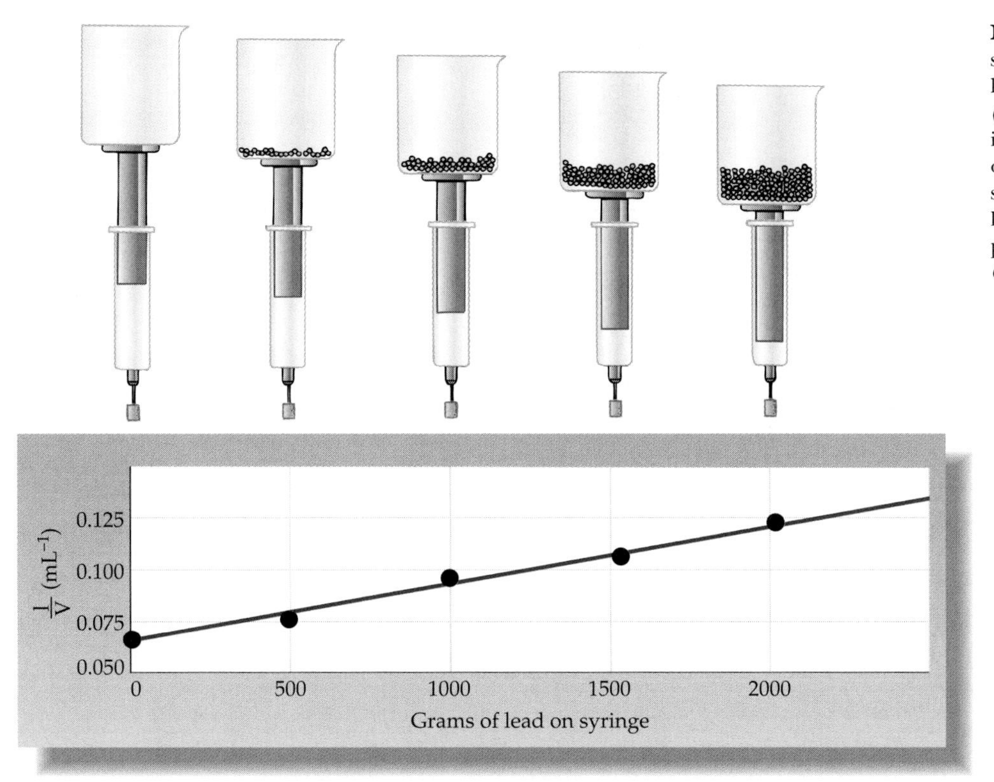

Figure 12.3 Demonstrating Boyle's law. A syringe containing air is sealed and then lead shot is added to the beaker at the top *(top)*. As the mass of lead increases, the air in the syringe is compressed *(bottom)*. A plot of $1/V$ versus the mass of lead gives a straight line, showing that $1/V$ and P (related to the mass of lead) are directly proportional. (See D. Davenport: *Journal of Chemical Education,* Vol. 39, p. 252, 1962.)

This form of Boyle's law expresses the fact that the product of the pressure and volume of a gas sample is a constant at a given temperature, where the constant C_B is determined by the quantity of gas (in moles) and its temperature (in kelvins). This means that if the pressure–volume product is known for one set of conditions (P_1 and V_1), then it is known for another set of conditions (P_2 and V_2). Under either set of conditions, the PV product is equal to C_B, so

$$P_1V_1 = P_2V_2$$

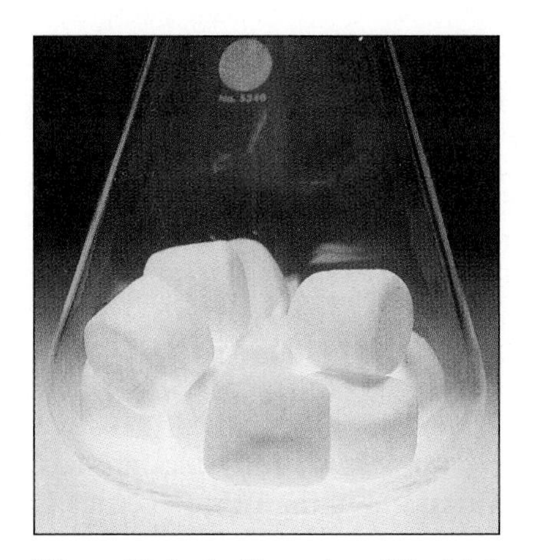

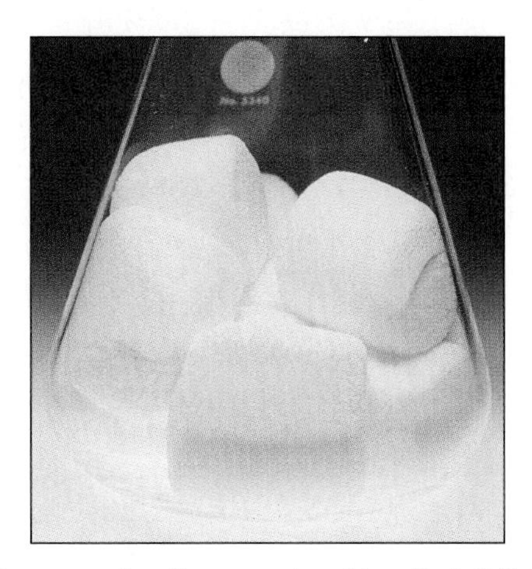

Figure 12.4 An illustration of Boyle's law. Some marshmallows are placed in a flask *(left)*. The flask is then evacuated (the air is withdrawn from the flask using a vacuum pump). The air in the marshmallows expands as the pressure is lowered, causing the marshmallows to expand *(right)*. *(C. D. Winters)*

Historical Perspective

Robert Boyle (1627–1691)

Robert Boyle was born in Ireland, in a home that still stands, as the 14th and last child of the first Earl of Cork. He first published his studies of gases in 1660, and his book, *The Sceptical Chymist,* was published in 1680. Boyle was the first to define elements in modern terms, although he had medieval views about what elements were. For example, Boyle thought that gold was not an element, but rather a metal that could be formed from other metals. Boyle was also a physiologist—he was the first to show that the healthy human body has a constant temperature. Not everyone applauded Boyle's work. Isaac Newton, a young man when Boyle was at the peak of his career, questioned the correctness of Boyle's ideas.

Robert Boyle (Oesper Collection in the History of Chemistry/University of Cincinnati)

This form of Boyle's law is useful when we want to know, for example, what happens to the volume of a given quantity of gas (at a given T) when the pressure changes.

Example 12.2 Boyle's Law

A sample of gaseous nitrogen in a 65.0-L automobile air bag has a pressure of 745 mm Hg. If this sample is transferred to a 25.0-L bag with the same temperature as before, what is the pressure of the gas in the new bag?

Solution It is often useful to make a table of the information provided.

Original Conditions	Final Conditions
$P_1 = 745$ mm Hg	$P_2 = ?$
$V_1 = 65.0$ L	$V_2 = 25.0$ L

You know that $P_1V_1 = P_2V_2$. Therefore,

$$P_2 = \left(\frac{P_1V_1}{V_2}\right) = \frac{(745 \text{ mm Hg})(65.0 \not{\text{L}})}{(25.0 \not{\text{L}})} = 1940 \text{ mm Hg}$$

Remember that the essence of Boyle's law is that P and V *change in opposite directions.* Because you know that the volume has decreased, you know the new pressure (P_2) must be greater than the original pressure (P_1); thus, P_1 must be multiplied by a volume factor that has a value *greater than 1* to reflect the fact that P_2 must be greater than P_1. Indeed, simply understanding Boyle's law leads us to the equation

$$P_2 = P_1 \cdot \left(\frac{65.0 \not{\text{L}}}{25.0 \not{\text{L}}}\right) = 1940 \text{ mm Hg}$$

Exercise 12.2 Boyle's Law

A sample of CO_2 has a pressure of 55 mm Hg in a volume of 125 mL. The sample is compressed so that the new pressure of the gas is 78 mm Hg. What is the new volume of the gas? (The temperature does not change in this process.)

The Effect of Temperature on Gas Volume: Charles's Law

In 1787, the French scientist Jacques Charles (1746–1823) discovered that the volume of a fixed quantity of gas at constant pressure increased with increasing temperature. An extreme example of Charles's law is shown in Figure 12.5.

Figure 12.6 illustrates how the volumes of two different gas samples change with temperature (at a constant pressure). When the plots of volume versus temperature are extended toward lower temperatures, they all reach zero volume at a common temperature, −273.15 °C. Of course gases will not actually reach zero volume; they liquefy well above that temperature.

In 1848, William Thomson (1824–1907), also known as Lord Kelvin, proposed a temperature scale in which the zero point was −273.15 °C. This temperature scale has been named for Lord Kelvin, and the units of the scale are known as kelvins. The kelvin, which has been adopted as the SI unit for temperature measurement, is equivalent in size to the Celsius degree. When the Kelvin temperature scale is used, the volume–temperature relationship, now known as **Charles's law,** is as follows: If a given quantity of gas is held at a constant pressure, its volume is directly proportional to the Kelvin temperature. This can be expressed by the relation

The Kelvin temperature scale was introduced on page 32.

$$V \propto T$$

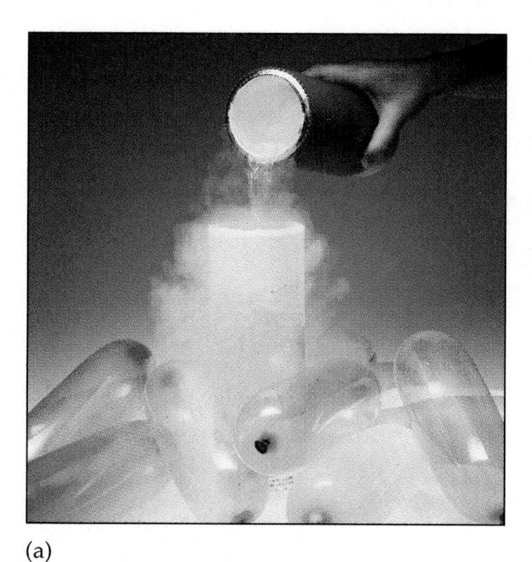

(a)

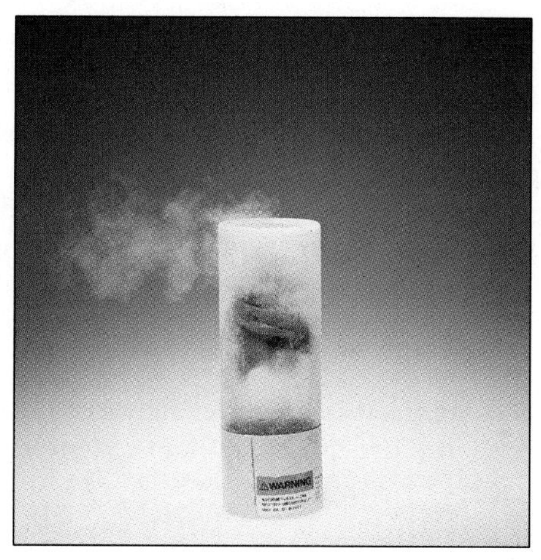

(b)

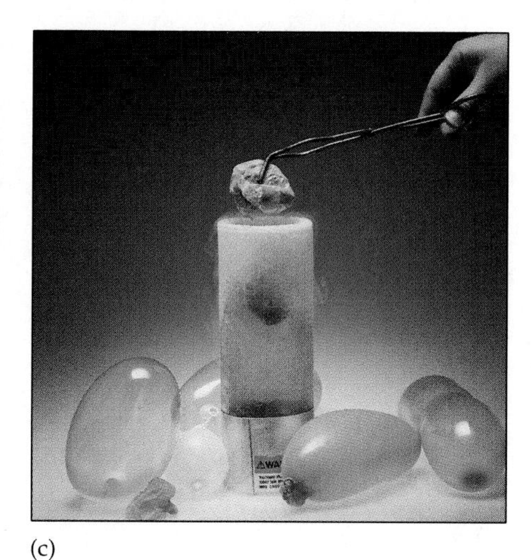

(c)

***Figure* 12.5 A dramatic illustration of Charles's law.** (a) Air-filled balloons are placed in liquid nitrogen (77 K). The volume of the gas in the balloons is dramatically reduced at this temperature. (b) After all of the balloons have been placed in the liquid nitrogen, (c) they are removed; as they warm to room temperature they reinflate to their original volume. *(C. D. Winters)*

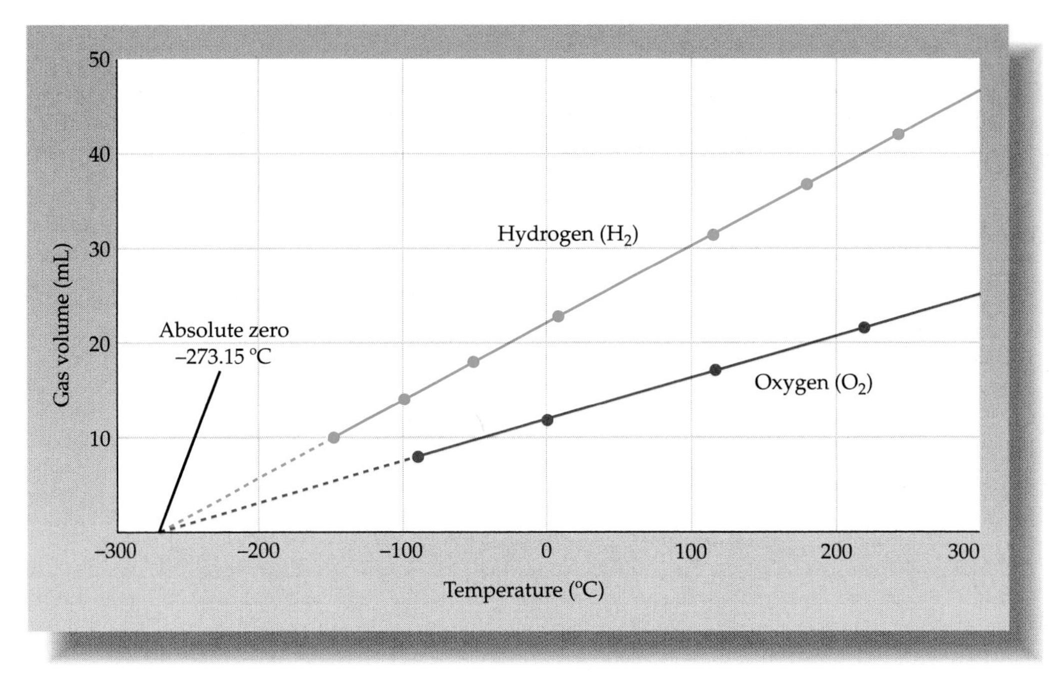

Figure 12.6 Charles's law. The solid lines represent the volumes of the samples of hydrogen and oxygen at different temperatures. The volumes decrease as the temperature is lowered (constant pressure). These lines, if extended, intersect the temperature axis at approximately −273 °C.

or, using a proportionality constant, C_C, we can write the equation

$$V = C_C \cdot T$$

or

$$C_C = \frac{V}{T} \quad [12.2]$$

The quotient V/T always has the same value for a given sample of gas at a specified pressure. Therefore, if we know the volume and temperature of a given quantity of gas (V_1 and T_1), we can find the volume, V_2, at some other temperature, T_2, using the equation

$$\frac{V_1}{T_1} = \frac{V_2}{T_2}$$

Calculations using Charles's law are illustrated by the following example and exercise. Be sure to notice that the temperature T *must always be expressed in kelvins.*

Example 12.3 Charles's Law

Suppose you have a sample of CO_2 in a gas-tight syringe (as in Figure 12.3). The gas volume is 25.0 mL at room temperature (20.0 °C). What is the final volume of the gas if you hold the syringe in your hand to raise its temperature to 37 °C?

Historical Perspective

***Jacques Alexandre César Charles* (1746–1823)**

The French chemist Jacques Charles was most famous in his lifetime for his experiments in ballooning. The first such flights were made by the Montgolfier brothers in June 1783, using a large spherical balloon made of linen and paper and filled with hot air. In August 1783, however, another group supervised by Jacques Charles tried a different approach. Exploiting his recent discoveries in the study of gases, Charles decided to inflate the balloon with hydrogen gas. Because hydrogen would escape easily from a paper bag, Charles made a bag of silk coated with a rubber solution. Inflating the bag to its final diameter took several days and required nearly 500 lb of acid and 1000 lb of iron. (Iron and sulfuric acid, for example, produce H_2 gas and Iron (II) sulfate.) A huge crowd watched the ascent on August 27, 1783. The balloon stayed aloft for almost 45 minutes and traveled about 15 miles. When it landed in a village, however, the people were so terrified that they tore it to shreds.

Jacques Charles (The Bettmann Archive)

Solution To organize the information, construct a table. Notice that the temperature *must* be converted to kelvins.

Original Conditions	Final Conditions
$V_1 = 25.0$ mL	$V_2 = ?$
$T_1 = 20.0 + 273 = 293$ K	$T_2 = 37 + 273 = 310.$ K

You know that $V_1/T_1 = V_2/T_2$. Substitute the known quantities into this equation and solve for V_2:

$$V_2 = T_2 \cdot \left(\frac{V_1}{T_1}\right) = (310.\ \text{K}) \cdot \left(\frac{25.0\ \text{mL}}{293\ \text{K}}\right) = 26.5\ \text{mL}$$

As expected the volume of the gas increased with a temperature increase. As in Example 12.2, this again suggests a logical approach to the problem. The new volume (V_2) must equal the original volume (V_1) multiplied by a temperature fraction that is greater than 1 to reflect the effect of the temperature increase. That is,

$$V_2 = V_1 \cdot \left(\frac{310.\ \text{K}}{293\ \text{K}}\right)$$

Exercise 12.3 Charles's Law

A balloon is inflated with helium to a volume of 45 L at room temperature (25 °C). If the balloon is cooled to −10 °C, what is the new volume of the balloon? Assume that the pressure does not change.

Combining Boyle's and Charles's Laws: The General Gas Law

Neither Boyle's law nor Charles's law depends on the identity of the gas being studied. These laws describe the behavior of any gaseous substance, regardless of its identity.

The volume of a gas is inversely proportional to its pressure at constant temperature (Boyle's law) and directly proportional to the Kelvin temperature at constant pressure (Charles's law). But what if we need to know what happens to a given amount of gas when two of the three parameters (P, V, and T) change? For example, what would happen to the pressure of a sample of nitrogen in an automobile air bag if the gas is placed in a smaller bag and heated to a higher temperature? You can deal with this kind of problem by combining the two equations that express Boyle's and Charles's laws.

$$\frac{P_1 V_1}{T_1} = \frac{P_2 V_2}{T_2} \qquad [12.3]$$

This equation is sometimes called the **general gas law,** or **combined gas law.** It applies specifically to situations in which the *quantity of gas does not change.* The next example and exercise illustrate its use.

A weather balloon is filled with helium. As it ascends into the troposphere, does the volume increase or decrease? *(NASA/Science Source/Photo Researchers, Inc.)*

Example 12.4 The General Gas Law

Helium-filled balloons are used to carry scientific instruments high into the atmosphere. Suppose that a balloon is launched when the temperature is 22.5 °C and the barometric pressure is 754 mm Hg. If the balloon's volume is 4.19×10^3 L (and no helium escapes from the balloon), what will the volume be at a height of 20 miles, where the pressure is 76.0 mm Hg and the temperature is −33.0 °C?

Solution It is a good idea to begin by setting out the information given in a table.

Initial Conditions	Final Conditions
$V_1 = 4.19 \times 10^3$ L	$V_2 = ?$ L
$P_1 = 754$ mm Hg	$P_2 = 76.0$ mm Hg
$T_1 = 22.5$ °C (295.7 K)	$T_2 = -33.0$ °C (240.2 K)

We can rearrange the general gas law (Equation 12.3) to calculate the new volume V_2:

$$V_2 = \left(\frac{T_2}{P_2}\right) \cdot \left(\frac{P_1 V_1}{T_1}\right) = V_1 \cdot \frac{P_1}{P_2} \cdot \frac{T_2}{T_1}$$

$$= 4.19 \times 10^3 \text{ L} \cdot \frac{754 \text{ mm Hg}}{76.0 \text{ mm Hg}} \cdot \frac{240.2 \text{ K}}{295.7 \text{ K}} = 3.38 \times 10^4 \text{ L}$$

The pressure decreased by almost a factor of 10, which should lead to about a 10-fold volume increase. This increase is partly offset by a drop in temperature that leads to a volume decrease. On balance, the volume increases because the pressure has dropped so substantially.

Exercise 12.4 The General Gas Law

You have a 22.-L cylinder of helium at a pressure of 150 atm and at 31 °C. How many balloons can you fill, each with a volume of 5.0 L, on a day when the atmospheric pressure is 755 mm Hg and the temperature is 22 °C?

As a final comment on the gas laws, notice that, for a given quantity of gas in a constant volume, the gas pressure is directly proportional to the absolute temperature:

$$P_2 = P_1 \cdot \frac{T_2}{T_1} \text{ at constant volume}$$

When you drive an automobile for some distance, you know the air pressure in the tires goes up. The reason for this is that friction warms the tires, and, since the tire volume is nearly constant, the pressure must increase.

The Laws of Gay-Lussac and Avogadro

At the beginning of the 19th century, the French chemist Joseph Gay-Lussac (1778–1850) found that the ratio of the volumes of gases in a reaction was always a small whole number, as long as the volumes are measured at the same temperature and pressure. This statement is now referred to as **Gay-Lussac's law of combining volumes.** As an example, this means that 100 mL of H_2 gas combines exactly with 50 mL of O_2 gas to give exactly 100 mL of H_2O vapor if all the gases are measured at the same T and P (Figure 12.7).

Gay-Lussac's law remained only a summary of experimental observations until it was explained by the Italian physicist and lawyer, Amedeo Avogadro (1776–1856). In 1811, Avogadro published his ideas, now known as **Avogadro's hypothesis,** that equal volumes of gases under the same conditions of temperature and pressure have equal numbers of molecules. Applied to the example above, this means that 100 mL of H_2 molecules in the gas phase must have twice the number of molecules as 50 mL of O_2, and the reaction between the gases should produce 100 mL of gaseous H_2O molecules.

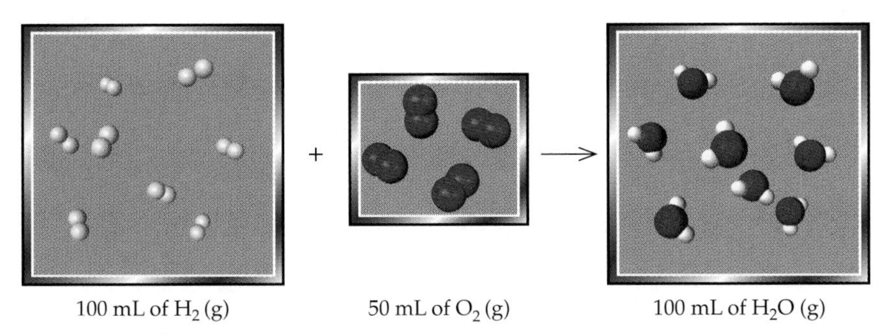

***Figure* 12.7 Illustration of Gay-Lussac's law.** This law relates volumes of gases in chemical reactions. This is understood based on Avogadro's hypothesis that the number of gas molecules in a sample is proportional to the volume.

A balloon inflates because as you blow into it you are introducing a greater and greater mass of air. The pressure also increases because as the balloon inflates, the skin of the balloon is stretched. *(C. D. Winters)*

Avogadro's law follows from Avogadro's hypothesis: The volume of a gas, at a given temperature and pressure, is directly proportional to the quantity of gas. Thus, V is proportional to n, the number of moles of gas, so

$$V \propto n \quad \text{or} \quad V = C_A \cdot n \qquad [12.4]$$

where C_A is the proportionality constant.

Example 12.5 Avogadro's Law

Ammonia can be made directly from the elements:

$$N_2(g) + 3\ H_2(g) \longrightarrow 2\ NH_3(g)$$

If you begin with 15.0 L of $H_2(g)$, what volume of $N_2(g)$ is required for complete reaction (both gases being at the same T and P)? What is the theoretical yield of NH_3, in liters, under the same conditions?

Solution Because gas volume is proportional to the number of moles of a gas, we can use volumes instead of moles and treat this just as we did stoichiometry problems.

Calculate the volumes of N_2 required and NH_3 produced (in liters) by multiplying the volume of H_2 available by a stoichiometric factor (also in units of liters) obtained from the chemical equation:

$$V(N_2 \text{ required}) = 15.0\ \cancel{L\ H_2} \text{ available} \cdot \frac{1\ L\ N_2}{3\ \cancel{L\ H_2}} = 5.00\ L\ N_2$$

$$V(NH_3 \text{ produced}) = 15.0\ \cancel{L\ H_2} \text{ available} \cdot \frac{2\ L\ NH_3}{3\ \cancel{L\ H_2}} = 10.0\ L\ NH_3$$

Exercise 12.5 Avogadro's Law

Methane burns in oxygen to give CO_2 and H_2O, according to the equation

$$CH_4(g) + 2\ O_2(g) \longrightarrow CO_2(g) + 2\ H_2O(g)$$

If 22.4 L of gaseous CH_4 is burned, what volume of O_2 is required for complete combustion? What volumes of CO_2 and H_2O are produced? Assume all gases have the same temperature and pressure.

12.3 THE IDEAL GAS LAW

See the *Saunders Interactive General Chemistry CD-ROM,* Screen 12.4, The Ideal Gas Law.

Four interrelated quantities can be used to describe a gas: pressure, volume, temperature, and quantity (moles). We know from experiments that three gas laws can be used to describe the relationship of these properties (Section 12.2).

Boyle's Law	Charles's Law	Avogadro's Law
$V \propto (1/P)$ (constant T, n)	$V \propto T$ (constant P, n)	$V \propto n$ (constant T, P)

If all three laws are combined, the result is

$$V \propto \frac{nT}{P}$$

This can be made into a mathematical equation by introducing a proportionality constant, labeled ***R*,** called the **gas constant.** The gas constant is a "universal constant," a number that you can use to interrelate the properties of any gas:

$$V = R\left(\frac{nT}{P}\right)$$

or

$$PV = nRT \qquad [12.5]$$

The equation $PV = nRT$ is called the **ideal gas law.** It describes the behavior of a so-called "ideal" gas. As you will learn in Section 12.8, there is no such thing as an "ideal" gas. However, real gases at pressures around an atmosphere or less and temperatures around room temperature usually behave close enough to ideality that $PV = nRT$ adequately describes their behavior.

To use the equation $PV = nRT$, we need a value for R. This is readily determined experimentally. By carefully measuring P, V, n, and T for a sample of gas we can calculate the value of R from these values using the ideal gas law equation. For example, under conditions of **standard temperature and pressure (STP)** (a gas temperature of 0 °C or 273.15 K and a pressure of 1 atm) 1 mol of gas occupies 22.414 L, a quantity called the **standard molar volume.** Substituting these values into the ideal gas law gives a value for R:

A gas is at **STP,** or **standard pressure and temperature,** when its pressure is 1 atm and its temperature is 0 °C or 273.15 K. Under these conditions, exactly 1 mol of a gas occupies 22.414 L.

$$R = \frac{PV}{nT} = \frac{(1.0000 \text{ atm})(22.414 \text{ L})}{(1.0000 \text{ mol})(273.15 \text{ K})} = 0.082057 \frac{\text{L} \cdot \text{atm}}{\text{K} \cdot \text{mol}}$$

With a value for R, we can now use the ideal gas law to perform some useful calculations.

Example 12.6 The Ideal Gas Law

The nitrogen gas in an air bag, with a volume of 65 L, exerts a pressure of 829 mm Hg at 25 °C. What quantity of N_2 gas (in moles) is in the air bag?

Solution First, let us write down the information provided.

$$V = 65 \text{ L} \qquad P = 829 \text{ mm Hg} \qquad T = 25 \text{ °C} \qquad n = ?$$

To use the ideal gas law with R having units of L · atm/K · mol), the pressure must be expressed in atmospheres and the temperature in kelvins. Therefore,

$$P = 829 \text{ mm Hg}\left(\frac{1 \text{ atm}}{760 \text{ mm Hg}}\right) = 1.09 \text{ atm}$$

$$T = 25 + 273 = 298 \text{ K}$$

Now substitute the values of P, V, T and R into this equation, and solve for the number of moles, n,

$$n = \frac{PV}{RT} = \frac{(1.09 \text{ atm})(65 \text{ L})}{(0.082057 \text{ L} \cdot \text{atm/K} \cdot \text{mol})(298 \text{ K})} = 2.9 \text{ mol } N_2$$

Notice that units of atmospheres, liters, and kelvins cancel to leave the answer in units of moles.

Exercise 12.6 The Ideal Gas Law

The balloon used by Charles in his historic flight in 1783 (see Historical Perspective, p. 547) was filled with about 1300 mol of H_2. If the temperature of the gas was 23 °C, and its pressure was 750 mm Hg, what was the volume of the balloon?

The Density of Gases

See the *Saunders Interactive General Chemistry CD-ROM,* Screen 12.5, Gas Density.

The density of a gas at a given temperature and pressure (Figure 12.8) is a useful quantity. Let us see how this is related to the ideal gas law. Because the number of moles (n) of any compound is given by its mass (m) divided by its molar mass (M), we can substitute m/M for n in the ideal gas equation.

$$PV = \left(\frac{m}{M}\right) RT$$

Density (d) is defined as mass divided by volume (m/V); therefore, if we rearrange this equation so that m/V appears on one side,

$$d = \frac{m}{V} = \frac{PM}{RT} \qquad [12.6]$$

we find the density of a gas is related to the pressure and temperature of the gas and to its molar mass. This fact is useful because a measurement of the density of a gas at a given pressure and temperature can be used to calculate its molar mass.

Figure 12.8 Gas density. The two balloons are filled with nearly equal quantities of gas at the same temperature and pressure. The blue balloon contains low-density hydrogen ($d = 0.090$ g/L), and the red balloon contains argon, a higher density gas ($d = 1.8$ g/L). In comparison, the density of dry air at 1 atm pressure and 25 °C is about 1.2 g/L. *(C. D. Winters)*

Example 12.7 Gas Density and Molar Mass

The density of an unknown gas is 1.23 g/L at STP. Calculate its molar mass.

Solution List the information given in a short table, recalling that "STP" is shorthand for standard conditions, a pressure of 1 atm and a temperature of 0 °C or 273.15 K:

$d = 1.23$ g/L

$T =$ standard temperature $= 0$ °C $= 273.15$ K

$P =$ standard pressure $= 1.00$ atm

$R = 0.082057$ L · atm/K · mol

$M = ?$

These values are substituted into the equation for gas density (Equation 12.6), which is then solved for molar mass (M):

$$M = \frac{dRT}{P} = \frac{(1.23 \text{ g/L})(0.082057 \text{ L} \cdot \text{atm/K} \cdot \text{mol})(273.15 \text{ K})}{1.00 \text{ atm}} = 27.6 \text{ g/mol}$$

Exercise 12.7 Gas Density Calculation

Calculate the density of dry air at 15.0 °C and 1.00 atm if its molar mass (average) is 28.96 g/mol.

Gas density has some practical implications. Among other things it is important to a hot-air balloonist (Figure 12.9). Perhaps you have seen hot air balloons floating across the sky in the early morning or evening. When the air inside the balloon is heated (usually with a propane heater), it expands. Rather than increasing the volume of the balloon, however, air is forced from the inside of the balloon. The result is that less air remains inside. The lower mass of air in the same balloon volume means that the gas inside the balloon has a lower density. Just as a balloon filled with hydrogen or helium will rise, so will a balloon filled with hot air.

Another way to look at explaining why a hot air balloon can rise starts with the equation for the density of a gas, $d = PM/RT$. From this you see that the density is inversely proportional to temperature. Raising the temperature of a gas results in a decrease in density. (This assumes the pressure remains the same in the balloon. This must be the case because gas can escape from the balloon, keeping the pressure inside equal to the pressure outside the balloon.)

From the equation $d = PM/RT$ we also see that the density of a gas is directly proportional to its molar mass. Dry air, which has an average molar mass of about 29 g/mol, has a density of about 1.2 g/L at 1 atm and 25 °C. Gases or vapors with molar masses greater than 29 g/mol have densities larger than 1.2 g/L under these same conditions (1 atm and 25 °C). Gases such as CO_2, SO_2, and gasoline vapor settle along the ground if released into the atmosphere. Conversely, gases such as H_2, He, CO, CH_4 (methane), and NH_3 rise if released into the atmosphere (Figure 12.10).

Figure 12.9 Hot air balloons. A hot air balloon rises because the heated air has a lower density. *(Greg Galowski/Dembinsky Photo Associates)*

The value for the average molar mass for dry air falls between the values for the primary components of the atmosphere, N_2 (28 g/mol) and O_2 (32 g/mol). The value is nearer 28 than 32 because nitrogen is the more abundant gas.

Figure 12.10 A fire extinguisher uses carbon dioxide. Carbon dioxide from fire extinguishers is denser than air, so it settles on top of a fire and smothers it. (When CO_2 gas is released from the tank, it expands and cools significantly. The white cloud is solid CO_2 and condensed moisture from the air.) *(C. D. Winters)*

The significance of gas density was tragically revealed in several recent events. One such instance occurred in the African country of Cameroon in 1984 when Lake Nyos expelled a huge bubble of CO_2 into the atmosphere. Because CO_2 is denser than air, the CO_2 cloud hugged the ground, killing 1700 people living in a nearby village. On December 3, 1984, in Bhopal, India, the accidental release of methyl isocyanate (CH_3NCO) vapor into the air from a chemical plant killed several hundred people and injured thousands more. The vapors of this toxic chemical are heavier than air and thus settled close to the ground, where they were inhaled.

Calculating the Molar Mass of a Gas from *P*, *V*, and *T* Data

The molar mass of a gas is now usually obtained by mass spectroscopy. See page 73.

When a new compound is isolated in the laboratory, one of the first things to be done is to determine its molar mass. If the compound is in the gas phase, a classical method of determining the molar mass is to measure the pressure and volume exerted by a given mass of the gas at a given temperature.

Example 12.8 Calculating the Molar Mass of a Gas from *P*, *V*, and *T* Data

See the *Saunders Interactive General Chemistry CD-ROM,* Screen 12.6, Using Gas Laws—Molar Mass.

Suppose you have done an experiment to determine the empirical formula of a compound now used to replace CFCs in air conditioners. Your results give a formula of CHF_2. Now you need the molar mass of the compound to find the molecular formula. You therefore do another experiment and find that a 0.100-g sample of the compound exerts a pressure of 70.5 mm Hg in a 256-mL container at 22.3 °C. What is the molar mass of the compound? What is its formula?

Solution Let us begin by organizing the data:

$$V = 256 \text{ mL, or } 0.256 \text{ L}$$

$$T = 22.3 \text{ °C, or } 295.5 \text{ K}$$

$$P = 70.5 \text{ mm Hg}\left(\frac{1 \text{ atm}}{760 \text{ mm Hg}}\right) = 0.0928 \text{ atm}$$

We can find the molar mass of the gas by first using the ideal gas law to calculate n, the number of moles equivalent to 0.100 g of the gas:

$$n = \frac{PV}{RT} = \frac{(0.0928 \text{ atm})(0.256 \text{ L})}{(0.082057 \text{ L} \cdot \text{atm/K} \cdot \text{mol})(295.5 \text{ K})} = 9.80 \times 10^{-4} \text{ mol}$$

Now you know that 0.100 g of gas is equivalent to 9.80×10^{-4} mol. Therefore,

$$\text{Molar mass} = \frac{0.100 \text{ g}}{9.80 \times 10^{-4} \text{ mol}} = 102 \text{ g/mol}$$

With this result, we can compare the experimentally determined molar mass with the mass of a mole of gas having the empirical formula CHF_2.

$$\frac{\text{Experimental molar mass}}{\text{Mass of 1 mol of } CHF_2} = \frac{102 \text{ g/mol}}{51.0 \text{ g/formula unit}} = 2 \text{ formula units of } CHF_2 \text{ per mole}$$

Therefore, the formula of the compound is $C_2H_2F_4$.

You may wish to use an alternative approach to calculate this formula. Recall that the density of a gas is related to the molar mass, temperature, and pressure by Equation 12.6. The density of the gas in question is the mass of the gas divided by the volume:

$$d = \frac{0.100 \text{ g}}{0.256 \text{ L}} = 0.391 \text{ g/L}$$

Use this value of density along with the values of pressure and temperature in the equation $d = PM/RT$ and solve for the molar mass (M):

$$M = \frac{dRT}{P} = \frac{(0.391 \text{ g/}\cancel{\text{L}})(0.082057 \cancel{\text{L}} \cdot \cancel{\text{atm}}/\cancel{\text{K}} \cdot \text{mol})(295.5 \cancel{\text{K}})}{0.0928 \cancel{\text{atm}}} = 102 \text{ g/mol}$$

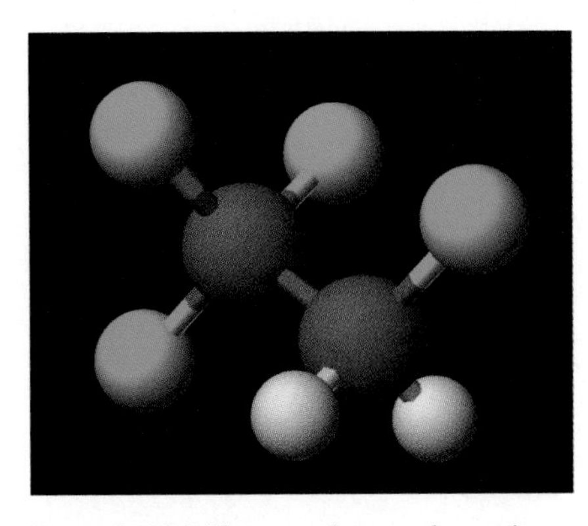

Example 12.8 illustrates how to determine the molar mass of a gas such as the compound known as HCFC-134, a fluorocarbon now being used in home and automobile air conditioners.

Exercise 12.8 Molar Mass from P, V, and T Data

A 0.105-g sample of a gaseous compound has a pressure of 561 mm Hg in a volume of 125 mL at 23.0 °C. What is its molar mass?

12.4 GAS LAWS AND CHEMICAL REACTIONS

See the *Saunders Interactive General Chemistry CD-ROM,* Screen 12.7, Gas Laws and Chemical Reactions.

Many important chemical reactions, such as the industrial production of ammonia or chlorine, involve gases:

$$N_2(g) + 3\ H_2(g) \longrightarrow 2\ NH_3(g)$$

$$2\ NaCl(aq) + 2\ H_2O \longrightarrow 2\ NaOH(aq) + H_2(g) + Cl_2(g)$$

It is important to deal with such reactions quantitatively. The examples that follow explore stoichiometry calculations involving gases in more detail. The scheme in Figure 12.11 connects these calculations for gas reactions with calculations done in Chapters 4 and 5.

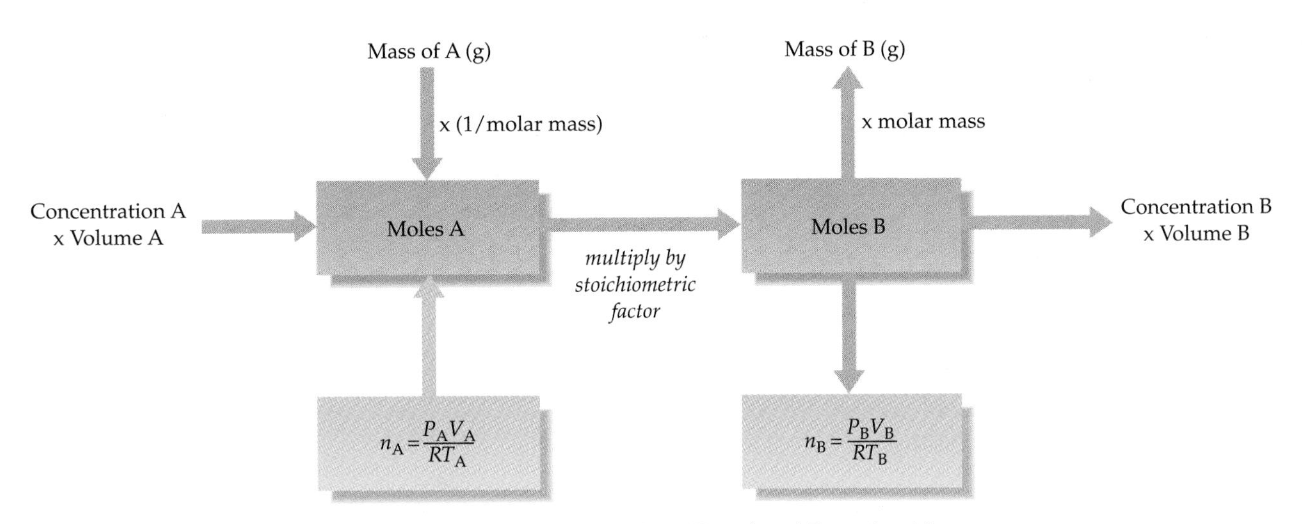

Figure 12.11 A scheme for performing stoichiometry calculations. Here A and B may be either reactants or products. The number of moles of A can be calculated from its mass in grams, from P, V, and T data by using the ideal gas law, or from the concentration and volume of a solution. Once the number of moles of B is determined, this value can be converted to mass or solution concentration or volume, or to a volume of gas at a given pressure and temperature.

Example 12.9 Gas Laws and Stoichiometry

Let us say you are asked to design an air bag for a car. You know that the bag should be filled with gas with a pressure higher than atmospheric pressure, say 829 mm Hg, at a temperature of 22.0 °C. The bag has a volume of 45.5 L. What quantity of sodium azide, NaN_3, should be used to generate the required quantity of gas? The gas-producing reaction is

$$2\ NaN_3(s) \longrightarrow 2\ Na(s) + 3\ N_2(g)$$

Solution The general logic to be followed here can be outlined as follows:

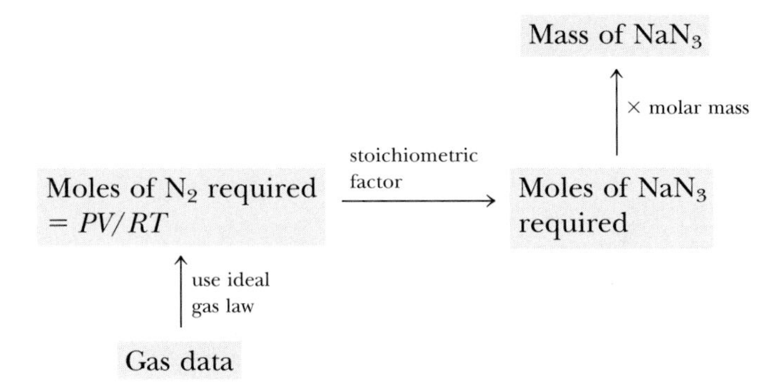

The first step is to find the number of moles of gas required so that this can be related to the quantity of sodium azide required:

$$P = 829 \text{ mmHg, or } 1.09 \text{ atm}$$
$$V = 45.5 \text{ L}$$
$$T = 22.0 \text{ °C, or } 295.2 \text{ K}$$

$$n = \text{moles of } N_2 \text{ required} = \frac{(1.09\ \cancel{\text{atm}})(45.5\ \cancel{\text{L}})}{(0.082057\ \cancel{\text{L}} \cdot \cancel{\text{atm}}/\cancel{\text{K}} \cdot \text{mol})(295.2\ \cancel{\text{K}})}$$
$$= 2.05 \text{ mol } N_2$$

Now that the required quantity of nitrogen has been calculated, we can calculate the quantity of sodium azide that will produce 2.05 mol of N_2 gas.

$$\text{Mass of } NaN_3 = 2.05\ \cancel{\text{mol } N_2}\left(\frac{2\ \cancel{\text{mol } NaN_3}}{3\ \cancel{\text{mol } N_2}}\right)\left(\frac{65.01 \text{ g}}{1\ \cancel{\text{mol } NaN_3}}\right) = 88.8 \text{ g } NaN_3$$

Lithium reacting with water produces lithium hydroxide, LiOH, and hydrogen gas. *(C. D. Winters)*

Example 12.10 Gas Laws and Stoichiometry

Suppose we wish to prepare some deuterium gas, D_2, for use in an experiment. One way to do this is to react heavy water, D_2O, with an active metal such as lithium.

$$2\ Li(s) + 2\ D_2O(\ell) \longrightarrow 2\ LiOD(aq) + D_2(g)$$

If we place 0.125 g of Li metal in 15.0 mL of D_2O ($d = 1.11$ g/mL), what quantity of D_2 (in moles) can be prepared? If dry D_2 gas is captured in a 1450-mL flask at 22.0 °C, what is the pressure of the gas in mm Hg? (Deuterium has a molar mass of 2.0147 g/mol.)

Solution We are combining two reactants with no guarantee that they are in the correct stoichiometric ratio. This reaction must therefore be approached as a *limiting reactant problem.* We shall have to find the number of moles of each substance and then see if one of them is present in a limited amount. Once the limiting reactant is known, we can find the quantity of D_2 produced and then calculate its pressure under the conditions given.

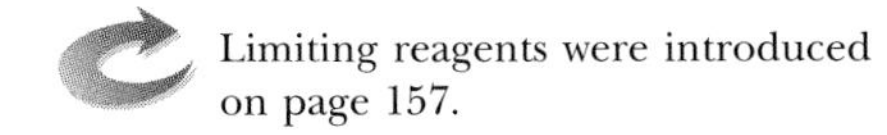

Limiting reagents were introduced on page 157.

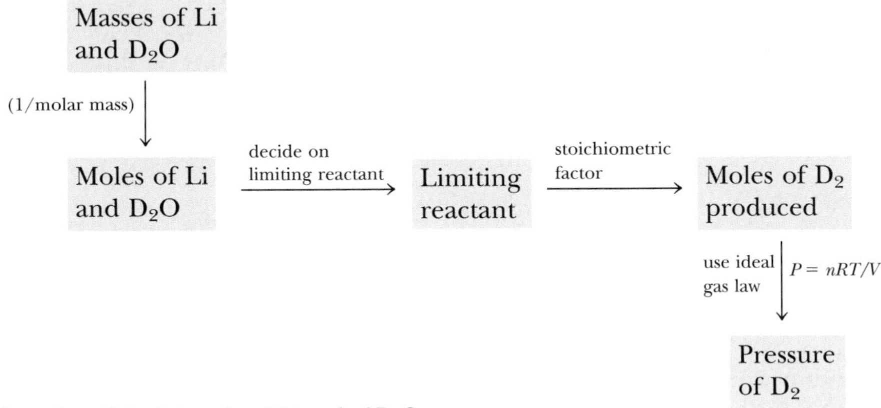

Step 1. *Calculate moles of Li and of D_2O:*

$$0.125 \text{ g Li}\left(\frac{1 \text{ mol Li}}{6.941 \text{ g Li}}\right) = 0.0180 \text{ mol Li}$$

$$15.0 \text{ mL } D_2O\left(\frac{1.11 \text{ g } D_2O}{\text{mL } D_2O}\right)\left(\frac{1 \text{ mol } D_2O}{20.03 \text{ g } D_2O}\right) = 0.831 \text{ mol } D_2O$$

Step 2. *Decide which reactant is the limiting reactant:*

$$\text{Ratio of moles of reactants available} = \frac{0.831 \text{ mol } D_2O}{0.0180 \text{ mol Li}} = \frac{46.2 \text{ mol } D_2O}{1 \text{ mol Li}}$$

The balanced equation shows that the ratio should be 1 mol of D_2O to 1 mol of Li. From the calculated values, we can see that D_2O is in large excess and Li is the limiting reactant. Therefore, further calculations are based on the moles of Li available.

Step 3. *Use the limiting reactant to calculate the quantity of D_2 produced:*

$$0.0180 \text{ mol Li}\left(\frac{1 \text{ mol } D_2 \text{ produced}}{2 \text{ mol Li}}\right) = 0.00900 \text{ mol } D_2 \text{ produced}$$

Step 4. *Calculate the pressure of D_2:*

$$P = ?$$
$$V = 1450 \text{ mL, or } 1.45 \text{ L}$$
$$T = 22.0 \text{ °C, or } 295.2 \text{ K}$$
$$n = 0.00900 \text{ mol } D_2$$

$$P = \frac{nRT}{V} = \frac{(0.00900 \text{ mol})(0.082057 \text{ L} \cdot \text{atm/K} \cdot \text{mol})(295.2 \text{ K})}{1.45 \text{ L}} = 0.150 \text{ atm}$$

$$0.150 \text{ atm}\left(\frac{760 \text{ mm Hg}}{1 \text{ atm}}\right) = 114 \text{ mm Hg}$$

Exercise 12.9 Gas Laws and Stoichiometry

Gaseous ammonia is synthesized by the reaction

$$N_2(g) + 3\ H_2(g) \xrightarrow[500\ °C]{\text{iron catalyst}} 2\ NH_3(g)$$

Assume that you take 355 L of H_2 gas at 25.0 °C and 542 mm Hg and combine it with excess N_2 gas. What quantity of NH_3 gas, in moles, is produced? If this amount of NH_3 gas is stored in a 125-L tank at 25.0 °C, what is the pressure of the gas?

Exercise 12.10 Gas Laws and Stoichiometry

Oxygen reacts with aqueous hydrazine to produce water and gaseous nitrogen according to the balanced equation

$$N_2H_4(aq) + O_2(g) \longrightarrow 2\ H_2O(\ell) + N_2(g)$$

If a solution contains 180 g of N_2H_4, what is the maximum volume of O_2 that will react with the hydrazine if the oxygen is measured at a pressure of 750 mm Hg and a temperature of 21 °C?

12.5 GAS MIXTURES AND PARTIAL PRESSURES

See the *Saunders Interactive General Chemistry CD-ROM,* Screen 12.8, Gas Mixtures and Partial Pressures.

The air you breathe is a blend of nitrogen, oxygen, carbon dioxide, water vapor, and small amounts of other gases (Table 12.2). Each of these gases is exerting pressure, and atmospheric pressure is the sum of the pressures exerted by each individual gas. The pressure of each gas in the mixture is called its **partial pressure.**

John Dalton (page 61) was the first to observe that the pressure of a mixture of gases is the sum of the partial pressures of the different components of the mixture. This observation is now known as **Dalton's law of partial pressures** (Figure 12.12). Mathematically, we can write Dalton's law of partial pressures as

$$P_{total} = P_1 + P_2 + P_3 + \cdots \quad [12.7]$$

where P_1, P_2, and P_3 are the pressures of the different gases in a mixture and P_{total} is the total pressure.

The pressure of a gas in a given volume and at a specific temperature depends only on the quantity of gas. There is no reference in the ideal gas law to the identity of the gas. It is also true that each gas behaves independently of any others in a mixture of ideal gases. In a gas mixture, therefore, we can consider the behavior of each gas separately. As an example let us take a mixture of three ideal gases, labeled A, B, and C. There are n_A moles of A, n_B moles of B, and n_C moles of C. Assume that the mixture ($n_{total} = n_A + n_B + n_C$) is contained in a given volume *(V)* at a given temperature *(T)*. We can calculate the pressure exerted by each from the ideal gas law equation:

$$P_AV = n_ART \qquad P_BV = n_BRT \qquad P_CV = n_CRT$$

Table 12.2 • Components of Atmospheric Dry Air

Constituent	Molar Mass*	Mole Percent	Partial Pressure at STP (atm)
N_2	28.01	78.08	0.7808
O_2	32.00	20.95	0.2095
CO_2	44.01	0.033	0.00033
Ar	39.95	0.934	0.00934

*The average molar mass of dry air = 28.960 g/mol.

According to Dalton's law, the total pressure exerted by the mixture is the sum of the pressures exerted by each component:

$$P_{\text{total}} = P_A + P_B + P_C = n_A\left(\frac{RT}{V}\right) + n_B\left(\frac{RT}{V}\right) + n_c\left(\frac{RT}{V}\right)$$

$$= (n_A + n_B + n_C)\left(\frac{RT}{V}\right)$$

$$P_{\text{total}} = n_{\text{total}}\left(\frac{RT}{V}\right)$$

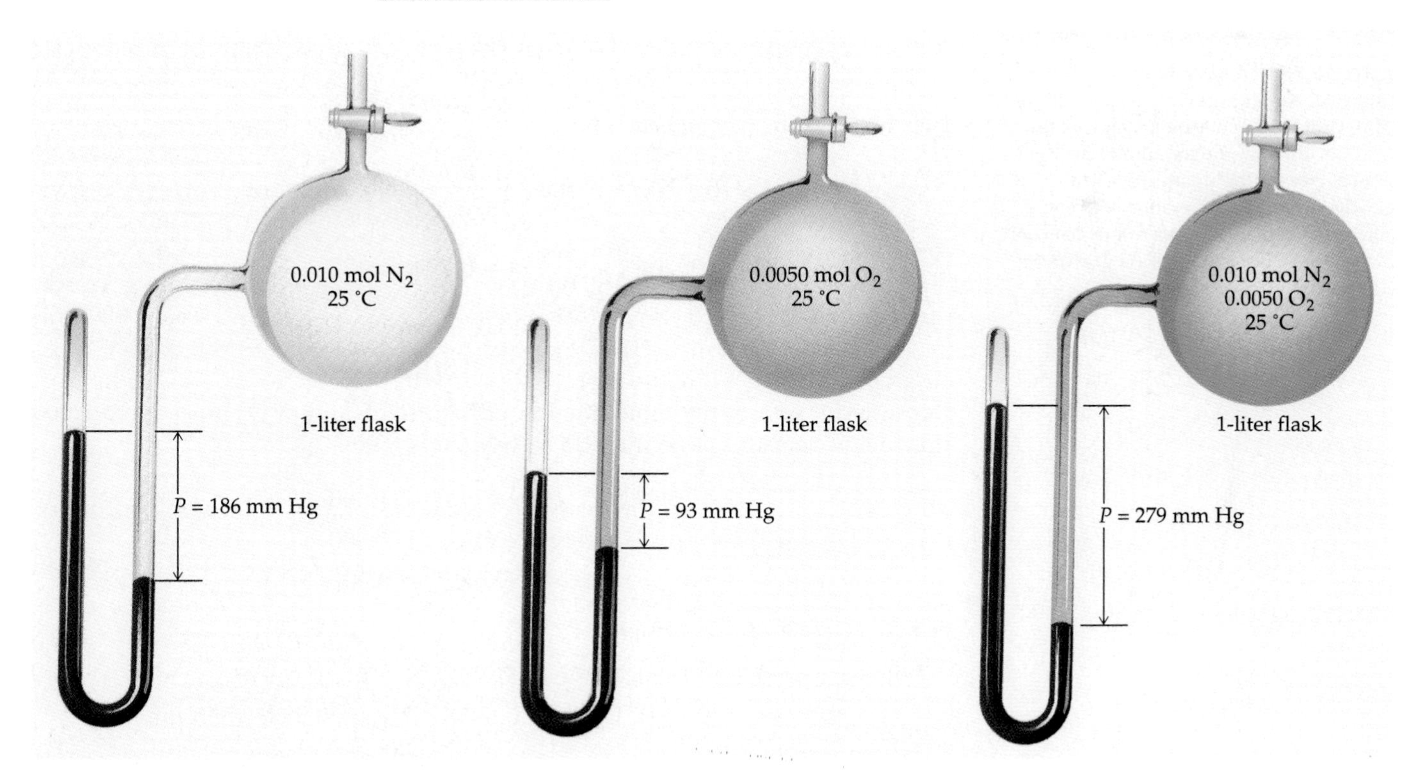

Figure 12.12 **Dalton's law.** In a 1.0-L flask at 25 °C, 0.010 mol of N_2 exerts a pressure of 186 mm Hg, and 0.0050 mol of O_2 in a 1.0-L flask at 25 °C exerts a pressure of 93 mm Hg *(left* and *middle).* The N_2 and O_2 samples are mixed in a 1.0-L flask at 25 °C *(right).* The total pressure, 279 mm Hg, is the sum of the pressures that each gas alone exerts in the flask.

Notice that the sum of the mole fractions of all components of a mixture is 1.0. In the example on this page, $X_A + X_B + X_C = 1.0$.

For mixtures of gases, it is convenient to introduce a quantity called the **mole fraction,** X, which is defined as the number of moles of a particular substance in a mixture divided by the total number of moles of all substances present. Mathematically, the mole fraction of a substance A in a mixture with B and C is expressed as

$$X_A = \frac{n_A}{n_A + n_B + n_C} = \frac{n_A}{n_{total}}$$

Now we can combine this equation ($n_{total} = n_A/X_A$) with the equations for P_A and P_{total}, and derive the very useful equation

$$P_A = X_A P_{total} \quad [12.8]$$

***Figure* 12.13 A gas-mixing manifold for anesthesia.** An anesthesiologist uses this device to prepare an anesthetic gas mixture. By proper mixing, the anesthetic gas can be added slowly to the breathing mixture. Near the end of the operation, the anesthetic gas is replaced by air of normal composition or by pure oxygen. *(C. D. Winters)*

Example 12.11 Partial Pressures of Gases

Halothane has the formula $C_2HBrClF_3$. It is a nonflammable, nonexplosive, and nonirritating gas that is a commonly used as an inhalation anesthetic (Figure 12.13). Suppose you mix 15.0 g of halothane vapor with 23.5 g of oxygen gas. If the total pressure of the mixture is 855 mm Hg, what is the partial pressure of each gas?

Solution One way to solve this problem is to recognize that the partial pressure of a gas is given by the total pressure of the mixture multiplied by the mole fraction of the gas. We therefore first calculate the mole fractions of halothane and of O_2.

Step 1. *Calculate mole fractions:*

$$\text{Moles } C_2HBrClF_3 = 15.0\,\text{g}\left(\frac{1\text{ mol}}{197.4\text{ g}}\right) = 0.0760\text{ mol}$$

$$\text{Moles } O_2 = 23.5\,\text{g}\left(\frac{1\text{ mol}}{32.00\text{ g}}\right) = 0.734\text{ mol}$$

$$\text{Mole fraction } C_2HBrClF_3 = \frac{0.0760\text{ mol } C_2HBrClF_3}{0.810\text{ total moles}} = 0.0938$$

Because the sum of the mole fraction of halothane and of O_2 must equal 1.000, this means that the mole fraction of oxygen is 0.906.

$$X_{halothane} + X_{oxygen} = 1.000$$
$$0.0938 + X_{oxygen} = 1.000$$
$$X_{oxygen} = 0.906$$

Step 2. *Calculate partial pressures:*

$$\text{Partial pressure of halothane} = P_{halothane} = X_{halothane} \cdot P_{total}$$
$$P_{halothane} = 0.0938 \cdot P_{total} = 0.0938\ (855\text{ mm Hg})$$
$$= 80.2\text{ mm Hg}$$

Because $P_{halothane} + P_{oxygen} = 855$ mm Hg, this means

$$P_{oxygen} = 855\text{ mm Hg} - 80.2\text{ mm Hg} = 775\text{ mm Hg}$$

Figure 12.14 Collecting a gas over water. Hydrogen gas is generated by the reaction of magnesium metal with aqueous HCl. The pressure of the generated gas forces water from inside the inverted beaker. The beaker will contain both hydrogen and water vapor, and the pressure inside the beaker (P_{total}) will equal the sum of the pressure exerted by hydrogen ($P_{hydrogen}$) and water vapor ($P_{water\ vapor}$). *(Photo, C. D. Winters; model, S. M. Young)*

Exercise 12.11 Partial Pressures

The halothane–oxygen mixture described in Example 12.10 is placed in a 5.00-L tank at 25.0 °C. What is the total pressure (in mm Hg) of the gas mixture in the tank? What are the partial pressures (in mm Hg) of the gases?

An application of Dalton's law occurs with laboratory experiments that involve collecting a gas by displacing water from a container (Figure 12.14). Here the total pressure in the collecting flask is the sum of the pressures of the gas collected plus the pressure of the water vapor. Water evaporates and the vapor exerts a pressure.

Vapor pressure is described in Section 13.3. The vapor pressure of water is given at different temperatures in Appendix G.

Example 12.12 Using Partial Pressures—Collecting a Gas over Water

Small quantities of H_2 gas can be prepared in the laboratory by the following reaction:

$$\mathrm{Mg(s)} + 2\ \mathrm{HCl(aq)} \longrightarrow \mathrm{MgCl_2(aq)} + \mathrm{H_2(g)}$$

Assume you carried out this experiment and collected 456 mL of gas as illustrated in Figure 12.14. The temperature of the gas mixture (H_2 + water vapor) is 22.0 °C, and the total pressure of gases in the flask is 742 mm Hg. How many moles of gas (hydrogen + water vapor) are in the flask? How many moles of H_2 did you prepare?

Solution

Step 1. *Calculate the total number of moles of gas:*

P = total pressure = 742 mm Hg = 0.976 atm

V = 0.456 L

T = 22.0 °C, or 295.2 K

n = total moles of gas = moles H_2 + moles H_2O vapor

$$= \frac{PV}{RT} = \frac{(0.976 \text{ atm})(0.456 \text{ L})}{(0.082057 \text{ L} \cdot \text{atm/K} \cdot \text{mol})(295.2 \text{ K})} = 0.0184 \text{ mol gas}$$

Step 2. *Calculate the number of moles of H_2 gas.* The total pressure in the collecting flask, 742 mm Hg, is the sum of the partial pressures of the two gases in the flask:

$$P_{\text{total}} = 742 \text{ mm Hg} = P_{\text{hydrogen}} + P_{\text{water vapor}}$$

To find the number of moles of H_2, you must know the partial pressure of H_2. The partial pressure of water vapor over liquid water can be obtained from Appendix G. Here $P_{\text{water vapor}}$ is 19.8 mm Hg, so P_{hydrogen} = (742 mm Hg − 19.8 mm Hg) = 722 mm Hg (or 0.950 atm). Now you can find the number of moles of this gas in the flask:

$$n_{H_2} = \frac{P_{H_2}V}{RT} = \frac{(0.950 \text{ atm})(0.456 \text{ L})}{(0.082057 \text{ L} \cdot \text{atm/K} \cdot \text{mol})(295.2 \text{ K})} = 0.0179 \text{ mol } H_2$$

Alternatively, you can use the ratio of P_{hydrogen} to the total pressure to find the mole fraction of H_2 and from that the number of moles of H_2:

$$\text{Mole fraction } H_2 = \frac{P_{H_2}}{P_{\text{total}}} = \frac{722 \text{ mm Hg}}{742 \text{ mm Hg}} = 0.973$$

$$\text{Moles } H_2 = X_{\text{hydrogen}} \cdot \text{total moles} = (0.973)(0.0184 \text{ mol}) = 0.0179 \text{ mol } H_2$$

Exercise 12.12 Partial Pressures of Gases

In an experiment similar to that in Figure 12.14, 352 mL of gaseous nitrogen is collected in a flask over water at a temperature of 24.0 °C. The total pressure of the gases in the flask is 742 mm Hg. What mass of N_2 is collected? (See Appendix G for water vapor pressure.)

12.6 THE KINETIC-MOLECULAR THEORY OF GASES

So far we have discussed the macroscopic properties of gases. Now we turn to the kinetic-molecular theory, a description of the behavior of gases at the molecular or atomic level. A qualitative introduction to kinetic-molecular theory was given in Section 1.1, where it was mentioned that this theory applies to liquids and solids as well as gases. In this section we shall discuss its application in describing the behavior of gases.

See the *Saunders Interactive General Chemistry CD-ROM,* Screen 12.9, The Kinetic-Molecular Theory of Gases.

If your friend is wearing a perfume or if a pizza smells good, how do you know it? In scientific terms, we know that molecules of perfume or the odor-causing molecules of food enter the gas phase and drift through space until they reach the cells of your body that react to odors. The same thing happens in the laboratory when bottles of aqueous ammonia and hydrochloric acid sit side by side (Figure 12.15). Molecules of the two compounds enter the gas phase and drift along until they encounter one another, at which time they react and form a cloud of tiny particles of solid ammonium chloride.

If you changed the temperature of the environment of the containers in Figure 12.15 and measured the time needed for the cloud of ammonium chloride to form, you would find that the time would be longer at lower temperatures. The speed at which molecules move depends on the temperature. In fact, it can be shown that the average kinetic energy, $\overline{KE}$, of a collection of gas molecules depends *only* on the Kelvin temperature, a concept expressed by the following relation

$$\overline{KE} \propto T \qquad [12.9]$$

where the horizontal bar over the symbol KE indicates an average value.

The kinetic energy of only one molecule is given by

$$KE = \frac{1}{2}\,(\text{mass})\,(\text{speed})^2 = \frac{1}{2}\,mu^2 \qquad [12.10]$$

where u is the speed of the molecule. There are trillions of molecules in a gas sample, however, and their speeds vary greatly. Mathematically, this situation is described by saying that some have a speed of u_1, some have a different speed u_2, and so on. The *average* speed, $\overline{u}$, of the molecules is therefore

$$\overline{u} = \frac{n_1u_1 + n_2u_2 + \cdots}{N} \qquad [12.11]$$

where n_1 is the number of molecules with the speed u_1, n_2 is the number of molecules with speed n_2, and so on. N is the total number of molecules $(n_1 + n_2 + \cdots)$. This means the *average* kinetic energy of many molecules, $\overline{KE}$, is related to the *average of the squares of their speeds*, $\overline{u^2}$ (called the "mean square speed"), by the equation

$$\overline{KE} = \frac{1}{2}\,m\overline{u^2} \qquad [12.12]$$

Now, because $\overline{KE}$ is proportional to T, $1/2\ m\overline{u^2}$ must be proportional to T as well, and so we can say that

$$\frac{1}{2}\,m\overline{u^2} = CT \qquad [12.13]$$

where C is a proportionality constant. This is a useful relationship and the basis for the discussion that follows.

Two characteristic properties of gases are that they completely occupy the volume available to them and that they are compressible. By contrast, it is difficult to squeeze solids or liquids to force them to occupy a smaller volume. These observations are explained by assuming *the distance between gas particles* (atoms or molecules) *is very large relative to the actual size of the particles* (see Figure 1.2). Compressing a gas forces the gas molecules together.

Gases can also be condensed to form liquids and solids when the temperature is lowered sufficiently (Figure 12.16). The formation of a liquid or solid

Figure 12.15 The movement of gas molecules. Open dishes of aqueous ammonia and hydrochloric acid were placed side by side. When molecules of NH_3 and HCl escape from solution to the atmosphere and encounter one another, solid ammonium chloride, NH_4Cl, is formed. *(C. D. Winters)*

Figure 12.16 Liquid nitrogen. The fact that a large volume of N_2 gas can be condensed to a small volume of liquid indicates that the distance between molecules in the gas phase is large and that there are forces of attraction between N_2 molecules, called intermolecular forces. *(C. D. Winters)*

Current Perspectives

Methane, A Greenhouse Gas

Methane, CH_4, is the simplest hydrocarbon. It is also one of the most important gases within the earth and in the atmosphere. We have already described how methane is found locked up as a hydrated compound at the bottom of the oceans (Chapter 11, page 482) and how methane can be converted to methanol (Chapter 11, page 506).

Methane is an important fuel. About 70% of the natural gas from Texas, for example, is methane, and most is used as a fuel in homes and industry. There are other reasons to be interested in this gas, however. Each year microbes produce about 4×10^9 metric tons of the gas, an enormous mass that has a profound effect on us and our environment.

Methane is produced in anaerobic (without oxygen) environments. These occur inside animals (the rumen of cows or in the guts of insects such as termites), in natural wetlands or rice paddies, or in human-made sites such as landfills and sewage digestors. As more and more people inhabit the earth, however, the methane content of the atmosphere increases, the increase being about 1%/year over the past decade.

Among questions of interest to scientists is the way that methane is generated in animals. Very recently the structure of methyl-coenzyme reductase, a key enzyme common to all methane-producing pathways, was revealed. (An enzyme is a biological catalyst, a substance that accelerates or promotes a chemical process that would otherwise occur only slowly or would produce different products.) Pathways to methane begin with the fermentation of cellulose and other complex molecules to acids such as acetic acid, CH_3CO_2H. Through a complex sequence of reactions, this substance is transformed into methane, the overall reaction being

$$CH_3CO_2H(aq) \longrightarrow CH_4(g) + CO_2(g)$$

It is known that the $—CH_3$ group of acetic acid is ultimately transferred to the Ni(I) ion in the active site in the enzyme where it adds a proton from another source and releases CH_4.

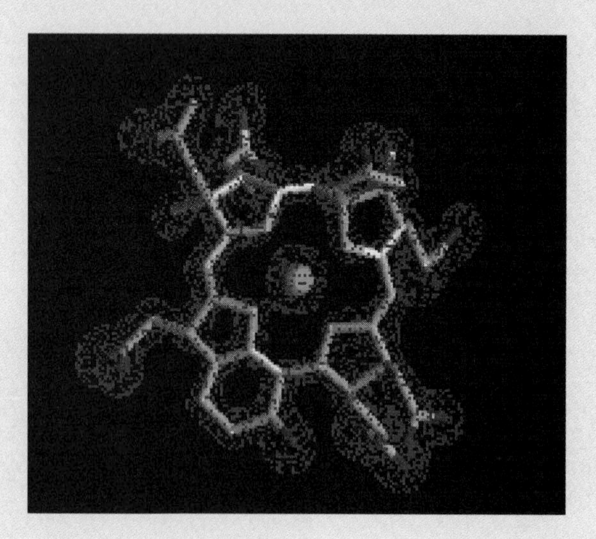

The structure of the active Ni-containing site in a methane-producing enzyme. The nickel atom is in the center (green). N atoms are at the blue points and O atoms at the red points. The remaining atoms are C atoms. (See U. Ermler, W. Grabarse, S. Shima, M. Goubeaud, R. K. Thauer; "Crystal structure of methyl-coenzyme M reductase: The key enzyme of biological methane formation." *Science*, Vol. 278, pp. 1457–1462, 1997.)

from a gas is evidence for the existence of forces of attraction between molecules, forces called **intermolecular forces.** However, the fact that gases completely occupy the volume available to them is evidence that intermolecular forces in the gas phase must be weak.

The experimental observations just described can guide us in formulating the principal features of the **kinetic-molecular theory** as applied to gases.

1. Gases consist of molecules whose separation is much greater than the size of the molecules themselves.
2. The molecules of a gas are in continual, random, and rapid motion.
3. The average kinetic energy of gas molecules is proportional to the gas temperature. *All gases, regardless of their molecular mass, have the same average kinetic energy at the same temperature.*
4. Gas molecules collide with one another and with the walls of their container, but they do so without loss of energy.

A second reason scientists are interested in methane is that it is an important "greenhouse" gas, along with CO_2, H_2O vapor, and O_3. Greenhouse gases behave like the glass panels of a greenhouse; they keep the heat from radiating to the outside. When sunlight strikes the earth, it warms the earth's surface. The warmed surfaces reradiate their energy, largely in the infrared region of the spectrum. The small molecules of the atmosphere, including methane, absorb this radiation, and the atmosphere is warmed. In general, this is a good effect, because the earth would otherwise be too cool for habitation. The problem lies in the fact that gases such as CO_2 and CH_4 are increasing in the atmosphere. Current thinking is that this will lead to an overall warming of the earth's atmosphere with potentially damaging effects. While much of the focus recently has been on the increase in atmospheric CO_2, there is considerable interest in lowering the CH_4 concentration. About 45 million metric tons of CH_4 escape to the troposphere annually, and a CH_4 molecule is far more effective than a CO_2 molecule at absorbing and radiating energy back to earth.

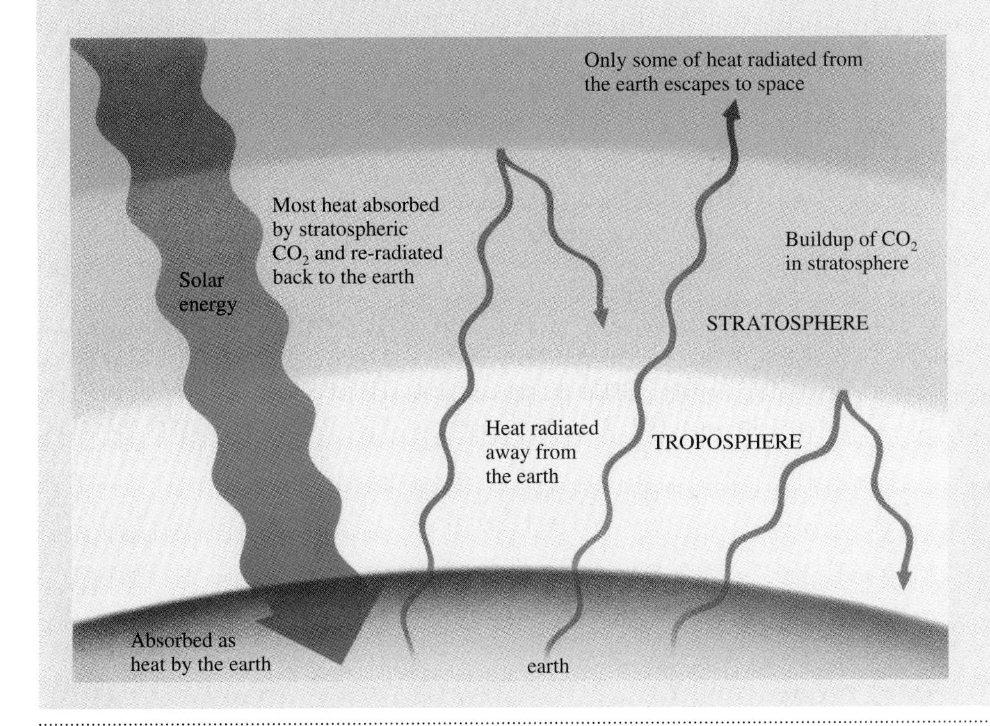

The greenhouse effect. Greenhouse gases (CO_2, CH_4, H_2O, and O_3) trap infrared radiation from the earth, and the troposphere and stratosphere are warmed. Increased concentrations of these gases could lead to excessive warming, with profound environmental effects.

Kinetic-Molecular Theory and the Gas Laws

The gas laws, which come from experiment, can be explained by the kinetic-molecular theory. The starting place is to describe how pressure arises from collisions of gas molecules with the walls of the container holding the gas (Figure 12.17). The force developed by these collisions depends on the number of collisions and the average force per collision. When the temperature of a gas is increased, the average force of a collision on the walls of its container increases because the average kinetic energy of the molecules increases. Also, because the speed increases with temperature, more collisions occur per second. Thus, the collective force per square centimeter is greater, and the pressure increases. Mathematically, this means P is proportional to T when n and V are constant.

Increasing the number of molecules of a gas at a fixed temperature and volume does not change the average collision force, but it does increase the num-

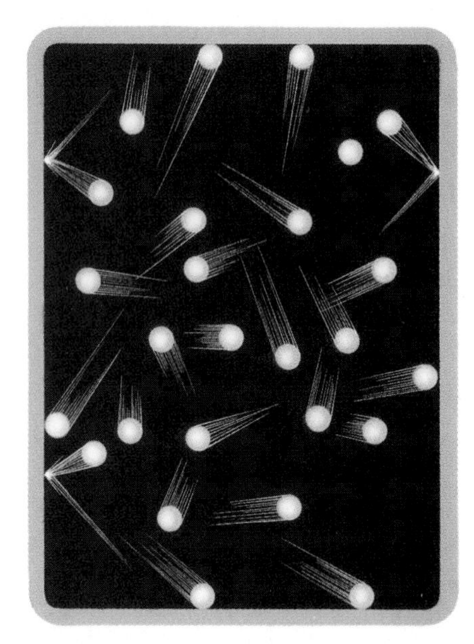

***Figure* 12.17 Gas pressure.** According to the kinetic-molecular theory, gas pressure is caused by gas molecules bombarding the container walls.

See the *Saunders Interactive General Chemistry CD-ROM,* Screen 12.10, Gas Laws and the Kinetic-Molecular Theory

Collisions between two bodies that occur without loss of energy are called *elastic collisions.* Collisions between billiard balls come close to being elastic, but basketballs have inelastic collisions.

ber of collisions occurring per second. Thus, the pressure increases, and we can say that P is proportional to n when V and T are constant.

If the pressure is not allowed to increase when either the number of molecules of gas or the temperature is increased, the volume of the container (and the area over which the collisions can take place) must increase. This is expressed by stating that V is proportional to nT when P is constant, a statement that is a *combination of Avogadro's law and Charles's law.*

Finally, if the temperature is constant, the average impact force of molecules of a given mass with the container walls must be constant. If n is kept constant while the volume of the container is made smaller, however, the number of collisions with the container walls per second must increase. This means the pressure increases, and so P is proportional to $1/V$ when n and T are constant, as stated by *Boyle's law.*

Distribution of Molecular Speeds

The relative number of molecules that have a given speed can be measured experimentally. Figure 12.18 is a graph of the number of molecules versus their speed; the higher a point on the curve, the greater the number of molecules having that speed.

Two important observations can be made about Figure 12.18. First, some molecules have high speeds (and thus high kinetic energy) and others have low speeds (low kinetic energy). The most common speed corresponds to the maximum in the distribution curve. For oxygen gas at 25 °C, for example, the max-

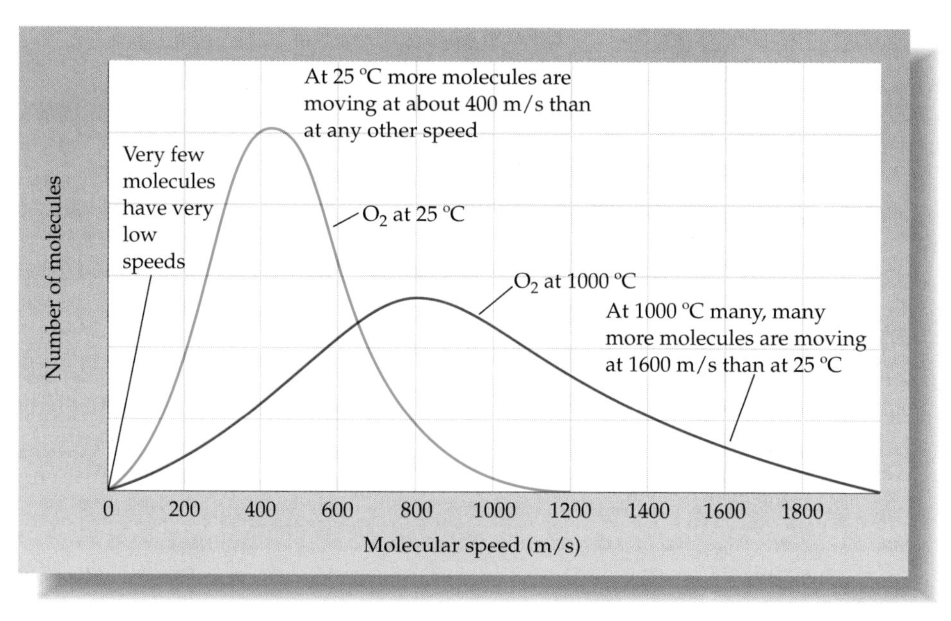

***Figure* 12.18 The distribution of molecular speeds.** A graph of the number of molecules with a given speed versus that speed shows the distribution of molecular speeds. The red curve shows the effect of increased temperature. Notice, however, that even though the curve for the higher temperature is "flatter" and broader than the one at a lower temperature, the areas under the curves are the same because the number of molecules in the sample is fixed.

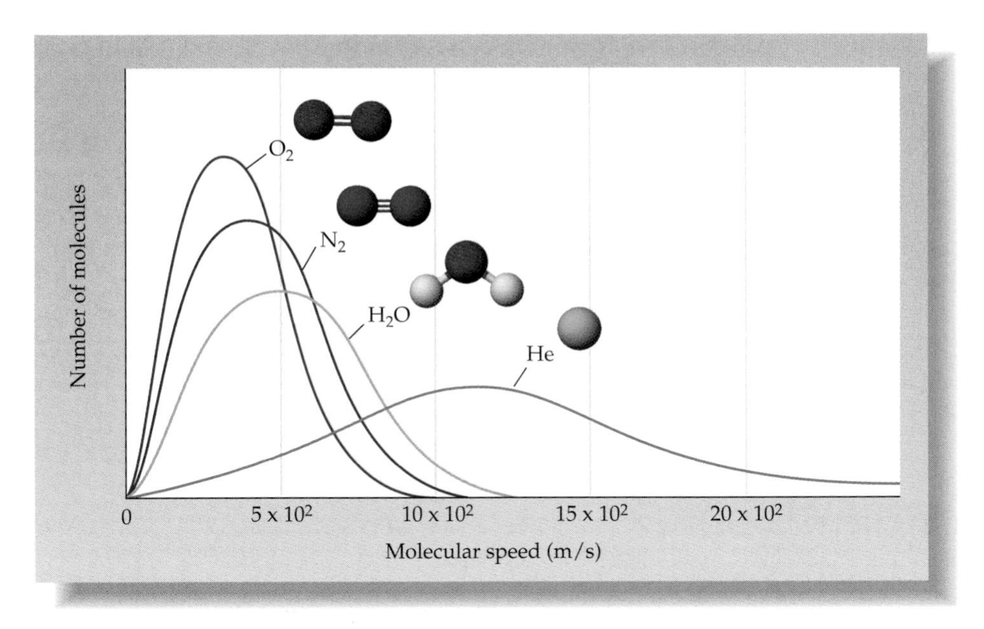

Figure 12.19 The effect of molecular mass on the distribution of speeds. At a given temperature, molecules with higher masses have lower speeds.

Curves of speed or energy versus number of molecules, such as those in Figure 12.18, are often called **Maxwell-Boltzmann distribution curves.** They are named after James Clerk Maxwell (1831–1879) and Ludwig Boltz581 mann (1844–1906).

imum is at a speed of about 400 m/s, and most of the molecules have speeds within the range from 200 m/s to 700 m/s. Notice that the curves are not symmetrical. A consequence of this is that the average speed is a little faster than the most common speed.

The second observation from Figure 12.18 is that, as the temperature is increased, the most common speed goes up, and the number of molecules traveling very fast goes up a great deal. There is an increase in average speed as we would have expected from Equation 12.13.

See the *Saunders Interactive General Chemistry CD-ROM,* Screen 12.11, Distribution of Molecular Speeds.

A relationship exists among molecular mass, average speed, and temperature. Because two gases with two different molecular masses must have the same average kinetic energy at the same temperature, the heavier gas molecules must have a lower average speed (Figure 12.19). Equation 12.14 expresses this idea in quantitative form. Here the square root of the mean square speed ($\sqrt{\overline{u^2}}$, called the **root-mean-square,** or **rms speed**), the temperature (T, in kelvins), and the molar mass (M) are related.

$$\sqrt{\overline{u^2}} = \sqrt{\frac{3RT}{M}} \qquad [12.14]$$

In this equation, sometimes called "Maxwell's equation" after James Clerk Maxwell (Section 7.1), R is the gas constant expressed in SI units; that is, $R = 8.314510\ \text{J/K} \cdot \text{mol}$.

The distribution of speeds (or kinetic energies) of molecules illustrated by Figure 12.18 is an important concept that is often used to explain chemical phenomena. You will see the concept several times more in the book; the next time will be in Chapter 13, in a discussion of the behavior of liquids.

Example 12.13 Molecular Speed

Calculate the rms speed of oxygen molecules at 25 °C.

The gas constant R can be expressed in different units. If P is measured in SI units of pascals (kg/m · s^2), and V is measured in cubic meters (m^3), then R has the value

$$R = \frac{(1.01325 \times 10^5 \text{ kg/m} \cdot \text{s}^2)(22.414 \times 10^{-2} \text{ m}^3)}{(1 \text{ mol})(273.15 \text{ K})}$$

$$= 8.3145 \text{ kg} \cdot \text{m}^2/\text{s}^2 \cdot \text{mol} \cdot \text{K}$$

Because 1 kg · m^2/s^2 is 1 joule, the gas constant can be expressed as

$$R = 8.3145 \text{ J/mol} \cdot \text{K}.$$

Solution We must use Equation 12.14 with M in units of kg/mol. The reason for this is that R is in units of J/K · mol, and 1 J = 1 kg · m^2/s^2. Thus, the molar mass of O_2 is 32.0 × 10^{-3} kg/mol.

$$\sqrt{\overline{u^2}} = \sqrt{\frac{3(8.3145 \text{ J/K} \cdot \text{mol})(298 \text{ K})}{32.0 \times 10^{-3} \text{ kg/mol}}} = \sqrt{2.32 \times 10^5 \text{ J/kg}}$$

To obtain the answer in meters per second, we use the relation 1 J = 1 kg · m^2/s^2. This means we have

$$\sqrt{\overline{u^2}} = \sqrt{2.32 \times 10^5 \text{ kg} \cdot \text{m}^2/\text{kg} \cdot \text{s}^2} = \sqrt{2.32 \times 10^5 \text{ m}^2/\text{s}^2}$$

$$= 482 \text{ m/s}$$

This speed is equivalent to about 1100 miles per hour!

***Exercise* 12.13** Molecular Speeds

Calculate the rms speeds of helium atoms and N_2 molecules at 25 °C.

12.7 DIFFUSION AND EFFUSION

When a pizza is brought into a room, the volatile aroma-causing molecules vaporize into the atmosphere where they mix with the oxygen, nitrogen, carbon dioxide, water vapor, and other gases present. Even if there were no movement of the air in the room caused by fans or people moving about, the smell would eventually reach everywhere in the room. This mixing of molecules of two or more gases due to their molecular motions is called gaseous **diffusion;** it results from the random molecular motion of all gases. Given time, the molecules of one component in a gas mixture will thoroughly and completely mix with all other components of the mixture (Figure 12.20).

Closely related to diffusion is **effusion,** which is the movement of gas through a tiny opening in a container into another container where the pressure is very low (Figure 12.21). Thomas Graham (1805–1869), a Scottish chemist, studied the effusion of gases and found that the rate of effusion of a gas—the quantity of material moving from one place to another in a given amount of time—was inversely proportional to the square root of its molar mass. Based on

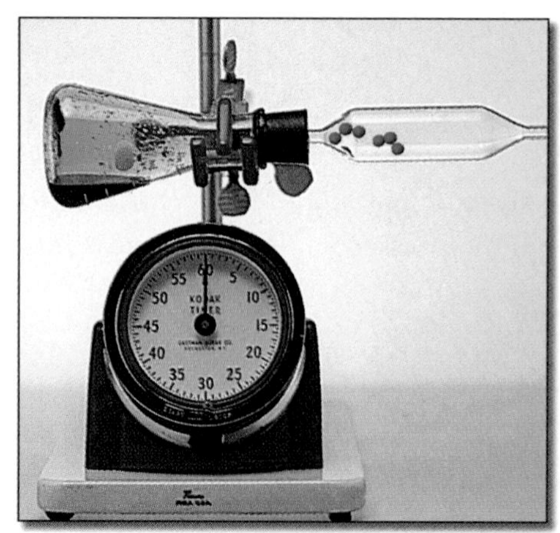

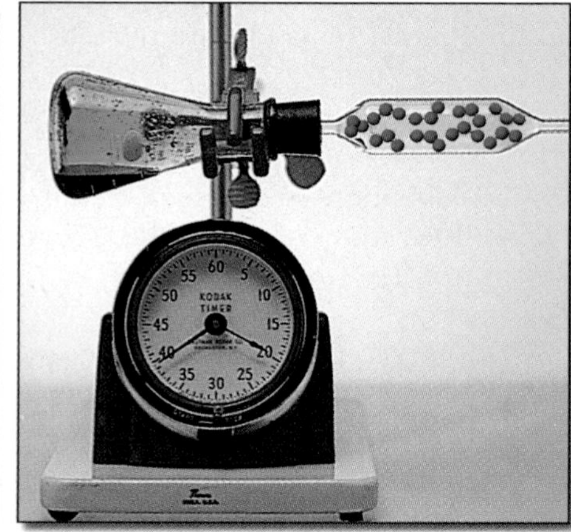

***Figure* 12.20 Diffusion.** Brown NO_2 gas diffuses out of the flask in which it was generated and into the attached tube within a few minutes. (The NO_2 is being made by the reaction of copper with nitric acid.) *(Photos, C. D. Winters; models, S. M. Young)*

Graham's experimental results, the rates of effusion of two gases can be compared:

$$\frac{\text{Rate of effusion of gas 1}}{\text{Rate of effusion of gas 2}} = \sqrt{\frac{\text{molar mass of gas 2}}{\text{molar mass of gas 1}}} \qquad [12.15]$$

The relationship in Equation 12.15, known as **Graham's law.** It is readily derived from Maxwell's equation by recognizing that the rate of effusion depends on the speed of the molecules. The ratio of the rms speeds is the same as the ratio of the effusion rates:

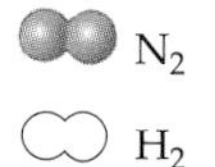

See the *Saunders Interactive General Chemistry CD-ROM,* Screen 12.12, Application of the Kinetic-Molecular Theory.

$$\frac{\text{Rate of effusion of gas 1}}{\text{Rate of effusion of gas 2}} = \frac{\sqrt{\overline{u^2} \text{ of gas 1}}}{\sqrt{\overline{u^2} \text{ of gas 2}}} = \sqrt{\frac{3RT/(M \text{ of gas 1})}{3RT/(M \text{ of gas 2})}}$$

Canceling out like terms gives the expression in Equation 12.15.

Figure 12.21 illustrates the relative rates of effusion of H_2 and N_2 molecules, and Graham's law allows a quantitative comparison of their relative rates:

$$\frac{\text{Rate of effusion of } H_2}{\text{Rate of effusion of } N_2} = \sqrt{\frac{M \text{ of } N_2}{M \text{ of } H_2}} = \sqrt{\frac{28.0 \text{ g/mol}}{2.02 \text{ g/mol}}} = \frac{3.72}{1}$$

This calculation tells you that H_2 molecules will effuse through the barrier 3.72 times faster than N_2 molecules.

Example 12.14 Graham's Law of Effusion

Tetrafluoroethylene, C_2F_4, effuses through a barrier at a rate of 4.6×10^{-6} mol/h. An unknown gas, consisting only of boron and hydrogen, effuses at the rate of 5.8×10^{-6} mol/h under the same conditions. What is the molar mass of the unknown gas?

Solution From Graham's law we know that a light molecule will effuse more rapidly than a heavier one. Because the unknown gas effuses more rapidly than C_2F_4 ($M = 100.0$ g/mol), the unknown must have a molar mass less than 100 g/mol. Substituting the experimental data into Graham's law (Equation 12.15), we have

$$\frac{\text{Rate of effusion of unknown gas}}{\text{Rate of effusion of } C_2F_4} = \sqrt{\frac{M \text{ of } C_2F_4}{M \text{ of unknown gas}}}$$

$$\frac{5.8 \times 10^{-6} \text{ mol/h}}{4.6 \times 10^{-6} \text{ mol/h}} = 1.3 = \sqrt{\frac{100.0 \text{ g/mol}}{M \text{ of unknown gas}}}$$

To solve for the unknown molar mass, square both sides of the equation and rearrange to find M for the unknown.

$$1.6 = \frac{100.0 \text{ g/mol}}{M \text{ of unknown gas}}$$

$$M = 63 \text{ g/mol}$$

A boron–hydrogen compound corresponding to this molar mass is B_5H_9, called pentaborane.

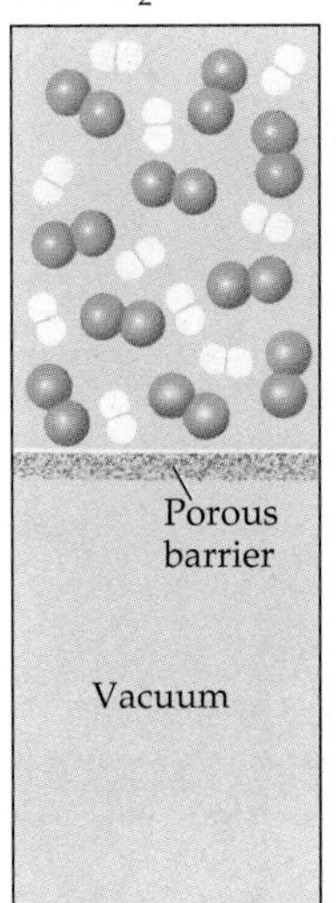

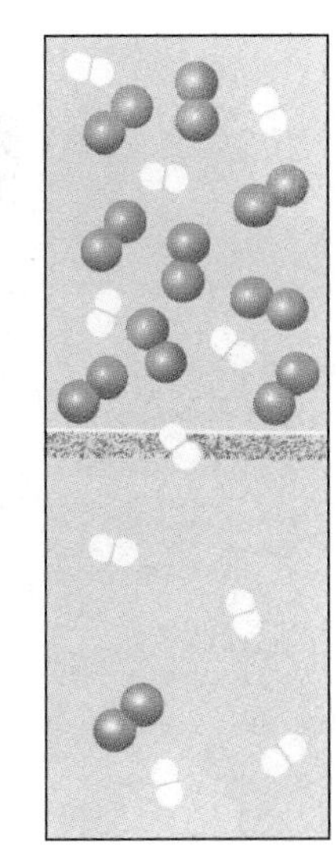

***Figure* 12.21 Effusion.** Gas molecules effuse through the pores of a porous barrier. Lighter molecules (H_2) with higher average speeds strike the barrier more often and pass more rapidly through it than heavier, slower molecules (N_2) at the same temperature.

Mylar film is now used for helium-filled balloons instead of rubber because effusion of helium through Mylar is less rapid. (Mylar is polyethylene terephthalate; see page 525.) *(C. D. Winters)*

Exercise 12.14 Graham's Law

A sample of pure methane, CH_4, is found to effuse through a porous barrier in 1.50 min. Under the same conditions, an equal number of molecules of an unknown gas effuses through the barrier in 4.73 min. What is the molar mass of the unknown gas?

12.8 SOME APPLICATIONS OF THE GAS LAWS AND KINETIC-MOLECULAR THEORY

Rubber Balloons and Why They Leak

Graham's law allows us to explain the earlier observation concerning the balloon used by Charles in 1783 (see page 547). In the brief biography of Charles, we mentioned that he filled the balloon with H_2 gas, but he had to go to extraordinary lengths to keep the gas in the balloon. Unlike the first hot air balloons that were made of paper, Charles had to use silk and coat it with rubber. The reason for this is simple: At any temperature at which H_2, N_2, and O_2 are gases, the lighter weight H_2 molecules have a higher average speed than the N_2 or O_2 molecules of air, so H_2 molecules would effuse rapidly through paper, a more porous material than silk coated with rubber.

For much the same reason that Charles used special materials in his hydrogen-filled balloon, you should be cautious when buying a helium-filled, rubber balloon at a carnival. Lightweight He atoms have a higher rms speed than the heavier N_2 or O_2 molecules of air, so He atoms can effuse rapidly through the balloon wall, and the balloon deflates. The newer balloons made of Mylar film, however, keep He gas enclosed much longer because the film has much smaller pores than rubber, and the He atoms cannot escape as easily.

Divers going to depths over about 130 feet are at risk from nitrogen narcosis. To prevent this, and oxygen toxicity, they breathe a mixture of oxygen (with a lower partial pressure than sea-level air) and helium. *(Tom & Therisa Stack/Tom Stack & Associates)*

Deep Sea Diving

Diving with a self-contained underwater breathing apparatus (SCUBA) is exciting. If you want to dive much beyond about 100 ft, however, you need to take special precautions.

When you breathe air from a SCUBA tank, the pressure of the gas in your lungs is equal to the pressure exerted on your body. At the surface, air has an oxygen concentration of 21%, so the partial pressure of O_2 is about 0.21 atm. If you are at a depth of about 33 ft the water pressure is 2 atm. This means the oxygen partial pressure is double the surface partial pressure, or about 0.4 atm. Similarly, the partial pressure of N_2, which is about 0.8 atm at the surface, doubles to about 1.6 atm at a depth of 33 ft. What is the problem?

Nitrogen narcosis, also called "rapture of the deep" or the "martini effect," results from the toxic effect of high N_2 pressure on nerve conduction. Its effect is comparable to drinking a martini on an empty stomach or taking nitrous oxide (N_2O) at the dentist; it makes you slightly giddy. In severe cases, it can impair a diver's judgment and even cause a diver to take the regulator out of his or her mouth and hand it to a fish! Some people can go as deep as 130 ft with no problem, but others experience nitrogen narcosis at 80 ft.

Another problem with breathing air at depths beyond 100 ft or so is oxygen toxicity. Our bodies are regulated for a partial pressure of O_2 of 0.21 atm. At a depth of 130 ft (of sea water), the partial pressure of O_2 is comparable to breathing 100% oxygen at sea level. These higher partial pressures can harm the lungs and cause central nervous system damage. Oxygen toxicity is the reason very deep dives are done not with compressed air but with gas mixtures with a much lower percentage of O_2, say about 10%.

Because of the risk of nitrogen narcosis, divers going beyond about 130 ft, such as those who work for offshore oil drilling companies, use a mixture of oxygen and helium. This solves the nitrogen narcosis problem, but it introduces another. If the diver has a voice link to the surface, the diver's speech sounds like Donald Duck! The reason for this is that the velocity of sound in helium is different from that in air, and the density of gas at several hundred feet is much higher than at the surface. This is the reason that a search of the Internet will turn up several papers from the military and from commercial diving companies on "helium speech descrambling" devices.

12.9 NONIDEAL BEHAVIOR: REAL GASES

If you are working with a gas at approximately room temperature and a pressure of 1 atm or less, the ideal gas law is remarkably successful in relating the quantity of gas and its pressure, volume, and temperature. At higher pressures or lower temperatures, however, deviations from the ideal gas law occur. The origin of these deviations is understood in terms of the breakdown of the assumptions used when describing ideal gases.

At standard temperature and pressure (STP), the volume occupied by a single molecule is *very* small relative to its share of the total gas volume. Recall that there are 6.02×10^{23} molecules in a mole and that a mole of gas occupies 22.4 L (22.4×10^{-3} m^3) at STP. The volume, V, that each molecule has to move around in is given by

$$V = \frac{22.4 \times 10^{-3}\ \text{m}^3}{6.023 \times 10^{23}\ \text{molecules}} = 3.72 \times 10^{-26}\ \text{m}^3/\text{molecule}$$

The volume of a sphere is given by $(4/3)\pi r^3$. Solving for r when $V = 3.72 \times 10^{-26}$ m^3,

$$r = \left[\frac{3V}{4\pi}\right]^{1/3} = \left[\frac{3(3.72 \times 10^{-26}\ \text{m}^3)}{4(3.14)}\right]^{1/3}$$

$$= 2.1 \times 10^{-9}\ \text{m, or 2100 pm}$$

If this volume is assumed to be a sphere, then the radius, r, of the sphere is about 2000 pm. The radius of the smallest gaseous substance, the helium atom, is 31 pm (see Figure 8.10), so the helium atom has about the same amount of space to move around in as a pea has inside a basketball.

Now suppose the pressure is increased significantly, to 1000 atm. The volume available to each molecule is a sphere with a radius of only about 200 pm, which means the situation is now like that of a pea inside a sphere a bit larger than a Ping-Pong ball. The important point is that the volume occupied by the gas molecules themselves is no longer negligible at higher pressures; this violates the first tenet of the kinetic-molecular theory. The kinetic-molecular theory and the ideal gas law are concerned with the volume available to the molecules to move about, not the volume of the molecules themselves. The experimentally determined volume, however, must include both. At high pressures, therefore, the measured volume is larger than predicted by $PV = nRT$.

Another assumption of the kinetic-molecular theory was that collisions between molecules are elastic, an assumption that implies that the atoms or mole-

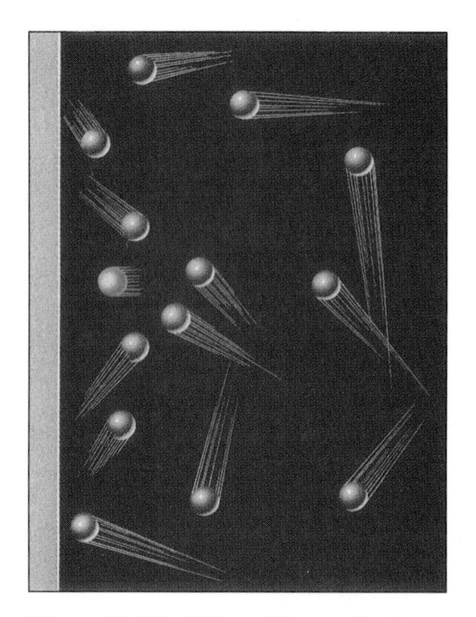

Figure 12.22 Effect of intermolecular forces. A gas molecule strikes the container wall with less force because of the attractive forces between it and its neighbors.

cules of the gas never stick to one another by some type of force. This is clearly not true as well. All gases can be liquefied—although some gases require a very low temperature—and the only way this can happen is if there are forces between the molecules (see Figure 12.16). When a molecule is about to hit the wall of its container, most other molecules are farther away from the wall and therefore pull it away from the wall (Figure 12.22). This attraction causes the molecule to hit the wall with less force—the collision is softer than if no attraction existed between the molecules. Because all collisions are softer, the observed gas pressure is less than that predicted by the ideal gas law. Although this effect can be observed in gases at pressures around 1 atm, it becomes particularly pronounced when the pressure is high, that is, when the molecules are close together.

The Dutch physicist Johannes van der Waals (1837–1923) studied the breakdown of the ideal gas law equation and developed an equation to correct for the errors arising from nonideality. This equation is known as the **van der Waals equation:**

$$\left[P + a\left(\frac{n}{V}\right)^2\right][V - bn] = nRT \qquad [12.16]$$

correction for intermolecular forces

correction for molecular volume

a and b = van der Waals constants

Although Equation 12.16 might seem complicated at first glance, the terms in square brackets are simply those of the ideal gas law, each corrected for the effects discussed previously. The pressure correction term, $a(n/V)^2$, corrects for intermolecular forces. Because the observed gas pressure is lower than the ideal pressure (as calculated using the equation $PV = nRT$) owing to intermolecular forces, $a(n/V)^2$ is added to the observed pressure. The constant a, which is determined experimentally, typically has values in the range 0.01 to 10 atm(L/mol)^2. The bn term corrects the observed volume, V_{observed}, to a smaller value, the volume actually available to the gas molecules. Here n is the number of moles of gas, and b is an experimental quantity that corrects for the molecular volume. Typical values of b range from 0.01 to 0.1 L/mol, roughly increasing with increasing molecular size (Table 12.3).

Table 12.3 • Some van der Waals Constants

Substance	a (atm · L^2/mol^2)	b (L/mol)
He	0.034	0.0237
Ar	1.34	0.0322
H_2	0.244	0.0266
N_2	1.39	0.0391
O_2	1.36	0.0318
Cl_2	6.49	0.0562
CO_2	3.59	0.0427
H_2O	2.25	0.0428

As an example of the importance of these corrections, consider a sample of 8.00 mol of chlorine gas, Cl_2, in a 4.00-L tank at 27.0 °C. The ideal gas law would lead you to expect a pressure of 49.2 atm. A more realistic estimate of the pressure, obtained from the van der Waals equation, is only 29.5 atm, about 20 atm less than the ideal pressure!

Exercise 12.15 Van der Waals's Equation

Using both the ideal gas law and van der Waals's equation, calculate the pressure expected for 10.0 mol of helium gas in a 1.00-L container at 25 °C.

CHAPTER HIGHLIGHTS

When you have finished studying this chapter, you should be able to

- Describe how pressure measurements are made (Section 12.1).
- Use the units of pressure, especially atmospheres (atm) and millimeters of mercury (mm Hg) (Section 12.1).

The **standard atmosphere** is the pressure capable of supporting a column of mercury 760 mm high; therefore, 1 atm is defined as exactly 760 mm Hg. Also 1 atm = 101.3 kPa or 1.013 bar. Conditions of **STP, standard temperature and pressure,** are 273 K and 1 atm.

- Understand the basis of the gas laws and how to use these laws (Section 12.2).

Boyle's law specifies that the pressure of a gas is inversely proportional to its volume ($P \propto 1/V$) at constant T and n. **Charles's law** tells us that gas volume is directly proportional to temperature ($V \propto T$) at constant n and P. **Avogadro's law** specifies that gas volume is directly proportional to its quantity ($V \propto n$) at constant T and P. The **general gas law** (or **combined gas law**), $P_1V_1/T_1 = P_2V_2/T_2$, is a combination of Boyle's and Charles's law. It is usually applied to find out what happens when a given quantity of gas undergoes changes in P, V, and T.

- Understand the **ideal gas law** and how to use this equation (Section 12.3).

The three gas laws are combined into one statement called the ideal gas law ($PV = nRT$). R, the gas law constant, = 0.082051 L · atm/K · mol.

- Calculate the molar mass of a compound from a knowledge of the pressure of a known quantity of a gas in a given volume at a known temperature (Section 12.3).
- Apply the gas laws to a study of the stoichiometry of reactions (Section 12.4).
- Use **Dalton's law** (Section 12.5).

Dalton's law of partial pressures specifies that the total pressure of a mixture of gases is the sum of the partial pressures of the individual gases (A, B, C, . . .) in the mixture:

$$P_{total} = P_A + P_B + P_C + \cdots$$

The **partial pressure** of a gas in a mixture (P_A) is given by its mole fraction (X_A) times the total pressure of the mixture (P_{total}), $P_A = X_A P_{total}$. The **mole fraction** of a component (A) of a mixture is defined as the number of moles of A divided by the total moles of all components.

- Apply the **kinetic-molecular theory,** a theory of gas behavior at the molecular level (Section 12.6).

The average kinetic energy of gas molecules ($\overline{KE}$) is proportional to the mass of the gas and to $\overline{u^2}$ (the average of the squares of the molecular speeds).

$$\overline{KE} = \frac{1}{2} m\overline{u^2}$$

Because $\overline{KE}$ is determined by temperature, heavier molecules (large m) must move with a slower average speed ($\overline{u}$) than lighter molecules (small m) at a given temperature. For a given gas, the only way to change its average kinetic energy is to change the temperature ($\overline{KE} \propto T$).

- Understand the phenomena of **diffusion** and **effusion** and how to use **Graham's law** (Section 12.6).

Graham's law of effusion states that the rates of effusion of two gases are inversely proportional to the square roots of their molar masses at the same temperature and pressure.

$$\frac{\text{Rate of effusion of gas 1}}{\text{Rate of effusion of gas 2}} = \sqrt{\frac{\text{molar mass of gas 2}}{\text{molar mass of gas 1}}} \qquad [12.15]$$

- Appreciate the fact that gases usually do not behave as ideal gases (Section 12.9). Deviations from ideal behavior are largest at high pressure and low temperature.

Ideal gas molecules are assumed to have no volume and to not interact with one another by intermolecular forces. These assumptions lead to the simple model of ideal gas behavior as expressed by $PV = nRT$. Because neither assumption is completely correct, a real gas has a more complex behavior than an ideal gas. One description of the behavior of real gases is found in the van der Waals equation (12.16).

STUDY QUESTIONS

REVIEW QUESTIONS

1. Name the three gas laws that interrelate *P*, *V*, and *T*. Explain the relationships in words and in equations.
2. What conditions are represented by STP? What is the volume of a mole of ideal gas under these conditions?
3. Show how to calculate the molar mass of a gas from *P*, *V*, and *T* measurements and other information.
4. State Avogadro's law. Relate your discussion to the formation of water vapor from its elements [$2\ H_2(g) + O_2(g) \longrightarrow 2\ H_2O(g)$]. For example, if 3 mol of H_2 is used at STP, how many liters of O_2 are required at STP and how many liters of H_2O vapor are produced at STP?
5. State Dalton's law. If the air you breathe is 78% N_2 and 22% O_2 (on a mole basis), what is the mole fraction of O_2? What is the partial pressure of O_2 when the atmospheric pressure is 748 mm Hg?
6. What are the basic assumptions of the kinetic-molecular theory? Explain Boyle's law on the basis of kinetic-molecular theory.
7. What assumptions are made in describing an ideal gas? Under what circumstances is the ideal gas law least accurate?
8. In van der Waals's equation, what properties of a real gas are accounted for by the constants *a* and *b*?
9. State Graham's law in words and in equation form. If an unknown gas effuses four times slower than H_2 gas at 25 °C, what is the molar mass of the unknown gas? What common gas has this molar mass?

NUMERICAL QUESTIONS

10. The pressure of a gas is 440 mm Hg. Express this pressure in units of (a) atmospheres, (b) bars, and (c) pascals.

11. The average barometric pressure at an altitude of 10 km is 210 mm Hg. Express this pressure in atmospheres, bars, and kilopascals.

12. If the mercury levels in a U-tube manometer are as illustrated here, express the gas pressure in millimeters of mercury, atmospheres, torrs, and kilopascals.

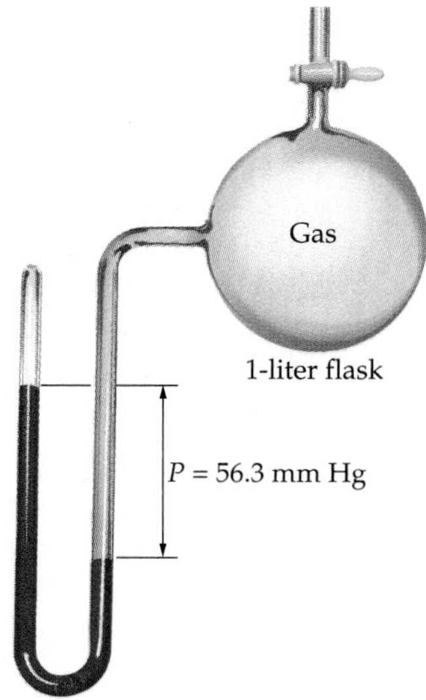

13. If the pressure of a gas is 95.0 kPa, what is the difference (in mm Hg) in the mercury levels in the U-tube manometer? (See the figure accompanying Study Question 12.)

The Gas Laws

14. A sample of nitrogen gas has a pressure of 67.5 mm Hg in a 500.-mL flask. What is the pressure of this gas sample when it is transferred to a 125-mL flask at the same temperature?

15. A sample of CO_2 gas has a pressure of 56.5 mm Hg in a 125-mL flask. The sample is transferred to a new flask where it has a pressure of 62.3 mm Hg at the same temperature. What is the volume of the new flask?

16. You have 3.5 L of NO at a temperature of 22.0 °C. What volume would the NO occupy at 37 °C? (Assume the pressure of the NO sample is constant.)

17. A 5.0-mL sample of CO_2 gas is enclosed in a gas-tight syringe (see Figure 12.3) at 22 °C. If the syringe is immersed in an ice bath (0 °C), what is the new gas volume, assuming that the pressure is held constant?

18. Water can be made by combining gaseous O_2 and H_2. If you begin with 3.6 L of H_2 gas at 380 mm Hg and 25 °C, how many liters of O_2 would you need for complete reaction if the O_2 gas is also measured at 380 mm Hg and 25 °C?

19. Ethane, C_2H_6, burns in air according to the equation

$$2\ C_2H_6(g) + 7\ O_2(g) \longrightarrow 4\ CO_2(s) + 6\ H_2O(g)$$

How many liters of O_2 are required for complete reaction with 5.2 L of C_2H_6? How many liters of H_2O vapor are produced? Assume all gases are measured at the same temperature and pressure.

20. You have a sample of gas in a flask with a volume of 250 mL. At 25.5 °C the pressure of the gas is 360 mm Hg. If you decrease the temperature to −5.0 °C, what is the gas pressure at the lower temperature?

21. A sample of gas occupies 135 mL at 22.5 °C; the pressure is 165 mm Hg. What is the pressure of the gas sample when it is placed in a 252-mL flask at a temperature of 0.0 °C?

22. You have a sample of CO_2 in a flask (A) with a volume of 25.0 mL. At 20.5 °C, the pressure of the gas is 436.5 mm Hg. To find the volume of another flask (B), you move the CO_2 to that flask and find that its pressure is now 94.3 mm Hg at 24.5 °C. What is the volume of flask B?

23. One of the cylinders of an automobile engine has a volume of 400. cm^3. The engine takes in air at a pressure of 1.00 atm and a temperature of 15 °C and compresses it to a volume of 50.0 cm^3 at 77 °C. What is the final pressure of the gas in the cylinder? (The ratio of before and after volumes, in this case 400:50 or 8:1, is called the compression ratio.)

24. A helium-filled balloon of the type used in long-distance flying contains 420,000 ft^3 (1.2×10^7 L) of helium. Let us say you fill the balloon with helium on the ground where the pressure is 737 mm Hg and the temperature is 16.0 °C. When the balloon ascends to a height of 2 miles where the pressure is only 600. mm Hg and the temperature is −33

°C, what volume is occupied by the helium gas? Assume the pressure inside the balloon matches the external pressure.

This balloon was built for a possible flight around the world. It is called a "Rozier" type after Jean Rozier. (On November 21, 1783, he became one of the first two humans to fly.) The balloon uses helium as well as hot air for lift. *(David Kipler/Washington University Photographic Services)*

25. A sample of gas is contained in a 245-mL flask at a temperature of 23.5 °C; the gas pressure is 48.5 mm Hg. The gas is moved to a new flask, which is then immersed in ice water, and which has a volume of 68 mL. What is the pressure of the gas in the smaller flask at the new temperature?

The Ideal Gas Law

26. A 1.25-g sample of CO_2 is contained in a 750.-mL flask at 22.5 °C. What is the pressure of the gas?

27. A balloon holds 30.0 kg of helium. What is the volume of the balloon if the final pressure is 1.20 atm and the temperature is 22 °C?

28. A flask is first evacuated so that it contains no gas at all. Then, 2.2 g of CO_2 is introduced into the flask. On warming to 22 °C, the gas exerts a pressure of 318 mm Hg. What is the volume of the flask?

29. A steel cylinder holds 1.50 g of ethanol, C_2H_5OH. What is the pressure of the ethanol vapor if the cylinder has a volume of 251 cm^3 and the temperature is 250 °C? (Assume all the ethanol is in the vapor phase at this temperature.)

30. A balloon for long-distance flying contains 1.2×10^7 L of helium. If the helium pressure is 737 mm Hg at 25 °C, what mass of helium (in grams) does the balloon contain? (See Study Question 24.)

31. What mass of helium, in grams, is required to fill a 5.0-L balloon to a pressure of 1.1 atm at 25 °C?

32. A 1.007-g sample of an unknown gas exerts a pressure of 715 mm Hg in a 452-mL container at 23 °C. What is the molar mass of the gas?

33. A 0.0125-g sample of a gas with an empirical formula of CHF_2 is placed in a 165-mL flask. It has a pressure of 13.7 mm Hg at 22.5 °C. What is the molecular formula of the compound?

34. Analysis of a gaseous chlorofluorocarbon, CCl_xF_y, shows it contains 11.79% C and 69.57% Cl. In another experiment you find that 0.107 g of the compound fills a 458-mL flask at 25 °C with a pressure of 21.3 mm Hg. What is the molecular formula of the compound?

35. There are five compounds in the family of sulfur–fluorine compounds with the general formula S_xF_y. One of these compounds is 25.23% S. If you place 0.0955 g of the compound in a 89-mL flask at 45 °C, the pressure of the gas is 83.8 mm Hg. What is the molecular formula of S_xF_y?

36. Forty miles above the earth's surface the temperature is 250 K and the pressure is only 0.20 mm Hg. What is the density of air (in grams per liter) at this altitude? (Assume the molar mass of air is 29 g/mol.)

37. Diethyl ether, $(C_2H_5)_2O$, vaporizes easily at room temperature. If the vapor exerts a pressure of 233 mm Hg in a flask at 25 °C, what is the density of the vapor?

38. A gaseous organofluorine compound has a density of 0.355 g/L at 17 °C and 189 mm Hg. What is the molar mass of the compound?

39. Chloroform is a common liquid used in the laboratory. It vaporizes readily. If the pressure of chloroform vapor in a flask is 195 mm Hg at 25.0 °C, and the density of the vapor is 1.25 g/L, what is the molar mass of chloroform?

40. If 12.0 g of O_2 is required to inflate a balloon to a certain size at 27 °C, what mass of O_2 is required to inflate it to the same size (and pressure) at 81 °C?

41. You have two gas-filled balloons, one containing He and the other H_2. The H_2 balloon is twice the size of the He balloon. The pressure of gas in the H_2 balloon is 1 atm, and that in the He balloon is 2 atm. The H_2 balloon is outside in the snow (−5 °C), and the He balloon is inside a warm building (23 °C).
   (a) Which balloon contains the greater number of molecules?
   (b) Which balloon contains the greater mass of gas?

42. Assume that a bicycle tire with an internal volume of 1.52 L contains 0.406 mol of air. The tire will burst if its internal pressure reaches 7.25 atm. To what temperature, in degrees Celsius, does the air in the tire need to be heated to cause a blowout?

43. The temperature of the atmosphere on Mars can be as high as 27 °C at the equator at noon, and the atmospheric pressure is about 8 mm Hg. If a spacecraft could collect 10. m^3 of this atmosphere, compress it to a small volume, and send it back to Earth, how many moles would the sample contain?

Gas Laws and Stoichiometry

44. Iron reacts with hydrochloric acid to produce iron(II) chloride and hydrogen gas:

$$Fe(s) + 2\,HCl(aq) \longrightarrow FeCl_2(aq) + H_2(g)$$

The H_2 gas from the reaction of 2.2 g of iron with excess acid is collected in a 10.0-L flask at 25 °C. What is the pressure of the H_2 gas in this flask?

45. Silane, SiH_4, reacts with O_2 to give silicon dioxide and water:

$$SiH_4(g) + 2\ O_2(g) \longrightarrow SiO_2(s) + 2\ H_2O(\ell)$$

A 5.20-L sample of SiH_4 gas at 356 mm Hg pressure and 25 °C is allowed to react with O_2 gas. What volume of O_2 gas, in liters, is required for complete reaction if the oxygen has a pressure of 425 mm Hg at 25 °C?

46. Sodium azide, the explosive compound in automobile air bags, decomposes according to the equation

$$2\ NaN_3(s) \longrightarrow 2\ Na(s) + 3\ N_2(g)$$

What mass of sodium azide is required to provide the nitrogen needed to inflate a 25.0-L bag to a pressure of 1.3 atm at 25 °C?

47. The hydrocarbon octane burns to give CO_2 and water vapor:

$$2\ C_8H_{18}(g) + 25\ O_2(g) \longrightarrow 16\ CO_2(g) + 18\ H_2O(g)$$

If a 0.095-g sample of octane burns completely in O_2, what will be the pressure of water vapor in a 4.75-L flask at 30.0 °C? If the O_2 gas needed for complete combustion was contained in a 4.75-L flask at 22 °C, what would its pressure be?

48. Hydrazine reacts with O_2 according to the equation

$$N_2H_4(g) + O_2(g) \longrightarrow N_2(g) + 2\ H_2O(\ell)$$

Assume the O_2 needed for the reaction is in a 450-L tank at 23 °C. What must the oxygen pressure be in the tank to have enough oxygen to consume 1.00 kg of hydrazine completely?

49. A self-contained breathing apparatus uses canisters containing potassium superoxide. The superoxide consumes the CO_2 exhaled by a person and replaces it with oxygen.

$$4\ KO_2(s) + 2\ CO_2(g) \longrightarrow 2\ K_2CO_3(s) + 3\ O_2(g)$$

What mass of KO_2, in grams, is required to react with 8.90 L of CO_2 at 22.0 °C and 767 mm Hg?

50. $Ni(CO)_4$ can be made by reacting finely divided nickel with gaseous CO. If you have CO in a 1.50-L flask at a pressure of 418 mm Hg at 25.0 °C, along with 0.450 g of Ni powder, what is the maximum number of grams of $Ni(CO)_4$ that can be made?

51. If you place 2.25 g of solid silicon in a 6.56-L flask that contains CH_3Cl with a pressure of 585 mm Hg at 25 °C, how many grams of $(CH_3)_2SiCl_2(g)$, dimethyldichlorosilane, can be formed?

$$Si(s) + 2\ CH_3Cl(g) \longrightarrow (CH_3)_2SiCl_2(g)$$

What pressure of $(CH_3)_2SiCl_2(g)$ would you expect in this same flask at 95 °C on completion of the reaction? (Dimethyldichlorosilane is one starting material used to make silicones, polymeric substances used as lubricants, antistick agents, and in water-proofing caulk.)

Gas Mixtures

52. What is the total pressure in atmospheres of a gas mixture that contains 1.0 g of H_2 and 8.0 g of Ar in a 3.0-L container at 27 °C? What are the partial pressures of the two gases?

53. A cylinder of compressed gas is labeled "Composition (mole %): 4.5% H_2S, 3.0% CO_2, balance N_2." The pressure gauge attached to the cylinder reads 46 atm. Calculate the partial pressure of each gas, in atmospheres, in the cylinder.

54. A halothane–oxygen mixture ($C_2HBrClF_3 + O_2$) can be used as an anesthetic. Assume that a tank containing such a mixture has the following partial pressures: P(halothane) = 170 mm Hg and $P(O_2)$ = 570 mm Hg.
(a) What is the ratio of the number of moles of halothane to the number of moles of O_2?
(b) If the tank contains 160 g of O_2, how many grams of $C_2HBrClF_3$ are present?

55. A collapsed balloon is filled with He to a volume of 12 L at a pressure of 1.0 atm. Oxygen (O_2) is then added so that the final volume of the balloon is 26 L with a total pressure of 1.0 atm. The temperature, constant throughout, is equal to 20 °C.
(a) How many grams of He does the balloon contain?
(b) What is the final partial pressure of He in the balloon?
(c) What is the partial pressure of O_2 in the balloon?
(d) What is the mole fraction of each gas?

56. A miniature volcano can be made in the laboratory with ammonium dichromate. When ignited it decomposes in a fiery display.

$$(NH_4)_2Cr_2O_7(s) \longrightarrow N_2(g) + 4\ H_2O(g) + Cr_2O_3(s)$$

If 5.0 g of ammonium dichromate is used, and if the gases from this reaction are trapped in a 3.0-L flask at 23 °C, what is the total pressure of the gas in the flask? What are the partial pressures of N_2 and H_2O?

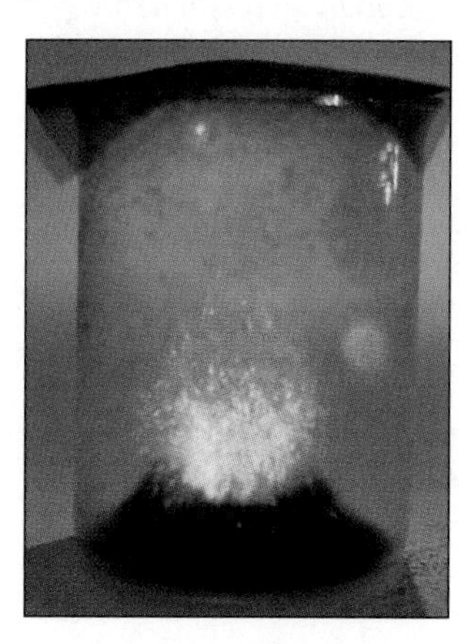

Ammonium dichromate decomposes exothermically to give nitrogen gas, water vapor, and green chromium(III) oxide. *(C. D. Winters)*

57. A study of climbers who reached the summit of Mt. Everest without supplemental oxygen showed that the partial pressures of O_2 and CO_2 in their lungs were 35 mm Hg and 7.5

mm Hg, respectively. The barometric pressure at the summit was 253 mm Hg. Assume the lung gases are saturated with moisture at a body temperature of 37 °C (so $P(H_2O)$ = 47.1 mm Hg). If you assume the lung gases are only O_2, N_2, CO_2, and H_2O, what is the partial pressure of N_2?

58. A sample of nitrogen gas is collected over water at 18 °C (see Figure 12.14). If the barometric pressure in the laboratory is 747 mm Hg, what is the partial pressure of the dry nitrogen gas in the sample? (See Appendix G for water vapor pressure data.)

59. You are given 1.56 g of a mixture of $KClO_3$ and KCl. When heated, the $KClO_3$ decomposes to KCl and O_2,

$$2\ KClO_3(s) \longrightarrow 2\ KCl(s) + 3\ O_2(g)$$

and 327 mL of O_2 is collected over water at 19 °C. The total pressure of the gas in the collection flask is 735 mm Hg. What is the weight percentage of $KClO_3$ in the sample? (See Appendix G for water vapor pressure data.)

Kinetic-Molecular Theory

60. You have two flasks of equal volume. Flask A contains H_2 at 0 °C and 1 atm pressure. Flask B contains CO_2 gas at 0 °C and 2 atm pressure. Compare these two gases with respect to each of the following:
(a) Average kinetic energy per molecule
(b) Average molecular velocity
(c) Number of molecules
(d) Mass of gas

61. Equal masses of gaseous N_2 and Ar are placed in separate flasks of equal volume, both gases being at the same temperature. Tell whether each of the following statements is true or false. Briefly explain your answer in each case.
(a) There are more molecules of N_2 present than atoms of Ar.
(b) The pressure is greater in the Ar flask.
(c) Ar atoms have a greater average speed than the N_2 molecules.
(d) The molecules of N_2 collide more frequently with the walls of the flask than do the atoms of Ar.

62. If the average speed of an oxygen molecule is 4.28×10^4 cm/s at 25 °C, what is the average speed of a CO_2 molecule at the same temperature?

63. Calculate the rms speed for CO molecules at 25 °C. What is the ratio of this speed to that of Ar atoms at the same temperature?

64. Place the following gases in order of increasing average molecular speed at 25 °C: (a) Ar, (b) CH_4, (c) N_2, and (d) CH_2F_2.

65. The reaction of SO_2 with Cl_2 gives dichlorine oxide, which is used to bleach wood pulp and to treat wastewater:

$$SO_2(g) + 2\ Cl_2(g) \longrightarrow OSCl_2(g) + Cl_2O(g)$$

All the compounds involved in the reaction are gases. List them in order of increasing average speed.

Diffusion and Effusion

66. Argon gas is ten times denser than helium gas at the same temperature and pressure. Which gas is predicted to effuse faster? How much faster?

67. In each pair of gases below, tell which will effuse faster:
(a) CO_2 or F_2
(b) O_2 or N_2
(c) C_2H_4 or C_2H_6
(d) two CFCs: $CFCl_3$ or $C_2Cl_2F_4$

68. A gas whose molar mass you wish to know effuses through an opening at a rate one-third as fast as that of helium gas. What is the molar mass of the unknown gas?

69. A sample of uranium fluoride is found to effuse at the rate of 17.7 mg/h. Under comparable conditions, gaseous I_2 effuses at the rate of 15.0 mg/h. What is the molar mass of the uranium fluoride? (*Caution*: Rates must be used in units of moles per time.)

Nonideal Gases

70. In the text it is stated that the pressure of 8.00 mol of Cl_2 in a 4.00-L tank at 27.0 °C should be 29.5 atm if calculated using van der Waals's equation. Verify this result and compare it with the pressure predicted by the ideal gas law.

71. You want to store 165 g of CO_2 gas in a 12.5-L tank at room temperature (25 °C). Calculate the pressure the gas would have using (a) the ideal gas law and (b) van der Waals's equation.

GENERAL QUESTIONS

72. Complete the following table:

	atm	mm Hg	kPa	bar
Standard atmosphere	____	____	____	____
Partial pressure of N_2 in the atmosphere	____	593	____	____
Tank of compressed H_2	____	____	____	133
Atmospheric pressure at the top of Mt. Everest	____	____	33.7	____

73. You want to fill a cylindrical tank with CO_2 gas at 865 mm Hg and 25 °C. The tank is 20.0 m long with a 10.0-cm radius. What mass of CO_2 (in grams) is required?

74. Acetaldehyde is a common liquid compound that vaporizes readily. The vapor has a pressure of 331 mm Hg in a 125-mL flask at 0.0 °C, and the density of the vapor is 0.855 g/L. What is the molar mass of acetaldehyde?

75. On combustion 1.0 L of a gaseous compound of hydrogen, carbon, and nitrogen gives 2.0 L of CO_2, 3.5 L of H_2O vapor, and 0.50 L of N_2 at STP. What is the empirical formula of the compound?

76. To what temperature, in degrees Celsius, must a 25.5-mL sample of oxygen at 90 °C be cooled for its volume to shrink to 21.5 mL? Assume the pressure and mass of the gas are constant.

77. You have a sample of helium gas at −33 °C, and you want to increase the average speed of helium atoms by 10.0%. To what temperature should the gas be heated to accomplish this?

78. The vapor pressure of water at 25 °C is 23.8 mm Hg. How many molecules of water per cubic centimeter exist in the vapor phase?

79. Which of the following gas samples contains the largest number of gas molecules? Which contains the smallest number? Which represents the largest mass of gas?
(a) 1.0 L of H_2 at STP
(b) 1.0 L of Ar at STP
(c) 1.0 L of H_2 at 27 °C and 760 mm Hg
(d) 1.0 L of He at 0 °C and 900 mm Hg

80. You are given a solid mixture of $NaNO_2$ and NaCl and are asked to analyze it for the amount of $NaNO_2$ present. To do so you allow it to react with sulfamic acid, HSO_3NH_2, in water according to the equation

$$NaNO_2(aq) + HSO_3NH_2(aq) \longrightarrow NaHSO_4(aq) + H_2O(\ell) + N_2(g)$$

What is the weight percentage of $NaNO_2$ in 1.232 g of the solid mixture if reaction with sulfamic acid produces 295 mL of N_2 gas? The gas was collected over water (see Figure 12.14) at a temperature of 21.0 °C and with the barometric pressure equal to 736.0 mm Hg.

81. The density of air 20 km above the earth's surface is 92 g/m^3. The pressure of the atmosphere is 42 mm Hg and the temperature is −63 °C.
(a) What is the average molar mass of the atmosphere at this altitude?
(b) If the atmosphere at this altitude is only O_2 and N_2, what is the mole fraction of each gas?

82. A 3.0-L bulb containing He at 145 mm Hg is connected by a valve to a 2.0-L bulb containing Ar at 355 mm Hg. (See the following figure.) Calculate the partial pressure of each gas and the total pressure after the valve between the flasks is opened.

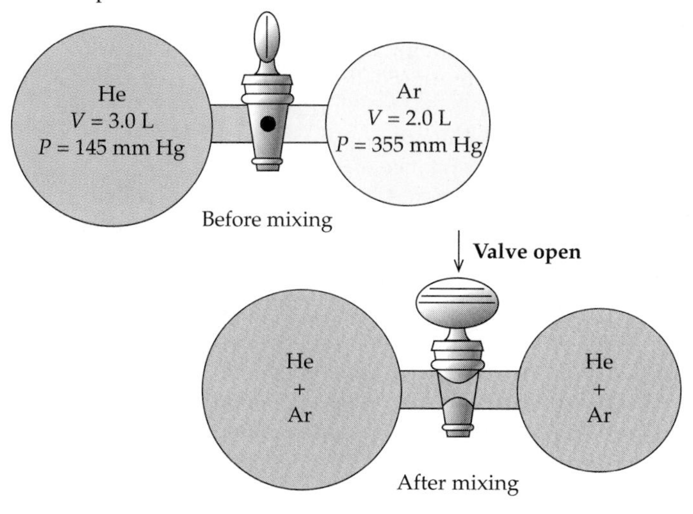

83. Phosphine gas, PH_3, is toxic when it reaches a concentration of 7×10^{-5} mg/L. To what pressure does this correspond at 25 °C?

84. Chlorine dioxide, ClO_2, reacts with fluorine to give a new gas that contains Cl, O, and F. In an experiment you find that 0.150 g of the gas has a pressure of 17.2 mm Hg in a 1850-mL flask at 21 °C. What is the identity of this unknown gas?

85. A xenon fluoride can be prepared by heating a mixture of Xe and F_2 gases to a high temperature in a pressure-proof container. Assume that xenon gas was added to a 0.25-L container until its pressure was 0.12 atm at 0.0 °C. Fluorine gas was then added until the total pressure was 0.72 atm at 0.0 °C. After the reaction was complete, the xenon was consumed completely and the pressure of the F_2 remaining in the container was 0.36 atm at 0.0 °C. What is the empirical formula of the xenon fluoride?

86. A balloon at the circus is filled with helium gas to a gauge pressure of 22 mm Hg at 25 °C. The volume of the gas is 305 mL, and the barometric pressure is 755 mm Hg. How many moles of helium are in the balloon? (Remember that gauge pressure = total pressure − barometric pressure. See page 541.)

87. Acetylene can be made by allowing calcium carbide to react with water:

$$CaC_2(s) + 2\ H_2O(\ell) \longrightarrow C_2H_2(g) + Ca(OH)_2(s)$$

Suppose you react 2.65 g of CaC_2 with excess water. If you collect the acetylene over water (see Figure 12.14), and find that the gas (acetylene and water vapor) has a volume of 795 mL at 25.2 °C at a barometric pressure of 735.2 mm Hg, what is the percent yield of acetylene?

CONCEPTUAL QUESTIONS

88. A 1.0-L flask contains 10.0 g each of O_2 and CO_2 at 25 °C.
(a) Which gas has the greater partial pressure, O_2 or CO_2, or are they the same?
(b) Which molecules have the greater average speed, or are they the same?
(c) Which molecules have the greater average kinetic energy, or are they the same?

89. If equal masses of O_2 and N_2 are placed in separate containers of equal volume at the same temperature, which of the following statements is true? If false, tell why it is false.
(a) The pressure in the N_2-containing flask is greater than that in the flask containing O_2.
(b) There are more molecules in the flask containing O_2 than in the one containing N_2.

90. Suppose you have two pressure-proof steel cylinders of equal volume, one containing 1.0 kg of CO and the other 1.0 kg of acetylene, C_2H_2.
(a) In which cylinder is the pressure greater at 25 °C?
(b) Which cylinder contains the greater number of molecules?

91. Two flasks, each with a volume of 1.00 L, contain O_2 gas with a pressure of 380 mm Hg. Flask A is at 25 °C, and flask B is at 0 °C. Which flask contains the greater number of O_2 molecules?

92. State whether each of the following samples of matter is a gas. If there is not enough information for you to decide, write "insufficient information."
(a) A material is in a steel tank at 100 atm pressure. When the tank is opened to the atmosphere, the material suddenly expands, increasing its volume by 10%.
(b) A 1.0-mL sample of material weighs 8.2 g.
(c) A material is transparent and pale green in color.
(d) One cubic meter of material contains as many molecules as 1 m^3 of air at the same temperature and pressure.

93. Does the effect of intermolecular attraction on the properties of a gas become more significant or less significant if (a) the gas is compressed to a smaller volume at constant temperature; (b) more gas is forced into the same volume at constant temperature; (c) the temperature of the gas is raised at constant pressure?

94. Each of the four tires of a car is filled with a different gas. Each tire has the same volume and each is filled to the same pressure, 3.0 atm, at 25 °C. One tire contains 116 g of air, another tire has 80.7 g of neon, another tire has 16.0 g of helium, and the fourth tire has 160. g of an unknown gas.
(a) Do all four tires contain the same number of gas molecules? If not, which one has the greatest number of molecules?
(b) How many times heavier is a molecule of the unknown gas than an atom of helium?
(c) In which tire do the molecules have the largest kinetic energy? The greatest average speed?

95. In the Chemical Puzzler at the beginning of this chapter, we asked if the operation of the bag depended on the temperature or the altitude of the car. How would you answer these questions now?

CHALLENGING QUESTIONS

96. You have a 550-mL tank of gas with a pressure of 1.56 atm at 24 °C. You thought the gas was pure carbon monoxide gas, CO, but you later found it was contaminated by small quantities of gaseous CO_2 and O_2. Analysis shows that the tank pressure is 1.34 atm (at 24 °C) if the CO_2 is removed. Another experiment shows that 0.0870 g of O_2 can be removed chemically. What are the masses of CO and CO_2 in the tank, and what are the partial pressures of each of the three gases at 25 °C?

97. Methane is burned in a laboratory Bunsen burner to give CO_2 and water vapor. Methane gas is supplied to the burner at the rate of 5.0 L/min (at a temperature of 28 °C and a pressure of 773 mm Hg). At what rate must oxygen be supplied to the burner (at a pressure of 742 mm Hg and a temperature of 26 °C)?

98. Iron forms a series of compounds of the type $Fe_x(CO)_y$. If you heat any of these compounds in air, they decompose to Fe_2O_3 and CO_2 gas. After heating a 0.142-g sample of $Fe_x(CO)_y$, you isolate the CO_2 in a 1.50-L flask at 25 °C. The pressure of the gas is 44.9 mm Hg. What is the formula of $Fe_x(CO)_y$?

99. Group 2A metal carbonates are decomposed to the metal oxide and CO_2 on heating:

$$MCO_3(s) \longrightarrow MO(s) + CO_2(g)$$

You heat 0.158 g of a white, solid carbonate of a Group 2A metal (M) and find that the evolved CO_2 has a pressure of 69.8 mm Hg in a 285-mL flask at 25 °C. Identify M.

100. Silane, SiH_4, reacts with O_2 to give silicon dioxide and water:

$$SiH_4(g) + 2\ O_2(g) \longrightarrow SiO_2(s) + 2\ H_2O(g)$$

If you mix SiH_4 with O_2 in the correct stoichiometric ratio, and if the total pressure of the mixture is 120 mm Hg, what are the partial pressures of SiH_4 and O_2? When the reactants have been completely consumed, what is the total pressure in the flask?

101. Chlorine trifluoride, ClF_3, is a valuable reagent because it can be used to convert metal oxides to metal fluorides:

$$6\ NiO(s) + 4\ ClF_3(g) \longrightarrow 6\ NiF_2(s) + 2\ Cl_2(g) + 3\ O_2(g)$$

(a) How many grams of NiO will react with ClF_3 gas if the gas has a pressure of 250 mm Hg at 20 °C in a 2.5-L flask?
(b) If the ClF_3 described in part (a) is completely consumed, what are the partial pressures of Cl_2 and of O_2 in the 2.5-L flask at 20 °C (in mm Hg). What is the total pressure in the flask?

Summary Questions

102. Chlorine gas (Cl_2) is used as a disinfectant in municipal water supplies, although chlorine dioxide (ClO_2) and ozone are becoming more widely used. ClO_2 is better than Cl_2 because it leads to fewer chlorinated byproducts, which are themselves pollutants.
(a) How many valence electrons are in ClO_2?
(b) The chlorite ion, ClO_2^-, is obtained by reducing ClO_2. Draw a possible electron dot structure for the ion ClO_2^-. (Cl is the central atom.)
(c) What is the hybridization of the central Cl atom in ClO_2^-? What is the shape of the ion?
(d) Which species do you suppose has the larger bond angle, O_3 or ClO_2^-? Explain briefly.
(e) Chlorine dioxide, ClO_2, a yellow-green gas, can be made by the reaction of chlorine with sodium chlorite:

$$2\ NaClO_2(s) + Cl_2(g) \longrightarrow 2\ NaCl(s) + 2\ ClO_2(g)$$

Assume you react 15.6 g of $NaClO_2$ with chlorine gas, which has a pressure of 1050 mm Hg in a 1.45-L flask at 22 °C. How many grams of ClO_2 can be produced? What is the pressure of the ClO_2 gas in a 1.25-L flask at 25 °C?

103. The sodium azide required for automobile air bags is made by the reaction of sodium metal with dinitrogen oxide in liquid ammonia:

$$3\ N_2O(g) + 4\ Na(s) + NH_3(\ell) \longrightarrow NaN_3(s) + 3\ NaOH(s) + 2\ N_2(g)$$

(a) You have 65.0 g of sodium and a 35.0-L flask containing N_2O gas with a pressure of 2.12 atm at 23 °C. What is the maximum possible yield (in grams) of NaN_3?
(b) Draw a Lewis structure for the azide ion. Include all possible resonance structures. Which resonance structure is most likely?
(c) What is the shape of the azide ion?

INTERACTIVE GENERAL CHEMISTRY CD-ROM

Screen 12.3 Gas Laws

(a) In the animation of Charles's law, it is stated that if we extrapolate to a temperature of absolute zero a gas has no volume. Why can this experiment not be performed?
(b) Consider the animation of Avogadro's law. If we instead combined 12 H_2 molecules and 4 N_2 molecules to produce NH_3, what would be the final result? What would the relative volume be?

Screen 12.4 The Ideal Gas Law

The "tool" on this screen allows you to explore the gas laws.

(a) If you did not already know the gas laws, describe how you could use this "tool" to discover them.
(b) Set the external pressure at 200 mm Hg and the amount of CO_2 gas at 180 mg. Change the temperature of the gas. What do you observe? What gas law does this illustrate?
(c) Verify the volume given on the screen for mass = 180 mg, $P = 200$ mm Hg, and temperature = 100 °C.

Screen 12.9 The Kinetic-Molecular Theory of Gases

Answer the questions after reading the screen and observing the animation.

(a) What are all the pictures on this screen (of fish, coffee, garlic, blue cheese, and an orange) intended to make you think about? What does this have to do with the kinetic theory of gases?
(b) If the absolute temperature of a gas doubles, by how much does the average speed of the gaseous molecules increase?
(c) Do the principles of the kinetic-molecular theory include the shape, size, or chemical properties of gas molecules?

Screen 12.10 Gas Laws and the Kinetic-Molecular Theory

(a) What is the nature of pressure on the molecular level assumed to be on this screen?
(b) This screen shows animations describing the following relationships on the molecular scale: P versus n, P versus T, and P versus V. Sketch out a molecular scale animation for the relationship between n and V.

Screen 12.11 Distribution of Molecular Speeds

Examine the Maxwell-Boltzmann distribution tool on this screen.

(a) What trends do you observe about gas speeds and the molar mass of the gas?
(b) What trends do you see for changes in temperature?

Screen 12.12 Application of the Kinetic-Molecular Theory

(a) The presentation on this screen shows that NH_3 molecules diffuse through air more quickly than do HCl molecules. Why does this happen?
(b) What compound is formed when the two gases, NH_3 and HCl, meet?

Chapter 13

Bonding and Molecular Structure: Intermolecular Forces, Liquids, and Solids

(C. D. Winters)

A CHEMICAL PUZZLER

The authors of this text can readily identify with this figure. When we were young, home delivery of milk was a regular feature of urban life, but, living in the northern part of the country in winter, your milk might freeze before it was brought inside. Milk, which is primarily water, freezes at temperatures below 0 °C (32 °F), and the freezing of water is accompanied by an increase of volume. Either the bottle broke when the milk froze, or the ice pushed the cap off the bottle. You know of many other observations that indicate that water expands in volume when it freezes and that ice is less dense than liquid water. Ice cubes float, and lakes freeze at the top and not at the bottom. What is interesting about this, however, is that this feature has tremendous importance to our lives. These properties of water are shared by very few other substances in the universe. They are so strange that we want to know why they happen.

Of the 112 known elements, the vast majority are solids at 25 °C and 1 atm of pressure. Only 11 elements occur as gases under these conditions (H_2, N_2, O_2, F_2, Cl_2, and the 6 noble gases), and only 2 elements occur as liquids (Hg and Br_2), although 2 others (Cs and Ga) melt at only slightly higher temperatures. Many common compounds are gases (such as CO_2 and CH_4), and liquids (H_2O), but as is the case with the elements, the largest number of compounds are also solids.

We will often categorize materials by their physical state because properties that distinguish the three states of matter are unambiguous, and an identification as solid, liquid, or gas is easily made. In doing this, however, we tend to restrict our thinking to this fairly specific set of conditions because these are the conditions common to the surface of the earth, and the conditions in which we are most likely to encounter the materials in question. Taking a broader perspective, however, we will see that materials may be found in different physical states under other conditions. For example, at temperatures below −196 °C, nitrogen is a liquid; below −210 °C, it is a solid. Sodium chloride, a crystalline solid at room temperature, is a liquid above 800 °C and a gas above 1413 °C.

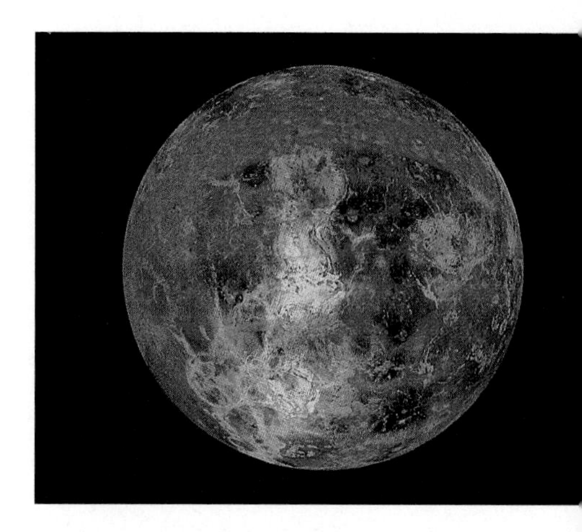

The temperature on the surface of Venus is many hundreds of degrees Celsius. If these were conditions under which we normally made observations, we would think of water as a gas rather than a liquid. *(JPL/Tsado/Tom Stack & Associates)*

Scientists will want to consider elements and compounds under a wide range of conditions (Figure 13.1). Thus, understanding the characteristics of the three states of matters takes on increased importance. The behavior of gases was explored in Chapter 12, so we now turn to the other phases of matter, liquids and solids.

The primary objective in this chapter is to evaluate the liquid and solid states of matter in more detail and, in particular, to gain an understanding of these states by looking at the particulate level, the level of atoms, molecules, and ions. You will find this a useful chapter because it explains, among other things, why your body is cooled when you sweat, how bodies of water can influence local climate, why one form of pure carbon (diamond) is hard and another (graphite) is slippery, and why many solid compounds form beautiful crystal shapes.

13.1 STATES OF MATTER AND THE KINETIC-MOLECULAR THEORY

The kinetic-molecular theory of gases (Section 12.6) assumes that gas molecules or atoms are widely separated and that these particles can be considered to be independent of one another. Consequently, we can relate the properties of gases under most conditions by a very simple mathematical equation, $PV = nRT$, the ideal gas law equation. Liquids and solids present a more complicated picture, however. In these states of matter, the particles are close together because forces of attraction occur between them, forces that we could mostly ignore when considering gases. As might be expected, when these features are introduced, it is not possible to create a simple "ideal liquid equation" or "ideal solid equation" to describe these states of matter.

Figure 13.1 Substances in different environments. Why should we want to know how elements and compounds behave under unusual conditions? Consider, for example, the background needed to create materials that will survive both the low temperatures and the extremely low pressures of outer space or the high temperatures experienced on reentry of a space vehicle into the atmosphere. *(David Ducros/Science Photo Library/Photo Researchers, Inc.)*

How different are the states of matter at the particulate level? We can get a sense of this by comparing volumes occupied by equal numbers of molecules of a material in different states. Figure 13.2 shows a flask containing about 300 mL of liquid nitrogen. If all this liquid were allowed to evaporate, the gaseous nitrogen would fill a very large balloon (>200 L) to a pressure of 1 atm at room temperature. A large amount of space exists between molecules in a gas, whereas

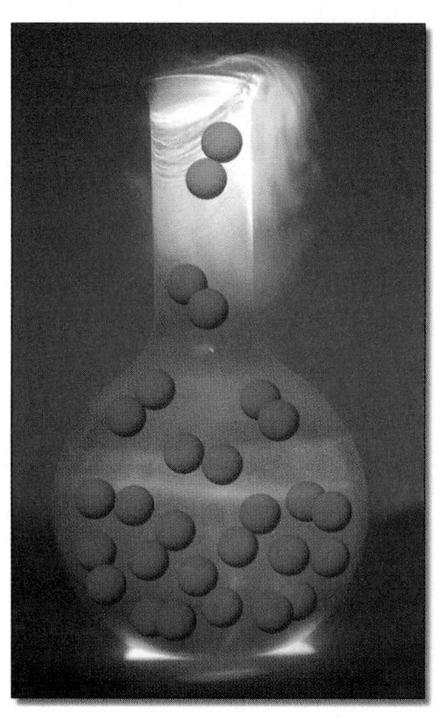

Figure 13.2 Liquid nitrogen. When this 300-mL sample of liquid nitrogen evaporates, it will produce over 200 L of gas at 25 °C and 1.0 atm. As a liquid, the molecules of N_2 are close together, but in the gas phase, they are far apart. *(Photo, C. D. Winters; model, S. M. Young)*

Figure 13.3 Liquid and solid benzene. The same volume of liquid benzene, C_6H_6, is placed in two test tubes, and one tube is cooled, freezing the liquid. The solid and liquid phases have almost the same volume, showing that the molecules are packed together almost as tightly in the liquid phase as they are in the solid. *(C. D. Winters)*

in liquids the molecules are assumed to be close together, in fact touching one another.

See the *Saunders Interactive General Chemistry CD-ROM,* Screen 12.2, Phases of Matter: The Kinetic-Molecular Theory.

The increase in volume when converting liquids to gases is strikingly large. In contrast, no dramatic change in volume occurs when a solid is converted to a liquid. Figure 13.3 shows the same amount of liquid and solid benzene side by side, and you see they are not appreciably different in volume. This means that the atoms in the liquid are packed together about as tightly as the atoms in the solid phase.

We know that gases are readily compressed, a process that involves forcing the gas molecules closer together. Molecules or atoms in the liquid or solid phase strongly resist forces that would push them closer together. This is reflected in the lack of compressibility of liquids and solids. The air–fuel mixture in your car's engine is routinely compressed by a factor of about 10 before it is ignited. In contrast, the volume of liquid water changes only by 0.005% per atmosphere of pressure applied to it.

Intermolecular forces are the attractive forces between molecules or between ions and molecules. Without such interactions, all substances would be ideal gases.

The kinetic-molecular theory of gases assumes that only weak forces of attraction exist between molecules (Section 12.6). Much stronger **intermolecular forces**—the forces between molecules or between ions and molecules—exist in liquids and solids. These forces hold the molecules together in these states and account for the fact that liquids (and solids if they are powdered) can be poured from one container to another and that water and other liquids form droplets. These forces need to be incorporated into a discussion of capillary action (in which a liquid moves up a narrow-diameter tube or channel) and the formation of a meniscus (a curved surface on the top of a liquid in a tube). Because of the importance of intermolecular forces in descriptions of the liquid and solid states, we will look next at this topic.

13.2 INTERMOLECULAR FORCES

The various types of intermolecular forces are listed in Table 13.1. They involve interactions between ions and polar molecules, between polar molecules, and between nonpolar molecules in which a dipole can be induced or created. It is interesting to note that all these forces arise from electrostatic attractions, that is, the attraction between positive and negative charges.

Intermolecular forces will be relevant to many discussions that follow. They are particularly important to an analysis of energies involved in converting between the various states of matter (solid, liquid, gas). They are directly related to properties such as melting point, boiling point, and enthalpy of vaporization, properties that involve the addition of energy to overcome forces of attraction between particles.

See the *Saunders Interactive General Chemistry CD-ROM,* Screens 13.3 and 13.4, Intermolecular Forces.

Intermolecular forces (excepting those involving ions) are known collectively as **van der Waals forces**, named for the physicist who developed the equation to describe the behavior of real gases (Section 12.9).

Table 13.1 • SUMMARY OF INTERMOLECULAR FORCES

Type of Interaction		Principal Factors Responsible for Interaction Energy
Ion (+)	Dipole (− +)	Ion charge; dipole moment
Dipole (− +) (including hydrogen bonding)	Dipole (− +)	Dipole moment
Dipole (− +)	Induced dipole (− +)	Dipole moment, polarizability
Induced dipole (− +)	Induced dipole (− +)	Polarizability

The magnitudes of intermolecular forces are also important in our discussions. In general, you should anticipate that *intermolecular* forces between molecules are not nearly as strong as *intramolecular* bonding forces (as evaluated in Chapters 9 and 10). Recall that most covalent bond energies are in the range of 100 to 400 kJ/mol (Table 9.9). The attractive forces between ions in ionic compounds are usually even higher, often in the range of 700 to 1100 kJ/mol. As a rough guideline, intermolecular forces are generally 15% (or less) of the values of bond energies.

Interactions Between Ions and Molecules with a Permanent Dipole

The distribution of bonding electrons in a molecule often results in a permanent dipole moment (see Section 9.8). Molecules with a dipole have positive and negative ends. If a polar molecule and an ionic compound are mixed, the negative end of the dipole will be attracted to a positive cation and, similarly, the positive end of the dipole will be attracted to a negative anion (see Table 13.1). Forces involved in the attraction between a positive or negative ion and polar molecules are less than for ion–ion attractions, but they are greater than other kinds of intermolecular forces. The **ion–dipole attraction** can be evaluated based on the equation for attraction between charges, Force $\propto$ $(n^+e)(n^-e)/d^2$ (Coulomb's law, Section 3.4). From this equation we can see that attractive forces depend on

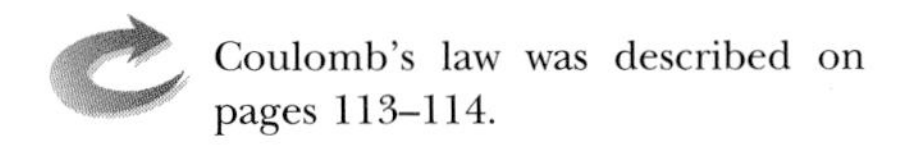

Coulomb's law was described on pages 113–114.

- *The distance between the ion and the dipole:* the closer the ion and dipole, the stronger the attraction.
- *The charge on the ion:* the higher the ion charge, the stronger the attraction.
- *The magnitude of the dipole:* the greater the magnitude of the dipole, the stronger the attraction.

Ion–dipole attractions are related to the sizes of atoms and ions (Section 8.6), to the concepts of formal charge and electronegativity (Section 9.6), and to molecular polarity (Section 9.8).

The formation of hydrated ions in aqueous solution is one of the most important examples of the interaction between an ion and a polar molecule. Water is a polar molecule with positive and negative electric poles (see Figure 9.8). When an ionic compound dissolves in water, the positive and negative ions are surrounded by polar water molecules (see Figures 5.1–5.3). The forces of attraction between the ions and the positive and negative ends of the dipole are strong.

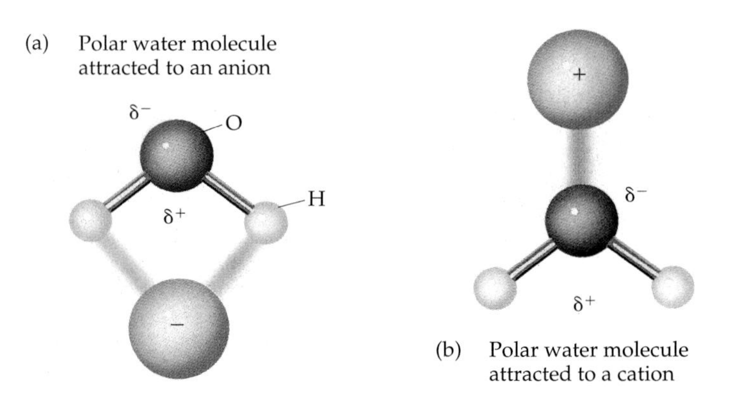

(a) Polar water molecule attracted to an anion

(b) Polar water molecule attracted to a cation

The energy associated with hydration of ions, called the **solvation energy**, or more specifically, the **enthalpy of hydration**, can be substantial. Solvation energy or enthalpy for an individual ion cannot be measured directly, but values can be estimated. The solvation, or hydration, of sodium ions is described by the reaction

$$Na^+(g) + x\,H_2O(\ell) \longrightarrow [Na(H_2O)_x]^+(aq) \quad (x \text{ probably} = 6)$$
$$\Delta H_{rxn} = -405 \text{ kJ}$$

Hydration is important in our watery world. Oceans, lakes, and rivers contain many different ions dissolved in the water (e.g., Na^+, Ca^{2+}, Cl^-, HCO_3^-, and CO_3^{2-}) (see Table 5.1). All are hydrated to a greater or lesser extent.

The energy of attraction depends on $1/d$, where d is the distance between the center of the ion and the oppositely charged "pole" of the dipole. Thus, as the ion radius becomes larger, the enthalpy of hydration becomes less exothermic. The trend in the enthalpy of hydration of the alkali metal cations illustrates this property.

Cation	Ion Radius (pm)	Enthalpy of Hydration (kJ/mol)
Li^+	90	−515
Na^+	116	−405
K^+	152	−321
Rb^+	166	−296
Cs^+	181	−263

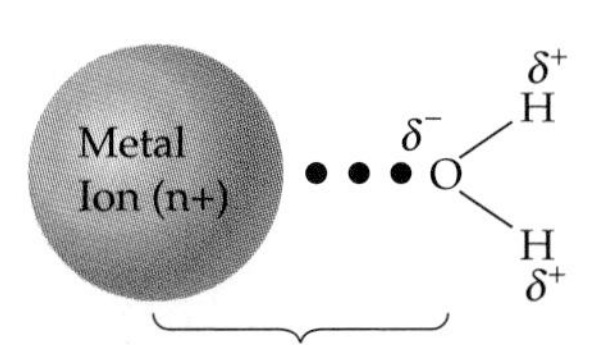

It is interesting to compare these values with the enthalpy of hydration of the H^+ ion, estimated to be −1090 kJ/mol. This extraordinarily large value is due to the tiny size of the H^+ ion. Because the H^+ ion is hydrated, it is commonly represented with the formula of the hydronium ion, H_3O^+, even though the structure is actually more complicated than this.

See the discussion of the hydronium ion on page 197.

Hydrated salts are often encountered in chemistry (Section 3.8). Their formulas are given by appending a specific number of water molecules at the end of the formula; two examples are $BaCl_2 \cdot 2\,H_2O$ or $CoCl_2 \cdot 6\,H_2O$. Sometimes the water molecules simply fill in empty spaces in a crystalline lattice, but often the cation in these salts is directly associated with water molecules. For example, the compound $CoCl_2 \cdot 6\,H_2O$ is better written as $[Co(H_2O)_4Cl_2] \cdot 2\,H_2O$ (Figure 13.4). Four of the six water molecules are associated with the Co^{2+} ion by ion–dipole attractive forces; the remaining two water molecules are in the lattice.

(a)

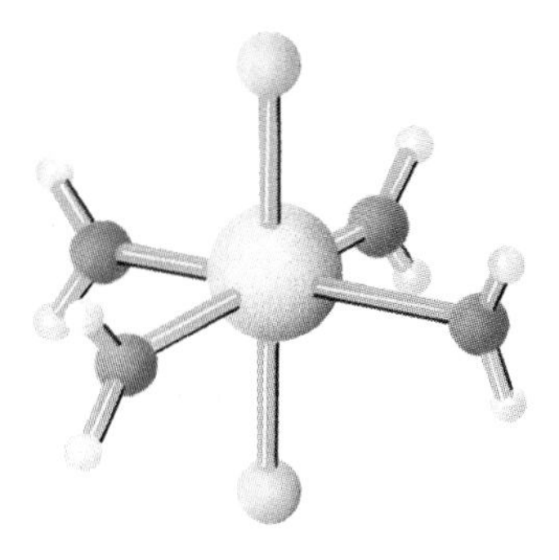
(b)

Figure 13.4 Hydrated cobalt(II) chloride, $CoCl_2 \cdot 6\,H_2O$. (a) The solid compound. (b) The structure of $[Co(H_2O)_4Cl_2] \cdot 2\,H_2O$. Four of the six water molecules are attracted to the Co^{2+} ion by ion–dipole forces. The compound is an example of a coordination compound, a class of compounds discussed in detail in Chapter 23. *(Photo, C. D. Winters)*

Example 13.1 Hydration Energy

Explain why the enthalpy of hydration of Na^+ (−405 kJ/mol) is somewhat more negative than that of Cs^+ (−263 kJ/mol), whereas that of Mg^{2+} is much more negative (−1922 kJ/mol) than that of either Na^+ or Cs^+.

Solution The strength of ion–dipole attractions depends directly on the size of the ion charge and the magnitude of the dipole and inversely on the distance between them. Here we are considering three different ions interacting with the same solvent, water, so only ion charge and size are important considerations. We can obtain ionic radii from Figure 8.14: Na^+ = 116 pm, Cs^+ = 181 pm, and Mg^{2+} = 86 pm. From these values we predict that the distances between the center of the positive charge on the metal ion and the negative side of the water dipole vary in this order: $Mg^{2+} < Na^+ < Cs^+$. The hydration energy varies in the reverse order (the hydration energy of Mg^{2+} being the most negative value).

Notice also that Mg^{2+} has a 2+ charge, whereas the other ions are 1+. The greater charge on Mg^{2+} leads to a greater force of ion–dipole attraction than for the other two ions, which only have a 1+ charge. This means the enthalpy of hydration for Mg^{2+} is much more negative than for the other two ions.

Exercise 13.1 Hydration Energy

Which should have the more negative enthalpy of hydration, F^- or Cl^-? Explain briefly.

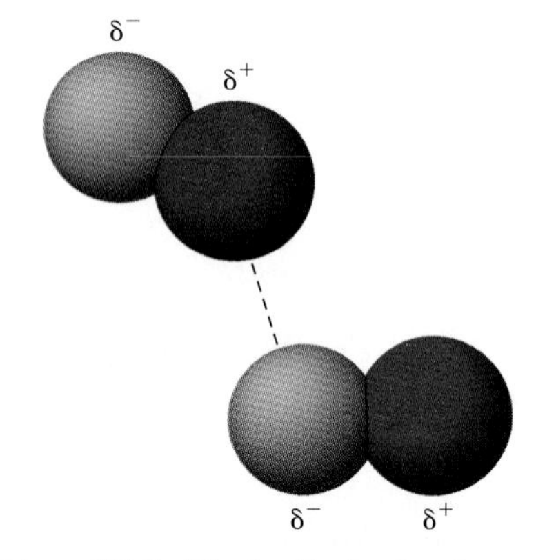

Figure 13.5 Dipole–dipole attractions. Two BrCl molecules are attracted to each other by dipole–dipole forces. Chlorine is more electronegative than bromine, so the bond between these atoms is polar. The negative end of one BrCl molecule is attracted to the positive end of a second molecule.

Interactions Between Molecules with Permanent Dipoles

When a polar molecule encounters another polar molecule, of the same or different kind, the two molecules can interact. The positive end of one molecule is attracted to the negative end of the other polar molecule (Figure 13.5 and Table 13.1). Many molecules have dipoles, and the occurrence of **dipole–dipole attractions** can have important effects on properties of substances.

We can see the importance of intermolecular forces of attraction when we look at the evaporation of a liquid or the condensation of a gas (Figure 13.6). In both processes an energy change occurs. Evaporation requires the addition of heat, specifically the enthalpy of vaporization (ΔH_{vap}) (see also Section 6.4). The value for the enthalpy of vaporization has a positive sign, that is, evaporation is an endothermic process. The enthalpy change for the condensation process—the reverse of evaporation—has a negative value because heat is transferred out of the system upon condensation.

The enthalpy of vaporization is a measure of the attractive forces among molecules in the liquid state. We can understand this by considering the process

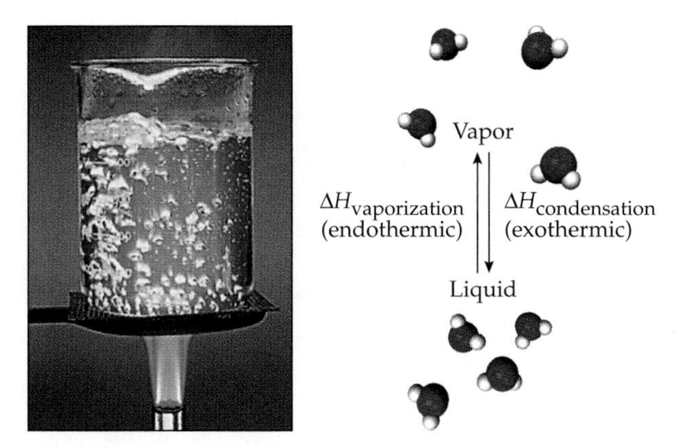

Figure 13.6 Evaporation at the molecular level. Energy must be supplied to separate molecules in the liquid state against intermolecular forces of attraction. *(Photo, C. D. Winters; model, S. M. Young)*

of vaporization at the molecular or particulate level (see Figure 13.6). Intermolecular forces allow molecules to associate with one another in the liquid state. To convert a substance to the gaseous state, a molecule must be separated from neighboring molecules in the liquid state against the forces of attraction in the liquid. This requires the addition of energy, in this case the enthalpy of vaporization.

See the *Saunders Interactive General Chemistry CD-ROM,* Screen 13.4, Intermolecular Forces (2).

The greater the forces of attraction between molecules in a liquid, the greater the energy that must be supplied to separate them. Thus, we expect polar compounds to have a higher value of the enthalpy of vaporization than nonpolar compounds with a similar molar mass. Comparisons between a few polar and nonpolar molecules that illustrate this are given in Table 13.2.

The boiling point of a liquid is also related to intermolecular forces of attraction. As the temperature of a substance is raised, its molecules gain kinetic energy. Eventually, when the boiling point is reached, the molecules have sufficient kinetic energy to escape the forces of attraction of their neighbors. The higher the forces of attraction, the higher the boiling point.

Intermolecular forces influence solubility. A qualitative observation on solubility is that "like dissolves like." This means polar molecules are likely to dissolve in a polar solvent, and nonpolar molecules are likely to dissolve in a non-

Table 13.2 • Molar Masses and Boiling Points of Nonpolar and Polar Substances

Nonpolar			Polar		
	M (g/mol)	bp (°C)		*M* (g/mol)	bp (°C)
N_2	28	−196	CO	28	−192
SiH_4	32	−112	PH_3	34	−88
GeH_4	77	−90	AsH_3	78	−62
Br_2	160	59	ICl	162	97

(a)

(b)

Figure 13.7 "Like dissolves like." (a) Ethylene glycol ($HOCH_2CH_2OH$), a polar compound used as antifreeze in automobiles, dissolves in water. (b) Nonpolar motor oil (a hydrocarbon) dissolves in nonpolar solvents such as gasoline or CCl_4. It will not dissolve in a polar solvent such as water, however. Commercial spot removers use nonpolar solvents to dissolve oil and grease from fabrics. *(Photos, C. D. Winters; models, S. M. Young)*

polar solvent (Figure 13.7). The converse is also true; that is, it is unlikely that polar molecules will dissolve in nonpolar solvents or that nonpolar molecules will dissolve in polar solvents.

Other factors besides energy are also important in controlling the mixing of substances. See Section 14.2.

For example, water and ethanol (C_2H_5OH) can be mixed in any ratio to give a homogeneous mixture. In contrast, water does not dissolve in gasoline to any appreciable extent. The difference in these two situations is that ethanol and water are polar molecules, whereas the hydrocarbon molecules in gasoline (e.g., octane, C_8H_{18}) are nonpolar. Water–ethanol interactions are strong enough that the energy expended in pushing water molecules apart to make room for ethanol molecules is made up for by the energy of attraction between the two polar molecules. In contrast, water–hydrocarbon attractions are weak. The hydrocarbon molecules cannot disrupt the stronger water–water attractions.

Hydrogen Bonding

Hydrogen fluoride and many other compounds with O—H and N—H bonds have exceptional properties. We can see this by examining the boiling points for hydrogen compounds of elements in Groups 4A through 7A (Figure 13.8). Generally, the boiling points of related compounds increase with molar mass. This trend is seen in the boiling points of the hydrogen compounds of Group 4A elements (CH_4, SiH_4, GeH_4, SnH_4). The same effect is also operating for the heavier molecules of the hydrogen compounds of elements of Groups 5A, 6A, and 7A. The boiling points of NH_3, H_2O, and HF, however, are greatly out of line with what might be expected based on molar mass alone. If we were to use the line from H_2Te to H_2Se to H_2S and extrapolate to the expected boiling point of water, water would be expected to boil around −90 °C. The boiling point of water is almost 200 °C higher than this value. Similarly, the boiling points of NH_3 and HF are much higher than would be expected based on molar mass.

See the *Saunders Interactive General Chemistry CD-ROM,* Screen 13.6, Hydrogen Bonding, and Screen 13.7, The Weird Properties of Water.

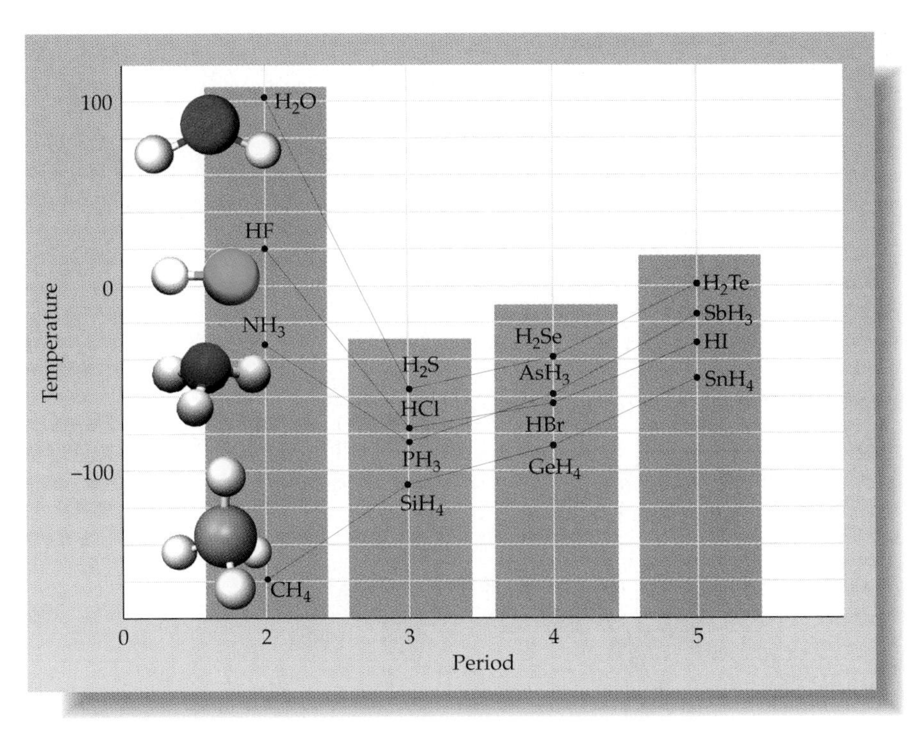

Figure 13.8 The boiling points of some simple hydrogen compounds. The effect of hydrogen bonding is apparent in the unusually high boiling points of H_2O, HF, and NH_3.

Because the temperature at which a substance boils depends on the attractive forces between molecules, the boiling points of H_2O, HF, and NH_3 clearly indicate strong intermolecular attractions. The unusually high boiling points in these compounds are due to **hydrogen bonding,** a special type of dipole–dipole interaction involving polar H—X bonds.

A bond dipole arises as a result of a difference in electronegativity between bonded atoms (see Section 9.6). The electronegativities of N (3.0), O (3.5), and F (4.0) are among the highest of all the elements, whereas the electronegativity of hydrogen is much lower (2.1). The large difference in electronegativity means that N—H, O—H, and F—H bonds are very polar. In bonds between H and N, O, or F, the more electronegative element takes on a partial negative charge, and the hydrogen atom acquires a partial positive charge.

In hydrogen bonding, there is an unusually strong attraction between an electronegative atom with a lone pair of electrons (an N, O, or F atom in another molecule or even in the same molecule) and the hydrogen atom of the N—H, O—H, or F—H bond. A hydrogen bond can be represented as

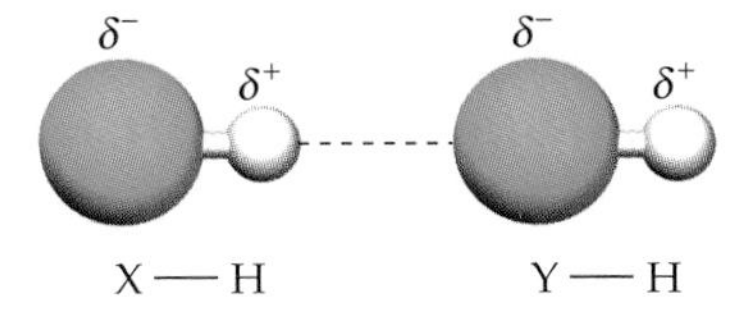

The hydrogen atom becomes a bridge between the two electronegative atoms X and Y, and the dotted line represents the hydrogen bond. The most pronounced effects of hydrogen bonding occur where X and Y are N, O, or F.

Energies associated with most hydrogen bonds involving these elements are in the range of 20 to 30 kJ/mol.

An important reason for the unusual strength of the hydrogen bond compared with dipole–dipole interactions is the small size of the atoms. The positive charge of the X—H dipole is concentrated in a very small volume, and the hydrogen can approach the small electronegative O, F, or N atoms closely. Hydrogen bonding X—H(δ^+) · · · Y(δ^-) can also occur if the element Y is Cl or S, but a much smaller effect is seen.

Types of Hydrogen Bonds

N—H···N—	O—H···N—	F—H···N—
N—H···O—	O—H···O—	F—H···O—
N—H···F—	O—H···F—	F—H···F—

Example 13.2 The Effect of Hydrogen Bonding

Ethanol, CH_3CH_2OH, and dimethyl ether, CH_3OCH_3, have the same formula but a different arrangement of atoms (they are *isomers*). Predict which of these compounds has the higher boiling point.

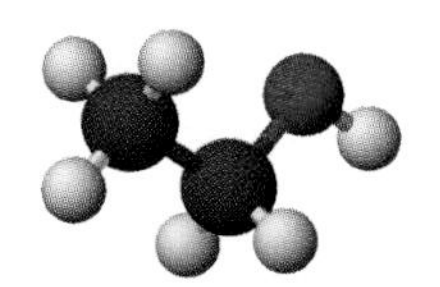

ethanol, CH_3CH_2OH

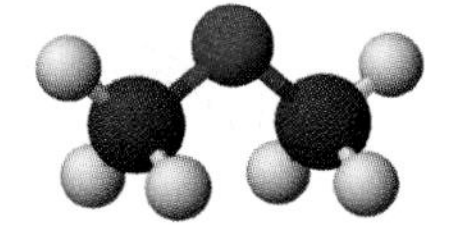

dimethyl ether, CH_3OCH_3

Solution Although these two compounds have identical masses they have very different structures. Ethanol possesses an O—H group, and thus hydrogen bonding is expected to be an important contribution to the intermolecular forces. In contrast, dimethyl ether, although a polar molecule, presents no opportunity for hydrogen bonding. We can predict, therefore, that intermolecular forces will be larger in ethanol than in dimethyl ether and that ethanol will have the higher boiling point. Indeed, ethanol boils at 78.3 °C, whereas dimethyl ether has a boiling point of −24.8 °C, more than 100 °C lower. Dimethyl ether is a gas, whereas ethanol is a liquid under standard conditions.

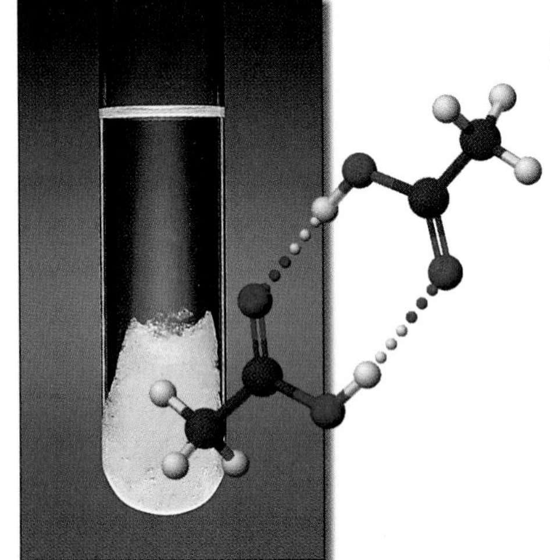

***Figure* 13.9 Hydrogen bonding.** Acetic acid molecules can interact through hydrogen bonds. This photo shows partly solid glacial acetic acid. Notice that the solid is denser than the liquid, a property shared by virtually all substances, the notable exception being water. *(Photo, C. D. Winters; model, S. M. Young)*

Hydrogen bonding has important implications for any property of a compound that is related to intermolecular forces of attraction. For example, it is important in determining structures of molecular solids, one example of which is solid acetic acid. In the solid state, two molecules of CH_3CO_2H are joined to one another by hydrogen bonding (Figure 13.9).

Another example of the effect of hydrogen bonding on the structure of a solid is ice (Figure 13.10). In ice, each hydrogen atom of a water molecule can form a hydrogen bond to a lone pair of electrons on the oxygen atom of an adjacent water molecule. In addition, because the oxygen atom in water has two lone pairs of electrons, it can form more hydrogen bonds with two hydrogen atoms from adjacent molecules (Figure 13.10a). The result is a tetrahedral arrangement for the hydrogen atoms around the oxygen, involving two covalently bonded hydrogens and two hydrogen-bonded hydrogens.

In both liquid and solid water, each water molecule hydrogen-bonds to other water molecules, as illustrated by Figure 13.10a. In liquid water, these interactions lead to a complicated, random structure. Solid structures, however, have

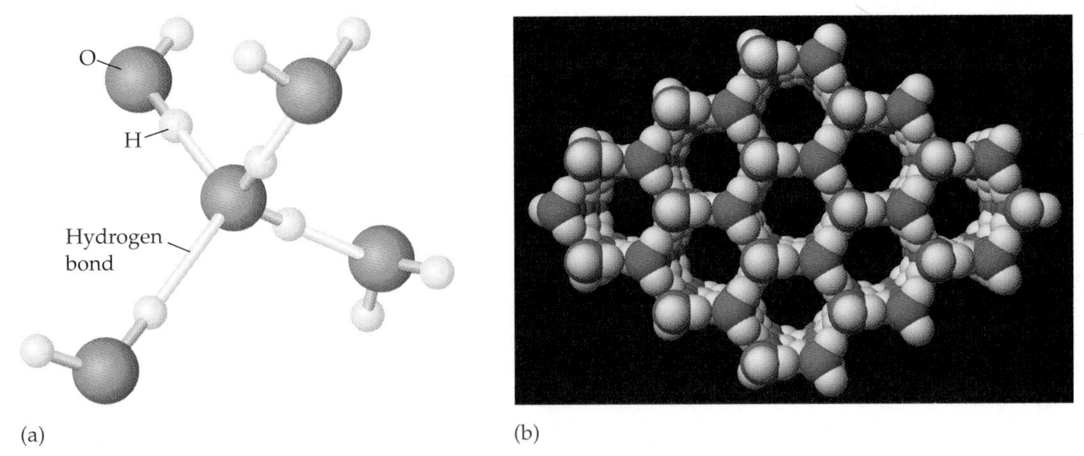

Figure 13.10 The structure of ice. (a) The oxygen atom of a water molecule attaches to four other water molecules. Notice that the four groups that surround an oxygen atom are arranged tetrahedrally. Each oxygen atom is covalently bound to two hydrogen atoms and hydrogen-bonded to hydrogen atoms from two other molecules. The hydrogen bonds are longer than the covalent bonds. (b) In ice, the structural unit shown in part (a) is repeated for each atom in the crystalline lattice. This computer-generated structure shows a small portion of the extensive lattice. Notice that six-member, hexagonal rings are formed. The corners of the hexagon are O atoms, and each side is composed of a normal O—H bond and a longer hydrogen bond. (The structure is on the *Saunders Interactive General Chemistry CD-ROM* and at the Web site for this book at http://www.saunderscollege.com.) *(b, S. M. Young)*

regular patterns. Ice is beautifully regular, a three-dimensional lattice of tetrahedral oxygen atoms with hydrogen atoms between them. A small segment of this lattice is shown in Figure 13.10b.

Hydrogen bonding is an important factor in the structure of some synthetic polymers including nylon, in which the N—H unit of the amide interacts with a carbonyl oxygen on an adjacent polymer chain (see Figure 11.20). Hydrogen bonding in Kevlar, also a polyamide (Figure 13.11), gives this material the exceptional strength-to-weight ratio needed for its use in making canoes, ski equipment, and bullet-proof vests.

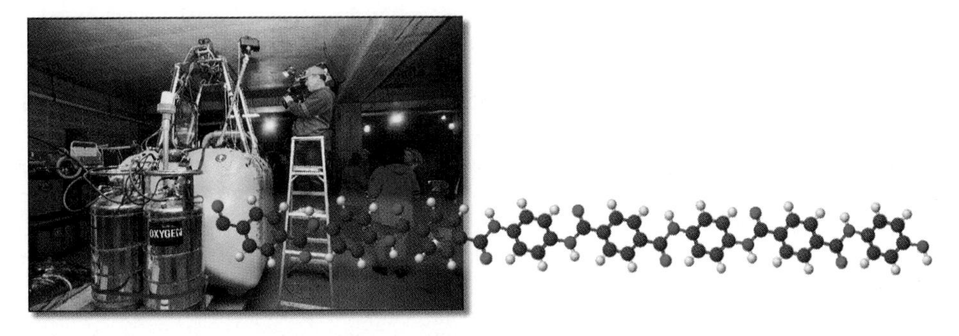

Figure 13.11 The structure of Kevlar. Kevlar is a condensation polymer of terephthalic acid and 1,4-benzenediamine. As in nylon, hydrogen bonding can occur between the N—H group of an amide and a carbonyl oxygen (see Figure 11.20). Kevlar is remarkably strong and is used in sails for racing boats, in bullet-proof vests, and in reinforcement fibers for automobile tires. The photo shows a basket made from Kevlar for a balloon built to circumnavigate the earth. *(Photo, David Kipler/Washington University Photographic Services; model, S. M. Young)*

A CLOSER LOOK

The Unusual Properties of Water: A Consequence of Hydrogen Bonding

One of the most striking differences between our planet and others in the solar system is the existence of liquid water on the earth. Three fourths of the globe is covered by oceans, the polar regions are vast ice fields, and even soil and rocks hold large amounts of water. Although we tend to take water for granted, almost no other substance behaves in a similar manner. Its unique features reflect the ability of water molecules to cling tenaciously to one another by hydrogen bonding.

To achieve the regular arrangement of water molecules linked by hydrogen bonding, ice has an open-cage structure with lots of empty space (see Figure 13.10). The result is a density about 10% less than that of liquid water, explaining why ice floats. We can also see in this structure that the oxygen atoms are arranged at the corners of puckered, six-sided rings, or hexagons. Snowflakes, as you know, are always based on six-sided figures, and this is a further reflection of the internal molecular structure of ice (see Figure 3.3).

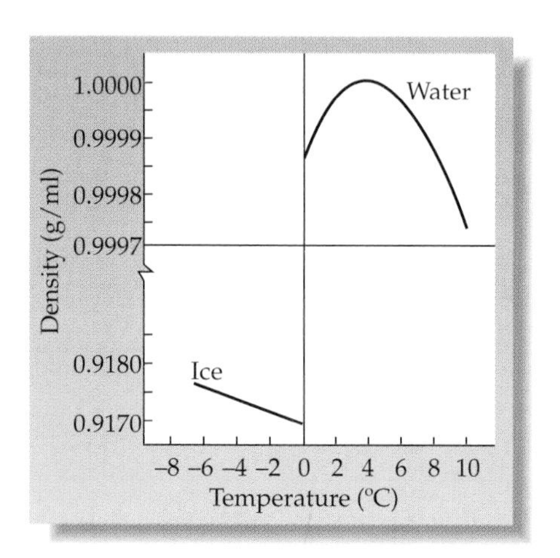

The temperature dependence of the densities of ice and pure water. In contrast, the density of sea water increases with falling temperature right down to the freezing point. This is a crucial distinction between freshwater and sea water, and it has an effect on oceanic circulation and on the formation of sea ice.

When ice melts at 0 °C, a relatively large increase in density occurs, a result of the breakdown of the regular structure imposed on the solid state by hydrogen bonding. As the temperature of liquid water is raised from 0 °C to 4 °C, another surprising thing occurs: the density of water increases. The general rule followed by almost every other substance we know of is that volume increases and the density decreases as the temperature is raised (and the kinetic energy of the water molecules increases). Once again, hydrogen bonding is the reason for water's behavior. At a temperature just above the melting point, some of the water molecules continue to be clustered in ice-like arrangements, which retain extra empty space. As the temperature is raised from 0 °C to 4 °C, the final vestiges of the ice structure disappear and the volume contracts further, giving rise to the increase in density. Water's density reaches a maximum at about 4 °C, and from this point the density declines with increasing temperatures in the normal fashion.

Because of the way that water density changes as the temperature approaches the freezing point, lakes do not freeze solidly from the bottom up in the winter. When lake water cools with the approach of winter, its density increases, the cooler water sinks, and the warmer water rises. This means the water and dissolved material turn over until all the water reaches 4 °C, the maximum density. (This process carries oxygen-rich water to the lake bottom to restore the oxygen used during the summer, and it brings nutrients to the top layers of the lake; see Figure 1.12.) As the water cools further, it stays on the top of the lake, because water cooler than 4 °C is less dense than water at 4 °C. With further heat loss, ice can then begin to form on the surface, floating there and protecting the underlying water and aquatic life from further heat loss.

Extensive hydrogen bonding is also the origin of the extraordinarily high heat capacity of water. Although liquid water does not have the regular structure of ice, hydrogen bonding still occurs. With a rise in temperature, the extent of hydrogen bonding diminishes. Disrupting hydrogen bonds requires heat. The high heat capacity of water is, in large part, why oceans and lakes have such an enormous effect on weather. In autumn, the temperature of the atmosphere drops. When the temperature of the air is lower than the temperature of the ocean or lake, the ocean or lake gives up heat to the atmosphere, moderating the drop in air temperature. Furthermore, so much heat must be given off for each degree drop in temperature (because so many hydrogen bonds form) that the decline in water temperature is gradual. For this reason the temperature of the ocean or a large lake is generally higher than the average air temperature until late in the fall.

A misty morning. A common observation on fall mornings is mist rising from a lake. The lake is giving up heat energy, through evaporation. Water vapor formed by evaporation of lake water condenses in the cold air, forming the small droplets of moisture that we call mist or fog. *(John Gerlach/Tom Stack & Associates)*

Dispersion Forces: Interactions Involving Induced Dipoles

Dispersion forces are found in all molecular substances. Such forces are electrostatic in nature and arise from attractions involving **induced dipoles.** It is these forces that can explain, for example, how some nonpolar molecules like iodine, I_2, can form solids at room temperature or how they can dissolve in water or ethanol.

See the *Saunders Interactive General Chemistry CD-ROM,* Screen 13.5, Intermolecular Forces (3).

Dispersion forces can range from very weak to quite strong. They are the major component of the forces between most molecules. In HCl, for example, about 20% of the intermolecular force is dipole–dipole attraction, and the remaining 80% is a dispersion force. Only in cases in which strong hydrogen bonding occurs are dispersion forces superseded by dipole forces.

Interactions Between Polar and Nonpolar Molecules

Oxygen dissolves in water to a very small extent; a typical concentration is about 10 ppm (or about 0.001% by weight). This is important because microorganisms use dissolved oxygen to convert organic substances dissolved in water to simpler compounds. The quantity of oxygen required to oxidize a given quantity of organic material is called the **biological oxygen demand (BOD).** Highly polluted water often has a high concentration of organic matter and so has a high BOD.

How can nonpolar O_2 molecules dissolve in a polar substance like water? The answer is that polar molecules such as water can induce, or create, a dipole in molecules that do not have a permanent dipole. To see how this can occur, picture a polar water molecule approaching a nonpolar molecule such as O_2 (Figure 13.12). The electron cloud of an isolated (gaseous) O_2 molecule is symmetrically distributed between the two oxygen atoms. As the negative end of the polar H_2O molecule approaches, however, the O_2 electron cloud distorts. In this process, the O_2 molecule itself becomes polar, that is, a dipole has been induced in the otherwise nonpolar O_2 molecule. The result is that H_2O and O_2 molecules are now attracted to one another, although only weakly. Oxygen can dissolve in water because a force of attraction exists between a permanent dipole and an induced dipole (dipole/induced dipole force).

The process of inducing a dipole is called **polarization,** and the degree to which the electron cloud of an atom (e.g., Ne or Ar) or a molecule (e.g., O_2, N_2, or I_2) can be distorted to induce a dipole depends on the **polarizability** of

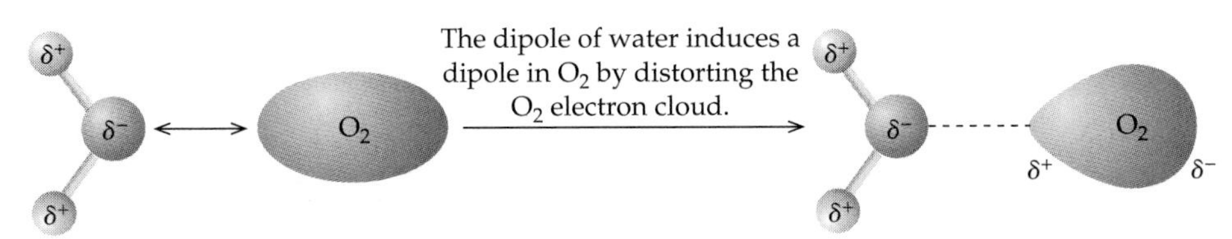

Figure 13.12 Dipole/induced dipole interaction. A polar molecule such as water can induce a dipole in nonpolar O_2.

Table 13.3 • The Solubility of Some Gases in Water*

Gas	Molar Mass (g/mol)	Solubility at 20 °C (g gas/100 g water)†
H_2	2.01	0.000160
N_2	28.0	0.000190
O_2	32.0	0.000434

*Data taken from J. A. Dean: *Lange's Handbook of Chemistry,* 14th ed., pp. 5.3–5.8. New York, McGraw-Hill, 1992.

†Pressure of gas + pressure of water vapor = 760 mmHg.

the atom or molecule. Although this property is difficult to measure experimentally, it makes sense that the valence electrons of atoms or molecules with large, extended electron clouds, such as I_2, can be polarized, or distorted, more readily than the electrons in a much smaller atom or molecule, such as He or H_2, in which the valence electrons are close to the nucleus and more tightly held. In general, for an analogous series of compounds, say the halogens or alkanes, *the higher the molar mass the greater the polarizability of the molecule.*

The table of the solubilities of common gases in water (see Table 13.3) illustrates the effect of interactions between a dipole and an induced dipole. Here you see a trend to higher solubility with increasing mass. As the molar mass of the gas increases, the polarizability of the electron cloud increases. Because a dipole is more readily induced as the polarizability increases, the strength of dipole/induced dipole interactions generally increases with mass. And finally, because the solubility of substances such as O_2 depends on the strength of the dipole/induced dipole interaction, the solubility of nonpolar substances in polar solvents generally increases with mass.

Interactions Between Nonpolar Molecules

Iodine, I_2, is a solid and not a gas at STP, proving that nonpolar molecules must also experience intermolecular forces. An estimate of these forces is the enthalpy of vaporization of the substance at its boiling point. You can see that they range from very weak forces (N_2, O_2, and CH_4 have very low boiling points and enthalpies of vaporization) to quite substantial (I_2 and benzene).

Enthalpy of vaporization was described in Chapter 6 (Section 6.4) and will be discussed again in Section 13.3 (page 599).

Compound	ΔH(vaporization) (kJ/mol)	Bp (°C)
N_2	5.58	−196
O_2	6.82	−183
I_2	41.95	185
CH_4 (methane)	8.2	−161.5
C_6H_6 (benzene)	30.7	80.1

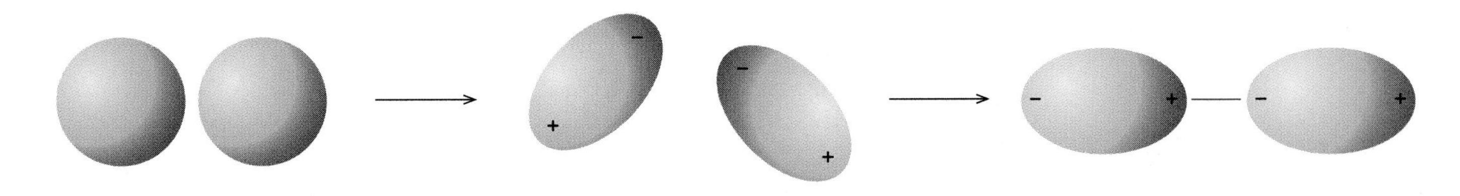

Two nonpolar atoms or molecules
(Time averaged shape is spherical)
Momentary attractions and repulsions between nuclei and electrons in neighboring molecules lead to induced dipoles.
Correlation of the electron motions between the two atoms or molecules (which are now dipolar) leads to a lower energy and stabilizes the system.

Figure 13.13 Induced dipole interactions. Momentary attraction and repulsion between nuclei and electrons creates induced dipoles and leads to a net stabilization due to attractive forces.

Induced dipole/induced dipole forces are sometimes referred to as London dispersion forces.

To understand how two nonpolar molecules can attract each other, remember that the electrons in atoms or molecules are in a state of constant motion. On average, the electron cloud around an atom is spherical (Figure 13.13). When two nonpolar atoms or molecules approach each other, however, attractions or repulsions between their electrons and nuclei can lead to distortions in their electron clouds. That is, dipoles can be induced momentarily in neighboring atoms or molecules, and these induced dipoles lead to intermolecular attraction. Thus, the intermolecular force of attraction in liquids and solids composed of nonpolar molecules is an *induced dipole/induced dipole force.*

Finally, it is important to reemphasize that dispersion forces are found in all molecules, both polar and nonpolar. Dispersion forces, however, are the only forces between nonpolar molecules. This point is made in Figure 13.14, a diagram that outlines the common intermolecular forces. Its purpose is to help you decide what types of forces are appropriate for a given set of molecules.

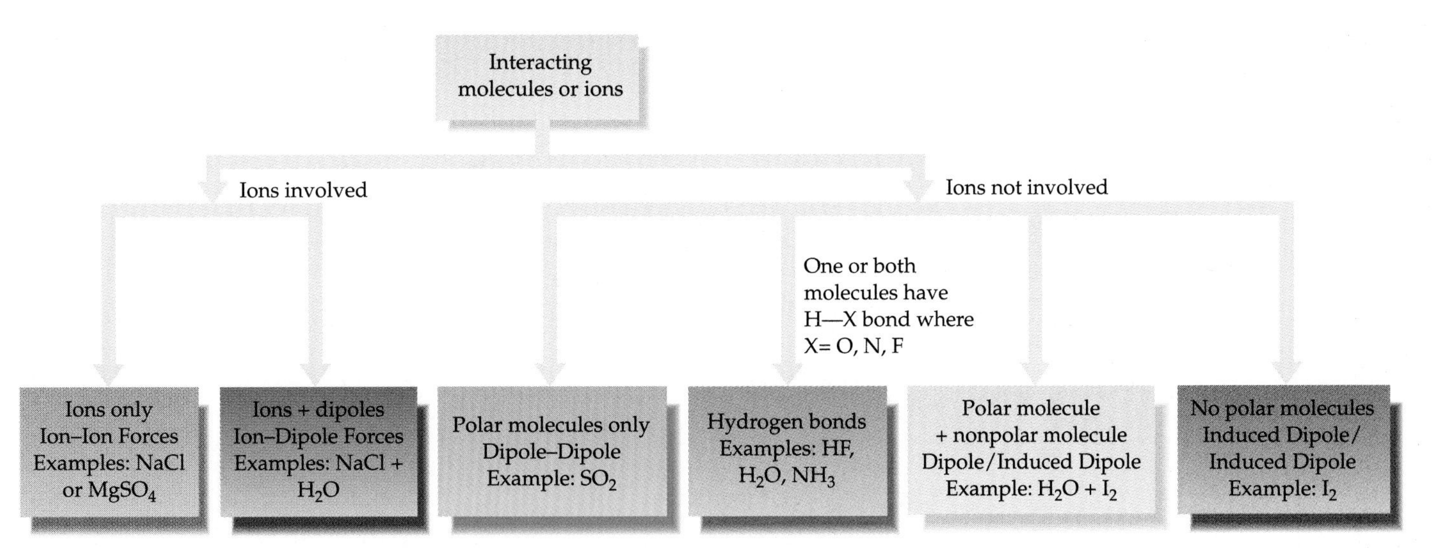

Figure 13.14 Deciphering intermolecular forces. A scheme to help decide what types of intermolecular forces are important in a given system.

Example 13.3 Intermolecular Forces

Decide what type of intermolecular force is involved in each of these examples, and place them in order of increasing strength of interaction: **(a)** liquid methane, CH_4; **(b)** a mixture of water and methanol, H_2O and CH_3OH; **(c)** a solution of lithium chloride, LiCl, in water.

Solution To answer this question it is helpful to follow the outline in Figure 13.14.

(a) Methane, CH_4, is a simple covalent molecule. No ions are involved. After drawing the Lewis structure you would conclude that it must be a tetrahedral molecule and that it cannot be polar. The only way methane molecules can interact is therefore through induced dipole forces.

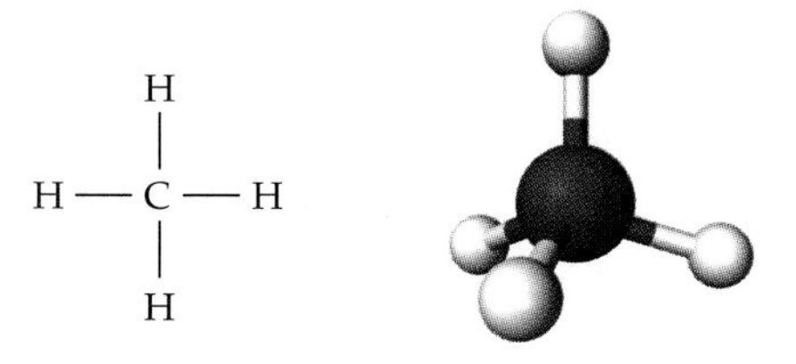

(b) Both water and methanol are covalently bonded molecules, both are polar, and both have an O—H bond. They therefore interact through the special dipole–dipole force called hydrogen bonding.

$$\overset{\delta+}{H}(\overset{\delta+}{H})\overset{\delta-}{O}\cdots\overset{\delta+}{H}-\overset{\delta-}{O}-CH_3 \quad \text{and} \quad \overset{\delta+}{H}(H_3C)\overset{\delta-}{O}\cdots\overset{\delta+}{H}-\overset{\delta-}{O}-H^{\delta+}$$

(c) LiCl is an ionic compound composed of Li^+ and Cl^- ions, and water is a polar molecule. Ion–dipole forces are therefore involved.

In order of increasing strength, the interactions are

$$\text{liquid } CH_4 < H_2O \text{ and } CH_3OH < \text{LiCl in } H_2O$$

Exercise 13.2 Intermolecular Forces

Decide what type of intermolecular force is involved in **(a)** liquid O_2; **(b)** hydrated $MgSO_4$; **(c)** O_2 dissolved in H_2O. Place the interactions in order of increasing strength.

13.3 PROPERTIES OF LIQUIDS

Of the three states of matter, liquids are the most complicated. The molecules in a gas under normal conditions are far apart and may be considered more or less independent of one another. The particles that make up solids—atoms, molecules, or ions—are close together and thus are influenced by their neighbors. The fact that they are in an orderly arrangement, however, makes a description easier. The particles of a liquid interact with their neighbors, like the particles in a solid, but unlike solids there is little order in their structure.

Still, we must turn to the particulate state to describe the characteristics of the liquid state. We can explain, at least qualitatively, such properties of liquids as vaporization, vapor pressure, and fluidity based on this perspective.

Enthalpy of Vaporization

Vaporization, or **evaporation,** the process in which a substance in the liquid state becomes a gas, is a well known phenomenon for liquids. In this process, molecules escape from the liquid surface and enter the gaseous state.

To understand the property of evaporation, we have to look at molecular energies. Molecules that make up a liquid have a range of energies (Figure 13.15) that closely resembles the distribution of energies for molecules of a gas (see Figure 12.18). As with gases, the average energy for molecules in a liquid depends only on temperature: the higher the temperature, the higher the average energy and the greater the relative number of molecules with high kinetic energy. In a sample of a liquid, at least a few molecules have very high energy, and some of the molecules in the liquid state are going to have more kinetic energy than the potential energy of the intermolecular attractive forces holding liquid molecules to one another. If such a high-energy molecule finds itself at the surface of the liquid, and if it is moving in the right direction, it can break free of its neighbors and enter the gas phase (Figure 13.16).

Vaporization is an endothermic process because energy must be expended to separate the molecules against the forces of attraction holding them together. The heat energy required to vaporize a sample is often given as the **molar enthalpy of vaporization,** $\boldsymbol{\Delta H^\circ_{vap}}$ (in units of kilojoules per mole; see Table 13.4).

$$\text{Liquid} \xrightarrow[\text{heat energy absorbed by liquid}]{\text{vaporization}} \text{Vapor}$$

See the *Saunders Interactive General Chemistry CD-ROM,* Screen 13.8, Properties of Liquids (1), Enthalpy of Vaporization.

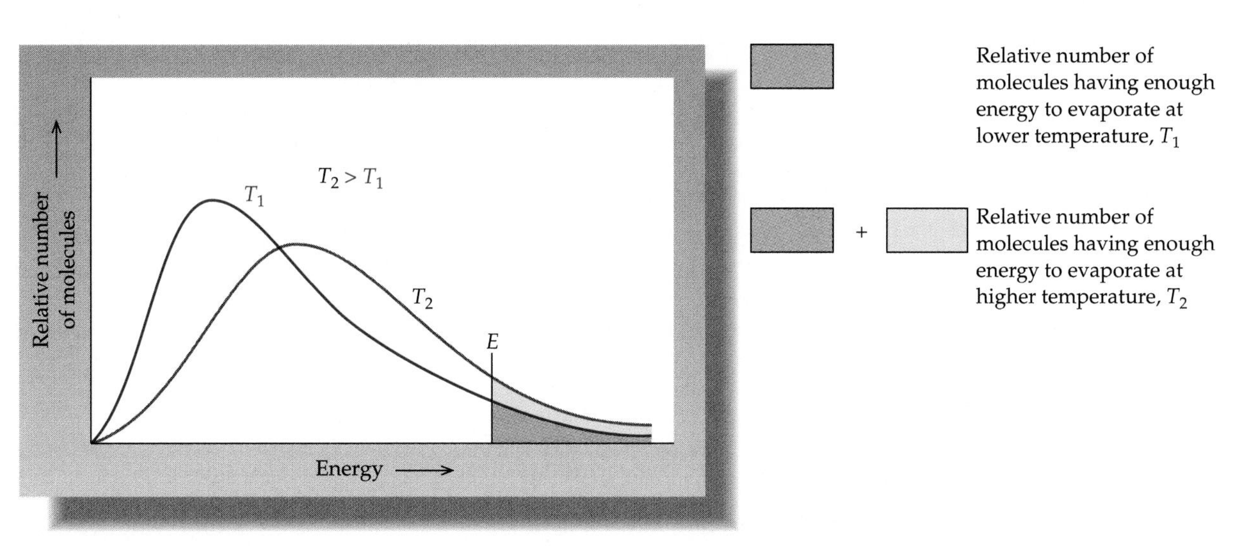

Figure 13.15 The distribution of energy among molecules in a liquid sample. T_2 is a higher temperature than T_1, and more molecules have an energy greater than the value marked E in the diagram at this temperature.

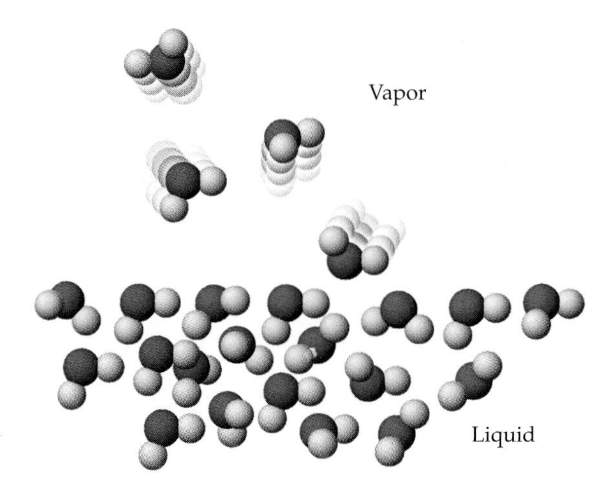

Figure 13.16 Evaporation. Some molecules at the surface of a liquid have enough energy to escape the attractions of their neighbors and enter the gaseous state. At the same time, some molecules in the gaseous state can reenter the liquid.

A molecule in the gas phase will eventually transfer some of its kinetic energy by colliding with slower gaseous molecules and solid objects. If this molecule comes in contact with the surface of the liquid again, it can reenter the liquid phase in a process called **condensation.**

$$\text{Vapor} \xrightarrow[\text{heat energy released by vapor}]{\text{condensation}} \text{Liquid}$$

Condensation is the opposite of vaporization. Condensation is exothermic; when condensation occurs, energy is transferred to the surroundings. The enthalpy change for condensation is equal but opposite in sign to the enthalpy of vaporization. For example, the enthalpy change for the vaporization of 1 mol of water at 100 °C is +40.7 kJ. On condensing 1 mol of water vapor to liquid water at 100 °C, the enthalpy change is −40.7 kJ.

In the discussion of intermolecular forces for nonpolar molecules you might have noticed there is a direct relationship between the ΔH°_{vap} values for various substances and the temperature at which they boil. The larger selection of substances in Table 13.4 confirms this. Both properties reflect the attractive forces between particles in the liquid. The boiling points of nonpolar liquids (e.g., the hydrocarbons, atmospheric gases, and the halogens) increase with increasing atomic or molecular mass, a reflection of increased intermolecular dispersion forces. The alkanes in this list show this trend clearly. Similarly, the boiling points and enthalpies of vaporization of the heavier hydrogen halides (HX, where X = Cl, Br, and I) increase with increasing molecular mass. For these molecules, hydrogen bonding is not as important as it is in HF, so dipole–dipole and dispersion forces take over, and the latter become increasingly important with increasing mass. Notice, however, the very high heat of vaporization of water and hydrogen fluoride that results from extensive hydrogen bonding.

Example 13.4 Enthalpy of Vaporization

You put 1.00 L of water (about 4 cupsful) in a pan at 100 °C, and it slowly evaporates. How much heat must have been supplied to vaporize the water?

Solution Three pieces of information are needed to solve this problem:

1. ΔH°_{vap} for water = +40.7 kJ/mol at 100 °C

Table 13.4 • Molar Enthalpy of Vaporization and Boiling Points for Common Compounds*

Compound	Molar Mass (g/mol)	ΔH°_{vap} (kJ/mol)†	Boiling Point (°C) (Vapor pressure = 760 mm Hg)
Polar Compounds			
HF	20.0	25.2	19.7
HCl	36.5	16.2	−84.8
HBr	80.9	19.3	−66.4
HI	127.9	19.8	−35.6
NH_3	17.0	23.3	−33.3
H_2O	18.0	40.7	100.0
SO_2	64.1	24.9	−10.0
Nonpolar Compounds			
CH_4 (methane)	16.0	8.2	−161.5
C_2H_6 (ethane)	30.1	14.7	−88.6
C_3H_8 (propane)	44.1	19.0	−42.1
C_4H_{10} (butane)	58.1	22.4	−0.5
He	4.0	0.08	−268.9
Ne	20.2	1.7	−246.1
Ar	39.9	6.4	−185.9
Xe	131.3	12.6	−108.0
H_2	2.0	0.90	−252.9
N_2	28.0	5.6	−195.8
O_2	32.0	6.8	−183.0
F_2	38.0	6.6	−188.1
Cl_2	70.9	20.4	−34.0
Br_2	159.8	30.0	58.8

*Data taken from D. R. Lide: *Basic Laboratory and Industrial Chemicals,* Boca Raton, FL, CRC Press, 1993.

†ΔH°_{vap} is measured at the normal boiling point of the liquid.

2. The density of water at 100 °C = 0.958 g/cm^3. (This is needed because ΔH°_{vap} has units of kilojoules per mole, so you first must find the mass of water and then the number of moles.)

3. Molar mass of water = 18.02 g/mol

Given the density of water, a volume of 1.00 L (or 1.00×10^3 cm^3) is equivalent to 958 g, and this mass is in turn equivalent to 53.2 mol of water.

$$1.00 \not{L}\left(\frac{1000 \not{mL}}{1 \not{L}}\right)\left(\frac{0.958 \not{g}}{1 \not{mL}}\right)\left(\frac{1 \text{ mol } H_2O}{18.02 \not{g}}\right) = 53.2 \text{ mol } H_2O$$

Therefore, the amount of energy required is

$$53.2 \not{mol}\left(\frac{40.7 \text{ kJ}}{\not{mol}}\right) = 2.16 \times 10^3 \text{ kJ}$$

$$= \text{heat energy required for vaporization}$$

2160 kJ is equivalent to about one quarter of the energy in your daily food intake.

Figure 13.17 Rainstorms release an enormous quantity of energy. When water vapor condenses, energy is evolved to the surroundings. The enthalpy of condensation of water is very large, so a large quantity of heat is released in a rainstorm. *(Mark A. Schneider/Dembinsky Photo Associates)*

Exercise 13.3 Enthalpy of Vaporization

The molar enthalpy of vaporization of methanol, CH_3OH, is 35.2 kJ/mol at 64.6 °C. How much energy is required to evaporate 1.00 kg of this alcohol?

Water is exceptional among the liquids listed in Table 13.4 in that an enormous amount of heat is required to convert liquid water to water vapor. This fact is important to your environment and your own physical well-being. When you exercise vigorously, your body responds by sweating to rid itself of the excess heat. Heat from your body is consumed in the process of evaporation, and your body is cooled.

Heats of vaporization and condensation of water also play an important role in our weather (Figure 13.17). For example, if enough water condenses from the air to fall as an inch of rain on an acre of ground, the heat released is over 2.0×10^8 kJ! This is equivalent to about 50 tons of exploded dynamite, that is, to the energy released by a small bomb.

Vapor Pressure

See the *Saunders Interactive General Chemistry CD-ROM,* Screen 13.9, Properties of Liquids (2), Vapor Pressure.

If you put some water in an open beaker, it will eventually evaporate completely. If you put water in a sealed flask (Figure 13.18), however, the liquid will evaporate only until the rate of vaporization equals the rate of condensation. At this point, no further change will be observed in the system. The situation is an example of what chemists call a **dynamic equilibrium.** A system at equilibrium is represented using a set of double arrows connecting the reactant and product or the two phases of the substance.

$$\text{Liquid} \rightleftharpoons \text{Vapor}$$

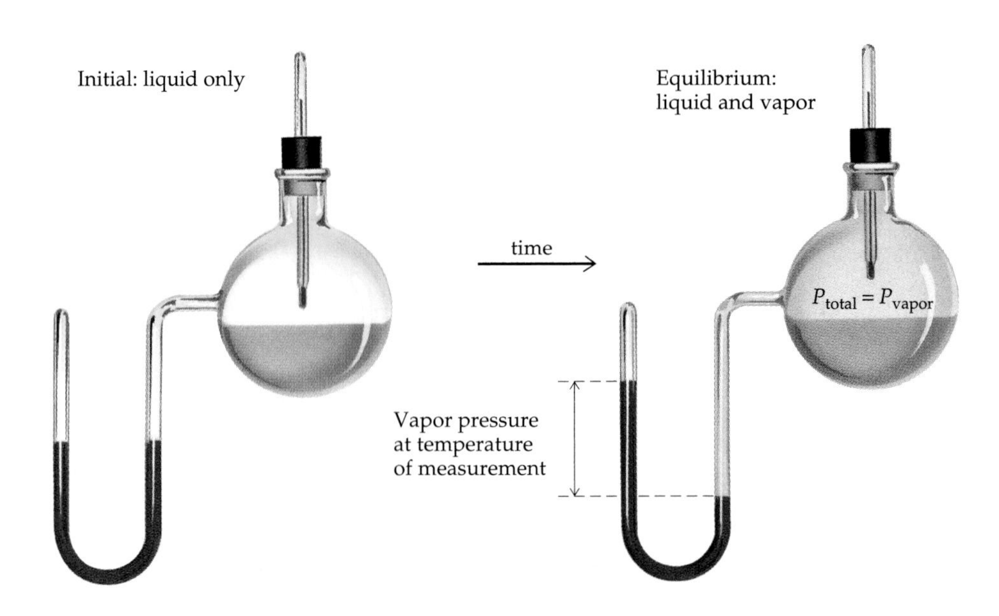

Figure 13.18 Vapor pressure. A volatile liquid is placed in an evacuated flask *(left)*. At the beginning, no molecules of the liquid are in the vapor phase. After a short time, however, some of the molecules have evaporated, and the gaseous molecules exert a vapor pressure. The pressure of the vapor measured when the liquid and vapor are in equilibrium is called the equilibrium vapor pressure *(right)*.

This indicates that two processes are going on at the same time. Molecules continuously move from the liquid to the vapor phase and from the vapor back to the liquid phase. Even though changes occur on the molecular level, no change can be detected on the macroscopic level. Furthermore, because the rate at which molecules move from liquid to vapor is the same as the rate at which they move from vapor to liquid, there is no net change in the masses of the two phases. In contrast to a closed flask, water in an open beaker does not reach an equilibrium with gas phase water molecules; air movement and gas diffusion remove the water vapor from the vicinity of the liquid surface so water molecules are not able to return to the liquid.

When a liquid–vapor equilibrium has been established, the pressure exerted by the water vapor is called the **equilibrium vapor pressure** (often called just *vapor pressure*). The equilibrium vapor pressure of any substance is a measure of the tendency of its molecules to escape from the liquid phase and enter the vapor phase at a given temperature. This tendency is referred to qualitatively as the **volatility** of the compound. The higher the equilibrium vapor pressure at a given temperature, the more volatile the compound.

Equilibrium is a concept used throughout chemistry and one we shall return to often. This situation is signaled by connecting the two states or the reactants and products by a set of double arrows ($\rightleftarrows$).

As described previously (see Figure 13.15), the average energy of molecules in the liquid phase is a function of temperature. At a higher temperature, more molecules have sufficient energy to escape the surface of the liquid. The equilibrium vapor pressure must therefore also increase with temperature (Figure 13.19). All points along the vapor-pressure-versus-temperature curve in Figure 13.19 represent conditions of pressure and temperature at which liquid and vapor are in equilibrium. For example, at 25 °C the equilibrium vapor pressure of water is 24 mm Hg, whereas at 60 °C it is 149 mm Hg. If water is placed in an evacuated flask that is maintained at 60 °C, liquid will evaporate until the pressure exerted by the vapor is 149 mm Hg (assuming there is enough water in the flask so some liquid remains when equilibrium is reached).

At the conditions of T and P given by any point on a curve in Figure 13.19, pure liquid and its vapor are in dynamic equilibrium. If T and P define a point not on the curve, the system is not at equilibrium.

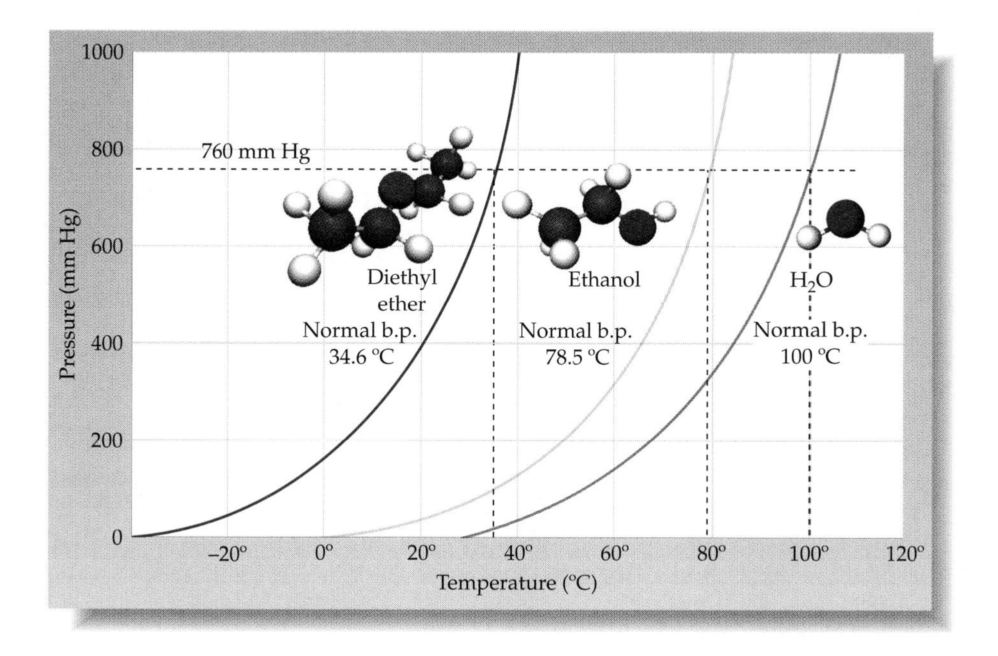

Figure 13.19 Vapor pressure curves for diethyl ether ($C_2H_5OC_2H_5$), ethanol (C_2H_5OH), and water. Each curve represents conditions of T and P at which the two phases, liquid and vapor, are in equilibrium. These compounds exist as liquids for temperatures and pressures to the left of the curve and as gases under conditions to the right of the curve.

Vapor pressures of water at various temperatures are given in Appendix G.

Example 13.5 Vapor Pressure

You place 2.00 L of water in your dormitory room, which has a volume of 4.33 × 10^4 L. You seal the room and wait for the water to evaporate. Will all of the water evaporate at 25 °C? (At 25 °C the density of water is 0.997 g/mL, and its vapor pressure is 23.8 mm Hg.)

Solution One approach to solving this problem is to calculate the quantity of water that must evaporate in order to exert a pressure of 23.8 mm Hg in a volume of 4.33 × 10^4 L at 25 °C. We shall use the ideal gas law to find the quantity of water vapor in moles:

$$n = \frac{PV}{RT} = \frac{\left(\frac{23.8 \text{ mm Hg}}{760 \text{ mm Hg/atm}}\right)(4.33 \times 10^4 \text{ L})}{0.082057 \text{ L} \cdot \text{atm} / \text{K} \cdot \text{mol})(298 \text{ K})} = 55.5 \text{ mol H}_2\text{O}$$

$$55.5 \text{ mol H}_2\text{O} \left(\frac{18.02 \text{ g}}{\text{mol}}\right) = 999 \text{ g H}_2\text{O}$$

$$999 \text{ g H}_2\text{O} \left(\frac{1 \text{ mL}}{0.997 \text{ g}}\right) = 1.00 \times 10^3 \text{ mL H}_2\text{O}$$

This calculation shows that only half of the available water needs to evaporate in order to achieve an equilibrium water vapor pressure of 23.8 mm Hg at 25 °C in the dorm room.

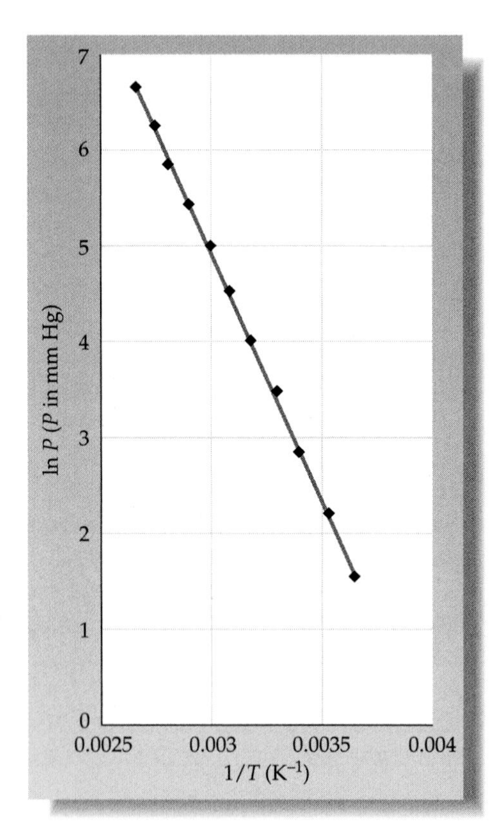

A graph of ln P vs $1/T$ for water. When the natural logarithm of the vapor pressure for water at various temperatures is plotted against $1/T$, a straight line is obtained. The slope of the line equals $-\Delta H_{vap}/R$. Values of T and P are from Appendix G.

Exercise 13.4 Vapor Pressure Curves

Look at the vapor pressure curve for ethanol in Figure 13.19.

(a) What is the approximate vapor pressure of ethanol at 40 °C?

(b) Are liquid and vapor in equilibrium when the temperature is 60 °C and the pressure is 600 mm Hg? If not, does liquid evaporate to form more vapor, or does vapor condense to form liquid?

Exercise 13.5 Vapor Pressure

If 0.50 g of pure water is sealed in an evacuated 5.0-L flask and the whole assembly is heated to 60 °C, will the pressure be equal to or less than the equilibrium vapor pressure of water at this temperature? What if you use 2.0 g of water? Under either set of conditions is any liquid water left in the flask, or does all of the water evaporate?

A mathematical relationship between temperature and the vapor pressure of a pure liquid shown in Figure 13.19 was determined by the German physicist R. Clausius (1822–1888) and the Frenchman B. P. E. Clapeyron (1799–1864) in the 19th century. They showed that vapor pressure and temperature are related by the mathematical equation

$$\ln P = -(\Delta H_{vap}/RT) + C$$

The term ln P is the natural logarithm of the vapor pressure, T is the Kelvin temperature at which P is measured, ΔH_{vap} is the enthalpy of vaporization of the liquid, R is the ideal gas constant, and C is a constant characteristic of the compound in question. This equation, now called the **Clausius-Clapeyron equation,** provides a method of obtaining values for ΔH_{vap}. The equilibrium vapor pressure of a liquid can be measured at several different temperatures, and these pressures are plotted versus $1/T$. The result is a straight line with a slope of $-\Delta H_{vap}/R$.

Boiling Point

See the *Saunders Interactive General Chemistry CD-ROM,* Screen 13.10, Properties of Liquids (3), Boiling Point.

If you have a beaker of water open to the atmosphere, the mass of the atmosphere is pressing down on the surface. As heat is added, more and more water evaporates, pushing the molecules of the atmosphere aside. If enough heat is added, a temperature is eventually reached at which the vapor pressure of the liquid equals the atmospheric pressure. At this point, bubbles of vapor begin to form in the liquid, and the liquid boils (Figure 13.20).

The boiling point of a liquid is the temperature at which its vapor pressure is equal to the external pressure, and, if the external pressure is 1 atm, the temperature is designated the **normal boiling point.** Normal boiling points of some liquids are listed in Table 13.4. Be sure to notice the direct relationship between normal boiling point, enthalpy of vaporization, and intermolecular forces.

The normal boiling point of water is 100 °C, and in a great many places in the United States, water boils at or very near this temperature. If you live at higher altitudes, however, such as in Salt Lake City, Utah, where the barometric pressure is about 650 mm Hg, water will boil at a noticeably lower temperature. The curve for the equilibrium vapor pressure of water in Figure 13.19 shows that a pressure of 650 mm Hg corresponds to a boiling temperature of about 95 °C. Cooks know that food has to be cooked a little longer in Salt Lake City or Denver to achieve the same result as in New York City at sea level.

To shorten cooking time, a pressure cooker can be used. This is a sealed pot that allows water vapor to build up to pressures slightly greater than the external or atmospheric pressure. The boiling point of the water increases, and foods cook faster.

Critical Temperature and Pressure

In a laboratory, it is possible to measure the vapor pressure of a liquid over a wide range of temperatures. Above the normal boiling point, the vapor pressure of a liquid will continue to increase. On first thought it might seem that vapor pressure–temperature curves (such as shown in Figure 13.19) should continue upward without limit, but this is not so. Instead, when a specific temperature and pressure are reached, the interface between the liquid and vapor disappears. This point is called the **critical point.** The temperature at which this occurs is the **critical temperature, T_c,** and the corresponding pressure is the **critical pressure, P_c** (Figure 13.21). The substance that exists under these conditions is called a **supercritical fluid.** It is like a gas under such a high pressure that its density resembles a liquid's, while its viscosity (ability to flow) remains close to that of a gas.

Figure 13.20 Vapor pressure and boiling. When the vapor pressure of the liquid equals the atmospheric pressure, bubbles of vapor begin to form within the body of liquid, and the liquid boils. *(Photo, C. D. Winters; model, S. M. Young)*

For most substances the critical point is at a very high temperature and pressure (see Table 13.5). Water, for instance, has a critical temperature of 374 °C and a critical pressure of 217.7 atm. Consider what the particulate level might look like under these conditions. At this high pressure, water molecules have been forced almost as close together as they are in the liquid state. The high

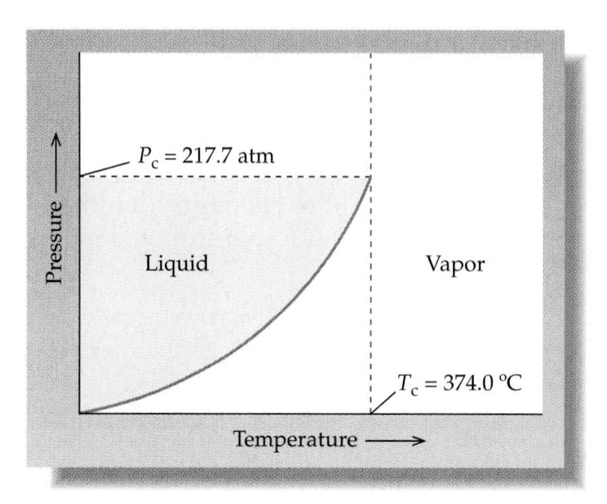

Figure 13.21 The vapor pressure curve for water. The curve representing equilibrium conditions for liquid and gaseous water ends at the critical point; above that temperature and pressure water becomes a supercritical fluid. Notice the scales for the two axes. The portion of the curve shown in Figure 13.19 is such a small part of this graph that it cannot be identified.

temperature, however, means that each of these molecules has enough kinetic energy to exceed the forces holding molecules together. So, the critical fluid has a tightly packed molecular arrangement like a liquid, but the intermolecular forces of attraction that characterize the liquid state are less than the kinetic energy of the particles.

Supercritical fluids like water and carbon dioxide take on unexpected properties, such as the ability to dissolve normally insoluble materials. Supercritical CO_2 is especially useful. This fluid does not dissolve water or polar compounds such as sugar, but it does dissolve nonpolar oils, which constitute many of the flavoring or odor-causing compounds in foods. As a result, food companies now use supercritical CO_2 to extract caffeine from coffee, for example.

Table 13.5 • Critical Temperatures and Pressures for Common Compounds*

Compound	T_c (°C)	P_c (atm)
CH_4 (methane)	−82.6	45.4
C_2H_6 (ethane)	32.3	49.1
C_3H_8 (propane)	96.7	41.9
C_4H_{10} (butane)	152.0	37.3
CCl_2F_2 (CFC-12)	111.8	40.9
NH_3	132.4	112.0
H_2O	374.0	217.7
CO_2	30.99	72.8
SO_2	157.7	77.8

*Data taken from D. R. Lide: *Basic Laboratory and Industrial Chemicals*, CRC Press, Boca Raton, FL, 1993.

Current Perspectives

Supercritical CO_2

Lots of us love a good cup of coffee, but sometimes we can't have a second cup because the caffeine will keep us awake at night. That's the reason the coffee industry has looked for ways to remove caffeine from coffee beans. Various solvents have been tried, but the best method now seems to be to use CO_2 in its supercritical state.

Carbon dioxide is cheap and widely available, essentially nontoxic, nonflammable, and the least expensive solvent after water. It is relatively easy to reach the supercritical state (its critical temperature of 30.99 °C and pressure of 72.8 atm are easily achieved). The material is also easy to handle.

To decaffeinate coffee, the beans are treated with steam to bring the caffeine to the surface. The beans are then immersed in supercritical CO_2, which selectively dissolves the caffeine but leaves intact the compounds that give flavor to coffee. (Decaffeinated coffee contains less than 3% of the original caffeine.) The supercritical CO_2 is poured off, and the CO_2 is recovered by evaporation and then reused in the process.

The fact that caffeine can be so easily dissolved by supercritical CO_2 suggests that it should be an ideal solvent for many purposes. This is important because more than 3.0×10^{10} pounds of organic and halogenated solvents are used worldwide every year as cleaning agents. Many have deleterious effects on the environment, so industry has long sought new methods. Although CO_2 seems an ideal solution, many materials have very low solubilities in it.

Progress is being made, however. A U.S.–Italian team of scientists, headed by University of North Carolina chemist Joseph DeSimone, has developed a surfactant to use with supercritical CO_2. Surfactants such as soaps and detergents enhance the ability of a solvent to dissolve substances. They allow greases and oils, nonpolar materials that are insoluble in water, to be taken into solution as emulsions. Surfactants are typically long-chain molecules that have polar and nonpolar ends. The nonpolar end has an affinity for nonpolar substances, and the polar end has an affinity for water.

A surfactant for supercritical CO_2 would greatly enhance its solvent properties. The problem, however, is to build a molecular substance in which one end of the molecule has an affinity for nonpolar CO_2. It turns out that fluorinated hydrocarbons fit this requirement. The surfactant developed by the DeSimone group is a polymer of styrene with a fluorinated acrylate at one end. Among other things, supercritical CO_2 with added surfactant dissolved polystyrene, a substance that would not otherwise dissolve. This useful property could allow supercritical CO_2 to be used to clean surfaces from small residues of the polymer left from manufacturing processes. (See Jocelyn Kaiser: "Supercritical solvent comes into its own." *Science,* Vol. 274, p. 2013, 1996.)

Supercritical CO_2 is used to remove caffeine, $C_8H_{10}N_4O_2$, from coffee beans.

(Photo, C. D. Winters; model, S. M. Young)

Surface Tension, Capillary Action, and Viscosity

Molecules at the surface of a liquid behave differently from those in the interior because molecules in the interior interact with molecules all around them (Figure 13.22). In contrast, surface molecules are affected only by those below the surface layer. This phenomenon leads to a net inward force of attraction on

See the *Saunders Interactive General Chemistry CD-ROM,* Screen 13.11, Properties of Liquids (4).

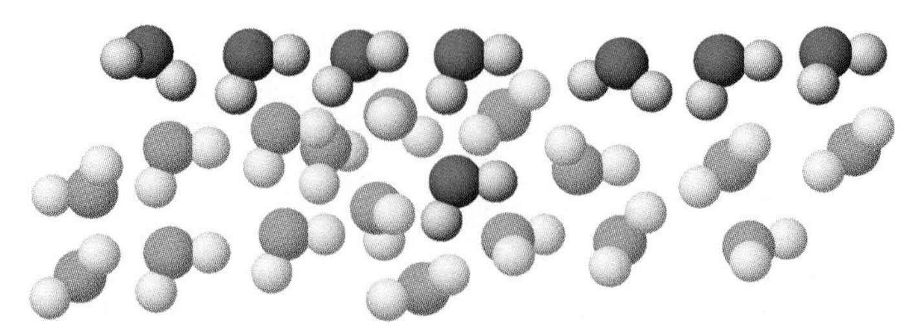

Figure 13.22 Intermolecular forces in a liquid. There is a difference in the forces acting on a molecule at the surface of a liquid and those acting on a molecule in the interior of a liquid.

the surface molecules, contracting the surface and making the liquid behave as though it had a "skin." The "toughness" of the skin of a liquid is measured by its **surface tension**—the energy required to break through the surface or to disrupt a liquid drop and spread the material out as a film. It is surface tension that causes water drops to be spheres and not little cubes, for example (Figure 13.23a), because the sphere has a smaller surface area than any other shape of the same volume.

Capillary action is closely related to surface tension. When a small-diameter glass tube is placed in water, the water rises in the tube, just as water rises in a piece of paper in water (Figure 13.23b). Because there are polar Si—O bonds on the surface of glass, polar water molecules are attracted by **adhesive forces** between the two different substances. These forces are strong enough that they can compete with the **cohesive forces** between the water molecules themselves. Thus, some water molecules can adhere to the walls, and others are attracted to these and build a "bridge" back into the liquid. The surface tension of the water (from cohesive forces) is great enough to pull the liquid up the tube, so the water level rises in the tube. The rise will continue until the various attractive forces—adhesion between water and glass, cohesion between water molecules—are balanced by the force of gravity on the water column. It is these forces that lead to the characteristic concave, or downward-curving, meniscus that you see for water in a drinking glass or in a laboratory test tube (Figure 13.23c).

Some liquids exist for which cohesive forces (high surface tension) are much greater than adhesive forces with glass. Mercury is one example. Mercury does not climb the walls of a glass capillary, and when it is in a tube mercury will form a convex or upward-curving meniscus (Figure 13.23c).

Finally, one additional but important property of liquids is their **viscosity,** the resistance of liquids to flow. When you turn over a glassful of water, it empties quickly. In contrast, it takes much more time to empty a glassful of honey or rubber cement. Although intermolecular forces play a significant role in determining viscosity, other factors are also clearly at work. For example, olive oil consists of molecules with very long chains of carbon and oxygen atoms (see Chapter 11), and it is about 70 times more viscous than ethanol, a small molecule with only two carbons and one oxygen. The long-chain molecules of natural oils are floppy and become entangled with one another; the longer the chain the greater the tangling and the greater the viscosity. Also, longer chains have greater intermolecular forces because there are more atoms to attract one another, each atom contributing to the total force.

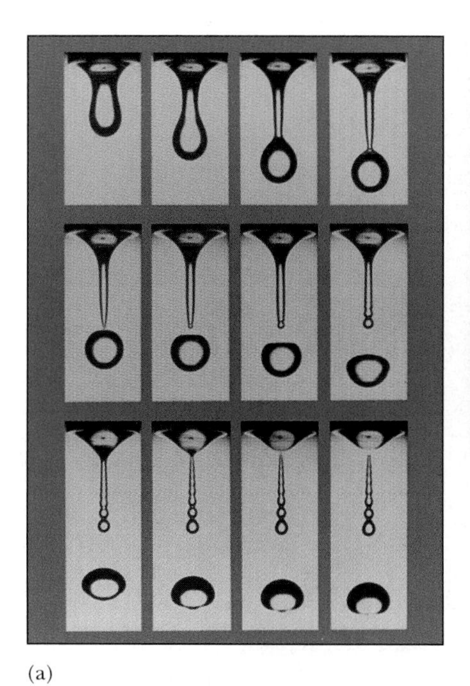

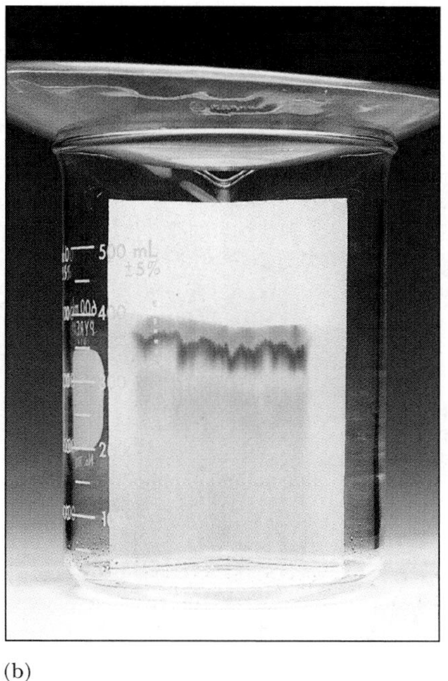

(a) (b) (c)

Figure 13.23 Intermolecular forces in water. (a) A series of photographs showing the different stages when a water drop falls. The drop was illuminated by a strobe light of 5-ms duration. The total time for this sequence was 0.05 s. (b) Capillary action. Polar water molecules are attracted to the O—H bonds in paper fibers, and water rises in the paper. If a line of ink is placed in the path of the rising water, the different components of the ink are attracted differently to the water and paper and are separated in a process called chromatography. (c) Water (top layer) forms a concave meniscus, while mercury (bottom layer) forms a convex meniscus. The different shapes are determined by the adhesive forces of the molecules of the liquid with the walls of the tube and the cohesive forces between molecules of the liquid. *(a, S. R. Nagel, James Frank Institute, University of Chicago; b and c, C. D. Winters)*

Example 13.6 Viscosity

Glycerol ($HOCH_2CHOHCH_2OH$) is used in cosmetics. Do you expect its viscosity to be larger or smaller than the viscosity of ethanol, CH_3CH_2OH? Why or why not?

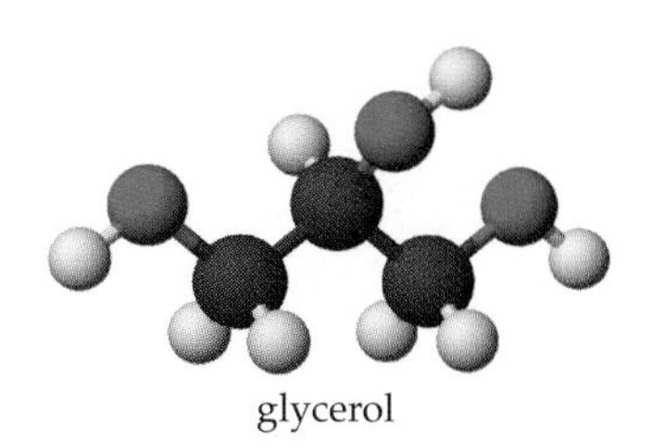

glycerol

13.4 METALLIC AND IONIC SOLIDS

See the *Saunders Interactive General Chemistry CD-ROM,* Screen 13.12, Crystalline and Amorphous Solids.

Many kinds of solids exist in the world around us (Figure 13.24). Solid-state chemistry is one of the booming areas of science, especially because it relates to the development of interesting new materials. As we describe various kinds of solids, we hope to provide a glimpse of the reason this area is exciting.

Solid-state chemistry can be organized by classifying the common types of solids (Table 13.6). This section will describe the solid-state structures of common metallic and ionic solids, with some information on a few other types as well.

Figure 13.24 Types of solids. The following types of solids are included in this photograph (counterclockwise, from bottom right): (a) Salt, NaCl, an ionic solid. (b) A metallic solid, aluminum. (c) The round bar is silicon, a network solid. (d) The plastic laboratory "squeeze bottle" is composed of polyethylene, an amorphous solid. *(Photos, C. D. Winters; models, S. M. Young)*

Crystal Lattices and Unit Cells of Metal Atoms

In both gases and liquids, molecules move continually and randomly, and rotate and vibrate as well. Because of this movement, orderly arrangement of molecules in the gas or liquid state is not possible. In solids, the molecules, atoms, or ions cannot move (although they vibrate and occasionally rotate). Thus, a regular, repeating pattern of atoms or molecules within the structure—a long-range order—is a characteristic of the solid state.

Table 13.6 • Structures and Properties of Various Types of Solid Substances

Type	Examples	Structural Units
Ionic	NaCl, K_2SO_4, $CaCl_2$, $(NH_4)_3PO_4$	Positive and negative ions; no discrete molecules
Metallic	Iron, silver, copper, other metals and alloys	Metal atoms (or positive metal ions surrounded by an electron sea)
Molecular	H_2, O_2, I_2, H_2O, CO_2, CH_4, CH_3OH, CH_3CO_2H	Molecules held together by covalent bonds
Network	Graphite, diamond, quartz, feldspars, mica	Atoms held in an infinite one-, two-, or three-dimensional network
Amorphous (glassy)	Glass, polyethylene, nylon	Covalently bonded networks with no long-range regularity

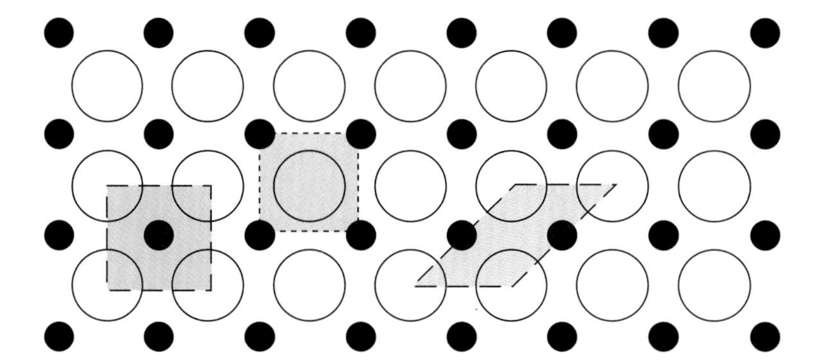

Figure 13.25 Unit cells for a flat, two-dimensional solid made from circular "atoms." Several unit cells are possible, two of the most obvious being squares. The lattice can be built by translating the unit cells throughout the plane of the figure. All unit cells contain a net of one black circle and one white circle.

The beautiful, external (macroscopic) regularity of a crystal of salt (see Figure 13.24 and page 114) suggests that it has an internal symmetry, a symmetry involving the ions that make up the solid. Structures of solids can be described as a three-dimensional lattice of atoms, ions, or molecules. To describe the structure, we can describe the unit cell. For a crystalline solid, a **unit cell** is the smallest repeating unit that has all of the symmetry characteristic of the way the atoms, ions, or molecules are arranged.

To help understand unit cells, let us use a two-dimensional lattice model, the repeating pattern of circles shown in Figure 13.25. The green square shaded in this figure is a unit cell. The complete lattice could be created from a group of unit cells identical to this by joining the unit cells edge to edge. It is also a requirement that a unit cell reflect the "stoichiometry" of the solid. Here the square unit cell clearly contains one fourth of each of four white circles and one black circle, giving a total of one white and one black circle per unit cell.

You will recognize that it is possible to draw other unit cells for the two-dimensional lattice. Another is a square (red) that fully encloses a single circle,

Forces Holding Units Together	Typical Properties
Ionic; attractions among charges on positive and negative ions	Hard; brittle; high melting point; poor electric conductivity as solid, good as liquid; often water-soluble
Metallic; electrostatic attraction among metal ions and electrons	Malleable; ductile; good electric conductivity in solid and liquid; good heat conductivity; wide range of hardness and melting points
Dispersion forces, dipole–dipole forces, hydrogen bonds	Low to moderate melting points and boiling points; soft; poor electric conductivity in solid and liquid
Covalent; directional electron-pair bonds	Wide range of hardnesses and melting points (three-dimensional bonding > two-dimensional bonding > one-dimensional bonding); poor electric conductivity, with some exceptions
Covalent; directional electron-pair bonds	Noncrystalline; wide temperature range for melting; poor electric conductivity, with some exceptions

Figure 13.26 "Cubic Space Division" by M. C. Escher. The artist has depicted three-dimensional space built up by stacking many, many cubes (with each corner of each cube itself a smaller cube). Each cube is a unit cell of the whole, because the whole can be made up from the smaller cubes. *(M. C. Escher's "Cubic Space Division" © 1998 Cordon Art B.V.—Baarn, The Netherlands. All rights reserved.)*

and still another is a parallelogram. It is conventional, however, to try to draw unit cells in which atoms or ions are placed at the **lattice points,** that is, at the corners of the cube or other geometric object that constitutes the unit cell.

Solids (three-dimensional lattices) can be built by assembling three-dimensional unit cells much like building blocks, as the artist M. C. Escher illustrated in a well-known drawing (Figure 13.26). The corners (or lattice points) defining each unit cell in simple solids represent identical environments for the ions in an **ionic solid,** metal atoms in a **metallic solid,** or molecules in a **molecular solid.** Such points are equivalent to one another, and collectively they define the **crystal lattice.**

To construct crystal lattices, nature uses seven, three-dimensional unit cells. These differ from one another in that their sides have different relative lengths and their edges meet at different angles (Figure 13.27). In addition, one of these lattices is differentiated from the others by being hexagonal. The simplest of the seven crystal lattices is the **cubic unit cell,** a cell with edges of equal length that meet at 90° angles. We shall look in detail only at this structure, not only because cubic unit cells are easily visualized but also because they are very common.

Within the cubic class, three important cell symmetries occur: **primitive** or **simple cubic (sc), body-centered cubic (bcc),** and **face-centered cubic (fcc)** (Figure 13.28). All three have eight identical atoms or ions at the corners of the cubic unit cell. The bcc and fcc arrangements, however, differ from the primitive cube because they have additional particles, of the same type as those at the

Unit cells for a number of metallic solids have been illustrated in the book:
- Figure 2.16, potassium (bcc)
- Figure 2.18, copper (fcc)
- Figures 1.2 and 2.18, iron (bcc)
- Figure 2.20, aluminum (fcc)
- Figures 2.17 and 4.3, magnesium (hcp)

Unit cells for ionic solids are
- Figure 1.4, iron pyrite
- Figure 1.9, salt and sodium
- Figure 2.17, MgO

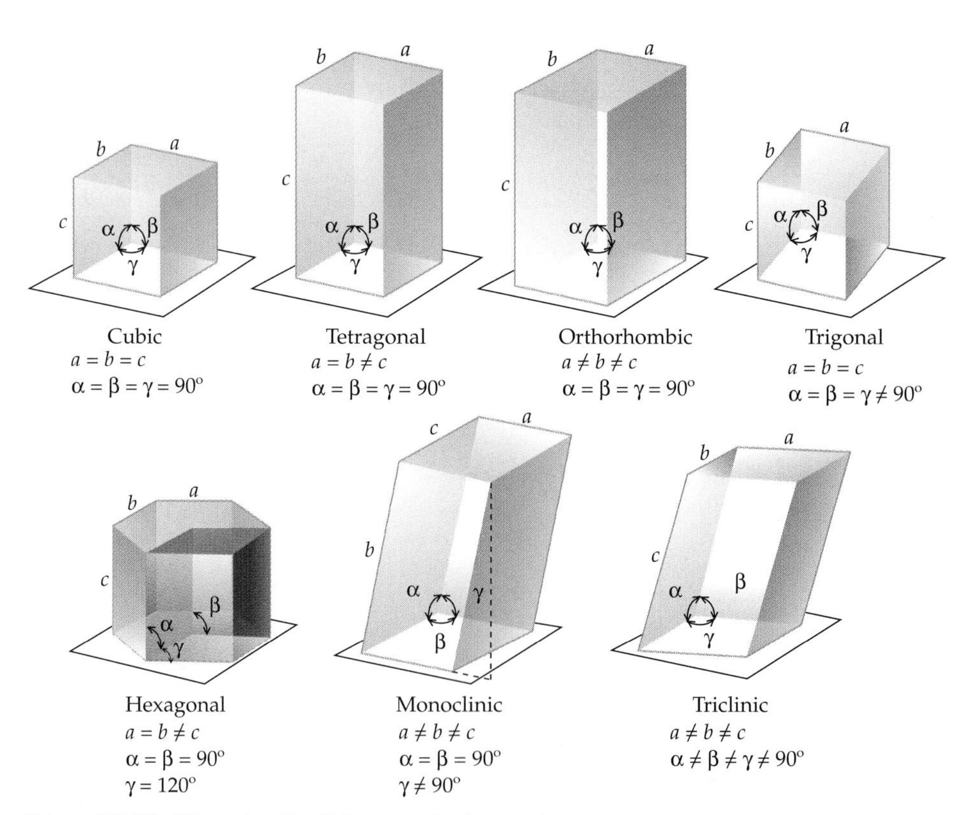

Figure 13.27 The unit cells of the seven basic crystal systems.

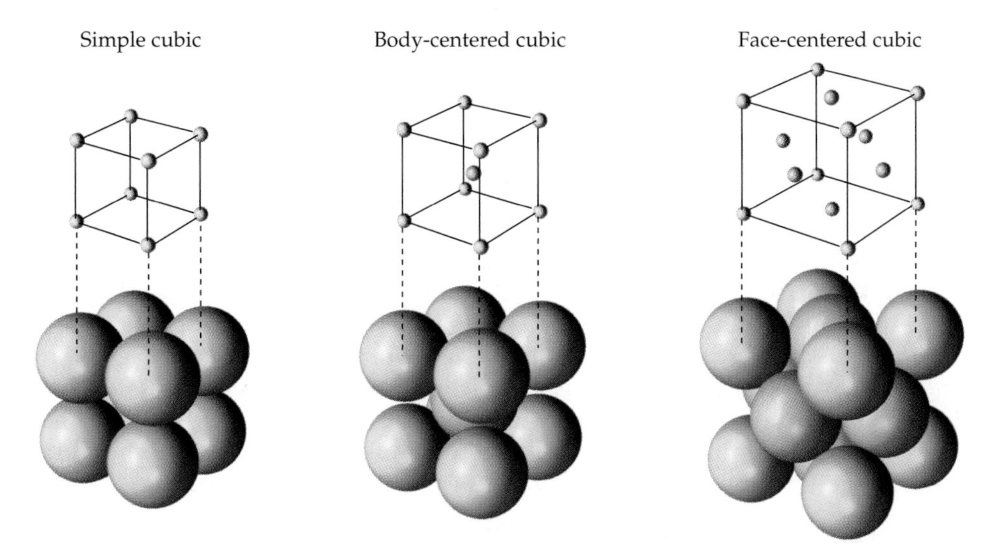

Figure 13.28 The three cubic unit cells. The top row shows the lattice points of the three cells, and the bottom row shows the same cells using space-filling spheres. The spheres in each figure represent identical atoms or ions centered on the lattice points; different colors were used in the drawing only to call attention to the spheres in the center of the cube or on the faces.

corners, at other locations. The bcc structure is called "body-centered" because it has an additional particle at the center or "body" of the cube. The fcc arrangement is called "face-centered" because it has, in the center of each of the six faces of the cube, a particle of the same type as the corner atoms. Metals may assume any of these structures. The alkali metals, for example, are body-centered cubic, whereas nickel, copper, and aluminum are face-centered cubic.

When the cubes pack together to make a three-dimensional crystal, the particle at each corner is shared among eight cubes (Figure 13.29a). Because of this, only one eighth of each corner particle is actually within a given unit cell.

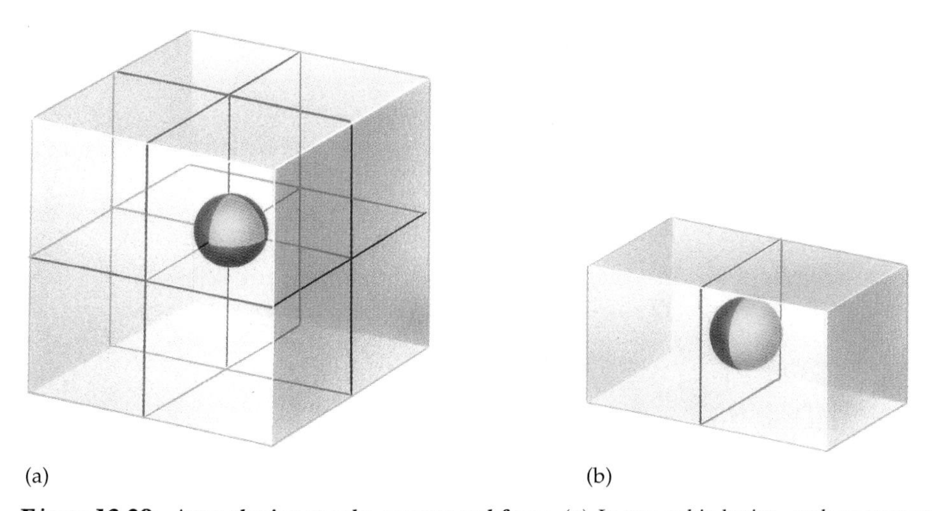

Figure 13.29 Atom sharing at cube corners and faces. (a) In any cubic lattice, each corner particle is shared equally among eight cubes, so one eighth of the particle is within a particular cubic unit cell. (b) In a face-centered lattice, each particle on a cube face is shared equally between two unit cells. One half of each particle of this type is within the given unit cell.

Furthermore, because a cube has eight corners, and because one eighth of the atom or ion at each corner "belongs to" a particular unit cell, the corner particles contribute a net of one particle to a given unit cell.

(8 corners of a cube) (1/8 of each corner particle within a unit cell)
= 1 net particle per unit cell

A primitive or simple cubic arrangement has one net particle within the unit cell.

In contrast to the simple cube, a body-centered cube has an additional particle, and it is wholly within the unit cell at the cube's center. The center particle is present in addition to those at the cube corners, so *a body-centered cubic arrangement has a net of two particles within the unit cell.*

In an fcc arrangement, there is a particle in each of the six faces of the cube in addition to those at the cube corners. One half of each particle on a face belongs to the given unit cell (Figure 13.29b). Three net particles are therefore contributed by the particles on the faces of the cube; that is,

(6 faces of a cube) (1/2 of a particle within a unit cell)
= 3 net face-centered particles within a unit cell

Thus, *the face-centered cubic arrangement has a net of four atom or ions within the unit cell,* one contributed by the corner atoms or ions and another three contributed by the ones centered in the six faces.

An experimental technique, x-ray crystallography, can be used to determine the structure of a crystalline substance (page 616). Once the structure is known, the information can be combined with other experimental information to calculate such useful parameters as the radius of an atom. This approach is outlined in Example 13.6.

A piece of aluminum metal and a face-centered cubic unit cell. *(Photo, C. D. Winters; model, S. M. Young)*

Example 13.6 Determination of an Atom Radius from Measurements of a Crystal Lattice

Aluminum has a density of 2.699 g/cm^3, and the atoms are packed into a face-centered cubic unit cell. Use this information to find the radius of an aluminum atom.

Solution Our strategy for solving this problem is as follows:

1. Find the mass of a unit cell from the knowledge that it is face-centered cubic.
2. Combine the density of the metal with the mass of the unit cell to find the unit cell volume.
3. Use the volume to find the length of an edge of the unit cell.
4. Calculate the radius of an Al atom from the edge dimension.

Step 1. *Calculate the mass of the unit cell.*

$$\text{Mass of unit cell} = \underbrace{\left(\frac{26.98 \text{ g}}{1 \text{ mol}}\right)}_{\text{Molar mass}}\underbrace{\left(\frac{1 \text{ mol}}{6.022 \times 10^{23} \text{ atoms}}\right)}_{\text{Avogadro's number}}\underbrace{\left(\frac{4 \text{ atoms}}{\text{unit cell}}\right)}_{\text{Number of atoms in fcc unit cell}} = 1.792 \times 10^{-22} \text{ g}$$

Step 2. *Calculate the volume of the unit cell.*

$$\text{Volume of unit cell} = \left(\frac{1.792 \times 10^{-22}\ \cancel{g}}{1\ \text{unit cell}}\right)\left(\frac{1\ \text{cm}^3}{2.699\ \cancel{g}}\right)$$

↑ Mass of unit cell ↑ Density of Al

$$= 6.640 \times 10^{-23}\ \text{cm}^3/\text{unit cell}$$

Step 3. *Calculate the length of a unit cell edge.* Now, recall that the volume of a cube = (length of edge)^3. To find the length of an edge, therefore, we need to find the cube root of the cube volume:

$$\text{Length of unit cell edge} = \sqrt[3]{6.640 \times 10^{-23}\ \text{cm}^3} = 4.049 \times 10^{-8}\ \text{cm}$$

Step 4. *Calculate the radius of the atom.* With the edge dimension known, the radius of an Al atom can be calculated. A face of a face-centered cube is shown here. Be sure to notice that, although the corner atoms do not touch one another, the center atom touches each of the corner atoms. The diagonal distance across a face of the cell is equal to four times the radius of each of the spheres. In this case

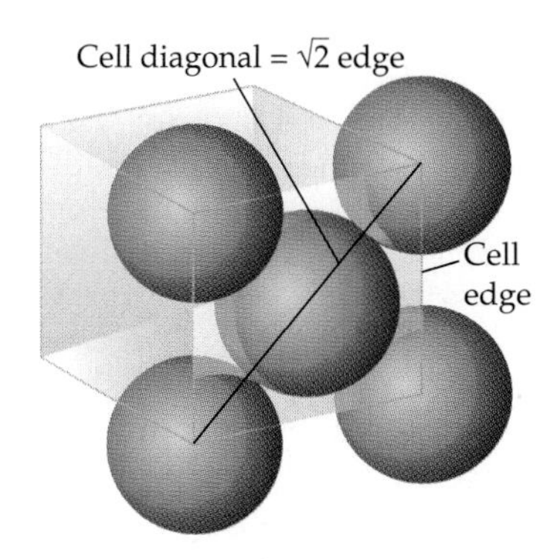

One face of a face-centered cubic unit cell.

$$\text{Cell face diagonal} = 4 \times (\text{Al atom radius})$$

So, our problem is to find the diagonal distance. For a right triangle, the sum of the squares of the sides equals the square of the hypotenuse—the hypotenuse is the face diagonal in this case. Therefore,

$$(\text{Diagonal distance})^2 = (\text{edge})^2 + (\text{edge})^2 = 2 \times (\text{edge})^2$$

Taking the square root of both sides, we have

$$\begin{aligned}\text{Diagonal distance} &= \sqrt{2} \times (\text{cell edge})\\ &= \sqrt{2} \times (4.049 \times 10^{-8}\ \text{cm}) = 5.727 \times 10^{-8}\ \text{cm}\end{aligned}$$

If this distance is divided by 4, we obtain the Al atom radius.

$$\text{Al atom radius} = \frac{(5.727 \times 10^{-8}\ \text{cm})}{4} = 1.432 \times 10^{-8}\ \text{cm}$$

Atomic dimensions are usually expressed in picometers. We apply the appropriate conversion factors to obtain a value for the radius of 143.2 pm.

$$1.432 \times 10^{-8}\ \cancel{\text{cm}}\left(\frac{1\ \cancel{\text{m}}}{100\ \cancel{\text{cm}}}\right)\left(\frac{1\ \text{pm}}{1 \times 10^{-12}\ \cancel{\text{m}}}\right) = 143.2\ \text{pm}$$

This is in excellent agreement with the radius given in Figure 8.10.

Example 13.7 The Structure of Solid Iron

Iron has a density of 7.8740 g/cm^3, and the radius of an iron atom is 126 pm. Verify that the structure of the solid is a body-centered cube.

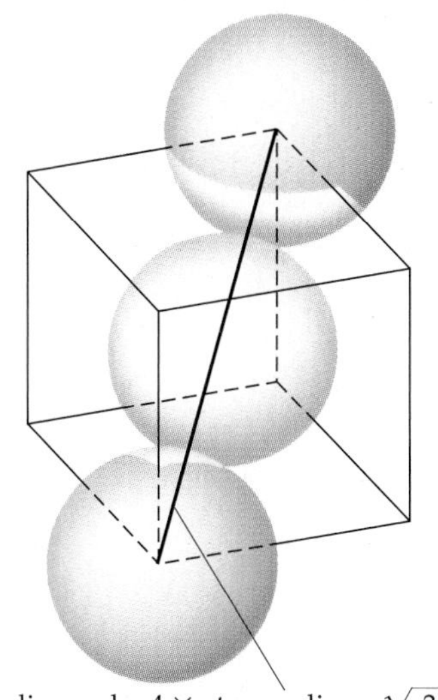

Body-centered cube with two opposite corner atoms and center atom.

Solution As outlined in the text, a body-centered cubic unit cell contains a net of two atoms. We can verify the bcc structure by using the density of iron and the atom radius to calculate the net number of atoms in the unit cell. If

A CLOSER LOOK *Using X-rays to Determine Crystal Structure*

How do chemists determine the distance between the atoms in a metal, the ions in a salt, or the atoms of a molecule? In 1912 Max von Laue (1879–1960), a German physicist, found that crystalline solids diffract x-rays. Somewhat later, the English scientists William (1862–1942) and Lawrence (1890–1971) Bragg (father and son), showed that x-ray diffraction by crystalline solids could be used to determine distances between atoms. X-ray crystallography, the science of determining atomic-scale crystal structure, is now a procedure used extensively by chemists.

To determine distances on an atomic scale we have to use a probe that can locate atoms or ions. The probe must have dimensions not much larger than the atoms, otherwise it would pass over them unperturbed. The probe must therefore be only a few picometers in size. X-rays meet this requirement.

Solids are generally studied because the atoms or molecules are in fixed positions, so they can be located with x-ray photons and because the ions, atoms, or molecules of the crystal must be arranged in an orderly manner to get a regular diffraction pattern.

To "see" radiation interact with atoms we have to rely on a change in the radiation as its passes through the crystal. The change is observed as a scattering, or "diffraction," of the photons of the radiation from their original path (Figure A).

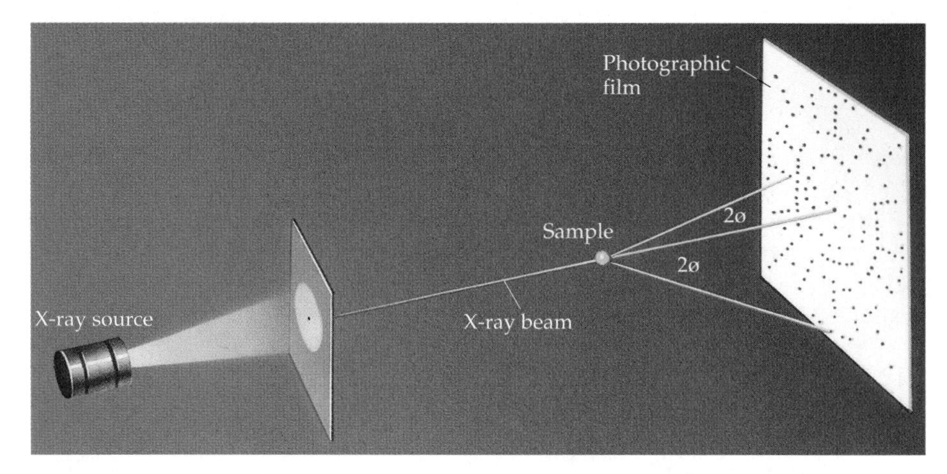

Figure A In the x-ray diffraction experiment, a beam of x-rays is directed at a crystalline solid. The photons of the x-ray beam are scattered by the atoms of the solid. The angle of scattering depends on the locations of the atoms in the crystal. The scattered x-rays are detected by a photographic film or an electronic detector.

there are two atoms, the unit cell is bcc. Our strategy can therefore be laid out as follows:

1. Use the atom radius to find the length of an edge of the cube.
2. Calculate the unit cell volume from the edge dimension.
3. Combine the unit cell volume with the experimental density to calculate the mass of the unit cell.
4. Compare the mass of one iron atom with the mass of the unit cell. The unit cell mass should be equal to twice the mass of one Fe atom.

Step 1. *Use atom radius to find length of cube's edge.*

In a body-centered cubic unit cell the atoms of the lattice touch only along the diagonal line running from a corner at the "top" of the cube to the opposite corner at the "bottom" of the cube. (As in the fcc structure in Example 13.6 the corner atoms do not touch one another.) The distance along this diagonal is equal to 4 times the radius of an atom and is also equal to $\sqrt{3}$ times the length of the edge of the cube. The reason for this is as follows:

$$\text{Diagonal distance across cube face} = \sqrt{2} \times (\text{cell edge})$$
$$(\text{Diagonal distance across the cube})^2 = (\text{cell edge})^2 + (\text{face diagonal})^2$$
$$= (\text{cell edge})^2 + [\sqrt{2} \times (\text{cell edge})]^2$$
$$= 3 \times (\text{cell edge})^2$$

The diffraction of x-rays produces a regularly arranged series of spots on a fluorescent screen or photographic film.

Structural information from x-ray crystallography comes from the fact that crystals scatter x-rays depending on the locations of the atoms in the crystal. We can understand this by considering the wave properties of photons. You know that waves have peaks and valleys at different points in space. If two waves meet at some point, and the peak of one wave meets the valley of another, the waves cancel each other at that point. If the waves meet peak-to-peak, however, they reinforce each other, and the radiation produces a detectable signal or spots on a photographic film (Figure B). This condition is met when the x-rays are scattered at special values of the angle ϕ, an angle related to the distances between atoms in the solid.

The experiment as it is really done is more complicated than we have described. Nonetheless, with modern instruments and computers, chemists and physicists can usually determine quite readily the location of atoms in a crystal and the distances between them. Indeed, the technique has provided so much structural information in the past 20 years or so that the science of chemistry has itself been revolutionized. Many of the structural models you have seen in this book are based on the results of x-ray crystallography.

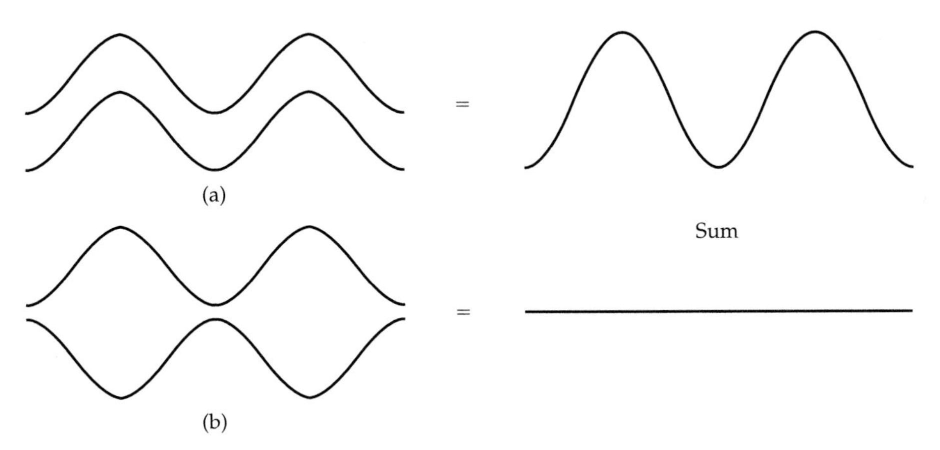

Figure B (a) Constructive interference occurs when two in-phase waves combine to produce a wave of greater amplitude. (b) Destructive interference results from the combination of two waves of equal magnitude that are exactly out of phase.

Taking the square root of both sides, we have

$$\text{Diagonal distance across the cube} = \sqrt{3} \times (\text{cell edge})$$

The cell diagonal distance is equal to 4 times the radius of an atom. Therefore,

$$4 \times \text{atom radius} = \sqrt{3} \times (\text{cell edge})$$

and so the length of an edge of the unit cell in this case is

$$\text{Cell edge} = \frac{4 \times \text{atom radius}}{\sqrt{3}} = \frac{4 \times 126 \text{ pm}}{\sqrt{3}} = 291 \text{ pm}$$

Expressing the length in centimeters, we have

$$291 \text{ pm} \left(\frac{1 \text{ m}}{1 \times 10^{12} \text{ pm}}\right)\left(\frac{100 \text{ cm}}{1 \text{ m}}\right) = 2.91 \times 10^{-8} \text{ cm}$$

Step 2. *Calculate the unit cell volume.*

$$\text{Unit cell volume} = (2.91 \times 10^{-8} \text{ cm})^3 = 2.46 \times 10^{-23} \text{ cm}^3$$

Step 3. *Calculate the mass of the unit cell from the cell volume and density.*

$$\text{Mass of unit cell} = (2.46 \times 10^{-23} \text{ cm}^3)\left(\frac{7.8740 \text{ g}}{\text{cm}^3}\right) = 1.94 \times 10^{-22} \text{ g}$$

Step 4. *Calculate the mass of an iron atom, and compare it with the mass of a unit cell.*

$$\text{Mass of one Fe atom} = \left(\frac{55.85 \text{ g}}{1 \text{ mol}}\right)\left(\frac{1 \text{ mol}}{6.022 \times 10^{23} \text{ atoms}}\right)$$
$$= 9.274 \times 10^{-23} \text{ g/atom}$$

$$\frac{1.94 \times 10^{-22} \text{ g/unit cell}}{9.274 \times 10^{-23} \text{ g/atom}} = 2.09 \text{ atoms per unit cell}$$

This is close to the value for two atoms per unit cell, the value required by a body-centered cubic cell. We can assume that rounding off this value to 2 is reasonable based on uncertainties in the radii used. The iron unit cell is therefore body-centered cubic.

Exercise 13.7 Determination of an Atomic Radius from Measurements of a Crystal Lattice

Gold is a face-centered cubic unit cell. The density of the solid is 19.32 g/cm^3. Calculate the radius of a gold atom.

See the *Saunders Interactive General Chemistry CD-ROM,* Screen 13.13, Ionic Solids.

The holes in the lattice are smaller than the ions used to create an ionic lattice. A lattice, therefore, is usually built out of the larger ions, and the smaller ions are placed into the holes in the lattice. For NaCl, for example, an fcc lattice is built out of the Cl^- ions (radius = 167 pm), and the smaller Na^+ cations (radius = 116 pm) are placed in appropriate holes in the lattice.

Structures and Formulas of Ionic Solids

The lattices of many ionic compounds are built by taking a simple cubic or face-centered cubic lattice of spherical ions of one type and placing ions of opposite charge in spaces between these ions (called holes) within the lattice. This produces a three-dimensional lattice of regularly placed ions. The smallest repeating unit in these structures is, by definition, the unit cell for the ionic compound.

The choice of the lattice and the number and location of the holes that are filled are the keys to understanding the relationship between the lattice structure and the formula of a salt. This is illustrated with the ionic compound cesium chloride, CsCl (Figure 13.30). The structure of CsCl has a primitive cubic unit cell of chloride ions. The cesium ion fits into a hole in the center of the cube. Notice that the Cs^+ ion has eight Cl^- ions as nearest neighbors in this lattice.

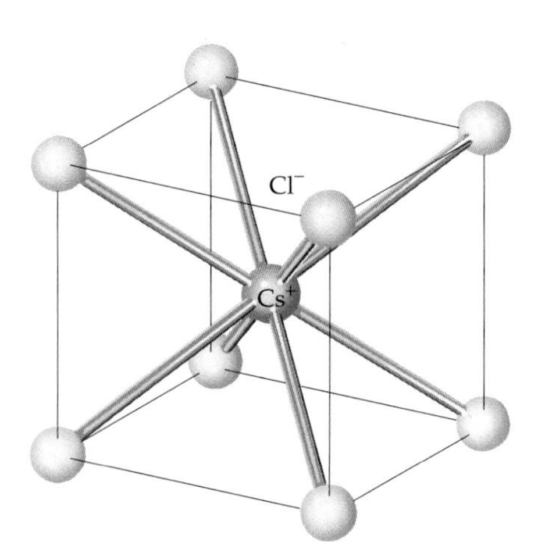

Figure 13.30 Cesium chloride (CsCl) crystal structure. This unit cell has Cl^- ions at the corners of a simple cube and a Cs^+ in the center of the cube.

Next consider the structure for NaCl. An extended view of the lattice and one unit cell are illustrated in Figure 13.31a and 13.31b. The Cl^- ions are arranged in a face-centered cubic unit cell, and the Na^+ ions are arranged in a regular manner between these ions. Notice that each Na^+ ion is surrounded by six Cl^- ions. An octahedral geometry is assumed by the ions surrounding an Na^+ ion so the Na^+ ions are said to be in **octahedral holes** (Figure 13.31c).

The formula for NaCl can be related to this structure by counting the number of cations and anions contained in one unit cell. A face-centered cubic lattice of Cl^- ions has a net of four Cl^- ions within the unit cell. There is one Na^+ in the center of the unit cell, contained totally within the unit cell. In addition, there are 12 Na^+ ions along the edges of the unit cell. Each of these Na^+ ions is shared among four unit cells, so each contributes one fourth of an Na^+ ion to the unit cell, giving three additional Na^+ ions within the unit cell. This accounts for all the ions contained in the unit cell: 4 Cl^- and 4 Na^+ ions. Thus, a unit cell of NaCl has a 1:1 ratio of Na^+ and Cl^- ions as the formula requires.

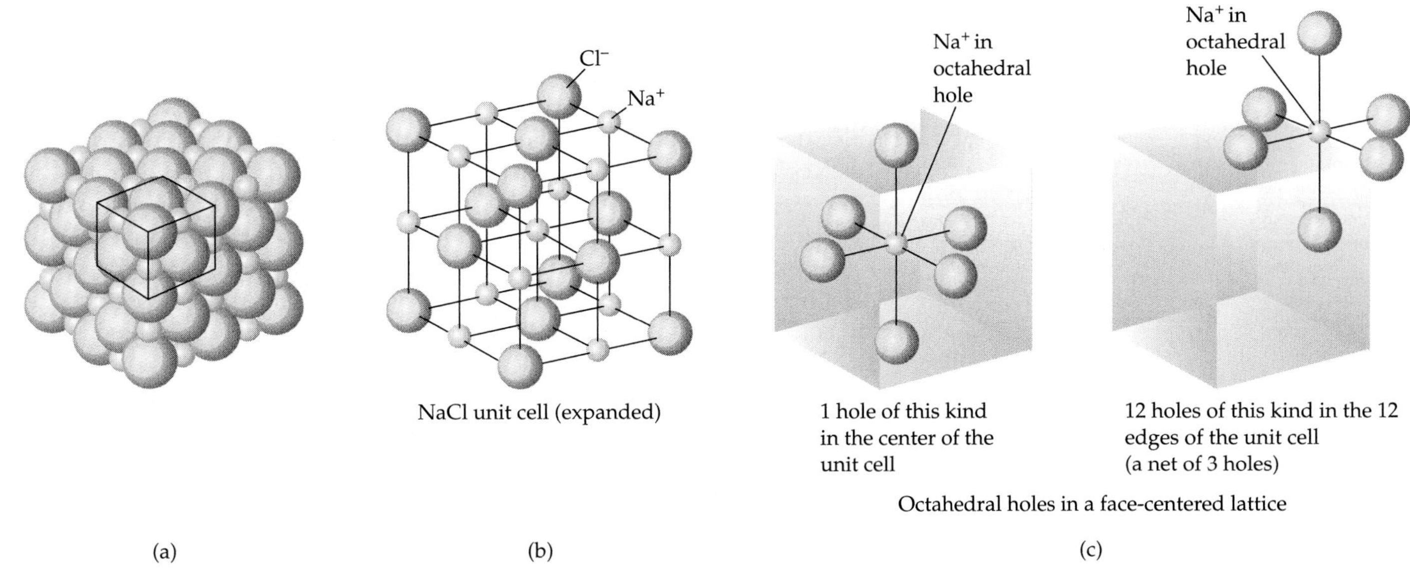

***Figure* 13.31 Sodium chloride.** (a) An extended view of a sodium chloride lattice. The smaller Na^+ ions (silver) are packed into a face-centered cubic lattice of larger Cl^- ions (yellow). One unit cell is outlined. (b) An expanded view of the NaCl face-centered cubic unit cell. The lines only represent the connections between lattice points. (c) A close-up view of the octahedral holes in the lattice.

Example 13.8 Ionic Structure and Formula

One unit cell of the mineral perovskite is illustrated here. The compound is composed of calcium, titanium, and oxygen. Based on the unit cell, what is the formula of perovskite? What is the oxidation number of the titanium ion?

Ca
Ti
O

Perovskite unit cell. See Example 13.8.

Solution The unit cell has Ca^{2+} ions at the corners of the cubic unit cell, a titanium ion in the center of the cell, and oxide ions in the face centers.

Number of Ca^{2+} ions: (8 Ca^{2+} ions at cube corners) $\times$ (1/8 of each ion inside unit cell) = 1 net Ca^{2+} ion

Number of titanium ions: One ion is in the cube center = 1 net titanium ion

Number of O^{2-} ions: (6 O^{2-} ions in cube faces) $\times$ (1/2 of each ion inside cell) = 3 net O^{2-} ions

This means the formula of perovskite is $CaTiO_3$.

The oxidation number of titanium can be figured out using the rules on page 207. Calcium ions are always +2, and oxygen atoms have a −2 oxidation number. Therefore, the oxidation number of titanium in this formula is +4.

Exercise 13.8 Ionic Structure and Formula

Cesium chloride, CsCl, has a cubic unit cell of Cl^- ions with Cs^+ ions in the cubic holes. Prove that the formula of the salt must have one Cs^+ ion per Cl^- ion.

In summary, compounds with the formula MX are commonly formed in either of two crystal structures: (1) M^{n+} ions occupying all the cubic holes of a simple cubic X^{n-} lattice, or (2) M^{n+} ions in all the octahedral holes in a face-centered cubic X^{n-} lattice. CsCl is a good example of (1) (Exercise 13.8), and NaCl is the usual example of (2). Indeed, chemists and geologists in particular have observed that the sodium chloride, or "rock salt," structure is adopted by many ionic compounds, most especially by all the alkali metal halides (except CsCl, CsBr, and CsI), all the oxides and sulfides of the alkaline earth metals, and all the oxides of formula MO of the transition metals of the fourth period. Finally, the formulas of compounds in general must be reflected in the structures of their unit cells (see Example 13.8); therefore, the formula can always be derived from the unit cell structure.

Atoms or ions have different radii, depending on their position in the periodic table (see Figures 8.10 and 8.11). Because metallic or ionic solids form by packing atoms or ions as closely as possible (see page 622), the size of the unit cell and the density of the solid depend on the radii of the atoms or ions involved.

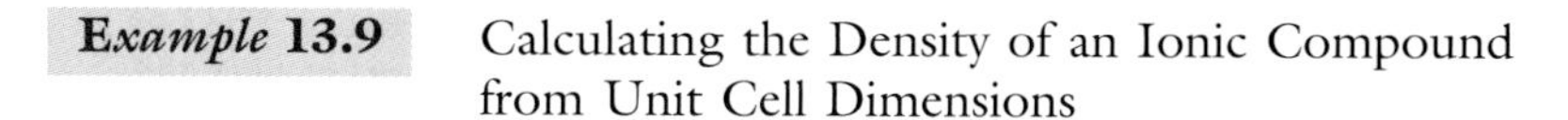

Example 13.9 Calculating the Density of an Ionic Compound from Unit Cell Dimensions

Magnesium oxide has a face-centered unit cell of oxide ions with magnesium ions in the octahedral holes. The radius of the Mg^{2+} ion is 86 pm, and the radius of O^{2-} is 126 pm. Calculate the density of MgO in grams per cubic centimeter.

Solution The density of a solid is the mass of a unit cell divided by the volume of the cell. Let us first find the volume of the unit cell and then its mass.

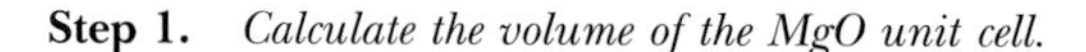

Step 1. *Calculate the volume of the MgO unit cell.*

One face of the face-centered MgO unit cell is shown in the margin. The O^{2-} ions define the lattice, and the Mg^{2+} and O^{2-} ions along each edge just touch one another. This means that one edge of the unit cell is equal to one O^{2-} ion radius plus twice the radius of Mg^{2+} plus another O^{2-} ion radius, or

$$\text{MgO unit cell edge} = 126\text{ pm} + 2\ (86\text{ pm}) + 126\text{ pm} = 424\text{ pm}$$

Because the crystal is cubic, the volume of the unit cell is the cube of the edge.

$$\text{Volume of unit cell} = (\text{edge})^3 = (424\text{ pm})^3 = 7.62 \times 10^7\text{ pm}^3$$

Converting this to cubic centimeters,

$$\text{Volume in cm}^3 = 7.62 \times 10^7\ \cancel{\text{pm}^3}\ (10^{-10}\text{ cm}/\cancel{\text{pm}})^3 = 7.62 \times 10^{-23}\text{ cm}^3$$

Step 2. *Calculate the mass of the MgO unit cell.*

As outlined previously in the text, the unit cell contains four Mg^{2+} and four O^{2-} ions. We can obtain the mass of one formula unit of MgO from the molar mass of MgO and Avogadro's number.

$$\left(\frac{40.30\text{ g}}{1\ \cancel{\text{mol MgO}}}\right)\left(\frac{1\ \cancel{\text{mol MgO}}}{6.022 \times 10^{23}\text{ formula units}}\right) = 6.692 \times 10^{-23}\text{ g/formula unit}$$

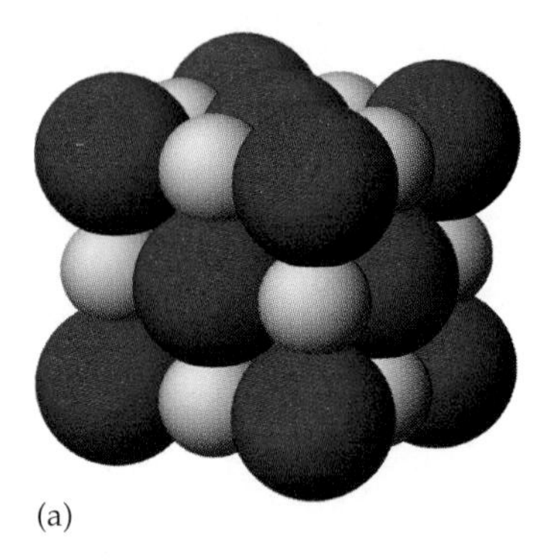

(a)

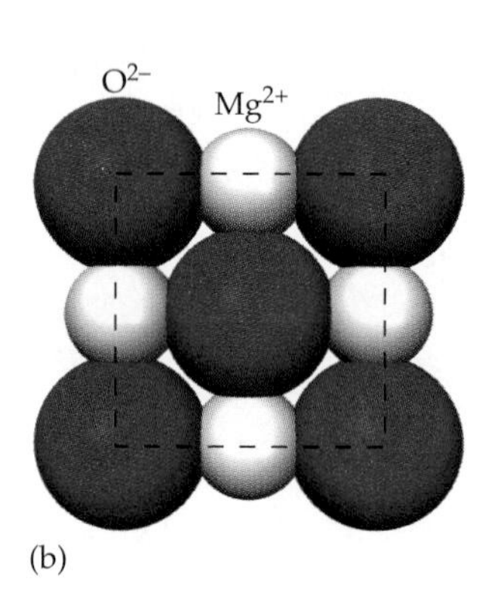

(b)

Magnesium oxide. (a) A unit cell showing oxide ions as a face-centered cubic lattice with magnesium ions in octahedral holes. (b) One face of the cell.

Because there are four MgO "formula units" per unit cell, the cell has a mass of

$$\left(\frac{6.692 \times 10^{-23}\ \text{g}}{\text{formula unit}}\right)\left(\frac{4\ \text{MgO formula units}}{1\ \text{unit cell}}\right) = 2.677 \times 10^{-22}\ \text{g/unit cell}$$

Step 3. *Calculate the density of MgO from the mass and volume of a unit cell.*

$$\text{Density} = \left(\frac{2.677 \times 10^{-22}\ \text{g}}{1\ \text{unit cell}}\right)\left(\frac{1\ \text{unit cell}}{7.62 \times 10^{-23}\ \text{cm}^3}\right) = 3.51\ \text{g/cm}^3$$

The experimental density is 3.6 g/cm^3. The small difference between the calculated and experimental values comes from using average Mg^{2+} and O^{2-} radii.

Exercise 13.9 Density from Unit Cell Dimensions

Potassium chloride, KCl, has the same crystal structure as NaCl. Using the ion sizes in Figure 8.14, calculate the density of KCl.

PROBLEM-SOLVING TIPS AND IDEAS

13.1 Calculations with Unit Cells

In Example 13.6 we used the density of a solid and a knowledge of the cell type to find the atom radius. In Example 13.7 this procedure was reversed. We verified the cell type from the density of the solid and the ionic radii. Then, in Example 13.8, we used the ionic radii for Mg^{2+} and O^{2-} and a knowledge of the structure of MgO to calculate the density of this compound. From these three examples, we conclude that the three pieces of information—particle radii, geometry of the crystal lattice, and density—are tied closely together. Provided with information on any two of these, we can calculate or determine the third.

The following observations may help to explain relationships between these pieces of information:

- The density is a macroscopic quantity, easily measurable in a laboratory. Bear in mind, however, that the density of a macroscopic sample must also be the density of the unit cell.
- Density equals mass divided by volume. The mass of a unit cell can be determined (using values of atomic weights) if we know how many atoms or ions are in the unit cell. The volume of the unit cell depends on the size of the atoms or ions and how they are arranged in the unit cell. So, both the mass and volume of the unit cell require knowing the crystal lattice structure.

13.5 OTHER KINDS OF SOLID MATERIALS

So far we have described the structures of metals and simple ionic solids. Next, we will look briefly at the other categories of solids: molecular solids, network solids, and amorphous solids.

A CLOSER LOOK

Packing Atoms and Ions into Unit Cells

It is a "rule" that nature does things as efficiently as possible. So, if we imagine the atoms or ions of a crystal lattice to be tiny, hard spheres, it is no surprise that they will be packed together to fill space as efficiently as possible. A simple cubic arrangement does not do this well; only 52% of the space is filled. The efficiency of packing in a body-centered unit cell is somewhat better, 68%. A face-centered cubic lattice, however, is the most efficient way to fill space; atoms or ions occupy 74% of the space within the fcc unit cell.

You can begin to appreciate the idea of efficient packing by looking at a two-dimensional analogy—how marbles can be arranged in a layer. In one arrangement (Figure Aa), the marbles are at the corners of a square and each touches four other spheres. In the other (Figure Ab), each marble touches six others at the corners of a hexagon. In both arrangements, space is left between atoms in the layer, but the amount of unfilled space is less in Ab than in Aa. Thus, Ab is the more efficient way of packing spheres in two dimensions.

(a)

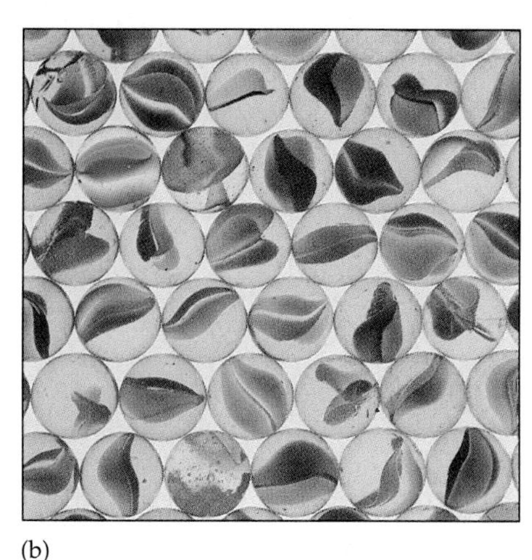

(b)

Figure A Packing of spheres in one layer. (a) Marbles are arranged in a square pattern, and each marble contacts four others. (b) Each marble contacts six others at the corners of a hexagon. Pattern (b) is a more efficient packing arrangement than (a). *(C. D. Winters)*

To fill three-dimensional space, layers of atoms are stacked one on top of another. If you start with the square arrangement (Figure Aa) and stack the next layer directly on top of the first, the arrangement resembles Escher's work (see Figure 13.26); the result is a solid with a simple cubic unit cell.

Molecular Solids

See the *Saunders Interactive General Chemistry CD-ROM,* Screen 13.16, Silicate Minerals.

Covalently bonded molecules such as H_2O and CO_2 are found in the solid state under appropriate conditions. In these compounds, molecules, rather than atoms or ions, line up in a regular fashion in a three-dimensional lattice. We have commented already on one such structure, the structure of ice (Section 13.3, Figure 13.10).

How molecules are arranged in a crystalline lattice depends on the shape of the molecules and the types of intermolecular forces. The molecules tend to pack in the most efficient manner, that is, they are arranged to fill space efficiently, to be as close together as possible, and to align in ways that maximize intermolecular forces of attraction. Thus, the water structure was established to gain the maximum intermolecular attraction through hydrogen bonding. As illustrated in Figure 13.9, organic acid molecules often assemble in the solid state as dimers, with two molecules linked by hydrogen bonding.

The greatest interest in structures of molecular solids centers on the structure of the molecules themselves. It is from structural studies on molecular solids that most of the information on molecular geometries, bond lengths, and bond angles discussed in Chapters 9 through 11 was assembled.

There are two more efficient ways of packing atoms or ions. They begin with a layer of atoms as in Figure Ab. Succeeding layers of atoms or ions are then stacked one on top of the other in two different ways. Depending on the stacking pattern, you will get either a **cubic close-packed (ccp)** or **hexagonal close-packed (hcp)** arrangement (Figure B). In the hcp arrangement (Figure Ba), additional layers of particles are placed above and below a given layer, fitting into the same depressions on either side of the middle layer. In a three-dimensional crystal, the layers repeat their pattern in the manner ABABAB. . . . Atoms in each A layer are directly above the ones in another A layer; the same holds for the B layers. In the ccp arrangement (Figure Bb), the atoms of the "top" layer rest in depressions in the middle layer, and those of the "bottom" layer are oriented opposite to those in the top layer. In a crystal, the pattern is repeated ABCABCABC. . . . By turning the whole crystal, you can see that the ccp arrangement is the face-centered cubic structure.

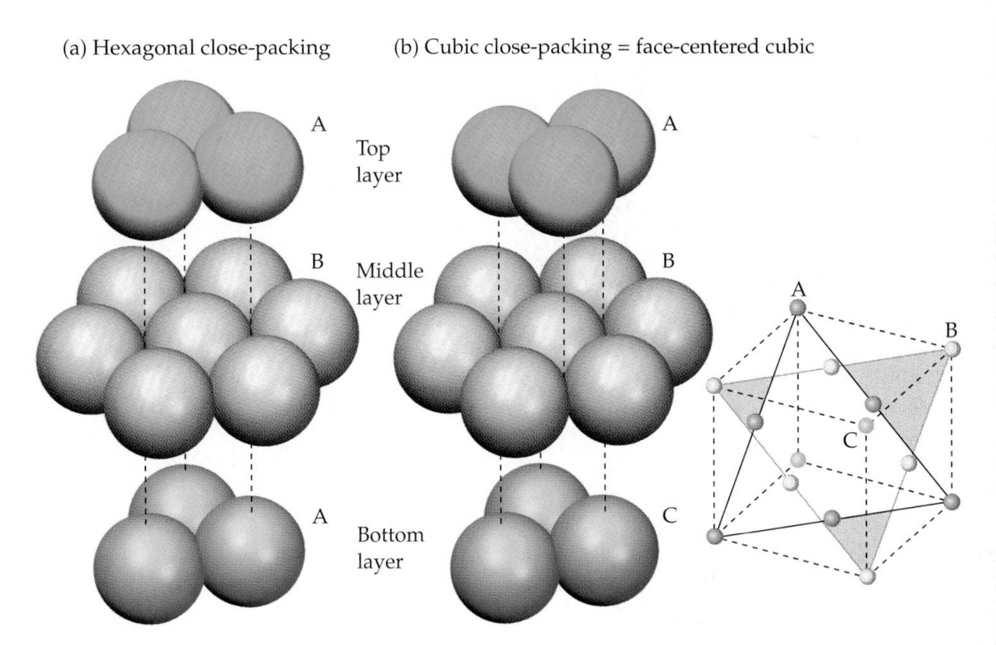

Figure B The most efficient ways to pack atoms or ions in crystalline materials, hexagonal close packing (hcp) and cubic close packing (ccp). The drawing on the right shows that cubic close packing results in the face-centered cubic lattice.

Network Solids

Network solids are substances composed of networks of covalently bonded atoms. Common examples include two allotropes of carbon: graphite and diamond (see Figure 2.21). Elemental silicon is a network solid with a diamond-like structure (Figure 2.23a), and silicon dioxide, SiO_2, is also a network solid (Figure 2.23b).

Diamonds have a low density ($d = 3.51\ g/cm^3$), but they are also the hardest material and the best conductor of heat known. They are transparent to visible light, as well as infrared and ultraviolet radiation. They are electrically insulating but behave as semiconductors with some advantages over silicon. What more could a scientist want in a material—except a cheap, practical way to make it!

In the 1950s, scientists at General Electric in Schenectady, New York, achieved something alchemists had sought for centuries, the synthesis of diamonds from carbon-containing materials, including wood or peanut butter. Their technique is to heat graphite to a temperature of 1500 °C in the presence of a metal, such as nickel or iron, and under a pressure of 50,000 to 65,000 atm. Under these conditions, the carbon dissolves in the metal and slowly forms di-

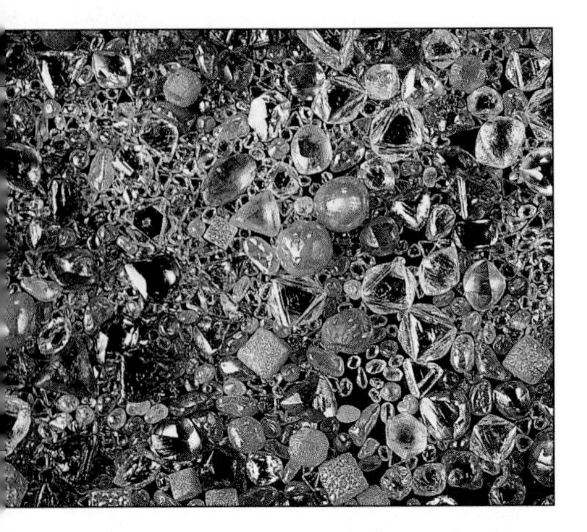

Figure 13.32 A mixture of natural and synthetic industrial diamonds. The colors of diamonds may range from colorless to yellow, brown, or black. Poorer-quality diamonds are used extensively in industry, mainly for cutting or grinding tools. Industrial-quality diamonds are produced synthetically at present by heating graphite, along with a metal catalyst, to 1200–1500 °C and a pressure of 65–90 kilobars. New techniques are being sought, however, to manufacture them less expensively and to be able to coat surfaces with a diamond film. *(Sinclair Stammers/Science Photo Library/Photo Researchers, Inc.)*

See the *Saunders Interactive General Chemistry CD-ROM,* Screen 13.16, Silicate Minerals.

amonds (Figure 13.32). Over $500 million worth of diamonds are made this way. Most are used for abrasives and diamond-coated cutting tools.

The high-pressure synthesis of diamonds is expensive, and the diamonds are not entirely pure nor crystalline enough for making semiconductor devices. Even from the beginning, therefore, the search was on for better ways to make diamonds. The most promising method so far seems to be a low-pressure process called "chemical vapor deposition" (CVD). A mixture of hydrogen and a carbon-containing gas such as methane (CH_4) is decomposed by heating at about 2200 °C with microwaves or a hot wire. Carbon atoms deposit on a silicon plate or other material in the chamber and slowly build up a film of carbon atoms in the diamond structure.

Silicates, compounds composed of silicon and oxygen, represent an enormous class of chemical compounds. You know them in the form of sand, quartz, talc, and mica, or as a major constituent of rocks such as granite (Figure 13.33). Details of several silicate structures are outlined in Chapter 22.

Most network solids are very hard and rigid and are characterized by high melting and boiling points. These characteristics simply reflect the fact that a great deal of energy must be provided to break the covalent bonds in the lattice. The contrasting properties of CO_2 and SiO_2, compounds formed by the lightest members of Group 4A, illustrate this point. Carbon dioxide is a gas at room temperature (although it condenses to solid "dry ice" when cooled to −78 °C), whereas silicon dioxide, a solid, melts above 1600 °C. This difference in properties is explained by their structures. Silicon dioxide consists of tetrahedral silicon atoms linked together through covalent bonds to oxygen in a giant three-dimensional lattice. A very high temperature is required to break the covalent bonds between silicon and oxygen and disrupt this very stable structure. In contrast, relatively weak dispersion forces hold the molecules of CO_2 together in the solid state, and low temperatures are required to separate the molecules from one another.

Amorphous Solids

A characteristic property of pure crystalline solids—metals, ionic solids, and molecular solids—is that they melt at a specific temperature. For example, water melts at 0 °C, aspirin at 135 °C, lead at 327.5 °C, and NaCl at 801 °C. Heated to 0 °C, ice begins to melt, and the temperature will remain at 0 °C until all of the solid has been converted to a liquid. Only after all the ice has melted will the temperature rise further. Because of their specific and reproducible values, melting points are often used as a means of identifying chemical compounds.

Another property of these crystalline solids is that they form well-defined crystals, with smooth flat faces. When a sharp force is applied to a crystal, it will most often cleave to give smooth flat faces, and the resulting solid particles are smaller versions of the macroscopic crystal (Figure 13.34).

Figure 13.33 Silicates. Naturally occurring silicates include clear quartz, sand, sheets of mica, light green talc, and sandstone. *(C. D. Winters)*

Many common solids, ones that we encounter everyday, do not have these properties, however. Glass is a good example. When glass is heated it softens over a wide temperature range, a property useful for artisans and craftsmen who can create beautiful and functional products for our enjoyment and use (see Figure 3.2). We also know of a property of glass that we would just as soon not have: When glass breaks it leaves randomly shaped jagged pieces without flat faces. Other materials that behave similarly, with respect to both melting and

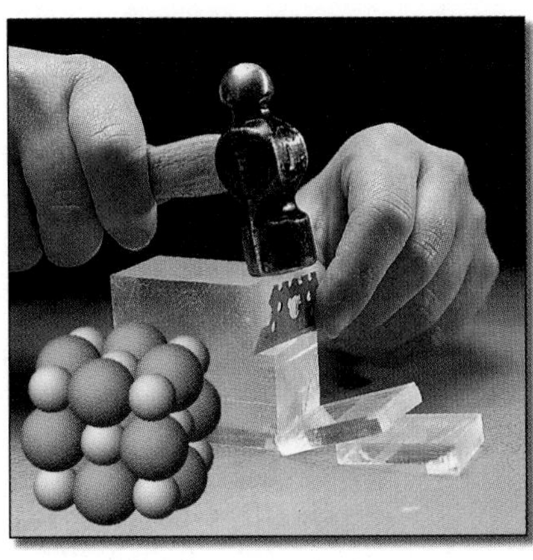

(a)

(b)

Figure 13.34 Crystalline and amorphous solids. (a) A salt crystal can be cleaved cleanly into smaller and smaller crystals that are duplicates of the larger crystal. This is evidence for the underlying lattice structure. (b) Glass is an amorphous solid composed of linked silicon–oxygen tetrahedra. It has, however, no long-range order as in crystalline quartz. *(Photos, C. D. Winters; models, S. M. Young)*

breaking, are the polymers used extensively in our society. Polyethylene, nylon, and other common plastics have the same basic properties as glass.

The characteristics of these **amorphous solids** relate to their molecular structure. At the particulate level, amorphous solids do not have a regular structure. This means that such substances look a lot like liquids, and, in many ways, this is an appropriate description. Unlike liquids, however, the forces of attraction are strong enough that movement of the molecules is restricted.

Scientists have made advances in understanding the properties of these kind of materials. One view is that small regions exist within the amorphous solid that have their own regular structures. The properties of a substance are interpreted based on the sum of behaviors of the different segments of the molecular structure.

It is not possible to identify a unit cell, a repeating structural unit that is reproduced in all directions, in the solid phase of an amorphous solid. Such solids do not have a regular, long-range molecular structure.

13.6 THE PHYSICAL PROPERTIES OF SOLIDS

The outward shape of a crystalline solid is a reflection of its internal structure. But what about the temperatures at which solids melt, their hardness, or their solubility in water? All of these and many other physical properties of solids are of interest to chemists, geologists, engineers, and others. A few such properties are examined here.

The melting point of a solid is the temperature at which the lattice collapses and the solid is converted to liquid. Just as in the case of the liquid-to-vapor transformation, melting requires energy, the **enthalpy of fusion** (in kilojoules per mole) (Chapter 6).

$$\text{Heat energy absorbed on melting} = \text{Enthalpy of fusion} = \Delta H_{\text{fusion}}\ (\text{kJ/mol})$$

$$\text{Heat energy evolved on freezing} = \text{Enthalpy of crystallization}$$
$$= -\Delta H_{\text{fusion}}\ (\text{kJ/mol})$$

Enthalpies of fusion can range from just a few thousand joules per mole to many thousands per mole (Table 13.7). A low melting temperature will certainly mean a low value for the enthalpy of fusion, whereas high melting points are associated with high enthalpies of fusion. Figure 13.35 shows the enthalpy of fusion of most of the metals of the periodic table relative to one another. From this you can conclude (1) that metals that have notably low melting points, such as the alkali metals and mercury (mp = −39 °C), also have very low enthalpies of fusion; and (2) that metals of the transition series have high heats of fusion, with those of the third transition series being extraordinarily high. This parallels melting points for these elements. Tungsten, which has the highest melting point of all the known elements except for carbon, also has the highest enthalpy of fusion among the transition metals. These properties relate to the uses of this metal. Tungsten, for example, is used for the filaments in light bulbs; no other material has been found to work better since the invention of the light bulb in 1908.

The melting temperature of a solid can convey a great deal of information. Table 13.7 gives you some data for several basic types of substances: metals, polar and nonpolar molecules, and ionic solids. In general, nonpolar substances

Table 13.7 • Melting Points and Enthalpies of Fusion of Some Solids

Compound	Melting Point (°C)	Enthalpy of Fusion (kJ/mol)	Type of Intermolecular Forces
		Metals	
Hg	−39	2.29	Metal bonding; see Section 10.4
Na	98	2.60	
Al	660	10.7	
Ti	1668	20.9	
W	3422	35.2	
		Molecular Solids: Nonpolar Molecules	
O_2	−219	0.440	Dispersion forces only
F_2	−220	0.510	
Cl_2	−102	6.41	
Br_2	−7.2	10.8	
		Molecular Solids: Polar Molecules	
HCl	−114	1.99	All three HX molecules have dipole–dipole forces and dispersion forces that increase with size and molar mass.
HBr	−87	2.41	
HI	−51	2.87	
H_2O	0	6.02	Hydrogen bonding
		Ionic Solids	
NaF	996	33.4	All ionic solids have extended ion–ion interactions. Note the general trend is the same as for lattice energies (see Section 9.3) and Figure 9.2.
NaCl	801	28.2	
NaBr	747	26.1	
NaI	660	23.6	

Historical Perspective

Dorothy Crowfoot Hodgkin (1910–1994)

A recent biography of Dorothy Crowfoot Hodgkin said that "more than any other scientist, she personified the transformation of crystallography from a black art into an indispensable scientific tool. . . . She made not one brilliant breakthrough but a series of them, deciphering the structure of one medically important substance after another."

Dorothy Crowfoot Hodgkin was born in Egypt, then a British colony, where her father was a supervisor of schools and ancient monuments. She went to school in England, however, and graduated from Oxford University in 1931. (At the time, Oxford University had only one woman student for every five men.) She was fascinated by the structures of crystals, and so went to Cambridge University where

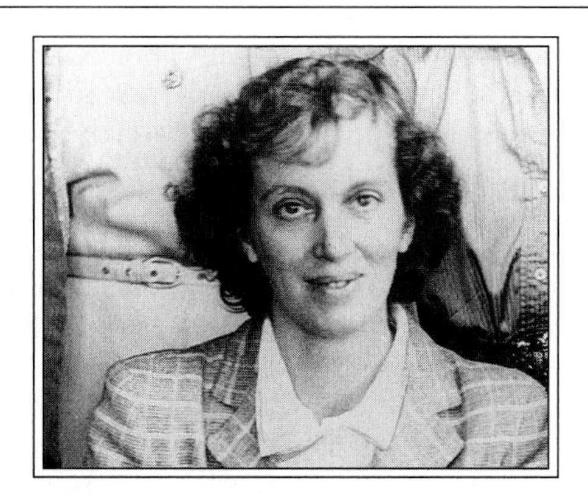

Dorothy Crowfoot Hodgkin. *(The Bettmann Archive)*

she worked with John D. Bernal of the Mineralogical Institute. After only a year, Oxford invited her back as a chemistry instructor, and she stayed there until her retirement in 1977.

Her first major achievement was the determination of the structure of penicillin in the early 1940s. This was followed by a determination of the structure of cephalosporin. After World War II she began the work that earned her the Nobel Prize in chemistry in 1964: the determination of the structure of vitamin B_{12}, the factor used to treat pernicious anemia. Later came another of her major accomplishments, the determination of the structure of insulin.

See S. B. McGrayne: *Nobel Prize Women in Science,* New York, Birch Lane Press, 1992, pp. 225–254.

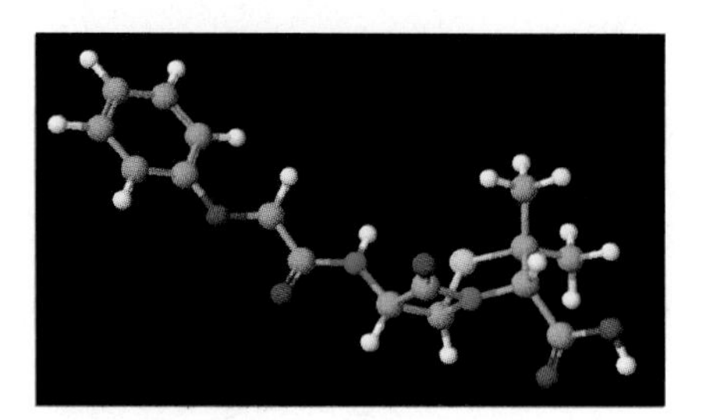

The structure of penicillin was first determined by Dorothy Crowfoot Hodgkin.

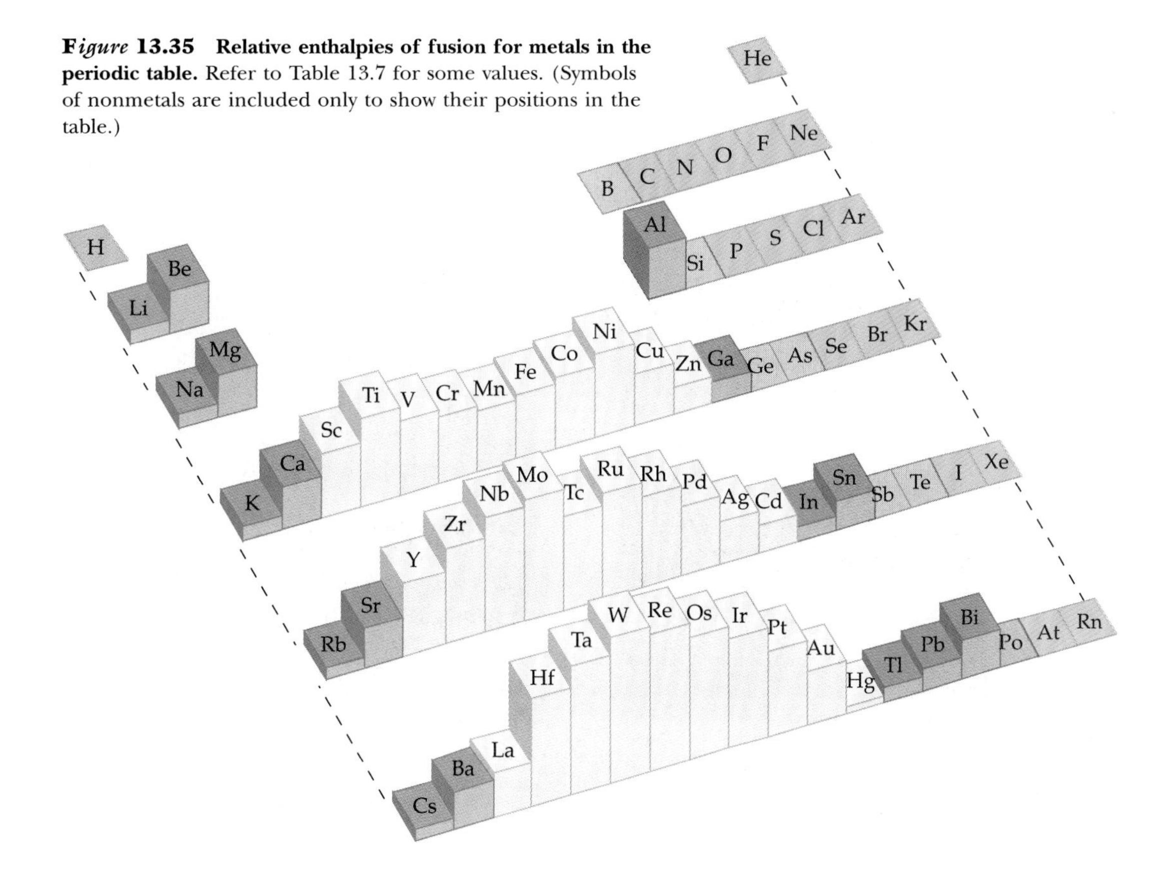

Figure 13.35 Relative enthalpies of fusion for metals in the periodic table. Refer to Table 13.7 for some values. (Symbols of nonmetals are included only to show their positions in the table.)

that form molecular solids have low melting points. Melting points increase in a series of related molecules, however, as the size and molar mass increase. This happens because dispersion forces are larger when the molar mass is larger. Thus, increasing amounts of energy are required to break down the intermolecular forces in the solid, a principle that is reflected in an increasing enthalpy of fusion.

The ionic compounds in Table 13.7 have higher melting points and higher enthalpies of fusion than molecular solids. This phenomenon is due to very strong ion–ion forces in ionic solids, forces that are reflected in high lattice energies (see Section 9.3). Because ion–ion forces depend on ion size (as well as ion charge), there is a good correlation between lattice energy and the position of the metal or halogen in the periodic table. That is, as the cation size increases from Li^+ to Cs^+, the lattice energy declines for compounds of a given halide ion. There is a similar decline in lattice energy as the halide ion size increases from F^- to I^- when salts of a given cation are considered. As illustrated by the data in Table 13.7, the decline in lattice energy with increasing ion size is accompanied by a decrease in melting point and enthalpy of fusion.

Molecules can escape directly from the solid to the gas phase by **sublimation** (Figure 13.36).

$$\text{Solid} \longrightarrow \text{gas} \qquad \text{heat energy required} = \Delta H_{\text{sublimation}}$$

Common substances that sublime at 1 atm pressure are I_2 and dry ice, solid CO_2.

Sublimation, like fusion and evaporation, is an endothermic process. The heat energy required is called the enthalpy of sublimation. Water, which has a molar enthalpy of sublimation of 51 kJ/mol, can be converted from solid ice to water vapor quite readily. One of the best examples of the use of this property is in a frost-free refrigerator. During certain times, the freezer compartment is warmed slightly. Molecules of water freed from the surface of the ice (the vapor pressure of ice is 4.60 mm Hg at 0 °C) are removed in a current of air blown through the freezer.

Figure 13.36 Sublimation. This process is the conversion of a solid directly to its vapor. Here, naphthalene ($C_{10}H_8$) sublimes when heated in warm water. *(Photo, C. D. Winters; model, S. M. Young)*

Example 13.10 Heat of Vaporization and Fusion

Rubbing alcohol, or 2-propanol, is an organic compound that freezes at −89.5 °C and boils at 82.3 °C. What quantity of heat is required to melt 10.0 g of 2-propanol at −89.5 °C, heat the resulting liquid to the boiling point, and then evaporate the liquid at this temperature? Required information is: ΔH_{fusion} = 5.37 kJ/mol; specific heat capacity of liquid 2-propanol = 2.60 J/g · K; and ΔH_{vap} = 39.85 kJ/mol.

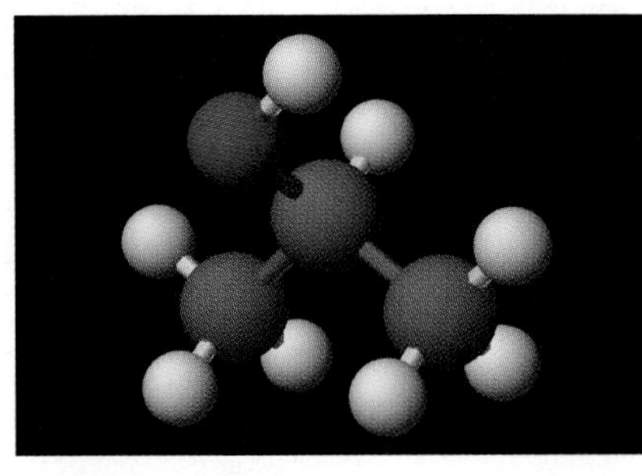

Common rubbing alcohol is the compound 2-propanol (or isopropanol).

Solution We shall break the calculation into the three steps in the transformation:

Step 1. *Heat required to melt the solid at −89.5 °C.*

$$10.0\,\text{g} \cdot \left(\frac{1\text{ mol}}{60.10\text{ g}}\right) \cdot \left(\frac{5.37\text{ kJ}}{\text{mol}}\right) = 0.894\text{ kJ}$$

Step 2. *Heat required to warm the liquid from −89.5 °C (183.7 K) to +82.3 °C (355.5 K).*

$$10.0\,\text{g} \cdot \left(\frac{2.60\text{ J}}{\text{g} \cdot \text{K}}\right) \cdot (355.5\text{ K} - 183.7\text{ K}) = 4.47 \times 10^3\text{ J, or } 4.47\text{ kJ}$$

Step 3. *Heat required to evaporate the liquid at 82.3 °C.*

$$10.0\,\text{g} \cdot \left(\frac{1\text{ mol}}{60.10\text{ g}}\right) \cdot \left(\frac{39.85\text{ kJ}}{\text{mol}}\right) = 6.63\text{ kJ}$$

Example 13.10 is an application of the discussion in Section 6.4 and Example 6.4.

Step 4. *Total heat energy required.*

Total heat required = 0.894 kJ + 4.47 kJ + 6.63 kJ = 11.99 kJ

Exercise 13.10 Heat of Fusion

Calculate the heat required to melt 100.0 g of water at its melting point. Compare this with the heat required to melt 100.0 g of the hydrocarbon octane (C_8H_{18}, melting point = −56.8 °C; ΔH_{fusion} = 20.65 kJ/mol). Is there a relation between the calculated heats and the types of intermolecular forces involved?

13.7 PHASE DIAGRAMS

See the *Saunders Interactive General Chemistry CD-ROM,* Screen 13.17, Phase Diagrams.

Depending on the conditions of temperature and pressure, a substance exists either as a gas, a liquid, or a solid. In addition, under certain specific conditions, two (or even three) states can coexist in equilibrium. It is possible to summarize this information in the form of a graph called a **phase diagram.** Phase diagrams are used to illustrate the relationship between phases of matter and the pressure and temperature. A phase diagram for water is shown in Figure 13.37.

The lines in a phase diagram identify the conditions under which two phases exist at equilibrium. (Conversely, all points that do not fall on the lines in the figure represent conditions under which only one state exists.) Line A-D represents the combinations of temperature and pressure for which there can be equilibrium between the liquid and vapor phases. Line A-B represents conditions for

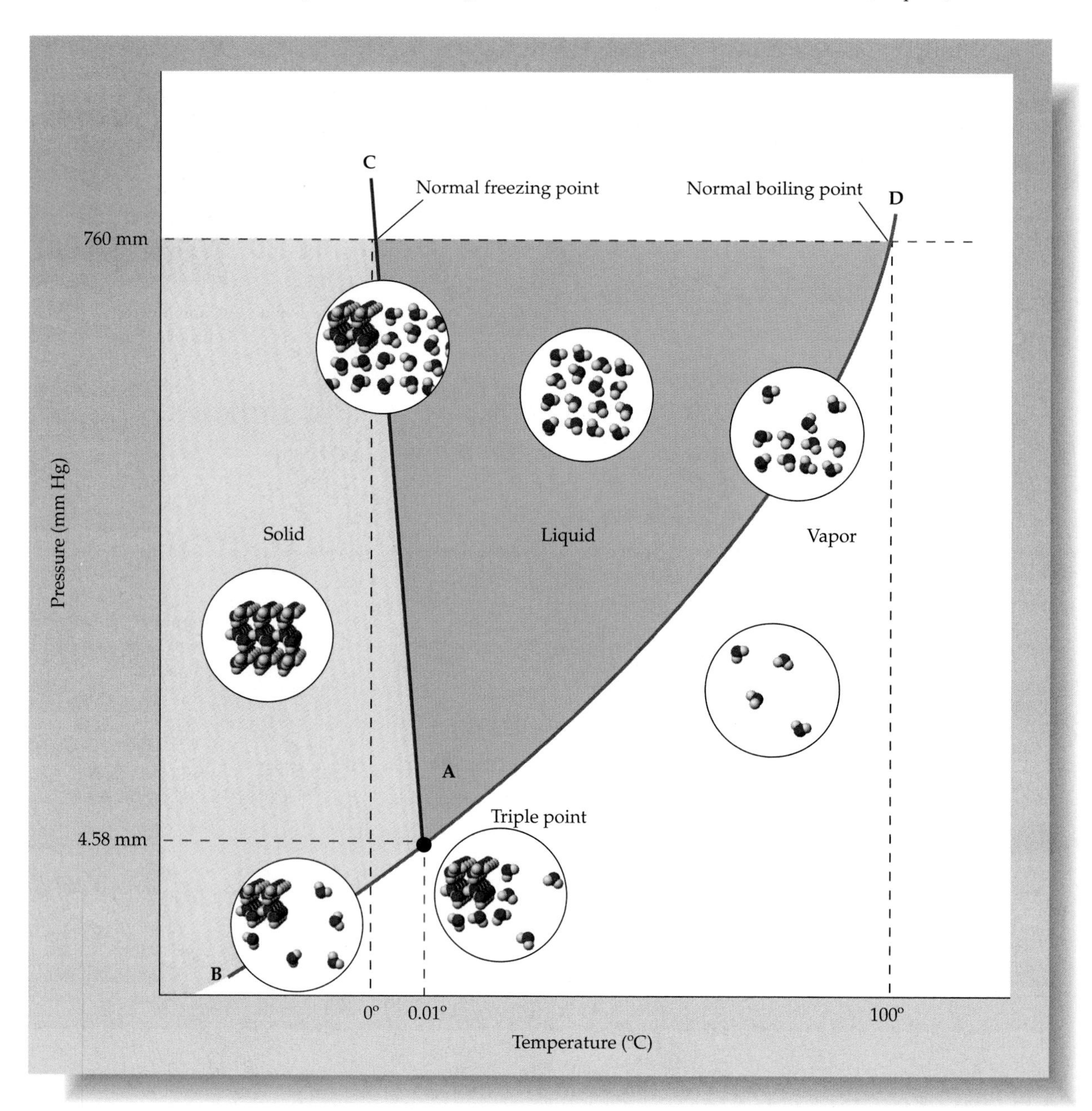

Figure 13.37 Phase diagram for water. The scale is intentionally skewed (nonlinear) to be able to show the triple point and the negative slope of the line representing the liquid–solid equilibrium.

solid–vapor equilibrium, and line A-C for liquid–solid equilibrium. The line from point A to point D, which represents the equilibrium between liquid and gaseous water, is the same curve plotted for water vapor pressure in Figure 13.18. Recall that the normal boiling point, 100 °C in the case of water, is the temperature at which the equilibrium vapor pressure is 760 mm Hg.

Point A represents conditions under which all three phases can coexist in equilibrium. This is aptly called the **triple point.** For water, the triple point is at $P = 4.58$ mm Hg and $T = 0.01$ °C.

The line A-C shows the conditions of pressure and temperature at which solid–liquid equilibrium exists. (Because no vapor pressure is involved here, the pressure referred to is the external pressure on the liquid.) For water, this line has a negative slope. That is, the higher the external pressure, the lower the melting point. The change for water is approximately 0.01 °C for each atmosphere increase in pressure. The negative slope of the water solid–liquid equilibrium line can be explained from our knowledge of the structure of water and ice. When pressure on an object is increased, common sense tells you the volume of the object will become smaller, giving the substance a higher density. Because ice is less dense than liquid water (due to the open lattice structure of ice), ice and water in equilibrium respond to increased pressure (at constant T) by melting ice to form more water because the same mass of water requires less volume (Figure 13.38a). This phenomenon is illustrated in Figure 13.37 using a solid–liquid equilibrium line with a greatly exaggerated slope.

Of the thousands of substances for which phase diagrams are known, only water, bismuth, and antimony have solid–liquid equilibrium lines with a negative slope.

Ice is slippery stuff, and it has long been assumed you could ski or skate on it because the surface melted slightly from the pressure of a skate blade or ski, or that surface melting occurred because of frictional heating. These have always seemed to be unsatisfying explanations, however, because it does not seem possible that just standing or sliding on a piece of ice could produce a pressure or temperature high enough to cause sufficient melting. Recently, though, surface chemists have studied ice surfaces and may have come up with a better explanation (Figure 13.38b). They have concluded that water molecules on the surface of ice are vibrating rapidly, that "the surface of ice is molten." In fact, they say that the outermost layer or two of water molecules is almost liquid-like and that this is the reason we can ski or skate on ice.

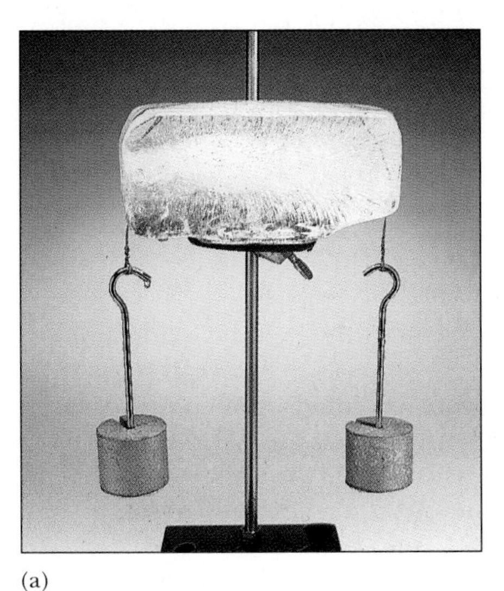

(a)

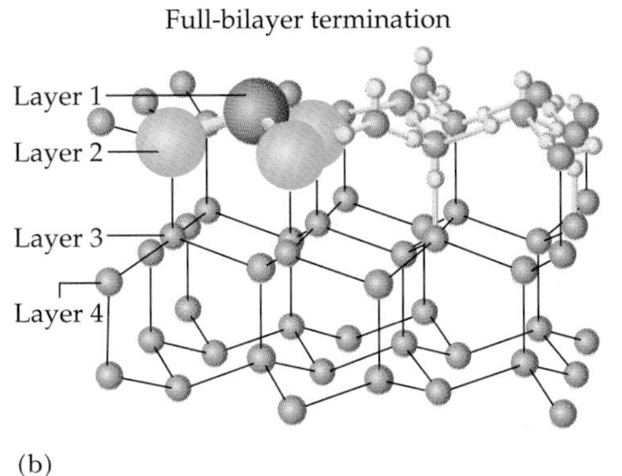

(b)

Figure 13.38 **Ice.** (a) Ice melts under pressure. Here a thin wire is passed over a block of ice, and weights are suspended at either end. With time, the pressure exerted on the ice by the wire causes the ice to melt, and the wire eventually melts through the ice. (b) A computer model of ice at its surface. (The larger spheres represent complete water molecules.) The molecules in Layer 1 are vibrating faster than those in Layer 2, which in turn are vibrating faster than those in the interior. The surface molecules behave more like water molecules in the liquid phase than in the solid phase. *(a, C. D. Winters; b, Permission of M. A. Van Hove & G. A. Somorjai/Lawrence Berkeley National Laboratory)*

Figure 13.39 Phase diagram for CO_2.

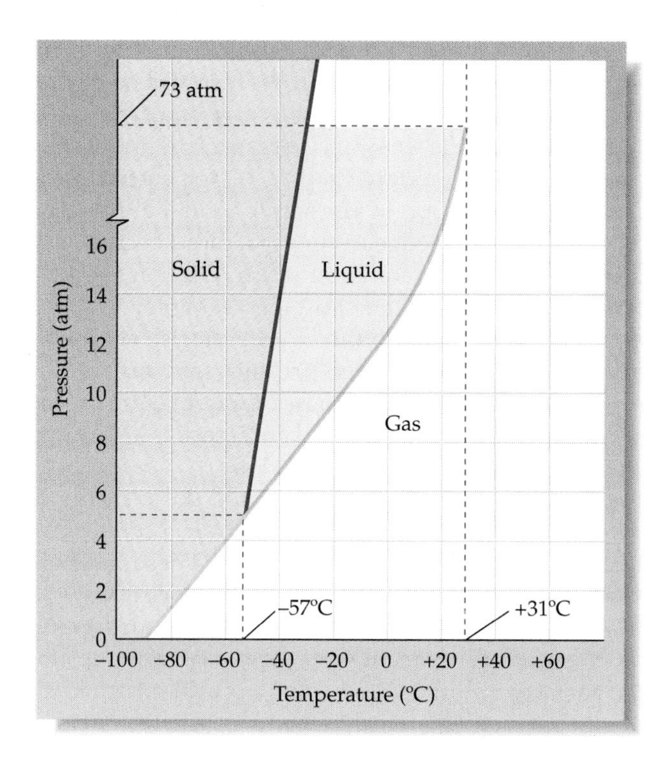

The basic features of the phase diagram for CO_2 (Figure 13.39) are the same as those for water, except that the CO_2 solid–liquid equilibrium line has a positive slope. Solid CO_2 is denser than the liquid, so solid CO_2 sinks to the bottom in a container of liquid CO_2. Notice also that solid CO_2 sublimes directly to CO_2 gas as it warms to room temperature (at 1 atm pressure). This is the reason CO_2 is called *dry ice*; it looks like water ice, but it does not melt.

CHAPTER HIGHLIGHTS

Having studied this chapter, you should be able to

- Use the kinetic-molecular theory to define the difference in solids, liquids, and gases (Section 13.1).
- Describe the different **intermolecular forces** in liquids and solids (Section 13.2).
- Identify examples of **ion–dipole attraction** (Section 13.2). Such forces depend on the distance between the ion and the dipole, the magnitude of the dipole, and the charge on the ion (Example 13.1).
- Tell when two molecules can interact through a **dipole–dipole attraction** or when **hydrogen bonding** may occur. The latter occurs most strongly when H is attached to O, N, or F (Section 13.2 and Example 13.2).
- Explain how hydrogen bonding affects the properties of water (Section 13.2).
- Identify instances in which molecules interact by **induced dipoles (dispersion forces)** (Section 13.2).
- Explain the process of evaporation of a liquid or condensation of its vapor, and use the **enthalpy of vaporization** in calculations (Section 13.3).

- Define and use the concept of the **equilibrium vapor pressure** of a liquid and its relation to the boiling point of a liquid (Section 13.3).

  Evaporation of a liquid continues at a given temperature until a **dynamic equilibrium** is established at which point the rate of evaporation equals the rate at which vapor molecules condense or reenter the liquid phase. At equilibrium, the pressure exerted by the vapor phase molecules is called the equilibrium vapor pressure. More **volatile** compounds have higher equilibrium vapor pressures at a given temperature. If the vapor pressure equals the atmospheric or external pressure, the liquid boils. When the vapor pressure is 1 atm, the temperature is the **normal boiling point.** The equilibrium vapor pressure of a liquid increases with temperature.
- Describe the phenomena of the **critical temperature, T_c,** and **critical pressure, P_c,** of a substance (Section 13.3).
- Describe how intermolecular interactions affect the **cohesive forces** between identical liquid molecules, the energy necessary to break through the surface of a liquid **(surface tension)**, and the resistance to flow, or **viscosity,** of liquids (Section 13.3).
- Characterize different types of solids: **metallic** (e.g., copper), ionic (e.g., NaCl and CaF_2), **molecular** (e.g., water and I_2), **network** (e.g., diamond), and **amorphous** (e.g., glass and many synthetic polymers) (Sections 13.4 and 13.5 and Table 13.6).
- Describe the three types of cubic unit cells: **primitive** or **simple cubic** (sc); **body-centered cubic** (bcc); and **face-centered cubic** (fcc) (Section 13.4).
- Perform calculations that relate the characteristics of solids and their unit cells (density, cell dimensions, ion or atom radius, and unit cell type) (Section 13.4 and Examples 13.6, 13.7, and 13.8).
- Define the **enthalpy of fusion** and use this in a calculation (Section 13.6).
- Identify the different points (triple point, normal boiling point, freezing point) and regions (solid, liquid, vapor) of a **phase diagram,** and use the diagram to evaluate the vapor pressure of a liquid or the relative densities of liquid and solid (Section 13.7).

STUDY QUESTIONS

The structures of many of the molecules and solids referred to in these questions are found on the Saunders Interactive General Chemistry CD-ROM *or at the Web site for this book (http://www.saunderscollege.com).*

REVIEW QUESTIONS

1. Name the types of forces that can be involved between two molecules and between ions and molecules.
2. Explain how a water molecule can interact with a molecule such as CO_2. What intermolecular force is involved?
3. What intermolecular interactions can occur in an aqueous solution of copper(II) chloride? Draw a simple, molecular level picture of these interactions.
4. Hydrogen bonding is most important when H is attached to certain very electronegative atoms. Which are those atoms? Why does this lead to strong hydrogen bonding?
5. Explain how hydrogen bonding leads to the decline in density of water from 4 °C to solid ice at 0 °C.
6. Why is the specific heat capacity of water so large compared with many other liquids? How does this affect weather?
7. When you exercise vigorously your body sweats. Why does this help your body cool down?
8. Explain, in terms of intermolecular forces, the trends in boiling points of the molecules EH_3, where E is a Group 5A element (see Figure 13.8).
9. Explain why the viscosity of an oil is so much greater than the viscosity of liquid benzene.
10. What are the three types of cubic unit cells? Explain their similarities and differences.
11. Sketch the body-centered cubic unit cell of potassium metal. What is the net number of potassium atoms within the unit cell?
12. Explain why a cubic salt crystal can be cleaved cleanly into smaller cubes, whereas glass shatters into randomly shaped pieces.
13. What types of solids tend to have very high melting points?

14. Sketch the phase diagram for water. Label the normal boiling point, melting point, and triple point and show what regions of temperature and pressure are appropriate to solid, liquid, and vapor.
15. Explain why the solid–liquid equilibrium line in the water phase diagram has a negative slope.

Intermolecular Forces

16. What intermolecular force(s) must be overcome to
   (a) melt ice
   (b) melt solid I_2
   (c) remove the water of hydration from $MnCl_2 \cdot 4\,H_2O$
   (d) convert liquid NH_3 to NH_3 vapor
17. When KCl dissolves in water, what type of attractive forces must be overcome in the liquid water? What type of forces must be overcome in the solid KCl? What type of attractive forces are important when KCl dissolves in liquid water?
18. What type of forces must be overcome within the solid I_2 when I_2 dissolves in methanol, CH_3OH? What type of forces must be disrupted between CH_3OH molecules when I_2 dissolves? What type of forces exist between I_2 and CH_3OH molecules in solution?
19. One example of a hydrated salt is $NiSO_4 \cdot 6\,H_2O$. What kind of attractive force is responsible for binding the water to the nickel ions of nickel sulfate?
20. What type of intermolecular force must be overcome in converting each of the following from a liquid to a gas?
   (a) liquid O_2 (c) CH_3I (methyl iodide)
   (b) mercury (d) CH_3CH_2OH (ethanol)
21. What type of intermolecular forces must be overcome in converting each of the following from a liquid to a gas?
   (a) CO_2 (c) $CHCl_3$
   (b) NH_3 (d) CCl_4
22. Rank the following atoms or molecules in order of increasing strength of intermolecular forces in the pure substance. Which do you think might exist as a gas at 25 °C and 1 atm?
   (a) Ne (c) CO
   (b) CH_4 (d) CCl_4
23. Rank the following in order of increasing strength of intermolecular forces in the pure substances. Which do you think might exist as a gas at 25 °C and 1 atm?
   (a) $CH_3CH_2CH_2CH_3$ (butane)
   (b) CH_3OH (methanol)
   (c) He
24. Which member of each of the following pairs of compounds has the higher boiling point?
   (a) O_2 or N_2 (c) HF or HI
   (b) SO_2 or CO_2 (d) SiH_4 or GeH_4
25. Place the following four compounds in order of increasing boiling point:
   (a) SCl_2 (c) CH_4
   (b) NH_3 (d) CO
26. Which of the following compounds would be expected to form intermolecular hydrogen bonds in the liquid state?
   (a) CH_3OCH_3 (dimethyl ether)
   (b) CH_4
   (c) HF
   (d) CH_3CO_2H (acetic acid)
   (e) Br_2
   (f) CH_3OH (methanol)
27. Which of the following compounds would be expected to form intermolecular hydrogen bonds in the liquid state?
   (a) H_2Se
   (b) HCO_2H (formic acid)
   (c) HI
   (d) acetone

$$H_3C-\overset{\overset{\displaystyle O}{\|}}{C}-CH_3$$

28. In each pair of ionic compounds, which is more likely to have the greater heat of hydration? Briefly explain your reasoning in each case.
   (a) LiCl or CsCl
   (b) $NaNO_3$ or $Mg(NO_3)_2$
   (c) RbCl or $NiCl_2$
29. When salts of Mg^{2+}, Na^+, and Cs^+ are placed in water, the positive ion is hydrated (as is the negative ion). Which of these three cations is most strongly hydrated? Which one is least strongly hydrated?

Liquids

30. Ethanol, CH_3CH_2OH, has a vapor pressure of 59 mm Hg at 25 °C. What quantity of heat energy is required to evaporate 125 mL of the alcohol at 25 °C? The enthalpy of vaporization of the alcohol at 25 °C is 42.32 kJ/mol. The density of the liquid is 0.7849 g/mL.
31. The enthalpy of vaporization of liquid mercury is 59.11 kJ/mol. What quantity of heat is required to vaporize 0.500 mL of mercury at 357 °C, its normal boiling point? The density of mercury is 13.6 g/mL.
32. Answer the following questions using Figure 13.19:
   (a) What is the approximate equilibrium vapor pressure of water at 60 °C? Compare your answer with the data in Appendix G.
   (b) At what temperature does water have an equilibrium vapor pressure of 600 mm Hg?
   (c) Compare the equilibrium vapor pressures of water and ethyl alcohol at 70 °C. Which is higher?
33. Answer the following questions using Figure 13.19:
   (a) What is the equilibrium vapor pressure of diethyl ether at room temperature (approximately 20 °C)?
   (b) Place the three compounds in Figure 13.19 in order of increasing intermolecular forces.
   (c) If the pressure in a flask is 400 mm Hg and if the temperature is 40 °C, which of the three compounds (diethyl ether, ethanol, and water) are predominantly liquids and which are gases?
34. Assume you seal 1.0 g of diethyl ether (see Figure 13.19) in an evacuated 100.-mL flask. (There are no molecules of any other gas in the flask.) If the flask is held at 30 °C, what is the approximate gas pressure in the flask? If the flask is

placed in an ice bath, does additional liquid ether evaporate or does some ether condense to a liquid?

35. Refer to Figure 13.19 as an aid in answering these questions:
   (a) You put some water at 60 °C in a plastic milk carton and seal the top very tightly so gas cannot enter or leave the carton. What happens when the water cools?
   (b) If you put a few drops of liquid diethyl ether on your hand, does it evaporate completely or remain a liquid?

36. Vapor pressure curves for CS_2 (carbon disulfide) and CH_3NO_2 (nitromethane) are drawn here.

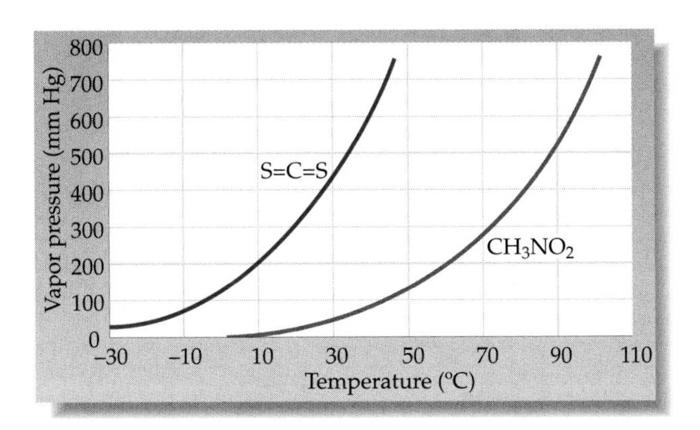

   (a) What are the approximate vapor pressures of CS_2 and CH_3NO_2 at 40 °C?
   (b) What type of intermolecular forces exist in the liquid phase of each compound?
   (c) What is the normal boiling point of CS_2? Of CH_3NO_2?
   (d) At what temperature does CS_2 have a vapor pressure of 600 mm Hg?
   (e) At what temperature does CH_3NO_2 have a vapor pressure of 60 mm Hg?

37. Answer each of the following questions with *increases, decreases,* or *does not change.*
   (a) If the intermolecular forces in a liquid increase, the normal boiling point of the liquid ______.
   (b) If the intermolecular forces in a liquid decrease, the vapor pressure of the liquid ______.
   (c) If the surface area of a liquid decreases, the vapor pressure ______.
   (d) If the temperature of a liquid increases, the equilibrium vapor pressure ______.

38. Carbon dioxide can be converted to a supercritical fluid. Its critical temperature and pressure are 304.2 K and 72.8 atm, respectively. Can carbon monoxide (T_c = 132.9 K; P_c = 34.5 atm) be liquefied at or above room temperature? Explain briefly.

39. The simple hydrocarbon methane (CH_4) cannot be liquefied at room temperature, no matter how high the pressure. Propane (C_3H_8), another compound in the series of simple alkanes, has a critical pressure of 42 atm and a critical temperature of 96.7 °C. Can this compound be liquefied at room temperature?

Metallic and Ionic Solids

40. Outline a two-dimensional unit cell for the pattern shown here. If the black squares are labeled A and the white squares are B, what is the simplest formula for a "compound" based on this pattern?

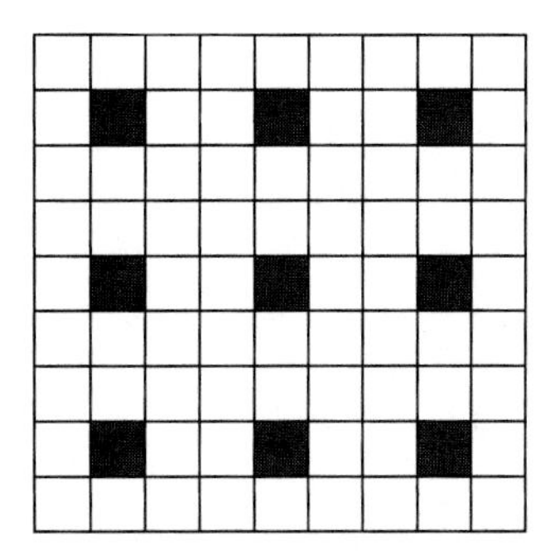

41. Outline a two-dimensional unit cell for the pattern shown here. If the black squares are labeled A and the white squares are B, what is the simplest formula for a "compound" based on this pattern?

42. The very dense metal iridium has a face-centered cubic unit cell and a density of 22.56 g/cm^3. Use this information to calculate the radius of an atom of the element.

43. Calcium metal crystallizes in a face-centered cubic unit cell. The density of the solid is 1.54 g/cm^3. What is the radius of a calcium atom?

44. The density of copper metal is 8.95 g/cm^3. If the radius of a copper atom is 127.8 pm, is the copper unit cell simple cubic, body-centered cubic, or face-centered cubic? (See Figure 2.18 for a model of the unit cell.)

45. Vanadium metal has a density of 6.11 g/cm^3. Assuming the vanadium atomic radius is 132 pm, is the vanadium unit cell simple cubic, body-centered cubic, or face-centered cubic?

46. Thallium(I) chloride, TlCl, crystallizes in either a simple cubic or a face-centered cubic unit cell of Cl^- ions with Tl^+ ions in the lattice holes. The density of the solid is 7.00 g/cm^3, and the edge of the unit cell is 385 pm. What is the unit cell geometry?

47. The metal hydride LiH has a density of 0.77 g/cm^3. The edge of the unit cell is 408.6 pm. If it is assumed that the H^- ions define the lattice points, does the compound have a face-centered cubic or a simple cubic unit cell?

48. One way of viewing the unit cell of perovskite was illustrated in Example 13.8. Another way is shown here. Prove that this view also leads to a formula of $CaTiO_3$.

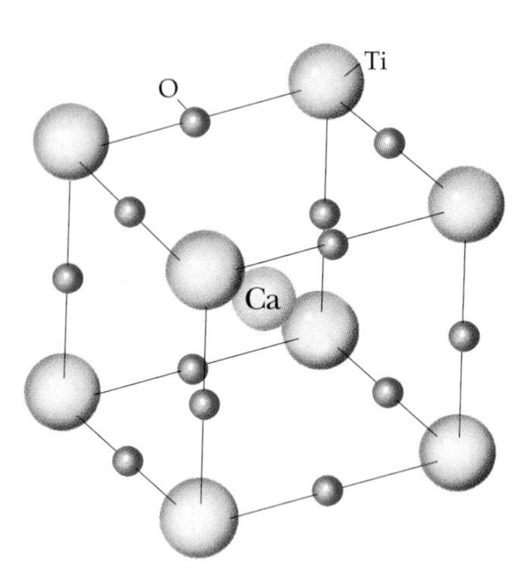

49. Rutile, TiO_2, crystallizes in a structure characteristic of many other ionic compounds. How many formula units of TiO_2 are in the unit cell illustrated here? (The oxide ions marked by an *x* are wholly within the cell; the others are in the cell faces.)

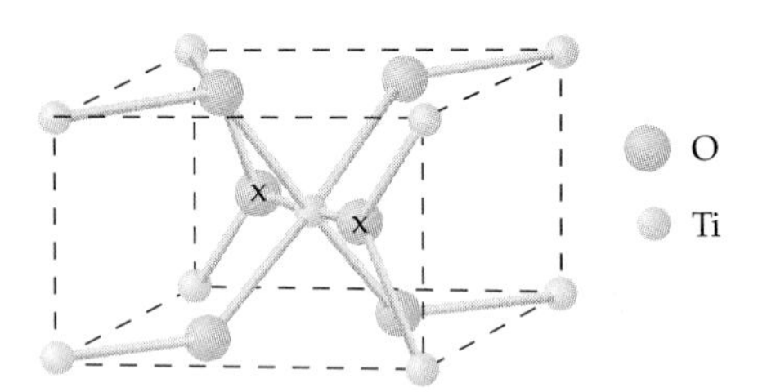

50. Based on the unit cell pictured here, what is the formula of zinc blende? Zinc blende is the main source of zinc. Many other compounds crystallize in this structure, including many important semiconductors such as GaAs.

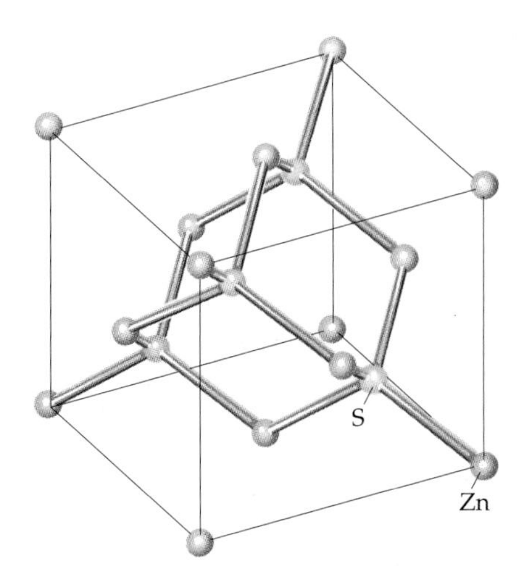

51. Cuprite is a semiconductor. Oxide ions are at the cube corners and in the cube center. Copper ions are wholly within the unit cell. What is the formula of cuprite? What is the oxidation number of copper?

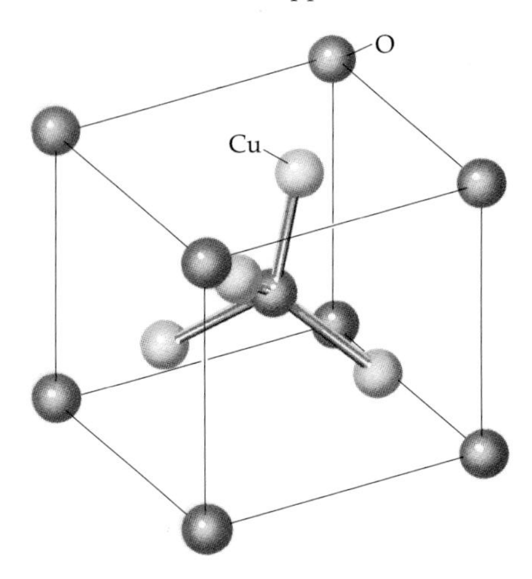

Molecular and Network Solids

52. A diamond unit cell is shown here.

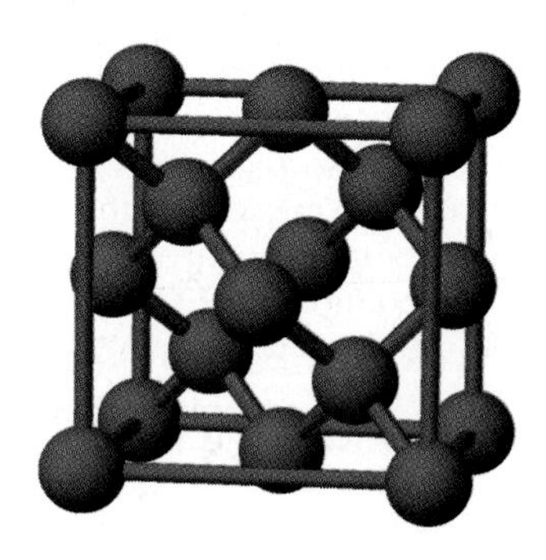

(a) How many carbon atoms are in one unit cell?
(b) What type of unit cell is this (sc, bcc, fcc)?
(c) If the density of diamond is 3.51 g/cm^3, what is the volume of the unit cell and the length of the edge?

53. The structure of graphite is given in Figure 2.21.
(a) What type of intermolecular bonding forces exist between the layers of six-member carbon rings?
(b) Account for the lubricating ability of graphite, that is, why does graphite feel slippery? Why does pencil lead (which is really graphite in clay) leave black marks on paper?

Phase Changes

54. Consider the phase diagram of CO_2 in Figure 13.39.
(a) Is the density of liquid CO_2 greater or less than that of solid CO_2?
(b) In what phase do you find CO_2 at 5 atm and 0 °C?
(c) What is the critical temperature of CO_2?

55. Use the phase diagram of xenon given here to answer the following questions:

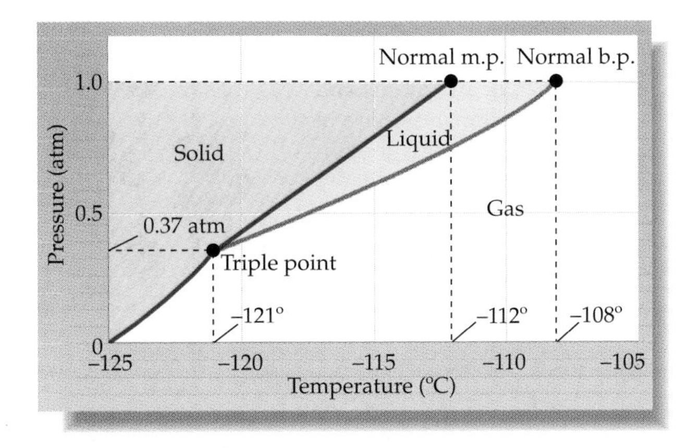

(a) In what phase is xenon found at room temperature and 1.0 atm pressure?
(b) If the pressure exerted on a xenon sample is 0.75 atm, and the temperature is −114 °C, in what phase does the xenon exist?
(c) If you measure the vapor pressure of a liquid xenon sample to be 380 mm Hg, what is the temperature of the liquid phase?
(d) What is the vapor pressure of the solid at −122 °C?
(e) Which is the denser phase—solid or liquid? Explain briefly.

56. Liquid ammonia, $NH_3(\ell)$, was once used in home refrigerators as the heat transfer fluid. The specific heat of the liquid is 4.7 J/g · K and that of the vapor is 2.2 J/g · K. The enthalpy of vaporization is 23.33 kJ/mol at the boiling point. If you heat 12 kg of liquid ammonia from −50.0 °C to its boiling point of −33.3 °C, allow it to evaporate, and then continue warming on to 0.0 °C, how much heat energy must you supply?

57. If your air conditioner is more than several years old, it may use the chlorofluorocarbon CCl_2F_2 as the heat transfer fluid. (CFCs such as this are being replaced as rapidly as possible by substances less harmful to the environment.) The normal boiling point of CCl_2F_2 is −29.8 °C, and the enthalpy of vaporization is 20.11 kJ/mol. The gas and the liquid have specific heats of 117.2 J/mol · K and 72.3 J/mol · K, respectively. How much heat is evolved when 20.0 g of CCl_2F_2 is cooled from +40 °C to −40 °C?

Physical Properties of Solids

58. Benzene, C_6H_6, is an organic liquid that freezes at 5.5 °C (see Figure 13.3) to beautiful, feather-like crystals. How much heat is evolved when 15.5 g of benzene freezes at 5.5 °C? (The heat of fusion of benzene is 9.95 kJ/mol.) If the 15.5-g sample is remelted, again at 5.5 °C, what quantity of heat is required to convert it to a liquid?

59. The specific heat of silver is 0.235 J/g · K. Its melting point is 962 °C, and its heat of fusion is 11.3 kJ/mol. What quantity of heat, in joules, is required to change 5.00 g of silver from solid at 25 °C to liquid at 962 °C?

GENERAL QUESTIONS

60. Rank the following substances in order of increasing strength of intermolecular forces: (a) Ar, (b) CH_3OH, and (c) CO_2.

61. What types of intermolecular forces are important in the liquid phase of (a) C_2H_6 and (b) $(CH_3)_2CHOH$.

62. Construct an approximate phase diagram for O_2 from the following information: normal boiling point, 90.18 K; normal melting point, 54.8 K; and triple point 54.34 K (at a pressure of 2 mm Hg). Very roughly estimate the vapor pressure of liquid O_2 at −196 °C, the lowest temperature easily reached in the laboratory. Is the density of liquid O_2 greater or less than that of solid O_2?

63. Cooking oil floats on top of water. From this observation, what conclusions can you draw regarding the polarity or hydrogen-bonding ability of molecules found in cooking oil?

64. Acetone, $(CH_3)_2C{=}O$, is a common laboratory solvent. It is usually contaminated with water, however. Why does acetone absorb water so readily? Draw molecular structures showing how water and acetone can interact. What intermolecular force(s) is (are) involved in the interaction?

65. A unit cell of cesium chloride is shown on page 618. The density of the solid is 3.99 g/cm^3, and the radius of the Cl^- ion is 167 pm. What is the radius of the Cs^+ ion in the center of the cell? (Although it is not quite true, for the sake of this calculation we shall assume that the Cs^+ ion touches all of the corner Cl^- ions and all ions in the face of the cell touch one another.)

66. If you place 1.0 L of ethanol (C_2H_5OH) in a room that is 3.0 m long, 2.5 m wide, and 2.5 m high, will all the alcohol evaporate? If some liquid remains, how much will there be? The vapor pressure of ethyl alcohol at 25 °C is 59 mm Hg, and the density of the liquid at this temperature is 0.785 g/cm^3.

67. Liquid methanol, CH_3OH, is placed in a glass tube. Is the meniscus of the liquid concave or convex?

68. Liquid ethylene glycol, $HOCH_2CH_2OH$, is one of the main ingredients in commercial antifreeze. Do you predict its viscosity to be greater or less than that of ethanol, CH_3CH_2OH?

69. Select the substance in each of the following pairs that should have the higher boiling point:
(a) Br_2 or ICl
(b) neon or krypton
(c) CH_3CH_2OH (ethanol) or C_2H_4O (ethylene oxide, structure below)

$H_2C{-}CH_2$ (ring closed through O)

70. If the lakes and oceans of another planet are filled with ammonia, what quantity of energy is evolved (in joules) when 1.00 mol of liquid ammonia cools from −33.3 °C (its boiling point) to −43.3 °C? (The specific heat capacity of liquid NH_3 is 4.70 J/g · K.) Compare this with the quantity of heat evolved by 1.00 mol of liquid water cooling by exactly 10 °C. Which evolves more heat on cooling 10 °C, liquid water or liquid ammonia?

71. Account for these facts:
   (a) Although ethanol (C_2H_5OH) (bp, 80 °C) has a higher molar mass than water (bp, 100 °C), the alcohol has a lower boiling point.
   (b) Mixing 50 mL of ethanol with 50 mL of water produces a solution with a volume slightly less than 100 mL.

72. Use the vapor pressure curves illustrated here to answer the questions that follow.

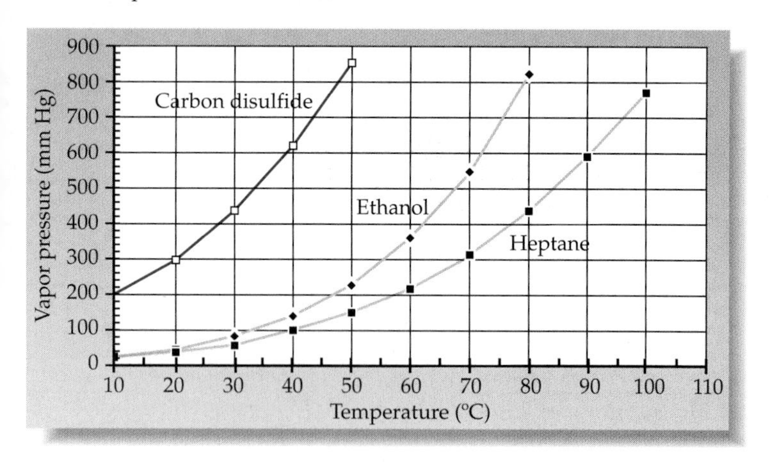

   (a) What is the vapor pressure of ethanol, C_2H_5OH, at 60 °C?
   (b) Considering only carbon disulfide (CS_2) and ethanol, which has the stronger intermolecular forces in the liquid state?
   (c) At what temperature does heptane (C_7H_{16}) have a vapor pressure of 500 mm Hg?
   (d) What are the approximate normal boiling points of each of the three substances?
   (e) Suppose the pressure is 400 mm Hg and the temperature is 70 °C. Is each substance a liquid, a gas, or a mixture of liquid and gas?

73. Silver crystallizes in the unit cell illustrated here. The density of silver is 10.5 g/cm^3. What is the radius of a silver atom?

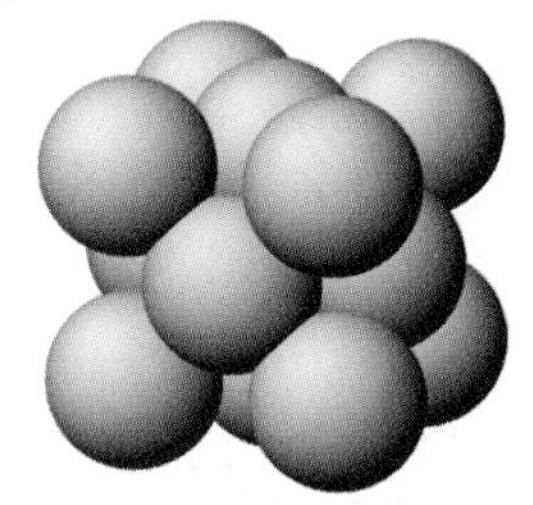

74. Tungsten crystallizes in the unit cell shown here.
   (a) What type of cell is this?
   (b) How many tungsten atoms occur per unit cell?
   (c) If the edge of the unit cell is 315.5 pm, what is the radius of a tungsten atom?

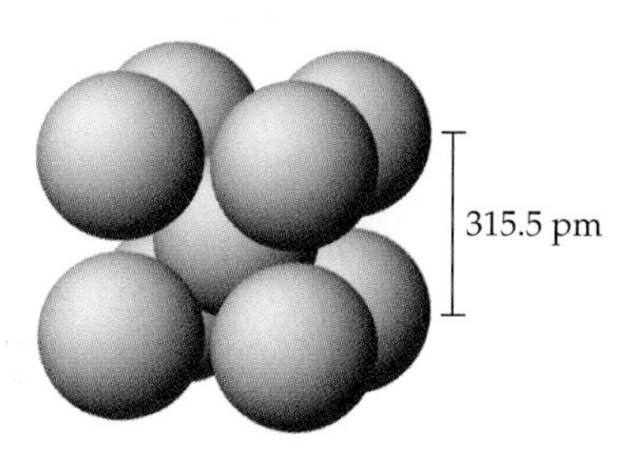

75. The noncubic unit cell shown here is for calcium carbide. How many calcium atoms and how many carbon atoms are in each unit cell? What is the formula of calcium carbide?

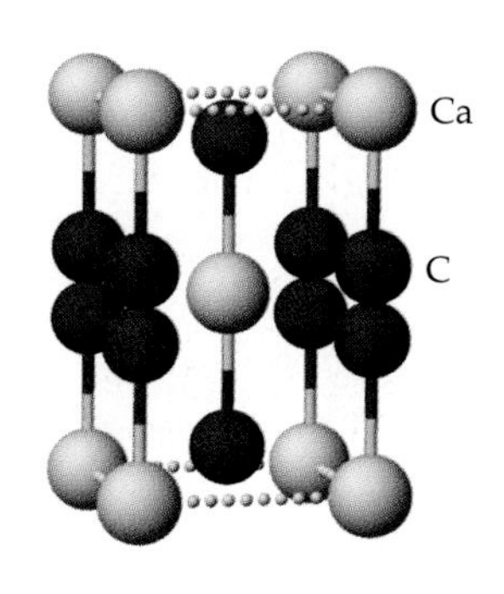

CONCEPTUAL QUESTIONS

76. Rationalize the observation that $CH_3CH_2CH_2OH$, 1-propanol, has a boiling point of 97.2 °C, whereas a compound with the same empirical formula, methyl ethyl ether ($CH_3CH_2—O—CH_3$) boils at 7.4 °C.

77. Many common salts are hydrated, for example, $CoCl_2 \cdot 6\ H_2O$. Which of the following two salts has the more exothermic heat of hydration, $BeSO_4$ or $BaSO_4$?

78. Can $CaCl_2$ have a unit cell like that of sodium chloride? Explain.

79. Cite two pieces of evidence to support the statement that water molecules in the liquid state exert considerable attractive force on one another.

80. Why is it not possible for a salt with the formula M_3X (Na_3PO_4, for example) to have a face-centered cubic lattice of X anions and M cations?

81. During thunderstorms in the Midwest, very large hailstones can fall from the sky. (Some are the size of golf balls!) To preserve some of these stones, we put them in the freezer compartment of a frost-free refrigerator. Our friend, who is a chemistry student, tells us to use an older model that is not frost-free. Why?

82. The photos illustrate an experiment you can do yourself. Place 10–20 mL of water in an empty soda can and heat the water to boiling. Using tongs or pliers, turn the can over in a pan of cold water, making sure the opening in the can is below the water level in the pan. Describe what happens and explain it in terms of the subject of this chapter.

(C. D. Winters)

CHALLENGING QUESTIONS

83. Calcium fluoride is the well-known mineral fluorite. It is known that each unit cell contains four Ca^{2+} ions and eight F^- ions and that the Ca^{2+} ions are arranged in an fcc lattice. The F^- ions fill all the so-called tetrahedral holes in a face-centered cubic lattice of Ca^{2+} ions. The edge of the CaF_2 unit cell is 5.46295×10^{-8} cm in length. The density of the solid is 3.1805 g/cm^3. Use this information to calculate Avogadro's number.

The mineral fluorite, CaF_2, and its unit cell. *(Photo, C. D. Winters; model, S. M. Young)*

84. Mercury and many of its compounds are dangerous poisons if breathed, swallowed, or even absorbed through the skin. The liquid metal has a vapor pressure of 0.00169 mm Hg at 24 °C. If the air in a small room is saturated with mercury vapor, how many atoms of mercury vapor occur per cubic meter? Assume the room is 4 m square and 2.5 m high (about 13 ft on a side with a ceiling at about 10 ft).

85. You can get some idea of how efficiently spherical atoms or ions are packed in a three-dimensional solid by seeing how well circular atoms pack in two dimensions. Using the drawings shown here, prove that B is a more efficient way to pack circular atoms than A. A unit cell of A contains portions of four circles and one hole. In B, packing coverage can be calculated by looking at a triangle that contains portions of three circles and one hole. Show that A fills about 80% of the available space, whereas B fills closer to 90% of the available space.

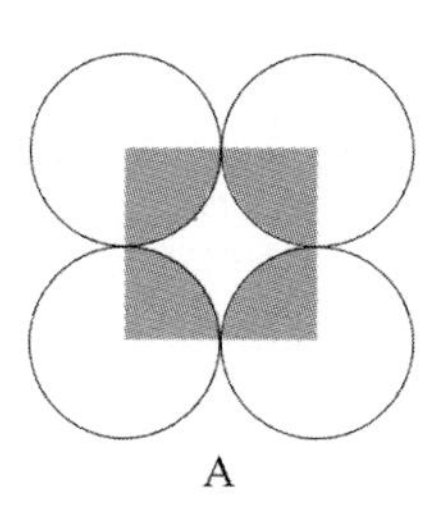

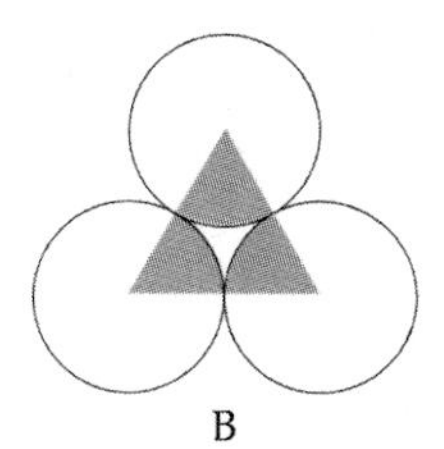

86. If a simple cubic unit cell is formed so that the spherical atoms or ions just touch one another along the edge, calculate the percentage of empty space within the unit cell. (Recall that the volume of a sphere is $4/3(\pi r^3)$, where r is the radius of the sphere.)

87. Two identical swimming pools are filled with uniform spheres of ice packed as closely as possible. The spheres in the first pool are the size of grains of sand; those in the second pool are the size of oranges. The ice in both pools melts. In which pool, if either, will the water level be higher? (Ignore any differences in filling space at the planes next to the walls and bottom.)

Summary Question

88. Sulfur dioxide, SO_2, is found in polluted air. It comes from the combustion of fossil fuels, and from industries that convert certain metal-containing ores to metals or metal oxides.

(a) Draw the Lewis structure for SO_2. From this tell the O—S—O angle and molecular geometry of the compound.

(b) What types of forces are responsible for binding SO_2 molecules to one another in the solid or liquid phase?

(c) Using the information in the table, place the compounds listed in order of increasing intermolecular forces.

Compound	Normal Boiling Point (°C)
SO_2	−10.05
NH_3	−33.33
CH_4	−161.48
H_2O	100.0

(d) Airborne sulfur dioxide is one of the compounds responsible for "acid rain." It is thought that the $SO_2(g)$ first oxidizes to $SO_3(g)$ in air. What is the standard enthalpy change for this process? What is the standard enthalpy change for the combination of $SO_3(g)$ and water to give $H_2SO_4(aq)$?

INTERACTIVE GENERAL CHEMISTRY CD-ROM

Screen 13.4 Intermolecular Forces (2)

Watch the animation when you study *Ion–Dipole Forces* on this screen.

(a) Why do you see an interaction between the Na^+ ions and the O atom of water molecules and not between Na^+ ions and the H atoms of water?
(b) Considering Coulomb's law, explain the trend in enthalpy of hydration values for Group 1A cations.
(c) Which ion would you expect to have a greater enthalpy of hydration: Na^+ or Mg^{2+}?

Click on *Dipole–Dipole Forces* and then on the table of *Molar Masses and Boiling Points.*

(a) What is the connection between the strength of a compound's intermolecular forces and its boiling point?
(b) What two factors control the strength of intermolecular forces, as reflected in the boiling point table?

Screen 13.5 Intermolecular Forces (3)

(a) What does the term "polarizable" mean?
(b) Rank the halogens F_2, Cl_2, and Br_2 in terms of their expected boiling points. Explain briefly.

Screen 13.12 Solid Structures (Crystalline and Amorphous Solids)

(a) What is the difference between crystalline and amorphous solids?
(b) Watch the simple cubic unit cell animation on this screen. How many unit cells make up the rotating structure?
(c) Consider the shapes of the following solids and predict if they are crystalline solids or amorphous solids: table salt, ice (think of snow flakes), glass, and wood.

Screen 13.13 Solid Structures (Ionic Solids)

(a) Watch the animation of a face-centered cubic (fcc) unit cell on this screen. Why is the sodium ion said to occupy "octahedral" holes?
(b) Examine the structure of CaO using the model in the *Molecular Models* folder (in the *SOLIDS* folder).
   1. Describe the structure. Why type of lattice is described by the Ca^{2+} ions? What type of hole does the O^{2-} occupy in the lattice of Ca^{2+} ions?
   2. How is the formula of CaO related to its unit cell structure? (How many Ca^{2+} ions and how many O^{2-} ions are in one unit cell?)
   3. How is the CaO structure related to the NaCl structure?
(c) Examine the structure of ZnS using the model in the *Molecular Models* folder (in the *SOLIDS* folder or directory). Describe the structure. What type of lattice is described by the Zn^{2+} ions (silver color)? What type of holes do the S^{2-} ions (yellow color) occupy in the lattice of Zn^{2+} ions? How is the formula of ZnS related to its unit cell structure?
(d) Lead sulfide, PbS (commonly called galena), has the same formula as ZnS. Does it have the same solid structure? (See the model in the *Molecular Models* folder.) If different, how it is different? How is its unit cell related to its formula?
(e) Examine the structure of CaF_2 using the model in the *Molecular Models* folder (in the *SOLIDS* folder or directory).
   1. Describe the structure. Why type of lattice is described by the Ca^{2+} ions? What type of holes do the F^- occupy in the lattice of Ca^{2+} ions?
   2. How is the formula of CaF_2 related to its unit cell structure? (How many Ca^{2+} ions and how many F^- ions are in one unit cell?)
   3. How is the CaF_2 structure related to the ZnS structure?

Chapter 14

Solutions and Their Behavior

(C. D. Winters)

A Chemical Puzzler

It has probably happened to us all at one time or another. You take a can of soda from the refrigerator and pop it open, only to be greeted by a spray of liquid as dissolved carbon dioxide bubbles out of the aqueous solution. A SCUBA diver rising to the surface too quickly is likely to have another unhappy experience—the bends—as bubbles of gaseous nitrogen form in the blood as the external gas pressure is decreased. How are these events related to an important medical technique, in which the life of a person with severe respiratory problems is saved when placed in a hyperbaric chamber under a pressure of oxygen greater than 1 atm? We will examine these examples later in this chapter when we discuss solutions of gases in liquids.

We come into contact with solutions every day: aqueous solutions of ionic salts, motor oil and gasoline with additives to improve their properties, and household cleaners such as ammonia in water. We purposely make solutions. Adding sugar, flavoring, and sometimes CO_2 to water produces a palatable soft drink. Athletes drink commercial beverages that are specially prepared solutions, called isotonic solutions, with dissolved salts to match precisely salt concentrations in body fluids, thus allowing the fluid to be taken into the body much more rapidly. In medicine, saline solutions (an aqueous solution with NaCl and other soluble salts) are infused into the body to replace lost fluids.

A **solution** is a homogeneous mixture of two or more substances in a single phase. It is usual to think of the component present in largest amount as the **solvent** and the other component as the **solute.** When you think of solutions, those that occur to you first probably involve a liquid as solvent. Some solutions, however, do not involve a liquid solvent at all; examples include the air you breathe (a solution of nitrogen, oxygen, carbon dioxide, water vapor, and other gases) and solid solutions—called *alloys*—such as 18 K gold, brass, bronze, and pewter. Although there are many types of solutions, the objective in this chapter is to develop an understanding of gases, liquids, and solids dissolved in liquid solvents.

Experience tells you that adding a solute to a pure liquid will change the properties of the liquid. Indeed, that is the reason some solutions are made. For instance, adding antifreeze to your car's radiator prevents the coolant from boiling in the summer and freezing in the winter. The changes that occur in the freezing and boiling points when a substance is dissolved in a pure liquid are two properties that we will examine in detail. These properties, as well as the osmotic pressure of a solution and changes in vapor pressure, are examples of colligative properties. **Colligative properties** are those that ideally depend only on the number of solute particles per solvent molecule and not on the nature (identity) of the solute.

Four major topics are covered in this chapter. First, because colligative properties depend on the relative number of solvent and solute particles in solution, convenient ways of describing solution concentration in these terms are required. Second, we consider how and why solutions form on the molecular level. This gives us some insight into the third topic, the colligative properties themselves. The chapter concludes with a brief discussion of colloids, mixtures that have properties intermediate between solutions and suspensions and that are important in many biological systems.

14.1 UNITS OF CONCENTRATION

See the *Saunders Interactive General Chemistry CD-ROM,* Screen 14.1, Solubility.

To analyze the colligative properties of a solution, we need ways of measuring solute concentrations that reflect the number of molecules or ions of solute per molecule of solvent. Molarity, the concentration unit useful in stoichiometry calculations, does not work when we are dealing with colligative properties. Recall that **molarity** (M) is defined as the number of moles of solute per liter of solution:

$$\text{Molar concentration of solute A} = \frac{\text{quantity of A (mol)}}{\text{volume of solution (L)}}$$

Using molarity, it is possible to determine the number of solute molecules, but it is not possible to identify the amount of solvent used to make the solution. This problem is illustrated in Figure 14.1. The flask on the right contains a 0.100 M aqueous solution of potassium chromate. It was made by adding enough water to 0.100 mol of K_2CrO_4 to make 1.000 L of solution. No mention is made of how much solvent (water) was actually added. If 1.000 L of water had been added to 0.100 mol of K_2CrO_4, as illustrated by the flask on the left side of Figure 14.1, the volume of solution would not be 1.000 L. It is slightly greater than a liter, a result that might have been anticipated because the solute molecules are expected to occupy some volume.

Several concentration units reflect the number of molecules or ions of solute per solvent molecule. Four common units are molality, mole fraction, weight percent, and parts per million.

The **molality** *(m)* of a solution is defined as the number of moles of solute per kilogram of solvent.

$$\text{Molality of solute } (m) = \frac{\text{quantity of solute (mol)}}{\text{kilograms of solvent}} \qquad [14.1]$$

Figure 14.1 Preparing 0.100 molal and 0.100 molar solutions. In the flask on the right, 0.100 mol (19.4 g) of K_2CrO_4 was mixed with enough water to make 1.000 L of solution. (The volumetric flask was filled to the mark on its neck, indicating the volume is exactly 1.000 L.) Exactly 1.000 kg of water was added to 0.100 mol of K_2CrO_4 in the flask on the left. Notice that the volume of solution is greater than 1.000 L. (The small pile of yellow solid in front of the flask is 0.100 mol of K_2CrO_4.) *(C. D. Winters)*

The symbol for molality is a small italicized *m*. The symbol for molarity is a regular capital M.

Recall that the mole fraction was first used in connection with gas mixtures (Section 12.5).

The concentration of the K_2CrO_4 solution in the flask on the left side of Figure 14.1 is 0.100 molal (0.100 *m*). It was prepared from 19.4 g (0.100 mol) of K_2CrO_4 and 1.00 kg (1.00 L × 1 kg/L) of water:

$$\text{Molality of } K_2CrO_4 = \frac{19.4 \cancel{\text{g}} (1 \text{ mol}/194 \cancel{\text{g}})}{1.00 \text{ kg water}} = 0.100 \ m$$

Notice that different quantities of water were used to make the 0.100 M (0.100 molar) and 0.100 *m* (0.100 molal) solutions of K_2CrO_4. This means the *molarity and molality of a given solution cannot be the same* (although the difference may be negligibly small when the solution is quite dilute).

The **mole fraction,** X, of a solution component is defined as the number of moles of a given component of a mixture divided by the total number of moles of all of the components of the mixture. Mathematically this is represented as

$$\text{Mole fraction of A } (X_A) = \frac{n_A}{n_A + n_B + n_C + \cdots} \qquad [14.2]$$

Consider a solution that contains 1.00 mol (46.1 g) of ethanol, C_2H_5OH, in 9.00 mol (162 g) of water. Here the mole fraction of alcohol is 0.100 and that of water is 0.900:

$$X_{\text{ethanol}} = \frac{1.00 \text{ mol ethanol}}{1.00 \text{ mol ethanol} + 9.00 \text{ mol water}} = 0.100$$

$$X_{\text{water}} = \frac{9.00 \text{ mol water}}{1.00 \text{ mol ethanol} + 9.00 \text{ mol water}} = 0.900$$

Notice that the sum of the mole fractions of the components in the solution equals 1.000, a fact that is true for the solute and solvent in all solutions:

$$X_{\text{water}} + X_{\text{ethanol}} = 1.000$$

The composition of many common products is often given in terms of weight percent. Here the label on household bleach indicates that it contains 5.25% sodium hypochlorite. *(C. D. Winters)*

Weight percent is the mass of one component divided by the total mass of the mixture (multiplied by 100):

$$\text{Weight \%A} = \frac{\text{mass of A}}{\text{mass of A} + \text{mass of B} + \text{mass of C} + \cdots} \times 100\% \quad [14.3]$$

The alcohol–water mixture has 46.1 g of ethanol and 162 g of water, so the total mass of solution is 208 g, and the weight % of alcohol is

$$\text{Weight \% ethanol} = \frac{46.1 \text{ g ethanol}}{46.1 \text{ g ethanol} + 162 \text{ g water}} \times 100\% = 22.2\%$$

Notice that if you know the weight percentage of a solute, you can also determine its mole fraction or molality (or vice versa) because the masses of solute and solvent are known.

Weight percent is a common unit in everyday life. Vinegar, for example, is an aqueous solution containing approximately 5% acetic acid and 95% water, and the label on a common household bleach lists its active ingredient as 5.25% sodium hypochlorite (NaOCl) and 94.75% inert ingredients.

Naturally occurring solutions are often very dilute. Environmental chemists, biologists, geologists, oceanographers and others frequently use the units milligrams per liter (mg/L) and **parts per million (ppm).** The unit ppm refers to relative amounts by weight; 1.0 ppm represents 1.0 g of a substance in sample with a total weight of 1.0×10^6 g. Because water at 25 °C has a density of 1.0 g/mL, a concentration of 1.0 mg/L is equivalent to 1.0 mg of solute in 1000 g of water or to 1.0 g of solute in 1,000,000 g of water; that is, units of ppm and mg/L are approximately equivalent:

$$\frac{1.0 \text{ mg solute}}{1.0 \text{ L solution}} = \frac{1.0 \text{ mg solute}}{1.0 \times 10^3 \text{ g water}} = \frac{1.0 \times 10^{-3} \text{ g solute}}{1.0 \times 10^3 \text{ g water}} = \frac{1.0 \text{ g solute}}{1.0 \times 10^6 \text{ g water}}$$

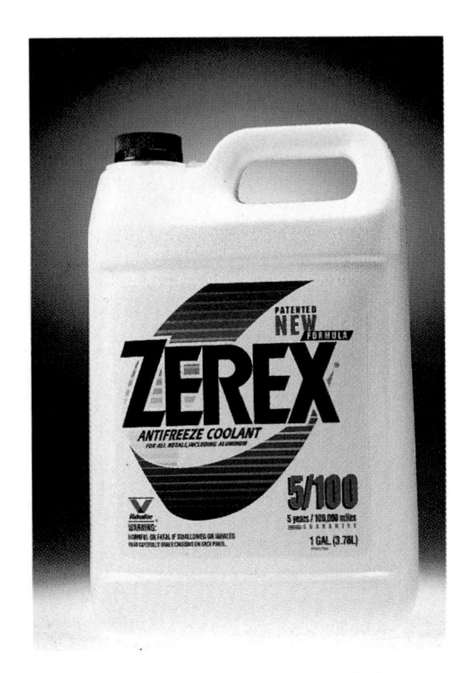

Commercial antifreeze contains ethylene glycol, $HOCH_2CH_2OH$, an organic alcohol that is readily soluble in water. *(Photo, C. D. Winters; model, S. M. Young)*

Example 14.1 Calculating Mole Fractions, Molality, and Weight Percent

Assume you add 1.2 kg of ethylene glycol, $HOCH_2CH_2OH$, as an antifreeze to 4.0 kg of water in the radiator of your car. What are the mole fraction, molality, and weight percent of the ethylene glycol?

Solution 1.2 kg of ethylene glycol (molar mass = 62.1 g/mol) is equivalent to 19 mol, and 4.0 kg of water represents 220 mol.

Mole fraction:

$$X_{\text{glycol}} = \frac{19 \text{ mol glycol}}{19 \text{ mol glycol} + 220 \text{ mol water}} = 0.080$$

Molality:

$$\text{Molality} = \frac{19 \text{ mol glycol}}{4.0 \text{ kg water}} = 4.8\ m$$

Weight percentage:

$$\text{Weight \%} = \frac{1.2 \times 10^3 \text{ g glycol}}{1.2 \times 10^3 \text{ g glycol} + 4.0 \times 10^3 \text{ g water}} \times 100\% = 23\%$$

Example 14.2 Parts per Million

You dissolve 560 g of $NaHSO_4$ in a swimming pool that contains 4.5×10^5 L of water at 25 °C. What is the sodium ion concentration in parts per million? (Sodium hydrogen sulfate is used to adjust the pH of the pool water. Recognize that the anion, HSO_4^-, can furnish $H^+(aq)$ to the solution.)

Solution As a first step, we calculate the quantity of sodium ions (in grams) in 560 g of $NaHSO_4$:

$$560 \text{ g NaHSO}_4 \left(\frac{1 \text{ mol NaHSO}_4}{120 \text{ g NaHSO}_4}\right)\left(\frac{1 \text{ mol Na}^+}{1 \text{ mol NaHSO}_4}\right)\left(\frac{23.0 \text{ g Na}^+}{1 \text{ mol Na}^+}\right) = 110 \text{ g Na}^+$$

Then, we can use the mass of sodium ions added to the pool to find the number of milligrams per liter, which is equivalent to parts per million:

$$\frac{110 \text{ g}(1000 \text{ mg}/1 \text{ g})}{4.5 \times 10^5 \text{ L}} = 0.24 \text{ mg/L, or } 0.24 \text{ ppm}$$

Exercise 14.1 Mole Fraction, Molality, and Weight Percentage

If you dissolve 10.0 g of sugar (sucrose), $C_{12}H_{22}O_{11}$ (about one heaping teaspoonful), in a cup of water (250. g), what are the mole fraction, molality, and weight percent of sugar?

Exercise 14.2 Parts per Million

Sea water has a sodium ion concentration of 1.08×10^4 ppm. If the sodium is present in the form of dissolved sodium chloride, how many grams of NaCl are in each liter of sea water? Sea water is denser than pure water because of dissolved salts. Assume that the density of sea water is 1.05 g/mL.

14.2 THE SOLUTION PROCESS

If solid salt (NaCl) is added to a beaker of water, the salt will begin to dissolve (Figures 1.5 and 5.1). The amount of solid diminishes, and the concentration of $Na^+(aq)$ and $Cl^-(aq)$ in solution increases. If we continue to add NaCl, however, a point will eventually be reached when no additional NaCl seems to dissolve. The concentrations of $Na^+(aq)$ and $Cl^-(aq)$ will not increase further, and any additional solid NaCl added after this point will simply remain as a solid at the bottom of the beaker. We say that the solution is **saturated.**

Although no change is observed on the macroscopic level, things are different on the particulate level. The process of dissolving is still going on, with $Na^+(aq)$ and $Cl^-(aq)$ ions leaving the solid state and entering solution. Concurrently, however, a second process is occurring; this is the formation of solid NaCl(s) from $Na^+(aq)$ and $Cl^-(aq)$. The rates at which NaCl is dissolving and reprecipitating are equal in a saturated solution, so that there is no net change on the macroscopic level.

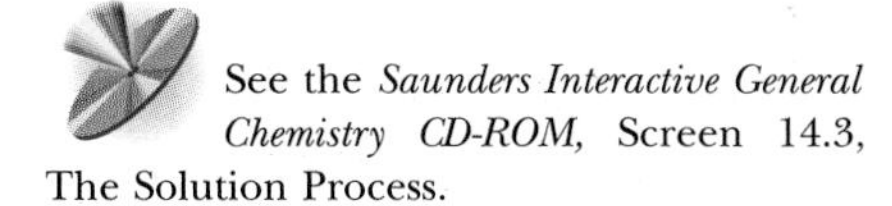

See the *Saunders Interactive General Chemistry CD-ROM,* Screen 14.3, The Solution Process.

The term **unsaturated** is used when referring to solutions with concentrations of solute that are less than that of a saturated solution.

A CLOSER LOOK Supersaturated Solutions

Although at first glance it may seem a contradiction, it is possible to have a solution in which there is *more* dissolved solute than the amount in a saturated solution. Such solutions are referred to as **supersaturated** solutions. Supersaturated solutions are unstable, and the excess solid eventually crystallizes from the solution until the equilibrium concentration of the solute is reached.

The solubility of substances often decreases if the temperature is lowered. Supersaturated solutions are usually made by preparing a saturated solution at a given temperature and then carefully cooling it. If the rate of crystallization is slow, the solid may not precipitate when the solubility is exceeded. Going to still lower temperatures results in a solution that has more solute than the amount defined by equilibrium conditions; it is supersaturated.

When disturbed in some manner, a supersaturated solution moves toward equilibrium by precipitating solute. This can occur rapidly and often with the evolution of heat. In fact, supersaturated solutions are used in "heat packs" to apply heat to injured muscles (see page 279). When crystallization of sodium acetate ($NaCH_3CO_2$) from a supersaturated solution in a heat pack is initiated, the temperature of the heat pack rises to close to 50 °C, and crystals of solid sodium acetate are detectable inside the bag.

When a supersaturated solution is disturbed, the dissolved salt (here sodium acetate, $NaCH_3CO_2$) rapidly crystallizes. *(C. D. Winters)*

This is an another example of equilibrium in chemistry, and we can describe the situation in terms of an equation with substances linked by a set of double arrows ($\rightleftarrows$):

$$NaCl(s) \overset{H_2O}{\rightleftarrows} Na^+(aq) + Cl^-(aq)$$

Recall that equilibrium systems were encountered earlier with changes of state (Chapter 13). In equilibria involving two states, the description was very similar to what we are seeing for a saturated solution: no changes were observable at the macroscopic level, but two opposing processes go on at the particulate level at the same rate.

A saturated solution gives us a way to define solubility precisely. **Solubility** is the concentration of solute in equilibrium with undissolved solute in a saturated solution. Sodium chloride's solubility, for example, is 35.7 g in 100 mL of water at 0 °C. If we added 50.0 g of NaCl to 100 mL of water at 0 °C, we can expect 35.7 g to dissolve, and 14.3 g of solid to remain.

Quantitative data on solubility for many compounds are listed in chemical handbooks. It is from these data that solubility rules such as those given in Table 5.4 were created.

Liquids Dissolving in Liquids

If two liquids mix to an appreciable extent to form a solution, they are said to be **miscible.** In contrast, **immiscible** liquids do not mix to form a solution; they exist in contact with each other as separate layers.

The polar compound ethanol (C_2H_5OH) dissolves in water, which is also a polar compound. Beer, wine, and other alcoholic beverages contain amounts of alcohol ranging from just a few percent to more than 50%. Ethanol commonly used in laboratories has a concentration of 95% ethanol and 5% water. In fact, ethanol and water are miscible in all proportions. The nonpolar liquids octane (C_8H_{18}) and carbon tetrachloride (CCl_4) are also miscible in all proportions. On the other hand, neither C_8H_{18} nor CCl_4 is miscible with water. Observations like these have led to a familiar rule of thumb: *like dissolves like.* That is, two or more nonpolar liquids frequently are miscible, just as are two or more polar liquids.

The miscibility of different liquids is illustrated in Figure 13.7.

What is the molecular basis for the "like dissolves like" guideline? In pure water and pure ethanol, the major force between molecules is hydrogen bonding involving O—H groups. When the two liquids are mixed, hydrogen bonding between ethanol and water molecules also occurs and assists in the solution process. Molecules of pure octane or pure CCl_4, both of which are nonpolar, are held together in the liquid phase by dispersion forces (Section 13.2). When these nonpolar liquids are mixed, the energy associated with these forces of attraction is similar in value to the energy due to the forces of attraction between octane and CCl_4 molecules. Thus, little or no energy change occurs when octane–octane and CCl_4–CCl_4 attractive forces are replaced with octane–CCl_4 forces. The solution process is expected to be nearly energy-neutral. So, why do they mix? The answer lies deeper in thermodynamics. As you shall see in Chapter 20, processes that move to less orderly arrangements tend to occur. The tendency is measured by a thermodynamic function called *entropy* (Figure 14.2).

In contrast, polar and nonpolar liquids usually do not mix to an appreciable degree; placed together in a container they separate into two distinct layers (Figure 14.3). The rationale for this behavior is complicated. Experimental data show that the enthalpy of mixing of dissimilar liquids is also near zero, so the energetics of the process is not the primary factor. Apparently interposing nonpolar molecules into a polar solvent such as water causes the internal structure of water at the molecular level to become more ordered. Forming a more ordered system disfavors the process of mixing.

Although the energetics of solution formation are important, it is generally accepted that the more important contributor is the entropy of mixing. As you shall see in Chapter 20, entropy favors the formation of less ordered systems such as solutions. See also T. P. Silverstein: "The real reason why oil and water don't mix." *Journal of Chemical Education,* Vol. 75, pp. 116–118, 1998.

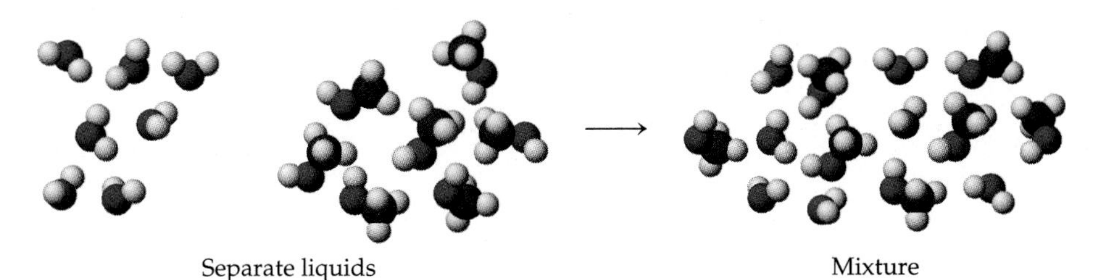

Figure 14.2 Like dissolves like. When two similar liquids—here water and methanol—are mixed, the molecules are intermingled. The mixture has a more disorderly arrangement of molecules than the separate liquids. It is this disordering process that largely drives solution formation. (The disordering of a system is measured by the thermodynamic function called *entropy.* It is discussed in Chapter 20.)

(a)

(b)

Figure 14.3 Miscibility. (a) The colorless, denser bottom layer is nonpolar carbon tetrachloride, CCl_4. The blue middle layer is a solution of $CuSO_4$ in water, and the colorless, less dense top layer is nonpolar octane (C_8H_{18}). This mixture was prepared by carefully layering one liquid on top of another, without mixing. (b) After stirring the mixture, the two nonpolar liquids form a homogeneous mixture. This layer of mixed liquids is under the water layer because the mixture of CCl_4 and C_8H_{18} has a greater density than water. *(Photo, C. D. Winters; model, S. M. Young)*

Solids Dissolving in Liquids

The "*like dissolves like*" guideline also holds for solids dissolving in liquids. For example, solid I_2 is a nonpolar molecular solid in which the I_2 molecules are held together by intermolecular dispersion forces. Iodine is expected to dissolve in a nonpolar liquid such as CCl_4 (Figure 14.4).

Sucrose, a sugar, is also a molecular solid. The molecule is capable of hydrogen bonding, however, because it contains numerous O—H groups. Sucrose is readily soluble in water, a fact that we know well because of its use to sweeten beverages. We might also have predicted its solubility in water because strong solute–solvent attractive forces are expected. Hydrogen bonding between sugar and water is strong enough that the energy evolved here can be supplied to disrupt the sugar–sugar and water–water interactions. On the other hand, sugar is not soluble in CCl_4 or other nonpolar liquids.

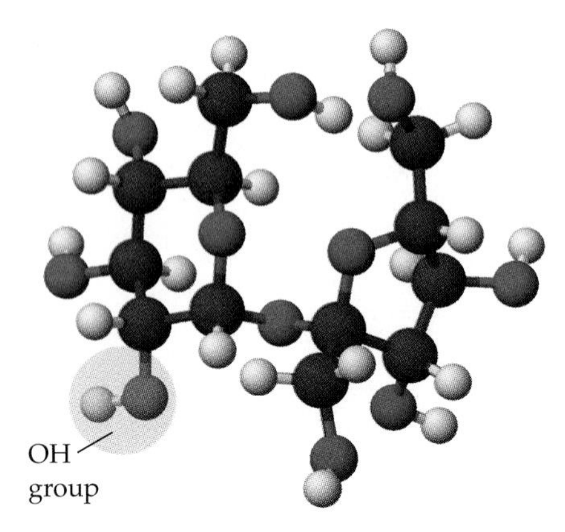

Sucrose, or cane sugar, has a number of —OH groups on each molecule, groups that allow it to form hydrogen bonds with water molecules.

Common ionic solids are not soluble to any appreciable extent in nonpolar solvents such as CCl_4. They are often soluble in water, but as we noted in discussions concerning solubility in Chapter 5, their solubility in this solvent varies greatly (Figure 5.4). For example, at 20 °C, 100 g of water dissolves 74.5 g of $CaCl_2$. In contrast, only 0.0014 g of limestone, $CaCO_3$, dissolves in 100 g of water. And $CaCO_3$ is not the least soluble substance known.

Network solids include graphite, diamond, and quartz sand (SiO_2), and experience and intuition tell you they do not dissolve readily in water. After all, where would all the beaches be if sand dissolved readily in water? What if the diamond in a diamond ring dissolved when you washed your hands? The normal covalent chemical bonding in network solids is simply too strong to be replaced by weaker hydrogen bonding attraction to water dipoles.

(a)

(b)

Figure 14.4 Water, carbon tetrachloride (CCl_4), and iodine. (a) Water (a polar molecule) and CCl_4 (a nonpolar molecule) are immiscible, and the less dense water layer is found on top of the denser CCl_4 layer. A small amount of iodine dissolves in water to give a brown solution. (b) The mixture in (a) has been stirred. The nonpolar molecule I_2 is extracted into nonpolar CCl_4 in which it is more soluble, producing a purple solution. Solvent extraction is one of the methods available to chemists to separate materials. *(Photo, C. D. Winters; model, S. M. Young)*

Heat of Solution

Both ammonium chloride and sodium hydroxide dissolve readily in water, but the solution becomes colder when NH_4Cl dissolves, and it warms up when NaOH dissolves. Why do energy changes occur when substances dissolve?

We can get a sense of this situation by examining the solution process at the particulate level (see Figure 5.1). When a solute is dissolved in a solvent, the solute particles are separated from one another and dispersed among molecules of solvent. In addition, the solvent molecules are pushed apart, making space for molecules of the solute. To understand the energetics of this process we focus on attractive forces (solute–solute, solvent–solvent, and solute–solvent) and the energies (enthalpies) related to these forces. Separating solute and solvent particles from one another requires the expenditure of energy. We know, however, that solute and solvent molecules will be attracted to one another and that this is a process that evolves energy. In the overall process of dissolving, the energy expended to separate solute and solvent will be partially or fully offset by the energy associated with the attraction between solute and solvent molecules.

The heat of solution, the overall enthalpy change for the solution process, is the sum of the enthalpy changes for these individual processes. If the energy

A CLOSER LOOK

Dissolving Ionic Solids

Alkali metal halides dissolve in water because strong ion-dipole forces promote ion hydration and help break down the cation–anion attraction in the crystal lattice. Analysis of the various energy changes occurring when a salt such as KF dissolves, for example, will give us greater insight into the process.

Let us imagine that the salt dissolves by first breaking up into ions in the gas phase and that these ions are then hydrated. The enthalpy change for the first part of the process can be estimated from the lattice energy of the salt (page 379). The enthalpy changes for the second stage are the enthalpies of hydration of the gaseous ions to form hydrated ions ($\Delta H_{hydration}$) (page 587).

From Hess's law (Chapter 6) you know the enthalpy of solution, $\Delta H_{solution}$, is the sum of the enthalpies of the processes along the path from reactants [KF(s)] to products [K^+(aq) + F^-(aq)]:

$$\Delta H_{solution} = -\text{lattice energy} + \text{enthalpy of hydration}$$

$$\Delta H_{solution} \text{ for KF(s)} = 821 \text{ kJ/mol} + (-819 \text{ kJ/mol}) = +2 \text{ kJ/mol}$$

ESTIMATED ENTHALPY OF SOLUTION AND WATER SOLUBILITY OF SOME IONIC COMPOUNDS

Compound	Lattice Energy (kJ/mol)	$\Delta H_{hydration}$ (kJ/mol)	$\Delta H_{solution}$ (kJ/mol)	Solubility in H_2O (g/100 mL)
AgCl	−912	−851	+61	0.000089 (10 °C)
NaCl	−786	−760	+26	35.7 (0 °C)
LiF	−1037	−1005	+32	0.3 (18 °C)
KF	−821	−819	+2	92.3 (18 °C)
RbF	−789	−792	−3	130.6 (18 °C)

related to solvent–solute interactions is the greater quantity, the solution process will occur with evolution of energy. This is the case when NaOH dissolves in water (Figure 14.5). If the energy derived from solute–solvent interactions is the smaller quantity, however, then the solution process is endothermic, as with NH_4Cl or NH_4NO_3 dissolving in water (page 279 and Figure 14.5).

Heats of solution are easily measured using a calorimeter. This is usually done in an open system, such as the coffee cup calorimeter described in Section 6.10, and for an experiment run under standard conditions the resulting mea-

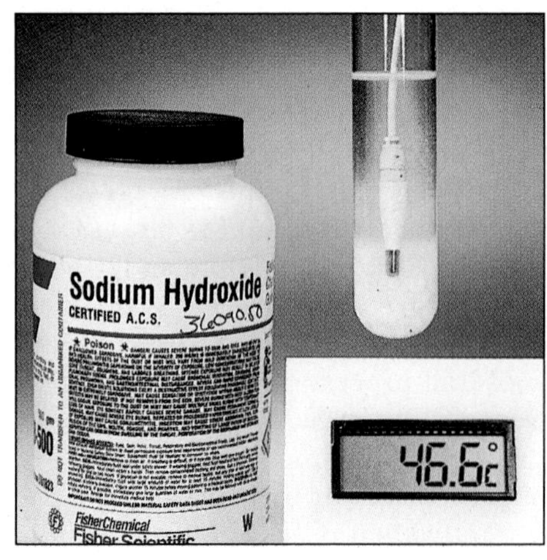

(a)

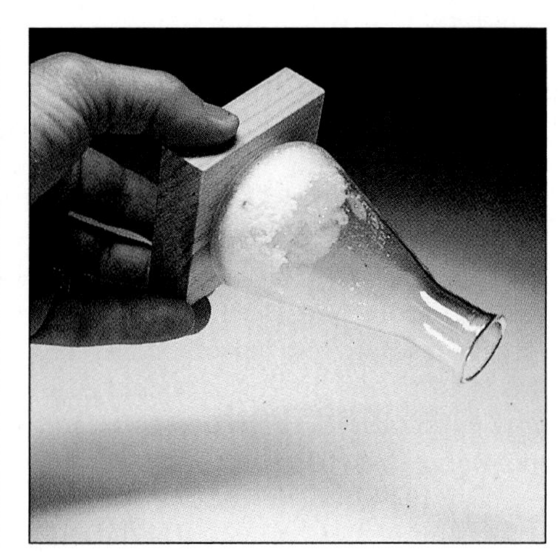

(b)

***Figure* 14.5 Heat of solution.** (a) Dissolving NaOH in water is a strongly exothermic process. (b) When NH_4NO_3 is dissolved in water, the temperature of the system drops. [Here the solids $Ba(OH)_2 \cdot 8 H_2O$ and NH_4NO_3 were mixed. These react to form $Ba(NO_3)_2$ and NH_3 and to release water. When excess NH_4NO_3 dissolves in this small amount of water, the temperature of the system drops because of the endothermic heat of solution of NH_4NO_3 in water.] If a flask containing the mixture is set on a few drops of water on a piece of wood, the water freezes, and the flask is stuck to the board. *(C. D. Winters)*

Using this approach we can estimate that the enthalpy change for dissolving KF is about zero. What is important here is to put this in the context of the data in the table, which lists the water solubility of some simple salts with their calculated enthalpy of solution. Here you see that there is a rough correlation between the estimated value of $\Delta H_{solution}$ and solubility in water: As $\Delta H_{solution}$ becomes more and more negative, the solubility increases. Rubidium fluoride is very soluble in water and has an exothermic enthalpy of solution. Silver chloride is quite insoluble and has a very positive value of $\Delta H_{solution}$. Solubility is favored when the energy required to break down the lattice of the solid (the lattice energy) is smaller or roughly equal to the energy given off when the ions are hydrated ($\Delta H_{hydration}$).

The approach outlined here for predicting water solubility is helpful, in part because it is an easily visualized process. Unfortunately, reasonably good values for lattice energy and $\Delta H_{hydration}$ are available for only a few salts. Thus, the approach that obtains the enthalpy of solution from the enthalpies of formation of solid and aqueous salts is more useful because these values are more generally available.

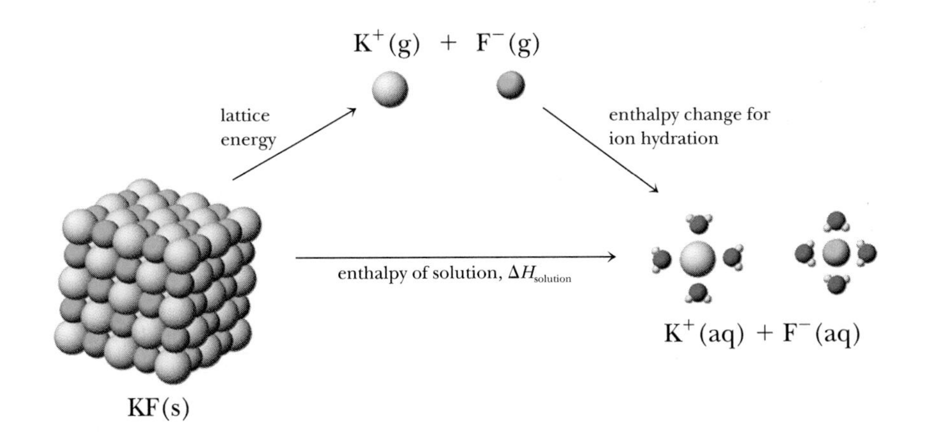

surement produces a value for the enthalpy of solution, ΔH°_{soln}. For solutions, standard conditions refer to a concentration of 1 *m*.

As we noted in Chapter 6, it is convenient to present thermodynamic data in a table of standard enthalpies of formation, and such tables often include values for the heats of formation of aqueous solutions of salts. For example, a value of ΔH°_f for NaCl(aq) of −407.3 kJ/mol is listed in Appendix L. This value refers to the formation of a 1 *m* solution of NaCl from the elements. It may be considered to involve the enthalpies of two steps: (1) the formation of NaCl(s) from the elements Na(s) and $Cl_2(g)$ in their standard states, and (2) the formation of a 1 *m* solution by dissolving solid NaCl in water:

Formation of NaCl(s) $\quad Na(s) + \frac{1}{2} Cl_2(g) \longrightarrow NaCl(s) \quad \Delta H^\circ_f = -411.2 \text{ kJ/mol}$

Dissolving NaCl $\quad NaCl(s) \longrightarrow NaCl(aq, 1\ m) \quad \Delta H^\circ_{soln} = +3.9 \text{ kJ/mol}$

Net process $\quad Na(s) + \frac{1}{2} Cl_2(g) \longrightarrow NaCl(aq, 1\ m) \quad \Delta H^\circ_f = -407.3 \text{ kJ/mol}$

As described in Section 13.2, when solutes are dissolved, they become solvated. If water is the solvent, the term "hydration" is used.

See the *Saunders Interactive General Chemistry CD-ROM,* Screen 14.4, Energetics of Solution Formation.

Heats of solution, ΔH°_{soln} (Table 14.1), can be calculated using ΔH°_f data. The solution process for NaCl(s), for example, is represented by the equation

$$NaCl(s) \longrightarrow NaCl(aq, m)$$

The energy of this process is calculated using Equation 6.6:

$$\begin{aligned}\Delta H^\circ_{soln} &= \sum[\Delta H^\circ_f(\text{products})] - \sum[\Delta H^\circ_f(\text{reactants})]\\ &= \Delta H^\circ_f[NaCl(aq)] - \Delta H^\circ_f[NaCl(s)]\\ &= -407.3 \text{ kJ/mol} - (-411.2 \text{ kJ/mol}) = +3.9 \text{ kJ/mol}\end{aligned}$$

Table 14.1 • DATA FOR CALCULATING ENTHALPY OF SOLUTION

Compound	ΔH_f°(s) (kJ/mol)	ΔH_f°(aq, 1 *m*) (kJ/mol)
LiF	−616.0	−611.1
NaF	−573.6	−572.8
KF	−526.3	−585.0
RbF	−557.7	−583.8
LiCl	−408.6	−445.6
NaCl	−411.2	−407.3
KCl	−436.7	−419.5
RbCl	−435.4	−418.3
NaOH	−425.6	−470.1
NH_4NO_3	−365.6	−339.9

Example 14.3 Calculating the Enthalpy of Solution

Use the data given in Table 14.1 to determine the heat of solution for NH_4NO_3, the compound used in cold packs (see page 279 and Figure 14.5).

Solution The solution process for NH_4NO_3 is represented by the equation

$$NH_4NO_3(s) \longrightarrow NH_4NO_3(aq)$$

The energy of this process is calculated using heats of formation given in Table 14.1:

$$\begin{aligned}\Delta H^\circ_{soln} &= \sum[\Delta H_f^\circ(\text{product})] - \sum[\Delta H_f^\circ(\text{reactant})]\\ &= \Delta H_f^\circ[NH_4NO_3(aq)] - \Delta H_f^\circ[NH_4NO_3(s)]\\ &= -339.9 \text{ kJ/mol} - (-365.6 \text{ kJ/mol}) = +25.7 \text{ kJ/mol}\end{aligned}$$

The process is endothermic as indicated by the fact that ΔH°_{soln} has a positive value.

Exercise 14.3 Calculating Enthalpy of Solution

Use the data in Table 14.1 to calculate the enthalpy of solution for NaOH.

Generally product-favored processes are accompanied by evolution of energy, so we can expect that if the solution process is exothermic a substance is likely to be soluble. This is not the whole story, however. As described earlier (page 647) another factor besides energy influences whether a substance dissolves. This is the driving force to form a mixture from pure substances, or more generally to move from a more ordered to a less ordered system. In Chapter 20 we will describe this when another thermodynamic quantity, called entropy, is introduced.

Factors Affecting Solubility: Pressure and Temperature

Biochemists and physicians, among others, are interested in the solubility of gases such as CO_2 and O_2 in water or body fluids, and scientists and engineers need to know about the solubility of solids in various solvents. Pressure and temperature are two external factors that influence such processes. Both affect the solubility of gases in liquids, whereas normally only temperature is an important factor in the solubility of solids in liquids.

Dissolving Gases in Liquids: Henry's Law

The solubility of a gas in a liquid is directly proportional to the gas pressure. As the gas pressure increases, so does its solubility. This is a statement of **Henry's law,**

$$S_g = k_H P_g \qquad [14.4]$$

where S_g is the gas solubility, P_g is the partial pressure of the gaseous solute, and k_H is Henry's law constant (Table 14.2), a constant characteristic of the solute and solvent.

Carbonated soft drinks illustrate Henry's law. These are packed under pressure in a chamber filled with carbon dioxide gas, some of which dissolves in the drink. When the can or bottle is opened, the partial pressure of CO_2 above the solution drops, which causes the concentration of CO_2 in solution to drop, and gas to bubble out of the solution (Figure 14.6). Sometimes this occurs very rapidly, as illustrated in the *Chemical Puzzler* at the beginning of the chapter.

Henry's law has important consequences in SCUBA diving (Figure 14.7) [SCUBA stands for self-contained underwater breathing apparatus and was invented by the famous oceanographer Jacques Cousteau (1910–1997)]. When you dive the pressure of the air you breathe must be balanced against the external pressure of the water. In deeper dives, the pressure of the gases in the SCUBA gear must be several atmospheres and, as a result, more gas dissolves in the blood. This can lead to a problem if you ascend too rapidly. You can experience a painful and potentially lethal condition referred to as "the bends," in which nitrogen gas bubbles form in the blood as the solubility of nitrogen decreases with decreasing pressure. In an effort to prevent the bends, divers may use a

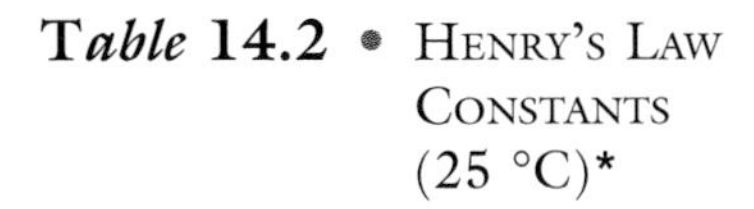

Table 14.2 • Henry's Law Constants (25 °C)*

Gas	k_H (M/mm Hg)
N_2	8.42×10^{-7}
O_2	1.66×10^{-6}
CO_2	4.48×10^{-5}

*From W. Stumm and J. J. Morgan: *Aquatic Chemistry,* p. 109. New York, Wiley, 1981.

See the *Saunders Interactive General Chemistry CD-ROM,* Screen 14.5, Henry's Law and Gas Pressure.

Henry's law holds quantitatively only for gases that do not interact chemically with the solvent. It does not work perfectly for NH_3, for example, which gives small concentrations of NH_4^+ and OH^- in water, or for CO_2, which reacts with water to form H_3O^+ and HCO_3^- ions.

Figure 14.6 An illustration of Henry's law. Carbonated beverages are bottled under CO_2 pressure. When the bottle is opened, the pressure is released and bubbles of CO_2 form within the liquid and rise to the surface. After some time, an equilibrium between dissolved CO_2 and atmospheric CO_2 is reached. Because CO_2 provides some of the taste in the beverage, the beverage tastes flat when most of its dissolved CO_2 is lost. *(Photo, C. D. Winters; model, S. M. Young)*

helium–oxygen mixture (rather than nitrogen–oxygen) because helium is not as soluble in aqueous media as nitrogen.

We can understand the effect of pressure on solubility by examining the system at the particulate level. The solubility of a gas is defined as the concentration of the dissolved gas in equilibrium with the substance in the gaseous state. At equilibrium, the rate at which solute (gas) molecules escape the solution and enter the gaseous state equals the rate at which gas molecules reenter the solution (see Figure 14.6). An increase in pressure results in more molecules of gas striking the surface of the liquid and entering solution in a given time. The solution eventually reaches equilibrium when the concentration of gas dissolved in the solvent is high enough that the rates of gas molecules escaping and entering the solution are the same.

Figure 14.7 SCUBA diving. If you dive to any appreciable depth using SCUBA gear, you will have to be concerned with the solubility of gases in your blood. Nitrogen is soluble in blood, so if you breathe a high-pressure mixture of O_2 and N_2 deep under water, the concentration of N_2 in your blood can be appreciable. If you ascend too rapidly, the nitrogen is released as the pressure decreases and forms bubbles in the blood. This affliction, called the "bends," is painful and can be fatal if blood forced out of capillaries and the brain is deprived of oxygen. In an effort to prevent the bends in very deep dives, deep sea divers sometimes use an $He–O_2$ mixture instead of $N_2–O_2$ because helium is less soluble in blood than nitrogen. See page 570 in Chapter 12. *(Susan Blanchet/Dembinsky Photo Associates)*

Example 14.4 Using Henry's Law

What is the concentration of O_2 in a fresh water stream in equilibrium with air at 25 °C and 1.0 atm? Express the answer in grams of O_2 per liter of water.

Solution Henry's law can be used to calculate the molar solubility of oxygen. First, however, the partial pressure of O_2 in air must be calculated. Recall from Chapter 12 that the mole fraction of O_2 in air is 0.21 (Table 12.2). If the total pressure is 1.0 atm, the partial pressure of O_2 is 160 mm Hg. Using the Henry's law constant for oxygen (Table 14.2) to calculate solubility, we have

$$\text{Solubility of } O_2 = \left(\frac{1.66 \times 10^{-6}\ \text{M}}{\text{mm Hg}}\right)(160\ \text{mm Hg}) = 2.66 \times 10^{-4}\ \text{M}$$

This concentration can be expressed in grams per liter using the molar mass of O_2:

$$\text{Solubility of } O_2 = \left(\frac{2.66 \times 10^{-4}\ \text{mol}}{\text{L}}\right)\left(\frac{32.0\ \text{g}}{\text{mol}}\right) = 0.00850\ \text{g/L}$$

This concentration of O_2 (8.50 mg/L) is quite low, but it is sufficient to provide the oxygen required by aquatic life.

Exercise 14.4 Using Henry's Law

Henry's law constant for CO_2 in water at 25 °C is 4.48×10^{-5} mol/mm Hg. What is the concentration of CO_2 in water when the partial pressure is 0.33 atm? (Although CO_2 reacts with water to give traces of H_3O^+ and HCO_3^-, the reaction occurs to such a small extent that Henry's law is obeyed at low CO_2 partial pressures.)

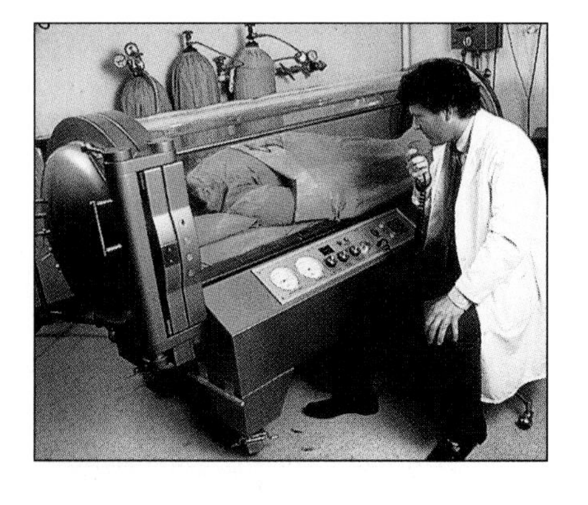

A hyperbaric chamber makes use of Henry's law. Persons with diseases such as emphysema find it difficult to breathe. In a hyperbaric chamber they are exposed to a higher pressure of oxygen, which is used to enhance the uptake of oxygen and the concentration of oxygen in the blood. *(Peter Arnold, Inc.)*

Temperature Effects on Solubility: Le Chatelier's Principle

The solubility of all gases in water decreases with increasing temperature. You may realize this from everyday observations such as the appearance of bubbles of gas that are seen as water is heated mildly (below the boiling point) (Figure 14.8), and from the fact that gas evolution is less vigorous from a bottle of a carbonated beverage when it is cold than when it is warm.

Decreased solubility of gases with increasing temperature has environmental consequences. Fish often seek lower depths in summer because the warmer surface layers of water have lower oxygen concentrations. Thermal pollution, resulting from the use of surface water as a coolant for various industries, can be a special problem for the marine life that requires oxygen to survive. Effluent water returned to a natural water source at a warmer temperature will be depleted of oxygen.

Figure 14.8 Effect of heat on solubility of a gas. A warm glass rod is placed in a glass of ginger ale. The heat energy of the rod is absorbed by a cold solution of CO_2 in water and causes the CO_2 to be less soluble. The Henry's law constant for CO_2 in water is 4.48×10^{-5} M/mm Hg at 25 °C; it drops 20% to 3.6×10^{-5} M/mm Hg at 100 °C. *(C. D. Winters)*

To understand the effect of temperature on the solubility of gases, let us reexamine the heat of solution. Gases that dissolve to an appreciable extent in water usually do so in an exothermic process. That is, the enthalpy of solution has a negative value:

$$\text{Gas} + \text{liquid solvent} \longrightarrow \text{saturated solution} + \text{heat energy}$$
$$\Delta H_{\text{solution}} < 0$$

The reverse process, loss of dissolved gas molecules from a solution, is the opposite of this process and heat is required. We can depict this situation in an equation with a double arrow ($\rightleftarrows$), indicating an equilibrium system:

$$\text{Gas} + \text{liquid solvent} \rightleftarrows \text{saturated solution} + \text{heat}$$

To understand how temperature affects solubility, we turn to **Le Chatelier's principle,** which states that a change in any of the factors determining an equilibrium causes the system to adjust to reduce or counteract the effect of the change. If a solution of a gas in a liquid is heated, the equilibrium will shift to absorb some of the added heat energy. That is, the reaction

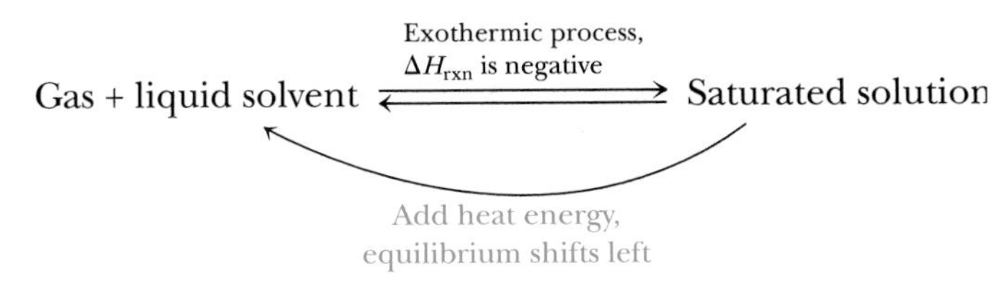

shifts back to the left if the temperature is raised because heat energy can be consumed in the process that gives free gas molecules and pure solvent. This shift corresponds to less gas dissolved, or a lower solubility, at higher temperature, the observed result.

The solubility of solids in water is also affected by temperature, but unlike the situation involving solutions of gases there is no general pattern of behavior. In Figure 14.9, the solubilities of several salts are plotted versus temperature. The solubility of many salts increases with increasing temperature, but there are notable exceptions. Predictions based on whether the heat of solution is positive or negative work most of the time, but exceptions do occur.

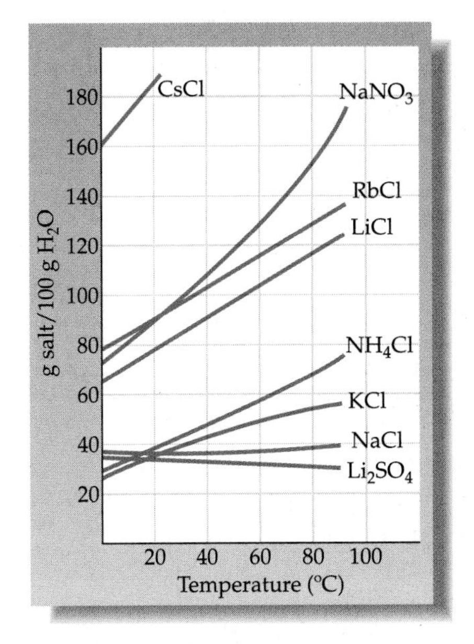

Figure 14.9 The temperature dependence of the solubility of some ionic compounds in water. Most compounds increase in solubility with increasing temperature.

The variation of solubility with temperature is used by chemists to purify compounds. An impure sample of a given compound is dissolved in a solvent at high temperature because the solubility is usually increased under these conditions. The solution is cooled to decrease the solubility, and, when the limit of

CURRENT PERSPECTIVES

Carbon Dioxide, Seashells, and Killer Lakes

Carbon dioxide is such a simple molecule, but it plays important roles in your body, in our environment, and in our economy.

Exhale—CO_2 is among the gases coming from your lungs. This is because CO_2 is produced in the "burning" of carbohydrates. In the following equation, CH_2O is used to represent a carbohydrate.

$$O_2(g) + CH_2O \longrightarrow CO_2(g) + H_2O(\ell)$$

Some of the CO_2 from this "combustion" is not exhaled but is dissolved in body fluids where it can react with water to produce the weak acid H_2CO_3. This in turn loses H^+ to give the hydrogen carbonate ion, HCO_3^-, in solution. These substances act together as a "buffer" (see Chapter 18) to control the pH of your blood and body tissues.

Carbon dioxide is also the end product of fermentation, the slow oxidation of carbohydrates. When beverages such as beer and champagne are made, the gas remains dissolved in the liquid. When a bottle of the beverage is opened, some of the dissolved CO_2 comes out of solution, and you see it as a foamy head on beer or bubbles in champagne. The presence of dissolved CO_2 in many beverages is also part of the reason they are acidic.

More rapid burning of carbohydrates or of hydrocarbons—as in the burning of forests or the combustion of fossil fuels—is the source of an ever-increasing amount of CO_2 in earth's atmosphere. The concentration of CO_2 has steadily increased since the beginning of the Industrial Revolution, and the increase is accelerating. Since 1957, the amount of CO_2 in the atmosphere has increased about 6%.

It has been estimated that the mean global surface temperature has increased about 0.5 °C since the late 19th century and that the mean sea level has risen 10 to 15 cm in that period, in part because of melting polar ice and in part because of the thermal expansion of surface water. An increase in the concentration of CO_2 in the atmosphere contributes to the warming of the planet because CO_2 is a "greenhouse" gas, that is, it traps heat from the earth that would normally be dissipated into space.

The increase in atmospheric CO_2 concentration has not been as large as predicted, however, because of "sinks," mechanisms in the environment that absorb the gas. Some CO_2 is used up in increased plant respiration. In fact, plant ecologists have found that doubling the CO_2 level can cause silver maple trees, for example, to grow 61% faster.

Much of the CO_2 on earth is tied up in the form of carbonates, such as $CaCO_3$, calcium carbonate, the primary constituent of seashells and corals. Indeed, about 85% of the carbon on earth is in the form of "carbonate sediments." As CO_2 enters the atmosphere, much of it finds it way to this sink, thereby limiting the increase in atmospheric CO_2 concentration.

Large amounts of carbon dioxide can dissolve in underground water, and people have enjoyed sparkling mineral wa-

solubility is reached at the lower temperature, crystals of the pure compound form. If the process is done slowly and carefully it is sometimes possible to obtain very large crystals (Figure 14.10).

Predicting the solubility of solids in nonaqueous solvents is less complicated. In general, solubility increases with increasing temperature. For more on the subject of solids dissolving in water, see R. S. Treptow: "Le Chatelier's principle applied to the temperature dependence of solubility." *Journal of Chemical Education,* Vol. 61, p. 499, 1984.

14.3 COLLIGATIVE PROPERTIES

When water contains dissolved sodium chloride, the vapor pressure of water over the solution is different from that of pure water, as is the freezing point of the solution, its boiling point, and its osmotic pressure. These colligative properties depend on the relative numbers of solute and solvent particles.

Changes in Vapor Pressure: Raoult's Law

The equilibrium vapor pressure at a particular temperature is the pressure of the vapor when the rate at which molecules escape the liquid and enter the gaseous state equals the rate at which gaseous molecules condense to re-form the liquid. Using this model as a starting point, let us evaluate the effect of a dissolved solute on vapor pressure.

Notice that it is the number of solvent and solute particles that is important in determining vapor pressure. Colligative properties are those properties that depend on the relative number of particles.

ter from natural springs for centuries. Dissolved CO_2 can be deadly, however. In the western African nation of Cameroon in 1986 a huge bubble of CO_2 gas escaped from Lake Nyos and moved down a river valley at a speed of 20 m/s (about 45 mph). Because CO_2 is denser than air, it hugged the ground and displaced the air in its path. More than 1700 people suffocated. The CO_2 came from springs of carbonated groundwater at the bottom of the lake. The lake is deep and the bottom layer became saturated with CO_2 under pressure. The CO_2 mixed little with the upper layers of water, but when this delicate situation changed, perhaps because of an earthquake or landslide, the CO_2 came out of the lake water just like it does when a can of soda is opened.

Carbon dioxide is one of the top 20 chemicals made in the United States, where it is used largely for refrigeration in the form of solid CO_2, "dry ice," and to carbonate beverages. You will see it as a product or reactant in many reactions in this book.

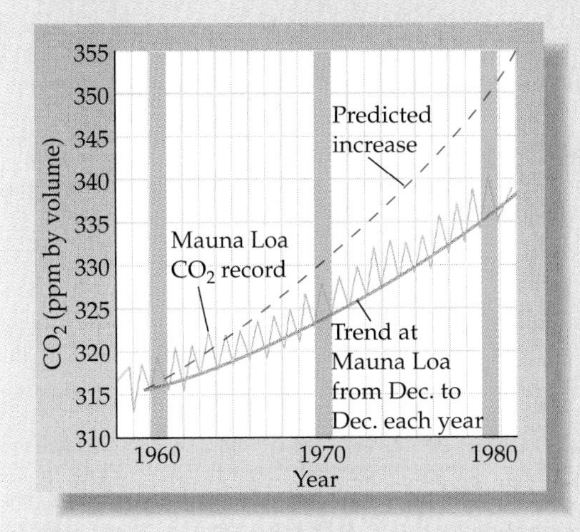

The concentration of atmospheric CO_2 since 1958. This plot is based on data collected at the Mauna Loa Observatory in Hawaii, which is located far from urban areas where carbon dioxide levels are high because of factories, power plants, and motor vehicles. There is a seasonal fluctuation because photosynthesis consumes CO_2 in the summer. The dotted line is the CO_2 concentration (in ppm) predicted on the basis of fossil fuel consumption.

Lake Nyos in Cameroon (western Africa), the site of a natural disaster. In 1986, a huge bubble of CO_2 escaped from the lake and asphyxiated more than 1700 people. *(Courtesy of George Kling)*

To escape the surface of the liquid and enter the gaseous state, a molecule must be at the surface and have sufficient kinetic energy to escape. This presents vaporization as a kind of statistical problem. A certain number of molecules exist at the surface of the liquid, and there is some probability that a given molecule has enough energy to escape. Dissolved solute alters this situation in two ways. First, the presence of solute molecules results in fewer solvent molecules being at the surface. Second, the solute introduces a new set of forces, those between solute and solvent molecules. Let us look at each of these separately.

At the surface of an aqueous solution (Figure 14.11), there are ions or molecules of solute as well as molecules of water. Not as many water molecules are present at the surface of a solution as in pure water because some of them have been displaced by dissolved ions or molecules. This means that not as many water molecules are available to leave the liquid surface. As a result, the vapor pressure is lower than the vapor pressure of pure water at a given temperature because the vapor pressure of the solvent, $P_{solvent}$, is proportional to the relative number of solvent molecules at the surface, that is, to the mole fraction of the solvent. For example, if only half as many solvent molecules are present at the surface of a solution as at the surface of the pure liquid, then the vapor pressure of the solvent in the solution is only half the value of the pure solvent at

***Figure* 14.10 Crystals of ionic solids.** Crystals of ionic compounds were grown on pieces of granite or on seashells. *Blue:* $CuSO_4 \cdot 5\,H_2O$. *Turquoise:* mixture of $(NH_4)_2Cu(SO_4)_2 \cdot 6\,H_2O$ and $(NH_4)_2Zn(SO_4)_2 \cdot 6\,H_2O$. *Red:* $K_3Fe(CN)_6$. *Green:* mixture of $(NH_4)_2Cu(SO_4)_2 \cdot 6\,H_2O$ and $K_2Ni(SO_4)_2 \cdot 6\,H_2O$. *White:* $NaKC_4H_4O_6$. (Crystals supplied by S. M. Young and students of Hartwick College. Method devised by Clementina Teixeira of the Instituto Superior Tecnico, Lisbon, Portugal.) *(C. D. Winters)*

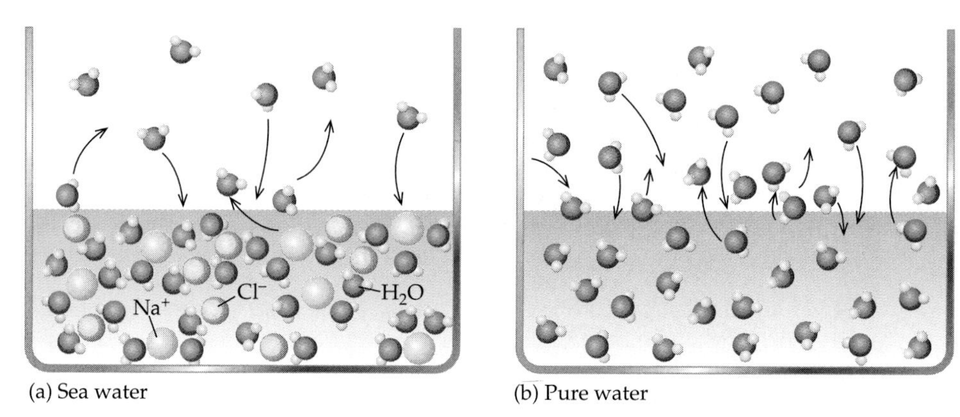

Figure 14.11 The effect of dissolved solute on vapor pressure. (a) Sea water is an aqueous solution of many salts, including sodium chloride. (b) The vapor pressure over an aqueous solution (with a nonvolatile solute) is not as large as the vapor pressure of water over pure water at the same temperature.

the same temperature. Because $P_{\text{solvent}} \propto X_{\text{solvent}}$, we can write the following equation for the equilibrium vapor pressure of the solvent over a solution:

$$P_{\text{solvent}} = X_{\text{solvent}} \cdot K \quad [14.5]$$

where K is a constant. A useful form of this equation can be developed if we first rewrite Equation 14.5 for a pure solvent. If $P^{\circ}_{\text{solvent}}$ is the vapor pressure of the pure solvent, then Equation 14.5 becomes

$$P^{\circ}_{\text{solvent}} = X_{\text{solvent}} \cdot K$$

Because X_{solvent} is 1 for the pure solvent, the constant K is just the vapor pressure of the pure solvent. Substituting $P^{\circ}_{\text{solvent}}$ for K in Equation 14.5, we arrive at the equation

$$P_{\text{solvent}} = X_{\text{solvent}} P^{\circ}_{\text{solvent}} \quad [14.6]$$

See the *Saunders Interactive General Chemistry CD-ROM,* Screen 14.7, Vapor Pressure and Raoult's Law.

This equation is called **Raoult's law,** named for Francois M. Raoult (1830–1901), a professor of chemistry at the University of Grenoble in France, who did the pioneering studies in this area.

Like the ideal gas law, Raoult's law describes a simplified, or *ideal,* model of a solution. Although few solutions are ideal, just as few gases are ideal, Raoult's law is a good approximation for solution behavior in many instances, especially at low solute concentration. We say that an ideal solution is one that obeys Raoult's law.

When will a solution not be ideal? This brings us to the second effect of dissolved solutes, the forces of attraction between solute and solvent molecules. For Raoult's law to hold, the forces between solute and solvent molecules must be the same as those between the solvent molecules in the pure solvent. This is frequently the case when molecules with similar structures are involved. Solutions of one hydrocarbon in another (hexane, C_6H_{14}, dissolved in octane, C_8H_{18}, for example) usually follow Raoult's law quite closely. If solvent–solute interactions are stronger than solvent–solvent interactions, the actual vapor pressure will be lower than calculated by Raoult's law. If the solvent–solute interactions are weaker than solvent–solvent interactions, the vapor pressure will be higher.

Example 14.5 Using Raoult's Law

Ethylene glycol, $HOCH_2CH_2OH$, is a common ingredient in automobile antifreeze (Figure 14.12 and page 644). If 651 g of ethylene glycol is dissolved in 1.50 kg of water (a 30.2% solution, a commonly used antifreeze solution for automobiles), what is the vapor pressure of the water over the solution at 90 °C? (The vapor pressure of pure water at 90 °C is 525.8 mm Hg, as given in Appendix G.) Assume ideal behavior for the solution.

Figure 14.12 **Antifreeze.** Adding antifreeze to water prevents the water from freezing. Here a jar of pure water *(left)* and a jar of water to which automobile antifreeze had been added *(right)* were kept overnight in the freezing compartment of a home refrigerator. *(C. D. Winters)*

Solution First, calculate the number of moles of water and ethylene glycol, and, from these, the mole fraction of water:

$$\text{Moles of water} = 1.50 \times 10^3 \text{ g}\left(\frac{1 \text{ mol}}{18.02 \text{ g}}\right) = 83.2 \text{ mol water}$$

$$\text{Moles ethylene glycol} = 651 \text{ g}\left(\frac{1 \text{ mol}}{62.07 \text{ g}}\right) = 10.5 \text{ mol glycol}$$

$$X_{water} = \frac{83.2 \text{ mol water}}{83.2 \text{ mol water} + 10.5 \text{ mol glycol}} = 0.888$$

Next apply Raoult's law, calculating the vapor pressure from the mole fraction of water and the vapor pressure of pure water:

$$P_{water} = X_{water}\, P^{\circ}_{water} = (0.888)(525.8 \text{ mm Hg}) = 467 \text{ mm Hg}$$

The dissolved solute decreases the vapor pressure by 59 mm Hg, or about 11%:

$$\Delta P_{water} = P_{water} - P^{\circ}_{water} = 467 \text{ mm Hg} - 525.8 \text{ mm Hg} = -59 \text{ mm Hg}$$

Ethylene glycol dissolves easily in water due to extensive hydrogen bonding. It is nonvolatile so it will not boil off, and it is noncorrosive. And it is relatively inexpensive. These features make it ideal for use as antifreeze.

Exercise 14.5 Using Raoult's Law

Assume you dissolve 10.0 g of sugar ($C_{12}H_{22}O_{11}$) in 225 mL (225 g) of water and warm the water to 60 °C. What is the vapor pressure of the water over this solution?

Adding a nonvolatile solute to a solvent lowers the vapor pressure of the solvent, and Raoult's law can be modified to calculate directly the lowering of the vapor pressure. The mole fraction of the solvent in any solution is always less than 1, so the vapor pressure of the solvent over an ideal solution ($P_{solvent}$) must be less than the vapor pressure of the pure solvent ($P^{\circ}_{solvent}$). The vapor pressure change, $\Delta P_{solvent}$, is given by

$$\Delta P_{solvent} = P_{solvent} - P^{\circ}_{solvent}$$

Substituting Raoult's law for $P_{solvent}$, we have

$$\Delta P_{solvent} = (X_{solvent} \cdot P^{\circ}_{solvent}) - P^{\circ}_{solvent} = -(1 - X_{solvent})P^{\circ}_{solvent}$$

In a solution that has only the volatile solvent and one nonvolatile solute, the sum of the mole fraction of solvent and solute must be 1:

$$X_{solvent} + X_{solute} = 1$$

Therefore, $1 - X_{solvent} = X_{solute}$, and the equation for $\Delta P_{solvent}$ can be rewritten as

$$\Delta P_{solvent} = -X_{solute}P^{\circ}_{solvent} \qquad [14.7]$$

Thus, the change in the vapor pressure of the solvent is *proportional to the mole fraction* (the relative number of particles) of solute.

Boiling Point Elevation

See the *Saunders Interactive General Chemistry CD-ROM,* Screen 14.8, Boiling Point and Freezing Point.

Suppose you have a solution of a nonvolatile solute in the volatile solvent benzene. If the solute concentration is 0.200 mol in 100. g of benzene (C_6H_6), this means that $X_{benzene} = 0.865$. Using Raoult's law, we can calculate that the vapor pressure of the solvent at 60 °C will drop from 400. mm Hg for the pure solvent to 346 mm Hg for the solution:

$$X_{benzene} = 0.865$$

$$P_{benzene} = X_{benzene}\ P^{\circ}_{benzene} = (0.865)(400.\text{ mm Hg}) = 346\text{ mm Hg}$$

This point is marked on the vapor pressure graph in Figure 14.13. Now, what is the vapor pressure when the temperature of the solution is raised another 10 °C? $P^{\circ}_{benzene}$ becomes larger with increasing temperature, so $P_{benzene}$ for the solution must also become larger. This new point, and additional ones calculated in the same way for other temperatures, define the vapor pressure curve for the solution (the lower curve in Figure 14.13).

An important observation we can make in Figure 14.13 is that the lowering of the vapor pressure caused by the nonvolatile solute leads to an increase in the boiling point. The normal boiling point of a liquid is the temperature at which its vapor pressure is equal to 1 atm or 760 mm Hg (see Section 13.3). From the figure we see that the normal boiling point of pure benzene is about 80 °C. Tracing the vapor pressure curve for the solution in Figure 14.13, we can see that the vapor pressure reaches 760 mm Hg at a temperature about 5 °C higher than this value. Thus, the *boiling point of a solution is raised relative to that of the pure solvent.*

The vapor pressure curve and increase in the boiling point shown in Figure 14.13 refer specifically to a 2.00 *m* solution. We might wonder at this point how the boiling point of the solution would vary with solute concentration, and it is possible to reason out the answer. Recall that the change in vapor pressure is directly proportional to the concentration of solute ($\Delta P_{benzene} = -X_{solute} P^{\circ}_{benzene}$, Equation 14.7). Concentrations of solute greater than 2.00 *m* lead to a larger decrease in vapor pressure and consequently to a higher boiling point. Conversely, solute concentrations less than 2.00 *m* show a smaller decrease in vapor pressure and less of an increase in boiling point. In fact, a simple relationship exists between boiling point elevation and molal concentration: the *boiling point elevation, Δt_{bp}, is directly proportional to the molality of the solute:*

As was the case with Raoult's law, the equation for boiling point elevation (Equation 14.8) assumes that the solution is "ideal." Most dilute solutions ($<$0.10 *m*) come close to meeting this requirement.

$$\text{Elevation in boiling point} = \Delta t_{bp} = K_{bp} \cdot m_{solute} \qquad [14.8]$$

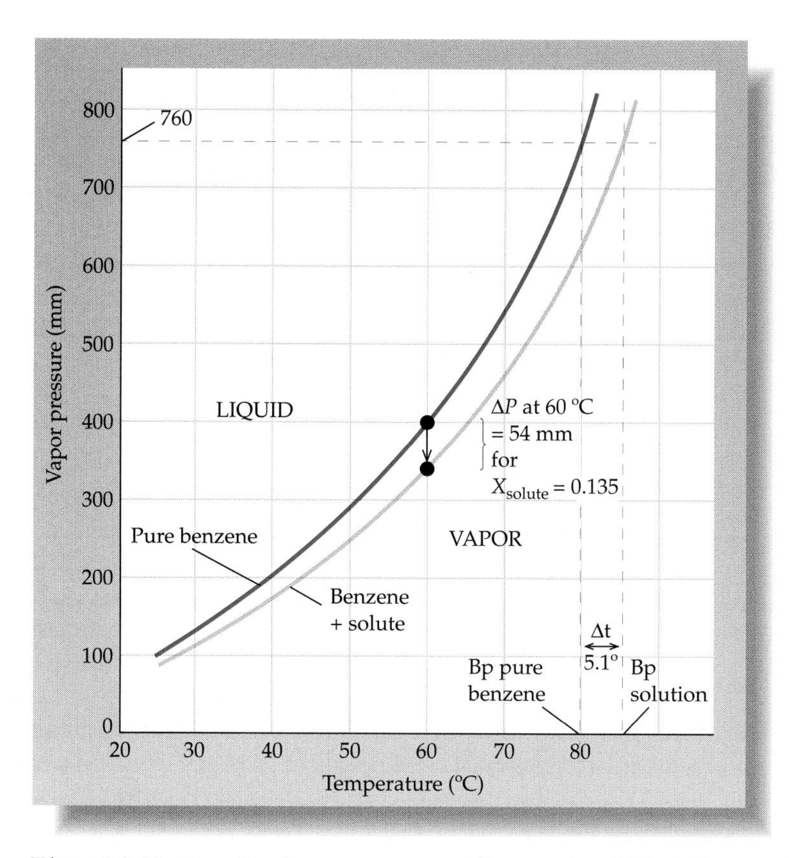

Figure 14.13 Lowering the vapor pressure of benzene by addition of a nonvolatile solute. The curve drawn in red represents the vapor pressure of pure benzene, and the curve below it, in blue, the vapor pressure of a solution containing 0.200 mol of a solute dissolved in 0.100 kg of solvent (2.00 *m*). This graph was created using a series of calculations such as those shown in the text. As an alternative, the graph could be created by measuring the values for the vapor pressure for the solution in a laboratory experiment.

In this equation, K_{bp} is a proportionality constant called the **molal boiling point elevation constant.** It has the units of "degrees/molal" (°C/*m*). Values for K_{bp} are determined experimentally, and different solvents have different values (Table 14.3). Formally, the value corresponds to the elevation in boiling point for a 1 *m* solution.

Table 14.3 • Some Boiling Point Elevation and Freezing Point Depression Constants

Solvent	Normal bp (°C) Pure Solvent	K_{bp} (°C/*m*)	Normal fp (°C) Pure Solvent	K_{fp} (°C/*m*)
Water	100.00	+0.5121	0.0	−1.86
Benzene	80.10	+2.53	5.50	−5.12
Camphor	207.4	+5.611	179.75	−39.7
Chloroform ($CHCl_3$)	61.70	+3.63	—	—

Example 14.6 Boiling Point Elevation

Eugenol, the active ingredient in cloves, has a formula of $C_{10}H_{12}O_2$ (page 130). What is the boiling point of a solution when 0.144 g of this compound is dissolved in 10.0 g of benzene?

Solution Equation 14.7 allows us to calculate the change in boiling point. This value must then be added to the boiling point of pure benzene to provide the answer.

Equation 14.7 requires values for K_{bp} and the molality of the solution. You can obtain the value for K_{bp} for benzene in Table 14.3, but you need to calculate the molality, m:

$$0.144 \text{ g eugenol} \left(\frac{1 \text{ mol eugenol}}{164.2 \text{ g}}\right) = 8.77 \times 10^{-4} \text{ mol eugenol}$$

$$\frac{8.77 \times 10^{-4} \text{ mol eugenol}}{0.0100 \text{ kg benzene}} = 8.77 \times 10^{-2}\ m \text{ eugenol}$$

Use the value for the molality to calculate the boiling point elevation and then the boiling point:

$$\Delta t_{bp} = (2.53\ °\text{C}/m)\ (0.0877\ m) = 0.222\ °\text{C}$$

Because the boiling point *rises* relative to that of the pure solvent, the boiling point of the solution is 80.10 °C + 0.222 °C = 80.32 °C.

Exercise 14.6 Boiling Point Elevation

What quantity of ethylene glycol, $HOCH_2CH_2OH$, must be added to 125 g of water to raise the boiling point by 1.0 °C? Express the answer in grams.

Colligative Properties and Molar Mass Determination

Early in this book you learned how to calculate a molecular formula from an empirical formula when given the molar mass. But how do you know the molar mass of an unknown compound? An experiment must be done to find this crucial piece of information, and one way to do this is to use a colligative property of a solution of the compound. If the compound is soluble in a solvent of appreciable vapor pressure and a known K_{bp} or K_{fp}, the molar mass can then be determined. All approaches use the same basic logic:

Change in vapor pressure, boiling point elevation, or freezing point depression, or osmotic pressure → Solution concentration —(use mass of solvent)→ Moles of solute —(g solute / mol solute)→ Molar mass

Example 14.7 Determining Molar Mass by Boiling Point Elevation

A solution prepared from 1.25 g of oil of wintergreen (methyl salicylate) in 99.0 g of benzene has a boiling point of 80.31 °C. Determine the molar mass of this compound.

Solution Calculations using colligative properties to determine a molar mass always follow the pattern outlined just above. We first use the boiling point elevation to calculate the solution concentration:

$$\text{Boiling point elevation } (\Delta t_{bp}) = 80.31\ °\text{C} - 80.10\ °\text{C} = 0.21\ °\text{C}$$

and then calculate the molality:

$$\text{Molality of solution} = \frac{\Delta t_{bp}}{K_{bp}} = \frac{0.21\ °\cancel{\text{C}}}{2.53\ °\cancel{\text{C}}/m} = 0.083\ m$$

The quantity of solute in the solution is calculated from the solution concentration:

$$\text{Moles of solute} = \left(\frac{0.083 \text{ mol}}{1.00\ \cancel{\text{kg}} \text{ solvent}}\right)(0.099\ \cancel{\text{kg}} \text{ solvent}) = 0.0082 \text{ mol solute}$$

Now we can combine the moles of solute with its mass:

$$\frac{1.25 \text{ g}}{0.0083 \text{ mol}} = 150 \text{ g/mol}$$

Methyl salicylate has the formula $C_8H_8O_3$ and a molar mass of 152.14 g/mol.

The structure of oil of wintergreen (methyl salicylate). Notice that it is based on the benzene ring. Also notice its relation to aspirin, acetylsalicylic acid (page 126 and Figure 1.14). *(Photo, C. D. Winters; model, S. M. Young)*

Exercise 14.7 Determining Molar Mass by Boiling Point Elevation

Crystals of the beautiful blue hydrocarbon, azulene (0.640 g), which has an empirical formula of C_5H_4, are dissolved in 99.0 g of benzene. The boiling point of the solution is 80.23 °C. What is the molecular formula of azulene?

The elevation of the boiling point of a solvent on adding a solute has many practical consequences. One of them is the summer protection your car's engine receives from "all-season" antifreeze. The main ingredient of commercial antifreeze is ethylene glycol, $HOCH_2CH_2OH$. The car's radiator and cooling system are sealed to keep the coolant under pressure, so that it will not vaporize at normal engine temperatures. When the air temperature is high in the summer, however, the radiator could still "boil over" if it were not protected with "antifreeze." By adding this nonvolatile liquid, the solution in the radiator has a higher boiling point than that of pure water.

Freezing Point Depression

Another consequence of dissolving a solute in a solvent is that the freezing point of the solution is lower than that of the pure solvent. For an ideal solution, the depression of the freezing point is given by an equation similar to that for the elevation of the boiling point:

$$\text{Freezing point depression, } \Delta t_{fp} = K_{fp} \cdot m_{solute} \quad [14.9]$$

where K_{fp} is the **freezing point depression constant** in degrees per molal (°C/*m*). Values of K_{fp} for a few common solvents are given in Table 14.3. The values are negative quantities, so that the result of the calculation is a negative value for Δt_{fp}, signifying a decrease in temperature.

The practical aspects of freezing point changes from pure solvent to solution are similar to those for boiling point elevation. The very name of the liquid you add to the radiator in your car, *antifreeze*, indicates its purpose (see Figure 14.12). The label on the container of antifreeze tells you, for example, to add 6 qt (5.7 L) of antifreeze to a 12-qt (11.4-L) cooling system in order to lower the freezing point to −34 °C and raise the boiling point to +109 °C.

Example 14.8 Freezing Point Depression

How many grams of ethylene glycol, $HOCH_2CH_2OH$, must be added to 5.50 kg of water to lower the freezing point of the water from 0.0 °C to −10.0 °C? (This is approximately the situation in your car.)

Solution The solution concentration and freezing point depression are related by Equation 14.9. The solute concentration (molality) in a solution with a freezing point depression of −10.0 °C is

$$\text{Solute concentration } (m) = \frac{\Delta t_{fp}}{K_{fp}} = \frac{-10.0\ \text{°C}}{-1.86\ \text{°C}/m} = 5.38\ m$$

Because the radiator contains 5.50 kg of water, we need 29.6 mol of glycol:

$$\left(\frac{5.38 \text{ mol glycol}}{1.00 \text{ kg water}}\right)(5.50 \text{ kg water}) = 29.6 \text{ mol glycol}$$

The molar mass of glycol is 62.07 g/mol, so the mass required is

$$29.6 \text{ mol glycol}\left(\frac{62.07 \text{ g}}{1 \text{ mol}}\right) = 1840 \text{ g glycol}$$

The density of ethylene glycol is 1.11 kg/L, so the volume of antifreeze to be added is 1.84 kg (1 L/1.11 kg) = 1.66 L.

Exercise 14.8 Freezing Point Depression

Some people have summer homes on a lake or in the woods. In the northern United States, these summer homes are usually closed up for the winter. When doing so the owners "winterize" the plumbing by putting antifreeze in the toilet tanks, for example. Will adding 525 g of $HOCH_2CH_2OH$ to 3.00 kg of water ensure that the water will not freeze at −25 °C?

Why Is a Solution Freezing Point Depressed?

Imagine the freezing process this way. When the temperature is held at the freezing point of a pure solvent, freezing begins with a few molecules clustering together to form a tiny amount of solid. More molecules of the liquid move to the surface of the solid, and the solid grows. Heat, the heat of fusion, is evolved; as long as this heat energy is removed, solidification continues. If the heat is not removed, however, the opposing processes of freezing and melting can come into equilibrium; at this point, the number of molecules moving from solid to liquid is the same as the number moving from liquid to solid in a given time.

But what happens when a solution freezes? Again, a few molecules of solvent cluster together to form some solid. More and more solvent molecules join them, and the solid phase, which is pure solid solvent, continues to grow as long as the heat of fusion is removed. At the same time some solvent molecules are returning to the liquid from the solid. The freezing and melting processes can come into equilibrium when the numbers of molecules moving in the two directions in a given time are the same. Notice, however, that the liquid layer next to the solid contains solute molecules or ions, whereas the solid is pure solvent. This is analogous to the situation at the solution–vapor interface in Figure 14.11. If the temperature is held at the normal freezing point of the pure solvent, the number of molecules of the solid (pure solvent) entering the liquid phase must be greater than the number of solvent molecules leaving the solution and depositing on the solid, in a given time. Why? For the same reason the vapor pressure of a solution is lower than that of the pure solvent: Solute molecules have replaced some solvent molecules in the liquid at the liquid–vapor or liquid–solid interface. In order to have the same number of solvent molecules moving in each direction (solid $\longrightarrow$ liquid and liquid $\longrightarrow$ solid) in a given time, the temperature must be lowered to slow down movement from solid to liquid. That is, the freezing temperature of a solution must be less than that of the pure liquid solvent.

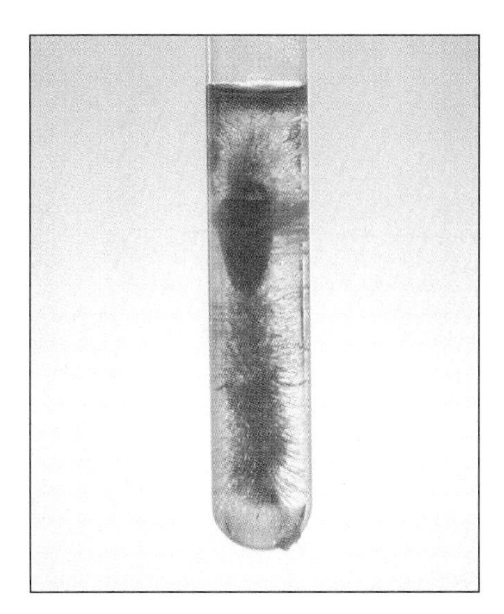

Figure 14.14 Freezing a solution. When a solution freezes, the solvent solidifies as the pure substance. To take this photo, a purple dye was dissolved in water, and the solution was frozen slowly. Pure ice formed along the walls of the tube, and the dye stayed in solution. The concentration of the solute increased as more and more solvent was frozen out, and the resulting solution had a lower and lower freezing point. At equilibrium, the system contains pure, colorless ice that had formed along the walls of the tube and a concentrated solution of dye in the center of the tube. *(C. D. Winters)*

As a solution freezes, solvent molecules are removed from the liquid phase and are deposited on the solid. The concentration of the solute in the liquid solution increases, and the solution freezing point declines further as a result. When reporting the freezing point of a solution, we usually take this to mean the point at which solid solvent crystals first begin to appear.

The fact that the solid is pure ice while the solution is more concentrated than before (Figure 14.14) can be put to practical use. Early Americans (and some contemporary ones as well) knew that a drink called "apple jack" can be made this way. Fermenting apple cider produces a small amount of alcohol. If the fermented cider is cooled, some of the water freezes to pure ice, leaving a solution higher in alcohol content. A modern adaptation is found in ice beer, whose alcoholic content is increased by freezing and separating out the ice crystals.

One way to purify a liquid solvent in the laboratory is to freeze it slowly, saving the solid (pure solvent) and throwing away any remaining solution (now more concentrated with impurities).

Colligative Properties of Solutions Containing Ions

In the northern United States it is common practice to scatter salt on snowy or icy roads or sidewalks. When the sun shines on the snow or patch of ice, a small amount is melted, and the water dissolves some of the salt. As a result of the dissolved solute, the freezing point of the solution is lower than 0 °C. The solution "eats" its way through the ice, breaking it up, and the icy patch is no longer dangerous for drivers or for people walking.

Salt (NaCl) is the most common substance used on roads because it is inexpensive and dissolves readily in water. Its relatively low molar mass means that the effect per gram is large. In addition, salt is especially effective because it is an electrolyte; it dissolves to give ions in solution:

$$NaCl(s) \longrightarrow Na^+(aq) + Cl^-(aq)$$

Remember that colligative properties depend not on what is dissolved but *only on the number of particles of solute per solvent particle.* When 1 mol of NaCl dissolves, 2 mol of ions are formed, which means that the effect on the freezing point of water should be twice as large as that expected for a mole of sugar. This peculiarity was discovered by Raoult in 1884 and studied in detail by Jacobus Henrikus van't Hoff (1852–1911) in 1887. Later in that same year Svante Arrhenius (1859–1927) provided the explanation for the behavior of electrolytes based on ions in solution. A 0.100 m solution of NaCl really contains two solutes, 0.100 m Na^+ and 0.100 m Cl^-. What we should use to estimate the freezing point depression is the *total* molality of solute particles:

$$m_{\text{total}} = m(Na^+) + m(Cl^-) = (0.100 + 0.100)\text{mol/kg} = 0.200 \text{ mol/kg}$$
$$\Delta t_{\text{fp}} = (-1.86\ °\text{C}/\cancel{\text{molal}})(0.200\ \cancel{\text{molal}}) = -0.372\ °\text{C}$$

To estimate the freezing point depression for an ionic compound, first find the molality of solute from the mass and molar mass of the compound. Then, multiply the number you get by the number of ions in the formula: two for NaCl, three for Na_2SO_4, four for $LaCl_3$, five for $Al_2(SO_4)_3$, and so on.

As it turns out, this model is a reasonable first guess at the effect of ionization of an electrolyte on colligative properties, but it is not exact. Let us look at some experimental data (Table 14.4), evaluating the effect of the dissociation of two ionic compounds, NaCl and Na_2SO_4, on the solution freezing point. The measured freezing point depression is larger than that calculated from Equation 14.9 assuming no ionization. As seen in the last column of the table, however,

Table 14.4 • Freezing Point Depression of Some Ionic Solutions

Mass %	m (mol/kg)	Δt_{fp} (measured, °C)	Δt_{fp} (calculated, °C)	Δt_{fp}, measured / Δt_{fp}, calculated
NaCl				
0.00700	0.0120	−0.0433	−0.0223	1.94
0.500	0.0860	−0.299	−0.160	1.87
1.00	0.173	−0.593	−0.322	1.84
2.00	0.349	−1.186	−0.649	1.83
Na_2SO_4				
0.00700	0.00493	−0.0257	−0.00917	2.80
0.500	0.0354	−0.165	−0.0658	2.51
1.00	0.0711	−0.320	−0.132	2.42
2.00	0.144	−0.606	−0.268	2.26

Δt_{fp} is not twice the value expected for NaCl, but only about 1.8 times larger. Likewise, Δt_{fp} for Na_2SO_4 approaches but does not reach a value that is 3 times larger than the value assuming no ionization. The ratio of the experimentally observed value of Δt_{fp} to the value calculated, assuming no ionization, is called the **van't Hoff factor** and is represented by *i*.

$$i = \frac{\Delta t_{fp}\text{, measured}}{\Delta t_{fp}\text{, calculated}} = \frac{\Delta t_{fp}\text{, measured}}{K_{fp} \cdot m}$$

or

$$\Delta t_{fp}\text{, measured} = K_{fp} \cdot m \cdot i \qquad [14.10]$$

The numbers in the last column of Table 14.4 are van't Hoff factors. These values can be used in calculations of any colligative property. Vapor pressure lowering, boiling point elevation, freezing point depression, and osmotic pressure are all larger for electrolytes than for nonelectrolytes of the same molality.

The van't Hoff factor approaches a whole number (2, 3, and so on) only in very dilute solutions. In more concentrated solutions, the experimental freezing point depressions tell us that there are fewer ions in solution than expected. This behavior, which is typical of all ionic compounds, is a consequence of the strong attractions between ions. The result is as if some of the positive and negative ions are paired, decreasing the total molality of particles. Indeed, in more concentrated solutions, and especially in solvents less polar than water, ions are extensively associated in ion pairs and in even larger clusters.

***Example* 14.9** Freezing Point and Ionic Solutions

A 0.00200 *m* aqueous solution of an ionic compound $Co(NH_3)_5(NO_2)Cl$ freezes at −0.00732 °C. How many moles of ions does 1 mol of the salt give on being dissolved in water?

Solution First, let us calculate the freezing-point depression expected for a 0.00200 *m* solution assuming that the salt does not dissociate into ions:

$$\Delta t_{fp}\text{, calculated} = K_{fp} \cdot m = (-1.86\ °\text{C}/m)(0.00200\ m) = -3.72 \times 10^{-3}\ °\text{C}$$

Now compare the calculated freezing point depression with the measured depression. This gives us the van't Hoff factor:

$$i = \frac{\Delta t_{fp}\text{, measured}}{\Delta t_{fp}\text{, calculated}} = \frac{-7.32 \times 10^{-3}\ °\text{C}}{-3.72 \times 10^{-3}\ °\text{C}} = 1.97 \approx 2$$

It appears that 1 mol of this compound gives 2 mol of ions. In this case, the ions are $[Co(NH_3)_5(NO_2)]^+$ and Cl^-.

***Exercise* 14.9** Freezing Point and Ionic Compounds

Calculate the freezing point of 525 g of water that contains 25.0 g of NaCl. Assume *i*, the van't Hoff factor, is 1.85 for NaCl.

See the *Saunders Interactive General Chemistry CD-ROM,* Screen 14.9, Osmosis.

Osmosis

Osmosis is the movement of solvent molecules through a semipermeable membrane from a region of lower to a region of higher solute concentration. This movement can be demonstrated with a simple experiment. The beaker in Figure 14.15 contains pure water, and a concentrated sugar solution is in the bag and tube. The liquids are separated by a semipermeable membrane, a thin sheet of material (such as a vegetable tissue or cellophane) through which only certain types of molecules can pass. Here, water molecules can pass through but larger sugar molecules (or hydrated ions) cannot (Figure 14.16). When the experiment is begun, the liquid levels in the beaker and the tube are the same. Over time, however, the level of the sugar solution inside the tube rises, the level of pure water in the beaker falls, and the sugar solution becomes steadily more dilute. After a while, no further net change occurs; equilibrium is reached.

From a molecular point of view, the semipermeable membrane does not present a barrier to the movement of water molecules, so they move through the membrane in both directions. When a solution contains large sugar molecules or hydrated ions, however, not as many water molecules strike the membrane in a given time on the solution side as on the pure water side. Thus, over a given time, more water molecules pass through the membrane from the pure

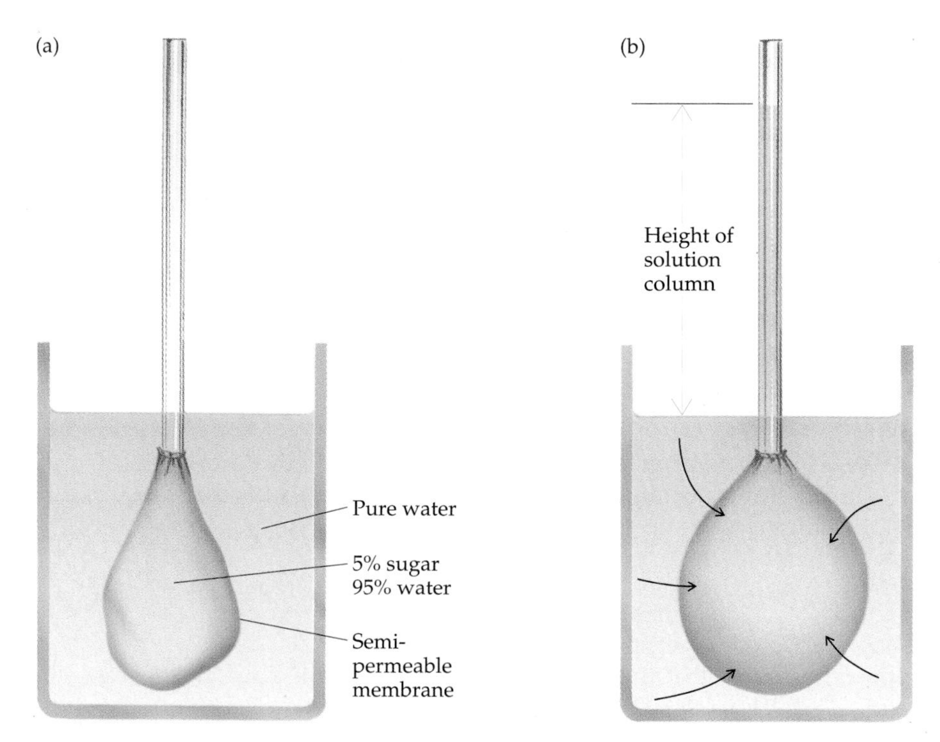

***Figure* 14.15 The process of osmosis.** (a) The bag attached to the tube contains a solution that is 5% sugar and 95% water. The beaker contains pure water. The bag is made of a material that is semipermeable, meaning that it allows water but not sugar molecules to pass through. (b) Over time, water flows from the region of low solute concentration (pure water) to the region of higher solute concentration (the sugar solution). Flow continues until the pressure exerted by the column of solution in the tube above the water level in the beaker is great enough to result in equal rates of passage of water molecules in both directions. The height of the column of solution (b) is a measure of the osmotic pressure, Π.

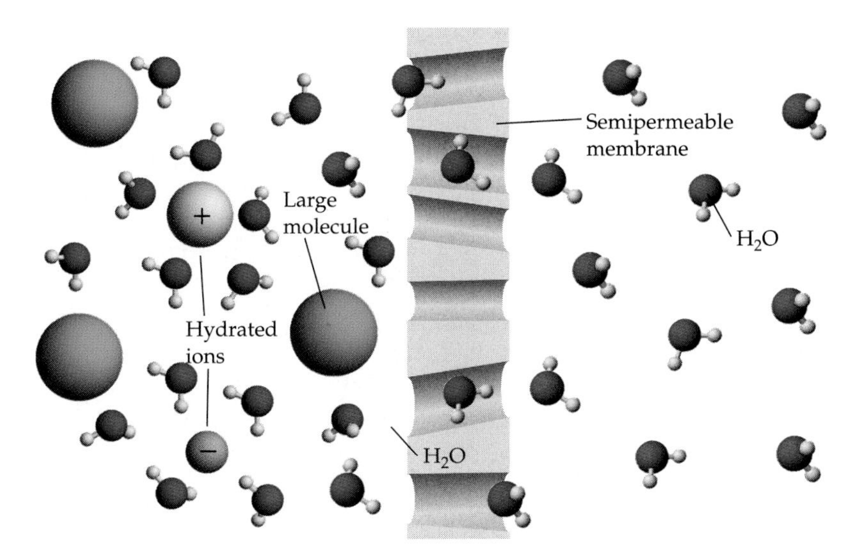

Figure 14.16 Osmosis at the particulate level. Osmotic flow through a membrane that is selectively permeable (semipermeable) to water. Dissolved substances such as hydrated ions or large sugar molecules cannot diffuse through the membrane. The membrane acts as a molecular sieve.

water side to the solution side than in the opposite direction. In effect, water molecules tend to move from regions of low solute concentration to regions of high solute concentration. The same is true for any solvent, as long as the membrane allows solvent molecules but not solute molecules to pass through.

Why does the system eventually reach equilibrium? It is evident that the solution in the tube in Figure 14.16 can never reach zero sugar or salt concentration, which would be required to equalize the number of water molecules moving through the membrane in each direction in a given time. The answer lies in the fact that the solution moves higher and higher in the tube as osmosis continues and water moves into the sugar solution. Eventually the pressure exerted by this height of solution counterbalances the pressure of the water moving through the membrane from the pure water side and no further net movement of water occurs. An equilibrium of forces is achieved. The pressure created by the column of solution for the system at equilibrium is called the **osmotic pressure.** A measure of this pressure is the difference in height between the solution in the tube and the level of pure water in the beaker.

The osmotic pressure of sea water, which contains about 35 g of dissolved salts per kilogram of sea water, is about 27 atm.

From experimental measurements on dilute solutions, it is known that osmotic pressure (given the symbol Π) and concentration (c) are related by the equation

$$\Pi = cRT \tag{14.11}$$

In this equation, the quantity c is the molar concentration (in moles per liter), R is the gas constant, and T is the absolute temperature (in kelvins). Using a value for the gas law constant of 0.082057 L · atm/K · mol allows calculation of the osmotic pressure Π in atmospheres. This equation is analogous to the ideal gas law $(PV = nRT)$, with Π taking the place of P and c being equivalent to n/V.

According to the osmotic pressure equation, the pressure exerted by a 0.10 M solution of particles at 25 °C is

$$\Pi = (0.10\ \cancel{\text{mol}}/\cancel{\text{L}})(0.0821\ \cancel{\text{L}} \cdot \text{atm}/\cancel{\text{K}} \cdot \cancel{\text{mol}})(298\ \cancel{\text{K}}) = 2.4\ \text{atm}$$

Current Perspectives

The Chemistry of Survival

You have always dreamed of sailing the South Pacific so you take a year out of college and hitch a ride on a 35-ft sailboat bound from San Francisco to Fiji. On the way, however, your small boat is rammed by a whale somewhere west of Hawaii. The boat sinks, but you do have time to climb into the life raft.

Now your thoughts turn to survival, and the main problem is water. Your body can lose as much as 2.5 L/day of water in exhaled breath, in sweat, and in urine and feces. This water has to be replenished. You can survive for only 12 days without water. If you drink a half liter a day, you can make it for 20 to 24 days, and 1 L of water a day will enable you to survive indefinitely, provided you can find some food.

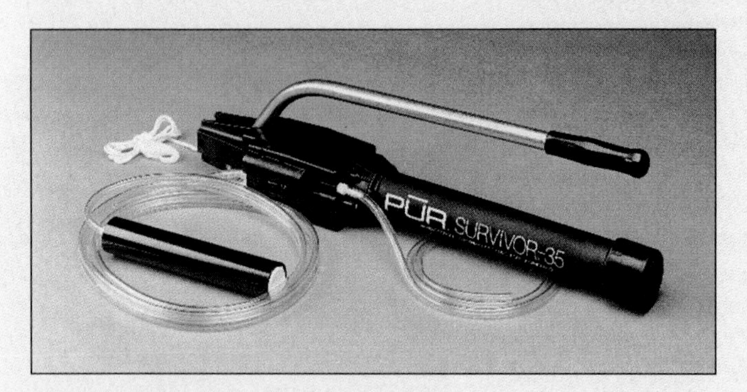

This hand-operated water desalinator works by reverse osmosis. It weighs just 7 lb and can produce 4.5 L/h of pure water from sea water. *(Courtesy of Recovery Engineering, Inc.)*

In Samuel Coleridge's epic poem *The Rime of the Ancient Mariner,* the mariner laments that, on the ocean, there is "Water, water everywhere, . . . Nor any drop to drink." The human body can use only fresh water. You cannot drink the salt water of the ocean because the salt concentration is higher than that of the cells in the body. Taken into the body, salt water will cause water to flow out of the cells into the bloodstream. This is the reason that one of the most important pieces of equipment in a life raft is a portable device for removing the salt from ocean water. The device works by the principle of reverse osmosis. If the osmotic pressure can be counterbalanced by an external pressure, then the osmotic flow of water can be reversed. The problem is that the osmotic pressure of sea water is about 27 atm. To obtain a useful amount of fresh water from sea water requires the application of at least twice that pressure to the semipermeable membrane.

The reverse osmosis device in the figure was designed so that an individual can apply enough pressure to generate about 4.5 L of fresh water in an hour. In fact, in 1989 William and Simonne Butler survived in a life raft in the Pacific using such a hand-pumped reverse osmosis water purifier. Adrift with only a fishing hook, a piece of line, and the reverse osmosis device, they were able to make about 2.5 L of water a day and caught fish for food. Although they lost some weight from their ordeal, they were in fairly good health when the Costa Rican Coast Guard rescued them after 66 days.

See J. Alper: "Survival at Sea," *Chem. Matters,* p. 4. October 1992

Because pressures on the order of 10^{-3} atm are easily measured, concentrations of about 10^{-4} M can be determined through measurements of osmotic pressure. Osmosis is an ideal method for measuring the molar masses of very large molecules, including polymers and compounds that are of biological importance.

Example 14.10 Osmotic Pressure and Molar Mass

Beta-carotene is the most important of the A vitamins. Its molar mass can be determined by measuring the osmotic pressure generated by a given mass of β-carotene dissolved in the solvent chloroform. Calculate the molar mass of β-carotene if 7.68 mg, dissolved in 10.0 mL of chloroform, gives an osmotic pressure of 26.57 mm Hg at 25.0 °C.

Solution The concentration of β-carotene in chloroform can be calculated from the osmotic pressure (Equation 14.11):

$$\text{Concentration (M)} = \frac{\Pi}{RT} = \frac{(26.57\ \text{mm Hg})(1\ \text{atm}/760\ \text{mm Hg})}{(0.082057\ \text{L}\cdot\text{atm/K}\cdot\text{mol})(298.15\ \text{K})}$$
$$= 1.429 \times 10^{-3}\ \text{mol/L}$$

Now the quantity of β-carotene dissolved in 10.0 mL of solvent can be calculated:

$$(1.429 \times 10^{-3}\ \text{mol/L})(0.0100\ \text{L}) = 1.429 \times 10^{-5}\ \text{mol}$$

This quantity of β-carotene (1.429×10^{-5} mol) is equivalent to 7.68 mg (7.68×10^{-3} g). This gives us a way to calculate the molar mass:

$$\frac{7.68 \times 10^{-3}\ \text{g}}{1.429 \times 10^{-5}\ \text{mol}} = 537\ \text{g/mol}$$

Beta-carotene is a hydrocarbon with the formula $C_{40}H_{56}$.

Exercise 14.10 Osmotic Pressure and Molar Mass

A 1.40-g sample of polyethylene, a common plastic, is dissolved in enough benzene to give exactly 100 mL of solution. The measured osmotic pressure of the solution is 1.86 mm Hg at 25 °C. Calculate the average molar mass of the polymer.

Osmosis is of great practical significance for people in the health professions. Patients who become dehydrated through illness often need to be given water and nutrients intravenously. Water cannot simply be dripped into a patient's vein, however. Rather, the intravenous solution must have the same overall solute concentration as the patient's blood: the solution must be isoosmotic or *isotonic.* If pure water was used, the inside of a blood cell would have a higher solute concentration (lower water concentration), and water would flow into the cell. This *hypotonic* situation would cause the red blood cells to burst (lyse) (Figure 14.17). The opposite situation, *hypertonicity*, occurs if the intravenous solution is more concentrated than the contents of the blood cell. In this case the cell would lose water and shrivel up (crenate). To combat this, a dehydrated patient is rehydrated in the hospital with a sterile saline solution that is 0.16 M NaCl, a solution that is isotonic with the cells of the body.

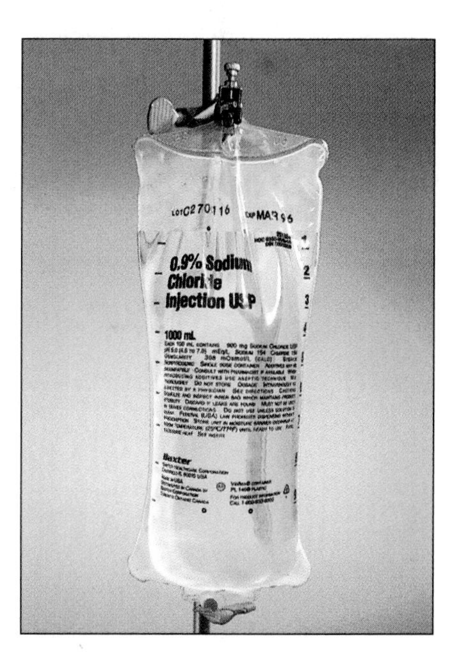

An isotonic saline solution. *(C. D. Winters)*

14.4 COLLOIDS

We defined a solution broadly as a homogeneous mixture of two or more substances in a single phase. To this we should add that, in a true solution, no settling of the solute should be observed and the solute particles should be in the form of ions or relatively small molecules. Thus, NaCl and sugar form true solutions in water. You are also familiar with suspensions, which result, for example, if a handful of fine sand is added to water and shaken vigorously. Sand particles are still visible and gradually settle to the bottom of the beaker or bottle.

See the *Saunders Interactive General Chemistry CD-ROM,* Screen 14.10, Colloids.

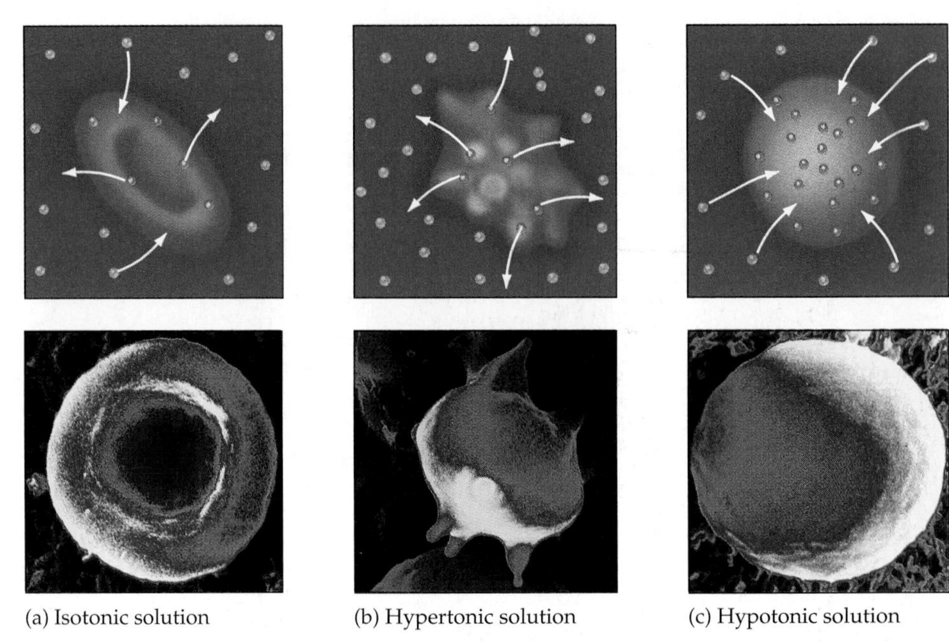

***Figure* 14.17 Osmosis and living cells.** (a) A cell placed in an isotonic solution. The net movement of water into and out of the cell is zero because the concentration of solutes inside and outside the cell is the same. (b) In a hypertonic solution, the concentration of solutes outside the cell is greater than that inside. There is a net flow of water out of the cell, causing the cell to dehydrate, shrink, and perhaps die. (c) In a hypotonic solution, the concentration of solutes outside the cell is less than that inside. There is a net flow of water into the cell, causing the cell to swell and perhaps to burst. *(David Phillips/Science Source/Photo Researchers, Inc.)*

Colloidal dispersions, also called **colloids,** represent a state intermediate between a solution and a suspension. Colloids include many of the foods you eat and the materials around you; among them are Jello, milk, fog, and porcelain (see Table 14.5).

Around 1860, the British chemist Thomas Graham (1805–1869) found that substances such as starch, gelatin, glue, and albumin from eggs diffused only very slowly when placed in water, compared with sugar or salt. In addition, the former substances differ significantly in their ability to diffuse through a thin

***Table* 14.5 • Types of Colloids**

Type	Dispersing Medium	Dispersed Phase	Examples
Aerosol	Gas	Liquid	Fog, clouds, aerosol sprays
Aerosol	Gas	Solid	Smoke, airborne viruses, automobile exhaust
Foam	Liquid	Gas	Shaving cream, whipped cream
Foam	Solid	Gas	Foam rubber, sponge, pumice
Emulsion	Liquid	Liquid	Mayonnaise, milk, face cream
Gel	Solid	Liquid	Jelly, cheese, butter
Sol	Liquid	Solid	Gold in water, milk of magnesia, mud
Solid sol	Solid	Solid	Milk glass, alloys (e.g., steel or brass)

(a) (b)

Figure 14.18 The Tyndall effect. Colloidal suspensions or dispersions scatter light, a phenomenon known as the Tyndall effect. (a) Dust in the air scatters the light coming through the trees in a forest along the Oregon coast. (b) A narrow beam of light from a laser is passed through an NaCl solution *(left)* and then a colloidal mixture of gelatin and water *(right)*. *(C. D. Winters)*

membrane: sugar molecules can diffuse through many membranes, but the very large molecules that make up starch, gelatin, glue, and albumin do not. Moreover, Graham found that he could not crystallize these latter substances, whereas he could crystallize sugar, salt, and other materials that form true solutions. Graham coined the word "colloid" (from the Greek meaning "glue") to describe this class of substances distinctly different from true solutions and suspensions.

We now know that it is possible to crystallize some colloidal substances, albeit with difficulty, so there really is no sharp dividing line between these classes based on this property. Colloids do, however, have the following distinguishing characteristics: (1) It is generally true that colloidal materials have very high molar masses; this is certainly true of human cells and of proteins such as hemoglobin that have molar masses in the thousands. (2) The particles of a colloid are relatively large (say 1000 nm in diameter), large enough that they scatter visible light when dispersed in a solvent, making the mixture appear cloudy (Figure 14.18). (3) Even though colloidal particles are large, they are not so large that they settle out.

Graham also gave us the words **sol** for a colloidal solution (a dispersion of a solid substance in a fluid medium) and **gel** for a dispersion that has a structure that prevents it from being mobile. Jello is a sol when the solid is first mixed with boiling water, but it becomes a gel when cooled. Other examples of gels are the gelatinous precipitates of $Al(OH)_3$, $Fe(OH)_3$, and $Cu(OH)_2$ (Figure 14.19).

Colloidal dispersions consist of finely divided particles that, as a result, have a very high surface area. For example, if you have one-millionth of a mole of colloidal particles, each assumed to be a sphere with a diameter of 200 nm, the total surface area of the particles would be on the order of 100 million cm^2, or the size of several football fields. It is not surprising, therefore, that many of the properties of colloids depend on the properties of surfaces.

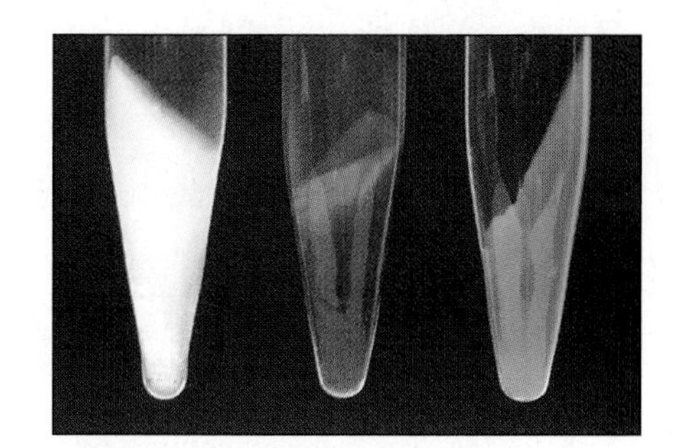

Figure 14.19 Gelatinous precipitates. *(left)* $Al(OH)_3$, *(center)* $Fe(OH)_3$, and *(right)* $Cu(OH)_2$. *(C. D. Winters)*

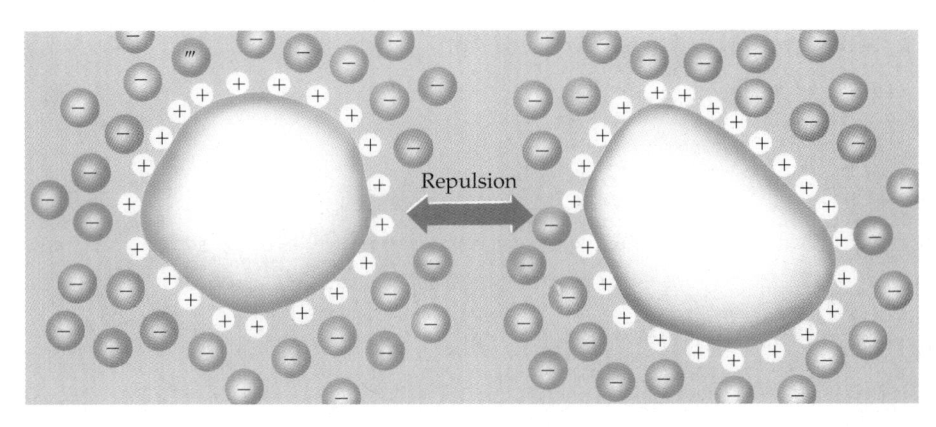

Figure 14.20 Hydrophobic colloids. A hydrophobic colloid is stabilized by positive ions absorbed onto each particle and a secondary layer of negative ions. Because the particles bear similar charges they repel one another and precipitation is prevented.

Types of Colloids

Colloids are classified according to the state of the dispersed phase and the dispersing medium. Table 14.5 lists several types of colloids and gives examples of each.

Colloids with water as the dispersing medium can be classified as **hydrophobic** (from the Greek, meaning "water-fearing") or **hydrophilic** ("water loving"). A hydrophobic colloid is one in which only weak attractive forces exist between the water and the surface of the colloidal particles. Examples are dispersions of metals and of nearly insoluble salts in water. When salts like AgCl precipitate (e.g., from the reaction of $AgNO_3$ and NaCl), the result is often a colloidal dispersion. The precipitation reaction occurs too rapidly for ions to gather from long distances and make large crystals, so the ions aggregate to form small particles.

Figure 14.21 Formation of silt. Silt forms at a river delta as colloidal soil particles come in contact with salt water in the ocean. Here the Ashley and Cooper Rivers empty into the Atlantic Ocean at Charleston, SC. The high concentration of ions in salt water causes the colloidal soil particles to coagulate. *(NASA/Peter Arnold, Inc.)*

Why do hydrophobic colloids exist? Why don't the particles come together (coagulate) and form larger particles? The answer seems to be that the colloidal particles carry electric charges. An AgCl particle, for example, will absorb Ag^+ ions if the ions are present in substantial concentration; an attraction occurs between Ag^+ ions in solution and Cl^- ions on the surface of the particle. The colloidal particles thus become positively charged, and attract a secondary layer of anions. The particles, now surrounded by layers of ions, repel one another and are prevented from coming together to form a precipitate (Figure 14.20).

A stable hydrophobic colloid can be made to coagulate by introducing ions into the dispersing medium. Milk is a colloidal suspension of hydrophobic particles. When milk ferments, lactose (milk sugar) is converted to lactic acid, which forms lactate ions and hydrogen ions. The protective charge on the surfaces of the colloidal particles is overcome, and the milk coagulates; the milk solids come together in clumps called "curds."

Soil particles are often carried by water in rivers and streams as hydrophobic colloids. When river water carrying large amounts of colloidal particles meets sea water with its high concentration of salts, the particles coagulate to form the silt seen at the mouth of the river (Figure 14.21). Municipal water treatment plants often add aluminum salts such as $Al_2(SO_4)_3$ to water. In aqueous solution, aluminum ions exist as $[Al(H_2O)_6]^{3+}$, which neutralize the charge on the hydrophobic colloidal soil particles, causing the soil particles to aggregate and settle out.

Hydrophilic colloids are strongly attracted to water molecules. They often have groups such —OH and $—NH_2$ on their surfaces (Figure 14.22). These groups form strong hydrogen bonds to water, thus stabilizing the colloid. Proteins and starch are important examples, and homogenized milk is the most familiar example.

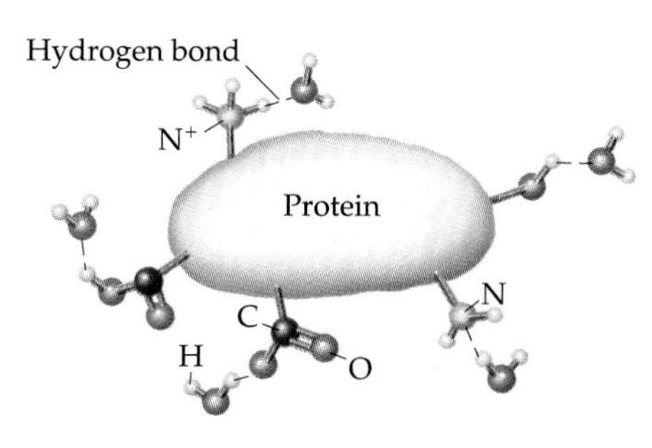

***Figure* 14.22 Hydrophilic colloids.** A hydrophilic colloidal particle is stabilized by hydrogen bonding to water.

Emulsions are colloidal dispersions of one liquid in another, such as oil or fat in water. You find emulsions in familiar items such as salad dressing, mayonnaise, and milk. If vegetable oil and vinegar are mixed to make a salad dressing, the mixture quickly separates into two layers because the nonpolar oil molecules do not interact with polar water and acetic acid molecules. So why are milk and mayonnaise apparently homogeneous mixtures that do not separate into layers? The answer is that they contain an **emulsifying agent** such as soap or a protein. Lecithin is a protein found in egg yolks, so mixing egg yolks with oil and vinegar stabilizes the colloidal dispersion known as mayonnaise. To understand how this works, we can look into the functioning of soaps and detergents, substances known as surfactants.

One of the best places to find out about the properties of food in terms of chemistry is in *The Cookbook Decoder,* a cookbook written by a chemist, A. Grosser (Beaufort Books, Inc., New York, 1981).

Surfactants

Soaps and detergents are emulsifying agents. Soap is made by heating a fat with sodium or potassium hydroxide (see Section 11.6).

$$\begin{array}{l} H_2C—O—C(=O)—(CH_2)_{16}CH_3 \\ \quad | \\ HC—O—C(=O)—(CH_2)_{16}CH_3 \\ \quad | \\ H_2C—O—C(=O)—(CH_2)_{16}CH_3 \end{array}$$

fat (tristearin or glyceryl tristearate)

$$\downarrow + 3\ NaOH$$

$$\begin{array}{l} H_2C—O—H \\ \quad | \\ HC—O—H \\ \quad | \\ H_2C—O—H \end{array} + 3\left[H_3C—(CH_2)_{16}—C(=O)O^- \quad Na^+\right]$$

glycerol

Hydrocarbon tail: Soluble in oil — Polar head: Soluble in water

sodium stearate, a soap

A sodium soap is a solid at room temperature, whereas potassium soaps are usually liquids.

The soap and detergent industry is enormous. Roughly 3×10^7 tons of household and toilet soaps, and synthetic and soap-based laundry detergents are produced annually worldwide.

The resulting fatty acid anion has a split personality: it has a nonpolar, hydrophobic hydrocarbon tail that is soluble in other similar hydrocarbons and a polar, hydrophilic head that is soluble in water.

Oil cannot be simply washed away from dishes or clothing with water because oil is nonpolar and thus insoluble in water. Instead, we must add soap to

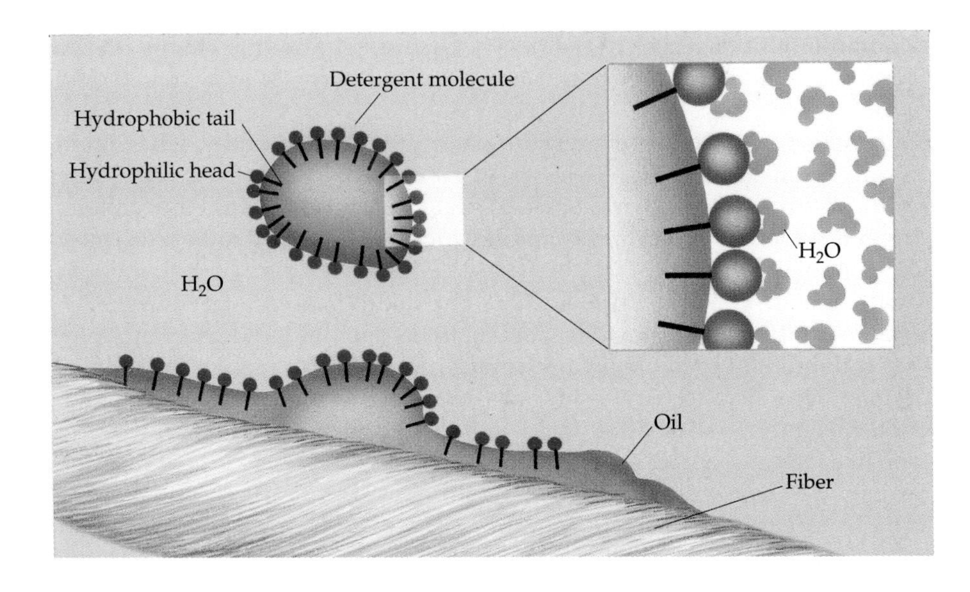

Figure 14.23 The cleaning action of soap. Soap molecules interact with water through the charged, hydrophilic end of the molecule. The long, hydrocarbon end of the molecule is hydrophobic but can bind through dispersion forces with hydrocarbons and other nonpolar substances.

the water to clean away the oil. The nonpolar molecules of the oil interact with the nonpolar hydrocarbon tails of the soap molecules, leaving the polar heads of the soap to interact with surrounding water molecules. The oil and water then mix (Figure 14.23). If the oily material on a piece of clothing or a dish also contains some dirt particles, the dirt can now be washed away.

See the *Saunders Interactive General Chemistry CD-ROM,* Screen 14.11, Surfactants.

Substances such as soaps that affect the properties of surfaces, and so affect the interaction between two phases, are called surface-active agents, or **surfactants,** for short. A surfactant that is used for cleaning has come to be called a **detergent.** One function of a surfactant is to lower the surface tension of water, which enhances the cleansing action of the detergent (Figure 14.24).

Figure 14.24 Effect of a detergent on the surface tension of water. Sulfur (density = 2.1 g/cm^3) is carefully placed on the surface of water (density 1.0 g/cm^3) *(left).* The surface tension of the water keeps the denser sulfur afloat. Several drops of detergent are then placed on the surface of the water *(right).* The surface tension of the water is reduced, and the sulfur sinks to the bottom of the beaker. *(C. D. Winters)*

Many detergents used in the home and industry are synthetic. One example is sodium lauryl benzenesulfonate, a biodegradable compound.

$$CH_3CH_2CH_2CH_2CH_2CH_2CH_2CH_2CH_2CH_2CH_2CH_2{-}C_6H_4{-}SO_3^-\ Na^+$$

sodium lauryl benzenesulfonate

In general, synthetic detergents use the sulfonate $-SO_3^-$ group as the polar head instead of the carboxylate group, $-CO_2^-$. The carboxylate anions form an insoluble precipitate with any Ca^{2+} or Mg^{2+} ions present in water. Because hard water is characterized by high concentrations of these ions, using soaps containing carboxylates produces bathtub rings and tattle-tale gray clothing. The synthetic sulfonate detergents have the advantage that they do not form such precipitates because their calcium salts are more soluble in water.

CHAPTER HIGHLIGHTS

When you have finished studying this chapter, you should be able to

- Define the terms **solution, solvent, solute,** and **colligative properties** (Section 14.1).
- Use the following concentration units: **molality, mole fraction, weight percent,** and **parts per million** (Section 14.1), and calculate solution concentrations using these units.
- Describe the process of dissolving a solute in a solvent, including the energy changes that may occur (Section 14.2).
- Understand the distinctions between **saturated, unsaturated,** and **supersaturated** solutions (Section 14.2).
- Define and illustrate the terms **miscible** and **immiscible** (Section 14.2).
- Understand the relation of **lattice energy** and **enthalpy of hydration** to the **enthalpy of solution** for an ionic solute (Section 14.2).
- Describe the effect of pressure and temperature on the solubility of a solute (Sections 14.2).
- Use **Henry's law** to calculate the solubility of a gas in a solvent (Section 14.2).
- Apply **Le Chatelier's principle** to the change in solubility of gases with pressure and temperature changes (Section 14.2).
- Calculate the mole fraction of a solute or solvent ($X_{solvent}$) and the effect of a solute on solvent vapor pressure ($P_{solvent}$) using **Raoult's law** (Section 14.3).

$$\text{Raoult's law: } P_{solvent} = X_{solvent} \cdot P^{\circ}_{solvent} \qquad [14.6]$$

(where $P^{\circ}_{solvent}$ is the vapor pressure of the pure solvent at the specified temperature).

- Calculate the boiling point elevation or freezing point depression caused by a solute in a solvent (Section 14.3).

$$\text{Elevation in boiling point, } \Delta t_{bp} = K_{bp} \cdot m_{solute} \qquad [14.8]$$

$$\text{Depression of freezing point, } \Delta t_{fp} = K_{fp} \cdot m_{solute} \qquad [14.9]$$

- Use colligative properties to determine the molar mass of a solute (Section 14.3).
- Give a molecular-level explanation for boiling point elevation and freezing point depression (Section 14.3).

- Characterize the effect of ionic solutes on colligative properties (Section 14.3).
- Use the **van't Hoff factor,** *i,* in calculations involving colligative properties (Section 14.3).
- Give a molecular-level explanation for osmosis (Section 14.3).
- Calculate the **osmotic pressure** (Π) for solutions, and use the equation defining osmotic pressure to determine the molar mass of a solute (Section 14.3).

$$\Pi = CRT \quad [14.11]$$

- Recognize the difference among a homogeneous solution, a suspension, and a **colloid** (or **colloidal dispersion**) (Section 14.4).
- Recognize **hydrophobic** and **hydrophilic** colloids (Section 14.4).
- Describe the action of a **surfactant** (Section 14.4).

STUDY QUESTIONS

REVIEW QUESTIONS

1. Is the following statement true or false? If false, change it to make it true. "Colligative properties depend on the nature of the solvent and solute and on the concentration of the solute."
2. Name the four colligative properties described in this chapter. Write the mathematical expression that describes each of these.
3. Define molality, and tell how it differs from molarity.
4. Name three effects that govern the solubility of a gas in water.
5. If you dissolve equal molar amounts of NaCl and $CaCl_2$ in water, the calcium salt lowers the freezing point of the water almost 1.5 times as much as the NaCl. Why?
6. Explain why a cucumber shrivels up when it is placed in a concentrated solution of salt.
7. Explain the differences among a colloidal dispersion, a suspension, and a true solution.
8. Explain the difference between a sol and a gel. Give an example of each.
9. Explain how a surfactant such as a soap or detergent functions.

NUMERICAL QUESTIONS

Concentration Units

10. Assume you dissolve 2.56 g of malic acid, $C_4H_6O_5$, in half a liter of water (500.0 g). Calculate the molarity, molality, mole fraction, and weight percentage of acid in the solution.

11. Camphor, a white solid with a pleasant odor, is extracted from the roots, branches, and trunk of the camphor tree. Assume you dissolve 45.0 g of camphor ($C_{10}H_{16}O$) in 425 mL of ethanol, C_2H_5OH. Calculate the molarity, molality, mole fraction, and weight percentage of camphor in this solution. (The density of ethanol is 0.785 g/mL.)

12. Fill in the blanks in the table. Aqueous solutions are assumed.

Compound	Molality	Weight Percentage	Mole Fraction
NaI	0.15	______	______
C_2H_5OH	______	5.0	______
$C_{12}H_{22}O_{11}$	0.15	______	______

13. Fill in the blanks in the table. Aqueous solutions are assumed.

Compound	Molality	Weight Percentage	Mole Fraction
KNO_3	______	10.0	______
CH_3CO_2H	0.0183	______	______
$HOCH_2CH_2OH$	______	18.0	______

14. You want to prepare a solution that is 0.200 *m* in Na_2CO_3. How many grams of the salt must you add to 125 g of water? What is the mole fraction of Na_2CO_3 in the resulting solution?

15. You want to prepare a solution that is 0.0512 *m* in $NaNO_3$. How many grams of the salt must you add to exactly 500 g of water? What is the mole fraction of $NaNO_3$ in the solution?

16. You wish to prepare an aqueous solution of glycerol, $C_3H_5(OH)_3$, in which the mole fraction of the solute is 0.093. How many grams of glycerol must you combine with 425 g of water to make this solution? What is the molality of the solution?

17. You want to prepare an aqueous solution of ethylene glycol, $HOCH_2CH_2OH$, in which the mole fraction of solute is 0.125. What mass of ethylene glycol, in grams, should you combine with 955 g of water? What is the molality of the solution?

18. Fill in the blanks in the following table:

Compound	Grams Compound	Grams Water	Molality	Mole Fraction of Compound
K_2CO_3	______	125	0.0125	______
C_2H_5OH	13.5	175	______	______
$NaNO_3$	______	555	______	0.0934

19. Fill in the blanks in the following table:

Compound	Grams Compound	Grams Water	Molality	Mole Fraction of Compound
$AgNO_3$	______	625	0.0245	______
$HOCH_2CH_2OH$	______	256	______	0.0545
$Pt(NH_3)_2Cl_2$	0.0075	225	______	______

20. Concentrated sulfuric acid has a density of 1.84 g/cm^3 and is 95.0% by weight H_2SO_4. What is the molality of this acid? What is its molarity?

21. Hydrochloric acid is sold as a concentrated aqueous solution. If the molarity of commercial HCl is 12.0 and its density is 1.18 g/cm^3, calculate
(a) The molality of the solution
(b) The weight percentage of HCl in the solution

22. A 10.7 m solution of NaOH has a density of 1.33 g/cm^3 at 20 °C. Calculate
(a) The mole fraction of NaOH
(b) The weight percentage of NaOH
(c) The molarity of the solution

23. Concentrated aqueous ammonia has a molarity of 14.8 and a density of 0.90 g/cm^3. What is the molality of the solution? Calculate the mole fraction and weight percentage of NH_3.

24. If you dissolve 2.00 g of $Ca(NO_3)_2$ in 750 g of water, what is the molality of $Ca(NO_3)_2$? What is the total molality of ions in solution? (Assume total dissociation of the ionic solid.)

25. If you want a solution that is 0.100 m in ions, how many grams of Na_2SO_4 must you dissolve in 125 g of water? (Assume total dissociation of the ionic solid.)

26. The average lithium ion concentration in sea water is 0.18 ppm. What is the molality of Li^+ in sea water?

27. Silver ion has an average concentration of 28 ppb (parts per billion) in U.S. water supplies.
(a) What is the molality of the silver ion?
(b) If you wanted 1.0×10^2 g of silver and could recover it chemically from water supplies, how many liters of water would you have to treat? (Assume the density of water is 1.0 g/cm^3.)

The Solution Process

28. Which pairs of liquids will be miscible?
(a) H_2O and $CH_3CH_2CH_2CH_3$
(b) C_6H_6 (benzene) and CCl_4
(c) H_2O and CH_3CO_2H

29. Acetone, $(CH_3)_2CO_2$ is quite soluble in water. Explain why this should be so.

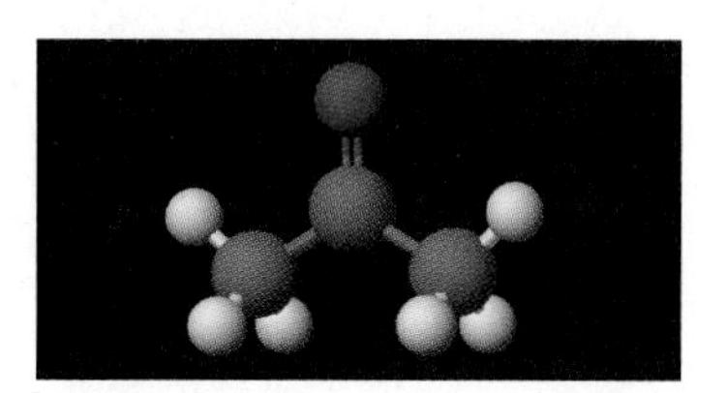

30. Use the data of Table 14.1 to calculate the enthalpy of solution of LiCl. How does the value differ from that for NaCl?

31. Use the following data to calculate the enthalpy of solution of sodium perchlorate, $NaClO_4$:

$$\Delta H_f^\circ\text{(s)} = -382.9 \text{ kJ/mol}$$
$$\Delta H_f^\circ\text{(aq, 1 } m) = -369.5 \text{ kJ/mol}$$

32. You make a saturated solution of NaCl at 25 °C. No solid is present in the beaker holding the solution. What can be done to increase the amount of dissolved NaCl in this solution? (See Figure 14.9)
(a) Add more solid NaCl.
(b) Raise the temperature of the solution.
(c) Raise the temperature of the solution and add some NaCl.
(d) Lower the temperature of the solution and add some NaCl.

33. Some lithium chloride, LiCl, is dissolved in 100 mL of water in one beaker and some Li_2SO_4 is dissolved in 100 mL of water in another beaker. Both are at 10 °C and both are saturated solutions; some solid remains undissolved in each beaker. Describe what you would observe as the temperature is raised. The following data are available to you from a handbook of chemistry:

Compound	Solubility (g/100 mL) 10 °C	40 °C
Li_2SO_4	35.5	33.7
LiCl	74.5	89.8

Henry's Law

34. The partial pressure of O_2 in your lungs varies from 25 mm Hg to 40 mm Hg. How much O_2 can dissolve in water at 25 °C if the partial pressure of O_2 is 40 mm Hg?

35. Henry's law constant for O_2 in water at 25 °C is 1.66×10^{-6} M/mm Hg. Which of the following is a reasonable constant when the temperature is 50 °C? Explain the reason for your choice.
(a) 8.80×10^{-7} M/mm Hg
(b) 3.40×10^{-6} M/mm Hg
(c) 1.66×10^{-6} M/mm Hg
(d) 8.40×10^{-5} M/mm Hg

36. An unopened soda can has an aqueous CO_2 concentration of 0.0506 M at 25 °C. What is the pressure of CO_2 gas in the can?

37. Hydrogen gas has a Henry's law constant of 1.07×10^{-6} M/mm Hg at 25 °C when dissolving in water. If the total pressure of gas (H_2 gas plus water vapor) over water is 1.0 atm, what is the concentration of H_2 in the water in grams per milliliter? (See Appendix G for the vapor pressure of water.)

Vapor Pressure Changes

38. Urea, $(NH_2)_2CO$, is widely used in fertilizers and plastics. The compound is quite soluble in water; 1.00 g can be dissolved in 1.00 mL of water. If you dissolve 9.00 g of urea in 10.0 mL of water, what is the vapor pressure of the solution at 24 °C? Assume the density of water is 1.00 g/mL.

39. A 35.0-g sample of ethylene glycol, $HOCH_2CH_2OH$, is dissolved in half a liter of water (500.0 g). The vapor pressure of water at 32 °C is 35.7 mm Hg. What is the vapor pressure of the water–ethylene glycol solution at 32 °C? (The glycol is nonvolatile.)

40. Pure ethylene glycol [$HOCH_2CH_2OH$] is added to 2.00 kg of water in the cooling system of a car. The vapor pressure of the water in the system when the temperature is 90 °C is 457 mm Hg. How many of grams of glycol are added? (Assume that ethylene glycol is not volatile at this temperature. See Appendix G for the vapor pressure of water.)

41. Pure iodine (105 g) is dissolved in 325 g of CCl_4 at 65 °C. Given that the vapor pressure of CCl_4 at this temperature is 531 mm Hg, what is the vapor pressure of the CCl_4–I_2 solution at 65 °C? (Assume that I_2 does not contribute to the vapor pressure.)

Boiling Point Elevation

42. Verify that 0.200 mol of a nonvolatile solute in 125 g of benzene (C_6H_6) produces a solution whose boiling point is 84.2 °C.

43. What is the boiling point of a solution composed of 15.0 g of urea, $(NH_2)_2CO$, in 0.500 kg of water?

44. What is the boiling point of a solution composed of 15.0 g of $CHCl_3$ and 0.515 g of the nonvolatile solute acenaphthalene, $C_{12}H_{10}$, a component of coal tar?

45. What is the boiling point of a solution composed of 0.755 g of caffeine ($C_8H_{10}O_2N_4$) in 95.6 g of benzene, C_6H_6?

46. Phenanthrene, $C_{14}H_{10}$, is an aromatic hydrocarbon. If you dissolve some phenanthrene in 50.0 g of benzene, the boiling point of the solution is 80.51 °C. How many grams of the hydrocarbon must have been dissolved?

47. A solution of glycerol, $C_3H_5(OH)_3$, in 735 g of water has a boiling point of 104.4 °C at a pressure of 760 mm Hg. How many grams of glycerol are in the solution? What is the mole fraction of the solute?

48. Arrange the following solutions in order of increasing boiling point:
(a) 0.10 *m* KCl
(b) 0.10 *m* sugar
(c) 0.080 *m* $MgCl_2$

49. Arrange the following aqueous solutions in order of increasing boiling point:
(a) 0.20 *m* ethylene glycol (nonvolatile, nonelectrolyte)
(b) 0.12 *m* $(NH_4)_2SO_4$
(c) 0.10 *m* $CaCl_2$
(d) 0.12 *m* KNO_3

50. You add 0.255 g of an orange, crystalline compound whose empirical formula is $C_{10}H_8Fe$ to 11.12 g of benzene. The boiling point of the benzene rises from 80.10 °C to 80.26 °C. What is the molar mass and molecular formula of the compound?

51. Butylated hydroxyanisole (BHA) is used as an antioxidant in margarine and other fats and oils; it prevents oxidation and prolongs the shelf life of the food. What is the molar mass of BHA if 0.640 g of the compound, dissolved in 25.0 g of chloroform, produces a solution whose boiling point is 62.22 °C?

52. Benzyl acetate is one of the active components of oil of jasmine. If 0.125 g of the compound is added to 25.0 g of chloroform ($CHCl_3$), the boiling point of the solution is 61.82 °C. What is the molar mass of benzyl acetate?

53. Anthracene is a hydrocarbon obtained from coal. The empirical formula of anthracene is C_7H_5. To find its molecular formula you dissolve 0.500 g in 30.0 g of benzene. The boiling point of the pure benzene is 80.10 °C, whereas the solution has a boiling point of 80.34 °C. What is the molecular formula of anthracene?

Freezing Point Depression

54. A mixture of ethanol (C_2H_5OH) and water has a freezing point of −16.0 °C.
(a) What is the molality of the alcohol?
(b) What is the weight percent of alcohol in the solution?

55. Some ethylene glycol, $HOCH_2CH_2OH$, is added to your car's cooling system along with 5.0 kg of water.
(a) If the freezing point of the water–glycol solution is −15.0 °C, how many grams of $HOCH_2CH_2OH$ must have been added?
(b) What is the boiling point of the solution?

56. If 52.5 g of LiF is dissolved in 306 g of water, what is the expected freezing point of the solution? (Assume the van't Hoff factor *i* for LiF is 2.)

57. If you have ever made homemade ice cream, you know that you cool the milk and cream by immersing the container in ice and a concentrated solution of rock salt (NaCl) in water. If you want to have a water–salt solution that freezes at −10. °C, how many grams of NaCl must you add to 3.0 kg of water? (Assume the van't Hoff factor *i* for NaCl is 1.85.)

58. An aqueous solution contains 0.180 g of an unknown, nonionic solute in 50.0 g of water. The solution freezes at −0.040 °C. What is the molar mass of the solute?

59. The organic compound called aluminon is used as a reagent to test for the presence of the aluminum ion in

aqueous solution. A solution of 2.50 g of aluminon in 50.0 g of water freezes at −0.197 °C. What is the molar mass of aluminon?

60. The melting point of pure biphenyl ($C_{12}H_{10}$) is found to be 70.03 °C. If 0.100 g of naphthalene is added to 10.0 g of biphenyl, the freezing point of the mixture is found to be 69.40 °C. If K_{fp} for biphenyl is −8.00 °C/*m*, what is the molar mass of naphthalene?

61. Phenylcarbinol is used in nasal sprays as a preservative. A solution of 0.52 g of the compound in 25.0 g of water has a melting point of −0.36 °C. What is the molar mass of phenylcarbinol?

62. List the following aqueous solutions in order of increasing melting point. (The last three are all assumed to dissociate completely into ions in water.)
(a) 0.1 *m* sugar
(b) 0.1 *m* NaCl
(c) 0.08 *m* $CaCl_2$
(d) 0.04 *m* Na_2SO_4

63. Arrange the following aqueous solutions in order of decreasing freezing point. (The last three are all assumed to dissociate completely into ions in water.)
(a) 0.20 *m* ethylene glycol (nonvolatile, nonelectrolyte)
(b) 0.12 *m* K_2SO_4
(c) 0.10 *m* $MgCl_2$
(d) 0.12 *m* KBr

Osmosis

64. An aqueous solution contains 3.00% phenylalanine ($C_9H_{11}NO_2$) by mass. Assume the phenylalanine is nonionic and nonvolatile. Find
(a) The freezing point of the solution
(b) The boiling point of the solution
(c) The osmotic pressure of the solution at 25 °C
In your view, which of these is most easily measurable in the laboratory?

65. Estimate the osmotic pressure of human blood at 37 °C. Assume blood is isotonic with a 0.16 M NaCl solution, and assume the van't Hoff factor *i* is 1.9 for NaCl.

66. An aqueous solution containing 1.00 g of bovine insulin (a protein, not ionized) per liter has an osmotic pressure of 3.1 mm Hg at 25 °C. Calculate the molar mass of bovine insulin.

67. Calculate the osmotic pressure of a 0.0120 M solution of NaCl in water at 0 °C. Assume the van't Hoff factor *i* is 1.94 for this solution.

Colloids

68. When solutions of $BaCl_2$ and Na_2SO_4 are mixed, the mixture becomes cloudy. After a few days, a white solid is observed on the bottom of the beaker with a clear liquid above it.
(a) Write a balanced equation for the reaction that occurs.
(b) Why is the solution cloudy at first?
(c) What happens during the few days of waiting?

69. The dispersed phase of a certain colloidal dispersion consists of spheres of diameter 1.0×10^2 nm.
(a) What is the volume ($V = \frac{4}{3}\pi r^3$) and surface area ($A = 4\pi r^2$) of each sphere?
(b) How many spheres are required to give a total volume of 1.0 cm^3? What is the total surface area of these spheres in square meters?

GENERAL QUESTIONS

70. Solution properties
(a) Which solution is expected to have the higher boiling point: 0.10 *m* Na_2SO_4 or 0.15 *m* sugar?
(b) For which aqueous solution is the vapor pressure of water higher: 0.30 *m* NH_4NO_3 or 0.15 *m* Na_2SO_4?

71. Arrange the following aqueous solutions in order of (i) increasing vapor pressure of water and (ii) increasing boiling point:
(a) 0.35 *m* $HOCH_2CH_2OH$ (a nonvolatile solute)
(b) 0.50 *m* sugar
(c) 0.20 *m* KBr (a strong electrolyte)
(d) 0.20 *m* Na_2SO_4 (a strong electrolyte)

72. Dimethylglyoxime [DMG, $(CH_3CNOH)_2$] is used as a reagent to precipitate nickel ion. Assume that 53.0 g of DMG has been dissolved in 525 g of ethanol (C_2H_5OH).

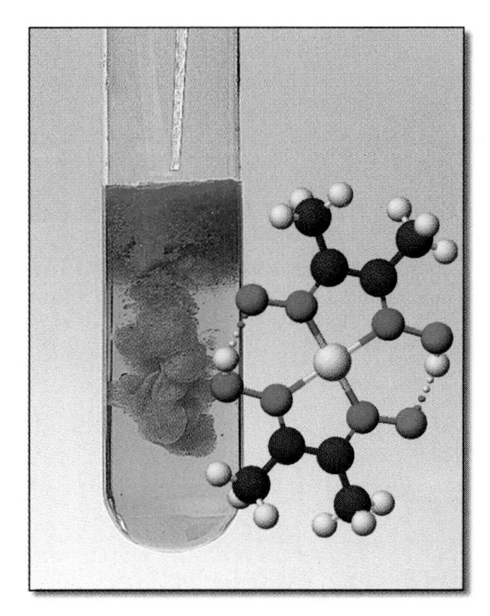

The red, insoluble compound formed between nickel(II) ion and dimethylglyoxime (DMG) is precipitated when DMG is added to a basic solution of Ni^{2+}(aq). *(Photo, C. D. Winters; model, S. M. Young)*

(a) What is the mole fraction of DMG?
(b) What is the molality of the solution?
(c) What is the vapor pressure of the ethanol over the solution at ethanol's normal boiling point of 78.4 °C?
(d) What is the boiling point of the solution? (DMG does not produce ions in solution.) (K_{bp} for ethanol = +1.22 °C/*m*)

73. Making homemade ice cream is one of life's great pleasures. Fresh milk and cream, sugar, and flavorings are churned in a bucket suspended in an ice–water mixture, the freezing point of which has been lowered by adding rock salt. One manufacturer of home ice cream freezers recommends adding 2.50 lb (1130 g) of rock salt (NaCl) to 16.0 lb of ice (7250 g) in a 4-qt freezer. For the solution when this mixture melts calculate
    (a) The weight percentage of NaCl
    (b) The mole fraction of NaCl
    (c) The molality of the solution

74. Consider the following aqueous solutions: (i) 0.20 *m* $HOCH_2CH_2OH$ (nonvolatile, nonelectrolyte); (ii) 0.10 *m* $CaCl_2$; (iii) 0.12 *m* KBr; and (iv) 0.12 *m* Na_2SO_4.
    (a) Which solution has the highest boiling point?
    (b) Which solution has the lowest freezing point?
    (c) Which solution has the highest water vapor pressure?

75. Solution properties:
    (a) Which solution is expected to have the higher boiling point: 0.20 *m* KBr or 0.30 *m* sugar?
    (b) Which aqueous solution has the lower freezing point: 0.12 *m* NH_4NO_3 or 0.10 *m* Na_2CO_3?

76. Instead of using NaCl to melt the ice on your sidewalk, you decide to use $CaCl_2$. (Like NaCl, $CaCl_2$ is a strong electrolyte.) If you add 35.0 g of $CaCl_2$ to 150. g of water, what is the freezing point of the solution? (Assume $i = 2.7$ for $CaCl_2$.)

77. The solubility of NaCl in water at 100 °C is 39.1 g/100. g of water. Calculate the boiling point of this solution. (Assume $i = 1.85$ for NaCl.)

78. The smell of ripe raspberries is due to *p*-hydroxyphenyl-2-butanone, which has the empirical formula C_5H_6O. To find its molecular formula, you dissolve 0.135 g in 25.0 g of chloroform, $CHCl_3$. The boiling point of the solution is 61.82 °C. What is the molecular formula of the solute?

79. Hexachlorophene is used in germicidal soap. What is its molar mass if 0.640 g of the compound, dissolved in 25.0 g of chloroform, produces a solution whose boiling point is 61.93 °C?

80. The solubility of ammonium formate, NH_4CHO_2, in water is 102 g in 100 g of water at 0 °C and 546 g/100 g of water at 80 °C. A solution is prepared by dissolving NH_4CHO_2 in 200 g of water until no more will dissolve at 80 °C. The solution is then cooled to 0 °C. How many grams of NH_4CHO_2 precipitate? (Assume no water evaporates and that the solution is not supersaturated.)

81. How much N_2 can dissolve in water at 25 °C if the N_2 partial pressure is 585 mm Hg?

82. Cigars are best stored in a "humidor" at 18 °C and 55% relative humidity. This means the pressure of water vapor should be 55% of the vapor pressure of pure water at the same temperature. The proper humidity can be maintained by placing a solution of glycerol [$C_3H_5(OH)_3$] and water in the humidor. Calculate the percentage by mass of glycerol that will lower the vapor pressure of water to the desired value. (The vapor pressure of glycerol is zero.)

83. An aqueous solution containing 10.0 g of starch per liter has an osmotic pressure of 3.8 mm Hg at 25 °C.
    (a) What is the molar mass of starch? (Because not all starch molecules are identical, the result will be an average.)
    (b) What is the freezing point of the solution? Would it be easy to determine the molecular weight of starch by measuring the freezing point depression? (Assume that the molarity and molality are the same for this solution.)

CONCEPTUAL QUESTIONS

84. If an egg is placed in dilute acetic acid (vinegar), the acid reacts with the calcium carbonate of the shell but the membrane around the egg remains intact (a). If the egg, without its shell, is placed in pure water, the egg swells (b). If, however, the egg is placed in a solution with a high solute concentration (a mixture of equal volumes of water and corn syrup), it shrivels dramatically (c). Explain these observations. (See Screen 14.9 of the *Saunders Interactive General Chemistry CD-ROM*.)

(a)

(b)

(c)

(a) A fresh egg is placed in dilute acetic acid. The acid reacts with the $CaCO_3$ of the shell but leaves the egg's membrane intact. (b) If the egg, with its shell removed, is placed in pure water, the egg swells. (c) Placed in a concentrated sugar solution, the egg shrivels. *(C. D. Winters)*

85. A protozoan (single-celled animal) that normally lives in the ocean is placed in fresh water. Will it shrivel or burst? Explain briefly.

86. A solution of 5.00 g of acetic acid in 100. g of benzene freezes at 3.37 °C. A solution of 5.00 g of acetic acid in 100. g of water freezes at −1.49 °C. Find the molar mass of acetic acid from each of these experiments. What can you conclude about the state of the acetic acid molecules dissolved in each of these solvents? Recall the discussion of hydrogen bonding in Section 13.2, and propose a structure for the species in benzene solution.

87. Solutions of salts have boiling points higher than those calculated using the equation

$$\Delta t_{bp} = K_{bp} \cdot m$$

(where m is the molality of the salt). Briefly explain this observation.

88. Account for the fact that alcohols such as methanol (CH_3OH) and ethanol (C_2H_5OH) are quite miscible with water, whereas an alcohol with a long carbon chain, such as octanol ($C_8H_{17}OH$), is poorly soluble in water.

89. Starch contains C—C, C—H, C—O, and O—H bonds. Hydrocarbons have only C—C and C—H bonds. Both starch and hydrocarbons can form colloidal dispersions in water. Which dispersion is classified as hydrophobic? Which is hydrophilic? Explain briefly.

90. How are the aqueous solubilities of NaCl, KCl, RbCl, and CsCl in Figure 14.9 related to their lattice energies (see Table 9.3)?

91. One theory to explain why oil and water are not miscible is that water would have to become *more* ordered as water molecules move aside to accommodate large oil molecules. Thinking in terms of hydrogen bonding, and what you learned about the structure of ice in Chapter 13, explain how water could become more ordered. Why is this unlikely to occur at constant temperature?

CHALLENGING PROBLEMS

92. Calculate the enthalpies of solution for Li_2SO_4 and K_2SO_4. Are the solution processes exothermic or endothermic? Compare them with LiCl and KCl. What similarities or differences do you find?

Compound	ΔH_f° (s) (kJ/mol)	ΔH_f° (aq, 1 m) (kJ/mol)
Li_2SO_4	−1436.4	−1464.4
K_2SO_4	−1437.7	−1414.0

93. Water at 25 °C has a density of 0.997 g/cm^3. Calculate the molality and molarity of pure water at this temperature.

94. If a volatile solute is added to a volatile solvent, both substances contribute to the vapor pressure over the solution. Assuming an ideal solution, the vapor pressure of each is given by Raoult's law, and the total vapor pressure is the sum of the vapor pressure of each component. A solution, assumed to be ideal, is made from 1.0 mol of toluene ($C_6H_5CH_3$) and 2.0 mol of benzene (C_6H_6). The vapor pressures of the pure solvents are 22 mm Hg and 75 mm Hg, respectively, at 20 °C. What is the total vapor pressure of the mixture? What is the mole fraction of each component in the liquid and in the vapor?

95. A solution is made by adding 50.0 mL of ethanol (C_2H_5OH) to 50.0 mL of water. What is the total vapor pressure over the solution at 20 °C? (See Study Question 94.) The vapor pressure of ethanol at 20 °C is 43.6 mm Hg.

96. A 2.0% (by mass) aqueous solution of novocainium chloride ($C_{13}H_{21}ClN_2O_2$) freezes at −0.237 °C. Calculate the van't Hoff factor i. How many moles of ions are in solution per mole of compound?

97. A solution is 4.00% (by mass) maltose and 96.00% water. It freezes at −0.229 °C.
   (a) Calculate the molar mass of maltose (which is not an ionic compound).
   (b) The density of the solution is 1.014 g/mL. Calculate the osmotic pressure of the solution.

98. The following table lists the concentrations of the principal ions in sea water:

Ion	Concentration (ppm)
Cl^-	1.95×10^4
Na^+	1.08×10^4
Mg^{2+}	1.29×10^3
SO_4^{2-}	9.05×10^2
Ca^{2+}	4.12×10^2
K^+	3.80×10^2
Br^-	67

   (a) Calculate the freezing point of water.
   (b) Calculate the osmotic pressure of sea water at 25 °C. What is the minimum pressure needed to purify sea water by reverse osmosis?

99. A tree is exactly 10 m tall.
   (a) What must be the total molarity of solutes if the sap rises to the top of the tree by osmotic pressure at 20 °C? Assume the groundwater outside the tree is pure water and that the density of the sap is 1.0 g/mL. (1 mm Hg = 13.6 mm H_2O)
   (b) If the only solute in the sap is sucrose, $C_{12}H_{22}O_{11}$, what is its percentage by mass?

100. A 2.00% solution of H_2SO_4 in water freezes at −0.796 °C.
   (a) Calculate the van't Hoff factor i.
   (b) Which of the following best represents sulfuric acid in a dilute aqueous solution: H_2SO_4, $H^+ + HSO_4^-$, or $2\ H^+ + SO_4^{2-}$?

101. A compound is known to be a potassium salt, KX. If 4.00 g of the salt is dissolved in exactly 100 g of water, the solution freezes at −1.28 °C. Which of the elements of Group 7A is X?

Summary Questions

102. A newly synthesized compound containing boron and fluorine is 22.1% boron. Dissolving 0.146 g of the compound in 10.0 g of benzene gives a solution with a vapor pressure of 94.16 mm Hg at 25 °C. (The vapor pressure of pure benzene at this temperature is 95.26 mm Hg.) In a separate ex-

periment, it is found that the compound does not have a dipole moment.
(a) What is the molecular formula for the compound?
(b) Draw a Lewis structure for the molecule, and suggest a possible molecular structure. Give the bond angles in the molecule and the hybridization of the boron atom.

103. In chemical research we often send newly synthesized compounds to commercial laboratories for analysis. These laboratories determine the weight percentage of C and H by burning the compound and collecting the evolved CO_2 and H_2O. They determine the molar mass by measuring the osmotic pressure of a solution of the compound. Calculate the empirical and molecular formulas of a compound, C_xH_yCr, given the following information:
(a) The compound contains 73.94% C and 8.27% H; the remainder is chromium.
(b) At 25 °C, the osmotic pressure of 5.00 mg of the unknown dissolved in exactly 100 mL of chloroform solution is 3.17 mm Hg.

INTERACTIVE GENERAL CHEMISTRY CD-ROM

Screen 14.2 Solubility

(a) Examine the video that plays when the "Unsaturated" button is clicked. What is the evidence that the final solution of nickel chloride is unsaturated?
(b) How can you test a solution to see if it is supersaturated?

Screen 14.3 The Solution Process

Examine the problem screen (in particular, the video sequence) associated with this screen.
(a) In which solvent is I_2 more soluble: H_2O or CCl_4? Could you have predicted this?
(b) What do you think will be found if the solubility of hexane (C_6H_{14}) is examined in these same two solvents: water and carbon tetrachloride? In which is hexane more soluble?

Screen 14.5 Henry's Law

Click on the button for the audio explanation of Henry's law. What is meant by a "dynamic equilibrium"?

Screen 14.7 Colligative Properties (1)

(a) Why is the vapor pressure of a liquid lowered by a dissolved solute?
(b) Why, on the molecular scale, is the vapor pressure of a solution proportional to the mole fraction of solvent?
(c) What substance would have the greatest influence on the vapor pressure of water when added to 1000 g of the liquid: 10.0 g of sucrose ($C_{12}H_{22}O_{11}$), 10.0 g of ethylene glycol [$HOCH_2CH_2OH$], or 10.0 g of AgCl?

Screen 14.8 Colligative Properties (2)

(a) Explain why the boiling point of a liquid is elevated on adding a solute. (Click on the animation for boiling point.)
(b) Which should lower the freezing point to a greater degree: 0.10 *m* NH_4NO_3 or 0.10 *m* $HOCH_2CH_2OH$?

Screen 14.9 Osmosis

(a) This screen explains the observations first seen on the *Chemical Puzzler* screen. Osmotic pressure is found to be responsible for the changes in the egg's size. What part of the egg acts as a semipermeable membrane?
(b) If the egg were put in concentrated salt water, what would happen to its size?
(c) Suppose you have two solutions separated by a semipermeable membrane. One contains 5.85 g of NaCl dissolved in 100 mL of solution and the other 8.88 g of KNO_3 dissolved in 100 mL of solution. In which direction will solvent flow: from the NaCl solution to the KNO_3 solution or from KNO_3 to NaCl? Explain briefly.

Screen 14.11 Surfactants

(a) Explain how surfactants act to help oil and water form a colloid.
(b) Explain how a fabric softener works.

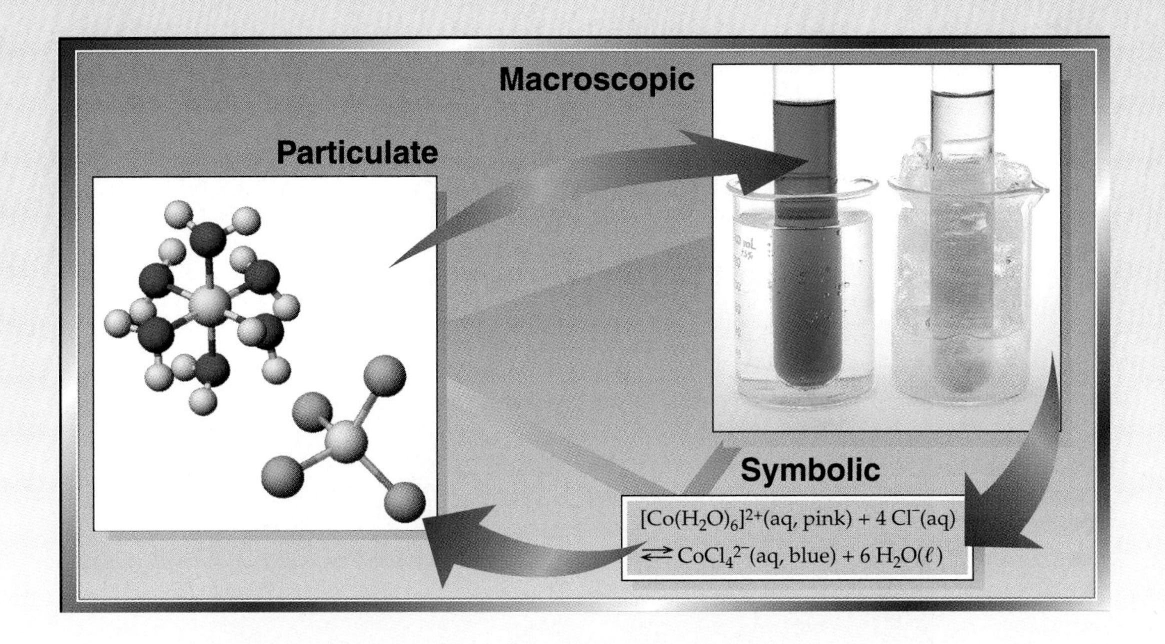

Part 4

The Control of Chemical Reactions

New compounds are created by allowing elements and compounds to react with one another. To many scientists, this is the essence of chemistry. But, if a reaction occurs, how rapidly does it occur and how does it occur? This is the subject of Chapter 15, a study of the rates and mechanisms of chemical reactions.

When a reaction occurs, it can reach a state of equilibrium. After defining chemical equilibria in Chapter 16, we shall proceed to study reactions in aqueous solution in particular. We are especially interested in reactions involving acids and bases (Chapters 17 and 18) because biochemical and other natural systems are so highly dependent on acid–base reactions. Furthermore, many environmental reactions involve the precipitation and dissolving of insoluble compounds, so this is the subject of Chapter 19.

To tie together the discussion of chemical equilibria, we shall further explore the science of thermodynamics (Chapter 20), and you will learn to predict if a given chemical reaction can occur.

As a final topic in this section, we shall describe in more detail another major class of reactions: oxidation–reduction reactions (Chapter 21). Because such reactions can result in the flow of an electric current, we shall describe the chemistry of batteries in this chapter. ● *(Photo, C. D. Winters; model, S. M. Young)*

Chapter 15

Principles of Reactivity: Chemical Kinetics

(Photo, C. D. Winters; model, S. M. Young)

A Chemical Puzzler

Automobiles in the United States and in other parts of the world must now be equipped with catalytic converters in their exhaust systems. Internal combustion engines produce gases such as NO and CO in addition to the usual products of combustion (CO_2 and H_2O). Because NO and CO contribute to air pollution, the exhaust gases are sent through a catalytic converter to change them to nonpolluting gases such as N_2, O_2, CO_2, and others. How does a catalytic converter work? In spite of the fact that the converter is the site of continuous chemical reactions when the engine is running, the converter does not need to be refilled. Chemicals in the catalytic converter are not used up. Why is this so?

Methane is the primary component of natural gas. The oxidation of methane to form CO_2 and H_2O is product-favored and is used to produce an important fraction of the energy required for heating of homes. Other petroleum products, including gasoline, kerosene, and fuel oil, react with oxygen in a similar fashion in product-favored reactions (Figure 15.1). Methane and other hydrocarbons, however, can coexist with atmospheric oxygen for a very long time without a reaction occurring. Without an external stimulus, such as a spark, these oxidation reactions are very slow.

Similarly, metals such as iron and aluminum are widely used in current technology. All metals react with oxygen in product-favored reactions to form metal oxides (see Figure 15.1). Fortunately these reactions are also slow; if the formation of metal oxides occurred rapidly this would present major problems.

When carrying out a chemical reaction, chemists are concerned with two issues: the *rate* at which the reaction proceeds and the *extent* to which the reaction is product-favored. Chapter 6 provided some clues to address the second question, and Chapters 16 and 20 will develop that topic further. In this chapter we turn to the other part of the equation, **chemical kinetics,** a study of the rates of chemical reactions.

See the *Saunders Interactive General Chemistry CD-ROM,* Screen 6.3, for a discussion of thermodynamics and kinetics.

The study of kinetics is divided into two parts. The first part is at the *macroscopic level,* which addresses rates of reactions: what reaction rate means, how to determine a reaction rate by doing experiments, and how factors such as the concentrations of reactants and temperature influence rates. The second part of this subject considers chemical reactions at the *particulate level.* Here, the concern is with the collision theory of reactions and with **mechanisms,** the detailed pathways taken by atoms and molecules as a reaction proceeds. The goal is to

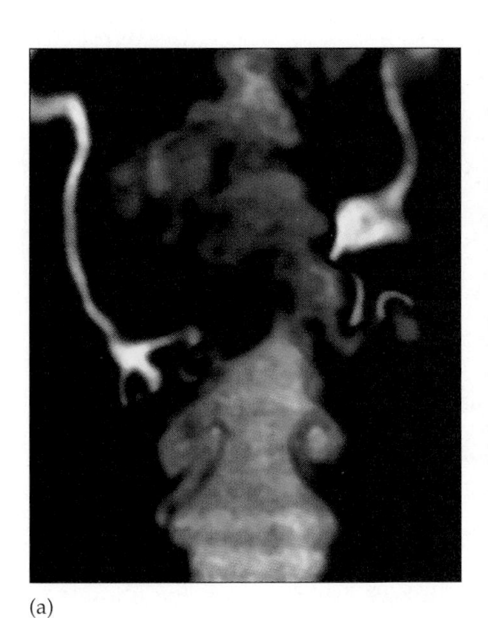

(a)

(b)

(c)

Figure **15.1 Oxidation reactions, both rapid and slow.** (a) Laser-induced fluorescence image of a hydrogen-air jet flame. The fuel is visualized using acetone (CH_3COCH_3) fluorescence (orange) simultaneously with fluorescence from OH radicals (white). (b) Iron dust, sprayed into a bunsen burner flame, oxidizes rapidly. (c) Iron also rusts in air, a much slower oxidation process. *(a, Courtesy Sandia National Laboratories; b and c, C. D. Winters)*

reconcile data in the macroscopic world of chemistry with an understanding of how and why chemical reactions occur at the particulate level—and then further to apply this information to control important reactions.

15.1 RATES OF CHEMICAL REACTIONS

See the *Saunders Interactive General Chemistry CD-ROM,* Screen 15.2, Rates of Chemical Reactions.

The concept of rate is encountered in many nonchemical circumstances. Common examples of rates are the speed of an automobile in terms of the distance traveled per unit time (e.g., kilometers per hour); the rate of flow of water from a faucet as volume per unit time (liters per minute); and the rate of growth of the population in terms of the number of births per day. In each case a change is measured over an interval of time. The **rate of a chemical reaction** refers to the change in concentration of a substance per unit of time. During a chemical reaction, the concentration of the reactants decreases with time, and the concentration of the products increases. It is possible to describe the rate of reaction based on either the increase in concentration of a product or the decrease of the concentration of a reaction per unit of time.

An easy way to gauge the speed of an automobile is to measure how far it travels during a time interval. Two measurements are made: distance traveled and time elapsed. The speed is the distance traveled divided by the time elapsed, or $\Delta(\text{distance})/\Delta(\text{time})$. If an automobile travels 2.4 miles in 4.5 min (0.075 h), its speed is (2.4 miles/0.075 h), or 32 mph.

The Greek letter Δ (delta) means that a change in some quantity has been measured. As usual, Δ = final − initial.

Chemical reaction rates are determined in a similar manner. Two quantities, concentration and time, must be measured. The rate of the reaction can then be described as the decrease in concentration of a reactant, or the increase in concentration of a product per unit time, $\Delta(\text{concentration})/\Delta(\text{time})$. (This is, of course, the average rate for the defined time interval. The rate at any given instant may be different. We will discuss this issue further shortly).

For a rate study, the concentration of a substance undergoing reaction can be determined by a variety of methods. Concentrations can sometimes be measured directly, using a pH meter, for example. Often, concentrations are obtained by measuring a property such as the absorbance of light that is related to concentration (Figure 15.2).

Consider the decomposition of N_2O_5 in a solvent, liquid carbon tetrachloride. This reaction occurs according to the following equation:

$$2\ N_2O_5(\text{solvent}) \longrightarrow 4\ NO_2(\text{solvent}) + O_2(g)$$

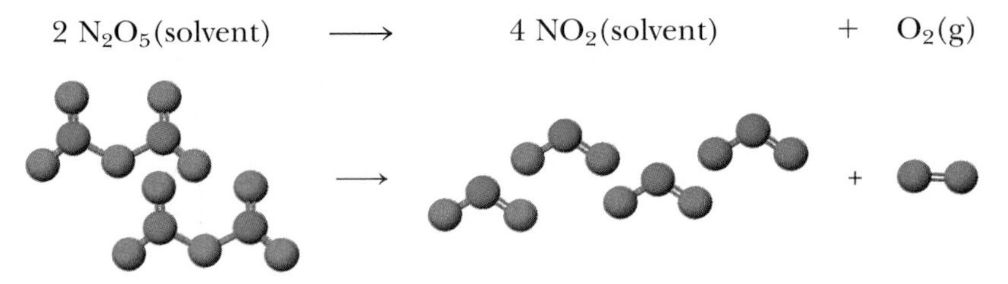

The progress of this reaction can be followed in a number of ways, one of which is by monitoring the increase in O_2 pressure. The number of moles of O_2 formed (calculated from measured values of P, V, and T), is related to the amount of N_2O_5 that has decomposed: for every 1 mol of O_2 formed, 2 mol of N_2O_5 decomposed. The N_2O_5 concentration in solution at a given time equals the initial concentration of N_2O_5 minus the amount decomposed. Data for a typical experiment done at 30.0 °C are presented as a graph of concentration of N_2O_5 versus time in Figure 15.3.

(a)

(b)

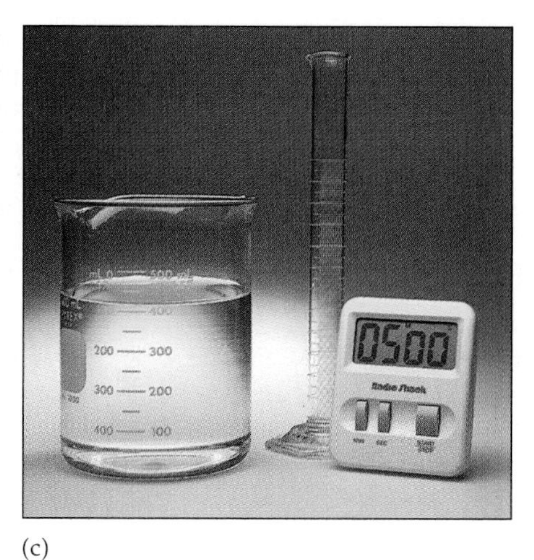

(c)

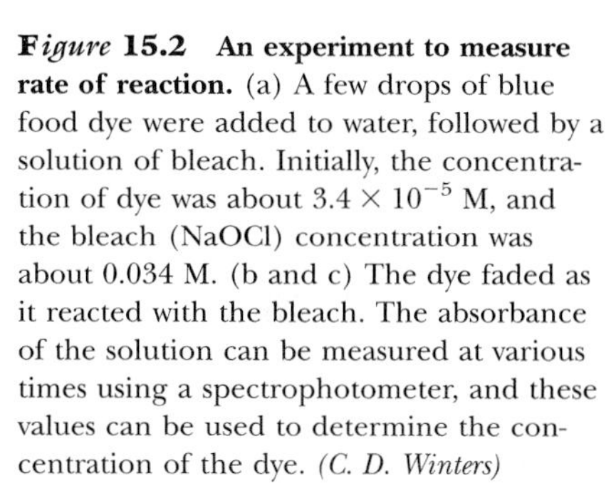

Figure 15.2 An experiment to measure rate of reaction. (a) A few drops of blue food dye were added to water, followed by a solution of bleach. Initially, the concentration of dye was about 3.4×10^{-5} M, and the bleach (NaOCl) concentration was about 0.034 M. (b and c) The dye faded as it reacted with the bleach. The absorbance of the solution can be measured at various times using a spectrophotometer, and these values can be used to determine the concentration of the dye. *(C. D. Winters)*

The rate of this reaction in any interval of time can be expressed as the change in concentration of N_2O_5 divided by the change in time,

$$\text{Rate of reaction} = \frac{\text{Change in } [N_2O_5]}{\text{Change in time}} = -\frac{\Delta[N_2O_5]}{\Delta t}$$

The minus sign is needed because the concentration of N_2O_5 decreases with time, and the rate is always expressed as a positive quantity.

The rate could also be expressed in terms of the rate of formation of NO_2 or the rate of formation of O_2. Rates expressed in these ways will have a positive sign because the concentration is increasing. The rate of formation of NO_2 is twice the rate of decomposition of N_2O_5 because the balanced chemical equation tells us that 2 mol of NO_2 is formed from 1 mol of N_2O_5. The rate of formation of O_2 is one half of the rate of decomposition of N_2O_5 because half a

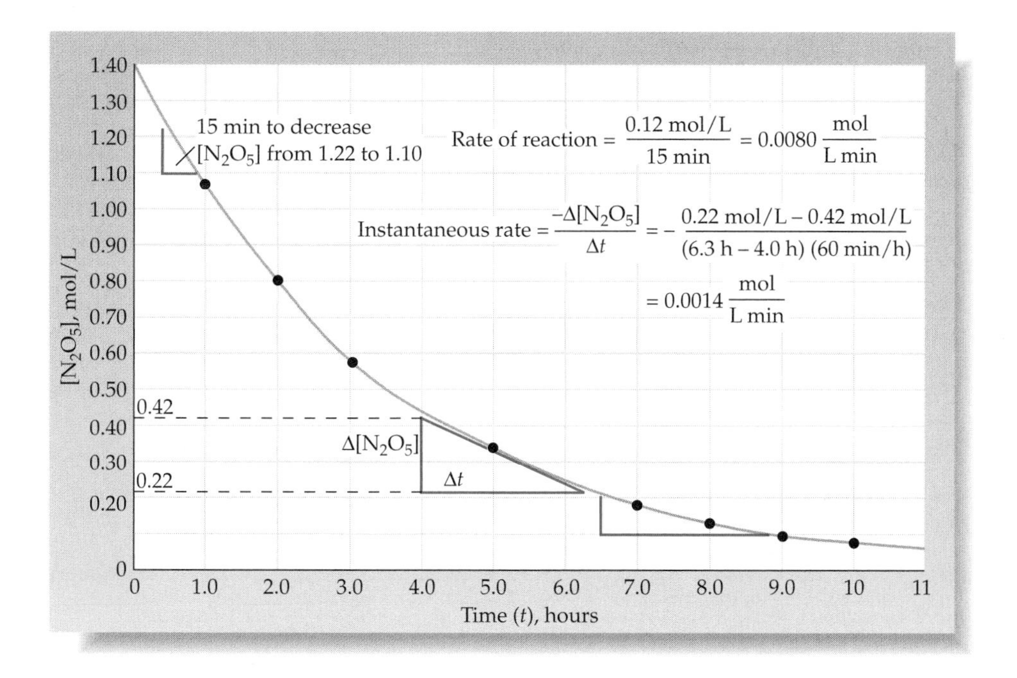

Figure 15.3 A plot of reactant concentration versus time for the decomposition of N_2O_5. The average rate for a 15-min interval from 45 min to 1 h is 0.0080 mol/L · min. The instantaneous rate calculated when $[N_2O_5] = 0.34$ M is 0.0014 mol/L · min.

mole of O_2 is formed per mole of N_2O_5 decomposed. For example, the rate of disappearance of N_2O_5 between 40 min and 55 min (see Figure 15.3) is given by

$$-\frac{\Delta[N_2O_5]}{\Delta t} = -\frac{(1.10 \text{ mol/L}) - (1.22 \text{ mol/L})}{55 \text{ min} - 40 \text{ min}} = +\frac{0.12 \text{ mol/L}}{15 \text{ min}}$$

$$= 0.0080 \frac{\text{mol } N_2O_5 \text{ consumed}}{\text{L} \cdot \text{min}}$$

Expressing the rate in terms of the rate of appearance of NO_2 produces a rate that is twice the rate of disappearance of N_2O_5.

$$\text{Rate} = \frac{\Delta[NO_2]}{\Delta t} = \frac{0.0080 \text{ mol } N_2O_5 \text{ consumed}}{\text{L} \cdot \text{min}} \cdot \frac{2 \text{ mol } NO_2 \text{ formed}}{1 \text{ mol } N_2O_5 \text{ consumed}}$$

$$= 0.016 \frac{\text{mol } NO_2 \text{ formed}}{\text{L} \cdot \text{min}}$$

In terms of the rate at which O_2 is formed the rate of the reaction is

$$\text{Rate} = \frac{\Delta[O_2]}{\Delta t} = \frac{0.0080 \text{ mol } N_2O_5 \text{ consumed}}{\text{L} \cdot \text{min}} \cdot \frac{\frac{1}{2} \text{ mol } O_2 \text{ formed}}{1 \text{ mol } N_2O_5 \text{ consumed}}$$

$$= 0.0040 \frac{\text{mol } O_2 \text{ formed}}{\text{L} \cdot \text{min}}$$

The graph of concentration versus time on Figure 15.3 is not a straight line because the rate of the reaction changes during the course of the reaction. The concentration of N_2O_5 decreases rapidly at the beginning of the reaction but more slowly near the end. We can verify this by comparing the rate of disappearance of N_2O_5 calculated previously (the concentration decreased by 0.12 mol/L in 15 min) with the rate of reaction calculated for the time interval from 6.5 h to 9.0 h (when the concentration drops by 0.12 mol/L in 2.5 h).

$$-\frac{\Delta[N_2O_5]}{\Delta t} = -\frac{(0.1 \text{ mol/L}) - (0.22 \text{ mol/L})}{540 \text{ min} - 390 \text{ min}} = +\frac{0.12 \text{ mol/L}}{150 \text{ min}}$$

$$= 0.00080 \frac{\text{mol}}{\text{L} \cdot \text{min}}$$

The rate in this later stage of this reaction has dropped to only one tenth of its previous value.

The procedure we used to calculate rate gives us the **average rate** over the chosen time interval. We might also ask what the **instantaneous rate** is at a single point in time. The instantaneous rate is determined by drawing a line tangent to the concentration–time curve at a particular time (see Figure 15.2) and obtaining the rate from the slope of this line. For example, when $[N_2O_5] = 0.34$ mol/L and $t = 5.0$ h, the rate is

$$\text{Rate when } [N_2O_5] \text{ is } 0.34 \text{ M} = -\frac{\Delta[N_2O_5]}{\Delta t}$$

$$= -\frac{(0.22 \text{ mol/L}) - (0.42 \text{ mol/K})}{(6.3 \text{ h} - 4.0 \text{ h})(60 \text{ min/h})} = +\frac{0.20 \text{ mol/L}}{140 \text{ min}} = 1.4 \times 10^{-3} \frac{\text{mol}}{\text{L} \cdot \text{min}}$$

This shows that N_2O_5 is being consumed, at that moment in time, at a rate of 0.0014 mol/L · min.

The difference between an average rate and an instantaneous rate has an analogy in the speed of an automobile. In the previous example, the car traveled 2.4 miles in 4.5 min for an average speed of 32 mph. At any instant in time, however, the car may have moved much slower or much faster. The instantaneous speed at any instant is indicated by the car's speedometer.

Example 15.1 Reaction Rates and Stoichiometry

Give the relative rates for disappearance of reactants and formation of products for the following reaction:

$$4\ PH_3(g) \longrightarrow P_4(g) + 6\ H_2(g)$$

Solution In this reaction 4 mol of PH_3 disappears when 1 mol of P_4 and 6 mol of H_2 are formed. To equate rates, we must divide $\Delta[\text{reagent}]/\Delta t$ by the stoichiometric coefficient in the balanced equation:

$$\text{Reaction rate} = -\frac{1}{4}\left(\frac{\Delta[PH_3]}{\Delta t}\right) = +\frac{\Delta[P_4]}{\Delta t} = +\frac{1}{6}\left(\frac{\Delta[H_2]}{\Delta t}\right)$$

$$= -\frac{1}{4}(\text{rate of change of } [PH_3])$$

$$= \text{rate of change of } [P_4]$$

$$= \frac{1}{6}(\text{rate of change of } [H_2])$$

Because 4 mol of PH_3 disappears for every mole of P_4 formed, the numerical value of the rate of formation of P_4 is only one fourth of the rate of disappearance of PH_3. Similarly, P_4 is formed at only one sixth of the rate that H_2 is formed.

Example 15.2 Rate of Reaction

Data collected on the concentration of dye as a function of time (see Figure 15.2) are given in the graph. What is the average rate of change of the dye concentration over the first 2 min? What is the average rate of change during the fifth minute (from $t = 4$ to $t = 5$)? Estimate the instantaneous rate at 4 min.

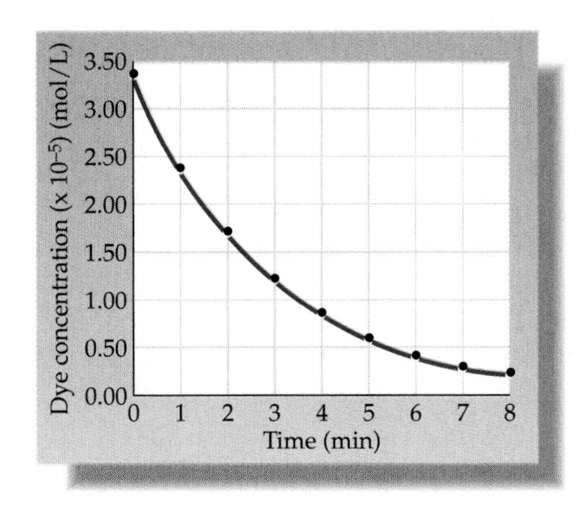

The decrease in concentration of dye as a function of time (see Figure 15.2).

Solution The concentration of dye decreases from 3.4×10^{-5} M at $t = 0$ min, to 1.7×10^{-5} M at $t = 2$ min. The *average rate* of the reaction in this interval of time is

$$-\frac{\Delta[\text{Dye}]}{\Delta t} = -\frac{(1.7 \times 10^{-5}\ \text{mol/L}) - (3.4 \times 10^{-5}\ \text{mol/L})}{2\ \text{min}} = \frac{8.5 \times 10^{-6}\ \text{mol}}{\text{L} \cdot \text{min}}$$

The concentration of dye decreases from 0.90×10^{-5} M at $t = 4$ min, to 0.60×10^{-5} M at $t = 5$ min. The *average rate* of the reaction in this interval of time is

$$-\frac{\Delta[\text{Dye}]}{\Delta t} = -\frac{(0.6 \times 10^{-5}\ \text{mol/L}) - (0.9 \times 10^{-5}\ \text{mol/L})}{1\ \text{min}} = \frac{3.0 \times 10^{-6}\ \text{mol}}{\text{L} \cdot \text{min}}$$

To find the instantaneous rate at 4 min, it is necessary to draw a line tangent to the line on the graph at this point. The slope of this line is the instantaneous rate. The approximate value is -3.5×10^{-6} mol/L · min.

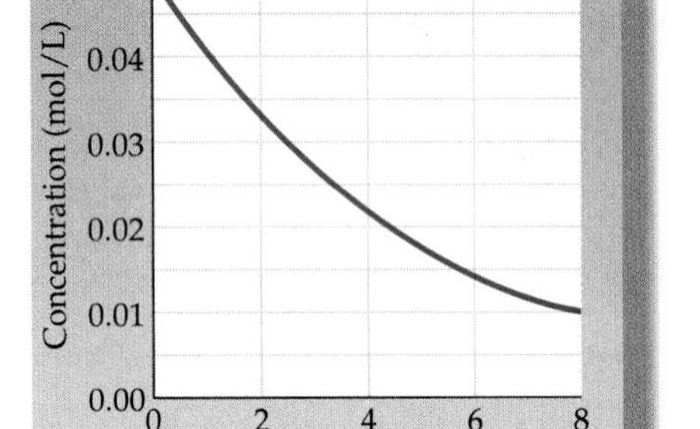

See Exercise 15.2.

Exercise 15.1 Reaction Rates and Stoichiometry

What are the relative rates of appearance or disappearance of each product and reactant, respectively, in the decomposition of nitrosyl chloride, NOCl?

$$2\ \text{NOCl(g)} \longrightarrow 2\ \text{NO(g)} + \text{Cl}_2\text{(g)}$$

Exercise 15.2 Rate of Reaction

Sucrose decomposes to fructose and glucose in acid solution. A plot of the concentration of sucrose as a function of time is given at left. What is the rate of change of the sucrose concentration over the first 2 h? What is the rate of change over the last 2 h? Estimate the instantaneous rate at 4 h.

15.2 REACTION CONDITIONS AND RATE

For a chemical reaction to occur, molecules of the reactants must come together so that atoms can be exchanged or rearranged. Atoms and molecules are mobile in the gas phase or in solution, and so reactions are often carried out using a mixture of gases or using solutions of reactants. Under these circumstances, several factors affect the speed of a reaction.

- *Concentrations of reactants.* Alka-Seltzer contains $NaHCO_3$ and citric acid and when a tablet is placed in water these compounds react to give CO_2 (Figure 15.4). The reaction is faster in pure water than in an ethanol–water mixture in which the concentration of water is low.
- *Temperature.* Cooking involves chemical reactions, and a higher temperature results in foods cooking faster. In the laboratory the reaction mixture is often heated to make reactions occur faster (Figure 15.5).
- *Catalysts.* Catalysts are substances that accelerate chemical reactions but are not themselves transformed. For example, hydrogen peroxide, H_2O_2, decomposes to water and oxygen,

$$2\ \text{H}_2\text{O}_2(\ell) \longrightarrow \text{O}_2\text{(g)} + 2\ \text{H}_2\text{O}(\ell)$$

Figure 15.4 Dependence of reaction rate on concentration. The rate of reaction depends on the concentrations of the reactants. Here an Alka-Seltzer tablet is placed in pure water *(right)* or in ethanol to which a trace of water has been added *(left)*. Water is a reactant, and the reaction rate is much greater when the concentration of water is high. *(C. D. Winters)*

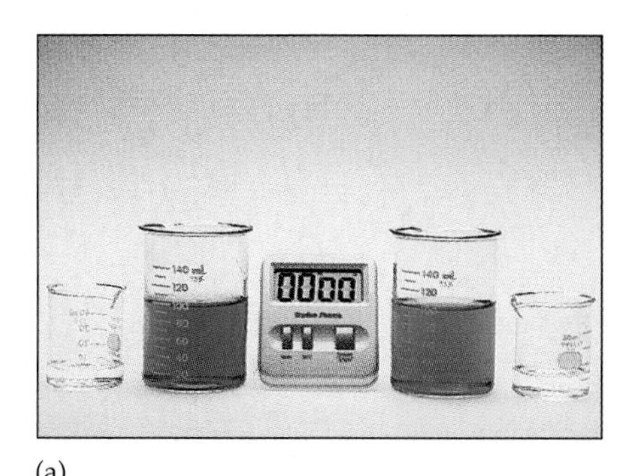

(a)

(b)

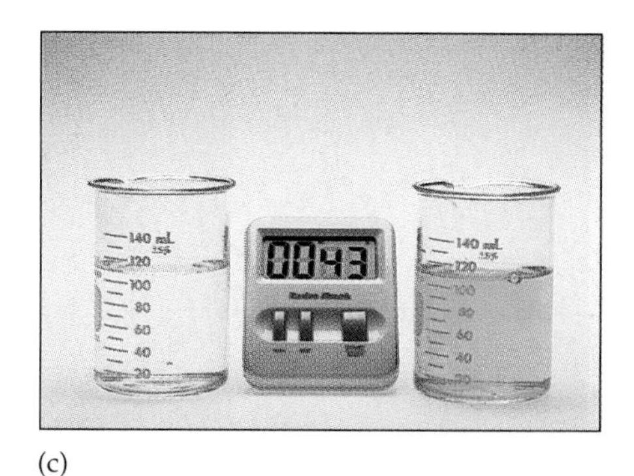

(c)

***Figure* 15.5 Dependence of reaction rate on temperature.** Bleach, at two different temperatures, is poured into water containing blue dye. (a) *(far left)* Bleach at 54 °C. *(far right)* Bleach at 22 °C. (b) The bleach is poured into the two solutions. (c) The dye treated with warm bleach *(left)* has turned almost colorless after 2 min, whereas the solution into which the cooler bleach was poured *(right)* is still blue. *(C. D. Winters)*

but a solution of H_2O_2 can be stored for many months because the rate of the decomposition reaction is extremely slow. Adding a manganese salt, an iodide-containing salt, or a biological substance called an *enzyme,* however, causes this reaction to occur rapidly, as shown by vigorous bubbling as gaseous oxygen escapes from the solution (Figure 15.6).

See the *Saunders Interactive General Chemistry CD-ROM,* Screens 15.3 and 15.4, Control of Reaction Rates.

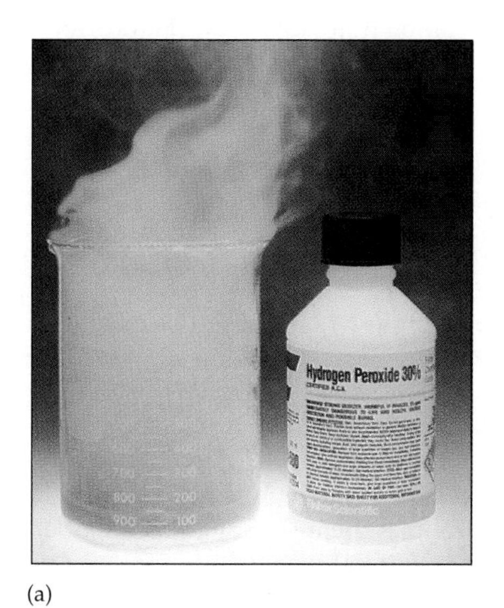

(a)

(b)

(c)

***Figure* 15.6 Catalyzed decomposition of H_2O_2.** (a) The rate of decomposition of hydrogen peroxide is increased by the catalyst MnO_2. Here a 30% solution of H_2O_2, poured onto the black solid MnO_2, rapidly decomposes to O_2 and H_2O. Steam forms because of the high heat of reaction. (b) A bombardier beetle uses the catalyzed decomposition of H_2O_2 as a defense mechanism. The heat of the reaction lets the insect eject steam and other irritating chemicals with explosive force. (c) A naturally occurring catalyst, called an enzyme, decomposes hydrogen peroxide. Here the enzyme found in a potato is used, and bubbles of O_2 gas are seen rising in the solution. *(a and c, C. D. Winters; b, Thomas Eisner with Daniel Aneshansley, Cornell University)*

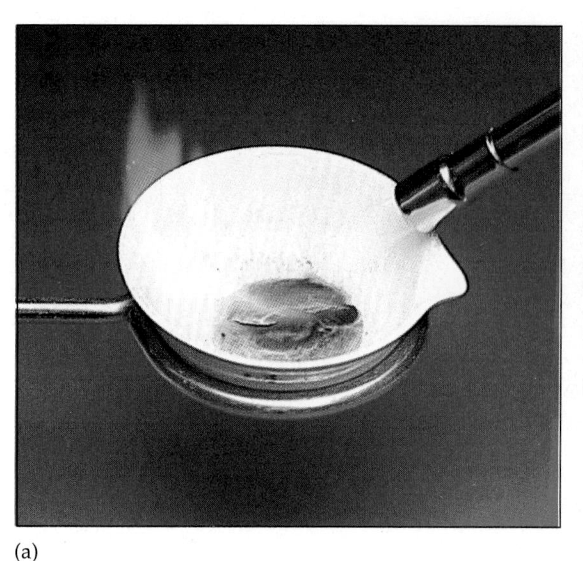

(a)

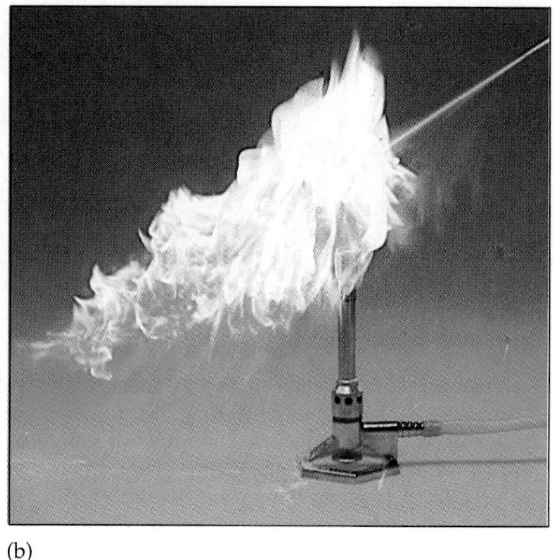

(b)

Figure 15.7 The combustion of lycopodium powder. (a) The spores of this common fern burn only with difficulty when piled in a dish. (b) If the spores are ground to a fine powder and sprayed into a flame, combustion is rapid. *(C. D. Winters)*

The surface area of a solid reactant can also affect the reaction rate. Only molecules at the surface of a solid can come in contact with other reactants. The smaller the particles of a solid, the more molecules there are on the surface. With very small particles the effect on rate can be quite dramatic (Figure 15.7). Farmers know that dust explosions (in an enclosed silo or at a feed mill) represent a major hazard.

15.3 EFFECT OF CONCENTRATION ON REACTION RATE

See the *Saunders Interactive General Chemistry CD-ROM,* Screen 15.4, Concentration Dependence.

It is often possible to change the rate of a reaction by changing the concentrations of reactants. One goal in studying kinetics is to determine how concentrations of reactants affect the reaction rate. The effect can be determined by evaluating the rate of a reaction using different concentrations of each reactant (with temperature held constant). Consider, for example, the decomposition of N_2O_5 to NO_2 and O_2. Figure 15.3 presented data on the concentration of N_2O_5 as a function of time. We previously calculated that the instantaneous rate of disappearance of N_2O_5 when $[N_2O_5] = 0.34$ mol/L is 0.0014 mol/L · min. An evaluation of the instantaneous rate of the reaction when $[N_2O_5] = 0.68$ mol/L shows a rate of 0.0028 mol/L · min. Doubling the concentration of N_2O_5 has clearly doubled the reaction rate. A similar exercise shows that if $[N_2O_5]$ is 0.17 mol/L, the reaction rate is halved. These results tell us that the reaction rate is directly proportional to reactant concentration for this reaction. That is,

$$\text{Rate of reaction} \propto [N_2O_5]$$

where the symbol $\propto$ means "proportional to."

Different relationships between reaction rate and reactant concentration are encountered in other reactions. For example, the reaction rate could be independent of concentration, or the rate may depend on the concentration of some reactant raised to some power (that is, $[\text{reactant}]^n$). Finally, if there are several reactants, the rate of reaction may depend on the concentrations of each reactant.

Rate Equations

The relationship between reactant concentration and reaction rate is expressed by an equation called a **rate equation,** or **rate law.** For the N_2O_5 decomposition reaction the rate equation is

$$\text{Rate of reaction} = k[N_2O_5]$$

where the proportionality constant, k, is called the **rate constant.** This rate equation tells us that the reaction rate is proportional to the concentration of the reactant.

In general, for a reaction such as

$$\text{a A} + \text{b B} \xrightarrow{\text{C}} \text{x X}$$

A catalyst does not appear as a reactant in the balanced, overall equation for the reaction, but it may appear in the rate expression. It is common practice to indicate catalysts by placing them above the reaction arrow, as shown in the examples.

where C is a homogeneous catalyst, the rate equation has the form

$$\text{Rate} = k[\text{A}]^m[\text{B}]^n[\text{C}]^p$$

The rate equation expresses the fact that the rate of reaction is proportional to the reactant concentrations (and perhaps the catalyst concentration), each concentration being raised to some power. It is important to recognize that the exponents, m, n, and p in this case, are *not necessarily the stoichiometric coefficients* for the balanced chemical equation. The exponents must be determined by experiment. They are often positive whole numbers, but they can be negative numbers, fractions, or zero.

Consider another example, the decomposition of hydrogen peroxide in the presence of a catalyst such as iodide ion.

$$2\ H_2O_2(\text{aq}) \xrightarrow{I^-(\text{aq})} 2\ H_2O(\ell) + O_2(\text{g})$$

Experiments show that the reaction has the following rate equation:

$$\text{Reaction rate} = k[H_2O_2][I^-]$$

Here the exponent on each concentration term is 1, even though the stoichiometric coefficient of H_2O_2 is 2 and I^- does not appear in the balanced equation.

The Rate Constant, k

The rate constant, k, is a proportionality constant that relates rate and concentration *at a given temperature.* It is an important quantity because it enables you to find the reaction rate for a new set of concentrations. To see how to use k, consider the substitution of Cl^- ion by water in the cancer chemotherapy agent cisplatin, $Pt(NH_3)_2Cl_2$. The rate expression for this reaction is

$$\text{Rate} = k[Pt(NH_3)_2Cl_2]$$

$$Pt(NH_3)_2Cl_2(aq) + H_2O(\ell) \longrightarrow [Pt(NH_3)_2(H_2O)Cl]^+(aq) + Cl^-(aq)$$

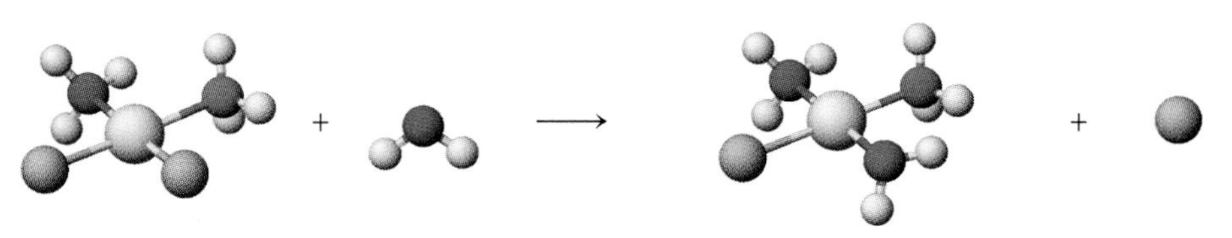

and the rate constant, k, is 0.090/h. A knowledge of k allows us to calculate the rate at a particular reactant concentration, for example when $[Pt(NH_3)_2Cl_2] =$ 0.018 mol/L:

$$\text{Rate} = (0.090/\text{h})(0.018 \text{ mol/L}) = 0.0016 \text{ mol/L} \cdot \text{h}$$

The Order of a Reaction

The order with respect to a particular reactant is the exponent of its concentration term in the rate expression, and the total reaction order is the sum of the exponents on all concentration terms. The rate equation for the decomposition of N_2O_5

$$2\ N_2O_5 \longrightarrow 4\ NO_2 + O_2$$
$$\text{Rate} = k[N_2O_5]$$

has an exponent of 1 on $[N_2O_5]$, which means the reaction is first order with respect to N_2O_5. If the concentration of N_2O_5 is doubled, the rate of reaction doubles. If the concentration of N_2O_5 decreases by half, the rate is half as fast.

Consider another example, the reaction of NO and Cl_2:

$$2\ NO(g) + Cl_2(g) \longrightarrow 2\ NOCl(g)$$

The rate equation for this reaction is

$$\text{Rate} = k\ [NO]^2[Cl_2]$$

The reaction is second order in NO, first order in Cl_2, and third order overall. Experimental data at 50 °C shown in the following table indicate that the reaction rate, $\Delta[NOCl]/\Delta t$, is 1.43×10^{-6} mol/L · s when $[NO] = [Cl_2] = 0.250$ mol/L.

Experiment	[NO] mol/L	$[Cl_2]$ mol/L	Rate mol/L · s
1	0.250	0.250	1.43×10^{-6}
2	0.500	0.250	5.72×10^{-6}
3	0.250	0.500	2.86×10^{-6}
4	0.500	0.500	11.4×10^{-6}

If $[Cl_2]$ is held constant and [NO] is doubled to 0.500 mol/L (Experiment 2), then the reaction rate increases by a factor of 4 (to 5.72×10^{-6} mol/L · s). If [NO] is held constant and $[Cl_2]$ is doubled to 0.500 mol/L (Experiment 3), the rate is doubled (to 2.86×10^{-6} mol/L · s). If both [NO] and $[Cl_2]$ are doubled (Experiment 4), then the rate is 8 times the original value (11.4×10^{-5} mol/L · s).

The decomposition of ammonia on a platinum surface at 856 °C is interesting because it is zero order:

$$2\ NH_3(g) \longrightarrow N_2(g) + 3\ H_2(g)$$

This means the reaction is independent of NH_3 concentration:

$$\text{Rate} = k[NH_3]^0 = k$$

The reaction order is important because it gives some insight into the most interesting question of all—how the reaction occurs. This is described further in Section 15.6.

Determination of the Rate Equation

See the *Saunders Interactive General Chemistry CD-ROM,* Screen 15.5, Determination of the Rate Equation.

The relation between rate and concentration must be determined experimentally. One way to do this is the method of initial rates. The **initial rate** is the instantaneous reaction rate at the start of the reaction (the rate at $t = 0$). An approximate value of the initial rate can be obtained by mixing the reactants and determining $\Delta[\text{product}]/\Delta t$ or $-\Delta[\text{reactant}]/\Delta t$ after 1% to 2% of the limiting reactant has been consumed. Measuring the rate during the initial stage of a reaction is convenient because initial concentrations are known, and this method avoids possible complications arising from interference by reaction products or the occurrence of other reactions.

As an example of the determination of a reaction rate by this method, let us look at the reaction of sodium hydroxide with methyl acetate to produce acetate ion and methyl alcohol.

Methyl acetate is an ester. For more on ester hydrolysis, see page 515.

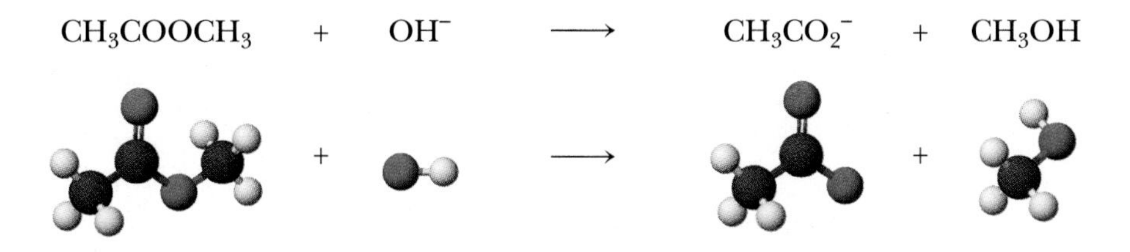

Data in the table were collected for several experiments at 25 °C:

Experiment	Initial Concentrations $[CH_3CO_2CH_3]$	$[OH^-]$	Initial Reaction Rate (mol/L · s) at 25 °C
1	0.050 M	0.050 M	0.00034
	↓ no change	↓ × 2	↓ × 2
2	0.050 M	0.10 M	0.00069
	↓ × 2	↓ no change	↓ × 2
3	0.10 M	0.10 M	0.00137

When the initial concentration of one reactant, either $CH_3CO_2CH_3$ or OH^-, is doubled, and the concentration of the other reactant is held constant, the initial reaction rate doubles. This rate doubling shows that the rate for the reaction is directly proportional to the concentration of *both* $CH_3CO_2CH_3$ and OH^- so the reaction is first order in each of these reactants. The rate equation that reflects these experimental observations is

$$\text{Rate} = k[CH_3CO_2CH_3][OH^-]$$

Using this equation we can predict that doubling *both* concentrations at the same time should cause the rate to go up by a factor of 4. What happens, however, if one concentration is doubled and the other halved? The rate equation tells us the rate should not change, and that is what is observed!

The value for k, the rate constant, can be found by substituting values of rate and concentration into the rate equation. To find k for the methyl acetate/hydroxide ion reaction, for example, data from one of the experiments are substituted into the rate equation. Using the data from the first experiment, we have

$$\text{Rate} = 0.00034 \text{ mol/L} \cdot \text{s} = k(0.050 \text{ mol/L})(0.050 \text{ mol/L})$$

$$k = \frac{0.00034 \cancel{\text{mol/L}} \cdot \text{s}}{(0.050 \text{ mol/L})(0.050 \cancel{\text{mol/L}})} = 0.14 \text{ L/mol} \cdot \text{s}$$

Example 15.3 Determining a Rate Equation

The rate of the reaction between CO and NO_2

$$CO(g) + NO_2(g) \longrightarrow CO_2(g) + NO(g)$$

was studied at 540 K starting with various concentrations of CO and NO_2 and the data in the table were collected. Determine the rate equation from these data. What is the value of the rate constant?

	Initial Concentrations		Initial Rate
Experiment	**[CO], mol/L**	**[NO_2], mol/L**	**(mol/L · h)**
1	5.10×10^{-4}	0.350×10^{-4}	3.4×10^{-8}
2	5.10×10^{-4}	0.700×10^{-4}	6.8×10^{-8}
3	5.10×10^{-4}	0.175×10^{-4}	1.7×10^{-8}
4	1.02×10^{-3}	0.350×10^{-4}	6.8×10^{-8}
5	1.53×10^{-3}	0.350×10^{-4}	10.2×10^{-8}

Solution In the first three experiments the concentration of CO is constant. In the second experiment the NO_2 concentration has been doubled, leading to an increase in rate of a factor of 2, and in the third experiment, decreasing [NO_2] to half its original value has caused the rate to decrease by half. These results tell us that the reaction is first order in [NO_2].

The data in Experiments 1 and 4 (with constant [NO_2]) show that doubling [CO] doubles the rate, whereas data from Experiments 1 and 5 show that tripling the concentration triples the rate. This means the reaction is also first order in [CO], and we now know the rate equation is

$$\text{Rate} = k[CO][NO_2]$$

The rate constant k can be found by inserting data for one of the experiments into the rate equation. Using data from Experiment 1, for example, Rate = 3.4×10^{-8} mol/L · h = $k(5.10 \times 10^{-4}$ mol/L$)(0.350 \times 10^{-4}$ mol/L$)$

$$k = 1.9 \text{ L/mol} \cdot \text{h}$$

A better value of k in Example 15.3 is obtained by calculating the value for each experiment and then averaging the values.

Example 15.4 Using a Rate Equation to Determine Rates

Using the rate equation and rate constant determined for the reaction of CO and NO_2 at 540 K in Example 15.3, determine the initial rate of the reaction when [CO] = 3.8×10^{-4} and $[NO_2] = 0.600 \times 10^{-4}$.

Solution Knowing the rate constant (k = 1.9 L/mol · h) and the concentration of each reactant, we can use the rate law to calculate the rate as follows:

$$\text{Rate} = k[\text{CO}][\text{NO}_2]$$
$$= 1.9 \text{ L/mol} \cdot \text{h}(3.8 \times 10^{-4} \text{ mol/L})(0.650 \times 10^{-4} \text{ mol/L})$$
$$= 4.7 \times 10^{-8} \text{ mol/L} \cdot \text{h}$$

It is sometimes useful to make an educated guess at this answer before carrying out the mathematical solution. In this case, the guess would act as a check on the result that we have just calculated. We know that the reaction is first order in both reactants. Comparing concentration values in this problem with the values of concentration in Experiment 1 in Example 15.3, we notice that [CO] is about three fourths of the value, whereas $[NO_2]$ is almost twice the value. The effects do not precisely offset each other, but we might predict that the difference in rates between this experiment and Experiment 1 will be fairly small, with the rate of this experiment being just a little greater. The calculated values bears this out.

Exercise 15.3 Determining a Rate Equation

The initial rate of the reaction of nitrogen monoxide and oxygen

$$2 \text{ NO(g)} + \text{O}_2\text{(g)} \longrightarrow 2 \text{ NO}_2\text{(g)}$$

was measured at 25 °C for various initial concentrations of NO and O_2. Data are collected in the table. Determine the rate equation from these data. What is the value of the rate constant and what are the appropriate units of k?

Experiment	Initial Concentrations (mol/L) [NO]	$[O_2]$	Initial Rate (mol/L · s)
1	0.020	0.010	0.028
2	0.020	0.020	0.057
3	0.020	0.040	0.114
4	0.040	0.020	0.227
5	0.010	0.020	0.014

Exercise 15.4 Using Rate Laws

The rate constant k is 0.090/h for the reaction

$$Pt(NH_3)_2Cl_2(aq) + H_2O(\ell) \longrightarrow [Pt(NH_3)_2(H_2O)Cl]^+(aq) + Cl^-(aq)$$

and the rate equation is

$$\text{Rate} = k[Pt(NH_3)_2Cl_2]$$

Calculate the rate of reaction when the concentration of $Pt(NH_3)_2Cl_2$ is 0.020 M. What is the rate of change in the concentration of Cl^- under these conditions?

15.4 CONCENTRATION–TIME RELATIONSHIPS: INTEGRATED RATE LAWS

See the *Saunders Interactive General Chemistry CD-ROM,* Screen 15.6, Concentration–Time Relationships.

It is sometimes useful or important to know how long a reaction must proceed to reach a predetermined concentration of some reagent, or what the reactant and product concentrations will be after some time has elapsed. One way to do this is to collect experimental data and construct graphs such as that shown in Figure 15.3. These methods can be inconvenient and time-consuming, however. It would be easier if we had a mathematical equation relating concentration and time, an equation in which concentration of the reactant and time were the only unknowns. Such an equation could then be used to calculate a concentration at any given time, or the length of time for a given amount of reactant to react.

In effect, we are asking the following question: Can mathematical equations be written that define relationships between the concentration of a reactant and time, such as that shown graphically in Figure 15.3? The answer is yes, and we will next proceed to develop and use several equations applying to several rate laws.

First-Order Reactions

Suppose the reaction "R $\longrightarrow$ products" is first order. This means the reaction rate is directly proportional to the concentration of R raised to the first power, or, mathematically,

$$\text{Rate} = -\frac{\Delta[R]}{\Delta t} = k[R]$$

Using the methods of calculus, this relationship can be transformed into a very useful equation called an **integrated rate equation** because integral calculus is used in its derivation:

$$\ln \frac{[R]_t}{[R]_0} = -kt \qquad [15.1]$$

Equation 15.1 can also be written $\ln [R]_t = -kt + \ln [R]_0$; see Appendix A (Section A.3) for a further discussion on using logarithms.

Here $[R]_0$ and $[R]_t$ are concentrations of the reactant at time $t = 0$ and at a later time, t ($t = 0$ does not need to correspond to the actual beginning of the experiment; it can be the time when instrument readings were started, for example.) The *ratio* of concentrations, $[R]_t/[R]_0$, is the fraction of reactant that *remains* after a given time has elapsed. In words, the equation says

$$\text{Natural logarithm}\left(\frac{[\text{R after some time}]}{[\text{R at start of experiment}]}\right) = -(\text{rate constant})(\text{elapsed time})$$

Notice the negative sign in the equation. The ratio $[R]_t/[R]_0$ is less than 1 because $[R]_t$ is always less than $[R]_0$. This means the logarithm of $[R]_t/[R]_0$ is negative, and so the other side of the equation must also bear a negative sign. Equation 15.1 is useful in three ways:

- If $[R]_t/[R]_0$ is measured in the laboratory after some amount of time has elapsed, then k can be calculated.
- If $[R]_0$ and k are known, the concentration of material expected to remain after a given amount of time can be calculated.
- If k is known, Equation 15.1 can be used to calculate the time elapsed until R reaches some predetermined concentration.

Finally, notice that the ratio $[R]_t/[R]_0$ is dimensionless, and so the product kt must have no dimensions. This means that k has units of 1/time (1/y or 1/s, for example) and that k for first-order reactions is independent of the units chosen for concentration. [R] can be expressed in any convenient quantity unit—moles per liter, moles, grams, number of atoms, number of molecules, or pressure.

The fraction 1/time can also be written as time^{-1}. For example, y^{-1} or s^{-1}.

Example 15.5 The First-Order Rate Equation

Cyclopropane, C_3H_6, has been used in a mixture with oxygen as an anesthetic. This practice has diminished greatly, however, because the compound is very flammable. When heated, this compound rearranges to propene:

$$\text{Rate} = k[\text{cyclopropane}] \qquad k = 5.4 \times 10^{-2}\ \text{h}^{-1}$$

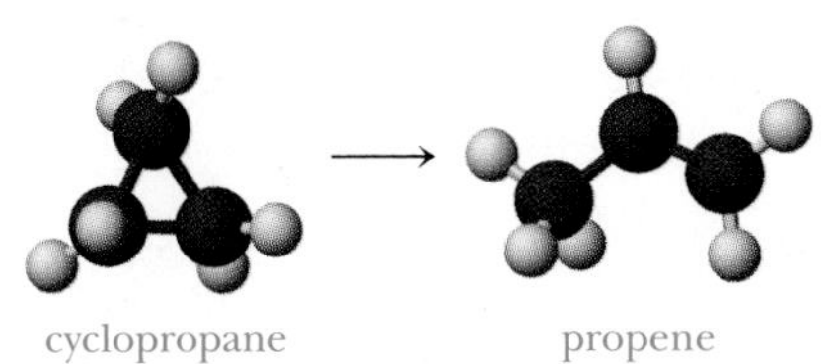

Cycloalkanes with fewer than five carbon atoms are strained because the C—C—C bond angles cannot match the preferred tetrahedral angle of 109.5°. Thus, the ring opens readily, forming propene in the case of cyclopropane (see page 491).

If the initial concentration of cyclopropane is 0.050 mol/L, how many hours must elapse for the concentration of the compound to drop to 0.010 mol/L?

Solution The first-order rate equation applied to this reaction is

$$\ln \frac{[\text{cyclopropane}]_t}{[\text{cyclopropane}]_0} = -kt$$

where $[\text{cyclopropane}]_t$, $[\text{cyclopropane}]_0$, and k are given:

$$\ln \frac{[0.010]}{[0.050]} = -(5.4 \times 10^{-2}\ \text{h}^{-1})t$$

$$\frac{-\ln(0.20)}{5.4 \times 10^{-2}\ \text{h}^{-1}} = t$$

$$\frac{-(-1.61)}{5.4 \times 10^{-2}\ \text{h}^{-1}} = t$$

$$t = 30.\ \text{h}$$

Example 15.6 Using the First-Order Rate Equation

Hydrogen peroxide decomposes in dilute sodium hydroxide at 20 °C in a first-order reaction:

$$2\ H_2O_2(aq) \longrightarrow 2\ H_2O(\ell) + O_2(g)$$

$$\text{Rate} = k[H_2O_2] \quad k = 1.06 \times 10^{-3}\ \text{min}^{-1}$$

If the initial concentration of H_2O_2 is 0.020 mol/L, what is the concentration of the peroxide after exactly 100 min? What is the fraction remaining after exactly 100 min?

Solution Here $[H_2O_2]_0$, k, and t are known, and we are asked to find $[H_2O_2]_t$. One approach is to write Equation 15.1 as

$$\ln\ [H_2O_2]_t - \ln\ [H_2O_2]_0 = -kt$$

Substituting into this equation, we have

$$\begin{aligned} \ln\ [H_2O_2]_t - \ln\ (0.020) &= -(1.06 \times 10^{-3}\ \text{min}^{-1})(100.\ \text{min}) \\ \ln\ [H_2O_2]_t - (-3.91) &= -(1.06 \times 10^{-1}) \\ \ln\ [H_2O_2]_t &= -3.91 - 0.106 \\ &= -4.02 \end{aligned}$$

Taking the antilogarithm of -4.02 (i.e., the inverse of ln (-4.02), or $e^{-4.02}$), we find the concentration of hydrogen peroxide after exactly 100 min is $[H_2O_2]_t =$ 0.018 mol/L.

With the final concentration of H_2O_2 after 100 min having already been calculated, we can calculate the fraction that remains. This is (0.018 mol/L)/(0.020 mol/L), or 0.90 (90%). We could also have calculated this value from Equation 15.1, recognizing that the answer is $[H_2O_2]_t/[H_2O_2]_0$

$$\begin{aligned} \ln \left\{\frac{[H_2O_2]_t}{[H_2O_2]_0}\right\} &= -kt \\ &= -(1.06 \times 10^{-3}\ \text{min}^{-1})(100\ \text{min}) \\ &= -0.106 \end{aligned}$$

Taking the antilog of both sides gives the answer:

$$\text{Fraction remaining after 100 min} = \frac{[H_2O_2]_t}{[H_2O_2]_0} = 0.90$$

Exercise 15.5 Using the First-Order Rate Equation

Sucrose, a sugar, decomposes in acid solution to glucose and fructose. The reaction is first order in sucrose, and the rate constant at 25 °C is $k = 0.21\ \text{h}^{-1}$. If the initial concentration of sucrose is 0.010 mol/L, what is its concentration after 5.0 h?

Exercise 15.6 Using the First-Order Rate Equation

Gaseous NO_2 decomposes when heated

$$2\ NO_2\ (g) \longrightarrow 2\ NO(g) + O_2(g)$$

This is a first-order reaction with $k = 3.6 \times 10^{-3}\ s^{-1}$ at 300 °C.
(a) A sample of gaseous NO_2 is placed in a flask and heated at 300 °C for 150 s. What fraction of the initial sample remains after this time?
(b) How long must a sample be heated so that 99% of the sample has decomposed?

Second-Order Reactions

Suppose the reaction "R ⟶ products" is second order. The rate equation is

$$\text{Rate} = -\frac{\Delta[R]}{\Delta t} = k[R]^2$$

Using the methods of calculus, this relationship can be transformed into the following equation that relates reactant concentration and time:

$$\frac{1}{[R]_t} - \frac{1}{[R]_0} = kt \qquad [15.2]$$

The same symbolism used with first-order reactions applies: $[R]_0$ is the concentration of reactant at the time $t = 0$, and $[R]_t$ is the concentration at a later time.

Example 15.7 Using the Second-Order Integrated Rate Equation

The gas phase decomposition of HI

$$HI(g) \longrightarrow \tfrac{1}{2}\ H_2(g) + \tfrac{1}{2}\ I_2(g)$$

has the rate equation

$$\text{Rate} = k[HI]^2$$

where $k = 30.$ L/mol · min at 443 °C. How much time does it take for the concentration of HI to drop from 0.010 mol/L to 0.0050 mol/L at 443 °C?

Solution Here $[R]_0 = 0.010$ mol/L and $[R]_t = 0.0050$ mol/L. Substituting into Equation 15.2, we have

$$\frac{1}{0.0050\ \text{mol/L}} - \frac{1}{0.010\ \text{mol/L}} = (30.\ \text{L/mol} \cdot \text{min})t$$

$$(2.0 \times 10^2\ \text{L/mol}) - (1.0 \times 10^2\ \text{L/mol}) = (30.\ \text{L/mol} \cdot \text{min})t$$

$$t = 3.3\ \text{min}$$

Exercise 15.7 Using the Second-Order Concentration/Time Equation

Using the rate constant for HI decomposition given in Example 15.7, calculate the concentration of HI after 12 min if $[HI]_0 = 0.010$ mol/L.

Zero-Order Reactions

For a zero-order reaction of the kind "R $\longrightarrow$ products," the rate equation is

$$\text{Rate} = -\frac{\Delta[\text{R}]}{\Delta t} = k[\text{R}]^0 = k$$

This equation leads to the integrated rate equation

$$[\text{R}]_0 - [\text{R}]_t = kt \qquad [15.3]$$

where the units of k are mol/(L · s).

Graphical Methods for Determining Reaction Order and the Rate Constant

See the *Saunders Interactive General Chemistry CD-ROM,* Screen 15.7, Determination of a Rate Equation: Graphical Methods.

Equations 15.1, 15.2, and 15.3 relating concentration and time for zero-, first-, and second-order reactions are very different. Nonetheless, they suggest a convenient way to determine the order of a reaction and its rate constant. Rearranged slightly, each of these equations has the form $y = a + bx$. This is the equation for a straight line, where b is the slope of the line and a is the y-intercept (the value of y when x is zero). As illustrated here, $x = t$ in each case.

Zero-order	First-order	Second-order
$\underset{y}{[\text{R}]_t} = \underset{a}{[\text{R}]_0} - \underset{bx}{kt}$	$\underset{y}{\ln [\text{R}]_t} = \underset{a}{\ln [\text{R}]_0} - \underset{bx}{kt}$	$\underset{y}{\frac{1}{[\text{R}]_t}} = \underset{a}{\frac{1}{[\text{R}]_0}} + \underset{bx}{kt}$

The decomposition of ammonia on a platinum surface was previously mentioned as a zero-order reaction,

$$2\ NH_3(g) \longrightarrow N_2(g) + 3\ H_2(g) \qquad \text{Rate} = k\ [NH_3]^0 = k$$

which means the reaction is independent of NH_3 concentration. The straight line, obtained when the concentration at time t, $[\text{R}]_t$, is plotted against time (Figure 15.8), is proof that this reaction is zero-order in NH_3 concentration. The rate constant, k, can be determined from the slope of the line. The slope of the line is found by selecting any two points on the line and reading off the coordinates as indicated by the arrows. The slope $= -k$, so in this case

$$-k = -1.5 \times 10^{-6}\ \text{mol/L} \cdot \text{s}$$
$$k = 1.5 \times 10^{-6}\ \text{mol/L} \cdot \text{s}$$

The intercept of the line at $t = 0$ is equal to $[\text{R}]_0$.

The plot of concentration versus time for a first-order reaction is a curved line (see Figure 15.3). Plotting ln [reactant] versus time, however, produces a straight line with a negative slope when the reaction is first order in that reactant. Consider the decomposition of hydrogen peroxide, a first-order reaction (see Example 15.6).

$$2\ H_2O_2(aq) \longrightarrow 2\ H_2O(\ell) + O_2(g)$$
$$\text{Rate} = k\ [H_2O_2]$$

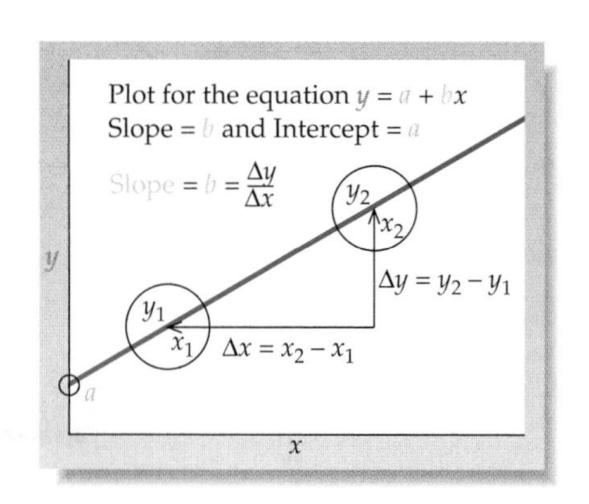

The line plotted here has a positive slope. A line slanting downward from the left has a negative slope.

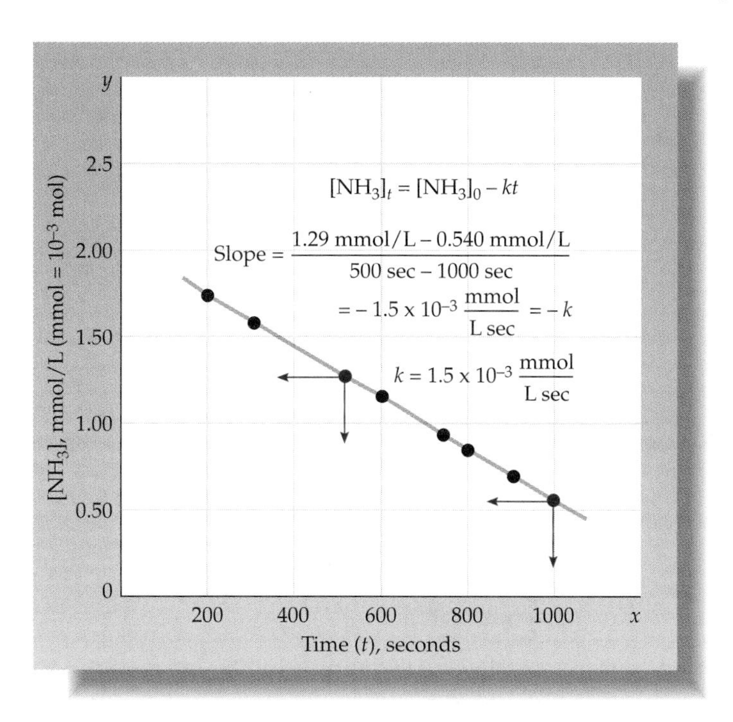

Figure 15.8 Plot of a zero-order reaction. A graph of the concentration of ammonia, $[NH_3]_t$, against time for the decomposition of NH_3 [$2\ NH_3(g) \longrightarrow N_2(g) + 3\ H_2(g)$] on a metal surface at 856 °C is a straight line, indicating that this is a zero-order reaction. The rate constant k for this reaction is found from the slope of the line; $k = -$slope. (The points chosen to calculate the slope are given in red.)

Values of concentration of H_2O_2 ($[H_2O_2]$) as a function of time for a typical experiment are given as the first two columns of numbers in Figure 15.9. The third column contains values of ln $[H_2O_2]$. A graph of ln $[H_2O_2]$ versus time produces a straight line, showing that the reaction is first order in H_2O_2. The negative slope of the line equals the rate constant for the reaction, 1.06×10^{-3} min^{-1}.

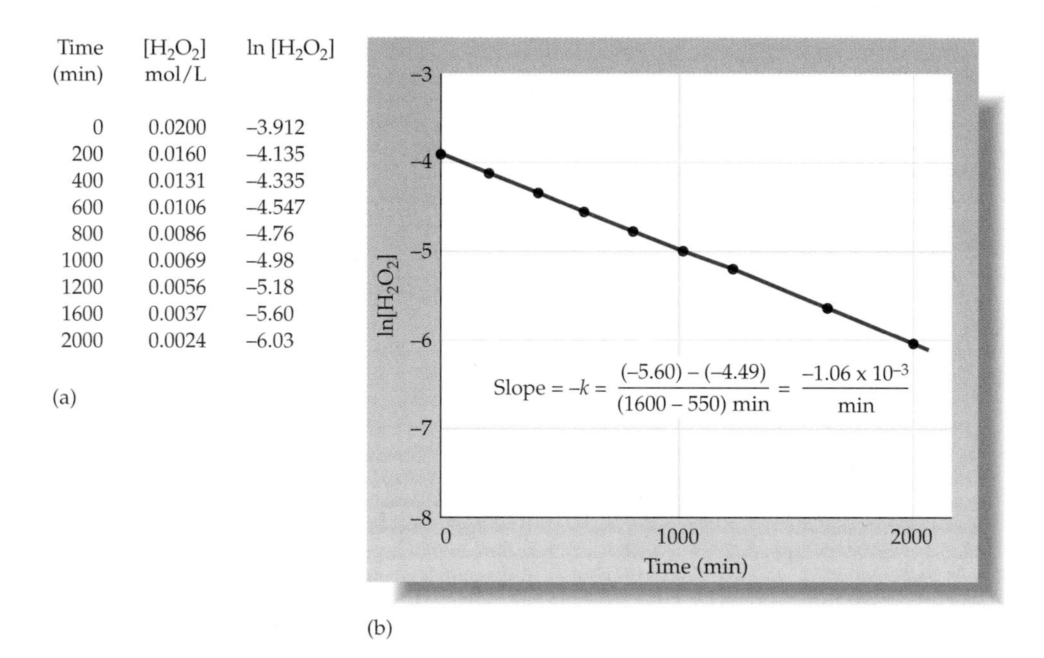

Time (min)	$[H_2O_2]$ mol/L	ln $[H_2O_2]$
0	0.0200	–3.912
200	0.0160	–4.135
400	0.0131	–4.335
600	0.0106	–4.547
800	0.0086	–4.76
1000	0.0069	–4.98
1200	0.0056	–5.18
1600	0.0037	–5.60
2000	0.0024	–6.03

(a)

(b)

Figure 15.9 The decomposition of H_2O_2. (a) Concentration-versus-time data for the decomposition of hydrogen peroxide [$2\ H_2O_2(aq) \longrightarrow 2\ H_2O(\ell) + O_2(g)$]. (b) A plot of ln $[H_2O_2]$ against time is a straight line with a negative slope, indicating a first-order reaction. The rate constant $k = -$slope.

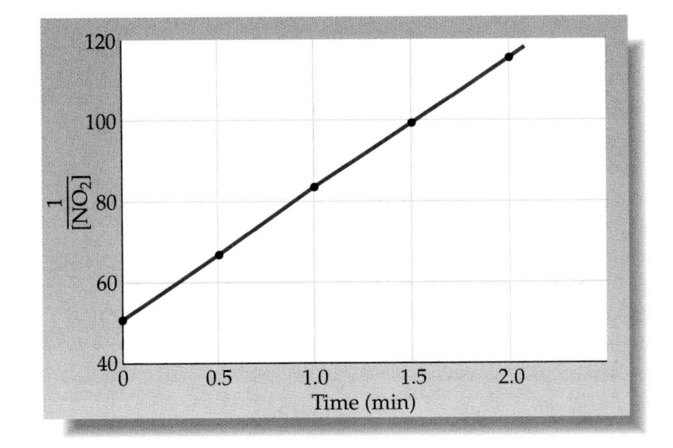

Figure 15.10 A second-order reaction. Concentration-versus-time curve for the decomposition of NO_2 [$2\ NO_2(g) \longrightarrow 2\ NO(g) + O_2(g)$]. A straight line for the plot of $1/[NO_2]$ versus time confirms that this is a second-order reaction. The slope of the line equals the rate constant for this reaction.

The decomposition of NO_2 is a second-order process,

$$NO_2(g) \longrightarrow NO(g) + \tfrac{1}{2}\ O_2(g)$$

$$\text{Rate} = k[NO_2]^2$$

This fact can be verified by showing that a plot of $1/[NO_2]$ versus time is a straight line (Figure 15.10). Here the slope of the line is equal to k.

Problem-Solving Tips and Ideas

15.1 Using Integrated Rate Laws

Integrated rate expressions provide a useful way to determine reaction order. Graphs based on these laws give a straight line *only* when

- [R] is plotted against time for a zero-order process (rate = k).
- ln [R] is plotted against time for a first-order process (rate = k[R]).
- 1/[R] is plotted against time for a second-order process with a rate equation of the type "rate = $k[R]^2$."

To determine the reaction order, therefore, a chemist will plot the experimental concentration–time data in different ways until a straight line plot is achieved. The mathematical relationships for zero-, first-, and second-order reactions are summarized in Table 15.1.

Table 15.1 • Characteristic Properties of Reactions of the Type R ⟶ Products

Order	Rate Equation	Integrated Rate Equation	Straight Line Plot	Slope	k Units
0	$k[R]^0$	$[R]_0 - [R]_t = kt$	$[R]_t$ vs. t	$-k$	mol/L · s
1	$k[R]^1$	$\ln([R]_t/[R]_0) = -kt$	$\ln [R]_t$ vs. t	$-k$	s^{-1}
2	$k[R]^2$	$(1/[R]_t) - (1/[R]_0) = kt$	$1/[R]_t$ vs. t	k	L/mol · s

Exercise 15.8 Using Graphical Methods

Data for the decomposition of N_2O_5 in a particular solvent at 45 °C are as follows:

$[N_2O_5]$, mol/L	t, min
2.08	3.07
1.67	8.77
1.36	14.45
0.72	31.28

A convenient graphing program is included on the *Saunders Interactive General Chemistry CD-ROM.*

Plot $[N_2O_5]$, ln $[N_2O_5]$, and $1/[N_2O_5]$ versus time, t. What is the order of the reaction? What is the rate constant for the reaction?

Half-Life and First-Order Reactions

See the *Saunders Interactive General Chemistry CD-ROM,* Screen 15.8, Half-Life.

The **half-life, $t_{1/2}$,** of a reaction is the time required for the concentration of a reactant to decrease to one-half its initial value. It indicates the rate at which a reactant is consumed in a chemical reaction; the longer the half-life, the slower the reaction. Half-life is used primarily when dealing with first-order processes.

If a reaction is first-order in a reactant R, $t_{1/2}$ is the time when

$$[R]_t = \frac{1}{2}[R]_0 \quad \text{or} \quad \frac{[R]_t}{[R]_0} = \frac{1}{2}$$

Again $[R]_0$ is the initial concentration, and $[R]_t$ is the concentration after the reaction is half completed. To evaluate $t_{1/2}$ we substitute $[R]_t/[R]_0 = 1/2$ and $t = t_{1/2}$ into the integrated first-order rate equation (Equation 15.1),

$$\ln \frac{[R]_t}{[R]_0} = -kt$$

$$\ln (1/2) = -kt_{1/2}$$

Rearranging this equation (and knowing that ln 2 = 0.693), we come to the very useful equation that relates half-life and the first-order rate constant:

$$t_{1/2} = \frac{0.693}{k} \qquad [15.4]$$

The significant feature of this equation is that $t_{1/2}$ is *independent* of concentration for a first-order reaction.

To illustrate the concept of half-life, consider the decomposition of H_2O_2:

$$2\ H_2O_2(aq) \longrightarrow 2\ H_2O(\ell) + O_2(g)$$

Earlier, this chemical reaction was identified as a first-order reaction (see Figure 15.9). The data provided in Figure 15.9 allowed us to determine that the rate constant k for this reaction is $1.06 \times 10^{-3}\ \text{min}^{-1}$. Using Equation 15.4, the half-life of H_2O_2 in this reaction can be calculated from the rate constant.

$$t_{1/2} = \frac{0.693}{k} = \frac{0.693}{1.06 \times 10^{-3}\ \text{min}^{-1}} = 654\ \text{min}$$

In Figure 15.11, the concentration of H_2O_2 has been plotted as a function of time. This shows that $[H_2O_2]$ decreases by half within each 650-min period. The initial concentration of H_2O_2 is 0.020 M, but this drops to 0.010 M after 650 min. The concentration drops again by half (to 0.0050 M) after another 650 min. That is, after 2 half-lives (1300 min) the concentration is $(1/2) \cdot (1/2) = (1/2)^2 = 1/4$, or 25% of the initial concentration. After 3 half-lives (1950 min), the concentration has now dropped to $(1/2) \cdot (1/2) \cdot (1/2) = (1/2)^3 = 1/8$, or 12.5% of the initial value; here $[H_2O_2] = 0.0025$ M.

It is hard to visualize whether a reaction is fast or slow from the value of the rate constant. Can you tell from the value of the rate constant, $k = 1.06 \times 10^{-3}$ min, whether the decomposition of H_2O_2 will require seconds, minutes, hours, or days to reach completion? Probably not, but this is easily assessed from the value of half-life for this reaction, 650 min. The half-life is just under 11 h, so you will have to wait several days for most of the H_2O_2 in a sample to decompose.

The concept of half-life is generally well known because it is a common term used when dealing with radioactive elements. Radioactive decay is a first-order process (see Chapter 24), and half-life is commonly used to describe how rapidly a radioactive element decays.

Although less widely used, half-lives of zero- and second-order processes are also easily defined. Substituting $t_{1/2}$ for t and $\frac{1}{2}[R]_0$ for $[R]_t$ in Equation 15.3 for a zero-order reaction gives the equation

$$[R]_0 - \frac{1}{2}[R]_0 = kt_{1/2}$$

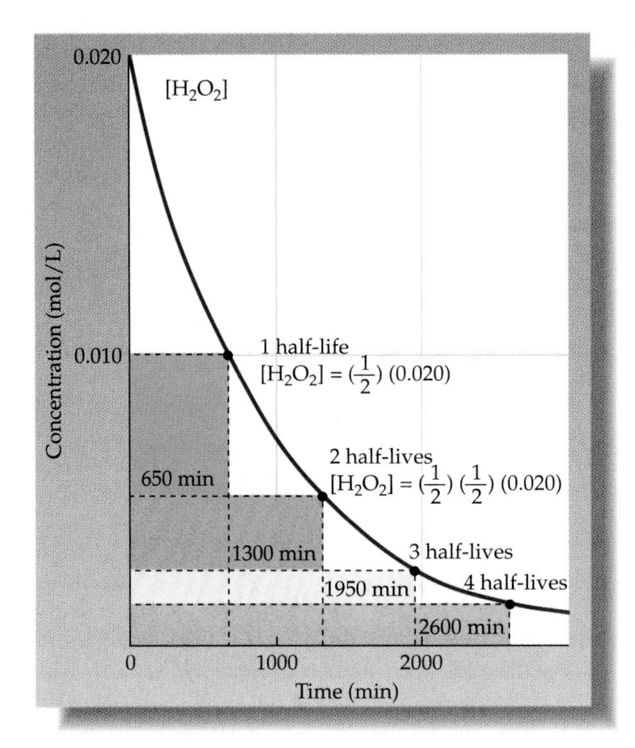

Figure 15.11 Half-life of a first-order reaction. The concentration-versus-time curve for the disappearance of H_2O_2 (where $k = 1.06 \times 10^{-3}$ min^{-1}). The concentration of H_2O_2 is halved every 650 min.

Solving for $t_{1/2}$

$$t_{1/2} = \frac{[\mathrm{R}]_0}{2k} \quad \text{for a zero-order reaction}$$

Making the same substitution in Equation 15.2 for a second-order reaction

$$\frac{1}{1/2[\mathrm{R}]_0} - \frac{1}{[\mathrm{R}]_0} = kt_{1/2}$$

leads to the following equation:

$$t_{1/2} = \frac{1}{k[\mathrm{R}]_0} \quad \text{for a second-order reaction}$$

Notice that the half-life expressions for zero- and second-order processes include the initial concentration, whereas that for a first-order process does not. This means that half-lives for zero- and second-order processes vary with initial concentration; each successive half-life requires a different amount of time than the one before it.

Example 15.8 Half-Life of a First-Order Process

Sucrose, $C_{12}H_{22}O_{11}$, decomposes to fructose and glucose in acid solution with the rate law

$$\text{Rate} = k[\text{sucrose}] \qquad k = 0.208\ \mathrm{h^{-1}}\ \text{at 25 °C}$$

Find the half-life of sucrose under these conditions. Calculate the time required for 87.5% of the initial concentration of sucrose to decompose.

Solution The half-life for the reaction is

$$t_{1/2} = \frac{0.693}{k} = \frac{0.693}{0.208\ \mathrm{h^{-1}}} = 3.33\ \mathrm{h}$$

The concentration of a reactant in a first-order process has dropped to 12.5% of its original value after 3 half-lives. This is the same as saying that 87.5% of the reactant has been consumed. The time required for sucrose to reach this concentration is $3 \times 3.33\ \mathrm{h} =$ 9.99 h.

Exercise 15.9 Half-Life of a First-Order Process

Americium is used in smoke detectors and in medicine for the treatment of certain malignancies. The isotope of americium with a mass number of 241, ^{241}Am, has a rate constant, k, for radioactive decay of 0.0016 $\mathrm{year^{-1}}$. In contrast, radioactive iodine-125, which is used for studies of thyroid functioning, has a rate constant for decay of 0.011 $\mathrm{day^{-1}}$. What are the half-lives of these isotopes? Which element decays faster?

15.5 A MICROSCOPIC VIEW OF REACTION RATES

The connection between reactions and their rate laws is often best understood from a particulate view.

Again and again throughout this book we have turned to the particulate level of chemistry to understand chemical phenomena. The rate of a reaction is no exception. Looking at the way reactions occur at the atomic and molecular level provides some insight into the various influences on rates of reactions.

Let us review the macroscopic observations that we have made so far concerning reaction rates. We know that the wide difference in rates of reaction relates to the specific compounds involved—from very fast reactions such as an explosion that occurs when hydrogen and oxygen are exposed to a spark or flame (see Figure 1.15), to slow reactions like the formation of rust that occur over days, weeks, or years (see Figure 15.1). For a specific reaction, a number of factors influence rate. These include the concentration of reactants, the temperature of the reaction system, and the presence of catalysts, substances that do not appear as reactants or products in a chemical reaction but that participate in the reaction in some way that causes the reaction to go faster. Let us next look at each of these influences in more depth.

See the *Saunders Interactive General Chemistry CD-ROM,* Screen 15.9, Collision Theory.

Concentration, Reaction Rate, and Collision Theory

An example will help to describe how concentrations of reactants affect reaction rate. Consider the gas phase reaction of nitric oxide and ozone,

$$NO(g) + O_3(g) \longrightarrow NO_2(g) + O_2(g)$$

The rate law for this product-favored reaction is first order in each reactant, that is Rate = $k[NO][O_3]$. How is this rate law rationalized?

Let us consider the reaction at the particulate level and imagine a flask containing a mixture of NO and O_3 molecules in the gas phase. Both kinds of molecules are in rapid and random motion within the flask. They strike the walls of the vessel and collide with other molecules. For this or any other reaction to occur, the **collision theory** of reaction rates states that three conditions must be met:

Point 1 is related to reactant concentrations. Point 2 is related to the effect of temperature on the reaction rate.

1. The reacting molecules must collide with one another.
2. The reacting molecules must collide with sufficient energy.
3. The molecules must collide in an orientation that can lead to rearrangement of the atoms.

We shall discuss each of these within the context of the effects of concentration and temperature on reaction rate.

To react, molecules must collide with one another. The rate of their reaction is primarily related to the number of collisions and that is related in turn to their concentrations (Figure 15.12). Doubling the concentration of one reagent in the NO + O_3 reaction, say NO, will lead to twice the number of molecular collisions. Figure 15.12a shows a single molecule of one of the reactants (NO) moving randomly among 16 O_3 molecules. In a given time period, it might collide with 2 O_3 molecules. The number of NO–O_3 collisions will double, however, if the concentration of NO molecules is doubled (to 2, as shown in Figure

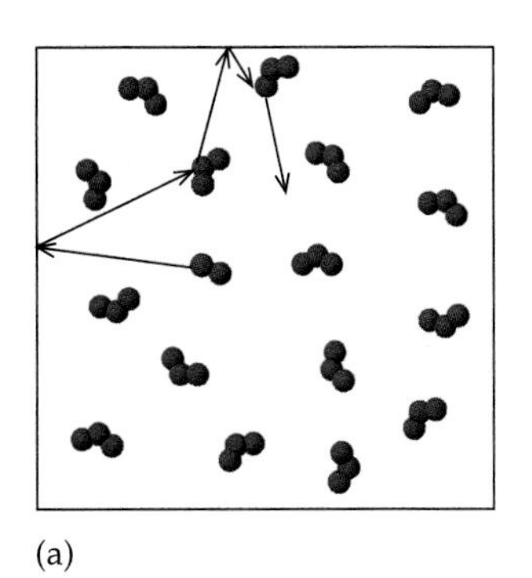
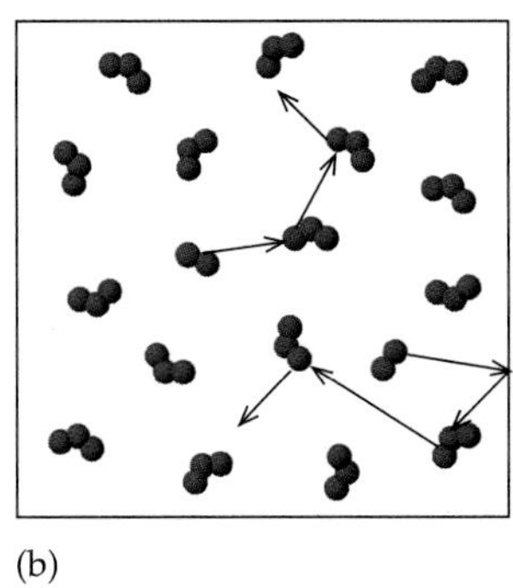
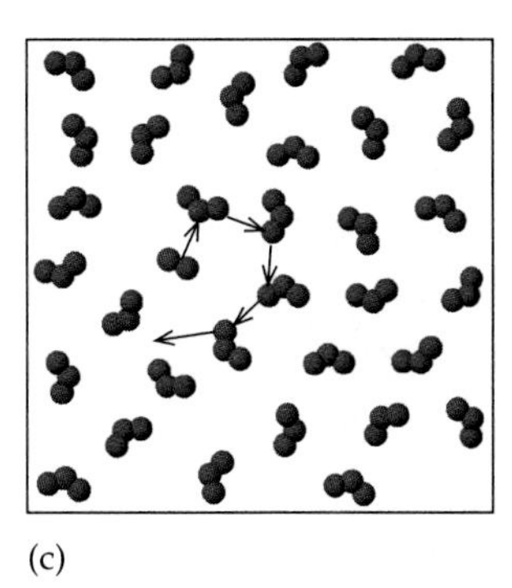

(a) (b) (c)

Figure 15.12 The effect of concentration on the frequency of molecular collisions. (a) A single NO molecule, moving among 16 O_3 molecules, is shown colliding with two of them per second. (b) If two NO molecules move among 16 O_3 molecules, we would predict that 4 NO–O_3 collisions would occur per second. (c) If the number of O_3 molecules is doubled (to 32), the frequency of NO–O_3 collisions is also doubled, to four per second.

15.12b) or if the number of O_3 molecules is doubled (to 32, as in Figure 15.12c). Thus we can explain the dependence of reaction rate on concentration: the number of collisions between the two reactant molecules is directly proportional to the concentrations of each reactant, and the rate of the reaction shows a first-order dependence on each reactant.

Temperature, Reaction Rate, and Activation Energy

See the *Saunders Interactive General Chemistry CD-ROM,* Screen 15.10, Activation Energy.

In a laboratory or in the chemical industry, chemical reactions are often carried out at elevated temperature because this allows the reaction to occur more rapidly. Conversely, it is sometimes desirable to lower the temperature to slow down a chemical reaction (to avoid an uncontrollable reaction or an explosion that could be dangerous). Chemists are very aware of the effect of temperature on the rate of a reaction. But how and why does temperature influence reaction rate?

A discussion of the effect of temperature on reaction rate builds further on collision theory and goes back to the concept of the distribution of energies for molecules in a sample of a gas or liquid. First, recall from your study of gases and liquids that the molecules in a sample have a wide range of energies, described earlier as a Boltzmann distribution of energies (see Figure 12.13 and Figure 13.15). That is, in any sample of a gas or liquid, some molecules have very low energies, others have very high energies, but most have some intermediate energy. As the temperature increases, the average energy of the molecules increases as does the fraction having higher energies (Figure 15.13).

Activation Energy

Now let us first consider the requirement that reacting molecules need some minimum energy. Chemists visualize this as an energy barrier that must be surmounted by the reactants for a reaction to occur (Figure 15.14). This energy, the energy required to mount the barrier, is called the **activation energy, E_a**. If the barrier is low, the kinetic energy required is low, and a high proportion of the molecules in a sample may have sufficient energy to react. The reaction

The value of E_a is controlled by, among other things, the strengths of the bonds in the reactants and products.

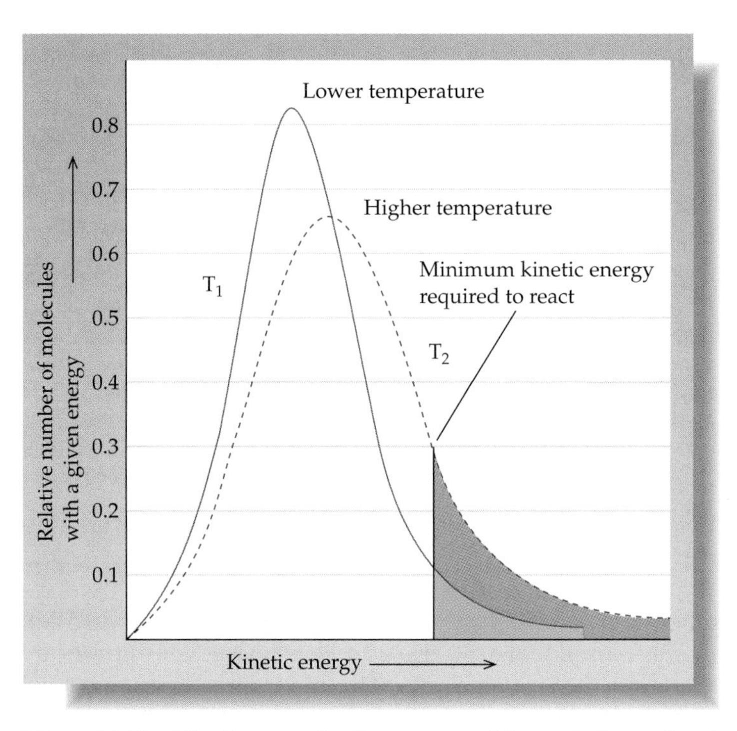

Figure 15.13 Kinetic-energy distribution curve. The vertical axis gives the relative number of molecules possessing the energy indicated on the horizontal axis. The graph indicates the minimum energy required for an arbitrary reaction. At a higher temperature, a larger fraction of the molecules have sufficient energy to react. (See Figure 12.18, the Boltzmann distribution function for a collection of gas molecules.)

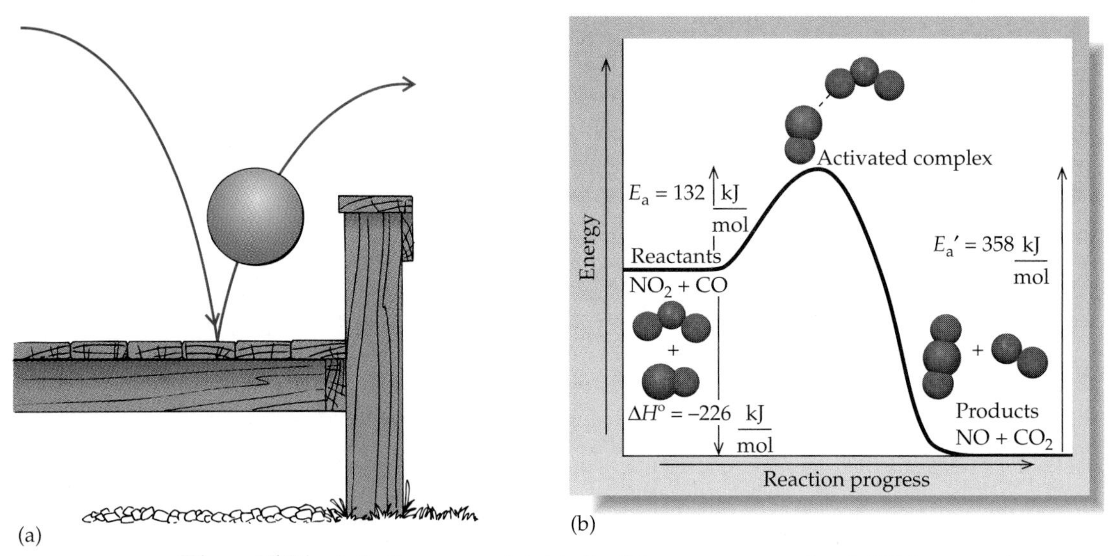

Figure 15.14 Activation energy. (a) A ball will bounce over a wall only if it has enough energy to bounce high enough to go over the wall. (b) The reaction of NO_2 and CO to produce NO and CO_2 requires 132 kJ/mol to surmount the activation energy barrier (E_a = 132 kJ/mol). The reaction is exothermic because the energy of the products is lower by 226 kJ/mol than the energy of the reactants ($\Delta H°$ = −226 kJ/mol). If the reaction is reversed, and NO and CO_2 give NO_2 and CO, 358 kJ/mol is required to surmount the activation energy barrier. That is, the activation energy in the reverse direction is greater than in the forward direction, and the reaction is endothermic in the reverse direction by 226 kJ/mol ($\Delta H°$ = +226 kJ/mol).

will be fast. If the barrier is high, the activation energy is high, and only a few reactant molecules in a sample may have sufficient energy. The reaction will be slow.

To illustrate the activation energy barrier for a reaction, consider a simple process, the conversion of *cis*-2-butene to *trans*-2-butene.

Cis-2-butene and *trans*-2-butene are isomers. They have the same formula, but the atoms fill space differently. The conversion of one isomer to the other is called an isomerization. See Section 11.3.

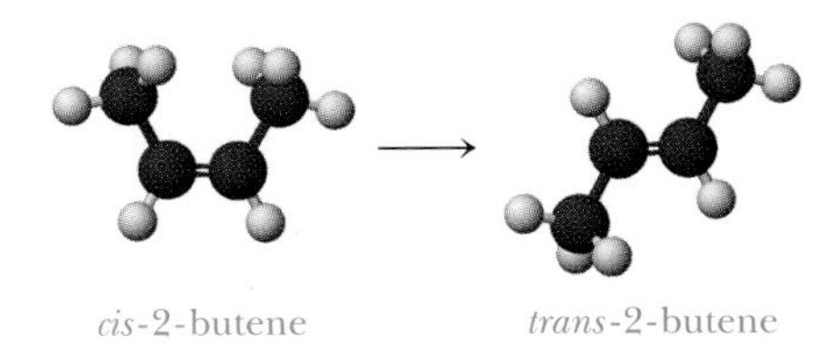

Heating either butene isomer to about 500 °C leads to a mixture of the two molecules. (There is only a very slight preference, about 4 kJ/mol, favoring the formation of *trans*-2-butene in the product mixture.)

At the molecular level we imagine that interconversion of the two isomers—called a *cis–trans* isomerization—is achieved by twisting one end of the molecule with respect to the other end. This requires considerable energy, however, because for rotation about a double bond to occur, the C═C π-bond must be broken (see p. 457).

We can look at this reaction from the viewpoint of the energy of the system (Figure 15.15). In order for *cis*-2-butene to undergo a twisting motion the energy of the system must increase, reaching a maximum when the twist is 90°. This maximum in the potential energy diagram represents the activation energy barrier. The activation energy for this reaction (the forward reaction, *cis* $\longrightarrow$ *trans*) is 262 kJ/mol.

See the *Saunders Interactive General Chemistry CD-ROM*, Screen 10.8, for an animation of this reaction.

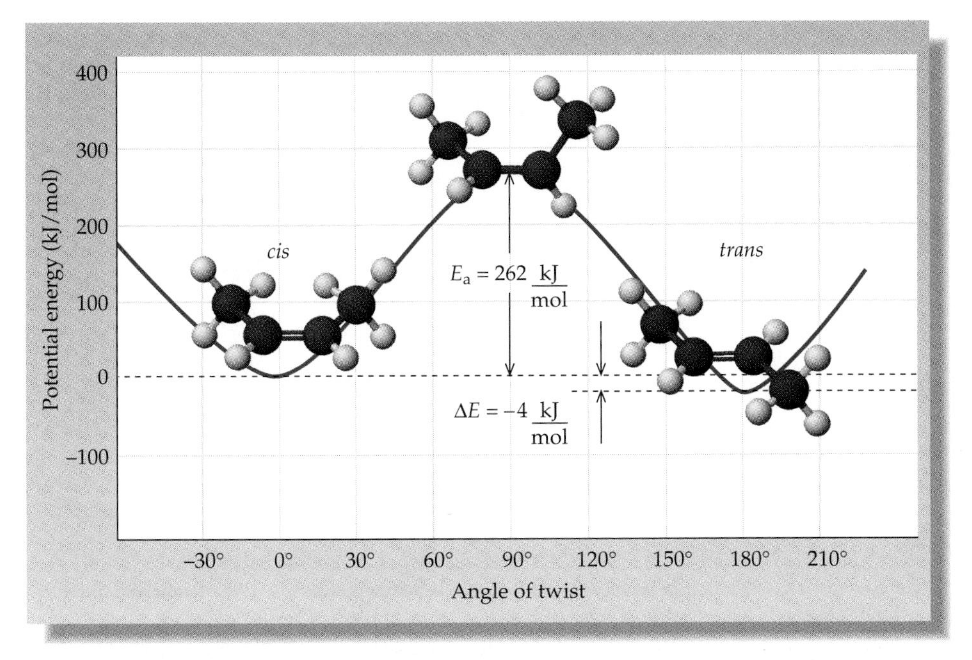

Figure 15.15 Energy profile for the isomerization of *cis*-2-butene to *trans*-2-butene. As the molecule twists around the central carbon–carbon bond, the energy of the system increases. The maximum on this curve is 262 kJ/mol higher than the energy of the reactant; this is the activation energy for the reaction of *cis*-2-butene to *trans*-2-butene. Notice that the energy of the product, the *trans* isomer, is lower than that of the reactant, so the reaction is exothermic.

A CLOSER LOOK Reaction Coordinate Diagrams

Energy diagrams such as those in Figures 15.14 and 15.15 are called *reaction coordinate diagrams.* They can convey a great deal of information. For example, the diagram for the butene isomerization shows that, with time, the reaction progresses smoothly in a single step from reactants to products. Another example is the substitution of a halogen atom of CH_3Cl by an ion such as F^-. Here the F^- ion attacks the molecule from the side opposite the Cl substituent. As F^- begins to form a bond to carbon, the C—Cl bond weakens and the CH_3 portion of the molecule changes shape. As time progresses, the products CH_3F and Cl^- are formed. The reaction would have a reaction coordinate diagram like that in Figure 15.14b.

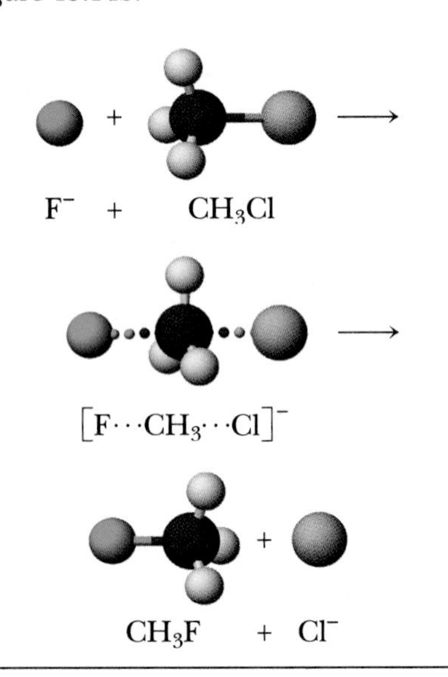

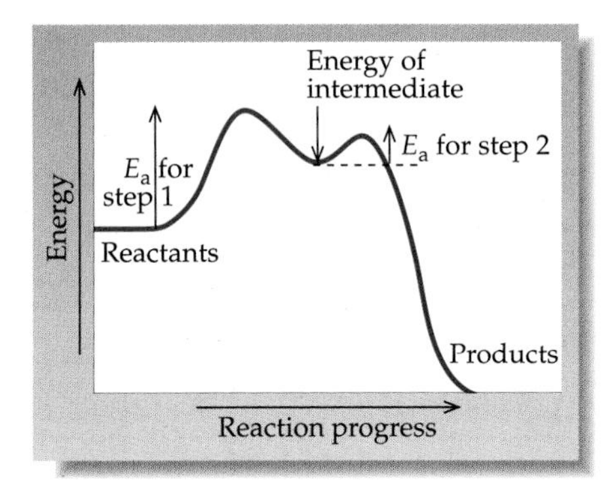

Figure A A reaction coordinate diagram for a two-step reaction, a process involving an intermediate.

The diagram in Figure A shows a two-step reaction that involves a reaction intermediate. An example would be the substitution of the —OH group on methanol by a halide ion in the presence of acid.

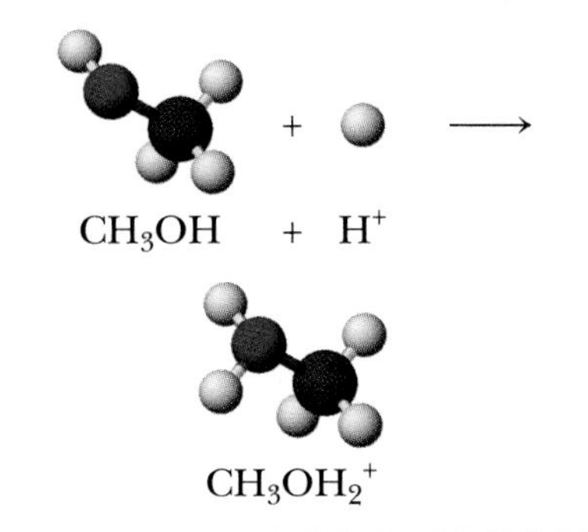

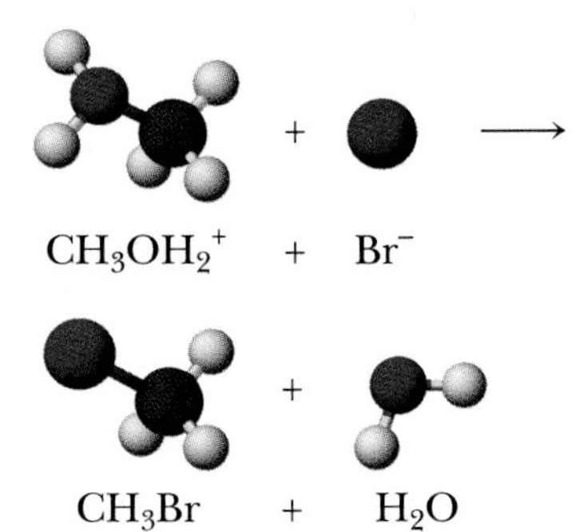

In the first step, an H^+ ion attaches to the O of the C—O—H group in a rapid, reversible reaction. Activation energy is required to reach this state. The energy of this protonated species, a reaction intermediate, is therefore higher than that of the reactants and is represented by the dip in the diagram. In the second step, a halide ion, say Br^-, attacks the intermediate in a process that requires further activation energy. The final result is methyl bromide, CH_3Br, and water.

Notice in Figure A, as in Figures 15.14 and 15.15, that the energy of the products is lower than the energy of the reactants. The reactions are exothermic.

Animations of these reaction coordinate diagrams are on Screen 15.10 of the *Saunders Interactive General Chemistry CD-ROM.*

The rate of isomerization of *cis*-2-butene is a first-order reaction; that is, "Rate = k[*cis*-2-butene]." In a large collection of *cis*-2-butene molecules, the probability that a molecule will isomerize is related to the fraction of molecules that have a high enough energy. The rate of such a reaction would have a first-order dependence on concentration: Doubling the concentration doubles the number of molecules and doubles the rate.

Effect of a Temperature Increase

The conversion of *cis*-2-butene to the *trans* isomer at room temperature is slow because only a small fraction of the butene molecules have enough energy to undergo this reaction. The rate can be increased, however, by heating the sample, which has the effect of increasing the fraction of molecules having higher energies. In general, the effect of an increase in temperature is to increase reaction rates by increasing the fraction of molecules with enough energy to surmount the activation energy barrier.

Effect of Molecular Orientation on Reaction Rate

Not only must the NO and O_3 molecules collide with sufficient energy, but they must come together in the correct orientation. Having a sufficiently high energy is necessary, but this is not sufficient to ensure that reactants will form products. For the reaction of NO and O_3, the N of NO must come together with one of the O atoms at the end of the O_3 molecule (Figure 15.16). This so-called "steric factor" is important in determining how fast the reaction is and a factor affecting the value of the rate constant k. The lower the probability of achieving the proper alignment, the lower the value of k, and the slower the reaction.

If there is a constraint based on molecular orientation in the NO/O_3 reaction, imagine what happens when two or more complicated molecules collide. To have them come together in exactly the correct relationship means only a tiny fraction of the collisions can be effective. No wonder some reactions are so slow. Conversely, it is amazing that so many are so fast!

The Arrhenius Equation

As we have said, reaction rates depend on the energy and frequency of collisions between reacting molecules, on the temperature, and on whether the collisions have the correct geometry. These requirements are summarized by the **Arrhenius equation**

$$k = \text{reaction rate constant} = Ae^{-E_a/RT} \qquad [15.5]$$

A: Frequency of collisions with correct geometry when reactant concentrations = 1 M.

$e^{-E_a/RT}$: Fraction of molecules with minimum energy for reaction

where R is the gas constant with a value of 8.314510×10^{-3} kJ/K · mol. The parameter A is called the *frequency factor*, and it has units of L/mol · s. It is related

See the *Saunders Interactive General Chemistry CD-ROM,* Screen 15.11, Temperature Dependence.

Svante Arrhenius (1859–1927) was a Swedish chemist who, among other things, derived the relation between the rate constant and temperature from experiment.

An interpretation of the significance of the term "A" in the Arrhenius equation goes beyond the level of this text. One aspect of "A" that is well to remember, however, is that it becomes smaller as the reactants become larger, a reflection of the "steric effect."

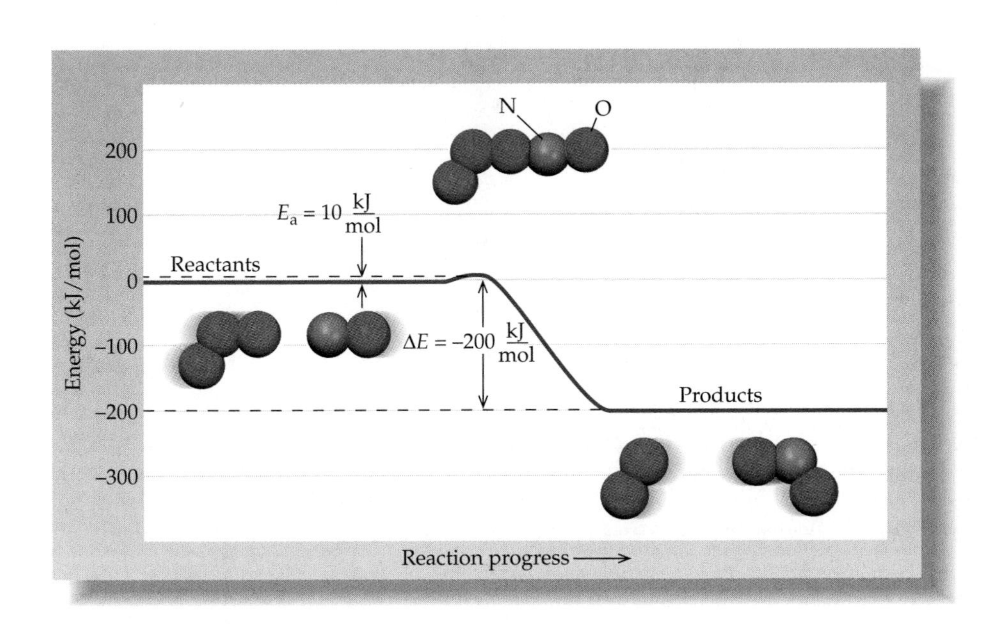

Figure 15.16 The exothermic reaction of NO and O_3. Reaction is possible only if ozone and NO approach each other in the proper orientation and with sufficient energy to cross over the activation energy barrier. If O_3 approaches the O atom of NO, or if NO approaches the central O atom of O_3, for example, reaction should not occur. Notice that the energy of the products is lower than that of the reactants, so the reaction is exothermic.

Temperature (K)	Value of $e^{-E_a/RT}$ for $E_a = 40$ kJ
298	9.7×10^{-8}
400	5.9×10^{-6}
600	3.3×10^{-4}

to the number of collisions and to the fraction of collisions that have the correct geometry. The factor $e^{-E_a/RT}$ is always less than 1, and it is interpreted as the *fraction of molecules having the minimum energy required for reaction.* As the table in the margin shows, the factor changes significantly with temperature.

The Arrhenius equation is valuable because it can be used to (1) calculate the value of the activation energy from the temperature dependence of the rate constant and (2) calculate the rate constant for a given temperature if the activation energy and A factor are known. Taking the natural logarithm of each side of Equation 15.5, we have

$$\ln k = \ln A - \left(\frac{E_a}{RT}\right)$$

and, if we rearrange this slightly, it becomes the equation for a straight line relating ln k to $(1/T)$:

$$\ln k = \ln A - \frac{E_a}{R}\left(\frac{1}{T}\right) \quad \longleftarrow \text{Arrhenius equation} \qquad [15.6]$$

$$y \quad = a \quad + bx \quad \longleftarrow \text{Equation for straight line}$$

This means that, if the natural logarithm of k (ln k) is plotted versus $1/T$, the result is a downward sloping line with a slope of $(-E_a/R)$ (Figure 15.17). So, now we have a means to calculate E_a from experimental values of k at several temperatures, a calculation illustrated in Example 15.9.

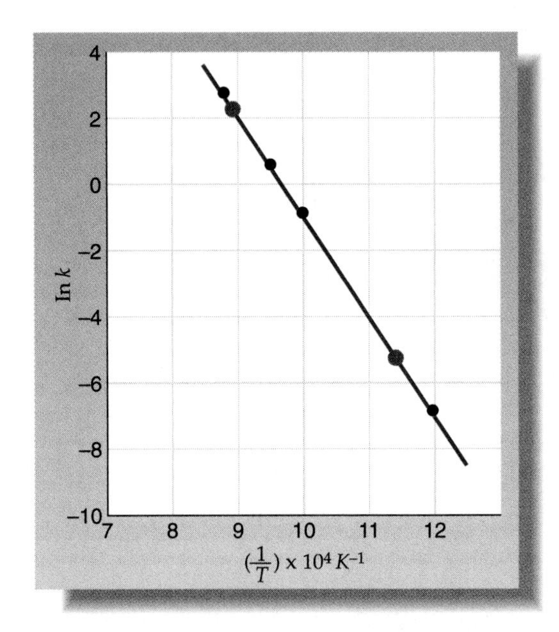

Figure 15.17 Determining the activation energy. A plot of ln k versus $1/T$ for the reaction

$$2\ N_2O(g) \longrightarrow 2\ N_2(g) + O_2(g)$$

The slope of the line gives E_a, as outlined in Example 15.9.

Example 15.9 Determination of E_a from the Arrhenius Equation

Using the experimental data shown in the table, calculate the activation energy E_a for the reaction

$$2\ N_2O(g) \longrightarrow 2\ N_2(g) + O_2(g)$$

Temperature (K)	***k* [L/mol · s)]**
1125	11.59
1053	1.67
1001	0.380
838	0.0011

Solution The first step is to find the reciprocal of the kelvin temperature and the natural logarithm of *k*.

$(1/T)/K^{-1}$	**ln *k***
8.889×10^{-4}	2.4501
9.497×10^{-4}	0.513
9.990×10^{-4}	−0.968
11.9×10^{-4}	−6.81

These data are then plotted as illustrated in Figure 15.17. Choosing the blue points on the graph, the slope is found to be

$$\text{Slope} = \frac{\Delta \ln k}{\Delta(1/T)} = \frac{2.0 - (-5.6)}{(9.0 - 11.5)(10^{-4})/\text{K}} = -\frac{7.6}{2.5 \times 10^{-4}}\ \text{K}$$
$$= -3.0 \times 10^4\ \text{K}$$

The activation energy is then evaluated from

$$\text{Slope} = -\frac{E_a}{R}$$

$$-3.0 \times 10^4\ \text{K} = -\frac{E_a}{8.31 \times 10^{-3}\ \text{kJ/K} \cdot \text{mol}}$$

$$E_a = 250\ \text{kJ/mol}$$

In addition to the graphical method for evaluating E_a used in Example 15.9, E_a can also be obtained algebraically. Knowing *k* at two different temperatures, we can write the equation for each of these conditions:

$$\ln k_2 = \ln A - \left(\frac{E_a}{RT_2}\right) \quad \text{and} \quad \ln k_1 = \ln A - \left(\frac{E_a}{RT_1}\right)$$

If one of these equations is subtracted from the other, we have

$$\ln k_2 - \ln k_1 = \ln \frac{k_2}{k_1} = -\frac{E_a}{R}\left[\frac{1}{T_2} - \frac{1}{T_1}\right] \qquad [15.7]$$

A good rule of thumb is that reaction rates double for every 10 °C rise in temperature in the vicinity of room temperature.

an equation from which E_a can be obtained. This equation now suggests an alternative use, however. The rate constant k_2 can be calculated for another temperature T_2 if E_a, k_1, and T_1 are known. The first of these situations is modeled in Example 15.10.

Example 15.10 Calculating E_a from the Temperature Dependence of k

Using values of k determined at two different temperatures, calculate the value of E_a for the decomposition of HI:

$$2\ HI(g) \longrightarrow H_2(g) + I_2(g)$$
$$k = 2.15 \times 10^{-8}\ L/(mol \cdot s)\ \text{at}\ 6.50 \times 10^2\ K$$
$$k = 2.39 \times 10^{-7}\ L/(mol \cdot s)\ \text{at}\ 7.00 \times 10^2\ K.$$

Solution Here we have k_1 at T_1 and k_2 at T_2, and so we use Equation 15.7:

$$\ln \frac{2.39 \times 10^{-7}\ L/(mol \cdot s)}{2.15 \times 10^{-8}\ L/(mol \cdot s)} = -\frac{E_a}{8.315 \times 10^{-3}\ kJ/K \cdot mol}\left[\frac{1}{7.00 \times 10^2\ K} - \frac{1}{6.50 \times 10^2\ K}\right]$$

$$E_a = 182\ kJ/mol$$

Exercise 15.10 Calculating E_a from the Temperature Dependence of k

The colorless gas N_2O_4 decomposes to the brown gas NO_2 (nitrogen dioxide) in a first-order reaction:

$$N_2O_4(g) \longrightarrow 2\ NO_2(g)$$

The rate constant $k = 4.5 \times 10^3\ s^{-1}$ at 274 K and $1.00 \times 10^4\ s^{-1}$ at 283 K. What is the energy of activation, E_a?

See the *Saunders Interactive General Chemistry CD-ROM,* Screen 15.14, Catalysts and Reaction Rate.

Effect of Catalysts on Reaction Rate

Catalysts are substances that speed up the rate of a chemical reaction, and we have seen several examples of catalysts in earlier discussions in this chapter: MnO_2 (see Figure 15.6), iodide ion (page 695), an enzyme in a potato (page 693) and hydroxide ion (page 702) catalyze the decomposition of hydrogen peroxide. In biological systems, catalysts called *enzymes* are especially significant. One such enzyme is catalase, and its function is to speed up the decomposition of hydrogen peroxide. This enzyme ensures that hydrogen peroxide, which is highly toxic, does not build up in the body.

Catalysts are not consumed in a chemical reaction. They are, however, intimately involved in the details of the reaction at the particulate level. Their function is to provide a different pathway with a lower activation energy for the reaction.

To illustrate how a catalyst participates in a reaction, let us again consider the conversion of *cis*-2-butene to *trans*-2-butene:

$$\underset{cis\text{-2-butene}}{\mathrm{(H_3C)(H)C{=}C(CH_3)(H)}}\ (g) \longrightarrow \underset{trans\text{-2-butene}}{\mathrm{(H)(H_3C)C{=}C(CH_3)(H)}}\ (g)$$

As noted earlier (page 713), this is a first-order reaction. The rate equation for the reaction is "Rate = k[*cis*-2-butene]," and the energy profile shows a high activation energy barrier associated with the breaking of the π bond in the molecule (see Figure 15.15). Because of the high activation energy this is a slow reaction, and rather high temperatures are required for it to occur at a reasonable rate.

This reaction is greatly speeded up in the presence of iodine, however. If a trace of gaseous molecular iodine, I_2, is added to *cis*-2-butene, the iodine catalyzes this reaction. The presence of iodine allows the isomerization reaction to be carried out at a temperature several hundred degrees lower than the uncatalyzed reaction. Iodine is not consumed (nor is it a product) in the overall reaction, and it does not appear in the overall balanced equation. It does appear in the reaction rate law, however; the rate of the reaction depends on the concentration of I_2 as the square root of the iodine concentration:

$$\text{Rate} = k\,[cis\text{-2-butene}]\,[\mathrm{I_2}]^{1/2}$$

The rate of the conversion of the *cis* isomer to *trans*-2-butene changes because the presence of I_2 changes the reaction mechanism (Figure 15.18). The best hypothesis is that iodine molecules first dissociate to form iodine atoms (Step 1). An I atom then adds to one of the C atoms of the C=C double bond

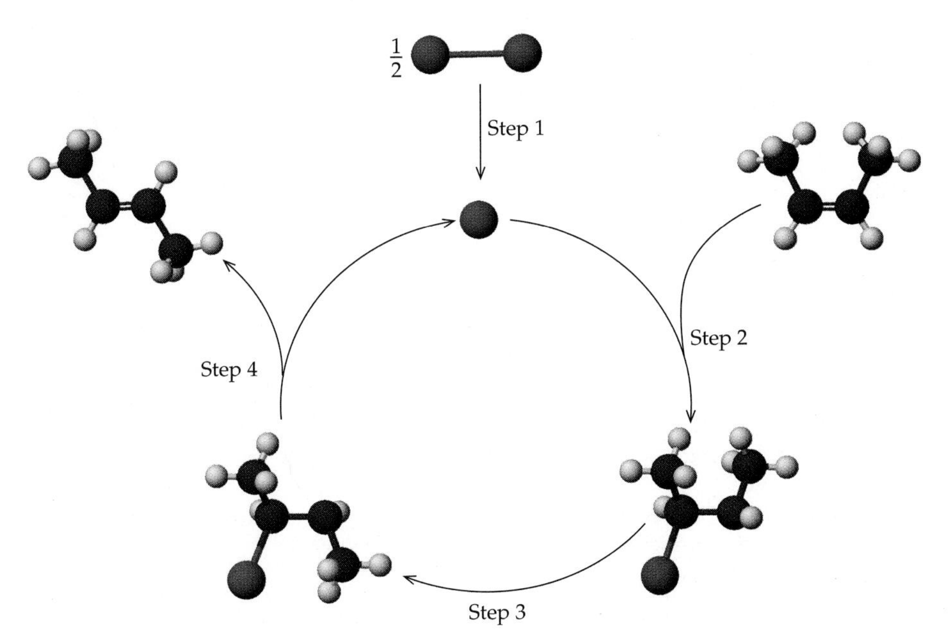

Figure 15.18 The mechanism of the iodine-catalyzed isomerization of *cis*-2-butene. *Cis*-2-butene is converted to *trans*-2-butene in the presence of a catalytic amount of iodine.

(Step 2). This converts the double bond between the carbon atoms to a single bond (the π bond is broken) and allows the ends of the molecule to twist freely relative to each other (Step 3). If the I atom then dissociates from the intermediate, the double bond re-forms, and the molecule may now be in the *trans* configuration (Step 4).

The iodine atom catalyzing the rotation is now free to add to another molecule of *cis*-2-butene. The result is a kind of chain reaction as one molecule of *cis*-2-butene after another is converted to the *trans* isomer. The chain is broken if the iodine atom recombines with another iodine atom to re-form molecular iodine.

An energy profile for the catalyzed reaction (Figure 15.19) contains several interesting features. First, the overall energy barrier has been greatly lowered from the situation in the uncatalyzed reaction. Second, the profile for the reaction includes several steps (five in all), representing each step in the reaction. This diagram includes a series of chemical species called **reaction intermediates,** which are species formed in one step of the reaction and consumed in a later step. Iodine atoms are intermediates, as are the free radical species formed when an iodine atom adds to *cis*-2-butene.

Five important points are associated with this mechanism.

- Iodine, I_2, molecules dissociate to atoms and then re-form. On the macroscopic level the concentration of I_2 is unchanged. Iodine is not involved in the balanced, stoichiometric equation even though it has appeared in the rate equation. This is generally true of catalysts.
- Both the catalyst I_2 and the reactant *cis*-2-butene are in the gas phase. If a catalyst is present in the same phase as the reacting substance it is called a **homogeneous catalyst.**

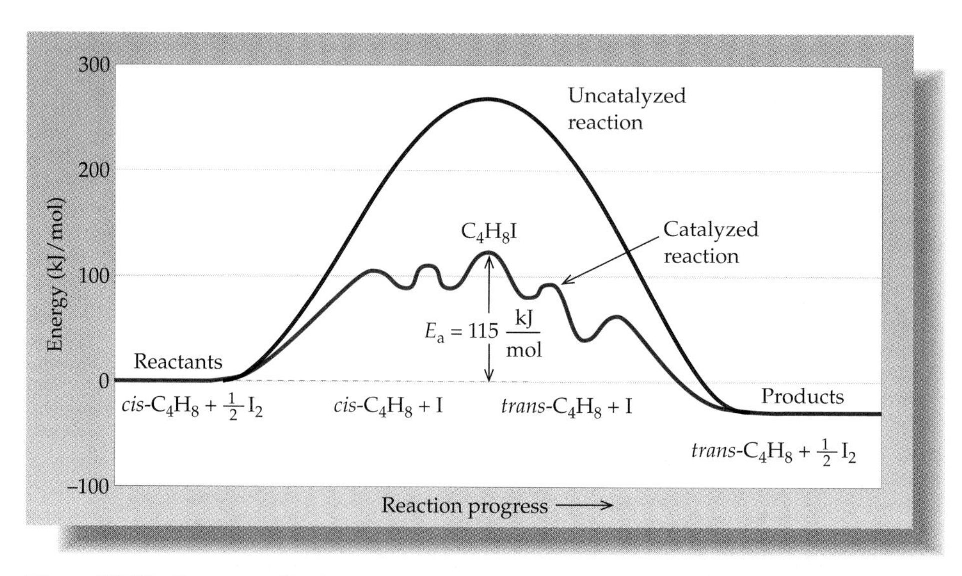

Figure 15.19 Energy profile for the iodine-catalyzed reaction of *cis*-2-butene. A catalyst accelerates a reaction by altering the mechanism so that the activation energy is lowered. With a smaller barrier to overcome, more reacting molecules have sufficient energy to surmount the barrier, and the reaction occurs more readily. The energy profile for the uncatalyzed conversion of *cis*-2-butene to *trans*-2-butene is shown by the black curve, and that for the iodine-catalyzed reaction is represented by the red curve. Notice that the shape of the barrier has changed because the mechanism has changed.

- Iodine atoms and the radical species formed by addition of I to 2-butene are intermediates.
- The activation energy barrier to reaction is significantly lower because the mechanism changed and the reaction rate increases. (Compare Figures 15.15 and 15.19.) In fact, dropping the activation energy from 262 kJ/mol for the uncatalyzed reaction to 115 kJ/mol for the catalyzed process makes the catalyzed reaction 10^{15} times faster!
- The diagram of energy-versus-reaction progress has five energy barriers (five humps appear in the curve). This feature in the diagram means that the reaction occurs in a series of five steps.

The fifth step is the recombination of I atoms to form I_2 molecules.

What we have described here is a *reaction mechanism.* The uncatalyzed isomerization reaction of *cis*-2-butene is a one-step reaction mechanism, whereas the catalyzed mechanism involves a series of steps. We shall discuss reaction mechanisms in more detail in Section 15.6.

Catalysis in Industry and the Environment

An expert in the field of industrial chemistry has said that "Every year more than a trillion dollars worth of goods is manufactured with the aid of synthetic catalysts. Without them, fertilizers, pharmaceuticals, fuels, synthetic fibers, solvents, and surfactants would be in short supply. Indeed, 90 percent of all manufactured items use catalysts at some stage of production." The major areas of catalyst use are petroleum refining, industrial production of chemicals, and environmental controls.

About 12×10^6 kg of polyethylene (see Section 11.7) is produced annually in the United States. The catalysts used to make these and similar polymers are often called Ziegler-Natta catalysts and were named for Karl Ziegler (1898–1973), a German chemist, and Guilio Natta (1903–1979), an Italian chemist. They shared the Nobel Prize in chemistry in 1963 for this work.

Almost all industrial-scale reactions use **heterogeneous catalysts,** catalysts that are present in a different phase than the reactants being catalyzed. Heterogeneous catalysts are used because they are more easily separated from the products and leftover reactants than are homogeneous catalysts. Catalysts for chemical processing are generally metal-based and often contain precious metals such as platinum and palladium. In the United States more than $600 million worth of such catalysts are employed annually by the chemical-processing industry, almost half of them in the preparation of polymers such as polyethylene (Section 11.7).

Almost 8×10^9 kg of nitric acid is made annually in the United States using the Ostwald process, the first step of which involves the controlled oxidation of ammonia over a Pt-containing catalyst (Figure 15.20):

$$4\ NH_3(g) + 5\ O_2(g) \xrightarrow{\text{Pt-containing catalyst}} 4\ NO(g) + 6\ H_2O(g) \qquad \Delta H^\circ_{rxn} = -905.5\ \text{kJ}$$

followed by further oxidation of NO to NO_2:

$$2\ NO(g) + O_2(g) \longrightarrow 2\ NO_2(g) \qquad \Delta H^\circ_{rxn} = -114.1\ \text{kJ}$$

In a typical plant, a mixture of air with 10% NH_3 is passed very rapidly over the catalyst at a high pressure and at about 850 °C. Roughly 96% of the ammonia is converted to NO_2, making this one of the most efficient industrial catalytic reactions. The final step is to absorb the NO_2 into water to give the acid (HNO_3) and NO, the latter being recycled into the process:

$$3\ NO_2(g) + H_2O(\ell) \longrightarrow 2\ HNO_3(aq) + NO(g) \qquad \Delta H^\circ_{rxn} = -138.2\ \text{kJ}$$

Figure 15.20 Catalyzed oxidation of ammonia. The platinum–rhodium gauze catalyst used for the oxidation of ammonia in the manufacture of nitric acid. *(Johnson Matthey)*

Acetic acid, CH_3CO_2H, has a place in the organic chemicals industry comparable to that of sulfuric acid in the inorganic chemicals industry; about 2.1×10^9 kg of acetic acid was made in the United States in 1995. Acetic acid is used widely in industry to make plastics and synthetic fibers, as a fungicide, and as the starting material for preparing many dietary supplements. One way of synthesizing the acid uses homogeneous catalysis. A rhodium-based compound catalyzes the combination of carbon monoxide and methyl alcohol, both inexpensive chemicals, to form acetic acid:

$$CH_3OH + CO \xrightarrow{\text{Rh-containing catalyst}} CH_3C(=O){-}OH$$

methanol acetic acid

The role of the rhodium-containing catalyst in this reaction is to bring the reactants together and allow them to rearrange to products. The first step in the process is the reaction of the alcohol with hydrogen iodide to give methyl iodide, CH_3I,

$$CH_3OH + HI \longrightarrow CH_3I + H_2O$$

which then reacts with the catalyst, a molecule containing a rhodium(I) ion and CO. This gives a new molecule with CH_3, I, and CO attached to the metal center. Acetic acid is the product after these fragments rearrange and the intermediate reacts with water:

$$\{Rh(CH_3)(CO)I\} + H_2O \longrightarrow \text{Rh catalyst} + HI + CH_3CO_2H$$

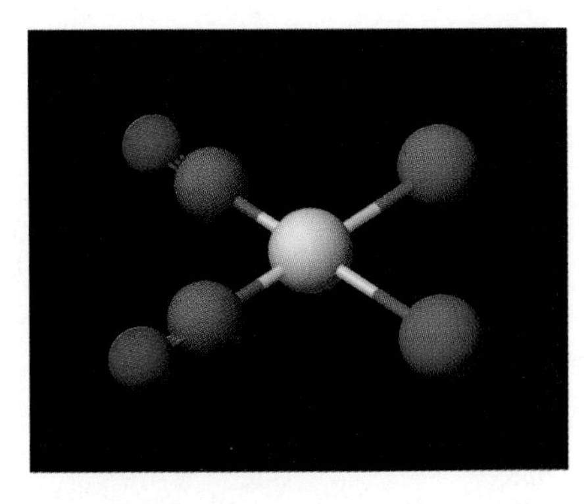

$[Rh(CO)_2I_2]^-$, the catalyst in the conversion of methanol to acetic acid.

Besides producing acetic acid, this final step regenerates the Rh-containing catalyst and produces HI, which is then available to react with more CH_3OH to begin a new catalytic cycle.

The largest recent growth in catalyst use is in emission controls for automobiles and power plants. This market consumes very large quantities of platinum group metals: platinum, palladium, rhodium, and iridium. Over 9000 kg of platinum, and 1300 kg each of palladium and rhodium, were sold in the United States in the first half of 1991 for automotive uses. In contrast, chemical processing used only about 1400 kg of all three metals, and the petroleum industry used about 1800 kg of platinum and rhodium.

The purpose of the catalysts in the exhaust system of an automobile is to ensure that the combustion of carbon monoxide and hydrocarbons is complete (see the *Chemical Puzzler* and Figure 15.21)

$$2\,CO(g) + O_2(g) \xrightarrow{\text{Pt-NiO catalyst}} 2\,CO_2\,(g)$$

$$\underset{\text{isooctane, a component of gasoline}}{2\,C_8H_{18}(g)} + 25\,O_2(g) \xrightarrow{\text{Pt-NiO catalyst}} 16\,CO_2(g) + 18\,H_2O(g)$$

and to convert nitrogen oxides to molecules less harmful to the environment. At the high temperatures in an internal combustion engine, some N_2 from air reacts with O_2 to give NO, a serious air pollutant. Nitrogen oxide is unstable and should revert to N_2 and O_2. Its rate of reversion is slow, but the catalysts in a catalytic converter greatly speed the reaction.

$$2\,NO(g) \xrightarrow{\text{catalyst}} N_2(g) + O_2(g)$$

The role of the heterogeneous catalyst in the preceding reactions is probably to weaken the bonds of the reactants and to assist in product formation.

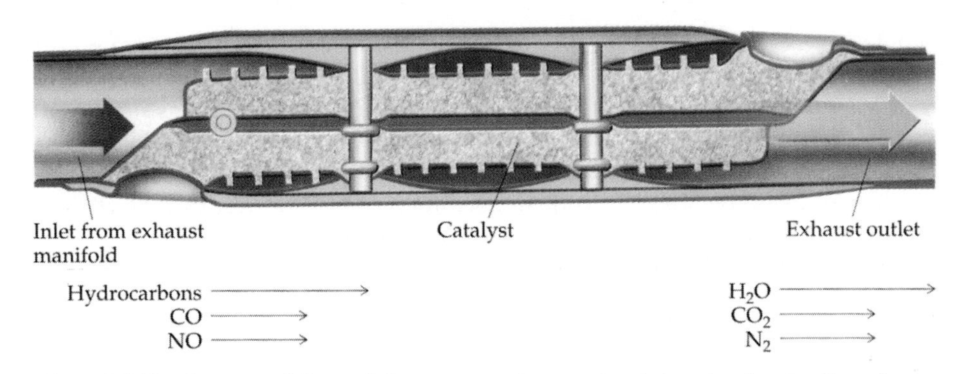

Figure 15.21 An automobile catalytic converter. Cross-sectional view showing the flow of gases through an automobile catalytic converter.

Exercise 15.11 Catalysis

Which of the following statements is (are) true? If any are false, change the wording to make them true.

1. The concentration of a homogeneous catalyst may appear in the rate equation.
2. A catalyst is always consumed in the overall reaction.
3. A catalyst must always be in the same phase as the reactants.

Rate laws are macroscopic observations. Mechanisms analyze how reactions occur at the particulate level.

See the *Saunders Interactive General Chemistry CD-ROM,* Screens 15.12 and 15.13, Reaction Mechanisms.

15.6 REACTION MECHANISMS

We come now to one of the most important reasons to study the rates of reactions: to understand **reaction mechanisms,** the sequence of bond-making and bond-breaking steps that occurs during the conversion of reactants to products. As you have seen in the previous discussion, the study of reaction mechanisms places you squarely in the realm of the particulate level of chemistry. We want to analyze the changes that atoms and molecules undergo when they react. And then, we want to relate this description back to the macroscopic world, to the experimental observations of reaction rates.

You have seen that the rate equation for a reaction is determined by experimentation. Based on the rate equation, and applying chemical intuition, chemists can often make an educated guess about the mechanism for a reaction. In some reactions, the conversion of reactants to products in a single step is envisioned. For example, nitric oxide and ozone react in a single-step reaction, with the reaction occurring as a consequence of a collision between reactant molecules:

$$NO(g) + O_3(g) \longrightarrow NO_2(g) + O_2(g)$$

The uncatalyzed isomerization of *cis*-2-butene to *trans*-2-butene was also best described as a single-step reaction.

Most chemical reactions occur in a sequence of steps, however. We saw an example of this with the iodine-catalyzed 2-butene isomerization reaction. Another example of a reaction that occurs in several steps is the reaction of bromine and NO:

$$Br_2(g) + 2\ NO(g) \longrightarrow 2\ BrNO(g)$$

A single-step reaction would require that three molecules collide simultaneously with the reactant molecules in just the right orientation. There is a pretty low probability that such an event would occur frequently. Thus, for this reaction it would be reasonable to look for a mechanism that involves a series of steps. A two-step mechanism is one possibility. In one possible mechanism, Br_2 and NO combine in an initial step to produce an intermediate species, Br_2NO (Figure 15.22). This intermediate then reacts with another NO to give the reaction products. The overall reaction is obtained by adding the equations for these two steps:

Step 1	$Br_2(g) + NO(g) \longrightarrow Br_2NO(g)$
Step 2	$Br_2NO(g) + NO(g) \longrightarrow 2\ BrNO(g)$
Overall Reaction	$Br_2(g) + 2\ NO(g) \longrightarrow 2\ BrNO(g)$

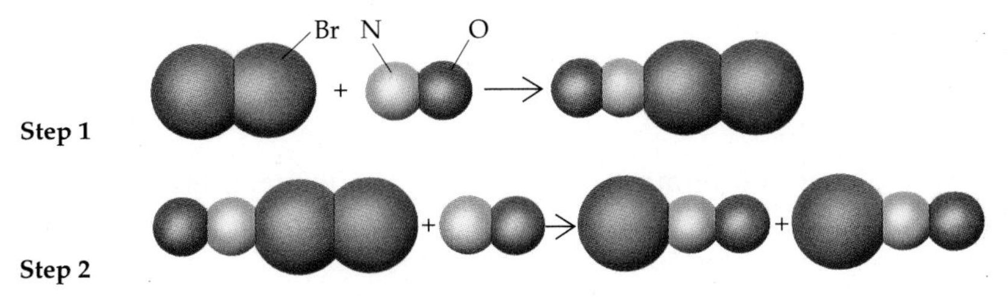

Figure 15.22 A reaction mechanism. A representation of the proposed two-step mechanism by which NO and Br_2 are converted to NOBr.

Each step in a multistep-reaction sequence is an **elementary step,** which is defined as a chemical equation that describes an assumed single molecular event such as the formation or rupture of a chemical bond or the displacement of atoms as a result of a molecular collision. Each step has its own activation energy barrier, E_a, and rate constant, k. The steps must add up to give the balanced equation for the overall reaction, and the time required to complete all of the steps defines the overall reaction rate. A series of steps that satisfactorily explain the kinetic properties of a chemical reaction constitutes a possible reaction mechanism.

Mechanisms of reactions are postulated starting with experimental data. To see how this is done, we first describe three types of elementary steps.

Molecularity of Elementary Steps

Elementary steps are classified by the number of reactant molecules (or ions, atoms, or free radicals) that come together. This number is called the **molecularity** of the elementary step. The order of a reaction can be a fractional number, whereas the molecularity of a step must be a whole, positive number. When one molecule is the only reactant in an elementary step, the reaction is a **unimolecular** process. A **bimolecular** elementary process involves two molecules. These may be identical molecules ($A + A \longrightarrow$ products) or different molecules ($A + B \longrightarrow$ products). For example, a two-step mechanism has been proposed for the decomposition of ozone:

Step 1	Unimolecular	$O_3(g) \longrightarrow O_2(g) + O(g)$
Step 2	Bimolecular	$O_3(g) + O(g) \longrightarrow 2\ O_2(g)$
Overall reaction		$2\ O_3(g) \longrightarrow 3\ O_2(g)$

The reaction is suggested to occur by an initial unimolecular step followed by a second bimolecular step.

An elementary step involving three molecules is called **termolecular.** This could involve three molecules of the same or different type ($3\ A \longrightarrow$ products; $2\ A + B \longrightarrow$ products; or $A + B + C \longrightarrow$ products). As might be suspected, the simultaneous collision of three molecules is not very likely, unless one of the molecules involved is in high concentration, such as a solvent molecule. In fact, most termolecular processes involve the reaction of two molecules; the function of the third particle is to absorb the excess energy produced when a new chemical bond is formed by the first two molecules in an exothermic step. For example, N_2 is unchanged in a termolecular reaction between oxygen molecules and oxygen atoms that produces ozone in the upper atmosphere:

$$O(g) + O_2(g) + N_2(g) \longrightarrow O_3(g) + \text{energetic } N_2(g)$$

The probability that four or more molecules will simultaneously collide with sufficient kinetic energy and proper orientation to react is so small that reaction molecularities greater than three are never proposed.

Rate Equations for Elementary Steps

As you have already seen, the experimentally determined rate equation for a reaction cannot be predicted from its overall stoichiometry. In contrast, *the rate equation for any elementary step is defined by the reaction stoichiometry.* The rate equation of an elementary step is given by the product of the rate constant and the concentrations of the reactants in that step. This means we can write the rate equation for any elementary step, as shown by examples in the table:

Elementary Step	Molecularity	Rate Equation
$A \longrightarrow$ product	unimolecular	Rate = $k[A]$
$A + B \longrightarrow$ product	bimolecular	Rate = $k[A][B]$
$A + A \longrightarrow$ product	bimolecular	Rate = $k[A]^2$
$2\,A + B \longrightarrow$ product	termolecular	Rate = $k[A]^2[B]$

For example, the rate laws for each of the two steps in the decomposition of ozone are

$$\text{Rate for (unimolecular) Step 1} = k[O_3]$$

$$\text{Rate for (bimolecular) Step 2} = k'\,[O_3][O]$$

When a reaction mechanism consists of two elementary steps, it is likely that the two steps are going to occur at different rates. The two rate constants (k and k' in this example) are not expected to have the same value (nor the same units, if the two steps have different molecularities).

Molecularity and Reaction Order

The molecularity of an elementary step and its order are the same. A unimolecular elementary step must be first order, a bimolecular elementary step must be second order, and a termolecular elementary step must be third order. Such a direct relation between molecularity and order is emphatically *not* true for the *overall* reaction. If you discover experimentally that a reaction is first order, you cannot conclude that it occurs in a single, unimolecular elementary step. Similarly, a second-order rate equation does not imply the reaction occurs in a single, bimolecular elementary step. An example illustrating this is the decomposition of N_2O_5:

$$2\,N_2O_5(g) \longrightarrow 4\,NO_2(g) + O_2(g)$$

Here the rate equation is "Rate = $k[N_2O_5]$," but chemists are fairly certain the mechanism involves a series of unimolecular and bimolecular steps.

To see how the experimentally observed rate equation for the *overall reaction* is connected with a possible mechanism or sequence of elementary steps requires some chemical intuition. We will provide only a glimpse of the subject in the next section.

Example 15.11 Elementary Steps

The hypochlorite ion undergoes self oxidation–reduction to give chlorate, ClO_3^-, and chloride ions:

$$3\ ClO^-(aq) \longrightarrow ClO_3^-(aq) + 2\ Cl^-(aq)$$

It is thought that the reaction occurs in two steps:

Step 1 $ClO^-(aq) + ClO^-(aq) \longrightarrow ClO_2^-(aq) + Cl^-(aq)$

Step 2 $ClO_2^-(aq) + ClO^-(aq) \longrightarrow ClO_3^-(aq) + Cl^-(aq)$

What is the molecularity of each step? Write the rate equation for each reaction step. Show that the sum of these reactions gives the equation for the net reaction.

Solution Because two ions are involved in each step, each step is bimolecular. The rate equation for any elementary step is the product of the concentrations of the reactants. In this case, the rate equations are

Step 1 $\text{Rate} = k[ClO^-]^2$

Step 2 $\text{Rate} = k[ClO^-][ClO_2^-]$

Finally, on adding the equations for the two elementary steps, we see that the ClO_2^- ion is a product of the first step and a reactant in the second step. It therefore cancels out, and we are left with the stoichiometric equation for the overall reaction:

Step 1 $ClO^-(aq) + ClO^-(aq) \longrightarrow \cancel{ClO_2^-}(aq) + Cl^-(aq)$

Step 2 $\cancel{ClO_2^-}(aq) + ClO^-(aq) \longrightarrow ClO_3^-(aq) + Cl^-(aq)$

Sum of steps $3\ ClO^-(aq) \longrightarrow ClO_3^-(aq) + 2\ Cl^-(aq)$

Exercise 15.12 Elementary Steps

Nitric oxide is reduced by hydrogen to give nitrogen and water,

$$2\ NO(g) + 2\ H_2(g) \longrightarrow N_2(g) + 2\ H_2O(g)$$

and one possible mechanism to account for this reaction is

$$2\ NO(g) \rightleftharpoons N_2O_2(g)$$
$$N_2O_2(g) + H_2(g) \longrightarrow N_2O(g) + H_2O(g)$$
$$N_2O(g) + H_2(g) \longrightarrow N_2(g) + H_2O(g)$$

What is the molecularity of each of the three steps? What is the rate equation for the third step? Show that the sum of these elementary steps is the net reaction.

CURRENT PERSPECTIVES

Depletion of Stratospheric Ozone

Much of our life in the United States today depends on refrigeration. We cool our homes, cars, offices, and shopping centers with air conditioners. We preserve our food and medicines with refrigerators. Until very recently all of these refrigeration units used *chlorofluorocarbons,* or CFCs, as the heat-exchanging fluid.

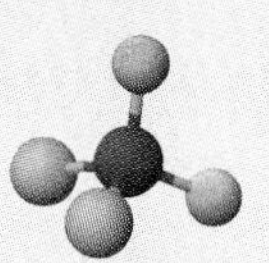

freon-12
dichlorodifluoromethane

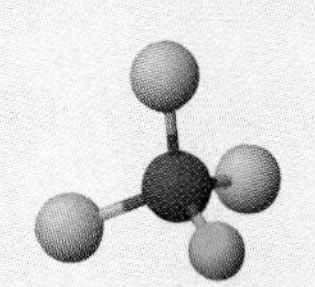

freon-11
trichlorofluoromethane

That situation has now changed dramatically, in part because of laboratory studies of the kinetics of the reactions that CFCs might undergo in the stratosphere.

Nonflammable, nontoxic CFCs such as CCl_2F_2 were discovered by scientists at the Frigidaire Division of General Motors in 1928. By 1988, the total worldwide consumption of CFCs was over 1×10^9 kilograms annually. In the United States thousands of businesses produced CFC-related goods and services worth more than $28 billion a year, and there were more than 700,000 CFC-related jobs. CFCs were used in the United States mostly as refrigerants, foam-blowing agents for polystyrene and polyurethane, aerosol propellants, and industrial solvents.

It is ironic that the very properties that led to the first use of CFCs are now causing worldwide concern. Once gaseous CFCs are released into the troposphere, that part of the earth's atmosphere ranging from the surface to an altitude of about 10 km, they persist for a long time because there is no mechanism for their destruction. Through atmospheric mixing they rise to the stratosphere where they are eventually destroyed by solar radiation—but with significant consequences to our environment, as first recognized by M. J. Molina and F. S. Rowland in 1974 (Figure A). From laboratory experiments, they predicted that continued use of CFCs would lead eventually to a significant depletion of the ozone layer around the earth. This is a serious concern because, for every 1% loss of ozone from the stratosphere, an additional 2% of the sun's high-energy ultraviolet radiation can reach earth's surface, resulting in increases in skin cancer, damage to plants, and possibly other effects that we do not even suspect at this time.

Ozone is produced in the stratosphere when high-energy ultraviolet radiation causes the photodissociation of oxygen to give O atoms, which react with O_2 molecules:

$$O_2(g) + \text{radiation } (\lambda < 280 \text{ nm}) \longrightarrow 2\,O(g)$$
$$O(g) + O_2(g) \longrightarrow O_3(g)$$

The ozone produced by this mechanism in the stratosphere is quite abundant (10 ppm), which is fortunate because O_3 is also photodissociated by sunlight,

$$O_3(g) \longrightarrow O(g) + O_2(g)$$

and the O atoms produced react with more O_2 to regenerate O_3. The process keeps 95% to 99% of the sun's ultraviolet radiation from reaching earth's surface.

The problem with CFCs is that they disrupt the protective ozone layer by a "chlorine catalytic cycle." The CFCs rise to the stratosphere where C—Cl bonds are broken by high-energy photons. The Cl atoms attack ozone to give ClO, chlorine oxide, and oxygen:

$$Cl(g) + O_3(g) \longrightarrow ClO(g) + O_2(g)$$

This would not necessarily be a problem, except that ClO can react with an O atom to give O_2 and regenerate a Cl atom:

$$ClO(g) + O(g) \longrightarrow Cl(g) + O_2(g)$$

The Cl atom can then destroy still another O_3, and so on and on in a "catalytic cycle." The net reaction is the destruction of a significant quantity of ozone. It is estimated that each Cl atom can destroy as many as 100,000 ozone molecules before the Cl atom is inactivated or returned to the troposphere (probably as HCl).

At least two other major kinds of reactions are believed to *interfere* with ozone loss. In one case ClO reacts with nitrogen monoxide, NO, to release a Cl atom and form NO_2. The NO_2 goes on to regenerate a molecule of ozone. In another reaction, ClO forms chlorine nitrate ($ClONO_2$), a compound that at least temporarily acts as a "chlorine reservoir." Eventually, though, this compound also breaks apart and frees Cl atoms to resume their ozone destruction.

If these interference reactions are important, CFCs might have only a minimal effect on earth's ozone layer. In the early spring in the southern hemisphere, however, the ozone layer over the Antarctic is significantly depleted, a fact clearly illustrated in satellite images of ozone concentration done in the early 1990s (Figure B). One theory used to explain this observation involves the high-altitude clouds that are common over the Antarctic continent in the winter. Chlorine nitrate could condense in these extremely cold clouds, and chlorine atoms could also be trapped as HCl. Rowland and Molina estimate that one out of every three or four collisions of $ClONO_2$ with HCl-containing ice crystals leads to reactions such as

$$ClONO_2 + HCl \longrightarrow HNO_3 + Cl_2$$

The first sunlight of spring that warms the clouds can trigger the release of atomic chlorine by photodissociating chlorine molecules. Because nitrogen oxides are trapped in the clouds

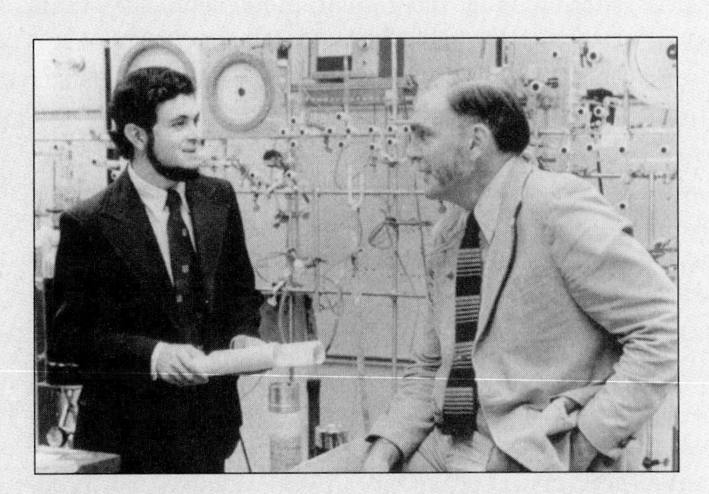

Figure A Mario J. Molina *(left)* and F. Sherwood Rowland *(right)*. These scientists first recognized the potential for the depletion of the earth's atmosphere by CFCs. Rowland is Professor of Chemistry at the University of California, Irvine, and Molina is Professor of Environmental Sciences at the Massachusetts Institute of Technology. Molina and Rowland shared the 1995 Nobel Prize in Chemistry with Paul Crutzen of Germany for their studies of the earth's ozone layer and the effect of pollutants on it. *(The Bettmann Archive)*

as nitric acid, the "chlorine catalytic cycle" can run unchecked for 5 or 6 weeks in the spring.

Whatever the theories for the springtime Antarctic ozone loss, the problem is real, and people around the world have taken steps to halt any further deterioration. Chemical companies in the United States have halted CFC production and are actively searching for substitutes. In January, 1989, 24 nations signed the *Montreal Protocol*, which calls for reductions in production and use of certain CFCs. Another meeting in Denmark in 1992 led to a complete ban on CFC production.

But there will be trade-offs. For example, CFC substitutes now available are less efficient as refrigerants, so it is estimated that appliances will use 3% more electricity in the United States, and this will increase consumer costs. Furthermore, because electricity is mostly generated by burning fossil fuels, the amount of CO_2 evolved will increase, which in turn will contribute to the "greenhouse" effect (page 563).

CFCs and their relation to ozone depletion is just one more example of the risks-and-benefits problem first described in the Introduction to this book (page 12). In this case scientists and citizens have concluded that the risks of CFCs outweigh the benefits.

Figure B This satellite image, acquired in September 1992, shows that the stratospheric ozone concentration is depleted near the South Pole. The image is color-coded to indicate relative concentrations, blue being low and red being high. For more images and information on ozone in the atmosphere go to Web site http://daac.gsfc.nasa.gov. *(NASA)*

Reaction Mechanisms and Rate Equations

Because rate equations are determined by experiment, the dependence of rate on concentration is an experimental fact. Mechanisms are constructs of our imagination, intuition, and good "chemical sense." To describe a mechanism we need to make a guess (hopefully a *good* guess) about how the reaction occurs at the particulate level. Often, several mechanisms can be proposed that correspond to the observed rate equation, and a postulated mechanism will often be wrong. A good mechanism is a worthy goal, however; it allows us to understand the chemistry better. A practical consequence of a good mechanism is that it allows us to predict important things, such as how to control a reaction better and how to design new and related experiments that expand our knowledge of chemistry.

Given the scope of chemical reactions that are known, it should not be surprising that the link between rate laws (the macroscopic world) and mechanisms (the particulate world) is very complex. We can only go a small way into this subject and enunciate a few basic principles.

One of the most important aspects of kinetics to understand is that products of a reaction can never be produced at a rate faster than the rate of the slowest step. The rate of the overall reaction is limited by, and is exactly equal to, the combined rates of all elementary steps up through the slowest step in the mechanism. The slowest elementary step of a sequence is called the **rate-determining step,** or rate-limiting step. You are already familiar with rate-determining steps. No matter how fast you shop in the supermarket, it always seems that the time it takes to finish is determined by the wait in the checkout line.

Imagine a reaction takes place with a mechanism involving two sequential steps and assume that we know the rates of both steps. The first step is slow, the second is fast:

Elementary Step 1 $$\mathrm{A + B \xrightarrow[\text{Slow, } E_a \text{ large; } 0.001 \text{ reactions/s}]{k_1} X + M}$$

Elementary Step 2 $$\mathrm{M + A \xrightarrow[\text{Fast, } E_a \text{ small; } 100 \text{ reactions/s}]{k_2} Y}$$

Overall Reaction $$\mathrm{2\,A + B \longrightarrow X + Y}$$

In the first step, A and B come together and slowly react to form one of the products (X) plus another reactive species, M. Almost as soon as M is formed, however, it is rapidly consumed by reaction with an additional molecule of A to form the second product Y. The products X and Y are the result of two elementary steps. The rate-determining elementary step is the first step. This step is bimolecular and so has the rate equation

$$\text{Rate} = k_1[\mathrm{A}][\mathrm{B}]$$

where k_1 is the rate constant for that step. The overall reaction is expected to follow this same second-order rate equation.

At this introductory level you cannot be expected to derive reaction mechanisms. Given a mechanism, however, you can decide if it is in agreement with experiment.

Let us apply these ideas to the mechanism of a real reaction. Experiment shows that the reaction of nitrogen dioxide with fluorine has a second-order rate equation:

Overall Reaction $2\ NO_2(g) + F_2(g) \longrightarrow 2\ FNO_2(g)$

$$\text{Rate} = k[NO_2][F_2]$$

The experimental rate equation immediately rules out the possibility that the reaction occurs in a single step. If the reaction was an elementary step, the rate law would have a second-order dependence on $[NO_2]$. Because a single-step reaction is ruled out, it follows that there must be at least two steps in the mechanism. We can also conclude that the rate-determining elementary step must involve NO_2 and F_2 in a 1:1 ratio. The simplest possible mechanism is identical with the hypothetical example:

Elementary Step 1 Slow $NO_2(g) + F_2(g) \xrightarrow{k_1} FNO_2(g) + F(g)$

Elementary Step 2 Fast $NO_2(g) + F(g) \xrightarrow{k_2} FNO_2(g)$

Overall Reaction $2\ NO_2(g) + F_2(g) \longrightarrow 2\ FNO_2(g)$

This proposed mechanism suggests that molecules of NO_2 and F_2 first react to produce one molecule of the product (FNO_2) plus an F atom. In a second step, the F atom produced in the first step reacts with additional NO_2 to give a second molecule of product. If we assume that the first, bimolecular step is rate-determining, its rate equation would be "Rate = $k_1[NO_2][F_2]$, the rate equation observed experimentally. The experimental rate constant is, therefore, the same as k_1.

The F atom in the first step of the NO_2/F_2 reaction is a reaction intermediate. It does not appear in the net stoichiometric equation describing the overall reaction. Reaction intermediates usually have a very fleeting existence, but they occasionally have long enough lifetimes to be observed. One of the tests of a proposed mechanism is the detection of an intermediate.

Example 15.12 Elementary Steps and Reaction Mechanisms

Oxygen atom transfer from nitrogen dioxide to carbon monoxide produces nitrogen monoxide and carbon dioxide:

$$NO_2(g) + CO(g) \longrightarrow NO(g) + CO_2(g)$$

It has the following rate equation at temperatures less than 500 K:

$$\text{Rate} = k[NO_2]^2$$

Can this reaction occur in one bimolecular step whose stoichiometry is the same as the overall reaction?

Solution If the reaction occurs simply by the collision of one NO_2 molecule with one CO molecule (i.e., the equation for the overall reaction and the equation for the single elementary step are the same), the rate equation would be

$$\text{Rate} = k[NO_2][CO]$$

This does not agree with experiment, so the mechanism must involve more than a single step. In fact, the reaction is thought to occur in two bimolecular steps:

Elementary Step 1 Slow, rate-determining $2\ NO_2(g) \longrightarrow NO_3(g) + NO(g)$

Elementary Step 2 Fast $NO_3(g) + CO(g) \longrightarrow NO_2(g) + CO_2(g)$

Overall Reaction $NO_2(g) + CO(g) \longrightarrow NO(g) + CO_2(g)$

The first or rate-determining step indeed has a rate equation that agrees with experiment.

Exercise 15.13 Elementary Steps and Reaction Mechanism

The Raschig reaction produces the industrially important reducing agent hydrazine, N_2H_4, from NH_3 and OCl^- in basic, aqueous solution. A proposed mechanism is

Step 1 Fast $NH_3(aq) + OCl^-(aq) \longrightarrow NH_2Cl(aq) + OH^-(aq)$

Step 2 Slow $NH_2Cl(aq) + NH_3(aq) \longrightarrow N_2H_5^+(aq) + Cl^-(aq)$

Step 3 Fast $N_2H_5^+(aq) + OH^-(aq) \longrightarrow N_2H_4(aq) + H_2O(\ell)$

1. What is the overall stoichiometric equation?
2. Which step of the three is rate-determining?
3. Write the rate equation for the rate-determining elementary step.
4. What reaction intermediates are involved?

Another common two-step reaction mechanism involves an initial fast reaction that produces an intermediate, followed by a slower second step in which the intermediate is converted to the final product. The rate of the reaction is determined by the second step, for which a rate law can be written. The rate of that step, however, depends on the concentration of the intermediate. An important thing to remember, though, is that the rate law must be written with respect to the *reactants only*. An intermediate, whose concentration will probably not be measurable, cannot appear as a term in the rate expression.

The reaction of nitric oxide and oxygen provides an example in this category to discuss in more detail:

$$2\,NO(g) + O_2(g) \longrightarrow 2\,NO_2(g)$$

$$\text{Rate} = k[NO]^2[O_2]$$

The experimentally determined rate law shows second-order dependence on NO_2 and first-order dependence on O_2. Although this rate law would be correct for a termolecular reaction, there is experimental evidence that there is an intermediate in this reaction. A possible two-step mechanism that proceeds through an intermediate is

The first step described in this mechanism involves a chemical equilibrium. We have encountered equilibrium situations in earlier chapters, in particulate level descriptions of vapor pressure, in phase changes, and in saturated solutions. Chemical equilibrium will be discussed in detail in Chapters 16 through 19.

Elementary Step 1	Fast, Equilibrium	$NO(g) + O_2(g) \underset{k_{-1}}{\overset{k_1}{\rightleftharpoons}} OONO(g)$	intermediate
Elementary Step 2	Slow, Rate-determining	$NO(g) + OONO(g) \xrightarrow{k_2} 2\,NO_2(g)$	
Overall Reaction		$2\,NO(g) + O_2(g) \longrightarrow 2\,NO_2(g)$	

The second step of this reaction is the slow step, and the overall rate depends on it. We can write a rate law for the second step:

$$\text{Rate} = k_2[NO][OONO]$$

This rate law cannot be compared directly with the experimental rate law because it contains the concentration of a transient intermediate, OONO. Recall that the experimental rate law must be written only in terms of compounds appearing in the overall equation. We therefore need to express the postulated rate law in a way that eliminates the intermediate. To do this, we must look at the very rapid first step in this reaction sequence.

At the beginning of the reaction NO and O_2 react rapidly and produce the intermediate OONO; the rate of formation can be defined by a rate law with a rate constant k_1:

$$\text{Rate of production of OONO} = k_1[NO][O_2]$$

Because the intermediate is consumed only very slowly in the second step, it is possible for the OONO to revert to NO and O_2 before it reacts further:

$$\text{Rate of reversion of OONO to NO and } O_2 = k_{-1}[OONO]$$

As NO and O_2 form OONO, their concentration drops, so the rate of the forward reaction decreases. At the same time, the concentration of OONO builds up, so the rate of the reverse reaction increases. Eventually, the rates of the forward and reverse reactions become the same, and the first elementary step reaches a *state of equilibrium.* The forward and reverse reactions in the first step are so much faster than the second elementary step that equilibrium is established before any significant amount of OONO is consumed by NO to give NO_2. The state of equilibrium for the first step remains throughout the lifetime of the overall reaction.

Because equilibrium is established when the rates of the forward and reverse reactions are the same, this means

$$\text{Rate of forward reaction} = \text{rate of reverse reaction}$$
$$k_1[NO][O_2] = k_{-1}[OONO]$$

Rearranging this equation, we find

$$\frac{k_1}{k_{-1}} = \frac{[OONO]}{[NO][O_2]} = K$$

Both k_1 and k_{-1} are constants (they will change only if the temperature changes). We can define a new constant K equal to the ratio of these two constants and equate K to the quotient $[OONO]/[NO][O_2]$. From this we can come up with an expression for the concentration of OONO:

$$[OONO] = K[NO][O_2]$$

If $K[NO][O_2]$ is substituted for [ONOO] in the rate law for the rate-determining elementary step, we have

$$\text{Rate} = k_2[NO][OONO] = k_2[NO]\{K[NO][O_2]\}$$
$$= k_2K[NO]^2[O_2]$$

Because both k_2 and K are constants, their product is another constant k', and we have

$$\text{Rate} = k'[NO]^2[O_2]$$

This is exactly the rate law that was derived from experiment, and so this is a reasonable mechanism to postulate for this reaction. It is *not* the only possible mechanism, however. As described in Example 15.13, at least one other mechanism is consistent with the same experimental rate law.

Example 15.13 Reaction Mechanisms Involving an Equilibrium Step

The NO + O_2 reaction described in the text could also occur by the following mechanism:

Elementary Step 1 Fast, Equilibrium $NO(g) + NO(g) \underset{k_{-1}}{\overset{k_1}{\rightleftharpoons}} N_2O_2(g)$ intermediate

Elementary Step 2 Slow, Rate-determining $N_2O_2(g) + O_2(g) \xrightarrow{k_2} 2\ NO_2(g)$

Overall Reaction $2\ NO(g) + O_2(g) \longrightarrow 2\ NO_2(g)$

Show that this mechanism also leads to the experimental rate law, Rate = $k[NO]^2[O_2]$.

Solution The rate law for the rate determining elementary step is

$$\text{Rate} = k_2[N_2O_2][O_2]$$

The compound N_2O_2 is an intermediate, which is consumed in the slow step and cannot appear in the final derived rate law. Again we need to express the postulated rate law in a way that eliminates the intermediate. As shown in the preceding discussion, we recognize that $[N_2O_2]$ is related to [NO] by the equilibrium constant, $K(= [N_2O_2]/[NO]^2)$.

If this is solved for $[N_2O_2]$, we have $[N_2O_2] = K[NO]^2$. When this is substituted into the derived rate law

$$\text{Rate} = k_2\{K[NO]^2\}[O_2]$$

the resulting equation is seen to be identical with the experimental rate law where $k_2K = k$.

The NO + O_2 reaction has an experimental rate law for which at least two mechanisms can be proposed. The challenge is to decide which is correct. In this case further experimentation detected the species OONO as a short-lived intermediate, confirming the mechanism involving this intermediate.

Exercise 15.14 Reaction Mechanisms Involving a Fast Initial Step

One possible mechanism for the decomposition of nitryl chloride, NO_2Cl, is

Elementary Step 1 Fast, Equilibrium $$NO_2Cl(g) \underset{k-1}{\overset{k_1}{\rightleftharpoons}} NO_2(g) + Cl(g)$$

Elementary Step 2 Slow $$NO_2Cl(g) + Cl(g) \xrightarrow{k_2} NO_2(g) + Cl_2(g)$$

What is the overall reaction? What rate law would be derived from this mechanism? What effect does increasing the concentration of the product NO_2 have on the reaction rate?

Problem-Solving Tips and Ideas

15.2 A Summary of the Principles of Rate Equations and Reaction Mechanisms

The connection between an experimental rate equation and the hypothesis of a reaction mechanism is important in chemistry.

1. Experiments must first be performed that define the effect of reactant concentrations on the rate of the reaction. This gives the experimental rate equation.
2. A mechanism for the reaction is proposed on the basis of the experimental rate equation, the principles of stoichiometry and molecular structure and bonding, general chemical experience, and intuition.
3. The *proposed* reaction mechanism is used to *derive* a rate equation. This rate equation must contain only those species present in the overall chemical reaction and not reaction intermediates. If the derived and experimental rate equations are the same, the postulated mechanism *may* be a reasonable hypothesis of the reaction sequence.
4. If more than one mechanism can be proposed, and they all predict derived rate equations in agreement with experiment, then more experiments must be done.

We have described only a few of many possible reaction mechanisms. Many reactions have quite complex mechanisms. As an example, you might wish to consider some of the reactions that are thought to lead to the much-discussed hole in the earth's ozone layer. Although many interrelated processes combine in this phenomenon, scientists now understand some of them reasonably well, and a few are described on page 728.

CHAPTER HIGHLIGHTS

Having studied this chapter, you should be able to

- Explain the concept of **reaction rate** (Section 15.1).
- Derive the **average** and **instantaneous rate** of a reaction from experimental information (Section 15.1).
- Describe the various conditions that affect reaction rate (i.e., reactant concentrations, temperature, presence of a catalyst, and the state of the reactants) (Section 15.2).
- Define the various parts of a **rate equation** and their significance (the **rate constant** and **order of reaction**) (Section 15.3).
- Derive a rate equation from experimental information (Section 15.3).
- Describe and use the relationships between reactant concentration and time for zero-order, first-order, and second-order reactions (Section 15.4 and Table 15.1).

  For the first-order reaction R $\longrightarrow$ products, the integrated rate law is

$$\ln \frac{[R]_t}{[R]_0} = -kt \qquad [15.1]$$

  where $[R]_0$ is the initial concentration of R, $[R]_t$ is the concentration after some time t has elapsed, and k is the rate constant.
- Apply graphical methods for determining reaction order and the rate constant from experimental data (Section 15.4 and Table 15.1).
- Use the concept of **half-life ($t_{1/2}$)**, especially for first-order reactions (Section 15.4). For a first-order reaction, the half-life is inversely proportional to k:

$$t_{1/2} = \frac{0.693}{k} \qquad [15.4]$$

- Describe the **collision theory** of reaction rates (Section 15.5).
- Appreciate the relation of **activation energy (E_a)** to the rate and thermodynamics of a reaction (Section 15.5).
- Use collision theory to describe the effect of reactant concentration on reaction rate (Section 15.5).
- Describe the functioning of a catalyst and its effect on the activation energy and mechanism of a reaction (Section 15.5).
- Define **homogeneous** and **heterogeneous catalysts** (Section 15.5).
- Understand the effect of molecular orientation on reaction rate (Section 15.5).
- Describe the effect of temperature on reaction rate using the theories of reaction rates and the **Arrhenius equation** (Equation 15.7 and Section 15.5).
- Use Equations 15.5, 15.6, and 15.7 to calculate the activation energy from experimental data (Section 15.5).

- Understand the concept of **reaction mechanism** (the sequence of bond-making and bond-breaking steps that occurs during the conversion of reactants to products) and the relation of the mechanism to the overall, stoichiometric equation for a reaction (Section 15.6).
- Describe the **elementary steps** of a mechanism and give their **molecularity** (Section 15.6).
- Define the **rate-determining step** in a mechanism and identify any **reaction intermediates** (Section 15.6).

STUDY QUESTIONS

REVIEW QUESTIONS

1. Which of the following can be used to determine the rate equation for a chemical reaction?
   (a) Theoretical calculations
   (b) Measuring the rate of the reaction as a function of the concentration of the reacting species
   (c) Measuring the rate of the reaction as a function of temperature
2. Describe four factors that influence the rate of a chemical reaction.
3. Refer to Figure 15.3. After 2.0 h, what is the concentration of NO_2? Of O_2?
4. Using the rate equation "Rate = $k[A]^2[B]$," define the order of the reaction with respect to A and B. What is the total order of the reaction?
5. A reaction has the experimental rate equation "Rate = $k[A]^2$." How will the rate change if the concentration of A is tripled? If the concentration of A is halved?
6. A reaction has the experimental rate equation "Rate = $k[A]^2[B]$." If the concentration of A is doubled, and the concentration of B is halved, what happens to the reaction rate?
7. Write the equation relating concentration of reactant and time for a first-order reaction. Define each term in the equation.
8. After 5 half-life periods for a first-order reaction, what fraction of reactant remains?
9. Plotting 1/[reactant] versus time for a reaction produces a straight line. What is the order of the reaction? If a straight line of negative slope is observed for a plot of ln [reactant] versus time, what is the order of the reaction?
10. Draw a reaction energy diagram for a single-step, exothermic process. Mark the activation energies of the forward and reverse processes. Identify the net energy change for the reaction on this diagram.
11. Explain how collision theory accounts for the temperature dependence of reaction rates.
12. What experimental information is required to use the Arrhenius equation to calculate the activation energy of a reaction?
13. Define what is meant by the term "mechanism" for a chemical reaction.
14. Define the terms "elementary step" and "rate-determining step" as applied to mechanisms.
15. What is a reaction intermediate? Give an example using one of the reaction mechanisms described in the text.
16. What is a catalyst? What is the effect of a catalyst on the mechanism of a reaction?
17. Explain the difference between a homogeneous and a heterogeneous catalyst. Give an example of each.

NUMERICAL QUESTIONS

Reaction Rates

18. Give the relative rates of disappearance of reactants and formation of products for each of the following reactions:
   (a) $2\ O_3(g) \longrightarrow 3\ O_2(g)$
   (b) $2\ HOF(g) \longrightarrow 2\ HF(g) + O_2(g)$

19. Give the relative rates of disappearance of reactants and formation of products for each of the following reactions:
   (a) $2\ NO(g) + Br_2(g) \longrightarrow 2\ NOBr(g)$
   (b) $N_2(g) + 3\ H_2(g) \longrightarrow 2\ NH_3(g)$

20. Experimental data are listed here for the reaction $A \longrightarrow 2\ B$.

Time (s)	**[B] (mol/L)**
0.00	0.000
10.0	0.326
20.0	0.572
30.0	0.750
40.0	0.890

   (a) Prepare a graph from these data, connect the points with a smooth line, and calculate the rate of change of [B] for each 10-s interval from 0.0 to 40.0 s. Does the rate of change decrease from one time interval to the next? Suggest a reason for this result.
   (b) How is the rate of change of [A] related to the rate of change of [B] in each time interval? Calculate the rate of change of [A] for the time interval from 10.0 to 20.0 s.
   (c) What is the instantaneous rate when [B] = 0.750 mol/L?

21. Phenyl acetate, an ester, reacts with water according to the equation

$$\underset{\text{phenyl acetate}}{CH_3\overset{\overset{O}{\parallel}}{C}—O—C_6H_5} + H_2O(\ell) \longrightarrow \underset{\text{acetic acid}}{CH_3\overset{\overset{O}{\parallel}}{C}—O—H(aq)} + \underset{\text{phenol}}{C_6H_5—OH(aq)}$$

The data in the table were collected for this reaction at 5 °C.

Time (s)	[Phenyl acetate] (mol/L)
0	0.55
15.0	0.42
30.0	0.31
45.0	0.23
60.0	0.17
75.0	0.12
90.0	0.085

(a) Plot the phenyl acetate concentration versus time, and describe the shape of the curve observed.
(b) Calculate the rate of change of the phenyl acetate concentration during the period 15.0 s to 30.0 s and also during the period 75.0 s to 90.0 s. Compare the values, and suggest a reason why one value is smaller than the other.
(c) What is the rate of change of the phenol concentration during the time period 60.0 s to 75.0 s?
(d) What is the instantaneous rate at 15.0 s?

Concentration and Rate Equations

22. The reaction between ozone and nitrogen dioxide at 231 K is first order in both $[NO_2]$ and $[O_3]$.

$$2\ NO_2(g) + O_3(g) \longrightarrow N_2O_5(s) + O_2(g)$$

(a) Write the rate equation for the reaction.
(b) If the concentration of NO_2 is tripled, what is the change in the reaction rate?
(c) What is the effect on reaction rate if the concentration of O_3 is halved?

23. Nitrosyl bromide, NOBr, is formed from NO and Br_2:

$$2\ NO(g) + Br_2(g) \longrightarrow 2\ NOBr(g)$$

Experiments show the reaction is second order in NO and first order in Br_2.
(a) Write the rate equation for the reaction.
(b) How does the reaction rate change if the concentration of Br_2 is changed from 0.0022 mol/L to 0.0066 mol/L?
(c) What is the change in the reaction rate if the concentration of NO is changed from 0.0024 mol/L to 0.0012 mol/L?

24. The data in the table are for the reaction of NO and O_2 at 660 K.

$$2\ NO(g) + O_2(g) \longrightarrow 2\ NO_2(g)$$

Reactant Concentration (mol/L)		Rate of Disappearance of NO
[NO]	$[O_2]$	(mol/L · s)
0.010	0.010	2.5×10^{-5}
0.020	0.010	1.0×10^{-4}
0.010	0.020	5.0×10^{-5}

(a) Determine the order of the reaction for each reactant.
(b) Write the rate equation for the reaction.
(c) Calculate the rate constant.
(d) Calculate the rate (in mol/L · s) at the instant when [NO] = 0.015 mol/L and $[O_2]$ = 0.0050 mol/L.
(e) At the instant when NO is reacting at the rate 1.0×10^{-4} mol/L · s, what is the rate at which O_2 is reacting and NO_2 is forming?

25. The reaction $2\ NO(g) + 2\ H_2(g) \longrightarrow N_2(g) + 2\ H_2O(g)$ was studied at 904 °C, and the data in the table were collected.

Reactant Concentration (mol/L)		Rate of Appearance of N_2
[NO]	$[H_2]$	(mol/L · s)
0.420	0.122	0.136
0.210	0.122	0.0339
0.210	0.244	0.0678
0.105	0.488	0.0339

(a) Determine the order of the reaction for each reactant.
(b) Write the rate equation for the reaction.
(c) Calculate the rate constant for the reaction at 904 °C.
(d) Find the rate of appearance of N_2 at the instant when [NO] = 0.350 mol/L and $[H_2]$ = 0.205 mol/L.

Concentration–Time Equations

26. The rate equation for the hydrolysis of sucrose to fructose and glucose

$$C_{12}H_{22}O_{11}(aq) + H_2O(\ell) \longrightarrow 2\ C_6H_{12}O_6\ (aq)$$

is "Rate = $k[C_{12}H_{22}O_{11}]$." After 2.57 h at 27 °C, 5.00 g/L of sucrose has decreased to 4.50 g/L. Find the rate constant k.

27. The decomposition of N_2O_5 in CCl_4 is a first-order reaction. If 2.56 mg of N_2O_5 is present initially, and 2.50 mg is present after 4.26 min at 55 °C, what is the value of the rate constant k?

28. The decomposition of SO_2Cl_2 is a first-order reaction:

$$SO_2Cl_2(g) \longrightarrow SO_2(g) + Cl_2(g)$$

The rate constant for the reaction is $2.8 \times 10^{-3}\ min^{-1}$ at 600 K. If the initial concentration of SO_2Cl_2 is 1.24×10^{-3} mol/L, how long will it take for the concentration to drop to 0.31×10^{-3} mol/L?

29. The conversion of cyclopropane to propene, described in Example 15.3, occurs with a first-order rate constant of $5.4 \times 10^{-2}\ h^{-1}$. How long will it take for the concentration of cyclopropane to decrease from an initial concentration 0.080 mol/L to 0.020 mol/L?

30. Ammonium cyanate, NH_4NCO, rearranges in water to give urea, $(NH_2)_2CO$:

$$NH_4NCO(aq) \longrightarrow (NH_2)_2CO(aq)$$

The rate equation for this process is "Rate $= k[NH_4NCO]^2$," where $k = 0.0113$ L/mol · min. If the original concentration of NH_4NCO in solution is 0.229 mol/L, how long will it take for the concentration to decrease to 0.180 mol/L?

31. The decomposition of nitrogen dioxide at a high temperature

$$NO_2(g) \longrightarrow NO(g) + \tfrac{1}{2}\,O_2(g)$$

is second order in this reactant. The rate constant for this reaction is 3.40 L/mol · min. Determine the time needed for the concentration of NO_2 to decrease from 2.00 mol/L to 1.50 mol/L.

Half-Life

32. The rate equation for the decomposition of N_2O_5 (giving NO_2 and O_2) is "Rate $= k[N_2O_5]$." For the reaction at a particular temperature, the value of k is $5.0 \times 10^{-4}\ s^{-1}$.
(a) Calculate the half-life of N_2O_5 in the reaction.
(b) How long does it take for the N_2O_5 concentration to drop to one tenth of its original value?

33. The decomposition of SO_2Cl_2

$$SO_2Cl_2(g) \longrightarrow SO_2(g) + Cl_2(g)$$

is first order in SO_2Cl_2, and the reaction has a half-life of 245 min at 600 K. If you begin with 3.6×10^{-3} mol of SO_2Cl_2 in a 1.0-L flask, how long will it take for the quantity of SO_2Cl_2 to decrease to 2.00×10^{-4} mol?

34. Gaseous azomethane, $CH_3N{=}NCH_3$, decomposes in a first-order reaction when heated:

$$CH_3N{=}NCH_3(g) \longrightarrow N_2(g) + C_2H_6(g)$$

The rate constant for this reaction at 425 °C is 40.8 min^{-1}. If the initial amount of azomethane in the flask is 2.00 g, how much remains after 0.0500 min? How many moles of N_2 are formed in this time?

35. The compound $Xe(CF_3)_2$ is unstable, decomposing to elemental Xe with a half-life of 30. min. If you place 7.50 mg of $Xe(CF_3)_2$ in a flask, how long must you wait until only 0.25 mg of $Xe(CF_3)_2$ remains?

36. The radioactive isotope ^{64}Cu is used in the form of copper(II) acetate to study Wilson's disease. The isotope has a half-life of 12.70 h. What fraction of radioactive copper(II) acetate remains after 64 h?

37. Radioactive gold-198 is used as the metal in the diagnosis of liver problems. The half-life of this isotope is 2.7 days. If you begin with a 5.6-mg sample of the isotope, how much of this sample remains after 1.0 day?

Graphical Analysis of Rate Equations and k

38. Common sugar, sucrose, breaks down in dilute acid solution to give the simpler sugars glucose and fructose. Both of the simple sugars have the same formula, $C_6H_{12}O_6$:

$$C_{12}H_{22}O_{11}(aq) + H_2O(\ell) \longrightarrow 2\ C_6H_{12}O_6(aq)$$

The rate of this reaction has been studied in acid solution, and the data in the table were obtained.

Time (min)	$[C_{12}H_{22}O_{11}]$ (mol/L)
0	0.316
39	0.274
80	0.238
140	0.190
210	0.146

(a) Plot the data in the table as ln [sucrose] versus time and 1/[sucrose] versus time. What is the order of the reaction?
(b) Write the rate equation for the reaction, and calculate the rate constant k.
(c) Estimate the concentration of sucrose after 175 min.

39. Data for the reaction of phenyl acetate with water are given in Study Question 21. Plot these data as ln [phenyl acetate] and 1/[phenyl acetate] versus time. Based on the appearance of the two graphs, what can you conclude about the order of the reaction with respect to phenyl acetate? Working from the data and the rate law, determine the rate constant for the reaction.

40. Data for the decomposition of dinitrogen oxide

$$2\ N_2O(g) \longrightarrow 2\ N_2(g) + O_2(g)$$

on a gold surface at 900 °C are given. Verify that the reaction is first order by preparing a graph of ln $[N_2O]$ versus time. Derive the rate constant from the slope of the line in this graph (see Table 15.1). Using the rate law and value of k, determine the decomposition rate at 900 °C when $[N_2O] = 0.035$ mol/L.

Time (min)	$[N_2O]$ (mol/L)
15.0	0.0835
30.0	0.0680
80.0	0.0350
120.0	0.0220

41. Ammonia decomposes when heated according to the equation

$$NH_3(g) \longrightarrow NH_2(g) + H(g)$$

The data in the table for this reaction were collected at 2000 K.

Time (h)	$[NH_3]$ (mol/L)
0	8.00×10^{-7}
25	6.75×10^{-7}
50	5.84×10^{-7}
75	5.15×10^{-7}

Prepare graphs of ln $[NH_3]$ versus time, and $1/[NH_3]$ versus time. Can you draw a conclusion from these graphs about the order of this reaction with respect to NH_3? Find the rate constant for the reaction from the slope of the line.

Kinetics and Energy

42. Calculate the activation energy, E_a, for the reaction

$$N_2O_5(g) \longrightarrow 2\ NO_2(g) + \tfrac{1}{2}\ O_2(g)$$

from the observed rate constants: k at 25 °C = 3.46×10^{-5} s^{-1} and k at 55 °C = 1.5×10^{-3} s^{-1}.

43. If the rate constant for a reaction triples in value when the temperature rises from 3.00×10^2 K to 3.10×10^2 K, what is the activation energy of the reaction?

44. When heated to a high temperature, cyclobutane, C_4H_8, decomposes to ethylene:

$$C_4H_8(g) \longrightarrow 2\ C_2H_4(g)$$

The activation energy, E_a, for this reaction is 260 kJ/mol. At 800 K, the rate constant $k = 0.0315$ s^{-1}. Determine the value of k at 850 K.

45. When heated to a high temperature, cyclopropane is converted to propene (see Example 15.3). Rate constants for this reaction at 470 °C and 510 °C are $k = 1.10 \times 10^{-4}$ s^{-1} and $k = 1.02 \times 10^{-3}$ s^{-1}, respectively. Determine the activation energy, E_a, from these data.

46. The reaction of H_2 molecules with F atoms

$$H_2(g) + F(g) \longrightarrow HF(g) + H(g)$$

has an activation energy of 8 kJ/mol and an energy change for the reaction of −133 kJ/mol. Draw a diagram similar to Figure 15.14b for this process. Indicate the activation energy and enthalpy of reaction on this diagram.

47. Answer questions (a) and (b) based on the accompanying reaction coordinate diagram.

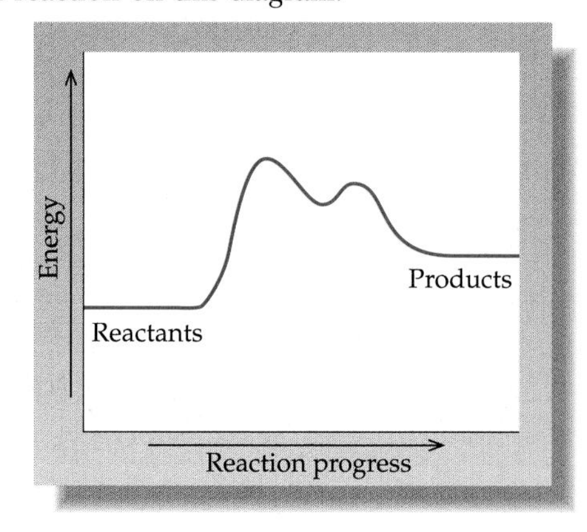

(a) Is the reaction exothermic or endothermic?

(b) Does the reaction occur in more than one step? If so, how many?

Mechanisms

48. What is the rate law for each of the following *elementary* reactions?

(a) $NO(g) + NO_3(g) \longrightarrow 2\ NO_2(g)$

(b) $Cl(g) + H_2(g) \longrightarrow HCl(g) + H(g)$

(c) $(CH_3)_3CBr(aq) \longrightarrow (CH_3)_3C^+(aq) + Br^-(aq)$

49. What is the rate law for the following *elementary* reactions?

(a) $Cl(g) + ICl(g) \longrightarrow I(g) + Cl_2(g)$

(b) $O(g) + O_3(g) \longrightarrow 2\ O_2(g)$

(c) $2\ NO_2(g) \longrightarrow N_2O_4(g)$

50. The reaction between chloroform, $CHCl_3$, and chlorine gas proceeds in a series of three elementary steps:

Step 1	Fast, reversible	$Cl_2(g) \rightleftharpoons 2\ Cl(g)$
Step 2	Slow	$CHCl_3(g) + Cl(g) \longrightarrow CCl_3(g) + HCl(g)$
Step 3	Fast	$CCl_3(g) + Cl(g) \longrightarrow CCl_4(g)$
Overall reaction		$CHCl_3(g) + Cl_2(g) \longrightarrow CCl_4(g) + HCl(g)$

(a) Which step is the rate-determining step?

(b) Write the rate equation for the rate-determining step.

51. Ozone, O_3, in the earth's upper atmosphere decomposes according to the equation $2\ O_3(g) \longrightarrow 3\ O_2(g)$. The mechanism of the reaction is thought to proceed through an initial fast, reversible step followed by a slow second step.

Step 1	Fast, reversible	$O_3(g) \rightleftharpoons O_2(g) + O(g)$
Step 2	Slow	$O_3(g) + O(g) \longrightarrow 2\ O_2(g)$

(a) Which of the steps is the rate-determining step?

(b) Write the rate equation for the rate-determining step.

52. The reaction of $NO_2(g)$ and $CO(g)$ is thought to occur in two steps:

Step 1	Slow	$NO_2(g) + NO_2(g) \longrightarrow NO(g) + NO_3(g)$
Step 2	Fast	$NO_3(g) + CO(g) \longrightarrow NO_2(g) + CO_2(g)$

(a) Show that the elementary steps add up to give the overall, stoichiometric equation.

(b) What is the molecularity of each step?

(c) For this mechanism to be consistent with kinetic data, what must be the experimental rate equation?

(d) Identify any intermediates in this reaction.

53. Iodide ion is oxidized in acid solution by hydrogen peroxide:

$$H_2O_2(aq) + 2\ H^+(aq) + 2\ I^-(aq) \longrightarrow I_2(aq) + 2\ H_2O(\ell)$$

A proposed mechanism is

Step 1	Slow	$H_2O_2(aq) + I^-(aq) \longrightarrow H_2O(\ell) + OI^-(aq)$

Step 2 Fast $H^+(aq) + OI^-(aq) \longrightarrow HOI(aq)$

Step 3 Fast $HOI(aq) + H^+(aq) + I^-(aq) \longrightarrow I_2(aq) + H_2O(\ell)$

(a) Show that the three elementary steps add up to give the overall, stoichiometric equation.
(b) What is the molecularity of each step?
(c) For this mechanism to be consistent with kinetic data, what must be the experimental rate equation?
(d) Identify any intermediates in the elementary steps in this reaction.

54. Hydrogen and carbon monoxide react to give formaldehyde under certain conditions:

$$H_2(g) + CO(g) \longrightarrow H_2CO(g)$$

The mechanism proposed for this reaction is

Step 1 Fast, reversible $H_2 \rightleftarrows 2\ H$

Step 2 Slow $H + CO \longrightarrow HCO$

Step 3 Fast $H + HCO \longrightarrow H_2CO$

What rate law would be derived from this mechanism?

55. The experimental rate equation for the reaction

$$H_2(g) + I_2(g) \rightleftarrows 2\ HI(g)$$

is "Rate = $k[H_2][I_2]$." Does the following mechanism satisfy the experimental rate equation?

Step 1 Fast, reversible $I_2 \rightleftarrows 2\ I$

Step 2 Fast, reversible $I + H_2 \rightleftarrows IH_2$

Step 3 Slow $IH_2 + I \longrightarrow 2\ HI$

Catalysis

56. Which of the following statements is (are) false? If the statement is incorrect, change it to make it correct.
(a) The concentration of a homogeneous catalyst may appear in the rate equation.
(b) A catalyst is consumed in the reaction.
(c) A catalyst must always be in the same phase as the reactants.

57. Which of the following statements is (are) false? If the statement is incorrect, change it to make it correct.
(a) A catalyst can change the course of a reaction so that different products are formed.
(b) A catalyst changes the rate-determining step in a reaction.
(c) The energy evolved in a reaction without a catalyst is different from the energy evolved in a reaction in which a catalyst is present.

58. Carbonic anhydrase is a biological catalyst, an enzyme that catalyzes the hydration of CO_2:

$$CO_2(g) + H_2O(\ell) \longrightarrow H_2CO_3(aq)$$

This is a critical reaction involved in the transfer of CO_2 from tissues to the lung via the bloodstream. One enzyme molecule hydrates 10^6 molecules of CO_2 per second. How many kilograms of CO_2 are hydrated in 1 h in 1 L containing 5×10^{-6} M enzyme?

59. Many biologically important reactions occur in the presence of catalysts called enzymes. The enzyme catalase (see Figure 15.6) catalyzes the decomposition of peroxides, reducing the activation energy from 72 kJ/mol (for an uncatalyzed process) to 28 kJ/mol at 298 K. By what factor does the rate constant k increase? (Assume A in the Arrhenius equation remains constant.)

GENERAL QUESTIONS

60. Data in the table were collected at 540 K for the following reaction:

$$CO(g) + NO_2(g) \longrightarrow CO_2(g) + NO(g)$$

(a) Derive the rate equation.
(b) Determine the reaction order with respect to each reactant.
(c) Calculate the rate constant, giving the correct units for k.

Initial Concentration (mol/L) [CO]	[NO_2]	**Initial Rate (mol/L · h)**
5.1×10^{-4}	0.35×10^{-4}	3.4×10^{-8}
5.1×10^{-4}	0.70×10^{-4}	6.8×10^{-8}
5.1×10^{-4}	0.18×10^{-4}	1.7×10^{-8}
1.0×10^{-3}	0.35×10^{-4}	6.8×10^{-8}
1.5×10^{-3}	0.35×10^{-4}	10.2×10^{-8}

61. Ammonium cyanate, NH_4NCO, rearranges in water to give urea, $(NH_2)_2CO$:

$$NH_4NCO(aq) \longrightarrow (NH_2)_2CO(aq)$$

Time (min)	**[NH_4NCO] (mol/L)**
0	0.458
4.50×10^1	0.370
1.07×10^2	0.292
2.30×10^2	0.212
6.00×10^2	0.114

Using the data in the table
(a) Decide if the reaction is first or second order.
(b) Calculate k for this reaction.
(c) Calculate the half-life of ammonium cyanate under these conditions.
(d) Calculate the concentration of NH_4NCO after 12.0 h.

62. Nitrogen oxides, NO_x (a mixture of NO and NO_2 collectively designated as NO_x), play an essential role in the production of pollutants found in photochemical smog. The NO_x in the atmosphere is slowly broken down to N_2 and O_2 in a first-order reaction. The average half-life of NO_x in the smokestack emissions in a large city during daylight is 3.9 h.

(a) Starting with 1.50 mg in an experiment, what quantity of NO_x remains after 5.25 h?
(b) How many hours of daylight must have elapsed to decrease 1.50 mg of NO_x to 2.50×10^{-6} mg?

63. At temperatures below 500 K the reaction between carbon monoxide and nitrogen dioxide

$$NO_2(g) + CO(g) \longrightarrow CO_2(g) + NO(g)$$

has the rate equation "Rate = $k[NO_2]^2$." Which of the three mechanisms suggested here best agrees with the experimentally observed rate equation?

Mechanism 1 *Single, elementary step*

$$NO_2 + CO \longrightarrow CO_2 + NO$$

Mechanism 2 *Two steps*

Slow $\quad NO_2 + NO_2 \longrightarrow NO_3 + NO$

Fast $\quad NO_3 + CO \longrightarrow NO_2 + CO_3$

Mechanism 3 *Two steps*

Slow $\quad NO_2 \longrightarrow NO + O$

Fast $\quad CO + O \longrightarrow CO_2$

64. Chlorine atoms contribute to the destruction of the earth's ozone layer by the following sequence of reactions:

$$Cl + O_3 \longrightarrow ClO + O_2$$
$$ClO + O \longrightarrow Cl + O_2$$

where the O atoms in the second step come from the decomposition of ozone by sunlight:

$$O_3(g) \longrightarrow O(g) + O_2(g)$$

What is the net equation on summing these three equations? Why does this lead to ozone loss in the stratosphere? What is the role played by Cl in this sequence of reactions? What name is given to species such as ClO?

65. Nitryl fluoride, an explosive compound, can be made by treating nitrogen dioxide with fluorine:

$$2\ NO_2(g) + F_2(g) \longrightarrow 2\ NO_2F(g)$$

Use the rate data in the table to do the following:
(a) Write the rate equation for the reaction.
(b) Indicate the order of reaction with respect to each component of the reaction.
(c) Indicate the numerical value of the rate constant k.

Experiment	Initial Concentrations (mol/L) $[NO_2]$	$[F_2]$	$[NO_2F]$	Initial Rate (mol/L · s)
1	0.001	0.005	0.001	2×10^{-4}
2	0.002	0.005	0.001	4×10^{-4}
3	0.006	0.002	0.001	4.8×10^{-4}
4	0.006	0.004	0.001	9.6×10^{-4}
5	0.001	0.001	0.001	4×10^{-5}
6	0.001	0.001	0.002	4×10^{-5}

66. Describe each of the following statements as true or false. If false, rewrite the sentence to make it correct.
(a) The rate-determining elementary step in a reaction is the slowest step in a mechanism.
(b) It is possible to change the rate constant by changing the temperature.
(c) As a reaction proceeds at constant temperature, the rate remains constant.
(d) A reaction that is third order overall must involve more than one step.

67. The decomposition of dinitrogen pentaoxide

$$2\ N_2O_5(g) \longrightarrow 4\ NO_2(g) + O_2(g)$$

has the rate equation "Rate = $k[N_2O_5]$." It has been found experimentally that the decomposition is 20% complete in 6.0 h at 300 K. Calculate the rate constant and the half-life at 300 K.

68. The data in the table give the temperature dependence of the rate constant for the reaction $N_2O_5(g) \longrightarrow 2\ NO_2(g) + \frac{1}{2}\ O_2(g)$. Plot these data in the appropriate way to derive the activation energy for the reaction.

T (K)	k (s^{-1})
338	4.87×10^{-3}
328	1.50×10^{-3}
318	4.98×10^{-4}
308	1.35×10^{-4}
298	3.46×10^{-5}
273	7.87×10^{-7}

69. The decomposition of gaseous dimethyl ether at ordinary pressures is first order. Its half-life is 25.0 min at 500 °C:

$$CH_3OCH_3(g) \longrightarrow CH_4(g) + CO(g) + H_2(g)$$

(a) Starting with 8.00 g of dimethyl ether, what mass remains (in grams) after 125 min and after 145 min?
(b) Calculate the time in minutes required to decrease 7.60 ng (nanograms) to 2.25 ng.
(c) What fraction of the original dimethyl ether remains after 150 min?

70. The decomposition of phosphine, PH_3, proceeds according to the equation

$$4\ PH_3(g) \longrightarrow P_4(g) + 6\ H_2(g)$$

It is found that the reaction has the rate equation "Rate = $k[PH_3]$." The half-life of PH_3 is 37.9 s at 120 °C.
(a) How much time is required for three fourths of the PH_3 to decompose?
(b) What fraction of the original sample of PH_3 remains after 1 min?

71. Three mechanisms are proposed for the gas phase reaction of NO with Br_2 to give BrNO:

Mechanism 1

$$NO(g) + NO(g) + Br_2(g) \longrightarrow 2\ BrNO(g)$$

Mechanism 2

Step 1 $NO(g) + Br_2(g) \longrightarrow Br_2NO(g)$

Step 2 $Br_2NO(g) + NO(g) \longrightarrow 2\ BrNO(g)$

Mechanism 3

Step 1 $NO(g) + NO(g) \longrightarrow N_2O_2(g)$

Step 2 $N_2O_2(g) + Br_2(g) \longrightarrow 2\ BrNO(g)$

(a) Write the balanced equation for the net reaction.
(b) What is the molecularity for each step in each mechanism?
(c) What are the intermediates formed in Mechanisms 2 and 3?

72. An energy diagram is given here for the adsorption and dissociation of O_2 on a platinum surface.

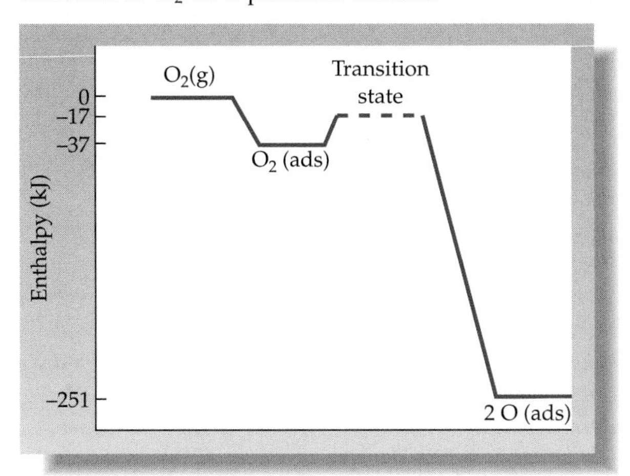

(a) What is the energy change for (1) $O_2(g) \longrightarrow O_2(\text{adsorbed})$; (2) $O_2(\text{adsorbed}) \longrightarrow 2\ O(\text{adsorbed})$; (3) $O_2(g) \longrightarrow 2\ O(\text{adsorbed})$?
(b) What is approximate activation energy for $O_2(\text{adsorbed}) \longrightarrow 2\ O(\text{adsorbed})$?

73. Radioactive iodine-131, which has a half-life of 8.04 days, is used in the form of sodium iodide to treat cancer of the thyroid. If you begin with 25.0 mg of $Na^{131}I$, what quantity of the material remains after 31 days?

74. The reaction between chloroform, $CHCl_3$, and chlorine gas is thought to proceed as discussed in Study Question 50. Show that the mechanism agrees with the experimental rate law "Rate = $k[CHCl_3][Cl_2]^{1/2}$."

75. The ozone in the earth's ozone layer decomposes according to the equation

$$2\ O_3(g) \longrightarrow 3\ O_2(g)$$

The mechanism of the reaction is thought to proceed through an initial fast equilibrium and a slow step:

Step 1 Fast, Reversible $O_3(g) \rightleftharpoons O_2(g) + O(g)$

Step 2 Slow $O_3(g) + O(g) \longrightarrow 2\ O_2(g)$

Show that the mechanism agrees with the experimental rate law "Rate = $k[O_3]^2/[O_2]$."

CONCEPTUAL QUESTIONS

76. Hydrogenation reactions, processes wherein H_2 is added to a molecule, are usually catalyzed. An excellent catalyst is a very finely divided metal suspended in the reaction solvent. Tell why finely divided rhodium, for example, is a much more efficient catalyst than a small block of the metal.

77. It is instructive to use a mathematical model in connection with Study Question 76. Suppose you have 1000 blocks, each of which is 1.0 cm on a side. If all 1000 of these blocks are stacked to give a cube that is 10. cm on a side, what fraction of the 1000 blocks have at least one surface on the outside surface of the cube? Now divide the 1000 blocks into eight equal piles of blocks and form them into eight cubes, 5.0 cm on a side. Now what fraction of the blocks have at least one surface on the outside of the cubes? How does this mathematical model pertain to Study Question 76?

78. Isotopes are often used as "tracers" to follow an atom through a chemical reaction, and the following is an example. Acetic acid reacts with methanol by eliminating a molecule of water and forming methyl acetate (see Chapter 11).

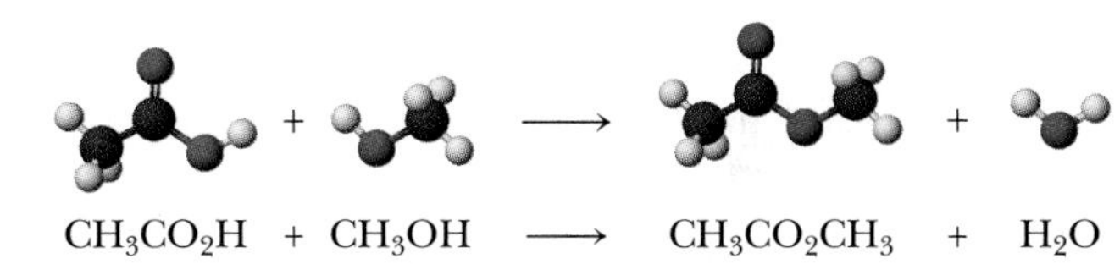

$$CH_3CO_2H + CH_3OH \longrightarrow CH_3CO_2CH_3 + H_2O$$

Explain how you could use the isotope ^{18}O to show whether the oxygen atom in the water comes from the —OH of the acid or the —OH of the alcohol.

79. Examine the reaction coordinate diagram given here.

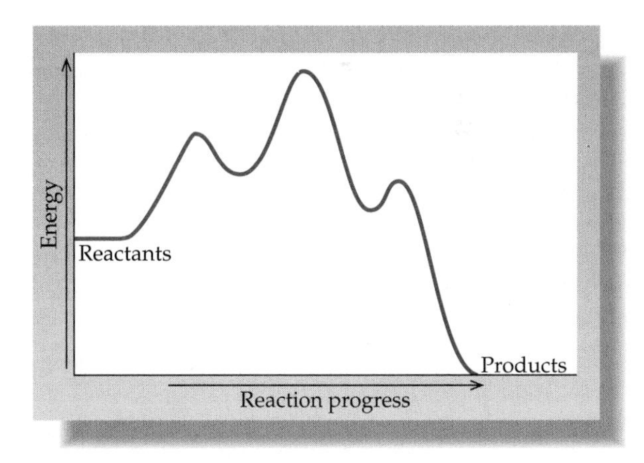

(a) How many steps are in the mechanism for the reaction described by this diagram?
(b) Which is the slowest step in the reaction?
(c) Is the reaction overall exothermic or endothermic?

CHALLENGING QUESTIONS

80. Gaseous radon is sometimes found in homes (see Chapter 24). The isotope radon-222 (^{222}Rn) is radioactive with a half-

life of 3.82 d. Assume there is radon gas in the basement of a home (with the dimensions 12 m × 7.0 m × 3.0 m) and that the gas has a partial pressure of 1.0×10^{-6} mm Hg.
(a) How many atoms of ^{222}Rn are in each liter of air in the basement?
(b) If the radon gas is not replenished in the basement, how many atoms of ^{222}Rn remain per liter of air after 1 month (31 days)?

81. Data for the following reaction

$$[Mn(CO)_5(CH_3CN)]^+ + NC_5H_5 \longrightarrow [Mn(CO)_5(NC_5H_5)]^+ + CH_3CN$$

appear in the table. Calculate E_a from a plot of ln k versus $1/T$.

k (min^{-1})	T (K)
0.0409	298
0.0818	308
0.157	318

82. Draw a reaction energy diagram for an exothermic reaction that occurs in a single step. Mark the activation energies, and identify the net energy change for the reaction on this diagram. Draw a second diagram that represents the same reaction in the presence of a catalyst. Identify the activation energy of this reaction and the energy change. Is the activation energy in the two drawings different? Does the energy evolved in the two reactions differ?

83. The gas phase reaction

$$2\ N_2O_5(g) \longrightarrow 4\ NO_2(g) + O_2(g)$$

has an activation energy of 103 kJ, and the rate constant is $0.0900\ min^{-1}$ at 328.0 K. Find the rate constant at 318.0 K.

84. Egg protein albumin is precipitated when an egg is cooked in boiling (100 °C) water. The E_a for this first-order reaction is 52.0 kJ/mol. Estimate the time to prepare a 3-min egg at an altitude at which water boils at 90 °C.

85. Many biochemical reactions are catalyzed by acids. A typical mechanism consistent with the experimental results (in which HA is the acid and X is the reactant) is

Step 1	Fast, Reversible	$HA \rightleftharpoons H^+ + A^-$
Step 2	Fast, Reversible	$X + H^+ \rightleftharpoons XH^+$
Step 3	Slow	$XH^+ \longrightarrow$ products

What rate law is derived from this mechanism? What is the order of the reaction with respect to HA? How would doubling the concentration of acid HA affect the reaction?

86. Hypofluorous acid, HOF, is very unstable, decomposing in a first-order reaction to give HF and O_2, with a half-life of only 30 min at room temperature:

$$HOF(g) \longrightarrow HF(g) + \tfrac{1}{2}\ O_2(g)$$

If the partial pressure of HOF in a 1.00-L flask is initially 1.00×10^2 mm Hg at 25 °C, what is the total pressure in the flask and the partial pressure of HOF after exactly 30 min? After 45 min?

87. We know that the decomposition of SO_2Cl_2 is first order in SO_2Cl_2,

$$SO_2Cl_2(g) \longrightarrow SO_2(g) + Cl_2(g)$$

with a half-life of 245 min at 600 K. If you begin with a partial pressure of SO_2Cl_2 of 25 mm Hg in a 1.0-L flask, what is the partial pressure of each reactant and product after 245 min? What is the partial pressure of each reactant after 12 h?

Summary Question

88. The substitution of CO in $Ni(CO)_4$ (in the nonaqueous solvents toluene and hexane) was studied some years ago and led to an understanding of some of the general principles that govern the chemistry of compounds having metal–CO bonds. (See J. P. Day, F. Basolo, and R. G. Pearson, *Journal of the American Chemical Society,* Vol. 90, p. 6927, 1968.)

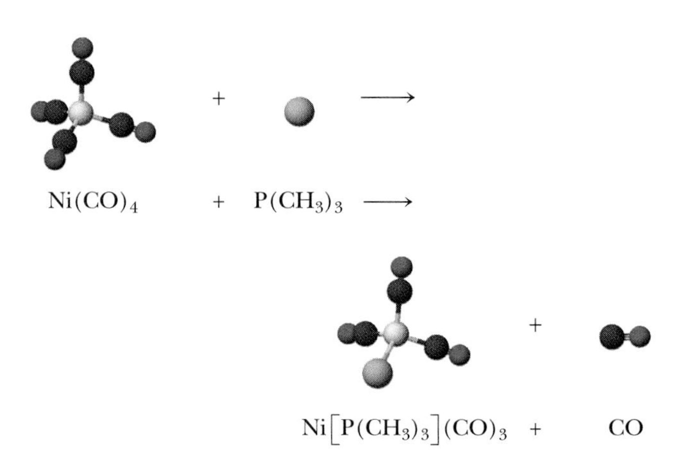

A detailed study of the kinetics of the reaction led to the following mechanism:

Slow	$Ni(CO)_4 \longrightarrow Ni(CO)_3 + CO$
Fast	$Ni(CO)_3 + L \longrightarrow Ni(CO)_3L$

L in this reaction is an electron pair donor such as $P(CH_3)_3$.
(a) What is the molecularity of each of the elementary reactions?
(b) It was found that doubling the concentration of $Ni(CO)_4$ led to an increase in reaction rate by a factor of 2. Doubling the concentration of L had no effect on the reaction rate. Based on this information, write the rate equation for the reaction. Does this agree with the mechanism described?
(c) The experimental rate constant for the reaction, when $L = P(C_6H_5)_3$, is $9.3 \times 10^{-3}\ s^{-1}$ at 20 °C. If the initial

concentration of $Ni(CO)_4$ is 0.025 M, what is the concentration of the product after 5.0 min?

(d) $Ni(CO)_4$ is formed by reacting nickel metal with carbon monoxide. If you have 750 mL of CO at a pressure of 1.50 atm at 22 °C, and the CO is combined with 0.125 g of nickel metal, how many grams of $Ni(CO)_4$ can be formed? If CO remains after reaction, what is its pressure in the 750-mL flask at 29 °C?

(e) An excellent way to make pure nickel metal is to decompose $Ni(CO)_4$ in a vacuum at a temperature slightly higher than room temperature. What is the enthalpy change for the reaction

$$Ni(CO)_4(g) \longrightarrow Ni(s) + 4\ CO(g)$$

if the molar enthalpy of formation of $Ni(CO)_4$ gas is −602.91 kJ/mol?

INTERACTIVE GENERAL CHEMISTRY CD-ROM

Screen 15.2 Rates of Chemical Reactions

(a) What is the difference between an instantaneous rate and an average rate?
(b) Observe the graph of food dye versus time on this screen. (Click on the "tool" icon on this screen.) The plot shows the concentration of dye as the reaction progresses. What does the steepness of the plot at any particular time tell you about the rate of the reaction at that time?
(c) As the reaction progresses, the concentration of dye decreases as it is consumed. What happens to the reaction rate as this occurs? What is the relationship between reaction rate and dye concentration?

Screen 15.4 Control of Reaction Rates (Concentration Dependence)

(a) Watch the video on this screen. How does an increase in HCl concentration affect the rate of the reaction of the acid with magnesium metal?
(b) On the second portion of this screen are data for the rate of decomposition of N_2O_5 (click on "More"). The initial reaction rate is given for three separate experiments, each beginning with a different concentration of N_2O_5. How is the initial reaction rate related to $[N_2O_5]$?

Screen 15.5 Determination of Rate Equation (Method of Initial Rates)

This screen describes how to determine experimentally a rate law using the method of initial rates.

(a) Why must the rate of reaction be measured at the very beginning of the process for this method to be valid?
(b) The first experiment shows that the initial rate of NH_4NCO degradation is 2.2×10^{-4} mol/L · s when $[NH_4NCO] =$ 0.14 M. Using the rate law determined on this screen, predict what the rate would be if $[NH_4NCO] = 0.18$ M.

Screen 15.7 Determination of a Rate Equation (Graphical Methods)

The rate law for the decomposition of NH_4NCO is "Rate = $k[NH_4NCO]$". If you wanted to calculate the concentration of NH_4NCO after some time had elapsed, which equation would you use?

Screen 15.8 Half-Life: First-Order Reactions

(a) Examine the graph and table of concentrations on the second portion of this screen.
   1. What will the concentration of H_2O_2 be after 3270 min? After 3924 min?
   2. What fraction of the original concentration of H_2O_2 remains after each of these times?

Screen 15.9 Microscopic View of Reactions (1)

(a) According to collision theory, what three conditions must be met for two molecules to react?
(b) Examine the animations that play when numbers 1 and 2 are selected. One of these occurs at a higher temperature than the other. Which one? Explain briefly.
(c) Examine the animations that play when numbers 2 and 3 are selected. Would you expect the reaction of O_3 with N_2

$$O_3(g) + N_2(g) \longrightarrow O_2(g) + ONN(g)$$

to be more or less sensitive to requiring a proper orientation for reaction than the reaction displayed on this screen? Explain briefly.

Screen 15.10 Microscopic View of Reactions

Examine the two animations on the *Reaction Coordinate Diagrams* sidebar to this screen. What is the difference between the ways in which the two reactions occur?

Screen 15.12 Reaction Mechanisms

Examine the two mechanisms on this screen. Why is the second step in each said to be bimolecular?

Screen 15.13 Reaction Mechanisms and Rate Equations

(a) What is the difference between an overall mechanism and an elementary step?
(b) What is the relationship between the stoichiometric coefficients of the reactants in an elementary step and the rate law for that step?
(c) What is the rate law for Step 2 of Mechanism 2?
(d) Examine the *Isotopic Labeling* sidebar to this screen. If the reaction occurred in a single step transfer of an oxygen atom from NO_2 to CO, would any $N^{16}O^{18}O$ be found if the reaction is started using a mixture of $N^{16}O_2$ and $N^{18}O_2$? Why or why not?

Screen 15.14 Catalysis and Reaction Rate

(a) Examine the mechanism for the iodide ion-catalyzed decomposition of H_2O_2. Explain how the mechanism shows that I^- is a catalyst.
(b) How does the reaction coordinate diagram show that the catalyzed reaction is expected to be faster than the uncatalyzed reaction?

Chapter **16**

Principles of Reactivity: Chemical Equilibria

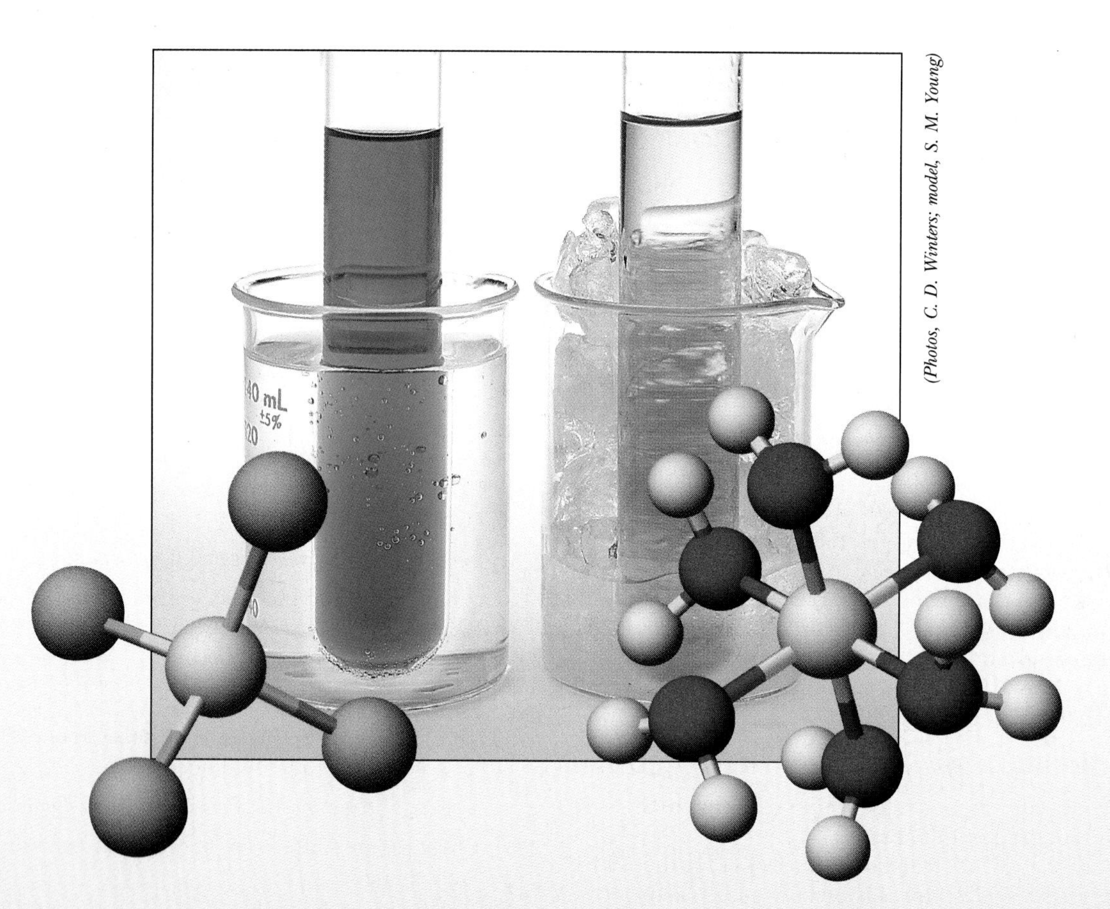

(Photos, C. D. Winters; model, S. M. Young)

A Chemical Puzzler

In the Chemical Puzzler for Chapter 3 you saw that heating pink $CoCl_2 \cdot 6\ H_2O$ caused the compound to lose some water, to give a blue compound. Here $CoCl_2 \cdot 6\ H_2O$ and HCl were dissolved in water to give a pink solution. When the solution was heated, the color turned blue *(left)*, whereas the pink color was restored when the solution was cooled *(right)*. The change is reversible. The color changes to pink when the heated solution is cooled, and the color changes to blue when the cold solution is warmed. The endothermic change occurring is

$$\underset{\text{pink}}{[Co(H_2O)_6]^{2+}}(aq) + 4\ Cl^-(aq) \rightleftharpoons \underset{\text{blue}}{CoCl_4{}^{2-}}(aq) + 6\ H_2O(\ell)$$

Why is this change reversible? Could we have predicted the effect of heat on this reaction?

The concept of equilibrium is fundamental in chemistry, and so this and the next three chapters explore the application of equilibrium to chemical reactions. The concept of equilibrium is not peculiar to chemistry, however. You participate in social situations and live in an economy that represent an equilibrium of competing forces. You and your family or your roommate have arrived at some arrangements that keep your personal relationships as smooth as possible. That is, you have achieved an equilibrium. Of course this equilibrium is easily upset when a stress is applied to the arrangement, as when another child is added to the family or another roommate moves in. In international affairs, countries achieve an equilibrium of competing interests, an equilibrium that can be upset if, for example, the currency of one country changes in value or if a third country joins the relationship. Chemical systems are rather analogous to our social and political arrangements, and chemical systems can also be forced to move to a new equilibrium position by outside influences.

Our goal in this chapter is to explore the consequences of the fact that chemical reactions are reversible, that in a closed system a state of equilibrium is eventually achieved between reactants and products, and that outside forces can affect that equilibrium. A major result of this exploration will be an ability to describe "chemical reactivity" in quantitative terms. As you will see, the concentrations of reactants and products at equilibrium are a measure of the intrinsic tendency of chemical reactions to proceed from reactants to products.

***Figure* 16.1 Cave chemistry.** Calcium carbonate stalactites cling to the roof of a cave, and stalagmites grow up from the cave floor. The chemistry producing these formations is a good example of the reversibility of chemical reactions. *(Arthur N. Palmer)*

16.1 THE NATURE OF THE EQUILIBRIUM STATE

If you ever visited a limestone cave, you were surely impressed with the beautiful limestone stalactites and stalagmites, which are made chiefly of calcium carbonate (Figure 16.1). How did these evolve?

See the *Saunders Interactive General Chemistry CD-ROM,* Screen 16.2, The Principle of Microscopic Reversibility and Screen 16.3, The Equilibrium State.

The key concept for describing the formation of stalactites and stalagmites is the reversibility of chemical reactions. Calcium carbonate is found in underground deposits in the form of limestone, a leftover from ancient oceans. If water seeping through the limestone contains dissolved CO_2, a reaction occurs in which the mineral dissolves, giving an aqueous solution of Ca^{2+} and HCO_3^- ions:

$$CaCO_3(s) + CO_2(aq) + H_2O(\ell) \longrightarrow Ca^{2+}(aq) + 2\ HCO_3^-(aq)$$

When the mineral-laden water reaches a cave, the reverse reaction occurs, with CO_2 being evolved into the cave and solid $CaCO_3$ being deposited.

$$Ca^{2+}(aq) + 2\ HCO_3^-(aq) \longrightarrow CaCO_3(s) + CO_2(g) + H_2O(\ell)$$

Dissolving and reprecipitating limestone can be illustrated by a laboratory experiment with some soluble salts containing the Ca^{2+} and HCO_3^- ions (say $CaCl_2$ and $NaHCO_3$). If you put them in an open beaker of water in the laboratory, you will soon see bubbles of CO_2 gas and a precipitate of solid $CaCO_3$ (Figure 16.2a). If you drop some dry ice (solid CO_2) into the solution, however, the solid $CaCO_3$ redissolves (Figure 16.2b and c). This experiment illustrates an important feature of chemical reactions: *in principle, all chemical reactions are reversible.*

The concept of reversibility was first introduced in Chapter 13 when discussing phase equilibria and was used in the discussion of the properties of solutions.

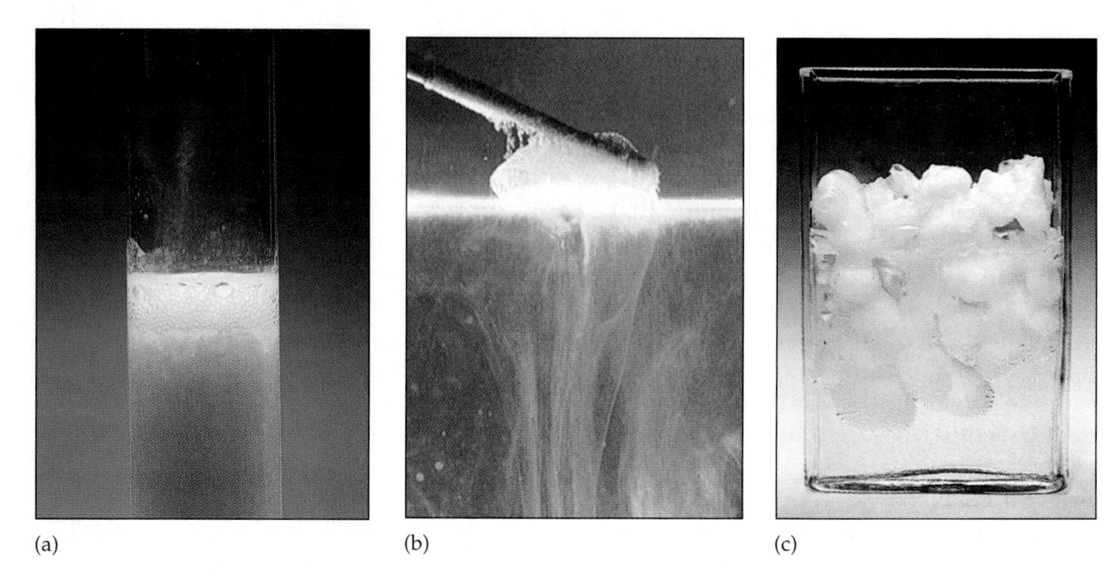

(a) (b) (c)

***Figure* 16.2 Equilibria in the $CO_2/Ca^{2+}/H_2O$ system.** (a) Adding calcium ions (as $CaCl_2$) to bicarbonate ions (as $NaHCO_3$) produces CO_2 gas and solid $CaCO_3$. (b) Adding dry ice (solid CO_2) to a saturated solution of $Ca(OH)_2$ produces solid $CaCO_3$. (c) If excess dry ice (the white solid) is added to the $CaCO_3$ precipitated in (b), the calcium carbonate dissolves to give $Ca^{2+}(aq)$ and $HCO_3^-(aq)$. *(C. D. Winters)*

Now let us carry out the reaction of calcium carbonate, water, and carbon dioxide in a different way. Suppose a solution of Ca^{2+} and HCO_3^- ions is placed in a closed container, from which the CO_2 gas cannot escape. The reaction producing $CaCO_3$ and CO_2 will continue for a while, but eventually no further change will occur. If we were to examine the reaction system, we would find that Ca^{2+}, HCO_3^-, $CaCO_3$, CO_2, and H_2O coexist in the system. No further observable change occurs. It looks as though the reaction has stopped, but this is not the case. Instead, the reaction has reached equilibrium:

The set of double arrows, $\rightleftharpoons$, in an equation indicates that the reaction is reversible and, in general chemistry courses, is a signal to you that the reaction will be studied using the concepts of chemical equilibria.

$$Ca^{2+}(aq) + 2\ HCO_3^-(aq) \rightleftharpoons CaCO_3(s) + CO_2(g) + H_2O(\ell)$$

How might the equilibrium condition be described at the molecular level? As we begin the reaction in a closed container, only Ca^{2+} and HCO_3^- are present, and they react to give the products ($CaCO_3$, CO_2, and H_2O) at some rate. As the reactants are used up, the rate of reaction slows. The reaction products, $CaCO_3$, CO_2, and H_2O, begin to combine, with a rate that increases as their concentration increases. Eventually the rate of the forward reaction, the formation of $CaCO_3$, and the rate of the reverse reaction, the redissolving of $CaCO_3$, become equal. With $CaCO_3$ being formed and redissolving at the same rate, no further *macroscopic* change is observed. The system is at equilibrium.

This description of events at the particulate level defines a basic principle of chemical equilibria. Chemical equilibria are *dynamic.* When the system is at equilibrium, the forward and reverse reactions continue, but they take place at equal rates.

To prove that equilibria are dynamic, let us look at a common introductory chemistry experiment, the study of the equilibrium that exists in the reaction of aqueous iron(III) ion and thiocyanate ion, $SCN^-(aq)$.

$$Fe(H_2O)_6^{3+}(aq) + SCN^-(aq) \rightleftharpoons Fe(H_2O)_5(SCN)^{2+}(aq) + H_2O(\ell)$$

nearly colorless — colorless — red-orange

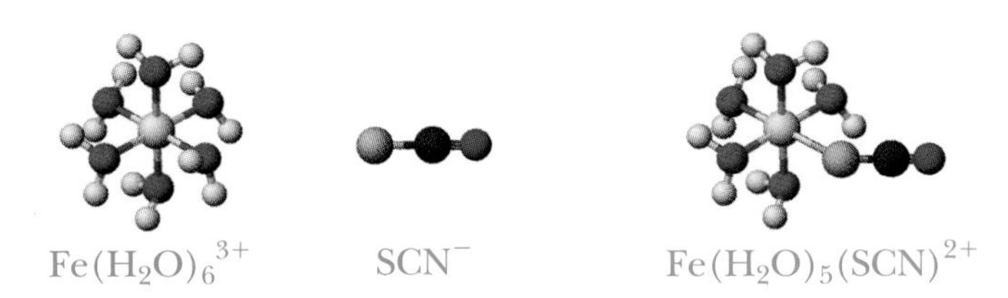

When colorless solutions of the reactant ions are mixed (Figure 16.3), the SCN^- ion rapidly replaces water on the $Fe(H_2O)_6^{3+}$ ion to give a red-orange ion in which the SCN^- ion is bonded to Fe^{3+}. As the concentration of $Fe(H_2O)_5(SCN)^{2+}$ begins to build up, this ion reacts with water to release SCN^- and revert to aqueous $Fe(H_2O)_6^{3+}$. Eventually, the rate at which SCN^- replaces H_2O on $Fe(H_2O)_6^{3+}$ to form $Fe(H_2O)_5(SCN)^{2+}$ (the "forward" reaction) becomes equal to the rate at which $Fe(H_2O)_5(SCN)^{2+}$ sheds the SCN^- ion to go back to the simple ions in solution (the "reverse" reaction). When the rates of the forward and reverse reactions become equal, equilibrium has been achieved.

***Figure* 16.3 The reaction of aqueous iron(III) ion and thiocyanate ion, SCN^-.** The colorless solutions are mixed to give the red-orange ion $Fe(H_2O)_5(SCN)^{2+}$. *(C. D. Winters)*

When equilibrium has been achieved in the $Fe(H_2O)_6^{3+} + SCN^-$ reaction, the forward and reverse reactions do not stop. To prove this, we could add a drop of an aqueous solution containing radioactive SCN^- ion to the solution. (The ion is made radioactive by using radioactive ^{14}C in place of normal, nonradioactive ^{12}C.) When the solution is sampled shortly after adding the radioactive SCN^- ion, we would observe that the radioactive ion is incorporated into $Fe(H_2O)_5(SCN)^{2+}$, an observation explained by reactions such as those described by the following, simplified equations:

$$Fe(H_2O)_5(SCN)^{2+}(aq) + H_2O(\ell) \rightleftharpoons Fe(H_2O)_6^{3+}(aq) + SCN^-(aq)$$
$$Fe(H_2O)_6^{3+}(aq) + S^{14}CN^-(aq) \rightleftharpoons Fe(H_2O)_5(S^{14}CN)^{2+}(aq) + H_2O(\ell)$$

The only way for radioactive $S^{14}CN^-$ ion to be incorporated into the red-orange $Fe(H_2O)_5(SCN)^{2+}$ ion is if the exchange reaction with water is dynamic and reversible, and continues even at equilibrium.

Not only are equilibrium processes dynamic and reversible, but, for a specific reaction, the nature of the equilibrium state is the same, no matter what the direction of approach. Let us say you measure the concentration at equilibrium of acetic acid and the ions that come from its ionization in aqueous solution.

$$CH_3CO_2H(aq) + H_2O(\ell) \rightleftharpoons CH_3CO_2^-(aq) + H_3O^+(aq)$$

acetic acid — acetate ion — hydronium ion

Because acetic acid is a weak acid, the concentrations of the acetate and hydronium ions are small. Now mix sodium acetate and hydrochloric acid:

$$NaCH_3CO_2(aq) + HCl(aq) \longrightarrow CH_3CO_2H(aq) + NaCl(aq)$$

Because HCl is a strong acid, the ensuing reaction has the net ionic equation

$$CH_3CO_2^-(aq) + H_3O^+(aq) \rightleftharpoons CH_3CO_2H(aq) + H_2O(\ell)$$

acetate ion — hydronium ion — acetic acid

If you begin with, say, 1 mol of acetic acid in the first experiment, and 1 mol each of sodium acetate and HCl in the second experiment (all in the same volume of solution), the concentrations of acetic acid, acetate ion, and hydronium ion at equilibrium will be identical!

16.2 THE EQUILIBRIUM CONSTANT

See the *Saunders Interactive General Chemistry CD-ROM,* Screen 16.4, The Equilibrium Constant.

One way to describe the equilibrium position of a chemical reaction is to give concentrations of the reactants and products at this point. These are usually expressed in terms of an **equilibrium constant expression,** which relates concentrations of reactants and products at equilibrium at a given temperature to a numerical constant.

The notion that equilibrium concentrations of reactants and products are related in a simple manner is easy to prove by experiments such as those for the reaction of hydrogen and iodine to produce hydrogen iodide:

$$H_2(g) + I_2(g) \rightleftharpoons 2\ HI(g)$$

A very large number of experiments have shown that *at equilibrium* the ratio of the square of the HI concentration to the product of the H_2 and I_2 concentrations

Square brackets around a chemical formula indicate molar concentrations.

$$\frac{[HI]^2}{[H_2][I_2]}$$

is always the same within experimental error for all experiments done at 425 °C. For example, assume enough H_2 and I_2 has been placed in a flask so that the concentration of each is 0.0175 mol/L at 425 °C. With time, the concentrations of H_2 and I_2 will decline and that of HI will increase; a state of equilibrium will be attained eventually. If the gases in the flask are then analyzed, the result would be $[H_2] = [I_2] = 0.0037$ mol/L and $[HI] = 0.0276$ mol/L.

INITIAL AND EQUILIBRIUM CONCENTRATION (MOL/L)

Equation	$H_2(g)$	+	$I_2(g)$	$\rightleftharpoons$	$2\ HI(g)$
Initial concentration	0.0175		0.0175		0
Change in concentration as reaction proceeds to equilibrium	−0.0138		−0.0138		+0.0276
Equilibrium concentration	0.0037		0.0037		0.0276

Putting these equilibrium concentration values into the preceding expression

$$\frac{[HI]^2}{[H_2][I_2]} = \frac{(0.0276)^2}{(0.0037)(0.0037)} = 56$$

Show that if you begin with 2.0 mol/L of H_2 and 2.0 mol/L of I_2, at equilibrium $[H_2] = [I_2] = 0.42$ mol/L and $[HI] = 3.16$ mol/L. See Example 16.7.

gives a ratio of about 56 (or 55.64 if the experimental information contains more significant figures). This ratio is always the same for all experiments at 425 °C, no matter from which direction the reaction is approached (mixing H_2 and I_2 or allowing HI to decompose) and no matter what the initial concentrations.

Hundreds of experiments on many different chemical systems have proved that equilibrium constants can be calculated from the following general expression. For the general reaction

$$a\,A + b\,B \rightleftharpoons c\,C + d\,D$$

the equilibrium concentrations of reactants and products are always related by the **equilibrium constant expression**

$$\text{Equilibrium constant} = K = \frac{[C]^c[D]^d}{[A]^a[B]^b} \quad [16.1]$$

(numerator: Product concentrations; denominator: Reactant concentrations)

Important features of Equation 16.1 are*

- Product concentrations appear in the numerator.
- Reactant concentrations appear in the denominator.
- Each concentration is raised to the power of its stoichiometric coefficient in the balanced equation.
- The value of the constant K depends on the particular reaction and on the temperature. Units are not given with K.

The value of an equilibrium constant for a given reaction provides information about the extent of reaction when equilibrium has been achieved. After learning a few characteristics of the equilibrium expression, we shall turn to that topic.

In Chapter 15 you learned about the rate at which reactions approach equilibrium and about some common mechanisms or pathways for reactions. This group of chapters (16–19) considers how we can define the position of equilibrium in terms of experimentally measured concentrations and the practical consequences of that action. Chapter 20 describes how we can predict whether the reaction mixture will consist mostly of reactants or of products when equilibrium has been achieved.

Writing Equilibrium Constant Expressions

Reactions Involving Solids and Water

You should know a few rules about writing equilibrium constant expressions for chemical reactions. For example, the oxidation of solid, yellow sulfur produces colorless sulfur dioxide gas (Figure 16.4).

$$\tfrac{1}{8}\,S_8(s) + O_2(g) \rightleftharpoons SO_2(g)$$

Following the general principle that products appear in the numerator and reactants in the denominator, you would write

$$K' = \frac{[SO_2]}{[S_8]^{1/8}[O_2]}$$

Because sulfur is a molecular solid and because the concentration of molecules within any solid is fixed, the sulfur concentration is not changed either by reaction or by addition or removal of some solid. Furthermore, it is an experimental fact that the equilibrium concentrations of O_2 and SO_2 are not changed by the amount of sulfur present, as long as there is some solid sulfur present at equilibrium. Therefore, chemists do not include the concentrations of *any solid*

Figure 16.4 Burning sulfur. Elemental sulfur burns in oxygen with a beautiful blue flame to give SO_2 gas. *(Photo, C. D. Winters; models, S. M. Young)*

*Accurate mathematical analysis of equilibria requires the use of *activities,* which may be thought of as *effective* concentrations (or pressures). Relating activities to concentrations is often quite tedious, however, and using concentrations leads to only small errors in the situations we deal with in introductory chemistry. Thus, we shall use concentrations (or partial pressures) in our calculations.

reactants and products in the equilibrium constant expression, and you should write the equilibrium expression for the combustion of sulfur as

$$K = \frac{[SO_2]}{[O_2]}$$

There are special considerations for reactions occurring in aqueous solution. Consider ammonia, which is a weak base owing to its interaction with water (see Figure 5.11).

$$NH_3(aq) + H_2O(\ell) \rightleftharpoons NH_4^+(aq) + OH^-(aq)$$

Because the water concentration is very high in dilute solutions, the concentration of water is essentially unchanged by the reaction. For this reason the *molar concentration of water,* like that of solids, is not included in the equilibrium constant expression.* Thus, we write

$$K = \frac{[NH_4^+][OH^-]}{[NH_3]}$$

Expressing Concentrations: K_c *and* K_p

See the *Saunders Interactive General Chemistry CD-ROM,* Screen 16.6, Writing Equilibrium Expressions.

The concentrations in the equilibrium constant expression are usually given in moles per liter (M), and so the symbol K is often written with a subscript c for "concentration," as in K_c. K expressions involving gases, however, can also be written in terms of partial pressures of reactants and products because the ideal gas law specifies that "gas concentration," (n/V), is equivalent to P/RT; that is, the partial pressure of a gas is proportional to its concentration $[P = (n/V)RT]$. If reactant and product concentrations are given in partial pressures, then K has a subscript p, as in K_p.

In some cases the numerical values of K_c and K_p may be the same, but they are often different. In this book we shall generally use equilibrium constants where concentrations are given in moles per liter, K_c. The box *A Closer Look: Equilibrium Constant Expressions for Gases,* however, shows you how to express them in terms of gas pressures for reactions involving gases and how K_c and K_p are related.

Exercise 16.1 Writing Equilibrium Constant Expressions

Write equilibrium constant expressions for each of the following reactions in terms of concentrations:

1. $PCl_5(g) \rightleftharpoons PCl_3(g) + Cl_2(g)$
2. $Cu(OH)_2(s) \rightleftharpoons Cu^{2+}(aq) + 2\ OH^-(aq)$
3. $Cu(NH_3)_4^{2+}(aq) \rightleftharpoons Cu^{2+}(aq) + 4\ NH_3(aq)$
4. $CH_3CO_2H(aq) + H_2O(\ell) \rightleftharpoons CH_3CO_2^-(aq) + H_3O^+(aq)$

---

*The equilibrium constant expression for aqueous ammonia could be written

$$K' = \frac{[NH_4^+][OH^-]}{[NH_3][H_2O]}$$

However, virtually all solutions that concern us are quite dilute. The concentration of water ($\approx$ 55.5 M) is always larger than the solute concentration and is essentially constant. The usual convention, therefore, is to include $[H_2O]$ in the equilibrium constant for aqueous solution equilibria:

$$K'[H_2O] = K = \frac{[NH_4^+][OH^-]}{[NH_3]}$$

A CLOSER LOOK — *Equilibrium Constant Expressions for Gases—K_c and K_p*

Many metal carbonates, such as limestone, decompose on heating to give the metal oxide and CO_2 gas:

$$CaCO_3(s) \rightleftharpoons CaO(s) + CO_2(g)$$

The equilibrium condition for this reaction can be expressed either in terms of the number of moles per liter of CO_2, $K_c = [CO_2]$, or in terms of the pressure of CO_2, $K_p = P_{CO_2}$. From the ideal gas law, you know that

$$P = (n/V)RT$$
$$= (\text{concentration in moles per liter}) \cdot RT$$

For this reaction, we can therefore say that $K_p = [CO_2]RT$. Because $K_c = [CO_2]$, the interesting conclusion is that $K_p = K_c(RT)$. That is, the *values* of K_p and K_c are *not* the same; K_p for the decomposition of calcium carbonate is the product of K_c and the factor RT.

The equilibrium constant in terms of partial pressures, K_p, is known for the reaction of N_2 and H_2 to produce ammonia:

$$N_2(g) + 3\ H_2(g) \rightleftharpoons 2\ NH_3(g)$$

$$K_p = \frac{(P_{NH_3})^2}{(P_{N_2})(P_{H_2})^3} = 5.8 \times 10^5 \text{ at } 25\ °C$$

Does K_c, the equilibrium constant in terms of concentrations, have the same value or a different value as K_p? We can answer this by substituting for each pressure in K_p the equivalent expression $[mol/L](RT)$. That is,

$$K_p = \frac{\{[NH_3](RT)\}^2}{\{[N_2](RT)\}\{[H_2](RT)\}^3} =$$

$$\frac{[NH_3]^2}{[N_2][H_2]^3} \cdot \frac{1}{(RT)^2} = \frac{K_c}{(RT)^2}$$

Solving for K_c, we find

$$K_p = 5.8 \times 10^5 = \frac{K_c}{[(0.08206)(298)]^2}$$

$$K_c = 3.5 \times 10^8$$

Once again you see that K_p and K_c are not the same but are related by some function of RT.

Looking carefully at these and other examples, we find that, in general,

$$K_p = K_c(RT)^{\Delta n} \qquad [16.2]$$

where Δn is the change in the number of moles of gas on going from reactants to products.

Δn = total moles of gaseous products— total moles of gaseous reactants

For the decomposition of $CaCO_3$,

$$\Delta n = 1 - 0 = 1$$

whereas the value of Δn for the ammonia synthesis is

$$\Delta n = 2 - 4 = -2$$

Be sure to notice that $K_p = K_c$ when $\Delta n = 0$. This occurs when the number of reactant and product molecules is the same, as in the reaction

$$2\ HI(g) \rightleftharpoons H_2(g) + I_2(g)$$

Manipulating Equilibrium Expressions

Chemical equations can be balanced using different sets of stoichiometric coefficients. For example, oxidation of carbon can give carbon monoxide:

$$C(s) + \tfrac{1}{2}\ O_2(g) \rightleftharpoons CO(g)$$

and the equilibrium constant expression for this reaction as written would be

$$K_1 = \frac{[CO]}{[O_2]^{1/2}} = 4.6 \times 10^{23} \text{ at } 25\ °C$$

You can write the chemical equation equally well, however, as

$$2\ C(s) + O_2(g) \rightleftharpoons 2\ CO(g)$$

and the equilibrium constant would now be

$$K_2 = \frac{[CO]^2}{[O_2]} = 2.1 \times 10^{47} \text{ at } 25\ °C$$

When you compare the two equilibrium expressions you find that $K_2 = (K_1)^2$; that is,

$$K_2 = \frac{[CO]^2}{[O_2]} = \left\{\frac{[CO]}{[O_2]^{1/2}}\right\}^2 = K_1^{\,2}$$

In general, when the stoichiometric coefficients of a balanced equation are multiplied by some factor, the equilibrium constant for the new equation (K_{new}) is the old equilibrium constant (K_{old}) *raised to the power of the multiplication factor.* In the case of the oxidation of carbon, the second equation was obtained by multiplying the first equation by 2. K_2 is therefore the *square* of K_1.

Closely related to the effect of using a new set of stoichiometric coefficients is what happens to K when a chemical equation is reversed. Compare the values of K for formic acid transferring an H^+ ion to water

$$HCO_2H(aq) + H_2O(\ell) \rightleftharpoons HCO_2^-(aq) + H_3O^+(aq)$$

$$K_1 = \frac{[HCO_2^-][H_3O^+]}{[HCO_2H]} = 1.8 \times 10^{-4} \text{ at } 25\ °C$$

with the opposite reaction, the gain of an H^+ ion by the formate ion, HCO_2^-.

$$HCO_2^-(aq) + H_3O^+(aq) \rightleftharpoons HCO_2H(aq) + H_2O(\ell)$$

$$K_2 = \frac{[HCO_2H]}{[HCO_2^-][H_3O^+]} = 5.6 \times 10^{3} \text{ at } 25\ °C$$

Here $K_2 = 1/K_1$. The equilibrium constants for a reaction and its reverse are the *reciprocals* of each other.

It is often useful to add two equations together to obtain the equation for a net process. As an example, consider the reactions that take place when silver chloride dissolves in water (to a very small extent). Ammonia is then added to the solution, and the ammonia reacts with the silver ion to form the polyatomic ion, $Ag(NH_3)_2^+$ (Figure 16.5). Adding the equation for the process of dissolving solid AgCl to the equation for the reaction of Ag^+ ion with ammonia gives the equation for the net reaction, the process of dissolving solid AgCl in the

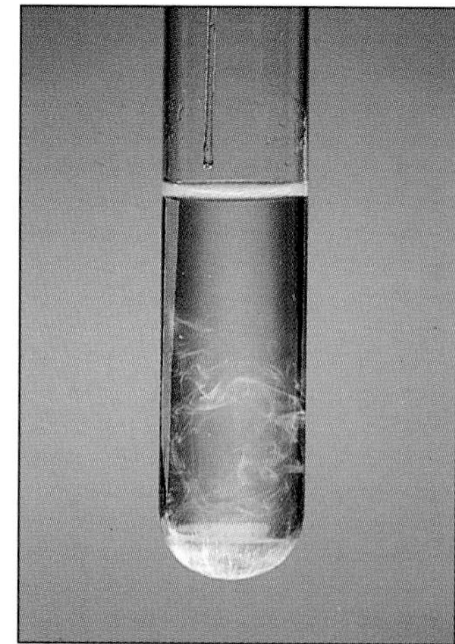
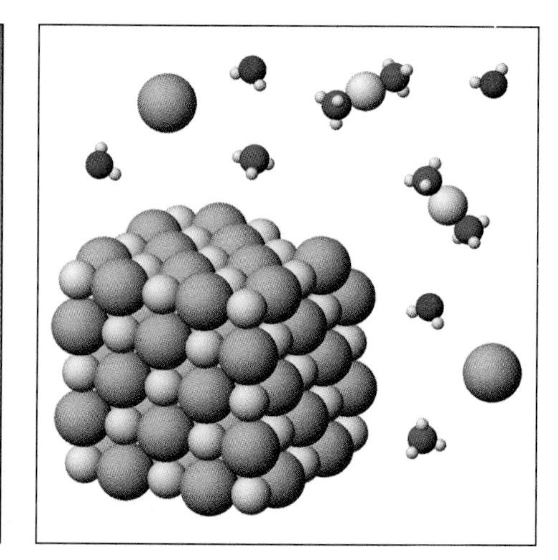

Figure 16.5 Dissolving AgCl in aqueous ammonia. *(left)* A precipitate of AgCl(s) is suspended in water. *(center)* When aqueous ammonia is added, the ammonia reacts with the trace of silver ion in solution, and the silver chloride dissolves. *(right)* A particulate level model of the dissolving process. See Figure 5.6 for the process of forming a precipitate of AgCl. *(Photos, C. D. Winters; model, S. M. Young)*

presence of aqueous ammonia. (All equilibrium constants are given at 25 °C.)

$$AgCl(s) \rightleftharpoons Ag^+(aq) + Cl^-(aq) \qquad K_1 = [Ag^+][Cl^-] = 1.8 \times 10^{-10}$$

$$Ag^+(aq) + 2\ NH_3(aq) \rightleftharpoons Ag(NH_3)_2{}^+(aq) \qquad K_2 = \frac{[Ag(NH_3)_2{}^+]}{[Ag^+][NH_3]^2} = 1.6 \times 10^7$$

Net Reaction

$$AgCl(s) + 2\ NH_3(aq) \rightleftharpoons Ag(NH_3)_2{}^+(aq) + Cl^-(aq)$$

Multiplying the equilibrium constants for the two reactions, K_1 and K_2, gives the equilibrium constant for the net reaction, K_{net}:

$$K_{net} = K_1 \cdot K_2 = [Ag^+][Cl^-] \cdot \frac{[Ag(NH_3)_2{}^+]}{[Ag^+][NH_3]^2} = \frac{[Ag(NH_3)_2{}^+][Cl^-]}{[NH_3]^2}$$

$$= K_1K_2 = 2.9 \times 10^{-3}$$

In general, when two or more equations are added to produce a net equation, the equilibrium constant for the net equation is the product of the equilibrium constants for the added equations.

***Example* 16.1** Manipulating Equilibrium Constant Expressions

A mixture of nitrogen, hydrogen, and ammonia is brought to equilibrium. When the equation is written using whole-number coefficients, the value of K_c is 3.5×10^8 at 25 °C.

Equation 1 $\quad N_2(g) + 3\ H_2(g) \rightleftharpoons 2\ NH_3(g) \qquad K_1 = 3.5 \times 10^8$

However, the equation can also be written

Equation 2 $\quad \frac{1}{2}\ N_2(g) + \frac{3}{2}\ H_2(g) \rightleftharpoons NH_3(g) \qquad K_2 = ?$

What is the value of K_2? What is the value of K_3, the equilibrium constant for the reverse of Equation 1, that is, the decomposition of ammonia to the elements?

Equation 3 $\quad 2\ NH_3(g) \rightleftharpoons N_2(g) + 3\ H_2(g) \qquad K_3 = ?$

Solution To see the relation between K_1 and K_2, first write the equilibrium constant expressions for these two balanced equations:

$$K_1 = \frac{[NH_3]^2}{[N_2][H_2]^3} = 3.5 \times 10^8 \qquad \text{and} \qquad K_2 = \frac{[NH_3]}{[N_2]^{1/2}[H_2]^{3/2}}$$

Writing these expressions makes it clear that K_1 is the square of K_2; that is, $K_1 = K_2{}^2$. The answer to our question is therefore

$$K_2 = \sqrt{K_1} = \sqrt{3.5 \times 10^8} = 1.9 \times 10^4$$

Equation 3 is the reverse of Equation 1, and its equilibrium constant expression is

$$K_3 = \frac{[N_2][H_2]^3}{[NH_3]^2}$$

In this case, K_3 is the reciprocal of K_1. That is, $K_3 = 1/K_1$:

$$K_3 = \frac{1}{K_1} = \frac{1}{3.5 \times 10^8} = 2.9 \times 10^{-9}$$

As a final comment, notice that the production of ammonia from the elements has a large equilibrium constant. As expected, the reverse reaction, the decomposition of ammonia to its elements, has a small equilibrium constant.

Exercise 16.2 Manipulating Equilibrium Constant Expressions

The conversion of oxygen to ozone has a very small equilibrium constant:

$$\tfrac{3}{2}\ O_2(g) \rightleftharpoons O_3(g) \qquad K_c = 2.5 \times 10^{-29}$$

1. What is the value of K_c when the equation is written using whole-number coefficients?

$$3\ O_2(g) \rightleftharpoons 2\ O_3(g)$$

2. What is the value of K_c for the conversion of ozone to oxygen?

$$2\ O_3(g) \rightleftharpoons 3\ O_2(g)$$

Problem-Solving Tips and Ideas

16.1 Writing and Manipulating Equilibrium Constant Expressions

This section is an important one. You should now know

1. How to write an equilibrium constant expression from the balanced equation, recognizing that the concentrations of solids and of liquids used as solvents do not appear in the expression.
2. That when the stoichiometric coefficients in a balanced equation are changed by a factor of n, then $K_{new} = (K_{old})^n$.
3. That when a balanced equation is reversed, then $K_{new} = 1/K_{old}$.
4. That when several balanced equations (each with its own equilibrium constant, K_1, K_2, etc.) are added to obtain a net, balanced equation, then $K_{net} = K_1 \cdot K_2 \cdot K_3 \cdot \ldots$.
5. That the value of K depends on the way the equilibrium concentrations are expressed (as moles per liter or P).

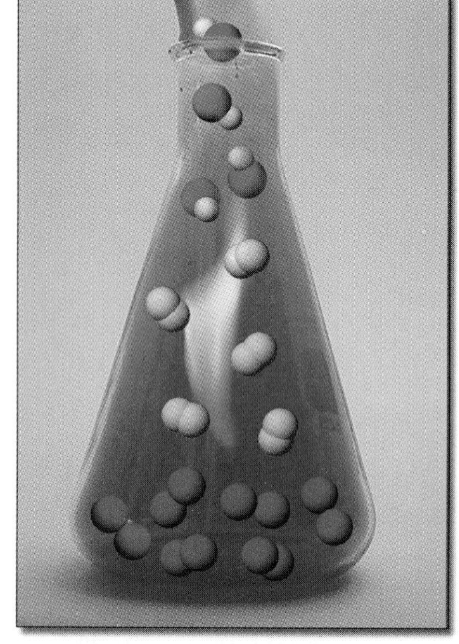

Covalent hydrogen compounds. Hydrogen burns in an atmosphere of bromine to give hydrogen bromide, HBr. Models of the reactants and products are included. *(Photo, C. D. Winters; model, S. M. Young)*

Exercise 16.3 Manipulating Equilibrium Constant Expressions

The following equilibrium constants are given at 500 K:

$$H_2(g) + Br_2(g) \rightleftharpoons 2\ HBr(g) \qquad K = 7.9 \times 10^{11}$$

$$H_2(g) \rightleftharpoons 2\ H(g) \qquad K = 4.8 \times 10^{-41}$$

$$Br_2(g) \rightleftharpoons 2\ Br(g) \qquad K = 2.2 \times 10^{-15}$$

Calculate K for the reaction of H and Br atoms to give HBr:

$$H(g) + Br(g) \rightleftharpoons HBr(g) \qquad K = ?$$

The Meaning of the Equilibrium Constant

The value of the equilibrium constant indicates whether a reaction is product- or reactant-favored. In addition, it can be used to calculate how much product will be present at equilibrium, which is valuable information to chemists and chemical engineers.

A large value of K means that reactants are converted largely to products when equilibrium has been achieved. That is, the products are strongly favored over the reactants at equilibrium. An example is the reaction of nitrogen monoxide and ozone.

See the *Saunders Interactive General Chemistry CD-ROM,* Screen 16.5, The Meaning of the Equilibrium Constant, and Screen 16.8, Determining an Equilibrium Constant.

$K \gg 1$: Reaction is product-favored; equilibrium concentrations of products are greater than equilibrium concentrations of reactants.

$$NO(g) + O_3(g) \rightleftharpoons NO_2(g) + O_2(g)$$

$$K_c = 6 \times 10^{34} \text{ at } 25\ °C = \frac{[NO_2][O_2]}{[NO][O_3]}$$

$$K_c \gg 1, \text{ so, at equilibrium, } [NO_2][O_2] \gg [NO][O_3]$$

The very large value of K indicates that, if stoichiometric amounts of NO and O_3 are mixed and allowed to come to equilibrium, virtually none of the reactants will be found. Essentially all will have been converted to NO_2 and O_2. A chemist would say "the reaction has gone to completion."

Conversely, a small K (as in the formation of ozone from oxygen) means that very little of the reactants have formed products when equilibrium has been achieved. In other words, the reactants are favored over the products at equilibrium.

$K \ll 1$: Reaction is reactant-favored; equilibrium concentrations of reactants are greater than equilibrium concentrations of products.

$$\tfrac{3}{2}\, O_2(g) \rightleftharpoons O_3(g)$$

$$K_c = 2.5 \times 10^{-29} \text{ at } 25\ °C = \frac{[O_3]}{[O_2]^{3/2}}$$

$$K_c \ll 1, \text{ this means } [O_3] \ll [O_2]^{3/2} \text{ at equilibrium}$$

The very small value of K indicates that, if O_2 is placed in a flask, very little O_2 will have been converted to O_3 when equilibrium has been achieved.

Equilibrium constant values for a few reactions are given in Table 16.1. These reactions occur to widely varying extents, as shown by the wide range of values of K.

Exercise 16.4 The Equilibrium Constant and Extent of Reaction

Solid AgCl and AgBr were each placed in 1.0 L of water in separate beakers. Are these reactions product- or reactant-favored? When equilibrium is achieved, in which beaker will the concentration of silver ion be larger?

$$AgCl(s) \rightleftharpoons Ag^+(aq) + Cl^-(aq) \qquad K_c = 1.8 \times 10^{-10}$$

$$AgBr(s) \rightleftharpoons Ag^+(aq) + Br^-(aq) \qquad K_c = 3.3 \times 10^{-13}$$

Table 16.1 • Selected Equilibrium Constants

Reaction	Equilibrium Constant K_c (at 25 °C)
Nonmetal Reactions	
$\frac{1}{8} S_8(s) + O_2(g) \rightleftharpoons SO_2(g)$	4.2×10^{52}
$2 H_2(g) + O_2(g) \rightleftharpoons 2 H_2O(g)$	3.2×10^{81}
$N_2(g) + 3 H_2(g) \rightleftharpoons 2 NH_3(g)$	3.5×10^{8}
$N_2(g) + O_2(g) \rightleftharpoons 2 NO(g)$	1.7×10^{-3} (at 2300 K)
Weak Acids and Bases	
$HCO_2H(aq) + H_2O(\ell) \rightleftharpoons HCO_2^-(aq) + H_3O^+(aq)$ formic acid	1.8×10^{-4}
$CH_3CO_2H(aq) + H_2O(\ell) \rightleftharpoons CH_3CO_2^-(aq) + H_3O^+(aq)$ acetic acid	1.8×10^{-5}
$H_2CO_3(aq) + H_2O(\ell) \rightleftharpoons HCO_3^-(aq) + H_3O^+(aq)$ carbonic acid	4.2×10^{-7}
$NH_3(aq) + H_2O(\ell) \rightleftharpoons NH_4^+(aq) + OH^-(aq)$ ammonia (weak base)	1.8×10^{-5}
"Insoluble" Solids	
$CaCO_3(s) \rightleftharpoons Ca^{2+}(aq) + CO_3^{2-}(aq)$	3.8×10^{-9}
$AgCl(s) \rightleftharpoons Ag^+(aq) + Cl^-(aq)$	1.8×10^{-10}

16.3 THE REACTION QUOTIENT

See the *Saunders Interactive General Chemistry CD-ROM,* Screen 16.9, Systems at Equilibrium.

In Chapter 13 you first encountered phase diagrams, graphic representations of the equilibria involved in transformations between the phases of matter. For any liquid, the plot of temperature-versus-equilibrium vapor pressure is a curved line (Figure 13.19). This line defines all the conditions of T and P at which equilibrium can exist. There can be no equilibrium in a system whose temperature and vapor pressure is not represented by a set of points on the line.

The equilibrium situation for a chemical reaction can also be represented graphically for some simple cases. For example, consider a reaction such as the transformation of butane to isobutane (2-methyl-propane).

$$\text{Butane} \rightleftharpoons \text{isobutane}$$

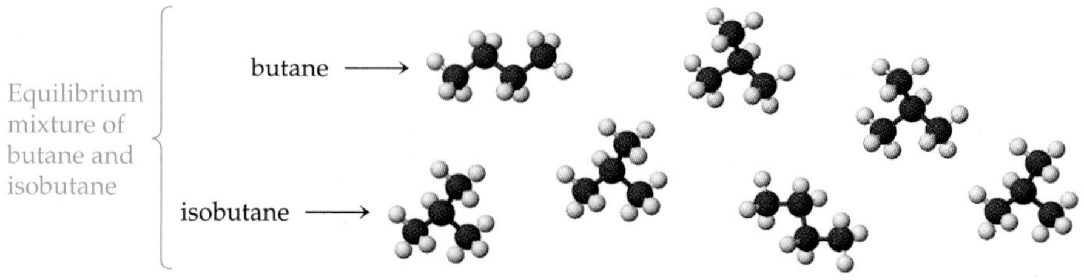

Equilibrium mixture of butane and isobutane

butane →

isobutane →

$$K_c = \frac{[\text{isobutane}]}{[\text{butane}]} = 2.50 \text{ at } 298 \text{ K}$$

If the concentration of one of the compounds is known, then there is only one value of the other concentration that will satisfy the equilibrium constant expression. For example, if [butane] is 1.0 mol/L, then the equilibrium concentration of isobutane, [isobutane], must be 2.5 mol/L. If [butane] were changed to 0.80 M, then [isobutane] at equilibrium must be

$$[\text{isobutane}] = K_c\,[\text{butane}] = 2.50\ (0.80\ \text{M}) = 2.0\ \text{M}$$

Choosing a number of possible values for [butane] and solving for the allowed value of [isobutane] would eventually lead to the points plotted in Figure 16.6. That is, the equilibrium expression K_c = [isobutane]/[butane] is just the equation of a straight line (with a slope equal to K_c).

Similar plots can be made for other types of equilibrium expressions, but they are more complicated.

The graphical treatment of the simple equilibrium expression K_c = [isobutane]/[butane] is useful for two reasons. First, Figure 16.6 illustrates that an infinite number of *sets* of equilibrium concentrations of reactant and product are possible (all the points along the equilibrium line). Second, sets of concentrations for butane and isobutane that do not lie along the line in Figure 16.6 do not satisfy the equilibrium condition (just as in the case of phase equilibria; Figure 13.19). If this is the case, the system will attempt to shift to a new set of concentrations that will satisfy the equilibrium condition.

Any point in Figure 16.6, whether on or off the line, can be defined by the ratio [isobutane]/[butane]. This ratio is give ı the general name of the **reaction quotient, Q,** and it is equal to the equilibrium constant, K_c, only when the reaction is at equilibrium. Suppose you have a system composed of 3 mol/L of butane and 4 mol/L of isobutane (at 298 K). This means that the ratio of concentrations, Q_c, is

$$Q_c = \frac{[\text{isobutane}]}{[\text{butane}]} = \frac{4.0}{3.0} = 1.3$$

a ratio given by a point in the tan portion of Figure 16.6. This set of concentrations clearly does *not* represent an equilibrium system because $Q_c < K_c$. To reach equilibrium, some of the butane must change into isobutane, thereby lowering [butane] and raising [isobutane]. Indeed, this transformation will continue until the ratio [isobutane]/[butane] = 2.5; that is, until $Q_c = K_c$, and the

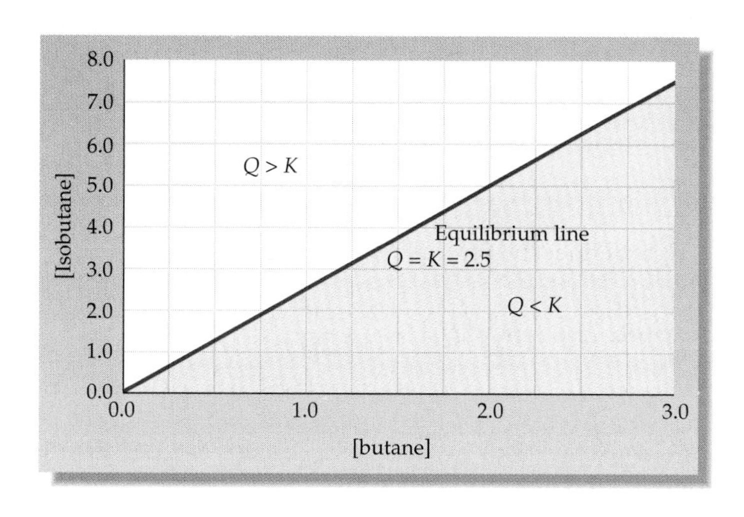

***Figure* 16.6 An equilibrium plot.** A plot of the concentrations of butane and isobutane that satisfies the expression Q = [isobutane]/[butane] = 2.50. When $Q = K = 2.50$, the system is at equilibrium. When $Q < K$ (the light tan portion of the diagram), the system is not at equilibrium, and reactants are further converted to products. When $Q > K$ (the white portion of the diagram), the system is also not at equilibrium, but products must revert to reactants to establish equilibrium.

ratio of concentrations in the system is represented by a point on the line in Figure 16.6. This is somewhat analogous to the changes that must occur when a liquid and its vapor are not in equilibrium.

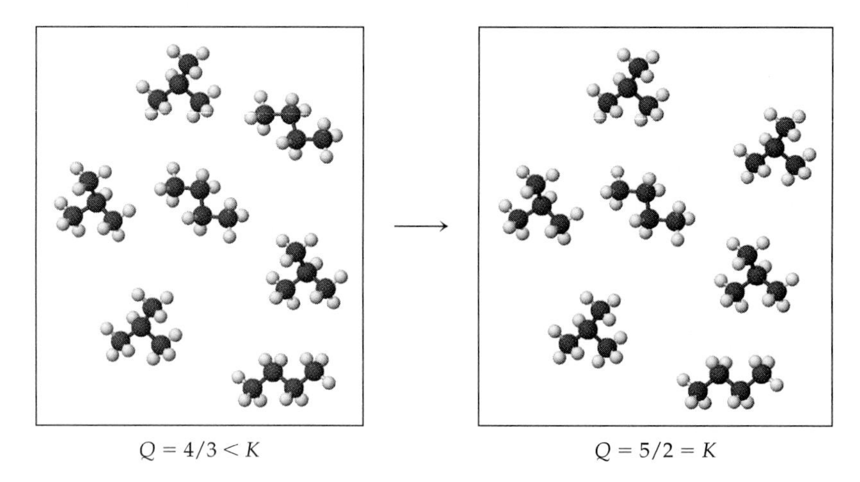

What happens when too much isobutane is in the system relative to the amount of butane? Suppose [isobutane] = 6.0 M but [butane] is only 1.0 M. Now the reaction quotient Q_c is greater than K_c ($Q_c > K_c$). This is represented by a point in the white portion of Figure 16.6. The system is again not at equilibrium, but it can proceed to an equilibrium state by converting isobutane to butane. Only when Q_c has a value of 2.5 ($Q_c = K_c$) is the system at equilibrium.

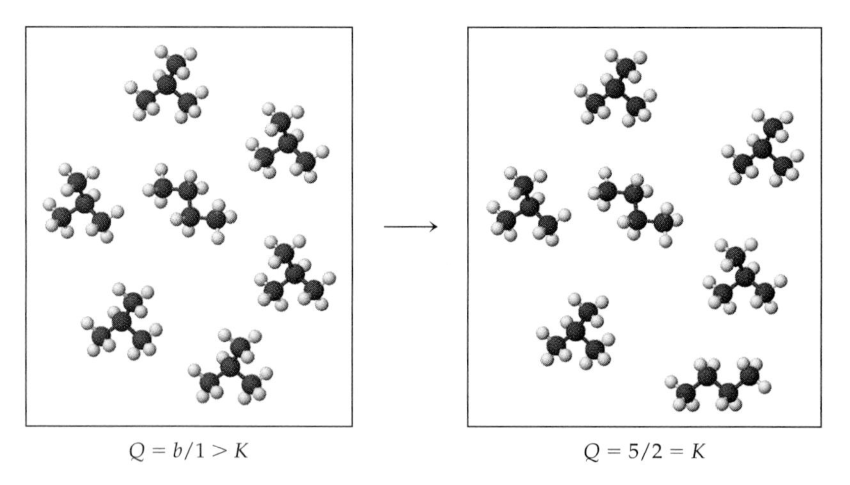

For any reaction

$$a\,\text{A} + b\,\text{B} \rightleftharpoons c\,\text{C} + d\,\text{D}$$

the **reaction quotient, *Q*** is defined by the equation

$$\text{Reaction quotient} = Q = \frac{[\text{C}]^c[\text{D}]^d}{[\text{A}]^a[\text{B}]^b}$$

The expression for Q has the *appearance* of the equilibrium expression, but Q differs from K in that the concentrations in the expression are not necessarily equilibrium concentrations. The following general statements can be made regarding the relationship of K and Q.

1. **If $Q < K$, the system is not at equilibrium and some reactants will be converted to products.**

The ratio of product concentrations to reactant concentrations is too small. More reactants must be converted to products (thus increasing Q) to achieve equilibrium (when $Q = K$).

2. **If $Q = K$, the system is at equilibrium.**
3. **If $Q > K$, the system is not at equilibrium and some products will be converted to reactants.**

The ratio of product concentrations to reactant concentrations is too large. To reach equilibrium, products must be converted to reactants (thus decreasing the value of Q until $Q = K$).

We will often make use of these ideas in our discussions on chemical equilibria (see Section 16.5, for example).

Example 16.2 The Reaction Quotient

The brown gas nitrogen dioxide, NO_2, is in equilibrium with the colorless gas dinitrogen tetraoxide, N_2O_4. $K_c = 170$ at 298 K.

$$2\ NO_2(g) \rightleftharpoons N_2O_4(g)$$

Suppose the concentration of NO_2 is 0.015 M, and the concentration of N_2O_4 is 0.025 M. Is Q_c larger than, smaller than, or equal to K_c? If the system is not at equilibrium, in which direction will the reaction proceed to achieve equilibrium?

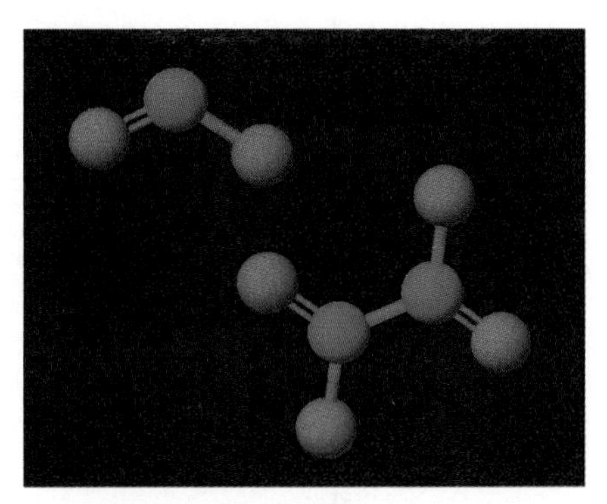

Nitrogen dioxide (*top*) and dinitrogen tetraoxide (*bottom*)

Solution The equilibrium constant expression for the reaction is

$$K_c = \frac{[N_2O_4]}{[NO_2]^2} = 170$$

If the reactant and product concentrations are substituted into the reaction quotient expression, we have

$$Q_c = \frac{(0.025)}{(0.015)^2} = 110$$

The value of Q_c is less than the value of K_c ($Q_c < K_c$), so the reaction is not at equilibrium. It must proceed to equilibrium by converting some NO_2 to N_2O_4, thus increasing $[N_2O_4]$ and decreasing $[NO_2]$ until $Q_c = K_c$.

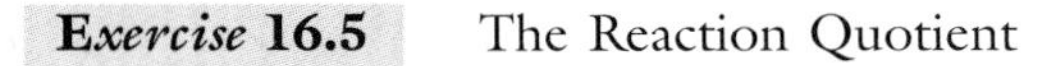

Exercise 16.5 The Reaction Quotient

At 2000 K the equilibrium constant, K_c, for the formation of NO(g)

$$N_2(g) + O_2(g) \rightleftharpoons 2\ NO(g)$$

is 4.0×10^{-4}. You have a container in which, at 2000 K, the concentration of N_2 is 0.50 mol/L, that of O_2 is 0.25 mol/L, and that of NO is 4.2×10^{-3} mol/L. Is the system at equilibrium? If not, predict which way the reaction will proceed to achieve equilibrium.

16.4 CALCULATING AN EQUILIBRIUM CONSTANT

When the values of the concentrations of all of the reactants and products are known *at equilibrium*, calculating an equilibrium constant simply involves substituting the data into the equilibrium constant expression. This is illustrated in the following example.

Example 16.3 Calculating an Equilibrium Constant

A mixture of SO_2, O_2, and SO_3 is allowed to reach equilibrium at 852 K. The equilibrium concentrations are $[SO_2] = 3.61 \times 10^{-3}$ mol/L, $[O_2] = 6.11 \times 10^{-4}$ mol/L, and $[SO_3] = 1.01 \times 10^{-2}$ mol/L. Calculate K_c, at 852 K, for the reaction

$$2\ SO_2(g) + O_2(g) \rightleftharpoons 2\ SO_3(g)$$

Solution First, write the equilibrium constant expression in terms of concentrations:

$$K_c = \frac{[SO_3]^2}{[SO_2]^2[O_2]}$$

Next, substitute the experimental information into the expression and calculate K_c:

$$K_c = \frac{(1.01 \times 10^{-2})^2}{(3.61 \times 10^{-3})^2(6.11 \times 10^{-4})} = 1.28 \times 10^4$$

More commonly, an experiment will only provide information on the initial quantities of reactants and the concentration at equilibrium of only one of the reactants or of one of the products. Equilibrium concentrations of the rest of the reactants and products must then be inferred from the balanced chemical equation. The remainder of this section describes calculations of this type.

Consider the oxidation of sulfur dioxide to sulfur trioxide:

$$2\ SO_2(g) \quad + \quad O_2(g) \rightleftharpoons \quad 2\ SO_3(g)$$

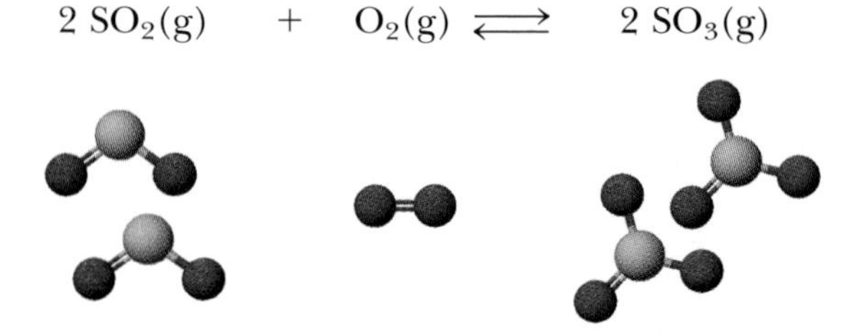

The expression for K_c was given in Example 16.3. Now let us suppose that, in an experiment to determine K_c for this reaction, you place 1.00 mol of SO_2 and 1.00 mol O_2, but no SO_3, in a 1.00-L flask. You cannot tell from this information how much of the SO_2 and O_2 will react on reaching equilibrium or how much SO_3 will then form. You only know that the system will contain a mixture of SO_2, O_2, and SO_3 at equilibrium.

The balanced chemical equation for the oxidation of SO_2 helps to define the situation at equilibrium. Let x be the number of moles of SO_2 consumed as the reaction proceeds to equilibrium. The balanced equation tells us that $x/2$ mol of O_2 must be consumed for each mole of SO_2 consumed. Because 1 mol of SO_3 is formed for each mole of SO_2 consumed, x moles of SO_3 are formed when x mol of SO_2 are consumed. The key idea is that all the changes can be expressed in terms of a single unknown (here we have used x) and the known coefficients in the balanced equation. These numbers can be displayed in the form of an *equilibrium table:*

Equation	**$2\ SO_2$**	**+**	**O_2**	**$\rightleftharpoons$**	**$2\ SO_3$**
Initial moles	1.00		1.00		0
Change on proceeding to equilibrium	$-x$		$-x/2$		$+x$
Moles at equilibrium	$1.00 - x$		$1.00 - x/2$		x
Equilibrium concentrations (mol/L)	$1.00 - x$		$1.00 - x/2$		x

Now that the equilibrium table is constructed, we only need to measure experimentally the equilibrium quantity of any one of the reactants or products. This can be used to find the value of x and therefore the equilibrium concentrations of all the substances in the reactions.

Example 16.4 Calculating an Equilibrium Constant

1.00 mol of SO_2 and 1.00 mol of O_2 are placed in a 1.00-L flask at 1000 K. When equilibrium has been achieved, 0.925 mol of SO_3 has formed. Calculate K_c, at 1000 K, for the reaction

$$2\ SO_2(g) + O_2(g) \rightleftharpoons 2\ SO_3(g)$$

Solution First, write the equilibrium constant expression in terms of concentrations:

$$K_c = \frac{[SO_3]^2}{[SO_2]^2[O_2]}$$

Next, refer to the equilibrium table described previously. The amount of SO_3 produced, 0.925 mol, is the value of x. Therefore, x mol, or 0.925 mol, of SO_2 must have been consumed along with $x/2$, or 0.925 mol/2 (= 0.463 mol), of O_2. The equilibrium table for this case is therefore

Equation	**$2\ SO_2$**	**+**	**O_2**	**$\rightleftharpoons$**	**$2\ SO_3$**
Initial moles	1.00		1.00		0
Change on proceeding to equilibrium	−0.925		−0.925/2		+0.925
Moles at equilibrium	1.00 − 0.925		1.00 − 0.925/2		0.925
Concentration at equilibrium (mol/L)	0.075		0.537		0.925

Remember that *concentrations* are needed in the equilibrium constant expression. The final line of the table gives the concentration of each substance at equilibrium. With these now known, it is possible to calculate K_c:

$$K_c = \frac{[SO_3]^2}{[SO_2]^2[O_2]} = \frac{(0.925)^2}{(0.075)^2(0.537)} = 2.8 \times 10^2$$

Example 16.5 Calculating an Equilibrium Constant

1.00 mol of ethanol and 1.00 mol of acetic acid are dissolved in water and kept at 100 °C. The volume of the solution is 250 mL. At equilibrium, 0.25 mol of acetic acid has been consumed in producing ethyl acetate. Calculate K_c at 100 °C for the reaction

$$\underset{\text{ethanol}}{C_2H_5OH(aq)} + \underset{\text{acetic acid}}{CH_3CO_2H(aq)} \rightleftharpoons \underset{\text{ethyl acetate}}{CH_3CO_2C_2H_5(aq)} + H_2O(\ell)$$

Solution At equilibrium, 0.25 mol of acid is consumed, so 0.25 mol of ethanol must also be consumed and 0.25 mol of ethyl acetate has formed. The equilibrium table is therefore

Equation	C_2H_5OH +	$CH_3CO_2H \rightleftharpoons$	$CH_3CO_2C_2H_5 + H_2O$
Initial moles	1.00	1.00	0
Change in moles	−0.25	−0.25	+0.25
Moles at equilibrium	0.75	0.75	0.25
Concentration at equilibrium	$\frac{0.75 \text{ mol}}{0.250 \text{ L}}$ = 3.0 M	$\frac{0.75 \text{ mol}}{0.250 \text{ L}}$ = 3.0 M	$\frac{0.25 \text{ mol}}{0.250 \text{ L}}$ = 1.0 M

The concentration of each substance at equilibrium is now known, and K_c can be calculated:

$$K_c = \frac{[CH_3CO_2C_2H_5]}{[C_2H_5OH][CH_3CO_2H]} = \frac{1.0}{(3.0)(3.0)} = 0.11$$

Notice that water, a reaction product, is not included in the equilibrium constant expression. Water is the solvent in the reaction and the small amount of additional water produced in the reaction does not affect the concentrations.

Example 16.6 Calculating an Equilibrium Constant

As described in *A Closer Look: Equilibrium Constant Expressions for Gases* (page 753), equilibrium quantities of reactants and products can be expressed in terms of partial pressures. Calculate K_p for the decomposition of H_2S

$$2\ H_2S(g) \rightleftharpoons 2\ H_2(g) + S_2(g)$$

from the following information: A tank initially contains H_2S with a pressure of 10.00 atm at 800 K. When the reaction has come to equilibrium, the partial pressure of S_2 vapor is 2.0×10^{-2} atm.

Solution The concentration of a gas at equilibrium can be expressed as moles per liter or as a partial pressure. Recall from our previous discussion (page 753) that partial pressures of gases are proportional to concentration [$P = (n/V)RT$]. The equilibrium expression that we want to evaluate is, therefore,

$$K_p = \frac{P^2_{H_2} \cdot P_{S_2}}{P^2_{H_2S}}$$

Let us set up an equilibrium table that expresses the equilibrium partial pressures of each gas in terms of a single unknown quantity, x.

Equation	**2 H_2S**	$\rightleftharpoons$	**2 H_2**	**+**	**S_2**
Initial pressure (atm)	10.0		0		0
Change	$-2x$		$+2x$		$+x$
Equilibrium partial pressure (in terms of x)	$10.00 - 2x$		$2x$		x
Equilibrium partial pressures (actual values) (atm)	$10.00 - 2(0.020)$		$2(0.020)$		0.020

Here we have designated the quantity of S_2 formed at equilibrium as x. The balanced equation informs us that for every mole (or atmosphere of pressure) of S_2 formed, 2 mol (or 2 atm) of H_2 are also formed, and 2 mol (or 2 atm) of H_2S are consumed. Because x is known from experiment to be 0.020 atm in this case, the partial pressure of each gas is known at equilibrium, and K_p can be calculated:

$$K_p = \frac{P^2_{H_2} \cdot P_{S_2}}{P^2_{H_2S}} = \frac{[2(0.020)]^2(0.020)}{[10.00 - 2(0.020)]^2} = 3.2 \times 10^{-7}$$

Exercise 16.6 Calculating an Equilibrium Constant

A mixture of H_2 (9.838×10^{-4} mol) and I_2 (1.377×10^{-3} mol) is sealed in a quartz tube and kept at 350 °C for a week. During this time the reaction

$$H_2(g) + I_2(g) \rightleftharpoons 2\,HI(g)$$

comes to equilibrium. The tube is broken open, and 4.725×10^{-4} mol of I_2 is found.

1. Calculate the number of moles of H_2 and HI present at equilibrium.
2. Assume the volume of the tube is 10.0 mL. Calculate K_c for the reaction.
3. Is your value of K_c different if the volume of the tube is 20.0 mL?

Exercise 16.7 Calculating an Equilibrium Constant

A solution is prepared by dissolving 0.050 mol of diiodocyclohexane, $C_6H_{10}I_2$, in the solvent CCl_4. The total solution volume is 1.00 L. When the reaction

$$C_6H_{10}I_2 \rightleftharpoons C_6H_{10} + I_2$$

has come to equilibrium at 35 °C, the concentration of I_2 is 0.035 mol/L.

1. What are the concentrations of $C_6H_{10}I_2$ and C_6H_{10} at equilibrium?
2. Calculate K_c, the equilibrium constant.

16.5 USING EQUILIBRIUM CONSTANTS IN CALCULATIONS

See the *Saunders Interactive General Chemistry CD-ROM,* Screen 16.10, Estimating Equilibrium Concentrations.

Another type of situation arises when you know the value of K as well as the initial number of moles, concentrations, or partial pressures of reactants or of reactants and products and need to find the quantities present at equilibrium. Once again we will solve these problems by constructing a table in which the unknown chemical quantities are expressed in terms of one unknown, say x.

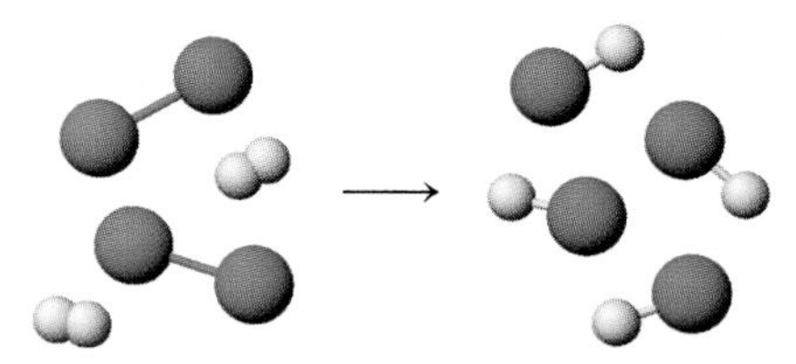

Hydrogen and iodine react to give hydrogen iodide.

Example 16.7 Calculating a Concentration from an Equilibrium Constant

At 425 °C, $K_c = 55.64$ for the reaction

$$H_2(g) + I_2(g) \rightleftharpoons 2\,HI(g)$$

If 1.00 mol each of H_2 and I_2 are placed in a 0.500-L flask at 425 °C, what are the equilibrium concentrations of H_2, I_2, and HI?

Solution Having written the balanced chemical equation, the next step is to write the equilibrium constant expression:

$$K_c = \frac{[HI]^2}{[H_2][I_2]} = 55.64$$

Now set up an equilibrium table to find a way to express the equilibrium concentrations of H_2, I_2, and HI in terms of a single unknown, x. Here we define x as the quantity of H_2 or of I_2 that is consumed in the reaction. Furthermore, $2x$ is the quantity of HI produced because the stoichiometric factor is (2 mol HI/1 mol H_2).

Equation	H_2	+	I_2	$\rightleftharpoons$ 2 HI
Initial moles	1.00		1.00	0
Initial concentrations (mol/L)	$\frac{1.00\text{ mol}}{0.500\text{ L}}$ = 2.00 M		$\frac{1.00\text{ mol}}{0.500\text{ L}}$ = 2.00 M	0
Change in concentrations	$-x$		$-x$	$+2x$
Equilibrium concentrations (mol/L)	$2.00 - x$		$2.00 - x$	$2x$

Now the equilibrium concentrations can be substituted into the equilibrium constant expression:

$$K_c = 55.64 = \frac{[HI]^2}{[H_2][I_2]} = \frac{(2x)^2}{(2.00 - x)(2.00 - x)} = \frac{(2x)^2}{(2.00 - x)^2}$$

In this case, the unknown quantity x can be found by taking the square root of both sides of the equation,

$$\sqrt{K_c} = 7.459 = \frac{2x}{2.00 - x}$$

and then solving for x:

$$7.459(2.00 - x) = 2x$$
$$14.9 - 7.459x = 2x$$
$$14.9 = 9.459x$$
$$x = 1.58$$

With x known, we can now solve for the equilibrium concentrations of the reactants and products:

$$[H_2] = [I_2] = 2.00 - x = 0.42 \text{ M}$$
$$[HI] = 2x = 3.16 \text{ M}$$

It is always wise to verify these values by substituting them back into the equilibrium expression and see if your calculated K_c agrees with the one given in the problem. In this case $(3.16)^2/(0.42)^2 = 57$. The slight discrepancy with the given value, $K_c = 55.64$, is due to the fact that we know $[H_2]$ and $[I_2]$ to only two significant figures.

Problem-Solving Tips and Ideas

16.2 How to Assign the Unknown x

In Example 16.7 we set up the equilibrium table shown here.

Equation	**H_2**	**+**	**I_2**	$\rightleftarrows$	**2 HI**
Initial concentrations	2.00 M		2.00M		0
Change in concentrations	$-x$		$-x$		$+2x$
Equilibrium concentrations	$2.00 - x$		$2.00 - x$		$2x$

Then we assigned the unknown x to the quantity of H_2 and I_2 that is consumed on proceeding to equilibrium. We could, however, also have made the unknown the quantity of HI formed at equilibrium (and give it the symbol y to avoid confusion). The table would have been

Equation	**H_2**	**+**	**I_2**	$\rightleftarrows$	**2 HI**
Initial concentrations	2.00 M		2.00 M		0
Change in concentrations	$-y/2$		$-y/2$		$+y$
Equilibrium concentrations	$2.00 - y/2$		$2.00 - y/2$		y

These can be substituted into the usual equilibrium expression:

$$K_c = 55.64 = \frac{[HI]^2}{[H_2][I_2]} = \frac{(y)^2}{(2.00 - y / 2)(2.00 - y / 2)}$$

Solving for y, you would find $[HI] = y = 3.16$, exactly the same result found in Example 16.7. We conclude that the unknown may be assigned in any of several ways when solving a problem. The important point is to define clearly the meaning of the unknown.

Example 16.8 Calculating a Concentration from an Equilibrium Constant

The reaction

$$N_2(g) + O_2(g) \rightleftharpoons 2\ NO(g)$$

contributes to air pollution whenever a fuel is burned in air at a high temperature, as in a gasoline engine. At 1500 K, $K_c = 1.0 \times 10^{-5}$. A sample of air is heated in a closed container to 1500 K. Before any reaction occurs, $[N_2] = 0.80$ mol/L and $[O_2] = 0.20$ mol/L. Calculate the equilibrium concentration of NO.

Solution As in previous problems, we shall write the equilibrium expression and then set up a table of equilibrium concentrations.

Equation	N_2	+	O_2	$\rightleftharpoons$	2 NO
Initial concentrations (mol/L)	0.80		0.20		0
Change in concentration	$-x$		$-x$		$+2x$
Equilibrium concentrations (mol/L)	$0.80 - x$		$0.20 - x$		$2x$

Next, the equilibrium concentrations are substituted into the equilibrium constant expression:

$$K_c = 1.0 \times 10^{-5} = \frac{(2x)^2}{(0.80 - x)(0.20 - x)}$$

This is a quadratic equation, which can be rearranged into the standard form ($ax^2 + bx + c = 0$) (see Appendix A).

$$(1.0 \times 10^{-5})(0.80 - x)(0.20 - x) = 4x^2$$

$$(1.0 \times 10^{-5})(0.16 - 1.00x + x^2) = 4x^2$$

$$(4 - 1.0 \times 10^{-5})x^2 + (1.0 \times 10^{-5})x - 0.16 \times 10^{-5} = 0$$

$$ax^2 \qquad\qquad bx \qquad\qquad c \qquad 0$$

Such equations can be solved by the quadratic formula given in Appendix A. Using this, you will find two roots to this equation:

$$x = 6.3 \times 10^{-4} \qquad \text{or} \qquad x = -6.3 \times 10^{-4}$$

Because x stands for the quantity of N_2 that reacts on proceeding to equilibrium, and $2x$ is the quantity of NO formed, the negative root is physically meaningless. Thus, we choose the positive root and use this to calculate the equilibrium concentrations of the reactants and products.

$$[N_2] = 0.80 - 6.3 \times 10^{-4} \approx 0.80\ M$$

$$[O_2] = 0.20 - 6.3 \times 10^{-4} \approx 0.20\ M$$

$$[NO] = 2x = 1.26 \times 10^{-3}\ M$$

This result brings us to an important point. Notice in this example that the quantity of NO formed is very small, as are the quantities of N_2 and O_2 consumed. Indeed, the concentrations of N_2 and O_2 have not changed (to two significant figures) on proceeding to equilibrium. In situations where the change in the reactant concentration is very small, equilibrium calculations can be simplified greatly by making an assumption. Here we assume that x is very small relative to 0.80 or 0.20. This means that $0.80 - x \approx 0.80$ and $0.20 - x \approx 0.20$. The equilibrium constant expression can now be written and solved as follows:

$$K_c = 1.0 \times 10^{-5} = \frac{(2x)^2}{(0.80)(0.20)}$$

$$1.6 \times 10^{-6} = 4x^2$$

$$x = 6.3 \times 10^{-4}$$

The value of x obtained using the assumption is the same as from the quadratic equation.

How do you know when you can make the assumption that the unknown quantity x is so small that you can write a simplified equation that does not require the quadratic formula to solve? In most equilibrium calculations, a quantity (x) may be ignored if it is less than 5% of the smallest quantity initially present (here 0.20). In this example, x from the approximate equation was 6.3×10^{-4}, and it is $[(6.3 \times 10^{-4})/0.20]100\% = 0.32\%$ of the smallest quantity initially present. In general, when solving equilibrium problems of this type you (1) make an assumption that the unknown (x) is small and solve the simplified equation. Then (2) compare it with the smallest quantity available. If the result is less than 5% of the smallest quantity, then there is no need to solve the full equation using the quadratic formula.

The 5% figure chosen here is arbitrary. It is chosen because it is close to the accuracy of the data used in the problems in this book. See Appendix A.

For a further discussion of such calculations, see Section 17.7 and Appendix A.

Exercise 16.8 Calculating a Concentration from an Equilibrium Constant

At some temperature, $K_c = 33$ for the reaction

$$H_2(g) + I_2(g) \rightleftharpoons 2\ HI(g)$$

H_2 and I_2 are initially present at equal concentrations, 6.00×10^{-3} mol/L for each. Find the concentration of each reactant and product at equilibrium.

Exercise 16.9 Calculating a Concentration from an Equilibrium Constant

Graphite and carbon dioxide are kept at constant volume at 1000 K until the reaction

$$C(graphite) + CO_2(g) \rightleftharpoons 2\ CO(g)$$

has come to equilibrium. At this temperature, $K_c = 0.021$. The initial concentration of CO_2 is 0.012 mol/L. Calculate the equilibrium concentration of CO.

16.6 DISTURBING A CHEMICAL EQUILIBRIUM: LE CHATELIER'S PRINCIPLE

See the *Saunders Interactive General Chemistry CD-ROM,* Screen 16.11, Le Chatelier's Principle, and Screen 16.12, Temperature Changes.

There are three common ways a chemical reaction at equilibrium may be disturbed: (1) a change in temperature, (2) a change in the concentration of a reactant or product, and (3) a change in volume (Table 16.2). If you try to raise the temperature of the system (by adding heat energy), a chemical reaction will occur that acts to cool the system (by using heat energy). If additional reactant or product is added to a reaction at equilibrium, the system will respond by using up some of what has been added. In a system where one or more of the reactants or products is a gas, decreasing the available volume will lead to a pressure increase; the system will react in a way that decreases the pressure. These outcomes can be predicted by **Le Chatelier's principle:** a change in any of the factors that determine the equilibrium conditions of a system will cause the system to change in such a manner as to reduce or counteract the effect of the change.

Effect of Temperature Changes on Equilibria

Temperature effects on phase equilibria and on the solubility of gases in water were described in Chapters 13 and 14. For example, an increase in temperature leads to an increase in the vapor pressure of a liquid (Section 13.2) and to a de-

Table 16.2 • EFFECTS OF DISTURBANCES ON EQUILIBRIUM AND K

Disturbance	Change as Mixture Returns to Equilibrium	Effect on Equilibrium	Effect on K
Addition of reactant	Some of added reactant is consumed	Shift to right	No change
Addition of product	Some of added product is consumed	Shift to left	No change
Decrease in volume, increase in pressure	Pressure decreases	Shift toward fewer gas molecules	No change
Increase in volume, decrease in pressure	Pressure increases	Shift toward more gas molecules	No change
Rise in temperature	Heat energy is consumed	Shift in the endothermic direction	Change
Drop in temperature	Heat energy is generated	Shift in exothermic direction	Change

crease in the solubility of most gases (Section 14.2). As in the case of the solubility of gases, you can make a qualitative prediction about the effect of a temperature change on the equilibrium position in a chemical reaction if you know whether the reaction is exothermic or endothermic. As an example, consider the endothermic reaction of N_2 with O_2 to give NO.

In Section 14.2 Le Chatelier's principle was applied to the solubility of substances.

$$N_2(g) + O_2(g) \rightleftarrows 2\ NO(g) \qquad \Delta H^\circ_{rxn} = +180.5\ kJ$$

Equilibrium Constant, K_c	**Temperature**
4.5×10^{-31}	298 K
6.7×10^{-10}	900 K
1.7×10^{-3}	2300 K

We are surrounded by N_2 and O_2, but you know that they do not react appreciably at room temperature. The position of equilibrium lies almost completely to the left. If a mixture of N_2 and O_2 is heated above 700 °C, however, as in an automobile engine, the equilibrium shifts toward NO, and a greater concentration of NO can exist in equilibrium with reactants. The experimental equilibrium constants given above show that [NO] increases and [N_2] and [O_2] decrease as the temperature increases at equilibrium.

To see why the value of *K* can change with temperature, consider the N_2/O_2 reaction further. The enthalpy change for the reaction is +180.5 kJ, so we might imagine heat as a "reactant." Le Chatelier's principle informs us that input of energy (as heat) causes the equilibrium to shift in a direction that counteracts this input. The way to counteract the energy input here is to use up some of the added heat by consuming N_2 and O_2 and producing more NO. An increase in temperature must therefore be accompanied by increased production of NO and consumption of N_2 and O_2. Because this raises the value of the numerator and lowers the value of the denominator in the *K* expression, *K* must also increase in value.

A more quantitative analysis of temperature effects on equilibria can be made using thermodynamics. See Chapter 20.

Two molecules of the brown gas NO_2 combine readily to form colorless N_2O_4, and an equilibrium between these compounds is readily achieved in a closed system (Figure 16.7):

$$2\ NO_2(g) \rightleftarrows N_2O_4(g) \qquad \Delta H^\circ = -57.2\ kJ$$

Equilibrium Constant, K_c	**Temperature**
1300	273 K
170	298 K

Here the reaction is exothermic, so we might imagine heat as a reaction "product." By lowering the temperature of the reaction, as in Figure 16.7, some heat is removed. According to Le Chatelier's principle, this consumption of heat can be counteracted if the reaction produces more heat by combination of NO_2 to give more N_2O_4. Thus, the equilibrium concentration of NO_2 declines, that of N_2O_4 increases, and the values of *K* become larger as the temperature declines.

Based on the examples in this section, it can be stated that *increasing* the temperature of a system at equilibrium causes a reaction in the direction that results in *absorption* of heat energy; *decreasing* the temperature causes a reaction to go in the direction that results in *evolution* of heat energy.

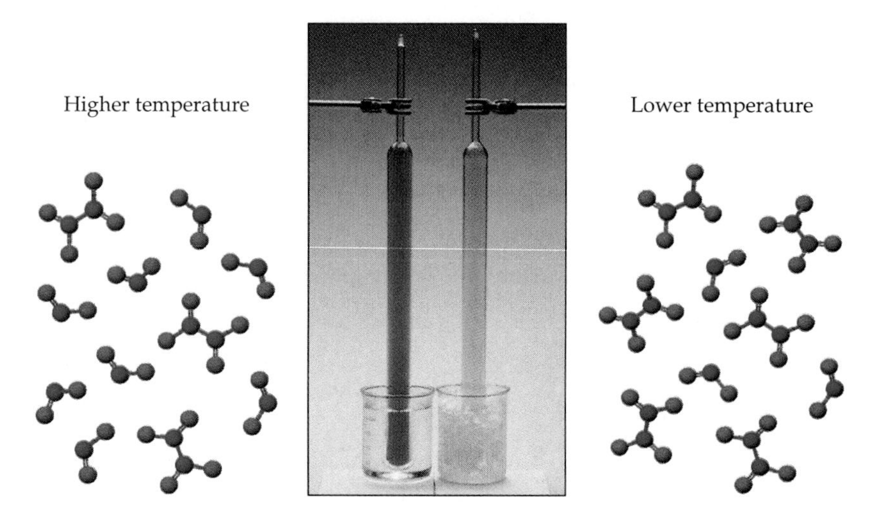

Figure 16.7 Effect of temperature on an equilibrium. The tubes in the photograph both contain gaseous NO_2 (brown) and N_2O_4 (colorless). As predicted by Le Chatelier's principle, the equilibrium favors colorless N_2O_4 at lower temperatures. This is clearly seen in the tube at the right, where the gas in the ice bath at 0 °C is only slightly brown because there is only a small partial pressure of the brown gas NO_2. At 50 °C *(the tube at the left)*, the equilibrium shifts, forming more NO_2, as indicated by the darker brown color. *(Photo, Marna G. Clarke; models, S. M. Young)*

Exercise 16.10 Le Chatelier's Principle

Consider the effect of temperature changes on the following equilibria.

1. A mixture of three gases is in equilibrium:

$$2\ NOCl(g) \rightleftharpoons 2\ NO(g) + Cl_2(g) \qquad \Delta H^\circ_{rxn} = +77.1\ kJ$$

   Does the concentration of NOCl increase or decrease at equilibrium as the temperature of the system is increased?

2. Does the concentration of SO_3 increase or decrease when the temperature increases?

$$2\ SO_2(g) + O_2(g) \rightleftharpoons 2\ SO_3(g) \qquad \Delta H^\circ_{rxn} = -198\ kJ$$

Effect of the Addition or Removal of a Reactant or Product

See the *Saunders Interactive General Chemistry CD-ROM,* Screen 16.13, Addition and Removal of a Reagent.

If the concentration of a reactant or product is changed from its equilibrium value *at a given temperature,* the reaction must shift to a new equilibrium position for which the reaction quotient still equals *K*. To illustrate this, let us return to the butane/isobutane equilibrium and Figure 16.8.

$$\text{butane} \rightleftharpoons \text{isobutane} \qquad K_c = 2.5$$

Suppose the equilibrium mixture consists of two molecules of butane and five of isobutane. The ratio of molecules is 5/2, or 2.5/1, the value of the equilibrium constant for the reaction. Now add 7 more molecules of isobutane to the mixture to give a ratio of 12 isobutane molecules to 2 of butane. The ratio or reaction quotient, *Q*, is 6/1. *Q* is greater than *K*, so the system will change to

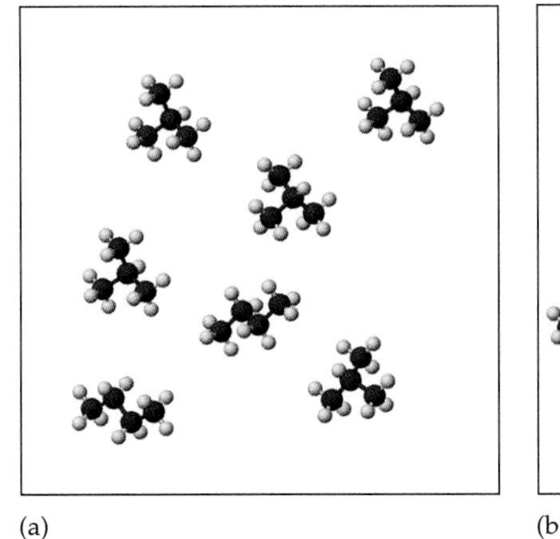
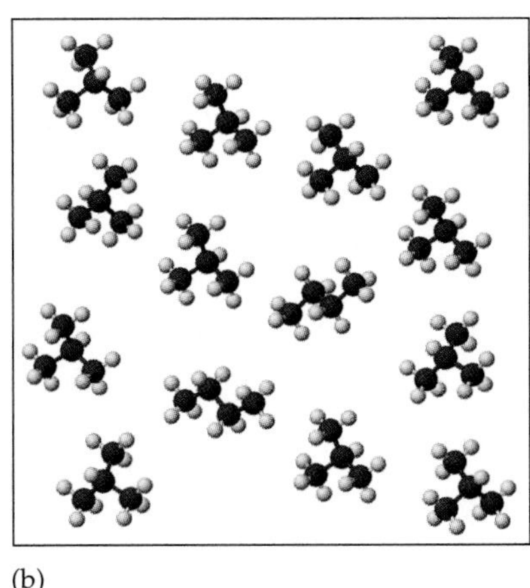
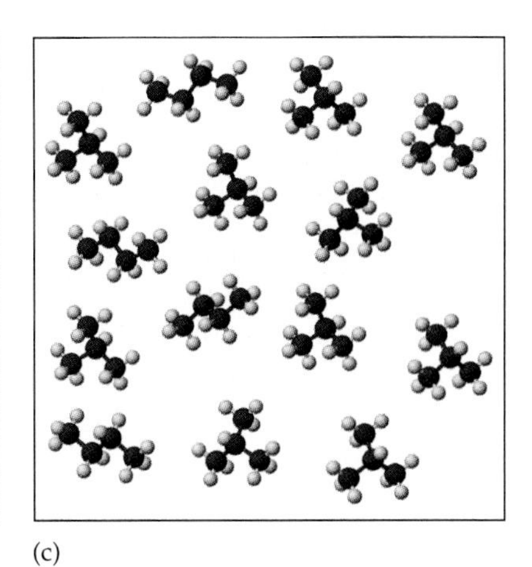

(a) (b) (c)

***Figure* 16.8 Changing concentrations and Le Chatelier's principle.** (a) An equilibrium mixture of five isobutane molecules and two butane molecules. (b) Seven isobutane molecules are added, so the system is no longer at equilibrium. (c) A net of two isobutane molecules have changed to butane molecules, to once again give an equilibrium mixture where the ratio of isobutane to butane is 5 to 2 (or 2.5/1).

reestablish equilibrium. The only way this can happen is for some molecules of isobutane to be changed into butane molecules, a process that continues until the ratio [isobutane]/[butane] is once again 5/2, or 2.5/1. In this particular case, it means that if 2 of the 12 isobutane molecules change to butane, the *ratio* of isobutane to butane at the new equilibrium position is again 2.5/1.

Example 16.9 The Effect of Concentration Changes on an Equilibrium

Here we wish to work out an algebraic solution to the problem of disturbing the butane/isobutane equilibrium. Assume equilibrium has been established in a 1.00-L flask with [butane] = 0.500 mol/L and [isobutane] = 1.25 mol/L. Then 1.50 mol/L of butane is added. What are the new equilibrium concentrations of butane and isobutane?

Solution Let us first organize the information in a table (where all concentrations are in moles per liter).

Equilibrium	Butane	$\rightleftarrows$	Isobutane
Initial concentration (M)	0.500		1.25
Concentration immediately on adding butane	0.500 + 1.50		1.25
Change in concentration on proceeding to new equilibrium position	$-x$		$+x$
Concentration at new equilibrium position (M)	$2.00 - x$		$1.25 + x$

The entries in this table were arrived at as follows:

1. The concentration of butane at the new equilibrium position will be the old equilibrium concentration plus what was added (1.50 mol/L) minus the concentration of butane that must be converted to isobutane in order to reestab-

CURRENT PERSPECTIVES

Kinetics, Equilibrium, and Ammonia

This book began (page 3) with a question: *Too much of a good thing?* This referred to the practice of adding nitrogen-containing substances to croplands to stimulate growth and to the recent finding that nitrogen-based fertilizers can lead to significant environmental changes. The main nitrogen-based substance used is ammonia, and its history is interesting.

Until the end of the 19th century nitrogen was obtained from naturally occurring salts such as $NaNO_3$ or from bird droppings. It was clear that this would not sustain the worldwide demand, so chemists sought ways to make some nitrogen-based compound cheaply. In 1909, the German chemist Fritz Haber (1868–1934) accomplished the feat of making ammonia from its elements, nitrogen and hydrogen:

$$N_2(g) + 3\ H_2(g) \rightleftharpoons 2\ NH_3(g)$$

At 25 °C, K_c (calc) = 3.5×10^8 and
$\Delta H_{rxn} = -92.2$ kJ/mol

At 450 °C, K_c (experiment) = 0.16 and
$\Delta H_{rxn} = -111.3$ kJ/mol

Now ammonia is made for pennies a kilogram and is consistently ranked in the top five chemicals produced in the United States. In 1995, approximately 17 billion kilograms were produced. Not only is it used as fertilizer but it is also a starting material for making nitric acid and ammonium nitrate, among other things.

The manufacture of ammonia illustrates the principles of kinetics and chemical equilibria.

- The reaction is exothermic and so is predicted to be product-favored ($K > 1$ at 25 °C). The reaction is slow, however, at 25 °C, so it is carried out at a higher temperature to increase the reaction rate.
- Although the reaction rate increases with temperature, the equilibrium constant declines, as predicted by Le Chatelier's principle. Thus, for a given concentration of starting material, the equilibrium concentration of NH_3 is smaller at higher temperatures.
- To increase the equilibrium concentration of NH_3, the reaction is carried out at a higher pressure. This does not change the value of K, but an increase in pressure can be compensated by converting 4 mol of reactants to 2 mol of product.

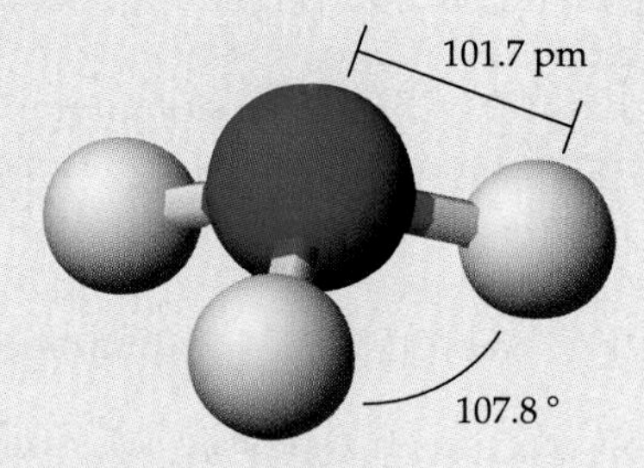

The ammonia molecule has a trigonal-pyramidal molecular geometry.

lish equilibrium. The quantity of butane that is converted is unknown as yet and so is designated as x.

2. The quantity of isobutane at the new equilibrium position is the quantity that was already present (1.25 mol/L) plus the quantity formed (x mol/L) on proceeding to the new equilibrium position.

Having defined [butane] and [isobutane] at the new equilibrium position, and remembering that K is a constant (= 2.50), we can write

$$K_c = \frac{[\text{isobutane}]_{\text{new}}}{[\text{butane}]_{\text{new}}} = 2.50 = \frac{1.25 + x}{2.00 - x}$$

$$2.50\ (2.00 - x) = 1.25 - x$$

$$3.75 = 3.50\ x$$

$$x = 1.07 \text{ mol/L}$$

- A catalyst is used to increase the reaction rate further. An effective catalyst for the Haber process is Fe_3O_4 mixed with KOH, SiO_2, and Al_2O_3 (all inexpensive chemicals). Because the catalyst is not effective below 400 °C, however, the optimum temperature is about 450 °C.

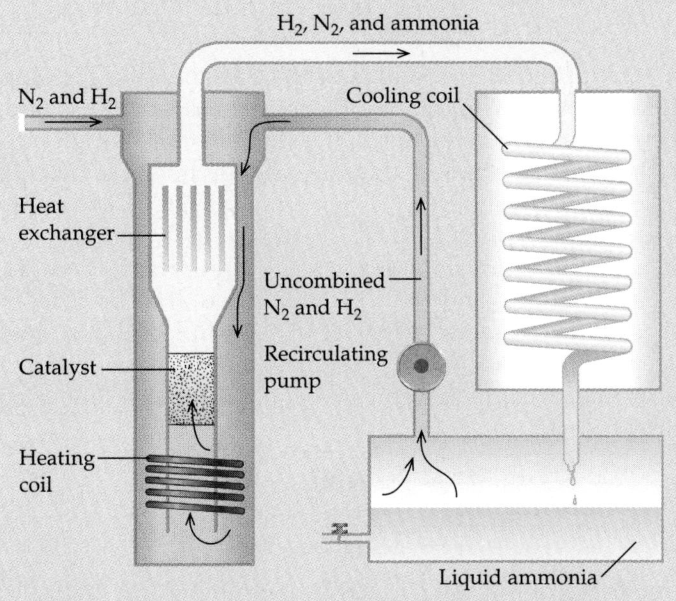

The Haber process for ammonia synthesis. A mixture of H_2 and N_2 is pumped over a catalytic surface. The NH_3 is collected as a liquid (at −33 °C), and unchanged reactants are recycled in the catalytic chamber.

Fritz Haber (1868–1934) received the Nobel Prize in chemistry in 1918 for the industrial ammonia synthesis. He is, however, also infamous because he was the director of the German Chemical Warfare Office during World War I. After the war he continued his work in thermodynamics (see the Born-Haber cycle, page 350). Because he had a Jewish background he had to leave Germany in 1933. He worked for a time in England but died in Switzerland in 1934. *(Oesper Collection in the History of Chemistry/University of Cincinnati)*

We now know that the new equilibrium position is established at

$$[\text{butane}] = 2.00 - x = 0.93 \text{ mol/L}$$

$$[\text{isobutane}] = 1.25 + x = 2.32 \text{ mol/L}$$

Be sure to verify that [isobutane]/[butane] = 2.32/0.93 = 2.5.

Exercise 16.11 The Effect of Concentration Changes on an Equilibrium

Equilibrium exists between butane and isobutane when [butane] = 0.20 M and [isobutane] = 0.50 M. What are the equilibrium concentrations of butane and isobutane if 2.00 mol/L of isobutane is added to the original mixture?

The Effect of Volume Changes on Gas Phase Equilibria

See the *Saunders Interactive General Chemistry CD-ROM,* Screen 16.14, Volume Changes.

For a reaction that involves gases, what happens to equilibrium concentrations or pressures if the size of the container is changed? (This occurs, for example, when fuel and air are compressed in an automobile engine.) To answer this question, recall that concentrations are in moles per liter. If the volume of a gas changes, the concentration must also change, and the equilibrium position changes. As an example, once again consider the equilibrium

$$\underset{\text{brown gas}}{2\ NO_2(g)} \rightleftharpoons \underset{\text{colorless gas}}{N_2O_4(g)} \qquad K_c = \frac{[N_2O_4]}{[NO_2]^2} = 170 \text{ at } 298\text{ K}$$

What happens to this equilibrium if the volume of the flask holding the gases is suddenly halved? Because the concentration of a gas increases as the volume available to the gas decreases, the immediate result is that the concentrations of both gases will double. This means, however, that the system is no longer at equilibrium because the quotient $[N_2O_4]/[NO_2]^2$ is not equal to 170. For example, assume equilibrium is established when $[N_2O_4]$ is 0.0280 mol/L and $[NO_2]$ is 0.0128 mol/L. If both concentrations double when the volume is halved, this means $[N_2O_4]$ is now 0.0560 mol/L and $[NO_2]$ is 0.0256 mol/L. The reaction quotient, Q, under these circumstances is $(0.0560)/(0.0256)^2 = 85.5$, a value clearly less than K. Because Q is less than K, the quantity of product must increase at the expense of the reactants to return to equilibrium, and the equilibrium will shift in favor of N_2O_4:

$$2\ NO_2(g) \rightleftharpoons N_2O_4(g)$$

$$\xrightarrow[\text{Equilibrium shifts right}]{\text{Decrease volume of container}}$$

This means that one molecule of N_2O_4 is formed by consuming two molecules of NO_2. The concentration of NO_2 decreases twice as fast as that of N_2O_4 increases until the reaction quotient, $[N_2O_4]/[NO_2]^2$, is once again equal to K.

The conclusions for the NO_2/N_2O_4 equilibrium can be generalized. For any reaction involving gases, the stress of a volume *decrease* (a pressure increase) is counterbalanced by a shift in the equilibrium to the side of the reaction with the *fewer number of molecules.* For a volume increase (a pressure decrease), the opposite situation results: The equilibrium shifts to the side of the reaction with the greater number of molecules.

Exercise 16.12 Effect of Concentration and Volume Changes on Equilibria

The formation of ammonia from its elements is an important industrial process:

$$N_2(g) + 3\ H_2(g) \rightleftharpoons 2\ NH_3(g)$$

1. Does the equilibrium shift to the left or the right when extra H_2 is added? When extra NH_3 is added?
2. What is the effect on the position of the equilibrium when the volume of the system is increased? Does the equilibrium shift to the left or to the right, or is the system unchanged?

16.7 IS THERE LIFE AFTER EQUILIBRIUM?

When the battery "runs down" in your calculator or car, you say that the battery is "dead." But it hasn't "died." As will be explained in Chapter 20, the reactions in the battery have just come to a state of equilibrium. It is for this reason that many of the chemical reactions that are interesting and important to us all occur under nonequilibrium conditions: the biochemical reactions in plants and animals, the production of elements in the stars, and the synthesis of useful materials. It is important for us to know how far away from equilibrium a system may be, how strong the tendency or drive is toward equilibrium and the rate of progress in that direction, how the system may be prevented from reaching equilibrium, and, perhaps, how to derive useful work from the system as it moves toward equilibrium. Thus, although it is useful to know about the state of a system when it achieves equilibrium, an understanding of systems not at equilibrium is also extremely useful.

The chemical reactions in living plants have not achieved equilbrium. When equilibrium is attained, the plant literally dies. *(J. Kotz)*

CHAPTER HIGHLIGHTS

Having studied this chapter, you should be able to

- Understand the nature and characteristics of the state of equilibrium: (a) chemical reactions are reversible; (b) equilibria are dynamic, and (c) the nature of the equilibrium state is the same, no matter what the direction of approach (Section 16.1).
- Write an equilibrium constant expression for any chemical reaction (Section 16.2). For the general reaction

$$a\,\mathrm{A} + b\,\mathrm{B} \rightleftharpoons c\,\mathrm{C} + d\,\mathrm{D}$$

  the equilibrium concentrations of reactants and products are always related by the **equilibrium constant expression,** Equation 16.1.

$$\text{Equilibrium constant} = K_c = \frac{\overbrace{[\mathrm{C}]^c[\mathrm{D}]^d}^{\text{Product concentrations}}}{\underbrace{[\mathrm{A}]^a[\mathrm{B}]^b}_{\text{Reactant concentrations}}}$$

- Recognize that the concentrations of solids and solvents (e.g., water) are not included in equilibrium constant expressions (Section 16.2).
- Appreciate the fact that equilibrium concentrations may be expressed in terms of reactant and product concentrations (expressed in moles per liter), and K is then designated as K_c. Alternatively, concentrations of gases may be represented by partial pressures, and K for such cases is designated K_p (Section 16.2).
- Know how K changes as different stoichiometric coefficients are used in a balanced equation or if the equation is reversed (Section 16.2).
- Know that, when two chemical equations are added to give a net equation, the value of K for the net equation is the product of the values of K for the summed equations (Section 16.2).

- Recognize that a large value of K ($K >> 1$) means the reaction is product-favored, and the product concentrations are greater than the reactant concentrations at equilibrium. A small value of K ($K << 1$) indicates a reactant-favored reaction in which the product concentrations are smaller than the reactant concentrations at equilibrium (Section 16.2).
- Apply the idea of the **reaction quotient** (Q) to decide if a reaction is at equilibrium ($Q = K$), or if there will be a net conversion of reactants to products ($Q < K$) or products to reactants ($Q > K$) to attain equilibrium (Section 16.3).
- Calculate an equilibrium constant given the reactant and product concentrations at equilibrium (Section 16.4).
- Use equilibrium constants to calculate the concentration of a reactant or product at equilibrium (Section 16.5).
- Apply **Le Chatelier's principle** to predict the effect of a disturbance on a chemical equilibrium: a change in temperature, a change in concentrations, or a change in volume or pressure for a reaction involving gases (Section 16.6 and Table 16.2).

STUDY QUESTIONS

REVIEW QUESTIONS

1. Name three important features of the equilibrium condition.
2. Decide if each of the following statements is true or false. If false, change the wording to make it true.
   (a) The magnitude of the equilibrium constant is always independent of temperature.
   (b) When two chemical equations are added to give a net equation, the equilibrium constant for the net equation is the product of the equilibrium constants of the summed equations.
   (c) The equilibrium constant for a reaction has the same value as K for the reverse reaction.
   (d) Only the concentration of CO_2 appears in the equilibrium constant expression for the reaction

$$CaCO_3(s) \rightleftharpoons CaO(s) + CO_2(g)$$

   (e) For the reaction $CaCO_3(s) \rightleftharpoons CaO(s) + CO_2(g)$, the value of K is the same whether the amount of CO_2 is expressed as moles per liter or as gas pressure.
3. Neither $PbCl_2$ nor PbF_2 is appreciably soluble in water. If solid $PbCl_2$ and solid PbF_2 are placed in equal amounts of water in separate beakers, in which beaker is the concentration of Pb^{2+} greater? Equilibrium constants for these solids dissolving in water are

$$PbCl_2(s) \rightleftharpoons Pb^{2+}(aq) + 2\ Cl^-(aq) \qquad K = 1.7 \times 10^{-5}$$
$$PbF_2(s) \rightleftharpoons Pb^{2+}(aq) + 2\ F^-(aq) \qquad K = 3.7 \times 10^{-8}$$

4. How does the reaction quotient for a reaction differ from the equilibrium constant for that reaction?
5. If the reaction quotient is smaller than the equilibrium constant for a reaction such as $A \rightleftharpoons B$, does this mean that reactant A continues to be consumed to form B, or does B form A, as the system moves to equilibrium?
6. The decomposition of calcium carbonate

$$CaCO_3(s) \rightleftharpoons CaO(s) + CO_2(g)$$

   is an endothermic process. Using Le Chatelier's principle, explain how increasing the temperature affects the equilibrium. If more $CaCO_3$ is added to a flask in which this equilibrium exists, how is the equilibrium affected? What if some additional CO_2 is placed in the flask?
7. The oxidation of NO to NO_2 is exothermic:

$$2\ NO(g) + O_2(g) \rightleftharpoons 2\ NO_2(g)$$

   Which way will the equilibrium shift when (a) the temperature is lowered and (b) the volume of the reaction flask is increased?

NUMERICAL QUESTIONS

Writing and Manipulating Equilibrium Constant Expressions

8. Write equilibrium constant expressions for the following reactions. For gases use either pressures or concentrations.
   (a) $2\ H_2O_2(g) \rightleftharpoons 2\ H_2O(g) + O_2(g)$
   (b) $CO(g) + \frac{1}{2}\ O_2(g) \rightleftharpoons CO_2(g)$
   (c) $C(s) + CO_2(g) \rightleftharpoons 2\ CO(g)$
   (d) $FeO(s) + CO(g) \rightleftharpoons Fe(s) + CO_2(g)$
9. Write equilibrium constant expressions for the following reactions. For gases use either pressures or concentrations.
   (a) $3\ O_2(g) \rightleftharpoons 2\ O_3(g)$
   (b) $Ni(s) + 4\ CO(g) \rightleftharpoons Ni(CO)_4(g)$
   (c) $(NH_4)_2CO_3(s) \rightleftharpoons 2\ NH_3(g) + CO_2(g) + H_2O(g)$
   (d) $BaSO_4(s) \rightleftharpoons Ba^{2+}(aq) + SO_4^{2-}(aq)$

10. Consider the following equilibria involving $SO_2(g)$ and their corresponding equilibrium constants:

$$SO_2(g) + \tfrac{1}{2}O_2(g) \rightleftharpoons SO_3(g) \qquad K_1$$
$$2\,SO_3(g) \rightleftharpoons 2\,SO_2(g) + O_2(g) \qquad K_2$$

Which of the following expressions relates K_1 to K_2?

(a) $K_2 = K_1^2$ (d) $K_2 = \dfrac{1}{K_1}$

(b) $K_2^2 = K_1$ (e) $K_2 = \dfrac{1}{K_1^2}$

(c) $K_2 = K_1$

11. The following process can be considered as the sum of two reactions, each with its own equilibrium constant, K_1 and K_2. How are K_1 and K_2 related to K_{net}?

$$Cu(OH)_2(s) + 4\,NH_3(aq) \rightleftharpoons Cu(NH_3)_4^{2+}(aq) + 2\,OH^-(aq) \qquad K_{net}$$
$$Cu(OH)_2(s) \rightleftharpoons Cu^{2+}(aq) + 2\,OH^-(aq) \qquad K_1$$
$$Cu^{2+}(aq) + 4\,NH_3(aq) \rightleftharpoons Cu(NH_3)_4^{2+}(aq) \qquad K_2$$

12. Calculate K_c for the reaction

$$SnO_2(s) + 2\,CO(g) \rightleftharpoons Sn(s) + 2\,CO_2(g)$$

given the following information:

$$SnO_2(s) + 2\,H_2(g) \rightleftharpoons Sn(s) + 2\,H_2O(g) \qquad K_c = 8.12$$
$$H_2(g) + CO_2(g) \rightleftharpoons H_2O(g) + CO(g) \qquad K_c = 0.771$$

13. Calculate K_c for the reaction

$$Fe(s) + H_2O(g) \rightleftharpoons FeO(s) + H_2(g)$$

given the following information:

$$H_2O(g) + CO(g) \rightleftharpoons H_2(g) + CO_2(g) \qquad K_c = 1.6$$
$$FeO(s) + CO(g) \rightleftharpoons Fe(s) + CO_2(g) \qquad K_c = 0.67$$

14. The equilibrium constant, K_c, for the reaction

$$H_2(g) + Cl_2(g) \rightleftharpoons 2\,HCl(g)$$

at 500 K is 4.8×10^{10}. Calculate K_c for
(a) $\tfrac{1}{2}H_2(g) + \tfrac{1}{2}Cl_2(g) \rightleftharpoons HCl(g)$
(b) $HCl(g) \rightleftharpoons \tfrac{1}{2}H_2(g) + \tfrac{1}{2}Cl_2(g)$

15. The equilibrium constant K_c for the reaction

$$CO_2(g) \rightleftharpoons CO(g) + \tfrac{1}{2}O_2(g)$$

is 6.66×10^{-12} at 1000 K. Calculate K_c for the reaction

$$2\,CO(g) + O_2(g) \rightleftharpoons 2\,CO_2(g)$$

The Reaction Quotient

16. $K_c = 5.6 \times 10^{-12}$ at 500 K for the reaction

$$I_2(g) \rightleftharpoons 2\,I(g)$$

A mixture kept at 500 K contains I_2 at a concentration of 0.020 mol/L and I at a concentration of 2.0×10^{-8} mol/L. Is the reaction at equilibrium? If not, which way must the reaction go to reach equilibrium?

17. A mixture of SO_2, O_2, and SO_3 at 1000 K contains the gases at the following concentrations: $[SO_2] = 5.0 \times 10^{-3}$ mol/L, $[O_2] = 1.9 \times 10^{-3}$ mol/L, and $[SO_3] = 6.9 \times 10^{-3}$ mol/L. Is the reaction at equilibrium?

$$2\,SO_2(g) + O_2(g) \rightleftharpoons 2\,SO_3(g)$$

If not, which way will the reaction proceed to reach equilibrium? K_c for the reaction is 279.

18. The reaction

$$2\,NO_2(g) \rightleftharpoons N_2O_4(g)$$

has an equilibrium constant, K_c, of 170 at 25 °C. If 2.0×10^{-3} mol of NO_2 is present in a 10.-L flask along with 1.5×10^{-3} mol of N_2O_4, is the system at equilibrium? If it is not at equilibrium, does the concentration of NO_2 increase or decrease as the system proceeds to equilibrium?

19. The equilibrium constant, K_c, for the reaction

$$2\,NOCl(g) \rightleftharpoons 2\,NO(g) + Cl_2(g)$$

is 3.9×10^{-3} at 300 °C. A mixture contains the gases at the following concentrations: $[NOCl] = 5.0 \times 10^{-3}$ mol/L, $[NO] = 2.5 \times 10^{-3}$ mol/L, and $[Cl_2] = 2.0 \times 10^{-3}$ mol/L. Is the reaction at equilibrium? If not, in which direction does the reaction move to come to equilibrium?

Calculating Equilibrium Constants

20. The reaction below was examined at 250 °C. At equilibrium $[PCl_5] = 4.2 \times 10^{-5}$ mol/L, $[PCl_3] = 1.3 \times 10^{-2}$ mol/L, and $[Cl_2] = 3.9 \times 10^{-3}$ mol/L. Calculate K_c for the reaction.

$$PCl_5(g) \rightleftharpoons PCl_3(g) + Cl_2(g)$$

21. An equilibrium mixture of SO_2, O_2, and SO_3 at 1000 K contains the gases at the following concentrations: $[SO_2] = 3.77 \times 10^{-3}$ mol/L, $[O_2] = 4.30 \times 10^{-3}$ mol/L, and $[SO_3] = 4.13 \times 10^{-3}$ mol/L. Calculate the equilibrium constant, K_c, for the reaction

$$2\,SO_2(g) + O_2(g) \rightleftharpoons 2\,SO_3(g)$$

22. The reaction

$$C(s) + CO_2(g) \rightleftharpoons 2\,CO(g)$$

occurs at high temperatures. At 700 °C, a 2.0-L flask contains 0.10 mol of CO, 0.20 mol of CO_2, and 0.40 mol of C at equilibrium.
(a) Calculate K_c for the reaction at 700 °C.
(b) Calculate K_c for the reaction, also at 700 °C, if the amounts at equilibrium in the 2.0-L flask are 0.10 mol of CO, 0.20 mol of CO_2, and 0.80 mol of C.

23. Hydrogen and carbon dioxide react at a high temperature to give water and carbon monoxide:

$$H_2(g) + CO_2(g) \rightleftharpoons H_2O(g) + CO(g)$$

(a) Laboratory measurements at 986 °C show that there is 0.11 mol each of CO and water vapor and 0.087 mol each of H_2 and CO_2 at equilibrium in a 1.0-L container. Calculate the equilibrium constant for the reaction at 986 °C.

(b) If there is 0.050 mol each of H_2 and CO_2 in a 2.0-L container at equilibrium at 986 °C, what amounts of CO(g) and H_2O(g), in moles, could be present?

24. At a very high temperature water vapor is 10.% dissociated into H_2(g) and O_2(g) (i.e., 10.% of the original water has been transformed into products, and 90.% remains):

$$H_2O(g) \rightleftharpoons H_2(g) + \tfrac{1}{2}\,O_2(g)$$

Assuming a water concentration of 2.0 mol/L before dissociation, calculate the equilibrium constant, K_c.

25. You place 2.0 mol of hydrogen iodide in a 1.0-L container at a certain temperature. The compound partially dissociates according to the equation $2\,HI(g) \rightleftharpoons H_2(g) + I_2(g)$. If 22% of the HI has dissociated at equilibrium, calculate K_c.

26. You place 3.00 mol of pure SO_3 in an 8.00-L flask at 1150 K. At equilibrium, 0.58 mol of O_2 has been formed. Calculate K_c for the reaction at 1150 K:

$$2\,SO_3(g) \rightleftharpoons 2\,SO_2(g) + O_2(g)$$

27. A mixture of CO and Cl_2 is placed in a reaction flask: [CO] = 0.0102 mol/L and [Cl_2] = 0.00609 mol/L. When the reaction has come to equilibrium at 600 K, [Cl_2] = 0.00301 mol/L.

$$CO(g) + Cl_2(g) \rightleftharpoons COCl_2(g)$$

(a) Calculate the concentrations of CO and $COCl_2$ at equilibrium.

(b) Calculate K_c.

28. Ammonium iodide dissociates reversibly to ammonia and hydrogen iodide if the compound is heated to a sufficiently high temperature:

$$NH_4I(s) \rightleftharpoons NH_3(g) + HI(g)$$

Some ammonium iodide is placed in a flask, and then heated to 400 °C. If the total pressure in the flask when equilibrium has been achieved is 705 mm Hg, what is the value of K_p (when partial pressures are in atmospheres)?

29. When solid ammonium carbamate sublimes, it dissociates completely into ammonia and carbon dioxide according to the equation

$$(NH_4)(H_2NCO_2)(s) \rightleftharpoons 2\,NH_3(g) + CO_2(g)$$

At 25 °C, experiment shows that the total pressure of the gases in equilibrium with the solid is 0.116 atm. What is the equilibrium constant, K_p?

Using Equilibrium Constants in Calculations

30. K_c for the interconversion of butane and isobutane is 2.5 at 25 °C.

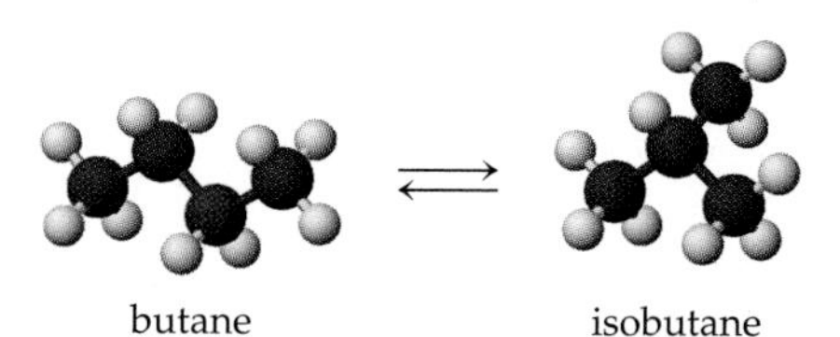

butane isobutane

If you place 0.017 mol of butane in a 0.50-L flask at 25 °C and allow equilibrium to be established, what will be the equilibrium concentrations of butane and isobutane?

31. Cyclohexane, C_6H_{12}, a hydrocarbon, can isomerize or change into methylcyclopentane, a compound of the same formula ($C_5H_9CH_3$) but with a different molecular structure:

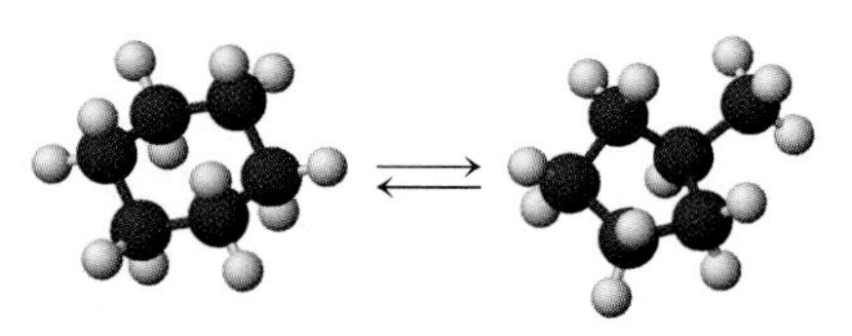

cyclohexane methylcyclopentane

The equilibrium constant is 0.12 at 25 °C. If you had originally placed 0.045 mol of cyclohexane in a 2.8-L flask, what would be the concentrations of cyclohexane and methylcyclopentane when equilibrium is established?

32. Carbonyl bromide, $COBr_2$, decomposes to CO and Br_2 with an equilibrium constant, K_c, of 0.190 at 73 °C:

$$COBr_2(g) \rightleftharpoons CO(g) + Br_2(g)$$

A 0.015-mol sample of $COBr_2$ was heated in a 2.5-L flask until equilibrium was attained. What are the concentrations of CO and Br_2 at equilibrium?

33. The equilibrium constant for the reaction, $K_p(= K_c)$,

$$Pb(\ell) + H_2O(g) \rightleftharpoons PbO(s) + H_2(g)$$

is 1.3×10^{-4} at 1000 K. A mixture of gases in a 1.0-L flask initially contains H_2O at 1.0 atm and H_2 at 1.0×10^{-3} atm at 1000 K. The mixture is kept in contact with molten lead and solid PbO at 1000 K until the reaction has come to equilibrium. What are the partial pressures of H_2O and H_2 at equilibrium? Which way did the reaction move to come to equilibrium?

34. The equilibrium constant, K_c, for the dissociation of iodine

$$I_2(g) \rightleftharpoons 2\,I(g)$$

is 3.76×10^{-3} at 1000 K. Suppose 0.105 mol of I_2 is placed in a 12.3-L flask at 1000 K. What are the concentrations of I_2 and I when the system comes to equilibrium?

35. The equilibrium constant, K_c, for the reaction

$$N_2O_4(g) \rightleftharpoons 2\ NO_2(g)$$

at 25 °C is 5.88×10^{-3}. Suppose 15.6 g of N_2O_4 is placed in a 5.00-L flask at 25 °C. Calculate:
(a) The number of moles of NO_2 present at equilibrium
(b) The percentage of the original N_2O_4 that is dissociated

36. The equilibrium constant K_p for $N_2O_4(g) \rightleftharpoons 2\ NO_2(g)$ is 0.15 at 25 °C. If the pressure of N_2O_4 at equilibrium is 0.85 atm, what is the total pressure of the gas mixture ($N_2O_4 + NO_2$) at equilibrium?

37. At 450 °C, 3.60 moles of ammonia is placed in a 2.00-L vessel and allowed to decompose to the elements:

$$2\ NH_3(g) \rightleftharpoons N_2(g) + 3\ H_2(g)$$

If the experimental value of K_c is 6.3 for this reaction at this temperature, calculate the equilibrium concentration of each reagent. What is the total pressure in the flask?

Disturbing a Chemical Equilibrium: Le Chatelier's Principle

38. Dinitrogen trioxide decomposes to NO and NO_2 in an endothermic process ($\Delta H = 40.5$ kJ/mol).

$$N_2O_3(g) \rightleftharpoons NO(g) + NO_2(g)$$

Predict the effect of the following changes on the position of the equilibrium; that is, state which way the equilibrium will shift (left, right, or no change) when each of the following changes is made:
(a) Adding more $N_2O_3(g)$
(b) Adding more $NO_2(g)$
(c) Increasing the volume of the reaction flask
(d) Lowering the temperature

39. K_p for the following reaction is 0.16 at 25 °C.

$$2\ NOBr(g) \rightleftharpoons 2\ NO(g) + Br_2(g)$$

The enthalpy change for the reaction at standard conditions is +16.1 kJ. Predict the effect of the following changes on the position of the equilibrium; that is, state which way the equilibrium will shift (left, right, or no change) when each of the following changes is made:
(a) Adding more $Br_2(g)$
(b) Removing some NOBr(g)
(c) Decreasing the temperature
(d) Increasing the container volume

40. Consider the isomerization of butane with an equilibrium constant of $K_c = 2.5$. (See Study Question 30.) The system is originally at equilibrium with [butane] = 1.0 M and [isobutane] = 2.5 M.
(a) If 0.50 mol/L of isobutane is suddenly added, and the system shifts to a new equilibrium position, what is the new equilibrium concentration of each gas?
(b) If 0.50 mol/L of butane is added, and the system shifts to a new equilibrium position, what is the equilibrium concentration of each gas?

41. K_c for the decomposition of ammonium hydrogen sulfide is 1.8×10^{-4} at 25 °C.

$$NH_4HS(s) \rightleftharpoons NH_3(g) + H_2S(g)$$

(a) When the pure salt decomposes in a flask, what are the equilibrium concentrations of NH_3 and H_2S?
(b) If NH_4HS is placed in a flask already containing 0.020 mol/L of NH_3 and then the system is allowed to come to equilibrium, what are the equilibrium concentrations of NH_3 and H_2S?

GENERAL QUESTIONS

42. The equilibrium constant, K_c, for the reaction

$$N_2(g) + O_2(g) \rightleftharpoons 2\ NO(g)$$

is 1.7×10^{-3} at 2300 K.
(a) What is K_p for this reaction?
(b) What is K_c for the reaction when written as

$$\tfrac{1}{2}\ N_2(g) + \tfrac{1}{2}\ O_2(g) \rightleftharpoons NO(g)$$

(c) What is K_c for the reaction

$$2\ NO(g) \rightleftharpoons N_2(g) + O_2(g)$$

43. If K_p for the formation of phosgene, $COCl_2$, is 6.5×10^{11} at 25 °C

$$CO(g) + Cl_2(g) \rightleftharpoons COCl_2(g)$$

what is the value of K_p for the dissociation of phosgene?

$$COCl_2(g) \rightleftharpoons CO(g) + Cl_2(g)$$

44. Suppose 0.086 mol of Br_2 is placed in a 1.26-L flask and heated to 1756 K, a temperature at which the halogen dissociates to atoms

$$Br_2(g) \rightleftharpoons 2\ Br(g)$$

If Br_2 is 3.7% dissociated at this temperature, calculate K_c.

45. The equilibrium constant, K_p, for the reaction

$$N_2(g) + O_2(g) \rightleftharpoons 2\ NO(g)$$

is 1.7×10^{-3} at 2300 K. The gases in this reaction have the following partial pressures in a reaction vessel at 2300 K: $P(N_2) = 0.50$ atm, $P(O_2) = 0.25$ atm, and $P(NO) = 4.2 \times 10^{-3}$ atm. Is the system at equilibrium? If not, which way does the reaction proceed?

46. Calculate the equilibrium constant, K_c, at 25 °C for the reaction

$$2\ NOCl(g) \rightleftharpoons 2\ NO(g) + Cl_2(g)$$

using the following information. In one experiment 2.00 mol of NOCl was placed in a 1.00-L flask, and the concentration of NO after equilibrium was achieved was 0.66 mol/L.

47. Equal numbers of moles of H_2 gas and I_2 vapor are mixed in a flask and heated to 700 °C. The initial concentration of

each gas is 0.0088 mol/L, and 78.6% of the I_2 is consumed when equilibrium is achieved according to the equation

$$H_2(g) + I_2(g) \rightleftharpoons 2\ HI(g)$$

Calculate K_c for this reaction.

48. The total pressure for a mixture of N_2O_4 and NO_2 is 1.5 atm. If K_p = 6.75 (at 25 °C), calculate the partial pressure of each gas in the mixture:

$$2\ NO_2(g) \rightleftharpoons N_2O_4(g)$$

49. Ammonium hydrogen sulfide decomposes on heating:

$$NH_4HS(s) \rightleftharpoons NH_3(g) + H_2S(g)$$

If K_p for this reaction is 0.11 at 25 °C (when the partial pressures are measured in atmospheres), what is the total pressure in the flask at equilibrium?

50. Two molecules of gaseous acetic acid can form a dimer through hydrogen bonds. The equilibrium constant K_c at 25 °C has been determined to be 3.2×10^4. Assume that acetic acid is present initially at a concentration of 5.4×10^{-4} mol/L at 25 °C and that no dimer is present initially.

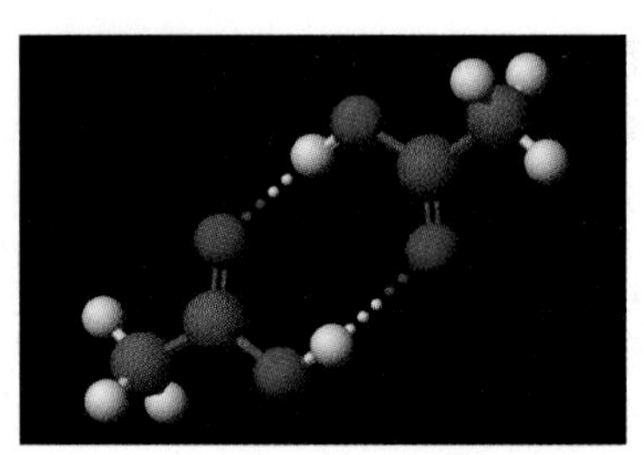

Acetic acid dimer

(a) What percentage of acetic acid is converted to dimer?
(b) As the temperature goes up, in which direction does the equilibrium shift? (Recall that hydrogen-bond formation is an exothermic process.)

51. The equilibrium constant for the butane/isobutane isomerization reaction is 2.5 at 25 °C. (See Study Question 30.) If 1.75 mol of butane and 1.25 mol of isobutane are mixed, is the system at equilibrium? If not, when it proceeds to equilibrium which reagent increases in concentration? Calculate the concentrations of the two compounds when the system reaches equilibrium.

52. The equilibrium constant, K_c, for the reaction

$$2\ SO_2(g) + O_2(g) \rightleftharpoons 2\ SO_3(g)$$

is 279 at 1000 K. Calculate K_p for the reaction.

53. The decomposition of NOCl

$$2\ NOCl(g) \rightleftharpoons 2\ NO(g) + Cl_2(g)$$

has K_p = 0.039 at 250 °C. Calculate K_c for the reaction.

54. Zinc carbonate dissolves very poorly in water ($K_c = 1.5 \times 10^{-11}$).

$$ZnCO_3(s) \rightleftharpoons Zn^{2+}(aq) + CO_3^{\ 2-}(aq)$$

If some solid $ZnCO_3$ is placed in water, what are the molar concentrations of Zn^{2+} and $CO_3^{\ 2-}$ when equilibrium has been achieved?

55. At 2300 K the equilibrium constant for the formation of NO(g) is $K_c = 1.7 \times 10^{-3}$.

$$N_2(g) + O_2(g) \rightleftharpoons 2\ NO(g)$$

Analysis shows that the concentrations of N_2 and O_2 are both 0.25 M, and that of NO is 0.0042 M under certain conditions.

(a) Is the system at equilibrium?
(b) If the system is not at equilibrium, in which direction does the reaction proceed?
(c) When the system is at equilibrium, what are the equilibrium concentrations?

CONCEPTUAL QUESTIONS

56. A sample of liquid water is sealed in a container. Over time some of the liquid evaporates, but equilibrium is reached eventually. At this point you can measure the equilibrium vapor pressure of the water. Is the process $H_2O(g) \rightleftharpoons H_2O(\ell)$ a dynamic equilibrium? Explain the changes that take place in reaching equilibrium in terms of the rates of the competing processes of evaporation and condensation.

57. An ice cube is placed in a beaker of water at 20 °C. The ice cube partially melts, and the temperature of the water is lowered to 0 °C. At this point, both ice and water are at 0 °C, and no further change is apparent. Is the system at equilibrium? Is this a dynamic equilibrium? That is, are events still occurring at the molecular level? Suggest an experiment to test whether this is so. (*Hint*: Consider using D_2O.)

CHALLENGING QUESTIONS

58. A reaction important in smog formation is

$$O_3(g) + NO(g) \rightleftharpoons O_2(g) + NO_2(g)$$
$$K_c = 6.0 \times 10^{34}$$

(a) If the initial concentrations are $[O_3] = 1.0 \times 10^{-6}$ M, $[NO] = 1.0 \times 10^{-5}$ M, $[NO_2] = 2.5 \times 10^{-4}$ M, and $[O_2] = 8.2 \times 10^{-3}$ M, is the system at equilibrium? If not, in which direction does the reaction proceed?
(b) If the temperature is increased, as on a very warm day, will the concentrations of the products increase or decrease? (*Hint*: You may have to calculate the enthalpy change for the reaction to find out if it is exothermic or endothermic.)

59. The equilibrium $N_2O_4(g) \rightleftharpoons 2\ NO_2(g)$ has been thoroughly studied (see Figure 16.7). If the total pressure in a flask containing NO_2 and N_2O_4 gas at 25 °C is 1.50 atm, and the value of K_p at this temperature is 0.148, what fraction of the N_2O_4 has dissociated to NO_2? What happens to the fraction dissociated if the volume of the container is increased so that the total pressure falls to 1.00 atm?

60. The ammonia complex of trimethylborane, $(NH_3)B(CH_3)_3$, dissociates at 100 °C to its components with $K_c = 0.15$.

$$(NH_3)B(CH_3)_3 \rightleftharpoons B(CH_3)_3 + NH_3$$

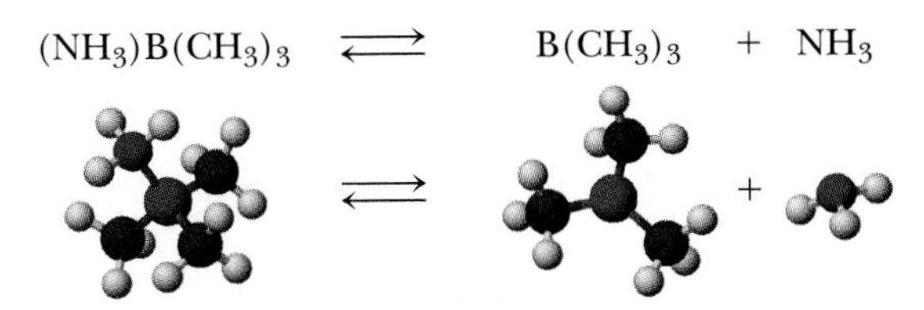

If NH_3 is substituted by some other molecule, a different equilibrium constant is obtained.

For $[(CH_3)_3P]B(CH_3)_3$	$K_c = 4.2 \times 10^{-3}$
For $[(CH_3)_3N]B(CH_3)_3$	$K_c = 1.5 \times 10^{-2}$
For $(NH_3)B(CH_3)_3$	$K_c = 0.15$

(a) If you begin an experiment by placing 0.010 mol of each complex in a flask, which would have the largest concentration of $B(CH_3)_3$ at 100 °C?
(b) If 0.73 g (0.010 mol) of $(NH_3)B(CH_3)_3$ is placed in a 100.-mL flask and heated to 100 °C, what is the concentration of each gas in the equilibrium mixture? What is the percent dissociation of $(NH_3)B(CH_3)_3$?

61. Sulfuryl chloride, SO_2Cl_2, a compound with very irritating vapors, is used as a reagent in the synthesis of organic compounds. When heated to a sufficiently high temperature it decomposes to SO_2 and Cl_2.

$$K_c = 0.045 \text{ at } 375 \text{ °C}$$

$$SO_2Cl_2(g) \rightleftharpoons SO_2(g) + Cl_2(g)$$

(a) A sample of 6.70 g of SO_2Cl_2 is placed in a 1.00-L flask and then heated to 375 °C. What is the concentration of each of the compounds in the system when equilibrium is achieved? What fraction of SO_2Cl_2 has dissociated?
(b) What are the concentrations of SO_2Cl_2, SO_2, and Cl_2 at equilibrium in the 1.00-L flask at 375 °C if you begin with a mixture of SO_2Cl_2 (6.70 g) and Cl_2 (1.00 atm)? What fraction of SO_2Cl_2 has dissociated?
(c) Does the fraction of SO_2Cl_2 in parts (a) and (b) agree with your expectation based on Le Chatelier's principle?

62. Hemoglobin (Hb) can form a complex with either O_2 or CO. For the reaction

$$HbO_2(aq) + CO(g) \rightleftharpoons HbCO(aq) + O_2(g)$$

at body temperature, K_c is about 200. If the ratio $[HbCO]/[HbO_2]$ comes close to one, death is probable. What partial pressure of CO in the air is likely to be fatal? Assume the partial pressure of O_2 is 0.2 atm.

63. At 1800 K, oxygen molecules dissociate very slightly into atoms:

$$O_2(g) \rightleftharpoons 2\,O(g) \qquad K_p = 1.2 \times 10^{-10}$$

If you place 1.0 mol of O_2 in a 10.-L vessel and heat it to 1800 K, how many O atoms are present in the flask?

Summary Question

64. Nitrosyl bromide, NOBr, is prepared by the direct reaction of NO and Br_2.

$$2\,NO(g) + Br_2(g) \longrightarrow 2\,NOBr(g)$$

but the compound dissociates readily at room temperature:

$$2\,NOBr(g) \rightleftharpoons 2\,NO(g) + Br_2(g)$$

(a) If you mix 3.50 g of NO and 9.67 g of Br_2, how many grams of NOBr can be prepared?
(b) If N is the central atom of nitrosyl bromide, draw the electron dot structure for the molecule.
(c) What is the electron pair geometry of NOBr? What is its molecular geometry? Is the molecule polar?
(d) Some NOBr is placed in a flask at 25 °C and allowed to dissociate. The total pressure at equilibrium is 190 mm Hg and the compound is found to be 34% dissociated. What is the value of K_p?

INTERACTIVE GENERAL CHEMISTRY CD-ROM

Screen 16.2 The Principle of Microscopic Reversibility

(a) Describe the experiment shown on this screen. What is the objective of the experiment?
(b) What is the principle of microscopic reversibility?

Screen 16.4 The Equilibrium Constant

(a) The experiment seen on Screen 16.3 is used again on this screen. Examine the table of initial concentrations and the beakers of solutions formed by mixing these solutions. The quantity of iron used in all the experiments is the same, but additional SCN^- is used going from left to right. What effect does using more SCN^- have on the quantity of product formed?
(b) Realizing that in each case the SCN^- is the limiting reactant, calculate the percent yield for each of the four reactions.

(c) Use two sets of experimental results (final concentrations) and calculate the equilibrium constant in each case. What is the effect on K_c of varying the initial concentrations?

Screen 16.5 The Meaning of the Equilibrium Constant

Would you expect the reaction $2\ NO_2(g) \rightleftharpoons 2\ NO(g) + O_2(g)$ to have a large or small equilibrium constant? Explain briefly.

Screen 16.11 Le Chatelier's Principle

(a) Examine the water tank animation on this screen. Describe how addition of water to the left-hand tank illustrates Le Chatelier's principle.
(b) Of the three potential changes to an equilibrium system described on the screen—*Temperature Change, Addition or Removal of a Reactant or Product,* and *Volume Changes in Gas Phase Equilibria*—which does the tank demonstration illustrate?
(c) What would you expect to occur in the tank equilibrium if the width of the left-hand tank were suddenly decreased to one half its present radius?

Screen 16.12 Disturbing an Equilibrium (Temperature Changes)

Watch the photographs shown on this screen. Describe the difference between the two states shown, both in terms of temperature and in terms of the concentrations of the species in the flask.

In contrast, aqueous ammonia (see Figure 5.11) and the carbonate ion produce only a very small concentration of OH^- ion and are classed as weak Brønsted bases:

$$NH_3(aq) + H_2O(\ell) \rightleftharpoons NH_4^+(aq) + OH^-(aq)$$
$$CO_3^{2-}(aq) + H_2O(\ell) \rightleftharpoons HCO_3^-(aq) + OH^-(aq)$$

Weak bases
$[OH^-] \ll$ initial concentration of base

In the Brønsted model, an acid donates a proton and produces a conjugate base. This model also informs us that, in general, *the stronger the acid, the weaker its conjugate base.* Aqueous HCl, for example, is a strong acid because it has a strong tendency to donate a proton to water and produce its conjugate base Cl^-. In this reaction water acts as a base and accepts the proton from HCl to produce H_3O^+, the conjugate acid of water. The reaction proceeds almost completely to the right; essentially no HCl molecules are present in the solution at equilibrium:

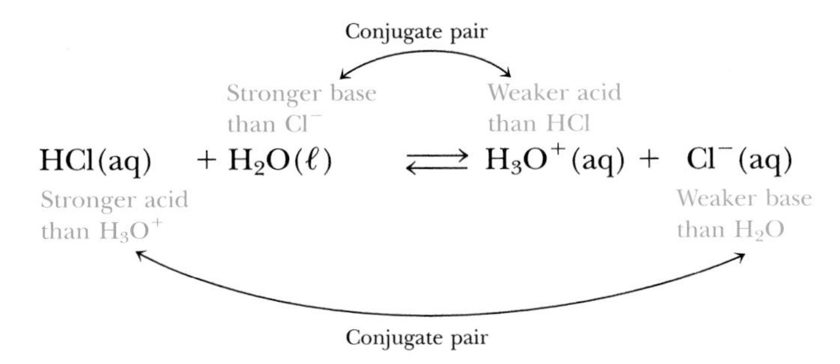

In the case of a strong acid, the acid and base on the left side of the balanced equation are stronger than the conjugate acid and base on the right. The stronger acid and base react to give predominantly the weaker acid and base. Of the two acids here, HCl and H_3O^+, HCl is better able to donate a proton. Of the two bases, H_2O and Cl^-, water must be the stronger base, and it wins out in the competition for the proton. The equilibrium lies far to the right.

Acetic acid, a weak acid, ionizes to a very small extent in water. Thus, of the two acids present in aqueous acetic acid (CH_3CO_2H and H_3O^+), the hydronium ion is the stronger. Of the two bases (H_2O and acetate ion, $CH_3CO_2^-$), the acetate ion must be the stronger. At equilibrium, the solution consists mostly of acetic acid with only a small concentration of acetate ion and hydronium ion. Again, the equilibrium lies toward the side of the reaction having the weaker acid and base:

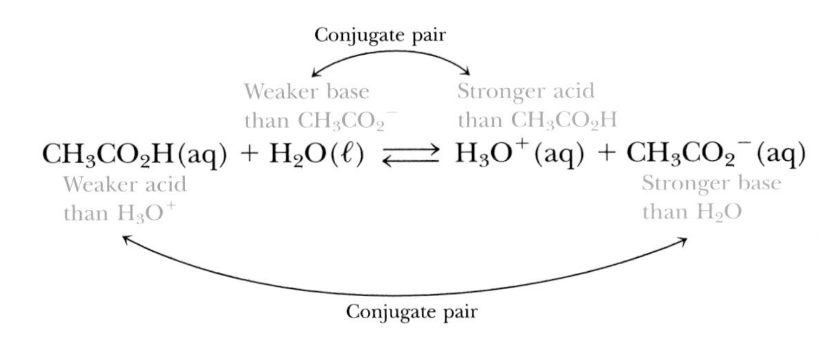

Table 17.3 • Relative Strengths of Acids and Bases

Conjugate Acid		Conjugate Base	
Name	**Formula**	**Formula**	**Name**
Perchloric acid	$HClO_4$	ClO_4^-	Perchlorate ion
Sulfuric acid	H_2SO_4	HSO_4^-	Hydrogen sulfate ion
Hydrochloric acid	HCl	Cl^-	Chloride ion
Nitric acid	HNO_3	NO_3^-	Nitrate ion
Hydronium ion	H_3O^+	H_2O	Water
Hydrogen sulfate ion	HSO_4^-	SO_4^{2-}	Sulfate ion
Phosphoric acid	H_3PO_4	$H_2PO_4^-$	Dihydrogen phosphate ion
Acetic acid	CH_3CO_2H	$CH_3CO_2^-$	Acetate ion
Hexaaquaaluminum ion	$[Al(H_2O)_6]^{3+}$	$[Al(H_2O)_5(OH)]^{2+}$	Pentaaquahydroxoaluminum ion
Carbonic acid	H_2CO_3	HCO_3^-	Hydrogen carbonate ion
Hydrogen sulfide	H_2S	HS^-	Hydrogen sulfide ion
Dihydrogen phosphate ion	$H_2PO_4^-$	HPO_4^{2-}	Hydrogen phosphate ion
Ammonium ion	NH_4^+	NH_3	Ammonia
Hydrocyanic acid	HCN	CN^-	Cyanide ion
Hydrogen carbonate ion	HCO_3^-	CO_3^{2-}	Carbonate ion
Phenol	C_6H_5OH	$C_6H_5O^-$	Phenoxide ion
Water	H_2O	OH^-	Hydroxide ion
Ethanol	C_2H_5OH	$C_2H_5O^-$	Ethoxide ion
Ammonia	NH_3	NH_2^-	Amide ion
Methylamine	CH_3NH_2	CH_3NH^-	Methylamide ion
Hydrogen	H_2	H^-	Hydride ion
Methane	CH_4	CH_3^-	Methide ion

(Increasing Acid Strength: upward arrow along the acid column; Increasing Base Strength: downward arrow along the base column.)

Figure 17.3 The basic properties of the hydride ion, H^-. Calcium hydride, CaH_2, is the source of H^- ion. This ion is such a powerful proton acceptor that it reacts vigorously with water, a proton donor, to give H_2 and hydroxide ion:

$$CaH_2(s) + 2\,H_2O(\ell) \longrightarrow 2\,H_2(g) + Ca^{2+}(aq) + 2\,OH^-(aq)$$

(C. D. Winters)

These two examples of the relative extent of acid–base reactions illustrate an important principle in Brønsted acid–base theory: All proton transfer reactions proceed *from the stronger* acid–base pair *to the weaker* acid–base pair.

Acids and bases can be ordered on the basis of their relative abilities to donate or accept protons in aqueous solution, and we have done so for a few Brønsted acids and bases in Table 17.3. At the top on the left are the stronger acids; those substances strongly donate protons, and their conjugate bases are extremely weak. The opposite is true for acids at the bottom left of the table. For example, the H_2 molecule can be considered an acid in the sense that it can conceivably donate a proton (H^+) and form the conjugate base H^-, the hydride ion. Hydrogen H_2, however, is an exceedingly weak acid. We know this because its conjugate base, the hydride ion (which is known in substances such as NaH, sodium hydride) is a *very* strong base. It reacts explosively with H^+ donors such as H_2O, to form H_2 (Figure 17.3).

Exercise 17.3 Relative Strengths of Acids and Bases

1. Which is the stronger Brønsted acid, HCO_3^- or NH_4^+? Which has the stronger conjugate base?
2. Which is the stronger Brønsted base, CN^- or SO_4^{2-}?

Using Relative Acid–Base Strengths to Predict the Direction of Acid–Base Reactions

The chart of acids and bases in Table 17.3 can be used to predict whether the equilibrium lies predominantly to the left or the right in an acid–base reaction. The examples that follow show you how to do this.

Example 17.1 Predicting the Direction of Acid–Base Reactions

Write a balanced equation for the reaction that occurs between each acid–base pair in water. Decide whether the equilibrium lies predominantly to the left or right.

1. Acetic acid, CH_3CO_2H, and sodium cyanide, NaCN. Is HCN, a poisonous acid, formed to a significant extent?
2. Ammonium chloride, NH_4Cl, and sodium carbonate, Na_2CO_3.

Solution

1. **Reaction of CH_3CO_2H and NaCN.** The conjugate base of acetic acid is the acetate ion, $CH_3CO_2^-$. The other reactant, NaCN, is a water-soluble salt that forms Na^+ and CN^- ions in water. Sodium ion is not listed in Table 17.3 because it does not react appreciably with water. The CN^- ion, however, is a base with HCN as its conjugate acid. The net ionic equation for the reaction between CH_3CO_2H and CN^- can therefore be written as

$$CH_3CO_2H(aq) + CN^-(aq) \rightleftharpoons HCN(aq) + CH_3CO_2^-(aq)$$

To decide to which side the equilibrium lies, left or right, we can compare the two acids (or the two bases) in the reaction. According to Table 17.3, HCN is a weaker acid than CH_3CO_2H, and $CH_3CO_2^-$ is a weaker base than CN^-. In a sense, we can view the situation as a competition between two bases for the available H^+ ion. The stronger base is expected to win out.

Because all Brønsted acid–base reactions move predominantly toward the weaker acid and base, the reaction favors the $HCN/CH_3CO_2^-$ pair at equilibrium:

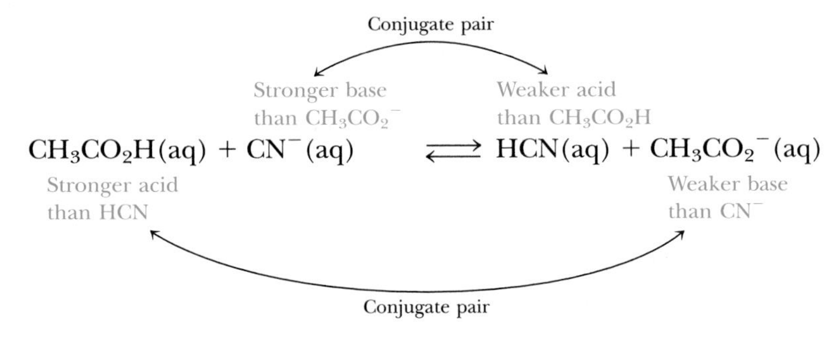

This result tells you that you certainly would *not* want to mix acetic acid and a cyanide ion salt because HCN is the product. The average fatal dose of HCN is 50 to 60 mg, or about 0.002 mol.

2. **Reaction of NH_4Cl and Na_2CO_3.** Ammonium chloride dissociates to form NH_4^+ and Cl^- ions in water. The ammonium ion is a weak acid

$$NH_4^+(aq) + H_2O(\ell) \rightleftharpoons H_3O^+(aq) + NH_3(aq)$$

and the chloride ion is a weak base:

$$Cl^-(aq) + H_2O(\ell) \rightleftharpoons HCl(aq) + OH^-(aq)$$

As you will see in Section 17.8, the base properties of Cl^-, ClO_4^-, and NO_3^- can be disregarded in aqueous solution. These ions do not contribute to the basicity of a solution.

As shown in Table 17.3, the OH^- ion is a *much* stronger base than Cl^-, and HCl is a much stronger acid than H_2O. The equilibrium lies so far to the left that we can assume this reaction does not occur. In general, any anion X^- that is a weaker base than OH^- *does not react appreciably with water* to give HX and OH^-. Thus, for NH_4Cl, we only need to be concerned with the NH_4^+ ion.

Sodium carbonate, Na_2CO_3, dissociates in water to give the ions Na^+ and CO_3^{2-}. As in the reaction of CH_3CO_2H and NaCN, we can ignore any contribution from the Na^+ ion. The carbonate ion, however, is a base (Table 17.3):

$$CO_3^{2-}(aq) + H_2O(\ell) \rightleftharpoons HCO_3^-(aq) + OH^-(aq)$$

Having decided the ammonium ion is an acid and the carbonate ion is a base, we can write the equation for their reaction:

$$\underset{\text{Acid}}{NH_4^+(aq)} + \underset{\text{Base}}{CO_3^{2-}(aq)} \rightleftharpoons \underset{\text{Conjugate acid}}{HCO_3^-(aq)} + \underset{\text{Conjugate base}}{NH_3(aq)}$$

Ammonium ion is a stronger acid than HCO_3^-, and CO_3^{2-} is a stronger base than NH_3. Some reaction does therefore occur. You know this because you can detect the characteristic odor of ammonia over the solution. (If you open a bottle of *solid* ammonium carbonate, $(NH_4)_2CO_3$, you can also smell ammonia.)

Exercise 17.4 Predicting the Direction of Acid–Base Reactions

For each of the following reactions, predict whether the equilibrium lies predominantly to the left or to the right:

1. $HSO_4^-(aq) + NH_3(aq) \rightleftharpoons NH_4^+(aq) + SO_4^{2-}(aq)$
2. $HCO_3^-(aq) + HS^-(aq) \rightleftharpoons H_2S(aq) + CO_3^{2-}(aq)$

Exercise 17.5 Writing Acid–Base Reactions

Ammonium chloride and sodium sulfate are mixed in water. Write a balanced, net ionic equation for the acid–base reaction that could, in principle, occur. Does the reaction in fact occur to any appreciable extent?

17.4 STRONG ACIDS AND BASES

See the *Saunders Interactive General Chemistry CD-ROM,* Screen 17.5, Strong Acids and Bases.

The acids in Table 17.3 are listed in descending order of their ability to donate a proton. *The hydronium ion is actually the strongest acid that can exist in water.* Notice, however, that several acids are listed even higher in the table than H_3O^+. How can this be? When they are placed in water, strong acids such as HCl, HBr, HI,

A CLOSER LOOK — A Base "In the Limelight"

When you have done something especially well, and all of your friends know about it, you are said to be in the "limelight." The word has its roots deep in chemistry.

Limestone, $CaCO_3$, has been called one of the six most important chemicals in the world of chemistry. The others are coal, oil, iron ore, sulfur, and salt. When heated strongly, limestone is converted to lime, or quicklime, CaO, and billions of kilograms of CaO are made annually this way:

$$CaCO_3(s) \longrightarrow CaO(s) + CO_2(g)$$

In a reaction typical of many metal oxides, CaO forms slaked lime, $Ca(OH)_2$, when treated with water.

$$CaO(s) + H_2O(\ell) \longrightarrow Ca(OH)_2(s)$$

Calcium hydroxide is poorly soluble in water, but saturated solutions have a pH in the range of 12 or so. As such it is a very useful base and is widely used in water treatment and manufacturing other chemicals. It is also used to make cement and in the paper and steel industries.

In the 1820s, Lt. Thomas Drummond (1797–1840), a member of the Royal Engineers, was involved in a survey of Great Britain. During the winters he attended the famous public chemistry lectures and demonstrations by the great chemist Michael Faraday at the Royal Institution in London. There he apparently heard about the very bright light that is emitted when a piece of lime is heated to a very high temperature. It occurred to him that it could be used to make distant surveying stations visible, especially at night. Soon he developed an apparatus in which a ball of lime was heated by an alcohol flame in a stream of oxygen gas. It was reported at the time that the light from a "ball of lime not larger than a boy's marble" could be seen at a distance of 70 miles! Such lights were soon adapted to lighthouses and became known as Drummond lights.

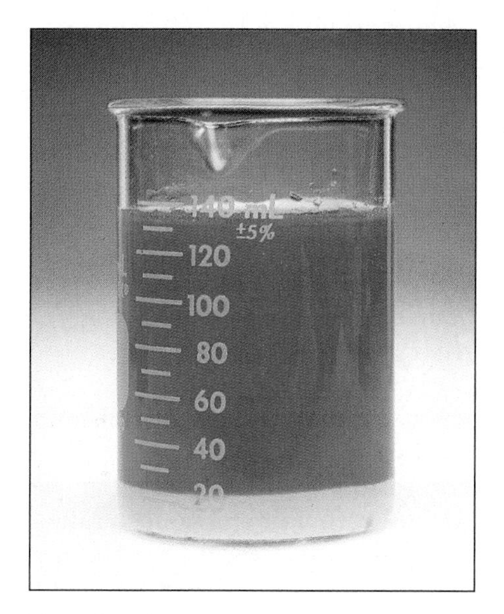

Lime, CaO, is not very soluble in water, but it does react with water to give $Ca(OH)_2$, which gives a solution with a pH of about 12. The lattice of CaO is a face-centered lattice of O^{2-} ions with Ca^{2+} in octahedral holes. As such it resembles the NaCl lattice (page 619). *(C. D. Winters)*

Metal oxides such as CaO and thorium(IV) oxide, ThO_2, emit a brilliant white light when heated to incandescence. Thorium oxide has been used in the mantles of camping lanterns such as the one shown here. *(C. D. Winters)*

Many inventions are soon adapted to warfare and that was the case with limelights. They were used to illuminate targets in the battle of Charleston, SC, during the United States Civil War in the 1860s. The public came to know about limelight, though, when it moved into theaters. Gaslights were used in the early 1800s to illuminate the stage, but they were clearly not adequate. Soon after Drummond's invention, though, actors trod the boards "in the limelight."

HNO_3, H_2SO_4, and $HClO_4$ (see Table 5.2) ionize *completely* to form H_3O^+ and their conjugate base by reacting with water (see Figure 17.2).

$$H_2SO_4(aq) + H_2O(\ell) \longrightarrow H_3O^+(aq) + HSO_4^-(aq)$$

Strong acid, 100% ionization for loss of first H atom as H_3O^+ — Weak conjugate base of H_2SO_4

Thus, H_2SO_4 molecules do not exist in water; rather, only H_3O^+ and HSO_4^- are present because H_3O^+ is the strongest acid species that can exist in aqueous solution. Acids that ionize 100% in aqueous solution are called **strong acids.**

Figure 17.4 Consumer products containing strong acids and bases. Muriatic acid is hydrochloric acid, and battery acid is sulfuric acid. Oven cleaners often contain sodium hydroxide. *(C. D. Winters)*

Because strong acids (Figure 17.4) are converted completely to H_3O^+ and the appropriate anion in water, they appear to be the same strength. We say that water "levels" their acidic character to the level of H_3O^+.

Just as no acids can be stronger in water than H_3O^+, no base can be stronger than OH^- in aqueous solution. The strong bases in Table 17.3 are species lower than OH^- in the right column: $C_2H_5O^-$, NH_2^-, H^-, and CH_3^- all react completely with water to produce OH^- (see Figure 17.3):

$$\underset{\text{amide ion, Strong base}}{NH_2^-(aq)} + H_2O(\ell) \longrightarrow \underset{\text{Weak conjugate acid of } NH_2^-}{NH_3(aq)} + OH^-(aq)$$

17.5 WEAK ACIDS AND BASES

Very few acids and bases strongly donate or accept protons, respectively. *The vast majority of acids and bases are weak.*

The relative strength of an acid or base can be expressed quantitatively with an **equilibrium constant.** For the general *weak acid* HA, for example, we can write

$$HA(aq) + H_2O(\ell) \rightleftharpoons H_3O^+(aq) + A^-(aq)$$

$$K_a = \frac{[H_3O^+][A^-]}{[HA]} \qquad [17.1]$$

See the *Saunders Interactive General Chemistry CD-ROM,* Screen 17.6, Weak Acids and Bases.

where K has a subscript "a" to indicate that it is an equilibrium constant for a weak acid in water. The value of K is less than 1 for a weak acid, indicating that the product of the equilibrium concentrations of the hydronium ion and the conjugate base of the weak acid is smaller than the equilibrium concentration of the weak acid.

Similarly, we can write the equilibrium expression for a *weak base* B in water. Here we label K with a subscript "b." Its value is also less than 1.

$$B(aq) + H_2O(\ell) \rightleftharpoons BH^+(aq) + OH^-(aq)$$

$$K_b = \frac{[BH^+][OH^-]}{[B]} \qquad [17.2]$$

Some common acids and bases are ordered by their relative abilities to donate or accept protons in aqueous solution in Table 17.3. This has been done for a larger number of substances in Table 17.4, where each acid and base is listed with its value of K_a or K_b, respectively. The following are important ideas concerning Table 17.4.

- Acids are listed at the left and their conjugate bases are on the right.
- A large value of K indicates products are strongly favored, whereas a small value of K indicates the reactants are favored.
- The strongest acids, at the upper left (see also Table 17.3), also have the largest K_a values. K_a values become smaller on descending the chart as the acid strength declines.

Table 17.4 • IONIZATION CONSTANTS FOR SOME ACIDS AND THEIR CONJUGATE BASES

Acid Name	Acid	K_a	Base	K_b	Base Name
Perchloric acid	$HClO_4$	large	ClO_4^-	very small	perchlorate ion
Sulfuric acid	H_2SO_4	large	HSO_4^-	very small	hydrogen sulfate ion
Hydrochloric acid	HCl	large	Cl^-	very small	chloride ion
Nitric acid	HNO_3	large	NO_3^-	very small	nitrate ion
Hydronium ion	H_3O^+	55.5	H_2O	1.8×10^{-16}	water
Sulfurous acid	H_2SO_3	1.2×10^{-2}	HSO_3^-	8.3×10^{-13}	hydrogen sulfite ion
Hydrogen sulfate ion	HSO_4^-	1.2×10^{-2}	SO_4^{2-}	8.3×10^{-13}	sulfate ion
Phosphoric acid	H_3PO_4	7.5×10^{-3}	$H_2PO_4^-$	1.3×10^{-12}	dihydrogen phosphate ion
Hexaaquairon(III) ion	$Fe(H_2O)_6^{3+}$	6.3×10^{-3}	$Fe(H_2O)_5OH^{2+}$	1.6×10^{-12}	pentaaquahydroxoiron(III) ion
Hydrofluoric acid	HF	7.2×10^{-4}	F^-	1.4×10^{-11}	fluoride ion
Nitrous acid	HNO_2	4.5×10^{-4}	NO_2^-	2.2×10^{-11}	nitrite ion
Formic acid	HCO_2H	1.8×10^{-4}	HCO_2^-	5.6×10^{-11}	formate ion
Benzoic acid	$C_6H_5CO_2H$	6.3×10^{-5}	$C_6H_5CO_2^-$	1.6×10^{-10}	benzoate ion
Acetic acid	CH_3CO_2H	1.8×10^{-5}	$CH_3CO_2^-$	5.6×10^{-10}	acetate ion
Propanoic acid	$CH_3CH_2CO_2H$	1.3×10^{-5}	$CH_3CH_2CO_2^-$	7.7×10^{-10}	propanoate ion
Hexaaquaaluminum ion	$Al(H_2O)_6^{3+}$	7.9×10^{-6}	$Al(H_2O)_5OH^{2+}$	1.3×10^{-9}	pentaaquahydroxoaluminum ion
Carbonic acid	H_2CO_3	4.2×10^{-7}	HCO_3^-	2.4×10^{-8}	hydrogen carbonate ion
Hexaaquacopper(II) ion	$Cu(H_2O)_6^{2+}$	1.6×10^{-7}	$Cu(H_2O)_5OH^+$	6.25×10^{-8}	pentaaquahydroxocopper(II) ion
Hydrogen sulfide	H_2S	1×10^{-7}	HS^-	1×10^{-7}	hydrogen sulfide ion
Dihydrogen phosphate ion	$H_2PO_4^-$	6.2×10^{-8}	HPO_4^{2-}	1.6×10^{-7}	hydrogen phosphate ion
Hydrogen sulfite ion	HSO_3^-	6.2×10^{-8}	SO_3^{2-}	1.6×10^{-7}	sulfite ion
Hypochlorous acid	$HOCl$	3.5×10^{-8}	ClO^-	2.9×10^{-7}	hypochlorite ion
Hexaaqualead(II) ion	$Pb(H_2O)_6^{2+}$	1.5×10^{-8}	$Pb(H_2O)_5OH^+$	6.7×10^{-7}	pentaaquahydroxolead(II) ion
Hexaaquacobalt(II) ion	$Co(H_2O)_6^{2+}$	1.3×10^{-9}	$Co(H_2O)_5OH^+$	7.7×10^{-6}	pentaaquahydroxocobalt(II) ion
Boric acid	$B(OH)_3(H_2O)$	7.3×10^{-10}	$B(OH)_4^-$	1.4×10^{-5}	tetrahydroxoborate ion
Ammonium ion	NH_4^+	5.6×10^{-10}	NH_3	1.8×10^{-5}	ammonia
Hydrocyanic acid	HCN	4.0×10^{-10}	CN^-	2.5×10^{-5}	cyanide ion
Hexaaquairon(II) ion	$Fe(H_2O)_6^{2+}$	3.2×10^{-10}	$Fe(H_2O)_5OH^+$	3.1×10^{-5}	pentaaquahydroxoiron(II) ion
Hydrogen carbonate ion	HCO_3^-	4.8×10^{-11}	CO_3^{2-}	2.1×10^{-4}	carbonate ion
Hexaaquanickel(II) ion	$Ni(H_2O)_6^{2+}$	2.5×10^{-11}	$Ni(H_2O)_5OH^+$	4.0×10^{-4}	pentaaquahydroxonickel(II) ion
Hydrogen phosphate ion	HPO_4^{2-}	3.6×10^{-13}	PO_4^{3-}	2.8×10^{-2}	phosphate ion
Water	H_2O	1.8×10^{-16}	OH^-	55.5	hydroxide ion
Hydrogen sulfide ion*	HS^-	1×10^{-19}	S^{2-}	1×10^{5}	sulfide ion
Ethanol	C_2H_5OH	very small	$C_2H_5O^-$	large	ethoxide ion
Ammonia	NH_3	very small	NH_2^-	large	amide ion
Hydrogen	H_2	very small	H^-	large	hydride ion
Methane	CH_4	very small	CH_3^-	large	methide ion

↑ Increasing Acid Strength (upward) — Increasing Base Strength ↓ (downward)

*The values of K_a for HS^- and K_b for S^{2-} are estimates.

- The strongest bases are at the lower right. They have the largest K_b values. K_b values become larger on descending the chart as base strength increases.
- The weaker the acid, the stronger its conjugate base. That is, the smaller the value of K_a, the larger the value of K_b.
- The K_a values suggest a way to classify acids and their conjugate bases.

Acid Strength	K_a	Conjugate Base Strength	K_b
Strong	>1	Very weak	$<10^{-16}$
Weak	1 to 10^{-16}	Weak	10^{-16} to 1
Very weak	$<10^{-16}$	Strong	>1

Some Weak Acids

Neutral Molecules as Acids

A few neutral molecules with an ionizable hydrogen atom are strong Brønsted acids, but the vast majority are weak. Several weak acids are pictured in Figure 17.5, and a number are listed in Table 17.4.

These models, and many others of weak acids and bases, are in the "Models" folder on the *Saunders Interactive General Chemistry CD-ROM* and on the Web site for this book.

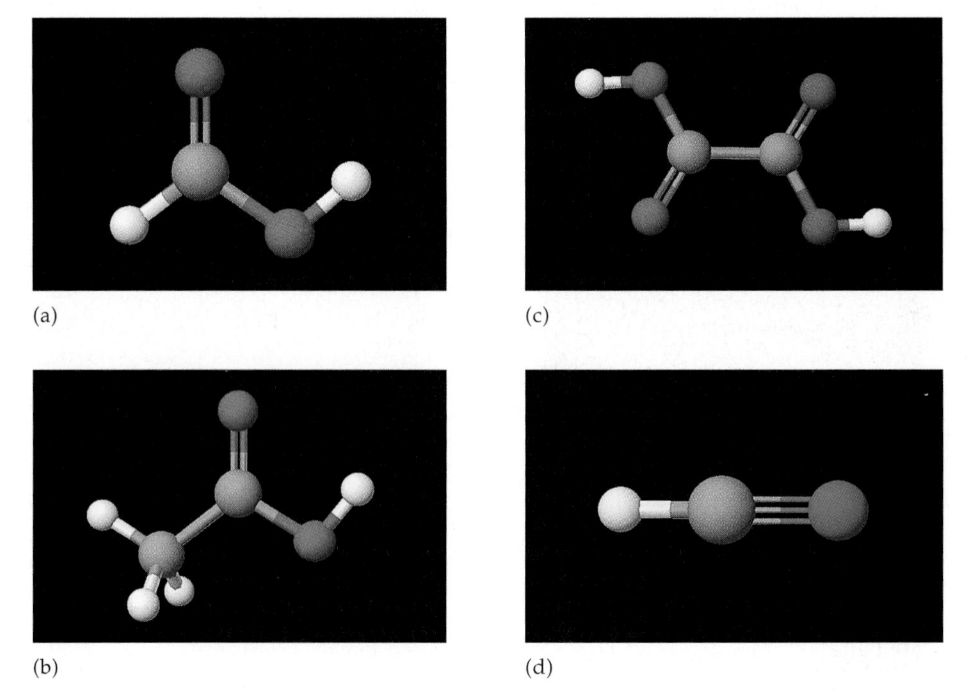

Figure 17.5 Some weak acids. (a) Formic acid, HCO_2H. (b) Acetic acid, CH_3CO_2H. (c) Oxalic acid, $H_2C_2O_4$. (d) Hydrocyanic acid, HCN. Note the similarity in structure of the organic acids. All have the carboxylic acid group, $—CO_2H$. (See page 512.)

Cations as Weak Acids

The ammonium ion is an example of a cation acting as a weak Brønsted acid:

$$NH_4^+(aq) + H_2O(\ell) \rightleftharpoons NH_3(aq) + H_3O^+(aq) \qquad K_a = 5.6 \times 10^{-10}$$

When the salt of a metal cation is placed in water, the metal ion becomes hydrated, as explained in Section 13.2. In fact, the interaction is usually sufficiently strong that the ion is surrounded by as many as six water molecules, $[M(H_2O)_6]^{n+}$, where M represents a metal ion with a charge of $n+$. Metal ions with a charge of 2+ or 3+, such as Al^{3+} and many transition metal ions, produce acidic, aqueous solutions:

$$[Cu(H_2O)_6]^{2+}(aq) + H_2O(\ell) \rightleftharpoons [Cu(H_2O)_5(OH)]^+(aq) + H_3O^+(aq) \qquad K_a = 1.6 \times 10^{-7}$$

With a K_a value of 1.6×10^{-7}, the aqueous Cu^{2+} ion is less acidic than acetic acid, but more acidic than $H_2PO_4^-$ (see Table 17.4).

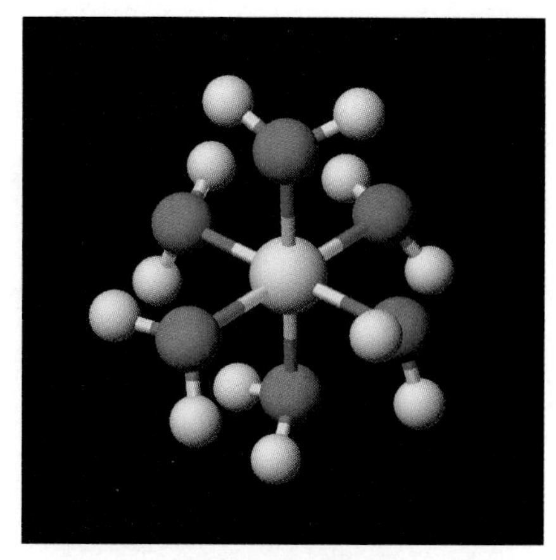

As described in Chapter 13 (page 587), metal ions in water are hydrated. They are surrounded directly by water molecules, often six at the corners of an octahedron. All are weak acids (except for the Group 1A cations). That is, common cations from Groups 1A and 2A, such as Na^+, K^+, Ca^{2+}, and Mg^{2+}, do not contribute to the acidity of solutions. See Example 17.1 and Section 17.8.

Anions as Weak Acids

Six anionic Brønsted acids are listed in Table 17.4. The dihydrogen phosphate anion, $H_2PO_4^-$, for example, is the acid in baking powder,

$$H_2PO_4^-(aq) + H_2O(\ell) \rightleftharpoons HPO_4^{2-}(aq) + H_3O^+(aq) \qquad K_a = 6.2 \times 10^{-8}$$

where its function is to provide a hydrogen ion to produce CO_2 from baking soda ($NaHCO_3$) (Figure 17.6).

$$\underset{\text{Weak acid}}{H_2PO_4^-(aq)} + \underset{\text{Weak base}}{HCO_3^-(aq)} \rightleftharpoons \underset{\text{Conjugate acid of bicarbonate ion}}{H_2CO_3(aq)} + \underset{\text{Conjugate base of } H_2PO_4^-}{HPO_4^{2-}(aq)}$$

$$H_2CO_3(aq) \rightleftharpoons CO_2(g) + H_2O(\ell)$$

(a)

(b)

***Figure* 17.6 Weak acids.** (a) Baking powder contains the weak acid calcium dihydrogen phosphate, $Ca(H_2PO_4)_2$. This can react with the basic hydrogen carbonate ion in baking soda (HCO_3^-) to form HPO_4^{2-} ion, carbon dioxide gas, and water. (b) Many foods and household products contain weak acids such as acetic acid (in vinegar) and citric acid (in fruit juices). The hydrogen sulfate ion (in sodium hydrogen sulfate, $NaHSO_4$) is used as a pH adjuster for swimming pools). *(C. D. Winters)*

Figure 17.7 Weak bases in consumer products. Examples of products containing weak bases are household ammonia, detergents, phosphates in fertilizers, and citrates and benzoates in soft drinks. The models shown here are the phosphate ion, the citrate ion ($C_3H_5O_7^{3-}$), and the benzoate ion ($C_6H_5CO_2^-$). *(Photo, C. D. Winters; models, S. M. Young)*

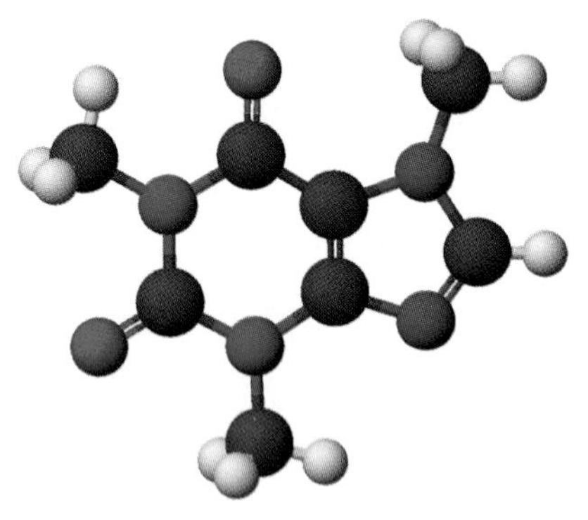

Caffeine is a weak base and can readily interact with protons from an acid.

Some Weak Bases

A variety of weak bases are important in chemistry and in consumer products, and some are listed in Table 17.4 and illustrated in Figure 17.7.

Neutral Molecules as Bases

Ammonia, NH_3, which is perhaps the best known weak base, produces a very small amount of hydroxide ion when it accepts a proton from water:

$$NH_3(aq) + H_2O(\ell) \rightleftharpoons NH_4^+(aq) + OH^-(aq) \qquad K_b = 1.8 \times 10^{-5}$$

The compound is the simplest member of a large series of compounds called amines, in which the H atoms of NH_3 are replaced by some other substituent. For example, if the substituent is the methyl group, CH_3, we have

$H_3C{-}\ddot{N}(H){-}H$	$H_3C{-}\ddot{N}(H){-}CH_3$	$H_3C{-}\ddot{N}(CH_3){-}CH_3$
methylamine	dimethylamine	trimethylamine
$K_b = 5.0 \times 10^{-4}$	$K_b = 7.4 \times 10^{-4}$	$K_b = 7.4 \times 10^{-5}$

The structures of these amines are found on page 508. Other examples of weak bases that are neutral molecules include nicotine (page 508) and caffeine.

Anions as Weak Bases

You have already seen examples of anions acting as Brønsted bases in aqueous solution. For example, the cyanide ion, CN^-, the conjugate base of the weak acid HCN, produces a measurable concentration of hydroxide ion in water,

$$\underset{\text{Base}}{CN^-(aq)} + \underset{\text{Acid}}{H_2O(\ell)} \rightleftharpoons \underset{\text{Conjugate acid of } CN^-}{HCN(aq)} + \underset{\text{Conjugate base of water}}{OH^-(aq)}$$

$$K_b = \frac{[HCN][OH^-]}{[CN^-]} = 2.5 \times 10^{-5}$$

Cyanide ion is a weaker base than hydroxide ion, so the equilibrium is predicted to lie to the left. Nonetheless, according to Table 17.4, CN^- is intermediate in base strength between two well-known bases, NH_3 and CO_3^{2-}, so the CN^- ion clearly reacts with water to a small extent to produce a basic solution. In general, *the conjugate base of a weak acid produces a basic solution in water.*

Exercise 17.6 Weak Acids and Bases

1. Lactic acid, $CH_3CHOHCO_2H$, has a value of K_a of 1.4×10^{-4}. Where does this fit in Table 17.4? Is lactic acid stronger or weaker than acetic acid?

$$CH_3CHOHCO_2H(aq) + H_2O(\ell) \rightleftharpoons H_3O^+(aq) + CH_3CHOHCO_2^-(aq)$$

2. Write a balanced equation for the lactate ion, $CH_3CHOHCO_2^-$, functioning as a Brønsted base in water.

Problem-Solving Tips and Ideas

17.1 Strong or Weak?

How can you tell whether an acid or base is weak? The easiest way is to remember those few that are strong (see Table 5.2 and the following short table), and assume that all others are probably weak. See also Sections 17.3 and 17.4. The common **strong acids** are

Hydrohalic acids: HCl, HBr, and HI

Nitric acid: HNO_3

Sulfuric acid: H_2SO_4 (for loss of first H^+ only)

Perchloric acid: $HClO_4$

Some common **strong bases** are

Group 1A hydroxides: LiOH, NaOH, KOH

Group 2A hydroxides: $Sr(OH)_2$ and $Ba(OH)_2$

17.6 WATER AND THE pH SCALE

The stronger a Brønsted acid or base, the larger the concentration of $H_3O^+(aq)$ or $OH^-(aq)$ for a given concentration of acid or base, respectively. If these concentrations could be measured quantitatively, we would have a way to compare acid and base strengths.

See the *Saunders Interactive General Chemistry CD-ROM,* Screen 17.3, Acid–Base Properties of Water, and Screen 17.4, pH Scale.

The Water Ionization Constant, K_w

Water autoionizes, transferring a proton from one water molecule to another and producing a hydronium ion and a hydroxide ion:

$$2\ H_2O(\ell) \rightleftharpoons H_3O^+(aq) + OH^-(aq)$$

Because the hydroxide ion is a much stronger base than water, and the hydronium ion is a much stronger acid than water (see Table 17.4), the equilibrium lies far to the left. In fact, in pure water at 25 °C only about two out of a billion water molecules are ionized at any instant. To express this idea more quantitatively, we can write the equilibrium constant expression

$$K = \frac{[H_3O^+][OH^-]}{[H_2O]^2}$$

Recall from Section 16.2 that in pure water or in dilute aqueous solutions (say 0.1 M solute or less), the concentration of water can be considered to be a constant (55.5 M). For this reason $[H_2O]^2$ is combined with the constant K, and the equilibrium constant expression becomes

$$K[H_2O]^2 = [H_3O^+][OH^-]$$

$$K_w = [H_3O^+][OH^-] \qquad [17.3]$$

This equilibrium constant is given a special symbol, $\boldsymbol{K_w}$, and is known as the **ionization constant for water.** In pure water, the transfer of a proton between two water molecules leads to one H_3O^+ and one OH^-. Because this is the only source of these ions, we know that $[H_3O^+]$ must equal $[OH^-]$ in pure water. Electrical conductivity measurements of pure water show that $[H_3O^+] = [OH^-] = 1.0 \times 10^{-7}$ M at 25 °C, and so

$$K_w = [H_3O^+][OH^-] = (1.0 \times 10^{-7})(1.0 \times 10^{-7}) = 1.0 \times 10^{-14}$$

The equation $K_w = [H_3O^+][OH^-]$ is valid in pure water and in any aqueous solution. K_w is temperature-dependent; the autoionization reaction is endothermic, so K_w increases with temperature.

°C	K_w
10	0.29×10^{-14}
15	0.45×10^{-14}
20	0.68×10^{-14}
25	1.01×10^{-14}
30	1.47×10^{-14}
50	5.48×10^{-14}

The hydronium ion and hydroxide ion concentrations in pure water are *both* 1.0×10^{-7} at 25 °C, and water is said to be **neutral.** If some acid or base is added to pure water, however, the equilibrium

$$2\ H_2O(\ell) \rightleftharpoons H_3O^+(aq) + OH^-(aq)$$

is disturbed. Adding acid raises the concentration of the H_3O^+ ions. To oppose this increase, Le Chatelier's principle (Section 16.6) predicts that a small fraction of the H_3O^+ ions reacts with OH^- ions to form water. This lowers $[OH^-]$ until the product of $[H_3O^+]$ and $[OH^-]$ is again equal to 1.0×10^{-14} at 25 °C. Similarly, adding a base to pure water raises the OH^- ion concentration. Le Chatelier's principle predicts that some of the added OH^- ions react with H_3O^+ ions present in the solution, thereby lowering $[H_3O^+]$ until the value of the product $[H_3O^+][OH^-]$ equals 1.0×10^{-14} at 25 °C.

Thus, for aqueous solutions at 25 °C, we can say that

- In a neutral solution $[H_3O^+] = [OH^-]$

  Both are equal to 1.0×10^{-7} M
- In an acidic solution $[H_3O^+] > [OH^-]$

  $[H_3O^+] > 1.0 \times 10^{-7}$ M and $[OH^-] < 1.0 \times 10^{-7}$ M
- In a basic solution $[H_3O^+] < [OH^-]$

  $[H_3O^+] < 1.0 \times 10^{-7}$ M and $[OH^-] > 1.0 \times 10^{-7}$ M

Example 17.2 Ion Concentrations in a Solution of a Strong Base

If you have a 0.0010 M aqueous solution of NaOH, what are the hydroxide and hydronium ion concentrations?

Solution NaOH, a strong base, is 100% dissociated into ions in water to give an initial concentration of OH^- of 0.0010 M. Before adding NaOH, the water initially contained a small concentration of OH^- (10^{-7} M) from the autoionization of water.

To solve this problem, we take the point of view that the water ionization reaction occurs only *after* excess OH^- has been added (in the form of NaOH). Water ionization gives equal concentrations of H_3O^+ and OH^-, and both are equal to the unknown quantity x. Thus, the *net* OH^- equilibrium concentration must be (0.0010 M + OH^- from water) = $(0.0010 + x)$.

Equation	$2\ H_2O(\ell) \rightleftarrows H_3O^+(aq)$	$+\ OH^-(aq)$
Before ionization (M)	0	0.0010
Change in concentrations on proceeding to equilibrium	$+x$	$+x$
After equilibrium is achieved (M)	x	$0010 + x$

We can solve for x from the expression for K_w:

$$K_w = 1.0 \times 10^{-14} = [H_3O^+][OH^-] = (x)(0.0010 + x)$$

Expanding this equation gives a quadratic equation that can be solved by the usual methods (see Appendix A and Example 16.8). A useful approximation, however, can be made that simplifies the calculation. According to Le Chatelier's principle, the water autoionization equilibrium is suppressed by the presence of the OH^- ion from NaOH. Thus, the value of x, the concentration of H_3O^+ and OH^- coming from water, must be *smaller* than 10^{-7}. This means that x in the term $(0.0010 + x)$ can be ignored. (Following the usual rules for significant figures, the sum of 0.0010 and 1.0×10^{-7} is 0.0010.) For this reason, we can say $[OH^-] = 0.0010$ M and $[H_3O^+] = x$ at equilibrium and we can write the following approximate expression:

$$K_w = 1.0 \times 10^{-14} = [H_3O^+][OH^-] \approx (x)(0.0010)$$

and so $x = [H_3O^+]$ in the presence of 0.0010 M NaOH $\approx$ 1.0×10^{-11} M.

As a final step, it is useful to check our approximation:

$$[H_3O^+][OH^-] = (1.0 \times 10^{-11})(0.0010 + 1.0 \times 10^{-11}) \approx 1.0 \times 10^{-14}$$

The calculated product of the H_3O^+ and OH^- ion concentrations equals K_w, so the approximation is valid.

Exercise 17.7 Hydronium Ion Concentration in a Solution of a Strong Acid

Gaseous HCl (0.0020 mol) is bubbled into 5.0×10^2 mL of water to make an aqueous HCl solution. What are the concentrations of H_3O^+ and OH^- in this solution?

The Connection Between the Ionization Constants for an Acid and Its Conjugate Base

Table 17.4 informs us that, for a series of acids, the strength of their conjugate bases increases as the acid strength decreases. Now that you have been introduced to K_w, the autoionization constant for water, this relationship can be expressed mathematically,

$$K_a \cdot K_b = K_w \qquad [17.4]$$

where K_a is the ionization constant for a weak acid, and K_b is the ionization constant for its conjugate base. As the value of K_a decreases, the value of K_b must increase because their product is a constant (at a given temperature).

To see the origin of Equation 17.4, suppose you add the equation for the ionization of a weak acid, say HCN, and the equation for the hydrolysis of its conjugate base, CN^-:

$$\begin{array}{rcll} HCN(aq) + H_2O(\ell) & \rightleftharpoons & H_3O^+(aq) + CN^-(aq) & K_a = 4.0 \times 10^{-10} \\ CN^-(aq) + H_2O(\ell) & \rightleftharpoons & HCN(aq) + OH^-(aq) & K_b = 2.5 \times 10^{-5} \\ \hline 2\ H_2O(\ell) & \rightleftharpoons & H_3O^+(aq) + OH^-(aq) & K_w = 1.0 \times 10^{-14} \end{array}$$

The result is the equation for the autoionization of water. Recall from Section 16.2 that the equilibrium constant for a reaction that is the sum of two others is the product of the equilibrium constants for the summed reactions. Therefore,

$$K_a \cdot K_b = \left(\frac{[H_3O^+][CN^-]}{[HCN]}\right)\left(\frac{[HCN][OH^-]}{[CN^-]}\right) = [H_3O^+][OH^-] = K_w$$

Equation 17.4 is useful because K_b can be calculated from a knowledge of K_a. For example, the value of K_b for the cyanide ion in Table 17.4 was calculated from the value of K_a for its conjugate acid, HCN.

$$K_b \text{ for } CN^- = \frac{K_w}{K_a \text{ for HCN}} = \frac{1.0 \times 10^{-14}}{4.0 \times 10^{-10}} = 2.5 \times 10^{-5}$$

Exercise 17.8 Using the Equation $K_a \cdot K_b = K_w$

K_a for lactic acid, $CH_3CHOHCO_2H$, is 1.4×10^{-4} (see Exercise 17.6). What is K_b for the conjugate base of this acid, $CH_3CHOHCO_2^-$? Where does this base fit in Table 17.4? Is it stronger or weaker than the acetate ion?

The pH Scale

A second way of expressing hydronium ion concentration avoids using very small numbers or exponential notation. This is the pH scale, a widely used method of expressing acidity. The **pH** of a solution is defined as the negative of the base-10 logarithm (log) of the hydronium ion concentration:

In general, $pX = -\log X$, where X can be any measurable quantity.

$$pH = -\log[H_3O^+] \qquad [17.5]$$

In a similar way, the pOH of a solution is defined as the negative of the base-10 logarithm of the hydroxide ion concentration:

$$pOH = -\log[OH^-]$$

In pure water, the hydronium and hydroxide ion concentrations are both 1.0×10^{-7} M. Therefore, for pure water at 25 °C

$$\begin{aligned} pH &= -\log(1.0 \times 10^{-7}) = -[\log(1.0) + \log(10^{-7})] \\ &= -[(0.00) + (-7)] \\ &= 7.00 \end{aligned}$$

In the same way, you can show that the pOH of pure water is also 7.00 at 25 °C.

If we take the negative logarithms of both sides of the expression $K_w = [H_3O^+][OH^-]$, we obtain another useful equation:

$$K_w = [H_3O^+][OH^-] = 1.0 \times 10^{-14}$$
$$-\log([H_3O^+][OH^-]) = -\log(1.0 \times 10^{-14})$$
$$(-\log[H_3O^+]) + (-\log[OH^-]) = 14.00$$

$$pH + pOH = 14.00 \quad [17.6]$$

The sum of the pH and pOH of a solution must be equal to 14.00 at 25 °C.

Example 17.3 Calculating pH

An aqueous solution contains 0.700 g of NaOH and has a volume of 485 mL. What is its pH?

Solution As described in Example 17.2, sodium hydroxide is totally dissociated in aqueous solution to Na^+ and OH^- ions. The first step is to find the concentration of the OH^- ions.

$$0.700 \text{ g NaOH} \left(\frac{1 \text{ mol NaOH}}{40.00 \text{ g NaOH}}\right) = 0.0175 \text{ mol NaOH}$$

$$[OH^-] = \frac{0.0175 \text{ mol}}{0.485 \text{ L}} = 0.0361 \text{ M}$$

Here the OH^- ion concentration is 0.0361 M from the NaOH *plus* an additional amount from the autoionization of water. As we concluded in Example 17.2, however, the contribution to the OH^- ion concentration from water is *exceedingly small,* so we assume $[OH^-] = 0.0361$ M. Substituting this into the K_w expression, we have

$$K_w = [H_3O^+][OH^-] = [H_3O^+](0.0361) = 1.0 \times 10^{-14}$$
$$[H_3O^+] = 2.8 \times 10^{-13} \text{ M}$$
$$pH = -\log[H_3O^+] = -\log(2.8 \times 10^{-13}) = -(-12.55) = 12.55$$

Notice that the same result is obtained if the pOH is calculated from the OH^- ion concentration:

$$pOH = -\log[OH^-] = -\log(0.0361) = -(-1.442) = 1.442$$

With the pOH known, pH can be found from Equation 17.6:

$$pH + pOH = 14.00$$
$$pH + 1.442 = 14.00$$
$$pH = 12.55$$

Example 17.4 pH and Hydronium Ion Conversions

1. Suppose the hydronium ion concentration in vinegar is 1.6×10^{-3} M. Calculate the pH.
2. The pH of sea water is 8.30. Calculate $[H_3O^+]$ and $[OH^-]$.

Solution

1. The pH of vinegar is found using Equation 17.5.

$$pH = -\log [H_3O^+] = -\log (1.6 \times 10^{-3}) = -(-2.80) = 2.80$$

2. The pH of sea water is 8.30. Therefore, we can find $[H_3O^+]$ by finding the antilog of the negative of the pH:

$$[H_3O^+] = 10^{-pH} = 10^{-8.30} = 5.0 \times 10^{-9} \text{ M}$$

There are two ways to calculate the concentration of the OH^- ion. You can use Equation 17.4

$$K_w = [H_3O^+][OH^-] = 1.0 \times 10^{-14} \text{ at } 25\ °C$$

and calculate $[OH^-]$ from it:

$$(5.0 \times 10^{-9})[OH^-] = 1.0 \times 10^{-14}$$
$$[OH^-] = 2.0 \times 10^{-6} \text{ M}$$

Alternatively, you can calculate the pOH using Equation 17.6, and then convert pOH to $[OH^-]$. Try this to see if you do indeed obtain 2.0×10^{-6} M.

Problem-Solving Tips and Ideas

17.2 Calculating and Using pH

- An alternative and useful form of the definitions of pH and pOH is

$$[H_3O^+] = 10^{-pH} \qquad [OH^-] = 10^{-pOH}$$

This is useful because you can key −pH (or −pOH) into your calculator, for example, and then press the 10^x key (or its equivalent on your calculator) to find the hydronium ion (or hydroxide ion) concentration.

- Significant figures and logarithms:

The digits to the left of the decimal point in a pH represent a power of 10. Only the digits to the right of the decimal are significant. In Example 17.3,

$$\begin{aligned} pH &= -\log (2.8 \times 10^{-13}) \\ &= -\log (2.8) + (-\log 10^{-13}) \\ &= -0.45 + 13.00 \\ &= 12.55 \end{aligned}$$

there are two digits to the right of the decimal in 12.55 because the H_3O^+ concentration had two significant figures.

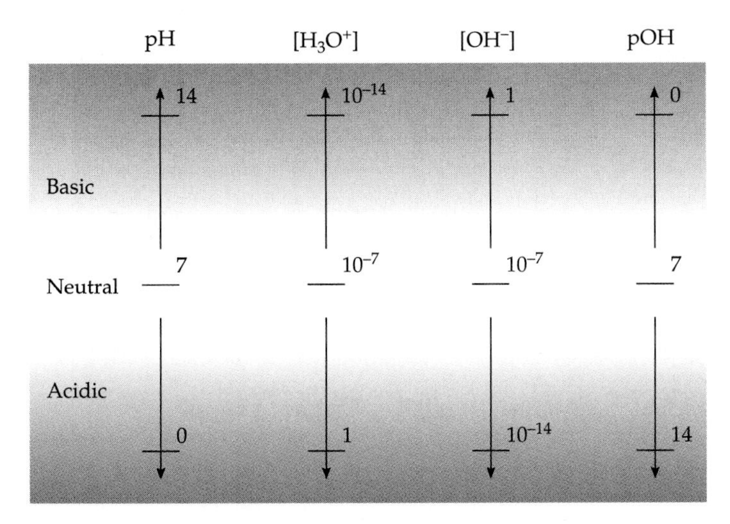

Figure 17.8 The relation between hydronium ion and hydroxide ion concentrations and pH and pOH.

Solutions with pH less than 7.00 (at 25 °C) are acidic; solutions with pH greater than 7.00 are basic. Solutions with pH = 7.00 at 25 °C are considered neutral. The relation between acidity, basicity, and pH or pOH is illustrated graphically in Figure 17.8. To give you a feeling for the pH scale, the approximate pH values for some common aqueous solutions are given in Figure 17.9.

Exercise 17.9 pH and Hydronium Ion Conversions

If the pH of a diet soda is 3.12 at 25 °C (Figure 17.10a), what are the hydronium and hydroxide ion concentrations in the soda?

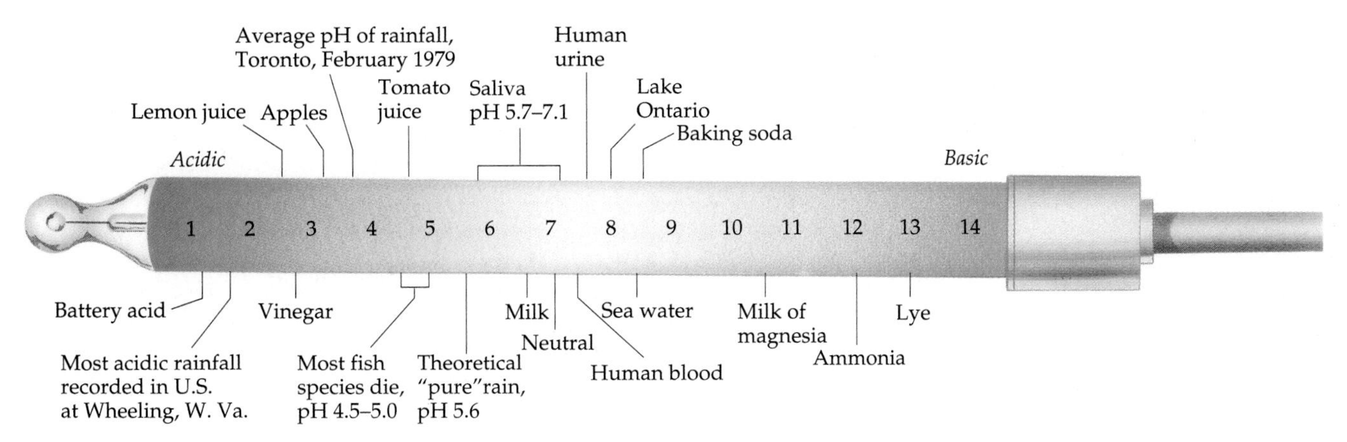

Figure 17.9 The pH of some common aqueous solutions. The scale is superimposed on the drawing of a pH electrode used in the measurement of pH by an instrumental method.

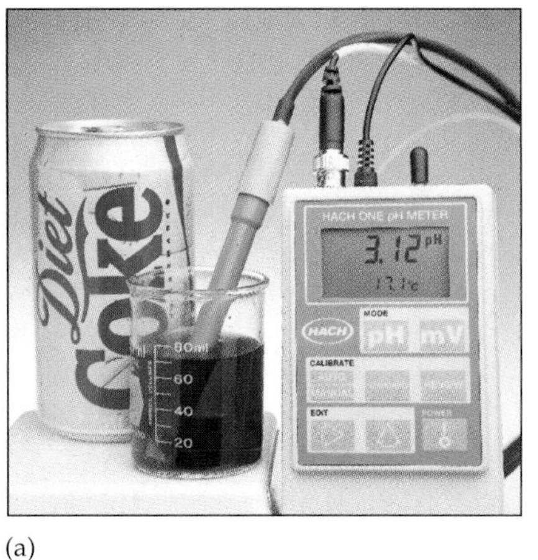

(a)

(b)

Figure 17.10 Determining pH. (a) The pH of a soda is measured with a modern pH meter. Soft drinks are often quite acidic owing to the dissolved CO_2 and other ingredients. (b) Some household products. Each solution contains a few drops of a chemical called a pH indicator (in this case a "universal indicator"). A color of yellow or red indicates a pH less than 7. A green to purple color indicates a pH greater than 7. *(C. D. Winters)*

Determining pH

The pH of a solution can be determined approximately using an **indicator,** a substance that changes color in some known pH range (Figures 17.10b and 17.11). Recall that acids were originally defined by the fact that they made certain vegetable dyes turn red. Indicators, such as phenolphthalein, are often large molecules derived from plants. These dyes, or indicators, can exist in conjugate

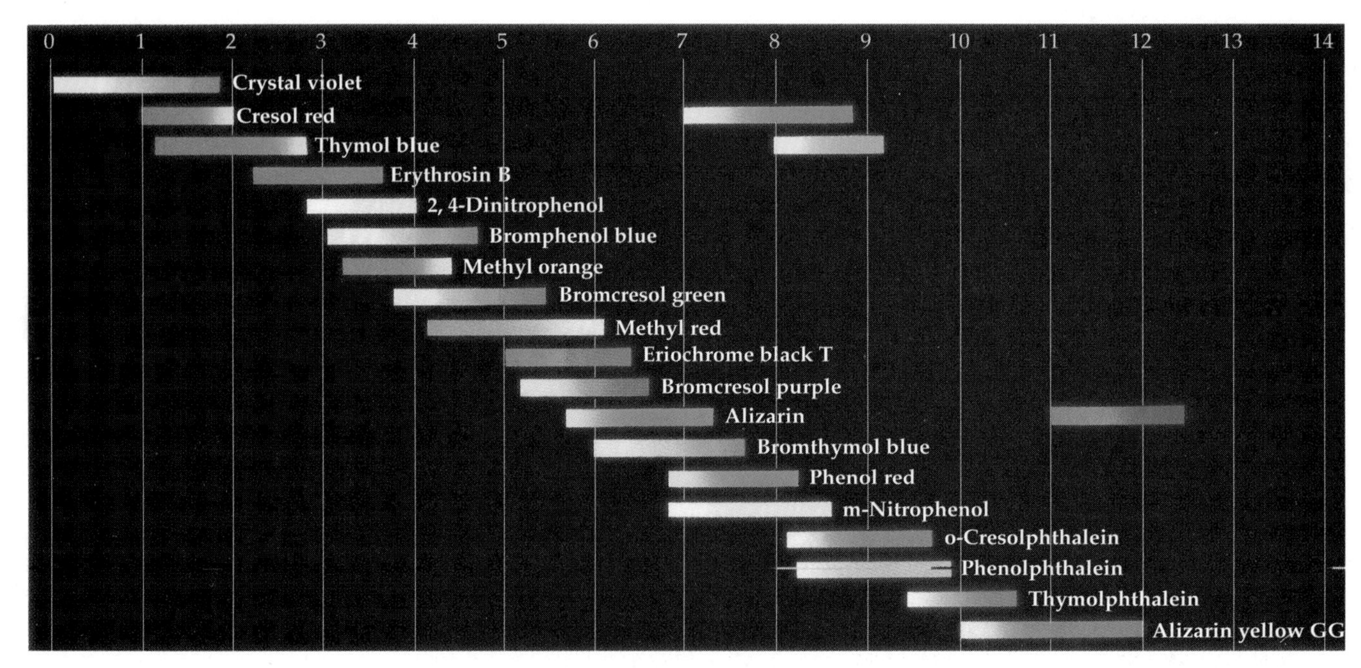

Figure 17.11 Some common acid–base indicators. The color changes occur over a range of pH values. Notice that a few indicators have two color changes over two different pH ranges. *(Hach Company)*

acid and base forms, and they have different colors depending on the form in which they are found. Common litmus paper is impregnated with a natural plant juice that is red in solutions more acidic than about pH = 5 but blue when the pH exceeds about 8.2 in basic solution.

Although an approximate pH can be obtained using an appropriate chemical indicator, a modern pH meter is far preferable for accurately determining pH (see Figure 17.10a).

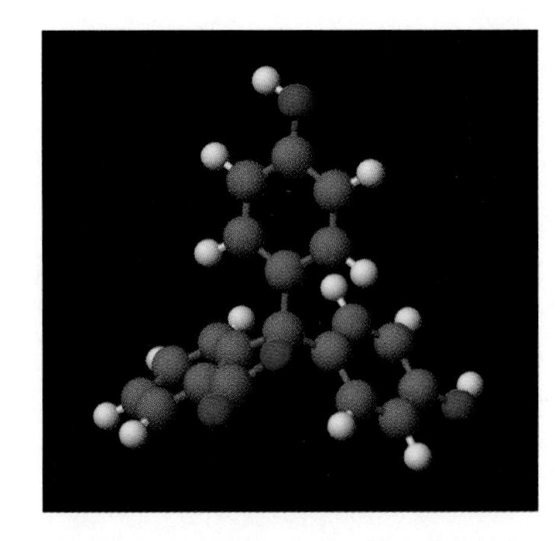

The common indicator phenolphthalein is a Brønsted acid and is colorless in that form. Its conjugate base is pink. (See Figure 5.9.) Recall that another plant, red cabbage, is a very good acid–base indicator (Figure 5.27, page 223).

17.7 EQUILIBRIA INVOLVING WEAK ACIDS AND BASES

Calculating K_a or K_b from Initial Concentrations and Measured pH

The K_a and K_b values in Table 17.4 and in the more extensive tables in Appendices H and I were all determined by experiment. There are several ways to determine the pH of the solution. The following example illustrates one method.

See the *Saunders Interactive General Chemistry CD-ROM,* Screen 17.7, Determining K_a and K_b values, and Screens 17.8 and 17.9, pH of Weak Acid Solutions and Weak Base Solutions.

Example 17.5 Calculating K_a from a Measured pH

Lactic acid is a monoprotic acid that occurs naturally in sour milk and arises from metabolism in the human body. (See Section 11.5 and *A Closer Look: Muscle Cramps and Hangovers,* page 514). A 0.10 M aqueous solution of lactic acid, $CH_3CHOHCO_2H$, has a pH of 2.43. What is the value of K_a for lactic acid?

Solution The equation for the ionization of lactic acid in water is

$$\underset{\text{lactic acid}}{CH_3CHOHCO_2H(aq)} + H_2O(\ell) \rightleftharpoons \underset{\text{lactate ion}}{CH_3CHOHCO_2^-(aq)} + H_3O^+(aq)$$

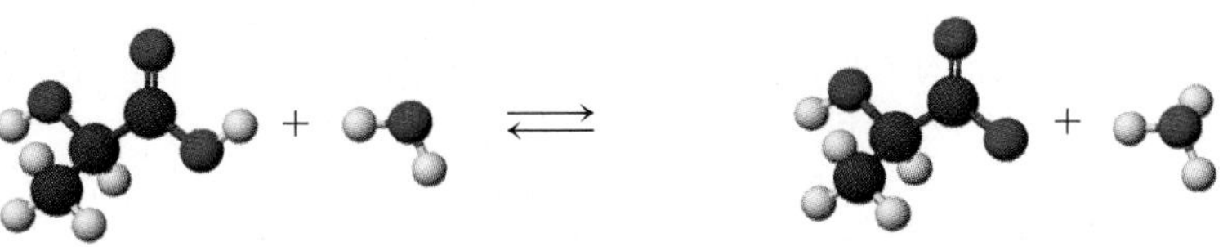

$$K_a \text{ (lactic acid)} = \frac{[H_3O^+][CH_3CHOHCO_2^-]}{[CH_3CHOHCO_2H]}$$

To calculate K_a, we need the equilibrium concentration of each species. The pH of the solution directly tells us the equilibrium concentration of $[H_3O^+]$, and we can derive the others ($[CH_3CHOHCO_2H]$ and $[CH_3CHOHCO_2^-]$) from this. We therefore begin by converting the pH to $[H_3O^+]$:

$$[H_3O^+] = 10^{-pH} = 10^{-2.43} = 3.7 \times 10^{-3}\ M$$

The following equilibrium table gives the initial concentrations of the important species in solution, describes what happens as the reaction proceeds to equilibrium, and then shows the equilibrium concentrations.

Equation	$CH_3CHOHCO_2H + H_2O$	$\rightleftharpoons$ H_3O^+	$+ CH_3CHOHCO_2^-$
Before ionization (M)	0.10	0	0
Change on proceeding to equilibrium	$-x$	$+x$	$+x$
After equilibrium is achieved (M)	$0.10 - x$	x	x

The following points can be made concerning the equilibrium table:

1. Hydronium ion, H_3O^+, is present in solution from lactic acid ionization *and* from water autoionization. Le Chatelier's principle informs us that the H_3O^+ added to the water by lactic acid suppresses the H_3O^+ coming from the water autoionization. Because measurement of the pH tells us the *total* $[H_3O^+]$ is 3.7×10^{-3} M, and because $[H_3O^+]$ from water must be less than 10^{-7} M, the pH is almost completely a reflection of H_3O^+ from lactic acid. (For weak acids and bases, this approximation can *almost* always be made. The exception is when x is near 10^{-7}; that is, when the equilibrium pH is in the 6 to 8 range. Think through each case to be certain.)
2. The quantity x represents the equilibrium concentrations of hydronium ion and lactate ion. That is, at equilibrium $[H_3O^+] \approx [C_3H_5O_3^-] = x = 3.7 \times 10^{-3}$ M.
3. By stoichiometry, x is also the quantity of acid that ionized on proceeding to equilibrium.

With these points in mind, we can calculate K_a for lactic acid:

$$K_a = \frac{[H_3O^+][CH_3CHOHCO_2^-]}{[CH_3CHOHCO_2H]} = \frac{(3.7 \times 10^{-3})(3.7 \times 10^{-3})}{0.10 - 0.0037} = 1.4 \times 10^{-4}$$

Comparing this value of K_a with others in Table 17.4, you see that lactic acid can be classed as a moderately weak acid.

Propanoic acid is used in making esters for fruit flavors and perfume bases (Section 11.5). Salts of the conjugate base, the propanoate ion (a model of which is shown here) are used as preservatives in baked goods (see Exercise 17.10). *(Photo, C. D. Winters; models, S. M. Young)*

Exercise 17.10 Calculating a K_a Value from a Measured pH

A solution prepared from 0.10 mol of propanoic acid dissolved in sufficient water to give 1.0 L of solution has a pH of 2.94. Determine K_a for propanoic acid. The acid ionizes according to the equation

$$CH_3CH_2CO_2H(aq) + H_2O(\ell) \rightleftharpoons H_3O^+(aq) + CH_3CH_2CO_2^-(aq)$$

There is an important point to notice in Example 17.5. The lactic acid concentration at equilibrium was given by "original acid concentration − quantity of acid ionized." In this example the expression was (0.10 − 0.0037). By the usual rules governing significant figures, (0.10 − 0.0037) is equal to 0.10. The acid is weak, so very little of it ionizes (approximately 4%), and the value of $[CH_3CHOHCO_2H]$ at equilibrium is essentially equal to its initial value. The approximation that the denominator in the K_a expression for a weak acid, HA,

$$HA(aq) + H_2O(\ell) \rightleftharpoons H_3O^+(aq) + A^-(aq)$$

$$K_a = \frac{[H_3O^+][A^-]}{[HA]_0 - [H_3O^+]} \approx \frac{[H_3O^+][A^-]}{[HA]_0}$$

is just $[HA]_0$, the initial acid concentration, is useful in calculations involving weak acids and bases. Error analysis shows that

The approximation that $[\text{acid}]_{\text{equilibrium}}$ is essentially equal to $[\text{acid}]_{\text{initial}}$ ($= [HA]_0$) is valid whenever $[HA]_0$ is greater than or equal to $100 \cdot K_a$.

The use of this approximation is illustrated in the next several examples.

Calculating Equilibrium Concentrations and pH from Initial Concentrations and K_a or K_b

Knowing values of the equilibrium constants for weak acids and bases enables us to calculate the pH of a solution of a weak acid or base.

Example 17.6 Calculating Equilibrium Concentrations and pH from K_a

$$C_6H_5-C(=O)-OH(aq) + H_2O(\ell) \rightleftharpoons H_3O^+(aq) + C_6H_5-C(=O)-O^-(aq)$$

benzoic acid ($C_6H_5CO_2H$) — benzoate ion ($C_6H_5CO_2^-$)

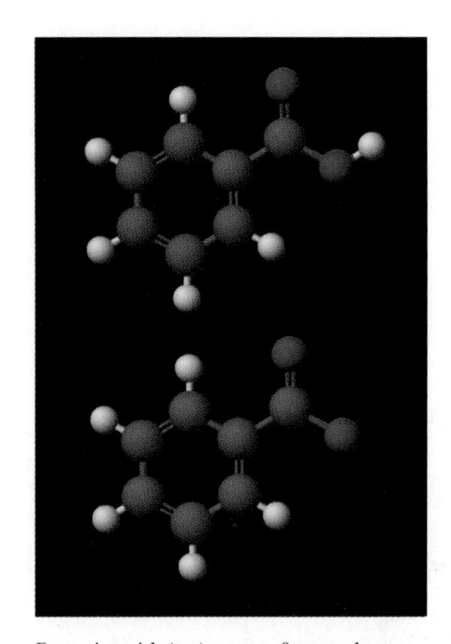

Benzoic acid *(top)* occurs free and combined in nature. Most berries, for example, contain up to 0.05% of the acid by weight. Sodium benzoate is often used as a preservative in soft drinks. (A model of the benzoate ion is at the bottom.)

Calculate the pH of a 0.020 M solution of benzoic acid if $K_a = 6.3 \times 10^{-5}$.

Solution As usual, we organize the information in a table.

Equation	$C_6H_5CO_2H + H_2O$	$\rightleftharpoons H_3O^+$	$+ C_6H_5CO_2^-$
Before ionization (M)	0.020	0	0
Change on proceeding to equilibrium	$-x$	$+x$	$+x$
After equilibrium is achieved (M)	$(0.020 - x)$	x	x

According to the reaction stoichiometry, $[H_3O^+] = [C_6H_5CO_2^-] = x$ at equilibrium. (Here we ignored any H_3O^+ that arises from water ionization; see Example 17.5.) Stoichiometry tells us that the quantity of acid ionized is also x. Thus, the benzoic acid concentration at equilibrium is

$$\begin{aligned}[C_6H_5CO_2H] &= \text{initial acid concentration} - \text{quantity of acid that ionized}\\ &= [C_6H_5CO_2H]_0 - x\\ &= 0.020 - x\end{aligned}$$

Substituting these equilibrium concentrations into the K_a expression, we have

$$K_a = \frac{[H_3O^+][C_6H_5CO_2^-]}{[C_6H_5CO_2H]} = \frac{(x)(x)}{(0.020 - x)} = 6.3 \times 10^{-5}$$

As described in the text just previous to this example, we know x is small compared with 0.020 (because $[HA]_0 \geq 100 \cdot K_a$). Therefore,

$$K_a = 6.3 \times 10^{-5} \approx \frac{x^2}{0.020}$$

This means that

$$x = \sqrt{K_a(0.020)} = 0.0011 \text{ M}$$

and we find that

$$[H_3O^+] = [C_6H_5CO_2^-] = 0.0011 \text{ M}$$

and

$$[C_6H_5CO_2H] = (0.020 - x) = 0.019 \text{ M}$$

Finally, the pH of the solution is found to be

$$\text{pH} = -\log (1.1 \times 10^{-3}) = 2.96$$

Now, let's think about the result. Because benzoic acid is weak, we made the approximation that $(0.020 - x) \approx 0.020$. If we do *not* make the approximation, the following equation results:

$$K_a(0.020 - x) = x^2$$
$$0.020\ K_a - K_a x = x^2$$
$$x^2 + K_a x - 0.020\ K_a = 0$$

This is a quadratic equation of the type $ax^2 + bx + c = 0$, where $a = 1$, $b = K_a$, and $c = -0.020K_a$. Such an expression can be solved by the quadratic formula or by the method of successive approximations (see Appendix A). Either method gives $x = [H_3O^+] = 1.1 \times 10^{-3}$, the same answer to two significant figures that we obtained from the "approximate" expression $x = \sqrt{K_a(0.020)}$.

Exercise 17.11 Calculating Equilibrium Concentrations and pH from K_a

What are the equilibrium concentrations of acetic acid, the acetate ion, and H_3O^+ for a 0.10 M solution of acetic acid ($K_a = 1.8 \times 10^{-5}$)? What is the pH of the solution?

In the previous example, we calculated the pH of an aqueous solution of a weak acid. You know it is weak because K_a was so small that $[H_3O^+] << [HA]_0$. Another way to say this is that only a small percentage of the acid is ionized in water to produce hydronium ion. Indeed, in the case of a 0.020 M solution of benzoic acid, the percentage ionization is less than 10%:

$$\text{Percentage ionized} = \frac{\text{quantity of acid ionized}}{\text{initial acid concentration}} \times 100\% = \frac{0.0011 \text{ M}}{0.020 \text{ M}} \times 100\% = 5.5\%$$

In general, the hydronium ion concentration obtained from this approximate expression is the same (to two significant figures) as that from a more exact solution to the problem whenever the weak acid is no more than about 10% ionized (and the approximation given on page 813 can be used).

Example 17.7 Calculating Equilibrium Concentrations and pH from K_a

If you have a 0.0010 M solution of formic acid (Figure 17.12), what is the pH of the solution? What is the concentration of formic acid at equilibrium? The acid is moderately weak, with $K_a = 1.8 \times 10^{-4}$.

$$HCO_2H(aq) + H_2O(\ell) \rightleftharpoons HCO_2^-(aq) + H_3O^+(aq)$$

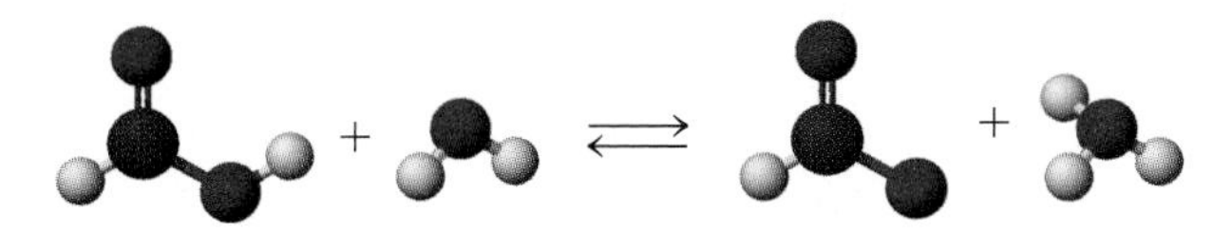

Solution The usual equilibrium table is given here. Notice that we have again made the reasonable approximation that H_3O^+ from water ionization can be ignored.

Equation	$HCO_2H + H_2O \rightleftharpoons$	H_3O^+	$+ HCO_2^-$
Before ionization (M)	0.0010	0	0
Change on proceeding to equilibrium	$-x$	$+x$	$+x$
After equilibrium is achieved (M)	$0.0010 - x$	x	x

Substituting the values in the table into the K_a expression, we have

$$K_a = \frac{[H_3O^+][HCO_2^-]}{[HCO_2H]} = \frac{(x)(x)}{(0.0010 - x)} = 1.8 \times 10^{-4}$$

Formic acid is a weak acid because it has a value of K_a much less than 1. In this situation, however, $[HA]_0$ is *less than* $100 \cdot (1.8 \times 10^{-4})$, so the usual approximation is not reasonable. If we do solve for the $[H_3O^+]$ using the approximate expression, $x \approx \sqrt{K_a(0.0010)}$, we obtain a value of x ($= [H_3O^+] = [HCO_2^-]$) of 4.2×10^{-4} M. But a check of the percentage ionization of the acid

$$\text{Percentage ionized} = \frac{4.2 \times 10^{-4}}{0.0010} \times 100\% = 42\%$$

shows clearly that 0.0010 M formic acid is *much* more than 10% ionized, so the answers from the approximate and exact solutions to the problem should differ significantly. This means we have to find the equilibrium concentrations by solving the "exact" expression:

$$K_a = \frac{[H_3O^+][HCO_2^-]}{[HCO_2H]} = \frac{(x)(x)}{(0.0010 - x)} = 1.8 \times 10^{-4}$$

$$x^2 + (1.8 \times 10^{-4})\,x + (-1.8 \times 10^{-4})(0.0010) = 0$$

Here $a = 1$, $b = 1.8 \times 10^{-4}$, and $c = -1.8 \times 10^{-7}$ in the quadratic expression $ax^2 + bx + c = 0$. Solving for x, we find

$$x = -(1.8 \times 10^{-4}) \pm \frac{\sqrt{(1.8 \times 10^{-4})^2 - 4(1)(-1.8 \times 10^{-7})}}{2(1)}$$

$$= 3.4 \times 10^{-4}\ \text{M} \quad \text{and} \quad -5.2 \times 10^{-4}\ \text{M}$$

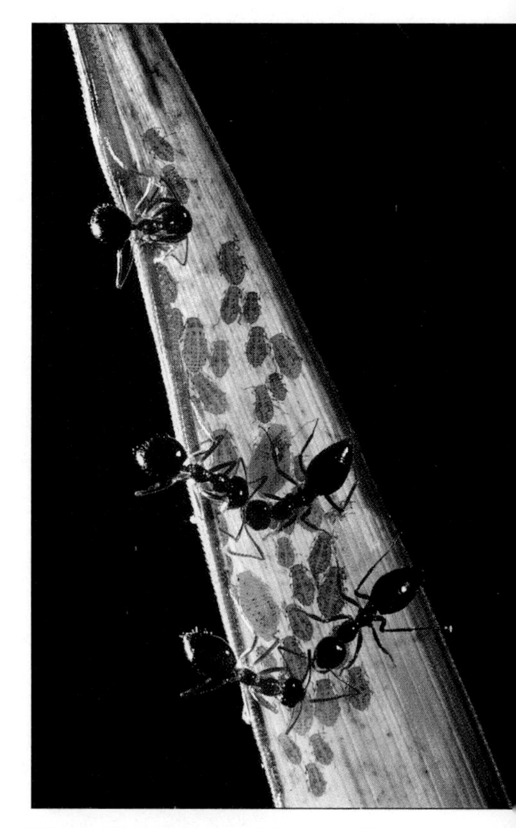

Figure **17.12 Ants and formic acid, HCO_2H.** The acid was first obtained in 1670 as a product of the destructive distillation of ants, whose Latin genus name is *Formica.* It is one component of the venom injected by stinging ants. Formic acid is a colorless, odorless substance whose main uses include the manufacture of other chemicals. *(Ted Nelson/Dembinsky Photo Associates)*

Because the negative root of the equation is chemically meaningless, we use the positive root and find that

$$[H_3O^+] = [HCO_2^-] = 3.4 \times 10^{-4}\ M$$

and so

$$[HCO_2H] = 0.0010 - x = 0.0007\ M$$

The pH of the formic acid solution is therefore

$$pH = -\log\ (3.4 \times 10^{-4}) = 3.47$$

It is important to notice in this problem that the approximate solution failed because (1) the acid concentration is small and (2) the acid is not all that weak. These made invalid the approximation that $[HA]_{equilibrium} \approx [HA]_{initial}$.

Exercise 17.12 Calculating Equilibrium Concentrations and pH from K_a

What are the equilibrium concentrations of HF, F^-, and H_3O^+ in a 0.015 M solution of HF?

The examples solved thus far have involved weak acids. Weak bases, however, are widely found in nature. Examples include caffeine (Figure 17.13) and nicotine (page 508). Caffeine, which is found in coffee, tea, and cola nuts, is a well-known stimulant. Pure nicotine, which is extracted from tobacco, is highly toxic. It is used as an insecticide in the form of its salt with sulfuric acid.

The next example describes the calculation of the pH of an aqueous solution of the weak base pyridine. Salts of pyridine derivatives are widely used in consumer products (Figure 17.13b).

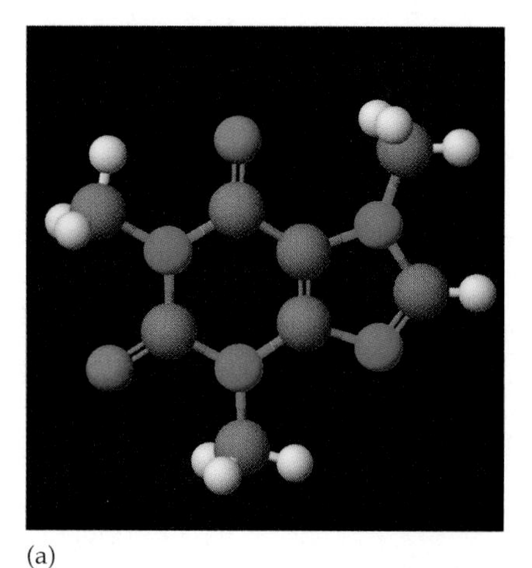

(a)

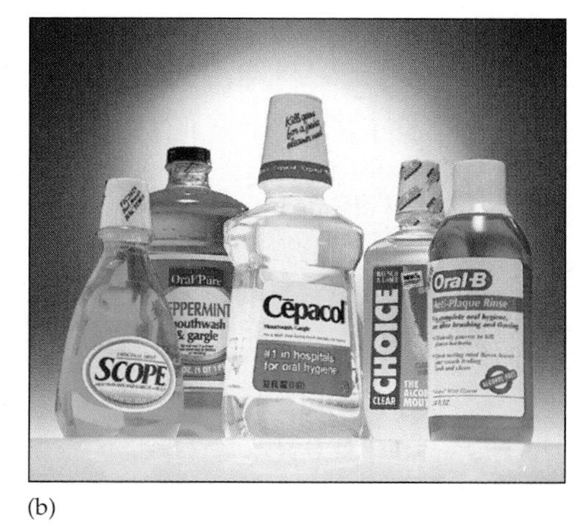

(b)

Figure 17.13 Weak bases. (a) Caffeine, $C_8H_{10}N_4O_2$, is a naturally occurring base. (b) The pyridinium ion is the conjugate acid of the weak base pyridine. Several over-the-counter mouthwashes contain a pyridinium chloride salt as an antibacterial agent. *(Photo, C. D. Winters; model, S. M. Young)*

Example 17.8 Calculating the pH of an Aqueous Solution of a Weak Base

Pyridine, a weak base, was discovered in coal tar in 1846:

$$C_5H_5N(aq) + H_2O(\ell) \rightleftharpoons C_5H_5NH^+(aq) + OH^-(aq) \qquad K_b = 1.5 \times 10^{-9}$$

pyridine (C_5H_5N)

pyridinium ion ($C_5H_5NH^+$)

What is the pH of a 0.010 M aqueous solution of pyridine?

Solution The equilibrium expression for this weak base is

$$K_b = 1.5 \times 10^{-9} = \frac{[C_5H_5NH^+][OH^-]}{[C_5H_5N]}$$

The concentrations of the base and ionic products are related in an equilibrium table:

Equation	C_5H_5N + H_2O ⇌	$C_5H_5NH^+$ +	OH^-
Before ionization (M)	0.010	0	0
Change on proceeding to equilibrium	$-x$	$+x$	$+x$
After equilibrium is achieved (M)	$0.010 - x$	x	x

If the expressions for the equilibrium concentrations are substituted into the K_b expression, we have

$$K_b = 1.5 \times 10^{-9} = \frac{(x)(x)}{0.010 - x}$$

an expression similar to the one used to find the hydronium ion concentration in a weak acid solution. Two points should be noticed here.

1. We ignored any OH^- ion that can arise from the autoionization of water. The reasons are similar to those for ignoring H_3O^+ from water in solutions of weak acids.
2. Pyridine is a very weak base, and it is probably a good assumption that it is less than 10% ionized in solution. We therefore use the approximate expression to solve for x,

$$K_b = 1.5 \times 10^{-9} = \frac{(x)(x)}{0.010}$$

and this gives a value of

$$x = [OH^-] = [C_5H_5NH^+] = 3.9 \times 10^{-6}\ M$$

The hydroxide ion concentration is indeed *much* less than the original concentration of pyridine. That is, the pyridine equilibrium concentration $(0.010\ M - 3.9 \times 10^{-6})$ is nearly equal to 0.010 M. Less than 1% of the base has reacted with water to give ions, and the approximate solution is valid.

3. This example shows that the "rule of thumb" that was used with weak acids can be applied to weak bases. That is, if $[Base]_0 \geq 100 \cdot K_b$, then the approximation that $[OH^-] = \sqrt{K_b[Base]_0}$ can be used.

The objective of the problem was to find the pH of the solution:

$$pOH = -\log [OH^-] = -\log (3.9 \times 10^{-6}) = 5.41$$
$$pH = 14.00 - pOH = 14.00 - 5.41 = 8.59$$

The solution is weakly basic.

Exercise 17.13 Calculating the pH of an Aqueous Solution of a Weak Base

What is the pH of a 0.025 M solution of ammonia, NH_3?

Problem-Solving Tips and Ideas

17.3 Using K_a and K_b

The mathematical approach used to this point can be summarized using the hypothetical weak acid HA. (The same ideas apply to weak bases.)

$$HA(aq) + H_2O(\ell) \rightleftharpoons H_3O^+(aq) + A^-(aq)$$

The equilibrium constant expression, *in terms of equilibrium concentrations,* is

$$K_a = \frac{[H_3O^+][A^-]}{[HA]}$$

where $[H_3O^+]$ is the hydronium ion concentration from the reaction of the weak acid with water and from the autoionization of water. When the pH falls outside the range between 6 and 8, we can assume that H_3O^+ ion from water autoionization does not contribute significantly. This also means that the equilibrium concentration of HA can be written $[HA] = [HA]_0 - [H_3O^+]$, where $[HA]_0$ is the original concentration of HA. Therefore, we can write K_a in a form that allows us to solve for $[H_3O^+]$ if we know $[HA]_0$.

$$K_a = \frac{[H_3O^+]^2}{[HA]_0 - [H_3O^+]}$$

Solving for $[H_3O^+]$ generally requires the use of the quadratic formula or the method of successive approximations (Appendix A). This is not necessary, however, if we make the approximation that $[HA]_0 - [H_3O^+] \approx [HA]_0$, an approximation that is valid when $[HA]_0 \geq 100 \cdot K_a$.

A CLOSER LOOK Stomach Acidity

If you think you have an "acid stomach," you're correct! The pH is about 1.5, due largely to hydrochloric acid, which is needed for the enzyme pepsin to catalyze the digestion of proteins in food. As soon as food reaches your stomach, acidic gastric juices are released by glands in the mucous lining of the stomach. Hydrogen ions are produced in blood plasma by the ionization of dissolved CO_2. These ions are transported through the stomach lining along with Cl^-, which provides charge balance. Because the stomach wall contains protein just like some foods, it is a wonder the stomach does not digest itself. In fact, it sometimes does, producing a hole—an ulcer. Most often, however, the stomach wall resists the attack of H_3O^+, thanks to a protective layer of mucous-producing cells. These cells prevent the H_3O^+ and Cl^- ions from diffusing back into the blood plasma. Ordinarily, these cells are continuously being sloughed off and replaced at the rate of about half a million cells per minute.

When you eat too much food, or when your stomach is irritated by very spicy food, it responds with an outpouring of acid, and the pH is lowered to the point where discomfort is felt. You know you can take an antacid to relieve an "acid stomach," and their chemistry was described on page 165.

The stomach wall can also be damaged by the action of aspirin. Aspirin, acetylsalicylic acid, is a weak carboxylic acid (Section 11.5) with $K_a = 3.2 \times 10^{-4}$.

$$C_6H_4(CO_2CH_3)CO_2H(aq) + H_2O(aq) \rightleftharpoons C_6H_4(CO_2CH_3)CO_2^-(aq) + H_3O^+(aq)$$

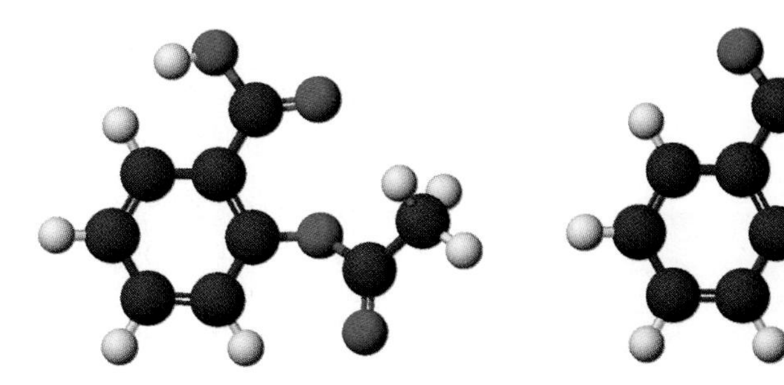

At the pH of the stomach, most of the conjugate base, the acetylsalicylate ion, reacts with H_3O^+, leaving aspirin largely un-ionized. The equilibrium lies to the left. In the neutral form aspirin molecules are able to penetrate the stomach wall, and, once inside, are in a region of lower acidity. They are therefore able to ionize and produce H_3O^+. The accumulation of H_3O^+ in the wall can cause bleeding, but the amount of blood lost per aspirin tablet is not generally harmful.

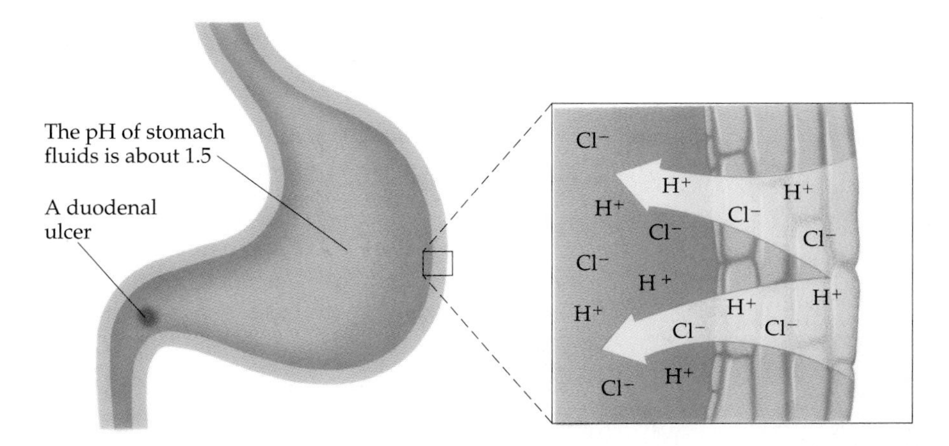

The lining of the stomach contains cells that secrete a solution of hydrochloric acid. The pH of the solution is about 1.5. Some people's stomachs produce more acid than is needed for the primary digestion of food, resulting in indigestion, among other things.

17.8 ACID–BASE PROPERTIES OF SALTS: HYDROLYSIS

Many compounds in nature and in consumer products are salts. A salt is an ionic compound that could have been formed by the reaction of an acid with a base; a salt's positive ions come from the base, and its negative ions come from the acid. Common table salt is one example; other examples include potassium ben-

See the *Saunders Interactive General Chemistry CD-ROM,* Screen 17.10, The Acid–Base Properties of Salts.

zoate ($KC_6H_5CO_2$) and potassium citrate ($K_3C_6H_5O_7$) in soft drinks, sodium bicarbonate in stomach antacids, and various phosphates in fertilizers (Figure 17.14).

The word "hydrolysis" literally means "breaking a substance apart" (-lysis) by water (hydro-).

What is important to us here is that many salts produce acidic or basic aqueous solutions by hydrolysis. A **hydrolysis** reaction is said to have occurred when a salt dissolves in water and leads to changes in the H_3O^+ and OH^- concentrations of the water. For example, you have already seen that the ammonium ion and some metal ions—most notably ions such as Al^{3+}, Pb^{2+}, and transition metal ions of 2+ and 3+ charge—increase the hydronium ion concentration in aqueous solution (see Section 17.5 and Figure 17.15). Furthermore, we described the fact that the anionic portion of a weak acid—the conjugate base of the weak acid—reacts with water to give a measurable concentration of hydroxide ion. For example, the benzoate ion, the conjugate base of the weak acid benzoic acid (Table 17.4 and page 813), hydrolyzes to give a basic solution and re-form the weak acid:

$$\underset{\substack{\text{benzoate ion} \\ \text{Weak base}}}{C_6H_5CO_2^-(aq)} + H_2O(\ell) \rightleftharpoons \underset{\substack{\text{benzoic acid} \\ \text{Conjugate acid}}}{C_6H_5CO_2H(aq)} + OH^-(aq) \qquad K_b = 1.6 \times 10^{-10}$$

Example 17.9 illustrates the calculation of the pH of an aqueous solution of a salt in which the anion is the conjugate base of a weak acid.

***Figure* 17.14 Salts.** Many common household products contain salts. Some are neutral in aqueous solution (NaCl) and some are acidic (such as ammonium salts and salts of amine-type bases). Many more, however, are basic. These include carbonates, bicarbonates, and benzoates. *(C. D. Winters)*

Example 17.9 Calculating the pH of an Aqueous Solution of the Salt of a Weak Acid

Sodium hypochlorite, NaOCl, is used as a source of chlorine in some laundry bleaches, swimming pool disinfectants, and water treatment plants. Estimate the pH of a 0.015 M solution of NaOCl. (Use Table 17.4 to obtain K_b for OCl^-.)

Solution Sodium hypochlorite consists of a sodium ion and a hypochlorite ion. The Na^+ ion does not hydrolyze in water (page 801), but the OCl^- ion is the conjugate base of a weak acid and so is expected to produce a basic solution:

$$\underset{\substack{\text{hypochlorite ion} \\ \text{Weak base}}}{OCl^-(aq)} + H_2O(\ell) \rightleftharpoons \underset{\substack{\text{hypochlorous acid} \\ \text{Conjugate acid of OCl}^-}}{HOCl(aq)} + OH^-(aq)$$

$$K_b = 2.9 \times 10^{-7} = \frac{[HOCl][OH^-]}{[OCl^-]}$$

The concentrations of the hypochlorite ion, hypochlorous acid, and hydroxide ion are given in the equilibrium table.

Equation	$OCl^- + H_2O$	$\rightleftharpoons$ $HOCl$	$+ OH^-$
Concentration before ionization (M)	0.015	0	0
Change in concentration on proceeding to equilibrium	$-x$	$+x$	$+x$
Concentration after equilibrium is achieved (M)	$0.015 - x$	x	x

Products that contain an aqueous solution of sodium hypochlorite, NaClO. The hypochlorite ion, ClO^-, is a weak base. See Example 17.9. *(C. D. Winters)*

The hypochlorite ion is a very weak base, as indicated by the very small value of K_b. It is therefore reasonable to assume x is small compared with 0.015 in $(0.015 - x)$, and so we can write

$$K_b = 2.9 \times 10^{-7} \approx \frac{x^2}{0.015}$$

Solving for x gives $x = 6.6 \times 10^{-5}$. Because $(0.015 - 6.6 \times 10^{-5}) \approx 0.015$, our assumption that x is negligible is justified. At equilibrium, therefore, the concentrations are

$$[HOCl] = [OH^-] = 6.6 \times 10^{-5}\ M \qquad \text{and} \qquad [OCl^-] = 0.015\ M$$

Finally, the pH of the solution is

$$K_w = 1.0 \times 10^{-14} = [H_3O^+][OH^-] = [H_3O^+]\ (6.6 \times 10^{-5})$$
$$[H_3O^+] = 1.5 \times 10^{-10}\ M$$
$$pH = 9.82$$

The pH indicates a basic solution, as expected for the conjugate base of a weak acid.

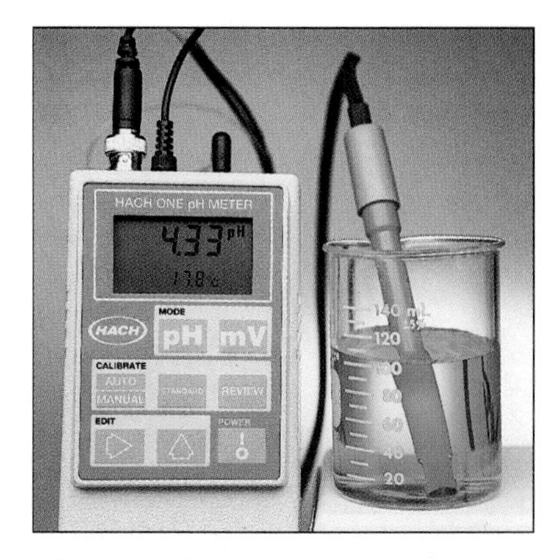

Figure 17.15 Acidity of ions. The blue solution is 0.10 M $CuSO_4$. Its pH is about 4.3, indicating that an aqueous solution of Cu^{2+} ion is acidic, as are aqueous solutions of many metal cations. *(C. D. Winters)*

Exercise 17.14 Calculating the pH of an Aqueous Salt Solution

The ammonium ion is the conjugate acid of the weak base ammonia:

$$NH_4^+(aq) + H_2O(\ell) \rightleftharpoons H_3O^+(aq) + NH_3(aq)$$

What is the pH of a 0.50 M solution of ammonium chloride?

With the knowledge you have gained in this chapter of the types of Brønsted acids and bases, you can predict with some confidence whether a given salt will be acidic or basic. To summarize, the bases listed above water on the right side of Table 17.4 are extremely weak and do not affect the pH of an aqueous solution. Acids listed below water on the left side of Table 17.4 are extremely weak; they do not make a solution acidic. This can be used to make the following predictions:

1. A salt such as $NaNO_3$ gives a neutral, aqueous solution because Na^+ does not hydrolyze to an appreciable extent and because NO_3^- is the conjugate base of a strong acid and also does not hydrolyze appreciably.
2. An aqueous solution of K_3PO_4 should be basic because PO_4^{3-} is the conjugate base of the weak acid HPO_4^{2-}, whereas K^+ does not hydrolyze appreciably.
3. An aqueous solution of $FeCl_2$ should be weakly acidic, because $Fe(H_2O)_6^{2+}$ hydrolyzes to give an acidic solution (Section 17.5), whereas Cl^- is the conjugate base of the strong acid HCl and so does not hydrolyze to an appreciable extent.

Some additional explanation is needed concerning salts of amphiprotic anions, such as HCO_3^- and $H_2PO_4^-$. Because they have an ionizable hydrogen, they can act as acids,

$$\underset{\text{Acid}}{HCO_3^-(aq)} + H_2O(\ell) \rightleftharpoons CO_3^{2-}(aq) + H_3O^+(aq) \qquad K_a = 4.8 \times 10^{-11}$$

but they are also the conjugate bases of weak acids:

$$\underset{\text{Base}}{HCO_3^-(aq)} + H_2O(\ell) \rightleftharpoons H_2CO_3(aq) + OH^-(aq) \qquad K_b = 2.4 \times 10^{-8}$$

Is a solution of sodium dihydrogen phosphate, NaH_2PO_4, weakly acidic or weakly basic?

Whether the solution is acidic or basic depends on the relative sizes of K_a and K_b. In the case of an aqueous solution of the hydrogen carbonate anion, K_b is larger than K_a, so $[OH^-]$ is larger than $[H_3O^+]$, and a solution of a salt such as $NaHCO_3$ is basic.

Finally, what happens if you have a salt based on an acidic cation and a basic anion? One example is ammonium fluoride. Here the ammonium ion decreases the pH, and the fluoride ion increases the pH:

$$NH_4^+(aq) + H_2O(\ell) \rightleftharpoons H_3O^+(aq) + NH_3(aq) \qquad K_a(NH_4^+) = 5.6 \times 10^{-10}$$

$$F^-(aq) + H_2O(\ell) \rightleftharpoons HF(aq) + OH^-(aq) \qquad K_b\ (F^-) = 1.4 \times 10^{-11}$$

Because $K_a(NH_4^+) > K_b(F^-)$, the ammonium ion is a stronger acid than fluoride ion is a base. The resulting solution should be slightly acidic.

In general, for a salt that has an acidic cation and a basic anion, the pH of the solution will be determined by the ion that is the stronger acid or base.

All the necessary information to determine the pH of a salt is in Table 17.4. We have also summarized it, however, in a concise manner in Table 17.5 and in the following statements:

Cation	**Anion**	**pH of the Solution**
From strong base (Na^+)	From strong acid (Cl^-)	$= 7$ (neutral)
From strong base (K^+)	From weak acid ($CH_3CO_2^-$)	> 7 (basic)
From weak base (NH_4^+)	From strong acid (Cl^-)	< 7 (acidic)
From any weak base (BH^+)	From any weak acid (A^-)	Depends on relative strengths of acid and base

Table 17.5 • ACID–BASE PROPERTIES OF TYPICAL IONS IN AQUEOUS SOLUTION

	Neutral		**Basic**			**Acidic**
Anions	Cl^- Br^- I^-	NO_3^- ClO_4^-	$CH_3CO_2^-$ HCO_2^- CO_3^{2-} S^{2-} F^-	CN^- PO_4^{3-} HCO_3^- HS^- NO_2^-	SO_4^{2-} HPO_4^{2-} SO_3^{2-} OCl^-	HSO_4^- $H_2PO_4^-$ HSO_3^-
Cations	Li^+ Na^+ K^+	Mg^{2+} Ca^{2+} Ba^{2+}	None			Al^{3+} NH_4^+ Transition metal ions

Exercise 17.15 Predicting the pH of Salt Solutions

For each of the following salts predict whether the pH will be greater than, less than, or equal to 7.

1. $NaCl$ **2.** $FeCl_3$ **3.** NH_4NO_3 **4.** Na_2HPO_4

17.9 POLYPROTIC ACIDS AND BASES

As described in Section 17.3 and Table 17.1, some important acids are capable of donating more than one proton and are therefore called polyprotic. Indeed, many occur in nature. Examples include oxalic acid, citric acid, malic acid (found in apples) and tartaric acid (Figure 17.16).

Phosphoric acid is commonly used in the food industry. It ionizes in three steps.

First ionization step.

$$H_3PO_4(aq) + H_2O(\ell) \rightleftharpoons H_3O^+(aq) + H_2PO_4^-(aq) \qquad K_{a1} = 7.5 \times 10^{-3}$$

Second ionization step.

$$H_2PO_4^-(aq) + H_2O(\ell) \rightleftharpoons H_3O^+(aq) + HPO_4^{2-}(aq) \qquad K_{a2} = 6.2 \times 10^{-8}$$

Third ionization step.

$$HPO_4^{2-}(aq) + H_2O(\ell) \rightleftharpoons H_3O^+(aq) + PO_4^{3-}(aq) \qquad K_{a3} = 3.6 \times 10^{-13}$$

Figure 17.16 Naturally occurring polyprotic acids. Among them is tartaric acid, $C_4H_6O_6$. It is found in grapes and other fruits, both free and as its salts. Notice that the acid has two carboxylic acid groups ($—CO_2H$). (Other polyprotic acids include oxalic acid, Figure 17.5c, and citric acid, page 194). *(Photo, C. D. Winters; model, S. M. Young)*

Notice that the K_a values for each successive step become smaller and smaller. It is more difficult to remove H^+ from a negatively charged ion, such as $H_2PO_4^-$, than from a neutral molecule, such as H_3PO_4. Furthermore, the larger the negative charge of the anionic acid, the more difficult it is to remove H^+.

For many inorganic acids, such as phosphoric acid, carbonic acid, and hydrogen sulfide, each successive loss of a proton is about 10^4 to 10^6 times more difficult than the previous ionization step. This means that the first ionization step of a polyprotic acid produces up to about a million times more H_3O^+ than the second step. A consequence is that, for many inorganic polyprotic acids, the pH of the solution depends primarily on the hydronium ion generated *in the first ionization step*; the hydronium ion produced in the second and any subsequent can be ignored.

The principles used to describe the behavior of polyprotic acids apply to their conjugate bases. This is illustrated by the calculation of the pH of a solution containing the carbonate ion (Example 17.10).

Example 17.10 Calculating the pH of the Solution of a Polyprotic Base

The carbonate ion, CO_3^{2-}, is important in the environment. It is a base in water, forming the hydrogen carbonate ion, which in turn can form carbonic acid.

$$CO_3^{2-}(aq) + H_2O(\ell) \rightleftharpoons HCO_3^-(aq) + OH^-(aq) \qquad K_{b1} = 2.1 \times 10^{-4}$$
$$HCO_3^-(aq) + H_2O(\ell) \rightleftharpoons H_2CO_3(aq) + OH^-(aq) \qquad K_{b2} = 2.4 \times 10^{-8}$$

What is the pH of a 0.10 M solution of Na_2CO_3?

Solution The fact that the second ionization constant, K_{b2}, is so much smaller than the first, K_{b1}, means that hydroxide ion concentration in the solution results almost entirely from the first step. So, let us assume that *only* the first step occurs and calculate the resulting concentrations.

Equation	CO_3^{2-} + H_2O ⇌	HCO_3^- +	OH^-
Before ionization (M)	0.10	0	0
Change on proceeding to equilibrium	$-x$	$+x$	$+x$
After equilibrium is achieved (M)	$0.10 - x$	x	x

The concentration of OH^- ($= x$) can be derived from the expression

$$K_{b1} = 2.1 \times 10^{-4} = \frac{[HCO_3^-][OH^-]}{[CO_3^{2-}]} = \frac{x^2}{0.10 - x}$$

Because K_{b1} is relatively small, it is reasonable to make the approximation that $(0.10 - x) \approx 0.10$, and so

$$x = [HCO_3^-] = [OH^-] \approx \sqrt{K_{b1}\,(0.10)} = 4.6 \times 10^{-3}\ M$$

Using this value of $[OH^-]$ to calculate the pH, we find

$$pOH = -\log\,(4.6 \times 10^{-3}) = 2.34$$

$$pH = 11.66$$

and the concentration of the hydrogen carbonate ion is

$$[HCO_3^-] = 0.10 - 0.0046 \approx 0.10\ M$$

This result shows that the approximation was indeed justified.

Next, let us turn to the problem of obtaining the concentration of carbonic acid, H_2CO_3, a product of the second ionization step.

Equation	HCO_3^- + H_2O ⇌	H_2CO_3 +	OH^-
Before ionization (M)	4.6×10^{-3}	0	4.6×10^{-3}
Change on proceeding to equilibrium	$-y$	$+y$	$+y$
After equilibrium is achieved (M)	$(4.6 \times 10^{-3} - y)$	y	$(4.6 \times 10^{-3} + y)$

Because K_{b2} is so small, the second step occurs to a *much* smaller extent than the first. This means the amount of H_2CO_3 and OH^- produced in the second step ($= y$) is *much* smaller than 10^{-3} M. It is therefore reasonable that both $[HCO_3^-]$ and $[OH^-]$ are very close to 4.6×10^{-3} M.

$$K_{b2} = 2.4 \times 10^{-8} = \frac{[H_2CO_3][OH^-]}{[HCO_3^-]} \approx \frac{(y)(4.6 \times 10^{-3})}{4.6 \times 10^{-3}}$$

Because $[HCO_3^-]$ and $[OH^-]$ have nearly identical values, they cancel from the expression, and we find that $[H_2CO_3]$ is simply equal to K_{b2}.

$$y = [H_2CO_3] = K_{b2} = 2.4 \times 10^{-8}\ M$$

Exercise 17.16 Calculating the pH of the Solution of a Polyprotic Acid

Suppose you have a 0.10 M solution of oxalic acid, $H_2C_2O_4$. What is the pH of the solution? What is the concentration of the oxalate ion, $C_2O_4^{2-}$?

PROBLEM-SOLVING TIPS AND IDEAS

17.4 Polyprotic Acids or Bases

For polyprotic acids, H_2A, in which K_{a1} and K_{a2} are different by 10^3 or more,

$$HA(aq) + H_2O(\ell) \rightleftharpoons H_3O^+(aq) + HA^-(aq) \qquad K_{a1}$$
$$HA^-(aq) + H_2O(\ell) \rightleftharpoons H_3O^+(aq) + A^{2-}(aq) \qquad K_{a2}$$

we can say that

- The hydronium ion comes largely from the first ionization step and is given by $[H_3O^+] = \sqrt{K_{a1} \cdot [H_2A]_0}$ (where $[H_2A]_0$ is the original concentration of the diprotic acid).
- The hydronium ion produced in the second ionization step is extremely small. The hydronium ion concentration of the solution is therefore effectively equal to $[HA^-]$. From this it follows that $[A^{2-}] = K_{a2}$.

The conclusions described for polyprotic acids can be applied to polyprotic bases such as the carbonate ion:

$$CO_3^{2-}(aq) + H_2O(\ell) \rightleftharpoons HCO_3^-(aq) + OH^-(aq) \qquad K_{b1} = 2.1 \times 10^{-4}$$
$$HCO_3^-(aq) + H_2O(\ell) \rightleftharpoons H_2CO_3(aq) + OH^-(aq) \qquad K_{b2} = 2.4 \times 10^{-8}$$

- The hydroxide ion present in the solution comes largely from the first step.
- The OH^- concentration is effectively equal to that of the conjugate acid of the base in the first step (here $[OH^-] \approx [HCO_3^-]$).

17.10 MOLECULAR STRUCTURE, BONDING, AND ACID–BASE BEHAVIOR

One of the most interesting aspects of chemistry is the correlation between molecular structure, bonding, and observed properties. Here it is useful to analyze the connection between the structure and bonding of some acids and their relative strengths.

When an acid HA dissociates in water, we can think of the process as the sum of a series of steps. (For simplicity, we ignore the solvation of the individual species, assuming this has *about* the same effect in analogous cases.)

Step 1. *H—A bond breaking (one bonding electron is retained by each partner):*

$$H—A \longrightarrow H\cdot + \cdot A$$

Step 2. *Loss of an electron by H to form H^+:*

$$H\cdot \longrightarrow H^+ + e^-$$

Step 3. *Gain of an electron by A to form A^-:*

$$A\cdot + e^- \longrightarrow A^-$$

Net reaction: $$H{-}A \longrightarrow H^+ + A^-$$

Because we are only interested in relative acidities, we can further simplify our analysis by ignoring the second step, one common to all acid dissociations. Thus, to get some insight into the relative strengths of several acids, we can compare the H—A bond strengths in the acids (Step 1) and the relative affinities of A for an electron (Step 3). In general, the more easily the H—A bond is broken and the greater the electron affinity of A, the greater the relative strength of the acid. These two effects can work together (a weak H—A bond and a high affinity of A for an electron lead to a strong acid). It is often observed, however, that they work in opposite directions, so the balance between the effects controls acidity.

Let us use this approach with the binary hydrogen halides, HA (Figure 17.17a). Their relative acid strengths are known to be in the order shown in the table.

	—Increasing acid strength⟶			
	HF	**HCl**	**HBr**	**HI**
H—A bond strength (kJ/mol)	569	431	368	297
Electron affinity of A (kJ/mol)	328	349	325	295

Here it is evident that H—A bond strengths largely control the relative acidities of these compounds. The strongest acid, HI, has the weakest H—A bond. If the affinity of A for an electron were the controlling factor, then HCl would have been the strongest acid.

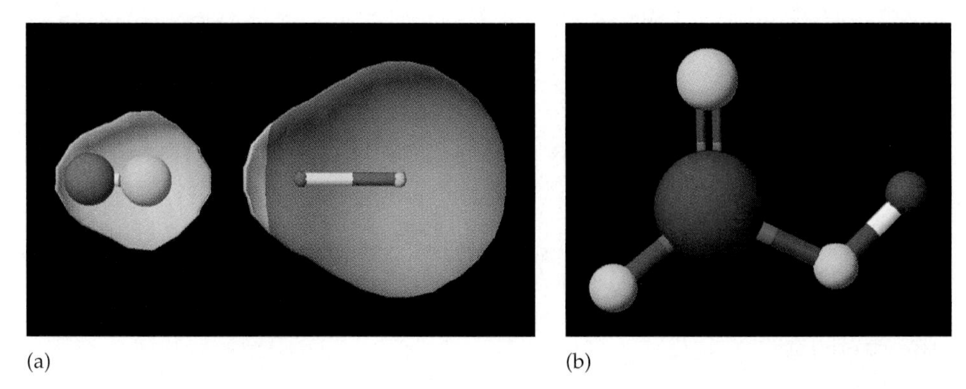

(a) (b)

F*igure* 17.17 Computer-generated models of the properties of acids. In all of these models red spheres represent positive atoms and yellow spheres represent negative atoms. The size of the sphere reflects the relative charge. (a) Models of HF (*left*) and HBr (*right*). In HF the H atom on the left is positive (+0.29) and the F atom on the right is negative (−0.29). There is a substantial charge on each atom, so the molecule is quite polar. In HBr, the partial electric charges are much smaller than in HF (as reflected by the much smaller spheres). Both molecules are surrounded by a surface that represents the electron density contour. (b) In this model of nitric acid, HNO_3, the central N atom is positively charged as is the acidic H atom. *(S. M. Young)*

Next, let us examine acids in which the proton is always bonded to the same element, but other changes are made in the molecule. Oxoacids fit this description, and two series are of interest:

—Increasing acid strength→

O=N—O—H (nitrous acid) O=N(=O)—O—H (nitric acid)

Cl—O—H (hypochlorous acid) O—Cl—O—H (chlorous acid) O—Cl(O)—O—H (chloric acid) O—Cl(O)(O)—O—H (perchloric acid)

From the trends seen in the nitrogen- and chlorine-based oxoacids, it is logical to conclude that the greater the number of O atoms attached to the central atom in the acid, the stronger the acid. This increase in acidity is due to the **inductive effect,** the attraction of electrons from adjacent bonds by a more electronegative atom. As more and more electronegative O atoms are attached, the inductive effect is stronger, the O—H bond is more strongly polarized ($^{\delta-}$O—H$^{\delta+}$), and the bond is more readily broken (Figure 17.17b).

$$HONO_2 \longrightarrow [NO_3]^- + H^+$$

Inductive effect of electronegative O atoms weakens the O—H bond.

Another interesting question is: Why do so few substances behave as Brønsted acids, even though hundreds of compounds have some type of E—H bond? Acetic acid, for example, loses the hydrogen of the —OH group (and not that attached to carbon) as H^+ when placed in water, and a negative charge is left on the oxygen atoms of the acetate ion:

$$H_3C—C(=O)—O—H \longrightarrow H^+ + H_3C—C(=O)—O^-$$

C—H bonds not broken in water

O—H bond is broken when molecule acts as a Brønsted acid.

Such cleavage is favorable because the oxygen atom can accommodate the negative charge; because oxygen has a high electronegativity, it can accept this charge relatively easily. In contrast, the C—H hydrogen of acetic acid (and many other molecules) is not dissociated as H^+ in the presence of water because the carbon atom is not sufficiently electronegative to accommodate the negative charge left if the bond breaks as C—H ⟶ C:$^-$ + H^+ (Figure 17.18).

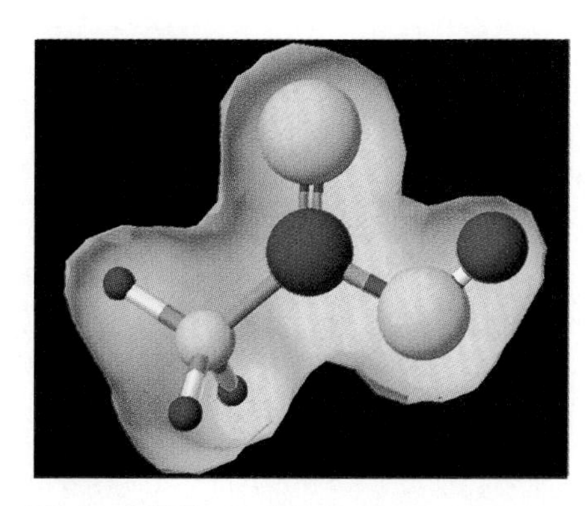

Figure 17.18 Computer-generated model of acetic acid. Molecular modeling methods support the notion that it is the —OH hydrogen atom that is donated when acetic acid, CH_3CO_2H, functions as an acid. (Here the molecule is enclosed within a surface that shows the distribution of electrons in the molecule. Red spheres represent positive atoms and yellow spheres represent negative atoms. The size of the spheres reflects the relative charge.) The —OH hydrogen bears a positive partial charge (+0.24), whereas the —CH hydrogen atoms are essentially uncharged. *(S. M. Young)*

Exercise 17.17 Relative Acid Strengths

In each pair of acids, tell which is the stronger and why. Verify your answers using Appendix H.

1. H_2SO_4 (sulfuric acid) and H_2SO_3 (sulfurous acid)
2. H_3AsO_3 (arsenious acid) and H_3AsO_4 (arsenic acid)

17.11 THE LEWIS CONCEPT OF ACIDS AND BASES

G. N. Lewis developed the first concepts of the electron pair bond. See page 375 for a short biography of this important American scientist.

See the *Saunders Interactive General Chemistry CD-ROM,* Screens 17.11, 17.12, and 17.13, Lewis Acids and Bases.

The theory of acid–base behavior advanced by Brønsted and Lowry in the 1920s works well for aqueous solutions. A more general theory, however, was developed by Gilbert N. Lewis (1875–1946) in the 1930s. His theory is based on the sharing of electron pairs between acid and base rather than the proton transfer idea of Brønsted and Lowry. A **Lewis acid** is a substance that can accept a pair of electrons from another atom to form a new bond, and a **Lewis base** is a substance that can donate a pair of electrons to another atom to form a new bond. This means that an acid–base reaction in the Lewis sense can occur if there is a molecule (or ion) with a pair of electrons that can be donated and a molecule (or ion) that can accept an electron pair:

$$\underset{\text{Acid}}{\mathrm{A}} + \underset{\text{Base}}{\mathrm{B:}} \longrightarrow \underset{\text{Adduct or complex}}{\mathrm{B:A}}$$

The result is often called an acid–base **adduct,** or complex. In Section 9.4 this type of chemical bond was called a **coordinate covalent bond.**

A simple example of a Lewis acid–base reaction is the formation of the hydronium ion from H^+ and water. The H^+ ion has no electrons in its valence, or 1*s*, orbital, and the water molecule has two unshared pairs of electrons (located in sp^3 hybrid orbitals). One of the pairs can be shared between H^+ and water, thus forming an O—H bond. A similar interaction occurs between H^+ and the base ammonia to form the ammonium ion:

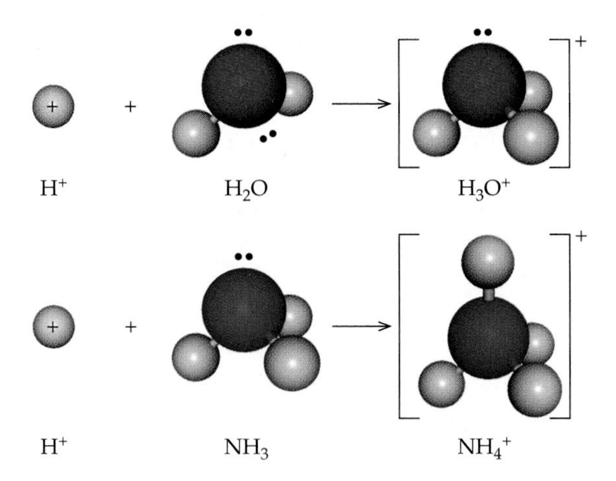

Such reactions are very common. In general, they involve Lewis acids that are cations or neutral molecules with an available, empty valence orbital and bases that are anions or neutral molecules with a lone electron pair.

Cationic Lewis Acids

All metal cations are potential Lewis acids. Not only are electron pairs attracted to their positive charge, but all have at least one empty orbital. This empty orbital can overlap the orbital bearing the electron pair of the base and can thereby form a two-electron chemical bond. Consider the beryllium ion, Be^{2+}, which can form four donor–acceptor bonds:

$$BeCl_2(s) + 4\ H_2O(\ell) \longrightarrow [Be(H_2O)_4]^{2+}(aq) + 2\ Cl^-(aq)$$

Water molecules are electron pair donors, or Lewis bases, so Be^{2+} and H_2O form an acid–base adduct, and the four empty orbitals on beryllium mean that up to four such bonds can form. Once formed, the 2+ charge of the beryllium ion means that the electrons of the $H_2O—Be^{2+}$ bond are very strongly attracted to the cation. As a result, the O—H bonds of the bound water molecules are also polarized, because the oxygen of water, in relinquishing electrons to Be^{2+}, demands more from its O—H bonds:

$$(H_2O)_3Be^{+2} \longleftarrow :\underset{\delta^-}{\ddot{O}}(-H^{\delta^+})—H^{\delta^+}$$

The net effect is that the O—H bond in a coordinated water molecule is weakened and the H atom is lost as a proton more readily than in an uncoordinated water molecule. Thus, the $[Be(H_2O)_4]^{2+}$ complex ion functions as a Brønsted acid, or proton donor:

$$[Be(H_2O)_4]^{2+}(aq) + H_2O(\ell) \rightleftharpoons [Be(H_2O)_3(OH)]^+(aq) + H_3O^+(aq)$$

Polarization of the O—H bond arises from the inductive effect described on page 827, and it explains why metal cations such as Al^{3+} and 2+ and 3+ transition metal cations generally form acidic aqueous solutions (see Figure 17.15 and Table 17.4).

The hydroxide ion, OH^-, is an excellent Lewis base and so binds readily to metal cations to give metal hydroxides. An important feature of the chemistry of some metal hydroxides is that they are **amphoteric.** An amphoteric metal hydroxide can behave as a Brønsted base and react with a Brønsted acid, or it can behave as a Lewis acid and react with a Lewis base (Table 17.6).

Table 17.6 • Some Common Amphoteric Metal Hydroxides

Hydroxide	Reaction as a Brønsted Base	Reaction as a Lewis Acid
$Al(OH)_3$	$Al(OH)_3(s) + 3\ H_3O^+(aq) \longrightarrow Al^{3+}(aq) + 6\ H_2O(\ell)$	$Al(OH)_3(s) + OH^-(aq) \longrightarrow [Al(OH)_4]^-(aq)$
$Zn(OH)_2$	$Zn(OH)_2(s) + 2\ H_3O^+(aq) \longrightarrow Zn^{2+}(aq) + 4\ H_2O(\ell)$	$Zn(OH)_2(s) + 2\ OH^-(aq) \longrightarrow [Zn(OH)_4]^{2-}(aq)$
$Sn(OH)_4$	$Sn(OH)_4(s) + 4\ H_3O^+(aq) \longrightarrow Sn^{4+}(aq) + 8\ H_2O(\ell)$	$Sn(OH)_4(s) + 2\ OH^-(aq) \longrightarrow [Sn(OH)_6]^{2-}(aq)$
$Cr(OH)_3$	$Cr(OH)_3(s) + 3\ H_3O^+(aq) \longrightarrow Cr^{3+}(aq) + 6\ H_2O(\ell)$	$Cr(OH)_3(s) + OH^-(aq) \longrightarrow [Cr(OH)_4]^-(aq)$

(a) (b) (c)

Figure 17.19 The amphoteric nature of $Al(OH)_3$. (a) Adding aqueous ammonia to a soluble salt of Al^{3+} leads to a precipitate of $Al(OH)_3$. (b) Adding a strong base (NaOH) to $Al(OH)_3$ dissolves the precipitate. Here aluminum hydroxide acts as a Lewis acid toward the Lewis base OH^- and forms the soluble sodium salt of the complex ion $Al(OH)_4^-$. (c) If we begin again with freshly precipitated $Al(OH)_3$, it is dissolved as strong acid (HCl) is added. In this case $Al(OH)_3$ acts as a Brønsted base and forms a soluble aluminum salt and water. *(C. D. Winters)*

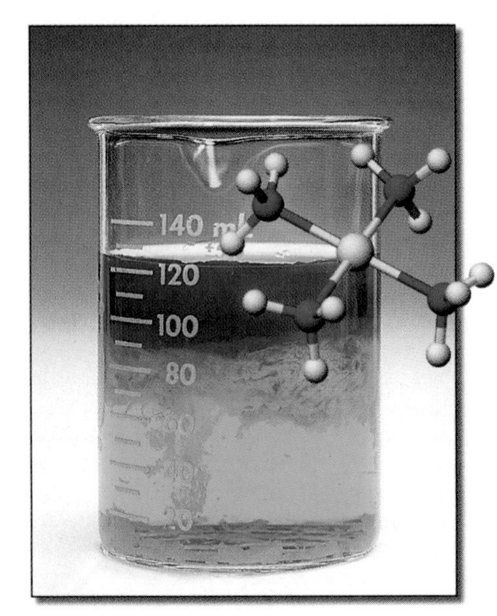

Figure 17.20 The Lewis acid–base complex ion $[Cu(NH_3)_4]^{2+}$. Here aqueous ammonia was added to aqueous $CuSO_4$ (the light blue solution at the bottom of the beaker). The small concentration of OH^- in $NH_3(aq)$ first formed insoluble blue-white $Cu(OH)_2$ (the solid in the middle of the beaker). With additional NH_3, however, the deep blue, soluble complex ion formed (the solution at the top of the beaker). The model shows the copper(II)–ammonia complex ion. *(Photo, C. D. Winters; model, S. M. Young)*

One of the best examples of amphoterism is aluminum hydroxide, $Al(OH)_3$ (Figure 17.19). Adding OH^- to a precipitate of $Al(OH)_3$ produces the water-soluble $[Al(OH)_4]^-$ ion. Here $Al(OH)_3$ acts as a Lewis acid, and the hydroxide ion is a Lewis base:

$$\underset{\text{Lewis acid}}{Al(OH)_3(s)} + \underset{\text{Lewis base}}{OH^-(aq)} \longrightarrow [Al(OH)_4]^-(aq)$$

If acid is added, the $Al(OH)_3$ precipitate again dissolves. This time, however, the aluminum hydroxide is a Brønsted base:

$$\underset{\text{Brønsted base}}{Al(OH)_3(s)} + \underset{\text{Brønsted acid}}{3\ H_3O^+(aq)} \longrightarrow Al^{3+}(aq) + 6\ H_2O(\ell)$$

Many metal ions also form Lewis acid–base complexes with the Lewis base ammonia. For example, silver ion readily forms a water-soluble, colorless complex ion in aqueous ammonia. Indeed, this complex ion is so stable that the insoluble compound AgCl can be dissolved in aqueous ammonia (Figure 16.5).

$$AgCl(s) + 2\ NH_3(aq) \longrightarrow [H_3N{-}Ag{-}NH_3]^+(aq) + Cl^-(aq)$$

Light blue aqueous copper(II) ions also react with ammonia to produce a beautiful, deep blue complex ion with four ammonia molecules surrounding each metal ion (Figure 17.20).

$$Cu^{2+}(aq) + 4\ NH_3(aq) \longrightarrow [Cu(NH_3)_4]^{2+}(aq)$$

Molecular Lewis Acids

Lewis's acid–base concept accounts nicely for the fact that oxides of nonmetals behave as acids (Section 5.3). Carbon dioxide is an important example. Because oxygen is electronegative, the C—O bonds in CO_2 are polarized away from carbon and toward oxygen. This causes the carbon atom to be slightly deficient in electrons and positive. The negatively charged Lewis base OH^- can therefore attack the carbon to give, ultimately, the bicarbonate ion:

$$\ddot{O}=C=\ddot{O} \; (\delta^-, \delta^+, \delta^-) + :\ddot{O}-H^- \longrightarrow \ddot{O}=C(-\ddot{O}-H)(-:\ddot{O}:^-)$$

This reaction also accounts for the precipitation of $CaCO_3$ when CO_2 is bubbled into a solution of $Ca(OH)_2$ (Figure 16.2),

$$\underset{\text{Lewis base}}{Ca(OH)_2(s)} + \underset{\text{Lewis acid}}{CO_2(aq)} \longrightarrow CaCO_3(s) + H_2O(\ell)$$

a reaction that we can properly call a Lewis acid–base reaction.

Exercise 17.18 Lewis Acids and Bases

Tell if each of the following is a Lewis acid or a Lewis base:

1. PH_3
2. BCl_3
3. H_2S
4. HS^-

CHAPTER HIGHLIGHTS

Having studied this chapter, you should be able to

- Define and use the three main acid–base theories (Sections 17.1, 17.3, and 17.11).

Concept	Acid	Base
Arrhenius	source of H^+	source of OH^-
Brønsted	H^+ donor	H^+ acceptor
Lewis	electron-pair acceptor	electron-pair donor

- Recognize common **monoprotic** and **polyprotic** acids and bases and write balanced equations for their ionization in water (Section 17.3).
- Appreciate when a substance can be **amphiprotic** (Section 17.3).
- Recognize the **Brønsted acid** and base in a reaction and identify the conjugate partner of each (Section 17.3).
- Use Tables 17.3 and 17.4 to decide on the relative strengths of acids and bases.
- Write equations for acid–base reactions and decide whether they are product- or reactant-favored (Section 17.3 and Table 17.3).
- Understand the concept of **water autoionization** and its role in Brønsted acid–base chemistry (Sections 17.2 and 17.3).

- Identify common strong acids and bases (Section 17.4).
- Recognize some common weak acids and understand that they can be neutral molecules (e.g., acetic acid), cations (e.g., NH_4^+), hydrated metal ions (e.g., $Fe(H_2O)_6^{2+}$), or anions (e.g., HCO_3^-) (Section 17.5 and Table 17.4).
- Recognize some common weak bases and understand that they can be neutral molecules (e.g., NH_3) or anions (e.g., CO_3^{2-}) (Section 17.5 and Table 17.4).
- Calculate the pH of a solution from a knowledge of the hydronium ion or hydroxide ion concentration (Section 17.6).
- Use the pH of a solution to calculate the hydronium ion or hydroxide ion concentration (Section 17.6).
- Calculate the equilibrium constant for a weak acid (K_a) or weak base (K_b) from experimental information (such as pH, $[H_3O^+]$, or $[OH^-]$) (Section 17.7).
- Use the equilibrium constant and other information to calculate the concentrations of dissolved molecules and ions and the pH for a solution of a weak acid or weak base (Section 17.7).
- Describe the acid–base properties of salts and calculate the pH of a solution of a salt of a weak acid or of a weak base (Section 17.8).
- Calculate the pH of a solution of a polyprotic acid (Section 17.9).
- Appreciate the connection between the structure of a compound and its acidity or basicity (Section 17.10).
- Characterize a compound as a Lewis base (an electron-pair donor) or Lewis acid (an electron-pair acceptor) (Section 17.11).

STUDY QUESTIONS

REVIEW QUESTIONS

1. Outline the main ideas of the Arrhenius, Brønsted, and Lewis theories of acids and bases. How does the Lewis theory relate to the Brønsted theory?
2. Write a balanced equation depicting the autoionization of water. What is the experimental evidence for this reaction?
3. Write balanced chemical equations showing that phosphoric acid is a polyprotic acid.
4. Write balanced chemical equations showing that the hydrogen sulfate ion, HSO_4^-, is amphiprotic.
5. Identify the Brønsted acid and the base on the left side of each of the following equations, and designate the conjugate partner of each on the right side:

$$HNO_2(aq) + H_2O(\ell) \rightleftharpoons H_3O^+(aq) + NO_2^-(aq)$$
$$NH_4^+(aq) + CN^-(aq) \rightleftharpoons NH_3(aq) + HCN(aq)$$

6. Show that water can be a Brønsted base and a Lewis base.
7. If you have 0.1 M solutions of each of the following Brønsted acids, in which solution is the hydronium ion concentration larger? In which solution is the pH higher?

$$NH_4^+(aq) + H_2O(\ell) \rightleftharpoons H_3O^+(aq) + NH_3(aq) \quad K_a = 5.6 \times 10^{-10}$$
$$HCO_2H(aq) + H_2O(\ell) \rightleftharpoons H_3O^+(aq) + HCO_2^-(aq) \quad K_a = 1.8 \times 10^{-4}$$

8. If you have 0.1 M solutions of each of the following Brønsted bases, in which solution is the hydroxide ion concentration larger? In which solution is the pH higher?

$$CN^-(aq) + H_2O(\ell) \rightleftharpoons HCN(aq) + OH^-(aq) \quad K_b = 2.5 \times 10^{-5}$$
$$CH_3NH_2(aq) + H_2O(\ell) \rightleftharpoons CH_3NH_3^+(aq) + OH^-(aq) \quad K_b = 4.2 \times 10^{-4}$$

9. Write a balanced equation to show how hydrolysis of the $[Ni(H_2O)_6]^{2+}$ ion leads to an acidic solution.
10. Define the term "amphoteric." Give an example.
11. Which should be the stronger acid and why: H_2SeO_4 or H_2SeO_3?

NUMERICAL AND OTHER QUESTIONS

The Brønsted Concept of Acids and Bases

12. Write the formula and give the name of the conjugate base of each of the following acids:
(a) HCN (b) HSO_4^- (c) HF (d) HNO_2 (e) HCO_3^-

13. Write the formula, and give the name of the conjugate acid of each of the following bases:
(a) NH_3 (b) HCO_3^- (c) HS^- (d) Br^- (e) HSO_4^-

14. What are the products for each of the following acid–base reactions? Indicate the acid and its conjugate base and the base and its conjugate acid.
(a) $HNO_3 + H_2O \longrightarrow$
(b) $HSO_4^- + H_2O \longrightarrow$
(c) $H_3O^+ + F^- \longrightarrow$

15. What are the products for each of the following acid–base reactions? Indicate the acid and its conjugate base and the base and its conjugate acid.
(a) $HClO_4 + H_2O \longrightarrow$
(b) $NH_4^+ + H_2O \longrightarrow$
(c) $NH_2^- + H_2O \longrightarrow$
(d) $HCO_3^- + OH^- \longrightarrow$

16. Dissolving K_2CO_3 in water gives a basic solution. Write a balanced equation showing how the carbonate ion is responsible for this effect.

17. Dissolving ammonium bromide in water gives an acidic solution. Write a balanced equation showing how this can occur.

18. Write balanced equations showing how the HPO_4^{2-} ion of sodium monohydrogen phosphate, Na_2HPO_4, can be a Brønsted acid or a Brønsted base.

19. Write balanced equations showing how the hydrogen oxalate ion, $HC_2O_4^-$, can be both a Brønsted acid and a Brønsted base.

20. In each of the following acid–base reactions, identify the Brønsted acid and base on the left, and their conjugate partners on the right:
(a) $HCO_2H(aq) + H_2O(\ell) \rightleftharpoons HCO_2^-(aq) + H_3O^+(aq)$
(b) $H_2S(aq) + NH_3(aq) \rightleftharpoons HS^-(aq) + NH_4^+(aq)$
(c) $HSO_4^-(aq) + OH^-(aq) \rightleftharpoons SO_4^{2-}(aq) + H_2O(\ell)$

21. In each of the following acid–base reactions, identify the Brønsted acid and base on the left, and their conjugate partners on the right.
(a) $CH_3CO_2H(aq) + C_5H_5N(aq) \rightleftharpoons CH_3CO_2^-(aq) + C_5H_5NH^+(aq)$
(b) $N_2H_4(aq) + HSO_4^-(aq) \rightleftharpoons N_2H_5^+(aq) + SO_4^{2-}(aq)$
(c) $[Al(H_2O)_6]^{3+}(aq) + OH^-(aq) \rightleftharpoons [Al(H_2O)_5OH]^{2+}(aq) + H_2O(\ell)$

22. Several acids are listed here with their respective equilibrium constants:

$$HF(aq) + H_2O(\ell) \rightleftharpoons H_3O^+(aq) + F^-(aq) \quad K_a = 7.2 \times 10^{-4}$$

$$HPO_4^-(aq) + H_2O(\ell) \rightleftharpoons H_3O^+(aq) + PO_4^{3-}(aq) \quad K_a = 3.6 \times 10^{-13}$$

$$CH_3CO_2H(aq) + H_2O(\ell) \rightleftharpoons H_3O^+(aq) + CH_3CO_2^-(aq) \quad K_a = 1.8 \times 10^{-5}$$

(a) Which is the strongest acid? Which is the weakest?
(b) What is the conjugate base of the acid HF?
(c) Which acid has the weakest conjugate base?
(d) Which acid has the strongest conjugate base?

23. Several acids are listed here with their respective equilibrium constants:

$$C_6H_5OH(aq) + H_2O(\ell) \rightleftharpoons H_3O^+(aq) + C_6H_5O^-(aq) \quad K_a = 1.3 \times 10^{-10}$$

$$HCO_2H(aq) + H_2O(\ell) \rightleftharpoons H_3O^+(aq) + HCO_2^-(aq) \quad K_a = 1.8 \times 10^{-4}$$

$$HC_2O_4^-(aq) + H_2O(\ell) \rightleftharpoons H_3O^+(aq) + C_2O_4^{2-}(aq) \quad K_a = 6.4 \times 10^{-5}$$

(a) Which is the strongest acid? Which is the weakest?
(b) Which acid has the weakest conjugate base?
(c) Which acid has the strongest conjugate base?

24. Several bases are listed here with their respective K_b values:

$$NH_3(aq) + H_2O(\ell) \rightleftharpoons NH_4^+(aq) + OH^-(aq) \quad K_b = 1.8 \times 10^{-5}$$

$$C_5H_5N(aq) + H_2O(\ell) \rightleftharpoons C_5H_5NH^+(aq) + OH^-(aq) \quad K_b = 1.5 \times 10^{-9}$$

$$N_2H_4(aq) + H_2O(\ell) \rightleftharpoons N_2H_5^+(aq) + OH^-(aq) \quad K_b = 8.5 \times 10^{-7}$$

(a) Which is the strongest base? Which is the weakest?
(b) What is the conjugate acid of C_5H_5N?
(c) Which base has the strongest conjugate acid? Which has the weakest?

25. Several bases are listed here with their respective K_b values:

$$HS^-(aq) + H_2O(\ell) \rightleftharpoons H_2S(aq) + OH^-(aq) \quad K_b = 1 \times 10^{-7}$$

$$CN^-(aq) + H_2O(\ell) \rightleftharpoons HCN(aq) + OH^-(aq) \quad K_b = 2.5 \times 10^{-5}$$

$$NO_2^-(aq) + H_2O(\ell) \rightleftharpoons HNO_2(aq) + OH^-(aq) \quad K_b = 2.2 \times 10^{-11}$$

(a) Which is the strongest base? Which is the weakest?
(b) What is the conjugate acid of HS^-?
(c) Which base has the strongest conjugate acid? Which has the weakest?

26. State which of the following ions or compounds has the strongest conjugate base and briefly explain your choice:
(a) HSO_4^-
(b) CH_3CO_2H
(c) HOCl

27. Which of the following ions or compounds has the strongest conjugate acid? Briefly explain your choice.
(a) CN^-
(b) NH_3
(c) SO_4^{2-}

Writing Acid–Base Reactions

28. Ammonium chloride and sodium dihydrogen phosphate, NaH_2PO_4, are mixed in water. Using Table 17.3, write a balanced equation for the acid–base reaction that could, in principle, occur. Does the reaction occur to a significant extent?

29. Acetic acid, CH_3CO_2H, and sodium hydrogen carbonate, $NaHCO_3$, are mixed in water. Using Table 17.3, write a balanced equation for the acid–base reaction that could, in principle, occur. Does the reaction occur to any significant extent?

30. For each reaction below, predict whether the equilibrium lies predominantly to the left or to the right. Explain your prediction briefly.
(a) $NH_4^+(aq) + Br^-(aq) \rightleftharpoons NH_3(aq) + HBr(aq)$
(b) $HPO_4^{2-}(aq) + CH_3CO_2^-(aq) \rightleftharpoons PO_4^{3-}(aq) + CH_3CO_2H(aq)$
(c) $NH_2^-(aq) + H_2O(\ell) \rightleftharpoons NH_3(aq) + OH^-(aq)$
(d) $Fe(H_2O)_6^{3+}(aq) + HCO_3^-(aq) \rightleftharpoons Fe(H_2O)_5(OH)^{2+}(aq) + H_2CO_3(aq)$

31. For each reaction below, predict whether the equilibrium lies predominantly to the left or to the right. Explain your prediction briefly.
(a) $H_2S(aq) + CO_3^{2-}(aq) \rightleftharpoons HS^-(aq) + HCO_3^-(aq)$
(b) $HCN(aq) + SO_4^{2-}(aq) \rightleftharpoons CN^-(aq) + HSO_4^-(aq)$
(c) $CN^-(aq) + NH_3(aq) \rightleftharpoons HCN(aq) + NH_2^-(aq)$
(d) $SO_4^{2-}(aq) + CH_3CO_2H(aq) \rightleftharpoons HSO_4^-(aq) + CH_3CO_2^-(aq)$

pH Calculations

32. A certain table wine has a pH of 3.40. What is the hydronium ion concentration of the wine? Is it acidic or basic?

33. A saturated solution of milk of magnesia, $Mg(OH)_2$, has a pH of 10.5. What is the hydronium ion concentration of the solution? What is the hydroxide ion concentration? Is the solution acidic or basic?

34. What is the pH of a 0.0013 M solution of HNO_3? What is the hydroxide ion concentration of the solution?

35. What is the pH of a 1.2×10^{-4} M solution of KOH? What is the hydronium ion concentration of the solution?

36. What is the pH of a 0.0015 M solution of $Ca(OH)_2$?

37. The pH of a solution of $Ba(OH)_2$, a strong base, is 10.66 at 25 °C. What is the hydroxide ion concentration in the solution? If the solution volume is 125 mL, how many grams of $Ba(OH)_2$ must have been dissolved?

38. Make the following conversions. In each case, tell whether the solution is acidic or basic.

pH	$[H_3O^+]$	$[OH^-]$
(a) 1.00	______	______
(b) 10.50	______	______
(c) ______	1.3×10^{-5} M	______
(d) ______	______	2.3×10^{-4} M

39. Make the following conversions. In each case, tell whether the solution is acidic or basic.

pH	$[H_3O^+]$	$[OH^-]$
(a) ______	6.7×10^{-8} M	______
(b) ______	______	2.2×10^{-7} M
(c) 5.25	______	______
(d) ______	2.5×10^{-2} M	______

Using pH to Calculate Ionization Constants

40. A 2.5×10^{-3} M solution of an unknown acid has a pH of 3.80 at 25 °C.
(a) What is the hydronium ion concentration of the solution?
(b) Is the acid a strong acid, a moderately weak acid (K_a of about 10^{-5}), or a very weak acid (K_a of about 10^{-10})?

41. A 0.015 M solution of an unknown base has a pH of 10.09.
(a) What are the hydroxide and hydronium ion concentrations of this solution?
(b) Is the base a strong base, a moderately weak base (K_b of about 10^{-5}), or a very weak base (K_b of about 10^{-10})?

42. A 0.015 M solution of hydrogen cyanate, HOCN, has a pH of 2.67.
(a) What is the hydronium ion concentration in the solution?
(b) What is the ionization constant, K_a, for the acid?

43. A 0.10 M solution of chloroacetic acid, $ClCH_2CO_2H$, has a pH of 1.95. Calculate K_a for the acid.

44. A 0.025 M solution of hydroxylamine has a pH of 9.11. What is the value of K_b for this weak base?

$$H_2NOH(aq) + H_2O(\ell) \rightleftharpoons H_3NOH^+(aq) + OH^-(aq)$$

45. Methylamine, CH_3NH_2, is a weak base.

$$CH_3NH_2(aq) + H_2O(\ell) \rightleftharpoons CH_3NH_3^+(aq) + OH^-(aq)$$

If the pH of a 0.065 M solution of the amine is 11.70, what is its K_b?

Using Ionization Constants to Calculate pH

46. The ionization constant of a very weak acid HA is 4.0×10^{-9}. Calculate the equilibrium concentrations of H_3O^+, A^-, and HA in a 0.040 M solution of the acid. What is the pH of the solution?

47. What are the equilibrium concentrations of hydronium ion, acetate ion, and acetic acid in a 0.20 M aqueous solution of acetic acid? What is the pH of the solution?

48. If you have a 0.025 M solution of HCN, what are the equilibrium concentrations of H_3O^+, CN^-, and HCN? What is the pH of the solution?

49. Phenol, C_6H_5OH, is a weak organic acid:

$$C_6H_5OH(aq) + H_2O(\ell) \rightleftharpoons C_6H_5O^-(aq) + H_3O^+(aq) \quad K_a = 1.3 \times 10^{-10}$$

Although somewhat toxic to humans, it is used as a disinfectant and in the manufacture of plastics. If you dissolve 0.195 g of the acid in enough water to make 125 mL of solution, what is the equilibrium hydronium ion concentration? What is the pH of the solution?

50. The hydrogen phthalate ion, $C_8H_5O_4^-$, is a weak acid with $K_a = 2.0 \times 10^{-7}$:

$$C_8H_5O_4^-(aq) + H_2O(\ell) \rightleftharpoons C_8H_4O_4^{2-}(aq) + H_3O^+(aq)$$

If you dissolve 1.28 g of potassium hydrogen phthalate, $KC_8H_5O_4$, in enough water to make 125 mL of solution, what is the pH of the solution?

51. Propanoic acid, $CH_3CH_2CO_2H$, ionizes in water according to the equation

$$CH_3CH_2CO_2H(aq) + H_2O(\ell) \rightleftharpoons CH_3CH_2CO_2^-(aq) + H_3O^+(aq)$$

If you dissolve 0.294 g of the acid in enough water to make 125 mL of solution, what is the equilibrium hydronium ion concentration? What is the pH of the solution?

52. Benzoic acid, $C_6H_5CO_2H$, has a K_a of 6.3×10^{-5}, while that of a derivative of this acid, 4-chlorobenzoic acid ($ClC_6H_4CO_2H$), is 1.0×10^{-4}.
(a) Which is the stronger acid?
(b) For a 0.010 M solution of each of these monoprotic acids, which will have the higher pH?

53. Place the following acids in order of (i) increasing strength and (ii) increasing pH. Assume you have a 0.10 M solution of each acid.
(a) 4-chlorobenzoic acid, $ClC_6H_4CO_2H$ $K_a = 1.0 \times 10^{-4}$
(b) Bromoacetic acid, $BrCH_2CO_2H$ $K_a = 1.3 \times 10^{-3}$
(c) Trimethylammonium ion, $(CH_3)_3NH^+$ $K_a = 1.6 \times 10^{-10}$

54. What are the equilibrium concentrations of NH_3, NH_4^+, and OH^- in a 0.15 M solution of aqueous ammonia? What is the pH of the solution?

55. A hypothetical weak base has $K_b = 5.0 \times 10^{-4}$. Calculate the equilibrium concentrations of the base, its conjugate acid, and OH^- in a 0.15 M solution of the base.

56. The weak base methylamine, CH_3NH_2, has $K_b = 4.2 \times 10^{-4}$. It reacts with water according to the equation

$$CH_3NH_2(aq) + H_2O(\ell) \rightleftharpoons CH_3NH_3^+(aq) + OH^-(aq)$$

Calculate the equilibrium hydroxide ion concentration in a 0.25 M solution of the base. What are the pH and pOH of the solution?

57. Calculate the pH of a 0.12 M aqueous solution of the base aniline, $C_6H_5NH_2$ ($K_b = 4.0 \times 10^{-10}$).

$$C_6H_5NH_2(aq) + H_2O(\ell) \rightleftharpoons C_6H_5NH_3^+(aq) + OH^-(aq)$$

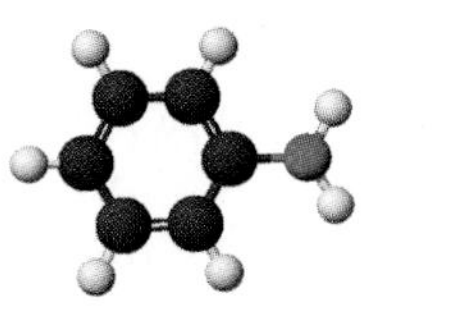
aniline

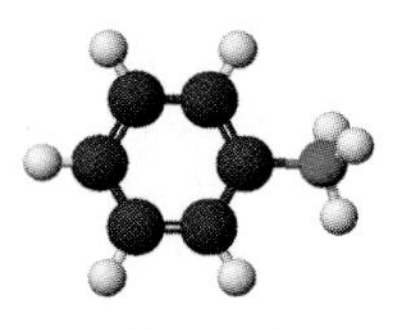
anilinium ion

58. Calculate the pH of a 0.0010 M aqueous solution of HF.

59. A solution of hydrofluoric acid, HF, has a pH of 2.30. Calculate the equilibrium concentrations of HF, F^-, and H_3O^+, and calculate the amount of HF originally dissolved per liter.

Acid-Base Properties of Salts

60. If each of the salts listed here were dissolved in water to give a 0.10 M solution, which solution would have the highest pH? Which would have the lowest pH?
(a) Na_2S (d) NaF
(b) Na_3PO_4 (e) $NaCH_3CO_2$
(c) NaH_2PO_4 (f) $AlCl_3$

61. Which of the following common food additives will give a basic solution when dissolved in water?
(a) $NaNO_3$ (used as a meat preservative)
(b) $NaC_6H_5CO_2$ (sodium benzoate; used as a soft drink preservative)
(c) Na_2HPO_4 (used as an emulsifier in the manufacture of pasteurized Kraft cheese)

62. Calculate the hydronium ion concentration and pH in a 0.20 M solution of the salt ammonium chloride, NH_4Cl.

63. Calculate the hydronium ion concentration and pH for a 0.015 M solution of the salt sodium acetate, $NaCH_3CO_2$.

64. Sodium cyanide is the salt of the weak acid HCN. Calculate the concentration of H_3O^+, OH^-, HCN, and Na^+ in a solution prepared by dissolving 10.8 g of NaCN in enough water to make 5.00×10^2 mL of solution at 25 °C.

65. The sodium salt of propanoic acid, $NaCH_3CH_2CO_2$, is used as an antifungus agent by veterinarians. Calculate the equilibrium concentrations of H_3O^+ and OH^-, and the pH, for a solution of 0.10 M $NaCH_3CH_2CO_2$. (See Table 17.4 for the K_b value of the $CH_3CH_2CO_2^-$ ion.)

Conjugate Acid–Base Pairs

66. The organic base aniline forms its conjugate acid, the anilinium ion, when treated with HCl. (See Study Question 57.)

$$\underset{\text{aniline}}{C_6H_5NH_2(aq)} + HCl(aq) \longrightarrow \underset{\text{anilinium ion}}{C_6H_5NH_3^+(aq)} + Cl^-(aq)$$

(a) If K_b for aniline is 4.0×10^{-10}, what is K_a for the anilinium ion?
(b) What is the pH of a 0.080 M solution of anilinium hydrochloride, $C_6H_5NH_3^+Cl^-$?

67. The local anesthetic Novocaine is the hydrogen chloride salt of an organic base, procaine:

$$C_{13}H_{20}N_2O_2(aq) + HCl(aq) \longrightarrow [HC_{13}H_{20}N_2O_2]^+ Cl^-(aq)$$

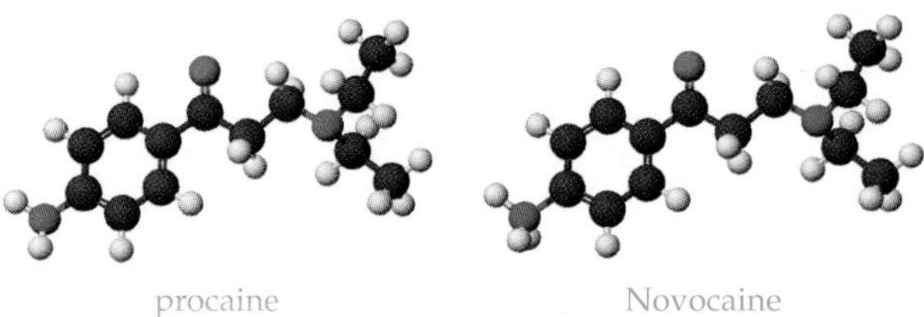
procaine Novocaine

(a) What is K_b for procaine if K_a for Novocaine is 1.4×10^{-9}?
(b) What is the pH of a 0.0015 M solution of Novocaine?

68. Saccharin ($HC_7H_4NO_3S$) is a weak acid with $K_a = 2.1 \times 10^{-12}$ at 25 °C. It is used in the form of sodium saccharide, $NaC_7H_4NO_3S$. What is the pH of a 0.10 M solution of sodium saccharide at 25 °C?

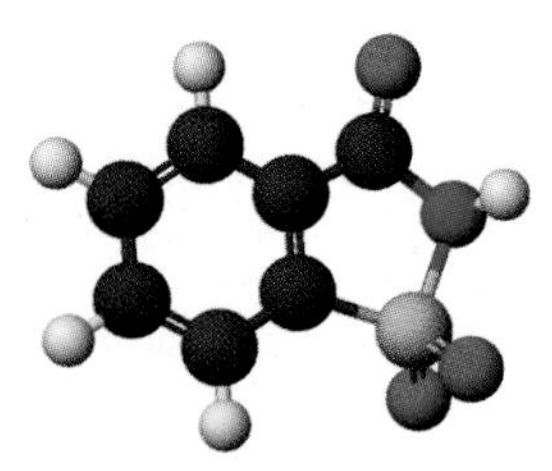
saccharin

69. Pyridine is a weak organic base and readily forms a salt with hydrochloric acid. (See Example 17.8.)

$$C_5H_5N(aq) + HCl(aq) \longrightarrow C_5H_5NH^+(aq) + Cl^-(aq)$$

pyridine — pyridinium ion

What is the pH of a 0.025 M solution of pyridinium hydrochloride, $[C_5H_5NH^+]\ Cl^-$?

Polyprotic Acids and Bases

70. Sulfurous acid, H_2SO_3, is a weak acid capable of providing two H^+ ions.
(a) What is the pH of a 0.45 M solution of H_2SO_3?
(b) What is the equilibrium concentration of the sulfite ion, SO_3^{2-}, in the 0.45 M solution of H_2SO_3?

71. Ascorbic acid (vitamin C, $C_6H_8O_6$) is a diprotic acid ($K_{a1} = 6.8 \times 10^{-5}$ and $K_{a2} = 2.7 \times 10^{-12}$). What is the pH of a solution that contains 5.0 mg of acid per milliliter of solution?

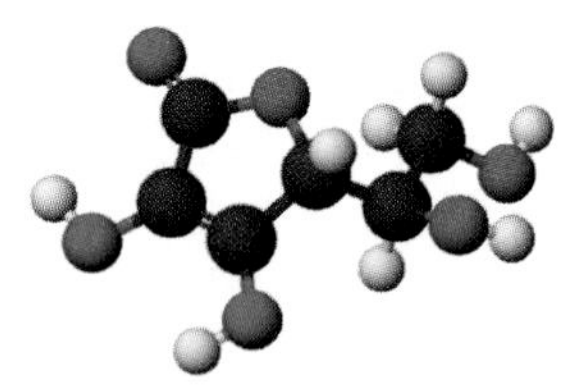
vitamin C

72. Hydrazine, N_2H_4, can interact with water in two steps:

$$N_2H_4(aq) + H_2O(\ell) \rightleftharpoons N_2H_5^+(aq) + OH^-(aq) \quad K_{b1} = 8.5 \times 10^{-7}$$

$$N_2H_5^+(aq) + H_2O(\ell) \rightleftharpoons N_2H_6^{2+}(aq) + OH^-(aq) \quad K_{b2} = 8.9 \times 10^{-16}$$

(a) What are the concentrations of OH^-, $N_2H_5^+$, and $N_2H_6^{2+}$ in a 0.010 M aqueous solution of hydrazine?
(b) What is the pH of the 0.010 M solution of hydrazine?

73. Ethylenediamine, $H_2NCH_2CH_2NH_2$, can interact with water in two steps, forming OH^- in each step (see Appendix I). If you have a 0.15 M aqueous solution of the amine, calculate the concentrations of $[H_3NCH_2CH_2NH_3]^{2+}$ and OH^-.

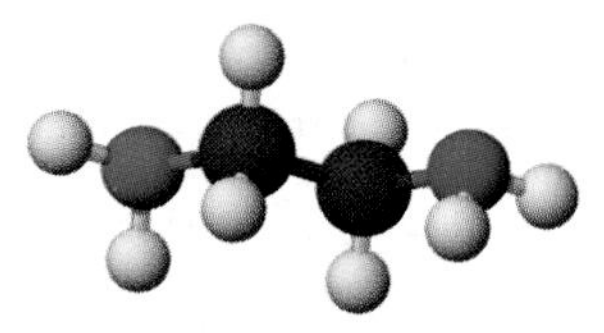
ethylenediamine

Lewis Acids and Bases

74. Decide if each of the following substances should be classified as a Lewis acid or base:
(a) Mn^{2+}
(b) CH_3NH_2
(c) H_2NOH in the reaction:
$H_2NOH(aq) + HCl(aq) \longrightarrow [H_3NOH]Cl(aq)$
(d) SO_2 in the reaction:
$SO_2(g) + BF_3(g) \rightleftharpoons O_2S \rightarrow BF_3(s)$
(e) $Zn(OH)_2$ in the reaction:
$Zn(OH)_2(s) + 2\ OH^-(aq) \rightleftharpoons Zn(OH)_4^{2-}(aq)$

75. Decide if each of the following substances should be classified as a Lewis acid or a Lewis base:
(a) BCl_3
(b) H_2NNH_2, hydrazine
(c) CN^- in the reaction:
$Au^+(aq) + 2\ CN^-(aq) \rightleftharpoons [Au(CN)_2]^-(aq)$
(d) Fe^{3+}

76. Trimethylamine, $(CH_3)_3N$, is a common reagent. It interacts readily with diborane gas, B_2H_6. The latter dissociates to BH_3, and this forms a complex with the amine, $(CH_3)_3N \rightarrow BH_3$. Is the BH_3 fragment a Lewis acid or a Lewis base?

77. Carbon monoxide forms complexes with low-valent metals. For example, $Ni(CO)_4$ and $Fe(CO)_5$ are well known. CO also forms complexes with the iron(II) ion in hemoglobin, which prevents the hemoglobin from taking up oxygen. Is CO a Lewis acid or a Lewis base?

78. Draw a Lewis dot structure of ICl_3. Does it function as a Lewis acid or base in reacting with Cl^- to form ICl_4^-? What are the likely structures of ICl_3 and ICl_4^-? What are the hybridizations used by I in ICl_3 and ICl_4^-?

79. CO_2 reacts with the oxide ion, O^{2-}, to form the carbonate ion. Show that the following reaction is a typical Lewis acid–base reaction:

$$CO_2 + O^{2-} \longrightarrow CO_3^{2-}$$

GENERAL PROBLEMS

80. If NaH, a salt of the hydride ion (H^-), is put in water, it reacts explosively with water to form $H_2(g)$ (see Figure 17.3):

$$H^- + H_2O \longrightarrow H_2 + OH^-$$

Specify the Brønsted acid and Brønsted base in this reaction. Is the resulting aqueous solution acidic or basic?

81. Liquid ammonia autoionizes just as water does.
(a) Write a balanced equation for the autoionization process of liquid ammonia.
(b) What is the conjugate acid of NH_3? The conjugate base?
(c) If $NaNH_2$ is dissolved in liquid ammonia, is it an acid or a base?

82. About this time, you may be wishing you had an aspirin. Aspirin is an organic acid (page 819) with a K_a of 3.27×10^{-4} for the reaction

$$HC_9H_7O_4(aq) + H_2O(\ell) \rightleftharpoons C_9H_7O_4^-(aq) + H_3O^+(aq)$$

If you have two tablets, each having 0.325 g of aspirin (mixed with a neutral "binder" to hold the tablet together) and dissolve them in a glass of water (225 mL), what is the pH of the solution?

83. Consider the following ions: CO_3^{2-}, Br^-, S^{2-}, and ClO_4^-.
   (a) Which of these anions gives a basic solution in water?
   (b) Which one is the strongest base?
   (c) Write a chemical equation for the reaction of each basic anion with water.

84. Hydrogen sulfide, H_2S, and sodium acetate, $NaCH_3CO_2$, are mixed in water. Using Table 17.3, write a balanced equation for the acid–base reaction that could, in principle, occur. Does the reaction occur to any significant extent?

85. For each reaction below, predict whether the equilibrium lies predominantly to the left or to the right. Explain your prediction briefly.
   (a) $HCN(aq) + CO_3^{2-}(aq) \rightleftharpoons CN^-(aq) + HCO_3^-(aq)$
   (b) $HCO_3^-(aq) + SO_4^{2-}(aq) \rightleftharpoons CO_3^{2-}(aq) + HSO_4^-(aq)$
   (c) $HSO_4^-(aq) + CH_3CO_2^-(aq) \rightleftharpoons SO_4^{2-}(aq) + CH_3CO_2H(aq)$
   (d) $Co(H_2O)_6^{2+}(aq) + HCO_3^-(aq) \rightleftharpoons Co(H_2O)_5(OH)^+(aq) + H_2CO_3(aq)$

86. The base $Ca(OH)_2$ is almost insoluble in water; only 0.50 g dissolves in 1.0 L of water at 25 °C. If the dissolved substance is completely dissociated into its constituent ions, what is the pH of a saturated solution?

87. A monoprotic acid HX has $K_a = 1.3 \times 10^{-3}$. Calculate the equilibrium concentration of HX and H_3O^+ and the pH for a 0.010 M solution of the acid.

88. *m*-Nitrophenol, a weak acid, can be used as a pH indicator because it is yellow at a pH above 8.6 and colorless at a pH below 6.8. If the pH of a 0.010 M solution of the compound is 3.44, calculate the K_a of the compound.

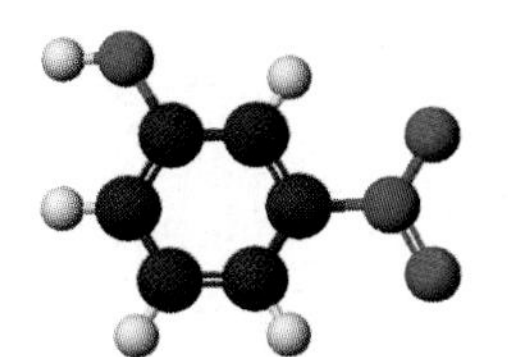

m-nitrophenol

89. The ionization constant for water at body temperature (37 °C), K_w, is 2.5×10^{-14}. What are the concentrations of H_3O^+ and OH^- at this temperature? How do these differ from the concentrations at 25 °C?

90. The butylammonium ion, $C_4H_9NH_3^+$, has a K_a of 2.3×10^{-11}:

$$C_4H_9NH_3^+(aq) + H_2O(\ell) \rightleftharpoons H_3O^+(aq) + C_4H_9NH_2(aq)$$

   (a) Calculate K_b for the conjugate base, $C_4H_9NH_2$ (butylamine).
   (b) Place the butylammonium ion and its conjugate base in Table 17.4. Name an acid weaker than $C_4H_9NH_3^+$ and a base stronger than $C_4H_9NH_2$.
   (c) What is the pH of a 0.015 M solution of the butylammonium ion?

91. The base ethylamine ($CH_3CH_2NH_2$) has a K_b of 4.3×10^{-4}. A closely related base, ethanolamine ($HOCH_2CH_2NH_2$), has a K_b of 3.2×10^{-5}.
   (a) If you have a 0.10 M solution of each, which has the higher pH?
   (b) Which of the two bases is stronger? Calculate the pH of its 0.10 M solution.

92. For each of the following salts, predict whether an aqueous solution has a pH less than, equal to, or greater than 7:

(a) $NaHSO_4$	(d) Na_2CO_3	(g) Na_2HPO_4
(b) NH_4Br	(e) $(NH_4)_2S$	(h) LiBr
(c) $KClO_4$	(f) $NaNO_3$	(i) $FeCl_3$

93. Chloroacetic acid, $ClCH_2CO_2H$, is a moderately weak acid ($K_a = 1.40 \times 10^{-3}$). If you dissolve 94.5 mg of the acid in water, giving a total volume of 125 mL, what is the pH of the solution?

94. Hydroxylamine, NH_2OH, has a K_b of 6.6×10^{-9}. What are the pH and pOH of a 0.051 M solution of the base?

95. To what volume should 1.00×10^2 mL of any weak acid HA with a concentration 0.20 M be diluted in order to double the percentage dissociation?

96. Identify each of the following solutions as acidic, base, or neutral.

(a) 0.1 M NH_3	(e) 0.1 M NH_4Cl
(b) 0.1 M Na_2CO_3	(f) 0.1 M $NaCH_3CO_2$
(c) 0.1 M NaCl	(g) 0.1 M $NH_4CH_3CO_2$
(d) 0.1 M CH_3CO_2H	

97. Arrange the following 0.1-M solutions in order of increasing pH:

(a) NaCl	(d) $NaCH_3CO_2$
(b) NH_4Cl	(e) KOH
(c) HCl	

98. Arrange the following 1.0-M solutions in order of increasing pH:

(a) NaCl	(d) HCl
(b) NH_3	(e) NaOH
(c) NaCN	(f) CH_3CO_2H

99. Nicotinic acid, $C_6H_5NO_2$, is found in minute amounts in all living cells, but appreciable amounts occur in liver, yeast, milk, adrenal glands, white meat, and corn. Whole wheat flour contains about 60. μg/g of flour. One gram of the acid dissolves in 60. mL of water and gives a pH of 2.70. What is the approximate value of K_a for the acid?

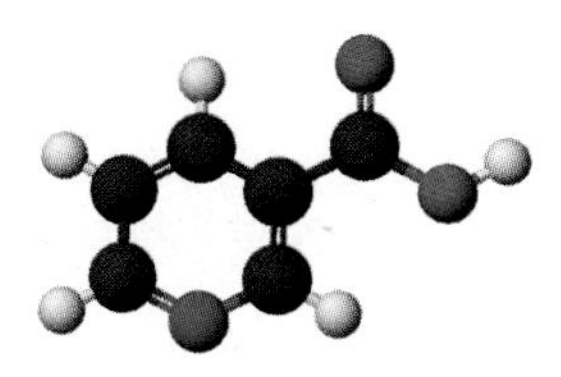

nicotinic acid

100. Oxalic acid is a relatively weak diprotic acid capable of losing two protons. Calculate the equilibrium constant for the overall reaction from K_{a1} and K_{a2} (see Appendix H for the required K_a values):

$$H_2C_2O_4(aq) + 2\ H_2O(\ell) \rightleftharpoons C_2O_4^{2-}(aq) + 2\ H_3O^+(aq)$$

101. Nicotine, $C_{10}H_{14}N_2$, has two basic nitrogen atoms (page 508), and both can react with water to give a basic solution:

$$Nic(aq) + H_2O(\ell) \rightleftharpoons NicH^+(aq) + OH^-(aq)$$
$$NicH^+(aq) + H_2O(\ell) \rightleftharpoons NicH_2^{2+}(aq) + OH^-(aq)$$

K_{b1} is 7.0×10^{-7} and K_{b2} is 1.1×10^{-10}. Calculate the approximate pH of 0.020 M solution.

102. Iodine, I_2, is much more soluble in an aqueous solution of potassium iodide, KI, than it is in pure water. The anion found in solution is I_3^-. Write an equation for this reaction, indicating the Lewis acid and Lewis base.

103. Sulfur trioxide reacts with O^{2-} to form the sulfate ion and with S^{2-} to form the thiosulfate ion, $S_2O_3^{2-}$. Each can be considered the reaction of Lewis acid and Lewis base.
   (a) Write balanced equations to depict the formation of sulfate and thiosulfate ions.
   (b) Write electron dot structures for each species, and show which act as Lewis acids and which as Lewis bases.

CONCEPTUAL QUESTIONS

104. Acetic acid is a weak Brønsted acid. It is interesting to compare the strength of this acid with a related series of acids where the H atoms of the CH_3 group in acetic acid are replaced by Cl.

Acid	K_a
CH_3CO_2H	1.8×10^{-5}
$ClCH_2CO_2H$	1.4×10^{-3}
Cl_2CHCO_2H	3.3×10^{-2}
Cl_3CCO_2H	2.0×10^{-1}

   (a) What trend in acid strength do you observe as H is successively replaced by Cl? Can you suggest a reason for this trend?
   (b) Suppose each of the acids in the table was present as a 0.10 M aqueous solution. Which would have the highest pH? The lowest pH?

105. The acid-base indicator phenolphthalein is a weak Brønsted acid (see Figure 5.9 and page 811):

$$\underset{\text{phenolphthalein colorless}}{C_{20}H_{14}O_4(aq)} + 2\ H_2O(\ell) \rightleftharpoons 2\ H_3O^+(aq) + \underset{\text{conjugate base pink}}{C_{20}H_{12}O_4^{2-}(aq)}$$

Using Le Chatelier's principle, explain why the dye exists as a colorless molecule in acidic solution, while it is deep red in basic solution.

106. Perchloric acid behaves as an acid even when it is dissolved in 100% sulfuric acid.
   (a) Write a balanced equation showing how perchloric acid can transfer a proton to sulfuric acid.
   (b) Draw a Lewis electron dot structure for sulfuric acid. How can sulfuric acid function as a base?

107. Discuss the validity of this statement: Strong acids leveled by water have conjugate bases weaker than water (see Section 17.4).

CHALLENGING QUESTIONS

108. A hydrogen atom in the organic base pyridine, C_5H_5N, can be substituted by various atoms or groups to give XC_5H_4N, where X is an atom such as Cl or a group such as CH_3. The following table gives K_a values for the conjugate acids of a variety of substituted pyridines:

$$XC_5H_4N(aq) + HCl(aq) \longrightarrow XC_5H_4NH^+(aq) + Cl^-(aq)$$

Substituted pyridine — Conjugate acid

Atom or Group X	**K_a of Conjugate Acid**
NO_2	5.9×10^{-2}
Cl	1.5×10^{-4}
H	6.8×10^{-6}
CH_3	1.0×10^{-6}

   (a) Suppose each conjugate acid is dissolved in sufficient water to give a 0.050 M solution. Which solution would have the highest pH? The lowest pH?
   (b) Which of the substituted pyridines is the strongest Brønsted base? Which one is the weakest Brønsted base?

109. Let us consider ammonium cyanide, a salt of a weak base and a weak acid. Both NH_4^+ and CN^- ion hydrolyze in aqueous solution, but the net reaction can be considered as a proton transfer from NH_4^+ to CN^-.

$$NH_4^+(aq) + CN^-(aq) \rightleftharpoons NH_3(aq) + HCN(aq)$$

   (a) Show that the equilibrium constant for this reaction, K_{total}, is

$$K_{total} = \frac{K_w}{K_a K_b}$$

   where K_a is the ionization constant for the weak acid HCN and K_b is the constant for the weak base NH_3.
   (b) Prove that the hydronium ion concentration in this solution must be given by

$$[H_3O^+] = \sqrt{\frac{K_w K_a}{K_b}}$$

(c) What is the pH of a 0.15 M solution of ammonium cyanide?

110. Tables of acid and base ionization constants are usually given in handbooks as pK_a values for weak acids (or for the corresponding conjugate acid of a weak base). The pK_a value is $-\log K_a$. The pK_a for the conjugate acid of caffeine ($C_8H_{10}N_4O_2$, page 802) is 10.4.

(a) What is the K_a value for the conjugate acid of caffeine, $C_8H_{11}N_4O_2$?

(b) What is K_b for caffeine, the free base?

111. Listed below are values of pK_a for some compounds or ions. (See Question 110.)

Acid	pK_a
Benzoic acid, $C_6H_5CO_2H$	4.20
Benzylammonium ion, $C_6H_5CH_2NH_3^+$	9.35
Chloroacetic acid, $ClCH_2CO_2H$	2.87
Conjugate acid of cocaine, $C_{17}H_{22}NO_4^+$	8.41
Thioacetic acid, $HSCH_2CO_2H$	3.33

(a) Which is the strongest acid?

(b) Which acid has the strongest conjugate base?

(c) List the acids in order of increasing strength.

Summary Questions

112. Equilibrium constants can be measured for the dissociation of Lewis acid–base complexes such as the dimethyl ether complex of BF_3, $(CH_3)_2O \rightarrow BF_3$. The value of K (here K_p) for the reaction is 0.17:

$$(CH_3)_2O{-}BF_3(g) \rightleftharpoons BF_3(g) + (CH_3)_2O(g)$$

(a) Tell which product is the Lewis acid and which is the Lewis base.

(b) What is the F—B—F angle in BF_3? In $(CH_3)_2O \rightarrow BF_3$?

(c) What is the hybridization of O in $(CH_3)_2O \rightarrow BF_3$? Of the boron atom?

(d) If you place 1.00 g of the complex in a 565-mL flask at 25 °C, what is the total pressure in the flask? What are the partial pressures of the Lewis acid, the Lewis base, and the complex?

113. Sulfanilic acid, which is used in making dyes, is made by reacting aniline with sulfuric acid:

$$H_2SO_4(aq) + \underset{\text{aniline}}{C_6H_5NH_2(aq)} \longrightarrow \underset{\text{sulfanilic acid}}{H_2N{-}C_6H_4{-}S(=O)(O){-}OH} + H_2O(\ell)$$

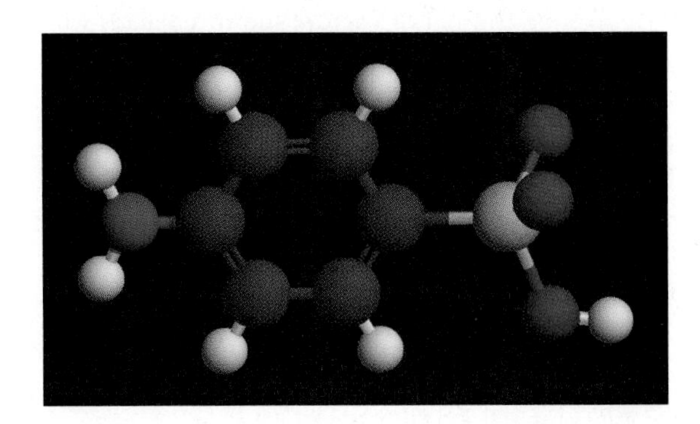

(a) If you want to prepare 150.0 g of sulfanilic acid, and you expect only an 85% yield, how many milliliters of aniline should you take as a starting material (density of aniline = 1.02 g/mL)?

(b) Give approximate values for the following bond angles in sulfanilic acid: C—N—H; H—N—H; O—S—C; and O—S—O.

(c) The acid has a K_a value of 5.9×10^{-4}. The sodium salt of the acid, $Na(H_2NC_6H_4SO_3)$, is quite soluble in water. If you dissolve 1.25 g of the salt in 125 mL of water, what is the pH of the solution?

INTERACTIVE GENERAL CHEMISTRY CD-ROM

Screen 17.2 Brønsted Acids and Bases

What is the difference between a Brønsted acid and a Brønsted base? How are the two related?

Screen 17.3 The Acid–Base Properties of Water

(a) Explain how water acts as both an acid and a base in its autoionization reaction.

(b) Write a balanced equation to show how ammonia could autoionize in liquid ammonia.

Screen 17.4 The pH Scale

How does a pH meter work? Do H^+ ions move in and out of the probe that is inserted into a solution?

Screen 17.5 Strong Acids and Bases

(a) What is the definition of a strong acid?

(b) What strong base is illustrated on this screen? What is its conjugate acid? Under ordinary conditions would you expect this acid to behave as a Brønsted acid?

Screen 17.6 Weak Acids and Bases

(a) What is the difference between a strong acid and a weak acid?
(b) The chemistry of aspirin and of digestion is described in the sidebar. Is aspirin a strong or weak acid? In what form does aspirin exist in your stomach, as the acid or its conjugate base?

Screen 17.7 Determining K_a and K_b Values

Why is trichloroacetic acid so much stronger an acid than acetic acid? (See the sidebar to this screen.)

Screen 17.8 Estimating the pH of Weak Acid Solutions

(a) Perform the experiment on this screen. What is the general effect of increasing acid concentration on solution pH?
(b) Determine the pH of each of the three acids when each is 0.003 M.
(c) Based on your pH measurements, rank the acids in order of increasing strength.

Screen 17.9 Estimating the pH of Weak Base Solutions

Measure the pH of solutions of the three bases listed here and then rank them in order of increasing base strength.

Screen 17.11 Lewis Acids and Bases

(a) What is the difference between a Lewis base and a Brønsted base?
(b) Why can BF_3 function as a Lewis acid?
(c) When BF_3 reacts with ammonia, what happens to the F—B—F angles in BF_3? What is the hybridization of the B atom in BF_3 before reaction? After reaction?

Screen 17.12 Cationic Lewis Acids

(a) Explain what happens as aqueous ammonia is added to a solution of Cu^{2+} ions. Why does $Cu(OH)_2$ dissolve in excess NH_3?
(b) Why do you asphyxiate if you breathe carbon monoxide? (See the sidebar to this screen.)

Screen 17.13 Puzzler

Why are most foods acidic? Why are most cleaners basic?

Chapter 18

Principles of Reactivity: Reactions Between Acids and Bases

(C. D. Winters)

A Chemical Puzzler

Because of severe anxiety, you can suffer from alkalosis. That is, the pH of your blood is too high. This can lead to serious medical problems such as muscle spasms and convulsions. One cure is to breathe into a paper bag. Knowing that the pH of your blood is maintained by the weak acid carbonic acid and its conjugate base, the hydrogen carbonate ion, why does this cure for alkalosis work?

Naturally occurring acid–base reactions. Volcanoes spew tons of hydrochloric acid, sulfur dioxide, and other acidic compounds into the earth's atmosphere where they react with natural bases such as carbonate ion (CO_3^{2-}). *(Pat and Tom Leeson/Photo Researchers, Inc.)*

Reactions of acids with bases are all around you and within you. The pH of the oceans and your blood is controlled by the chemistry of carbonic acid. Well over 50 million tons of fertilizers such as ammonium nitrate, made by reaction of the base ammonia with nitric acid, are manufactured every year (page 774). The acid in "acid rain" comes from the interaction of gaseous, nonmetal oxides, such as CO_2, SO_2, and nitrogen oxides, with water, and this acid rain falls to the earth where it is neutralized by reaction with minerals.

This chapter continues the exploration of acid–base chemistry with particular emphasis on the results of acid–base reactions, the control of such reactions, and some reactions of practical concern.

18.1 ACID–BASE REACTIONS

Using tables such as Table 17.3, the direction of an acid–base reaction can be predicted. For example, acetic acid (CH_3CO_2H) should react to a significant extent with ammonia in water:

$$CH_3CO_2H(aq) + NH_3(aq) \rightleftharpoons NH_4^+(aq) + CH_3CO_2^-(aq)$$

Remember that reactions always proceed in the direction of the weaker acid–base pair, and you find here that NH_4^+ is a weaker acid than CH_3CO_2H and that $CH_3CO_2^-$ is a weaker base than NH_3. We can verify this because the overall reaction is the sum of three other reactions whose equilibrium constants are known.

The formation of NH_4^+ and $CH_3CO_2^-$ is driven by coupling the $CH_3CO_2H + NH_3$ reaction with another, strongly driven reaction. This happens because some of the products of one reaction are reactants in the second reaction—in this case the formation of H_2O from H_3O^+ and OH^-—thereby removing these products as they are formed. These types of coupled systems are very common in living systems, such as in the role played by adenosine triphosphate (ATP) in our bodies.

Ionization of the acid:

$$CH_3CO_2H(aq) + H_2O(\ell) \rightleftharpoons H_3O^+(aq) + CH_3CO_2^-(aq) \qquad K_a = 1.8 \times 10^{-5}$$

Ionization of the base:

$$NH_3(aq) + H_2O(\ell) \rightleftharpoons NH_4^+(aq) + OH^-(aq) \qquad K_b = 1.8 \times 10^{-5}$$

Union of hydronium ion and hydroxide ion:

$$H_3O^+(aq) + OH^-(aq) \rightleftharpoons 2\ H_2O(\ell) \qquad K = \frac{1}{K_w} = 1.0 \times 10^{14}$$

Net reaction:

$$CH_3CO_2H(aq) + NH_3(aq) \rightleftharpoons NH_4^+(aq) + CH_3CO_2^-(aq)$$

$$K_{net} = \frac{K_a K_b}{K_w} = 3.2 \times 10^4$$

Recall that when equations are added to find a net equation, K_{net} is the product of the K values for each of the summed equations (see Section 16.1). In this case K_{net} is large because, although the acid and base are both weak, the H_3O^+ and OH^- ions that they produce are very strong and are "swept up" by the water formation reaction with its extraordinarily high equilibrium constant of 10^{14}. Thus, as suggested by Le Chatelier's principle, the acid and base ionization reactions shift much farther to the right than either would go if it occurred alone, and the overall process has a very large equilibrium constant. It is a product-favored reaction.

The reaction of acetic acid and ammonia is a good example of a weak acid reacting with a weak base, and so is the reaction of citric acid and bicarbonate ion in Figure 18.1. These reactions, however, are examples of only one of four possible types (Type 4) of acid–base reaction in aqueous solution.

Type of Acid–Base Reaction	Strength of Acid	Strength of Base
1	Strong	Strong
2	Strong	Weak
3	Weak	Strong
4	Weak	Weak

Figure 18.1 Reaction of an acid with a base. The bubbles coming from the tablet are carbon dioxide, which comes from the reaction of a weak Brønsted acid (citric acid) with a weak Brønsted base (HCO_3^-). The reaction is driven to completion by gas evolution. *(C. D. Winters)*

The outcome of each of these combinations is considered in turn.

The Reaction of a Strong Acid with a Strong Base

Strong acids and bases are 100% ionized in aqueous solution (see Section 17.4); therefore, if we mix HCl and NaOH, we can write the equation

$$H_3O^+(aq) + Cl^-(aq) + Na^+(aq) + OH^-(aq) \longrightarrow 2\,H_2O(\ell) + Na^+(aq) + Cl^-(aq)$$

which leads to the *net ionic equation*

$$H_3O^+(aq) + OH^-(aq) \longrightarrow 2\,H_2O(\ell) \qquad K = \frac{1}{K_w} = 1.0 \times 10^{14}$$

The net ionic equation for the reaction of any strong base with any strong acid is always simply the union of hydronium ion and hydroxide ion to give water (page 198). The enormous value of K for the reaction shows that it is, for all practical purposes, quantitatively complete. Thus, if equal numbers of moles of NaOH and HCl are mixed, the result is just a solution of NaCl in water. Because the constituents of NaCl, Na^+ and Cl^- ions, arise from a strong base and a strong acid, respectively, they produce a *neutral* aqueous solution (see Table 17.5). For this reason reactions of strong acids and bases have come to be called "neutralization reactions."

The Reaction of a Strong Acid with a Weak Base

In the reaction between HCl and ammonia we assume that HCl is 100% ionized in solution and that the H_3O^+ ion produced by the acid reacts with the OH^- ion from the weak base:

$$NH_3(aq) + H_2O(\ell) \rightleftharpoons NH_4^+(aq) + OH^-(aq) \qquad K_b = 1.8 \times 10^{-5}$$

$$H_3O^+(aq) + OH^-(aq) \rightleftharpoons 2\,H_2O(\ell) \qquad K = \frac{1}{K_w} = 1.0 \times 10^{14}$$

$$H_3O^+(aq) + NH_3(aq) \rightleftharpoons H_2O(\ell) + NH_4^+(aq) \qquad K_{net} = \frac{K_b}{K_w} = 1.8 \times 10^{9}$$

The overall equilibrium constant is quite large, and the reaction proceeds essentially to completion. After reaction of equal molar quantities of HCl and NH_3, the solution contains the salt ammonium chloride, NH_4Cl. The Cl^- ion is the conjugate base of the strong acid HCl and so has no effect on the solution pH (see Table 17.5). The NH_4^+ ion, however, is the conjugate acid of the weak base NH_3, so it produces an acidic solution:

$$NH_4^+(aq) + H_2O(\ell) \rightleftharpoons H_3O^+(aq) + NH_3(aq)$$

$$K_a \text{ for } NH_4^+ \text{ (from Table 17.4)} = 5.6 \times 10^{-10}$$

Two important conclusions can be drawn from this example.

- Mixing equal molar quantities of a strong acid and a weak base gives an acidic solution.
- The overall reaction is the reverse of the reaction that defines K_a for the conjugate acid of the weak base, so its K is just $1/K_a$ for the conjugate acid of the weak base in the reaction. Here $K_a = 5.6 \times 10^{-10}$, so $K_{net} = 1/(5.6 \times 10^{-10}) = 1.8 \times 10^9$.

The point at which the moles of OH^- supplied by a base is exactly equal to the moles of H_3O^+ supplied by the acid is called the *equivalence point*. See Section 5.9.

Example 18.1 pH at the Equivalence Point of a Strong Acid/Weak Base Reaction

Suppose you mix exactly 100 mL of 0.10 M HCl with exactly 50 mL of 0.20 M NH_3. What is the pH of the resulting solution?

Solution The balanced, net ionic equation for the HCl/NH_3 reaction is given in the text preceding this example:

$$H_3O^+(aq) + NH_3(aq) \rightleftharpoons H_2O(\ell) + NH_4^+(aq)$$

Ammonium ion is a weak Brønsted acid, which means that the pH of the solution depends on K_a for this ion and on its concentration.

Here we are mixing 0.010 mol of HCl (0.100 L × 0.10 M) with 0.010 mol of NH_3 (0.050 L × 0.20 M NH_3); 0.010 mol of NH_4Cl is produced. Because the solution volume after reaction is the total of the volumes of the reacting solutions, exactly 150 mL, the product ion concentrations are

$$[NH_4^+] = [Cl^-] = 0.010 \text{ mol}/0.150 \text{ L} = 0.067 \text{ M}$$

Equation	$H_3O^+(aq)$ +	$NH_3(aq)$ ⇌	$H_2O(\ell)$ +	$NH_4^+(aq)$
Concentrations before reaction (M)	0.10	0.20		0
Quantity before reaction	0.010 mol	0.010 mol		0
Quantity after reaction	0	0		0.010 mol
Concentrations after reaction (M)	0	0		0.067

Now we can solve for the hydronium ion concentration of the solution using the value of K_a for ionization of NH_4^+.

Equation	$NH_4^+(aq)$	+ $H_2O(\ell)$ ⇌	$H_3O^+(aq)$ +	$NH_3(aq)$
Concentrations before NH_4^+ hydrolysis (M)	0.067		0	0
Change in concentrations on proceeding to equilibrium	$-x$		$+x$	$+x$
Concentrations at equilibrium (M)	$(0.067 - x)$		x	x

Substituting the equilibrium concentrations into the equation for K_a, we have

$$K_a = 5.6 \times 10^{-10} = \frac{[NH_3][H_3O^+]}{[NH_4^+]} = \frac{(x)(x)}{0.067 - x}$$

Let us make the approximation that $(0.067 - x) \approx 0.067$, and so the hydronium ion concentration is

$$[H_3O^+] = 6.1 \times 10^{-6}\ M$$

and the pH is 5.21. The solution is clearly acidic, as anticipated for a solution containing the conjugate acid of a weak base.

Note that the approximation $(0.067 - x) \approx 0.067$ is valid. Subtracting 6.1×10^{-6} from 0.067 indeed makes no difference in the final answer.

Exercise 18.1 pH at the Equivalence Point of a Strong Acid/Weak Base Reaction

Aniline, $C_6H_5NH_2$, is a weak organic base first discovered in 1826. If you mix exactly 50 mL of 0.20 M HCl with 0.93 g of aniline, are the acid and base completely consumed? What is the pH of the resulting solution? (The K_a for $C_6H_5NH_3^+$ is 2.4×10^{-5}).

$$HCl(aq) + C_6H_5NH_2(aq) \rightleftharpoons C_6H_5NH_3^+(aq) + Cl^-(aq)$$

Aniline, $C_6H_5NH_2$, is a weak base that was the basis of the first synthetic dyes. Its main use at the present time is as a starting material in the synthesis of polymers. Here it is shown reacting with formaldehyde to produce a polymer. *(Photo, C. D. Winters; model, S. M. Young)*

The Reaction of a Weak Acid with a Strong Base

Consider the reaction of the weak acid formic acid, HCO_2H, with the strong base NaOH:

$$HCO_2H(aq) + H_2O(\ell) \rightleftharpoons H_3O^+(aq) + HCO_2^-(aq) \qquad K_a = 1.8 \times 10^{-4}$$

$$H_3O^+(aq) + OH^-(aq) \rightleftharpoons 2\ H_2O(\ell) \qquad K = \frac{1}{K_w} = 1.0 \times 10^{14}$$

$$HCO_2H(aq) + OH^-(aq) \rightleftharpoons H_2O(\ell) + HCO_2^-(aq) \qquad K_{net} = \frac{K_a}{K_w} = 1.8 \times 10^{10}$$

Once again the equilibrium constant for the overall process is large, so the net reaction goes essentially to completion. Assuming that equal molar quantities of acid and base were mixed, the final solution contains only sodium formate ($NaHCO_2$), a salt that is 100% dissociated in water. The Na^+ ion is the cation of a strong base and so gives a neutral solution. The formate ion, however, is the conjugate base of a weak acid (see Table 17.4). The solution is basic.

$$H_2O(\ell) + HCO_2^-(aq) \rightleftharpoons HCO_2H(aq) + OH^-(aq) \qquad K_b = 5.6 \times 10^{-11}$$

Again there are two important conclusions:

- Mixing equal molar quantities of a strong base with a weak acid produces a salt whose anion is the conjugate base of the weak acid. The solution is *basic*, with the pH depending on K_b for the anion.
- The net reaction between a strong base and a weak acid is the reverse of the reaction that defines K_b for the conjugate base of the weak acid. K for the net reaction is $1/K_b$ for the conjugate base of the weak acid (here $K_{net} = 1/K_b = 1/5.6 \times 10^{-11} = 1.8 \times 10^{10}$).

In Example 18.2 $[OH^-]$ is given as 0 before equilibrium is established. It is actually 10^{-7} M, but as described in Chapter 17, OH^- and H_3O^+ ions from water autoionization can generally be ignored. This type of simplification is made in the other examples in this chapter.

Example 18.2 pH at the Equivalence Point of a Strong Base/Weak Acid Reaction

Suppose you mix exactly 50 mL of 0.10 M NaOH with exactly 50 mL of 0.10 M formic acid, HCO_2H. What is the pH of the resulting solution?

Solution The net ionic equation for the acid–base reaction that occurs is

$$HCO_2H(aq) + OH^-(aq) \rightleftharpoons H_2O(\ell) + HCO_2^-(aq)$$

Here we mix 0.0050 mol of HCO_2H (0.050 L × 0.10 M) with 0.0050 mol of NaOH; 0.0050 mol of $NaHCO_2$ is produced. Because 50 mL of each reagent was mixed, $NaHCO_2$ is dissolved in 1.0×10^2 mL of water. The concentration of this salt is therefore $[NaHCO_2] = 0.050$ M.

With the salt concentration known, we can solve for the pH of the solution. From the discussion preceding this example we know the final solution is basic due to the hydrolysis of the conjugate base of formic acid.

Equation	$H_2O(\ell) + HCO_2^-(aq)$	$\rightleftharpoons HCO_2H(aq)$	$+ OH^-(aq)$
Concentrations before HCO_2^- hydrolysis (M)	0.050	0	0
Change in concentrations on proceeding to equilibrium	$-x$	$+x$	$+x$
Concentrations at equilibrium (M)	$(0.050 - x)$	x	x

We therefore can write

$$K_b = 5.6 \times 10^{-11} = \frac{[HCO_2H][OH^-]}{[HCO_2^-]} = \frac{(x)(x)}{0.050 - x}$$

Let us make the approximation that $(0.050 - x) \approx 0.050$, and so the hydroxide ion concentration is

$$[OH^-] = 1.7 \times 10^{-6} \text{ M}$$

This gives a H_3O^+ concentration of 6.0×10^{-9} M and a pH of 8.23. The solution is basic, as predicted.

As a final comment, note that the approximation that $(0.050 - x) \approx 0.050$ is valid. Subtracting 1.7×10^{-6} from 0.050 indeed makes no difference in the final answer.

Exercise 18.2 pH at the Equivalence Point of a Strong Base/Weak Acid Reaction

What volume of 0.100 M NaOH, in milliliters, is required to react completely with 0.976 g of the weak, monoprotic acid benzoic acid ($C_6H_5CO_2H$)? What is the pH of the solution after reaction? (See Table 17.4 for the K_b for the benzoate ion, $C_6H_5CO_2^-$.)

Problem-Solving Tips and Ideas

18.1 The pH of the Solution at the Equivalence Point of an Acid–Base Reaction

Finding the pH at the equivalence point for an acid–base reaction always involves three calculation steps. There are no shortcuts. Consider the reaction of a weak acid, HA, with a strong base as an example. (The same principles apply to other acid–base reactions.)

$$HA(aq) + OH^-(aq) \longrightarrow A^-(aq) + H_2O(\ell)$$

Step 1. *Stoichiometry problem.* At the equivalence point, the acid and base have been consumed completely to leave a solution of A^-, the conjugate base of the weak acid. Use the principles of stoichiometry to calculate (a) the quantity of base added, (b) the quantity of acid consumed, and (c) the quantity of conjugate base formed.

Step 2. *Calculate the concentration of conjugate base,* $[A^-]$. Recognize that the volume of the solution is the sum of the original volume of acid solution plus the volume of base solution added.

Step 3. *pH at the equivalence point.* Use the concentration of conjugate base from Step 2 and the value of K_b for the base (A^-) to calculate the concentration of OH^- in the solution:

$$H_2O(\ell) + A^-(aq) \rightleftharpoons HA(aq) + OH^-(aq) \qquad K = K_b \text{ for } A^-$$

The Reaction of a Weak Acid with a Weak Base

If acetic acid, a weak acid, is mixed with ammonia, a weak base, the following reaction occurs (see page 842).

$$CH_3CO_2H(aq) + NH_3(aq) \rightleftharpoons NH_4^+(aq) + CH_3CO_2^-(aq) \qquad K_{net} = 3.2 \times 10^4$$

If equal molar quantities of acid and base are mixed, the resulting solution contains only ammonium acetate, $NH_4CH_3CO_2$. Is this solution acidic or basic? Ammonium ion is the conjugate acid of a weak base, and so should contribute to the solution acidity:

$$NH_4^+(aq) + H_2O(\ell) \rightleftharpoons H_3O^+(aq) + NH_3(aq) \qquad K_a = 5.6 \times 10^{-10}$$

On the other hand, acetate ion ($CH_3CO_2^-$) is the conjugate base of a weak acid; it should make the solution basic:

$$CH_3CO_2^-(aq) + H_2O(\ell) \rightleftharpoons CH_3CO_2H(aq) + OH^-(aq) \qquad K_b = 5.6 \times 10^{-10}$$

In Section 17.8 you learned that the pH of a solution of the conjugate acid and base of a weak base and acid depends on the relative values of K_a and K_b. Here $K_a = K_b$. This tells us that equal quantities of H_3O^+ and OH^- are formed, so the solution is expected to be neutral. When equal molar amounts of a weak acid and a weak base react, the pH of the solution depends on the *relative K values of the conjugate base and acid.*

Table 18.1 • CHARACTERISTICS OF ACID–BASE REACTIONS

Type	Example	Net Ionic Equation	K	Species Present After Equal Molar Amounts Are Mixed; pH
Strong acid + strong base	$HCl + NaOH$	$H_3O^+(aq) + OH^-(aq) \rightleftharpoons 2\ H_2O(\ell)$	1.0×10^{14}	Cl^-, Na^+, pH = 7, neutral
Strong acid + weak base	$HCl + NH_3$	$H_3O^+(aq) + NH_3(aq) \rightleftharpoons NH_4^+(aq) + H_2O(\ell)$	1.8×10^{9}	Cl^-, NH_4^+, pH < 7, acidic
Weak acid + strong base	$HCO_2H + NaOH$	$HCO_2H(aq) + OH^-(aq) \rightleftharpoons HCO_2^-(aq) + H_2O(\ell)$	1.8×10^{10}	HCO_2^-, Na^+, pH > 7, basic
Weak acid + weak base	$HCO_2H + NH_3$	$HCO_2H(aq) + NH_3(aq) \rightleftharpoons HCO_2^-(aq) + NH_4^+(aq)$	3.2×10^{4}	HCO_2^-, NH_4^+, pH depends on K_b and K_a of conjugate base and acid

A Summary of Acid–Base Reactions

When writing an acid–base reaction, we first pay attention to whether the reactants are strong or weak. When one reactant is strong and the other one weak, *the pH of the solution after mixing equal molar amounts of acid and base is controlled by the conjugate partner of the weak acid or weak base.* This means the equilibrium constant for the acid–base reaction is the reciprocal of the K for ionization of the conjugate base or acid. These relationships are summarized in Table 18.1.

Exercise 18.3 pH of a Solution of a Salt of a Weak Base and Weak Acid

Suppose you mix exactly 50 mL of 0.10 M acetic acid and 50 mL of 0.10 M pyridine (C_5H_5N), a weak base. (K_a for the pyridinium ion, $C_5H_5NH^+$, is 6.7×10^{-6}, and K_b for acetate ion is 5.6×10^{-10}.) Is the solution acidic or basic?

18.2 THE COMMON ION EFFECT

One objective of this chapter is to follow the course of events during an acid–base titration. Before doing this, though, we have to explore the nature of solutions of acids and bases that contain another solute, in particular an ion that is "common" to the acid or base equilibrium.

Lactic acid is a weak acid found in sour milk, apples and other fruit, beer and wine, and several plants (see page 514):

$$H_3C{-}CH(OH){-}C({=}O){-}O{-}H(aq) + H_2O(\ell) \rightleftharpoons H_3O^+(aq) + H_3C{-}CH(OH){-}C({=}O){-}O^-(aq)$$

lactic acid ($C_3H_6O_3$)
$K_a = 1.3 \times 10^{-4}$

lactate ion ($C_3H_5O_3^-$)

Suppose a sample containing lactic acid is analyzed by titrating the acid with the strong base NaOH:

$$\underset{\text{lactic acid}}{H_3CHOHCO_2H(aq)} + OH^-(aq) \rightleftharpoons \underset{\text{lactate ion}}{CH_3CHOHCO_2^-(aq)} + H_2O(\ell)$$

At the equivalence point in the titration, the acid has been consumed and converted completely to its conjugate base, the lactate ion. *Before* the equivalence point, however, some lactate ion is present in solution along with as yet unreacted lactic acid. Thus, before the equivalence point, the *weak acid is present* along with *some amount of its conjugate base.* If we halt the titration at some intermediate stage before the equivalence point, we observe that the ionization of the remaining lactic acid is affected by the lactate ion present. This phenomenon is called the **common ion effect** because an ion (here the lactate ion) "common" to the ionization of the acid is present in an amount greater than that produced by simple acid ionization. In an acid ionization, the presence of the conjugate base in solution limits the extent to which the acid ionizes and thus affects the pH of the solution (Figure 18.2).

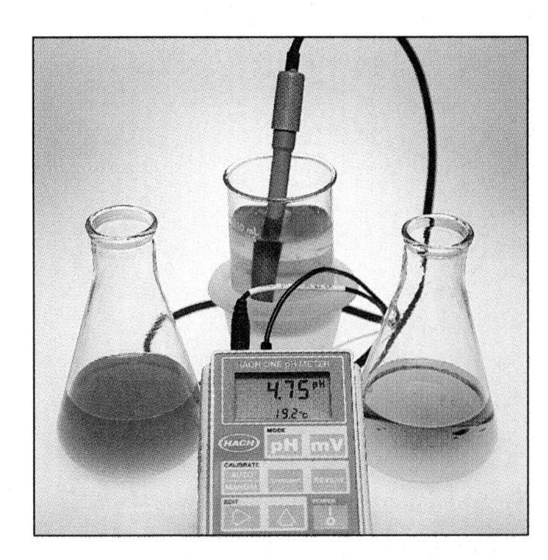

***Figure* 18.2 The common ion effect.** An acid (0.25 M acetic acid, *left*) is mixed with a base (0.10 M sodium acetate, *right*). The pH meter shows the resulting solution (*center*) has a lower hydronium ion concentration (a pH of about 5) than the acetic acid solution (pH is about 2.7). (Each solution contains universal indicator. This dye is red in low pH, yellow in slightly acidic media, and green in neutral to weakly basic media.) *(C. D. Winters)*

Example 18.3 The Common Ion Effect

What is the pH of 0.25 M aqueous acetic acid? What is the pH of the solution after adding sodium acetate to make the solution 0.10 M in the salt? (Ignore any change in volume on adding the sodium acetate.)

Solution Let us first determine the pH of the 0.25 M acid solution using the approach outlined in Chapter 17. Then we can turn to the effect of adding a "common ion."

Step 1. *Determine the pH of 0.25 M acetic acid.*

Equation	$CH_3CO_2H(aq) + H_2O(\ell) \rightleftharpoons$	$H_3O^+(aq)$	$+ CH_3CO_2^-(aq)$
Concentration before CH_3CO_2H ionizes (M)	0.25	0	0
Change in concentration on proceeding to equilibrium	$-x$	$+x$	$+x$
Concentration at equilibrium (M)	$0.25 - x$	x	x

The appropriate equilibrium expression is

$$K_a = 1.8 \times 10^{-5} = \frac{[CH_3CO_2^-][H_3O^+]}{[CH_3CO_2H]} = \frac{(x)(x)}{0.25 - x}$$

Let us make the following approximation: $(0.25 - x) \approx 0.25$, and so the hydronium ion concentration and pH are

$$[H_3O^+] = 2.1 \times 10^{-3}\ M \qquad pH = 2.67$$

(The approximation is valid because subtracting 2.1×10^{-3} from 0.25 indeed makes no difference in the final answer.)

Step 2. *Effect of added "common ion."* Sodium acetate, $NaCH_3CO_2$, is very soluble in water and is 100% dissociated into its ions, Na^+ and $CH_3CO_2^-$. As you learned in Chapter 17, Na^+ ion has no effect on the pH of a solution (see Table 17.5). On the other hand, $CH_3CO_2^-$ is the conjugate base of the weak acid CH_3CO_2H, so it should contribute some excess OH^- ions to the solution,

$$\underset{\text{acetate ion}}{CH_3CO_2^-(aq)} + H_2O(\ell) \rightleftharpoons \underset{\text{acetic acid}}{CH_3CO_2H(aq)} + OH^-(aq)$$

and thus reduce the hydronium ion concentration below 2.1×10^{-3} M, the concentration found if the solution contains only acetic acid.

We solve for the net hydronium ion concentration in the following way. First, imagine the solution contains only the weak acid (CH_3CO_2H, 0.25 M) and its conjugate base ($CH_3CO_2^-$, 0.10 M) and that neither has yet reacted with water to generate H_3O^+ or OH^-, respectively.

Concentrations in the acetic acid/sodium acetate mixture ***before*** *ionization of the acid:*

$$\underset{0.25\text{ M}}{CH_3CO_2H(aq)} + H_2O(\ell) \rightleftharpoons \underset{0}{H_3O^+(aq)} + \underset{0.10\text{ M}}{CH_3CO_2^-(aq)}$$

Now imagine that the acid ionizes to give H_3O^+ and $CH_3CO_2^-$, both in the amount y, in the presence of the added $NaCH_3CO_2$.

Concentrations in the acetic acid/sodium acetate mixture ***after*** *ionization of the acid:*

$$\underset{(0.25 - y)\text{ M}}{CH_3CO_2H(aq)} + H_2O(\ell) \rightleftharpoons \underset{y\text{ M}}{H_3O^+(aq)} + \underset{(0.10 + y)\text{ M}}{CH_3CO_2^-(aq)}$$

Without the added salt $NaCH_3CO_2$, which provides the "common ion" $CH_3CO_2^-$, ionization of the acid would have produced H_3O^+ and $CH_3CO_2^-$, both in the amount $x = 2.1 \times 10^{-3}$ M. We know from Le Chatelier's principle, however, that y must be less than x ($y < x$). The concentration table for the ionization of acetic acid in the presence of the "common ion" acetate ion is therefore

Equation	$CH_3CO_2H(aq) + H_2O(\ell) \rightleftharpoons$	$H_3O^+(aq)$ +	$CH_3CO_2^-(aq)$
Concentration before CH_3CO_2H ionizes (M)	0.25	0	0.10
Change in concentration on proceeding to equilibrium	$-y$	$+y$	$+y$
Concentration at equilibrium (M)	$(0.25 - y)$	y	$0.10 + y$

and the appropriate equilibrium expression is

$$K_a = 1.8 \times 10^{-5} = \frac{[H_3O^+][CH_3CO_2^-]}{[CH_3CO_2H]} = \frac{(y)(0.10 + y)}{0.25 - y}$$

We know that the amount of $CH_3CO_2^-$ formed by ionization in a 0.25 M solution of acetic acid is 0.0021 M, and Le Chatelier's principle predicts that the acid produces even less acetate ion in the presence of the "common ion." It is therefore reasonable to assume that $(0.10 + y)$ M ≈ 0.10 M and that $(0.25 - y)$ M ≈ 0.25 M. Solving the "approximate" expression, we find that

$$y = [H_3O^+] = [CH_3CO_2^-] \text{ added to solution by } CH_3CO_2H \text{ ionization}$$
$$\approx 4.5 \times 10^{-5} \text{ M}$$

$$pH = 4.35$$

This is indeed a very small value for $[H_3O^+]$ and for the concentration of $CH_3CO_2^-$ that is added to the solution by acetic acid ionization. Our approximations are reasonable.

Finally, compare the situations before and after adding acetate ion, the common ion. The pH before adding sodium acetate was 2.67. The acidity decreased (the pH increased to 4.35), however, after adding the weak base sodium acetate.

Exercise 18.4 The Common Ion Effect

Assume you have a 0.30 M solution of formic acid (HCO_2H) and add enough sodium formate ($NaHCO_2$) to make the solution 0.10 M in the salt. Calculate the pH of the formic acid solution before and after adding sodium formate.

Problem-Solving Tips and Ideas

18.2 Solving Common Ion Problems

The key idea of common ion problems is that the weak acid HA, for example, ionizes to produce *less* H_3O^+ in the presence of the common ion A^- than it would have in the absence of A^-. Thus, adding A^- to a solution of the weak acid HA causes the pH to be higher than in the absence of A^-. This is easy to remember because adding a base to any solution raises the pH of the solution.

18.3 BUFFER SOLUTIONS

The normal pH of human blood is 7.4. Experiment clearly shows that the addition of a small quantity of strong acid or base to blood, say 0.01 mol to a liter of blood, leads to a change in pH of only about 0.1 pH unit (Figure 18.3). In comparison, if you add 0.01 mol of HCl to 1.0 L of pure water, the pH drops from 7 to 2, whereas addition of 0.01 mL of NaOH increases the pH from 7 to 12. Blood, and many other body fluids, are said to be buffered; that is, its pH is resistant to change on addition of a strong acid or base.

In general, two species are required in a **buffer solution,** one (an acid) capable of reacting with added OH^- ions and another (a base) that can consume added H_3O^+ ions. An additional requirement is that the acid and base do not react with each another. This means that a buffer is usually prepared from roughly equal quantities of a *conjugate acid–base pair:* (a) a weak acid and its conjugate base (acetic acid and acetate ion, for example), or (b) a weak base and its conjugate acid (ammonia and ammonium ion, for example). Some systems commonly used in the laboratory are given in Table 18.2.

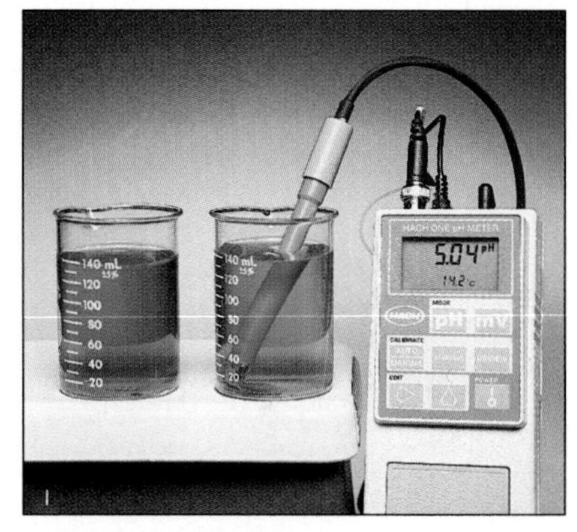

(a)

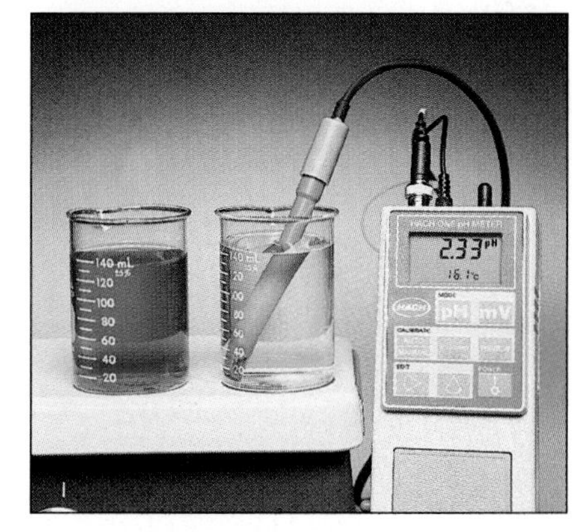

(b)

Figure 18.3 Buffer solutions. (a) The pH electrode is indicating the pH of water that contains a trace of acid (and bromphenol blue indicator; see Figure 17.11). The solution at the left is a buffer solution with a pH of about 7. (It also contains bromphenol blue dye.) (b) When 5 mL of 0.10 M HCl is added to each solution, the pH of the water drops several units, whereas the pH of the buffer stays constant as implied by the fact that the indicator color did not change. *(C. D. Winters)*

Table 18.2 • SOME COMMONLY USED BUFFER SYSTEMS

Weak Acid	Conjugate Base	Acid K_a	Useful pH Range
Phthalic acid $C_6H_4(CO_2H)_2$	Hydrogen phthalate ion $C_6H_4(CO_2H)(CO_2)^-$	1.3×10^{-3}	1.9–3.9
Acetic acid CH_3CO_2H	Acetate ion $CH_3CO_2^-$	1.8×10^{-5}	3.7–5.8
Dihydrogen phosphate ion $H_2PO_4^-$	Hydrogen phosphate ion HPO_4^{2-}	6.2×10^{-8}	6.2–8.2
Hydrogen phosphate ion HPO_4^{2-}	Phosphate ion PO_4^{3-}	3.6×10^{-13}	11.3–13.3

To see how a buffer works, let us consider an acetic acid/acetate ion buffer. Acetic acid, the weak acid, will consume any added hydroxide ion:

$$CH_3CO_2H(aq) + OH^-(aq) \longrightarrow CH_3CO_2^-(aq) + H_2O(\ell) \qquad K = 1.8 \times 10^9$$

The equilibrium constant for the reaction is very large because OH^- is a much stronger base than acetate, $CH_3CO_2^-$ (Table 17.4). This means that any OH^- entering the solution from an outside source is consumed completely. In a similar way, any hydronium ion added to the solution reacts with the acetate ion present in the buffer:

$$H_3O^+(aq) + CH_3CO_2^-(aq) \longrightarrow H_2O(\ell) + CH_3CO_2H(aq) \qquad K = 5.6 \times 10^4$$

The equilibrium constant for this reaction is also quite large because H_3O^+ is a much stronger acid than CH_3CO_2H.

Now that we have established that a buffer solution should effectively remove small amounts of added acid or base, work through the following example, which shows more quantitatively how the pH of a solution can be maintained.

Example 18.4 A Buffer Solution

Calculate the pH change that occurs when 1.00 mL of 1.0 M HCl is added to (1) 1.00 L of pure water and to (2) 1.00 L of acetic acid/sodium acetate buffer with $[CH_3CO_2H] = 0.700$ M and $[CH_3CO_2^-] = 0.600$ M.

Solution

Part 1. *Adding Acid to Pure Water.* 1.00 mL of 1.00 M HCl represents 0.00100 mol of acid. If this is added to 1.00 L of pure water, the hydrogen ion concentration is 1.00×10^{-3} M. This means the H_3O^+ concentration is raised from 10^{-7} to 10^{-3}, so the pH falls from 7 to 3.

Part 2. *pH of Acetic Acid/Acetate Ion Buffer Solution.* We first need to determine the pH of the buffer solution before any HCl is added. The balanced equation connecting the species in solution is just the ionization of acetic acid.

Equation	$CH_3CO_2H(aq)$	$+ H_2O(\ell) \rightleftarrows$	$H_3O^+(aq)$	$+ CH_3CO_2^-(aq)$
Concentrations before CH_3CO_2H ionizes (M)	0.700		0	0.600
Change in concentrations on proceeding to equilibrium	$-x$		$+x$	$+x$
Concentrations at equilibrium (M)	$0.700 - x$		x	$0.600 + x$

and the appropriate equilibrium expression is

$$K_a = 1.8 \times 10^{-5} = \frac{[H_3O^+][CH_3CO_2^-]}{[CH_3CO_2H]} = \frac{(x)(0.600 + x)}{0.700 - x}$$

The value of x is very small with respect to 0.700 or 0.600, so we can use the "approximate expression" to find x, the hydronium ion concentration. On rearranging the "approximate" expression, we have

$$[H_3O^+] = x = \frac{[CH_3CO_2H]}{[CH_3CO_2^-]} K_a \approx \frac{(0.700)}{(0.600)} (1.8 \times 10^{-5}) = 2.1 \times 10^{-5} \text{ M}$$

$$\text{pH} = 4.68$$

Part 3. *Add HCl to the Acetic Acid/Acetate Ion Buffer Solution.* HCl is a strong acid that is 100% ionized in water. Therefore, we are only concerned with the fact that it supplies H_3O^+ and reacts *completely* with the base in the solution according to the following equation:

$$H_3O^+(aq) + CH_3CO_2^-(aq) \rightleftarrows H_2O(\ell) + CH_3CO_2H(aq)$$

	H_3O^+ from added HCl	$CH_3CO_2^-$ from buffer	CH_3CO_2H from buffer
Moles before reaction	0.00100	0.600	0.700
Change in moles from complete reaction	−0.00100	−0.001	+0.001
Moles after reaction	0	0.599	0.701
Concentrations after reaction (M)	0	0.598	0.700

Because the added HCl reacts *completely* with acetate ion to produce acetic acid, the solution is once again a buffer containing only the weak acid and its salt. We only need to consider the equilibrium for the ionization of the weak acid in the presence of its common ion, $CH_3CO_2^-$, as in the previous part:

Equation	$CH_3CO_2H(aq)$	$+ H_2O(\ell) \rightleftarrows$	$H_3O^+(aq)$	$+ CH_3CO_2^-(aq)$
Concentrations before CH_3CO_2H ionizes (M)	0.700		0	0.598
Change in concentrations on proceeding to equilibrium	$-y$		$+y$	$+y$
Concentrations at equilibrium (M)	$(0.700 - y)$		y	$0.598 + y$

As usual, we make the approximation that y, the amount of H_3O^+ formed by ionizing acetic acid in the presence of acetate ion, is small compared with 0.700 M or 0.598 M. We can therefore write

$$[H_3O^+] = y = \frac{[CH_3CO_2H]}{[CH_3CO_2^-]} K_a \approx \frac{(0.700)}{(0.598)} (1.8 \times 10^{-5}) = 2.1 \times 10^{-5} \text{ M}$$

$$pH = 4.68$$

Within the number of significant figures allowed, the pH *does not change* in the buffer solution after adding HCl, even though it changed by 4 units when 1 mL of 1.00 M HCl was added to 1.00 L of pure water. Buffer solutions do indeed "buffer." In this case the acetate ion consumed the added hydronium ion, and the solution thus resisted a change in pH.

Exercise 18.5 Buffer Solutions

Calculate the pH of 0.500 L of a buffer solution composed of 0.50 M formic acid (HCO_2H) and 0.70 M sodium formate ($NaHCO_2$) before and after adding 10.0 mL of 1.0 M HCl.

General Expressions for Buffer Solutions

In Example 18.4 we solved for the hydronium ion concentration of the acetic acid/acetate ion buffer solution by rearranging the K_a expression to give

$$[H_3O^+] = \frac{[CH_3CO_2H]}{[CH_3CO_2^-]} K_a$$

This result can be generalized for any buffer. For a buffer solution based on a *weak acid and its conjugate base,* the hydronium ion concentration is given by

$$[H_3O^+] = \frac{[\text{acid}]}{[\text{conjugate base}]} K_a \qquad [18.1]$$

When the usual expression for K_b is rearranged, the hydroxide ion concentration in a buffer composed of a *weak base and its conjugate acid* (e.g., NH_3 and NH_4^+) is

$$[OH^-] = \frac{[\text{base}]}{[\text{conjugate acid}]} K_b \qquad [18.2]$$

The Henderson-Hasselbalch Equation

The general forms for calculating the hydronium ion or hydroxide ion concentrations of buffer solutions are useful in finding the pH of such solutions and in deciding how to prepare buffers. They are often written in a different form, however. If we take the negative logarithm of each side of the equation for $[H_3O^+]$, for example, we have

$$-\log [H_3O^+] = \left\{ -\log \frac{[\text{acid}]}{[\text{conjugate base}]} \right\} + (-\log K_a)$$

You know that $-\log [H_3O^+]$ is defined as pH, and from Chapter 17 you recall that the definition can be extended to other quantities. Thus, $-\log K_a$ is equivalent to pK_a. Furthermore, because

$$-\log \frac{[\text{acid}]}{[\text{conjugate base}]} = +\log \frac{[\text{conjugate base}]}{[\text{acid}]}$$

the preceding equation can be rewritten as

$$\text{pH} = pK_a + \log \frac{[\text{conjugate base}]}{[\text{acid}]} \qquad [18.3]$$

This equation, known as the **Henderson-Hasselbalch equation,** shows clearly that the pH of the solution of a weak acid and its conjugate base is controlled primarily by the strength of the acid (as expressed by pK_a). The "fine" control of the pH is then given by the relative amounts of conjugate base and acid. When the concentrations of conjugate base and acid are the same in a solution, the ratio of [conjugate base]/[acid] is 1. The log of 1 is zero, so pH = pK_a under these circumstances. If there is more of the conjugate base in the solution than acid, for example, then pH > pK_a.

The Henderson-Hasselbalch equation is valid when the ratio of [conjugate base]/[acid] is no larger than 10 and no smaller than 0.1. Many handbooks of chemistry list ionization constants in terms of pK_a values, so approximate pH values of possible buffer solutions are readily apparent.

Example 18.5 Using the Henderson-Hasselbalch Equation

You dissolve 2.00 g of benzoic acid ($C_6H_5CO_2H$) and 2.00 g of sodium benzoate ($NaC_6H_5CO_2$) in enough water to make 1.00 L of solution. Calculate the pH of the solution using the Henderson-Hasselbalch equation.

Solution Our first objective is to calculate the pK_a of the acid. Appendix H (and Table 17.4) gives K_a for benzoic acid as 6.3×10^{-5}. Therefore,

$$-\log (6.3 \times 10^{-5}) = pK_a = 4.20$$

Next, we need the concentrations of the acid (benzoic acid) and its conjugate base (benzoate ion):

$$2.00 \text{ g benzoic acid} \left(\frac{1 \text{ mol}}{122.1 \text{ g}}\right) = 0.0164 \text{ mol benzoic acid}$$

$$2.00 \text{ g sodium benzoate} \left(\frac{1 \text{ mol}}{144.1 \text{ g}}\right) = 0.0139 \text{ mol sodium benzoate}$$

Because the solution volume is 1.00 L, the concentrations are [benzoic acid] = 0.0164 M and [sodium benzoate] = 0.0139 M. Therefore,

$$\text{pH} = 4.20 + \log \frac{0.0139}{0.0164}$$

$$= 4.20 + \log (0.848) = 4.13$$

Notice that the pH is less than the pK_a because the ratio of conjugate base to acid concentration was less than 1.

Exercise 18.6 Using the Henderson-Hasselbalch Equation

Suppose you dissolve 15.0 g of $NaHCO_3$ and 18.0 g of Na_2CO_3 in enough water to make 1.00 L of solution. Use the Henderson-Hasselbalch equation to calculate the pH of the solution. (Consider this as a solution of the weak acid HCO_3^- and its conjugate base, the CO_3^{2-} ion.)

Preparing Buffer Solutions

A buffer solution has two obvious requirements. First, it should have the capacity to control the pH after the addition of reasonable amounts of acid and base. That is, the concentration of acetic acid in an acetic acid/acetate ion buffer, for example, must be large enough to consume all the hydroxide ion that may be added and still control the pH (see Example 18.4). Buffers are usually prepared from 0.10 M to 1.0 M solutions of reagents. Any buffer, however, loses its capacity if too much strong acid or base is added.

The second requirement for a buffer solution is that it should control the pH at the desired value. The equation for hydronium ion concentration of an acid buffer

$$[H_3O^+] = \frac{[\text{acid}]}{[\text{conjugate base}]} K_a$$

shows us how to prepare a buffer solution of given pH. First, we want [acid] ≈ [conjugate base] so that the solution can buffer equal amounts of added acid or base. Given this, we want to choose an acid whose K_a is near the $[H_3O^+]$ we want. The exact value of $[H_3O^+]$ is then achieved by adjusting the acid/conjugate base ratio. The following example illustrates this approach.

Example 18.6 Preparing a Buffer Solution

Suppose you wish to prepare a buffer solution to maintain the pH at 4.30. A list of possible acids (and their conjugate bases) is shown in the table.

Acid	Conjugate Base	K_a
HSO_4^-	SO_4^{2-}	1.2×10^{-2}
CH_3CO_2H	$CH_3CO_2^-$	1.8×10^{-5}
HCN	CN^-	4.0×10^{-10}

Which combination should be selected, and what should the ratio of acid to conjugate base be?

Solution The desired hydronium ion concentration is 5.0×10^{-5} M,

$$pH = 4.30, \quad \text{so} \quad [H_3O^+] = 10^{-pH} = 10^{-4.30} = 5.0 \times 10^{-5} \text{ M}$$

Of the acids given, only acetic acid (CH_3CO_2H) has a K_a value close to that of the desired $[H_3O^+]$. You therefore need only adjust the ratio $[CH_3CO_2H]/[CH_3CO_2^-]$ to achieve the desired hydronium ion concentration:

$$[H_3O^+] = 5.0 \times 10^{-5} = \frac{[CH_3CO_2H]}{[CH_3CO_2^-]}(1.8 \times 10^{-5})$$

To satisfy the previous expression, the ratio $[CH_3CO_2H]/[CH_3CO_2^-]$ must be 2.8 to 1:

$$\frac{[CH_3CO_2H]}{[CH_3CO_2^-]} = \frac{5.0 \times 10^{-5}}{1.8 \times 10^{-5}} = \frac{2.8}{1}$$

Therefore, if you add 0.28 mol of acetic acid and 0.10 mol of sodium acetate (or any other pair of molar quantities in the ratio 2.8/1) to enough water to make 1 L of solution, the buffer solution will have a pH of 4.30.

Exercise 18.7 Preparing a Buffer Solution

Using an acetic acid/sodium acetate buffer solution, what ratio of acid to conjugate base is necessary to maintain the pH at 5.00? Explain how you would make such a solution.

Example 18.6 raises one more important point concerning buffer solutions. The hydronium ion concentration depends in part on the ratio of [acid]/[conjugate base] (or the hydroxide ion concentration depends to some extent on [base]/[conjugate acid]). Although we write these ratios in terms of reagent concentrations, it is not concentrations that are important. Rather, *the relative number of moles of acid and conjugate base is important.* Because both reagents are dissolved in the same solution, their concentrations depend on the same solution volume. In Example 18.6, the ratio 2.8/1 for acetic acid and sodium acetate implied that 2.8 mol of the acid and 1.0 mol of sodium acetate were dissolved per liter:

$$\frac{[CH_3CO_2H]}{[CH_3CO_2^-]} = \frac{2.8 \text{ mol/L}}{1.0 \text{ mol/L}} = \frac{2.8 \text{ mol}}{1.0 \text{ mol}}$$

Alternatively, 1.4 mol of acetic acid and 0.50 mol of sodium acetate, or any reasonably large amount of these reagents that give a ratio of 2.8/1, could have been used. Furthermore, because the actual concentration is not important, the acid and its conjugate base could have been dissolved in any reasonable amount of water. The mole ratio of acid to its conjugate base is maintained, no matter what the solution volume may be. This is the reason that commercially available buffer solutions are sold as the premixed dry ingredients. To use them you only need to mix the ingredients in some volume of pure water (Figure 18.4).

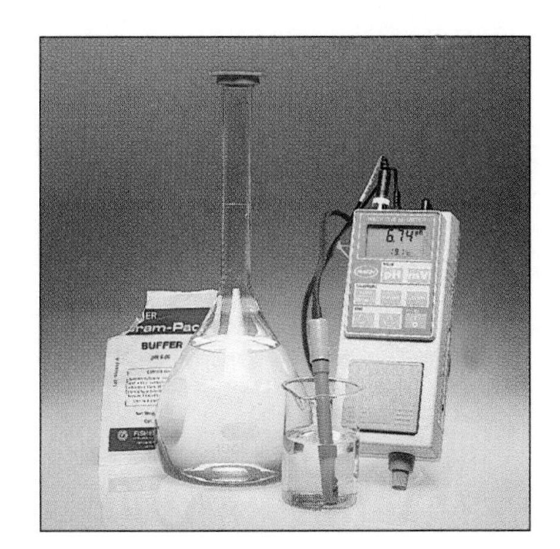

Figure 18.4 A commercial buffer solution. The solid acid and conjugate base in the packet are mixed with water to give a solution with the indicated pH. The quantity of water used does not matter because the ratio [acid]/[conjugate base] does not depend on the solution volume. *(C. D. Winters)*

The pH of a buffer depends on the pK_a of the acid and the ratio of moles of weak acid and its conjugate base. The pH does not change, therefore, if a buffer solution is diluted.

Problem-Solving Tips and Ideas

18.3 The pH of a Buffer Solution

Suppose you wish to calculate the pH of a buffer solution composed of 0.10 M NH_3 and 0.050 M NH_4Cl. The result is the same whether you consider this a solution of a weak base (NH_3) and its conjugate acid (NH_4^+)

$$NH_3(aq) + H_2O(\ell) \rightleftharpoons NH_4^+(aq) + OH^-(aq)$$

$$[OH^-] = \frac{[NH_3]}{[NH_4^+]} K_b = \frac{0.10}{0.050}(1.8 \times 10^{-5}) = 3.6 \times 10^{-5}$$

$$pH = 9.55$$

or as a solution of a weak acid (NH_4^+) and its conjugate base (NH_3):

$$NH_4^+(aq) + H_2O(\ell) \rightleftharpoons NH_3(aq) + H_3O^+(aq)$$

$$[H_3O^+] = \frac{[NH_4^+]}{[NH_3]} K_a = \frac{0.050}{0.10}(5.6 \times 10^{-10}) = 2.8 \times 10^{-10}$$

$$pH = 9.55$$

All buffer solutions may be treated in a similar manner.

18.4 ACID–BASE TITRATION CURVES

In Section 5.9 we described acid–base titrations, a method for the accurate analysis of the amount of acid or base in a sample. In the current chapter you learned something more about acid–base titrations. For example, you know now that the pH at the equivalence point is 7, and the solution is truly "neutral," only when a strong acid is titrated with a strong base and vice versa. If one of the substances being titrated is weak, then the pH at the equivalence point is not 7. We found that (see Table 18.1)

Titrations of a weak acid with a weak base are generally not done because the equivalence point cannot be judged accurately.

1. Weak acid + strong base $\longrightarrow$ pH > 7 at equivalence point due to hydrolysis of the conjugate base of the weak acid
2. Strong acid + weak base $\longrightarrow$ pH < 7 at equivalence point due to hydrolysis of the conjugate acid of the weak base.

To give us more insight into the properties of weak acids and bases, we want to describe how the pH of a solution of an acid or base changes as it is being titrated with a base or acid. To illustrate this, the following examples explore two of the most common situations: (1) the titration of a strong acid with a strong base and (2) the titration of a weak acid with a strong base.

Example 18.7 The Titration of a Strong Acid with a Strong Base

You begin with 50.0 mL of a 0.100 M solution of HCl. Calculate the pH of the solution as 0.100 M NaOH is slowly added. Plot the results with the pH of the resulting solution on the vertical axis and milliliters of base added on the horizontal axis.

A CLOSER LOOK Buffers and Biochemistry

Maintenance of pH is vital to the cells of all living organisms because enzyme activity is influenced by pH. The primary protection against harmful pH changes in cells is provided by buffer systems. The intracellular pH of most cells is maintained in a range between 6.9 and 7.4, and there are three important biological buffer systems that can control pH in this range: the phosphate system ($HPO_4^{2-}/H_2PO_4^-$), the bicarbonate/carbonic acid system (HCO_3^-/H_2CO_3), and the histidine system.

Phosphate ions are abundant in cells, both as the ions themselves and as important substituents on organic molecules. Most importantly, the pK_a for the $H_2PO_4^-$ ion is 7.20, which is very close to the pH at the high end of the normal range.

$$H_2PO_4^-(aq) + H_2O(\ell) \rightleftharpoons H_3O^+(aq) + HPO_4^{2-}(aq)$$

This means that if the buffer is to control the pH at about 7.4, the ratio of HPO_4^{2-} to $H_2PO_4^-$ must be 1.58.

$$pH = pK_a + \log \frac{[HPO_4^{2-}]}{[H_2PO_4^-]}$$

$$7.4 = 7.20 + \log \frac{[HPO_4^{2-}]}{[H_2PO_4^-]}$$

$$\frac{[HPO_4^{2-}]}{[H_2PO_4^-]} = 1.58$$

If the total phosphate concentration, $[HPO_4^{2-}] + [H_2PO_4^-]$, is 2.0×10^{-2} M, a typical concentration, you can calculate that $[HPO_4^{2-}]$ should be about 1.2×10^{-2} M and $[H_2PO_4^-]$ should be about 7.7×10^{-3} M.

The bicarbonate/carbonic acid buffer is important in blood plasma. Three equilibria are important here:

$$CO_2(g) \rightleftharpoons CO_2(dissolved)$$

$$CO_2(dissolved) + H_2O(\ell) \rightleftharpoons H_2CO_3(aq)$$

$$H_2CO_3(aq) + H_2O(\ell) \rightleftharpoons H_3O^+(aq) + HCO_3^-(aq)$$

The overall equilibrium constant for the second and third steps is p$K_{overall}$ = 6.3 at 37 °C, the temperature of the human body. Thus,

$$7.4 = 6.3 + \log \frac{[HCO_3^-]}{[CO_2(dissolved)]}$$

Although the value of p$K_{overall}$ is about 1 pH unit away from the blood pH, the natural partial pressure of CO_2 in the aveoli of the lungs (about 40 mm Hg) is sufficient to keep $[CO_2(dissolved)]$ at about 1.2×10^{-3} M and $[HCO_3^-]$ at about 1.5×10^{-2} M.

Histidine is one of the 20 naturally occurring amino acids. The pK_a value for the loss of the H^+ on the N atom of the five-member imidazole ring is 6.04.

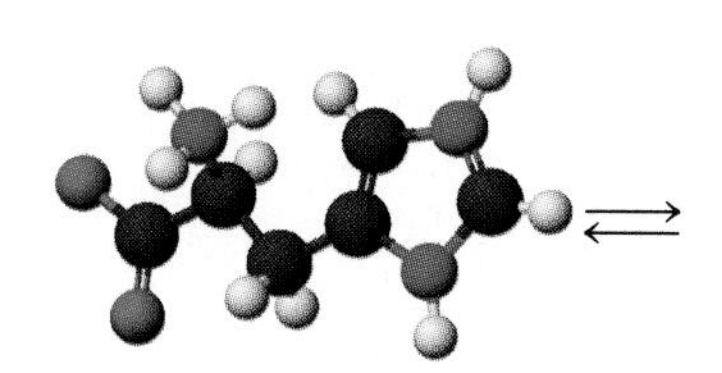

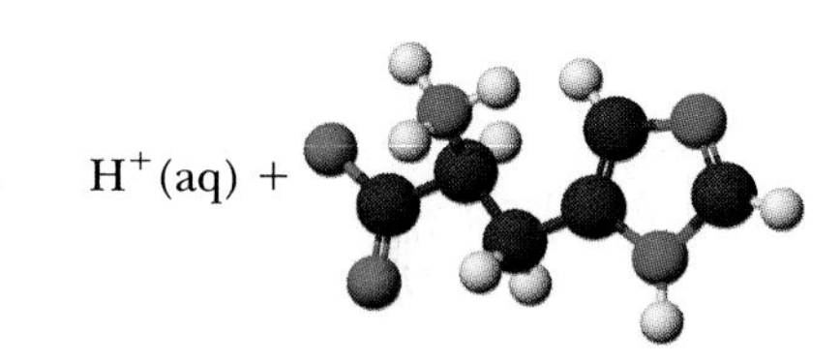

Histidine not only occurs as the free amino acid in cells, but it also occurs as part of proteins or peptides (see Chapter 11, page 528). When attached to another molecule, its pK_a value can change and move into the range required to act as a buffering agent.

Solution

Step 1. *pH before adding base.* HCl is a strong acid, so $[H_3O^+] = 0.100$ M, which means the pH in the beginning is 1.000.

Step 2. *pH after adding 10.0 mL of 0.100 M NaOH.* To find this answer, we must know the number of moles of H_3O^+ in the solution *after* the base is added. We therefore calculate the quantity of acid in the beginning and subtract from it the quantity of acid that reacts with the base according to the net ionic equation

$$H_3O^+(aq) + OH^-(aq) \longrightarrow 2\ H_2O(\ell)$$

Initially, the number of moles of $H_3O^+ = (0.0500\ L) \cdot (0.100\ M) = 0.00500$ mol. The number of moles of OH^- added $= (0.0100\ L) \cdot (0.100\ M) = 0.00100$ mol.

Equation	Moles H_3O^+	Moles OH^-
Before reaction	0.00500 M in 50.0 mL solution	0.00100 mol in 10.0 mL solution
Change on reaction	−0.00100 mol	−0.00100 mol
After reaction	0.00400 mol in 60.0 mL solution	0 mol
Concentrations after reaction	0.0667 M	0 M

Here you see that 0.00100 mol of NaOH consumed 0.00100 mol of HCl to leave 0.00400 mol of HCl in the solution. Because the solutions were combined, the total volume after reaction is the sum of the combining volumes, 60.0 mL in this case. The hydronium ion concentration after reaction is therefore

$$[H_3O^+] = \frac{0.00400 \text{ mol}}{0.0600 \text{ L}} = 0.0667 \text{ M}$$

and the pH of the solution is 1.176.

Step 3. *pH after any volume of base is added, up to the equivalence point.* In general, the hydronium ion concentration up to the equivalence point can be calculated from the equation

$$[H_3O^+] = \frac{\text{original moles acid} - \text{total moles base added}}{\text{volume acid} + \text{volume base added}}$$

For instance, after 45.0 mL of NaOH has been added, the acid concentration is

$$[H_3O^+] = \frac{0.00500 \text{ mol } H_3O^+ - (0.0450 \text{ L})(0.100 \text{ M})}{0.0500 \text{ L} + 0.0450 \text{ L}} = 0.00526 \text{ M}$$

and the pH is 2.279.

When the volume of added base has reached 49.5 mL, only 0.50 mL away from the equivalence point, the pH indicates the solution is still quite acidic:

$$[H_3O^+] = \frac{0.00500 \text{ mol } H_3O^+ - (0.0495 \text{ L})(0.100 \text{ M})}{0.0500 \text{ L} + 0.0495 \text{ L}} = 5.03 \times 10^{-4} \text{ M}$$

$$\text{pH} = 3.299$$

Step 4. *pH at the equivalence point.* This is the reaction of a strong acid with a strong base. The pH is therefore 7.00 at the equivalence point (at 25 °C).

Step 5. *pH beyond the equivalence point.* No acid remains in the solution after the equivalence point. NaOH is added to a solution that contains only NaCl, a neutral salt. The pH therefore depends only on the concentration of OH^- from added NaOH. For example, after only 1.0 mL of base has been added beyond the equivalence point (total volume of added base = 51.0 mL), the pH is 11.00:

$$[OH^-] = \frac{\text{moles excess base}}{\text{total volume}} = \frac{(0.0010 \text{ L})(0.100 \text{ M})}{0.0500 \text{ L acid} + 0.0510 \text{ L base}} = 9.9 \times 10^{-4} \text{ M}$$

$$\text{pOH} = 3.00 \quad \text{and so the} \quad \text{pH} = 11.00$$

The results of these calculations are summarized in the table and titration curve in Figure 18.5.

The equivalence point in an acid–base titration is the midpoint of the vertical portion of the pH curve.

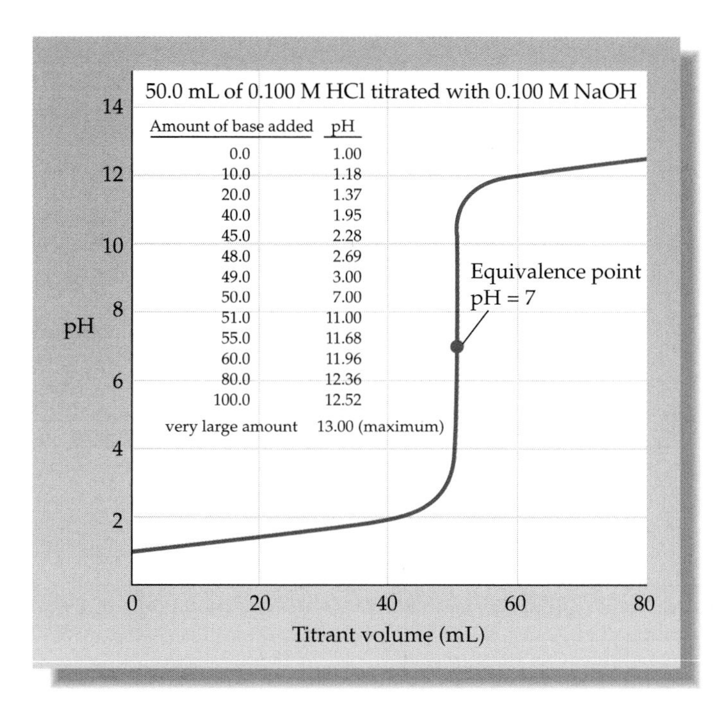

***Figure* 18.5 The change in pH as a strong acid is titrated with a strong base.** In this case 50.0 mL of 0.100 M HCl is titrated with 0.100 M NaOH. The pH at the equivalence point is 7.0 (at 25 °C), as is always the case for the reaction of a strong acid with a strong base.

***Exercise* 18.8** Titration of a Strong Acid with a Strong Base

For the titration outlined in Example 18.7, verify the pH calculated for the addition of (1) 40.0 mL of NaOH and (2) 60.0 mL of NaOH. What is the pH after 49.9 mL of NaOH is added?

The pH at each point in the titration of the strong acid HCl with the strong base NaOH is illustrated in Figure 18.5. As calculated in Example 18.7, the pH rises slowly until very close to the equivalence point. Then the pH increases very rapidly, rising 7 units (the H_3O^+ concentration decreases by a factor of 10 million!) when only a very small amount of base (perhaps a drop or two) is added. After the equivalence point is passed, only a small further rise in pH is seen.

The titration of a weak acid with a strong base is somewhat different from the strong acid/strong base titration just described. To illustrate, let us look carefully at the titration curve for the titration of 100.0 mL of 0.100 M acetic acid with 0.100 M NaOH (Figure 18.6):

$$CH_3CO_2H(aq) + NaOH(aq) \longrightarrow NaCH_3CO_2(aq) + H_2O(\ell)$$

Three points, or regions, on this curve are especially important

1. The pH before titration begins
2. The pH at the midpoint of the titration
3. The pH at the equivalence point

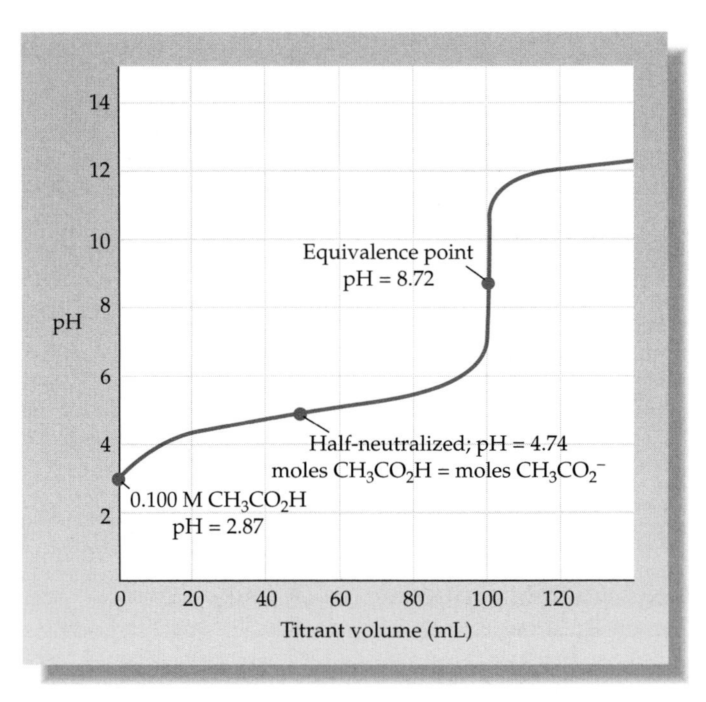

***Figure* 18.6 The change in pH during the titration of a weak acid with a strong base.** Here 100.0 mL of 0.100 M acetic acid is titrated with 0.100 M NaOH. Note especially that (a) acetic acid is a weak acid so the pH of the original solution is 2.87. (b) The pH at the point at which half the acid has reacted with base is equal to the pK_a for the acid (pH = pK_a = 4.74), a general observation in the titration of a weak acid with a strong base. (c) At the equivalence point the solution consists of acetate ion, a weak base. The solution is basic with a pH of 8.72.

The pH before any base is added (2.87) is found in the usual way (Example 17.6) for the ionization of a weak acid. At the equivalence point of the titration, the solution contains sodium acetate, and so the pH (8.72) is controlled by the acetate ion, the conjugate base of acetic acid (see Example 18.2).

Now let us determine the pH at the midpoint of the titration, that is, at the point at which half of the acid has been consumed. As NaOH is added to the acetic acid, the base is consumed and sodium acetate is produced. This means that at every point between the beginning of the titration (when only pure acetic acid is present) and the equivalence point (where only sodium acetate is present) the solution contains *both* acetic acid *and* its salt, sodium acetate. These are the components of a buffer solution, and the hydronium ion concentration can be found from the following expression:

$$[H_3O^+] = \frac{[CH_3CO_2H]}{[CH_3CO_2^-]} K_a$$

Thus, at any point in the titration (before the equivalence point), the hydronium ion concentration can be calculated from

$$[H_3O^+] = \frac{[CH_3CO_2H \text{ remaining}]}{[CH_3CO_2^- \text{ formed}]} K_a$$

Now, what happens when exactly half of the acid has been consumed by base? At this point half of the acid has been converted to the conjugate base, $CH_3CO_2^-$. Half of the acid remains. Therefore, $[CH_3CO_2H] = [CH_3CO_2^-]$, and

$$[H_3O^+] \text{ at the midpoint of the titration} = K_a$$

In the particular case of the titration of acetic acid with strong base, $[H_3O^+] = 1.8 \times 10^{-5}$ M at the midpoint, and so the pH is 4.74. This result is of general importance, however, because we can say to a good approximation that

- At the midpoint in any weak acid/strong base titration, $[H_3O^+] = K_a$ (or $pH = pK_a$) of the weak acid.
- At the midpoint in any weak base/strong acid titration, $[OH^-] = K_b$ (or $pOH = pK_b$) of the weak base.

You can see the value of these conclusions. If you perform an acid–base titration in which one of the components is weak and record the pH values at each step, you can readily determine K for the weak acid or base.

As a final point concerning the titration of a weak acid with a strong base, we might wish to know the pH *after* the equivalence point. In the case of the CH_3CO_2H/NaOH titration, the solution consists of sodium acetate *plus* any NaOH that was added in excess after passing the equivalence point. Both acetate ion and hydroxide ion are bases, so the total concentration of OH^- is the sum of that from excess NaOH *plus* that produced by the hydrolysis of the acetate ion:

$$CH_3CO_2^-(aq) + H_2O(\ell) \rightleftharpoons CH_3CO_2H(aq) + OH^-(aq)$$

The OH^- concentration from the hydrolysis reaction is very small, however, compared with that from excess NaOH, so, after the equivalence point

$$[OH^-] = \frac{\text{moles excess OH}^- \text{ from NaOH}}{\text{total volume in liters}}$$

Example 18.8 Titration of Acetic Acid with Sodium Hydroxide

What is the pH of the solution when 90.0 mL of 0.100 M NaOH has been added to 100.0 mL of 0.100 M acetic acid (see Figure 18.6)?

Solution The pH is desired at a point after the midpoint of the titration (pH = 4.74) but before the equivalence point (pH = 8.72). The solution is a buffer solution, containing some unreacted acetic acid plus some sodium acetate formed in the reaction of the acid with sodium hydroxide. Let us first calculate the quantities of reactants before reaction (= concentration × volume) and then use the principles of stoichiometry to calculate the quantities of reactants and products after reaction.

Equation	$CH_3CO_2H(aq)$ +	$OH^-(aq)$ ⟶	$CH_3CO_2^-(aq) + H_2O(\ell)$
Quantity before reaction (mol)	0.0100	0.00900	0
Change occurring on reaction	−0.00900	−0.00900	+0.00900
Quantity after reaction (mol)	0.0010	0	0.00900

This is a buffer solution, whose hydronium ion concentration can be found from the equation

$$[H_3O^+] = \frac{[CH_3CO_2H \text{ remaining}]}{[CH_3CO_2^- \text{ formed}]} K_a$$

Recall that the ratio of moles of acid and conjugate base is the same as the ratio of their concentrations. Therefore,

$$[H_3O^+] = \frac{0.0010 \text{ mol}}{0.00900 \text{ mol}} \times 1.8 \times 10^{-5} = 2.0 \times 10^{-6} \text{ M}$$

and this gives a pH of 5.70, in agreement with Figure 18.6.

Exercise 18.9 Titration of a Weak Acid with a Strong Base

The titration of 100.0 mL of 0.100 acetic acid with 0.100 M NaOH is described in the text. What is the pH of the solution when 35.0 mL of the base has been added?

Problem-Solving Tips and Ideas

18.4 Calculating the pH at Various Stages of an Acid–Base Reaction

Finding the pH at some point in an acid–base reaction always involves three or four calculation steps. Consider *the titration of a weak base (B) with a strong acid* as an example. (The same principles apply to other acid–base titrations.)

$$H_3O^+(aq) + B(aq) \longrightarrow BH^+(aq) + H_2O(\ell)$$

Step 1. *Stoichiometry problem.* Up to the equivalence point, acid is consumed completely to leave a solution containing some base (B) and its conjugate acid (BH^+). Use the principles of stoichiometry to calculate (a) the quantity of acid added, (b) the quantity of base consumed, (c) the quantity of base remaining, and (d) the quantity of conjugate acid (BH^+) formed.

Step 2. *Calculate the concentrations of base, [B], and conjugate acid, [BH^+].* Recognize that the volume of the solution at any point is the sum of the original volume of the base solution plus the volume of acid solution added.

Step 3. *Calculate pH before the equivalence point.* At any point before the equivalence point, the solution is a buffer solution because the base and its conjugate acid are present. Calculate [OH^-] using the concentrations from Step 2 and the value of K_b for the weak base (see Equation 18.2).

Step 4. *pH at the equivalence point.* At the equivalence point, calculate the concentration of the conjugate acid using the procedure of Steps 1 and 2 and use this value with the value of K_a for this acid to calculate the concentration of H_3O^+ in the solution.

The titrations illustrated thus far involve a monoprotic acid (HA) reacting with a base, such as NaOH or NH_3. Many acids and bases are polyprotic, however. One example is oxalic acid, $H_2C_2O_4$:

$$H_2C_2O_4(aq) + H_2O(\ell) \rightleftharpoons HC_2O_4^-(aq) + H_3O^+(aq) \qquad K_{a1} = 5.9 \times 10^{-2}$$
$$HC_2O_4^-(aq) + H_2O(\ell) \rightleftharpoons C_2O_4^{2-}(aq) + H_3O^+(aq) \qquad K_{a2} = 6.4 \times 10^{-5}$$

Figure 18.7 illustrates a titration of 100. mL of 0.100 M oxalic acid with 0.100 M NaOH. The first significant rise in pH is experienced after 100 mL of base has been added, indicating that the first proton of the acid has been titrated:

$$H_2C_2O_4(aq) + OH^-(aq) \rightleftharpoons HC_2O_4^-(aq) + H_2O(\ell)$$

The rise in pH between 50 to 150 mL of added base is not well defined, however. The reason for this is that the solution contains not only $HC_2O_4^-$ but also the dianion $C_2O_4^{2-}$. These ions are, respectively, a weak acid and its conjugate base,

$$HC_2O_4^-(aq) + OH^-(aq) \rightleftharpoons C_2O_4^{2-}(aq) + H_2O(\ell)$$

in other words, the components of a buffer. Thus, the species in the solution resist a rapid rise in pH. Only when the second proton of oxalic acid is titrated,

$$HC_2O_4^-(aq) + OH^-(aq) \rightleftharpoons C_2O_4^{2-}(aq) + H_2O(\ell)$$

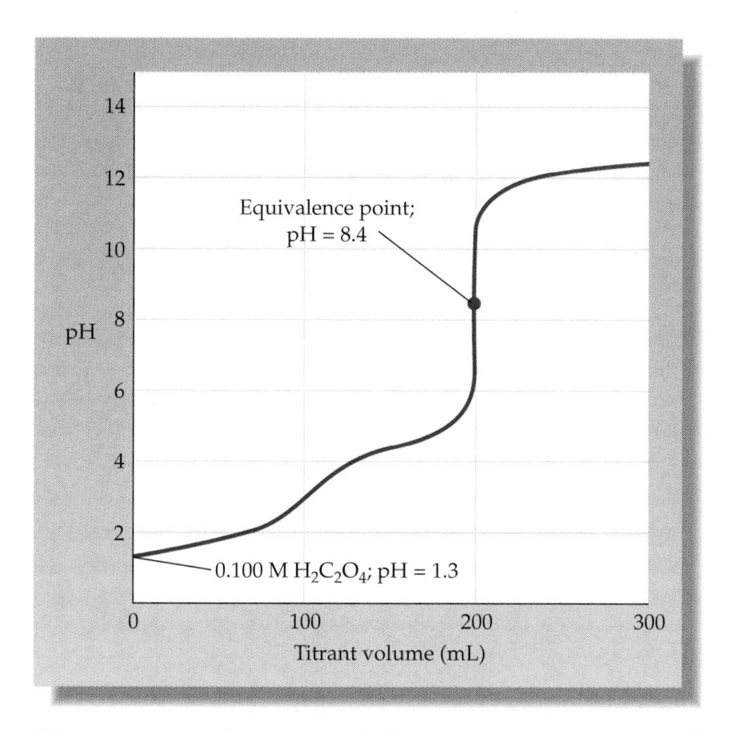

Figure 18.7 Titration curve for a diprotic acid. The curve for the titration of 100.0 mL of 0.100 M oxalic acid ($H_2C_2O_4$, a weak diprotic acid) with 0.100 M NaOH. The first rise in pH (at 100 mL) occurs when the first H atom of $H_2C_2O_4$ is titrated, and the second rise (at 200 mL) occurs at the completion of the reaction

$$HC_2O_4^-(aq) + OH^-(aq) \longrightarrow C_2O_4^{2-}(aq) + H_2O(\ell)$$

does the pH increase significantly. The pH at this second equivalence point is controlled by the K_b value for $C_2O_4^{2-}$, the conjugate base of the hydrogen oxalate ion:

$$C_2O_4^{2-}(aq) + H_2O(\ell) \rightleftharpoons HC_2O_4^-(aq) + OH^-(aq)$$

$$K_b = \frac{K_w}{K_{a2}} = 1.6 \times 10^{-10}$$

Calculation of the pH at the equivalence point indicates that it should be about 8.5, as observed.

Finally, it is useful to consider the titration of a weak base with a strong acid. Figure 18.8 illustrates the case for the titration of 100.0 mL of 0.100 M NH_3 with 0.100 M HCl:

$$NH_3(aq) + HCl(aq) \longrightarrow NH_4^+(aq) + Cl^-(aq)$$

Notice that the pH of the original 0.100 M NH_3 solution is 11.12, as expected for a weak base. At the midpoint of the titration, half of the original ammonia has been converted to ammonium chloride, and half of the ammonia remains. The hydroxide ion concentration at the midpoint is therefore

$$[OH^-] \text{ at midpoint} = \frac{[NH_3]}{[NH_4^+]} K_b$$

$$[OH^-] = K_b = 1.8 \times 10^{-5} \text{ M}$$

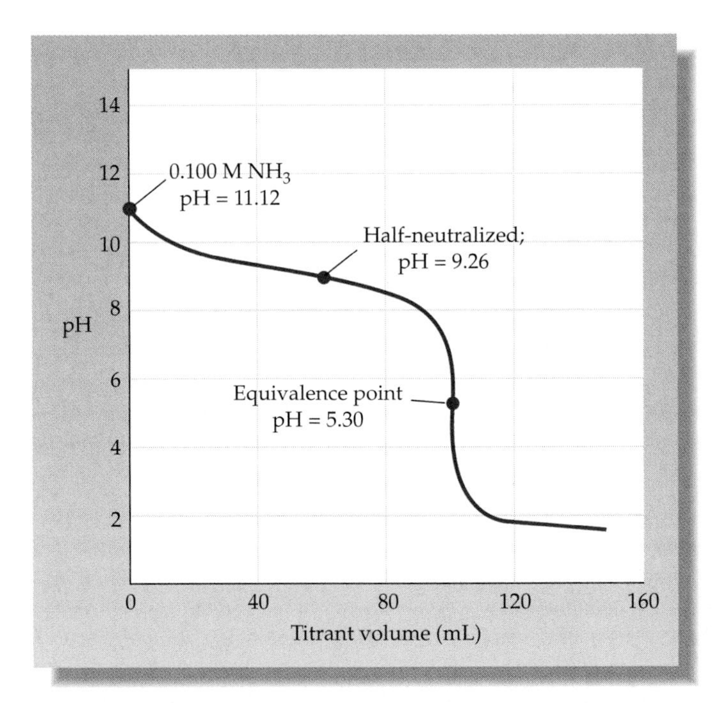

***Figure* 18.8 Titration of a weak base with a strong acid.** The change in pH during the titration of 100.0 mL of 0.100 M NH_3 with 0.100 M HCl. The pH at the half-neutralization point is equal to the pK_a for the conjugate acid (NH_4^+) of the weak base (NH_3) (pH = pK_a = 9.26). At the equivalence point the solution contains the NH_4^+ ion, a weak acid, so the pH is about 5.

The pOH is 4.74 and so the pH is 9.26 (see Figure 18.8). As the addition of strong acid continues, the pH is relatively constant because of the buffering action of the NH_3/NH_4^+ combination. Near the equivalence point, however, the pH drops rapidly. At the equivalence point, the solution contains only ammonium chloride, a weak Brønsted acid, and the solution is weakly acidic.

At the midpoint in the titration of a weak base with a strong acid, the pH is equivalent to the pK_a of the conjugate acid of the weak base.

Exercise 18.10 Titration of a Weak Base with a Strong Acid

Calculate the pH at the equivalence point in the titration of 100.0 mL of 0.100 M NH_3 with 0.100 M HCl, as described in the text.

18.5 ACID–BASE INDICATORS

The goal of an acid–base titration is to add a substance in an amount that is stoichiometrically exactly equivalent to the substance with which it reacts. This condition is achieved at the equivalence point. One way to estimate when you have reached the equivalence point is to observe some change in the solution. Using a pH meter (see Figure 18.3) you can see when the pH of the solution changes by a number of units, or you can add a few drops of an indicator, a dye that has different colors in solutions of different pH (see Figure 17.11). The dye's color change is said to occur at the *end point* of the titration, and we try to make the difference between the end and equivalence points negligible.

An acid–base indicator (HInd) is usually an organic dye that is itself a weak acid:

$$\text{HInd(aq)} + H_2O(\ell) \rightleftharpoons H_3O^+(\text{aq}) + \text{Ind}^-(\text{aq})$$

$$K_a \text{ for the indicator} = \frac{[H_3O^+][\text{Ind}^-]}{[\text{HInd}]}$$

The acid form of the compound (HInd) has one color, and the conjugate base (Ind^-) has another (Figure 17.11). According to Le Chatelier's principle, addition of H_3O^+ or OH^- to an indicator causes one color or the other to appear, depending on whether HInd or Ind^- is the predominant species. The idea is to try to choose an indicator with a K_a near that of the acid being titrated and one whose color changes strongly at the equivalence point. If the color change is strong on going from an acidic to a basic medium, we need add only a tiny bit of the indicator to the solution and then titrate both acids at the same time.

If the K_a expression for the indicator is rearranged

$$[H_3O^+] = \frac{[\text{HInd}]}{[\text{Ind}^-]} K_a \quad \text{or} \quad \frac{[H_3O^+]}{K_a} = \frac{[\text{HInd}]}{[\text{Ind}^-]}$$

it is apparent that a tiny amount of an indicator can in fact reveal the pH of a solution by changing the $[\text{HInd}]/[\text{Ind}^-]$ ratio. The ratio of $[\text{HInd}]/[\text{Ind}^-]$ is controlled by the hydronium ion concentration of the solution. As $[H_3O^+]$ changes in the course of a titration, the ratio of $[\text{HInd}]/[\text{Ind}^-]$ must also change in order to maintain the equality in the preceding equation. When the H_3O^+ concentration is high, [HInd] must therefore be large and $[\text{Ind}^-]$ small. The indicator color is that of HInd. In the case of the common indicator phenolphthalein, the dye is colorless at high concentrations of H_3O^+ (or low pH), whereas the indicator thymol blue is red under these conditions (Figure 17.11). When the pH increases, and $[H_3O^+]$ declines, the ratio $[HInd]/[Ind^-]$ must also decrease. Thus, the concentration of Ind^- increases, and the color of Ind^- is seen. Phenolphthalein, for example, is red above pH 10, whereas thymol blue is yellow between pH 3 and 8 and purple at a pH of about 9.

In principle, the color of the indicator changes when $[H_3O^+] = K_a$ of the indicator (because $[HInd] = [Ind^-]$ at this point), and so you might think that you can accurately determine $[H_3O^+]$ by carefully observing the color change. In practice, your eyes are not quite that good. Usually, you see the color of HInd when $[HInd]/[Ind^-]$ is about 10/1, and you see the color of Ind^- when $[HInd]/[Ind^-]$ is about 1/10. This means the color change is observed over a hydronium ion concentration interval of about 100, which corresponds to 2 pH units. This is not really a problem, however, as you can see in Figures 18.5 and 18.6; on passing through the equivalence point of these titrations the pH changes by as many as 7 units.

A variety of indicators is available, each changing color in a different pH range (see Figure 17.11). If you are analyzing a weak acid or base by titration, you must choose an indicator that changes color in a range that includes the pH to be observed at the equivalence point. This means that an indicator that changes color at or near pH 7 should be used for a strong acid/strong base titration. On the other hand, the pH at the equivalence point in the titration of a weak acid with a strong base is greater than 7; therefore, you should use an indicator that changes color at a pH of about 8.

Exercise 18.11 Indicators

Use Figure 17.11 to decide which indicator is best to use in the titration of NH_3 with HCl in Figure 18.8.

CHAPTER HIGHLIGHTS

When you have finished reading this chapter, you should be able to

- Predict the pH of an acid–base reaction at its equivalence point (Section 18.1).

Acid	Base	pH at equivalence point
Strong	Strong	= 7 (neutral)
Strong	Weak	< 7 (acidic)
Weak	Strong	> 7 (basic)
Weak	Weak	Depends on K values of conjugate base and acid

- Calculate the pH at the equivalence point in the reaction of a strong acid with a weak base, or in the reaction of a strong base with a weak acid (see Examples 18.1 and 18.2).
- Predict the effect of the presence of a "common ion" on the pH of the solution of a weak acid or base (Section 18.2).

- Describe the composition of a **buffer solution** and explain how buffers work (Section 18.3).
- Calculate the pH of a buffer solution before and after adding excess acid or base (Section 18.3 and Example 18.4).
- Use the **Henderson-Hasselbalch equation** (Equation 18.3) to calculate the pH of a buffer solution of given composition (Section 18.3).
- Use the Henderson-Hasselbalch equation to predict the change in pH when the composition of the buffer changes.
- Describe how a buffer solution of a given pH can be prepared (Section 18.3).
- Calculate the pH at any point in an acid–base titration (Section 18.4).
- Understand the differences between the titration curves for a strong acid/strong base titration versus cases in which one of the substances is weak (Section 18.4).
- Describe how an **indicator** functions in an acid–base titration (Section 18.5).

STUDY QUESTIONS

REVIEW QUESTIONS

1. What is the equivalence point of an acid–base reaction?
2. Decide if the pH is equal to 7, less than 7, or greater than 7 at equivalence point in each of the following titrations:
   (a) A weak base with a strong acid
   (b) A strong base with a strong acid
   (c) A strong base with a weak acid
3. Sketch the general shape of the titration curve when (a) a strong base is titrated with a strong acid and (b) a weak base is titrated with a strong acid. In each case indicate if the pH at the equivalence point is less than 7, equal to 7, or greater than 7.
4. Briefly describe how a buffer solution controls the pH when excess strong acid is added. Use the NH_3/NH_4Cl buffer as an example.
5. Briefly describe how a buffer solution controls the pH of a solution when excess strong base is added. Use a solution of acetic acid and sodium acetate as an example.
6. What is the difference between the equivalence point of a titration and its end point?
7. Use the Henderson-Hasselbalch equation to decide how the pH of a buffer solution changes
   (a) when the weak acid concentration decreases relative to the concentration of its conjugate base.
   (b) when the weak acid/conjugate base pair is changed to a stronger acid and its conjugate base.

NUMERICAL AND OTHER QUESTIONS

Acid–Base Reactions

8. Calculate the value of the equilibrium constant for the reaction of benzoic acid, $C_6H_5CO_2H$, with NaOH. Does the equilibrium lie predominantly to the left or to the right side of the reaction?

$$C_6H_5CO_2H(aq) + OH^-(aq) \rightleftharpoons H_2O(\ell) + C_6H_5CO_2^-(aq)$$

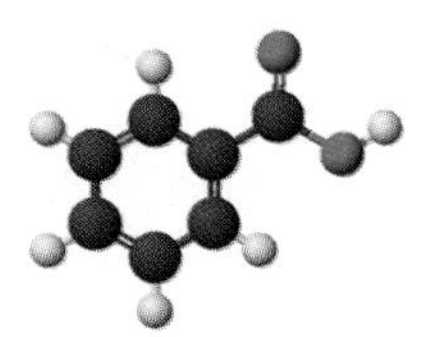

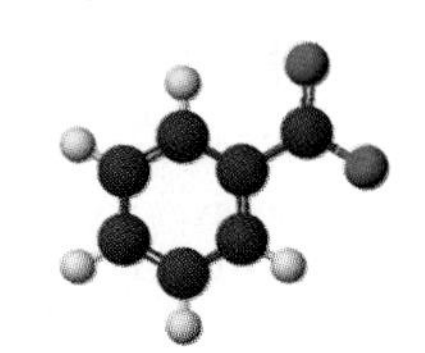

9. What is the value of the equilibrium constant for the reaction of the weak base aniline, $C_6H_5NH_2$, with HCl? Does the equilibrium lie predominantly to the left or to the right side of the reaction?

$$C_6H_5NH_2(aq) + H_3O^+(aq) \rightleftharpoons C_6H_5NH_3^+(aq) + H_2O(\ell)$$

10. Calculate the hydronium ion concentration and pH of the solution that results when 22.0 mL of 0.15 M acetic acid, CH_3CO_2H, is mixed with 22.0 mL of 0.15 M NaOH.

11. Calculate the hydronium ion concentration and the pH when 50.0 mL of 0.40 M NH_3 is mixed with 50.0 mL of 0.40 M HCl.

12. For each of the following cases, decide whether the pH is less than 7, equal to 7, or greater than 7:
   (a) Equal volumes of 0.10 M acetic acid, CH_3CO_2H, and 0.10 M KOH are mixed.
   (b) 25 mL of 0.015 M NH_3 is mixed with 25 mL of 0.015 M HCl.
   (c) 150 mL of 0.20 M HNO_3 is mixed with 75 mL of 0.40 M NaOH.

13. For each of the following cases, decide whether the pH is less than 7, equal to 7, or greater than 7:
   (a) 25 mL of 0.45 M H_2SO_4 is mixed with 25 mL of 0.90 M NaOH.
   (b) 15 mL of 0.050 M formic acid, HCO_2H, is mixed with 15 mL of 0.050 M NaOH.

(c) 25 mL of 0.15 M $H_2C_2O_4$ (oxalic acid) is mixed with 25 mL of 0.30 M NaOH. (Both H^+ ions of oxalic acid are titrated.)

14. Phenol, C_6H_5OH, is a weak organic acid that has been used as a disinfectant. Suppose 0.515 g of the compound is dissolved in water to give exactly 125 mL of solution. The resulting solution is titrated with 0.123 M NaOH:

$$C_6H_5OH(aq) + OH^-(aq) \longrightarrow C_6H_5O^-(aq) + H_2O(\ell)$$

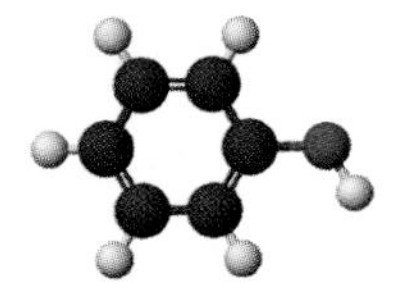

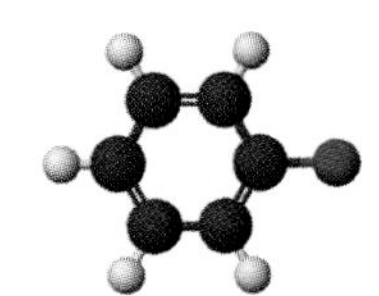

What are the concentrations of all of the following ions at the equivalence point: Na^+, H_3O^+, OH^-, and $C_6H_5O^-$? What is the pH of the solution?

15. Assume you dissolve 0.235 g of the weak acid benzoic acid, $C_6H_5CO_2H$, in enough water to make 1.00×10^2 mL of solution and then titrate the solution with 0.108 M NaOH:

$$C_6H_5CO_2H(aq) + OH^-(aq) \longrightarrow C_6H_5CO_2^-(aq) + H_2O(\ell)$$

What are the concentrations of the following ions at the equivalence point: Na^+, H_3O^+, OH^-, and $C_6H_5CO_2^-$? What is the pH of the solution?

16. You require 36.78 mL of 0.0105 M HCl to reach the equivalence point in the titration of 25.0 mL of aqueous ammonia. What are the concentrations of H_3O^+, OH^-, and NH_4^+ at the equivalence point? What is the pH of the solution at the equivalence point? What was the concentration of NH_3 in the original ammonia solution?

17. A solution of the weak base aniline, $C_6H_5NH_2$, in 25.0 mL of water requires 25.67 mL of 0.175 M HCl to reach the equivalence point:

$$C_6H_5NH_2(aq) + H_3O^+(aq) \rightleftharpoons C_6H_5NH_3^+(aq) + H_2O(\ell)$$

What are the concentrations of H_3O^+, OH^-, and $C_6H_5NH_3^+$ at the equivalence point? What is the pH of the solution at the equivalence point? What was the concentration of aniline in the original solution?

The Common Ion Effect and Buffer Solutions

18. Does the pH of the solution increase, decrease, or stay the same when you
(a) Add solid ammonium chloride to a dilute aqueous solution of NH_3?
(b) Add solid sodium acetate to a dilute aqueous solution of acetic acid?
(c) Add solid NaCl to a dilute aqueous solution of NaOH?

19. Does the pH of the solution increase, decrease, or stay the same when you
(a) Add solid sodium oxalate, $Na_2C_2O_4$, to 50.0 mL of 0.015 M oxalic acid, $H_2C_2O_4$?
(b) Add solid ammonium chloride to 75 mL of 0.016 M HCl?
(c) Add 20.0 g of NaCl to 1.0 L of 0.10 M sodium acetate, $NaCH_3CO_2$?

20. Does the pH of the solution increase, decrease, or stay the same when you
(a) Add 10.0 mL of 0.10 M HCl to 25.0 ml of 0.10 M NH_3?
(b) Add 25.0 mL of 0.050 M NaOH to 50.0 mL of 0.050 M acetic acid?

21. Does the pH of the solution increase, decrease, or stay the same when you
(a) Add 25.0 mL of 0.050 M NaOH to 25.0 mL of 0.075 M oxalic acid, $H_2C_2O_4$?
(b) Add 1.0 mL of 0.10 M HCl to 25.0 mL of 0.10 M pyridine, a weak base?

22. What is the pH of a buffer solution that is 0.20 M with respect to ammonia, NH_3, and 0.20 M with respect to ammonium chloride, NH_4Cl?

23. What is the pH of a buffer solution if 100.0 mL of the solution is 0.15 M with respect to acetic acid and contains 1.56 g of sodium acetate, $NaCH_3CO_2$?

24. What is the pH of the buffer solution that results when 2.2 g of NH_4Cl, ammonium chloride, is added to 250 mL of 0.12 M NH_3? Is the final pH lower or higher than the pH of the original ammonia solution?

25. Lactic acid ($CH_3CHOHCO_2H$) is found in sour milk, in sauerkraut, and in muscles after activity (see page 514). If 2.75 g of sodium lactate, $NaCH_3CHOHCO_2$, is added to 5.00×10^2 mL of 0.100 M lactic acid, what is the pH of the resulting buffer solution? Is the final pH lower or higher than the pH of the lactic acid solution? (K_a for lactic acid = 1.4×10^{-4}.)

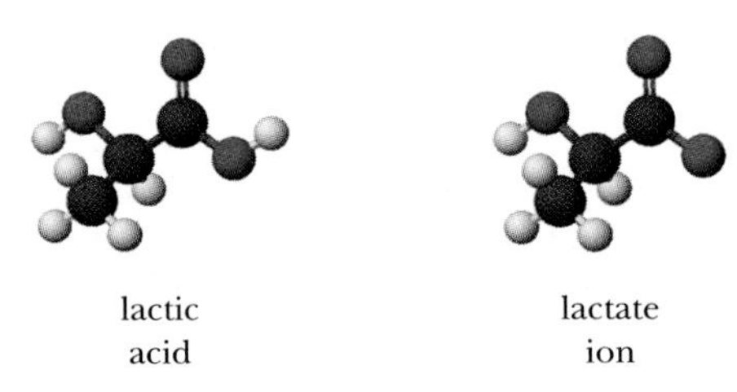

lactic acid   lactate ion

26. What mass of sodium acetate, $NaCH_3CO_2$, must be added to 1.00 L of 0.10 M acetic acid to give a solution with a pH of 4.5?

27. What mass of ammonium chloride, NH_4Cl, must be added to exactly 5.00×10^2 mL of 0.10 M NH_3 solution to give a solution with a pH of 9.0?

28. A buffer solution is made of 12.2 g of benzoic acid ($C_6H_5CO_2H$) and 7.20 g of sodium benzoate ($NaC_6H_5CO_2$) in 2.50×10^2 mL of solution. What is the pH of the solution? If the solution is diluted to 5.00×10^2 mL with pure water, what is the new pH of the solution?

29. What is the pH of a buffer solution prepared from 5.15 g of NH_4NO_3 and 0.10 L of 0.15 M NH_3? What is the new pH if the solution is diluted with pure water to a volume of 5.00×10^2 mL?

30. Which of the following combinations would be the best to buffer a solution at a pH of approximately 9?
(a) HCl/NaCl
(b) Ammonia/ammonium nitrate [NH_3/NH_4Cl]
(c) Acetic acid/sodium acetate [$CH_3CO_2H/NaCH_3CO_2$]

31. Many natural processes are studied in the laboratory in an environment of controlled pH. Which of the following combinations would be the best choice to buffer the pH at approximately 7?
(a) H_3PO_4/NaH_2PO_4
(b) NaH_2PO_4/Na_2HPO_4
(c) Na_2HPO_4/Na_3PO_4

32. A buffer solution was prepared by adding 4.95 g of sodium acetate, $NaCH_3CO_2$, to 2.50×10^2 mL of 0.150 M acetic acid, CH_3CO_2H.
(a) What is the pH of the buffer?
(b) What is the pH of 1.00×10^2 mL of the buffer solution if you add 82 mg of NaOH to the solution?

33. You dissolve 0.425 g of NaOH in 2.00 L of a solution that has $[H_2PO_4^-] = [HPO_4^{2-}] = 0.132$ M. What was the pH of the solution before adding NaOH? After adding NaOH?

34. A buffer solution is prepared by adding 0.125 mol of ammonium chloride to 5.00×10^2 mL of 0.500 M solution of ammonia.
(a) What is the pH of the buffer solution?
(b) If 0.0100 mol of HCl gas is bubbled into 5.00×10^2 mL of the buffer, what is the new pH of the solution?

35. What will be the pH change when 20.0 mL of 0.100 M NaOH is added to 80.0 mL of a buffer solution consisting of 0.169 M NH_3 and 0.183 M NH_4Cl?

Using the Henderson-Hasselbalch Equation

36. What is the value of pK_a for acetic acid? Use the Henderson-Hasselbalch equation to calculate the pH of a solution that has an acetic acid concentration of 0.050 M and a sodium acetate concentration of 0.075 M.

37. What is the value of pK_a for the ammonium ion? Use the Henderson-Hasselbalch equation to calculate the pH of a solution that has an ammonium chloride concentration of 0.050 M and an ammonia concentration of 0.045 M.

38. Assume a buffer solution is prepared using formic acid and sodium formate.
(a) What is the pK_a for formic acid?
(b) What is the pH of a solution that has a formic acid concentration of 0.050 M and a sodium formate concentration of 0.035 M?
(c) What must the ratio of acid to conjugate base be in order to increase the pH by 0.5?

39. A buffer solution is composed of 1.360 g of KH_2PO_4 and 5.677 g of Na_2HPO_4.
(a) What is the pK_a of the weak acid, the $H_2PO_4^-$ ion?
(b) What is the pH of the buffer solution?

Titration Curves and Indicators

40. Without doing detailed calculations, sketch the curve for the titration of 30.0 mL of 0.10 M NaOH with 0.10 M HCl. Indicate the approximate pH at the beginning of the titration and at the equivalence point. What is the total solution volume at the equivalence point?

41. Without doing detailed calculations, sketch the curve for the titration of 50 mL of 0.050 M pyridine, C_5H_5N (a weak base), with 0.10 M HCl. Indicate the approximate pH at the beginning of the titration and at the equivalence point. What is the total solution volume at the equivalence point?

42. You titrate 25.0 mL of 0.10 M NH_3 with 0.10 M HCl.
(a) What is the pH of the NH_3 solution before the titration begins?
(b) What is the pH at the equivalence point?
(c) What is the pH at the midpoint of the titration?
(d) What indicator in Figure 17.11 would be best to detect the equivalence point?
(e) Calculate the pH of the solution after adding 5.00, 15.0, 20.0, 22.0, and 30.0 mL of the acid. Combine this information with that from the preceding parts of this study question, and plot the titration curve.

43. Construct a rough plot of pH versus volume of base for the titration of 0.050 M HCN with 0.075 M NaOH.
(a) What is the pH before any NaOH is added?
(b) What is the pH at the half-neutralization point?
(c) What is the pH when 95% of the required NaOH has been added?
(d) What volume of base, in milliliters, is required to reach the equivalence point?
(e) What is the pH at the equivalence point?
(f) What indicator would be most suitable? (See Figure 17.11.)
(g) What is the pH when 105% of the required base has been added?

44. The weak base ethanolamine, $HOCH_2CH_2NH_2$, can be titrated with HCl:

$$HOCH_2CH_2NH_2(aq) + H_3O^+(aq) \longrightarrow HOCH_2CH_2NH_3^+(aq) + H_2O(\ell)$$

Assume you have 25.0 mL of a 0.010 M solution of ethanolamine and titrate it with 0.0095 M HCl. (K_b for ethanolamine is 3.2×10^{-5})
(a) What is the pH of the ethanolamine solution before the titration begins?
(b) What is the pH at the equivalence point?
(c) What is the pH at the midpoint of the titration?
(d) What indicator in Figure 17.11 would be best to detect the equivalence point?
(e) Calculate the pH of the solution after adding 5.00, 10.0, 20.0, and 30.0 mL of the acid. Combine this information with that from the preceding parts of this study question, and plot an approximate titration curve.

45. Aniline hydrochloride, $[C_6H_5NH_3]Cl$, is a weak acid. [Its conjugate base is the weak base aniline, $C_6H_5NH_2$ (page 845)]. The acid can be titrated with a strong base such as NaOH:

$$C_6H_5NH_3^+(aq) + OH^-(aq) \longrightarrow C_6H_5NH_2(aq) + H_2O(\ell)$$

Assume 50.0 mL of 0.100 M aniline hydrochloride is titrated with 0.185 M NaOH. (K_a for aniline hydrochloride is 2.4×10^{-5}.)
(a) What is the pH of the $[C_6H_5NH_3]Cl$ solution before the titration begins?
(b) What is the pH at the equivalence point?
(c) What is the pH at the midpoint of the titration?
(d) What indicator in Figure 17.11 would be best to detect the equivalence point?
(e) Calculate the pH of the solution after adding 10.0, 20.0, and 30.0 mL of the base. Combine this information with that from the preceding parts of this study question, and plot an approximate titration curve.

46. Using Figure 17.11, suggest an indicator to use in each of the following titrations:
(a) The weak base pyridine is titrated with HCl
(b) Formic acid is titrated with NaOH
(c) Hydrazine, a weak diprotic base, is titrated with HCl

47. Using Figure 17.11, suggest an indicator to use in each of the following titrations:
(a) Na_2CO_3 is titrated to HCO_3^- with HCl.
(b) Hypochlorous acid is titrated with NaOH.
(c) Trimethylamine is titrated with HCl.

GENERAL QUESTIONS

48. You dissolve 1.00 mol of propanoic acid and 0.40 mol of NaOH in enough water to make 1.00 L of solution.

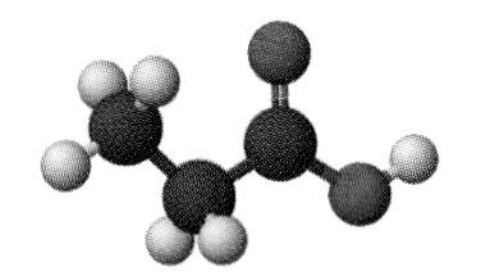

propanoic acid propanoate ion

(a) Write a balanced equation to depict the reaction that occurs.
(b) How many moles of acid and of its conjugate base, the propanoate ions, are present after the reaction?
(c) Calculate the pH of the solution.
(d) Would the pH increase, decrease, or remain the same if 0.40 g of NaOH is added to the solution?

49. If you add 12.5 mL of 4.15 M acetic acid to 25.0 mL of 1.00 M NaOH, what are the hydronium ion concentration and pH of the resulting solution?

50. Calculate the concentrations of NH_4^+, OH^-, NH_3, and Na^+ in a solution that contains 0.040 M NaOH and 0.20 M NH_3.

51. A titration of 25.0 mL of aqueous formic acid, HCO_2H, requires 25.67 mL of 0.275 M NaOH to reach the equivalence point. What are the concentrations of H_3O^+, OH^-, and HCO_2^- at the equivalence point? What is the pH of the solution at the equivalence point? What was the concentration of formic acid in the original solution?

52. What is the pH of a 0.160 M acetic acid solution? If 56.8 g of sodium acetate is added to 1.50 L of the 0.160 M acetic acid solution, what is the new pH of the solution?

53. You dissolve 0.515 g of phenol, C_6H_5OH, a weak organic acid, in enough water to make 1.00×10^2 mL of solution. The resulting solution is titrated with 0.123 M NaOH:

$$C_6H_5OH(aq) + OH^-(aq) \longrightarrow C_6H_5O^-(aq) + H_2O(\ell)$$

What are the concentrations of the following ions at the equivalence point: Na^+, H_3O^+, OH^-, and $C_6H_5O^-$? What is the pH at the equivalence point?

54. Calculate the pH at the equivalence point in a titration of 25.0 mL of 0.120 M formic acid, HCO_2H, with 0.105 M NaOH.

55. A 5.00×10^2-mL solution contains 0.150 mol of $NaNO_2$ and 0.200 mol of the weak acid HNO_2. How many more grams of $NaNO_2$ must be added to this solution to have a pH of 4.00?

56. Arrange the following solutions in order of increasing pH (all reagents are 0.10 M):
(a) NaCl
(b) NH_3
(c) $CH_3CO_2H/NaCH_3CO_2$
(d) HCl
(e) NH_3/NH_4Cl
(f) CH_3CO_2H

57. A buffer solution has an acetic acid concentration of 0.50 M and a sodium acetate concentration of 0.88 M.
(a) What is the pH of the buffer solution?
(b) If 0.10 mol of NaOH is added to 1.00 L of the buffer solution, what is the pH after addition?

58. A 0.30 M solution of a weak acid HA has a pH of 2.25. What is the pH of an equimolar solution of HA and the Na^+ salt of the conjugate base, NaA?

59. You titrate 25.0 mL of a 0.0256 M aqueous solution of the weak base aniline, $C_6H_5NH_2$, with 0.0195 M HCl.
(a) What is the pH of the aniline solution before the titration begins?
(b) What is the pH at the equivalence point of the titration?
(c) What is the pH at the midpoint of the titration?
(d) What indicator in Figure 17.11 would be best to detect the equivalence point?
(e) Calculate the pH of the solution after adding 5.00, 10.0, 15.0, 20.0, 24.0, and 30.0 mL of the acid. Combine this information with that from the preceding parts of this study question, and plot the titration curve.

60. The pH of human blood is controlled by several buffer systems, among them the reaction

$$H_2PO_4^-(aq) + H_2O(\ell) \rightleftharpoons H_3O^+(aq) + HPO_4^{2-}(aq)$$

Calculate the ratio $[H_2PO_4^-]/[HPO_4^{2-}]$ in normal blood having a pH of 7.40.

61. You dissolve 0.221 g of the weak base trimethylamine, $(CH_3)_3N$, in water to give 50.0 mL of solution. This solution is titrated with 0.100 M HCl.
(a) What is the pH of the original amine solution?
(b) What is the pH of the solution at the midpoint of the titration?
(c) What is the pH at the equivalence point?
(d) Sketch a rough titration curve for the titration and decide on a suitable indicator (Figure 17.11).

62. Hydroxylamine is a weak base that readily forms salts such as $[NH_3OH]Cl$, hydroxylamine hydrochloride. This compound is used as a reducing agent in photography and as an antioxidant in soaps. Assume you have 25.0 mL of a 0.155 M solution of $[NH_3OH]Cl$ and titrate it with 0.108 M NaOH:

$$NH_3OH^+(aq) + OH^-(aq) \longrightarrow NH_2OH(aq) + H_2O(\ell)$$

(a) What is the pH of the $[NH_3OH]Cl$ solution before the titration begins?
(b) What is the pH at the equivalence point?
(c) What is the pH at the midpoint of the titration?

63. Three titration curves are pictured here. Each represents the titration of 100 mL of 0.050 M acid (or base) with 0.050 M base (or acid). Four types of titrations are possible: (a) a strong acid titrated with a strong base; (b) a strong base titrated with a strong acid; (c) a weak acid titrated with a strong base; or (d) a weak base titrated with a strong acid. Indicate for each diagram the type of titration and briefly describe your reasoning.

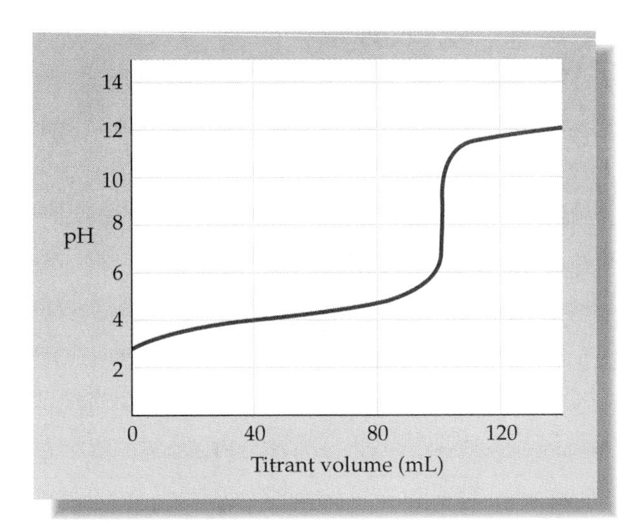

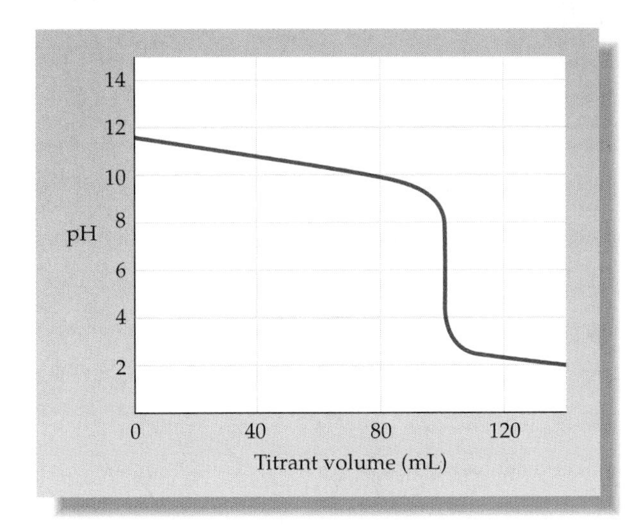

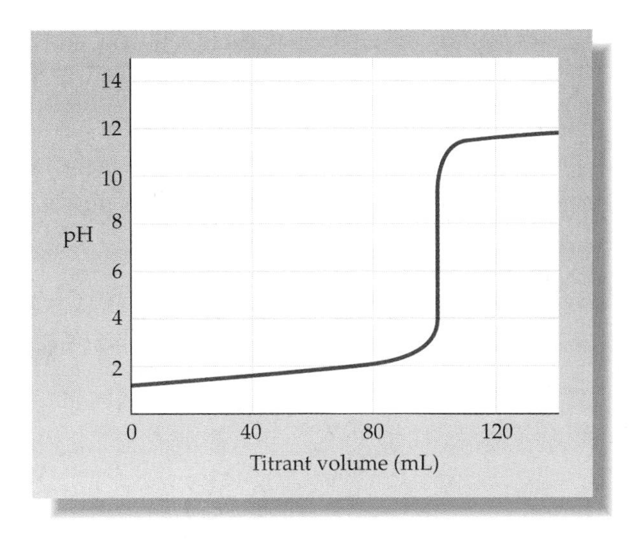

CONCEPTUAL QUESTIONS

64. Composition diagrams, commonly known as "alpha plots," are often used to represent the species in a solution of an acid or base as the pH varies. The diagram for 0.100 M acetic acid is shown here. The plot shows how the mole fraction [= alpha (α)] of acetic acid and its conjugate base, $\{[CH_3CO_2H]/[CH_3CO_2^-]\}$, changes as the pH increases. It is another way of viewing the relative concentrations of acetic acid and acetate ion as a strong base is added to a solution of acetic acid in the course of a titration.

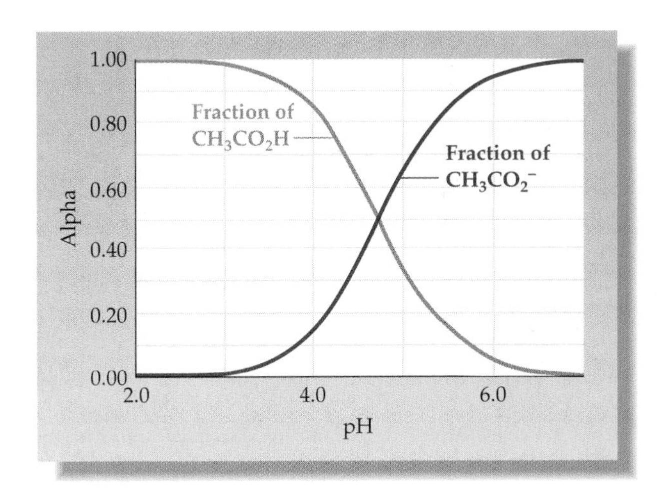

(a) Explain why the fraction of acetic acid declines and that of acetate ion increases as the pH increases.
(b) Which species predominates at a pH of 4, acetic acid or acetate ion? What is the situation at a pH of 6?
(c) Consider the point where the two lines cross. The fraction of acetic acid in the solution is 0.5, and so is that of acetate ion. That is, the concentrations of acid and conjugate base are equal. At this point the graph shows the pH is 4.75. Explain why the pH at this point is 4.75.

65. The composition diagram, or alpha plot, for the important acid–base system of carbonic acid, H_2CO_3, is illustrated here. (See Study Question 64 for more information on such diagrams.)

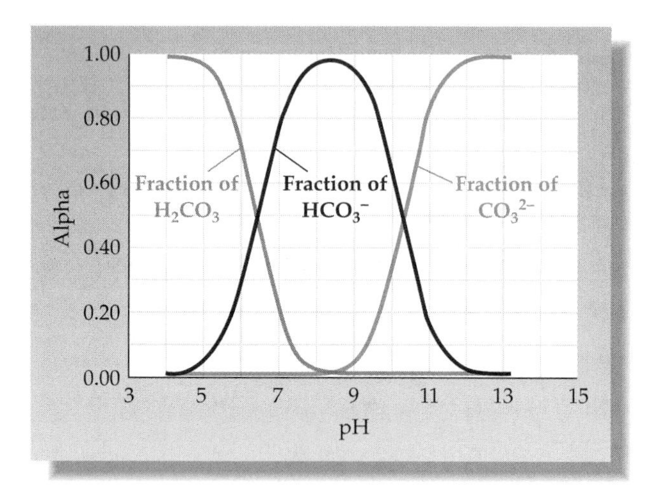

(a) Explain why the fraction of bicarbonate ion, HCO_3^-, rises and then falls as the pH increases.
(b) What is the composition of the solution when the pH is 6.0? When the pH is 10.0?
(c) If you wanted to buffer a solution at a pH of 11.0, what should be the ratio of HCO_3^- to CO_3^{2-}?

66. Two acids, each approximately 10^{-2} M in concentration, are titrated separately with a strong base. The acids show the following pH values at the equivalence point: HA, pH = 9.5 and HB, pH = 8.5.
(a) Which is the stronger acid, HA or HB?
(b) Which conjugate base, A^- or B^-, is stronger?

67. The chapter-opening *Chemical Puzzler* described a cure for alkalosis, an elevation of blood pH. The equilibria involved in blood pH control are

$$CO_2(g) + H_2O(\ell) \rightleftarrows H_2CO_3(aq)$$
$$H_2CO_3(aq) + H_2O(\ell) \rightleftarrows H_3O^+(aq) + HCO_3(aq)$$

(a) Explain why breathing into a paper bag lowers the blood pH.
(b) Acidosis is the opposite of alkalosis; it is a depression of blood pH. In this case you would want to take deeper breaths or breathe more rapidly to exhale more CO_2. Explain why this would cure acidosis.

CHALLENGING QUESTIONS

68. Prove that the pH at the midpoint in the titration of a weak base with a strong acid is pH = 14.00 + log K_b (at 25 °C).

69. When 16.0 mL of 0.20 M benzoic acid, a weak acid, is mixed with 32.0 mL of 0.10 M NH_3, what is the pH of the resulting solution? See Study Question 17-109 for a method of calculating the hydronium ion concentration.

70. The pH at the second equivalence point in the titration of 0.100 M oxalic acid with 0.100 M NaOH was given as approximately 8.4 on page 865. Perform suitable calculations to verify this result.

71. During active exercise, lactic acid ($K_a = 1.4 \times 10^{-4}$) is produced in the muscle tissues (see page 514 and Study Question 18.25). At the pH of the body (pH = 7.4), which form will be primarily present: unionized lactic acid [$CH_3CH(OH)CO_2H$] or the lactate ion [$CH_3CH(OH)CO_2^-$]?

72. How many moles of HCl must be added to 1.00 L of a buffer made from 0.150 M NH_3 and 10.0 g of NH_4Cl to decrease the pH by one unit?

73. Buffer capacity is defined as the number of moles of a strong acid or strong base that are required to change the pH of 1 L of the buffer solution by one unit. What is the buffer capacity of a solution that is 0.10 M in acetic acid and 0.10 M in sodium acetate?

74. Amino acids are the building blocks of proteins, and the simplest of these is glycine, $NH_2CH_2CO_2H$. It can exist in three forms in equilibrium with one another.

$$^+NH_3CH_2CO_2H(aq) + H_2O(\ell) \rightleftarrows {}^+NH_3CH_2CO_2^-(aq) + H_3O^+(aq)$$
$$K_a = 4.5 \times 10^{-3}$$

$$^+NH_3CH_2CO_2^-(aq) + H_2O(\ell) \rightleftarrows NH_2CH_2CO_2^-(aq) + H_3O^+(aq)$$
$$K_a = 1.7 \times 10^{-10}$$

If the pH of a natural system is 7.2, in what form is the amino acid found? Sketch an alpha plot (see Study Question 18.64) for glycine from pH = 1 to pH = 12.

Summary Question

75. The chemical name for aspirin is acetylsalicylic acid. It is believed that the analgesic and other desirable properties of aspirin are due not to the aspirin but to the simpler compound salicylic acid, $C_6H_4(OH)CO_2H$, that results from the breakdown of aspirin in the stomach:

O
‖
C—OH
OH

salicylic acid, $K_a = 1.1 \times 10^{-3}$

(a) Give approximate values for the following bond angles in the acid: (i) C—C—C in the ring; (ii) O—C=O, (iii) either of the C—O—H angles; and (iv) C—C—H.
(b) What is the hybridization of the C atoms of the ring? Of the C atom in the $—CO_2H$ group?
(c) Experiment shows that 1.00 g of the acid will dissolve in 460 mL of water. What is the pH of this solution?
(d) If you have salicylic acid in your stomach, and if the pH of gastric juice is 2.0, calculate the percentage of salicylic acid that will be present in the stomach in the form of the salicylate ion, $C_6H_4(OH)CO_2^-$.
(e) Assume you have 25.0 mL of a 0.014 M solution of salicylic acid and titrate it with 0.010 M NaOH. What is the pH at the midpoint of the titration? What is the pH at the equivalence point?

INTERACTIVE GENERAL CHEMISTRY CD-ROM

Screen 18.2 Acid–Base Reactions

(a) What are the four types of acid–base reactions?
(b) What pH is expected for each of these reactions when an acid has been completed consumed by a base? That is, is the final solution acidic, neutral, or basic?

Screen 18.8 Buffer Solutions

(a) What are the main components of a buffer solution?
(b) What buffer system is used in the kidneys? What is its purpose? Explain how this system works.

Screen 18.9 Buffer Solutions

(a) An explanation of the Henderson-Hasselbalch equation is obtained by pressing the *General Rule* button on this screen. Which is more important in controlling the pH of a solution, the pK_a of the weak acid or the relative concentrations of the acid and its conjugate base?
(b) Examine the sidebar on this screen. Suggest a reason the human body does not use an acetic acid/acetate ion buffer for holding blood near pH 7.4.

Screen 18.12 Titration Curves

Use the *Titration Tool* on this screen to titrate 50.0 mL of 0.1 M acetic acid ($pK_a = 4.75$) with 0.2 M NaOH in 0.10-mL increments.

(a) What should the pH of the solution be before NaOH is added?
(b) What quantity of 0.2 M NaOH is required to reach the equivalence point?
(c) Proceed to the point in the titration where one half of the acid has been neutralized. (This is called the half-neutralization point or midpoint.) What is the pH at this point in the titration? What is its relation to the pK_a of the acid?
(d) What is the pH at the equivalence point? What are the species in solution at this point? Which of these determines the pH at the equivalence point? Explain briefly.

Examine the alpha plot that is displayed following the acetic acid titration.

(e) What is depicted by the line that begins at an alpha of 1.0 and slopes downward as the pH increases?
(f) What is depicted by the line that begins at an alpha of 0.0 and slopes upward?
(g) What is the composition of the solution when the two lines cross? How is this related to the titration of acetic acid with NaOH? How is this related to the pK_a of the acid?
(h) How are alpha plots related to buffer solutions?
(i) What is the pH of a solution that contains 0.90 M acetic acid and 0.10 M acetate ion?
(j) Which species predominates in a solution that has a pH of 5, acetic acid or acetate ion? Explain briefly.

Screen 18.13 Return to the Puzzler

(a) Explain why hyperventilating can cause the pH of blood to become dangerously high.
(b) Why does the loss of $CO_2(g)$ during hyperventilation change blood pH even though CO_2 contains no H atoms?
(c) What is acidosis? What is alkalosis?

Chapter 19

Principles of Reactivity: Precipitation Reactions

(C. D. Winters and Arthur Palmer)

A CHEMICAL PUZZLER

Some caves are formed by dissolving limestone, $CaCO_3$, by organic acids and carbon dioxide in water that trickles through the ground. The stalactites and stalagmites you see, however, are formed by the reprecipitation of the limestone. How are these processes connected? What is their chemistry? (This composite photo shows limestone stalactites and stalagmites in a cave (about 10 m high; Carlsbad Caverns, NM) and the precipitation of calcium carbonate from a solution of calcium hydroxide in contact with a piece of solid CO_2, or dry ice.) *(Cave photo Arthur N. Palmer, dry ice photo C. D. Winters)*

In Chapter 5 you were introduced to several types of reactions, among them precipitation reactions. A precipitation reaction is an exchange reaction in which one of the products is an insoluble compound,

$$MA(aq) + BX(aq) \longrightarrow MX(s) + BA(aq)$$
$$CaCl_2(aq) + Na_2CO_3(aq) \longrightarrow CaCO_3(s) + 2\ NaCl(aq)$$

that is, a compound having a solubility of less than about 0.01 mol of dissolved material per liter of solution. If you stir calcium carbonate (as the mineral calcite) into pure water, only about 6 mg (6×10^{-5} mol) will dissolve per liter at 25 °C. No wonder that corals or sea shells, which are mostly calcium carbonate, do not dissolve appreciably in the sea (Figure 19.1). On the other hand, you know that "sea salt," NaCl, is very soluble in water.

***Figure* 19.1 Corals.** Corals and sea shells do not dissolve appreciably in sea water. They are composed mostly of insoluble calcium carbonate. *(Brian Parker/Tom Stack & Associates)*

How do you know when to predict an insoluble compound as the product of a reaction—and why should you care? In Chapter 5 we described some guidelines for making predictions (Figure 5.4), and we also discussed some important minerals that are insoluble compounds—compounds formed by precipitation reactions. In this chapter we want to make our estimates of solubility more quantitative, and we want to explore the conditions under which some compounds precipitate and others do not. Finally, we want to discuss some more practical aspects of chemical equilibria, this time applied to insoluble substances.

19.1 THE SOLUBILITY OF SALTS

See the *Saunders Interactive General Chemistry CD-ROM,* Screen 19.2, Precipitation Reactions, and Screen 19.3, Solubility.

If we define an insoluble compound as one having a solubility in water of less than about 0.01 mol/L, then many, many compounds fit that description. Black smokers at the bottom of the ocean spew out insoluble metal sulfides, caves are lined with insoluble limestone, and eggs are protected by an insoluble calcium carbonate shell (Figure 19.2).

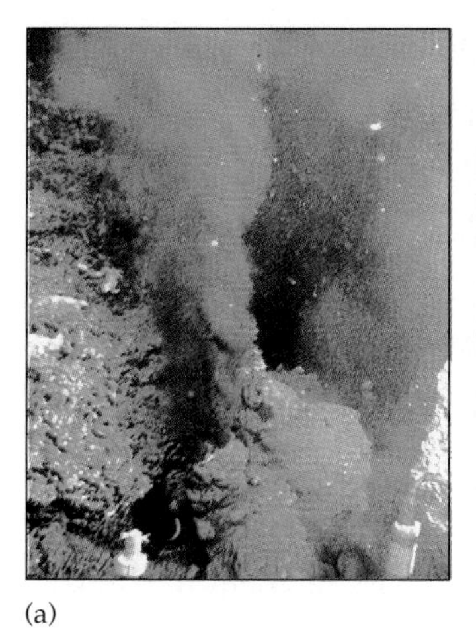

(a)

(b)

(c)

***Figure* 19.2 Insoluble substances in nature.** (a) At the bottom of the ocean there are vents that spew out minerals and other compounds from deep within the earth. As the broth encounters cold water, metal ions and sulfide ions form metal sulfides, which are usually black. (See also page 190.) (b) Stalactites with water drops, from which CO_2 gas is being lost (Jewel Cave, South Dakota). (c) Egg shells are calcium carbonate. *(a, Dudley Foster/Woods Hole Oceanographic Institution; b, Arthur N. Palmer; c, Charles D. Winters.)*

Table 19.1 • SOME COMMON MINERALS

Common Names (and Uses)	Formula	Chemical Name
Calcite, aragonite, iceland spar (source of lime, CaO)	$CaCO_3$	Calcium carbonate
Azurite (one source of copper)	$2\ CuCO_3 \cdot Cu(OH)_2$	—
Cinnabar (source of mercury)	HgS	Mercury(II) sulfide
Zinc blende (one source of zinc)	ZnS	Zinc sulfide
Galena (one source of lead)	PbS	Lead(II) sulfide
Iron pyrite, fool's gold	FeS_2	—
Apatite (source of phosphate for phosphoric acid and fertilizers)	$Ca_{10}(PO_4)_6(OH)_2$	—
Fluorite or fluorspar (source of HF and other inorganic fluorides)	CaF_2	Calcium fluoride
Gypsum	$CaSO_4 \cdot 2\ H_2O$	Calcium sulfate
Magnetite (one source of iron)	Fe_3O_4	—
Rutile (one source of titanium)	TiO_2	Titanium(IV) oxide
Bauxite (source of Al)	Al_2O_3	Aluminum oxide

Insoluble compounds are important because our economy depends on metals and chemicals produced from minerals found in the earth. These minerals are often found in particular geographic regions and are usually insoluble compounds (Table 19.1). Notice that all of the minerals listed in Table 19.1 fit the guidelines for solubility in Figure 5.4.

Because so many important minerals are sulfides, oxides, or carbonates, methods had to be found to recover the metal from the mineral. The refining of nickel is an example of the use of precipitation reactions. Nickel oxide occurs in very low concentrations in iron-bearing rock. To recover the nickel, it is "leached," or extracted, from the rock with sulfuric acid to give soluble nickel(II) sulfate and water, an exchange reaction driven by the formation of water:

$$NiO(s) + H_2SO_4(aq) \longrightarrow NiSO_4(aq) + H_2O(\ell)$$

Then, in another exchange reaction, the nickel(II) ion is recovered by precipitating it with sulfide ion to give black nickel(II) sulfide.

$$NiSO_4(aq) + H_2S(aq) \longrightarrow NiS(s) + H_2SO_4(aq)$$

The recovered nickel, in the form of its sulfide, is then converted to nickel metal or other nickel-containing compounds.

Exercise 19.1 Solubility of Salts

Using the guidelines of Figure 5.4, predict whether each of the following is likely to be soluble or insoluble in water:

1. $AgBr$
2. K_2CrO_4
3. $SrCO_3$
4. $Ba(NO_3)_2$

Exercise 19.2 Precipitation Reactions

Which of the following combinations should lead to a precipitate when 0.10 M solutions are mixed? Give the formula of the precipitate.

1. $NaI(aq) + AgNO_3(aq)$
2. $KCl(aq) + Pb(NO_3)_2(aq)$
3. $CuCl_2(aq) + Mg(NO_3)_2(aq)$

19.2 THE SOLUBILITY PRODUCT CONSTANT, K_{sp}

See the *Saunders Interactive General Chemistry CD-ROM,* Screen 19.4, Solubility Product Constant.

Silver bromide, AgBr, is used in photographic film. This water-insoluble salt can be made by adding a water-soluble silver salt ($AgNO_3$) to an aqueous solution of a bromide-containing salt (KBr) (see also Figure 5.6). The *net ionic equation* for the reaction is

$$Ag^+(aq) + Br^-(aq) \longrightarrow AgBr(s)$$

If some of the precipitated AgBr is placed in pure water, a very small amount of the salt eventually dissolves, and an equilibrium is established:

$$AgBr(s) \rightleftharpoons Ag^+(aq,\ 5.7 \times 10^{-7}\ M) + Br^-(aq,\ 5.7 \times 10^{-7}\ M)$$

When the AgBr has dissolved to the greatest extent possible, the solution is said to be *saturated* (see Section 14.2), and experiment shows that the concentrations of the silver and bromide ions in the solution are each about 5.7×10^{-7} M at 25 °C. The extent to which an insoluble salt dissolves can be expressed in terms of the equilibrium constant for the dissolving process. In this case,

$$K = [Ag^+(aq)][Br^-(aq)]$$

Recall from Chapter 16 that the "concentration" of a solid does not appear in the equilibrium constant expression.

When silver bromide dissolves in water, one mole of Ag^+ ions and one mole of Br^- ions are produced for every mole of AgBr dissolved.* The solubility of silver bromide can therefore be determined if the concentration of *either* the silver ion *or* the bromide ion is measured. This means that the previous equilibrium expression tells us that the *product* of the two concentrations measures the *solubility* of the solid compound and is a *constant.* Hence, this constant has come to be called the **solubility product constant,** and it is often designated by K_{sp}.

$$K_{sp} = [Ag^+(aq)][Br^-(aq)]$$

Because the concentrations of both Ag^+ and Br^- are 5.7×10^{-7} M when silver bromide is in equilibrium with its ions, K_{sp} is

$$K_{sp} = [5.7 \times 10^{-7}][5.7 \times 10^{-7}] = 3.3 \times 10^{-13} \text{ (at 25 °C)}$$

As with any equilibrium constant, K_{sp} values change with temperature.

*Dissolving ionic solids is a complex process. It almost always involves more than just the simple equilibria shown here. As a result, you should realize that simple K_{sp} calculations can be inaccurate. In particular, the calculated solubility can be incorrect, especially for salts having ions with charges larger than 1+ or 1−. It has been noted that "the solubility product [is more properly] a description, not of solubility, but simply of the ionic concentrations in the saturated solution." (Meites, L., Pode, J. S. F., and Thomas, H. C.: "Are solubilities and solubility products related?" *Journal of Chemical Education,* Vol. 43, p. 667, 1966.)

Table 19.2 • K_{sp} VALUES FOR SOME INSOLUBLE SALTS

Compound	K_{sp} at 25 °C
$CaCO_3$	3.8×10^{-9}
$SrCO_3$	9.4×10^{-10}
$BaCO_3$	8.1×10^{-9}
$BaSO_4$	1.1×10^{-10}
CaF_2	3.9×10^{-11}
CdS	1.0×10^{-27}
PbS	3.2×10^{-28}
CuS	7.9×10^{-37}
HgS	2.0×10^{-53}
AgCl	1.8×10^{-10}
AgBr	3.3×10^{-13}
AgI	1.5×10^{-16}
Ag_2CrO_4	9.0×10^{-12}
$PbCl_2$	1.7×10^{-5}
$PbCrO_4$	1.8×10^{-14}
Hg_2Cl_2	1.1×10^{-18}

The solubility product constant K_{sp} always has the form

$$A_xB_y(s) \rightleftharpoons xA^{y+}(aq) + yB^{x-}(aq) \qquad K_{sp} = [A^{y+}]^x[B^{x-}]^y$$

This means that you write K_{sp} expressions for the dissolving of other salts as follows

$$CaF_2(s) \rightleftharpoons Ca^{2+}(aq) + 2\ F^-(aq) \qquad K_{sp} = [Ca^{2+}][F^-]^2 = 3.9 \times 10^{-11}$$

$$Ag_2SO_4(s) \rightleftharpoons 2\ Ag^+(aq) + SO_4^{2-}(aq) \quad K_{sp} = [Ag^+]^2[SO_4^{2-}] = 1.7 \times 10^{-5}$$

The numerical values of K_{sp} for a few salts are given in Table 19.2, and many more values are collected in Appendix J. Notice that all are given for a temperature of 25 °C.

Finally, be sure not to confuse *solubility* with the *solubility product*. The solubility of a salt is the quantity present in a unit amount of a saturated solution, expressed in moles per liter, grams per 100 mL, or other units. The *solubility product* is an equilibrium constant. Nonetheless, there is a connection between them: if either is known, the other can be calculated.

***Exercise* 19.3** Writing K_{sp} Expressions

Write K_{sp} expressions for the following insoluble salts, and look up numerical values for the constant in Appendix J.

1. $BaSO_4$
2. BiI_3
3. Ag_2CO_3

19.3 DETERMINING K_{sp} FROM EXPERIMENTAL MEASUREMENTS

See the *Saunders Interactive General Chemistry CD-ROM,* Screen 19.5, Determining K_{sp}.

In practice, solubility product constants are determined by careful laboratory measurements using various chemical and spectroscopic methods. We shall not go into these methods but shall assume that the given ion concentrations have been measured experimentally.

***Example* 19.1** K_{sp} from Solubility Measurements

The solubility of silver iodide, AgI, is 1.22×10^{-8} mol/L at 25 °C. Calculate K_{sp} for AgI.

Solution When silver iodide dissolves in water, the following reaction and K_{sp} expression are appropriate:

$$AgI(s) \rightleftharpoons Ag^+(aq) + I^-(aq) \qquad K_{sp} = [Ag^+][I^-]$$

We know from experiment that the solubility of AgI is 1.22×10^{-8} mol/L at 25 °C. Because each mole of AgI that dissolves provides a mole of Ag^+ ions and a mole of I^- ions, the concentration of each ion is 1.22×10^{-8} M. Therefore,

$$K_{sp} = [Ag^+][I^-] = (1.22 \times 10^{-8})(1.22 \times 10^{-8}) = 1.49 \times 10^{-16}$$

Example 19.2 K_{sp} from Solubility Measurements

Lead(II) chloride dissolves to a slight extent in water:

$$PbCl_2(s) \rightleftharpoons Pb^{2+}(aq) + 2\ Cl^-(aq)$$

Calculate K_{sp} if the lead ion concentration is found to be 1.62×10^{-2} mol/L.

Solution The balanced equation for dissolving $PbCl_2$ shows that the concentration of Cl^- ion must be twice the Pb^{2+} ion concentration. If

$$[Pb^{2+}] = 1.62 \times 10^{-2}\ M \quad \text{then} \quad [Cl^-] = 2 \times [Pb^{2+}] = 3.24 \times 10^{-2}\ M$$

This means the solubility product constant is

$$K_{sp} = [Pb^{2+}][Cl^-]^2 = (1.62 \times 10^{-2})(3.24 \times 10^{-2})^2 = 1.70 \times 10^{-5}$$

Exercise 19.4 K_{sp} from Solubility Measurements

The barium ion concentration, $[Ba^{2+}]$, in a saturated solution of barium fluoride is 7.5×10^{-3} M. Calculate the K_{sp} for BaF_2:

$$BaF_2(s) \rightleftharpoons Ba^{2+}(aq) + 2\ F^-(aq)$$

19.4 ESTIMATING SALT SOLUBILITY FROM K_{sp}

The K_{sp} values for many insoluble salts have been determined through a variety of laboratory measurements (Appendix J). These are an invaluable aid because they can be used to estimate the solubility of a solid salt or to determine if a solid will precipitate if solutions of its anion and cation are mixed. We turn now to methods that can be used to estimate the solubility of a salt from its K_{sp} value. Later we shall see how to use these predictions to plan the separation of ions that are mixed in solution (see Section 19.7).

See the *Saunders Interactive General Chemistry CD-ROM,* Screen 19.6, Estimating Salt Solubility.

Example 19.3 Solubility from K_{sp}

The K_{sp} for $CaCO_3$ (as the mineral calcite) is 3.8×10^{-9} at 25 °C. Calculate the solubility of calcium carbonate in pure water in (1) moles per liter and (2) grams per liter.

Solution The equation for the solubility of $CaCO_3$ is

$$CaCO_3(s) \rightleftharpoons Ca^{2+}(aq) + CO_3^{2-}(aq)$$

and the equilibrium expression is $K_{sp} = [Ca^{2+}][CO_3^{2-}]$. When 1 mol of calcium carbonate dissolves, 1 mol of Ca^{2+} and 1 mol of CO_3^{2-} ions are produced. Thus, the solubility of $CaCO_3$ can be measured by determining the concentration of *either* Ca^{2+} *or* CO_3^{2-}. Let us denote the solubility of $CaCO_3$ (in moles per liter) by x; that is, x mol of $CaCO_3$ dissolves per liter. Both $[Ca^{2+}]$ and $[CO_3^{2-}]$ must therefore also equal x at equilibrium.

Equilibria involving the carbonate ion are actually more complex than can be presented here. The solubility of $CaCO_3$ is best defined only in terms of $[Ca^{2+}]$ because hydrolysis of CO_3^{2-} actually makes $[CO_3^{2-}]$ slightly less than $[Ca^{2+}]$. See Section 19.11.

Equation	$CaCO_3(s) \rightleftharpoons$	$Ca^{2+}(aq)$ +	$CO_3^{2-}(aq)$
Initial concentration (M)		0	0
Change on proceeding to equilibrium		$+x$	$+x$
Equilibrium concentration (M)		x	x

Because K_{sp} is the product of the calcium and carbonate ion concentrations, K_{sp} is the square of the solubility x,

$$K_{sp} = [Ca^{2+}][CO_3^{2-}] = 3.8 \times 10^{-9} = (x)(x) = x^2$$

and so the value of x is

$$x = \sqrt{3.8 \times 10^{-9}} = 6.2 \times 10^{-5} \text{ M}$$

The solubility of $CaCO_3$ in pure water is 6.2×10^{-5} mol/L. To find its solubility in grams per liter, we need only multiply by the molar mass of $CaCO_3$.

$$\text{Solubility in grams per liter} = (6.2 \times 10^{-5} \text{ mol/L})(100. \text{ g/mol}) = 0.0062 \text{ g/L}$$

Example 19.4 Solubility from K_{sp}

Knowing that the K_{sp} of CaF_2 is 3.9×10^{-11}, calculate the solubility of the salt (Figure 19.3) in (1) moles per liter and (2) grams per liter.

Solution As usual, we begin by writing the equilibrium equation and the K_{sp} expression:

$$CaF_2(s) \rightleftharpoons Ca^{2+}(aq) + 2\,F^-(aq) \qquad K_{sp} = [Ca^{2+}][F^-]^2$$

The next problem is to define the salt solubility in terms that allow us to solve the K_{sp} expression for this value. From the balanced equation we know that, if 1 mol of CaF_2 dissolves, 1 mol of Ca^{2+} and 2 mol of F^- appear in the solution. This means that the CaF_2 solubility is equivalent to the concentration of Ca^{2+} in the solution. If the solubility of CaF_2 is given by the unknown quantity x, then $[Ca^{2+}] = x$ and $[F^-] = 2x$.

Figure **19.3** **Fluorite.** The mineral fluorite (or fluorospar), CaF_2, has a face-centered cubic lattice of calcium ions with fluoride ions in the tetrahedral holes in the lattice. Millions of tons are used on a worldwide basis, most coming from Mexico, Mongolia, China, and the former Soviet Union. It is used to make hydrogen fluoride, which is used in the manufacture of other chemicals such as fluorocarbons and aluminum. *(Photo, C. D. Winters; model, S. M. Young)*

Equation	$CaF_2(s) \rightleftharpoons$	$Ca^{2+}(aq)$ +	$2\,F^-(aq)$
Initial concentration (M)		0	0
Change on proceeding to equilibrium		$+x$	$+2x$
Equilibrium concentration (M)		x	$2x$

Substituting these values into the K_{sp} expression, we find

$$K_{sp} = [Ca^{2+}][F^-]^2 = (x)(2x)^2 = 4x^3$$

Solving the equation for x,

$$x = \sqrt[3]{\frac{K_{sp}}{4}} = \sqrt[3]{\frac{3.9 \times 10^{-11}}{4}} = 2.1 \times 10^{-4}\ \text{M}$$

we find that 2.1×10^{-4} mol of CaF_2 dissolves per liter to give

$$[Ca^{2+}] = x = 2.1 \times 10^{-4}\ \text{M}$$
$$[F^-] = 2x = 2 \times (2.1 \times 10^{-4}) = 4.2 \times 10^{-4}\ \text{M}$$

Finally, the solubility of CaF_2 in grams per liter is

$$(2.1 \times 10^{-4}\ \text{mol/L})(78.1\ \text{g/mol}) = 0.016\ \text{g}\ CaF_2/\text{L}$$

Exercise 19.5 Salt Solubility from K_{sp}

Using the values of K_{sp} in Appendix J, calculate the solubility of CuI and of $Mg(OH)_2$ in moles per liter.

Problem-Solving Tips and Ideas

19.1 Seeing Double?

Problems like Example 19.4 (and Example 19.2) often provoke such questions as "Aren't you counting things twice when you multiply x by 2 and then square it as well?" in the expression $K_{sp} = (x)(2x)^2$. The answer is no. The quantities that belong in the equation are the *concentration* of Ca^{2+} and the *concentration* of F^-.

$$K_{sp} = (\text{concentration of } Ca^{2+})(\text{concentration of } F^-)^2$$

From the formula of the compound we know that the concentration of F^- must be *twice* that of calcium: if $[Ca^{2+}] = x$ then $[F^-]$ must equal $2x$. Having defined the ion concentration, we then square the concentration of F^- as the equilibrium expression demands.

The relative solubilities of salts can often be deduced by comparing values of solubility product constants, but you must be careful! For example, the K_{sp} for silver chloride is

$$AgCl(s) \rightleftharpoons Ag^+(aq) + Cl^-(aq) \qquad K_{sp} = 1.8 \times 10^{-10}$$

whereas that for silver chromate is

$$Ag_2CrO_4(s) \rightleftharpoons 2\,Ag^+(aq) + CrO_4^{2-}(aq) \qquad K_{sp} = 9.0 \times 10^{-12}$$

In spite of the fact that silver chromate has a numerically smaller K_{sp} value, it is about ten times more soluble than silver chloride. If you solve for their solubilities as in the previous examples, you find the solubility S for AgCl is 1.3×10^{-5} mol/L, whereas $S(Ag_2CrO_4) = 1.3 \times 10^{-4}$ mol/L. Direct comparisons of solubility on the basis of K_{sp} values can *only* be made for salts having the same ion ratio. This means, for example, that you can directly compare solubilities of 1:1 salts such as the silver halides by comparing their K_{sp} values. The K_{sp} values for the compounds of Ag^+ with Cl^-, Br^-, and I^- are in the order

$$AgI\ (K_{sp} = 1.5 \times 10^{-16}) < AgBr\ (K_{sp} = 3.3 \times 10^{-13}) < AgCl\ (K_{sp} = 1.8 \times 10^{-10})$$

and so their solubilities are: $S(AgI) < S(AgBr) < S(AgCl)$. Similarly, you can compare 1:2 salts such as the lead halides:

$$PbI_2\ (K_{sp} = 8.7 \times 10^{-9}) < PbBr_2\ (K_{sp} = 6.3 \times 10^{-6}) < PbCl_2\ (K_{sp} = 1.7 \times 10^{-5})$$

$$\xrightarrow[\text{Increase in } K_{sp} \text{ and increase in solubility}]{S(PbI_2) < S(PbBr_2) < S(PbCl_2)}$$

Exercise 19.6 Comparing Salt Solubilities

Using K_{sp} values, tell which salt in each pair is more soluble in water.

1. AgCl or AgCN
2. $Mg(OH)_2$ or $Ca(OH)_2$
3. $MgCO_3$ or $CaCO_3$

An interesting example of the practical consequences of relative solubilities of insoluble salts occurs in the sea. One group of marine animals, the pteropods, forms the mineral aragonite, one type of $CaCO_3$. Most marine organisms, however, form calcite, another type of calcium carbonate. Measurements at 2 °C and at depths of several thousand meters show that aragonite is about 1.5 times more soluble than calcite. This suggests why pteropod shells are not found at all beyond depths of a few hundred meters in the Pacific Ocean, whereas substantial calcite deposits are found to depths of 3500 m. The solubility of $CaCO_3$ in sea water in either mineral form is given by the reaction

$$CaCO_3(s) \rightleftharpoons Ca^{2+}(aq) + CO_3^{2-}(aq)$$

Because this is a dynamic equilibrium, aragonite (the more soluble form) dissolves, and the less soluble form (calcite) precipitates.

A cluster of aragonite needles formed by precipitation in Jewel Cave, Wind Cave National Park (South Dakota). Aragonite is a form of calcium carbonate. *(Arthur N. Palmer)*

See the *Saunders Interactive General Chemistry CD-ROM,* Screen 19.7, Can a Precipitation Reaction Occur?

19.5 PRECIPITATION OF INSOLUBLE SALTS

Metal-bearing ores often contain the metal in the form of an insoluble salt (Figures 19.3 and 19.4), and, to complicate matters, the ores often contain several such metal salts. Virtually all industrial methods for separating metals from their ores involve dissolving the metal salts to obtain the metal ion or ions in solution. The solution is then usually concentrated in some manner, and a precipitating agent is added to precipitate selectively only one type of metal ion as an insoluble salt. In the case of nickel, the ion can be precipitated as insoluble nickel(II) sulfide or nickel(II) carbonate:

$$Ni^{2+}(aq) + S^{2-}(aq) \rightleftharpoons NiS(s) \qquad K = \frac{1}{K_{sp}} = 3.3 \times 10^{20}$$

$$Ni^{2+}(aq) + CO_3{}^{2-}(aq) \rightleftharpoons NiCO_3(s) \qquad K = \frac{1}{K_{sp}} = 1.5 \times 10^{8}$$

The final step in obtaining the metal is to reduce the ion to the metal either chemically or electrochemically (see Chapter 21).

The goal of this section is to work out methods for determining if a precipitate forms under a given set of conditions. For example, if Ag^+ and Cl^- are present at some given concentrations, does AgCl precipitate from the solution?

Figure 19.4 Minerals are often insoluble Compounds. Besides iron pyrite (FeS_2, Figure 1.4), This group includes such minerals as (left) goethite, a mixture of iron (III) oxide and iron (III) hydroxide, (center) fluorite (calcium fluoride, CaF_2), and (right) hematite [iron (III) oxide, Fe_2O_3]. *(C. D. Winters)*

K_{sp} and the Reaction Quotient Q

Silver chloride, like silver bromide, is used in photographic films. It dissolves to a very small extent in water and has a correspondingly small value of K_{sp}:

$$AgCl(s) \rightleftharpoons Ag^+(aq) + Cl^-(aq) \qquad K_{sp} = [Ag^+][Cl^-] = 1.8 \times 10^{-10}$$

But let us look at the problem from the other direction: If a solution contains Ag^+ and Cl^- ions does AgCl precipitate from solution? This is the same question we asked in Section 16.3 when we wanted to know if a given mixture of reactants and products was an equilibrium mixture or if the reactants continued to form products or if products reverted to reactants. The solution there was to calculate the **reaction quotient Q.**

The "reaction quotient" was introduced in Section 16.3. Some chemists prefer to call the reaction quotient the "ion product."

Recall that the reaction quotient expression is the same as that for the equilibrium constant. For silver chloride, Q is given by

$$Q = [Ag^+][Cl^-]$$

The difference between Q and K is that the concentrations required in the reaction quotient expression may or may not be those at equilibrium.

Based on the discussion in Section 16.3, we can reach the following important conclusions for a slightly soluble salt such as AgCl:

1. **If $Q = K_{sp}$, the system is at equilibrium.** When the product of the ion concentrations is equal to K_{sp}, the solution is **saturated.** No more solid AgCl will dissolve.
2. **If $Q < K_{sp}$, the system is not at equilibrium; the solution is *not* saturated.** This means one of two things: (a) If solid AgCl is present, more will dissolve until equilibrium is achieved (when $Q = K_{sp}$). (b) If solid AgCl is not already present, more $Ag^+(aq)$ or more $Cl^-(aq)$ (or more of both) can be added to the solution until precipitation of solid AgCl begins (when $Q = K_{sp}$).
3. **If $Q > K_{sp}$, the system is not at equilibrium; the solution is *super*saturated.** The concentrations of Ag^+ and Cl^- in solution are too high, and AgCl will precipitate (until $Q = K_{sp}$).

Example 19.5 Solubility and the Reaction Quotient

Some solid AgCl has been placed in a beaker of water. After some time, experiment shows that the concentrations of Ag^+ and Cl^- are each 1.2×10^{-5} mol/L.

Has the system reached equilibrium? That is, is the solution saturated? If not, will more AgCl dissolve?

Solution To solve the problem compare Q and K_{sp}.

$$Q = [Ag^+][Cl^-] = (1.2 \times 10^{-5})(1.2 \times 10^{-5}) = 1.4 \times 10^{-10}$$

$$K_{sp} = 1.8 \times 10^{-10}$$

Here Q is less than K_{sp}. The solution is therefore not yet saturated, and AgCl will continue to dissolve until $Q = K_{sp}$, at which point $[Ag^+] = [Cl^-] = 1.3 \times 10^{-5}$ M. That is, an additional 0.1×10^{-5} mol of AgCl (about 0.14 mg) will dissolve per liter.

Lead(II) iodide is precipitated by mixing solutions of lead(II) nitrate and potassium iodide. See Exercise 19.7. (The model shows solid PbI_2.) *(Photo, C. D. Winters; model, S. M. Young)*

Exercise 19.7 Solubility and the Reaction Quotient

Solid PbI_2 ($K_{sp} = 8.7 \times 10^{-9}$) is placed in a beaker of water. After some time, the lead(II) concentration is measured and found to be 1.1×10^{-3} M. Has the system reached equilibrium? That is, is the solution saturated? If not, will more PbI_2 dissolve?

K_{sp} and Precipitation Reactions

Knowing the reaction quotient for a precipitation reaction, we can decide (1) if a precipitate forms when the ion concentrations are known or (2) what concentrations of ions are required to begin the precipitation of an insoluble salt.

Example 19.6 Deciding Whether a Precipitate Forms

Suppose the concentration of aqueous nickel(II) ion in a solution is 1.5×10^{-6} M. If enough Na_2CO_3 is added to make the solution 6.0×10^{-4} M in carbonate ion, CO_3^{2-}, does precipitation of $NiCO_3$ occur? If not, does it occur if the concentration of CO_3^{2-} is raised by a factor of 100?

Solution The insoluble salt $NiCO_3$ dissolves according to the balanced equation

$$NiCO_3(s) \rightleftharpoons Ni^{2+}(aq) + CO_3^{2-}(aq) \quad K_{sp} = 6.6 \times 10^{-9}$$

and the solubility product expression is $K_{sp} = [Ni^{2+}][CO_3^{2-}]$. When the concentrations of nickel(II) and carbonate ion are those stated earlier,

$$Q = [Ni^{2+}][CO_3^{2-}] = (1.5 \times 10^{-6})(6.0 \times 10^{-4}) = 9.0 \times 10^{-10}$$
$$Q\ (9.0 \times 10^{-10}) < K_{sp}\ (6.6 \times 10^{-9})$$

It is evident that the reaction quotient is less than the value of K_{sp}. The solution is therefore not saturated, and precipitation does *not* occur.

If $[CO_3^{2-}]$ is increased by a factor of 100, it becomes 6.0×10^{-2} M. Under these circumstances, Q is 9.0×10^{-8}.

$$Q = [Ni^{2+}][CO_3^{2-}] = (1.5 \times 10^{-6})(6.0 \times 10^{-2}) = 9.0 \times 10^{-8}$$

a value larger than K_{sp}. Precipitation of $NiCO_3$ occurs and continues until the Ni^{2+} and CO_3^{2-} concentrations have declined to the point at which their product is equal to K_{sp}.

The mineral celestite is largely strontium sulfate. Its clear, light blue color led to its name, which was derived from the Latin *caelestis,* meaning celestial. See Exercise 19.8. *(C. D. Winters)*

Exercise 19.8 Deciding Whether a Precipitate Forms

If the concentration of strontium ion is 2.5×10^{-4} M, does precipitation of $SrSO_4$ occur when enough of the soluble salt Na_2SO_4 is added to make the solution 2.5×10^{-4} M in SO_4^{2-}? K_{sp} for $SrSO_4$ is 2.8×10^{-7}.

Now that we know how to decide if a precipitate forms when the concentration of each ion is known, let us turn to the problem of deciding how much of the precipitating agent is required to begin the precipitation of an ion at a given concentration level.

Example 19.7 Ion Concentrations Required to Begin Precipitation

The concentration of barium ion, Ba^{2+}, in a solution is 0.010 M.

1. What concentration of sulfate ion, SO_4^{2-}, is required to just begin precipitating $BaSO_4$ (Figure 19.5)?
2. When the concentration of sulfate ion in the solution reaches 0.015 M, what concentration of barium ion remains in solution?

(a)

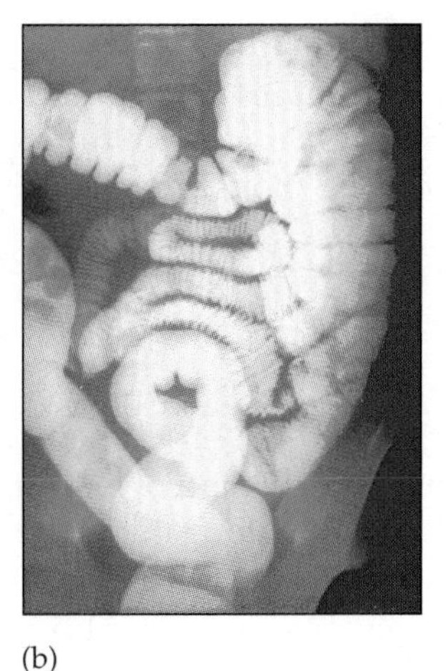

(b)

Figure 19.5 **Barium sulfate.** Barium sulfate, a white solid, is quite insoluble in water ($K_{sp} = 1.1 \times 10^{-10}$) (see Example 19.7). (a) A sample of the mineral barite, which is mostly barium sulfate. (b) Barium sulfate is opaque to x-rays; therefore, if you drink a "cocktail" containing $BaSO_4$ it does not dissolve in your stomach or intestines. Its progress through your digestive organs can be followed by x-ray analysis. This photo is an x-ray film of a person's gastrointestinal tract after the person ingested barium sulfate. It is fortunate that $BaSO_4$ is so insoluble, because water- and acid-soluble barium salts are toxic. *(a, C. D. Winters; b, Susan Leavines/Science Source/Photo Researchers, Inc.)*

Solution As usual, let us begin by writing the balanced equation for the equilibrium that exists when $BaSO_4$ precipitates and the K_{sp} expression:

$$BaSO_4(s) \rightleftharpoons Ba^{2+}(aq) + SO_4^{2-}(aq) \qquad K_{sp} = [Ba^{2+}][SO_4^{2-}] = 1.1 \times 10^{-10}$$

1. From the K_{sp} expression, we know that, when the product of the ion concentrations exceeds 1.1×10^{-10}, that is, when $Q > K_{sp}$, precipitation occurs. The Ba^{2+} ion concentration is known (0.010 M), so the SO_4^{2-} ion concentration that leads to precipitation can be calculated:

$$K_{sp} = 1.1 \times 10^{-10} = [0.010][SO_4^{2-}]$$

$$[SO_4^{2-}] = \frac{K_{sp}}{[Ba^{2+}]} = \frac{1.1 \times 10^{-10}}{0.010} = 1.1 \times 10^{-8}\ M$$

The result tells us that if the sulfate ion is just *slightly* greater than 1.1×10^{-8} M, $BaSO_4$ begins to precipitate; $Q = [Ba^{2+}][SO_4^{2-}]$ would then be greater than K_{sp}.

2. If the sulfate ion concentration is increased to 0.015 M, the maximum concentration of Ba^{2+} ion that can exist in solution (in equilibrium with $BaSO_4$) is

$$[Ba^{2+}] = \frac{K_{sp}}{[SO_4^{2-}]} = \frac{1.1 \times 10^{-10}}{0.015} = 7.3 \times 10^{-9}\ M$$

The fact that the barium ion concentration is so small under these circumstances means that the Ba^{2+} ion has been essentially completely removed from solution. (It began at 0.010 M and has dropped by a factor of about 10 million.) For all practical purposes, the Ba^{2+} ion precipitation is complete.

Exercise 19.9 Ion Concentrations Required to Begin Precipitation

What is the minimum concentration of I^- that can cause precipitation of PbI_2 from a 0.050 M solution of $Pb(NO_3)_2$? K_{sp} for PbI_2 is 8.7×10^{-9}. What concentration of Pb^{2+} ions remains in solution when the concentration of I^- is 0.0015 M?

Example 19.8 K_{sp} and Precipitations

Suppose you mix 100.0 mL of 0.0200 M $BaCl_2$ with 50.0 mL of 0.0300 M Na_2SO_4. Does $BaSO_4$ precipitate ($K_{sp} = 1.1 \times 10^{-10}$) (see Figure 19.5)?

Solution To answer this question, we need to know the concentrations of the Ba^{2+} and SO_4^{2-} ions. We know these concentrations in the original solutions but not in the solution after they have been mixed. We therefore first calculate these new concentrations in the total solution volume of 150.0 mL:

$$[Ba^{2+}] = \frac{(0.0200\ mol/L)(0.1000\ \cancel{L})}{0.1500\ \cancel{L}} = 0.0133\ M \text{ after mixing}$$

$$[SO_4^{2-}] = \frac{(0.0300\ mol/L)(0.0500\ \cancel{L})}{0.1500\ \cancel{L}} = 0.0100\ M \text{ after mixing}$$

Now the reaction quotient can be calculated:

$$Q = [Ba^{2+}][SO_4^{2-}] = (0.0133)(0.0100) = 1.33 \times 10^{-4}$$

Because Q is greater than K_{sp}, precipitation of $BaSO_4$ occurs.

Exercise 19.10 K_{sp} and Precipitations

You have 100.0 mL of 0.0010 M silver nitrate. Does AgCl precipitate if you add 5.0 mL of 0.025 M HCl?

PROBLEM-SOLVING TIPS AND IDEAS

19.2 Does a Precipitate Form?

One difficulty students have when solving a problem such as "If the concentrations of the ions Ag^+ and Cl^- are 1.0×10^{-5} M, does a precipitate of AgCl form?" is the temptation to write the equation

$$Ag^+(aq) + Cl^-(aq) \rightleftharpoons AgCl(s)$$

Instead, approach *all* problems involving "insoluble" salts from the viewpoint of the salt *dissolving* in water,

$$AgCl(s) \rightleftharpoons Ag^+(aq) + Cl^-(aq)$$

and use the K_{sp} expression with the appropriate ion concentrations.

19.6 SOLUBILITY AND THE COMMON ION EFFECT

The test tube on the left in Figure 19.6 contains a precipitate of silver acetate, $AgCH_3CO_2$, in water. The solution is saturated, so that silver ions and acetate ions occur in solution in equilibrium with solid silver acetate:

$$AgCH_3CO_2(s) \rightleftharpoons Ag^+(aq) + CH_3CO_2^-(aq)$$

But what happens if some excess silver ions are added, say by adding silver nitrate? Le Chatelier's principle (Section 16.6) suggests that more precipitate should form because a product ion has been added, thus causing the equilibrium position to shift to the left. This is indeed observed, as illustrated by the test tube on the right in Figure 19.6.

The ionization of weak acids and bases is affected by the presence of an ion common to the equilibrium process (Section 18.2), and the effect of adding silver ion to the saturated silver acetate solution is another example of the *common ion effect.* In the case of slightly soluble salts, the effect is always to lower the salt solubility.

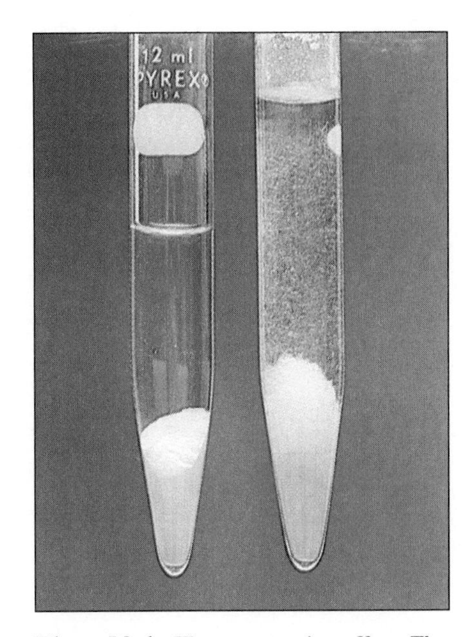

***Figure* 19.6 The common ion effect.** The tube at the left contains a saturated solution of silver acetate, $AgCH_3CO_2$. When 1 M $AgNO_3$ is added to the tube, the equilibrium

$$AgCH_3CO_2(s) \rightleftharpoons Ag^+(aq) + CH_3CO_2^-(aq)$$

shifts to the left, as evidenced by the tube at the right, where more solid silver acetate has formed. *(C. D. Winters)*

See the *Saunders Interactive (General Chemistry CD-ROM,* Screen 19.8, The Common Ion Effect.

Example 19.9 The Common Ion Effect and Salt Solubility

The solubility of AgCl is 1.3×10^{-5} M (0.0019 g/L). If some AgCl is placed in 1.00 L of a 0.55 M solution of NaCl, how many grams of AgCl dissolve?

Solution In *pure* water, the solubility of AgCl is equal to either $[Ag^+]$ or $[Cl^-]$.

$$\text{Solubility of AgCl in pure water} \equiv [Ag^+] \text{ or } [Cl^-] = \sqrt{K_{sp}} = 1.3 \times 10^{-5} \text{ M}$$

In water already containing the Cl^- ion, however, Le Chatelier's principle predicts that the solubility is *less than* 1.3×10^{-5} mol/L. In addition, the solubility can now be expressed only in terms of the concentration of the silver ion.

$$\text{Solubility of AgCl in presence of added } Cl^- \equiv [Ag^+] = x < 1.3 \times 10^{-5} \text{ M}$$

$$AgCl(s) \rightleftharpoons Ag^+(aq) + Cl^-(aq)$$

⟵ Excess Cl^- shifts equilibrium to the left—

To solve for the solubility of AgCl in the presence of added Cl^- ion, set up the usual table, which shows the concentrations of Ag^+ and Cl^- when equilibrium is attained.

Equation	**$AgCl(s) \rightleftharpoons$ $Ag^+(aq)$**	**+ $Cl^-(aq)$**
Initial concentration (M)	0	0.55
Change on proceeding to equilibrium	$+x$	$+x$
Equilibrium concentration (M)	x	$(x + 0.55)$

Some AgCl dissolves in the presence of chloride ion and produces Ag^+ and Cl^- ion concentrations of x mol/L. Some chloride ion was already present, however, so the total chloride ion concentration is the amount coming from AgCl (equals x) plus what was already there (0.55 M).

Now the K_{sp} expression can be written using the equilibrium concentrations from the table:

$$K_{sp} = 1.8 \times 10^{-10} = [Ag^+][Cl^-] = (x)(x + 0.55)$$

and rearranged to

$$x^2 + 0.55x - K_{sp} = 0$$

This is a quadratic equation and can be solved by the methods in Appendix A. The easier approach, however, is to assume that x is *very* small with respect to 0.55; that is, we assume it makes only a negligible difference to the answer if we assume that $(x + 0.55) \approx 0.55$. This is a very reasonable assumption because we know that the solubility equals 1.3×10^{-5} M *without* the common ion Cl^-, and it is even smaller in the presence of added Cl^-. Solving the equation with the approximation,

$$K_{sp} = 1.8 \times 10^{-10} = (x)(0.55)$$

$$x = [Ag^+] = 3.3 \times 10^{-10} \text{ M} \quad \text{or} \quad 4.7 \times 10^{-8} \text{ g AgCl/L}$$

As predicted by Le Chatelier's principle, the solubility of AgCl in the presence of added Cl^- is less (3.3×10^{-10} M) than in pure water (1.3×10^{-5} M).

As a final step, let us check the approximation we made by substituting the value of x into the exact expression $K_{sp} = (x)(x + 0.55)$. Then, if the product $(x)(x + 0.55)$ is the same as the given value of K_{sp}, the approximation is valid:

$$K_{sp} = (x)(x + 0.55) = (3.3 \times 10^{-10})(3.3 \times 10^{-10} + 0.55) = 1.8 \times 10^{-10}$$

Exercise 19.11 The Common Ion Effect and Salt Solubility

Calculate the solubility of $BaSO_4$ (1) in pure water and (2) in the presence of 0.010 M $Ba(NO_3)_2$. K_{sp} for $BaSO_4$ is 1.1×10^{-10}.

Problem-Solving Tips and Ideas

19.3 Solubility in the Presence of a Common Ion

To calculate the solubility of a slightly soluble salt in the presence of an ion common to the equilibrium, let us say $BaSO_4$ in the presence of added Ba^{2+} ion, proceed in the following way:

1. The concentration of the common ion, here Ba^{2+}, is known.
2. Assume some of the slightly soluble salt dissolves, producing Ba^{2+} and SO_4^{2-} in this case. The equilibrium constant (K_{sp}) for the dissolving process is known.
3. To solve the K_{sp} expression, the concentration of the common ion is assumed to be equal to that of the added, soluble salt. The concentration of the ion other than the common ion is a measure of the solubility of the salt. For the $BaSO_4$ case

$$K_{sp} = [Ba^{2+}][SO_4^{2-}] = [\text{common ion}][\text{ion that reflects salt solubility}]$$

Note two important ideas from Example 19.9. First, the solubility of AgCl in the presence of a common ion was reduced by a factor of about 10^5, in accordance with Le Chatelier's principle. Second, we made the approximation that the amount of common ion added to the solution was very large compared with the amount of that ion coming from the insoluble salt, which allowed us to simplify our calculations. This is almost always the case, but you should *always* check the approximation.

Example 19.10 The Common Ion Effect and Salt Solubility

Calculate the solubility of silver chromate, Ag_2CrO_4, at 25 °C in (1) pure water and (2) in the presence of 0.0050 M K_2CrO_4 solution.

Solution The balanced equation defining silver chromate's solubility is

$$Ag_2CrO_4(s) \rightleftharpoons 2\ Ag^+(aq) + CrO_4^{2-}(aq)$$

$$K_{sp} = [Ag^+]^2[CrO_4^{2-}] = 9.0 \times 10^{-12}$$

Part 1. *Solubility in pure water.* Because there is a 1 : 1 stoichiometric ratio for moles of $CrCrO_4^{2-}$ in solution to moles of Ag_2CrO_4 dissolved, we define the solubility of this salt in terms of $[CrO_4^{2-}] = x$. Thus, x is not only the concentration of CrO_4^{2-} in solution but is also the number of moles per liter of Ag_2CrO_4 dissolved.

Equation	$Ag_2CrO_4(s) \rightleftharpoons$ **2** $Ag^+(aq)$ +	$CrO_4^{2-}(aq)$
Before Ag_2CrO_4 begins to dissolve (M)	0	0
Change on proceeding to equilibrium	$+2x$	$+x$
After equilibrium is achieved (M)	$2x$	x

This must mean that $[Ag^+] = 2x$, and so the K_{sp} expression can be written as

$$K_{sp} = 9.0 \times 10^{-12} = [Ag^+]^2[CrO_4^{2-}] = (2x)^2(x) = 4x^3$$
$$x^3 = 2.3 \times 10^{-12}\text{ M}$$
$$x = [CrO_4^{2-}] = \sqrt[3]{2.3 \times 10^{-12}} = 1.3 \times 10^{-4}\text{ M}$$

This means the ion concentrations at equilibrium in pure water are

$$[Ag^+] = 2x = 2.6 \times 10^{-4}\text{ M} \qquad \text{and} \qquad [CrO_4^{2-}] = x = 1.3 \times 10^{-4}\text{ M}$$

We know now that 1.3×10^{-4} mol of Ag_2CrO_4 dissolves per liter.

Do not be confused about the use of $(2x)^2$, where the factor 2 appears both as the multiplier of x and the exponent of $(2x)$. This occurs because $[CrO_4^{2-}]$ is defined as x and the stoichiometry demands that $[Ag^+]$ be defined as $2x$. This is substituted into the K_{sp} expression in place of $[Ag^+]$. The K_{sp} expression then demands that $[Ag^+]$ or the equivalent expression $2x$ be squared. See *Problem-Solving Tips and Ideas 19.1.*

Part 2. *Solubility in a solution containing chromate ion.* In the presence of excess chromate ion from dissolved K_2CrO_4, the concentration of Ag^+ is less than in pure water. To calculate this concentration, let us again assume the solubility of Ag_2CrO_4 is equivalent to the amount of CrO_4^{2-} ion that appears in solution from the dissolving process. Now, however, this quantity is assigned a different unknown value, y. Thus the concentration of Ag^+ ion at equilibrium is $2y$ and that of CrO_4^{2-} ion is y *plus* the amount of CrO_4^{2-} already in the solution.

Equation	$Ag_2CrO_4(s) \rightleftharpoons Ag^+(aq)$ +	$CrO_4^{2-}(aq)$
Before Ag_2CrO_4 begins to dissolve (M)	0	0.0050
Change on proceeding to equilibrium	$+2y$	$+y$
After equilibrium is achieved (M)	$2y$	$(y + 0.0050)$

Substituting the equilibrium amounts into the K_{sp} expression, we have

$$K_{sp} = 9.0 \times 10^{-12} = [Ag^+]^2[CrO_4^{2-}] = (2y)^2(y + 0.0050)$$

Again make the approximation that y is very small with respect to 0.0050, and so $(y + 0.0050) \approx 0.0050$. (This is probably reasonable because $[CrO_4^{2-}]$ is 0.00013 M *without* added chromate ion, and it is certain that y is even smaller in the presence of extra chromate ion.) The approximate expression is therefore

$$K_{sp} = 9.0 \times 10^{-12} = [Ag^+]^2[CrO_4^{2-}] = (2y)^2(0.0050)$$

and so y is

$$y = 2.1 \times 10^{-5}\text{ M} = [CrO_4^{2-}]\text{ from }Ag_2CrO_4$$

This means the silver ion concentration in the presence of the common ion is

$$[Ag^+] = 2y = 4.2 \times 10^{-5}\ M$$

This silver ion concentration is less than its value in pure water (2.6×10^{-4}), and again you see the result of adding an ion "common" to the equilibrium.

As a final step, check the approximation. Substitute y back into the expression $K_{sp} = (2y)^2(0.0050 + y)$ to see if K_{sp} calculated from the expression agrees with the given K_{sp}:

$$K_{sp} = [2(2.1 \times 10^{-5})]^2(0.0050 + 0.000021) = 8.9 \times 10^{-12}$$

Here you see that the calculated K_{sp} is only about 1% different from the given K_{sp}. The approximation is valid.

Exercise 19.12 The Common Ion Effect and Salt Solubility

Calculate the solubility of $Zn(CN)_2$ at 25 °C (1) in pure water and (2) in the presence of 0.10 M KCN. K_{sp} for $Zn(CN)_2$ is 8.0×10^{-12}.

19.7 SOLUBILITY, ION SEPARATIONS, AND QUALITATIVE ANALYSIS

In many courses in introductory chemistry a portion of the laboratory work is devoted to the qualitative analysis of aqueous solutions, the identification of anions and metal cations. The purpose of such laboratory work is (1) to introduce you to some basic chemistry of various ions and (2) to illustrate how the principles of chemical equilibria can be applied.

Assume you have an aqueous solution that contains some or all of the following metal ions: Ag^+, Pb^{2+}, Cd^{2+}, and Cu^{2+}. Your objective is to separate the ions from one another so that each type of ion ends up in a separate test tube; the presence or absence of the ion can then be established. As a first step in this process, you want to find one reagent that forms a precipitate with one or more of the cations and leaves the others in solution. This is done by comparing K_{sp} values for salts of cations with various anions (say S^{2-}, OH^-, Cl^-, or SO_4^{2-}), looking for an anion that gives insoluble salts for some cations but not others.

Looking over the list of solubility products in Appendix J, you notice that all of the ions in our example solution form very insoluble sulfides (Ag_2S, PbS, CdS, and CuS). However, only two of them form insoluble chlorides: AgCl and $PbCl_2$. Thus, your "magic reagent" for partial cation separation could be aqueous HCl, which forms precipitates with two of the ions, leaving the other two in solution (Figure 19.7). Now you are left with the task of separating precipitated AgCl and $PbCl_2$ from each other and aqueous Cd^{2+} and Cu^{2+} from each other (see Figure 19.7). The separation of AgCl and $PbCl_2$ is not difficult because $PbCl_2$ dissolves in hot water, but AgCl remains insoluble.

• Some Insoluble Sulfides and Chlorides

Compound	K_{sp} at 25 °C
Ag_2S	1.0×10^{-49}
PbS	3.2×10^{-28}
CdS	1.0×10^{-27}
CuS	7.9×10^{-37}
AgCl	1.8×10^{-10}
$PbCl_2$	1.7×10^{-5}

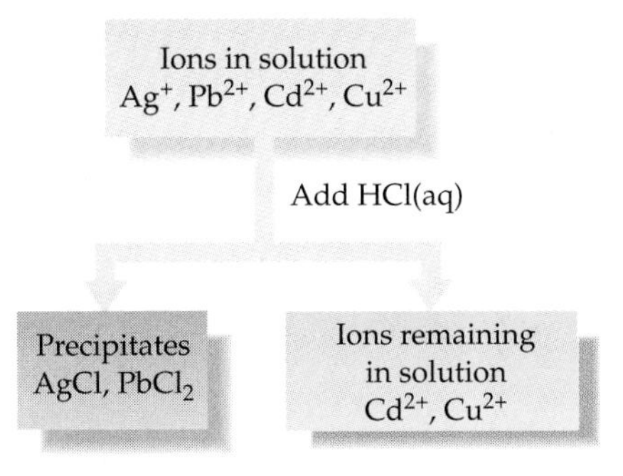

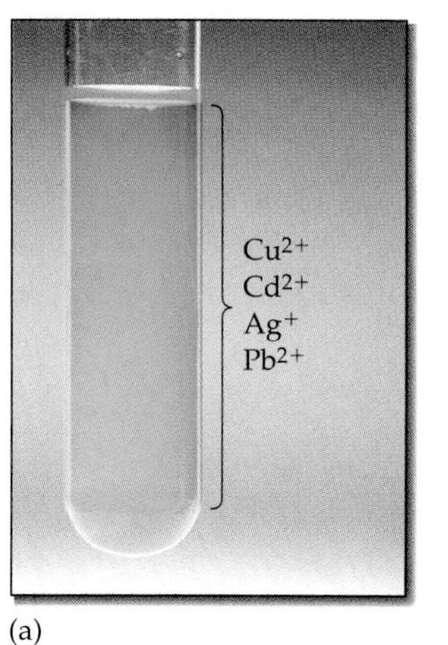

(a)

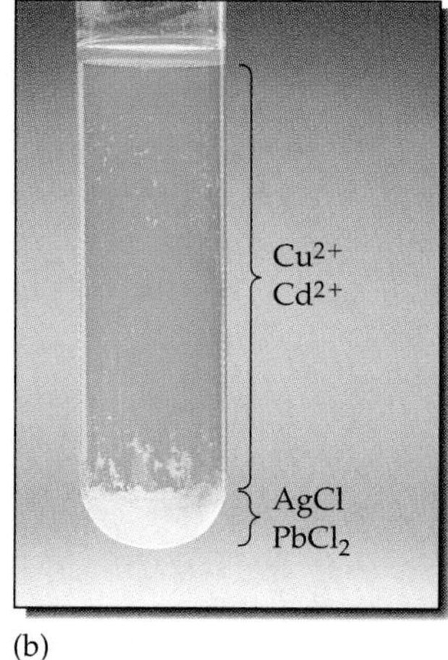

(b)

(c)

***Figure* 19.7 Ion separations by solubility difference.** (a) The solution contains nitrate salts of Ag^+, Pb^{2+}, Cd^{2+}, and Cu^{2+}. (The Cu^{2+} ion in water is light blue; the others are colorless.) (b) Aqueous HCl is added in an amount sufficient to precipitate completely AgCl and $PbCl_2$ (both white solids). (c) The blue solution, now containing only Cu^{2+} and Cd^{2+} ions, is poured into another test tube, leaving white, solid AgCl (top model) and $PbCl_2$ (bottom model) in the first test tube. *(Photos, C. D. Winters; models, S. M. Young)*

***Example* 19.11** Separation of Two Ions by Difference in Solubility

A solution contains Cl^- (as NaCl, 0.010 M) and CrO_4^{2-} (as K_2CrO_4, 0.0010 M). A solution of $AgNO_3$ is added slowly (without changing the volume of the original solution appreciably).

1. Which precipitates first, AgCl or Ag_2CrO_4?
2. We find that AgCl precipitates before Ag_2CrO_4. What is the Cl^- ion concentration when Ag_2CrO_4 begins to precipitate?

See the *Saunders Interactive General Chemistry CD-ROM,* Screen 19.9, Using Solubility.

Solution As usual, we first write the equations for dissolving the insoluble salts and look up their K_{sp} values:

$$AgCl(s) \rightleftharpoons Ag^+(aq) + Cl^-(aq) \qquad K_{sp} = 1.8 \times 10^{-10}$$
$$Ag_2CrO_4(s) \rightleftharpoons 2\,Ag^+(aq) + CrO_4^{2-}(aq) \qquad K_{sp} = 9.0 \times 10^{-12}$$

1. To discover which anion precipitates first, calculate the concentration of Ag^+ required to just begin the precipitation of each salt.

   (a) To begin to precipitate AgCl when $[Cl^-] = 0.010$ M

$$K_{sp} = [Ag^+][Cl^-] = 1.8 \times 10^{-10}$$
$$[Ag^+] = \frac{1.8 \times 10^{-10}}{0.010} = 1.8 \times 10^{-8}\ \text{M}$$

   (b) To begin to precipitate Ag_2CrO_4 when $[CrO_4^{2-}] = 0.0010$ M

$$K_{sp} = [Ag^+]^2[CrO_4^{2-}] = 9.0 \times 10^{-12}$$
$$[Ag^+] = \sqrt{\frac{9.0 \times 10^{-12}}{0.0010}} = 9.5 \times 10^{-5}\ \text{M}$$

These calculations show that much less silver ion is needed to begin the precipitation of AgCl than to begin the precipitation of Ag_2CrO_4. Therefore, AgCl precipitates before Ag_2CrO_4 as $AgNO_3$ is added slowly to the solution.

2. Ag_2CrO_4 begins to precipitate when $[Ag^+] = 9.5 \times 10^{-5}$. The concentration of Cl^- when the Ag^+ ion concentration has this value can be calculated using the K_{sp} expression for AgCl:

$$[Cl^-] = \frac{K_{sp}}{[Ag^+]} = \frac{1.8 \times 10^{-10}}{9.5 \times 10^{-5}} = 1.9 \times 10^{-6}\ M$$

This means the percentage of Cl^- ion still in solution when Ag_2CrO_4 just begins to precipitate is

$$\frac{1.9 \times 10^{-6}\ M}{0.010\ M} \times 100\% = 0.019\%$$

Thus, if the AgCl precipitate is removed from the reaction mixture (by filtration) just before the Ag_2CrO_4 begins to precipitate, 99.98% of the Cl^- ion is by then separated from the CrO_4^{2-} ion (Figure 19.8).

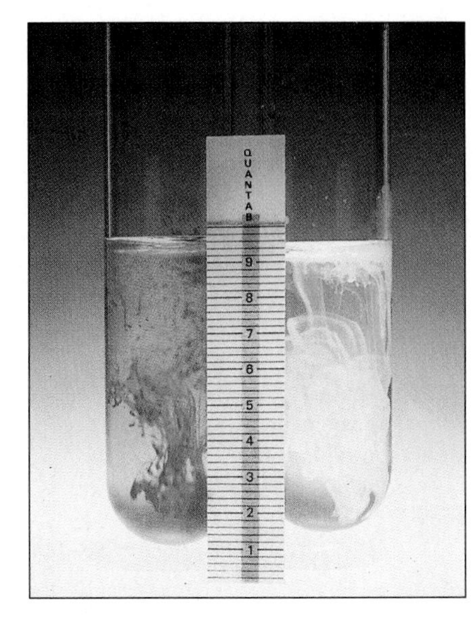

Figure 19.8 Selective precipitation and water testing. Red-orange Ag_2CrO_4 is precipitated in the test tube on the left and white AgCl in the test tube on the right. In between is a "test strip" to test water for the presence of chloride ion. The strip consists of a film of Ag_2CrO_4. When water containing chloride ion enters the strip, Ag_2CrO_4 is converted to the less soluble salt AgCl, and the strip turns white. *(C. D. Winters)*

***Exercise* 19.13** Separation of Two Ions by Difference in Solubility

Under the right circumstances, Ag^+ can be separated from Pb^{2+} in aqueous solution based on the difference in the solubilities of their chloride salts, AgCl and $PbCl_2$.

1. If you begin with both metal ions having a concentration of 0.0010 M, which ion precipitates first as an insoluble chloride on adding HCl?
2. What is the concentration of the metal ion that precipitates first, just before the second metal chloride begins to precipitate?

***Exercise* 19.14** Schemes for Ion Separation

The cations of each of the following pairs appear together in one solution:

1. Ag^+ and Bi^{3+} **2.** Fe^{2+} and K^+

You may add only one reagent to precipitate one cation and not the other. Consult the solubility product table in Appendix J to decide whether to use Cl^-, S^{2-}, or OH^- as the precipitating ion in each case. (The precipitating ions are introduced in the form of HCl, $(NH_4)_2S$, or NaOH, for example.)

19.8 SIMULTANEOUS EQUILIBRIA

In many instances two or more reactions occur at the same time in a solution, all of them being described as equilibrium processes. Chemists characterize such situations as examples of **simultaneous equilibria.**

One example of simultaneous equilibria is the case in which a reagent is added to a saturated solution of an insoluble salt, converting the salt to another, even less soluble salt (see Figure 19.8). Consider two common lead compounds, lead(II) chloride and lead(II) chromate.

$$PbCl_2,\ K_{sp} = 1.7 \times 10^{-5} \qquad \text{and} \qquad PbCrO_4,\ K_{sp} = 1.8 \times 10^{-14}$$

See the *Saunders Interactive General Chemistry CD-ROM,* Screen 19.10, Simultaneous Equilibria.

A CLOSER LOOK

Equilibria Involving Sulfide Ions

Equilibria involving sulfide ions have long presented a difficult problem, especially in view of the hydrolysis of the strongly basic sulfide ion, and the uncertainty in the equilibrium constant for this reaction. It has been argued, however, that the solubility of metal sulfides should be treated as in the PbS example in the text (see page 897). That is, a metal sulfide such as zinc sulfide dissolves according to the equation

$$ZnS(s) + H_2O(\ell) \rightleftharpoons Zn^{2+}(aq) + OH^-(aq) + HS^-(aq)$$

and that the equilibrium constant for this reaction, is actually

$$K_{sp} = [Zn^{2+}][OH^-][HS^-] = 2 \times 10^{-20}$$

In practice, we are interested in dissolving metal sulfides in acid solution. In the presence of a strong acid both OH^- and HS^- are protonated to form H_2O and H_2S, respectively. In strong acid, therefore, zinc sulfide dissolves to some extent to give zinc ions, H_2O, and H_2S,

$$ZnS(s) + 2\ H_3O^+(aq) \rightleftharpoons Zn^{2+}(aq) + H_2S(aq) + 2\ H_2O(\ell)$$

and the equilibrium constant for the reaction is called K_{spa}, for K_{sp} in acid.

$$K_{spa} = \frac{[Zn^{2+}][H_2S]}{[H_3O^+]^2} = 2 \times 10^{-4}$$

The value of K_{spa} is considerably larger than K_{sp} because the equilibrium is shifted strongly to the right due to the protonation of the two basic anions, OH^- and HS^-. Indeed, it is generally true that K_{spa} values are about 10^{21} larger than K_{sp} values for metal sulfides, and sulfides for which K_{spa} is greater than about 10^{-2} are readily soluble in strong acids.

Metal Sulfide	K_{sp}	K_{spa}
HgS (black)	2×10^{-53}	2×10^{-32}
CuS	7.9×10^{-37}	6×10^{-16}
PbS	3.2×10^{-28}	3×10^{-7}
ZnS	2.0×10^{-25}	2×10^{-4}
FeS	4.9×10^{-18}	6×10^{2}
MnS	5.1×10^{-15}	3×10^{7}

For more details see R. J. Myers: "The new low value for the second dissociation constant for H_2S." *Journal of Chemical Education,* Vol. 63, p. 687, 1986.

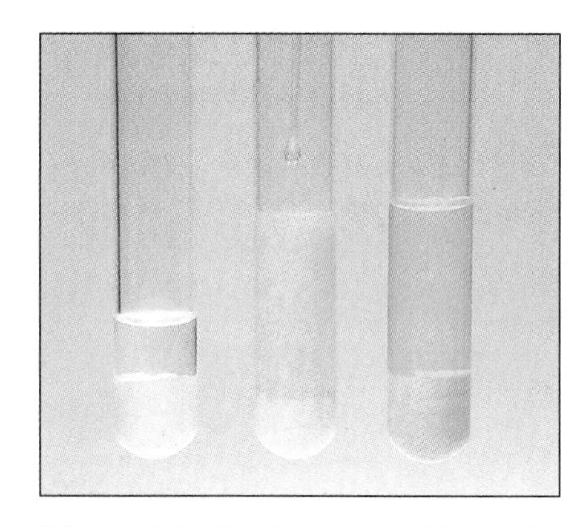

Figure 19.9 Simultaneous equilibria. The test tube at the left contains a precipitate of white $PbCl_2$. In the middle test tube, a solution of yellow K_2CrO_4 has been added to insoluble white $PbCl_2$. The test tube at the right shows what happens when the $PbCl_2$ precipitate is stirred in the presence of K_2CrO_4. The white solid $PbCl_2$ has been transformed into less soluble yellow $PbCrO_4$. *(C. D. Winters)*

If you add a few drops of K_2CrO_4 to a small amount of a white precipitate of $PbCl_2$ and shake the mixture, the solid changes to yellow $PbCrO_4$ (Figure 19.9). That this is possible is evident from the equilibrium constant, K_{net}, for the process

$$PbCl_2(s) + CrO_4^{2-}(aq) \rightleftharpoons PbCrO_4(s) + 2\ Cl^-(aq) \qquad K_{net} = 9.4 \times 10^8$$

This reaction is the sum of two reactions whose equilibrium constants are known:

$$PbCl_2(s) \rightleftharpoons Pb^{2+}(aq) + 2\ Cl^-(aq) \qquad K_1 = K_{sp} = 1.7 \times 10^{-5}$$

$$Pb^{2+}(aq) + CrO_4^{2-}(aq) \rightleftharpoons PbCrO_4(s) \qquad K_2 = \frac{1}{K_{sp}} = \frac{1}{1.8 \times 10^{-14}}$$

The equilibrium constant for the overall reaction is the product of the equilibrium constants for the summed reactions. That is, $K_{net} = K_1 \cdot K_2 = 9.4 \times 10^8$. This very large value indicates that the reaction should proceed from left to right.

Simultaneous equilibria are very important in many operations in the laboratory and in nature. The next sections explore more examples.

Exercise 19.15 Simultaneous Equilibria

Silver forms many insoluble salts. Which is more soluble, AgCl or AgBr? If you add sufficient bromide ion to an aqueous suspension of AgCl(s), can AgCl be converted to AgBr? To answer this, derive the value of the equilibrium constant for

$$AgCl(s) + Br^-(aq) \rightleftharpoons AgBr(s) + Cl^-(aq)$$

19.9 SOLUBILITY AND pH

The next time you are tempted to wash a supposedly insoluble salt down the kitchen or laboratory drain, stop and consider the consequences. Many metal ions, such as lead, chromium, and mercury, are *toxic* in the environment. Even if the so-called insoluble salt does not appear to dissolve, its solubility in water may be greater than you think, in part owing to the possibility of hydrolysis of the anion of the salt.

See the *Saunders Interactive General Chemistry CD-ROM,* Screen 19.11, Solubility and pH.

Lead sulfide, PbS, is found in nature as the mineral galena (Figure 19.10). Let us consider what happens if a trace of lead(II)-sulfide dissolves in water. First, the insoluble salt dissolves to an extremely small extent.

Lead(II) sulfide dissolves particularly well in HNO_3, in part because this acid oxidizes the sulfide ion to elemental sulfur.

Step 1. *The solubility of PbS in water.*

$$PbS(s) \rightleftharpoons Pb^{2+}(aq) + S^{2-}(aq)$$

One product of the reaction is a sulfide ion, an anion even more basic than the OH^- ion (see Table 17.4). The sulfide ion is therefore extensively hydrolyzed.

Step 2. *The hydrolysis of sulfide ion.*

$$S^{2-}(aq) + H_2O(\ell) \rightleftharpoons HS^-(aq) + OH^-(aq) \qquad K_b = 1 \times 10^5$$

The overall process is the sum of two simultaneous equilibria,

$$PbS(s) + H_2O(\ell) \rightleftharpoons Pb^{2+}(aq) + HS^-(aq) + OH^-(aq)$$

which has an equilibrium constant of 3.2×10^{-23}. Thus, because of the sulfide ion hydrolysis, metal sulfides are more soluble in water than expected from the simple ionization of the salt.

The lead sulfide example leads to the general observation that any salt containing an anion that is the conjugate base of a weak acid dissolves in water to a greater extent than given by K_{sp}. This means that salts of phosphate, acetate, carbonate, and cyanide, as well as sulfide, can be affected, because all of these anions undergo the general hydrolysis reaction

$$X^-(aq) + H_2O(\ell) \rightleftharpoons HX(aq) + OH^-(aq)$$

in aqueous solution.

Figure 19.10 Lead sulfide. This and other metal sulfides dissolve in water to a greater extent than expected, in part because the sulfide ion reacts with water to form the very stable species HS^- and OH^-. This shifts the solubility equilibrium toward products, including aqueous Pb^{2+} ion. *(Photo, C. D. Winters; model, S. M. Young)*

The possibility that the anion of an insoluble salt undergoes hydrolysis leads to yet another useful, general conclusion: If a strong acid is added to a water-insoluble salt such as $CaCO_3$, OH^- ion from X^- ion hydrolysis is removed (by formation of water). This shifts the X^- ion hydrolysis reaction further to the right; the weak acid HX is formed, and the salt dissolves (Figure 19.11). For example, calcium carbonate dissolves readily when hydrochloric acid is added, and some of the reactions occurring are

$$\begin{array}{lll}
CaCO_3(s) \rightleftharpoons Ca^{2+}(aq) + CO_3^{2-}(aq) & & K = K_{sp} = 3.8 \times 10^{-9} \\
CO_3^{2-}(aq) + H_2O(\ell) \rightleftharpoons HCO_3^-(aq) + OH^-(aq) & & K = K_b = 2.1 \times 10^{-4} \\
OH^-(aq) + H_3O^+(aq) \rightleftharpoons 2\ H_2O(\ell) & & K = 1/K_w = 1/1.0 \times 10^{-14} \\
\hline
CaCO_3(s) + H_3O^+(aq) \rightleftharpoons Ca^{2+}(aq) + HCO_3^-(aq) + H_2O(\ell) & & K_{net} = K_{sp}(K_b)(1/K_w) = 80.0
\end{array}$$

In the presence of strong acid, the hydrogen carbonate ion reacts further with acid to give H_2CO_3, which then reverts to CO_2 gas and water.

$$H_2CO_3(aq) \rightleftharpoons CO_2(g) + H_2O(\ell) \qquad K \approx 10^5$$

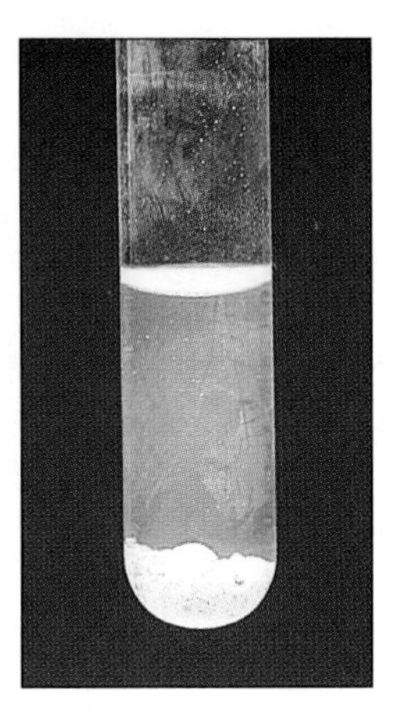

Figure 19.11 Selectively dissolving precipitates in acid. *(left)* A precipitate of AgCl (white) and Ag_3PO_4 (yellow). *(right)* Adding a strong acid dissolves Ag_3PO_4 (and leaves insoluble AgCl) because the basic PO_4^{3-} anion reacts with acid to give HPO_4^{2-} and $H_2PO_4^-$ ions. *(C. D. Winters)*

The CO_2 bubbles out of solution, and the equilibrium is moved even further to the right (see Figure 5.13).

Carbonates are generally soluble in strong acids, and so are many metal sulfides:

$$FeS(s) + 2\ H_3O^+(aq) \rightleftharpoons Fe^{2+}(aq) + H_2S(aq) + 2\ H_2O(\ell)$$

and metal hydroxides:

$$Mg(OH)_2(s) + 2\ H_3O^+(aq) \rightleftharpoons Mg^{2+}(aq) + 4\ H_2O(\ell)$$

In general, the solubility of a salt containing the conjugate base of a weak acid is increased by *addition of a stronger acid* to the solution. In contrast, the salts are not soluble in strong acid if the anion is the conjugate base of a strong acid. For example, AgCl is not soluble in strong acid.

$$AgCl(s) \rightleftharpoons Ag^+(aq) + Cl^-(aq) \qquad K_{sp} = 1.8 \times 10^{-10}$$
$$H_3O^+(aq) + Cl^-(aq) \rightleftharpoons HCl(aq) + H_2O(\ell) \qquad K \ll 1$$

Because Cl^- is a *very* weak base (Table 17.3) it is not removed by the strong acid H_3O^+. You can see that this same conclusion applies to insoluble salts of Br^- and I^-.

19.10 SOLUBILITY AND COMPLEX IONS

See the *Saunders Interactive General Chemistry CD-ROM,* Screen 19.12, Complex Ion Formation and Solubility.

Other mentions of complex ions in the book occur on page 746 [$Co(H_2O)_6^{2+}$ and $CoCl_4^{2-}$], Figure 16.3 [$Fe(H_2O)_5SCN^-$], and Figure 17.20 [$Cu(NH_3)_4^{2+}$].

The metal ions we have been discussing exist in aqueous solution as complex ions. That is, they consist of the metal ion and water molecules, bound into a single entity. The negative end of the polar water molecule, the oxygen atom, is attracted to positive metal ions (Section 17.11). Indeed, any negative ion or Lewis base, such as ammonia, can be attracted to a metal ion (Figure 19.12).

As you shall see in Chapter 23, complex ions are important in chemistry. They are the basis of such biologically important substances as hemoglobin and vitamin B_{12}. For our present purposes, they are important in that a water-soluble complex ion of a metal ion can often be formed in preference to an insoluble salt. For example, adding sufficient ammonia to a precipitate of AgCl causes the insoluble salt to dissolve because the water-soluble complex ion $Ag(NH_3)_2^+$ is preferentially formed (see Figures 16.5 and 19.13).

$$AgCl(s) + 2\ NH_3(aq) \rightleftharpoons Ag(NH_3)_2^+(aq) + Cl^-(aq)$$

We can consider dissolving AgCl(s) in this way as a two-step process. First, AgCl dissolves minimally in water, giving $Ag^+(aq)$ and $Cl^-(aq)$ ions. Then, the $Ag^+(aq)$ ion combines with NH_3 to give the ammonia complex. Lowering the $Ag^+(aq)$ concentration through complexation with NH_3 shifts the solubility equilibrium to the right, and more solid AgCl dissolves:

$$AgCl(s) \rightleftharpoons Ag^+(aq) + Cl^-(aq) \qquad K_{sp} = 1.8 \times 10^{-10}$$

$$Ag^+(aq) + 2\ NH_3(aq) \rightleftharpoons Ag(NH_3)_2^+(aq) \qquad K = 1.6 \times 10^7$$

This is another example of simultaneous equilibria. This is similar to the case in which an even more insoluble precipitate is formed from a more soluble one or hydrolysis of an anion of an insoluble salt leads to greater than expected solubility (page 897).

The equilibrium constant for the formation of a complex ion is called a formation constant, $K_{formation}$. (Values of formation constants are given in Appendix K.) For the silver–ammonia complex, the value of the constant is

$$Ag^+(aq) + 2\ NH_3(aq) \rightleftharpoons Ag(NH_3)_2^+(aq) \qquad K_{formation} = 1.6 \times 10^7$$

The large value of the equilibrium constant means that the equilibrium lies well to the right and provides the driving force for dissolving AgCl. If we combine $K_{formation}$ with K_{sp} to give the net equilibrium constant for dissolving AgCl, we obtain

$$AgCl(s) + 2\ NH_3(aq) \rightleftharpoons Ag(NH_3)_2^+(aq) + Cl^-(aq)$$

$$K_{net} = K_{sp} \cdot K_{formation} = (1.8 \times 10^{-10})(1.6 \times 10^7) = 2.9 \times 10^{-3}$$

AgCl is much more soluble in the presence of ammonia than in pure water.

Figure 19.12 Complex ions. The green solution contains soluble $Ni(H_2O)_6^{2+}$ ions where water molecules are bound to nickel(II) ions by ion–dipole forces. This ion, which gives the solution its green color, is referred to as a "complex ion" because it is formed by a combination of a metal ion and molecules of water. The red, insoluble solid is the dimethylglyoxime $[(CH_3CNOH)_2]$ complex of nickel(II) ion. Formation of this beautiful red solid is the classical test for the presence of aqueous nickel(II) ion. *(Photo, C. D. Winters; models, S. M. Young)*

Figure 19.13 Forming and dissolving precipitates. (a) AgCl is precipitated by adding NaCl(aq) to $AgNO_3(aq)$. (b) The precipitate of AgCl dissolves on adding aqueous NH_3 to give water-soluble $Ag(NH_3)_2^+$ (see Figure 16.5). (c) The silver–ammonia complex ion is changed to insoluble AgBr on adding NaBr(aq). (d) The precipitate of AgBr is dissolved on adding $Na_2S_2O_3(aq)$. The product is the water-soluble complex ion $Ag(S_2O_3)_2^{3-}$. *(C. D. Winters)*

Formation constants have been measured for the formation of many other complex ions, making it possible to compare the stabilities of various complex ions by comparing values of their formation constants. For silver(I) ion, a few other values are

Formation Equilibrium	$K_{formation}$
$Ag^+(aq) + 2\ Cl^-(aq) \rightleftharpoons AgCl_2^-(aq)$	2.5×10^5
$Ag^+(aq) + 2\ S_2O_3^{2-}(aq) \rightleftharpoons Ag(S_2O_3)_2^{3-}(aq)$	2.0×10^{13}
$Ag^+(aq) + 2\ CN^-(aq) \rightleftharpoons Ag(CN)_2^-(aq)$	5.6×10^{18}

Figure 19.13 shows what happens as complex ions are formed. Beginning with a precipitate of AgCl, adding aqueous ammonia dissolves the precipitate to give the soluble complex ion $Ag(NH_3)_2^+$. Silver bromide is even more stable than $Ag(NH_3)_2^+$, so AgBr forms in preference to the complex ion on adding bromide ion. If thiosulfate ion, $S_2O_3^{2-}$, is then added, however, the large value for $K_{formation}$ for $Ag(S_2O_3)_2^{3-}$ shows that it is very stable, and this complex ion forms in preference to AgBr.

Another conclusion that comes from the table of formation constants is that K_{sp} calculations are not very accurate. One reason is that such calculations ignore complex ion formation. For example, in the presence of excess Cl^- ion, the solubility of AgCl is larger than expected because the complex ion $AgCl_2^-$ is formed.

Example 19.12 Dissolving Precipitates Using Complex Ion Formation

How many moles of ammonia must be added to dissolve 0.050 mol of AgCl suspended in 1.0 L of water?

Solution As described in the preceding text, the reaction that occurs is

$$AgCl(s) + 2\ NH_3(aq) \rightleftharpoons Ag(NH_3)_2^+(aq) + Cl^-(aq) \qquad K_{net} = 2.9 \times 10^{-3}$$

If 0.050 mol/L of AgCl is to be dissolved, 0.050 mol/L of the complex ion $Ag(NH_3)_2^+$ and 0.050 mol/L of Cl^- ion are formed. If we wish to have these concentrations at equilibrium, what must the concentration of ammonia be?

$$K_{net} = 2.9 \times 10^{-3} = \frac{[Ag(NH_3)_2^+][Cl^-]}{[NH_3]^2}$$

Solving for $[NH_3]$, we find it is 0.93 mol/L. To dissolve AgCl, enough ammonia must therefore be added to form the complex ion with Ag^+ ion (2×0.050 mol = 0.10 mol), and additional ammonia must then be added to bring the concentration to 0.93 mol/L. Therefore, a total of (0.10 mol + 0.93 mol), or 1.03 mol/L, of NH_3 must be added.

Exercise 19.16 Dissolving Precipitates Using Complex Ion Formation

Does 1.00×10^2 mL of 4.0 M aqueous ammonia completely dissolve 0.010 mol of AgCl suspended in 1.0 L of pure water?

19.11 EQUILIBRIA IN THE ENVIRONMENT: CARBON DIOXIDE AND CARBONATES

Our biosphere is a complicated mixture of carbon-containing compounds, some being created, some being transformed, and others being decomposed at any moment. The compound that links these processes and their materials or products is carbon dioxide, which is produced in the biological process of respiration and consumed by photosynthetic organisms. Although CO_2 constitutes only about 0.0325% of the atmosphere, a recent estimate is that the earth's atmosphere contains 700 billion tons of carbon in the form of the gas.

Recall the passage from Primo Levi (from his book *The Periodic Table*) about CO_2 and its role in our lives (see page 60).

Our atmosphere normally contains about 0.0003 atm of CO_2, and when this is in equilibrium with CO_2 dissolved in water, the concentration of aqueous H_2CO_3 is about 10^{-5} M. Although this solubility is small, the oceans are thought to hold roughly 60 times as much CO_2 as the atmosphere. In the oceans, however, the CO_2 is in the form of (1) dissolved gas or carbonic acid; (2) one of the ionization products of H_2CO_3; (3) solid metal carbonates, such as $CaCO_3$ or $MgCO_3$ (see Figure 19.1); and (4) organic matter.

Carbonate equilibria are also illustrated in Figure 16.2.

Because H_2CO_3 is a weak acid, solutions of CO_2 in pure water are slightly acidic, and, because K_{a2} is much smaller than K_{a1}, the pH can be estimated from the first equilibrium reaction (Section 17.9). Using $[H_2CO_3] \approx 10^{-5}$ M, the hydronium ion concentration is about 2.2×10^{-6} M, and the pH is therefore about 5.6. This means that the rain that falls even in a nonpolluted environment should be slightly acidic (Figure 19.14a).

(a)

(b)

Figure 19.14 Natural limestone formations. (a) A limestone wall in Wind Cave National Park (South Dakota). The gaps in the wall were formed as rainwater seeped into cracks in the wall. Because rainwater is slightly acidic (because of dissolved CO_2), it dissolved the limestone. As the water seeped deeper into the wall, however, its acidity dropped, less limestone dissolved, and the gaps narrowed. (b) Much of the Grand Canyon in Arizona is "red-wall limestone," so-called because iron in the Supai formation higher in the canyon walls has washed down over the lower red-wall formation. *(a, Arthur N. Palmer; b, James Cowlin/Image Enterprises)*

The oceans of the earth contain enormous quantities of calcium carbonate produced by various sea creatures. There is, therefore, the additional equilibrium of calcium carbonate to consider:

$$CaCO_3(s) \rightleftharpoons Ca^{2+}(aq) + CO_3^{2-}(aq) \qquad K_{sp} = 3.8 \times 10^{-9}$$

When this is included, sea water saturated with carbon dioxide is found to have a pH of 8.2 ± 0.3. The slightly alkaline character of sea water arises from the fact that the carbonate ion is the conjugate base of a weak acid, the bicarbonate ion, and so carbonate hydrolysis produces OH^-:

$$CO_3^{2-}(aq) + H_2O(\ell) \rightleftharpoons HCO_3^-(aq) + OH^-(aq) \qquad K_b = 2.1 \times 10^{-4}$$

The most important aspect of the carbonate equilibrium system is that it buffers the pH of sea water. The addition of acids by undersea volcanic activity or other natural processes is countered primarily by carbonic acid formation (and subsequent loss of carbon dioxide to the atmosphere):

$$HCO_3^-(aq) + H_3O^+(aq) \rightleftharpoons H_2CO_3(aq) + H_2O(\ell)$$

and secondarily by bicarbonate formation:

$$H_3O^+(aq) + CO_3^{2-}(aq) \rightleftharpoons HCO_3^-(aq) + H_2O(\ell)$$

On the other hand, an increase in hydroxide ion concentration is counteracted by the reaction

$$OH^-(aq) + HCO_3^-(aq) \rightleftharpoons H_2O(\ell) + CO_3^{2-}(aq)$$

This reaction of course leads to an increase in the carbonate concentration. If the sea water is above the saturation limit of calcium carbonate, limestone is precipitated (Figure 19.14b).

The greenhouse effect was described in Chapter 11 (page 564).

Carbon dioxide is being generated in ever-increasing amounts, due in part to the increase in the population of the earth, in part to the clearing of forests (and thus to less use of CO_2 in photosynthesis), and in part to increased combustion of fossil fuels. Indeed, there is a fear that this could lead to a global warming trend caused by the "greenhouse effect."

Fortunately, the amount of CO_2 in the atmosphere is not increasing as rapidly as might be expected, largely because the ocean is a great CO_2 sink. As the partial pressure of CO_2 increases, CO_2 solubility increases, and it is estimated that the sea has absorbed roughly half of the increase in CO_2. Although this should lead in turn to an increase in hydronium ion concentrations, it can be controlled through reaction with carbonate ion. Furthermore, as the upper layers of sea water containing dissolved CO_2 are mixed with the lower layers in contact with carbonate-containing sediments, hydronium ion can be removed by a reaction such as

$$H_3O^+(aq) + CaCO_3(s) \rightleftharpoons HCO_3^-(aq) + Ca^{2+}(aq) + H_2O(\ell)$$

The problem is that ocean mixing is relatively slow, requiring times on the order of 1000 years for complete mixing.

19.12 CHEMICAL EQUILIBRIA: AN EPILOGUE

The chemistry of CO_2 and carbonates in the environment, which is similar to their chemistry in the human body, illustrates the vastness of the subject of chemical equilibria and their importance. We are reminded again, however, that many of the chemical reactions that are interesting and important to us all occur under *nonequilibrium conditions*: the biochemical reactions in plants and animals, the production of elements in the stars, and the synthesis of useful materials. Although many processes in living entities can be modeled using the concepts of equilibrium, it is the movement toward a state of equilibrium that is the driving force of life.

CHAPTER HIGHLIGHTS

Now that you have finished studying this chapter, you should be able to

- Recall the solubility guidelines (see Figure 5.4) and write balanced equations for precipitation reactions (Section 19.1).
- Write the equilibrium constant expression—the **solubility product constant, K_{sp}**—for any insoluble salt (Section 19.2).
- Calculate K_{sp} values from experimental data (Section 19.3).
- Estimate the solubility of a salt from the value of K_{sp} (Section 19.4).
- Use K_{sp} values to decide the order of precipitation of two or more insoluble salts (Section 19.5).
- Decide if a precipitate forms when the ion concentrations are known (Section 19.6).
- Calculate the ion concentrations that are required to begin the precipitation of an insoluble salt (Section 19.5).
- Calculate the solubility of a salt in the presence of a common ion (Section 19.6).
- Use K_{sp} values to devise a method of separating ions in solution from one another (Section 19.7).
- Calculate the equilibrium constant for the net reaction for a situation in which two or more equilibrium processes are occurring in solution (Section 19.8).
- Understand that hydrolysis increases the solubility of a salt when the anion is the conjugate base of a weak acid (Section 19.9).
- Recognize that the solubility of insoluble salts containing basic anions is affected by the pH of the solution (Section 19.9).
- Understand that the formation of a complex ion can increase the solubility of an insoluble salt (Section 19.10).

STUDY QUESTIONS

REVIEW QUESTIONS

1. Explain why $[CaCO_3]$ does not appear in the K_{sp} expression for

$$CaCO_3(s) \rightleftharpoons Ca^{2+}(aq) + CO_3^{2-}(aq)$$

2. What is a "reaction quotient" and how does it differ from an equilibrium constant? Use the following equilibrium in your discussion:

$$AgCl(s) \rightleftharpoons Ag^{+}(aq) + Cl^{-}(aq)$$

3. Find two salts in Appendix J that have K_{sp} values less than 1×10^{-4}.
   (a) Write balanced equations to show the equilibria existing when the compounds dissolve in water.
   (b) Write the K_{sp} expression for each compound.
4. Explain the terms "saturated," "not saturated" ("unsaturated" or "undersaturated"), and "supersaturated." Use an equilibrium such as

$$CaF_2(s) \rightleftharpoons Ca^{2+}(aq) + 2\ F^{-}(aq)$$

to illustrate your answer.

5. What is the common ion effect? Use the equilibrium

$$Fe(OH)_2(s) \rightleftharpoons Fe^{2+}(aq) + 2\ OH^-(aq)$$

in your discussion.

6. Explain why the solubility of Ag_3PO_4 can be greater in water than is calculated from the K_{sp} of the salt.
7. Silver chloride is insoluble in water. If aqueous ammonia is added, however, the precipitate of AgCl dissolves. Explain.

NUMERICAL AND OTHER QUESTIONS

Solubility Guidelines

The following questions are a review of the solubility guidelines discussed in Section 5.1.

8. Name two insoluble salts of each of the following ions:
(a) Cl^-
(b) Zn^{2+}
(c) Fe^{2+}

9. Name two insoluble salts of each of the following ions:
(a) SO_4^{2-}
(b) Ni^{2+}
(c) Br^-

10. Using the table of solubility guidelines (Figure 5.4), predict whether each of the following is insoluble or soluble in water:
(a) $(NH_4)_2CO_3$ (c) NiS
(b) $ZnSO_4$ (d) $BaSO_4$

11. Using the table of solubility guidelines (Figure 5.4), predict whether each of the following is insoluble or soluble in water:
(a) $Pb(NO_3)_2$ (c) $ZnCl_2$
(b) $Fe(OH)_3$ (d) CuS

12. Will a precipitate form when 0.10 M solutions of the following compounds are mixed? Give the formula of the precipitate.
(a) $NaBr(aq) + AgNO_3(aq)$
(b) $KCl(aq) + Pb(NO_3)_2(aq)$
(c) $ZnCl_2(aq) + Mg(NO_3)_2(aq)$

13. Each of the following pairs of compounds is soluble in water. State whether a precipitate will form if you mix equal volumes of 0.1 M solutions of these compounds. Give the formula of the precipitate.
(a) $Na_2SO_4 + AgNO_3$ (c) KOH + LiCl
(b) $NaOH + CaCl_2$ (d) $KBr + Al(NO_3)_3$

Writing Solubility Product Constant Expressions

14. For each of the following insoluble salts, (i) write a balanced equation showing the equilibrium occurring when the salt is added to water, and (ii) write the K_{sp} expression. Give the value for the K_{sp} of each salt.
(a) AgCN
(b) $NiCO_3$
(c) $AuBr_3$

15. For each of the following insoluble salts, (i) write a balanced equation showing the equilibrium occurring when the salt is added to water, and (ii) write the K_{sp} expression. Give the value for the K_{sp} of each salt.
(a) $PbSO_4$
(b) BaF_2
(c) Ag_3PO_4

Calculating K_{sp}

16. When 1.55 g of solid thallium(I) bromide is added to 1.00 L of water, the salt dissolves to a small extent:

$$TlBr(s) \rightleftharpoons Tl^+(aq) + Br^-(aq)$$

The thallium(I) and bromide ions in equilibrium with TlBr each have a concentration of 1.8×10^{-3} M. What is the value of K_{sp} for TlBr?

17. When 250 mg of CaF_2, calcium fluoride, is added to 1.00 L of water, a very small amount dissolves:

$$CaF_2(s) \rightleftharpoons Ca^{2+}(aq) + 2\ F^-(aq)$$

At equilibrium, the concentration of Ca^{2+} is found to be 2.1×10^{-4} M. What is the value of K_{sp} for CaF_2?

18. At 20 °C, a saturated aqueous solution of silver acetate, $AgCH_3CO_2$, contains 1.0 g dissolved in 100.0 mL of solution. Calculate K_{sp} for silver acetate:

$$AgCH_3CO_2(s) \rightleftharpoons Ag^+(aq) + CH_3CO_2^-(aq)$$

19. Calcium hydroxide, $Ca(OH)_2$, dissolves in water to the extent of 0.93 g per liter. What is the K_{sp} of $Ca(OH)_2$?

$$Ca(OH)_2(s) \rightleftharpoons Ca^{2+}(aq) + 2\ OH^-(aq)$$

20. You place 2.75 g of barium fluoride in 1.00 L of pure water at 25 °C. After equilibrium has been established,

$$BaF_2(s) \rightleftharpoons Ba^{2+}(aq) + 2\ F^-(aq)$$

the fluoride ion concentration is 0.0150 M. What is the K_{sp} of BaF_2?

21. At 25 °C, 34.9 mg of Ag_2CO_3 will dissolve in 1.0 L of pure water:

$$Ag_2CO_3(s) \rightleftharpoons 2\ Ag^+(aq) + CO_3^{2-}(aq)$$

What is the solubility product constant for this salt?

22. You add 0.979 g of $Pb(OH)_2$ to 1.00 L of pure water at 25 °C, and it partially dissolves to give a saturated solution. The pH is 8.92. Estimate the value of K_{sp} for $Pb(OH)_2$.

23. You place 1.234 g of solid $Ca(OH)_2$ in 1.00 L of pure water at 25 °C and it partially dissolves. The pH of the solution is 12.40. Estimate the K_{sp} for $Ca(OH)_2$.

Estimating Salt Solubility from K_{sp}

The K_{sp} values required for the problems that follow are found in Table 19.2 or in Appendix J.

24. Estimate the solubility of silver cyanide in (a) moles per liter and (b) grams per liter in pure water at 25 °C:

$$AgCN(s) \rightleftharpoons Ag^+(aq) + CN^-(aq)$$

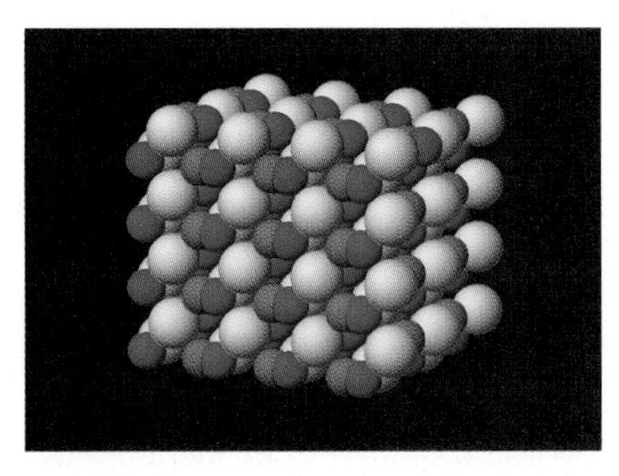

A model of the structure of silver cyanide, AgCN.

25. What is the molar concentration of $Au^+(aq)$ in a saturated solution of AuCl in pure water at 25 °C?

$$AuCl(s) \rightleftharpoons Au^+(aq) + Cl^-(aq)$$

26. The K_{sp} of radium sulfate, $RaSO_4$, is 4.2×10^{-11}. If 25 mg of radium sulfate is placed in 1.00×10^2 mL of water, how many milligrams of the salt dissolve?

27. If 55 mg of lead(II) sulfate is placed in 250 mL of pure water, what mass of the lead compound remains undissolved?

28. Estimate the solubility of magnesium fluoride, MgF_2, in (a) moles per liter and (b) grams per liter of pure water:

$$MgF_2(s) \rightleftharpoons Mg^{2+}(aq) + 2\ F^-(aq)$$

29. Estimate the solubility of lead(II) bromide in (a) moles per liter and (b) grams per liter of pure water.

30. Use K_{sp} values to decide which compound in each of the following pairs is the more soluble.
(a) $PbCl_2$ or $PbBr_2$ (c) BiI_3 or $Bi(OH)_3$
(b) HgS or FeS (d) $Fe(OH)_2$ or $Zn(OH)_2$

31. Use K_{sp} values to decide which compound in each of the following pairs is the more soluble.
(a) AgBr or AgSCN (c) AgI or HgI_2
(b) $SrCO_3$ or $SrSO_4$ (d) MgF_2 or CaF_2

32. Rank the following compounds in order of increasing solubility in water: Na_2CO_3, $BaCO_3$, and Ag_2CO_3.

33. Rank the following compounds in order of increasing solubility in water: AgI, HgS, PbI_2, $PbSO_4$, NH_4NO_3.

34. If you place 5.0 mg of $NiCO_3$ in 1.0 L of pure water, does all the salt dissolve before equilibrium can be established, or does some salt remain undissolved?

35. If you place the amounts of the compounds given here in pure water, does the salt dissolve before equilibrium can be established, or does some salt remain undissolved?
(a) 5.0 mg of MgF_2 in 125 mL of pure water
(b) 0.50 g of CaF_2 in 95 mL of pure water

Precipitations

36. You have a solution that has a lead(II) concentration of 0.0012 M. If enough soluble chloride-containing salt is added so that the Cl^- concentration is 0.010 M, does $PbCl_2$ precipitate? The equilibrium process involved is

$$PbCl_2(s) \rightleftharpoons Pb^{2+}(aq) + 2\ Cl^-(aq)$$

37. Sodium carbonate is added to a solution in which the concentration of Ni^{2+} ion is 0.0024 M. Does $NiCO_3$ precipitate (a) when the concentration of the carbonate ion is 1.0×10^{-6} M or (b) when it is 100 times greater (or 1.0×10^{-4} M)? The equilibrium process is

$$NiCO_3(s) \rightleftharpoons Ni^{2+}(aq) + CO_3^{\ 2-}(aq)$$

38. If the concentration of Zn^{2+} in exactly 10 mL of water is 1.6×10^{-4} M, does zinc hydroxide, $Zn(OH)_2$, precipitate when 4.0 mg of NaOH is added?

39. You have 95 mL of a solution that has a lead(II) concentration of 0.0012 M. Does $PbCl_2$ precipitate when 1.20 g of solid NaCl is added?

40. A sample of hard water contains about 2.0×10^{-3} M Ca^{2+}. A soluble fluoride-containing salt such as NaF is added to "fluoridate" the water (to aid in the prevention of dental caries). What is the maximum concentration of F^- that can be present *without* precipitating CaF_2?

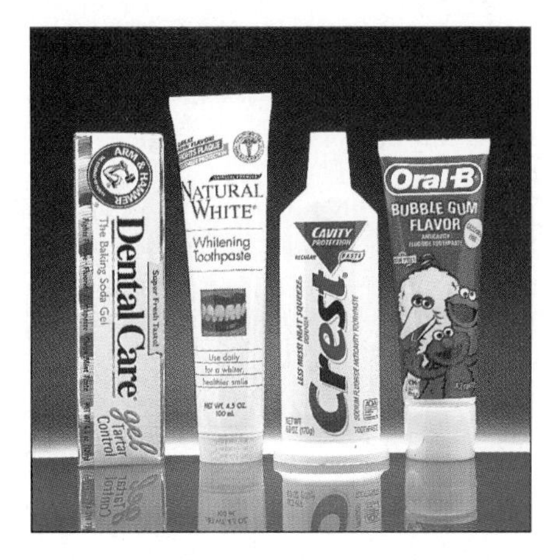

Adding fluoride ion to drinking water (or toothpaste) prevents the formation of dental caries. *(C. D. Winters)*

41. If the concentration of Mg^{2+} in sea water is 1350 mg/L, what OH^- concentration is required to precipitate $Mg(OH)_2$?

42. Does a precipitate of $Mg(OH)_2$ form when 25.0 mL of 0.010 M NaOH is combined with 75.0 mL of a 0.10 M solution of magnesium chloride?

43. If you mix 48 mL of 0.0012 M $BaCl_2$ with 24 mL of 1.0×10^{-6} M H_2SO_4, does a precipitate of $BaSO_4$ form? The equilibrium process involved is

$$BaSO_4(s) \rightleftharpoons Ba^{2+}(aq) + SO_4^{\ 2-}(aq)$$

44. If you mix 10.0 mL of 0.0010 M $Pb(NO_3)_2$ with 5.0 mL of 0.015 M HCl, does $PbCl_2$ precipitate?

45. If 1.00×10^2 mL of 0.10 M H_2SO_4 is added to 50.0 mL of 0.00013 M $BaCl_2$, does $BaSO_4$ precipitate?

46. The cations Ba^{2+}, Sr^{2+}, and Pb^{2+} can all be precipitated as very insoluble sulfates. If you add sodium sulfate to a solution containing all of these metal cations, each with a concentration of 0.1 M, in what order are the insoluble sulfates precipitated?

47. Hydrogen iodide, HI, is added slowly to a solution 0.10 M in each of the following ions: Pb^{2+}, Ag^+, and Hg_2^{2+}. The insoluble salts PbI_2, AgI, and Hg_2I_2 eventually form. In what order do these salts precipitate?

48. You often work with salts of Fe^{3+}, Pb^{2+}, and Al^{3+} in the laboratory. (All are found in nature, and all are important economically.) If you have a solution containing these three ions, each at a concentration of 0.1 M, what is the order in which their hydroxides precipitate as aqueous NaOH is slowly added?

49. Alkaline earth metal ions can be precipitated as insoluble carbonates. If you have a solution of Mg^{2+}, Ca^{2+}, Sr^{2+}, and Ba^{2+} ions, all with the same concentration, what is the order in which their carbonates are precipitated as sodium carbonate is slowly added?

Common Ion Effect

50. Calculate the molar solubility of silver thiocyanate, AgSCN, in pure water and in water containing 0.010 M NaSCN.

51. Calculate the solubility of silver carbonate, Ag_2CO_3, in moles per liter, in pure water. Compare this with the molar solubility of Ag_2CO_3 in 225 mL of water to which 0.15 g of Na_2CO_3 has been added.

52. Calculate the solubility, in milligrams per milliliter, of silver phosphate, Ag_3PO_4, in (a) pure water and (b) in water that is 0.020 M in $AgNO_3$.

53. What is the solubility, in milligrams per milliliter, of BaF_2, (a) in pure water and (b) in water containing 5.0 mg/mL of KF?

Separations

54. To separate Ca^{2+} and Mg^{2+} ions from one another, ammonium oxalate, $(NH_4)_2C_2O_4$, is added to a solution that is 0.020 M in both metal ions. If the concentration of the oxalate ion is adjusted properly, the metal oxalates can be precipitated separately. In this case CaC_2O_4 precipitates before MgC_2O_4. The chemical equilibria involved are

$$MgC_2O_4(s) \rightleftharpoons Mg^{2+}(aq) + C_2O_4^{2-}(aq)$$
$$CaC_2O_4(s) \rightleftharpoons Ca^{2+}(aq) + C_2O_4^{2-}(aq)$$

(a) What concentration of oxalate ion, $C_2O_4^{2-}$, precipitates the maximum amount of Ca^{2+} ion without precipitating Mg^{2+}?

(b) What concentration of Ca^{2+} remains in solution when Mg^{2+} just begins to precipitate?

55. In principle, the ions Ba^{2+} and Ca^{2+} can be separated by the difference in solubility of their fluorides, BaF_2 and CaF_2. If you have a solution that is 0.10 M in both Ba^{2+} and Ca^{2+}, CaF_2 begins to precipitate first as fluoride ion is added slowly to the solution.

(a) What concentration of fluoride ion precipitates the maximum amount of Ca^{2+} ion without precipitating BaF_2?

(b) What concentration of Ca^{2+} remains in solution when BaF_2 just begins to precipitate?

56. A solution contains 0.10 M iodide ion, I^-, and 0.10 M carbonate ion, CO_3^{2-}.

(a) If solid $Pb(NO_3)_2$ is slowly added to the solution, which salt precipitates first, PbI_2 or $PbCO_3$?

(b) What is the concentration of the first ion that precipitates (CO_3^{2-} or I^-) when the second or more soluble salt begins to precipitate?

57. A solution contains Ca^{2+} and Pb^{2+} ions, both at a concentration of 0.010 M. You wish to separate the two ions from each other as completely as possible by precipitating one but not the other using aqueous Na_2SO_4 as the precipitating agent.

(a) Which precipitates first as sodium sulfate is added, $CaSO_4$ or $PbSO_4$?

(b) What is the concentration of the first ion that precipitates (Ca^{2+} or Pb^{2+}) when the second, or more soluble, salt begins to precipitate?

58. Each of the following pairs of ions is found together in aqueous solution. Using the table of solubility product constants in Appendix J, devise a way to separate these ions by precipitating one of them as an insoluble salt and leaving the other in solution.

(a) Ba^{2+} and Na^+

(b) Bi^{3+} and Cd^{2+}

59. Each of the following pairs of ions is found together in aqueous solution. Using the table of solubility product constants in Appendix J, devise a way to separate these ions by adding one reagent to precipitate one of them as an insoluble salt and leave the other in solution.

(a) Cu^{2+} and Ag^+

(b) Al^{3+} and Fe^{3+}

Simultaneous Equilibria and Complex Ions

60. Solid gold(I) chloride, AuCl, is dissolved when excess cyanide ion, CN^-, is added to give a water-soluble complex ion:

$$AuCl(s) + 2\ CN^-(aq) \rightleftharpoons Au(CN)_2^-(aq) + Cl^-(aq)$$

Show that this equation is the sum of two other equations, one for dissolving AuCl to give its ions and the other for the formation of the $Au(CN)_2^-$ ion from Au^+ and CN^-. Calculate K_{net} for the overall reaction. (See Appendices J and K for the required equilibrium constants.)

61. Solid silver bromide, AgBr, can be dissolved by adding concentrated aqueous ammonia to give the water-soluble silver–ammonia complex ion.

$$AgBr(s) + 2\ NH_3(aq) \rightleftharpoons [Ag(NH_3)_2]^+(aq) + Br^-(aq)$$

Show that this equation is the sum of two other equations, one for dissolving AgBr to give its ions and the other for the formation of the $[Ag(NH_3)_2]^+$ ion from Ag^+ and NH_3. Calculate K_{net} for the overall reaction.

62. Calculate the equilibrium constant for the following reaction:

$$AgCl(s) + I^-(aq) \rightleftharpoons AgI(s) + Cl^-(aq)$$

Does the equilibrium lie predominantly to the left or right? Does AgI form if iodide ion, I^-, is added to a saturated solution of AgCl?

63. Calculate the equilibrium constant for the following reaction:

$$Zn(OH)_2(s) + 2\ CN^-(aq) \rightleftharpoons Zn(CN)_2(s) + 2\ OH^-(aq)$$

Does the equilibrium lie predominantly to the left or right? Can zinc hydroxide be transformed into zinc cyanide by adding a soluble salt of the cyanide ion? (See Appendix J for the required equilibrium constants.)

64. Can 5.0 mL of 2.5 M NH_3 dissolve 1.0×10^{-4} mol of AgBr suspended in 1.0 L of pure water?

65. Can you completely dissolve 15.0 mg of AuCl in 100.0 mL of pure water if you add 15.0 mL of 6.00 M NaCN? (See Study Question 60.)

Solubility and pH

66. Which of the following barium compounds should be soluble in a strong acid such as HCl: $Ba(OH)_2$, $BaSO_4$, or $BaCO_3$?

67. Which of the following silver salts should be soluble in a strong acid: AgI, Ag_2CO_3, or Ag_3PO_4?

68. Which of the following salts should be soluble in a strong acid: BiI_3, $BaCO_3$, or $BiPO_4$?

69. Suggest a method for separating a precipitate consisting of a mixture of CuS and $Cu(OH)_2$.

GENERAL QUESTIONS

70. A saturated solution of Li_3PO_4 has $[PO_4^{3-}] = 3.3 \times 10^{-3}$ M.
(a) What is the value of K_{sp} for Li_3PO_4?
(b) Does Li_3PO_4 become more soluble or less soluble on adding aqueous HCl?

71. If you place 1.00×10^2 mg of $BaSO_4$ in 1.00 L of water, approximately how many milligrams of $BaSO_4$ are dissolved at 25 °C?

72. The alkaline earth metal ions can be precipitated from aqueous solution by addition of fluoride ion.
(a) If you have a 0.015 M solution of Ba^{2+}, what is the minimum concentration of F^- ion necessary to begin precipitation of BaF_2?
(b) If you start with 100.0 mL of the 0.015 M Ba^{2+}-containing solution, how many milligrams of NaF are required to just begin precipitation?
(c) After adding 0.50 g of NaF to 100.0 mL of 0.015 M Ba^{2+}-containing solution, what is the concentration of Ba^{2+} remaining in solution?

73. The alkaline earth cations Mg^{2+}, Ca^{2+}, and Ba^{2+} can be precipitated as their sulfate salts. If you have a solution of these three ions, each with a concentration of 0.001 M, in what order do these ions precipitate as their insoluble sulfates as you slowly add sulfuric acid to the solution?

74. To make up an unknown solution for the students in your laboratory, you dissolve 1.0×10^{-5} mol of $AgNO_3$ in 1.0 L of water. Carelessly, you used tap water instead of distilled water. If the chloride ion concentration in tap water is 2.0×10^{-4} M, is your error revealed by the formation of a white precipitate of AgCl?

75. The solubilities of lead and zinc compounds are often similar. For example, the K_{sp} of $Pb(OH)_2$ is 2.8×10^{-16} and that of $Zn(OH)_2$ is 4.5×10^{-17}. If Pb^{2+} and Zn^{2+} each have a concentration of 1.0×10^{-6} M in a solution, and if the solution has a pH of 8.0, do either $Pb(OH)_2$ or $Zn(OH)_2$ precipitate?

76. Predict the order in which insoluble iodide salts precipitate as sodium iodide is slowly added to an aqueous solution containing Ag^+, Bi^{3+}, and Pb^{2+}. (The metal ions are initially each 0.10 M.)

77. Suppose you mix 15.0 mL of 0.010 M $CaCl_2$ and 25.0 mL of 0.0010 M NaOH. Does $Ca(OH)_2$ precipitate? If so, how many grams of $Ca(OH)_2$ are formed?

78. $Zn(OH)_2$ is a relatively insoluble base. A saturated solution has a pH of 8.65. Calculate the K_{sp} for $Zn(OH)_2$.

79. Although it is a violent poison if swallowed, mercury(II) cyanide, $Hg(CN)_2$, has been used as a topical (skin) antiseptic.
(a) What is the molar solubility of this salt in pure water?
(b) How many milligrams of $Hg(CN)_2$ dissolve per liter of pure water?
(c) How many milliliters of water are required to dissolve 1.0 g of the salt?

80. A saturated solution of $Mg(OH)_2$ has a pH of 10.49. Use this information to calculate the K_{sp} of $Mg(OH)_2$.

81. If 0.581 g of solid $Mg(OH)_2$ is added to 1.00 L of pure water, what quantity dissolves? If the solution is buffered at a pH of 5.00, does more of the hydroxide dissolve? If so, how much?

82. Suppose you mix 2.00 g of $AgNO_3$ and 3.00 g of K_2CrO_4 in enough water to make 50.0 mL of solution. What is the concentration of the ions—Ag^+, NO_3^-, K^+, and CrO_4^{2-}—in the final solution? The K_{sp} for Ag_2CrO_4 is 9.0×10^{-12}.

83. The Ca^{2+} ion in hard water is often precipitated as $CaCO_3$ by adding soda ash, Na_2CO_3. If the calcium ion concentration in hard water is 0.010 M, and if the Na_2CO_3 is added until the carbonate ion concentration is 0.050 M, what percentage of the calcium ion will be removed from the water? (You may ignore hydrolysis of the carbonate ion.)

84. Photographic film is coated with crystals of AgBr suspended in gelatin. Light exposure leads to the reduction of some of the silver ions to metallic silver. Unexposed AgBr is dissolved with sodium thiosulfate in the fixing step:

$$AgBr(s) + 2\ S_2O_3^{2-}(aq) \rightleftharpoons [Ag(S_2O_3)_2]^{3-}(aq) + Br^-(aq)$$

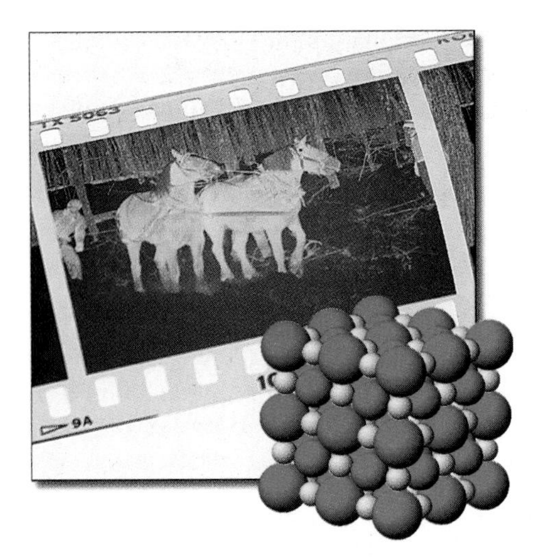

Photographic film is coated with silver bromide. Notice that the crystal structure of AgBr is identical to that of AgCl (Figure 19.7). See also Figure 19.13. *(Photo, C. D. Winters; models, S. M. Young)*

(a) Using the appropriate K_{sp} and $K_{formation}$ values in Appendices J and K, calculate the equilibrium constant for the dissolving process.
(b) If you want to dissolve 1.0 g of AgBr in 1.0 L of solution, how many grams of $Na_2S_2O_3$ must be added?

85. Can 5.0 mL of 2.5 M NH_3 dissolve 25 mg of AgI suspended in 1.0 L of pure water?

CONCEPTUAL QUESTIONS

86. Calcium carbonate, $CaCO_3$, can exist in two mineral forms: calcite and aragonite. In pure water, calcite has a K_{sp} of 3.8×10^{-9}, whereas the K_{sp} of aragonite is 6.0×10^{-9}. Which is the more soluble in pure water?

87. Which of the following reagents would you add to increase the water solubility of magnesium hydroxide?

$$Mg(OH)_2(s) \rightleftharpoons Mg^{2+}(aq) + 2\ OH^-(aq)$$

(a) NaCl (d) HCl
(b) $MgCl_2$ (e) H_2O
(c) NaOH

88. Explain how changing pH affects the solubility of $Ni(OH)_2$.

89. Explain how changing pH affects the solubility of Ag_3PO_4.

90. The formula for insoluble mercury(I) chloride is Hg_2Cl_2. Suppose a student incorrectly assumed its formula was HgCl and that the following equilibrium occurred in aqueous solution

$$HgCl(s) \rightleftharpoons Hg^+(aq) + Cl^-(aq)$$

instead of the correct reaction

$$Hg_2Cl_2(s) \rightleftharpoons Hg_2^{2+}(aq) + 2\ Cl^-(aq)$$

If the student determined $[Cl^-]$ in a saturated solution of mercury(I) chloride, would the correct K_{sp} value be calculated? If not, would the calculated value be too high or too low?

91. Explain why $PbCO_3$ dissolves in strong acid but $PbCl_2$ cannot.

92. Barium carbonate dissolves to some extent in the presence of carbon dioxide:

$$BaCO_3(s) + CO_2(g) + H_2O(\ell) \rightleftharpoons Ba^{2+}(aq) + 2\ HCO_3^-(aq) \quad K = 4.5 \times 10^{-5}$$

(a) How is the solubility of barium carbonate affected by the pressure of CO_2?
(b) How is the solubility of barium carbonate affected by a decrease in the pH?

93. Two common constituents of kidney stones are calcium phosphate, $Ca_3(PO_4)_2$, and hydrated calcium oxalate, $Ca(C_2O_4)_2 \cdot H_2O$. Are kidney stones containing these salts more likely to form when the urine is acidic or when it is basic?

CHALLENGING QUESTIONS

94. On the basis of the following facts and experimental observations, list $Al(OH)_3$, AgSCN, AgCl, $Fe(OH)_3$, and Ag_3PO_4 in order of decreasing solubility in water.
(a) A dilute solution of NaSCN is added to white, solid AgCl. The white solid dissolves and pale, amber AgSCN is formed. K_{sp} for AgSCN is 1.0×10^{-12}.
(b) A saturated solution of $Fe(OH)_3$ has a lower pH than a saturated solution of $Al(OH)_3$. K_{sp} for $Al(OH)_3$ is 1.9×10^{-33}.
(c) The concentration of Ag^+ ion in equilibrium with solid Ag_3PO_4 is four times larger than $[Ag^+]$ in equilibrium with solid AgCl.

95. Explain why each of the following four facts is consistent with the other three based on values of K_{sp} and formation constants for complex ions.
(a) AgI is less soluble in water than AgCl.
(b) The value of K_{sp} for AgCl is larger than K_{sp} for AgI.
(c) AgCl is soluble in 15 M NH_3(aq). Solid AgI is not soluble under these conditions.
(d) If solid AgCl is stirred in an aqueous solution of KI, AgCl dissolves and AgI precipitates.

96. A test tube contains insoluble ZnS and 0.10 M H_2S. A strong acid such as HCl(aq) is added. What is the concentration of the Zn^{2+} ion when the pH of the solution is adjusted to 1.50? [See *A Closer Look: Equilibria Involving Sulfide Ions* (page 896) for the equilibrium constant expression for ZnS dissolving in acid.]

97. A test tube contains 5.0 mg of FeS in exactly 10 mL of water. Assume the solution also contains H_2S with a concentration of 0.10 M. What should the pH be if all of the FeS is to dissolve? (See *A Closer Look: Equilibria Involving Sulfide Ions,* page 896.)

Summary Question

98. Aluminum chloride reacts with phosphoric acid to give aluminum phosphate, $AlPO_4$. The solid exists in many of the same crystal forms as SiO_2 and is used industrially as the basis of adhesives, binders, and cements. ($AlPO_4$ is often referred to by the same name as a popular pet food company.)

(a) Write a balanced equation for the reaction of aluminum chloride and phosphoric acid.

(b) If you begin with 152 g of aluminum chloride, and 3.0 L of 0.750 M phosphoric acid, how many grams of $AlPO_4$ can be isolated?

(c) If you place 25.0 g of $AlPO_4$ in enough pure water to have a volume of exactly one liter, what are the concentrations of Al^{3+} and PO_4^{3-} at equilibrium?

(d) Does the solubility of $AlPO_4$ increase or decrease on adding HCl? Explain briefly.

(e) If you mix 1.50 L of 0.0025 M Al^{3+} (in the form of $AlCl_3$) with 2.50 L of 0.035 M Na_3PO_4, does a precipitate of $AlPO_4$ form? If so, how many grams of $AlPO_4$ precipitate?

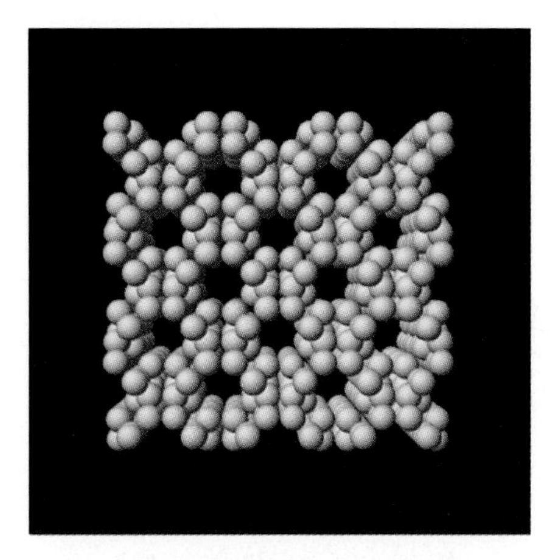

The structure of an aluminum phosphate, $AlPO_4$. Each sphere represents a tetrahedral aluminum or phosphorus center. An O atom is shared between each Al and P. See http://www-iza-sc.csb.yale.edu/IZA-SC/ for information on the structures of aluminum phosphates and related zeolite structures.

INTERACTIVE GENERAL CHEMISTRY CD-ROM

Screen 19.5 Determining K_{sp}

(a) In the example on this screen, we determine the K_{sp} value for $PbCl_2$ by measuring the concentration of Pb^{2+} ions in solution. Why do we not need to measure the concentration of Cl^- ions in this case?

(b) Watch the video on this screen describing the use of an atomic absorption spectrometer. Explain how the spectrometer works.

Screen 19.6 Estimating Salt Solubility

This screen describes how the solubility of $BaSO_4$ is estimated using its K_{sp} value. The calculation assumes that the ions produced, Ba^{2+} and SO_4^{2-}, do not react further in solution. What would happen to the solubility of $BaSO_4$ if some of the sulfate ion reacted with acid in solution?

$$SO_4^{2-}(aq) + H_3O^+(aq) \longrightarrow HSO_4^-(aq) + H_2O(\ell)$$

Screen 19.7 Can a Precipitation Reaction Occur?

Three cases are given on this screen: $Q < K_{sp}$, $Q > K_{sp}$, and $Q = K_{sp}$.

(a) Which of these represents a system at equilibrium?

(b) Examine the $Q > K_{sp}$ video and animation. Describe the silver chloride precipitation at the particulate level. (See also Figure 5.6.)

Screen 19.8 The Common Ion Effect

The animation on this screen illustrates the common ion effect for the case of adding extra chloride ion to an equilibrium system containing $PbCl_2(s)$, $Pb^{2+}(aq)$, and $Cl^-(aq)$. Explain the changes you see in terms of the solubility product constant expression for this system.

Screen 19.9 Using Solubility

Examine the flowchart for the separation of the ions Ag^+, Pb^{2+}, and Cu^{2+} in aqueous solution.

(a) What is the initial step in the separation, and why does it work?

(b) In what way does the solubility of $PbCl_2$ differ from that of AgCl?

(c) How do we know that Pb^{2+} ions are in solution?

Screen 19.10 Simultaneous Equilibria

(a) The video on this screen shows that the white solid $PbCl_2$ can be converted into the yellow solid $PbCrO_4$. It might appear that the reaction occurs in the solid state, even though it actually occurs in aqueous solution. Write chemical equations to show how it occurs in solution.

(b) Why is the experiment on this screen good evidence that chemical equilibria are dynamic as opposed to static?

Screen 19.11 Solubility and pH

(a) Explain how the example on this screen, the solubility of $Co(OH)_2(s)$, illustrates Le Chatelier's principle.

(b) Notice that the solubility of the compound increases as the pH decreases. What effect do you think a pH decrease would have on the solubility of $MgCO_3$, considering the fact that carbonate ion, CO_3^{2-}, is a weak base?

(c) Examine the pH and Solubility Table. Explain why the solubility of $Co(OH)_2$ increases by 100 for each 1.0 decrease in pH.

Screen 19.12 Complex Ion Formation and Solubility

Examine the sidebar to this screen. How does the chemistry of floor wax support the idea that reactions can be reversible?

Chapter 20

Principles of Reactivity: Entropy and Free Energy

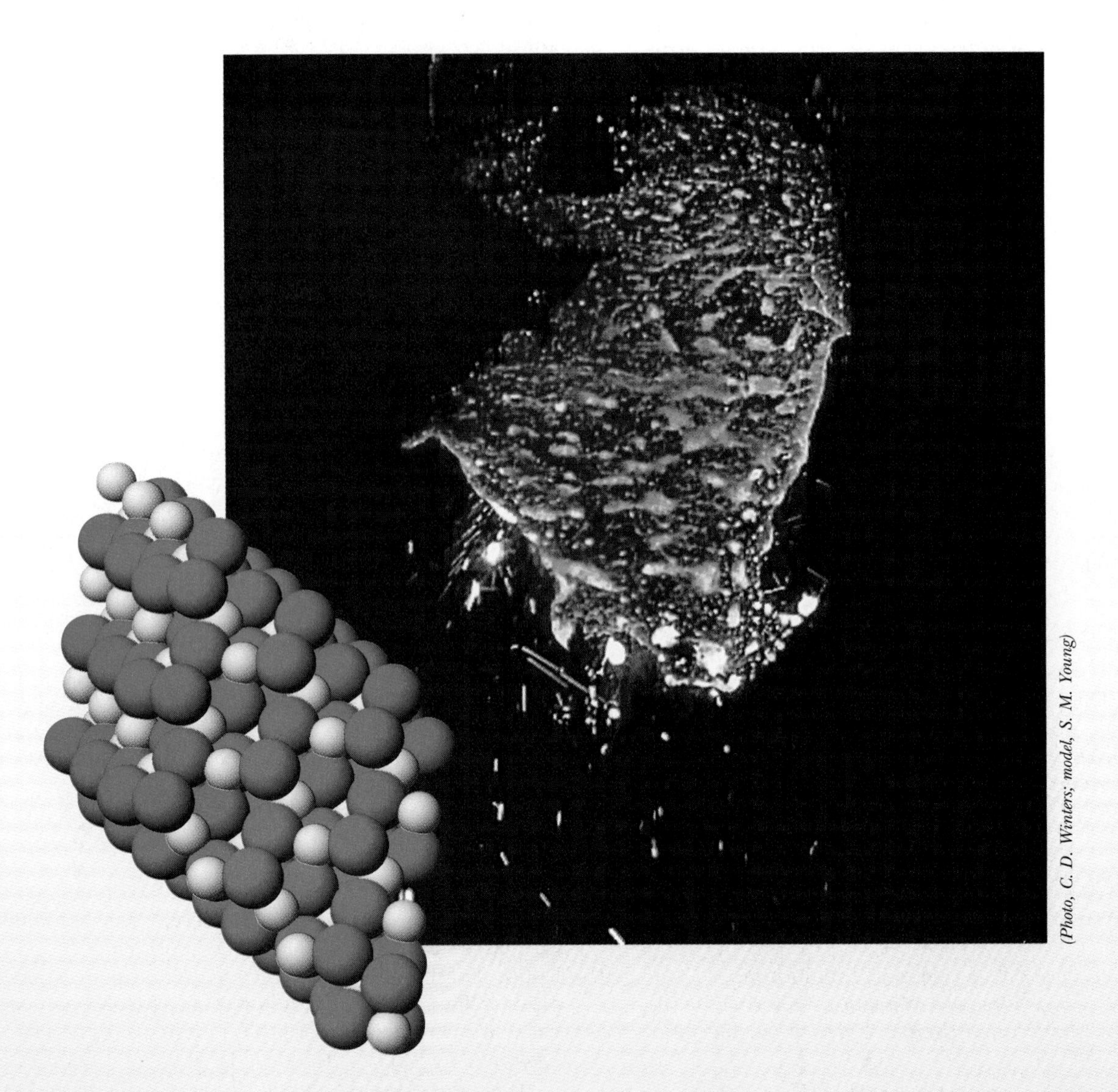

(Photo, C. D. Winters; model, S. M. Young)

A Chemical Puzzler

This reaction of iron wool with oxygen is highly product-favored. How might we have predicted that this would be the case? [The model is iron(III) oxide.]

Some chemical and physical changes take place by themselves, given enough time. If you stretch a rubber band, it snaps back spontaneously and quickly. If you put a spoonful of sugar in your coffee or tea, it dissolves, and the molecules distribute themselves evenly throughout the liquid. When placed in a flask of chlorine gas, hot copper reacts readily (Figure 20.1). These changes are all said to be spontaneous, or product-favored, although they do proceed at different speeds.

To control chemical reactions in order to produce new pharmaceuticals or new polymers or to prevent unwanted reactions in our environment, we must understand why some chemical reactions are product-favored and others are not. We must understand how to predict when a reaction will be product-favored and how to control that tendency, if necessary. The objective of this chapter is to describe a way to predict when reactions will be product- or reactant-favored.

Figure 20.1 Reaction of copper metal and chlorine. Copper wool burns in chlorine gas to give copper(II) chloride in a product-favored reaction.[The model is copper(II) chloride] *(Photo, C. D. Winters; model, S. M. Young)*

20.1 SPONTANEOUS REACTIONS AND SPEED: THERMODYNAMICS VERSUS KINETICS

As described in Chapter 6, a product-favored chemical reaction is one in which most of the reactants can eventually be converted to products, given sufficient time. The reactions of most elements with halogens are examples of product-favored reactions, and the reaction of copper with chlorine is no exception (see Figure 20.1):

$$\xrightarrow{\text{Product-favored}}$$
$$Cu(s) + Cl_2(g) \longrightarrow CuCl_2(s)$$

The contrasting term, *reactant-favored reaction,* may be misleading. One of the most widely discussed reactions in the past few years is the decomposition of ozone, O_3, to O atoms and O_2 molecules, a reaction that can occur by the action of photons of sunlight in the 280- to 310-nm range of wavelengths.

$$\xleftarrow{\text{Reactant-favored}}$$
$$O_3(g) \longrightarrow O(g) + O_2(g)$$

This reaction is reactant-favored. This does not mean it does not occur at all. What it does mean is that, when equilibrium is achieved, not many molecules of O_3 will have broken down into products. The equilibrium constant is much less than 1.

$$K_c = 0.0063 \text{ at } 25\ ^\circ\text{C} = \frac{[O][O_2]}{[O_3]}$$

The oxidation of H_2 by O_2 to form water is another product-favored reaction, but it occurs only if you ignite the mixture (see Figure 1.15). The mere presence of oxygen is not enough; H_2 gas can stay in contact with air a long time if you are careful not to set off the reaction. That is, there is a difference between the speed of a reaction and whether it is product-favored. **Thermodynamics,** the subject of this chapter, is the science of energy transfer, and it will help us predict whether a reaction can occur given enough time. Thermodynamics, however, can tell us nothing about the speed of the reaction. The study of the rates of reactions, and why some are fast and others are slow, is called **kinetics,** and it was the topic of Chapter 15.

You have seen a number of product-favored reactions in this book, examples of which include:

- $Al + Br_2$, page 2
- $Na + H_2O$, page 6
- $K + H_2O$, pages 18, 83
- $H_2 + O_2$, page 35
- $Mg + O_2$, pages 84, 151
- $Fe + Cl_2$, page 195
- $Na + Cl_2$, page 91

See the *Saunders Interactive General Chemistry CD-ROM,* Screen 20.2, Reaction Spontaneity: Thermodynamics and Kinetics.

A CLOSER LOOK — A Review of Definitions of Thermodynamics

System: The part of the universe (the specific atoms, molecules, or ions) under study.

Surroundings: The rest of the universe exclusive of the system.

Exothermic: Thermal energy transfer proceeds from the system to the surroundings.

Endothermic: Thermal energy transfer proceeds from the surroundings to the system.

First law of thermodynamics: The law of the conservation of energy, $\Delta E = q + w$. The change in the energy of a system is equal to the heat transferred to the system and the work done on the system.

Enthalpy change: The thermal energy transferred at constant pressure.

State function. A quantity whose changed value is determined only by the initial and final values.

Standard conditions: $P = 1$ bar. Material in solution has a concentration of 1 molal. (Note that in a dilute solution $m \approx$ M.)*

Enthalpy of formation: The enthalpy change occurring when a compound is formed from its elements, ΔH_f°. Elements in their standard states have $\Delta H_f^\circ = 0$.

*Note that standard pressure for thermodynamic quantities is 1 bar, where 1 bar = 0.98692 atm. Most data are quoted at 298 K. See Chapters 6 and 12.

Beginning with Chapter 6, we often used ideas of energy differences and energy transfer. In Chapter 6 we described the energy involved in chemical reactions, in Chapters 7 through 10 the concern was with energy on the atomic and molecular level, and in Chapters 13 and 14 we discussed the energy involved in changes in physical state. Now, we want to define more completely the energy changes that cause some chemical and physical changes to be product-favored, regardless of their speed.

Before beginning this survey, a word about nomenclature is important. Chemists often use the term *spontaneous* to refer to a product-favored reaction. A *nonspontaneous* process is reactant-favored. The word "spontaneous," however, often brings with it the idea that something not only happens but happens quickly, thus mixing up the concepts of thermodynamics and kinetics. The terms "product"- and "reactant-favored," therefore, are generally used in this book to separate clearly the concept of reaction probability from that of reaction speed.

20.2 DIRECTIONALITY OF REACTIONS: ENTROPY

See the *Saunders Interactive General Chemistry CD-ROM,* Screen 20.3, Directionality of Reactions.

In Chapter 6 we stated that at room temperature most exothermic reactions are product-favored and therefore are useful for carrying out chemical transformations. As we shall now see, this is true because such reactions lead to the dispersal of stored chemical potential energy over a larger number of atoms and molecules and because the matter concentrated in the reactants is dispersed. Let us persue these ideas further.

Dispersal of Energy and Matter

After an exothermic reaction, the chemical potential energy that had been stored in relatively few atoms and molecules (the reactants) is always distributed more randomly than it was before. The dispersal of energy over a much larger number of atoms or molecules can be partly or largely responsible for a reaction being product-favored.

To understand this, let us describe a simple model, the interaction of a Cl atom with an ozone molecule:

$$Cl(g) + O_3(g) \longrightarrow ClO(g) + O_2(g)$$

The Cl/O_3 reaction occurs in earth's ozone layer and is partly responsible for the loss of tropospheric ozone. The Cl atoms come from the action of sunlight on chlorofluorocarbons, or CFCs, compounds long used in refrigeration equipment and for many other purposes.

Take a very small sample of matter, say two Cl atoms and two O_3 molecules. The two Cl atoms have some amount of energy, but we shall suppose that the O_3 molecules are at a lower temperature and so have a lower kinetic energy. When these atoms and molecules are placed in a container together, there are six possible ways for them to come together (if we can differentiate the Cl and O_3 atoms and molecules). Two Cl atoms can come together or two O_3 molecules can come together. Chlorine atom A, however, can interact with O_3 molecule A or O_3 molecule B. For Cl atom B, interactions are also possible with each of the two O_3 molecules.

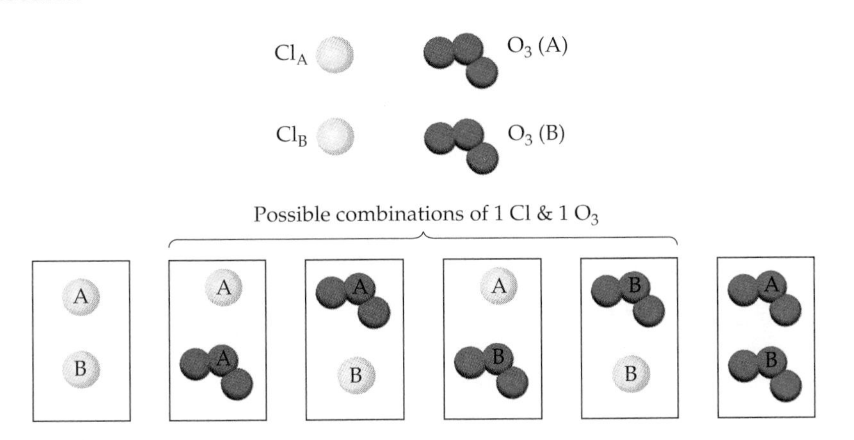

On page 244 we stated that energy always transfers from a hotter object to a colder one. Energy therefore transfers from the "hotter" Cl atoms to the "colder" O_3 molecules. Presumably no energy will be transferred from one Cl atom to another because both have the same energy, and the situation is the same for two O_3 molecules. Four of the six encounters that happen in our mixture do lead to energy transfer from Cl to O_3, however. That is, there is a high probability that the kinetic energy of the Cl atoms is dispersed over the system.

The transfer of energy from hotter Cl atoms to cooler O_3 molecules results ultimately in the destruction of the ozone and the formation of ClO radicals and O_2 molecules. Some portion of the kinetic energy originally possessed by the Cl atoms is now dispersed over the products.

A situation in which energy is concentrated in only a few atoms or molecules is unlikely. In fact, this is something we value highly because a substance in which a large quantity of chemical potential energy is concentrated among relatively few atoms or molecules is an energy resource. Examples are coal, oil, wood, and natural gas.

Just as there is a tendency for highly concentrated energy to disperse, highly concentrated matter also tends to disperse. For example, suppose a few crystals of $KMnO_4$ are placed in some water. You know from the experience of putting salt or sugar in water, and you can see in Figure 20.2, that the solute disperses throughout the water in time.

To model the dispersal of matter, let us turn again to a simple model. Suppose two O_3 molecules are placed in a flask. This flask is then opened to a second flask, which has the same volume but contains no molecules. We observed in a gas diffusion experiment (see Figure 12.20) that molecules rapidly diffuse from one place to another. Similarly, in Figure 20.3 you see that it is highly probable that the two O_3 molecules will be dispersed over the two flasks.

Figure 20.2 Dispersal of matter. (a) A small quantity of $KMnO_4$ is placed in water. (b) In time, the ionic solid dissolves and the highly colored MnO_4^- ions (and K^+ ions) are dispersed throughout the water. *(C. D. Winters)*

It is certainly possible that both molecules will stay in the original flask, and it is possible that both of them will move to the second flask. It is twice as likely, however, that one will be in each flask rather than having both in one flask.

In general, the probability that all of the molecules remain in the original flask of a two flask system is $(\frac{1}{2})^n$, where n is the number of molecules.

If you now put one more O_3 molecule in the original flask, you can show there are eight possible ways to distribute three molecules between two flasks, only one of which corresponds to having all three O_3 molecules in the original flask. For ten molecules, there is only one chance in 1.0×10^3 that all will be in the original flask! This makes it obvious that if we have a gas sample of the size we usually work with, say 0.1 mol, then the probability that matter is dispersed is overwhelming.

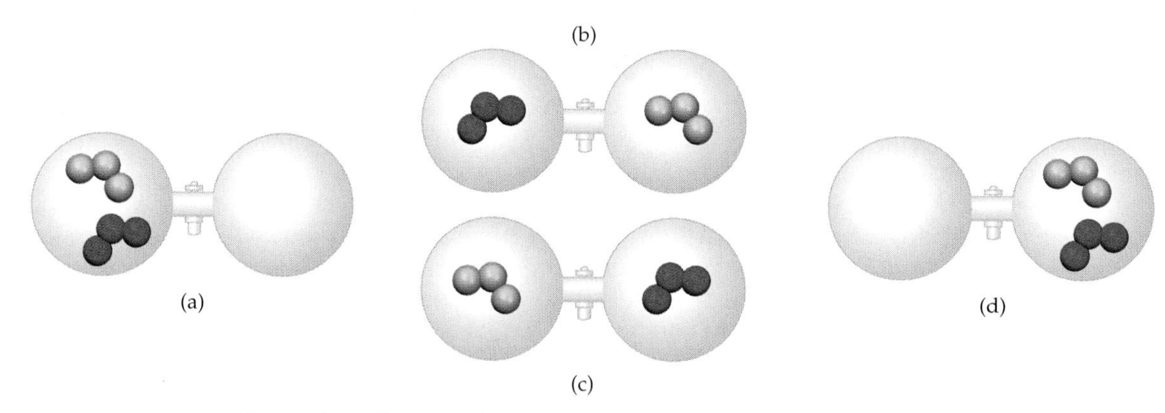

Figure 20.3 The expansion of a gas is a highly probable process. Two O_3 molecules were originally in one flask (a). When the valve to the other flask (of equal volume) is opened, the two molecules distribute themselves between the two flasks. For two molecules and two flasks, four arrangements are possible (a, b, c, and d). Two of them are less likely (a and d). In general, the probability that the molecules will stay in the original flask is $(1/2)^n$, where n is the number of molecules.

No wonder the $KMnO_4$ sample in Figure 20.2 eventually disperses over the entire container. Furthermore, the process occurs on its own—it is spontaneous. It is also evident from these examples that if we wanted to put all of the O_3 molecules back into one flask, or recover the $KMnO_4$ crystals, we would have to intervene in some way. We could use a pump to force all of the O_3 molecules from one side to the other or we could lower the temperature drastically (Figure 20.4). In the latter case the molecules move "downhill" energetically to a place of very low kinetic energy.

In summary, the final state of a system can be more probable than the initial one in one of two ways: (1) energy can be dispersed over a greater number of atoms and molecules and (2) the atoms and molecules can be more disordered.

- If energy and matter are both dispersed in a reaction, it is definitely product-favored because both the products and the distribution of energy are more probable.
- If only energy or matter is dispersed, then quantitative information is needed to decide which effect is greater. (As you will see, energy dispersal is more important at room temperature than matter dispersal, and most exothermic reactions are product-favored. At high temperatures the opposite is true—matter dispersal becomes more important.)
- If neither matter nor energy is more spread out after a process occurs, then that process will be reactant-favored—the initial substances will remain no matter how long we wait.

(a)

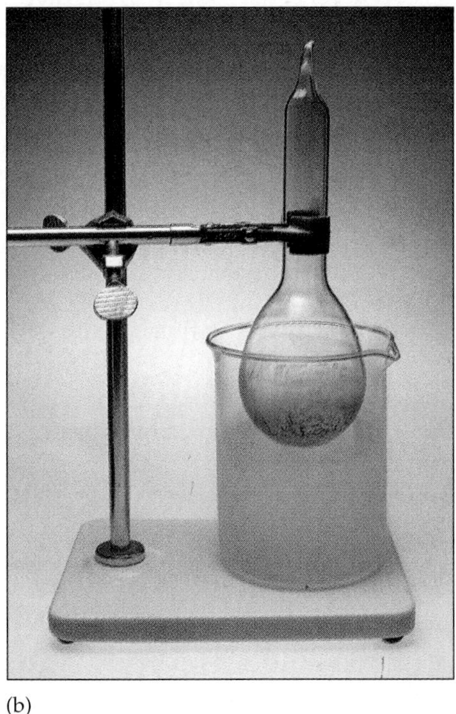

(b)

***Figure* 20.4** **Reversing the process of matter dispersal.** (a) Brown NO_2 gas is dispersed evenly throughout the flask. (b) If the flask is immersed in liquid nitrogen at −196 °C, the kinetic energy of the NO_2 molecules is reduced to the point that they become a solid and collect on the cold walls of the flask. *(C. D. Winters)*

Historical Perspective

Ludwig Boltzmann (1844–1906)

Ludwig Boltzmann was an Austrian mathematician and physicist who gave us a useful interpretation of entropy (and who also did much of the work on the kinetic theory of gases). Engraved on his tombstone in Vienna is his equation relating entropy and "chaos," $S = k \log W$ (where k is a fundamental constant of nature, now called Boltzmann's constant). Boltzmann said the symbol W was related to the number of ways that atoms or molecules can be arranged in a given state, always keeping their total energy fixed. His equation tells us, therefore, that if the atoms of a substance can only be arranged in a few ways—that is, if we can put our atoms or molecules in only a few places—then the entropy is low. On the other hand, the entropy is high if many arrangements are possible ($W >> 1$), that is, if the level of chaos is high.

The tombstone of Boltzmann with the equation $S = k \log W$ engraved on it. *(Oesper Collection in the History of Chemistry/University of Cincinnati)*

Entropy: A Measure of Matter Dispersal or Disorder

See the *Saunders Interactive General Chemistry CD-ROM.* Screen 20.4, Entropy.

The third law of thermodynamics states that the entropy of each element in some crystalline state is taken as zero at 0 K. Based on this assumption, every substance has a positive entropy.

Although it is impossible to cool anything to absolute zero, it is possible to come very close. Methods exist for estimating the disorder that is in a substance near 0 K, so accurate entropy values can be obtained for many substances.

Additional values of entropy are given in Appendix L.

The dispersal or disorder in a sample of matter can be measured with a calorimeter (see Figure 6.17), the same instrument needed to measure the enthalpy change when a reaction occurs. The result is a thermodynamic function called **entropy** and symbolized by S. Measurement of entropy depends on the assumption that in a perfect crystal at the absolute zero of temperature (0 K, or −273.15 °C) all translational motion ceases and there is no disorder; this sets the zero of the entropy scale.

When energy is transferred to matter in very small increments, so that the temperature change is very small, the entropy change can be calculated as $\Delta S = q/T$, the heat absorbed divided by the absolute temperature at which the change occurs. By starting as close as possible to absolute zero and repeatedly introducing small quantities of energy, an entropy change can be determined for each small increase of temperature. These entropy changes can then be added to give the total (or absolute) entropy of a substance at any desired temperature.

The results of such measurements for several substances at 298 K are given in Table 20.1. These are standard molar entropy values, and so they apply to 1 mol of each substance at the standard pressure of 1 bar (= 0.98692 atm) and are expressed in units of joules per kelvin per mole (J/K · mol).

Some interesting and useful generalizations can be drawn from the data given in Table 20.1.

- *When comparing the same or very similar substances, entropies of gases are much larger than those of liquids, which are larger than for solids.* In a solid the particles can only vibrate around lattice positions. When a solid melts, its particles are freer to move around, and molar entropy increases. When a liq-

(a)

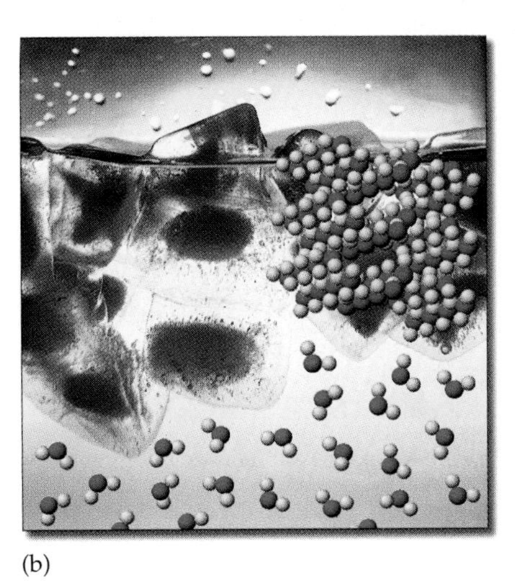

(b)

Figure 20.5 Entropy and phases. (a) The entropy of $Br_2(\ell)$ is 152.2 J/K · mol and that for more disordered bromine vapor is 175.0 J/K · mol. (b) The entropy of ice, which has an orderly structure, is smaller than for disordered liquid water. *(Photos, C. D. Winters; models, S. M. Young)*

uid vaporizes, restrictions due to forces between the particles nearly disappear, and another large entropy increase occurs. For example, the entropies (in J/K · mol) of I_2 (solid), Br_2 (liquid), and Cl_2(gas) are 116.1, 152.2, and 223.0, respectively, and the entropies of $Br_2(\ell)$ and $Br_2(g)$ are 152.2 and 175.0, respectively (Figure 20.5).

- *Entropies of more complex molecules are larger than those of simpler molecules, especially in a series of closely related compounds.* In a more complicated molecule there are more ways for the atoms to be arranged in three-dimensional space and hence greater entropy. Entropies (in J/K · mol) for methane (CH_4), ethane (CH_3CH_3), and propane ($CH_3CH_2CH_3$) are 186.3, 229.6, and 269.9, respectively. For atoms or molecules of similar molar mass, we have Ar, CO_2, and $CH_3CH_2CH_3$, with entropies of 154.7, 213.7, and 269.9 J/K · mol, respectively.
- *Entropies of ionic solids become larger as the attractions among the ions become weaker.* The weaker the forces between ions, the easier it is for the ions to vibrate about their lattice positions. Examples are MgO(s) and NaF(s) with entropies of 26.9 and 51.5 J/K · mol, respectively; the 2+ and 2− charges on the magnesium ions and oxide ions result in greater attractive forces and hence lower entropy.
- *Entropy usually increases when a pure liquid or solid dissolves in a solvent.* Because Table 20.1 refers only to pure substances, no example values are available; however, matter usually becomes more dispersed or disordered when a substance dissolves and different kinds of molecules mix together (Figure 20.6).

Table 20.1 • Some Standard Molar Entropy Values at 298 K

Compound or Element	Entropy, $S°$ (J/K · mol)
C(graphite)	5.7
C(g)	158.1
$CH_4(g)$	186.3
$C_2H_6(g)$	229.6
$C_3H_8(g)$	269.9
$CH_3OH(\ell)$	126.8
$CO_2(g)$	213.7
Ca(s)	41.4
Ar(g)	154.7
$H_2(g)$	130.7
$O_2(g)$	205.1
$N_2(g)$	191.6
$H_2O(g)$	188.8
$H_2O(\ell)$	69.9
HCl(g)	186.9
$F_2(g)$	202.8
$Cl_2(g)$	223.1
$Br_2(\ell)$	152.2
$I_2(s)$	116.1
NaF(s)	51.5
MgO(s)	26.9
$CaCO_3(s)$	92.9

From *The NBS Tables of Chemical Thermodynamic Properties,* New York, American Institute of Physics, 1982 or http://webbook.nist.gov

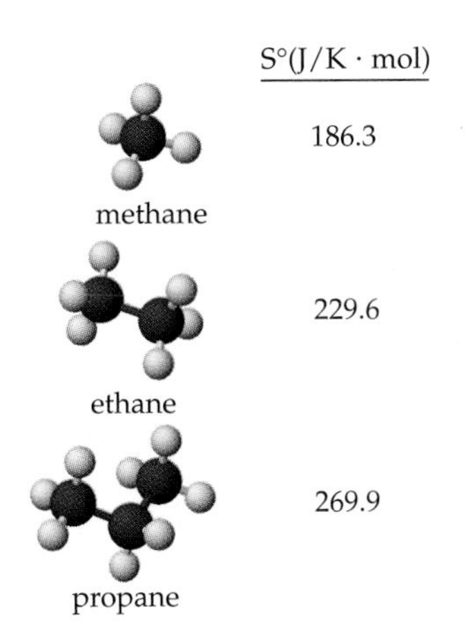

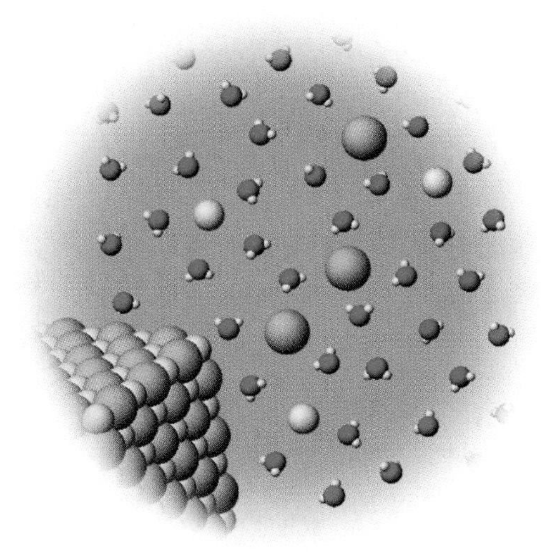

Figure 20.6 Entropy change on dissolving. A large increase in entropy occurs when a highly ordered substance such as crystalline NaCl dissolves in water. *(Photo, C. D. Winters; model, S. M. Young)*

- *Entropy increases when a dissolved gas escapes from a solution.* Although gas molecules are dispersed among solvent molecules in solution, the very large entropy increase that occurs on changing from the liquid to the gas phase results in a higher entropy for separated gas and liquid than for the mixture.

The previous generalizations can be used to predict whether the disorder of the substances increases when reactants are converted to products. (Such predictions are much easier to make for entropy changes than for enthalpy changes.) For the processes $H_2O(s) \longrightarrow H_2O(\ell)$ and $H_2O(\ell) \longrightarrow H_2O(g)$, we expect an entropy increase in each case because water molecules in the solid are more ordered than in the liquid and much more ordered than in the gas (see Figure 20.5). This is confirmed by entropy measurements. At 273.15 K, for example, ice can be converted to water very slowly, and no temperature change occurs. The quantity of energy transferred as heat is the heat of fusion (6020 J/mol). Moreover, the transfer of energy is into the water, so $q = +6020$ J/mol. This gives an entropy change of +22 J/K · mol.

$$\Delta S \text{ for } H_2O(s) \longrightarrow H_2O(\ell) = \frac{q}{T} = \frac{+6020 \text{ J/mol}}{273.15 \text{ K}} = +22.0 \text{ J/K} \cdot \text{mol}$$

Similarly, boiling water requires that energy be transferred into the water, so q must again be positive, and so must ΔS. Here $q = +40.7$ kJ/mol (the heat of vaporization of water at the boiling point), and ΔS is +109 J/K · mol.

$$\Delta S \text{ [for } H_2O(\ell) \longrightarrow H_2O(g)] = \frac{q}{T} = \frac{+40{,}700 \text{ J/mol}}{373.15 \text{ K}} = +109 \text{ J/K} \cdot \text{mol}$$

For the decomposition of calcium carbonate,

$$CaCO_3(s) \longrightarrow CaO(s) + CO_2(g)$$

we would also predict an increase in entropy, because 1 mol of gaseous CO_2 is present in the products, whereas the reactant is a solid. Because gases usually have much higher entropies than solids or liquids, gaseous substances are most important in determining entropy changes. Indeed, $\Delta S^\circ = +160.6$ J/K in this case.

An example in which a decrease in entropy would be predicted is

$$2\ CO(g) + O_2(g) \longrightarrow 2\ CO_2(g)$$

Here there is 3 mol of gaseous substances (2 of CO and 1 of O_2) at the beginning of the reaction but only 2 mol of gaseous substances at the end. Two moles of gas (of whatever kind) contain less entropy than 3 mol of gas, and so ΔS is negative. The actual value is $\Delta S^\circ = -173.0$ J/K.

Another example of decreased entropy is the process

$$Ag^+(aq) + Cl^-(aq) \longrightarrow AgCl(s)$$

Here the reactant ions are free to move about among water molecules in aqueous solution, but those same ions are held in a crystal lattice in the solid, a situation with much greater constraint. As a result, $\Delta S^\circ = -33.1$ J/K.

Exercise 20.1 Calculating ΔS Values for Phase Changes

The enthalpy of vaporization of benzene (C_6H_6) is 30.9 kJ/mol at the boiling point of 80.1 °C. Calculate the entropy change for benzene going from liquid to vapor and from vapor to liquid at 80.1 °C.

Exercise 20.2 Entropy

For each of the following processes, predict whether you expect entropy to be greater for the products than for the reactants. Explain how you arrived at your prediction.

1. $CO_2(g) \longrightarrow CO_2(s)$
2. $KCl(s) \longrightarrow KCl(aq)$
3. $MgCO_3(s) + \text{heat} \longrightarrow MgO(s) + CO_2(g)$

Entropy and the Second Law of Thermodynamics

Experience with many, many chemical reactions and other processes in which energy is transferred has led to the **second law of thermodynamics,** which states that *the total entropy of the universe is continually increasing.* Whenever anything happens, matter, energy, or both become more dispersed or disordered. This means that a *product-favored reaction* is accompanied by an *increase in the entropy* of the universe; $\Delta S_{universe}$ is positive. Evaluating whether this will happen during a proposed chemical reaction allows us to predict whether or not reactants form appreciable quantities of products.

Such an evaluation can be made using tables of data if we consider a carefully specified situation—a set of standard conditions as defined on page 268. Once a prediction has been made, corrections can be applied to account for differences from the standard conditions. Predicting whether a reaction is product-favored involves two steps: (1) calculating how much entropy is created by dispersal of matter; and (2) calculating how much entropy is created by dispersal of energy. Both calculations are done assuming that reactants at standard conditions are converted completely to products at standard conditions.

As an example of predicting the product favorability of a reaction, let us consider a process that might be used to manufacture liquid methanol for use as automobile fuel:

$$CO(g) + 2\ H_2(g) \longrightarrow \underset{\text{methanol}}{CH_3OH(\ell)}$$

We base our prediction on having 1 mol CO(g) and 2 mol $H_2(g)$ as reactants, each at a pressure of 1 bar. We further assume that the product is 1 mol of liquid methanol, also at a pressure of 1 bar. If the total entropy is predicted to be higher after the product has been produced, then the reaction is product-favored under these conditions and might be useful. If not, perhaps some other conditions could be used, or perhaps we should consider some other methanol-producing reaction altogether.

Calculating the Entropy Change for a System

See the *Saunders Interactive General Chemistry CD-ROM,* Screen 20.5, Calculating ΔS.

To calculate the entropy change due to dispersal of matter in the course of a reaction, ΔS_{system}, we assume that each reactant and each product is present in the amount required by its stoichiometric coefficient. All substances are assumed to be at standard pressure and at the temperature specified, so that values from Table 20.1 (or Appendix L) apply. Then we can add up all the entropies of the products and subtract the entropies of the reactants to see whether entropy increases or decreases:

$$\Delta S^\circ_{system} = \sum S^\circ(\text{products}) - \sum S^\circ(\text{reactants}) \qquad [20.1]$$

Notice that the equation for calculating ΔS°_{system} has the same form as that for calculating ΔH° for a reaction (Equation 6.6).

The entropy change for the reacting system, ΔS°_{system}, is the change occurring when reactants in their standard states are converted completely to products in their standard states. Thus, for the conversion of 1 mol of CO(g) and 2 mol of $H_2(g)$ to 1 mol of $CH_3OH(\ell)$, we have

$$CO(g) + 2\ H_2(g) \longrightarrow CH_3OH(\ell)$$

$$\begin{aligned}\Delta S^\circ_{system} &= S^\circ[CH_3OH(\ell)] - \{S^\circ[CO(g)] + 2\ S^\circ[H_2(g)]\} \\ &= 1\text{ mol }(126.8\text{ J/K}\cdot\text{mol}) - \{1\text{ mol }(197.6\text{ J/K}\cdot\text{mol}) + 2\text{ mol }(130.7\text{ J/K}\cdot\text{mol})\} \\ &= -332.2\text{ J/K}\end{aligned}$$

Notice that this calculation gives the entropy change for the reaction *system.* In this case the decrease in entropy is large because 3 mol of gaseous material is converted to only 1 mol of a liquid–phase product.

Example 20.1 Calculating an Entropy Change for a Chemical Reaction

Nitrogen dioxide is formed from nitrogen monoxide and oxygen in a product-favored reaction at 25 °C (Figure 20.7). Determine the standard entropy change, $\Delta S°$, for the reaction, $\Delta S°_{rxn}$ ($= \Delta S°_{system}$).

Solution As in any problem involving a chemical reaction, first write the balanced equation:

$$2\ NO(g) + O_2(g) \longrightarrow 2\ NO_2(g)$$

Next, subtract the entropies of the reactants from the entropies of the products (Appendix L), carefully scaling each entropy by the number of moles of reactant or product involved:

$$\Delta S°_{rxn} = (2\ mol\ NO_2)(240.1\ J/K \cdot mol) - [(2\ mol\ NO)(210.8\ J/K \cdot mol) + (1\ mol\ O_2)(205.1\ J/K \cdot mol)]$$

$$= -146.5\ J/K$$

or -73.25 J/K for the formation of 1 mol of NO_2.

Notice that the sign of the entropy change is negative. This is largely due to the fact that the chemical reaction began with 3 mol of gaseous reactants and ended with 2 mol of gaseous product.

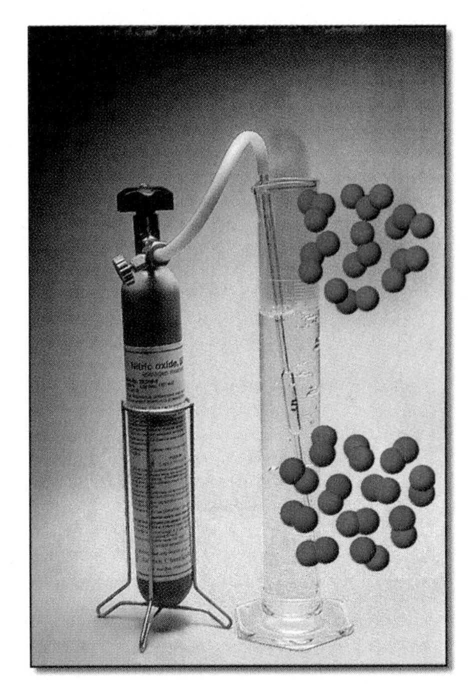

Figure 20.7 The reaction of NO with O_2. The entropy of the system decreases in this reaction because 3 mol of reactants ($2\ NO + O_2$) produce only 2 mol of products ($2\ NO_2$). *(Photo, C. D. Winters; model, S. M. Young)*

Exercise 20.3 Calculating an Entropy Change for a Chemical Reaction

The active ingredient in a popular antacid remedy is $CaCO_3$ (more familiar as the main component of chalk or limestone). Using the entropy values in Table 20.1, calculate the entropy change for the formation of $CaCO_3(s)$ from the elements. (Use graphite as the standard state of carbon.) Is the sign of the entropy change for the formation of $CaCO_3(s)$ positive or negative? Account for the increase or decrease in entropy in the formation of this compound in terms of the disorder in this chemical system.

Calculating the Entropy Change in the Surroundings

Predicting the product favorability of a reaction involves calculating the entropy created by the dispersal of matter *and* of energy (page 920). We have done the first of these calculations for the formation of $CH_3OH(\ell)$ from CO and H_2O:

$$CO(g) + 2\ H_2(g) \longrightarrow CH_3OH(\ell) \qquad \Delta S°_{system} = -332.2\ J/K$$

Next, the entropy created by the dispersal of energy by this chemical reaction—the entropy change for the surroundings—can be evaluated by calculating $\Delta H°$ for the reaction (from tables such as Table 6.2 or Appendix L) and by assuming that this quantity of energy is transferred to or from the surroundings. If the process is reversible and occurs at a constant temperature, the entropy change for the surroundings can be calculated as

$$\Delta S°_{surroundings} = \frac{q_{surroundings}}{T} = -\frac{\Delta H_{system}}{T}$$

The minus sign in this equation comes from the fact that $-\Delta H_{system}$ is the energy transferred *out* of the system and hence equals the energy transferred *into* the surroundings, $q_{surroundings}$.

The equation states that for an exothermic reaction (ΔH_{system} has a negative value) there will be an increase in entropy of the surroundings. For the proposed methanol-producing reaction $\Delta H^\circ_{system} = -128.14$ kJ (calculated from Table 6.2), and so the entropy change is

$$\Delta S^\circ_{surroundings} = \frac{-(-128.14\ \text{kJ})(1000\ \text{J/kJ})}{298\ \text{K}} = +430.\ \text{J/K}$$

Calculating the Total Entropy Change for System and Surroundings

The entropy changes for the system and its surroundings have been calculated for the reaction

$$CO(g) + 2\ H_2(g) \longrightarrow CH_3OH(\ell)$$
$$\Delta S^\circ_{system} = -332.2\ \text{J/K} \qquad \Delta S^\circ_{surroundings} = +430.\ \text{J/K}$$

The *total* entropy change for a process, referred to as $\Delta S^\circ_{universe}$ (the entropy change for the universe), is the sum of the entropy change for the system (ΔS°_{system}) and the entropy change for the surroundings ($\Delta S^\circ_{surroundings}$). (We assume that nothing else but our reaction happens, and so there are no other entropy changes.) For the formation of methanol from CO(g) and $H_2O(g)$, this total entropy change is

$$\Delta S^\circ_{universe} = \Delta S^\circ_{system} + \Delta S^\circ_{surroundings} = (-332.2 + 430.)\ \text{J/K} = 98\ \text{J/K}$$

See the *Saunders Interactive General Chemistry CD-ROM,* Screen 20.6, The Second Law.

The reaction is accompanied by an *increase* in the entropy of the universe. It follows from the second law of thermodynamics that, if we had CO(g) and $H_2(g)$, each at 1 bar pressure and in contact with one another, they would react to form $CH_3OH(\ell)$. The process is product-favored and might be useful for manufacturing methanol.

Predictions of the sort we have just made by calculating $\Delta S^\circ_{universe}$ can also be made qualitatively, without calculations, if we know whether or not a reaction is exothermic and if we can predict whether matter is dispersed when the reaction takes place. A reaction is sure to be product-favored if it is *exothermic* and proceeds from a state of order to one of disorder *(entropy increases)*. Also, a reaction is certainly *not* product-favored if it is *endothermic* and *entropy decreases* for the system. Two other possible cases are indicated in Table 20.2, but they are more difficult to predict without quantitative information.

As examples of such predictions, consider gas-producing reactions of solids. They are product-favored because they are exothermic and produce highly disordered gases:

$$CaCO_3(s) + 2\ HCl(aq) \longrightarrow CaCl_2(aq) + H_2O(\ell) + CO_2(g) + \text{heat}$$
$$\Delta H^\circ_{rxn} = -15.2\ \text{kJ} \qquad \Delta S^\circ_{rxn} = 137.6\ \text{J/K}$$

Similarly, combustion reactions are product-favored because they are exothermic and produce a larger number of product molecules from a few reactant molecules:

$$\underset{\text{butane}}{2\ C_4H_{10}(g)} + 13\ O_2(g) \longrightarrow 8\ CO_2(g) + 10\ H_2O(g) + \text{heat}$$
$$\Delta H^\circ_{rxn} = -5315.1\ \text{kJ} \qquad \Delta S^\circ_{rxn} = 2362.8\ \text{J/K}$$

Table 20.2 • PREDICTING WHETHER A REACTION IS PRODUCT-FAVORED

ΔH_{system}	ΔS_{system}	**Product-Favored**
−, exothermic	+, less order	Yes
−, exothermic	−, more order	Depends on T and relative magnitudes of ΔH and ΔS for the system but generally product-favored at lower T.
+, endothermic	+, less order	Depends on T and relative magnitudes of ΔH and ΔS for the system but generally product-favored at higher T.
+, endothermic	−, more order	No

But what about a reaction such as the production of ethylene, C_2H_4, from ethane, C_2H_6? The reaction is very endothermic, although the entropy change is predicted to be positive:

$$C_2H_6(g) \longrightarrow H_2(g) + C_2H_4(g)$$
$$\Delta H^\circ_{rxn} = +136.94 \text{ kJ} \qquad \Delta S^\circ_{rxn} = +120.6 \text{ J/K}$$

So, the enthalpy change suggests the reaction should not be product-favored, whereas the entropy change suggests the opposite. Which is the more important? Calculating $\Delta S^\circ_{surroundings}$ involves dividing the enthalpy change by temperature, and ΔH° does not change much as temperature increases; therefore, the higher the temperature the less important the ΔH° term is. At room temperature ΔH° is usually the deciding factor, and so the ethylene-producing reaction is not expected to be product-favored. This is indeed the case at 25 °C. To make this reaction work in industry, chemical engineers have designed a special process at about 1000 °C. At this higher temperature the ΔS° term is more important, and more products can be produced.

One goal of chemists is often to take small molecules and assemble them into larger molecules that can be sold for much more than the cost of the reactants. An example is the system described earlier, assembling CO and H_2 into methanol, CH_3OH,

$$CO(g) + 2\ H_2(g) \longrightarrow CH_3OH(\ell)$$
$$\Delta H^\circ_{rxn} = -128.14 \text{ kJ} \qquad \Delta S^\circ_{system} = -332.2 \text{ J/K} \qquad \Delta S^\circ_{universe} = 98 \text{ J/K}$$

and then turning the methanol into gasoline (i.e., into molecules such as octane, C_8H_{18}). The problem with this is that we are fighting a losing battle with entropy. The way to get this to work, however, is to increase the entropy somewhere else in the universe. Indeed, Roald Hoffmann (1937–), who shared the 1981 Nobel Prize in chemistry, has said that "One amusing way to describe synthetic chemistry, the making of molecules that is at the intellectual and economic center of chemistry, is that it is the local defeat of entropy."*

*R. Hoffmann: "Unstable." *American Scientist,* Vol. 75, Nov.–Dec. 1987, pp. 619–621.

Exercise 20.4 Is a Reaction Product- or Reactant-Favored?

Classify the following reactions as one of the four types of reactions summarized in Table 20.2:

Reaction	ΔH°_{rxn} (298 K) kJ	ΔS°_{system} (298 K) J/K
1. $CH_4(g) + 2\ O_2(g) \longrightarrow 2\ H_2O(\ell) + CO_2(g)$	−890.3	−243.0
2. $2\ Fe_2O_3(s) + 3\ C(graphite) \longrightarrow 4\ Fe(s) + 3\ CO_2(g)$	+467.9	+560.3
3. $C(graphite) + O_2(g) \longrightarrow CO_2(g)$	−393.5	+2.9
4. $N_2(g) + 3\ F_2(g) \longrightarrow 2\ NF_3(g)$	−248.6	−278.7

Exercise 20.5 Is a Reaction Product- or Reactant-Favored?

Is the direct reaction of hydrogen and chlorine to give hydrogen chloride gas predicted to be product-favored or reactant-favored?

$$H_2(g) + Cl_2(g) \longrightarrow 2\ HCl(g)$$

Answer the question by calculating the values for ΔS°_{system} and $\Delta S^\circ_{surroundings}$ (at 298 K) and then summing them to determine $\Delta S^\circ_{universe}$.

20.3 GIBBS FREE ENERGY

See the *Saunders Interactive General Chemistry CD-ROM,* Screen 20.7, Gibbs Free Energy.

Calculations of the sort done in the previous section would be simpler if we did not have to evaluate separately the entropy change of the surroundings from a table of ΔH°_f values and the entropy change of the system from a table of S° values. Another thermodynamic function, defined by J. Willard Gibbs (1839–1903), a professor at Yale University, solved this dilemma. In Gibbs's honor this function is now called the **Gibbs free energy** and given the symbol *G.*

In the previous section we showed that the total entropy change accompanying a chemical reaction carried out slowly at constant temperature and pressure is

$$\Delta S_{universe} = \Delta S_{surroundings} + \Delta S_{system}$$
$$= -\frac{\Delta H_{system}}{T} + \Delta S_{system}$$

Multiplying through this equation by $-T$, the result is

$$-T\,\Delta S_{universe} = \Delta H_{system} - T\,\Delta S_{system}$$

Gibbs defined the free energy function so that $-T\,\Delta S_{universe}$ is equal to the change in the free energy of the system, ΔG_{system}. That is,

$$\Delta G_{system} = -T\,\Delta S_{universe} = \Delta H_{system} - T\,\Delta S_{system}$$

Under standard conditions, the equation becomes

$$\Delta G^\circ_{system} = \Delta H^\circ_{system} - T\,\Delta S^\circ_{system} \quad [20.2]$$

State functions were described on page 267.

This equation, called the **Gibbs free energy equation,** is one of the important equations in science. Gibbs free energy, G, is a state function, like H (enthalpy) and E (internal energy), functions that depend only on initial and final states. Furthermore, it is a property of the system; it carries no reference to the universe. The change in Gibbs free energy specifically relates to changes in a system; that is, ΔG is evaluated based on properties—ΔH, S, and T—that are measurable for the system being studied. Because ΔG equals $-T\Delta S_{\text{universe}}$, a negative value of ΔG ($\Delta G < 0$ corresponding to $T\Delta S_{\text{universe}} > 0$) refers to a product-favored (or spontaneous) process. Conversely, a positive value of ΔG refers to a reactant-favored (or nonspontaneous) process.

The Gibbs free energy equation (Equation 20.2) allows us to answer the question unanswered in Table 20.2: What happens when both $\Delta H^\circ_{\text{rxn}}$ and $\Delta S^\circ_{\text{rxn}}$ have the same sign?

- If the reaction is exothermic (negative $\Delta H^\circ_{\text{system}} = \Delta H^\circ_{\text{rxn}}$) and if the entropy of the system increases (positive $\Delta S^\circ_{\text{system}} = \Delta S^\circ_{\text{rxn}}$), then $\Delta G^\circ_{\text{system}}$ ($= \Delta G^\circ_{\text{rxn}}$) must be negative and the reaction is product-favored.
- If ΔH° and ΔS° have the same sign, then the temperature T and the relative magnitudes of the enthalpy and entropy changes determine whether ΔG° is negative and the reaction is product-favored.
- If ΔH° is positive and ΔS° is negative, then ΔG° must be positive, and the reaction cannot be product-favored under standard conditions and at the temperature for which the data were tabulated.

These conclusions are the same ones previously tabulated in Table 20.2. The Gibbs function is useful because it allows us to make a decision about the favorability of a reaction, especially in cases in which both the enthalpy change and entropy change have the same algebraic sign. Let us turn to the various ways this valuable tool can be used.

J. Willard Gibbs (1839–1903). *(Burndy Library/Courtesy AIP Emilo Segre Visual Archives)*

Calculating $\Delta G^\circ_{\text{rxn}}$, the Free Energy Change for a Reaction

Enthalpy and entropy changes can be calculated for chemical reactions using values of ΔH°_f and S° for substances in the reaction. Then, $\Delta G^\circ_{\text{rxn}}$ ($= \Delta G^\circ_{\text{system}}$) can be found from the resulting values of $\Delta H^\circ_{\text{rxn}}$ and $\Delta S^\circ_{\text{rxn}}$ using Equation 20.2, as illustrated in the following example and exercise.

Example 20.2 Calculating $\Delta G^\circ_{\text{rxn}}$ from $\Delta H^\circ_{\text{rxn}}$ and $\Delta S^\circ_{\text{rxn}}$

Calculate the standard free energy change, ΔG°, for the formation of methane at 298 K:

$$\text{C(graphite)} + 2\,\text{H}_2\text{(g)} \longrightarrow \text{CH}_4\text{(g)}$$

Solution The following values for ΔH°_f and S° are provided in Appendix L.

	C(graphite)	H_2(g)	CH_4(g)
ΔH°_f (kJ/mol)	0	0	−74.8
S° (J/K · mol)	5.7	130.7	186.3

From these values, we can find both $\Delta H°$ and $\Delta S°$ for the reaction:

$$\begin{aligned}\Delta H°_{rxn} &= \Delta H°_f\,[CH_4(g)] - \{\Delta H°_f\,[C(graphite)] + 2\,\Delta H°_f\,[H_2(g)]\}\\ &= -74.8\text{ kJ} - (0 + 0)\\ &= -74.8\text{ kJ}\end{aligned}$$

$$\begin{aligned}\Delta S°_{rxn} &= S°\,[CH_4(g)] - \{S°\,[C\,(graphite)] + 2\,S°\,[H_2(g)]\}\\ &= 186.3\text{ J/K}\cdot\text{mol} - [1\text{ mol}(5.7\text{ J/K}\cdot\text{mol}) + 2\text{ mol}\,(130.7\text{ J/K}\cdot\text{mol})]\\ &= -80.8\text{ J/K}\end{aligned}$$

Both the enthalpy change and the entropy change for this reaction are negative. This is a case when the reaction is predicted to be product-favored at "low temperature" (see Table 20.2). These values alone do not tell us if the temperature is low enough, however. By combining them in the Gibbs equation, and calculating $\Delta G°_{rxn}$ for a temperature of 25 °C, we can predict with certainty the outcome of the reaction:

$$\begin{aligned}\Delta G°_{rxn} &= \Delta H°_{rxn} - T\,\Delta S°_{rxn}\\ &= -74.8\text{ kJ} - (298\text{ K})(-80.8\text{ J/K})(1\text{ kJ}/1000\text{ J})\\ &= -74.8\text{ kJ} - (-24.1\text{ kJ})\\ &= -50.7\text{ kJ}\end{aligned}$$

An enthalpy-driven reaction has a relatively large and negative ΔH that can overcome a small, negative ΔS.

$\Delta G°_{rxn}$ is negative at 298 K, so the reaction is predicted to be product-favored.

In this case the product $T\,\Delta S°$ is negative and smaller than $\Delta H°_{rxn}$ because the entropy change is relatively small. Chemists call this an "enthalpy-controlled reaction" because the exothermic nature of the reaction overcomes the decrease in entropy of the system.

Exercise 20.6 Calculating $\Delta G°_{rxn}$ from $\Delta H°_{rxn}$ and $\Delta S°_{rxn}$

Using values of $\Delta H°_f$ and $S°$ to find $\Delta H°_{rxn}$ and $\Delta S°_{rxn}$, respectively, calculate the free energy change, $\Delta G°$, for the formation of 1 mol of $NH_3(g)$ from the elements at standard conditions (and 25 °C): $\frac{1}{2}\,N_2(g) + \frac{3}{2}\,H_2(g) \longrightarrow NH_3(g)$.

Standard Free Energy of Formation, $\Delta G°_f$

Recall that the standard enthalpy of formation, $\Delta H°_f$, is defined as the enthalpy change for the formation of 1 mol of a compound from the elements, with both reactants and products in their standard states (Section 6.9 and page 268). The **Gibbs free energy of formation, $\Delta G°_f$,** can be defined in a similar manner: $\Delta G°_f$ is the change in Gibbs free energy when 1 mol of a compound is formed from the elements, the reactants and products being in their standard states. The values of $\Delta G°_{rxn}$ calculated in Example 20.2 and Exercise 20.6 are in fact values of $\Delta G°_f$ for $CH_4(g)$ and $NH_3(g)$. Yet another example is *A Chemical Puzzler,* on page 910:

$$2\,Fe(s) + \tfrac{3}{2}\,O_2(g) \longrightarrow Fe_2O_3(s)$$

Here $\Delta G°_f$ for $Fe_2O_3(s)$ is -742.2 kJ/mol under standard conditions (1 bar pressure and 298 K). These and a few other values of $\Delta G°_f$ are given in Table 20.3, and many others are listed in Appendix L. Notice that $\Delta G_f^\circ = 0$ for elements in their standard states, for the same reason they have ΔH_f° values of 0.

Just as ΔH_{rxn}° can be calculated from standard enthalpies of formation, the free energy change for any reaction (at standard conditions) can also be found from values of ΔG_f° by the general equation

$$\Delta G_{reaction}^\circ = \sum \Delta G_f^\circ\,(\text{products}) - \sum \Delta G_f^\circ\,(\text{reactants}) \qquad [20.3]$$

The example and exercises that follow illustrate how this is done.

Table 20.3 • STANDARD MOLAR FREE ENERGIES OF FORMATION FOR SOME SUBSTANCES AT 298 K

Element or Compound	ΔG_f° (kJ/mol)
$H_2(g)$	0
$O_2(g)$	0
$N_2(g)$	0
C(graphite)	0
C(diamond)	2.9
$CO(g)$	−137.2
$CO_2(g)$	−394.4
$CH_4(g)$	−50.7
$H_2O(g)$	−228.6
$H_2O(\ell)$	−237.1
$NH_3(g)$	−16.5
$Fe_2O_3(s)$	−742.2

Example 20.3 Calculating ΔG_{rxn}° from ΔG_f°

Calculate the free energy change for the combustion of methane from the standard free energies of formation of the products and reactants.

Solution We first write the balanced equation for the reaction and then find the value of ΔG_f° for each reactant and product:

$$CH_4(g) + 2\,O_2(g) \longrightarrow 2\,H_2O(g) + CO_2(g)$$

ΔG_f° (kJ/mol): −50.7, 0, −228.6, −394.4

The free energy of formation values are given for 1 mol, so each value must be multiplied by the number of moles involved:

$$\begin{aligned}\Delta G_{rxn}^\circ &= 2\,\Delta G_f^\circ[H_2O(g)] + \Delta G_f^\circ[CO_2(g)] - \{\Delta G_f^\circ[CH_4(g)] + 2\,\Delta G_f^\circ[O_2(g)]\}\\ &= 2\text{ mol }(-228.6\text{ kJ/mol}) + 1\text{ mol }(-394.4\text{ kJ/mol}) - [1\text{ mol }(-50.7\text{ kJ/mol}) + 2\text{ mol }(0\text{ kJ/mol})]\\ &= -800.9\text{ kJ}\end{aligned}$$

ΔG_{rxn}° has a large, negative value, indicating that the reaction is product-favored under standard conditions.

Exercise 20.7 Calculating ΔG_{rxn}° from ΔG_f°

1. Write a balanced chemical equation depicting the formation of gaseous carbon dioxide (CO_2) from its elements.
2. What is the standard free energy of formation of 1.00 mol of CO_2 gas?
3. What is the standard free energy change for the reaction when 2.5 mol of CO_2 gas is formed from the elements?

Exercise 20.8 Calculating ΔG_{rxn}° from ΔG_f°

Calculate the standard free energy change for the combustion of 1.00 mol of benzene, $C_6H_6(\ell)$, to give $CO_2(g)$ and $H_2O(g)$.

PROBLEM-SOLVING TIPS AND IDEAS

20.1 Using ΔG_f°

Equation 20.3 is simply a shortcut that can be used to calculate the free energy change of a reaction when the free energies of formation (ΔG_f°) of the reactants and products are known. Do not, however, let this obscure the fact that it is the balance of the enthalpy and entropy changes, as well as the temperature, that determines the value of ΔG_{rxn}° as expressed by the Gibbs equation (Equation 20.2).

As a final thought in this section, we might ask what is meant by the term "free" energy? Does it mean that we can get something for nothing? No, the free energy change of a reaction is a measure of the *maximum magnitude of the net useful work* that can be obtained from a reaction. Consider the formation of methane again:

$$C(graphite) + 2\ H_2(g) \longrightarrow CH_4(g)$$

$$\Delta H_{rxn}^\circ = -74.8\ kJ \quad \text{and} \quad \Delta S_{rxn}^\circ = -80.8\ J/K$$

$$\Delta G_{rxn}^\circ = -74.8\ kJ - (298\ K)(-80.8\ J/K)(1\ kJ/1000\ J)$$

$$= -74.8\ kJ + 24.1\ kJ = -50.7\ kJ$$

The reaction is exothermic, with an enthalpy change of -74.8 kJ. Part of this thermal energy, though, is used to bring order to the system (and so the entropy declines). The amount of thermal energy diverted to this is $T\Delta S^\circ = 24.1$ kJ. Only 50.7 kJ of energy is therefore "free," or available for useful work.

Product-Favored or Reactant-Favored?

The enthalpy and entropy changes for a chemical reaction depend on the reactants and products and may be positive or negative. These in turn influence the sign of the free energy change and whether the reaction is product- or reactant-favored (see Table 20.2 and Examples 20.2 and 20.3). Let us look at some further predictions that are based on the calculation of the free energy change for a reaction (Figure 20.8).

If the free energy of the system decreases in a process (ΔG_{system} is negative), then the process is product-favored. Because $\Delta G^\circ = \Delta H^\circ - T\Delta S^\circ$, a process is certainly product-favored if the enthalpy of the system decreases *and* its entropy increases (Line 1 in Figure 20.8). This is the case for the reaction of potassium with water illustrated by Figure 20.9 or for the combustion of carbon:

	C(graphite)	**+ O_2(g)**	**⟶ CO_2(g)**	**Overall**
ΔH_f° (kJ/mol)	0	0	−393.5	−393.5
ΔS° (J/K · mol)	5.7	205.1	213.7	2.9
ΔG_f° (kJ/mol)	0	0	−394.4	−394.4

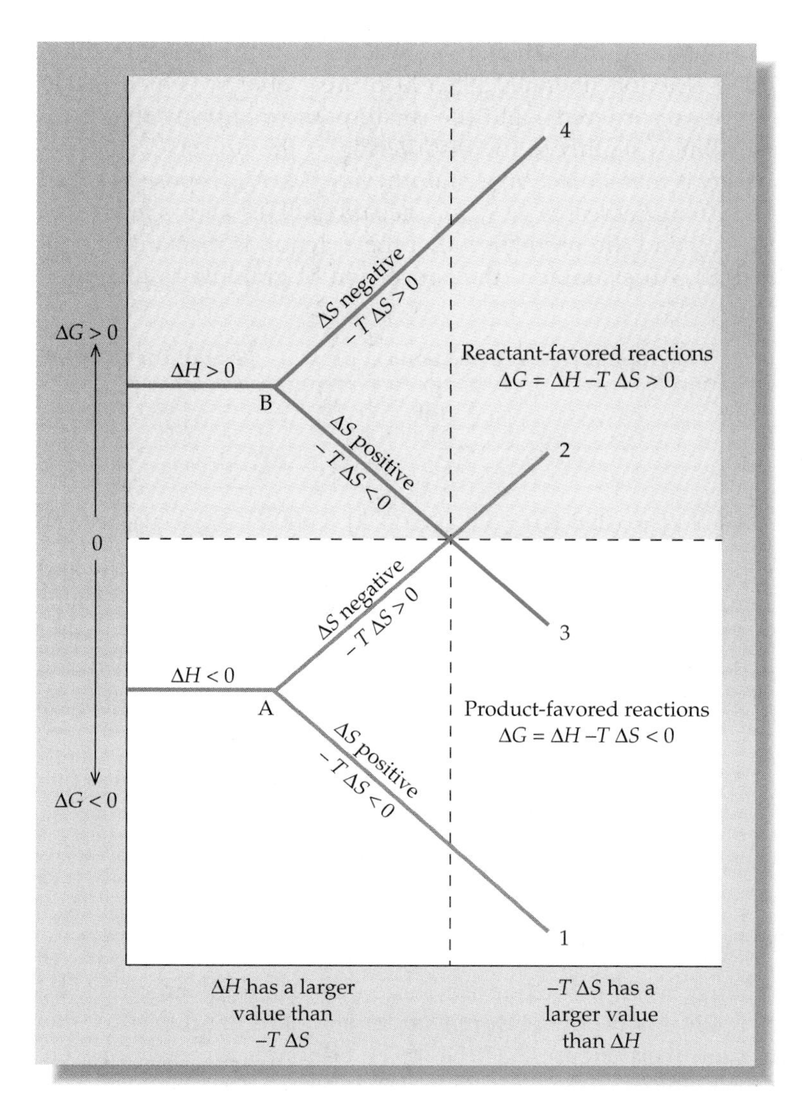

Figure 20.8 Changes in ΔG as a function of ΔH and ΔS. *Line 1*: If ΔH has a negative value and ΔS has a positive value for a reaction, the reaction is predicted to be product-favored. *Line 4*: If ΔH has a positive value and ΔS has a negative value, the reaction is predicted to be reactant-favored. *Lines 2 and 3*: Whether the reaction is product-favored or reactant-favored depends on the relative values of ΔH and ΔS and on temperature.

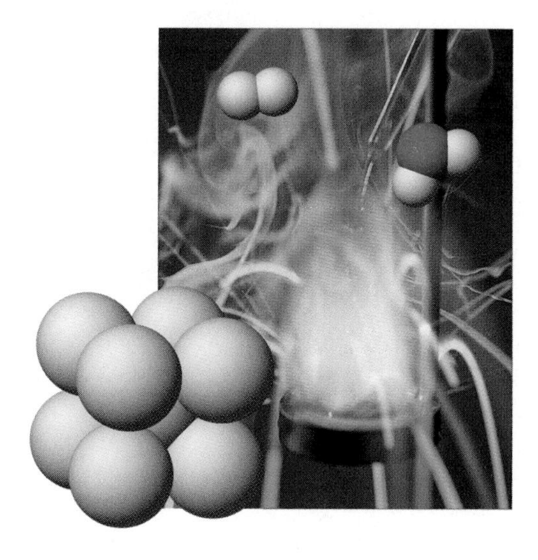

Figure 20.9 The product-favored reaction of potassium with water. The reaction

$$K(s) + H_2O(\ell) \longrightarrow KOH(aq) + \tfrac{1}{2}\,H_2(g)$$

is driven by both enthalpy and entropy changes, and so the free energy change for the reaction is less than zero. ($\Delta H^\circ_{rxn} = -196.5$ kJ; $\Delta S^\circ_{rxn} = 22.9$ J/K; and $\Delta G^\circ_{rxn} = -203.4$ kJ.) *(Photo, C. D. Winters; model, S. M. Young)*

The free energy change for this reaction can be calculated in either of two ways. You can add up all of the ΔG°_f values for the products and subtract the sum of those for the reactants:

$$\begin{aligned}\Delta G^\circ_{rxn} &= \sum \Delta G^\circ_f\,(\text{products}) - \Sigma\,\Delta G^\circ_f\,(\text{reactants}) = -394.4 \text{ kJ} - (0 + 0)\\ &= -394.4 \text{ kJ}\end{aligned}$$

Alternatively, you can calculate the enthalpy and entropy changes for the reaction, ΔH°_{rxn} and ΔS°_{rxn}, and combine them using the Gibbs equation:

$$\begin{aligned}\Delta G^\circ_{rxn} &= \Delta H^\circ_{rxn} - T\,\Delta S^\circ_{rxn}\\ &= -393.5 \text{ kJ} - (298\text{ K})(2.9\text{ J/K})(1\text{ kJ}/1000\text{ J})\\ &= -393.5 \text{ kJ} - 0.86 \text{ kJ}\\ &= -394.4 \text{ kJ}\end{aligned}$$

Both the enthalpy and the entropy changes contribute to making this reaction product-favored. The reaction liberates 393.5 kJ of heat energy (which makes the surroundings more disordered), and the small, positive entropy change of the system also contributes slightly to the disordering of the universe.

Another possibility is that a reaction is endothermic (ΔH°_{rxn} is positive), and the entropy of the system decreases (ΔS°_{rxn} is negative). This always leads to a positive value of ΔG°_{rxn}, and the prediction that the reaction is reactant-favored (Line 4 in Figure 20.8). An example is the conversion of graphite to diamond:

The conversion of graphite to diamond is thermodynamically unfavorable at all temperatures (at 1 bar pressure). The process can be carried out, however, at high pressure in the presence of a catalyst.

	C(graphite) $\longrightarrow$	**C(diamond)**	**Overall for Reaction**
ΔH°_f (kJ/mol)	0	1.9	+1.9
S° (J/K · mol)	5.7	2.4	−3.3

$$\Delta G^\circ_{rxn} = +1.9 \text{ kJ} - (298 \text{ K})(-3.3 \text{ J/K})(1 \text{ kJ}/1000 \text{ J}) = 2.9 \text{ kJ}$$

If the entropy decreases (ΔS° is negative), $-T\Delta S^\circ$ is a positive quantity and the reaction can be spontaneous only if ΔH° is large and negative and outweighs the positive ($-T\ \Delta S^\circ$) term (Line 2 in Figure 20.8). Such cases are called *enthalpy-driven* reactions, an example of which is the formation of NaCl from the elements:

	2 Na(s) +	**Cl_2(g)** $\longrightarrow$	**2 NaCl(s)**	**Overall for Reaction**
ΔH°_f (kJ/mol)	0	0	2(−411.2)	−822.4
S° (J/K · mol)	2(51.2)	223.1	2(72.1)	−181.3

$$\Delta G^\circ_{rxn} = -822.4 \text{ kJ} - (298 \text{ K})(-181.3 \text{ J/K})(1 \text{ kJ}/1000 \text{ J}) = -768.4 \text{ kJ}$$

The free energy change for the reaction is a large negative number, clearly indicating that the reaction is product-favored under standard conditions, something you already knew from Figure 2.27. However, although the reaction generates 822.4 kJ of heat energy (which is used to disorder the surroundings), only 768.4 kJ is available to disorder the universe. That is, 54 kJ of energy at 25 °C ($= T\Delta S^\circ$) is not "free," having been used to create some order in the system. Thus, the formation of salt is *enthalpy-driven,* as are many such reactions.

Now let us consider a case in which both the enthalpy and entropy changes are positive (Line 3 in Figure 20.8). The only way the reaction can be spontaneous is for $-T\Delta S^\circ$ to be large enough that ΔG° is negative. This situation often arises when salts, particularly those with low ion charges, dissolve in water. Because salts are highly ordered solids they have a low entropy. When they dissolve in water, however, the final state of the system is a jumble of ions with a high entropy.

	NH_4NO_3(s) $\longrightarrow$	**NH_4NO_3(aq, 1 *m*)**	**Overall for Process**
ΔH°_f (kJ/mol)	−365.6	−339.9	+25.7
S° (J/K · mol)	151.1	259.8	+108.7

$$\Delta G^\circ_{rxn} = +25.7 \text{ kJ} - (298 \text{ K})(+108.7 \text{ J/K})(1 \text{ kJ}/1000 \text{ J}) = -6.7 \text{ kJ}$$

The reaction is product-favored, and the enthalpy and entropy changes for the reaction clearly show that the reaction is *entropy-driven.* That is, although the process is endothermic (and you would feel an obvious cooling effect if you held your hand on a beaker or cold pack containing dissolving ammonium nitrate, see page 279), this is outweighed by the large increase in entropy of the system, and the reaction is product-favored under standard conditions. The disordering of the system when the solid dissolves is a very potent driving force.

Entropy is often the "*force*" that drives the mixing of any two substances, liquid or gas, with each other. Chapter 14 described the formation of ideal solutions, ones in which the forces between solute molecules and between solvent molecules are the same as between solute and solvent molecules. Because the energies of the two kinds of molecules *cannot* change on forming an ideal solution, it is the increase in entropy experienced by the solute and solvent molecules as they mix that provides the driving force (see Figure 14.2).

Exercise 20.9 Predicting the Outcome of a Reaction

Using free energies of formation in Appendix L, calculate ΔG°_{rxn} for (1) the formation of $CaCO_3(s)$ from $CO_2(g)$ and $CaO(s)$ and (2) the decomposition of $CaCO_3(s)$ to give $CO_2(g)$ and $CaO(s)$. Which reaction is product-favored under standard conditions?

Free Energy and Temperature

See the *Saunders Interactive General Chemistry CD-ROM,* Screen 20.8, Free Energy and Temperature.

If a reaction has a positive enthalpy change *and* a positive entropy change, the only way it can be product-favored under standard conditions is if $-T\,\Delta S^\circ$ is large enough to outweigh ΔH°. This can happen in two ways: the entropy change can be positive and large (as is the case in dissolving NH_4NO_3 and many other ionic solids in water), or the entropy change can be positive and the temperature high. The latter is in fact one of the reasons reactions are often carried out at high temperatures. Let us consider an example of this case.

Our economy is based in large measure on the production of iron, and we can think of at least three different ways to reduce iron(III) oxide to metallic iron. One way is to heat iron(III) oxide and hope it decomposes to iron and oxygen:

$$Fe_2O_3(s) \longrightarrow 2\,Fe(s) + \tfrac{3}{2}\,O_2(g) \qquad \Delta G^\circ_{rxn} = +742\ \text{kJ}$$

We can obtain some idea of the feasibility of this by calculating ΔG°_{rxn}. Using the data in Appendix L, we find that ΔH°_{rxn} is +824.2 kJ and ΔS°_{rxn} is +275 J/K. The enthalpy change means the reaction could be reactant-favored, whereas the entropy change means it could be product-favored. Unfortunately, ΔH°_{rxn} is so positive that it cannot be outweighed by $-T\,\Delta S^\circ$ (= −82 kJ at 25 °C) at any reasonable temperature, and ΔG°_{rxn} at room temperature is +742 kJ at 298 K!

Another way to reduce iron(III) oxide is the so-called **thermite reaction.** Here the reactant-favored decomposition of $Fe_2O_3(s)$ to $Fe(s)$

$$Fe_2O_3(s) \longrightarrow 2\,Fe(s) + \tfrac{3}{2}\,O_2(g) \qquad \Delta G^\circ_{rxn} = +742\ \text{kJ}$$

is coupled with the highly product-favored oxidation of aluminum to Al_2O_3:

$$2\,Al(s) + \tfrac{3}{2}\,O_2(g) \longrightarrow Al_2O_3(s) \qquad \Delta G^\circ_{rxn} = -1582\ \text{kJ}$$

Figure 20.10 The thermite reaction, a product-favored process. After starting the reaction with a fuse of burning magnesium wire, iron(III) oxide reacts with aluminum powder to give aluminum oxide and iron. The Fe_2O_3/Al reaction generates so much heat that the iron is produced in the molten state. (See also Figure 5.17.) *(C. D. Winters)*

In practice, Fe_2O_3 is not reduced directly with C. Instead, the carbon is burned to give CO, which then acts as the reducing agent. See Chapter 23.

The sum of these reactions is

$$Fe_2O_3(s) + 2\ Al(s) \longrightarrow 2\ Fe(s) + Al_2O_3(s) \qquad \Delta G^\circ_{rxn} = -840.\ kJ$$
$$\Delta H^\circ_{rxn} = -851.5\ kJ \qquad \Delta S^\circ_{rxn} = -37.5\ J/K$$

This is certainly a product-favored process (Figure 20.10). The unfavorable ΔS°_{rxn} at 298 K ($-T\Delta S^\circ = +11.2$ kJ) is swamped by a very large and negative enthalpy change. That means that a large amount of heat is produced, so much in fact that the products are raised to the melting point of iron (1530 °C), and white hot, molten metal streams out of the reaction. The reaction has been applied to welding procedures, but it is unfortunately not practical for the production of iron on a large scale. The cost of the aluminum to use as a reducing agent is much larger than the value of the iron produced.

The usual method of reducing iron(III) oxide to iron metal is to use carbon or carbon monoxide, both inexpensive reducing agents. The decomposition of Fe_2O_3 is coupled with the highly product-favored combustion of carbon or carbon monoxide:

$$C(graphite) + O_2(g) \longrightarrow CO_2(g)$$
$$\Delta H^\circ_{rxn} = -393.5\ kJ \qquad S^\circ_{rxn} = -2.9\ J/K \qquad \Delta G^\circ_{rxn} = -394.4\ kJ$$

The overall process for reduction by carbon can be thought of as the sum of two reactions:

$$\begin{array}{rcll} 2\ Fe_3O_3(s) & \longrightarrow & 4\ Fe(s) + 3\ O_2(g) & 2(\Delta G^\circ_{rxn} = +742\ kJ) \\ 3\ C(graphite) + 3\ O_2(graphite) & \longrightarrow & 3\ CO_2(g) & 3(\Delta G^\circ_{rxn} = -394.4\ kJ) \\ \hline 2\ Fe_2O_3(s) + 3\ C(graphite) & \longrightarrow & 4\ Fe(s) + 3\ CO_2(s) & \Delta G^\circ_{rxn} = +300.8\ kJ \end{array}$$

$$\Delta H^\circ_{rxn} = +467.9\ kJ \qquad \Delta S^\circ_{rxn} = +560.3\ J/K$$

Even though the entropy is large, $-T\Delta S^\circ$ ($= -167$ kJ at 25 °C) is not large enough to offset the very unfavorable (positive) enthalpy change at 25 °C. Why then is this process used in industry? Precisely because the large, positive entropy change allows the reaction to become spontaneous at higher temperature. Let us calculate the *minimum* temperature T at which ΔG°_{rxn} is no longer positive, that is, the temperature at which it is zero:

$$\Delta G^\circ_{rxn} = \Delta H^\circ_{rxn} - T\Delta S^\circ_{rxn}$$
$$0 = +467.9\ kJ - T(0.5603\ kJ/K)$$
$$T = 835\ K \text{ or } 562\ °C$$

The free energy change for the reaction becomes 0 at 562 °C, and at higher temperatures it is negative. Thus, the reaction can become product-favored by raising the temperature to a point easily reached in an industrial furnace.*

*To calculate ΔG° at another temperature, we used the enthalpy and entropy values appropriate for 25 °C. This is not entirely correct, because ΔH° and ΔS° are somewhat temperature-dependent. Their dependency on T, however, is *much* smaller than that of ΔG° (as long as the temperature change does not include a phase change). Thus, our calculation provides an estimate of the point at which the reaction becomes spontaneous.

Exercise 20.10 Temperature and Free Energy Change

Is the reduction of magnesia, MgO, with carbon a product-favored process at 25 °C? If not, at what temperature does it become so?

$$MgO(s) + C(graphite) \longrightarrow Mg(s) + CO(g)$$

20.4 THERMODYNAMICS AND THE EQUILIBRIUM CONSTANT

See the *Saunders Interactive General Chemistry CD-ROM,* Screen 20.9, Thermodynamics and the Equilibrium Constant.

The free energy change for a reaction, $\Delta G°$, is the increase or decrease in free energy as the reactants in their standard states are converted *completely* to the products in their standard states. But complete conversion is not often observed in practice. A product-favored reaction proceeds largely to products, but some reactants may remain at equilibrium. A reactant-favored reaction proceeds only partially to products before achieving equilibrium. The question now is how $\Delta G°$ is related to the conditions at equilibrium.

To answer this, let us consider what happens as a reaction proceeds from reactants to products at constant temperature and pressure. When the reactants are mixed, the system becomes more disordered, and the entropy of the system increases. Then, as the reaction proceeds, heat energy might be evolved, further increasing the entropy of the system and its surroundings. When the system reaches equilibrium, products or reactants may predominate. Rarely, however, would reactants be converted *completely* to products. Under these conditions the free energy change for the reaction is not equal to $\Delta G°$ but to ΔG (without the superscript ° that signifies standard conditions). The relationship between $\Delta G°$ and ΔG is

$$\Delta G = \Delta G° + RT \ln Q$$

where R is the universal gas constant, T is the temperature in kelvins, and Q is the reaction quotient (see Section 16.3). That is, for the general reaction of A and B giving products C and D

$$a\,A + b\,B \rightleftharpoons c\,C + d\,D$$

the reaction quotient, Q, is

$$Q = \frac{[C]^c[D]^d}{[A]^a[B]^b}$$

You learned in Chapter 16 that if $Q = K$, then the system is at equilibrium. The amounts of reactants and products no longer change at *equilibrium,* and $\Delta G = 0$.

Equilibrium is characterized by the inability to do work.

Because $\Delta G = 0$ and $Q = K$ at equilibrium, the equation $\Delta G = \Delta G° + RT \ln Q$ leads to

$$0 = \Delta G° + RT \ln K \qquad \text{(at equilibrium)}$$

Rearranging the last equation leads to a useful relationship between the standard free energy change for a reaction and the **thermodynamic equilibrium constant, *K*.**

$$\Delta G° = -RT \ln K \qquad [20.4]$$

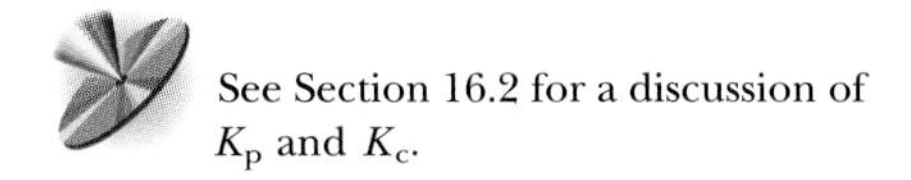

See Section 16.2 for a discussion of K_p and K_c.

For equilibria involving only gases, the thermodynamic equilibrium constant is K_p. For those that involve compounds in solution, it is equal to K_c.

The relationship between free energy and equilibrium is illustrated by Figure 20.11. The left side of each diagram gives the total free energy of the reactants, and the right side gives the total free energy of the products. The difference between the free energy of the pure reactants and the pure products is $\Delta G°$, which, like K, depends only on temperature and is a constant for a given reaction.

The relationship between $\Delta G°$ and K in Equation 20.4 informs us that, when $\Delta G°$ is negative, K must be greater than 1. Furthermore, the more negative the value of $\Delta G°$, the larger the equilibrium constant. We say that products are favored over reactants or that products are more stable than reactants. This situation is illustrated by Figure 20.11a. The opposite case is illustrated by Figure 20.11b. Here $\Delta G°$ is positive, the reaction is reactant-favored, and K must be less than 1 (when $K < 1$, ln K is negative, and the term $-RT \ln K$ is positive). Now the reactants are more stable than the products. Finally, it is possible that $\Delta G°$ is 0, so K would be equal to 1. This extremely rare situation would mean that $[C]^c[D]^d = [A]^a[B]^b$ (for the reaction $a\,A + b\,B \rightleftarrows c\,C + d\,D$) at equilibrium. These relationships can be summarized as follows:

$\Delta G°$	K	**Product Formation**
$\Delta G° < 0$	$K > 1$	Products favored over reactants at equilibrium
$\Delta G° = 0$	$K = 1$	At equilibrium when $[C]^c[D]^d = [A]^a[B]^b$, very rare
$\Delta G° > 0$	$K < 1$	Reactants favored over products at equilibrium

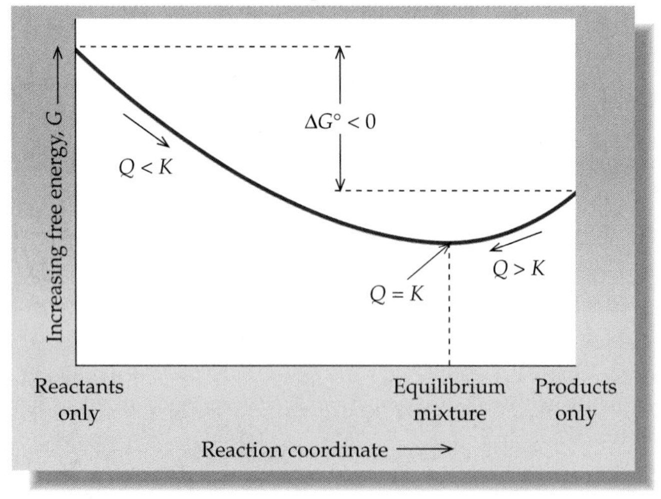

(a)

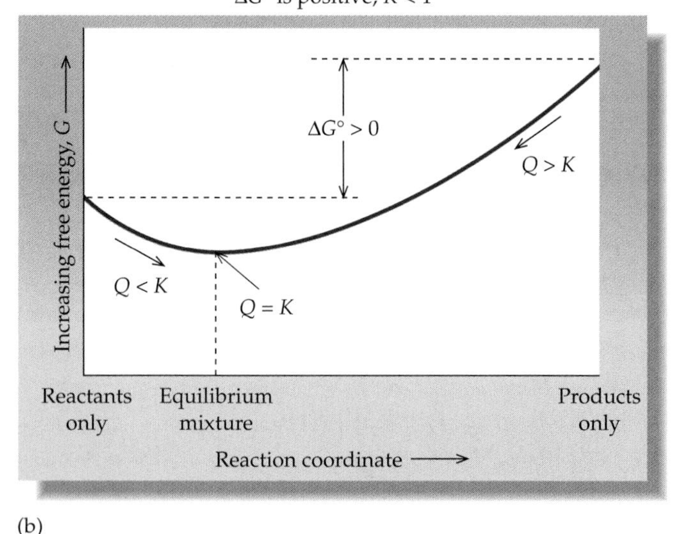

(b)

Figure 20.11 The variation in free energy for a reversible reaction carried out at constant temperature. The standard free energy change for a reaction, $\Delta G°_{rxn}$, is the change in free energy for the complete conversion of reactants in their standard states to products in their standard states. In (a) $\Delta G°_{rxn}$ is negative, the reaction is product-favored, and $K > 1$. In (b) $\Delta G°_{rxn}$ is positive, the reaction is reactant-favored, and $K < 1$. In both cases, the mixture of reactants and products at equilibrium is more stable than pure products or pure reactants. The position of the equilibrium depends on the $\Delta G°_{rxn}$ and T. Comparing Q and K at any point gives the direction of approach to equilibrium. If Q is not equal to K, the reaction runs "downhill" until $Q = K$.

In Section 16.3 we described what happens when a reaction is not at equilibrium.

- When $Q < K$ the reaction continues to proceed from reactants to products until equilibrium is achieved.
- When $Q > K$ the reaction proceeds from products to reactants until equilibrium is achieved.

These situations are also illustrated by Figure 20.11. If $Q < K$, equilibrium is approached from left to right on each of the curves in the figure. If $Q > K$, equilibrium is approached from right to left on each of the curves. The composition of the equilibrium mixture, and whether or not the reaction is product- or reactant-favored, depends on the value of $\Delta G°$, whereas the direction of approach to equilibrium depends on the value of Q relative to K.

Finally, Figure 20.11 also illustrates the fact that the mixture of reactants and products at equilibrium is more stable than either the pure reactants or pure products. Thus, in either a product-favored or reactant-favored reaction, the reaction runs "downhill" along the reaction coordinate until $Q = K$. The position of the equilibrium depends on the $\Delta G°$ and T.

Two applications of Equation 20.4 are (1) the calculation of equilibrium constants from values of $\Delta G_f°$ for reactants and products and (2) the evaluation of $\Delta G°_{rxn}$ from an experimental determination of K. These applications are explored in the following examples.

Example 20.4 Calculating K_p from $\Delta G°_{rxn}$

The standard free energy change for the reaction, $\Delta G°_{rxn}$,

$$N_2(g) + 3\ H_2(g) \rightleftharpoons 2\ NH_3(g)$$

is -32.9 kJ. Calculate the equilibrium constant for this reaction at 25 °C.

Solution In this case, we need only substitute the appropriate values into Equation 20.4, taking care that the units of $\Delta G°_{rxn}$ are the same as those of RT:

$$\begin{aligned}\Delta G°_{rxn} &= -RT \ln K_p \\ (-32.9\ \text{kJ})(1000\ \text{J/kJ}) &= -(8.3145\ \text{J/K}\cdot\text{mol})(298\ \text{K}) \ln K_p \\ \ln K_p &= 13.3 \\ K_p &= e^{13.3} \\ &= 6 \times 10^5\end{aligned}$$

To find K_p using a calculator, enter 13.3 and then strike the key labeled "e^x" or "inv(erse) ln x." See Appendix A for more information.

The equilibrium constant has a very large value, which means the equilibrium position lies very far to the product side at 25 °C.

Example 20.5 Calculating $\Delta G°_{rxn}$ from K_p

Calculate $\Delta G°_{rxn}$ for the decomposition of ammonium chloride (Figure 20.12) at 25 °C from $K_p = 1.1 \times 10^{-16}$.

$$NH_4Cl(s) \rightleftharpoons NH_3(g) + HCl(g)$$

Figure 20.12 The decomposition of ammonium chloride. Heating NH_4Cl produces gaseous NH_3 and HCl in a reactant-favored reaction. The NH_3 and HCl recombine in the vapor phase to give a cloud of NH_4Cl. (The model of NH_4Cl is at the right. Notice that it has a NaCl-like structure. That is, the Cl^- ions form a face-centered cubic lattice with NH_4^+ ions in the octahedral holes. See Chapter 13 and the cover of this book.) *(Photo, C. D. Winters; models, S. M. Young)*

Solution Here we substitute the value of K_p into Equation 20.4 at a temperature of 25 °C (= 298 K).

$$\Delta G^\circ_{rxn} = -RT \ln K_p = -(8.3145\ \text{J/K}\cdot\text{mol})(298\ \text{K})\ \ln\ (1.1 \times 10^{-16})$$
$$= 9.1 \times 10^4\ \text{J}$$
$$= 91\ \text{kJ}$$

Using values of ΔG°_f from Appendix L you can verify that the free energy change is indeed +91 kJ.

Example 20.6 Uses of ΔH°_f, S°, ΔG°_f, and the Calculation of K

In Chapter 5 you studied in a qualitative way the driving forces of chemical reactions. One of these is the formation of a gaseous product that can escape when the reaction is open to the surroundings. Let us look at such a reaction from the quantitative, thermodynamic point of view:

$$MgCO_3(s) \longrightarrow MgO(s) + CO_2(g)$$

1. Is the reaction product-favored at room temperature?
2. Is the reaction enthalpy-driven or entropy-driven?
3. What is the value of K_p at 25 °C?
4. At what temperature does $K_p = 1$?
5. Does high temperature make the reaction more or less product-favored?

Solution To answer the first two questions we need to know ΔG°_{rxn} and its sign, as well as ΔH°_{rxn}. Using data from Appendix L, we have

$$\Delta H^\circ_{rxn} = \Delta H^\circ_f\ [MgO(s)] + \Delta H^\circ_f\ [CO_2(g)] - \Delta H^\circ_f\ [MgCO_3(s)]$$
$$= 1\ \text{mol}\ (-601.7\ \text{kJ/mol}) + 1\ \text{mol}\ (-393.5\ \text{kJ/mol}) - 1\ \text{mol}\ (-1095.8\ \text{kJ/mol})$$
$$= +100.6\ \text{kJ}$$

$$\Delta S^\circ_{rxn} = S^\circ[MgO(s)] + S^\circ[CO_2(g)] - S^\circ[MgCO_3(s)]$$
$$= 1\ \text{mol}\ (26.9\ \text{J/K}\cdot\text{mol}) + 1\ \text{mol}\ (213.7\ \text{J/K}\cdot\text{mol}) - 1\ \text{mol}\ (65.7\ \text{J/K}\cdot\text{mol})$$
$$= +174.9\ \text{J/K}$$

Knowing the enthalpy and entropy changes for reaction, we can combine them to calculate the free energy change for the reaction:

$$\Delta G^\circ_{rxn} = \Delta H^\circ_{rxn} - T\Delta S^\circ_{rxn} = 100.6\ \text{kJ} - (298\ \text{K})(174.9\ \text{J/K})(1\ \text{kJ}/1000\ \text{J})$$
$$= 48.5\ \text{kJ}$$

The decomposition of magnesium carbonate to give carbon dioxide is reactant-favored at 298 K. What little reaction that does occur at 298 K is entropy-driven because the entropy change is positive. The enthalpy change is positive and too large to be offset by the increase in entropy to make the reaction product-favored at this temperature.

Having found ΔG°_{rxn}, K_p can now be calculated at 298 K from

$$\Delta G^\circ_{rxn} = -RT \ln K_p$$

or,

$$\ln K_p = -\frac{\Delta G^\circ_{rxn}}{RT} = -\frac{48{,}500 \text{ J/mol}}{(8.3145 \text{ J/K}\cdot\text{mol})(298 \text{ K})} = -19.6$$

$$K_p = e^{-19.6} = 3 \times 10^{-9}$$

This means that $K_p = P_{CO_2} = 3 \times 10^{-9}$, or that the partial pressure of CO_2 is 3×10^{-9} bar at equilibrium at 25 °C. The partial pressure is extremely small because ΔG°_{rxn} has such a large, positive value.

At what temperature does $K_p = 1$? From the preceding analysis you know that ΔS°_{rxn} drives the reaction. This means that, at a sufficiently high temperature, the negative value of $-T\,\Delta S^\circ$ could become large enough to outweigh the inhibiting effect of a positive ΔH°_{rxn}, and ΔG°_{rxn} would become negative. A balance between ΔH° and $-T\,\Delta S^\circ$ is reached when $\Delta G^\circ = 0$, that is, when $K = 1$. The breakeven temperature is

$$T = \frac{\Delta H^\circ}{\Delta S^\circ} = \frac{100.6 \text{ kJ}}{0.1749 \text{ kJ/K}} = 575.2 \text{ K (or 302.0 °C)}$$

At approximately 300 °C the equilibrium constant is 1; that is, $P_{CO_2} = 1$ bar. At still higher temperatures, the $-T\Delta S^\circ$ term increasingly dominates the ΔH° term, and the yield increases because K increases. This means the answer to question 5 is "more product-favored," and this is true for any reaction with $\Delta S^\circ_{rxn} > 0$ as illustrated by line 3 in Figure 20.8.

Exercise 20.11 Calculating K_p from the Free Energy Change

Calculate K_p at 298 K from the value of ΔG°_{rxn} for the reaction

1. $S(s) + O_2(g) \longrightarrow SO_2(g)$
2. $CaCO_3(s) \longrightarrow CaO(s) + CO_2(g)$

Exercise 20.12 Coupling Chemical Reactions

Tin(IV) oxide can be reduced to tin metal using carbon as the reducing agent:

$$SnO_2(s) + C(s) \longrightarrow Sn(s) + CO_2(g)$$

Show that this process is the sum of two reactions: the oxidation of carbon to give CO_2, and the decomposition of SnO_2 to give Sn and O_2. What are the values of ΔG°_{rxn} for the separate reactions? What is ΔG°_{rxn} for the overall process? Is the loss of oxygen by tin more or less product-favored when carbon is used as the reducing agent?

20.5 THERMODYNAMICS AND TIME

With this chapter, we brought together the **three laws of thermodynamics.**

First law: The total energy of the universe is a constant.

Second law: The total entropy of the universe is always increasing.

Third law: The entropy of a pure, perfectly formed crystalline substance at absolute zero is zero.

Some cynic long ago paraphrased the first two laws into simpler statements. The first law is a statement that "You can't win!," and the second law tells you that "You can't break even either!" Yet another interpretation of the second law is Murphy's Law that "Things always tend to go wrong."

The second law tells us the entropy of the universe increases in a product-favored process. A snowflake spontaneously melts in a warm room, but you never see a glassful of water molecules spontaneously reassembling into snowflakes (at temperatures greater than 0 °C). Molecules of your perfume or cologne spontaneously diffuse throughout a room, but they don't spontaneously collect again on your body. With time, all natural processes result in chaos. This is what scientists mean when they say that the second law is an expression in physical—as opposed to psychological—form of what we call *time.* In fact, entropy has been called "time's arrow."

The second law of thermodynamics demands that disorder increases with time. Because all natural processes take place as time progresses and result in increased disorder, it is apparent that entropy and time "point" in the same direction.

Neither of the first two laws of thermodynamics has ever been or can be proven. It is just that there never has been a single example showing otherwise. No less a scientist than Albert Einstein once remarked that thermodynamic theory " . . . is the only physical theory of the universe content [which], within the framework of applicability of its basic concepts, will never be overthrown."

Einstein's statement does not mean that people have not tried (and are continuing to try) to disprove the laws of thermodynamics. Someone is always claiming to have invented a machine that performs useful work without expending energy—a perpetual motion machine (which should be a violation of the second law). Although such a machine was actually granted a patent recently by the U.S. Patent Office (presumably the patent examiner had not had a course in thermodynamics), no workable perpetual motion machine has ever been demonstrated. The laws of thermodynamics are safe.

If you are interested in the theories of the origin of the universe, and in "time's arrow," read Stephen W. Hawking's *A Brief History of Time, From the Big Bang to Black Holes,* New York, Bantam Books, 1988.

CHAPTER HIGHLIGHTS

For many practical reasons chemists would like to be able to predict if a reaction favors the products or the reactants. That is the main objective of this chapter. To achieve that, we have introduced the concept of entropy and have described the interplay of enthalpy and entropy changes for chemical reactions. A summary of the concepts in this chapter is given in Figure 20.8 and the following table:

ΔH°_{rxn}	ΔS°_{rxn}	Sign of ΔG°_{rxn}	K	Reaction Outcome
−, exothermic	+, less order	−	>1	Product-favored at all T
−, exothermic	−, more order	− or +	Depends on T	Depends on T and on relative magnitude of ΔH°_{rxn} and ΔS°_{rxn}. Generally product-favored at *lower* T.
+, endothermic	+, less order	+ or −	Depends on T	Depends on T and on relative magnitude of ΔH°_{rxn} and ΔS°_{rxn}. Generally product-favored at *higher* T.
+, endothermic	−, more order	+	<1	Reactant-favored at all T.

When you have finished studying this chapter, you should be able to

- Describe the difference between the information provided by kinetics and thermodynamics (Section 20.1).
- Understand that **entropy** is a measure of matter and energy dispersal (Section 20.2).
- Predict the sign of the entropy change for a reaction or change in state (Section 20.2).
- Calculate the entropy change for a change of state or for a chemical reaction (Section 20.2 and Equation 20.1).
- Use entropy and enthalpy changes to predict whether a reaction is product- or reactant-favored (Section 20.2 and Table 20.2).
- Understand the connection between enthalpy and entropy changes for a reaction and the **Gibbs free energy** change (Section 20.3, Table 20.2, and the summary table on page 938).
- Calculate the change in free energy for a reaction from the enthalpy and entropy changes (Equation 20.2) or from the standard free energy of formation of reactants and products, ΔG_f° (Equation 20.3) (Section 20.3).
- Describe the relationship between the free energy change for a reaction and its equilibrium constant (Section 20.4 and Equation 20.4).
- Show that a reactant-favored reaction can become product-favored if coupled with another reaction that is strongly product-favored (Section 20.4).

STUDY QUESTIONS

Many of these questions require thermodynamic data. If the required data are not given in the question, consult the tables in this chapter or Appendix L.

REVIEW QUESTIONS

1. State the three laws of thermodynamics.
2. What is meant by a "product-favored chemical reaction?"
3. Criticize the following statements:
   (a) The entropy increases in all product-favored reactions.
   (b) A reaction with a negative free energy change ($\Delta G^\circ_{rxn} < 0$) is predicted to be product-favored with rapid transformation of reactants to products.
   (c) All product-favored processes are exothermic.
   (d) Endothermic reactions are never product-favored.
4. Decide if each of the following statements is true or false. If false, rewrite to make it true.
   (a) The entropy of a substance increases on going from the liquid to the vapor state at any temperature.
   (b) An exothermic reaction is always product-favored.
   (c) Reactions with a positive ΔH°_{rxn} and a positive ΔS°_{rxn} can never be product-favored.
   (d) Reactions with $\Delta G^\circ_{rxn} < 0$ have an equilibrium constant greater than 1.
   (e) When the equilibrium constant of a reaction is less than 1, then ΔG°_{rxn} is less than 0.
5. Explain why the entropy of the system increases on dissolving solid NaCl in water {$S^\circ[NaCl(s)] = 72.1$ J/K · mol and $S^\circ[NaCl(aq)] = 115.5$ J/K · mol}.

NUMERICAL AND OTHER QUESTIONS

Entropy

6. Which substance has the higher entropy in each of the following pairs?
   (a) A sample of pure silicon (to be used in a computer chip) or a piece of silicon containing a trace of some other atoms such as B or P.
   (b) An ice cube or liquid water, both at 0 °C.
   (c) A sample of pure solid I_2 or iodine vapor, both at room temperature.

7. Which substance has the higher entropy in each of the following pairs?
   (a) A sample of Dry Ice (solid CO_2) at −78 °C or CO_2 vapor at 0 °C.
   (b) Sugar, as a solid or dissolved in a cup of tea.
   (c) Two 100-mL beakers, one containing pure water and the other containing pure alcohol, or a beaker containing a mixture of water and alcohol.

8. By comparing the formulas or states for each pair of compounds, decide which is expected to have the higher entropy at the same temperature.
   (a) KCl(s) or $AlCl_3$(s)
   (b) $CH_3I(\ell)$ or $CH_3CH_2I(\ell)$
   (c) NH_4Cl(s) or NH_4Cl(aq)

9. By comparing the formulas or states for each pair of compounds, decide which is expected to have the higher entropy at the same temperature.

(a) $NaCl(s)$ or $MgCl_2(s)$
(b) $CH_3NH_2(g)$ or $(CH_3)_2NH(g)$
(c) $Au(s)$ or $Hg(\ell)$

10. Calculate the entropy change, $\Delta S°$, for each of the following changes and comment on the sign of the change:
(a) C(diamond) $\longrightarrow$ C(graphite)
(b) $Na(g) \longrightarrow Na(s)$
(c) $Br_2(\ell) \longrightarrow Br_2(g)$

11. Calculate the entropy change, $\Delta S°$, for each of the following changes and comment on the sign of the change:
(a) $NH_4Cl(s) \longrightarrow NH_4Cl(aq)$
(b) $C_2H_5OH(\ell) \longrightarrow C_2H_5OH(g)$
(c) $CCl_4(g) \longrightarrow CCl_4(\ell)$

12. Calculate the entropy change, $\Delta S°$, for the vaporization of ethanol, C_2H_5OH, at the normal boiling point of the pure alcohol, 78.0 °C. The enthalpy of vaporization of the alcohol is 39.3 kJ/mol.

13. The enthalpy of vaporization of liquid diethyl ether, $(C_2H_5)_2O$, is 26.0 kJ/mol at the boiling point of 35.0 °C. Calculate $\Delta S°$ for (a) liquid to vapor and (b) vapor to liquid at 35.0 °C.

Reactions and Entropy Change

14. Calculate the standard entropy change for the formation of 1.0 mol of gaseous propane (C_3H_8) at 25 °C:

$$3\ C(\text{graphite}) + 4\ H_2(g) \longrightarrow C_3H_8(g)$$

15. Calculate the standard entropy change for the formation of 1.0 mol of silicon hydride (silane) at 25 °C:

$$Si(s) + 2\ H_2(g) \longrightarrow SiH_4(s)$$

16. Calculate the standard entropy change for the formation of 1.0 mol of each of the following compounds from the elements at 25 °C:
(a) $H_2O(\ell)$
(b) $Mg(OH)_2(s)$
(c) $PbCl_2(s)$

17. Calculate the standard entropy change for the formation of 1.0 mol of each of the following compounds from the elements at 25 °C:
(a) $ICl(g)$
(b) $COCl_2(g)$
(c) $CaCO_3(s)$

18. Calculate the standard entropy change for each of the following reactions at 25 °C:
(a) $Mg(s) + 2\ H_2O(\ell) \longrightarrow Mg(OH)_2(s) + H_2(g)$
(b) $Na_2CO_3(s) + 2\ HCl(aq) \longrightarrow 2\ NaCl(aq) + H_2O(\ell) + CO_2(g)$

19. Calculate the standard entropy change for each of the following reactions at 25 °C:
(a) $2\ Al(s) + 3\ Cl_2(g) \longrightarrow 2\ AlCl_3(s)$
(b) $C_2H_5OH(\ell) + 3\ O_2(g) \longrightarrow 2\ CO_2(g) + 3\ H_2O(g)$

20. Almost 5.0×10^9 kg of benzene, C_6H_6, is made each year. It is a starting material for many other compounds and a solvent (although it is also a carcinogen, and its use is restricted) (page 499). One compound that can be made from benzene is cyclohexane, C_6H_{12}:

$$C_6H_6(\ell) + 3\ H_2(g) \longrightarrow C_6H_{12}(\ell)$$

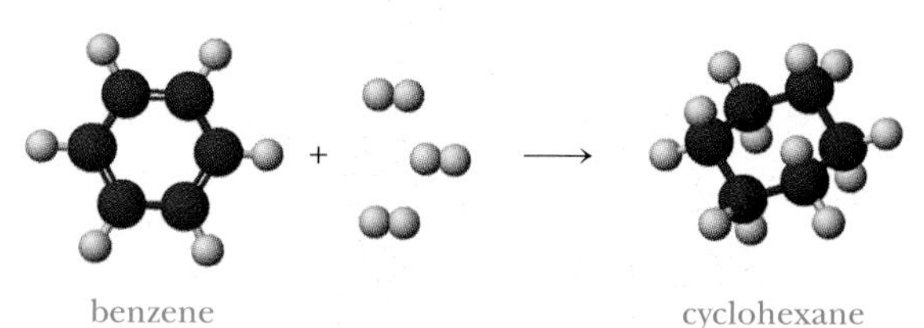

The enthalpies of formation are

$$\Delta H_f^\circ[C_6H_6(\ell)] = +49.0\text{ kJ/mol}$$
$$\Delta H_f^\circ[C_6H_{12}(\ell)] = -156.4\text{ kJ/mol}$$

Is this reaction likely to be product-favored or reactant-favored? Explain your reasoning briefly.

21. What are the signs of the enthalpy and entropy changes for the splitting of water to give gaseous hydrogen and oxygen, a process that requires considerable energy? Is this reaction likely to be product-favored? Explain your answer briefly.

22. Classify each of the following reactions as one of the four reaction types summarized in Table 20.2:
(a) $Fe_2O_3(s) + 2\ Al(s) \longrightarrow 2\ Fe(s) + Al_2O_3(s)$

$$\Delta H° = -851.5\text{ kJ} \qquad \Delta S° = -363.12\text{ J/K}$$

(b) $N_2(g) + 2\ O_2(g) \longrightarrow 2\ NO_2(g)$

$$\Delta H° = 64.4\text{ kJ} \qquad \Delta S° = -120\text{ J/K}$$

23. Classify each of the following reactions as one of the four reaction types summarized in Table 20.2:
(a) $C_6H_{12}O_6(s) + 6\ O_2(g) \longrightarrow 6\ CO_2(g) + 6\ H_2O(\ell)$

$$\Delta H° = -673\text{ kJ} \qquad \Delta S° = 60.4\text{ J/K}$$

(b) $MgO(s) + C(\text{graphite}) \longrightarrow Mg(s) + CO(g)$

$$\Delta H° = 491.18\text{ kJ} \qquad \Delta S° = 197.67\text{ J/K}$$

24. Is the combustion of ethane, C_2H_6, likely to be a product-favored reaction?

$$C_2H_6(g) + \tfrac{7}{2}\ O_2(g) \longrightarrow 2\ CO_2(g) + 3\ H_2O(g)$$

Answer the question by calculating the value of $\Delta S_{\text{universe}}$. Required values of ΔH_f° and $S°$ are in Appendix L. Does your calculated answer agree with your preconceived idea of this reaction?

25. In the U. S. military, packaged meals can be heated in the field. The reaction of magnesium with water provides the heat to warm so-called "meals ready to eat" (MREs).

$$Mg(s) + 2\ H_2O(\ell) \longrightarrow Mg(OH)_2(s) + H_2(g)$$

Is this reaction in fact predicted to be product-favored? Answer the question by calculating the value of $\Delta S^\circ_{\text{universe}}$. Required values of ΔH_f° and $S°$ are in Appendix L. Does your calculated answer agree with your preconceived idea of this reaction?

Soldier using an MRE, a meal heated by the reaction of magnesium and water. (*U.S. Army Natick Research and Development Center*)

Free Energy

26. Using values of ΔH_f° and S°, and calculate ΔG°_{rxn} for each of the following reactions:
(a) $Cu(s) + Cl_2(g) \longrightarrow CuCl_2(s)$ (see Figure 20.1)
(b) $NH_3(g) + HCl(g) \longrightarrow NH_4Cl(s)$ (see book's cover)
Which of the values of ΔG°_{rxn} that you have just calculated corresponds to a standard free energy of formation, ΔG_f°? In those cases, compare your calculated values with the values of ΔG_f° tabulated in Appendix L. Which of these reactions is (are) predicted to be product-favored? Are the reactions enthalpy- or entropy-driven?

27. Using values of ΔH_f° and S°, calculate ΔG°_{rxn} for each of the following reactions:
(a) $Ca(s) + 2\ H_2O(\ell) \longrightarrow Ca(OH)_2(aq) + H_2(g)$
(b) $6\ C(graphite) + 3\ H_2(g) \longrightarrow C_6H_6(\ell)$
Which of the values of ΔG°_{rxn} that you have just calculated corresponds to a standard free energy of formation, ΔG_f°? In those cases, compare your calculated values with the values of ΔG_f° tabulated in Appendix L. Which of these reactions is (are) predicted to be product-favored? Are the reactions enthalpy- or entropy-driven?

28. Using values of ΔH_f° and S°, calculate the standard molar free energy of formation, ΔG_f°, for each of the following compounds:
(a) $CS_2(g)$ (b) $N_2H_4(\ell)$ (c) $COCl_2(g)$
Compare your calculated values of ΔG_f° with those listed in Appendix L. Which reactions are predicted to be product-favored?

29. Using values of ΔH_f° and S°, calculate the standard molar free energy of formation, ΔG_f°, for each of the following compounds:
(a) $Mg(OH)_2(s)$ (b) $NOCl(g)$ (c) $Na_2CO_3(s)$
Compare your calculated values of ΔG_f° with those listed in Appendix L. Which reactions are predicted to be product-favored?

30. Hydrazine is used to remove dissolved oxygen from the water in hot water heating systems:

$$N_2H_4(\ell) + O_2(g) \longrightarrow 2\ H_2O(\ell) + N_2(g)$$

What is the value of ΔG°_{rxn} when 1.00 mol of N_2H_4 is oxidized? What is the value of ΔG°_{rxn} for the oxidation of 1.00 kg of hydrazine?

31. Write a balanced equation that depicts the formation of 1 mol of $Fe_2O_3(s)$ from its elements (see *A Chemical Puzzler*, page 910). What is the standard free energy of formation of 1.00 mol of $Fe_2O_3(s)$? What is the value of ΔG°_{rxn} when 454 g (1 lb) of $Fe_2O_3(s)$ is formed from the elements?

32. Using values of ΔG_f° calculate ΔG°_{rxn} for each of the following reactions. Which are predicted to be product-favored?
(a) $Ca(s) + Cl_2(g) \longrightarrow CaCl_2(s)$
(b) $2\ HgO(s) \longrightarrow 2\ Hg(\ell) + O_2(g)$
(c) $NH_3(g) + 2\ O_2(g) \longrightarrow HNO_3(\ell) + H_2O(\ell)$

33. Using values of ΔG_f° calculate ΔG°_{rxn} for each of the following reactions. Which are predicted to be product-favored?
(a) $HgS(s) + O_2(g) \longrightarrow Hg(\ell) + SO_2(g)$
(b) $2\ H_2S(g) + 3\ O_2(g) \longrightarrow 2\ H_2O(g) + 2\ SO_2(g)$
(c) $SiCl_4(g) + 2\ Mg(s) \longrightarrow 2\ MgCl_2(s) + Si(s)$

34. What is the value of ΔG_f° for $BaCO_3(s)$? You know that $\Delta G^\circ_{rxn} = +218.1$ kJ for the reaction

$$BaCO_3(s) \longrightarrow BaO(s) + CO_2(g)$$

and other data are available in Appendix L.

35. What is the value of ΔG_f° for $TiCl_2(s)$? You know that $\Delta G^\circ_{rxn} = -272.8$ kJ for

$$TiCl_2(s) + Cl_2(g) \longrightarrow TiCl_4(\ell)$$

and other data are available in Appendix L.

36. Hydrogenation, the addition of hydrogen to an organic compound, is a reaction of considerable industrial importance. Calculate ΔH°, ΔS°, and ΔG° at 25 °C for the hydrogenation of 1-octene, C_8H_{16}, to give octane, C_8H_{18}. Is the reaction product- or reactant-favored under standard conditions?

$$C_8H_{16}(g) + H_2(g) \longrightarrow C_8H_{18}(g)$$

The following information is required, in addition to data in Appendix L.

Compound	ΔH_f° (kJ/mol)	S°(J/K · mol)
1-Octene	−81.4	462.5
Octane	−208.6	466.7

37. If gaseous hydrogen can be produced cheaply, it can be burned directly as a fuel or converted to another fuel, methane (CH_4), for example:

$$3\ H_2(g) + CO(g) \longrightarrow CH_4(g) + H_2O(g)$$

Calculate ΔH°, ΔS°, and ΔG° at 25 °C for this reaction. Is it predicted to be product- or reactant-favored under standard conditions?

Thermodynamics and Equilibrium Constants

38. The formation of NO(g) from its elements,

$$\tfrac{1}{2}\ N_2(g) + \tfrac{1}{2}\ O_2(g) \longrightarrow NO(g)$$

has a standard free energy change, ΔG_f°, of +86.55 kJ/mol at 25 °C. Calculate K_p at this temperature. Comment on the connection between the sign of ΔG° and the magnitude of K_p.

39. Methanol, CH_3OH, is now widely used as a fuel in race cars such as those that compete in the Indianapolis 500 (see Chapter 6):

$$C(\text{graphite}) + \tfrac{1}{2}\ O_2(g) + 2\ H_2(g) \longrightarrow CH_3OH(\ell)$$

Calculate K_p for the formation of methanol from the elements at 25 °C. Comment on the connection between the sign of ΔG° and the magnitude of K_p.

40. Ethylene reacts with hydrogen to produce ethane:

$$H_2C{=}CH_2(g) + H_2(g) \longrightarrow H_3C{-}CH_3(g)$$

(a) Using the data in Appendix L, calculate ΔG° for the reaction at 25 °C. Is the reaction predicted to be product-favored under standard conditions?

(b) Calculate K_p from ΔG°_{rxn}. Comment on the connection between the sign of ΔG° and the magnitude of K_p.

41. Use the data in Appendix L to calculate ΔG° and K_p at 25 °C for the reaction

$$2\ HBr(g) + Cl_2(g) \rightleftharpoons 2\ HCl(g) + Br_2(\ell)$$

Comment on the connection between the sign of ΔG° and the magnitude of K_p.

GENERAL QUESTIONS

42. Sodium reacts violently with water according to the equation

$$Na(s) + H_2O(\ell) \longrightarrow NaOH(aq) + \tfrac{1}{2}\ H_2(g)$$

First predict the signs of ΔH° and ΔS° for the reaction and then verify your prediction with a calculation.

43. Calculate the entropy change involved in the formation of 1.0 mol of each of the following gaseous hydrocarbons under standard conditions from carbon and hydrogen. (Use graphite as the standard state of carbon.)

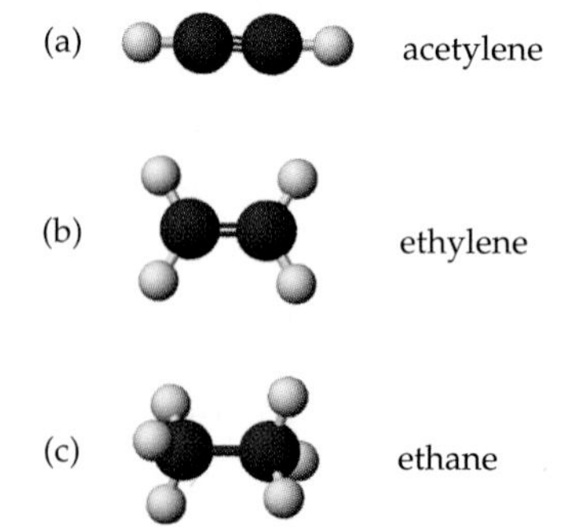

What trend do you see in these values? Does ΔS° increase or decrease on adding H atoms?

44. Yeast produces ethanol by the fermentation of glucose, the basis for the production of most alcoholic beverages:

$$\underset{\text{glucose}}{C_6H_{12}O_6(aq)} \longrightarrow 2\ \underset{\text{ethanol}}{C_2H_5OH(\ell)} + 2\ CO_2(g)$$

Calculate ΔH°, ΔS°, and ΔG° for the reaction. Is the reaction product- or reactant-favored? (In addition to the thermodynamic values in Appendix L, you will need the following for $C_6H_{12}O_6$(aq): $\Delta H_f^\circ = -1260.0$ kJ/mol; $S^\circ = 289$ J/K · mol; and $\Delta G_f^\circ = -918.8$ kJ/mol.)

45. Elemental boron, in the form of thin fibers, can be made by reducing a boron halide with H_2:

$$BCl_3(g) + \tfrac{3}{2}\ H_2(g) \longrightarrow B(s) + 3\ HCl(g)$$

The standard enthalpy of formation of BCl_3 is −403.8 kJ/mol, and its entropy, S°, is 290 J/K · mol. The entropy, S°, for B(s) is 5.86 J/K · mol. Calculate ΔH°, ΔS°, and ΔG° at 25 °C for this reaction. Is it predicted to be product-favored under standard conditions? If product-favored, is it enthalpy-driven or entropy-driven?

46. The equilibrium constant, K_p, for $N_2O_4(g) \rightleftharpoons 2\ NO_2(g)$ is 0.14 bar at 25 °C. Calculate ΔG° from this constant, and compare your calculated value with that determined from the ΔG_f° values in Appendix L.

47. Nitric oxide and chlorine combine to produce nitrosyl chloride, NOCl:

$$NO(g) + \tfrac{1}{2}\ Cl_2(g) \longrightarrow NOCl(g)$$

Calculate the equilibrium constant, K_p, for the reaction. Is the reaction product- or reactant-favored?

48. The equilibrium constant for the butane $\rightleftharpoons$ isobutane equilibrium at 25 °C is 2.5. Calculate ΔG°_{rxn} at this temperature in kilojoules per mole.

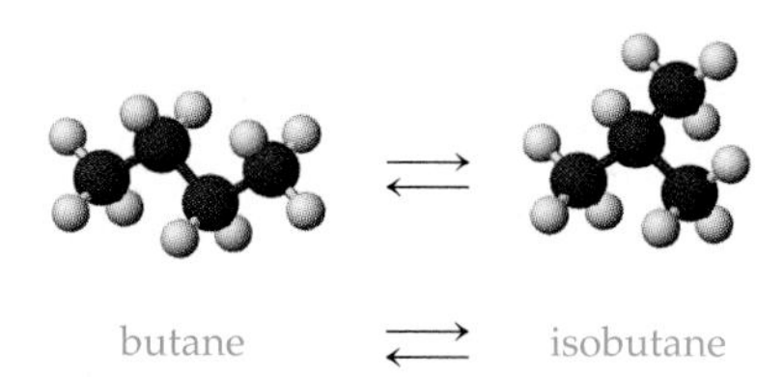

49. Some metal oxides can be decomposed to the metal and oxygen under reasonable conditions. Is the decomposition of silver(I) oxide product-favored at 25 °C?

$$2\ Ag_2O(s) \longrightarrow 4\ Ag(s) + O_2(g)$$

If not, can it become so if the temperature is raised? At what temperature is the reaction product-favored?

50. Iodine, I_2, dissolves readily in carbon tetrachloride with an enthalpy change that is approximately zero:

$$I_2(s) \longrightarrow I_2\ (\text{in } CCl_4 \text{ solution}) \qquad \Delta H^\circ_{rxn} \approx 0$$

What is the sign of ΔG°_{rxn}? Is the dissolving process entropy-driven or enthalpy-driven? Explain briefly.

51. A crucial reaction for the production of synthetic fuels is the conversion of coal to H_2 with steam:

$$C(s) + H_2O(g) \longrightarrow CO(g) + H_2(g)$$

   (a) Calculate ΔG°_{rxn} for this reaction at 25 °C assuming C(s) is graphite.
   (b) Calculate K_p for the reaction at 25 °C.
   (c) Is the reaction predicted to be product-favored under standard conditions? If not, at what temperature will it become so?

52. Calculate ΔG°_{rxn} for the decomposition of sulfur trioxide to sulfur dioxide and oxygen:

$$2\ SO_3(g) \longrightarrow 2\ SO_2(g) + O_2(g)$$

   (a) Is the reaction product-favored under standard conditions at 25 °C?
   (b) If the reaction is not product-favored at 25 °C, is there a temperature at which it will become so?
   (c) What is the equilibrium constant for the reaction at 1500 °C? (Make the assumption that ΔH° and ΔS° are not temperature-dependent.)

53. Methanol is relatively inexpensive to produce. Some consideration has been given to using it as a precursor to other fuels such as methane, which could be obtained by the decomposition of the alcohol:

$$CH_3OH(\ell) \longrightarrow CH_4(g) + \tfrac{1}{2}\ O_2(g)$$

   (a) What are the sign and magnitude of the entropy change for the reaction? Does the sign of ΔS° agree with your expectation? Explain briefly.
   (b) Is the reaction product-favored under standard conditions at 25 °C? Use thermodynamic values to prove your answer.
   (c) If not product-favored at 25 °C, at what temperature does the reaction become so?

54. A cave in Mexico was recently discovered to have some interesting chemistry (see Chapter 22). Hydrogen sulfide, H_2S, reacts with oxygen in the cave to give sulfuric acid, which drips from the ceiling in droplets with a pH less than 1. If the reaction occurring is

$$H_2S(g) + 2\ O_2(g) \longrightarrow H_2SO_4(\ell)$$

calculate ΔH°, ΔS°, and ΔG°. Is the reaction product-favored? Is it enthalpy- or entropy-driven?

55. Wet limestone is used to scrub SO_2 gas from the exhaust gases of power plants. One possible reaction gives hydrated calcium sulfite,

$$CaCO_3(s) + SO_2(g) + \tfrac{1}{2}\ H_2O(\ell) \longrightarrow CaSO_3 \cdot \tfrac{1}{2}\ H_2O(s) + CO_2(g)$$

and another reaction gives hydrated calcium sulfate:

$$CaCO_3(s) + SO_2(g) + \tfrac{1}{2}\ H_2O(\ell) + \tfrac{1}{2}\ O_2(g) \longrightarrow CaSO_4 \cdot \tfrac{1}{2}\ H_2O(s) + CO_2(g)$$

Which is the more product-favored reaction? Use the data in the table and any other needed in Appendix L.

	$CaSO_3 \cdot \frac{1}{2}\ H_2O(s)$	$CaSO_4 \cdot \frac{1}{2}\ H_2O(s)$
ΔH°_f (kJ/mol)	−1311.7	−1574.65
S° (J/K · mol)	121.3	134.8

CONCEPTUAL QUESTIONS

56. Calculate the entropy change for dissolving HCl gas in water. Is the sign of ΔS° what you expected? Why or why not?

57. Why is the value of S° of Br_2 greater than that of I_2 at 25 °C?

58. Sulfur undergoes a phase transition between 80 °C and 100 °C:

$$S_8(\text{rhombic}) \longrightarrow S_8(\text{monoclinic})$$

$$\Delta H^\circ_{rxn} = 3.213\ \text{kJ/mol} \qquad \Delta S^\circ_{rxn} = 8.7\ \text{J/K}$$

   (a) Estimate ΔG° for the phase transition at 80 °C and 100 °C. What do these results tell you about the stability of the two forms of sulfur at each of these temperatures?
   (b) Calculate the temperature at which $\Delta G^\circ = 0$. What is the significance of this temperature?

59. For each of the following processes, give the algebraic sign of ΔH°, ΔS°, and ΔG°. No calculations are necessary; use your common sense.
   (a) The splitting of liquid water to give gaseous oxygen and hydrogen, a process that requires a considerable amount of energy.
   (b) The explosion of dynamite, a mixture of nitroglycerin, $C_3H_5N_3O_9$, and diatomaceous earth. The explosive decomposition gives gaseous products such as water, CO_2, and others; much heat is evolved.
   (c) The combustion of gasoline in the engine of your car, as exemplified by the combustion of octane:

$$2\ C_8H_{18}(g) + 25\ O_2(g) \longrightarrow 16\ CO_2(g) + 18\ H_2O(g)$$

CHALLENGING PROBLEMS

60. Titanium(IV) chloride, $TiCl_4$, is produced by the reaction of carbon and chlorine with TiO_2:

$$TiO_2(s) + C(s) + 2\ Cl_2(g) \longrightarrow TiCl_4(\ell) + CO_2(g)$$

The reaction can be thought of as occurring in two steps: the reduction of TiO_2 with carbon

$$TiO_2(s) + C(s) \longrightarrow Ti(s) + CO_2(g)$$

and the oxidation of titanium metal by chlorine to form the product:

$$Ti(s) + 2\ Cl_2(g) \longrightarrow TiCl_4(\ell)$$

   (a) Calculate ΔG°_{rxn} and K_p for each of the two steps.
   (b) Calculate ΔG°_{rxn} and K_p for the overall process. How are these related to the values of the free energy change and K_p for the two steps in the process?
   (c) Is the overall reaction enthalpy- or entropy-driven?

61. Insoluble silver chloride dissolves in the presence of excess chloride ion. The equation for the overall reaction is

$$AgCl(s) + Cl^-(aq) \rightleftharpoons AgCl_2^-(aq)$$

(a) Show that the overall reaction is the sum of two others: the ionization of AgCl(s) to give silver(I) and chloride ions, and the formation of $AgCl_2^-(aq)$ from $Ag^+(aq)$ and $Cl^-(aq)$ ions.
(b) Calculate the equilibrium constant for the overall process from the equilibrium constants for the two steps. (Constants for the two steps are found in Appendices J and K.)
(c) Calculate the free energy change for each step and for the overall reaction. Required values are in Appendix L or in the following list:

Species	**ΔG_f° (kJ/mol)**
$Ag^+(aq)$	+77.1
$Cl^-(aq)$	−131.2
$AgCl_2^-(aq)$	−215.4

How is $K_{overall}$ related to $\Delta G^\circ_{overall}$? Is the overall reaction product-favored?

62. Silver(I) oxide can be formed by the reaction of silver metal and oxygen:

$$4\ Ag(s) + O_2(g) \longrightarrow 2\ Ag_2O(s)$$

(a) Calculate ΔH°_{rxn}, ΔS°_{rxn}, and ΔG°_{rxn} for the reaction.
(b) What is the pressure of O_2 (in bars) in equilibrium with Ag and Ag_2O at 25 °C?
(c) At what temperature is the pressure of O_2 in equilibrium with Ag and Ag_2O equal to 1.00 bar?

63. Mercury vapor is dangerous because it can be ingested into the lungs. We wish to estimate the vapor pressure of mercury at two different temperatures from the data in the table.

Substance	**ΔH_f°, kJ/mol**	**ΔG_f°, kJ/mol**	**S°, J/K · mol**
$Hg(\ell)$	0	0	76.02
$Hg(g)$	61.317	31.85	174.85

Estimate the temperature at which K_p for the process $Hg(\ell) \longrightarrow Hg(g)$ is equal to (a) 1.00 bar and (b) (1/760) bar. What is the vapor pressure at each of these temperatures? (Experi-mental vapor pressures are 1 atm at 356.6 °C and 1 mm Hg at 126.2 °C.) (*Note*: The temperature at which P = 1.00 bar can be calculated from thermodynamic data. To find the other temperature you will need to use the temperature for P = 1.00 atm and the Clausius-Clapeyron equation on page 604.)

Summary Question

64. Phenol, C_6H_5OH, is widely used to manufacture phenol-formaldehyde polymers (known as phenolics).

Phenol–formaldehyde polymer. When a mixture of phenol and formaldehyde (HCHO) is treated with concentrated HCl, a polymer begins to grow. These polymers are thermosetting polymers. Bakelite is an example of a phenol–formaldehyde polymer. *(C. D. Winters)*

(a) Suppose phenol can be made from the reaction of benzene and water:

$$C_6H_6(\ell) + H_2O(\ell) \longrightarrow C_6H_5OH(s) + H_2(g)$$
$$\Delta H_f^\circ = -165.1\ kJ/mol$$
$$S^\circ = 144.0\ J/K \cdot mol$$

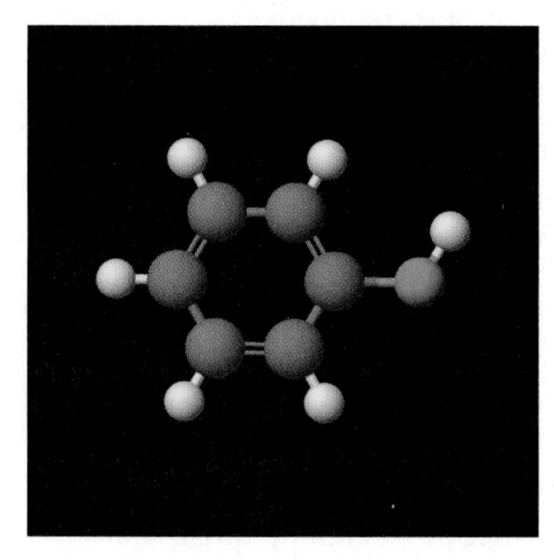

A model of phenol, C_6H_5OH.

Is this reaction predicted to be product-favored under standard conditions at 25 °C? Is the reaction enthalpy- or entropy-driven?
(b) What is the C—O—H angle in phenol?
(c) What is the hybridization of the C atoms in phenol? The O atom?
(d) Phenol is a weak acid in aqueous solution. What is the pH of a 0.012 M solution?

INTERACTIVE GENERAL CHEMISTRY CD-ROM

Screen 20.2 Reaction Spontaneity (Thermodynamics and Kinetics)

This screen is an expansion of Screen 6.3. You might wish to review that screen.

(a) What is the relationship between the terms "spontaneous" and "product-favored?"
(b) If gaseous H_2 and O_2 are carefully mixed and left alone, they will remain intact for millions of years. Is this "stability" a function of thermodynamics or of kinetics?

Screen 20.3 Directionality of Reactions

(a) Which of the dispersal mechanisms involves the concept of enthalpy change?
(b) Is energy dispersed during the process described in the *Chemical Puzzler* (Screen 20.1)?

$$NH_4NO_3(s) \longrightarrow NH_4NO_3(aq)$$

Screen 20.4 Entropy

(a) Which of the dispersal mechanisms described on Screen 20.3 involves a change in a system's entropy?
(b) Why does the entropy of a substance increase with temperature?

Screen 20.6 The Second Law of Thermodynamics

What is the second law of thermodynamics? How is it illustrated by the reaction of H_2 with O_2?

Screen 20.7 Gibbs Free Energy

Why are Gibbs free energy changes calculated for reactions?

Screen 20.8 Free Energy and Temperature

(a) Are reactions that occur only at high temperature but not at low temperature enthalpy-favored, entropy-favored, or both? Or, can't you tell?
(b) Are reactions that occur only at low temperature but not at high temperature enthalpy-favored, entropy-favored, or both? Or, can't you tell?

Screen 20.9 Thermodynamics and the Equilibrium Constant

What is the relationship between $\Delta G°$ for a reaction and the reaction's equilibrium constant, K? Does K increase or decrease as $\Delta G°$ becomes a more negative number?

Chapter 21

Principles of Reactivity: Electron Transfer Reactions

(C. D. Winters)

A Chemical Puzzler

Batteries are used in a wider and wider variety of appliances and are beginning to be used for automobiles. Fuel cells, a related technology, have been used extensively in space vehicles and are now being developed for automotive use. How do batteries and fuel cells work? What do they have in common with the reaction of zinc and hydrochloric acid in Figure 21.9, page 965?

In 1994, the California Air Resources Board determined that, by 1998, 2% of new cars and light trucks in California are to be zero-emissions vehicles (ZEVs). That is, they should give off absolutely no volatile organic compounds, nitrogen oxides, or carbon monoxide. By 2001, 5% of the vehicles should be ZEVs, and 10% must meet that requirement by 2003. About a dozen states—including such populous states as Maryland, Massachusetts, New Jersey, and New York—later indicated that they would follow California's lead. How will this be accomplished? It is likely that electric vehicles (EVs) will be the only type of vehicle capable of meeting the California standards in the near future, but in the longer term cars with fuel cells may come into use (Figure 21.1). As a result, the California standards have given impetus to auto manufacturers and technology companies around the world to develop new methods of supplying electric power. This can only be done by batteries or fuel cells, which in turn depend on electron transfer reactions and electrochemistry—the subject of this chapter.

21.1 OXIDATION–REDUCTION REACTIONS

See the *Saunders Interactive General Chemistry CD-ROM,* Screen 21.2, Redox Reactions.

Oxidation–reduction reactions—also called *redox reactions*—occur by electron transfer and constitute a major class of chemical reactions (see Section 5.7). Because examples of redox reactions occur everywhere, you experience their consequences daily. Corrosion, for example, occurs by redox reactions. The iron and steel in cars, bridges, and buildings can oxidize to rust, and aluminum structures can corrode (Section 21.7). Many biological processes depend on electron transfer reactions. For example, the oxygen you take in as you breathe is converted ultimately to water and carbon dioxide. The oxidation number of the oxygen in the product molecules (H_2O and CO_2) is -2, so electrons must have been transferred to O_2 molecules to cause their reduction. Where did the electrons come from? At least in the final step they are transferred to O_2 from cytochromes, large iron-containing molecules (see Figure 2.19). Other biological electron transfer processes include the conversion of water to O_2 in green plants

(a)

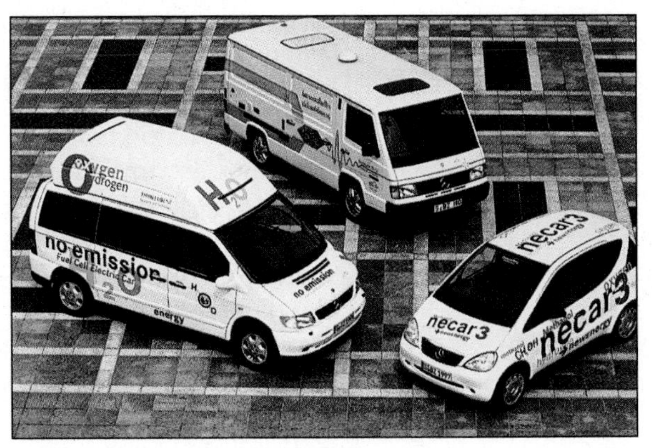

(b)

***Figure* 21.1 Battery- or fuel cell–powered vehicles.** (a) A battery-powered automobile now on sale in certain states. (b) Vehicles powered by fuel cells. *(a, General Motors; b, Ballard Power Systems)*

by photosynthesis, and the conversion of atmospheric N_2 by bacteria to a usable form of nitrogen, such as NH_4^+:

$$N_2(g) + 8\ H_3O^+(aq) + 6\ e^- \longrightarrow 2\ NH_4^+(aq) + 8\ H_2O\ (\ell)$$

A **battery** is an **electrochemical cell,** or a collection of such cells, that produces a current, or flow of electrons, at a constant voltage as a result of an electron transfer reaction. The power to run your calculator or computer or to start your car, or even to propel your ZEV, comes from a battery.

Redox reactions are important in manufacturing chemicals and producing metals. For example, many metals are commercially prepared or purified by the direct application of electricity in a process called **electrolysis,** the use of electric energy to produce a chemical change such as the reduction of copper ions to copper metal:

$$Cu^{2+}(aq) + 2\ e^- \longrightarrow Cu(s)$$

Other metals, such as iron, are prepared using chemical reducing agents (see page 932).

$$Fe_2O_3(s) + 3\ CO(g) \longrightarrow 2\ Fe(s) + 3\ CO_2(g)$$

You may wish to review Section 5.7 and recall the definitions of terms such as "oxidation," "reduction," "reducing agent," "oxidizing agent," and "oxidation number."

Chemists have only recently begun to understand the way in which electrons are transferred from one site to another in redox reactions. Moreover, the prevention of corrosion, the construction of fuel cells and more powerful batteries, and the plating of metals using electricity are now better understood. The general subject is fascinating because it applies to so many problems of practical interest.

To understand the general subject of electron transfer reactions, and electrochemistry in particular, we organize the subject as follows: Equations for electron transfer reactions can appear complicated at first glance, so we first describe some special techniques for balancing such equations. This is followed by a description of electrochemical cells in which an electric current is produced by an electron transfer reaction—cells that are sometimes called **voltaic cells,** or **galvanic cells.** This then leads to a discussion of batteries and fuel cells and then to the process of corrosion and its prevention. Finally, we describe electrolysis and some important industrial processes.

Voltaic cells are named for Count Alessandro Volta (1745–1827), who studied current-producing reactions, and galvanic cells are named for Dr. Luigi Galvani (1737–1798), who made early studies of animal electricity and produced electricity chemically.

Balancing Equations for Oxidation–Reduction Reactions

See the *Saunders Interactive General Chemistry CD-ROM,* Screen 21.3, Balancing Equations for Redox Reactions.

Balancing equations for redox reactions may appear to be a formidable task in some cases. Fortunately, though, there are systematic ways of doing so. One approach is illustrated by the equation for the reaction of aqueous copper(II) ions and metallic zinc (Figure 21.2). Here a piece of zinc is immersed in an aqueous solution of copper sulfate. After a time, the blue color of the aqueous Cu^{2+} ion fades, and copper metal "plates out," or forms a coating on the zinc strip. In addition, the zinc strip slowly disappears. What happened?

The fate of the Cu^{2+} ion in the test tube in Figure 21.2 is probably obvious; the blue color of aqueous copper(II) ion fades as it is reduced to copper metal. We depict this by the equation

$$Cu^{2+}(aq) + 2\ e^- \longrightarrow Cu(s)$$

Cu^{2+} gains electrons, is reduced, and is the oxidizing agent.

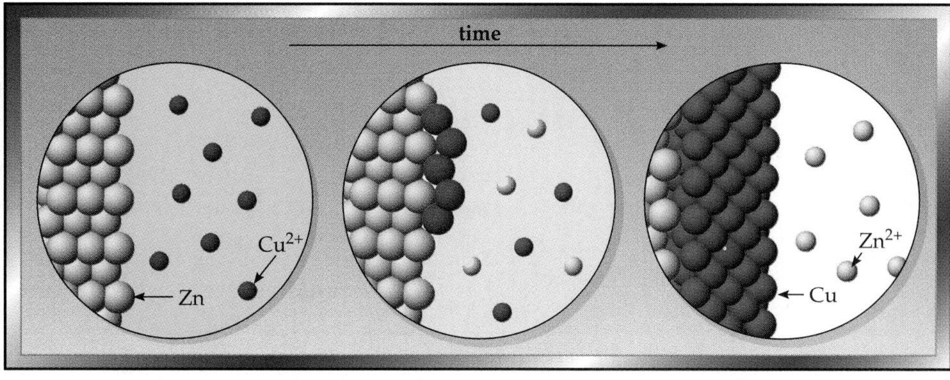

(a) (b)

Figure 21.2 An oxidation–reduction reaction. (a) A strip of zinc metal was placed in a solution of copper(II) sulfate *(left)*, and the zinc reacts with the copper(II) ions to give copper metal and zinc ions in solution:

$$Zn(s) + Cu^{2+}(aq) \longrightarrow Zn^{2+}(aq) + Cu(s)$$

Copper metal accumulates on the zinc strip, and the blue color of aqueous copper(II) ions fades as copper(II) ions disappear from solution *(middle and right)*. (b) Electrons are transferred from the zinc atoms to copper(II) ions. The oxidized zinc, now in the form of Zn^{2+} ions, enters solution, and the reduced copper atoms form a film of solid copper metal on the surface. *(Photo, C. D. Winters; model, S. M. Young)*

Because we observe that the zinc metal disappears, it must be the source of the electrons that cause reduction of Cu^{2+}, and we depict this by the equation

$$Zn(s) \longrightarrow Zn^{2+}(aq) + 2\,e^-$$

Zn loses electrons, is oxidized, and is the reducing agent.

Zinc is the reducing agent because it donates electrons and forms aqueous Zn^{2+} in the process.

The equation for the net chemical reaction occurring in the test tube is the sum of the equations for the two **half-reactions:** one for the oxidation of zinc and one for the reduction of copper(II) ion.

$$\begin{array}{lrcl} \textit{Oxidation:} & Zn(s) & \longrightarrow & Zn^{2+}(aq) + \cancel{2\,e^-} \\ \textit{Reduction:} & Cu^{2+}(aq) + \cancel{2\,e^-} & \longrightarrow & Cu(s) \\ \hline \textit{Net reaction:} & Zn(s) + Cu^{2+}(aq) & \longrightarrow & Cu(s) + Zn^{2+}(aq) \end{array}$$

Notice that the equation for each half-reaction is balanced for mass: one atom of each kind appears on each side of the equation. There is also a *charge balance,* that is, the algebraic sum of charges on one side of the equation equals the algebraic sum of the charges on the other side. (Here both sides have a net charge of 0.) Because the equation for the net reaction is the sum of the balanced equations for the half-reactions, the net equation is likewise balanced for mass and charge.

The preceding redox reaction illustrates the general approach to balancing equations for oxidation–reduction reactions, which is explored further in the following examples.

Example 21.1 Balancing an Equation for an Oxidation–Reduction Reaction

Balance the equation for the reaction of silver(I), Ag^+, with copper (see Figure 5.15):

$$Cu(s) + Ag^+(aq) \longrightarrow Ag(s) + Cu^{2+}(aq)$$

Solution

Step 1. *Recognize the reaction as an oxidation–reduction process.* Here the oxidation number for silver changes from +1 to 0, and that for copper changes from 0 to +2.

Step 2. *Separate the process into half-reactions.*

Reduction: $Ag^+(aq) \longrightarrow Ag(s)$ Ag^+ is the oxidizing agent.
Oxidation: $Cu(s) \longrightarrow Cu^{2+}(aq)$ Cu is the reducing agent.

Step 3. *Balance each half-reaction for mass.* Both half-reactions are already balanced for mass; the same number of atoms of each element appears on each side.

Step 4. *Balance each half-reaction for charge.* The equation is balanced for charge by adding electrons to the more positive side of the half-reaction:

$Ag^+(aq) + e^- \longrightarrow Ag(s)$ Each Ag^+ ion acquires an electron.

$Cu(s) \longrightarrow Cu^{2+}(aq) + 2\ e^-$ Each Cu atom loses two electrons.

In each case the net electric charge on each side of the equation is 0.

Step 5. *Multiply each half-reaction by an appropriate factor.* The reducing agent must donate as many electrons as the oxidizing agent acquires. Here one atom of copper produces two electrons, whereas one Ag^+ ion acquires one electron. Because two Ag^+ ions are required to consume the two electrons produced by a Cu atom, the Ag^+/Ag half-reaction is multiplied by 2.

$$2\ Ag^+(aq) + 2\ e^- \longrightarrow 2\ Ag(s)$$

Step 6. *Add the half-reactions to produce the overall balanced equation.*

$$Cu(s) \longrightarrow Cu^{2+}(aq) + \cancel{2\ e^-}$$
$$2\ Ag^+(aq) + \cancel{2\ e^-} \longrightarrow 2\ Ag(s)$$
$$Cu(s) + 2\ Ag^+(aq) \longrightarrow 2\ Ag(s) + Cu^{2+}(aq)$$

Step 7. *Check the overall equation to ensure that both mass and charge balance.* Here two silver atoms or ions and one copper atom or ion appear on each side, and the net charge on each side is 2+. The equation is balanced.

Exercise 21.1 Balancing an Equation for an Oxidation–Reduction Reaction

Balance the equation

$$Cr^{2+}(aq) + I_2(aq) \longrightarrow Cr^{3+} + I^-(aq)$$

Write the balanced half-reactions and the balanced net ionic equation. Identify the oxidizing agent, the reducing agent, the substance oxidized, and the substance reduced.

A problem often encountered when balancing equations for reactions in aqueous solution is that water and either hydrogen ion or hydroxide ion can enter into the reaction. In acidic conditions, H^+ may be a reactant or a product; in basic conditions OH^- ions can also participate. It may also be necessary to use the *pair* of species H^+ and H_2O to balance equations for reactions in acid solution. Similarly, the *pair* OH^- and H_2O may be needed to balance equations for reactions in basic solution. Examples 21.2 through 21.4 show how to determine when these species are needed and how to place them in the equation.

We balance equations using H^+ rather than H_3O^+ because it is much simpler. If desired, the equation can be adjusted later to include hydronium ions by adding the same number of H_2O molecules to each side of the equation.

Example 21.2 Balancing Equations for Oxidation–Reduction Reactions in Acid Solution

Balance the net ionic equation for the reaction of the dioxovanadium(V) ion, VO_2^+, with zinc in acid solution. (The entire sequence of reactions occurring when this ion is reduced by zinc is illustrated in Figure 21.3.)

$$VO_2^+(aq) + Zn(s) \longrightarrow VO^{2+}(aq) + Zn^{2+}(aq)$$

Solution

Step 1. *Recognize the reaction as an oxidation–reduction process.* The oxidation number of V changes from +5 in VO_2^+ to +4 in VO^{2+}. The oxidation number of Zn changes from 0 in the metal to +2 in Zn^{2+}.

Step 2. *Separate the overall process into half-reactions.*

Oxidation: $Zn(s) \longrightarrow Zn^{2+}(aq)$ Zn is the reducing agent.
Reduction: $VO_2^+(aq) \longrightarrow VO^{2+}(aq)$ VO_2^+ is the oxidizing agent.

(a)

(b)

(c)

(d)

***Figure* 21.3 The reduction of the vanadate ion, VO_2^+, by zinc.** (a) Ammonium metavanadate, $(NH_4)_2VO_3$, is dissolved in water. On adding sulfuric acid, an acid–base reaction occurs to give the yellow VO_2^+ ion:

$$VO_3^-(aq) + 2\ H_3O^+(aq) \longrightarrow VO_2^+(aq) + 3\ H_2O(\ell)$$

(b) When zinc metal is added, the VO_2^+ ion is reduced to the blue VO^{2+} ion. See Example 21.2 for the balanced equation. (c) Further reduction occurs to give the green V^{3+} ion and finally (d) to produce the violet V^{2+} ion. *(C. D. Winters)*

Step 3. *Balance each half-reaction for mass.* Begin by balancing all atoms except H and O. The atoms H and O are always the last to be balanced because they often appear in more than one reactant or product.

Zinc half-reaction: $Zn(s) \longrightarrow Zn^{2+}(aq)$

This half-reaction is already balanced for mass.

Vanadium half-reaction: $VO_2^{+}(aq) \longrightarrow VO^{2+}(aq)$

The V atoms in this half-reaction are already balanced. An oxygen-containing species must be added to the right side to achieve an O atom balance, however.

$$VO_2^{+}(aq) \longrightarrow VO^{2+}(aq) + \text{(need 1 O atom)}$$

In acid solution, add H_2O to the side requiring O atoms, one H_2O molecule for each O atom required:

$$VO_2^{+}(aq) \longrightarrow VO^{2+}(aq) + H_2O(\ell)$$

This means there are now two unbalanced H atoms on the right. Because the reaction occurs in acidified solution, H^+ ions are present and can be used to balance the equation. In acid solution, a mass balance for H may be achieved by adding H^+ to the side of the equation deficient in H atoms. Here two H^+ ions are added to the left side:

$$2\,H^{+}(aq) + VO_2^{+}(aq) \longrightarrow VO^{2+}(aq) + H_2O(\ell)$$

Step 4. *Balance the half-reactions for charge.*

Zinc half-reaction: $Zn(s) \longrightarrow Zn^{2+}(aq) + 2\,e^-$

The mass-balanced VO_2^+ equation has a net charge of 3+ on the left side and 2+ on the right. Therefore, 1 e^- is added to the more positive left side.

Vanadium half-reaction: $1\,e^- + 2\,H^{+}(aq) + VO_2^{+}(aq) \longrightarrow VO^{2+}(aq) + H_2O(\ell)$

Step 5. *Multiply the half-reactions by appropriate factors so that the reducing agent donates as many electrons as the oxidizing agent consumes.* Here the reducing agent supplies 2 mol of electrons and the oxidizing agent consumes 1 mol of electrons. Therefore, the oxidizing half-reaction must be multiplied by 2. Now 2 mol of the oxidizing agent (VO_2^+) are available to consume the 2 mol of electrons provided per mole of the reducing agent (Zn):

$$Zn(s) \longrightarrow Zn^{2+}(aq) + 2\,e^-$$
$$2[1\,e^- + 2\,H^{+}(aq) + VO_2^{+}(aq) \longrightarrow VO^{2+}(aq) + H_2O(\ell)]$$

Step 6. *Add the half-reactions to give the balanced, overall equation.*

$$Zn(s) \longrightarrow Zn^{2+}(aq) + \cancel{2\,e^-}$$
$$\cancel{2\,e^-} + 4\,H^{+}(aq) + 2\,VO_2^{+}(aq) \longrightarrow 2\,VO^{2+}(aq) + 2\,H_2O(\ell)$$
$$\overline{Zn(s) + 4\,H^{+}(aq) + 2\,VO_2^{+}(aq) \longrightarrow Zn^{2+}(aq) + 2\,VO^{2+}(aq) + 2\,H_2O(\ell)}$$

Step 7. *Check the final result to ensure mass and charge balance.*

Mass balance: 1 Zn, 2 V, 4 H, and 4 O
Charge balance: Each side has a net charge of 6+.

Example 21.3 Balancing Equations for Oxidation–Reduction Reactions in Acid Solution

Balance the net ionic equation for the reaction of the organic compound ethanol, C_2H_5OH, with the dichromate ion in acid solution (see Figure 5.20).

$$C_2H_5OH(aq) + Cr_2O_7^{2-}(aq) \longrightarrow CH_3CO_2H(aq) + Cr^{3+}(aq)$$

ethanol; dichromate ion; orange-red; acetic acid; chromium(III) ion; green

Solution

Step 1. *Recognize the reaction as an oxidation–reduction process.* Here Cr changes from +6 to +3 (so $Cr_2O_7^{2-}$ is reduced), and C changes from −2 to 0 (and so C_2H_5OH is oxidized).

Step 2. *Break the overall equation into half-reactions.*

Oxidation: $C_2H_5OH(aq) \longrightarrow CH_3CO_2H(aq)$ — C_2H_5OH is the reducing agent.
Reduction: $Cr_2O_7^{2-}(aq) \longrightarrow Cr^{3+}(aq)$ — $Cr_2O_7^{2-}$ is the oxidizing agent.

Step 3. *Balance each half-reaction for mass.* Here we start with the C_2H_5OH half-reaction. The C atoms are balanced, but O and H are not. The first step is therefore to add H_2O to the O-deficient left side,

$$C_2H_5OH(aq) + H_2O(\ell) \longrightarrow CH_3CO_2H(aq)$$

and then to balance H by adding H^+ to the right side of the equation. There are 8 H atoms on the left and 4 on the right. Therefore, 4 H^+ ions are added to the right.

$$C_2H_5OH(aq) + H_2O(\ell) \longrightarrow CH_3CO_2H(aq) + 4\,H^+(aq)$$

Next, turn to the $Cr_2O_7^{2-}$ half-reaction. The Cr atoms should be balanced first,

$$Cr_2O_7^{2-}(aq) \longrightarrow 2\,Cr^{3+}(aq)$$

and then H_2O is added to the O-deficient right side:

$$Cr_2O_7^{2-}(aq) \longrightarrow 2\,Cr^{3+}(aq) + 7\,H_2O(\ell)$$

Finally, the H atoms are balanced by placing H^+ on the H-deficient left side:

$$14\,H^+(aq) + Cr_2O_7^{2-}(aq) \longrightarrow 2\,Cr^{3+}(aq) + 7\,H_2O(\ell)$$

Step 4. *Balance the half-reactions for charge.*

$$C_2H_5OH(aq) + H_2O(\ell) \longrightarrow CH_3CO_2H(aq) + 4\,H^+(aq) + 4\,e^-$$
$$6\,e^- + 14\,H^+(aq) + Cr_2O_7^{2-}(aq) \longrightarrow 2\,Cr^{3+}(aq) + 7\,H_2O(\ell)$$

Addition of electrons to each half-reaction confirms that C_2H_5OH is the reducing agent (electron donor) and $Cr_2O_7^{2-}$ is the oxidizing agent (electron acceptor).

Step 5. *Multiply the balanced half-reactions by appropriate factors.*

$$3[C_2H_5OH(aq) + H_2O(\ell) \longrightarrow CH_3CO_2H(aq) + 4\,H^+(aq) + 4\,e^-]$$
$$2[6\,e^- + 14\,H^+(aq) + Cr_2O_7^{2-}(aq) \longrightarrow 2\,Cr^{3+}(aq) + 7\,H_2O(\ell)]$$

Three moles of ethanol produces 12 mol of electrons, which are then consumed by 2 mol of dichromate ion.

Step 6 *Add the balanced half-reactions.*

$$3\ C_2H_5OH(aq) + 3\ H_2O(\ell) \longrightarrow 3\ CH_3CO_2H(aq) + 12\ H^+(aq) + \cancel{12\ e^-}$$
$$\cancel{12\ e^-} + 28\ H^+(aq) + 2\ Cr_2O_7^{2-}(aq) \longrightarrow 4\ Cr^{3+}(aq) + 14\ H_2O(\ell)$$

$$3\ C_2H_5OH(aq) + 3\ H_2O(\ell) + 28\ H^+(aq) + 2\ Cr_2O_7^{2-}(aq) \longrightarrow 3\ CH_3CO_2H(aq) + 12\ H^+(aq) + 4\ Cr^{3+}(aq) + 14\ H_2O(\ell)$$

Step 7. *Eliminate common reactants and products.* Water and H^+ ions appear on both sides of the overall equation in Step 6. The equation can therefore be simplified by recognizing that a *net* of 11 H_2O appears on the right, and a *net* of 16 H^+ is on the left. The final, balanced net ionic equation is

$$3\ C_2H_5OH(aq) + 16\ H^+(aq) + 2\ Cr_2O_7^{2-}(aq) \longrightarrow 3\ CH_3CO_2H(aq) + 4\ Cr^{3+}(aq) + 11\ H_2O(\ell)$$

Step 8. *Check the final result for mass and charge balance.*

Mass balance: 6 C, 17 O, 34 H, and 4 Cr
Charge balance: 12+ on both sides

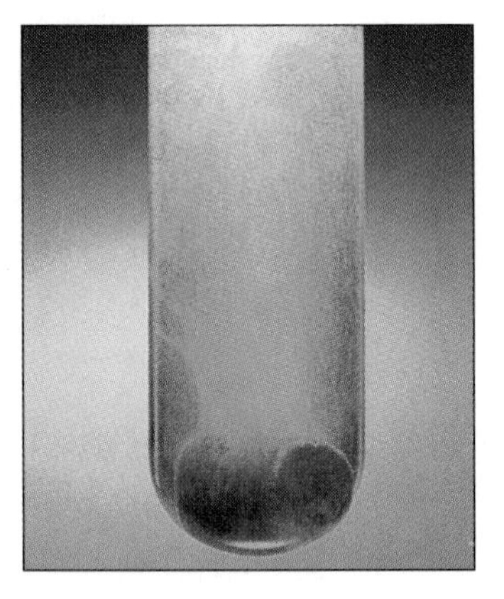

Metallic cobalt reacts with nitric acid to give pink $Co(NO_3)_3$, H_2, and nitrogen oxides—among them NO_2. (See Exercise 21.2) *(C. D. Winters)*

Exercise 21.2 Balancing Equations for Oxidation–Reduction Reactions in Acid Solution

Cobalt metal reacts with nitric acid to give a cobalt(III) salt and NO_2 gas. The unbalanced net ionic equation is

$$Co(s) + NO_3^-(aq) \longrightarrow Co^{3+}(aq) + NO_2(g)$$

1. Balance the equation for the reaction in acid solution.
2. Identify the oxidizing and reducing agents and the substance oxidized and the substance reduced.

Examples 21.2 and 21.3 illustrate the technique of balancing equations for redox reactions that occur in acid solution. Under those conditions H^+ ions or the H^+/H_2O pair can be used to achieve balanced equations. Conversely, in basic solution, only OH^- ions or the OH^-/H_2O pair can be used.

Example 21.4 Balancing Equations for Oxidation–Reduction Reactions in Basic Solution

As an introduction to balancing redox reactions in basic solutions, let us balance the half-reaction for the oxidation of SnO_2^{2-} in basic solution:

$$SnO_2^{2-}(aq) \longrightarrow SnO_3^{2-}(aq)$$

Solution

Step 1. *Verify that the reaction is an oxidation–reduction process.* The oxidation number of tin changes from +2 to +4.

Step 2. *Separate the overall reaction into half-reactions.*

$$SnO_2^{2-}(aq) \longrightarrow SnO_3^{2-}(aq)$$

A CLOSER LOOK — *Redox Reactions in Acidic and Basic Solution*

Why are some redox reactions involving metal oxoanions carried out in acidic aqueous solutions (Figure 21.3) whereas others are done in basic solution (Figure 21.4)? The answer is apparent when you think about some oxidizing agents or reducing agents as effectively releasing or requiring oxide ions. Understanding this idea will also help when balancing equations for such reactions.

Metal oxoanions such as MnO_4^- and $Cr_2O_7^{2-}$ are excellent oxidizing agents in acid solution. You can think of these ions as essentially shedding oxide ions as the metal ion is reduced:

$$MnO_4^-(aq) \longrightarrow Mn^{2+}(aq) + [4\ O^{2-}] + 5\ e^-$$

Because the oxide ion is such a strong base that it cannot exist in aqueous solution (see Table 17.4), the function of the acid is to remove oxide ion as water:

$$[O^{2-}] + 2\ H^+(aq) \longrightarrow H_2O(\ell)$$

Thus, when oxoanions are used as oxidizing agents, the reactions are usually done in acidic solutions.

When metals or metal ions are used as reducing agents, their oxidation number increases. Because transition metal ions with higher oxidation numbers are more stable in the form of oxoanions, the reaction must effectively supply oxide ions. For example, vanadium(II) ion can be a reducing agent, but oxide ion must be supplied to stabilize it as a vanadium(IV) or (V) ion:

$$V^{2+}(aq) + 2\ [O^{2-}] \longrightarrow VO_2^+(aq) + 3\ e^-$$

Again, however, oxide ions do not exist in aqueous solution, so hydroxide ions act as the source of the required oxide ions:

$$2\ OH^-(aq) \longrightarrow H_2O(\ell) + [O^{2-}]$$

Step 3. *Balance each half-reaction for mass.* The Sn atoms are balanced, but the left side is deficient in oxygen:

$$(1\ \text{O atom required}) + SnO_2^{2-}(aq) \longrightarrow SnO_3^{2-}(aq)$$

In basic solution, there are two O-containing species, OH^- and H_2O. However, OH^- is oxygen-rich compared with H_2O. (Half of the atoms are O in OH^-, whereas only one third of them are O in H_2O). Therefore, OH^- can be thought of as a supplier of oxide ion (and also of water), as suggested by the following *hypothetical,* balanced equation:

$$2\ OH^- \longrightarrow [O^{2-}] + H_2O$$

Two OH^- ions are therefore added to the O-deficient side to supply the one O atom needed, and H_2O is added to the other side to balance the H atoms:

$$2\ OH^-(aq) + SnO_2^{2-}(aq) \longrightarrow SnO_3^{2-}(aq) + H_2O(\ell)$$

Notice that two OH^- ions are required to supply one O atom and that one H_2O molecule must be added to the other side for every O atom supplied by OH^-.

Step 4. *Balance the half-reactions for charge:*

$$2\ OH^-(aq) + SnO_2^{2-}(aq) \longrightarrow SnO_3^{2-}(aq) + H_2O(\ell) + 2\ e^-$$

This confirms that SnO_2^{2-} is a reducing agent (electron donor).

Step 5. *Check the final results.*

Mass balance: 1 Sn, 4 O, and 2 H
Charge balance: 4− on both sides

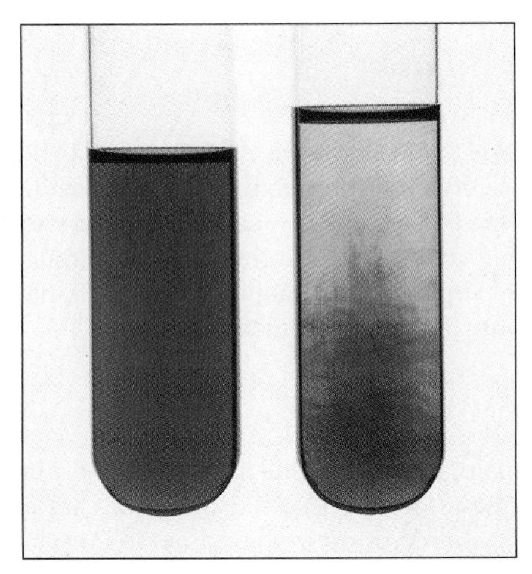

Figure 21.4 The reduction of permanganate ion by basic hydrogen peroxide. The purple permanganate ion, MnO_4^- *(left)*, is reduced to green manganate ion, MnO_4^{2-} *(right)*, by hydrogen peroxide in basic solution. *(C. D. Winters)*

Example 21.5 Balancing Equations for Oxidation–Reduction Reactions in Basic Solution

The purple permanganate ion, MnO_4^-, is reduced to the green manganate ion, MnO_4^{2-}, by hydrogen peroxide in basic solution. (Hydrogen peroxide, H_2O_2, exists as the HO_2^- ion in basic solution. See Figure 21.4.)

$$MnO_4^-(aq) + HO_2^-(aq) \longrightarrow MnO_4^{2-}(aq) + O_2(g)$$

The fact that the reaction occurs in basic solution means that OH^- ions and H_2O molecules are available to achieve a mass balance in O and H.

Solution

Step 1. *Verify that the reaction is an oxidation–reduction process.* The oxidation number of Mn changes from +7 to +6 (MnO_4^- is reduced) and that for O changes from −1 to 0 (the peroxide is oxidized).

Step 2. *Separate the overall reaction into half-reactions.*

Reduction half-reaction: $MnO_4^-(aq) \longrightarrow MnO_4^{2-}(aq)$
Oxidation half-reaction: $HO_2^-(aq) \longrightarrow O_2(g)$

Step 3. *Balance each half-reaction for mass.* The reaction involving the permanganate ion is already balanced for mass:

Reduction half-reaction: $MnO_4^-(aq) \longrightarrow MnO_4^{2-}(aq)$

In the peroxide reaction, however, the right side is deficient in H. In basic solution we have H_2O and OH^- available to balance an equation. The H_2O molecule is "richer" in H, so place that on the H-deficient side of the equation and OH^- on the other side:

Oxidation half-reaction: $OH^-(aq) + HO_2^-(aq) \longrightarrow O_2(g) + H_2O(\ell)$

In doing so, we have balanced not only the O atoms but also the H atoms.

Another way of balancing equations in basic solution is to treat the half-reaction as though it occurred in acid solution (using H^+ and H_2O to balance the equation). As a final step, add OH^- ions to both sides of the equation to "neutralize" any added H^+ ions. See *Problem-Solving Tips and Ideas 21.1.*

Step 4. *Balance the half-reactions for charge.*

Reduction half-reaction: $e^- + MnO_4^-(aq) \longrightarrow MnO_4^{2-}(aq)$
Oxidation half-reaction: $OH^-(aq) + HO_2^-(aq) \longrightarrow O_2(g) + H_2O(\ell) + 2\,e^-$

This confirms that MnO_4^- is the oxidizing agent (electron acceptor) and HO_2^- is the reducing agent (electron donor).

Step 5. *Multiply the half-reactions by appropriate factors to balance the number of electrons donated and accepted:*

$$2[e^- + MnO_4^-(aq) \longrightarrow MnO_4^{2-}(aq)]$$
$$OH^-(aq) + HO_2^-(aq) \longrightarrow O_2(g) + H_2O(\ell) + 2\,e^-$$

Step 6. *Add the half-reactions:*

$$\cancel{2\,e^-} + 2\,MnO_4^-(aq) \longrightarrow 2\,MnO_4^{2-}(aq)$$
$$OH^-(aq) + HO_2^-(aq) \longrightarrow O_2(g) + H_2O(\ell) + \cancel{2\,e^-}$$

$$OH^-(aq) + HO_2^-(aq) + 2\,MnO_4^-(aq) \longrightarrow 2\,MnO_4^{2-}(aq) + O_2(g) + H_2O(\ell)$$

Step 7. *Simplify by eliminating reactants and products common to both sides.* Not required here.

Step 8. *Check the final result.*

Mass balance: 2 Mn, 11 O, and 2 H
Charge balance: 4− on both sides

Exercise 21.3 Balancing Equations for Oxidation–Reduction Reactions in Basic Solution

Batteries based on the reduction of sulfur appear very promising. (See D. Peramunage and S. Licht: "A solid sulfur cathode for aqueous batteries," *Science*, Vol. 261, p. 1029, 1993.) One being studied currently involves the reaction of sulfur with aluminum in base:

$$Al(s) + S(s) \longrightarrow Al(OH)_3(s) + HS^-(aq)$$

1. Balance the equation, showing each balanced half-reaction (in basic solution).
2. Identify the oxidizing and reducing agents, the substance oxidized, and the substance reduced.

The preceding examples illustrate one method for balancing equations for redox reactions. The steps you go through are always the same, but *many variations* in detail exist, especially in the mass-balancing step, Step 3. Only a few of these variations can be illustrated here, so the best way to learn to balance redox equations is to practice.

Problem-Solving Tips and Ideas

21.1 Balancing Equations for Oxidation–Reduction Reactions

- Never just add O^{2-} ions, O atoms, or O_2 molecules to balance oxygen in an equation. (Molecular O_2 should only appear if it is known to be a reactant or product.)
- Oxide ion cannot exist in water. Nonetheless, it often appears as though O^{2-} is the product of a reduction half-reaction in an acid solution or is required in an oxidation reaction in a basic solution. To balance such half-reactions, you can treat them as though (a) the product O^{2-} ion is converted to H_2O in acid solution by reaction with H^+ ion ($O^{2-} + 2\ H^+ \longrightarrow H_2O$) or (b) the required O^{2-} ion is produced by OH^- in basic solution, with water as another product ($2\ OH^- \longrightarrow O^{2-} + H_2O$).
- You may only use the pairs H^+/H_2O in acid or OH^-/H_2O in base (or H^+ or OH^- alone if appropriate).
- Another method of handling reactions in basic solution is to treat the half-reaction as if it had occurred in acid and then add OH^- to both sides of the equation to "neutralize" the H^+. For example, for the half-reaction $SnO_2^{2-}(aq) \longrightarrow SnO_3^{2-}(aq)$, first balance the equation in acid solution,

$$H_2O(\ell) + SnO_2^{2-}(aq) \longrightarrow SnO_3^{2-}(aq) + 2\ H^+(aq) + 2\ e^-$$

and then add enough OH^- ions to both sides of the equation so that the H^+ ions are converted to water. Here two OH^- ions are added and the equation is balanced after eliminating excess water molecules:

$$H_2O(\ell) + 2\ OH^-(aq) + SnO_2^{2-}(aq) \longrightarrow SnO_3^{2-}(aq) + 2\ H^+(aq) + 2\ OH^-(aq) + 2\ e^-$$

$$2\ OH^-(aq) + SnO_2^{2-}(aq) \longrightarrow SnO_3^{2-}(aq) + H_2O(\ell) + 2\ e^-$$

- Never add H atoms or H_2 molecules to balance hydrogen. (Molecular H_2 should only appear if it is known to be a reactant or product.) Hydrogen balance is achieved with H^+ and H_2O in acid or with H_2O and OH^- in basic solution.
- Be sure to write all the charges on any ions involved. Failing to include the correct charge is the most common error seen on student papers.

21.2 CHEMICAL CHANGE LEADING TO AN ELECTRIC CURRENT

See the *Saunders Interactive General Chemistry CD-ROM,* Screen 21.4, Electrochemical Cells.

Metallic zinc reacts readily with aqueous copper(II) ion to produce aqueous zinc(II) ion and copper metal (see Figure 21.2).

$$Zn(s) + Cu^{2+}(aq) \longrightarrow Zn^{2+}(aq) + Cu(s)$$

This may be interesting, but it would also be useful if the electrons transferred from zinc to copper(II) ion could be employed to power a computer or car. The problem with just dropping a piece of zinc into a solution of copper(II) ions is that the electrons provided by the zinc move directly to the aqueous Cu^{2+} ions on contact. In order to use the reaction as the basis of a battery, zinc metal and Cu^{2+} ions must be placed in separate containers (Figure 21.5). Electrons can then pass from the zinc **electrode,** a conductor of electrons, out of the solution into the external wire, through the device to be powered, and onto an electrode dipping into the solution of Cu^{2+} ions. Copper metal is then "plated out" onto the electrode in the beaker containing $Cu^{2+}(aq)$.

An electrode conducts electrons into and out of a solution. It is most often a metal plate or wire or a piece of graphite.

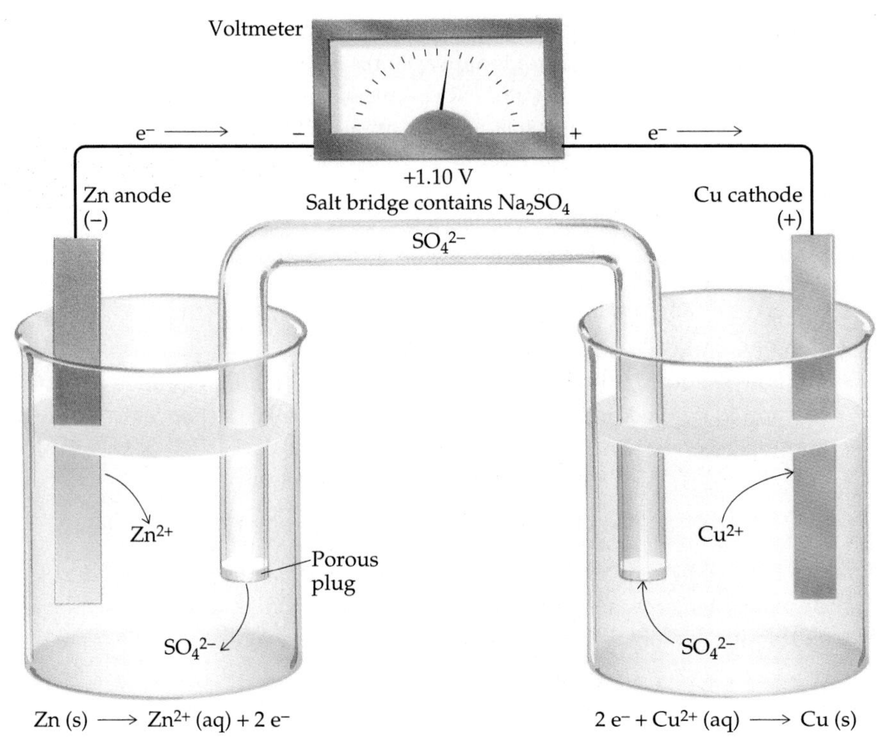

Figure **21.5 A voltaic cell using $Cu^{2+}(aq)/Cu(s)$ and $Zn^{2+}(aq)/Zn(s)$ half-cells.** A voltage of 1.10 V is generated if the cell is set up under the conditions shown. Electrons flow through the external wire from the Zn electrode (anode) to the Cu electrode (cathode). A salt bridge provides a connection between the half-cells for ion flow; thus, SO_4^{2-} ions flow from the copper to the zinc compartment.

The arrangement we just described works *only* if a **salt bridge,** a device for maintaining a balance of ion charges in the cell compartments, is also included. When the Zn electrode provides electrons to the wire, Zn^{2+}(aq) enters the solution in the Zn compartment (see Figure 21.5), and negative ions must be found to balance these newly generated positive charges. Similarly, the loss of Cu^{2+} ions in the copper compartment leaves behind negative ions that were associated with Cu^{2+}. Some way must be found for these negative ions, now in excess, to leave the solution. Thus, to achieve a balance of ion charges in each compartment, the negative ion concentration must *decrease* in the copper compartment and *increase* in the zinc compartment.

The salt bridge also allows cation flow. In Figure 21.5 cations (Na^+) move into the copper compartment (to replenish the Cu^{2+} ions consumed in the electrode reaction) and out of the zinc compartment (because Zn^{2+} cations are being added to the solution by the electrode reaction).

The function of the salt bridge is to allow ions to pass freely from the compartment where cations are being lost to the compartment where cations are being generated. The salt bridge contains an aqueous solution of Na_2SO_4, allowing $SO_4{}^{2-}$ to transfer (and Na^+ to transfer out of the compartment where the cation concentration is increasing to the one where it is decreasing). If the salt bridge were removed, ion flow would cease as would electron flow. The voltage indicated on the meter would be zero.

An oxidizing agent and a reducing agent arranged so they can react only if electrons flow through an outside conductor is called an electrochemical cell, a voltaic cell, or a battery. In *all* electrochemical cells **oxidation** occurs at the **anode,** and **reduction** occurs at the **cathode.** In a flashlight or car battery the anode is marked "−" because oxidation produces electrons that make the electrode negative. Conversely, the cathode is marked "+" because reduction consumes electrons, leaving the metal electrode positive. Important terms are summarized in Figure 21.6. Figure 21.7 illustrates an electrochemical cell built of substances that give rise to a product-favored electron transfer reaction and of an **electrolyte** that allows ion movement between electrodes.

Strictly speaking, a battery is a collection of cells that provides a current or flow of electrons at a constant voltage as a result of an oxidation–reduction reaction.

To remember that anode and cathode are paired with oxidation and reduction, note the alphabetic orders:

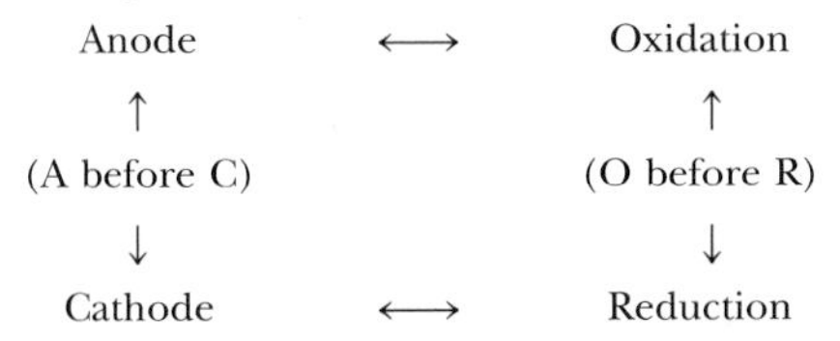

Finally, it is useful to point out that we often talk about electric circuits for the simple reason that electric charges—electrons or ions—always flow in a "circle." As shown in Figures 21.5 and 21.6 electrons move through the external circuit from reducing agent (Zn) to oxidizing agent (Cu^{2+}); the negative ions complete the circle of negative charge movement through the salt bridge (from Cu^{2+} to Zn^{2+}).

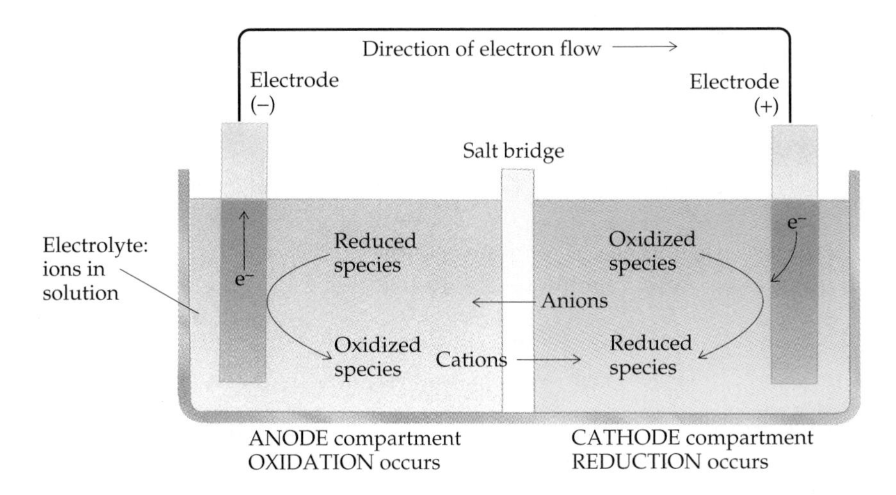

***Figure* 21.6 A summary of the terminology used in voltaic cells.** Notice that negative charge moves in a circle through the cell and external wire. Electrons move from the negative electrode (anode, site of oxidation) to the positive electrode (cathode, site of reduction) in the external wire, and anions move from the cathode compartment to the anode compartment in the cell. Ions of the electrolyte carry charge from one electrode to the other.

(a)

(b)

***Figure* 21.7 Simple electrochemical cells.** (a) An electrochemical cell can be made by inserting copper and zinc electrodes into almost any conductive material. (b) A cell made from a piece of Zn and a graphite rod in aqueous NH_4Cl produces 1.1 V. Neither could be used as a battery because the current is very small. *(C. D. Winters)*

Example 21.6 Electrochemical Cells

A simple voltaic cell has been assembled with Ni(s) and $Ni(NO_3)_2$(aq) in one compartment and Cd(s) and $Cd(NO_3)_2$(aq) in the other. An external wire connects the two electrodes, and a salt bridge containing $NaNO_3$ connects the two solutions. The net reaction is

$$Cd(s) + Ni^{2+}(aq) \longrightarrow Ni(s) + Cd^{2+}(aq)$$

What half-reaction occurs at each electrode? Which is the anode, and which is the cathode? What is the direction of electron flow in the external wire and of anion flow in the salt bridge?

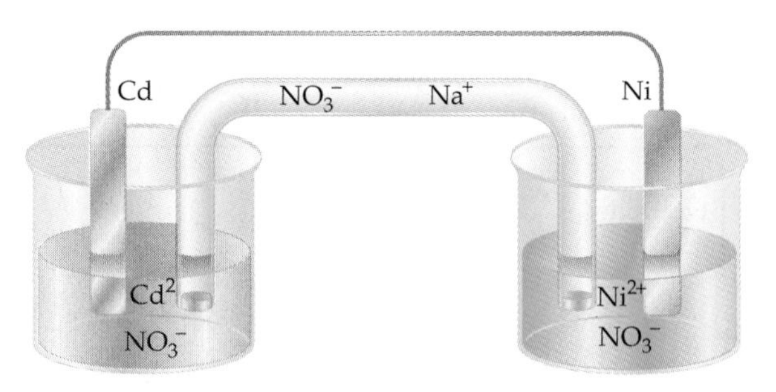

Solution Based on the net ionic equation we know Cd(s) is the reducing agent, and Ni^{2+}(aq) is the oxidizing agent. Thus, the half-reactions are

Anode, oxidation: $Cd(s) \longrightarrow Cd^{2+}(aq) + 2\,e^-$

Cathode, reduction: $Ni^{2+}(aq) + 2\,e^- \longrightarrow Ni(s)$

Electrons flow from their source (the oxidation of Cd) through the wire to the electrode where they are used to reduce Ni^{2+}(aq).

Because Cd^{2+} ions are formed in the anode compartment, anions must move into that compartment from the salt bridge. The Ni^{2+} concentration in the cathode compartment is decreasing, so anions move out of that compartment into the salt bridge. The "circle" of flow of negative charge is complete: Electrons flow from Cd to Ni, and negative ions move from Ni to Cd.

Although we have described the salt bridge as a device that allows anion flow to maintain electric neutrality, it also allows cation flow. Cations (here Na^+ from the salt bridge) move into the nickel compartment (to take the place of the Ni^{2+} cations consumed in the electrode reaction) and out of the cadmium compartment (because Cd^{2+} cations are being added to the solution by the electrode reaction).

Exercise 21.4 Electrochemical Cells

A voltaic cell has been assembled with the net reaction

$$Ni(s) + 2\ Ag^+(aq) \longrightarrow Ni^{2+}(aq) + 2\ Ag(s)$$

Give the half-reactions for this electron transfer process, indicating whether each is an oxidation or reduction and deciding which happens at the anode and which at the cathode. What is the direction of electron flow in an external wire connecting the two electrodes? If a salt bridge connecting the cell compartments contains KNO_3, what is the direction of flow of the nitrate ions?

21.3 ELECTROCHEMICAL CELLS AND POTENTIALS

See the *Saunders Interactive General Chemistry CD-ROM,* Screen 21.5, Electrochemical Cells and Potentials.

Electrons generated at the site of oxidation (the anode) of a cell are thought to be "driven," or "pushed," toward the cathode by an **electromotive force,** or **emf**. This force is due to the difference in electric potential energy of an electron at the two electrodes. Just as a ball rolls downhill in response to a difference in gravitational potential energy, an electron moves from an electrode of higher electric potential energy to one of lower electric potential energy. A moving ball can do work and so can moving electrons.

The quantity of electric work done is proportional to the number of electrons (the quantity of electric charge) that go from higher to lower potential energy and to the size of the potential energy difference:

$$\text{Electric work} = \text{charge} \times \text{potential energy difference}$$

Charge is measured in coulombs. A **coulomb (C)** is the quantity of charge that passes a point in an electric circuit when a current of one ampere flows for one second. The charge on a single electron is very small (1.6022×10^{-19} C), so it takes 6.24×10^{18} electrons to produce just one coulomb of charge. Electric potential energy difference is measured in volts. The volt is defined so that one joule of work is performed when one coulomb of charge passes through a potential difference of one volt:

$$\text{Volt} = \frac{1\ \text{joule}}{1\ \text{coulomb}} \quad \text{or} \quad \text{Work (joule)} = 1\ \text{volt} \times 1\ \text{coulomb}$$

Figure 21.8 Dry cell batteries. These dry cell batteries all produce 1.5 V. Some are larger than others, however, because they contain more oxidizing and reducing agent and so can produce more electrical work. *(C. D. Winters)*

The maximum work that can be accomplished by an electrochemical cell is equal to the product of the cell potential and the charge passing through the circuit. The cell potential depends in turn on the substances that make up the cell, whether they are gases or solutes in solution, and on their concentration. The quantity of charge depends on the quantity of reactants consumed. You know that if two batteries produce the same voltage, the one with the greater quantity of reactants is capable of doing more work (Figure 21.8).

The standard state for substances in solution is a concentration of 1.0 molal. However, for dilute solutions we use the approximation that 1.0 m $\approx$ 1.0 M. Recall that 1 atm = 1.013 bar.

Because the potential of a cell depends on the concentrations of reactants and products, **standard conditions** have been defined for electrochemical measurements. These are the same as for standard enthalpies and free energies of formation (see Sections 6.9 and 20.3). When the reactants and products are present as pure solids or in solution at a concentration of 1.0 M, or as gases at 1.0 bar, the measured cell potential is the **standard electrode potential, $E°$.** Cell potentials for product-favored electrochemical reactions have *positive* values, and, unless specified otherwise, are given at 25 °C (298 K). For example, the cell illustrated in Figure 21.3 has a potential of +1.10 V at 25 °C.

$E°$ and $\Delta G°$

The standard potential $E°$ is a quantitative measure of the tendency of the reactants in their standard states to proceed to products in their standard states and so is the standard free energy change for a reaction, $\Delta G°_{rxn}$. The exact relation between $E°$ and $\Delta G°_{rxn}$ is

$$\Delta G°_{rxn} = -nFE° \qquad [21.1]$$

The Faraday constant is named in honor of Michael Faraday, the first person to investigate quantitatively the relationship between chemistry and electricity.

where n is the number of moles of electrons transferred between oxidizing and reducing agents in a balanced redox reaction, and F is the **Faraday constant,** 9.6485309×10^4 J/V · mol.

The reaction of Zn(s) and Cu^{2+}(aq) produces a current, so you readily conclude that it is a product-favored reaction as written

$$Zn(s) + Cu^{2+}(aq) \longrightarrow Zn^{2+}(aq) + Cu(s) \qquad E° = +1.10 \text{ V}$$

Because product-favored reactions have a negative free energy change, $\Delta G°_{rxn}$, the negative sign in Equation 21.1 is in agreement with the fact that *all product-favored* electron transfer reactions have a *positive* $E°$.

Historical Perspective

Michael Faraday (1791–1867)

The terms "anion," "cation," "electrode," and "electrolyte" originated with Michael Faraday, one of the most influential men in the history of chemistry. Faraday was apprenticed to a bookbinder in London (England) when he was 13. This suited him, however, as he enjoyed reading the books sent to the shop for binding. By chance, one of these was a small book on chemistry, and his appetite for science was whetted. He soon began performing experiments on electricity, and in 1812 a patron of the shop invited Faraday to accompany him to a lecture at the Royal Institution by one of the most famous chemists of his day, Sir Humphry Davy (1778–1829). Faraday was so intrigued by Davy's lecture that he wrote to ask Davy for a position as an assistant. Faraday was accepted and began work in 1813. His work was very fruitful, and Faraday was so talented that he was made the director of the laboratory of the Royal Institution about 12 years later.

Michael Faraday *(Oesper Collection in the History of Chemistry/University of Cincinnati)*

It has been said that Faraday's contributions are so enormous that, had there been Nobel Prizes when he was alive, he would have received at least six. These could have been for discoveries such as

- Electromagnetic induction, which led to the invention of the first transformer and electric motor
- The laws of electrolysis
- The discovery of the magnetic properties of matter
- The discovery of benzene and other organic chemicals (which led to the development of important chemical industries)
- The discovery of the "Faraday effect" (the rotation of the plane of polarized light by a magnetic field)
- The introduction of the concept of electric and magnetic fields

In addition to making discoveries that had profound effects on science, Faraday was an educator. He wrote and spoke about his work in memorable ways, especially in lectures to the general public that helped to popularize science.

If the direction of the reaction is reversed, the sign of ΔG°_{rxn} and so the sign of E° are reversed. Thus, if we write the equation for the reduction of Zn^{2+} by Cu,

When a reaction is reversed, the magnitude of ΔG° and E° remains the same, but their signs are reversed.

$$\mathrm{Cu(s)} + \mathrm{Zn^{2+}(aq)} \longrightarrow \mathrm{Cu^{2+}(aq)} + \mathrm{Zn(s)} \qquad E^\circ = -1.10\ \mathrm{V}$$

this equation depicts a reaction that is reactant-favored. It must have a positive ΔG°_{rxn} and a negative E°. *Reactant-favored* reactions have a *negative* E°.

Example 21.7 The Relation Between E° and ΔG°_{rxn}

The reaction of zinc metal with copper(II) ions (see Figure 21.5) has a standard cell potential E° of +1.10 V at 25 °C. Calculate ΔG°_{rxn} for the reaction:

$$\mathrm{Zn(s)} + \mathrm{Cu^{2+}(aq)} \longrightarrow \mathrm{Zn^{2+}(aq)} + \mathrm{Cu(s)}$$

Solution To obtain ΔG°_{rxn} we substitute the given E° value into Equation 21.1:

$$\Delta G^\circ_{rxn} = -(2.00\ \text{mol electrons transferred})\left(\frac{9.65 \times 10^4\ \mathrm{J}}{\mathrm{V \cdot mol}}\right)(1.10\ \mathrm{V})\left(\frac{1\ \mathrm{kJ}}{1000\ \mathrm{J}}\right)$$

$$= -212\ \mathrm{kJ}$$

See the *Saunders Interactive General Chemistry CD-ROM,* Section 21.6, Standard Potentials.

Exercise 21.5 The Relation Between $E°$ and $\Delta G°_{rxn}$

The following reaction has an $E°$ value of -0.76 V. Calculate $\Delta G°_{rxn}$, and decide whether the reaction is product-favored or reactant-favored:

$$H_2(g) + 2\ H_2O(\ell) + Zn^{2+}(aq) \longrightarrow Zn(s) + 2\ H_3O^+(aq)$$

Calculating the Potential $E°$ of an Electrochemical Cell

An oxidation-reduction reaction is the sum of two half-reactions, one for oxidation and the other for reduction (see Section 21.1). For the $Zn(s)/Cu^{2+}(aq)$ reaction in Figure 21.5, for example,

Anode, oxidation:	$Zn(s)$	$\longrightarrow$	$Zn^{2+}(aq) + 2\ e^-$
Cathode, reduction:	$Cu^{2+}(aq) + 2\ e^-$	$\longrightarrow$	$Cu(s)$
Net process:	$Zn(s) + Cu^{2+}(aq)$	$\longrightarrow$	$Cu(s) + Zn^{2+}(aq)$

It would be very helpful to be able to *predict* the standard potential for this reaction, and it makes sense that it might be the sum of the potentials of the half-reactions. The problem is that potentials for isolated half-reactions cannot be obtained directly because the potential measures the potential energy difference for electrons in two different chemical environments. However, we can always measure the *standard* potential for any half-reaction in combination with some other standard half-reaction. Indeed, chemists decided that the standard half-cell reaction against which all others are measured is the **standard hydrogen electrode,**

$$2\ H_3O^+(aq) + 2\ e^- \longrightarrow H_2(g,\ 1\ bar) + 2\ H_2O(\ell) \qquad E° = 0.00\ V$$

and a *potential of 0.00 V* has been assigned to this half-reaction. (This value has no physical meaning in itself, just as the half-reaction alone has no meaning.) To measure the potential for any half-reaction, we make that reaction one side of an electrochemical cell and the H_2/H_3O^+ half-reaction the other side. Molecular hydrogen, H_2, is a reducing agent, and $H_3O^+(aq)$ is an oxidizing agent. Thus, when the standard hydrogen electrode is paired with another half-cell in an electrochemical cell, the H_2/H_3O^+ half-reaction can be either an oxidation or a reduction. If the other half-cell contains a better reducing agent than H_2, then H_3O^+ is reduced to H_2:

H_3O^+ reduced:

$$2\ H_3O^+(aq,\ 1\ M) + 2\ e^- \longrightarrow H_2(g,\ 1\ bar) + 2\ H_2O(\ell) \qquad E° = 0.00\ V$$

If the other half-cell contains a better oxidizing agent than H_3O^+, then H_2 is oxidized:

H_2 oxidized:

$$H_2(g,\ 1\ bar) + 2\ H_2O(\ell) \longrightarrow 2\ H_3O^+(aq,\ 1\ M) + 2\ e^- \qquad E° = 0.00\ V$$

In either direction, the H_2/H_3O^+ half-cell has a potential of 0.00 V. The measured potential of the electrochemical cell is then *assigned* as the potential of the half-cell being studied. To demonstrate the strategy for determining half-cell potentials, consider the following examples.

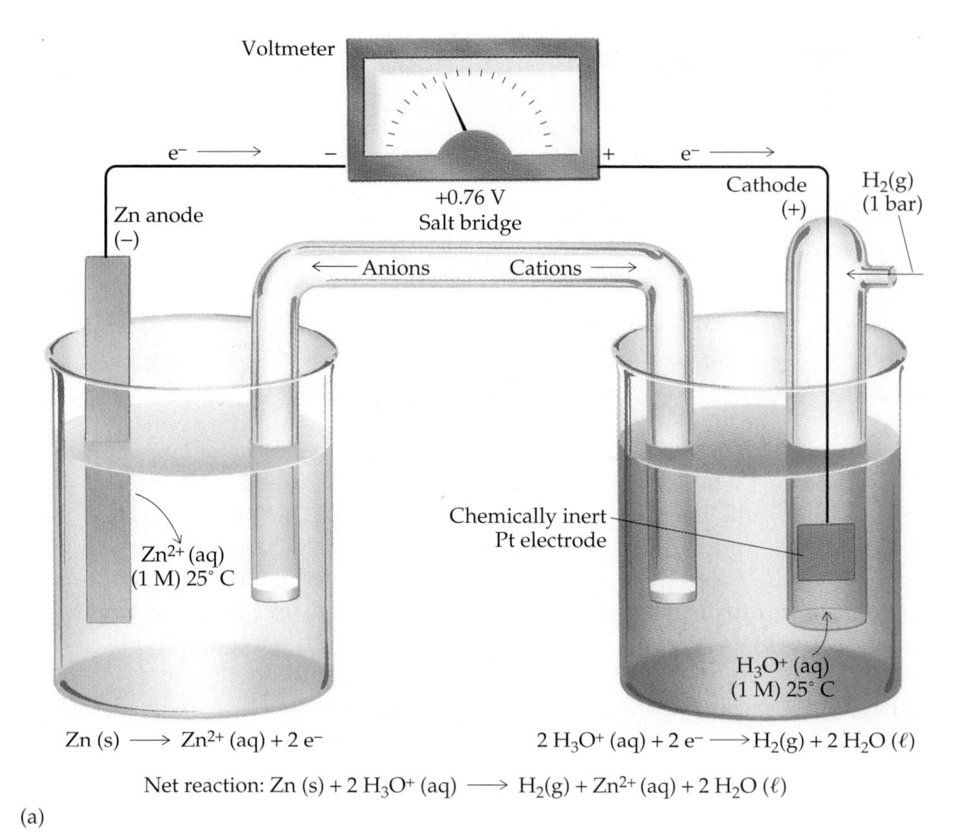

Figure 21.9 An electrochemical cell using Zn^{2+}(aq)/Zn(s) and H_3O^+(aq)/H_2(g) half-cells. (a) A voltage of +0.76 V is generated when the cell is set up under the conditions shown at 25 °C. Electrons flow from the Zn electrode (anode) to the H_3O^+(aq)/H_2(g) electrode (cathode) to produce Zn^{2+}(aq) and H_2(g). Zinc is the reducing agent, and H_3O^+(aq) is the oxidizing agent. (Because the species in the H_3O^+(aq)/H_2(g) half-reaction cannot be fashioned into a solid electrode, electrons are transferred using a piece of platinum foil.) (b) Zinc metal reacts with 1 M HCl in a product-favored reaction. *(b, C. D. Winters)*

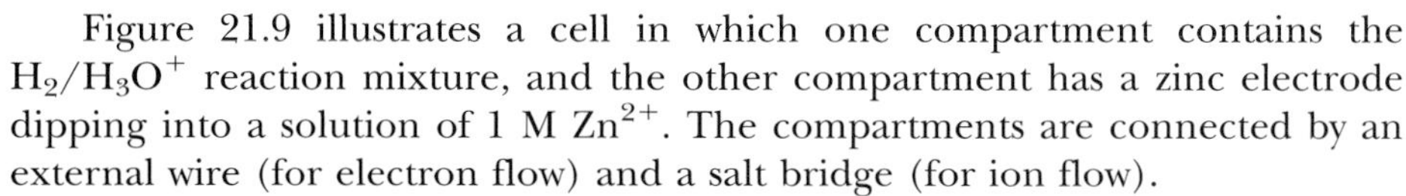

Figure 21.9 illustrates a cell in which one compartment contains the H_2/H_3O^+ reaction mixture, and the other compartment has a zinc electrode dipping into a solution of 1 M Zn^{2+}. The compartments are connected by an external wire (for electron flow) and a salt bridge (for ion flow).

The *measured potential* of an electrochemical cell is *always positive.* The device used to measure potentials, a voltmeter, is designed to give a positive potential only when the positive terminal (+) of the voltmeter is connected to the positive electrode and the negative terminal (−) to the negative electrode. In this way we not only measure the potential, but we find the sign of each of the electrodes. Thus, when a voltmeter is attached to the cell shown in Figure 21.9, it is found that the H_2/H_3O^+ electrode is positive, the zinc electrode is negative, and the measured potential is $E° = +0.76$ V. Although both Zn and H_2 can potentially function as reducing agents, the observation that the Zn electrode is negative here means that Zn is the source of electrons, and so we know that Zn is a *better reducing agent* than H_2 gas. We therefore observe that the zinc electrode

When a redox reaction involves an ion or compound that cannot be made into a solid electrode, a chemically inert electric conductor can be used. Platinum and gold are frequently used for this purpose.

The potential produced by an electrochemical cell is the sum of the potentials of the oxidizing half-reaction and the reducing half-reaction.

dissolves as zinc metal is oxidized to Zn^{2+} ions, and H_2 gas is formed from the reduction of H_3O^+:

$$\begin{array}{rcll} Zn(s) & \longrightarrow & Zn^{2+}(aq,\ 1\ M) + 2\ e^- & E° = ?\ V \\ 2\ H_3O^+(aq,\ 1\ M) + 2\ e^- & \longrightarrow & H_2(g,\ 1\ bar) + 2\ H_2O(\ell) & E° = 0.00\ V \\ \hline Zn(s) + 2\ H_3O^+(aq,\ 1\ M) & \longrightarrow & Zn^{2+}(aq) + H_2(g,\ 1\ bar) + 2\ H_2O(\ell) & E°_{net} = +0.76\ V \end{array}$$

The + sign of $E°_{net}$ correctly reflects the fact that the overall reaction is product-favored as written (and we already knew this from the fact that zinc reacts readily with acid [see Figure 21.9b]). Because the potential for the H_2/H_3O^+ half-cell is 0.00 V, $E°$ for the Zn/Zn^{2+} half-cell must be +0.76 V:

$$Zn(s) \longrightarrow Zn^{2+}(aq,\ 1\ M) + 2\ e^- \qquad E° = +0.76\ V$$

What is $E°$ for the Cu/Cu^{2+} half-cell in our original electrochemical cell in Figure 21.5? In Figure 21.10 this half-cell is shown coupled with the standard H_2/H_3O^+ half-cell. The measured cell potential is +0.34 V, the H_2/H_3O^+ half-cell is negative, and the Cu/Cu^{2+} electrode is positive. Furthermore, the concentration of Cu^{2+} ions declines, and metallic copper forms. These experimental observations confirm that Cu^{2+} is being reduced and that H_2 must be the reducing agent, the source of electrons. Because the reducing agent, H_2, is oxidized to H_3O^+, the H_2 electrode must be the anode (and is negatively charged). Furthermore, because Cu^{2+} ions are the acceptors of electrons (the oxidizing agent), this electrode is the cathode (and is positively charged). Most impor-

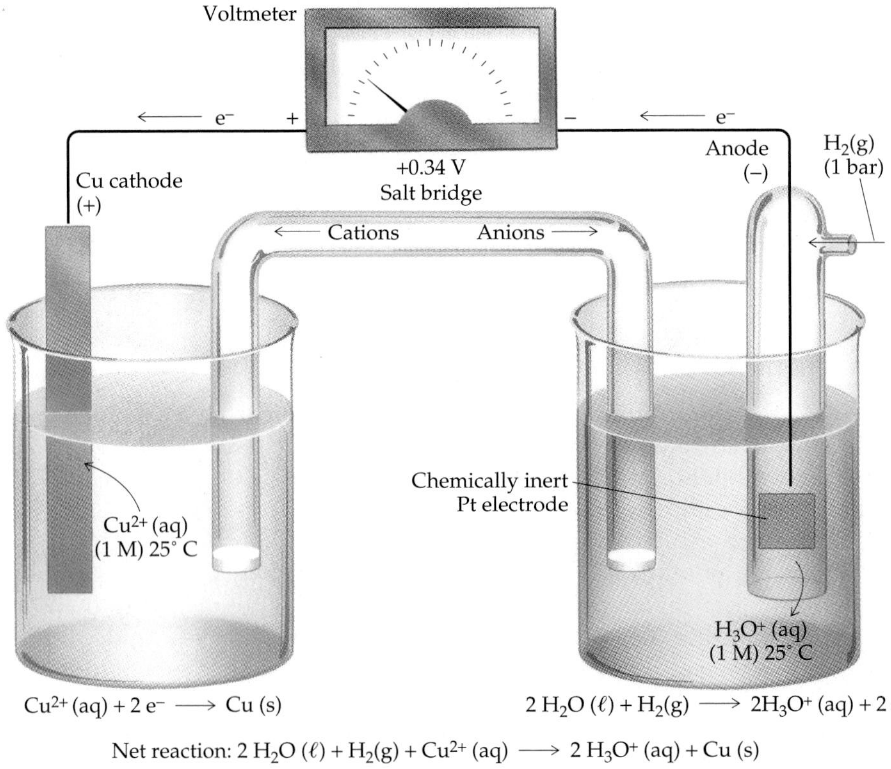

Figure 5.19 illustrates the reduction of copper(II) oxide with H_2 to form copper metal and water. Although not done under standard electrochemical conditions, this experiment shows H_2 is a better reducing agent than copper metal.

$$Cu^{2+}(aq) + 2\ e^- \longrightarrow Cu(s) \qquad 2\ H_2O(\ell) + H_2(g) \longrightarrow 2H_3O^+(aq) + 2\ e^-$$

Net reaction: $2\ H_2O(\ell) + H_2(g) + Cu^{2+}(aq) \longrightarrow 2\ H_3O^+(aq) + Cu(s)$

***Figure* 21.10 An electrochemical cell using $Cu^{2+}(aq)/Cu(s)$ and $H_3O^+(aq)/H_2(g)$ half-cells.** A voltage of +0.34 V is generated when the cell is set up as pictured at 25 °C. Electrons flow from the $H_3O^+(aq)/H_2(g)$ electrode (the anode) to the copper electrode (cathode) to produce copper metal and hydronium ions. Consequently, $H_2(g)$ is the reducing agent, and $Cu^{2+}(aq)$ is the oxidizing agent.

tantly, it is apparent that $H_2(g)$ is a better reducing agent than Cu(s), and the appropriate half-reactions and net reaction are

Anode, oxidation:	$H_2(g, 1\ bar) + 2\ H_2O(\ell) \longrightarrow 2\ H_3O^+(aq, 1\ M) + 2\ e^-$	$E° = 0.00$ V
Cathode, reduction:	$Cu^{2+}(aq, 1\ M) + 2\ e^- \longrightarrow Cu(s)$	$E° = ?$ V
Net reaction:	$Cu^{2+}(aq, 1\ M) + H_2(g, 1\ bar) + 2\ H_2O(\ell) \longrightarrow Cu(s) + 2\ H_3O^+(aq, 1\ M)$	$E°_{net} = +0.34$

The half-cell potential for $Cu^{2+}(aq, 1\ M) + 2\ e^- \longrightarrow Cu(s)$ must be +0.34 V at 25 °C.

The potential of a cell in which Zn(s) reduces $Cu^{2+}(aq)$ to Cu(s) can now be calculated, because we have $E°$ values for the half-reactions involved:

Anode, oxidation:	$Zn(s) \longrightarrow Zn^{2+}(aq, 1\ M) + 2\ e^-$	$E° = +0.76$ V
Cathode, reduction:	$Cu^{2+}(aq, 1\ M) + 2\ e^- \longrightarrow Cu(s)$	$E° = +0.34$ V
Net reaction:	$Cu^{2+}(aq, 1\ M) + Zn(s) \longrightarrow Cu(s) + Zn^{2+}(aq, 1\ M)$	$E°_{net} = +1.10$ V

This is an important result. Two half-cell potentials were measured independently against the same standard H_2/H_3O^+ half-cell, and the sum of the two potentials is equal to the experimentally measured $E°$ (see Figure 21.5.) Now we can use a similar technique with any of the preceding half-cells (Cu/Cu^{2+}, Zn/Zn^{2+}, or H_2/H_3O^+) as reference cells and determine $E°$ values for hundreds of other possible half-cells.

Example 21.8 Determining a Half-Reaction Potential

The cell illustrated here has a potential of $E° = +0.51$ V at 25 °C. The net ionic equation for the cell reaction is

$$Zn(s) + Ni^{2+}(aq, 1\ M) \longrightarrow Zn^{2+}(aq, 1\ M) + Ni(s)$$

Which electrode is the anode, and which is the cathode? What are the polarities of the electrodes? What is the value of $E°$ for the half-cell $Ni^{2+}(aq) + 2\ e^- \longrightarrow Ni(s)$?

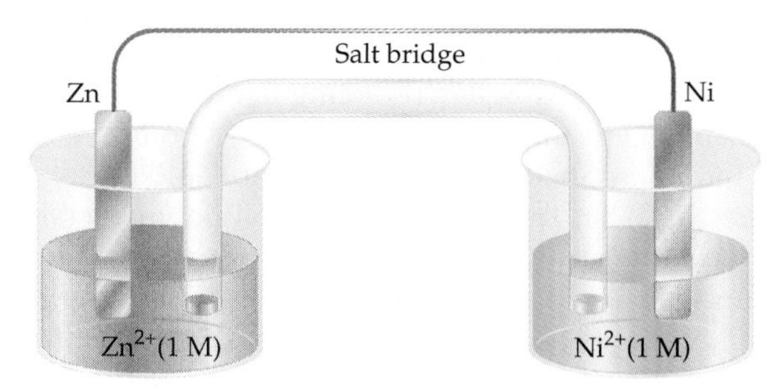

Solution Zn(s) is oxidized to $Zn^{2+}(aq)$, so the Zn electrode is the anode and is negatively charged because it is the source of electrons. Nickel(II) ions are reduced to Ni metal at the Ni electrode, so this is the cathode, and it is positive.

Because the overall cell potential is known, and the potential for the $Zn(s)/Zn^{2+}(aq)$ half-cell is known, $E°$ for the nickel half-cell can be calculated:

Anode, oxidation:	$Zn(s) \longrightarrow Zn^{2+}(aq) + 2\ e^-$	$E° = +0.76$ V
Cathode, reduction:	$Ni^{2+}(aq) + 2\ e^- \longrightarrow Ni(s)$	$E° = ?$ V
Net reaction:	$Zn(s) + Ni^{2+}(aq) \longrightarrow Zn^{2+}(aq) + Ni(s)$	$E°_{net} = +0.51$

At 25 °C the value of $E°$ for the reaction $Ni^{2+}(aq) + 2\ e^- \longrightarrow Ni(s)$ is −0.25 V.

Exercise 21.6 Determining a Half-Reaction Potential

Given that the reduction of aqueous copper(II) with iron metal has an E°_{net} value of +0.78 V, what is E° for the half-cell $Fe(s) \longrightarrow Fe^{2+}(aq, 1\ M) + 2\ e^-$?

$$Fe(s) + Cu^{2+}(aq, 1\ M) \longrightarrow Fe^{2+}(aq, 1\ M) + Cu(s)$$

21.4 USING STANDARD POTENTIALS

The standard potential E° for the reduction of Ni^{2+} to metallic nickel is −0.25 V (see Example 21.18). What does this mean? This tells us that if Ni^{2+} is used as an oxidizing agent, coupled with H_2 as the reducing agent,

Cathode, reduction:	$Ni^{2+}(aq, 1\ M) + 2\ e^- \longrightarrow Ni(s)$	$E^\circ = -0.25$ V
Anode, oxidation:	$H_2(g, 1\ bar) + 2\ H_2O(\ell) \longrightarrow 2\ H_3O^+(aq, 1\ M) + 2\ e^-$	$E^\circ = 0.00$ V
Net reaction:	$Ni^{2+}(aq, 1\ M) + H_2(g, 1\ bar) + 2\ H_2O(\ell) \longrightarrow Ni(s) + 2\ H_3O^+(aq, 1\ M)$	$E^\circ_{net} = -0.25$ V

the reaction is *reactant-favored* under standard conditions; E°_{net} is negative so ΔG° is positive. From thermodynamics, you know that if a reaction is not product-favored in the direction written (ΔG° is positive), the reverse reaction is product-favored (ΔG° is negative). Therefore, the reaction

$$Ni(s) + 2\ H_3O^+(aq, 1\ M) \longrightarrow Ni^{2+}(aq, 1\ M) + H_2(g, 1\ bar) + 2\ H_2O(\ell) \qquad E^\circ_{net} = +0.25\ V$$

is product-favored, and E°_{net} for the reaction is positive. Just as acid–base reactions always move in the direction of the weaker acid–base pair (see Section 17.3), redox reactions move toward the weaker oxidizing agent/reducing agent pair. In the Ni/H_3O^+ reaction, there are two possible reducing agents: Ni(s) and H_2. Because nickel metal reduces hydronium ion in a product-favored reaction, nickel metal must be a better reducing agent than H_2, and H_3O^+ must be a better oxidizing agent than aqueous Ni^{2+}.

Thus far we have described four elements that could act as reducing agents: Zn, Cu, H_2, and Ni. What are their *relative* abilities to act in this manner? Conversely, what are the relative abilities of their ions (Zn^{2+}, Cu^{2+}, H_3O^+, and Ni^{2+}) to act as electron acceptors or oxidizing agents? Writing half-reactions involving these elements and their ions in the form

$$\text{Oxidized form} + n\ e^- \longrightarrow \text{Reduced form}$$

and listing them in order of descending E° values, places the oxidizing agents in descending order of their ability to attract electrons:

Increasing strength as oxidizing agent ↑		Increasing strength as reducing agent ↓
$Cu^{2+}(aq, 1\ M) + 2\ e^- \longrightarrow Cu(s)$		$E^\circ = +0.34$ V
$2\ H_3O^+(aq, 1\ M) + 2\ e^- \longrightarrow H_2(g, 1\ atm) + 2\ H_2O(\ell)$		$E^\circ = 0.00$ V
$Ni^{2+}(aq, 1\ M) + 2\ e^- \longrightarrow Ni(s)$		$E^\circ = -0.25$ V
$Zn^{2+}(aq, 1\ M) + 2\ e^- \longrightarrow Zn(s)$		$E^\circ = -0.76$ V

The value of $E°$ becomes more negative down the series. This means that Cu^{2+}(aq) is the best oxidizing agent of the substances on the left; that is, Cu^{2+} shows the greatest tendency to be reduced. Conversely, Zn^{2+} is the worst oxidizing agent; it has the least tendency to be reduced. Of the substances on the right, Zn(s) is the best reducing agent (best electron donor), because $E°$ for the half-reaction

$$Zn(s) \longrightarrow Zn^{2+}(aq, 1\ M) + 2\ e^- \qquad E° = +0.76$$

has the most positive value. By the same reasoning, Cu is the worst reducing agent.

The preceding table of half-reaction potentials tells us that, at standard conditions, the following reactions are product-favored:

Increasing oxidizing ability ↑	Increasing reducing ability ↓
Cu^{2+} can oxidize H_2, Ni, and Zn	H_2 can reduce Cu^{2+} (but not Ni^{2+} or Zn^{2+})
H_3O^+ can oxidize Ni and Zn (but not Cu)	Ni can reduce H_3O^+ and Cu^{2+} (but not Zn^{2+})
Ni^{2+} can oxidize Zn (but not H_2 or Cu)	Zn can reduce Ni^{2+}, H_3O^+, and Cu^{2+}

Each of these reactions has a positive $E°_{net}$ (and a negative $\Delta G°$). The reduction of Cu^{2+}(aq) with Ni further illustrates the point:

Oxidation:	$Ni(s) \longrightarrow$	$Ni^{2+}(aq) + 2\ e^-$	$E° = +0.25$ V
Reduction:	$Cu^{2+}(aq) + 2\ e^- \longrightarrow$	$Cu(s)$	$E° = +0.34$ V
Net reaction:	$Ni(s) + Cu^{2+}(aq) \longrightarrow$	$Ni^{2+}(aq) + Cu(s)$	$E°_{net} = +0.59$ V

We have just created a small portion of a **table of standard reduction potentials** (Table 21.1). Like the table of conjugate acids and bases (Table 17.4), Table 21.1 is *very* useful. Some important points concerning this table are

A larger table of $E°$ values is given in Appendix M.

- The $E°$ values are for reactions written in the form "oxidized form + electrons $\longrightarrow$ reduced form." The species on the left side of the reaction is an oxidizing agent, and the species on the right is a reducing agent. All potentials are therefore for reduction reactions.
- When writing the reaction "reduced form $\longrightarrow$ oxidized form + electrons," the sign of $E°$ is reversed, but the value of $E°$ is unaffected. Thus,

$$Li(s) \longrightarrow Li^+(aq) + e^- \qquad E° = +3.045\ V$$

- All the half-reactions are reversible. For example, aqueous H_3O^+ is reduced to H_2(g) in Figure 21.9, whereas H_2(g) is oxidized to H_3O^+ in Figure 21.10.
- The more positive the value of $E°$ for the reactions in Table 21.1, the better the oxidizing ability of the ion or compound on the left side of the reaction. This means *F_2(g) is the best oxidizing agent in the table:*

$$F_2(g, 1\ atm) + 2\ e^- \longrightarrow 2\ F^-(aq, 1\ M) \qquad E° = +2.87\ V$$

The ion at the bottom left corner of the table, Li^+(aq), is the poorest oxidizing agent because its $E°$ is the most negative. The oxidizing agents in the table (ions, elements, and compounds at the left) *increase* in strength *from the bottom to the top of the table.*

Table 21.1 • Standard Reduction Potentials in Aqueous Solution at 25 °C*

Reduction Half-Reaction		$E°$ (V)
$F_2(g) + 2\ e^-$	$\longrightarrow 2\ F^-(aq)$	+2.87
$H_2O_2(aq) + 2\ H_3O^+(aq) + 2\ e^-$	$\longrightarrow 4\ H_2O(\ell)$	+1.77
$PbO_2(s) + SO_4^{2-}(aq) + 4\ H_3O^+(aq) + 2\ e^-$	$\longrightarrow PbSO_4(s) + 6\ H_2O(\ell)$	+1.685
$MnO_4^-(aq) + 8\ H_3O^+(aq) + 5\ e^-$	$\longrightarrow Mn^{2+}(aq) + 12\ H_2O(\ell)$	+1.52
$Au^{3+}(aq) + 3\ e^-$	$\longrightarrow Au(s)$	+1.50
$Cl_2(g) + 2\ e^-$	$\longrightarrow 2\ Cl^-(aq)$	+1.360
$Cr_2O_7^{2-}(aq) + 14\ H_3O^+(aq) + 6\ e^-$	$\longrightarrow 2\ Cr^{3+}(aq) + 21\ H_2O\ (\ell)$	+1.33
$O_2(g) + 4\ H_3O^+(aq) + 4\ e^-$	$\longrightarrow 6\ H_2O(\ell)$	+1.229
$Br_2(\ell) + 2\ e^-$	$\longrightarrow 2\ Br^-(aq)$	+1.08
$NO_3^-(aq) + 4\ H_3O^+(aq) + 3\ e^-$	$\longrightarrow NO(g) + 6\ H_2O(\ell)$	+0.96
$OCl^-(aq) + H_2O(\ell) + 2\ e^-$	$\longrightarrow Cl^-(aq) + 2\ OH^-(aq)$	+0.89
$Hg^{2+}(aq) + 2\ e^-$	$\longrightarrow Hg(\ell)$	+0.855
$Ag^+(aq) + e^-$	$\longrightarrow Ag(s)$	+0.80
$Hg_2^{2+}(aq) + 2\ e^-$	$\longrightarrow 2\ Hg(\ell)$	+0.789
$Fe^{3+}(aq) + e^-$	$\longrightarrow Fe^{2+}(aq)$	+0.771
$I_2(s) + 2\ e^-$	$\longrightarrow 2\ I^-(aq)$	+0.535
$O_2(g) + 2\ H_2O(\ell) + 4\ e^-$	$\longrightarrow 4\ OH^-(aq)$	+0.40
$Cu^{2+}(aq) + 2\ e^-$	$\longrightarrow Cu(s)$	+0.337
$Sn^{4+}(aq) + 2\ e^-$	$\longrightarrow Sn^{2+}(aq)$	+0.15
$2\ H_3O^+(aq) + 2\ e^-$	$\longrightarrow H_2(g) + 2\ H_2O(\ell)$	0.00
$Sn^{2+}(aq) + 2\ e^-$	$\longrightarrow Sn(s)$	−0.14
$Ni^{2+}(aq) + 2\ e^-$	$\longrightarrow Ni(s)$	−0.25
$V^{3+}(aq) + e^-$	$\longrightarrow V^{2+}(aq)$	−0.255
$PbSO_4(s) + 2\ e^-$	$\longrightarrow Pb(s) + SO_4^{2-}(aq)$	−0.356
$Cd^{2+}(aq) + 2\ e^-$	$\longrightarrow Cd(s)$	−0.40
$Fe^{2+}(aq) + 2\ e^-$	$\longrightarrow Fe(s)$	−0.44
$Zn^{2+}(aq) + 2\ e^-$	$\longrightarrow Zn(s)$	−0.763
$2\ H_2O(\ell) + 2\ e^-$	$\longrightarrow H_2(g) + 2\ OH^-(aq)$	−0.8277
$Al^{3+}(aq) + 3\ e^-$	$\longrightarrow Al(s)$	−1.66
$Mg^{2+}(aq) + 2\ e^-$	$\longrightarrow Mg(s)$	−2.37
$Na^+(aq) + e^-$	$\longrightarrow Na(s)$	−2.714
$K^+(aq) + e^-$	$\longrightarrow K(s)$	−2.925
$Li^+(aq) + e^-$	$\longrightarrow Li(s)$	−3.045

Increasing strength of oxidizing agents (↑)

Increasing strength of reducing agents (↓)

*In volts (V) versus the standard hydrogen electrode

- The more negative the value of the reduction potential $E°$, the less likely the reaction occurs as a reduction and the more likely the reverse reaction occurs (as an oxidation). Thus, Li(s) is the strongest reducing agent in the table, and F^- is the weakest reducing agent. The reducing agents in the table (the ions, elements, or compounds at the right) *increase* in strength *from the top to the bottom.*
- The reaction between any substance on the left in this table (an oxidizing agent) with any substance lower than it on the right (a reducing agent) is product-favored under standard conditions.
- The algebraic sign of the half-reaction potential is the sign of the electrode when it is attached to the H_2/H_3O^+ standard cell. (See Figures 21.9 and 21.10.)

- Electrochemical potentials depend on the nature of the reactants and products and their concentrations, not on the quantities of material used. Changing the stoichiometric coefficients for a half-reaction therefore does not change the value of $E°$. For example, the reduction of Fe^{3+} has an $E°$ of +0.771 V, whether the reaction is written as

$$Fe^{3+}(aq, 1\ M) + e^- \longrightarrow Fe^{2+}(aq, 1\ M) \qquad E° = +0.771\ V$$

or as

$$2\ Fe^{3+}(aq, 1\ M) + 2\ e^- \longrightarrow 2\ Fe^{2+}(aq, 1\ M) \qquad E° = +0.771\ V$$

The volt is defined as "energy per charge." Multiplying a reaction by some number causes both the energy and the charge to be multiplied by that number. Thus, the ratio "energy per charge = volt" does not change.

Example 21.9 Predicting $E°$ and the Direction of a Redox Reaction

Decide if each of the following reactions is product-favored in the direction written, and calculate $E°_{net}$:

1. $2\ Al(s) + 3\ Sn^{4+}(aq) \longrightarrow 2\ Al^{3+}(aq) + 3\ Sn^{2+}(aq)$
2. $2\ Br^-(aq) + I_2(s) \longrightarrow 2\ I^-(aq) + Br_2(\ell)$

Solution Reaction (1) is predicted to be product-favored as written (and thus to have a positive $E°$). As indicated in Table 21.1, Al(s) is a stronger reducing agent than Sn^{2+}, and Sn^{4+} is a stronger oxidizing agent than Al^{3+}. This is verified by the positive value of $E°$:

Oxidation:	$2\ [Al(s) \longrightarrow Al^{3+}(aq) + 3\ e^-]$	$E° = +1.66\ V$
Reduction:	$3\ [Sn^{4+}(aq) + 2\ e^- \longrightarrow Sn^{2+}(aq)]$	$E° = +0.15\ V$
Net reaction:	$2\ Al(s) + 3\ Sn^{4+}(aq) \longrightarrow 3\ Sn^{2+}(aq) + 2\ Al^{3+}(aq)$	$E°_{net} = +1.81\ V$

Reaction (2) is predicted to be reactant-favored. Based on the reduction potentials in Table 21.1, we know that I_2 is a weaker oxidizing agent than Br_2 and Br^- is a weaker reducing agent than I^-. The negative sign of $E°_{net}$ for the reaction as written confirms this:

Oxidation:	$2\ Br^-(aq) \longrightarrow Br_2(\ell) + 2\ e^-$	$E° = -1.08\ V$
Reduction:	$I_2(s) + 2\ e^- \longrightarrow 2\ I^-(aq)$	$E° = +0.535\ V$
Net reaction:	$2\ Br^-(aq) + I_2(s) \longrightarrow 2\ I^-(aq) + Br_2(\ell)$	$E°_{net} = -0.55\ V$

The reaction written in the opposite direction,

$$2\ I^-(aq) + Br_2(\ell) \longrightarrow I_2(s) + 2\ Br(aq) \qquad E°_{net} = +0.55\ V$$

is product-favored. Indeed, it can be observed that bromine dissolved in water oxidizes iodide ion to iodine (see Figure 21.11).

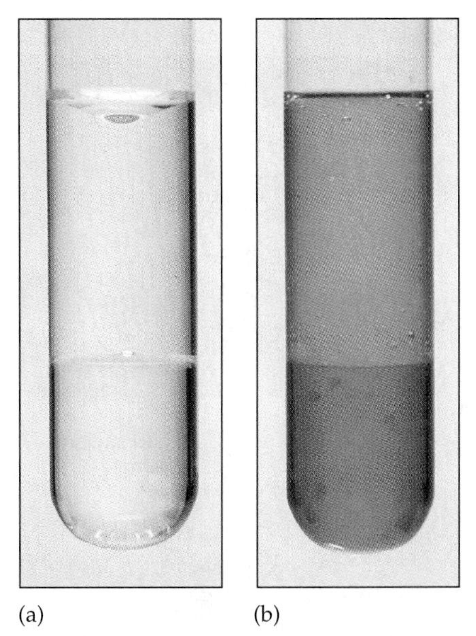

***Figure* 21.11 The reaction of bromine and iodide ion.** The reaction of Br_2 with I^- to give Br^- and I_2 is product-favored with $E° = +0.545$ V. (a) The test tube contains an aqueous solution of KI (top layer) and immisible CCl_4 (bottom layer). (b) A few drops of Br_2 in water are added. The iodine produced in the reaction collects in the bottom CCl_4 layer and gives it a purple color (see also Figure 14.4). (The top layer contains excess Br_2 in water.) This proves that I_2 is in fact a product of the reaction and that Br_2 is a better oxidizing agent than I_2. *(C. D. Winters)*

Exercise 21.7 Predicting $E°$ and the Direction of a Redox Reaction

Is the following reaction product-favored under standard conditions? What is the value of $E°_{net}$?

$$H_2O_2(aq) + 2\ H_3O^+(aq) + 2\ Br^-(aq) \longrightarrow Br_2(\ell) + 4\ H_2O(\ell)$$

Current Perspectives

Dental Amalgams and Electrochemistry

Amalgams are commonly used to restore single teeth, and they are unique substances. A small amount of mercury is mixed with a nearly equal mass of powdered alloy of silver, tin, copper, and zinc. A few seconds after mixing, the material is very plastic and can be packed into a prepared cavity. After 3 min the material becomes solid enough to be carved, and after about 2 h it is sufficiently hard that the patient can bite down on the tooth. The final product resists corrosion and does not react to produce toxic or soluble salts.

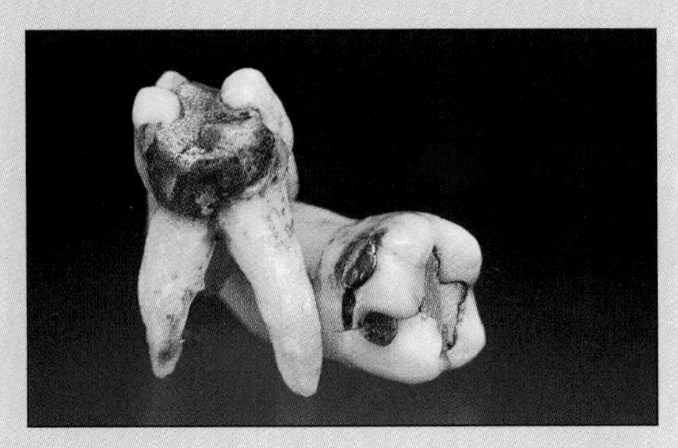

Teeth with fillings of dental amalgam. *(C. D. Winters)*

If you have dental fillings, however, you know that one thing can cause a problem. If you bite a piece of aluminum foil from a gum or candy wrapper with a filled tooth—ouch! Dentists say the same thing can happen when two fillings come into contact. The sensation is caused by the nerves in the teeth detecting the electron flow from an electrochemical reaction. Dental amalgam consists of several phases, some having compositions approximating Ag_2Hg, Ag_3Sn, and Sn_xHg (where x is 7 to 9). All three of these may undergo electrochemical reactions. For example,

$$3\ Hg_2^{2+}(aq) + 4\ Ag(s) + 6\ e^- \longrightarrow 2\ Ag_2Hg_3(s) \qquad E° = +0.85\ V$$

$$Sn^{2+}(aq) + 3\ Ag(s) + 2\ e^- \longrightarrow Ag_3Sn(s) \qquad E° = -0.05\ V$$

Aluminum is a much better reducing agent than any of the solid solutions ($E°$ for $Al(s) \longrightarrow Al^{3+}(aq) + 3\ e^-$ is +1.66 V). If a piece of aluminum therefore comes into contact with a dental filling, the saliva and gum tissue act as a salt bridge, and a reaction occurs. The tiny current from the reaction produces a jolt of pain.

Example 21.10 Constructing an Electrochemical Cell

Using the half-reactions $Fe(s)/Fe^{2+}(aq)$ and $Cu(s)/Cu^{2+}(aq)$, construct an electrochemical cell, and predict its standard potential $E°$.

Solution First, let us decide which is the reducing agent and which is the oxidizing agent. From Table 21.1 we conclude that Fe(s) is a better reducing agent than Cu(s), and $Cu^{2+}(aq)$ is a better oxidizing agent than $Fe^{2+}(aq)$. Therefore, Fe(s) reduces $Cu^{2+}(aq)$ to copper metal:

Oxidation:	$Fe(s) \longrightarrow Fe^{2+}(aq) + 2\ e^-$	$E° = +0.44\ V$
Reduction:	$Cu^{2+}(aq) + 2\ e^- \longrightarrow Cu(s)$	$E° = +0.34\ V$
Net reaction:	$Fe(s) + Cu^{2+}(aq) \longrightarrow Fe^{2+}(aq) + Cu(s)$	$E°_{net} = +0.78\ V$

For a cell that operates under standard conditions, we need to have a solid iron electrode dipping into a 1.0 M solution of $Fe^{2+}(aq)$. In another compartment we can have a solid copper electrode dipping into a solution of 1.0 M $Cu^{2+}(aq)$. An external wire connects the two electrodes, and a salt bridge allows for ion flow. (Because virtually all metal ions form soluble nitrate salts, we can use $Cu(NO_3)_2$ and $Fe(NO_3)_2$ in the cell compartments and $NaNO_3$ in the salt bridge.) When the cell is assembled, electrons flow from the Fe(s) anode to the Cu(s) cathode, and anions (NO_3^- in this case) flow from the $Cu^{2+}(aq)$ compartment to the $Fe^{2+}(aq)$ compartment.

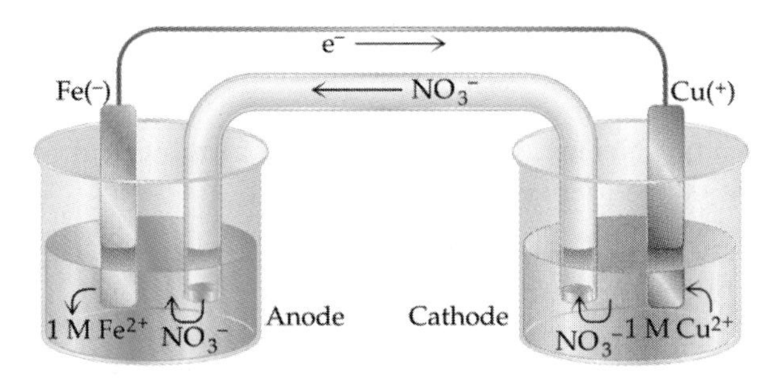

Exercise 21.8 Constructing an Electrochemical Cell

Draw a diagram of an electrochemical cell using the half-cells $Zn(s)/Zn^{2+}(aq)$ and $Al(s)/Al^{3+}(aq)$. Decide first on the net reaction, and predict its $E°$ value. Show the direction of electron flow in the external wire and the directions of ion flow in the salt bridge. Tell which compartment is the anode and which is the cathode.

21.5 ELECTROCHEMICAL CELLS AT NONSTANDARD CONDITIONS

Oxidation–reduction reactions in the real world are rarely carried out under standard conditions. Even if the cell started out with all dissolved species at 1 M concentration, these would change as the reaction progressed; reactant concentrations decrease and those of the products increase. How can we define the potential of cells under *non*standard conditions?

See the *Saunders Interactive General Chemistry CD-ROM,* Screen 21.7, Electrochemical Cells at Nonstandard Conditions.

The Nernst Equation

The standard cell potential, $E°$, is the potential measured under standard conditions, that is, with all dissolved substances having a concentration of 1.0 mol/L. These are almost never the conditions in a real electrochemical cell, however, so how can we predict the potential under nonstandard conditions, E? The answer is that the standard potential, $E°$, can be corrected by a factor that includes the temperature of the reaction, the number of moles of electrons transferred between oxidizing and reducing agents in a balanced redox equation (n), and the concentrations of reactants and products. This relationship is called the **Nernst equation** after Walther Nernst (1864–1941), the German physical chemist who also discovered the third law of thermodynamics:

$$E = E° - (RT/nF) \ln Q$$

where Q is the reaction quotient (page 759), F is the Faraday constant (9.6485309×10^4 J/V · mol), and R is the gas constant (8.314510 J/K · mol). When T is 298 K, we can write a modified form of the Nernst equation that we find useful in practical chemical applications.

$$E = E° - \frac{0.0257 \text{ V}}{n} \ln Q \quad \text{at } 25\ °\text{C} \qquad [21.2]$$

The Nernst equation is often written using the base-10 logarithm of Q:

$$E = E° - \frac{0.0592 \text{ V}}{n} \log Q$$

Walther Nernst (1864–1941) was a German physicist and chemist known for his work relating to the third law of thermodynamics. *(Francis Simon, AIP Niels Bohr Library)*

This equation allows us to find the potential produced by a cell under nonstandard conditions or to find the concentration of a reactant or product by measuring the potential produced by a cell.

Before using the Nernst equation to derive a numerical answer, let us explore some of its consequences. Taking the reaction

$$Zn(s) + Ni^{2+}(1.0\ M, aq) \longrightarrow Zn^{2+}(1.0\ M, aq) + Ni(s) \qquad E^\circ_{net} = +0.51\ V$$

what happens to the cell potential if, for example, $[Ni^{2+}]$ is 1.0 M, whereas $[Zn^{2+}]$ is only 0.0010 M? In this case Q is much smaller than 1:

$$Q = \frac{[Zn^{2+}]}{[Ni^{2+}]} = \frac{0.0010}{1.0} = 0.0010$$

When you take the logarithm of a number less than 1, the result is a negative number ($\ln 0.001 = -6.91$). Because the "correction factor" in the Nernst equation is subtracted from E°, the combination of negative signs means that E is more positive than E° in this case. When the concentrations of products are low relative to the reactant concentrations in a product-favored reaction, the cell potential is more positive than E°; the reaction becomes even more product-favored.

Example 21.11 Using the Nernst Equation

Determine the cell potential at 25 °C for

$$Fe(s) + Cd^{2+}(aq) \longrightarrow Fe^{2+}(aq) + Cd(s)$$

when (1) $[Fe^{2+}] = 0.010$ M and $[Cd^{2+}] = 1.0$ M and (2) $[Fe^{2+}] = 1.0$ M and $[Cd^{2+}] = 0.010$ M.

Solution To calculate a nonstandard potential, we first need the standard potential for the cell reaction, E°_{net}.

Oxidation:	$Fe(s) \longrightarrow Fe^{2+}(aq) + 2\ e^-$	$E^\circ = +0.44$ V
Reduction:	$Cd^{2+}(aq) + 2\ e^- \longrightarrow Cd(s)$	$E^\circ = -0.40$ V
Net reaction:	$Fe(s) + Cd^{2+}(aq) \longrightarrow Fe^{2+}(aq) + Cd(s)$	$E^\circ_{net} = +0.04$ V

Next, let us substitute E° and the conditions for solution (1) into the Nernst equation:

$$E = E^\circ - \frac{0.0257\ V}{n} \ln \frac{[Fe^{2+}]}{[Cd^{2+}]}$$

As in chemical equilibrium calculations (see Chapter 16), the concentration of a solid is not part of the expression. Now, using $n = 2$ (the number of moles of electrons transferred), and the ion concentrations given earlier,

$$E = +0.04\ V - \frac{0.0257\ V}{2} \ln \frac{0.010}{1.0} = +0.10\ V$$

The cell potential E is more positive than E°_{net}, so the tendency to transfer electrons from Fe(s) to Cd^{2+}(aq) is greater than under standard conditions.

For the condition described by (2), the Nernst equation is

$$E = +0.04 \text{ V} - \frac{0.0257 \text{ V}}{2} \ln \frac{1.0}{0.010} = -0.02 \text{ V}$$

The cell potential E is now negative. Although the reaction under standard conditions is product-favored, the nonstandard conditions defined here lead to the reaction becoming reactant-favored. Under these conditions, cadmium metal reduces iron(II) ion:

$$Fe^{2+}(aq, 1.0 \text{ M}) + Cd(s) \longrightarrow Fe(s) + Cd^{2+}(aq, 0.010 \text{ M}) \qquad E_{net} = +0.02 \text{ V}$$

Exercise 21.9 Using the Nernst Equation

Calculate E_{net} for the following reaction:

$$2\ Ag^{+}(aq, 0.80 \text{ M}) + Hg(\ell) \longrightarrow 2\ Ag(s) + Hg^{2+}(0.0010 \text{ M}, aq)$$

Is the reaction product-favored or reactant-favored under these conditions? How does this compare with the reaction under standard conditions?

$E°$ and the Equilibrium Constant

The cell potential, and even the reaction direction, can change when the concentrations of products and reactants change (see Example 21.11). Thus, as reactants are converted to products in a product-favored reaction, the value of E_{net} must decline from its initial positive value to eventually reach zero. A *potential of zero* means that no *net* reaction is occurring; it is an indication that the cell has reached *equilibrium.* Thus, when $E_{net} = 0$, the Q term in the Nernst equation is equivalent to the equilibrium constant K for the reaction. So, when equilibrium has been attained, Equation 21.2 can be rewritten as

$$E = 0 = E° - \frac{0.0257 \text{ V}}{n} \ln K$$

which rearranges to

$$\ln K = \frac{nE°}{0.0257 \text{ V}} \quad \text{at 25 °C} \qquad [21.3]$$

This is an extremely useful equation, because it allows us to obtain the equilibrium constant for a reaction from a calculation or measurement of $E°_{net}$.

Example 21.12 $E°$ and Equilibrium Constants

Calculate the equilibrium constant for the reaction

$$Fe(s) + Cd^{2+}(aq) \longrightarrow Fe^{2+}(aq) + Cd(s) \qquad E°_{net} = +0.04 \text{ V}$$

What are the equilibrium concentrations of the Fe^{2+} and Cd^{2+} ions if each began with a concentration of 1.0 M?

Solution The reaction was found to have a value of E°_{net} of +0.04 V in Example 21.11. Substituting into Equation 21.3, we therefore have

$$\ln K = \frac{(2.00)(0.04 \text{ V})}{0.0257 \text{ V}} = 3.1$$

$$K = 20$$

The concentration of Fe^{2+} and Cd^{2+} when the cell has reached equilibrium are given by the equilibrium expression.

$$K = 20 = \frac{[Fe^{2+}]}{[Cd^{2+}]}$$

Because the cell began at standard conditions, the original concentrations of both ions were 1.0 M. As the reaction proceeded to equilibrium, x mol/L of Cd^{2+} was consumed and x mol/L of Fe^{2+} was produced. Therefore,

$$K = 20 = \frac{1.0 + x}{1.0 - x}$$

Solving this we find $x = 0.9$ M. Thus, the equilibrium concentrations are

$$[Fe^{2+}] = 1.0 + x = 1.9 \text{ M} \quad \text{and} \quad [Cd^{2+}] = 1.0 - x = 0.10 \text{ M}$$

Exercise 21.10 E° and Equilibrium Constants

In Exercise 21.9 you calculated E_{net} for

$$2\ Ag^+(aq, 0.80\ M) + Hg(\ell) \longrightarrow 2\ Ag(s) + Hg^{2+}(0.0010\ M, aq)$$

What is the equilibrium constant for this reaction?

21.6 BATTERIES AND FUEL CELLS

See the *Saunders Interactive General Chemistry CD-ROM,* Screen 21.8, Batteries.

The voltaic cells we described to this point can produce a useful potential, but the potential declines rapidly as the reactant concentrations decline. For this reason, there has been great interest over the years in the design of usable batteries, voltaic cells that deliver current at a constant potential. As we depend more and more on portable computers, cellular phones, personal information devices, and pagers—and if there is any hope of producing a usable electric car—it is ever more important that lightweight, long-lived batteries be developed.

Batteries can be classified as primary and secondary. **Primary batteries** use oxidation–reduction reactions that cannot be reversed easily, so when the cell reaction reaches equilibrium, the battery is "dead" and is discarded. **Secondary batteries** are often called **storage batteries**, or **rechargeable batteries**. The reactions in these batteries can be reversed; the battery can be "recharged."

Fuel cells are another type of electrochemical device. In a battery the oxidizing and reducing agents are held within a closed container. In contrast, the reactants in a fuel cell are supplied from an outside source.

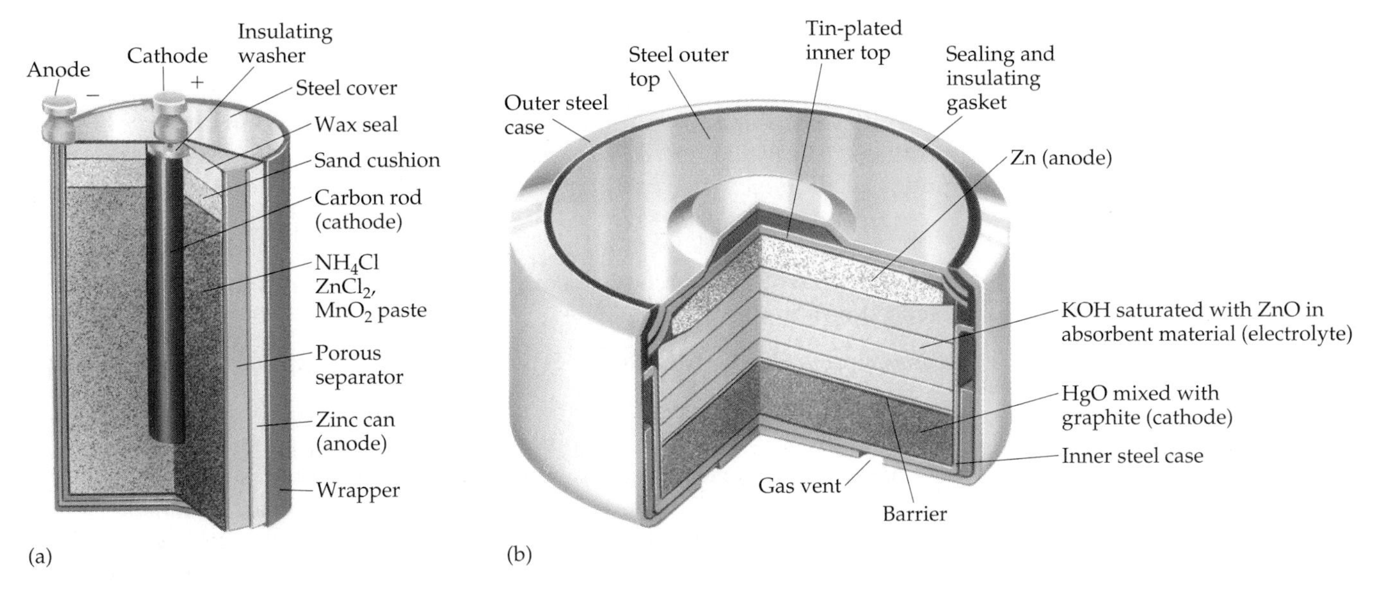

Figure 21.12 Primary batteries. (a) The Leclanché dry cell. (b) A mercury battery. The reducing agent is zinc, and the oxidizing agent is mercury(II) oxide.

Primary Batteries

The common **dry cell battery**, invented by Georges Leclanché (1839–1882) in 1866, is the energy source in toys, flashlights, and remote controllers for TVs, among other things. This type of battery contains a carbon rod electrode inserted into a moist paste of NH_4Cl, ZnCl, and MnO_2 in a zinc can that serves as the anode (Figure 21.12).

Anode, oxidation: $Zn(s) \longrightarrow Zn^{2+}(aq) + 2\ e^-$

The electrons reduce the ammonium ion to ammonia and hydrogen at the carbon cathode.

Cathode, reduction: $2\ NH_4^+(aq) + 2\ e^- \longrightarrow 2\ NH_3(g) + H_2(g)$

The products of the cathode reaction are gases and would cause the sealed dry cell to explode if they were not removed. This is the reason for the manganese(IV) oxide, an oxidizing agent that consumes the hydrogen,

$$2\ MnO_2(s) + H_2(g) \longrightarrow Mn_2O_3(s) + H_2O(\ell)$$

and the ammonia is taken up by zinc(II) ion:

$$Zn^{2+}(aq) + 2\ NH_3(g) + 2\ Cl^-(aq) \longrightarrow Zn(NH_3)_2Cl_2(s)$$

All these reactions lead to the following net process with a potential of 1.5 V:

Net reaction: $2\ MnO_2(s) + 2\ NH_4Cl(s) + Zn(s) \longrightarrow Mn_2O_3(s) + H_2O(\ell) + Zn(NH_3)_2Cl_2(s)$

Unfortunately, there are at least two disadvantages to this battery. If current is drawn from the battery rapidly, the gaseous products cannot be consumed rapidly enough, and the potential drops. Furthermore, a slow reaction between the zinc electrode and ammonium ion leads to further deterioration, and the battery has a poor "shelf life."*

The somewhat more expensive "alkaline" battery is now commonly used because it avoids some of the problems of dry cell batteries. An **alkaline battery** produces 1.54 V, and the key reaction is again the oxidation of zinc, this time under alkaline or basic conditions. The oxidation or anode reaction is

Anode, oxidation: $Zn(s) + 2\ OH^-(aq) \longrightarrow ZnO(s) + H_2O(\ell) + 2\ e^-$

and the electrons produced are consumed by reduction of manganese(IV) oxide at the cathode:

Cathode, reduction: $2\ MnO_2(s) + H_2O(\ell) + 2\ e^- \longrightarrow Mn_2O_3(s) + 2\ OH^-(aq)$

In contrast to the Leclanché battery, no gases are formed in the alkaline battery, and there is no decline in potential under high current loads.

Mercury batteries are a close relative of alkaline batteries (see Figure 21.12) and are typically used in calculators, cameras, watches, heart pacemakers, and other devices in which a small battery is required. As in dry cells and alkaline batteries; the anode is metallic zinc, but the cathode is mercury(II) oxide:

Anode, oxidation: $Zn(s) + 2\ OH^-(aq) \longrightarrow ZnO(s) + H_2O(\ell) + 2\ e^-$
Cathode, reduction: $HgO(s) + H_2O(\ell) + 2\ e^- \longrightarrow Hg(\ell) + 2\ OH^-(aq)$

These materials are tightly compacted powders separated by a moist paste of HgO containing some NaOH or KOH. Moistened paper serves as the "salt bridge," and the battery produces 1.35 V. These batteries are widely used, but, because they contain mercury, they can lead to some environmental problems. Mercury and its compounds are poisonous, so, if possible, mercury cells should be reprocessed to recover the metal when the battery is no longer useful.

The **lithium battery** has become popular because of its light weight. Lithium, which has a lower density ($d = 0.534\ g/cm^3$) than zinc ($d = 7.14$ g/cm^3), is used as the anode. Another advantage is that lithium is a stronger reducing agent than zinc (see Table 21.1), so lithium batteries produce about 3 V.

Secondary Batteries

The Leclanché cell and alkaline and mercury batteries no longer produce a current when the chemicals inside have reached equilibrium conditions. At that point they must be discarded. In contrast, secondary, or storage, batteries can be recharged, some of them hundreds of times. The original reactant concentrations can be restored by reversing the net cell reaction using an external source of electric energy.

An automobile battery—the **lead storage battery**—is perhaps the best example. Such batteries are used to supply the energy to the engine starter of a car, but once the engine is running, the battery is recharged by current from the car's alternator. There are two types of electrodes in a lead storage battery (Figure 21.13): one made of porous lead (the reducing agent) and the other of compressed, insoluble lead(IV) oxide (the oxidizing agent). The electrodes, arranged alternately in a stack and immersed in aqueous sulfuric acid, are sep-

*The shelf life of a dry cell is also affected by temperature. You can double or triple the shelf life of a battery if you store it in a refrigerator at about 4 °C.

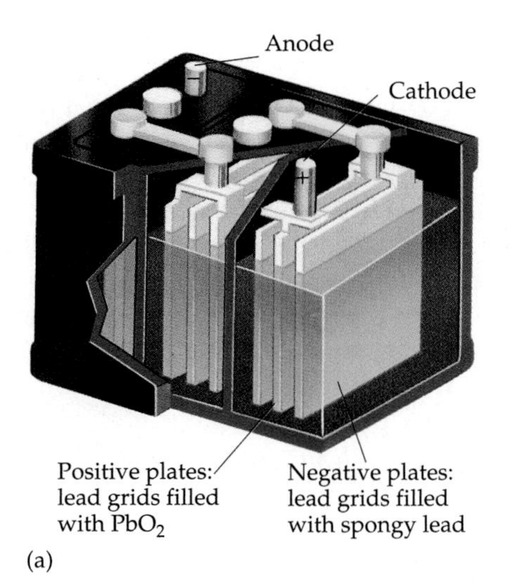

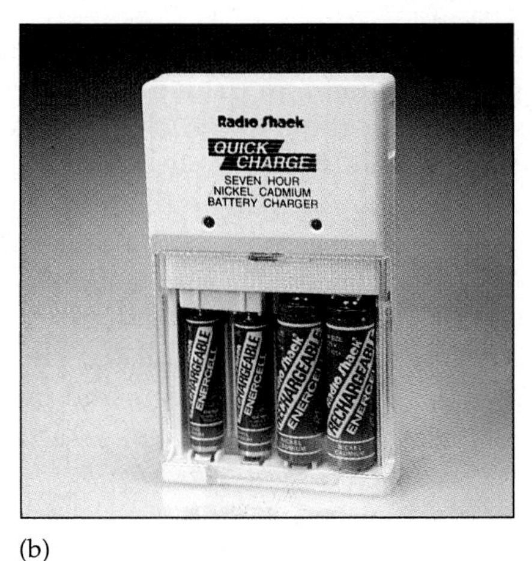

Figure 21.13 Secondary batteries. These are also called storage, or rechargeable, batteries. (a) The lead storage battery. (b) Nickel–cadmium or "ni–cad" batteries. *(b, C. D. Winters)*

arated by thin fiberglass sheets. When the cell acts as a supplier of electric energy, the lead electrode is oxidized to insoluble lead(II) sulfate:

Anode, oxidation:
$$Pb(s) + HSO_4^-(aq) + H_2O(\ell) \longrightarrow PbSO_4(s) + H_3O + 2\,e^- \qquad E° = +0.356\text{ V}$$

The electrons move through the external circuit to the lead(IV) oxide electrode where they cause reduction of PbO_2:

Cathode, reduction:
$$PbO_2(s) + 3\,H_3O^+(aq) + HSO_4^-(aq) + 2\,e^- \longrightarrow PbSO_4(s) + 5\,H_2O(\ell) \qquad E° = +1.685\text{ V}$$

The net reaction supplies electric energy but leaves both electrodes coated with an adhering film of white lead(II) sulfate and consumes sulfuric acid:

Net process:
$$Pb(s) + PbO_2(s) + 2\,HSO_4^-(aq) + 2\,H_3O^+(aq) \longrightarrow 2\,PbSO_4(s) + 4\,H_2O(\ell) \qquad E° = +2.041\text{ V}$$

A lead storage battery can usually be recharged by supplying electric energy to reverse the net process. The film of $PbSO_4$ is converted back to metallic lead and PbO_2, and sulfuric acid is regenerated.

Common car batteries have six cells in series, producing 12 V.

Lead storage batteries are large and heavy and produce a relatively low power for their mass, so many attempts have been and are being made to improve on them. Nonetheless, they do produce a relatively constant 2 V and a large initial current. This combination has been hard to beat, and they remain in widespread use.

Rechargeable, lightweight **nickel–cadmium** ("ni–cad") **batteries** (see Figure 21.13) are used in a variety of cordless appliances, such as telephones and video camcorders. These have the advantage that the oxidizing and reducing agent can be regenerated easily when recharged, and they produce a nearly constant potential:

Anode, oxidation:
$$Cd(s) + 2\,OH^-(aq) \longrightarrow Cd(OH)_2(s) + 2\,e^-$$
Cathode, reduction:
$$NiO(OH)(s) + H_2O(\ell) + e^- \longrightarrow Ni(OH)_2(s) + OH^-(aq)$$

Like mercury batteries, ni–cad batteries should be disposed of with care because cadmium and its compounds are toxic.

Fuel Cells

The first fuel cell was demonstrated in 1839 to the Royal Institution in London by William Grove. (This was during the time that Michael Faraday was director of the Institution.) Grove (1811–1896) was a barrister but was interested in electrochemistry and energy.

A **fuel cell** is also an electrochemical device for converting chemical energy, provided by a fuel and an oxidant, into electricity (Figure 21.14). In contrast to a storage battery, though, a fuel cell does not need to involve a reversible reaction; the reactants are supplied to the cell as needed from an external source.

Fuel cells are actually rather old technology. A hydrogen–oxygen fuel cell was first demonstrated in 1839, but it was not until the space program began in the 1960s that they came into use. Fuel cells have been used for some years in the Gemini, Apollo, and Space Shuttle programs to provide electric power, heat, and fresh water for space vehicles. They are now being tested by several companies for use in buses and cars and in stationary power plants.

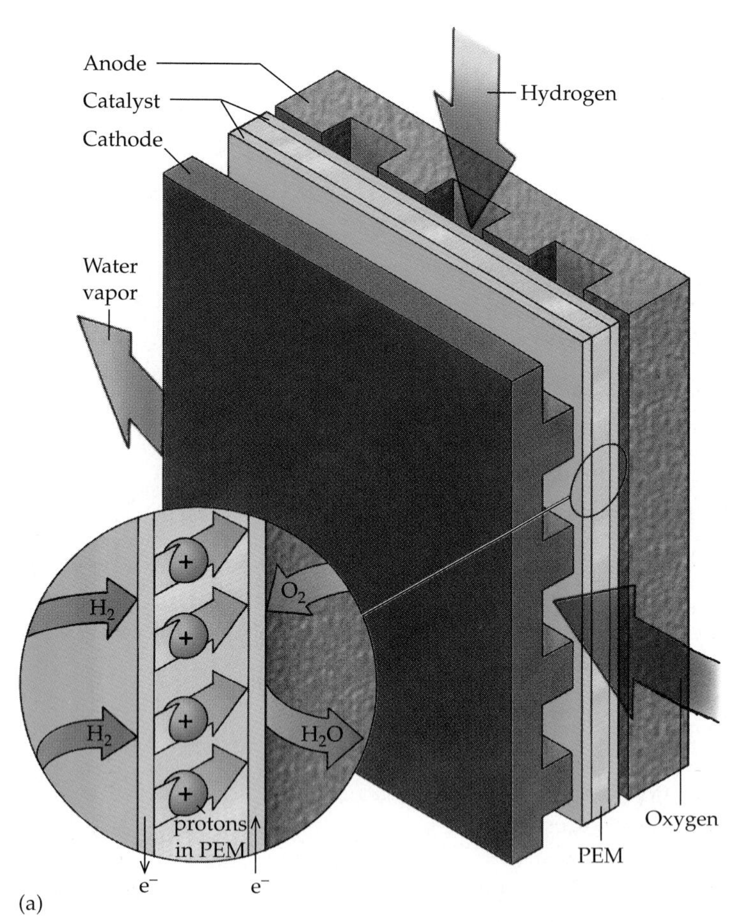

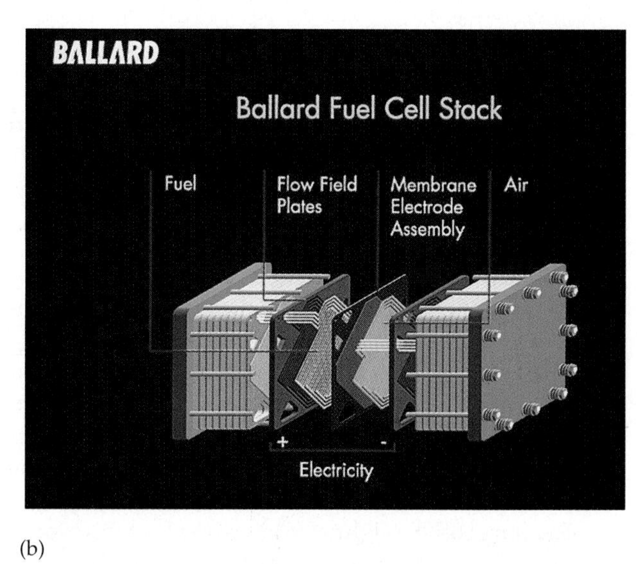

Figure 21.14 A hydrogen–oxygen fuel cell. (a) At the center of this fuel cell is a proton-conducting plastic foil called a "proton exchange membrane," or PEM. The foil is coated with a platinum catalyst and an electrode made of gas-permeable graphite paper. This device is sandwiched between graphite plates that have channels that carry the gas to the electrode. Hydrogen molecules are oxidized to H^+ ions at the catalytic surface of the anode. The H^+ ions move to the cathode through the PEM, which functions as the electrolyte in the cell. The H^+ ions join with reduced oxygen, O^{2-} ions, in the cathode compartment to produce water. (b) Detail of a fuel cell "stack." Here a number of cells, as diagrammed in (a), are connected in series. (The stack is pulled apart to see more of the detail.) Stacking cells in series increases the voltage produced. Increasing the cells' surface area increases the current produced. *(Ballard Power Systems)*

The net cell reaction of the hydrogen–oxygen cell is the oxidation of hydrogen with oxygen to give water (and provide electric energy):

Anode, oxidation:	$2\ H_2(g) + 4\ H_2O(\ell) \longrightarrow 4\ H_3O^+(aq) + 4\ e^-$
Cathode, reduction:	$O_2(g) + 4\ H_3O^+(aq) + 4\ e^- \longrightarrow 6\ H_2O(\ell)$
Net reaction:	$2\ H_2(g) + O_2(g) \longrightarrow 2\ H_2O(\ell)$
	$E^\circ_{net} = 1.23$ V

Gaseous H_2 is pumped onto the anode of the cell, and O_2 or air is directed to the cathode. The product, water, is swept out of the cell as a vapor and can be purified for drinking purposes. Electrons flowing in the external circuit can be used to provide power for an automobile, for example. The fuel cells on board the Space Shuttle deliver the same power as batteries weighing ten times as much would provide. On a typical seven-day mission, the Shuttle fuel cells consume 1500 lb of hydrogen and generate 190 gal of drinking water.

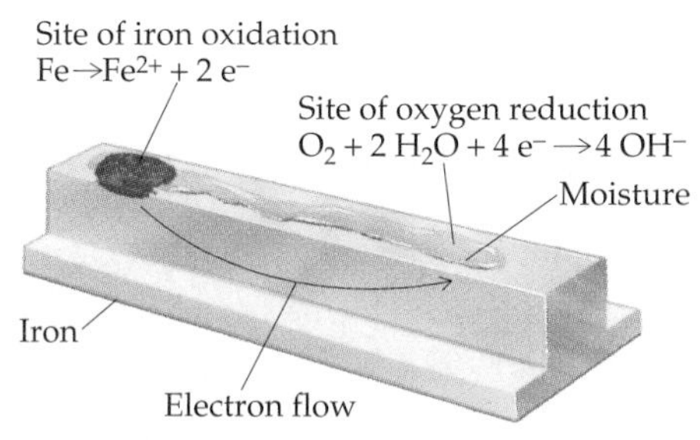

Figure 21.15 The reactions that occur when iron corrodes in an aqueous environment with oxygen present. The site of oxidation of iron may be different from the site of oxygen reduction because electrons can flow through the metal.

21.7 CORROSION: REDOX REACTIONS IN THE ENVIRONMENT

See the *Saunders Interactive General Chemistry CD-ROM,* Screen 21.9, Corrosion.

Corrosion is the deterioration of metals, usually with loss of metal to a solution in some form, by a product-favored oxidation–reduction reaction. The corrosion of iron, for example, is the conversion of the metal to red-brown rust, hydrated iron(III) oxide $[Fe_2O_3 \cdot H_2O]$, and other products. This is significant because an estimated 25% of the annual steel production in the United States is used to replace material lost to corrosion.

For the corrosion to occur at the surface of a metal, there must be anodic areas where the metal can be oxidized to metal ions as electrons are produced,

Anode, oxidation: $M(s) \longrightarrow M^{n+}(aq) + n\ e^-$

and cathodic areas where the electrons are consumed by half-reactions such as

Cathode, reduction:

$$2\ H_3O^+(aq) + 2\ e^- \longrightarrow H_2(g) + 2\ H_2O(\ell)$$
$$2\ H_2O(\ell) + 2\ e^- \longrightarrow H_2(g) + 2\ OH^-(aq)$$
$$O_2(g) + 2\ H_2O(\ell) + 4\ e^- \longrightarrow 4\ OH^-(aq)$$

Anodic areas occur at cracks in the oxide coating of the metal, at boundaries between phases, or around impurities. The cathodic areas occur at the metal oxide coating, at less reactive metallic impurity sites, or around other metal compounds such as sulfides.

The other requirements for corrosion are (1) an electric connection between the anode and cathode and (2) an electrolyte with which both anode and cathode are in contact. Both requirements are easily fulfilled, as seen in Figures 21.15 and 21.16.

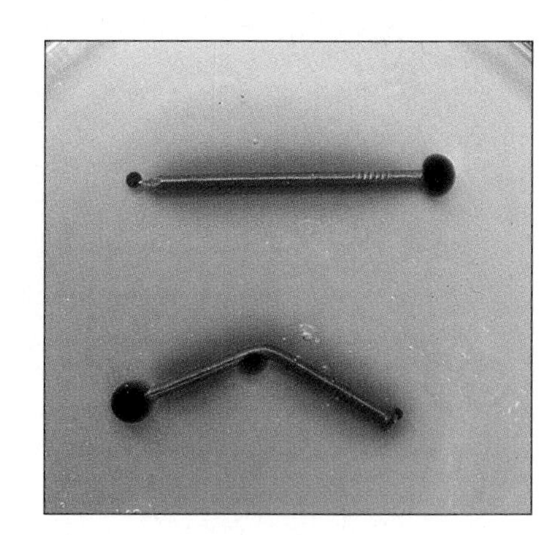

Figure 21.16 Corroding iron nails. Two nails were placed in an agar gel, which also contained the indicator phenolphthalein and $[Fe(CN)_6]^{3-}$. The nails began to corrode and gave Fe^{2+} ions at the tip or where the nail is bent. That these points are the anode is indicated by the blue-green color of Prussian blue, a complex of $[Fe(CN)_6]^{3-}$ and Fe^{2+}, the latter from oxidized iron. The remainder of the nail is the cathode, because oxygen is reduced in water to give OH^- (see Figure 21.15). The presence of the OH^- ion is indicated by the red color of the indicator. *(C. D. Winters)*

If the relative rates of the anodic and cathodic corrosion reactions could be measured independently, the anodic reaction would be found to be the faster. When the two reactions are coupled to each other as in a corroding metal, however, the overall rate can only be that of the slower process. This generally means that corrosion is controlled by the rate of the cathodic process, a fact that helps

CURRENT PERSPECTIVES

Your Next Car?

With some state governments now mandating the availability of vehicles with low emissions of pollutants, automobile manufacturers and many other companies are exploring ways to bring this about. One way is to make gasoline or diesel engines emission-free, a very difficult engineering problem. Another way is to use electric vehicles powered by batteries or fuel cells.

The large car manufacturers have already designed prototype electric cars (page 947), and most use tried-and-true lead-acid storage batteries. The problem with these batteries, however, is that they consume 321 g of reactants per mole of electrons:

$$Pb(s) + PbO_2(s) + 2\ H_2SO_4(aq) \longrightarrow 2\ PbSO_4(s) + 2\ H_2O(\ell)$$

In fact, lead storage batteries deliver the least power per kilogram of battery weight. But the table and photograph show another problem—the power available from any type of battery is much less than is available from an equivalent mass of gasoline.

AMOUNT OF ENERGY STORED PER KILOGRAM OF BATTERY MASS

Chemical System	**Watt-hour/kg (1 W · h = 3600 J)**
Lead-acid battery	18–56
Nickel–cadmium battery	33–70
Sodium–sulfur battery	80–140
Lithium polymer battery	150
Gasoline–air combustion engine	12,200

Fuel cells are another possible solution (see Figure 21.14). Hydrogen–oxygen fuel cells are well developed and have a maximum efficiency of 83%. In particular, hydrogen–oxygen fuels cells with a "proton exchange membrane" (PEM) as the electrolyte are promising for automotive use. They operate at room temperature or slightly above, start rapidly, have high specific power, and can tolerate impurities in the reactants. Cars and buses equipped with fuel cells are now being tested (see Figure 21.1).

One problem with fuel cells is how to transport the hydrogen gas needed by the cell, and how to "fill up" the hydrogen tank. (The bulge on top of the van in Figure 21.1b is a hydrogen tank.) The latest idea is to generate the hydrogen in the car by combining gasoline (or another fuel such as methanol or ethanol) and air over an appropriate catalyst to give CO and H_2 gas:

$$C_8H_{18}(g) + O_2(g) \longrightarrow 8\ CO(g) + 9\ H_2(g)$$

The CO is converted to CO_2 in a reaction with steam, which produces still more hydrogen:

$$CO(g) + H_2O(g) \longrightarrow CO_2(g) + H_2(g)$$

The first reaction is a catalyzed partial oxidation of a hydrocarbon and the second is the "water gas shift" reaction. Both are old technology. What is new is that they were recently demonstrated on a scale appropriate for automotive use. The drawing (opposite) is of a "concept" car from a major U.S. manufacturer.

The gasoline engines used in cars and trucks today are the product of years of engineering development. Auto companies, however, are investing heavily in fuel cell technology in the hope of producing a "green" car within the next ten years.

to explain the chemistry of corroding systems and how to prevent corrosion.

In the corrosion of iron, the anodic reaction is the oxidation of iron. Which of the possible cathodic reactions is the fastest depends, in part, on the acidity of the surrounding solution and the amount of oxygen present. When little or no oxygen is present—as when a piece of iron is buried in moist clay—hydrogen ion and water are reduced. As indicated in the preceding equations, $H_2(g)$ and hydroxide ions are the products. Iron(II) hydroxide can form, and because it is relatively insoluble, it can precipitate on the metal surface and inhibit the further formation of Fe^{2+} at the anodic site:

Anode, oxidation: $$Fe(s) \longrightarrow Fe^{2+}(aq) + 2\ e^-$$

Cathode, reduction: $$2\ H_2O(\ell) + 2\ e^- \longrightarrow H_2(g) + 2\ OH^-(aq)$$

$$Fe^{2+}(aq) + 2\ OH^-(aq) \longrightarrow Fe(OH)_2(s)$$

Net reaction: $$Fe(s) + 2\ H_2O(\ell) \longrightarrow H_2(g) + Fe(OH)_2(s)$$

Corrosion under oxygen-free conditions is slow for two reasons. First, H_2O re-

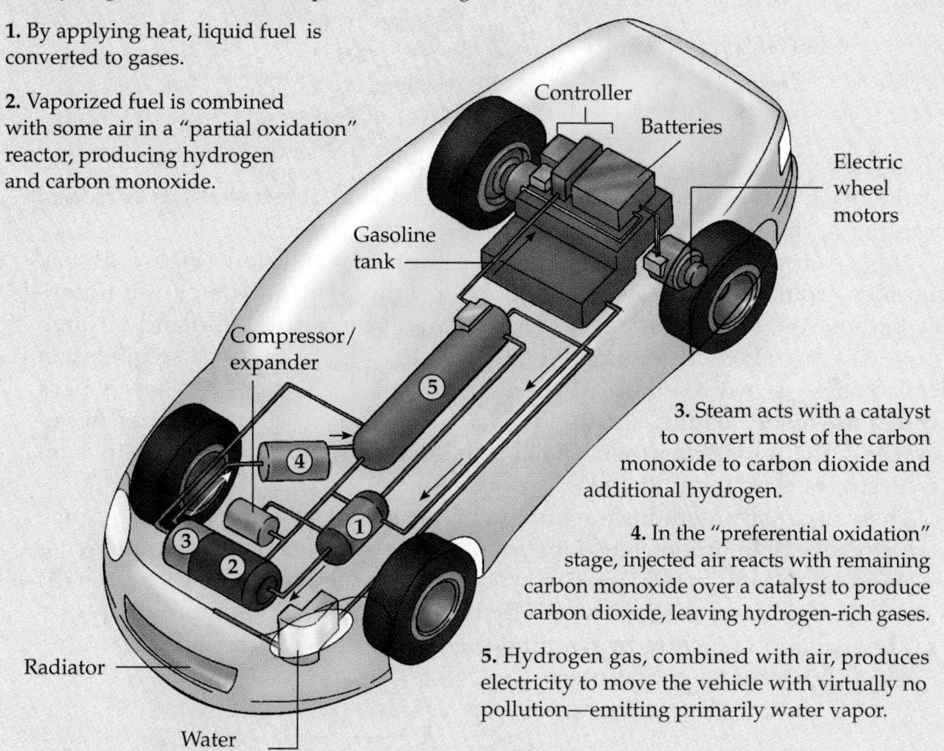

The fuel system of a hypothetical car. Liquid fuel (gasoline, ethanol, or methanol) is first vaporized (Step 1). In Step 2 the fuel is converted to CO and H_2, and in Step 3 the CO is combined with steam to produce more H_2 and CO_2. Any remaining CO is converted to CO_2 in Step 4. Finally, the H_2 is combined with air in the fuel cell stack (see Figure 21.14), and the electricity generated is used by the batteries to propel the car.

Design for a car using a fuel cell.
The hydrogen for the fuel cell is produced from gasoline in the car.
1. By applying heat, liquid fuel is converted to gases.
2. Vaporized fuel is combined with some air in a "partial oxidation" reactor, producing hydrogen and carbon monoxide.
Controller
Batteries
Electric wheel motors
Gasoline tank
Compressor/ expander
3. Steam acts with a catalyst to convert most of the carbon monoxide to carbon dioxide and additional hydrogen.
4. In the "preferential oxidation" stage, injected air reacts with remaining carbon monoxide over a catalyst to produce carbon dioxide, leaving hydrogen-rich gases.
5. Hydrogen gas, combined with air, produces electricity to move the vehicle with virtually no pollution—emitting primarily water vapor.
Radiator
Water reservoir

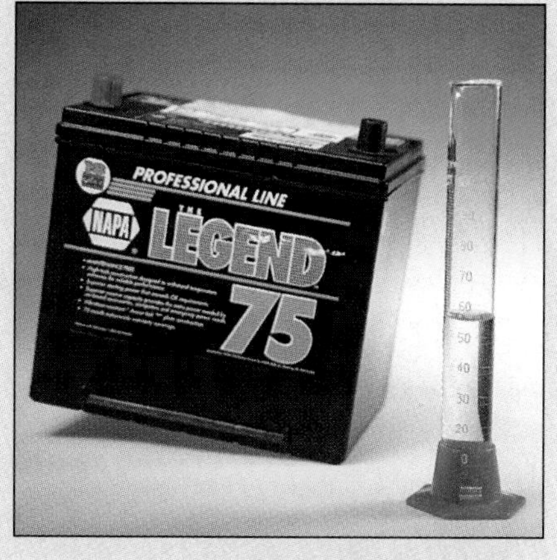

Gram for gram, gasoline packs more energy than a fully charged lead storage battery. A 15-kg battery has the same amount of stored energy as 59 mL of gasoline.

duction is slow. Second, the coating of insoluble $Fe(OH)_2$ inhibits further formation of Fe^{2+} at the anodic site.

If both water and O_2 are present, the chemistry of iron corrosion is somewhat different, and the corrosion reaction is about 100 times faster than without oxygen:

Anode, oxidation: $2\,[Fe(s) \longrightarrow Fe^{2+}(aq) + 2\,e^-]$

Cathode, reduction: $O_2(g) + 2\,H_2O(\ell) + 4\,e^- \longrightarrow 4\,OH^-(aq)$

$Fe^{2+}(aq) + 2\,OH^-(aq) \longrightarrow Fe(OH)_2(s)$

Net reaction: $2\,Fe(s) + 2\,H_2O(\ell) + O_2(g) \longrightarrow 2\,Fe(OH)_2(s)$

If the iron object has free access to oxygen and water, as in the open or in flowing water, red-brown iron(III) oxide forms:

$$4\,Fe(OH)_2(s) + O_2(g) \longrightarrow \underset{\text{red-brown}}{2\,Fe_2O_3 \cdot H_2O(s)} + 2\,H_2O(\ell)$$

This is the familiar rust you see on cars and buildings and the substance that colors the water red in some mountain streams or in your home. On the other hand, if oxygen is not freely available, further oxidation of the iron(II) hydroxide is limited to the formation of magnetic iron oxide (which can be thought of as a mixed oxide of Fe_2O_3 and FeO):

$$6\ Fe(OH)_2(s) + O_2(g) \longrightarrow \underset{\text{green hydrated magnetite}}{2\ Fe_3O_4 \cdot H_2O(s)} + 4\ H_2O(\ell)$$

$$Fe_3O_4 \cdot H_2O(s) \longrightarrow H_2O(\ell) + \underset{\text{black magnetite}}{Fe_3O_4(s)}$$

It is the black magnetite that you find coating an iron object that has corroded by resting in moist soil.

Other substances in air and water can assist in corrosion (Figure 21.17). Chlorides, from sea air or from salt spread on the roads in winter, are notorious. Because the chloride ion is relatively small, it can diffuse into and through a protective metal oxide coating. Metal chlorides, which are more soluble than metal oxides or hydroxides, can then form. These chloride salts leach back through the oxide coating, and a path is now open for oxygen and water to further attack the underlying metal. This is the reason you often see small pits on the surface of a corroded metal.

There are many methods for stopping a metal object from corroding, some more effective than others, but none totally successful. The general approaches are (1) to inhibit the anodic process, (2) to inhibit the cathodic process, or (3) to do both. The usual method is **anodic inhibition,** attempting to directly prevent the oxidation reaction by painting the metal surface or by allowing a thin

(a)

(b)

***Figure* 21.17 Reduction of copper(II) ions by aluminum metal.** (a) A ball of aluminum foil is added to a solution of copper(II) nitrate and sodium chloride. Normally, the coating of chemically inert Al_2O_3 on the surface of aluminum protects the metal from further oxidation. (b) In the presence of the Cl^- ion, however, the coating of Al_2O_3 is breached, and aluminum reduces copper(II) ions to copper metal. The reaction is rapid and is so exothermic that the water can boil on the surface of the foil. (Notice that the blue color of copper(II) ions has faded as these ions are consumed in the reaction.) (See Screen 21.9 of the *Saunders Interactive General Chemistry CD-ROM.*) *(C. D. Winters)*

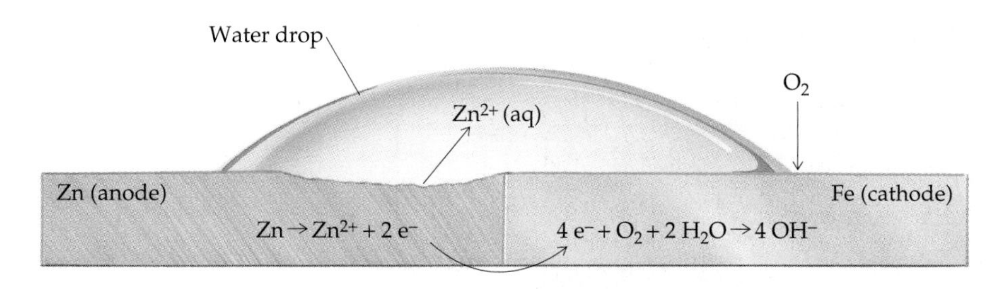

Figure 21.18 Cathodic protection of an iron-containing object. The iron is coated with a film of zinc, a metal more easily oxidized than iron. The zinc acts as an anode and forces iron to become the cathode, thereby preventing the corrosion of the iron.

oxide film to form. More recently developed methods are illustrated by the following reaction:

$$2\ Fe(s) + 2\ Na_2CrO_4(aq) + 2\ H_2O(\ell) \longrightarrow Fe_2O_3(s) + Cr_2O_3(s) + 4\ NaOH(aq)$$

An iron surface is oxidized by a chromium(VI) salt to give iron(III) and chromium(III) oxides. These form a coating impervious to O_2 and water, and further atmospheric oxidation is inhibited.

There are several other ways to inhibit metal oxidation, one of which is to force the metal to become the cathode, instead of the anode, in an electrochemical cell. Hence, this is called **cathodic protection.** This is usually done by attaching another, more readily oxidized metal. The best example of this is **galvanized iron,** iron that has been coated with a thin film of zinc (Figure 21.18). The $E°$ for zinc oxidation is considerably more positive than $E°$ for iron oxidation (see Table 21.1), so the zinc metal film is oxidized before any of the iron. The zinc coating forms what is called a **sacrificial anode.** Another reason for using a zinc coating on iron is that, when the zinc is corroded, $Zn(OH)_2$ forms on the surface and inhibits further oxidation.

21.8 ELECTROLYSIS: CHEMICAL CHANGE FROM ELECTRIC ENERGY

See the *Saunders Interactive General Chemistry CD-ROM,* Screen 21.10, Electrolysis.

An electric current can be produced by a chemical change. Equally important, however, is the opposite process, **electrolysis,** the use of an electric current to bring about chemical change.

What happens if a pair of inert electrodes, which are connected to a battery, are inserted into a bath of molten NaCl (Figure 21.19)? The external battery, or other source of electric potential, acts as an "electron pump," and electrons flow from this source into one of the electrodes, thereby giving it a negative charge. Sodium ions are attracted to this negative electrode and are reduced when electrons from the electrode are accepted, making the electrode the cathode. The battery simultaneously draws electrons from the other electrode, giving it a positive electric charge. Chloride ions are attracted to this electrode and surrender electrons. Because oxidation has occurred, this is the anode. Thus, the following reactions have occurred in molten NaCl:

Anode, oxidation: $2\ Cl^- \longrightarrow Cl_2(g) + 2\ e^-$

Cathode, reduction: $2\ Na^+ + 2\ e^- \longrightarrow 2\ Na(\ell)$

Net reaction: $2\ Cl^- + 2\ Na^+ \longrightarrow 2\ Na(\ell) + Cl_2(g) \qquad E_{net} \approx -4\ V$

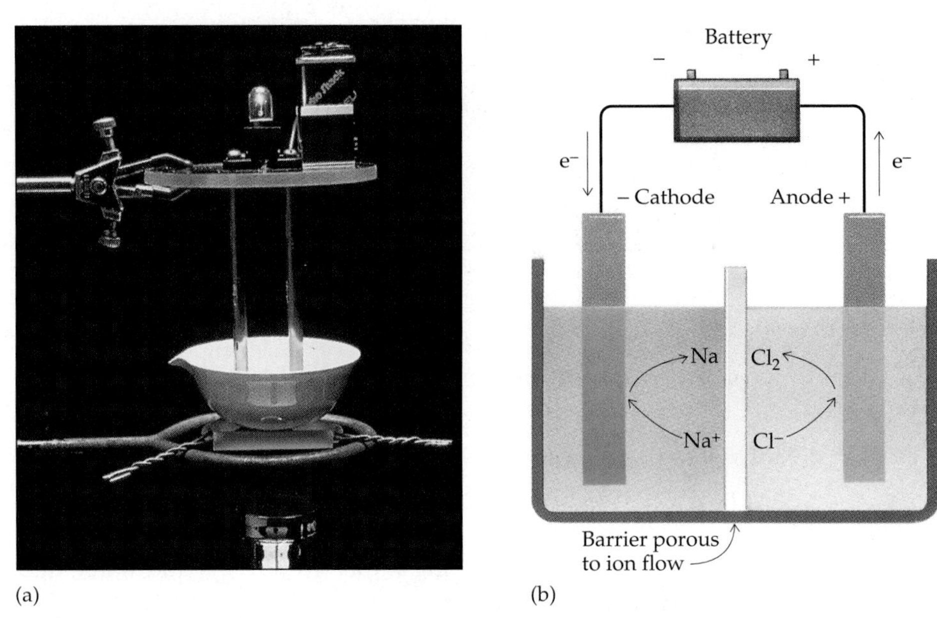

Figure 21.19 Electrolysis of molten sodium chloride. (a) When melted, a salt can conduct electricity because the ions are free to move in the molten salt. Here a small light shows the circuit is conducting when electrodes are immersed in the molten salt. (b) In the molten state, sodium cations migrate to the negative cathode where they are reduced to sodium metal. Chloride ions migrate to the positive anode where they are oxidized to elemental chlorine, Cl_2. *(a, C. D. Winters)*

The half-cell potentials in Table 21.1 apply only to aqueous solutions. If we use these to *estimate* the potential for the preceding reaction, we obtain a value of about −4 V. The reaction is clearly not product-favored in the direction written, and this is the reason that an external battery has been attached. The battery, with a potential greater than 4 V, forces the reactant-favored reaction to occur by "pumping" electrons in the proper direction.

Problem-Solving Tips and Ideas

21.2 A Summary of Electrochemical Terminology

Whether you are describing a voltaic cell or an electrolysis cell, the terms "anode" and "cathode" always refer to the electrodes at which oxidation and reduction occur, respectively. The polarity of the electrodes is reversed, however, in a voltaic or electrolysis cell.

Type of Cell	Electrode	Function	Polarity
Voltaic	Anode	Oxidation	−
	Cathode	Reduction	+
Electrolysis	Anode	Oxidation	+
	Cathode	Reduction	−

In a voltaic cell, the negative electrode is the one at which electrons are produced. In an electrolysis cell, the negative electrode is the one onto which the external source is "pumping" the electrons.

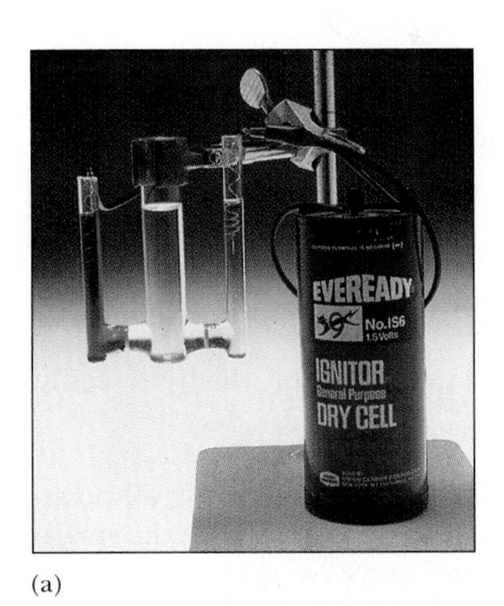

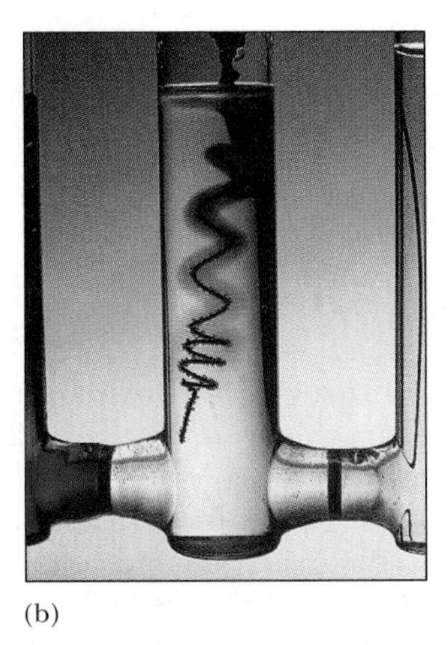

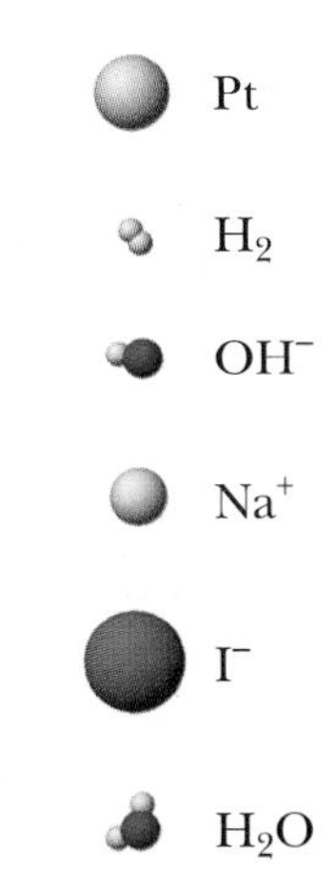

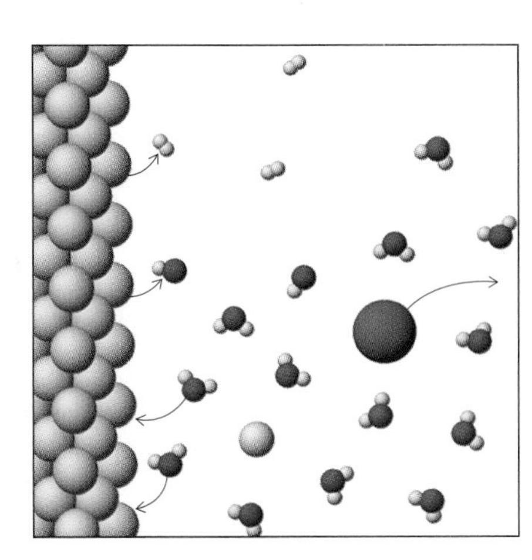

(a) (b) (c)

Figure 21.20 The electrolysis of aqueous sodium iodide. Aqueous NaI is contained in all three compartments of the cell, and both electrodes are platinum. (a) At the positive electrode, or anode *(right)*, the I^- ion is oxidized to iodine, I_2, which gives the solution a yellow-brown color. At the negative electrode, or cathode *(left)*, water is reduced, and the presence of the OH^- ion is indicated by the red color of the acid–base indicator phenolphthalein. (b) In a close-up of the cathode, bubbles of H_2 and evidence of OH^- being generated at the electrode are clearly seen. (c) A model of the process. Water molecules are reduced at the electrode surface, producing H_2 gas and aqueous OH^- ions. *(Photos, C. D. Winters; models, S. M. Young)*

What if an aqueous solution of a salt, say sodium iodide, is used instead of a molten salt (Figure 21.20)? With water now present, are Na^+ and I^- ions reduced and oxidized, respectively, or is water involved? Possible *reduction reactions* are

$$Na^+(aq) + e^- \longrightarrow Na(s) \qquad E^\circ = -2.71\ V$$
$$2\ H_2O(\ell) + 2\ e^- \longrightarrow H_2(g) + 2\ OH^-(aq) \qquad E^\circ = -0.83\ V$$
$$2\ H_3O^+(aq) + 2\ e^- \longrightarrow 2\ H_2O(\ell) + H_2(g) \qquad E^\circ = 0.00\ V$$

When several reactions at an electrode are possible, the cathode in this case, we must consider not only which is the most easily reduced (best oxidizing agent) but also which is reduced most *rapidly*. Complications usually occur when currents are large—as in a commercial electrolysis cell—and when reactant concentrations are small. In this case H_2 and OH^- are clearly observed as products. This is reasonable because water is certainly reduced more readily than sodium. Furthermore, because the hydronium ion concentration is only about 10^{-7} M in an NaI solution, the middle reaction is the best description of the net change occurring at the cathode.*

For the oxidation processes possible in aqueous sodium iodide, we need to compare the two reactions

$$2\ I^-(aq) \longrightarrow I_2(s) + 2\ e^- \qquad E^\circ = -0.535\ V$$
$$6\ H_2O(\ell) \longrightarrow O_2(g) + 4\ H_3O^+(aq) + 4\ e^- \qquad E^\circ = -1.23\ V$$

* All of the electrode potentials for the reactions discussed here should be corrected to nonstandard conditions. Furthermore, one should take into account the kinetic phenomenon of *overpotential.* Despite these complications, E° values do allow us to make some *estimates* of the reactions that occur under electrolysis conditions.

(A third possibility, the oxidation of OH^- ion, is not considered due to the very low concentration of the ion.) The iodide ion is the more easily oxidized species. The best description of the chemistry that occurs on electrolysis of aqueous NaI is therefore the reactions

Anode, oxidation:	$2\ I^-(aq) \longrightarrow I_2(s) + 2\ e^-$	$E° = -0.535$ V
Cathode, reduction:	$2\ H_2O(\ell) + 2\ e^- \longrightarrow H_2(g) + 2\ OH^-(aq)$	$E° = -0.83$ V
Net reaction:	$2\ I^-(aq) + 2\ H_2O(\ell) \longrightarrow H_2(g) + 2\ OH^-(aq) + I_2(s)$	$E°_{net} = -1.37$ V

The products are hydrogen, hydroxide ion, and iodine, all of which are easily identified in the experiment in Figure 21.20.

What happens if an aqueous solution of some other metal halide such as $SnCl_2$ is electrolyzed? As before, consult Table 21.1, and, considering all possible reactions, find the oxidation and reduction reactions that require the smallest potential. In this case aqueous Sn^{2+} ion is *much* more easily reduced ($E° = -0.14$ V) than water ($E° = -0.83$ V) at the cathode, so tin metal is produced. At the anode, two oxidations are possible: $Cl^-(aq)$ to $Cl_2(g)$ and $H_2O(\ell)$ to $O_2(g)$. Experiments show that chloride ion is generally oxidized more rapidly than water, so the reactions occurring on electrolysis of aqueous tin(II) chloride are (Figure 21.21)

Anode, oxidation:	$2\ Cl^-(aq) \longrightarrow Cl_2(g) + 2\ e^-$	$E° = -1.36$ V
Cathode, reduction:	$Sn^{2+}(aq) + 2\ e^- \longrightarrow Sn(s)$	$E° = -0.14$ V
Net reaction:	$Sn^{2+}(aq) + 2\ Cl^-(aq) \longrightarrow Sn(s) + Cl_2(g)$	$E°_{net} = -1.50$ V

Again, this process is of obvious commercial importance. Tin is used to coat food containers ("tin" cans) and, in alloys with other metals, to make solder, pewter, and bronze.

A useful general principle can be derived from the preceding discussion. If an electric current is passed through a solution, the electrode reactions are most likely those requiring the least potential (and those that occur most rapidly). In water, this means that a substance is reduced if it has a reduction potential *less negative* than about -0.8 V, the potential for the reduction of water (containing 1 M OH^-). A check of Table 21.1 and Appendix M shows that this includes commercially useful metals, such as Sn, Pt, Cu, Ag, Au, and Cd. Indeed, electrolysis is used to coat, or "plate," other materials with these metals. If a substance has a reduction potential *more negative* than about -0.8 V, then only water is reduced. Substances falling into this category include Na, K, Mg, and Al. To produce these

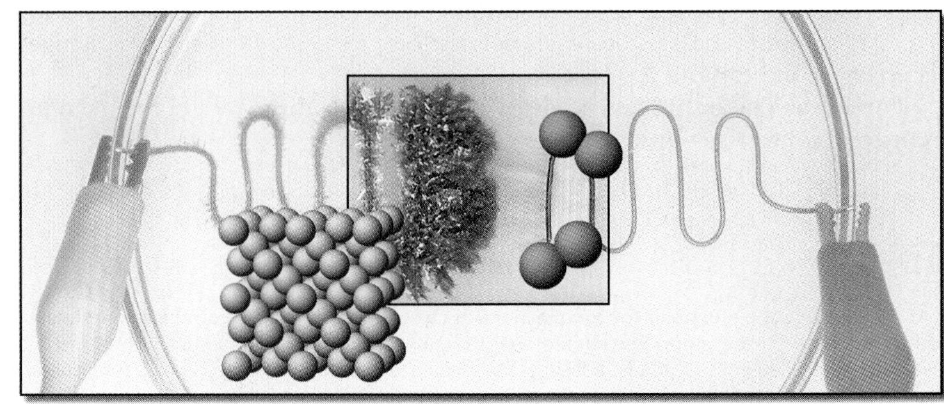

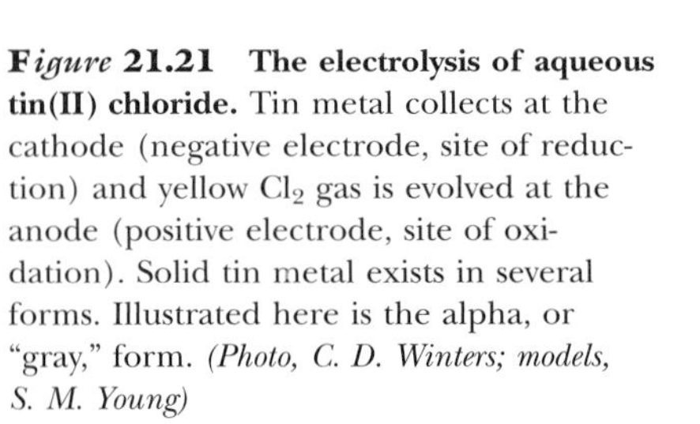

***Figure* 21.21 The electrolysis of aqueous tin(II) chloride.** Tin metal collects at the cathode (negative electrode, site of reduction) and yellow Cl_2 gas is evolved at the anode (positive electrode, site of oxidation). Solid tin metal exists in several forms. Illustrated here is the alpha, or "gray," form. *(Photo, C. D. Winters; models, S. M. Young)*

metals requires methods other than the reduction of their ions in aqueous solution.

Example 21.13 Electrolysis of Aqueous NaOH

Predict what happens when an electric current is passed through aqueous sodium hydroxide.

Solution First, list all the species in solution. In this case they are Na^+, OH^-, and H_2O. Next, use Table 21.1 to decide which of the species can be oxidized and which can be reduced, and note the potential of each possible reaction:

Reductions:

$$Na^+(aq) + e^- \longrightarrow Na(s) \qquad E° = -2.71\ V$$

$$2\ H_2O(\ell) + 2\ e^- \longrightarrow H_2(g) + 2\ OH^-(aq) \qquad E° = -0.83\ V$$

Oxidation:

$$4\ OH^-(aq) \longrightarrow O_2(g) + 2\ H_2O(\ell) + 4\ e^- \qquad E° = -0.40\ V$$

It is evident that water is reduced to H_2 at the cathode, and OH^- is oxidized to O_2 at the anode. The cell reaction is $2\ H_2O(\ell) \longrightarrow 2\ H_2(g) + O_2(g)$, and the potential under standard conditions is -1.23 V.

Exercise 21.11 Electrolysis of Salts

Predict the results of passing an electric current through each of the following solutions:

1. Molten NaBr
2. Aqueous NaBr
3. Aqueous $CuCl_2$

21.9 COUNTING ELECTRONS

Metallic silver is produced at the cathode in the electrolysis of aqueous $AgNO_3$, the reaction being $Ag^+(aq) + e^- \longrightarrow Ag(s)$. One mole of electrons is required to produce 1 mol of silver from 1 mol of silver ions. In contrast, 2 mol of electrons is required to produce 1 mol of tin.

$$Sn^{2+}(aq) + 2\ e^- \longrightarrow Sn(s)$$

See the *Saunders Interactive General Chemistry CD-ROM,* Screen 21.11, Coulometry.

It follows that, if the number of moles of electrons flowing through the electrolysis cell could be measured, the number of moles of silver or tin produced could be calculated. Conversely, if the amount of silver or tin produced is known, then the number of moles of electrons used could be calculated.

The number of moles of electrons consumed or produced in an electron transfer reaction is usually obtained by measuring the current flowing in the external electric circuit in a given time. The **current** flowing in an electric circuit is the charge (in units of coulombs) passed per unit time, and the usual unit for current is the **ampere.**

$$\text{Current, } I \text{ (amperes, A)} = \frac{\text{electric charge (coulombs, C)}}{\text{time (seconds, s)}} \qquad [21.4]$$

Michael Faraday (1791–1867) first explored the quantitative aspects of electricity (page 963). In his honor scientists have defined the Faraday constant, 96,485.31 C/mol, as the *charge carried by 1 mol of electrons* (see page 962). The current passing through an electrochemical cell, and the time the current flowed, are easily measured with modern instruments. The charge that passed through the cell can therefore be obtained by multiplying the current (in amperes) by the time (in seconds). Knowing the charge, and using the Faraday constant as a conversion factor, the number of moles of electrons that passed through an electrochemical cell can be calculated.

Example 21.14 Using the Faraday Constant

A current of 1.50 A is passed through a solution containing silver ions for 15.0 min. The voltage is such that silver is deposited at the cathode. What mass of silver, in grams, is deposited?

$$Ag^+(aq) + e^- \longrightarrow Ag(s)$$

Solution From the half-reaction we know that if 1 mol of electrons passed through the cell, then 1 mol of silver was deposited. To find the number of moles of electrons, we need to know the total electric charge passed through the cell. The charge can be calculated from experimental measurements of the current (in amperes) and the time the current flowed. Thus, the logic of the calculation is

$$\text{Time (s)} \times \text{current (A)} \longrightarrow \text{Charge (C)} \longrightarrow \text{Mol } e^- \longrightarrow \text{Mol Ag} \longrightarrow \text{Mass Ag}$$

1. Calculate the charge (number of coulombs) passed in 15.0 min:

$$\begin{aligned}\text{Charge (coulombs, C)} &= \text{current (A)} \times \text{time (s)} \\ &= 1.50 \text{ A } (15.0 \text{ min})(60.0 \text{ s/min}) \\ &= 1.35 \times 10^3 \text{ C}\end{aligned}$$

2. Calculate the number of moles of electrons:

$$1.35 \times 10^3 \text{ C}\left(\frac{1 \text{ mol } e^-}{9.65 \times 10^4 \text{ C}}\right) = 1.40 \times 10^{-2} \text{ mol } e^-$$

3. Calculate the number of moles of silver and then the mass of silver deposited:

$$1.40 \times 10^{-2} \text{ mol } e^-\left(\frac{1 \text{ mol Ag}}{1 \text{ mol } e^-}\right)\left(\frac{107.9 \text{ g Ag}}{1 \text{ mol Ag}}\right) = 1.51 \text{ g Ag}$$

Example 21.15 Using the Faraday Constant

One of the half-reactions occurring in the lead storage battery is

$$Pb(s) + HSO_4^-(aq) + H_2O(\ell) \longrightarrow PbSO_4(s) + H_3O^+(aq) + 2\ e^-$$

If a battery delivers 1.50 A, and if its lead electrode contains 427 g of lead, how long can current flow before the lead in the electrode is consumed?

Solution Here we begin with current and the mass of material used in the reaction, and we want to calculate a time. This is the reverse of Example 21.13:

Mass Pb ⟶ Mol Pb ⟶ Mol e^- ⟶ Charge (C) ⟶ Time (s)

1. Calculate the quantity of lead available (mol):

$$427\ \text{g}\left(\frac{1\ \text{mol}}{207.2\ \text{g}}\right) = 2.06\ \text{mol Pb}$$

2. Calculate the number of moles of electrons passed through the cell, assuming all the lead was consumed:

$$2.06\ \text{mol Pb}\left(\frac{2\ \text{mol } e^-}{1\ \text{mol Pb}}\right) = 4.12\ \text{mol } e^-$$

3. Calculate the charge carried by the electrons that passed through the cell.

$$4.12\ \text{mol } e^-\left(\frac{9.65 \times 10^4\ \text{C}}{1\ \text{mol } e^-}\right) = 3.98 \times 10^5\ \text{C}$$

4. Use the charge passed through the cell and the measured current to calculate the time.

$$3.98 \times 10^5\ \text{C} = 1.50\ \text{A} \times \text{time (s)}$$
$$\text{Time} = 2.65 \times 10^5\ \text{s}\ \ (\text{or } 73.6\ \text{h})$$

***Exercise* 21.12** Using the Faraday Constant

In the commercial production of sodium by electrolysis, the cell operates at 7.0 V and a current of 25×10^3 A. How many grams of sodium can be produced in one hour?

21.10 THE COMMERCIAL PRODUCTION OF CHEMICALS BY ELECTROCHEMICAL METHODS

Aluminum

Aluminum is the third most abundant element in the earth's crust and has many important uses in our economy. You probably know it best in its use in the kitchen to wrap food, a use demonstrating its excellent formability. Just as importantly, aluminum has a low density and excellent corrosion resistance; the latter property comes from the fact that a transparent, chemically inert film of aluminum oxide, Al_2O_3, clings tightly to the metal's surface. It is these properties that have led to the many other uses of aluminum in aircraft parts, ladders, and automobile parts, for example.

There is an interesting history to the development of a practical method for aluminum production. Aluminum was originally made in the 19th century by reducing $AlCl_3$ with sodium,

$$3\ Na(s) + AlCl_3(s) \longrightarrow Al(s) + 3\ NaCl(s)$$

Charles Martin Hall was only 22 years old in 1886 when he discovered the electrolytic process for extracting aluminum from Al_2O_3 in a woodshed behind his family's home in Oberlin, Ohio. Following his discovery, Hall went on to found the company that eventually became Aluminum Corporation of America. He died a millionaire in 1914. *(Oesper Collection in the History of Chemistry/University of Cincinnati)*

but only at a very high cost. It was therefore considered a precious metal, chiefly used in jewelry. In fact, in the 1855 Paris Exposition, some of the first aluminum metal produced was exhibited along with the crown jewels of France. Napoleon III saw its possibilities for military use, however, and commissioned studies on improving its production. The French had a ready source of aluminum-containing ore, bauxite, and in 1886 a 23-year-old Frenchman, Paul Héroult (1863–1914), conceived the electrochemical method in use today. In an interesting coincidence, an American, Charles Hall (1863–1914), who was only 22 at the time, announced his invention of the identical process in the same year. Hence, the commercial process is now known as the Hall-Héroult process.

The essential features of the Hall-Héroult process are illustrated in Figure 21.22. The aluminum-containing ore, chiefly in the form of Al_2O_3, is mixed with cryolite, Na_3AlF_6. The mixture is melted at about 980 °C and electrolyzed using graphite electrodes. Aluminum is produced at the cathode and oxygen at the anode. The cells operate at the very low potential of 4.0 to 5.5 V, but at a current of 50,000 to 150,000 A. Each kilogram of aluminum requires 13 to 16 kilowatt-hours (kwh) of energy, exclusive of that required to heat the furnace. This is the reason there is so much interest in recycling soft drink cans and other aluminum objects. The recycled metal can be purified and made into new materials at a fraction of the cost of making aluminum from the ore.

Chlorine and Sodium Hydroxide

Chlorine is used to treat water and sewage and in the production of organic chemicals, such as pesticides and vinyl chloride, the building block of plastics

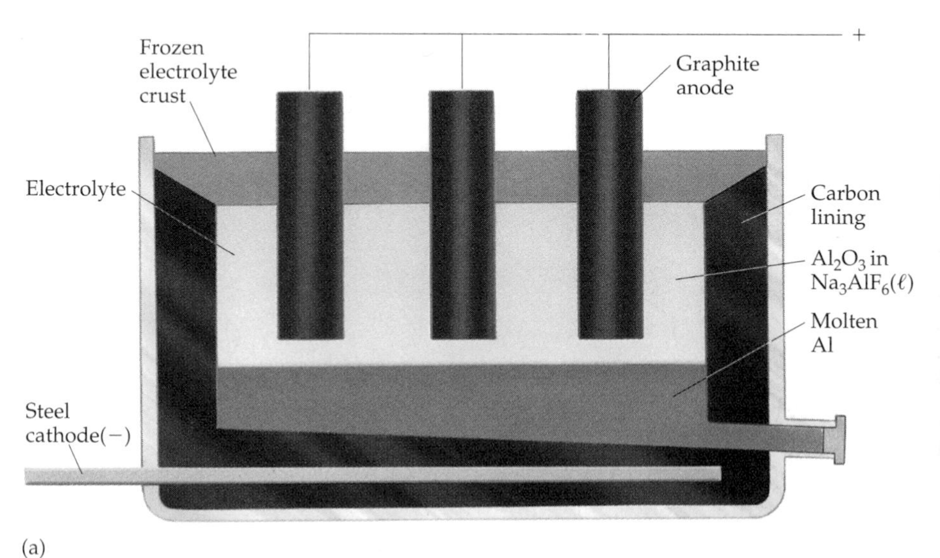

(a)

(b)

Figure 21.22 Industrial production of aluminum by electrolysis. (a) Purified aluminum-containing ore (bauxite), essentially Al_2O_3, is mixed with cryolite (Na_3AlF_6) to give a mixture that melts at a lower temperature than Al_2O_3 alone. The aluminum-containing substances are reduced at the graphite cathode to give molten aluminum. Oxygen is produced at the carbon anode, and the gas reacts slowly with the carbon to give CO_2, leading to eventual loss of the electrode. (b) Molten aluminum alloy, produced from recycled metal, at 760 °C, in 1.6×10^4-kg capacity crucibles. *(b, Courtesy, Allied Metal Company, Chicago, IL)*

called PVCs (polyvinyl chloride). In 1995, chlorine was tenth on the list of chemicals produced in the largest amounts in the United States, with about 1.1×10^{10} kg having been made. Almost all Cl_2 is made by electrolysis, with 95% coming from the electrolysis of brine, a saturated aqueous solution of NaCl. The other product coming from these cells, NaOH, is equally valuable, and almost 1.2×10^{10} kg was produced in the United States in 1995.

Three different electrolysis methods are used to convert aqueous NaCl to Cl_2 and NaOH: the mercury process, the diaphragm process, and the membrane process. The mercury process uses liquid mercury as the cathode. Although this process is still used, largely because it produces high-purity sodium hydroxide and Cl_2 gas, the problems of mercury pollution from these cells are enormous. As a consequence, mercury electrolysis cells are being phased out in favor of other methods, chiefly the membrane process.

A very simple diagram of a membrane electrolysis cell is shown in Figure 21.23. The anode reaction is the oxidation of chloride ion to Cl_2 gas

$$2\ Cl^-(aq) \longrightarrow Cl_2(g) + 2\ e^-$$

and the cathode process is the reduction of water:

$$2\ H_2O(\ell) + 2\ e^- \longrightarrow H_2(g) + 2\ OH^-(aq)$$

Activated titanium is used for the anode, and stainless steel or nickel is preferred for cathodes. In the membrane cell, brine is introduced into the anode chamber, and chloride ion oxidation occurs. The membrane separating the anode and cathode is not permeable to water, but it does allow ions to pass; that is, the membrane is just a salt bridge between anode and cathode. To maintain charge balance in the cell, therefore, sodium ions pass through the membrane. Because

One of the greatest difficulties in building economical membrane cells has been to find membrane material that is reasonable in cost, is stable at high salt concentrations, and is stable to a high pH and to the strong oxidant Cl_2.

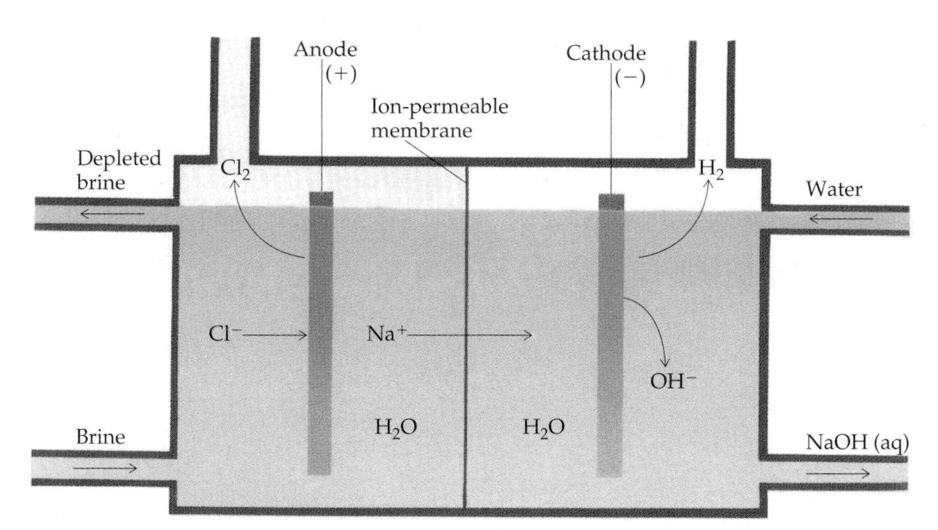

Figure 21.23 A membrane cell for the production of NaOH and Cl_2 gas from a saturated, aqueous solution of NaCl (brine). Here the anode and cathode compartments are separated by a water-impermeable but ion-conducting membrane. A widely used membrane is made of Nafion, a fluorine-containing polymer that is a relative of polytetrafluoroethylene (Teflon). Pure brine is fed into the anode compartment and dilute sodium hydroxide or water into the cathode compartment. Overflow pipes carry the evolved gases and NaOH away from the chambers of the electrolysis cell.

A chlor-alkali plant. Each membrane cell shown here contains many anodes and cathodes arranged in series. *(OxyTech Systems)*

the reduction of water at the cathode produces hydroxide ion, the product in the cathode chamber is aqueous sodium hydroxide with a concentration of 20% to 35% by weight. The energy consumption of these cells is in the range of 2000 to 2500 kwh/ton of NaOH produced.

Example 21.16 Chlorine Production

Assume an electrolysis cell that produces chlorine from aqueous sodium chloride (called "brine") operates at 4.6 V (with a current of 3.0×10^5 A). Calculate the number of kilowatt-hours of energy required to produce 1.00 kg of chlorine.

Solution The *watt* is a unit of electric power. It describes the *rate* of energy consumption or production; that is, the watt has units of energy per time and is defined as 1 watt = 1 joule/second. From page 961 you know that 1 joule = 1 volt · coulomb. Therefore,

$$1 \text{ joule} = 1 \text{ watt} \cdot \text{second} = 1 \text{ volt} \cdot \text{coulomb}$$

A kilowatt-hour is the expenditure of 1000 W for 1 h, so

$$(1000 \text{ W})(1 \text{ h})\left(\frac{3600 \text{ s}}{\text{h}}\right)\left(\frac{1 \text{ J/s}}{\text{W}}\right) = 3.60 \times 10^6 \text{ J/kwh}$$

To calculate the number of kilowatt-hours, therefore, we need to calculate the energy involved (in joules), and the chain of calculations leading to this is

Mass of Cl_2 ⟶ Moles of Cl_2 ⟶ Charge required ⟶ Energy

The reaction producing chlorine gas is

$$2\ Cl^-(aq) \longrightarrow Cl_2(g) + 2\ e^-$$

To produce 1.00 kg of Cl_2, 2.72×10^6 C is required:

$$1.00 \times 10^3 \text{ g}\left(\frac{1 \text{ mol } Cl_2}{70.91 \text{ g}}\right)\left(\frac{2 \text{ mol } e^-}{1 \text{ mol } Cl_2}\right)\left(\frac{9.65 \times 10^4 \text{ C}}{1 \text{ mol } e^-}\right) = 2.72 \times 10^6 \text{ C}$$

The energy required is

$$\text{Energy (J)} = (4.6 \text{ V})(2.72 \times 10^6 \text{ C}) = 1.3 \times 10^7 \text{ J}$$

Finally, the power required is

$$1.3 \times 10^7 \text{ J}\left(\frac{1 \text{ kwh}}{3.60 \times 10^6 \text{ J}}\right) = 3.5 \text{ kwh}$$

Exercise 21.13 Producing Sodium

Sodium metal is produced by electrolysis from molten sodium chloride. The cell operates at 7.0 V with a current of 25×10^3 A. How many kilowatt-hours of electricity is used to produce 1.00 kg of sodium metal?

CHAPTER HIGHLIGHTS

Having studied this chapter, you should be able to

- Define and use the terms **battery,** or **electrochemical cell; fuel cell; electrolysis; electrode; electrolyte; salt bridge; anode;** and **cathode.**
- Balance equations for oxidation–reduction reactions in acidic or basic solutions using the half-reaction approach (Section 21.1).
- Explain the workings of an electrochemical cell (which half-reaction occurs at the anode, which at the cathode, the polarity of the electrodes, the direction of electron flow in the external connection, and the direction of ion flow in the salt bridge) (Section 21.2).
- Appreciate the meaning of the **standard electrode potential $E°$,** and its connection to the free energy change $\Delta G°$ (Section 21.3).
- Recognize that product-favored reactions have a positive $E°$, whereas reactant-favored reactions have negative $E°$ (Section 21.3 and Equation 21.1).
- Know that the **standard hydrogen electrode** ($E° = 0.00$ V) is the standard against which all half-reaction potentials are measured (Section 21.3).
- Appreciate the method by which standard potentials of half-reactions can be determined (Section 21.3).
- Use Table 21.1, the **table of standard reduction potentials** (Section 21.4).
- Understand that when a half-reaction or net electrochemical reaction is reversed, the sign of $E°$ is reversed but its value does not change (Section 21.4).
- Recognize that, as the value of $E°$ for a reduction half-reaction becomes less negative, the ion or molecule becomes a better oxidizing agent (the substance on the left is more readily reduced) (Section 21.4).
- Apply the idea that the reaction between any substance on the left in Table 21.1 (an oxidizing agent) and any substance lower than it on the right (a reducing agent) is product-favored under standard conditions (Section 21.4).
- Recognize that electrochemical potentials depend on the nature of the reactants and products and their concentrations, not on the quantities of material used (Sections 21.4 and 21.5).
- Predict the sign and value of $E°_{net}$ for a redox reaction (Section 21.4).
- Use the **Nernst equation** (Equation 21.2) to calculate the cell potential under nonstandard conditions (Section 21.5).
- Calculate the equilibrium constant for a reaction from the value of $E°$ (Equation 21.3 and Section 21.5).
- Recognize the difference between **primary** and **secondary batteries** (Section 21.6).
- Appreciate the chemistry and advantages and disadvantages of dry cells, alkaline batteries, mercury batteries, lithium batteries, lead storage batteries, and ni–cad batteries (Section 21.6).
- Understand the difference between batteries and **fuel cells** (Section 21.6).
- Understand reactions involved in **corrosion** and how **anodic inhibition** and **cathodic protection** can inhibit corrosion (Section 21.7).
- Describe the difference between electrolysis of an electrolyte and the operation of a galvanic or voltaic cell (Section 21.8).
- Identify the reactions that occur in the electrolysis of a molten salt (Section 21.8).
- Characterize the reactions occurring on electrolysis of an aqueous solution of electrolyte (Section 21.8).

- Use the relationship between current (amperes, A), electric charge (coulombs, C), and time (seconds, s) (Equation 21.4) and use the Faraday constant (9.65×10^4 C/mol e^-) (Section 21.9).
- Describe electrochemical methods for the production of aluminum, chlorine, and sodium hydroxide (Section 21.10).

STUDY QUESTIONS

REVIEW QUESTIONS

1. In each of the following reactions, identify the substance oxidized and the substance reduced. Identify the oxidizing agent and the reducing agent.
   (a) $2\ Al(s) + 3\ Cl_2(g) \longrightarrow 2\ AlCl_3(s)$
   (b) $FeS(s) + 3\ NO_3^-(aq) + 4\ H_3O^+(aq) \longrightarrow 3\ NO(g) + SO_4^{2-}(aq) + Fe^{3+}(aq) + 6\ H_2O(\ell)$
2. Explain the function of a salt bridge in an electrochemical cell.
3. Decide if each of the following statements is true or false. If false, rewrite it to make it a correct statement.
   (a) Oxidation always occurs at the anode of an electrochemical cell.
   (b) The anode of a battery is the site of reduction and is negative.
   (c) Standard conditions for electrochemical cells are a concentration of 1.0 M for dissolved species and 1 bar pressure for gases.
   (d) The potential of a cell does not change with temperature.
   (e) All product-favored oxidation–reduction reactions have a negative E°_{net}.
4. Decide which phrase (a through d) best completes the sentence: A product-favored oxidation–reduction reaction has (a) a positive ΔG° and a positive E°, (b) a negative ΔG° and a positive E°, (c) a positive ΔG° and a negative E°, or (d) a negative ΔG° and a negative E°.
5. Decide which phrase (a through c) best completes the sentence: According to Table 21.1, Zn(s) can (a) oxidize Fe(s) and Cd(s), (b) reduce Al^{3+} and Mg^{2+}, or (c) reduce Cd^{2+} and Ag^+.
6. Decide if each of the following statements is true or false. If false, rewrite it to make it a correct statement.
   (a) The value of the electrode potential, E°, for ($2\ Li^+ + 2\ e^- \longrightarrow 2\ Li$) is twice that for ($Li^+ + e^- \longrightarrow Li$).
   (b) Al is the strongest reducing agent listed in Table 21.1.
   (c) The equilibrium constant for an oxidation–reduction reaction can be calculated using the Nernst equation.
   (d) Changing the concentrations of dissolved substances does not change the potential observed for an electrochemical cell.
7. What are the advantages and disadvantages of lead storage batteries?
8. How does a fuel cell differ from a battery?
9. Explain why the products of electrolysis of molten NaCl differ from those obtained by electrolysis of aqueous NaCl.
10. Describe the electrochemical method for the manufacture of Cl_2 and NaOH.
11. What is the difference between anodic and cathodic protection against corrosion? Explain how each works.

NUMERICAL AND OTHER QUESTIONS

Balancing Equations for Redox Reactions

Balance equations for the half-reactions in Questions 12 through 17. Describe the reactant as an oxidizing or reducing agent, and decide if the overall process is an oxidation or reduction reaction. Unless noted otherwise, all are carried out in acid solution, meaning that H^+ or H^+ and H_2O may be used to balance the equation.

12. (a) $Cr(s) \longrightarrow Cr^{3+}(aq)$
   (b) $AsH_3(g) \longrightarrow As(s)$
   (c) $VO_3^-(aq) \longrightarrow V^{2+}(aq)$
13. (a) $Br_2(aq) \longrightarrow Br^-(aq)$
   (b) $OsO_4(s) \longrightarrow Os^{4+}(aq)$
   (c) $U^{4+}(aq) \longrightarrow UO_2^+(aq)$
14. (a) $Cr_2O_7^{2-}(aq) \longrightarrow Cr^{3+}(aq)$
   (b) $CH_3CHO(aq) \longrightarrow CH_3CO_2H(aq)$
   (c) $Bi^{3+}(aq) \longrightarrow HBiO_3(aq)$
15. (a) $HOI(aq) \longrightarrow I^-(aq)$
   (b) $NO(g) \longrightarrow HNO_2(aq)$
   (c) $C_6H_5CH_3(aq) \longrightarrow C_6H_5CO_2H(aq)$
16. The half-reactions here are in basic solution. You may need to use OH^- or the OH^-/H_2O pair to balance the equation.
   (a) $Sn(s) \longrightarrow Sn(OH)_4^{2-}(aq)$
   (b) $MnO_4^-(aq) \longrightarrow MnO_2(s)$
   (c) $ClO^-(aq) \longrightarrow Cl^-(aq)$
17. The half-reactions here are in basic solution. You may need to use OH^- or the OH^-/H_2O pair to balance the equation.
   (a) $CrO_2^-(aq) \longrightarrow CrO_4^{2-}(aq)$
   (b) $Br_2(aq) \longrightarrow BrO_3^-(aq)$
   (c) $Ni(OH)_2(s) \longrightarrow NiO_2(s)$

Use the half-reaction method to balance the equations in Questions 18 through 23. For reactions in acid solution you may need to use H^+ or H^+ and H_2O to balance the equation.

18. Reaction (a) is in neutral solution. Reactions (b) and (c) are in acid solution.
(a) $Cl_2(aq) + Br^-(aq) \longrightarrow Br_2(\ell) + Cl^-(aq)$
(b) $Sn(s) + H^+(aq) \longrightarrow Sn^{2+}(aq) + H_2(g)$
(c) $Zn(s) + VO_2^+(aq) \longrightarrow Zn^{2+}(aq) + V^{2+}(aq)$

19. The reactions here are in acid solution.
(a) $Hg^{2+}(aq) + Cu(s) \longrightarrow Cu^{2+}(aq) + Hg(\ell)$
(b) $MnO_2(s) + Cl^-(aq) \longrightarrow Mn^{2+}(aq) + Cl_2(g)$
(c) $Zn(s) + NO_3^-(aq) \longrightarrow Zn^{2+}(aq) + N_2O(g)$

20. The reactions here are in acid solution.
(a) $Ag^+(aq) + HCHO(aq) \longrightarrow Ag(s) + HCO_2H(aq)$
(b) $H_2S(aq) + Cr_2O_7^{2-}(aq) \longrightarrow S(s) + Cr^{3+}(aq)$
(c) $H_2C_2O_4(aq) + MnO_4^-(aq) \longrightarrow CO_2(g) + Mn^{2+}(aq)$

21. The reactions here are in acid solution.
(a) $MnO_4^-(aq) + HSO_3^-(aq) \longrightarrow Mn^{2+}(aq) + SO_4^{2-}(aq)$
(b) $Cr_2O_7^{2-}(aq) + Fe^{2+}(aq) \longrightarrow Cr^{3+}(aq) + Fe^{3+}(aq)$
(c) $Ag(s) + NO_3^-(aq) \longrightarrow NO_2(g) + Ag^+(aq)$

22. The reactions here are in basic solution. You may need to add OH^- or OH^- and H_2O to balance the equation.
(a) $Zn(s) + ClO^-(aq) \longrightarrow Zn(OH)_2(s) + Cl^-(aq)$
(b) $ClO^-(aq) + CrO_2^-(aq) \longrightarrow Cl^-(aq) + CrO_4^{2-}(aq)$
(c) $Br_2(\ell) \longrightarrow Br^-(aq) + BrO_3^-(aq)$
(This is a disproportionation reaction: One substance, Br_2, functions both as the reducing and the oxidizing agent.)

23. The reactions here are in basic solution. You may need to add OH^- or OH^- and H_2O to balance the equation.
(a) $Fe(OH)_2(s) + CrO_4^{2-}(aq) \longrightarrow Fe_2O_3(s) + Cr(OH)_4^-(aq)$
(b) $PbO_2(s) + Cl^-(aq) \longrightarrow ClO^-(aq) + Pb(OH)_3^-(aq)$
(c) $Al(s) + OH^-(aq) \longrightarrow Al(OH)_4^-(aq) + H_2(g)$

Electrochemical Cells and Cell Potentials

24. In principle, the reaction of chromium and iron(II) ion can be used to build an electrochemical cell:

$$2\ Cr(s) + 3\ Fe^{2+}(aq) \longrightarrow 2\ Cr^{3+}(aq) + 3\ Fe(s)$$

(a) Write the half-reactions involved.
(b) Which half-reaction is an oxidation and which is a reduction?
(c) Which half-reaction occurs in the anode compartment and which in the cathode compartment?

25. Chlorine gas, Cl_2, can oxidize zinc metal in a reaction that has been suggested as the basis of a battery.
(a) Write the half-reactions involved.
(b) Which half-reaction is an oxidation and which is a reduction? Which half-reaction would occur in the anode compartment and which would occur in the cathode compartment?

26. Suppose the half-cells $Cu^{2+}(aq)/Cu(s)$ and $Sn^{2+}(aq)/Sn(s)$ are to be used as the basis of a battery. Assume the polarity of the copper electrode is positive and that of the tin electrode is negative. Write the half-reactions that occur in each half-cell. Decide which is the oxidation and which is the reduction. Identify which half-reaction occurs at the anode and which at the cathode.

27. In principle, the half cells $Fe^{2+}(aq)/Fe(s)$ and $O_2(g)/H_2O$ (in acid solution) could be used as the basis of a battery. Assuming the polarity of the iron electrode is negative, write the half-reactions that occur in the cell. Identify the oxidation and reduction half-reactions. Which half-reaction occurs at the anode and which at the cathode?

28. The standard potential for the reaction of Mg(s) with $I_2(s)$ is +2.91 V. What is the standard free energy change, $\Delta G°$, for the reaction?

29. The standard potential, $E°$, for the reaction of Zn(s) and $Cl_2(g)$ is +2.12 V (Study Question 25). What is the standard free energy change, $\Delta G°$, for the reaction?

30. Calculate the value of $E°$ for each of the following reactions. Decide if each is product-favored in the direction written.
(a) $Zn^{2+}(aq) + 2\ I^-(aq) \longrightarrow Zn(s) + I_2(s)$
(b) $Zn(s) + Ni^{2+}(aq) \longrightarrow Zn^{2+}(aq) + Ni(s)$
(c) $Cu(s) + Cl_2(g) \longrightarrow Cu^{2+}(aq) + 2\ Cl^-(aq)$

31. Calculate the value of $E°$ for each of the following reactions. Decide if each is product-favored in the direction written.
(a) $Mg(s) + Br_2(\ell) \longrightarrow Mg^{2+}(aq) + 2\ Br^-(aq)$
(b) $Sn^{2+}(aq) + 2\ Ag^+(aq) \longrightarrow Sn^{4+}(aq) + 2\ Ag(s)$
(c) $2\ Zn(s) + O_2(g) + 2\ H_2O(\ell) \longrightarrow 2\ Zn(OH)_4^{2-}(aq)$

32. Balance each of the following unbalanced equations, then calculate the standard potential, $E°$, and decide whether each is product-favored as written. (Half-reaction potentials are found in Appendix M.)
(a) $Sn^{2+}(aq) + Ag(s) \longrightarrow Sn(s) + Ag^+(aq)$
(b) $Zn(s) + Sn^{4+}(aq) \longrightarrow Sn^{2+}(aq) + Zn^{2+}(aq)$
(c) $I_2(s) + Br^-(aq) \longrightarrow I^-(aq) + Br_2(\ell)$

33. Balance each of the following unbalanced equations, then calculate the standard potential, $E°$, and decide whether each is product-favored as written. (Half-reaction potentials are found in Appendix M.)
(a) $Ce^{4+}(aq) + Cl^-(aq) \longrightarrow Ce^{3+}(aq) + Cl_2(g)$
(b) $Cu(s) + NO_3^-(aq) + H_3O^+(aq) \longrightarrow Cu^{2+}(aq) + NO(g) + H_2O(\ell)$
(c) $Fe^{2+}(aq) + Cr_2O_7^{2-}(aq) + H_3O^+(aq) \longrightarrow Fe^{3+}(aq) + Cr^{3+}(aq) + H_2O(\ell)$

34. Consider the following half-reactions:

Half-Reaction	***E*° (V)**
$Cl_2(g) + 2\ e^- \longrightarrow 2\ Cl^-(aq)$	+1.36
$I_2(s) + 2\ e^- \longrightarrow 2\ I^-(aq)$	+0.535
$Pb^{2+}(aq) + 2\ e^- \longrightarrow Pb(s)$	−0.126
$V^{2+}(aq) + 2\ e^- \longrightarrow V(s)$	−1.18

(a) Which is the weakest oxidizing agent in the list?
(b) Which is the strongest oxidizing agent?
(c) Which is the strongest reducing agent?
(d) Which is the weakest reducing agent?
(e) Can Pb(s) reduce $V^{2+}(aq)$ to V(s)?
(f) Can $I^-(aq)$ reduce $Cl_2(g)$ to $Cl^-(aq)$?
(g) Name the elements or ions that can be reduced by Pb(s).

35. Consider the following half-reactions:

Half-Reaction	$E°$ (V)
$Ce^{4+}(aq) + e^- \longrightarrow Ce^{3+}(aq)$	+1.61
$Ag^+(aq) + e^- \longrightarrow Ag(s)$	+0.80
$Hg_2^{2+}(aq) + 2\ e^- \longrightarrow 2\ Hg(\ell)$	+0.79
$Sn^{2+}(aq) + 2\ e^- \longrightarrow Sn(s)$	−0.14
$Ni^{2+}(aq) + 2\ e^- \longrightarrow Ni(s)$	−0.25
$Al^{3+}(aq) + 3\ e^- \longrightarrow Al(s)$	−1.66

(a) Which is the weakest oxidizing agent in the list?
(b) Which is the strongest oxidizing agent?
(c) Which is the strongest reducing agent?
(d) Which is the weakest reducing agent?
(e) Can Sn(s) reduce $Ag^+(aq)$ to Ag(s)?
(f) Can $Hg(\ell)$ reduce $Sn^{2+}(aq)$ to Sn(s)?
(g) Name the ions that can be reduced by Sn(s).
(h) What metals can be oxidized by $Ag^+(aq)$?

36. Use the list of half-reactions in Study Question 34 to answer the following questions:
(a) What two half-reactions, when combined in an electrochemical cell, will result in the largest positive standard potential?
(b) Combine the $I_2(s)/I^-(aq)$ half-cell with the $V^{2+}(aq)/V(s)$ half-cell, and write a balanced equation for the product-favored reaction that occurs. What is the value of $E°_{net}$ for the reaction?

37. Use the list of half-reaction potentials in Study Question 35 to answer the following questions:
(a) What reaction between an oxidizing agent and a reducing agent on the list leads to the maximum positive standard potential?
(b) If the $Ni^{2+}(aq)/Ni(s)$ half-cell is combined with the $Hg_2^{2+}(aq)/Hg(\ell)$ half-cell write a balanced equation for the product-favored reaction that occurs. What is its standard potential?
(c) Write a balanced equation for the product-favored reaction that occurs when the half-reactions $Ag^+(aq)/Ag(s)$ and $Ce^{4+}(aq)/Ce^{3+}(aq)$ are combined. What is the value of $E°_{net}$ for this reaction?

38. Assume that you assemble an electrochemical cell based on the half-reactions $Zn^{2+}(aq)/Zn(s)$ and $Ag^+(aq)/Ag(s)$.
(a) Write a balanced equation for the product-favored reaction that occurs in the cell and calculate $E°_{net}$.
(b) Which electrode is the anode and which is the cathode?
(c) Diagram the components of the cell.
(d) If you use a silver wire as an electrode, is it the anode or cathode?
(e) Do electrons flow from the Zn electrode to the Ag electrode, or vice versa?
(f) If a salt bridge containing $NaNO_3$ connects the two half-cells, in which direction do the nitrate ions move, from the zinc to the silver compartment, or vice versa?

39. An electrochemical cell uses Al(s) and $Al^{3+}(aq)$ in one compartment and Ag(s) and $Ag^+(aq)$ in the other.
(a) Write a balanced equation for the product-favored reaction that occurs in this cell, and calculate $E°_{net}$.
(b) Which is the better reducing agent, Ag or Al?
(c) Which is the anode and which is the cathode? Indicate the polarity of each electrode.
(d) Sodium nitrate is in the salt bridge connecting the two half-cells. In which direction do the nitrate ions flow, from Al to Ag or from Ag to Al?

Cells Under Nonstandard Conditions, E and K

40. Calculate the voltage delivered by an electrochemical cell using the following reaction if all dissolved species are 0.10 M:

$$2\ Fe^{3+}(aq) + 2\ I^-(aq) \longrightarrow 2\ Fe^{2+}(aq) + I_2(s)$$

Does the voltage increase or decrease relative to $E°_{net}$, the standard voltage for the reaction?

41. Calculate the voltage delivered by an electrochemical cell using the following reaction if all dissolved species are 0.015 M:

$$2\ Fe^{2+}(aq) + H_2O_2(aq) + 2\ H_3O^+(aq) \longrightarrow 2\ Fe^{3+}(aq) + 4\ H_2O(\ell)$$

Does the voltage increase or decrease relative to $E°_{net}$, the standard voltage for the reaction?

42. An electrochemical cell is constructed of one half-cell in which a silver wire dips into an aqueous solution of $AgNO_3$. The other half-cell consists of an inert platinum wire in an aqueous solution of Fe^{2+} and Fe^{3+}.
(a) Write a balanced equation for the product-favored reaction that occurs in this cell under standard conditions.
(b) What is $E°_{net}$?
(c) If $[Ag^+] = 0.10$ M, but $[Fe^{2+}]$ and $[Fe^{3+}]$ are 1.0 M, what is the value of E_{net}? Is the net cell reaction still that in part (a)? If not, what is the net cell reaction under the new conditions?

43. An electrochemical cell is constructed of one half-cell in which a silver wire dips into a 1.0 M aqueous solution of $AgNO_3$. The other half-cell consists of a zinc electrode in a solution of 1.0 M $Zn(NO_3)_2$.
(a) Write a balanced equation for the product-favored reaction that occurs in this cell under standard conditions.
(b) What is $E°_{net}$?
(c) What is the polarity of the Zn electrode?
(d) Do electrons flow from Zn to Ag or from Ag to Zn?
(e) If Ag^+ has a concentration of 0.50 M, but Zn^{2+} is held at 1.0 M, what is the voltage of the cell? Compare this with $E°_{net}$ and comment on the effect of the change in silver concentration.

44. Calculate equilibrium constants for the following reactions:
(a) $2\ Fe^{3+}(aq) + 2\ I^-(aq) \longrightarrow 2\ Fe^{2+}(aq) + I_2(s)$
(b) $I_2(s) + 2\ Br^-(aq) \longrightarrow 2\ I^-(aq) + Br_2(\ell)$

45. Calculate equilibrium constants for the following reactions:
(a) $Zn^{2+}(aq) + Ni(s) \longrightarrow Zn(s) + Ni^{2+}(aq)$
(b) $Cu(s) + 2\ Ag^{+}(aq) \longrightarrow Cu^{2+}(aq) + 2\ Ag(s)$

Electrolysis, Electrical Energy, and Power

46. If you wish to convert 1.00 g of $Au^{3+}(aq)$ ion to Au(s) in a "gold-plating" process, how long must you electrolyze a solution if the current passing through the circuit is 2.00 A?

47. If you wish to produce 0.085 g of nickel metal, how long must you electrolyze a solution containing $Ni^{2+}(aq)$ if the current passed through the cell is 0.15 A?

48. A current of 2.50 A is passed through a solution of $Ni(NO_3)_2$ for 2.00 h. What mass of nickel is deposited at the cathode?

49. A current of 0.0125 A is passed through a solution of $CuCl_2$ for 96 min. What mass of copper is deposited at the cathode, and what mass of chlorine gas is produced at the anode?

50. An old method of measuring the current flowing in a circuit was to use a "silver coulometer." The current passed first through a solution of $Ag^{+}(aq)$ and then into another solution containing an electroactive species. The amount of silver metal deposited at the cathode was weighed, and, if the time was noted, the current could be calculated. If 0.052 g of Ag was deposited during a 450-s experiment, what was the current flowing in the circuit?

51. As noted in Study Question 50, a "silver coulometer" was used in the past to measure the current flowing in an electrochemical cell. Suppose you found that the current flowing through an electrolysis cell deposited 0.089 g of Ag metal at the cathode after exactly 10 min. If this same current then passed through a cell containing gold(III) ion in the form of $[AuCl_4]^-$, how much gold was deposited at the cathode in that electrolysis cell? The half-reaction for the production of gold is

$$[AuCl_4]^-(aq) + 3\ e^- \longrightarrow Au(s) + 4\ Cl^-(aq)$$

52. The basic reaction occurring in the cell in which Al_2O_3 and aluminum salts are electrolyzed is $Al^{3+} + 3\ e^- \longrightarrow Al(s)$. If the cell operates at 5.0 V and 1.0×10^5 A, how many grams of aluminum metal can be produced in an 8.0-h day?

53. The vanadium(II) ion can be produced by electrolysis of a vanadium(III) salt in solution. How long must you carry out an electrolysis if you wish to convert completely 0.125 L of 0.015 M $V^{3+}(aq)$ to $V^{2+}(aq)$ if the current is 0.268 A?

54. The reactions occurring in a lead storage battery are given in Section 21.6. A typical battery might be rated at "50 ampere-hours (A-h)." This means it has the capacity to deliver 50. A for 1.0 h (or 1.0 A for 50. h). How many grams of lead would be consumed to deliver 1.0 A for 50. h?

55. A lead storage battery (Section 21.6) operates at 12.0 V and is rated at "100 A-h." This term means the battery can deliver 1.0 A of current for 100 h. What is the power rating of the battery in watts?

56. Electrolysis of brine leads to chlorine gas. In 1995, 1.138×10^{10} kg of Cl_2 was manufactured. How many kilowatt-hours of energy must have been used to produce this amount of chlorine from NaCl? Assume the electrolysis cells operate at 4.6 V and 3.0×10^5 A.

57. Electrolysis of molten NaCl is done in cells operating at 7.0 V and 4.0×10^4 A. How much Na(s) and $Cl_2(g)$ can be produced in one day in such a cell? What is the energy consumption in kilowatt-hours?

58. An aqueous solution of $NiCl_2$ is placed in a beaker with two inert platinum electrodes. When the cell is attached to an external battery, electrolysis occurs.
(a) Write an equation for the half-reaction occurring at the anode. What is the polarity of the electrode?
(b) Write an equation for the half-reaction occurring at the cathode. What is the polarity of the electrode?
(c) If 0.0250 A flows through the cell for 1.25 h, how many grams of each product are expected?

59. An aqueous solution of NaI is placed in a beaker with two inert platinum electrodes. When the cell is attached to an external battery, electrolysis occurs.
(a) Write an equation for the half-reaction occurring at the anode. What is the polarity of the electrode?
(b) Write an equation for the half-reaction occurring at the cathode. What is the polarity of the electrode?
(c) If 0.050 A pass through the cell for 5.0 h, how many grams of each product are expected?

60. Describe what happens at the anode and at the cathode when electrolyzing each of the following solutions:
(a) KBr(aq)
(b) NaF(molten)
(c) NaF(aq)

61. Describe what happens at the anode and at the cathode when electrolyzing each of the following solutions:
(a) $NiBr_2(aq)$
(b) KI(aq)
(c) $CdCl_2(aq)$

GENERAL QUESTIONS

62. Balance the following equations for redox reactions that occur in acid solution:
(a) $I^-(aq) + Br_2(\ell) \longrightarrow IO_3^-(aq) + Br^-(aq)$
(b) $U^{4+}(aq) + MnO_4^-(aq) \longrightarrow Mn^{2+}(aq) + UO_2^+(aq)$
(c) $I^-(aq) + MnO_2(s) \longrightarrow Mn^{2+}(aq) + I_2(s)$

63. Balance the following equations for redox reactions that occur in basic solution:
(a) $CN^-(aq) + CrO_4^{2-}(aq) \longrightarrow NCO^-(aq) + Cr(OH)_4^-(aq)$
(b) $Co^{2+}(aq) + OCl^-(aq) \longrightarrow Co(OH)_3(s) + Cl^-(aq)$

64. You are told to assemble an electrochemical cell with one half-cell being $Cl_2(g)/Cl^-(aq)$. The other half-cell could be $Al^{3+}(aq)/Al(s)$, $Mg^{2+}(aq)/Mg(s)$, or $Zn^{2+}(aq)/Zn(s)$. Which of the metal ion/metal combinations would you choose to produce the largest possible positive E°_{net}? Write a balanced equation for the reaction you have chosen.

65. The total charge that can be delivered by a large dry cell battery before its voltage drops too low is usually about 35 A-h. (One amp-hour [A-h] is the charge that passes through a circuit when 1 amp flows for 1 h.) What mass of Zn is consumed when 35 A-h of charge are drawn from the cell?

66. The standard potential for the lead storage battery is 2.04 V.
(a) What is the equilibrium constant for the reaction occurring in the battery under standard conditions?
(b) What is the potential of the battery at 25 °C when the sulfuric acid concentration is 6.00 M?

67. Mendelevium, Md, was the first actinide element found to exist as a stable 2+ ion in aqueous solution. The equilibrium constant for the reaction

$$V^{3+}(aq) + Md^{3+}(aq) \longrightarrow Md^{2+}(aq) + V^{2+}(aq)$$

is 15 at 25 °C. Use this to calculate E° for the half-reaction $Md^{3+}(aq) + e^- \longrightarrow Md^{2+}(aq)$.

68. A magnesium bar with a mass of 5.0 kg is attached to a buried iron pipe to protect the pipe from corrosion.
(a) Explain how the magnesium protects the pipe.
(b) If a current of 0.030 A flows between the bar and the pipe, how many years will elapse before the magnesium is entirely consumed?

69. A proposed automobile battery involves the reaction of Zn(s) and $Cl_2(g)$ to give $ZnCl_2$. If you want such a battery to operate 12 h and deliver 1.5 A of current, what is the minimum mass of zinc that the anode must contain?

70. In principle, a battery could be made from aluminum metal and chlorine gas.
(a) Write a balanced equation for the reaction that would occur in a battery using $Al^{3+}(aq)/Al(s)$ and $Cl_2(g)/Cl^-(aq)$ half-reactions.
(b) Tell which half-reaction occurs at the anode and which at the cathode. What are the polarities of these electrodes?
(c) Calculate the standard potential, E°_{net}, for the battery.
(d) If the pressure of $Cl_2(g)$ is only 0.50 atm, how is the voltage of the battery affected? Is it more positive or more negative than E°_{net}? Calculate E_{net}.
(e) If you want the battery to deliver a current of 0.75 A, how long can it operate if the aluminum electrode contains 30.0 g of Al? (Assume an unlimited supply of chlorine.)

71. A battery can be built using the reaction between Al metal and O_2 from the air. If the Al anode of this battery consists of 84 g of aluminum, how many hours can the battery produce 1.0 A of electricity (assuming an unlimited supply of O_2)?

72. A silver–zinc battery could be a lighter, but more expensive, alternative to the lead storage battery:

$$Ag_2O(s) + Zn(s) + H_2O(\ell) \longrightarrow Zn(OH)_2(s) + 2\ Ag(s)$$

The electrolyte is 40% KOH, and silver/silver oxide electrodes are separated from zinc/zinc hydroxide electrodes by a plastic sheet permeable to hydroxide ion. Under normal operating conditions, the battery has a potential of 1.59 V.
(a) How much energy can be produced per gram of reactants in the silver/zinc battery? Assume the battery produces a current of 0.10 A.
(b) How much energy can be produced per gram of reactants in the standard lead storage battery? Assume the battery produces a current of 0.10 A at 2.0 V.
(c) Which battery produces the greater energy per gram of reactants?

73. Fluorinated organic compounds are important commercially, as they are used as herbicides, flame retardants, and fire extinguishing agents, among other things. A reaction such as

$$CH_3SO_2F + 3\ HF \longrightarrow CF_3SO_2F + 3\ H_2$$

is actually carried out electrochemically in liquid HF as the solvent.
(a) If you electrolyze 150 g of CH_3SO_2F, what mass of HF is required and what mass of each product can be isolated?
(b) Is H_2 produced at the anode or cathode of the electrolysis cell?
(c) A typical electrolysis cell operates at 8.0 V and a current such as 250 A. How many kilowatt-hours of energy does one such cell consume in 24 h?

CONCEPTUAL QUESTIONS

74. Four metals, A, B, C, and D, exhibit the following properties:
(a) Only A and C react with 1.0 M hydrochloric acid to give $H_2(g)$.
(b) When C is added to solutions of the ions of the other metals, metallic B, D, and A are formed.
(c) Metal D reduces B^{n+} to give metallic B and D^{n+}.
Based on the preceding information, arrange the four metals in order of increasing ability to act as reducing agents.

75. Which of the following is the best way to store partly used steel wool? Which is the most conducive to rusting?
(a) Keep it immersed in plain tap water.
(b) Keep it immersed in soapy water (a basic solution).
(c) Just leave it in the kitchen sink.

76. Why does iron corrode more rapidly in solutions with a high concentration of dissolved CO_2?

77. Two factors are important in determining the potential of an electrochemical cell: the nature of the reactants and the concentration of those reactants. Which plays the more important role? Explain your answer in terms of the Nernst equation.

CHALLENGING QUESTIONS

78. The half-cells Ni(s)/Ni^{2+}(aq) and Cd(s)/Cd^{2+}(aq) are assembled into a battery.

(a) Write a balanced equation for the reaction occurring in the cell.

(b) What is oxidized and what is reduced? What is the reducing agent, and what is the oxidizing agent?

(c) Which is the anode, and which is the cathode? What is the polarity of the Cd electrode?

(d) What is E°_{net} for the cell?

(e) What is the direction of electron flow in the external wire?

(f) If the salt bridge contains KNO_3, toward which compartment do the NO_3^- ions migrate?

(g) Calculate the equilibrium constant for the net reaction.

(h) If the concentration of Cd^{2+} is reduced to 0.010 M, and $[Ni^{2+}]$ = 1.0 M, what is the voltage produced by the cell? Is the net reaction still the reaction given in part (a)?

(i) If 0.050 A is drawn from the battery, how long can it last if you begin with 1.0 L of each of the solutions, and each was initially 1.0 M in dissolved species? The electrodes each weigh 50.0 g in the beginning.

79. Suppose you use Cu^{2+}(aq)/Cu(s) and Sn^{2+}(aq)/Sn(s) half-cells as the basis of an electrochemical cell. If the cell starts at standard conditions, and 0.400 A of current flows for 48.0 h, what are the concentrations of the dissolved species at this point? What is the potential of the cell after 48.0 h? (Assume there is 1.00 L of solution.)

80. A current of 0.0100 A is passed through a solution of rhodium sulfate. The only reaction at the cathode is the deposition of rhodium metal. After 3.00 h, 0.038 g of Rh has been deposited. What is the charge on the rhodium ion, Rh^{n+} in rhodium sulfate?

81. Batteries are listed by their "cranking power," which is the amount of current the battery can produce for 30 s; a typical value is 450 A. How many coulombs flow through the battery in exactly 30 s? If this is a lead storage battery, how much lead (Pb) is consumed in exactly 30 s?

82. A hydrogen–oxygen fuel cell operates on the simple reaction

$$H_2(g) + \tfrac{1}{2} O_2(g) \longrightarrow H_2O(\ell)$$

If the cell is designed to produce 1.5 A of current, and if the hydrogen is contained in a 1.0-L tank at 200. atm pressure at 25 °C, how long can the fuel cell operate before the hydrogen runs out? (Assume there is an unlimited supply of O_2.)

83. Living organisms derive energy from the oxidation of food, typified by glucose, $C_6H_{12}O_6$:

$$C_6H_{12}O_6(aq) + 6\,O_2(g) \longrightarrow 6\,CO_2(g) + 6\,H_2O(\ell)$$

Electrons in this redox process are transferred from glucose to oxygen in a series of at least 25 steps. It is interesting to calculate the total daily current flow in a typical organism and the rate of energy expenditure (power). (See T. P. Chirpich: "Electrochemistry in organisms." *Journal of Chemical Education,* Vol. 52, page 99, 1975.)

(a) The molar enthalpy of combustion of glucose is −2800 kJ. If you are on a typical daily diet of 2400 Cal (kilocalories), how many moles of glucose must be consumed in a day if glucose is assumed to be the only source of energy? How many moles of O_2 must be consumed in the oxidation process?

(b) How many moles of electrons must be supplied to reduce the amount of O_2 calculated in part (a)?

(c) Based on the answer in part (b), calculate the current flowing, per second, in your body from the combustion of glucose.

(d) If the average standard potential in the electron transport chain is 1.0 V, what is the rate of energy expenditure in watts?

Summary Question

84. *Nicotinamide adenine dinucleotide,* abbreviated NAD^+, is an oxidizing agent and one of the central agents for electron transfer in biological systems. The center for its redox chemistry is nicotinamide. In NAD^+, nicotinamide is bound to a ribose sugar through the N atom. (Completing NAD^+ are two phosphate units, another ribose, and the organic base adenine.) Let us depict NAD^+ here as nicotinamide with an R group representing the ribose unit.

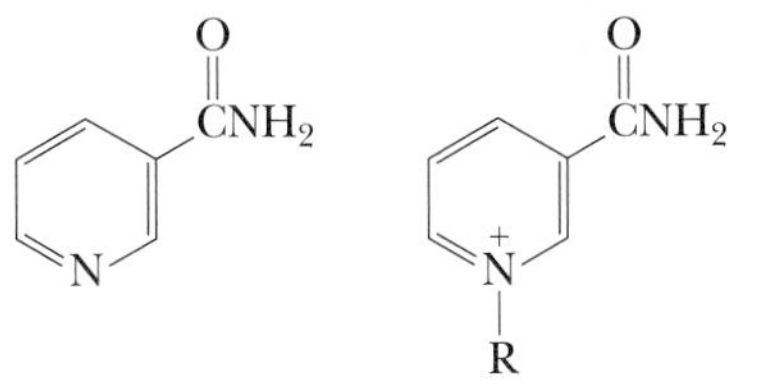

Nicotinamide NAD^+

NAD^+ can oxidize secondary alcohols to ketones and aldehydes to carboxylic acids in a two-electron process. Its half-reaction is

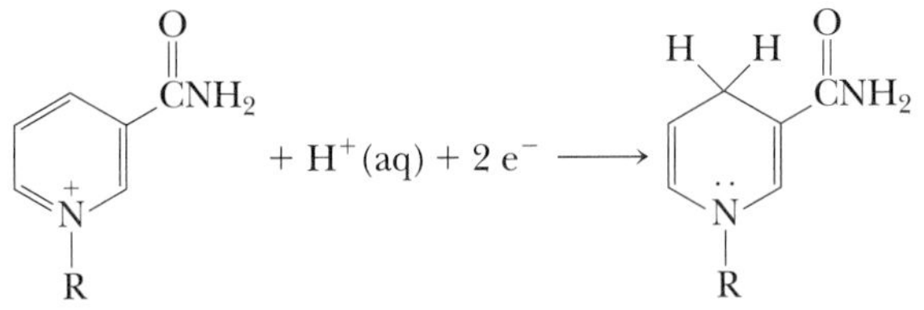

NAD^+ $+ H^+(aq) + 2\,e^- \longrightarrow$ NADH

(a) If E° for the reduction of NAD^+ is −0.320 V, will NAD^+ oxidize iron(II) to iron(III) under standard conditions?

(b) When NAD^+ is reduced, describe what happens to the structure of the C_5N ring. What is the electron pair geometry around the N atom in the ring? Around the C atom *para* to the N atom? Do the hybridizations of these atoms change?

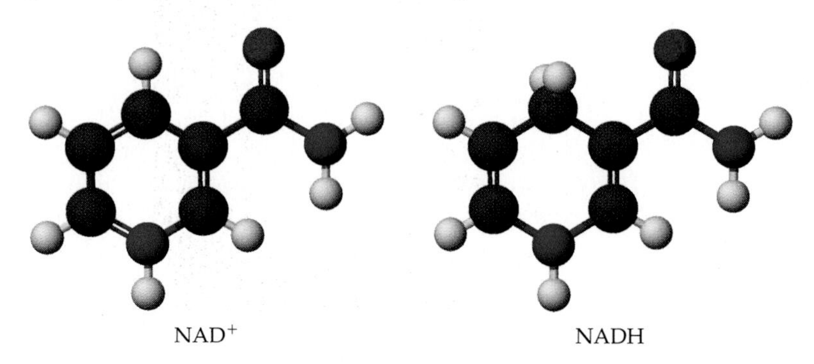

INTERACTIVE GENERAL CHEMISTRY CD-ROM

Screen 21.2 Redox Reactions

What is the difference between a direct redox reaction and an indirect redox reaction?

Screen 21.5 Electrochemical Cells and Potentials

(a) When the contents of the half-cells in the photos are changed, the standard potentials for the cells also change. What does this imply about the tendency of different metals to either hold onto or release electrons?
(b) Rank the three metals used on this screen, Ag, Cu, and Zn, in increasing tendency to release electrons.
(c) Rank the three metal ions used in the reactions on this screen, Ag^+, Cu^{2+}, and Zn^{2+}, in increasing tendency to accept electrons.
(d) What is the connection between the two rankings you have just made?

Screen 21.8 Batteries

(a) What common feature is shared by the dry cell battery, the alkaline battery, and the mercury battery?
(b) What is the anode material in a ni–cad battery?

Screen 21.9 Corrosion

(a) What are the half-reactions in the corrosion of iron?
(b) What is responsible for the fact that aluminum does not corrode, thermodynamics or kinetics?

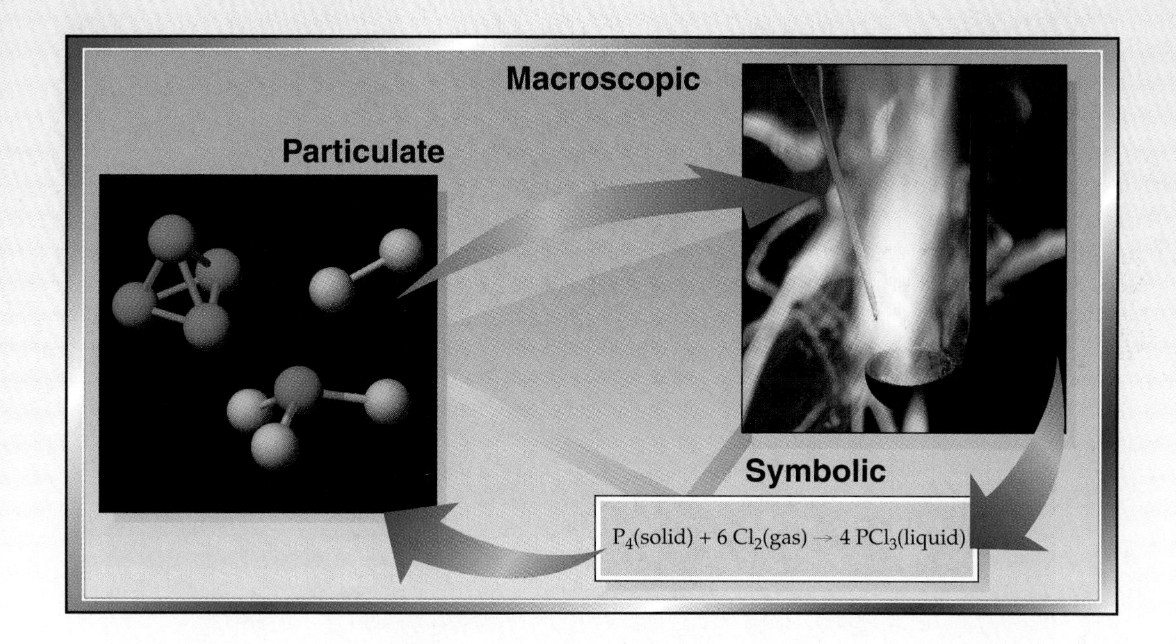

Part 5

The Chemistry of the Elements

The goal of this section is to lay out, in a reasonably systematic way, something of the chemistry of the elements, especially those of economic and biochemical importance. In this very short space we can give you only a brief introduction to the subject. For this reason, we will dwell on the chemistry you may find interesting and useful in everyday life, rather than dig into every nook and cranny of all 112 elements.

We shall first look into the chemistry of the A-group elements (Chapter 22). These groups are, of course, dominated by metals, although we will discuss the chemistry of the common nonmetals as well.

The transition elements, which include the lanthanides and actinides, make up a huge portion of the known elements, and their chemistry is considered in Chapter 23.

Finally, we want to give you a brief introduction to the chemistry of the radioactive elements in Chapter 24. This is always a subject of great interest because radioactive materials are used so widely in medicine and because such topics as food irradiation and nuclear waste disposal are frequently in the news. ● *(Photo, C. D. Winters; models, S. M. Young)*

Chapter 22

The Chemistry of the Main Group Elements

(Photo, C. D. Winters; models, S. M. Young)

A Chemical Puzzler

Aluminum *(right)* does not react with nitric acid, whereas copper *(left)* reacts vigorously to give NO_2 gas and $Cu(NO_3)_2$. Why does aluminum metal not react with strongly oxidizing nitric acid?

The abundances of the elements in the solar system are shown plotted against atomic number in Figure 22.1. What do you see? Hydrogen and helium are clearly the most abundant—they make up about 97% of the mass of the universe. Indeed, most of the mass of our solar system is in the sun, and its primary components are hydrogen and helium. Lithium, beryllium, and boron are very low in abundance, but carbon is very high. From that point on, the abundances generally decline as the atomic number increases. The elements of the A groups of the periodic table—often referred to as the main group, or representative, elements—are more abundant, on the whole, than the transition elements. Except for the high abundance of iron and nickel, the trend continues downward.

On earth, analysis to determine composition is limited to the crust, the outer shell of the planet, and its atmosphere. Here, oxygen, silicon, and aluminum together make up over 80% of the mass. Altogether eight main group elements are among the ten most abundant elements in earth's crust. Oxygen and nitrogen are the primary components of the atmosphere, and water is highly abundant on the surface, underground, and as vapor in the atmosphere. Most rocks and minerals are compounds of the main group elements. The chemistry of limestone ($CaCO_3$), which contains calcium, carbon, and oxygen, was described in Section 19.11. Sand and quartz (Figure 2.23) are composed of silicon and oxygen, and many other common minerals, such as $CaSO_4 \cdot 2\ H_2O$ and fluorite, CaF_2 (Figure 3.11), are composed of main group elements.

As a further indication of the importance of the main group elements, we need only look at the top ten chemicals produced by the U. S. chemical industry in 1995 (see inside back cover). All are main group elements or their compounds.

Because main group elements and their compounds are of such great economic importance—and have an interesting chemistry—we devote this chapter to a brief tour of these elements.

Abundances in the Earth's Crust

O	49.5%	Na	2.6%
Si	25.7%	K	2.4%
Al	7.4%	Mg	1.9%
Fe	4.7%	H	0.9%
Ca	3.4%	Ti	0.6%

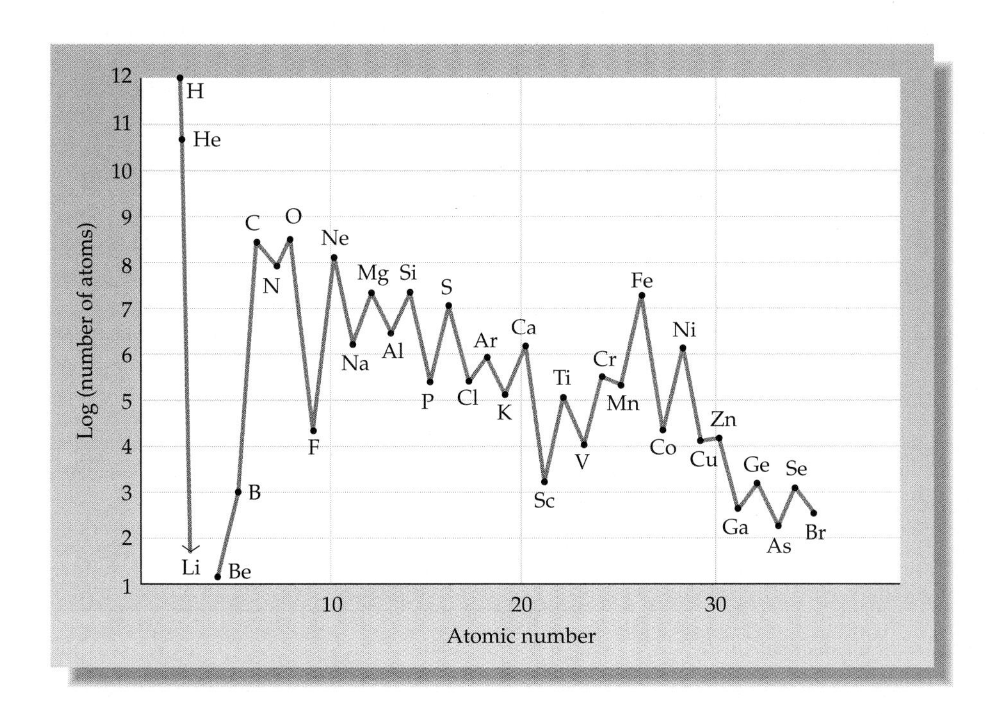

***Figure* 22.1 The abundances of the elements in the solar system.** The scale is relative to the number of H atoms.

22.1 THE PERIODIC TABLE—A GUIDE TO THE ELEMENTS

See the periodic table in the "toolbox" on the *Saunders Interactive General Chemistry CD-ROM.* There is a photograph of each element in the periodic table along with a listing of its properties.

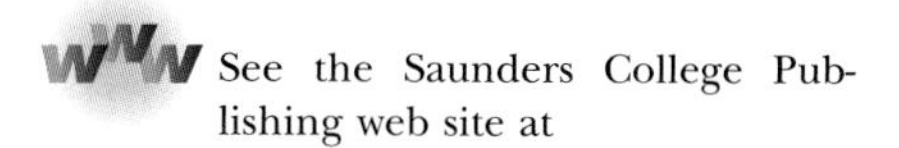

See the Saunders College Publishing web site at

http://www.saunderscollege.com

This will link you to various periodic tables and information on the elements on the World Wide Web.

The many similarities among the properties of elements in a given group include the common oxidation states of the elements and the formulas of their common oxides, oxoanions, and hydrides (Table 22.1). These similarities guided Mendeleev in creating the periodic table. As illustrated in Figure 2.15, Mendeleev placed elements in groups based on the stoichiometries of their common oxygen and hydrogen compounds.

A knowledge of trends in metallic character is particularly useful as we survey these elements. The Group 1A elements, the alkali metals, are the most metallic elements in the periodic table. In contrast, elements on the far right are nonmetals, and in between are the metalloids. Metallic character increases from the top of a group to the bottom. This is particularly well illustrated in Group 4A; carbon at the top of the group is a nonmetal, silicon and germanium are classic metalloids, and tin and lead are metals (Figure 22.2).

Valence Electrons

As mentioned on page 79, the groups in the periodic table can be numbered in two ways. In this book we retained the traditional representation, using the designation A for main group elements and B for transition elements. This approach makes it easier to identify the number of valence electrons.

The *ns* and *np* electrons are the valence electrons for the main group elements (Section 8.4), and you learned in Chapters 3 and 8 that these can be determined from the position of an element in the periodic table.

The elements of Group 8, known as the **noble,** or **rare gases,** have complete electron subshells. Helium has an electron configuration $1s^2$; the others have ns^2np^6 valence electron configurations. These elements are generally unreactive. Indeed, the first three elements in the group form no isolable compounds. The other three elements are now known to have limited chemistry, however, and the discovery of xenon compounds ranks as one of the most interesting developments in modern chemistry (see XeF_2 on page 396). For the moment, we merely observe that an element or ion with a valence electron configuration with filled *s* and *p* subshells is very stable—a fact that guides predictions of the chemical behavior of other elements that often react in a way to achieve a "noble gas configuration."

Figure 22.2 Group 4A elements. These elements illustrate the trend to increasing metallic character on descending a periodic group: Nonmetallic carbon; silicon, a metalloid; and the metals tin and lead. (Carbon is shown here in the form of a graphite crucible, and silicon is a round bar. Lead is present as bullets, a toy, and a sphere, and tin is present as chips of the metal.) *(C. D. Winters)*

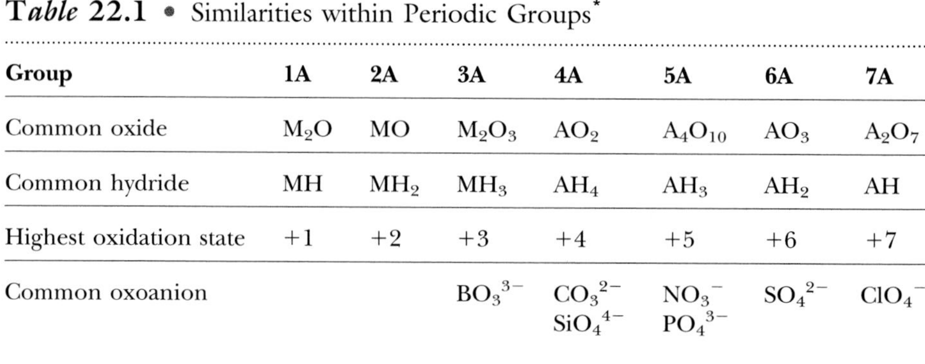

Table 22.1 • Similarities within Periodic Groups*

Group	1A	2A	3A	4A	5A	6A	7A
Common oxide	M_2O	MO	M_2O_3	AO_2	A_4O_{10}	AO_3	A_2O_7
Common hydride	MH	MH_2	MH_3	AH_4	AH_3	AH_2	AH
Highest oxidation state	+1	+2	+3	+4	+5	+6	+7
Common oxoanion			BO_3^{3-}	CO_3^{2-} SiO_4^{4-}	NO_3^- PO_4^{3-}	SO_4^{2-}	ClO_4^-

*M denotes a metal and A denotes a nonmetal.

Ionic Compounds of Main Group Elements

The elements in Groups 1A and 2A invariably form positively charged ions whose electron configuration is the same as that for the previous noble gas. Thus, the commonly encountered compounds of these elements (e.g., NaCl, $CaCO_3$) can immediately be identified as ionic. We may expect all compounds of these elements to have typical ionic properties: they will be crystalline solids with high melting points and conduct electricity in the molten state (see Figure 21.19).

The elements in Groups 6A and 7A can reach a noble gas configuration by adding electrons. Thus, in many reactions, the Group 7A elements (halogens) form anions with a 1− charge (halide ions, F^-, Cl^-, Br^-, I^-), and the Group 6A elements form anions with a 2− charge (O^{2-}, S^{2-}, Se^{2-}, Te^{2-}). Unlike the Group 1A and 2A metals, however, other avenues are open to the nonmetallic elements. Covalent compounds are formed in reactions between two or more nonmetals. Thus, a description of the chemistry of main group elements is in two parts: first, the formation of ionic compounds from metals; second, the formation of covalent compounds from nonmetals.

A majority of compounds of Group 3A elements contain 3+ ions. For example, many compounds of aluminum contain the Al^{3+} ion. In Group 5A chemistry, ions are also encountered; for example, the nitride ion, N^{3-}, has a neon electron configuration. The energy required to form highly charged cations and anions is large (Section 8.7), which means that their formation is often unfavorable relative to other possible modes of behavior.

Some covalent chemistry also occurs in Group 3A, especially with boron, and nitrogen chemistry is dominated by covalent compounds. Consider ammonia, NH_3; the ammonium ion, NH_4^+; the various nitrogen oxides; nitric acid, HNO_3; and the nitrate ion, NO_3^-, all of which involve covalent bonding from nitrogen to other nonmetallic elements.

With this brief overview, we can now predict the results of reactions between elements of Groups 1A, 2A, and 3A and elements of Group 6A or 7A (Table 22.2). In general, we expect the metal to be oxidized and the nonmetal reduced to form ionic compounds. The structures of such compounds will be an infinite lattice of positive and negative ions.

Liquid BBr_3 *(left)* and solid BI_3 *(right)*. Formed from a metalloid and a nonmetal, both are covalent compounds. Included are models of these trigonal-planar molecules. *(Photo, C. D. Winters; models, S. M. Young)*

Table 22.2 • Some Reactions of Group 1A and 2A Metals

Metal	Nonmetal	Product
K(s), Group 1A	$Br_2(\ell)$, Group 7A	KBr(s), ionic
Ba(s), Group 2A	Cl_2(g), Group 7A	$BaCl_2$(s), ionic
Al(s), Group 3A	F_2(g), Group 7A	AlF_3(s), ionic
Na(s), Group 1A	S_8(s), Group 6A	Na_2S(s), ionic
Mg(s), Group 2A	O_2(g), Group 6A	MgO(s), ionic

Example 22.1 Reactions of Group 1A and 2A elements

Give the formula and name for the product in each of the following reactions. Write a balanced chemical equation for the reaction.

1. $Ca(s) + S_8(s)$
2. $Rb(s) + I_2(s)$
3. Lithium and chlorine
4. Aluminum and oxygen

Solution Group 1A elements form 1+ ions, Group 2A elements form 2+ ions, and Group 3A elements are converted to 3+ ions. In their reactions with metals, halogen atoms add a single electron to give anions with a 1− charge, whereas Group 6A elements are predicted to add two electrons to form anions with a 2− charge (although sometimes they do not react in this simple fashion, as we shall see in Section 22.3.) These predictions are based on the assumption that ions are formed with electron configurations of the nearest noble gas.

Balanced Equation	Product Name
1. $8\ Ca(s) + S_8(s) \longrightarrow 8\ CaS(s)$	Calcium sulfide
2. $2\ Rb(s) + I_2(s) \longrightarrow 2\ RbI(s)$	Rubidium iodide
3. $2\ Li(s) + Cl_2(g) \longrightarrow 2\ LiCl(s)$	Lithium chloride
4. $4\ Al(s) + 3\ O_2(g) \longrightarrow 2\ Al_2O_3(s)$	Aluminum oxide

Exercise 22.1 Main Group Element Chemistry

Write a balanced chemical equation for the formation of the following compounds from the elements:

1. $NaBr$
2. $CaSe$
3. $AlCl_3$
4. K_2O

Covalent Compounds and Electron Configurations

You have already seen many reactions between two nonmetals such as that between carbon and excess oxygen to form CO_2. The products of these reactions are compounds in which the atoms share electron pairs.

The valence electron configuration of an element controls the stoichiometry of its compounds as well as the charge on ions. Involving all the valence electrons in the formation of compounds is a frequent and reasonable occurrence in main group element chemistry. We should not be surprised to discover halogen compounds whose stoichiometries reflect the fact that the core element has the highest possible oxidation number. For example, the very electronegative element fluorine forms ionic compounds with elements of Groups 1A through 3A, whereas the compounds for Groups 4A through 7A are covalent with highly polar bonds. In every case the central element has its maximum oxidation number.

FLUORINE COMPOUNDS FORMED BY MAIN GROUP ELEMENTS

Group	Compound	Bonding
1A	NaF	Ionic
2A	MgF_2	Ionic
3A	AlF_3	Ionic
4A	SiF_4	Covalent
5A	PF_5	Covalent
6A	SF_6	Covalent
7A	IF_7	Covalent
8A	XeF_4	Covalent

Example 22.2 Predicting Formulas for Compounds of Main Group Elements

What formula is predicted for each of the following:

1. The product of a reaction between germanium and excess oxygen.
2. The product of the reaction of arsenic and fluorine.
3. A compound formed from phosphorus and excess chlorine.
4. An anion of selenic acid.

Solution

1. Germanium is in Group 4A. We predict that this element will form an oxide of formula GeO_2, in which germanium has an oxidation number of +4.
2. Arsenic, in Group 5A, reacts vigorously with fluorine to form AsF_5, a compound in which arsenic has an oxidation number of +5.
3. PCl_5 is the product formed when phosphorus reacts with excess chlorine.
4. Selenium is below sulfur in the periodic table, and their chemistry is similar. Thus, Se is oxidized to SeO_3 with excess oxygen. This species is an acid anhydride (like SO_3) and forms selenic acid, H_2SeO_4, with water. The anion of this acid, the selenate ion, has the formula SeO_4^{2-}. Selenium's reaction with oxygen, the acid properties of the oxide, and the formula for selenate ion all have analogues in sulfur chemistry.

Exercise 22.2 Predicting Formulas for Main Group Compounds

Write formulas for

1. Hydrogen telluride
2. Sodium arsenate
3. Selenium hexachloride
4. Perbromic acid

Expect many similarities among elements in the same periodic group. Such analogies allow us to extrapolate from simple compounds that are easily recognized to analogous compounds of less common elements. You already know examples of this. Water, H_2O, is the simplest hydrogen compound of oxygen, and you expect to find hydrogen compounds with similar stoichiometries with other elements in Group 6A. Indeed, H_2S, H_2Se, and H_2Te are well known.

Example 22.3 Predicting Formulas

Predict formulas for

1. A compound of hydrogen and phosphorus.
2. The hypobromite ion.
3. Germane (the simplest hydrogen compound of germanium).

Solution

1. Phosphine, PH_3, has a stoichiometry similar to ammonia, NH_3.
2. Hypobromite ion, OBr^-, is similar to hypochlorite ion, OCl^-, the anion of hypochlorous acid.
3. GeH_4 is analogous to CH_4 and SiH_4.

Exercise 22.3 Predicting Formulas

Identify a second-period compound or ion that has a formula and Lewis structure analogous to each of the following:

1. PH_4^+
2. S_2^{2-}
3. P_2H_4
4. PF_3

Example 22.4 Recognizing Incorrect Formulas

One formula is incorrect in each of the following groups. Pick out the incorrect formula and indicate why it is incorrect.

1. $CsSO_4$, KCl, $NaNO_3$, Li_2O
2. MgO, CaI_2, Ba_2SO_4, $CaCO_3$
3. CO, CO_2, CO_3
4. PF_3, PF_4^+, PF_2, PF_6^-

Solution

1. $CsSO_4$. Sulfate ion has a 2− charge, so this formula would require a Cs^{2+} ion. Cesium, in Group 1A, forms only 1+ ions.
2. Ba_2SO_4. This formula requires a Ba^+ ion. The ion charge does not equal the group number.
3. CO_3. Given that O has an oxidation number of −2, carbon would have an oxidation number of +6. Carbon is in Group 4A, however, and can have a maximum oxidation state of +4.
4. PF_2. This species has an odd number of electrons.

Exercise 22.4 Recognizing Incorrect Formulas

Explain why compounds with the following formulas do not exist: ClO, Na_2Cl, $CaCH_3CO_2$, C_3H_7.

Problem-Solving Tips and Ideas

22.1 Nonexistent Compounds?

It may seem that the array of known chemical compounds is without limit. Not all combinations of elements are acceptable, however. It will help if you recognize when a formula is incorrect or when a combination of elements would not be found together in a compound. Here are some useful hints about what compounds do *not* exist as stable entities. All are based on the ideas outlined in Chapter 8.

- The three lightest noble gases do not form any compounds; among the noble gases only Xe has an extensive chemistry.
- Group 1A elements never form 2+ or 3+ ions; Group 2A metals never form 1+ ions.
- Very few covalent compounds contain an odd number of electrons. The same is true of polyatomic ions. (You are likely to encounter four exceptions in this text:

NO and NO_2, O_2^-, and ClO_2). Count the electrons in a formula; if the number is odd, then the formula is probably wrong.

- The maximum oxidation state of an element in a compound is equal to the number of the group in which it is found.

22.2 HYDROGEN

Hydrogen was first prepared in 1660 by Robert Boyle, from the reaction of iron with "oil of vitriol" (sulfuric acid). But it was not until 1766 that the British scientist Henry Cavendish (1731–1810) prepared a pure sample of this gas. Cavendish called hydrogen "inflammable air"; he found that it did not dissolve in water, has the lowest density of any gas, and produced water when it was burned. Shortly after Cavendish did his experiments hydrogen was recognized as an element and given its name, which means "water former."

Hydrogen is ninth in abundance in the earth's crust, 0.9% by mass, where it occurs primarily in water and in fossil fuels. In its earliest history, H_2 was mainly used as a fuel. In the middle of the 19th century it was found that heating soft coal (in the absence of air) gave a gas that could be used for cooking and lighting. This gas, called coal gas, contains about 20% H_2 along with several lightweight hydrocarbons.

A modern plant for the production of syngas, located in South Africa. Coal, oxygen, and steam are heated under pressure to form a mixture of hydrogen and carbon monoxide called "syngas." After purification, the syngas is used to manufacture various hydrocarbons. *(Fluor Engineers, Inc.)*

Because coal gas was useful, new methods were sought for its production. It was found that injecting water into a bed of red hot coke produces a mixture of H_2 and CO. This mixture is known as water gas or synthesis gas (syngas):

$$\underset{\text{coke}}{C(s)} + H_2O(g) \longrightarrow \underbrace{H_2(g) + CO(g)}_{\text{water gas}}$$

Water gas burns cleanly and can be handled readily. The amount of heat produced, however, is only about half that from the combustion of coal gas, and its flame is nearly invisible. Furthermore, carbon monoxide is highly toxic and has no odor. Despite its hazards, water gas was used as a cooking gas until about 1950, but only after adding another material to make the flame luminous and a malodorous compound so that leaks would be detected. Although no longer used as a fuel, there is renewed interest in syngas because recent chemical research has shown that it can be used as a raw material to manufacture hydrocarbons.

Synthesis of Hydrogen Gas

About 300 billion L (STP) of hydrogen gas is produced worldwide in a year, and virtually all is used immediately in other processes. The largest quantity of hydrogen is produced by the catalytic steam re-formation of hydrocarbons. This process uses methane, CH_4, as the primary starting material. Methane reacts with steam at high temperature to give H_2 and CO:

$$CH_4(g) + H_2O(g) \longrightarrow 3\,H_2(g) + CO(g) \qquad \Delta H^\circ_{rxn} = +206 \text{ kJ}$$

The reaction of steam and a hydrocarbon has been scaled to run in an automobile to provide hydrogen for a fuel cell. See page 982.

The reaction is rapid in the 900 °C to 1000 °C range and goes nearly to completion. More hydrogen can be obtained in a second step in which the CO that

See O. Khaselev and J. A. Turner, "A monolithic photovoltaic-photoelectrochemical device for hydrogen production via water splitting," *Science*, Vol. 280, p. 425, 1998.

is formed reacts with more water. This so-called water gas shift reaction is run at 400 °C to 500 °C and is slightly exothermic:

$$H_2O(g) + CO(g) \longrightarrow H_2(g) + CO_2(g) \qquad \Delta H^\circ = -41\ kJ$$

The CO_2 formed in the process is removed by reaction with CaO (to give $CaCO_3$), thus leaving fairly pure hydrogen.

Electrolysis of water is the cleanest method of H_2 production (Figure 22.3), and it provides a valuable byproduct, high-purity O_2. Electric energy is quite expensive, however, so this method is not generally used commercially.

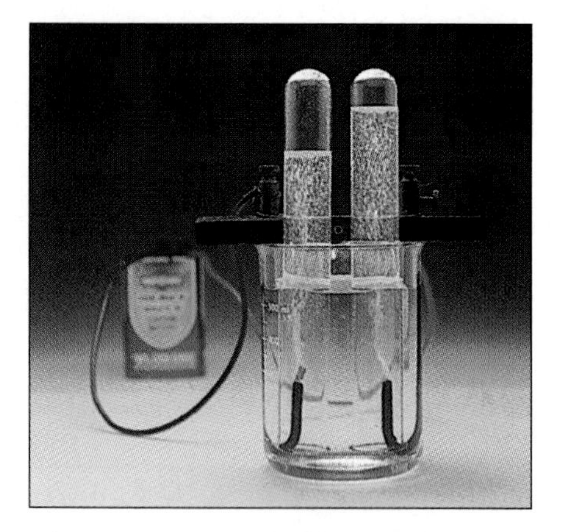

Figure 22.3 Electrolysis. Electrolysis of a dilute aqueous solution of H_2SO_4, to give H_2 (*left*) and O_2 (*right*). *(C. D. Winters)*

Because hydrogen is such a valuable commodity, there is considerable interest in finding a way to split water into hydrogen and oxygen using thermal energy. At temperatures over 1000 °C, H_2O, H_2, and O_2 exist at equilibrium. Achieving such high temperatures is quite costly, however, so many approaches have been suggested for achieving this result at lower temperature using a sequence of reactions. One example is the following:

Step 1, *at 750 °C:* $CaBr_2 + H_2O \longrightarrow 2\ HBr + CaO$

Step 2, *at 100 °C:* $Hg + 2\ HBr \longrightarrow HgBr_2 + H_2$

Step 3, *at 25 °C:* $HgBr_2 + CaO \longrightarrow HgO + CaBr_2$

Step 4, *at 500 °C:* $HgO \longrightarrow Hg + \frac{1}{2}\ O_2$

Net reaction: $H_2O(\ell) \longrightarrow H_2(g) + \frac{1}{2}\ O_2(g)$

Even though water is essentially free, thermal processes like this to produce hydrogen are not yet economical relative to those using natural gas or coal as a starting material.

A number of reactions can be used in the laboratory to form hydrogen (Table 22.3). One of the simplest is the reaction of a metal with an acid (see Figure 5.9). In 1783, Jacques Charles (of Charles's law) used the reaction of sulfuric acid with iron to produce the hydrogen for a lighter-than-air balloon.

The reaction of aluminum with NaOH (Figure 22.4) also generates hydrogen as one product. During World War II, this method was used to obtain hydrogen to inflate small balloons for weather observation and to raise radio antennas. Metallic aluminum was plentiful because it came from damaged aircraft. Finally, reaction (3) in Table 22.3 is perhaps the most efficient way to syn-

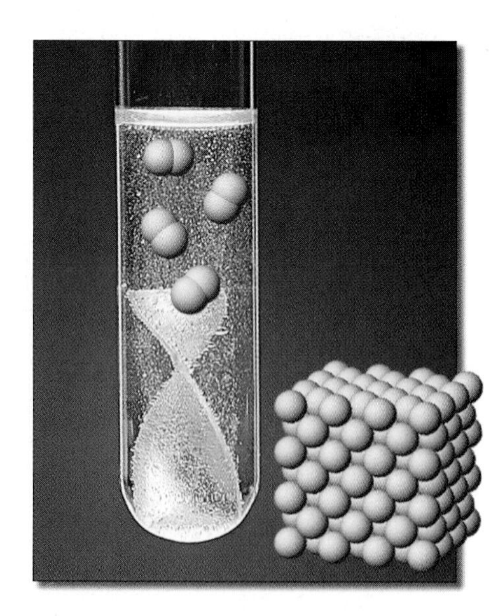

Figure 22.4 Oxidation of aluminum in aqueous base. The reaction of aluminum with aqueous NaOH produces hydrogen and $Na[Al(OH)_4]$. Included are models of aluminum metal and diatomic hydrogen. *(Photo, C. D. Winters; models, S. M. Young)*

Table 22.3 • Methods for Preparing H_2 in the Laboratory

1. Metal + Acid $\longrightarrow$ metal salt + H_2

$Mg(s) + 2\ HCl(aq) \longrightarrow MgCl_2(aq) + H_2(g)$

2. Metal + H_2O or base $\longrightarrow$ metal hydroxide or oxide + H_2

$2\ Na(s) + 2\ H_2O(\ell) \longrightarrow 2\ NaOH(aq) + H_2(g)$
$2\ Fe(s) + 3\ H_2O(\ell) \longrightarrow Fe_2O_3(s) + 3\ H_2(g)$
$2\ Al(s) + 2\ KOH(aq) + 6\ H_2O(\ell) \longrightarrow 2\ KAl(OH)_4(aq) + 3\ H_2(g)$

3. Metal hydride + H_2O $\longrightarrow$ metal hydroxide + H_2

$CaH_2(s) + 2\ H_2O(\ell) \longrightarrow Ca(OH)_2(s) + 2\ H_2(g)$

thesize H_2 in the laboratory (Figure 17.4). It is also useful for removing traces of water from liquid compounds that do not have a reactive —OH group.

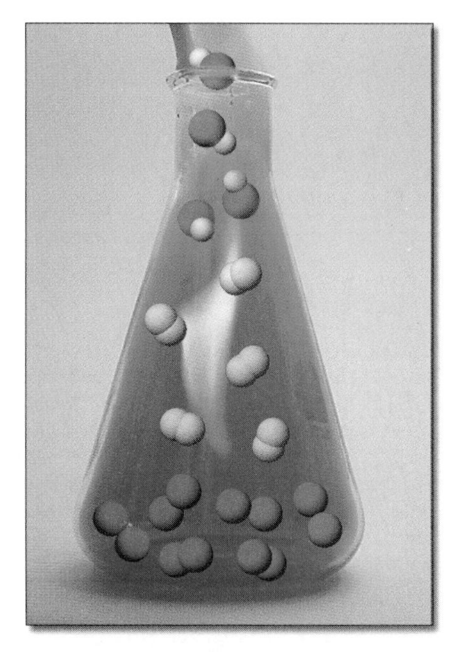

Covalent hydrogen compounds. Hydrogen burns in an atmosphere of bromine to give hydrogen bromide, HBr. Models of the reactants and products are included. *(Photo, C. D. Winters; model, S. M. Young)*

Properties of Hydrogen

Under standard conditions, hydrogen is a colorless gas. Its very low boiling point, 20.7 K, reflects its nonpolar character and low molar mass. It is, of course, the least dense gas known.

Hydrogen combines chemically with virtually every other element except the noble gases. Three different types of binary hydrogen-containing compounds are known.

Ionic Metal Hydrides

Ionic metal hydrides form when H_2 reacts with Group 1A and 2A metals:

$$2\ Na(s) + H_2(g) \longrightarrow 2\ NaH(s)$$
$$Ca(s) + H_2(g) \longrightarrow CaH_2(s)$$

These ionic compounds contain the hydride ion, H^-, in which hydrogen has −1 oxidation number.

Covalent Hydrides

Covalent hydrides are formed with electronegative elements, such as the nonmetals carbon, nitrogen, oxygen, and fluorine. Here the oxidation number of the hydrogen atom is +1:

$$N_2(g) + 3\ H_2(g) \longrightarrow 2\ NH_3(g)$$
$$F_2(g) + H_2(g) \longrightarrow 2\ HF(g)$$

See the previous discussion of the "hydrogen economy," page 278.

Interstitial Hydrides

Hydrogen is absorbed by many metals forming interstitial hydrides, in which hydrogen atoms reside in the spaces between metal atoms (called interstices) in the crystal lattice. For example, when a piece of palladium metal is used as an electrode for the electrolysis of water, the metal can soak up a thousand times its volume of hydrogen (at STP). Most interstitial hydrides are nonstoichiometric, that is, the ratio of metal and hydrogen is not a whole number. When interstitial hydrides are heated, H_2 can be driven out. Thus, these materials can be used to store H_2, the same as a sponge can store water. This is one way to store hydrogen for use as a fuel in automobiles (page 278).

Methanol is often added to gasoline to prevent "gas line freeze" in automobiles in the winter. The alcohol forms hydrogen bonds with water, thus preventing the water from freezing and clogging the gas line. *(Photo, C. D. Winters; model, S. M. Young)*

Some Uses of Hydrogen

By far the largest use of H_2 gas is in the production of ammonia, NH_3, by the Haber process (see Section 16.6):

$$N_2(g) + 3\ H_2(g) \longrightarrow 2\ NH_3(g)$$

A large amount is also used to make methanol, CH_3OH (page 506).

$$2\ H_2(g) + CO(g) \longrightarrow CH_3OH(\ell) \qquad \Delta H^\circ_{rxn} = -128.2\ \text{kJ}$$

Almost 5.1×10^9 kg of methanol was produced in 1995. Methanol is used as an additive in gasoline because oxygen-containing compounds cause gasoline to burn more cleanly. In addition, methanol is often added to gasoline in cold weather to prevent "fuel line freeze"; in this use, its function is to dissolve traces of water that often contaminate gasoline.

22.3 SODIUM AND POTASSIUM

The name "sodium" comes from caustic soda, NaOH. The Latin name for NaOH, *natrium*, furnished the symbol Na. The name "potassium" is derived from the word "potash" (literally, the ashes from a fire pot). Potash (K_2CO_3) can be extracted from ashes with water. The Latin word for potash is *kalium*, and hence we get the symbol K for this element.

Sodium and potassium are the sixth and seventh most abundant elements in earth's crust, 2.6% and 2.4%, respectively, by mass. Both metals, as well as the other Group 1A elements, are highly reactive with oxygen, water, and other oxidizing agents (see Figure 2.16). In all cases, compounds of the Group 1A metals contain the element as a 1+ ion.

The solubility guidelines indicate that most sodium and potassium compounds are water-soluble (see Figure 5.4), so it is not surprising that sodium and potassium salts are found on earth either in the oceans or in underground deposits that are the residue of ancient seas. To a much smaller extent, these elements are also found in minerals, like Chilean saltpeter ($NaNO_3$) and borax ($Na_2B_4O_7 \cdot 10\ H_2O$) (see Figure 9.5).

Within the earth's crust, sodium and potassium are about equally abundant; however, sea water contains about 2.8% NaCl but only about 0.8% KCl. Why this great difference, given that compounds of these elements have similar solubilities? The answer lies in the fact that potassium is an important factor in plant growth. Much of the potassium in ground water is taken up by plants. Most plants contain four to six times as much potassium as sodium. You may be aware that potassium is one of the three key elements in most fertilizers (nitrogen and phosphorus being the other two).

Some NaCl is essential in the diet of humans and animals because many biological functions are controlled by the concentrations of Na^+ and Cl^- ions. Animals travel great distances to reach a "salt lick," and farmers often place large blocks of salt in fields for cattle. The fact that salt has been important for a long time is evident in surprising ways. We are paid a "salary" for work done; this word is derived from the Latin word *salarium*, which meant "salt money" because Roman soldiers were paid in salt. We still talk about "salting away" money for a rainy day, a term related to the practice of preserving meat by salting it.

All animals, including humans, need a certain amount of salt in their diet. Sodium ion is important in maintaining electrolyte balance and in regulating osmotic pressure. *(C. D. Winters)*

Preparation and Properties of Sodium and Potassium

The pure metals were prepared first by the English scientist Sir Humphry Davy (1778–1829) in 1807 by electrolyzing the molten carbonates Na_2CO_3 and K_2CO_3. Sodium is still produced by electrolysis (Figure 22.5), although molten NaCl is now used (Chapter 21). Common chemical species are not strong enough reducing agents to convert Na^+ to the metal, so electrolysis is the only viable method of preparation. Potassium can be made by electrolysis also, but there are problems with this method, not the least of which is that molten potassium is soluble in the molten salt, making separation difficult. The preferred preparation of potassium uses the reaction of sodium vapor with molten KCl:

$$Na(g) + KCl(\ell) \longrightarrow K(g) + NaCl(\ell)$$

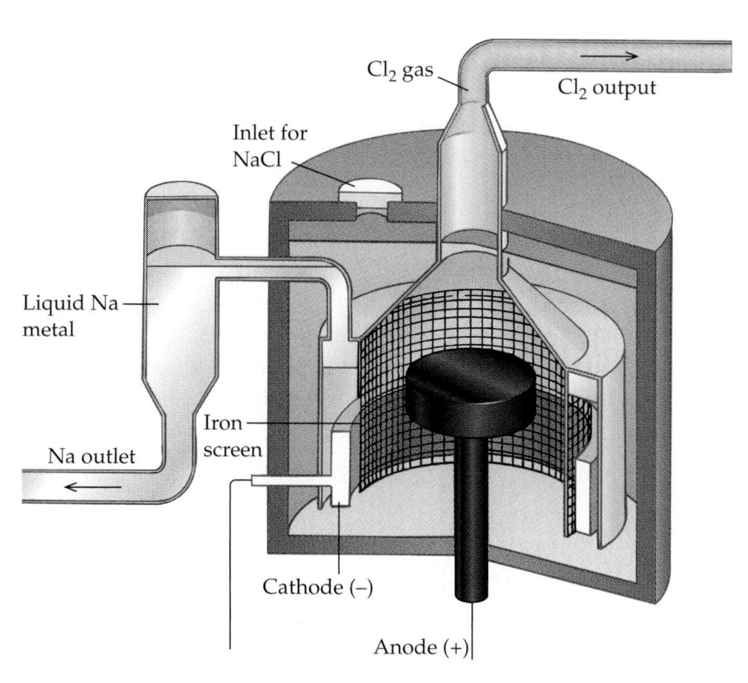

Figure 22.5 Downs cell for producing sodium metal by electrolysis. A circular iron cathode is separated from the graphite anode by an iron screen. Because the cell operates at about 600 °C, sodium is produced at the cathode as a liquid. The liquid metal floats on top of the molten salt. Chlorine gas, produced at the anode, bubbles out of the cell and is collected.

This reaction is a good example of the importance of understanding chemical equilibria. The equilibrium constant is less than 1, indicating that reactants are favored. The potassium vapor, however, is removed continuously from the reaction, shifting the equilibrium to the right.

Both sodium and potassium are silvery metals that are soft and easily cut with a knife (see Figure 2.13). Their densities are just a bit less than the density of water. Melting points for both elements are quite low, 93.5 °C for sodium and 65.65 °C for potassium. As with all alkali metals, these elements are highly reactive. When exposed to air, the metal surface is quickly coated with an oxide film. Consequently, the metals must be stored in a way that avoids contact with air; this is often done by placing them in kerosene or mineral oil.

Cesium (mp 28 °C), gallium (mp 29.8 °C), and mercury (mp −38.9 °C) are the only metals that are liquid at or near room temperature.

The Group 1A metals are highly reactive. Their reaction with water generates an aqueous solution of the metal hydroxide and hydrogen (pages 6 and 18 and Figure 2.16):

$$2\ Na(s) + 2\ H_2O(\ell) \longrightarrow 2\ Na^+(aq) + 2\ OH^-(aq) + H_2(g)$$

and reaction with any of the halogens yields a metal halide (Figure 2.27):

$$2\ Na(s) + Cl_2(g) \longrightarrow 2\ NaCl(s)$$
$$2\ K(s) + Br_2(\ell) \longrightarrow 2\ KBr(s)$$

Chemistry also produces some surprises. Group 1A metal oxides, M_2O, are known, but they are not the principal products of reactions between the Group 1A elements and oxygen. The primary product of the reaction of sodium and oxygen is sodium peroxide, Na_2O_2, and not sodium oxide, Na_2O. The principal product from the reaction of potassium and oxygen is KO_2, potassium superoxide:

$$2\ Na(s) + O_2(g) \longrightarrow Na_2O_2(s)$$
$$K(s) + O_2(g) \longrightarrow KO_2(s)$$

Both Na_2O_2 and KO_2 are ionic, with Group 1A cations paired with either the peroxide ion (O_2^{2-}) or the superoxide ion O_2^-. These compounds are not just laboratory curiosities. They are used in oxygen generation devices in places where people are confined, such as submarines, aircraft, and spacecraft. When a person breathes, for every liter of O_2 inhaled, 0.82 L of CO_2 is exhaled. A requirement of an O_2 generation system is that it should produce a larger volume of O_2 than the volume of CO_2 taken in. This requirement is met with peroxides and superoxides. With KO_2 the reaction is

Recall that molecular orbital theory (Section 10.3) was needed to describe the bonding in O_2^+, O_2, O_2^-, and O_2^{2-}. The bond orders for these four species are 2.5, 2.0, 1.5, and 1.0, respectively.

$$4\ KO_2(s) + 2\ CO_2(g) \longrightarrow 2\ K_2CO_3(s) + 3\ O_2(g)$$

Sodium Compounds of Commercial Importance

Electrolysis of aqueous sodium chloride (Chapter 21) is the basis of the chlor-alkali industry, one of the largest chemical industries in the United States. The major commercial products from this process are chlorine and sodium hydroxide:

$$2\ NaCl(aq) + 2\ H_2O(\ell) \longrightarrow 2\ NaOH(aq) + H_2(g) + Cl_2(g)$$

In 1995, 11.9×10^9 kg of NaOH and 11.4×10^9 kg of Cl_2 were produced in the United States.

Sodium carbonate is another commercially important compound. In 1995, in the United States 10.1×10^9 kg of Na_2CO_3 was produced, making it the 11th ranked industrial chemical. This compound has the common name *soda ash,* or *washing soda,* and it has been obtained, since prehistoric times, from naturally occurring deposits of $Na_2CO_3 \cdot 10\ H_2O$. *Trona,* $Na_2CO_3 \cdot NaHCO_3 \cdot 2\ H_2O$, however, estimated at 6×10^{10} tons, was discovered in Wyoming, and virtually all Na_2CO_3 in the United States now comes from that source. About 40% of the soda ash is used in the manufacture of glass, but large amounts are also used in water treatment, in pulp and paper manufacture, and in cleaning materials.

Soda ash obtained from Searles Lake, California, saline deposits. *(Jack Dermid/Photo Researchers, Inc.)*

As will be described on page 1038 the use of chlorine has come under attack from environmental groups; therefore, considerable interest has arisen in manufacturing sodium hydroxide without the coproduct chlorine. This has led to a revival of the old "soda-lime process," which produces NaOH from inexpensive lime (CaO) and soda (Na_2CO_3):

$$Na_2CO_3(aq) + CaO(s) + H_2O(\ell) \longrightarrow 2\ NaOH(aq) + CaCO_3(s)$$

The calcium carbonate byproduct is filtered off and is recycled into the process by heating it (calcining) to recover lime:

$$CaCO_3(s) \longrightarrow CaO(s) + CO_2(g)$$

Sodium bicarbonate, $NaHCO_3$, is formed in small amounts from soda ash. You are probably more aware of the bicarbonate under the common name **baking soda.** Not only is $NaHCO_3$ used in cooking, but it is also added in small amounts to table salt. This is because NaCl is often contaminated with small amounts of $MgCl_2$. The magnesium salt is hygroscopic; that is, it picks water up from the air and, in doing so, causes the NaCl to clump. Adding $NaHCO_3$ converts $MgCl_2$ to magnesium carbonate, a nonhygroscopic salt:

$$MgCl_2(s) + 2\ NaHCO_3(s) \longrightarrow MgCO_3(s) + 2\ NaCl(s) + H_2O(\ell) + CO_2(g)$$

22.4 CALCIUM AND MAGNESIUM

The Group 2A elements are called *alkaline earths.* The "earth" part of the group name is left over from the days of medieval alchemy. To alchemists, any solid substance that did not melt and was not changed by fire into another substance was called an "earth." Compounds of Group 1A and 2A elements, such as NaOH and CaO, were alkaline according to the experimental tests of the alchemists: they had a bitter taste and could be shown to neutralize acids. Group 1A compounds, however, melted in a fire or combined with the clay containers in which they were heated. Melting points of most Group 2A compounds are very high, due to the strong attractive forces between the M^{2+} cation with the anions present (page 381). For example, CaO melts at 2572 °C, a temperature well beyond the range of an ordinary fire.

The rocks in the darker band shown in this photograph of one section of the Grand Canyon are largely dolomite, a mixture of $MgCO_3$ and $CaCO_3$. *(James Cowlin/Image Enterprises)*

Calcium compounds such as lime (CaO) were known and used in ancient times. Calcium metal, however, was first prepared in 1808 by Sir Humphry Davy, who also prepared magnesium, strontium, and barium in the same year. As with sodium and potassium, Davy's preparation of these elements was accomplished by electrolysis of a molten salt.

The great abundance of calcium and magnesium on earth leads to their occurrence in plants and animals, and both elements form many commercially important compounds. It is on this chemistry that we want to focus our attention.

Like the Group 1A elements, the Group 2A elements are very reactive, so they are only found in nature combined with other elements. Unlike Group 1A metals, however, many compounds of the Group 2A elements have low water solubility. This explains their common occurrence in various minerals. Limestone ($CaCO_3$), gypsum ($CaSO_4 \cdot 2\ H_2O$), and fluorite (CaF_2) are examples of common calcium-containing minerals. Magnesite ($MgCO_3$), talc or soapstone ($3\ MgO \cdot 4\ SiO_2 \cdot H_2O$), and asbestos ($3\ MgO \cdot 4\ SiO_2 \cdot 2\ H_2O$) are magnesium-containing minerals, whereas dolomite is a compound with both elements ($MgCO_3 \cdot CaCO_3$).

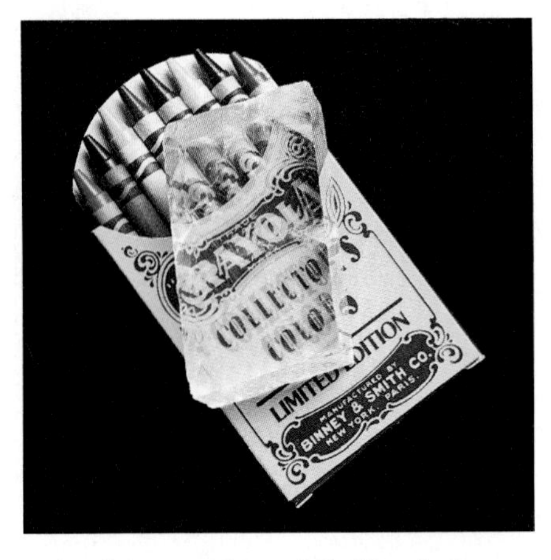

Icelandic spar, a form of $CaCO_3$, displays birefringence, a property in which a double image is formed as light passes through the crystal. *(C. D. Winters)*

Limestone, a sedimentary rock, is found widely on the surface of the earth (Figure 19.14). These deposits are the fossilized remains of marine life and are the most common of the *calcite* forms of the compound. You also know other forms of calcite, however. Marble is fairly pure calcite formed by crystallization of $CaCO_3$ under high pressure. High-quality deposits of marble are found in Italy and in the United States in Vermont, Georgia, and Colorado. Another form of calcite is *Icelandic spar,* which forms large, clear crystals.

Properties of Calcium and Magnesium

Calcium and magnesium are fairly high melting, silvery metals. The chemical properties of Group 2A elements present few surprises. All (except beryllium) are oxidized by a wide range of oxidizing agents to form ionic compounds that contain the M^{2+} ion. For example, these elements combine with halogens to form MX_2, and with oxygen or sulfur to form MO or MS (see Figure 2.17 for the reaction of Mg and O_2). All except beryllium react with water to form hydrogen and the metal hydroxide, $M(OH)_2$. With acids, hydrogen is evolved and a salt of the metal and the anion of the acid results.

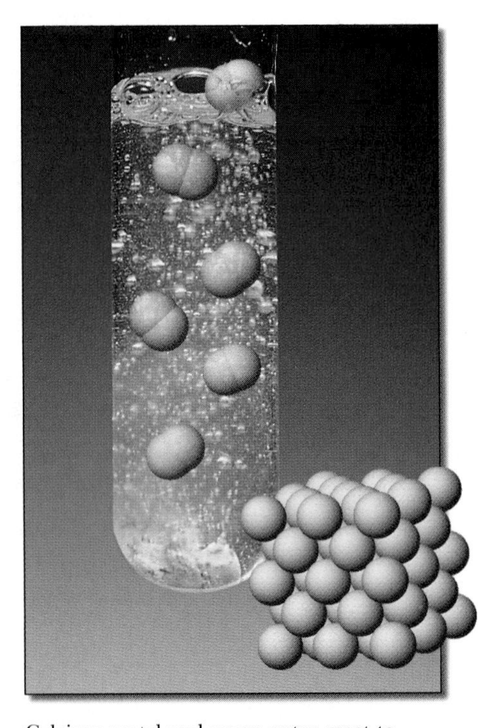

Calcium metal and warm water react to form hydrogen gas and calcium hydroxide. Included are models of hexagonal close-packed calcium metal and diatomic hydrogen. *(Photo, C. D. Winters; models, S. M. Young)*

Metallurgy of Magnesium

Several hundred thousand tons of magnesium are produced annually. The low density of the metal (1.74 g/cm^3) makes it useful in lightweight alloys. Most aluminum used today contains about 5% magnesium to improve its mechanical properties and to make it more resistant to corrosion. Other alloys having more magnesium than aluminum are used when a high strength-to-weight ratio is needed and when corrosion resistance is important, such as in aircraft and automotive parts and in lightweight tools.

Although there are many magnesium-containing minerals, most magnesium is obtained from sea water, in which it is present in a concentration of about 0.05 M. To obtain the element, magnesium is first precipitated from sea water as the relatively insoluble hydroxide (K_{sp} for $Mg(OH)_2 = 1.5 \times 10^{-11}$). The base used in this reaction, CaO, is prepared from sea shells, $CaCO_3$. Heating $CaCO_3$ gives CO_2 and CaO, and addition of water to CaO gives calcium hydroxide. When $Ca(OH)_2$ is added to sea water, $Mg(OH)_2$ precipitates:

$$Mg^{2+}(aq) + Ca(OH)_2(s) \longrightarrow Mg(OH)_2(s) + Ca^{2+}(aq)$$

Magnesium hydroxide is isolated by filtration and then neutralized with hydrochloric acid:

$$Mg(OH)_2(s) + 2\ HCl(aq) \longrightarrow MgCl_2(aq) + 2\ H_2O(\ell)$$

After evaporating the water, solid magnesium chloride is left. The anhydrous $MgCl_2$ melts at 708 °C, and the molten salt is electrolyzed to give the metal and chlorine:

$$MgCl_2(\ell) \longrightarrow Mg(s) + Cl_2(g)$$

Calcium Compounds of Commercial Importance

Apatite, a mineral with the general formula $3\ Ca_3(PO_4)_2 \cdot CaX_2$ (X = F, Cl, OH). *(C. D. Winters)*

The most important fluoride of the alkaline earth metals is fluorite, CaF_2, but fluorapatite $[CaF_2 \cdot 3\ Ca_3(PO_4)_2]$ is becoming increasingly important as a commercial source of fluorine.

Almost half of the CaF_2 mined is used in the steel industry where it is added to the mixture of materials that are melted to make crude iron. The CaF_2 acts to remove some impurities and improves the separation of molten metal from slag, the layer of silicate impurities and byproducts that comes from reducing iron ore to the metal (Chapter 23).

The second major use of fluorite is in the manufacture of hydrofluoric acid by reaction of the mineral with concentrated sulfuric acid:

$$CaF_2(s) + H_2SO_4(\ell) \longrightarrow 2\ HF(g) + CaSO_4(s)$$

Hydrogen fluoride is very reactive and is an extremely important chemical. It is used to make cryolite, Na_3AlF_6, a material needed in aluminum production and in the manufacture of fluorocarbons such as tetrafluoroethylene, which is used to make polytetrafluoroethylene (Teflon) (see Table 11.15).

Apatites are collectively referred to as phosphate rock. Over 1.0×10^8 tons are mined annually; Florida alone accounts for about one third of the world's output. Much of this rock is converted to phosphoric acid by reaction with sulfuric acid. Phosphoric acid is used to manufacture a multitude of products, including fertilizer and detergents. Its reaction products are found in baking powder, in frozen fish, and in many other food products:

$$\underset{\text{fluorapatite}}{CaF_2 \cdot 3\ Ca_3(PO_4)_2(s)} + 10\ H_2SO_4(aq) \longrightarrow 10\ CaSO_4(s) + 6\ H_3PO_4(aq) + 2\ HF(g)$$

Because of their economic importance, calcium carbonate and calcium oxide are of special interest. The thermal decomposition of $CaCO_3$ to lime, CaO, is one of the oldest chemical reactions known. Among industrial chemicals produced today, lime is fifth, with 18.7×10^9 kg produced in 1995. Most limestone and lime are used in the chemicals industry, and one third of the lime is used to make steel by the basic oxygen process (Chapter 23).

Limestone has been used in agriculture for centuries. It is spread on fields to neutralize acidic compounds in the soil and to supply the essential nutrient Ca^{2+}. Because magnesium carbonate also is often present in limestone, "liming" a field also supplies Mg^{2+}, another important plant nutrient.

For several thousand years, lime has been used as mortar (a lime, sand, and water paste) to secure stones to one another in building houses, walls, and roads. The Chinese used it in setting stones in the Great Wall. The Romans perfected its use, and the fact that many of their constructions still stand is testament both to their skill and the usefulness of lime. In 312 BC, the famous Appian Way, a Roman highway stretching from Rome to Brindisi (a distance of about 350 miles) was begun, and lime mortar was used between several layers of its stones.

The utility of mortar depends on some simple chemistry. Mortar consists of one part lime to three parts sand, with water added to make a thick paste. The first reaction that occurs is the formation of $Ca(OH)_2$, the process referred to as "slaking," the product being called *slaked lime.* When the mortar is placed

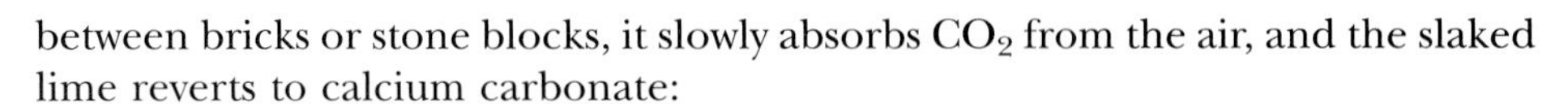

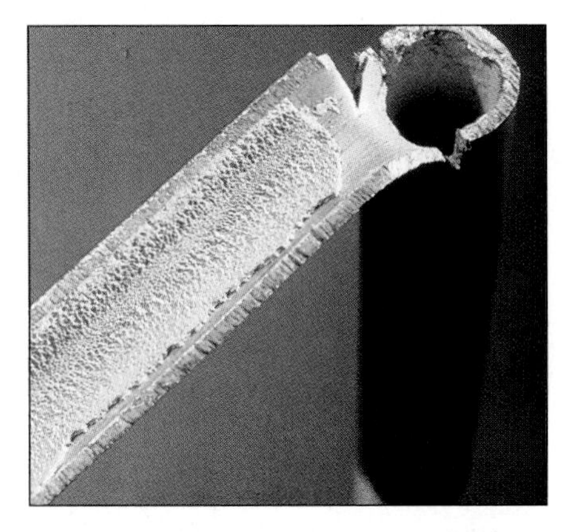

A section of water pipe that has been coated on the inside with $CaCO_3$ deposited from hard water. *(Courtesy of Betz Dearborn)*

between bricks or stone blocks, it slowly absorbs CO_2 from the air, and the slaked lime reverts to calcium carbonate:

$$Ca(OH)_2(s) + CO_2(g) \rightleftharpoons CaCO_3(s) + H_2O(\ell)$$

Although sand mixed into mortar is chemically inert, the grains are bound together by the particles of calcium carbonate, and a hard material results.

"Hard water" contains dissolved ions, chiefly Ca^{2+} and Mg^{2+}. Water containing dissolved CO_2 reacts with limestone:

$$CaCO_3(s) + H_2O(\ell) + CO_2(g) \rightleftharpoons Ca^{2+}(aq) + 2\ HCO_3^-(aq)$$

This reaction is reversible, however. When hard water is heated, the solubility of CO_2 drops, and the equilibrium shifts to the left. If this happens in a heating system or steam-generating plant, the walls of the hot water pipes can become coated or even blocked with solid $CaCO_3$. In your house, you may notice a coating of calcium carbonate on the inside of cooking pots.

The previous equation describes the chemistry of caves as well. The acidic oxide CO_2 reacts with $Ca(OH)_2$ to produce white, solid $CaCO_3$. When further CO_2 is available, however, the $CaCO_3$ can redissolve because of the formation of aqueous Ca^{2+} and HCO_3^- ions. (See Sections 16.1 and 19.11.)

22.5 ALUMINUM

Aluminum is the third most abundant element in earth's crust (7.4%). You are familiar with the metal because it is widely used in packaging (aluminum foil and aluminum cans) and as a structural material.

Pure aluminum is soft and weak; moreover, it loses strength rapidly above 300 °C. What you recognize as aluminum is actually aluminum alloyed with small amounts of other elements, which strengthen the metal and improves its properties. A large passenger plane may use more than 50 tons of aluminum alloy. A typical alloy may contain about 4% copper with smaller amounts of silicon, magnesium, and manganese. To make a softer, more corrosion-resistant alloy for window frames, furniture, highway signs, and cooking utensils, however, only manganese may be included.

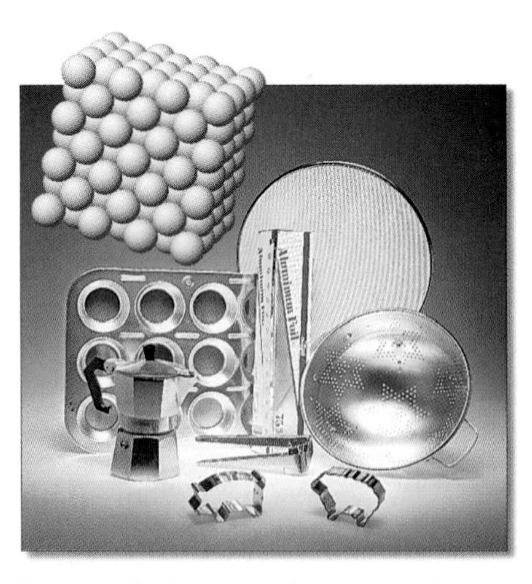

Because aluminum metal is an attractive and durable metal, it is widely used. Notice that aluminum metal has a face-centered unit cell. *(Photo, C. D. Winters; models, S. M. Young)*

Aluminum is readily oxidized, as seen from its position in a table of standard reduction potentials (see Table 21.1):

$$Al^{3+}(aq) + 3\ e^- \longrightarrow Al(s) \qquad E° = -1.66\ V$$

Aluminum's resistance to corrosion is therefore unexpected. We know, however, that this corrosion resistance is due to the formation of a thin, tough, and transparent skin of oxide, Al_2O_3, that adheres to the metal surface:

$$4\ Al(s) + 3\ O_2(g) \longrightarrow 2\ Al_2O_3(s) \qquad \Delta H° = -3351.4\ kJ$$

An important feature of the protective oxide layer is that it rapidly self-repairs. If you scratch the surface coating, a new layer of oxide immediately forms over the damaged area. This feature explains the lack of reactivity of aluminum toward nitric acid, an oxidizing acid, shown in the *Chemical Puzzler* for this chapter.

Metallurgy of Aluminum

Aluminum is found in varying amounts in nature as aluminosilicates, minerals such as clay that are based on aluminum, silicon, and oxygen. As these minerals are weathered, they gradually break down to various forms of hydrated aluminum oxide, $Al_2O_3 \cdot n\,H_2O$, called *bauxite.*

Aluminum is obtained by electrolysis of bauxite (see Figure 21.22). For this process it is first necessary to purify the ore, separating Al_2O_3 from iron and silicon oxides. Purification is done by the *Bayer process,* which uses the amphoteric, basic, or acidic nature of the various oxides. Silica, SiO_2, is an acidic oxide, Al_2O_3 is amphoteric, and Fe_2O_3 is a basic oxide. The first two oxides therefore dissolve in a hot concentrated solution of caustic soda (NaOH), leaving insoluble Fe_2O_3 to be filtered out:

$$Al_2O_3(s) + 2\,NaOH(aq) + 3\,H_2O(\ell) \longrightarrow 2\,Na[Al(OH)_4](aq)$$
$$SiO_2(s) + 2\,NaOH(aq) + 2\,H_2O(\ell) \longrightarrow Na_2[Si(OH)_6](aq)$$

By treating the solution containing aluminum and silicon anions with CO_2, Al_2O_3 is precipitated, and the silicate ion remains in solution. Recall that CO_2 forms the weak acid H_2CO_3 in water, so Al_2O_3 precipitation is an acid–base reaction

$$H_2CO_3(aq) + 2\,Na[Al(OH)_4](aq) \longrightarrow Na_2CO_3(aq) + Al_2O_3(s) + 5\,H_2O(\ell)$$

Properties of Aluminum and Its Compounds

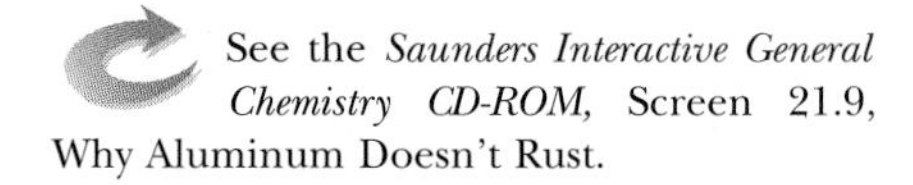
See the *Saunders Interactive General Chemistry CD-ROM,* Screen 21.9, Why Aluminum Doesn't Rust.

Aluminum dissolves in HCl(aq) but not in nitric acid. The latter is a powerful oxidizing agent and a source of oxygen atoms, so it oxidizes the surface of aluminum rapidly, and the film of Al_2O_3 protects the metal from further attack. In fact, nitric acid is often shipped in aluminum tanks.

Various salts of aluminum dissolve in water, giving the hydrated Al^{3+}(aq) ion. As described in Section 17.11, these solutions are acidic, due to the following equilibrium:

$$[Al(H_2O)_6]^{3+}(aq) + H_2O(\ell) \rightleftharpoons [Al(H_2O)_5(OH)]^{2+}(aq) + H_3O^+(aq)$$

The addition of acid shifts the equilibrium to the left, whereas base causes the equilibrium to shift to the right. Eventually, with further hydroxide, the hydrated oxide $Al_2O_3 \cdot 3\,H_2O$ precipitates.

Aluminum oxide, Al_2O_3, which can be formed by dehydrating $Al(OH)_3$, is quite insoluble in water and generally resistant to chemical attack. In the crystalline form, aluminum oxide is known as *corundum.* This material is extraordinarily hard, a property that leads to its use as the abrasive in grinding wheels, "sandpaper," and toothpaste.

Some gems are impure aluminum oxide. Rubies, beautiful red crystals prized for jewelry and used in some lasers, are Al_2O_3 contaminated with a small amount of Cr^{3+} (Figure 3.8). The Cr^{3+} ions replace some of the Al^{3+} ions in the crystal lattice. Blue sapphires occur when Fe^{2+} and Ti^{4+} impurities are present in Al_2O_3. Synthetic rubies were first made in 1902, and the worldwide capacity is now about 200,000 kg/year; much of this production is used for jewel bearings in watches and instruments.

Synthetic rubies. These are crystals of Al_2O_3 containing a few Cr^{3+} ions in place of Al^{3+} ions in the lattice. The model shows a portion of the aluminum oxide lattice. *(Photo, K. Nassau; model, S. M. Young)*

22.6 SILICON

Elemental silicon has the same structure as diamond. Tetrahedral Si atoms are linked by Si—Si bonds in a network solid. See Figure 2.23.

Silicon is the second most abundant element in the earth's crust. It is almost inevitable, therefore, that silicon compounds would be important in the development of society. Pottery, made of silicon-based natural materials, was made at least 6000 years ago in the Middle East, and sophisticated techniques for making porcelain were developed by the Chinese 5000 years ago.

The name "silicon" is derived from the Latin word *silex,* meaning flint, a silicate mineral often used by prehistoric people to make knives and other tools. Today we are surrounded by silicon-containing materials: bricks, pottery, porcelain, lubricants, sealants, computer chips, and solar cells. Silicon-based semiconductors have fueled the computer revolution of the past decade.

Reasonably pure silicon is made in large quantities by heating pure silica sand with purified coke to approximately 3000 °C in an electric furnace:

$$SiO_2(s) + 2\ C(s) \longrightarrow Si(\ell) + 2\ CO(g)$$

The molten silicon is drawn off the bottom of the furnace and allowed to cool to a shiny blue-gray solid (Figures 2.23a and 22.2). For applications in the electronics industry, extremely high purity silicon is needed. The crude silicon is first chlorinated to form silicon tetrachloride, a liquid with a boiling point of 57.6 °C.

$$Si(s) + 2\ Cl_2(g) \longrightarrow SiCl_4(\ell)$$

The volatile tetrachloride is purified by distillation and then reduced to silicon using very pure magnesium or zinc:

$$SiCl_4(g) + 2\ Mg(s) \longrightarrow 2\ MgCl_2(s) + Si(s)$$

The magnesium chloride is washed out with water. The silicon is then remelted and cast into bars. A final purification is carried out by *zone refining,* a process in which a special heating device is used to melt a narrow segment of the silicon rod. The heater is then moved slowly down the rod. Impurities contained in the silicon tend to remain in the liquid phase because the melting point of a mixture is lower than that of the pure element (Chapter 14). The silicon that crystallizes above the heated zone is therefore of a higher purity (Figure 22.6).

Figure **22.6 Making pure silicon.** A rod of pure silicon has just been withdrawn from a vat of molten silicon. To form the rod, a seed crystal was mechanically rotated in molten silicon. Pure silicon formed on the seed crystal as the rod was slowly withdrawn from the vat. *(C. Bruce Forster, Wacker Siltronic, Portland, OR)*

Silicon Dioxide

The simplest oxide of silicon is SiO_2, commonly called silica. Silica is a major constituent of many rocks, such as granite and sandstone. Quartz is a pure crystalline form of silica. The introduction of impurities into quartz produces gemstones such as amethyst (Figure 22.7).

The fact that SiO_2 is a high-melting solid (quartz melts at 1610 °C) should catch your attention because CO_2, the oxide of the element above silicon in the periodic table, is a gas at room temperature and 1 atm. This great disparity in properties arises from the different structures of CO_2 and SiO_2. Carbon dioxide is a molecular species with the carbon atom linked to each oxygen by a double bond. In contrast, SiO_2 has silicon and oxygen atoms bonded together, giving a giant network (see Figure 13.33). This structure is preferred over a simple molecular structure because the energy of two Si═O double bonds is much less than the energy of four Si—O single bonds. In fact, the contrast between

(a)

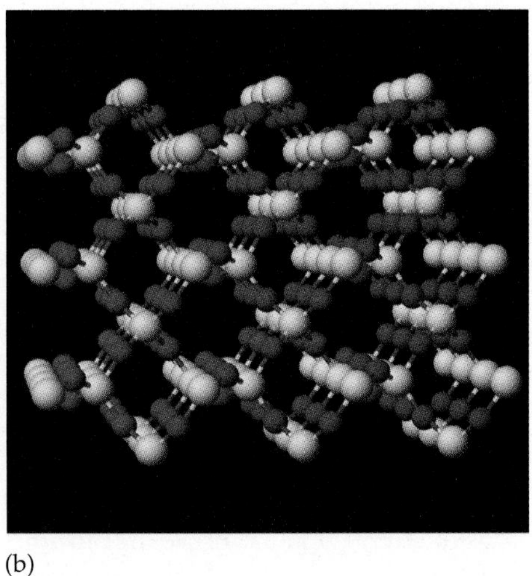

(b)

Figure 22.7 Various forms of quartz. (a) Pure, colorless quartz was used as an ornamental material as early as the Stone Age; now, quartz is used for its electric properties in such consumer products as phonographs, watches, and radios. Purple amethyst is the most highly prized variety of quartz. It ranges in color from pale lilac to a deep, royal purple. The name comes from the Greek *amethustos,* meaning "not drunken"; it was believed that an amethyst wearer could never become intoxicated. The blue and brown objects in the photo are synthetic quartz. The blue color comes from the inclusion of some cobalt, and the brown color is from the inclusion of iron. (b) A model of the structure of quartz. Notice the six-member rings of Si and O atoms. Also notice that each Si atom is surrounded tetrahedrally by O atoms. *(Photo, C. D. Winters; model, S. M. Young)*

SiO_2 and CO_2 exemplifies a more general phenomenon. Multiple bonds, often encountered between second-period elements, are rare among elements in the lower periods.

Other examples of the preference of third-period elements to form single bonds rather than multiple bonds are seen in the structures of S_8 versus O_2 and P_4 versus N_2.

Crystalline quartz is used to control the frequency of almost all radio and television transmissions. These and related applications use so much quartz that there is not enough natural quartz to fulfill demand, so quartz is synthesized. Noncrystalline, or vitreous, quartz, made by melting pure silica sand, is placed in a steel "bomb" and dilute aqueous NaOH is added. A "seed" crystal is placed in the mixture, just as you might place a seed crystal in a hot sugar solution to grow rock candy. When the mixture is heated above the critical temperature of water (above 400 °C and 1700 atm) over a period of days, pure quartz crystallizes (Figure 22.7a).

Silica is resistant to attack by all acids except HF, with which it reacts to give SiF_4 and H_2O. It does dissolve slowly in hot, molten NaOH or Na_2CO_3 to give Na_4SiO_4:

$$SiO_2(s) + 4\ HF(\ell) \longrightarrow SiF_4(g) + 2\ H_2O(\ell)$$
$$2\ Na_2CO_3(\ell) + SiO_2(s) \longrightarrow Na_4SiO_4(s) + 2\ CO_2(g)$$

After the molten mixture has cooled, hot water under pressure is added. This partially dissolves the material to give a solution of sodium silicate. After filtering off insoluble sand or glass, the solvent is evaporated, leaving sodium silicate, called *water glass.* The biggest single use of this material is in household and industrial detergents. A sodium silicate solution maintains pH by its buffering ability and can degrade animal and vegetable fats and oils. Sodium silicate is also used in various adhesives and binders, especially for gluing corrugated cardboard boxes.

Current Perspectives

Lead Pollution, Old and New

Anchoring the bottom of Group 4A is the element lead. One of a handful of elements known since ancient times, it is also a metal with modern uses. It is an essential commodity in the modern industrial world, ranking fifth in consumption behind iron, copper, aluminum, and zinc. The major uses of the metal and its compounds are in storage batteries (page 978), pigments, ammunition, solders, plumbing, and bearings. Its chief source is lead sulfide, PbS, known commonly as galena (Figure 19.10).

Lead and its compounds are cumulative poisons, particularly in children. At a blood level as low as about 50 ppb (parts per billion), blood pressure is elevated; intelligence is affected at about 100 ppb; and blood levels higher than about 800 ppb can lead to coma and possible death. Health experts believe that more than 200,000 children become ill from lead poisoning annually. This is caused chiefly by eating paint containing lead pigments. Older homes often contain lead-based paint because white lead [$2\ PbCO_3 \cdot Pb(OH)_2$] was the pigment used in white paint until about 40 years ago when it was replaced by TiO_2. Lead salts have a sweet taste, which may contribute to the tendency of children to chew on painted objects.

Uses of lead. The primary use of lead is in lead storage batteries (where the metal is used as the anode, and PbO_2 is used as the cathode.) Other uses include as plumbers' lead and as a component of solder and pewter. Lead is also used to frame the glass pieces in stained glass windows. *(C. D. Winters)*

Until just a few years ago a major use of lead was as tetraethyllead, $Pb(C_2H_5)_4$. This was added to gasoline to improve the burning properties of fuel. The compound has been phased out, however, because of the hazards from tons of lead compounds being spewed into the environment. All gasoline sold in the United States now bears the label "unleaded" or "no lead."

Lead is often implicated as one cause of the collapse of the Roman Empire. The Romans used lead to make pipes to carry water, and they cooked in lead vessels. Because many foods are acidic, and lead reacts slowly with acid, lead could be incorporated into food.

Scientists have recently obtained evidence that extensive lead mining and smelting operations during the Roman Empire contributed to atmospheric pollution on a global scale. Drilling in the Greenland ice cap produced an ice core, a cylinder of ice 9938 feet long. The core contains trace remnants of the atmosphere for the last 9000 years, and it is possible to obtain very accurate dates on the various segments of this core.

Among the trace elements in the polar ice core was lead, in concentrations in the parts per trillion (ppt) range. Samples of lead have isotope ratios (such as $^{208}Pb/^{206}Pb$) that depend on the source of the element. (Analysis of isotope ratios is carried out using sensitive mass spectrometric techniques; see Figure 2.11.) In the segment of the ice core between 300 BC and 600 AD, samples had lead isotope ratios identical to the lead from Roman mines in southwestern Spain! Lead polluted the earth's atmosphere 2000 years ago.

If sodium silicate is treated with acid, a gelatinous precipitate of noncrystalline, or amorphous, SiO_2, called *silica gel,* is obtained. Washed and dried, silica gel is a very porous material with dozens of uses. Because it can absorb up to 40% of its own weight of water, you may know it as a drying agent. (Small packets of silica gel are often placed in packing boxes of merchandise during storage. It is frequently stained with $(NH_4)_2CoCl_4$, a humidity detector that is pink when hydrated, but remains blue when dry.) Finally, silicates are used to clarify beer; passage through a bed of silica gel removes minute particles that make the brew cloudy.

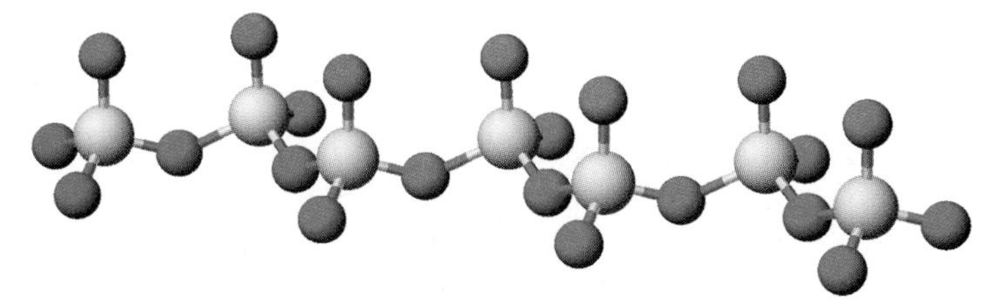

Figure 22.8 **Pyroxene.** Each SiO_4 unit shares two O atoms with neighboring units. The ratio of Si atoms to O atoms in this chain structure is 1 to 3.

The Silicate Minerals

The **silicate minerals** are a world in themselves. All silicates are built from tetrahedral SiO_4 units. The greatly differing properties of silicate materials are determined by the way that these tetrahedral SiO_4 units link together.

The simplest silicates, *orthosilicates,* contain SiO_4^{4-} anions. The 4− charge of the anion can be balanced by four M^+ ions, two M^{2+} ions, or a combination of ions. For example, calcium orthosilicate, Ca_2SiO_4, is one of the components of *Portland cement. Olivine,* an important mineral in earth's mantle, contains Mg^{2+}, Fe^{2+}, and Mn^{2+}; the Fe^{2+} ion gives olivine its characteristic olive color.

A group of minerals called *pyroxenes* have as their basic structural unit an extended chain of linked SiO_4 tetrahedra (Figure 22.8). If two such chains link by sharing oxygen atoms, the result is an *amphibole,* of which the *asbestos* minerals are one example. As a result of its chain structure, asbestos is a fibrous material. The asbestos minerals have a very low thermal conductivity, which has led to their use in insulation and fireproofing.

Linking many silicate chains together produces a sheet of SiO_4 tetrahedra (Figure 22.9). Substances in this category include the mineral mica and clays. The molecular sheet of SiO_4 tetrahedra leads to the characteristic appearance

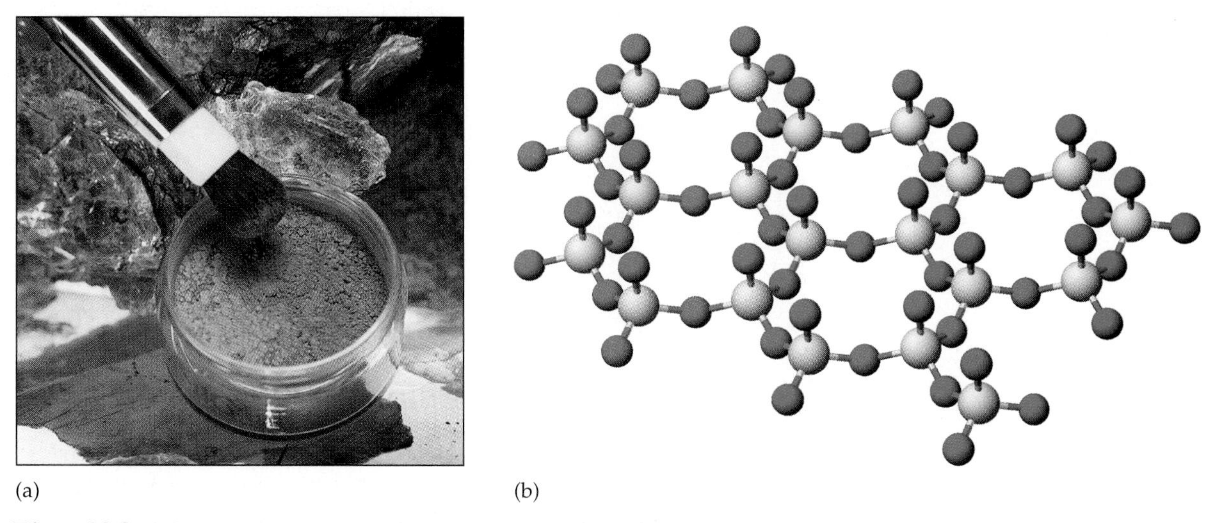

(a) (b)

Figure 22.9 **Mica.** (a) Mica is a layered silicate, and this photo shows the sheet-like structure of the mineral. Also shown in this photo is a cosmetic containing a *small* amount of mica. It is said to make the wearer "glow." (b) A model of a portion of a silicate sheet. Chains of SiO_4 (as in the pyroxene structure) are joined by sharing O atoms. *(Photo, C. D. Winters; model, S. M. Young)*

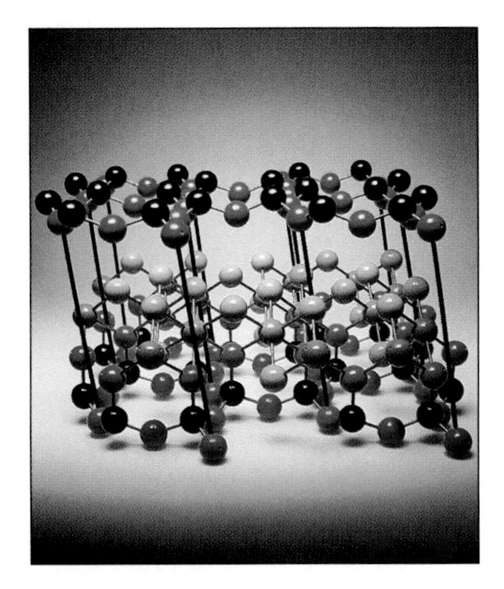

Figure 22.10 A model of kaolinite, an aluminosilicate. Each Si atom (black) is surrounded tetrahedrally by O atoms (red) to give rings consisting of six Si atoms and six O atoms. The layer of Al ions (light blue) is attached through O atoms to the Si—O rings. Hydroxide ions (light green) act as bridges between Al ions. The net result is a layered structure that gives clays their slipperiness and workability when wet. *(C. D. Winters)*

of mica, which is often found as "books" of thin, silicate sheets. Mica is used in furnace windows and as insulation, and flecks of mica give the glitter to "metallic" paints.

Clay minerals are essential components of soils, and are the raw material for pottery, bricks, and tiles. Clays come from the weathering and decomposition of igneous rocks. Specifically, the aluminosilicate *kaolinite* comes from the weathering of feldspar, an idealized version of the reaction being

$$\underset{\text{feldspar}}{2\ KAlSi_3O_8(s)} + CO_2(g) + 2\ H_2O(\ell) \longrightarrow \underset{\text{kaolinite}}{Al_2(OH)_4Si_2O_5(s)} + 4\ SiO_2(s) + K_2CO_3(aq)$$

Its structure consists of layers of SiO_4 tetrahedra linked into sheets, as in mica, but these sheets are interleaved with six-coordinate Al^{3+} ions. The aluminum atoms are positioned in the lattice to be surrounded octahedrally by O atoms of the silicon–oxygen sheets and OH^- ions (Figure 22.10).

China clay is primarily kaolinite. It is practically free of iron (a common impurity), and so it is colorless, making it particularly valuable. The predominant use for kaolin in the United States is for paper filling and coating, but some is also used for china, crockery, and earthenware.

All clays are aluminosilicates in that they contain both aluminum and silicon. In the clay minerals some Si^{4+} ions are also replaced by Al^{3+} ions. To make up for the "missing" positive charge, 1+ for every Si^{4+} replaced by an Al^{3+}, nature adds positive ions such as Na^+ and Mg^{2+}, which are located between the aluminosilicate sheets. This gives these materials interesting properties, one of which is their use in medicines (Figure 22.11). Several remedies for the relief of upset stomach contain highly purified clays, which absorb excess stomach acid as well as potentially harmful bacteria and their toxins by exchanging the intersheet cations in the clays for the toxins, which are often organic cations. Indeed, this is a remedy learned from other cultures, where clay has long been eaten for medicinal purposes.

Other aluminosilicates include *feldspars* (among the most common minerals; they make up about 60% of earth's crust) and *zeolites.* Both materials are again composed of SiO_4 tetrahedra with some of the silicon atoms replaced by Al atoms. Because the silicon atoms formally bear a 4+ charge and are replaced by Al^{3+} ions, other positive ions are present for charge balance. Typically, alkali and alkaline earth ions serve this purpose. For example, the synthetic zeolite "Linde A" has the formula $Na_{12}(Al_{12}Si_{12}O_{48}) \cdot 27\ H_2O$.

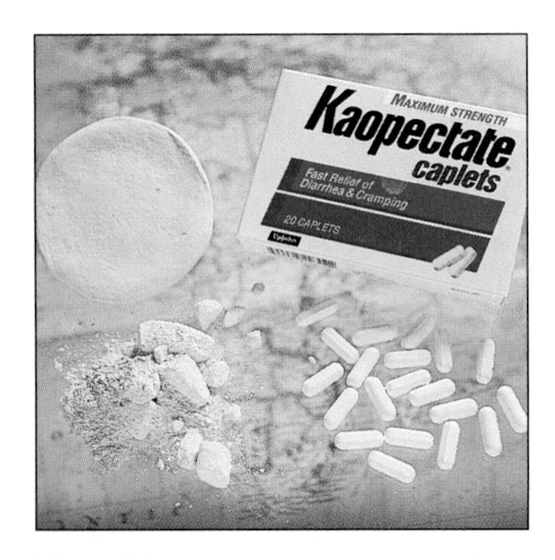

Figure 22.11 A commercial remedy for diarrhea contains one type of clay. The off-white objects in the photo are pieces of clay purchased in a market in Ghana, West Africa. This clay is made to be eaten as a remedy for stomach ailments. The practice of eating clay is widespread among many cultures of the world. *(C. D. Winters)*

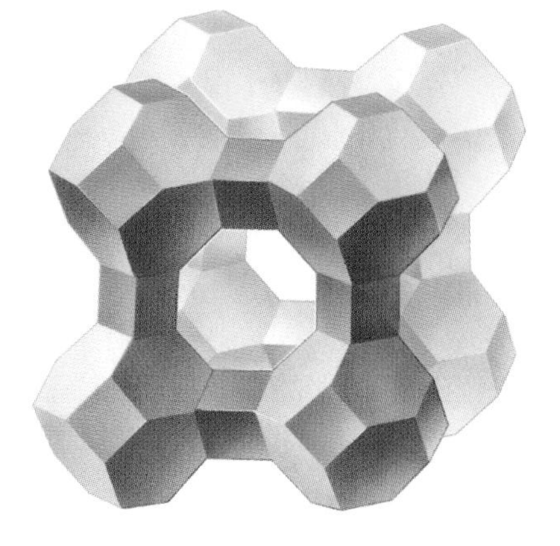

Naturally occurring zeolites are aluminosilicates. Unlike mica and clays, the aluminum, silicon, and oxygen atoms are linked into polyhedral frameworks. In this structure, each vertex is either an Si or Al atom, and each edge consists of an Si—O—Si, Al—O—Si, or Al—O—Al bond. The channels in the framework can be used to selectively capture small molecules or act as catalytic sites. See the sidebar on the *Saunders Interactive General Chemistry CD-ROM,* Screen 9.14.

The main feature of zeolite structures is their regularly shaped tunnels and cavities. Hole diameters are between 300 and 1000 pm, and small molecules such as water can fit into the cavities of the zeolite. As a result, zeolites can be used as drying agents to selectively absorb water from air or a solvent. Small amounts are sealed into multipane windows to keep the air dry between the panes.

Zeolites are also used as catalysts. Mobil Oil Corporation, for example, has patented a process in which methanol, CH_3OH, is converted to gasoline in the presence of specially tailored zeolites. Finally, zeolites are used as water-softening agents in detergents, because the sodium ions of the zeolite can be exchanged for Ca^{2+} ions in hard water, effectively removing Ca^{2+} from the water.

Silicone Polymers

Just as silicon reacts readily with chlorine to produce $SiCl_4$, a similar reaction occurs between silicon and methyl chloride, CH_3Cl. This reaction forms $(CH_3)_2SiCl_2$; the structure of this compound has two methyl groups (CH_3) and two Cl atoms bound to tetrahedral silicon:

$$Si(s) + 2\ CH_3Cl(g) \xrightarrow{\text{Cu powder catalyst/300 °C}} (CH_3)_2SiCl_2(\ell)$$

Unlike compounds with C—Cl bonds, halides of the other Group 4A elements hydrolyze readily. For $(CH_3)_2SiCl_2$, this reaction with water initially produces $(CH_3)_2Si(OH)_2$; however, the molecules condense together, eliminating water:

$$(CH_3)_2SiCl_2 + 2\ H_2O \longrightarrow (CH_3)_2Si(OH)_2 + 2\ HCl$$
$$n\ (CH_3)_2Si(OH)_2 \longrightarrow -\!\!\left[(CH_3)_2SiO\right]\!\!-_n + n\ H_2O$$

The product of this reaction is a *polymer* called polydimethylsiloxane, a member of the *silicone* polymer family.

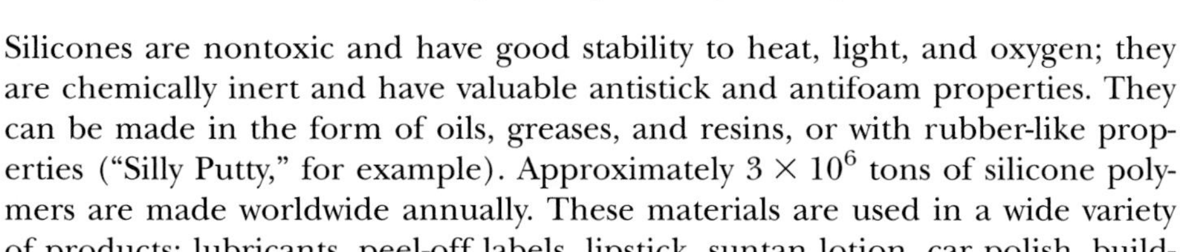

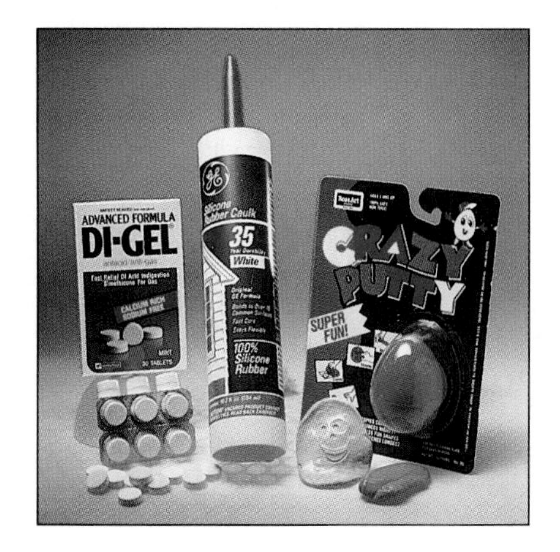

Some examples of products containing silicones, polymers with repeating —R_2Si—O— units. *(C. D. Winters)*

Silicones are nontoxic and have good stability to heat, light, and oxygen; they are chemically inert and have valuable antistick and antifoam properties. They can be made in the form of oils, greases, and resins, or with rubber-like properties ("Silly Putty," for example). Approximately 3×10^6 tons of silicone polymers are made worldwide annually. These materials are used in a wide variety of products: lubricants, peel-off labels, lipstick, suntan lotion, car polish, building caulk, and remedies for upset stomachs.

22.7 NITROGEN AND PHOSPHORUS

Nitrogen is found primarily as N_2 in the atmosphere, where it constitutes 78.1% by volume (or 75.5% by weight). Interestingly, neither nitrogen nor phosphorus is among the 10 most common elements in earth's crust, but both are essential to life on this planet. Nitrogen and its compounds, including ammonia, nitric acid, ammonium nitrate, and urea, play a key role in our economy, and

Mining phosphate rock near Ft. Meade, Florida. Phosphates are used for fertilizers and to prepare phosphoric acid. *(Runk/Schoenberger, from Grant Heilman)*

phosphoric acid is an important commodity chemical. The major use of all of these chemicals is in fertilizers.

Over 200 different phosphorus-containing minerals are known; all are *orthophosphates,* that is, they contain the tetrahedral PO_4^{3-} ion or a derivative of this ion. By far the largest source of this element is the *apatite* mineral family, members of which have the general formula $CaX_2 \cdot 3\ Ca_3(PO_4)_2$ (X = F, Cl, OH) (see page 1019).

Both phosphorus and nitrogen are part of every living organism. The element phosphorus is contained in biochemicals called nucleic acids and phospholipids, and nitrogen occurs in proteins and nucleic acids. Nitrogen and phosphorus constitute about 3% and 1.2% by weight, respectively, of the human body.

The Elements: Nitrogen and Phosphorus

Nitrogen (N_2) is a colorless gas that liquifies at 77 K (−196 °C) (see Figure 2.13). Its most notable feature is its reluctance to react with other elements or compounds. This comes about because the N≡N triple bond has a large dissociation energy (945.4 kJ/mol) and because the molecule is nonpolar. Nitrogen does react, however, with hydrogen to give ammonia (page 774) and with a few metals to give nitrides, compounds with the N^{3-} ion such as Mg_3N_2:

$$3\ Mg(s) + N_2(g) \longrightarrow \underset{\text{magnesium nitride}}{Mg_3N_2(s)}$$

Elemental nitrogen, N_2, is a very useful material. The largest quantity of this gas is used to provide a nonoxidizing atmosphere for packaged foods and wine, and to pressurize electric cables and telephone wires. Liquid nitrogen is used as a coolant in a variety of ways, including freezing soft materials such as rubber so they can be ground to a powder and preserving biological samples (e.g., blood and semen).

Phosphorus was first isolated from an animal source rather than a mineral one. In 1669, an alchemist obtained phosphorus from the distillation of purified urine.

Elemental phosphorus is produced in large quantities by the reduction of phosphate minerals:

$$2\ Ca_3(PO_4)_2(s) + 10\ C(s) + 6\ SiO_2(s) \longrightarrow P_4(g) + 6\ CaSiO_3(s) + 10\ CO(g)$$

The allotrope of phosphorus that is white or yellowish (see Figure 2.25) has the molecular formula P_4, and consists of a tetrahedron of phosphorus atoms.

Nitrogen Compounds

One of the most interesting features of nitrogen is the wide diversity of its compounds (Figure 22.12). Compounds with nitrogen in all oxidation states between −3 and +5 are known. Several of these compounds are especially significant.

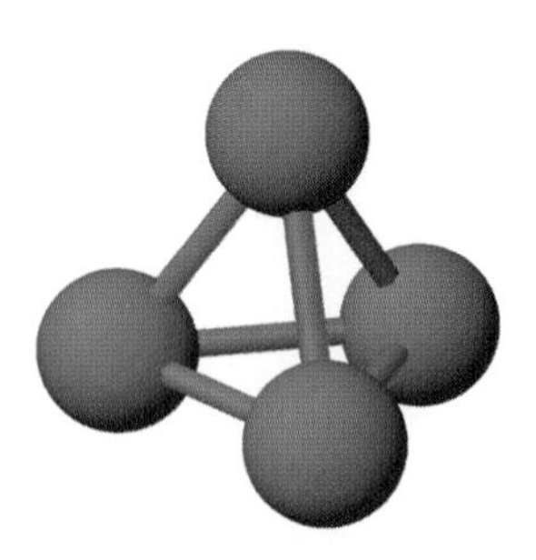

Yellow phosphorus is a tetrahedron of phosphorus atoms. See also Figure 2.25.

Ammonia and Nitrogen Fixation

Nitrogen gas, N_2, cannot be used by plants until it is "fixed," that is, converted into a form that can be used by living systems. Nitrogen fixation is done naturally by organisms such as blue-green algae. A few field crops, such as alfalfa and soybeans, fix nitrogen; this is actually done by nitrogen-fixing bacteria that have a symbiotic relationship with the plant. Because most plants cannot fix N_2, however, it is necessary that the nitrogen be provided by an external source. This is especially true of the new varieties of wheat, corn, and rice that grow fast or that have been bred to provide higher levels of protein. Thus, the production of nitrogen-containing fertilizers is a huge part of the chemical industry today.

Prior to World War I, the majority of commercial nitrogen fertilizer came from deposits of Chilean saltpeter ($NaNO_3$) and from guano (the excrement of sea birds and bats).

The Haber Process

A feasible process for fixing nitrogen in the form of NH_3 was devised by Fritz Haber (1868–1934). The Haber process, as it is called, produces ammonia by direct combination of nitrogen and hydrogen:

$$N_2(g) + 3\ H_2(g) \rightleftharpoons 2\ NH_3(g)$$

Nitrogen from the air is free, but H_2 must be made, the most common route being from natural gas by steam re-forming (page 1011). The Haber process is so efficient that the cost of ammonia is now almost entirely the cost of the hydrogen consumed in making the NH_3.

Ammonia is a gas at room temperature and pressure with a very penetrating odor. It condenses to a liquid at −33 °C under 1 atm pressure. Solutions in water, often referred to as ammonium hydroxide, are basic, due to the reaction of ammonia with water (Section 17.5 and Figure 5.11).

The Haber process was described from the viewpoint of chemical equilibria on page 774.

$$NH_3(aq) + H_2O(\ell) \rightleftharpoons NH_4^+(aq) + OH^-(aq) \qquad K_b = 1.8 \times 10^{-5} \text{ at } 25\ °C$$

Hydrazine

Another nitrogen-hydrogen compound is hydrazine, N_2H_4, a colorless fuming liquid with an ammonia-like odor (mp, 2.0 °C; bp, 113.5 °C). About 9.1×10^6 kg of hydrazine is produced annually by the *Raschig process*—the oxidation of ammonia with alkaline sodium hypochlorite in the presence of gelatin:

$$2\ NH_3(aq) + NaOCl(aq) \longrightarrow N_2H_4(aq) + NaCl(aq) + H_2O(\ell)$$

Not surprisingly, hydrazine is a base:

$$N_2H_4(aq) + H_2O(\ell) \rightleftharpoons N_2H_5^+(aq) + OH^-(aq) \qquad K_b = 8.5 \times 10^{-7}$$

It is also a strong reducing agent, as reflected in the $E°$ value in basic solution:

$$N_2(g) + 4\ H_2O(\ell) + 4\ e^- \longrightarrow N_2H_4(aq) + 4\ OH^-(aq) \qquad E° = -1.16\ V$$

This ability is exploited in the use of hydrazine to treat waste water from chemical plants. It removes ions such as CrO_4^{2-} by reducing them and thus preventing them from entering the environment. A related use is the treatment of water boilers in large electric-generating plants. Oxygen dissolved in the water is a serious problem in these plants because the dissolved gas can oxidize the metal of the boiler and pipes and lead to corrosion. Hydrazine reduces the dissolved oxygen to water:

$$N_2H_4(aq) + O_2(g) \longrightarrow N_2(g) + 2\ H_2O(\ell)$$

Oxides of Nitrogen

Nitrogen is unique among elements in the number of binary oxides it forms. The most common of these are listed in Table 22.4. It is interesting to note that all of these oxides are thermodynamically unstable with respect to decomposition to N_2 and O_2; all have positive ΔG_f° values. Most are slow to decompose and are said to be kinetically stable.

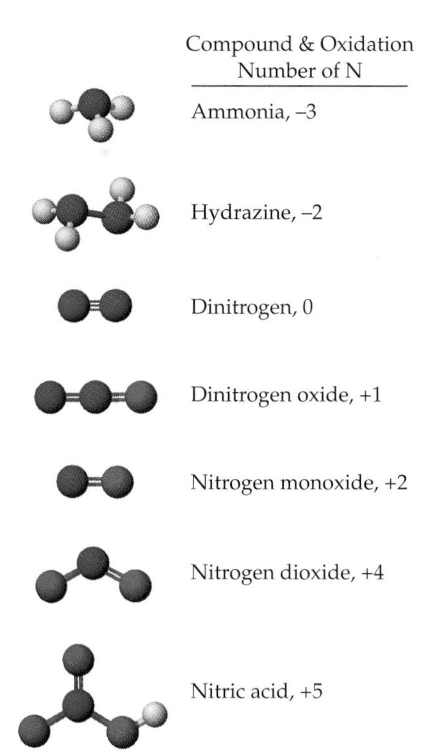

Figure 22.12 Some compounds of nitrogen. The N atom in nitrogen-containing compounds can have oxidation states ranging from −3 to +5.

Table 22.4 • Some Oxides of Nitrogen

Formula	Name	Structure	Nitrogen Oxidation Number	Description
N_2O	Dinitrogen oxide (nitrous oxide)	:N≡N—Ö: linear	+1	Colorless gas (laughing gas)
NO	Nitrogen monoxide (nitric oxide)	*	+2	Colorless gas, odd-electron molecule (paramagnetic)
N_2O_3	Dinitrogen trioxide	O=N—N(=O)—O planar	+3	Blue solid (mp, −100.7 °C), reversibly dissociates to NO and NO_2
NO_2	Nitrogen dioxide	O—N=O	+4	Brown, paramagnetic gas
N_2O_4	Dinitrogen tetraoxide	(O=)(O—)N—N(=O)(—O) planar	+4	Colorless liquid/gas, dissociates to NO_2 (see Figure 16.7)
N_2O_5	Dinitrogen pentaoxide	(O=)(O—)N—O—N(=O)(—O)	+5	Colorless solid

*It is not possible to draw a Lewis structure that accurately represents the electronic structure of NO. See Chapter 10.

Dinitrogen Oxide N_2O is a nontoxic, odorless, and tasteless gas having nitrogen with the lowest oxidation number (+1) in the series of nitrogen oxides. It can be made by the careful decomposition of ammonium nitrate at 250 °C:

$$NH_4NO_3(s) \longrightarrow N_2O(g) + 2\ H_2O(g)$$

Dinitrogen oxide gas is used as an anesthetic in minor surgery and has come to be called "laughing gas" because of its effects. Because it is soluble in vegetable fats, the largest commercial use of N_2O is as a propellent and aerating agent in cans of whipped cream.

See "NO—A Small Molecule with a Big Role," Chapter 9, page 400 and Screen 9.8 of the *Saunders Interactive General Chemistry CD-ROM.*

Nitrogen Monoxide NO is a simple odd-electron molecule. On a laboratory scale, it can be synthesized conveniently by the reduction of nitrate or nitrite ions using a mild reducing agent in acid solution:

$$KNO_2(aq) + KI(aq) + H_2SO_4(aq) \longrightarrow NO(g) + K_2SO_4(aq) + H_2O(\ell) + \tfrac{1}{2}\ I_2(aq)$$

Nitrogen monoxide, NO, has recently been the subject of intense research because it has been found to be important in a number of biochemical processes (page 400).

Nitrogen Dioxide The brown gas you see when a bottle of nitric acid is allowed to stand in the sunlight is NO_2, nitrogen dioxide:

$$2\ HNO_3(aq) \longrightarrow 2\ NO_2(g) + H_2O(\ell) + \tfrac{1}{2}\ O_2(g)$$

Nitrogen dioxide is also a culprit in air pollution. Nitrogen monoxide is present in urban polluted air, but it reacts rapidly with oxygen to form NO_2 (page 400):

$$2\ NO(g) + O_2(g) \longrightarrow 2\ NO_2(g)$$

You may have noticed the dioxide as a brown haze hovering over a city.

Dinitrogen Tetraoxide Like NO, the dioxide NO_2 is an odd-electron molecule, but unlike NO, NO_2 molecules can dimerize to give a species that obeys the octet rule. Two molecules of NO_2 combine to form N_2O_4, dinitrogen tetraoxide, a molecule with an N—N single bond:

$$\underset{\text{deep brown gas}}{2\ NO_2(g)} \rightleftharpoons \underset{\text{colorless (mp, −11.2 °C)}}{N_2O_4(g)}$$

When N_2O_4 is frozen (mp −11.2 °C), the solid consists entirely of N_2O_4 molecules; as the solid melts and the temperature increases to the boiling point, dissociation to NO_2 begins to occur. At the boiling point (21.5 °C) and 1 atm pressure, the distinctly brown gas phase consists of 15.9% NO_2 and 84.1% N_2O_4 (page 771).

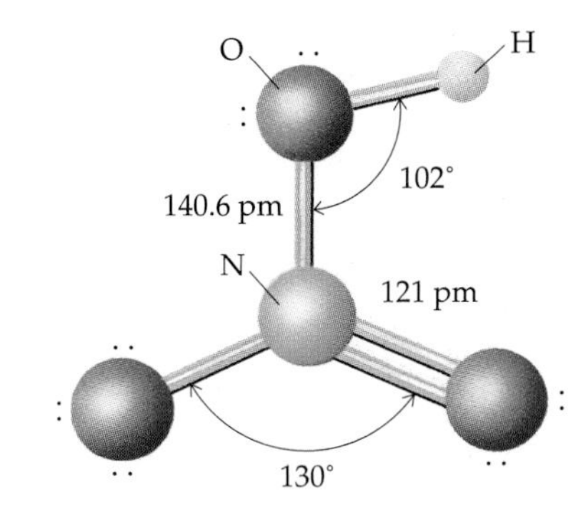

The structure of nitric acid, HNO_3.

Nitric Acid Nitrogen dioxide and N_2O_4 react with water to form nitric acid, HNO_3, so the moist gases are not only toxic but highly corrosive as well:

$$N_2O_4(g) + H_2O(\ell) \longrightarrow \underset{\text{nitric acid}}{HNO_3(aq)} + \underset{\text{nitrous acid}}{HNO_2(aq)}$$

Nitric acid has been known for centuries and is still one of the most important compounds in our economy. The oldest way to make the acid is to treat Chilean saltpeter, $NaNO_3$, with sulfuric acid (Figure 22.13):

$$2\ NaNO_3(s) + H_2SO_4(aq) \longrightarrow 2\ HNO_3(aq) + Na_2SO_4(s)$$

The anhydride of nitric acid is dinitrogen pentaoxide, N_2O_5. This compound is made by dehydrating nitric acid:

$$4\ HNO_3 + P_4O_{10} \longrightarrow 2\ N_2O_5 + 4\ HPO_3$$

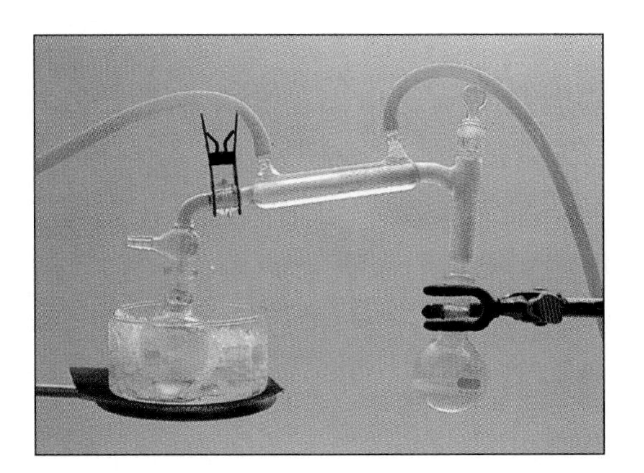

***Figure* 22.13 The preparation of nitric acid by the reaction of sulfuric acid and sodium nitrate.** Pure HNO_3 is colorless, but some acid decomposes to give brown NO_2, and it is this gas that fills the apparatus and colors the liquid in the distillation flask. *(C. D. Winters)*

Enormous quantities of nitric acid are now produced, however, from ammonia in the multistep *Ostwald process,* which was described in Section 15.5 as an example of industrial catalysis. Roughly 20% of the ammonia produced every year is converted to nitric acid. The acid has many uses, but by far the greatest amount is turned into ammonium nitrate by the reaction of nitric acid and ammonia.

Nitric acid is a powerful oxidizing agent, as the large, positive $E°$ values for the following half-reactions illustrate:

$$NO_3^-(aq) + 4\ H_3O^+(aq) + 3\ e^- \longrightarrow NO(g) + 6\ H_2O(\ell) \qquad E° = +0.96\ V$$
$$NO_3^-(aq) + 2\ H_3O^+(aq) + e^- \longrightarrow NO_2(g) + 3\ H_2O(\ell) \qquad E° = +0.80\ V$$

Concentrated nitric acid attacks and oxidizes almost all metals. In this process, the nitrate ion is usually reduced to one of the nitrogen oxides. Which oxide is formed depends on the metal and on reaction conditions. With copper, for example, either NO or NO_2 (page 209) is produced, depending on the concentration of the acid:

In dilute acid:

$$3\ Cu(s) + 8\ H_3O^+(aq) + 2\ NO_3^-(aq) \longrightarrow 3\ Cu^{2+}(aq) + 12\ H_2O(\ell) + 2\ NO(g)$$

In concentrated acid:

$$Cu(s) + 4\ H_3O^+(aq) + 2\ NO_3^-(aq) \longrightarrow Cu^{2+}(aq) + 6\ H_2O(\ell) + 2\ NO_2(g)$$

At least four metals are not attacked by nitric acid: Au, Pt, Rh, and Ir. These came to be known as the "noble metals." The alchemists of the 14th century, however, knew that if you mixed HNO_3 with HCl in a ratio of about 1:3, this *aqua regia,* or "kingly water," would attack even the noblest of metals. The reaction of platinum in aqua regia is

$$5\ Pt(s) + 4\ NO_3^-(aq) + 30\ Cl^-(aq) + 24\ H_3O^+(aq) \longrightarrow 5\ [PtCl_6]^{2-}(aq) + 2\ N_2(g) + 36\ H_2O(\ell)$$

Some common sulfur-containing minerals: golden iron pyrite (FeS_2); black cubes of galena (PbS); black needles of stibnite (Sb_2S_3); and yellow orpiment (As_2S_3). *(C. D. Winters)*

22.8 OXYGEN AND SULFUR

Oxygen is by far the most abundant element in the earth's crust, representing just under 50% by weight. It appears as elemental oxygen in the atmosphere or combined, in water and in many minerals. Scientists believe that elemental oxygen did not appear on this planet until about 2 billion years ago. The theory is that oxygen was formed on the planet by plants in the process of photosynthesis.

Sulfur is 15th in abundance in earth's crust. It too is found in its elemental form in nature, but only in certain concentrated deposits. Generally, it exists in the form of sulfur-containing compounds in natural gas and oil and as metal sulfide minerals, including cinnabar, HgS, galena, PbS, and iron pyrite, FeS_2, or "fool's gold," a salt of the disulfide ion S_2^{2-}. Sulfur also occurs as sulfate ion (for example, in gypsum, $CaSO_4 \cdot 2\ H_2O$, Figure 3.15). Sulfur oxides (SO_2 and SO_3) are also found in nature, primarily as products of volcanic activity (Figure 22.14).

Recall the boxed discussion of "Black Smokers and the Origin of Life," page 190. Hydrogen sulfide escapes from vents in the ocean floor and forms clouds of insoluble metal sulfides.

In the United States, most sulfur, about 10 million tons per year, is obtained from deposits along the Gulf of Mexico. These occur in the caprock over subterranean salt domes, typically at a depth of 150 to 750 m below the surface in layers perhaps 30 m thick. The theory explaining their existence is that the sulfur was formed by anaerobic ("without air") bacteria acting on sedimentary sulfate deposits such as gypsum ($CaSO_4 \cdot 2\ H_2O$).

Figure 22.14 Sulfur. Most sulfur comes from deep within the earth. In this photo the sulfur has been forced to the surface by volcanic activity deposited on the rim of a steam vent. (About 10% of the sulfur used in the United States is obtained from underground deposits.) *(David Cavagnaro)*

Preparation and Properties of the Elements

Pure oxygen is obtained by fractionation of air and is third among industrial chemicals produced in the United States. Very pure oxygen can be made in the laboratory by electrolysis of water and by the catalyzed decomposition of metal chlorates such as $KClO_3$:

$$2\ KClO_3(s) \longrightarrow 2\ KCl(s) + 3\ O_2(g)$$

At room temperature and pressure oxygen is a colorless gas, but it is pale blue when condensed to the liquid at −183 °C (Figure 10.20). As described in Section 10.3, diatomic oxygen is paramagnetic because it has two unpaired electrons.

Ozone, O_3, is a second, less stable, allotrope of oxygen. It is a blue, diamagnetic gas with an odor so strong that it can be detected in concentrations as low as 0.05 ppm. Ozone is conveniently synthesized by passing O_2 through an electric discharge, or by irradiation of O_2 with ultraviolet light. The gas is in the news constantly because of the realization that the earth's protective layer of ozone is being disrupted by chemicals such as the chlorofluorocarbons (see page 728).

Sulfur has many allotropes. The most common and most stable allotrope is the yellow, orthorhombic form, which consists of S_8 molecules with the sulfur atoms arranged in a crown-shaped ring (see Figures 2.26 and 22.15). Less stable allotropes have rings of 6 to 20 sulfur atoms. Another form of sulfur has a molecular structure with chains of sulfur atoms.

(a)

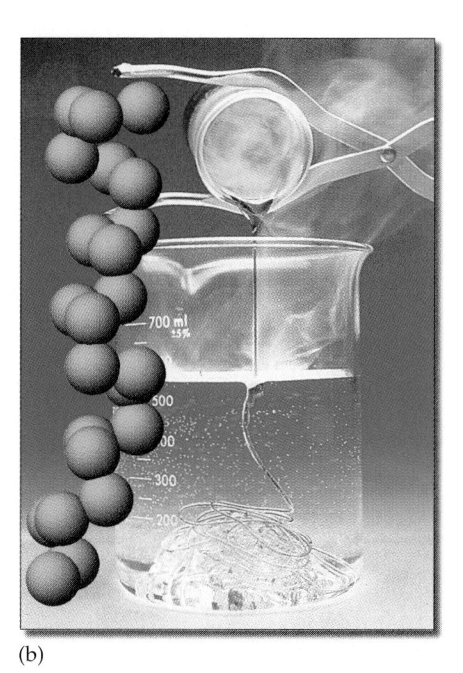

(b)

Figure 22.15 Sulfur allotropes. (a) At room temperature sulfur exists as a bright yellow solid composed of S_8 rings. (b) When heated, the rings break open, and eventually form chains of S atoms in a material described as "plastic sulfur." *(Photos, C. D. Winters; models, S. M. Young)*

Some common household products containing sulfur or sulfur-based compounds. *(C. D. Winters)*

Sulfur is obtained from underground deposits by a process developed by Herman Frasch (1851–1914) about 1900. Superheated water (at 165 °C) and then air are forced into the deposit; the sulfur melts (mp, 113 °C) and is forced to the surface as a frothy, yellow stream where it solidifies (see Figure 22.14).

The largest use of sulfur by far is the production of *sulfuric acid,* H_2SO_4, the compound produced in largest quantity by the chemical industry (page 196). In the United States, roughly 70% of the sulfuric acid is used to manufacture "super-phosphate" fertilizer. Smaller amounts are used in the conversion of ilmenite, a titanium-bearing ore, to TiO_2, which is then used as a white pigment in paint, plastics, and paper. The acid is also used to make iron and steel, petroleum products, synthetic polymers, and paper.

Sulfur Compounds

Hydrogen sulfide, H_2S, has a structure resembling that of water. Unlike water, however, hydrogen sulfide is a gas under standard conditions (mp, −85.6 °C; bp, −60.3 °C) because only very weak hydrogen bonding occurs between molecules compared with the strong hydrogen bonding in water (Figure 13.8). Hydrogen sulfide is a deadly poison, comparable in toxicity to hydrogen cyanide, but it fortunately has a terrible odor and is detected in concentrations as low as 0.02 ppm. You must be careful with H_2S, however, because it has an anesthetic effect, and your nose rapidly loses its ability to detect H_2S. Death occurs at H_2S concentrations of 100 ppm.

Sulfur is often found as the sulfide ion in conjunction with metals. The recovery of metals from their sulfide ores usually begins by heating, or *roasting,* the ore in air. The outcome of this process is the conversion of the metal sulfide to either a metal oxide or the metal itself, the sulfur appearing as SO_2:

$$2\ ZnS(s) + 3\ O_2(g) \longrightarrow 2\ ZnO(s) + 2\ SO_2(g)$$
$$2\ PbO(s) + PbS(s) \longrightarrow 3\ Pb(s) + SO_2(g)$$

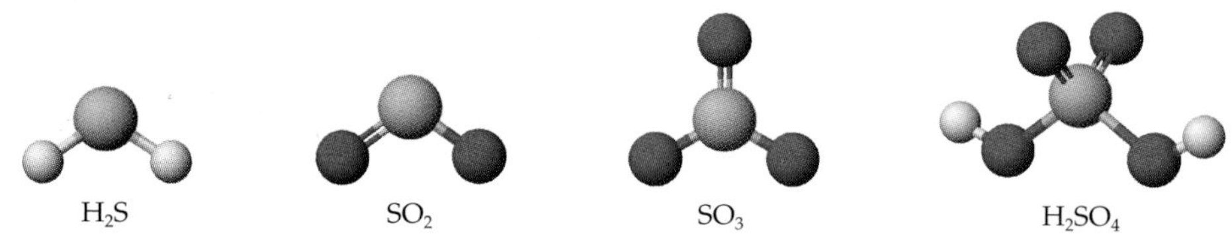

Models of some common sulfur-containing molecules: H_2S, SO_2, SO_3, and H_2SO_4.

Sulfur dioxide (SO_2) and trioxide (SO_3) are the most important oxides of the element. The former is produced on an enormous scale by the combustion of sulfur and by roasting sulfide ores in air. The combustion of sulfur in sulfur-containing coal and fuel oil creates particularly large environmental problems. It has been estimated that about 2.0×10^8 tons of sulfur oxides are released into the atmosphere each year by human activities, primarily in the form of SO_2; this is more than half of the total emitted by all other natural sources of sulfur in the environment.

Sulfur dioxide is a colorless, toxic gas with a choking odor. It readily dissolves in water. The most important reaction of SO_2 is oxidation to the trioxide:

$$SO_2(g) + \tfrac{1}{2}\ O_2(g) \longrightarrow SO_3(g) \qquad \Delta H^\circ = -98.9\ \text{kJ/mol}$$

Sulfur trioxide is extremely reactive and is very difficult to handle; it is almost always converted directly to sulfuric acid by reaction with water (page 196).

Current Perspectives

Snot-Tites

(By Arthur N. Palmer, Earth Sciences Department, SUNY-Oneonta)

Recent discoveries on the deep ocean floor and in limestone caves have revealed biological communities whose primary source of energy is the oxidation of sulfur compounds, thriving in environments that seem to us very inhospitable (page 190). Caves are especially fertile sites for study because they provide us with a first-hand look at these bizarre ecosystems.

The solution process that forms limestone caves is usually performed by carbonic acid derived from carbon dioxide (page 901). About 10% of all known caves are produced by sulfuric acid, however, which results from the oxidation of hydrogen sulfide that rises from sources deep beneath the surface. Of those sulfuric-acid caves with accessible entrances, fewer than a dozen are known to be actively forming.

In the jungles of southern Mexico is a spring that is milky white with suspended particles of sulfur. In the cliff above, hydrogen sulfide in stifling amounts spews from a dark hole—the Cueva de Villa Luz. This cave can be followed downward to a large underground stream and a maze of actively enlarging cave passages. Water rises into the cave from underlying sulfur-bearing strata, releasing hydrogen sulfide at concentrations up to 150 ppm. Yellow sulfur crystallizes on the cave walls around the inlets. Sulfur and sulfuric acid are produced by the following reactions:

$$2\ H_2S(g) + O_2(g) \longrightarrow 2\ S\ (\text{solid}) + 2\ H_2O(\ell)$$
$$2\ S(s) + 2\ H_2O(\ell) + 3\ O_2(g) \longrightarrow 2\ H_2SO_4(aq)$$

The sulfuric acid reacts with the limestone bedrock, which consists mostly of calcium carbonate and replaces the limestone with gypsum ($CaSO_4 \cdot 2\ H_2O$).

$$H^+(aq) + HSO_4^-(aq) + 2\ CaCO_3(s) + 2\ H_2O(\ell) \longrightarrow CaSO_4 \cdot 2\ H_2O(s) + Ca^{2+}(aq) + 2\ HCO_3^-(aq)$$

Some of the gypsum is removed by the cave stream, which is undersaturated with that mineral, and this is the main process that enlarges the cave.

The cave atmosphere is poisonous to humans, so that gas masks are essential. But surprisingly, the cave is teeming with life. Sulfur-oxidizing bacteria—called chemoautotrophs—speed the chemical reactions and thrive on the large amounts of energy released by them, even in the absence of all other food sources. They use the chemical energy to obtain carbon for their bodies from calcium carbonate and carbon dioxide, both of which are abundant in the cave. Bacterial filaments hang from the walls and ceilings in bundles. (Because the filaments look like something coming from a runny nose, the cave explorers refer to them as "snot-tites.") Other microbes feed on the bacteria, and so on up the food chain—which includes spiders, gnats, and pygmy snails—all the way to sardine-like fish that swim in the cave stream. At the top of the chain, the local Zoque Indians hold occasional festivals at the cave entrance that culminate in a fish fry. This entire ecosystem, except the humans, is supported by the sulfur reactions within the cave.

But this is no ordinary cave environment. Droplets of water seeping in from the surface absorb both hydrogen sulfide and oxygen from the cave air. As illustrated by the reactions shown previously, these two dissolved gases react to produce sulfuric acid. This depletes the concentration of both gases in the droplets, allowing more to be absorbed from the air. Meanwhile the droplets grow more acidic. The longer the droplets remain on the ceiling, the lower their pH becomes, until some reach the strength of battery acid. The droplets clinging to the bacterial filaments in the photo had average pH values of 1.4, with some as low as zero! Drops that landed on explorers burned their skin and disintegrated their clothing. Fortunately the cave stream has a nearly neutral pH and provides a refreshing bath.

Filaments of sulfur-oxidizing bacteria (dubbed "snot-tites") hang from the ceiling of a Mexican cave containing an atmosphere rich in hydrogen sulfide. The bacteria thrive on the energy released by oxidation of the hydrogen sulfide, forming the base of a complex food chain. Droplets of sulfuric acid on the filaments have the pH of battery acid. *(A. N. Palmer)*

See C. Petit: "The walls are alive." *U. S. News and World Report,* page 59, February 9, 1998.

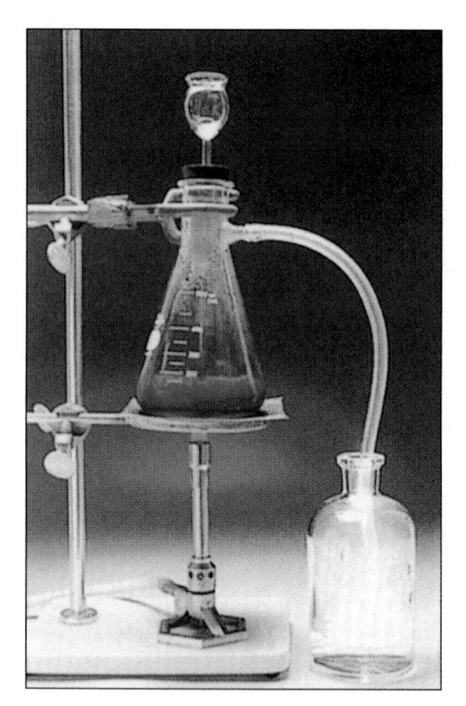

Figure 22.16 Chlorine preparation. Chlorine is prepared by oxidation of chloride ion using a strong oxidizing agent. Here, oxidation of NaCl is accomplished using $K_2Cr_2O_7$ in H_2SO_4. *(C. D. Winters)*

22.9 CHLORINE

Yellow-green chlorine gas was first made by the Swedish chemist Karl Wilhelm Scheele (1742–1786) in 1774 by a reaction still used today as a laboratory source of Cl_2 (Figure 22.16):

$$2\ NaCl(s) + 2\ H_2SO_4(aq) + MnO_2(s) \longrightarrow Na_2SO_4(aq) + MnSO_4(aq) + 2\ H_2O(\ell) + Cl_2(g)$$

Chlorine exists in the earth's crust as chloride ion in sea water and in brine wells. It is the 11th ranked element in abundance and is the most abundant halogen. Today, chlorine is made in enormous quantities by electrolysis of brine (Chapter 21) and ranks eighth among industrial chemicals. Almost 70% of the chlorine manufactured is used for the production of organic chemicals including vinyl chloride, the precursor to polyvinyl chloride (PVC), a plastic used in many consumer items (see Section 11.6).

One of the first properties of Cl_2 that Scheele recognized was its ability to bleach textiles and paper. Soon thereafter, chlorine's ability to act as a disinfectant in water was also recognized. Today, these two uses consume about 20% of the Cl_2 produced.

Chlorine Compounds

Hydrogen Chloride

Hydrogen chloride, in the form of hydrochloric acid, ranks 26th among industrial chemicals. The gas can be prepared by the reaction of hydrogen and chlorine, but the rapid, exothermic reaction is difficult to control. The classical method of making laboratory quantities of HCl uses the reaction of NaCl and sulfuric acid. This procedure takes advantage of the fact that HCl is a gas:

$$2\ NaCl(s) + H_2SO_4(\ell) \longrightarrow Na_2SO_4(s) + 2\ HCl(g)$$

Gaseous hydrogen chloride has a sharp, irritating odor. It dissolves in water to give hydrochloric acid, a strong acid. Gaseous HCl reacts with metals, metal oxides, and metal hydrides to give metal chlorides and, depending on the reactant, water or hydrogen:

$$Mg(s) + 2\ HCl(g) \longrightarrow MgCl_2(s) + H_2(g)$$
$$ZnO(s) + 2\ HCl(g) \longrightarrow ZnCl_2(s) + H_2O(g)$$
$$NaH(s) + HCl(g) \longrightarrow NaCl(s) + H_2(g)$$

Oxoacids of Chlorine

Oxoacids of chlorine range from HOCl, in which chlorine has an oxidation number of +1, to $HClO_4$, in which the chlorine oxidation number is equal to the group number, +7. All are strong oxidizing agents.

Acid	Name	Anion	Name
HOCl	Hypochlorous	OCl^-	Hypochlorite
HOClO	Chlorous	ClO_2^-	Chlorite
$HOClO_2$	Chloric	ClO_3^-	Chlorate
$HOClO_3$	Perchloric	ClO_4^-	Perchlorate

Hypochlorous acid, HOCl, which forms when chlorine dissolves in water, was discovered over two centuries ago in the original work on chlorine. In this reaction, part of the chlorine is oxidized to hypochlorite ion and part is reduced to chloride ion.

A reaction in which an element or compound is simultaneously oxidized and reduced is called a *disproportionation reaction*. Here Cl_2 is oxidized to OCl^- and reduced to Cl^-.

$$Cl_2(g) + 2\ H_2O(\ell) \rightleftharpoons H_3O^+(aq) + HOCl(aq) + Cl^-(aq)$$

Chlorine, chloride ion, and hypochlorous acid exist in equilibrium; a low pH favors Cl_2, whereas a high pH favors the products.

If, instead of dissolving Cl_2 in pure water, the element is added to cold aqueous NaOH, hypochlorite ion and chloride ion form:

$$Cl_2(g) + 2\ OH^-(aq) \rightleftharpoons OCl^-(aq) + Cl^-(aq) + H_2O(\ell)$$

The resulting alkaline solution is the "liquid bleach" used in home laundries. Note that the reaction equation bears a close resemblance to the equation for the reaction of Cl_2 with water. With basic conditions the equilibrium has shifted far to the right. The bleaching action of this solution is a result of the oxidizing ability of OCl^-. Most dyes are colored organic compounds, and hypochlorite ion oxidizes dyes to colorless products.

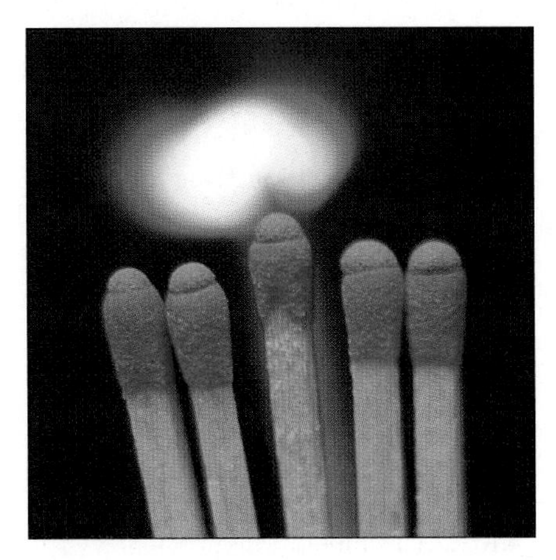

The head of a "strike anywhere" match contains, among other things, P_4S_3, and an oxidizing agent, potassium chlorate. Safety matches have sulfur (3–5%) and $KClO_3$ (45–55%) in the head and red phosphorus (page 225) in the striking strip. *(C. D. Winters)*

When calcium hydroxide is used for this reaction in place of sodium hydroxide, solid $Ca(OCl)_2$ is the product. This compound is easily handled and is the "chlorine" that is sold for swimming pool sterilization.

When a basic solution of hypochlorite ion is heated, disproportionation occurs, forming chlorate ion and chloride ion:

$$3\ OCl^-(aq) \longrightarrow ClO_3^-(aq) + 2\ Cl^-(aq)$$

Sodium and potassium chlorates are made by this reaction in large quantities. The sodium salt can be reduced to ClO_2, which is used for bleaching paper pulp. Some $NaClO_3$ is also converted to potassium chlorate, $KClO_3$, the preferred oxidizer in fireworks and a component of safety matches.

Perchlorates

Perchlorates, salts containing ClO_4^-, are the most stable oxochlorine compounds, although they are powerful oxidants. Pure perchloric acid, $HClO_4$, is a colorless liquid that explodes if shocked. It explosively oxidizes organic materials and rapidly oxidizes silver and gold. Dilute aqueous solutions of the acid are safe to handle, however.

Perchlorate salts of most metals exist. Although many are relatively stable, they are unpredictable. *Great care should be used when handling any perchlorate salt.* Ammonium perchlorate, for example, bursts into flame if heated above 200 °C.

$$2\ NH_4ClO_4(s) \longrightarrow N_2(g) + Cl_2(g) + 2\ O_2(g) + 4\ H_2O(g)$$

This property of the ammonium salt accounts for its use as the oxidizer in the solid booster rockets for the Space Shuttle. The solid propellent in these rockets is largely NH_4ClO_4, the remainder being the reducing agent, powdered aluminum. Each Shuttle launch requires about 750 tons of ammonium perchlorate, and more than half of the sodium perchlorate currently manufactured is converted to the ammonium salt. The process for doing this is an exchange reaction that takes advantage of the fact that ammonium perchlorate is less soluble in water than sodium perchlorate:

The solid-fuel booster rockets of the Space Shuttle are fueled with a mixture of NH_4ClO_4 and aluminum powder. *(NASA)*

$$NaClO_4(aq) + NH_4Cl(aq) \longrightarrow NaCl(aq) + NH_4ClO_4(s)$$

CURRENT PERSPECTIVES

Banning Chlorine

Elemental chlorine is produced in enormous quantities in the industrialized world. As depicted in the "chlorine family tree," it is the starting point for the production of at least 10,000 compounds that contain chlorine and that are important in our economy. The major uses of chlorine and chlorine-containing compounds are in water purification, as solvents for the preparation of pharmaceuticals, in making plastics, and in pulp and paper manufacturing.

Chlorine-containing compounds are also used as bactericides and pesticides. Dichlorodiphenyltrichloroethane, commonly called DDT, was widely used to eradicate mosquitoes that carried malaria. It was so effective that the World Health Organization once regarded shortages of the chemical as a threat to public health. Our experience with DDT, however, is one reason that environmental groups want to ban chlorine and chlorine-containing organic compounds.

Most chlorine-containing compounds are very stable. Although this property makes them valuable in some applications, it may also mean that they persist in the environment. Chlorine-containing compounds accumulate in animals, reaching higher and higher concentrations the farther up the food chain you go. Biologists found that DDT accumulated in bald eagles and that this was a cause of the decline in the bald eagle population in the United States in the 1960s. The pesticide apparently interferes with the calcium carbonate structure of the eggs, creating thin shells. This caused the ea-

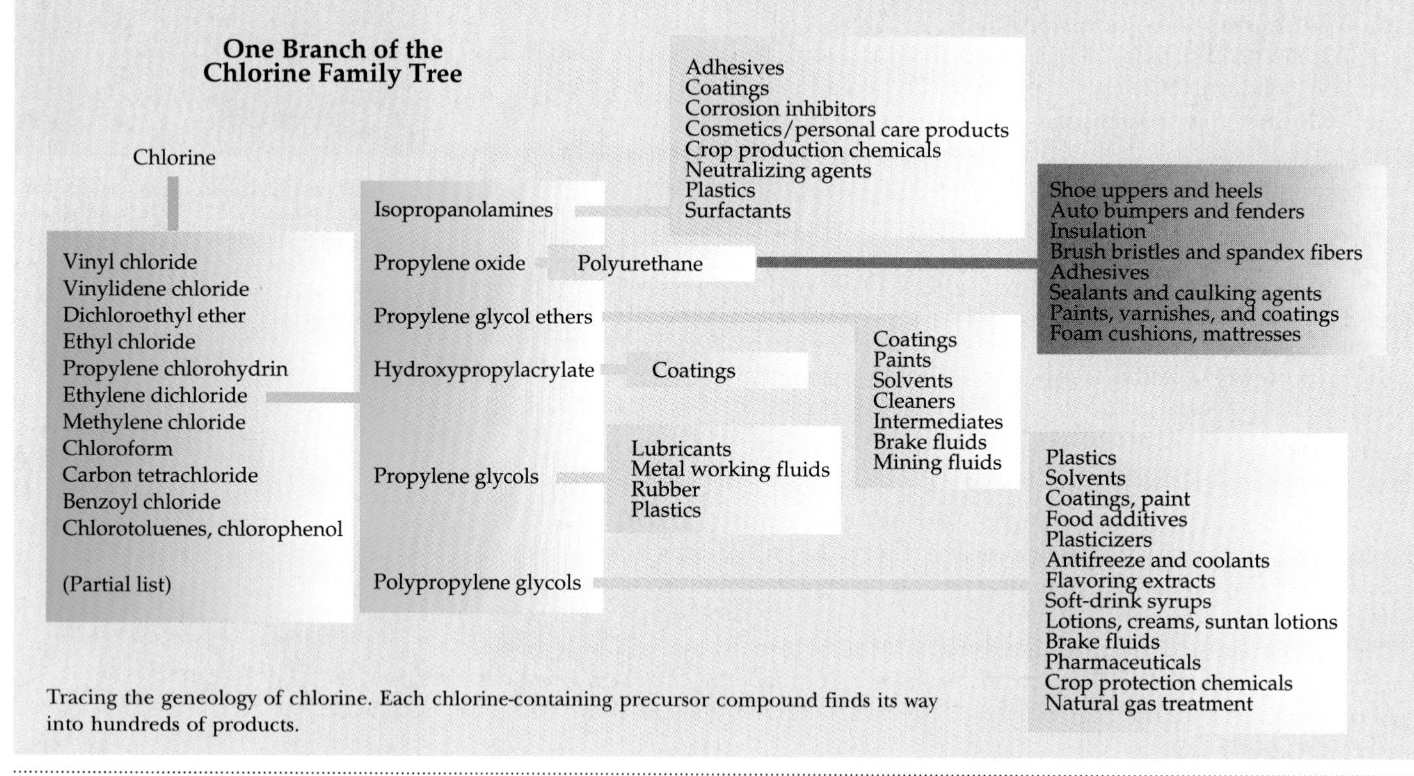

Tracing the geneology of chlorine. Each chlorine-containing precursor compound finds its way into hundreds of products.

CHAPTER HIGHLIGHTS

When you have finished studying this chapter you should be able to

- Predict general types of chemical behavior of the Group A elements (Section 22.1).
- Predict similarities and differences among the elements in a given group, based on periodic properties (Section 22.1).
- Know which reactions produce ionic compounds, and predict formulas for

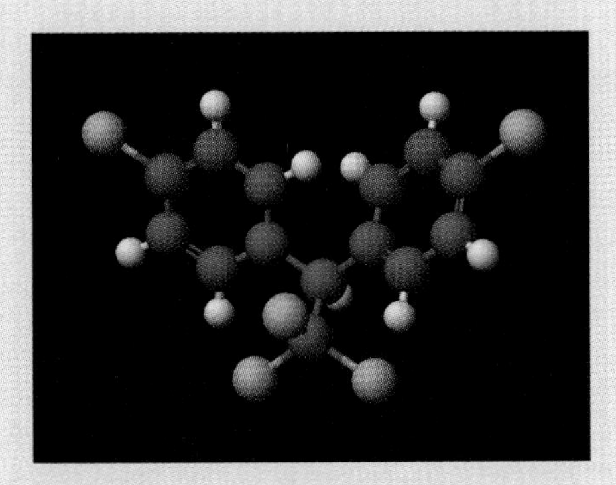

Model of dichlorodiphenyltrichloroethane, commonly called DDT.

gles' reproductive rate to drop severely; only 417 nesting pairs were counted in the lower 48 states in 1972. After DDT was banned for use in the United States in 1972, the eagle population began to recover, and 3747 nesting pairs were estimated in 1992.

Other environmental problems arise from the formation of toxic chlorine-containing compounds in the environment. For example, the largest use of chlorine gas is in bleaching paper. If the used chlorine is dumped into the effluent of a paper mill, it can combine with naturally occurring compounds to turn them into toxins. For this reason the paper industry has been working for some years to substitute processes that rely on oxygen, ozone, or peroxides to do the job that chlorine once did.

Some environmental organizations advocate a total ban on the use of chlorine in any form, arguing that alternatives exist for some chlorine-containing products. Mario Molina, who, with Sherwood Rowland, first recognized the effect of CFCs on the earth's ozone layer (page 728), believes, however, that a total ban on chlorine would go too far and would deprive society of many useful materials while removing deleterious ones. As evidence of this, it has been noted that of the nearly 400 new drugs approved for therapeutic use in humans since 1984, more than 60 are organochlorine compounds. Two of the ten most prescribed pharmaceuticals, Ceclor (cefaclor) and Xanax (alprazolam), contain chlorine as do Vancocin (vancomycin), the only antibiotic effective against some *Staphylococcus* infections, and Lorstan, a new drug used to treat hypertension.*

Mario Molina has said that "The underlying need is to make reasonable assessments of risk, and then tackle the problems that matter." Chlorine and its compounds are an example of the discussion of risks and benefits of chemistry in the Introduction to this book. A spokesperson for the chlorine industry said that "The stakes are high in this discussion. We must balance the risks of chlorine chemistry against not having access to chlorine and look carefully at the alternatives because they may have adverse impacts of their own."

*G. W. Gribble: a letter to the editor, *Chemical and Engineering News,* April 18, p. 18, 1994.

See also I. Amato, "The crusade against chlorine." *Science, 193,* Vol. 261, pp. 152–154; and R. Carson: *Silent Spring.* New York, Houghton-Mifflin, 1962.

Some chlorine-containing products. The white material is PVC (polyvinyl chloride), which is used to make water and waste pipes in homes. *(C. D. Winters)*

common ions and common ionic compounds based on electron configurations (Section 22.1).

- Recognize when a formula is incorrectly written, based on general principles governing electron configurations (Section 22.1).
- Be able to summarize briefly a series of facts about the most common compounds of main group elements (ionic or covalent structure, color, solubility, simple reaction chemistry) (Sections 22.2–22.9).
- Identify uses of common elements and compounds and understand the chemistry that relates to their usage (Sections 22.2–22.9).

STUDY QUESTIONS

REVIEW QUESTIONS

1. The periodic table is one of the most useful sources of information available to a chemist. List at least three types of information that can be obtained from the periodic table.
2. Give formulas for the following common acids: nitric acid, sulfuric acid, hydrobromic acid, perchloric acid, carbonic acid. What is the oxidation number of the central atom in each of these compounds? How are these related to the periodic group number of the element?
3. Define the term "amphoteric." Write chemical equations to illustrate the amphoteric character of $Al(OH)_3$.
4. Give examples of two basic oxides. Write equations illustrating the formation of each oxide from its component elements. Write another chemical equation that illustrates the basic character of each oxide.
5. Give examples of two acidic oxides. Write equations illustrating the formation of each oxide from its component elements. Write another chemical equation that illustrates the acidic character of each oxide.
6. Write complete electron configurations for each of the elements in the second period. How many valence electrons are present on each atom?
7. Write complete electron configurations for the main group elements in the third period. How many valence electrons are present on each atom?
8. What is the general electron configuration of the elements in Group 4A in the periodic table? Relate this to the formulas of some compounds formed by these elements.
9. Give the name and symbol of each element having the valence configuration [noble gas]ns^2np^1.
10. Give symbols and names for four monatomic ions that have the same electron configuration as argon.
11. Describe the structure of NaCl. What are the particles making up this structure? What forces hold these particles together?

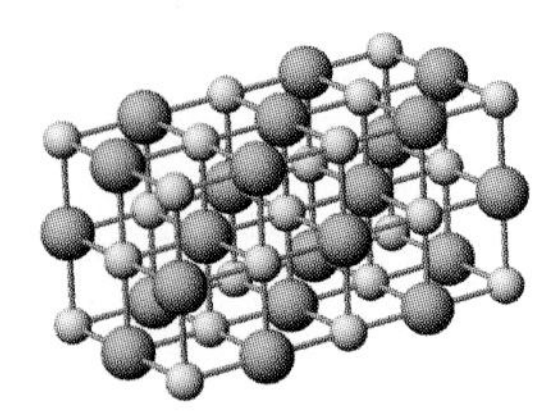

Model of the NaCl lattice.

12. Select one of the alkali metals and write a balanced chemical equation for its reaction with chlorine. Is the reaction likely to be exothermic or endothermic? Is the product ionic or covalent?
13. Select one of the alkaline earth metals and write a balanced chemical equation for its reaction with oxygen. Is the reaction likely to be exothermic or endothermic? Is the product ionic or covalent?
14. For the product of the reaction you wrote for Question 12, predict the following physical properties: color, state of matter (s, ℓ, or g), solubility in water.
15. For the product of the reaction you wrote for Question 13, predict the following physical properties: color, state of matter (s, ℓ, or g), solubility in water.
16. List, in order, the ten most abundant elements in earth's crust. Which of these elements are main group elements? Identify one or more chemical species that occur in earth's crust containing these main group elements.
17. Would you expect to find calcium occurring naturally in the earth's crust as a free element? Why or why not?
18. Which of the first ten elements in the periodic table are found as the free element in the earth's crust? Which elements in this group occur in the earth's crust only as part of a chemical compound?
19. Place the following oxides in order of increasing basicity: CO_2, SiO_2, and SnO_2.
20. Place the following oxides in order of increasing basicity: Na_2O, Al_2O_3, SiO_2, and SO_3.
21. Complete and balance equations for the following reactions. [Assume an excess of oxygen for (d).]
(a) $Na(s) + Br_2(\ell)$
(b) $Mg(s) + O_2(g)$
(c) $Al(s) + F_2(g)$
(d) $C(s) + O_2(g)$
22. Complete and balance equations for the following reactions:
(a) $K(s) + I_2(g)$
(b) $Ba(s) + O_2(g)$
(c) $Al(s) + S_8(s)$
(d) $Si(s) + Cl_2(g)$

Hydrogen

23. Write balanced chemical equations for the reaction of hydrogen gas with oxygen, chlorine, nitrogen.
24. Write an equation for the reaction of potassium and hydrogen. Name the product. Is it ionic or covalent? Predict one physical property and one chemical property of this compound.
25. A method recently suggested for the preparation of hydrogen (and oxygen) from water proceeds as follows:
(a) Sulfuric acid and hydrogen iodide are formed from sulfur dioxide, water, and iodine.
(b) The sulfuric acid from the first step is decomposed by heat to water, sulfur dioxide, and oxygen.
(c) The hydrogen iodide from the first step is decomposed with heat to hydrogen and iodine.
Write a balanced equation for each of these steps and show that their sum is the decomposition of water to form hydrogen and oxygen.
26. Write a balanced chemical equation for the preparation of H_2 by the reaction of CH_4 and water. Using data in Appendix L, calculate ΔH°, ΔG°, and ΔS° for this reaction.
27. Using data in Appendix L, calculate ΔH°, ΔG°, and ΔS° for the reaction of carbon and water to give CO and H_2 (water gas).

28. You are given a stoppered flask containing a gas and told that the gas is either hydrogen, nitrogen, or oxygen. Suggest an experiment to identify the gas.

29. Hydrogen is one of the more important chemicals produced by the chemical industry. How is it made? What is the major use of hydrogen gas? Name several other uses of hydrogen.

Alkali Metals

30. Write equations for the reaction of sodium with each of the halogens. Predict several physical properties that are common to all of the alkali metal halides.

31. Sodium peroxide is the primary product when sodium metal is burned in oxygen. Write an equation for this reaction. What ions make up this compound?

32. Write balanced equations for the reaction of lithium, sodium, and potassium with O_2. Specify which metal largely forms an oxide, which one forms a peroxide, and which one forms a superoxide.

33. A piece of sodium catches on fire in the laboratory! Why shouldn't you try to extinguish the fire using water? Suggest several better alternatives.

34. The electrolysis of aqueous NaCl gives NaOH, Cl_2, and H_2.
(a) Write a balanced equation for the process.
(b) In 1995 in the United States, 1.19×10^{10} kg of NaOH and 1.14×10^{10} kg of Cl_2 were produced. Does the ratio of masses of NaOH and Cl_2 produced agree with the ratio of masses expected from the balanced equation? If not, what does this tell you about the way in which NaOH or Cl_2 are actually produced? Is the electrolysis of aqueous NaCl the only source of these chemicals?

35. Write equations for the half-reactions that occur at the cathode and the anode when an aqueous solution of KCl is electrolyzed. What chemical species is oxidized, and what chemical species is reduced in this reaction? Predict the products formed if an aqueous solution of CsI is electrolyzed.

Alkaline Earths

36. When magnesium burns in air, it forms both an oxide and a nitride. Write balanced equations for the formation of both compounds.

37. Calcium reacts with hydrogen gas at elevated temperatures to form a hydride. This compound reacts readily with water, so it is an excellent drying agent for organic solvents.
(a) Write a balanced equation showing the formation of calcium hydride from Ca and H_2.
(b) Write a balanced equation for the reaction of calcium hydride with water (see Figure 17.4).

38. Name three uses of limestone. Write a balanced equation for the reaction of limestone with CO_2 in water.

39. Explain what is meant by "hard water." What causes hard water, and what problems are associated with hard water?

40. Calcium oxide, CaO, is used to remove SO_2 from power plant exhaust. These two compounds react to give solid $CaSO_3$. How many grams of SO_2 can be removed using 1.2×10^3 kg of CaO?

41. $Ca(OH)_2$ has a K_{sp} of 7.9×10^{-6}, whereas that for $Mg(OH)_2$ is 1.5×10^{-11}. Calculate the equilibrium constant for the reaction,

$$Ca(OH)_2(s) + Mg^{2+}(aq) \rightleftharpoons Ca^{2+}(aq) + Mg(OH)_2(s)$$

and explain why this reaction can be used in the commercial isolation of magnesium from sea water.

Aluminum

42. Write an equation for the reactions of aluminum with HCl(aq), Cl_2, and O_2.

43. Write an equation for the reaction of Al and $H_2O(\ell)$ to produce H_2 and Al_2O_3. Using thermodynamic data in Appendix L, calculate $\Delta H°$, $\Delta G°$, and $\Delta S°$ for this reaction. Do these data indicate that the reaction should favor the products? Why is aluminum metal unaffected by water?

44. Aluminum dissolves readily in hot aqueous NaOH to give the aluminate ion, $Al(OH)_4^-$, and H_2. Write a balanced equation for this reaction. If you begin with 13.2 g of Al, what volume of H_2 gas (in milliliters) is produced when the gas is measured at 735 mm Hg and 22.5 °C?

45. Alumina, Al_2O_3, is amphoteric. Among examples of its amphoteric character are the reactions that occur when Al_3O_3 is heated strongly or "fused" with acidic oxides and basic oxides.
(a) Write a balanced equation for the reaction of alumina with silica, an acidic oxide, to give aluminum metasilicate, $Al_2(SiO_3)_3$.
(b) Write a balanced equation for the reaction of alumina with the basic oxide CaO to give calcium aluminate, $Ca(AlO_2)_2$.

46. Aluminum sulfate (1995 worldwide production of about 3×10^9 kg) is the most commercially important aluminum compound after aluminum oxide and aluminum hydroxide. Write a balanced equation for the reaction of aluminum oxide with sulfuric acid to give aluminum sulfate. In order to manufacture 1.00 kg of aluminum sulfate, how many kilograms of aluminum oxide and sulfuric acid must be used?

47. Gallium hydroxide, like aluminum hydroxide, is amphoteric. Write balanced equations for the reaction of solid $Ga(OH)_3$ with solutions of HCl and NaOH. What volume of 0.0112 M HCl is needed to react completely with 1.25 g of $Ga(OH)_3$?

48. Halides of the Group 3A elements are excellent Lewis acids. When a Lewis base such as Cl^- interacts with $AlCl_3$, the ion $AlCl_4^-$ is formed. Write a Lewis electron dot structure for this ion. What structure is predicted for $AlCl_4^-$? What hybridization is assigned to the aluminum atom in $AlCl_4^-$?

49. "Aerated" concrete bricks are widely used building materials. They are obtained by mixing gas-forming additives with a moist mixture of lime, sand, and possibly cement. Industrially, the following reaction is important:

$$2\,Al(s) + 3\,Ca(OH)_2(s) + 6\,H_2O(\ell) \longrightarrow [3\,CaO \cdot Al_2O_3 \cdot 6\,H_2O](s) + 3\,H_2(g)$$

Assume that the mixture of reactants contains 0.56 g of Al for each brick. What volume of hydrogen gas do you expect at 26 °C and atmospheric pressure (745 mm Hg)?

Silicon

50. Describe the structures of SiO_2 and CO_2. Explain why SiO_2 has a very high melting point, whereas CO_2 is a gas.

51. Describe how ultrapure silicon can be produced from sand.

52. One material needed to make silicones is dichlorodimethylsilane, $(CH_3)_2SiCl_2$. It is made by treating silicon powder at about 300 °C with CH_3Cl in the presence of a copper-containing catalyst.
   (a) Write a balanced equation for the reaction.
   (b) Assume you carry out the reaction on a small scale with 2.65 g of silicon. To measure the CH_3Cl gas, you fill a 5.60-L flask at 24.5 °C. What pressure of CH_3Cl gas must you have in the flask in order to have the stoichiometrically correct amount of the compound?
   (c) What mass of $(CH_3)_2SiCl_2$ is produced? (Assume 100% yield.)

53. Describe the structure of pyroxenes (see Figure 22.8). What is the ratio of silicon to oxygen in this type of compound?

Nitrogen and Phosphorus

54. Consult the data in Appendix L. Are any of the nitrogen oxides listed there stable with respect to N_2 and O_2?

55. Use data in Appendix L to calculate the enthalpy change for the reaction

$$2\ NO_2(g) \rightleftharpoons N_2O_4(g)$$

Is this reaction exothermic or endothermic?

56. Use data in Appendix L to calculate the enthalpy change for the reaction

$$2\ NO(g) + O_2(g) \longrightarrow 2\ NO_2(g)$$

Is the reaction exothermic or endothermic?

57. The overall reaction involved in the industrial synthesis of nitric acid is

$$NH_3(g) + 2\ O_2(g) \longrightarrow HNO_3(aq) + H_2O(\ell)$$

Calculate $\Delta G°$ for the reaction and then the equilibrium constant, at 25 °C.

58. A major use of hydrazine, N_2H_4, is in steam boilers in power plants.
   (a) The reaction of hydrazine with O_2 dissolved in water gives N_2 and water. Write a balanced equation for this reaction.
   (b) O_2 dissolves in water to the extent of 3.08 cm^3 (gas at STP) in 100. mL of water at 20 °C. In order to consume all of the dissolved O_2 in 3.00×10^4 L of water (enough to fill a small swimming pool), how many grams of N_2H_4 are needed?

59. Before hydrazine came into use to remove dissolved oxygen in the water in steam boilers, Na_2SO_3 was commonly used:

$$2\ Na_2SO_3(aq) + O_2(aq) \longrightarrow 2\ Na_2SO_4(aq)$$

How many grams of Na_2SO_3 are required to remove O_2 from 3.00×10^4 L of water as outlined in Study Question 58?

60. A common analytical method for hydrazine involves its oxidation with iodate ion, IO_3^-. In the process, hydrazine is a four-electron reducing agent:

$$N_2(g) + 5\ H_3O^+(aq) + 4\ e^- \longrightarrow N_2H_5^+(aq) + 5\ H_2O(\ell) \qquad E° = -0.23\ V$$

Write the balanced equation for the reaction of hydrazine in acid solution ($N_2H_5^+$) with $IO_3^-(aq)$ to give N_2 and I_2. Calculate $E°$ for this reaction.

61. The steering rockets in the Space Shuttle use N_2O_4 and a derivative of hydrazine, 1,1-dimethylhydrazine. This mixture is called a *hypergolic fuel* because it ignites when the reactants come into contact:

$$H_2NN(CH_3)_2(\ell) + 2\ N_2O_4(\ell) \longrightarrow 3\ N_2(g) + 4\ H_2O(g) + 2\ CO_2(g)$$

   (a) Identify the oxidizing agent and the reducing agent in this reaction.
   (b) The same propulsion system was used by the Lunar Lander on moon missions in the 1970s. If the Lander used 4100 kg of $H_2NN(CH_3)_2$, how many kilograms of N_2O_4 were required to react with it? How many kilograms of each of the reaction products were generated?

62. Unlike carbon, which can form extended chains of atoms, nitrogen can form chains of very limited length. Draw the Lewis electron dot structure of the azide ion, N_3^-.

63. $CaHPO_4$ is used as an abrasive in toothpaste. Write a balanced equation showing a possible preparation for this compound.

Oxygen and Sulfur

64. In the "contact process" for making sulfuric acid, sulfur is first burned to SO_2. Environmental restrictions allow no more than 0.30% of this SO_2 to be vented to the atmosphere.
   (a) If enough sulfur is burned in a plant to produce 1.80×10^6 kg of pure, anhydrous H_2SO_4 per day, what is the maximum amount of SO_2 that is allowed to be exhausted to the atmosphere?
   (b) One way to prevent any SO_2 from reaching the atmosphere is to "scrub" the exhaust gases with slaked lime, $Ca(OH)_2$:

$$Ca(OH)_2(s) + SO_2(g) \longrightarrow CaSO_3(s) + H_2O(\ell)$$
$$2\ CaSO_3(s) + O_2(g) \longrightarrow 2\ CaSO_4(s)$$

What mass of $Ca(OH)_2$ (in kilograms) is needed to remove the SO_2 calculated in part (a)?

65. A sulfuric acid plant produces an enormous amount of heat. To keep costs as low as possible, much of this heat is used to make steam to generate electricity. Some of the electricity is used to run the plant, and the excess is sold to the local electrical utility. Three reactions are important in sulfuric acid production: (1) burning S to SO_2; (2) oxidation of SO_2 to SO_3; and (3) reaction of SO_3 with H_2O:

$$SO_3(g) + H_2O\ (\text{in 98\% } H_2SO_4) \longrightarrow H_2SO_4(\ell)$$

The enthalpy change of the third reaction is −130 kJ/mol. Estimate the total heat produced per mole of H_2SO_4 produced. How much heat is produced per ton (907 kg) of H_2SO_4?

66. In addition to the sulfide ion, S^{2-}, there are polysulfides, S_n^{2-}. (Such ions are chains of sulfur atoms, not rings.) Draw a Lewis electron dot structure for the S_2^{2-} ion. The S_2^{2-} ion is the disulfide ion, an analogue of the peroxide ion. It occurs in iron pyrites, FeS_2.

67. Sulfur forms a range of compounds with fluorine. Draw Lewis electron dot structures for S_2F_2 (connectivity is FSSF), SF_2, SF_4, SF_6, and S_2F_{10}. What is the formal oxidation number of sulfur in each of these compounds?

Chlorine

68. The halogen oxides and oxoanions are good oxidizing agents. For example, the reduction of bromate ion has an $E°$ value of 1.44 V in acid solution:

$$BrO_3^-(aq) + 6\ H_3O^+(aq) + 6\ e^- \longrightarrow Br^-(aq) + 9\ H_2O(\ell)$$

Can you oxidize aqueous 1.0 M Mn^{2+} to aqueous MnO_4^- with 1.0 M bromate ion?

69. The hypohalite ions, OX^-, are the anions of weak acids. Calculate the pH of a 0.10 M solution of NaOCl. What is the concentration of HOCl in this solution?

70. Bromine is obtained from sea water. The process involves treating water containing bromide ion with Cl_2 and extracting the Br_2 from the solution using an organic solvent. Write a balanced equation for the reaction of Cl_2 and Br^-. What are the oxidizing and reducing agents in this reaction? Using the table of standard reduction potentials (Appendix M), verify that this is a product-favored reaction.

71. To prepare chlorine from chloride ion a strong oxidizing agent is required. The dichromate ion, $Cr_2O_7^{2-}$, is one example (see Figure 22.16). Consult the table of standard reduction potentials (Appendix M) and identify several other oxidizing agents that may be suitable. Write balanced equations for the reactions of these substances with chloride ion.

GENERAL QUESTIONS

72. For each of the third-period elements (Na through Ar), identify the following:
(a) Whether the element is a metal, nonmetal, or metalloid
(b) The color and appearance of the element
(c) The state of the element (s, ℓ, or g) under standard conditions

For aid in this question, consult Figure 2.13 or use the periodic table "tool" on the *Saunders Interactive General Chemistry CD-ROM*. The latter provides a picture of each element and a listing of its properties.

73. For each of the second-period elements (Li through Ne), identify the following:
(a) Whether the element is a metal, nonmetal, or metalloid
(b) The color and appearance of the element
(c) The state of the element (s, ℓ, or g) under standard conditions

For aid in this question, consult Figure 2.13 or use the periodic table "tool" on the *Saunders Interactive General Chemistry CD-ROM*. The latter provides a picture of each element and a listing of its properties.

74. Let us consider the chemistry of the elements sodium, magnesium, aluminum, silicon, phosphorus, and sulfur.
(a) Write a balanced chemical equation to depict the reaction of each of the elements with elemental chlorine.
(b) Describe the bonding in each of the products of the reactions with chlorine as ionic or covalent.
(c) Draw Lewis electron dot structures for the products of the reactions of silicon and phosphorus with chlorine. What are their electron-pair and molecular geometries?

75. Let us consider the chemistry of the elements of Group 4A.
(a) Write a balanced chemical equation to depict the reaction of each of the elements with elemental chlorine.
(b) Describe the bonding in each of the products of the reactions with chlorine as ionic or covalent.

(You have not seen reactions of some of these elements in the text, but you have been given enough information to be able to predict the reactions that can occur.)

76. One of the pieces of evidence for the hydride ion in metal hydrides comes from electrochemistry. Predict the reactions that occur at each electrode when molten LiH is electrolyzed.

77. To store 2.88 kg of gasoline with an energy equivalence of 1.43×10^8 J requires a volume of 4.1 L. In comparison, 1.0 kg of H_2 has the same energy equivalence. What volume is required if this quantity of H_2 is to be stored at 25 °C and 1.0 atm of pressure?

78. Using data in Appendix L and those given in the table shown here, calculate $\Delta G°$ values for the decomposition of MCO_3 to MO and CO_2 where M = Mg, Ca, and Ba. What is the relative tendency of these carbonates to decompose?

Compound	$\Delta G_f°$ (kJ/mol)
$MgCO_3$	−1012.1
$BaCO_3$	−1137.6
BaO	−525.1

79. Ammonium perchlorate is used as the oxidizer in the solid fuel booster rockets of the Space Shuttle. If one launch requires 700 tons (6.35×10^5 kg) of the salt, and the salt decomposes according to the equation on page 1037, what mass of water is produced? What mass of O_2 is produced? If the O_2 produced is assumed to react with the powdered aluminum present in the rocket engine, how much aluminum is necessary to use up all of the O_2, and how much Al_2O_3 is produced?

80. Metals react with hydrogen halides (such as HCl) to give the metal halide and hydrogen:

$$M(s) + n\ HX(g) \longrightarrow MX_n(s) + \frac{n}{2}\ H_2(g)$$

The free energy change for the reaction is

$$\Delta G°_{rxn} = \Delta G_f°\ (MX_n) - n\ \Delta G_f°\ [HX(g)]$$

(a) ΔG_f° for HCl(g) is −95.3 kJ/mol. What must be the value for ΔG_f° (MX_n) in order for the reaction to be product-favored?
(b) Which of the following metals is (are) predicted to react with HCl(g): Ba, Pb, Hg, or Ti?

81. The boron atom in boric acid, $B(OH)_3$, is bonded to three —OH groups. (In the solid state, the —OH groups are in turn hydrogen-bonded to —OH groups in neighboring molecules.)
(a) Draw the Lewis structure for boric acid.
(b) What is the hybridization of the boron in the acid?
(c) Sketch a picture showing how hydrogen bonding can occur between neighboring molecules.
(d) Boric acid is an acid because it reacts with water to give the borate and hydronium ions:

$$B(OH)_3(aq) + 2\,H_2O(\ell) \longrightarrow B(OH)_4^-(aq) + H_3O^+(aq)$$

Given the nature of bonding in boric acid, why do you believe this reaction is possible? Is boric acid acting as a Lewis acid or Lewis base in this reaction?

82. The structure of nitric acid is illustrated on pages 390 and 1031.
(a) Why are the N—O bonds the same length, and why are both shorter than the N—OH bond length?
(b) Rationalize the bond angles in the molecule.
(c) What is the hybridization of the central N atom? What orbitals overlap to form the N—O π bond?

83. Like many metals, aluminum reacts readily with halogens. Aluminum bromide has the structure illustrated here in the solid and gaseous phases. That is, it is a dimer of the monomer units $AlBr_3$ with a Br atom bridging from one Al atom to another. (This form of halogen bridging is fairly common in chemistry.) The Br—Al—Br angle is 115° and the Al—Br—Al angle is 87°.

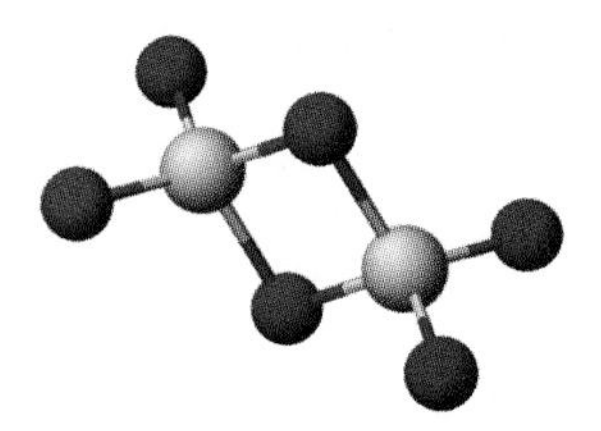

Model of aluminum bromide dimer.

(a) What is the hybridization of the Al atoms?
(b) Draw the Lewis structure for an $AlBr_3$ monomer.
(c) How might a Br atom form a bridge from one Al atom to another? Describe the bonding in the Al_2Br_6 molecule.

CHALLENGING QUESTIONS

84. Identify the lettered compounds in the following reaction scheme. When 1.00 g of a white solid A is strongly heated, you obtain another white solid, B, and a gas. (An experiment is carried out on the gas, showing that it exerts a pressure of 209 mm Hg in a 450-mL flask at 25 °C.) Bubbling the gas into a solution of $Ca(OH)_2$ gives another white solid, C. If the white solid B is added to water, the resulting solution turns red litmus paper blue. To the solution of B, you add dilute, aqueous HCl and evaporate to dryness to yield a white solid D. When D is placed in a Bunsen burner flame, it colors the flame green. Finally, if the aqueous solution of B is treated with sulfuric acid, a white precipitate, E, forms.

85. You have a 1.0-L flask that contains a mixture of argon and hydrogen. The pressure inside the flask is 745 mm Hg and the temperature is 22 °C. Describe an experiment that you could use to determine the percentage of hydrogen in this mixture.

86. Use ΔH_f° data in Appendix L to calculate the enthalpy change of the reaction

$$2\,N_2(g) + 5\,O_2(g) + 2\,H_2O(\ell) \longrightarrow 4\,HNO_3(aq)$$

Speculate on whether such a reaction could be used to "fix" nitrogen. Would research to find ways to accomplish this reaction be a good endeavor?

87. You are given air and water for starting materials, along with whatever laboratory equipment you need. Describe how you could synthesize ammonium nitrate from these reagents.

Summary Question

88. Magnesium chemistry.
(a) Magnesium is obtained from sea water. If sea water is 0.050 M in magnesium ion, what volume of sea water (in liters) must be treated to obtain 1.00 kg of magnesium metal? What mass of lime (CaO, in kilograms) must be used to precipitate the magnesium in this volume of sea water?
(b) When 1.2×10^3 kg of molten $MgCl_2$ is electrolyzed to produce magnesium, how many kilograms of metal are produced at the cathode? What is produced at the anode? How many kilograms of the other product are produced? What is the total number of faradays of electricity used in the process?
(c) One industrial process has an energy consumption of 8.4 kwh/lb of Mg. How many joules are therefore required per mole? How does this energy compare with the energy of the process

$$MgCl_2(s) \longrightarrow Mg(s) + Cl_2(g)?$$

Chapter **23**

The Transition Elements

(David Heald © The Solomon R. Guggenheim Foundation, New York)

A Chemical Puzzler

The new Guggenheim Museum recently opened in Bilbao, Spain. The entire exterior of the very striking building is covered with a fourth-period transition metal. The metal has a density of 4.5 g/cm^3 and a melting point of about 1900 K. It is most commonly found with oxidation numbers of +2, +3, and +4 in its compounds, and those with +2 and +3 are paramagnetic. It resists corrosion due to an oxide layer on the metal. In fact, the oxide is a common pigment in paint. Finally, the metal is unaffected by most acids and alkalis. What is the metal?

The **transition elements,** the large block of elements in the central portion of the periodic table, form a bridge between the *s*-block elements at the left and the *p*-block metals, metalloids, and nonmetals on the right (Figure 23.1). These elements are also referred to as the ***d*-block elements** because their occurrence in the periodic table coincides with the filling of the *d* orbitals. Contained within this group of elements are two subgroups sometimes called inner transition elements; these are the **lanthanide elements** that occur between La and Hf, and the **actinide elements** that occur between Ac and element 104. Because these subgroups arise as the *f* orbitals are filled, they are also called ***f*-block elements.**

For historical reasons the lanthanide elements are sometimes called "rare earths" although many of these elements are not actually rare. Cerium, for example, is half as abundant in the earth's crust as chlorine and about five times more abundant than lead.

In this chapter, the primary focus is on the *d*-block elements, and within this group we concentrate mainly on the elements in the fourth period, that is, the elements of the first transition series, the group from scandium to zinc.

23.1 PROPERTIES OF THE TRANSITION ELEMENTS

The *d*-block elements include the most common metal used in construction and manufacturing (iron), metals that are valued for their beauty (gold, silver, and platinum), metals used in coins (nickel, copper), and metals used in modern technology (titanium). Copper, silver, gold, and iron were known and used in early civilization. This group of elements contains the densest elements (osmium, $d = 22.49$ g/cm^3, and iridium, $d = 22.41$ g/cm^3), the metals with the highest and lowest melting points (tungsten, mp = 3410 °C, and mercury, mp = −38.9 °C), and one of two radioactive elements with atomic number less than 83 [technetium (Tc), atomic number 43; the other is promethium (Pm), atomic number 61, in the *f*-block].

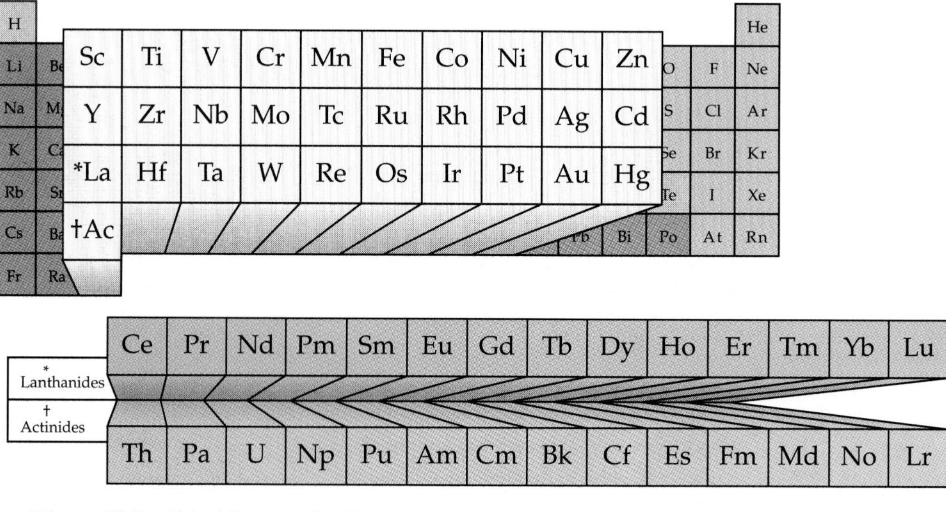

Figure 23.1 Transition metals. The periodic table showing the *d*-block, or transition, elements (red), and the *f*-block elements, the lanthanides (blue) and the actinides (lavender).

Certain *d*-block elements are particularly important in living organisms. For example, cobalt is the crucial element in vitamin B_{12}, a compound that acts as a catalyst in the metabolism of carbohydrates, fats, and proteins. Hemoglobin and myoglobin, compounds in biochemical oxidation–reduction processes, contain iron. Molybdenum and iron, together with sulfur, form the reactive portion of nitrogenase, a biological catalyst used by nitrogen-fixing organisms to convert atmospheric nitrogen into ammonia. Copper and zinc are important in other biological catalysts.

Some "bad actors" also appear in this group of elements. Mercury, for example, is toxic and is a threat in the environment. Other metals are toxic as well, so that disposal of "heavy-metal" wastes is generally a significant problem.

Figure 23.2 Paint pigments. Pigments often contain transition metal compounds: Yellow, CdS; Green, Cr_2O_3; white, TiO_2 and ZnO; purple, $Mn_3(PO_4)_2$; blue, cobalt and aluminum oxides; and ochre, Fe_2O_3. *(C. D. Winters)*

Many transition metal compounds are highly colored, which makes them useful as pigments in paints and dyes. Prussian blue, $Fe_4[Fe(CN)_6]_3$, is the "bluing agent" in laundry bleach and in engineering blueprints. A common pigment (artist's cadmium-yellow) contains cadmium sulfide (CdS), and the color in most white paints is provided by titanium(IV) oxide (Figure 23.2).

The presence of transition metal ions in crystalline silicates or alumina causes a common material to be transformed into a gemstone. Iron(II) is the cause of the yellow color in citrine, whereas chromium(III) causes the red color of a ruby. Transition metal complexes in small quantities add color to glass (Figure 23.3). For example, blue glass is made by adding a small amount of a cobalt(III) oxide, and green glass is made by adding Cr_2O_3. Old window panes sometimes take on a purple color over time as a consequence of oxidation of traces of manganese(II) ion present in the glass to purple permanganate (MnO_4^-).

The black coating of tarnish that develops on silver is the result of oxidation. Similarly, for copper, both the blackening of the surface and the attractive green "patina" that is sometimes seen on copper roofs are the result of oxidation.

Most transition elements are solids with relatively high melting and boiling points. They have a metallic sheen, and are conductors of electricity and heat. Metals typically undergo oxidation reactions, forming ionic compounds. But there are some exceptions. Mercury is a liquid. Whereas the oxidation of iron (rust) is well known, and a problem that we go to some lengths to prevent, silver and gold are used in coins and jewelry because they are resistant to oxidation.

Let us look more closely at the properties of the transition elements, concentrating especially on the underlying principles that govern the characteristics of these elements.

Electron Configurations

Because chemical behavior is related to electronic structure, it is important to know the electron configurations of the *d*-block elements and their common ions (see Section 8.4). Recall that the configuration of these metals has the general form [noble gas core]$ns^a(n-1)d^b$, that is, valence electrons for the transition elements reside in the *ns* and $(n-1)d$ subshells (see Tables 8.2 and 8.4).

On oxidizing a transition metal, the *s* electrons are lost first and, in some instances, one or more *d* electrons are lost as well. The resulting ions have the electron configuration [noble gas core]$(n-1)d^x$. In contrast to ions formed by main group elements, transition metal ions do not have noble gas configurations, and their compounds often possess unpaired electrons, leading to paramagnetism (see p. 336), a behavior that will be discussed later in this section.

Figure 23.3 Colored glass. Colored glass can be made by adding small amounts of metal oxides to clear glass. Blue glass often contains cobalt(III) oxide, copper or chromium oxides give green glass, nickel or cobalt oxides give a purple color, copper or selenium oxides give red, and an iridescent green color is due to uranium oxide. *(C. D. Winters)*

Current Perspectives

Metal with a Memory

In the early 1960s, William J. Buehler, a metallurgical engineer at the Naval Ordnance Laboratory in White Oak, Maryland, was experimenting with binary alloys, that is, alloys made up of two metals. He was looking for a material that was resistant to impact and heat because it was to be used in the nose cone of a Navy missile. It was also important that the material be fatigue-resistant, that is, it should not lose its desirable properties when heated or handled. An alloy of nickel and titanium appeared to have some desirable properties, so Buehler prepared long, thin strips of this alloy to demonstrate that it could be folded and unfolded many times without breaking. At a meeting to discuss this material, a colleague wanted to see what happened when the strip was heated. He held a pipe lighter to a folded-up piece of metal and was amazed to observe that the metal strip immediately unfolded to its original shape. Thus, memory metal was discovered. This unusual alloy is now called nitinol, a name constructed out of *ni*ckel, *ti*tanium, *N*aval *O*rdnance *L*ab.

The shape that NiTi "remembers" is established by heating the alloy to between 500 and 550 °C for about an hour and then allowing it to cool. At the low temperature, the alloy is fairly soft and may be bent and twisted out of shape. When warmed, the metal returns to its original shape. The temperature at which the change in shape occurs varies with small differences in the NiTi ratio. Depending on composition, materials that change shape at temperatures ranging from −125 °C to about 70 °C are possible, greatly increasing the number of possible uses for this intriguing material.

Memory metal never made it into missile nose cones, but it has found a wide variety of other uses, and some of the most interesting are in medicine. For example, bone anchors can be made of nitinol; on warming to body temperature the alloy expands and locks the anchor in place. It is possible to thread a filament of nitinol into a vein and have the filament rearrange itself to form a fine-mesh screen to filter out blood clots. Nitinol can also be used in orthodontics; braces made of nitinol remember their shape and apply a steady constant pressure to move teeth into position. Eyeglass frames of nitinol can be twisted into odd shapes only to return to their original shape when warmed to body temperature.

What is this remarkable alloy, and how does it work? When heated above 500 °C the alloy crystallizes in a structure with eight atoms of nickel at the corners of a cube and a titanium atom in the center. Extended in three dimensions, the crystal creates interpenetrating cubic lattices of nickel and titanium atoms.

The key to understanding the behavior of this material is knowing that solid nitinol undergoes a transition between two solid phases. Solid phases differ in important ways at the atomic level. For example, a solid-solid phase transition you have heard about is the change between the diamond and graphite phases of carbon.

At higher temperatures, nitinol exists in what is called the austenite phase, but at temperatures lower than the transition temperature the atomic arrangement distorts slightly as the material shifts to the martensite phase. There is no visible change in the shape of the metal when the phase change occurs; the change occurs entirely at the atomic level. The distortion of the atomic structure, however, imposes a strain on the system. When the metal is heated, the process reverses and the metal atoms move back into their original positions to relieve the strain.

For more information, see G. B. Kauffman and I. Mayo: "Memory metal." *Chem Matters,* October, 1993, p. 4; and "The metal with a memory." *Invention and Technology,* Fall, 1993, p. 18.

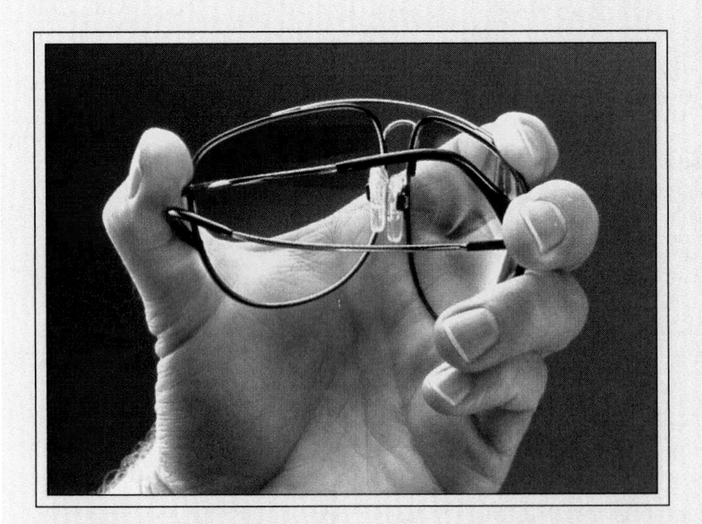

These sunglass frames are made of nitinol, so they snap back to the proper fit even after being twisted like a pretzel. The nitinol used in these frames has a critical temperature below room temperature, so the metal readily returns to its "memorized" shape. Similar alloys are used for wires in dental braces and surgical anchors, which cannot be heated after insertion. *(NASA/Science Source/Photo Researchers, Inc.)*

Table 23.1 • COMMON OXIDATION PRODUCTS OF ELEMENTS IN THE FIRST Transition Series

Element	Reaction with O_2	Reaction with Cl_2	Reaction with Aqueous HCl
Scandium	Sc_2O_3	$ScCl_3$	$Sc^{3+}(aq)$
Titanium	TiO_2	$TiCl_4$	$Ti^{3+}(aq)$
Vanadium	V_2O_5	VCl_4	NR*
Chromium	Cr_2O_3	$CrCl_3$	$Cr^{2+}(aq)$
Manganese	Mn_3O_4	$MnCl_2$	$Mn^{2+}(aq)$
Iron	Fe_2O_3	$FeCl_3$	$Fe^{2+}(aq)$
Cobalt	Co_3O_4	$CoCl_2$	$Co^{2+}(aq)$
Nickel	NiO	$NiCl_2$	$Ni^{2+}(aq)$
Copper	CuO	$CuCl_2$	NR
Zinc	ZnO	$ZnCl_2$	$Zn^{2+}(aq)$

*NR = no reaction.

Oxidation Numbers

Oxidation numbers of +2 and +3 are commonly observed in compounds of the metals in the first transition series. Examples of oxidation reactions of transition metals include those with oxygen to form metal oxides, with halogens to form metal halides, and with aqueous acid to form the hydrated metal ion (Table 23.1 and Figure 23.4). With iron, for example, oxidation processes usually convert Fe ($[Ar]3d^64s^2$) to Fe^{2+} ($[Ar]3d^6$) or to Fe^{3+} ($[Ar]3d^5$). The rusting of iron is a well-known example of oxidation. Rusting is actually a complicated process

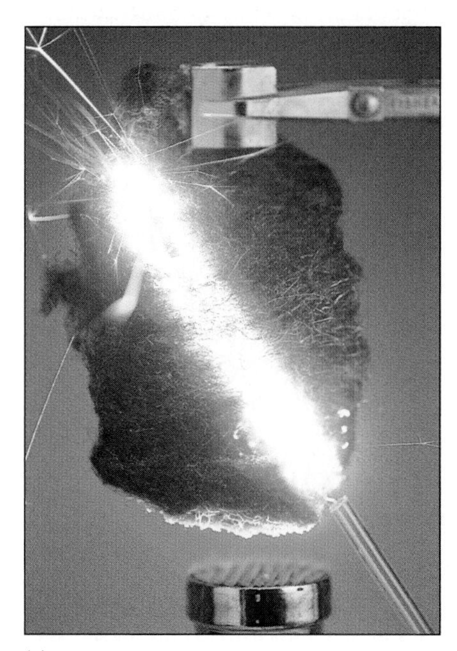

(a)

(b)

(c)

Figure 23.4 Typical reactions of transition metals. These metals react with oxygen, with halogens, and with acids under appropriate conditions. (a) Here steel wool reacts with O_2, (b) with chlorine gas, Cl_2, and (c) iron filings react with aqueous HCl. *(C. D. Winters)*

Table 23.2 • Standard Aqueous Reduction Potentials of Elements of the First Transition Series

Element	Half-Reaction	$E°$ (V)
Scandium	$Sc^{3+} + 3\ e^- \longrightarrow Sc$	−2.08
Titanium	$Ti^{2+} + 2\ e^- \longrightarrow Ti$	−1.63
Vanadium	$V^{2+} + 2\ e^- \longrightarrow V$	−1.2
Chromium	$Cr^{2+} + 2\ e^- \longrightarrow Cr$	−1.18
Manganese	$Mn^{2+} + 2\ e^- \longrightarrow Mn$	−0.91
Iron	$Fe^{2+} + 2\ e^- \longrightarrow Fe$	−0.44
Cobalt	$Co^{2+} + 2\ e^- \longrightarrow Co$	−0.28
Nickel	$Ni^{2+} + 2\ e^- \longrightarrow Ni$	−0.23
Copper	$Cu^{2+} + 2\ e^- \longrightarrow Cu$	+0.34
Zinc	$Zn^{2+} + 2\ e^- \longrightarrow Zn$	−0.76

For more on the process of iron corrosion, see pages 981–985.

that requires both oxygen and water, and the product (rust) is a hydrated iron(III) oxide. Iron reacts with chlorine to give $FeCl_3$, and it reacts with $H_3O^+(aq)$ to produce $Fe^{2+}(aq)$ and H_2.

Recall from Chapter 21 that a table of electrochemical potentials is a source of useful information on oxidations and reductions. Standard reduction potentials for the elements of the first transition series are shown in Table 23.2. All of these metals except copper can be oxidized by $H_3O^+(aq)$.

Despite the preponderance of +2 and +3 compounds of these elements, the range of oxidation states is considerably broader (Figures 23.5 and 23.6). In examples earlier in this text, we encountered chromium with a +6 oxidation number (CrO_4^{2-}, $Cr_2O_7^{2-}$), manganese with +7 (MnO_4^-), and silver and copper as 1+ ions. The most common oxidation state of titanium is +4, seen for example in the common minerals of this element: TiO_2 (rutile) and ilmenite ($FeTiO_3$).

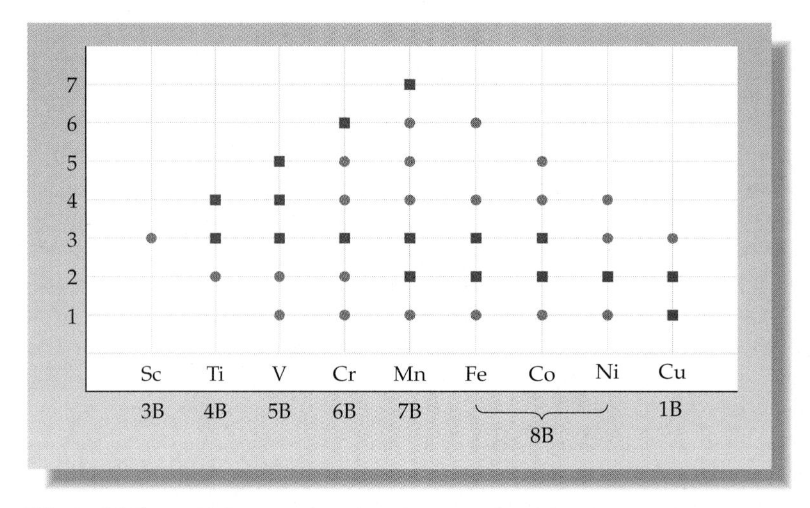

Figure 23.5 Oxidation numbers for elements of the first transition series. Oxidation numbers for the elements Sc through Cu. The more common oxidation numbers are given in red. Less commonly encountered ones are in blue.

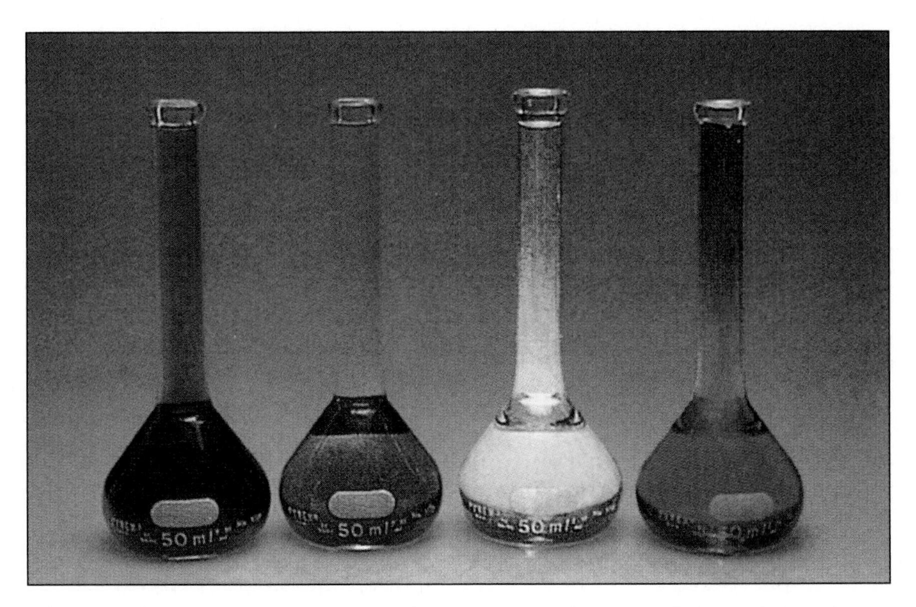

***Figure* 23.6 Some common compounds of chromium.** Aqueous solutions of chromium compounds in which the oxidation number of the metal is +3 [in $Cr(NO_3)_3$ (violet) and $CrCl_3$ (green)] and +6 [in K_2CrO_4 (yellow) and $K_2Cr_2O_7$ (orange)]. *(C. D. Winters)*

Higher oxidation states are more common in compounds of the elements in the second and third transition series, whereas +2 and +3 ions are less often encountered. For example, the naturally occurring sources of molybdenum and tungsten are the ores molybdenite (MoS_2) and wolframite (WO_3). In contrast, the principal ore of chromium is chromite, $FeO \cdot Cr_2O_3$. This general trend is carried over in the *f*-block. The lanthanides form 3+ ions, whereas actinide elements usually also have higher oxidation numbers, such as +4 and even +6 in typical compounds. For example, UO_3 is a common oxide of uranium, and UF_6 is a compound important in processing uranium fuel for nuclear reactors (see Sections 24.6 and 24.7).

Several of the lanthanides form compounds with metal oxidation numbers of +2 or +4.

Metal Atom Radii

Figure 8.11 shows the variation in atomic radii for the first 12 elements in the fourth, fifth, and sixth periods. A rapid decrease in size is seen among the first three elements in the period. In contrast, the radii of the transition elements vary over a fairly narrow range, dropping to a minimum around the middle of this group of elements and then rising slowly. This variation can be understood based on electron configurations. Atom size is determined by the radius of the outermost orbital, which for these elements is the *ns* orbital ($n = 4$, 5, or 6). As we progress from left to right in the periodic table, the effect caused by the increasing nuclear charge is mostly canceled by the increasing number of electrons in the $(n - 1)d$ orbitals. In the first half of the transition series the *ns* electrons experience a small increase in net nuclear attraction across the series and decrease slightly in size. The small increase in radius in the second half of this group of elements is due to increased electron–electron repulsions as the *d* subshell is completed.

Information on each of the transition metals is found in the periodic table on the *Saunders Interactive General Chemistry CD-ROM.*

Table 23.3 • Radii of Group 3A Elements

Element	Period	Radius (pm)
Boron	2	85
Aluminum	3	143
Gallium	4	135
Indium	5	167
Thallium	6	170

The overall decrease in metal radii across the fourth-period transition metals has a noticeable effect on the radii of main group elements that follow (Table 23.3). Instead of the normal increase in size down a periodic group, we see that gallium is actually smaller than aluminum.

Radii of the transition elements in the fifth and sixth periods are almost identical. The reason for this is that the lanthanide elements are inserted into the table just before the third series of transition elements. The filling of $4f$ orbitals through the lanthanide elements is accompanied by a steady contraction in size (not unlike the overall size decrease across each series of transition elements). At the point where the $5d$ orbitals begin to fill again, the radii have decreased to a size similar to that of elements in the previous period. Because this is such a noticeable effect with significant consequences, it is given a specific name, the **lanthanide contraction.**

The effect of the lanthanide contraction carries over into the p block as well. Thus, the increase in radius from indium to thallium is smaller than expected (see Table 23.3).

The similar radii of the atoms of the transition metals and of their ions affect their chemistry. For example, the "platinum group metals" (Ru, Os, Rh, Ir, Pd, and Pt) form similar compounds. Thus, it is not surprising that minerals containing these metals are found in the same geological zones in the earth.

Density

A consequence of the variation in metal radii is that the densities of the metals first increase and then decrease across a period (Figure 23.7). This variation reflects the slow increase in atomic mass and the change in volume. Although the overall change in radii among these elements is small, the effect is magnified because the volume is actually changing with the cube of the radius ($V = \frac{4}{3}\pi r^3$).

The lanthanide contraction is the reason that elements in the sixth period have the highest density. The relatively small radii of sixth-period transition metals, combined with the fact that their atomic mass is considerably larger than their counterparts in the fifth period, causes sixth-period metal densities to be very large.

Melting Point

The periodic variation of melting points of the transition elements is illustrated in Figure 23.8. The metals with the highest melting points occur in the middle of each series.

The melting point of any substance reflects the forces of attraction between elementary particles. With metals, we are dealing with forces of attraction between atoms of metals, which we could also call metallic bonding (see Section 10.4). Again, electron configurations provide us with an explanation. The variation in melting point indicates the strongest metallic bonds occur when the d subshell is about half-filled. This is also the point at which the largest number of unpaired electrons occurs in the isolated atoms. We can conclude that the d electrons are playing an important role in metallic bonding in the solid metal.

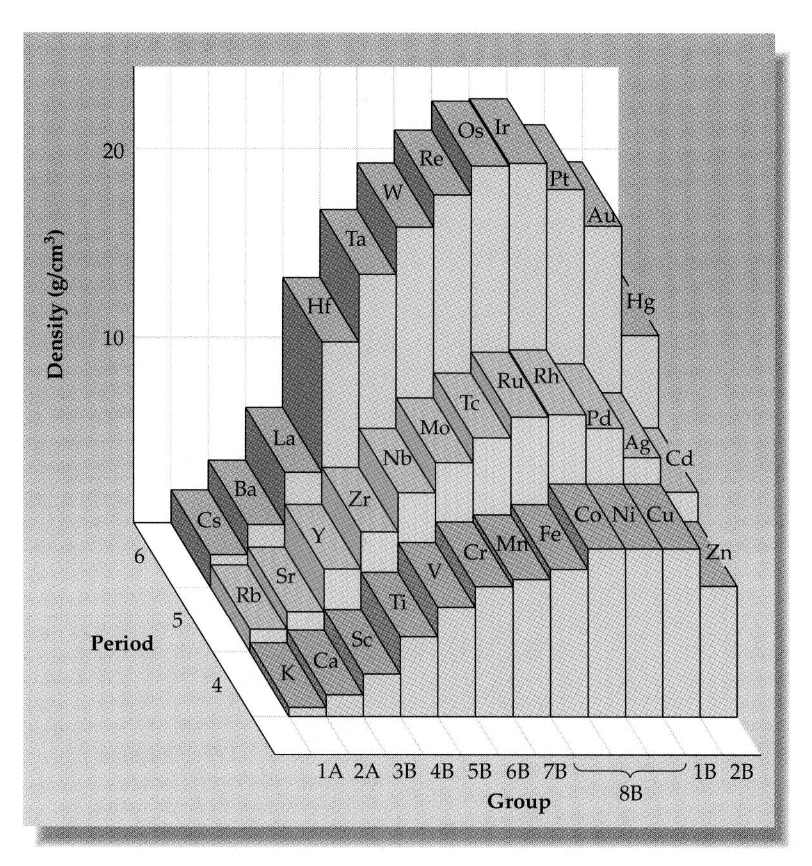

Figure 23.7 Density of the *d*-block elements. Densities of the transition metals (and the preceding *s*-block elements) as a function of periodic group.

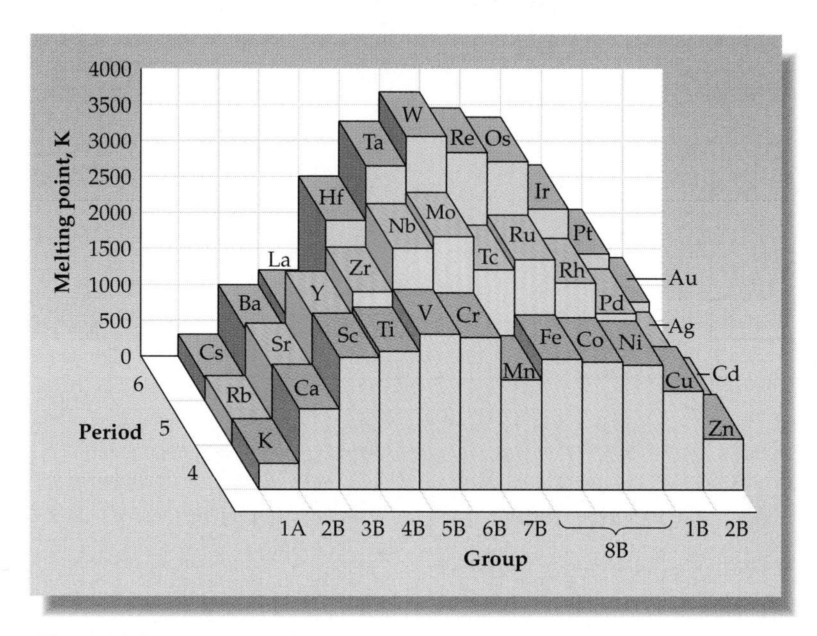

Figure 23.8 Melting points of the *d*-block elements. Melting points of the transition metals as a function of periodic group.

Magnetism

Between zero and five unpaired electrons occur in the *d* orbitals of transition metal ions, and up to seven unpaired electrons are in *f* orbitals of the lanthanide and actinide metal ions in their compounds. If a species contains unpaired electrons, it is **paramagnetic** and attracted by a magnet (see Section 8.1). As we will discover in Section 23.5, paramagnetism of transition metal compounds is an important feature that must be accounted for in a description of the species.

Recall that unpaired electrons can act as subatomic bar magnets. If, in a solid metal, the atomic magnets of a group of atoms (called a domain) interact so that their magnets are all oriented in the same direction, the magnitude of the magnetic effect is much larger than paramagnetism. We call this effect **ferromagnetism** (see page 336). Iron, cobalt, and nickel metals are ferromagnetic. Ferromagnetic materials are also unique in that, once the electron magnets are aligned by an external magnetic field, the metal is permanently magnetized. In such a case, the magnetism can be eliminated by heating or vibrating the metal to rearrange the electron spin domains.

Antiferromagnetism occurs when electron spins on adjacent atoms in a domain line up in opposite directions.

23.2 COMMERCIAL PRODUCTION OF TRANSITION METALS

Metals are found in nature as oxides, sulfides, halides, carbonates, or other ionic compounds (Figure 23.9). Some metal-containing mineral deposits are of little value, either because the deposit has too low a concentration of the desired metal or because the metal is difficult to separate from impurities in the ore. The relatively few minerals from which elements can be obtained profitably are called ores.

Very few ores are chemically pure substances. They are usually mixtures of the desired mineral and large quantities of impurities such as sand and clay, called **gangue** (pronounced "gang"). Generally, the first step in obtaining the

SOME ORES OF IMPORTANT METALS

Metal	Ore
Iron	magnetite, Fe_3O_4
	hematite, Fe_2O_3
Titanium	ilmenite, $FeTiO_3$
	rutile, TiO_2
Copper	chalcocite, Cu_2S
	chalcopyrite, $CuFeS_2$
Molybdenum	molybdenite, MoS_2
Zinc	zinc blende, ZnS
Lead	galena, PbS
Mercury	cinnabar, HgS

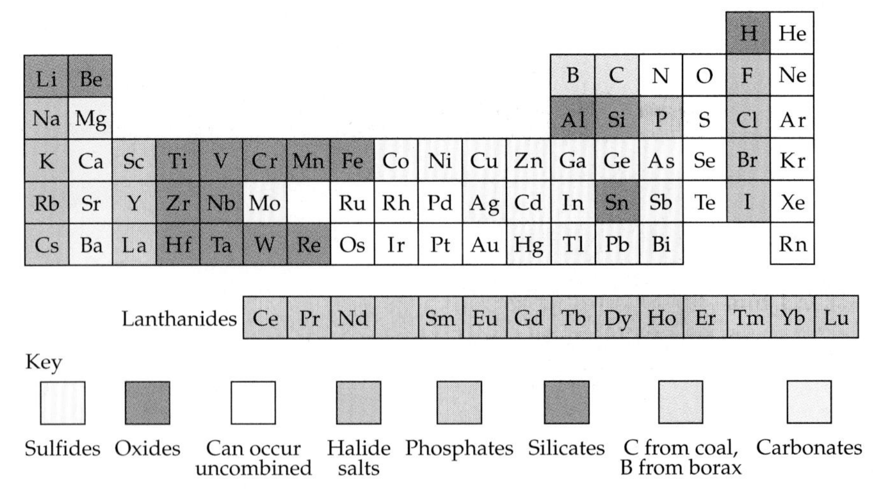

Figure 23.9 Sources of the elements. The transition metals are found naturally as oxides or sulfides or, in a few cases such as copper, silver, and gold, found uncombined. The blank spaces are for elements that do not occur in nature. See also Table 19.1.

desired metal is to separate the mineral from the gangue. The second major step involves converting the ore to the metal. Pyrometallurgy and hydrometallurgy are two methods of recovering metals from their ores. **Pyrometallurgy,** a high-temperature method, is illustrated by iron production. **Hydrometallurgy** uses aqueous solutions at relatively low temperatures for the extraction of metals, such as copper, zinc, tungsten, and gold.

Iron Production

The production of iron from its ores involves oxidation–reduction reactions carried out in a blast furnace (Figure 23.10). The furnace is charged at the top with a mixture of ore (usually hematite, Fe_2O_3), coke (which is primarily carbon), and limestone ($CaCO_3$), and a blast of hot air is forced in at the bottom. The coke burns with such an intense heat that the temperature at the bottom is almost 2000 °C, and a temperature of about 200 °C is attained at the top of the furnace. The quantity of oxygen used is controlled so that carbon monoxide is the primary product:

In 1993, 9.8×10^7 tons of raw steel and 5.3×10^7 tons of iron were produced in the United States.

$$2\ C(s,\ coke) + O_2(g) \longrightarrow 2\ CO(g) + heat$$

Coke is made by heating coal in a tall narrow oven that is sealed to keep out oxygen. Heat drives off volatile chemicals, including benzene and ammonia, leaving nearly pure carbon.

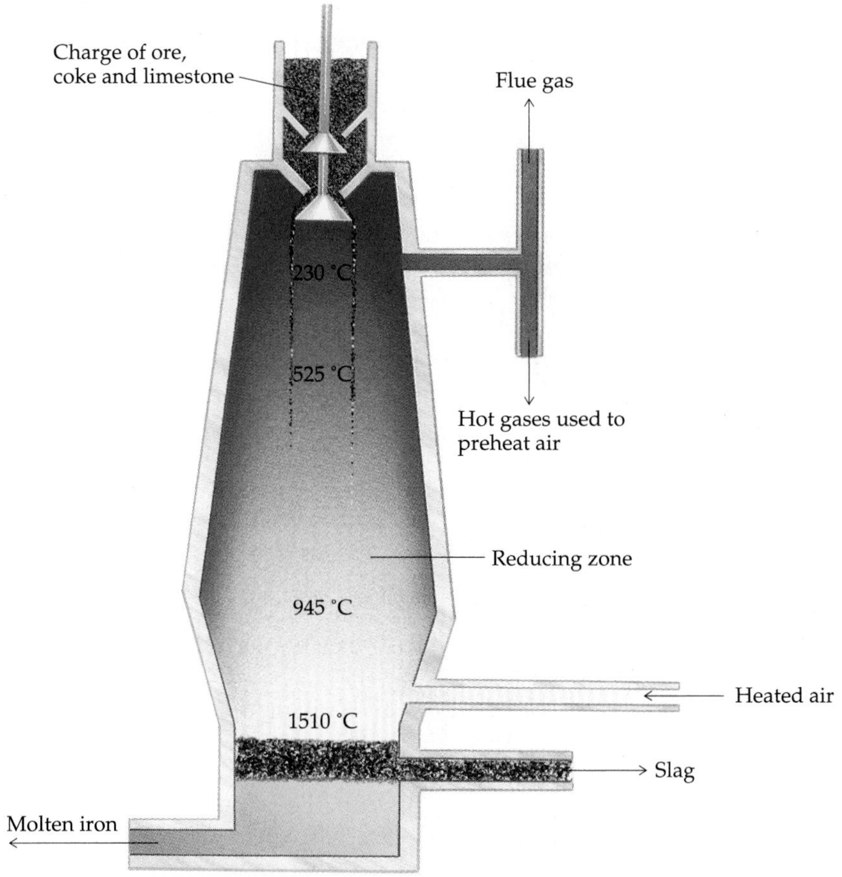

F*igure* 23.10 A blast furnace used for the reduction of iron ore to iron. The largest modern furnaces have hearths 14 m in diameter and can produce up to 10,000 tons of iron per day.

Both carbon and carbon monoxide participate in the reduction of iron(III) oxide to give impure metal:

$$Fe_2O_3(s) + 3\ C(s) \longrightarrow 2\ Fe(\ell) + 3\ CO(g)$$
$$Fe_2O_3(s) + 3\ CO(g) \longrightarrow 2\ Fe(\ell) + 3\ CO_2(g)$$

Much of the carbon dioxide formed in the reduction process (and from heating limestone) is itself reduced on contact with unburned coke and produces more reducing agent:

$$CO_2(g) + C(s) \longrightarrow 2\ CO(g)$$

The molten iron flows down through the furnace and collects at the bottom, where it can be tapped off through an opening in the side. When cooled, the impure iron is called "cast iron" or "pig iron." Usually, the material is either brittle or soft (rather undesirable properties for most uses), due to the presence of small amounts of impurities, such as elemental carbon, phosphorus, and sulfur.

Production of 1 ton of pig iron requires about 1.7 ton of iron ore, 0.5 ton of coke, 0.25 ton of limestone, and 2 tons of air.

Iron ores generally contain silicate minerals and silicon dioxide. Lime, formed from added limestone, reacts with these materials to give calcium silicate, which is molten at the temperature of the blast furnace:

$$SiO_2(s) + CaO(s) \longrightarrow CaSiO_3(\ell)$$

This is an acid–base reaction because CaO is a basic oxide and SiO_2 is an acidic oxide. The calcium silicate, being less dense than molten iron, floats on the iron as a separate layer. Other nonmetal oxides that may also be present dissolve in this layer, and the mixture is called a "slag." At the interface, or boundary, between molten iron and slag, ionic impurities tend to move into the slag layer, whereas other elements present (either other metals or nonmetals) are concentrated in the iron. The floating slag layer is easily removed and frees the iron from some of the original impurities.

The impure pig iron coming from the bottom of the blast furnace is purified to remove the nonmetal impurities. Several technologies can be used, but the most important is currently the **basic oxygen furnace** (Figure 23.11). In this furnace, oxygen is blown into the molten pig iron. This process oxidizes phosphorus, sulfur, and most of the excess carbon:

$$P_4(s) + 5\ O_2(g) \longrightarrow P_4O_{10}(g)$$
$$S_8(s) + 8\ O_2(g) \longrightarrow 8\ SO_2(g)$$
$$C(s) + O_2(g) \longrightarrow CO_2(g)$$

The oxides either escape as gases or react with basic oxides (such as CaO) that are added or are used to line the furnace. For example,

$$P_4O_{10}(g) + 6\ CaO(s) \longrightarrow 2\ Ca_3(PO_4)_2(\ell)$$

These salts form a floating layer of slag, which can be poured off to free the denser molten iron layer of impurities.

Pig iron may contain up to 4.5% carbon, 1.7% manganese, 0.3% phosphorus, 0.04% sulfur, and as much as 15% silicon. The final purification step removes much of the P, S, and Si and reduces the carbon content to about 1.3%. The result is ordinary *carbon steel.* Almost any degree of flexibility, hardness, strength, and malleability can be achieved in carbon steel by proper cooling, reheating, and tempering. The drawbacks to this material, however, are that it corrodes easily and loses its properties when heated strongly.

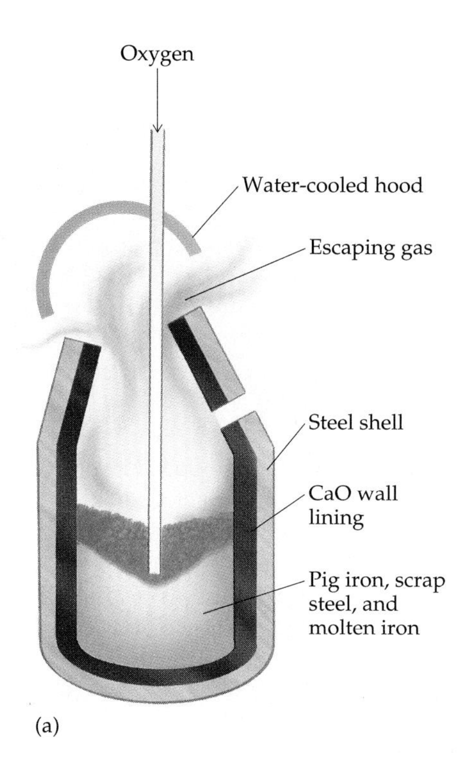

Figure 23.11 A "basic oxygen furnace." (a) Much of the steel produced today is made by blowing oxygen through a furnace charged with scrap and molten iron from a blast furnace. Measured amounts of alloying elements determine the particular steel produced. (b) Molten steel being poured from a basic oxygen furnace. *(b, Bethlehem Steel Corporation)*

During the processing of steel, other transition metals, such as chromium, manganese, and nickel, can be added to produce alloys that have specific physical, chemical, and mechanical properties. An important type of steel, *stainless steel,* typically contains 18% to 20% Cr and 8% to 12% Ni. Stainless steel's usefulness results from its resistance to corrosion. Another alloy of iron is *alnico V.* This alloy, used in loudspeaker magnets because of its permanent magnetism, contains five elements: Al (8%), Ni (14%), Co (24%), Cu (3%), and Fe (51%).

Copper Production

Copper-bearing minerals include chalcopyrite ($CuFeS_2$), chalcocite (Cu_2S), and covellite (CuS). Because the ores containing these minerals generally have a low percentage of copper, enrichment is necessary. This is carried out by a process known as *flotation.* First, the ore is finely powdered. Oil is then added and the mixture agitated with soapy water in a large tank (Figure 23.12). At the same time, compressed air is forced through the mixture, and the lightweight, oil-covered copper sulfide particles are carried to the top as a frothy mixture. The heavier gangue settles to the bottom of the tank, and the copper-laden froth is skimmed off.

In the pyrometallurgy of copper, the enriched ore is **roasted** with enough air to convert iron to its oxide while leaving copper(II) sulfide:

$$2\ CuFeS_2(s) + 3\ O_2(g) \longrightarrow 2\ CuS(s) + 2\ FeO(s) + 2\ SO_2(g)$$

This mixture of copper sulfide and iron oxide is mixed with ground limestone, sand, and some fresh concentrated ore and then heated to 1100 °C. As in the blast furnace, limestone, $CaCO_3$, is converted to lime, CaO; then lime and SiO_2

Native copper and two minerals of copper. Azurite [$2\ CuCO_3 \cdot Cu(OH)_2$] is blue and malachite [$CuCO_3 \cdot Cu(OH)_2$] is green. *(C. D. Winters)*

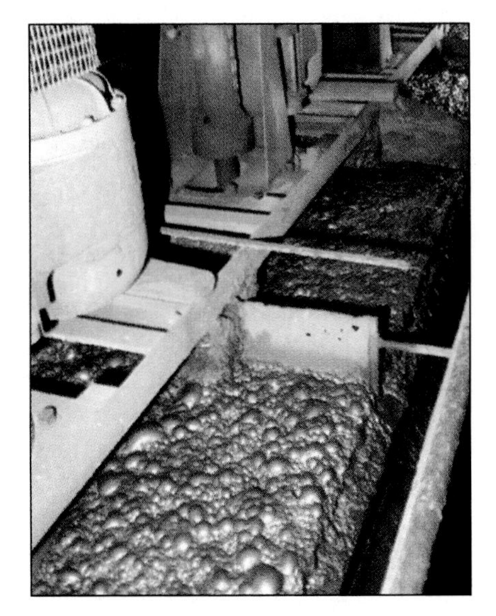

Figure 23.12 Enriching copper ore by the flotation process. The less dense particles of copper sulfide are trapped in soap bubbles and float on water. The denser gangue settles to the bottom.

react to form calcium silicate. Iron oxide reacts similarly with SiO_2, so the slag is a mixture of iron and calcium silicates:

$$FeO(s) + SiO_2(s) \longrightarrow FeSiO_3(\ell)$$
$$CaO(s) + SiO_2(s) \longrightarrow CaSiO_3(\ell)$$

At the same time, copper(II) sulfide, CuS, is reduced to copper(I) sulfide, Cu_2S, which melts and flows to the bottom of the furnace. The iron-containing slag is less dense than molten Cu_2S, so the Cu_2S and slag can be separated easily. The Cu_2S, called *copper matte,* is tapped off and run into another furnace (the converter), where it is "blown" with air. This converts the sulfur to SO_2 and produces impure copper metal, which is further refined in yet another furnace:

$$Cu_2S(\ell) + O_2(g) \longrightarrow 2\ Cu(\ell) + SO_2(g)$$

In the pyrometallurgical recovery of copper, each ton of copper produced is accompanied by about 1.5 tons of iron silicate slag and 2 tons of SO_2. These byproducts must be disposed of, not a simple task. One solution for SO_2 disposal is to convert it to sulfuric acid (see page 196).

Hydrometallurgy avoids some of the energy costs and pollution problems of pyrometallurgy. One method of copper recovery now in use in Arizona leaches, or dissolves, the impurities from the ore by treating the ore with a solution of copper(II) chloride and iron(III) chloride.

$$CuFeS_2(s) + 3\ CuCl_2(aq) \longrightarrow 4\ CuCl(s) + FeCl_2(aq) + 2\ S(s)$$
$$CuFeS_2(s) + 3\ FeCl_3(aq) \longrightarrow CuCl(s) + 4\ FeCl_2(aq) + 2\ S(s)$$

Copper is obtained in the form of copper(I) chloride by this reaction. To return copper to solution, sodium chloride is added because the soluble complex ion $[CuCl_2]^-$ is formed in the presence of excess chloride ion:

$$CuCl(s) + Cl^-(aq) \longrightarrow [CuCl_2]^-(aq)$$

The anion $[CuCl_2^-]$ is another example of a coordination complex; these species are discussed further in Section 23.3.

Copper(I) compounds are unstable with respect to Cu(0) and Cu(II), so the $[CuCl_2]^-$ ion disproportionates to the metal and $CuCl_2$; the latter is used to continue the leaching process:

$$2\ [CuCl_2]^-(aq) \longrightarrow Cu(s) + CuCl_2(aq) + 2\ Cl^-(aq)$$

Approximately 10% of the copper produced in the United States is obtained by using bacteria. Acidified water is sprayed onto copper-mining wastes that contain low levels of copper. As the water trickles down through the crushed rock, the bacterium *Thiobacillus ferrooxidans,* which thrives in the presence of acid and sulfur, breaks down the iron sulfides in the rock and converts iron(II) to iron(III). The iron(III) ion in turn oxidizes the sulfide ion of copper sulfide, leaving copper(II) ion in the water. Then the copper(II) ion is reduced to metallic copper by reaction with iron:

$$Cu^{2+}(aq) + Fe(s) \longrightarrow Cu(s) + Fe^{2+}(aq)$$

Whatever method is used to recover copper from its ores, the final step in the refining process is purification by electrolysis. Thin sheets of pure copper metal and slabs of impure copper are immersed in a solution containing $CuSO_4$ and H_2SO_4 (Figure 23.13). The pure copper sheets are the cathode of an electrolysis cell, and the impure slabs are the anode. Copper is oxidized to copper(II) ions at the anode; at the cathode, copper(II) ions in solution are reduced to pure copper.

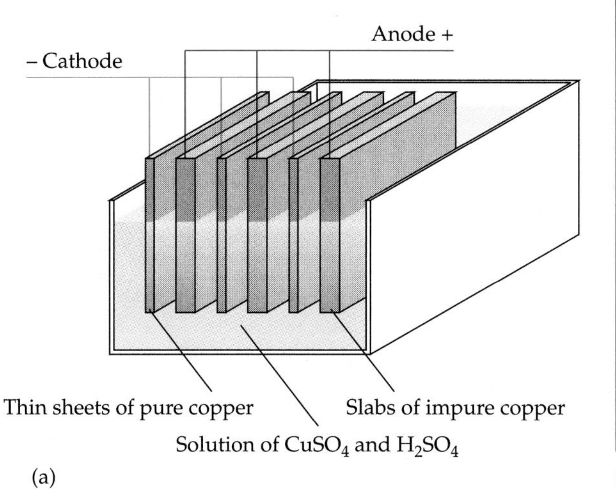

Figure 23.13 Electrolytic refining of copper. (a) Slabs of impure copper, called "blister copper," form the anode and pure copper is deposited at the cathode. (b) Photograph of electrolysis cells for refining copper.

23.3 COORDINATION COMPOUNDS

When a metal salt dissolves in water, water molecules cluster around the ions. We commonly indicate this by adding "(aq)" to the formula of these ions. The negative end of the polar water molecules is attracted to the positively charged metal ion, and the positive end of the water molecule is attracted to the anion (Figure 23.14). The energy of the ion-solvent interaction is partly responsible for the solution process (see page 650).

When crystallized from aqueous solution, compounds often retain one or more water molecules. Chemists indicate this by appending the appropriate number of water molecules to the formula for the salt. For example, hydrated nickel(II) chloride is written $NiCl_2 \cdot 6\ H_2O$. This formula indicates that six water molecules are present in the crystalline structure of the green solid.

Species in which a metal cation is associated with a number of anions were encountered earlier in the book. One example is the deep red ferricyanide ion, $[Fe(CN)_6]^{3-}$, which exists in solutions containing Fe^{3+} and CN^-, and in salts such as potassium ferricyanide, $K_3Fe(CN)_6$ (see Figures 14.10 and 23.15). In the metallurgy of copper, the solubility of copper(I) chloride in aqueous NaCl was due to formation of a complex ion $[CuCl_2]^-$. These are examples of a very large group of substances known as **coordination compounds** in which a metal atom or ion is associated with a group of neutral molecules or anions (see Figure 23.15).

Figure 23.14 Dissolving nickel chloride in water. When this transition metal salt dissolves in water, polar water molecules are attracted to the metal cation as well as to the halide anion. The models illustrate $[Ni(H_2O)_6]^{2+}$ and $NiCl_2(s)$. *(Photo, C. D. Winters; model, S. M. Young)*

Complexes and Ligands

Ammonia and nickel(II) chloride form a lilac-colored coordination compound with the formula $NiCl_2 \cdot 6\ NH_3$. This formula identifies the compound's composition, but it fails to give information about its structure. The usual method

(a)

(b)

(c)

Figure 23.15 Coordination compounds. (a) In aqueous solutions, transition metal ions are surrounded by a number of water molecules, usually six. Here the solution contains the cobalt(II) ion as the red cation $[Co(H_2O)_6{}^{2+}]$. (b) Crystals of $K_3[Fe(CN)_6]$, which contain the iron(III) ion as the deep red anion $[Fe(CN)_6]^{3-}$. (c) Crystals of the neutral compound $Cr(CH_3COCHCOCH_3)_3$. *(Photos, C. D. Winters; models, S. M. Young)*

of writing the formula for this or any other coordination compound places the metal atom or ion and the molecules or anions directly bonded to the metal within brackets. Thus, it is preferable to write the formula for the nickel(II)–ammonia compound as $[Ni(NH_3)_6]Cl_2$ to show that the nickel(II) ion and six ammonia molecules are a single structural unit, the cation $[Ni(NH_3)_6]^{2+}$. The two chloride ions are not part of this structural unit but are present as counterions to balance the charge in this ionic compound. The ion $[Ni(NH_3)_6]^{2+}$ is called a **coordination complex,** or **complex ion** (Figure 23.16). Coordination complexes can be cations, as in this example; anions (like $[Fe(CN)_6]^{3-}$); or neutral species (see Figure 23.15).

The molecules or ions attached to the metal are called **ligands,** from the Latin verb *ligare,* "to bind." Ligands have at least one atom having a lone pair of electrons, and it is this lone pair that gives a ligand the ability to bond to the metal. In the classic description of bonding in coordination complexes the lone pair of electrons on the ligand is shared with the metal ion. A ligand is there-

Sum of metal ion and ligand charges

Coordination complex

2+

H

N

Ni^{2+}

Coordinated metal ion

Ligand

©George V. Kelvin

$[Ni(NH_3)_6]^{2+}$

Figure 23.16 A coordination complex ion. In this complex, $[Ni(NH_3)_6]^{2+}$, the ligands are NH_3 molecules. They are bound to the Ni^{2+} ion by coordinate covalent bonds. Because the metal ion has a 2+ charge, and the ligands have no charge, the charge on the complex is 2+.

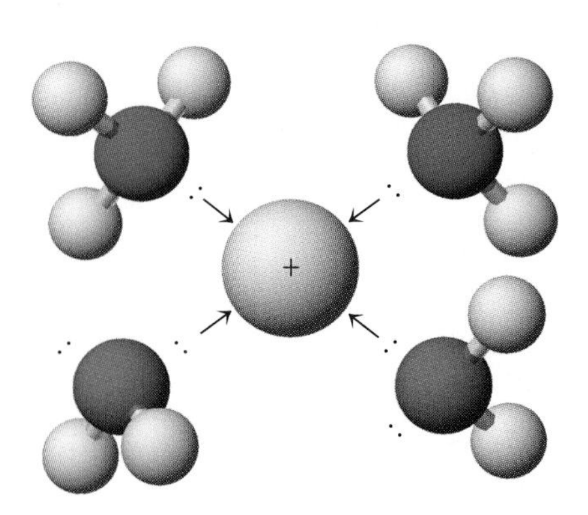

fore a Lewis base because it furnishes the electron pair, the metal ion is a Lewis acid (see Section 17.11), and the bond between ligand and metal is the result of a Lewis acid/Lewis base interaction.

The number of ligands attached to a metal is called the **coordination number.** Coordination complexes have a definite geometry or structure. The nickel ion in the complex $[Ni(NH_3)_6]^{2+}$ (see Figure 23.16) has a coordination number of 6, and the six ligands are in a regular octahedral geometry around the central metal ion.

A bond in which one of the atoms formally contributes the pair of electrons is a "coordinate covalent bond." The term "coordination complex" derives from this.

Ligands like H_2O and NH_3, with only a single Lewis base atom, are termed **monodentate.** The word "dentate" comes from the Latin word *dentis* meaning "tooth," so NH_3 is a "one-toothed" ligand. Other ligands, however, have more than one donor atom. When several atoms separate the Lewis base sites, it is possible for two or more atoms in the same ligand to bind to one metal atom. Ligands that contain two or more atoms attached to the metal are called **polydentate ligands.** Ethylenediamine (1,2-diaminoethane), $NH_2CH_2CH_2NH_2$, often abbreviated as *en*; the oxalate ion, $C_2O_4^{2-}$; the acetylacetonate ion (abbreviated $acac^-$); and phenanthroline (phen) are examples of **bidentate** ligands (Figures 23.17 and 23.18). Ethylenediaminetetraacetate ion ($EDTA^{4-}$) is an example of a **hexadentate** ligand (Figure 23.18).

Ligands that are capable of being bidentate or polydentate can act as monodentate ligands. These ligands can also "bridge" two or more metal atoms.

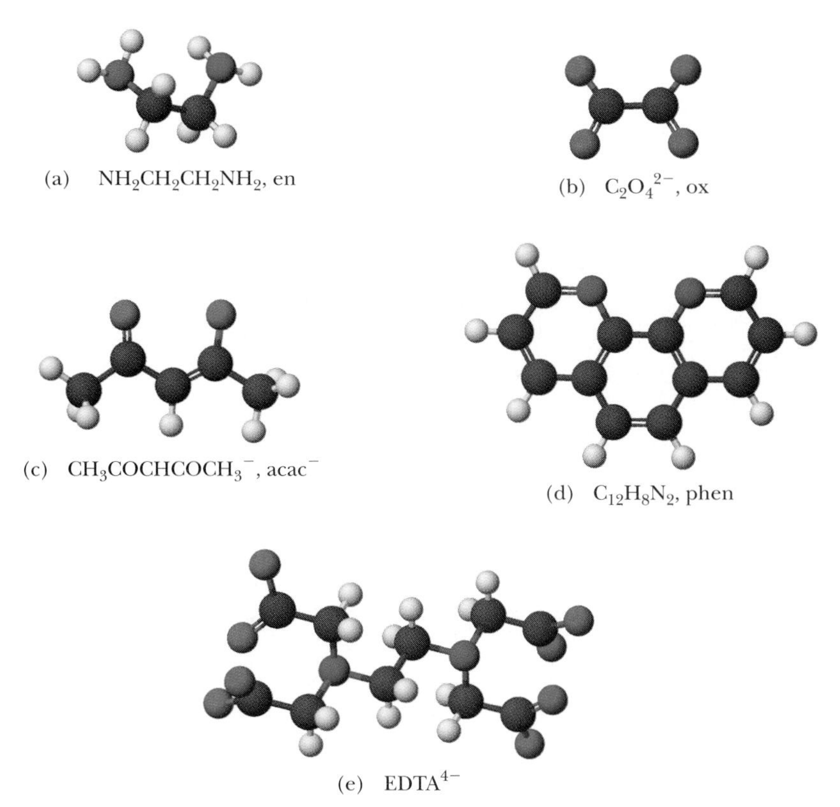

Figure 23.17 Polydentate ligands. (a) Bidentate ethylenediamine, $NH_2CH_2CH_2NH_2$ (*en*). (b) The bidentate oxalate ion, $C_2O_4^{2-}$ (*ox*). (c) The bidentate acetylacetonate anion, $CH_3COCHCOCH_3^-$ ($acac^-$). (d) The bidentate ligand phenanthroline. (d) The hexadentate ethylenediaminetetraacetate anion, $EDTA^{4-}$.

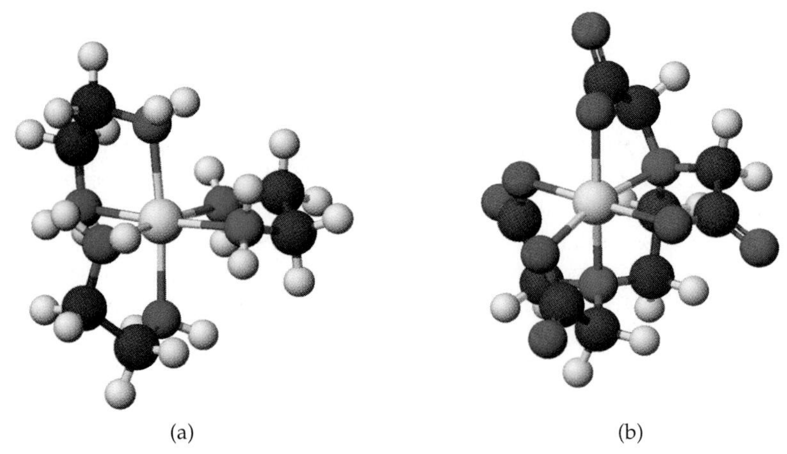

Figure 23.18 Coordination compounds with polydentate ligands. (a) The structure of $[Co(NH_2CH_2CH_2NH_2)_3]^{2+}$. (b) The structure of $[Co(EDTA)]^-$. (See also Figure 23.15c.)

A regular pentagon has angles of 110°, very close to the tetrahedral angle of 109.5° preferred by sp^3 hybridized carbon and nitrogen.

An important structural characteristic of a molecule or ion that allows it to act as a polydentate ligand is the ability to form five- and six-member rings. Ethylenediamine bonds to a metal, forming a five-member ring having two carbon atoms, two nitrogen atoms, and the metal ion (see Figure 23.18a). Structures containing five-member and six-member rings allow the atoms to achieve their normal bond angles and distances.

Polydentate ligands are also called **chelating ligands,** or chelates (pronounced "key-lates"). The name derives from the Greek word *chele* meaning "claw." Because two bonds must be broken to separate a chelated ligand from the metal, complexes having chelated ligands have greater stability. Chelated complexes are important in everyday life. One way to clean the rust out of water-cooled automobile engines and steam boilers, for example, is to add a solution of oxalic acid. Iron oxide dissolves in the presence of the acid to give a water-soluble iron oxalate complex:

$$Fe_2O_3(s) + 6\ H_2C_2O_4(aq) + 3\ H_2O(\ell) \longrightarrow 2\ [Fe(C_2O_4)_3]^{3-}(aq) + 6\ H_3O^+(aq)$$

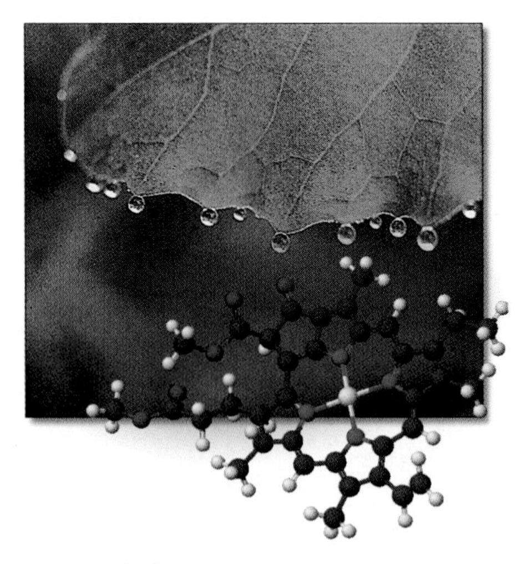

Figure 23.19 The structure of chlorophyll. Here an Mg^{2+} ion is coordinated to four N atoms in a square-planar geometry. *(Photo, C. D. Winters; model, S. M. Young)*

$EDTA^{4-}$ is an excellent chelating ligand because it encapsulates and firmly binds metal ions. It is often added to commercial salad dressing to remove traces of free metal ions from solution because these metal ions can otherwise act as catalysts for the oxidation of the oils in the dressing. Without $EDTA^{4-}$, the dressing would quickly become rancid. Another use is in bathroom cleansers, in which $EDTA^{4-}$ removes deposits of $CaCO_3$ and $MgCO_3$ left by hard water. The $EDTA^{4-}$ coordinates to Ca^{2+} or Mg^{2+}, creating a soluble complex ion. Figure 23.18b shows the structure of $[Co(EDTA)]^-$. Note how the ligand is oriented to allow carbon, nitrogen, oxygen, and cobalt atoms to achieve desired bond angles.

Complexes with polydentate ligands have a particularly important role in biochemistry. A chelating ligand encloses Mg^{2+} in plant chlorophyll (Figure 23.19). Similar ligands bonded to iron are part of the structures of the oxygen-carrying proteins hemoglobin and myoglobin (see page 324).

See *A Closer Look: Coordination Compounds in Biochemistry.*

It is useful to be able to predict the formula of a coordination complex, given the metal ion and ligands, and to derive the oxidation number of the coordinated metal ion. The following examples explore these questions.

A CLOSER LOOK Coordination Compounds in Biochemistry

The field of bioinorganic chemistry, the union of biochemistry and inorganic chemistry, has come into prominence in the last decade because so many biochemical reactions are dependent on metal centers. Specifically, many reactions involve coordination compounds.

You have already encountered coordination compounds such as the cancer chemotherapy agent *cis*-platin (page 11) and the oxygen-carrying iron complexes myoglobin (page 324) and heme (page 85). Finally, the center for biological methane formation was recently found to be a nickel(I) coordination compound (page 563).

In medicine *cis*-platin has become the treatment of choice for certain types of cancers. Another example of the medical use of a coordination compound is auranofin, a rheumatoid arthritis drug in which gold(I) is coordinated by sulfur and phosphorus (Figure A). Cardiolyte, a complex of the synthetic element technetium, is used for heart imaging (part *b*).

Among the most interesting biochemical coordination compounds are the iron-sulfur proteins (Figure B). In these complexes iron(II) and iron(III) ions are bound to sulfide ions or the S atom of amino acids such as cysteine. The structures are fascinating because they vary from a simple tetrahedron of an iron ion and four S-ligands to two iron ions bridged by sulfurs to give clusters of three or four iron ions bridged by sulfur. These iron–sulfur proteins are chiefly implicated in electron transfer reactions. One example is the enzyme nitrogenase, which assists in converting atmospheric nitrogen and a source of hydrogen ions to ammonia:

$$N_2(g) + 8\ H^+(aq) + 8\ e^- \longrightarrow 2\ NH_3(aq) + H_2(g)$$

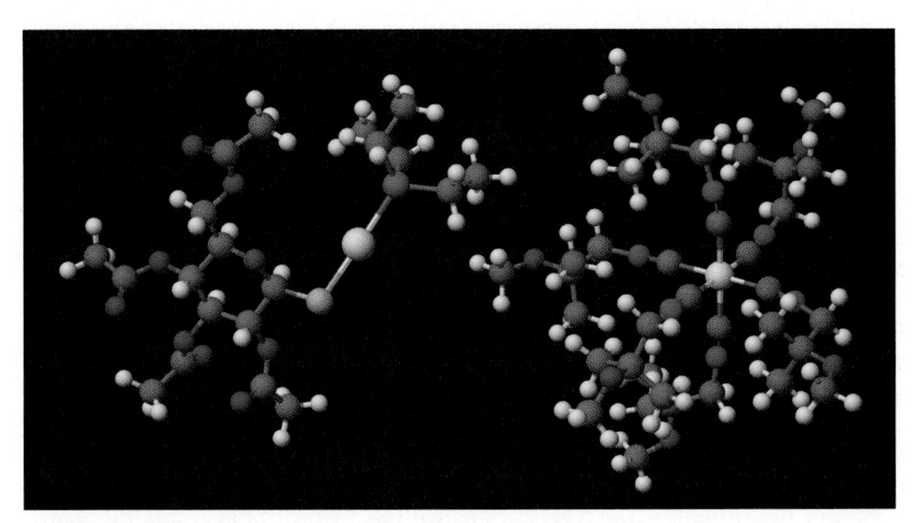

Figure A A gold-containing drug for rheumatoid arthritis (auranofin, *left*) and a technetium-containing coordination compound used in heart-imaging studies (cardiolyte, *right*). The metal atoms are silver-colored in each case. In the arthritis drug, a gold(I) ion is attached to a sulfur atom (yellow) and a phosphorus atom (purple). In the technetium-based molecule, a Tc ion is surrounded by six ligands. The coordinating molecules are isonitriles, :C≡N—R.

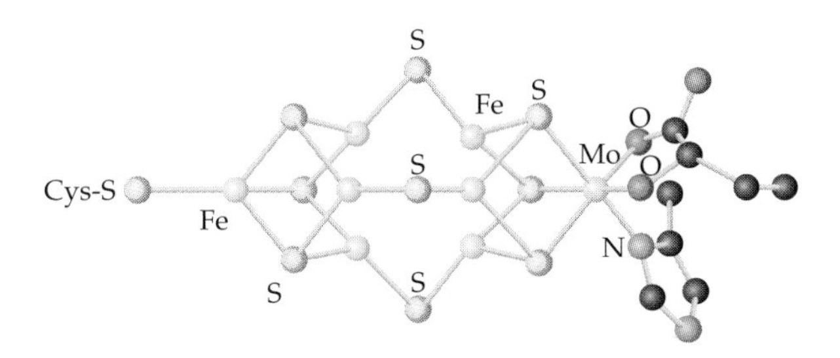

Figure B Iron–sulfur proteins commonly function as electron transfer agents. Shown here is a schematic structure of the cofactor cluster from nitrogenase. Notice that this also includes a coordinated molybdenum ion. (Redrawn with permission, R. H. Holm, et al., *Science*, Vol. 277, p. 655, 1997.)

Example 23.1 Coordination Complexes

Write the formula of the coordination complex in which the metal ion is coordinated to six Lewis base sites.

1. One Ni^{2+} ion is bound to two water molecules and two bidentate oxalate ions.
2. A Co^{3+} ion is bound to one Cl^- ion, one ammonia molecule, and two bidentate ethylenediamine molecules.

Solution

1. In the nickel(II) complex, the ligands consist of two neutral molecules and two with 2− charges. When these are combined with Ni^{2+}, the net charge is 2−.

$$Ni^{2+} + 2\ H_2O + 2\ C_2O_4^{2-} \longrightarrow [Ni(C_2O_4)_2(H_2O)_2]^{2-}$$

2. In the cobalt(III) complex there are two neutral en molecules, one neutral NH_3 molecule, and one Cl^- ion. When these are combined with Co^{3+}, the net charge is 2+.

$$Co^{3+} + 2\ NH_2CH_2CH_2NH_2 + NH_3 + Cl^- \longrightarrow [Co(NH_2CH_2CH_2NH_2)_2(NH_3)Cl]^{2+}$$

Example 23.2 Formulas of Coordination Compounds

Give the oxidation number of the metal ion in each of the following complexes:

1. $[Co(en)_2(NO_2)_2]^+$
2. $Pt(NH_3)_2(C_2O_4)$

Solution

1. In this cobalt complex there are two neutral, bidentate ethylenediamine molecules and two nitrite ions, NO_2^-. Because the overall charge on the ion is 1+, the cobalt ion must be Co^{3+}.

$$Co^{3+} + 2\ NH_2CH_2CH_2NH_2 + 2\ NO_2^- \longrightarrow [Co(NH_2CH_2CH_2NH_2)_2(NO_2)_2]^+$$

2. Platinum is coordinated to two neutral ammonia molecules and one bidentate oxalate ion. Thus, platinum is in the form of the Pt^{2+} ion.

$$Pt^{2+} + 2\ NH_3 + C_2O_4^{2-} \longrightarrow Pt(NH_3)_2(C_2O_4)$$

Exercise 23.1 Formulas of Coordination Complexes

1. Give the oxidation number of platinum in $Pt(NH_3)_2Cl_2$.
2. What is the formula of a complex assembled from one Co^{3+} ion, five ammonia molecules, and one monodentate carbonate ion?

Naming Coordination Compounds

Just as there are rules for naming simple inorganic and organic compounds, coordination compounds are named according to an established system. For example, the following compounds are named according to the rules outlined in the following list.

Compound	Systematic Name
$[Ni(H_2O)_6]SO_4$	Hexaaquanickel(II) sulfate
$[Cr(en)_2(CN)_2]Cl$	Dicyanobis(ethylenediamine)chromium(III) chloride
$K[Pt(NH_3)Cl_3]$	Potassium amminetrichloroplatinate(II)

As you read through the rules, notice how they apply to the examples in the preceding table:

1. In naming a coordination compound that is a salt, name the cation first and then the anion. (This is how all salts are commonly named.)
2. When giving the name of the complex ion or molecule, name the ligands first, in alphabetical order, followed by the name of the metal.
   - (a) If a ligand is an anion whose name ends in *-ite* or *-ate,* the final *e* is changed to *o* (as in sulfate ⟶ sulfato or nitrite ⟶ nitrito).
   - (b) If the ligand is an anion whose name ends in *-ide,* the ending is changed to *o* (as in chloride ⟶ chloro or cyanide ⟶ cyano).
   - (c) If the ligand is a neutral molecule, its common name is usually used. The important exceptions to this rule are water, which is called *aqua,* ammonia, which is called *ammine,* and CO, called *carbonyl.*
   - (d) When there is more than one of a particular monodentate ligand with a simple name, the number of ligands is designated by the appropriate prefix: *di, tri, tetra, penta,* or *hexa.* If the ligand name is complicated (whether monodentate or bidentate), the prefix changes to *bis, tris, tetrakis, pentakis,* or *hexakis,* followed by the ligand name in parentheses.
3. If the complex ion is an anion, the suffix *-ate* is added to the metal name.
4. Following the name of the metal, the oxidation number of the metal is given in Roman numerals.

Complexes can be considerably more complicated than those described in this chapter; then, even more rules of nomenclature must be applied. The brief rules just outlined, however, are sufficient for most complexes.

Example 23.3 Naming Coordination Compounds

Name the following compounds:

1. $[Cu(NH_3)_4]SO_4$
2. $K_2[CoCl_4]$
3. $Co(phen)_2Cl_2$
4. $[Co(en)_2(H_2O)Cl]Cl_2$

Solution

1. The sulfate ion has a 2− charge, so the complex ion has a 2+ charge (i.e., $[Cu(NH_3)_4]^{2+}$). Because NH_3 is a neutral molecule, the copper ion is Cu^{2+}. The compound's name is therefore tetraamminecopper(II) sulfate.
2. Two K^+ ions are part of the formula, so the complex ion has a 2− charge ($[CoCl_4]^{2-}$). Because four Cl^- ions occur in the complex ion, the cobalt center is Co^{2+}. Thus, the name of the compound is potassium tetrachlorocobaltate(II).

3. This is a neutral compound. Because two Cl^- ions and two neutral phen (phenanthroline) ligands are bonded to a cobalt ion, the metal ion must be Co^{2+}. The compound name is dichlorobis(phenanthroline)cobalt(II).
4. Here the complex ion has a 2+ charge because it is associated with two uncoordinated Cl^- ions. The cobalt ion must be Co^{3+} because it is bonded to two neutral en (ethylenediamine) ligands, one neutral water, and one Cl^-. The name is aquachlorobis(ethylenediamine)cobalt(III) chloride.

Exercise 23.2 Naming Coordination Compounds

Name the compounds $[Ru(phen)_2(H_2O)CN]Cl$ and $Pt(NH_3)_2Cl_2$.

23.4 STRUCTURES OF COORDINATION COMPOUNDS AND ISOMERS

Common Geometries

The geometry of a coordination complex is defined by the arrangement of donor atoms of the ligands around the central metal ion. The metal ion in a coordination compound may have a coordination number between 2 and 12. Only complexes with coordination numbers of two, four, and six are common, however, so we will concentrate on species such as $[ML_2]^{n\pm}$, $[ML_4]^{n\pm}$, and $[ML_6]^{n\pm}$, where M is the metal ion and L is a monodentate ligand.

Complexes with the General Formula $[ML_2]^{n\pm}$

This stoichiometry is often encountered with metal ions with a 1+ charge. One example, the copper complex $[CuCl_2]^-$, was mentioned earlier in this chapter. Another example is the complex ion that forms when AgCl(s) dissolves in aqueous ammonia (see page 754 and Section 19.9):

$$AgCl(s) + 2\ NH_3(aq) \longrightarrow [Ag(NH_3)_2]^+(aq) + Cl^-(aq)$$

In all cases, complexes of this stoichiometry have a linear geometry, that is, the two ligands are on opposite sides of the metal, with an L—M—L bond angle of 180°.

Complexes with the General Formula $[ML_4]^{n\pm}$

The VSEPR theory of Chapter 9 might lead us to expect tetrahedral structures for $[ML_4]^{n\pm}$ complexes. Indeed, this is observed for such complexes as $TiCl_4$, $[CoCl_4]^{2-}$, $[NiCl_4]^{2-}$, and $[Zn(NH_3)_4]^{2+}$. A large number of $[ML_4]^{n\pm}$ complexes, however, are *square-planar;* the four ligands lie in a plane surrounding the metal atom in the center with L—M—L bond angles of 90°. Square-planar complexes are particularly common with metal ions that have the electron configuration [noble gas]$(n-1)d^8$. This includes many complexes of Pt^{2+} and Pd^{2+}. It is interesting, however, that nickel(II) complexes can be either tetrahedral (as in $[NiCl_4]^{2-}$) or square-planar (as in $[Ni(CN)_4]^{2-}$):

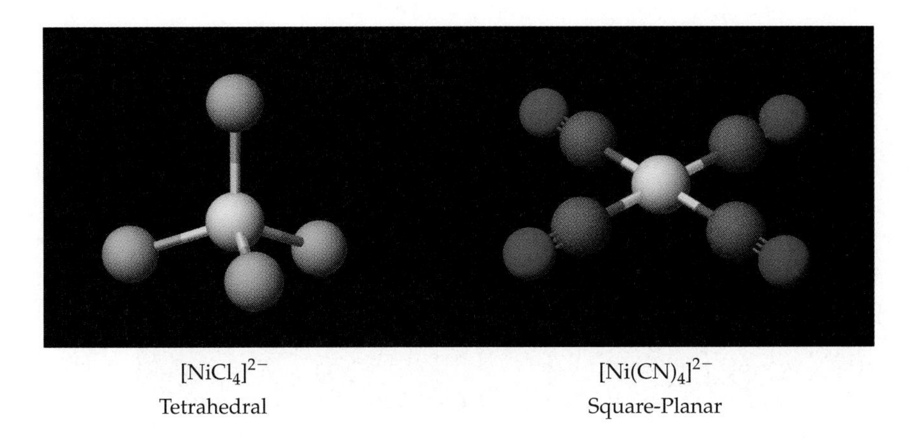

$[NiCl_4]^{2-}$ Tetrahedral

$[Ni(CN)_4]^{2-}$ Square-Planar

Complexes with the General Formula $[ML_6]^{n\pm}$

With *very* rare exceptions, the six ligands in an $[ML_6]^{n\pm}$ complex are arranged at the corners of an octahedron, with the metal at the center. The complexes have octahedral geometry as pictured in Figures 23.15, 23.16, and 23.18.

The examples given here were illustrated using complexes with monodentate ligands. The structural principles apply equally well to complexes with polydentate ligands, however. The structural geometry is defined by the metal and the donor atoms attached to it.

Isomerism

Isomerism is one of the most interesting aspects of molecular structure. Molecules that have the same molecular formula but different bonding arrangements of atoms are called **structural isomers.** We already encountered structural isomers in organic chemistry (see Section 11.2). Compounds like butane and 2-methylpropane are structural isomers (page 487). In a second type of isomers, **stereoisomerism,** the atom-to-atom bonding sequence is the same, but the atoms differ in their arrangement in space. Two types of stereoisomerism occur. One is **geometric isomerism,** in which the atoms making up a molecule are arranged in different geometrical relationships (see Section 11.3). *Cis*-2-butene and *trans*-2-butene are geometric isomers (page 457). The second type of stereoisomerism is **optical isomerism,** which arises when a molecule and its mirror image do not superimpose. Both geometric and optical isomers are encountered in coordination chemistry.

Geometric Isomerism

Geometric isomers result when the atoms bonded *directly* to the metal have a different spatial arrangement (different bond angles). The simplest example of geometric isomerism is *cis–trans* isomerism, which occurs in both square-planar and octahedral complexes. We can illustrate *cis–trans* isomerism with the square-planar complex $Pt(NH_3)_2Cl_2$ (Figure 23.20). This complex is formed from Pt^{2+}, two NH_3 molecules, and two Cl^- ions. The two Cl^- ions, for example, can be either adjacent to each other (*cis*) or on opposite sides of the complex (*trans*). The *cis* isomer is effective in the treatment of testicular, ovarian, bladder, and osteogenic sarcoma cancers, but the *trans* isomer has no effect on these diseases.

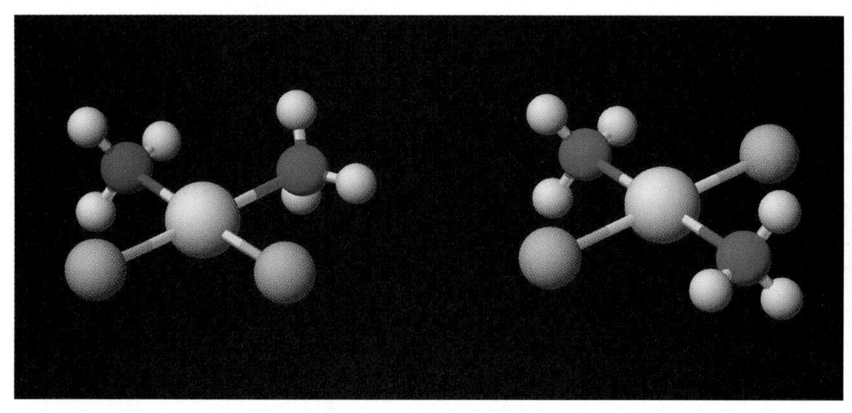

(a) *cis* isomer (b) *trans* isomer

Figure 23.20 ***Cis* and *trans* isomers of diamminedichloroplatinum(II).** In the *cis* isomer the Cl—Pt—Cl angle is 90° (the Cl^- ligands are next to one another), whereas in the *trans* isomer the angle is 180° (the Cl^- ligands are on opposite sides of the metal ion).

Cis–trans isomerism in an octahedral complex is illustrated with $[Co(NH_2CH_2CH_2NH_2)_2Cl_2]^+$, an octahedral complex with two bidentate ethylenediamine ligands and two chloride ligands. In this complex, the two Cl^- ions occupy positions that are either adjacent (*cis* isomer) or opposite (*trans* isomer) (Figure 23.21). It is interesting that these isomers have different colors: the *cis* isomer is purple, and the *trans* isomer is green.

Cis–trans isomerism is *not possible* for tetrahedral complexes. All L—M—L angles in tetrahedron geometry are 109.5°, and all positions are equivalent in this three-dimensional structure.

Another example of geometric isomerism, called *mer–fac* isomerism, occurs for octahedral complexes with the general formula MX_3Y_3. In a *fac* isomer, three similar ligands lie at the corners of a triangular face of the octahedron (*fac* = *facial*), whereas in the *mer* isomer, the ligands follow a meridian (*mer* = *meridional*). *Fac* and *mer* isomers of $Cr(NH_3)_3Cl_3$ are shown in Figure 23.22.

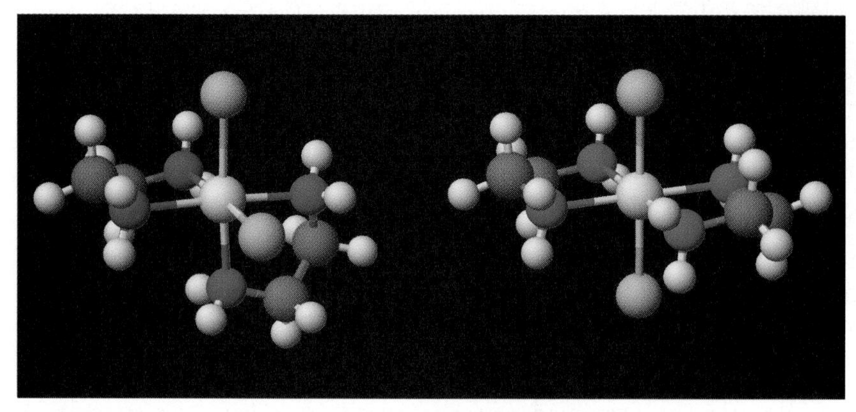

(a) *cis* isomer (b) *trans* isomer

Figure 23.21 ***Cis* and *trans* isomers of $[Co(en)_2Cl_2]^+$.** In the *cis* isomer the Cl—Co—Cl angle is 90° (the Cl^- ligands are next to one another), whereas in the *trans* isomer the angle is 180° (the Cl^- ligands are on opposite sides of the metal ion).

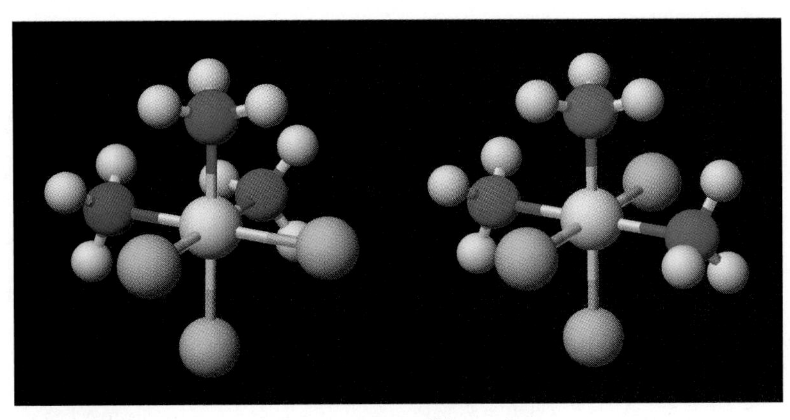

(a) *fac* isomer (b) *mer* isomer

Figure 23.22 ***Fac* and *mer* isomers of $Cr(NH_3)_3Cl_3$.** (a) In the *fac* isomer the three NH_3 ligands (and the three Cl^- ligands) are arranged on one octahedral face. (b) In the *mer* isomer, the three NH_3 ligands (and the three Cl^- ligands) are arranged around the meridian on the molecule.

Optical Isomerism

Everyone has, at one time or another, tried to put a left shoe on a right foot, or a left-handed glove on a right hand. It doesn't work very well. Even though our two hands and two feet appear generally similar, there is a very important distinction between them. Left hands and feet are mirror images of right hands and feet, and, most importantly, these mirror images cannot be superimposed.

Certain molecules have the same characteristic as gloves and hands: A given structure and its mirror image cannot be superimposed. Molecules (and other objects) that have nonsuperimposable mirror images are termed **chiral,** and objects with superimposable mirror images are **achiral.** Nonsuperimposable molecules are known as **enantiomers.** Many common objects have this property. For example, some seashells are chiral, and wood screws and machine bolts are also chiral, distinguished by left-handed or right-handed threads (Figure 23.23).

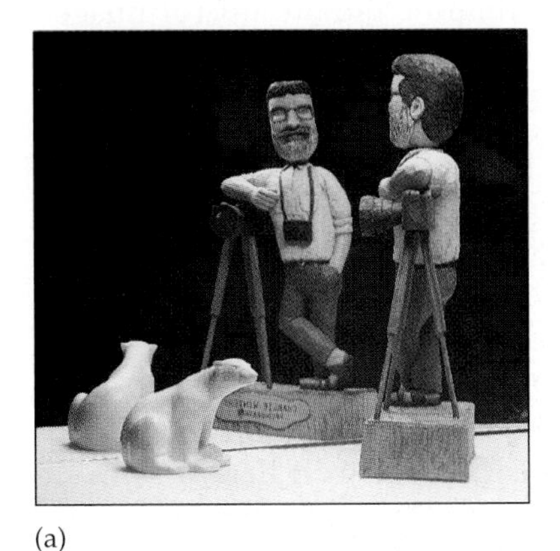

(a)

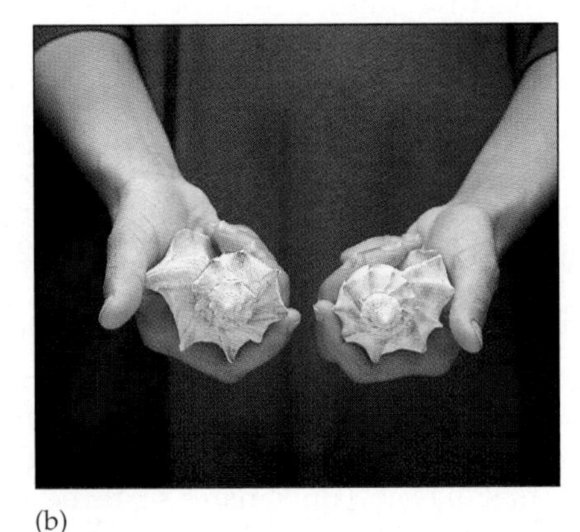

(b)

Figure 23.23 Mirror images. (a) Mirror images of two wood carvings. The mirror image cannot be superimposed on the actual statue. The man's right arm is resting on the camera in the mirror image, but in the actual statue the man's left arm is resting on the camera. (b) Left- and right-handed sea shells. If you cup a right-handed shell in your right hand, with your thumb pointing from the narrow end to the wide end, the opening will be on the right. *(C. D. Winters)*

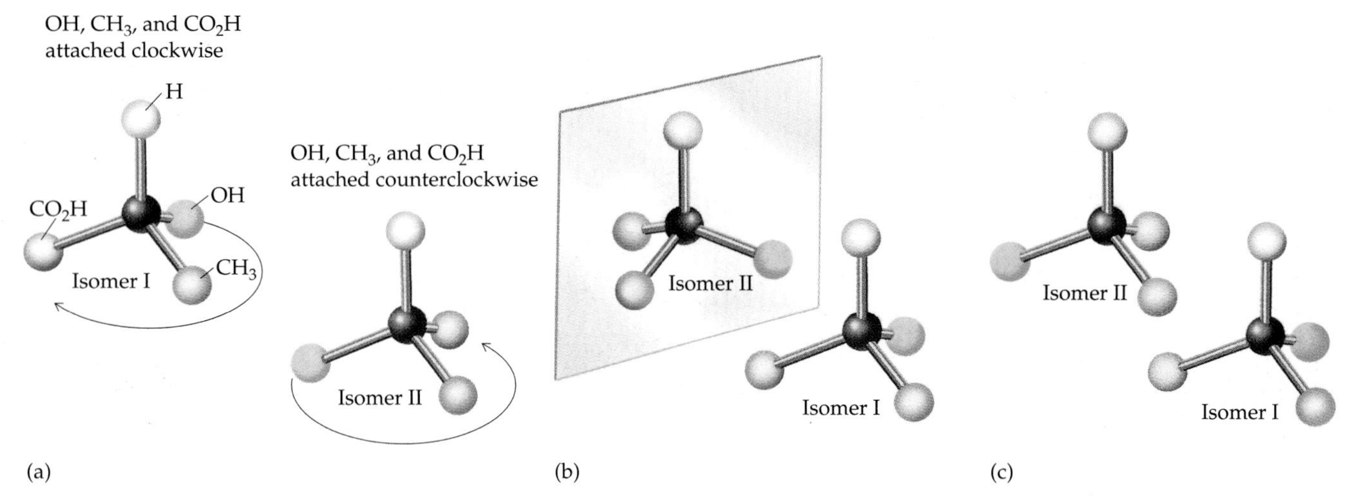

Figure 23.24 The two enantiomeric forms of lactic acid, $CH_3CH(OH)(CO_2H)$. (a) In Isomer I the groups —OH, —CH_3, and —CO_2H are attached in a clockwise manner. In Isomer II the groups —OH, —CH_3, and —CO_2H are attached in a counterclockwise manner. (b) Isomer I is placed in front of a mirror, and its mirror image is isomer II. (c) The isomers are nonsuperimposable.

Enantiomers have the same stoichiometry *and* the same atom-to-atom bonding sequences, but they differ in the details of the arrangement of atoms in space. The most common type of chirality that occurs in chemistry arises if a carbon atom is bonded to four different groups. An example of such a compound is lactic acid, $CH_3CH(OH)CO_2H$. The enantiomers, or mirror images, of lactic acid are shown in Figure 23.24.

Although we have chosen to describe this phenomenon in a discussion of metal complexes, it is important to realize that this is a particularly important aspect in structures of organic and biochemical molecules.

To see that the enantiomers of lactic acid differ, imagine that you are looking down the H—C bond from the "top" of the molecule. In one of the enantiomers the three other groups (CH_3, OH, and CO_2H) are arranged in a clockwise order, whereas in the second enantiomer these groups appear in counterclockwise order. It is impossible to superimpose the two mirror images; if you rotate the molecule, two groups of the four groups never line up.

The two enantiomers of a compound have the same physical properties, such as melting point, boiling point, density, and solubility in common solvents. They differ in a very significant way, however: When a beam of plane-polarized light is passed through a solution of a pure enantiomer, the plane of polarization is twisted in one direction (Figure 23.25). The two enantiomers rotate polarized light to an equal extent, *but in opposite directions.* Because this effect involves light, chiral compounds are called **optical isomers,** and chiral compounds are said to be **optically active.**

A mixture of equal quantities of two enantiomers is called a **racemic** mixture. Racemic mixtures do not rotate polarized light because the effects of the two optical isomers cancel.

The possibility for chirality arises for a number of coordination compounds based on octahedral geometry, only one of which is described here. In this situation a metal atom coordinates to three bidentate ligands, as for instance in $[Co(en)_3]^{3+}$, which has three bidentate ethylenediamine ligands coordinated to a cobalt(III) ion. The structure of this complex is portrayed schematically in Figure 23.26. Because of the way that the five-member chelate rings are arranged, mirror images of this molecule do not superimpose. Solutions of each optical isomer rotate polarized light in opposite directions.

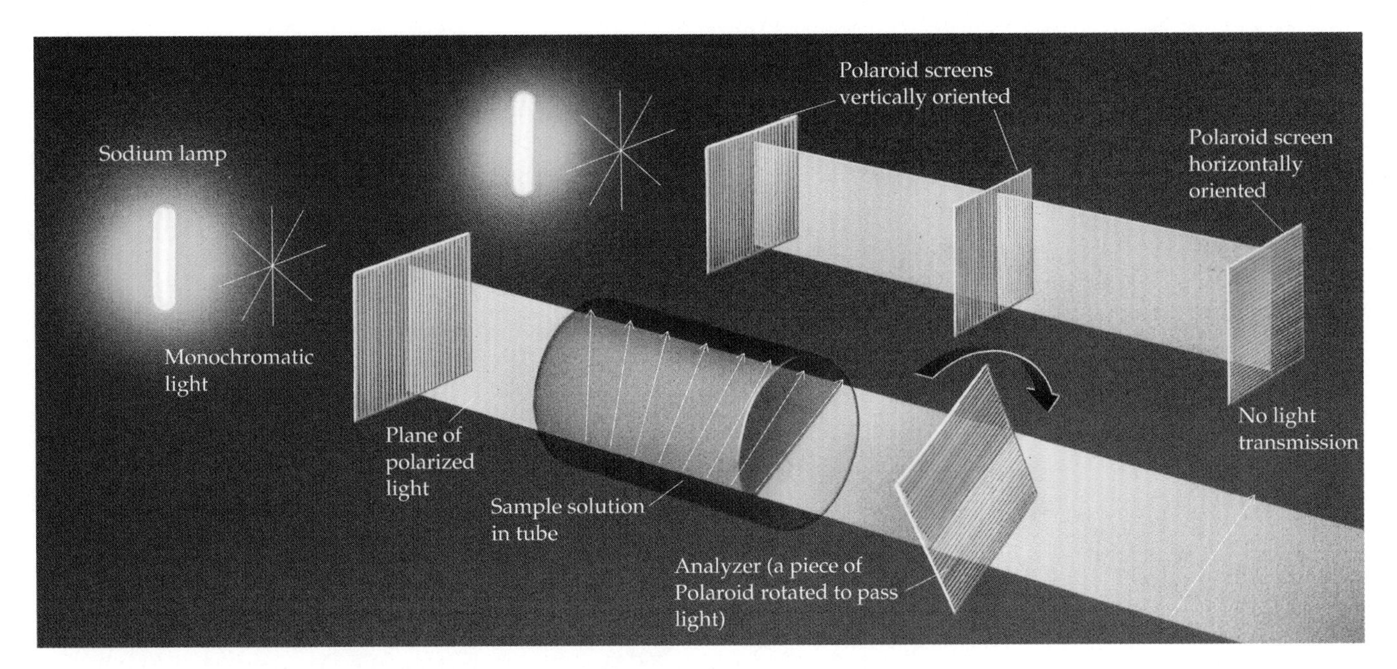

Figure 23.25 Rotation of plane-polarized light by an optical isomer. (*Top*) Monochromatic light (light of only one wavelength) is produced by a sodium lamp. After it passes through a polarizing filter, the light is vibrating in only one direction—it is polarized. Polarized light will pass through a second polarizing filter if this filter is parallel to the first filter, but not if the second filter is perpendicular. (*Bottom*) A solution of an optical isomer placed between the first and second polarizing filters causes rotation of the plane of polarized light. The angle of rotation can be determined by rotating the second filter until maximum light transmission occurs. The magnitude and direction of rotation are unique physical properties of the optical isomer being tested.

Square-planar complexes are incapable of optical isomerism that is based at the metal; mirror images are always superimposable. Although optical isomers of tetrahedral complexes are possible, no examples of stable complexes with a metal bonded tetrahedrally to four different monodentate ligands are known.

Other complexes in which optical activity can occur include *cis*-$[M(bidentate)_2X_2]^{n+}$ and *fac*-$[ML_3XYZ]^{n+}$.

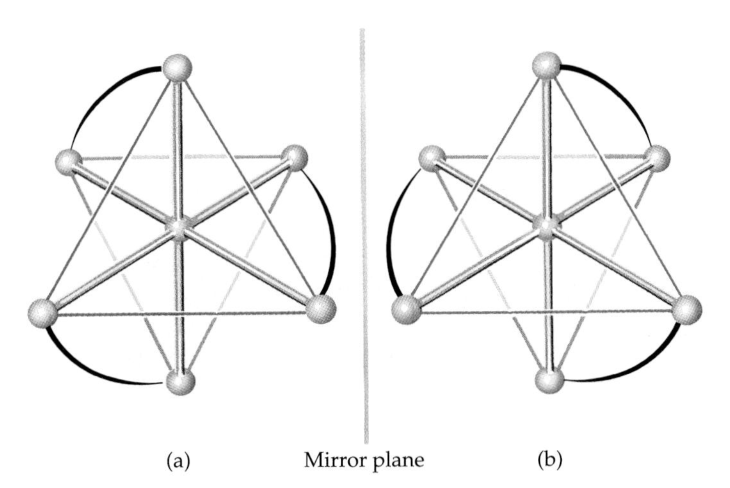

Figure 23.26 Optical isomerism in complexes of the type $M(bidentate)_3$. The three chelate rings are arranged so that the complex resembles a three-bladed propeller. (a) One of the complexes twists clockwise and (b) the other twists counterclockwise. The mirror images cannot be superimposed.

Example 23.4 Isomerism

What type of isomerism (geometric or optical isomerism or both) is exhibited by each of the following compounds?

1. $[Co(NH_3)_4Cl_2]^+$
2. $[Ru(phen)_3]Cl_2$
3. $[Pt(CN)_2Cl_2]^{2-}$

Solution

1. This Co^{3+} complex has an octahedral structure, and two geometric isomers are possible. One isomer has two Cl^- ions in *cis* positions; in the other isomer the Cl^- ligands are *trans*.

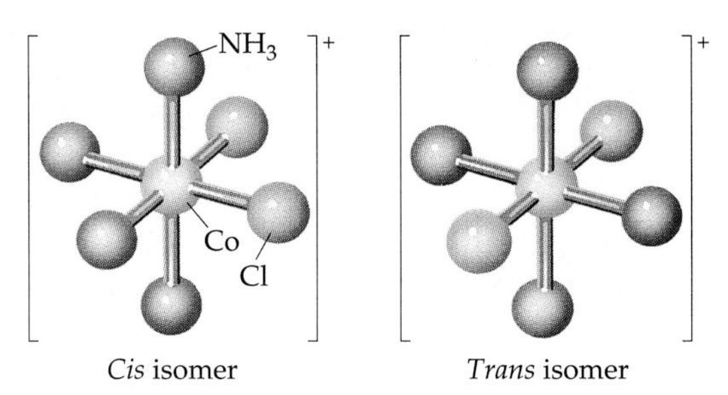

Cis isomer *Trans* isomer

2. In $[Ru(phen)_3]^{2+}$, the Ru^{2+} ion is surrounded by three bidentate phen ligands. Two optical isomers are possible for this complex. The curved lines represent the phenanthroline ligand.

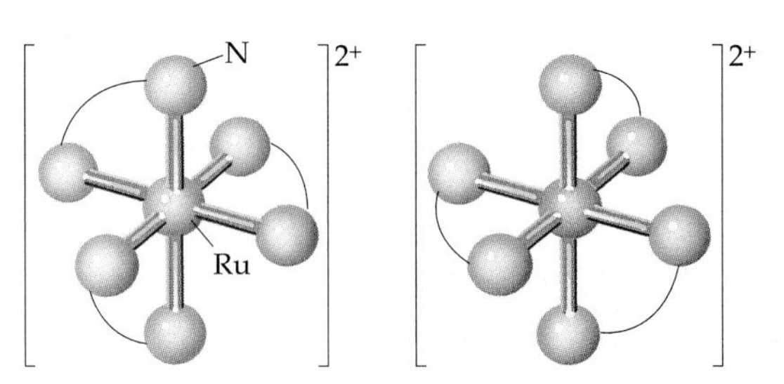

3. This platinum(II) complex has a square-planar geometry. *Cis* and *trans* isomers are possible.

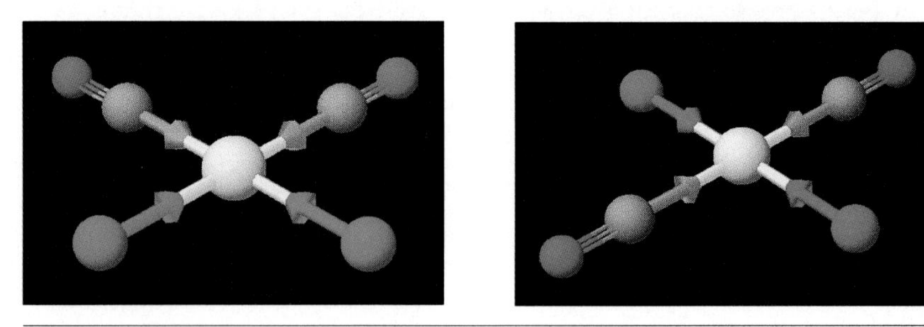

Exercise 23.3 Identifying Isomers

What type of isomers are possible for $[Co(en)_2(CN)_2]Br$?

23.5 BONDING IN COORDINATION COMPOUNDS

Metal–ligand bonding in a coordination complex was described earlier as the interaction between a Lewis acid (the metal ion) and a Lewis base (the ligand). This valence bond picture represents the ligand–metal bond as covalent, with an electron pair shared between the metal and the ligand donor atom. This model is frequently used, but it is not adequate to explain properties of complexes such as color (see Figures 23.6 and 23.15) and magnetism. As a result, other bonding models have largely superseded the valence bond model. Currently, the bonding in coordination complexes is usually described by either **molecular orbital theory** or by **crystal field theory.**

Molecular orbital theory and crystal field theory approach metal–ligand bonding using metal *d* orbitals and ligand lone pair orbitals. As ligands approach the metal to form bonds, two effects occur: (a) the metal and ligand orbitals overlap, and (b) the electrons of the metal repulse the electrons of the ligand. Molecular orbital theory takes both effects into account, whereas the crystal field model focuses on metal–ligand electron repulsion. The molecular orbital model assumes that metal and ligand bond through the molecular orbitals formed by atomic orbital overlap between metal and ligand. In contrast, the crystal field model assumes that the positive metal ion and negative ligand lone pair are attracted *electrostatically,* that is, the bond arises from the attractive force between a positively charged metal ion and a negative ion or the negative end of a polar molecule. Both the molecular orbital and crystal field models ultimately produce the same *qualitative* results regarding color and magnetic behavior. We will focus on the crystal field approach.

Crystal field theory was developed to explain the properties of metal ions in crystalline ionic solids. Solid ionic compounds contain metal ions surrounded by a group of negative ions. The negative ions create a "field" of charge in a specific geometric arrangement around the metal.

d-Orbital Energies in Coordination Compounds

To understand crystal field theory, let us look at the *d* orbitals in more detail. Our particular interest is in the orientation of the *d* orbitals relative to the positions of ligands in a metal complex. The five *d* orbitals (Figure 23.27) can be grouped into two sets: the $d_{x^2-y^2}$ and d_{z^2} orbitals in one set and the d_{xy}, d_{xz}, and d_{yz} orbitals in another. The $d_{x^2-y^2}$ and d_{z^2} orbitals are directed *along the x-, y-, and z-axes,* whereas the orbitals of the second group are aligned *between these axes.* The *d* orbitals are divided into these two sets because we have chosen to assign the ligands in square-planar and octahedral complexes to lie along the *x*-, *y*-, and *z*-axes.

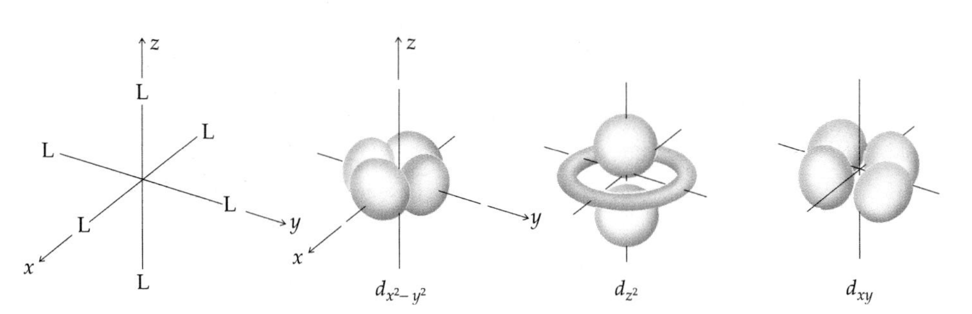

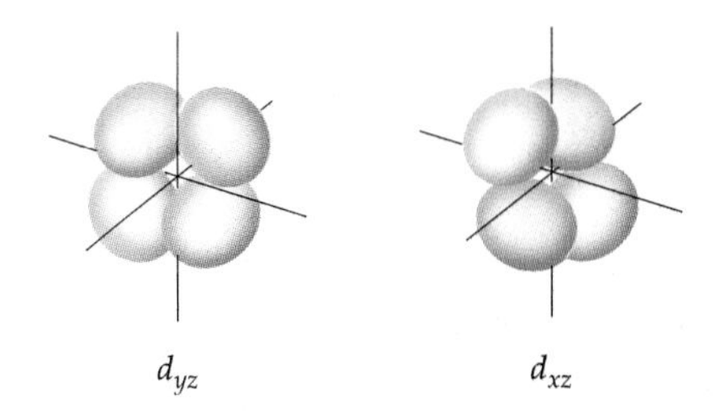

Figure 23.27 The *d* orbitals. The five *d* orbitals and their spatial relation to ligands on the *x*-, *y*-, and *z*-axes.

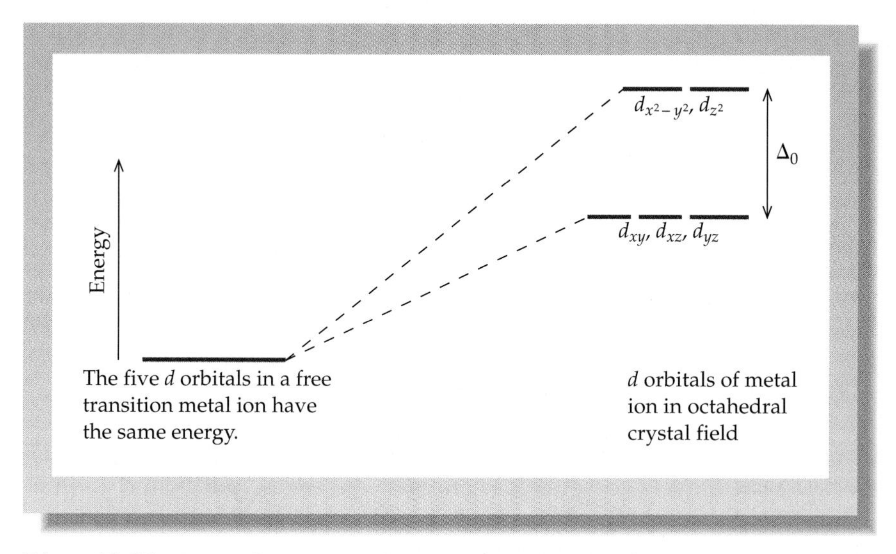

Figure 23.28 Crystal field splitting in an octahedral complex. The *d* orbital energy changes as six ligands approach the metal ion along the *x*-, *y*-, and *z*-axes. The energy difference between the two sets of *d* orbitals is labeled Δ_0.

In an isolated atom or ion, the *d* orbitals are degenerate, that is, they have the same energy. For a metal atom or ion in a coordination complex, however, the *d* orbitals have different energies. According to the crystal field model, repulsion between *d* electrons and the electron pairs of the ligands destabilizes the *d* orbitals (causes their energy to become higher). Electrons in the various *d* orbitals are not affected equally, however, because of their orientation in space relative to the position of the ligand lone pairs. In an octahedral complex, the *d* orbitals have different energies (Figure 23.28). Electrons in the $d_{x^2-y^2}$ and d_{z^2} orbitals experience a larger repulsion because these orbitals point directly at the incoming ligand electron pairs. A smaller repulsive effect is experienced by electrons in the d_{xy}, d_{xz}, and d_{yz} orbitals. The difference in degree of repulsion means an energy difference exists between the two sets of orbitals. This difference, called the *crystal field splitting* and given the symbol Δ_0, is a function of the metal and the ligands and varies predictably from one complex to another.

A different splitting pattern is encountered with square-planar complexes (Figure 23.29). If the four ligands are in the *xy*-plane, the $d_{x^2-y^2}$ orbital is at highest energy. The d_{z^2} energy is shifted upward to a lesser extent, whereas the d_{xy} orbital (which lies in the *xy*-plane) is found at higher energy than the d_{xz} and d_{yz} orbitals, both of which are partially pointing in the *z*-direction.

Magnetic Properties of Coordination Compounds

The *d*-orbital splitting in coordination complexes provides the means to explain both the magnetic behavior and color of these species. To understand these properties, we must first understand how to assign electrons to the various orbitals in square-planar and octahedral complexes.

A gaseous Cr^{2+} ion has the electron configuration $[Ar]3d^4$. (The term "gaseous" in this context is used by scientists to denote a single, isolated atom or ion with all other particles an infinite distance away.) In such an ion, the five

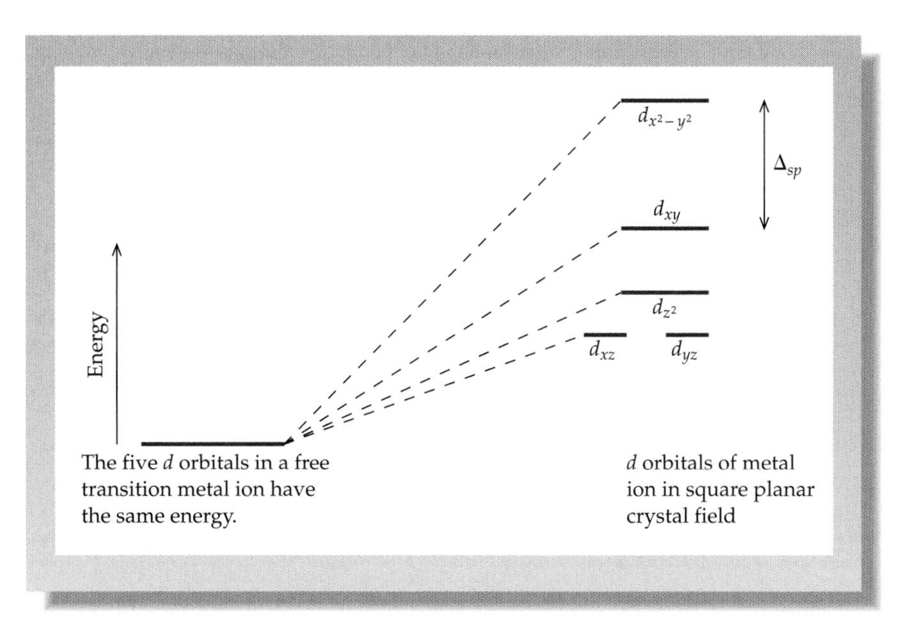

***Figure* 23.29 Splitting of the *d* orbitals in a square-planar complex.** Notice that the five d orbitals are split into four groups of orbitals when the metal ion is the center of a square-planar complex.

$3d$ orbitals have the same energy. The four electrons reside singly in different d orbitals, according to Hund's rule, and the Cr^{2+} ion has four unpaired electrons.

Cr^{2+} ion [Ar] [↑|↑|↑|↑|] ($3d$) [ ] ($4s$)

When the Cr^{2+} ion is part of an octahedral complex, however, the five d orbitals do not have identical energies. These orbitals divide into two sets with the d_{xy}, d_{xz}, and d_{yz} orbitals at a lower energy than the $d_{x^2-y^2}$ and d_{z^2} orbitals. Having two sets of orbitals means that two different electron configurations are possible. Three of the four d electrons in Cr^{2+} are assigned to the lower energy d_{xy}, d_{xz}, and d_{yz} orbitals. The fourth electron, however, can either be assigned to an orbital in the higher energy $d_{x^2-y^2}$ and d_{z^2} set (Configuration A) or pair up with an electron already in the lower energy set (Configuration B). The first arrangement is called **high spin** because it has the maximum number of unpaired electrons, four. In contrast, the second arrangement is called **low spin** because it has the minimum number of unpaired electrons possible.

Configuration A high spin: ↑ in d_{z^2}, $d_{x^2-y^2}$; ↑ ↑ ↑ in d_{xy}, d_{xz}, d_{yz}

Configuration B low spin: d_{z^2}, $d_{x^2-y^2}$ empty; ↑↓ ↑ ↑ in d_{xy}, d_{xz}, d_{yz}

At first glance, a high-spin configuration appears to contradict conventional thinking. It seems logical that the most stable situation would occur when electrons occupy the lowest energy orbitals. A second factor intervenes, however. When two electrons pair up in a single orbital they are constrained to the same region of space. Because all electrons are negatively charged, repulsion increases when they are in the same orbital. This is a destabilizing effect and bears the name **pairing energy (P).** The preference for an electron to be in the lowest energy orbital and the pairing energy have opposing effects.

Low-spin complexes arise when the splitting of the d orbitals by the crystal field is large, that is, there is a large value of Δ_O. In this situation, the energy

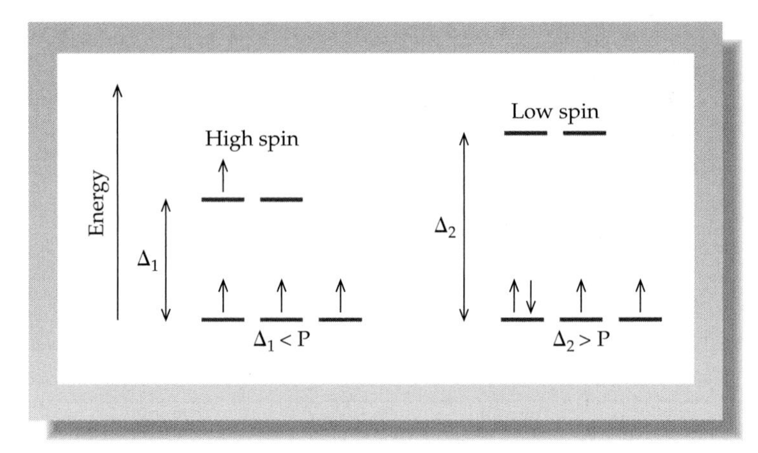

Figure 23.30 High- and low-spin cases for $[Cr(L)_6]^{2+}$ complexes. When the crystal field splitting is small, the electrons prefer to remain unpaired and the complex has four unpaired electrons. When the crystal field splitting is large, then the four electrons are in the lowest energy orbitals, and the complex has two unpaired electrons.

gained by putting all the electrons in the lowest energy level is the dominant effect. Conversely, high-spin complexes occur with small values of Δ_0, as illustrated in Figure 23.30.

For octahedral complexes, there is a choice between high and low spin only for configurations d^4 through d^7 (Figure 23.31). Complexes of the d^6 metal ion, Fe^{2+}, for example, can have either high spin or low spin. The complex formed when the ion is placed in water, $[Fe(H_2O)_6]^{2+}$, is high spin, whereas $[Fe(CN)_6]^{4-}$ is low spin.

It is possible to tell the difference between low- and high-spin complexes by determining the magnetic behavior of the substance. The high-spin complex $[Fe(H_2O)_6]^{2+}$ has four unpaired electrons and is *paramagnetic* (attracted by a

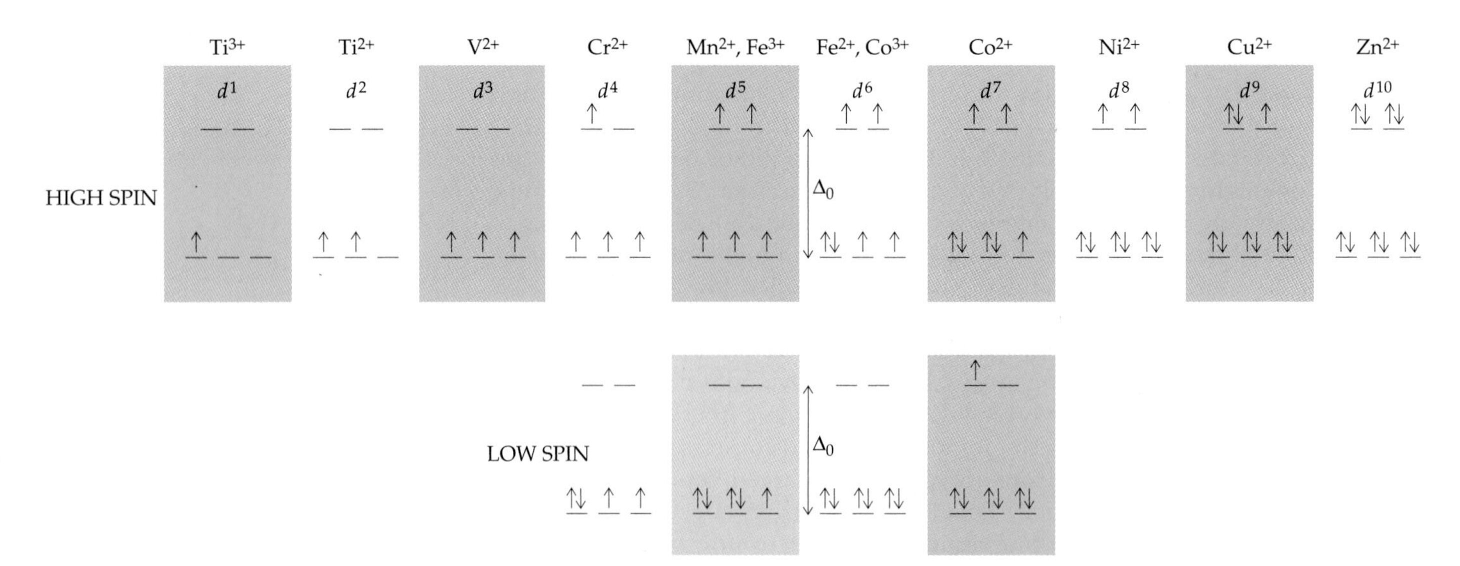

Figure 23.31 High- and low-spin complexes. Electron configurations for octahedral complexes of metal ions having from d^1 to d^{10} configurations. Only the d^4 through d^7 cases have both high-spin and low-spin configurations.

magnet), whereas the low-spin $[Fe(CN)_6]^{4-}$ complex has no unpaired electrons and is *diamagnetic* (repelled by a magnet) (see Section 8.1):

Electron configuration for Fe^{2+} in an octahedral complex

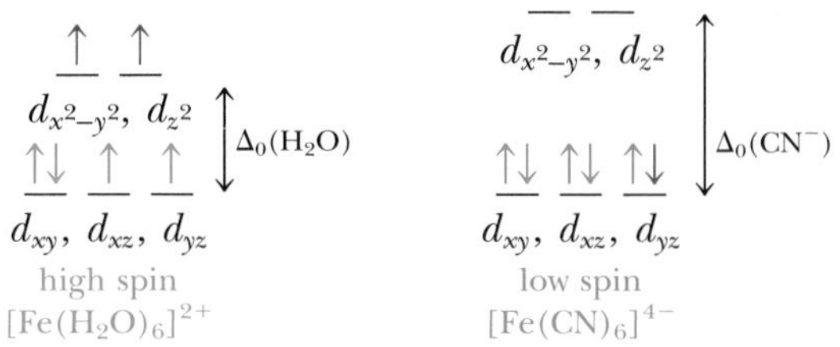

high spin $[Fe(H_2O)_6]^{2+}$	low spin $[Fe(CN)_6]^{4-}$
$d_{x^2-y^2}, d_{z^2}$: ↑ ↑	$d_{x^2-y^2}, d_{z^2}$: — —
$\Delta_o(H_2O)$	$\Delta_o(CN^-)$
d_{xy}, d_{xz}, d_{yz}: ↑↓ ↑ ↑	d_{xy}, d_{xz}, d_{yz}: ↑↓ ↑↓ ↑↓

The Ni^{2+}, Pd^{2+}, and Pt^{2+} ions have the electron configuration [noble gas]$(n-1)d^8$ and are likely to form square-planar complexes. In a square-planar complex there are four sets of orbitals (see Figure 23.29). Although high- and low-spin configurations may appear to be possible, only low-spin complexes are known:

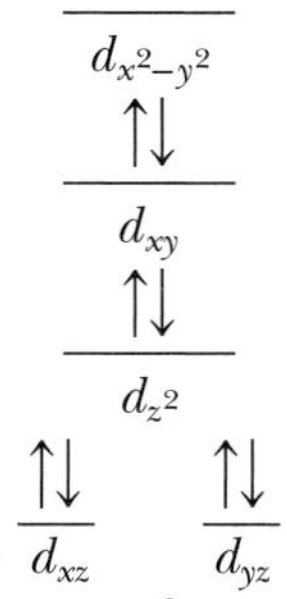

$d_{x^2-y^2}$: —
d_{xy}: ↑↓
d_{z^2}: ↑↓
d_{xz}, d_{yz}: ↑↓ ↑↓

Electron configuration in d^8 square-planar complexes

Example 23.5 High- and Low-Spin Complexes and Magnetism

Depict the electron configurations for each of the following complexes, and tell how many unpaired electrons are present in each. Describe each complex as paramagnetic or diamagnetic.

1. Low-spin $[Co(NH_3)_6]^{3+}$ **2.** High-spin $[CoF_6]^{3-}$

Solution

1. This is an octahedral complex because there are six ligands surrounding cobalt. Furthermore, because the NH_3 ligands are neutral molecules and because the overall charge on the complex is 3+, this complex is based on the Co^{3+} ion. The cobalt(III) ion has an electron configuration of $[Ar]3d^6$. To obtain the low-spin configuration for the complex ion, the lower energy set of orbitals is filled entirely. This d^6 complex ion has no unpaired electrons and so is diamagnetic:

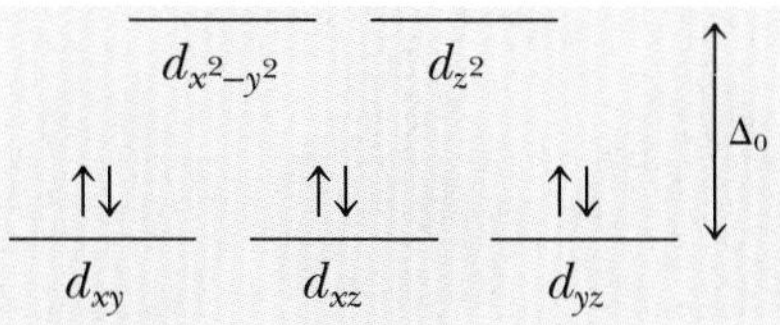

$d_{x^2-y^2}$, d_{z^2}: — —
Δ_o
d_{xy}, d_{xz}, d_{yz}: ↑↓ ↑↓ ↑↓

Electron configuration of low-spin, octahedral $[Co(NH_3)_6]^{3+}$

2. Again, this is a cobalt(III) complex with six d electrons. Experiment now shows this complex to be high spin, which means the crystal field splitting (Δ_O) is smaller than in $[Co(NH_3)_6]^{3+}$. To obtain the electron configuration for the d^6 Co^{3+} metal ion in $[CoF_6]^{3-}$, place one electron in each of the five d orbitals, and then place the sixth electron in one of the lower energy orbitals. The complex has four unpaired electrons and is paramagnetic:

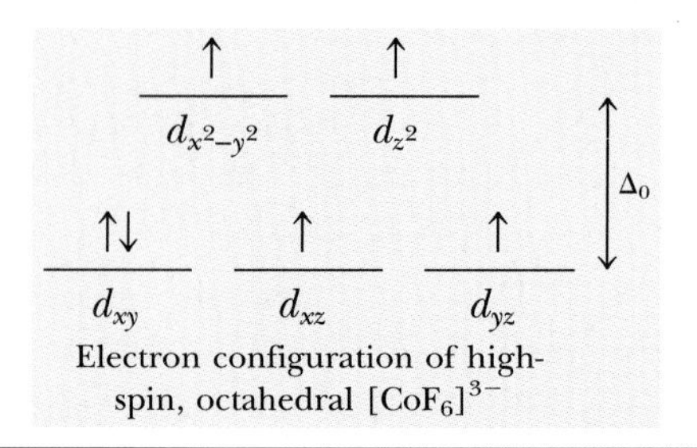

Electron configuration of high-spin, octahedral $[CoF_6]^{3-}$

Exercise 23.4 High- and Low-Spin Configurations and Magnetism

For each of the following complex ions, give the oxidation number of the metal ion, depict the low- and high-spin configurations, give the number of unpaired electrons in each state, and tell whether each is paramagnetic or diamagnetic.

1. $[Ru(H_2O)_6]^{2+}$
2. $[Ni(NH_3)_6]^{2+}$

23.6 THE COLORS OF COORDINATION COMPOUNDS

One of the most interesting properties of the transition elements is that their compounds are usually colored, whereas compounds of main group metals are usually colorless (see Figures 23.6, 23.15, and 23.32). With an understanding of d-orbital splitting, we can now explain the origin of the colors for complexes. First, however, let us look more closely at what we mean by color.

Figure 23.32 Aqueous solutions of some transition metal ions. Compounds of the transition elements are often colored, whereas those of main group elements are usually colorless. Pictured, from left to right, are aqueous solutions of nitrate salts of Fe^{3+}, Co^{2+}, Ni^{2+}, Cu^{2+}, and Zn^{2+}. *(C. D. Winters)*

Color

Visible light radiation with wavelengths from 400 nm to 700 nm (see Section 7.1) represents a very small portion of the electromagnetic spectrum. Within this region of the spectrum are all the colors you see when white light is passed through a prism: red, orange, yellow, green, blue, indigo, and violet (ROYGBIV). Each color is identified with a narrow wavelength range.

Red Green Blue

700 600 500 400 nm

R O Y G B I V

The blue in ROYGBIV is actually cyan according to now accepted color industry standards. The highlights in this book are printed in cyan. Note also that magenta doesn't have its own wavelength region, it is a superposition of B and R, and orange is simply yellow with some red content.

Isaac Newton did experiments with light and established that the mind's *perception* of color requires only three colors! When we see white light, we are seeing a mixture of all of the colors as the superposition of red, green, and blue. If one or more of these colors is absorbed, the light of the other colors can then pass through to your eyes. Your mind then perceives and interprets the color.

To understand the perceived colors of compounds, we divide the visible spectral range into three broad regions: red, green, and blue (Figure 23.33). The *primary* colors—red, green, and blue—appear at the corners of the triangle superimposed on the color discs, and the *secondary* colors—yellow, cyan, and magenta—appear at the edges of the triangle.

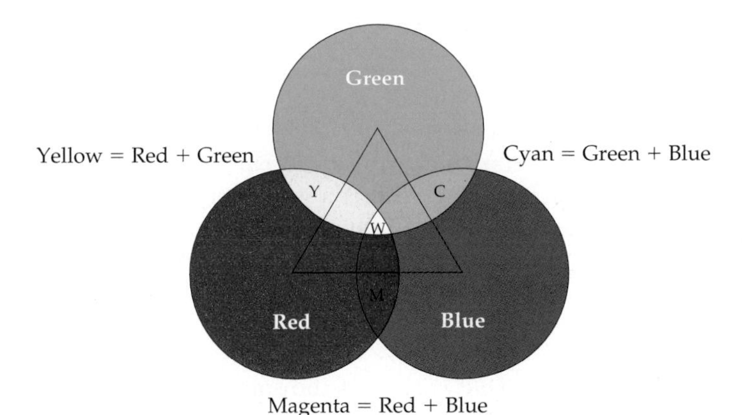

Figure 23.33 Using color discs to represent colors. Color discs can be used to explain the perception of color. The three primary colors are red, green, and blue. Light of two primary colors can add together to give a third color. For example, cyan arises when green and blue light are added, and yellow arises from addition of green and red. Alternatively, we can think of color as arising from the subtraction of a light from white light. For example, if red light is removed from white light, the color cyan results.

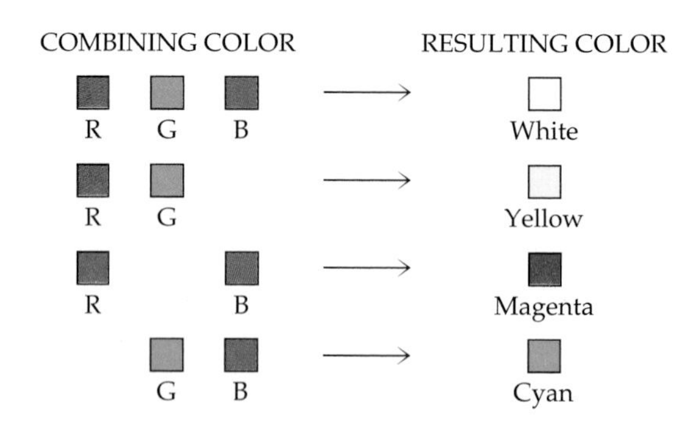

Figure 23.34 Another way to view colors. The three secondary colors shown at the right—cyan, yellow, and magenta—are derived from addition of the indicated primary colors. Alternatively, a secondary color results if one of the three primary colors is subtracted. For example, yellow is viewed as arising either from the addition of red and green or by subtraction (or absence) of blue light from white light.

Figure 23.34 presents this information in another way. A color arises in two ways: by *addition* of two other colors or by the *subtraction* of light of a particular color from white light. For example, adding blue and green light leads to a color called cyan. Alternatively, if red light is subtracted from white light, cyan light remains.

Now let us apply these ideas to transition metal complexes. A solution of $[Ni(H_2O)_6]^{2+}$ is green. We know that this color is the result of removing red (R) and blue (B) light from white light. As white light passes through an aqueous solution of Ni^{2+}, red and blue light are absorbed but green light is allowed to pass, and so this is the color we perceive. Similarly, the $[Co(NH_3)_6]^{3+}$ ion (see Figure 23.32) is yellow-orange because blue (B) light has been selectively absorbed; that is, the solution allows red (R) and green (G) light to pass.

The qualitative conclusions that we have drawn concerning colors and absorption of light are confirmed in the laboratory using a scientific instrument called a *spectrophotometer.* A schematic drawing of a spectrophotometer is shown in Figure 23.35. White light from a glowing filament is first passed through a device (a prism or diffraction grating) that divides the light according to its respective frequencies. The instrument selects a specific frequency to pass through a solution of the compound to be studied. If light of a given frequency is not absorbed, its intensity is unchanged when it emerges from the sample. On the other hand, if light of this frequency is absorbed, the light emerging from the sample has a lower intensity.

Modern spectrometers are designed to make absorption measurements over a range of frequencies. An **absorption spectrum** is a graph of the frequency or wavelength of the light against the intensity of light absorbed at that frequency or wavelength. When light in a certain frequency range is absorbed, the plot shows an *absorption band.* For example, $[Co(NH_3)_6]^{3+}$ absorbs in the blue region, leaving its complementary color yellow (Y = R + G) to pass through to the eye.

Scientists are not limited to the visible region of the spectrum when making measurements on absorption of electromagnetic radiation. Ultraviolet and infrared spectrometers are common pieces of scientific apparatus in a chemistry laboratory. They differ from the visible spectrometer depicted in Figure 23.35 only in that the light source must generate radiation in the correct region of the electromagnetic spectrum, and the detector must be able to detect this radiation.

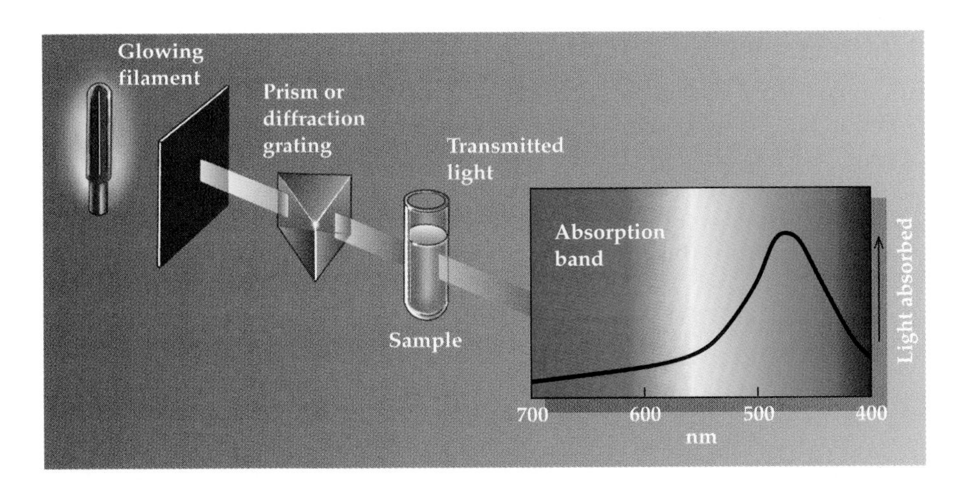

***Figure* 23.35 Obtaining a spectrum.** White light is passed through a solution of a transition metal complex ion, say $[Co(NH_3)_6]^{3+}$, and light is absorbed. The transmitted light is the color observed.

The Absorption of Light by Coordination Complexes

The color of a transition metal complex results from the absorption of light in the visible region of the spectrum. Although the details of the process by which light is absorbed are beyond the scope of this book, we can describe this process in qualitative terms.

Chapter 7 described atomic spectra, which are obtained when electrons are excited from one energy level to another. The absorptions correspond to specific wavelengths, and the energy of the light absorbed or emitted is related to the energy levels of the atom or ion under study. The concept that light is absorbed when electrons move between energy levels applies to all substances, and this is the basis for the spectra (and hence the colors) for transition metal coordination complexes. Because most transition metal complexes are colored, we conclude that the energy levels in these complexes are spaced so that visible light is absorbed.

In coordination complexes, the splitting between *d* orbitals often corresponds to the energy of visible light, so light in the visible region of the spectrum is absorbed when electrons move from a lower energy *d* orbital to a higher energy *d* orbital. This transition of an electron between two orbitals in a complex is labeled a *d-to-d transition.* Qualitatively, such a transition for $[Co(NH_3)_6]^{3+}$ might be represented using an energy level diagram such as that shown here:

$$\begin{array}{ccc} \_\_\_ & \_\_\_ & \\ d_{x^2-y^2} & d_{z^2} & \\ \uparrow\downarrow & \uparrow\downarrow & \uparrow\downarrow \\ d_{xy} & d_{xz} & d_{yz} \end{array} \quad \Delta_0 \xrightarrow[\text{(light absorbed)}]{+\text{ energy }(\Delta_0)} \quad \begin{array}{ccc} \downarrow & \_\_\_ & \\ d_{x^2-y^2} & d_{z^2} & \\ \uparrow\downarrow & \uparrow\downarrow & \uparrow \\ d_{xy} & d_{xz} & d_{yz} \end{array}$$

Ground state of low-spin, octahedral Co^{3+} complex — Excited state

The Spectrochemical Series of Ligands

Experiments with coordination complexes have revealed that, for a given metal ion, some ligands cause a small energy separation of the *d* orbitals, whereas others cause a large separation. In other words, some ligands create a small crystal field, and others create a large one.

Spectroscopic data for several cobalt(III) complexes are presented in Table 23.4. We can see from this information why different complexes have the indicated colors.

In Table 23.4, notice the connection between the wavelength of the absorbed light, its relative energy, and the color of the complex.

- Both $[Co(NH_3)_6]^{3+}$ and $[Co(en)_3]^{3+}$ are yellow-orange, because they absorb light in the blue portion of the visible spectrum. Note, by the way, that these compounds have very similar spectra, which is to be expected because both have six amine-type donor atoms.
- Although $[Co(CN)_6]^{3-}$ does not have an absorption band in the visible region, it is pale yellow. This is due to the fact that the absorption in the ultraviolet is broad and extends at least minimally into the visible region, resulting in the absorption of a small amount of blue light.
- $[Co(C_2O_4)_3]^{3-}$ and $[Co(H_2O)_6]^{3+}$ have similar absorptions, in the yellow and violet regions. Their colors are shades of green with a small difference due to the relative amount of light of each color being absorbed.

The absorption maxima among the listed complexes ranges from 700 nm for $[CoF_6]^{3-}$ to 310 nm for $[Co(CN)_6]^{3-}$. The ligands change from member to member in this series, and we can conclude that the energy of the light absorbed by the complex is related to different crystal field splittings, Δ_0, caused by the different ligands. Fluoride ion causes the smallest splitting of the *d* orbitals among the complexes listed, whereas cyanide caused the largest splitting.

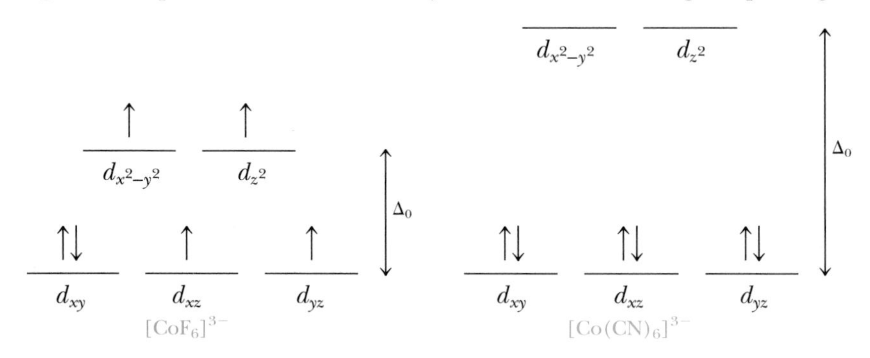

Spectra of complexes of other metals provide similar results. Based on this information, it is possible to list ligands in order of their ability to split the *d* orbitals. This list, called the **spectrochemical series** because it was determined by spectroscopy, is shown here. (The series contains many more ligands than are listed.)

Spectrochemical Series

$$\text{Halides} < C_2O_4^{2-} < H_2O < NH_3 = \text{en} < \text{phen} < CN^-$$

small orbital splitting / small Δ_0 / weak field ligands — large orbital splitting / large Δ_0 / strong field ligands

The spectrochemical series is applicable to a wide range of metal complexes and ligands. The ability of crystal field theory to explain differences in the color of transition metal complexes is one of the strengths of this theory.

Table 23.4 • THE COLORS OF SOME COMPLEXES OF THE Co^{3+} ION*

Complex Ion	Wavelength of Light Absorbed (nm)	Color of Light Absorbed	Color of Complex
$[CoF_6]^{3-}$	700	Red	Green
$[Co(C_2O_4)_3]^{3-}$	600, 420	Yellow, violet	Dark green
$[Co(H_2O)_6]^{3+}$	600, 400	Yellow, violet	Blue-green
$[Co(NH_3)_6]^{3+}$	475, 340	Blue, ultraviolet	Yellow-orange
$[Co(en)_3]^{3+}$	470, 340	Blue, ultraviolet	Yellow-orange
$[Co(CN)_6]^{3-}$	310	Ultraviolet	Pale yellow

*The complex with fluoride ion, $[CoF_6]^{3-}$, is high spin and has one absorption band. The other complexes are low spin and have two absorption bands. In all but one case, one of these absorptions is in the visible region of the spectrum. The wavelengths are measured at the top of that absorption band.

From the relative position of the ligand in the series, you can now also make predictions about a compound's magnetic behavior. Recall that d^4, d^5, d^6, and d^7 complexes can be high or low spin, depending on the crystal field splitting Δ_0. Complexes formed with ligands near the left end of the spectrochemical series are expected to have small Δ_0 values and thus are likely to be high spin. In contrast, complexes with ligands near the right end are expected to have large Δ_0 values and have low-spin configurations. The complex $[CoF_6]^{3-}$ is high spin, whereas $[Co(NH_3)_6]^{3+}$ and the others shown in Table 23.4 are low spin.

Example 23.6 Spectrochemical Series

An aqueous solution of $[Fe(H_2O)_6]^{2+}$ is light blue-green. Do you expect the d^6 Fe^{2+} ion in this complex to have a high- or low-spin configuration?

Solution Using the color wheel in Figure 23.30, we see that the blue-green color arises when the complex absorbs red light. The d-orbital splitting must be small, meaning that $[Fe(H_2O)_6]^{2+}$ has a good chance of being a high-spin complex. The presence of four unpaired electrons can be verified experimentally.

CHAPTER HIGHLIGHTS

When you finish studying this chapter, you should be able to

- Identify the **transition elements** (d-block elements) and the **lanthanide** and **actinide** (f-block) **elements,** and predict properties of these elements (Section 23.1).
- Describe a metal atom or ion as **paramagnetic** or diamagnetic (Section 23.1).
- Describe the metallurgy for iron and copper, two metals that have major uses in today's technology (Section 23.2).
- Describe the difference between **pyrometallurgy** (high-temperature processes) and **hydrometallurgy** (techniques using water) (Section 23.2).

- Identify and describe important features of **coordination complexes: ligands (monodentate** and **polydentate),** the charge on the complex ion, and the oxidation number of the central metal ion (Section 23.3).
- Provide the systematic name for a coordination complex (Section 23.3).
- Recognize examples of common **coordination numbers** and their geometries (2, linear; 4, tetrahedral and square-planar; 6, octahedral) (Section 23.4).
- Recognize and draw isomers of coordination complexes (Section 23.4).
- Know that there are two common types of isomers of coordination compounds: **geometric** and **optical isomers** (Section 23.4).
- Recognize that square-planar complexes may exist as geometric isomers but not optical isomers. Octahedral complexes can exhibit both forms of isomerism (Section 23.4).
- Explain the bonding in coordination compounds using **crystal field theory** (Section 23.5).
- Rationalize the existence of **high-** and **low-spin** complexes (Section 23.5).
- Understand why complexes are colored and how the color of a complex can be explained using the crystal field model of bonding (Section 23.6).

STUDY QUESTIONS

REVIEW QUESTIONS

1. What feature of electronic structure distinguishes transition elements, lanthanide elements, and actinide elements from the other elements and from one another?
2. Write out electron configurations for each element in the first transition series.
3. Write out electron configurations for the 2+ ion of each metal in the first transition series.
4. Identify two physical properties and two chemical properties for each element in the first transition series.
5. Identify a common use for each of the following elements: Ti, Cr, Mn, Fe, Ni, and Cu.
6. What are the two major types of processes for the recovery of metals from their ores? Give an example of the use of each type of process.
7. What transition elements are commonly used in the manufacture of permanent magnets?
8. Define the following terms: (a) transition element, (b) coordination compound, (c) complex ion, (d) ligand, (e) chelate, (f) bidentate. Give an example to illustrate each word or phrase.
9. Define the terms "diamagnetic" and "paramagnetic." What feature of electronic structure distinguishes these properties?
10. What are three common metal coordination numbers encountered in coordination chemistry, and what structure or structures are possible for each?
11. Give an example of a complex that displays (a) geometric isomerism and (b) optical isomerism.
12. According to the crystal field model, what is the origin of the splitting of metal d orbitals into two sets in an octahedral complex?
13. What factors determine whether a complex will be high- or low-spin?

NUMERICAL AND OTHER QUESTIONS

Configurations and Physical Properties

14. The metal atoms in the chromium compounds pictured in Figure 23.6 have two different oxidation numbers. Using an orbital box diagram, show the electron configuration for chromium in each oxidation state. In which of the two oxidation states is chromium expected to be paramagnetic?

15. Give the electron configuration for each of the following ions, and tell whether it is paramagnetic or diamagnetic.
(a) Y^{3+} (d) V^{2+}
(b) Pt^{2+} (e) Ce^{4+}
(c) Rh^{3+} (f) U^{4+}

16. Identify two transition metal ions with the following electron configurations:
(a) $[Ar]3d^6$ (c) $[Ar]3d^{10}$
(b) $[Ar]3d^5$ (d) $[Ar]3d^8$

17. Identify another ion of a first series transition metal that is isoelectronic with each of the following:
(a) Fe^{3+} (c) Zn^{2+}
(b) Fe^{2+} (d) Cr^{3+}

Metallurgy

18. The following equations represent various ways of obtaining transition metals from their compounds. Balance each equation.
(a) $Cr_2O_3(s) + Al(s) \longrightarrow Al_2O_3(s) + Cr(s)$
(b) $TiCl_4(\ell) + Mg(s) \longrightarrow Ti(s) + MgCl_2(s)$
(c) $[Ag(CN)_2]^-(aq) + Zn(s) \longrightarrow Ag(s) + Zn^{2+}(aq) + 2\ CN^-(aq)$

19. In the first step in the recovery of copper, an ore such as chalcopyrite, $CuFeS_2$, is roasted in air to give CuS, FeS, and SO_2. If you begin with one ton (908 kg) of chalcopyrite, what mass of SO_2 is produced?

20. Titanium is the seventh most abundant metal in earth's crust. It is strong, lightweight, and resistant to corrosion; these properties lead to its use in aircraft engines. To obtain metallic titanium, ilmenite ($FeTiO_3$), an ore of titanium, is first treated with sulfuric acid to form $FeSO_4$ and $Ti(SO_4)_2$. After separating these compounds, the latter substance is converted to TiO_2 in basic solution:

$$FeTiO_3(s) + 3\ H_2SO_4(aq) \longrightarrow FeSO_4(aq) + Ti(SO_4)_2(aq) + 3\ H_2O(\ell)$$

$$Ti^{4+}(aq) + 4\ OH^-(aq) \longrightarrow TiO_2(s) + 2\ H_2O(\ell)$$

What volume of 18.0 M H_2SO_4 is required to react completely with 1.00 kg of ilmenite? What mass of TiO_2 can theoretically be produced by this sequence of reactions?

21. In the process described in Study Question 20, ilmenite ore is leached with sulfuric acid. This leads to the significant environmental problem of disposal of the iron(II) sulfate (which, in its hydrated form, is commonly called "copperas"). To avoid this, it has been suggested that HCl be used to leach ilmenite so that the iron-containing product is $FeCl_2$. This can be treated with water and air to give commercially useful iron(III) oxide and regenerate HCl by the reaction

$$2\ FeCl_2(aq) + 2\ H_2O(\ell) + \tfrac{1}{2}\ O_2(g) \longrightarrow Fe_2O_3(s) + 4\ HCl(aq)$$

(a) Write a balanced equation for the treatment of ilmenite with aqueous HCl to give iron(II) chloride, titanium(IV) oxide, and water.
(b) If the equation written in part (a) is combined with the preceding equation for oxidation of $FeCl_2$ to Fe_2O_3, is the HCl used in the first step recovered in the second step?
(c) How many grams of iron(III) oxide can be obtained from one ton (908 kg) of ilmenite in this process?

Ligands and Formulas of Complexes

22. Which of the following ligands are expected to be monodentate and which are multidentate?
(a) CH_3NH_2 (e) en
(b) CH_3CN (f) Br^-
(c) N_3^- (g) phen
(d) $C_2O_4^{2-}$

23. Only one of the following nitrogen compounds or ions, NH_4^+, NH_3, or NH_2^-, is incapable of serving as a ligand. Identify this species, and explain your answer.

24. Give the oxidation number of the metal ion in each of the following compounds:
(a) $[Mn(NH_3)_6]SO_4$ (c) $[Co(NH_3)_4Cl_2]Cl$
(b) $K_3[Co(CN)_6]$ (d) $Mn(en)_2Cl_2$

25. Give the formula of a complex constructed from one Ni^{2+} ion, one ethylenediamine ligand, three ammonia molecules, and one water molecule. Is the complex neutral or is it charged? If charged, give the charge.

Naming

26. Write formulas for the following ions or compounds:
(a) Dichlorobis(ethylenediamine)nickel(II)
(b) Potassium tetrachloroplatinate(II)
(c) Potassium dicyanocuprate(I)
(d) Diaquatetraammineiron(II)

27. Write formulas for the following ions or compounds:
(a) Diamminetriaquahydroxochromium(II) nitrate
(b) Hexaammineiron(III) nitrate
(c) Pentacarbonyliron(0)
(d) Ammonium tetrachlorocuprate(II)

28. Name the following ions or compounds:
(a) $[Ni(C_2O_4)_2(H_2O)_2]^{2-}$ (c) $[Co(en)_2(NH_3)Cl]^{2+}$
(b) $[Co(en)_2Br_2]^+$ (d) $Pt(NH_3)_2(C_2O_4)$

29. Name the following ions or compounds:
(a) $[Co(H_2O)_4Cl_2]^+$ (c) $[Pt(NH_3)Br_3]^-$
(b) $Co(H_2O)_3F_3$ (d) $[Co(en)_2(NH_3)Cl]^{2+}$

30. Give the name or formula for each ion or compound, as appropriate:
(a) Hydroxopentaaquairon(III) ion
(b) $K_2[Ni(CN)_4]$
(c) $K[Cr(C_2O_4)_2(H_2O)_2]$
(d) Ammonium tetrachloroplatinate(II)

31. Give the name or formula for each ion or compound, as appropriate:
(a) Dichlorotetraaquachromium(III) chloride
(b) $[Cr(NH_3)_5SO_4]Cl$
(c) Sodium tetrachlorocobaltate(II)
(d) $[Fe(C_2O_4)_3]^{3-}$

Isomerism

32. Draw all possible geometric isomers of
(a) $Fe(NH_3)_4Cl_2$
(b) $Pt(NH_3)_2(NCS)(Br)$ (NCS is bonded to Pt^{2+} through S)
(c) $Co(NH_3)_3(NO_2)_3$ (NO_2 is bonded to Co^{3+} through N)
(d) $[Co(en)Cl_4]^-$

33. Which of the following complexes can have geometric isomers? (If isomers are possible, draw their structures and label them as *cis* or *trans*, or as *fac* or *mer* if appropriate.)
(a) $[Co(H_2O)_4Cl_2]^+$ (c) $[Pt(NH_3)Br_3]^-$
(b) $Co(H_2O)_3F_3$ (d) $[Co(en)_2(NH_3)Cl]^{2+}$

34. Determine whether the following molecules possess a chiral center:
(a) CH_2Cl_2 (c) $ClCH(OH)CH_2Cl$
(b) $H_2NCH(CH_3)CO_2H$ (d) $CH_3CH_2CH{=}CHC_6H_5$

35. Decide whether each of the following molecules has an enantiomer.

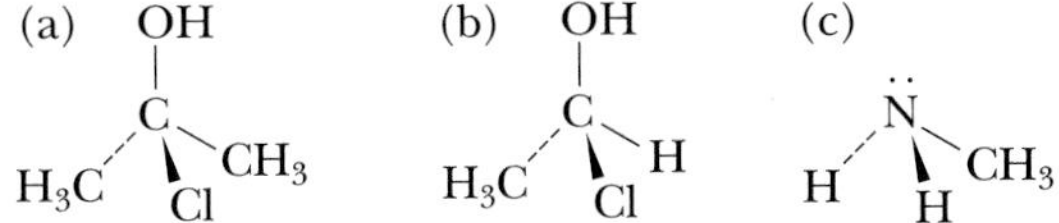

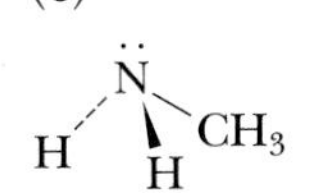

36. Four isomers are possible for $[Co(en)(NH_3)_2(H_2O)Cl]^{2+}$. (Two of the four have optical isomers and so each has a non-superimposable mirror image.) Draw the structures of the four isomers.

37. Draw all possible isomers (geometric and optical) for the ion $[Cr(C_2O_4)_2(H_2O)_2]^-$.

Magnetism of Coordination Complexes

38. What *d* electron configurations exhibit both high and low spin in octahedral complexes?

39. Can $[Cr(NH_3)_6]^{3+}$ form both low-spin and high-spin complexes? Explain your answer.

40. Depict high- and low-spin configurations for each of the complexes below. Tell whether each is diamagnetic or paramagnetic. Give the number of unpaired electrons for the paramagnetic cases.
(a) $[Fe(CN)_6]^{4-}$ (c) $[Fe(H_2O)_6]^{3+}$
(b) $[Co(NH_3)_6]^{3+}$ (d) $[CrF_6]^{4-}$

41. From experiment we know that $[CoF_6]^{3-}$ is paramagnetic and $[Co(NH_3)_6]^{3+}$ is diamagnetic. Using the crystal field model, depict the electron configuration for each ion. What can you conclude about the effect of the ligand on the magnitude of Δ_0?

Color

42. In water, the titanium(III) ion, $[Ti(H_2O)_6]^{3+}$, is violet. (Its broad absorption band occurs at about 500 nm.) What color light is absorbed by the ion?

43. The chromium(II) ion in water, $[Cr(H_2O)_6]^{2+}$, absorbs light with a wavelength of about 700 nm. What color is the solution?

GENERAL QUESTIONS

44. For the complex ion $[Fe(H_2O)_6]^{2+}$ identify
(a) The coordination number of iron
(b) The coordination geometry for iron
(c) The oxidation number of iron
(d) The number of unpaired electrons, assuming high spin
(e) Whether the complex is diamagnetic or paramagnetic

45. For the complex ion $[Co(en)(NH_3)_2Cl_2]^+$ identify
(a) The coordination number of cobalt
(b) The coordination geometry for cobalt
(c) The oxidation number of cobalt
(d) The number of unpaired electrons, assuming low spin
(e) Whether the complex is diamagnetic or paramagnetic

46. Predict whether each complex below is high- or low-spin and whether each is paramagnetic or diamagnetic, based on the position of the ligands in the spectrochemical series. If paramagnetic, give the number of unpaired electrons. Use the crystal field model to find the electron configuration of each ion.
(a) $[Fe(CN)_6]^{4-}$ (c) $[Cr(en)_3]^{3+}$
(b) $[MnF_6]^{4-}$ (d) $[Cu(phen)_3]^{2+}$

47. Which element in each of the following pairs is denser? Explain each answer briefly.
(a) Ti or Fe (c) Ti or Zr
(b) Ti or Os (d) Zr or Hf

48. In $Pt(NH_3)_2(C_2O_4)$ the metal ion is surrounded by a square plane of coordinating atoms. Draw a structure for this molecule. Give the oxidation number of the platinum and the name of the compound.

49. Worldwide production of nickel is 750,000 tons/year; most comes from Canada. The common mineral containing nickel is pentlandite, NiS. Roasting of NiS in air produces NiO, which is reduced to the metal using H_2:

$$NiO(s) + H_2(g) \longrightarrow Ni(s) + H_2O(g)$$

The Mond process was, at one time, the method used to purify the metal. At a temperature of about 50 °C, and at atmospheric pressure, the impure metal reacts with CO to give a volatile compound tetracarbonylnickel(0), $Ni(CO)_4$. If this compound is passed into another part of the reactor and heated to 250 °C, it reverts to pure Ni and CO:

$$Ni(s) + 4\,CO(g) \rightleftharpoons Ni(CO)_4$$

If you wish to produce one ton (908 kg) of pure nickel, how much NiS, H_2, and CO are required?

50. Give the formula of a complex formed from one Co^{3+} ion, two ethylenediamine molecules, one water molecule, and one chloride ion. Is the complex neutral or charged? If charged, give the net charge on the ion.

51. Name the following ligands:
(a) OH^- (c) I^-
(b) O^{2-} (d) $C_2O_4^{2-}$

52. Determine whether the following complexes have a chiral metal center:
(a) $[Fe(en)_3]^{2+}$
(b) $[Co(en)_2Br_2]^+$
(c) $[Co(en)(H_2O)Cl_3]^+$
(d) $Pt(NH_3)(H_2O)(Cl)(NO_2)$ (square-planar Pt)

53. From experiment we know that $[Mn(H_2O)_6]^{2+}$ has five unpaired electrons, whereas $[Mn(CN)_6]^{4-}$ has only one. Using the crystal field model, depict the electron configuration for each ion. What can you conclude about the effect of the different ligands on the magnitude of Δ_0?

54. Arrange the following ligands in order of increasing crystal field splitting:
(a) CN^- (b) NH_3 (c) F^- (d) H_2O

55. Experiments show that $K_4[Cr(CN)_6]$ is paramagnetic and has two unpaired electrons. In contrast, the related complex $K_4[Cr(SCN)_6]$ is paramagnetic with four unpaired electrons. Account for these differences using the crystal field model. Predict where the SCN^- ion occurs in the spectrochemical series relative to CN^-.

CONCEPTUAL QUESTIONS

56. Comment on the fact that although an aqueous solution of cobalt(III) sulfate is diamagnetic, the solution becomes paramagnetic when a large excess of fluoride ion is added.

57. It is usually observed that stability of analogous complexes $[ML_6]^{n+}$ is in the order $Mn^{2+} < Fe^{2+} < Co^{2+} < Ni^{2+} < Cu^{2+} > Zn^{2+}$. (This order of ions is called the *Irving-Williams series*.) Look up the values of formation constants for am-

monia complexes of Co^{2+}, Ni^{2+}, Cu^{2+}, and Zn^{2+} in Appendix K and verify this statement. (See also Section 19.9.)

58. Describe an experiment that would determine the following:
 (a) Whether the cation in $[Fe(H_2O)_6]Cl_2$ is a low-spin or a high-spin metal ion
 (b) Whether nickel in $K_2[NiCl_4]$ is square-planar or tetrahedral

59. How many geometric isomers of the complex $[Cr(dmen)_3]^{3+}$ can exist? Note that dmen is the bidentate ligand 1,1-dimethylethylenediamine, $(CH_3)_2NCH_2CH_2NH_2$.

60. Diethylenetriamine (dien), $H_2NCH_2CH_2NHCH_2CH_2NH_2$, is capable of serving as a tridentate ligand.
 (a) Draw the structure of *fac*-Cr(dien)Cl_3 and *mer*-Cr(dien)Cl_3.
 (b) Two different geometric isomers of *mer*-Cr(dien)Cl_2Br are possible. Draw the structure for each.
 (c) Three different geometric isomers are possible for $[Cr(dien)_2]^{3+}$. Two have the dien ligand in a *fac* configuration, and one has the ligand in the *mer* orientation. Draw the structure of each isomer.

61. The square-planar complex Pt(en)Cl_2 has chloride ligands in a *cis* configuration. No *trans* isomer is known. Based on the bond lengths and bond angles of carbon and nitrogen in the ethylenediamine ligand, explain why the *trans* compound is not possible.

62. The complex ion $[Co(CO_3)_3]^{3-}$, an octahedral complex with bidentate carbonate ions as ligands, has one absorption in the visible region of the spectrum at 640 nm. From this information
 (a) Predict the color of this complex, and explain your reasoning.
 (b) Place the carbonate ion in the proper place in the spectrochemical series.
 (c) Predict whether $[Co(CO_3)_3]^{3-}$ will be paramagnetic or diamagnetic.

63. Iron–sulfur proteins are important electron transfer agents in biochemistry. (See *A Closer Look: Coordination Compounds in Biochemistry,* page 1063.) What is the coordination geometry of the iron(III) ion at the left end of the molecule? Of the molybdenum ion?

CHALLENGING QUESTIONS

64. A complex ion formed from a cobalt(III) ion, five ammonia molecules, a bromide ion, and a sulfate ion exists in two forms, one dark violet (A) and the other violet-red (B). The dark violet form (A) gives a precipitate with $BaCl_2$ but none with $AgNO_3$. Form B behaves in the opposite manner. This tells you that one form is $[Co(NH_3)_5Br]SO_4$ and the other is $[Co(NH_3)_5(SO_4)]Br$. Which compound is A and which is B? (*Note:* Only when ions such as Br^- or SO_4^{2-} are not directly coordinated to the metal ion can they form free ions in aqueous solution.)

65. A platinum-containing compound, known as Magnus's green salt, has the formula $[Pt(NH_3)_4][PtCl_4]$ (in which both platinum ions are Pt^{2+}). Name the compound.

66. Give the formula of a complex ion formed from one Pt^{2+} ion, one nitrite ion (NO_2^-, which binds to Pt^{2+} through N), one chloride ion, and two ammonia molecules. Are isomers possible? If so, draw the structure of each isomer, and tell what type of isomerism is observed.

67. In this question, we wish to explore the differences between metal coordination by monodentate and bidentate ligands. Formation constants, K_f, for $[Ni(NH_3)_6]^{2+}$(aq) and $[Ni(en)_3]^{2+}$(aq) are shown here:

$$Ni^{2+}(aq) + 6\ NH_3(aq) \rightleftharpoons [Ni(NH_3)_6]^{2+}(aq) \qquad K_f = 10^8$$
$$Ni^{2+}(aq) + 3\ en(aq) \rightleftharpoons [Ni(en)_3]^{2+}(aq) \qquad K_f = 10^{18}$$

The difference in K_f between these complexes, reflecting a large increase in stability of the chelated complex, is caused by the *chelate effect.* Recall that K is related to the standard free energy of the reaction by $\Delta G° = -RT \ln K$ and $\Delta G° = \Delta H° - T\Delta S°$. Here we know from experiment that $\Delta H°$ for the NH_3 reaction is -109 kJ/mol, and $\Delta H°$ for the en reaction is -117 kJ/mol. Is the difference in $\Delta H°$ sufficient to account for the 10^{10} difference in K_f? Comment on the role of entropy in the second reaction.

68. A 0.213-g sample of uranyl(VI) nitrate, $UO_2(NO_3)_2$, is dissolved in 20.0 mL of 1.0 M H_2SO_4 and shaken with Zn. The zinc reduces the uranyl ion, UO_2^{2+}, to an ion with a lower oxidation number, U^{n+} ($n < 6$). The uranium-containing solution, after reduction, is titrated with 0.0173 M $KMnO_4$. The potassium permanganate oxidizes the uranium back to the +6 oxidation state. Given that 12.47 mL of the potassium permanganate is required for titration to a permanent pink color of the equivalence point, calculate the oxidation number of the uranium after the uranyl(VI) nitrate is reduced with zinc. Write a balanced, net ionic equation for the oxidation of U^{n+} (in which you now know n) by MnO_4^- in acid solution to give UO_2^{2+} and Mn^{2+}.

69. The glycinate ion, $H_2NCH_2CO_2^-$ (formed by deprotonation of the amino acid glycine), can function as a bidentate ligand, as pictured below.

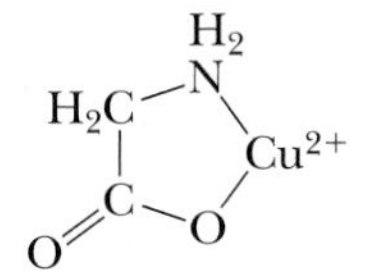

H_2N, H_2C, Cu^{2+}, C, O, O

Draw all of the isomers that can be formed by the coordination complex $Cu(H_2NCH_2CO_2)_2(H_2O)_2$.

70. You have a sample of an alloy of copper and aluminum and wish to determine the weight percentage of each element in the mixture. A 2.1309 g sample was first dissolved in a mixture of HCl and HNO_3. The resulting solution was made basic with excess ammonia, and the $Al(OH)_3$ that precipitated was collected and dried in a furnace to give 3.8249 g of Al_2O_3. What is the weight percentage of Al and Cu in the alloy?

71. Three different compounds of chromium(III) with water and chloride ion have the same composition: 19.51% Cr, 39.92% Cl, and 40.57% H_2O. One of the compounds is violet and dissolves in water to give a complex ion with a 3+ charge and three chloride ions. All three chloride ions precipitate immediately as AgCl on adding $AgNO_3$. Draw the structure of the complex ion and name the compound.

Chapter 24

Nuclear Chemistry

(C. D. Winters)

A Chemical Puzzler

About a decade ago, a new environmental hazard became big news: radon! It is often found lurking in basements, and this substance posed a threat to your health. Science and technology responded, with detection devices and the means of addressing the problem. But, what is this material, radon? Why is it a problem? And how did it get into your basement? Pictured here: a commercial kit to test for radon gas in the home.

A recent report issued by the National Research Council stated that "The future vigor and prosperity of American medicine, science, technology, and national defense clearly depend on continued use and development of nuclear techniques and use of radioactive isotopes." Nuclear chemistry, a subject that bridges many areas of science, has a significant influence on our society. Radioactive isotopes are now widely used in medicine, and some of the latest diagnostic techniques, such as positron emission tomography (PET) scans, depend on radioactivity. Similarly, your home may be protected with a smoke detector that contains a radioactive element, and research in all fields of science uses radioactive elements and their compounds. The national security of the United States since World War II has depended on nuclear weapons, and a number of nations around the world depend on nuclear reactors as a source of electricity. No matter what your reason for taking a college course in chemistry—to prepare for a career in one of the sciences or simply to gain knowledge as a concerned citizen—you should know something about nuclear chemistry. This chapter, therefore, considers changes in the atomic nucleus and the effects of those changes, the fissioning and fusion of nuclei and the energy that can be derived from such changes, the units used to measure radioactivity, and the uses of radioactive isotopes.

24.1 THE NATURE OF RADIOACTIVITY

In the late 19th century Ernest Rutherford (1871–1937) and J. J. Thomson (1856–1940) were studying the radiation from uranium and thorium. Rutherford found that "There are present at least two distinct types of radiation—one that is readily absorbed, which will be termed for convenience α [alpha] radiation, and the other of a more penetrative character, which will be termed β [beta] radiation." **Alpha radiation,** he discovered, was composed of particles, which, when passed through an electric field, were attracted to the negative side of the field (see Figure 2.2); indeed, his later studies showed that these particles are helium nuclei, which are ejected at high speeds from a radioactive element (Table 24.1). As might be expected, such massive particles have limited penetrating power and can be stopped by several sheets of ordinary paper or clothing (Figure 24.1).

The discovery of radioactivity by Henri Becquerel (1852–1908), and the subsequent chemical studies by Marie Curie (1867–1934) that isolated radium and polonium from pitchblende, a uranium ore, were described in Chapter 2. See the short biographies of Marie Curie (page 64) and Ernest Rutherford (page 72).

In the same experiment, Rutherford also found that **beta radiation** was attracted to the electrically positive plate. Thus, β radiation must be composed of negatively charged particles. Becquerel's work showed that these particles have an electric charge and mass equal to those of an electron. Thus, β (beta) par-

Table 24.1 • CHARACTERISTICS OF α, β, AND γ EMISSIONS

Name	Symbols	Charge	Mass (g/particle)
Alpha	${}^{4}_{2}He$, ${}^{4}_{2}\alpha$	2+	6.65×10^{-24}
Beta	${}^{0}_{-1}e$, ${}^{0}_{-1}\beta$	1−	9.11×10^{-28}
Gamma	${}^{0}_{0}\gamma$, γ	0	0

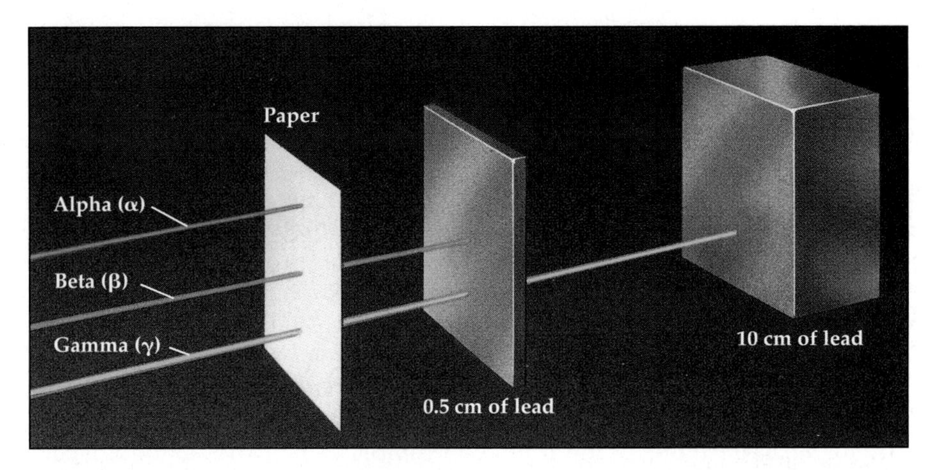

Figure 24.1 The relative penetrating ability of the three major types of nuclear radiation. Heavy, highly charged α particles interact with matter most strongly and are stopped by a piece of paper or a layer of skin. Beta particles and positrons are lighter, have a lower charge, and so interact to a lesser extent with matter; they are stopped by about 0.3 cm of aluminum. Gamma rays are uncharged, massless photons and are the most penetrating.

ticles are electrons ejected at high speeds from some radioactive nuclei. They are more penetrating than α particles. At least a 0.3-cm piece of aluminum is necessary to stop β particles, and they can penetrate several millimeters of living bone or tissue.

Nuclei formed after α and β emission are usually in an excited state. To return to the ground state γ radiation is emitted.

Rutherford hedged his bets when he said there were *at least* two types of radiation. Indeed, a third type was later discovered by Paul Villard (1860–1934), a Frenchman, who named it **γ (gamma) radiation,** using the third letter in the Greek alphabet in keeping with Rutherford's scheme. Unlike α and β radiation, which are particulate in nature, γ radiation is a form of electromagnetic radiation like x-radiation, although γ rays are even more energetic than x-rays (see Figure 7.3). Furthermore, γ rays have no electric charge and are not affected by an electric field (see Figure 2.2). Finally, γ radiation is the most penetrating; it can pass completely through the human body. Thick layers of lead or concrete are required to shield the body from this radiation.

24.2 NUCLEAR REACTIONS

Equations for Nuclear Reactions

Ernest Rutherford found that radium not only emits α particles but that it also produces the radioactive gas radon in the process. Rutherford and Frederick Soddy (1877–1956), in 1903, proposed the revolutionary theory that radioactivity is the result of a natural change of the isotope of one element into the isotope of a *different* element. In such changes, called **nuclear reactions,** or *transmutations,* an unstable nucleus emits radiation and is converted into a more stable nucleus of a different element. Thus, a nuclear reaction results in a change in atomic number and often a change in mass number as well. For example, the reaction studied by Rutherford can be written as

The product in this equation, ${}^{4}_{2}He$, a helium nucleus, has a 2+ charge. By convention, however, ion charges are not shown in balanced equations for nuclear reactions.

$${}^{226}_{88}Ra \longrightarrow {}^{4}_{2}He + {}^{222}_{86}Rn$$

Historical Perspective

The Nuclear Age Dawns

On August 2, 1939, as the world was on the brink of World War II, Albert Einstein sent a letter to President Franklin D. Roosevelt. In this letter, which profoundly changed the course of history, Einstein called attention to work being done on the physics of the atomic nucleus. He said he and others believed this work suggested the possibility that "uranium may be turned into a new and important source of energy . . . and [that it was] conceivable . . . that extremely powerful bombs of a new type may thus be constructed. . . . "

Powerful indeed! Einstein's letter was the beginning of the Manhattan Project, the project that led to the detonation of the first atomic bomb at 5:30 AM on July 16, 1945, in the desert of New Mexico. The rest of the world would learn the truth of the power locked in the atomic nucleus a few weeks later, on August 6 and August 9, when the United States used atomic weapons against Japan. J. Robert Oppenheimer, the director of the atomic bomb project, is said to have recalled the following words from the sacred Hindu epic, Bhagavad-Gita, at the moment of the explosion of the first atomic bomb:

The first nuclear bomb detonation in July, 1945, at White Sands, New Mexico. The Doomsday Clock, from *The Bulletin of Atomic Scientists,* appears at lower right. *(Los Alamos National Laboratory, Bulletin of Atomic Scientists)*

> If the radiance of a thousand suns
> Were to burst at once upon the sky,
> That would be like the splendor of the Mighty One . . .
> I am become Death,
> The shatterer of worlds.

In the 50 years since the first—and thankfully only—use of atomic weapons in war, more powerful weapons have been developed and stockpiled by a number of nations. With the end of the Cold War, fears of a nuclear holocaust are fading, but they are being replaced to some extent by the concern that Third World nations have developed nuclear weapons. The respected magazine *The Bulletin of Atomic Scientists* has used for many years the symbol of a clock with its hands near midnight, illustrating the danger faced by the world from atomic weapons. Even with the end of the Cold War, the hands have moved back only a little.

See R. Rhodes: *The Making of the Atomic Bomb,* New York, Simon and Schuster, 1986. This is a comprehensive, very readable history of nuclear physics and the events leading up to the development of the atomic bomb.

In this balanced equation the subscripts are the atomic numbers and the superscripts are the mass numbers.

Atoms in molecules and ions are rearranged in a chemical change; they are not created or destroyed. The number of atoms remains the same. Similarly, in nuclear reactions the total number of nuclear particles, or **nucleons** (protons and neutrons), remains the same. The essence of nuclear reactions, however, is that one nucleon can change into a different nucleon. A proton can change to a neutron or a neutron can change to a proton, but the total number of nucleons remains the same. Therefore, the sum of the mass numbers of reacting nuclei must *equal* the sum of the mass numbers of the nuclei produced. Furthermore, to maintain charge balance, the sum of the atomic numbers of the products must equal the sum of the atomic numbers of the reactants. These principles may be verified for the preceding nuclear equation:

	$^{226}_{88}Ra$	$\longrightarrow$	$^{4}_{2}He$	+	$^{222}_{86}Rn$
	radium-226		α particle		radon-222
Mass number: (protons + neutrons)	226	$\longrightarrow$	4	+	222
Atomic number: (protons)	88	$\longrightarrow$	2	+	86

Nuclear Reactions Involving α and β Particles

One way a radioactive isotope can disintegrate or decay is to eject an α particle from the nucleus as illustrated by the conversion of radium to radon and by the following reaction:

	${}^{234}_{92}\text{U}$	$\longrightarrow$	${}^{4}_{2}\text{He}$	+	${}^{230}_{90}\text{Th}$
	uranium-234		α particle		thorium-230
Mass number	234	$\longrightarrow$	4	+	230
Atomic number	92	$\longrightarrow$	2	+	90

Notice that in α emission the *atomic number decreases by two* units and the *mass number decreases by four* units for each α particle emitted.

Emission of a β particle is another way for an isotope to decay. For example, loss of a β particle by uranium-239 is represented by

	${}^{239}_{92}\text{U}$	$\longrightarrow$	${}^{0}_{-1}\beta$	+	${}^{239}_{93}\text{Np}$
	uranium-239		β particle		neptunium-239
Mass number	239	$\longrightarrow$	0	+	239
Atomic number	92	$\longrightarrow$	-1	+	93

Because a β particle has a charge of -1, electric balance requires that the atomic number of the product be one unit *greater* by one than that of the reacting nucleus. The mass number does not change, however. The mass number of 0 for the electron is due to the small mass of the particle (only 1/1836 the mass of a proton).

How does a nucleus, composed only of protons and neutrons, eject an electron? It is generally accepted that a series of steps is involved, but the net process is

$$\underset{\text{neutron}}{{}^{1}_{0}\text{n}} \longrightarrow \underset{\text{electron}}{{}^{0}_{-1}\beta} + \underset{\text{proton}}{{}^{1}_{1}\text{p}}$$

where we use the symbol p for a proton. The ejection of a β particle always means that a new element is formed with an atomic number *one unit greater* than the decaying nucleus.

Radioactive Decay Series

In some instances, nuclear reactions result in the formation of an isotope that is also radioactive. When this happens, the initial nuclear reaction is followed by a second, and if the situation is repeated, a third and a fourth, and so on, until finally a stable nonradioactive isotope is formed to end the series. A sequence of reactions like this is called a **radioactive decay series.** Radioactive decay series are particularly notable because they occur naturally in uranium ores.

Let us select ${}^{238}_{92}\text{U}$, the most abundant of three naturally occurring uranium isotopes, to illustrate a natural radioactive decay series. Like most uranium isotopes, $^{238}_{92}U$ is an α particle emitter. The loss of an α particle from $^{238}_{92}U$ results in formation of $^{234}_{90}Th$, but this product is also radioactive. Thorium-234 decomposes by β emission to $^{234}_{91}Pa$, and $^{234}_{91}Pa$ undergoes β emission to give $^{234}_{92}U$. Following this comes a series of further α and β emissions, eventually ending in the nonradioactive isotope, $^{206}_{82}Pb$. Radioactive decay series like this are often portrayed in a graph plotting atomic number and mass number (Figure 24.2). In all, the radioactive decay series converting $^{238}_{92}U$ to $^{206}_{82}Pb$ involves 14 sequential reactions and the emission of a total of eight α and six β particles.

An equation can be written for each step in this decay sequence. The first four steps in the ^{238}U radioactive decay series are:

Step 1. $$^{238}_{92}U \longrightarrow {}^{234}_{90}Th + {}^{4}_{2}\alpha$$

Step 2. $$^{234}_{90}Th \longrightarrow {}^{234}_{91}Pa + {}^{0}_{-1}\beta$$

Step 3. $$^{234}_{91}Pa \longrightarrow {}^{234}_{92}U + {}^{0}_{-1}\beta$$

Step 4. $$^{234}_{92}U \longrightarrow {}^{230}_{90}Th + {}^{4}_{2}\alpha$$

In spontaneous nuclear reactions such as those shown here, the reactant nucleus is often called the "parent" and the product is called the "daughter."

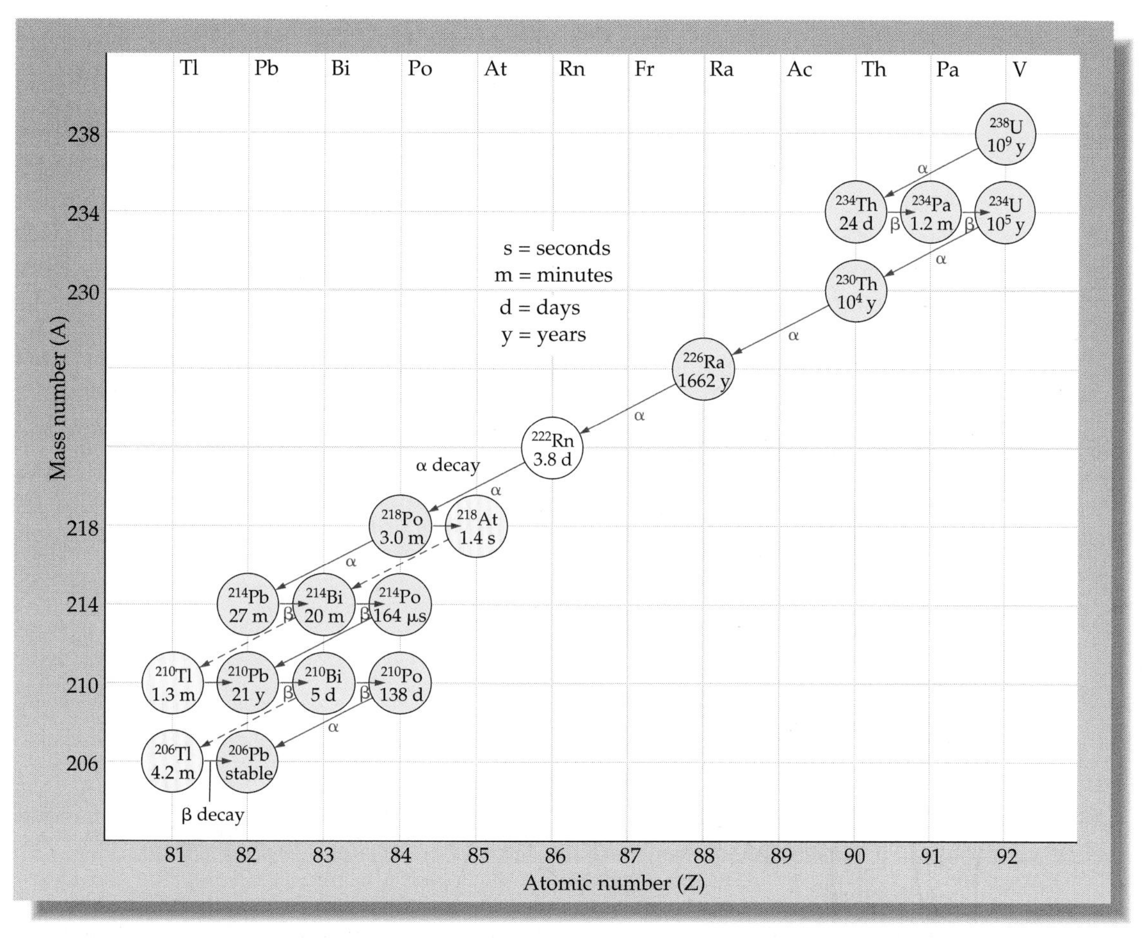

Figure 24.2 The ^{238}U radioactive decay series. This radioactive decay series begins with $^{238}_{92}U$ and ends with $^{206}_{82}Pb$. In the first step, $^{238}_{92}U$ emits an α particle to give $^{234}_{90}Th$. This radioactive isotope then emits a β particle to give $^{234}_{91}Pa$. Further α and β emissions occur as shown to reach the final product, $^{206}_{82}Pb$. The half-life is given for each isotope (see Section 24.4).

As shown in Figure 24.2, a series of radioactive products is formed as intermediates in this reaction series, including $^{226}_{88}Ra$ and several polonium isotopes. None of these isotopes have very long half-lives, so they will be present in uranium samples only in small amounts. It is a credit to Marie Curie's abilities as a chemist that she was able to extract sufficient radium and polonium from uranium ore to identify these new elements.

Marie Curie (page 64) succeeded in isolating 1 g of radium from 7 tons of pitchblende (a uranium ore).

The $^{238}_{92}U$ radioactive decay series is also the source of the environmental hazard radon, mentioned in the *Chemical Puzzler.* Radon is being formed continuously in our environment by decay of ^{238}U. It is one of the noble gases, and it is radioactive, with a half-life of 3.82 days. Because radon is chemically inert, it is not trapped by chemical processes in the soil or water and is free to seep up from the ground and into underground mines or into homes through pores in block walls, cracks in the basement floor or walls, or around pipes. When inhaled by humans, radon-222 decays inside the lungs to give polonium, a radioactive element that is not a gas and is not chemically inert:

$$^{222}_{86}Rn \longrightarrow {}^{4}_{2}He + {}^{218}_{84}Po \qquad t_{1/2} = 3.82 \text{ days}$$

$$^{218}_{84}Po \longrightarrow {}^{4}_{2}He + {}^{214}_{82}Pb \qquad t_{1/2} = 3.11 \text{ minutes}$$

Polonium-218 lodges in body tissues where it undergoes α decay to give lead-214, itself a radioactive isotope. The range of an α particle is quite small, perhaps 0.7 mm (about the thickness of a sheet of paper). This is, however, approximately the thickness of the epithelial cells of the lungs, so the radiation can damage these tissues and induce lung cancer.

Virtually every home in the United States is believed to have some level of radon gas. To test for the presence of the gas, you can purchase testing kits, such as the one shown in the *Chemical Puzzler* (page 1088). There is currently a great deal of controversy over the level of radon that is considered "safe." The U. S. Environmental Protection Agency has set a standard of 4 pCi/L of air as an "action level." Some people believe 1.5 pCi is close to the average level. It is estimated that only about 2% of homes contain over 8 pCi/L. If your home shows higher levels of radon gas, you should probably have it tested further and perhaps take corrective actions, such as sealing cracks around the foundation and in the basement. Keep in mind the relative risks involved, however (see page 12). A 1.5 pCi/L level of radon leads to a lung cancer risk about the same as the risk of your dying in an accident in your home.

The abbreviation pCi stands for picocurie, 1×10^{-12} Ci (curies). A curie is a unit of radioactivity. It is further described on page 1120.

Example 24.1 Radioactive Decay Series

Another radioactive decay series begins with $^{235}_{92}U$ and ends with $^{207}_{82}Pb$.

1. What is the total number of α and β particles emitted in this series?
2. The first three steps of this series involve (in order) α, β, and α emissions. Write nuclear equations for each of these steps.

Solution

1. Mass declines by 28 mass units (235 − 207) in this series. Because a decrease in mass can only occur with α emission, we conclude that seven α particles must be emitted. For each α emission, the atomic number must decrease by 2, so emission of seven α particles causes the atomic number to decrease by

14. The actual decrease in atomic number is 10, however (92 − 82). Four β particles cause the atomic number to increase by 4. This radioactive decay sequence involves loss of seven α and four β particles.

2. **Step 1.** $^{235}_{92}U \longrightarrow ^{231}_{90}Th + ^{4}_{2}He$

Step 2. $^{231}_{90}Th \longrightarrow ^{231}_{91}Pa + ^{0}_{-1}\beta$

Step 3. $^{231}_{91}Pa \longrightarrow ^{227}_{89}Ac + ^{4}_{2}He$

***Exercise* 24.1** Radioactive Decay Series

1. Six α and 4 β particles are emitted in the thorium-232 radioactive decay series. What is the final product in this series?
2. The first three steps in the thorium-232 decay series are α, β, and β emissions, respectively. Write a nuclear equation for each step.

Other Types of Radioactive Decay

In addition to radioactive decay by emission of α, β, or γ radiation, other decay processes are now known. Some nuclei decay, for example, by emission of a **positron,** $^{0}_{+1}\beta$. As its symbol suggests, a positron has a mass equal to the mass of an electron, and a 1+ charge. Positron emission by polonium-207 leads to the formation of bismuth-207, for example.

Positrons were discovered by Carl Anderson (1905–1971) in 1932. The positron is one of a group of particles that are known as "antimatter." If matter and antimatter particles collide, mutual annihilation occurs, with energy being emitted.

	$^{207}_{84}$**Po**	⟶	$^{0}_{+1}\beta$	+	$^{207}_{83}$**Bi**
	polonium-207		positron		bismuth-207
Mass number	207	⟶	0	+	207
Atomic number	84	⟶	+1	+	83

In contrast to β decay, positron decay leads to a *decrease* in the atomic number.

The atomic number is also reduced by one when **electron capture** occurs. In this process an inner-shell electron is captured by the nucleus.

	$^{7}_{4}$**Be**	+	$^{0}_{-1}$**e**	⟶	$^{7}_{3}$**Li**
	beryllium-7		electron		lithium-7
Mass number	7	+	0	⟶	7
Atomic number	4	+	−1	⟶	3

In an old nomenclature used in atomic physics the innermost electron shell was called the K shell, so the electron capture decay mechanism is sometimes called *K capture.*

In summary, a radioactive nucleus can decay in four common ways, as summarized in Figure 24.3.

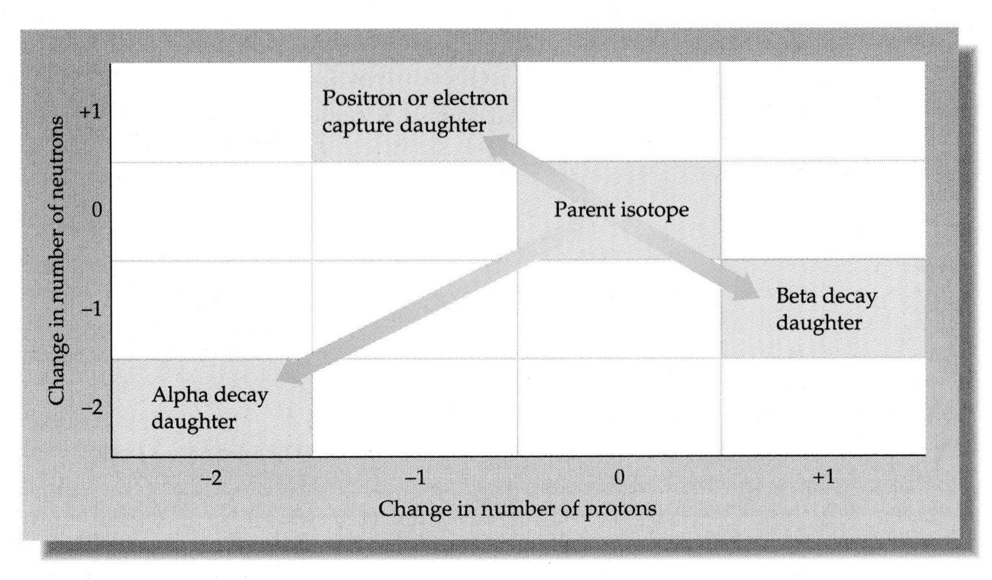

Figure 24.3 Predicting products of nuclear reactions. The change in mass number and atomic number for α emission, β emission, and positron ($_{+1}^{0}\beta$) emission, and electron capture are portrayed.

Example 24.2 Nuclear Reactions

Complete the following equations. Give the symbol, mass number, and atomic number of the species indicated by the question mark (?).

1. $_{18}^{37}\text{Ar} + _{-1}^{0}\text{e} \longrightarrow ?$
2. $_{6}^{11}\text{C} \longrightarrow _{5}^{11}\text{B} + ?$
3. $_{16}^{35}\text{S} \longrightarrow _{17}^{35}\text{Cl} + ?$
4. $_{15}^{30}\text{P} \longrightarrow _{+1}^{0}\beta + ?$

Solution

1. This is an electron capture reaction. The product has a mass number of 37 + 0 = 37 and an atomic number of 18 − 1 = 17. The symbol for the product is $_{17}^{37}\text{Cl}$.
2. This reaction is recognized as positron ($_{+1}^{0}\beta$) emission. By choosing this particle, the sum of the atomic numbers (6 = 5 + 1) and the mass numbers (11) on either side of the reaction are equal.
3. Beta ($_{-1}^{0}\beta$) emission is required to balance the mass numbers (35) and atomic numbers (16 = 17 − 1) on both sides of the equation.
4. The product nucleus is $_{14}^{30}\text{Si}$. This balances the mass numbers (30) and atomic numbers (15 = 1 + 14) on both sides of the equation.

Exercise 24.2 Nuclear Reactions

Complete the following nuclear equations. Indicate the symbol, the mass number, and the atomic number of "?".

1. $_{7}^{13}\text{N} \longrightarrow _{6}^{13}\text{C} + ?$

2. $^{41}_{20}Ca + ^{0}_{-1}e \longrightarrow ?$
3. $^{90}_{38}Sr \longrightarrow ^{90}_{39}Y + ?$
4. $^{22}_{11}Na \longrightarrow ? + ^{0}_{+1}\beta$

24.3 STABILITY OF ATOMIC NUCLEI

The fact that some nuclei are unstable (radioactive), and others are stable (nonradioactive), leads us to consider the reasons for stability. Figure 24.4 shows the naturally occurring isotopes of the elements from hydrogen to bismuth. It is quite astonishing that there are so few. Why not hundreds more?

In its simplest and most abundant form, hydrogen has only one nuclear particle, the proton. In addition, the element has two other well-known isotopes: nonradioactive deuterium, with one proton and one neutron ($^{2}_{1}H$ = D), and radioactive tritium, with one proton and two neutrons ($^{3}_{1}H$ = T). Helium, the next element, has two protons and two neutrons in its most stable isotope. At the end of the actinide series is element 103, lawrencium, one isotope of which has a mass number of 257 and 154 neutrons. From hydrogen to lawrencium, except for $^{1}_{1}H$ and $^{3}_{2}He$, the mass numbers of stable isotopes are always *at least twice as large* as the atomic number. Except for $^{1}_{1}H$ and $^{3}_{2}He$, every isotope of every element has a nucleus containing *at least* one neutron for every proton. Apparently

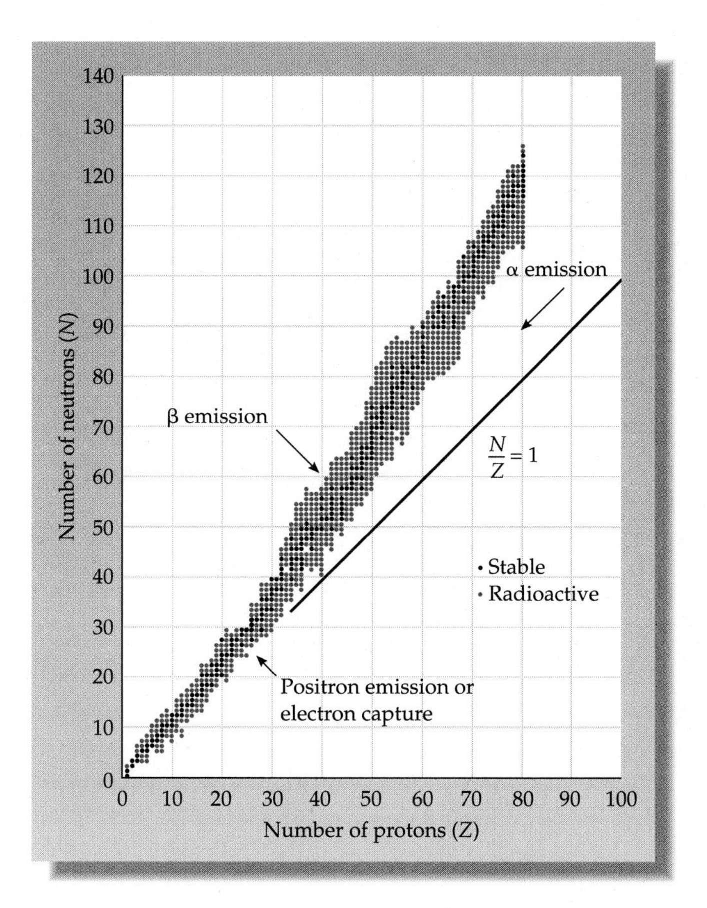

Figure 24.4 Stable and unstable isotopes. A plot of the number of neutrons (N) versus the number of protons (Z) for stable (black circles) and radioactive (red circles) isotopes from hydrogen ($Z = 1$) through bismuth ($Z = 83$). The stable isotopes lie on a narrow "peninsula of stability." Radioactive isotopes above the band of stable isotopes decay by β emission. Radioactive isotopes below this band decay by either positron emission or electron capture. These modes of decay lead to formation of daughter nuclei that are closer to or within the peninsula of stability.

the tremendous *repulsive* forces between the positively charged protons in the nucleus are moderated by the presence of neutrons with no electric charge.

1. For elements up to Ca ($Z = 20$), the stable isotopes usually have equal numbers of protons and neutrons, or perhaps one more neutron than protons. Examples include $^{7}_{3}Li$, $^{12}_{6}C$, $^{16}_{8}O$, and $^{32}_{16}S$.
2. Beyond calcium the neutron–proton ratio becomes increasingly greater than 1. The band of stable isotopes deviates more and more from the line $N = Z$. It is evident that more neutrons are needed for nuclear stability in the heavier elements. For example, whereas one stable isotope of Fe has 26 protons and 30 neutrons, one of the stable isotopes of platinum has 78 protons and 117 neutrons.
3. Beyond bismuth (83 protons and 126 neutrons) all isotopes are unstable and radioactive. There is apparently no nuclear "superglue" strong enough to hold heavy nuclei together. Furthermore, the rate of disintegration becomes greater the heavier the nucleus. For example, half of a sample of $^{238}_{92}U$ disintegrates in 4.5 billion years, whereas half of a sample of $^{257}_{103}Lr$ is gone in only 8 s.
4. A careful look at Figure 24.4 shows other interesting features. First, elements of even atomic number have more stable isotopes than do those of odd atomic number. Second, stable isotopes generally have an *even* number of neutrons. For elements of odd atomic number, most stable isotopes have an even number of neutrons. To emphasize these points, of the more than 300 stable isotopes represented in Figure 24.4, roughly 200 have an even number of neutrons *and* an even number of protons. Only about 120 have an odd number of either protons *or* neutrons. Only five isotopes ($^{2}_{1}H$, $^{6}_{3}Li$, $^{10}_{5}B$, $^{14}_{7}N$, and $^{180}_{73}Ta$) have odd numbers of *both* protons and neutrons.

The Band of Stability and Type of Radioactive Decay

The black dots in Figure 24.4 represent stable isotopes. All other isotopes (the red dots) are unstable. The unstable isotopes decay by processes that give stable products. The chart can help us predict what type of decay will be observed.

All elements beyond Bi ($Z = 83$) are unstable—that is, radioactive—and many decay by ejecting an α particle. For example, americium, the radioactive element used in smoke alarms, decays in this manner:

$$^{243}_{95}Am \longrightarrow ^{4}_{2}He + ^{239}_{93}Np$$

Beta emission occurs in isotopes that have too many neutrons to be stable, that is, isotopes *above* the peninsula of stability in Figure 24.4. During β decay the atomic number increases by one and the mass number remains constant:

$$^{60}_{27}Co \longrightarrow ^{0}_{-1}\beta + ^{60}_{28}Ni$$

Isotopes that have too few neutrons—isotopes *below* the peninsula of stability—attain stability by positron emission or by electron capture. Both processes lead to a daughter nucleus with a lower atomic number and the same mass number:

$$^{13}_{7}N \longrightarrow ^{0}_{+1}\beta + ^{13}_{6}C$$

$$^{41}_{20}Ca + ^{0}_{-1}e \longrightarrow ^{41}_{19}K$$

Example 24.3 Nuclear Stability

Identify the probable mode or modes of decay, and write the symbol for the product formed when the following unstable isotopes undergo radioactive decay:

1. Oxygen-15, $^{15}_{8}O$
2. Uranium-234, $^{234}_{90}U$
3. Fluorine-20, $^{20}_{9}F$
4. Manganese-56, $^{56}_{25}Mn$

Solution Each of the isotopes has been identified as unstable. Uranium-234, with an atomic number of 92, undergoes α emission, a common mode of decomposition for elements of high mass number and atomic number. For the remaining isotopes, the problem is to determine whether there are too many or too few neutrons in the nucleus. To do this, compare the mass number against the atomic weight of the element. If the mass number of the isotope is lower than the atomic weight, then there are probably too few neutrons; if the mass number is higher than the atomic weight, then there are too many neutrons. From this information, we can predict the possible mode of decomposition as either $^{0}_{-1}\beta$ emission, for too many neutrons, or $^{0}_{+1}\beta$ emission or electron capture, for too few neutrons. (It is not possible to choose between the latter two modes of decay.)

1. Oxygen-15 has too few neutrons and is expected to decay either by $^{0}_{+1}\beta$ emission or electron capture. Both processes give nitrogen-15 as the product. (In this instance, the actual process is $^{0}_{+1}\beta$ emission.)
2. Uranium-234 is expected to decay by α emission to give thorium-230.
3. With an excess of neutrons, fluorine-20 is predicted to decay by $^{0}_{-1}\beta$ emission, giving neon-20.
4. Manganese-56 has an excess of neutrons, and is thus expected to decay by $^{0}_{-1}\beta$ emission to form iron-56.

Exercise 24.3 Nuclear Stability

Write an equation for the probable mode of decay for each of the following unstable isotopes.

1. Silicon-32, $^{32}_{14}Si$
2. Titanium-45, $^{45}_{22}Ti$
3. Plutonium-239, $^{239}_{94}Pu$
4. Potassium-42, $^{42}_{19}K$

Binding Energy

As proved by Ernest Rutherford's experiments (see Chapter 2), the nucleus of an atom is extremely small. Yet the nucleus can contain up to 83 protons before becoming unstable. This is evidence that a very strong short-range binding force must be able to overcome the electrostatic repulsive forces between protons. A measure of the force holding the nucleus together is the **nuclear binding energy.** This energy (E_b) is defined as the negative of the energy change (ΔE) that would occur if a nucleus were formed directly from its component protons and neutrons. For example, if a mole of protons and a mole of neutrons directly formed a mole of deuterium nuclei, the energy change would be more than 2×10^8 kJ! Or, put in the opposite way, you would need to add 2.15×10^8 kJ/mol to separate the protons and neutrons in a mole of deuterium nuclei:

$$^{1}_{1}\mathrm{H} + {}^{1}_{0}\mathrm{n} \longrightarrow {}^{2}_{1}\mathrm{H} \qquad \Delta E = -2.15 \times 10^8 \text{ kJ/mole}$$

$$\text{Binding energy} = -\Delta E = E_b = +2.15 \times 10^8 \text{ kJ/mole}$$

This nuclear synthesis reaction is highly exothermic (and so E_b is very positive), an indication of the strong attractive forces holding the nucleus together. The deuterium nucleus is more stable than an isolated proton and an isolated neutron, just as the H_2 molecule is more stable than two isolated H atoms. Recall, however, that the energy released when a mole of H—H covalent bonds form is only 436 kJ. This is a tiny fraction of the energy released when protons and neutrons coalesce to form a nucleus.

To understand the enormous energy released during the formation of an atomic nucleus, we turn to an experimental observation and a theory. The experimental observation is that the mass of a nucleus is always less than the sum of the masses of its constituent protons and neutrons:*

$$\underset{1.007825\text{ g/mol}}{^{1}_{1}\mathrm{H}} \quad + \quad \underset{1.008665\text{ g/mol}}{^{1}_{0}\mathrm{n}} \quad \longrightarrow \quad \underset{2.01410\text{ g/mol}}{^{2}_{1}\mathrm{H}}$$

$$\begin{aligned}\text{Change in mass} = \Delta m &= \text{mass of product} - \text{sum of masses of reactants}\\ &= 2.01410\text{ g/mol} - 2.016490\text{ g/mol}\\ &= -0.00239\text{ g/mol}\end{aligned}$$

The quantity Δm is called the mass defect.

The theory is that the "missing mass," Δm, has been converted to energy, and this is the energy we described as the binding energy.

The relation between mass and energy is contained in Albert Einstein's 1905 theory of special relativity, which holds that mass and energy are simply different manifestations of the same quantity. Einstein stated that the energy of a body is equivalent to its mass times the square of the speed of light, $E = mc^2$. So, to calculate the energy change in a process in which the mass has changed, the equation becomes

$$\Delta E = (\Delta m)\,c^2 \qquad [24.1]$$

We can calculate ΔE in joules if the change in mass is given in kilograms and the velocity of light is in meters per second (because $1\text{ J} = 1\text{ kg} \cdot \text{m}^2/\text{s}^2$). For

*Nuclear binding energy is formally defined based on the masses of the nuclear particles. We can use atomic masses, however, in calculating the changes in mass. In the example here, the mass of $^{1}_{1}\mathrm{H}$ is used instead of the mass of a proton, and the mass of $^{2}_{1}\mathrm{D}$ is used instead of the mass of a deuterium nucleus. Atomic masses can easily be obtained from tables. By using atom mass we are including the mass of the extranuclear electrons in the masses of the reactants and products in the calculation.

the formation of one mole of deuterium nuclei from one mole of protons and one mole of neutrons, we have

$$\Delta E = (-2.39 \times 10^{-6}\ \text{kg/mole})(3.00 \times 10^{8}\ \text{m/s})^2 = -2.15 \times 10^{11}\ \text{J/mole}$$
$$E_b = 2.15 \times 10^{8}\ \text{kJ/mol of } {}^{2}_{1}\text{H nuclei}$$

This is the value of ΔE given at the beginning of this section for the change in energy when a mole of protons and a mole of neutrons form a mole of deuterium nuclei.

A helium nucleus is composed of two protons and two neutrons. As expected, the binding energy, E_b, is very large, even larger than for deuterium:

$$2\ {}^{1}_{1}\text{H} + 2\ {}^{1}_{0}\text{n} \longrightarrow {}^{4}_{2}\text{He} \qquad E_b = +2.73 \times 10^{9}\ \text{kJ/mol of helium nuclei}$$

When comparing nuclear stabilities, scientists generally calculate the **binding energy per nucleon.** For helium-4 this is

$$\frac{E_b}{\text{mol nucleons}} = \frac{2.73 \times 10^{9}\ \text{kJ/mol of } {}^{4}_{2}\text{He}}{4\ \text{mol nucleons/mol of } {}^{4}_{2}\text{He}} = 6.83 \times 10^{8}\ \text{kJ/mol nucleons}$$

The greater the binding energy per nucleon, the greater is the stability of the nucleus. Scientists have calculated the binding energies of a great number of nuclei and have plotted them as a function of mass number (Figure 24.5). It is very interesting—and important—that the point of maximum stability occurs in the vicinity of iron-56, ${}^{56}_{26}\text{Fe}$. This means that *all elements are thermodynamically unstable with respect to iron.* That is, very heavy nuclei may split, or undergo **fission,** with the release of enormous quantities of energy, to give more stable nuclei with atomic numbers nearer iron. In contrast, two very light nuclei may come together and undergo **fusion** exothermically to form heavier nuclei.

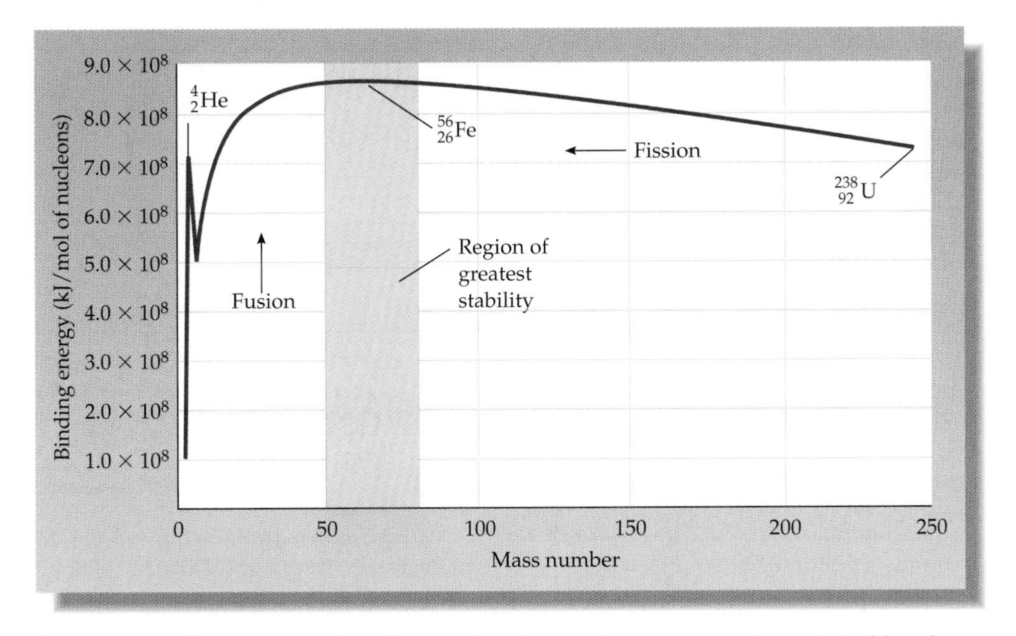

Figure 24.5 The relative stability of nuclei. This "curve of binding energy" was derived by calculating the binding energy per nucleon (in kilojoules per mol of nucleons) for the most abundant isotope of the elements from hydrogen to uranium.

Example 24.4 Nuclear Binding Energy

Calculate the binding energy (in kJ/mole) and the binding energy per nucleon (in kJ/mole nucleon) for carbon-12.

Solution The following reaction results in formation of carbon-12:

$$6\ {}^{1}_{1}\text{H} + 6\ {}^{1}_{0}\text{n} \longrightarrow {}^{12}_{6}\text{C}$$

The mass of ${}^{1}_{1}H$ is 1.00783 g/mol and the mass of ${}^{1}_{0}n$ is 1.00867 g/mol. Carbon-12, ${}^{12}_{6}C$, is the standard for the atomic masses in the periodic table, and its mass is defined as exactly 12.000 . . . g/mol. To determine binding energy we must first determine the difference in mass of the products and reactants in this reaction:

$$\begin{aligned}\Delta m &= 12.000000 - [(6 \times 1.00783) + (6 \times 1.00867)]\\ &= -9.9000 \times 10^{-2}\ \text{g/mol}\end{aligned}$$

The binding energy is calculated using Equation 24.1. Using the mass in kilograms, and the speed of light in meters per/second gives an answer for the binding energy in joules:

$$\begin{aligned}E_b &= -(\Delta m)c^2 = -(-9.9 \times 10^{-5}\ \text{kg/mol})(3.00 \times 10^{8}\ \text{m/s})^2\\ &= 8.91 \times 10^{12}\ \text{J/mol} = 8.91 \times 10^{9}\ \text{kJ/mol}\end{aligned}$$

The binding energy per nucleon is determined by dividing the binding energy by the number of nucleons, which in this instance is 12.

$$\begin{aligned}\frac{E_b}{\text{mol nucleons}} &= \frac{8.91 \times 10^{9}\ \text{kJ mol}^{-1}}{12\ \text{nucleons mol}^{-1}}\\ &= 7.43 \times 10^{8}\ \text{kJ/mol nucleons}\end{aligned}$$

Exercise 24.4 Binding Energy

Calculate the binding energy, in kilojoules per mole, for the formation of lithium-6:

$$3\ {}^{1}_{1}\text{H} + 3\ {}^{1}_{0}\text{n} \longrightarrow {}^{6}_{3}\text{Li}$$

The necessary masses are ${}^{1}_{1}H = 1.00783$ g/mol, ${}^{1}_{0}n = 1.00867$ g/mol, and ${}^{6}_{3}Li =$ 6.015125 g/mol.

24.4 RATES OF DISINTEGRATION REACTIONS

Half-Life

Cobalt-60 is used as a source of β particles and γ rays to treat malignancies in the human body. Although the isotope is radioactive, its decomposition is slow; only half of a sample of cobalt-60 decays in a little over 5 years. On the other hand, copper-64, which is used in the form of copper acetate to detect brain tumors, decays much more rapidly; half of the radioactive copper decays in slightly less than 13 h.

The half-lives for radioactive isotopes cover a wide range. Uranium-238 has one of the longer half-lives, 4.51×10^9 years. This length of time is close to the age of the earth, estimated to be $4.5–4.6 \times 10^9$ years. We can conclude that roughly half of the uranium-238 present when the planet was formed is still around. At the other end of the scale is element-112, whose discovery was reported in 1996. Two atoms of this element were created by nuclear reactions (see *A Closer Look, The Search for Element 114*, page 1112). Element-112 decomposed with a half-life of 280 μs (microseconds, $1\ \mu s = 1 \times 10^{-6}$ s).

The half-life is typically reported whenever a radioactive element is mentioned. You will have noticed this already in Figure 24.2, describing the radioactive decay series that begins with uranium-238. In that figure, a half-life was indicated for each of the isotopes formed in the decay series. Half-life can be used to assess how long a radioactive element will persist before its activity becomes negligible. It can also be used to estimate the amount of a radioactive substance that remains after a given time (Figure 24.6). The following example and exercise illustrate this point further.

The half-life for a radioactive isotope is a constant, independent of temperature and of the number of nuclei present.

Example 24.5 Using Half-Life

Tritium (^3_1H), a radioactive isotope of hydrogen, has a half-life of 12.3 years.

1. If you begin with 1.5 mg of this isotope, how many milligrams remain after 49.2 years?
2. How long will it take for a sample of tritium to decay to one eighth of its activity?
3. Estimate the length of time necessary for the sample to decay to 1% of its original activity.

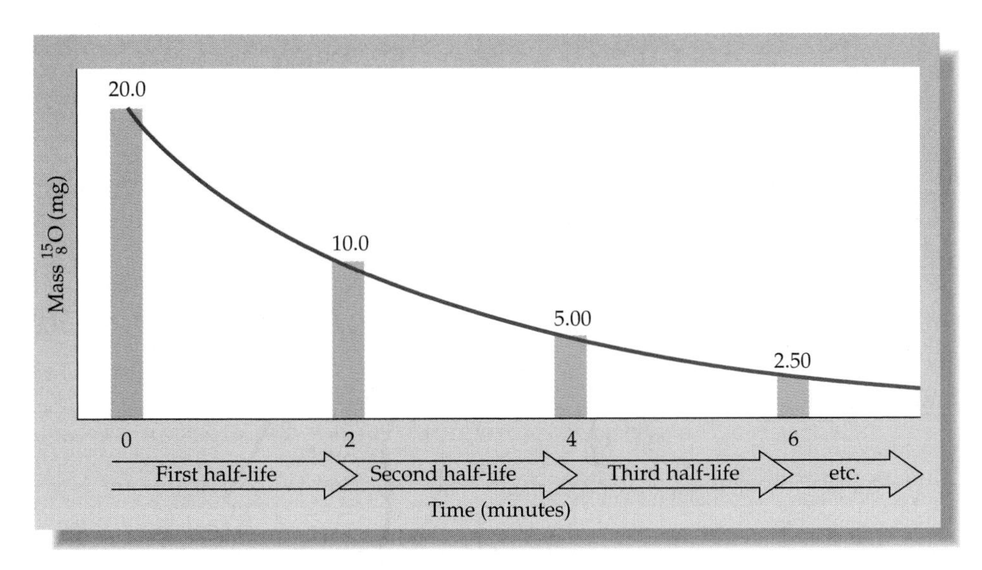

***Figure* 24.6** **Decay of 20 mg of oxygen-15.** After each half-life period of 2.0 min, the quantity present at the beginning of the period is reduced by half.

Solution

1. First we find the number of half-lives in 49.2 years.

$$\text{Number of half-lives} = 49.2 \text{ years} \times \left(\frac{1 \text{ half-life}}{12.3 \text{ years}}\right) = 4.0$$

This means that the initial quantity of $^{3}_{1}H$ is reduced four times by 1/2:

$$1.5 \times \frac{1}{2} \times \frac{1}{2} \times \frac{1}{2} \times \frac{1}{2} = 1.5 \text{ mg} \left(\frac{1}{2}\right)^4 = 0.094 \text{ mg}$$

2. To reach one eighth of its original activity, the sample must decay over a period of 3 half-lives; 3×12.3 years, or 36.9 years, is required.
3. During each half-life, the amount of the sample decreases by 1/2. After 6 half-lives, the sample will be 1/64th (0.0156 times) its original size. After 7 half-lives, the amount will be 1/128 (0.0078 times) its original size. One percent (0.01) lies between 6 and 7 half-lives; between 74 and 86 years must elapse before the sample reaches 1% of its original value. The exact value is 82 years; its calculation will be considered next.

***Exercise* 24.5** Using Half-Life

Strontium-90, $^{90}_{38}Sr$, is a radioactive isotope produced when an atomic bomb explodes. It is a particular hazard because of its long half-life and because it is taken up by the body and concentrates in bone and bone marrow (it replaces Ca^{2+} in bone). The half-life of strontium-90 is 28.1 years.

1. A sample of this isotope emits 2000 β particles per minute. How many years are required for the level of radioactivity to be reduced to 250 β particles per minute?
2. Estimate the length of time required for the radioactivity of strontium-90 to decrease to 1/16th (about 6.2%) of its original value.

Rate of Radioactive Decay

To determine the half-life of a radioactive element, the *rate of decay* must be measured. That is, we must measure the number of atoms that disintegrate per second or per hour or per year.

The rate of nuclear decay is often described in terms of the **activity** *(A)* of the sample, the number of disintegrations observed per unit time. The activity is *proportional* to the number of radioactive atoms present *(N)*.

$$\text{Rate of radioactive decay} \equiv \text{activity } (A) \propto \text{number of radioactive atoms present } (N)$$

This proportionality can also be expressed in the form

$$A = kN \qquad [24.2]$$

$$\frac{\text{Disintegrations}}{\text{Time}} = \frac{\text{disintegrations}}{(\text{number of atoms})(\text{time})} \times \text{number of atoms}$$

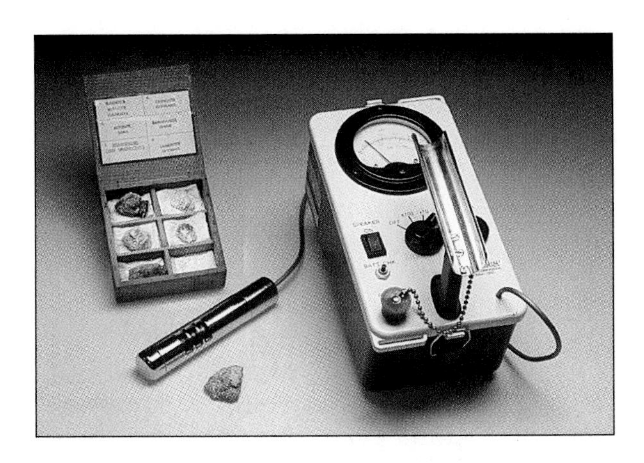

***Figure* 24.7 A Geiger Müller counter.** A charged particle (e.g., an α or β particle) enters the gas-filled tube and ionizes the gas. These gaseous ions migrate to electrically charged plates and thereby give rise to a "pulse," or momentary flow of electric current. The current is amplified and used to operate a counter. A sample of carnotite, a mineral containing uranium oxide, is also shown in the figure. *(C. D. Winters)*

where k is the proportionality constant, or *decay constant.* In the language of kinetics, Equation 24.2 is simply a rate law that is first order in the number of atoms in the sample, and k is the rate constant.

The activity of a sample can be measured with a device such as a Geiger-Müller counter (Figure 24.7). Let us say the activity is measured at some time t_0 and then measured again after a few minutes, hours, or days. If the initial activity is A_0 at t_0, then a second measurement gives a smaller activity A at a later time t. From Equation 24.2, you can see that the ratio of the activity A at some time t to the activity at the beginning of the experiment (A_0) must be equal to the ratio of the number of radioactive atoms N present at time t to the number present at the beginning of the experiment (N_0).

$$\frac{A}{A_0} = \frac{kN}{kN_0}$$

or

$$\frac{A}{A_0} = \frac{N}{N_0}$$

This means that experimental information is related directly to the fraction of radioactive atoms remaining in a sample after some time has passed.

Equation 24.3 relates the period over which a sample is observed (t) to the fraction of radioactive atoms present after that amount of time has passed:

$$\ln \frac{N}{N_0} = -kt \qquad [24.3]$$

Equation 24.3 is similar to the integrated rate equation for a first-order reaction. It differs from Equation 15.1 only in the fact that the equation refers to numbers of atoms (N and N_0) instead of concentrations.

Equation 24.3 is useful in three ways:

- If A/A_0 (and thus N/N_0) is measured in the laboratory over some period t, then k can be calculated. The decay constant k can then be used to determine the half-life of the sample, as illustrated in Example 24.6.
- If k is known, the fraction of a radioactive sample still present after some time t has elapsed can be calculated.
- If k is known for a radioactive isotope, you can calculate the time required for that isotope to decay to a fraction of the original activity.

Recognizing Equation 24.3 as a first-order integrated rate law, we are in a position to see how the half-life of a radioactive isotope $t_{1/2}$ is determined. The half-life is the time needed for half of the material present at the beginning of the experiment (N_0) to disappear. Thus, when time = $t_{1/2}$, then $N = 1/2N_0$. This means that

$$\ln \frac{1/2N_0}{N_0} = -kt_{1/2}$$

Solving for $t_{1/2}$ gives the equation that connects the half-life and decay constant:

$$t_{1/2} = \frac{0.693}{k} \quad [24.4]$$

Example 24.6 Determination of Half-Life

A sample of radon initially undergoes 7.0×10^4 α particle disintegrations per second (dps). After 6.6 days, it undergoes only 2.1×10^4 α particle dps. What is the half-life of this isotope of radon?

Solution Experiment has provided us with both A and A_0.

$$A = 2.1 \times 10^4 \text{ dps} \qquad A_0 = 7.0 \times 10^4 \text{ dps}$$

and the time ($t = 6.6$ days). We can therefore find the value of k. Because $N/N_0 = A/A_0$,

$$\ln \left(\frac{2.1 \times 10^4}{7.0 \times 10^4}\right) = -k(6.6 \text{ days})$$

$$\ln (0.30) = -k(6.6 \text{ days})$$

$$k = -\frac{\ln (0.30)}{6.6 \text{ days}} = -\frac{(-1.20)}{6.6 \text{ days}} = 0.18 \text{ day}^{-1}$$

and from k we can obtain $t_{1/2}$:

$$t_{1/2} = \frac{0.693}{k} = \frac{0.693}{0.18 \text{ day}^{-1}} = 3.8 \text{ days}$$

Example 24.7 Time and Radioactivity

Some high-level radioactive waste with a half-life $t_{1/2}$ of 200. years is stored in underground tanks. What time is required to reduce an activity of 6.50×10^{12} disintegrations per minute (dpm) to a fairly harmless activity of 3.00×10^{-3} dpm?

Solution The data give you the initial activity ($A_0 = 6.50 \times 10^{12}$ dpm) and the activity after some elapsed time ($A = 3.00 \times 10^{-3}$ dpm). To find the elapsed time t, you must first find k from the half-life:

$$k = \frac{0.693}{t_{1/2}} = \frac{0.693}{200. \text{ years}} = 0.00347 \text{ year}^{-1}$$

With k known, the time t can be calculated:

$$\ln\left(\frac{3.00 \times 10^{-3}}{6.50 \times 10^{12}}\right) = -[0.00347 \text{ year}^{-1}]t$$

$$-35.312 = -[0.00347 \text{ year}^{-1}]t$$

$$t = \frac{-35.312}{-[0.00347 \text{ year}^{-1}]}$$

$$= 1.02 \times 10^4 \text{ years}$$

Exercise 24.6 Determination of Half-Life

Technetium-99m, a γ-ray emitter, is used in medical imaging. A sample of this isotope emits 3.28×10^5 photons/s. After 1.0 h, the γ-ray emission has dropped to 2.92×10^5 photons/s. Calculate the half-life of technetium-99m.

Exercise 24.7 Rate of Radioactive Decay

Gallium citrate, containing the radioactive isotope gallium-67, is used medically as a tumor-seeking agent. It has a half-life of 77.9 h. How much time is needed for a sample of gallium citrate to decay to 10% of its original activity?

Radiochemical Dating

Willard Libby received the 1960 Nobel Prize in chemistry for developing carbon-14 dating techniques.

In certain situations the age of a material can be determined based on the rate of decay of a radioactive isotope. The best known example of this is the carbon-14 (or radiocarbon) dating technique that was developed in 1946 by Willard Libby (1908–1980), a noted nuclear chemist. Carbon-14 dating is widely used in such fields as anthropology to date historical artifacts.

Most of the carbon on earth is composed of two stable isotopes, carbon-12 and carbon-13 with isotopic abundances of 98.9 and 1.1%, respectively. In addition, however, natural carbon contains traces of a third isotope, carbon-14. This isotope is radioactive; it is a β emitter with a half-life of 5730 years. Carbon-14 is present on earth to the extent of about 1 out of 10^{12} atoms. A 1-g sample of carbon will show about 14 dpm (decompositions per minute), not a lot of radioactivity but easily detectable by modern methods.

You might wonder why carbon-14 is present at all because it is radioactive and its half-life is short, at least on the geological time scale. The reason is that carbon-14 is continually being formed in the upper atmosphere by nuclear reactions initiated by cosmic radiation:

$$^{14}_{7}\text{N} + ^{1}_{0}\text{n} \longrightarrow ^{14}_{6}\text{C} + ^{1}_{1}\text{H}$$

It is estimated that about 7.5 kg of carbon-14 is produced per year from cosmic radiation. Once formed, carbon-14 is oxidized to $^{14}CO_2$ and this becomes part of the carbon cycle, circulating worldwide through the atmosphere, oceans, and biosphere. The amount of carbon-14 on earth is fairly constant, in an amount that is related to the rates of its formation and decay.

Willard Libby and his apparatus for carbon-14 dating. *(Oesper Collection in the History of Chemistry/University of Cincinnati)*

The usefulness of carbon-14 for dating comes about in the following way. Plants absorb CO_2 and convert it to organic compounds, thereby incorporating the carbon-14 into living tissue. As long as the plant is alive, this process continues, and the amount of carbon-14 in the plant equals the amount in the atmosphere, about 14 dpm/g of carbon. When the plant dies, however, additional carbon-14 is no longer incorporated. In effect, the clock starts to tick, and the activity decreases over time. In 5730 years, the activity is 7 dpm/g, in 11,460 years it is 3.5 dpm/g. By measuring the activity, and knowing the half-life of carbon-14, it is possible to calculate when the plant died.

Carbon-14 dating is not without its limitations. The procedure assumes that the amount of carbon-14 has remained constant over time; that is, the amount of carbon-14 in the atmosphere thousands of years ago is the same as it is now. In addition, the procedure has some significant limitations in the range of applicable times. It is not possible to date an object that is less than about 100 years old; the radiation level from carbon-14 has not changed enough to be detected. The accuracy of any measurement, in fact, is only about ±100 years. Neither is it possible to determine ages much past about 40,000 years. At that point (nearly 7 half-lives), the radiation has decayed virtually to zero. But for the span of time between 100 and 40,000 years, the use of this technique has been highly valuable and has provided important information for scientists (Figure 24.8).

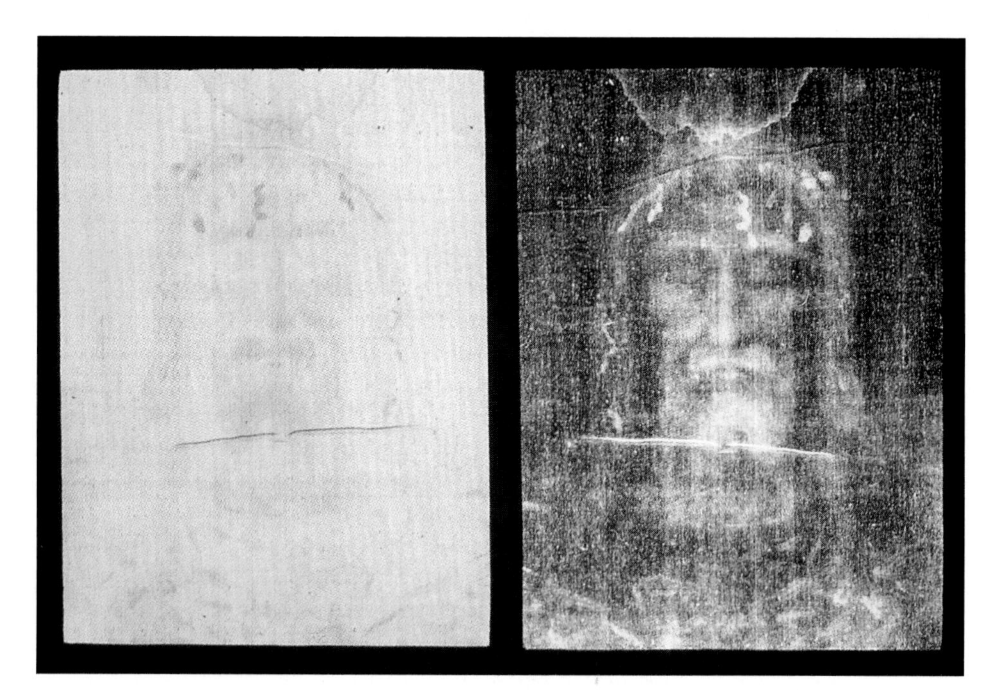

***Figure* 24.8 Dating using carbon-14.** The Shroud of Turin is a linen cloth over 4 m long. It bears a faint, straw-colored image of an adult male of average build who had apparently been crucified. Although reliable records of the shroud date to about 1350 AD, for the past 600 years it has been alleged to be the burial shroud of Jesus Christ. In 1987–1988, the age of the cloth was determined by radiocarbon dating. This study showed that the flax from which the linen was made was grown between 1260 and 1390 AD. See *Time,* April 20, 1998, pages 52–61. *(Santi Visalli/The Image Bank)*

Example 24.8 Radiochemical Dating

The Dead Sea Scrolls, Hebrew manuscripts of the books of the Old Testament, were found in 1947. The activity of carbon-14 in the linen wrappings of the book of Isaiah is about 11 dpm/g. Calculate the approximate age of the linen.

Solution We use Equation 24.3

$$\ln\left(\frac{N}{N_0}\right) = -kt$$

where N is proportional to the activity at the present time (11 dpm/g) and N_0 is proportional to the activity of carbon-14 in the living material (14 dpm/g). To calculate the time elapsed since the linen wrappings were part of a living plant, we first need k, the rate constant. From the text you know that $t_{1/2}$ is 5.73×10^3 years, so

$$k = \frac{0.693}{t_{1/2}} = \frac{0.693}{5.73 \times 10^3 \text{ years}} = 1.21 \times 10^{-4} \text{ year}^{-1}$$

Now everything is in place to calculate t:

$$\ln\left(\frac{11 \text{ dpm/g}}{14 \text{ dpm/g}}\right) = -(1.21 \times 10^{-4} \text{ year}^{-1})t$$

$$t = \frac{\ln 0.79}{-(1.21 \times 10^{-4} \text{ year}^{-1})}$$

$$= \frac{-0.24}{-(1.21 \times 10^{-4} \text{ year}^{-1})}$$

$$= 2.0 \times 10^3 \text{ years}$$

The linen is therefore about 2000 years old.

Exercise 24.8 Radiochemical Dating

A wooden Japanese temple guardian statue of the Kamakura period (AD 1185–1334) had a carbon-14 activity of 12.9 dpm/g in 1990. What is the approximate age of the statue? The initial activity of carbon-14 was 14 dpm/g, and $t_{1/2} = 5.73 \times 10^3$ years.

24.5 ARTIFICIAL NUCLEAR REACTIONS

Scientists have amassed information on a great many isotopes. This fact will be apparent on a brief review of Figure 24.4. Over 300 stable isotopes (the black circles on the graph) exist, and in addition there is a considerably greater number of radioactive nuclei. All of the stable isotopes are found in the earth's crust. Only a few radioactive isotopes are found in nature, however; this includes uranium and thorium, which are fairly abundant, along with very small amounts of the decomposition products of these elements. We mentioned radioactive carbon-14, formed in a nuclear reaction. This accounts for a very small fraction of known radioactive isotopes. Where did information on the rest of the unstable isotopes come from?

The answer is that scientists are able to synthesize a great many unstable isotopes. This is done by carrying out nuclear reactions, or nuclear transformations, processes in which one isotope is converted into another.

The first nuclear reaction was identified almost 80 years ago. In the course of his experiments, Ernest Rutherford found in 1919 that bombardment of nitrogen atoms with α particles *produced protons.* Quite correctly he concluded that the α particles had knocked a proton out of the nitrogen nucleus and that an isotope of another element had been produced. Nitrogen had undergone a *transmutation* to oxygen:

$$^{4}_{2}\text{He} + ^{14}_{7}\text{N} \longrightarrow ^{17}_{8}\text{O} + ^{1}_{1}\text{H}$$

During the next decade, Rutherford and his coworkers discovered several other nuclear transformations using α particles. Only limited progress was made in the area of artificial nuclear reactions using these techniques, however, because in most instances, α particles were simply scattered by target nuclei. The failure of particles to react with the nucleus of a target atom is due to the strong electrostatic repulsive forces between the α particle and the positively charged nucleus. In 1932, however, two new discoveries were made that greatly influenced nuclear reaction chemistry and led to rapid growth in this area. The first was the use of highly accelerated high-energy particles as projectiles. The second was the use of neutrons as the bombarding particles for nuclear reactions.

Alpha particles used in the early studies on nuclear reactions were from naturally radioactive materials such as uranium. The energies of these α particles depended on their source, and it was apparent that these energies were not high enough to overcome the electrostatic barrier imposed by the positive charge of the nucleus. Seeking higher energy particles, J. D. Cockcroft (1897–1967) and E. T. S. Walton (1903–1995), in Rutherford's laboratory in Cambridge, England, turned to protons. Protons are formed when hydrogen atoms ionize in a cathode-ray tube, and it was known that they could be accelerated to higher energy by applying a high voltage. These scientists allowed the energetic protons to strike a lithium target, and they observed the following reaction:

$$^{7}_{3}\text{Li} + ^{1}_{1}\text{p} \longrightarrow 2\ ^{4}_{2}\text{He}$$

Common particle accelerators include the cyclotron and linear accelerator. Both operate on the principle that when a charged particle is placed between charged plates it accelerates to a higher speed and thus has a higher energy.

This was the first example of reaction initiated by a particle that had been artificially accelerated to high energy. Since that time, this technique has been further developed, and the use of particle accelerators in carrying out nuclear reactions is now commonplace. Modern examples of this process are seen in the synthesis of the transuranium elements, several of which are described in more detail in *A Closer Look: The Search for Element 114* (page 1112).

The first experiments involving neutrons as the bombarding particles were carried out in both the United States and Great Britain in 1932. When nitrogen, oxygen, fluorine, and neon were bombarded with energetic neutrons, α particles were detected among the products. Using neutrons in these reactions made sense; because neutrons have no charge, it was reasoned that these particles would not be repelled by the nucleus as were positively charged α particles and protons. Thus neutrons did not need high energies to react.

Neutrons had been predicted to exist for over a decade before they were identified in 1932 by James Chadwick (1891–1974). Chadwick produced neutrons in a nuclear reaction between α particles and beryllium: $^{4}_{2}\alpha + ^{9}_{4}\text{Be} \longrightarrow ^{12}_{6}\text{C} + ^{1}_{0}\text{n}$.

In 1934, Enrico Fermi (1901–1954), one of the great nuclear physicists of the time, and his coworkers showed that in fact the reverse was true, that nuclear reactions using neutrons were more favorable if the neutrons had low energy. When its energy was low, the neutron was simply captured by the nucleus, giving a product nucleus in which the mass number was increased by one unit. The product nucleus didn't fragment in these reactions but it was usually produced in an excited state, and when the nucleus returned to the ground state a γ ray was emitted. These reactions, in which a neutron is captured and a γ ray given off, are now called (n, γ) reactions.

Enrico Fermi (1901–1954), the Italian physicist who was an early contributor in the development of neutron capture reactions, first experimentally observed nuclear fission and demonstrated a nuclear chain reaction. *(AIP/Niels Bohr Library)*

Virtually all nuclei undergo this kind of reaction. Neutron capture (n, γ) reactions are now the source of most of the radioisotopes used in medicine and chemistry. Preparation of a radioactive isotope involves irradiating a sample of an element with neutrons, often in a nuclear reactor. Radioactive phosphorus, $^{32}_{15}\text{P}$, which is used in chemical studies tracing the uptake of phosphorus in the body, is made by this route:

$$^{31}_{15}\text{P} + {}^{1}_{0}\text{n} \longrightarrow {}^{32}_{15}\text{P} + \gamma$$

The products of (n, γ) reactions are usually radioactive. With an excess of neutrons, these nuclei decompose by β emission. This secondary reaction provides a route to yet other new isotopes. For example, iodine-131 is a radioactive isotope commonly used to treat hyperthyroidism, a condition in which the thyroid gland is too active. A "radioactive cocktail" containing iodine-131 as iodide ion is ingested. The iodide ion migrates specifically to the thyroid gland, where the radioactivity causes the thyroid to be partially deactivated. Iodine-131 is made in the following nuclear reaction sequence:

$$^{130}_{52}\text{Te} + {}^{1}_{0}\text{n} \longrightarrow {}^{131}_{52}\text{Te} \qquad (t_{1/2} \text{ for } {}^{131}_{52}\text{Te} = 25 \text{ min})$$
$$^{131}_{52}\text{Te} \longrightarrow {}^{131}_{53}\text{I} + {}^{0}_{-1}\beta$$

The first examples of transuranium elements (elements with an atomic number >92) were discovered in this kind of nuclear reaction sequence. In 1940, scientists in Berkeley, California, bombarded uranium-238 with neutrons. Among the products they identified neptunium-239 and plutonium-239, which formed by the following reaction sequence:

Neptunium, plutonium, and americium were unknown prior to their preparation via nuclear reactions. Later, however, these elements were found to be present in trace quantities in uranium ores. The half-lives of these elements are not long enough for them to occur independently in the earth's crust.

$$^{238}_{92}\text{U} + {}^{1}_{0}\text{n} \longrightarrow {}^{239}_{92}\text{Up}$$
$$^{239}_{92}\text{U} \longrightarrow {}^{239}_{93}\text{Np} + {}^{0}_{-1}\beta$$
$$^{239}_{93}\text{Np} \longrightarrow {}^{239}_{94}\text{Pu} + {}^{0}_{-1}\beta$$

Four years later a similar reaction was used to make americium-241. Plutonium-239 was found to add two neutrons; the resulting plutonium-241 decayed by β emission to give americium-241.

Example 24.9 Nuclear Reactions

Write nuclear equations for the following reactions:

1. The product of an (n, γ) reaction of fluorine-19 is radioactive. Predict how it will decay, and write an equation for this reaction.
2. A commonly used neutron source is a plutonium–beryllium alloy. Plutonium-239 is an α emitter. Beryllium-9 (the only stable isotope of beryllium) reacts with α particles formed by plutonium, and neutrons are ejected. Write equations for both reactions.

Historical Perspective

Glenn Seaborg and the Transuranium Elements

Among the most significant contributors to the modern periodic chart is Nobel laureate Glenn Seaborg (born 1912). Thanks to his insights, it is now very well established that the transuranium elements (atomic numbers greater than 92), a number of which he either discovered or helped to discover in the course of the Manhattan Project during World War II, are members of the actinide series.

Until Seaborg offered his version of the periodic table, chemists were convinced that Th, Pa, and U belonged in the main body of the table, Th under Hf, Pa under Ta, and U under W. Seaborg proposed that Th was the beginning of the actinides and that the transuranium elements belonged as a group under the lanthanides. Some prominent inorganic chemists, many of them Seaborg's friends, tried to discourage his publication of this finding in the open literature. One very prominent inorganic chemist felt that Seaborg would ruin his scientific reputation. Nevertheless Seaborg, strongly convinced, persisted and, as a result, on Seaborg's expansion of the periodic table it was possible to predict accurately the properties of many of the as yet undiscovered transuranium elements. Subsequent preparation of these elements in atomic accelerators proved him right, and it was fitting that he was awarded the Nobel Prize in 1951 for his work. The name seaborgium for synthetic element number 106 was recently accepted by IUPAC, the international society responsible for naming elements. No element has ever been named for a living person. Indeed, his daughter was said to have remarked that her father must surely be dead when she heard of the naming of the element.

Glenn Theodore Seaborg (1912–) began his college education as a literature major, but changed to science in his junior year at the University of California. He shared the Nobel Prize in 1951 with Edward M. McMillan (1907–1991), who started Seaborg in this area of research. Seaborg is the only scientist honored by having an element named for him while he was alive (Seaborgium, element 106). *(Lawrence Berkeley Laboratory)*

Solution

1. Fluorine-19 adds a neutron to form fluorine-20. This isotope is radioactive (notice that there is an odd number of protons and neutrons, a good predictor of instability), and the addition of a neutron has resulted in an excess of neutrons. Thus, ${}^{20}_{9}\mathrm{F}$ is predicted to be a β emitter. The equations for these reactions are

$$
\begin{aligned}
{}^{19}_{9}\mathrm{F} + {}^{1}_{0}\mathrm{n} &\longrightarrow {}^{20}_{9}\mathrm{F} \\
{}^{20}_{9}\mathrm{F} &\longrightarrow {}^{20}_{10}\mathrm{Ne} + {}^{\;\,0}_{-1}\beta
\end{aligned}
$$

2. Plutonium-239 emits α particles. These particles add to the beryllium-9, and a neutron is ejected, producing carbon-12. This product is determined from the nuclear equation, which is written so that there is a balance in the mass number and atomic number:

$$
\begin{aligned}
{}^{239}_{94}\mathrm{Pu} &\longrightarrow {}^{235}_{92}\mathrm{U} + {}^{4}_{2}\mathrm{He} \\
{}^{4}_{2}\alpha + {}^{9}_{4}\mathrm{Be} &\longrightarrow {}^{12}_{6}\mathrm{C} + {}^{1}_{0}\mathrm{n}
\end{aligned}
$$

A CLOSER LOOK *The Search for Element 114*

The discovery of new elements has long been a goal of scientists. Through the second half of the 19th century, guided first by Mendeleev's predictions and later by atomic theory, all the stable isotopes were identified. The last nonradioactive element to be identified was rhenium in 1925. In addition, several radioactive elements that existed in nature were also characterized during this period, including radium and polonium, which were isolated by Marie Curie. By 1936, all of the elements from atomic number 1 through 92 were known, with the exception of technetium and promethium. Progress toward finding new elements at that point required a new avenue of discovery. This was found in the rapidly rising star of nuclear science.

Technetium and promethium (radioactive elements, and the last two elements with atomic numbers less than 92 to be discovered) were identified as products of nuclear reactions in 1937 and 1942, respectively. About the same time, in 1940 to be precise, the first two transuranium elements (elements with atomic numbers higher than uranium) were also reported. The race to discover elements of even higher atomic number was on. The competitors were American (at the Lawrence Berkeley National Laboratory), Russian (at the Joint Institute for Nuclear Research at Dubna, near Moscow), and the Europeans (at the Institute for Heavy Ion Research at Darmstadt, in Germany). With victory, besides the scientific acclaim, comes the right to propose a name for the new element.

All syntheses of new transuranium elements have a common methodology. An element of fairly high atomic number is subjected to bombardment by a beam of particles, initially neutrons, later helium nuclei and then larger nuclei such as ^{11}B and ^{12}C, and most recently nuclei of atoms like chromium, cobalt, and zinc. The incoming particle fuses with the nucleus of the target atoms, forming a compound nucleus, which lasts for a short time before it decomposes. A new element is detected by its decomposition, a signature of particles with specific energies. Particle accelerator and detector technology has improved with time and, by the end of the 1970s, elements through 106 were known.

The tunnel housing the four-mile-long particle accelerator at Fermilab, Batavia, Illinois *(Fermilab Visual Media Services, Batavia, IL)*

New elements were progressively harder to make, however, because each successive step produced elements with shorter lifetimes. Element 105, now named dubnium, had a half-life of 34 s; atoms of element 106, seaborgium, last only 20 s until they decompose.

Until 1970, the favored technique was "hot fusion." This process used high-energy particles that fused with the target nucleus and then shed off excess energy by expelling neutrons. A more subtle methodology was needed, and Russian scientists provided the theory. The idea, called "cold fusion," seeks to match precisely the energy of the bombarding particle with the energy required to fuse the nuclei. The idea was picked up in Darmstadt, and in 1981 element 107 was discovered. Within three more years elements 108 and 109 were also reported from the same lab.

Elements 110 and 111 were reported in 1994, and 112 was reported in 1996, but the experimental difficulties continued to grow. Lifetimes of these elements were in the millisecond range (for element 112, the lifetime was 280 μs). Nuclear stability had clearly emerged as the limiting factor.

But on the horizon is a ray of hope. Scientist have long known that certain isotopes are more stable than others. This fact can be seen in the numbers of isotopes with certain "magic numbers" of neutrons and protons. The magic numbers have a theoretical basis; they correspond to filled shells in the nucleus and can be regarded as more or less analogous to the filled shells for electronic structure. Elements with 2, 8, 20, 50, and 82 protons are in this category, as are elements with 126 neutrons. The next "magic numbers" according to theoreticians are 114 protons and 184 neutrons. The isotope $^{298}_{114}X$ matches these requirements. So element 114 may be out there, with a longer half-life, waiting to be discovered. Watch for announcements as we reach the end of the 20th century.

Exercise 24.9 Nuclear Reactions

Complete the following nuclear equations, indicating the symbol, the mass number, and the atomic number of the remaining product.

1. $^{13}_{6}C + ^{1}_{0}n \longrightarrow ^{4}_{2}He + ?$
2. $^{14}_{7}N + ^{4}_{2}He \longrightarrow ^{1}_{0}n + ?$
3. $^{253}_{99}Es + ^{4}_{2}He \longrightarrow ^{1}_{0}n + ?$
4. $^{27}_{13}Al + ^{1}_{1}H \longrightarrow ?$

24.6 NUCLEAR FISSION

In 1938, the radiochemists Otto Hahn (1879–1968) and Fritz Strassman (1902–1980) found some barium in a sample of uranium that had been bombarded with neutrons. Further work by Lise Meitner (1878–1968), Otto Frisch (1904–1979), Niels Bohr (1885–1962), and Leo Szilard (1898–1964) confirmed that a uranium-235 nucleus had captured a neutron to form uranium-236 and that this heavier isotope had then undergone **nuclear fission;** that is, the nucleus had split in two (Figure 24.9)

$$^{235}_{92}U + ^{1}_{0}n \longrightarrow ^{236}_{92}U \longrightarrow ^{141}_{56}Ba + ^{92}_{36}Kr + 3\ ^{1}_{0}n \qquad \Delta E = -2 \times 10^{10}\ kJ/mol$$

Later, other elements were found among the fission products. This is only one of many reactions that occur.

The fact that the fission reaction produces more neutrons than are required to begin the process is important. In this nuclear reaction, bombardment with a single neutron produces 3 neutrons capable of inducing 3 more fission reactions, which release 9 neutrons to induce 9 more fissions, from which 27 neutrons are obtained, and so on. This sequence of reactions is called a **chain reaction.** If the amount of uranium-235 is small, few neutrons are captured by ^{235}U nuclei, and the chain reaction cannot be sustained. In an atomic bomb,

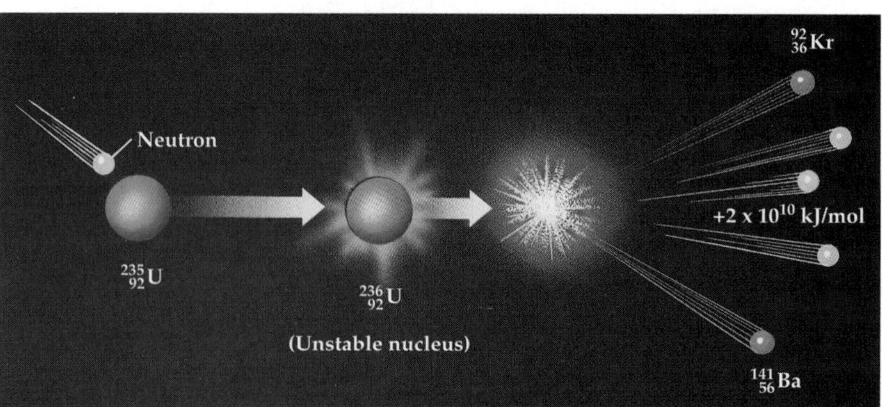

Figure 24.9 **Nuclear fission.** Neutron capture by $^{235}_{92}U$ produces $^{236}_{92}U$. This isotope is unstable and falls apart into several atomic fragments of intermediate mass along with several neutrons. The neutrons that are formed can initiate further nuclear reactions by adding to other $^{235}_{92}U$ nuclei. The result is a chain reaction. The nuclear reaction is highly energetic, producing about 2×10^{10} kJ/mol of energy.

Historical Perspective

Lise Meitner (1878–1968)

Element number 109 is named meitnerium to honor the contributions of Lise Meitner. She was born in November, 1878, in Vienna, Austria, the third of eight children. Fortunately, she had the support crucial to the success of women in science: her father demonstrated a strong interest in her education and accomplishments, and her mother supported him in this. At the turn of the century young women were not prepared academically to attend a university. Nonetheless, she had private tutors, and was finally allowed to take the university entrance examinations when she was almost 23 years old. She began as a student at the University of Vienna shortly thereafter to study science. According to one biographer, "women university students were widely regarded as freaks," but Lise Meitner found a mentor, Ludwig Boltzmann, one of the giants of science in the 20th century. He was an enthusiastic and emotional lecturer who drew her to the study of physics. She earned her Ph.D. in physics in 1905, only the second woman to be awarded a doctorate in physics in the university's 500-year history.

Lise Meitner (1878–1968) *(AIP-Emilio Segré Visual Archives, Herzfeld Collection)*

Boltzmann took his own life in 1906, so Meitner moved to Berlin. There she began working with a young chemist, Otto Hahn, in the new field of radiochemistry. The problem was that Hahn worked in an institute directed by Emil Fischer (Nobel laureate, 1902), and Fischer absolutely forbade women in his laboratory. They reached a compromise, however: Meitner could set up a laboratory in a damp basement carpentry shop as long as she did not come upstairs. And she could use the restroom in a neighboring hotel! Later, Meitner was taken on as an assistant by Max Planck (Nobel laureate, 1918) who became one of her greatest supporters.

Working with Hahn, Meitner discovered protactinium (Pa, element 91) in 1918 and carried out many significant studies on radioactive elements. Her greatest contribution to 20th-century science, however, was to explain the process of nuclear fission. She and her nephew, Otto Frisch, also a physicist, published a paper in 1939 that first used the term "nuclear fission." When she did this, in 1938–1939, she was living in exile in Sweden. She had fled Germany because of her Jewish ancestry.

Meitner was suggested for a Nobel Prize for many years, and her coworker Hahn received the chemistry prize in 1944 for his work on nuclear fission. Lise Meitner never did receive that accolade. Nonetheless, the leader of the team that discovered element 109 in Germany recently said that "She should be honored as the most significant woman scientist of this century."

See R. L. Sime: "Lise Meitner and the discovery of nuclear fission," *Scientific American,* Vol. 278, p. 80, 1998.

two small pieces of uranium, neither capable of sustaining a chain reaction, are brought together to form one larger piece capable of supporting a chain reaction, and an explosion results.

Rather than allow a fission reaction to run away explosively, engineers can slow it by limiting the number of neutrons available, and energy can be derived safely and used as a heat source in a power plant (Figure 24.10). In a **nuclear,** or **atomic, reactor,** the rate of fission is controlled by inserting cadmium rods or other "neutron absorbers" into the reactor. The rods absorb the neutrons that cause fission reactions; by withdrawing or inserting the rods, the rate of the fission reaction can be increased or decreased.

Few nuclei undergo fission on colliding with a low-energy neutron, but ^{235}U and ^{239}Pu are two isotopes for which fission is possible. Natural uranium contains an average of only 0.72% of the fissionable 235 isotope; more than 99% of the natural element is uranium-238, which is fissionable only with high-energy neutrons. Because the percentage of natural ^{235}U is too small to sustain a chain

If the uranium-235 content is over 90%, a sample is considered to be weapon quality.

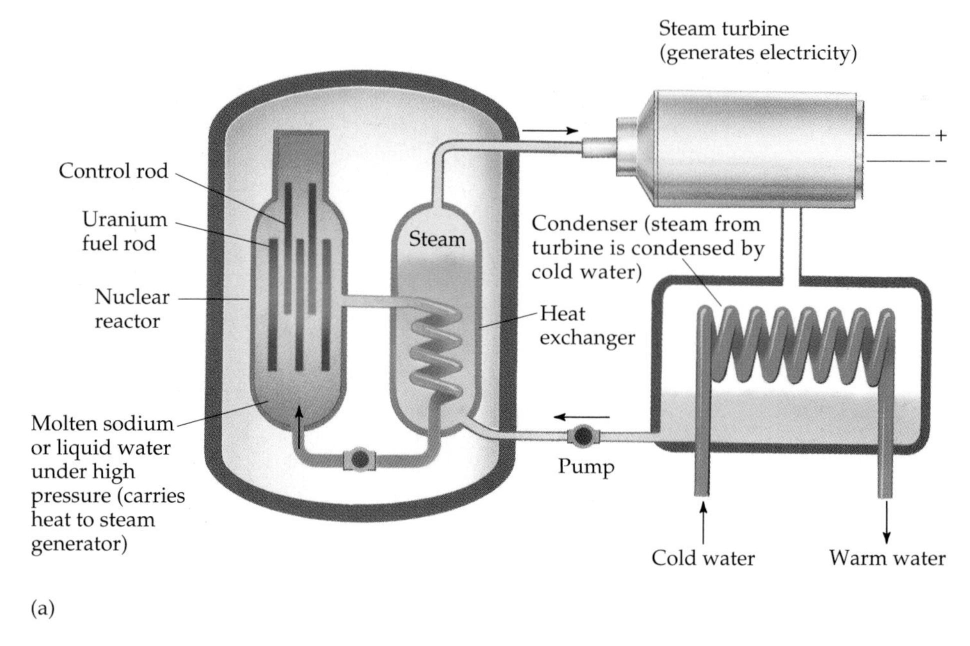

(a)

(b)

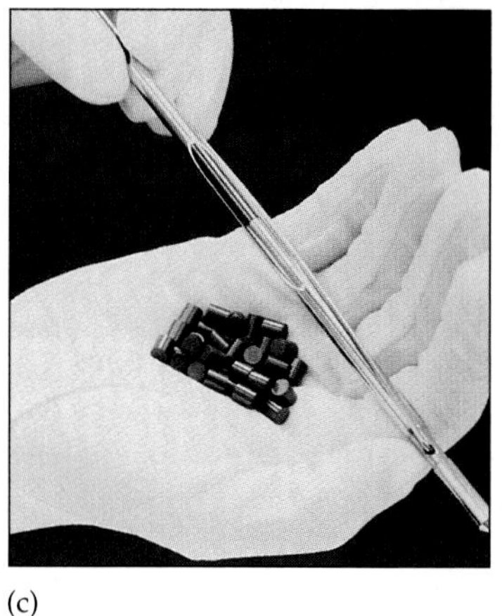

(c)

Figure 24.10 Nuclear power plants. (a) Liquid water (or liquid sodium) is circulated through the reactor, where it is heated to about 325 °C by the energy generated in the nuclear fission reaction. When this hot liquid is circulated through a steam generator, water in the generator is turned to steam, which in turn drives a steam turbine. After passing through the turbine, the steam is converted back to liquid water and is recirculated through the steam generator. Enormous quantities of cooling water from rivers or lakes are necessary to condense the steam. (This basic system is the same as in nonnuclear power plants, except that the water or circulating liquid is heated initially by coal, gas, or oil-fired burners.) (b) A nuclear power plant at Indian Pointe, New York. (c) Uranium pellets used in the reactor fuel rods. *(b, Joe Azzara/The Image Bank; c, D. O. E. / Science Source/Photo Researchers, Inc.)*

reaction, uranium for nuclear power fuel must be enriched. To accomplish this, some of the ^{238}U isotope in a sample is separated, thereby raising the concentration of ^{235}U. One way to do this is by gaseous diffusion (see Section 12.7).

Controversy surrounds the use of nuclear power plants, particularly in the United States. Proponents regard nuclear power to be an essential part of an advancing, technologically dependent society. Our standard of living depends on inexpensive, reliable, and safe sources of energy. Many believe nuclear power plants should be built to meet the demand. Nuclear power plants are capable of supplying these demands, and they can be the source of "clean" energy in that they do not pollute the atmosphere with ash, smoke, or oxides of sulfur, nitrogen, or carbon. In addition, they help to ensure that our supplies of fossil fuels will not be depleted in the near future, and they free us of dependence on such fuels from other countries. There are currently more than 100 operating plants in the United States, and more than 350 worldwide. The nuclear plants in the United States supply about 20% of the nation's electric energy; only coal-fired plants contribute a greater share (57%).

There are *no* new nuclear power plants now under construction in the United States because these plants do have disadvantages. One problem is presented by the reactor fission products. Because these products are highly radioactive, their disposal poses an enormous problem. Perhaps the most reasonable suggestion is that radioactive wastes can be converted to a glassy material having a volume of about 2 m^3 per reactor per year; this relatively small volume of material can then be stored underground in geological formations, such as salt deposits, that are expected to be stable for hundreds of millions of years.

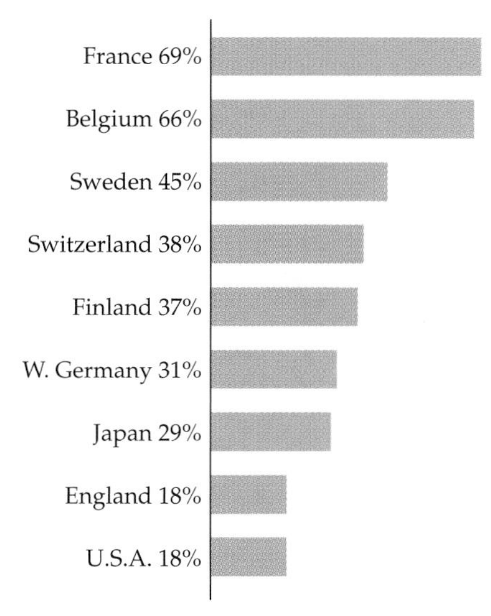

The approximate share of electricity generated by nuclear power in various countries.

24.7 NUCLEAR FUSION

Tremendous amounts of energy are generated when light nuclei combine to form heavier nuclei. Such a reaction is called **nuclear fusion,** and one of the best examples is the fusion of hydrogen nuclei (protons) to give helium nuclei:

$$4\ {}^{1}_{1}\text{H} \longrightarrow {}^{4}_{2}\text{He} + 2\ {}^{\ 0}_{+1}\beta \qquad \Delta E = -2.5 \times 10^{9}\ \text{kJ/mol}$$

This reaction is the source of the energy from our sun and other stars, and it is the beginning of the synthesis of the other elements in the universe. Temperatures of 10^6 to 10^7 K, found in the interior of the sun, are required to bring the positively charged nuclei together with enough kinetic energy to overcome nuclear repulsions.

Deuterium—heavy hydrogen—can also be fused to give helium-3,

$${}^{2}_{1}\text{H} + {}^{2}_{1}\text{H} \longrightarrow {}^{3}_{2}\text{He} + {}^{1}_{0}\text{n} \qquad \Delta E = -3.2 \times 10^{8}\ \text{kJ/mol}$$

or deuterium can be fused with tritium(${}^{3}_{1}\text{H}$), a radioactive isotope of hydrogen, to give helium-4:

$${}^{2}_{1}\text{H} + {}^{3}_{1}\text{H} \longrightarrow {}^{4}_{2}\text{He} + {}^{1}_{0}\text{n} \qquad \Delta E = -1.7 \times 10^{9}\ \text{kJ/mol}$$

Both of these reactions evolve an enormous quantity of energy, so it has been the dream of nuclear physicists to try to harness them to provide power for the nations of the world.

At the very high temperatures needed for fusion reactions to occur, atoms do not exist as such; instead, a **plasma** is created consisting of unbound nuclei and electrons. To achieve the high temperatures required for the fusion reaction of the hydrogen bomb, a fission bomb (atomic bomb) is first set off. One type of hydrogen bomb depends on the production of tritium (${}^{3}_{1}\text{H}$) in the bomb.

Nuclear power plants generate highly radioactive wastes. Some nuclear waste products have very long half-lives, up to tens of thousands of years. The storage of nuclear wastes is a formidable problem. They are now stored in large double-walled tanks buried in the ground, but better long-term solutions must be found. *(U. S. Department of Energy)*

In this type, lithium-6 deuteride (LiD, a solid salt) is placed around an ordinary $^{235}_{92}U$ or $^{239}_{94}Pu$ fission bomb, and the fission is set off in the usual way. A $^{6}_{3}Li$ nucleus absorbs one of the neutrons produced and splits into tritium and helium:

$$^{6}_{3}Li + ^{1}_{0}n \longrightarrow ^{3}_{1}H + ^{4}_{2}He$$

The temperature reached by the fission of uranium or plutonium is high enough to bring about the fusion of tritium and deuterium and the release of 1.7×10^{9} kJ/mol of ^{3}He. A 20-megaton bomb usually contains about 300 lb of lithium deuteride, as well as a considerable amount of plutonium and uranium.

The possibility of "cold fusion" of deuterium was announced in a press conference in 1989, with much fanfare. This work has been discredited, for the most part, in the scientific community although some research efforts continue. The release of this announcement via a press conference, rather than through the normal mode of publication in a scientific journal in which the work can be scrutinized and checked, is one aspect of this case condemned by the scientific community. See J. R. Huizenga: *Cold Fusion, The Scientific Fiasco of the Century.* New York, Oxford University Press, 1993.

Controlling a nuclear fusion reaction in order to harness it for peaceful uses is extraordinarily difficult and has not yet been achieved. Three critical requirements must be met for controlled fusion. First, the temperature must be high enough for fusion to occur. The fusion of deuterium and tritium, for example, requires a temperature of 1.0×10^{8} degrees or more. Second, the plasma must be confined long enough to release a net output of energy. Third, the energy must be recovered in some usable form.

In spite of the problems in controlling fusion, a number of attractive features encourage research on nuclear fusion. For example, the hydrogen fuel is cheap and abundant. Furthermore, most radioisotopes produced by fusion have short half-lives and so are a serious radiation hazard for only a short time.

24.8 RADIATION EFFECTS AND UNITS OF RADIATION

All three types of radiation (α, β and γ) disrupt normal cell processes in living organisms, and the potential for serious radiation damage to humans is well known. The biological effects of the atomic bombs exploded at Hiroshima and Nagasaki, Japan, at the close of World War II in 1945 have been well documented. Controlled exposure, however, can be beneficial in destroying unwanted tissue, as in the radiation therapy used in treating some types of cancer.

To quantify radiation and its effects, particularly on humans, several units have been developed. The **röntgen,** or roentgen (R), is a measure of radiation exposure and is proportional to the amount of ionization produced in air by x-rays and γ rays. A normal chest x-ray exposes you to about 0.1 R.

The **rad** (short for "radiation absorbed dose") measures the radiation dose to tissue rather than to air. The röntgen and rad are similar in size. One rad represents a dose of 1.00×10^{-5} J absorbed per gram of material. In more meaningful terms, a whole-body dose of 450 rad would be fatal to about 50% of the population.

Different types of radiation have different biological effects. A rad of α particles can produce 10 to 20 times as much of an effect as a rad of x-rays, for example. To take these differences into account, a unit called the **rem** (standing for *r*öntgen *e*quivalent *m*an) is used. The dose in rems is the product of the absorbed dose in rads times a "quality factor." The "quality factor" is 1 for γ and β radiation, 5 for low-energy neutrons and protons, and 10 to 20 for α particles and high-energy neutrons and protons. Because most radiation doses are fairly small, the millirem, or mrem, is commonly used, where 1 mrem = 10^{-3} rem.

CURRENT PERSPECTIVES

Assessing Your Exposure to Radiation*

The Committee on Biological Effects of Ionizing Radiation of the National Academy of Sciences issued a report in 1980 that contained a survey for individual evaluation of exposure to ionizing radiation. The following table is adapted from this report. By adding up your exposure, you can compare your annual dose to the United States annual average of 180 to 200 mrem.

(Adapted from A. R. Hinrichs: *Energy,* pp. 335–336. Philadelphia, Saunders College Publishing, 1992.)

	Common Sources of Radiation	Your Annual Dose (mrem)
Where You Live	**Location:** Cosmic radiation at sea level	26
	For your elevation (in feet), add this number of mrem	____
	Elevation 1000, **mrem** 2; **Elevation** 4000, **mrem** 15; **Elevation** 7000, **mrem** 40	
	Elevation 2000, **mrem** 5; **Elevation** 5000, **mrem** 21; **Elevation** 8000, **mrem** 53	
	Elevation 3000, **mrem** 9; **Elevation** 6000, **mrem** 29; **Elevation** 9000, **mrem** 70	
	Ground: U. S. average	26
	House construction: For stone, concrete, or masonry building, add 7	____
What You Eat, Drink, and Breathe	**Food, water, air:** U. S. average	24
	Weapons test fallout	4
How You Live	**X-ray and radiopharmaceutical diagnosis**	
	Number of chest x-rays ____ × 10	____
	Number of lower gastrointestinal tract x-rays ____ × 500	____
	Number of radiopharmaceutical examinations ____ × 300	____
	(Average dose to total U. S. population = 92 mrem)	
	Jet plane travel: For each 2500 miles add 1 mrem	____
	TV viewing: Number of hours per day ____ × 0.15	____
How Close You Live to a Nuclear Plant	**At site boundary:** average number of hours per day ____ × 0.2	____
	One mile away: average number of hours per day ____ × 0.02	____
	Five miles away: average number of hours per day ____ × 0.002	____
	Over 5 miles away: none	____
	Note: Maximum allowable dose determined by "as low as reasonably achievable" (ALARA) criteria established by the U. S. Nuclear Regulatory Commission. Experience shows that your actual dose is substantially less than these limits.	
	Your total annual dose in mrem	____

Compare your annual dose to the U. S. annual average of 180 mrem.
One mrem per year is a risk equal to increasing your diet by 4%, or taking a 5-day vacation in the Sierra Nevada (CA) mountains.

*Based on the "BEIR Report III"—National Academy of Sciences, Committee on Biological Effects of Ionizing Radiation: *The Effects on Populations of Exposure to Low Levels of Ionizing Radiation,* National Academy of Sciences, Washington, D.C., 1980.

Finally, the **curie** (Ci) is commonly used as a unit of activity. One curie represents the quantity of any radioactive isotope that undergoes 3.7×10^{10} dps.

We are constantly exposed to natural and artificial **background radiation,** estimated to be about 200 mrem/year (Table 24.2). More than half of this is from natural background radiation sources: cosmic radiation and radioactive elements and minerals found naturally in the earth and air.

Cosmic radiation, emitted by the sun and other stars, continually bombards the earth and accounts for about 40% of background radiation. The remainder comes from radioactive isotopes such as ^{40}K. Because potassium (which is present to the extent of about 0.3 g/kg of soil) is essential to all living organisms, we all carry some radioactive potassium. Other radioactive isotopes found in some abundance on the earth are thorium-232, uranium-238, and radium-226. Thorium, for example, is found to the extent of 12 g/1000 kg of soil.

Burning fossil fuels also releases traces of radioactive isotopes into the atmosphere, which has also added significantly to background radiation in recent years.

Roughly 17% of our annual exposure comes from medical procedures such as diagnostic x-rays and the use of radioactive compounds to trace the body's functions. Finally, another 17% comes from such sources as the radioactive products from testing nuclear explosives in the atmosphere, x-ray generators, televisions, nuclear power plants and their wastes, nuclear weapons manufacture, and nuclear fuel processing.

Table 24.2 • RADIATION EXPOSURE FOR ONE YEAR FROM NATURAL AND ARTIFICIAL SOURCES*

	Millirem/Year	Percentage
Natural Sources		
Cosmic radiation	50.0	25.8
The earth	47.0	24.2
Building materials	3.0	1.5
Inhaled from the air	5.0	2.6
Elements found naturally in human tissues	21.0	10.8
Subtotal	126.0	64.9
Medical Sources		
Diagnostic x-rays	50.0	25.8
Radiotherapy	10.0	5.2
Internal diagnosis	1.0	0.5
Subtotal	61.0	31.5
Other Artificial Sources		
Nuclear power industry	0.85	0.4
Luminous watch dials, TV tubes, industrial wastes	2.0	1.0
Fallout from nuclear testing	4.0	2.1
Subtotal	6.9	3.5
Total	193.9	99.9

*From J. R. Amend, B. P. Mundy, and M. T. Arnold: *General, Organic, and Biochemistry,* 2nd ed., p. 356. Philadelphia, Saunders College Publishing, 1993.

24.9 APPLICATIONS OF RADIOACTIVITY

Food Irradiation

Irradiated strawberries. *(MDS Nordian)*

Although uncontrolled radioactivity may be harmful, the radiation from radioisotopes can be put to beneficial use. For example, consider the importance of killing pests that would destroy food during storage. In some parts of the world stored-food spoilage may claim up to 50% of the food crop. In our society, refrigeration, canning, and chemical additives lower this figure considerably. Still, there are problems with food spoilage, and food protection costs amount to a sizable fraction of the final cost of food. Food irradiation with γ rays from sources such as ^{60}Co and ^{137}Cs is commonly used in European countries, Canada, and Mexico. Some irradiated foods are sold in the United States as well. Foods may be pasteurized by irradiation to retard the growth of organisms, such as bacteria, molds, and yeasts. This irradiation prolongs shelf life under refrigeration in much the same way that heat pasteurization protects milk. Chicken normally has a three-day refrigerated shelf life; after irradiation, it may have a three-week refrigerated shelf life.

The FDA may soon permit irradiation up to 100 krad for the pasteurization of foods. Radiation levels in the 1- to 5-Mrad (1 Mrad = 1×10^6 rad) range sterilize; that is, every living organism is killed. Foods irradiated at these levels will keep indefinitely when sealed in plastic or aluminum-foil packages. Indeed, radiation-sterilized foods can last three to seven years without refrigeration. Ham, beef, turkey, and corned beef sterilized by radiation have been used on many Space Shuttle flights. An astronaut said that "The beautiful thing was that it didn't disturb the taste, which made the meals much better than the freeze-dried and other types of foods we had."

More than 40 classes of foods are already irradiated in 24 countries. In the United States, in contrast, only a small number of foods may be irradiated (Table 24.3).

Recent studies indicate that there may be harmful health effects from several common agricultural fumigants; irradiation of fruits and vegetables could be an effective alternative to some chemical fumigants. The agricultural products may be picked, packed, and readied for shipment. After that, the entire shipping container can be passed through a building containing a strong source of radiation. This type of sterilization offers greater worker safety because it lessens chances of exposure to harmful chemicals, and it protects the environment by avoiding contamination of water supplies with these toxic chemicals.

Table 24.3 • EXAMPLES OF IRRADIATED FOODSTUFFS

Food	Purpose	Status
Potatoes	Retardation of sprouts	FDA approved
Wheat	Insect disinfection	FDA approved
Wheat flour	Insect disinfection	FDA approved
Spices	Retardation of microbe growth	FDA approved
Grapefruit	Mold control	Approved for export
Strawberries	Mold control	Approved for export
Fish	Microbe control	Approved for export
Shrimp	Microbe control	Approved for export

CURRENT PERSPECTIVES

Protect Your Home with a Smoke Detector

Radioactive isotopes are used in many ways, one of which may already be familiar to you. That is, some home smoke alarms use the disintegration of a radioactive element, americium, as a way to detect smoke particles in the air. As shown here, a weak radioactive source in the smoke alarm ionizes the air, thus setting up a small current in an electric circuit. If smoke is present, the ions become attached to the smoke particles. The slower movement of the heavier, charged smoke particles reduces the current in the circuit and sets off the alarm.

Americium-241 was prepared in 1944 by G. T. Seaborg, R. A. James, L. O. Morgan, and A. Ghiorso at what is now the Argonne National Laboratory. It arises from successive neutron capture by ^{239}Pu, followed by β decay. The isotope is now obtained in kilogram quantities as a byproduct of processing plutonium produced in nuclear reactors. The isotope ^{241}Am has a half-life of 432.7 years.

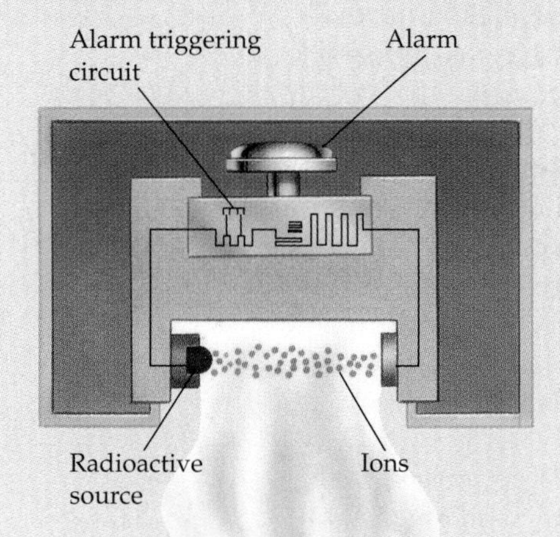

(Left) Diagram of a smoke alarm. *(Right)* A household smoke alarm. (Photo, *C. D. Winters)*

Radioactive Tracers

The chemical behavior of a radioisotope is almost identical to that of the nonradioactive isotopes of the same element, because the energies of the valence electrons are nearly the same in both atoms. Chemists can therefore use radioactive isotopes as **tracers** in chemical reactions and biological processes. To use a tracer, a chemist prepares a reactant compound in which one of the elements consists of both radioactive and stable isotopes, and introduces it into the reaction (or feeds it to an organism). After the reaction, the chemist measures the radioactivity of the products (or determines which parts of the organism contain the radioisotope) by using a Geiger-Müller counter or similar instrument. Several radioisotopes commonly used as tracers are listed in Table 24.4.

For example, plants are known to take up phosphorus-containing compounds from the soil through their roots. The use of the radioactive phosphorus isotope $^{32}_{15}$P, a β emitter, presents a way not only of detecting the uptake of phosphorus by a plant but also of measuring the speed of uptake under various

Table 24.4 • RADIOISOTOPES USED AS TRACERS

Isotope	Half-Life	Use
^{14}C	5730 years	CO_2 for photosynthesis research
^{3}H	12.26 years	Tag hydrocarbons
^{35}S	87.2 days	Tag pesticides, measure air flow
^{32}P	14.23 days	Measure phosphorus uptake by plants

conditions. Plant biologists can grow hybrid strains of plants that can absorb phosphorus quickly, and they can test this ability with the radioactive phosphorus tracer. This type of research leads to faster maturing crops, better yields per acre, and more food or fiber at less expense.

Important characteristics of a pesticide can be measured by tagging the pesticide with radioisotopes that have short half-lives and then applying it to a test field. Following the tagged pesticide can provide information on its tendency to accumulate in the soil, to be taken up by the plant, and to accumulate in runoff surface water. This is done with a high degree of accuracy by counting the disintegrations of the radioactive tracer. After these tests are completed, the radioactive isotopes in the tagged pesticides decay to a harmless level in a few days or a few weeks because of the short half-lives of the species used.

Medical Imaging

Radioactive isotopes are used in **nuclear medicine** in both diagnosis and therapy. In the diagnosis of internal disorders such as tumors, physicians need information on the locations of abnormal tissue. This is done by **imaging,** a technique in which the radioisotope, either alone or combined with some other chemical, accumulates at the site of the disorder. There, acting like a homing device, the radioisotope disintegrates and emits its characteristic radiation, which is detected. Modern medical diagnostic instruments construct an image of the area within the body where the radioisotope is concentrated.

Four of the most common diagnostic radioisotopes are given in Table 24.5. All are made in a particle accelerator. Each of these radioisotopes produces γ

Table 24.5 • DIAGNOSTIC RADIOISOTOPES

Radioisotope	Name	Half-Life (hours)	Uses
$^{99m}Tc^{*}$	Technetium-99m	6.0	To the thyroid, brain, kidneys
^{201}Tl	Thallium-201	73.1	To the heart
^{123}I	Iodine-123	13.3	To the thyroid
^{67}Ga	Gallium-67	78.25	To various tumors and abscesses

*The technetium-99m isotope is the one most commonly used for diagnostic purposes. The *m* stands for "metastable," a term explained in the text.

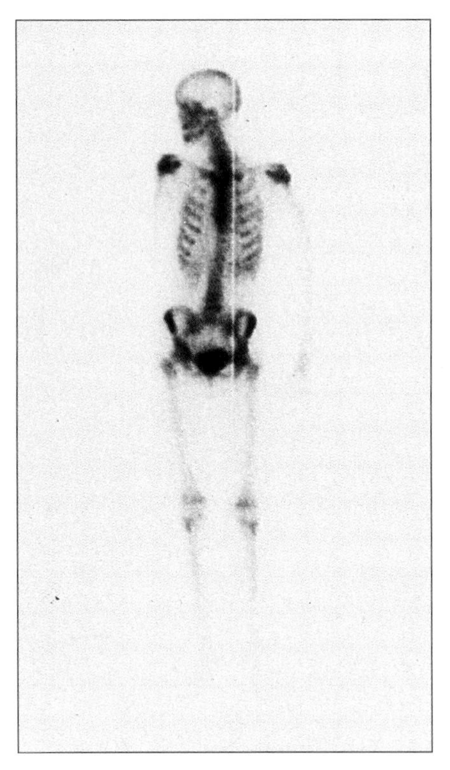

Figure 24.11 A whole-body scan. Phosphate with technetium-99m was injected into the blood and then absorbed by the bones and kidneys. This picture was taken 3 hours after injection. *(SUNY Upstate Medical Center)*

radiation, which in low doses is less harmful to the tissue than ionizing radiations, such as β or α particles. By the use of special carrier compounds, these radioisotopes can be made to accumulate in specific areas of the body. For example, the pyrophosphate ion, $P_4O_7^{4-}$, can bond to the technetium-99m radioisotope; together they accumulate in the skeletal structure where abnormal bone metabolism is occurring (Figure 24.11). The technetium-99m radioisotope is metastable, as denoted by the letter *m*; this term means that the nucleus loses energy by emitting a γ ray

$$^{99m}\text{Tc} \longrightarrow {}^{99}\text{Tc} + \gamma$$

and the γ rays are detected. Such investigations often pinpoint bone tumors.

Positron emission tomography (PET) is a form of nuclear imaging that uses **positron emitters,** such as carbon-11, fluorine-18, nitrogen-13, or oxygen-15. All these radioisotopes are neutron-deficient, have short half-lives, and therefore must be prepared in a cyclotron immediately before use. When these radioisotopes decay, the positron travels less than a few millimeters before it encounters an electron and undergoes antimatter–matter annihilation.

$$_{+1}^{0}\beta + {}_{-1}^{0}\beta \longrightarrow 2\ \gamma$$

The annihilation event produces two γ rays that radiate in opposite directions and are detected by two scintillation detectors located 180° apart in the PET scanner. By detecting several million annihilation γ rays within a circular slice around the subject over approximately 10 min, the region of tissue containing the radioisotope can be imaged with computer signal-averaging techniques (Figure 24.12).

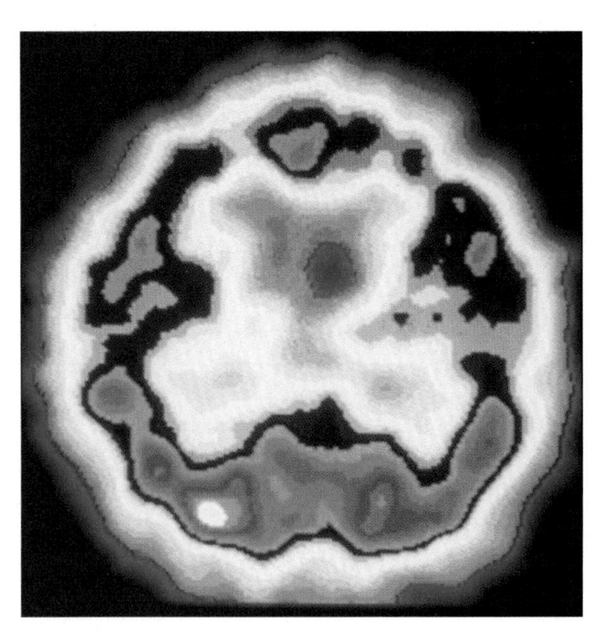

Figure 24.12 A PET scan of a normal human brain. *(CEA-ORSAY/CNRI/Science Photo Library/Photo Researchers, Inc.)*

CHAPTER HIGHLIGHTS

Having studied this chapter, you should be able to

- Characterize the three major types of radiation observed in natural radioactive decay: α, β, and γ (Section 24.1).
- Write a balanced equation for a nuclear reaction (Section 24.2).
- Decide whether a particular radioactive isotope will decay by α, β, or positron emission or by electron capture (Sections 24.2 and 24.3).
- Calculate the **binding energy** for a particular isotope and understand what this energy means in terms of nuclear stability (Section 24.3).
- Use the equation $\ln (N/N_0) = -kt$ (Equation 24.2), which relates (through the decay constant k) the number of radioactive atoms present at the beginning (N_0) and end (N) of the period to time elapsed (Section 24.4).
- Calculate the half-life of a radioactive isotope ($t_{1/2}$) from the activity of a sample, or use the half-life to find the time required for an isotope to decay to a particular activity (Section 24.4).
- Describe nuclear chain reactions, **nuclear fission,** and **nuclear fusion** (Sections 24.6 and 24.7).
- Describe some sources of background radiation and the units used to measure radiation (Section 24.8).
- Relate some uses of radioisotopes (Section 24.9).

STUDY QUESTIONS

REVIEW QUESTIONS

1. Some very important milestones in scientific history contributed greatly to the development of nuclear chemistry. Among them are the discoveries listed here. Describe, briefly, each of these discoveries, identify prominent scientists who contributed to them, and comment on the significance of each to the development of this field.
   (a) 1896, the discovery of radioactivity
   (b) 1898, the isolation of radium and polonium
   (c) 1918, the first artificial nuclear reaction
   (d) 1932, (n, γ) reactions
   (e) 1939, fission reactions
2. In Chapter 2, the law of conservation of mass was introduced as an important principle in chemistry. The discovery of nuclear reactions has forced scientists to modify this law. Explain why this is so, and give an example illustrating that mass is not conserved in a nuclear reaction.
3. A graph of binding energy per nucleon is shown in the text as Figure 24.5. Explain how the data used to construct this graph were obtained.
4. How is Figure 24.4 used to predict the type of decomposition for unstable (radioactive) isotopes?
5. Outline how a nuclear reaction is carried out in the laboratory. Describe one example in which an artificial nuclear reaction has been used to make an element with an atomic number greater than 92.
6. Describe how a nuclear reactor works. What are the advantages and disadvantages of nuclear power in today's technological world?
7. What are the mathematical equations that govern the rates of decay for radioactive elements?
8. Explain how carbon-14 is used to obtain dates of archeological artifacts. What are the limitations on the use of this technique?
9. Explain the concept of half-life for nuclear decay, and indicate how it is used.
10. What is a radioactive decay series? Explain why radium and polonium are found in uranium ores.
11. Food irradiation is a subject of considerable current interest. Use the World Wide Web to research this topic. (See the site for the United States Food and Drug Administration, for example, at http://www.fda.gov.) What foods are currently being treated with radiation? What advantage does the irradiation method offer?

Nuclear Reactions

12. Complete the following nuclear equations. Write the mass number and atomic number for the remaining particle, as well as its symbol.

(a) ${}^{54}_{26}Fe + {}^{4}_{2}He \longrightarrow 2\,{}^{1}_{0}H + ?$
(b) ${}^{27}_{13}Al + {}^{4}_{2}He \longrightarrow {}^{30}_{15}P + ?$
(c) ${}^{32}_{16}S + {}^{1}_{0}n \longrightarrow {}^{1}_{1}H + ?$
(d) ${}^{96}_{42}Mo + {}^{2}_{1}H \longrightarrow {}^{1}_{0}n + ?$
(e) ${}^{96}_{42}Mo + {}^{1}_{0}n \longrightarrow {}^{99}_{43}Tc + ?$

13. Complete the following nuclear equations. Write the mass number and atomic number for the remaining particle, as well as its symbol.

(a) ${}^{9}_{4}Be + ? \longrightarrow {}^{6}_{3}Li + {}^{4}_{2}He$
(b) $? + {}^{1}_{0}n \longrightarrow {}^{24}_{11}Na + {}^{4}_{2}He$
(c) ${}^{40}_{20}Ca + ? \longrightarrow {}^{40}_{19}K + {}^{1}_{1}H$
(d) ${}^{241}_{95}Am + {}^{4}_{2}He \longrightarrow {}^{243}_{97}Bk + ?$
(e) ${}^{246}_{96}Cm + {}^{12}_{6}C \longrightarrow 4\,{}^{1}_{0}n + ?$
(f) ${}^{238}_{92}U + ? \longrightarrow {}^{249}_{100}Fm + 5\,{}^{1}_{0}n$

14. Complete the following nuclear equations. Write the mass number and atomic number for the remaining particle, as well as its symbol.

(a) ${}^{111}_{47}Ag \longrightarrow {}^{111}_{48}Cd + ?$
(b) ${}^{87}_{36}Kr \longrightarrow {}^{0}_{-1}\beta + ?$
(c) ${}^{231}_{91}Pa \longrightarrow {}^{227}_{89}Ac + ?$
(d) ${}^{230}_{90}Th \longrightarrow {}^{4}_{2}He + ?$
(e) ${}^{82}_{35}Br \longrightarrow {}^{82}_{36}Kr + ?$
(f) $? \longrightarrow {}^{24}_{12}Mg + {}^{0}_{-1}\beta$

15. Complete the following nuclear equations. Write the mass number and atomic number for the remaining particle, as well as its symbol.

(a) ${}^{19}_{10}Ne \longrightarrow {}^{0}_{+1}\beta + ?$
(b) ${}^{59}_{26}Fe \longrightarrow {}^{0}_{-1}\beta + ?$
(c) ${}^{40}_{19}K \longrightarrow {}^{0}_{-1}\beta + ?$
(d) ${}^{37}_{18}Ar + {}^{0}_{-1}e$ (electron capture) $\longrightarrow ?$
(e) ${}^{55}_{26}Fe + {}^{0}_{-1}e$ (electron capture) $\longrightarrow ?$
(f) ${}^{26}_{13}Al \longrightarrow {}^{25}_{12}Mg + ?$

16. The uranium-235 radioactive decay series, beginning with ${}^{235}_{92}U$ and ending with ${}^{207}_{82}Pb$, occurs in the following sequence: α, β, α, β, α, α, α, α, β, β, α. Write an equation for each step in this series.

17. The thorium-232 radioactive decay series, beginning with ${}^{232}Th$ and ending with ${}^{208}Pb$, occurs in the following sequence: α, β, β, α, α, α, α, β, β, α. Write an equation for each step in this series.

18. What particle is emitted in the following nuclear reactions? Write an equation for each reaction.

(a) Gold-198 decays to mercury-198.
(b) Radon-222 decays to polonium-218.
(c) Cesium-137 decays to barium-137.
(d) Indium-110 decays to cadmium-110.

19. What is the product of the following nuclear decay processes? Write an equation for each process.

(a) Gallium-67 decays by electron capture.
(b) Potassium-38 decays with positron emission.
(c) Technetium-99m decays with γ emission.
(d) Manganese-56 decays by β emission.

Nuclear Stability

20. Boron has two stable isotopes, ${}^{10}B$ (abundance = 19.78%) and ${}^{11}B$ (abundance = 80.1%). Calculate the binding energies per nucleon of these two nuclei and compare their stabilities:

$$5\,{}^{1}_{1}H + 5\,{}^{1}_{0}n \longrightarrow {}^{10}_{5}B$$
$$5\,{}^{1}_{1}H + 6\,{}^{1}_{0}n \longrightarrow {}^{11}_{5}B$$

The required masses (in grams per mole) are: ${}^{1}_{1}H = 1.00783$; ${}^{1}_{0}n = 1.00867$; ${}^{10}_{5}B = 10.01294$; and ${}^{11}_{5}B = 11.00931$.

21. Calculate the binding energy in kilojoules per mole of nucleons of P for the formation of ${}^{30}_{15}P$

$$15\,{}^{1}_{1}H + 15\,{}^{1}_{0}n \longrightarrow {}^{30}_{15}P$$

and for the formation of ${}^{31}_{15}P$:

$$15\,{}^{1}_{1}H + 16\,{}^{1}_{0}n \longrightarrow {}^{31}_{15}P$$

The required masses (in grams per mole) are ${}^{1}_{1}H = 1.00783$; ${}^{1}_{0}n = 1.00867$; ${}^{30}_{15}P = 29.97832$; and ${}^{31}_{15}P = 30.97376$.

22. Calculate the binding energy per nucleon for calcium-40, and compare your result with the value for calcium-40 in Figure 24.5. Masses needed for this calculation are ${}^{1}_{1}H = 1.00783$, ${}^{1}_{0}n = 1.00867$, and ${}^{40}_{20}Ca = 39.96259$.

23. Calculate the binding energy per nucleon for iron-56. Masses needed for this calculation are ${}^{1}_{1}H = 1.00783$, ${}^{1}_{0}n = 1.00867$, ${}^{56}_{26}Fe = 55.9349$. Compare the result of your calculation to the value for iron-56 in the graph in Figure 24.5, page 1101.

24. Calculate the binding energy per mole of nucleons for ${}^{15}_{8}O$, ${}^{16}_{8}O$, and ${}^{17}_{8}O$. Masses needed for these calculations are ${}^{1}_{1}H = 1.00783$, ${}^{1}_{0}n = 1.00867$, ${}^{16}_{8}O = 15.99492$, ${}^{15}_{8}O = 15.003065$, and ${}^{17}_{8}O = 16.999132$.

25. Calculate the binding energy per nucleon for nitrogen-14. The mass of nitrogen-14 is 14.003074.

Rates of Disintegration Reactions

26. Copper-64 is used in the form of copper acetate to study brain tumors. It has a half-life of 12.8 h. If you begin with 15.0 mg of ${}^{64}Cu$-labeled copper acetate, what mass in milligrams remains after 64 h?

27. Gold-198 is used as the metal in the diagnosis of liver problems. The half-life of ${}^{198}Au$ is 2.69 days. If you begin with 5.6 mg of this gold isotope, what mass remains after 10.8 days?

28. Iodine-131 is used in the form of sodium iodide to treat cancer of the thyroid.

(a) The isotope decays by ejecting a β particle. Write a balanced equation to show this process.
(b) The isotope has a half-life of 8.04 days. If you begin with 25.0 mg of radioactive Na ${}^{131}I$, what mass remains after 32.2 days (about a month)?

29. Phosphorus-32 is used in the form of Na_2HPO_4 in the treatment of chronic myeloid leukemia, among other things.

(a) The isotope decays by emitting a β particle. Write a balanced equation to show this process.

(b) The half-life of ^{32}P is 14.3 days. If you begin with 9.6 mg of radioactive Na_2HPO_4, what mass remains after 28.6 days (about one month)?

30. Gallium-67 ($t_{1/2}$ = 78.25 h) is used in the medical diagnosis of certain kinds of tumors. If you ingest a compound containing 0.15 mg of this isotope, what mass in milligrams remains in your body after 13 days? (Assume none is excreted.)

31. Radioisotopes of iodine are widely used in medicine. For example, iodine-131 ($t_{1/2}$ = 8.05 days) is used to treat thyroid cancer. If you ingest a sample of NaI containing ^{131}I, how much time is required for the isotope to fall to 5.0% of its original activity?

32. The noble gas radon has been the focus of much attention recently because it can be found in homes. Radon-222 emits α particles and has a half-life of 3.82 days.
(a) Write a balanced equation to show this process.
(b) How long does it take for a sample of radon to decrease to 10.0% of its original activity?

33. A sample of wood from a Thracian chariot found in an excavation in Bulgaria has a ^{14}C activity of 11.2 dpm/g. Estimate the age of the chariot and the year it was made. ($t_{1/2}$ for ^{14}C is 5.73×10^3 years and the activity of ^{14}C in living material is 14.0 dpm/g.)

34. A piece of charred bone found in the ruins of an American Indian village has a ^{14}C to ^{12}C ratio of 0.72 times that found in living organisms. Calculate the age of the bone fragment. (See Study Question 33 for required data on carbon-14.)

35. Strontium-90 is often mentioned as one of the more hazardous radioactive isotopes that resulted from atmospheric nuclear testing. A sample of strontium-90 is found to have an activity of 1.0×10^3 dpm. One year later the activity of this sample is 975 dpm.
(a) Calculate the half-life of strontium-90 from this information.
(b) How long will it take for the activity of this sample to drop to 1.0% of the initial value?

36. Radioactive cobalt-60 is used extensively in nuclear medicine as a γ-ray source. It is made by a neutron capture reaction from cobalt-59, and it is a β emitter; β emission is accompanied by strong γ radiation. The half-life of cobalt-60 is 5.27 years.
(a) How long will it take for a cobalt-60 source to decrease to one eighth of its original activity?
(b) What fraction of the activity of a cobalt-60 source remains after 1.0 year?

37. A technique to date geological samples uses rubidium-87, a long-lived radioactive isotope of rubidium ($t_{1/2} = 4.9 \times 10^{10}$ years). Rubidium-87 decays by β emission to strontium-87. If the rubidium-87 is part of a rock or mineral, then strontium-87 will remain trapped within the crystalline structure of the rock. The age of the rock dates back to the time when the rock solidified. Chemical analysis of the rock gives the amounts of ^{87}Rb and ^{87}Sr, and from this the fraction of ^{87}Rb that remains can be calculated.

Analysis of a stony meteorite determined that 1.8 μmol of ^{87}Rb and 1.6 μmol of ^{87}Sr were present (1 μmol = 1×10^{-6} mol). Estimate the age of the meteorite. (*Hint*: The amount of ^{87}Rb at t_0 is moles ^{87}Rb + moles ^{87}Sr.)

38. The age of minerals can sometimes be determined by measuring the amounts of ^{206}Pb and ^{238}U in a sample. This determination assumes that all of the ^{206}Pb in the sample comes from the decay of ^{238}U. The date obtained relates to the time when the rock solidified. Assume that the ratio of ^{206}Pb to ^{238}U in an igneous rock sample is 0.33. Calculate the age of the rock. ($t_{1/2}$ for ^{238}U is 4.5×10^9 years)

39. Scandium occurs in nature as a single isotope, scandium-45. Neutron irradiation of scandium produces scandium-46, a radioactive isotope (a β emitter) that has a half-life of 83.8 days. Assume you have a sample that contains scandium-46 that has a rate of decay of 6.2×10^4 dpm. Draw a graph showing disintegrations per minute as a function of time during a period of one year.

40. Sodium-23 (in a sample of NaCl) is subjected to neutron bombardment in a nuclear reactor. When removed from the reactor, the sample is radioactive with β activity of 2.54×10^4 dpm. The decrease in radioactivity over time was studied, producing the following data:

Activity (dpm)	Time (h)
2.54×10^4	0
2.42×10^4	1
2.31×10^4	2
2.00×10^4	5
1.60×10^4	10
1.01×10^4	20

(a) Write equations for the neutron capture reaction and for the reaction in which the product of this reaction decays by β emission.
(b) Determine the half-life, $t_{1/2}$, for the radioactive product from the data.

41. The isotope of polonium most likely isolated by Madame Curie in her pioneering studies is polonium-210. A sample of this element was prepared in a nuclear reaction. Initially its activity (α emission) was 7840 dpm. The decrease in radioactivity over time was studied, producing the following data:

Activity (dpm)	Time (days)
7840	0
7570	7
7300	14
5920	56
5470	72

Determine the half-life, $t_{1/2}$, for polonium-210 from these data.

Nuclear Transmutations

42. There are two isotopes of americium, both with half-lives sufficiently long to allow the handling of massive quantities. Americium-241, for example, has a half-life of 248 years as an α emitter, and it is used in gauging the thickness of materials and in smoke detectors. The isotope is formed from ^{239}Pu by absorption of two neutrons followed by emission of a β particle. Write a balanced equation for this process.

43. Americium-240 is made by bombarding a plutonium-239 atom with an α particle. In addition to ^{240}Am, the products are a proton and two neutrons. Write a balanced equation for this process.

44. To synthesize the heavier transuranium elements, a lighter nucleus must be bombarded with a relatively large particle. If you know the products are californium-246 and four neutrons, with what particle would you bombard uranium-238 atoms?

45. The element with the highest known atomic number is 112. It is thought that still heavier elements are possible, especially with $Z = 114$ and $N = 184$. To this end, serious attempts have been made to force calcium-40 and curium-248 to merge. What would be the atomic number of the element formed?

46. Deuterium nuclei (^2_1H) were found to be particularly effective as bombarding particles to carry out nuclear reactions. Complete the following equations:

(a) $^{114}_{48}Cd + ^2_1H \longrightarrow ? + ^1_1H$
(b) $^6_3Li + ^2_1H \longrightarrow ? + ^1_0n$
(c) $^{40}_{20}Ca + ^2_1H \longrightarrow ^{38}_{19}K + ?$
(d) $? + ^2_1H \longrightarrow ^{65}_{30}Zn + \gamma$

47. Some of the reactions explored by Rutherford and others are listed below. Identify the unknown species in each reaction:

(a) $^{14}_7N + ^4_2He \longrightarrow ^{17}_8O + ?$
(b) $^9_4Be + ^4_2He \longrightarrow ? + ^1_0n$
(c) $? + ^4_2He \longrightarrow ^{30}_{15}P + ^1_0n$
(d) $^{239}_{94}Pu + ^4_2He \longrightarrow ? + ^1_0n$

48. Boron is used for control rods in nuclear reactors because it is an effective absorber of neutrons. It controls nuclear fission by absorbing some of the neutrons given off when fission occurs. When boron-10 adds a neutron, an α particle is emitted. Write an equation for this nuclear reaction.

49. Tritium, ^3_1H, is one of the nuclei used in fusion reactions. This isotope is radioactive with a short half-life, 12.3 years. Like carbon-14, tritium is formed in the upper atmosphere from cosmic radiation, and it is found in trace amounts on earth. In amounts required for a fusion reaction, however, it must be made via a nuclear reaction. The reaction of ^6_3Li with neutrons is used to produce tritium, with α particles being given off. Write a equation for this nuclear reaction.

GENERAL QUESTIONS

50. Complete the following nuclear equations. Write the mass number and atomic number for the remaining particle, as well as its symbol.

(a) $^{13}_6C + ? \longrightarrow ^{14}_6C$
(b) $^{40}_{18}Ar + ? \longrightarrow ^{43}_{19}K + ^1_1H$
(c) $^{250}_{98}Cf + ^{11}_5B \longrightarrow 4\ ^1_0n + ?$
(d) $^{53}_{24}Cr + ^4_2He \longrightarrow ? + ^{56}_{26}Fe$
(e) $^{212}_{84}Po \longrightarrow ^{208}_{82}Pb + ?$
(f) $^{122}_{53}I \longrightarrow ^0_{+1}\beta + ?$
(g) $? \longrightarrow ^{23}_{11}Na + ^0_{-1}\beta$
(h) $^{137}_{53}I \longrightarrow ^1_0n + ?$

51. The following reaction sequence involves four unknown elements, Q, Δ, Σ, and Π. Based on the reactions in the sequence, identify the unknown elements.

$$\begin{aligned} Pb + Cr &\longrightarrow Q \\ Cf + O &\longrightarrow Q \\ Q &\longrightarrow \alpha + \Delta \\ \Delta &\longrightarrow \alpha + \Sigma \\ \Sigma &\longrightarrow \alpha + \Pi \end{aligned}$$

52. The oldest known fossil found in South Africa has been dated by the reaction

$$^{87}Rb \longrightarrow ^{87}Sr + ^0_{-1}\beta \qquad t_{1/2} = 4.8 \times 10^{10} \text{ years}$$

If the ratio of the present quantity of ^{87}Rb to the original quantity is 0.951, calculate the age of the fossil.

53. Balance the following reactions used for the synthesis of transuranium elements:

(a) $^{238}_{92}U + ^{14}_7N \longrightarrow ? + 5\ ^1_0n$
(b) $^{238}_{92}U + ? \longrightarrow ^{249}_{100}Fm + 5\ ^1_0n$
(c) $^{253}_{99}Es + ? \longrightarrow ^{256}_{101}Md + ^1_0n$
(d) $^{246}_{96}Cm + ? \longrightarrow ^{254}_{102}No + 4\ ^1_0n$
(e) $^{252}_{98}Cf + ? \longrightarrow ^{257}_{103}Lr + 5\ ^1_0n$

54. In June, 1972, *natural* fission reactors, which operated billions of years ago, were discovered in Oklo, Gabon. At present, natural uranium contains 0.72% ^{235}U. How many years ago did natural uranium contain 3.0% ^{235}U, the amount needed to sustain a natural reactor? ($t_{1/2}$ for ^{235}U is 7.04×10^8 years.)

55. Predict the probable mode of decay for each of the following radioactive isotopes and write an equation to show the products of decay:

(a) Radon-218
(b) Californium-240
(c) Cobalt-61
(d) Carbon-11

56. The average energy output of a good grade of coal is 2.6×10^7 kJ/ton. Fission of 1 mol of ^{235}U releases 2.1×10^{10} kJ. Find the number of tons of coal needed to produce the same energy as 1 lb of ^{235}U. (See Appendix C for conversion factors.)

57. A concern in the nuclear power industry is that, if nuclear power becomes more widely used, there may be serious shortages in worldwide supplies of fissionable uranium. One solution is to build "breeder" reactors that manufacture more fuel than they consume. One such cycle works as follows:

(a) A "fertile" ^{238}U nucleus collides with a neutron to produce ^{239}U.

(b) ^{239}U decays by β emission ($t_{1/2}$ = 23.5 min) to give an isotope of neptunium.
(c) This neptunium isotope decays by β emission to give a plutonium isotope.
(d) The plutonium isotope is fissionable. On collision of one of these plutonium isotopes with a neutron, fission occurs with energy, at least two neutrons, and other nuclei as products.

Write an equation for each of the steps, and explain how this process can be used to breed more fuel than the reactor originally contained and still produce energy.

58. When an electron and a positron collide they are converted into two γ rays. In the process, their masses are converted completely into energy.
(a) Calculate the energy of this process, in joules.
(b) Using Planck's equation (page 300) determine the frequency of the γ rays emitted in this process.

59. When a neutron is captured by an atomic nucleus, energy is released as γ radiation. This energy can be calculated based on the change in mass in converting reactants to products. For the following nuclear reaction:

$$^{6}_{3}Li + ^{1}_{0}n \longrightarrow ^{7}_{3}Li + \gamma$$

(a) Calculate the energy evolved in this reaction (per atom). Masses needed for this calculation are $^{6}_{3}Li$ = 6.01512, $^{1}_{0}n$ = 1.00867, and $^{7}_{3}Li$ = 7.01600.
(b) Use the answer in part (a) to calculate the wavelength of the γ rays emitted in the reaction.

60. In order to measure the volume of the blood system of an animal, the following experiment was done. A 1.0-mL sample of an aqueous solution containing tritium with an activity of 2.0×10^6 dps was injected into the bloodstream. After time was allowed for complete circulatory mixing, a 1.0-mL blood sample was withdrawn and found to have an activity of 1.5×10^4 dps. What was the volume of the circulatory system? (The half-life of tritium is 12.3 years, so this experiment assumes that only a negligible amount of tritium has decayed in the time of the experiment.)

61. Radioactive isotopes are often used as "tracers" to follow an atom through a chemical reaction, and the following is an example. Acetic acid reacts with methanol, CH_3OH, by eliminating a molecule of H_2O to form methyl acetate, $CH_3CO_2CH_3$. Explain how you would use the radioactive isotope ^{15}O to show whether the oxygen atom in the water product comes from the —OH of the acid or the —OH of the alcohol:

$$CH_3\overset{\overset{\large O}{\|}}{C}OH + HOCH_3 \longrightarrow CH_3\overset{\overset{\large O}{\|}}{C}OCH_3 + H_2O$$

acetic acid; methanol; methyl acetate

CONCEPTUAL QUESTION

62. Radioactive decay series always begin with a very long-lived isotope. For example, the half-life of ^{238}U is 4.5×10^9 years. Each series is identified by the name of the long-lived parent isotope of highest mass.
(a) The uranium-238 radioactive decay series is sometimes referred to as the $4n + 2$ series because masses of all 13 members of this series can be expressed by the equation $m = 4n + 2$, where n is an integer. Explain why the masses correlate in this way.
(b) Two other radioactive decay series identified in minerals in the earth's crust are the thorium-232 series and the uranium-235 series. Do the masses of the isotopes in these series conform to a simple mathematical equation? If so, identify that equation.
(c) Identify the radioactive decay series to which each of the following isotopes belong: $^{226}_{88}Ra$, $^{215}_{86}At$, $^{228}_{90}Th$, $^{210}_{83}Bi$.
(d) Evaluation reveals one series of elements, the $4n + 1$ series, is missing in the earth's crust. Speculate why this is so.

CHALLENGING QUESTIONS

63. You might wonder how it is possible to determine the half-life of long-lived radioactive isotopes like ^{238}U. With a half-life of more than 10^9 years, the radioactivity of a sample of uranium will not measurably change in your lifetime. The answer is that you can calculate the half-life using the mathematics governing first-order reactions.

It can be shown that a 1.0-mg sample of ^{238}U decays at the rate of 12 α emissions per second. Set up a mathematical equation for the rate of decay, $\Delta N/\Delta t = kN$, where N is the number of nuclei in the 1.0-mg sample and $\Delta N/\Delta t$ is 12 dps. Next, solve this equation for the rate constant for this process and then relate the rate constant to the half-life of the reaction. Carry out this calculation, and compare your result to the literature value, 4.5×10^9 years.

64. The last unknown element between bismuth and uranium was discovered by Lise Meitner (1878–1968) and Otto Hahn (1879–1968) in 1918. They obtained ^{231}Pa by chemical extraction of pitchblende, in which it is present in an amount of 1 ppm. This isotope has a half-life of 3.27×10^4 years. What radioactive decay series (the uranium-235, uranium-238, or thorium-232 series) contains ^{231}Pa as a member? What sequence of nuclear reactions produces this isotope?

List of Appendices

Appendix A

Some Mathematical Operations

The mathematical skills required in this introductory course are basic skills in algebra and a knowledge of (1) exponential (or scientific) notation, (2) logarithms, and (3) quadratic equations. This appendix reviews each of the final three topics.

A.1 ELECTRONIC CALCULATORS

The directions for calculator use in this section are given for calculators using "algebraic" logic. Such calculators are the most common type used by students in introductory courses. The procedures differ slightly for calculators using RPN logic (such as those made by Hewlett-Packard).

The advent of inexpensive electronic calculators a few years ago has made calculations in introductory chemistry much more straightforward. You are well advised to purchase a calculator that has the capability of performing calculations in scientific notation, has both base-10 and natural logarithms, and is capable of raising any number to any power and of finding any root of any number. In the following discussion, we point out how these functions of your calculator can be used.

Although electronic calculators have greatly simplified calculations, they have also forced us to focus again on significant figures. A calculator easily handles eight or more significant figures, but real laboratory data are never known to this accuracy. You are therefore urged to review Section 1.7 on handling numbers.

A.2 EXPONENTIAL (SCIENTIFIC) NOTATION

In exponential, or scientific, notation, a number is expressed as a product of two numbers: $N \times 10^n$. The first number, N, is the so-called *digit term* and is a number between 1 and 10. The second number, 10^n, the *exponential term*, is some integer power of 10. For example, 1234 is written in scientific notation as 1.234×10^3, or 1.234 multiplied by 10 three times:

$$1234 = 1.234 \times 10^1 \times 10^1 \times 10^1 = 1.234 \times 10^3$$

Conversely, a number less than 1, such as 0.01234, is written as 1.234×10^{-2}. This notation tells us that 1.234 should be divided twice by 10 to obtain 0.01234:

$$0.01234 = \frac{1.234}{10^1 \times 10^1} = 1.234 \times 10^{-1} \times 10^{-1} = 1.234 \times 10^{-2}$$

Some other examples of scientific notation are

$10000 = 1 \times 10^4$	$12345 = 1.2345 \times 10^4$
$1000 = 1 \times 10^3$	$1234 = 1.234 \times 10^3$
$100 = 1 \times 10^2$	$123 = 1.23 \times 10^2$
$10 = 1 \times 10^1$	$12 = 1.2 \times 10^1$
$1 = 1 \times 10^0$	(any number to the zero power = 1)
$1/10 = 1 \times 10^{-1}$	$0.12 = 1.2 \times 10^{-1}$
$1/100 = 1 \times 10^{-2}$	$0.012 = 1.2 \times 10^{-2}$
$1/1000 = 1 \times 10^{-3}$	$0.0012 = 1.2 \times 10^{-3}$
$1/10000 = 1 \times 10^{-4}$	$0.00012 = 1.2 \times 10^{-4}$

When converting a number to scientific notation, notice that the exponent n is positive if the number is greater than 1 and negative if the number is less than 1. The value of n is the number of places by which the decimal is shifted to obtain the number in scientific notation:

$$1\ 2\ 3\ 4\ 5. = 1.2345 \times 10^4$$

Decimal shifted 4 places to the left. Therefore, n is positive and equal to 4.

$$0.0\ 0\ 1\ 2 = 1.2 \times 10^{-3}$$

Decimal shifted 3 places to the right. Therefore, n is negative and equal to 3.

If you wish to convert a number in scientific notation to the usual form, the procedure is simply reversed:

$$6\ .\ 2\ 7\ 3 \times 10^2 = 627.3$$

Decimal point moved 2 places to the right, since n is positive and equal to 2.

$$0\ 0\ 6.273 \times 10^{-3} = 0.006273$$

Decimal point shifted 3 places to the left, since n is negative and equal to 3.

Two final points must be made concerning scientific notation. First, if you are used to working on a computer, you may be in the habit of writing a number such as 1.23×10^3 as 1.23E3 or 6.45×10^{-5} as 6.45E-5. Second, some electronic calculators allow you to convert numbers readily to the scientific notation. If you have such a calculator, you can change a number shown in the usual form to scientific notation simply by pressing the EE or EXP key and then the " = " key.

1. Adding and Subtracting Numbers

When adding or subtracting two numbers, first convert them to the same powers of 10. The digit terms are then added or subtracted as appropriate:

$$(1.234 \times 10^{-3}) + (5.623 \times 10^{-2}) = (0.1234 \times 10^{-2}) + (5.623 \times 10^{-2}) = 5.746 \times 10^{-2}$$

$$(6.52 \times 10^2) - (1.56 \times 10^3) = (6.52 \times 10^2) - (15.6 \times 10^2) = -9.1 \times 10^2$$

2. Multiplication

The digit terms are multiplied in the usual manner, and the exponents are added algebraically. The result is expressed with a digit term with only one nonzero digit to the left of the decimal:

$$(1.23 \times 10^3)(7.60 \times 10^2) = (1.23)(7.60) \times 10^{3+2}$$
$$= 9.35 \times 10^5$$

$$(6.02 \times 10^{23})(2.32 \times 10^{-2}) = (6.02)(2.32) \times 10^{23-2}$$
$$= 13.966 \times 10^{21}$$
$$= 1.40 \times 10^{22} \text{ (answer in three significant figures)}$$

3. Division

The digit terms are divided in the usual manner, and the exponents are subtracted algebraically. The quotient is written with one nonzero digit to the left of the decimal in the digit term:

$$\frac{7.60 \times 10^3}{1.23 \times 10^2} = \frac{7.60}{1.23} \times 10^{3-2} = 6.18 \times 10^1$$

$$\frac{6.02 \times 10^{23}}{9.10 \times 10^{-2}} = \frac{6.02}{9.10} \times 10^{(23)-(-2)} = 0.662 \times 10^{25} = 6.62 \times 10^{24}$$

4. Powers of Exponentials

When raising a number in exponential notation to a power, treat the digit term in the usual manner. The exponent is then multiplied by the number indicating the power:

$$(1.25 \times 10^3)^2 = (1.25)^2 \times 10^{3\times 2}$$
$$= 1.5625 \times 10^6 = 1.56 \times 10^6$$

$$(5.6 \times 10^{-10})^3 = (5.6)^3 \times 10^{(-10)\times 3}$$
$$= 175.6 \times 10^{-30} = 1.8 \times 10^{-28}$$

Electronic calculators usually have two methods of raising a number to a power. To square a number, enter the number and then press the "x^2" key. To raise a number to any power, use the "y^x" key. For example, to raise 1.42×10^2 to the fourth power,

1. Enter 1.42×10^2.
2. Press "y^x".
3. Enter 4 (this should appear on the display).
4. Press " = " and 4.0659×10^8 appears on the display.

As a final step, express the number in the correct number of significant figures (4.07×10^8) in this case.

5. Roots of Exponentials

Unless you use an electronic calculator, the number must first be put into a form in which the exponential is exactly divisible by the root. The root of the digit term is found in the usual way, and the exponent is divided by the desired root:

$$\sqrt{3.6 \times 10^7} = \sqrt{36 \times 10^6} = \sqrt{36} \times \sqrt{10^6} = 6.0 \times 10^3$$
$$\sqrt[3]{2.1 \times 10^{-7}} = \sqrt[3]{210 \times 10^{-9}} = \sqrt[3]{210} \times \sqrt[3]{10^{-9}} = 5.9 \times 10^{-3}$$

To take a square root on an electronic calculator, enter the number and then press the "$\sqrt{x}$" key. To find a higher root of a number, such as the fourth root of 5.6×10^{-10},

1. Enter the number.
2. Press the "$\sqrt[x]{y}$" key. (On most calculators, the sequence you actually use is to press "2ndF" and then "$\sqrt[x]{y}$." Alternatively, you press "INV" and then "y^x.")
3. Enter the desired root, 4 in this case.
4. Press " = ". The answer here is 4.8646×10^{-3}, or 4.9×10^{-3}.

A general procedure for finding any root is to use the "y^x" key. For a square root, x is 0.5 (or $\frac{1}{2}$), whereas it is 0.33 (or $\frac{1}{3}$) for a cube root, 0.25 (or $\frac{1}{4}$) for a fourth root, and so on.

A.3 LOGARITHMS

Two types of logarithms are used in this text: (1) common logarithms (abbreviated log) whose base is 10 and (2) natural logarithms (abbreviated ln) whose base is e (=2.71828):

$$\log x = n, \text{ where } x = 10^n$$
$$\ln x = m, \text{ where } x = e^m$$

Most equations in chemistry and physics were developed in natural, or base e, logarithms, and we follow this practice in this text. The relation between log and ln is

$$\ln x = 2.303 \log x$$

Despite the different bases of the two logarithms, they are used in the same manner. What follows is largely a description of the use of common logarithms.

A common logarithm is the power to which you must raise 10 to obtain the number. For example, the log of 100 is 2, since you must raise 10 to the second power to obtain 100. Other examples are

$$\begin{aligned} \log 1000 &= \log (10^3) = 3 \\ \log 10 &= \log (10^1) = 1 \\ \log 1 &= \log (10^0) = 0 \\ \log 0.1 &= \log (10^{-1}) = -1 \\ \log 0.0001 &= \log (10^{-4}) = -4 \end{aligned}$$

To obtain the common logarithm of a number other than a simple power of 10, you must resort to a log table or an electronic calculator. For example,

$$\log 2.10 = 0.3222, \text{ which means that } 10^{0.3222} = 2.10$$
$$\log 5.16 = 0.7126, \text{ which means that } 10^{0.7126} = 5.16$$
$$\log 3.125 = 0.49485, \text{ which means that } 10^{0.49485} = 3.125$$

To check this on your calculator, enter the number, and then press the "log" key. When using a log table, the logs of the first two numbers can be read directly from the table. The log of the third number (3.125), however, must be interpolated. That is, 3.125 is midway between 3.12 and 3.13, so the log is midway between 0.4942 and 0.4955.

To obtain the natural logarithm ln of the numbers shown here, use a calculator having this function. Enter each number and press "ln:"

$$\ln 2.10 = 0.7419, \text{ which means that } e^{0.7419} = 2.10$$
$$\ln 5.16 = 1.6409, \text{ which means that } e^{1.6409} = 5.16$$

To find the common logarithm of a number greater than 10 or less than 1 with a log table, first express the number in scientific notation. Then find the log of each part of the number and add the logs. For example,

$$\log 241 = \log (2.41 \times 10^2) = \log 2.41 + \log 10^2$$
$$= 0.382 + 2 = 2.382$$

$$\log 0.00573 = \log (5.73 \times 10^{-3}) = \log 5.73 + \log 10^{-3}$$
$$= 0.758 + (-3) = -2.242$$

Significant Figures and Logarithms

LOGARITHMS AND NOMENCLATURE: The number to the left of the decimal in a logarithm is called the **characteristic,** and the number to the right of the decimal is the **mantissa.**

Notice that the mantissa has as many significant figures as the number whose log was found. (So that you could more clearly see the result obtained with a calculator or a table, this rule was not strictly followed until the last two examples.)

Obtaining Antilogarithms

If you are given the logarithm of a number, and find the number from it, you have obtained the "antilogarithm," or "antilog," of the number. Two common procedures used by electronic calculators to do this are:

Procedure A
1. Enter the log or ln.
2. Press 2ndF.
3. Press 10^x or e^x.

Procedure B
1. Enter the log or ln.
2. Press INV.
3. Press log or ln x.

Test one or the other of these procedures with the following examples:

1. Find the number whose log is 5.234:

Recall that $\log x = n$, where $x = 10^n$. In this case $n = 5.234$. Enter that number in your calculator, and find the value of 10^n, the antilog. In this case,

$$10^{5.234} = 10^{0.234} \times 10^5 = 1.71 \times 10^5$$

Notice that the characteristic (5) sets the decimal point; it is the power of 10 in the exponential form. The mantissa (0.234) gives the value of the number x. Thus, if you use a log table to find x, you need only look up 0.234 in the table and see that it corresponds to 1.71.

2. Find the number whose log is -3.456:

$$10^{-3.456} = 10^{0.544} \times 10^{-4} = 3.50 \times 10^{-4}$$

Notice here that -3.456 must be expressed as the sum of -4 and $+0.544$.

Mathematical Operations Using Logarithms

Because logarithms are exponents, operations involving them follow the same rules used for exponents. Thus, multiplying two numbers can be done by adding logarithms:

$$\log xy = \log x + \log y$$

For example, we multiply 563 by 125 by adding their logarithms and finding the antilogarithm of the result:

$$\begin{aligned} \log 563 &= 2.751 \\ \log 125 &= \underline{2.097} \\ \log xy &= 4.848 \\ xy &= 10^{4.848} = 10^4 \times 10^{0.848} = 7.05 \times 10^4 \end{aligned}$$

One number (x) can be divided by another (y) by subtraction of their logarithms:

$$\log \frac{x}{y} = \log x - \log y$$

For example, to divide 125 by 742,

$$\begin{aligned} \log 125 &= 2.097 \\ -\log 742 &= \underline{2.870} \\ \log \frac{x}{y} &= -0.773 \\ \frac{x}{y} &= 10^{-0.773} = 10^{0.227} \times 10^{-1} = 1.68 \times 10^{-1} \end{aligned}$$

Similarly, powers and roots of numbers can be found using logarithms.

$$\log x^y = y(\log x)$$

$$\log \sqrt[y]{x} = \log x^{1/y} = \frac{1}{y} \log x$$

As an example, find the fourth power of 5.23. We first find the log of 5.23 and then multiply it by 4. The result, 2.874, is the log of the answer. Therefore, we find the antilog of 2.874:

$$\begin{aligned} (5.23)^4 &= ? \\ \log (5.23)^4 &= 4 \log 5.23 = 4(0.719) = 2.874 \\ (5.23)^4 &= 10^{2.874} = 748 \end{aligned}$$

As another example, find the fifth root of 1.89×10^{-9}:

$$\sqrt[5]{1.89 \times 10^{-9}} = (1.89 \times 10^{-9})^{1/5} = ?$$

$$\log (1.89 \times 10^{-9})^{1/5} = \frac{1}{5} \log (1.89 \times 10^{-9}) = \frac{1}{5}(-8.724) = -1.745$$

The answer is the antilog of -1.745:

$$(1.89 \times 10^{-9})^{1/5} = 10^{-1.745} = 1.80 \times 10^{-2}$$

A.4 QUADRATIC EQUATIONS

Algebraic equations of the form $ax^2 + bx + c = 0$ are called **quadratic equations.** The coefficients a, b, and c may be either positive or negative. The two roots of the equation may be found using the *quadratic formula*:

$$x = \frac{-b \pm \sqrt{b^2 - 4ac}}{2a}$$

As an example, solve the equation $5x^2 - 3x - 2 = 0$. Here $a = 5$, $b = -3$, and $c = -2$. Therefore,

$$\begin{aligned} x &= \frac{3 \pm \sqrt{(-3)^2 - 4(5)(-2)}}{2(5)} \\ &= \frac{3 \pm [2(5)/\sqrt{9 - (-40)}]}{10} = \frac{3 \pm \sqrt{49}}{10} = \frac{3 \pm 7}{10} \\ &= 1 \text{ and } -0.4 \end{aligned}$$

How do you know which of the two roots is the correct answer? You have to decide in each case which root has physical significance. It is *usually* true in this course, however, that negative values are not significant.

When you have solved a quadratic expression, you should always check your values by substitution into the original equation. In the previous example, we find that $5(1)^2 - 3(1) - 2 = 0$ and that $5(-0.4)^2 - 3(-0.4) - 2 = 0$.

The most likely place you will encounter quadratic equations is in the chapters on chemical equilibria, particularly in Chapters 16 through 18. Here you will often be faced with solving an equation such as

$$1.8 \times 10^{-4} = \frac{x^2}{0.0010 - x}$$

This equation can certainly be solved using the quadratic equation (to give $x = 3.4 \times 10^{-4}$). You may find the *method of successive approximations* to be especially convenient, however. Here we begin by making a reasonable approximation of x. This approximate value is substituted into the original equation, and this is solved to give what is hoped to be a more correct value of x. This process is repeated until the answer converges on a particular value of x, that is, until the value of x derived from two successive approximations is the same.

Step 1: First assume that x is so small that $(0.0010 - x) \approx 0.0010$. This means that

$$\begin{aligned} x^2 &= 1.8 \times 10^{-4}(0.0010) \\ x &= 4.2 \times 10^{-4} \text{ (to 2 significant figures)} \end{aligned}$$

Step 2: Substitute the value of x from Step 1 into the denominator of the original equation, and again solve for x:

$$\begin{aligned} x^2 &= 1.8 \times 10^{-4}(0.0010 - 0.00042) \\ x &= 3.2 \times 10^{-4} \end{aligned}$$

Step 3: Repeat Step 2 using the value of x found in that step:

$$x = \sqrt{1.8 \times 10^{-4}(0.0010 - 0.00032)} = 3.5 \times 10^{-4}$$

Step 4: Continue repeating the calculation, using the value of x found in the previous step:

$$x = \sqrt{1.8 \times 10^{-4}(0.0010 - 0.00035)} = 3.4 \times 10^{-4}$$

Step 5: $$x = \sqrt{1.8 \times 10^{-4}(0.0010 - 0.00034)} = 3.4 \times 10^{-4}$$

Here we find that iterations after the fourth step give the same value for x, indicating that we have arrived at a valid answer (and the same one obtained from the quadratic formula).

Here are several final thoughts on using the method of successive approximations. First, in some cases the method does not work. Successive steps may give answers that are random or that diverge from the correct value. In Chapters 16 through 18, you confront quadratic equations of the form $K = x^2/(C - x)$. The method of approximations works as long as $K < 4C$ (assuming one begins with $x = 0$ as the first guess, that is, $K \approx x^2/C$). This is always going to be true for weak acids and bases (the topic of Chapters 17 and 18), but it may *not* be the case for problems involving gas phase equilibria (Chapter 16), where K can be quite large.

Second, values of K in the equation $K = x^2/(C - x)$ are usually known only to two significant figures. We are therefore justified in carrying out successive steps until two answers are the same to two significant figures.

Finally, we highly recommend this method of solving quadratic equations, especially those in Chapters 17 and 18. If your calculator has a memory function, successive approximations can be carried out easily and rapidly.

Appendix B*

Some Important Physical Concepts

B.1 MATTER

The tendency to maintain a constant velocity is called inertia. Thus, unless acted on by an unbalanced force, a body at rest remains at rest, and a body in motion remains in motion with uniform velocity. Matter is anything that exhibits inertia; the quantity of matter is its mass.

B.2 MOTION

Motion is the change of position or location in space. Objects can have the following classes of motion:

- Translation occurs when the center of mass of an object changes its location. Example: a car moving on the highway.
- Rotation occurs when each point of a moving object moves in a circle about an axis through the center of mass. Examples: a spinning top, a rotating molecule.
- Vibration is a periodic distortion of and then recovery of original shape. Examples: a struck tuning fork, a vibrating molecule.

B.3 FORCE AND WEIGHT

Force is that which changes the velocity of a body; it is defined as

$$\text{Force} = \text{mass} \times \text{acceleration}$$

The SI unit of force is the **newton,** N, whose dimensions are kilograms times meter per second squared ($kg \cdot m/s^2$). A newton is therefore the force needed to change the velocity of a mass of 1 kilogram by 1 meter per second in a time of 1 second.

Because the earth's gravity is not the same everywhere, the weight corresponding to a given mass is not a constant. At any given spot on earth gravity is constant, however, and therefore weight is proportional to mass. When a balance tells us that a given sample (the "unknown") has the same weight as another sample (the "weights," as given by a scale reading or by a total of coun-

*Adapted from F. Brescia, J. Arents, H. Meislich, et al.: *General Chemistry*, 5th ed. Philadelphia, Harcourt Brace, 1988.

terweights), it also tells us that the two masses are equal. The balance is therefore a valid instrument for measuring the mass of an object independently of slight variations in the force of gravity.

B.4 PRESSURE*

Pressure is force per unit area. The SI unit, called the pascal, Pa, is

$$1 \text{ pascal} = \frac{1 \text{ newton}}{\text{m}^2} = \frac{1 \text{ kg} \cdot \text{m/s}^2}{\text{m}^2} = \frac{1 \text{ kg}}{\text{m} \cdot \text{s}^2}$$

The International System of Units also recognizes the bar, which is 10^5 Pa and which is close to standard atmospheric pressure (Table 1).

Table 1 • PRESSURE CONVERSIONS

From	To	Multiply By
Atmosphere	mm Hg	760 mm Hg/atm (exactly)
Atmosphere	lb/in.^{2}	14.6960 lb/(in.^{2} atm)
Atmosphere	kPa	101.325 kPa/atm
Bar	Pa	10^5 Pa/bar (exactly)
Bar	lb/in.^{2}	14.5038 lb/(in.^{2} bar)
mm Hg	torr	1 torr/mm Hg (exactly)

Chemists also express pressure in terms of the heights of liquid columns, especially water and mercury. This usage is not completely satisfactory, because the pressure exerted by a given column of a given liquid is not a constant but depends on the temperature (which influences the density of the liquid) and the location (which influences gravity). Such units are therefore not part of the SI, and their use is now discouraged. The older units are still used in books and journals, however, and chemists must still be familiar with them.

The pressure of a liquid or a gas depends only on the depth (or height) and is exerted equally in all directions. At sea level, the pressure exerted by the earth's atmosphere supports a column of mercury about 0.76 m (76 cm, or 760 mm) high.

One **standard atmosphere** (atm) is the pressure exerted by exactly 76 cm of mercury at 0 °C (density 13.5951 g/cm^3) and at standard gravity, 9.80665 m/s^2. The **bar** is equivalent to 1.01325 atm. One **torr** is the pressure exerted by exactly 1 mm of mercury at 0 °C and standard gravity.

B.5 ENERGY AND POWER

The SI unit of energy is the product of the units of force and distance, or kilograms times meter per second squared (kg · m/s^2) times meters (× m), which

*See Section 12.1

is $kg \cdot m^2/s^2$; this unit is called the **joule,** J. The joule is thus the work done when a force of 1 newton acts through a distance of 1 meter.

Work may also be done by moving an electric charge in an electric field. When the charge being moved is 1 coulomb (C), and the potential difference between its initial and final positions is 1 volt (V), the work is 1 joule. Thus,

$$1 \text{ joule} = 1 \text{ coulomb volt (CV)}$$

Another unit of electric work that is not part of the International System of Units but is still in use is the **electron volt,** eV, which is the work required to move an electron against a potential difference of 1 volt. (It is also the kinetic energy acquired by an electron when it is accelerated by a potential difference of 1 volt.) Because the charge on an electron is 1.602×10^{-19} C, we have

$$1 \text{ eV} = 1.602 \times 10^{-19} \text{ CV} \cdot \frac{1 \text{ J}}{1 \text{ CV}} = 1.602 \times 10^{-19} \text{ J}$$

If this value is multiplied by Avogadro's number, we obtain the energy involved in moving 1 mole of electron charges (1 faraday) in a field produced by a potential difference of 1 volt:

$$1 \frac{\text{eV}}{\text{particle}} = \frac{1.602 \times 10^{-19} \text{ J}}{\text{particle}} \cdot \frac{6.022 \times 10^{23} \text{ particles}}{\text{mol}} \cdot \frac{1 \text{ kJ}}{1000 \text{ J}} = 96.49 \text{ kJ/mol}$$

Power is the amount of energy delivered per unit time. The SI unit is the watt, W, which is a joule per second. One kilowatt, kW, is 1000 W. Watt hours and kilowatt hours are therefore units of energy (Table 2). For example, 1000 watts, or 1 kilowatt, is

$$1.0 \times 10^3 \text{ W} \cdot \frac{1 \text{ J}}{1 \text{ W} \cdot \text{s}} \cdot \frac{3.6 \times 10^3 \text{ s}}{1 \text{ h}} = 3.6 \times 10^6 \text{ J}$$

Table 2 • ENERGY CONVERSIONS

From	To	Multiply By
Calorie (cal)	joule	4.184 J/cal (exactly)
Kilocalorie (kcal)	cal	10^3 cal/kcal (exactly)
Kilocalorie	joule	4.184×10^3 J/kcal (exactly)
Liter atmosphere (L · atm)	joule	101.325 J/L · atm
Electron volt (eV)	joule	1.60218×10^{-19} J/eV
Electron volt per particle	kilojoules per mole	96.485 kJ · particle/eV · mol
Coulomb volt (CV)	joule	1 CV/J (exactly)
Kilowatt hour (kWh)	kcal	860.4 kcal/kWh
Kilowatt hour	joule	3.6×10^6 J/kWh (exactly)
British thermal unit (Btu)	calorie	252 cal/Btu

Appendix C

Abbreviations and Useful Conversion Factors

Table 3 • Some Common Abbreviations and Standard Symbols

Term	Abbreviation	Term	Abbreviation
Activation energy	E_a	Faraday constant	F
Ampere	A	Gas constant	R
Aqueous solution	aq	Gibbs free energy	G
Atmosphere, unit of pressure	atm	Standard free energy	$G°$
Atomic mass unit	amu	Standard free energy of formation	ΔG°_f
Avogadro constant	N_A	Free energy change for reaction	ΔG°_{rxn}
Bar, unit of pressure	bar	Half-life	$t_{1/2}$
Body-centered cubic	bcc	Heat	q
Bohr radius	a_0	Hertz	Hz
Boiling point	bp	Hour	h
Celsius temperature, °C	t	Joule	J
Charge number of an ion	z	Kelvin	K
Coulomb, electric charge	C	Kilocalorie	kcal
Curie, radioactivity	Ci	Liquid	ℓ
Cycles per second, hertz	Hz	Logarithm, base 10	log
Debye, unit of electric dipole	D	Logarithm, base e	ln
Electron	e^-	Minute	min
Electron volt	eV	Molar	M
Electronegativity	χ	Molar mass	M
Energy	E	Mole	mol
Enthalpy	H	Osmotic pressure	Π
Standard enthalpy	$H°$	Planck's constant	h
Standard enthalpy of formation	ΔH°_f	Pound	lb
Standard enthalpy of reaction	ΔH°_{rxn}	Pressure	
Entropy	S	Pascal, unit of pressure	Pa
Standard entropy	$S°$	In atmospheres	atm
Entropy change for reaction	ΔS°_{rxn}	In millimeters of mercury	mm Hg
Equilibrium constant	K	Proton number	Z
Concentration basis	K_c	Rate constant	k
Pressure basis	K_p	Simple cubic (unit cell)	sc
Ionization weak acid	K_a	Standard temperature and pressure	STP
Ionization weak base	K_b	Volt	V
Solubility product	K_{sp}	Watt	W
Formation constant	K_{form}	Wavelength	λ
Ethylenediamine	en		
Face-centered cubic	fcc		

C.1 FUNDAMENTAL UNITS OF THE SI SYSTEM

The metric system was begun by the French National Assembly in 1790 and has undergone many modifications. The International System of Units or *Système International* (SI), which represents an extension of the metric system, was adopted by the 11th General Conference of Weights and Measures in 1960. It is constructed from seven base units, each of which represents a particular physical quantity (Table 4).

Table 4 • SI Fundamental Units

Physical Quantity	Name of Unit	Symbol
Length	meter	m
Mass	kilogram	kg
Time	second	s
Temperature	kelvin	K
Amount of substance	mole	mol
Electric current	ampere	A
Luminous intensity	candela	cd

The first five units listed in Table 4 are particularly useful in general chemistry and are defined as follows:

1. The *meter* was redefined in 1960 to be equal to 1,650,763.73 wavelengths of a certain line in the emission spectrum of krypton-86.
2. The *kilogram* represents the mass of a platinum–iridium block kept at the International Bureau of Weights and Measures at Sèvres, France.
3. The *second* was redefined in 1967 as the duration of 9,192,631,770 periods of a certain line in the microwave spectrum of cesium-133.
4. The *kelvin* is 1/273.15 of the temperature interval between absolute zero and the triple point of water.
5. The *mole* is the amount of substance that contains as many entities as there are atoms in exactly 0.012 kg of carbon-12 (12 g of ^{12}C atoms).

C.2 PREFIXES USED WITH TRADITIONAL METRIC UNITS AND SI UNITS

Decimal fractions and multiples of metric and SI units are designated by using the prefixes listed in Table 5. Those most commonly used in general chemistry appear in italics.

Table 5 • TRADITIONAL METRIC AND SI PREFIXES

Factor	Prefix	Symbol	Factor	Prefix	Symbol
10^{12}	tera	T	10^{-1}	*deci*	d
10^{9}	giga	G	10^{-2}	*centi*	c
10^{6}	mega	M	10^{-3}	*milli*	m
10^{3}	*kilo*	k	10^{-6}	micro	μ
10^{2}	hecto	h	10^{-9}	*nano*	n
10^{1}	deka	da	10^{-12}	*pico*	p
			10^{-15}	femto	f
			10^{-18}	atto	a

C.3 DERIVED SI UNITS

In the International System of Units, all physical quantities are represented by appropriate combinations of the base units listed in Table 4. A list of the derived units frequently used in general chemistry is given in Table 6.

Table 6 • DERIVED SI UNITS

Physical Quantity	Name of Unit	Symbol	Definition
Area	square meter	m^2	
Volume	cubic meter	m^3	
Density	kilogram per cubic meter	kg/m^3	
Force	newton	N	$kg \cdot m/s^2$
Pressure	pascal	Pa	N/m^2
Energy	joule	J	$kg \cdot m^2/s^2$
Electric charge	coulomb	C	$A \cdot s$
Electric potential difference	volt	V	$J/(A \cdot s)$

Table 7 • COMMON UNITS OF MASS AND WEIGHT

1 Pound = 453.39 Grams
1 kilogram = 1000 grams = 2.205 pounds
1 gram = 10 decigrams = 100 centigrams = 1000 milligrams
1 gram = 6.022×10^{23} atomic mass units
1 atomic mass unit = 1.6605×10^{-24} gram
1 short ton = 2000 pounds = 907.2 kilograms
1 long ton = 2240 pounds
1 metric tonne = 1000 kilograms = 2205 pounds

Table 8 • Common Units of Length

1 Inch = 2.54 Centimeters (Exactly)

1 mile = 5280 feet = 1.609 kilometers
1 yard = 36 inches = 0.9144 meter
1 meter = 100 centimeters = 39.37 inches = 3.281 feet = 1.094 yards
1 kilometer = 1000 meters = 1094 yards = 0.6215 mile
1 Ångstrom = 1.0×10^{-8} centimeter = 0.10 nanometer = 100 picometers
= 1.0×10^{-10} meter = 3.937×10^{-9} inch

Table 9 • Common Units of Volume

1 quart = 0.9463 liter
1 liter = 1.0567 quarts

1 liter = 1 cubic decimeter = 1000 cubic centimeters = 0.001 cubic meter
1 milliliter = 1 cubic centimeter = 0.001 liter = 1.056×10^{-3} quart
1 cubic foot = 28.316 liters = 29.924 quarts = 7.481 gallons

Appendix D

Physical Constants

Table 10

Quantity	Symbol	Traditional Units	SI Units
Acceleration of gravity	g	980.6 cm/s	9.806 m/s
Atomic mass unit (1/12 the mass of ^{12}C atom)	amu or u	1.6605×10^{-24} g	1.6605×10^{-27} kg
Avogadro's number	N	6.0221367×10^{23} particles/mol	6.0221367×10^{23} particles/mol
Bohr radius	a_0	0.52918 Å 5.2918×10^{-9} cm	5.2918×10^{-11} m
Boltzmann constant	k	1.3807×10^{-16} erg/K	1.3807×10^{-23} J/K
Charge-to-mass ratio of electron	e/m	1.7588×10^{8} C/g	1.7588×10^{11} C/kg
Electronic charge	e	1.6022×10^{-19} C 4.8033×10^{-10} esu	1.6022×10^{-19} C
Electron rest mass	m_e	9.1094×10^{-28} g 0.00054858 amu	9.1094×10^{-31} kg
Faraday constant	F	96,485 C/mol e^- 23.06 kcal/V · mol e^-	96,485 C/mol e^- 96,485 J/V · mol e^-
Gas constant	R	$0.08206 \dfrac{L \cdot atm}{mol \cdot K}$ $1.987 \dfrac{cal}{mol \cdot K}$	$8.3145 \dfrac{Pa \cdot dm^3}{mol \cdot K}$ 8.3145 J/mol · K
Molar volume (STP)	V_m	22.414 L/mol	22.414×10^{-3} m^3/mol 22.414 dm^3/mol
Neutron rest mass	m_n	1.67495×10^{-24} g 1.008665 amu	1.67493×10^{-27} kg
Planck's constant	h	6.6261×10^{-27} erg · s	$6.6260755 \times 10^{-34}$ J · s
Proton rest mass	m_p	1.6726×10^{-24} g 1.007276 amu	1.6726×10^{-27} kg
Rydberg constant	R_α	3.289×10^{15} cycles/s 2.1799×10^{-11} erg	1.0974×10^{7} m^{-1} 2.1799×10^{-18} J
Velocity of light (in a vacuum)	c	2.9979×10^{10} cm/s (186,282 miles/s)	2.9979×10^{8} m/s

$\pi = 3.1416$
$e = 2.7183$
$\ln X = 2.303 \log X$

Table 11 • Specific Heats and Heat Capacities for Some Common Substances at 25 °C

Substance	Specific Heat J/g · K	Molar Heat Capacity J/mol · K
Al(s)	0.902	24.3
Ca(s)	0.653	26.2
Cu(s)	0.385	24.5
Fe(s)	0.451	24.8
Hg(ℓ)	0.138	27.7
H_2O(s), ice	2.06	37.7
H_2O(ℓ), water	4.18	75.3
H_2O(g), steam	2.03	36.4
C_6H_6(ℓ), benzene	1.74	136
C_6H_6(g), benzene	1.04	81.6
C_2H_5OH(ℓ), ethanol	2.46	113
C_2H_5OH(g), ethanol	0.954	420
$(C_2H_5)_2O$(ℓ), diethyl ether	3.74	172
$(C_2H_5)_2O$(g), diethyl ether	2.35	108

Table 12 • Heats of Transformation and Transformation Temperatures of Several Substances

Substance	MP (°C)	Heat of Fusion		BP (°C)	Heat of Vaporization	
		J/g	*kJ/mol*		*J/g*	*kJ/mol*
*Elements**						
Al	660	395	10.7	2467	12083	294
Ca	839	230	9.3	1493	3768	151
Cu	1083	205	13.0	2567	4799	305
Fe	1535	267	14.9	2750	6285	351
Hg	−38.8	11	2.3	357	294	59.0
Compounds						
H_2O	0.00	333	6.02	100.0	2260	40.7
CH_4	−182	58.6	0.92	−164	—	—
C_2H_5OH	−117	109	5.02	78.0	855	39.3
C_6H_6	5.48	127	9.92	80.1	395	30.8
$(C_2H_5)_2O$	−116	97.9	7.66	35	351	26.0

*Data for the elements are taken from J. Emsley, *The Elements*, 2nd ed. Oxford University Press, New York, 1991; *Saunders Interactive General Chemistry CD-ROM*.

Appendix E

Naming Organic Compounds

It seems a daunting task—to devise a systematic procedure that gives each organic compound a unique name—but that is what has been done. A set of rules was developed to name organic compounds by the International Union of Pure and Applied Chemistry, IUPAC. The IUPAC nomenclature allows chemists to write a name for any compound based on its structure or to identify the formula and structure for a compound from its name. In this book, we have generally used the IUPAC nomenclature scheme when naming compounds.

In addition to the systematic names, many compounds also have common names. The common names came into existence before the nomenclature rules were developed, and they have continued in use. For some compounds, these names are so well entrenched that they are used most of the time. One such compound is acetic acid, which is almost always referred to by that name and not by its systematic name, ethanoic acid.

The general procedure for systematic naming of organic compounds begins with nomenclature for hydrocarbons. Other organic compounds are then named as derivatives of hydrocarbons. Nomenclature rules for simple organic compounds are given in the following section.

E.1 HYDROCARBONS

Alkanes

The names of alkanes end in "-ane." When naming a specific alkane, the root of the name identifies the longest carbon chain in a compound. Specific substituent groups attached to this carbon chain are identified by name and position.

Alkanes with chains of from one to ten carbon atoms are given in Table 11.3. After the first four compounds, the names derive from Latin numbers—pentane, hexane, heptane, octane, nonane, decane—and this regular naming continues for higher alkanes. For substituted alkanes, the substituent groups on a hydrocarbon chain must be identified both by a name and by the position of substitution; this information precedes the root of the name. The position is indicated by a number that refers to the carbon atom to which the substituent is attached. (Numbering of the carbon atoms in a chain should begin at the end of the carbon chain that allows the substituent groups to have the lowest numbers.)

Names of hydrocarbon substituents are derived from the name of the hydrocarbon. The group $—CH_3$, derived by taking a hydrogen from methane, is called the methyl group; the C_2H_5 group is the ethyl group. The nomenclature scheme is easily extended to derivatives of hydrocarbons with other substituent groups like -Cl (chloro), $-NO_2$ (nitro), -CN (cyano), -D (deuterio), and so on (Table 13). If two or more of the same substituent groups occur, the prefixes di-, tri-, and tetra- are added. When different substituent groups are present, they are generally listed in alphabetical order.

Table 13 • NAMES OF COMMON SUBSTITUENT GROUPS

Formula	Name	Formula	Name
$—CH_3$	methyl	—D	deuterio
$—C_2H_5$	ethyl	—Cl	chloro
$—CH_2CH_2CH_3$	1-propyl (*n*-propyl)	—Br	bromo
$—CH(CH_3)_2$	2-propyl (isopropyl)	—F	fluoro
$—CH{=}CH_2$	ethenyl (vinyl)	—CN	cyano
$—C_6H_5$	phenyl	$—NO_2$	nitro
—OH	hydroxo		
$—NH_2$	amino		

Example:

$$\begin{array}{c} \quad\;\; CH_3 \quad C_2H_5 \\ \quad\;\; | \qquad\;\; | \\ CH_3CH_2CHCH_2CHCH_2CH_3 \end{array}$$

Step	Information to include	Contribution to name
1.	An alkane	name will end in "-ane"
2.	Longest chain is 7 carbons	name as a *heptane*
3.	$—CH_3$ group at carbon 3	3-*methyl*
4.	$—C_2H_5$ group at carbon 5	5-*ethyl*

Name: 5-ethyl-3-methylheptane

Cycloalkanes are named based on the ring size adding the prefix "cyclo"; for example, the cycloalkane with a six-member ring of carbons is called cyclohexane.

Alkenes

Alkenes have names ending in "-ene." The name of an alkene must specify the length of the carbon chain, the position of the double bond (and when appropriate, the configuration, either *cis* or *trans*). As with alkanes, both identity and position of substituent groups must be given. The carbon chain is numbered from the end that gives the double bond the lowest number.

Compounds with two double bonds are called dienes and they are named similarly—specifying the positions of the double bonds and the name and position of any substituent groups.

For example, the compound $H_2C{=}C(CH_3)CH(CH_3)CH_2CH_3$ has a five-carbon chain with a double bond between carbon atoms 1 and 2 and methyl groups on carbon atoms 2 and 3. Its name using IUPAC nomenclature is **2,3-dimethyl-1-pentene.** The compound $CH_3CH{=}CHCCl_3$ with a *cis* configuration around the double bond is named **1,1,1-trichloro-*cis*-2-butene.** The compound $H_2C{=}C(Cl)CH{=}CH_2$ is **2-chloro-1,3-butadiene.**

Alkynes

The naming of alkynes is similar to the naming of alkenes, except that *cis–trans* isomerism isn't a factor. The ending "-yne" on a name identifies a compound as an alkyne.

Benzene Derivatives

The carbon atoms in the six-member ring are numbered 1 through 6, and the name and position of substituent groups are given. The two examples shown here are **1-ethyl-3-methylbenzene** and **1,4-diaminobenzene.**

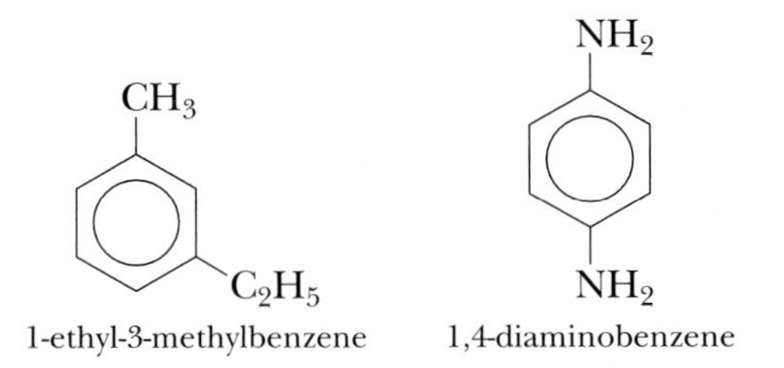

1-ethyl-3-methylbenzene 1,4-diaminobenzene

E.2 DERIVATIVES OF HYDROCARBONS

The names for alcohols, aldehydes, ketones, and acids are based on the name of the hydrocarbon with an appropriate suffix to denote the class of compound, as follows:

- **Alcohols:** Substitute "-ol" the final "-e" in the name of the hydrocarbon, and designate the position of the —OH group by the number of the carbon atom. For example, $CH_3CH_2CHOHCH_3$ is named as a derivative of the 4-carbon hydrocarbon butane. The —OH group is attached to the second carbon, so the name is 2-butanol.
- **Aldehydes:** Substitute "-al" for the final "-e" in the name of the hydrocarbon. The carbon atom of an aldehyde is, by definition, carbon-1 in the hydrocarbon chain. For example, the compound $CH_3CH(CH_3)CH_2CH_2CHO$ contains a 5-carbon chain with the aldehyde functional group being carbon 1 and the $—CH_3$ group at position 4; thus the name is **4-methylpentanal.**
- **Ketones:** Substitute "-one" for the final "-e" in the name of the hydrocarbon. The position of the ketone functional group (the carbonyl group) is

indicated by the number of the carbon atom. For example, the compound $CH_3COCH_2CH(C_2H_5)CH_2CH_3$ has the carbonyl group at the 2 position and an ethyl group at the 4 position of a 6-carbon chain; its name is **4-ethyl-2-hexanone.**

- **Carboxylic acids (organic acids):** Substitute "-oic" for the final "-e" in the name of the hydrocarbon. The carbon atoms in the longest chain are counted beginning with the carboxylic carbon atom. For example, *trans*-$CH_3CH{=}CHCH_2CO_2H$ is named as a derivative of *trans*-3-pentene, that is, ***trans*-3-pentenoic acid.**

An **ester** is named as a derivative of the alcohol and acid from which it is made. The name of an ester is obtained by splitting the formula RCO_2R' into two parts, the RCO_2— portion and the —R′ portion. The —R′ portion comes from the alcohol and is identified by the hydrocarbon group name; derivatives of ethanol, for example, are called *ethyl* esters. The acid part of the compound is named by dropping the "-oic" ending for the acid and replacing it by "-oate." The compound $CH_3CH_2CO_2CH_3$ is named **methyl propanoate.**

Notice that an anion derived from a carboxylic acid by loss of the acidic proton is named the same way. Thus, $CH_3CH_2CO_2^-$ is the **propanoate anion,** and the sodium salt of this anion, $Na(CH_3CH_2CO_2)$, is **sodium propanoate.**

Appendix F

Values for the Ionization Energies and Electron Affinities of the Elements

1A (1)	2A (2)	3B (3)	4B (4)	5B (5)	6B (6)	7B (7)	8B (8,9,10)			1B (11)	2B (12)	3A (13)	4A (14)	5A (15)	6A (16)	7A (17)	8 (18)
H 1312																	He 2371
Li 520	Be 899											B 801	C 1086	N 1402	O 1314	F 1681	Ne 2081
Na 496	Mg 738											Al 578	Si 786	P 1012	S 1000	Cl 1251	Ar 1521
K 419	Ca 599	Sc 631	Ti 658	V 650	Cr 652	Mn 717	Fe 759	Co 758	Ni 757	Cu 745	Zn 906	Ga 579	Ge 762	As 947	Se 941	Br 1140	Kr 1351
Rb 403	Sr 550	Y 617	Zr 661	Nb 664	Mo 685	Tc 702	Ru 711	Rh 720	Pd 804	Ag 731	Cd 868	In 558	Sn 709	Sb 834	Te 869	I 1008	Xe 1170
Cs 377	Ba 503	La 538	Hf 681	Ta 761	W 770	Re 760	Os 840	Ir 880	Pt 870	Au 890	Hg 1007	Tl 589	Pb 715	Bi 703	Po 812	At 890	Rn 1037

T*able* 14 • ELECTRON AFFINITY VALUES FOR SOME ELEMENTS (KJ/MOL)*

H						
−72.77						
Li	Be	B	C	N	O	F
−59.63	0†	−26.7	−121.85	0	−140.98	−328.0
Na	Mg	Al	Si	P	S	Cl
−52.87	0	−42.6	−133.6	−72.07	−200.41	−349.0
K	Ca	Ga	Ge	As	Se	Br
−48.39	0	−30	−120	−78	−194.97	−324.7
Rb	Sr	In	Sn	Sb	Te	I
−46.89	0	−30	−120	−103	−190.16	−295.16
Cs	Ba	Tl	Pb	Bi	Po	At
−45.51	0	−20	−35.1	−91.3	−180	−270

*Data taken from H. Hotop and W.C. Lineberger: *Journal of Physical Chemistry, Reference Data,* Vol. 14, p. 731, 1985. (This paper also includes data for the transition metals.) Some values are known to more than two decimal places.

†Elements with an electron affinity of zero indicate that a stable anion A^- of the element does not exist in the gas phase.

Appendix G

Vapor Pressure of Water at Various Temperatures

Table 15 • VAPOR PRESSURE OF WATER AT VARIOUS TEMPERATURES

Temperature °C	Vapor Pressure torr	Temperature °C	Vapor Pressure torr	Temperature °C	Vapor Pressure torr	Temperature °C	Vapor Pressure torr
−10	2.1	21	18.7	51	97.2	81	369.7
−9	2.3	22	19.8	52	102.1	82	384.9
−8	2.5	23	21.1	53	107.2	83	400.6
−7	2.7	24	22.4	54	112.5	84	416.8
−6	2.9	25	23.8	55	118.0	85	433.6
−5	3.2	26	25.2	56	123.8	86	450.9
−4	3.4	27	26.7	57	129.8	87	468.7
−3	3.7	28	28.3	58	136.1	88	487.1
−2	4.0	29	30.0	59	142.6	89	506.1
−1	4.3	30	31.8	60	149.4	90	525.8
0	4.6	31	33.7	61	156.4	91	546.1
1	4.9	32	35.7	62	163.8	92	567.0
2	5.3	33	37.7	63	171.4	93	588.6
3	5.7	34	39.9	64	179.3	94	610.9
4	6.1	35	42.2	65	187.5	95	633.9
5	6.5	36	44.6	66	196.1	96	657.6
6	7.0	37	47.1	67	205.0	97	682.1
7	7.5	38	49.7	68	214.2	98	707.3
8	8.0	39	52.4	69	223.7	99	733.2
9	8.6	40	55.3	70	233.7	100	760.0
10	9.2	41	58.3	71	243.9	101	787.6
11	9.8	42	61.5	72	254.6	102	815.9
12	10.5	43	64.8	73	265.7	103	845.1
13	11.2	44	68.3	74	277.2	104	875.1
14	12.0	45	71.9	75	289.1	105	906.1
15	12.8	46	75.7	76	301.4	106	937.9
16	13.6	47	79.6	77	314.1	107	970.6
17	14.5	48	83.7	78	327.3	108	1004.4
18	15.5	49	88.0	79	341.0	109	1038.9
19	16.5	50	92.5	80	355.1	110	1074.6
20	17.5						

Appendix H

Ionization Constants for Weak Acids at 25 °C

Table 16 • IONIZATION CONSTANTS FOR WEAK ACIDS AT 25 °C

Acid	Formula and Ionization Equation	K_a
Acetic	$CH_3CO_2H \rightleftharpoons H^+ + CH_3CO_2^-$	1.8×10^{-5}
Arsenic	$H_3AsO_4 \rightleftharpoons H^+ + H_2AsO_4^-$	$K_1 = 2.5 \times 10^{-4}$
	$H_2AsO_4^- \rightleftharpoons H^+ + HAsO_4^{2-}$	$K_2 = 5.6 \times 10^{-8}$
	$HAsO_4^{2-} \rightleftharpoons H^+ + AsO_4^{3-}$	$K_3 = 3.0 \times 10^{-13}$
Arsenous	$H_3AsO_3 \rightleftharpoons H^+ + H_2AsO_3^-$	$K_1 = 6.0 \times 10^{-10}$
	$H_2AsO_3^- \rightleftharpoons H^+ + HAsO_3^{2-}$	$K_2 = 3.0 \times 10^{-14}$
Benzoic	$C_6H_5CO_2H \rightleftharpoons H^+ + C_6H_5CO_2^-$	6.3×10^{-5}
Boric	$H_3BO_3 \rightleftharpoons H^+ + H_2BO_3^-$	$K_1 = 7.3 \times 10^{-10}$
	$H_2BO_3 \rightleftharpoons H^+ + HBO_3^{2-}$	$K_2 = 1.8 \times 10^{-13}$
	$HBO_3^{2-} \rightleftharpoons H^+ + BO_3^{3-}$	$K_3 = 1.6 \times 10^{-14}$
Carbonic	$H_2CO_3 \rightleftharpoons H^+ + HCO_3^-$	$K_1 = 4.2 \times 10^{-7}$
	$HCO_3^- \rightleftharpoons H^+ + CO_3^{2-}$	$K_2 = 4.8 \times 10^{-11}$
Citric	$H_3C_6H_5O_7 \rightleftharpoons H^+ + H_2C_6H_5O_7^-$	$K_1 = 7.4 \times 10^{-3}$
	$H_2C_6H_5O_7^- \rightleftharpoons H^+ + HC_6H_5O_7^{2-}$	$K_2 = 1.7 \times 10^{-5}$
	$HC_6H_5O_7^{2-} \rightleftharpoons H^+ + C_6H_5O_7^{3-}$	$K_3 = 4.0 \times 10^{-7}$
Cyanic	$HOCN \rightleftharpoons H^+ + OCN^-$	3.5×10^{-4}
Formic	$HCO_2H \rightleftharpoons H^+ + HCO_2^-$	1.8×10^{-4}
Hydrazoic	$HN_3 \rightleftharpoons H^+ + N_3^-$	1.9×10^{-5}
Hydrocyanic	$HCN \rightleftharpoons H^+ + CN^-$	4.0×10^{-10}
Hydrofluoric	$HF \rightleftharpoons H^+ + F^-$	7.2×10^{-4}
Hydrogen peroxide	$H_2O_2 \rightleftharpoons H^+ + HO_2^-$	2.4×10^{-12}
Hydrosulfuric	$H_2S \rightleftharpoons H^+ + HS^-$	$K_1 = 1 \times 10^{-7}$
	$HS^- \rightleftharpoons H^+ + S^{2-}$	$K_2 = 1 \times 10^{-19}$
Hypobromous	$HOBr \rightleftharpoons H^+ + OBr^-$	2.5×10^{-9}
Hypochlorous	$HOCl \rightleftharpoons H^+ + OCl^-$	3.5×10^{-8}
Nitrous	$HNO_2 \rightleftharpoons H^+ + NO_2^-$	4.5×10^{-4}
Oxalic	$H_2C_2O_4 \rightleftharpoons H^+ + HC_2O_4^-$	$K_1 = 5.9 \times 10^{-2}$
	$HC_2O_4^- \rightleftharpoons H^+ + C_2O_4^{2-}$	$K_2 = 6.4 \times 10^{-5}$
Phenol	$C_6H_5OH \rightleftharpoons H^+ + C_6H_5O^-$	1.3×10^{-10}
Phosphoric	$H_3PO_4 \rightleftharpoons H^+ + H_2PO_4^-$	$K_1 = 7.5 \times 10^{-3}$
	$H_2PO_4^- \rightleftharpoons H^+ + HPO_4^{2-}$	$K_2 = 6.2 \times 10^{-8}$
	$HPO_4^{2-} \rightleftharpoons H^+ + PO_4^{3-}$	$K_3 = 3.6 \times 10^{-13}$
Phosphorus	$H_3PO_3 \rightleftharpoons H^+ + H_2PO_3^-$	$K_1 = 1.6 \times 10^{-2}$
	$H_2PO_3 \rightleftharpoons H^+ + HPO_3^{2-}$	$K_2 = 7.0 \times 10^{-7}$

continued

Table 16 • *continued*

Acid	Formula and Ionization Equation	K_a
Selenic	$H_2SeO_4 \rightleftharpoons H^+ + HSeO_4^-$	K_1 = very large
	$HSeO_4^- \rightleftharpoons H^+ + SeO_4^{2-}$	$K_2 = 1.2 \times 10^{-2}$
Selenous	$H_2SeO_3 \rightleftharpoons H^+ + HSeO_3^-$	$K_1 = 2.7 \times 10^{-3}$
	$HSeO_3^- \rightleftharpoons H^+ + SeO_3^{2-}$	$K_2 = 2.5 \times 10^{-7}$
Sulfuric	$H_2SO_4 \rightleftharpoons H^+ + HSO_4^-$	K_1 = very large
	$HSO_4^- \rightleftharpoons H^+ + SO_4^{2-}$	$K_2 = 1.2 \times 10^{-2}$
Sulfurous	$H_2SO_3 \rightleftharpoons H^+ + HSO_3^-$	$K_1 = 1.7 \times 10^{-2}$
	$HSO_3^- \rightleftharpoons H^+ + SO_3^{2-}$	$K_2 = 6.4 \times 10^{-8}$
Tellurous	$H_2TeO_3 \rightleftharpoons H^+ + HTeO_3^-$	$K_1 = 2 \times 10^{-3}$
	$HTeO_3^- \rightleftharpoons H^+ + TeO_3^{2-}$	$K_2 = 1 \times 10^{-8}$

Appendix I

Ionization Constants for Weak Bases at 25 °C

Table 17 • IONIZATION CONSTANTS FOR WEAK BASES AT 25 °C

Base	Formula and Ionization Equation	K_b
Ammonia	$NH_3 + H_2O \rightleftharpoons NH_4^+ + OH^-$	1.8×10^{-5}
Aniline	$C_6H_5NH_2 + H_2O \rightleftharpoons C_6H_5NH_3^+ + OH^-$	4.0×10^{-10}
Dimethylamine	$(CH_3)_2NH + H_2O \rightleftharpoons (CH_3)_2NH_2^+ + OH^-$	7.4×10^{-4}
Ethylenediamine	$H_2NCH_2CH_2NH_2 + H_2O \rightleftharpoons H_2NCH_2CH_2NH_3^+ + OH^-$	$K_1 = 8.5 \times 10^{-5}$
	$H_2NCH_2CH_2NH_3^+ + H_2O \rightleftharpoons H_3NCH_2CH_2NH_3^{2+} + OH^-$	$K_2 = 2.7 \times 10^{-8}$
Hydrazine	$N_2H_4 + H_2O \rightleftharpoons N_2H_5^+ + OH^-$	$K_1 = 8.5 \times 10^{-7}$
	$N_2H_5^+ + H_2O \rightleftharpoons N_2H_6^{2+} + OH^-$	$K_2 = 8.9 \times 10^{-16}$
Hydroxylamine	$NH_2OH + H_2O \rightleftharpoons NH_3OH^+ + OH^-$	6.6×10^{-9}
Methylamine	$CH_3NH_2 + H_2O \rightleftharpoons CH_3NH_3^+ + OH^-$	5.0×10^{-4}
Pyridine	$C_5H_5N + H_2O \rightleftharpoons C_5H_5NH^+ + OH^-$	1.5×10^{-9}
Trimethylamine	$(CH_3)_3N + H_2O \rightleftharpoons (CH_3)_3NH^+ + OH^-$	7.4×10^{-5}

Appendix J

Solubility Product Constants for Some Inorganic Compounds at 25 °C

Table 18 • SOLUBILITY PRODUCT CONSTANTS FOR SOME INORGANIC COMPOUNDS AT 25 °C

Substance	K_{sp}	Substance	K_{sp}
Aluminum compounds		$CaCO_3$	3.8×10^{-9}
$AlAsO_4$	1.6×10^{-16}	$CaCrO_4$	7.1×10^{-4}
$Al(OH)_3$	1.9×10^{-33}	$CaC_2O_4 \cdot H_2O$*	2.3×10^{-9}
$AlPO_4$	1.3×10^{-20}	CaF_2	3.9×10^{-11}
Antimony compounds		$Ca(OH)_2$	7.9×10^{-6}
Sb_2S_3	1.6×10^{-93}	$CaHPO_4$	2.7×10^{-7}
Barium compounds		$Ca(H_2PO_4)_2$	1.0×10^{-3}
$Ba_3(AsO_4)_2$	1.1×10^{-13}	$Ca_3(PO_4)_2$	1.0×10^{-25}
$BaCO_3$	8.1×10^{-9}	$CaSO_3 \cdot 2\ H_2O$*	1.3×10^{-8}
$BaC_2O_4 \cdot 2\ H_2O$*	1.1×10^{-7}	$CaSO_4 \cdot 2\ H_2O$*	2.4×10^{-5}
$BaCrO_4$	2.0×10^{-10}	*Chromium compounds*	
BaF_2	1.7×10^{-6}	$CrAsO_4$	7.8×10^{-21}
$Ba(OH)_2 \cdot 8\ H_2O$*	5.0×10^{-3}	$Cr(OH)_3$	6.7×10^{-31}
$Ba_3(PO_4)_2$	1.3×10^{-29}	$CrPO_4$	2.4×10^{-23}
$BaSeO_4$	2.8×10^{-11}	*Cobalt compounds*	
$BaSO_3$	8.0×10^{-7}	$Co_3(AsO_4)_2$	7.6×10^{-29}
$BaSO_4$	1.1×10^{-10}	$CoCO_3$	8.0×10^{-13}
Bismuth compounds		$Co(OH)_2$	2.5×10^{-16}
$BiOCl$	7.0×10^{-9}	$CoS\ (\alpha)$	5.9×10^{-21}
$BiO(OH)$	1.0×10^{-12}	$Co(OH)_3$	4.0×10^{-45}
$Bi(OH)_3$	3.2×10^{-40}	*Copper compounds*	
BiI_3	8.1×10^{-19}	$CuBr$	5.3×10^{-9}
$BiPO_4$	1.3×10^{-23}	$CuCl$	1.9×10^{-7}
Bi_2S_3	1.6×10^{-72}	$CuCN$	3.2×10^{-20}
Cadmium compounds		$Cu_2O\ (Cu^+ + OH^-)$†	1.0×10^{-14}
$Cd_3(AsO_4)_2$	2.2×10^{-32}	CuI	5.1×10^{-12}
$CdCO_3$	2.5×10^{-14}	Cu_2S	1.6×10^{-48}
$Cd(CN)_2$	1.0×10^{-8}	$CuSCN$	1.6×10^{-11}
$Cd_2[Fe(CN)_6]$	3.2×10^{-17}	$Cu_3(AsO_4)_2$	7.6×10^{-36}
$Cd(OH)_2$	1.2×10^{-14}	$CuCO_3$	2.5×10^{-10}
CdS	1.0×10^{-27}	$Cu_2[Fe(CN)_6]$	1.3×10^{-16}
Calcium compounds		$Cu(OH)_2$	1.6×10^{-19}
$Ca_3(AsO_4)_2$	6.8×10^{-19}	CuS	7.9×10^{-37}

Substance	K_{sp}	Substance	K_{sp}
Gold compounds		Hg_2S	5.8×10^{-44}
$AuBr$	5.0×10^{-17}	$Hg(CN)_2$	3.0×10^{-23}
$AuCl$	2.0×10^{-13}	$Hg(OH)_2$	2.5×10^{-26}
AuI	1.6×10^{-23}	HgI_2	4.0×10^{-29}
$AuBr_3$	4.0×10^{-36}	HgS	2.0×10^{-53}
$AuCl_3$	3.2×10^{-25}	*Nickel compounds*	
$Au(OH)_3$	1×10^{-53}	$Ni_3(AsO_4)_2$	1.9×10^{-26}
AuI_3	1.0×10^{-46}	$NiCO_3$	6.6×10^{-9}
Iron compounds		$Ni(CN)_2$	3.0×10^{-23}
$FeCO_3$	3.5×10^{-11}	$Ni(OH)_2$	2.8×10^{-16}
$Fe(OH)_2$	7.9×10^{-15}	NiS	3.0×10^{-21}
FeS	4.9×10^{-18}	*Silver compounds*	
$Fe_4[Fe(CN)_6]_3$	3.0×10^{-41}	Ag_3AsO_4	1.1×10^{-20}
$Fe(OH)_3$	6.3×10^{-38}	$AgBr$	3.3×10^{-13}
Fe_2S_3	1.4×10^{-88}	Ag_2CO_3	8.1×10^{-12}
Lead compounds		$AgCl$	1.8×10^{-10}
$Pb_3(AsO_4)_2$	4.1×10^{-36}	Ag_2CrO_4	9.0×10^{-12}
$PbBr_2$	6.3×10^{-6}	$AgCN$	1.2×10^{-16}
$PbCO_3$	1.5×10^{-13}	$Ag_4[Fe(CN)_6]$	1.6×10^{-41}
$PbCl_2$	1.7×10^{-5}	Ag_2O $(Ag^+ + OH^-)^{\dagger}$	2.0×10^{-8}
$PbCrO_4$	1.8×10^{-14}	AgI	1.5×10^{-16}
PbF_2	3.7×10^{-8}	Ag_3PO_4	1.3×10^{-20}
$Pb(OH)_2$	2.8×10^{-16}	Ag_2SO_3	1.5×10^{-14}
PbI_2	8.7×10^{-9}	Ag_2SO_4	1.7×10^{-5}
$Pb_3(PO_4)_2$	3.0×10^{-44}	Ag_2S	1.0×10^{-49}
$PbSeO_4$	1.5×10^{-7}	$AgSCN$	1.0×10^{-12}
$PbSO_4$	1.8×10^{-8}	*Strontium compounds*	
PbS	3.2×10^{-28}	$Sr_3(AsO_4)_2$	1.3×10^{-18}
Magnesium compounds		$SrCO_3$	9.4×10^{-10}
$Mg_3(AsO_4)_2$	2.1×10^{-20}	$SrC_2O_4 \cdot 2\ H_2O$*	5.6×10^{-8}
$MgCO_3 \cdot 3\ H_2O$*	4.0×10^{-5}	$SrCrO_4$	3.6×10^{-5}
MgC_2O_4	8.6×10^{-5}	$Sr(OH)_2 \cdot 8\ H_2O$*	3.2×10^{-4}
MgF_2	6.4×10^{-9}	$Sr_3(PO_4)_2$	1.0×10^{-31}
$Mg(OH)_2$	1.5×10^{-11}	$SrSO_3$	4.0×10^{-8}
$MgNH_4PO_4$	2.5×10^{-12}	$SrSO_4$	2.8×10^{-7}
Manganese compounds		*Tin compounds*	
$Mn_3(AsO_4)_2$	1.9×10^{-11}	$Sn(OH)_2$	2.0×10^{-26}
$MnCO_3$	1.8×10^{-11}	SnI_2	1.0×10^{-4}
$Mn(OH)_2$	4.6×10^{-14}	SnS	1.0×10^{-28}
MnS	5.1×10^{-15}	$Sn(OH)_4$	1.0×10^{-57}
$Mn(OH)_3$	$\approx 1 \times 10^{-36}$	SnS_2	1.0×10^{-70}
Mercury compounds		*Zinc compounds*	
Hg_2Br_2	1.3×10^{-22}	$Zn_3(AsO_4)_2$	1.1×10^{-27}
Hg_2CO_3	8.9×10^{-17}	$ZnCO_3$	1.5×10^{-11}
Hg_2Cl_2	1.1×10^{-18}	$Zn(OH)_2$	4.5×10^{-17}
Hg_2CrO_4	5.0×10^{-9}	$Zn(CN)_2$	8.0×10^{-12}
Hg_2I_2	4.5×10^{-29}	$Zn_3(PO_4)_2$	9.1×10^{-33}
$Hg_2O \cdot H_2O$ $(Hg_2^{2+} + 2\ OH^-)^{*\dagger}$	1.6×10^{-23}	$Zn_3[Fe(CN)_6]_2$	4.1×10^{-16}
Hg_2SO_4	6.8×10^{-7}	ZnS	2.0×10^{-25}

*Since $[H_2O]$ does not appear in equilibrium constants for equilibria in aqueous solution in general, it does *not* appear in the K_{sp} expressions for hydrated solids.

†Very small amounts of oxides dissolve in water to give the ions indicated in parentheses. Solid hydroxides are unstable and decompose to oxides as rapidly as they are formed.

Appendix K

Formation Constants for Some Complex Ions in Aqueous Solution

Table 19 • Formation Constants for Some Complex Ions in Aqueous Solution

Formation Equilibrium	*K*
$Ag^+ + 2\ Br^- \rightleftharpoons [AgBr_2]^-$	1.3×10^7
$Ag^+ + 2\ Cl^- \rightleftharpoons [AgCl_2]^-$	2.5×10^5
$Ag^+ + 2\ CN^- \rightleftharpoons [Ag(CN)_2]^-$	5.6×10^{18}
$Ag^+ + 2\ S_2O_3^{2-} \rightleftharpoons [Ag(S_2O_3)_2]^{3-}$	2.0×10^{13}
$Ag^+ + 2\ NH_3 \rightleftharpoons [Ag(NH_3)_2]^+$	1.6×10^7
$Al^{3+} + 6\ F^- \rightleftharpoons [AlF_6]^{3-}$	5.0×10^{23}
$Al^{3+} + 4\ OH^- \rightleftharpoons [Al(OH)_4]^-$	7.7×10^{33}
$Au^+ + 2\ CN^- \rightleftharpoons [Au(CN)_2]^-$	2.0×10^{38}
$Cd^{2+} + 4\ CN^- \rightleftharpoons [Cd(CN)_4]^{2-}$	1.3×10^{17}
$Cd^{2+} + 4\ Cl^- \rightleftharpoons [CdCl_4]^{2-}$	1.0×10^4
$Cd^{2+} + 4\ NH_3 \rightleftharpoons [Cd(NH_3)_4]^{2+}$	1.0×10^7
$Co^{2+} + 6\ NH_3 \rightleftharpoons [Co(NH_3)_6]^{2+}$	7.7×10^4
$Cu^+ + 2\ CN^- \rightleftharpoons [Cu(CN)_2]^-$	1.0×10^{16}
$Cu^+ + 2\ Cl^- \rightleftharpoons [CuCl_2]^-$	1.0×10^5
$Cu^{2+} + 4\ NH_3 \rightleftharpoons [Cu(NH_3)_4]^{2+}$	6.8×10^{12}
$Fe^{2+} + 6\ CN^- \rightleftharpoons [Fe(CN)_6]^{4-}$	7.7×10^{36}
$Hg^{2+} + 4\ Cl^- \rightleftharpoons [HgCl_4]^{2-}$	1.2×10^{15}
$Ni^{2+} + 4\ CN^- \rightleftharpoons [Ni(CN)_4]^{2-}$	1.0×10^{31}
$Ni^{2+} + 6\ NH_3 \rightleftharpoons [Ni(NH_3)_6]^{2+}$	5.6×10^8
$Zn^{2+} + 4\ OH^- \rightleftharpoons [Zn(OH)_4]^{2-}$	2.9×10^{15}
$Zn^{2+} + 4\ NH_3 \rightleftharpoons [Zn(NH_3)_4]^{2+}$	2.9×10^9

Appendix L

Selected Thermodynamic Values*

Table 20 • SELECTED THERMODYNAMIC VALUES*

Species	ΔH_f° (298.15 K) kJ/mol	S° (298.15 K) J/K · mol	ΔG_f° (298.15 K) kJ/mol
Aluminum			
Al(s)	0	28.3	0
$AlCl_3$(s)	−704.2	110.67	−628.8
Al_2O_3(s)	−1675.7	50.92	−1582.3
Barium			
$BaCl_2$(s)	−858.6	123.68	−810.4
BaO(s)	−553.5	70.42	−525.1
$BaSO_4$(s)	−1473.2	132.2	−1362.2
Beryllium			
Be(s)	0	9.5	0
$Be(OH)_2$(s)	−902.5	51.9	−815.0
Bromine			
Br(g)	111.884	175.022	82.396
Br_2(ℓ)	0	152.2	0
Br_2(g)	30.907	245.463	3.110
BrF_3(g)	−255.60	292.53	−229.43
HBr(g)	−36.29	198.70	−53.45
Calcium			
Ca(s)	0	41.42	0
Ca(g)	178.2	158.884	144.3
Ca^{2+}(g)	1925.90	—	—
CaC_2(s)	−59.8	69.96	−64.9
$CaCO_3$(s, calcite)	−1206.92	92.9	−1128.79
$CaCl_2$(s)	−795.8	104.6	−748.1
CaF_2(s)	−1219.6	68.87	−1167.3
CaH_2(s)	−186.2	42	−147.2
CaO(s)	−635.09	39.75	−604.03
CaS(s)	−482.4	56.5	−477.4
$Ca(OH)_2$(s)	−986.09	83.39	−898.49
$Ca(OH)_2$(aq)	−1002.82	−74.5	−868.07
$CaSO_4$(s)	−1434.11	106.7	−1321.79

continued

Table 20 • *continued*

Species	ΔH_f° (298.15 K) kJ/mol	S° (298.15 K) J/K · mol	ΔG_f° (298.15 K) kJ/mol
Carbon			
C(s, graphite)	0	5.740	0
C(s, diamond)	1.895	2.377	2.900
C(g)	716.682	158.096	671.257
$CCl_4(\ell)$	−135.44	216.40	−65.21
$CCl_4(g)$	−102.9	309.85	−60.59
$CHCl_3(\ell)$	−134.47	201.7	−73.66
$CHCl_3(g)$	−103.14	295.71	−70.34
CH_4(g, methane)	−74.81	186.264	−50.72
C_2H_2(g, acetylene)	226.73	200.94	209.20
C_2H_4(g, ethylene)	52.26	219.56	68.15
C_2H_6(g, ethane)	−84.68	229.60	−32.82
C_3H_8(g, propane)	−103.8	269.9	−23.49
C_6H_6(ℓ, benzene)	49.03	172.8	124.5
CH_3OH(ℓ, methanol)	−238.66	126.8	−166.27
CH_3OH(g, methanol)	−200.66	239.81	−161.96
C_2H_5OH(ℓ, ethanol)	−277.69	160.7	−174.78
C_2H_5OH(g, ethanol)	−235.10	282.70	−168.49
CO(g)	−110.525	197.674	−137.168
$CO_2(g)$	−393.509	213.74	−394.359
$CS_2(g)$	117.36	237.84	67.12
$COCl_2(g)$	−218.8	283.53	−204.6
Cesium			
Cs(s)	0	85.23	0
$Cs^+(g)$	457.964	—	—
CsCl(s)	−443.04	101.17	−414.53
Chlorine			
Cl(g)	121.679	165.198	105.680
$Cl^-(g)$	−233.13	—	—
$Cl_2(g)$	0	223.066	0
HCl(g)	−92.307	186.908	−95.299
HCl(aq)	−167.159	56.5	−131.228
Chromium			
Cr(s)	0	23.77	0
$Cr_2O_3(g)$	−1139.7	81.2	−1058.1
$CrCl_3(s)$	−556.5	123.0	−486.1
Copper			
Cu(s)	0	33.150	0
CuO(s)	−157.3	42.63	−129.7
$CuCl_2(s)$	−220.1	108.07	−175.7
Fluorine			
$F_2(g)$	0	202.78	0
F(g)	78.99	158.754	61.91
$F^-(g)$	−255.39	—	—
$F^-(aq)$	−332.63	−13.8	−278.79
HF(g)	−271.1	173.779	−273.2
HF(aq)	−332.63	−13.8	−278.79
Hydrogen			
$H_2(g)$	0	130.684	0
H(g)	217.965	114.713	203.247

Species	**ΔH_f° (298.15 K) kJ/mol**	**S° (298.15 K) J/K · mol**	**ΔG_f° (298.15 K) kJ/mol**
$H^+(g)$	1536.202	—	—
$H_2O(\ell)$	−285.830	69.91	−237.129
$H_2O(g)$	−241.818	188.825	−228.572
$H_2O_2(\ell)$	−187.78	109.6	−120.35
Iodine			
$I_2(s)$	0	116.135	0
$I_2(g)$	62.438	260.69	19.327
$I(g)$	106.838	180.791	70.250
$I^-(g)$	−197	—	—
$ICl(g)$	17.78	247.551	−5.46
Iron			
$Fe(s)$	0	27.78	0
$FeO(s)$	−272	—	—
Fe_2O_3(s, hematite)	−824.2	87.40	−742.2
Fe_3O_4(s, magnetite)	−1118.4	146.4	−1015.4
$FeCl_2(s)$	−341.79	117.95	−302.30
$FeCl_3(s)$	−399.49	142.3	−344.00
FeS_2(s, pyrite)	−178.2	52.93	−166.9
$Fe(CO)_5(\ell)$	−774.0	338.1	−705.3
Lead			
$Pb(s)$	0	64.81	0
$PbCl_2(s)$	−359.41	136.0	−314.10
PbO(s, yellow)	−217.32	68.70	−187.89
$PbS(s)$	−100.4	91.2	−98.7
Lithium			
$Li(s)$	0	29.12	0
$Li^+(g)$	685.783	—	—
$LiOH(s)$	−484.93	42.80	−438.95
$LiOH(aq)$	−508.48	2.80	−450.58
$LiCl(s)$	−408.701	59.33	−384.37
Magnesium			
$Mg(s)$	0	32.68	0
$MgCl_2(s)$	−641.32	89.62	−591.79
$MgCO_3(s)$	−1095.8	65.7	−1012.1
$MgO(s)$	−601.70	26.94	−569.43
$Mg(OH)_2(s)$	−924.54	63.18	−833.51
$MgS(s)$	−346.0	50.33	−341.8
Mercury			
$Hg(\ell)$	0	76.02	0
$HgCl_2(s)$	−224.3	146.0	−178.6
HgO(s, red)	−90.83	70.29	−58.539
HgS(s, red)	−58.2	82.4	−50.6
Nickel			
$Ni(s)$	0	29.87	0
$NiO(s)$	−239.7	37.99	−211.7
$NiCl_2(s)$	−305.332	97.65	−259.032
Nitrogen			
$N_2(g)$	0	191.61	0
$N(g)$	472.704	153.298	455.563
$NH_3(g)$	−46.11	192.45	−16.45
$N_2H_4(\ell)$	50.63	121.21	149.34

continued

Table 20 • *continued*

Species	ΔH_f° (298.15 K) kJ/mol	S° (298.15 K) J/K · mol	ΔG_f° (298.15 K) kJ/mol
$NH_4Cl(s)$	−314.43	94.6	−202.87
$NH_4Cl(aq)$	−299.66	169.9	−210.52
$NH_4NO_3(s)$	−365.56	151.08	−183.87
$NH_4NO_3(aq)$	−339.87	259.8	−190.56
$NO(g)$	90.9	210.76	86.55
$NO_2(g)$	33.18	240.06	51.31
$N_2O(g)$	82.05	219.85	104.20
$N_2O_4(g)$	9.16	304.29	97.89
$NOCl(g)$	51.71	261.8	66.08
$HNO_3(\ell)$	−174.10	155.60	−80.71
$HNO_3(g)$	−135.06	266.38	−74.72
$HNO_3(aq)$	−207.36	146.4	−111.25
Oxygen			
$O_2(g)$	0	205.138	0
$O(g)$	249.170	161.055	231.731
$O_3(g)$	142.7	238.93	163.2
Phosphorus			
P_4(s, white)	0	41.09	0
P_4(s, red)	−17.6	22.80	−12.1
$P(g)$	314.64	163.193	278.25
$PH_3(g)$	5.4	210.23	13.4
$PCl_3(g)$	−287.0	311.78	−267.8
$P_4O_{10}(s)$	−2984.0	228.86	−2697.7
$H_3PO_4(\ell)$	−1279.0	110.5	−1119.1
Potassium			
$K(s)$	0	64.18	0
$KCl(s)$	−436.747	82.59	−409.14
$KClO_3(s)$	−397.73	143.1	−296.25
$KI(s)$	−327.90	106.32	−324.892
$KOH(s)$	−424.764	78.9	−379.08
$KOH(aq)$	−482.37	91.6	−440.50
Silicon			
$Si(s)$	0	18.33	0
$SiBr_4(\ell)$	−457.3	277.8	−443.9
$SiC(s)$	−65.3	16.61	−62.8
$SiCl_4(g)$	−657.01	330.73	−616.98
$SiH_4(g)$	34.3	204.62	56.9
$SiF_4(g)$	−1614.94	282.49	−1572.65
SiO_2(s, quartz)	−910.94	41.84	−856.64
Silver			
$Ag(s)$	0	42.55	0
$Ag_2O(s)$	−31.05	121.3	−11.20
$AgCl(s)$	−127.068	96.2	−109.789
$AgNO_3(s)$	−124.39	140.92	−33.41
Sodium			
$Na(s)$	0	51.21	0
$Na(g)$	107.32	153.712	76.761
$Na^+(g)$	609.358	—	—
$NaBr(s)$	−361.02	86.82	−348.983
$NaCl(s)$	−411.153	72.13	−384.138

Species	ΔH_f° (298.15 K) kJ/mol	S° (298.15 K) J/K · mol	ΔG_f° (298.15 K) kJ/mol
NaCl(g)	−176.65	229.81	−196.66
NaCl(aq)	−407.27	115.5	−393.133
NaOH(s)	−425.609	64.455	−379.494
NaOH(aq)	−470.114	48.1	−419.150
Na_2CO_3(s)	−1130.68	134.98	−1044.44
Sulfur			
S(s, rhombic)	0	31.80	0
S(g)	278.805	167.821	238.250
S_2Cl_2(g)	−18.4	331.5	−31.8
SF_6(g)	−1209	291.82	−1105.3
H_2S(g)	−20.63	205.79	−33.56
SO_2(g)	−296.830	248.22	−300.194
SO_3(g)	−395.72	256.76	−371.06
$SOCl_2$(g)	−212.5	309.77	−198.3
H_2SO_4(ℓ)	−813.989	156.904	−690.003
H_2SO_4(aq)	−909.27	20.1	−744.53
Tin			
Sn(s, white)	0	51.55	0
Sn(s, gray)	−2.09	44.14	0.13
$SnCl_4$(ℓ)	−511.3	258.6	−440.1
$SnCl_4$(g)	−471.5	365.8	−432.2
SnO_2(s)	−580.7	52.3	−519.6
Titanium			
Ti(s)	0	30.63	0
$TiCl_4$(ℓ)	−804.2	252.34	−737.2
$TiCl_4$(g)	−763.2	354.9	−726.7
TiO_2(s)	−939.7	49.92	−884.5
Zinc			
Zn(s)	0	41.63	0
$ZnCl_2$(s)	−415.05	111.46	−369.398
ZnO(s)	−348.28	43.64	−318.30
ZnS(s, sphalerite)	−205.98	57.7	−201.29

*Taken from *The NBS Tables of Chemical Thermodynamic Properties,* U.S. Government Printing Office, Washington, D.C., 1982.

Appendix M

Standard Reduction Potentials in Aqueous Solution at 25 °C

Table 21 • STANDARD REDUCTION POTENTIALS IN AQUEOUS SOLUTION AT 25 °C

Acidic Solution	Standard Reduction Potential, $E°$ (volts)
F_2 (g) + 2 e^- $\longrightarrow$ 2 F^- (aq)	2.87
Co^{3+} (aq) + e^- $\longrightarrow$ Co^{2+} (aq)	1.82
Pb^{4+} (aq) + 2 e^- $\longrightarrow$ Pb^{2+} (aq)	1.8
H_2O_2 (aq) + 2 H^+ (aq) + 2 e^- $\longrightarrow$ 2 H_2O	1.77
NiO_2 (s) + 4 H^+ (aq) + 2 e^- $\longrightarrow$ Ni^{2+} (aq) + 2 H_2O	1.7
PbO_2 (s) + SO_4^{2-} (aq) + 4 H^+ (aq) + 2 e^- $\longrightarrow$ $PbSO_4$ (s) + 2 H_2O	1.685
Au^+ (aq) + e^- $\longrightarrow$ Au (s)	1.68
2 HClO (aq) + 2 H^+ (aq) + 2 e^- $\longrightarrow$ Cl_2 (g) + 2 H_2O	1.63
Ce^{4+} (aq) + e^- $\longrightarrow$ Ce^{3+} (aq)	1.61
$NaBiO_3$ (s) + 6 H^+ (aq) + 2 e^- $\longrightarrow$ Bi^{3+} (aq) + Na^+ (aq) + 3 H_2O	≈ 1.6
MnO_4^- (aq) + 8 H^+ (aq) + 5 e^- $\longrightarrow$ Mn^{2+} (aq) + 4 H_2O	1.51
Au^{3+} (aq) + 3 e^- $\longrightarrow$ Au (s)	1.50
ClO_3^- (aq) + 6 H^+ (aq) + 5 e^- $\longrightarrow$ $\frac{1}{2}$ Cl_2 (g) + 3 H_2O	1.47
BrO_3^- (aq) + 6 H^+ (aq) + 6 e^- $\longrightarrow$ Br^- (aq) + 3 H_2O	1.44
Cl_2 (g) + 2 e^- $\longrightarrow$ 2 Cl^- (aq)	1.36
$Cr_2O_7^{2-}$ (aq) + 14 H^+ (aq) + 6 e^- $\longrightarrow$ 2 Cr^{3+} (aq) + 7 H_2O	1.33
$N_2H_5^+$ (aq) + 3 H^+ (aq) + 2 e^- $\longrightarrow$ 2 NH_4^+ (aq)	1.24
MnO_2 (s) + 4 H^+ (aq) + 2 e^- $\longrightarrow$ Mn^{2+} (aq) + 2 H_2O	1.23
O_2 (g) + 4 H^+ (aq) + 4 e^- $\longrightarrow$ 2 H_2O	1.229
Pt^{2+} (aq) + 2 e^- $\longrightarrow$ Pt (s)	1.2
IO_3^- (aq) + 6 H^+ (aq) + 5 e^- $\longrightarrow$ $\frac{1}{2}$ I_2 (aq) + 3 H_2O	1.195
ClO_4^- (aq) + 2 H^+ (aq) + 2 e^- $\longrightarrow$ ClO_3^- (aq) + H_2O	1.19
Br_2 (ℓ) + 2 e^- $\longrightarrow$ 2 Br^- (aq)	1.08
$AuCl_4^-$ (aq) + 3 e^- $\longrightarrow$ Au (s) + 4 Cl^- (aq)	1.00
Pd^{2+} (aq) + 2 e^- $\longrightarrow$ Pd (s)	0.987
NO_3^- (aq) + 4 H^+ (aq) + 3 e^- $\longrightarrow$ NO (g) + 2 H_2O	0.96
NO_3^- (aq) + 3 H^+ (aq) + 2 e^- $\longrightarrow$ HNO_2 (aq) + H_2O	0.94
2 Hg^+ (aq) + 2 e^- $\longrightarrow$ Hg_2^{2+} (aq)	0.920
Hg^{2+} (aq) + 2 e^- $\longrightarrow$ Hg (ℓ)	0.855
Ag^+ (aq) + e^- $\longrightarrow$ Ag (s)	0.7994
Hg_2^{2+} (aq) + 2 e^- $\longrightarrow$ 2 Hg (ℓ)	0.789

Acidic Solution	Standard Reduction Potential, $E°$ (volts)
Fe^{3+} (aq) + e^- ⟶ Fe^{2+} (aq)	0.771
$SbCl_6^-$ (aq) + 2 e^- ⟶ $SbCl_4^-$ (aq) + 2 Cl^- (aq)	0.75
$[PtCl_4]^{2+}$ (aq) + 2 e^- ⟶ Pt (s) + 4 Cl^- (aq)	0.73
O_2 (g) + 2 H^+ (aq) + 2 e^- ⟶ H_2O_2 (aq)	0.682
$[PtCl_6]^{2-}$ (aq) + 2 e^- ⟶ $[PtCl_4]^{2-}$ (aq) + 2 Cl^- (aq)	0.68
H_3AsO_4 (aq) + 2 H^+ (aq) + 2 e^- ⟶ H_3AsO_3 (aq) + H_2O	0.58
I_2 (s) + 2 e^- ⟶ 2 I^- (aq)	0.535
TeO_2 (s) + 4 H^+ (aq) + 4 e^- ⟶ Te (s) + 2 H_2O	0.529
Cu^+ (aq) + e^- ⟶ Cu (s)	0.521
$[RhCl_6]^{3-}$ (aq) + 3 e^- ⟶ Rh (s) + 6 Cl^- (aq)	0.44
Cu^{2+} (aq) + 2 e^- ⟶ Cu (s)	0.337
$HgCl_2$ (s) + 2 e^- ⟶ 2 Hg (ℓ) + 2 Cl^- (aq)	0.27
AgCl (s) + e^- ⟶ Ag (s) + Cl^- (aq)	0.222
SO_4^{2-} (aq) + 4 H^+ (aq) + 2 e^- ⟶ SO_2 (g) + 2 H_2O	0.20
SO_4^{2-} (aq) + 4 H^+ (aq) + 2 e^- ⟶ H_2SO_3 (aq) + H_2O	0.17
Cu^{2+} (aq) + e^- ⟶ Cu^+ (aq)	0.153
Sn^{4+} (aq) + 2 e^- ⟶ Sn^{2+} (aq)	0.15
S (s) + 2 H^+ + 2 e^- ⟶ H_2S (aq)	0.14
AgBr (s) + e^- ⟶ Ag (s) + Br^- (aq)	0.0713
2 H^+ (aq) + 2 e^- ⟶ H_2 (g) (reference electrode)	0.0000
N_2O (g) + 6 H^+ (aq) + H_2O + 4 e^- ⟶ 2 NH_3OH^+ (aq)	−0.05
Pb^{2+} (aq) + 2 e^- ⟶ Pb (s)	−0.126
Sn^{2+} (aq) + 2 e^- ⟶ Sn (s)	−0.14
AgI (s) + e^- ⟶ Ag (s) + I^- (aq)	−0.15
$[SnF_6]^{2-}$ (aq) + 4 e^- ⟶ Sn (s) + 6 F^- (aq)	−0.25
Ni^{2+} (aq) + 2 e^- ⟶ Ni (s)	−0.25
Co^{2+} (aq) + 2 e^- ⟶ Co (s)	−0.28
Tl^+ (aq) + e^- ⟶ Tl (s)	−0.34
$PbSO_4$ (s) + 2 e^- ⟶ Pb (s) + SO_4^{2-} (aq)	−0.356
Se (s) + 2 H^+ (aq) + 2 e^- ⟶ H_2Se (aq)	−0.40
Cd^{2+} (aq) + 2 e^- ⟶ Cd (s)	−0.403
Cr^{3+} (aq) + e^- ⟶ Cr^{2+} (aq)	−0.41
Fe^{2+} (aq) + 2 e^- ⟶ Fe (s)	−0.44
2 CO_2 (g) + 2 H^+ (aq) + 2 e^- ⟶ $H_2C_2O_4$ (aq)	−0.49
Ga^{3+} (aq) + 3 e^- ⟶ Ga (s)	−0.53
HgS (s) + 2 H^+ (aq) + 2 e^- ⟶ Hg (ℓ) + H_2S (g)	−0.72
Cr^{3+} (aq) + 3 e^- ⟶ Cr (s)	−0.74
Zn^{2+} (aq) + 2 e^- ⟶ Zn (s)	−0.763
Cr^{2+} (aq) + 2 e^- ⟶ Cr (s)	−0.91
FeS (s) + 2 e^- ⟶ Fe (s) + S^{2-} (aq)	−1.01
Mn^{2+} (aq) + 2 e^- ⟶ Mn (s)	−1.18
V^{2+} (aq) + 2 e^- ⟶ V (s)	−1.18
CdS (s) + 2 e^- ⟶ Cd (s) + S^{2-} (aq)	−1.21
ZnS (s) + 2 e^- ⟶ Zn (s) + S^{2-} (aq)	−1.44
Zr^{4+} (aq) + 4 e^- ⟶ Zr (s)	−1.53
Al^{3+} (aq) + 3 e^- ⟶ Al (s)	−1.66
Mg^{2+} (aq) + 2 e^- ⟶ Mg (s)	−2.37
Na^+ (aq) + e^- ⟶ Na (s)	−2.714
Ca^{2+} (aq) + 2 e^- ⟶ Ca (s)	−2.87
Sr^{2+} (aq) + 2 e^- ⟶ Sr (s)	−2.89

continued

Table 21 • *continued*

Basic Solution	Standard Reduction Potential, $E°$ (volts)
Ba^{2+} (aq) + 2 e^- ⟶ Ba (s)	−2.90
Rb^+ (aq) + e^- ⟶ Rb (s)	−2.925
K^+ (aq) + e^- ⟶ K (s)	−2.925
Li^+ (aq) + e^- ⟶ Li (s)	−3.045
ClO^- (aq) + H_2O + 2 e^- ⟶ Cl^- (aq) + 2 OH^- (aq)	0.89
OOH^- (aq) + H_2O + 2 e^- ⟶ 3 OH^- (aq)	0.88
2 NH_2OH (aq) + 2 e^- ⟶ N_2H_4 (aq) + 2 OH^- (aq)	0.74
ClO_3^- (aq) + 3 H_2O + 6 e^- ⟶ Cl^- (aq) + 6 OH^- (aq)	0.62
MnO_4^- (aq) + 2 H_2O + 3 e^- ⟶ MnO_2 (s) + 4 OH^- (aq)	0.588
MnO_4^- (aq) + e^- ⟶ MnO_4^{2-} (aq)	0.564
NiO_2 (s) + 2 H_2O + 2 e^- ⟶ $Ni(OH)_2$ (s) + 2 OH^- (aq)	0.49
Ag_2CrO_4 (s) + 2 e^- ⟶ 2 Ag (s) + CrO_4^{2-} (aq)	0.446
O_2 (g) + 2 H_2O + 4 e^- ⟶ 4 OH^- (aq)	0.40
ClO_4^- (aq) + H_2O + 2 e^- ⟶ ClO_3^- (aq) + 2 OH^- (aq)	0.36
Ag_2O (s) + H_2O + 2 e^- ⟶ 2 Ag (s) + 2 OH^- (aq)	0.34
2 NO_2^- (aq) + 3 H_2O + 4 e^- ⟶ N_2O (g) + 6 OH^- (aq)	0.15
N_2H_4 (aq) + 2 H_2O + 2 e^- ⟶ 2 NH_3 (aq) + 2 OH^- (aq)	0.10
$[Co(NH_3)_6]^{3+}$ (aq) + e^- ⟶ $[Co(NH_3)_6]^{2+}$ (aq)	0.10
HgO (s) + H_2O + 2 e^- ⟶ Hg (ℓ) + 2 OH^- (aq)	0.0984
O_2 (g) + H_2O + 2 e^- ⟶ OOH^- (aq) + OH^- (aq)	0.076
NO_3^- (aq) + H_2O + 2 e^- ⟶ NO_2^- (aq) + 2 OH^- (aq)	0.01
MnO_2 (s) + 2 H_2O + 2 e^- ⟶ $Mn(OH)_2$ (s) + 2 OH^- (aq)	−0.05
CrO_4^{2-} (aq) + 4 H_2O + 3 e^- ⟶ $Cr(OH)_3$ (s) + 5 OH^- (aq)	−0.12
$Cu(OH)_2$ (s) + 2 e^- ⟶ Cu (s) + 2 OH^- (aq)	−0.36
S (s) + 2 e^- ⟶ S^{2-} (aq)	−0.48
$Fe(OH)_3$ (s) + e^- ⟶ $Fe(OH)_2$ (s) + OH^- (aq)	−0.56
2 H_2O + 2 e^- ⟶ H_2 (g) + 2 OH^- (aq)	−0.8277
2 NO_3^- (aq) + 2 H_2O + 2 e^- ⟶ N_2O_4 (g) + 4 OH^- (aq)	−0.85
$Fe(OH)_2$ (s) + 2 e^- ⟶ Fe (s) + 2 OH^- (aq)	−0.877
SO_4^{2-} (aq) + H_2O + 2 e^- ⟶ SO_3^{2-} (aq) + 2 OH^- (aq)	−0.93
N_2 (g) + 4 H_2O + 4 e^- ⟶ N_2H_4 (aq) + 4 OH^- (aq)	−1.15
$[Zn(OH)_4]^{2-}$ (aq) + 2 e^- ⟶ Zn (s) + 4 OH^- (aq)	−1.22
$Zn(OH)_2$ (s) + 2 e^- ⟶ Zn (s) + 2 OH^- (aq)	−1.245
$[Zn(CN)_4]^{2-}$ (aq) + 2 e^- ⟶ Zn (s) + 4 CN^- (aq)	−1.26
$Cr(OH)_3$ (s) + 3 e^- ⟶ Cr (s) + 3 OH^- (aq)	−1.30
SiO_3^{2-} (aq) + 3 H_2O + 4 e^- ⟶ Si (s) + 6 OH^- (aq)	−1.70

Appendix N

Answers to Exercises

CHAPTER 1

Chemical Puzzler: Solid potassium reacts with liquid water to produce gaseous hydrogen (H_2) and potassium hydroxide dissolved in water. One can observe the evolution of energy in the form of light during this chemical change.

1.1. The beaker on the left holds a homogeneous solution, whereas the beaker on the right contains a heterogeneous mixture of sand and iron chips.

1.2. (a) Sodium, chlorine, chromium
(b) Zn, Ni, K

1.3. Iron: Lustrous solid, metallic, conducts heat and electricity, malleable, ductile
Water: Colorless liquid at room temperature, melting point is 0 °C and boiling point is 100 °C
Salt: White, water-soluble solid, high melting point
Oxygen: Colorless gas at room temperature

1.4. $$15.5 \text{ g}\left(\frac{1 \text{ cm}^3}{1.12 \times 10^{-3} \text{ g}}\right) = 1.38 \times 10^4 \text{ cm}^3$$

1.5. $77 \text{ K} - 273.15 = -196\ °\text{C}$

1.6. The plastic will float in CCl_4 because plastic has a lower density than the liquid. Aluminum metal, with a greater density than CCl_4, will sink in the liquid.

1.7. Chemical changes: The fuel in the campfire burns in air (combustion).
Physical changes: Water boils.
Energy is evolved in the combustion and transferred to the water, to the water container, and to the surrounding air.

1.8. $$25.3 \text{ cm}\left(\frac{1 \text{ m}}{100 \text{ cm}}\right) = 0.253 \text{ m}$$

$$25.3 \text{ cm}\left(\frac{10 \text{ mm}}{1 \text{ cm}}\right) = 253 \text{ mm}$$

$$(25.3 \text{ cm})(21.6 \text{ cm}) = 546 \text{ cm}^2$$

$$(546 \text{ cm}^2)\left(\frac{1 \text{ m}}{100 \text{ cm}}\right)^2 = 0.0546 \text{ m}^2$$

1.9. Area of sheet $= (2.50 \text{ cm})^2 = 6.25 \text{ cm}^2$

$$\text{Volume of piece} = 1.656 \text{ g}\left(\frac{1 \text{ cm}^3}{21.45 \text{ g}}\right) = 0.07720 \text{ cm}^3$$

$$\text{Thickness} = \frac{\text{volume}}{\text{area}} = 0.0124 \text{ cm}$$

$$0.0124 \text{ cm}\left(\frac{10 \text{ mm}}{1 \text{ cm}}\right) = 0.124 \text{ mm}$$

1.10. (a) $$750 \text{ mL}\left(\frac{1 \text{ L}}{1000 \text{ mL}}\right) = 0.750 \text{ L}$$
(b) $2.0 \text{ qt} = 0.50 \text{ gal}$

$$0.50 \text{ gal}\left(\frac{3.786 \text{ L}}{1 \text{ gal}}\right) = 1.9 \text{ L} = 1.9 \text{ dm}^3$$

1.11. (a) $$5.00 \times 10^2 \text{ mg}\left(\frac{1 \text{ g}}{1000 \text{ mg}}\right) = 0.500 \text{ g}$$

$$0.500 \text{ g}\left(\frac{1 \text{ kg}}{1000 \text{ g}}\right) = 5.00 \times 10^{-4} \text{ kg}$$

(b) Length $= 4.5 \text{ m} (100 \text{ cm}/1 \text{ m}) = 450 \text{ cm}$
Width $= 1.5 \text{ m} (100/1 \text{ m}) = 150 \text{ cm}$
Volume $= (450 \text{ cm})(150 \text{ cm})(25 \text{ cm}) = 1.7 \times 10^6 \text{ cm}^3$

$$\text{Mass} = 1.7 \times 10^6 \text{ cm}^3\left(\frac{7.874 \text{ g}}{\text{cm}^3}\right)\left(\frac{1 \text{ kg}}{1000 \text{ g}}\right) = 1.3 \times 10^4 \text{ kg}$$

1.12. Student A: average $= -0.1$ °C; deviation $= 0.2$ °C; error $= -0.1$ °C
Student B: average $= 273.16$ K; deviation $= 0.02$ K; error $= +0.01$ K
Student B is more accurate and more precise.

1.13. (a) 11.24, has two places to the right of the decimal as in 11.19. Product is 0.60, with two significant figures.
(b) 2.4×10^3, has two significant figures.

1.14. Change dimensions to centimeters: 7.6 m = 760 cm; 2.74 m = 274 cm; and 0.13 mm = 0.013 cm.
Volume of paint = (760 cm)(274 cm)(0.013 cm) = $2.7 \times 10^3 \text{ cm}^3$
Volume (L) $= 2.7 \times 10^3 \text{ cm}^3 (1 \text{ L}/10^3 \text{ cm}^3) = 2.7 \text{ L}$
Mass $= (2.7 \times 10^3 \text{ cm}^3)(0.914 \text{ g/cm}^3) = 2.5 \times 10^3 \text{ g}$

1.15. $$15 \text{ g fertilizer}\left(\frac{15 \text{ g N}}{100 \text{ g fertilizer}}\right) = 2.3 \text{ g N}$$

$$15 \text{ g fertilizer}\left(\frac{30 \text{ g } P_2O_5}{100 \text{ g fertilizer}}\right) = 4.5 \text{ g } P_2O_5$$

CHAPTER 2

Chemical Puzzler: Begin with the disk at the back and left and move clockwise: A wafer for computer chips made of silicon; a flask containing orange bromine vapor; copper wool; a tube containing liquid mercury; pieces of aluminum wire; an iron rod; and yellow, powdered sulfur.

2.1. (1) Molecules of water leave the clothes and enter the atmosphere.
(2) Molecules of water from the air condense on a cold glass.
(3) Molecules of sugar break away from one another in the solid and mix with molecules of water to form a solution.
(4) Molecules move faster as the temperature increases; therefore, sugar molecules more rapidly break away from the solid and mix with water molecules.

2.2. An atom is about 10^5 times larger than the nucleus. The nucleus, therefore, has a radius of $(100\text{ m})(1/10^5) = 0.001\text{ m}$, or about 1 mm. A pencil lead has a diameter of about 1 mm.

2.3. (1) A (for Cu) = 34 n + 29 p = 63

(2) $62.9396\ \cancel{\text{amu}}\left(\frac{1.661 \times 10^{-24}\text{ g}}{\cancel{\text{amu}}}\right) = 1.045 \times 10^{-22}\text{ g}$

(3) Nickel-59 has 28 protons, 28 electrons, and (59 − 28) = 31 neutrons.

2.4. ${}^{28}_{14}Si$, ${}^{29}_{14}Si$, ${}^{30}_{14}Si$

2.5. (0.7577)(34.96885 amu) + (0.2423)(36.96590 amu) = 35.45 amu

2.6. Eight elements in the third period: sodium (Na), magnesium (Mg), and aluminum (Al) are metals. Silicon (Si) is a metalloid. Phosphorus (P), sulfur (S), chlorine (Cl), and argon (Ar) are nonmetals.

CHAPTER 3

Chemical Puzzler: The compound used to make "disappearing ink" is hydrated cobalt(II) chloride. When painted onto a sheet of paper, you cannot see the faint pink solution. If the paper is heated, however, the compound is dehydrated to form blue $CoCl_2$, which is visible. See Section 3.8 (and Figure 3.15) for a discussion of hydrated compounds.

3.1. Styrene = C_8H_8

3.2. Glycine formula = $C_2H_5NO_2$. You will often see its formula written as $NH_2CH_2CO_2H$ to emphasize the different parts of the molecule.

3.3. (1) K^+; (2) Se^{2-}; (3) Ba^{2+}; (4) Cs^+

3.4. 1. (a) 1 Na^+ and 1 F^- ion (b) 1 Cu^{2+} and 2 NO_3^- ions (c) 1 Na^+ and 1 $CH_3CO_2^-$ ion
2. $FeCl_2$ and $FeCl_3$
3. Na_2S, Na_3PO_4, BaS, $Ba_3(PO_4)_2$

3.5. 1. (a) NH_4NO_3 (d) V_2O_3
(b) $CoSO_4$ (e) $Ba(CH_3CO_2)_2$
(c) $Ni(CN)_2$ (f) $Ca(OCl)_2$
2. (a) Magnesium bromide
(b) Lithium carbonate
(c) Potassium hydrogen sulfite
(d) Potassium permanganate
(e) Ammonium sulfide
(f) Copper(I) chloride and copper(II) chloride

3.6. 1. (a) CO_2 (d) BF_3
(b) PI_3 (e) O_2F_2
(c) SCl_2 (f) XeO_3
2. (a) Dinitrogen tetrafluoride
(b) Hydrogen bromide
(c) Sulfur tetrafluoride
(d) Boron trichloride
(e) Tetraphosphorus decaoxide
(f) Chlorine trifluoride

3.7. (1) $(1.5\ \cancel{\text{mol Si}}) \cdot \left(\frac{28.1\text{ g Si}}{\cancel{\text{mol Si}}}\right) = 42\text{ g Si}$

(2) $(454\ \cancel{\text{g S}}) \cdot \left(\frac{1\text{ mol}}{32.06\ \cancel{\text{g S}}}\right) = 14.2\text{ mol}$

$(14.2\ \cancel{\text{mol}}) \cdot \left(\frac{6.022 \times 10^{23}\text{ atoms S}}{\cancel{\text{mol}}}\right) = 8.53 \times 10^{24}\text{ atom S}$

3.8. $2.6 \times 10^{24}\ \cancel{\text{atoms Au}}\left(\frac{1\ \cancel{\text{mol Au}}}{6.022 \times 10^{23}\ \cancel{\text{atoms Au}}}\right)\left(\frac{196\text{ g Au}}{1\ \cancel{\text{mol Au}}}\right) = 850\text{ g Au}$

$850\ \cancel{\text{g Au}}\left(\frac{1\text{ cm}^3}{19.32\ \cancel{\text{g Au}}}\right) = 44\text{ cm}^3$

$\text{Area of Au piece} = \frac{44\text{ cm}^3}{0.10\ \cancel{\text{cm}}} = 440\text{ cm}^2$

$\text{Length of edge} = \sqrt{440\text{ cm}^2} = 21\text{ cm}$

3.9. 1. (a) $CaCO_3 = \frac{100.1\text{ g}}{\text{mol}}$; (b) cysteine = 121.2 g/mol

2. $454\ \cancel{\text{g CaCO}_3}\left(\frac{1\text{ mol}}{100.1\ \cancel{\text{g CaCO}_3}}\right) = 4.54\text{ mol}$

3. $2.50 \times 10^{-3}\ \cancel{\text{mol cysteine}}\left(\frac{121.2\text{ g}}{1\ \cancel{\text{mol cysteine}}}\right) = 0.303\text{ g cysteine}$

3.10. 1. NaCl has 23.0 g of Na (39.3%) and 35.5 g of Cl (60.7%) in 1.00 mol.
2. C_6H_{14} has 72.07 g of C (83.63%) and 14.11 g of H (16.37%) in 1.00 mol (86.18 g/mol).
3. $(NH_4)_2SO_4$ has a molar mass of 132.15 g/mol. It has 28.0 g of N (21.2%), 8.06 g of H (6.10%), 32.1 g of S (24.3%), and 64.0 g of O (48.4%) in 1.00 mol.

3.11. $88.17\ \cancel{\text{g C}}\left(\frac{1\text{ mol C}}{12.011\ \cancel{\text{g C}}}\right) = 7.341\text{ mol C}$

$11.83\ \cancel{\text{g H}}\left(\frac{1\text{ mol H}}{1.008\ \cancel{\text{g H}}}\right) = 11.74\text{ mol H}$

$$\frac{11.74\text{ mol H}}{7.341\text{ mol C}} = \frac{1.599\text{ H}}{1\text{ C}} = \frac{1\ (3/5)\text{ H}}{1\text{ C}} = \frac{(8/5)\text{ H}}{1\text{ C}} = \frac{8\text{ H}}{5\text{ C}}$$

$$\text{Empirical formula mass} = \frac{68.11\text{ g}}{\text{formula unit}} = \text{molar mass}$$

Empirical formula and molecular formula are both C_5H_8.

3.12. $$47.35 \text{ g C}\left(\frac{1 \text{ mol C}}{12.011 \text{ g C}}\right) = 3.942 \text{ mol C}$$

$$10.60 \text{ g H}\left(\frac{1 \text{ mol H}}{1.008 \text{ g H}}\right) = 10.52 \text{ mol H}$$

$$42.05 \text{ g O}\left(\frac{1 \text{ mol O}}{16.00 \text{ g O}}\right) = 2.628 \text{ mol O}$$

$$\frac{10.52 \text{ mol H}}{2.628 \text{ mol O}} = \frac{4 \text{ H}}{1 \text{ O}}$$

$$\frac{3.942 \text{ mol C}}{2.628 \text{ mol O}} = \frac{1.5 \text{ C}}{1 \text{ O}} = \frac{3 \text{ C}}{2 \text{ O}}$$

Empirical formula = $C_3H_8O_2$

$$\text{Empirical formula mass} = \frac{76.10 \text{ g}}{\text{formula unit}} = \text{molar mass}$$

Empirical formula and molecular formula are both $C_3H_8O_2$.

3.13. $$0.586 \text{ g K}\left(\frac{1 \text{ mol K}}{39.10 \text{ g K}}\right) = 0.0150 \text{ mol K}$$

$$0.480 \text{ g O}\left(\frac{1 \text{ mol O}}{16.00 \text{ g O}}\right) = 0.0300 \text{ mol O}$$

Empirical formula = KO_2

3.14. Formula of iron(III) oxide is Fe_2O_3 (molar mass = 159.7 g/mol)

$$50.0 \text{ g Fe}\left(\frac{1 \text{ mol Fe}}{55.85 \text{ g Fe}}\right) = 0.895 \text{ mol Fe}$$

$$0.895 \text{ mol Fe}\left(\frac{1 \text{ mol } Fe_2O_3}{2 \text{ mol Fe}}\right] = 0.448 \text{ mol } Fe_2O_3$$

$$0.448 \text{ mol } Fe_2O_3\left(\frac{159.7 \text{ g } Fe_2O_3}{1 \text{ mol } Fe_2O_3}\right) = 71.5 \text{ g } Fe_2O_3$$

3.15. Mass of water = 0.235 g − 0.185 g = 0.500 g H_2O

$$0.500 \text{ g } H_2O\left(\frac{1 \text{ mol } H_2O}{18.02 \text{ g } H_2O}\right) = 0.00278 \text{ mol } H_2O$$

$$0.185 \text{ g } CuCl_2\left(\frac{1 \text{ mol } CuCl_2}{134.4 \text{ g } CuCl_2}\right) = 0.00138 \text{ mol } CuCl_2$$

$$\text{Mole ratio} = \frac{0.00278 \text{ mol } H_2O}{0.00138 \text{ mol } CuCl_2} = \frac{2.01 \text{ mol } H_2O}{1.00 \text{ mol } CuCl_2}$$

Formula = $CuCl_2 \cdot 2\ H_2O$

CHAPTER 4

Chemical Puzzler: The reaction ceases when the baking soda ($NaHCO_3$) has been consumed. This is a situation that chemists call a reaction with a "limiting reactant." As long as baking soda is present, adding vinegar (acetic acid) will produce CO_2. Eventually, however, the baking soda will be consumed. No more will be available to react with additional vinegar. Initially, the limiting reactant is vinegar. After the baking soda has been consumed, the soda is the limiting reactant.

4.1. (1) Stoichiometric coefficients: 4 for Fe, 3 for O_2, and 2 for Fe_2O_3
(2) 8000 atoms of Fe require (3/4) × 8000 = 6000 molecules of O_2

4.2. (1) $C_5H_{12}(g) + 8\ O_2(g) \longrightarrow 5\ CO_2(g) + 6\ H_2O(\ell)$
(2) $2\ Pb(C_2H_5)_4(\ell) + 27\ O_2(g) \longrightarrow 2\ PbO(s) + 16\ CO_2(g) + 20\ H_2O(\ell)$

4.3. $$454 \text{ g } O_2\left(\frac{1 \text{ mol } O_2}{32.00 \text{ g } O_2}\right) = 14.2 \text{ mol } O_2$$

$$14.2 \text{ mol } O_2\left(\frac{2 \text{ mol C}}{1 \text{ mol } O_2}\right)\left(\frac{12.01 \text{ g C}}{1 \text{ mol C}}\right) = 341 \text{ g C}$$

$$14.2 \text{ mol } O_2\left(\frac{2 \text{ mol CO}}{1 \text{ mol } O_2}\right)\left(\frac{28.01 \text{ g CO}}{1 \text{ mol CO}}\right) = 795 \text{ g CO}$$

4.4. $S_8(s) + 8\ O_2(g) \longrightarrow 8\ SO_2(g)$

$$20.0 \text{ g } S_8\left(\frac{1 \text{ mol } S_8}{256.5 \text{ g } S_8}\right)2 = 0.0780 \text{ mol } S_8$$

$$160.\text{ g } O_2\left(\frac{1 \text{ mol } O_2}{32.00 \text{ g } O_2}\right) = 5.00 \text{ mol } O_2$$

$$\frac{5.00 \text{ mol } O_2}{0.0780 \text{ mol } S_8} = \frac{64.1 \text{ mol } O_2}{1 \text{ mol } S_8}$$

S_8 is the limiting reactant.

$$0.0780 \text{ mol } S_8\left(\frac{8 \text{ mol } SO_2}{1 \text{ mol } S_8}\right) = 0.624 \text{ mol } SO_2 \text{ produced}$$

$$0.624 \text{ mol } SO_2\left(\frac{64.07 \text{ g}}{\text{mol } SO_2}\right) = 40.0 \text{ g } SO_2$$

Some O_2 remains after reaction.

$$0.0780 \text{ mol } S_8\left(\frac{8 \text{ mol } O_2 \text{ required}}{1 \text{ mol } S_8}\right) = 0.624 \text{ mol } O_2 \text{ required}$$

Excess O_2 = 5.00 mol − 0.624 mol = 4.38 mol O_2

4.5. $$125 \text{ g } Cl_2\left(\frac{1 \text{ mol } Cl_2}{70.91 \text{ g } Cl_2}\right) = 1.76 \text{ mol } Cl_2$$

$$125 \text{ g C}\left(\frac{1 \text{ mol C}}{12.01 \text{ g C}}\right) = 10.4 \text{ mol C}$$

$$\frac{10.4 \text{ mol C}}{1.76 \text{ mol } Cl_2} = \frac{5.91 \text{ mol C}}{1.00 \text{ mol } Cl_2}$$

The balanced equation requires 1 mol C to 2 mol Cl_2. The ratio of moles available is 11.8 mol C to 2 mol Cl_2. This means Cl_2 is the limiting reactant.

$$1.76 \text{ mol } Cl_2\left(\frac{1 \text{ mol } TiCl_4}{2 \text{ mol } Cl_2}\right)\left(\frac{189.7 \text{ g } TiCl_4}{1 \text{ mol } TiCl_4}\right) = 167 \text{ g } TiCl_4$$

4.6. Theoretical yield of hydrogen:

$$125 \text{ g } CH_3OH\left(\frac{1 \text{ mol } CH_3OH}{32.04 \text{ g}}\right)\left(\frac{2 \text{ mol } H_2}{1 \text{ mol } CH_3OH}\right)\left(\frac{2.016 \text{ g } H_2}{1 \text{ mol } H_2}\right) = 15.7 \text{ g } H_2$$

$$\text{Percentage yield of hydrogen: } \frac{13.6 \text{ g } H_2}{15.7 \text{ g } H_2} \cdot 100\% = 86.5\%$$

4.7. Mass of water = 2.357 g − 2.108 g = 0.249 g

$$0.249\ \text{g H}_2\text{O}\left(\frac{1\ \text{mol H}_2\text{O}}{18.02\ \text{g H}_2\text{O}}\right) = 0.0138\ \text{mol H}_2\text{O}$$

$$0.0138\ \text{mol H}_2\text{O}\left(\frac{1\ \text{mol BaCl}_2\cdot 2\ \text{H}_2\text{O}}{2\ \text{mol H}_2\text{O}}\right)$$

$$\left(\frac{244.3\ \text{g BaCl}_2\cdot 2\ \text{H}_2\text{O}}{1\ \text{mol BaCl}_2\cdot 2\ \text{H}_2\text{O}}\right) = 1.688\ \text{g BaCl}_2\cdot 2\ \text{H}_2\text{O}$$

$$\frac{1.688\ \text{g BaCl}_2\cdot 2\ \text{H}_2\text{O}}{2.357\ \text{g sample}}\cdot 100\% = 71.6\%\ \text{BaCl}_2\cdot 2\ \text{H}_2\text{O}$$

4.8. $$1.612\ \text{g}\left(\frac{1\ \text{mol CO}_2}{44.01\ \text{g}}\right)\left(\frac{1\ \text{mol C}}{1\ \text{mol CO}_2}\right) = 0.03663\ \text{mol C}$$

$$0.7425\ \text{g H}_2\text{O}\left(\frac{1\ \text{mol H}_2\text{O}}{18.01\ \text{g H}_2\text{O}}\right)\left(\frac{2\ \text{mol H}}{1\ \text{mol H}_2\text{O}}\right) = 0.08245\ \text{mol H}$$

$$\frac{0.08245\ \text{mol H}}{0.03663\ \text{mol C}} = \frac{2.25\ \text{H}}{1\ \text{C}} = \frac{9\ \text{H}}{4\ \text{C}}$$

The empirical formula is C_4H_9, which has a molar mass of 57 g/mol. The molecular formula is $(C_4H_9)_2$, or C_8H_{18}.

4.9. $$0.600\ \text{g}\left(\frac{1\ \text{mol CO}_2}{44.01\ \text{g}}\right)\left(\frac{1\ \text{mol C}}{1\ \text{mol CO}_2}\right) = 0.0136\ \text{mol C}$$

$$0.0136\ \text{mol C}\left(\frac{12.01\ \text{g C}}{\text{mol C}}\right) = 0.163\ \text{g C}$$

$$0.0163\ \text{g H}_2\text{O}\left(\frac{1\ \text{mol H}_2\text{O}}{18.01\ \text{g H}_2\text{O}}\right)\left(\frac{2\ \text{mol H}}{1\ \text{mol H}_2\text{O}}\right) = 0.0181\ \text{mol H}$$

$$0.0181\ \text{mol H}\left(\frac{1.008\ \text{g H}}{\text{mol H}}\right) = 0.0182\ \text{g H}$$

Mass of O in unknown =
0.400 g − mass of C − mass of H= 0.219 g O

$$0.219\ \text{g O}\left(\frac{1\ \text{mol O}}{16.00\ \text{g O}}\right) = 0.0137\ \text{mol O}$$

The ratio of C to O is 1 C/1 O. The ratio of H to C is

$$\frac{0.0181\ \text{mol H}}{0.0136\ \text{mol C}} = \frac{1.33\ \text{H}}{1\ \text{C}} = \frac{4\ \text{H}}{3\ \text{C}}$$

The empirical formula is $C_3H_4O_3$.

4.10. $$0.452\ \text{g}\left(\frac{1\ \text{mol CO}_2}{44.01\ \text{g}}\right)\left(\frac{1\ \text{mol C}}{1\ \text{mol CO}_2}\right) = 0.0103\ \text{mol C}$$

$$0.0103\ \text{mol C}\left(\frac{12.01\ \text{g C}}{\text{mol C}}\right) = 0.123\ \text{g C}$$

$$0.0924\ \text{g H}_2\text{O}\left(\frac{1\ \text{mol H}_2\text{O}}{18.01\ \text{g H}_2\text{O}}\right)\left(\frac{2\ \text{mol H}}{1\ \text{mol H}_2\text{O}}\right) = 0.0103\ \text{mol H}$$

$$0.0103\ \text{mol H}\left(\frac{1.008\ \text{g H}}{\text{mol H}}\right) = 0.0103\ \text{g H}$$

Mass of Cr in unknown =
0.178 g − mass of C − mass of H
= 0.0447 g Cr

$$0.0447\ \text{g Cr}\left(\frac{1\ \text{mol Cr}}{52.00\ \text{g Cr}}\right) = 8.60\times 10^{-4}\ \text{mol Cr}$$

The ratio of C to H is 1 C/1 H. The ratio of C to Cr is 12 to 1; therefore, the empirical formula is $C_{12}H_{12}Cr$.

CHAPTER 5

Chemical Puzzler: The acid in the tablet (citric acid) reacts with the bicarbonate ion (or hydrogen carbonate ion) to give water and bubbles of CO_2 gas. This is a typical gas-forming reaction of an acid with a metal carbonate or bicarbonate.

5.1. Epsom salt is an electrolyte, whereas methanol is a nonelectrolyte.

5.2. (1) KNO_3 is soluble and gives K^+ and NO_3^- ions.
(2) $CaCl_2$ is soluble and gives Ca^{2+} and Cl^- ions.
(3) CuO is not water-soluble.
(4) $NaCH_3CO_2$ is soluble and gives Na^+ and $CH_3CO_2^-$ ions.

5.3. (1) $Na_2CO_3(aq) + NiCl_2(aq) \longrightarrow 2\ NaCl(aq) + NiCO_3(s)$
(2) No reaction; no insoluble compound is produced.
(3) $AlCl_3(aq) + 3\ KOH(aq) \longrightarrow Al(OH)_3(s) + 3\ KCl(aq)$

5.4. (1) $BaCl_2(aq) + Na_2SO_4(aq) \longrightarrow BaSO_4(s) + 2\ NaCl(aq)$
$Ba^{2+}(aq) + SO_4^{2-}(aq) \longrightarrow BaSO_4(s)$
(2) $FeCl_3(aq) + 3\ KOH(aq) \longrightarrow Fe(OH)_3(s) + 3\ KCl(aq)$
$Fe^{3+}(aq) + 3\ OH^-(aq) \longrightarrow Fe(OH)_3(s)$
(3) $Pb(NO_3)_2(aq) + 2\ KCl(aq) \longrightarrow PbCl_2(s) + 2\ KNO_3(aq)$
$Pb^{2+}(aq) + 2\ Cl^-(aq) \longrightarrow PbCl_2(s)$

5.5. (1) $HClO_4$ in water gives $H^+(aq)$ and $ClO_4^-(aq)$.
(2) $Ba^{2+}(aq) + 2\ OH^-(aq)$

5.6. (1) SeO_2 is an acidic oxide (like SO_2).
(2) MgO is a basic oxide.
(3) P_4O_{10} is an acidic oxide.

5.7. $Mg(OH)_2(s) + 2\ HCl(aq) \longrightarrow MgCl_2(aq) + 2\ H_2O(\ell)$
$Mg(OH)_2(s) + 2\ H^+(aq) \longrightarrow Mg^{2+}(aq) + 2\ H_2O(\ell)$

5.8. (1) $BaCO_3(s) + 2\ HNO_3(aq) \longrightarrow Ba(NO_3)_2(aq) + CO_2(g) + H_2O(\ell)$
Barium carbonate and nitric acid produce barium nitrate, carbon dioxide, and water.
(2) $(NH_4)_2SO_4(aq) + 2\ NaOH(aq) \longrightarrow NH_3(g) + Na_2SO_4(aq) + 2\ H_2O(\ell)$

Ammonium sulfate and sodium hydroxide produce ammonia, sodium sulfate, and water.

5.9. (1) Gas-forming reaction

$$CuCO_3(s) + H_2SO_4(aq) \longrightarrow CuSO_4(aq) + H_2O(\ell) + CO_2(g)$$

(2) Acid–base reaction

$$Ba(OH)_2(s) + 2\ HNO_3(aq) \longrightarrow Ba(NO_3)_2(aq) + 2\ H_2O(\ell)$$

(3) Precipitation reaction

$$ZnCl_2(aq) + (NH_4)_2S(aq) \longrightarrow ZnS(s) + 2\ NH_4Cl(aq)$$

5.10. (1) +3 (3) +4
(2) +5 (4) +5

5.11. The dichromate ion is the oxidizing agent and is reduced (Cr with a +6 oxidation number gives Cr with a +3 oxidation number). Ethanol is the reducing agent and is oxidized. (The C atom in ethanol has an oxidation number of −2, whereas it is 0 in acetic acid.)

5.12. (1) This is an acid–base reaction.
(2) This is an oxidation–reduction because Cu is oxidized (oxidation number changes from 0 to +2 in $CuCl_2$), and Cl_2 is reduced (oxidation number changes from 0 to −1). Cu is the reducing agent and Cl_2 is the oxidizing agent.
(3) This is a gas-forming reaction.
(4) This is an oxidation–reduction because S in $S_2O_3^{2-}$ is oxidized (oxidation number changes from +2 to +2.5 in $S_4O_6^{2-}$), and I_2 is reduced (oxidation number changes from 0 to −1). $S_2O_3^{2-}$ is the reducing agent and I_2 is the oxidizing agent.

5.13. $26.3\ \cancel{g}(1\ mol\ NaHCO_3/84.01\ \cancel{g}) = 0.313\ mol\ NaHCO_3$
$0.313\ mol\ NaHCO_3/0.200\ L = 1.57\ M$

5.14. $[Na^+] = 1.00\ M$, $[SO_4^{2-}] = 0.500\ M$

5.15. Mass of $AgNO_3$ required = (0.0200 M)(0.250 L) = 5.00×10^{-3} mol

$$5.00 \times 10^{-3}\ \cancel{mol}\left(\frac{169.9\ g}{\cancel{mol}}\right) = 0.850\ g\ AgNO_3$$

You should weigh out 0.850 g $AgNO_3$, dissolve it in a small amount of water in the volumetric flask. After the solid is dissolved, fill the flask to the mark.

5.16. $(2.00\ M)V_{conc} = (1.00\ M)(0.250\ L)$
$V_{conc} = 0.125\ L$
Take 125 mL of the 2.00 M NaOH solution and dilute it to a total volume of 250 mL with distilled water.

5.17. $(0.15\ M)(0.0060\ L) = (0.0100\ L)C_{dilute}$
$C_{dilute} = 0.090\ M$

5.18. $(0.0500\ L)(0.450\ M) = 0.0225\ mol\ HCl$

$$(0.0225\ \cancel{mol\ HCl})\left(\frac{2\ mol\ NaCl}{2\ \cancel{mol\ HCl}}\right) = 0.0225\ mol\ NaCl$$

$$(0.0225\ \cancel{mol\ NaCl})\left(\frac{58.44\ g\ NaCl}{\cancel{mol\ NaCl}}\right) = 1.31\ g\ NaCl$$

5.19. $(0.02833\ L)(0.953\ M) = 0.0270\ mol\ NaOH$
$(0.0270\ \cancel{mol\ NaOH})(1\ mol\ CH_3CO_2H/1\ \cancel{mol\ NaOH}) = 0.0270\ mol\ CH_3CO_2H$

$$0.0270\ \cancel{mol\ CH_3CO_2H}\left(\frac{60.05\ g\ CH_3CO_2H}{\cancel{mol\ CH_3CO_2H}}\right) = 1.62\ g\ CH_3CO_2H$$

$0.0270\ mol\ CH_3CO_2H/0.0250\ L = 1.08\ M$

5.20. $(0.02967\ L)(0.100\ M) = 0.00297\ mol\ HCl$

$$0.00297\ \cancel{mol\ HCl}\left(\frac{1\ mol\ NaOH}{1\ \cancel{mol\ HCl}}\right) = 0.00297\ mol\ NaOH$$

$$\frac{0.00297\ mol\ NaOH}{0.0250\ L} = 0.119\ M\ NaOH$$

5.21. $(0.02030\ L)(0.196\ M\ Na_2S_2O_3) = 0.00398\ mol\ Na_2S_2O_3$

$0.00398\ \cancel{mol\ Na_2S_2O_3}\ (1\ mol\ I_2/2\ \cancel{mol\ Na_2S_2O_3}) = 0.00199\ mol\ I_2$

0.00199 mol I_2 was not used in reaction with ascorbic acid.

I_2 originally added = (0.05000 L)(0.0520 M) = 0.00260 mol

I_2 used in reaction with ascorbic acid = 0.00260 mol − 0.00199 mol = 6.1×10^{-4} mol I_2

$$(6.1 \times 10^{-4}\ \cancel{mol\ I_2})\left(\frac{1\ \cancel{mol\ C_6H_8O_6}}{1\ \cancel{mol\ I_2}}\right)\left(\frac{176.1\ g\ C_6H_8O_6}{\cancel{mol\ C_6H_8O_6}}\right) = 0.11\ g\ C_6H_8O_6$$

CHAPTER 6

Chemical Puzzler: This question is answered thoroughly on the *Saunders Interactive General Chemistry CD-ROM*, Screen 6.19 (where the answer is shown to be 39 peanuts). The quantity of thermal energy that can be derived from a peanut can be determined with a calorimeter (Section 6.10). We also know from experiment the quantity of heat required to heat water to its boiling point and then to evaporate the water (see Sections 6.3 and 6.4). Knowing the quantity of heat required to heat and then evaporate a cup of water, and the quantity of heat available from one burning peanut, we could then calculate the number of peanuts required.

6.1. The chemical potential energy of a battery can be converted to work (to run a motor), heat (an electric space heater), and light (a light bulb). See the *Saunders Interactive General Chemistry CD-ROM*, Screen 6.6.

6.2. (1) $250\ \cancel{Cal}\ (1000\ \cancel{cal}/\cancel{Cal})(4.184\ J/\cancel{cal}) = 1.1 \times 10^6\ J$
(2) $(75\ W)(3.0\ \cancel{h})(3600\ s/\cancel{h}) = 8.1 \times 10^5\ Ws = 8.1 \times 10^5\ J$
(3) $(16\ \cancel{kJ})(1\ kcal/4.184\ \cancel{kJ}) = 3.8\ kcal$

6.3. $q = 24.1 \times 10^3\ J = (250.\ g)(0.902\ J/g \cdot K)(T_{final} - 5.0\ °C)$
$T_{final} = 112\ °C$

6.4. $(15.5\ g)(C_{metal})(18.9\ °C - 100.0\ °C) =$

$$-(55.5\ g)(4.184\ J/g \cdot K)(18.9\ °C - 16.5\ °C)$$

$$C_{metal} = \frac{0.44\ J}{g \cdot K}$$

6.5. $(400.\ g\ iron)(0.451\ J/g \cdot K)(32.8\ °C - T_{initial}) =$

$$-(1000.\ g)(4.184\ J/g \cdot K)(32.8\ °C - 20.0\ °C)$$

$T_{initial} = 330.\ °C$

6.6. $(25.0\ g\ CH_3OH)(2.53\ J/g \cdot K)(64.6\ °C - 25.0\ °C) = 2.50 \times 10^3\ J$

$$(25.0\ g\ CH_3OH)\left(\frac{2.00 \times 10^3\ J}{g}\right) = 5.00 \times 10^4\ J$$

Total heat energy $= 5.25 \times 10^4\ J$

6.7. $(1.32\ g\ ice)\left(\frac{333\ J}{g\ ice}\right) = 440.\ J$

$$440.\ J = -(9.85\ g)(C_{metal})(0.0\ °C - 100.0\ °C)$$

$$C_{metal} = \frac{0.446\ J}{g \cdot K}$$

6.8. (a) $(10.0\ g)\left(\frac{1\ mol\ I_2}{253.8\ g}\right)\left(\frac{62.4\ kJ}{mol}\right) = 2.46\ kJ$

(b) The process is exothermic.

$$(3.45\ g)\left(\frac{1\ mol\ I_2}{253.8\ g}\right)\left(\frac{62.4\ kJ}{mol}\right) = 0.848\ kJ$$

6.9. $(12.6\ g\ H_2O)\left(\frac{1\ mol}{18.02\ g}\right)\left(\frac{285.8\ kJ}{mol}\right) = 2.00 \times 10^2\ kJ$

6.10. $2\ CO_2(g) + 3\ H_2O(\ell) \longrightarrow C_2H_6(g) + \frac{7}{2}\ O_2(g)$ $\quad \Delta H = +1559.7\ kJ$

$$2\ C(s) + 2\ O_2(g) \longrightarrow 2\ CO_2(g) \quad \Delta H = 2\ (-393.5\ kJ)$$
$$3\ H_2(g) + \tfrac{3}{2}\ O_2(g) \longrightarrow 3\ H_2O(\ell) \quad \Delta H = 3\ (-285.8\ kJ)$$
$$2\ C(s) + 3\ H_2(g) \longrightarrow C_2H_6(g) \quad \Delta H = -84.7\ kJ$$

6.11. $\Delta H_{overall} = -413.7\ kJ + 106.8\ kJ = \frac{-306.9\ kJ}{mol}$

$$(454\ g)\left(\frac{1\ mol\ PbS}{239.3\ g}\right)\left(\frac{-306.9\ kJ}{mol}\right) = -582\ kJ$$

Reaction is exothermic.

6.12. $2\ C(s) + 3\ H_2(g) + \frac{1}{2}\ O_2(g) \longrightarrow C_2H_5OH(\ell)$ $\quad \Delta H_f^\circ = -277.7\ kJ/mol$

$Cu(s) + \frac{1}{8}\ S_8(s) + 2\ O_2(g) \longrightarrow CuSO_4(s)$ $\quad \Delta H_f^\circ = -771.4\ kJ/mol$

$$\Delta H^\circ_{rxn} = (-277.7\ kJ/mol)(1.5\ mol) = -4.2 \times 10^2\ kJ$$

6.13. $\Delta H^\circ_{rxn} = 6\ \Delta H_f^\circ[CO_2(g)] + 3\ \Delta H_f^\circ[H_2O(\ell)] - \{\Delta H_f^\circ[C_6H_6(\ell)] + 15/2\ \Delta H_f^\circ[O_2(g)]\} =$

$6\ mol(-393.5\ kJ/mol) + 3\ mol\ (-285.8\ kJ/mol) - 1\ mol\ (+49.0\ kJ/mol) - 0 = -3267.4\ kJ$

6.14. Heat transferred to calorimeter water $=$ $(1.50 \times 10^3\ g)(4.184\ J/g \cdot K)(2.32\ K) = 14.6 \times 10^3\ J$

Heat transferred to calorimeter bomb $=$ $(837\ J/K)(2.32\ K) = 1.94 \times 10^3\ J$

Total heat transferred by 1.00 g sucrose $= -16.5\ kJ$

Heat transferred per mole $= (-16.5\ kJ/g)\left(\frac{342.2\ g}{mol}\right) = -5650\ kJ/mol$

6.15. Mass of final solution $= 400.\ g$

$q = (4.20\ J/g \cdot K)(400.\ g)(26.60 - 25.10)°C = 2.52 \times 10^3\ J$ for 0.0800 mol HCl

2.52 kJ/0.0800 mol HCl = 31.5 kJ/mol of HCl

ΔH of neutralization $= -31.5\ kJ$

CHAPTER 7

Chemical Puzzler: When heated, metal ions absorb energy and move from the ground state to an excited state. (You can picture what happens by imagining an electron moving from an orbital in its highest, filled subshell to an orbital in a still higher energy subshell.) When the metal ion returns to the ground state, it loses energy, and this energy appears as light. This occurs for atoms in the gas phase in a "neon" sign and for atoms excited by entraining them in a burning fuel. See Section 7.3 for a discussion of this in the case of the H atom.

7.1. (1) Highest frequency = violet; lowest frequency = red
(2) FM radio has lower frequency than microwave oven.
(3) Wavelength of x-rays is shorter than that of ultraviolet light.

7.2. (1) 10 cm
(2) 6.67 cm
(3) 2 waves; 5 nodes (1 at each end and 3 in the middle)

7.3. Blue light: $4.00 \times 10^2\ nm = 4.00 \times 10^{-7}\ m$

$$\nu = \frac{(2.998 \times 10^8\ m/s)}{(4.00 \times 10^{-7}\ m)} = 7.50 \times 10^{14}/s$$

$$E = (6.626 \times 10^{-34}\ J \cdot s/photon)(7.50 \times 10^{14}/s)(6.022 \times 10^{23}\ photons/mol)$$
$$= 2.99 \times 10^5\ J/mol$$

Microwaves: $E = (6.626 \times 10^{-34}\ J \cdot s/photon)(2.45 \times 10^9/s)(6.022 \times 10^{23}\ photon/mol) = 0.978\ J/mol$

$$\frac{E\ (blue\ light)}{E\ (microwaves)} = 3.1 \times 10^5$$

Blue light is almost half a million times more energetic than microwaves.

7.4. Energy for $n = 3 = \frac{-Rhc}{n^2} = \frac{-Rhc}{9}$

If $-Rhc = -2.179 \times 10^{-18}$ J/atom (see Example 7.3), then $-Rhc/9 = -2.421 \times 10^{-19}$ J/atom

$(-2.421 \times 10^{-19}\ \cancel{J}/\cancel{atom})(6.022 \times 10^{23}\ \cancel{atom}/mol)$

$(1\ kJ/1000\ \cancel{J}) = 145.8\ kJ/mol$

7.5. Use Equation 7.5 where $n_{final} = 1$ and $n_{initial} = 2$.

$$\Delta E = -Rhc\left[\frac{1}{1^2} - \frac{1}{2^2}\right]$$

$$= -(2.179 \times 10^{-18}\ J/atom)\left(\frac{3}{4}\right) = 1.634 \times 10^{-18}\ J/atom$$

$$\lambda = hc/\Delta E$$

$$= (6.626 \times 10^{-34}\ \cancel{J} \cdot s/\cancel{atom})\ \frac{2.998 \times 10^8\ \cancel{m/s}}{1.634 \times 10^{-18}\ \cancel{J}/\cancel{atom}}$$

$= 1.216 \times 10^{-7}$ m, or 121.6 nm (see Figure 7.12), and frequency $= 2.466 \times 10^{15}$/s

7.6. First calculate the velocity of the neutron:

$$v = \left[\frac{2E}{m}\right]^{1/2}$$

$$= [2(6.21 \times 10^{-21}\ \cancel{kg} \cdot m^2/s^2)]/(1.675 \times 10^{-27}\ \cancel{kg})]^{1/2}$$

$$= 2723\ m/s$$

With the velocity known, the wavelength can be calculated:

$$\lambda = \frac{h}{mv} = \frac{6.626 \times 10^{-34}\ (kg \cdot m^2/s^2)s}{(1.675 \times 10^{-31}\ kg)(2723\ m/s)}$$

$$\lambda = 1.45 \times 10^{-10}\ m$$

7.7. (1) $\ell = 0$ and 1
(2) $m_\ell = +1$, 0, and -1; subshell label $= p$
(3) d subshell
(4) $\ell = 0$ and $m_\ell = 0$ for an s subshell.
(5) 3 orbitals in a p subshell
(6) f subshell has 7 values of m_ℓ and 7 orbitals.

7.8. (1)

Orbital	n	ℓ
$6s$	6	0
$4p$	4	1
$5d$	5	2
$4f$	4	3

(2) A $4p$ orbital has one nodal plane, and a $6d$ orbital has two nodal planes.

CHAPTER 8

Chemical Puzzler: Iron is a good reducing agent because, like all metals, its ionization energy is relatively low (Section 8.6). Chlorine is a good oxidizing agent because, like many nonmetals, it has a relatively large affinity for electrons (Section 8.6). The iron ion in $FeCl_3$ is Fe^{3+}, which has unpaired electrons, making it paramagnetic (Sections 8.1 and 8.5). (See the discussion of ferromagnetism on page 336.)

8.1. (1) 4s ($n + \ell = 4$) filled before $4p$ ($n + \ell = 5$)
(2) $6s$ ($n + \ell = 6$) filled before $5d$ ($n + \ell = 7$)
(3) $5s$ ($n + \ell = 5$) filled before $4f$ ($n + \ell = 7$)

8.2. (1) Cl
(2) Cl has the spectroscopic notation given in part 1.

[Ne] $3s$ [↑↓] $3p$ [↑↓][↑↓][↑]

(3) Ca has two valence electrons, the $4s$ electrons.
For the first electron, $n = 4$, $\ell = 0$, $m_\ell = 0$, and $m_s = +\frac{1}{2}$
For the second electron, $n = 4$, $\ell = 0$, $m_\ell = 0$, and $m_s = -\frac{1}{2}$

8.3. See Table 8.2.

8.4. V^{2+}: [Ar] $4s$ [ ] $3d$ [↑][↑][↑][ ][ ]

V^{3+}: [Ar] $4s$ [ ] $3d$ [↑][↑][ ][ ][ ]

Co^{3+}: [Ar] $4s$ [ ] $3d$ [↑↓][↑][↑][↑][↑]

All three ions are paramagnetic, with 3, 2, and 4 unpaired electrons, respectively.

8.5. Radii are in the order C < Si < Al. See Figure 8.10.

8.6. 1. H—O distance = 37 pm + 73 pm = 110 pm
H—S distance = 37 pm + 103 pm = 140 pm
2. Br has a radius of 228 pm/2 = 114 pm. The Br—Cl distance is 114 pm + 100 pm = 214 pm

8.7. 1. Radii are in the order C (77 pm) < B (85 pm) < Al (143 pm). See Figure 8.10.
2. Ionization energies: Al < B < C. See Figure 8.12 and Appendix F.
3. C should have the most positive EA. See Figure 8.13.

8.8. The ion radii are in the order $N^{3-} > O^{2-} > F^-$. All three ions have 10 electrons, but N has only 7 protons, O has 8, and F has 9.

8.9. $MgCl_3$ would have a Mg^{3+} ion. The ionization energy of Mg^{3+} is very large (page 357), making it improbable that the ion can be formed in the reaction.

CHAPTER 9

Chemical Puzzler: See "Why Are Sweeteners Sweet?" on page 423.

9.1. ·Ba· Ba, Group 2A, has 2 valence electrons.
·Äs· As, Group 5A, has 5 valence electrons.
·Br: Br, Group 7A, has 7 valence electrons.

9.2. Enthalpy of formation of Na(g) = +107.3 kJ/mol

Enthalpy of formation of I(g) = +106.8 kJ/mol

Formation of Na^+(g) (= IE) = +496 kJ/mol

Formation of I^- (= −EA) = −295.2 kJ/mol

Formation of NaI(s) from ions (= lattice energy) = −702 kJ/mol

Sum = calculated ΔH of formation = -287 kJ/mol (The enthalpy of formation of NaI(s) is -287.78 kJ/mol.)

9.3. $[NH_4]^+$, :C≡O:, $[S{=}C{=}N]^-$, $[SO_4]^{2-}$ (Lewis structures)

9.4. H₃C—Ö—H (methanol); H—N(H)—O—H (hydroxylamine)

methanol   hydroxylamine

9.5. $[H_2PO_4]^-$: H—O—P(O)(O)—O—H (Lewis structure)

9.6. NO_2^- has 18 valence electrons. Ozone, O_3, is an uncharged, triatomic molecule with 18 valence electrons. Notice that N^- is isoelectronic with O. Hydroxide ion, OH^-, is isoelectronic with HF.

9.7. Resonance structures for the nitrate ion:

$[NO_3]^- \longleftrightarrow [NO_3]^- \longleftrightarrow [NO_3]^-$ (three resonance structures, with the N=O double bond in a different position in each)

Lewis electron dot structure for nitric acid, HNO_3:

:Ö—H bonded to N, with N also bonded to two O atoms (one N=O double bond)

9.8. [:F̤—C̤l̤—F̤:]⁻ ClF_2^+, 2 bond pairs and 2 lone pairs

[:F̤—C̤l̤—F̤:]⁻ ClF_2^-, 2 bond pairs and 3 lone pairs

9.9. C—N (order = 1) > C=N (order = 2) > C≡N (order = 3)

$[:\ddot{O}{-}\ddot{N}{=}\ddot{O}]^- \longleftrightarrow [\ddot{O}{=}\ddot{N}{-}\ddot{O}:]^-$

The bond order in NO_2^- is 1.5. Therefore, the NO bond length (124 pm) should be between the length of a N—O single bond (136 pm) and a N=O double bond (115 pm).

9.10. $CH_4(g) + 2\ O_2(g) \longrightarrow CO_2(g) + 2\ H_2O(g)$

Break 4 C—H bonds and 2 O=O bonds =
4 mol (413 kJ/mol) + 2 mol (498 kJ/mol) = 2648 kJ

Make 2 C=O bonds and 4 H—O bonds =
2 mol (732 kJ/mol) + 4 mol (463 kJ/mol) = 3316 kJ

ΔH°_{rxn} = 2648 kJ − 3316 kJ = −668 kJ

(Calculated from ΔH°_f, the value of ΔH°_{rxn} is −802.3 kJ)

9.11. (1) CN^-, C = −1 and N = 0
(2) SO_3, S = +2 and O = −2/3

9.12. (1) The H atom is the positive atom in each case. H—F ($\Delta\chi = 1.9$) is more polar than H—I $\Delta\chi = 0.4$).
(2) B—F ($\Delta\chi = 2.0$) is much more polar than B—C ($\Delta\chi = 0.5$).
(3) C—Si is more polar ($\Delta\chi = 0.7$) than C—S (which is not polar at all, $\Delta\chi = 0$)

9.13.
−1 +1 0   0 +1 −1
:Ö—S̈=Ö ⟷ Ö=S̈—Ö:

The S—O bonds are polar, with the negative end being the O atom. (The O atom is more electronegative than the S atom.) Formal charges show that these bonds are in fact polar, with the O atom being the more negative atom.

9.14. The molecule is tetrahedral in shape with a Cl—C—Cl bond angle of about 109°. (There are four bonds around the central C atom.)

9.15. BF_3 is a planar, trigonal molecule because there are only three bonds (and no lone pairs) around the central atom. The BF_4^- ion, however, has four B—F bonds and so has a tetrahedral shape. Adding an F^- ion to BF_3 changes the molecular shape.

9.16. The electron-pair geometry around the I atom is trigonal-bipyramidal. The geometry of the ion is linear.

$[ICl_2]^-$: Cl—I—Cl with three lone pairs on I (structure)

9.17. 1. Phosphate ion has a tetrahedral electron-pair geometry, and the shape of the ion is likewise tetrahedral.

$[PO_4]^{3-}$ (Lewis structure)

2. The sulfite ion has a tetrahedral electron-pair geometry, and the shape of the ion is pyramidal.

$[SO_3]^{2-}$ (Lewis structure)

3. The IF_5 ion has an octahedral electron geometry and a square-pyramidal molecular shape. (Lone pairs on the F atoms are not shown.)

IF_5: I bonded to five F atoms, with one lone pair on I (structure)

9.18. 1. $BFCl_2$, polar, negative side is the F atom because F is the most electronegative atom in the molecule.

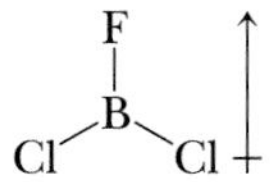

2. NH_2Cl, polar, negative side is the Cl atom.

3. SCl_2, polar, Cl atoms are on the negative side.

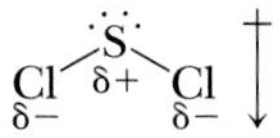

CHAPTER 10

Chemical Puzzler: The properties of the O_2 molecule are not satisfactorily explained by the simple electron dot picture. A more sophisticated bonding model—molecular orbital theory (Section 10.3)—leads to a better understanding of the O_2 molecule and many others as well.

10.1. (a) Hydronium ion, H_3O^+. This ion is isoelectronic and isostructural with NH_3 and so has an sp^3 hybridized O atom.

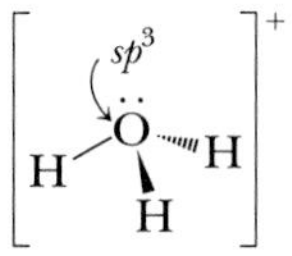

(b) Methylamine is also related to ammonia in that an H atom of ammonia is replaced by a CH_3 (methyl) group. The N atom is sp^3 hybridized as is the C of the CH_3 group.

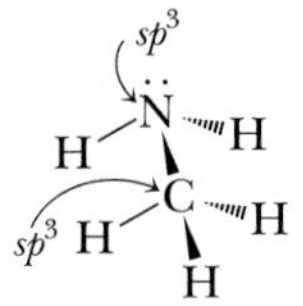

10.2. XeF_4 has an octahedral electron-pair geometry and a square-planar molecular geometry. The Xe atom hybridization is sp^3d^2.

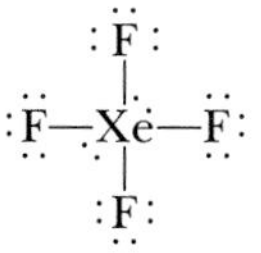

Electron dot structure 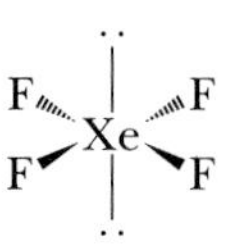 Electron-pair geometry

10.3. (a) BH_4^-, tetrahedral electron-pair geometry, sp^3
(b) SF_5^-, octahedral electron-pair geometry, sp^3d^2
(c) OSF_4, trigonal-bipyramidal electron-pair geometry, sp^3d
(d) ClF_3, trigonal-bipyramidal electron-pair geometry, sp^3d
(e) BCl_3, trigonal-planar electron-pair geometry, sp^2
(f) XeO_6^{4-}, octahedral electron-pair geometry, sp^3d^2

10.4. The two CH_3 carbon atoms are sp^3-hybridized. Three of these hybrid atomic orbitals are used to form C—H bonds and the fourth is used to form the C—C bond. The C atom of the C=O group is sp^2-hybridized. Two of the sp^2 hybrid orbitals are used to form C—C bonds, and a third is used in C—O bond formation. The unhybridized p orbital on the C atom is used in π bond formation with an O atom p orbital.

10.5. The bonding in N_2 is identical to that in CO. That is, there is one σ bond, two π bonds, and one lone pair on each N atom. The N atoms are sp-hybridized: one hybrid orbital accommodating a lone pair and the other used in sigma bond formation. The two unhybridized p atomic orbitals on each N atom are used in π bond formation.

10.6. Bond angles: H—C—H = 109°; H—C—C = 109°; and C—C—N = 180°. Bonding analysis: The CH_3 carbon has a tetrahedral electron-pair geometry and is sp^3-hybridized. The CN carbon has a linear electron pair geometry and is sp-hybridized. The N atom has a linear electron-pair geometry (the triple bond and lone pair are 180° apart) and can be considered sp-hybridized.

10.7. MO configuration of H_2^+: $(\sigma_{1s})^1$. The ion therefore has a bond order of $\frac{1}{2}$, and the ion is expected to exist. This is the same bond order as in the ions He_2^+ and H_2^-, whose molecular orbital configuration is $(\sigma_{1s})^2\ (\sigma^*_{1s})^1$.

10.8. The anion Li_2^- has the configuration $(\sigma_{1s})^2\ (\sigma^*_{1s})^2\ (\sigma_{2s})^2\ (\sigma^*_{2s})^1$. This gives a bond order of $\frac{1}{2}$, implying that the anion might be stable enough to prepare a salt such as $NaLi_2$.

10.9. O_2^+: [core electrons]$(\sigma_{2s})^2(\sigma^*_{2s})^2(\pi_{2p})^4(\sigma_{2p})^2(\pi^*_{2p})^1$. The net bond order is 2.5, a higher bond order than in O_2 and thus a stronger bond. The ion is paramagnetic to the extent of one electron.

CHAPTER 11

Chemical Puzzler: See pages 524–525.

11.1. Isomers of C_7H_{16}

$CH_3CH_2CH_2CH_2CH_2CH_2CH_3$ heptane

$CH_3CH_2CH_2CH_2CH(CH_3)CH_3$ 2-methylhexane

$CH_3CH_2CH_2CH(CH_3)CH_2CH_3$ 3-methylhexane

$CH_3CH_2CH(CH_3)CH(CH_3)CH_3$ 2,3-dimethylpentane

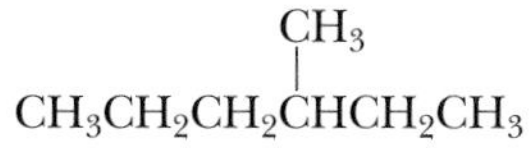

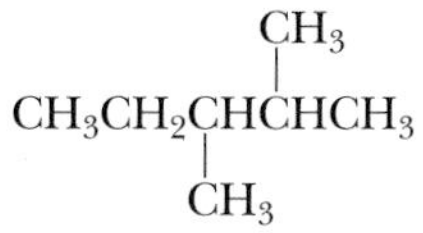

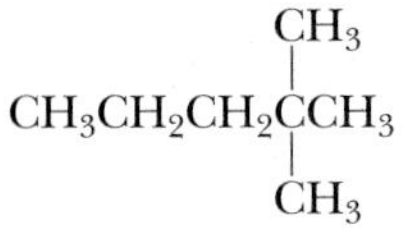

$CH_3CH_2CH_2C(CH_3)_2CH_3$ 2,2-dimethylpentane

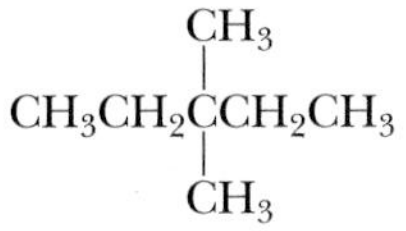

$CH_3CH_2C(CH_3)_2CH_2CH_3$ 3,3-dimethylpentane

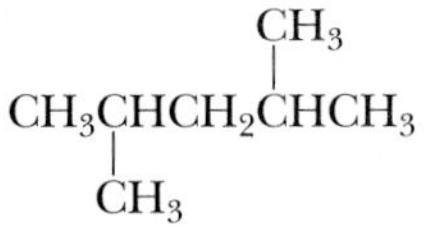

$CH_3CH(CH_3)CH_2CH(CH_3)CH_3$ 2,4-dimethylpentane

2-ethylpentane is pictured on page 490

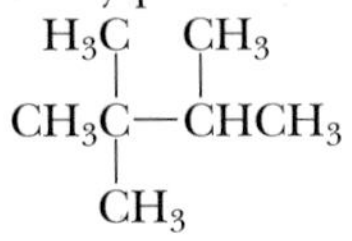

$CH_3C(CH_3)_2—CH(CH_3)CH_3$ 2,2,3-trimethylbutane

11.2. Some isomers of C_6H_{12} (in which the longest chain has six C atoms):

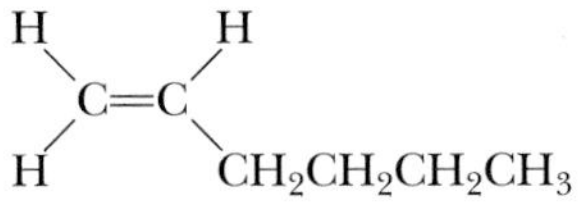

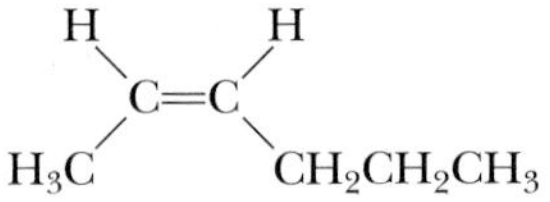

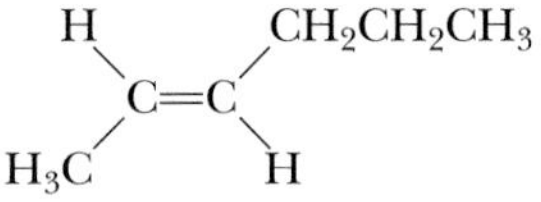

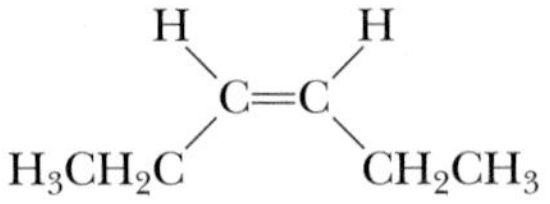

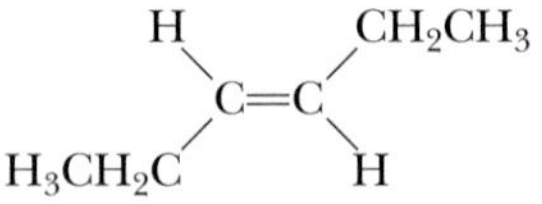

$H_2C=CHCH_2CH_2CH_2CH_3$

cis-$H_3CCH=CHCH_2CH_2CH_3$ trans-$H_3CCH=CHCH_2CH_2CH_3$

cis-$H_3CH_2CCH=CHCH_2CH_3$ trans-$H_3CH_2CCH=CHCH_2CH_3$

11.3. Reactions of alkenes:

(a) $H—CH_2—CH_2—Br$ (b) $H_3C—CHBr—CHBr—CH_3$

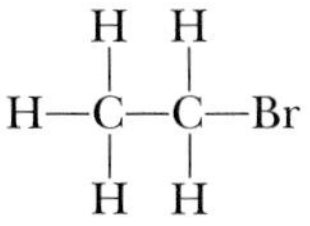

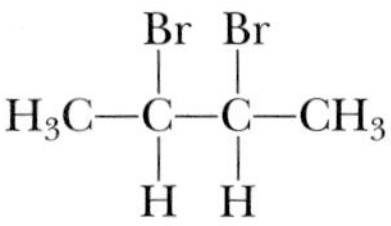

2,3-dibromobutane

11.4. p-dichlorobenzene

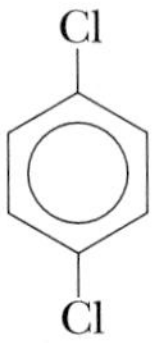

1,4-dichlorobenzene

11.5. $CH_3CH_2CH_2CH_2OH$ 1-butanol

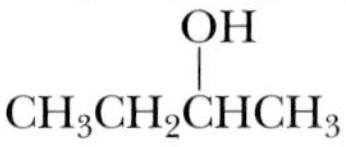

$CH_3CH_2CH(OH)CH_3$ 2-butanol

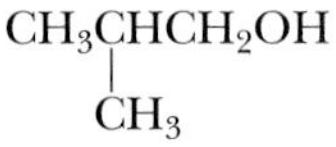

$CH_3CH(CH_3)CH_2OH$ 2-methyl-1-propanol

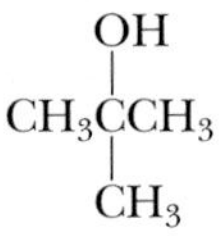

$CH_3C(OH)(CH_3)CH_3$ 2-methyl-2-propanol

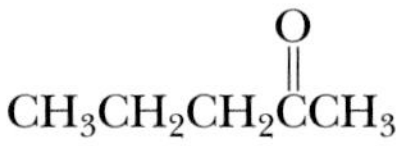

11.6. $CH_3CH_2CH_2C(=O)CH_3$ 2-pentanone

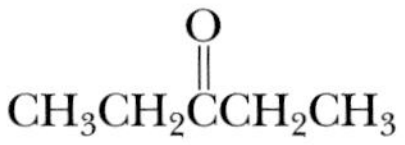

$CH_3CH_2C(=O)CH_2CH_3$ 3-pentanone

$CH_3CH_2CH_2CH_2CH(=O)$ pentanal

$CH_3CH(CH_3)CH_2CH(=O)$ 3-methylbutanal

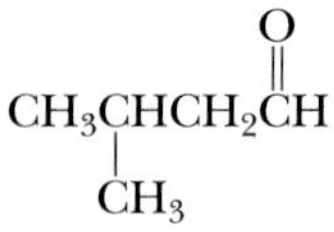

11.7. (1) 1-butanol gives butanal $CH_3CH_2CH_2CH(=O)$

(2) 2-butanol gives butanone $CH_3CH_2C(=O)CH_3$

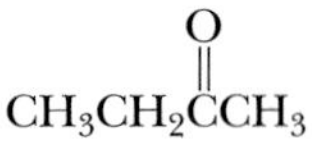

(3) 2-methyl-1-propanol gives 2-methylpropanal $CH_3CH(CH_3)—CH(=O)$

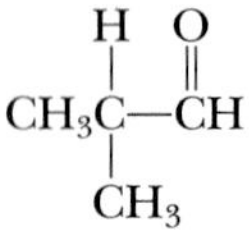

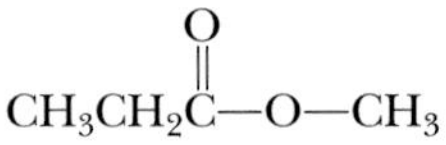

11.8. (1) $CH_3CH_2C(=O)—O—CH_3$ methyl propanoate

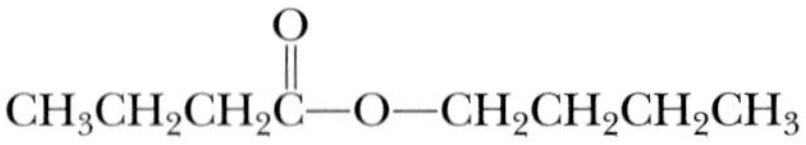

(2) $CH_3CH_2CH_2C(=O)—O—CH_2CH_2CH_2CH_3$ butyl butanoate

(3) $CH_3CH_2CH_2CH_2CH_2C(=O)—O—CH_2CH_3$ ethyl hexanoate

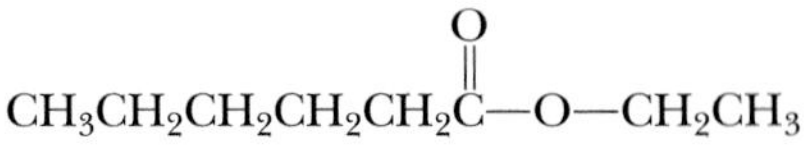

11.9. (1) Propyl acetate from acetic acid and propanol

$CH_3C(=O)OH + CH_3CH_2CH_2OH$

(2) 3-Methylpentyl benzoate from benzoic acid and 3-methylpentanol

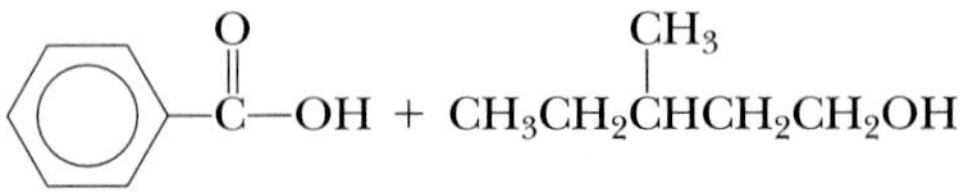

$C_6H_5—C(=O)—OH + CH_3CH_2CH(CH_3)CH_2CH_2OH$

(3) Ethyl salicylate from salicylic acid and ethanol

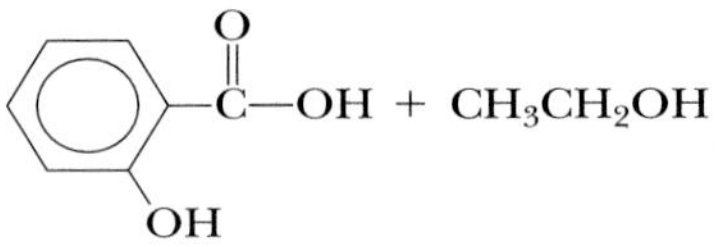

$HOC_6H_4—C(=O)—OH + CH_3CH_2OH$

11.10.

$$\begin{array}{l} \quad\;\; H \\ \quad\;\; | \\ H\!-\!C\!-\!O_2C(CH_2)_{14}CH_3 \\ \quad\;\; | \\ H\!-\!C\!-\!O_2C(CH_2)_{14}CH_3 \\ \quad\;\; | \\ H\!-\!C\!-\!O_2C(CH_2)_{14}CH_3 \\ \quad\;\; | \\ \quad\;\; H \end{array}$$

The products of saponification are glycerol and sodium palmitate, $NaCH_3(CH_2)_{14}CO_2$.

11.11. A polymer made from propylene glycol and maleic acid

$$\left(\begin{array}{c} H_3C \;\; H \qquad\quad O \;\; H \;\; H \;\; O \\ -C-C-O-\overset{\|}{C}-C{=}C-\overset{\|}{C}-O- \\ H \;\;\; H \end{array} \right)_n$$

CHAPTER 12

Chemical Puzzler: The chemistry of air bags is described on pages 539 and 556. The volume of an air bag can be affected by the temperature and pressure of the surrounding air. The effect of T and P on gas volume is described in this chapter.

12.1. 0.83 bar (0.82 atm) > 75 kPa (0.74 atm) > 0.63 atm > 250. mm Hg (0.329 atm)

12.2. P_1 = 55 mm Hg and V_1 = 125 mL
P_2 = 78 mm Hg and V_2 = ?
$$V_2 = \frac{P_1V_1}{P_2} = 88 \text{ mL}$$

12.3. T_1 = 298 K and V_1 = 45 L
T_2 = 263 K and V_2 = ?
$$V_2 = V_1\left(\frac{T_2}{T_1}\right) = 4.0 \times 10^1 \text{ L}$$

12.4. Calculate new volume (V_2) under new set of conditions.
V_1 = 22 L, P_1 = 150 atm, T_1 = 304 K
V_2 = ?, P_2 = 0.998 atm, T_1 = 295 K
$$V_2 = V_1\left(\frac{P_1}{P_2}\right)\left(\frac{T_2}{T_1}\right) = 3200 \text{ L}$$
At 5.0 L per balloon, one can fill 640 balloons.

12.5. $$22.4 \text{ L } CH_4 \left(\frac{2 \text{ L } O_2}{1 \text{ L } CH_4}\right) = 44.8 \text{ L } O_2 \text{ required}$$
44.8 L of H_2O and 22.4 L of CO_2 are produced.

12.6. n = 1300 mol, P = (750/760) atm, T = 296 K
$$V = \frac{nRT}{P} = 3.2 \times 10^4 \text{ L}$$

12.7. M = 28.96 g/mol, P = 1.00 atm, T = 288 K
$$d = \frac{PM}{RT} = 1.23 \text{ g/L}$$

12.8. P = 0.738 atm, V = 0.125 L, T = 296.2 K
$$n = \frac{PV}{RT} = 3.80 \times 10^{-3} \text{ mol}$$
$$\text{Molar mass} = \frac{0.105 \text{ g}}{3.80 \times 10^{-3} \text{ mol}} = 27.7 \text{ g/mol}$$

12.9. $$n\,(H_2) = \frac{PV}{RT} = 10.4 \text{ mol}$$
when P = 0.713 atm, V = 355 L, and T = 298.2 K.
$$10.4 \text{ mol } H_2 \left(\frac{2 \text{ mol } NH_3}{3 \text{ mol } H_2}\right) = 6.90 \text{ mol } NH_3$$
$$P\,(NH_3) = \frac{nRT}{V} = 1.35 \text{ atm}$$
when n = 6.90 mol, T = 298.2 K, and V = 125 L.

12.10. $$180 \text{ g}\left(\frac{1 \text{ mol } N_2H_4}{32.0 \text{ g}}\right) = 5.6 \text{ mol } N_2H_4$$
$$5.6 \text{ mol } N_2H_4 \left(\frac{1 \text{ mol } O_2}{1 \text{ mol } N_2H_4}\right) = 5.6 \text{ mol } O_2$$
$$V(O_2) = \frac{nRT}{P} = 140 \text{ L}$$
when n = 5.6 mol, T = 294 K, and P = 0.99 atm.

12.11. $$P_{\text{halothane}} = \frac{0.0760 \text{ mol})(0.082057 \text{ L} \cdot \text{atm/K} \cdot \text{mol})(298.2 \text{ K})}{5.00 \text{ L}}$$
$P_{\text{halothane}}$ = 0.372 atm (or 283 mm Hg)
$P(O_2)$ = 3.59 atm (or 2730 mm Hg)
P_{total} = 3.96 atm (or 3010 mm Hg)

12.12. $P\,(N_2)$ = 742 mm Hg − vapor pressure H_2O
= (742 − 22.4)mm Hg = 720. mm Hg
= 0.947 atm
$$n = \frac{PV}{RT} = 0.0137 \text{ mol } N_2 = 0.383 \text{ g } N_2$$
when V = 0.352 L and T = 297.2 K.

12.13. Using Equation 12.14 with M = 4.00 g/mol (or 4.00 × 10^{-3} kg/mol), T = 298 K, and R = 8.314 J/K · mol, a root mean square speed of 1360 m/s for He results. In contrast, N_2 molecules have a much smaller rms speed (515 m/s) owing to their greater mass.

12.14. The molar mass of CH_4 is 16.0 g/mol. Therefore,
$$\frac{\text{Rate for } CH_4}{\text{Rate for unk}} = \frac{n \text{ molecules/1.50 min}}{n \text{ molecules/4.73 min}} = \sqrt{\frac{M_{\text{unk}}}{16.0}}$$
M_{unk} = 159 g/mol

12.15. For n = 10.0 mol, V = 1.00 L, T = 298 K:
(a) Ideal gas law: $P = \frac{nRT}{V} = 245$ atm
(b) Van der Waals's equation (where a = 0.034 and b = 0.0237): P = 320 atm

CHAPTER 13

Chemical Puzzler: Ice is less dense than liquid water because of the open structure of ice (Figure 13.10), a result of hydrogen bonding. This means that a given mass of water occupies a larger volume when frozen than in the liquid state. This latter phenomenon means that expansion occurs when water freezes, and it can expand out of its container.

13.1. F^- should have the more negative hydration energy because its radius is much less than that of Cl^- (see Figure 8.14).

13.2. (a) O_2 interactions occur by induced-dipole/induced-dipole forces, generally the weakest of all intermolecular forces.

(b) The common, hydrated salt $MgSO_4 \cdot 7\ H_2O$ (epsomite) is widely used in agriculture and medicine. $MgSO_4$ consists of the ions Mg^{2+} and SO_4^{2-}, so there are ion–dipole forces involved in binding water to the ions.

(c) Dipole/induced-dipole forces exist between H_2O and O_2. Order of strength is $O_2—O_2 < O_2—H_2O < MgSO_4—H_2O$.

13.3. $(1.00 \times 10^3\ g)(1\ mol/32.04\ g)(35.2\ kJ/mol) = 1.10 \times 10^3\ kJ$

13.4. (a) Vapor pressure of ethanol at 40 °C is about 120 mm Hg.

(b) The equilibrium vapor pressure at 60 °C is about 370 mm Hg. At 60 °C and 600 mm Hg pressure, ethanol exists in the liquid state.

13.5. $P = \dfrac{nRT}{V}$

$= [(0.028\ mol)(0.0821\ L \cdot atm/K \cdot mol)(333\ K)]/5.0\ L$

$= 0.15\ atm = 120\ mm\ Hg$

Appendix G gives a vapor pressure of H_2O at 60 °C of 149 mm Hg. The calculated pressure of water in the flask is smaller than this, so all the water (0.50 g) evaporates. With 2.0 g, however, the calculated pressure (460 mm Hg) is much larger than the vapor pressure of water at 60 °C, so only enough water can evaporate to give an equilibrium pressure of 149 mm Hg.

13.6. Glycerol has three —OH groups per molecule that can be used in hydrogen bonding, as compared with only one for ethanol. The viscosity of glycerol should be greater than that of ethanol.

13.7. Mass of unit cell $= \left(\dfrac{4\ \text{Au atoms}}{\text{unit cell}}\right)(\text{mass of one atom})$

$= \left(\dfrac{4\ \text{Au atoms}}{1\ \text{unit cell}}\right)\left(\dfrac{197.97\ g}{mol}\right)\left(\dfrac{1\ mol}{6.022 \times 10^{23}\ \text{atoms}}\right)$

$= 1.308 \times 10^{-21}$ g/unit cell

Volume of unit cell = (mass of 1 cell) · (1/density)

$= \left(\dfrac{1.308 \times 10^{-21}\ g}{\text{unit cell}}\right)\left(\dfrac{1\ cm^3}{19.32\ g}\right)$

$= 6.772 \times 10^{-23}\ cm^3$/unit cell

Unit cell edge $= (\text{volume})^{1/3} = 4.076 \times 10^{-8}$ cm

Atom radius $= \left(\dfrac{1}{4}\right)(\text{diagonal distance}) = \left(\dfrac{1}{4}\right)(\sqrt{2})(\text{edge})$

Atom radius $= 1.441 \times 10^{-8}$ cm, or 144.1 pm

13.8. (8 corner Cl^- ions)(1/8 ion per corner) = 1 net Cl^- ion in the unit cell. Because there is 1 Cs^+ ion in the center of the unit cell, the formula of the salt must be CsCl.

13.9. Cube edge = 2(radius of K^+) + (radius of Cl^-)

= 2(152.0 pm) + 2(167.0 pm)

= 638.0 pm (or 6.38×10^{-8} cm)

Volume $= (\text{edge})^3 = (6.38 \times 10^{-8}\ cm)^3 = 2.60 \times 10^{-22}\ cm^3$

Mass of KCl unit cell

$= (4\ KCl/\text{cell})(74.55\ g/mol)(1\ mol/6.022 \times 10^{23}\ KCl)$

$= 4.95 \times 10^{-22}$ g/cell

Density $= (4.95 \times 10^{-22}\ g/\text{cell})/(2.60 \times 10^{-22}\ cm^3/\text{cell}) = 1.91\ g/cm^3$

(Literature density of KCl = 1.99 g/cm^3)

13.10. (a) $100.0\ g\ H_2O\ (1\ mol/18.02\ g)(6.02\ kJ/mol) = 33.4\ kJ$

(b) $100.0\ g\ C_8H_{18}\ (1\ mol/114.23\ g)(20.65\ kJ/mol) = 18.08\ kJ$

The intermolecular forces are weaker in the nonpolar hydrocarbon octane, so solid octane requires less energy to melt than an equal mass of water.

CHAPTER 14

Chemical Puzzler: This is an illustration of Henry's law (page 653). The solubility of a gas such as CO_2 depends largely on its pressure, the higher the pressure, the higher the solubility.

14.1. 10.0 g sugar = 0.0292 mol and 250. g water = 13.9 mol

$$X_{sugar} = \frac{0.0292\ \text{mol sugar}}{0.0292\ \text{mol sugar} + 13.9\ \text{mol water}} = 0.00210$$

$$\frac{0.0292\ \text{mol sugar}}{0.250\ \text{kg}} = 0.117\ \text{molal}$$

$$\text{Percentage sugar} = \frac{10.0\ \text{g sugar}}{260.\ \text{g solution}} \times 100\% = 3.85\%$$

14.2. 1.08×10^4 ppm Na $= 1.08 \times 10^4$ mg Na/1000 g solution

$(1.08 \times 10^4$ mg Na/1000 g solution) · (1050 g solution/1 L) $= 1.13 \times 10^4$ mg Na/L

$(1.13 \times 10^4$ mg Na/L) · (1 g/1000 mg) = 11.3 g Na/L

(11.3 g Na/L) · (58.45 g NaCl/23.0 g Na) = 28.7 g NaCl/L

14.3. $\Delta H°(\text{solution}) = \Delta H_f°(\text{aq}, 1\ m) - \Delta H_f°(\text{s})$

$= (-470.1\ kJ/mol) - (-425.6\ kJ/mol)$

$= -44.5\ kJ$

14.4. Solubility(CO_2) $= (4.48 \times 10^{-5}$ M/mm Hg)(251 mm Hg) $= 1.12 \times 10^{-2}$ M

14.5. Solution consists of sucrose (0.0292 mol) and water (12.5 mol).

$$X_{water} = \frac{12.5\ \text{mol}\ H_2O}{12.5\ \text{mol}\ H_2O + 0.0292\ \text{mol sugar}} = 0.998$$

$P_{water} = (0.998)(149.4\ \text{mm Hg}) = 149\ \text{mm Hg}$

Even with 10.0 g of sugar, the vapor pressure of water has changed very little.

14.6. $$\text{Concentration}\ (m) = \frac{\Delta t_{bp}}{K_{bp}} = \frac{1.0\ °C}{(+0.512\ °C/m)} = 2.0\ m$$

(2.0 mol/kg) · (0.125 kg) = 0.25 mol

(0.25 mol) · (62.07 g/mol) = 16 g glycol

14.7. $\Delta t_{bp} = 80.23\ °C - 80.10\ °C = 0.13\ °C$

$$\text{Concentration } (m) = \frac{\Delta t_{bp}}{K_{bp}} = 0.13\ °C/(2.53\ °C/m)$$
$$= 0.051\ m$$
$$(0.051\ \text{mol/kg}) \cdot (0.0990\ \text{kg}) = 0.0051\ \text{mol}$$
$$\frac{0.640\ \text{g}}{0.0051\ \text{mol}} = 130\ \text{g/mol (to 2 significant figures)}$$

(Answer is based on an outcome of 126 g/mol; numbers are not rounded until the end of the calculation.) (Azulene has the formula $C_{10}H_8$ with a molar mass of 128.2 g/mol.)

14.8. Molality of $HOCH_2CH_2OH$ (ethylene glycol) =
$$\frac{8.47\ \text{mol}}{3.00\ \text{kg}} = 2.82\ m$$
$$\Delta t_{fp} = K_{fp} \cdot m = (-1.86\ °C/\text{molal})(2.82\ m) = -5.24\ °C$$
525 g of glycol is not sufficient to keep the plumbing from freezing at −25 °C.

14.9. 25.0 g NaCl = 0.428 mol
$$\text{Concentration } (m) = \frac{0.428\ \text{mol}}{0.525\ \text{kg}} = 0.815\ m$$
$$\Delta t_{fp} = K_{fp} \cdot m \cdot i = (-1.86\ °C/\text{m})(0.815\ m)(1.85) = -2.80\ °C$$

14.10. $$M = \frac{\Pi}{RT} = \frac{(1.86\ \text{mm Hg} / 760\ \text{mm Hg atm}^{-1})}{(0.0821\ \text{L} \cdot \text{atm} / \text{K} \cdot \text{mol})(298\ \text{K})}$$
$$= 1.00 \times 10^{-4}\ \text{M}$$
$$(1.00 \times 10^{-4}\ \text{mol / L})(0.100\ \text{L}) = 1.00 \times 10^{-3}\ \text{mol}$$
$$\text{Molar mass} = \frac{1.40\ \text{g}}{1.00 \times 10^{-3}} = 1.4 \times 10^4\ \text{g/mol}$$
Assuming the polymer is composed of CH_2 units (molar mass = 14 g / mol), the polymer is about 100 units long.

CHAPTER 15

Chemical Puzzler: The catalyst is involved in the chemical reactions in the converter, but it is not consumed by those reactions. This is a property of all catalysts, as described in Section 15.5. See page 718 in particular.

15.1. $$-\left(\frac{1}{2}\frac{\Delta[\text{NOCl}]}{\Delta t}\right) = \frac{1}{2}\left(\frac{\Delta[\text{NO}]}{\Delta t}\right) = \frac{\Delta[\text{Cl}_2]}{\Delta t}$$

15.2. For first 2 h:
$$\frac{\Delta[\text{sucrose}]}{\Delta t} = \frac{(0.034 - 0.050)\text{mol / L}}{2\text{h}}$$
$$= \frac{-0.0080\ \text{mol / L}}{\text{h}}$$
For last 2 h:
$$\frac{\Delta[\text{sucrose}]}{\Delta t} = \frac{(0.010 - 0.015)\text{mol / L}}{2\text{h}}$$
$$= \frac{-0.0025\ \text{mol / L}}{\text{h}}$$
Notice that the rate is much slower over the last 2 h; less reactant is present, so the rate of reaction is smaller.

Instantaneous rate at 4 h = −0.0045 mol/L · h

Notice that this rate is intermediate between the rate over the first 2 h and that over the last 2 h.

15.3. (1) Second order with respect to NO and first order with respect to O_2. That is, Rate = $k[NO_2]^2[O_2]$.
(2) $k = 7.0 \times 10^3\ \text{L}^2/\text{mol}^2 \cdot \text{s}$

15.4. Rate of reaction = (0.090/h)(0.020 mol/L)
= 0.0018 mol/L · h
Cl^- appears at a rate of 0.0018 mol/L · h

15.5. $$\ln\left(\frac{[\text{sucrose}]}{[\text{sucrose}]_0}\right) = -kt$$
$$\ln\left(\frac{[\text{sucrose}]}{(0.010)}\right) = -(0.21/\text{h})(5.00\ \text{h}) = -1.05\ \text{h}$$
ln[sucrose] − ln(0.010) = −1.05 h
ln[sucrose] − (−4.61) = −1.05
ln[sucrose] = −5.66
[sucrose] after 5.00 h = 0.0035

15.6. (a) Fraction remaining after 150 s
$$\ln\frac{[\text{NO}_2]}{[\text{NO}_2]_0} = -(3.6 \times 10^{-3}\ \text{s}^{-1})(150\ \text{s})$$
$$\frac{[\text{NO}_2]}{[\text{NO}_2]_0} = 0.58 = \text{fraction remaining}$$
(b) Time for 99% to decompose
$$\ln(0.010) = -(3.6 \times 10^{-3}\ \text{s}^{-1})t$$
$$t = 1280\ \text{s}$$

15.7. $$\left(\frac{1}{[\text{HI}]}\right) - \left(\frac{1}{[\text{HI}]_0}\right) = kt$$
When $[HI]_0$ = 0.010 M, k = 30. L/(mol · min), and t = 12 min, [HI] = 0.0022 M.

15.8.

Concentration versus time

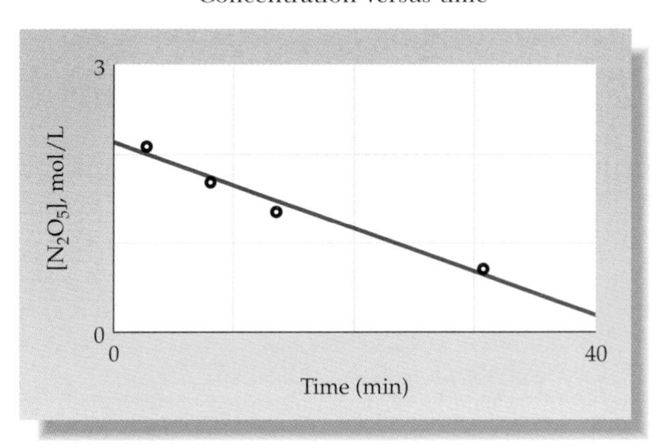

ln $[N_2O_5]$ versus time

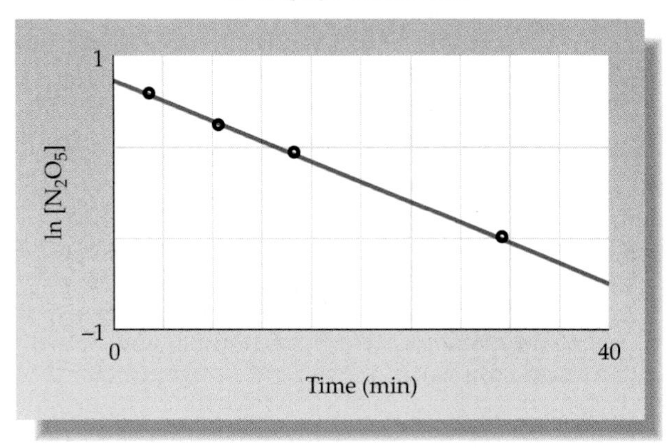

$1/[N_2O_5]$ versus time

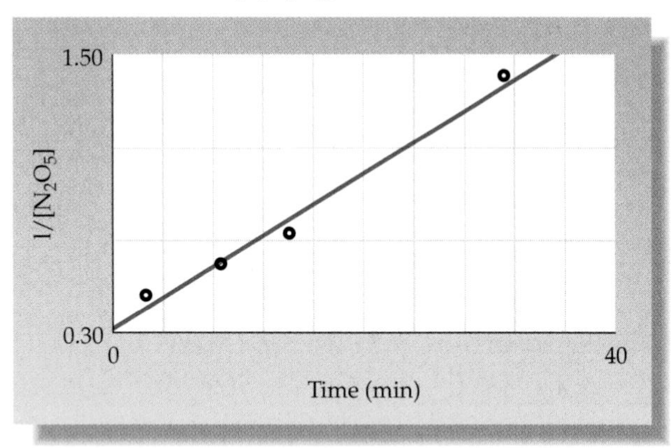

Only the plot of $\ln[N_2O_5]$ versus time is truly linear; the reaction is first order in N_2O_5. The slope of the line is -0.038 for $k = 0.038/\text{min}$.

15.9. For Am-241: $t_{1/2} = 430$ y
For I-125: $t_{1/2} = 63$ d
Iodine-125 decays faster.

15.10. Substituting the given values into Equation 15.7, we have

$$\ln \frac{100 \times 10^4\ \text{s}^{-1}}{4.5 \times 10^3\ \text{s}^{-1}} = -\frac{E_a}{8.31 \times 10^{-3}\ \text{kJ / K} \cdot \text{mol}}\left[\frac{1}{283\ \text{K}} - \frac{1}{274\ \text{K}}\right]$$

$E_a = 57$ kJ

15.11. (1) True
(2) A catalyst is never consumed in the overall reaction.
(3) A catalyst may be in the same phase or in a different phase than the reactants.

15.12. All three steps are bimolecular.
Rate for step 3 = $k_3\ [N_2O][H_2]$
N_2O_2, the product of the first step, is used in the second step, and N_2O, a product of the second step, is consumed in the third step. Adding the three reactions therefore gives the equation for the overall process.

15.13. (1) $2\ NH_3(aq) + OCl^-(aq) \longrightarrow N_2H_4(aq) + Cl^-(aq) + H_2O(\ell)$
(2) Step 2 is rate-determining.
(3) Rate = $k[NH_2Cl][NH_3]$
(4) NH_2Cl, $N_2H_5^+$, OH^-

15.14. *Overall equation*: $2\ NO_2Cl(g) \longrightarrow 2\ NO_2(g) + Cl_2(g)$
The rate law for the rate-determining step is
Rate = $k\ [NO_2Cl][Cl]$
Substituting for [Cl] using $K = [NO_2][Cl]/[NO_2Cl]$, we have

$$\text{Rate} = \frac{k'\ [NO_2Cl]^2}{[NO_2]}, \text{ where } k' = k_2 \cdot K = k_2\left(\frac{k_1}{k_{-1}}\right)$$

Because the rate is proportional to $1/[NO_2]$, an increase in NO_2 concentration will slow the reaction.

CHAPTER 16

Chemical Puzzler: This is an example of a chemical equilibrium, a reversible chemical process. The reactants are converted to products at the same time that products are converted back to reactants. All chemical equilibria are affected by changes in temperature. (Here the process interconverts the blue, tetrahedral $[CoCl_4]^{2-}$ ion and the pink, octahedral ion $[Co(H_2O)_6]^{2+}$. The blue ion is favored at higher temperature and the pink one at lower temperatures.) If we could know if the reaction was exothermic or endothermic, we could use Le Chatelier's principle (Section 16.6) to predict the effect of a temperature change.

16.1. (1) $K = \dfrac{[PCl_3][Cl_2]}{[PCl_5]}$

(2) $K = [Cu^{2+}][OH^-]^2$

(3) $K = \dfrac{[Cu^{2+}][NH_3]^4}{[Cu(NH_3)_4{}^{2+}]}$

(4) $K = \dfrac{[CH_3CO_2{}^-][H_3O^+]}{[CH_3CO_2H]}$

16.2. (1) $K_{new} = [K_{old}]^2 = 6.3 \times 10^{-58}$
(2) $\dfrac{K = 1}{6.3 \times 10^{-58}} = 1.6 \times 10^{57}$

16.3. $$K = \left[(7.9 \times 10^{11})\left(\frac{1}{4.8 \times 10^{-41}}\right)\left(\frac{1}{2.2 \times 10^{-15}}\right)\right]^{1/2} = 2.7 \times 10^{33}$$

16.4. Both reactions are reactant-favored. Concentration of Ag^+ in the AgCl beaker (1.3×10^{-5} M) is greater than in the AgI beaker (5.7×10^{-7}).

16.5. $$Q = \frac{[NO]^2}{[N_2][O_2]} = \frac{(4.2 \times 10^{-3})^2}{(0.50)(0.25)} = 1.4 \times 10^{-4}$$
$Q < K$, so the reaction is not at equilibrium. The reaction consumes N_2 and O_2 and produces NO on proceeding to equilibrium.

16.6. The equilibrium quantities are calculated starting with the amount of I_2 at equilibrium.

Equation	H_2	+	I_2	$\rightleftarrows$	2 HI
Initial (mol)	9.838×10^{-4}		13.77×10^{-4}		0
Change (mol)	-9.05×10^{-4}		-9.05×10^{-4}		$+18.1 \times 10^{-4}$
Equilibrium (mol)	7.93×10^{-5}		4.725×10^{-4}		18.1×10^{-4}
Equilibrium (M)	0.00793		0.04725		0.181

K (for volume of 10.0 mL) = 87.4

K (for volume of 20.0 mL) = 87.4

Notice that the volume of the container does not affect the value of K in this case.

16.7.

Equation	$C_6H_{10}I_2$	$\rightleftarrows$	C_6H_{10}	+	I_2
Initial (M)	0.050		0		0
Change (M)	−0.035		+0.035		+0.035
Equilibrium (M)	0.015		0.035		0.035

$K = 0.082$

16.8.

Equation	H_2	+	I_2	$\rightleftarrows$	2 HI
Initial (M)	6.00×10^{-3}		6.00×10^{-3}		0
Change (M)	$-x$		$-x$		$+2x$
Equilibrium (M)	$0.00600 - x$		$0.00600 - x$		$+2x$

$$K_c = 33 = \frac{(2x)^2}{(0.00600 - x)^2}$$

$x = 0.0045$ M, so $[H_2] = [I_2] = 0.0015$ M and $[HI] = 0.0090$ M

16.9.

Equation	C(s)	+	CO_2(g)	$\rightleftarrows$	2 CO(g)
Initial (M)	0		0.012		0
Change (M)			$-x$		$+2x$
Equilibrium (M)			$0.012 - x$		$2x$

$$K_c = 0.021 = \frac{(2x)^2}{(0.012 - x)}$$

$x = [CO_2] = 0.0057$ M and $2x = [CO] = 0.011$ M

16.10. (1) The reaction is endothermic. As the temperature increases, the equilibrium shifts to the right, and [NOCl] decreases.

(2) The reaction is exothermic. As the temperature increases, the equilibrium shifts to the left, and $[SO_3]$ decreases.

16.11.

Equation	butane	$\rightleftarrows$	isobutane
Initial (M)	0.20		0.50
After adding 2.0 M more isobutane	0.20		2.0 + 0.50
Change (M)	$+x$		$-x$
Equilibrium (M)	$0.20 + x$		$2.50 - x$

$$K = \frac{[\text{isobutane}]}{[\text{butane}]} = \frac{(2.50 - x)}{(0.20 + x)}.$$

Solving for x gives $x = 0.57$ M. Therefore, [isobutane] = 1.93 M and [butane] = 0.77 M.

16.12. (1) Added H_2 shifts the equilibrium right, and added NH_3 shifts it left.

(2) Increasing the volume decreases all the concentrations. The equilibrium shifts to the left, toward the side with the greater number of molecules.

CHAPTER 17

Chemical Puzzler: Soft drinks are usually acidic, aspirin is a weak acid and so is the citric acid found in citrus fruits such as lemons. Dish detergents are basic, and oven cleaner is often quite basic. Fertilizers contain basic salts such as phosphates and acidic metal cations.

17.1. (1) $H_3PO_4(aq) + H_2O(\ell) \longrightarrow H_3O^+(aq) + H_2PO_4^-(aq)$

(2) $H_2O(\ell) + CN^-(aq) \longrightarrow OH^-(aq) + HCN(aq)$
CN^- is a Brønsted base.

(3) $H_2C_2O_4(aq) + H_2O(\ell) \longrightarrow H_3O^+(aq) + HC_2O_4^-(aq)$

$HC_2O_4^-(aq) + H_2O(\ell) \longrightarrow H_3O^+(aq) + C_2O_4^{2-}(aq)$

17.2. (1) HBr is an acid, and Br^- is its conjugate base. NH_3 is a base, and its conjugate acid is NH_4^+.

(2) The conjugate base of H_2S is HS^-, and that for NH_4^+ is NH_3.

(3) The conjugate acid for NO_3^- is HNO_3, and that for HPO_4^{2-} is $H_2PO_4^-$.

17.3. (1) The NH_4^+ ion is a stronger Brønsted acid than HCO_3^-. This means that HCO_3^- has the stronger conjugate base.

(2) The CN^- ion is a stronger Brønsted base than SO_4^{2-}.

17.4. (1) HSO_4^- is a stronger acid than NH_4^+, so the equilibrium lies predominantly to the right. (We arrive at the same answer by knowing that NH_3 is a stronger base than SO_4^{2-}.)

(2) H_2S is a stronger acid than HCO_3^-, so the equilibrium lies predominantly to the left. (We arrive at the same answer by knowing that CO_3^{2-} is a stronger base than HS^-.)

17.5. The NH_4^+ ion is an acid, and NH_3 is its conjugate base. The SO_4^{2-} ion is a base, and HSO_4^- is its conjugate acid. Thus, the net ionic equation is

$$NH_4^+(aq) + SO_4^{2-}(aq) \rightleftarrows NH_3(aq) + HSO_4^-(aq)$$

The HSO_4^- ion is a stronger acid than NH_4^+, and NH_3 is a stronger base than SO_4^{2-}. The equilibrium therefore lies predominantly to the left.

17.6. 1. A K_a of 1.4×10^{-4} places lactic acid between formic acid and benzoic acid in strength. Lactic acid is a stronger acid than acetic acid.

2. $CH_3CHOHCO_2^-(aq) + H_2O(\ell) \rightleftarrows CH_3CHOHCO_2H(aq) + OH^-(aq)$

17.7. $$[H_3O^+] = \frac{0.0020 \text{ mol}}{0.50 \text{ L}} = 0.0040 \text{ M}$$

$$[OH^-] = \frac{1.0 \times 10^{-14}}{[H_3O^+]} = 2.5 \times 10^{-12}\ M$$

17.8. $$K_b = \frac{K_w}{K_a} = \frac{1.0 \times 10^{-14}}{1.4 \times 10^{-4}} = 7.1 \times 10^{-11}$$

This places the lactate ion between formate ion and benzoate in base strength. The lactate ion is a weaker base than the acetate ion.

17.9. pH = 3.12; $[H_3O^+] = 10^{-pH} = 10^{-3.12} = 7.6 \times 10^{-4}$ M

pOH = 14.00 − 3.12 = 10.88

$[OH^-] = 10^{-pOH} = 10^{-10.88} = 1.3 \times 10^{-11}$ M

17.10. $[H_3O^+] = 10^{-pH} = 10^{-2.94} = 1.1 \times 10^{-3}$ M

$$K_a = \frac{[H_3O^+][CH_3CH_2CO_2^-]}{[CH_3CH_2CO_2H]} = \frac{(1.1 \times 10^{-3})(1.1 \times 10^{-3})}{0.10 - (1.1 \times 10^{-3})}$$

$K_a = 1.2 \times 10^{-5}$

17.11. $$K_a = 1.8 \times 10^{-5} = \frac{[H_3O^+][CH_3CO_2^-]}{[CH_3CO_2H]} = \frac{(x)(x)}{0.10 - x}$$

$x = [H_3O^+] = [CH_3CO_2^-] = 1.3 \times 10^{-3}$ M; pH = 2.87

$[CH_3CO_2H] = 0.10 - x \approx 0.10$ M

17.12. $$K_a = 7.2 \times 10^{-4} = \frac{[H_3O^+][F^-]}{[HF]} = \frac{(x)(x)}{0.015 - x}$$

x is found by solving the quadratic equation or by using the method of successive approximations (Appendix A).

$x = [H_3O^+] = 2.9 \times 10^{-3}$ M; pH = 2.53

$[F^-] = 2.9 \times 10^{-3}$ M; [HF] = 0.012 M

17.13. $NH_3(aq) + H_2O(\ell) \rightleftharpoons NH_4^+(aq) + OH^-(aq)$

$$K_b = 1.8 \times 10^{-5} = \frac{[NH_4^+][OH^-]}{[NH_3]} = \frac{(x)(x)}{0.025 - x}$$

$x = [OH^-] = 6.7 \times 10^{-4}$ M; pOH = 3.17 and pH = 10.83

17.14. $$K_a = 5.6 \times 10^{-10} = \frac{[NH_3][H_3O^+]}{[NH_4^+]} = \frac{(x)(x)}{0.50 - x}$$

$x = [H_3O^+] = 1.7 \times 10^{-5}$ M; pH = 4.78

17.15.
1. NaCl is neutral and the pH is equal to 7.
2. $FeCl_3$ is acidic (because Fe^{3+} produces acidic solutions), and the pH is less than 7.
3. NH_4NO_3 is acidic (because NH_4^+ produces acidic solutions), and the pH is less than 7.
4. Na_2HPO_4 is basic, and the pH is greater than 7. (The HPO_4^{2-} ion is a stronger base than it is an acid.)

17.16. $H_2C_2O_4(aq) + H_2O(\ell) \longrightarrow H_3O^+(aq) + HC_2O_4^-(aq)$

(The K_a values are found in Appendix G)

$$K_{a1} = 5.9 \times 10^{-2} = \frac{[H_3O^+][HC_2O_4^-]}{[H_2C_2O_4]} = \frac{(x)(x)}{0.10 - x}$$

x is found by solving the quadratic equation or by using the method of successive approximations (Appendix A).

$x = [H_3O^+] = [HC_2O_4^-] = 5.3 \times 10^{-2}$ M; pH = 1.28

$[C_2O_4^{2-}] = K_{a2} = 6.4 \times 10^{-5}$ M

17.17. (1) H_2SO_4 (strong acid) > H_2SO_3 ($K_{a1} = 1.7 \times 10^{-2}$)
(2) H_3AsO_4 ($K_{a1} = 2.5 \times 10^{-4}$) > H_3AsO_3($K_{a1} = 6.0 \times 10^{-10}$)

In both cases the stronger acid has more O atoms attached to the central atom (S or As).

17.18. (1) PH_3 is a Lewis base. There is a lone pair of electrons on the P atom.
(2) BCl_3 is a Lewis acid. The central B atom is surrounded by only three bond pairs of electrons (and no lone pairs). It can therefore accept a pair of electrons from an electron-pair donor.
(3) H_2S is a Lewis base because there are two lone pairs of electrons on the S atom.
(4) HS^- is analogous with OH^-, an excellent Lewis base.

CHAPTER 18

Chemical Puzzler: If your blood pH is too high, it can be lowered by increasing the blood CO_2 concentration. This leads to additional hydronium ion in the blood, thus lowering the pH.

$$CO_2(g) + H_2O(\ell) \rightleftharpoons H_2CO_3(aq)$$
$$H_2CO_3(aq) + H_2O(\ell) \rightleftharpoons H_3O^+(aq) + HCO_3^-(aq)$$

By breathing into a paper bag, the CO_2 in your exhaled breath is captured and you breathe air that is higher in CO_2 concentration.

18.1. Moles of HCl = (0.050 L)(0.20 M) = 0.010 mol

Moles of aniline = (0.93 g)(1 mol/93.1 g) = 0.010 mol

Because 1 mol of aniline requires 1 mol of HCl and because they were present initially in equal molar quantities, the reaction completely consumes the acid and base. The solution after reaction contains 0.010 mol of $C_6H_5NH_3^+$ in 0.050 L of solution, so its concentration is 0.20 M. It is the conjugate acid of the weak base aniline, so

$$C_6H_5NH_3^+(aq) + H_2O(\ell) \rightleftharpoons C_6H_5NH_2(aq) + H_3O^+(aq)$$

$$K_a = 2.4 \times 10^{-5} = \frac{[H_3O^+][C_6H_5NH_2]}{[C_6H_5NH_3^+]} = \frac{(x)(x)}{0.2 - x}$$

$x = [H_3O^+] = 2.2 \times 10^{-3}$ M, which gives a pH of 2.66.

18.2. (0.976 g)(1 mol/122.1 g) = 0.00799 mol benzoic acid

Moles of NaOH required = 0.00799 mol; (0.00799 mol NaOH)$\left(\frac{1\ L}{0.100\ mol}\right)$ = 0.0799 L or 79.9 mL NaOH

Reaction gives 0.00799 mol of benzoate ion, $C_6H_5CO_2^-$.

$$\text{Concentration benzoate ion} = \frac{0.00799 \text{ mol}}{0.0799 \text{ L}} = 0.100 \text{ M}$$

Hydrolysis of benzoate ion:

$$C_6H_5CO_2^-(aq) + H_2O(\ell) \rightleftharpoons C_6H_5CO_2H(aq) + OH^-(aq)$$

$$K_b = 1.6 \times 10^{-10} = \frac{[C_6H_5CO_2H][OH^-]}{[C_6H_5CO_2^-]} = \frac{(x)(x)}{0.100 - x}$$

$[OH^-] = 4.0 \times 10^{-6}$ M; pOH = 5.40 and pH = 8.60

18.3. Equal numbers of moles of the acid and base were mixed, so reaction was complete to produce $CH_3CO_2^-$ (a weak base) and $C_5H_5NH^+$ (a weak acid).

$$CH_3CO_2H(aq) + C_5H_5N(aq) \longrightarrow CH_3CO_2^-(aq) + C_5H_5NH^+(aq)$$

K_b for $CH_3CO_2^- = 5.6 \times 10^{-10}$ and K_a for $C_5H_5NH^+ = 6.7 \times 10^{-6}$

The acid is stronger than the base here, so the solution containing the two ions is slightly acidic.

18.4. (a) pH of 0.30 M formic acid

$$K_a = 1.8 \times 10^{-4} = \frac{[H_3O^+][HCO_2^-]}{[HCO_2H]} = \frac{(x)(x)}{0.30 - x}$$

$x = [H_3O^+] = 7.3 \times 10^{-3}$ M, which gives a pH of 2.14.

(b) pH of 0.30 M formic acid + 0.010 M $NaHCO_2$

Equation	HCO_2H	$\rightleftharpoons$ H_3O^+	+ HCO_2^-
Initial concentrations (M)	0.30	0	0.10
Change (M)	$-x$	$+x$	$+x$
At equilibrium (M)	$0.30 - x$	x	$0.10 + x$

$$K_a = 1.8 \times 10^{-4} = \frac{[H_3O^+][HCO_2^-]}{[HCO_2H]} = \frac{(x)(0.1 + x)}{0.30 - x}$$

Assuming that x is small compared with 0.10 or 0.30, we find that $x = [H_3O^+] = 5.4 \times 10^{-4}$ M, which gives a pH of 3.27. As expected on adding a base (HCO_2^-), the pH has increased.

18.5. (1) Find pH of buffer before adding HCl.

$$[H_3O^+] \text{ before adding HCl} = \frac{[\text{acid}]}{[\text{conjugate base}]} \cdot K_a = \frac{0.50}{0.70} \cdot (1.8 \times 10^{-4}) = 1.3 \times 10^{-4} \text{ M}$$

pH of 3.89.

(2) Add 10.0 mL of 1.0 M HCl (= 0.010 mol). The reaction occurring is

$$H_3O^+(aq) + HCO_2^-(aq) \rightleftharpoons HCO_2H(aq)$$

The molar quantities and concentrations at this stage are as follows:

Concentrations	$[H_3O^+]$ from HCl	$[HCO_2^-]$ from buffer	$[HCO_2H]$ from buffer
Before reaction (mol)	0.010	0.35	0.25
Change when HCl reaction occurs (mol)	−0.010	−0.010	+0.010
After reaction (mol)	0	0.34	0.26
After reaction (mol/L in 0.510 L solution)	0	0.67	0.51

To find $[H_3O^+]$ at this stage, we again use the chemical equation for the ionization of formic acid.

$$HCO_2H(aq) + H_2O(\ell) \rightleftharpoons HCO_2^-(aq) + H_3O^+(aq)$$

$$[H_3O^+] \text{ after adding HCl} = \frac{[\text{acid}]}{[\text{conjugate base}]} \cdot K_a = \frac{0.51}{0.67} \cdot (1.8 \times 10^{-4}) = 1.4 \times 10^{-4} \text{ M}$$

From $[H_3O^+] = 1.4 \times 10^{-4}$ M the pH is 3.86. Only a small change in pH (3.89 $\longrightarrow$ 3.86) occurred on adding a concentrated acid.

18.6. K_a for $HCO_3^- = 4.8 \times 10^{-11}$ so $pK_a = 10.32$

15.0 g $NaHCO_3$ = 0.179 mol and 18.0 g Na_2CO_3 = 0.170 mol

$$\text{pH} = 10.32 + \log\left(\frac{0.170 \text{ mol}}{0.179 \text{ mol}}\right) = 10.30$$

18.7. A pH of 5.00 corresponds to $[H_3O^+] = 1.0 \times 10^{-5}$ M. This means that

$$[H_3O^+] = 1.0 \times 10^{-5} = \frac{[CH_3CO_2H]}{[CH_3CO_2^-]}(1.8 \times 10^{-5})$$

The ratio $[CH_3CO_2H]/[CH_3CO_2^-]$ must be 1/1.8 to achieve the correct hydronium ion concentration (1.0×10^{-5} M). Therefore, 1.0 mol of CH_3CO_2H is mixed with 1.8 mol of a salt of $CH_3CO_2^-$ (say $NaCH_3CO_2$) in some amount of water. (The volume of water is not critical; only the relative amounts of acid and conjugate base are important.)

18.8. (1) Using the expression in Example 18.7 to find $[H_3O^+]$ before the equivalence point, we have

$$[H_3O^+] = \frac{0.00500 \text{ mol HCl} - (0.0400 \text{ L NaOH})(0.100 \text{ M})}{0.0500 \text{ L HCl} + 0.0400 \text{ L NaOH}} = 0.0111 \text{ M}$$

pH = 1.954

(2) After 60.0 mL of base have been added, the HCl has been completely consumed, and we have added 10.0 mL of base in excess of the equivalence point.

$$[OH^-] = \frac{\text{moles excess base}}{\text{total volume}}$$
$$= \frac{(0.0100\ L)(0.100\ M)}{0.050\ L\ \text{acid} + 0.060\ L\ NaOH}$$
$$= 9.1 \times 10^{-3}\ M$$

pOH = 2.04, which gives a pH of 11.96

(3) When 49.9 mL of NaOH has been added, we are still 0.1 mL short of the equivalence point. Therefore, $[H_3O^+]$ is calculated as in part (a). This gives $[H_3O^+] = 1.00 \times 10^{-4}$ M, or a pH of 4.000. Even this close to the equivalence point the solution is still relatively acidic.

18.9. The equation for the reaction occurring here is

$$NaOH(aq) + CH_3CO_2H(aq) \rightleftharpoons NaCH_3CO_2(aq) + H_2O(\ell)$$

$$(0.0350\ L\ NaOH)(0.100\ M) = 0.00350\ mol\ NaOH$$

This leads to 0.00350 mol of $NaCH_3CO_2$.

Acid remaining = 0.0100 mol acid − 0.00350 mol NaOH
= 0.00650 mol

$$[H_3O^+] = \frac{[\text{acid remaining}]}{[\text{conjugate base formed}]} K_a$$
$$= \frac{0.0065\ mol}{0.00350}(1.8 \times 10^{-5}) = 3.3 \times 10^{-5}\ M$$

pH = 4.48

18.10. Quantity of NH_3 titrated = (0.100 L)(0.100 M NH_3) = 0.0100 mol NH_3

At the equivalence point, the reaction has produced 0.0100 mol NH_4^+. The titration required 100.0 mL of 0.100 M HCl, so the total solution volume at the equivalence point is 200. mL. Thus, $[NH_4^+] = 0.0500$ M.

$$K_a \text{ for } NH_4^+ = 5.6 \times 10^{-10} = \frac{[H_3O^+][NH_3]}{[NH_4^+]}$$

When $[NH_4^+] = 0.0500$ M, $[H_3O^+] = 5.3 \times 10^{-6}$ M and pH = 5.28.

18.11. The equivalence point occurs at pH of about 5.2 to 5.3. A suitable indicator might be methyl orange, which is yellow in a basic solution but red to red orange in an acidic solution (Figure 17.11). The color change occurs around pH 4.5–5.0. Other choices of indicator would include bromcresol green, bromphenol blue, and methyl red.

CHAPTER 19

Chemical Puzzler: The chemistry of caves is described on pages 747 and 901 as well as in Figures 16.1 and 19.1.

19.1. (1) AgBr, insoluble
(2) K_2CrO_4, soluble
(3) $SrCO_3$, insoluble
(4) $Ba(NO_3)_2$, soluble

19.2. (1) $NaI(aq) + AgNO_3(aq) \longrightarrow NaNO_3(aq) + AgI(s)$
(2) $2\ KCl(aq) + Pb(NO_3)_2(aq) \longrightarrow 2\ KNO_3(aq) + PbCl_2(s)$
(3) No reaction. No insoluble precipitate obtained. [Both $Cu(NO_3)_2$ and $MgCl_2$ are water-soluble.]

19.3. (1) $BaSO_4(s) \rightleftharpoons Ba^{2+}(aq) + SO_4^{2-}(aq)$

$$K_{sp} = [Ba^{2+}][SO_4^{2-}] = 1.1 \times 10^{-10}$$

(2) $BiI_3(s) \rightleftharpoons Bi^{3+}(aq) + 3\ I^-(aq)$

$$K_{sp} = [Bi^{3+}][I^-]^3 = 8.1 \times 10^{-19}$$

(3) $Ag_2CO_3(s) \rightleftharpoons 2\ Ag^+(aq) + CO_3^{2-}(aq)$

$$K_{sp} = [Ag^+]^2[CO_3^{2-}] = 8.1 \times 10^{-12}$$

19.4. $[Ba^{2+}] = 7.5 \times 10^{-3}$ M and $[F^-] = 2 \times [Ba^{2+}] = 1.5 \times 10^{-2}$ M

$$K_{sp} = [Ba^{2+}][F^-]^2 = (7.5 \times 10^{-3})(1.5 \times 10^{-2})^2 = 1.7 \times 10^{-6}$$

19.5. K_{sp} for CuI $= 5.1 \times 10^{-12} = [Cu^+][I^-] = (x)(x)$

Solubility of CuI $= x = (5.1 \times 10^{-12})^{1/2} = 2.2 \times 10^{-6}$ M

K_{sp} for $Mg(OH)_2 = 1.5 \times 10^{-11} = [Mg^{2+}][OH^-]^2 = (x)(2x)^2$

Solubility of $Mg(OH)_2 = [Mg^{2+}] = x = 1.6 \times 10^{-4}$ M

19.6. (1) Solubility of AgCl ($K_{sp} = 1.8 \times 10^{-10}$) is greater than the solubility of AgCN ($K_{sp} = 1.2 \times 10^{-16}$).
(2) Solubility of $Ca(OH)_2$ ($K_{sp} = 7.9 \times 10^{-6}$) is greater than the solubility of $Mg(OH)_2$($K_{sp} = 1.5 \times 10^{-11}$).
(3) Solubility of $MgCO_3$ ($K_{sp} = 4.0 \times 10^{-5}$) is greater than the solubility of $CaCO_3$($K_{sp} = 3.8 \times 10^{-9}$).

19.7. If $[Pb^{2+}] = 1.1 \times 10^{-3}$ M, then $[I^-] = 2\ (1.1 \times 10^{-3}\ M) = 2.2 \times 10^{-3}$ M

$$Q = [Pb^{2+}][I^-]^2 = (1.1 \times 10^{-3}\ M)(2.2 \times 10^{-3}\ M)^2 = 5.3 \times 10^{-9}$$

The value of Q is less than K_{sp} (8.7×10^{-9}), so PbI_2 can dissolve to a greater extent.

19.8. $Q = [Sr^{2+}][SO_4^{2-}] = (2.5 \times 10^{-4}\ M)(2.5 \times 10^{-4}\ M) = 6.3 \times 10^{-8}$

Q is less than K_{sp} so the solution is not yet saturated.

19.9. $$[I^-] = \left\{\frac{K_{sp}}{[Pb^{2+}]}\right\}^{1/2} = \left[\frac{8.7 \times 10^{-9}}{0.050\ M}\right]^{1/2}$$
$$= 4.2 \times 10^{-4}\ M$$

The I^- concentration needed to precipitate the lead(II) ion is 4.2×10^{-4} M. The lead(II) ion concentration remaining in solution when I^- reaches 0.0015 M is

$$[Pb^{2+}] = \frac{K_{sp}}{[I^-]^2} = \frac{8.7 \times 10^{-9}}{(0.0015)^2} = 3.9 \times 10^{-3}\ M$$

19.10. $[Ag^+]$ after mixing $= \dfrac{(0.0010 \text{ M})(0.1000 \text{ L})}{0.1050 \text{ L}}$

$= 9.5 \times 10^{-4}$ M

$[Cl^-]$ after mixing $= \dfrac{(0.025 \text{ M})(0.0050 \text{ L})}{0.1050 \text{ L}}$

$= 1.2 \times 10^{-3}$ M

$Q = [Ag^+][Cl^-] = (9.5 \times 10^{-4} \text{ M})(1.2 \times 10^{-3} \text{ M}) = 1.1 \times 10^{-6}$

Q is greater than K_{sp} (1.8×10^{-10}), so precipitation occurs.

19.11. Solubility in pure water:

$$[Ba^{2+}] = [SO_4^{\;2-}] = (K_{sp})^{1/2} = 1.0 \times 10^{-5} \text{ M}$$

Solubility with added Ba^{2+} ion:

$$[SO_4^{\;2-}] = \frac{K_{sp}}{(x + 0.010 \text{ M})}$$

Assuming x is very small relative to 0.010, we have

Solubility $= 1.1 \times 10^{-8}$ M

19.12. Solubility in pure water:
$[Zn^{2+}] = x$. Because $[CN^-] = 2x$, we have $K_{sp} = 8.0 \times 10^{-12} = (x)(2x)^2$ and $x = 1.3 \times 10^{-4}$ M.
Solubility in the presence of 0.10 M CN^-:
$[Zn^{2+}] = x$. Because $[CN^-] = 2x + 0.10$, we have $K_{sp} = 8.0 \times 10^{-12} = (x)(2x + 0.10)^2 \approx (x)(0.10)^2$. Solving, $x = 8 \times 10^{-10}$ M.

19.13. (1) $[Cl^-]$ needed to precipitate AgCl $= \dfrac{K_{sp}}{[Ag^+]}$

$= \dfrac{1.8 \times 10^{-10}}{0.0010 \text{ M}} = 1.8 \times 10^{-7}$ M

$[Cl^-]$ needed to precipitate $PbCl_2 = \left\{\dfrac{K_{sp}}{[Pb^{2+}]}\right\}^{1/2}$

$= \left(\dfrac{1.7 \times 10^{-5}}{0.0010 \text{ M}}\right)^{1/2} = 0.13$ M

Silver chloride precipitates before $PbCl_2$.

(2) When $[Cl^-] = 0.13$ M, $[Ag^+] = K_{sp}/(0.13) = 1.4 \times 10^{-9}$ M

19.14. (1) Can be separated by adding Cl^- to give AgCl(s) and leave Bi^{3+} in solution.
(2) Can be separated by adding S^{2-} or OH^- to give FeS(s) or $Fe(OH)_2$(s) and leave K^+ in solution because both K_2S and KOH are water-soluble.

19.15. K for the overall reaction $= K_{sp}$ for AgCl $\cdot$ $(1/K_{sp}$ for AgBr)

$= \dfrac{(1.8 \times 10^{-10})}{(3.3 \times 10^{-13})} = 550$

AgCl can be converted to AgBr.

19.16. 0.010 mol AgCl will give 0.010 M $Ag(NH_3)_2^{\;+}$ and 0.010 M Cl^- when dissolved completely in NH_3. Solving for $[NH_3]$ as in Example 19.12 gives $[NH_3] = 0.19$ M. The quantity of NH_3 available $= (0.100 \text{ L})(4.0 \text{ M}) = 0.40$ mol. Only 0.020 mol of NH_3 is required to form 0.010 M $Ag(NH_3)_2^{\;+}$ and to achieve a concentration of 0.19 M. Therefore, there is sufficient NH_3 to dissolve the AgCl completely.

CHAPTER 20

Chemical Puzzler: The objective of this chapter is to explore a new thermodynamic function, the Gibbs free energy, which allows one to predict if a reaction will be product- or reactant-favored.

20.1. Entropy change for liquid to vapor:

$$\Delta S° = \frac{30{,}900 \text{ J/mol}}{353.3 \text{ K}} = +87.5 \text{ J/K} \cdot \text{mol}$$

$\Delta S°$ for vapor $\longrightarrow$ liquid $= -87.5$ J/K $\cdot$ mol

20.2. (1) $S°$ for solid CO_2 is less than for CO_2 vapor. For a given compound, molecules in the vapor state always have higher entropy because the vapor state is more disordered than the solid state.
(2) KCl dissolved in water forms ions separated by water molecules (see Figure 20.6), a more highly disordered state than solid KCl. Thus, the entropy of the dissolved KCl ($S° = 159.0$ J/K $\cdot$ mol) is larger than that of solid KCl ($S° = 82.6$ J/K $\cdot$ mol).
(3) In this reaction a mole of solid has given rise to a mole of solid compound and a mole of gas. The entropy has increased.

20.3. $Ca(s) + C(s) + \frac{3}{2} O_2(g) \longrightarrow CaCO_3(s)$

$\Delta S° = S°[CaCO_3(s)] - \{S°[Ca(s)] + S°[C(\text{graphite})] + \frac{3}{2} S°[O_2(g)]\}$

$= (1 \text{ mol})(92.9 \text{ J/K} \cdot \text{mol}) - [(1 \text{ mol})(41.4 \text{ J/K} \cdot \text{mol}) + (1 \text{ mol})(5.7 \text{ J/K} \cdot \text{mol}) + (\frac{3}{2} \text{ mol})(205.1 \text{ J/K} \cdot \text{mol})]$

$= -261.9$ J/K

Here two solid and one gaseous reactant have been converted to a solid product. The entropy of the system has declined.

20.4. (1) Both $\Delta H°$ and $\Delta S°$ are negative, so the outcome depends on T.
(2) Both $\Delta H°$ and $\Delta S°$ are positive, so the outcome depends on T.
(3) Reaction is exothermic ($\Delta H°$ is negative), and $\Delta S°$ is positive. Reaction is product-favored.
(4) Both $\Delta H°$ and $\Delta S°$ are negative, so the outcome depends on T.

20.5. $\Delta H°_{rxn} = 2\,\Delta H°_f\,[HCl(g)] = -184.6$ kJ

$\Delta S°_{surroundings} = -\left[\dfrac{-184.6 \times 10^3 \text{ J}}{298 \text{ K}}\right] = +620.$ J/K

$\Delta S°_{rxn} = 2\,S°[HCl(g)] - \{S°[H_2(g)] + S°[Cl_2(g)]\}$
$= 2 \text{ mol } (186.9 \text{ J/K} \cdot \text{mol}) - [1 \text{ mol } (130.7 \text{ J/K} \cdot \text{mol}) + 1 \text{ mol } (223.1 \text{ J/K} \cdot \text{mol}]$
$= +20.0$ J/K

$\Delta S°_{universe} = +640.$ J/K

The reaction is predicted to be product-favored.

20.6. $\Delta H^\circ_{rxn} = \Delta H^\circ_f[NH_3(g)] - \{\frac{1}{2}\Delta H^\circ_f[N_2(g)] + \frac{3}{2}\Delta H^\circ_f[H_2(g)]\}$
$= (1\ mol)(-46.11\ kJ/mol) - [\frac{1}{2}\ mol(0) + \frac{3}{2}\ mol(0)]$
$= -46.11\ kJ$
$\Delta S^\circ_{rxn} = S^\circ[NH_3(g)] - \{\frac{1}{2}S^\circ[N_2(g)] + \frac{3}{2}S^\circ[H_2(g)]\}$
$= -99.38\ J/K$

$\Delta G^\circ_{rxn} = \Delta H^\circ_{rxn} - T\Delta S^\circ_{rxn} = -46.11\ kJ - (298\ K)(-99.38\ J/K)(1\ kJ/1000\ J) = -16.5\ kJ$

20.7. (1) $C(graphite) + O_2(g) \longrightarrow CO_2(g)$
(2) -394.4 kJ/mol
(3) $(2.5\ mol)(-394.4\ kJ/mol) = -990\ kJ$

20.8. $C_6H_6(\ell) + \frac{15}{2}O_2(g) \longrightarrow 6\ CO_2(g) + 3\ H_2O(\ell)$

$\Delta G^\circ_{rxn} = 6\ \Delta G^\circ_f[CO_2(g)] + 3\ \Delta G^\circ_f[H_2O(\ell)] - \{\Delta G^\circ_f[C_6H_6(\ell)] + \frac{15}{2}\Delta G^\circ_f[O_2(g)]\}$
$= (6\ mol)(-394.359\ kJ/mol) + (3\ mol)(-237.129\ kJ/mol) - [(1\ mol)(124.5\ kJ/mol) + (\frac{15}{2}\ mol)(0)]$
$= -3202.0\ kJ$

20.9. (1) $CaO(s) + CO_2(g) \longrightarrow CaCO_3(s)$

$\Delta G^\circ_{rxn} = \Delta G^\circ_f[CaCO_3(s)] - \{\Delta G^\circ_f[CaO(s)] + \Delta G^\circ_f[CO_2(g)]\}$
$= (1\ mol)(-1128.79\ kJ/mol) - [(1\ mol)(-604.03\ kJ/mol) + (1\ mol)(-394.359\ kJ/mol)]$
$= -130.40\ kJ$

(2) $CaCO_3(s) \longrightarrow CaO(s) + CO_2(g)$
ΔG° for the decomposition $= -\Delta G^\circ$ for the formation $= +130.40$ kJ The formation reaction is product-favored under standard conditions.

20.10. $MgO(s) + C(graphite) \longrightarrow Mg(s) + CO(g)$

$\Delta H^\circ_{rxn} = +491.18$ kJ and $\Delta S^\circ_{rxn} = 197.67$ J/K
$T = \Delta H^\circ_{rxn}/\Delta S^\circ_{rxn} = 491.18\ kJ/(0.198\ kJ/K) = 2480\ K$ (or about 2200 °C)

20.11. (1) $\Delta G^\circ_{rxn} = \Delta G^\circ_f[SO_2(s)] = -300.194\ kJ/mol$

$(-300.194\ kJ)(1000\ J/kJ) = -(8.314510\ J/K\cdot mol)\ (298\ K)\ \ln K_p$
$\ln K_p = 1.21 \times 10^2$ and so $K_p = 4.15 \times 10^{52}$

(2) $\Delta G^\circ_{rxn} = +130.4$ kJ from Exercise 20.9.

$\ln K_p = 52.6$ and so $K_p = 1.39 \times 10^{-23}$

20.12.

$SnO_2(s) \longrightarrow Sn(s) + O_2(g)$	$\Delta G^\circ_f = +519.6\ kJ$
$C(s) + O_2(g) \longrightarrow CO_2(g)$	$\Delta G^\circ_f = -394.4\ kJ$
$SnO_2(s) + C(s) \longrightarrow Sn(s) + CO_2(g)$	$\Delta G^\circ_f = +125.2\ kJ$

The decomposition of SnO_2 is more product-favored in the presence of carbon. Indeed, in practice tin(IV) oxide is readily reduced by carbon.

CHAPTER 21

Chemical Puzzler: Batteries and fuel cells are described in Section 21.6. The devices produce energy in the form of electricity, using product-favored oxidation–reduction reactions. The reaction of hydrochloric acid and zinc is a product-favored oxidation–reduction reaction, but in contrast to batteries and fuel cells, the energy from this reaction is released as heat.

21.1. *Reduction half-reaction:*
$I_2(s) \longrightarrow 2\ I^-(aq) + 2\ e^-$
Oxidation half-reaction:
$2\ [Cr^{2+}(aq) \longrightarrow Cr^{3+}(aq) + e^-]$
Net reaction: $2\ Cr^{2+}(aq) + I_2(s) \longrightarrow Cr^{3+}(aq) + 2\ I^-(aq)$

I_2 is the oxidizing agent, and Cr^{2+} is the reducing agent. Cr^{2+} is oxidized by I_2, and I_2 is reduced by Cr^{2+}.

21.2. (1) $Co(s) + 3\ NO_3^-(aq) + 6\ H^+(aq) \longrightarrow Co^{3+}(aq) + 3\ NO_2(g) + 3\ H_2O(\ell)$
(2) Cobalt metal is oxidized by nitric acid, and nitric acid is reduced by cobalt metal.

21.3. *Reduction half-reaction:*
$3\ [S(s) + H_2O(\ell) + 2\ e^- \longrightarrow HS^-(aq) + OH^-(aq)]$
Oxidation half-reaction:
$2\ [Al(s) + 3\ OH^-(aq) \longrightarrow Al(OH)_3(s) + 3\ e^-]$
Net: $2\ Al(s) + 3\ S(s) + 3\ OH^-(aq) + 3\ H_2O(\ell) \longrightarrow 2\ Al(OH)_3(s) + 3\ HS^-(aq)$

Aluminum metal, the reducing agent, is oxidized. Sulfur, the oxidizing agent, is reduced.

21.4. *Anode half-reaction (oxidation):*
$Ni(s) \longrightarrow Ni^{2+}(aq) + 2\ e^-$
Cathode half-reaction (reduction):
$e^- + Ag^+(aq) \longrightarrow Ag(s)$
Electrons in the external circuit flow from the anode (Ni) to the cathode (Ag). The NO_3^- ions in the salt bridge flow from the silver half-cell compartment to the anode compartment.

21.5. $\Delta G^\circ_{rxn} = -(2.00\ mol\ e^-)(9.65 \times 10^4\ J/V\cdot mol)(-0.76\ V)(1.0\ kJ/1000\ J) = +150\ kJ$
The reaction is not product-favored as written.

21.6.

$Fe(s) \longrightarrow Fe^{2+}(aq) + 2\ e^-$	$E^\circ = +0.44\ V$
$Cu^{2+}(aq) + 2\ e^- \longrightarrow Cu(s)$	$E^\circ = +0.34\ V$
$Fe(s) + Cu^{2+}(aq) \longrightarrow Fe^{2+}(aq) + Cu(s)$	$E^\circ_{net} = +0.78\ V$

21.7.

$H_2O_2(aq) + 2\ H_3O^+(aq) + 2\ e^- \longrightarrow 4\ H_2O(\ell)$	$E^\circ = +1.77\ V$
$2\ Br^-(aq) \longrightarrow Br_2(\ell) + 2\ e^-$	$E^\circ = -1.08\ V$
$H_2O_2(aq) + 2\ H_3O^+(aq) + 2\ Br^-(aq) \longrightarrow 4\ H_2O(\ell) + Br_2(\ell)$	$E^\circ_{net} = +0.69\ V$

The reaction is product-favored.

21.8.

$$\begin{array}{ll} 2\ Al(s) \longrightarrow 2\ Al^{3+}(aq) + 6\ e^- & E° = +1.66\ V \\ 3\ Zn^{2+}(aq) + 6\ e^- \longrightarrow 3\ Zn(s) & E° = -0.76\ V \\ \hline 2\ Al(s) + 3\ Zn^{2+}(aq) \longrightarrow 2\ Al^{3+}(aq) + 3\ Zn(s) & E°_{net} = +0.90\ V \end{array}$$

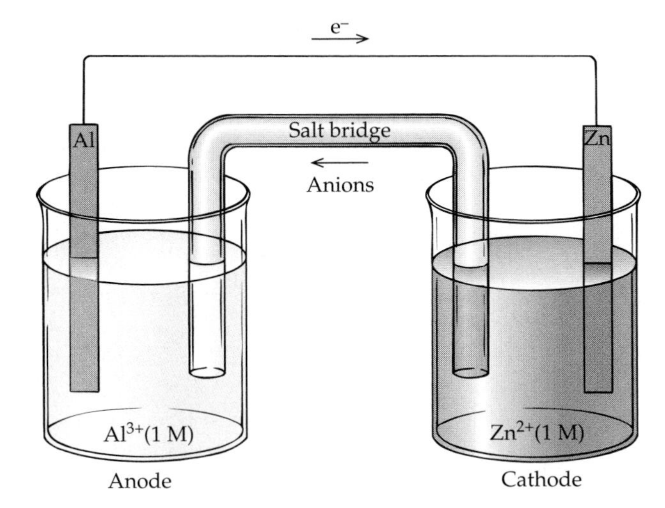

Electrons flow through the external wire from Al (anode) to Zn (cathode). Anions flow through the salt bridge from the Zn^{2+}-containing compartment to the Al^{3+}-containing compartment.

21.9.

$$\begin{array}{ll} 2\ [Ag^+(aq) + e^- \longrightarrow Ag(s)] & E° = +0.800\ V \\ Hg(\ell) \longrightarrow Hg^{2+}(aq) + 2\ e^- & E° = -0.855\ V \\ \hline 2\ Ag^+(aq) + Hg(\ell) \longrightarrow 2\ Ag(s) + Hg^{2+}(aq) & E°_{net} = -0.055\ V \end{array}$$

The negative sign for $E°_{net}$ indicates that reaction is reactant-favored under standard conditions.

$$E_{net} = E°_{net} - \frac{0.0257}{n} \ln \frac{[Hg^{2+}]}{[Ag^+]^2}$$

$$= -0.055 - \frac{0.0257}{2} \ln \frac{[0.0010]}{[0.80]^2}$$

$$= -0.55\ V + 0.083\ V = 0.028\ V$$

The reaction is product-favored under the new conditions.

21.10. $\ln K = \dfrac{nE°}{0.0257\ V} = \dfrac{(2)(-0.055\ V)}{0.0257\ V}$

$\ln K = -4.3$, and so $K = 0.014$. Notice that K is less than 1, as expected for a reaction with a negative $E°$.

21.11. (1) $2\ NaBr(\ell)$ electricity $\longrightarrow 2\ Na(\ell) + Br_2(\ell)$

(2) $NaBr(aq)$ + electricity $\longrightarrow OH^-(aq)$ and $H_2(g)$ from water + $Br_2(\ell)$ from $Br^-(aq)$

(3) $CuCl_2(aq)$ + electricity $\longrightarrow Cu(s) + Cl_2(g)$

21.12. $(25 \times 10^3\ A)(3600\ s/h) = 9.0 \times 10^7\ C/h$

$(9.0 \times 10^7\ C/h)(1\ mol\ e^-/96500\ C) = 9.3 \times 10^2\ mol\ e^-/h$

$(9.3 \times 10^2\ mol\ e^-/h)\left(\dfrac{1\ mol\ Na}{1\ mol\ e^-}\right)\left(\dfrac{23.0\ g}{mol}\right) = 2.2 \times 10^4\ g\ Na/h$

21.13. (See Exercise 21.12.)

$(9.0 \times 10^7\ C)(7.0\ V) = 6.3 \times 10^8\ J$

$(6.3 \times 10^8\ J)\left(\dfrac{1\ kwh}{3.60 \times 10^6\ J}\right) = 180\ kwh$

CHAPTER 22

Chemical Puzzler: Based on standard reduction potentials, aluminum is more easily oxidized than copper. Nonetheless, copper is oxidized by nitric acid, but aluminum is not. Aluminum does not react with nitric acid because the surface of the metal is protected by an unreactive coating of aluminum oxide (see Section 22.5). Copper has no such inert coating.

22.1. (1) $2\ Na(s) + Br_2(\ell) \longrightarrow 2\ NaBr(s)$

(2) $Ca(s) + Se(s) \longrightarrow CaSe(s)$

(3) $2\ Al(s) + 3\ Cl_2(g) \longrightarrow 2\ AlCl_3(s)$ (See the photos of the reaction of aluminum with bromine, page 2.)

(4) $4\ K(s) + O_2(g) \longrightarrow 2\ K_2O(s)$ (K_2O is one of the possible products of this reaction. The primary product from the reaction of potassium and oxygen is KO_2, potassium superoxide (see page 1016).

22.2. (1) H_2Te

(2) Na_2AsO_3

(3) $SeCl_6$

(4) $HBrO_4$

22.3. (1) NH_4^+ (ammonium ion)

(2) O_2^{2-} (peroxide ion)

(3) N_2H_4 (hydrazine)

(4) NF_3 (nitrogen trifluoride)

22.4. (1) ClO is an odd-electron molecule with Cl having the unlikely oxidation number of +2.

(2) Na_2Cl has chlorine with the unlikely charge of 2− (to balance the two positive charges of two Na^+ ions). The ionization energy of Cl (Chapter 8) is high, so loss of two electrons is unlikely.

(3) This compound would require either calcium ion to have the formula Ca^+ or acetate ion to be $CH_3CO_2^{2-}$. In all of its compounds, calcium occurs as the Ca^{2+} ion. The acetate ion, formed from acetic acid by loss of H^+, has a 1− charge.

(4) C_3H_7 is unlikely. No structure can be drawn that has seven H atoms and four bonds to each C atom (see Chapter 11). Notice that this formula has an odd number of electrons.

CHAPTER 23

Chemical Puzzler: The metal is titanium, and the positive identification can be made by comparing the data for density and melting point with literature values. Titanium's electron configuration is $[Ar]3d^2\ 4s^2$ and in its common compounds titanium occurs in the +2, +3, and +4 oxidation states. The properties of titanium, including in particular its inertness toward chemical reagent, make it an ideal structural material. Titanium (IV) oxide, TiO_2, is the common pigment in white paint.

23.1. (1) Platinum in $Pt(NH_3)_2Cl_2$ has an oxidation number of +2.

(2) $[Co(NH_3)_5CO_3]^+$

23.2. (1) Aquacyanobis(phenanthroline)ruthenium(II) chloride

(2) Diamminedichloroplatinum(II)

23.3. The complex $[Co(en)_2(CN)_2]^+$ exists as *cis* and *trans* geometric isomers. The *cis* isomer has two enantiomeric forms (two optical isomers, mirror images that do not superimpose).

(1) *cis–trans* isomers.

trans *cis*

(2) Optical isomers

23.4. (1) Ruthenium is in the +2 oxidation state. The Ru^{2+} ion has a d^6 configuration. The high-spin state has four unpaired electrons, and the low-spin state has zero unpaired electrons. Low-spin Ru^{2+} is diamagnetic; high-spin Ru^{2+} is paramagnetic.

high-spin Ru^{2+}: $d_{x^2-y^2}$ ↑, d_{z^2} ↑; d_{xy} ↑↓, d_{xz} ↑, d_{yz} ↑

low-spin Ru^{2+}: $d_{x^2-y^2}$ —, d_{z^2} —; d_{xy} ↑↓, d_{xz} ↑↓, d_{yz} ↑↓

(2) Ni^{2+} has a d^8 configuration. Thus, there is only one way to arrange electrons in the two sets of *d* orbitals created by octahedral geometry. This electronic configuration has two unpaired electrons; consequently, the Ni^{2+} ion is paramagnetic.

Ni^{2+} ion (d^8): $d_{x^2-y^2}$ ↑, d_{z^2} ↑; d_{xy} ↑↓, d_{xz} ↑↓, d_{yz} ↑↓

CHAPTER 24

Chemical Puzzler: Radon, element 86, is one of the products of radioactive decay of uranium. Radon is one of the noble gases and like uranium, it is also radioactive. It is an α-particle emitter, and its radioactivity poses a health hazard. An unreactive substance (as are all the noble gases), it is not trapped by chemical processes. It seeps through the soil, and it can enter your basement through cracks in the floor or walls. Because it is quite a bit denser than air, it tends to accumulate there. If the concentration becomes too high, it is important to take steps to remove it.

24.1. (1) The final product in this radioactive decay series is $^{208}_{82}Pb$, which is a stable (nonradioactive) isotope.

(2) $^{232}_{90}Th \longrightarrow ^{228}_{88}Ra + ^{4}_{2}He$

$^{228}_{88}Ra \longrightarrow ^{228}_{89}Ac + ^{0}_{-1}\beta$

$^{228}_{89}Ac \longrightarrow ^{228}_{90}Th + ^{0}_{-1}\beta$

24.2. (1) $^{13}_{7}N \longrightarrow ^{13}_{6}C + ^{0}_{+1}\beta$

(2) $^{41}_{20}Ca + ^{0}_{-1}e \longrightarrow ^{41}_{19}K$

(3) $^{90}_{38}Sr \longrightarrow ^{90}_{39}Y + ^{0}_{-1}\beta$

(4) $^{22}_{11}Na \longrightarrow ^{22}_{10}Ne + ^{0}_{+1}\beta$

24.3. (1) β decay: $^{32}_{14}Si \longrightarrow ^{0}_{-1}\beta + ^{32}_{15}P$

(2) Positron emission: $^{45}_{22}Ti \longrightarrow ^{0}_{+1}\beta + ^{45}_{21}Sc$

(3) α emission: $^{239}_{94}Pu \longrightarrow ^{4}_{2}He + ^{235}_{92}U$

(4) β decay: $^{42}_{19}K \longrightarrow ^{0}_{-1}\beta + ^{42}_{20}Ca$

24.4. Mass defect = $\Delta m = -0.03438$ g/mol

$\Delta E = (-3.438 \times 10^{-5}$ kg/mol$)(2.998 \times 10^8$ m/s$)^2$

$= -3.090 \times 10^{12}$ J/mol

E_b/mol of nucleons = 5.150×10^8 kJ/mol of nucleons

24.5. (1) Disintegrations are reduced to 1000 after 1 half-life, to 500 after 2 half-lives, to 250 after 3 half-lives. Thus, $3 \times 28 = 84$ years is required.

(2) To reach 1/16th of the original activity will require 4 half-lives, $(1/2)^4 = 1/16$; thus $4 \times 28 = 112$ years will be required.

24.6. $k = 0.693/t_{1/2} = 8.90 \times 10^{-3}\ h^{-1}$

$\ln\left(\frac{2.92 \times 10^5}{3.28 \times 10^5}\right) = -kt = -k(1.0\ h)$; solving, $k = 0.116\ h^{-1}$

$$t_{1/2} = \frac{0.693}{k} = \frac{0.693}{(0.116\ h^{-1})} = 5.97\ h$$

24.7. $$k = \frac{0.693}{t_{1/2}} = \frac{0.693}{78.3\ h} = 8.85 \times 10^{-3}\ h^{-1}$$

$\ln\left(\frac{0.10}{1.00}\right) = -(8.85 \times 10^{-3}\ h^{-1})t$; $t = 2.60 \times 10^2$ h

24.8. $\ln\left(\frac{12.9}{14.0}\right) = -(1.21 \times 10^{-4}/y)t$

$t = 676$ years. Because the accuracy of this dating method is around ± 100 years, the artifact is approximately 700 years old.

24.9. (1) $^{13}_{6}C + ^{1}_{0}n \longrightarrow ^{4}_{2}He + ^{10}_{4}Be$

(2) $^{14}_{7}N + ^{4}_{2}He \longrightarrow ^{1}_{0}n + ^{17}_{9}F$

(3) $^{253}_{99}Es + ^{4}_{2}He \longrightarrow ^{1}_{0}n + ^{256}_{101}Md$

(4) $^{27}_{13}Al + ^{1}_{1}H \longrightarrow ^{28}_{14}Si$

Appendix O

Answers to Study Questions

CHAPTER 1

1.4. Mercury and water are both liquids. Liquid mercury is the densest substance here: d (mercury) = 13.55 g/cm^3; d (copper) = 8.96 g/cm^3; d (water) = 1.0 g/cm^3.

1.6. (a) Physical
(b) Chemical
(c) Chemical
(d) Physical
(e) Physical
(f) Physical

1.8. (a) 3; (b) 3; (c) 3; (d) 3

1.10. Qualitative: clear, colorless
Quantitative: density, mass, length
Extensive: mass and length
Intensive: density, color, transparency

1.12. 557 g

1.14. 499 g; 0.499 kg

1.16. 2.68 g/cm^3. The metal is most likely aluminum.

1.18. 0.865 g/cm^3

1.20. 298 K

1.22. (a) 289 K; (b) 97 °C; (c) 233 K

1.24. Body temperature is approximately 98.6 °F, or about 37 °C. This is higher than the melting temperature of gallium.

1.26. (a) Carbon
(b) Sodium
(c) Chlorine
(d) Phosphorus
(e) Magnesium
(f) Calcium

1.28. (a) Li
(b) Ti
(c) Fe
(d) Si
(e) Pb
(f) Zn

1.30. 1.5 km; 1.5×10^5 cm

1.32. 5.3 cm^2; 5.3×10^{-4} m^2

1.34. 250. cm^3; 0.250 L; 2.50×10^{-4} m^3

1.36. 2520 g

1.38. 22 cm; 28 cm; 6.0×10^2 cm^2

1.40. (a) Method A: average density = 2.4 g/cm^3; average deviation = 0.2 g/cm^3
Method B: average density = 3.480 g/cm^3; average deviation = 1.166 g/cm^3. It appears that the person using method B misread the micrometer or miscalculated the density in the fourth trial. That data point is so far from the other three data points that it should be excluded. Without this data point, density is 2.703 ± 0.001g.
(b) Method A: error = −0.3 g/cm^3
Method B: error = +0.001 g/cm^3 after excluding the fourth data point.
(c) Method B is more precise (smaller deviation) after omitting the fourth data point. Method B is also more accurate (closer to the accepted value of density) after omitting the fourth data point.

1.42. 6.6×10^4 cm^3

1.44. 0.236

1.46. 0.294

1.48. 0.0854 cm^3

1.50. 1.54×10^{-10} m; 1.54 Å

1.52. 22 g; $290

1.54. Calculated density = density of nickel

1.56. 600 g of water occupies a greater volume (600 cm^3) than 600 g of lead (50 cm^3).

1.58. The density at 20 °C should be midway between 0.99913 and 0.99707 g/cm^3. Thus, a density of 0.99910 g/cm^3 is not reasonable.

1.60. 0.995 g platinum

1.62. 255 mL

1.64.

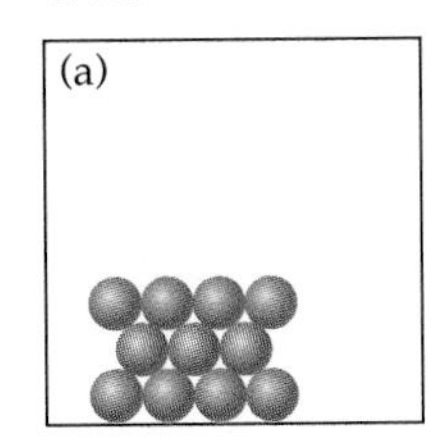

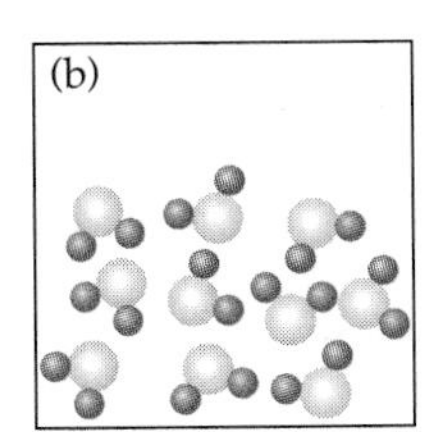

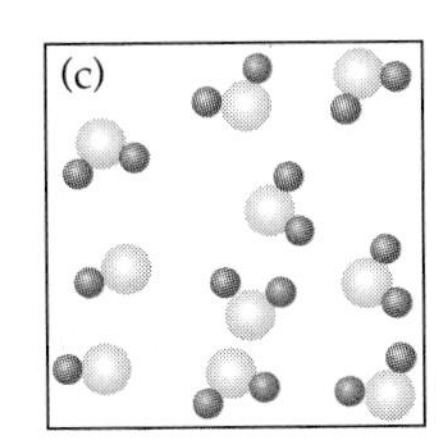

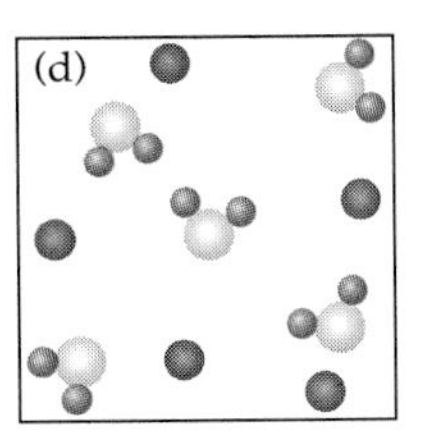

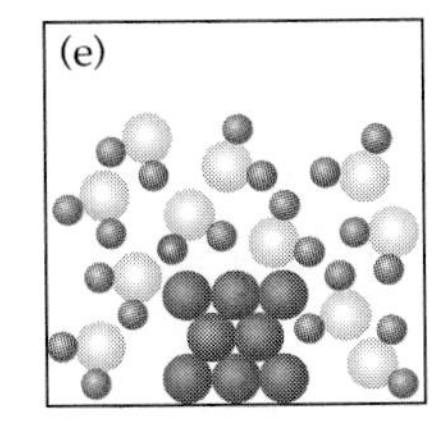

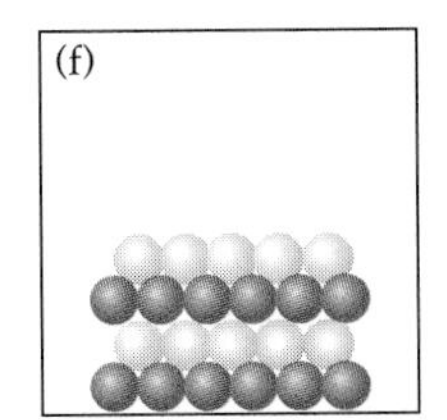

1.66. One could check for an odor, calculate the density, or perhaps check the boiling and freezing points of the liquid. If the density is approximately 1 g/cm^3 at room temperature, the liquid could be water. If it boils at about 100 °C and freezes at about 0 °C, that would be consistent with water. To check for the presence of salt, boil the liquid away. If a substance remains, it could be salt, but further testing would be required.

1.68.

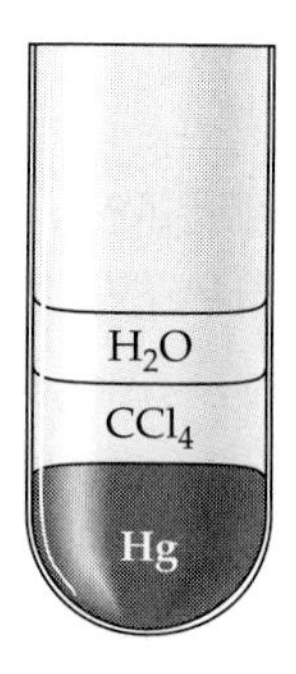

1.70. A copper-colored metal could be copper, but it may also be an alloy of copper, for example, brass or bronze. Testing the material's density and melting temperature would be one way to find out if it is copper.

1.72. 1.8×10^{-2} mm

1.74. Assume 57 kg: length = 9.0×10^1 m

1.76. Thickness = 2×10^{-7} cm. If the oil molecules spread out to form a single layer, the thickness could be the length of the oil molecule.

1.78. (a) Although not completely obvious in the photo, water is being dropped onto solid potassium metal. It produces hydrogen gas and the chemical compound potassium hydroxide. Steam is also seen because the heat of the chemical reaction vaporizes the water.
(b) Chemical changes: potassium metal reacts with water (to produce hydrogen gas and potassium hydroxide).
Physical changes: water is converted to steam. (It is also likely that the hydrogen gas produced in the reaction reacts with the oxygen in the air to produce water vapor.)
(c) Potassium reacts violently with water. The reaction produces heat and light.
(d) At the particulate level, potassium atoms are pictured as spheres stacked in a very regular arrangement.

CHAPTER 2

2.20. (a) 23; (b) 48; (c) 72

2.22. (a) $^{39}_{19}K$; (b) $^{39}_{18}Ar$; (c) $^{60}_{27}Co$

2.24. (a) 12 p, 12 n, 12 e^-
(b) 50 p, 69 n, 50 e^-
(c) 94 p, 150 n, 94 e^-

2.26.

Symbol	^{58}Ni	^{33}S	^{20}Ne	^{55}Mn
Number of protons	28	16	10	25
Number of neutrons	30	17	10	30
Number of electrons in the neutral atom	28	16	10	25
Name of element	nickel	sulfur	neon	manganese

2.28. 43 p, 56 n, 43 e^-

2.30. ^{57}Co, ^{58}Co, ^{60}Co

2.32.
$$\text{Atomic mass} = \left(\frac{7.50}{100}\right)(6.015121 \text{ amu}) + \left(\frac{92.50}{100}\right)(7.016003 \text{ amu})$$
$$= 0.451 \text{ amu} + 6.490 \text{ amu}$$
$$= 6.941 \text{ amu}$$

2.34. ^{69}Ga is 60.12% abundant and ^{71}Ga is 39.88% abundant.

2.36. Thallium has an average atomic mass of 204.3833 amu. This lies closer to 205, so ^{205}Tl must be more abundant. (^{205}Tl has a natural abundance of 70.48%.)

2.38. Group 5A has five elements
Nitrogen, N, nonmetal
Phosphorus, P, nonmetal
Arsenic, As, metalloid
Antimony, Sb, metalloid
Bismuth, Bi, metal

2.40. Two periods have 8 elements, two periods have 18 elements, and two have (or can have) 32 elements.

2.42. In Group 2A, the common compound lime is CaO, so we expect the formula of barium oxide to be BaO.

2.44. The average atomic mass of potassium is 39.0983 amu. This is very close to the mass of ^{39}K, so this isotope must be significantly more abundant than ^{40}K.

2.46. 1 = S, 2 = N, 3 = B, and 4 = I

2.48. (a) Co, Ni, and Cu have the highest density (about 9 g/cm^3). All are metals.
(b) In the second period, boron has the highest density. In the third period, aluminum, another Group 3A element, has the highest density.
(c) Of the first 36 elements, the elements that do not appear on this chart are gases: hydrogen, helium, nitrogen, oxygen, fluorine, neon, chlorine, argon, and krypton.

2.50. (a) Group 2A: could be beryllium, magnesium, calcium, strontium, barium, or radium.
(b) Third period: the period begins with sodium (Na) and ends with argon (Ar).
(c) Carbon
(d) Sulfur
(e) Iodine
(f) Magnesium
(g) Krypton
(h) Sulfur
(i) Germanium or arsenic

2.52. Because elements in the same group react similarly, one would expect the formula of the Group 2A oxides would be MO, where M is Be, Mg, Ca, Sr, Ba, or Ra.

2.54. In general, element abundance decreases as atomic number increases, with the exception of the transition metals Sc through Fe. Excluding some of the lightest elements (H through B), elements of even atomic number are more abundant than elements of odd atomic number.

2.56. (a) Solid element such as Fe = 5
(b) Liquid such as mercury = 4
(c) Oxygen (O_2) gas = 2
(d) Bronze, a solid mixture of Cu and Zn = 3
(e) Mixture of N_2 and Ne gases = 1

2.58. Ozone has three O atoms and is a bent (nonlinear) molecule. The O—O distances are both the same (116 pm). The O—O—O angle is about 121°.

2.60. The structures of the diamond allotrope of carbon and of elemental silicon are very similar. Both are extended networks of C or Si atoms. Each atom of C or Si is surrounded by a tetrahedron of other atoms. The tetrahedra are arranged in six-member rings. The C—C distance in diamond is 235 pm, whereas it is 357 pm in silicon.

2.62. 1.704 g of P_4O_{10} consists of 0.744 g P and 0.960 g O. If 4 "units" of P have a mass of 0.744 g, then each "unit" has a mass of 0.186 g. Likewise, for O, it has a mass of 0.0960 g per "unit." If 0.0960 g of O represents a mass of 16.000 amu, then 0.186 g must represent a unit of P. Thus, the mass of P must be 31.0 amu, very close to the "official" average atomic mass of P.

$$0.186 \text{ g P}/0.0960 \text{ g O} = x \text{ amu of P}/16.000 \text{ amu of O}$$

$$x \text{ amu} = 31.0$$

2.64. (a) 1.59×10^{-19} C
(b) Drop 1 = 1 e^-; 2 = 7 e^-; 3 = 6 e^-; 4 = 10 e^-; 5 = 4 e^-;
(c) Average deviation = 0.00; error = 0.6%

CHAPTER 3

3.2. $Pt(NH_3)_2Cl_2$: Each molecule has 2 N atoms and 6 H atoms. In 1 mol of the compound there are (6 H atoms/molecule) · (6.022×10^{23} molecules/mol) = 3.61×10^{24} H atoms/mol. The molar mass is 300.0 g/mol.

3.4. Each Sr atom has 38 electrons. An Sr atom can lose 2 electrons to form Sr^{2+}. The Sr^{2+} ion has the same number of electrons as a Kr atom (36 e^-).

3.6. (a) 40 g of vanillin ($C_8H_8O_3$, molar mass = 152.1 g/mol) represents 0.26 mol of the compound, whereas 3.0×10^{23} molecules represents about 0.5 mol.
(b) All have a C_6 ring.

3.8. 0.5 mol Si

3.10. C_2H_6 is the molecular formula of ethane, whereas its empirical formula is CH_3.

3.12. (a) B_4H_{10}; (b) $C_6H_8O_6$; (c) $C_{14}H_{18}N_2O_5$

3.14. (a) 1 calcium atom, 2 carbon atoms, and 4 oxygen atoms
(b) 8 carbon atoms and 8 hydrogen atoms
(c) 2 copper atoms, 1 carbon atom, 5 oxygen atoms, and 2 hydrogen atoms
(d) 1 cobalt atom, 6 nitrogen atoms, 18 hydrogen atoms, and 3 chlorine atoms
(e) 4 potassium atoms, 1 iron atom, 6 carbon atoms, and 6 nitrogen atoms

3.16. Alanine, $C_3H_7NO_2$; lactic acid, $C_3H_6O_3$

3.18. Sulfuric acid, H_2SO_4, is not flat. Check the model on the *Saunders Interactive General Chemistry CD-ROM* (see *Models* and then *Inorganic*), and see that the S atom and two of the O atoms are in a plane. The other two O atoms (and attached H atoms) are above or below that plane.

3.20. (a) Mg^{2+}; (b) Zn^{2+}; (c) Fe^{2+} or Fe^{3+}; (d) Ga^{3+}

3.22. (a) Sr^{2+} (e) S^{2-}
(b) Ti^{4+} (f) ClO_4^-
(c) Al^{3+} (g) Co^{2+}
(d) HCO_3^- (h) NH_4^+

3.24. Potassium loses one electron to form a K^+ ion, which has the same number of electrons Ar, argon.

3.26. Ba, Group 2A, should form Ba^{2+}. Br, Group 7A, should form Br^-. Barium bromide has the formula $BaBr_2$.

3.28. (a) Two K^+ ions and one S^{2-} ion
(b) One Ni^{2+} ion and one SO_4^{2-} ion
(c) Three NH_4^+ ions and one PO_4^{3-} ion
(d) One Ca^{2+} ion and two ClO^- ions
(e) One K^+ ion and one MnO_4^- ion

3.30. Co^{2+} forms CoO and Co^{3+} forms Co_2O_3

3.32. (a) Al, Group 3A, forms a 3+ ion, and the Group 7A atom Cl should form Cl^-. The formula should therefore be $AlCl_3$.
(b) F, Group 7A, forms a 1- ion, and the Group 1A atom Na forms Na^+. The formula should therefore be NaF.
(c) Ga_2O_3 is a correct formula for the Group 3A metal Ga.
(d) MgS is the correct formula for an element of Group 2A (Mg) and one from Group 6A (S).

3.34. The force of attraction is stronger in NaF than in NaI because the distance between ion centers is smaller in NaF (235 pm) than in NaI (322 pm).

3.36. (a) Potassium sulfide
(b) Nickel(II) sulfate
(c) Ammonium phosphate
(d) Calcium hypochlorite

3.38. (a) $(NH_4)_2CO_3$
(b) CaI_2
(c) $CuBr_2$
(d) $AlPO_4$
(e) $AgCH_3CO_2$

3.40. With K^+ ion: K_2CO_3, potassium carbonate; KBr, potassium bromide; KNO_3, potassium nitrate.
With Ba^{2+}: $BaCO_3$, barium carbonate; $BaBr_2$, barium bromide; $Ba(NO_3)_2$, barium nitrate.

With NH_4^+: $(NH_4)_2CO_3$, ammonium carbonate; NH_4Br, ammonium bromide; NH_4NO_3, ammonium nitrate.

3.42. (a) Nitrogen trifluoride
(b) Hydrogen iodide
(c) Boron tribromide
(d) Phosphorus pentafluoride

3.44. (a) SCl_2
(b) N_2O_5
(c) $SiCl_4$
(d) B_2O_3

3.46. (a) 27 g B
(b) 6.98×10^{-2} g Fe
(c) 0.48 g O_2
(d) 2610 g He

3.48. (a) 1.9998 mol Cu
(b) 3.1×10^{-4} mol K
(c) 2.1×10^{-5} mol Am
(d) 0.250 mol Al

3.50. 9.42×10^{-5} mol Kr; 5.67×10^{19} atoms

3.52. 1.0552×10^{-22} g Cu per atom

3.54. 2.97 cm^3; 1.44 cm

3.56. (a) 159.7 g/mol
(b) 117.2 g/mol
(c) 290.8 g/mol
(d) 176.1 g/mol

3.58. (a) 0.0166 mol; (b) 0.00555 mol;
(c) 0.00406 mol

3.60. 60.9 mol CH_3CN

3.62. (a) 0.00180 mol aspirin; 0.02266 mol $NaHCO_3$; 0.005205 mol citric acid
(b) 1.08×10^{21} molecules of aspirin

3.64. (a) 86.60% Pb and 13.40% S
(b) 81.71% C and 18.29% H
(c) 35.00% N, 5.037% H, and 59.96% O

3.66. (a) 305.4 g/mol
(b) 1.8×10^{-4} mol
(c) 70.79% C, 8.912% H, 4.587% N, and 15.71% O
(d) 39 mg carbon

3.68. $C_4H_6O_4$

3.70. Empirical formula = CH; molecular formula = C_2H_2

3.72. N_2O_3

3.74. Both the empirical and molecular formulas are $C_8H_8O_3$

3.76. Empirical formula = C_2H_6As; molecular formula = $C_4H_{12}As_2$

3.78. $KAl(SO_4)_2 \cdot 12\ H_2O$

3.80. XeF_2

3.82. 2.31 kg PbS

3.84. 75.1 kg $Ca_3(PO_4)_2$

3.86. 2×10^{21} molecules of water

3.88. 100. mg of iron(II) sulfate contains 36.8 mg of iron, whereas 100. mg of iron(II) gluconate contains 12.5 mg of iron.

3.90. If 17.46% Fe, then empirical formula is $Fe(C_2O_4)_3$. If 66.60% Fe, then empirical formula is FeC_2O_4.

3.92. 6.00 g H_2O

3.94. (a) Unlikely; two nonmetals
(b) Unlikely; two nonmetals
(c) Li_2S, lithium sulfide
(d) In_2O_3, indium oxide
(e) Unlikely; noble gas argon not likely to form compounds of any type
(f) Unlikely; two nonmetals
(g) MgF_2, magnesium fluoride

3.96. (a) NaClO, ionic
(b) BI_3, not ionic; two nonmetals
(c) $Al(ClO_4)_3$, ionic
(d) $Ca(CH_3CO_2)_2$, ionic
(e) $KMnO_4$, ionic
(f) $(NH_4)_2SO_3$, ionic
(g) KH_2PO_4, ionic
(h) S_2Cl_2, not ionic; two nonmetals
(i) ClF_3, not ionic; two nonmetals
(j) PF_3, not ionic; two nonmetals

3.98. Empirical formula = C_5H_4; molecular formula = $C_{10}H_8$

3.100. (a) Empirical formula = ICl_3; molecular formula = I_2Cl_6

3.102. 7.35 kg Fe

3.104. (d) is impossible. 3.43×10^{-27} mol of S_8 would be 0.00207 molecules, an impossibility.

3.106. M has a molar mass of 47.9 g/mol and so is most likely Ti.

3.108. V_2S_3

3.110. Al^{3+} has the largest positive charge and so should most strongly attract water (relative to Mg^{2+} and Na^+). As described in Chapter 8, Al^{3+} also has the smallest radius, which will also lead to the strongest force of attraction.

3.112. Their calculation would lead to a value of 0.77 mol of water to 1.0 mol of $CaCl_2$ (if they assume 0.739 g is the mass of pure $CaCl_2$). Of course, the correct value is 2.0 mol of water to 1.0 mol of $CaCl_2$. They therefore should reheat the crucible to drive off more water.

3.114. All of these acids have one (formic and acetic acid) or two (oxalic acid) groups of the type —CO_2H, called a carboxylic acid group. (On page 102 we referred to these as functional groups.) The O—C—O angle is about 120° in each one.

3.116. Z molar mass = 18 g/mol and A molar mass = 26 g/mol

3.118. (a) 1.0028×10^{23} C atoms; (b) 1.0028×10^{23} C atoms

3.120. (a) 0.0130 mol Ni
(b) NiF_2
(c) nickel(II) fluoride

CHAPTER 4

4.2. (c) $N_2(g) + 3\ H_2(g) \longrightarrow 2\ NH_3(g)$

4.4. 4 mol N_2 (2 mol NH_3/1 mol N_2)

4.6. (a) $4\ Cr(s) + 3\ O_2(g) \longrightarrow 2\ Cr_2O_3(s)$
(b) $Cu_2S(s) + O_2(g) \longrightarrow 2\ Cu(s) + SO_2(g)$

(c) $C_6H_5CH_3(\ell) + 9\,O_2(g) \longrightarrow 4\,H_2O(g) + 7\,CO_2(g)$

4.8. (a) $Fe_2O_3(s) + 3\,Mg(s) \longrightarrow 3\,MgO(s) + 2\,Fe(s)$
Mg, magnesium oxide; Fe, iron
(b) $AlCl_3(s) + 3\,H_2O(\ell) \longrightarrow Al(OH)_3(s) + 3\,HCl(g)$
$AlOH_3$, aluminum hydroxide; HCl, hydrogen chloride
(c) $2\,NaNO_3(s) + H_2SO_4(\ell) \longrightarrow Na_2SO_4(s) + 2\,HNO_3(g)$
Na_2SO_4, sodium sulfate; HNO_3, nitric acid
(d) $NiCO_3(s) + 2\,HNO_3(aq) \longrightarrow Ni(NO_3)_2(aq) + CO_2(g) + H_2O(\ell)$
$Ni(NO_3)_2$, nickel(II) nitrate; CO_2, carbon dioxide; H_2O, water

4.10. (a) $CO_2(g) + 2\,NH_3(g) \longrightarrow NH_2CONH_2(s) + H_2O(\ell)$
(b) $UO_2(s) + 4\,HF(aq) \longrightarrow UF_4(s) + 2\,H_2O(aq)$
$UF_4(s) + F_2(g) \longrightarrow UF_6(s)$
(c) $TiO_2(s) + 2\,Cl_2(g) + 2\,C(s) \longrightarrow TiCl_4(\ell) + 2\,CO$
$TiCl_4(\ell) + 2\,Mg(s) \longrightarrow Ti(s) + 2\,MgCl_2(s)$

4.12. 4.5 mol O_2; 310 g Al_2O_3

4.14. (a) CO_2 and water
(b) $CH_4(g) + 2\,O_2(g) \longrightarrow CO_2(g) + 2\,H_2O(g)$
(c) 63.99 g O_2
(d) 80.03 g products

4.16. 22.7 g Br_2; 25.3 g Al_2Br_6

4.18. (a) $4\,Fe(s) + 3\,O_2(g) \longrightarrow 2\,Fe_2O_3(s)$
(b) 3.83 g Fe_2O_3
(c) 1.15 g O_2

4.20. 71.2 mg acetone

4.22. (a) Cl_2 is the limiting reagent
(b) 5.08 g $AlCl_3$ produced
(c) 1.67 g Al remains

4.24. (a) CO is the limiting reagent
(b) 1.3 g H_2 in excess
(c) Theoretical yield = 85.2 g methanol

4.26. (a) 68.0 g NH_3
(b) 10. g NH_4Cl remains

4.28. 75.7% yield

4.30. 74.0 g $ZnCl_2$; 88.1% yield

4.32. 91.9% hydrate

4.34. 83.9%

4.36. Empirical formula = CH

4.38. Empirical formula = $C_{10}H_{20}O$

4.40. SiH_4

4.42. 0.133 g O_2 required; 0.183 g CO_2 and 0.075 g H_2O produced

4.44. (a) HCl is the limiting reactant
(b) 10.1 g H_2O formed

4.46. (a) $2\,Fe(s) + 3\,Cl_2(g) \longrightarrow 2\,FeCl_3(s)$
(b) 19.0 g Cl_2 required; 29.0 g $FeCl_3$ produced
(c) 63.8% yield of $FeCl_3$

4.48. (a) $TiCl_4$, titanium(IV) chloride; H_2O, water; TiO_2, titanium(IV) oxide; HCl, hydrogen chloride
(b) 4.60 g H_2O
(c) 10.2 g TiO_2; 18.6 g HCl

4.50. The balanced equation shows 1 mol of I_2 requires 3 mol of Cl_2 and forms 3 mol of ICl_3. Because there are 4 mol of I_2 in the original mixture, 3 mol remains. There was 4 mol of Cl_2 in the original mixture, so only 1 mol remains after using 3 mol. The net result of the reaction is represented by (b).

4.52. Initially the limiting reactant is baking soda; however, as more and more vinegar (acetic acid) is added, the baking soda is consumed. When all of the soda has been consumed, any additional vinegar will not produce more CO_2 gas. 5.0 g of $NaHCO_3$ requires 3.57 g of acetic acid for complete reaction.

4.54. Metal is most likely copper, Cu

4.56. Ti_2O_3

4.58. (a) $2\,Ag(s) + 2\,H_2SO_4(aq) \longrightarrow Ag_2SO_4(s) + SO_2(g) + 2\,H_2O(\ell)$
(b) $Au(s) + HNO_3(aq) + 4\,HCl(aq) \longrightarrow HAuCl_4(aq) + NO(g) + 2\,H_2O(\ell)$
(c) $Zn(s) + 2\,NaOH(aq) + 2\,H_2O(\ell) \longrightarrow Na_2Zn(OH)_4(aq) + H_2(g)$

4.60. (a) In the reactions represented by the sloping portion of the graph, Fe is the limiting reactant. At the point at which the yield of product begins to be constant (2.0 g Fe), the reactants are present in stoichiometric amounts. That is, 10.6 g of product contains 2.0 g Fe and 8.6 g Br_2.
(b) 2.0 g Fe = 0.036 mol Fe; 8.6 g Br_2 = 0.054 mol Br_2. The mol ratio is 1.5 mol Br_2 to 1.0 mol Fe.
(c) A mol ratio of 1.5 Br_2/1.0 mol Fe = 3 Br/1 Fe. The empirical formula is $FeBr_3$.
(d) $2\,Fe(s) + 3\,Br_2(\ell) \longrightarrow 2\,FeBr_3(s)$
(e) iron(III) bromide
(f) Statement (i) is correct.

CHAPTER 5

5.6. Soluble: $Ni(NO_3)_2$, $NiCl_2$. Insoluble: $NiCO_3$, $Ni_3(PO_4)_2$

5.8. (a) CuS, copper(II) sulfide
(b) $CaCO_3$, calcium carbonate
(c) silver iodide, AgI

5.12. Br^- has been oxidized to Br_2, and Cl_2 has been reduced to Cl^-. The oxidizing agent is Cl_2, and the reducing agent is Br^-.

5.16. 0.20 M $BaCl_2$ has an ion concentration of 0.60 M, whereas 0.25 M NaCl has an ion concentration of only 0.50 M.

5.18. (a) $FeCl_2$
(b) $AgNO_3$
(c) All three soluble in water

5.20. (a) LiCl, NaCl, KCl
(b) Any transition metal hydroxide [e.g., $Fe(OH)_3$] or such compounds as $Al(OH)_3$.
(c) $CaCO_3$ (or almost any transition metal carbonate)
(d) $NaNO_3$ or NH_4NO_3

5.22. (a) K^+ and OH^-
(b) K^+ and SO_4^{2-}
(c) Na^+ and NO_3^-

(d) NH_4^+ and SO_4^{2-}

5.24. (a) Soluble, Na^+ and CO_3^{2-}
(b) Soluble, Cu^{2+} and SO_4^{2-}
(c) Not soluble
(d) Soluble, Ca^{2+} and Br^-

5.26. Soluble: $Cu(NO_3)_2$, $CuCl_2$ (or other anions such as Br^-, SO_4^{2-}, ClO_4^-)
Insoluble: $CuCO_3$, $Cu_3(PO_4)_2$ (or other anions such as OH^-, O^{2-}, S^{2-})

5.28. $CdCl_2(aq) + 2\ NaOH(aq) \longrightarrow Cd(OH)_2(s) + 2\ NaCl(aq)$
$Cd^{2+}(aq) + 2\ OH^-(aq) \longrightarrow Cd(OH)_2(s)$

5.30. (a) $NiCl_2(aq) + (NH_4)_2S(aq) \longrightarrow NiS(s) + 2\ NH_4Cl(aq)$
(b) $3\ Mn(NO_3)_2(aq) + 2\ Na_3PO_4(aq) \longrightarrow Mn_3(PO_4)_2(s) + 6\ NaNO_3(aq)$

5.32. $Pb(NO_3)_2(aq) + 2\ KOH(aq) \longrightarrow Pb(OH)_2(s) + 2\ KNO_3(aq)$
Lead(II) nitrate and potassium hydroxide produce lead(II) hydroxide and potassium nitrate.

5.34. (a) $(NH_4)_2CO_3(aq) + Cu(NO_3)_2(aq) \longrightarrow CuCO_3(s) + 2\ NH_4NO_3(aq)$
$CO_3^{2-}(aq) + Cu^{2+}(aq) \longrightarrow CuCO_3(s)$
(b) $Pb(OH)_2(s) + 2\ HCl(aq) \longrightarrow PbCl_2(s) + 2\ H_2O(\ell)$
$Pb(OH)_2(s) + 2\ H^+(aq) + 2\ Cl^-(aq) \longrightarrow PbCl_2(s) + 2\ H_2O(\ell)$
(c) $BaCO_3(s) + 2\ HCl(aq) \longrightarrow BaCl_2(aq) + H_2O(\ell) + CO_2(g)$
$BaCO_3(s) + 2\ H^+(aq) \longrightarrow Ba^{2+}(aq) + H_2O(\ell) + CO_2(g)$

5.36. (a) $ZnCl_2(aq) + 2\ KOH(aq) \longrightarrow 2\ KCl(aq) + Zn(OH)_2(s)$
$Zn^{2+}(aq) + 2\ OH^-(aq) \longrightarrow Zn(OH)_2(s)$
(b) $AgNO_3(aq) + KI(aq) \longrightarrow AgI(s) + KNO_3(aq)$
$Ag^+(aq) + I^-(aq) \longrightarrow AgI(s)$
(c) $2\ NaOH(aq) + FeCl_2(aq) \longrightarrow Fe(OH)_2(s) + 2\ NaCl(aq)$
$2\ OH^-(aq) + Fe^{2+}(aq) \longrightarrow Fe(OH)_2(s)$

5.38. $HNO_3(aq) \longrightarrow H^+(aq) + NO_3^-(aq)$

5.40. $H_2C_2O_4(aq) \longrightarrow H^+(aq) + HC_2O_4^-(aq)$
$HC_2O_4^-(aq) \longrightarrow H^+(aq) + C_2O_4^{2-}(aq)$

5.42. $MgO(s) + H_2O(\ell) \longrightarrow Mg(OH)_2(s)$

5.44. (a) $2\ CH_3CO_2H(aq) + Mg(OH)_2(s) \longrightarrow Mg(CH_3CO_2)_2(aq) + 2\ H_2O(\ell)$
Acetic acid and magnesium hydroxide react to produce magnesium acetate and water.
(b) $HClO_4(aq) + NH_3(aq) \longrightarrow NH_4ClO_4(aq)$
Perchloric acid and ammonia produce ammonium perchlorate

5.46. $Ba(OH)_2(s) + 2\ HNO_3(aq) \longrightarrow Ba(NO_3)_2(aq) + 2\ H_2O(\ell)$

5.48. $FeCO_3(s) + 2\ HNO_3(aq) \longrightarrow Fe(NO_3)_2(aq) + CO_2(g) + H_2O(\ell)$
Iron(II) carbonate and nitric acid react to produce iron(II) nitrate, carbon dioxide, and water.

5.50. (a) $Ba(OH)_2(s) + 2\ HCl(aq) \longrightarrow BaCl_2(aq) + 2\ H_2O(\ell)$
Acid–base reaction
(b) $2\ HNO_3(aq) + CoCO_3(s) \longrightarrow Co(NO_3)_2(aq) + H_2O(\ell) + CO_2(g)$
Gas-forming reaction
(c) $2\ Na_3PO_4(aq) + 3\ Cu(NO_3)_2(aq) \longrightarrow Cu_3(PO_4)_2(s) + 6\ NaNO_3(aq)$
Precipitation reaction

5.52. (a) $MnCl_2(aq) + Na_2S(aq) \longrightarrow MnS(s) + 2\ NaCl(aq)$
$Mn^{2+}(aq) + S^{2-}(aq) \longrightarrow MnS(s)$
Precipitation reaction
(b) $K_2CO_3(aq) + ZnCl_2(aq) \longrightarrow ZnCO_3(s) + 2\ KCl(aq)$
$CO_3^{2-}(aq) + Zn^{2+}(aq) \longrightarrow ZnCO_3(s)$
Precipitation reaction
(c) $K_2CO_3(aq) + 2\ HClO_4(aq) \longrightarrow 2\ KClO_4(aq) + CO_2(g) + H_2O(\ell)$
$CO_3^{2-}(aq) + 2\ H^+(aq) \longrightarrow CO_2(g) + H_2O(\ell)$
Gas-forming reaction

5.54. (a) Br = +5 and O = −2
(b) C = +3 and O = −2
(c) F = 0
(d) Ca = +2 and H = −1
(e) H = +1, Si = +4, and O = −2
(f) S = +6 and O = −2

5.56. (a) Oxidation–reduction. Zn metal is oxidized to Zn^{2+} ions (and the metal is the reducing agent). The NO_3^- ion is an oxidizing agent and is reduced to NO_2. (N changes in oxidation number from +5 to +4.)
(b) Acid–base reaction
(c) Oxidation–reduction. Ca is oxidized to Ca^{2+} [in $Ca(OH)_2$] and is the reducing agent. Water is reduced (and is the oxidizing agent) to H_2 gas.

5.58. (a) Mg is oxidized (from an oxidation number of 0 to +2) and is the reducing agent. Oxygen is the oxidizing agent; O is reduced from an oxidation number of 0 to −2.
(b) Oxygen is the oxidizing agent; O is reduced from an oxidation number of 0 in O_2 to −2 in CO_2 and water. This means C_2H_4 is the reducing agent and so is oxidized. Carbon is oxidized from an oxidation number of −2 in C_2H_4 to +4 in CO_2.
(c) Chlorine is the oxidizing agent. Cl is reduced from an oxidation number of 0 in Cl_2 to −1 in $SiCl_4$. Silicon is therefore the reducing agent. It is oxidized from 0 in Si to +4 in $SiCl_4$.

5.60. $[Na_2CO_3]$ = 0.254 M; $[Na^+]$ = 0.508 M; $[CO_3^{2-}]$ = 0.254 M

5.62. 0.494 g $KMnO_4$

5.64. 5.08×10^3 mL

5.66. 0.0100 M $CuSO_4$

5.68. Method (a) gives 0.125 M H_2SO_4, the correct concentration. Method (b), however, leads to an incorrect concentration of 0.150 M.

5.70. (a) 0.50 M NH_4^+ and 0.25 M SO_4^{2-}

(b) 0.056 M H^+ and 0.056 M NO_3^-
(c) 0.246 M Na^+ and 0.123 M CO_3^{2-}
(d) 0.00124 M K^+ and 0.00124 M ClO_4^-

5.72. 121 mL

5.74. 60. g NaOH and 53 g of Cl_2

5.76. 193 mL $Na_2S_2O_3$

5.78. 1.50×10^3 mL $Pb(NO_3)_2$

5.80. HCl is the limiting reactant, so some $CaCO_3$ will remain after the reaction. Reaction can produce 1.73 g of $CaCl_2$.

5.82. 40.9 mL of HCl is required.

5.84. 42.5 mL of HCl is required.

5.86. 0.0219 g citric acid per 100 mL of soft drink

5.88. Citric acid. Calculate the mass of acid required by 29.1 mL of 0.513 M NaOH. Alternatively, calculate the molar mass of the acid; it will correspond to citric acid.

5.90. 0.500 g vitamin C

5.92. $MgCO_3(s) + 2\ H^+(aq) \longrightarrow Mg^{2+}(aq) + CO_2(g) + H_2O(\ell)$
The chloride ion, Cl^-, is a spectator ion in this gas-forming reaction.

5.94. (a) $(NH_4)_2S(aq) + Hg(NO_3)_2(aq) \longrightarrow HgS(s) + 2\ NH_4NO_3(aq)$
(b) Ammonium sulfide and mercury(II) nitrate produce mercury(II) sulfide and ammonium nitrate.
(c) Precipitation reaction

5.96. (a) Gas-forming reaction
(b) $C_6H_8O_7(aq) + 3\ NaHCO_3(aq) \longrightarrow Na_3C_6H_5O_7(aq) + 3\ CO_2(g) + 3\ H_2O(\ell)$
131 mg $NaHCO_3$ is contained in the tablet.

5.98. $[H^+] = [Cl^-] = 0.102$ M

5.100. 96.8% $Na_2S_2O_3$

5.102. (a) Combine $Zn(OH)_2$ (base) with HCl(acid) and then evaporate to dryness to obtain solid $ZnCl_2$.
(b) Combine $ZnCO_3$ with HCl to produce $ZnCl_2$, CO_2, and H_2O. Evaporate to dryness to obtain solid $ZnCl_2$.
(c) Combine Zn metal with HCl in an oxidation–reduction reaction,
$Zn(s) + 2\ HCl(aq) \longrightarrow ZnCl_2(aq) + H_2(g)$
or combine Zn metal with Cl_2 gas, also in an oxidation–reduction reaction.
$Zn(s) + Cl_2(g) \longrightarrow ZnCl_2(s)$

5.104. They will both calculate the same concentration of HCl. Diluting samples with pure water does not change the moles of HCl present, so each student should use the same volume of NaOH in the titration.

5.106. Calculate volume = 7.6×10^4 L. (This is a *very* large hot tub; more like a swimming pool.)

5.108. Mass of $Fe(OH)_3$ precipitated = 0.625 g. NaOH is in excess, and has a concentration of 0.025 M after the reaction.

5.110. $Co(NH_3)_6Cl_3$

5.112. (a) $(NH_4)_2PtCl_4(aq) + 2\ NH_3(aq) \longrightarrow Pt(NH_3)_2Cl_2(aq) + 2\ NH_4Cl(aq)$
(b) 15.54 g $(NH_4)_2PtCl_4$ and 667 mL NH_3
(c) $x = 2$; $Pt(NH_3)_2Cl_2(C_5H_5N)_2$

CHAPTER 6

6.10. 5.0×10^6 J

6.12. 2.4×10^4 kJ/day; 6.6 kwh/day

6.14. 2.90 kJ

6.16. Water requires more energy. $q(H_2O) = 13000$ J and $q(Al) = 11000$ J.

6.18. 101 kJ

6.20. Water requires more energy.

6.22. 279 K or 6 °C

6.24. $T_{initial} = 311$ K, or 37 °C

6.26. $C_{metal} = 0.24$ J/g · K

6.28. 3.30×10^2 kJ

6.30. 181 kJ

6.32. 48.2 kJ

6.34. (a) The enthalpy change is negative, so the reaction is exothermic.
(b) $q = -2.38$ kJ (heat energy evolved)

6.36. 3.29×10^4 kJ

6.38. $\Delta H^\circ_{rxn} = -763$ kJ/mol of CH_3OH, or -1.53×10^3 kJ for the reaction as written (for 2 mol of CH_3OH).

6.40. $6\ C(s) + 3\ H_2(g) \longrightarrow C_6H_6(\ell)$, $\Delta H^\circ_f[(C_6H_6(\ell)] = +49.0$ kJ/mol

6.42. $q = -260$ kJ

6.44. $2\ Cr(s) + \frac{3}{2}\ O_2(g) \longrightarrow Cr_2O_3(s)$, $\Delta H^\circ_f[Cr_2O_3(s)] = -1139.7$ kJ/mol

6.46. (a) Negative sign for ΔH°_f indicates an exothermic process.
(b) $6\ C(graphite) + 6\ H_2(g) + 3\ O_2(g) \longrightarrow C_6H_{12}O_6(s)$

6.48. $Pb(s) + Cl_2(g) \longrightarrow PbCl_2(s)$, $\Delta H^\circ_{rxn} = \Delta H^\circ_f[PbCl_2(s)] = -359.4$ kJ/mol

6.50. $\Delta H^\circ_{rxn} = -98.9$ kJ. Reaction is exothermic.

6.52. (a) $\Delta H^\circ_{rxn} = -905.2$ kJ. Reaction is exothermic.
(b) $q = -133$ kJ

6.54. (a) $\Delta H^\circ_{rxn} = -14.5$ kJ. Reaction is exothermic.
(b) $q = -62.5$ J

6.56. $\Delta H^\circ_f[C_{10}H_8(s)] = +77.9$ kJ/mol

6.58. $q = -6.62$ kJ

6.60. $q = -394$ kJ/mol

6.62. $\Delta H = -56$ kJ/mol

6.64. $C_{metal} = 0.489$ J/g · K

6.66. 100. g of ethanol transfer more heat. $q(H_2O) = -8.4$ kJ and $q(ethanol) = -9.8$ kJ.

6.68. (a) $\Delta H^\circ_{rxn} = -3.529 \times 10^5$ J
(b) 0.43 g Mg

6.70. $\Delta H^\circ_f[N_2H_4(\ell)] = 50.$ kJ/mol

6.72. (a) $\Delta H^\circ_{rxn} = +131.293$ kJ
(b) 1.0931×10^7 kJ

6.74. (a) $\Delta H^\circ_{rxn} = -35.9$ kJ
(b) $q = -3.59 \times 10^3$ kJ

6.76. Both sugars have the same balanced equation for combustion, so the difference in ΔH°_{rxn} depends only on ΔH°_f for the sugar. Fructose has the less negative enthalpy of formation, so its enthalpy of combustion will be more exothermic.

6.78. $C_{metal} = 0.162$ J/g · K

6.80. All of the molar heat capacities are about the same, that is, about 25 J/mol · K. Assuming this value for Ag, we would calculate $C_{Ag} = (25\text{ J/mol}\cdot\text{K}) \cdot (1\text{ mol}/107.9\text{ g}) = 0.23\text{ J/g}\cdot\text{K}$.

6.82. We know the enthalpy of combustion of both graphite and diamond.

$$C(graphite) + O_2(g) \longrightarrow CO_2(g) \quad \Delta H^\circ_{rxn} = -393.5\text{ kJ}$$

$$CO_2(g) \longrightarrow C(diamond) + O_2(g) \quad \Delta H^\circ_{rxn} = +395.4\text{ kJ}$$

Combining these gives +1.9 kJ for the conversion of graphite to diamond.

6.84. The condenser and motor create evolve more heat to cool the refrigerator's interior than the amount of cooling available when the door is first opened.

6.86. About 2 g

6.88. Isooctane produces more heat per gram ($\Delta H^\circ_{rxn} = -47.81$ kJ/g) than methanol ($\Delta H^\circ_{rxn} = -22.67$ kJ/g).

6.90. Butene energy diagram

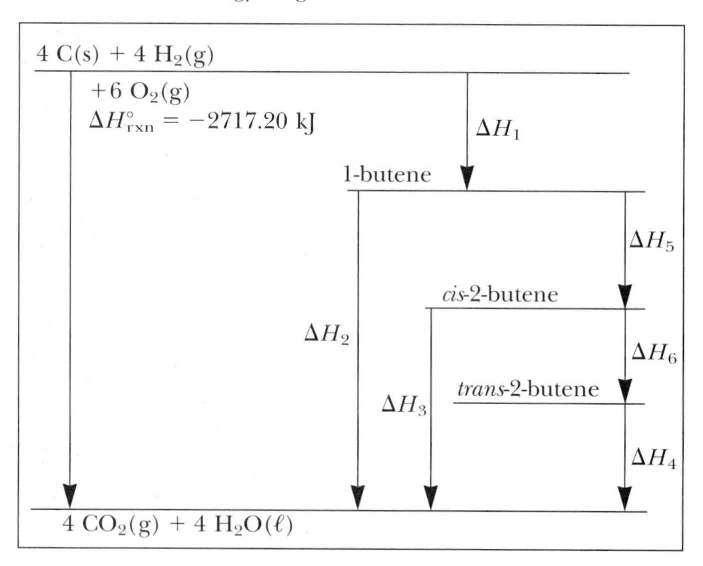

ΔH_1 is known to be −20.5 kJ/mol.

ΔH_2 = enthalpy of combustion of 1-butene = −2696.7 kJ

ΔH_3 = enthalpy of combustion of *cis*-2-butene = −2687.5 kJ

ΔH_4 = enthalpy of combustion of *trans*-2-butene = −2684.2 kJ

ΔH_5 = difference in enthalpies of combustion of 1-butene and *cis*-2-butene = $\Delta H_2 - \Delta H_3$ = −9.20 kJ

$\Delta H_6 = \Delta H$ for *cis*-2-butene to *trans*-2-butene = −3.3 kJ

ΔH°_f of *cis*-2-butene = $\Delta H_1 + \Delta H_5$ = −29.7 kJ

ΔH°_f of *trans*-2-butene = $\Delta H_1 + \Delta H_5 + \Delta H_6$ = −33.0 kJ

6.92. $\Delta H^\circ_f[B_2H_6(g)] = +35.6$ kJ/mol

6.94. $\Delta H^\circ_{rxn} = -305.3$ kJ

6.95. (a) 1.7×10^8 g SO_2 and 2.6×10^6 mol SO_2
(b) 1.2×10^{13} g MgO required and 3.6×10^{13} g $MgSO_4$ produced
(c) $\Delta H^\circ_{rxn} = -1919.0$ kJ
(d) 3.65×10^9 kJ/day

CHAPTER 7

7.20. (a) Red, orange, and yellow light
(b) Blue photons
(c) Blue light

7.22. 3.37 m

7.24. Frequency = $7.3 \times 10^{14}\text{ s}^{-1}$. Energy of 1 photon = 4.8×10^{-19} J. Energy of 1 mol of photons = 290 kJ. Energy of violet radiation is about 1.7 times that of red radiation.

7.26. Frequency = $7.5676 \times 10^{14}\text{ s}^{-1}$. Energy of 1 photon = 5.0144×10^{-19} J. Energy of 1 mol of photons = 302 kJ

7.28. (a) Line with shortest wavelength has the highest energy; 253.652 nm.
(b) Frequency = $1.18190 \times 10^{15}\text{ s}^{-1}$. Energy of 1 photon = 7.83139×10^{-19} J.
(c) The lines at 404 nm and 436 nm are visible in Figure 7.10. These are at the violet end of the spectrum.

7.30. FM < microwave < yellow light < x-rays

7.32. Calculated wavelength = 600 nm. In the visible portion of the spectrum, approximately corresponding to orange light.

7.34. (a) 10 emission lines
(b) Lowest frequency: $n = 5$ to $n = 4$
(c) Shortest wavelength: $n = 5$ to $n = 1$
(d) Highest energy: $n = 5$ to $n = 1$

7.36. $n = 2$ to $n = 4$ and $n = 3$ to $n = 5$

7.38. 102.6 nm; ultraviolet

7.40. $\Delta E = +2.093 \times 10^{-18}$ J

7.42. Wavelength = 2.9×10^{-10} m, or 0.29 nm

7.44. Wavelength = 2.2×10^{-25} nm; frequency = 1.2×10^{-21} m/s

7.46. (a) Orbital size: n
(b) Relative orbital orientation: m_ℓ
(c) Orbital shape: ℓ

7.48. (a) For $n = 4$: $\ell = 0, 1, 2, 3$
(b) For $\ell = 2$: $m_\ell = -2, -1, 0, +1, +2$
(c) For $4s$: $n = 4$, $\ell = 0$, $m_\ell = 0$
(d) For $4f$: $n = 4$, $\ell = 3$, $m_\ell = -3, -2, -1, 0, +1, +2, +3$

7.50. Three sets of quantum numbers are possible: $n = 4$ and $\ell = 1$, and then $m_\ell = -1, 0, +1$.

7.52. 4 subshells: *s*, *p*, *d*, and *f*

7.54. (a) ℓ cannot be equal to n
(b) m_ℓ can be no larger than $\pm\ell$.
(c) Same as (b)

7.56. (a) None, because m_ℓ can be no larger than $\pm\ell$.
(b) 3 orbitals. When $\ell = 1$, 3 p orbitals exist.
(c) 11 orbitals. When $\ell = 5$, 11 orbitals exist.
(d) 1 orbital. This set of three quantum numbers specifies an orbital.

7.58. (a) No planar nodes because $\ell = 0$.

(b) Two, because $\ell = 2$
(c) Three, because $\ell = 3$

7.60. (a) $2d$, because a d orbital must have $\ell = 2$, and ℓ cannot be larger than $n - 1$
(b) $3f$, for the same reason as (a)

7.62. (a) $2p$: $n = 2$, $\ell = 1$, and $m_\ell = -1, 0,$ or $+1$
(b) $3d$: $n = 3$, $\ell = 2$, and $m_\ell = -2, -1, 0, +1,$ or $+2$
(c) $4f$: $n = 4$, $\ell = 3$, and $m_\ell = -3, -2, -1, 0, +1, +2,$ or $+3$.

7.64. s orbital

7.66. 392 kJ

7.68. 130 mol photons

7.70. 260 s

7.72. Ionization energy for He^+ = 5248 kJ/mol. This is four times the values for the H atom.

7.74. (a) Smallest energy change: $n = 7$ to $n = 6$
(b) Highest frequency (largest energy change): $n = 7$ to $n = 1$
(c) Shortest wavelength (highest energy): $n = 7$ to $n = 1$

7.76. Energy depends only on n: $1s < 2s/2p < 3s/3p/3d < 4s$

7.78. (a) Size or energy
(b) Shape: ℓ
(c) Less energy
(d) 7 f orbitals
(e) One orbital
(f) (*left*) p orbital; (*right*) d orbital
(g) For $n = 5$, $\ell = 0, 1, 2, 3,$ and 4
(h) 16 orbitals ($= n^2$)

7.80. 9 orbitals

7.82. Light (radiation) has both wave and particle properties. In our modern view of atomic structure, this duality also applies to particles of very small mass such as protons, electrons, and neutrons.

7.84. (b), (e–g), (h), (i), (j)

7.86. When salts are placed in a burning organic liquid (here methanol), very small quantities of ions appear in the vapor phase, where they are excited by the heat of combustion. These excited ions emit radiation. Similarly, the atoms in the gas phase in a "neon" sign are excited electrically and emit radiation. The mechanism of emission is the same in both cases.

7.88. An electron in a $2p_z$ orbital has no probability of being in the xy-plane. That is, there is no electron density along the x-axis.

7.90. (a) For an s electron, the probability of it being a given distance from the nucleus is the same no matter what the direction in space around the nucleus.
(b) A $2p_x$ electron has its maximum probability along the x-axis. The yz-plane is a nodal plane. Thus, there is no probability of finding the electron at $y = 0.001$ nm.

CHAPTER 8

8.10. Aluminum and phosphorus are both in the third period of the periodic table and so are filling the $n = 3$ valence shell. Al is in Group 3A and so has three electrons beyond the nearest noble gas in $n = 3$, whereas P, which is in Group 5A, has five valence electrons in $n = 3$.

Al: $1s^2 2s^2 2p^3 3s^2 3p^1$

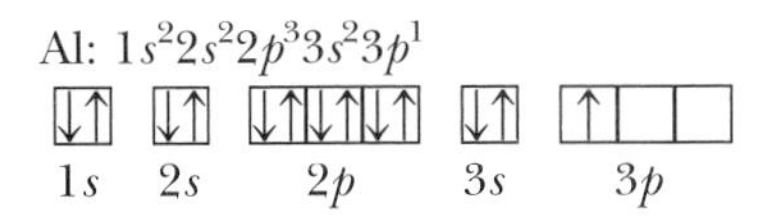

P: $1s^2 2s^2 2p^3 3s^2 3p^3$

[↓↑] [↓↑] [↓↑|↓↑|↓↑] [↓↑] [↑|↑|↑]
1s 2s 2p 3s 3p

8.12. V: $1s^2 2s^2 2p^6 3s^2 3p^6 3d^3 4s^2$

8.14. As: $[Ar]3d^{10}4s^2 4p^3$

[Ar] [↓↑|↓↑|↓↑|↓↑|↓↑] [↓↑] [↑|↑|↑]
3d 4s 4p

8.16. (a) Sr: $[Kr]5s^2$
(b) Zr: $[Kr]3d^1 5s^2$
(c) Rh: $[Kr]3d^7 5s^2$
(d) Sn: $[Kr]3d^{10} 5s^2 5p^2$

8.18. (a) Sm: $[Xe]4f^5 5d^1 6s^2$ (predicted) or $[Xe]4f^6 6s^2$ (actual)
(b) Yb: $[Xe]4f^{13} 5d^1 6s^2$ (predicted) or $[Xe]4f^{14} 6s^2$ (actual)

8.20. (a) Pu: $[Rn]5f^5 6d^1 7s^2$ (predicted) or $[Rn]5f^6 7s^2$ (actual)
(b) Cm: $[Rn]5f^7 6d^1 7s^2$ (predicted and actual)

8.22. (a) Mg^{2+}: [Ne] = [↓↑] [↓↑] [↓↑|↓↑|↓↑]
1s 2s 2p

(b) K^+: [Ar] = [↓↑] [↓↑] [↓↑|↓↑|↓↑] [↓↑] [↓↑|↓↑|↓↑]
1s 2s 2p 3s 3p

(c) Cl^-: [Ar] = [↓↑] [↓↑] [↓↑|↓↑|↓↑] [↓↑] [↓↑|↓↑|↓↑]
1s 2s 2p 3s 3p

(d) O^{2-}: [Ne] = [↓↑] [↓↑] [↓↑|↓↑|↓↑]
1s 2s 2p

8.24. Configurations of vanadium and its ions

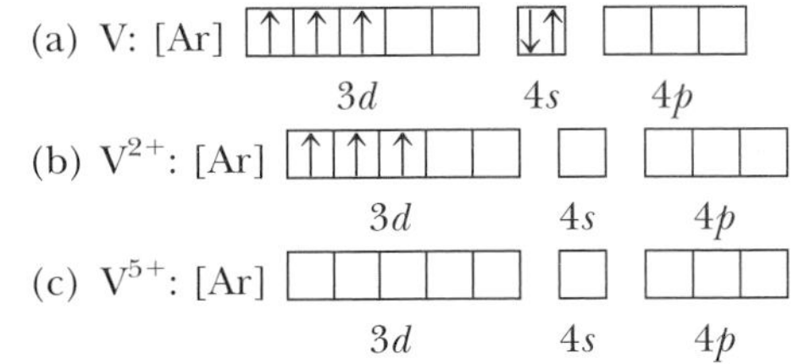

Both V and V^{2+} are paramagnetic.

8.26. (a) Mn: [Ar] [↑|↑|↑|↑|↑] [↓↑] [| |]
3d 4s 4p

(b) Mn^{2+}: [Ar] [↑|↑|↑|↑|↑] [ ] [| |]
3d 4s 4p

Mn^{2+} is paramagnetic, with five unpaired electrons in the $3d$ orbitals.

8.28. (a) Ru: Predicted configuration is $[Kr]4d^6 5s^2$. The actual configuration, however, is $[Kr]4d^7 5s^1$, showing that the details of the electron configurations of the transition metals can be different from the ones predicted by their position in the period table.
(b) Ru^{3+}: $[Kr]4d^5 5s^0$. Cations are always "formed" from the atom by first removing electrons of highest n ($5s^2$ or $5s^1$) and then electrons of next higher n (one or two electrons from $4d$). Thus, it makes no difference which atom configuration one begins with.

8.30. Uranium configuration: $[Rn]5f^3 6d^1 7s^2$, paramagnetic

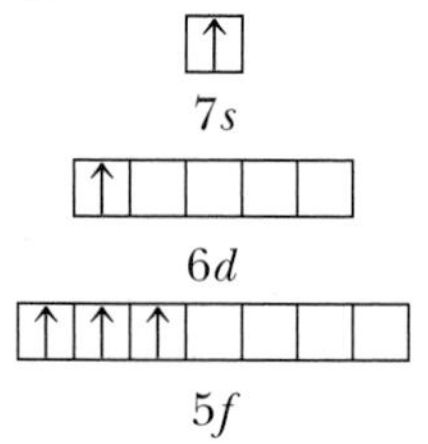

Uranium(IV) configuration: $[Rn]5f^2 6d^0 7s^0$, paramagnetic

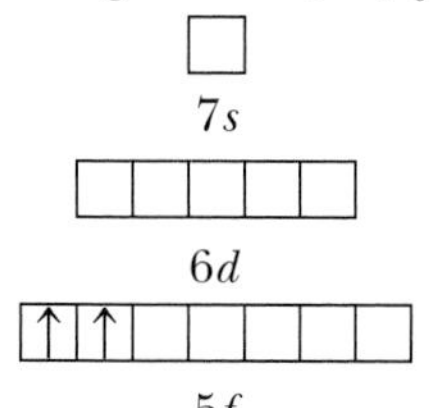

8.32. Ti^{2+} has 2 unpaired electrons {$[Ar]3d^2$}, whereas Mn^{3+} has 4 unpaired electron {$[Ar]3d^4$}. Both are paramagnetic.

8.34. Ti^{2+}: $[Ar]3d^2$; 2 unpaired electrons
V^{2+}: $[Ar]3d^3$; 3 unpaired electrons
Cr^{2+}: $[Ar]3d^4$; 4 unpaired electrons
Mn^{2+}: $[Ar]3d^5$; 5 unpaired electrons
Fe^{2+}: $[Ar]3d^6$; 4 unpaired electrons
Co^{2+}: $[Ar]3d^7$; 3 unpaired electrons
Ni^{2+}: $[Ar]3d^8$; 2 unpaired electrons
Cu^{2+}: $[Ar]3d^9$; 1 unpaired electron
Zn^{2+}: $[Ar]3d^{10}$; 0 unpaired electrons
Manganese(II) has the greatest number of unpaired electrons. Only Zn^{2+} in diamagnetic.

8.36. Magnesium electron configuration

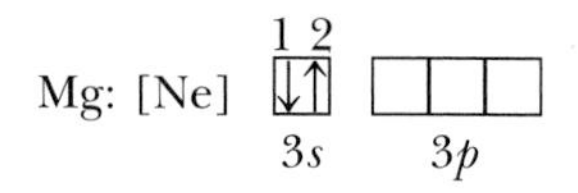

Electron 1: $n = 3$, $\ell = 0$, $m_\ell = 0$, $m_s = -\frac{1}{2}$
Electron 2: $n = 3$, $\ell = 0$, $m_\ell = 0$, $m_s = +\frac{1}{2}$

8.38. Gallium electron configuration

Ga: [Ar] ↓↑↓↑↓↑↓↑↓↑ ↓↑ ↑
3d 4s 4p

For the $4p$ electron: $n = 4$, $\ell = 1$, $m_\ell = +1$, $m_s = +\frac{1}{2}$ (Other possibilities for m_ℓ include 0 and -1 and m_s could also be $-\frac{1}{2}$.)

8.40. (a) 7 (because $\ell = 3$ indicates f orbitals)
(b) 2 (because this set indicates one of the three possible $6p$ orbitals)
(c) 0 (because ℓ cannot be equal to n)
(d) 1 (because this set indicates an electron occupying a $2p$ orbital)

8.42. (a) m_s cannot be 0
(b) m_ℓ cannot have an absolute value larger than ℓ
(c) ℓ cannot be equal to (or larger than) n

8.44. 203 pm

8.46. C < B < Al < Na < K

8.48. (a) Cl^-; (b) Al; (c) In

8.50. (c) is correct

8.52. K < Li < C < N

8.54. (a) Largest radius, Na
(b) O, most positive electron affinity
(c) Increasing ionization energy: Na < Mg < P < O

8.56. (a) Increasing ionization energy: S < O < F
(b) Largest ionization energy: O
(c) Most positive electron affinity: Cl
(d) Largest radius: O^{2-}

8.58. (a) Atomic number = 20 (Ca: $1s^2 2s^2 2p^6 3s^2 3p^6 4s^2$)
(b) s electrons = 8
(c) p electrons = 12
(d) d electrons = 0

8.60. (b), because ℓ cannot be equal to (or larger than) n.

8.62. Element 71 is lutetium, Lu {$[Xe]4f^{14} 5d^1 6s^2$}, so electron shells $n = 1, 2, 3,$ and 4 are complete.

8.64. (a) P
(b) Be
(c) N
(d) Tc
(e) Cl
(f) Zn

8.66. Element A is Rb, and element B is Se.
(a) A is a metal.
(b) B (= Se) has the greater ionization energy.
(c) B (= Se) has a more positive electron affinity.
(d) A (= Rb) has the larger radius.

8.68. In^{4+} (would remove an electron from the filled $n = 3$ shell)
Sn^{5+} (would remove an electron from the filled $n = 4$ shell)
Fe^{6+} (although this is plausible because the $4s$ and four of the $3d$ electrons could be removed, it represents a highly unlikely oxidation number for Fe because the charge is so high)

8.70. (a) Se
(b) Cl^-
(c) Na
(d) N
(e) N^{3-}

8.72. (a) Largest radius, Na
(b) Largest electron affinity, C
(c) Ionization energy: Na < Al < B < C

8.74. (a) Co, cobalt
(b) paramagnetic
(c) A 3+ ion has 4 unpaired electrons

8.76. Effective nuclear charge increases across the periodic table.

8.78. As the transition metals add electrons to the $(n-1)d$ orbitals, they repel the *ns* electrons and partially offset the effect of increasing nuclear charge. Consequently, the *ns* electrons experience only a slightly increasing nuclear attraction, and there is only a slight decline in radius.

8.80. Li. The first electron is removed from the 2*s* orbital (to form Li^+). Removing the second electron from Li (to form Li^{2+}) means breaking into a much lower energy electron shell (1*s*), and considerably more energy is required.

8.82. Cs^+ and Se^{2-}. Both would attain a noble gas configuration. (Cs^+ has a Xe configuration, whereas Se^{2-} is equivalent to Kr.)

8.84. Ca is smaller than K because the effective nuclear charge of Ca is larger than that for K. The second ionization of K is so much larger than that for Ca because, to create K^{2+}, we must break into a much lower energy electron shell. The first electron is removed from a 4*s* orbital of K, but the second must be removed from 3*p*.

8.86. This reaction implies that Ca has a 3+ charge. To form Ca^{3+}, one has to remove the two, 4*s* valence electrons plus a 3*p* electron. Too much energy is required.

8.88. Across this series of elements the mass is increasing but the atomic radius is decreasing. Radius determines the volume of the element, so the ratio "mass/volume," that is, the density of the element, must increase on proceeding from K to V.

8.90. The general trend for ionization energies across a period is inversely related to atomic size: the smaller the atom, the larger the ionization energy. Here, however, the ionization energy of S is smaller than expected (Figure 8.12). The reason for this is electron–electron repulsions make it easier to remove one of the paired electron in S {[Ne]$3s^2 3p^4$}. (Pairs also exist in Cl, but in Cl the effective nuclear charge is larger than in S, and the effect of electron–electron repulsions is more than offset.

8.92. (a) [↓↑] [↓↑] [↓↑|↓↑|↓↑] [↓↑] [↓↑|↑|↑]
1*s* 2*s* 2*p* 3*s* 3*p*
(b) $n = 3$, $\ell = 1$, $m_\ell = -1$, $m_s = +\frac{1}{2}$
(c) Smallest ionization energy: S; smallest radius, O
(d) S is smaller than S^{2-}.
(e) 804 g Cl_2
(f) Cl_2 is the limiting reactant. The theoretical yield of $OSCl_2$ is 16.8 g.
(g) $\Delta H_f^\circ[OSCl_2(g)] = -212.5$ kJ/mol

CHAPTER 9

9.14. (a) N—O bond order in NO_3^-, = 1.33

9.16. Bond dissociation energy is the energy required to break a given chemical bond. As such, it is always a positive quantity.

9.18. Need the O—O bond energy for O_2 (498 kJ), the H—H bond (2 mol of H—H at 436 kJ/mol). The total energy expended to break bonds will be 1370 kJ. When 2 mol H_2O is formed, 4 mol of O—H bonds forms (463 kJ/mol of O—H bonds), or 1850 kJ evolves. Thus, the total energy involved in the reaction is

ΔH = energy expended to break bonds + energy evolved on forming bonds
= +1370 kJ + (−1850 kJ) = −482 kJ

9.20. Electron affinity values are determined for isolated atoms in the gas phase, whereas electronegativity refers to an atom in a molecule.

9.22. The formal charge on each atom should be as close to 0 as possible. The following resonance structure is unlikely because +1 or −1 charges appear on the O atoms.

−1 +1
:Ö—C≡O:

9.24. Water has four electron pairs at the corners of a tetrahedron. The electron-pair geometry refers to the location of these pairs, so the electron-pair geometry is tetrahedral. The electron-pair geometry determines the molecular geometry, which describes the location of the atoms of the molecule. In water, only two of the electron pairs are bonds, so the H—O—H atoms describe a bent system.

9.26. Four electron pair give a pyramidal molecule (e.g., NH_3, H—N—H angle = 107°) if there are three bond pairs and one lone pair. A bent molecule (e.g., water, H—O—H = 105°) results if there are two bond pairs and two lone pairs.

9.28.

Element	**N**	**B**	**Na**	**Mg**	**F**	**S**
Group	5A	3A	1A	2A	7A	6A
Number of valence electrons	5	3	1	2	7	6

9.30. Group 1A, one bond; Group 2A, two bonds; Group 3A, three bonds; Group 4A, four bonds; Group 5A, three bonds; Group 6A, two bonds; Group 7A, one bond

9.32. Most negative, MgS. Least negative, NaCl

9.34. RbI < LiI < NaF < CaO

9.36. Because lattice energy is inversely related to the distance between ion centers, smaller ions should lead to a more negative lattice energy, requiring a larger energy input to disrupt the crystalline array. The melting point should increase as the distance between cation and anion decreases.

9.38. (a) NF_3, 26 valence electrons

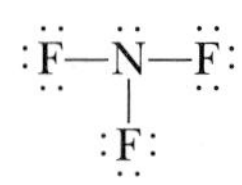

(b) ClO_3^-, 26 valence electrons

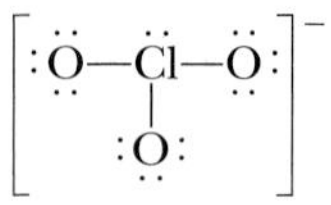

[:O—Cl—O: with :O: bonded to Cl]$^-$

(c) HOBr, 14 valence electrons

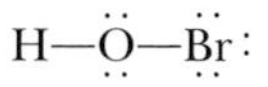

H—O—Br:

(d) SO_3^{2-}, 26 valence electrons

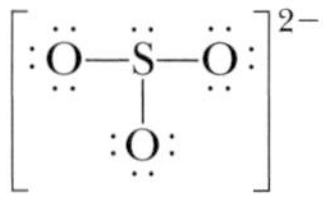

[:O—S—O: with :O: bonded to S]$^{2-}$

9.40. (a) $CHClF_2$, 26 valence electrons

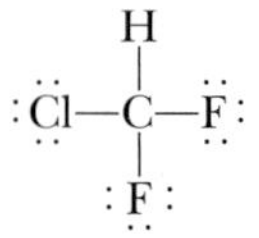

:Cl—C—F: (C bonded to H and :F:)

(b) CH_3CO_2H, 24 valence electrons

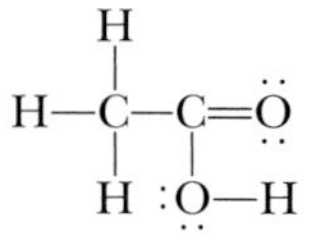

H—C—C=O (first C bonded to two H; second C bonded to :O—H)

(c) CH_3CN, 16 valence electrons

H—C—C≡N: (first C bonded to two more H)

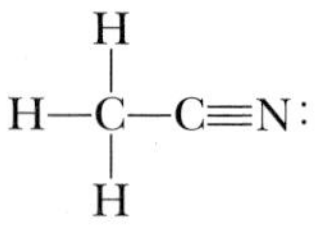

(d) CH_3OH, 14 valence electrons

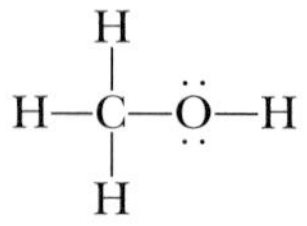

H—C—O—H (C bonded to two more H)

9.42. (a) SO_2, 18 valence electrons

:O—S=O ⟷ O=S—O:

(b) NO_2^-, 18 valence electrons

[:O—N=O]$^-$ ⟷ [O=N—O:]$^-$

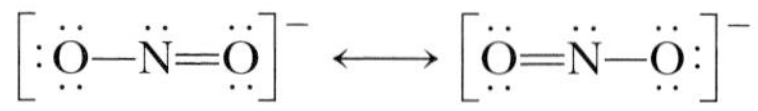

(c) SCN^-, 16 valence electrons

[S=C=N]$^-$ ⟷ [:S≡C—N:]$^-$ ⟷ [:S—C≡N:]$^-$

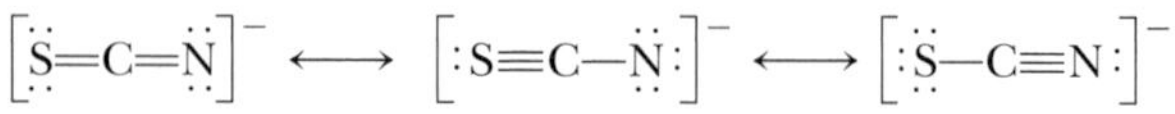

9.44. (a) BrF_3, 28 valence electrons

Br—F: (Br also bonded to two :F:)

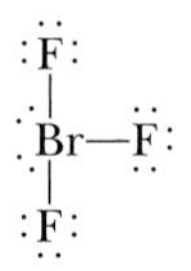

(b) I_3^-, 22 valence electrons

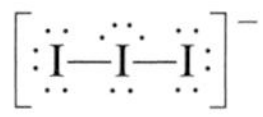

[:I—I—I:]$^-$

(c) XeO_2F_2, 34 valence electrons

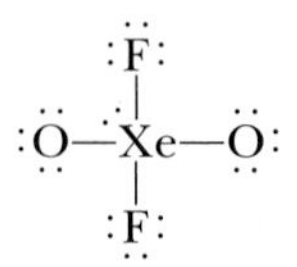

:O—Xe—O: (Xe also bonded to two :F:)

(d) XeF_3^+, 28 valence electrons

[Xe—F: (Xe also bonded to two :F:)]$^+$

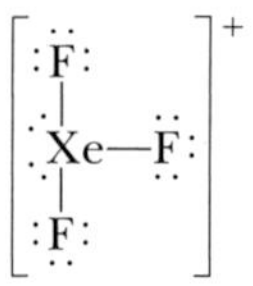

9.46. (a) 2 C—H bonds, bond order 1; 1 C=O bond, bond order 2
(b) 3 S—O single bonds, bond order 1
(c) 2 nitrogen–oxygen double bonds, bond order 2
(d) 1 N=O bond, bond order 2; 1 N—Cl bond, bond order 1

9.48. (a) B—Cl
(b) C—O
(c) P—O
(d) C=O

9.50. The CO bond in carbon monoxide is a triple bond, so it is both shorter and stronger than the CO double bond in H_2CO.

9.52. NO bond orders: 2 in NO_2^+; 1.5 in NO_2^-; 1.33 in NO_3^-. The NO bond is longest in NO_3^- and shortest in NO_2^+.

9.54. $\Delta H^\circ_{rxn} = -134$ kJ

9.56. $\Delta H^\circ_{rxn} = -96$ kJ

9.58. O—F bond dissociation energy = 192 kJ/mol

9.60. (a) N = 0; H = 0
(b) P = +1; O = −1
(c) B = −1; H = 0
(d) All are zero

9.62. (a) N = +1; O = 0
(b) The central N is 0, one O atom is −1, and the other is 0.

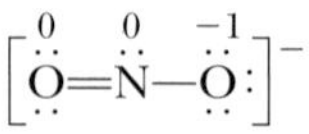

[O=N—O:]$^-$ (formal charges 0, 0, −1)

(c) N = 0; F = 0
(d) The central N is +1, one of the O atoms is −1, and the other two O atoms are both 0.

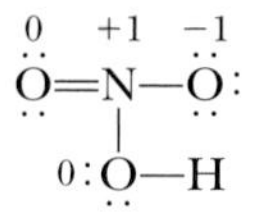

O=N—O: (formal charges 0, +1, −1), N bonded to :O—H (formal charge 0)

9.64. (a) $\overrightarrow{\text{C—O}}$ ($+\delta$ $-\delta$) $\overrightarrow{\text{C—N}}$ ($+\delta$ $-\delta$)

CO is more polar

(b) $\overrightarrow{\text{P—Cl}}$ ($+\delta$ $-\delta$) $\overrightarrow{\text{P—Br}}$ ($+\delta$ $-\delta$)

PCl is more polar

(c) $\overrightarrow{\text{B—O}}$ $\overrightarrow{\text{B—S}}$
$+\delta\ -\delta$ $+\delta\ -\delta$
BO is more polar

(d) $\overrightarrow{\text{B—F}}$ $\overrightarrow{\text{B—I}}$
$+\delta\ -\delta$ $+\delta\ -\delta$
BF is more polar

9.66. (a) CH and CO bonds are polar.
(b) The CO bond is the most polar, and O is the most negative atom.

9.68. (a) Even though the formal charge on B is −1 and on F is 0, F is much more electronegative. The four F atoms therefore likely bear the −1 charge of the ion, and the bonds are polar with the F atom the negative end.
(b) Even though the formal charge on B is −1 and on H is 0, H is slightly more electronegative. The four H atoms therefore likely bear the −1 charge of the ion. The BH bonds are polar with the H atom the negative end.
(c) The formal charge on O is −1 and on H it is 0. This conforms with the relative electronegativities. The bond is polar with O the negative end.
(d) The CH and CO bonds are polar. The negative charge of the CO bonds lies on the O atoms.

9.70.

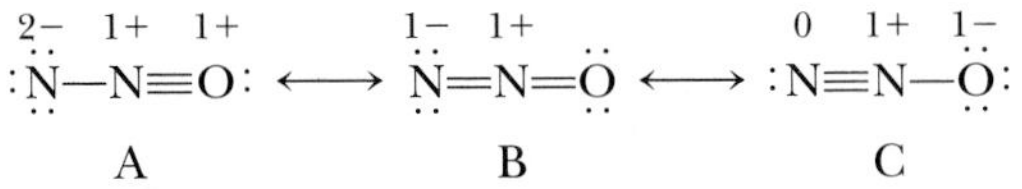

A	B	C
2− 1+ 1+	1− 1+	0 1+ 1−
:N̈—N≡O: ⟷	N̈=N=Ö ⟷	:N≡N—Ö:

Structure C is the most reasonable: The charges are as small as possible and the negative charge resides on the more electronegative atom.

9.72. [1− 0 0 :Ö—N̈=Ö]⁻ ⟷ [0 0 1− Ö=N̈—Ö:]⁻

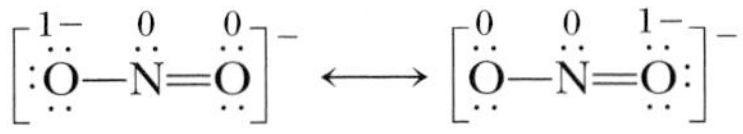

9.74. [1− 0 :C≡N:]⁻

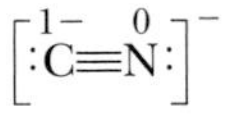

The negative (formal) charge resides on C, so the H^+ should attack that atom and form H—CN.

9.76. (a) 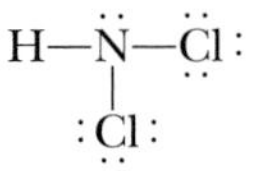 H—N̈—C̈l: with :C̈l: on N — Electron-pair geometry around N is tetrahedral. Molecular geometry is pyramidal.
(b) 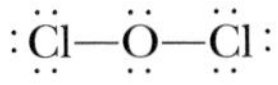 :C̈l—Ö—C̈l: — Electron-pair geometry around O is tetrahedral. Molecular geometry is bent.
(c) 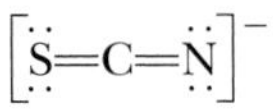 [S̈=C=N̈]⁻ — Electron-pair geometry around C is linear. Molecular geometry is linear.
(d) 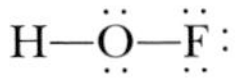 H—Ö—F̈: — Electron-pair geometry around O is tetrahedral. Molecular geometry is bent.

9.78. (a) 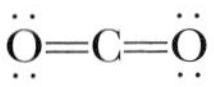 Ö=C=Ö — Electron-pair geometry around C is linear. Molecular geometry is linear.
(b) 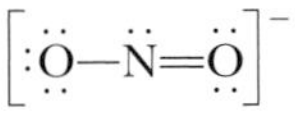 [:Ö—N̈=Ö]⁻ — Electron-pair geometry around O is trigonal-planar. Molecular geometry is bent.
(c) 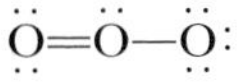 Ö=Ö—Ö: — Electron-pair geometry around O is trigonal-planar. Molecular geometry is bent.
(d) 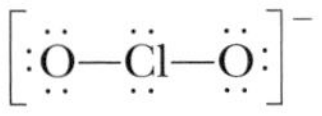 [:Ö—C̈l—Ö:]⁻ — Electron-pair geometry around O is tetrahedral. Molecular geometry is bent.

All have two atoms attached to the central atom. As the bond and lone pairs vary, the molecular geometries vary from linear to bent.

9.80. (a) 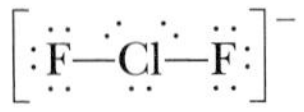 [:F̈—C̈l—F̈:]⁻ — Electron-pair geometry is trigonal-bipyramidal. Molecular geometry is linear.
(b) 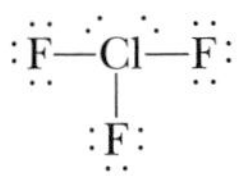 :F̈—C̈l—F̈: with :F̈: on Cl — Electron-pair geometry is trigonal-bipyramidal. Molecular geometry is T-shaped.
(c) 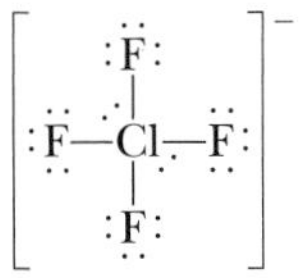 [ClF₄ with four F atoms around Cl]⁻ — Electron-pair geometry is octahedral. Molecular geometry is square-planar.
(d) 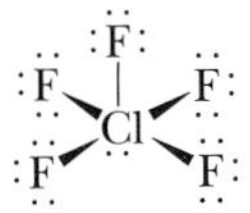 ClF₅ with five F atoms around Cl — Electron-pair geometry is octahedral. Molecular geometry is square-pyramidal.

9.82. (a) Ideal O—S—O angle = 120°
(b) 120°
(c) H—C—H = 109° and C—C—N angle = 180°
(d) 120°

9.84. 1 = 120°; 2 = 109°; 3 = 120°; 4 = 109°; 5 = 109°
The electron-pair geometry around the middle C atom is tetrahedral, so the chain of atoms cannot be linear.

9.86. (a) 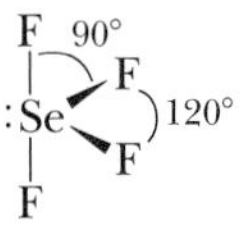 SeF_4: F—Se—F angles 90°, 120°
(b) 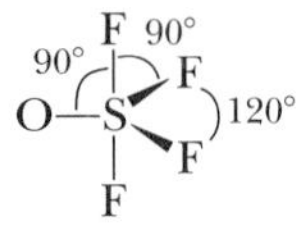 OSF_4: angles 90°, 90°, 120°
(c) 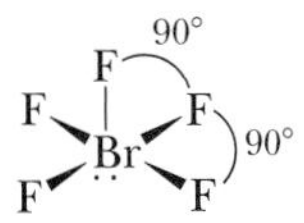 BrF_5: angles 90°, 90°
(d) 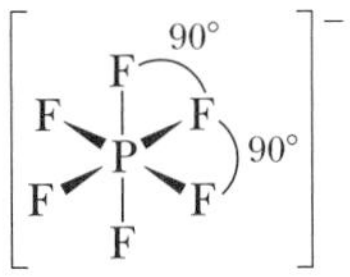 $[PF_6]^-$: angles 90°, 90°

9.88. NO_2^- has a larger bond angle (about 120°) than NO_2^+ (180°). The former has a trigonal-planar electron-pair geometry, whereas the latter is linear.

9.90. (i) Most polar bonds are in H_2O (O and H have the largest difference in electronegativity).
(ii) Not polar: CO_2, CCl_4
(iii) F

9.92. (a) Not polar; linear molecule
(b) HBF_2, polar, trigonal-planar molecule with F atoms negative end of dipole and H atom the positive end
(c) CH_3Cl, polar, Cl atom is the negative end and the 3 H atoms are on the positive side of the molecule.
(d) SO_3, not polar, trigonal-planar molecule

9.94. All of the molecules in the series have 16 valence electrons and all are linear.
(a) O=C=O ⟷ :O—C≡O: ⟷ :O≡C—O:
(b) $[N{=}N{=}N]^-$ ⟷ $[:N{-}N{\equiv}N:]^-$ ⟷ $[:N{\equiv}N{-}N:]^-$
(c) $[O{=}C{=}N]^-$ ⟷ $[:O{-}C{\equiv}N:]^-$ ⟷ $[:O{\equiv}C{-}N:]^-$

9.96. (a) 1 = 120°, 2 = 180°, 3 = 120°
(b) The C=C double bond is shorter than the C—C single bond.
(c) The C=C double bond is stronger than the C—C single bond.
(d) The CN triple bond is slightly more polar than the CH bonds.

9.98. (a) BF_3 (Lewis structure) Electron-pair geometry is trigonal-planar. Molecular geometry is trigonal-planar.
(b) CF_4 (Lewis structure) Electron-pair geometry is tetrahedral. Molecular geometry is tetrahedral.
(c) PF_3 (Lewis structure) Electron-pair geometry is tetrahedral. Molecular geometry is pyramidal.
(d) OF_2 (Lewis structure) Electron-pair geometry is tetrahedral. Molecular geometry is bent.

9.100. (a) The three lone pairs of XeF_2 occupy the equatorial positions. There the angles between lone pairs are 120°, so there is less lone pair/lone pair repulsion with this arrangement.
(b) The two lone pairs on the Cl atom occupy the equatorial positions. The reasoning is the same as for XeF_2.

9.102. +107 kJ

9.104. (a)
$[:C{-}N{\equiv}O:]^-$ (formal charges −3, +1, +1) ⟷ $[C{=}N{=}O]^-$ (formal charges −2, +1) ⟷ $[:C{\equiv}N{-}O:]^-$ (formal charges −1, +1, −1)
(b) The structure on the right seems most reasonable. It has the smallest charges on each atom.
(c) Carbon, the least electronegative element in the ion, has a negative charge. In addition, all three resonance structures have an unfavorable charge distribution. Compare with OCN^- on page 412.

9.106. The resonance structure for the acetamide molecule pictured here has formal charges of 1+ and 1− on the O and N atoms, respectively. These charges are 0 in the other resonance structure.

$H_3C{-}C(:O:^{-1}){=}N^{+1}H_2$ (Lewis structure)

9.108. (a) Both S and O have 0 formal charge.
(b) A = 109°; B = 120°; C = 120°; D = 109°
(c) The two C=C double bonds
(d) C—O bond is most polar.
(e) O atom is most electronegative.
(f) O atom is in the ring plane.
(g) The molecule is polar.

9.110. (a) $Cl_2(g) + 3\ F_2(g) \longrightarrow 2\ ClF_3(g)$
(b) F_2 is the limiting reagent. Theoretical yield of product is 1.62 g.
(c) ClF_3 (Lewis structure)
(d) The electron-pair geometry is trigonal-bipyramidal and the molecular geometry is T-shaped.

CHAPTER 10

10.16. Electron-pair geometry is tetrahedral.
Molecular geometry is pyramidal.
The N atom uses sp^3 hybrid orbitals. Three are used to form N—F σ bonds, and the remaining sp^3 hybrid accommodates the lone pair of electrons.

NF_3 (Lewis structure)

10.18. Electron-pair geometry is tetrahedral.
Molecular geometry is tetrahedral.
The C atom uses sp^3 hybrid orbitals. Three are used to form C—Cl σ bonds, and the remaining sp^3 hybrid forms the C—H bond.

$HCCl_3$ (Lewis structure)

10.20. (a) sp^2
(b) sp

(c) sp^3
(d) sp^2

10.22. (a) C, sp^3; O, sp^3
(b) CH_3, sp^3; middle C, sp^2; CH_2, sp^2
(c) CH_2, sp^3; CO_2H, sp^2; N, sp^3

10.24. (a) Electron-pair geometry is octahedral. Molecular geometry is octahedral. Se: sp^3d^2
(b) Electron-pair geometry is trigonal-bipyramidal. Molecular geometry is seesaw. Se: sp^3d
(c) Electron-pair geometry is trigonal-bipyramidal. Molecular geometry is linear. I: sp^3d
(d) Electron-pair geometry is octahedral. Molecular geometry is square-planar. Xe: sp^3d^2

10.26. HSO_3F, 32 valence electrons. The molecular geometry is tetrahedral, and the S atom is sp^3-hybridized.

SO_3F^-, 32 valence electrons. The molecular geometry is tetrahedral, and the S atom is sp^3-hybridized.

10.28. The C atom is sp^2-hybridized. Two of the sp^2 hybrid orbitals are used to form C—Cl σ bonds. The third is used to form the C—O σ bond. The p orbital not used in the C atom hybrid orbitals is used to form the CO π bond.

10.30. H_2^+ ion: $(\sigma_{1s})^1$. Bond order is 0.5. The bond in H_2^+ is weaker (bond order = 0.5) than in H_2 (bond order = 1).

10.32. MO diagram for C_2^{2-} ion.

σ^*_{2p} (empty)
π^*_{2p} (empty, empty)
σ_{2p} ↑↓
π_{2p} ↑↓ ↑↓
σ^*_{2s} ↑↓
σ_{2s} ↓↑

The ion has 10 valence electrons (isoelectronic with N_2). There are 1 net σ bond and 2 net π bonds, for a bond order of 3. The bond order increases by 1 on going from C_2 to C_2^{2-}. The ion is not paramagnetic.

10.34. (a) CO has 10 valence electrons:
$(\sigma_{2s})^2(\sigma^*_{2s})^2(\pi_{2p})^4(\sigma_{2p})^2$
(b) HOMO, σ_{2p}
(c) Diamagnetic
(d) σ bonds, 1; π bonds, 2; bond order, 3

10.36. (a) sp^2
(b) sp^2
(c) sp^3
(d) sp^3

10.38. The N atom is sp^2-hybridized. The three hybrid orbitals are used to form N—O sigma bonds. The N atom p orbital not used in hybrid formation overlaps with an O atom p orbital to form a NO π bond.

10.40. (a) C1, sp^2; C2, sp^2
(b) A, 120°; B, 120°; C, 120°

10.42. (a) A, 120°; B, 109°; C, 109°; D, 120°
(b) C1, sp^2; C2, sp^2; C3, sp^3

10.44. (a) N1, sp^3; C2, sp^3; C3, sp^2; N4; sp^2; C5, sp^2
(b) A, 109°; B, 120°; C, 109°

10.46. Valence bond theory: bond order is 1

$[:\ddot{O}-\ddot{O}:]^{2-}$

MO theory: [core electrons]
$(\sigma_{2s})^2(\sigma^*_{2s})^2(\pi_{2p})^4(\sigma_{2p})^2(\pi^*_{2p})^4$.
There is a net σ bond, just as predicted by valence bond theory.

10.48. The N atom has sp^3 hybridization in both NH_3 and NH_4^+. The H—N—H angle in NH_3 is close to 107°, whereas it is 109° in NH_4^+.

10.50. (a) Electron-pair geometry is tetrahedral. Molecular geometry is tetrahedral. I: sp^3
(b) Electron-pair geometry is trigonal-bipyramidal. Molecular geometry is trigonal-bipyramidal. I: sp^3d

10.52. See Table 10.1 on page 466.
(a) Paramagnetic diatomic molecules: B_2, O_2
(b) Bond order of 1: Li_2, B_2, F_2
(c) Bond order of 2: C_2, O_2
(d) Highest bond order is N_2

10.54. (a) Sulfamate ion

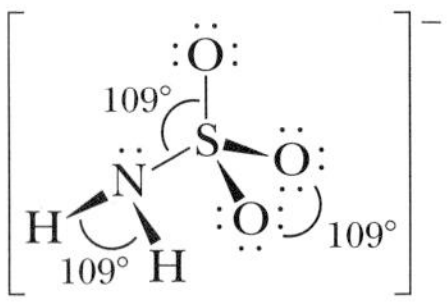

[Structure of sulfamate ion: H₂N–SO₃⁻ with angles 109°, 109°, 109°]

(b) N hybridization remains sp^3, but the S atom hybridization changes from sp^2 in SO_3 to sp^3 in the sulfamate ion.

10.56. CN, 9 valence electrons:
[core electrons] $(\sigma_{2s})^2(\sigma^*_{2s})^2(\pi_{2p})^4(\sigma_{2p})^1$
(a) HOMO, σ_{2p}
(b) Bond order = 2.5 (0.5 σ bond and 2 π bonds)
(c) Paramagnetic

10.58. (a) Ring carbons, sp^2; side-chain carbons, sp^3; N atom, sp^3
(b) A, 120°; B, 109°; C, 109°
(c) 23 σ bonds and 3 π bonds
(d) Polar
(e) To the N atom lone electron pair

10.60. (a) C, sp^2; N, sp^3
(b) The amide link has two resonance structures (shown here with formal charges on the O and N atoms). Structure B is less favorable owing to the separation of charges.

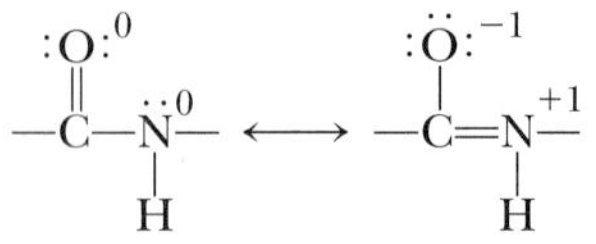

$$\underset{\text{A}}{-\overset{:\text{O}:^{0}}{\overset{\|}{\text{C}}}-\underset{\text{H}}{\overset{\cdot\cdot\,0}{\text{N}}}-} \longleftrightarrow \underset{\text{B}}{-\overset{:\ddot{\text{O}}:^{-1}}{\overset{|}{\text{C}}}=\underset{\text{H}}{\overset{+1}{\text{N}}}-}$$

(c) The fact that the amide link is planar indicates that structure B has some importance.

10.62. (a) The *keto* and *enol* forms are not resonance structures of each other because they do not have the same pattern of bonding. (The OH in the *enol* form becomes a CH in the *keto* form.) In constructing resonance structures one can only move electron pairs, not atoms.
(b) The hybridization of the central C atom changes from sp^2 to sp^3 on going from the *enol* to the *keto* form. (One O atom can also be thought of as changing from sp^3 to sp^2.)

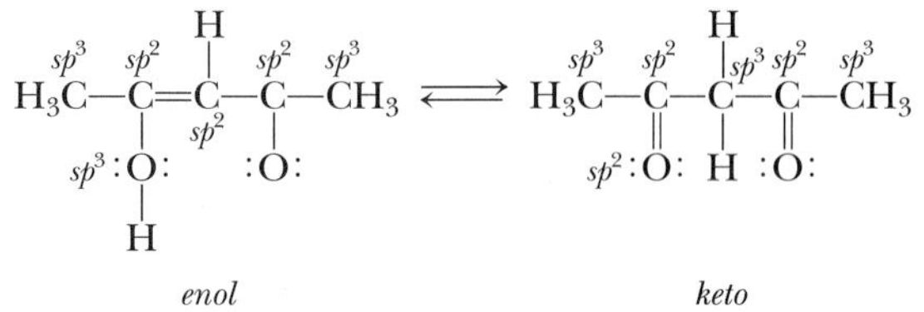

$$\underset{enol}{H_3C-C(OH)=CH-C(=O)-CH_3} \rightleftharpoons \underset{keto}{H_3C-C(=O)-CH_2-C(=O)-CH_3}$$

(enol: sp^3, sp^2, sp^2, sp^2, sp^3 carbons; O of OH sp^3. keto: sp^3, sp^2, sp^3, sp^2, sp^3 carbons; C=O oxygen sp^2)

(c) The CH_3 C atoms have tetrahedral electron-pair and molecular geometries. The two C atoms with O atoms attached have trigonal-planar electron-pair and molecular geometries. The electron-pair and molecular geometry of the central C atom changes from trigonal-planar in the *enol* form to tetrahedral in the *keto* form.

(d)

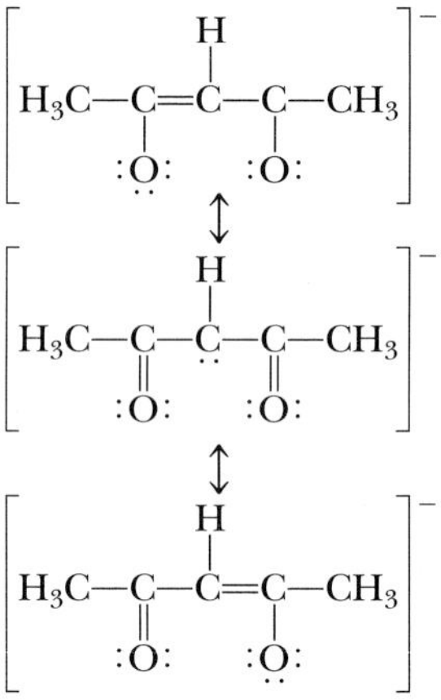

$$[H_3C-C(-\ddot{O}:^-)=CH-C(=O)-CH_3]^- \leftrightarrow [H_3C-C(=O)-\ddot{C}H-C(=O)-CH_3]^- \leftrightarrow [H_3C-C(=O)-CH=C(-\ddot{O}:^-)-CH_3]^-$$

(e) 6.80 g $CrCl_3$; 12.9 g $C_5H_8O_2$; 5.15 g NaOH

CHAPTER 11

11.2. $C_{12}H_{26}$

11.4. 2,2,4-trimethylpentane and two other isomers

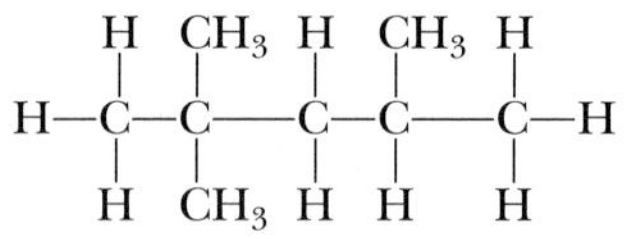

$(CH_3)_3C-CH_2-CH(CH_3)-CH_3$
2,2,4-trimethylpentane

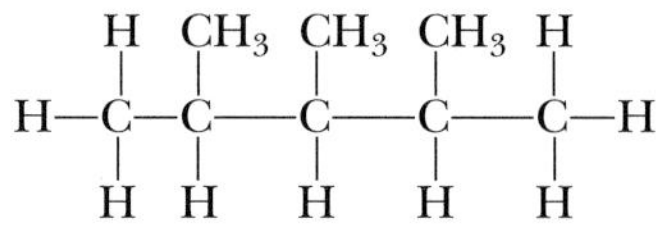

$CH_3-CH(CH_3)-CH(CH_3)-CH(CH_3)-CH_3$
2,3,4-trimethylpentane

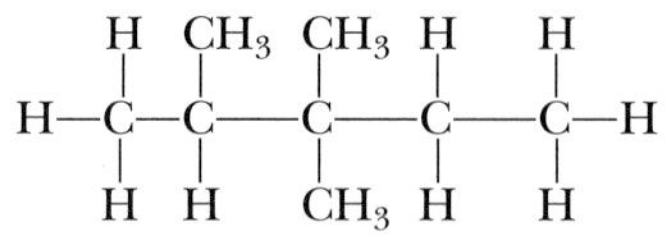

$CH_3-CH(CH_3)-C(CH_3)_2-CH_2-CH_3$
2,3,3-trimethylpentane

11.6. Illustrated isomer: 2,5-dimethylheptane

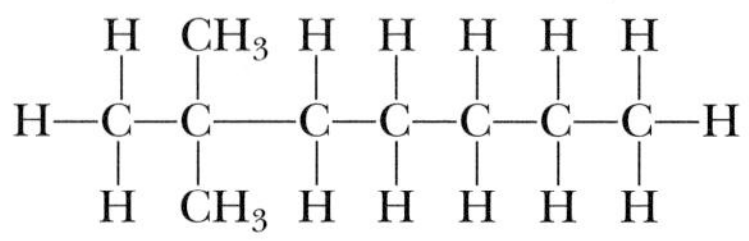

$(CH_3)_3C-CH_2-CH_2-CH_2-CH_2-CH_3$
2,2-dimethylheptane

11.8. The C—C—C angles is slightly greater than 109°, so the ring cannot be planar.

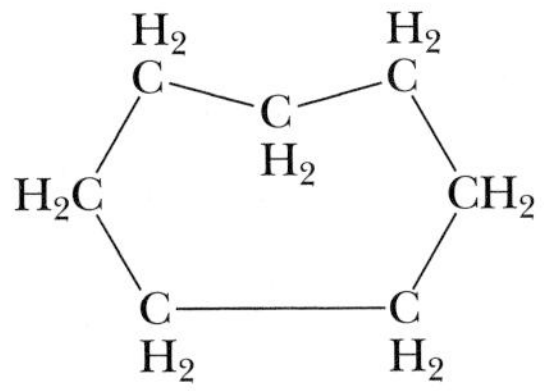

C_7H_{14}, cycloheptane

11.10. Hexene isomers:

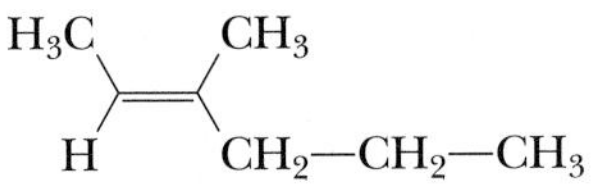

trans-3-methyl-2-hexene

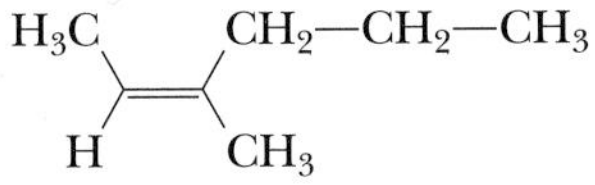

cis-3-methyl-2-hexene

11.12. (a) Alkenes with the formula C_5H_{10}

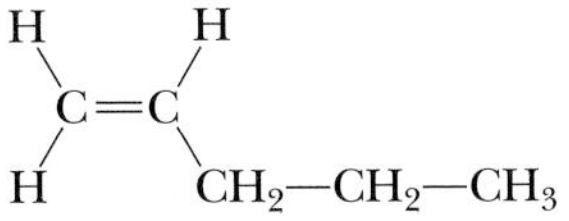

1-pentene

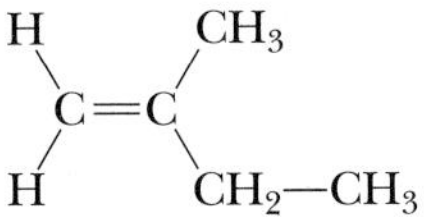

2-methyl-1-butene

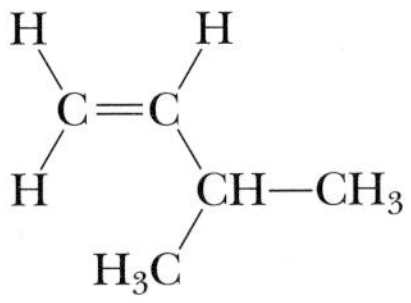

3-methyl-1-butene

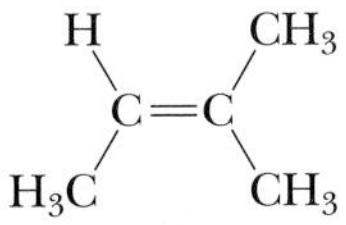

2-methyl-2-butene

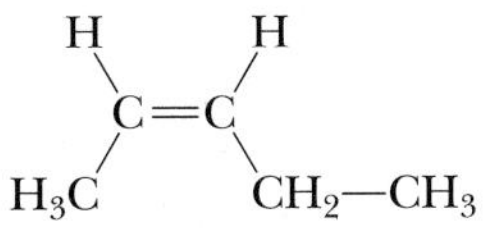

cis-2-pentene

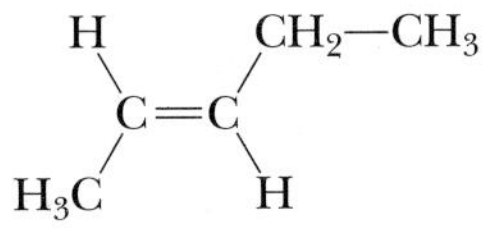

trans-2-pentene

(b) Cyclopentane

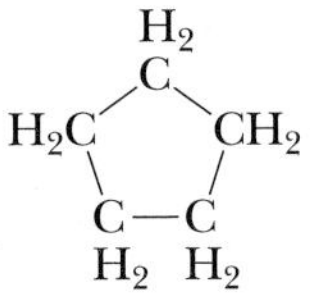

11.14. (a) 1,2-dibromopropane: $CH_3CHBrCH_2Br$
(b) $CH_3CH_2CH_2CH_2CH_3$

11.16. 1-butene: $CH_3CH_2CH{=}CH_2$

11.18.

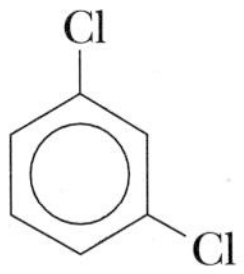

m-dichlorobenzene

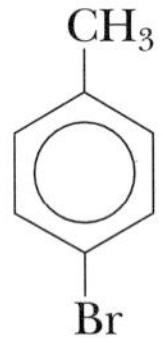

p-bromotoluene

11.20.

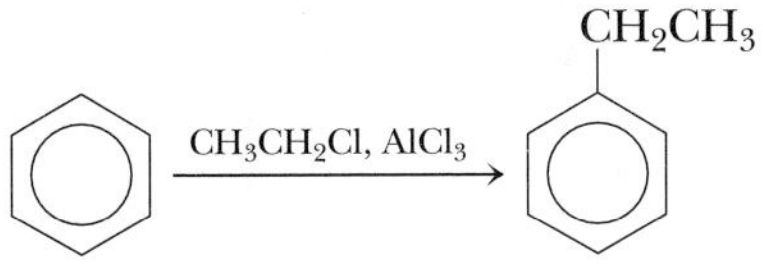

$$C_6H_6 \xrightarrow{CH_3CH_2Cl,\ AlCl_3} C_6H_5CH_2CH_3$$

11.22. (a) 1-propanol, primary alcohol
(b) 1-butanol, primary alcohol
(c) 2-methyl-2-propanol, secondary alcohol
(d) 2-methyl-2-butanol

11.24. (a) ethylamine, $CH_3CH_2NH_2$

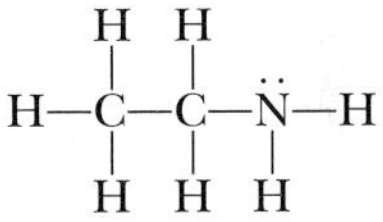

(b) dipropylamine, $(CH_3CH_2CH_2)_2NH$

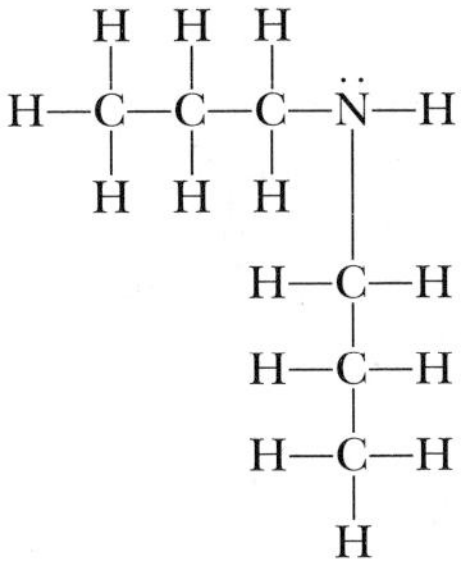

(c) butyldimethylamine, $CH_3CH_2CH_2CH_2N(CH_3)_2$

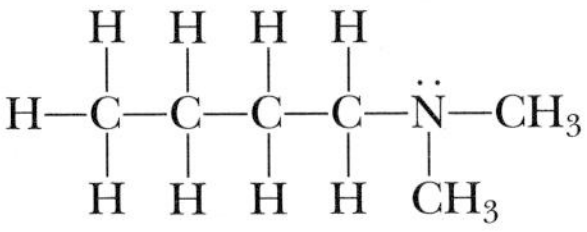

11.26.

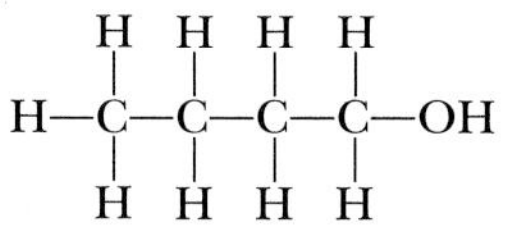

```
    H  H  H  H
    |  |  |  |
 H—C—C—C—C—OH
    |  |  |  |
    H  H  H  H
```

1-butanol

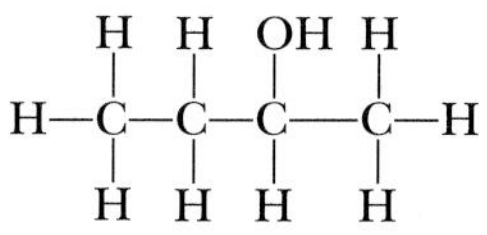

```
    H  H  OH  H
    |  |  |   |
 H—C—C—C——C—H
    |  |  |   |
    H  H  H   H
```

2-butanol

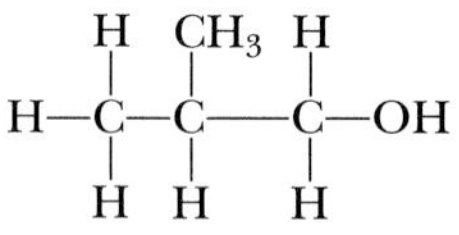

```
    H  CH3  H
    |  |    |
 H—C—C———C—OH
    |  |    |
    H  H    H
```

2-methyl-1-propanol

```
    H  CH3  H
    |  |    |
 H—C—C———C—H
    |  |    |
    H  OH   H
```

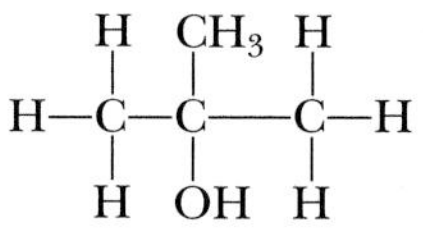

2-methyl-2-propanol

11.28. (a) $CH_3CH_2OCH_2CH_3 + H_2O$
(b) $[CH_3CH_2CH_2O]Na + \frac{1}{2} H_2$

11.30. (a) 2-pentanone

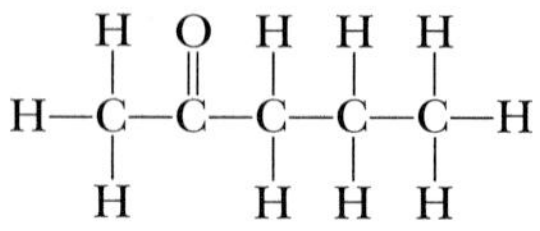

```
    H  O  H  H  H
    |  ‖  |  |  |
 H—C—C—C—C—C—H
    |     |  |  |
    H     H  H  H
```

(b) hexanal

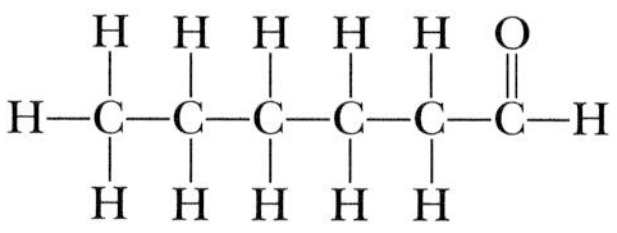

```
    H  H  H  H  H  O
    |  |  |  |  |  ‖
 H—C—C—C—C—C—C—H
    |  |  |  |  |
    H  H  H  H  H
```

(c) pentanoic acid

```
    H  H  H  H  O
    |  |  |  |  ‖
 H—C—C—C—C—C—H
    |  |  |  |
    H  H  H  H
```

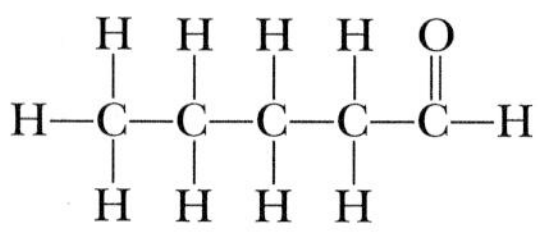

11.32. (a) 3-methylpentanoic acid
(b) methyl propanoate
(c) butyl acetate (or butyl ethanoate)
(d) *p*-bromobenzoic acid

11.34. (a) pentanoic acid (see Question 11.30c)
(b) 1-pentanol

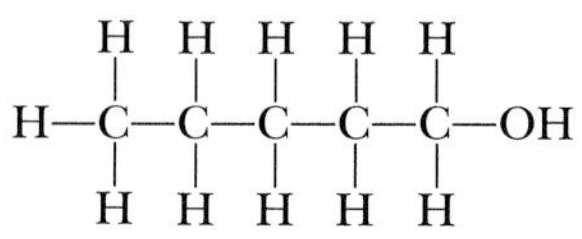

```
    H  H  H  H  H
    |  |  |  |  |
 H—C—C—C—C—C—OH
    |  |  |  |  |
    H  H  H  H  H
```

(c) 2-octanol

```
    H  H  H  H  H  H  OH  H
    |  |  |  |  |  |  |   |
 H—C—C—C—C—C—C—C——C—H
    |  |  |  |  |  |  |   |
    H  H  H  H  H  H  H   H
```

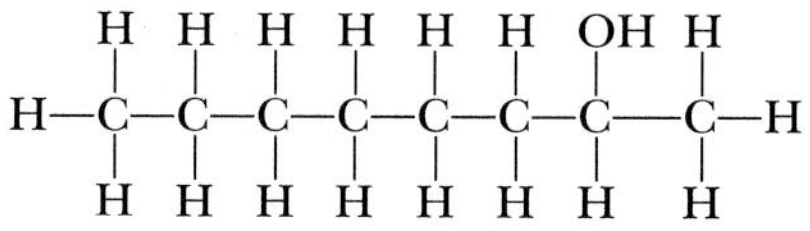

(d) No reaction. Ketone is not oxidized by $KMnO_4$.

11.36. Prepare propyl propanoate:
Step 1: oxidize 1-propanol to propanoic acid

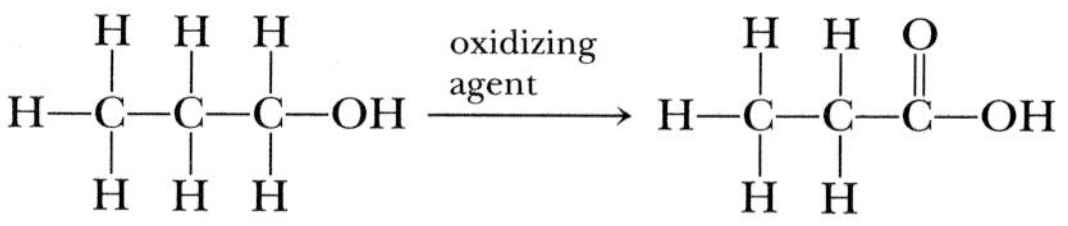

```
    H  H  H      oxidizing          H  H  O
    |  |  |      agent              |  |  ‖
 H—C—C—C—OH  ————————→   H—C—C—C—OH
    |  |  |                         |  |
    H  H  H                         H  H
```

Step 2: combine propanoic acid and 1-propanol

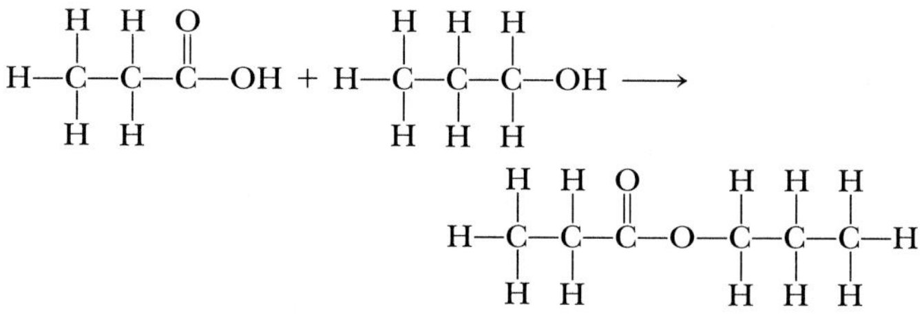

```
    H  H  O           H  H  H
    |  |  ‖           |  |  |
 H—C—C—C—OH + H—C—C—C—OH ——→
    |  |              |  |  |
    H  H              H  H  H

                 H  H  O     H  H  H
                 |  |  ‖     |  |  |
              H—C—C—C—O—C—C—C—H
                 |  |        |  |  |
                 H  H        H  H  H
```

11.38. Sodium acetate ($NaCH_3CO_2$) and 1-butanol, $CH_3CH_2CH_2CH_2OH$

11.40. (a) Prepare polyvinyl acetate (PVA) from vinyl acetate:

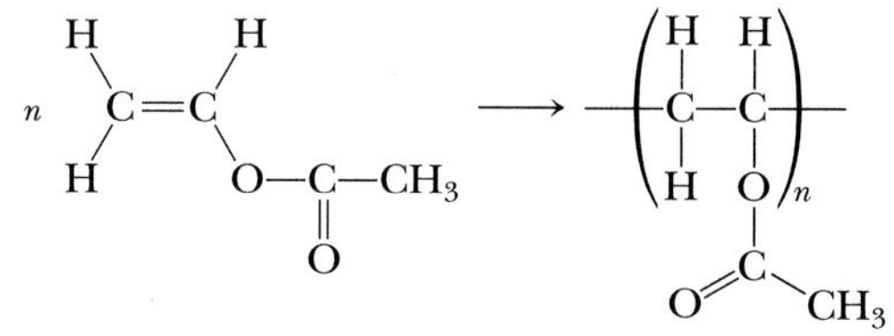

```
     H      H                       ⎛H  H⎞
      \    /                        ⎜|  |⎟
 n     C=C              ——→     —⎜C—C⎟—
      /    \                        ⎜|  |⎟
     H      O—C—CH3                 ⎝H  O⎠n
              ‖                          |
              O                          C
                                       //  \
                                      O     CH3
```

(b) Three units of PVA:

```
 ⎛H  H        H  H        H  H⎞
 ⎜|  |        |  |        |  |⎟
—⎜C—C————C—C————C—C⎟——
 ⎜|  |        |  |        |  |⎟
 ⎝H  O        H  O        H  O⎠
      |            |            |
      C            C            C
    //  \        //  \        //  \
   O    CH3     O    CH3     O    CH3
```

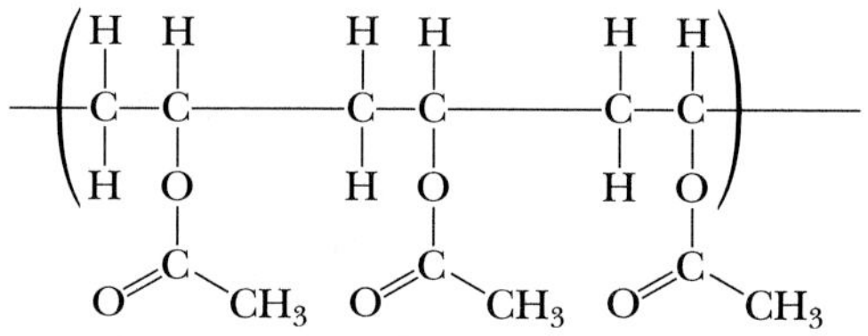

(c) Treat the polyvinyl acetate with base (NaOH).

11.42. Illustrated here is a segment of a copolymer composed of two units of 1,1-dichloroethene and two of chloroethene:

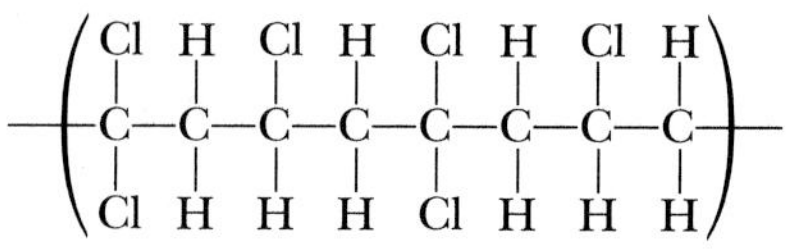

```
  ⎛Cl H  Cl H  Cl H  Cl H⎞
  ⎜|  |  |  |  |  |  |  |⎟
 —⎜C—C—C—C—C—C—C—C⎟—
  ⎜|  |  |  |  |  |  |  |⎟
  ⎝Cl H  H  H  Cl H  H  H⎠
```

11.44. (a) 2,2-dimethylpentane

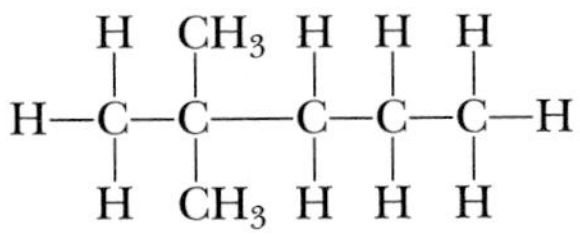

```
    H  CH3  H  H  H
    |   |   |  |  |
H—C—C———C—C—C—H
    |   |   |  |  |
    H  CH3  H  H  H
```

(b) 3,3-diethylpentane

```
          CH3
           |
    H  H  CH2  H  H
    |  |   |   |  |
H—C—C—C———C—C—H
    |  |   |   |  |
    H  H  CH2  H  H
           |
          CH3
```

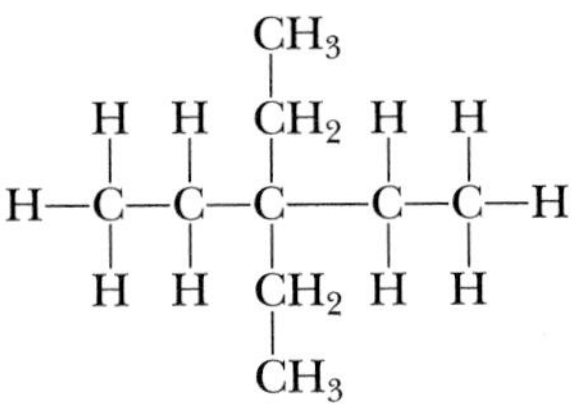

(c) 2-methyl-3-ethylpentane

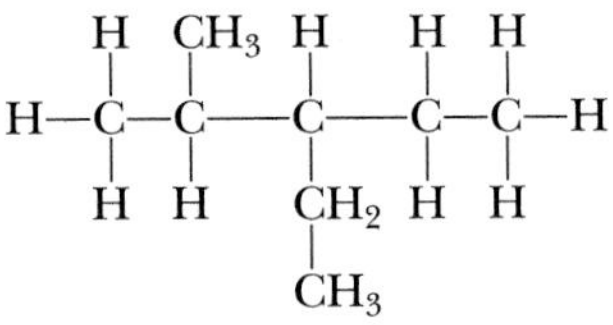

```
    H  CH3  H     H  H
    |   |   |     |  |
H—C—C———C———C—C—H
    |   |   |     |  |
    H   H  CH2    H  H
            |
           CH3
```

(d) 4-ethylheptane

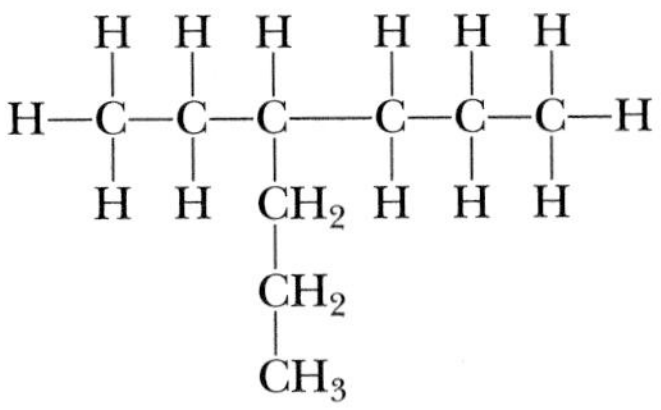

```
    H  H  H      H  H  H
    |  |  |      |  |  |
H—C—C—C———C—C—C—H
    |  |  |      |  |  |
    H  H CH2     H  H  H
          |
         CH2
          |
         CH3
```

11.46.

```
    Cl H  H             Cl Cl H
    |  |  |             |  |  |
H—C—C—C—H         H—C—C—C—H
    |  |  |             |  |  |
    Cl H  H             H  H  H
```

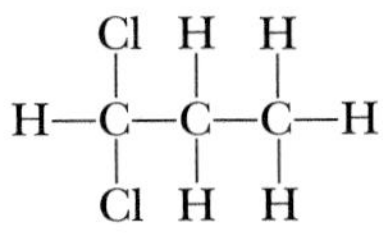

1,1-dichloropropane 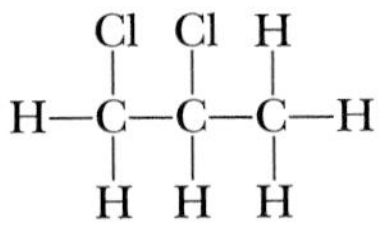 1,2-dichloropropane

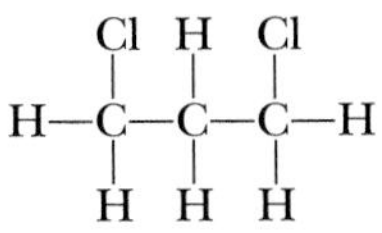

```
    Cl H  Cl            H  Cl H
    |  |  |             |  |  |
H—C—C—C—H         H—C—C—C—H
    |  |  |             |  |  |
    H  H  H             H  Cl H
```

1,3-dichloropropane 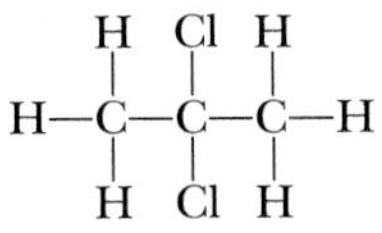 2,2-dichloropropane

11.48.

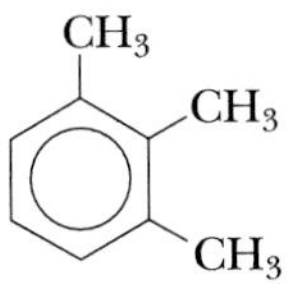

1,2,3-trimethylbenzene 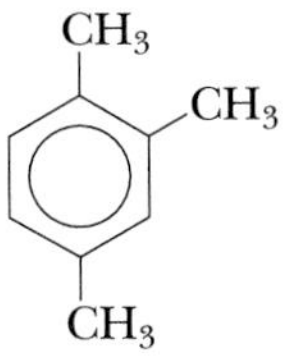 1,2,4-trimethylbenzene 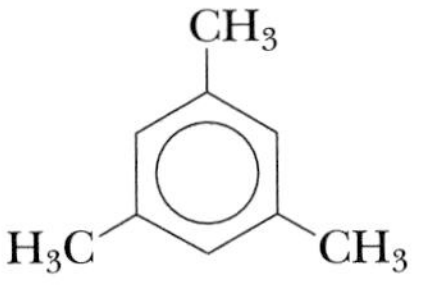 1,3,5-trimethylbenzene

11.50. Replace the carboxylic acid group with an H atom.

11.52.

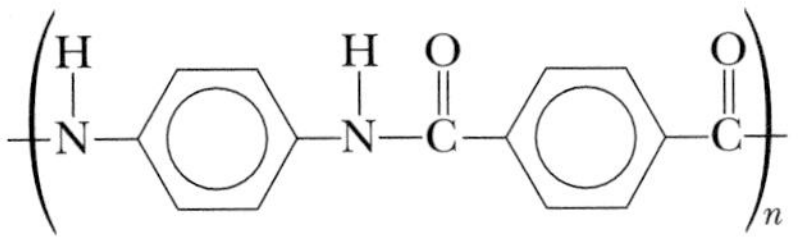

$$\left(\begin{array}{c} H \\ | \\ -N- \end{array} -C_6H_4- \begin{array}{c} H \quad O \\ | \quad \| \\ -N-C- \end{array} -C_6H_4- \begin{array}{c} O \\ \| \\ -C- \end{array} \right)_n$$

11.54.

```
    H     O
    |     ||
H—C—O—C—(CH2)10CH3                      H
    |     O                                 |
    |     ||                           H—C—OH
H—C—O—C—(CH2)10CH3   —NaOH→   H—C—OH + 3 CH3(CH2)10CO2Na
    |     O                            H—C—OH
    |     ||                                |
H—C—O—C—(CH2)10CH3                      H
    |
    H
```

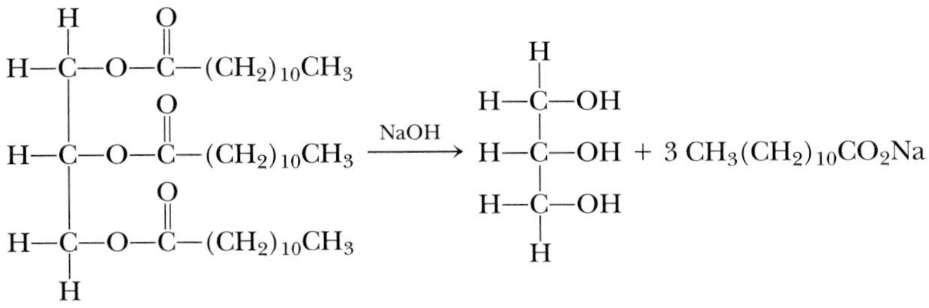

gylceryl trilaurate glycerol sodium laurate

11.56. Pyridine is isoelectronic with benzene. A CH in benzene has been replaced by an N atom in pyridine. Both benzene and pyridine have two resonance structures:

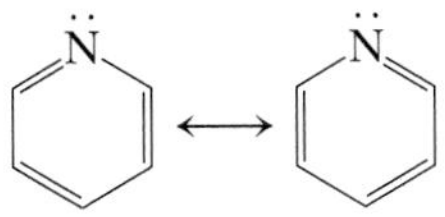

11.58. The alkene undergoes an addition reaction with bromine, whereas cyclopentane does not. When a solution of bromine in carbon tetrachloride is added to the alkene, the color of the bromine rapidly disappears.

11.60. (a) For rotation to occur, the CC π bond must be broken, and this requires energy.

(b) Rotation of one end of the molecule about the center C—C bond brings the H atoms on those two C atoms into proximity, causing a slight amount of repulsion between the electrons about those H atoms.

11.62. (a) CHO

(b) $C_4H_4O_4$

(c) Lewis structure, 44 valence electrons

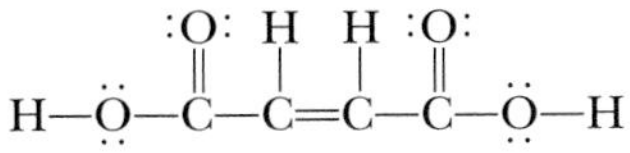

```
         :O: H   H :O:
          ||  |   |  ||
H—Ö—C—C=C—C—Ö—H
```

(d) All C atoms are sp^2-hybridized.

(e) All bond angles around the C atoms are 120°.

CHAPTER 12

12.10. (a) 0.58 atm; (b) 0.59 bar; (c) 5.9×10^4 pascals

12.12. 56.3 mm Hg; 56.3 torr; 0.0741 atm; 7.51 kPa

12.14. 2.70×10^2 mm Hg

12.16. 3.7 L

12.18. 1.8 L

12.20. 3.20×10^2 mm Hg

12.22. Volume of B = 117 mL

12.24. 1.2×10^7 L; the decrease in temperature is offset by the decrease in pressure.

12.26. $P = 0.919$ atm
12.28. Volume = 2.9 L
12.30. 1.9×10^6 g He
12.32. Molar mass = 57.5 g/mol
12.34. Empirical formula = CCl_2F; molar mass = 204 g/mol; molecular formula = $C_2Cl_4F_2$
12.36. Density = 3.7×10^{-4} g/L
12.38. Molar mass = 34.0 g/mol
12.40. 10.2 g O_2
12.42. 58 °C
12.44. 0.0394 mol H_2; 0.096 atm; 73 mm Hg
12.46. 58 g NaN_3
12.48. 1.7 atm O_2
12.50. Ni is the limiting reactant; 1.31 g $Ni(CO)_4$
12.52. 4.1 atm H_2; 1.6 atm Ar
12.54. (a) 0.30 mol halothane/1 mol O_2
(b) 290 g halothane
12.56. $P(N_2) = 0.16$ atm; $P(H_2O) = 0.64$ atm; P_{total} = 0.80 atm
12.58. 732 mm Hg
12.60. (a) They have the same average kinetic energy.
(b) Average speed of H_2 molecules > average speed of CO_2 molecules.
(c) There are twice as many CO_2 molecules as H_2 molecules.
(d) Mass of CO_2 is greater than mass of H_2.
12.62. Average speed of CO_2 molecule = 3.65×10^4 cm/s
12.64. Average speed (and molar mass) increases in the order $CH_2F_2 < Ar < N_2 < CH_4$
12.66. He diffuses faster than Ar (rate He = $3.2 \times$ rate Ar).
12.68. 36 g/mol
12.70. P from van der Waals = 29.5 atm; P(ideal) = 49.2 atm
12.72. (a) Standard atmosphere: 1 atm; 760 mm Hg; 101.325 kPa; 1.013 bar
(b) N_2 partial pressure: 0.780 atm; 593 mm Hg; 79.1 kPa; 0.791 bar
(c) H_2 pressure: 131 atm; 9.98×10^4 mm Hg; 1.33×10^4 kPa; 133 bar
(d) P(air): 0.333 atm; 253 mm Hg; 33.7 kPa; 0.337 bar
12.74. Molar mass = 44.0 g/mol
12.76. 310 K
12.78. 7.71×10^{17} molecules/cm^3
12.80. 64.6% $NaNO_2$
12.82. P(He) = 87 mm Hg; P(Ar) = 140 mm Hg; P_{total} = 230 mm Hg
12.84. ClO_2F
12.86. 0.0125 mol H_2
12.88. (a) 10.0 g O_2 represents more molecules than 10.0 g CO_2.
(b) Average speed of O_2 molecules [32 g/mol] > average speed CO_2 molecules [44 g/mol].
(c) The gases are at the same temperature so they have the same average kinetic energy.
12.90. (a) $P(C_2H_2) > P(CO)$
(b) There are more molecules in the C_2H_2 container than in the CO container.
12.92. (a) Not a gas; gas would expand to an infinite volume.
(b) Not a gas. Density of 8.2 g/mL is typical of a metal.
(c) Insufficient information
(d) Gas
12.94. (a) Because P, V, and T are the same for each tire, each contains the same number of gas molecules?
(b) The unknown gas molecules are 10.0 times heavier than He molecules.
(c) The kinetic energies of the gases are the same, but He has the largest average speed.
12.96. Mass of CO = 0.771 g and P(CO) = 1.22 atm
Mass of CO_2 = 0.22 g and $P(CO_2)$ = 0.22 atm
12.98. $Fe(CO)_5$
12.100. $P(SiH_4)$ = 40. mm Hg and $P(O_2)$ = 80. mm Hg
When reaction is complete, $P_{total} = P(H_2O)$ = 80. mm Hg
12.102. (a) 19 valence electrons
(b) $[:\ddot{O}-\ddot{Cl}-\ddot{O}:]^-$
(c) Cl is sp^3-hybridized, and the ion is bent.
(d) O_3 has an ideal bond angle of 120°, whereas the O—Cl—O angle in ClO_2^- is about 109°.
(e) Cl_2 is the limiting reactant, so 3.24 atm is the pressure of ClO_2.

CHAPTER 13

13.16. (a) Dipole–dipole interactions (and hydrogen bonds)
(b) Induced-dipole/induced-dipole forces
(c) Ion–dipole force
(d) Dipole–dipole interactions (and hydrogen bonds)
13.18. Induced-dipole/induced-dipole forces must be overcome in solid I_2. Dipole–dipole forces (and hydrogen bonds) exist between methanol molecules. In the $CH_3OH—I_2$ solution dipole/induced-dipole forces occur.
13.20. (a) Induced-dipole/induced-dipole forces
(b) Induced-dipole/induced-dipole forces
(c) Dipole–dipole forces
(d) Dipole–dipole forces (and hydrogen bonding)
13.22. Increasing strength: $Ne < CH_4 < CO < CCl_4$. Neon (Ne) exists as a gas.
13.24. (a) O_2 (has a higher mass than N_2)
(b) SO_2 (has a dipole moment, whereas CO_2 is nonpolar)
(c) HF (exhibits strong hydrogen bonding)
(d) GeH_4 (has a greater mass than SiH_4)
13.26. (c) HF; (d) acetic acid; (f) CH_3OH
13.28. (a) LiCl
(b) $Mg(NO_3)_2$
(c) $NiCl_2$
13.30. q = +90.1 kJ
13.32. Refer to Figure 13.19:
(a) Water vapor pressure is about 150 mm Hg at 60 °C.
(b) 600 mm Hg at about 93 °C
(c) At 70 °C, ethanol has a vapor pressure of about 520 mm Hg, whereas that of water is about 225 mm Hg.

13.34. At 30 °C the vapor pressure of ether is about 590 mm Hg. (This pressure requires 0.23 g of ether in the vapor phase at the given conditions, so there is sufficient ether in the flask.) At 0 °C the vapor pressure is about 160 mm Hg, so some ether condenses when the temperature declines.

13.36. (a) CS_2, about 620 mm Hg; CH_3NO_2, about 80 mm Hg
(b) CS_2, induced-dipole/induced-dipole forces; CH_3NO_2, dipole–dipole forces
(c) CS_2, about 46 °C; CH_3NO_2, about 100 °C
(d) About 39 °C
(e) About 34 °C

13.38. Because the critical pressure of CO is −140.3 °C, no amount of pressure applied at or above room temperature will liquefy CO.

13.40. Two possible unit cells are illustrated here.

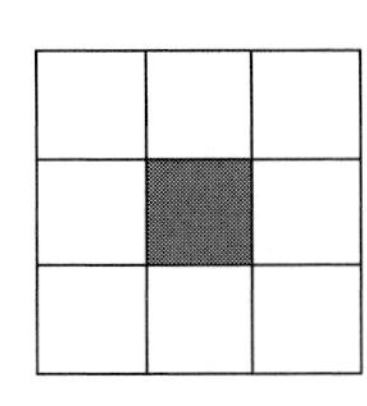

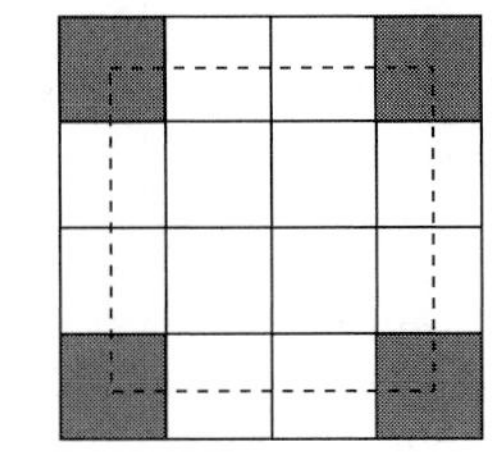

13.42. 1.356×10^{-8} cm (literature value is 1.357×10^{-8} cm)

13.44. Copper is face-centered cubic. (One approach to solving the problem is to use the copper radius to calculate the metal density, assuming each different cell type. Only when fcc is assumed do the calculated and experimental densities match.)

13.46. TlCl has a unit cell like that of CsCl (page 618), that is, there is 1 TlCl per unit cell. (One approach to solving this is similar to that in Question 13.44.)

13.48. Unit cell has 1 Ca atom, 1 Ti atom (each corner of the unit cell has 1/8 Ti atom), and 3 O atoms (each edge has 1/4 O atom). The formula is $CaTiO_3$.

13.50. ZnS. Each cell corner accounts for 1/8 Zn^{2+} ion (8 corners × 1/8 ion) and there are 8 S^{2-} ions within the cell.

13.52. (a) 8 C atoms per unit cell
(b) Unit cell could be described as fcc with 4 additional C atoms in tetrahedral holes within the lattice.
(c) Volume of unit cell = 4.55×10^{-23} cm^3. Edge = 357 pm

13.54. (a) Density of liquid CO_2 is less than that of solid CO_2.
(b) CO_2 is a gas at 5 atm and 0 °C.
(c) Critical temperature = 31 °C

13.56. q (to heat liquid) = 9.4×10^2 kJ
q (to vaporize NH_3) = 1.6×10^4 kJ
q (to heat the vapor) = 8.8×10^2 kJ
$q_{total} = 1.8 \times 10^4$ kJ

13.58. q (for fusion) = −1.97 kJ. q (for melting) = +1.97 kJ

13.60. $Ar < CO_2 < CH_3OH$

13.62. O_2 phase diagram. (i) Note the slight positive slope of the solid–liquid equilibrium line. This indicates the density of solid O_2 is greater than that of liquid O_2. (ii) Using the diagram here, the vapor pressure of O_2 at 74 K is between 150 mm Hg and 200 mm Hg.

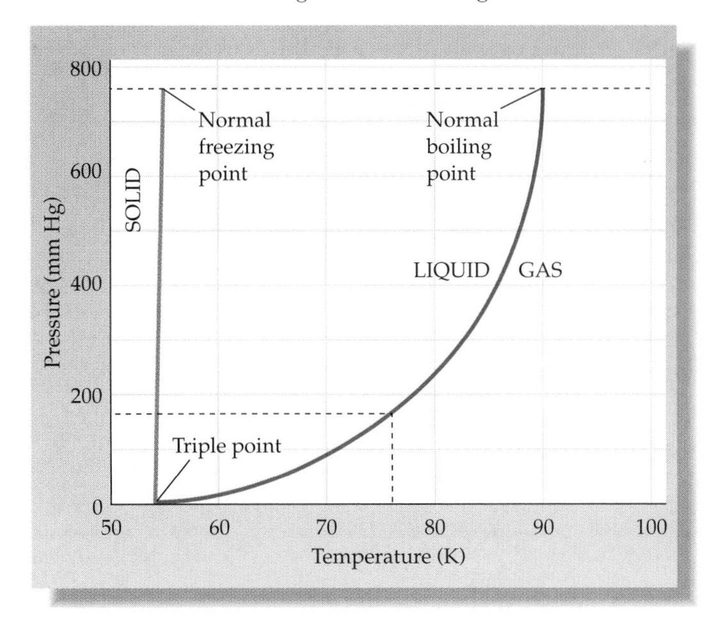

13.64. Acetone and water can interact by hydrogen bonding.

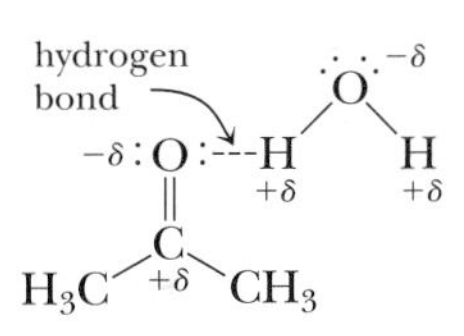

13.66. All the ethanol will evaporate. To achieve the equilibrium vapor pressure in the room 60. mol of ethanol is required, but only 17 mol is available.

13.68. Glycol's viscosity will be greater than ethanol's owing to the greater hydrogen bonding capacity of glycol.

13.70. $q(NH_3) = -8.00 \times 10^2$ J and $q(H_2O) = -7.54 \times 10^2$ J

13.72. (a) About 350 mm Hg
(b) Ethanol has stronger intermolecular forces than CS_2, resulting in a higher boiling point for ethanol.
(c) About 84 °C
(d) CS_2, 46 °C; ethanol, 78 °C; heptane, 99 °C

13.74. (a) Body-centered cubic
(b) 2 W atoms per unit cell
(c) 136.6 pm

13.76. Strong hydrogen bonding can occur between propanol molecules, whereas the dipole–dipole forces that exist between ether molecules are somewhat weaker.

13.78. NaCl has a 1:1 ratio of cations and anions in the unit cell, whereas the unit cell of $CaCl_2$ must have a 1:2 ratio of cations to anions.

13.80. A face-centered lattice of X anions with M cations would have a 1:1 ratio of anions to cations (and a formula of MX). The unit cell of Na_3PO_4 would require a ratio of 1 anion to 3 cations, an arrangement not possible in a NaCl-like unit cell.

13.82. When the can is inverted in cold water, the water vapor pressure in the can, that was approximately 760 mm Hg, drops rapidly, say to 9 mm Hg at 10 °C. This creates a partial vacuum in the can, and the can is crushed because of the difference in pressure inside the can and the pressure of the atmosphere pressing down on the outside of the can.

13.84. 5.5×10^{19} mercury atoms per cubic meter

13.86. Empty space = 47.6%

13.88. (a) The O—S—O angle is about 120° because the electron-pair geometry is trigonal-planar. The molecular geometry is bent.

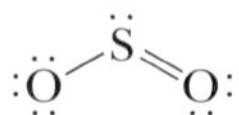

(b) Dipole–dipole forces
(c) $CH_4 < NH_3 < SO_2 < H_2O$
(d) $\Delta H^\circ_{rxn} = -98.89$ kJ for the oxidation of SO_2 to SO_3
$\Delta H^\circ_{rxn} = -227.72$ kJ for the formation of H_2SO_4

CHAPTER 14

14.10. Molarity, 3.82×10^{-2} M; molality, 3.82×10^{-2} *m*; mole fraction = 6.87×10^{-4}; weight percentage = 0.509%

14.12. NaI: 0.15 *m*; 2.2%; X = 2.7×10^{-3}
CH_3CH_2OH: 1.1 *m*; 5.0%; X = 0.02
$C_{12}H_{22}O_{11}$: 0.15 m; 4.9%; X = 2.7×10^{-3}

14.14. 2.65 g Na_2CO_3; $X(Na_2CO_3) = 3.59 \times 10^{-3}$

14.16. 220 g glycol; 5.7 *m*

14.18. K_2CO_3: 0.216 g; 125 g; 0.125 *m*; X = 2.25×10^{-4}
C_2H_5OH: 13.5 g; 175 g; 1.67 *m*; X = 0.0293
$NaNO_3$: 2.70×10^2 g; 555 g; 5.72 *m*; X = 0.0934

14.20. 2.00×10^2 *m*; 17.8 M

14.22. (a) X(NaOH) = 0.162
(b) 30.0% NaOH
(c) 9.97 M

14.24. Molality $[Ca(NO_3)_2] = 0.016$ *m*; Molality of ions = 3×0.016 *m* = 0.048 *m*

14.26. Molality = 2.6×10^{-5} *m* (assuming that 1 kg of sea water is equivalent to 1 kg of solvent)

14.28. (b) and (c)

14.30. $\Delta H^\circ_{solution}$ for LiCl = −37.0 kJ/mol. This is an exothermic heat of solution, as compared with the very slightly endothermic value for NaCl.

14.32. Above about 40 °C the solubility increases with temperature; therefore, add more NaCl and raise the temperature.

14.34. Solubility of $O_2 = 7 \times 10^{-5}$ M

14.36. 1130 mm Hg or 1.49 atm

14.38. 0.150 mol urea; 0.555 mol H_2O; 17.6 mm Hg water vapor

14.40. $X(H_2O) = 0.870$; 16.7 mol glycol; 1.04×10^3 g glycol

14.42. Calculated boiling point = 84.2 °C

14.44. $\Delta t_{bp} = 0.808$ °C; Solution boiling point = 62.51 °C

14.46. Molality = 0.162 *m*; 0.00810 mol solute; 1.4 g solute

14.48. 0.10 *m* sugar < 0.10 *m* KCl < 0.080 *m* $MgCl_2$. Both KCl and $MgCl_2$ are ionic and give 0.20 *m* and 0.24 *m* ions, respectively. Sugar consists of molecules, so the solute concentration is 0.10 *m*.

14.50. Molar mass = 360 g/mol; $C_{20}H_{16}Fe_2$

14.52. Molar mass = 150 g/mol

14.54. Molality = 8.60 *m*; 28.5%

14.56. Freezing point = −24.6 °C

14.58. Molar mass = 170 g/mol

14.60. Molar mass = 130 g/mol

14.62. 0.080 *m* $CaCl_2$ < 0.10 *m* NaCl < 0.040 *m* Na_2SO_4 < 0.10 sugar

14.64. (a) $\Delta t_{fp} = -0.348$ °C; fp = −0.348 °C
(b) $\Delta t_{bp} = +0.0959$ °C; bp = 100.0959 °C
(c) $\Pi = 4.58$ atm
The osmotic pressure is large and can be measured with a small experimental error.

14.66. Molar mass = 6.0×10^3 g/mol

14.68. (a) $BaCl_2(aq) + Na_2SO_4(aq) \longrightarrow BaSO_4(s) + 2\ NaCl(aq)$
(b) Initially the $BaSO_4$ particles form a colloidal suspension.
(c) Over time the particles of $BaSO_4(s)$ grow and precipitate.

14.70. (a) Na_2SO_4 dissociates in water to give three ions ($2\ Na^+$ and SO_4^{2-}), so its effective concentration is 0.30 *m*, greater than the sugar concentration.
(b) 0.15 *m* Na_2SO_4

14.72. (a) 0.456 mol DMG and 11.4 mol ethanol; X(DMG) = 0.0385
(b) 0.869 *m*
(c) VP ethanol over the solution at 78.4 °C = 730.7 mm Hg
(d) bp = 79.5 °C

14.74. (a) 0.12 *m* Na_2SO_4
(b) 0.12 *m* Na_2SO_4
(c) 0.20 *m* $HOCH_2CH_2OH$

14.76. Freezing point = −11 °C

14.78. Molar mass = 160 g/mol; $C_{10}H_{12}O_2$

14.80. 888 g NH_4CHO_2 precipitate

14.82. 81%

14.84. When the egg is placed in water, it swells because the concentration of solute is higher inside the egg than outside. There is therefore a net flow of water into the egg. The situation is opposite when the egg is placed in corn syrup. Water passes out of the egg from a solution of "low" solute concentration to one of relatively higher concentration.

14.86. Molar mass in benzene = 1.20×10^2 g/mol; molar mass in water = 62.4 g/mol. The actual molar mass of acetic acid is 60.1 g/mol. In benzene the molecules of acetic acid form "dimers." That is, two molecules form a single unit through hydrogen bonding. See Figure 13.9 on page 592.

14.88. Although the —OH group of octanol can interact with water by hydrogen bonds, the long, C_8 chain counters that effect. See page 507 for models of methanol and butanol and an explanation of their differing solubility in water.

14.90. Lattice energies decline as the cation ion size increases (NaCl, −786 kJ/mol; KCl, −717 kJ/mol), so we expect those of RbCl and CsCl to be smaller than −717 kJ/mol. The smaller the lattice energies, the larger the solubility.

14.92. $\Delta H^\circ_{solution}$ $[Li_2SO_4]$ = −28.0 kJ/mol
$\Delta H^\circ_{solution}$ [LiCl] = −37.0 kJ/mol
$\Delta H^\circ_{solution}$ $[K_2SO_4]$ = +23.7 kJ/mol
$\Delta H^\circ_{solution}$ [KCl] = +17.2 kJ/mol
Both lithium compounds have exothermic heats of solution, whereas both potassium compounds have endothermic values. Consistent with this is the fact that lithium salts (LiCl) are often more water-soluble than potassium salts (KCl) (see Figure 14.9).

14.94. X(benzene in solution) = 0.67 and X(toluene in solution) = 0.33

$$P_{total} = P_{benzene} + P_{toluene} = 7.3 \text{ mm Hg} + 50. \text{ mm Hg} = 57 \text{ mm Hg}$$

$$\text{X(benzene in vapor)} = \frac{7.3 \text{ mm Hg}}{57 \text{ mm Hg}} = 0.13$$

$$\text{X(toluene in vapor)} = \frac{50. \text{ mm Hg}}{57 \text{ mm Hg}} = 0.87$$

14.96. $i = 1.7$. That is, there is 1.7 mol of ions in solution per mol of compound.

14.98. (a) Calculate the number of moles of ions in 10^6 g H_2O: 550. mol Cl^-; 470. mol Na^+; 53.1 mol Mg^{2+}; 9.42 mol SO_4^{2-}; 10.3 mol Ca^{2+}; 9.72 mol K^+; 0.839 mol Br^-. Total moles of ions = 1.10×10^3 per 10^6 g water. This gives Δt_{fp} of −2.05 °C.
(b) Π = 27.0 atm. This means that a minimum pressure of 27 atm would have to be used in a reverse osmosis device.

14.100. (a) $i = 2.06$
(b) There are approximately 2 particles in solution, so $H^+ + HSO_4^{2-}$ best represents H_2SO_4 in aqueous solution.

14.102. (a) Molar mass = 98.9 g/mol. Empirical formula = BF_2, and molecular formula = B_2F_4.
(b)

CHAPTER 15

15.18. (a) $-\frac{1}{2}\frac{\Delta[O_3]}{\Delta t} = \frac{1}{3}\frac{\Delta[O_2]}{\Delta t}$
(b) $-\frac{1}{2}\frac{\Delta[HOF]}{\Delta t} = \frac{1}{2}\frac{\Delta[HF]}{\Delta t} = \frac{\Delta[O_2]}{\Delta t}$

15.20. (a) The graph of [B] (product concentration) versus time shows [B] increasing from zero. The line is curved, indicating the rate changes with time; this means that the rate depends on concentration. Rates for the four 10-s intervals are: from 0–10 s, 0.0326 mol/L s; from 10–20 s, 0.0246 mol/L s; from 20–30 s, 0.0178 mol/L s; from 30–40 s, 0.0140 mol/L s.
(b) $-\frac{\Delta[A]}{\Delta t} = \frac{1}{2}\frac{\Delta[B]}{\Delta t}$ throughout the reaction
In the interval from 10–20 s, $\frac{\Delta[A]}{\Delta t} = -0.0123 \frac{\text{mol}}{\text{L} \cdot \text{s}}$
(c) Instantaneous rate when [B] = 0.750 mol/L = $\frac{\Delta[B]}{\Delta t} = 0.0163 \frac{\text{mol}}{\text{L} \cdot \text{s}}$

15.22. (a) Rate = $k[NO_2][O_3]$
(b) If $[NO_2]$ is tripled, the rate triples.
(c) If $[O_3]$ is halved, the rate is halved.

15.24. (a) The reaction is second order in [NO] and first order in $[O_2]$.
(b) $\frac{\Delta[NO]}{\Delta t} = -k[NO]^2[O_2]$
(c) $k = 12.5\ L^2/mol^2 \cdot s$
(d) $\frac{\Delta[NO]}{\Delta t} = -1.4 \times 10^{-5}$ mol/L · s
(e) When $\Delta[NO]/\Delta t = -1.0 \times 10^{-4}$ mol/L · s, $\Delta[O_2]/\Delta t = 5.0 \times 10^{-5}$ mol/L · s and $\Delta[NO_2]/\Delta t = 1.0 \times 10^{-4}$ mol/L · s.

15.26. $k = 0.0410\ h^{-1}$

15.28. 5.0×10^2 min

15.30. 105 min

15.32. (a) $t_{1/2}$ = 1400 s
(b) 4600 s

15.34. 4.48×10^{-3} mol (0.260 g) of azomethane remain; 0.0300 mol N_2 is formed.

15.36. Fraction of ^{64}Cu = 0.030 (or expressed as a percent, 3.0 %)

15.38. (a) A graph of ln[sucrose] versus time produces a straight line, indicating that the reaction is first order in [sucrose]
(b) $\Delta[\text{sucrose}]/\Delta t = -k[\text{sucrose}]$; $k = 3.68 \times 10^{-3}$ min^{-1}
(c) At 175 min, [sucrose] = 0.167 M

15.40. The straight line obtained in a graph of $\ln[N_2O]$ versus time indicates a first-order reaction.
k = (− slope) = 0.0127 min^{-1}
The rate when $[N_2O]$ = 0.035 mol/L is 4.4×10^{-4} mol/L min.

15.42. 102 kJ/mol

15.44. 0.3 s^{-1}

15.46.

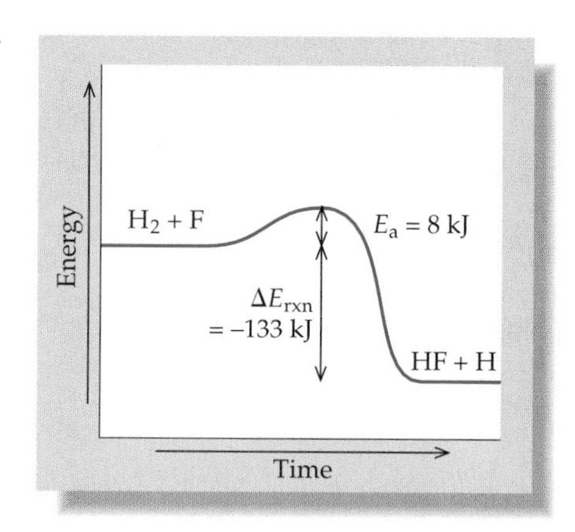

15.48. (a) Rate = $k[NO_3][NO]$
(b) Rate = $k[Cl][H_2]$
(c) Rate = $k[(CH_3)_3CBr]$

15.50. (a) Step 2
(b) Rate = $k[CHCl_3][Cl]$

15.52. (a) The sum of the two reactions is $NO_2(g) + CO(g) \longrightarrow CO_2(g) + NO(g)$.
(b) Both steps are bimolecular reactions.
(c) Rate = $k[NO_2]^2$.
(d) $NO_3(g)$

15.54. Rate = $k[H_2]^{1/2}[CO]$

15.56. (a) True
(b) False. A catalyst is not consumed in the reaction.
(c) False. Catalysts can be in the same phase or in a different phase from the reactants.

15.58. 800 kg CO_2/h

15.60. (a) Rate = $k[NO_2][CO]$
(b) The reaction is first order with respect to both CO and NO_2.
(c) $k = 1.9$ L/mol · h

15.62. (a) 0.59 mg NO_2
(b) 75 h

15.64. The sum of the three reactions is: $2\ O_3(g) \longrightarrow 3\ O_2(g)$, with two molecules of ozone decomposing to three O_2 molecules. Chlorine atoms catalyze the decomposition, and ClO is an intermediate in the reaction.

15.66. (a) True
(b) True
(c) False. Except when a reaction is zero order, the reaction rate decreases as the reactant concentrations decrease.
(d) False. Third-order reactions are not common, but they are known.

15.68. $E_a = 103$ kJ/mol

15.70. (a) 75.8 s
(b) One third of the PH_3 remains after 1 min.

15.72. (a) (1) $\Delta H = -37$ kJ ; (2) $\Delta H = -214$ kJ; (3) $\Delta H = -251$ kJ
(b) $\Delta H_{act} = 20$ kJ

15.74. For equation 1: rate forward = rate reverse; $k_1[Cl_2] = k_{-1}[Cl]^2$. Solving for [Cl] gives $[Cl] = K[Cl_2]^{1/2}$. Substituting this into the rate equation for the second (rate-determining) step gives Rate = $k_2K[CHCl_3][Cl_2]^{1/2}$

15.76. A finely powdered metal has a much larger surface area on which the reaction can occur.

15.78. Carry out the reaction with ^{18}O-labeled methanol. If the product, methyl acetate, contains ^{18}O, this will prove that the oxygen in the ester linkage comes from the alcohol. The water would therefore come from the —OH group of the acid.

15.80. (a) Use the ideal gas law to calculate the number of moles of Rn, use this (with Avogadro's number and the room volume) to calculate that 3.2×10^{13} atoms/L of Rn are present.
(b) 1.2×10^{11} atoms/L

15.82. The activation energies of the uncatalyzed (E_a) and catalyzed (E'_a) reactions are different. The same amount of energy (ΔE_{rxn}) is evolved in the two reactions.

Uncatalyzed

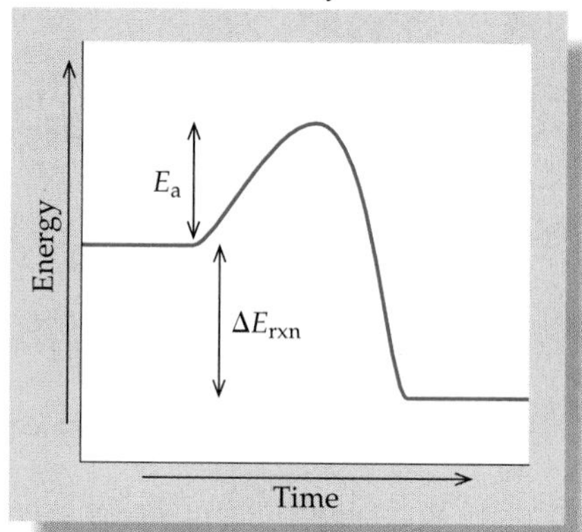

Catalyzed

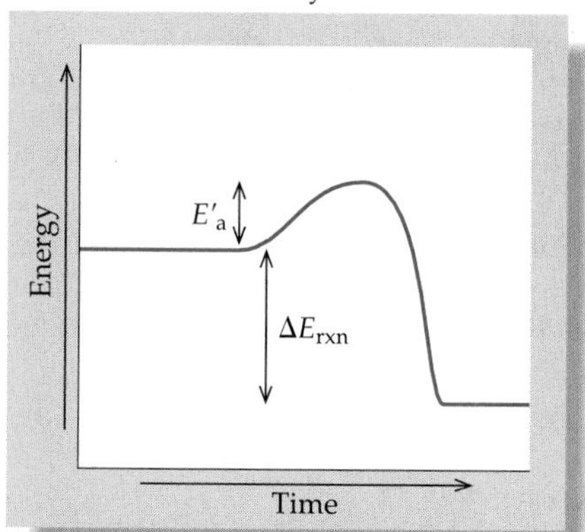

15.84. 4.3 min

15.86. After 30 min, $P(HOF) = 50.0$ mm Hg and $P_{total} = 125$ mm Hg. After 45 min, $P(HOF) = 40.0$ mm Hg and $P_{total} = 130$ mm Hg.

15.88. (a) Step 1 is unimolecular, step 2 is bimolecular.
(b) Rate = $k[Ni(CO)_4]$, in agreement with the mechanism.

(c) 1.5×10^{-3} M
(d) Ni is the limiting reagent. Mass of $Ni(CO)_4$ = 0.364 g and $P_{CO} = 1.25$ atm
(e) 160.8 kJ

CHAPTER 16

16.8. (a) $K_c = \frac{[H_2O]^2[O_2]}{[H_2O_2]}$ (b) $K_c = \frac{[CO_2]}{[CO][O_2]^{1/2}}$
(c) $K_c = \frac{[CO]^2}{[CO_2]}$ (d) $K_c = \frac{[CO_2]}{[CO]}$

16.10. Answer e, $K_2 = \frac{1}{[K_1]^2}$

16.12. $K_c = 13.7$

16.14. (a) $K_c = 2.2 \times 10^5$
(b) $K_c = 4.6 \times 10^{-6}$

16.16. Not at equilibrium; to reach equilibrium, the reaction proceeds to the right, forming more iodine atoms.

16.18. Not at equilibrium; to reach equilibrium, the reaction proceeds to the left, forming more NO_2.

16.20. $K_c = 1.2$

16.22. (a) $K_c = 0.025$
(b) $K_c = 0.025$; the amount of solid does not affect the equilibrium.

16.24. $K_c = 0.035$

16.26. $K_c = 0.029$

16.28. $K_p = 0.215$

16.30. [isobutane] = 0.024 M; [butane] = 0.010 M

16.32. $[CO] = [Br_2] = 0.034$ M

16.34. $[I_2] = 6.14 \times 10^{-3}$ M; $[I] = 4.79 \times 10^{-3}$ M

16.36. $P_{total} = 1.21$ atm

16.38. (a) Equilibrium shifts right.
(b) Equilibrium shifts left.
(c) Equilibrium shifts right.
(d) Equilibrium shifts left.

16.40. (a) [isobutane] = 2.9 M; [butane] = 1.1 M
(b) [isobutane] = 2.9 M; [butane] = 1.1 M

16.42. (a) $K_p = K_c = 1.7 \times 10^{-3}$
(b) $K_c = (1.7 \text{ x } 10^{-3})^{1/2} = 4.1 \times 10^{-2}$
(c) $K_c = \frac{1}{(1.7 \times 10^{-3})} = 5.9 \times 10^2$

16.44. $K_c = 3.9 \times 10^4$

16.46. $K_c = 0.080$

16.48. $P(NO_2) = 0.40$ atm; $P(N_2O_4) = 1.1$ atm

16.50. (a) 84%
(b) The equilibrium shifts to the left.

16.52. $K_p = 3.40$

16.54. $[Zn^{2+}] = [CO_3^{2-}] = 3.9 \times 10^{-6}$ M

16.56. This is a dynamic equilibrium. Initially, the rate of evaporation is greater than the rate of condensation. At equilibrium, these two rates are equal.

16.58. (a) The system is not at equilibrium. It must shift right to reach equilibrium.
(b) $\Delta H^\circ_{rxn} = -199.8$ kJ; raising the temperature shifts the equilibrium to the left, increasing the concentration of the reactants.

16.60. (a) $(NH_3)B(CH_3)_3$
(b) $P(NH_3) = P\{B(CH_3)_3\} = 2.1$ atm; 68% dissociated

16.62. $P(CO) = 1.0 \times 10^{-3}$ atm

16.64. (a) 12.8 g NOBr (NO is the limiting reactant.)
(b) O=N–Br (Lewis structure with lone pairs)
(c) The electron-pair geometry around nitrogen is trigonal-planar. The molecule has a bent molecular geometry.
(d) For the reaction as written: $K_p = 9.6 \times 10^{-3}$

CHAPTER 17

17.12. (a) CN^-; (b) SO_4^{2-}; (c) F^-; (d) NO_2^-; (e) CO_3^{2-}

17.14. (a) $H_3O^+(aq) + NO_3^-(aq)$; $H_3O^+(aq)$ is the conjugate acid of H_2O, and $NO_3^-(aq)$ is the conjugate base of HNO_3.
(b) $H_3O^+(aq) + SO_4^{2-}(aq)$; $H_3O^+(aq)$ is the conjugate acid of H_2O, and $SO_4^{2-}(aq)$ is the conjugate base of H_2SO_4.
(c) H_2O + HF; H_2O is the conjugate base of H_3O^+, and HF is the conjugate acid of F^-.

17.16. $CO_3^{2-}(aq) + H_2O(\ell) \longrightarrow HCO_3^-(aq) + OH^-(aq)$

17.18. As an acid: $HPO_4^{2-}(aq) + H_2O(\ell) \longrightarrow PO_4^{3-}(aq) + H_3O^+(aq)$
As a base: $HPO_4^{2-}(aq) + H_2O(\ell) \longrightarrow H_2PO_4^-(aq) + OH^-(aq)$

17.20.

	Acid (A)	Base (B)	Conjugate base of A	Conjugate acid of B
(a)	HCO_2H	H_2O	HCO_2^-	H_3O^+
(b)	H_2S	NH_3	HS^-	NH_4^+
(c)	HSO_4^-	OH^-	SO_4^{2-}	H_2O

17.22. (a) HF is the strongest acid, and HPO_4^{2-} is the weakest acid; (b) F^-; (c) HF; (d) HPO_4^{2-}

17.24. (a) NH_3 is the strongest base, C_5H_5N is the weakest.
(b) $C_5H_5NH^+$
(c) C_5H_5N has the strongest conjugate acid; NH_3 has the weakest conjugate base.

17.26. Answer is (c). HOCl, the weakest acid in this list (Table 17.4), has the strongest conjugate base.

17.28. $NH_4^+(aq) + H_2PO_4^-(aq) \rightleftharpoons NH_3(aq) + H_3PO_4(aq)$
The reaction does not occur to a significant extent. Equilibria favor the weaker acid and base, in this case the reactants.

17.30. Each equilibrium favors the weaker acid and base. The equilibrium positions are therefore: (a) left; (b) left; (c) right; (d) right

17.32. $[H_3O^+] = 4.0 \times 10^{-4}$ M

17.34. pH = 2.89; $[OH^-] = 7.7 \times 10^{-12}$ M

17.36. 11.48

17.38.

	pH	$[H_3O^+]$	$[OH^-]$	acid/base
(a)	1.00	0.10 M	1.0×10^{-13} M	acidic
(b)	10.50	3.2×10^{-11} M	3.2×10^{-4} M	basic
(c)	4.89	1.3×10^{-51} M	7.7×10^{-10} M	acidic
(d)	10.36	4.3×10^{-11} M	2.3×10^{-4} M	basic

17.40. (a) 1.6×10^{-4} M
(b) Moderately weak, $K_a = 1.1 \times 10^{-5}$

17.42. (a) 2.1×10^{-3} M
(b) $K_a = 3.6 \times 10^{-4}$

17.44. 6.6×10^{-9}

17.46. $[H_3O^+] = [A^-] = 1.3 \times 10^{-5}$ M; [HA] = 0.040 M

17.48. $[H_3O^+] = [CN^-] = 3.2 \times 10^{-6}$ M; [HCN] = 0.025 M; pH = 5.50

17.50. pH = 4.00

17.52. (a) 4-Chlorobenzoic acid is stronger.
(b) Benzoic acid, the weaker acid, has the higher pH.

17.54. $[NH_4^+] = [OH^-] = 1.64 \times 10^{-3}$ M; $[NH_3] = 0.15$ M; pH = 11.22

17.56. $[OH^-] = 0.0102$ M; pH = 1.99; pOH = 12.01

17.58. pH = 3.25

17.60. Highest pH, Na_2S; lowest pH, $AlCl_3$ (which gives the weak acid $[Al(H_2O)_6]^{3+}$ in solution.)

17.62. $[H_3O^+] = 1.1 \times 10^{-5}$ M; pH = 4.98

17.64. $[HCN] = [OH^-] = 3.3 \times 10^{-3}$ M; $[H_3O^+] = 3.0 \times 10^{-12}$ M; $[Na^+] = 0.441$ M

17.66. (a) $K_b = 2.5 \times 10^{-5}$
(b) pH = 2.85

17.68. pH = 12.29

17.70. (a) pH = 1.17; $[SO_3^{2-}] = 6.2 \times 10^{-8}$ M

17.72. (a) $[OH^-] = [N_2H_5^+] = 9.2 \times 10^{-5}$ M; $[N_2H_6^{2+}] = 8.9 \times 10^{-16}$ M
(b) pH = 9.96

17.74. (a) Lewis acid; (b) Lewis base; (c) Lewis base; (d) Lewis base; (e) Lewis acid

17.76. BH_3 is a Lewis acid.

17.78. ICl_3 is a Lewis acid. In ICl_3, I is sp^3d-hybridized; in ICl_4^- it is hybridized sp^3d^2.

$ICl_3 \xrightarrow{+Cl^-} ICl_4^-$

ICl_3 is T-shaped ICl_4^- is square planar

17.80. This reaction is best viewed as a Brønsted acid–base reaction; H^- is a Brønsted base and H_2O is a Brønsted acid.

17.82. pH = 2.672

17.84. $H_2S(aq) + CH_3CO_2^-(aq) \rightleftharpoons CH_3CO_2H(aq) + HS^-(aq)$
The equilibrium favors the reactants.

17.86. pH = 12.13

17.88. $K_a = 1.4 \times 10^{-5}$

17.90. (a) $K_b = 4.3 \times 10^{-4}$
(b) The $C_4H_9NH_3^+$ ion is a stronger acid than HPO_4^{2-}. $C_4H_9NH_2$ is a weaker base than $[Ni(H_2O)_5OH]^+$. $[Ni(H_2O)_6]^{2+}$ and all acids above it are stronger than $C_4H_9NH_3^+$ and PO_4^{3-}, and bases below it are stronger bases than $C_4H_9NH_2$.
(c) pH = 6.2

17.92. (a), (b), and (i) are less than 7; (c), (f), and (h) equal 7; and (d), (e), and (g) are greater than 7

17.94. pH = 9.26 and pOH = 4.74

17.96. Acidic: (d) and (e); basic: (a), (b), and (f); neutral: (c) and (g)

17.98. $HCl < CH_3CO_2H < NaCl < NH_3 < NaCN < NaOH$

17.100. $K_{net} = K_{a1}K_{a2} = 3.8 \times 10^{-6}$

17.102. $I_2(s) + I^-(aq) \rightleftharpoons I_3^-(aq)$
$I_2(s)$ is a Lewis acid, $I^-(aq)$ a Lewis base.

17.104. (a) Acid strength increases with increased chlorine substitution. This increase is due to the inductive effect, the attraction of electrons in adjacent bonds, by an electronegative atom. Bonding electrons are attracted to the electronegative chlorine atoms, increasing the polarity of the O—H bond and making it easier to separate an H^+ ion from the acid. See Section 17.10 and especially page 827.
(b) Cl_3CCO_2H has the lowest pH, CH_3CO_2H the highest.

17.106. (a) $HClO_4 + H_2SO_4 \rightleftharpoons ClO_4^- + H_3SO_4^+$
(b) $HO{-}SO_2{-}OH \xrightarrow{H^+} [HO{-}S(O)(OH){-}OH]^+$

17.108. (a) Highest pH, $CH_3C_5H_4NH^+$; lowest pH $O_2NC_5H_4NH^+$
(b) Strongest base is $CH_3C_5H_4N$; weakest is $O_2NC_5H_4N$.

17.110. (a) $K_a = 4 \times 10^{-11}$
(b) $K_b = 3 \times 10^{-4}$

17.112. (a) Lewis acid is BF_3; Lewis base is $(CH_3)_2O$.
(b) In BF_3, 120°; in $(CH_3)_2O{\rightarrow}BF_3$, 109.5°
(c) Both O and B are sp^3-hybridized.
(d) $P_{total} = 0.56$ atm

CHAPTER 18

18.8. $K = 6.3 \times 10^9$; equilibrium favors the right side (products).

18.10. (a) $[H_3O^+] = 1.5 \times 10^{-9}$ M; pH = 8.81

18.12. (a) Greater than 7
(b) Less than 7
(c) Equal to 7

18.14. $[Na^+] = [C_6H_5O^-] = 3.23 \times 10^{-2}$ M; $[C_6H_5OH] = [OH^-] = 1.5 \times 10^{-3}$ M; $[H_3O^+] = 6.5 \times 10^{-12}$ M; pH 11.19

18.16. At the equivalence point, $[NH_4^+] = 6.25 \times 10^{-3}$ M, $[H_3O^+] = 1.9 \times 10^{-6}$ M, $[OH^-] = 5.3 \times 10^{-9}$ M; pH = 5.73. In the original solution, $[NH_3] = 1.54 \times 10^{-2}$ M.

18.18. (a) Decrease pH
(b) Increase pH
(c) No change in pH

18.20. (a) pH decreases.
(b) pH increases.

18.22. pH = 9.26

18.24. pH = 9.11; pH of buffer is lower than the pH of the original solution of NH_3.

18.26. 5 g (*Note:* pH has one significant figure.)

18.28. pH = 3.90; dilution will not change the pH.

18.30. $NH_3 + NH_4Cl$ (answer b)

18.32. (a) pH = 4.95
(b) pH = 5.05

18.34. (a) pH = 9.55
(b) pH = 9.50

18.36. pK_a for acetic acid = 4.92; pH of buffer = 4.92

18.38. (a) $pK_a = 3.74$
(b) pH = 3.59
(c) $\frac{[HCO_2H]}{[HCO_2^-]} = 0.4$

18.40. The titration curve begins at pH = 13.00 and drops slowly as HCl is added. Just before the equivalence point (when 30.0 mL of acid has been added), the curve falls steeply. The pH at the equivalence point is exactly 7. Just after the equivalence point, the curve flattens again and begins to approach the final pH of just over 1.0.

18.42. (a) Starting pH = 11.15
(b) pH at equivalence point = 5.27
(c) pH at midpoint (half-neutralization point) = 9.26
(d) Methyl red, bromcresol green
(e) **Acid Added**

(mL)	**pH**
5.00	9.91
15.0	9.18
20.0	8.83
22.0	8.65
30.0	2.34

18.44. (a) pH = 10.75
(b) pH = 5.91
(c) pH = 9.51
(d) Eriochrome black T, bromcresol purple, alizarin
(e) **Acid Added**

(mL)	**pH**
5.00	10.13
10.0	9.72
20.0	9.00
30.0	3.20

18.46. (a) Methyl red
(b) Phenolphthalein
(c) Cresol red, thymol blue

18.48. (a) $CH_3CH_2CO_2H(aq) + OH^-(aq) \rightleftharpoons CH_3CH_2CO_2^-(aq) + H_2O(\ell)$
(b) 0.60 mol $CH_3CH_2CO_2H$, 0.40 mol $CH_3CH_2CO_2^-$
(c) pH = 4.71
(d) The mixture of acid and anion is a buffer, so there is very little change in pH (pH increases by 0.02 units.)

18.50. $[NH_4^+] = 9.0 \times 10^{-5}$ M; $[Na^+] = [OH^-] = 0.040$ M; $[NH_3] = 0.20$ M

18.52. pH of acetic acid = 2.77; pH after adding sodium acetate = 5.20

18.54. pH = 8.25

18.56. $HCl < CH_3CO_2H < CH_3CO_2H/NaCH_3CO_2 < NaCl < NH_3/NH_4Cl < NH_3$

18.58. pH = 3.97

18.60. $\frac{[H_2PO_4^-]}{[HPO_4^{2-}]} = 0.65$

18.62. (a) pH = 3.31
(b) pH = 9.31
(c) pH = 5.82

18.64. (a) To increase the pH, base must be added. The added base reacts with acetic acid to form more acetate in the mixture. As acid is converted to anion, the ratio $[CH_3CO_2H]/[CH_3CO_2^-]$ changes and the pH rises.
(b) At pH = 4 the acid predominates (85% acid to 15% acetate); at pH = 6 the acetate ion predominates (95% acetate to 5% acid).
(c) At the point where the lines cross $[CH_3CO_2H] = [CH_3CO_2^-]$. At this point pH = pK_a; pK_a for acetic acid is 4.75.

18.66. (a) HB is a stronger acid than HA.
(b) A^- is a stronger base than B^-.

18.68. At the midpoint of the titration of a weak base, $[OH^-] = K_b$. Given that $[H_3O^+][OH^-] = K_w$, we see that $[H_3O^+] \cdot K_b = K_w$. Take the negative log of both sides and recognizing that $pK_w = 14$, we get: pH + $(-\log K_b) = 14$. Rearranging this gives the answer.

18.70. The volume of base added to reach the equivalence point is twice the volume of the acid. At the equivalence point, $[C_2O_4^{2-}]$ is 0.033 M (the number of moles of $C_2O_4^{2-}$ equals the number of moles of acid, but the volume at this point is three times as large.) The problem can then be restated: what is the pH of a solution containing 0.033 M $C_2O_4^{2-}$ solution. Consider only the first hydrolysis step, $C_2O_4^{2-}(aq) + H_2O(\ell) \rightleftharpoons HC_2O_4^-(aq) + OH^-(aq)$, for which $K_b = K_w/K_{a2} = 1.6 \times 10^{-10}$, and solve using a standard weak base calculation.

18.72. Add 0.125 mol HCl.

CHAPTER 19

19.8. (a) Silver chloride, AgCl, and lead chloride, $PbCl_2$
(b) Zinc carbonate, $ZnCO_3$, zinc sulfide, ZnS
(c) Iron(II) carbonate, $FeCO_3$, iron (II) oxalate, FeC_2O_4

19.10. (a) and (b) are soluble, (c) and (d) are insoluble.

19.12. (a) AgBr will precipitate.
(b) $PbCl_2$ will precipitate.
(c) No precipitate

19.14. (a) $AgCN(s) \rightleftharpoons Ag^+(aq) + CN^-(aq)$; $K_{sp} = [Ag^+][CN^-] = 1.2 \times 10^{-16}$
(b) $NiCO_3(s) \rightleftharpoons Ni^{2+}(aq) + CO_3^{2-}(aq)$; $K_{sp} = [Ni^{2+}][CO_3^{2-}]^3 = 6.6 \times 10^{-9}$
(c) $AuBr_3(s) \rightleftharpoons Au^{3+}(aq) + 3\ Br^-(aq)$; $K_{sp} = [Au^{3+}][Br^-]^3 = 4.0 \times 10^{-36}$

19.16. $K_{sp} = 3.2 \times 10^{-6}$

19.18. $K_{sp} = 3.6 \times 10^{-3}$

19.20. $K_{sp} = 1.69 \times 10^{-6}$

19.22. $K_{sp} = 2.9 \times 10^{-16}$

19.24. (a) 1.1×10^{-8} mol/L; (b) 1.5×10^{-6} g/L

19.26. 0.21 mg $RaSO_4$

19.28. (a) 1.2×10^{-3} mol/L; (b) 7.3×10^{-2} g/L

19.30. (a) $PbCl_2$; (b) FeS; (c) BiI_3; (d) $Fe(OH)_2$

19.32. $BaCO_3 < Ag_2CO_3 < Na_2CO_3$

19.34. All $NiCO_3$ will dissolve.

19.36. $Q < K_{sp}$; no precipitate.

19.38. $Q > K_{sp}$; $Zn(OH)_2$ will precipitate.

19.40. 1.4×10^{-4} M

19.42. $Q > K_{sp}$; $Mg(OH)_2$ will precipitate

19.44. $Q < K_{sp}$; no precipitate

19.46. $BaSO_4$ precipitates first, then $PbSO_4$, then $SrSO_4$.

19.48. $Fe(OH)_3$ precipitates first, then $Al(OH)_3$, then $Pb(OH)_2$.

19.50. Solubility in water = 1.0×10^{-6} mol/L; solubility in 0.010 M SCN^- = 1.0×10^{-10} mol/L

19.52. (a) Solubility in water = 2.0×10^{-3} mg/mL
(b) Solubility (in 0.020 M Ag^+) = 6.8×10^{-13} mg/mL

19.54. (a) $[C_2O_4^{2-}] = 4.3 \times 10^{-3}$ M; (b) $[Ca^{2+}] = 5.3 \times 10^{-7}$ M

19.56. (a) $PbCO_3$ precipitates first; (b) $[CO_3^{2-}] = 1.7 \times 10^{-7}$ M

19.58. (a) Add H_2SO_4, precipitating $BaSO_4$ and leaving $Na^+(aq)$ in solution.
(b) Add NaOH; the concentration of OH^- needed to precipitate $Bi(OH)_3$ is about 10^6 less than that needed to precipitate $Cd(OH)_2$.

19.60.
$$AuCl(s) \rightleftharpoons Au^+(aq) + Cl^-(aq)$$
$$Au^+(aq) + 2\ CN^-(aq) \rightleftharpoons Ag(CN)_2^-(aq)$$

Net: $AuCl(s) + 2\ CN^-(aq) \rightleftharpoons Au(CN)_2^-(aq) + Cl^-(aq)$
$K_{net} = K_{sp} \cdot K_f = 4.0 \times 10^{25}$

19.62. $K = 1.2 \times 10^6$; yes, AgI forms

19.64. No. AgBr will not dissolve completely. (K_{net} for the equilibrium $AgBr(s) + 2\ NH_3(aq) \rightleftharpoons Ag(NH_3)_2^+(aq) + Br^-(aq)$ is 5.3×10^{-6}.) If all the AgBr dissolved, $[Ag(NH_3)_2^+]$ and $[Br^-]$ would be 1.0×10^{-4} M and $[NH_3]$ would have to be 0.043 M. However, the concentration of NH_3 is only 0.012 M.

19.66. Soluble in HCl: $Ba(OH)_2$ and $BaCO_3$

19.68. Soluble in strong acid: $BaCO_3$ and $BiPO_4$

19.70. $K_{sp} = 3.2 \times 10^{-9}$

19.72. (a) 1.1×10^{-2} M
(b) 45 mg
(c) 1.2×10^{-4} M

19.74. $Q > K_{sp}$; AgCl will precipitate.

19.76. AgI precipitates first, then BiI_3, then PbI_2.

19.78. $K_{sp} = 4.5 \times 10^{-17}$

19.80. $K_{sp} = 1.5 \times 10^{-11}$

19.82. $[Ag^+] = 6.9 \times 10^{-6}$ M; $[NO_3^-] = 0.235$ M; $[K^+] = 0.618$ M; $[CrO_4^{2-}] = 0.191$ M

19.84. (a) $K_{net} = 6.6$
(b) 2.0 g $Na_2S_2O_3$

19.86. Aragonite, having the larger K_{sp}, is more soluble.

19.88. The solubility increases as the pH decreases. Consider the equilibrium: $Ni(OH)_2(s) \rightleftharpoons Ni^{2+}(aq) + 2\ OH^-(aq)$. Adding OH^-, a common ion, shifts the equilibrium to the left, decreasing the solubility. Adding H_3O^+ lowers the pH; added H_3O^+ reacts with $OH^-(aq)$ to form water lowering $[OH^-]$ and causing the equilibrium to shift to the right.

19.90. The K_{sp} calculated (incorrectly) by the student is too high.

19.92. (a) The solubility increases with increased CO_2 pressure.
(b) The solubility increases if the pH is decreased.

19.94. $Ag_3PO_4 > AgCl > AgSCN > Al(OH)_3 > Fe(OH)_3$

19.96. $[Zn^{2+}] = 2 \times 10^{-6}$ M

19.98. (a) $AlCl_3(aq) + H_3PO_4(aq) \rightleftharpoons AlPO_4(s) + 3\ HCl(aq)$
(b) 139 g
(c) $[Al^{3+}] = [PO_4^{3-}] = 1.1 \times 10^{-10}$ M
(d) Solubility will increase. Added H_3O^+ reacts with PO_4^{3-} to form HPO_4^{2-} and shift the solubility equilibrium to the right.
(e) $Q > K_{sp}$; $AlPO_4$ will precipitate. (0.46 g. of $AlPO_4$ will precipitate.)

CHAPTER 20

20.6. (a) Silicon with trace elements present
(b) $H_2O(\ell)$
(c) $I_2(g)$

20.8. (a) $AlCl_3(s)$
(b) $CH_3CH_2I(\ell)$
(c) $NH_4Cl(aq)$

20.10. (a) +3.363 J/K; Graphite is less ordered than diamond, so entropy favors this process.
(b) −102.50 J/K; there is a decrease in order when a gas condenses to form a solid.

(c) $+93.3$ J/K; there is an increase in disorder when a liquid is converted to a gas.

20.12. $\Delta S° = +112$ J/k mol

20.14. $\Delta S° = -270.1$ J/K

20.16. (a) $\Delta S° = -163.34$ J/K
(b) $\Delta S° = -305.32$ J/K
(c) $\Delta S° = -151.9$ J/K

20.18. (a) $\Delta S° = +21.36$ J/K
(b) $\Delta S° = +266.7$ J/K

20.20. Product-favored. $\Delta H°_{rxn} = -205.4$ kJ/mol. The reaction is very favorable based on enthalpy, outweighing the unfavorable entropy (going from 3 mol gases to no moles gases).

20.22. (a) Exothermic (enthalpy-favored), and entropy-disfavored. Product-favored at lower temperatures.
(b) Endothermic (enthalpy-disfavored), and entropy-disfavored. Reactant-favored under all conditions.

20.24. $\Delta S°_{sys} = 46.37$ J/K; $\Delta S°_{surr} = 4788.8$ J/K; $\Delta S°_{universe} = 4835.2$ J/K. The reaction is product-favored, as expected for oxidation of a hydrocarbon.

20.26. (a) $\Delta G° = -175.9$ kJ
(b) $\Delta G° = -91.10$ kJ
(Both reactions are product-favored, and both are enthalpy-driven.)

20.28. (a) $\Delta G_f° = +67.12$ kJ/mol
(b) $\Delta G_f° = +149.55$ kJ/mol
(c) $\Delta G_f° = -204.5$ kJ/mol

20.30. $\Delta G°_{rxn} = -623.60$ kJ/mol; to oxidize 1.00 kg of hydrazine, $\Delta G°_{rxn} = -1.95 \times 10^4$ kJ

20.32. (a) $\Delta G°_{rxn} = -748.1$ kJ; product-favored.
(b) $\Delta G°_{rxn} = +117.078$ kJ; reactant-favored
(c) $\Delta G°_{rxn} = -301.39$ kJ; product-favored

20.34. $\Delta G_f° = -1137.6$ kJ/mol

20.36. $\Delta G°_{rxn} = -89.5$ kJ; product-favored

20.38. $K_p = 7 \times 10^{-16}$; a positive $\Delta G°_{rxn}$ indicates $K < 1$ and a reactant-favored reaction.

20.40. (a) $\Delta G°_{rxn} = -100.97$ kJ; product-favored
(b) $K_p = 5 \times 10^{17}$; the negative sign of $\Delta G°_{rxn}$ and the large value of K_p both indicate a product-favored reaction.

20.42. We predict that the enthalpy will be negative (violent reaction) and the entropy will be positive (formation of a gas). Calculations show $\Delta H°_{rxn} = -184.284$ kJ and surprisingly, $\Delta S°_{rxn} = -7.7$ J/K.

20.44. $\Delta H°_{rxn} = -82.4$ kJ; $\Delta S°_{rxn} = 460$ J/K; $\Delta G°_{rxn} = -219.4$ kJ; the reaction is product-favored.

20.46. Calculated from K_p: $\Delta G°_{rxn} = +4.87$ kJ; Calculated from $\Delta G_f°$ valued in Appendix L: $\Delta G°_{rxn} = 4.73$ kJ

20.48. $\Delta G°_{rxn} = -2.3$ kJ/mol

20.50. Iodine readily dissolves so the process is favorable and $\Delta G°$ must be less than zero. Because $\Delta H°_{soln}$ is approximately zero, this process must be entropy-driven.

20.52. (a) $\Delta G°_{rxn} = +141.73$ kJ so the reaction is reactant-favored at 25 °C.
(b) The reaction will be favorable above about 1050 K.
(c) $K_p = 1 \times 10^4$

20.54. $\Delta H°_{rxn} = -793.36$ kJ; $\Delta S°_{rxn} = -459.16$ J/K; $\Delta G°_{rxn} = -656.46$ kJ; the reaction is product-favored and enthalpy-driven.

20.56. $\Delta S°_{rxn} = -200.3$ J/K. The negative entropy change is expected because the number of moles of gases decreases.

20.58. (a) At 80 °C, $\Delta G° = +100$ J/mol; at 100 °C, $\Delta G° = -30$ J/mol. This means that at 80 °C the rhombic form is more stable. At 100 °C and higher, the monoclinic form is more stable.
(b) $\Delta G° = 0$ J/mol at 370 K. At this temperature rhombic and monoclinic sulfur exist in equilbrium.

20.60. (a) Step 1: $\Delta G°_{rxn} = +490.1$ kJ; $K_p = 1 \times 10^{-86}$
Step 2: $\Delta G°_{rxn} = -737.2$ kJ; $K_p = 1 \times 10^{129}$
(b) *Overall:* $\Delta G°_{net} = -247.1$ kJ; $K_p = 2 \times 10^{43}$
(c) The overall reaction is enthalpy-driven. ($\Delta H°_{net} = -258.0$ kJ; $\Delta S°_{net} = -35.7$ J/K.)

20.62. (a) $\Delta H°_{rxn} = -62.10$ kJ; $\Delta S°_{rxn} = -132.74$ J/K; $\Delta G°_{rxn} = -22.52$ kJ; at 298 K, the reaction is product-favored.
(b) $K_p = 8.79 \times 10^3$; $P(O_2) = 1.1 \times 10^{-4}$ bar
(c) $T = 468$ K

20.64. (a) $\Delta H°_{rxn} = +71.7$ kJ; $\Delta S°_{rxn} = +32.0$ J/K; $\Delta G°_{rxn} = +62.2$ kJ; the reaction is reactant-favored and enthalpy-driven.
(b) The electron-pair geometry is tetrahedral, so the C—O—H angle is about 109.5°
(c) Hybridization of C, sp^2; of oxygen, sp^3
(d) pH = 5.90

CHAPTER 21

21.12. (a) $Cr(s) \longrightarrow Cr^{3+}(aq) + 3\,e^-$
Cr is a reducing agent; this is an oxidation reaction.
(b) $AsH_3(g) + 3\,H_2O(\ell) \longrightarrow As(s) + 3\,H_3O^+(aq) + 3\,e^-$
AsH_3 is a reducing agent; this is an oxidation reaction.
(c) $VO_3^-(aq) + 6\,H_3O^+(aq) + 3\,e^- \longrightarrow V^{2+}(aq) + 9\,H_2O(\ell)$
$VO_3^-(aq)$ is an oxidizing agent; this is a reduction reaction.

21.14. (a) $Cr_2O_7^{2-}(aq) + 14\,H_3O^+ + 6\,e^- \longrightarrow 2\,Cr^{3+}(aq) + 21\,H_2O(\ell)$
$Cr_2O_7^{2-}(aq)$ is an oxidizing agent; this is a reduction reaction.
(b) $CH_3CHO(aq) + 3\,H_2O(\ell) \longrightarrow CH_3CO_2H(aq) + 2\,H_3O^+(aq) + 2\,e^-$
$CH_3CHO(aq)$ is a reducing agent; this is an oxidation reaction.
(c) $Bi^{3+}(aq) + 8\,H_2O(\ell) \longrightarrow HBiO_3(aq) + 5\,H_3O^+(aq) + 2\,e^-$
$Bi^{3+}(aq)$ is a reducing agent; this is an oxidation reaction.

21.16. (a) $Sn(s) + 4\,OH^-(aq) \longrightarrow [Sn(OH)_4]^{2-}(aq) + 2\,e^-$
Sn(s) is a reducing agent; this is an oxidation reaction.
(b) $MnO_4^-(aq) + 2\,H_2O(\ell) + 3\,e^- \longrightarrow MnO_2(s) + 4\,OH^-(aq)$
$MnO_4^-(aq)$ is an oxidizing agent; this is a reduction reaction.

(c) $ClO^-(aq) + H_2O(\ell) + 2\ e^- \longrightarrow Cl^-(aq) + 2\ OH^-(aq)$

$ClO^-(aq)$ is an oxidizing agent; this is a reduction reaction.

21.18. (a)

$$\begin{array}{r} Cl_2(g) + 2\ e^- \longrightarrow 2\ Cl^-(aq) \\ 2\ Br^-(aq) \longrightarrow Br_2(\ell) + 2\ e^- \\ \hline Cl_2(g) + 2\ Br^-(aq) \longrightarrow 2\ Cl^-(aq) + Br_2(\ell) \end{array}$$

(b)

$$\begin{array}{r} 2\ H_3O^+(aq) + 2\ e^- \longrightarrow 2\ H_2(g) + 2\ H_2O(\ell) \\ Sn(s) \longrightarrow Sn^{2+}(aq) + 2\ e^- \\ \hline 2\ H_3O^+(aq) + Sn(s) \longrightarrow H_2(g) + 2\ H_2O(\ell) + Sn^{2+}(aq) \end{array}$$

(c)

$$\begin{array}{r} 2\ \{VO_2^+(aq) + 4\ H_3O^+(aq) + 3\ e^- \longrightarrow V^{2+}(aq) + 6\ H_2O(\ell)\} \\ 3\ \{Zn(s) \longrightarrow Zn^{2+}(aq) + 2\ e^-\} \\ \hline 2\ VO_2^+(aq) + 8\ H_3O^+(aq) + 3\ Zn(s) \longrightarrow 2\ V^{2+}(aq) + 3\ Zn^{2+}(aq) + 12\ H_2O(\ell)\} \end{array}$$

21.20. (a)

$$\begin{array}{r} 2\ \{Ag^+(aq) + e^- \longrightarrow Ag(s)\} \\ HCHO(aq) + 3\ H_2O(\ell) \longrightarrow HCO_2H(aq) + 2\ H_3O^+(aq) + 2\ e^- \\ \hline 2\ Ag^+(aq) + HCHO(aq) + 3\ H_2O(\ell) \longrightarrow 2\ Ag(s) + HCO_2H(aq) + 2\ H_3O^+(aq) \end{array}$$

(b)

$$\begin{array}{r} 3\ \{H_2S(aq) + 2\ H_2O(\ell) \longrightarrow S(s) + 2\ H_3O^+(aq) + 2\ e^-\} \\ Cr_2O_7^{2-}(aq) + 14\ H_3O^+ + 6\ e^- \longrightarrow 2\ Cr^{3+}(aq) + 21\ H_2O(\ell) \\ \hline 3\ H_2S(aq) + Cr_2O_7^{2-}(aq) + 8\ H_3O^+(aq) \longrightarrow 3\ S(s) + 15\ H_2O(\ell) + 2Cr^{3+}(aq) \end{array}$$

(c)

$$\begin{array}{r} 5\ \{H_2C_2O_4(aq) + 2\ H_2O(\ell) \longrightarrow 2\ CO_2(g) + 2\ H_3O^+(aq) + 2\ e^-\} \\ 2\ \{MnO_4^-(aq) + 8\ H_3O^+(aq) + 5\ e^- \longrightarrow Mn^{2+}(aq) + 4\ H_2O(aq)\} \\ \hline 5\ H_2C_2O_4(aq) + 2\ MnO_4^-(aq) + 6\ H_3O^+(aq) \longrightarrow 10\ CO_2(g) + 2\ Mn^{2+}(aq) + 14\ H_2O(\ell) \end{array}$$

21.22. (a)

$$\begin{array}{r} Zn(s) + 2\ OH^-(aq) \longrightarrow Zn(OH)_2(s) + 2\ e^- \\ ClO^-(aq) + H_2O(\ell) + 2\ e^- \longrightarrow Cl^-(aq) + 2\ OH^-(aq) \\ \hline Zn(s) + ClO^-(aq) + H_2O(\ell) \longrightarrow Zn(OH)_2(s) + Cl^-(aq) \end{array}$$

(b)

$$\begin{array}{r} 3\ \{ClO^-(aq) + H_2O(\ell) + 2\ e^- \longrightarrow Cl^-(aq) + 2\ OH^-(aq)\} \\ 2\ \{CrO_2^-(aq) + 4\ OH^-(aq) \longrightarrow CrO_4^{2-}(aq) + 2\ H_2O(\ell) + 3\ e^-\} \\ \hline 3\ ClO^-(aq) + 2\ CrO_2^-(aq) + 2\ OH^-(aq) \longrightarrow 3\ Cl^-(aq) + 2\ CrO_4^{2-}(aq) + H_2O(\ell) \end{array}$$

(c)

$$\begin{array}{r} 5\ \{Br_2(\ell) + 2\ e^- \longrightarrow 2\ Br^-(aq)\} \\ Br_2(\ell) + 12\ OH^-(aq) \longrightarrow 2\ BrO_3^-(aq) + 6\ H_2O(\ell) + 10\ e^- \\ \hline 6\ Br_2(\ell) + 12\ OH^-(aq) \longrightarrow 10\ Br^-(aq) + 2\ BrO_3^-(aq) + 6\ H_2O(\ell) \end{array}$$

21.24. (a–c) $Cr(s) \longrightarrow Cr^{3+}(aq) + 3\ e^-$

Oxidation, anode compartment

$Fe^{2+}(aq) + 2\ e^- \longrightarrow Fe(s)$

Reduction, cathode compartment

21.26. The tin electrode is negative, so it furnishes electrons to the external circuit. The oxidation reaction $Sn(s) \longrightarrow Sn^{2+}(aq) + 2\ e^-$ is the source of these electrons and the tin electrode is the anode. Reduction occurs at the copper electrode, the cathode, and the half-reaction is $Cu^{2+}(aq) + 2\ e^- \longrightarrow Cu(s)$.

21.28. $\Delta G° = -561$ kJ

21.30. (a) $E°_{net} = -1.298$ V; the reaction is not product-favored.

(b) $E°_{net} = +0.51$ V; the reaction is product-favored.

(c) $E°_{net} = +1.023$ V; the reaction is product-favored.

21.32. (a) $Sn(s) + 2\ Ag^+(aq) \longrightarrow Sn^{2+}(aq) + 2\ Ag(s)$

$E° = -0.94$ V, the reaction is not product-favored.

(b) $Zn(s) + Sn^{4+}(aq) \longrightarrow Zn^{2+}(aq) + Sn^{2+}(aq)$

$E° = +\ 0.91$ V, the reaction is product-favored.

(c) $I_2(s) + 2\ Br^-(aq) \longrightarrow 2\ I^-(aq) + Br_2(\ell)$

$E° = -0.55$ V, the reaction is not product-favored.

21.34. (a) The weakest oxidizing agent is $V^{2+}(aq)$.

(b) The strongest oxidizing agent is $Cl_2(g)$.

(c) The strongest reducing agent is V(s).

(d) The weakest reducing agent is $Cl^-(aq)$.

(e) No, Pb(s) is not a strong enough reducing agent.

(f) Yes, Cl_2 can oxidize I^- to I_2.

(g) I_2 and Cl_2.

21.36. (a) The chlorine and vanadium half-cells, when combined produce the largest voltage, 2.54 V.

(b)

$$\begin{array}{r} I_2(s) + 2\ e^- \longrightarrow 2\ I^-(aq) \\ V(s) \longrightarrow V^{2+}(aq) + 2\ e^- \\ \hline I_2(s) + V(s) \longrightarrow V^{2+}(aq) + 2\ I^-(aq) \end{array}$$

For this voltaic cell, $E° = 1.72$ V.

21.38. (a) $Zn(s) + 2\ Ag^+(aq) \longrightarrow Zn^{2+}(aq) + 2\ Ag(s)$; $E°_{net} = 1.56$ V

(b–f)

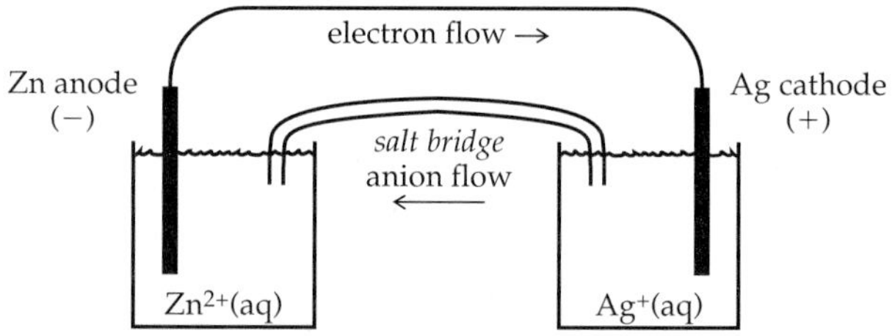

21.40. $E = 0.177$ V, the voltage has decreased by 0.060 V.

21.42. (a) $Fe^{2+}(aq) + Aq^+(aq) \longrightarrow Ag(s) + Fe^{3+}(aq)$

(b) $E°_{net} = 0.028$ V

(c) $E = -0.030$ V. The negative sign indicates that the reaction must occur in the reverse direction under these nonstandard conditions. The equation for the reaction is: $Ag(s) + Fe^{3+}(aq) \longrightarrow Fe^{2+}(aq) + Ag^+(aq)$

21.44. (a) $K = 9.5 \times 10^7$

(b) $K = 4 \times 10^{-19}$

21.46. 735 s (or 12.2 min)

21.48. 5.47 g Ni

21.50. 0.10 A

21.52. 2.7×10^5 g Al

21.54. 190 g Pb

21.56. 4.0×10^{10} kwh

21.58. (a) (Positive electrode polarity) $2\ Cl^-(aq) \longrightarrow Cl_2(g) + 2\ e^-$

(b) (Negative electrode polarity) $Ni^{2+}(aq) + 2\ e^- \longrightarrow Ni(s)$

(c) 0.0413 g Cl_2; 0.0342 g Ni

21.60. (a) Bromide ion is oxidized to Br_2 at the anode, H_2O is reduced to H_2 and $OH^-(aq)$ at the cathode.

(b) Fluoride ion is oxidized to F_2 at the anode, Na^+ is reduced to $Na(\ell)$ at the cathode.

(c) Water is oxidized to $H_3O^+(aq)$ and $O_2(g)$ at the anode, H_2O is reduced to H_2 and $OH^-(aq)$ at the cathode.

21.62. (a) $I^-(aq) + 3\ Br_2(\ell) + 9\ H_2O(\ell) \longrightarrow IO_3^-(aq) + 6\ Br^-(aq) + 6\ H_3O^+(aq)$

(b) $5\ U^{4+}(aq) + MnO_4^-(aq) + 18\ H_2O(\ell) \longrightarrow 5\ UO_2^+(aq) + Mn^{2+}(aq) + 12\ H_3O^+(aq)$

(c) $2\ I^-(aq) + MnO_2(s) + 4\ H_3O^+(aq) \longrightarrow I_2(s) + Mn^{2+}(aq) + 6\ H_2O(\ell)$

21.64. $Cl_2(g) + Mg(s) \longrightarrow Mg^{2+}(aq) + 2\ Cl^-(aq)$ $E^\circ_{net} = 3.73$ V

21.66. (a) $K = 8.8 \times 10^{68}$

(b) $E = 2.09$ V

21.68. (a) In effect, the magnesium and iron constitute an electrochemical cell, with the magnesium, the more easily oxidized element, being the anode and iron being the cathode. The magnesium is a sacrificial anode and is oxidized in preference to the iron.

(b) 42 years

21.70. (a) $3\ Cl_2(g) + 2\ Al(s) \longrightarrow 2\ Al^{3+}(aq) + 6\ Cl^-(aq)$

(b) *Anode reaction:* $Al(s) \longrightarrow Al^{3+}(aq) + 3\ e^-$

Cathode reaction: $Cl_2(g) + 2\ e^- \longrightarrow 2Cl^-(aq)$

(c) $E^\circ_{net} = 3.02$ V

(d) $E = 3.01$ V

(e) 119 h

21.72. (a) 5.0×10^{-4} W/g

(b) 3.1×10^{-4} W/g

(c) The Ag/Zn battery produces more energy per gram.

21.74. Conclusions: (a) A and C are stronger reducing agents than H_2; B, and D are weaker. (b) C is the strongest reducing agent. (c) D is a stronger reducing agent than B. Ordered relative to strength as a reducing agent: $B < D < A < C$.

21.76. CO_2 is an acid anhydride, and dissolved in water it produces a slightly acidic solution:

$$CO_2(aq) + 2\ H_2O(\ell) \rightleftharpoons HCO_3^-(aq) + H_3O^+(aq)$$

Corrosion, the oxidation of Fe(s) and reduction $H_3O^+(aq)$ to $H_2(g)$, is more favorable in acid solution. In addition, acid reacts with any surface layer of FeO(s) or $Fe(OH)_2(s)$, providing a clean Fe surface for the reaction to continue. Finally, the reaction of CO_2 and H_2O provide additional ions facilitating the migration of electric charge in the solution.

21.78. (a) $Cd(s) + Ni^{2+}(aq) \longrightarrow Cd^{2+}(aq) + Ni(s)$

(b) Cd(s) is oxidized and is the reducing agent; $Ni^{2+}(aq)$ is reduced and is the oxidizing agent.

(c) Cd is the anode, Ni is the cathode.

(d) $E^\circ_{net} = 0.15$ V

(e) Electrons flow from the anode (Cd) to the cathode (Ni).

(f) $NO_3^-(aq)$ ions migrate toward the Cd compartment.

(g) $K = 1.2 \times 10^5$

(h) $E = 0.21$ V

(i) 480 h

21.80. Rhodium is present in solution as Rh^{3+}

21.82. 290 hr

21.84. (a) No. (A calculation shows that $E^\circ_{net} = -1.091$ V)

(b) In NAD^+, the carbons and nitrogen in the C_5N ring are sp^2-hybridized and lie in a plane. In the NADH form, the C_5N ring remains planar. N is now sp^3-hybridized, however, and its hydrogen (and also the N lone pair) are no longer in the plane defined by the ring.

CHAPTER 22

22.2.

Name	Formula	Oxidation state of central atom
Nitric acid	HNO_3	N is +5
Sulfuric acid	H_2SO_4	S is +6
Hydrobromic acid	HBr	Br is −1
Perchloric acid	$HClO_4$	Cl is +7
Carbonic acid	H_2CO_3	C is +4

22.4. CaO

$2\ Ca(s) + O_2(g) \longrightarrow CaO(s)$

$CaO(s) + 2\ H_3O^+(aq) \longrightarrow Ca^{2+}(aq) + 3\ H_2O(\ell)$

BaO

$2\ Ba(s) + O_2(g) \longrightarrow 2\ BaO(s)$

$BaO(s) + 2\ H_3O^+(aq) \longrightarrow Ba^{2+}(aq) + 3\ H_2O(\ell)$

22.6.

Li	$1s^22s^1$	1
Be	$1s^22s^2$	2
B	$1s^22s^22p^1$	3
C	$1s^22s^22p^2$	4
N	$1s^22s^22p^3$	5
O	$1s^22s^22p^4$	6
F	$1s^22s^22p^5$	7
Ne	$1s^22s^22p^6$	8

22.8. [noble gas core] ns^2np^2

These elements most often share their four valence electrons to form compounds that have an octet of electrons around the central atom. Examples: CH_4, $SiCl_4$, GeO_2

22.10. Sulfide ion, S^{2-}; chloride ion, Cl^-; potassium ion, K^+; calcium ion, Ca^{2+}

22.12. $2\ Na(s) + Cl_2(g) \longrightarrow 2\ NaCl(s)$
The reaction is very exothermic, and the product, NaCl, is ionic, made up of Na^+ and Cl^- ions.

22.14. White, solid, soluble in water

22.16. In order of abundance: O > Si > Al > Fe > Ca > Na > K > Mg > H > Ti. All but Fe and Ti are main group elements. Ten compounds containing these elements: H_2O, SiO_2, Al_2O_3, Fe_2O_3, $CaCO_3$, CaF_2, NaCl, KCl, $MgCO_3$, CH_4, TiO_2

22.18. Among the first ten elements only carbon is present in the earth's crust in appreciable amounts. Oxygen (O_2), nitrogen (N_2), and the two noble gases, He and Ne, are found in nature in the elemental state, in the atmosphere.

22.20. In order of increasing basicity: $SO_3 < SiO_2 < Al_2O_3 < Na_2O$

22.22. (a) $2\ K(s) + I_2(s) \longrightarrow 2\ KI(s)$
(b) $2\ Ba(s) + O_2(g) \longrightarrow 2\ BaO(s)$
(c) $16\ Al(s) + 3\ S_8(s) \longrightarrow 8\ Al_2S_3(s)$
(d) $Si(s) + 2\ Cl_2(g) \longrightarrow SiCl_4(\ell)$

22.24. $2\ K(s) + H_2(g) \longrightarrow 2\ KH(s)$
Potassium hydride is an ionic compound, with a structure containing K^+ and H^- ions. It has a high melting point, typical of an ionic compound. It reacts violently with water:
$2\ KH(s) + 2\ H_2O(\ell) \longrightarrow H_2(g) + 2\ KOH(s)$
This is regarded as a Brønsted acid–base reaction.

22.26. $CH_4(g) + H_2O(g) \longrightarrow 3\ H_2(g) + CO(g)$

$$\Delta H^\circ_{rxn} = +206.10\ kJ;\ \Delta S^\circ_{rxn} = +214.64\ J/K;\ \Delta G^\circ_{rxn} = +142.14\ kJ$$

22.28. Insert a glowing splint into the gas. Hydrogen will ignite (burning in air). If the gas is oxygen, the splint will burst into flame. If it is nitrogen, the splint will cease to glow.

22.30. $Na(s) + F_2(g) \longrightarrow 2\ NaF(s)$
$Na(s) + Cl_2(g) \longrightarrow 2\ NaCl(s)$
$Na(s) + Br_2(\ell) \longrightarrow 2\ NaBr(s)$
$Na(s) + I_2(s) \longrightarrow 2\ NaI(s)$
The alkali metal halides are white, crystalline solids. They have high melting and boiling points, and are soluble in water.

22.32. $4\ Li(s) + O_2(g) \longrightarrow 2\ Li_2O(s)$ lithium oxide
$2\ Na(s) + O_2(g) \longrightarrow 2\ Na_2O_2(s)$ sodium peroxide
$K(s) + O_2(g) \longrightarrow KO_2(s)$ potassium superoxide

22.34. $2\ Cl^-(aq) + 2\ H_2O(\ell) \longrightarrow Cl_2(g) + H_2(g) + OH^-(aq)$
If this were the only process used to produce chlorine, the mass of Cl_2 reported for industrial production would be 0.88 times the mass of NaOH produced. (2 mol NaCl, 117 g, would yield 2 mol NaOH, 80 g, and 1 mol of Cl_2, 70 g). The amounts quoted indicate a Cl_2 to NaOH mass ratio of 0.96. So, chlorine is presumably also prepared by other routes than this.

22.36. $2\ Mg(s) + O_2(g) \longrightarrow 2\ MgO(s)$
$3\ Mg(s) + N_2(g) \longrightarrow Mg_3N_2(s)$

22.38. $CaCO_3$ is used in agriculture to neutralize acidic soil, to prepare CaO for use in mortar, and in steel production.

$$CaCO_3(s) + H_2O(\ell) + CO_2(g) \longrightarrow Ca^{2+}(aq) + 2\ HCO_3^-(aq)$$

22.40. 1.4×10^6 g

22.42. $2\ Al(s) + 6\ HCl(aq) \longrightarrow 2\ Al^{3+}(aq) + 6\ Cl^-(aq) + 3\ H_2(g)$
$2\ Al(s) + Cl_2(g) \longrightarrow 2\ AlCl_3(s)$
$4\ Al(s) + 3\ O_2(g) \longrightarrow 2\ Al_2O_3(s)$

22.44. $2\ Al(s) + 2\ OH^-(aq) + 6\ H_2O(\ell) \longrightarrow 2\ Al(OH)_4^-(aq) + 3\ H_2(g)$
18.4 L H_2

22.46. $Al_2O_3(s) + 3\ H_2SO_4(aq) \longrightarrow Al_2(SO_4)_3(s) + 3\ H_2O(\ell)$
0.298 kg Al_2O_3

22.48. $[AlCl_4]^-$: Lewis structure with Al bonded to four :Cl: atoms, in brackets with charge −
The ion has tetrahedral geometry. Aluminum is sp^3-hybridized.

22.50. SiO_2 is a network solid, with tetrahedral silicon atoms covalently bonded to four oxygens in an infinite array; CO_2 consists of individual molecules, with oxygen atoms double bonded to carbon. Melting SiO_2 requires breaking very stable Si—O bonds. Weak intermolecular forces of attraction between CO_2 molecules result in this substance being a gas at ambient conditions.

22.52. (a) $2\ CH_3Cl(g) + Si(s) \longrightarrow (CH_3)_2SiCl_2(\ell)$
(b) 0.823 atm
(c) 12.2 g

22.54. None of the four nitrogen oxides listed (N_2O, NO, NO_2, N_2O_4) are stable relative to the elements; for all of these compounds ΔG°_f is positive.

22.56. Exothermic, −114.14 kJ/mol

22.58. (a) $N_2H_4(aq) + O_2(g) \longrightarrow N_2(g) + 2\ H_2O(\ell)$
(b) 1.32×10^3 g

22.60. $5\ N_2H_5^+ + (aq) + 4\ IO_3^-(aq) \longrightarrow 5\ N_2(g) + 2\ I_2(aq) + H_3O^+(aq) + 11\ H_2O(\ell)$
$E^\circ_{net} = 1.43$ V

22.62. $:\ddot{N}{=}N{=}\ddot{N}: \longleftrightarrow :N{\equiv}N{-}\ddot{\underset{..}{N}}: \longleftrightarrow :\ddot{\underset{..}{N}}{-}N{\equiv}N:$

22.64. (a) 3.5×10^6 g
(b) 4.1×10^6 g

22.66. $:\ddot{\underset{..}{S}}{-}\ddot{\underset{..}{S}}:$

22.68. Not using standard conditions. (E°_{net} is −0.07 V for the net reaction that combines the $BrO_3^- \longrightarrow Br^-$ reduction half-reaction with the $Mn^{2+} \longrightarrow MnO_4^-$ oxidation half-reaction.)

22.70. $Cl_2(aq) + 2\ Br^-(aq) \longrightarrow 2\ Cl^-(aq) + Br_2(aq)$
Cl_2 is the oxidizing agent, Br^- is the reducing agent; $E^\circ_{net} = 0.292$ V

22.72.

Element	Appearance	State
Na, Mg, Al	silvery metals	solids
Si	black, shiny metalloid	solid
P	white, red, and black allotropes; nonmetal	solid
S	yellow nonmetal	solid
Cl	pale green, nonmetal	gas
Ar	colorless, nonmetal	gas

22.74. (a) $2\ Na(s) + Cl_2(g) \longrightarrow 2\ NaCl(s)$
$Mg(s) + Cl_2(g) \longrightarrow MgCl_2(s)$
$2\ Al(s) + 3\ Cl_2(g) \longrightarrow 2\ AlCl_3(s)$
$Si(s) + 2\ Cl_2(g) \longrightarrow SiCl_4(\ell)$
$P_4(s) + 10\ Cl_2(g) \longrightarrow 4\ PCl_5(s)$ (excess Cl_2)
$S_8(s) + 16\ Cl_2(g) \longrightarrow 8\ SCl_2(s)$
(b) NaCl and $MgCl_2$ are ionic, the other products are covalent.
(c) $SiCl_4$ is tetrahedral, PCl_5 is trigonal-bipyramidal.

:Cl:
:Cl—Si—Cl:
:Cl:

:Cl:
:Cl—P (Cl, Cl)
:Cl:

22.76. *Cathode:* $Li^+(\ell) + e^- \longrightarrow Li(\ell)$
Anode: $2\ H^-(\ell) \longrightarrow H_2(g) + 2\ e^-$
Formation of H_2 at the anode is evidence for the presence of H^-.

22.78. Mg: $\Delta G^\circ_{rxn} = +48.3$ kJ
Ca: $\Delta G^\circ_{rxn} = +130.40$ kJ
Ba: $\Delta G^\circ_{rxn} = +218.1$ kJ
Relative tendency to decompose: $MgCO_3 > CaCO_3 > BaCO_3$

22.80. (a) ΔG°_f should be more negative than $(-95.3\ \text{kJ}) \times n$
(b) Ba, Pb, and Ti

22.82. (a) We rationalize the equivalence of the two N—O bond lengths by writing two resonance structures for this molecule.
(b) The central atom, nitrogen, has three sets of bonding electrons in its valence shell; VSEPR predicts that this atom is trigonal-planar. Oxygen, in the —OH group, has four electron pairs in its valence shell, arranged tetrahedrally. Two are bonding pairs, defining the bent molecular geometry.
(c) sp^2. There is an empty p orbital on N that is perpendicular to the plane of the molecule; this can overlap with p orbitals on the two terminal oxygens to for a delocalized π bond.

22.84. A through E, in order: $BaCO_3$; BaO; $CaCO_3$; $BaCl_2$; $BaSO_4$

22.86. $\Delta H^\circ_{rxn} = -257.78$ kJ. This reaction is entropy-disfavored, however, with $\Delta S^\circ_{rxn} = -963$ J/K because of the decrease in the number of moles of gases. Combining these values gives $\Delta G^\circ_{rxn} = +29.34$ kJ, indicating that under standard conditions at 298 K the reaction is not favorable. (The reaction has a favorable ΔG°_{rxn} below 268 K, indicating that further research on this system might be worthwhile. Note, however, that at that temperature water is a solid.)

22.88. (a) 820 L sea water; 2.3 kg CaO
(b) 310 kg Mg, 890 kg Cl_2, 2.5×10^4 F
(c) 1600 kJ/mol. This is more than the energy required for $MgCl_2(s) \longrightarrow Mg(s) + Cl_2(g)$, for which 640 kJ/mol is required. The difference arises because of the need to run the electrolysis at high temperature.

CHAPTER 23

23.14. The Cr^{3+} ion is paramagnetic (3 unpaired electrons).

		3*d*	4*s*
Cr^{3+}	[Ar]	[↑][↑][↑][ ][ ]	[ ]
Cr^{6+}	[Ar]	[ ][ ][ ][ ][ ]	[ ]

23.16. (a) Fe^{2+} and Co^{3+}
(b) Mn^{2+} and Fe^{3+}
(c) Zn^{2+} and Cu^+
(d) Ni^{2+} and Cu^{3+}

23.18. (a) $Cr_2O_3(s) + 2\ Al(s) \longrightarrow Al_2O_3(s) + 2\ Cr(s)$
(b) $TiCl_4(\ell) + 2\ Mg(s) \longrightarrow Ti(s) + 2\ MgCl_2(s)$
(c) $2\ Ag(CN)_2^- + Zn(s) \longrightarrow 2\ Ag(s) + Zn^{2+}(aq) + 4\ CN^-(aq)$

23.20. 527 g

23.22. Monodentate: a, b, c, f; bidentate: d, e, g

23.24. (a) +2; (b) +3; (c) +3; (d) +2

23.26. (a) $Ni(en)_2Cl_2$ (en = $H_2NCH_2CH_2NH_2$)
(b) $K_2[PtCl_4]$
(c) $K[Cu(CN)_2]$
(d) $[Fe(NH_3)_4(H_2O)_2]^{2+}$

23.28. (a) Diaquabis(oxalato)nickelate(II) ion
(b) Dibromobis(ethylenediamine)cobalt(III) ion
(c) Ammine chlorobis(ethylenediamine)cobalt(III) ion
(d) Diammineoxalatoplatinum(II)

23.30. (a) $[Fe(H_2O)_5OH]^{2+}$
(b) Potassium tetracyanonickelate(II)
(c) Potassium diaquabis(oxalato)chromate(III)
(d) $(NH_4)_2[PtCl_4]$

23.30. (a)
cis: Fe with H_3N, NH_3, Cl, H_3N, NH_3, Cl
trans: Fe with H_3N, Cl, NH_3, H_3N, Cl, NH_3

(b)
cis: Pt with H_3N, SCN, H_3N, Br
trans: Pt with H_3N, SCN, Br, NH_3

(c)
fac: Co with O_2N, NH_3, NH_3, O_2N, NO_2, NH_3
mer: Co with H_3N, NH_3, NO_2, O_2N, NO_2, NH_3

(d)

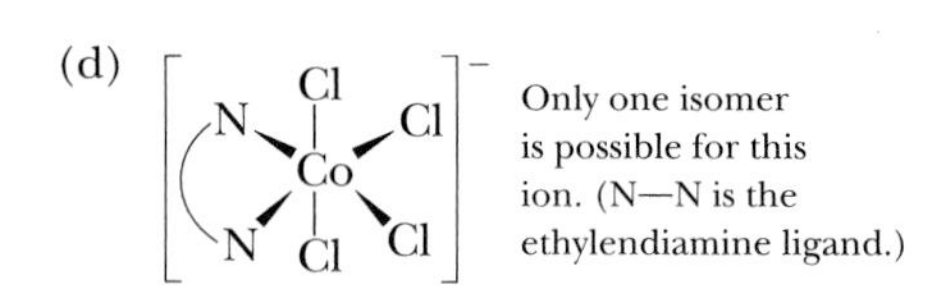

Only one isomer is possible for this ion. (N—N is the ethylendiamine ligand.)

23.34. (a) No chiral center
(b) Has a chiral center
(c) Has a chiral center
(d) No chiral center

23.36. Isomers of $[Co(en)(NH_3)_2(H_2O)Cl]^{2+}$, ethylenediamine (en) is depicted by N⌒N

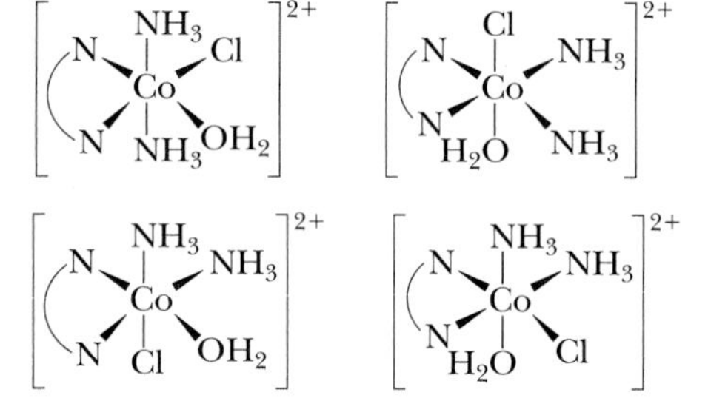

23.38. d^4, d^5, d^6, d^7

23.40. (a) $[Fe(CN)_6]^{4-}$: d^6, low spin is diamagnetic, high spin is paramagnetic (4 unpaired electrons)
(b) $[Co(NH_3)_6]^{3+}$: d^6, low spin is diamagnetic, high spin is paramagnetic (4 unpaired electrons)
(c) $[Fe(H_2O)_6]^{3+}$: d^5, both low spin (1 unpaired electron) and high spin (5 unpaired electrons) are paramagnetic.
(d) $[CrF_6]^{4-}$: d^4, both low spin (2 unpaired electrons) and high spin (4 unpaired electrons) are paramagnetic.
(a) and (b) d^6 configurations

low spin high spin

(c) d^5 configuration

low spin high spin

(d) d^4 configuration

low spin high spin

23.42. Green light is absorbed.

23.44. (a) 6; (b) octahedral; (c) +2; (d) 4 unpaired electrons; (e) paramagnetic

23.46. (a) d^6 low spin, diamagnetic, no unpaired electron
(b) d^5 high spin, paramagnetic, 5 unpaired electrons
(c) d^3 only one configuration possible, paramagnetic with 3 unpaired electrons
(d) d^9, only one configuration possible, paramagnetic with 1 unpaired electron.

23.48. Diammineoxalatoplatinum(II), Pt (+2)

NH_3 NH_3 Pt O C O C O O

23.50. $[Co(en)_2(H_2O)Cl]^{2+}$

23.52. (a) Iron is a chiral center.
(b) In the *cis* isomer, the cobalt is chiral. The *trans* isomer is not chiral.
(c) Neither of the two possible isomers is chiral.
(d) No. Square-planar compounds are never chiral.

23.54. $F^- < H_2O < NH_3 < CN^-$

23.56. When $Co_2(SO_4)_3$ dissolves in water it forms $[Co(H_2O)_6]^{3+}$; addition of fluoride converts this to $[CoF_6]^{3-}$. The hexaaquo complex is low spin (diamagnetic, no unpaired electrons), and the fluoride complex is high spin (paramagnetic, 4 unpaired electrons.) Notice that fluoride is a weaker field ligand than water.

23.58. (a) Determine the magnetism; a low-spin Fe^{2+} complex is diamagnetic, a high-spin complex is paramagnetic.
(b) Determine the magnetism; square-planar Ni^{2+} complexes are diamagnetic, tetrahedral Ni^{2+} complexes are paramagnetic.

23.60. (a)

fac *mer*

(b)

trans chlorides *cis* chlorides

(c)

23.62. (a) The complex absorbs in the red region. Blue and green are not absorbed and the complex will be cyan colored.
(b) Δ_o is small, so CO_3^{2-} is a weak field ligand and should be placed in the lower half of the series.
(c) A small Δ_o would lead to a prediction of high spin, so $[Co(CO_3)_3]^{3-}$ should be paramagnetic (4 unpaired electrons).

23.64. A, the dark violet isomer: $[Co(NH_3)_5Br]SO_4$; B, the violet-red isomer: $[Co(NH_3)_5(SO_4)]Br$

23.66. $Pt(NH_3)_2(NO_2)Cl$. Two geometric isomers (*cis* and *trans*) are possible.

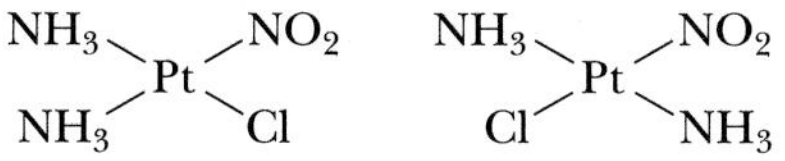

cis *trans*

23.68. Reduction product is $U^{4+}(aq)$
$5\ U^{4+}(aq) + 2\ MnO_4^-(aq) + 6\ H_2O \longrightarrow$
$5\ UO_2^{2+} + 4\ H_3O^+(aq) + 2\ Mn^{2+}(aq)$

23.70. 95.000% Al; 5.000% Cu

CHAPTER 24

24.12. (a) ${}^{56}_{26}Fe$; (b) ${}^{1}_{0}n$; (c) ${}^{32}_{15}P$; (d) ${}^{97}_{43}Tc$; (e) ${}^{0}_{-1}\beta$

24.14. (a) ${}^{0}_{-1}\beta$; (b) ${}^{87}_{37}Rb$; (c) ${}^{4}_{2}He$; (d) ${}^{226}_{88}Ra$; (e) ${}^{0}_{-1}\beta$; (f) ${}^{24}_{11}Na$

24.16. ${}^{235}_{92}U + {}^{4}_{2}He \longrightarrow {}^{231}_{90}Th$
${}^{231}_{90}Th + {}^{0}_{-1}\beta \longrightarrow {}^{231}_{91}Pa$
${}^{231}_{91}Pa + {}^{4}_{2}He \longrightarrow {}^{227}_{89}Ac$
${}^{227}_{89}Ac + {}^{0}_{-1}\beta \longrightarrow {}^{227}_{90}Th$
${}^{227}_{90}Th + {}^{4}_{2}He \longrightarrow {}^{223}_{88}Ra$
${}^{223}_{88}Ra + {}^{4}_{2}He \longrightarrow {}^{219}_{86}Rn$
${}^{219}_{86}Rn + {}^{4}_{2}He \longrightarrow {}^{215}_{84}Po$
${}^{215}_{84}Po + {}^{4}_{2}He \longrightarrow {}^{211}_{82}Pb$
${}^{211}_{82}Pb + {}^{0}_{-1}\beta \longrightarrow {}^{211}_{83}Bi$
${}^{211}_{83}Bi + {}^{0}_{-1}\beta \longrightarrow {}^{211}_{84}Po$
${}^{211}_{84}Po + {}^{4}_{2}He \longrightarrow {}^{207}_{82}Pb$

24.18. (a) ${}^{198}_{79}Au \longrightarrow {}^{198}_{80}Hg + {}^{0}_{-1}\beta$
(b) ${}^{222}_{86}Rn \longrightarrow {}^{218}_{84}Po + {}^{4}_{2}He$
(c) ${}^{137}_{55}Cs \longrightarrow {}^{137}_{56}Ba + {}^{0}_{-1}\beta$
(d) ${}^{110}_{49}In \longrightarrow {}^{110}_{50}Cd + {}^{0}_{-1}\beta$

24.20. Binding energy per mole of nucleons for ${}^{11}B = 4.779 \times 10^{-8}$ kJ
Binding energy per mole of nucleon for ${}^{10}B = 6.700 \times 10^{-8}$ kJ

24.22. 8.267×10^8 kJ/mol of nucleons

24.24. ${}^{15}O$: 7.216×10^8 kJ/mol of nucleons
${}^{16}O$: 7.707×10^8 kJ/mol of nucleons
${}^{17}O$: 7.493×10^8 kJ/mol of nucleons

24.26. 0.472 mg

24.28. (a) ${}^{131}_{53}I \longrightarrow {}^{131}_{54}Xe + {}^{0}_{-1}\beta$
(b) 1.56 mg

24.30. 9.5×10^{-3} mg

24.32. (a) ${}^{222}_{86}Rn \longrightarrow {}^{218}_{84}Po + {}^{4}_{2}He$
(b) 12.7 days

24.34. About 2700 years old

24.36. (a) 15.8 years
(b) 87.7%

24.38. 1.9×10^9 years

24.40. (a) ${}^{23}_{11}Na + {}^{1}_{0}n \longrightarrow {}^{24}_{11}Na$
${}^{24}_{11}Na \longrightarrow {}^{24}_{12}Mg + {}^{0}_{-1}\beta$
(b) $t_{1/2} = 15.1$ h

24.42. ${}^{239}_{94}Pu + 2\ {}^{1}_{0}n \longrightarrow {}^{241}_{94}Pu$
${}^{241}_{94}Pu \longrightarrow {}^{241}_{95}Am + {}^{0}_{-1}\beta$

24.44. ${}^{12}_{6}C$

24.46. (a) ${}^{115}_{48}Cd$ (b) ${}^{7}_{4}Be$ (c) ${}^{4}_{2}He$ (d) ${}^{63}_{29}Cu$

24.48. ${}^{10}_{5}B + {}^{1}_{0}n \longrightarrow {}^{7}_{3}Li + {}^{4}_{2}He$

24.50. (a) ${}^{1}_{0}n$ (b) ${}^{4}_{2}He$ (c) ${}^{257}_{103}Lw$ (d) ${}^{1}_{0}n$ (e) ${}^{4}_{2}He$ (f) ${}^{122}_{52}Te$ (g) ${}^{23}_{10}Ne$ (h) ${}^{136}_{53}I$

24.52. 3.6×10^9 years

24.54. 1.5×10^9 years

24.56. 1.6×10^3 tons

24.58. $\Delta m = 1.822 \times 10^{-27}$ g so $E = 1.6396 \times 10^{-10}$ J/event
2 photons; frequency $= 1.237 \times 10^{23}$ Hz (each takes half of the energy)

24.60. 130 mL

24.62. (a) The mass decreases by 4 units (with an ${}^{4}_{2}\alpha$ emission) or is unchanged (with a ${}^{0}_{-1}\beta$ emission) so the only masses possible are 4 units apart.
(b) ${}^{232}Th$ series, $m = 4n$; ${}^{235}U$ series $m = 4n + 3$
(c) ${}^{226}Ra$ and ${}^{210}Bi$, $4n + 2$ series; ${}^{215}At$, $4n + 3$ series; ${}^{228}Th$, $4n$ series)
(d) Each series is headed by a long-lived isotope (in the order of 10^9 years, the age of the earth.) The $4n + 1$ series is missing because there is no long-lived isotope in this series. Over geologic time, all the members of this series have decayed completely.

24.64. Uranium-235

Index/Glossary

Italicized page numbers indicate pages containing illustrations, and those followed by "t" indicate tables. Glossary terms, printed in boldface, are defined here as well as in the text.

Standard Atomic Weights of the Elements 1997 • Based on relative atomic mass of $^{12}C = 12$, where ^{12}C is a neutral atom in its nuclear and electronic ground state.†

Name	Symbol	Atomic Number	Atomic Weight	Name	Symbol	Atomic Number	Atomic Weight
Actinium*	Ac	89	(227)	Mercury	Hg	80	200.59(2)
Aluminium	Al	13	26.981538(2)	Molybdenum	Mo	42	95.94(1)
Americium*	Am	95	(243)	Neodymium	Nd	60	144.24(3)
Antimony	Sb	51	121.760(1)	Neon	Ne	10	20.1797(6)
Argon	Ar	18	39.948(1)	Neptunium*	Np	93	(237)
Arsenic	As	33	74.92160(2)	Nickel	Ni	28	58.6934(2)
Astatine*	At	85	(210)	Niobium	Nb	41	92.90638(2)
Barium	Ba	56	137.327(7)	Nitrogen	N	7	14.00674(7)
Berkelium*	Bk	97	(247)	Nobelium*	No	102	(259)
Beryllium	Be	4	9.012182(3)	Osmium	Os	76	190.23(3)
Bismuth	Bi	83	208.98038(2)	Oxygen	O	8	15.9994(3)
Bohrium	Bh	107	(264)	Palladium	Pd	46	106.42(1)
Boron	B	5	10.811(7)	Phosphorus	P	15	30.973762(4)
Bromine	Br	35	79.904(1)	Platinum	Pt	78	195.078(2)
Cadmium	Cd	48	112.411(8)	Plutonium*	Pu	94	(244)
Cesium	Cs	55	132.90545(2)	Polonium*	Po	84	(210)
Calcium	Ca	20	40.078(4)	Potassium	K	19	39.0983(1)
Californium*	Cf	98	(251)	Praseodymium	Pr	59	140.90765(2)
Carbon	C	6	12.0107(8)	Promethium*	Pm	61	(145)
Cerium	Ce	58	140.116(1)	Protactinium*	Pa	91	231.03588(2)
Chlorine	Cl	17	35.4527(9)	Radium*	Ra	88	(226)
Chromium	Cr	24	51.9961(6)	Radon*	Rn	86	(222)
Cobalt	Co	27	58.933200(9)	Rhenium	Re	75	186.207(1)
Copper	Cu	29	63.546(3)	Rhodium	Rh	45	102.90550(2)
Curium*	Cm	96	(247)	Rubidium	Rb	37	85.4678(3)
Dubnium	Db	105	(262)	Ruthenium	Ru	44	101.07(2)
Dysprosium	Dy	66	162.50(3)	Rutherfordium	Rf	104	(261)
Einsteinium*	Es	99	(252)	Samarium	Sm	62	150.36(3)
Erbium	Er	68	167.26(3)	Scandium	Sc	21	44.955910(8)
Europium	Eu	63	151.964(1)	Seaborgium	Sg	106	(266)
Fermium*	Fm	100	(257)	Selenium	Se	34	78.96(3)
Fluorine	F	9	18.9984032(5)	Silicon	Si	14	28.0855(3)
Francium*	Fr	87	(223)	Silver	Ag	47	107.8682(2)
Gadolinium	Gd	64	157.25(3)	Sodium	Na	11	22.989770(2)
Gallium	Ga	31	69.723(1)	Strontium	Sr	38	87.62(1)
Germanium	Ge	32	72.61(2)	Sulfur	S	16	32.066(6)
Gold	Au	79	196.96655(2)	Tantalum	Ta	73	180.9479(1)
Hafnium	Hf	72	178.49(2)	Technetium*	Tc	43	(98)
Hassium	Hs	108	(269)	Tellurium	Te	52	127.60(3)
Helium	He	2	4.002602(2)	Terbium	Tb	65	158.92534(2)
Holmium	Ho	67	164.93032(2)	Thallium	Tl	81	204.3833(2)
Hydrogen	H	1	1.00794(7)	Thorium*	Th	90	232.0381(1)
Indium	In	49	114.818(3)	Thulium	Tm	69	168.93421(2)
Iodine	I	53	126.90447(3)	Tin	Sn	50	118.710(7)
Iridium	Ir	77	192.217(3)	Titanium	Ti	22	47.867(1)
Iron	Fe	26	55.845(2)	Tungsten	W	74	183.84(1)
Krypton	Kr	36	83.80(1)	Ununnilium	Uun	110	(269)
Lanthanum	La	57	138.9055(2)	Unununium	Uuu	111	(272)
Lawrencium*	Lr	103	(262)	Ununbium	Uub	112	(277)
Lead	Pb	82	207.2(1)	Uranium*	U	92	238.0289(1)
Lithium	Li	3	6.941(2)	Vanadium	V	23	50.9415(1)
Lutetium	Lu	71	174.967(1)	Xenon	Xe	54	131.29(2)
Magnesium	Mg	12	24.3050(6)	Ytterbium	Yb	70	173.04(3)
Manganese	Mn	25	54.938049(9)	Yttrium	Y	39	88.90585(2)
Meitnerium	Mt	109	(268)	Zinc	Zn	30	65.39(2)
Mendelevium*	Md	101	(258)	Zirconium	Zr	40	91.224(2)

†The atomic weights of many elements can vary depending on the origin and treatment of the sample. This is particularly true for Li; commercially available lithium-containing materials have Li atomic weights in the range of 6.96 and 6.99. The uncertainties in atomic weight values are given in parentheses following the last significant figure to which they are attributed.

*Elements with no stable nuclide; the value given in parentheses is the atomic mass number of the isotope of longest known half-life. However, three such elements (Th, Pa, and U) have a characteristic terrestial isotopic composition, and the atomic weight is tabulated for these.

Top 25 Chemicals in the United States, 1995 • DATA ARE FROM *CHEMICAL & ENGINEERING NEWS*, JUNE 24, 1996

Rank	Name	Production (billions of kg)	How Made	End Uses
1	Sulfuric acid	43.3	Burning sulfur to SO_2, oxidation of SO_2 to SO_3, reaction with water. Also recovered from metal smelting.	Fertilizers, petroleum refining, manufacture of metals and chemicals.
2	Nitrogen	30.9	Separated from liquid air	Blanketing atmospheres for metals, electronics, and freezing agent for foods; ammonia production.
3	Oxygen	24.3	Separated from liquid air	Steel production, metal fabricating, chemical processing.
4	Ethylene	21.3	Cracking hydrocarbons from oil and natural gas.	Plastics, antifreeze production, fibers, solvents.
5	Calcium oxide (lime)	18.7	Heating limestone ($CaCO_3$)	Steel production, water treatment, refractories, pulp and paper.
6	Ammonia	16.2	Catalytic reaction of nitrogen, air, and hydrogen.	Fertilizers, plastics, fibers, resins.
7	Phosphoric acid	11.9	Sulfuric acid reacted with phosphate rock; elemental phosphorus burned and dissolved in water.	Fertilizers, detergents, water-treating compounds.
8	Sodium hydroxide	11.9	Electrolysis of aqueous NaCl	Chemicals, pulp and paper, aluminum, textiles, oil refining
9	Propylene	11.7	Cracking oil and oil products	Plastics, fibers, and solvents
10	Chlorine	11.4	Electrolysis of NaCl, recovery from HCl users	Chemical production, plastics, solvents, pulp and paper
11	Sodium carbonate	10.1	Trona ore, made from NaCl and limestone with ammonia	Glass, chemicals, pulp and paper
12	Methyl-*tert*-butyl ether	8.0	Acid-catalyzed reaction of methanol with isobutene	Gasoline additive
13	Ethylene dichloride	8.0	Chlorination of ethylene	Production of vinyl chloride
14	Nitric acid	7.8	Oxidation of ammonia to nitrogen dioxide, which is then dissolved in water	Ammonium nitrate and phosphate fertilizers, nitro explosives, plastics, dyes, and lacquers
15	Ammonium nitrate	7.3	Reaction of ammonia and nitric acid	Explosives, fertilizer, source of laughing gas, matches
16	Benzene	7.2	From oil and coal tar	Polystyrene, resins, nylon, rubber
17	Urea, $(NH_2)_2CO$	7.1	Reaction of NH_3 and CO_2 under pressure	Fertilizers, animal feeds, adhesives, plastics
18	Vinyl chloride	6.8	Dehydrochlorination of ethylene dichloride	Polymers, films, coatings, moldings
19	Ethylbenzene	6.2	Alkylation of benzene	Production of styrene
20	Styrene	5.2	Dehydrogenation of ethylbenzene.	Polymers, rubber, polyesters
21	Methanol	5.1	From natural gas. Methane oxidized to CO and H_2; catalytic conversion to alcohol	Polymers and adhesives, fuel
22	Carbon dioxide	4.9	Burn hydrocarbons; heat limestone	Production of urea, sodium carbonate, beverages; fire extinguishers
23	Xylene	4.3	Catalytic reforming of petroleum hydrocarbons	High-octane gasolines, urethane resins, solvents, intermediate to making TNT, benzaldehyde, and benzoic acid
24	Formaldehyde	3.7	Oxidation of methanol	Adhesives, plastics
25	Terephthalic acid	3.6	Air oxidation of *p*-xylene in acetic acid	Polyester fibers, films, resins